SOLID GEOMETRY

Sphere

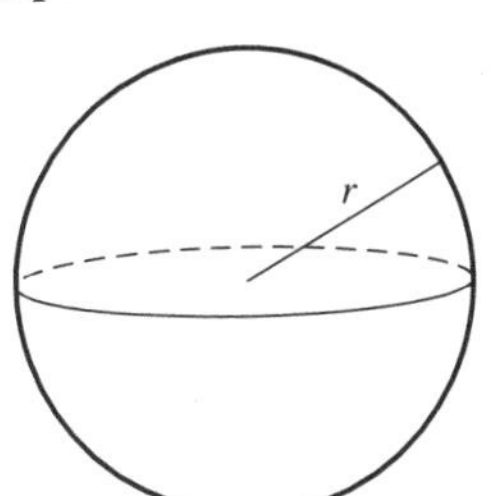

Volume $= \frac{4}{3}\pi r^3$
Surface Area $= 4\pi r^2$

Rectangular Parallelepiped

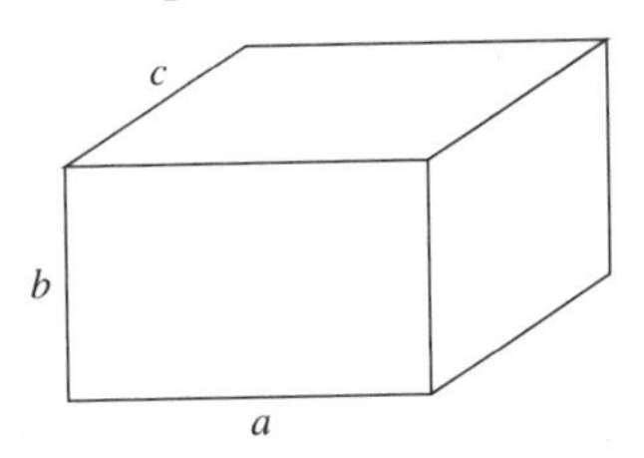

Volume $= abc$
Surface Area $= 2ab + 2ac + 2bc$

Cylinder

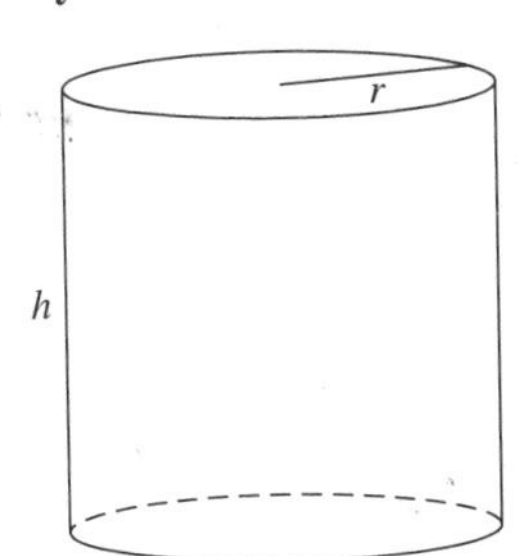

Volume $= \pi r^2 h$
Lateral Surface Area $= 2\pi rh$

Cone

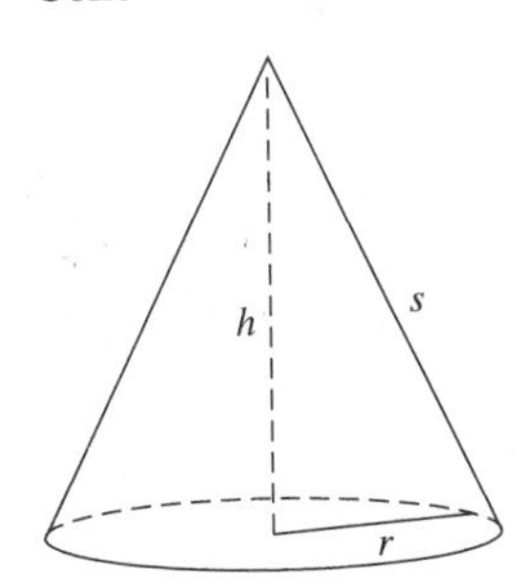

Volume $= \frac{1}{3}\pi r^2 h$
Lateral Surface Area $= \pi rs$

TRIGONOMETRY

Right Triangle

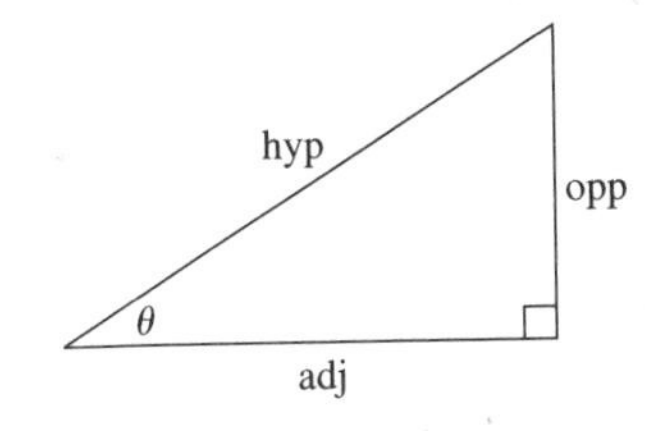

$\sin\theta = \dfrac{\text{opp}}{\text{hyp}}, \qquad \cos\theta = \dfrac{\text{adj}}{\text{hyp}}$

Unit Circle

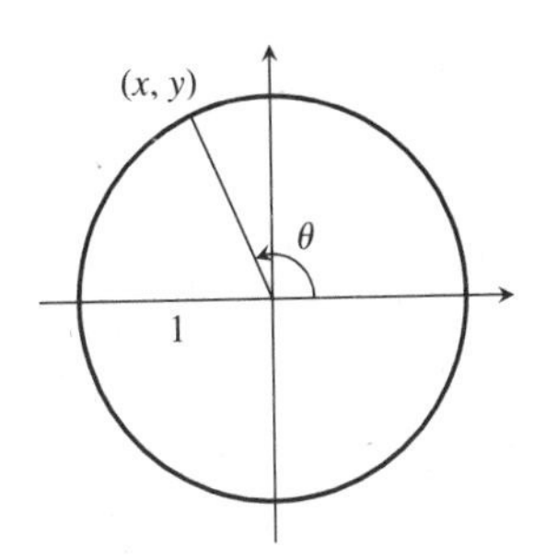

$\sin\theta = y, \qquad \cos\theta = x$

Other Trigonometric Functions

$\tan\theta = \dfrac{\sin\theta}{\cos\theta} \qquad \cot\theta = \dfrac{\cos\theta}{\sin\theta} \qquad \sec\theta = \dfrac{1}{\cos\theta} \qquad \csc\theta = \dfrac{1}{\sin\theta}$

30-60-90 Triangle

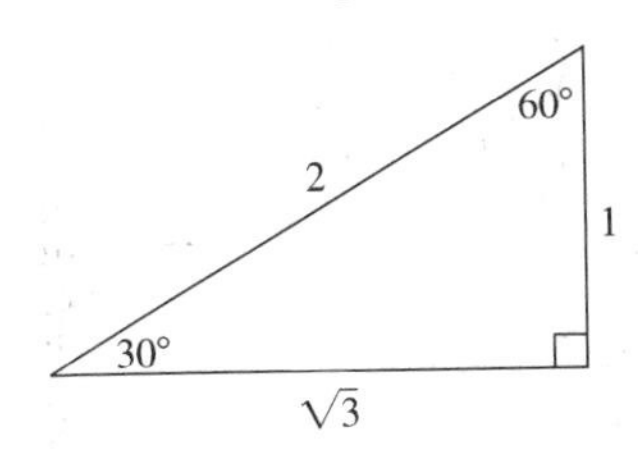

$\sin 30° = \cos 60° = \dfrac{1}{2}$

$\cos 30° = \sin 60° = \dfrac{\sqrt{3}}{2}$

45-45-90 Triangle

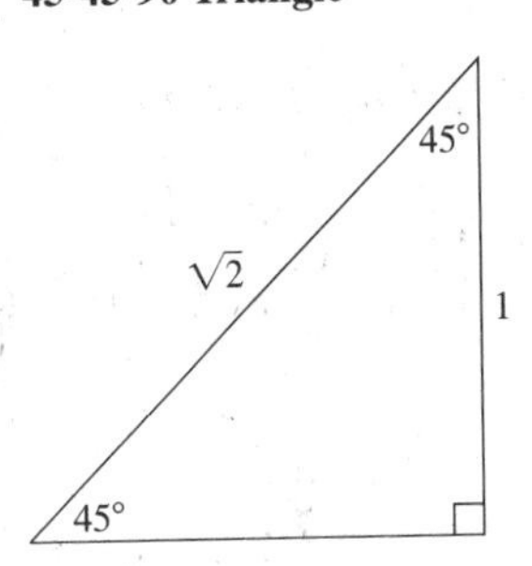

$\sin 45° = \cos 45° = \dfrac{\sqrt{2}}{2}$

Radian Measure

$360° = 2\pi$ radians

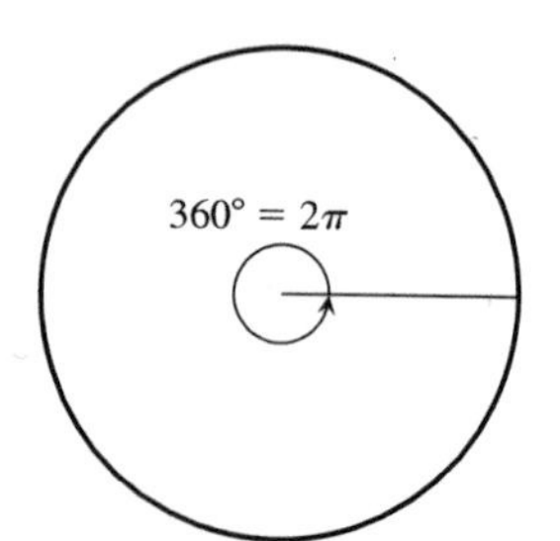

The Law of Sines

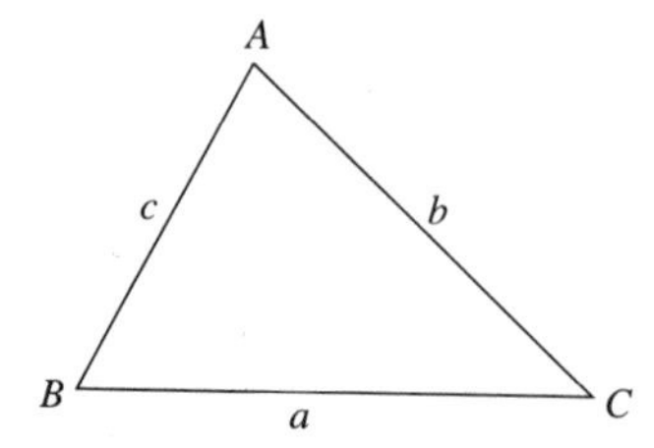

$$\frac{\sin A}{a} = \frac{\sin B}{b} = \frac{\sin C}{c}$$

Trigonometric Identities

$\cos^2\theta + \sin^2\theta = 1$

$\sin(-\theta) = -\sin\theta$

$\sin(\alpha + \beta) = \sin\alpha\cos\beta + \cos\alpha\sin\beta$

$\sin 2\theta = 2\sin\theta\cos\theta$

$$\sin^2\theta = \frac{1 - \cos 2\theta}{2}$$

$\sin\alpha\sin\beta = \frac{1}{2}(\cos(\alpha - \beta) - \cos(\alpha + \beta))$

$\cos(-\theta) = \cos\theta$

$\cos 2\theta = \cos^2\theta - \sin^2\theta$

$\sin\alpha\cos\beta = \frac{1}{2}(\sin(\alpha + \beta) + \sin(\alpha - \beta))$

$1 + \tan^2\theta = \sec^2\theta$

$\tan(-\theta) = -\tan\theta$

$\cos(\alpha + \beta) = \cos\alpha\cos\beta - \sin\alpha\sin\beta$

$$\tan 2\theta = \frac{2\tan\theta}{1 - \tan^2\theta}$$

$$\cos^2\theta = \frac{1 + \cos 2\theta}{2}$$

$\cos\alpha\cos\beta = \frac{1}{2}(\cos(\alpha + \beta) + \cos(\alpha - \beta))$

HYPERBOLIC FUNCTIONS

Hyperbolic Sine and Cosine

$$\cosh x = \frac{e^x + e^{-x}}{2} \qquad \sinh x = \frac{e^x - e^{-x}}{2}$$

Other Hyperbolic Functions

$$\tanh x = \frac{\sinh x}{\cosh x} \qquad \coth x = \frac{\cosh x}{\sinh x} \qquad \operatorname{sech} x = \frac{1}{\cosh x} \qquad \operatorname{csch} x = \frac{1}{\sinh x}$$

Some Hyperbolic Identities

$\cosh^2 x - \sinh^2 x = 1$

$\sinh(-x) = -\sinh x$

$\sinh 2x = 2\sinh x\cosh x$

$\cosh(-x) = \cosh x$

$\cosh 2x = \cosh^2 x + \sinh^2 x$

$1 - \tanh^2 x = \operatorname{sech}^2 x$

$\tanh(-x) = -\tanh x$

$$\tanh 2x = \frac{2\tanh x}{1 + \tanh^2 x}$$

PHYSICS FORMULAS AND CONSTANTS

D = RT (distance = (rate or speed) (time))

Average speed = (change in distance)/(change in time)

Average acceleration = (change in velocity)/(change in time)

Boyle's Law: pressure = k(temperature/volume), where k is a constant

Gravitational Constant: $G \approx 6.67259 \times 10^{-11}\ \mathrm{N \cdot m^2/kg^2}$

Mass of the Earth: $M_E \approx 5.97 \times 10^{24}$ kg

Radius of the Earth: $R_E \approx 6.378 \times 10^6$ m

Acceleration due to gravity near Earth's surface: $\approx 9.8\ \mathrm{m/s^2} \approx 32\ \mathrm{ft/s^2}$

Calculus

ELGIN H. JOHNSTON

Iowa State University

JEROLD C. MATHEWS

Iowa State University

Boston San Francisco New York
London Toronto Sydney Tokyo Singapore Madrid
Mexico City Munich Paris Cape Town Hong Kong Montreal

To Gail, for her patience, support, and love—EHJ
To Ellie, with gratitude for her steady support and love—JCM

Executive Editor: Laurie Rosatone
Marketing Manager: Michael Boezi
Senior Marketing Coordinator: Sara Anderson
Project Editor: Ellen Keohane
Development Editor: David Chelton
Managing Editor: Karen Guardino
Senior Production Supervisor: Peggy McMahon
Manufacturing Buyer: Evelyn Beaton
Senior Prepress Supervisor: Caroline Fell
Technical Art Supervisor: Joe Vetere
Production Services: Elm Street Publishing Services, Inc.
Composition: The Beacon Group, Inc.
Technical Art Illustration: Techsetters, Inc.
Senior Designer: Barbara T. Atkinson
Cover Photograph: ©2001 David Binder
Cover Design: Barbara T. Atkinson
Text Design: The Davis Group
Software Project Supervisor (TestGen): David Malone

PHOTO CREDITS:
p. 1, ©Eye Ubiquitous/Corbis; pp. 3, 70, 71, 87, 153, 154, 191, 202, 217, 249, 331, 343, 347, 369, 394, 443, 444, 541, 635, 667, 748, 843, PhotoDisc; p. 66, Kevin Fleming/CORBIS; p. 130, Professor Jim Watson, University of Southhampton/Science Photo Library/Photo Researchers Inc.; p. 179, Vince Streano/CORBIS; p. 189, Ron Kimball; p. 192, Courtesy of Fermilab/Visual Media Services; pp. 241, 452, Bettmann/CORBIS; p. 261, James L. Amos/CORBIS; p. 267, William James Warren/CORBIS; p. 297, Bob Rowan: Progressive Image/CORBIS; pp. 386, 1025, CORBIS; p. 493, FOXTROT ©1994 Bill Amend. Reprinted with permission of UNIVERSAL PRESS SYNDICATE. All rights reserved; p. 538, Wolfgang Kaehler/CORBIS; p. 588, Globus Bros./Corbis Stock Market; p. 636, SuperStock; p. 679, Morton Beebe, S.F./CORBIS; p. 703, JPL/NASA; p. 729, NRAO/AIU; p. 730, U.S. Geological Survey; p. 737, Philip Gould/CORBIS; p. 741, George B. Thomas, Jr., CALCULUS AND ANALYTIC GEOMETRY 4/e, figure page 355. Copyright ©1968. Reprinted by permission of Pearson Education, Inc.; p. 764, Map scanned from USGS topographic base map, Mt. Lassen National Park, by Dr. G. Kent Colbath, Cerritos Community College; p. 814, Teale Data Center/GIS Solutions Group; p. 870, Cordaiy Photo Library Ltd./CORBIS; p. 939, Francis G. Mayer/CORBIS; p. 946, Courtesy of Jerold Mathews; p. 973, Image created by Paul Bourke, Swinburne University of Technology; p. 974, Photofest; p. 999, David Young-Wolff/Photo Edit; p. 1100, Mobius Strip II by M.C. Escher ©2001 Cordon Art-Baarn-Holland. All rights reserved; pp. 1129, 1134, Tom Pantages. Back cover photo courtesy of the Central Artery/Tunnel Project and HNTB Corporation.

Library of Congress Cataloging-in-Publication Data
Johnston, Elgin H.
Calculus / Elgin H. Johnston, Jerold C. Mathews.—1st ed.
p. cm.
Includes index.
ISBN 0-321-00682-8
ISBN 0-321-02045-6 (pt. 1)—ISBN 0-321-02046-4 (pt. 2)
1. Calculus. I. Mathews, Jerold C. II. Title.
QA303 .J593 2001
515—dc21 2001031644
Calculus ISBN 0-321-00682-8

2 3 4 5 6 7 8 9 1 0—QWV—0403

Preface

Introduction

Since 1970, the content, teaching, and learning of calculus have been affected by the increasing importance of mathematics in the sciences; the availability of powerful, inexpensive computing technologies; and the needs of the more diverse student populations entering calculus. These developments make teaching calculus an exciting challenge. This book, written for students majoring in science, engineering, or the mathematical sciences, is our response to this challenge.

With the help of grants from the National Science Foundation, we have written a text intended to help students learn calculus more effectively so that they are better prepared for subsequent technical courses and possess more experience in solving realistic, multistep problems. Our intent is to do this within a fairly traditional syllabus.

We have provided for a more effective learning and problem-solving experience in several ways. Among the skills of good problem solvers is flexibility, including the use of numerical, graphical, and symbolic representations of a problem or its solution. We often use such representations and ask students to model our practice in their work.

In the first three chapters we discuss approximation and error, trigonometric and exponential functions, and vectors and parametric equations, all of which are important in the engineering, mathematical, and physical sciences. We return to these ideas in subsequent chapters, thus giving students repeated practice with important concepts. We have included as many realistic problems as possible, both in section exercises and in the student projects at the end of each chapter.

Approach

Our text was written for a three-semester sequence, with a provision for "early vectors."

Why Early Vectors? We have provided an option for early study of vectors and parametric equations in the plane for several reasons. Students taking engineering physics during their first two semesters often experience positive feedback from topics common to their calculus and physics courses. In addition, parametric equations provide a uniform framework for the study of many important topics in calculus including polar coordinates, arc length, surface area, and graphing. Throughout our text we often use ideas related to motion to motivate new topics. Discussions of motion lead naturally to parametric equations and vectors. By considering only planar vectors in Calculus I, students get a chance to understand vectors, parametric equations, and motion in familiar two-dimensional settings before they must deal with the additional complications of working in three dimensions. In addition, by introducing vectors early, students have more time to become familiar with vectors, parametric equations, and parametric surfaces before studying the Divergence Theorem and Stokes' Theorem towards the end of multivariable calculus. Finally, students in some engineering curricula require only one year of calculus. In recent years some engineering groups have expressed a desire that their students get some exposure to multivariable calculus during this year. Exposure to vectors in the first semester sets the stage for some multivariable calculus during the second semester.

Early Transcendentals We have followed the trend toward the early introduction of the transcendental functions and their derivatives. This makes possible more practice with these functions and more realistic examples and exercises. Along with many other contemporary calculus books, we complete the study (basic properties and differentiation) of trigonometric functions, exponentials, logarithms, and inverse trigonometric functions in the first semester.

Linear Approximation We have used the idea of linear approximation as a unifying theme throughout the book. The idea of linear approximation lies behind many concepts in calculus and can be part of a framework on which a student can build their understanding of calculus. We introduce the idea of linear approximation when we zoom in on the graphs of functions to find the rate of change at a point. The linear approximation theme continues in various applications including Newton's method and Euler's method. In later chapters we generalize linear approximation to higher-order approximations through Taylor polynomials and lay the groundwork for multivariable approximations by studying linear functions in two and three dimensions. We use linear approximations for functions of several variables to develop many of the ideas of multivariable calculus, including differentiability, the chain rule, and the change-of-variables formula for multiple integrals.

Flexibility for a Variety of Courses

The following flowcharts outline several ways of using our text. The sequence labeled "EV" (for early vectors) addresses a growing demand by engineering and physical science programs for exposure to vectors and some topics from multivariable calculus during the first year. Even more multivariable material can be included in the second semester by moving Chapter 7 (Infinite Series, Sequences, and Approximations) to the end of the third semester and including Chapter 9 (Functions of Several Variables) in the second semester.

The majority of courses will likely follow the standard sequence, designated "S" in the chart. In this sequence we defer work with vectors until Calculus II. Section 3.1 (Motion along a Line) is included in Calculus I. The remaining sections of Chapter 3, indicated by 3* in the chart, can be covered in Calculus II after Chapter 7 (Infinite Series, Sequences, and Approximations). We have written Chapters 4 (Applications of the Derivative) and 5 (The Integral) so that they are independent of the vector material in Chapter 3. In Chapter 6 (Applications of the Integral), Sections 6.1 (Volumes by Cross Section), 6.2 (Volumes by Shells), 6.3 (Polar Coordinates and Parametric Equations), 6.4 (Arc Length and Unit Tangent Vectors), 6.5 (Areas of Regions Described by Polar Equations), and 6.9 (Improper Integrals) can be covered before the vector material in Chapter 3. In Section 6.4 we use parametric representations of planar curves; we briefly review this topic in Section 6.3. Some applications in Section 6.6 (Work), 6.7 (Center of Mass), and 6.8 (Curvature, Acceleration, and Kepler's Second Law) depend upon material in Sections 3.2, 3.3, 3.4, and 3.5.

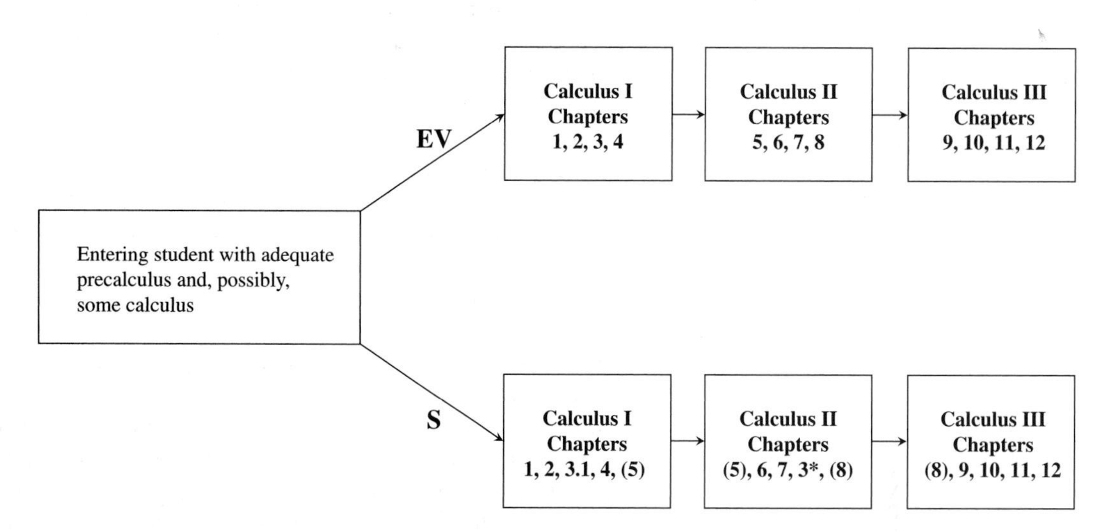

For transferring or entering students with credit for a full year of calculus, we have sequence "M" (for multivariable). Such a class can be taught from *Calculus, Part II* (ISBN 0-321-02046-4), a version of this text that contains the multivariable material of Chapters 8–12 and Chapter R, which contains some review of polar coordinates and most of the material on vectors from Calculus I and II. Depending on the backgrounds and needs of the students, the instructor can pick and choose as needed from Chapters R, 8, and 9 to fill in the material that is needed to study differentiability, multiple integrals, and vector function in Chapters 10, 11, and 12.

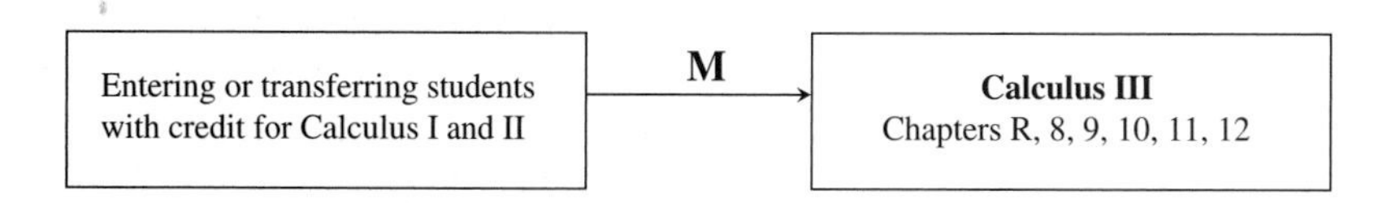

We also offer a version of the text for those students taking only one year of calculus, starting with Calculus I. It contains Chapters 1–8 of the full text.

Features

To better prepare for advanced courses in mathematics or physics or courses in their majors, students need to work actively at solving problems, studying examples, and investigating troublesome points through graphical, numerical, and analytic representations. Toward these ends we have provided a number of features to help students explore, learn, and understand calculus.

Technology

We assume that each student has convenient access to at least a graphics calculator. We feel that such calculators, as well as computer algebra systems, can be a valuable teaching and learning tool. One goal of the text is for students to become proficient at investigating problems using an appropriate combination of graphical, numerical, or symbolic tools. We have tried to model calculator and computer usage in the text and to provide students with opportunities to practice solving problems with the aid of a calculator or computer. Throughout the text, exercises requiring substantial calculator or computer algebra system (CAS) use are identified by the icon T. We comment further on exercises with and without this icon in the section labeled "Exercises," which follows shortly.

Precalculus Review

Students enter calculus with a wide variety of backgrounds in precalculus, and even well-prepared students may find their skills a little rusty after a few months without practice. Thus many students may need, and even welcome, some review of precalculus material. On the other hand, calculus is a new and exciting subject, so it is important to use and enhance this excitement from the first day of class. We have tried to provide for needed review while still bringing out new ideas or looking at old material in fresh ways. In Sections 1.1 and 1.2 we review the definition of function and operations on functions. In these sections we stress ways of representing functions (numerically, graphically, and symbolically) and, for each representation, how operations on functions can be interpreted. In studying composition of functions we work with graphics calculators and see that such calculators do not always perform functions composition correctly. This illustrates early the need to look at a problem in several ways and to think critically before trusting output from a calculator or CAS. Other review is built into the development of the derivative. In Chapters 1 and 2 many examples and exercises are designed not only to illustrate new concepts, but also to give students practice with algebraic and trigonometric skills. Finally, we have provided for more review in the Appendix, where we have summarized and, in some cases derived results from, algebra, geometry, and trigonometry that are useful in calculus.

Graphs and Tables

When every student has access to a graphics calculator or computer algebra system, the figures and tables in a text can play a larger role. To encourage more active reading, we often urge students to verify table entries or to reproduce graphs that accompany investigations and examples. In most cases, students can obtain qualitatively similar graphs with a graphics calculator.

Differential Equations

We tell students that calculus is the language of science and engineering. This statement is very well illustrated by the wide range of physical and biological phenomena that are modeled with differential equations. At the end of Chapter 2 we introduce differential equations. Our objective here is not to learn to solve differential equations, but rather to begin to understand what is meant by a solution to a differential equation and to get an idea of how to interpret and build equations that describe changing situations. As part of Section 3.6 on Newton's laws and as a prelude to the Fundamental Theorem of Calculus in Chapter 5, we use simple antidifferentiation to infer the position and velocity of objects in motion from Newton's second law. In an optional section of Chapter 4 we discuss Euler's method for solving differential equations. In Chapter 5 we discuss separable equations and how to solve them by integration. In Chapter 12 we touch on exact equations as a by-product of our work with potential functions and the Fundamental Theorem of Line Integrals.

Exercises

The problem sets at the end of each section contain a variety of exercises. Many of the problems reflect skills or problem-solving techniques encountered in the section. To encourage close reading of the examples, a few of these problems ask students to fill in some details in an example or to rework an example with different numerical data. Every exercise set also contains problems whose solution method is not covered in an example. In these problems students may be required to work a little beyond the material discussed in the text or to think about and use the text concepts in ways not illustrated in the examples. All of these problems can be solved using skills the student should already have mastered. Students find problems of this sort challenging and sometimes frustrating. However, one of the best ways for them to develop good problem-solving skills is to practice on problems that encourage them to think in new directions. For those who use calculus, it is also important to communicate ideas and conclusions. Thus, in some problems, students are asked to write a sentence or two explaining their result in words, as in Exercise 22 on page 77 or Exercise 32 on page 201. In others, we ask students to look at several examples illustrating a situation, then write a paragraph describing a conclusion that might be conjectured based on these examples. For example, see Exercises 38 and 39 on page 19 and Exercise 31 on page 528.

Most of the exercises in our text can be done by hand. At the same time, we have assumed that each student has convenient access to at least a graphics calculator. We expect that students will sometimes use their calculator or CAS in solving exercises involving routine graphing of functions or in evaluating a function or expression. We have not marked such exercises as requiring technology. We have marked with the icon T exercises requiring something beyond routine graphing or function evaluation. We have, for example, marked with T exercises involving graphical determination of error for an approximation or graphs of surfaces described parametrically. As students progress through the text they should become better at recognizing when the use of technology is appropriate and when it is not. Sound judgments of this sort are an important part of problem solving in today's engineering and science fields.

We have tried to denote decimal approximations to answers of Exercises by using the approximately equal to symbol $\approx$ and often have restricted the number of significant figures in an answer to that justified by the initial data of the Exercise. Usually, we have retained full calculator or CAS accuracy in a calculation, but have recorded intermediate and final results to six significant figures.

Investigations

We sometimes use the context of an "investigation" to introduce new topics. These investigations, often based on physical, numerical, graphical, or symbolic ideas, introduce students to the main ideas in a topic in advance of the formal definitions and development. Following each investigation, we use examples to show how one may solve the more or less standard kinds of problems associated with the topic. For example, see the Investigations on pages 543–544 (Section 7.1, Taylor Polynomials) or page 445 (Section 6.1, Volumes by Cross Section).

Chapter-End Materials

Chapter Review Materials Each chapter ends with review materials, beginning with a brief summary of the chapter highlights. Following these highlights is a review grid that summarizes key concepts with a graph, a formula, and a short example. Following the review grid are chapter review exercises that draw on material from the entire chapter.

Student Projects This book grew from an NSF grant to write projects designed to give students significant experience in solving multistep problems, writing, using appropriate computing technology, and working in groups. These projects have been class-tested extensively over a period of years. One or more of these projects are included at the end of each chapter, for example, see pages 189 and 537. Maple worksheets, *Mathematica* notebooks, and MATLAB M-files for these projects are available on the Web site located at *www.aw.com/johnston.*

Chapter Summaries

Chapter 1 Rates of Change, Limits, and the Derivative

In this chapter we lay the groundwork for our study of calculus and its applications. The ideas described in this chapter—functions, limits, and rates of change—are the foundation for the rest of the text. We begin by studying functions and how they can be presented or interpreted in symbolic, graphical, or tabular form. We then introduce the rate of change as the slope of the "line" we see when we zoom in on the graph of a function. As we modify this to get a symbolic definition of rate of change, we are led to define and investigate limits. Once we have limits, we refine our definition of rate of change and obtain the definition of the derivative.

Chapter 2 Finding the Derivative

In this chapter we approach the derivative more mechanically. Our goal is to develop quick, efficient procedures for producing the derivative of a function. Students are encouraged to recreate heuristic arguments in support of differentiation rules (product, quotient, chain rule). We also work with trigonometric functions and the unit circle and use this context to find the derivative of the sine function. We discuss the special role of the number e as the base of the exponential function and are then led to the natural logarithm. We then discuss inverse functions in more generality and investigate the notion of restriction of domain to obtain a one-to-one function for purpose of defining an inverse. With this preparation we then define the inverse sine, tangent, and cosine functions and discuss their derivatives. We conclude this chapter with a section on modeling, in which we give examples of some of the diverse and interesting situations that can be described by using derivatives.

Chapter 3 Motion, Vectors, and Parametric Equations

We work towards an understanding of vectors through the modeling of forces, velocities, and displacements. We model motion on lines and circles using vector/parametric equations and relate these motions to Newton's laws. The chapter includes vector arithmetic and geometric interpretations, including the dot product and the projection of a vector in a given direction; the calculation of the position, velocity, and speed of an object; tangent vectors and slope for curves described by parametric equations; and elementary antidifferentiation in connection with Newton's second law.

Chapter 4 Applications of the Derivative

We study such applications of the derivative as the $\Delta f \approx df$ approximation, Newton's method for solving equations like $f(x) = 0$, implicit differentiation, the analysis of the graph of a function by classifying it as increasing or decreasing or concave up or concave down, optimizing a process by locating the maximum or minimum of a function, the calculation of certain limits by applying l'Hôpital's Rule, and Euler's method, the simplest numerical method for solving a differential equation.

Chapter 5 The Integral

The goal of the chapter is to help students understand the integral as the limit of a Riemann sum associated with geometric or physical objects. The Fundamental Theorem of Calculus is discussed through a combination of position/velocity and area approaches. The integral is used to calculate simple areas and solve simple separable differential equations. We study the usual techniques of integration (substitution, integration by parts, and integration of rational functions through partial fractions) using a combination of symbolic calculation and tables. We discuss the trapezoid, midpoint, and Simpson's rules for numerical integration.

Chapter 6 Applications of the Integral

The integral is used to model and calculate the volumes of solids, lengths of curves, work done by force fields on objects, and masses and the centers of mass of objects. We also study curvature, the tangential and normal acceleration vectors of moving objects, and the classification and evaluation of "improper integrals." A major theme in this chapter is the use of "elements" of area, volume, arc length, work, and mass. Such elements increase understanding and decrease the need for memorized formulas by making direct use of underlying geometric concepts or physical principles. Applications involving vectors, which can be postponed or skipped if vectors are to be covered later, are as follows: Work (Section 6.6), Center of Mass (Section 6.7), and Curvature, Acceleration, and Kepler's Second Law (Section 6.8).

Chapter 7 Infinite Series, Sequences, and Approximations

In an effort to find approximations that do a better job than the tangent line approximation we are led to the Taylor polynomials. To better understand such approximations, we next study sequences and infinite series. We examine the dynamics of increasing sequences and the possible behaviors of such sequences. This in turn leads to results about the behavior of infinite series, including the comparison, integral, and ratio tests. This study of series leads us to a power series, and then back to Taylor polynomials and Taylor series.

Chapter 8 Vectors and Linear Functions

Chapter 8 lays the groundwork for the study of functions of several variables. Most of the ideas in this chapter are generalizations to three dimensions of ideas developed when we studied functions of one variable and vectors and motion in two dimensions. These ideas include vectors, planes, and motion in three dimensions. We also look at several simple but important ideas concerning linear functions in two and three dimensions. These ideas allow us to continue our theme of linear approximation and are the foundation of later results about differentiability of functions of several variables, the chain rule, and change of variables in multiple integrals.

Chapter 9 Functions of Several Variables

We start with a discussion of conic sections, because a sound understanding of graphs of conics is needed to understand many surface graphs. We then develop techniques to help students understand and visualize functions of several variables. Once these techniques are in place, we define and study rates of change for functions of several variables. Partial derivatives and the gradient vector are introduced, two fundamental concepts that are used in the remainder of this text.

Chapter 10 Differentiable Functions of Several Variables

In this chapter, we extend the ideas of differentiability, linear approximation, and the chain rule to higher dimensions by using the idea of a linear function and its associated matrix (see Chapter 8). The conceptual and notational connections with single variable calculus are emphasized. The chain rule is applied to express partial differential equations in new coordinate systems. Optimization is studied with and without constraints.

Chapter 11 Multiple Integrals

The goals of the chapter are for students to understand double and triple integrals as the limit of a Riemann sum associated with geometric or physical objects and to understand the process of evaluating multiple integrals through reduction to several single-dimensional integrals. Multiple integrals are used to calculate volumes of three-dimensional regions, areas of surfaces, masses, and centers of mass. We discuss the change-of-variables formula for multiple integrals with an emphasis on a change from rectangular to cylindrical or spherical coordinates. A major theme in this chapter is the use of "elements" of volume, surface area, and mass. Such elements increase understanding and decrease the need for memorized formulas by making direct use of underlying geometric concepts or physical principles.

Chapter 12 Line and Surface Integrals

In this chapter we study functions defined on curves or surfaces in space. We start by extending the definition of the definite integral to real- and vector-valued functions defined on curves. We use integrals to calculate the mass of a wire and the work done by a force and to study the flow of a fluid in the plane. In the second half of the chapter, we define the integral of real- and vector-valued functions defined on surfaces. With these integrals we study such phenomena as three-dimensional fluid flow and heat flux. Many of the integrals introduced in this chapter can be more easily evaluated by converting them to single, double, or triple integrals. To do this, we apply the Fundamental Theorem of Line Integrals, Green's Theorem, the Divergence Theorem, and Stokes' Theorem.

Supplements for the Instructor

Annotated Instructor's Edition

ISBN 0-201-75303-0
This specially bound version of the student edition contains answers to both even- and odd-numbered exercises at the back of the text. This edition also contains suggestions for the incorporation of technology and how aspects of the Web site can be used to enhance the presentation of chapter topics.

Calculus, Part I

ISBN 0-321-02045-6
This version of the student edition covers all of the material needed for a single-variable calculus course. Includes Chapters 1–8.

Calculus, Part II

ISBN 0-321-02046-4
This version of the student edition covers all of the material needed for a multivariable calculus course. Includes Chapters 8–12. Also contains an additional review chapter R on vectors and parametric equations for students entering the course without any prior instruction on these topics.

Instructor's Solutions Manual

Part I (Chapters 1–8), ISBN 0-321-09314-3
Part II (Chapters R, 8–12), ISBN 0-321-09311-9
The *Instructor's Solutions Manual* contains worked-out solutions to all of the exercises in the text. A PowerPoint CD is also included in the back of this solutions manual. It contains selected key figures from each chapter of the text for classroom demonstration purposes.

TestGen-EQ CD Package

ISBN 0-321-40073-9
Windows and MAC
TestGen-EQ is a computerized test generator with algorithmically defined problems organized specifically for this textbook. Its user-friendly graphical interface enables instructors to select, view, edit, and add test items, then print tests in a variety of fonts and forms. It offers a built-in question editor giving the user the power to create graphs, import graphics, insert mathematical symbols and templates, and insert variable numbers or text. It also has an "Export to HTML" feature and lets instructors create practice tests that can be posted to a Web site. Tests created with TestGen-EQ can be used with QuizMaster-EQ, which enables students to take exams on a computer network. QuizMaster-EQ automatically grades exams, stores results on disk, and allows the instructor to view or print a variety of reports for individual students, classes, or courses.

Video Series

ISBN 0-201-77071-7
These tutorial videos are designed to focus on key topics that students have trouble with in calculus. Experienced math instructors discuss material from the text in a clear and methodical presentation.

Interact MathXL Instructor Coupon

ISBN 0-201-71111-7
This Web-based diagnostic testing and tutorial system allows students to take practice tests correlated to the textbook and receive customized study plans based on their results. Each time a student takes a practice test, the resulting study plan identifies areas for improvement and links to appropriate practice exercises and tutorials. A course-management feature allows instructors to view students' test results, study plans, and practice work.

MyMathLab.com

MyMathLab.com is a complete, on-line course for Addison-Wesley mathematics textbooks that integrates interactive, multimedia instruction correlated to the textbook content. MyMathLab is easily customizable to suit the needs of students and instructors and provides a comprehensive and efficient on-line course-management system that allows for diagnosis, assessment, and tracking of students' progress.

MyMathLab Features

- Fully interactive multimedia textbooks are built in CourseCompass, a version of Blackboard™ designed specifically for Addison-Wesley.
- Chapter and section folders from the textbook contain a wide range of instructional content: videos, software tools, audio clips, animations, and electronic supplements.
- Hyperlinks take you directly to on-line testing, diagnosis, tutorials, and gradebooks in MathXL—Addison-Wesley's tutorial and testing system for mathematics and statistics.
- Instructors can create, copy, edit, assign, and track all tests for their course as well as track student tutorial and testing performance.
- With push-button ease, instructors can remove, hide, or annotate Addison-Wesley preloaded content, add their own course documents, or change the order in which material is presented.
- Using the communication tools found in MyMathLab, instructors can hold on-line office hours, host a discussion board, create communication groups within their class, send e-mail, and maintain a course calendar.
- Print supplements are available on-line, side by side with their textbooks.

For more information, visit our Web site at www.mymathlab.com or contact your Addison-Wesley sales representative for a live demonstration.

Supplements for the Student

Student's Solutions Manual

Part I (Chapters 1–8), ISBN 0-321-09313-5
Part II (Chapters R, 8–12), ISBN 0-321-09312-7
These manuals are designed for the student and contain carefully worked-out solutions to all of the odd-numbered exercises in the text.

Interact MathXL Student Coupon (24-month registration)

ISBN 0-201-76728-7
This Web-based diagnostic testing and tutorial system allows students to take practice tests correlated to the textbook and receive customized study plans based on their results. Each time a student takes a practice test, the resulting study plan identifies areas for improvement and links to appropriate practice exercises and tutorials generated by InterAct Math. A course-management feature allows instructors to view students' test results, study plans, and practice work.

Just-in-Time Algebra and Trigonometry for Students of Calculus, Second Edition

ISBN 0-201-66974-9
Sharp algebra and trigonometry skills are critical to mastering calculus, and *Just-in-Time Algebra and Trigonometry for Students of Calculus,* second edition, by

Guntram Mueller and Ronald I. Brent is designed to bolster these skills while students study calculus. As students make their way through calculus, this text is with them every step of the way, showing them the necessary algebra or trigonometry topics and pointing out potential problem spots. The easy-to-use table of contents has algebra and trigonometry topics arranged in the order in which students will need them as they study calculus.

Addison-Wesley Math Tutor Center

The Addison-Wesley Math Tutor Center is staffed by qualified mathematics instructors who provide students with tutoring on examples and exercises from the textbook. Tutoring is available via toll-free telephone, fax, e-mail, or whiteboard technology and is available five days a week, seven hours a day. www.awl.com/tutorcenter

Digital Video Tutor

ISBN 0-201-77070-9
The videos for this text are also available on CD-ROM. The complete set of digitized videos, affordable and portable for student use at home or on campus, is ideal for distance learning or supplemental instruction.

Web Site

The Web site for this text, www.aw.com/johnston, features several additional student and instructor resources.

Java Applets

There are more than 30 unique, interactive Java applets developed by Tom Leathrum of Jacksonville State University. Each is easy to use, with little syntax or special languages to learn. Students manipulate equations and graphs in real time. By bringing these applets into classroom demonstration and discussion, laboratory and homework assignments, or independent study, teachers and students can explore the mathematics of time and motion. These applets are designed to build a clear understanding of key concepts when they are first encountered and to help students over the hurdles of abstraction that have often confused them in the past. Web-site icons mark the locations in the text where material related to these Java applets is covered. WEB

Just-in-Time Online Algebra and Trigonometry

Compiled by Ronald I. Brent and Guntram Mueller of the University of Massachusetts, Lowell, this interactive Web-based testing and tutorial system allows students to practice the algebra and trigonometry skills that are critical to mastering calculus. *Just-in-Time Online* tracks student progress and provides personalized study plans to help students succeed. The registration coupon at the back of this text provides access to this feature of the Web site.

Downloadable Technology Resources for Specific Computer Algebra Systems and Graphing Calculators

Each manual provides detailed guidance for integrating *Mathematica,* Maple, MATLAB, or TI-graphing calculators throughout the course, including syntax and commands. These manuals are available as downloadable PDF files on the Web site.

Student Projects

The student projects at the end of each chapter are also available on the Web site for downloading in the form of *Mathematica* notebooks, Maple worksheets, or MATLAB M-files.

Streaming Videos

The tutorial videos to accompany this text are also available as streaming videos on the Web site.

Selected Solutions

Carefully worked-out solutions to odd-numbered problems from Chapters 1 and 2 are available to students as downloadable PDF files.

On-line Discussions

This suite of on-line communication tools includes a messageboard and iChat. These tools can be used to deliver courses in a distance learning environment. The messageboard allows users to post messages and check back periodically for responses. Students can also use the messageboard to obtain peer support for study guide activities, freeing up instructor time. iChat is a perfect arena for instructor-led live discussions with groups of students.

Acknowledgments

Many friends, students, and colleagues were of great help to us in writing this textbook. We were encouraged and guided by our own calculus students at Iowa State University as we tested these materials in our classrooms. The students were always willing to tell us what was working, what was not, and how things might be improved. Many of our colleagues at Iowa State University also contributed by class-testing some materials and refining many of our student projects. They include Roger Alexander, Clifford Bergman, Brian Cain, A. M. Fink, Alan Heckenbach, Irwin Hentzel, Fritz Keinert, Justin Peters, Richard Tondra, and Bruce Wagner. In addition, we'd like to express our sincere thanks to our many other colleagues and students.

During the summers of 1992, 1993, and 1996 we offered three workshops in calculus reform. The 1992 workshop was at the University of Wisconsin at La Crosse, the 1993 workshop was given at Idaho State University in Pocatello, and the 1996 workshop was at Iowa State University in Ames, Iowa. Although we went there to present our ideas and to demonstrate some of our student projects, we learned a lot by listening to participants' ideas on the directions calculus should take. We thank all of the workshop participants for their insights and ideas.

The University of Wisconsin at La Crosse Workshop:
Carrie Ash-Mott, John Bruha, Robert Coffman, Wesley Day, Ronald Dettmers, G. S. Gill, David Hardy, Marian Harty, Linda H. Host, Erna Jensen, Clement Jeske, Charles Kolsrud, Larry Krajewski, Robert Kreczner, Don Leake, Steve Leth, Franklyn Lightfoot, Rich Maresh, Dan Nicol, Gerald W. Niedfeldt, David Oakland, Marlene Pinzka, Kay Strangman, Gordon Sundberg, Jack Unbehaun, Calvin Van Niewaal, Paul Williams, Randy Wills, J. D. Wine, and Elizabeth Wood.

The Idaho State University at Pocatello Workshop:
Sam Berney, Janet Burgoyne, John C. Eilers, Chaitan Gupta, Joseph Hwang, Jim Brennan, Bob Davis, Bob Firman, Roger Higdem, Ken Meerdink, Tom Misseldine, Eric Rowley, Madeline Schaal, Peter Wildman, Mike Prophet, Dan Schaal, Don Shimamoto, Suzanne Wisner, and Larry Ford.

The Iowa State University at Ames Workshop:
Carrie Ash-Mott, Joel Bundt, Frank Carver, Wesley R. Day, Marlin Deweerdt, Loren Flater, Dan Fuchs, Peg Griffey, Ruth A. Hartman, Robert Hoile, Jenelle Jarnagin, Jim Jefson, Steven D. Klassen, David Kofoed, Virgil E. Larsen, Sergio Loch, Wanda Long, Herbert C. Lyon, Ronald M. Mathson, Cheryl Ooten, Constantin Pirvulescu, Jerry Schmidt, Virginia Swenson, Melvern K. Taylor, and James W. Van Ark.

We are grateful to the instructors who have class-tested and reviewed drafts of this book. Comments from these reviewers were invaluable in suggesting ways to improve our manuscript. Our sincere thanks to each of them, including those listed here:

Gregory T. Adams, *Bucknell University*

James C. Alexander, *Case Western Reserve University*

Johan G. F. Belinfante, *Georgia Institute of Technology*

Mark Bridger, *Northeastern University*

Donatella Danielli, *Johns Hopkins University*

John M. Erdman, *Portland State University*

Earl Hamilton, *North Seattle Community College*

Weimin Han, *University of Iowa*

Phillip Johnson, *University of North Carolina, Charlotte*

William J. Keane, *Boston College*

Joseph D. Lakey, *New Mexico State University*

Ronald M. Mathsen, *North Dakota State University*

Joe Miles, *University of Illinois, Urbana-Champaign*

Andrew Nestler, *Santa Monica College*

Scott Pauls, *Dartmouth College*

Roy Rakestraw, *Oral Roberts University*

Lynn M. Siedenstrang, *Grays Harbor College*

William L. Siegmann, *Rensselaer Polytechnic Institute*

Jennifer Slimowitz, *Rice University*

Harvey E. Wolff, *University of Toledo*

Many other people have been very helpful in finding materials on which to base problems and discussions. These include Susan North of the Ames Public Library, the reference staff at the Iowa State University Library, the reference staff at the Alaska State Library, the U.S. Geological Survey, and Michael Graff of the Iowa State University Computation Center.

Many people were involved in ensuring the accuracy of this text. We wish to express our thanks to those people who helped accuracy-check this text, which went through not one or two, but three rounds of checking: Jon Booze, Dr. Lyle Cochran, Paul Lorczak, Dr. Tim Mogill, Patricia Nelson, Jeffrey D. Oldham, Daniel Pick, Jeff Suzuki, Daniel P. Thompson, Marie Vanisko, Stephen Whalen, Dr. Yuri Zhorov, Dr. Holly Zullo, Dr. Mark Parker, Becky Cointin, and Tom Wegleitner.

We also wish to express our thanks to Laurel Technical Services for writing the Instructor's and Student's Solutions Manuals. Thanks to Tom Leathrum of Jacksonville State University for his extensive contributions to our Web site.

We hope that instructors and students find this text instructive and enlightening. We invite suggestions from readers—students, professors, and others—for improvements to the text. We will do our best to incorporate appropriate suggestions in future printings. Please write to us at Addison-Wesley or contact us by electronic mail.

Elgin H. Johnston
Iowa State University
ehjohnst@iastate.edu

Jerold C. Mathews
Iowa State University
mathews@iastate.edu

Contents

Chapter 1 Rates of Change, Limits, and the Derivative 1

Chapter 2 Finding the Derivative 87

Sections shaded in BLUE can be postponed.

Sections shaded in BLUE can be postponed.

Sections shaded in YELLOW contain some applications involving vectors which can be skipped or postponed.

1

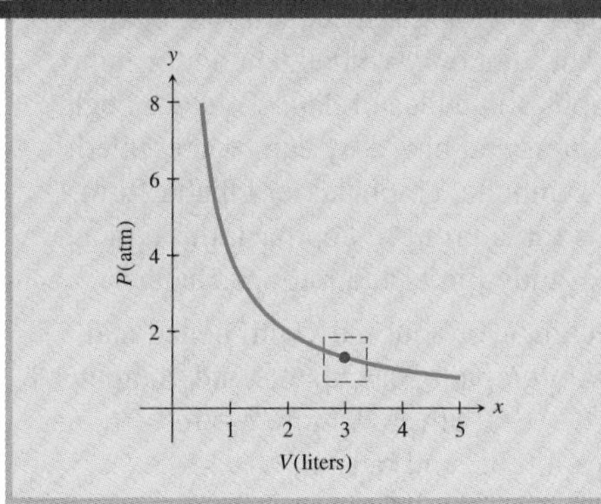

Rates of Change, Limits, and the Derivative

As the pistons move back and forth in the cylinders of an engine, the pressure and volume of the gas in the cylinder changes. With calculus, we can describe and study these changing conditions.

See the Investigation on page 24 for further explanation.

Change occurs all around us. We experience examples of change every day: the change in the position of a car moving down the street, the change in the temperature of our shower as we adjust the controls, or the change in volume of sound as we turn the knob on a stereo. Change is also present in such things as the stresses on a beam in a skyscraper, the shape of a loop in a roller coaster, or the marketing strategy for a new product. Diverse as these phenomena are, all can be described and analyzed with one powerful tool—calculus.

Credit for the invention of calculus usually goes to two men: Sir Isaac Newton (1642–1727), an English scientist and mathematician, and Gottfried Wilhelm Leibniz (1646–1716), a German philosopher and mathematician. The work of Leibniz and Newton rested on the work of many other scientists and mathematicians, including Archimedes, who used ideas foreshadowing integral calculus; Fermat, who devised a method for finding tangents to curves; and Kepler, who found formulas for the volumes of many solids.

In this chapter we lay the groundwork for our study of calculus and its applications. The ideas described in this chapter—functions, limits, and rates of change—are the foundation for the rest of the text. These concepts occur throughout our study of calculus, though often in new situations and applied to different problems.

In this chapter (and the rest of the text) much of our approach is driven by three realizations.

- People who use calculus (physicists, engineers, chemists, biologists, sociologists, economists, mathematicians, etc.) work not only with algebraic formulas, but also with graphs and tables of data. Thus we begin by studying functions and how they can be presented or interpreted in symbolic, graphical, or tabular form. We stick with this three-pronged approach throughout the chapter as we study limits and rates of change.
- People using calculus work not only with pencil and paper, but also with graphing calculators and computer algebra systems (CASs). Thus, we, too, use these tools to study functions, limits, and rates of change. As a student of calculus, you should use these tools at every opportunity to verify examples, to experiment and gather data about a problem, and to check the answers to your own work.
- Often, when people use calculus, they need to explain their answers to other people. A problem is not necessarily "finished" when an answer is produced or a graph is drawn. The answer usually has to be explained and interpreted in the context of the problem. In many of the examples in this text, we take a few sentences to explain the meaning of an answer. We also ask that you do this in many of the exercises.

So, do you have a pencil and some paper? Is your graphing calculator or computer algebra system (CAS) close at hand? If so, you're ready to start.

1.1 Functions

As students of science, engineering, and mathematics, you deal with functions every day. You are accustomed to using functions defined by formulas. For example, the function that gives the volume V of a sphere with radius r is described by

$$V = V(r) = \frac{4}{3}\pi r^3.$$

However, functions can be described in other ways, too. Graphs and tables that appear in newspapers and magazines may also represent functions, though we usually don't think "function" when we see such things. As an example, look at the graph in Fig. 1.1. This graph shows the level of the Skunk River as measured in Ames, Iowa, on July 8, 9, and 10 of 1993. This graph describes a function that tells

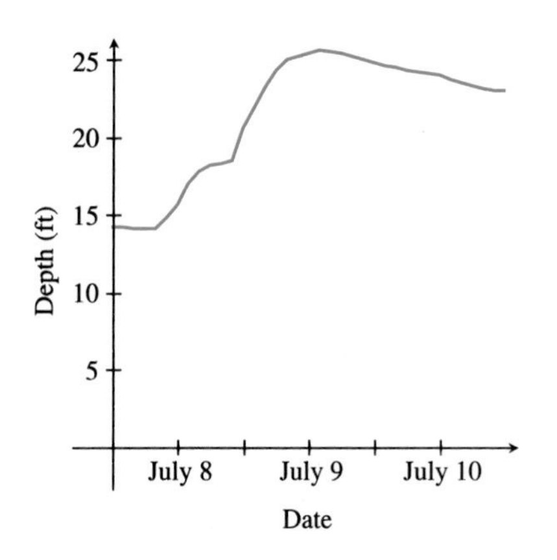

FIGURE 1.1 **Skunk River levels during the flood of 1993.**

(Data from the United States Department of the Interior Geological Society.)

the level of the river at any given time during these three days. We do not have an equation for this function, but the graph is nonetheless the graph of a function. So what exactly is a function? A function is not a formula, nor is it a graph, though it can be represented by either one. In this section we find out what a function is and what it is not.

The next time you use a word processor, take a moment to think about what is happening. You press the key marked D, and the letter d appears on the computer screen. You press <shift> K, and the letter K appears. Of course, you are seeing a marvelous piece of technology at work. You are also confident that what you type is what you will see on the screen. Why? Because the computer and word-processing software are designed to evaluate the function that assigns a particular "output" on the screen to a given "input" on the keyboard.

The word-processor function.

A function must have a **domain,** or collection of inputs. For the word-processor function, the domain is the set of possible keystrokes. Some elements of this domain are single keystrokes. For example, hit the key marked K to obtain k. Some of the elements of the domain are double keystrokes, as when you depress <shift> and K to get a K. We use $\mathcal{D}_W$ to denote the set of keystrokes that result in some output on the screen. A function must also have a **range,** or collection of outputs. In the case of the word-processor function, the range is the set of letters, numbers, and other symbols that can appear on the computer screen. We denote this set by $\mathcal{R}_W$. The function we are discussing is the rule that matches each keystroke input with the corresponding output on the screen. We call this rule W (for *word processor*). You should check that W satisfies the definition of *function:*

DEFINITION Function

Let $\mathcal{D}$ and $\mathcal{R}$ be sets. A function with domain $\mathcal{D}$ and range in $\mathcal{R}$ is a rule that assigns to each element in $\mathcal{D}$ one element in $\mathcal{R}$. The range of the function is the set of elements in $\mathcal{R}$ that are assigned to one or more elements of $\mathcal{D}$.

In the definition of *function,* we introduced a set $\mathcal{R}$ that *contains the range* of the function. It is not necessary to know the range of a function to define the function, but it is important to know some set $\mathcal{R}$ that contains the range. For many of the functions that we will work with, the range is very difficult to determine, though it is usually no trouble to give a set that contains the range.

What makes W a function? Given that the domain and a set containing the range are specified, W is a function because it is **well defined.** This means that the rule W assigns to each keystroke input in the domain one and only one screen output in the range. In plainer language, this means that every time you strike the key labeled V (and no other key), you will get v on the screen (assuming your computer is working properly!). You won't get v sometimes, with an occasional a or b—you will always get v. That's reassuring.

Sometimes it is helpful to think of a function as a machine that takes an object as input and returns an output, as illustrated in Fig. 1.2. For a function, the input is an element of the domain. The output is the corresponding element of the range. For example, if the domain item keystroke R is input into the word-processor function W, the output is r on the computer screen. As shorthand for this input-output relation, we write

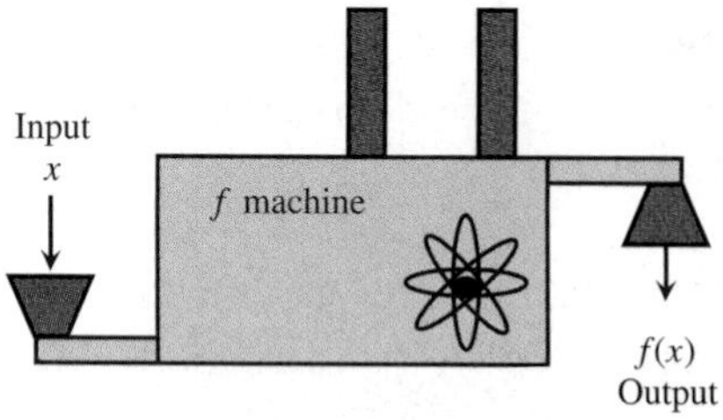

$$W(\text{keystroke } R) = r \text{ on computer screen.}$$

FIGURE 1.2 The f machine returns output $f(x)$ for input x.

In general, if f is a function and x is something in the domain of f, then we write $f(x)$ to denote the range element that corresponds to x. The input is x and the output is $f(x)$. This output is often called the value of $f(x)$ or the value of f at x. Sometimes we want to give this output a shorter name, such as y, and write

$$y = f(x).$$

We say that y **is a function of** x.

EXAMPLE 1 Let f be the function defined for all real numbers x by

$$f(x) = |x^2 - 2x - 10|. \tag{1}$$

Find the value of $f(6.2)$.

Solution

To find the value of $f(6.2)$, replace x by 6.2 on the right side of (1). Thus

$$f(6.2) = |6.2^2 - 2 \cdot 6.2 - 10| = |38.44 - 12.4 - 10| = |16.04| = 16.04.$$

Representing Functions According to our discussion, a function is a rule for linking elements of one set (the domain) to elements of another set (the range). Thus a function is a rule and not a formula, or a graph, or a table. However, formulas, graphs, and tables are very useful ways to describe functions. When a function is described in one of these ways, explicit mention of the domain and range of the function might be omitted. In these cases, it is usually assumed that the person working with the function can determine its domain and range from experience. For most of the functions that we work with, the domain and range are subsets of the real numbers. We will also work with functions that have sets of ordered pairs or ordered triples of real numbers as their domain or range. In these cases, if the domain is not given, we can try to determine the "likely" domain by using knowledge of algebra and trigonometry.

EXAMPLE 2 Discuss the domain and range of the function g described by the equation $g(x) = \sqrt{x}$.

Solution

Because we've all heard that "you can't take the square root of a negative number," it seems natural that the domain of g is the set $[0, \infty) = \{x : x \geq 0\}$ of nonnegative real numbers. We may take the range of the square root function to be the set of nonnegative real numbers. Thus if x is a nonnegative real number, then $\sqrt{x}$ denotes the nonnegative square root of x. Calculators and computer algebra systems follow this convention.

Sometimes a CAS or graphing calculator can help in determining the domain of a function represented by an equation. The graph in Fig. 1.3 was produced by having a CAS graph $y = \sqrt{x}$ for $-5 \leq x \leq 5$. The CAS responded that $\sqrt{x}$ is not a real number for $x < 0$, and then produced the graph for those x values for which $\sqrt{x}$ is real. The graph produced is consistent with our assumption that the domain of g is $x \geq 0$.

However, be careful when you use graphics packages to help determine the domain of a function. Sometimes the results are deceptive. Always think about the calculator or CAS result and interpret it in light of the problem that you are trying to solve. (See Exercises 37 and 38.)

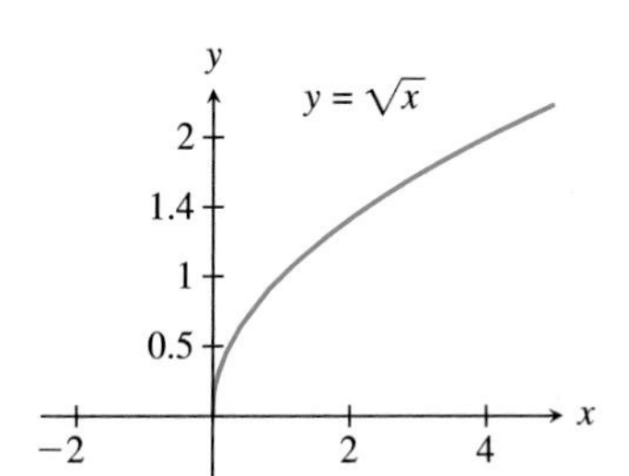

FIGURE 1.3 A CAS-produced graph of $y = \sqrt{x}$ for $-5 \leq x \leq 5$.

EXAMPLE 3 The Iditarod Trail sled dog race is an annual event to commemorate the courage of Leonhard Seppala and other dog mushers who delivered badly needed diphtheria serum to Nome, Alaska, in 1925. The 1100-mile course of the race runs from Anchorage to Nome. The first race was run in 1973. Table 1.1 gives the times of the winners for the races through 2000. Time is given as days:hours:minutes:seconds. This table represents a function whose domain is the set of years in which the race has been run and whose range is the set of winning times. Discuss this function.

Solution

Let I (for Iditarod) be the function that takes as input the year of the race and, as output, gives the winning time for that year. For example,

$$I(1980) = 14{:}07{:}11{:}51 \quad \text{and} \quad I(1994) = 10{:}13{:}02{:}39.$$

The domain of I is the set of years that the race has been run:

$$\{1973, 1974, 1975, \ldots, 2000\}.$$

The range of I is the set of winning times. A graphical representation of this function can be constructed by first forming the ordered pairs $(t, I(t))$, for t in the domain of I,

$$(1973, 20{:}00{:}49{:}41), (1974, 20{:}15{:}02{:}07), \ldots, (2000, 9{:}00{:}58{:}06).$$

A graph representing I is shown in Fig. 1.4. Note that the graph is not a curve, but consists of several points. The graph of I is simply the collection

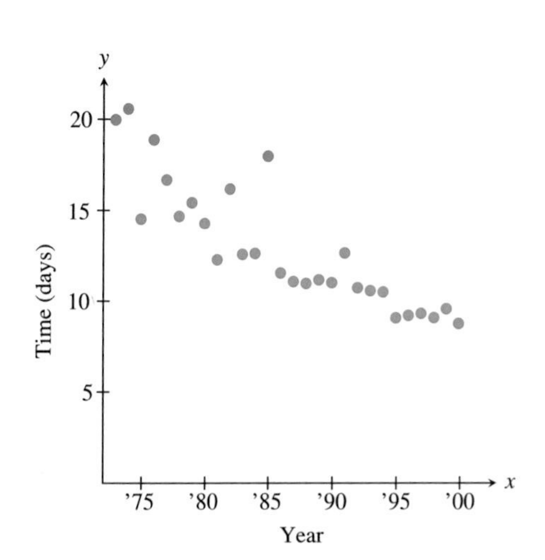

FIGURE 1.4 Winning times for the Iditarod Trail sled dog race.

TABLE 1.1 Iditarod winners.

Year	Name	Time	Year	Name	Time
1973	Dick Wilmarth	20:00:49:41	1987	Susan Butcher	11:02:05:13
1974	Carl Huntington	20:15:02:07	1988	Susan Butcher	11:11:41:40
1975	Emmitt Peters	14:14:43:45	1989	Joe Runyan	11:05:24:34
1976	Jerry Riley	18:22:58:17	1990	Susan Butcher	11:01:53:23
1977	Rick Swenson	16:27:13:00	1991	Rick Swenson	12:16:35:39
1978	Dick Mackey	14:18:52:24	1992	Martin Buser	10:19:17:15
1979	Rick Swenson	12:08:45:02	1993	Jeff King	10:15:38:15
1980	Joe May	14:07:11:51	1994	Martin Buser	10:13:02:39
1981	Rick Swenson	12:08:45:02	1995	Doug Swingley	9:02:42:19
1982	Rick Swenson	16:04:40:10	1996	Jeff King	9:05:43:13
1983	Rick Mackey	12:14:10:44	1997	Martin Buser	9:08:30:45
1984	Dean Osmar	12:15:07:33	1998	Jeff King	9:05:52:26
1985	Libby Riddles	18:00:20:17	1999	Doug Swingley	9:14:31:07
1986	Susan Butcher	11:15:06:00	2000	Doug Swingley	9:00:58:06

of points with coordinates of the form $(t, I(t))$, for t in the domain of I. Because the domain of I has 28 elements, there are only 28 such points. These points make up the graph of I. What are some other functions suggested by the table?

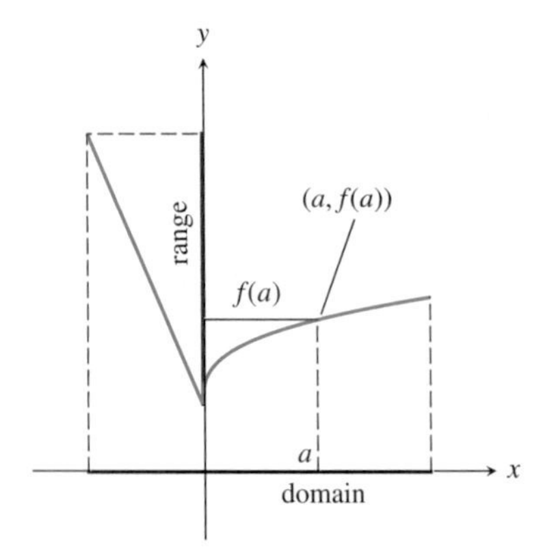

FIGURE 1.5 Graphical representation of a function.

When a function f is represented graphically, we assume, unless told otherwise, that the domain is represented by some part of the horizontal axis and the range by some part of the vertical axis. The graph of f consists of all points with coordinates $(x, f(x))$ where x is in the domain of f. The domain of the function contains the set of points on the horizontal axis that lie above or below the graph. The range of the function contains the set of points on the vertical axis that lie to the left or right of the graph. Given an a in the domain (on the horizontal axis), we find the corresponding element in the range by extending a dotted line from a, either up or down, until it meets the graph at the point $(a, f(a))$. From this point, extend the dotted line horizontally until it meets the vertical axis at $f(a)$. See Fig. 1.5. Graphs are a very useful way to study and represent functions. However, as the next example shows, graphs may not give complete information about a function.

EXAMPLE 4 Discuss the domain and range of the function f represented by the graph in Fig. 1.6.

Solution

The graph appears to lie above or below the interval $[-2, 3] = \{x: -2 \leq x \leq 3\}$ on the horizontal axis. Thus the domain of f contains the interval $[-2, 3]$. We cannot be certain that this is the entire domain of f because the portion of the graph shown does not tell us whether f is defined for real numbers outside of

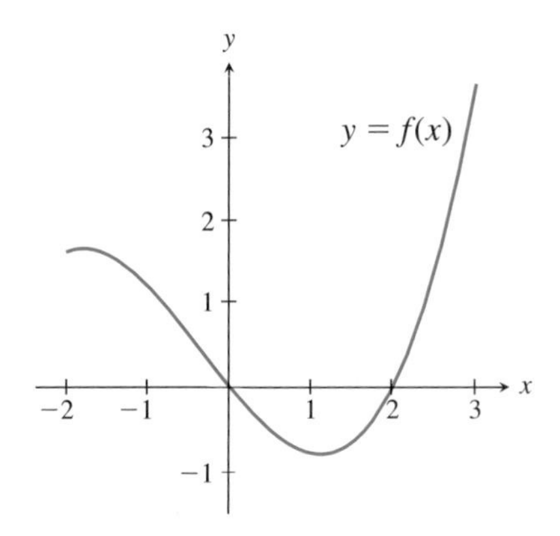

FIGURE 1.6 Graphical representation of f.

$[-2, 3]$; that is, the picture may show the graph for only part of the domain of f. The graph appears to lie to the left and right of (approximately) the interval $[-0.7, 3.8] = \{y: -0.7 \le y \le 3.8\}$; that is, all y for which $-0.7 \le y \le 3.8$ on the vertical axis. Hence, we conclude that the range of f contains this interval.

Interval Notation In the previous example, we used the notation $[-2, 3]$ to denote the set of x values between -2 and 3, with -2 and 3 included. Square brackets [ ] are used to indicate that an endpoint is to be included in the interval described. Parentheses ( ) are used when an endpoint is not to be included. Thus, $(-2, 3]$ is the set of x values between -2 and 3, not including -2 but including 3.We will also use this notation to describe intervals of infinite length. For example,

$$(-\infty, 6) \text{ describes the set of real numbers } x \text{ such that } x < 6,$$

and

$$\left[\sqrt{2}, \infty\right) \text{ describes the set of real numbers } x \text{ such that } x \ge \sqrt{2}.$$

Combining Functions Usually we work with functions that are defined by equations, as in Examples 1 and 2. Although such functions may be complicated, they are usually constructed from simple pieces with which we are already familiar. These pieces are often referred to as the **elementary functions.** They include

- The constant functions, given by $f(x) = c$, where c is a real number;
- the absolute value function, $|x|$;
- the power functions: $x, x^2, x^3, \ldots$;
- the reciprocal function defined by $f(x) = 1/x$;
- the root functions: $\sqrt{x}, \sqrt[3]{x}, \sqrt[4]{x}, \ldots$;
- the trigonometric functions: $\sin x, \cos x, \tan x, \sec x, \ldots$;
- the inverse trigonometric functions: $\arcsin x, \arctan x, \ldots$;
- the logarithmic functions: $\ln x, \log_{10} x, \log_2 x, \ldots$;
- the exponential functions: $e^x, 10^x, 2^x, \ldots$.

Though you are probably familiar with these functions, we will review many of them in later sections. In the remainder of this section, we review a few methods for combining functions to build new functions.

FIGURE 1.7 The graphs of $y = f(x) = \sqrt{x - 1}$ and $y = g(x) = \sqrt{x + 2}$.

EXAMPLE 5 Discuss the domains of the sum and product of the functions f and g defined by $f(x) = \sqrt{x - 1}$ and $g(x) = \sqrt{x + 2}$. The graphs of these functions are shown in Fig. 1.7.

Solution

The sum of f and g is denoted $f + g$ and is defined by

$$(f + g)(x) = f(x) + g(x) \tag{2}$$

for all x in the domain of $f + g$. The product function is denoted $f \cdot g$ and is defined by

$$(f \cdot g)(x) = f(x) \cdot g(x) \tag{3}$$

for all x in the domain of $f \cdot g$.

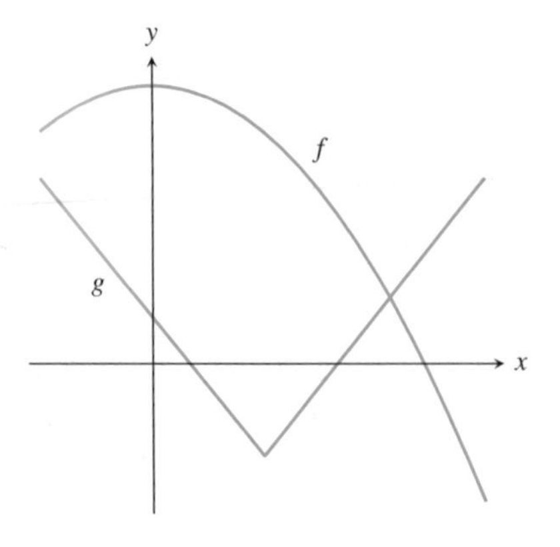

FIGURE 1.8 The graphs of f and g.

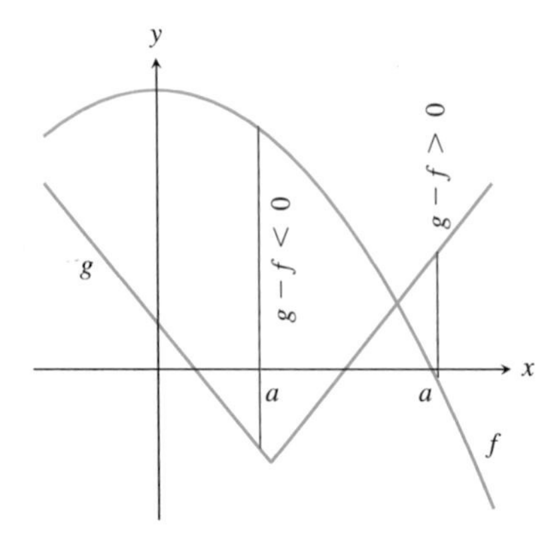

FIGURE 1.9 On a straightedge, mark the vertical distance between $f(a)$ and $g(a)$.

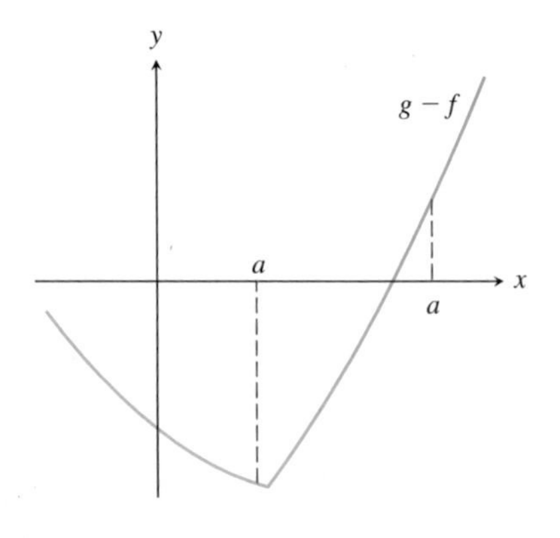

FIGURE 1.10 The graph of $g - f$.

The function f has domain $\mathcal{D}_f = [1, \infty)$ because this is the set of x values for which $x - 1$ is nonnegative. The function g has domain $\mathcal{D}_g = [-2, \infty)$. See Fig. 1.7. The domains of $f + g$ and $f \cdot g$ are the set of x values for which both (2) and (3) are defined. Hence, x must be in the domain of both f and g. Therefore, the domain of $f + g$ and the domain of $f \cdot g$ are the same and are equal to the intersection of the domains of f and g,

$$\mathcal{D} = \mathcal{D}_f \cap \mathcal{D}_g = [1, \infty). \tag{4}$$

An equation representing $f \cdot g$ can be written as

$$(f \cdot g)(x) = f(x)g(x) = \sqrt{x - 1}\,\sqrt{x + 2}.$$

We might be tempted to rewrite this last expression as

$$\sqrt{x - 1}\,\sqrt{x + 2} = \sqrt{(x - 1)(x + 2)} = \sqrt{x^2 + x - 2}$$

from which we might conclude that

$$(f \cdot g)(x) = \sqrt{x^2 + x - 2}. \tag{5}$$

But representing $f \cdot g$ by (5) is deceiving. Note that $x^2 + x - 2$ is nonnegative for all x in $(-\infty, -2]$ as well as for x in $[1, \infty)$. Thus $\sqrt{x^2 + x - 2}$ is defined on $(-\infty, -2]$. However, this interval is *not* part of the domain of $f \cdot g$, as seen from (4). Thus the function described by (5) is different from $f \cdot g$ because it has a different domain. This shows that we have to be careful when combining functions.

EXAMPLE 6 Figure 1.8 shows the graphs of two functions, f and g. Sketch the graph of the difference function, $g - f$.

Solution

The domains of f and g both contain the portion of the horizontal axis shown in the figure. Hence the domain of $g - f$ also contains this portion of the horizontal axis. We graph $g - f$ over this same interval.

The graph of the difference (or sum) of two functions can be found from the graphs of the two functions with the help of a straightedge. At a point a in the domain of $g - f$, simply mark on the straightedge the vertical distance d between $f(a)$ and $g(a)$. See Fig. 1.9. On the axes on which the difference is to be graphed, find a on the horizontal axis. If $g(a) \geq f(a)$ (i.e., g is above f at a), then

$$(g - f)(a) = g(a) - f(a) \geq 0,$$

and we mark a point at distance d above a. If $g(a) < f(a)$, then

$$(g - f)(a) = g(a) - f(a) < 0,$$

so we mark a point at distance d below a. Do this for as many points as needed to sketch a graph for $g - f$. Then sketch the appropriate curve through the plotted points. This is illustrated in Fig. 1.10.

Exercises 1.1

Exercises 1–6: Evaluate $f(a)$ for the given f and a.

1. $f(x) = -3x^2 + 2x + 7, \quad a = -2$

2. $f(s) = 2s^2 + 4s + 8, \quad a = 0.45$

3. $f(t) = 3t^2 - t + 1, \quad a = 1 + \sqrt{3}$

4. $f(x) = x^2 + x - 3, \quad a = t + h$

5. $f(t) = (2t + 1)^3, \quad a = x + h$

6. $f(z) = \dfrac{2z + 1}{5z - 3}, \quad a = \dfrac{x}{y}$

Exercises 7–12: State the domain of the function defined by the given equation.

7. $f(x) = \left(|x - 3| + 1\right)^2$

8. $g(t) = \dfrac{t + 3}{t + 4}$

9. $y = \sqrt{x(x + 1)}$

10. $h(r) = \dfrac{1}{\sqrt{(r - a)(r - b)}}, \quad$ where $b > a$

11. $H(r) = \dfrac{1}{\sqrt{r - a}} \dfrac{1}{\sqrt{r - b}}, \quad$ where $b > a$

12. $r = \dfrac{1}{\sin t}$

Exercises 13–17: State the domain and range of each function. A graph may be helpful in determining the range.

13. $f(x) = 4$

14. $y = -3t + 2$

15. $h(t) = \dfrac{1}{t^2 + 1}$

16. $g(x) = \dfrac{x^2 - 3x + 2}{x - 2}$

17. $y = \dfrac{|x|}{x}$

18. Sketch a graph for a function with domain $[-1, 3] = \{x: -1 \le x \le 3\}$ and range $[0, 2]$.

19. Write an equation for a function f whose natural domain is $[-1, 3]$.

20. Write an equation for a function whose natural domain is $(-\infty, -1) \cup (1, \infty)$. That is, the domain should be all real numbers *except* those in $[-1, 1]$.

Exercises 21–22: The graphs of two functions, f and g, are shown. Sketch the graphs of $f + g$ and $f - g$.

21.

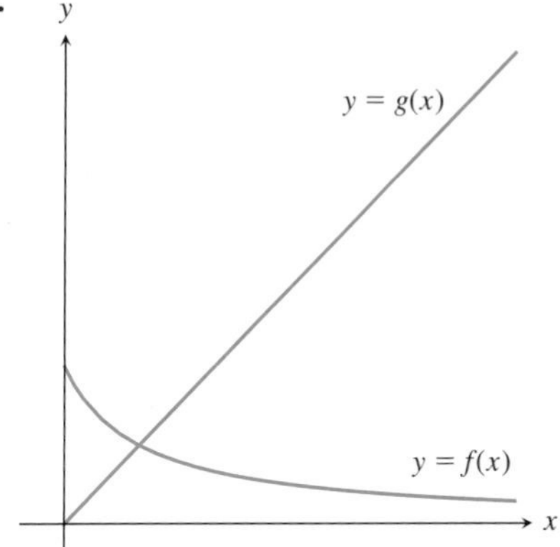

22.

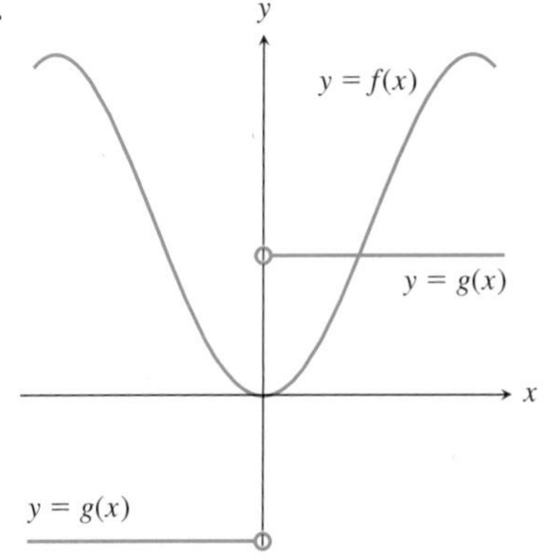

23. Let $f(x) = x^2$. Simplify the expression

$$\frac{f(x + h) - f(x)}{h}.$$

24. Let $g(t) = (t + 1)^2$. Simplify the expression

$$\frac{g(t) - g(x)}{t - x}.$$

25. A cylindrical pail contains water to a depth of 2 inches. At 1:00 P.M. it starts raining, and the water in the pail rises at a steady rate of $1/2$ inch per hour for three hours. At 4:00 it stops raining. Write a formula that gives the depth of water in the pail t hours after the rain starts. Graph the water depth as a function of time.

26. A parachutist is 5000 feet above the ground and falling at a steady rate of 12 feet per second. When will she reach the ground? Find a formula for her height above ground t seconds after the time at which she is 5000 feet above the ground.

27. A bicyclist rides a distance of 54 kilometers in four hours. What is the bicyclist's average speed during the ride?

28. A bicyclist rides from 10:00 A.M. until 3:00 P.M. His average speed for the ride is 12 km per hour, but during the first two hours he rides at a steady speed of 16 km per hour. What was his average speed during the last three hours of his ride?

29. The *vertical line test* is a method for deciding if a graph represents a function. Let G be a graph in the (x, y)-plane. If every vertical line intersects G at most once, then the graph represents y as a function of x. Explain why this test works.

30. Every year the owners of the Negative Outlook photography studio seek to enlarge their profits through newspaper ads. Records of dollars spent on advertising and corresponding profits for the last six years are summarized in the accompanying table. What function(s) are represented by this table? Does it appear that profits can be substantially increased by spending more on advertising?

Year	Ads($)	Profits($)
1	1000	163.93
2	2000	2105.26
3	3000	5744.68
4	4000	8101.26
5	5000	9124.08
6	6000	9557.52

Table for Exercise 30.

31. In some modern supercomputers, addition and multiplication "cost" the same amount, but exponentiations (powers) cost twice as much. (Assume that cost is directly proportional to computation time.) Hence, for large computations it can be profitable to replace exponentiations by multiplications. For example, computing

$$ax^3 + bx^2 + cx + d \tag{6}$$

in the normal way involves two exponentiations, three multiplications, and three additions. If we rewrite the polynomial as

$$x(x(ax + b) + c) + d, \tag{7}$$

we get by with three multiplications and three additions. Suppose that additions and multiplications cost one unit and exponentiations cost two units. Then the computation suggested by (6) costs

$$2 \cdot 2 + 3 + 3 = 10 \text{ units},$$

while the computation suggested by (7) costs

$$3 + 3 = 6 \text{ units}.$$

Make a table showing the costs of each computation method for polynomials of degree 1, 2, 3, 4, and 5. As the degree of the polynomial grows, what appears to happen to the ratio of the costs of the two methods?

32. At public swimming pools, the concentration of chlorine is checked regularly. The accompanying table shows the chlorine levels (in parts per million) in a pool for a 6-day period during which no chlorine was added. What are the domain and range of the function described by the table? If no chlorine is added, what do you predict the level will be in 10 days? Give reasons for your answer.

Day	Chlorine (ppm)
1	3
2	2.55
3	2.16
4	1.84
5	1.56
6	1.33

Table for Exercise 32.

33. The accompanying table gives the standings for Big Ten men's basketball teams in mid-January 1998. Describe two different functions suggested by the table. For each of your functions, give the domain, the range, and a *careful* description of the rule that determines the output for a given input from the domain.

	Conference		All	
Team	W	L	W	L
Illinois	3	0	11	5
Michigan	3	1	13	4
Iowa	2	1	13	2
Purdue	2	1	14	3
Michigan State	2	1	9	4
Indiana	2	2	11	5
Penn State	1	2	8	5
Northwestern	1	2	7	5
Wisconsin	1	2	8	7
Ohio State	0	2	7	8
Minnesota	0	3	7	7

Table for Exercise 33.

34. The accompanying bar graph gives information about soybean prices (in dollars per bushel) for the period from mid-June to mid-August 1999. Describe the function defined by the chart. Give the domain and range of the function and *carefully* describe the rule that determines the output for a given input from the domain.

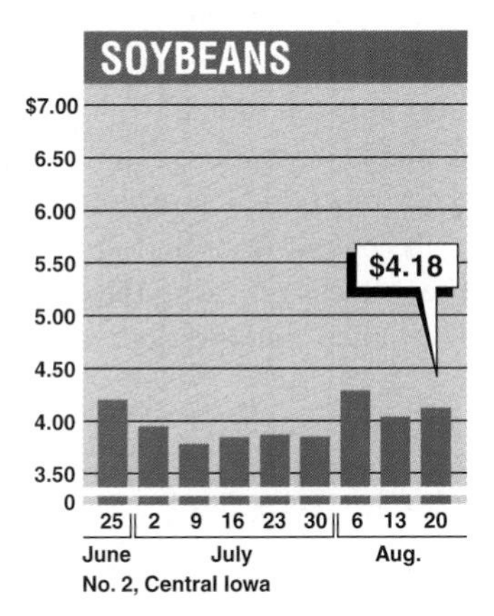

Graph for Exercise 34.

35. The accompanying weather map appeared in the *Des Moines Register* on June 15, 2000. Describe a function suggested by the map. Give the domain and range of the function and *carefully* describe the rule that determines the output for a given input from the domain.

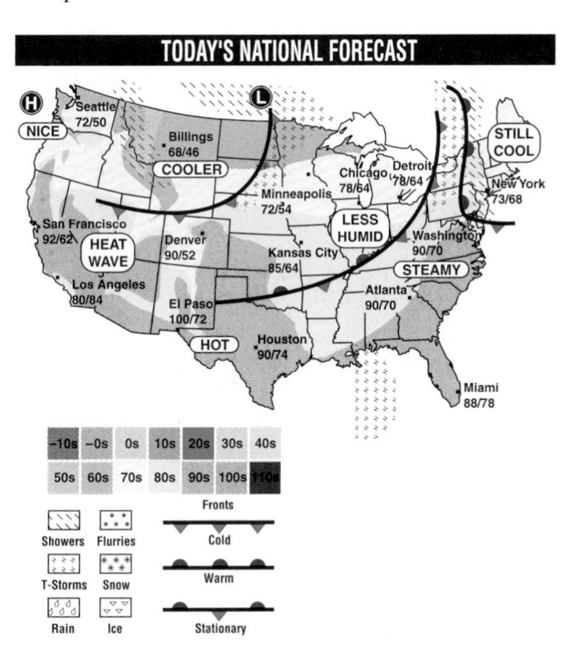

Map for Exercise 35.

36. The accompanying table can be used to find the driving distances between pairs of some major cities. Describe the function defined by the table. Give the domain and range of the function and *carefully* describe the rule that determines the output for a given input from the domain.

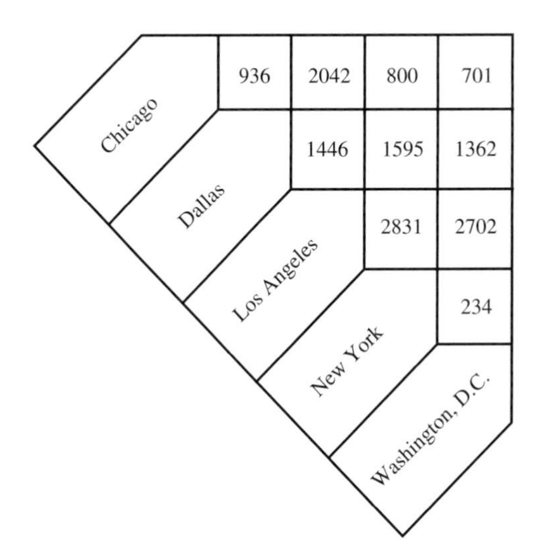

	Dallas	Los Angeles	New York	Washington, D.C.
Chicago	936	2042	800	701
Dallas		1446	1595	1362
Los Angeles			2831	2702
New York				234

Table for Exercise 36.

Exercises 37–38: An equation for a function f is given. The accompanying graph of $y = f(x)$ for $-5 \le x \le 5$ was produced by a CAS. Decide if the graph shown is the actual graph of the function f given. If not, state what is wrong with the graph. Try graphing f on your own calculator or CAS. Is the graph produced correct or not?

37.

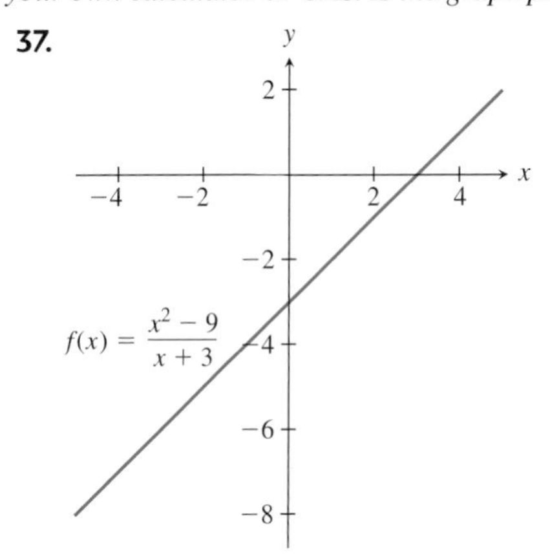

38. $f(x) = (2 - \sqrt{x + 1})^2$

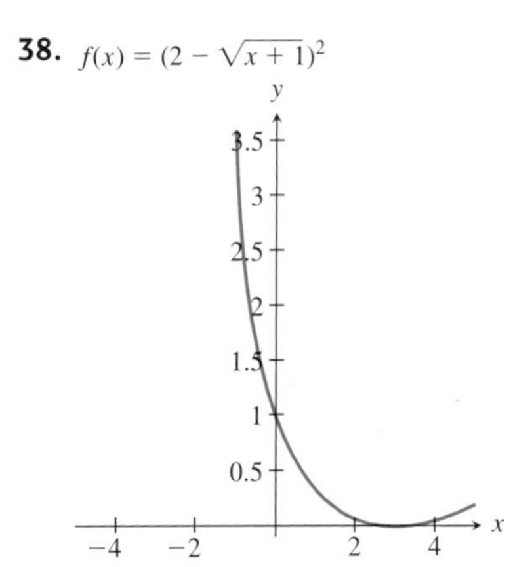

1.2 Compositions of Functions

Use your calculator (set in radian mode) to find the approximate value of $\sqrt{\sin 0.67}$. You may have done this in two steps. For the first step you calculated $\sin 0.67 \approx 0.620986$, and for the second step you found $\sqrt{0.620986} \approx 0.788027$. In performing these two steps, you evaluated two functions. The first was the sine function and the second was the square root function. The sine function was evaluated for the input value 0.67 radians, then the output from the sine function was used as input for the square root function. The process of taking the output from one function as the input for another function is called *function composition.*

DEFINITION Composition of Functions

Let f and g be functions. The composition $f \circ g$ of f with g is defined by

$$(f \circ g)(x) = f(g(x)), \tag{1}$$

where x is in the domain of g and $g(x)$ is in the domain of f. The domain of $f \circ g$ is

$$\mathcal{D}_{f \circ g} = \text{the set of } x \text{ such that } x \text{ is in } \mathcal{D}_g \text{ and } g(x) \text{ is in } \mathcal{D}_f.$$

EXAMPLE 1 Find $f \circ g$ and $g \circ f$ for the functions defined by the equations

$$f(x) = (x-1)^2 \quad \text{and} \quad g(x) = \sqrt{x-4}.$$

Give the domain of each composition.

Solution

The domain of f is $(-\infty, \infty)$, the set of all real numbers. The domain of g is $[4, \infty)$.

First consider $f \circ g$. This function is given by

$$(f \circ g)(x) = f(g(x)) = f\left(\sqrt{x-4}\right) = \left(\sqrt{x-4} - 1\right)^2,$$

and is defined when $\sqrt{x-4}$ makes sense. Hence $f \circ g$ has domain $[4, \infty)$.

The function $g \circ f$ is given by

$$(g \circ f)(x) = g(f(x)) = g((x-1)^2) = \sqrt{(x-1)^2 - 4}.$$

This function is defined only when $f(x) = (x-1)^2$ is in the domain of g. Thus we must have

$$(x-1)^2 \geq 4, \quad \text{that is} \quad \begin{cases} x - 1 \leq -2 \\ \text{or} \\ x - 1 \geq 2. \end{cases}$$

Solving these inequalities, we find $x \leq -1$ or $x \geq 3$. Hence the domain of $g \circ f$ is $(-\infty, -1] \cup [3, \infty)$.

EXAMPLE 2 Functions r and s are defined by Tables 1.2 and 1.3. Construct a table for $r \circ s$.

TABLE 1.2

t	$r(t)$
-2	-1
-1	3
0	4
1	2
2	-1

TABLE 1.3

x	$s(x)$
-3	0
-2	-2
0	0
1	5
2	2
4	3

Solution

The only possible inputs for $(r \circ s)(x) = r(s(x))$ are the allowable inputs for s, that is $x = -3, -2, 0, 1, 2, 4$. We try each of these values as input in $r(s(x))$. First try $x = -3$. Using the tables for s and r,

$$r(s(-3)) = r(0) = 4.$$

We record the fact that $(r \circ s)(-3) = 4$ in the table that represents $r \circ s$ (Table 1.4.) For $x = -2$,

$$r(s(-2)) = r(-2) = -1,$$

so $(r \circ s)(-2) = -1$. Similar calculations show that $(r \circ s)(0) = 4$. When we try $x = 1$, we find

$$r(s(1)) = r(5),$$

but $r(5)$ is undefined. Thus $x = 1$ is not in the domain of $r \circ s$. Continuing through the domain of s we find that $(r \circ s)(2) = -1$ and that $(r \circ s)(4)$ is undefined. Thus $(r \circ s)(x)$ is defined only for $x = -3, -2, 0, 2$. The value of $(r \circ s)(x)$ for each of these inputs is given in Table 1.4.

TABLE 1.4

x	$(r \circ s)(x)$
-3	4
-2	-1
0	4
2	-1

EXAMPLE 3 Let f be given by $f(x) = x^2$ and g by $g(x) = \sqrt{x-2}$. Sketch the graph of $f \circ g$.

Solution

The function g is defined for $x \geq 2$. Because f is defined for all real numbers, the domain of $f \circ g$ is $[2, \infty)$. For x in this domain,

$$(f \circ g)(x) = \left(\sqrt{x-2}\right)^2 = x - 2.$$

The graph of $y = (f \circ g)(x)$, shown in Fig. 1.11, is the part of the line of slope 1 and y-intercept -2 for which $x \geq 2$.

Be careful if you do this composition with a CAS or graphing calculator. Sometimes these devices simplify beyond what is proper and as a result may

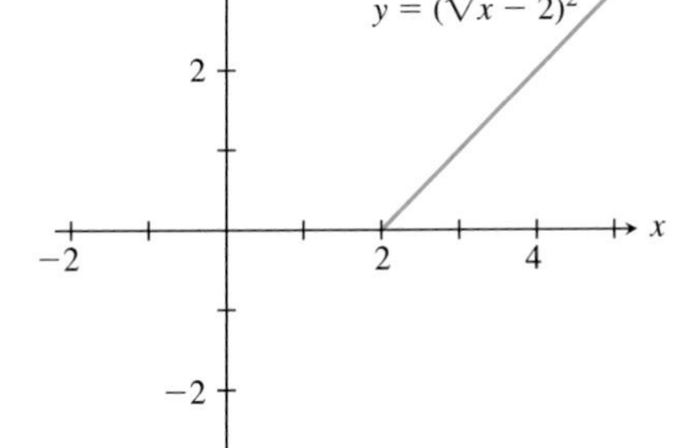

FIGURE 1.11 The function $f \circ g$ has domain $x \geq 2$.

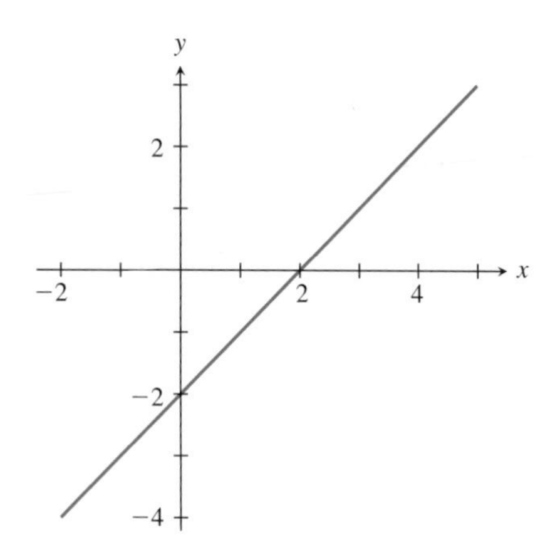

FIGURE 1.12 Sometimes a CAS may not produce the correct graph of $f \circ g$.

not give a true picture of the graph. The graph in Fig. 1.12 was produced when a CAS was used to graph $y = (f \circ g)(x)$ for f and g as defined above. Anyone looking at such a graph without thinking might be fooled into saying the domain for $f \circ g$ is all real numbers. There are two important lessons to be learned from this example:

1. Keep on thinking when using a calculator or computer.
2. Get to know the limitations and quirks of your calculator or CAS.

As you study calculus, you will encounter functions that can be defined as the composition of two or more "simple" functions. It may help your understanding to display a function with a complicated formula as a composition of simple functions.

EXAMPLE 4 The function f is given by

$$f(x) = 10^{\cos(x^2+1)}.$$

Describe f as a composition of "simpler" functions.

Solution

Imagine the process you might use to evaluate f for a given x. As a first step you might find the value of $x^2 + 1$. Next you would use the value of $x^2 + 1$ as input for the cosine function. Finally, you would use the output from the cosine function as the exponent for the power of 10. The three steps of this process describe three functions, g, h, and k, that can be composed to obtain f. These functions are described by

$$g(x) = x^2 + 1, \quad h(x) = \cos x, \quad \text{and} \quad k(x) = 10^x.$$

We then have

$$f(x) = (k \circ h \circ g)(x) = k(h(g(x))). \tag{2}$$

Let's check to see that the right side of (2) does indeed give $f(x)$. For a given input x, the innermost function, g, in (2) gives $g(x) = x^2 + 1$. The value of $g(x)$ is used as input for h to get $h(g(x)) = \cos(x^2 + 1)$. This result is then used as input for k to get

$$k(h(g(x))) = k(\cos(x^2 + 1)) = 10^{\cos(x^2+1)} = f(x).$$

Sometimes it is useful to display a composition in another format. Let $y = f(x) = 10^{\cos(x^2+1)}$ so y is a function of x. We can then write

$$\begin{aligned} y &= 10^u \\ u &= \cos v \\ v &= x^2 + 1. \end{aligned} \tag{3}$$

To find the value of y for a given x, work from the bottom up. An input x is used in the bottom line of (3) to get $v = x^2 + 1$. This value of v is used in the second line of (3) to get $u = \cos v$, then the value of u is used in the first line to get y. You should convince yourself that both (2) and (3) describe f as a composition of three simpler functions.

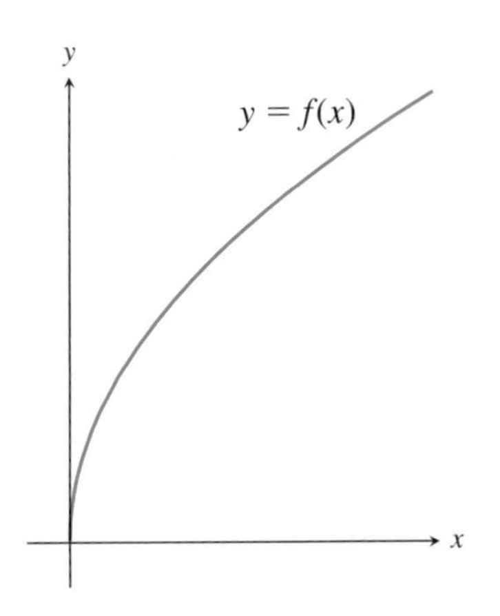

FIGURE 1.13 The graph of $y = f(x) = \sqrt{x}$.

Compositions and Graphs The graph of $y = f(x) = \sqrt{x}$ is shown in Fig. 1.13. Define a new function g by

$$g(x) = f(x - 3) = \sqrt{x - 3}. \tag{4}$$

What does the graph of $y = g(x)$ look like? We could produce the graph of g simply by plotting several points $(x, g(x))$ and then connecting the points with a "reasonable curve." This is how a calculator or CAS would produce the graph. On the other hand, the equations describing f and g are very closely related. The function g described in (4) is the composition of f with the linear function ℓ where $\ell(x) = x - 3$. It is not unusual to encounter two functions, one of which is obtained by composing the other with a linear function. When two functions are related in this way, the graph of one function can be obtained by translating, magnifying, shrinking, and/or reflecting the graph of the other.

Suppose we wish to plot the point $(a, g(a))$ on the graph of $y = g(x)$. Because $g(a) = f(a - 3)$, we see that the points $(a - 3, f(a - 3))$ and $(a, g(a))$ have the same y-coordinate, so they are at the same distance above or below the x-axis. See Fig. 1.14. In addition, the point $(a, g(a))$ is three units to the right of $(a - 3, f(a - 3))$. This means that the graph of $y = g(x)$ can be obtained by moving the graph of $y = f(x)$ three units to the right.

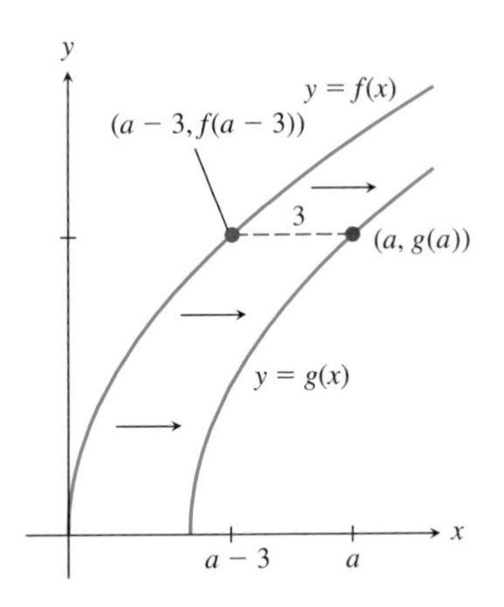

FIGURE 1.14 The points $(a - 3, f(a - 3))$ and $(a, g(a))$ have the same y-coordinates.

EXAMPLE 5 The graph of a function f is shown in Fig. 1.15.

a) Define g by $g(x) = 2f(x)$. Sketch the graph of $y = g(x)$.
b) Define h by $h(x) = f(2x)$. Sketch the graph of $y = h(x)$.

Solution

a) The domain of g is the same as the domain of f, so the graph of $y = g(x)$ lies above or below the same portion of the x-axis as the graph of $y = f(x)$. If a is in the domain of g, then $(a, g(a))$ is a point on the graph of $y = g(x)$ and $(a, f(a))$ is a point on the graph of $y = f(x)$. Because $g(a) = 2f(a)$, the y-coordinate of $(a, g(a))$ is twice the y-coordinate of the point $(a, f(a))$. It follows that the graph of $y = g(x)$ can be obtained by doubling the distance above or below the x-axis of each point on the graph of $y = f(x)$. See Fig. 1.16.

b) If a is in the domain of f, then $(a, f(a))$ is on the graph of $y = f(x)$. We also have

$$f(a) = f\left(2\frac{a}{2}\right) = h\left(\frac{a}{2}\right).$$

This means that the point $(a/2, h(a/2))$ is on the graph of $y = h(x)$ and has the same y-coordinate as the point $(a, f(a))$. Thus $(a/2, h(a/2))$ is the midpoint of the horizontal segment from $(a, f(a))$ to the y-axis. See Fig. 1.17. The graph of $y = h(x)$ can be obtained by halving the horizontal displacement from the y-axis of each point on the graph of $y = f(x)$.

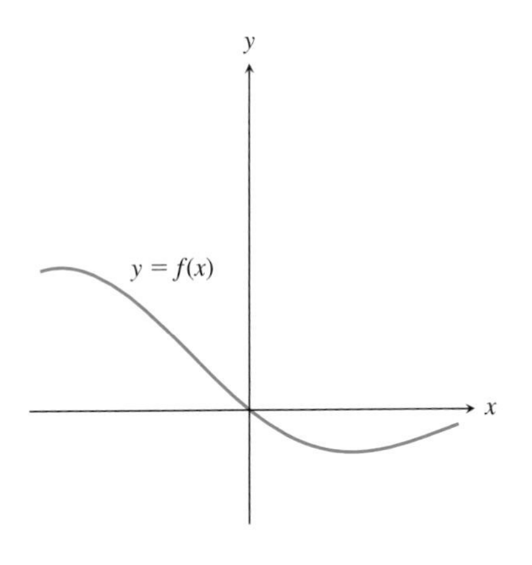

FIGURE 1.15 The graph of $y = f(x)$.

FIGURE 1.16 The graph of $y = 2f(x)$ is found by doubling the y-coordinate of each point on the graph of $y = f(x)$.

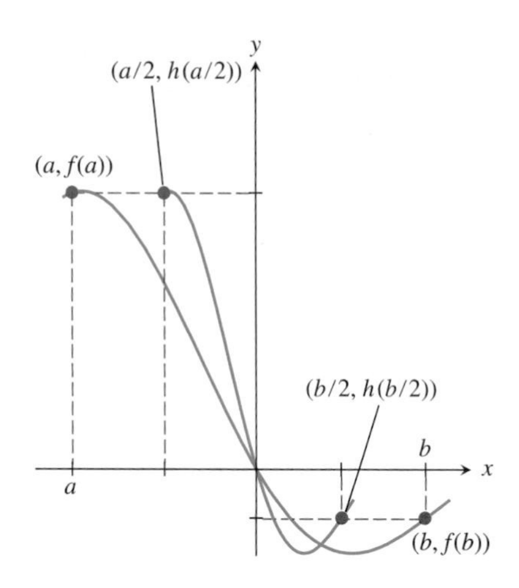

FIGURE 1.17 The graph of $y = f(2x)$ is found by halving the x-coordinate of each point on the graph of $y = f(x)$.

Graphical Composition When the graphs of two functions, f and g, are given, it is possible to use the graphs to find points on the graph of the composition $f \circ g$. We illustrate a method for doing this in the next example. In recent years this technique of "graphical composition" has been widely studied because it provides a means to illustrate chaos. See the project at the end of this chapter if you would like to explore some of the connections between compositions and chaos.

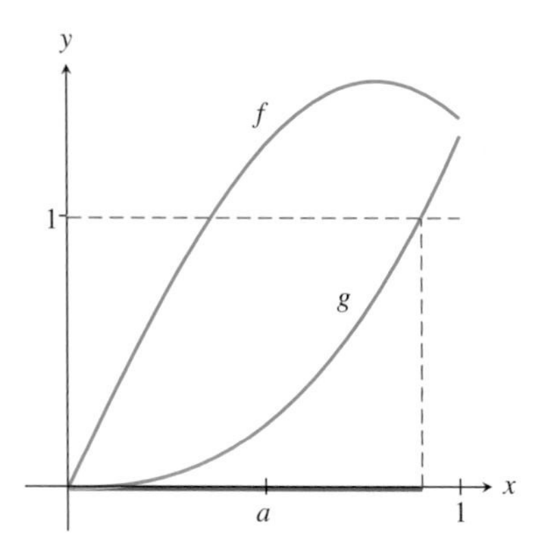

FIGURE 1.18 The domain of $f \circ g$ includes those x for which $0 \leq g(x) \leq 1$.

EXAMPLE 6 Figure 1.18 shows the graphs of two functions f and g, both of which have domain $[0, 1]$. A real number a is marked on the x-axis. Plot the point with coordinates $(a, (f \circ g)(a))$.

Solution

We first determine the domain of $f \circ g$. Because the domain of f is $[0, 1]$, the expression $f(g(x))$ will make sense if and only if $0 \leq g(x) \leq 1$. Thus any x values for which $g(x) > 1$ or $g(x) < 0$ are not in the domain of $f \circ g$. In Fig. 1.18 we see that $0 < g(a) < 1$, so a is in the domain of $f \circ g$. We now use the following four steps to find the point $(a, (f \circ g)(a))$ on the graph of the composition.

Step 1: From a on the horizontal axis draw a dashed line vertically to the graph of g. This line hits the graph at the point $(a, g(a))$.

Step 2: Think of the y-coordinate of the point $(a, g(a))$ as the "output" from the function g. To find $f(g(a))$, this output must become "input" for the function f. This is done by drawing a horizontal line from $(a, g(a))$ to the point $(g(a), g(a))$ on the line $y = x$.

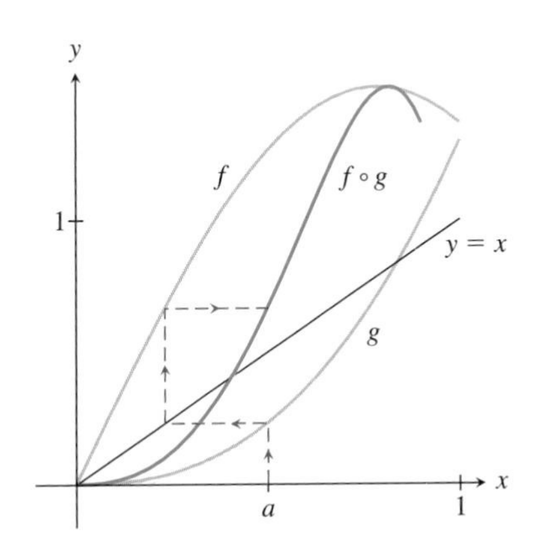

FIGURE 1.19 The domain of $f \circ g$ includes those x for which $0 \leq g(x) \leq 1$.

Step 3: To treat $g(a)$ as input for f, move vertically from this point to meet the graph of $y = f(x)$ at the point $(g(a), f(g(a)))$.

Step 4: From this point, extend the line horizontally until it hits the vertical line $x = a$. This final point has coordinates $(a, f(g(a)))$ and so is a point on the graph of $f \circ g$.

The four segments illustrating these four steps are shown in Fig. 1.19. The graph of $y = (f \circ g)(x)$ is also shown in the figure. Try the process just outlined for different starting values of a in Fig. 1.19. If you are careful, you should end up at other points on the graph of $y = (f \circ g)(x)$.

Exercises 1.2

Exercises 1–6: Find an equation for the composition and state the domain of the composition.

1. $(f \circ g)(x)$ with $f(x) = \dfrac{1}{x}$ and $g(x) = x^2 - 6x + 8$

2. $(r \circ r)(u)$ with $r(u) = \sqrt{1 - u}$

3. $(f \circ g)(t)$ with $f(t) = \dfrac{|t|}{t}$ and $g(t) = \sin t$

4. $(g \circ h)(z)$ with $g(x) = \dfrac{1}{x^{1/2}}$ and $h(z) = \dfrac{1}{z^{1/3}}$

5. $(h \circ k)(x)$ with $h(x) = \sqrt{x}$ and $k(x) = (x - 1)(x - 2)(x - 3)^2(x - 4)$

6. $(k \circ h)(x)$ with $h(x) = \sqrt{x}$ and $k(x) = (x - 1)(x - 2)(x - 3)^2(x - 4)$

Exercises 7–12: Decompose each function into a composition of simpler functions.

7. $G(x) = 5(x^3 - 3x + 7)^{10}$

8. $f(x) = \dfrac{2}{x^3 - 3x^2 + 1}$

9. $H(x) = \sin\left(\sqrt{x + 5}\right)$

10. $h(x) = f\left(\dfrac{1}{g(x)}\right)$

11. $r(x) = \left(\dfrac{1 - 2\sin x}{1 + 2\sin x}\right)^{1/2}$

12. $f(x) = 10^{10^x}$

Exercises 13–14: An equation for a function f is given. The accompanying graph of $y = f(x)$ for $-5 \leq x \leq 5$ was produced by a CAS. Decide if the graph shown is the actual graph of the f given. If not, state what is wrong with the graph. Try graphing f on your own calculator or CAS. Is the graph produced correct or not?

13. $f(x) = g(h(x))$ where $g(x) = x^2 + 1$ and $h(x) = \dfrac{1}{\sqrt{x - 3}}$.

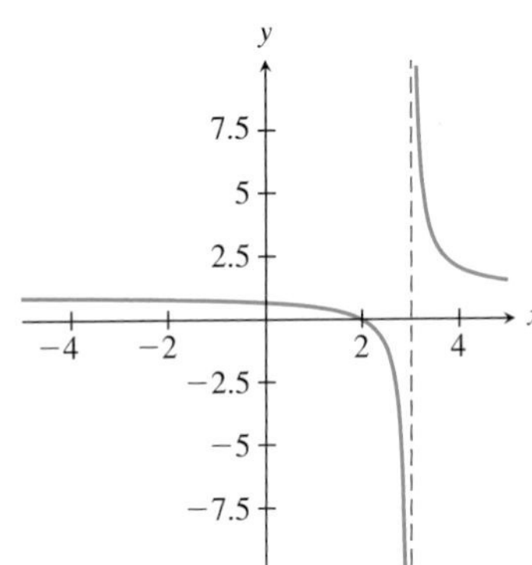

Figure for Exercise 13.

14. $f(x) = g(h(x))$ where $g(x) = \cos 3x$ and $h(x) = \sqrt{x}$.

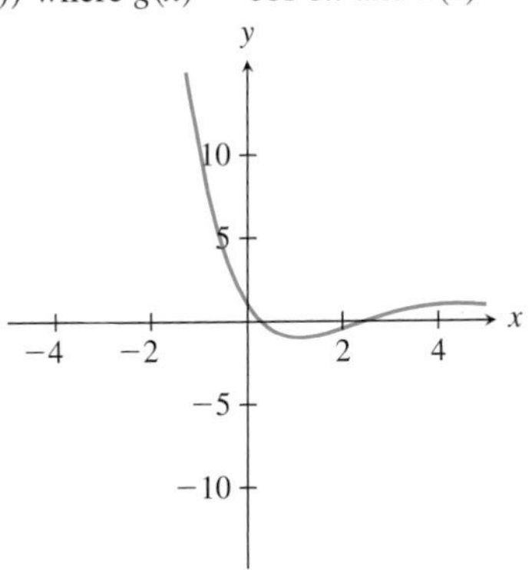

Figure for Exercise 14.

15. Referring to Example 2, construct a table for $s \circ r$.

16. Referring to Example 2, construct a table for $r \circ r$.

17. Referring to Example 2, construct a table for $r \circ (r \circ s)$.

18. Suppose that the domain of f is $[-1, 3]$ and that $g(x) = 2x - 1$. What is the domain of $f \circ g$?

19. Suppose that the domain of f is $[-1, 4]$ and that $g(x) = x^2$. What is the domain of $f \circ g$?

20. Suppose that the domain of h is $[0, \infty)$ and that $k(x) = 1/(x + 1)$. What can be said about the domain of $k \circ h$?

21. Suppose that the domain of F is $[-10, -8]$ and that $G(x) = x^3 + 1$. What can be said about the domain of $F \circ G$?

22. Tables A and B represent two functions.

x	
1	3
2	5
3	8
4	8
5	7
6	2

Table A.

x	$(f \circ g)(x)$
1	2
2	7
4	7
5	5
6	5

Table B.

a. If Table A represents the function f and Table B represents the function $f \circ g$, construct a table to represent g.

b. If Table A represents g and Table B represents the function $f \circ g$, construct a table to represent f.

23. The accompanying table defines a function f.

x	$f(x)$
-2	3
-1	1
0	0
1	-6
2	-5
3	3

Table for Exercise 23.

a. Let $g(x) = 2x - 1$ for all real numbers x. Give the domain and range of $f \circ g$. Also give the domain and range of $g \circ f$.

b. Let $h(x) = \sqrt{x + 1}$ for all real numbers $x \geq -1$. Give the domain and range of $f \circ h$. Also give the domain and range of $h \circ f$.

24. Let $f(x) = x^2 - 2x$.

a. On the same set of axes plot $y = f(x)$ and $y = f(x + 2)$. Carefully explain how one graph can be obtained from the other.

b. On the same set of axes plot $y = f(x)$ and $y = f(x) + 2$. Carefully explain how one graph can be obtained from the other.

25. Let $f(x) = x2^x$.

a. Use a calculator or computer to plot $y = f(x)$ and $y = f(-x)$ on the same set of axes. Carefully explain how one graph can be obtained from the other.

b. Use a calculator or computer to plot $y = f(x)$ and $y = -f(x)$ on the same set of axes. Carefully explain how one graph can be obtained from the other.

26. Let $f(x) = \sqrt{x^2 + 1}$ and let $\ell(x) = 2x + 1$.

a. Use a calculator or computer to plot $y = f(x)$ and $y = f(\ell(x))$ on the same set of axes. Carefully explain how one graph can be obtained from the other.

b. Use a calculator or computer to plot $y = f(x)$ and $y = \ell(f(x))$ on the same set of axes. Carefully explain how one graph can be obtained from the other.

27. Let $f(x) = \sin x$ and let $\ell(x) = -2x + \pi$.

a. Use a calculator or computer to plot $y = f(x)$ and $y = f(\ell(x))$ on the same set of axes. Carefully explain how one graph can be obtained from the other.

b. Use a calculator or computer to plot $y = f(x)$ and $y = \ell(f(x))$ on the same set of axes. Carefully explain how one graph can be obtained from the other.

Exercises 28–33: The graph of a function f is shown. Sketch the graph of $y = g(x)$.

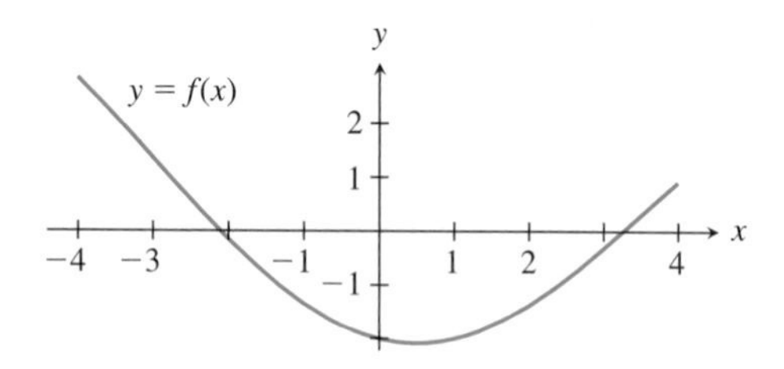

28. $g(x) = f(x + 1)$

29. $g(x) = f\left(\dfrac{x}{2}\right)$

30. $g(x) = -f(-x)$

31. $g(x) = |f(x)|$

32. $g(x) = f(|x|)$

33. $g(x) = \dfrac{|f(x)|}{f(x)}$

Exercises 34–35: Points a and b are shown on the x-axis. Plot the points $(a, (f \circ g)(a))$ and $(b, (f \circ g)(b))$. The scales on the x- and y-axes are the same.

34.

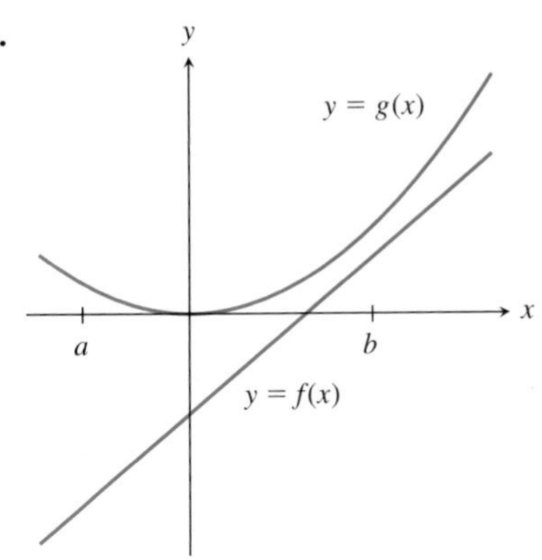

35.

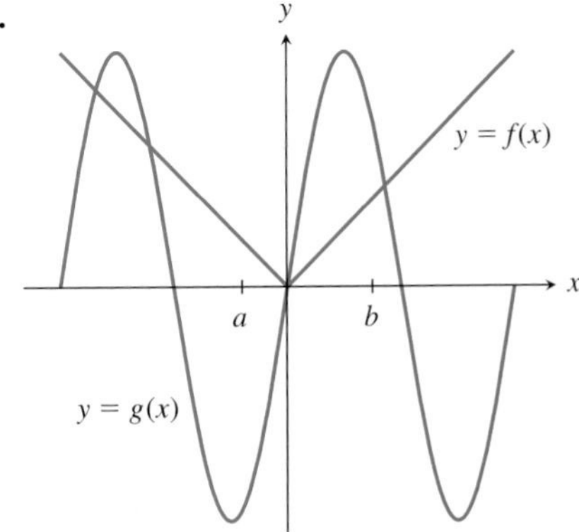

Exercises 36–37: The graph of a function f is shown along with the dotted line $y = x$. Graphically trace the progress of $f(a)$, $f(f(a))$, $f(f(f(a))), \ldots$ for the a shown on the x-axis. Describe what happens as the composition "deepens."

36.

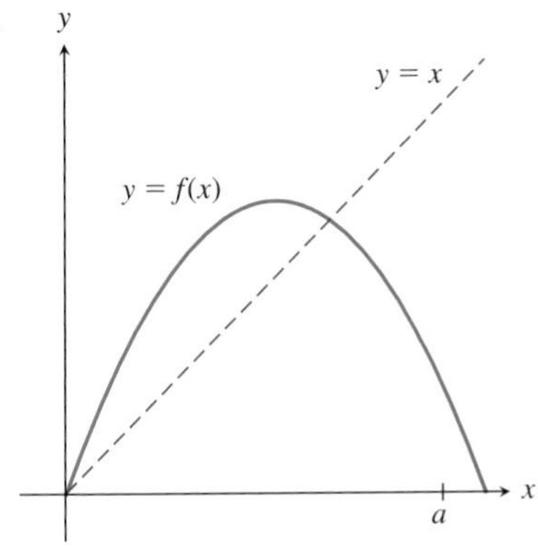

37.

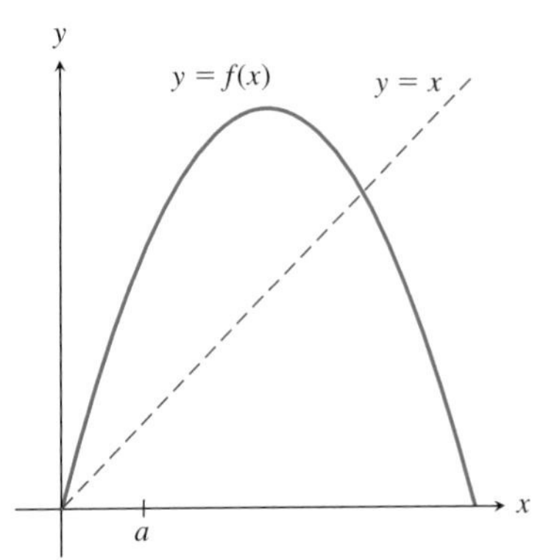

38. Plot $y = \cos x$, $y = \cos(\cos x)$, and $y = \cos(\cos(\cos(\cos x)))$. Describe what appears to be happening to the graphs as you graph "deeper and deeper" compositions of the cosine function. Try even longer cosine compositions to check your answer.

39. Plot $y = \sin x$, $y = \sin(\sin x)$, and $y = \sin(\sin(\sin(\sin x)))$. Describe what appears to be happening to the graphs as you graph "deeper and deeper" compositions of the sine function. Try even longer sine compositions to check your answer.

40. The accompanying figure shows the graph of a function f with domain and range both equal to the set $[0, 1] = \{x : 0 \le x \le 1\}$.

a. Sketch the graph of $y = (f \circ f)(x)$.

b. Sketch the graph of $y = (f \circ f \circ f)(x)$.

c. What do you think the graph of $y = (f \circ f \circ f \circ f \circ f)(x)$ would look like?

d. Verify that

$$f(x) = 1 - 2\left|x - \frac{1}{2}\right|, \quad 0 \le x \le 1.$$

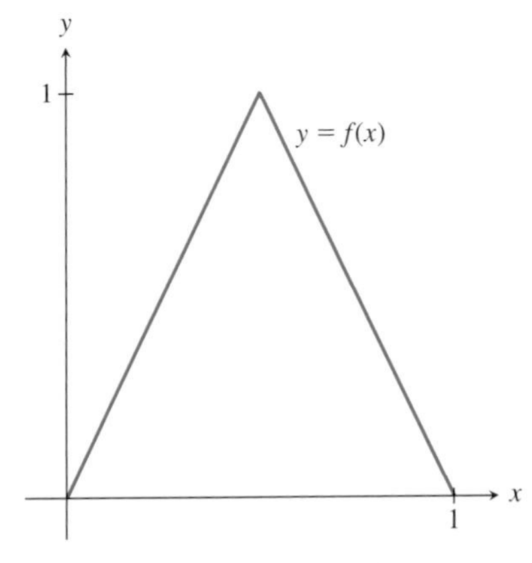

Figure for Exercise 40.

1.3 Slope as a Rate of Change

Equations of lines are the most important equations in science, engineering, and mathematics. Many real-life phenomena are described (or modeled) by lines or equations of lines. The speed of an object under constant acceleration, the work you do in climbing a ladder, and the profit a theater owner makes by showing a movie can all be described by an equation for a line. Even situations that cannot be described exactly by lines can often be accurately approximated by such equations. In fact, approximation by lines is one of the fundamental ideas behind the differential calculus.

In this section we begin our study of change by noting that the slope of a line can be interpreted as the rate of change of one quantity with respect to another. We use this idea to motivate a definition of rate of change for quantities that are related by nonlinear functions.

Lines and Slope

WEB In algebra you learned that a nonvertical line in the Cartesian plane, or (x, y)-plane, can be described by an equation of the form

$$y = mx + b.$$

The constant m is the *slope* of the line and tells us that when x increases by 1, y changes by m. The value of y increases if $m > 0$, decreases if $m < 0$, and is unchanged if $m = 0$. See Fig. 1.20. The number b is called the y-intercept of the line and tells us that the point $(0, b)$ is on the graph of the line.

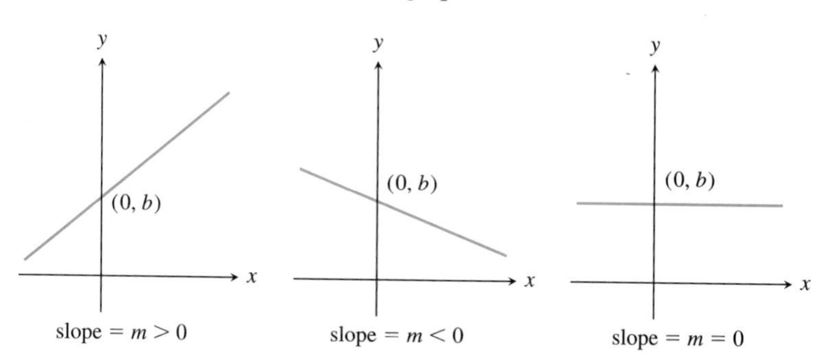

FIGURE 1.20 Slopes of lines.

EXAMPLE 1 Graphing a line

The equation

$$-3x + 4y - 8 = 0 \quad (1)$$

is the equation of a line. Find the slope and y-intercept of the line and sketch its graph.

Solution

Solving (1) for y we have

$$y = \frac{3}{4}x + 2. \quad (2)$$

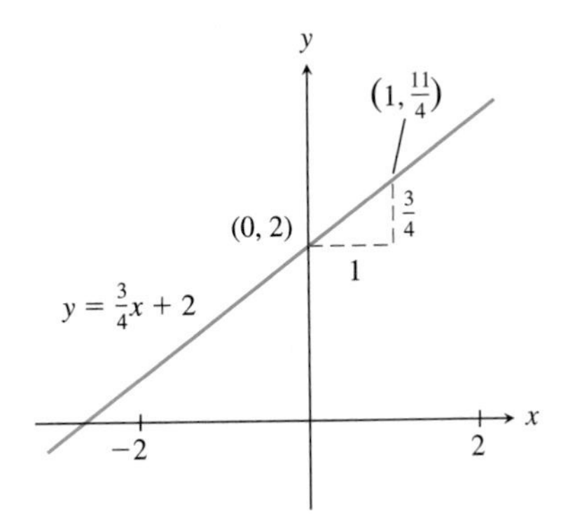

FIGURE 1.21 The line of slope 3/4 and y-intercept 2.

From this form of the equation we see that the line has slope 3/4 and y-intercept 2. To sketch the graph, we first plot the y-intercept, $(0, 2)$. To find a second point on the line, we use the slope. Since the slope is 3/4, we can start at $(0, 2)$, move horizontally to the right 1 unit (i.e., increase x by 1), and then move up 3/4 units (i.e., increase y by 3/4). We are now at the point $(1, 11/4)$, which is another point on the line. To complete the graph, draw the line determined by the two points that we have plotted. See Fig. 1.21.

Most problems in science, mathematics, and engineering can be solved in more than one way. One way of interpreting this statement is: *There is no one right way to solve a problem. Almost every problem has many equally valid solutions.* Get into the habit of thinking about different ways to solve a problem. Even if you have already solved a problem in one way, looking at alternative solutions is a good way to check your solution and almost always adds to your understanding and insight. We can even apply this philosophy to the previous example.

EXAMPLE 2 Another solution

Do Example 1 in another way.

Solution

Because two points determine a line, we use (1) to find the coordinates of two points on the line. One point can be the y-intercept, which is found by letting $x = 0$ in (1):

$$-3 \cdot 0 + 4y - 8 = 0.$$

Solving this equation, we find $y = 2$. Hence $(0, 2)$ is the y-intercept and is on the graph of the line. To find a second point on the line, pick a different x value, say $x = -4$, and substitute it into (1) to get

$$(-3)(-4) + 4y = 8.$$

The solution to this equation is $y = -1$. Hence the point $(-4, -1)$ is also on the line. Plot the points $(0, 2)$ and $(-4, -1)$ and draw the line determined by these two points to get the graph, as shown in Fig. 1.22. Because we know two points on the line, we can determine the slope, m, by

$$m = \frac{\text{change in } y}{\text{change in } x} = \frac{2 - (-1)}{0 - (-4)} = \frac{3}{4}.$$

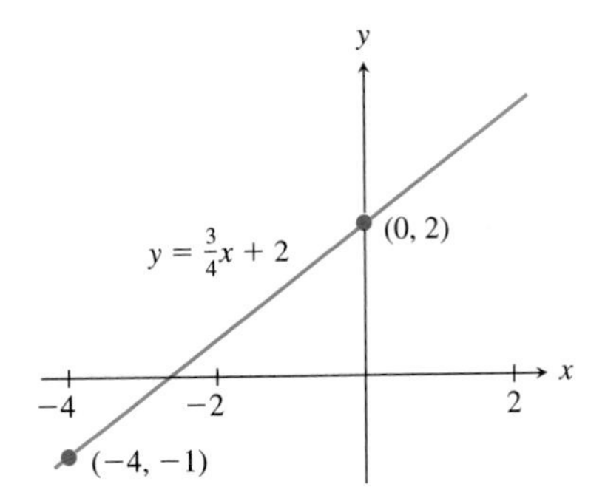

FIGURE 1.22 Two points determine a line.

Try using your graphing calculator or CAS to solve this example a third way. First produce the graph of the line on your calculator or CAS. Some graphics packages may take the equation as given in (1), but for most others you will have to enter the equation in a form similar to (2). Once the graph is drawn, use the cursor to read the coordinates of the y-intercept and the coordinates of some other points on the line. Use this information to compute the slope. How do your answers compare with the results of Examples 1 and 2?

Units for Slope In science and engineering, equations are used to model real-life situations, so the variables often have associated units (e.g., meters, minutes, dollars). If the equation is the equation of a line, these units can help us give a physical interpretation of the slope.

EXAMPLE 3 Predicting the flood stage of a river

During heavy rains, the water level in a local river must be closely monitored. At 1:00 P.M. a county agent measures the depth of a river and finds it to be 12 feet. Three hours later she takes another measurement and finds the depth is 14 feet. She assumes that the river is "rising steadily," so the depth as a function of time can be described by a line. She also knows that the river is at flood stage when its depth is 22 feet. The agent must prepare a report for the 6:00 news. The report is to include the rate at which the river is rising and the time at which the river will reach flood stage. Help the agent prepare her report.

Solution

Let t (hours) be time, with $t = 0$ corresponding to 12:00 noon. Let $d = d(t)$ (feet) be the depth of the river at time t. (We write $d = d(t)$ to remind us that d is a function of t.) Because we are assuming that $d(t)$ can be described by a line, we have

$$d = mt + b.$$

The agent's data give us two points on this line: $(t, d) = (1 \text{ hr}, 12 \text{ ft})$ and $(t, d) = (4 \text{ hr}, 14 \text{ ft})$. Hence the slope of the line is

$$m = \frac{14 \text{ ft} - 12 \text{ ft}}{4 \text{ hr} - 1 \text{ hr}} = \frac{2}{3} \text{ ft/hr}.$$

The slope tells us that the river is rising at a rate of 2/3 ft/hr. This is one of the pieces of information needed by the agent. Because the point $(1, 12)$ is on the line and we know the slope, we may use the point-slope form for the equation of a line to write

$$(d - 12) = \frac{2}{3}(t - 1), \quad \text{from which} \quad d = \frac{2}{3}t + \frac{34}{3}. \tag{3}$$

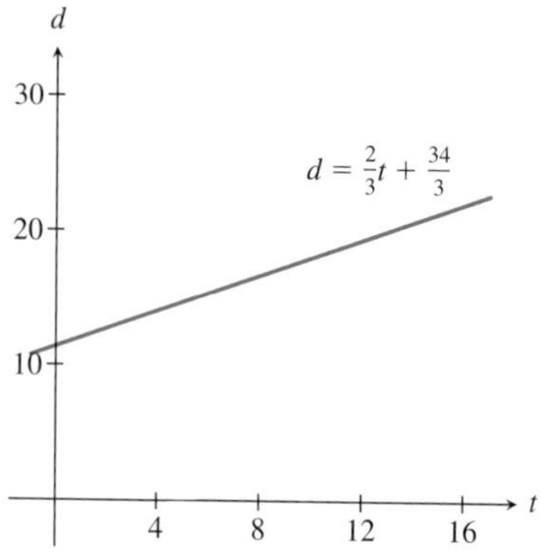

FIGURE 1.23 The river is rising!

To predict the time when the river will reach flood level, we need to know when $d = 22$. We can estimate this time from the graph in Fig. 1.23, or use the cursor and zoom-in feature on a calculator to see what t value corresponds to the point on the graph with a d value of 22. (Try it. What do you get?) This would certainly be accurate enough for broadcast purposes. We can also use the depth equation (3). Setting $d = 22$ in the equation,

$$22 = \frac{2}{3}t + \frac{34}{3}.$$

Solving this equation for t gives $t = 16$. Because the units are hours, and $t = 0$ corresponds to 12:00 noon, $t = 16$ corresponds to 4:00 A.M. the next morning.

Slope as a Rate of Change In Example 3 we found that the slope m of the line $d = mt + b$ could be interpreted as the rate at which the river is rising. In other words, m was the rate at which the depth of the river was changing with respect to time. We can interpret the slope of any nonvertical line in this way.

DEFINITION Rate of Change for a Linear Function

If y is related to x by the equation

$$y = mx + b$$

for a line, then the **rate of change of y with respect to x** is m. If x and y have associated units, then the units of m are

$$\text{units of } m = \frac{\text{units of } y}{\text{units of } x},$$

which is read as "units of y per unit of x."

Thus for lines, slope and rate of change are the same. Knowing the units for the rate of change is important in giving the slope the correct physical interpretation in real-life situations.

EXAMPLE 4 Expansion due to temperature increase

A steel girder is 15 meters long when it is at a temperature of $-20°$C. When in use the girder will be subjected to temperatures between $-20°$C and $40°$C. As the temperature changes, the steel expands or contracts, so the length of the girder changes. When the girder is at temperature T its length L is

$$L = L(T) = 15.0036 + 0.00018T, \quad -20 \leq T \leq 40. \tag{4}$$

Graph this equation and calculate the rate of change of L with respect to T.

Solution

When $T = -20$ is substituted into (4), we obtain $L = 15.0000$. When $T = 40$ is used, we obtain $L = 15.0108$. Thus we can graph the equation of the line by plotting the two points $(-20, 15.0000)$ and $(40, 15.0108)$ and then drawing the line they determine. Because the domain of the function described by (4) is $-20 \leq T \leq 40$, the graph consists only of that part of the line that lies above or below the interval $[-20, 40]$ on the T-axis. See Fig. 1.24. The slope of the line given in (4) is 0.00018 and carries units of

$$\frac{\text{units of } L}{\text{units of } T} = \text{meters/°C}.$$

Thus the rate of change of L with respect to T is 0.00018 meters/°C. This means that for each 1°C rise in temperature, the length of the beam increases by 0.00018 meters.

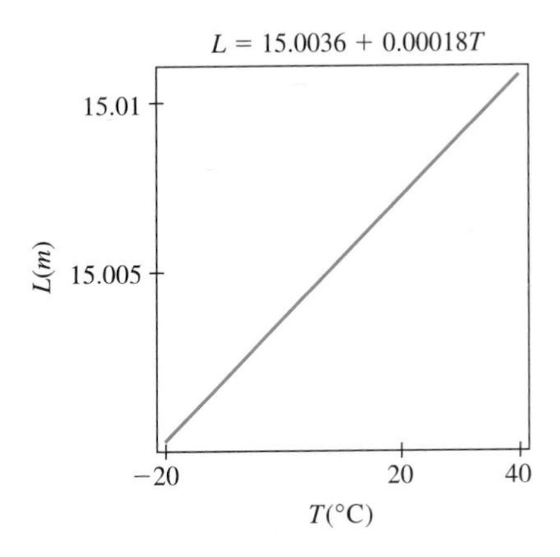

FIGURE 1.24 The graph describes the length of the girder as a function of temperature.

Rate of Change

In the preceding examples we saw that interpreting the slope of a line as a rate of change can provide important information. We use this idea to develop a definition for the rate of change (or slope) for a nonlinear function. The notion of rate of change for nonlinear functions is one of the most important ideas in calculus. We investigate this concept in the following Investigation, using a physical situation to illustrate the practical importance of this idea. We will examine the rate of change further in the next section.

INVESTIGATION

Change of Pressure with Volume in a Cylinder

A cylindrical chamber fitted with a movable piston contains 10^{23} molecules of an ideal gas at 0°C. As the piston moves, the volume V of the chamber varies. According to Boyle's law, the pressure P of the gas is

$$P = P(V) = \frac{4}{V}, \tag{5}$$

where V is measured in liters and P is measured in units of atmospheric pressure (atm). When the piston is moved outward, the volume V increases and the pressure changes. See Fig. 1.25. At what rate is the pressure changing with respect to volume when $V = 3$ liters?

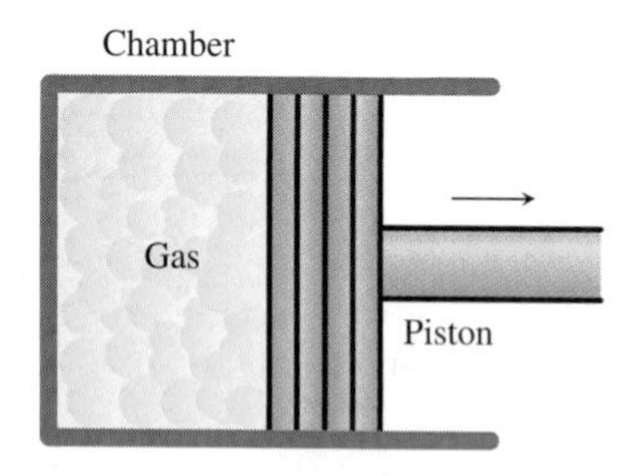

FIGURE 1.25 As the piston moves out, the available volume for the gas increases.

The graph of (5) is shown in Fig. 1.26. The units on the horizontal axis are liters and those on the vertical axis are atmospheres of pressure. This means that the rate of change has units as follows:

$$\frac{P\text{ units}}{V\text{ units}} = \text{atm/liter}.$$

These are appropriate units because we are looking for the rate of change of pressure with respect to volume. If the graph of this function were a line, we could find the rate of change by finding the slope of the line. Even though the graph is not a line, we see that a small piece of the graph looks very much like a line. In fact, if we take a small enough piece, it is almost impossible to distinguish it from a line. With the aid of a graphing calculator or computer, it is easy to "zoom in" on a small piece of the graph.

Because we are interested in the rate of change when

$$V = 3 \quad \text{and} \quad P = P(3) = 4/3,$$

we look at a piece of the graph containing the point $(3, 4/3)$. In Fig. 1.26 there is a square of side 1 centered at $(3, 4/3)$. Enlarge this piece of the graph by using the zoom feature on your calculator or CAS or by replotting (5) for $2.5 \le V \le 3.5$. The result should be something like Fig. 1.27. This portion of the graph looks straighter than the part shown in Fig. 1.26, but it still doesn't look like a line. We can take a closer look by zooming in on the small rectangle shown in Fig. 1.27. When you do this yourself, you should see something like the graph in Fig. 1.28.

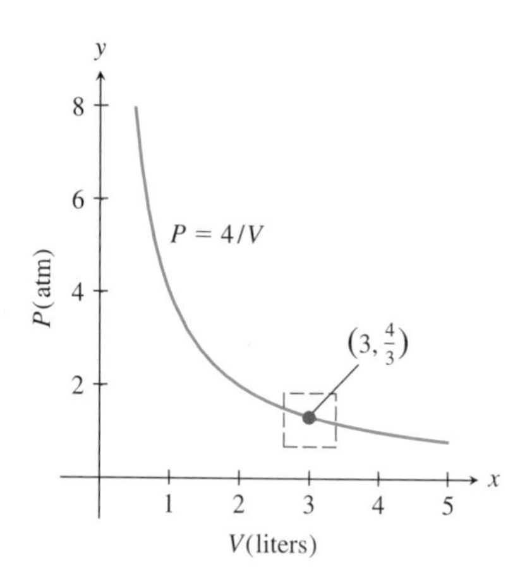

FIGURE 1.26 The graph of $P = 4/V$.

This piece of the graph looks very much like a line. (It isn't really. Hold your book up so the page is flat and level with your eye. Look in the direction of the graph in Fig. 1.28 and you will see a slight curve.) Reading from the vertical and horizontal scales of this graph, the slope appears to be about

$$\frac{1.38\text{ atm} - 1.28\text{ atm}}{2.9\text{ liters} - 3.1\text{ liters}} = -0.5\text{ atm/liter}.$$

This means that when the volume of the piston chamber is 3 liters and the piston is moved out to make the volume bigger, the pressure appears to be changing at a rate of about -0.5 atm/liter. The negative sign in front of this rate of change tells us that the pressure decreases as we make the piston chamber bigger. Is this consistent with common sense?

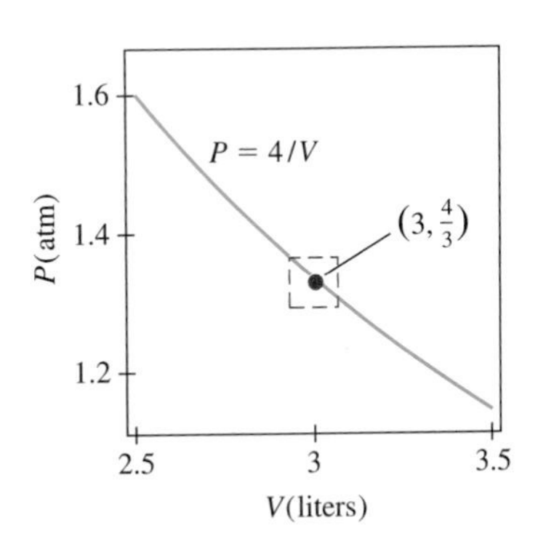

FIGURE 1.27 Zooming in on the graph near $(3, 4/3)$.

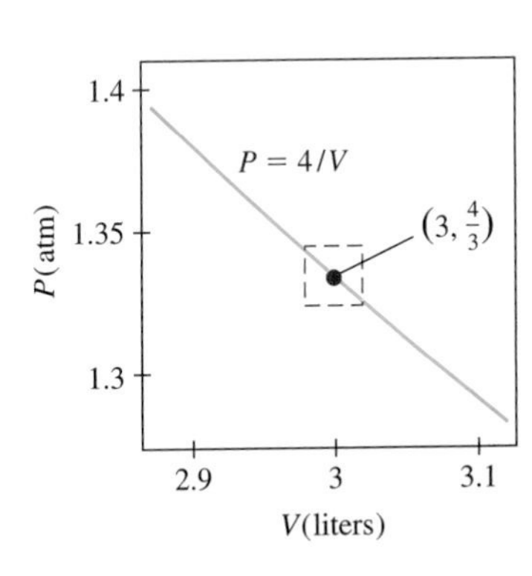

FIGURE 1.28 Graph of $P = 4/V$ for $2.9 \leq V \leq 3.1$. The slope of the "line" is approximately -0.5 atm/liter.

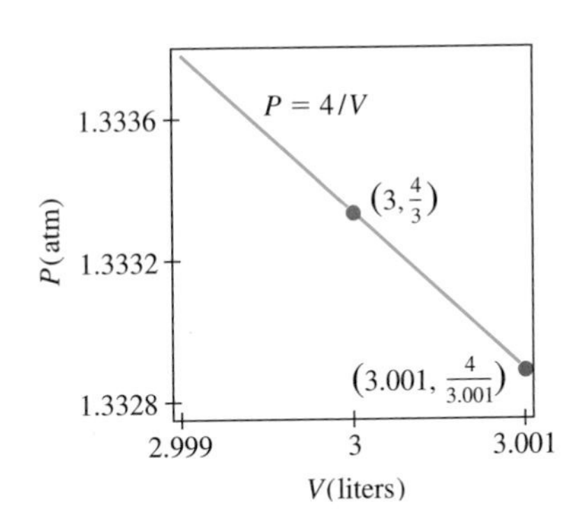

FIGURE 1.29 Graph of $P = 4/V$ for $2.999 \leq V \leq 3.001$. The slope of the "line" is ≈ -0.45 atm/liter.

How accurate is the rate of change? Can we do better? Let's try zooming in even closer by enlarging the portion of the graph in the square shown in Fig. 1.28. The result is shown in Fig. 1.29. In this figure the curve cannot be distinguished from a line.

Again using the vertical and horizontal coordinates shown on the axes, we estimate the slope as

$$\frac{1.3338 \text{ atm} - 1.3329 \text{ atm}}{2.999 \text{ liters} - 3.001 \text{ liters}} = -0.45 \text{ atm/liter}.$$

Based on these estimates from graphs, we conclude that when the piston chamber has volume 3 liters and the piston is pulled out to make the volume larger, the pressure of the gas in the chamber changes at a rate of about -0.45 atm/liter.

Looking closely at the graph like this can help us to estimate the rate of change of P with respect to V when $V = 3$. However, because it is difficult to get precise coordinate readings from a graph, our slope calculations may not be as accurate as we would like. We can do better by using equation (5) to actually find the coordinates of two points on the "line" of Fig. 1.29. Two such points are

$$(3, P(3)) = (3, 4/3) \quad \text{and} \quad (3.001, P(3.001)) = (3.001, 4/3.001).$$

Using these two points to compute the slope of the "line," our estimate for the rate of change is

$$\frac{\dfrac{4}{3} - \dfrac{4}{3.001}}{3 - 3.001} \text{ atm/liter} \approx -0.44430 \text{ atm/liter}.$$

WEB

Because we used coordinates of points on the graph for this last calculation, we might have more confidence that this number really represents the slope of the "line" in Fig. 1.29. This also suggests an easier way to approximate the rate of change at $V = 3$. Find the coordinates of a point R on the graph very near the point

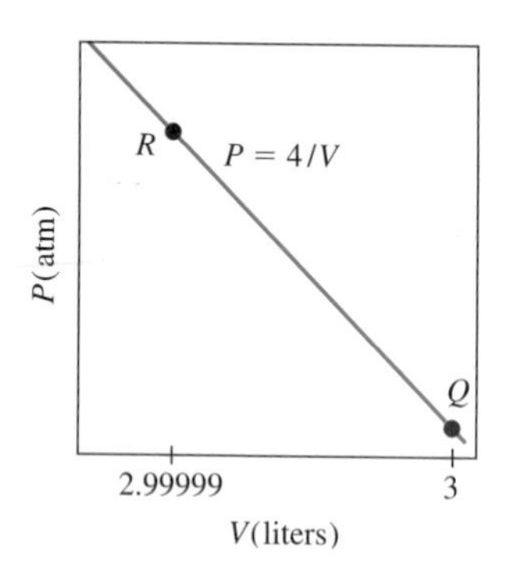

FIGURE 1.30 Computing slope using points $Q = (3, 4/3)$ and $R = (2.99999, 4/2.99999)$, the slope of the "line" is ≈ -0.444446 atm/liter.

$Q = (3, 4/3)$. The portion of the graph between Q and R is almost a line. The slope of the segment QR is an approximation of the slope of the line. Since the graph appears more "linelike" as we zoom in closer and closer, our calculations give better approximations of the rate of change by taking R closer to Q.

As an example, take $R = (2.99999, P(2.99999))$, a point on the graph very close to Q. As seen in Fig. 1.30, the graph between Q and R is again almost a line. The slope of the segment joining Q and R is

$$\frac{P(3) - P(2.99999)}{3 - 2.99999} \text{ atm/liter} = \frac{\frac{4}{3} - \frac{4}{2.99999}}{3 - 2.99999} \text{ atm/liter} \tag{6}$$
$$\approx -0.444446 \text{ atm/liter.}$$

In the next section, we streamline the process of finding the rate of change.

Exercises 1.3

1. What are the slope and y-intercept of the line with equation $y = 3x - 4$? Sketch a graph of the line.
2. Find the slope and y-intercept of the line with equation $4x - 7y + 5 = 0$. Sketch a graph of the line.
3. Find an equation for the line through the points $(-2, 5)$ and $(4, 1)$. Sketch a graph of the line.
4. Find an equation for the line through the points $(0, 8)$ and $(\sqrt{2}, 8)$. Sketch a graph of the line.
5. Find an equation for the line with slope -2 and y-intercept 5. Sketch a graph of the line.
6. Find an equation for the line that has slope $1/2$ and passes through the point $(-2, 3)$. Sketch a graph of the line.
7. Find an equation for the line that has slope 10 and passes through the point $(-10, 10)$. Sketch a graph of the line.
8. Find an equation for the line that passes through the point $(-1, 3)$ and makes an angle of $60°$ with the x-axis.
9. Sketch the graph of the line $y = 2x - 5$. At what angle does the line cross the x-axis?
10. A cylindrical pail with the base radius 3 inches contains water to a depth of 2 inches. At 1:00 P.M. it starts raining, and the water in the pail rises at a steady rate of $1/2$ inch per hour for three hours. At 4:00 it stops raining. Write a formula that gives the volume of water in the pail t hours after the rain starts. What is the rate of change of volume with respect to time while it is raining?
11. A parachutist is 10,000 feet above the ground and falling at a steady rate of 10 feet per second. When will she reach the ground? Find a formula for her height above ground t seconds after the time she is 10,000 feet above the ground. What is the rate of change of her height with respect to time?
12. A car accelerates down the highway in such a way that at time t seconds its velocity $v(t)$, in feet/second, is given by the linear function

$$v = v(t) = 6t + 15.$$

What are the units for the constants 6 and 15 in this formula? What is the rate of change of v with respect to t? (Don't forget the units.) This rate of change of velocity with respect to time has a familiar name. What is it?

13. The county agent featured in Example 3 decides she would like to give the rate at which the water level of the river is rising in inches/minute. What is the rate of rise in this case? In these units, what is the linear equation that gives the height of the river at time t?
14. (T) Referring to the Investigation, suppose that the piston is positioned so that the chamber has a volume of 5 liters and that the piston is being pulled outward. Approximate the rate of change of P with respect to V:
    a. graphically, by estimating the slope for small sections of the graph of $P(V)$ near $V = 5$.
    b. numerically, by computing the slope of the segment determined by the point $(5, P(5))$ and some nearby point of the graph.
15. An empty, cylindrical can, open at the top, has radius 8 cm and height 20 cm. Water is poured into the can at a rate of 10 cm^3/s.

a. Find a formula for the height $h(t)$ of water in the can after t seconds.

b. What is the rate of change of the height of water with respect to time?

c. When will the can be full?

16. The equation used to convert degrees Fahrenheit (F) to degrees Celsius (C) is the equation of a line. A temperature of 32°F corresponds to 0°C, and a temperature of 212°F corresponds to 100°C.

a. Use these data in finding a formula for converting a Fahrenheit temperature to Celsius.

b. Sketch a graph of the equation found in part **a**. Label your axes.

c. What is the rate of change of degrees Celsius with respect to degrees Fahrenheit?

d. Find a formula for converting a Celsius temperature to Fahrenheit.

e. Sketch a graph of the equation found in part **d**. Label your axes.

f. What is the rate of change of degrees Fahrenheit with respect to degrees Celsius?

17. In Example 4 we found that for the steel girder under consideration the rate of change of length, L, with respect to temperature T is 0.00018 m/°C. For this girder, what is the rate of change of *temperature* with respect to *length?* Explain what this rate of change means.

18. If the temperature of the gas in the Investigation is raised to 273°C and held there, the pressure of the gas in the chamber will double:

$$P(V) = \frac{8}{V} = 2\frac{4}{V}.$$

How will the rate of change of P with respect to V be affected when $V = 3$ liters? Explain your answer.

19. Sketch the graph of $y = |x|$ and zoom in on the graph at the point $(0, 0)$. If you zoom in close enough, will you ever see a straight line? Why?

T 20. Sketch the graph of the function $y = x^{1/3}$. Use graphical and numerical techniques to investigate the rate of change of y with respect to x:

a. at $x = 8$.

b. at $x = 0$.

21. Fill in the blanks in the following sentence: The speedometer in a car measures the rate of change of ________ with respect to ________ .

22. A calculus student might think of a speedometer as a "slope meter." Explain why this terminology is appropriate.

23. As seen in many of the figures in this section, computers and graphing calculators often produce graphs in which the horizontal and vertical scales are different. Care is needed when working with such graphs since the appearance of the graph of a function may change as the horizontal and vertical scales change. The accompanying figure shows the graph of a line $y = mx$ for $0 \le x \le 1$. Fill in the appropriate y-axis labels at the tick marks if:

a. $m = 1$. b. $m = 3$. c. $m = 1/4$.

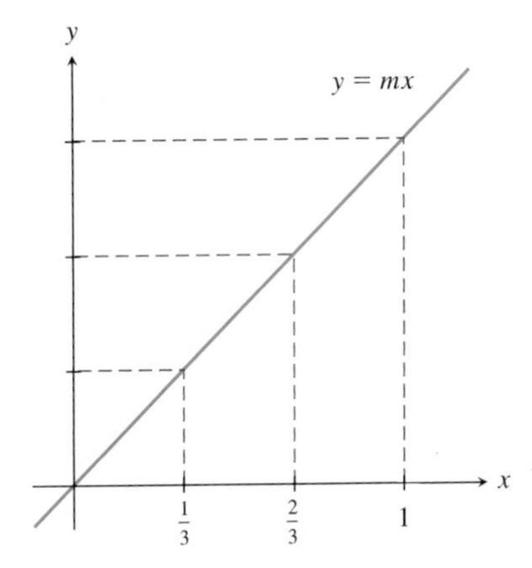

Figure for Exercise 23.

24. The line $y = \frac{1}{6}x$ and the line $y = mx + 2$ were graphed on the same set of axes with a computer graphics package. The graph is shown in the accompanying figure. In the resulting graph the unit in the vertical direction is three times as long as the unit in the horizontal direction, and the two lines cross at right angles. Find m.

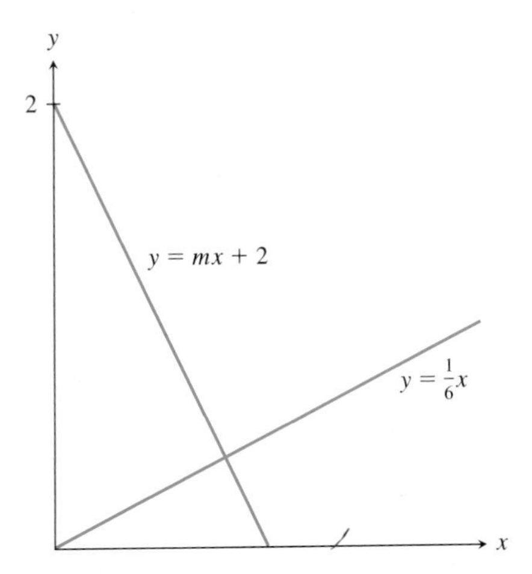

Figure for Exercise 24.

Exercises 25–28: A picture of a container is shown. Water is poured into the container at a constant rate (say, 1 liter/s). As the water is poured in, the height changes. Draw a graph of the height of water as a function of time. Discuss your graph, telling which features in the container relate to which features on the graph.

25.

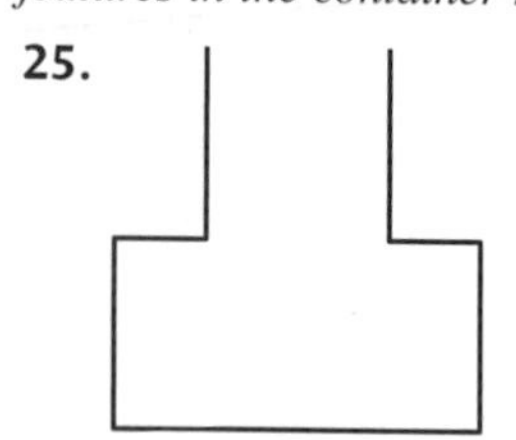

26.

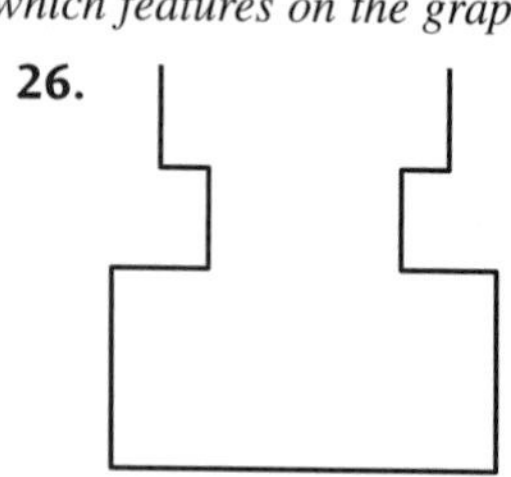

27.

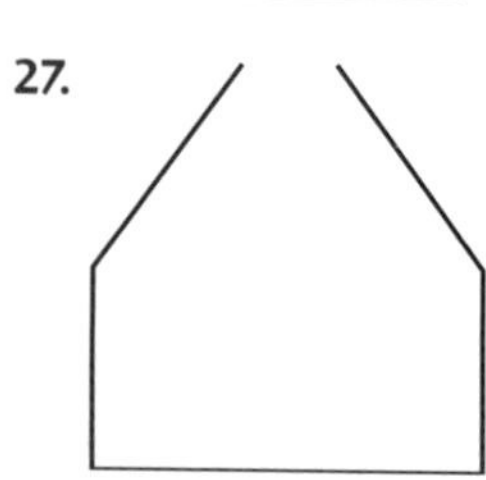

28.

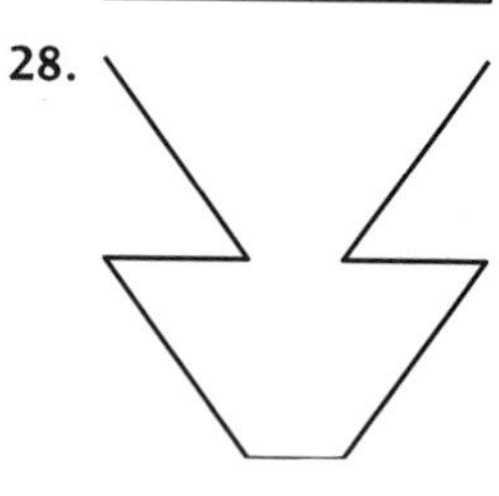

Exercises 29–32: Water is poured into a container at a constant rate (say, 1 liter/s). The graph shown is the graph of the height of water as a function of time. Draw a picture of the container that possibly resulted in the given graph.

29. h

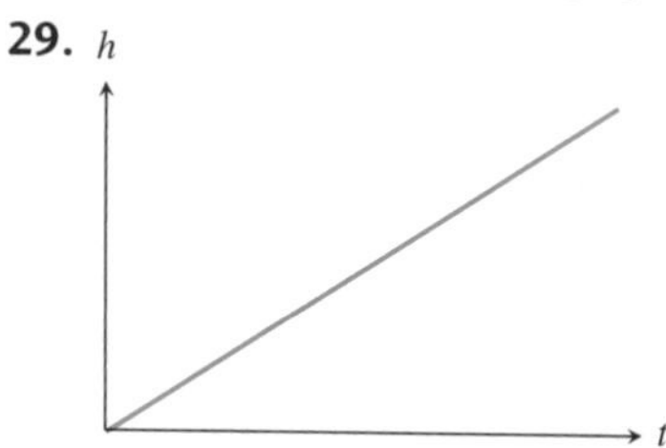

30. h

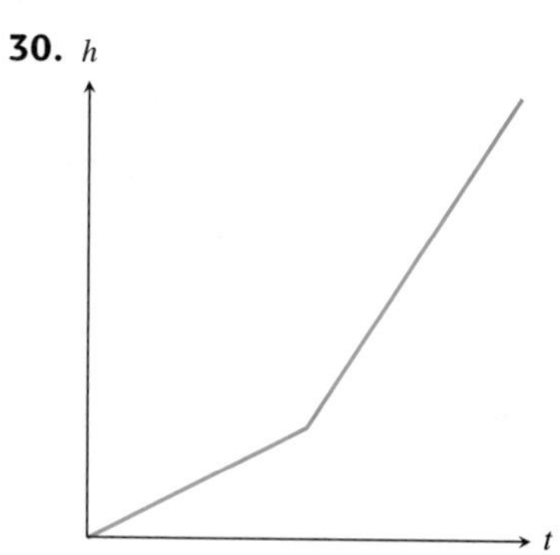

31. h

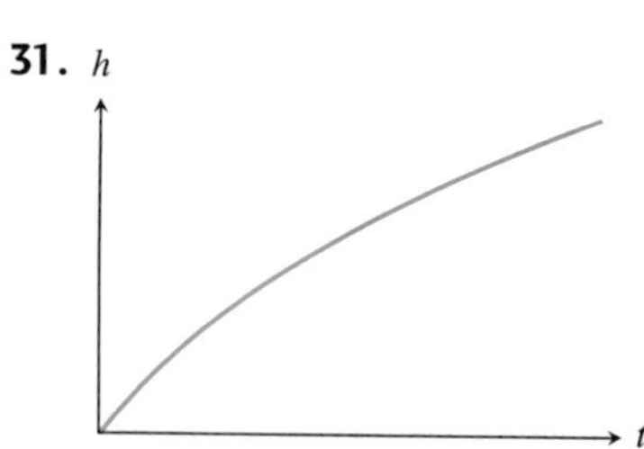

32. h

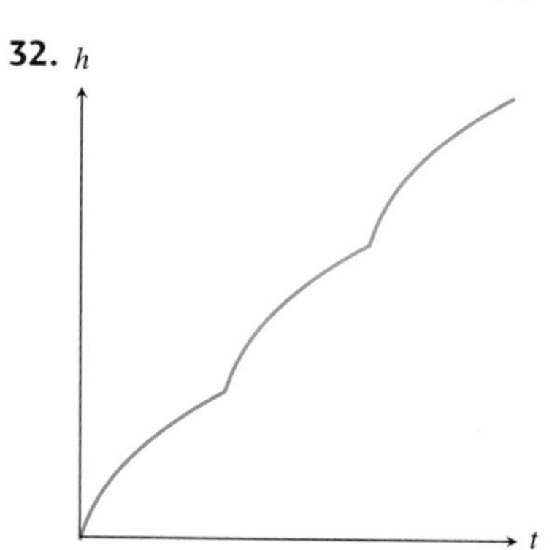

1.4 Calculating Rates of Change

In the preceding section we worked with the formula

$$P = P(V) = \frac{4}{V} \tag{1}$$

and looked at two ways to estimate the rate of change of P with respect to V when $V = 3$. The first method was a graphical approach. We saw that as we zoomed in at the point $(3, P(3))$, the graph of (1) looked more and more like a straight line. We estimated the "slope" of this "almost line" by reading the horizontal and vertical changes from the coordinate axes. The second method used a numerical calculation. We found the coordinates of two points on the graph and then found the slope of the segment joining them. One point was $Q = (3, P(3))$ and the second point, R, was taken close to Q. When R was close enough to Q, segment QR was almost indistinguishable from the portion of the graph between Q and R. We used the slope of QR as an estimate of the rate of change of P with respect to V when $V = 3$.

These two methods are not efficient if we are required to continually refine our estimate of the rate of change at $V = 3$. Each new estimate requires either that we zoom in closer or that we calculate the slope of QR for a new value of R. Furthermore, we can always get a better estimate by zooming in closer or by taking R closer to Q. When are we done? How can this process lead to a definite number that we can call *the* rate of change?

In this section we investigate a symbolic approach to finding the rate of change and see that this approach leads to a number that we can call the rate of change.

Refining Rate-of-Change Calculations

INVESTIGATION

Another Look at the Rate of Change

As in the Investigation of Section 1.3, let $P = 4/V$. We refine the rate-of-change calculations of the previous section and show that

$$-\frac{4}{9} \text{ atm/liter}$$

is a reasonable value for the rate of change of P with respect to V when $V = 3$.

Let $Q = (3, P(3)) = (3, 4/3)$, and let R be another point on the graph of

$$P = P(V) = \frac{4}{V}.$$

If R is close to Q, then we can express R as

$$R = (3 + h, P(3 + h)) = \left(3 + h, \frac{4}{3 + h}\right),$$

where $|h|$ is small, with $h > 0$ if R is to the right of Q, and $h < 0$ if R is to the left of Q. See Fig. 1.31, where the case $h > 0$ is illustrated.

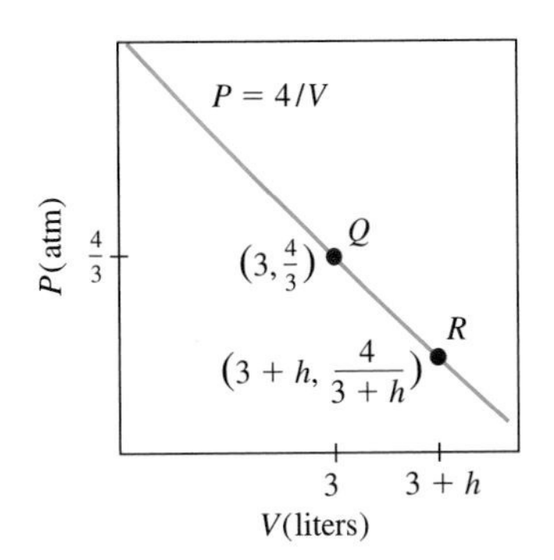

FIGURE 1.31 The graph of $P = 4/V$ near $Q = (3, 4/3)$. Note $h > 0$ since R is to the right of Q.

The slope of segment QR is

$$\begin{aligned}\frac{P(3 + h) - P(3)}{(3 + h) - 3} &= \frac{\frac{4}{3 + h} - \frac{4}{3}}{h} = \frac{4 \cdot 3 - 4(3 + h)}{(3 + h)3h} \\ &= -\frac{4h}{(3 + h)3h} = -\frac{4}{3(3 + h)}\end{aligned} \qquad (2)$$

Because (2) makes sense for any small, nonzero h, we can use this expression to streamline the calculations done in the last section. For example, in equation (6) of Section 1.3 we calculated the slope of QR with $R = (2.99999, P(2.99999))$. This R has the form $(3 + h, P(3 + h))$ with $h = -0.00001$. Thus the result in (6) of Section 1.3 can be found by substituting -0.00001 for h into (2):

$$-\frac{4}{3(3 + h)} = -\frac{4}{3(3 - 0.00001)} \approx -0.44445 \text{ atm/liter}.$$

More importantly, we can use (2) to see what happens to our rate-of-change calculations as we move R closer and closer to Q. As $R = (3 + h, P(3 + h))$ is taken closer and closer to Q, it must be true that h gets closer and closer to 0. But then the rate-of-change estimate (2) gets closer and closer to

$$-\frac{4}{3(3 + 0)} = -\frac{4}{9} = -0.44444\ldots.$$

This leads us to take $-(4/9)$ atm/liter as *the* rate of change of P with respect to V when $V = 3$. We interpret this to mean that if the piston is moving to expand the chamber, then at the instant when the chamber volume is 3 liters, the pressure P is changing at a rate of $-(4/9)$ atm/liter. The minus sign in front of the rate of change tells us that as V increases, P decreases.

This procedure can be used to find rates of change in many other situations.

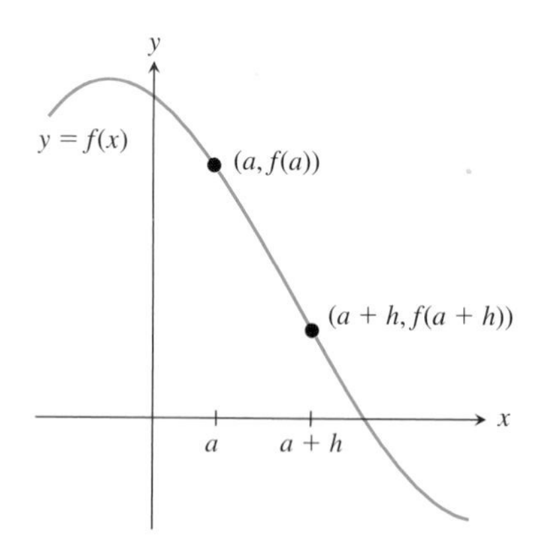

FIGURE 1.32 Find the slope of the segment determined by $(a, f(a))$ and $(a + h, f(a + h))$.

Calculating the Rate of Change

Let $y = f(x)$, where f is a function, and let $x = a$ be in the domain of f. To find the rate of change of y with respect to x at $x = a$:

STEP 1. Form and simplify the expression

$$\frac{f(a + h) - f(a)}{h}, \quad h \neq 0. \tag{3}$$

This is the slope of the segment determined by the points $(a, f(a))$ and $(a + h, f(a + h))$ on the graph of $y = f(x)$. When h is small, (3) can be used as an estimate of the rate of change of y (or f) with respect to x when $x = a$. See Fig. 1.32.

STEP 2. Investigate the behavior of (3) as h gets close to 0. If (3) gets close to a single number as h gets close to 0, then this number is the *rate of change of y (or f) with respect to x at $x = a$.*

We refer to the process just described as the rate-of-change algorithm. (By *algorithm* we mean a step-by-step procedure that can be used to solve a particular type of problem.) In the remaining sections of this chapter we will study the rate-of-change algorithm and introduce new ideas and terminology that allow us to express the algorithm in more precise language. This will add to our understanding of the rate of change and its role in describing the world around us. Meanwhile, we can use the rate-of-change algorithm as stated to find rates of change for some simple functions.

EXAMPLE 1 Calculating a rate of change

Let $y = f(x) = x^2$.

a) Find the rate of change of y with respect to x at $x = -1$.
b) Find the rate of change of y with respect to x at the arbitrary point $x = a$.

Solution

a) Before using the rate-of-change algorithm to find the answer, we try zooming in to get an estimate for the rate of change. (You should try this on your own calculator or CAS.) Figure 1.33 shows the graph of $y = x^2$. The small square in the figure is centered at the point of interest, $Q = (-1, (-1)^2) = (-1, 1)$. When we zoom in on the piece of the graph in the square we see something like Fig. 1.34.

The graph in Fig. 1.34 looks like the graph of a line. Using the tick marks on the axes we see that the y-coordinate of the "line" drops from 1.02 to 0.98 as the x-coordinate grows from -1.01 to -0.99. Hence the slope is about

$$\frac{1.02 - 0.98}{-1.01 - (-0.99)} = -2.$$

Thus we guess that when $x = -1$, the rate of change of y with respect to x is close to -2.

To apply the rate-of-change algorithm, use (3) with $f(x) = x^2$ and $a = -1$. We have

$$\begin{aligned} \frac{f(a+h) - f(a)}{h} &= \frac{(-1+h)^2 - (-1)^2}{h} \\ &= \frac{-2h + h^2}{h} = -2 + h. \end{aligned} \tag{4}$$

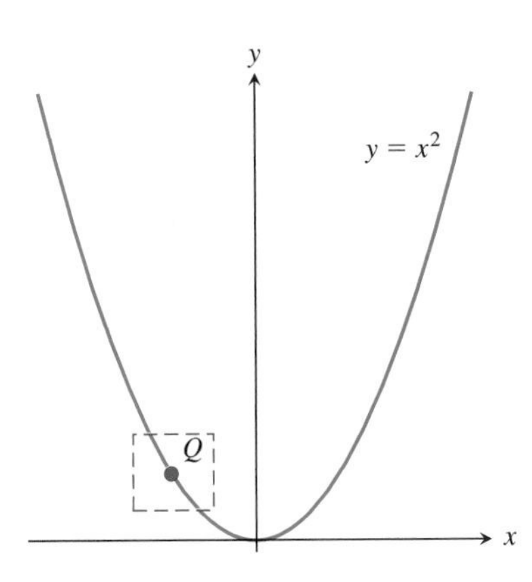

FIGURE 1.33 The graph of $y = x^2$. The small square is centered at $(-1, 1)$.

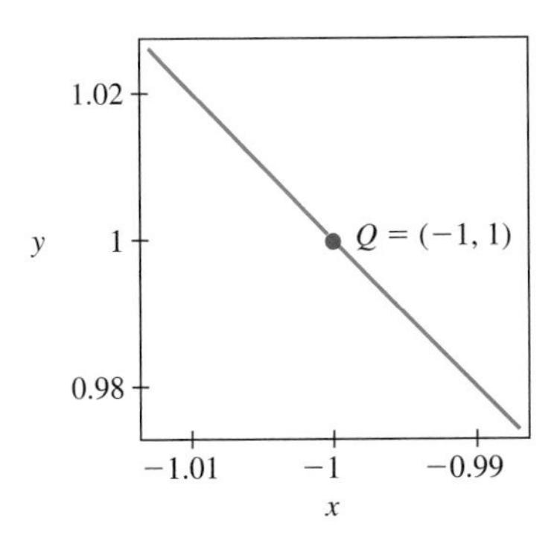

FIGURE 1.34 Zooming in on $y = x^2$ near $Q = (-1, 1)$.

This last expression is the slope of the segment determined by

$$Q = (-1, 1) \quad \text{and} \quad R = (-1 + h, (-1 + h)^2),$$

as seen in Fig. 1.35. For small h, (4) is an approximation of the rate of change of y with respect to x at $x = -1$. As we move R closer and closer to Q, we see that h becomes closer and closer to 0, and $-2 + h$ gets closer and closer to -2. We express this by writing

$$-2 + h \rightarrow -2 \text{ as } h \rightarrow 0,$$

and say "$-2 + h$ approaches -2 as h approaches 0." Thus when $x = -1$, the rate of change of y with respect to x is -2. Remember that the answer is a rate of change and so has units of y units/x units.

b) Using (3), the rate of change of y with respect to x when $x = a$ is approximated by

$$\frac{f(a+h) - f(a)}{h} = \frac{(a+h)^2 - a^2}{h} = \frac{2ah + h^2}{h} = 2a + h$$

when h is small. As h gets closer and closer to 0, this last expression approaches $2a$, that is,

$$2a + h \rightarrow 2a \text{ as } h \rightarrow 0.$$

Thus when $x = a$, the rate of change of y with respect to x is

$$2a \frac{y \text{ units}}{x \text{ units}}.$$

If we set $a = -1$ we get a rate of change of -2, as found in part **a)**.

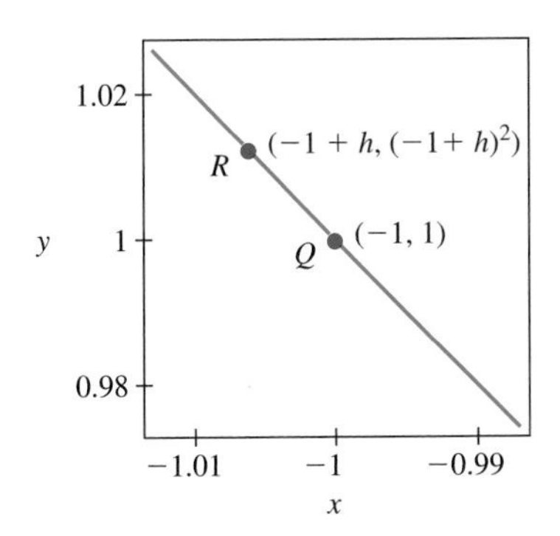

FIGURE 1.35 Take a point R close to Q.

The rate-of-change algorithm worked well in finding rates of change for the relatively simple functions $P = 4/V$ and $y = x^2$ because for these cases (3) could be simplified sufficiently to determine its behavior as h got close to 0. When such simplification is difficult or takes too much time we can use graphical or numerical techniques to gain insight into the behavior of (3) as h gets close to 0.

EXAMPLE 2 A rate of change for the sine function

Let $y = g(\theta) = \sin\theta$, where θ is measured in radians. Investigate the rate of change of y with respect to θ when $\theta = \pi/3$.

Solution

Figure 1.36 shows the graph of $y = \sin\theta$ for $0 < \theta < \pi$. In Fig. 1.37 we have zoomed in on the graph near the point

$$Q = (\pi/3, \sin(\pi/3)) = \left(\pi/3, \sqrt{3}/2\right) \approx (1.0472, 0.8660).$$

FIGURE 1.36 The graph of $y = \sin\theta$ for $0 < \theta < \pi$.

The graph in Fig. 1.37 is almost a line, and the slope appears to be about 0.5. Thus at $\theta = \pi/3$ the rate of change of y with respect to θ is approximately 0.5.

We can also estimate the rate of change by using (3). We have

$$\frac{g\left(\frac{\pi}{3}+h\right)-g\left(\frac{\pi}{3}\right)}{h} = \frac{\sin\left(\frac{\pi}{3}+h\right)-\sin\left(\frac{\pi}{3}\right)}{h} = \frac{\sin\left(\frac{\pi}{3}+h\right)-\frac{\sqrt{3}}{2}}{h}. \tag{5}$$

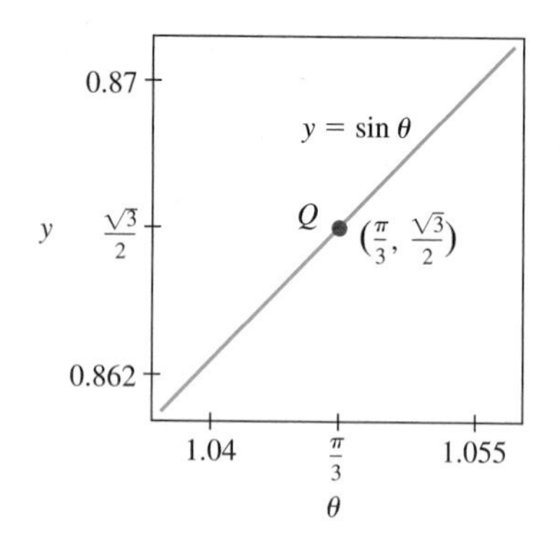

FIGURE 1.37 Zooming in near the point $Q = \left(\pi/3, \sqrt{3}/2\right)$.

We need to know what happens to this expression as h gets close to 0. In previous examples we were able to simplify the rate-of-change expressions to forms in which we could simply observe the behavior as $h \to 0$. This is not the case with (5). There are no obvious simplifications, and as $h \to 0$ both the numerator and the denominator approach 0. Another approach is to gather numerical data by trying several values of h close to 0. With the aid of a calculator or computer, we can generate data like those in Table 1.5. You should verify some of the numbers in this table with your own calculator, and then try some additional h values even closer to 0.

TABLE 1.5

h	$\dfrac{\sin\left(\frac{\pi}{3}+h\right)-\frac{\sqrt{3}}{2}}{h}$
0.1	0.455902
−0.1	0.542432
0.05	0.478146
−0.05	0.521438
0.001	0.499567
−0.001	0.500433
0.000005	0.499998
−0.000005	0.500002

Using data from the table we guess that the rate-of-change estimate in (5) is getting close to 0.5 as h gets close to 0, that is,

$$\frac{\sin\left(\frac{\pi}{3}+h\right)-\frac{\sqrt{3}}{2}}{h} \to 0.5 \text{ as } h \to 0.$$

We can also investigate the behavior of (5) by graphing this expression for h near 0. Let

$$R(h) = \frac{\sin\left(\frac{\pi}{3}+h\right)-\frac{\sqrt{3}}{2}}{h}$$

be the function defined by the right side of (5). The graph of $R = R(h)$ is shown in Fig. 1.38. Keep in mind that this graph is the graph of R (which is a function of h) and *not* the graph of $y = \sin\theta$. Although (5) (and hence the graph in Fig. 1.38) is not defined for $h = 0$, we can see from the graph that the value of the expression is closing in on 0.5 as h approaches 0.

The results of these investigations suggest that when $\theta = \pi/3$, the rate of change of $y = \sin\theta$ with respect to θ is close to (or perhaps equal to) 0.5.

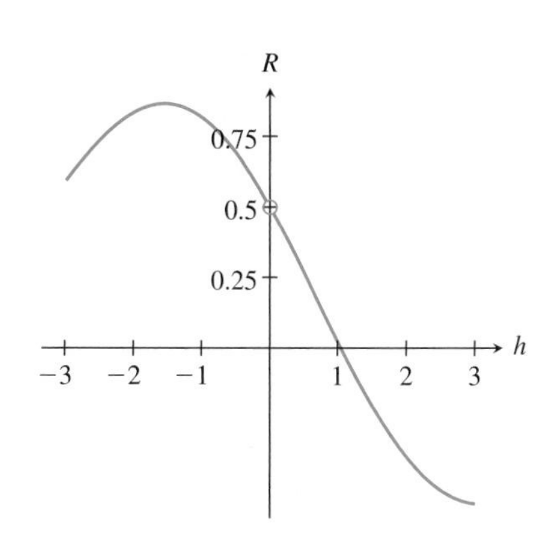

FIGURE 1.38 What appears to happen to $\left(\sin(\pi/3+h)-\sqrt{3}/2\right)/h$ as h gets close to 0?

Functions without Formulas

Scientists do not always have a formula to express a relationship between two quantities. Sometimes these relationships are given only graphically or as numerical data. Furthermore, as the next example shows, the information might be presented in an unusual format. In such cases we can still estimate rates of change, though it may be necessary to think carefully about what the information represents and how it is presented.

EXAMPLE 3 Temperature change in an autoclave

An autoclave is a piece of equipment used for high-temperature sterilization of laboratory and medical equipment. For sterilization to be successful, it is necessary that a high temperature be maintained for a certain period of time. Thus many autoclaves are linked to a device that produces a graph of temperature as a function of time during the sterilization process. One such graph is shown in the accompanying figure. This graph was produced on a circular piece of graph paper (we show only the portion of the paper that contains the graph). On the graph, temperature in °C is indicated by distance from the center of the paper, and time by marks around the circumference. As the autoclave runs, the paper rotates slowly and a pen driven by a thermostat produces the graph. This graph shows the output for a sterilization procedure that ran between 11:00 A.M. and 1:00 P.M. What was the rate of change of temperature with respect to time at 11:45 A.M.? At 12:15 P.M.?

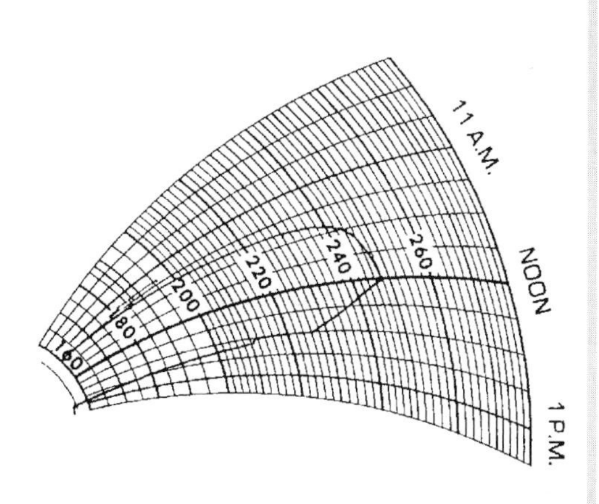

Solution

Let $T = T(t)$ be the temperature of the autoclave at time t. From the graph we read $T(11{:}45) \approx 247$°C. To estimate the rate of change of temperature with respect to time at 11:45, we must determine the behavior of

$$\frac{T(11{:}45 + h) - T(11{:}45)}{h} \tag{6}$$

when h is small. Because of the scale of the graph, we take $h = 15$ minutes to be small. Using this value of h in (6) we obtain

$$\frac{T(11{:}45 + {:}15) - T(11{:}45)}{15} = \frac{T(12{:}00) - T(11{:}45)}{15} \approx \frac{248 - 247}{15}$$
$$\approx 0.067°\text{C/min}.$$

As the graph shows, and the positive sign on our answer indicates, the temperature was rising at 11:45 A.M.

At 12:15 P.M. the temperature in the autoclave was 237°C. The rate of change of temperature with respect to time at that time was approximately

$$\frac{T(12{:}15 + h) - T(12{:}15)}{h} \tag{7}$$

where h is small. Setting $h = -15$ minutes in (7) we obtain

$$\frac{T(12{:}15 + (-{:}15)) - T(12{:}15)}{-15} = \frac{T(12{:}00) - T(12{:}15)}{-15} \approx \frac{248 - 237}{-15}$$
$$\approx -0.73°\text{C/min}.$$

The negative sign tells us that the temperature was decreasing at 12:15 P.M. You can see that this is true by looking at the graph.

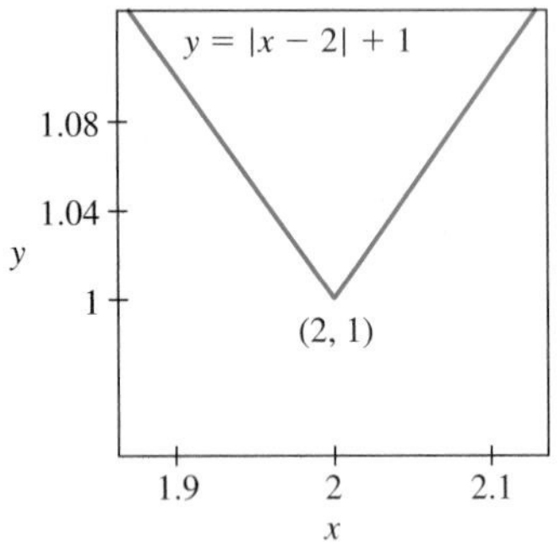

Functions without a Rate of Change

There are some functions for which the rate of change is not defined at one or more values. The rate-of-change algorithm can be used to help us identify some of these points for a given function.

EXAMPLE 4 Let

$$y = f(x) = |x - 2| + 1. \tag{8}$$

Investigate the rate of change of y with respect to x when $x = 2$.

Solution

Try zooming in on the graph of (8) near the point (2, 1). (See Fig. 1.39.) We find that no matter how close we zoom in, the portion of the graph we see is never going to look like a straight line because the corner at (2, 1) will not straighten out at any magnification. Thus we cannot approximate a rate of change at $x = 2$ by estimating the slope of an "almost line."

Next, try the rate-of-change algorithm. We investigate the expression

$$\frac{f(2 + h) - f(2)}{h} = \frac{\big(|(2 + h) - 2| + 1\big) - 1}{h} = \frac{|h|}{h} \tag{9}$$

for h close to 0. Table 1.6 shows values for (9) for several values of h close to 0. The data in the table indicate that the value of (9) does not settle down to one and only one number as h gets close to 0. This means that the rate of change of $y = |x - 2| + 1$ with respect to x is undefined at $x = 2$.

FIGURE 1.39 Zooming in near the point (2, 1) on the graph of $y = |x - 2| + 1$.

TABLE 1.6

h	$\frac{\lvert h\rvert}{h}$
0.01	1
−0.01	−1
0.001	1
−0.001	−1
0.0005	1
−0.0005	−1

Exercises 1.4

Exercises 1–10: Use the rate-of-change algorithm to find the rate of change for the given value a.

1. $f(x) = 2x^2 - 3, \quad a = 1$
2. $f(x) = -x^2 + 4, \quad a = -2$
3. $y = 5x - 7, \quad a = 4$
4. $s = -2t + 8, \quad a = c$
5. $g(x) = x^3, \quad a = 2$
6. $g(x) = x^3, \quad a = c$
7. $F(x) = 2/x, \quad a = 2$
8. $r(t) = 3/t^2, \quad a = 1$
9. $f(x) = \sqrt{x}, \quad a = 9$
10. $y = \sqrt{x}, \quad a = c, \quad c > 0$
11. Let $y = f(x) = mx + b$ where m and b are constants. Use the rate-of-change algorithm to show that for any value of x, the rate of change of y with respect to x is m.
12. **T** Let $y = \sin x$. Use both graphical and numerical methods, as in Example 2, to investigate the rate of change of y with respect to x when $x = 0$ and when $x = 2\pi/3$.
13. **T** Let $y = 2^x$. Use both graphical and numerical methods, as in Example 2, to investigate the rate of change of y with respect to x when $x = 0$, $x = 1$, $x = 2$, and $x = 4$.
14. Let $r = r(t) = |2t - 1|$. For what t values is the rate of change of r with respect to t undefined? What is the rate of change at those t values for which the rate of change is defined?
15. Let $g = g(x) = |3 + x| - 2|4 - 3x|$. For what x values is the rate of change of g with respect to x undefined? What is the rate of change at those x values for which the rate of change is defined?
16. When a heavy metal ball is dropped from the top of a 500 foot building, the height (in feet) of the ball above the ground after t seconds is

$$s(t) = 500 - 16t^2. \tag{10}$$

   a. Given the conditions described in the problem, for what t values does (10) make physical sense?

   b. What are the units for the rate of change of s with respect to t? What does this rate of change tell us about the falling ball?

   c. What is the rate of change of s with respect to t at any given time during the fall?

   d. When does the ball hit the ground? What is the rate of change of s with respect to t at the time of impact? What does this tell us about the motion of the ball at the time of impact?

17. The accompanying graph shows air pressure (in millibars) as a function of time in Ames, Iowa, from 4:00 A.M. Wednesday, December 11, 1991, to 2:00 P.M. Saturday, December 14, 1991. What was the rate of change of air pressure at 2:00 A.M. Thursday? Was the pressure rising or falling? At what time did the air pressure appear to be rising most rapidly? Falling most rapidly?

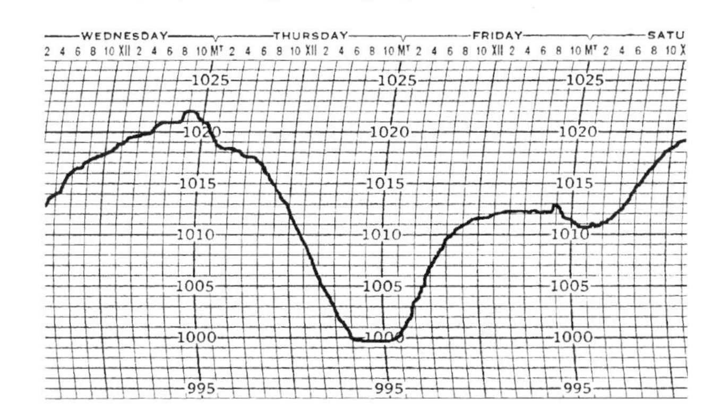

18. Let g be a function. The graph of $s = g(t)$ is shown.

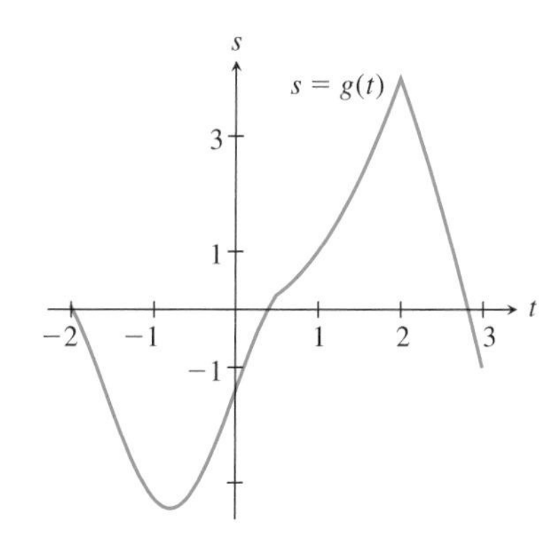

Figure for Exercise 18.

   a. Indicate with an asterisk those points on the graph where the rate of change of s with respect to t is undefined.

b. Indicate with a 0 those points where the rate of change is 0.

c. Indicate with a 1 those points where the rate of change is 1.

Give reasons for all answers.

19. Let f be a function with $f(-2) = 4$ and let $y = f(x)$. Suppose the rate of change of y with respect to x at $x = -2$ is 0. If we zoom in very close to the graph of $y = f(x)$ near the point $(-2, 4)$, what will we see? Explain your answer.

20. Let f and g be two functions, with $f(1) = 3$ and $g(1) = 3$. Suppose that at $x = 1$ the rate of change of f with respect to x is 2 and the rate of change of g with respect to x is 1. If $y = f(x)$ and $y = g(x)$ are graphed on the same set of axes and we then zoom in on the point $(1, 3)$, what will we see? Draw a sketch and explain.

21. Consider the functions f and g defined by $f(x) = x^2$ and $g(x) = x^2 + 5$. At a given $x = a$, how do the rates of change of the two functions compare?

22. Let F be a function and c a constant. How are the rates of change of $y = F(x)$ and $y = F(x) + c$ related? Explain your answer.

23. The accompanying table gives U.S. census data for the years 1790–1990. Estimate the rate of change of the U.S. population with respect to time for the years 1800, 1900, and 1980 (choose appropriate units). Explain how you got your estimates, and explain why someone else might arrive at different estimates.

Year	Pop. (millions)	Year	Pop. (millions)
1790	4	1900	76
1800	5	1910	92
1810	7	1920	106
1820	10	1930	123
1830	13	1940	132
1840	17	1950	151
1850	23	1960	178
1860	31	1970	203
1870	40	1980	227
1880	50	1990	247
1890	63		

Table for Exercise 23.

24. The accompanying map is a topographical map of the region surrounding Jackstraw Mountain on the west side of Rocky Mountain National Park. The contour lines on the map indicate altitudes above sea level. The change in elevation between adjacent lines is 80 feet. A backpacker hiking off-trail passes through the point labeled P on the map. Estimate the rate of change of the backpacker's elevation as he passes through P for each of the following directions of travel. (Your answers should be in units of vertical feet/horizontal mile. You will need the scale of miles accompanying the map.)

a. Along the path labeled N and in the direction of the arrow.

b. Along the path labeled N and in the direction opposite the arrow.

c. Along the path labeled W and in the direction of the arrow.

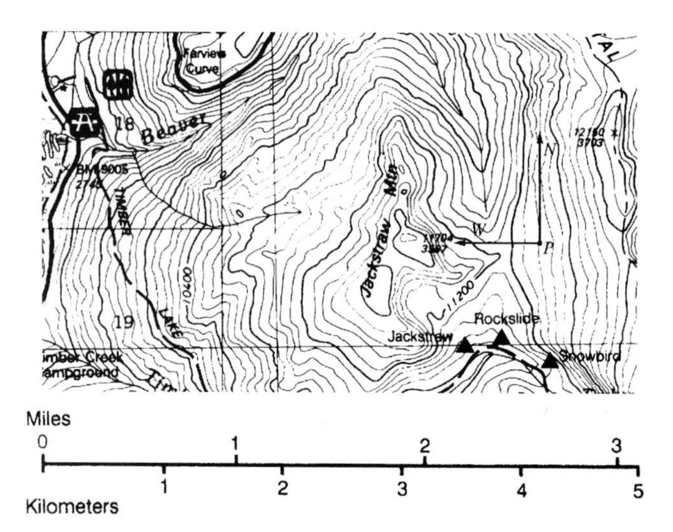

25. A car on a freeway is moving at exactly 60 miles per hour as it passes a white mark on the road. Estimate how far the car will be from the mark in 1 hour, in 1 minute, in 2 seconds, and in 0.01 second. Which of your four answers is likely to be the most accurate? Least accurate? Explain.

26. Let f be a function and $x = a$ a point in the domain of f. Explain why the rate of change of f with respect to x at $x = a$ can be found by considering the behavior of the expression

$$\frac{f(x) - f(a)}{x - a}$$

as x gets close to a. In particular, tell why this method always gives the same results as the rate-of-change algorithm.

1.5 Limits

The 1997 Nobel Prize in physics was awarded to Steven Chu of Stanford University, William D. Phillips of the National Institute of Standards and Technology, and Claude Cohen-Tannoudji of the Collège de France and École Normale Supérieure in recognition of their development of a technique for slowing and trapping atoms.

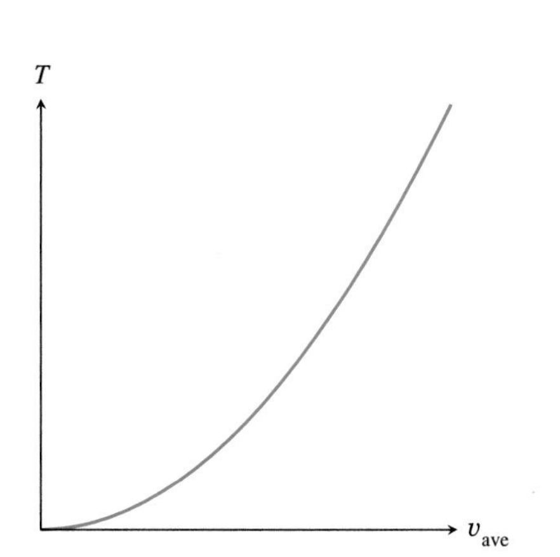

FIGURE 1.40 Temperature T, measured in Kelvins, is a function of average molecular speed, v_{ave}.

At room temperature, atoms and molecules in the air move with speeds of up to 1000 meters per second. Using the methods developed by the three Nobel laureates, scientists have been able to use lasers to slow individual atoms to speeds of less than 1 meter per second and to keep the slowed atoms confined using blasts of laser light and magnetic traps. When the atoms of a gas move at such low speeds, the temperature of the gas is very close to absolute zero (0 Kelvin). See Fig. 1.40. In fact, using the techniques described, temperatures of just a few millionths of a degree above absolute zero have been attained.

Scientists know that a temperature of absolute zero can never be achieved, but by studying slowed atoms they can learn more about the properties of matter at low temperatures, inferring what might happen if even lower temperatures were achieved. In mathematical language, we might say that physicists are investigating "the limit of properties of matter as temperature approaches absolute zero."

When we compute the rate of change of a function f at a point $x = a$ in its domain, we are in a situation similar to that of the physicist studying properties of matter at low temperatures. The rate of change is found by investigating the behavior of the quotient

$$\frac{f(a+h) - f(a)}{h} \tag{1}$$

when h is close to 0. We cannot simply set $h = 0$ in (1) because the expression is undefined for such h. Instead, we investigate the value of (1) for real numbers h close to 0. We might do so numerically (with a table of values), graphically, algebraically, or by combining two or more of these approaches. In doing this, our goal is to obtain a number or expression that we call the rate of change of f with respect to x at $x = a$. This number or expression is *the limit of* (1) *as h approaches 0.*

The definition of *limit* is motivated by the need to study the behavior of expressions like (1) near a value of h where the expression may not be defined.

DEFINITION Limit of a Function

Let g be a function and let a and L be real numbers. If we can make $g(x)$ as close to L as we like by taking x close to a, but not equal to a, then we say

$g(x)$ has **limit** L as x approaches a.

We denote this by

$$\lim_{x \to a} g(x) = L. \tag{2}$$

Equation (2) is read "the limit of $g(x)$ as x approaches a equals L."

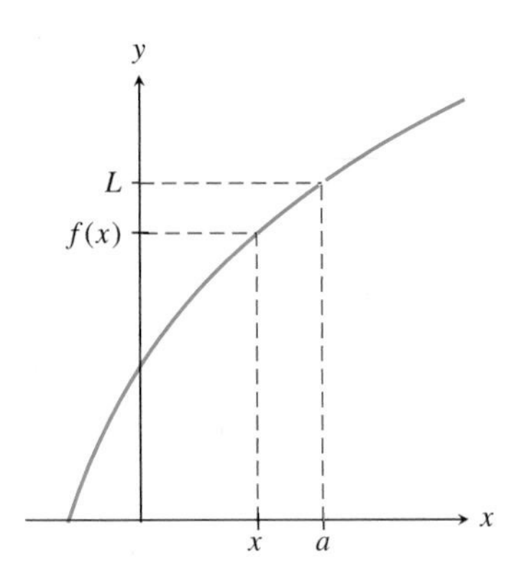

FIGURE 1.41 When x is close to a, $f(x)$ is close to L. We write $\lim_{x\to a} f(x) = L$.

This definition of *limit* is meant to be a "working" definition. By thinking of $L = \lim_{x\to a} g(x)$ as the number that $g(x)$ is close to when x is close to a, as illustrated in Fig. 1.41, we can evaluate most limits, including limits of expressions like (1). However, we will need to do more with limits than simply evaluate limits of functions. We will use limits to study rates of change, approximations, areas, and many other things. For this we will need a more precise definition than the one given. In the next section, we will see that mathematicians mean something very special when they say "close to." Once "close to" is properly defined, we will have a definition of *limit* that is good for more than just evaluation of limits.

Many limits can be evaluated using little more than common sense and knowledge about some familiar functions.

EXAMPLE 1 Evaluate $\lim_{x\to 5}(2x^2 + 4)$.

Solution

We know that when x is close to 5, x^2 is close to $5^2 = 25$. It follows that when x is close to 5, $2x^2 + 4$ is close to $2 \cdot 5^2 + 4 = 54$. Hence

$$\lim_{x\to 5}(2x^2 + 4) = 2 \cdot 5^2 + 5 = 54. \tag{3}$$

The graph of $y = 2x^2 + 4$ near $x = 5$ is shown in Fig. 1.42. From the graph we see that when x is close to 5, the value of $y = 2x^2 + 4$ is close to 54. This is consistent with (3).

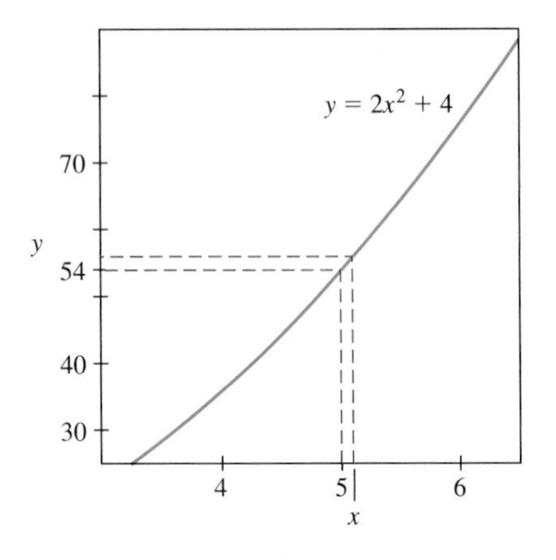

FIGURE 1.42 When x is close to 5, $2x^2 + 4$ is close to 54.

EXAMPLE 2 Find the value of $\lim_{\theta\to\pi/4} \cos\theta$.

Solution

When θ is close to $\pi/4$, $\cos\theta$ is close to $\cos \pi/4 = \sqrt{2}/2$. Therefore

$$\lim_{\theta\to\pi/4} \cos\theta = \cos(\pi/4) = \frac{\sqrt{2}}{2}.$$

This is illustrated by looking at the graph of $y = \cos\theta$ near $\theta = \pi/4$. In Fig. 1.43 we see that if θ is close to $\pi/4$, then $\cos\theta$ is close to $\sqrt{2}/2$.

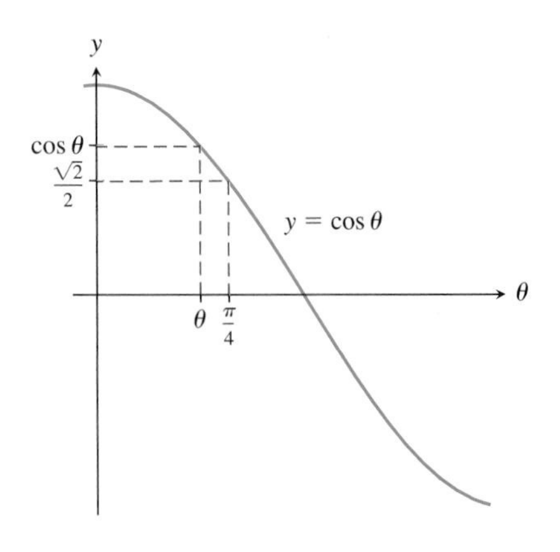

FIGURE 1.43 When θ is close to $\pi/4$, $\cos\theta$ is close to $\sqrt{2}/2$.

More care is needed in evaluating a limit near a point where the function is not defined. In these cases the limit cannot be determined by a simple substitution as in the previous examples.

EXAMPLE 3 Evaluate

$$\lim_{t\to 3} \frac{2t^2 - 4t - 6}{t^2 - 9}. \tag{4}$$

Solution

We demonstrate three techniques for exploring this limit: numerical, graphical, and analytical. First, however, let's see why the "commonsense" approach used in the two previous examples fails here. If t is close to but not equal to 3, then the numerator in (4) is close to 0 because

$$2t^2 - 4t - 6 \approx 2(3)^2 - 4(3) - 6 = 0.$$

Similarly, if t is close to 3, then the denominator in (4) is also close to 0 because

$$t^2 - 9 \approx 3^2 - 9 = 0.$$

Hence the numerator and denominator in (4) are both small, nonzero numbers. This means that when t is close to 3, the quotient in (4) is a ratio of two small, nonzero numbers. A problem arises because we cannot say for certain what happens when we divide one small number by another. The result might be relatively small, relatively large, or somewhere in between; for example,

$$0.0001/0.001 = 0.1, \quad 0.001/0.0001 = 10, \quad 0.002/0.001 = 2.$$

Thus simply knowing that the limit in (4) is close to the ratio of two small numbers is not enough information for us to determine what number, if any, is the limit. See Exercises 23, 24, and 25.

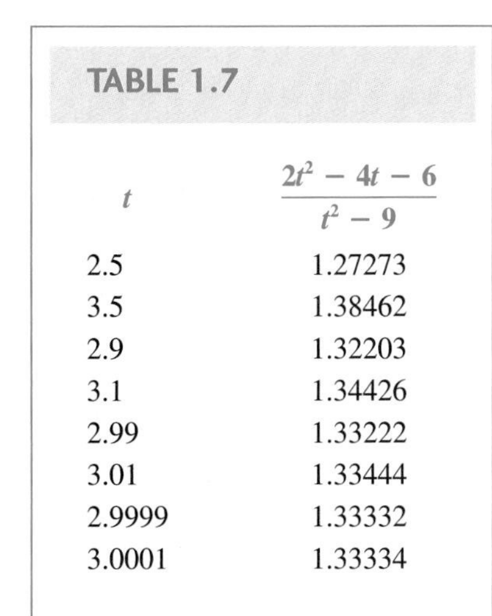

TABLE 1.7

t	$\dfrac{2t^2 - 4t - 6}{t^2 - 9}$
2.5	1.27273
3.5	1.38462
2.9	1.32203
3.1	1.34426
2.99	1.33222
3.01	1.33444
2.9999	1.33332
3.0001	1.33334

NUMERICAL APPROACH

With a calculator or computer, we can generate a list of values of

$$\frac{2t^2 - 4t - 6}{t^2 - 9} \tag{5}$$

for t values close to 3. A list of such values is shown in Table 1.7. (You should verify some of the values in this table, and add one or two new lines to the table.) The numerical data seem to indicate that the value of the quotient in (5) gets close to 1.3333 as t nears 3. Thus we have evidence that

$$\lim_{t \to 3} \frac{2t^2 - 4t - 6}{t^2 - 9} \approx 1.333.$$

GRAPHICAL APPROACH

Figure 1.44 shows the graph of

$$y = \frac{2t^2 - 4t - 6}{t^2 - 9}$$

for $2 < t < 3$ and $3 < t < 4$. (You should produce a similar graph on your computer or calculator.) Let t be close to 3 on the horizontal axis. Move vertically from t to the graph and from the graph over to the vertical axis. We meet the vertical axis at a point close to 1.33. This means that when t is close to 3, the value of (5) is close to 1.33. This suggests that

$$\lim_{t \to 3} \frac{2t^2 - 4t - 6}{t^2 - 9} \approx 1.33.$$

To get a better approximation of the limit, zoom in on the graph and investigate the graph for t values even closer to 3.

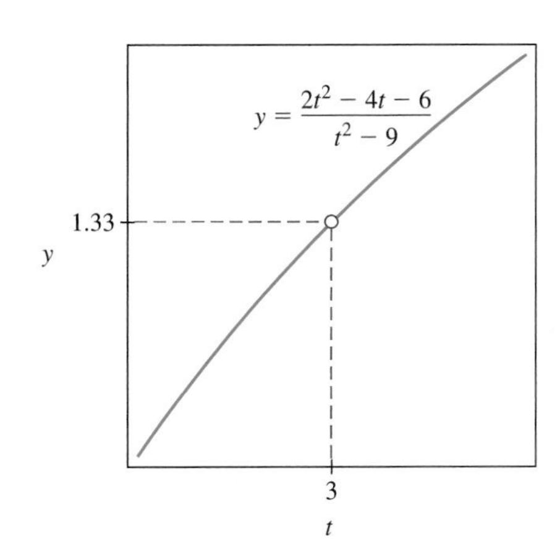

FIGURE 1.44 The graph indicates that $\lim_{t \to 3} \dfrac{2t^2 - 4t - 6}{t^2 - 9} \approx 1.3.$

ANALYTICAL APPROACH

After simplifying (5) algebraically, we can get very precise information about the limit. If t is close to 3 (but $t \neq 3$), then the expression in (5) can be simplified:

$$\frac{2t^2 - 4t - 6}{t^2 - 9} = \frac{2(t-3)(t+1)}{(t-3)(t+3)} = \frac{2(t+1)}{t+3}. \tag{6}$$

By canceling the factor of $t - 3$, we have eliminated the factor that causes the numerator and denominator of (5) to approach 0 as t approaches 3. When t is close to 3, the numerator of the last expression is close to $2(3 + 1) = 8$, and the denominator is close to $(3 + 3) = 6$. Thus when t is close to 3, the expression in (6) is close to

$$\frac{2(3+1)}{(3+3)} = \frac{4}{3}.$$

This means that

$$\lim_{t \to 3} \frac{2t^2 - 4t - 6}{t^2 - 9} = \lim_{t \to 3} \frac{2(t+1)}{(t+3)} = \frac{4}{3} \approx 1.33333. \tag{7}$$

Limits and the Rate of Change

In Section 1.4 we worked with the rate-of-change algorithm. This algorithm gives guidelines for determining the rate of change of a function f at a point a. We also saw that the rate-of-change algorithm can be used to identify instances in which the rate of change does not exist. According to the algorithm, the rate of change of f at a is determined by the behavior of the ratio, or "difference quotient"

$$\frac{f(a+h) - f(a)}{h}, \tag{8}$$

for values of h close to 0. Hence the rate of change, if it exists, is the limit of (8) as h tends to 0. We restate the rate-of-change algorithm in this more compact form.

The Rate-of-Change Algorithm

Let f be a function and a a real number in the domain of f. If the rate of change of f at a exists, then it is equal to

$$\lim_{h \to 0} \frac{f(a+h) - f(a)}{h}. \tag{9}$$

EXAMPLE 4 Calculating a rate of change

Let $y = f(x) = 2x^2 - 3x + 1$. Find the rate of change of y with respect to x at $x = -2$.

Solution

Using (9) with $a = -2$, the rate of change is

$$\lim_{h\to 0}\frac{f(-2+h)-f(-2)}{h} = \lim_{h\to 0}\frac{(2(-2+h)^2-3(-2+h)+1)-15}{h}$$
$$= \lim_{h\to 0}\frac{-11h+2h^2}{h}$$
$$= \lim_{h\to 0}(-11+2h) = -11.$$

EXAMPLE 5 Population growth rate

A population grows in such a way that at time t the population is

$$P = P(t) = 3\cdot 2^{t/40}, \tag{10}$$

where t is measured in years and population in billions of individuals. Find the rate of change of population with respect to time at time $t = 50$.

Solution

Using (9) with $a = 50$, the rate of change at time $t = 50$ is

$$\lim_{h\to 0}\frac{P(50+h)-P(50)}{h} = \lim_{h\to 0}\frac{3\cdot 2^{(50+h)/40}-3\cdot 2^{50/40}}{h}$$
$$= \lim_{h\to 0} 3\cdot\frac{2^{50/40}2^{h/40}-2^{50/40}}{h}$$
$$= \lim_{h\to 0} 3\cdot 2^{50/40}\frac{2^{h/40}-1}{h}.$$

Noting that the factor $3\cdot 2^{50/40}$ is $P(50)$, we have

$$\lim_{h\to 0}\frac{P(50+h)-P(50)}{h} = \lim_{h\to 0} P(50)\frac{2^{h/40}-1}{h}.$$

We approximate this last limit numerically by calculating the value of

$$P(50)\frac{2^{h/40}-1}{h}$$

for several values of h close to 0. These calculations are shown in Table 1.8. The values in the right column seem to be close to

$$P(50)\cdot 0.017329 \approx 0.123647. \tag{11}$$

Because the units for t are years and the units for P are billions of individuals, the rate of change has units of billions of people/year. Hence for $t = 50$, the rate of change of population is approximated by

0.123647 billion individuals per year.

This means that in year $t = 50$, the population grows by about 123,647,000 individuals. This is about 1.7 percent of the year 50 population.

TABLE 1.8

h	$P(50)\frac{2^{h/40}-1}{h}$
0.1	$0.017344\cdot P(50)$
−0.1	$0.017314\cdot P(50)$
0.005	$0.017329\cdot P(50)$
−0.005	$0.017328\cdot P(50)$
0.0001	$0.017329\cdot P(50)$
−0.0001	$0.017329\cdot P(50)$

Working with Limits

Sums, Products, and Quotients Many of the limits that we work with can be evaluated by algebraically combining simpler limits. For example, in (7) we evaluated

$$\lim_{t \to 3} \frac{2(t+1)}{(t+3)}$$

by noting that the numerator has limit 8 and the denominator has limit 6, and then reasoning that the quotient must have limit 8/6. This is an application of (15) in the following list of rules for limits.

Rules for Combining Limits

Let g and h be two functions with

$$\lim_{x \to a} g(x) = L \quad \text{and} \quad \lim_{x \to a} h(x) = M,$$

and let c be a real number. Then,

$$\lim_{x \to a}(c \cdot g(x)) = c\left(\lim_{x \to a} g(x)\right) = cL, \tag{12}$$

$$\lim_{x \to a}(g(x) + h(x)) = \left(\lim_{x \to a} g(x)\right) + \left(\lim_{x \to a} h(x)\right) = L + M, \tag{13}$$

$$\lim_{x \to a}(g(x)h(x)) = \left(\lim_{x \to a} g(x)\right)\left(\lim_{x \to a} h(x)\right) = LM. \tag{14}$$

If $M \neq 0$, then

$$\lim_{x \to a} \frac{g(x)}{h(x)} = \frac{\lim_{x \to a} g(x)}{\lim_{x \to a} h(x)} = \frac{L}{M}. \tag{15}$$

Although these results can all be proved, we will accept these statements without proof because they seem to follow from common sense. For example, to explain (14), we can reason as follows:

> When x is close to a, $f(x)$ is close to L and $g(x)$ is close to M. Thus when x is close to a, $f(x) \cdot g(x)$ is close to $L \cdot M$. This suggests that
>
> $$\lim_{x \to a}(g(x)h(x)) = LM.$$

Those who are interested in careful proofs for the limit rules should see Exercises 27 and 28 in Section 1.6.

Polynomials and Trigonometric Functions To use the limit rules effectively, we must be able to break complicated limits into simple ones and then evaluate the simple limits. Here are some of the more common "simple limits."

Limits of Polynomials, Sines, and Cosines

Let a be a real number. Then for any polynomial

$$p(x) = c_n x^n + c_{n-1} x^{n-1} + \cdots + c_1 x + c_0,$$
$$\lim_{x \to a} p(x) = p(a). \tag{16}$$

Also, for any real a,

$$\lim_{x \to a} \sin x = \sin a \quad \text{and} \quad \lim_{x \to a} \cos x = \cos a. \tag{17}$$

We also accept these results without proof because they seem reasonable given our experience with polynomials and trigonometric functions. We have already applied (16) and (17) in working Examples 1 and 2. Can you see where? When we apply the rules for combining limits and for finding limits of polynomials, sines, and cosines, we usually do not mention the application. However, we give one example to illustrate how we might document uses of these rules.

EXAMPLE 6 Evaluate

$$\lim_{t\to-1}(t^2 - 2t + 7)\tan t. \tag{18}$$

Solution

First recall that $\tan t = \sin t/\cos t$. By (17),

$$\lim_{t\to-1}\sin t = \sin(-1) \quad \text{and} \quad \lim_{t\to-1}\cos t = \cos(-1).$$

Because $\cos(-1) \neq 0$, it follows from (15) that

$$\lim_{t\to-1}\tan t = \lim_{t\to-1}\frac{\sin t}{\cos t} = \frac{\lim_{t\to-1}\sin t}{\lim_{t\to-1}\cos t} = \frac{\sin(-1)}{\cos(-1)} = \tan(-1).$$

Next, because $t^2 - 2t + 7$ is a polynomial, we use (16) to see that

$$\lim_{t\to-1}(t^2 - 2t + 7) = (-1)^2 - 2(-1) + 7 = 10.$$

We now know that the limit of each factor of (18) exists, so we apply (14) to get

$$\begin{aligned}\lim_{t\to-1}(t^2 - 2t + 7)\tan t &= \left(\lim_{t\to-1}(t^2 - 2t + 7)\right)\left(\lim_{t\to-1}\tan t\right)\\ &= 10\tan(-1) \approx -15.5741\end{aligned} \tag{19}$$

The graph of $y = (t^2 - 2t + 7)\tan t$ for t near -1 is shown in Fig. 1.45. From the graph we see that when t is close to -1, the value of $(t^2 - 2t + 7)\tan t$ is close to -15. This is consistent with (19).

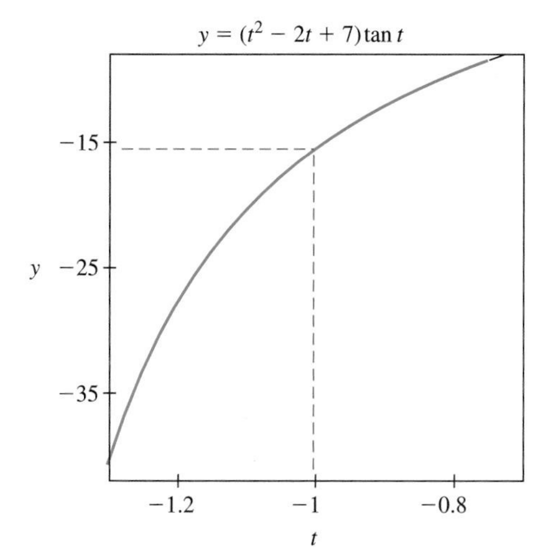

FIGURE 1.45 The graph of $y = (t^2 - 2t + 7)\tan t$ for t near -1.

Compositions With (12), (13), (14), and (15) we can find the limits of functions that are sums, products, or quotients of other functions whose limits we know. Because composition is another important way of combining functions, we give a rule for limits of compositions. This rule seems to be a natural consequence of the definition of limit, so we will accept it without proof. We discuss some aspects of this rule in Exercises 23 and 24 of Section 1.6.

Limits and Compositions

Let h and g be functions. Suppose that

$$\lim_{x\to a} g(x) = L$$

and that h is defined at L and

$$\lim_{x\to L} h(x) = h(L).$$

Then

$$\lim_{x\to a} h(g(x)) = h\left(\lim_{x\to a} g(x)\right) = h(L). \tag{20}$$

Most uses of (20) seem consistent with common sense, so we usually do not mention applications of the result. However, the following example is one in which we do point out the use of (20).

EXAMPLE 7 Find the value of

$$\lim_{\theta\to 1/2} \sin(\theta^2).$$

Solution

Let

$$p(\theta) = \theta^2.$$

Because $p(\theta)$ is a polynomial, we can apply (16) to obtain

$$\lim_{\theta\to 1/2} p(\theta) = p(1/2) = \left(\frac{1}{2}\right)^2 = \frac{1}{4}.$$

Next note that $\sin t$ is defined for $t = 1/4$ and that by (17)

$$\lim_{t\to 1/4} \sin t = \sin\left(\frac{1}{4}\right).$$

Now apply (20) to get

$$\begin{aligned}\lim_{\theta\to 1/2} \sin(\theta^2) &= \lim_{\theta\to 1/2} \sin(p(\theta)) \\ &= \sin\left(\lim_{\theta\to 1/2} p(\theta)\right) = \sin\left(\frac{1}{4}\right) \approx 0.247404\end{aligned} \tag{21}$$

The graph of $y = \sin(\theta^2)$ for θ near $1/2$ is shown in Fig. 1.46. Is the answer in (21) consistent with the graph?

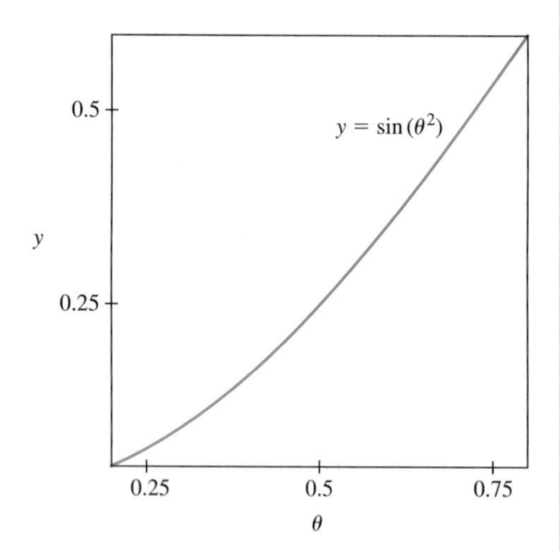

FIGURE 1.46 The graph of $y = \sin(\theta^2)$ for θ near $1/2$.

A Trigonometric Limit

In Chapter 2 we will calculate the rates of change of many different functions. When we find the rates of change of the sine and cosine functions, we will need to know the value of

$$\lim_{h\to 0} \frac{\sin h}{h}.$$

We evaluate this limit in the next example. In doing so, we see more techniques for working with limits and to review some trigonometry. For a more complete review, see Section A.2 in the Appendix.

EXAMPLE 8 Evaluate $\lim_{h\to 0} \dfrac{\sin h}{h}$, where h is in radians.

Solution

We first get an estimate for the limit by graphing

$$y = \frac{\sin h}{h} \tag{22}$$

for h values close to 0. The graph is shown in Fig. 1.47 and illustrates that the expression in (22) gets close to 1 as h gets close to 0. Hence we guess that

$$\lim_{h\to 0} \frac{\sin h}{h} \approx 1.$$

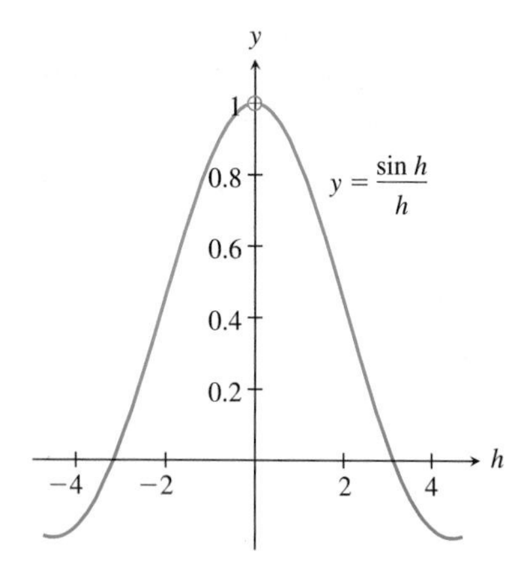

FIGURE 1.47 The graph of $y = (\sin h)/h$ suggests that $\lim_{h\to 0}(\sin h)/h = 1$.

We could also have found this estimate by gathering numerical data. See Exercise 13.

To verify analytically that the value of the limit is 1, we first show that for h close to 0,

$$\cos h \le \frac{\sin h}{h} \le \frac{1}{\cos h}. \tag{23}$$

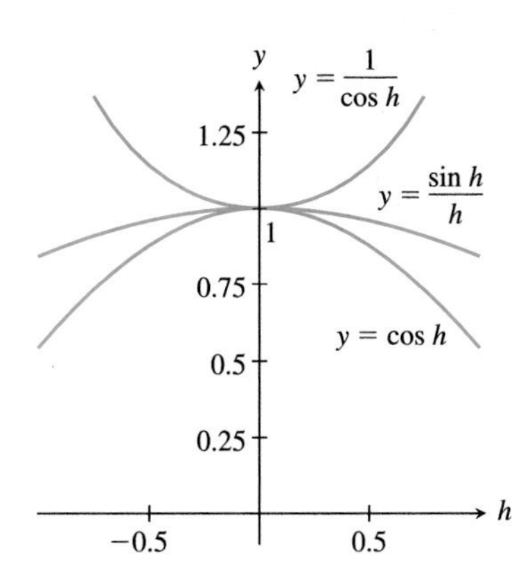

FIGURE 1.48 When h is small, $\sin h/h$ is between $\cos h$ and $1/\cos h$.

In Fig. 1.48 we show the graphs of

$$y = \cos h, \quad y = \frac{\sin h}{h}, \quad \text{and} \quad y = \frac{1}{\cos h},$$

and note that this figure does suggest (23). We start by establishing this inequality for small, positive values of h. Referring to Fig. 1.49, consider the circle of center $O = (0,0)$ and radius 1. Let $S = (1,0)$ and label P on the upper half of the circle so that $\angle POS$ has (radian) measure h. Then

$$P = (\cos h, \sin h).$$

From P draw a segment perpendicular to the x-axis. This segment intersects the axis at $R = (\cos h, 0)$. Let the line perpendicular to the x-axis at S meet OP at Q. Because triangle OSQ is a right triangle with right angle at S and $OS = 1$, it follows that

$$QS = \frac{QS}{1} = \frac{QS}{OS} = \frac{\text{opposite}}{\text{adjacent}} = \tan h.$$

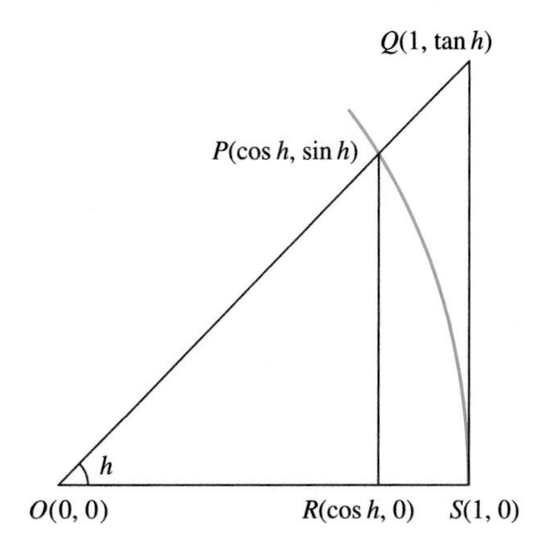

FIGURE 1.49 A right triangle in the circle of center $(0,0)$ and radius 1.

From Fig. 1.49,

$$\text{area}(\triangle ORP) < \text{area}(\text{sector } OSP) < \text{area}(\triangle OSQ). \tag{24}$$

Sector OSP has central angle h radians. The area of sector OSP is

$$\text{area}(\text{sector } OSP) = \frac{1}{2} 1^2 \cdot h = \frac{1}{2} h. \tag{25}$$

(See Section A.2 in the Appendix, where it is proved that in a circle of radius a, the area of a sector with central angle θ is $\frac{1}{2}a^2\theta$.) In addition,

$$\text{area}(\triangle ORP) = \frac{1}{2} \cos h \sin h \quad \text{and} \quad \text{area}(\triangle OSQ) = \frac{1}{2} \tan h. \tag{26}$$

Combining (24), (25), and (26), we have

$$\frac{1}{2} \cos h \sin h \le \frac{1}{2} h \le \frac{1}{2} \frac{\sin h}{\cos h} \tag{27}$$

Because we have assumed that h is a small, positive number, $\sin h > 0$. Multiplying inequality (27) by $2/\sin h$ gives

$$\cos h \le \frac{h}{\sin h} \le \frac{1}{\cos h}. \tag{28}$$

We now take reciprocals of the expressions in (28), recalling that if $0 < a < b$, then $0 < 1/b < 1/a$. We obtain

$$\cos h \le \frac{\sin h}{h} \le \frac{1}{\cos h}.$$

This establishes (23) for h positive and close to 0. In Exercise 44 we ask you to show that inequality is also true if h is small and negative. Hence (23) is true for all small, nonzero h.

We now use (23) to show that $\lim_{h \to 0} \frac{\sin h}{h} = 1$. First observe that

$$\lim_{h \to 0} \cos h = \cos 0 = 1 \quad \text{and} \quad \lim_{h \to 0} \frac{1}{\cos h} = \frac{1}{\lim_{h \to 0} \cos h} = 1.$$

From (23) we see that for small, nonzero h, the expression $(\sin h)/h$ is "trapped" between $\cos h$ and $1/\cos h$. See Fig. 1.48. Since $\cos h$ and $1/\cos h$ both tend to 1 as h approaches 0, it must be true that $(\sin h)/h$ also approaches 1 as h tends to 0, that is

$$\lim_{h \to 0} \frac{\sin h}{h} = 1.$$

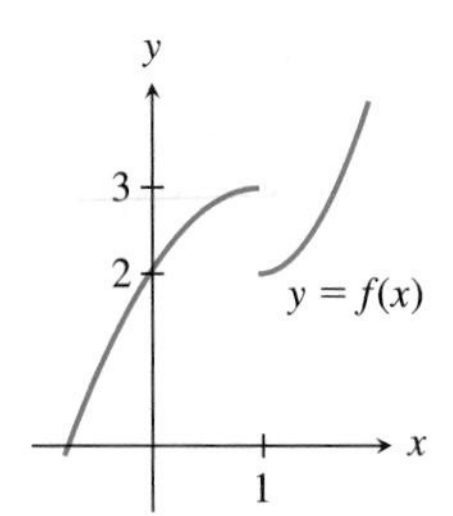

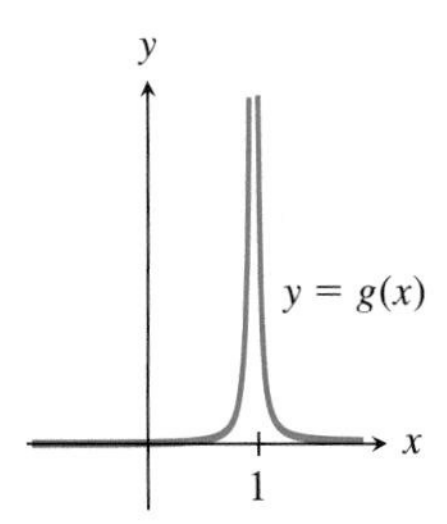

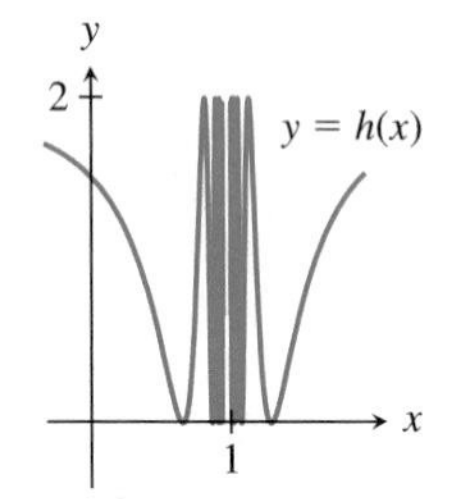

FIGURE 1.50 Sometimes a limit does not exist.

Existence of Limits

Figure 1.50 shows the graphs of three functions, f, g, and h, near $x = 1$. The graphs of these three functions look very different near $x = 1$. From the graph of f we see that if $x < 1$ and close to 1, then $f(x)$ is close to 3, but if $x > 1$ and close to 1, then $f(x)$ is close to 2.

From the graph of g we see that as x gets close to 1, the values of $g(x)$ get bigger and bigger. They seem to be growing to infinity, so the line $x = 1$ may be an asymptote to the graph of g.

The graph of h oscillates wildly between the horizontal lines $y = 0$ and $y = 2$ as x gets close to 1.

Though these graphs look very different, they do have one thing in common. Each graph indicates that *as x gets close to 1, the function values do not approach one and only one finite real number.* More specifically, as x approaches 1, $f(x)$ gets close to two different numbers, $g(x)$ does not get close to any finite number, and $h(x)$, as it oscillates, gets close to every number between 0 and 2. For each of these three functions, the limit as x approaches 1 does not exist.

When Limits Do Not Exist

Let F be a function and a a real number. If there is no single finite number L such that $F(x)$ gets close to L (and only L) as x gets close to a, then

$$\lim_{x \to a} F(x) \text{ does not exist.}$$

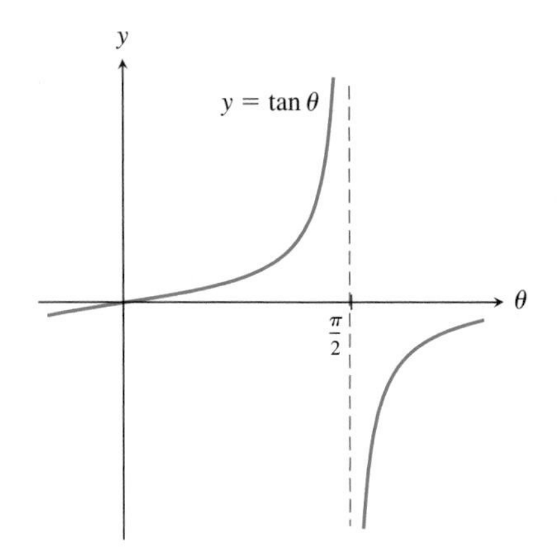

FIGURE 1.51 The graph indicates that $\lim_{\theta \to \pi/2} \tan \theta$ does not exist.

EXAMPLE 9 Show that $\lim_{\theta \to \pi/2} \tan \theta$ does not exist.

Solution

When evaluating a limit of a function, it is often a good idea to look at a graph of the function. In Fig. 1.51 we see the graph of $y = \tan \theta$ for θ near

TABLE 1.9

θ	$\tan\theta$
$\pi/2 + 0.5$	-1.83049
$\pi/2 - 0.5$	1.83049
$\pi/2 + 0.1$	-9.96664
$\pi/2 - 0.1$	9.96664
$\pi/2 + 0.01$	-99.9967
$\pi/2 - 0.01$	99.9967
$\pi/2 + 0.001$	-1000.0
$\pi/2 - 0.001$	1000.07

$\pi/2$. The graph suggests that the line $\theta = \pi/2$ is a vertical asymptote to the graph. When $\theta > \pi/2$ and close to $\pi/2$, the values of $\tan\theta$ are "large and negative." (This means that $\tan\theta < 0$ and $|\tan\theta|$ is large.) Thus it appears that the function values are heading toward $-\infty$ as θ closes in on $\pi/2$ from the right. When $\theta < \pi/2$ and close to $\pi/2$, the values of $\tan\theta$ are large and positive, and it appears that the function values are approaching $+\infty$ as θ approaches $\pi/2$ from the left. Since $\tan\theta$ is not getting close to a single number as θ gets close to $\pi/2$, we conclude that

$$\lim_{\theta\to\pi/2} \tan\theta \text{ does not exist.}$$

We can also investigate the behavior of the tangent function near $\theta = \pi/2$ by computing $\tan\theta$ for several θ values near $\pi/2 \approx 1.5708$. See Table 1.9. The data in the table reflect the behavior of the graph and so also suggest that

$$\lim_{\theta\to\pi/2} \tan\theta \text{ does not exist.}$$

Exercises 1.5

Exercises 1–12: Determine the limit, if it exists. Use analytical methods, then check your answer using graphical or numerical methods.

1. $\lim_{x\to 2}(x^2 - 3x)/(x + 1)$

2. $\lim_{t\to -1/2} \sin \pi t$

3. $\lim_{h\to 0}(3h^3 - 7h^2 + h)/h$

4. $\lim_{x\to -4} (2x^2 + 7x - 4)/(x + 4)$

5. $\lim_{t\to 1/2} \dfrac{2t + 1}{2t - 1}$

6. $\lim_{r\to\sqrt{2}} \dfrac{r^2 - 2}{r^3 + 2r^2 - 2r - 4}$

7. $\lim_{h\to 0} \dfrac{g(4 + h) - g(4)}{h}$, where $g(t) = t^3 - 3t$

8. $\lim_{h\to 0} \dfrac{G(-8 + h) - G(-8)}{h}$, where $G(s) = 2/(s + 1)$

9. $\lim_{x\to -1/3} \dfrac{\sqrt{6x + 3} - 1}{6x + 2}$

10. $\lim_{x\to -1} f(x)$, where

$$f(x) = \begin{cases} x + 1 & (x < -1) \\ 2 - 2x^2 & (-1 \le x \le 2) \\ 0 & (x > 2) \end{cases}$$

11. $\lim_{x\to 2} f(x)$ for f as given in Exercise 10.

12. $\lim_{h\to 0} \dfrac{f(-1 + h) - f(-1)}{h}$ for f as given in Exercise 10.

T **13.** Use numerical techniques to investigate

$$\lim_{h\to 0} \frac{\sin h}{h}.$$

T **14.** Investigate

$$\lim_{\theta\to 0} \frac{1 - \cos\theta}{\theta^2}.$$

Use graphical and numerical techniques.

T **15.** Investigate

$$\lim_{\theta\to 0} \frac{\tan\theta}{\theta}.$$

Use graphical and numerical techniques.

T **16.** Investigate

$$\lim_{\theta\to 0} \cos\left(\frac{1}{\theta}\right).$$

Use graphical and numerical techniques.

T **17.** Investigate

$$\lim_{\theta\to\pi/4} \frac{\tan\theta - 1}{\theta - \pi/4}.$$

Use graphical and numerical techniques.

18. The accompanying figure shows the graph of a function f.

a. For what values of c does $\lim_{x \to c} f(x) = 1$?

b. For what values of c does $\lim_{x \to c} f(x)$ not exist?

c. Evaluate $\lim_{x \to -1} f(x)$. d. Evaluate $\lim_{x \to -1} f(x^2)$.

e. Evaluate $\lim_{x \to -1} f(\sqrt{x})$.

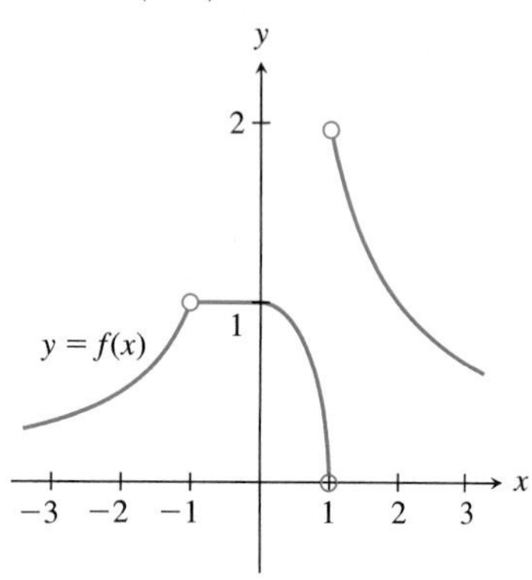

Figure for Exercise 18.

19. The accompanying figure shows the graph of a function g.

a. For what values of c does $\lim_{x \to c} g(x)$ fail to exist?

b. Evaluate $\lim_{x \to 2} g(x)$.

c. Evaluate $\lim_{x \to 2} (g(x))^2$.

d. Evaluate $\lim_{x \to 2} (g(x))^3$.

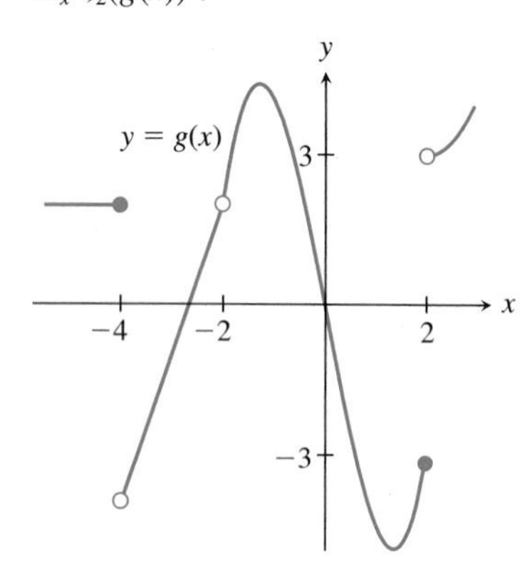

Figure for Exercise 19.

20. From the graph in Fig. 1.40, what is the temperature as the average molecular speed approaches 0?

T 21. Investigate the following limits both graphically and numerically:

a. $\lim_{h \to 0} \dfrac{2^h - 1}{h}$ b. $\lim_{h \to 0} \dfrac{3^h - 1}{h}$

c. $\lim_{h \to 0} \dfrac{(1/2)^h - 1}{h}$ d. $\lim_{h \to 0} \dfrac{(1/3)^h - 1}{h}$

T 22. Let a be a fixed positive number. Consider the limit

$$\lim_{h \to 0} \frac{a^h - 1}{h}.$$

By using numerical methods, find an approximate value of a for which this limit is 1.

23. Find real numbers a and b with $-0.01 \le a, b \le 0.01$ and:

a. $a/b = 3$.

b. $a/b = -500$.

c. $a/b = 10^9$.

d. $a/b = -10^{-9}$.

24. Let a and b be two real numbers. Find the largest and smallest possible values of a/b if:

a. $3.9 \le a \le 4.1$ and $5 \le b \le 5.2$.

b. $-3.1 \le a \le -2.9$ and $5 \le b \le 5.2$.

c. $-0.1 \le a \le 0.1$ and $9.9 \le b \le 10.2$.

25. Find functions f and g such that

$$\lim_{x \to 3} f(x) = \lim_{x \to 3} g(x) = 0$$

and:

a. $\lim_{x \to 3} \dfrac{f(x)}{g(x)} = 5$. b. $\lim_{x \to 3} \dfrac{f(x)}{g(x)} = -\sqrt{3}$.

c. $\lim_{x \to 3} \dfrac{f(x)}{g(x)}$ does not exist.

26. Let $0 < c < \pi/2$ and $0 < d < \pi/2$. If $\lim_{\theta \to c} \sin \theta = 0.3$ and $\lim_{\theta \to d} \sin \theta = 0.9$, find:

a. $\lim_{\theta \to c+d} \sin \theta$. b. $\lim_{\theta \to c-d} \cos \theta$.

c. $\lim_{\theta \to 2d} \cos \theta$.

27. The graph of a function f is shown in the accompanying figure. On the same set of axes draw the graph of a function g so that $\lim_{x \to -2} (f(x) + g(x))$ does not exist, and $\lim_{x \to 0} (f(x) + g(x)) = 1$.

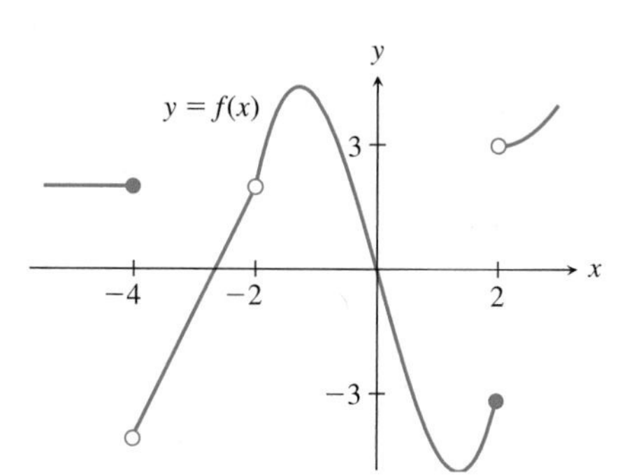

Figure for Exercise 27.

28. Let f be a function and assume that $\lim_{x \to a} f(x) = L$. Is it then true that

$$\lim_{x \to a} |f(x) - L| = 0?$$

Give reasons for your answer.

Exercises 29–36: Find the rate of change of y with respect to x for $x = a$.

29. $y = -2x + 3, \quad a = 4$

30. $y = x^2 - 3x, \quad a = -1$

31. $y = \dfrac{1}{x + 1}, \quad a = 1$

32. $y = x^3, \quad a = -2$

33. $y = x^3, \quad a = 0$

34. $y = -3x^2 + 4x + 2, \quad a = -1$

35. $y = \dfrac{2}{x - 2}, \quad a = 0$

36. $y = \dfrac{x}{x + 1}, \quad a = 2$

37. Let g be a function and assume that $\lim_{x \to a} |g(x) - M| = 0$. What is the value of

$$\lim_{x \to a} g(x)?$$

Give reasons for your answer.

T 38. Let $g(t) = t^3 - 4t - 4$.

a. Verify that $\lim_{t \to -2} g(t) = -4$.

b. Answer the following question by looking at a graph: If t is within 0.01 of -2, then what is the largest $|g(t) - (-4)|$ can be?

c. By looking at a graph, find a positive number r that satisfies the following: If t is within r of -2, then $|g(t) - (-4)| < 0.05$.

T 39. Let $f(x) = x^2 + 2x + 2$.

a. Verify that $\lim_{x \to 1} f(x) = 5$.

b. Answer the following question by looking at a graph: If x is within 0.05 of 1, then what is the largest $|f(x) - 5|$ can be?

c. By looking at a graph, find a positive number r that satisfies the following: If x is within r of 1, then $|f(x) - 5| < 0.1$.

T 40. Let $f(x) = (\sqrt{x} - 2)/(x - 4)$.

a. Verify that $\lim_{x \to 4} f(x) = 1/4$.

b. Answer the following question by looking at a graph: If x is within 0.05 of 4, then what is the largest $|f(x) - 1/4|$ can be?

c. By looking at a graph, find a positive number r that satisfies the following: If x is within r of 4, then $|f(x) - 1/4| < 0.1$.

T 41. Let $r(\theta) = (\sin 2\theta)/\theta$.

a. Verify that $\lim_{\theta \to 0} r(\theta) = 2$.

b. Answer the following question by looking at a graph: If θ is within 0.1 of 0, then what is the largest $|r(\theta) - 2|$ can be?

c. By looking at a graph, find a positive number r that satisfies the following: If θ is within r of 0, then $|r(\theta) - 2| < 0.05$.

42. Charles' law The accompanying table shows how the volume of a gas varies with temperature when the pressure of the gas is kept constant. Plot these data, and then find a line approximating the data. With the aid of this line, find the limit of the temperature as the volume approaches 0. (These data were generated by computer as part of a chemistry lab to investigate methods of finding the Celsius temperature of absolute zero.)

Volume (cm^3)	Temperature (°C)
230	9.4
233.9	14.3
238.1	19.4
242.9	25.3
246.7	30.0
250.9	35.1

Table for Exercise 42.

43. Use Fig. 1.49 to give a geometric argument that $|\sin h| \le |h|$.

44. Show that (23) also holds when h is a small negative number. (*Hint:* Recall that $\sin(-h) = -\sin h$ for all real numbers h.)

45. In Example 8 we found the limit of $(\sin h)/h$ as h approaches 0 by trapping $(\sin h)/h$ between two functions with known limits. In doing so, we illustrated an application of the squeeze theorem:

Let f, g, and h be three functions, each defined for x values close to a but possibly not defined at $x = a$. Assume that

$$\lim_{x \to a} f(x) = L = \lim_{x \to a} h(x)$$

and that for x near a,

$$f(x) \le g(x) \le h(x).$$

Then

$$\lim_{h \to a} g(x) = L.$$

a. Draw a good picture illustrating the squeeze theorem.

b. Write a paragraph explaining why the squeeze theorem is true.

c. Explain how the squeeze theorem was used in Example 8.

46. With the aid of the accompanying figure, give a geometric argument that:

a. $|\sin\,\theta - \sin\,\alpha| \le |\theta - \alpha|$.

b. $|\cos\,\theta - \cos\,\alpha| \le |\theta - \alpha|$.

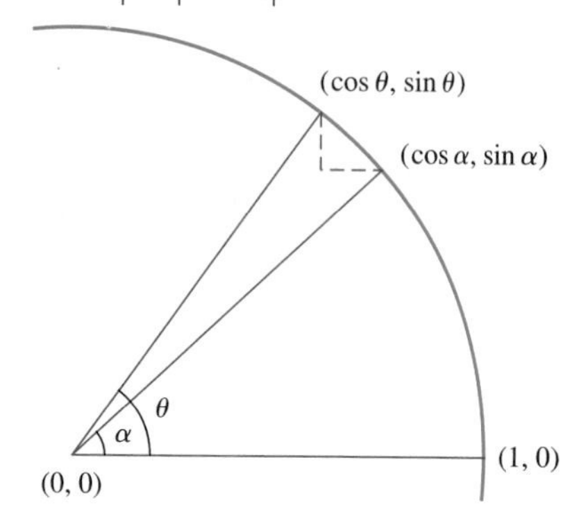

Figure for Exercise 46.

47. Use the squeeze theorem of Exercise 45 and the inequalities developed in Exercise 46 to show that:

a. $\lim_{\theta\to\alpha} \sin\,\theta = \sin\,\alpha$.

b. $\lim_{\theta\to\alpha} \cos\,\theta = \cos\,\alpha$.

How does this relate to (17)?

48. The world-record mile The accompanying table shows the history of the world record for the 1-mile run since 1875. These data are plotted in the accompanying figure.

a. Use any method to extrapolate the mile-run records to the year 2010.

b. Calculate the average of the times and the average of the dates using the table. Add the point for the average year and average time to the graph. (Such a point is sometimes called a *centroid.*) Put a nail through the centroid. Place a ruler against the nail and try to position the ruler so that if you were to draw a line with the ruler in this position, the line would "best fit" the data. Use this best-fit line to estimate what the record will be in 2010.

c. From these data, does it appear that there is a "limit" to how fast humans can run the mile? Give reasons for your answer, and discuss what you mean by "limit" in this situation.

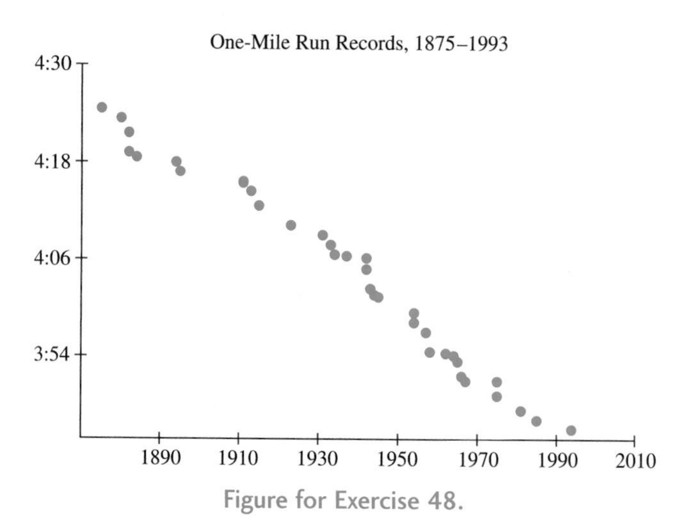

Figure for Exercise 48.

Year	Name (Country)	Time	Year	Name (Country)	Time
1875	Walter Slade (Britain)	4:24.5	1942	Gunder Haegg (Sweden)	4:04.6
1880	Walter George (Britain)	4:23.2	1943	Arne Andersson (Sweden)	4:02.6
1882	Walter George (Britain)	4:21.4	1944	Arne Andersson (Sweden)	4:01.6
1882	Walter George (Britain)	4:19.4	1945	Gunder Haegg (Sweden)	4:01.4
1884	Walter George (Britain)	4:18.4	1954	Roger Bannister (Britain)	3:59.4
1894	Fred Bacon (Scotland)	4:18.2	1954	John Landy (Australia)	3:58.0
1895	Fred Bacon (Scotland)	4:17.0	1957	Derek Ibbotson (Britain)	3:57.2
1911	Thomas Connett (U.S.)	4:15.6	1958	Herb Elliot (Australia)	3:54.5
1911	John Paul Jones (U.S.)	4:15.4	1962	Peter Snell (New Zealand)	3:54.4
1913	John Paul Jones (U.S.)	4:14.6	1964	Peter Snell (New Zealand)	3:54.1
1915	Norman Tauber (U.S.)	4:12.6	1965	Michel Jazy (France)	3:53.6
1923	Paavo Nurmi (Finland)	4:10.4	1966	Jim Ryun (U.S.)	3:51.3
1931	Jules Ladoumegue (France)	4:09.2	1967	Jim Ryun (U.S.)	3:51.1
1933	Jack Lovelock (New Zealand)	4:07.6	1975	Filbert Bayi (Tanzania)	3:51.0
1934	Glenn Cunningham (U.S.)	4:06.8	1975	John Walker (New Zealand)	3:49.4
1937	Sydney Wooderson (Britain)	4:06.4	1981	Sebastian Coe (Britain)	3:47.33
1942	Gunder Haegg (Sweden)	4:06.2	1985	Steve Cram (Britain)	3:46.30
1942	Arne Andersson (Sweden)	4:06.2	1993	Noureddine Morceli (Algeria)	3:44.39

Table for Exercise 48.

1.6 More Work with Limits

In the preceding section, we gave a "working" definition for *limit:*

DEFINITION Limit of a Function

Let g be a function and let a and L be real numbers. If we can make $g(x)$ as close to L as we like by taking x close to a, but not equal to a, then we say

$g(x)$ has limit L as x approaches a.

We denote this by

$$\lim_{x \to a} g(x) = L. \tag{1}$$

Equation (1) is read "the limit of $g(x)$ as x approaches a equals L."

For some work with limits, we need to be more precise about the phrase "close to" used in this definition. When we talk about the number $g(x)$ being close to the number L, it is natural to ask, "how close?" To answer this question, we look at

$$|g(x) - L|, \tag{2}$$

that is, the distance between $g(x)$ and L. If $g(x)$ is close to L, then the distance $|g(x) - L|$ between them should be small. We will measure how close $g(x)$ is to L by measuring how small (2) is. We do this with a number. For example, to say that $g(x)$ is within 0.01 of L is the same as saying that

$$|g(x) - L| \leq 0.01.$$

The definition of *limit* says: We can make $g(x)$ as close to L as we like by taking x close to a. This means that we can make (2) as small as we like by making $|x - a|$ sufficiently small. We illustrate this with some examples.

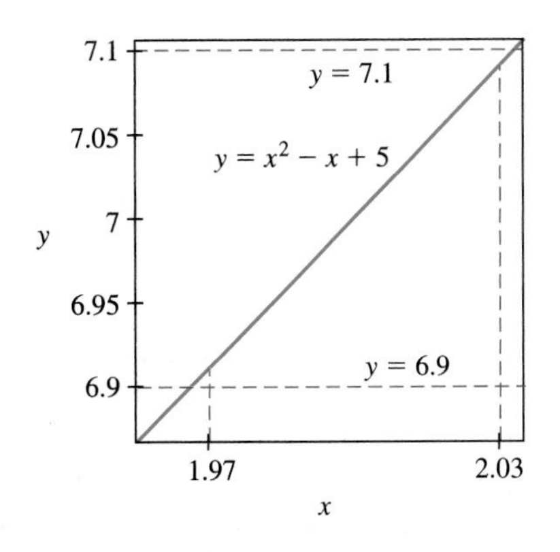

FIGURE 1.52 When $1.97 < x < 2.03$ we have $6.9 < x^2 - x + 5 < 7.1$.

EXAMPLE 1 Let $g(x) = x^2 - x + 5$ and consider the statement

$$\lim_{x \to 2} g(x) = 7.$$

a) Show that if $|x - 2| < 0.03$, then $g(x)$ is within 0.1 of 7.

b) Suppose we want

$$|g(x) - 7| < 0.001. \tag{3}$$

Find a positive number d so that (3) is true if $|x - 2| < d$.

Solution

a) We verify this graphically. First note that

$$|x - 2| < 0.03 \quad \text{is equivalent to} \quad 1.97 < x < 2.03.$$

The graph of $y = g(x)$ on this interval is shown in Fig. 1.52. We see that for $1.97 < x < 2.03$, the points on the graph lie between the lines

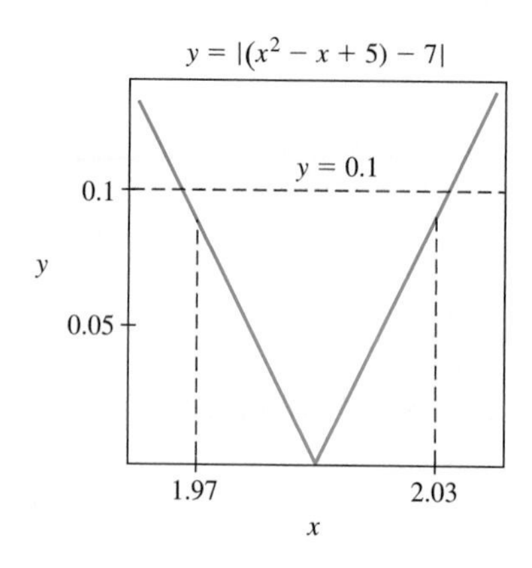

FIGURE 1.53 If $|x - 2| < 0.03$ then have $|(x^2 - x + 5) - 7| < 0.1$.

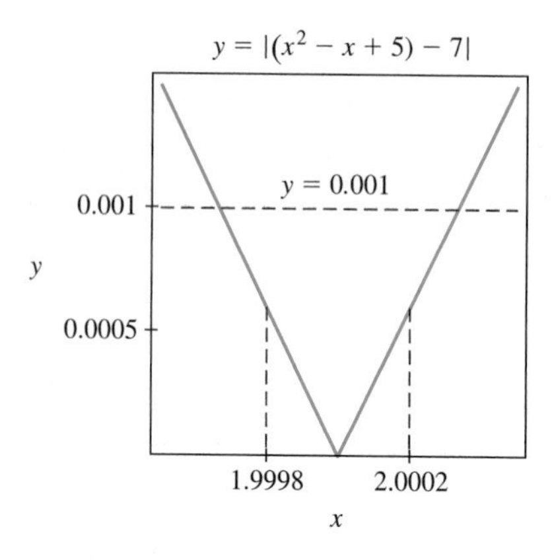

FIGURE 1.54 When $|x - 2| < 0.0002$ then $|(x^2 - x + 5) - 7| < 0.001$.

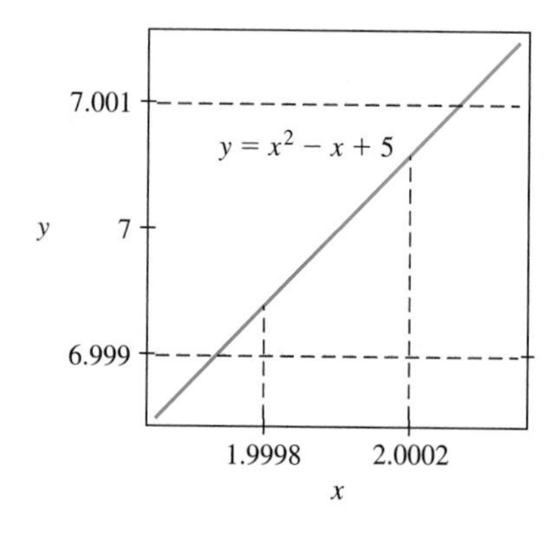

FIGURE 1.55 When $1.9998 < x < 2.0002$ we have $6.999 < g(x) < 7.001$.

$y = 6.9$ and $y = 7.1$. This means that if

$$1.97 < x < 2.03 \quad \text{then} \quad 6.9 < g(x) < 7.1.$$

Hence if $|x - 2| < 0.03$, then $g(x) = x^2 - x + 5$ is within 0.1 of 7.

For a different approach, graph $y = |g(x) - 7|$ for $1.97 < x < 2.03$. See Fig. 1.53. For x in this interval, the points on the graph lie below the line $y = 0.1$. This means that

$$|g(x) - 7| < 0.1 \quad \text{if} \quad 1.97 < x < 2.03.$$

Thus $g(x) = x^2 - x + 5$ is within 0.1 of 7 when x is within 0.03 of 2.

b) Because we want $|g(x) - 7| < 0.001$ for x close to 2, draw the graph of $y = |g(x) - 7|$ on some small interval centered at 2. In Fig. 1.54 we see the graph for $1.9995 < x < 2.0005$. If we include the line $y = 0.001$ on the graph, we can then see that

$$y = |g(x) - 7| < 0.001 \quad \text{for} \quad 1.9998 < x < 2.0002,$$

that is, when

$$|x - 2| < 0.0002.$$

Hence we may take $d = 0.0002$.

For another solution, see Fig. 1.55. In this figure we show the graph of $y = g(x)$ for $|x - 2| < 0.0002$. Note that for x values in this interval, the points of the graph lie between the lines $y = 6.999$ and $y = 7.001$. This again shows that 0.0002 is an acceptable value for d. Are there other acceptable values of d? How many other values? It is important to realize that graphs like those shown in Figs. 1.54 and 1.55 are usually not produced on the first attempt. The authors tried several graphs, with various parts of the domain and range specified, before producing pictures that show the needed details.

EXAMPLE 2 In the preceding section we showed that

$$\lim_{\theta \to 0} \frac{\sin \theta}{\theta} = 1.$$

Find $d > 0$ so that

$$\text{if } 0 < |\theta - 0| < d, \quad \text{then} \quad \left| \frac{\sin \theta}{\theta} - 1 \right| < 0.05. \tag{4}$$

Solution

In Figure 1.56 we show the graph of $y = (\sin \theta)/\theta$ for $-1 < \theta < 1$, $\theta \neq 0$. Because we want

$$0.95 < \frac{\sin \theta}{\theta} < 1.05,$$

we include the lines $y = 0.95$ and $y = 1.05$ in the figure. The graph of $y = (\sin \theta)/\theta$ lies between these two horizontal lines when $-0.5 < \theta < 0$ or

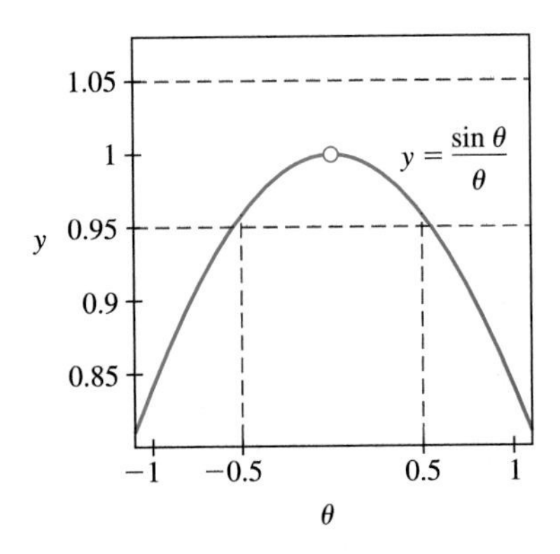

FIGURE 1.56 When $-0.5 < x < 0.5$ we have $0.95 < (\sin \theta)/\theta < 1.05$.

$0 < \theta < 0.5$. Thus we can take $d = 0.5$. For this value of d, statement (4) is true. Explain why any positive $d < 0.5$ would also work.

We could also find an acceptable value of d by looking at numerical data. Table 1.10 gives the value of $(\sin \theta)/\theta$ for several θ values close to 0. The data in Table 1.10 suggest that

$$\left|\frac{\sin \theta}{\theta} - 1\right| < 0.05 \quad \text{when} \quad 0 < |x| < 0.4.$$

(It would be a good idea to verify this graphically. Why?) Based on the data, we can take $d = 0.4$.

TABLE 1.10

θ	$\frac{\sin \theta}{\theta}$
−0.8	0.897
−0.6	0.941
−0.4	0.974
−0.2	0.993
0.2	0.993
0.4	0.974
0.6	0.941
0.8	0.897

The preceding examples illustrate the meaning of "close to" in the definition of *limit.* In this definition, the phrase "we can make $g(x)$ as close to L as we like" means that given any small number ϵ (pronounced ep′-si-lon), we can make $g(x)$ within ϵ of L, that is, we can make

$$|g(x) - L| < \epsilon. \tag{5}$$

We make $g(x)$ close to L by taking x close to a but not equal to a. This means that if ϵ is given, we can find another number δ (pronounced del′-ta) so that (5) is true when $0 < |x - a| < \delta$. That is, we must find δ so that

$$0 < |x - a| < \delta \quad \text{implies} \quad |g(x) - L| < \epsilon.$$

In Example 1b we were given $\epsilon = 0.001$ and found $\delta = d = 0.0002$. In Example 2 we were given $\epsilon = 0.05$ and found $\delta = d = 0.5$. We also showed that $\delta = d = 0.4$ was acceptable.

With this new understanding of "close to," we restate the definition of *limit.*

DEFINITION Limit of a Function

Let a and L be real numbers and assume that g is defined throughout an interval centered at a, except possibly at the point $x = a$. We say that $g(x)$ has limit L as x approaches a if for each $\epsilon > 0$ there is a corresponding $\delta > 0$ such that

$$\text{if } 0 < |x - a| < \delta, \quad \text{then} \quad |g(x) - L| < \epsilon. \tag{6}$$

See Fig. 1.57. We denote this by

$$\lim_{x \to a} g(x) = L.$$

If there is no real number L for which this is true, then we say $g(x)$ has no limit as x approaches a, or that $\lim_{x \to a} g(x)$ is undefined.

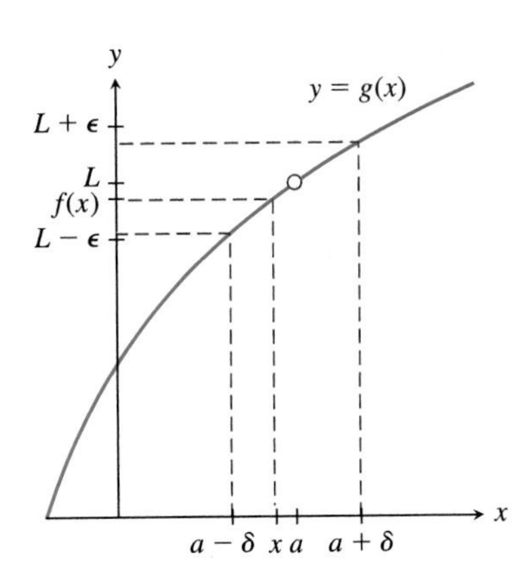

FIGURE 1.57 First we are given $\epsilon > 0$. There is a $\delta > 0$ so that when $0 < |x - a| < \delta$ we have $|g(x) - L| < \epsilon$.

According to this definition, a may or may not be in the domain of g. From (6) we see that we are concerned only with $g(x)$ values for x close to a but not equal to a. Hence whether a is in the domain of g or not has nothing to do with the value of $\lim_{x \to a} g(x)$.

EXAMPLE 3

a) Evaluate

$$\lim_{t\to 3}\frac{\sqrt{3t-5}-2}{t-3}. \tag{7}$$

b) Let L be the value of the limit in part **a)** and let $\epsilon = 0.035$. Find a number $\delta > 0$ so that

$$\text{if} \quad 0 < |t-3| < \delta, \quad \text{then} \quad \left|\frac{\sqrt{3t-5}-2}{t-3} - L\right| < \epsilon.$$

Use graphical methods.

Solution

a) Notice that when $t = 3$, the numerator and denominator of the fraction

$$\frac{\sqrt{3t-5}-2}{t-3} \tag{8}$$

are both 0. Thus we cannot hope to evaluate (7) by substituting $t = 3$ in (8). To simplify (8), rationalize the numerator by multiplying the numerator and denominator of the expression by $\sqrt{3t-5}+2$:

$$\begin{aligned}\frac{\sqrt{3t-5}-2}{t-3} &= \frac{\sqrt{3t-5}-2}{t-3}\cdot\frac{\sqrt{3t-5}+2}{\sqrt{3t-5}+2}\\ &= \frac{\left(\sqrt{3t-5}\right)^2-2^2}{(t-3)\left(\sqrt{3t-5}+2\right)}\\ &= \frac{3(t-3)}{(t-3)\left(\sqrt{3t-5}+2\right)}\\ &= \frac{3}{\sqrt{3t-5}+2}.\end{aligned}$$

Thus

$$\lim_{t\to 3}\frac{\sqrt{3t-5}-2}{t-3} = \lim_{t\to 3}\frac{3}{\sqrt{3t-5}+2} = \frac{3}{\sqrt{3\cdot 3-5}+2} = \frac{3}{4}.$$

b) With a calculator or CAS, sketch the graph of

$$y = \left|\frac{\sqrt{3t-5}-2}{t-3} - \frac{3}{4}\right|$$

for t values close to 3. See Fig. 1.58. We see for $t \neq 3$ and $2.8 < t < 3.2$, the graph lies below the line $y = \epsilon = 0.035$. Thus we can take $\delta = 0.2$ and see that

$$0 < |t-3| < 0.2 = \delta$$

implies

$$\left|\frac{\sqrt{3t-5}-2}{t-3} - \frac{3}{4}\right| < 0.035 = \epsilon.$$

Note that any positive number smaller than 0.2 would also be an acceptable value for δ.

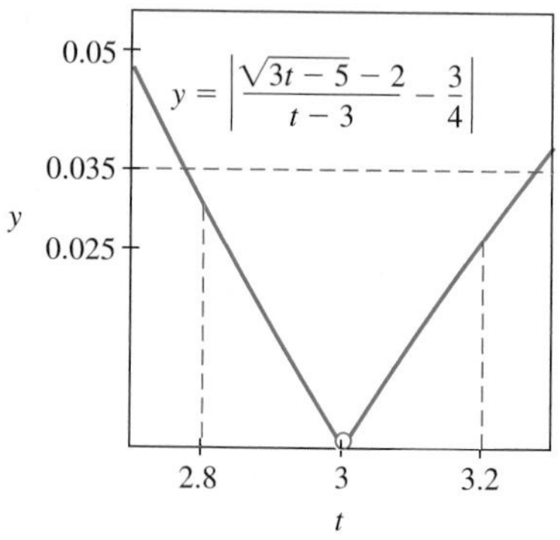

FIGURE 1.58 If $0 < |t-3| < 0.2$, then $\left|\frac{\sqrt{3t-5}-2}{t-3} - \frac{3}{4}\right| < 0.035$.

Limits and Approximation

When we use a calculator to find the value of sin 1 (where the 1 is in radians), we see the answer 0.841471 on the display. This decimal expression is not the *exact* value of sin 1, but we assume that the answer we see is correct to the number of digits shown. How does the calculator obtain this answer?

In calculators and computers the calculation of the values of square roots, trigonometric functions, logarithms, and exponentials of numbers is based on algorithms and formulas that approximate these functions. To guarantee that the calculator answer is correct to the number of digits shown, the people designing the calculator must know how accurate the approximation is. Limits are important in finding good approximations and in studying the accuracy of an approximation.

The statement

$$\lim_{x \to a} f(x) = L$$

means that $f(x)$ is close to L when x is close to a. Suppose that for a given x value we need an approximation of $f(x)$. Would it be acceptable to say

$$f(x) \approx L? \tag{9}$$

We need a lot more information before we can answer this question. For our value of x, how good is the approximation given in (9)? How good an approximation do we need? The answer to the second question depends on what we want to do with the approximation. To answer the first question, we need to be able to say something about the *error in the approximation,*

$$E(x) = f(x) - L.$$

We use the error, $E(x)$, to measure how good the approximation (9) is. If $E(x)$ is small enough, the approximation is a good one and may serve our needs. If $E(x)$ is large, the approximation may not be of any use. It is usually impossible or impractical to calculate $E(x)$ exactly, but we can often find a number larger than $|E(x)|$. If this larger number is small, then the error $E(x)$ is even smaller (and could be either positive or negative). In Examples 2 and 3 we learned techniques for finding δ given ϵ. These same techniques can be used to analyze $E(x)$. In fact, the E in $E(x)$ not only stands for *error,* but also reminds us of ϵ.

EXAMPLE 4 Because $\lim_{x \to 3} 2^x = 8$, we consider using the approximation

$$2^x \approx 8$$

when x is close to 3. For what values of x is this approximation within 0.01 of the actual value of 2^x?

Solution

We need to find x values such that

$$|E(x)| = |2^x - 8| < 0.01.$$

In the language of limits, we are given $\epsilon = 0.01$ and need to find $\delta > 0$ so that

$$\text{if} \quad |x - 3| < \delta, \quad \text{then} \quad |E(x)| = |2^x - 8| < \epsilon = 0.01.$$

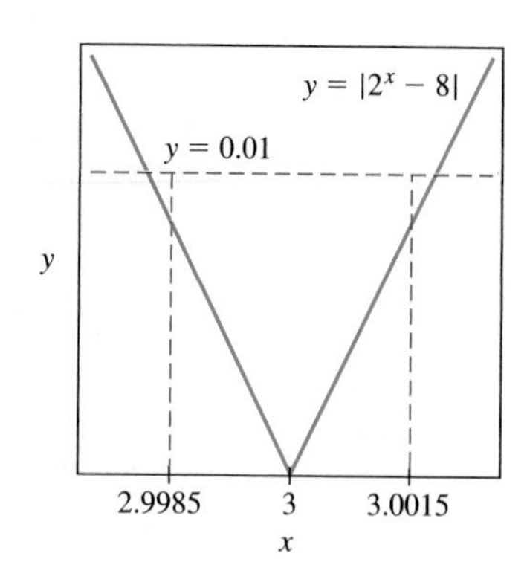

FIGURE 1.59 When $2.9985 < x < 3.0015$ we have 2^x within 0.01 of 8.

On your calculator or CAS graph $y = |2^x - 8|$ for x near 3. (Remember, the first graph you produce may not be what you want. The authors used their CAS to plot six different graphs of $y = |2^x - 8|$ before settling on the one in Fig. 1.59.) The graph of $y = |2^x - 8|$ lies below the line $y = 0.01$ for $2.9985 < x < 3.0015$. Hence for x in this interval we can say

$$2^x \approx 8$$

with an error of at most 0.01.

Sometimes we can manipulate a limit result to get a useful approximation. For example, in the preceding section we showed that

$$\lim_{x \to 0} \frac{\sin x}{x} = 1.$$

We use this result to find a good approximation to $\sin x$ for small x. Let

$$E(x) = \frac{\sin x}{x} - 1. \tag{10}$$

Because

$$\lim_{x \to 0} E(x) = \lim_{x \to 0} \left(\frac{\sin x}{x} - 1 \right) = 1 - 1 = 0,$$

we know that $E(x)$ can be made small by taking x close to 0. Rewrite (10) as

$$\frac{\sin x}{x} = 1 + E(x).$$

Then multiply by x to get

$$\sin x = x + x \cdot E(x),$$

or

$$\sin x - x = x \cdot E(x). \tag{11}$$

Hence

$$\sin x \approx x$$

because the error $\sin x - x = x \cdot E(x)$ is small when x is small. We say more about this error in the next example.

EXAMPLE 5 Discuss the error in the approximation

$$\sin x \approx x \tag{12}$$

for $-0.5 < x < 0.5$. (Make sure your calculator is in radian mode.)

Solution

In Example 2 we showed that

$$|E(x)| = \left| \frac{\sin x}{x} - 1 \right| < 0.05$$

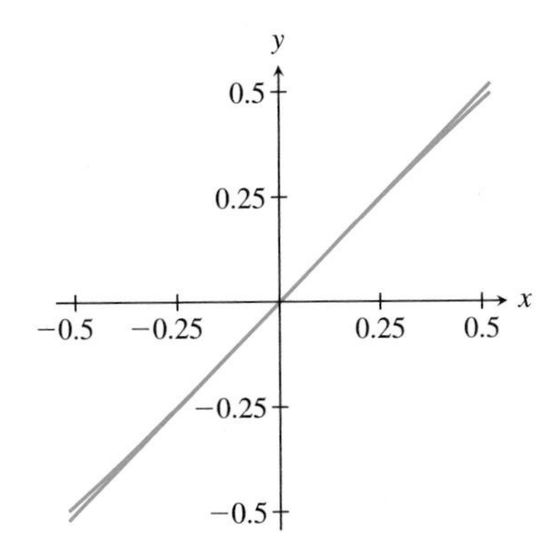

FIGURE 1.60 The graphs of $y = \sin x$ and $y = x$ are close for $-0.5 < x < 0.5$.

if $-0.5 < x < 0.5$. Using (11), we see that for x in this interval, the error in the approximation $\sin x \approx x$ is

$$|\sin x - x| = |x \cdot E(x)| = |x| \cdot |E(x)| \le 0.05|x| \le 0.05 \cdot 0.5 = 0.025. \quad (13)$$

Equation (13) tells us that if we use the approximation $\sin x \approx x$ for $-0.5 < x < 0.5$, then the error will be no more than 0.025 in magnitude. From (13) we also see that for $-0.5 < x < 0.5$,

$$|\sin x - x| \le 0.05|x|.$$

This means that for such x, the error in the approximation (12) is never more than 5 percent of $|x|$. Thus the error gets smaller as $|x|$ gets smaller, and the error is small compared to $|x|$.

To further understand the error, we graph $y = \sin x$ and $y = x$ for $-0.5 < x < 0.5$ on the same set of coordinate axes. The graph is shown in Fig. 1.60. The two graphs are close, which means that the numbers $\sin x$ and x are close for $-0.5 < x < 0.5$.

For another look at the error, graph the equation $y = |E(x)| = |\sin x - x|$ on $-0.5 < x < 0.5$. The graph is shown in Fig. 1.61. This graph shows plainly that the error in the approximation gets small as $|x|$ gets small.

For a more specific illustration, we compare $\sin x$ and x for a small value of x, say $x = 0.3$. A calculator gives $\sin(0.3) \approx 0.295520$, a value close to $x = 0.3$.

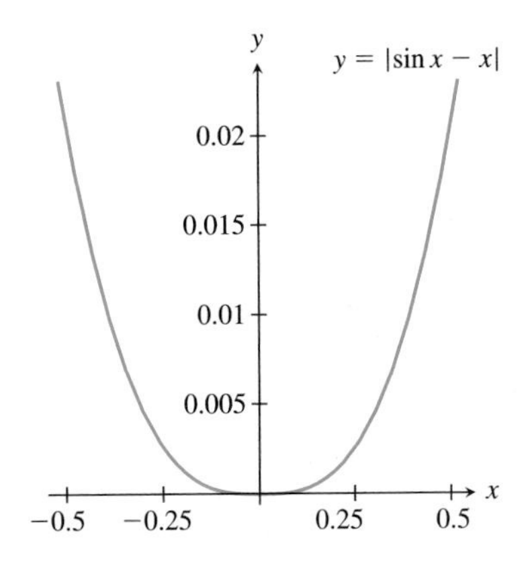

FIGURE 1.61 The graph of $y = |E(x)| = |\sin x - x|$ shows the error is small for $-0.5 < x < 0.5$.

EXAMPLE 6 Evaluate

$$\lim_{x \to 1} \frac{\sqrt{x} - 1}{x - 1}$$

and use the result to find a good approximation to $\sqrt{x}$ for x close to 1. Discuss the error in the approximation for x in the interval $0.9 \le x \le 1.1$.

Solution

For $x \ne 1$ we have

$$\frac{\sqrt{x} - 1}{x - 1} = \frac{\sqrt{x} - 1}{(\sqrt{x})^2 - 1^2} = \frac{\sqrt{x} - 1}{(\sqrt{x} - 1)(\sqrt{x} + 1)} = \frac{1}{\sqrt{x} + 1}.$$

Therefore

$$\lim_{x \to 1} \frac{\sqrt{x} - 1}{x - 1} = \lim_{x \to 1} \frac{1}{\sqrt{x} + 1} = \frac{1}{\sqrt{1} + 1} = \frac{1}{2}.$$

Now let

$$\frac{\sqrt{x} - 1}{x - 1} - \frac{1}{2} = E(x). \quad (14)$$

Solve this equation for $\sqrt{x}$. To do this, add $\frac{1}{2}$ to both sides of (14), then multiply both sides of the result by $x - 1$, and then add 1 to both sides. We have

$$\sqrt{x} = 1 + \frac{1}{2}(x - 1) + E(x)(x - 1). \quad (15)$$

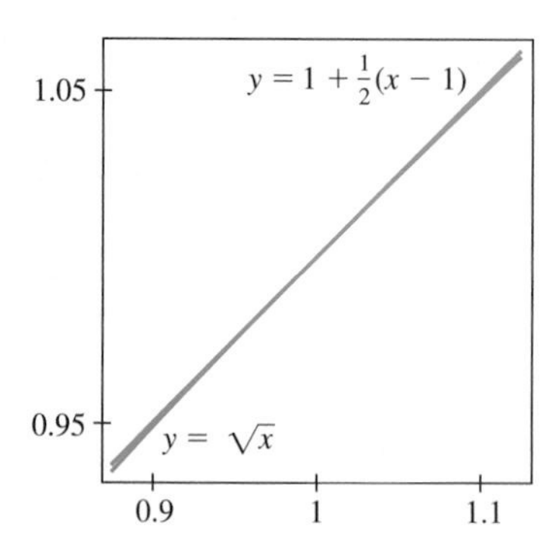

FIGURE 1.62 The graphs of $y = \sqrt{x}$ and $y = 1 + \frac{1}{2}(x - 1)$ are so close we see little difference.

When x is close to 1, we know that $E(x)$ is close to 0 and that $(x - 1)$ is small. Hence for x close to 1, $E(x)(x - 1)$ is small. If we drop this small term from (15), we get an approximation for $\sqrt{x}$:

$$\sqrt{x} \approx 1 + \frac{1}{2}(x - 1). \tag{16}$$

Figure 1.62 shows the graphs of $y = \sqrt{x}$ and $y = 1 + \frac{1}{2}(x - 1)$ on the interval $[0.9, 1.1]$. Apparently the approximation (16) is quite good.

From (15), the absolute value of the error in this approximation is

$$\left|\sqrt{x} - \left(1 + \frac{1}{2}(x - 1)\right)\right| = |(x - 1)E(x)|. \tag{17}$$

To estimate this error for $0.9 \le x \le 1.1$, recall (14) and graph

$$y = |E(x)| = \left|\frac{\sqrt{x} - 1}{x - 1} - \frac{1}{2}\right|$$

on this interval. The graph appears in Fig. 1.63 and shows that $|E(x)| < 0.015$ when $0.9 \le x \le 1.1$. Use this overestimate for $|E(x)|$ in (17) to see that

$$\left|\sqrt{x} - \left(1 + \frac{1}{2}(x - 1)\right)\right| \le 0.015|x - 1|.$$

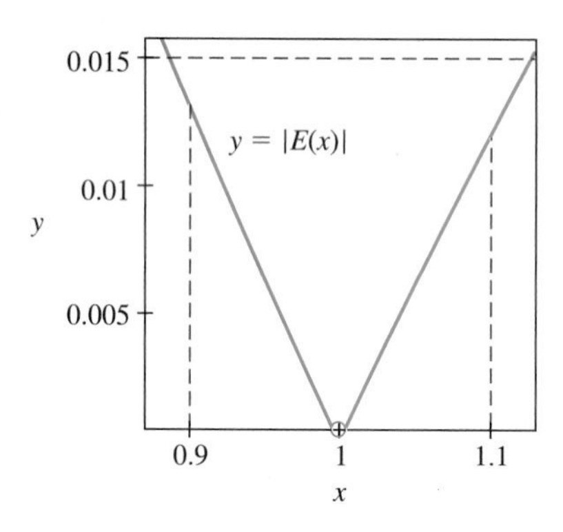

FIGURE 1.63 The graph shows that $y = |E(x)| = \left|\frac{\sqrt{x} - 1}{x - 1} - \frac{1}{2}\right| < 0.015$ when $0.9 < x < 1.1$.

Hence the error in the approximation (16) is at most 1.5% of the value of $|x - 1|$. Furthermore, for $0.9 \le x \le 1.1$, we have

$$|(x - 1)E(x)| \le (0.1)(0.015) = 0.0015,$$

so the error in (16) is no more than 0.0015. In Fig. 1.64 we show the graph of $y = |(x - 1)E(x)|$. This graph shows that on the interval $0.9 \le x \le 1.1$ the error in the approximation is less than 0.0015. For a more specific illustration, we use (16) to approximate $\sqrt{0.95}$. Putting $x = 0.95$ into (16),

$$\sqrt{0.95} \approx 1 + \frac{1}{2}(0.95 - 1) = 0.975,$$

while a calculator gives a value of 0.974679. Not bad! In Exercise 40 we do more work with this example to obtain a better approximation to the square root function.

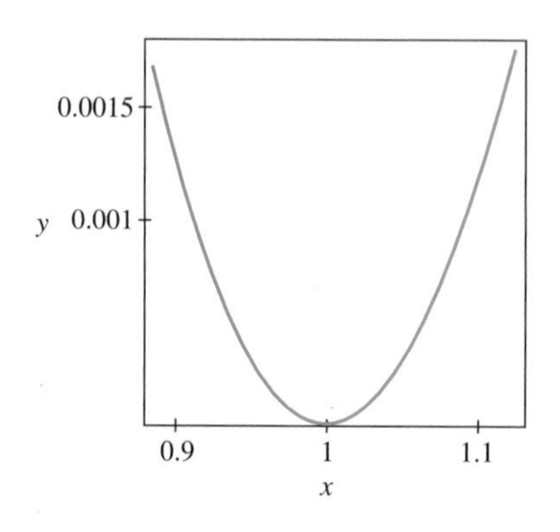

FIGURE 1.64 The graph of $y = \left|\sqrt{x} - \left(1 + \frac{1}{2}(x - 1)\right)\right|$ shows that the error is very small for $0.9 < x < 1.1$.

One-Sided Limits

According to Einstein's special theory of relativity, the length of a rod measured at rest is different from the length measured when the rod is in motion. If you measure an arrow on your desk and find it has length L_0, and then measure the same arrow as it flies by at speed v, the length will be

$$L = L_0\sqrt{1 - \left(\frac{v}{c}\right)^2}, \tag{18}$$

where $c = 2.998 \times 10^{10}$ cm/s is the speed of light. At the everyday speeds v of cars and planes, this effect is not noticeable, but if v is close to c, the difference between L_0 and L can be significant.

Although an arrow can never travel at the speed of light, physicists are interested in studying objects at speeds close to c. This suggests taking the limit of (18) as v approaches c. However, because (18) is defined for $0 < v < c$ but not for $v > c$, we really want to study (18) only for speeds v close to but less than c. In this case, we say that we are evaluating the limit of (18) as v approaches c from the left.

DEFINITION Left-hand Limit

Let $g(x)$ be defined in an interval $a - r < x < a$ for some $r > 0$. We say that $g(x)$ has **left-hand limit** L as x approaches a if we can make $g(x)$ as close to L as we like by taking $x < a$ and close to a. We write

$$\lim_{x \to a^-} g(x) = L.$$

This is read "the limit of $g(x)$ as x approaches a from the left is L."

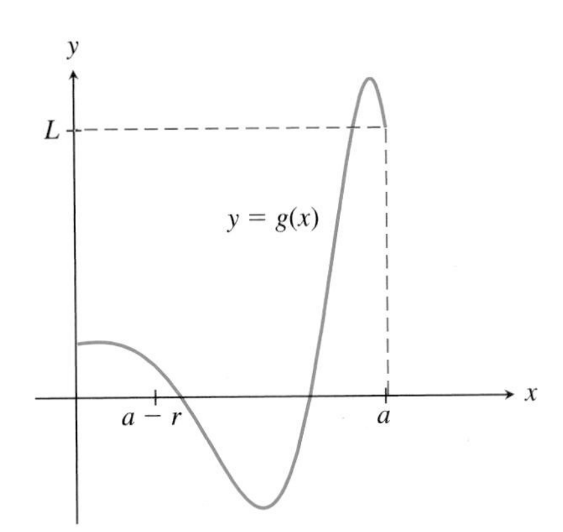

FIGURE 1.65 When $x < a$ and x is close to a, $g(x)$ is close to L. We write $\lim_{x \to a^-} g(x) = L$.

We illustrate the idea of the left-hand limit in Fig. 1.65. The right-hand limit, $\lim_{x \to b^+} g(x)$, is defined similarly. See Exercise 26.

EXAMPLE 7 Find the limit of the length of the flying arrow as v approaches c from the left.

Solution

When $v < c$ and close to c, expression (18) is defined. For such values of v,

$$L = L_0\sqrt{1 - \left(\frac{v}{c}\right)^2} \quad \text{is close to} \quad L_0\sqrt{1 - \left(\frac{c}{c}\right)^2} = 0.$$

Hence

$$\lim_{v \to c^-} L_0\sqrt{1 - \left(\frac{v}{c}\right)^2} = 0.$$

This means that as the speed of the arrow approaches the speed of light, the length of the arrow approaches 0.

EXAMPLE 8 The function f is defined by

$$f(x) = \begin{cases} -2x + 1 & x < 1 \\ 3 & x = 1 \\ 2x^2 & x > 1. \end{cases}$$

Evaluate

$$\lim_{x \to 1^-} f(x), \quad \lim_{x \to 1^+} f(x), \quad \text{and} \quad \lim_{x \to 1} f(x).$$

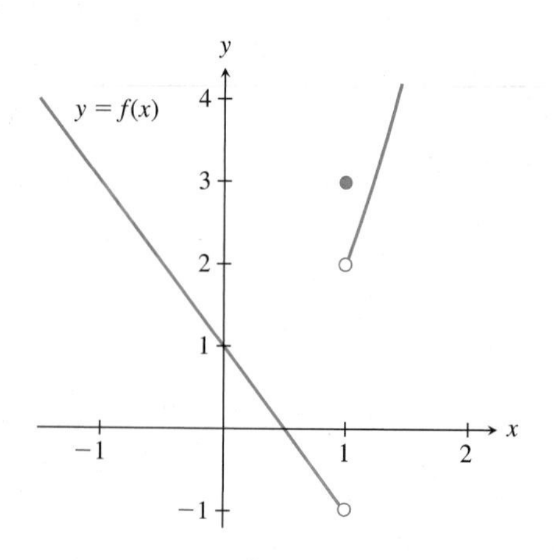

FIGURE 1.66 The graph of $y = f(x)$.

Solution

The graph of $y = f(x)$ is shown in Fig. 1.66. From the graph we see that if x is close to 1 and less than 1, then $f(x)$ is close to -1. In fact, when $x < 1$, we have $f(x) = -2x + 1$, so

$$\lim_{x \to 1^-} f(x) = \lim_{x \to 1^-}(-2x + 1) = -2 \cdot 1 + 1 = -1.$$

When $x > 1$, we have $f(x) = 2x^2$. Hence

$$\lim_{x \to 1^+} f(x) = \lim_{x \to 1^+} 2x^2 = 2 \cdot 1^2 = 2.$$

This can also be seen by looking at the graph for $x > 1$ and close to 1. This discussion also shows that when x is close to 1, then $f(x)$ might be close to -1 or close to 2, depending on whether $x < 1$ or $x > 1$. Because the values of $f(x)$ do not get close to one and only one number L as x approaches 1, we conclude that although the one-sided limits exist at $x = 1$,

$$\lim_{x \to 1} f(x) \quad \text{does not exist.}$$

Continuous Functions

In Section 1.5 we saw that if $p(x)$ is a polynomial and c is a real number, then

$$\lim_{x \to c} p(x) = p(c).$$

In fact, we have seen many examples with

$$\lim_{x \to c} f(x) = f(c).$$

Functions with such "well-behaved" limits are very important.

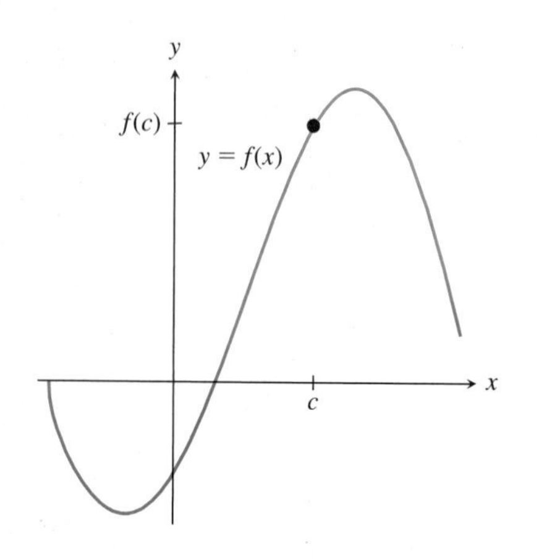

FIGURE 1.67 If $\lim_{x \to c} f(x) = f(c)$, then f is continuous at c.

DEFINITION Continuous Function

The function f is **continuous at** c if f is defined at c and

$$\lim_{x \to c} f(x) = f(c).$$

See Fig. 1.67. If f is continuous at every point in an interval $a < x < b$, we say that f is **continuous on the interval** $a < x < b$. If f is continuous at every point in its domain, then we say that f is **continuous.**

When we look closely at this definition, we see that three things must be true for a function f to be continuous at c:

a) $f(x)$ must be defined at $x = c$.
b) $\lim_{x \to c} f(x)$ must exist.
c) The value of the limit in **b)** is $f(c)$.

In Example 1 of Section 1.5 we showed that

$$\lim_{x \to 5}(2x^2 + 4) = 2 \cdot 5^2 + 4.$$

Hence the function defined by $2x^2 + 4$ is continuous at $x = 5$. In Example 2 of Section 1.5 we showed that

$$\lim_{\theta \to \pi/4} \cos\theta = \cos(\pi/4).$$

This means that the cosine function is continuous at $\theta = \pi/4$.

If one or more of the conditions **a), b), c)** is not satisfied, then $f(x)$ is **discontinuous** (or not continuous) at $x = c$. See Figs. 1.68 and 1.69. We have already seen many examples of discontinuity. For example, we have seen that

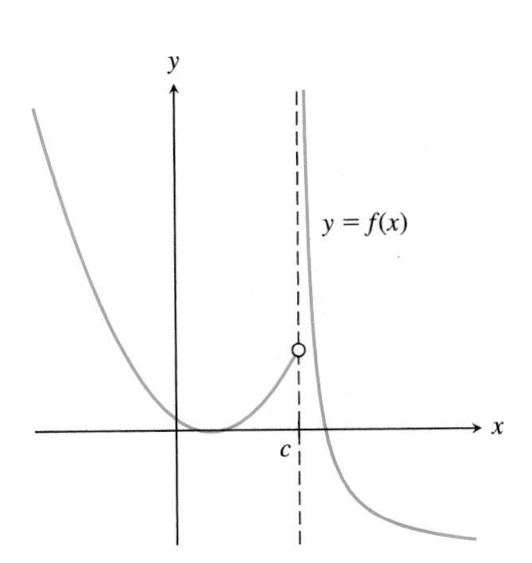

FIGURE 1.68 If f is not defined at c or $\lim_{x\to c} f(x)$ does not exist, then f is discontinuous at c.

$$\lim_{x \to 0} \frac{\sin x}{x} = 1.$$

However, the function defined by $(\sin x)/x$ is not continuous at $x = 0$ because the function is not defined for this value of x.

The function f of Example 8 is defined at $x = 1$ but is not continuous at this point because

$$\lim_{x \to 1} f(x) \quad \text{does not exist.}$$

As mentioned earlier, if p is any polynomial function

$$p(x) = a_n x^n + a_{n-1}x^{n-1} + \cdots + a_1 x + a_0,$$

then for any real number c,

$$\lim_{x \to c} p(x) = p(c).$$

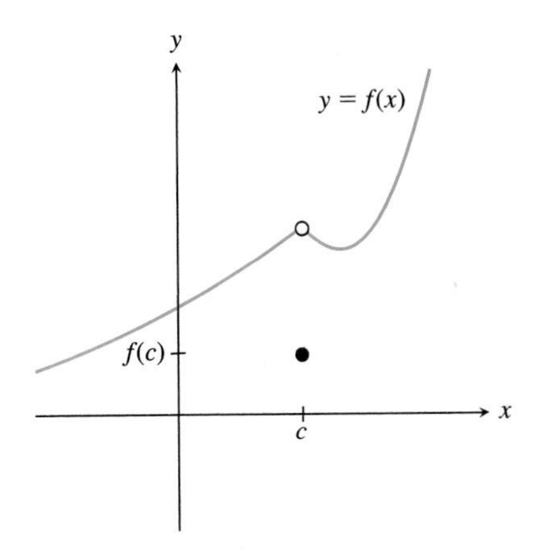

FIGURE 1.69 If $\lim_{x\to c} f(x)$ exists but is not equal to $f(c)$, then f is discontinuous at c.

This means that a polynomial is continuous at every point in its domain. Hence any polynomial is a continuous function.

In Section 1.5 we also saw that for any real number c

$$\lim_{x \to c} \sin x = \sin c \quad \text{and} \quad \lim_{x \to c} \cos x = \cos c.$$

Thus the sine function and the cosine function are continuous.

In addition, the rules for combining limits given in the previous section imply that most combinations of continuous functions are continuous. Likewise, the result on limits and compositions shows that compositions of continuous functions are also continuous.

Continuous functions have many important and interesting properties. In one sense, the graph of a continuous function is well behaved.

Graphs of Continuous Functions

Let f be continuous on an interval $a < x < b$. Then the graph of $y = f(x)$ on $a < x < b$ is an unbroken curve.

According to this statement, the graph of a function continuous on an interval cannot have jumps or breaks like those seen in Fig. 1.68, Fig. 1.69, or Fig. 1.50. The converse of this result is "almost" true. A precise statement of the converse would take some terminology not usually encountered in calculus courses. However, for our purposes, the following statement is enough.

Graphs Representing Continuous Functions

For a function f, suppose that the graph $y = f(x)$ for $a < x < b$ is an unbroken curve and the graph exhibits no oscillatory behavior like that seen in the third graph in Fig. 1.50. Then f is continuous on the interval $a < x < b$.

Thus the graph in Fig. 1.67 is the graph of a continuous function.

Exercises 1.6

1. Let $f(x) = 3x^2 - 2x + 1$.

T **a.** Find the value of $\lim_{x \to -1} f(x)$.

b. Let L be the value of the limit found in part **a** and let $\epsilon = 0.1$. Use graphical methods to find a $\delta > 0$ so that if $0 < |x - (-1)| < \delta$, then $|f(x) - L| < \epsilon$.

2. Let $f(x) = x + 2 \cos x$.

T **a.** Find the value of $\lim_{x \to \pi} f(x)$.

b. Let L be the value of the limit found in part **a** and let $\epsilon = 0.01$. Use graphical methods to find a $\delta > 0$ so that if $0 < |x - \pi| < \delta$, then $|f(x) - L| < \epsilon$.

3. Let $r(\theta) = 6$.

a. Find the value of $\lim_{\theta \to \sqrt{5}} r(\theta)$.

b. Let L be the value of the limit found in part **a** and let $\epsilon = 0.1$. Use graphical methods to find a $\delta > 0$ so that if $0 < |\theta - \sqrt{5}| < \delta$, then $|r(\theta) - L| < \epsilon$.

4. Let $g(t) = 2^t$.

T **a.** Find the value of $\lim_{t \to 1} g(t)$.

b. Let L be the value of the limit found in part **a** and let $\epsilon = 0.05$. Use graphical methods to find a $\delta > 0$ so that if $0 < |t - 1| < \delta$, then $|g(t) - L| < \epsilon$.

5. Let $H(r) = \dfrac{r^3 + 8}{r^2 - 4}$.

T **a.** Find the value of $\lim_{r \to -2} H(r)$.

b. Let L be the value of the limit found in part **a** and let $\epsilon = 0.1$. Use graphical methods to find a $\delta > 0$ so that if $0 < |r - (-2)| < \delta$, then $|H(r) - L| < \epsilon$.

6. Let $t(\theta) = \dfrac{\cos \theta - 1}{2\theta^2}$.

T **a.** Find the value of $\lim_{\theta \to 0} t(\theta)$.

b. Let L be the value of the limit found in part **a** and let $\epsilon = 0.005$. Use graphical methods to find a $\delta > 0$ so that if $0 < |\theta - 0| < \delta$, then $|t(\theta) - L| < \epsilon$.

Exercises 7–13: A function f and a number a are given.

a. Find the value of $\lim_{x \to a} f(x)$. Use analytical, numerical, or graphical means.

b. Let L be the value of the limit found in part **a**. Find an interval containing the point $x = a$ on which $f(x) \approx L$ with an error of at most 0.01.

T **7.** $f(x) = -4x^2 + 2x - 1, \quad a = 2$

T **8.** $f(x) = \dfrac{1}{\sqrt{x + 1}}, \quad a = 1$

T **9.** $f(x) = \cos(2x), \quad a = 0$

T **10.** $f(x) = \dfrac{10^x - 1}{x}, \quad a = 0$

T **11.** $f(x) = \dfrac{x - 5}{\sqrt{x} - \sqrt{5}}, \quad a = 5$

T **12.** $f(x) = \dfrac{\tan x - 1}{x - \pi/4}, \quad a = \pi/4$

T **13.** $f(x) = \dfrac{1/(x + 1) - \frac{1}{2}}{x - 1}, \quad a = 1$

T **14. a.** Evaluate

$$\lim_{x \to 2} \frac{3x^2 - 2x - 8}{x - 2}.$$

b. Use the result of part **a** to find an approximation to $3x^2 - 2x - 8$ for x values near 2.

c. Find an interval containing 2 on which the error in this approximation is less than 0.001.

d. For $x = 1.98$, compare the actual value of $3x^2 - 2x - 8$ with the value given by the approximation.

T **15. a.** Evaluate (or estimate)

$$\lim_{r \to -2} \frac{\sqrt{2r^2 + 1} - 3}{r + 2}.$$

b. Use the result of part **a** to find an approximation to $\sqrt{2r^2 + 1}$ for r values near -2.

c. Find an interval containing -2 on which the error in this approximation is less than 0.001.

d. For $r = -1.98$, compare the actual (calculator) value of $\sqrt{2r^2 + 1}$ with the value given by the approximation.

T **16.** a. Use graphical or numerical means to find the value of

$$\lim_{t \to 2} \frac{3^t - 3^2}{t - 2}.$$

b. Use the result of part **a** to find an approximation to 3^t for t values near 2.

c. Find an interval containing 2 on which the error in this approximation is less than 0.01.

d. For $t = 2.015$, compare the actual (calculator) value of 3^t with the value given by the approximation.

T **17.** a. Use graphical or numerical means to find the value of

$$\lim_{x \to \pi/4} \frac{\tan x - 1}{x - \pi/4}.$$

b. Use the result of part **a** to find an approximation to $\tan x$ for x values near $\pi/4$.

c. Find an interval containing $\pi/4$ on which the error in this approximation is less than 0.001.

d. For $x = 0.75$, compare the actual value of $\tan x$ with the value given by the approximation.

18. In Fig. 1.60, which is the graph of $y = x$ and which is the graph of $y = \sin x$? Give reasons for your answer.

19. Let $h(t) = t/|t|$.

a. Evaluate $\lim_{t \to 0^+} h(t)$ and $\lim_{t \to 0^-} h(t)$.

b. What can be said about $\lim_{t \to 0} h(t)$?

c. Is $h(t)$ continuous at $t = 0$? Why or why not?

20. Let $P(w)$ be the cost of first-class postage for a letter that weighs w ounces. If $0 < w < 1$, then $P(w) = 34¢$. If $w > 1$, the cost is 34¢ plus 22¢ for each ounce or fraction of an ounce above 1.

a. Sketch a graph of the function $C = P(w)$ for $0 < w < 5.5$.

b. Evaluate $\lim_{w \to 2^-} P(w)$ and $\lim_{w \to 2^+} P(w)$.

c. Tell why $P(w)$ is discontinuous for $w = 1, 2, 3, 4, \ldots$.

21. The graph of a function f is shown in the accompanying figure.

a. Evaluate $\lim_{x \to -1^+} f(x)$ and $\lim_{x \to -1^-} f(x)$.

b. Evaluate $\lim_{x \to 0^+} f(x)$ and $\lim_{x \to 0^-} f(x)$.

c. Evaluate $\lim_{x \to 1^+} f(x)$ and $\lim_{x \to 1^-} f(x)$.

d. For what values of x between -3 and 3 is f discontinuous?

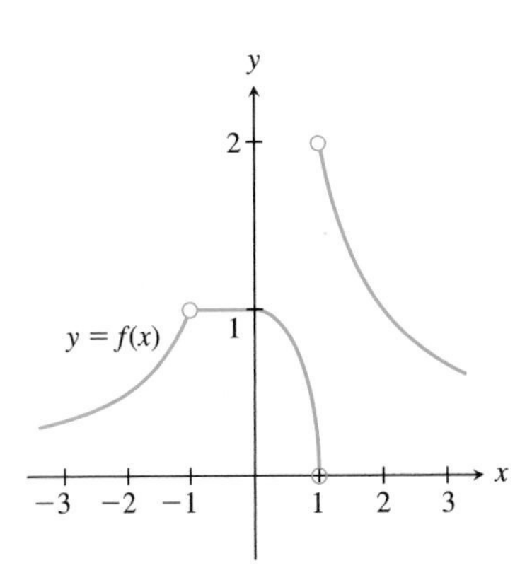

Figure for Exercise 21.

22. The graph of a function g is shown in the accompanying figure.

a. Evaluate $\lim_{x \to -4^+} g(x)$ and $\lim_{x \to -4^-} g(x)$.

b. Evaluate $\lim_{x \to -2^+} g(x)$ and $\lim_{x \to -2^-} g(x)$.

c. Evaluate $\lim_{x \to 2^+} g(x)$ and $\lim_{x \to 2^-} g(x)$.

d. For what values of x between -5 and 3 is g discontinuous?

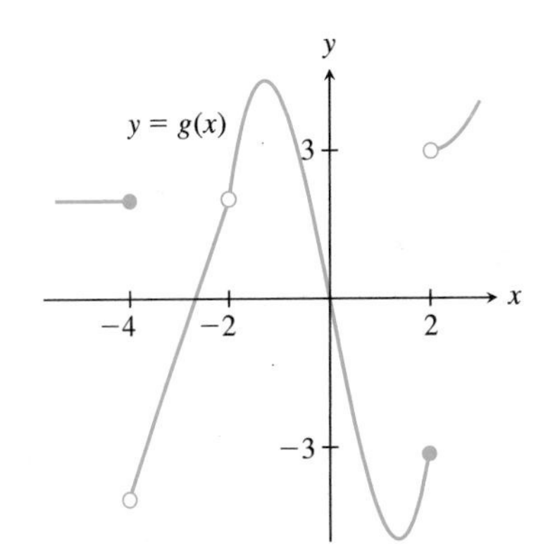

Figure for Exercise 22.

23. Restate the rule for limits and compositions in Section 1.5 in such a way that the conditions on the function h are expressed in terms of the continuity of h at L.

24. a. For $x \neq 0$ let

$$g(x) = x \sin\left(\frac{1}{x}\right).$$

Show that $\lim_{x \to 0} g(x) = 0$.

b. For $x \neq 0$ let

$$h(x) = \frac{\sin x}{x}.$$

Note that $h(x)$ is not defined for $x = 0$ but that $\lim_{x\to 0} h(x) = 1$. Show, however, that $\lim_{x\to 0} h(g(x))$ does not exist.

c. Carefully describe how this example relates to the requirement that h be defined at L in the rule for limits and compositions on page 43.

25. Give an ϵ, δ definition for the left-hand limit. Model it on the ϵ, δ definition for the limit given in this section.

26. Write a definition for the right-hand limit of a function. Model your definition on the definition of the left-hand limit given in this section.

27. In this exercise we "verify" equation (13) from the rules for combining limits in Section 1.5. Let g and h be functions and suppose that

$$\lim_{x\to a} g(x) = 3, \quad \lim_{x\to a} h(x) = 5.$$

Take $\epsilon = 0.1$. Use the following two steps as a guide to argue why there must be a $\delta > 0$ so that $0 < |x - a| < \delta$ implies

$$|(g(x) + h(x)) - (3 + 5)| < \epsilon.$$

a. First tell why there must be a $\delta > 0$ so that when $0 < |x - a| < \delta$ we have

$$|g(x) - 3| < 0.05$$

and

$$|h(x) - 5| < 0.05.$$

b. Let δ have the value discussed in part a. Show that $0 < |x - a| < \delta$ implies

$$|(g(x) + h(x)) - 8| < 0.1 = \epsilon.$$

28. In this exercise we "verify" equation (15) from the rules for combining limits in Section 1.5. Let g and h be functions and suppose that

$$\lim_{x\to a} g(x) = 3, \quad \lim_{x\to a} h(x) = 5.$$

Take $\epsilon = 0.1$. Use the following two steps as a guide to argue why there must be a $\delta > 0$ so that $0 < |x - a| < \delta$ implies

$$\left|\frac{g(x)}{h(x)} - \frac{3}{5}\right| < \epsilon.$$

a. First tell why there must be a $\delta > 0$ so that when $0 < |x - a| < \delta$ we have

$$|g(x) - 3| < 0.25 \quad \text{and} \quad |h(x) - 5| < 0.25.$$

b. Let δ have the value discussed in part a. Show that $0 < |x - a| < \delta$ implies

$$\left|\frac{g(x)}{h(x)} - \frac{3}{5}\right| < 0.1 = \epsilon.$$

29. Let

$$p(x) = x^3 - 4x + 6.$$

The following steps outline a method for using the squeeze theorem (see Exercise 45 of Section 1.5) to show that $\lim_{x\to 1} p(x) = p(1) = 3$.

a. Note that

$$p(x) - p(1)$$

takes the value 0 when $x = 1$. Hence the polynomial $p(x) - p(1)$ has a factor of $x - 1$. Verify that

$$p(x) - p(1) = (x - 1)(x^2 + x - 3).$$

b. For $|x - 1| < 1$ (i.e., $0 < x < 2$) show that

$$|x^2 + x - 3| < 9.$$

c. Show that for $|x - 1| < 1$,

$$|p(x) - 3| \le 9|x - 1|.$$

d. Use the squeeze theorem and part **c** to show that

$$\lim_{x\to 1} p(x) = p(1) = 3.$$

Relate this result to the rule for limits of polynomials, sines, and cosines on page 42 in Section 1.5.

30. Use the technique outlined in the previous problem to show that

$$\lim_{x\to 2}(-2x^3 + 3x^2 + x - 3) = -5.$$

31. Let $p(x)$ be a polynomial and a a real number. Based on the outline given in Exercise 29, discuss a method for showing that $\lim_{x\to a} p(x) = p(a)$. Note that this gives a method of proving equation (16) of the rule for limits of polynomials, sines, and cosines on page 42 in Section 1.5.

32. Define $r = r(c)$ to be the number of distinct real zeros of the polynomial

$$x^2 - 3x + c.$$

For example, $r(-4) = 2$ because $x^2 - 3x - 4 = 0$ has two different solutions, $x = 4$ and $x = -1$. However, $r(5) = 0$ because the equation $x^2 - 3x + 5 = 0$ has no real solutions. For what values of c is $r(c) = 2$? When is $r(c) = 1$? When is $r(c) = 0$? Graph the function $r = r(c)$ and list the values of c where the function r is not continuous.

33. Define $r = r(b)$ to be the number of distinct real zeros of the polynomial

$$x^2 + bx + 4.$$

For example, $r(3) = 0$ because $x^2 + 3x + 4 = 0$ has no real solutions, and $r(5) = 2$ because the equation $x^2 + 5x + 4 = 0$ has two real solutions. For what values of b is $r(b) = 2$? When is $r(b) = 1$? When is $r(b) = 0$? Graph the function $r = r(b)$ and list the values of b where the function r is not continuous.

34. Define $r(d)$ to be the number of distinct real zeros of the polynomial

$$2x^3 + 3x^2 - 12x + d.$$

For example, $r(0) = 3$ because $2x^3 + 3x^2 - 12x + 0 = 0$ has three real solutions, as shown in the accompanying figure. For what values of d is $r(d) = 3$? When is $r(d) = 2$? When is $r(d) = 1$? When is $r(d) = 0$? Graph the function $r = r(d)$ and list the values of d for which r is not continuous.

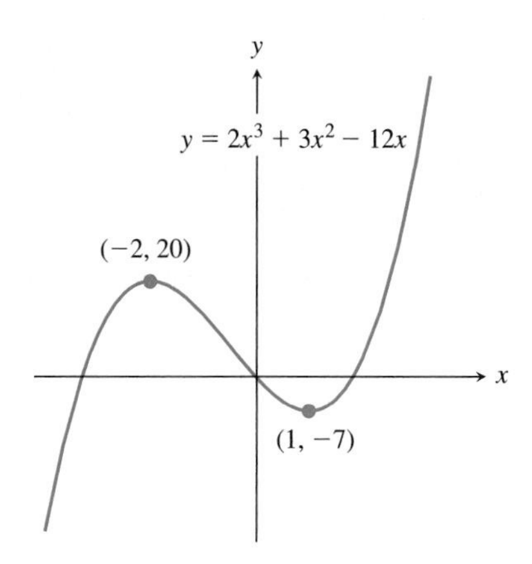

Figure for Exercise 34.

T 35. Find the value of

$$\lim_{x\to 8} \frac{\sqrt[3]{x} - 2}{x - 8}.$$

Use this result to develop an approximation to $\sqrt[3]{x}$ for x values close to 8. Find an interval containing 8 on which the error in the approximation is less than 1 percent of the value of $|x - 8|$.

T 36. Use graphical or numerical means to find the value of

$$\lim_{x\to 3} \frac{2^x - 8}{x - 3}.$$

Use this result to develop an approximation to 2^x for x values close to 3. Find an interval containing 3 on which the error in the approximation is less than 1 percent of the value of $|x - 3|$.

T 37. Find the value of

$$\lim_{x\to -2} \frac{\dfrac{1}{\sqrt{x^2+5}} - \dfrac{1}{3}}{x + 2}.$$

Use this result to develop an approximation to $1/\sqrt{x^2 + 5}$ for x values close to -2. Find an interval containing -2 on which the error in the approximation is less than 1 percent of the value of $|x - (-2)|$.

T 38. Use graphical or numerical means to find the value of

$$\lim_{\theta\to 0} \frac{\cos\theta - 1}{\theta^2}$$

(or see Exercise 14 in Section 1.5). Use this result to find an approximation to $\cos\theta$ for θ values close to 0. Find an interval containing 0 on which the error in the approximation is less than 1 percent of the value of θ^2. On this interval, compare $\cos\theta$ and the approximation graphically.

T 39. Use graphical or numerical means to find the value of

$$\lim_{\theta\to 0} \frac{\sin\theta - \theta}{\theta^3}.$$

Use this result to find an approximation to $\sin\theta$ for θ values close to 0. Find an interval containing 0 on which the error in the approximation is less then 1 percent of the value of $|\theta|^3$. Compare $\sin\theta$ and the approximation graphically.

40. In Example 6 we showed that

$$\sqrt{x} \approx 1 + \frac{1}{2}(x - 1).$$

We use this result to derive a better approximation to the square root function.

a. Verify that

$$\lim_{x\to 1} \frac{\sqrt{x} - \left(1 + \frac{1}{2}(x-1)\right)}{(x-1)^2} = -\frac{1}{8}.$$

b. Use the result of part **a** to derive an approximation for $\sqrt{x}$.

T c. Discuss the error in this approximation on the interval $0.9 \le t \le 1.1$ and compare the error to the approximation found in Example 6.

41. Suppose that $\lim_{x\to c} f(x) = L$. What can be said about

$$\lim_{x\to c^-} f(x) \quad\text{and}\quad \lim_{x\to c^+} f(x)?$$

42. Suppose that

$$\lim_{x\to c^-} f(x) = L \quad\text{and}\quad \lim_{x\to c^+} f(x) = M.$$

Under what conditions does $\lim_{x\to c} f(x)$ exist?

1.7 The Derivative

In his autobiography *Surely You're Joking, Mr. Feynman!* (Norton Publishing, 1985), physicist Richard Feynman tells the story of how he once pitted his pencil-and-paper arithmetic abilities against the skills of a merchant with an abacus.

They started with an addition problem, and the merchant was first with the answer. Next came a problem in multiplication. Again the merchant gave the answer first, but just barely ahead of Feynman. The next round tested division skills, and the result was a tie. For the last problem, the merchant suggested cube roots. Finding cube roots on an abacus is very difficult, but the merchant was evidently proud of his ability to do so. The party supplying the numbers challenged the competitors to find the cube root of 1729.03. Onlookers were amazed to see Feynman think a moment, then write the number 12.002 well before the merchant gave his answer of 12.

Feynman writes that the number 1729.03 was lucky for him. As a physicist he knew that $12^3 = 1728$, because this gives the number of cubic inches in a cubic foot. So $\sqrt[3]{1729.03}$ must be a little larger than 12... but how much larger? Feynman knew from calculus that for real numbers x close to 12^3,

$$\sqrt[3]{x} \approx 12 + \frac{1}{432}(x - 1728). \tag{1}$$

With $x = 1729.03$, this gave $\sqrt[3]{1729.03} \approx 12 + 1.03/432$. Feynman was able to quickly do the mental division to see that $1/432 \approx 0.002$. Let's try Feynman's approximation (1) for a few other numbers. Letting $c(x) = 12 + \dfrac{1}{432}(x - 1728)$ we find

$$\begin{aligned}
c(1728) &= 12 && \text{and} && \sqrt[3]{1728} = 12\\
c(1725) &\approx 11.9931 && \text{and} && \sqrt[3]{1725} \approx 11.9931\\
c(1740) &\approx 12.0278 && \text{and} && \sqrt[3]{1740} \approx 12.0277\\
c(5000) &\approx 19.5741 && \text{and} && \sqrt[3]{5000} \approx 17.0998
\end{aligned}$$

The first three examples support Feynman's statement that $c(x)$ is close to $\sqrt[3]{x}$ when x is close to 1728. On the other hand, the last entry shows that $c(x)$ may not be close to $\sqrt[3]{x}$ for *all* values of x. These numerical examples raise many interesting questions. Why should $c(x)$ be close to $\sqrt[3]{x}$ for any values of x close to 12^3? Suppose that we want $\sqrt[3]{x}$ accurate to three decimal places. For what values of x can we use $c(x)$? For x not near 12^3, is there another approximation that we can use, and if so, how do we find it? We can answer many of these questions after we take a closer look at the rate of change. (See also Problem 38.)

The Derivative

In Sections 1.3, 1.4, and 1.5 we discussed the rate of change of the pressure in a piston chamber with respect to volume, the rate of change of temperature in an autoclave with respect to time, and the rate of change of a population with respect to time. These rates of change are related in that they ask how a variable y changes with respect to a variable x, given that x and y are related by an equation of the form $y = f(x)$. We now define the *derivative* of a function f at a point $x = a$ of its domain.

DEFINITION Derivative

Let f be a function and suppose that f is defined at the point a (i.e., $f(a)$ is defined). The **derivative of f at a** is the number

$$f'(a) = \lim_{h \to 0} \frac{f(a+h) - f(a)}{h}, \tag{2}$$

provided that this limit exists. If this limit does not exist, we say f has no derivative at $x = a$.

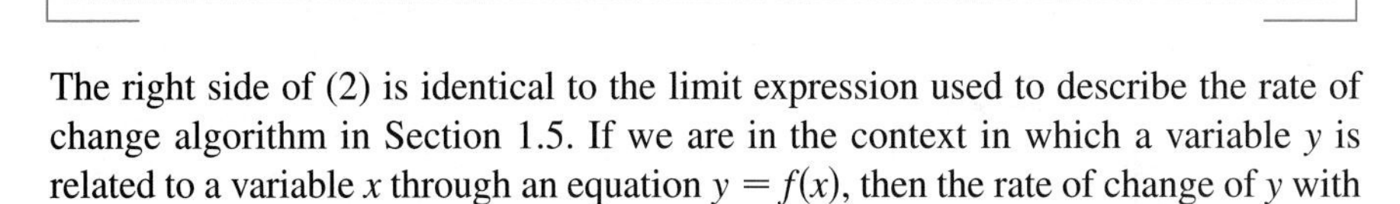

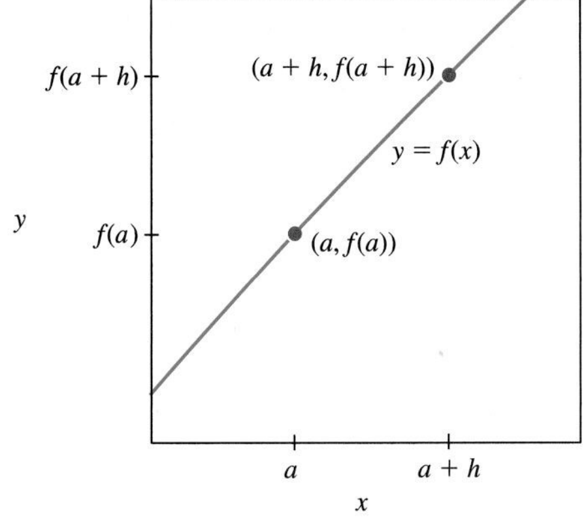

FIGURE 1.70 We see a "line" of slope $f'(a)$ when we zoom in on the graph of $y = f(x)$ near $(a, f(a))$.

The right side of (2) is identical to the limit expression used to describe the rate of change algorithm in Section 1.5. If we are in the context in which a variable y is related to a variable x through an equation $y = f(x)$, then the rate of change of y with respect to x when $x = a$ is $f'(a)$, the derivative of f at a.

If we are in a context in which our interest is primarily in the graph of $y = f(x)$ and, in particular, the slope of the "line" we see when we zoom in on the graph of f near $(a, f(a))$, then we may interpret the quotient

$$\frac{f(a+h) - f(a)}{h}$$

as the slope of the line joining the points $(a, f(a))$ and $(a + h, f(a + h))$ on the graph of $y = f(x)$. See Fig. 1.70. We expect that as $h \to 0$, the slope of this line approaches the slope of the graph of f at a.

Notation for the Derivative

There are many different notations for the derivative (or rate of change). The notation used often depends on how the function f is presented or on how the derivative is to be used. When a function is defined by a formula $y = f(x)$, the following are all notations for the derivative of f (or y) with respect to x at $x = a$:

$$f'(a), \quad \left.\frac{dy}{dx}\right|_{x=a}, \quad \left.\frac{df}{dx}\right|_{x=a}, \quad D_x f(a).$$

We shall use the first three of these notations.

The Derivative as a Function

If x is equal to a specific number, say $x = a$, then the derivative of f at a is the number denoted by $f'(a)$. If x is variable, we can think of the values $f'(x)$ as defining a function f'.

The Derivative as a Function

Let f be a function. The derivative function f' is defined by the equation

$$f'(x) = \lim_{h \to 0} \frac{f(x+h) - f(x)}{h}, \tag{3}$$

and is defined for all x for which this limit exists.

The derivative of f at a, $f'(a)$, is the value of the function f' at the value $x = a$. When $y = f(x)$, we will also use

$$y', \quad \frac{dy}{dx}, \quad \frac{df}{dx}, \quad \text{and} \quad \frac{d}{dx}f(x)$$

to denote the function f'. The $\frac{d}{dx}$ notation was introduced by Gottfried Liebnitz who, along with Isaac Newton, shares credit for the invention of calculus. The symbol $\frac{d}{dx}$ is often called an *operator;* it operates on a function f to give the derivative of f. Thus we can write

$$\frac{d}{dx}f(x) = f'(x).$$

EXAMPLE 1 Let $y = f(x) = x^2$. Find

a) $f'(1)$ **b)** $f'(x)$.

Solution

a) Setting $a = 1$ in (2) we have

$$f'(1) = \lim_{h \to 0} \frac{f(1+h) - f(1)}{h} = \lim_{h \to 0} \frac{(1+h)^2 - 1^2}{h}$$

$$= \lim_{h \to 0} \frac{(1 + 2h + h^2) - 1}{h} = \lim_{h \to 0}(2 + h) = 2.$$

b) Using (3),

$$f'(x) = \lim_{h \to 0} \frac{f(x+h) - f(x)}{h} = \lim_{h \to 0} \frac{(x+h)^2 - x^2}{h}$$

$$= \lim_{h \to 0} \frac{(x^2 + 2xh + h^2) - x^2}{h} = \lim_{h \to 0}(2x + h) = 2x.$$

Thus $f'(x) = 2x$. Setting $x = 1$ in this result we find $f'(1) = 2 \cdot 1 = 2$, as found in part **a)**.

Δx and Δy Notation

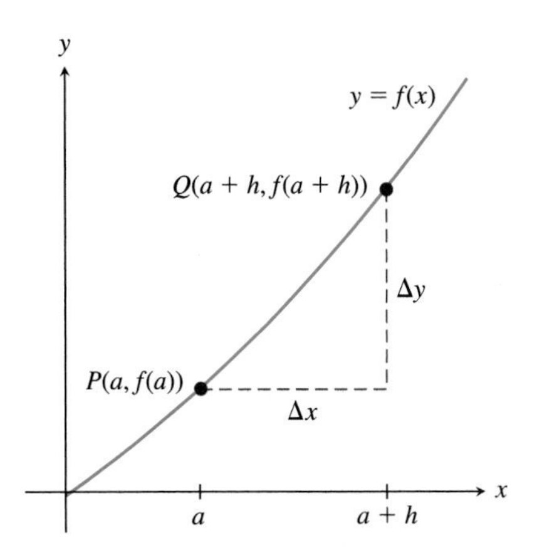

FIGURE 1.71 $Q(a+h, f(a+h))$ is near $P(a, f(a))$ on the graph of $y = f(x)$.

In Sections 1.3 and 1.4 we estimated the rate of change of a function f at a point $x = a$ by calculating the slope of the line determined by the point $P = (a, f(a))$ and a nearby point $Q = (a+h, f(a+h))$ on the graph of $y = f(x)$. Because Q was close to P, we knew that h had to be a small number. (See Fig. 1.71.) The slope of the line through P and Q was found by first calculating

$$\Delta y = \text{change in } y = f(a+h) - f(a)$$

and

$$\Delta x = \text{change in } x = (a+h) - a = h,$$

and then forming the quotient

$$\frac{\text{change in } y}{\text{change in } x} = \frac{\Delta y}{\Delta x} = \frac{f(a+h) - f(a)}{h}.$$

In view of the definition of *derivative,* the rate of change of f with respect to x at $x = a$ is then

$$\left.\frac{df}{dx}\right|_{x=a} = f'(a) = \lim_{\Delta x \to 0} \frac{\Delta y}{\Delta x} = \lim_{\Delta x \to 0} \frac{f(a+\Delta x) - f(a)}{\Delta x}. \tag{4}$$

Notations such as Δy and Δx are often used to represent small changes in quantities. This notation reminds us that the rate of change (or derivative) can be estimated by calculating the quotient $\frac{\Delta y}{\Delta x}$ for small Δx. This quotient also looks very much like the Leibnitz notation, $\frac{dy}{dx}$, for the derivative. The Leibnitz notation reminds us that the derivative is a limit of a ratio of small quantities.

Another Formula for the Derivative

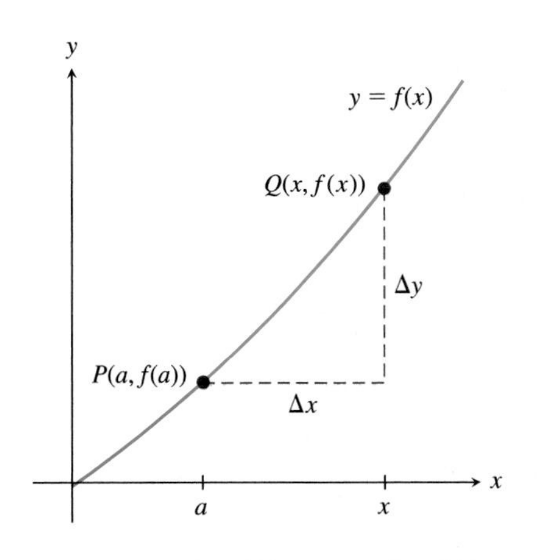

FIGURE 1.72 $Q(x, f(x))$ is near $P(a, f(a))$ on the graph of $y = f(x)$.

In the preceding discussion we used $Q = (a+h, f(a+h))$ to denote a point close to $P = (a, f(a))$. We could just as easily have denoted this nearby point $(x, f(x))$ with the understanding that x is close to a. This is illustrated in Fig. 1.72. Substituting x for $a + h$ in (2) and noting that $h \to 0$ is equivalent to $x \to a$ gives us another form for the definition of the derivative of f at a,

$$f'(a) = \lim_{x \to a} \frac{f(x) - f(a)}{x - a}. \tag{5}$$

Either (2) or (5) can be used to compute a derivative. Depending on the function f, calculations with the two formulas can involve different amounts of algebra.

Interpreting the Derivative

To effectively use the derivative as a tool, it is important to know how to interpret it in different contexts. Remember that the derivative (or rate of change) of $y = f(x)$ carries the units

$$\frac{\text{units of } y}{\text{units of } x}.$$

Knowing these units often helps in determining the meaning of the derivative.

EXAMPLE 2 A high diver jumps from a tower into a pool of water 100 feet below. After she has fallen for t seconds, her height (in feet) above the pool is $H(t) = 100 - 16t^2$. Find the derivative of H and discuss the meaning of the derivative. How long after she jumps does the diver reach the pool? What is the derivative at this time, and what does it tell us?

Solution

From (3), the derivative is

$$H'(t) = \lim_{h \to 0} \frac{H(t+h) - H(t)}{h}$$
$$= \lim_{h \to 0} \frac{(100 - 16(t+h)^2) - (100 - 16t^2)}{h}.$$

We cannot evaluate this limit by simply substituting $h = 0$ because this leads to $0/0$. Instead we do some algebra to find an equivalent expression whose limit we can determine. After simplifying the numerator in the last expression, we can factor and cancel an h from the numerator and denominator to get

$$H'(t) = \lim_{h \to 0} \frac{-32th - 16h^2}{h} = \lim_{h \to 0}(-32t - 16h) = -32t.$$

Since the height $H(t)$ is in feet and time t is in seconds, the units for the derivative are feet/second. Thus at time t the rate of change of the diver's height with respect to time is $-32t$ ft/s. Because the units are units of velocity, the derivative tells us how fast the diver is moving at time t. The minus sign tells us that when we zoom in on the graph of $H = H(t)$ near a point $(t, H(t))$, we will see a "line" of negative slope. Hence $H(t)$ decreases as time t increases. (This makes sense because as the diver falls, her height above the water is decreasing.)

When the diver reaches the water, her height $H(t)$ above the pool will be 0. Solving

$$H(t) = 100 - 16t^2 = 0 \quad \text{we find} \quad t = \pm 2.5 \text{ seconds}.$$

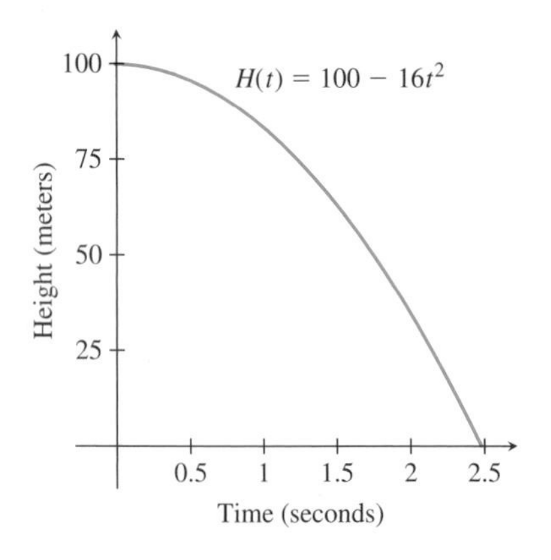

FIGURE 1.73 The graph illustrates the height of the diver as a function of time.

Because the t we seek must be positive, we conclude that the diver hits the water 2.5 seconds after jumping off the tower. See Fig. 1.73. The derivative at this time is

$$H'(2.5) = -32(2.5) = -80 \text{ ft/s}.$$

Thus the diver is moving downward at 80 ft/s when she hits the water.

EXAMPLE 3 According to Newton's law of gravitation, a 100-kg object r meters above the surface of Earth is subject to a gravitational force of

$$F(r) = \frac{100GM_e}{(R_e + r)^2} \text{ newtons (N)},$$

where R_e is the radius of the Earth ($\approx 6.37 \times 10^6$ m), M_e is the mass of Earth ($\approx 5.98 \times 10^{24}$ kg), and G is the gravitational constant ($\approx 6.67 \times 10^{-11}$ Nm^2/kg^2). Find the rate of change of F with respect to r as the object moves away from the surface of Earth along a ray from

the center of Earth. What is the rate of change when the object is 10^7 m above the surface of Earth?

Solution

The rate of change is the derivative. Using (5), we have

$$F'(r) = \lim_{t \to r} \frac{F(t) - F(r)}{t - r} = \lim_{t \to r} \frac{\dfrac{100GM_e}{(R_e + t)^2} - \dfrac{100GM_e}{(R_e + r)^2}}{t - r}.$$

Adding the fractions in the numerator of this expression gives

$$F'(r) = \lim_{t \to r} \frac{100GM_e((R_e + r)^2 - (R_e + t)^2)}{(R_e + r)^2(R_e + t)^2(t - r)}.$$

Next factor $(R_e + r)^2 - (R_e + t)^2$ as a difference of squares to obtain

$$\begin{aligned} F'(r) &= \lim_{t \to r} \frac{100GM_e(r - t)(2R_e + r + t)}{(R_e + r)^2(R_e + t)^2(t - r)} \\ &= \lim_{t \to r} \frac{-100GM_e(2R_e + r + t)}{(R_e + r)^2(R_e + t)^2} \\ &= \frac{-100GM_e(2R_e + 2r)}{(R_e + r)^4} \\ &= \frac{-200GM_e}{(R_e + r)^3}. \end{aligned}$$

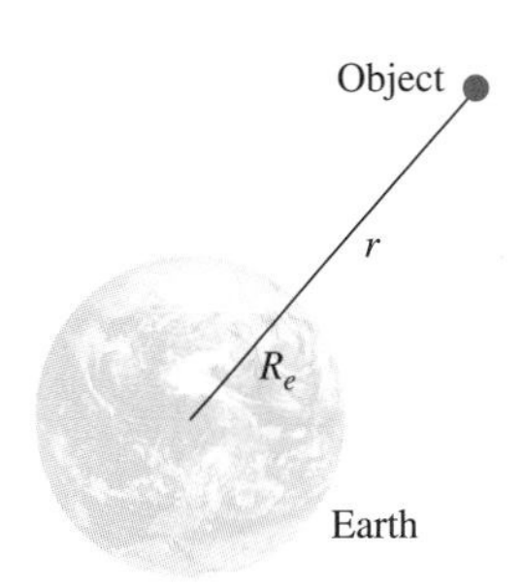

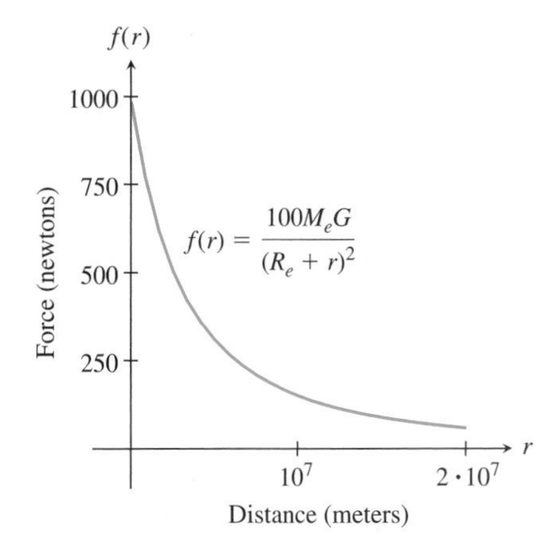

FIGURE 1.74 The graph illustrates the force on the object as a function of the distance from the Earth's surface.

The units for the derivative are newtons/meter. When $r = 10^7$ m, the derivative has value

$$F'(10^7) = \frac{-200GM_e}{(R_e + 10^7)^3} \approx -1.82 \times 10^{-5} \text{ N/m}.$$

When the object is 10^7 m above Earth's surface, the gravitational force on the object is $F(10^7) \approx 148.8$ N. See Fig. 1.74. The derivative tells us that the force on the object decreases by about 1.82×10^{-5} N if the object moves 1 m farther from the surface of Earth.

EXAMPLE 4 The owner of the Good Lookin' Glass Company asks a group of engineers and accountants to reflect on the cost of manufacturing several one-way mirrors. They report that the cost of producing $N > 0$ mirrors will be about

$$C(N) = 3000 + 5N + 2\sqrt{N} \quad \text{dollars.} \tag{6}$$

Compute the derivative of the function C and discuss the significance of the derivative.

Solution

We first discuss the meaning of (6). The glass company cannot manufacture $100\frac{1}{2}$ or $\sqrt{2}$ mirrors, but only a positive whole number of mirrors. Thus $C(N)$ is only defined for $N = 1, 2, 3, \ldots$. Thus, the graph of $C(N)$ shows just the

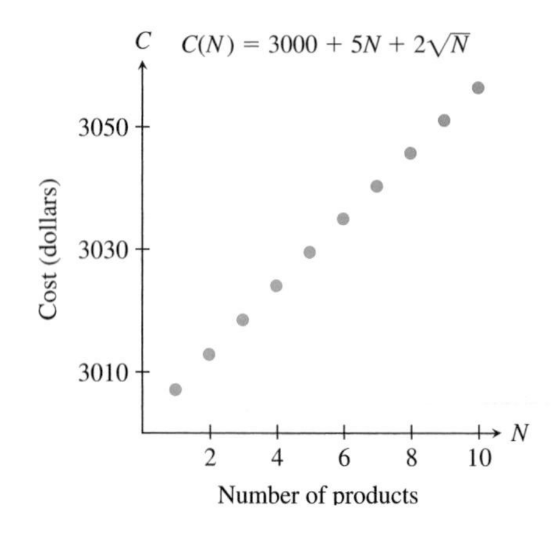

FIGURE 1.75 The cost function is defined only for nonnegative integers N.

individual points $(1, C(1))$, $(2, C(2))$, $(3, C(3))$, ... (see Fig. 1.75). Such a function does not have a derivative. When we zoom in on this graph, we will never see a straight line. This suggests that the rate of change (or derivative) of $C(N)$ with respect to N does not exist. Indeed, if the derivative of C did exist, it would be given by

$$\lim_{t\to N} \frac{C(t) - C(N)}{t - N}. \tag{7}$$

However, the quotient $(C(t) - C(N))/(t - N)$ is not defined for t in intervals on either side of N because any such interval contains noninteger values of t. This means that the conditions for the existence of the limit (7) are not satisfied. Hence the limit (7) does not exist.

Even though (6) may not be meaningful from a manufacturing point of view for nonintegers N, useful information can be obtained by assuming $C(N)$ is defined for all positive N. We can then compute $C'(N)$ for any positive N:

$$\begin{aligned} C'(N) &= \lim_{t\to N} \frac{C(t) - C(N)}{t - N} \\ &= \lim_{t\to N} \frac{\left(3000 + 5t + 2\sqrt{t}\right) - \left(3000 + 5N + 2\sqrt{N}\right)}{t - N} \\ &= \lim_{t\to N} \left(\frac{5(t - N)}{t - N} + \frac{2\left(\sqrt{t} - \sqrt{N}\right)}{\left(\sqrt{t} - \sqrt{N}\right)\left(\sqrt{t} + \sqrt{N}\right)} \right) \\ &= \lim_{t\to N} \left(5 + \frac{2}{\sqrt{t} + \sqrt{N}} \right) \\ &= 5 + \frac{1}{\sqrt{N}}. \end{aligned}$$

When $C(N)$ is the cost of manufacturing N units of a product, $C'(N)$ has units of dollars/product and is called the **marginal cost.** The marginal cost can be used as an estimate of the cost of producing a unit of the product after N units have already been manufactured. That is, $C'(N)$ is used as an estimate of $C(N + 1) - C(N)$. To see that this is reasonable, note that

$$C'(N) = \lim_{h\to 0} \frac{C(N + h) - C(N)}{h}.$$

Assuming that $h = 1$ is "small" in this context, we have

$$C'(N) \approx \frac{C(N + 1) - C(N)}{1} = C(N + 1) - C(N).$$

In manufacturing processes, where the average cost per product goes down as more products are produced, this approximation is often a good one. To check this, assume that 1000 mirrors have been produced. The extra cost to produce mirror 1001 is

$$\begin{aligned} C(1001) - C(1000) &= \left(3000 + 5 \cdot 1001 + 2\sqrt{1001}\right) \\ &\quad - \left(3000 + 5 \cdot 1000 + 2\sqrt{1000}\right) \\ &\approx \$5.031615. \end{aligned}$$

The approximation to this value given by the marginal cost is

$$C(1001) - C(1000) \approx C'(1000) = 5 + 1/\sqrt{1000} \approx \$5.031623.$$

Pretty close!

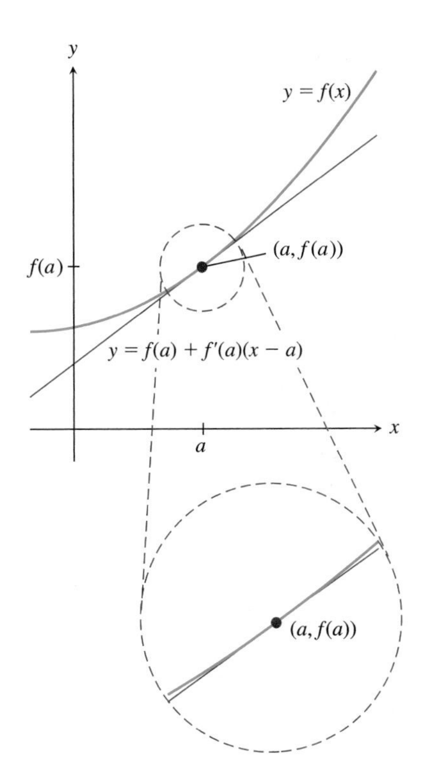

FIGURE 1.76 The graph of the tangent line is close to the graph of the function near the point of tangency.

The Tangent Line

Let f be a function with derivative $f'(a)$ at $x = a$. When we zoom in on the graph of $y = f(x)$ near the point $(a, f(a))$, the graph appears to be a straight line with slope $f'(a)$. See Fig. 1.76. This motivates the following definition.

DEFINITION Tangent Line

Let f have derivative $f'(a)$ at $x = a$. The line with slope $f'(a)$ through $(a, f(a))$ is the **tangent line** to the graph of $y = f(x)$ at the point $(a, f(a))$.

Because the line tangent to $y = f(x)$ at the point $(a, f(a))$ has slope $f'(a)$, we can easily write down an equation for the tangent line:

$$y - f(a) = f'(a)(x - a)$$

or

$$y = f(a) + f'(a)(x - a).$$

EXAMPLE 5 Find the equation of the line tangent to the graph of $y = x^2$ at the point $(-2, 4)$. On the same set of axes, sketch the graph of $y = x^2$ and the graph of the tangent line.

Solution

Let $f(x) = x^2$. In Example 1 we showed that $f'(x) = 2x$. Thus the desired tangent line has slope $f'(-2) = 2(-2) = -4$ and contains the point $(-2, 4)$. The equation for this line is

$$y - 4 = (-4)(x - (-2)).$$

This simplifies to $y = -4x - 4$. The graphs of $y = x^2$ and the tangent line are shown in Fig. 1.77.

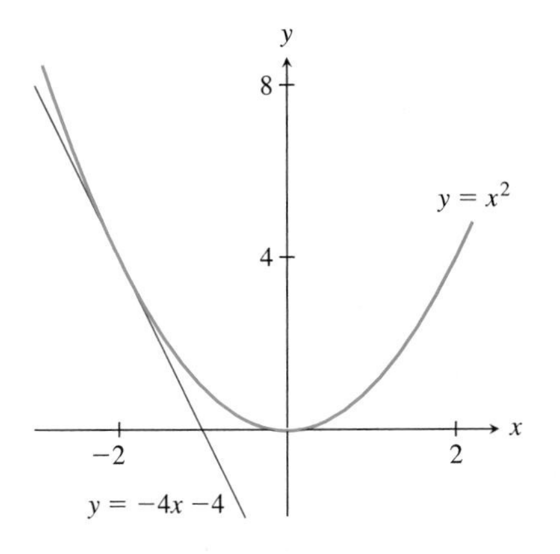

FIGURE 1.77 The line tangent to the graph of $y = x^2$ at the point $(-2, 4)$ has slope -4.

The Tangent Line as an Approximation If we draw the line tangent to the graph of $y = f(x)$ at the point $(a, f(a))$, then zoom in on the graph near this point, the graph and the tangent line will be almost indistinguishable. This is illustrated in Fig. 1.76.

If the graphs of the functions

$$y = f(x) \quad \text{and} \quad y = f(a) + f'(a)(x - a)$$

are almost indistinguishable, then the values of these two expressions must be close. Thus for x close to a,

$$f(x) \approx f(a) + f'(a)(x - a). \tag{8}$$

WEB Expression (8) says that when x is close to a, the function value $f(x)$ is approximated by the corresponding y-value for the tangent line.

EXAMPLE 6 Find the equation of the line tangent to

$$f(x) = 1/(x+1)$$

at $(2, 1/3)$ and use this equation to approximate $f(1.93) = 1/(1.93+1)$.

Solution

The tangent line passes through the point of tangency, $(2, 1/3)$. The slope of the tangent line is the derivative of f at $x = 2$,

$$f'(2) = \lim_{x\to 2} \frac{f(x) - f(2)}{x - 2} = \lim_{x\to 2} \frac{\dfrac{1}{x+1} - \dfrac{1}{3}}{x - 2}.$$

Add the fractions in the numerator and simplify the resulting complex fraction. We then have

$$f'(2) = \lim_{x\to 2} \frac{2 - x}{3(x+1)(x-2)} = \lim_{x\to 2} \frac{-1}{3(x+1)} = -\frac{1}{9}.$$

The equation of the line tangent to $y = 1/(x+1)$ at $(2, 1/3)$ is

$$y = f(2) + f'(2)(x-2) = \frac{1}{3} - \frac{1}{9}(x-2).$$

When x is close to 2, the y coordinates of the graph of $y = 1/(x+1)$ and the tangent line should be close. See Fig. 1.78. Hence for x near 2,

$$\frac{1}{x+1} \approx \frac{1}{3} - \frac{1}{9}(x-2).$$

Substitute $x = 1.93$ into this expression to get

$$f(1.93) = \frac{1}{1.93+1} \approx \frac{1}{3} - \frac{1}{9}(1.93 - 2) \approx 0.341111.$$

How does this compare with the value of $(1.93+1)^{-1}$ given by your calculator?

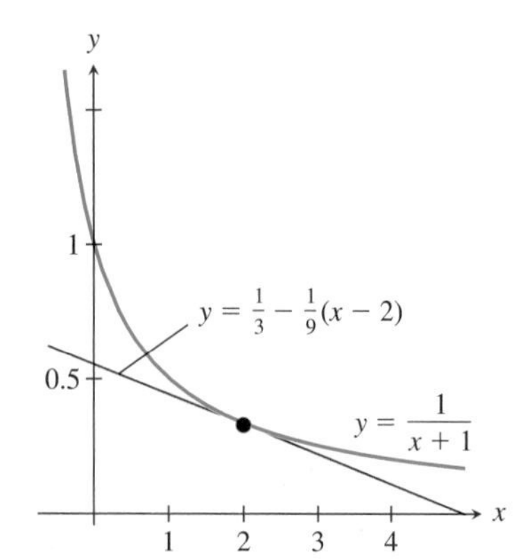

FIGURE 1.78 The line tangent to $y = 1/(x+1)$ at $(2, 1/3)$.

Error in the Tangent Line Approximation We can use the techniques developed in Section 1.6 to estimate the error in the tangent line approximation. Let

$$\frac{f(x) - f(a)}{x - a} - f'(a) = E(x). \tag{9}$$

Because

$$\lim_{x\to a} E(x) = \lim_{x\to a}\left(\frac{f(x) - f(a)}{x - a} - f'(a)\right) = f'(a) - f'(a) = 0,$$

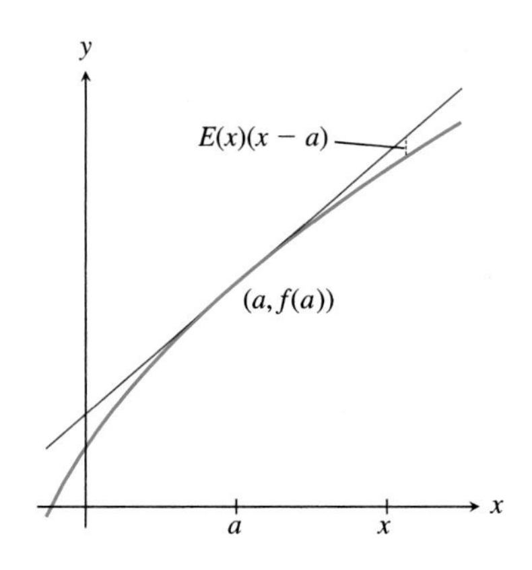

FIGURE 1.79 The difference between $f(x)$ and the y-coordinate of the tangent line is $E(x)\,(x - a)$.

we can make $E(x)$ small by taking x close to a. Solving (9) for $f(x)$, we find

$$f(x) = f(a) + f'(a)\,(x - a) + E(x)\,(x - a). \tag{10}$$

Hence the difference between the function value $f(x)$ and the value $f(a) + f'(a)\,(x - a)$ of the tangent line expression is

$$f(x) - (f(a) + f'(a)\,(x - a)) = E(x)\,(x - a). \tag{11}$$

From (11) we see that this difference is small compared to $|x - a|$ when $E(x)$ is small, that is, when x is close to a. The error is illustrated in Fig. 1.79.

EXAMPLE 7 The line tangent to $y = \sin x$ at $(0, 0)$ is used to approximate $\sin x$ for $-0.5 < x < 0.5$. Discuss the error in the approximation.

Solution

We first find the tangent line. Using (2) with $f(x) = \sin x$ and $a = 0$, the slope of the tangent line is

$$\lim_{h\to 0} \frac{\sin h - \sin 0}{h} = \lim_{h\to 0} \frac{\sin h}{h} = 1.$$

(The value of this limit was found in Example 8 of Section 1.5.) Because the tangent line has slope 1 and contains the point $(0, 0)$, its equation is $y = x$.

We now need to discuss the error in the approximation

$$\sin x \approx x$$

for $-0.5 \le x \le 0.5$. In Example 5 of Section 1.6 we showed that

$$|\sin x - x| \le 0.05|x| \le 0.025$$

for such x. Thus if we use the approximation $\sin x \approx x$ for x in the interval $(-0.5, 0.5)$, the error is no bigger than 0.025 in absolute value. Alternatively, the error is no bigger than 5% of $|x|$. The graph of $y = \sin x$ and the tangent line are shown in Fig. 1.80.

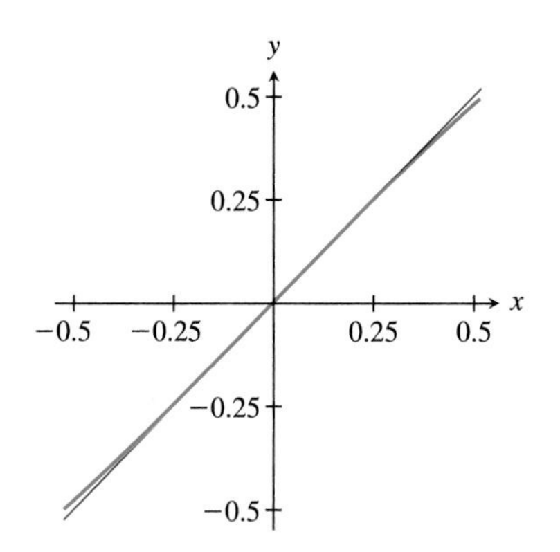

FIGURE 1.80 The line $y = x$ is tangent to the graph of $y = \sin x$ at $(0, 0)$.

EXAMPLE 8 The line tangent to $y = \sqrt{x}$ at $(1, 1)$ is used to approximate $\sqrt{x}$ for $0.9 < x < 1.1$. Discuss the error in the approximation.

Solution

The slope of the tangent line is the derivative of $\sqrt{x}$ at $x = 1$. Using (5) with $f(x) = \sqrt{x}$ and $a = 1$ the derivative is

$$\lim_{x\to 1} \frac{\sqrt{x} - \sqrt{1}}{x - 1} = \lim_{x\to 1} \frac{\sqrt{x} - 1}{(\sqrt{x} - 1)(\sqrt{x} + 1)} = \frac{1}{2}.$$

The tangent line is the line of slope $1/2$ that contains the point $(1, 1)$. This line has equation

$$y = \frac{1}{2}x + \frac{1}{2}.$$

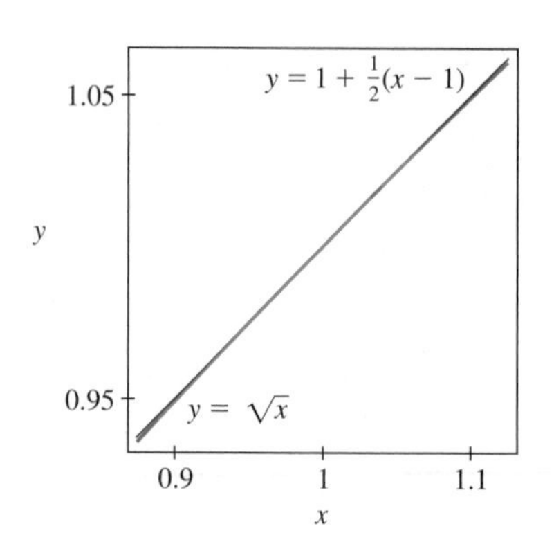

FIGURE 1.81 The line $y = \frac{1}{2}x + \frac{1}{2}$ is tangent to the graph of $y = \sqrt{x}$ at (1, 1).

We now need to consider the error in the approximation

$$\sqrt{x} \approx \frac{1}{2}x + \frac{1}{2} \tag{12}$$

for $0.9 < x < 1.1$. In Example 6 of Section 1.6 we showed that

$$\left|\sqrt{x} - \left(\frac{1}{2}x + \frac{1}{2}\right)\right| \leq 0.015|x - 1| \leq (0.015)(0.01) = 0.0015$$

for such x. Thus if we use the approximation (12) for $0.9 < x < 1.1$, the error is no more than 0.0015 in absolute value. In addition, the error is no bigger than 1.5% of the value of $|x - 1|$. The graph of $y = \sqrt{x}$ and the tangent line are shown in Fig. 1.81.

Derivatives and Continuous Functions As another consequence of (10), we obtain a necessary condition for a function f to have a derivative at a point $x = a$.

Continuity of Differentiable Functions

If the function f has a derivative at $x = a$, then f is continuous at $x = a$; that is,

$$\lim_{x \to a} f(x) = f(a).$$

To justify this statement, take the limit as $x \to a$ of both sides of (10):

$$\lim_{x \to a} f(x) = \lim_{x \to a}(f(a) + f'(a)(x - a) + E(x)(x - a)).$$

Applying results from Section 1.5 for limits of sums and products of functions, we obtain

$$\begin{aligned}
\lim_{x \to a} f(x) &= \lim_{x \to a} f(a) + \lim_{x \to a}(f'(a)(x - a)) + \lim_{x \to a}(E(x)(x - a)) \\
&= f(a) + f'(a)\lim_{x \to a}(x - a) + \left(\lim_{x \to a} E(x)\right)\left(\lim_{x \to a}(x - a)\right) \\
&= f(a) + f'(a) \cdot 0 + 0 \cdot 0 \\
&= f(a).
\end{aligned}$$

Because $\lim_{x \to a} f(x) = f(a)$, the function f is continuous at $x = a$.

Exercises 1.7

Exercises 1–10: Determine the derivative in each case. Use (2) for some of the problems and (5) for others.

1. $f(x) = x^2 - 3$
2. $y = 2x^2 - 6$
3. $h(t) = -3t^2 + 4t - \sqrt{2}$
4. $g(u) = \dfrac{2}{u + 2}$
5. $y = \dfrac{4}{2 - x}$
6. $s(t) = t - 2\sqrt{t}$

7. $y = 2t + 3\sqrt{t+9}$

8. $r(s) = as^2 + bs + c$ where a, b, c are constants.

9. $y = \dfrac{1}{ax+b}$ where a and b are constants.

10. $h(t) = \sqrt{at+b}$ where a and b are constants.

Exercises 11–17: Find an equation for the line tangent to the graph of the function at the specified point.

11. $y = x^2 - 3x$ at $(1, -2)$

12. $s(t) = -3t^2 - 2t + 1$ at $(-2, -7)$

13. $h(r) = 4r + 2$ at $(1, 6)$

14. $y = \dfrac{1}{x}$ at $(a, 1/a)$, $a \neq 0$

15. $g(u) = \dfrac{2}{u^2 - 1}$ at $(2, 2/3)$

16. $w = \sqrt{2v-1}$ at $\left(b, \sqrt{2b-1}\right)$, $b > 1/2$

17. $f(x) = 2|x|$ at $(2, 4)$

18. Use the equation for the line tangent to $y = x^2$ at $(4, 16)$ to approximate $(4.14)^2$ and $(3.91)^2$. In each case, compute the error in the approximation.

19. Use the equation for the line tangent to $y = 1/x$ at $(3, 1/3)$ to approximate $1/(3.15)$ and $1/(2.85)$. In each case, compute the error in the approximation.

20. In Chapter 2 we will see that the derivative of $\sin x$ is $\cos x$.

a. Find an equation for the line tangent to the graph of $y = \sin x$ at $\left(\pi/3, \sqrt{3}/2\right)$.

b. Use the result of part a to find an approximation to $\sin(\pi/3 + 0.1)$. (All angle measures are in radians.) Use your calculator to check the accuracy of the approximation.

21. In Chapter 2 we will see that the derivative of $\tan x$ is $\sec^2 x$.

a. Find an equation for the line tangent to the graph of $y = \tan x$ at $(\pi/4, 1)$.

b. Use the result of part a to find an approximation to $\tan(\pi/4 - 0.05)$. (All angle measures are in radians.) Use your calculator to check the accuracy of the approximation.

22. The planet Jupiter has mass $M \approx 1.9 \times 10^{27}$ kg and radius $R \approx 6.98 \times 10^7$ m. Suppose the 100-kg mass of Example 3 is 10^7 m above the surface of Jupiter. Find the rate of change of the gravitational force as the object moves away from Jupiter along a ray through the center of the planet. Discuss the meaning of this rate of change.

23. Repeat Exercise 22 for the moon. The moon has mass $M \approx 7.34 \times 10^{22}$ kg and radius $R \approx 1.74 \times 10^6$ m.

24. The accountants at the Seed & Sod Turf Company estimate that the cost of a grassroots advertising campaign that will reach $N \times 10^5$ consumers is roughly $C(N) = 5000 + 1000\sqrt{N} + 0.05(N-1)^2$ dollars.

a. Find the derivative of $C(N)$ and discuss the meaning of this derivative as a marginal cost.

b. Use the derivative to estimate the extra cost of a campaign to reach a total of 1,100,000 consumers over that of a campaign to reach 1,000,000 consumers.

25. A function f has derivative

$$f'(x) = \frac{1}{x^2+1}.$$

Given that $f(0) = 2$:

a. Find an equation for the line tangent to the graph of $y = f(x)$ at the point $(0, 2)$.

b. Find approximations to $f(0.05)$ and $f(-0.1)$.

26. A function h has derivative

$$h'(u) = u \sin(2u).$$

Given that $h(3\pi/4) = -3$, find an approximation to

$$h\left(\frac{3\pi}{4} + 0.15\right).$$

27. The line $y = 5x - 4$ is tangent to the graph of $y = g(x)$ at the point $(-1, -9)$. Find $g'(-1)$ and an approximation to $g(-0.88)$.

28. The line $y = -2t + 3$ is tangent to the graph of $y = h(t)$ at the point $(3, -3)$. Find $h'(3)$ and an approximation to $h(3.1)$.

29. Let $f(x) = x^3$.

a. Find an equation for the line tangent to the graph of $y = f(x)$ at the point $(-2, -8)$.

T **b.** Let $y = t(x)$ be an equation for the tangent line of part **a**. Suppose we wish to use the approximation

$$f(x) \approx t(x)$$

for x values near -2. Find r so that when $|x - (-2)| < r$, the error in the approximation is

$$|f(x) - t(x)| < 0.01|x - (-2)|.$$

(See Example 6 in Section 1.6.)

30. Let $f(x) = 1/\sqrt{x+1}$.

a. Find an equation for the line tangent to the graph of $y = f(x)$ at the point $(3, 1/2)$.

T **b.** Let $y = t(x)$ be an equation for the tangent line of part **a**. Suppose we wish to use the approximation

$$f(x) \approx t(x)$$

for x near 3. Find r so that when $|x - 3| < r$, the error in the approximation is

$$|f(x) - t(x)| < 0.01|x - 3|.$$

(See Example 6 in Section 1.6.)

31. Let f be a function with $f(4) = 3$ and $f'(4) = 7$. Define $h(x) = f(-x)$.

a. Tell why the point $(-4, 3)$ is on the graph of $y = h(x)$.

b. What is the value of $h'(-4)$? Justify your answer by comparing the graphs of $y = f(x)$ and $y = h(x) = f(-x)$.

c. By arguing graphically, tell why if the derivative of $h(x)$ at $x = a$ is defined, then $h'(a) = -f'(-a)$.

32. Let f be a function with $f(4) = 3$ and $f'(4) = 7$. Define $h(x) = -f(x)$.

a. Tell why the point $(4, -3)$ is on the graph of $y = h(x)$.

b. What is the value of $h'(4)$? Justify your answer by comparing the graphs of $y = f(x)$ and $y = h(x) = -f(x)$.

c. By arguing graphically, tell why if the derivative of $h(x)$ at $x = a$ is defined, then $h'(a) = -f'(a)$.

33. Let f be a function with $f(4) = 3$ and $f'(4) = 7$. Define $h(x) = 8f(x)$.

a. Tell why the point $(4, 24)$ is on the graph of $y = h(x)$.

b. What is the value of $h'(4)$? Justify your answer by comparing the graphs of $y = f(x)$ and $y = h(x) = 8f(x)$.

c. Express $h'(a)$ in terms of the derivative of f. Justify your answer with the aid of graphs.

34. Let f be a function with $f(4) = 3$ and $f'(4) = 7$. Define $h(x) = f(x + 7)$.

a. Tell why the point $(-3, 3)$ is on the graph of $y = h(x)$.

b. What is the value of $h'(-3)$? Justify your answer by comparing the graphs of $y = f(x)$ and $y = h(x) = f(x + 7)$.

c. Express $h'(a)$ in terms of the derivative of f. Justify your answer with the aid of graphs.

35. The limit statement

$$\lim_{x \to 2} \frac{(x^9 - 4x^7 + 3x - 2) - 4}{x - 2} = 515$$

is a statement about the derivative of some function f at some value $x = a$. What are f, a, and $f'(a)$?

36. The limit statement

$$\lim_{h \to 0} \frac{\sec(\pi/4 + h) - \sqrt{2}}{h} = \sqrt{2}$$

is a statement about the derivative of some function g at some $t = a$. What are g, a, and $g'(a)$?

37. A function f satisfies

$$f(1) = f(2) = f(3) = f(4) = f(5) = 0,$$
$$f'(1) = f'(3) = f'(5) = 1,$$

and

$$f'(2) = f'(4) = -1.$$

Use this information to sketch a possible graph for $y = f(x)$.

38. In this problem we discuss Richard Feynman's approximation (1) to the cube root function.

a. Show that

$$\lim_{x \to 1728} \frac{\sqrt[3]{x} - 12}{x - 1728} = \frac{1}{432}.$$

It will be helpful to keep in mind that $12^3 = 1728$. The factorization

$$b^3 - a^3 = (b - a)(b^2 + ba + a^2)$$

may also prove useful.

b. Use the result of part **a** to obtain the approximation (1).

T c. Use graphical means to investigate the error in the approximation for $1720 \leq x \leq 1736$.

Review of Key Concepts

We began this chapter with a review of functions. We saw that functions can be represented in a number of ways: symbolically (with a formula), graphically, or numerically (with a table of values). Functions represented in these ways arise regularly in science, engineering, and mathematics. It is important to know how to work with functions in all of these forms—not only with paper and pencil, but also with the aid of a calculator or computer algebra system.

Following our review of functions, we interpreted the slope of a line as a rate of change and used this to motivate a definition for the rate of change of an arbitrary function. For functions whose graphs look like lines as we zoom in, we defined the rate of change as the slope of this line. From this graphical definition we developed a numerical understanding of the rate of change and an analytic (or symbolic) definition for the rate of change.

The rate of change definition led to the idea of limit. Once limit was defined, we reformulated our definition of rate of change in a more precise form. We also saw that limits are very closely related to approximation and error. Because approximation is a very important idea in modern science, engineering, and mathematics, we spent some time discussing approximations and errors.

We concluded the chapter by defining the derivative of a function and noted that the derivative is really just another name for the rate of change. As one geometric application of the derivative, we defined the tangent line and noted that it is the "line" we see when we zoom in on a point of a graph. We saw that tangent lines can be used to obtain a good approximation to a function.

In the Chapter Summary that follows, we summarize many of these ideas in table form. For each concept, we present a general definition and a specific example and support both with a graph.

Chapter Summary

Representing Functions

A function can be represented by a graph:

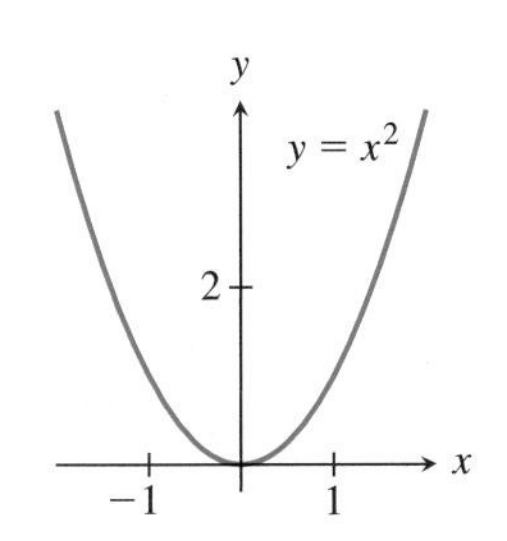

A function f can be represented by an equation:

$$f(x) = x^2.$$

A function f can be represented by a table of values:

x	$f(x)$
-2	4
-0.9	0.81
0	0
1	1
3/2	9/4
3	9

Rate of Change

The rate of change at $x = a$ is the slope of the "line" we see when we zoom in on the point $(a, f(a))$.

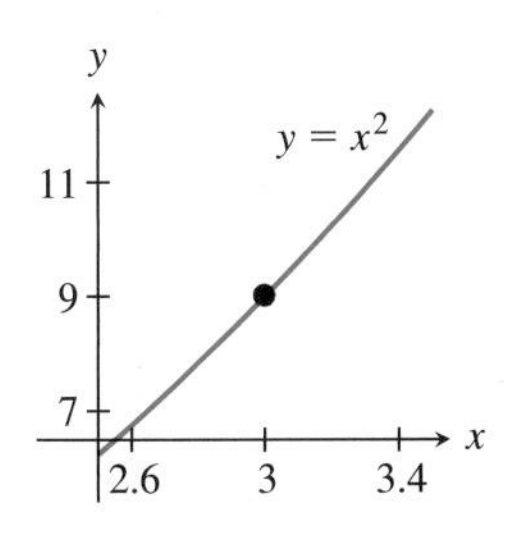

The rate of change of $y = f(x)$ with respect to x at $x = a$ is

$$\lim_{h\to 0} \frac{f(a+h) - f(a)}{h}.$$

If $y = f(x) = x^2$, the rate of change of y with respect to x at $x = 3$ is

$$\lim_{h\to 0} \frac{(3+h)^2 - 3^2}{h} = 6.$$

Limit of $f(x)$ at a

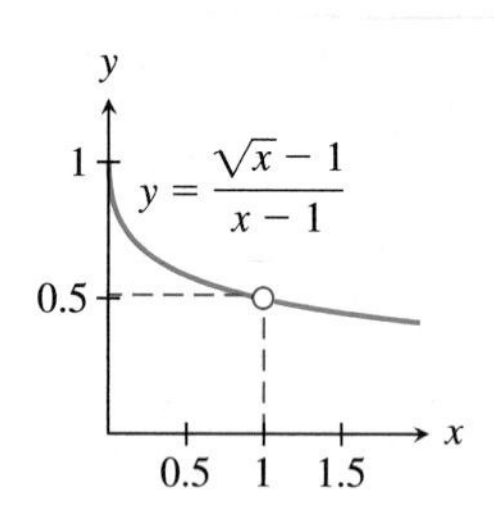

If we can make $f(x)$ as close to L as we like by taking x close to but not equal to a then

$$\lim_{x\to a} f(x) = L.$$

$$\lim_{x\to 1} \frac{\sqrt{x}-1}{x-1} = \frac{1}{2}$$

As x gets close to 1, $\dfrac{\sqrt{x}-1}{x-1}$ gets close to 1/2:

x	$(\sqrt{x}-1)/(x-1)$
1.1	0.488088
0.9	0.513167
1.005	0.499377

Limit of $f(x)$ at a

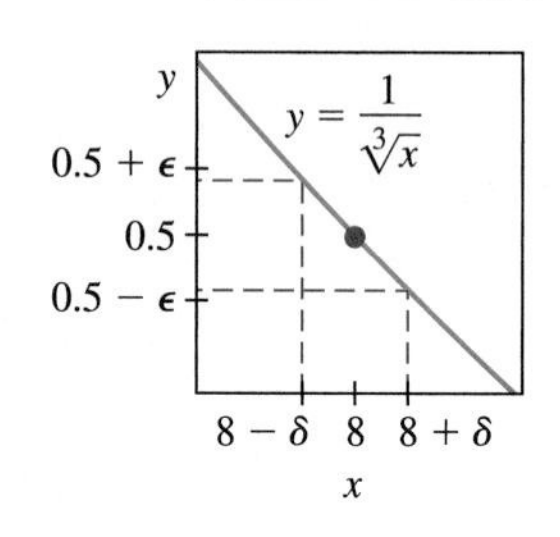

$$\lim_{x\to a} f(x) = L$$

means that if $\epsilon > 0$ is given, then there is a number $\delta > 0$ so that

$$|f(x) - L| < \epsilon$$

when

$$0 < |x - a| < \delta.$$

$$\lim_{x\to 8}\left(1/\sqrt[3]{x}\right) = \frac{1}{2}.$$

Let $\epsilon = 0.05$. Note (by graphing) that when

$$0 < |x - 8| < 0.2$$

we have

$$\left|1/\sqrt[3]{x} - 1/2\right| < 0.05 = \epsilon.$$

Thus for $\epsilon = 0.05$ we can take $\delta = 0.2$.

Continuous Function

A continuous function has an unbroken graph.

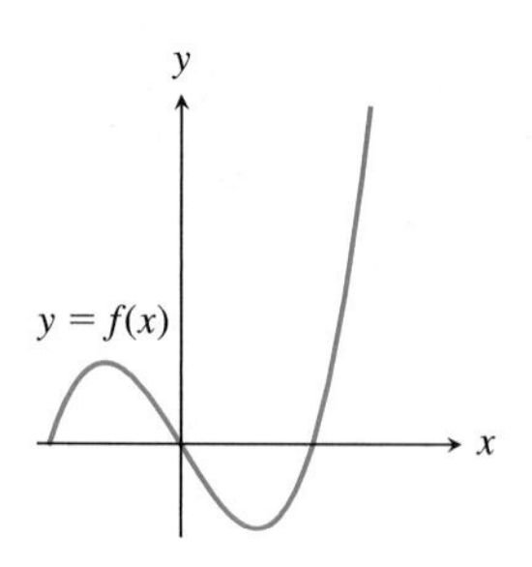

A function f is continuous at $x = a$ if

$$\lim_{x\to a} f(x) = f(a).$$

If a function f is continuous at every point in its domain, then we say that f is continuous.

Let c be a real number and $p(x)$ a polynomial. Then

$$\lim_{x\to c} p(x) = p(c).$$

Hence every polynomial is continuous on the real line. The sine and cosine functions are also continuous on the real line because for every real number c,

$$\lim_{\theta\to c} \sin\theta = \sin c$$

and

$$\lim_{\theta\to c} \cos\theta = \cos c.$$

The Derivative

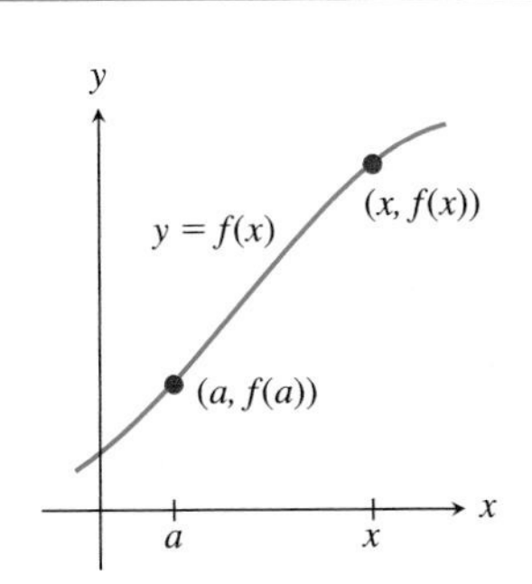

The derivative of f at a is

$$f'(a) = \lim_{h \to 0} \frac{f(a + h) - f(a)}{h}$$

provided this limit exists. The derivative at a is also given by

$$f'(a) = \lim_{x \to a} \frac{f(x) - f(a)}{x - a}.$$

Let $f(x) = x^2$. The derivative of f is the function f' defined by

$$f'(x) = \lim_{t \to x} \frac{t^2 - x^2}{t - x} = 2x.$$

The Line Tangent to a Graph

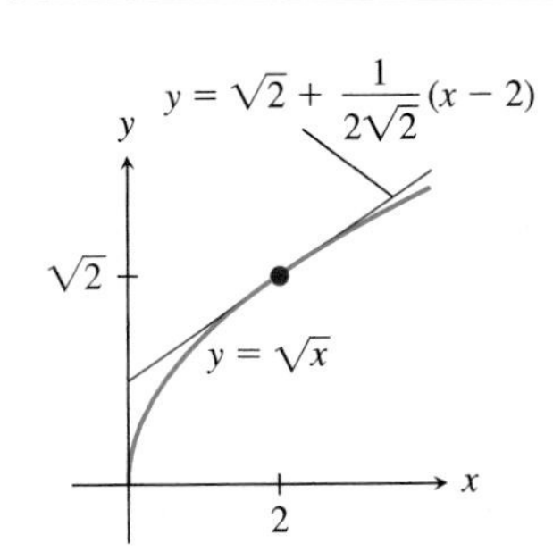

Suppose that the function f has derivative $f'(a)$ at $x = a$. The line tangent to the graph of $y = f(x)$ at the point $(a, f(a))$ is the line of slope $f'(a)$ that passes through $(a, f(a))$. This line has equation

$$y = f(a) + f'(a)(x - a).$$

The derivative of $f(x) = \sqrt{x}$ at $x = 2$ is

$$f'(2) = \frac{1}{2\sqrt{2}}.$$

The line tangent to the graph of $y = \sqrt{x}$ at the point $(2, \sqrt{2})$ has equation

$$y = \sqrt{2} + \frac{1}{2\sqrt{2}}(x - 2).$$

The Tangent Line Approximation

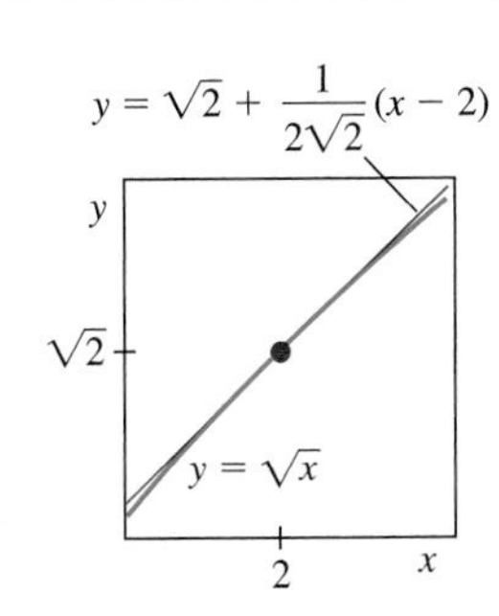

When x is close to a the approximation

$$f(x) \approx f(a) + f'(a)(x - a)$$

can be used. The error is

$$|f(x) - (f(a) + f'(a)(x - a))| = |E(x)(x - a)|$$

where $E(x) = \dfrac{f(x) - f(a)}{x - a} - f'(a)$.

Note that $E(x)$ is small when x is close to a.

The error in the approximation

$$\sqrt{x} \approx \sqrt{2} + \frac{1}{2\sqrt{2}}(x - 2)$$

is no more than 1% of $|x - 2|$ when x is within 0.2 of 2. This is because when $0 < |x - 2| < 0.2$ we have

$$|E(x)| = \left| \frac{\sqrt{x} - \sqrt{2}}{x - 2} - \frac{1}{2\sqrt{2}} \right| < 0.01.$$

Chapter Review Exercises

Exercises 1–6: The two functions f and g are defined by the accompanying tables.

1. State the domain of f and the domain of g.
2. What is the range of f? Of g?
3. What is the domain of the function $f + g$? Construct a table for $f + g$.
4. What is the domain of the function $\sqrt{g}$? Construct a table for $\sqrt{g}$.

5. What is the domain of the function f/g? Construct a table for f/g.

6. What is the domain of the function $f \circ g$? Construct a table for $f \circ g$.

x	$f(x)$
0	-3
1	π
2	sin 1
3	0
4	$(1, 3)$
5	12.4

x	$g(x)$
-3	0
-0.5	4.5
0	4
2	$-\pi$
3	3

Exercises 7–14: The two functions f and g are defined by the accompanying graphs.

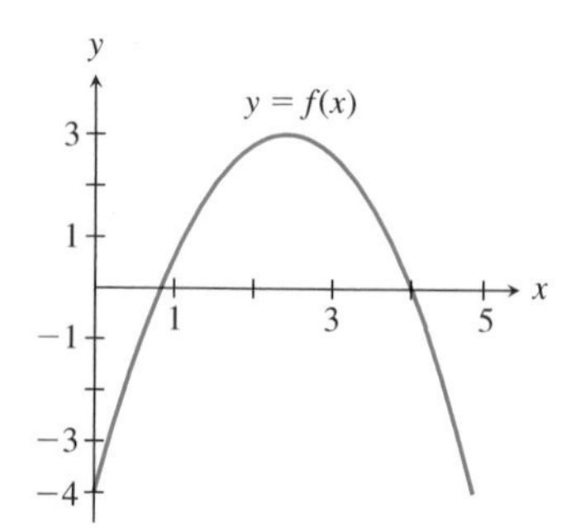

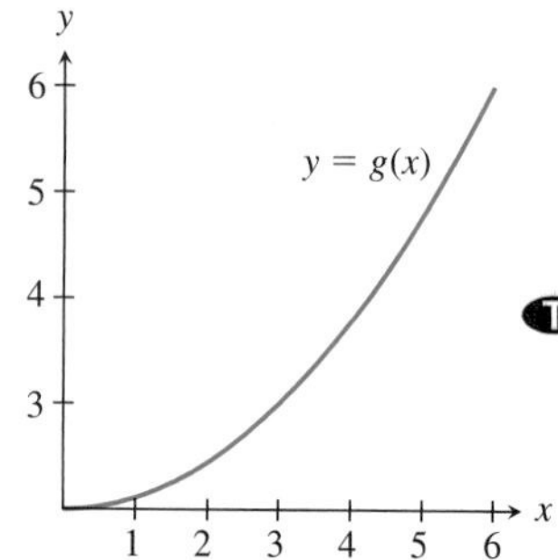

7. State the domain of f and the domain of g.

8. What is the range of f? Of g?

9. What is the domain of the function $f + g$? Draw a graph of $y = (f + g)(x)$.

10. What is the domain of the function $\sqrt{f}$? Draw a graph of $y = \sqrt{f(x)}$.

11. What is the domain of the function g/f?

12. What is the domain of the function $f \circ g$? Estimate the value of $(f \circ g)(1)$. Of $(f \circ g)(2)$.

13. Estimate the rate of change of f with respect to x at $x = 1$. Explain how you arrived at your answer.

14. Estimate the rate of change of g with respect to x at $x = 3$. Explain how you arrived at your answer.

15. In Section 1.6 we discussed the formula

$$L = L_0\sqrt{1 - \left(\frac{v}{c}\right)^2},$$

which gives the length L of an object moving at speed v if its length at rest is L_0. Assume that L is measured in meters and v in centimeters/second. What are the units of the rate of change of L with respect to v? Explain carefully the physical interpretation of this rate of change.

16. The speed of sound in air at 1 atm pressure and 0°C is approximately 1090 ft/s. A hiker yells and waits to hear her echo from a mountain d feet across a valley.

a. Find a formula for the length of time T it takes for the echo to return to the hiker.

b. What is the rate of change of T with respect to d? What are the units of this rate of change?

c. Explain carefully the physical interpretation of this rate of change.

17. The inflation rate tells how much the average cost of goods changes in a year. For example, if the rate of inflation is 3 percent for a year, then a product that cost \$1.00 at the beginning of the year would cost, on average, \$1.03 at the end of the year. At the end of every month, the United States government publishes the *annual* inflation rate for that month. But these annual rates often change from month to month! For example, in March 1994 the annual inflation rate was 3.5 percent, in April it was 3.2 percent, and in May it was 4.1 percent. Interpret these figures as rates of change and explain how the *annual* inflation rate can change every month.

T *Exercises 18–21: Investigate the limit (a) numerically, (b) graphically, and (c) analytically.*

18. $\lim_{x \to 2}(x^2 - 3x + 2)$

19. $\lim_{x \to -1} \dfrac{x^2 - 1}{x + 1}$

20. $\lim_{t \to 3} \dfrac{\sqrt{t + 3} - \sqrt{6}}{t - 3}$

21. $\lim_{r \to 2} \dfrac{r^3 - 8}{r^2 - 4}$

T *Exercises 22–25: Investigate the limit (a) numerically and (b) graphically.*

22. $\lim_{\theta \to \pi/3} \dfrac{\cos\theta - 1/2}{\theta - \pi/3}$

23. $\lim_{r \to 3} \dfrac{10^r - 1000}{r - 3}$

24. $\lim_{x \to 10} \dfrac{\log_{10} x - 1}{x - 10}$

25. $\lim_{t \to 0} \dfrac{\tan 3t}{5t}$

T **26.** Let L be the value of the limit in Exercise 19. Find a $\delta > 0$ so that when

$$0 < |x - (-1)| < \delta$$

we have $\left|\dfrac{x^2 - 1}{x + 1} - L\right| < 0.01.$

27. Let L be the value of the limit in Exercise 22. Find a $\delta > 0$ so that when

$$0 < |\theta - \pi/3| < \delta$$

we have

$$\left|\frac{\cos\theta - 1/2}{\theta - \pi/3} - L\right| < 0.01.$$

28. Let L be the value (or estimate) of the limit in Exercise 24. Find a $\delta > 0$ so that when

$$0 < |x - 10| < \delta$$

we have

$$\left|\frac{\log_{10} x - 1}{x - 10} - L\right| < 0.01.$$

Exercises 29–32: Find $f'(x)$ for the given f.

29. $f(x) = 2x - 5$

30. $f(x) = -3x^3 + 4x$

31. $f(x) = x + \dfrac{1}{x}$

32. $f(x) = (x + a)^2$, a constant.

33. Find the equation for the line tangent to $y = \sqrt{x + 2}$ at the point $(7, 3)$. Find $r > 0$ so that for $7 - r < x < 7 + r$ the error in the tangent line approximation to $\sqrt{x + 2}$ is no more than 0.001 in absolute value.

34. Use graphical or numerical methods to find the derivative of $f(x) = \cos x$ at $x = \pi/2$. Find an equation for the line tangent to $y = \cos x$ at the point $(\pi/2, 0)$. Find $r > 0$ so that for $\pi/2 - r < x < \pi/2 + r$ the error in the tangent line approximation to $\cos x$ is no more than 0.005 in absolute value.

35. Write a short paragraph that clearly explains what is meant by the statement

$$\lim_{x \to a} f(x) \quad \text{does not exist.}$$

Illustrate your explanation with some graphs.

36. In Section 1.7 we showed that if a function f has a derivative at $x = a$, then it is continuous at $x = a$. Is the converse true? That is, if a function f is continuous at $x = a$, must it have a derivative at $x = a$? Justify your answer.

37. In Section 1.7 we showed that if a function f has a derivative at $x = a$, then it is continuous at $x = a$. Explain why if f is not continuous at $x = a$, then we will never see a "straight line" when we zoom in on the graph of $y = f(x)$ at the point $(a, f(a))$.

STUDENT PROJECT

EXPLORING CHAOS

What Is Chaos?

Meteorologists say that "a butterfly flapping its wings in China can cause a tornado in Kansas several weeks later." Although it is very unlikely that the truth of this statement will ever actually be demonstrated, the statement does illustrate a problem with long-range weather prediction. Very small disturbances in the atmosphere can, over time, lead to large, noticeable phenomena. To accurately predict the weather a month from today, meteorologists would need precise information (pressure, humidity, velocity, temperature, etc.) now, and they would need this information about every cubic foot of the atmosphere. And even if it were possible to get information on this scale, small errors in some of these measurements might be enough to make weather patterns differ drastically from predictions.

The atmosphere is an example of a *chaotic system.* A chaotic system is one in which very small changes in one part of the system can lead to very big changes in another part. If the butterfly in China does not flap its wings at 2:00 P.M. on April 21, 2002, then it will be a nice day in Kansas on May 30. But if the butterfly does flap its wings, this small change in the atmosphere could propagate and eventually lead to a tornado on May 30. In recent years, mathematicians, scientists, and engineers have come to realize that many natural phenomena once thought to be well understood are really chaotic systems.

Compositions and Chaos In this project we will look at a fascinating chaotic system that arises by repeatedly composing the polynomial

$$p(x) = cx(1 - x)$$

with itself. Pick a number c with $0 < c < 4$. The system we examine will be the list of numbers we get by setting $x = 0.3$ and computing

$$p(0.3), p(p(0.3)), p(p(p(0.3))), p(p(p(p(0.3)))), \ldots. \tag{1}$$

(The choice $x = 0.3$ is arbitrary. Any x with $0 < x < 1$ would work as well.) In constructing (1) we use $x = 0.3$ as input for the polynomial p. We get output $p(0.3)$. We then use this output as input for p, and so get as a next output $p(p(0.3))$. Now we use this output as input for p, and so on. Repeat this process several times. What happens to the list (1)? We will see that the answer depends on the value of c, and that very small changes in c can cause substantial changes in the list (1).

PROBLEM 1 Let $c = 2.1$, so

$$p(x) = 2.1x(1 - x).$$

Use a calculator or CAS to compute the first 10 terms of the list (1). Describe what happens to the numbers in the list. Next let $c = 2.5$, so

$$p(x) = 2.5x(1 - x).$$

Compute the first 15 terms of the list (1). Describe what happens to the numbers in the new list. How is the second list similar to the first list? How is it different?

PROBLEM 2 The number 0.52381 should be familiar from Problem 1. Again setting $c = 2.1$, compute $p(0.52381)$. The number 0.52381 is called a *fixed point* for $p(x)$. Explain why this terminology is appropriate. The exact value of the fixed point can be found by solving the equation

$$p(x) = x.$$

Explain why a solution of this equation is a fixed point of p. Find the exact value of the fixed point that is approximated by 0.52381. Repeat for the number 0.6 with $c = 2.5$.

PROBLEM 3 Now let $c = 3.4$ so

$$p(x) = 3.4x(1 - x).$$

Use a calculator or CAS to compute the first 100 terms of the list (1). Describe what happens to the numbers in the list. Repeat with $c = 3.5$, $c = 3.83$, $c = 3.845$, and $c = 3.862$. How do the lists obtained differ?

PROBLEM 4 In the previous problem you found that the values in the list eventually start to repeat. For $c = 3.4$ the list (1) eventually settled down to alternate between the two values 0.451963 and 0.842154. Explain why these two numbers are fixed points of the polynomial $p(p(x))$. Explain why these two numbers must be (approximate) solutions to the equation

$$p(p(x)) = x.$$

For $c = 3.5$, $c = 3.83$, and $c = 3.845$, list (1) also settled down to cycle repeatedly through a few values. What are these values? For which polynomial are these values fixed points?

Where's the Chaos? At the time of this writing, there does not seem to be a universally accepted definition of chaos. However, most of the proposed definitions state that if there is chaos, then it must be true that small changes in the input to a system can result in large changes in the output. We saw hints of such behavior in the previous problems. In working these problems we found that the behavior of list (1) changes as c changes. Sometimes the list settles down to one number; sometimes it cycles repeatedly between two, three, four, or six numbers; and sometimes there appears to be no eventual pattern. In addition, we saw that in some cases a very small change in c substantially changed the behavior of the list. Thus the list (1) exhibits chaos. In particular, very small changes in the number c can change the list from one that cycles through a few numbers to one with no apparent pattern. Indeed, it can be shown that for any positive integer n, there is a value of c between 0 and 4 such that the list (1) eventually settles down to cycle through n different values. In addition, as c gets closer and closer to 4 (but remains less than 4), it takes smaller and smaller changes in c to effect substantial changes in (1).

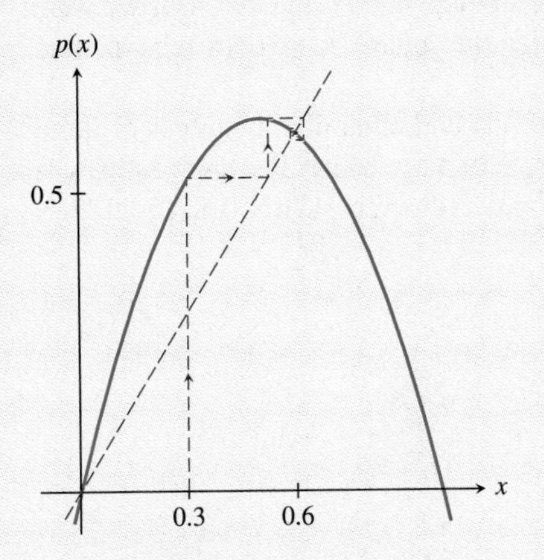

FIGURE 1.82 When $c = 2.5$, the list of compositions quickly closes in on 0.6.

Graphing the Composition List By using graphs, we can obtain a different view of how list (1) differs for different values of c. Since (1) consists of compositions, we can graph the compositions as we did in Section 1.1. For example, take $c = 2.5$, and then draw the graph of $y = p(x)$ and the line $y = x$. See Fig. 1.82. Starting from 0.3 on the x-axis, move up to the graph of p. We meet the graph of p at the point $(0.3, p(0.3))$. Change the output $p(0.3)$ to input by moving horizontally to the line $y = x$. We meet the line in the point $(p(0.3), p(0.3))$. Next move vertically to the graph of p again. We meet the graph in a point with y-coordinate $p(p(0.3))$. Continue the process just described, moving horizontally to the line $y = x$, and then vertically to the graph of p. Each time we meet p, the y-coordinate is the next element in the list (1). A programmable graphing calculator or CAS can be programmed to produce the resulting composition diagram very quickly. See Figs. 1.82 and 1.83.

For another way to investigate the behavior of (1), form the collection of ordered pairs

$$(0, 0.3), (1, p(0.3)), (2, p(p(0.3))), (3, p(p(p(0.3)))), \ldots,$$

where the first coordinate of an ordered pair tells how many times p was used in obtaining the second coordinate. When we plot several of these points (say, the first 200), we can get an idea of whether or not (1) settles down to cycle through some values. See Figs. 1.84 and 1.85.

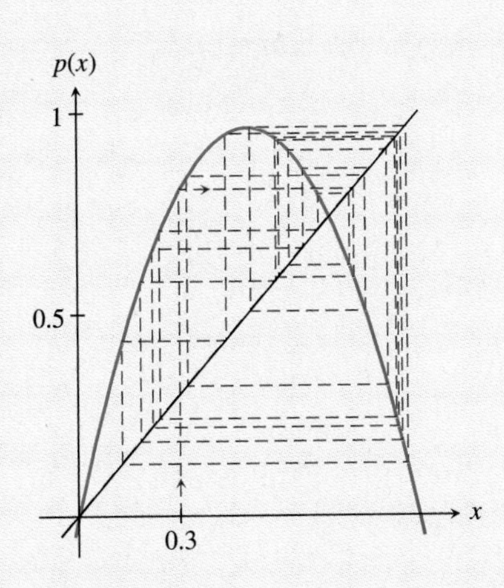

FIGURE 1.83 When $c = 3.862$, the list of compositions does not seem to enter a cycle.

PROBLEM 5 Generate graphs like those shown in Figs. 1.82, 1.83, 1.84, and 1.85 for three other values of c.

A Picture of Chaos To get a better picture of the chaotic nature of (1), we collect data on the behavior of the list for many values of c. For a given value of c, compute the list (1) until it seems to start cycling. For each number b in a cycle, construct the ordered pair (c, b). For example, when $c = 3.5$, list (1) eventually cycles through the

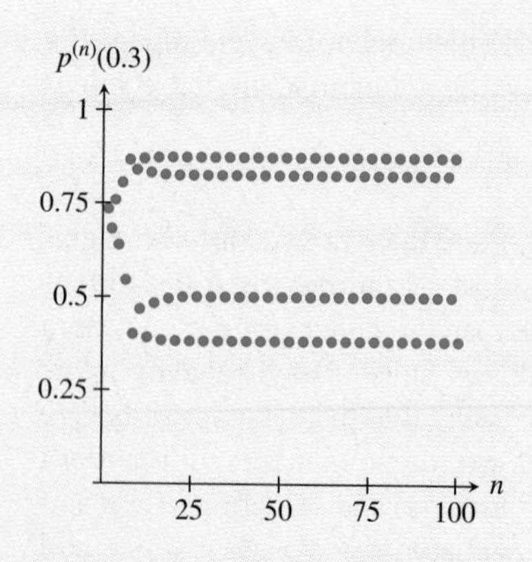

FIGURE 1.84 When $c = 3.5$, the list of compositions soon enters a cycle of length four.

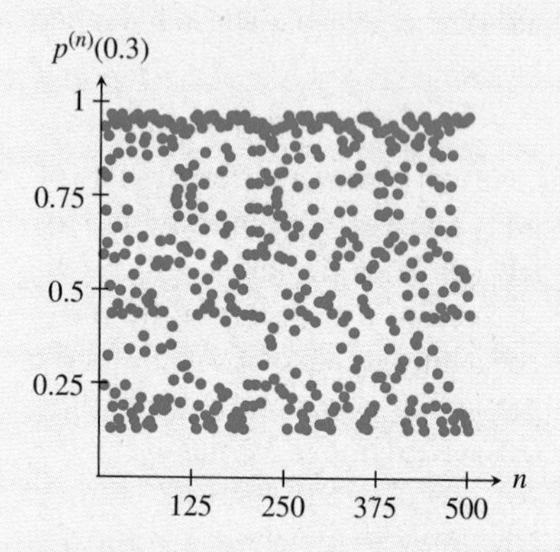

FIGURE 1.85 When $c = 3.862$, the list of compositions does not appear to enter a cycle.

four numbers 0.38282, 0.826941, 0.500884, and 0.874997. Thus we collect the four ordered pairs

$$(3.5, 0.38282), (3.5, 0.826941), (3.5, 0.500884), \text{ and } (3.5, 0.874997).$$

Do this for hundreds of values of c between 2 and 4 and then plot all of the points collected. Of course, there are many values of c for which (1) does not cycle, and others for which the cycle values become apparent only after many, many iterations. However, if we carry the list (1) to 1000 places for each of several hundred values of c, collect the last 100 entries in each list, and then use these 100 entries as the second coordinate in ordered pairs with first coordinate c, we obtain the graph shown in Fig. 1.86.

In this graph the c values run along the horizontal axis. For a given c value, the point or points above indicate entries 901 through 1000 in (1). In this picture we can see evidence of c values that result in cycles of length 2, 4, 8, and 3 as well as other c values that suggest more erratic behavior. This intriguing picture is also a fractal. Draw any small square in the graph with its right edge on the line $c = 4$. If we were to magnify the small portion of the graph inside the square, we would see a picture very similar to the original graph.

For more about the interesting behavior of (1) and about the dynamics of iterations of simple maps, see *Chaos and Fractals,* edited by Robert Devaney and Linda Keen and published by the American Mathematical Society, ISBN:0-8218-0137-6.

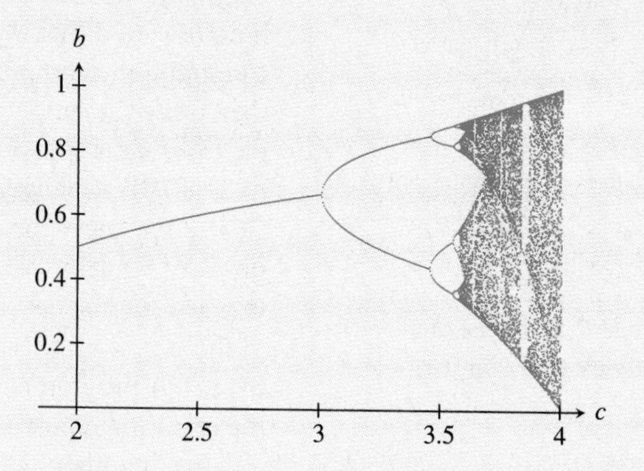

FIGURE 1.86. The graph indicates the cycle values for some values of c.

2

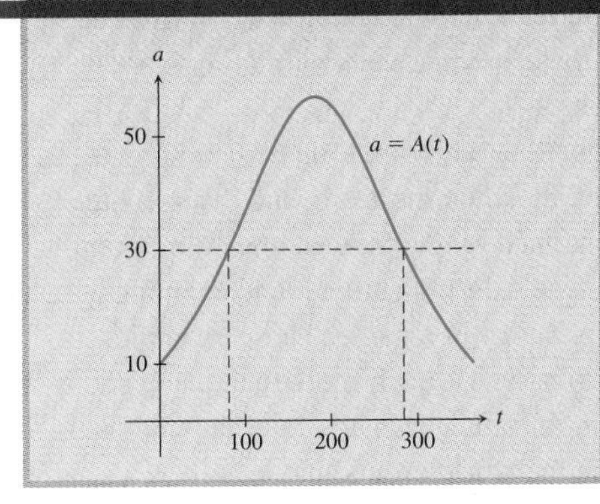

Finding the Derivative

In high northern latitudes, the sun does not rise far above the horizon during the winter. Pre-twentieth century navigators used information about the sun's height in calculating their position. Conversely, given their position and the height of the sun, they could calculate the date. This is an example of a function/inverse function relationship.

See page 154 for further explanation.

In Chapter 1 we looked at the derivative $f'(a)$ of f at $x = a$ in several ways. We saw that $f'(a)$ is the slope of the line we see when we zoom in on the graph of $y = f(x)$ at the point $(a, f(a))$. We also saw that $f'(a)$ is close to the number

$$\frac{f(x) - f(a)}{x - a} \tag{1}$$

for values of x close to a. This led to the interpretation of $f'(a)$ as a rate of change of one variable with respect to another.

In this chapter we approach the derivative more mechanically. Our goal is to develop quick, efficient procedures for producing the derivative of a function.

In Sections 2.1–2.4 we learn how to find the derivative of functions that are combinations of functions whose derivatives we know. For example, if we know the derivatives of f and g, what is the derivative of $f + g$? Of $f \cdot g$? Of $f \circ g$?

In Sections 2.5–2.9 we find the derivatives of the elementary functions. Using these results we can find the derivatives of almost all of the functions that we will encounter. The elementary functions are the power functions, the sine and cosine functions, the exponential and logarithm functions, and the inverse trigonometric functions. Once we know (i.e., memorize!) the derivatives of the elementary functions and learn how to find the derivatives of functions built from these pieces, we can find the derivative of most functions.

In Section 2.10 we look at several mathematical models. These models illustrate the many and diverse situations that can be described by derivatives, including population growth, blood levels during surgery, and even the progress of combat in war or in a game. These examples remind us that the derivative is much more than just a set of rules to manipulate formulas. It is an important and useful tool for describing the world around us.

Computer algebra systems and many calculators have procedures for finding derivatives. But with practice, you will find the calculation of the derivatives to be a straightforward mechanical process that in many cases is done more quickly by hand than by machine. Nonetheless, get to know the differentiation routines on your computer or calculator. These procedures can be very useful when working through a long calculation with the aid of technology. In this chapter we will concentrate on finding derivatives by hand. However, as part of getting to know your calculator or computer, get in the habit of checking some of your differentiation answers by machine.

2.1 Derivatives of Polynomials

Polynomial functions are among the most widely used functions in science, engineering, and mathematics. Every nonvertical line can be described by a first-degree polynomial. The force of gravity and the intensity of a light source are described by inverse square laws, and so both involve second-degree polynomials. In Chapter 7 we will show that nonpolynomial functions such as the sine function and the logarithm function can be closely approximated by polynomial functions.

Polynomials

A polynomial function p can be described by

$$p(x) = a_n x^n + a_{n-1}x^{n-1} + a_{n-2}x^{n-2} + \cdots + a_1 x + a_0,$$

where n is a nonnegative integer and the coefficients $a_n, a_{n-1}, a_{n-2}, \ldots, a_1, a_0$ are real numbers. The degree of p is the largest exponent on x. Hence, if $a_n \neq 0$, then p is a polynomial of degree n. Some examples are

$$q(x) = -4x^5 + 3x^4 - 7x^2 + x - 2 \quad \text{and} \quad r(x) = -x^3 + \sqrt{2}x^2 + \pi x - 13.$$

The polynomial function q has degree 5 and r has degree 3.

The basic building blocks of polynomials are the power functions, described by

$$1, x, x^2, x^3, \ldots.$$

Polynomials are formed by multiplying one or more of these power functions by constants and adding the resulting products. We can find the derivative of any polynomial if we know three things:

a) The derivatives of the power functions.
b) How to find the derivative of a constant multiple of a function, given the derivative of the function.
c) How to find the derivative of a sum of functions, given the derivative of each function in the sum.

The Derivative of x^n

In Chapter 1 we saw that the derivative of a function f at a point $x = a$ is the slope of the "line" that we see as we zoom in on the graph of f near the point $(a, f(a))$. If f is a constant function,

$$f(x) = c \tag{1}$$

where c is some real number, then we always see a horizontal line (slope 0) when we zoom in on the graph of $y = f(x)$. See Fig. 2.1. This means that the constant function (1) has derivative 0:

FIGURE 2.1 The graph of a constant function is a line of slope 0.

$$f'(x) = \frac{dy}{dx} = \frac{d}{dx}c = 0.$$

If g is the function defined by

$$g(x) = x,$$

then we always see a line of slope 1 when we zoom in on the graph at any point. See Fig. 2.2. Hence

$$g'(x) = \frac{dy}{dx} = \frac{d}{dx}x = 1.$$

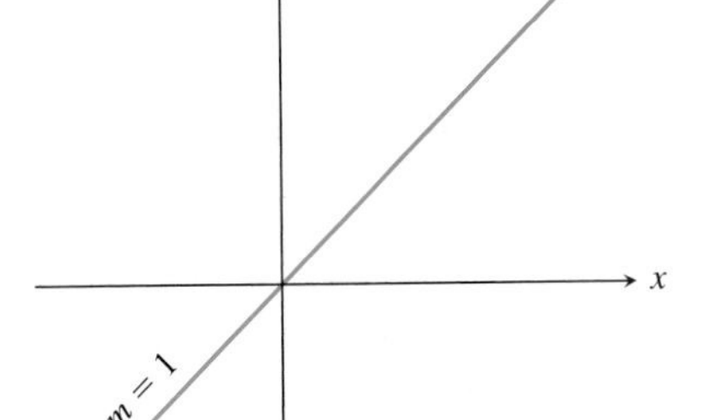

FIGURE 2.2 The graph $y = x$ is a line of slope 1.

We summarize these results:

Derivatives of c and x

The constant function f defined for all x by $y = f(x) = c$ has derivative

$$f'(x) = \frac{d}{dx}c = 0.$$

The function g defined for all x by $y = g(x) = x$ has derivative

$$g'(x) = \frac{d}{dx}x = 1.$$

We can also write these results as

$$c' = 0 \quad \text{and} \quad x' = 1.$$

To calculate the derivative of $y = x^n$ for $n \geq 2$, we recall the definition of derivative given in Section 1.7:

DEFINITION Derivative

Let f be a function and let x be in the domain of f. The derivative of f at the point x is

$$f'(x) = \lim_{t \to x} \frac{f(t) - f(x)}{t - x}, \tag{2}$$

provided that this limit exists.

WEB Applying (2) to $f(x) = x^n$, we have

$$\frac{d}{dx} x^n = f'(x) = \lim_{t \to x} \frac{t^n - x^n}{t - x}. \tag{3}$$

As t approaches x, both the numerator and the denominator in (3) approach 0. This suggests that we try to simplify the quotient in (3) before evaluating the limit. To do this, we recall some factorization formulas. We are familiar with the difference-of-squares factorization

$$t^2 - x^2 = (t - x)(t + x) \tag{4}$$

and the difference-of-cubes factorization

$$t^3 - x^3 = (t - x)(t^2 + tx + x^2). \tag{5}$$

It is not hard to verify a difference-of-fourth-powers factorization

$$t^4 - x^4 = (t - x)(t^3 + t^2x + tx^2 + x^3). \tag{6}$$

If n is a positive integer, then the polynomial $t^n - x^n$ is 0 when $t = x$. This means that the polynomial has a factor of $t - x$. From the pattern established in (4), (5), and (6), we can make a good guess at the other factor:

$$t^n - x^n = (t - x)\underbrace{(t^{n-1} + t^{n-2}x + t^{n-3}x^2 + \cdots + tx^{n-2} + x^{n-1})}_{n \text{ terms}}. \tag{7}$$

This is the difference-of-nth-powers factorization formula. This formula can be verified by multiplication of the terms on the right. See Exercises 21 and 22.

Using (7) to factor the numerator in (3), we have

$$\begin{aligned} f'(x) &= \lim_{t \to x} \frac{(t - x)(t^{n-1} + t^{n-2}x + t^{n-3}x^2 + \cdots + tx^{n-2} + x^{n-1})}{t - x} \\ &= \lim_{t \to x} (t^{n-1} + t^{n-2}x + t^{n-3}x^2 + \cdots + tx^{n-2} + x^{n-1}). \end{aligned}$$

Given that x is fixed in this limit, we may regard

$$t^{n-1} + t^{n-2}x + \cdots + tx^{n-2} + x^{n-1}$$

as a polynomial in the variable t. Hence we can evaluate this last limit by substituting x for t to get

$$\begin{aligned} f'(x) &= \underbrace{(x^{n-1} + x^{n-2}x + x^{n-3}x^2 + \cdots + xx^{n-2} + x^{n-1})}_{n \text{ terms}} \\ &= nx^{n-1}. \end{aligned}$$

We have proved the following important result.

Derivative of the Power Function x^n

If n is a positive integer and f is defined for all x by $f(x) = x^n$, then

$$\frac{d}{dx}x^n = f'(x) = (x^n)' = nx^{n-1}.$$

Multiplication by a Constant

When a function f is multiplied by a constant c, the derivative of the product, cf, is the product of c and the derivative of f.

Derivative of a Constant Multiple of a Function

If f is a function and c a real number, then

$$\frac{d}{dx}(cf(x)) = c\frac{d}{dx}f(x). \tag{8}$$

When this result is written using the "prime" notation, we have

$$(cf(x))' = cf'(x).$$

Almost any formula concerning derivatives can be justified by using the definition of derivative and applying one or more limit operations. To justify the constant multiple rule, apply (2) to the function cf. We get

$$\frac{d}{dx}(cf(x)) = \lim_{t \to x} \frac{cf(t) - cf(x)}{t - x}.$$

We factor the constant c from the expression on the right to get

$$\frac{d}{dx}(cf(x)) = c \lim_{t \to x} \frac{f(t) - f(x)}{t - x} = cf'(x).$$

where we have used (2) to recognize the limit as the definition of the derivative of f.

With this result we can quickly write down the derivative of any function of the form $f(x) = cx^n$ when n is a positive integer.

EXAMPLE 1 Find the derivative of $f(x) = -7x^{20}$.

Solution

The function defined by $-7x^{20}$ is the power function x^{20} multiplied by the constant -7. Thus by (8),

$$f'(x) = \frac{d}{dx}(-7x^{20}) = -7\frac{d}{dx}(x^{20}) = -7(20x^{19}) = -140x^{19}.$$

EXAMPLE 2 The point $(-0.2, 0.35)$ is on the graph of $y = f(x)$. The slope of the tangent to the graph at this point is 1.6. Find the equation for the line tangent to the graph of $y = 3f(x)$ at the point where $x = -0.2$.

Solution

Because $(-0.2, 0.35)$ is on the graph of f, we have $f(-0.2) = 0.35$. Thus

$$(3f)(-0.2) = 3f(-0.2) = 3(0.35) = 1.05,$$

so $(-0.2, 1.05)$ is on the graph of $y = 3f(x)$. The slope of the line tangent to the graph at this point is

$$(3f)'(-0.2) = 3f'(-0.2) = 3(1.6) = 4.8.$$

Because the tangent line contains the point of tangency, $(-0.2, 1.05)$, an equation for the line is

$$y - 1.05 = 4.8(x + 0.2).$$

The Derivative of a Sum

When functions f and g are added, the derivative of the sum $f + g$ is the sum of the derivative of f and the derivative of g.

Derivative of a Sum

Let f and g be differentiable functions. Then

$$\frac{d}{dx}(f + g)(x) = \frac{d}{dx}f(x) + \frac{d}{dx}g(x). \qquad (9)$$

The derivative of the difference of f and g is

$$\frac{d}{dx}(f - g)(x) = \frac{d}{dx}f(x) - \frac{d}{dx}g(x).$$

We will justify (9) in two ways. We do this to demonstrate that a problem can have more than one solution and to illustrate different ways of interpreting the derivative. First, we relate the formula to a physical situation.

Two hoses, a big one and a little one, are used to fill a swimming pool. The water is turned on at the time $t = 0$. Let $V = H(t)$, in ft^3, be the volume of water that flows from the big hose during the first t hours, and let $V = h(t)$ be the volume of water that flows from the little hose in the first t hours. Then

$$H'(t) \text{ has units of ft}^3/\text{hr}$$

and is the rate at which water is flowing from the big hose at time t. A similar statement holds for $h'(t)$.

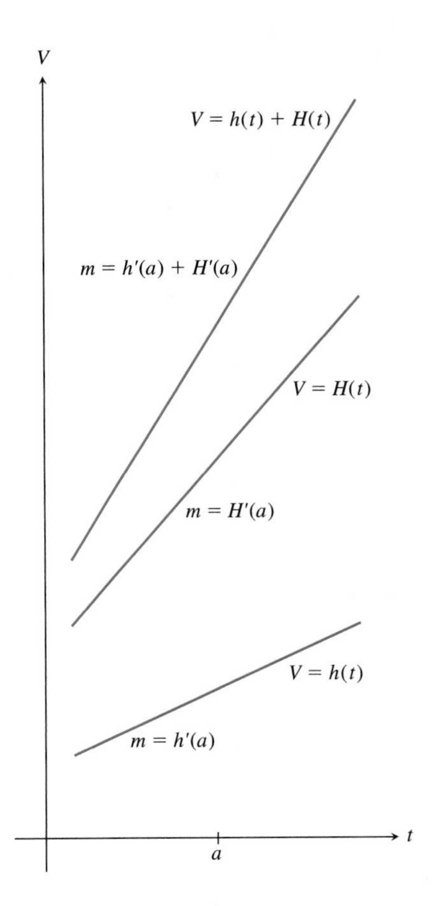

FIGURE 2.3 The slope m at time $t = a$ is the rate at which water flows into the pool at time $t = a$.

If the big hose alone is used to fill the pool, then at time t the pool has $H(t)$ ft^3 of water and the amount of water in the pool is changing at a rate of $H'(t)$ ft^3/hr. If the little hose alone is used, then at time t the pool has $h(t)$ ft^3 of water and the volume of water in the pool is changing at a rate of $h'(t)$ ft^3/hr. See Fig. 2.3.

If both hoses are used simultaneously, then at time t the volume of water in the pool is

$$(H + h)(t) = (H(t) + h(t)) \text{ ft}^3. \tag{10}$$

The derivative of (10),

$$(H + h)'(t) \text{ ft}^3/\text{hr}, \tag{11}$$

is the rate at which the volume of water in the pool is changing at time t. On the other hand, experience tells us that the rate of change of water in the pool is simply the sum of the rates at which water enters from the two hoses,

$$(H'(t) + h'(t)) \text{ ft}^3/\text{hr}. \tag{12}$$

Because (11) and (12) represent the same rate of change, the two expressions are equal. Hence

$$(H + h)'(t) = H'(t) + h'(t).$$

For a second way of justifying (9), we use the definition of derivative. Because f and g are differentiable, we have

$$\frac{d}{dx}f(x) = \lim_{t \to x} \frac{f(t) - f(x)}{t - x} \quad \text{and} \quad \frac{d}{dx}g(x) = \lim_{t \to x} \frac{g(t) - g(x)}{t - x}. \tag{13}$$

The derivative of $f + g$ is given by

$$\frac{d}{dx}(f + g)(x) = \lim_{t \to x} \frac{(f + g)(t) - (f + g)(x)}{t - x} \tag{14}$$

provided this limit exists. To evaluate this expression, we use the limit theorems for manipulating limits given in Section 1.5. Using these results, we obtain

$$\begin{aligned}
\lim_{t \to x} \frac{(f + g)(t) - (f + g)(x)}{t - x} &= \lim_{t \to x} \frac{(f(t) + g(t)) - (f(x) + g(x))}{t - x} \\
&= \lim_{t \to x} \left(\frac{f(t) - f(x)}{t - x} + \frac{g(t) - g(x)}{t - x} \right) \\
&= \lim_{t \to x} \frac{f(t) - f(x)}{t - x} + \lim_{t \to x} \frac{g(t) - g(x)}{t - x} \\
&= \frac{d}{dx}f(x) + \frac{d}{dx}g(x),
\end{aligned}$$

where the last equality follows from (13). This argument shows that the limit in (14) exists and is equal to $f'(x) + g'(x)$. This again verifies (9).

This argument can be extended to sums of three or more functions. Thus

$$\frac{d}{dx}(f(x) + g(x) + h(x)) = \frac{d}{dx}f(x) + \frac{d}{dx}g(x) + \frac{d}{dx}h(x),$$

and analogous formulas are true for sums of four, five, or more functions. For another proof of the derivative of a sum result, see Exercise 32.

The results of this section may seem obvious, but they are nonetheless important and will be used over and over. As an immediate application, we can now differentiate any polynomial.

EXAMPLE 3 Find the derivative of $p(x) = 4x^5 + 7x^3 - 3x^2 + 6$.

Solution

First write the derivative as a sum of derivatives:

$$p'(x) = (4x^5 + 7x^3 - 3x^2 + 6)' = (4x^5)' + (7x^3)' + (-3x^2)' + (6)'.$$

Next factor the constant multiples out of the derivatives:

$$p'(x) = 4(x^5)' + 7(x^3)' - 3(x^2)' + (6)'.$$

Now use the formulas for the derivative of a power function and the fact that a constant function has derivative 0:

$$p'(x) = 4(5x^4) + 7(3x^2) - 3(2x^1) + 0 = 20x^4 + 21x^2 - 6x.$$

EXAMPLE 4 True or false: The rate of change of the average of several functions is equal to the average of their rates of change.

Solution

We show that the statement is true for three functions f, g, and h. We assume that all derivatives exist on a common domain. The average of the three functions is given by

$$A(x) = \frac{f(x) + g(x) + h(x)}{3}.$$

The rate of change of the average is

$$A'(x) = \frac{d}{dx}A(x) = \frac{d}{dx}\left(\frac{1}{3}f(x) + \frac{1}{3}g(x) + \frac{1}{3}h(x)\right).$$

Because the derivative of a sum of functions is the sum of the derivatives of the functions, this becomes

$$A'(x) = \frac{d}{dx}\left(\frac{1}{3}f(x)\right) + \frac{d}{dx}\left(\frac{1}{3}g(x)\right) + \frac{d}{dx}\left(\frac{1}{3}h(x)\right).$$

Next factor the constant $1/3$ from each term to get

$$\begin{aligned} A'(x) &= \frac{1}{3}\frac{d}{dx}f(x) + \frac{1}{3}\frac{d}{dx}g(x) + \frac{1}{3}\frac{d}{dx}h(x) \\ &= \frac{f'(x) + g'(x) + h'(x)}{3}. \end{aligned}$$

This is the average of the rates of change.

In the solutions to Examples 3 and 4 we took great care to indicate where we were applying the results about derivatives of sums and derivatives of constant multiples. Though it is important to recognize when we are applying these results, in the future we will usually omit the details of these applications.

Exercises 2.1

Exercises 1–8: Find the derivative with respect to the implied variable (e.g., for the function $g(t)$, find the derivative with respect to t; for $h(r)$, with respect to r, etc.).

1. $f(x) = 4x - 7$

2. $g(t) = 3t^3 - 2t + \frac{1}{2}$

3. $y = \frac{1}{4}x^4 + \frac{1}{3}x^3 + \frac{1}{2}x^2 + x + 1$

4. $h(r) = ar^3 + br^2 + cr + d$, where a, b, c, d are constants.

5. $H(s) = \pi^2$

6. $s = t^{n+3}$, where n is a nonnegative integer.

7. $f(x) = (x + 1)(2x + 1)$

8. $y = 16(t + 2)^3$

9. a. Find the derivative of f where $f(x) = \frac{1}{3}x^3$.

b. Find a different function with the same derivative as f.

10. a. Find the derivative of g where $g(x) = -3x + 8$.

b. Find a different function with the same derivative as g.

11. a. Let $y = x^4$. Find the derivative of y with respect to x.

b. Find a function that has derivative $\frac{1}{3}x^3$.

12. a. Let $y = x^{n+1}$ where n is a positive integer. Find y'.

b. Find two different functions that have derivative x^n.

Exercises 13–16: Let f, g, and h be functions with $f'(3) = -4$, $g'(3) = 7$, and $h'(3) = 1$. Find $s'(3)$.

13. $s(x) = 3f(x) - 2g(x) + \frac{1}{2}h(x)$.

14. $f(x) = 2(g(x) + h(x)) - 3s(x)$.

15. $f(x) + g(x) + h(x) + s(x) = 20\sqrt{3}$.

16. $3f(x) + 2g(x) - 2h(x) + 5s(x) = 100$.

Exercises 17–20: Write an equation for the line tangent to the graph of the function at the given point.

17. $f(x) = 2x^2 - 3x + 4$ at $(2, 6)$

18. $h(t) = 2\sqrt{2}$ at $\left(-\sqrt{5}, 2\sqrt{2}\right)$

19. $g(x) = ax^2 + bx + c$ at $(d, g(d))$ (a, b, c constants)

20. $f(x) = x^n$ at (a, a^n) (n a positive integer)

21. Verify the factorizations given in (4), (5), and (6).

22. Verify the formula for the factorization of the difference of nth powers given in equation (7).

23. Let f be a differentiable function and c a real number. Use ideas about how multiplication by a constant affects slope to show that

$$(cf)' = cf'.$$

24. Is it true that

$$(p(x)q(x))' = p'(x)q'(x)$$

for every pair of polynomials p and q? Try some examples before answering.

25. Let $f(x) = 1/x^n$, where n is a positive integer. Use the definition of derivative and the difference-of-nth-powers factorization formula given in (7) to show that

$$f'(x) = \frac{d}{dx}\frac{1}{x^n} = \frac{d}{dx}x^{-n} = -nx^{-n-1}.$$

Exercises 26–29: Use the result of Exercise 25 to find the derivative of the given function.

26. $f(x) = \dfrac{2}{x^3}$

27. $r(\theta) = \theta + 6 + \dfrac{1}{\theta}$

28. $s(t) = 3t^3 - 4t + 7 - 6t^{-1} + \sqrt{2}t^{-5}$

29. $h(r) = \left(1 + \dfrac{2}{r^2}\right)^2$

30. Two balloonists, Sir Bass and Madam Alto, leave the ground in their balloons at the same instant. The height of Sir Bass's balloon at time t minutes is $h(t)$ meters, while the height of Madam Alto's balloon at time t is $2h(t)$ meters. At any given time, what can be said about the speeds at which the two balloons are ascending or descending? Give reasons for your answer.

31. Two gymnasts, Matt and Bart, climb a rope. At time $t = 0$, Matt starts climbing from ground level while Bart starts from 10 feet above ground. It is observed that at all times during the climb, Matt is ascending twice as fast as Bart. At time $t = 20$ minutes, Bart is 200 feet above the ground. How high is Matt at this time? Give reasons for your answer. Interpret the problem in terms of derivatives.

32. In this problem we outline another proof for (9), the rule for the derivative of a sum. If $f'(a)$ exists, then, as seen in (10) of Section 1.7,

$$f(t) = f(a) + f'(a)(t - a) + E_f(t)(t - a)$$

where $\lim_{t\to a} E_f(t) = 0$. Similarly,

$$g(t) = g(a) + g'(a)(t - a) + E_g(t)(t - a)$$

where $\lim_{t\to a} E_g(t) = 0$. Use these expressions to show that

$$(f + g)'(a) = \lim_{t\to a} \frac{(f + g)(t) - (f + g)(a)}{t - a} = f'(a) + g'(a).$$

2.2 Derivatives of Products and Quotients

The chain block hoist, shown in Fig. 2.4, is an ingenious mechanism that allows people to hoist loads of three or four tons by hand. This device is often found in garages, where it is used to move engines into and out of cars. When the chain is pulled at point P to rotate wheel A through one revolution, wheel B is raised $\pi(R - r)$ units. To lift an object of weight W, one has to pull at P with a force of

$$F = W\frac{R - r}{2Re} \tag{1}$$

where e is the mechanical efficiency of the mechanism (usually about 0.30 because of friction). Note that when R and r are close, it takes very little force to raise the weight, though in this case, the weight is not lifted very far. To understand the fascinating physics behind mechanisms like the chain block, we need to be able to study the derivative of expressions such as (1). See Exercises 29 and 30. Such expressions often involve products and quotients of functions with known derivatives. In this section we see how to differentiate product or quotient functions of the form

$$u(x)v(x) \quad \text{or} \quad \frac{u(x)}{v(x)},$$

provided that we already know the derivatives of u and v.

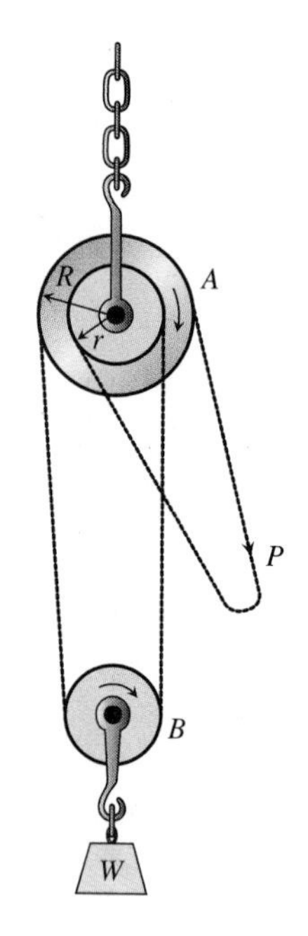

FIGURE 2.4 The chain block. Pull at P to raise the object below wheel B.

The Product Rule

Suppose that we know the rates of change of the functions u and v. What can we say about the rates of change of the product uv? The product rule tells us how the derivative of the product is related to the derivatives of the factors.

The Product Rule

If u and v are differentiable functions, then

$$\frac{d}{dx}(uv) = \frac{du}{dx}v + u\frac{dv}{dx} \qquad (2)$$

This can also be written $(uv)' = u'v + uv'$.

As (2) shows, the formula for differentiating a product is more complicated than the rule for differentiating a sum. The product rule reflects the way that the rates of change of the factors affect the rate of change of the product. This is illustrated by a very familiar process: that of changing the size of a window on a computer screen.

We change the size of a computer window by "grabbing" a corner and then "dragging" the mouse. The length and height of the window change in response to how we move the mouse. As a result, the area of the window changes. The rate of change of the area of the window depends on the rates of change of the height and length. This provides an illustration of the product rule.

Let $L(t)$, in centimeters, be the length of the window at time t seconds. The rate of change of the length at time t is

$$L'(t) \text{ cm/s}.$$

A short time later, at time $t + \Delta t$, the length of the window is $L(t + \Delta t)$. Because Δt is small, we know that

$$L'(t) \approx \frac{L(t + \Delta t) - L(t)}{\Delta t}.$$

Solving for $L(t + \Delta t)$, we find

$$L(t + \Delta t) \approx L(t) + L'(t)\Delta t.$$

Thus the length of the window has changed by approximately

$$L'(t)\Delta t \text{ cm}$$

during the time interval from t to $t + \Delta t$. Similarly, if $H(t)$ is the height of the window at time t, then the height changes by approximately

$$H'(t)\Delta t \text{ cm}$$

during this time interval.

The area of the window at time t is $A(t) = L(t)H(t)$. Now compare the windows at times t and $t + \Delta t$. See Fig. 2.5. The shaded region in the figure shows how the rates of change of the factors affect different parts of the area. The change in the area is the area of this shaded region,

$$A(t + \Delta t) - A(t) \approx L'(t)\Delta t \cdot H(t) + L(t) \cdot H'(t)\Delta t + L'(t)\Delta t \cdot H'(t)\Delta t.$$

Dividing by Δt, we find that the rate of change of the area is

$$\begin{aligned} A'(t) = (L(t)H(t))' &\approx \frac{A(t + \Delta t) - A(t)}{\Delta t} \\ &\approx L'(t)H(t) + L(t)H'(t) + L'(t)H'(t)\Delta t \\ &\approx L'(t)H(t) + L(t)H'(t). \end{aligned}$$

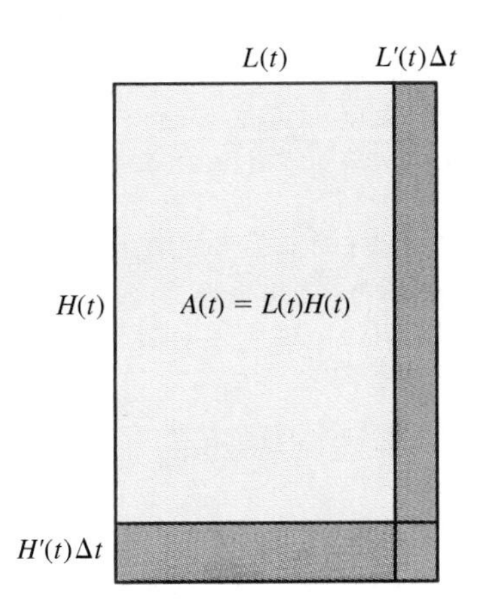

FIGURE 2.5 The shaded region indicates the change in the window size from time t to time $t + \Delta t$.

In the last line we have dropped the term $L'(t)H'(t)\Delta t$ because this expression is small when Δt is small.

The product rule can also be derived directly from the definition of derivative. Though this argument gives less insight into the product rule than the window/mouse argument, it gives us a chance to use some of the limit results from Chapter 1.

Deriving the Product Rule from the Definition of Derivative Let $w = uv$ and assume that u and v are both differentiable at $x = a$. Then

$$w'(a) = \lim_{x\to a} \frac{w(x) - w(a)}{x - a} = \lim_{x\to a} \frac{u(x)v(x) - u(a)v(a)}{x - a}.$$

Adding and subtracting $u(a)v(x)$ in the numerator, we have

$$w'(a) = \lim_{x\to a} \frac{\big(u(x)v(x) - u(a)v(x)\big) + \big(u(a)v(x) - u(a)v(a)\big)}{x - a}.$$

Separating this fraction into two fractions and using the fact that the limit of a sum is the sum of limits,

$$w'(a) = \lim_{x\to a} \frac{u(x)v(x) - u(a)v(x)}{x - a} + \lim_{x\to a} \frac{u(a)v(x) - u(a)v(a)}{x - a}.$$

Finally, because the limit of a product is the product of the limits,

$$w'(a) = \lim_{x\to a} \frac{u(x) - u(a)}{x - a} \lim_{x\to a} v(x) + u(a) \lim_{x\to a} \frac{v(x) - v(a)}{x - a}. \tag{3}$$

By the definition of derivative, we know

$$\lim_{x\to a} \frac{u(x) - u(a)}{x - a} = u'(a) \quad \text{and} \quad \lim_{x\to a} \frac{v(x) - v(a)}{x - a} = v'(a).$$

In addition, because $v(x)$ has a derivative at a, it is also continuous at a. (See Section 1.7.) Hence

$$\lim_{x\to a} v(x) = v(a).$$

Substituting these results into (3),

$$w'(a) = (uv)'(a) = u'(a)v(a) + u(a)v'(a).$$

See Exercise 37 for another way to justify the product rule.

EXAMPLE 1 If $y = (2x^3 - 4x + 2)(x^2 - x - 1)$, find $\dfrac{dy}{dx}$ by

a) using the product rule.

b) first expanding the product and then differentiating the resulting polynomial.

Solution

a) Using the product rule formula given in (2), we have

$$\begin{aligned}\frac{dy}{dx} &= (\underbrace{(2x^3 - 4x + 2)}_{u(x)}\underbrace{(x^2 - x - 1)}_{v(x)})' \\ &= \underbrace{(2x^3 - 4x + 2)'}_{u'(x)}\underbrace{(x^2 - x - 1)}_{v(x)} + \underbrace{(2x^3 - 4x + 2)}_{u(x)}\underbrace{(x^2 - x - 1)'}_{v'(x)} \\ &= (6x^2 - 4)(x^2 - x - 1) + (2x^3 - 4x + 2)(2x - 1).\end{aligned}$$

b) Expanding the product in the expression for y, we find

$$y = 2x^5 - 2x^4 - 6x^3 + 6x^2 + 2x - 2.$$

We can differentiate any polynomial in this form by using the techniques developed in Section 2.1. The result is

$$\frac{dy}{dx} = 10x^4 - 8x^3 - 18x^2 + 12x + 2.$$

Are the answers found in **a** and **b** the same?

EXAMPLE 2 The square root rule

Let $u(x) \geq 0$ and let

$$w(x) = \sqrt{u(x)}. \tag{4}$$

Assuming that $u(x)$ and $w(x)$ are both differentiable, find a formula for $w'(x)$.

Solution

Squaring both sides of (4), we have

$$w(x)w(x) = u(x).$$

Differentiate both sides of this equation, using the product rule on the left side. We obtain

$$w'(x)w(x) + w(x)w'(x) = u'(x).$$

Solving this equation for $w'(x)$ and then substituting $w(x) = \sqrt{u(x)}$ gives

$$w'(x) = \frac{u'(x)}{2w(x)} = \frac{1}{2}\frac{u'(x)}{\sqrt{u(x)}}. \tag{5}$$

There is no need to assume that w is differentiable in Example 2. If $u(x) > 0$ and $u'(x)$ exists, then $w'(x)$ must also exist. See Exercise 36.

EXAMPLE 3 Let $y = \sqrt{4x^2 - 3x + 7}$. Find dy/dx.

Solution

Let $w(x) = \sqrt{4x^2 - 3x + 7}$ and use (5) with

$$u(x) = 4x^2 - 3x + 7 \quad \text{and} \quad u'(x) = 8x - 3.$$

We then have

$$\frac{dy}{dx} = w'(x) = \frac{u'(x)}{2\sqrt{u(x)}} = \frac{8x - 3}{2\sqrt{4x^2 - 3x + 7}}.$$

Differentiating a Quotient

We can find a formula for differentiating a quotient of two functions by using the product rule.

EXAMPLE 4 The quotient rule

Assuming that $u(x)$, $v(x)$, and $w(x)$ are differentiable and that

$$w(x) = \frac{u(x)}{v(x)}, \tag{6}$$

find a formula for $w'(x)$.

Solution

Multiply both sides of (6) by $v(x)$ to get

$$w(x)v(x) = u(x).$$

Differentiate both sides of this equation, using the product rule on the left side. We obtain

$$w'(x)v(x) + w(x)v'(x) = u'(x).$$

Solve this equation for $w'(x)$ and substitute $\dfrac{u(x)}{v(x)}$ for $w(x)$ to get

$$\begin{aligned} w'(x) &= \frac{u'(x) - w(x)v'(x)}{v(x)} \\ &= \frac{u'(x) - \left(\dfrac{u(x)}{v(x)}\right)v'(x)}{v(x)} = \frac{u'(x)v(x) - u(x)v'(x)}{(v(x))^2}. \end{aligned} \tag{7}$$

There is no need to assume that w is differentiable in Example 4. If u and v are differentiable at x and $v(x) \neq 0$, then $w = u/v$ is differentiable at x. See Exercise 35. We will often use the result established in Example 4, because it gives an easy, direct way to differentiate a quotient.

EXAMPLE 5 Let $y = 1/x^n$, where n is a positive integer. Find dy/dx. (See also Exercise 25 in Section 2.1.)

Solution

Let $w(x) = 1/x^n$, then take $u(x) = 1$ and $v(x) = x^n$ in (7). We then have

$$\frac{dy}{dx} = w'(x) = \left(\frac{1}{x^n}\right)' = \frac{(1)'x^n - 1(x^n)'}{(x^n)^2} = \frac{0 \cdot x^n - nx^{n-1}}{x^{2n}} = -nx^{-n-1}.$$

Restated, this says that if n is a positive integer, then $\dfrac{d}{dx}x^{-n} = -nx^{-n-1}$.

Combining this result with the formula for the derivative of the power function found in the previous section, we have the following:

The Derivative of x^k for $k = 0, \pm 1, \pm 2, \ldots$

If k is an integer, then

$$\frac{d}{dx}x^k = (x^k)' = kx^{k-1}.$$

EXAMPLE 6 A bin buster

Good weather in the Midwest often results in large corn harvests called "bin busters" because there is not enough room in grain elevators for all of the corn harvested. At such times it is not uncommon to see large, conical piles of grain on the ground near the elevators. One such pile at a local elevator was in the shape of a right circular cone of height 15 ft. (The height was kept at 15 ft since corn was dumped from a chute 20 ft above ground level. Machines moved the grain in the pile around to keep the height at 15 ft during dumping.) Late in the harvest season this pile had a base radius of 100 ft and corn was being dumped at a rate of 2000 ft^3/day. How fast was the radius of the base of the pile changing at this time?

Solution

Let $r(t)$ be the radius of the pile at time t, where t is measured in days. (See Fig. 2.6.) We need to find $r'(t)$ because this derivative is the rate of change of r. Let $V(t)$ be the volume of corn in the pile at time t. At the time in question, the volume is changing (increasing) at a rate of 2000 ft^3/day, so

$$V'(t) = 2000 \text{ ft}^3/\text{day}$$

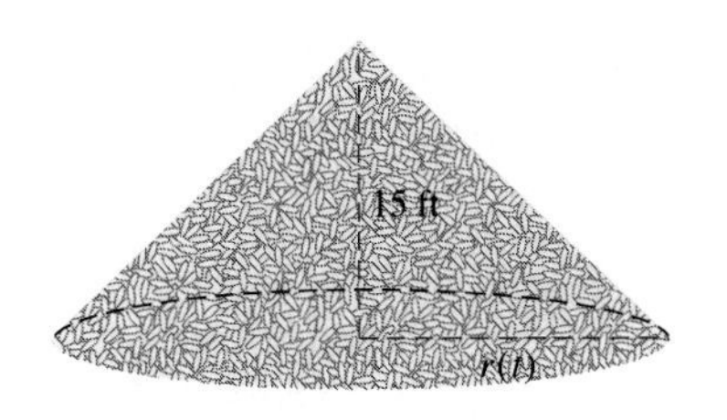

FIGURE 2.6 A bin buster!!!!

at this time. Because the corn is piled in a cone of (constant) height 15 ft and base radius $r(t)$, we have

$$V(t) = \frac{1}{3}\pi \cdot (15 \text{ ft})\,(r(t))^2.$$

Differentiate both sides of this equation, using the product rule on the right to differentiate $(r(t))^2 = r(t)r(t)$. We obtain

$$V'(t) = \frac{\pi}{3}(15\text{ ft})(r'(t)r(t) + r(t)r'(t)) = (10\pi\text{ ft})r(t)r'(t). \tag{8}$$

At the time in question we have $V'(t) = 2000\text{ ft}^3/\text{day}$ and $r(t) = 100$ ft. Putting these values, with units, into (8), we have

$$2000\text{ ft}^3/\text{day} = (10\pi\text{ ft})(100\text{ ft})r'(t).$$

Solving for $r'(t)$, we find that the radius of the pile is increasing at a rate of

$$r'(t) = \frac{2}{\pi}\text{ ft/day} \approx 0.63662\text{ ft/day}.$$

Exercises 2.2

Exercises 1–14: Find the derivative of the function represented by the given equation. Use the product, quotient, or square root rules developed in this section.

1. $y = (x^2 - 3x + 1)(2x + 4)$

2. $y = 2x(3x^3 - 3x + \pi)$

3. $H(t) = (-9t^5 + 6t^4 - 21t^2 + 8t - 1)^2$

4. $q(t) = \dfrac{2t - 1}{2t + 1}$

5. $f(x) = \dfrac{2x^2 - 3x + 10}{x^2 - 7}$

6. $p(x) = -3x^2 + 2x - 7 + \dfrac{4}{x} + \dfrac{21}{x^{21}}$

7. $s(r) = \left(1 - \dfrac{1}{r} + \dfrac{2}{r^2}\right)^2$

8. $g(u) = \dfrac{(u^2 - 2u + 3)(4u^2 - u + 1)}{u^3 + 8}$

9. $r = (\theta^2 - 2)(\theta^3 - 3)(\theta^4 - 4)$

10. $f(z) = z(z - z^{-1})(2z^2 - 4 + 2z^{-2})$

11. $p(r) = \sqrt{r^2 + 2r + 7}$

12. $w(v) = \sqrt{\dfrac{v^2 - 3v + 1}{2v + 5}}$

13. $F(s) = \sqrt{2s - 1}\,\sqrt{6s + 7}$

14. $y = \sqrt{\sqrt{x} + 1}$

Exercises 15–18: Find $f'(2)$, given that $u(2) = 1$, $v(2) = 2$, $w(2) = 3$, $u'(2) = 4$, $v'(2) = 5$, and $w'(2) = 6$.

15. $f(x) = u(x)v(x)$

16. $f(x) = \dfrac{u(x)v(x)}{w(x)}$

17. $f(x) = \sqrt{2w(x) - 3u(x)}$

18. $v(x) = \sqrt{2f(x) + 5}$

19. If $y = (2x^2 - 1)(3x + 7)$, find the coordinates (x, y) of all points where $dy/dx = 0$.

20. If $y = \sqrt{3x^2 + 2x + 7}$, find the coordinates (x, y) of all points where $dy/dx = 0$.

21. Find an equation for the line tangent to

$$y = \frac{2x^2 - 7}{7x + 1}$$

at the point where $x = 2$.

22. Find an equation for the line tangent to

$$g(t) = (t + 1)(t + 2)(t + 3)$$

at the point where $t = -2$.

23. A function w is expressed as a product of two functions and differentiated using the product rule. The result is

$$w'(x) = (2x - 3)(4x^3 + 3x - 1) + (x^2 - 3x + 8)(12x^2 + 3).$$

What is $w(x)$?

24. A function w is expressed as the square root of another function and differentiated using the "square root" rule of Example 2. The result is

$$w'(x) = \frac{3x^2 - 2x + 1}{\sqrt{x^3 - x^2 + x - 32}}.$$

What is $w(x)$?

25. Let $u(x)$ be a differentiable function. Find and simplify the derivative of each of the following:

a. $(u(x))^2$

b. $(u(x))^3$

c. $(u(x))^4$

d. $(u(x))^5$

You may wish to use the results of **a** and **b** in doing **c** and **d**. Based on your answers, what would you expect for the derivative of $(u(x))^n$?

26. Let n be a positive integer and $u(x)$ a differentiable function. Show that

$$u(t)^n - u(x)^n = (u(t) - u(x))\,(u(t)^{n-1} + u(t)^{n-2}u(x) + \cdots + u(t)u(x)^{n-2} + u(x)^{n-1})$$

and then use this factorization to find a formula for the derivative of $(u(x))^n$. Compare with your answers to Exercise 25.

27. Differentiate the identity $|x| = \sqrt{x^2}$ to find a formula for the derivative of $|x|$. For what values of x is this derivative defined? Use a graph of $y = |x|$ to check that your answer is correct.

28. Use the idea introduced in Exercise 27 to find the derivative of:

a. $|2x - 1|$.

b. $|x^2 - 4|$.

29. Consider the chain block shown in Fig. 2.4. Show that if you pull at P so that wheel A turns through one rotation, then wheel B will rise a distance of $\pi(R - r)$.

30. We can derive equation (1) for the force F needed to raise the weight W by equating the work done in applying the force F with the work needed to raise the weight W. Referring to Fig. 2.4, first suppose that force F is applied to pull at point P so that wheel A makes one full turn. Then a length of rope or chain equal to $2\pi R$ must be pulled "past" P. This is equivalent to applying the force F through a distance of $2\pi R$, so the total work done by the force is

$$\text{Work}_F = (\text{Force})\,(\text{Distance}) = 2\pi RF.$$

Now consider the object of weight W that is raised a distance of $\pi(R - r)$ during this procedure. Because the weight of an object *is* the force with which gravity acts on the object, the work done to raise the object must be

$$\text{Work}_W = \pi(R - r)W.$$

Because the force applied at point P acts with efficiency e on the weight, we have

$$e(\text{Work}_F) = \text{Work}_W.$$

Show that this leads to (1).

31. A substance is produced by a certain chemical process. Both the volume, $V = V(t)$ (in cubic meters), and the density, $D = D(t)$ (in units of kilograms per cubic meter), of the material produced vary with time. Thus the mass, $M(t) = D(t)V(t)$, of material produced is also a function of time. Calculate the rate of change of the mass at a time when the density is 1200 kg/m^3, the volume is 0.01 m^3, the rate of change of density is 0.001 kg/m^3/minute, and the rate of change of volume is 0.0005 m^3/minute.

32. An ideal gas is confined to a piston-cylinder system and kept at constant temperature. The pressure and volume of the gas at any time t are given by $P(t)$ and $V(t)$. According to the ideal gas law, $P(t)V(t) = k$, where k is a constant. Show that at any time t, the ratio of volume to pressure is equal to the ratio $-V'(t)/P'(t)$.

33. Oil spills from a ruptured tanker, forming a circular oil slick on the surface of the ocean. Concerned observers in a helicopter note that the oil slick is 2 km in radius and that the radius seems to be increasing at 10 m per hour. How fast is the area of the slick increasing at this time? If the oil slick is 0.5 cm thick, how fast is oil spilling from the tanker? (Be sure to include units with your answers.)

34. During a recent newspaper drive, volunteers collected papers and stacked them in a rectangular pile. At mid-morning the pile of papers measured 10 ft on the west side, 15 ft on the north side, and 5 ft high. At this time some people were stacking papers on the west side, causing this side to grow westward at 2 ft/hr. Others were stacking papers on the north side, causing this side to creep northward at $\frac{1}{2}$ ft/hr, and a third group was stacking papers on top, causing the height to increase by 3 ft/hr. How fast was the volume of the pile increasing at this time? How fast should papers be hauled away to keep the volume of the pile constant?

35. Derive the quotient rule (see Example 4) using the definition of derivative. Assume that $u(x)$ and $v(x)$ are differentiable at all points in the domain of $w(x) = u(x)/v(x)$. First show that

$$w'(x) = \lim_{t \to x} \frac{(u(t)/v(t)) - (u(x)/v(x))}{t - x}$$

$$= \lim_{t \to x} \frac{u(t)v(x) - u(x)v(t)}{t - x} \frac{1}{v(t)v(x)}.$$

Now use manipulations similar to those used in deriving the product rule from the definition of derivative. Use the final form of the quotient rule to guide you in these manipulations.

36. Use the definition of derivative to derive the rule for differentiating functions of the form $w(x) = \sqrt{u(x)}$. See

Example 2. Assume that $u(x)$ is differentiable at all points where $u(x) > 0$. Start by showing that

$$w'(x) = \lim_{t\to x} \frac{\sqrt{u(t)} - \sqrt{u(x)}}{t - x}$$
$$= \lim_{t\to x} \left(\frac{u(t) - u(x)}{t - x} \frac{1}{\sqrt{u(t)} + \sqrt{u(x)}} \right).$$

Now use the definition of the derivative of u and the continuity of u to obtain the final formula.

37. In this problem we outline another proof for the product rule. If $u'(a)$ exists, then, as seen in (10) of Section 1.7,

$$u(t) = u(a) + u'(a)(t - a) + E_u(t)(t - a)$$

where $\lim_{t\to a} E_u(t) = 0$. Similarly,

$$v(t) = v(a) + v'(a)(t - a) + E_v(t)(t - a)$$

where $\lim_{t\to a} E_v(t) = 0$. Use these expressions to show that

$$(uv)'(a) = \lim_{t\to a} \frac{u(t)v(t) - u(a)v(a)}{t - a}$$
$$= u'(a)v(a) + u(a)v'(a).$$

Exercises 38–41: The following exercises review some ideas about function composition in preparation for material in the next section.

38. Write

$$f(x) = \frac{1}{\sin \sqrt{x} - 3}$$

as a composition of simple functions.

39. Write

$$h(t) = \sqrt[3]{1 + \sqrt[4]{t^2 - 1}}$$

as a composition of simple functions.

40. Given a numerical value for x, explain to a friend the steps needed in using a calculator to find the value of

$$\tan\left(\frac{1 + \sqrt{x}}{2 - \sqrt{x}}\right).$$

41. Given a numerical value for t, explain to a friend the steps needed in using a calculator to find the value of

$$2^{2^{t^2+1}}.$$

2.3 Differentiating Compositions

Many important functions can be expressed as a composition of other functions. As an example, if you put \$1000 in a savings account that pays r percent annual interest compounded quarterly, then at the end of 20 years your account will contain

$$A(r) = 1000(1 + 0.25r)^{80} \text{ dollars.}$$

The function A is a composition of two simple functions. One of these is the linear function

$$l(r) = 1 + 0.25r,$$

and the other is the power function

$$p(x) = 1000x^{80}.$$

In Section 2.1 we learned how to find the derivatives of l and p. In this section we see how we can use the derivatives of l and p to find the derivative of

$$A(r) = p(l(r)).$$

We will find that the technique for computing such derivatives is not only a valuable practical tool, but also provides a useful means for studying rates of change of inverse functions (Section 2.8) and of implicitly defined functions (Section 2.4).

Differentiation of Compositions

The formula for differentiating compositions is often called the *chain rule.*

The Chain Rule

Let f and g be differentiable functions and let $h = f \circ g$. Then

$$\frac{d}{dx}h(x) = h'(x) = (f \circ g)'(x) = f'(g(x))g'(x). \tag{1}$$

We use the definition of derivative to obtain (1). According to the definition,

$$(f \circ g)'(x) = \lim_{t \to x} \frac{(f \circ g)(t) - (f \circ g)(x)}{t - x} = \lim_{t \to x} \frac{f(g(t)) - f(g(x))}{t - x}. \tag{2}$$

To evaluate this limit, we work with the approximation results developed in Sections 1.6 and 1.7. Let c be a number such that $f'(c)$ exists, and let z be close to c. Then

$$f(z) = f(c) + f'(c)(z - c) + E(z)(z - c) \tag{3}$$

where

$$E(z) = \frac{f(z) - f(c)}{z - c} - f'(c).$$

Because

$$\lim_{z \to c} \frac{f(z) - f(c)}{z - c} = f'(c),$$

we know that $E(z)$ approaches 0 as t approaches c.

Now assume that $g'(x)$ exists and that $f'(g(x))$ exists (i.e., that $g(x)$ is in the domain of f'). Substituting $g(x)$ for c and $g(t)$ for z in (3), we obtain

$$f(g(t)) = f(g(x)) + f'(g(x))(g(t) - g(x)) + E(g(t))(g(t) - g(x)), \tag{4}$$

where $E(g(t))$ approaches 0 as $g(t)$ approaches $g(x)$. Because $g'(x)$ exists, g is continuous at $t = x$ and so

$$g(t) \to g(x) \quad \text{as} \quad t \to x.$$

Hence

$$E(g(t)) \to 0 \quad \text{as} \quad t \to x.$$

Now substitute (4) into (2) to obtain

$$\begin{aligned}(f \circ g)'(x) &= \lim_{t \to x} \frac{\big(f(g(x)) + f'(g(x))(g(t) - g(x)) + E(g(t))(g(t) - g(x))\big) - f(g(x))}{t - x} \\ &= \lim_{t \to x}\left(f'(g(x))\frac{g(t) - g(x)}{t - x} + E(g(t))\frac{g(t) - g(x)}{t - x}\right) \\ &= f'(g(x)) \lim_{t \to x} \frac{g(t) - g(x)}{t - x} + \lim_{t \to x} E(g(t)) \cdot \lim_{t \to x} \frac{g(t) - g(x)}{t - x}\end{aligned} \tag{5}$$

By the definition of *derivative,*

$$\lim_{t \to x} \frac{g(t) - g(x)}{t - x} = g'(x).$$

Also, as noted above,

$$\lim_{t \to x} E(g(t)) = 0.$$

Using these results in (5), we obtain the chain rule,

$$(f \circ g)'(x) = f'(g(x))g'(x).$$

EXAMPLE 1 Find $\frac{dy}{dx}$ if $y = (-6x^8 + 3x^4 + x - 7)^{12}$.

Solution

We first note that $y = (f \circ g)(x)$, where g and f are given by

$$g(x) = -6x^8 + 3x^4 + x - 7 \quad \text{and} \quad f(x) = x^{12}.$$

We next find

$$g'(x) = -48x^7 + 12x^3 + 1 \quad \text{and} \quad f'(x) = 12x^{11}.$$

Thus,

$$\begin{aligned} \frac{dy}{dx} &= (f \circ g)'(x) = f'(g(x))g'(x) \\ &= 12(g(x))^{11}(-48x^7 + 12x^3 + 1) \\ &= 12(-6x^8 + 3x^4 + x - 7)^{11}(-48x^7 + 12x^3 + 1). \end{aligned}$$

In the preceding example we found the derivative of a power function. The result is a special case of the following formula for differentiation of power functions.

EXAMPLE 2 Let u be a differentiable function and n an integer. Find $h'(x)$ if $h(x) = (u(x))^n$.

Solution

We can write $h(x) = (f \circ u)(x)$ with $f(x) = x^n$. Because $f'(x) = nx^{n-1}$, application of the chain rule gives

$$h'(x) = (f \circ u)'(x) = f'(u(x))u'(x) = nu(x)^{n-1}u'(x).$$

Hence,

$$\frac{d}{dx}[u(x)^n] = nu(x)^{n-1}u'(x). \tag{6}$$

EXAMPLE 3 Find dy/dx if $y = (-3x^2 + 4)^8(x^3 - 3x + 1)^{-5}$.

Solution

Because the expression for y involves a product and compositions, we will need to use the product rule and the chain rule. By the product rule,

$$\frac{dy}{dx} = ((-3x^2 + 4)^8)'(x^3 - 3x + 1)^{-5} + (-3x^2 + 4)^8((x^3 - 3x + 1)^{-5})'. \tag{7}$$

To calculate $((-3x^2 + 4)^8)'$, use (6) with $n = 8$ and $u(x) = -3x^2 + 4$ to get

$$\begin{aligned}((-3x^2 + 4)^8)' &= 8(-3x^2 + 4)^7(-3x^2 + 4)' \\ &= 8(-3x^2 + 4)^7(-6x) = -48x(-3x^2 + 4)^7.\end{aligned} \tag{8}$$

To find $((x^3 - 3x + 1)^{-5})'$, apply (6) with $n = -5$ and $u(x) = x^3 - 3x + 1$. This gives

$$\begin{aligned}((x^3 - 3x + 1)^{-5})' &= (-5)(x^3 - 3x + 1)^{-6}(x^3 - 3x + 1)' \\ &= -5(x^3 - 3x + 1)^{-6}(3x^2 - 3) \\ &= (15 - 15x^2)(x^3 - 3x + 1)^{-6}.\end{aligned} \tag{9}$$

Substitute (8) and (9) into (7) to obtain the derivative

$$\frac{dy}{dx} = -48x(-3x^2 + 4)^7(x^3 - 3x + 1)^{-5} + (-3x^2 + 4)^8(15 - 15x^2)(x^3 - 3x + 1)^{-6}.$$

Differentiation of Rational Powers With the chain rule, we can extend our rule for differentiation of x^n to the case where n is a rational number.

EXAMPLE 4 Let $h(x) = x^{p/q}$, where p and q are integers and $q \neq 0$. Assuming $h(x)$ is differentiable for $x > 0$, find $h'(x)$.

Solution

Raise both sides of the equation $h(x) = x^{p/q}$ to the qth power to get

$$(h(x))^q = (x^{p/q})^q = x^p. \tag{10}$$

Next, differentiate both sides of this equation. Using (6) to differentiate the left side and the power rule for the right side,

$$q(h(x)^{q-1})h'(x) = px^{p-1}.$$

Solve this equation for $h'(x)$, then replace $h(x)$ by $x^{p/q}$. The result is

$$h'(x) = \frac{px^{p-1}}{q(h(x)^{q-1})} = \frac{p}{q}\frac{x^{p-1}}{(x^{p/q})^{q-1}} = \frac{p}{q}x^{(p/q)-1}.$$

Actually, there is no need to assume $x^{p/q}$ is differentiable to get this result. See Exercise 32 in Section 2.7.

Combining the result of Example 4 with the rules for differentiating power functions from Section 2.2, we have:

The Derivative of the Power Function

Let $y = f(x) = x^r$, where r is a rational number. Then

$$\frac{dy}{dx} = rx^{r-1}. \tag{11}$$

The Chain Rule Reformulated In some situations the chain rule is more easily applied if we use the Leibnitz notation for the derivative.

The Chain Rule

Suppose that $y = (f \circ g)(x)$. We can express this composition as a *composition chain:*

$$y = f(u)$$
$$u = g(x).$$

With this notation the chain rule can be written as

$$\frac{dy}{dx} = \frac{dy}{du}\frac{du}{dx}. \tag{12}$$

Since $dy/du = f'(u) = f'(g(x))$ and $du/dx = g'(x)$, we see that (12) is equivalent to (1). Hence (12) is another way of stating the chain rule.

EXAMPLE 5 Find $\frac{dy}{dx}$ if

$$y = \sqrt{3x^2 - 4x + 1/2}. \tag{13}$$

Solution

Begin by writing a composition chain that displays the given function as a composition of simple pieces:

$$y = \sqrt{u}$$
$$u = 3x^2 - 4x + \frac{1}{2}.$$

Because $y = u^{1/2}$, we apply (11) to get

$$\frac{dy}{du} = \frac{1}{2}u^{-1/2}. \tag{14}$$

Because u is a polynomial, we can immediately write down its derivative:

$$\frac{du}{dx} = 6x - 4. \tag{15}$$

Substituting (14) and (15) into (12), we have

$$\frac{dy}{dx} = \frac{dy}{du}\frac{du}{dx} = \frac{1}{2}u^{-1/2}(6x - 4).$$

Because (13) was given in terms of x, our answer should be in terms of x only. Thus we replace u by $3x^2 - 4x + \frac{1}{2}$. This leads to the result.

$$\frac{dy}{dx} = \frac{1}{2}(3x^2 - 4x + 1/2)^{-1/2}(6x - 4).$$

We will often work with functions that are compositions of three or more functions. The chain rule also applies in these cases.

EXAMPLE 6 Find a formula for dy/dx if $y = f(g(h(x)))$.

Solution

We write a composition chain for the function,

$$y = f(u)$$
$$u = g(v)$$
$$v = h(x).$$

Using the version of the chain rule given in (12), we have

$$\frac{dy}{dx} = \frac{dy}{du}\frac{du}{dx} = \frac{dy}{du}\frac{du}{dv}\frac{dv}{dx}. \tag{16}$$

Because

$$\frac{dy}{du} = f'(u) = f'(g(v)) = f'(g(h(x)))$$

$$\frac{du}{dv} = g'(v) = g'(h(x))$$

$$\frac{dv}{dx} = h'(x),$$

the derivative in (16) can be rewritten as

$$\frac{dy}{dx} = f'(g(h(x)))g'(h(x))h'(x).$$

EXAMPLE 7 Find $\frac{dy}{dx}$ if

$$y = \frac{3}{(\sqrt{x^2 + 2} + 2)^2}.$$

Solution

We use the Leibnitz notation as in (16). In this case the composition chain consists of three simple functions:

$$\begin{aligned} y &= \frac{3}{u^2} = 3u^{-2} \\ u &= \sqrt{v} + 2 = v^{1/2} + 2 \\ v &= x^2 + 2. \end{aligned} \tag{17}$$

By (11), we have

$$\frac{dy}{du} = -6u^{-3}, \quad \frac{du}{dv} = \frac{1}{2}v^{-1/2}, \quad \text{and} \quad \frac{dv}{dx} = 2x.$$

By the chain rule,

$$\frac{dy}{dx} = \frac{dy}{du}\frac{du}{dv}\frac{dv}{dx} = (-6u^{-3})\left(\frac{1}{2}v^{-1/2}\right)(2x).$$

Now substitute the expressions for u and v given in (17) to get

$$\begin{aligned} \frac{dy}{dx} &= \left(-6\left(\sqrt{v} + 2\right)^{-3}\right)\left(\frac{1}{2}v^{-1/2}\right)(2x) \\ &= \left(-6\left(\sqrt{x^2+2} + 2\right)^{-3}\right)\left(\frac{1}{2}(x^2+2)^{-1/2}\right)(2x) \\ &= \frac{-6x}{\left(\sqrt{x^2+2} + 2\right)^3\sqrt{x^2+2}}. \end{aligned}$$

Exercises 2.3

Exercises 1–18: Find the derivative of the given function.

1. $f(x) = 2x^{13/3} + 2$
2. $y = \dfrac{-3}{x^{5/2}}$
3. $y = (2x^2 - 3x + 1)^{23}$
4. $r = (\theta^3 - 3\theta^2 + \theta)^{2000}$
5. $g(x) = \left(x^3 - 5x + \dfrac{1}{2}\right)^{5/4}$
6. $H(s) = \left(3s^4 - 6s + \dfrac{1}{s}\right)^{7/3}$
7. $f(t) = \sqrt{2t - 7}$
8. $s = \sqrt[5]{u - 7}$
9. $r = (2\theta - 1)^6(4\theta + \pi)^8$
10. $y = (4x - 3)^{10}\sqrt{x^2 + 1}$
11. $y = \sqrt{1 + \sqrt{1 + \sqrt{x + 1}}}$
12. $y = ((2 - 3x^2)^{10} + 44)^{5/3}$
13. $h(u) = \sqrt[3]{\dfrac{4u - 1}{3u + 7}}$
14. $G(z) = \dfrac{(3z^2 - z + 1)^3}{(z^2 + 3z - 1)^4}$
15. $H(z) = (3z^2 - 1)^4(4z^{-3} + 2z^{-2} + 4)\sqrt{1 - \dfrac{1}{z}}$
16. $l(x) = \dfrac{1}{(1 + x + x^2)^{10}}$
17. $y = f(f(x))$ where $f(x) = x^2 - 3x + 1$
18. $h(t) = (g \circ g \circ g)(t)$ where $g(t) = \dfrac{1}{t^2} + 1$

Exercises 19–24: Find an expression for dy/dx.

19. $y = u^2 + 1,\ u = x^{-3/2} + 1$

20. $y = w^{-5/3},\ w = x^3 - 2x + 4$

21. $y = \sqrt{w},\ w = \dfrac{1}{v},\ v = 2x - 1$

22. $y = f(u),\ u = -2x^3 + x + 1$, where $f'(u) = 2\sin u$.

23. $y = H(t^2 + 1),\ t = (3x - 1)$, where $H'(v) = 1/v$.

24. $y = f\left(\sqrt{v+1}\right),\ v = w^3 + 3,\ w = 3x + 1$, with $f'(z) = \tan z$.

Exercises 25–26: In Exercise 27 of Section 2.2 we showed that $(d/dx)|x| = x/|x|$ for $x \neq 0$. Use this formula as needed to find the derivative of the given function.

25. $f(x) = |(x - 2)(x + 1)|$

26. $h(t) = \dfrac{2t^2 - 3}{|t^2 - 6|}$

27. $y = (|x^3 - x| + 2)^3$

Exercises 28–32: Sketch the graphs of $y = g(x)$ and $y = g'(x)$. The accompanying figures show the graph of $y = f(x)$ and the graph of $y = f'(x)$.

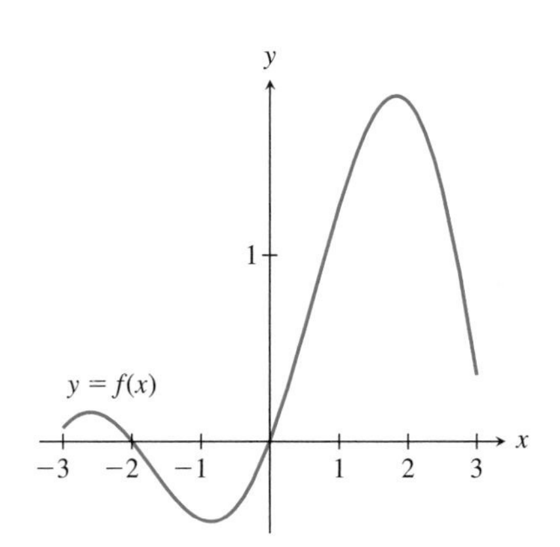

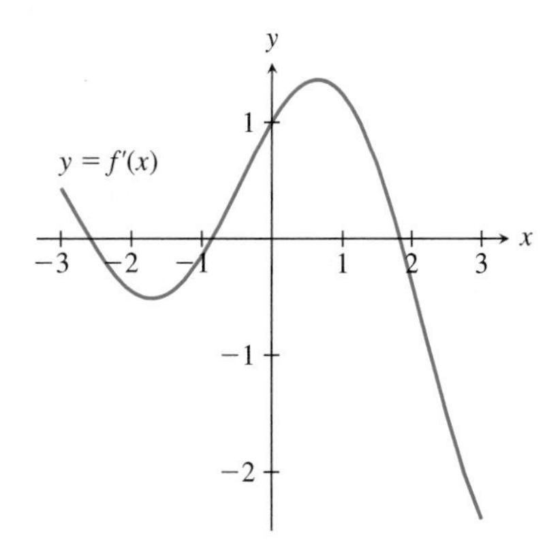

Figures for Exercises 28 – 32

28. $g(x) = f(-x)$

29. $g(x) = f(x - 2)$

30. $g(x) = f(|x|)$

31. $g(x) = |f(x)|$

32. $g(x) = \dfrac{|f(x)|}{f(x)}$

33. Find the equation of the line tangent to the graph of $y = 1/\sqrt{x^2 + 1}$ at the point $\left(1, 1/\sqrt{2}\right)$.

34. Find the equation of the line tangent to the graph of $y = x^{1/3}$ at the point $(0, 0)$.

35. Find the equation of the line tangent to the graph of $y = (2x - 3)^3/(-x^2 + x)^5$ at the point $(2, -1/32)$.

36. Let f and g be two functions. Assume that the line tangent to the graph of $y = f(x)$ at the point $(3, -1)$ has equation $y = 5x - 16$, and that the line tangent to the graph of $y = g(x)$ at $(1, 3)$ has equation $y = -2x + 5$. Find the equation of the line tangent to the graph of $y = (f \circ g)(x)$ at the point on the graph with $x = 1$.

37. The temperature F in degrees Fahrenheit at any point $y \geq 0$ on the y-axis is $F = F(y)$. An object moves on the positive y-axis so that at time $t \geq 0$ seconds its position is $y = y(t)$. Determine the rate of change of the temperature of the object with respect to time at $t = 15$ if the following data are known: $F(25.5) = -20$, $y(15) = 25.5$, $F'(25.5) = -1.9$, and $y'(15) = 4.25$. Give a brief physical interpretation of each of the four known data points.

38. Fill in reasons and complete the following argument to give another proof of the chain rule. Let $h(x) = f(g(x))$. Then

$$h'(x) = \lim_{t \to x} \frac{f(g(t)) - f(g(x))}{t - x}$$

$$= \lim_{t \to x} \frac{f(g(t)) - f(g(x))}{g(t) - g(x)} \frac{g(t) - g(x)}{t - x}.$$

Set $u = g(t)$ and note that as $t \to x$, $u = g(t) \to g(x)$. Hence

$$h'(x) = \lim_{u \to g(x)} \frac{f(u) - f(g(x))}{u - g(x)} \lim_{t \to x} \frac{g(t) - g(x)}{t - x}.$$

2.4 Implicit Differentiation

Implicitly Defined Functions

In Section 1.1 we remarked that when an equation is used to define a function, we often rely on our experience to supply a natural domain and range for the function. However, for many equations, the domain and range of the function are not at all obvious. As an example, consider the equation

$$y^3 - 3xy + x^2 - 1 = 0. \tag{1}$$

We can view this equation as defining y in terms of x, because if we put an x value into the equation, we can solve for the corresponding y value. For example, when $x = 1$, we find that $y = \sqrt{3}$ solves (1). So a function defined by this equation might map the domain element 1 to the range element $\sqrt{3}$. However, we also note that when $x = 1$, the y values 0 and $-\sqrt{3}$ also work in the equation. Which should we choose as the range element corresponding to $x = 1$? If we want to use (1) to define a function, we need to decide on a domain and then decide which y value is to correspond to a given x in the domain.

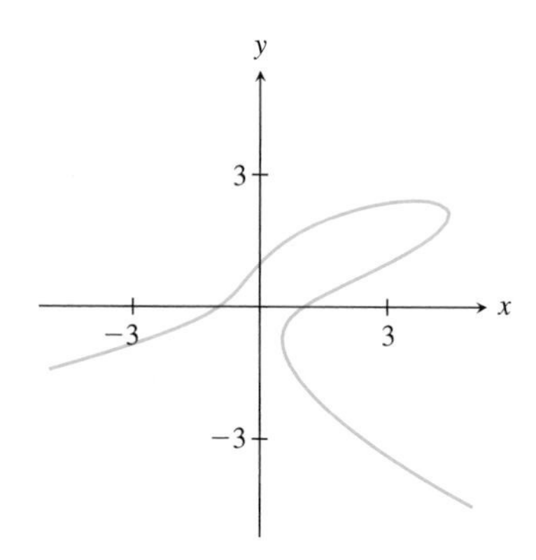

FIGURE 2.7 A graph of the equation $y^3 - 3xy + x^2 - 1 = 0$.

One way to gain insight into an equation such as (1) is by graphing; that is, plotting the set of all ordered pairs (x, y) that satisfy (1). Graphs of such equations can be very difficult to draw by hand. However, many computer algebra systems and plotting packages have commands to quickly draw graphs of the solutions to such equations.

The graph of (1) is shown in Fig. 2.7. From the graph we see that given a value x on the horizontal axis, there may be one, two, or three y values that correspond to x. To define a function using (1), we need to specify a domain for the function and a way of deciding which y to choose for each x in the domain. Suppose, for example, that we want (1) to define a function f so that $y = \sqrt{3}$ corresponds to the domain value $x = 1$, that is, so that $f(1) = \sqrt{3}$. We express this by saying:

Equation (1) defines y as a function of x near $\left(1, \sqrt{3}\right)$.

We may describe such a function f by using a graph like that shown in Fig. 2.7. To do this, we take a piece of the graph of the equation (1), a piece that contains $\left(1, \sqrt{3}\right)$, and identify this piece as the graph of a function f. We need only be sure that the piece of the graph chosen contains $\left(1, \sqrt{3}\right)$ and that for each x value that corresponding to a point on the graph there is only one corresponding y value; that is, the piece of graph we choose must satisfy the *vertical line test* (see Exercise 29 in Section 1.1). We have indicated a portion of the graph in Fig. 2.8. This darker piece of the graph defines a function with domain something like $\{x: -5 \leq x \leq 4.5\}$ and range something like $\{y: -1.5 \leq y \leq 2.5\}$.

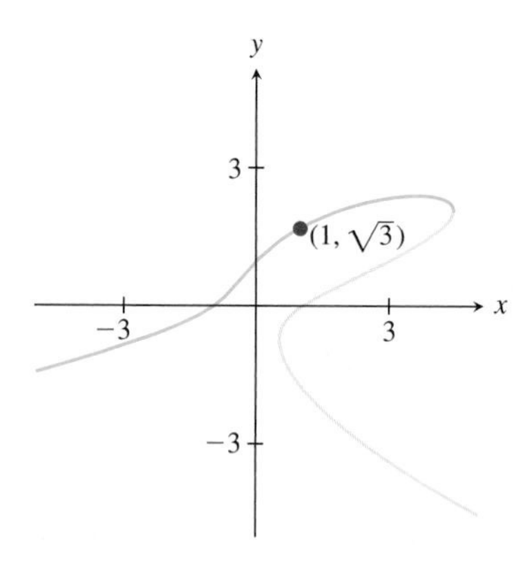

FIGURE 2.8 The darker portion gives the graph of a function f for which $f(1) = \sqrt{3}$.

To evaluate $f(x)$ for an x in the domain, we can put the x value into (1) and solve the resulting equation for y. To check that we have the right y value for the given x, we can make sure that the point (x, y) lies on the darker portion of the graph in Fig. 2.8. Can we find an explicit formula for $f(x)$? Yes, but not easily. Furthermore, if (1) were more complicated, it might be impossible to find such a formula. When a function f is "indirectly" defined through an equation such as (1), we say that f is defined **implicitly**.

EXAMPLE 1 The expression

$$x^2 + xy + y^2 = 3 \tag{2}$$

is satisfied if $x = 1$ and $y = -2$. Assume that this expression defines y as a function of x near $(1, -2)$. Write a formula for such a function.

Solution

The graph of $x^2 + xy + y^2 = 3$ is shown in Fig. 2.9. From this figure we see that the function g whose graph is the lower half of the graph is a function with $g(1) = -2$. Because (2) is quadratic in y, we can solve for y using the quadratic formula. Rewriting (2) as

$$y^2 + xy + (x^2 - 3) = 0,$$

we find

$$y = \frac{-x \pm \sqrt{x^2 - 4(x^2 - 3)}}{2} = \frac{-x \pm \sqrt{12 - 3x^2}}{2}.$$

If $x = 1$, we get $y = -2$ by using the negative sign in front of the square root. Thus the function g given by

$$g(x) = \frac{-x - \sqrt{12 - 3x^2}}{2}$$

is one function that is implicitly defined by (2) and has $g(1) = -2$.

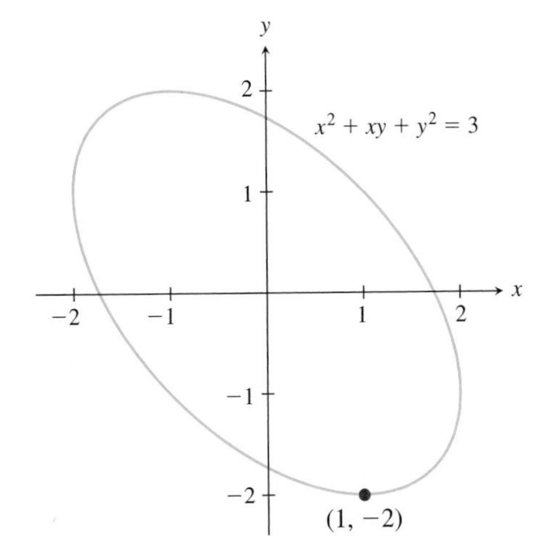

FIGURE 2.9 The graph of $x^2 + xy + y^2 = 3$.

Finding the Derivative of Implicitly Defined Functions

INVESTIGATION

Differentiating an Implicitly Defined Function

As in our earlier discussion, let

$$y^3 - 3xy + x^2 - 1 = 0 \tag{3}$$

define y as a function f of x near $(1, \sqrt{3})$. We assume that f is differentiable. As mentioned earlier, a formula for $y = f(x)$ is difficult to find. Surprisingly, though, it is not difficult to find an expression for f'. The process for finding this derivative is called **implicit differentiation.**

For a given x in the domain of f, we know that the pair of numbers

$$(x, y) = (x, f(x))$$

satisfies (3). In particular, because $y = f(x)$, equation (3) is true if we replace y with $f(x)$, that is,

$$(f(x))^3 - 3xf(x) + x^2 - 1 = 0. \tag{4}$$

This equation holds for all x in the domain of f. Differentiate both sides of (4) with respect to x. The derivative of the right side is 0. Why? The expression on the left side involves some products and compositions, so we will use the product rule and the chain rule. First break the derivative of the left side into the sum of several derivatives:

$$((f(x))^3)' - 3(xf(x))' + (x^2)' - 1' = 0'.$$

Apply the product rule to the second term of this expression to get

$$((f(x))^3)' - 3(x'f(x) + xf'(x)) + 2x = 0. \tag{5}$$

Next differentiate $(f(x))^3$. Using the result of Example 2 in Section 2.3, we have

$$((f(x))^3)' = 3(f(x))^2 f'(x). \tag{6}$$

Substituting (6) into (5) and replacing x' by 1, we have

$$3(f(x))^2 f'(x) - 3(f(x) + xf'(x)) + 2x = 0.$$

Expanding and regrouping this last equation, we see that $f'(x)$ can be factored from two terms, giving

$$f'(x)[3(f(x))^2 - 3x] + (2x - 3f(x)) = 0.$$

Solving for $f'(x)$, we get

$$f'(x) = \frac{3f(x) - 2x}{3(f(x))^2 - 3x}.$$

Replacing $f(x)$ with y, we have an expression for the desired derivative:

$$\frac{dy}{dx} = f'(x) = \frac{3y - 2x}{3y^2 - 3x}. \tag{7}$$

When $x = 1$, we have $y = \sqrt{3}$. The rate of change of y with respect to x at $(1, \sqrt{3})$ is found by substituting these x and y values into (7):

$$\left.\frac{dy}{dx}\right|_{x=1,y=\sqrt{3}} = \frac{3\sqrt{3} - 2\cdot 1}{3\sqrt{3}^2 - 3\cdot 1} = \frac{3\sqrt{3} - 2}{6} \approx 0.532692.$$

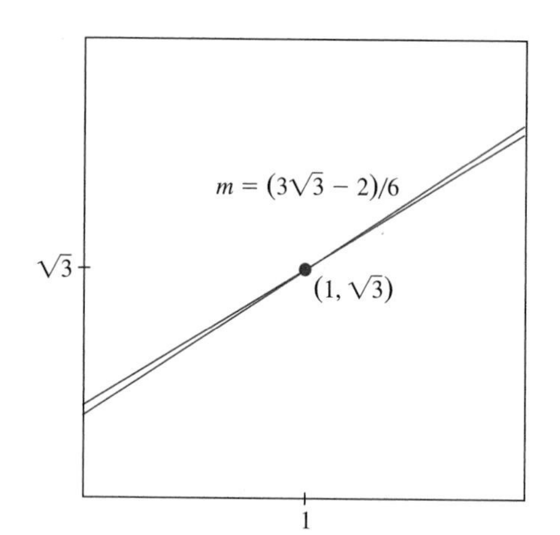

FIGURE 2.10 The line tangent to $y^3 - 2xy^2 + x^2 - 1 = 0$ at the point $(1, \sqrt{3})$ has slope $(3\sqrt{3} - 2)/6$.

In Fig. 2.10 we zoom in on the graph of $y^3 - 3xy + x^2 - 1 = 0$ near the point $(1, \sqrt{3})$ and include the graph of the line tangent to the graph at this point. What is the equation for this line?

Note that we can evaluate (7) for a given x value only if we know the corresponding y value. So even though dy/dx is not hard to find, it may be difficult to evaluate the derivative for a given x value!

EXAMPLE 2 Find the equation of the line tangent to the graph of the circle

$$x^2 + y^2 = 25 \tag{8}$$

at the point $(-3, 4)$. Show that the radial segment from $(0, 0)$ to the point $(-3, 4)$ is perpendicular to the tangent line.

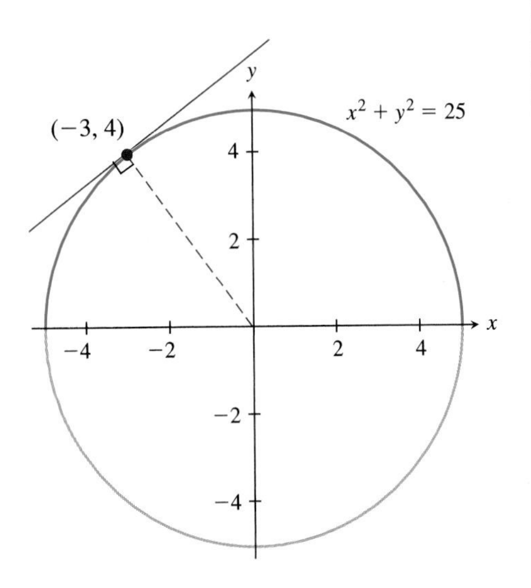

FIGURE 2.11 The graph of $x^2 + y^2 = 25$.

Solution

Equation (8) defines y as a function of x near the point $(-3, 4)$. Let f be such a function. Then the graph of f might be the upper half of the circle, the darker curve in Fig. 2.11. Because $y = f(x)$, we replace y with $f(x)$ in (8) to get

$$x^2 + (f(x))^2 = 25.$$

Differentiate both sides of this expression with respect to x to get

$$2x + 2f(x)f'(x) = 0.$$

Solve for $f'(x)$ to obtain

$$\frac{dy}{dx} = f'(x) = -\frac{x}{f(x)} = -\frac{x}{y}.$$

At the point $(-3, 4)$, the value of the derivative is

$$\frac{dy}{dx} = -\frac{-3}{4} = \frac{3}{4}.$$

The line tangent to the circle at the point $(-3, 4)$ has slope $3/4$ and equation

$$y - 4 = \frac{3}{4}(x + 3).$$

In Fig. 2.11 we show the tangent line and the radius from the center of the circle to $(-3, 4)$. The radial segment has slope $-4/3$. Because $(-4/3)(3/4) = -1$, the tangent line is perpendicular to the radius.

In practice, it is not necessary to replace y by $f(x)$ before differentiating. This substitution was just an aid to help us remember that y is a function of x.

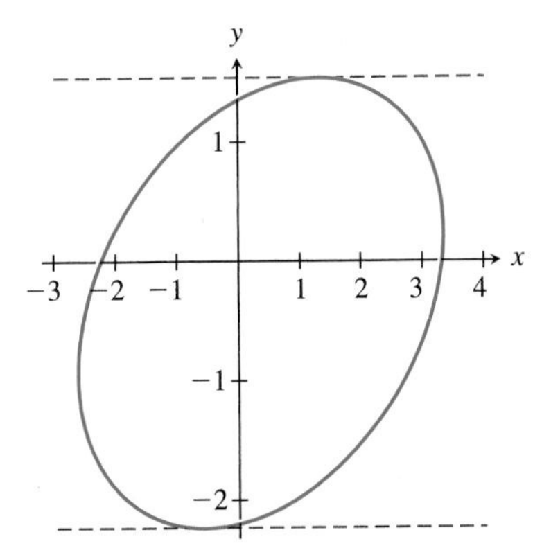

FIGURE 2.12 The graph of $(x - 2y - 1)^2 + (x + y)^2 = 16$.

EXAMPLE 3 Find the coordinates of the points on the graph of

$$(x - 2y - 1)^2 + (x + y)^2 = 16 \qquad (9)$$

where the tangent line is horizontal.

Solution

To get an idea of what the answer should be, we look at the graph of (9), shown in Fig. 2.12. From the graph, there appear to be two places where the tangent line is horizontal. The coordinates of the points of tangency are roughly $(1.5, 1.5)$ and $(-0.75, -2.2)$.

The tangent to the graph is horizontal at points where $\dfrac{dy}{dx}$ is equal to 0.

We find dy/dx by differentiating (9) implicitly, keeping in mind that the equation defines y as a function of x. Differentiate both sides of (9) with respect to x and apply the chain rule (as seen in Example 2 of Section 2.3) to get

$$2(x - 2y - 1)\frac{d}{dx}(x - 2y - 1) + 2(x + y)\frac{d}{dx}(x + y) = 0.$$

Writing the derivative of y with respect to x as $\frac{dy}{dx}$, this last equation becomes

$$2(x - 2y - 1)\left(1 - 2\frac{dy}{dx}\right) + 2(x + y)\left(1 + \frac{dy}{dx}\right) = 0.$$

Expand, and group the terms involving $\frac{dy}{dx}$ to get

$$\frac{dy}{dx}(-2x + 10y + 4) + (4x - 2y - 2) = 0.$$

Solve for $\frac{dy}{dx}$ to find

$$\frac{dy}{dx} = \frac{-2x + y + 1}{-x + 5y + 2}.$$

At a point where the tangent is horizontal (has slope 0), this derivative is 0. Hence at the point (x, y) where the tangent line is horizontal,

$$\frac{-2x + y + 1}{-x + 5y + 2} = 0. \tag{10}$$

A fraction is equal to 0 when its numerator is 0 and the denominator is nonzero. Hence from (10) we get

$$-2x + y + 1 = 0. \tag{11}$$

If (x, y) is a point on the graph where the tangent has slope 0, then this point must also be on the graph shown in Fig. 2.12, and hence must satisfy (9). With (9) and (11) we have a system of two equations and two unknowns,

$$\begin{aligned} (x - 2y - 1)^2 + (x + y)^2 &= 16 \\ -2x + y + 1 &= 0. \end{aligned} \tag{12}$$

Solve the second equation for y,

$$y = 2x - 1. \tag{13}$$

Substituting this result into the first equation in (12), we have

$$(x - 2(2x - 1) - 1)^2 + (x + (2x - 1))^2 = 16.$$

This equation can be solved using a calculator, computer, or the quadratic formula. The solutions are

$$x = \frac{1 + 2\sqrt{2}}{3} \approx 1.27 \quad \text{and} \quad x = \frac{1 - 2\sqrt{2}}{3} \approx -0.609.$$

Substituting these values into (13), we find the corresponding y values,

$$y = \frac{-1 + 4\sqrt{2}}{3} \approx 1.55 \quad \text{and} \quad y = \frac{-1 - 4\sqrt{2}}{3} \approx -2.22.$$

Thus, the two points on the graph where the tangent is horizontal are

$$\left(\frac{1 + 2\sqrt{2}}{3}, \frac{-1 + 4\sqrt{2}}{3}\right) \approx (1.27, 1.55)$$

and

$$\left(\frac{1 - 2\sqrt{2}}{3}, \frac{-1 - 4\sqrt{2}}{3}\right) \approx (-0.609, -2.22).$$

How do these answers compare with the estimates read from the graph in Fig. 2.12?

In the next example we use implicit differentiation to analyze the angle of intersection of two graphs.

EXAMPLE 4 Consider the graph of all ordered pairs (x, y) that satisfy the equation.

$$\frac{x}{x^2 + y^2} = 4, \tag{14}$$

and the graph of the ordered pairs that satisfy the equation

$$\frac{y}{x^2 + y^2} = -2. \tag{15}$$

Show that these two graphs are *orthogonal*; that is, at any point where the graphs intersect, the tangents to the graphs are at right angles to each other.

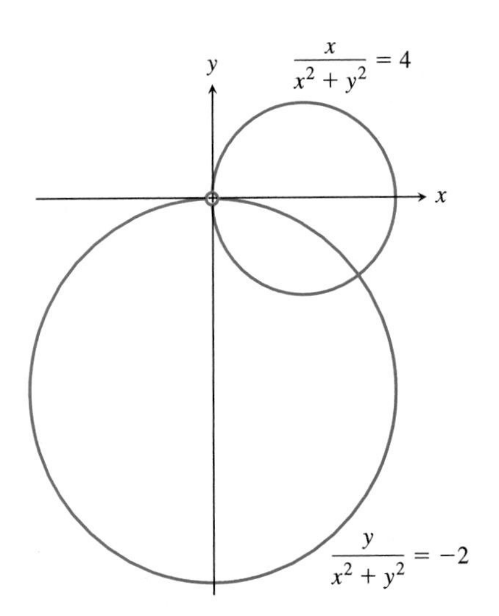

FIGURE 2.13 At the point of intersection, the lines tangent to the curves are perpendicular.

Solution

The graphs of (14) and (15) are shown in Fig. 2.13. It appears that each of the graphs is a circle. In fact, with a little algebraic manipulation, it can be shown that (14) describes all points except the origin on the circle of center $(1/8, 0)$ and radius $1/8$, and that (15) describes all points except the origin on the circle of center $(0, -1/4)$ and radius $1/4$. With a little more algebra, we could solve for the point of intersection of the two graphs. (See Exercises 23 and 24.) However, we will see that we do not need to know the intersection point to prove that the graphs are orthogonal at this point.

Let (x, y) be a point on the graph of (14). The slope of the line tangent to the graph at this point is the value of the derivative $\dfrac{dy}{dx}$. We find this derivative by implicitly differentiating (14). Assume that y is a function of x and differentiate both sides of (14) with respect to x:

$$\frac{d}{dx}\left(\frac{x}{x^2+y^2}\right) = \frac{d}{dx}4 = 0.$$

We apply the quotient rule to evaluate the derivative on the left side of this equation. We have

$$0 = \frac{d}{dx}\left(\frac{x}{x^2+y^2}\right) = \frac{\left(\frac{d}{dx}x\right)(x^2+y^2) - x\left(\frac{d}{dx}(x^2+y^2)\right)}{(x^2+y^2)^2}$$

$$= \frac{1(x^2+y^2) - x\left(2x + 2y\frac{dy}{dx}\right)}{(x^2+y^2)^2}$$

$$= \frac{x^2+y^2-2x^2-2xy\frac{dy}{dx}}{(x^2+y^2)^2}.$$

A fraction is 0 only if its numerator is 0. Thus we have

$$x^2+y^2-2x^2-2xy\frac{dy}{dx} = 0.$$

Solving for dy/dx, we find

$$\frac{dy}{dx} = \frac{y^2-x^2}{2xy}. \tag{16}$$

Differentiating both sides of (15), we find

$$\frac{\frac{dy}{dx}(x^2+y^2) - y\left(2x+2y\frac{dy}{dx}\right)}{(x^2+y^2)^2} = 0.$$

Setting the numerator equal to 0 and solving the resulting equation for dy/dx, we find

$$\frac{dy}{dx} = \frac{2xy}{x^2-y^2}. \tag{17}$$

Now let (x, y) be the coordinates of a point of intersection of the two curves. The lines tangent to the graphs of (14) and (15) at this point have slopes of (16) and (17), respectively. The product of these slopes is

$$\frac{y^2-x^2}{2xy}\,\frac{2xy}{x^2-y^2} = -1.$$

Because the products of the slopes is -1, we know that the two tangent lines are perpendicular. Thus the curves described by (14) and (15) are orthogonal. See Exercise 25 for another way to approach the implicit differentiation.

Orthogonal curves, such as those seen in Example 4, show up in many areas of engineering and physics. For example, suppose that different points of the plane are at different temperatures, and that at the point $(x, y) \neq (0, 0)$, the temperature is given by

$$T(x, y) = \frac{x}{x^2 + y^2}. \tag{18}$$

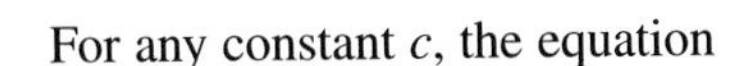

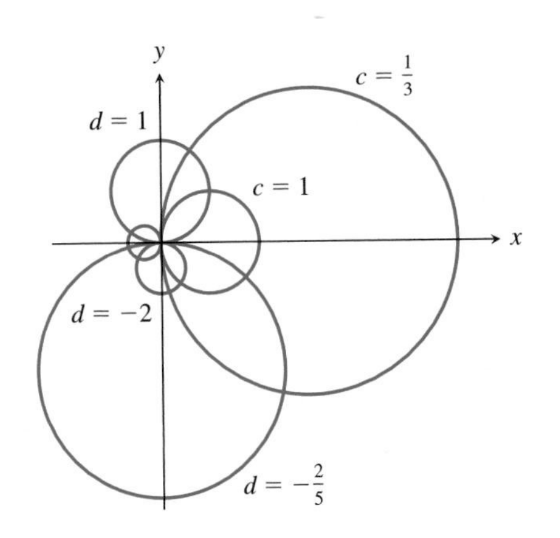

FIGURE 2.14 The curves $x/(x^2 + y^2) = c$ and $y/(x^2 + y^2) = d$ for some values of c and d.

For any constant c, the equation

$$\frac{x}{x^2 + y^2} = c \tag{19}$$

describes the set of points in the plane at which the temperature is c. The graph of all such points is called an *isotherm* for (18). As seen in Example 4, the curves described by equations of the form

$$\frac{y}{x^2 + y^2} = d \tag{20}$$

are orthogonal to the isotherms. The curves described by (20) are called the *heat flow lines* for the temperature distribution and are useful in analyzing the flow of heat from warm regions to cooler areas. A few of the curves defined by (19) and (20) are shown in Fig. 2.14.

Exercises 2.4

Exercises 1–10: Assume that the equation defines y as a function of x. Use implicit differentiation to find dy/dx.

1. $3x^2 - 4y^2 = 3$

2. $3x^2 + 5xy - 6y^2 = 15\sqrt{2}$

3. $3x - 2y + \dfrac{y}{x} - 7 = 0$

4. $x^3 + y^3 - x + 2y = 3$

5. $2x^2 - 3xy + y^2 = 2$

6. $x^3y + 3x^2y^2 - 4xy^3 = 12$

7. $(x - 2y)^2 - xy^2 = -2xy + 4$

8. $\sqrt{x^2 + y^2} - x^2y^3 = x + y$

9. $\dfrac{3x^2 + 4y}{2x - 6y^2} = x + y$

10. $(2x - 37)^2 - (4x + 7y)^3 = xy - 8$

Exercises 11–14: Check that the point P is on the graph of the equation. Assuming that the equation defines y as a function of x near the given point P, find the value of dy/dx at P.

11. $4y^2 - xy + 2x - 3y = 3, \quad P = (2, 1)$

12. $3x^2 - 4xy + 3y^3 = 12, \quad P = (-2, 0)$

13. $\dfrac{1}{x + y} + 3 - 4(x + y)^2 = -2, \quad P = (1, -2)$

14. $\sqrt{y + x} - \sqrt{y - x} = 2, \quad P = (10, 26)$

Exercises 15–18: Find the equation for the line tangent to the graph of the equation at the point Q.

15. $xy - x + 3y = 3, \quad Q = (1, 1)$

16. $(x + 2y)^3 - (2x + 4y)^3 = -56, \quad Q = (2, 0)$

17. $\dfrac{x^2 + y^2}{2x + 3y} = -2x, \quad Q = (-1, 1)$

18. $\sqrt{y + x} - \sqrt{y - x} = 2, \quad Q = (10, 26)$

Exercises 19–22: The graph of the solutions to an equation $f(x, y) = c$ is shown with one portion of the graph darkened. Solve the equation for y to find an expression of the form $y = f(x)$ whose graph is the darkened portion of the graph.

19.

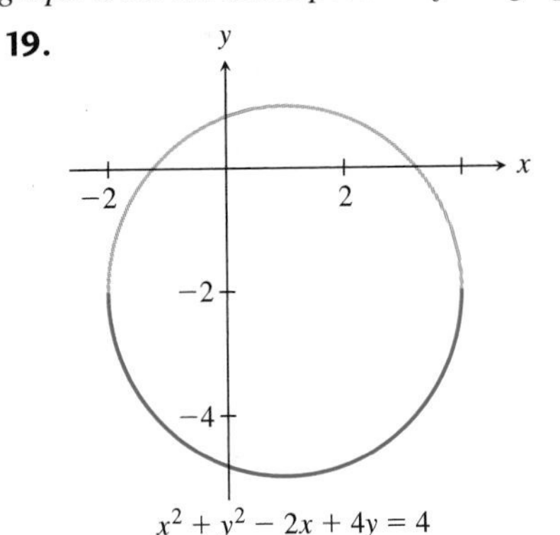

20.

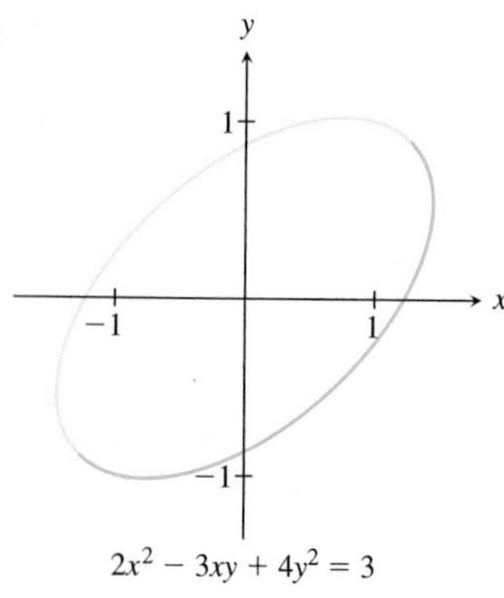

$2x^2 - 3xy + 4y^2 = 3$

21.

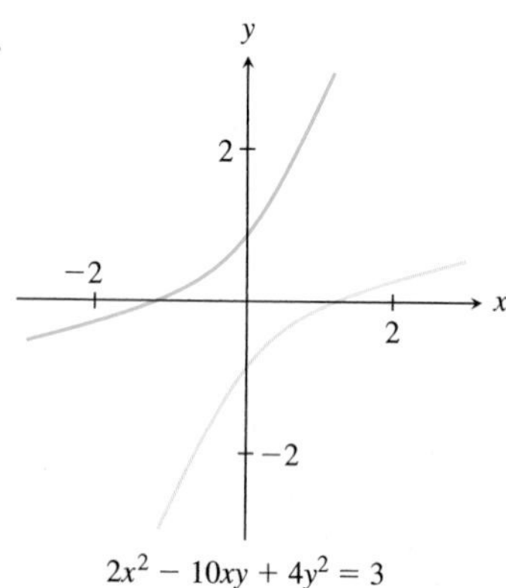

$2x^2 - 10xy + 4y^2 = 3$

22.

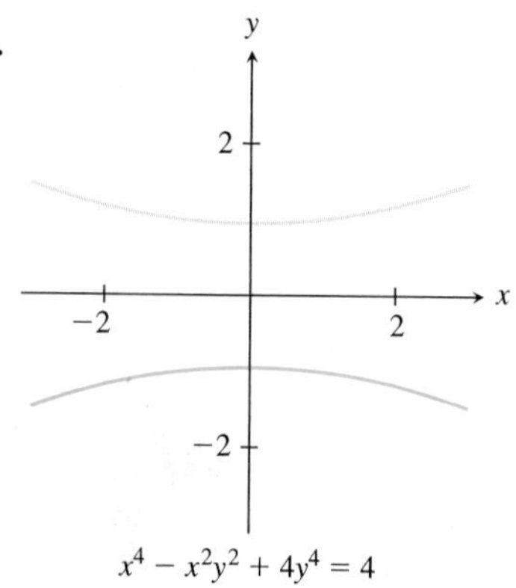

$x^4 - x^2y^2 + 4y^4 = 4$

23. Show that the equation

$$\frac{x}{x^2 + y^2} = 4$$

describes the points other than the origin on the circle of center $(1/8, 0)$ and radius $1/8$.

24. Find the point of intersection of the curve described by $\dfrac{x}{x^2 + y^2} = 4$ and the curve described by $\dfrac{y}{x^2 + y^2} = -2$.

25. Equation (14) can be written in the form

$$4x^2 + 4y^2 - x = 0.$$

Use implicit differentiation on this equation to find an expression for y'. Show that the expression you find is equivalent to that given in (16).

26. Consider the curves defined by

$$x = c \quad \text{and} \quad y = d,$$

where c and d are constants. Sketch these curves for a few values of c and d. Prove that the family of curves defined by $x = c$ is orthogonal to the family defined by $y = d$.

27. Consider the curves defined by

$$x^2 - y^2 = c \quad \text{and} \quad xy = d,$$

where c and d are constants. Sketch these curves for a few values of c and d. Prove that the family of curves defined by $x^2 - y^2 = c$ is orthogonal to the family defined by $xy = d$.

28. Let c and d be constants. Show that the family of curves defined by

$$x^3 - 3xy^2 = c$$

is orthogonal to the family defined by

$$3x^2y - y^3 = d.$$

29. Consider the equation

$$2xy^3 + x^2y^2 - 3y + x = 1.$$

a. Assume that this equation defines y as a function of x. Find an expression for dy/dx.

b. Assume that the equation defines x as a function of y. Find an expression for dx/dy.

c. Show that $(dy/dx)\,(dx/dy) = 1$. Explain this graphically by interpreting the derivative as the slope of a tangent.

30. Let the equation $y = f(x)$ define x as a function of y. Show that $dx/dy = 1/(f'(x))$. Interpret this result graphically by interpreting the derivative as the slope of a tangent.

31. The accompanying figure shows the graph of the ellipse with equation

$$x^2 - xy + \frac{3}{4}y^2 = 7.$$

The ellipse is centered at the origin but rotated from "standard position."

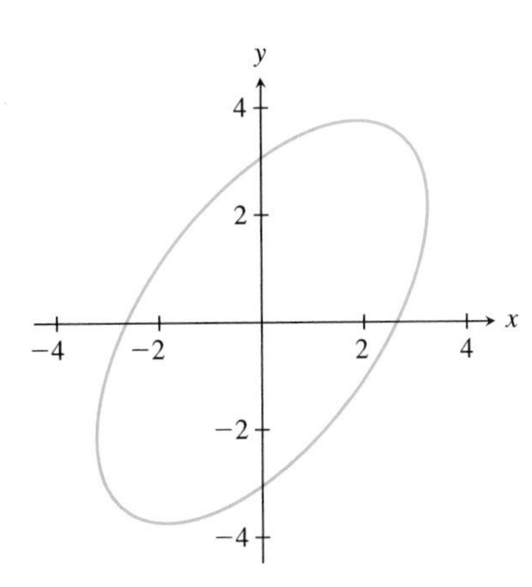

Figure for Exercise 31.

a. Find the coordinates of all points where the ellipse crosses the coordinate axes.

b. Find the coordinates of all points where the tangent line is horizontal.

c. Find the coordinates of all points where the tangent line is vertical.

32. Consider the graph of all points (x, y) that satisfy the equation

$$2x^2 - 3xy + y^2 = -2.$$

a. Find the coordinates of all points on the graph where the tangent line is horizontal.

b. Find the coordinates of all points on the graph where the tangent line is vertical.

c. Show that there are no points on the graph where the tangent line has slope 1.

2.5 Trigonometric Functions

If you are ever at a carnival near noon on a sunny day, go look at the Ferris wheel. When the sun is directly overhead, the shadow of the Ferris wheel will be on the ground below the wheel. Imagine watching the shadow of one of the cars on the Ferris wheel. The shadow will move back and forth across the ground as the wheel turns. The moving shadow illustrates two very important ideas. First, the moving shadow is an example of *periodic motion*. The back-and-forth motion of the shadow repeats regularly: The shadow moves from left to right under the Ferris wheel, then from right to left, then from left to right, and so on. Periodic phenomena are everywhere in our world. They are seen in the tides, the orbit of Earth about the sun, the phases of the moon, and the motion of a piston in an internal combustion engine.

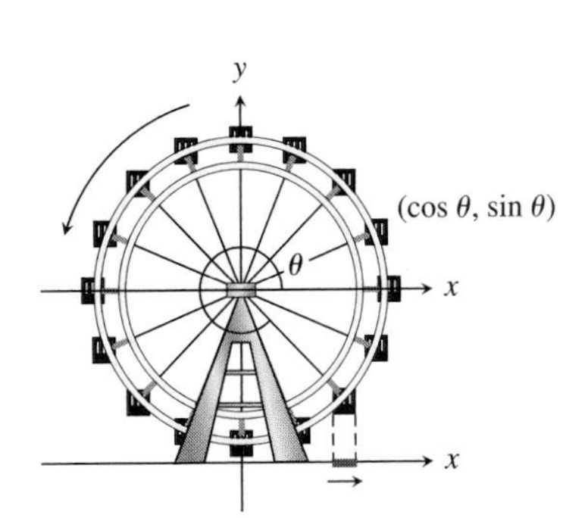

FIGURE 2.15 The shadow of a car on a Ferris wheel moves back and forth along the ground.

The moving shadow also illustrates the cosine function. Let the Ferris wheel have radius 1, and put a coordinate system in the plane of the wheel with the axle at the origin and the y-axis perpendicular to the ground. See Fig. 2.15. Let θ be the angle the ray from the origin through the car makes with the positive x-axis. The x-coordinate of the car is $\cos\theta$. Because the position of the car's shadow on the ground is just this x-coordinate, you are seeing an animation of the cosine function. This example illustrates the close relationship between periodic phenomena and trigonometric functions.

Now look more closely at the motion of the shadow. You will notice that the shadow moves very slowly near the ends of its back-and-forth motion, and it moves faster when it is under the center of the wheel. Because velocity is the rate of change of position with respect to time, the speed of the shadow is related to the derivative of the cosine function.

WEB

The Derivatives of Sine and Cosine

We use the geometry of the unit circle to find the derivative of the sine function. While the argument is not "mathematically rigorous," it does give some geometric insight into the relationship between the sine function and its derivative. See

Exercise 43 for an alternative argument using the definition of derivative and some of the limit results found in Chapter 1. Keep in mind that all angles are measured in radians. As we shall see later in this section, the calculus of trigonometric functions is easier if we use radians instead of degrees. Please refer to the Appendix if you need to review some basic trigonometry.

INVESTIGATION

The Derivative of the Sine Function

The derivative of the sine function at a value θ is the slope of the "line" we see as we zoom in on the graph of $y = \sin\theta$ near the point $(\theta, \sin\theta)$. Figure 2.16 shows $(\theta, \sin\theta)$ and a nearby point $(\theta + h, \sin(\theta + h))$. To find the derivative of the sine function at θ, we calculate the slope

$$\frac{\sin(\theta + h) - \sin\theta}{h} = \frac{\Delta y}{h} \tag{1}$$

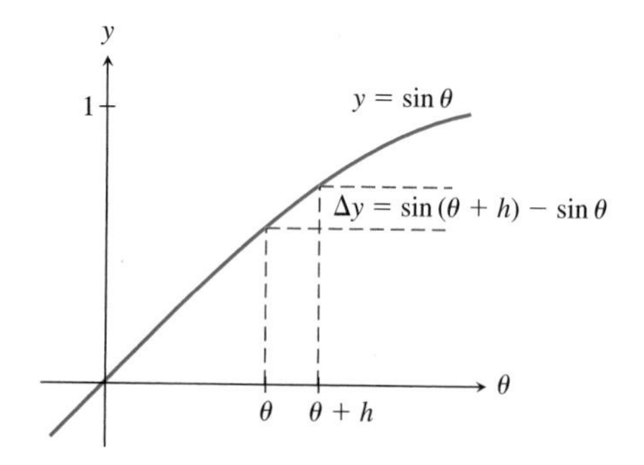

FIGURE 2.16 Finding the "slope" of the sine graph.

of the segment determined by these two points, then evaluate the limit of this slope as h approaches 0. In (1) we have set $\Delta y = \sin(\theta + h) - \sin\theta$ to denote the difference in the y-coordinates of the two points.

To calculate the limit of (1), we first approximate the numerator Δy. Consider the points $P = (\cos\theta, \sin\theta)$ and $Q = (\cos(\theta + h), \sin(\theta + h))$ on the unit circle, as shown in Fig. 2.17. Note that the vertical distance between P and Q is $\Delta y = \sin(\theta + h) - \sin\theta$. In the zoom-view of the line segment joining P and Q, we show a right triangle with one side Δy. We argue that the hypotenuse of this triangle is approximately h and the angle at Q is approximately θ.

Because h is small and angle POQ measures h radians,

$$\text{length of segment } PQ \approx \text{length of arc } PQ = h \cdot 1 = h. \tag{2}$$

For small h, segment PQ is very nearly tangent to the circle at P. Referring to the lower sketch in Fig. 2.17, it follows that angle OPQ is very nearly a right angle, and that the angle at Q is very nearly θ. Hence

$$\cos\theta \approx \frac{\Delta y}{h}. \tag{3}$$

Solving for Δy in this last equation, we see

$$\Delta y \approx h\cos\theta.$$

Substituting this result into (1) and letting $h \to 0$, we have

$$\frac{d}{d\theta}\sin\theta = \lim_{h\to 0}\frac{\sin(\theta + h) - \sin\theta}{h} = \lim_{h\to 0}\frac{\Delta y}{h} = \lim_{h\to 0}\frac{h\cos\theta}{h} = \cos\theta.$$

Once we have the derivative of the sine function, the derivatives of the other trigonometric functions follow easily.

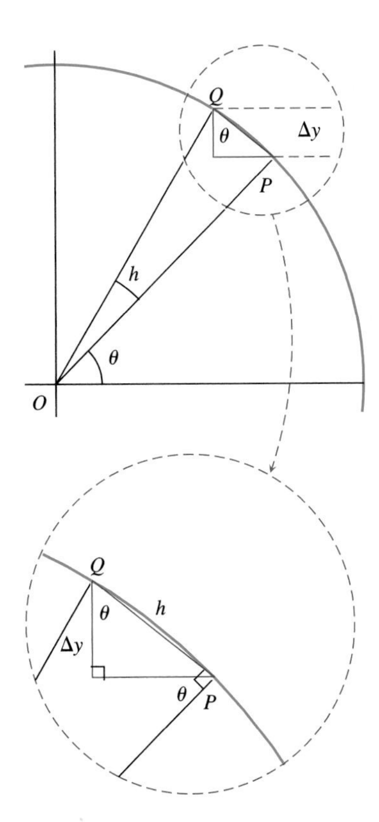

FIGURE 2.17 Finding Δy on the unit circle.

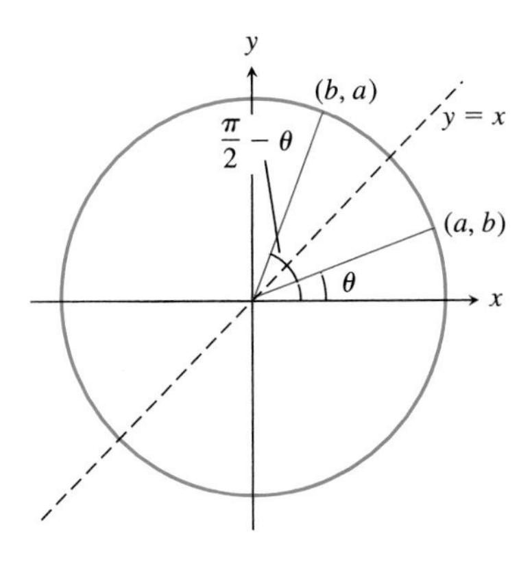

FIGURE 2.18 $\cos\left(\frac{\pi}{2} - \theta\right) = b = \sin\theta$
$\sin\left(\frac{\pi}{2} - \theta\right) = a = \cos\theta$

EXAMPLE 1 Find $\dfrac{d}{dx}\cos x$.

Solution

Let $w = \cos x$ and recall that

$$w = \cos x = \sin\left(\frac{\pi}{2} - x\right).$$

See Fig. 2.18 for a pictorial proof of this identity, or see the review of trigonometry in the Appendix. This equation expresses $\cos x$ as a composition of the sine function and the function given by $\dfrac{\pi}{2} - x$. As a composition chain this is

$$w = \sin u$$

$$u = \frac{\pi}{2} - x.$$

By the chain rule,

$$\frac{dw}{dx} = \frac{dw}{du}\frac{du}{dx} = (\cos u)(-1) = -\cos\left(\frac{\pi}{2} - x\right) = -\sin x.$$

Summarizing the last two results, we have:

Derivatives of the Sine and Cosine

$$\frac{d}{dx}\sin x = \cos x \quad \text{and} \quad \frac{d}{dx}\cos x = -\sin x \tag{4}$$

EXAMPLE 2 Find $\dfrac{d}{dx}\tan x$.

Solution

Because

$$\tan x = \frac{\sin x}{\cos x}$$

we can use the quotient rule developed in Example 4 of Section 2.2. We have

$$\frac{d}{dx}\tan x = \left(\frac{\sin x}{\cos x}\right)' = \frac{(\sin x)'(\cos x) - (\sin x)(\cos x)'}{(\cos x)^2}$$

$$= \frac{(\cos x)(\cos x) - (\sin x)(-\sin x)}{(\cos x)^2} = \frac{\cos^2 x + \sin^2 x}{\cos^2 x}$$

$$= \frac{1}{\cos^2 x} = \sec^2 x.$$

Any of the differentiation techniques studied in earlier sections may be needed to find the derivatives of the functions that involve trigonometric functions.

EXAMPLE 3 Find $\frac{dy}{dx}$ if $y = \sqrt{x \cos x}$.

Solution

First decompose the function into a composition chain:

$$y = \sqrt{u}$$
$$u = x \cos x$$

We then have

$$\frac{dy}{du} = \frac{d}{du}u^{1/2} = \frac{1}{2}u^{-1/2} = \frac{1}{2\sqrt{u}},$$

and by the product rule

$$\frac{du}{dx} = (x)' \cos x + x(\cos x)' = 1 \cdot \cos x + x(-\sin x) = \cos x - x \sin x.$$

By the chain rule,

$$\frac{dy}{dx} = \frac{dy}{du}\frac{du}{dx} = \left(\frac{1}{2\sqrt{u}}\right)(\cos x - x \sin x).$$

Substituting $u = x \cos x$ in the last expression, we obtain the derivative,

$$\frac{dy}{dx} = \frac{\cos x - x \sin x}{2\sqrt{x \cos x}}.$$

EXAMPLE 4 Find the equation of the line tangent to the graph of

$$y = \cos(\pi x^2) \tag{5}$$

at the point $\left(1/2, 1/\sqrt{2}\right)$. Use the tangent line to approximate the value of $\cos(\pi(0.55)^2)$.

Solution

To find the slope of the tangent line, we first need to calculate dy/dx. A composition chain for (5) is

$$y = \cos u$$
$$u = \pi x^2.$$

By (4),

$$\frac{dy}{du} = -\sin u.$$

By the chain rule,

$$\frac{dy}{dx} = \frac{dy}{du}\frac{du}{dx} = (-\sin u)(2\pi x) = -2\pi x \sin(\pi x^2).$$

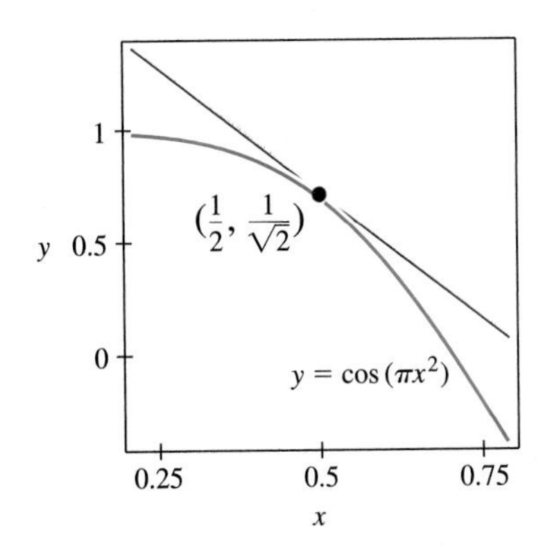

FIGURE 2.19 The line $y = \frac{1}{\sqrt{2}} - \frac{\pi}{\sqrt{2}}\left(x - \frac{1}{2}\right)$ is tangent to the graph of $y = \cos(\pi x^2)$ at the point $(1/2, 1/\sqrt{2})$.

The slope of the line tangent to the graph at $(1/2, 1/\sqrt{2})$ is the value of the derivative at this point:

$$\text{slope} = -2\pi(1/2)\sin(\pi(1/2)^2) = -\pi\sin\left(\frac{\pi}{4}\right) = -\frac{\pi}{\sqrt{2}}.$$

Because the tangent line has slope $-\pi/\sqrt{2}$ and contains the point $(1/2, 1/\sqrt{2})$, an equation for the line is

$$y - \frac{1}{\sqrt{2}} = -\frac{\pi}{\sqrt{2}}\left(x - \frac{1}{2}\right),$$

or

$$y = \frac{1}{\sqrt{2}} - \frac{\pi}{\sqrt{2}}\left(x - \frac{1}{2}\right). \tag{6}$$

The graph of $y = \cos(\pi x^2)$ and the graph of the tangent line are shown in Fig. 2.19. Because 0.55 is relatively close to $1/2 = 0.5$, we may approximate $\cos(\pi(0.55)^2)$ with the y value of the tangent line equation at $x = 0.55$. Thus

$$\cos(\pi(0.55)^2) \approx \frac{1}{\sqrt{2}} - \frac{\pi}{\sqrt{2}}\left(0.55 - \frac{1}{2}\right) = \frac{1}{\sqrt{2}} - \frac{\pi}{\sqrt{2}}(0.05) \approx 0.596.$$

How does this compare with the calculator value of $\cos(\pi(0.55)^2)$?

EXAMPLE 5 The equation

$$x = \tan y \tag{7}$$

implicitly defines y as a function of x. Find a formula for $\frac{dy}{dx}$.

Solution

Display (7) as a composition chain:

$$x = \tan y$$
$$y = y(x),$$

where we have written $y = y(x)$ to remind ourselves that y is a function of x. By the chain rule,

$$\frac{dx}{dx} = \frac{dx}{dy}\frac{dy}{dx}. \tag{8}$$

Because $\frac{dx}{dx} = 1$ and $\frac{dx}{dy} = \frac{d}{dy}\tan y = \sec^2 y$, equation (8) becomes

$$1 = \sec^2 y \frac{dy}{dx}.$$

Solving for $\frac{dy}{dx}$, we find

$$\frac{dy}{dx} = \frac{1}{\sec^2 y}. \tag{9}$$

There is nothing wrong with this answer, but we can put it in a form that might be easier to work with. Noting that

$$\sec^2 y = 1 + \tan^2 y = 1 + x^2,$$

we see that (9) can be written as

$$\frac{dy}{dx} = \frac{1}{1 + x^2}.$$

Derivatives of the Other Trigonometric Functions

We present here, for convenience, the derivatives of the other trigonometric functions. See Exercise 36.

Derivatives of Tangent, Cotangent, Secant, and Cosecant

$$\frac{d}{d\theta}\tan\theta = \sec^2\theta \qquad \frac{d}{d\theta}\cot\theta = -\csc^2\theta$$

$$\frac{d}{d\theta}\sec\theta = \sec\theta\tan\theta \qquad \frac{d}{d\theta}\csc\theta = -\csc\theta\cot\theta.$$

Why We Work in Radians

As mentioned earlier in this section, the calculus of trigonometric functions is easier to manage if angles are measured in radians. This is because of the relationship between central angle measure and arc length on the unit circle. A central angle of α radians subtends an arc of length α on the unit circle, while a central angle α *degrees* subtends an arc of length $\pi\alpha/180$. See Fig. 2.20. Now suppose that in the Investigation we had worked in degrees instead of radians. Then, with h measured in degrees, (2) would have been

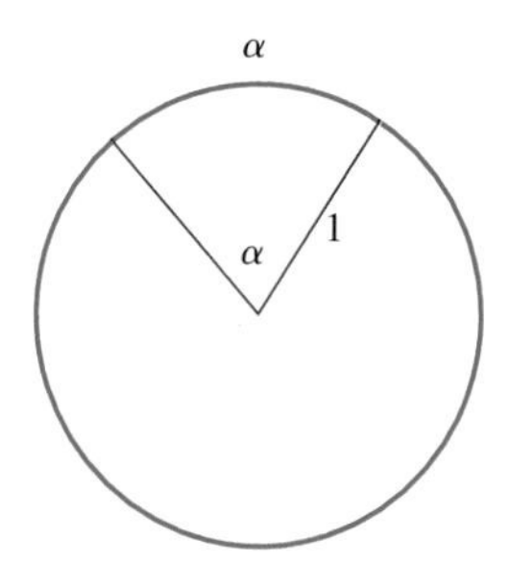

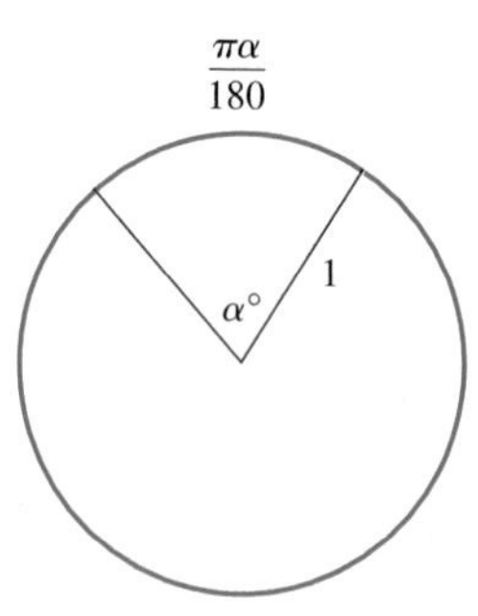

FIGURE 2.20 Arc length and the central angle

$$\text{length of segment } PQ \approx \text{length of arc } PQ = \frac{\pi h}{180}. \tag{10}$$

Then (3) would have been

$$\cos\theta \approx \frac{\Delta y}{\pi h/180}, \tag{11}$$

leading to

$$\Delta y \approx h\frac{\pi}{180}\cos\theta.$$

Substituting this result into (1) would have resulted in

$$\frac{d}{d\theta}\sin\theta = \lim_{h\to 0}\frac{\sin(\theta + h) - \sin\theta}{h} = \lim_{h\to 0}\frac{\Delta y}{h} = \lim_{h\to 0}\frac{h\frac{\pi}{180}\cos\theta}{h} = \frac{\pi}{180}\cos\theta.$$

This result illustrates a reason for working in radians rather than degrees. When we use radian measure, the derivative of the sine function is "cleaner." In degrees, the derivative carries an inconvenient constant multiplier of $\pi/180$.

Exercises 2.5

Exercises 1–20: Find the derivative.

1. $y = \sin(2x + 3)$
2. $r = \cos^2\theta$
3. $f(t) = \tan(3t^2 - 4t + 1)$
4. $q(t) = \sec(2t)$
5. $f(x) = -\dfrac{\cos x}{x}$
6. $w = \dfrac{1 - \sin 2u}{1 + \sin 2u}$
7. $T(\theta) = \tan\left(\dfrac{1 + \theta}{1 - \theta}\right)$
8. $h(t) = \cot 3t$
9. $f(x) = 2\sin(x^2 + 1)$
10. $y = -x^2 \csc(x^2)$
11. $g(t) = \dfrac{\cot(3t)}{1 + \csc^2(3t)}$
12. $s(v) = \csc 4v$
13. $z = \sin r \cos r$
14. $r = \tan\sqrt{x}$
15. $y = \sin(x \cos x)$
16. $F(w) = \left(2 - \sqrt{\sin w}\right)^5$
17. $f(x) = 1 + \tan^2(2x)$
18. $H(w) = \sqrt{\sec^2 w - 3\sec w + 4}$
19. $y = q(\cot t)$ where $q(t) = t \sin t$
20. $g(t) = p(\sin t)$, where $p(x) = 3x^4 - 2x^3 + x - 7$

Exercises 21–26: The given equation defines y as a function of x. Find the value of $\dfrac{dy}{dx}$ at point P.

21. $\sin y = x$ at $P = \left(\dfrac{\sqrt{2}}{2}, \dfrac{3\pi}{4}\right)$
22. $\cos y = x$ at $P = \left(-\dfrac{\sqrt{2}}{2}, \dfrac{3\pi}{4}\right)$
23. $xy\cos(x + y) = \pi^2$ at $P = (\pi, \pi)$
24. $x\sin(x + y) = y + \dfrac{\pi}{2}$ at $P = \left(\dfrac{\pi}{2}, 0\right)$
25. $\sin x + \cos y = 1$ at $P = \left(\dfrac{\pi}{6}, \dfrac{\pi}{3}\right)$
26. $\tan^2 x - \sec^2 y = -1$ at $P = \left(-\dfrac{\pi}{12}, -\dfrac{\pi}{12}\right)$
27. Find the equation of the line tangent to the graph of $y = \tan x$ at the point $(\pi/4, 1)$. Use the tangent line equation to find an approximate value for $\tan(\pi/4 + 0.03)$. How does the approximation compare with the calculator value for $\tan(\pi/4 + 0.03)$?
28. Find the equation of the line tangent to the graph of $y = \sec x$ at the point $(2\pi/3, -2)$. Use the tangent line equation to find an approximate value for $\sec(2\pi/3 - 0.002)$. How does the approximation compare with the calculator value for $\sec(2\pi/3 - 0.002)$?
29. Let $f(x) = \cos(\pi x)$. Find the equation of the line tangent to the graph of $y = f(x)$ at the point $\left(1/4, 1/\sqrt{2}\right)$. Use the tangent line equation to find an approximate value for $f(0.265)$. How does the approximation compare with the calculator value for $f(0.265)$?
30. Let $h(x) = x\sin(\pi x)$. Find the equation of the line tangent to the graph of $y = f(x)$ at the point $(3/2, -3/2)$. Use the tangent line equation to find an approximate value for $f(1.45)$. How does the approximation compare with the calculator value for $f(1.45)$?
31. Let $f(x) = \sin x$, and suppose that $f(a) = 0.6$. What are the possible values for $f'(a)$?
32. Let $g(x) = \sec x$, and suppose that $g(a) = 3$. What are the possible values for $g'(a)$?
33. Let $F(\theta) = \sin(2\theta)$, and suppose that $F(a) = -0.5$. What are the possible values for $F'(a)$?
34. Let $G(t) = 2\cot(2t)$, and suppose that $G(a) = 1$. What are the possible values for $G'(a)$?
35. Let $h(r) = 4\tan(2r + 1)$, and suppose that $h(a) = -2$. What are the possible values for $h'(a)$?
36. Use the derivative formulas given in (4) and the product, quotient, and/or chain rules, as needed, to derive the formulas for the derivative of the secant, cosecant, and cotangent functions.
37. Take a stick of length $\pi/2$ and fasten a pencil to each end. Holding the stick parallel to the x-axis, use the point of the left pencil to draw the graph of $y = \sin x$. What graph is drawn by the pencil on the right?

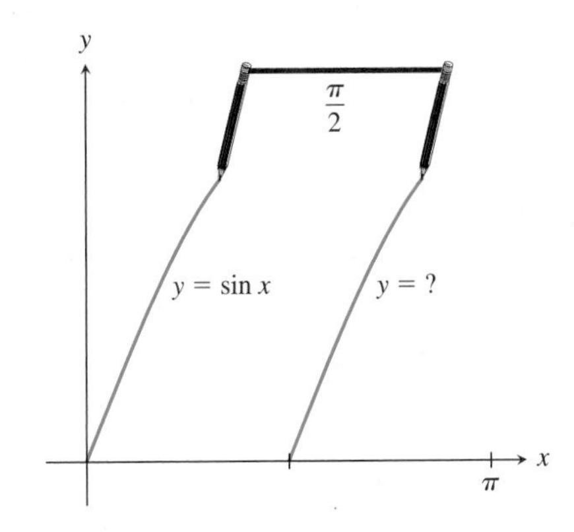

Figure for Exercise 37.

38. If the stick in the previous exercise is held at an angle of $\pi/4$ with respect to the x-axis, and the left pencil again draws the graph of $y = \sin x$, what graph does the right pencil draw?

39. Show that

$$\frac{d}{dx}\sin^2 x = \sin(2x)$$

and that

$$\frac{d}{dx}\cos^2 x = -\sin(2x).$$

40. Referring to Fig. 2.17, suppose that $\theta = \pi/4$. Calculate the differences between the length of PQ and h for $h = 0.01$, $h = 0.001$, and $h = 0.0001$.

41. What is the angle between the lines tangent to the sine and cosine graphs where they cross at $\left(\pi/4, \sqrt{2}/2\right)$?

42. The graphs of $y = \tan x$ and $y = \cos x$ cross at a point (a, b) with $0 < a < \pi/2$.

a. Show that

$$\sin a = \frac{-1 + \sqrt{5}}{2}.$$

b. What is the angle between the lines tangent to the graphs of $y = \tan x$ and $y = \cos x$ at this crossing point?

43. In Section 1.3, we showed that

$$\lim_{h \to 0} \frac{\sin h}{h} = 1$$

and

$$\lim_{h \to 0} \frac{\cos h - 1}{h} = 0.$$

The following is the beginning of an argument that $\frac{d}{d\theta}\sin\theta = \cos\theta$. Fill in the justification for each step and complete the argument.

$$\begin{aligned}\frac{d}{d\theta}\sin\theta &= \lim_{h \to 0} \frac{\sin(\theta + h) - \sin\theta}{h} \\ &= \lim_{h \to 0} \frac{(\sin\theta\cos h + \cos\theta\sin h) - \sin\theta}{h} \\ &= \lim_{h \to 0}\left(\sin\theta\frac{\cos h - 1}{h}\right) \\ &\quad + \lim_{h \to 0}\left(\cos\theta\frac{\sin h}{h}\right).\end{aligned}$$

44. The accompanying figure shows a metal rod of length 2 attached to the rim of a wheel of radius 1. The rod is attached to the wheel by a pin that allows the rod to pivot freely about the point of attachment. As the wheel rotates, the free end of the rod moves back and forth in a horizontal channel. The wheel turns counterclockwise at a constant rate of two revolutions per second. Assume at time $t = 0$ the free end of the rod was at position $x = 3$.

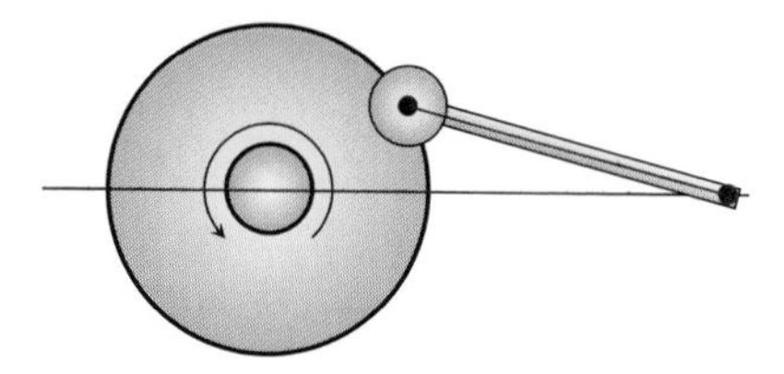

Figure for Exercise 44.

a. On a set of axes with the horizontal axis representing time and the vertical axis representing position on the x-axis, sketch a graph showing the position of the free end of the rod at any time t.

b. Find an equation that gives the position of the free end of the rod at any time t.

c. Find the derivative with respect to t of the function found in part **b**. Interpret the derivative as a rate of change. What information does the derivative give us?

45. A cork floating in the ocean is lifted up and down by the waves. The height difference between the low point of the cork and the high point is 10 m, and the cork rises from its low point to its high point and then falls back to its low point six times per minute.

a. On a set of axes with the horizontal axis representing time and the vertical axis representing height, sketch a graph showing the height of the cork at time t. What does height 0 mean in your graph? What is the significance of time $t = 0$ on your graph?

b. Write an equation (using a trigonometric function) to describe the height of the cork at any time t given in minutes. Include a discussion of all assumptions and decisions you made in coming up with your equation.

c. Using your answer to **b**, find the rate of change of height with respect to time. What does this rate of change tell you?

46. A piston moves back and forth in a cylindrical chamber that is filled with gas and has a radius of 4 centimeters. See the accompanying figure. At time t seconds, the end of the piston is $5 + 2\sin 2t$ centimeters from the closed end of the chamber.

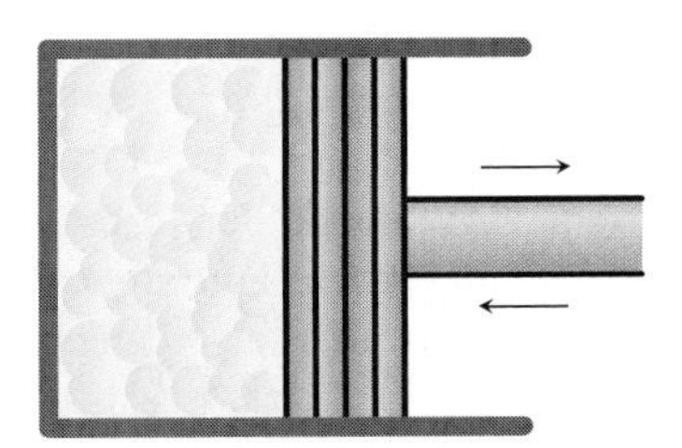

Figure for Exercise 46.

a. At time t, what is the volume of the gas in the chamber?

b. What is the rate of change of the volume of the gas at time $t = 3\pi/8$? Include units with your answer.

c. Assume that the temperature of the gas is held constant. Then the product of the volume of the gas and the pressure of the gas is a constant, say k. Assume that the pressure is measured in atmospheric pressures (atm). Find the rate of change of the pressure of the gas at time $t = 3\pi/8$. Include units with your answer.

47. Let $A = (\cos a, \sin a)$ and $B = (\cos b, \sin b)$ be distinct points on the unit circle, with neither one equal to the point $P = (1, 0)$ and so that the segment AB is not vertical. See the accompanying figure.

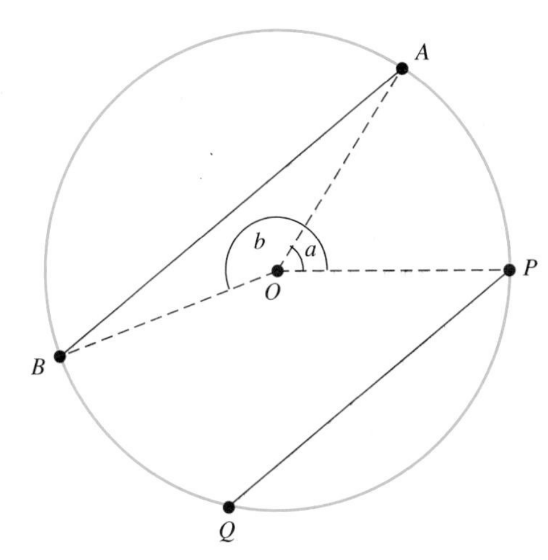

Figure for Exercise 47.

a. Find the slope of segment AB.

b. Write down an equation for the line through P and parallel to $\overline{AB}$.

c. Find the other point, Q, where this line intersects the unit circle.

d. Show by geometry that

$$Q = (\cos(a + b), \sin(a + b)).$$

e. Combine **c** and **d** to obtain angle addition formulas for the sine and cosine.

48. This exercise outlines a geometric argument that the derivative of $\tan\theta$ is $\sec^2\theta$ (at least for first-quadrant angles θ). In the accompanying diagram, $\angle DOA = \theta$, $\angle DOB = \theta + h$, and $\overline{AC}$ is perpendicular to $\overline{OA}$.

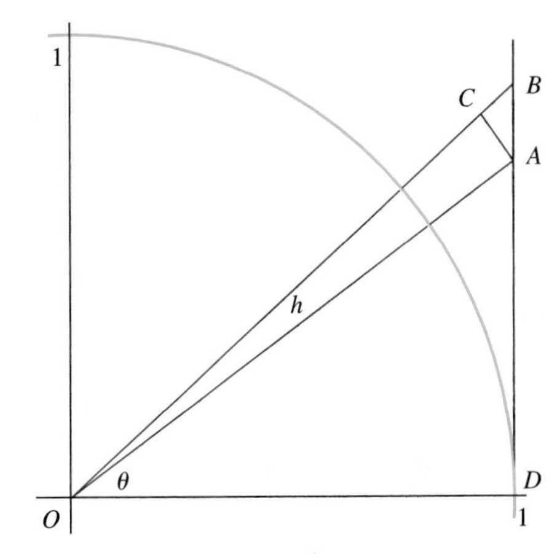

Figure for Exercises 48 and 49.

a. Show that $\overline{AB}$ has length

$$\tan(\theta + h) - \tan\theta$$

and that $\overline{OA}$ has length $\sec\theta$.

b. Show that $\angle BAC = \theta$.

c. Justify the statement $AC \approx h\sec\theta$.

d. Combine **a**, **b**, and **c** to conclude that

$$\frac{\tan(\theta + h) - \tan\theta}{h} \approx \sec^2\theta.$$

49. In the figure used in the preceding exercise, show that

$$BC = \sec(\theta + h) - \sec\theta.$$

Use this to show that

$$\frac{\sec(\theta + h) - \sec\theta}{h} \approx \tan\theta\sec\theta.$$

50. At the beginning of this section, we remarked that the speed of the shadow of the car of a Ferris wheel illustrates the rate of change of the cosine function. Check that the statements made about the shadow's motion are consistent with the fact that

$$\frac{d}{d\theta}\cos\theta = -\sin\theta.$$

2.6 Exponential Functions

The world's population doubles every 25 years. A star of the first magnitude is 100 times brighter than a star of the sixth magnitude. For safety, nuclear waste must be stored safely for 25,000 years. A dollar invested at 4 percent interest per year compounded daily will provide your descendants 1000 years later with over ten thousand trillion dollars. We have all read items like these in the news or in books of "amazing facts." Diverse as they may seem, these facts all have one thing in common. Each is described by an exponential function. Exponential functions are also used in describing more subtle aspects of our daily lives, from the warming of a can of soda to the swaying of tall skyscrapers in the wind.

INVESTIGATION

Bacteria and Uranium

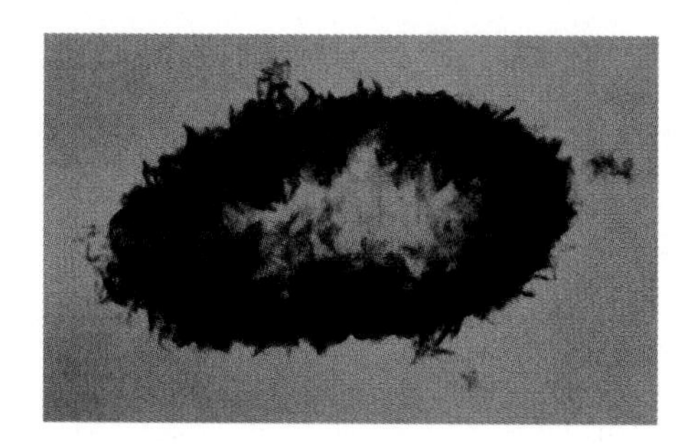

In the March 1992 issue of *Applied and Environmental Microbiology*, Derek Lovely and Elizabeth Phillips of the U.S. Geological Survey presented data on the role of bacterium *Desulfovibrio desulfuricans* in the reduction of uranium in solution. Uranium was added to a solution containing the bacteria. The amount of uranium (in micromoles, μM) in the solution was measured every hour for four hours. The purpose of the research was to study how bacteria contribute to the reduction of uranium in sedimentary environments. (One possible benefit of such research is a way to use microorganisms to help recover uranium from contaminated water.) Data from the experiment are summarized in Table 2.1 and Fig. 2.21. In this Investigation we approximate the rate of change of the amount of uranium with respect to time and discuss the significance of the result.

The graph in Fig. 2.21 shows that the amount of uranium decreases with time. The graph decreases rapidly near $t = 0$ and decreases slowly near $t = 4$. This corresponds to a rapid decrease in the amount of uranium at the beginning of the experiment and a more gradual decrease later. To better understand how the amount of uranium changes with time, we investigate the rate of change. Let $U(t)$ be the amount of uranium at time t. We use difference quotients to approximate the derivative of U at each hour:

TABLE 2.1

t hours	$U(t)$ μM
0	0.8
1	0.4
2	0.21
3	0.11
4	0.06

$$U'(0) \approx \frac{U(1) - U(0)}{1 - 0} = \frac{0.4 - 0.8}{1} = -0.40\ \mu\text{M/h}$$

$$U'(1) \approx \frac{U(2) - U(1)}{2 - 1} = \frac{0.21 - 0.4}{1} = -0.19\ \mu\text{M/h}$$

$$U'(2) \approx \frac{U(3) - U(2)}{3 - 2} = \frac{0.11 - 0.21}{1} = -0.10\ \mu\text{M/h}$$

$$U'(3) \approx \frac{U(4) - U(3)}{4 - 3} = \frac{0.06 - 0.11}{1} = -0.05\ \mu\text{M/h}.$$

These results are summarized in Table 2.2. In this table we have added a third column, showing the value of $U'(t)/U(t)$ for $t = 0$, 1, 2, and 3. The values in this third column do not change much and may suggest that

$$\frac{U'(t)}{U(t)} \approx -0.48 \quad \text{or} \quad U'(t) \approx -0.48U(t). \tag{1}$$

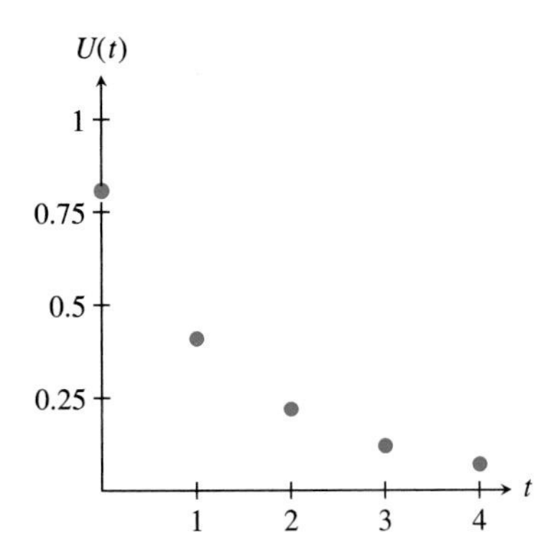

FIGURE 2.21 Uranium in solution.

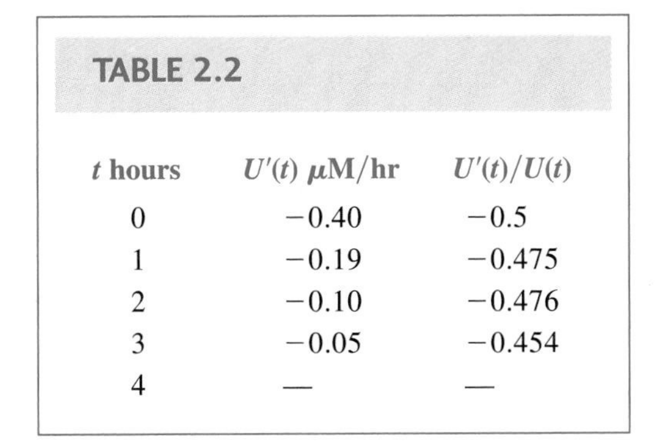

TABLE 2.2

t hours	$U'(t)$ μM/hr	$U'(t)/U(t)$
0	−0.40	−0.5
1	−0.19	−0.475
2	−0.10	−0.476
3	−0.05	−0.454
4	—	—

We interpret (1) as saying

> The rate of change in the amount of uranium is proportional to the amount of uranium.

This sounds reasonable. When there is a lot of uranium in solution, there is a lot to be removed, so the decrease is large. When there is not very much uranium, only a little bit is removed, so the decrease is small.

We can use (1) to say even more about $U(t)$. From

$$\frac{U(t+1) - U(t)}{(t+1) - t} \approx U'(t) \approx -0.48U(t),$$

solve for $U(t+1)$ to get

$$U(t+1) \approx U(t) - 0.48U(t) = 0.52U(t). \tag{2}$$

Put $t = 0$ into (2) and we have

$$U(1) \approx 0.52U(0) = 0.8(0.52). \tag{3}$$

Next set $t = 1$ in (2) and use (3) to get

$$U(2) \approx 0.52U(1) \approx 0.8(0.52)^2. \tag{4}$$

Continue this process by taking $t = 2$, and then $t = 3$ in (2). We then find

$$U(3) \approx 0.52U(2) \approx 0.8(0.52)^3 \tag{5}$$

and

$$U(4) \approx 0.52U(3) = 0.8(0.52)^4. \tag{6}$$

From (3), (4), (5), and (6) it appears reasonable to guess that

$$U(t) \approx 0.8(0.52)^t. \tag{7}$$

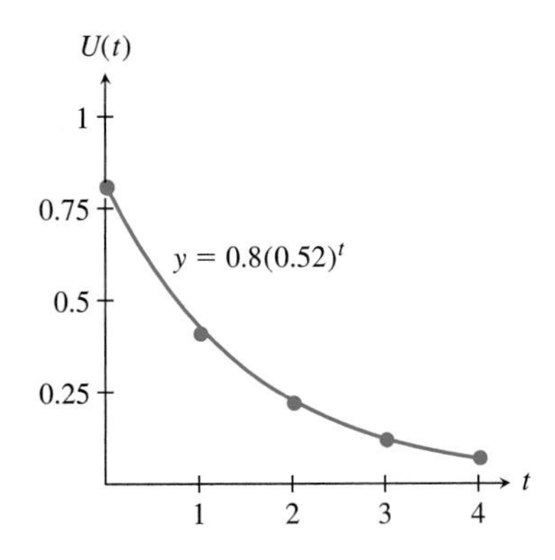

FIGURE 2.22 The graph of $U(t) = 0.8(0.52)^t$ fits the uranium data.

How good is this guess? In Fig. 2.22 we show the graph of (7) and the data points from Fig. 2.21. Pretty good agreement!

The function described by the right side of (7) is an example of an exponential function.

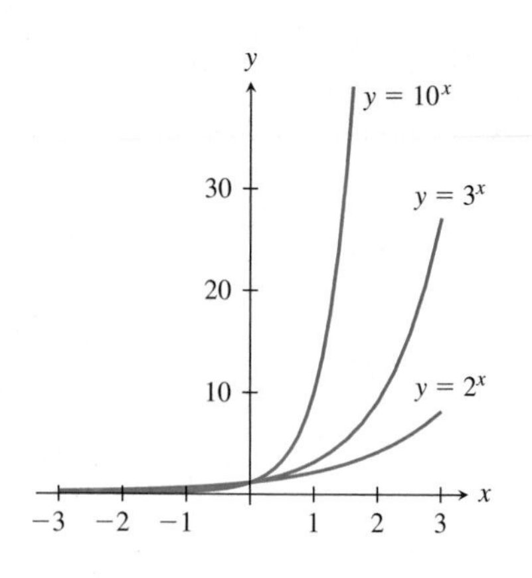

FIGURE 2.23 Graphs of $y = b^x$ for some $b > 1$.

Exponential Function

Let b be a positive real number with $b \neq 1$. The exponential function of base b is the function f defined by

$$f(x) = b^x. \tag{8}$$

The domain of the exponential function is the set of real numbers. (However, see Exercises 36 and 37.)

Use a calculator or computer to graph $y = b^x$ for various values of the base b. The graphs illustrate some of the important features of exponential functions. First graph $y = b^x$ for several values of $b > 1$. See Fig. 2.23. All of the graphs pass through the point $(0, 1)$ and all of the graphs rise as we move from left to right. The graphs appear to rise more rapidly for larger values of b. As we look far to the left, where x is large and negative, we see that the graph is very close to the x-axis. This suggests that the graph of $y = b^x$ is asymptotic to the negative x-axis.

Now graph $y = b^x$ for several values of b with $0 < b < 1$. See Fig. 2.24. These graphs also pass through the point $(0, 1)$. The graphs fall as we move from left to right, and appear to fall more rapidly for smaller values of b. The graphs also suggest that for $0 < b < 1$, the graph of $y = b^x$ is asymptotic to the positive x-axis.

The graphs suggest another very important feature of the exponential functions.

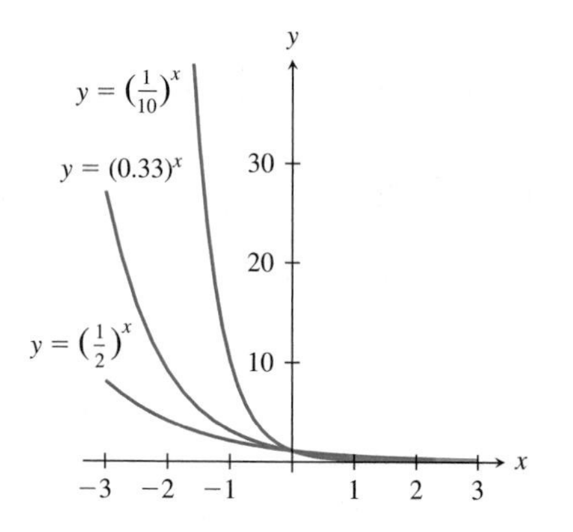

FIGURE 2.24 Graphs of $y = b^x$ for some $b < 1$.

For all real x, $b^x > 0$. An exponential function is never 0.

Exponential functions b^x are different from the power functions x^n in many ways. Although both 2^x and x^2 use exponential notation, these two functions are very different. For example, the graph of the exponential function $f(x) = 2^x$ is always rising as we move from left to right, whereas the graph of the power function $g(x) = x^2$ is falling for $x < 0$ but rising for $x > 0$. See Fig. 2.25. The exponential function has an asymptote, but the power function does not. The exponential function is never 0, but the power function does take on the value 0. The differences in the two functions are further illustrated by their derivatives.

WEB

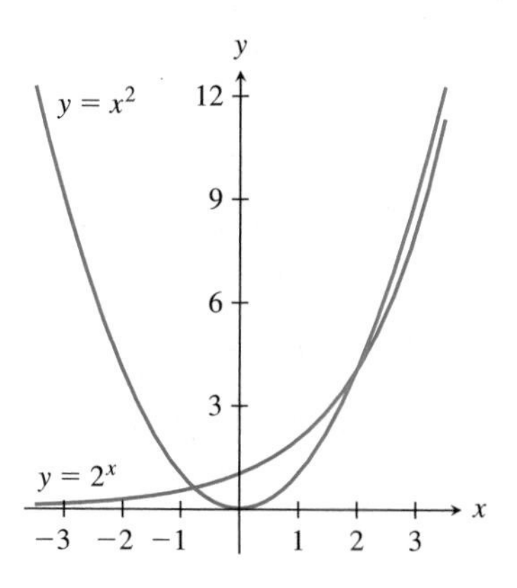

FIGURE 2.25 Graphs of $y = 2^x$ and $y = x^2$.

INVESTIGATION

The Derivative of an Exponential Function

Calculate $\frac{d}{dx}b^x$. By the definition of derivative, we have

$$\frac{d}{dx}b^x = \lim_{h\to 0}\frac{b^{x+h} - b^x}{h}. \tag{9}$$

From rules for working with exponents, $b^{x+h} = b^x b^h$. Substituting this into (9), then factoring b^x out of the numerator, gives

$$\frac{d}{dx}b^x = \lim_{h\to 0}\frac{b^x b^h - b^x}{h} = \lim_{h\to 0} b^x \frac{b^h - 1}{h} = b^x \lim_{h\to 0}\frac{b^h - 1}{h}. \tag{10}$$

In the last step we were able to factor b^x from the limit because b^x is independent of h. Assuming that this limit exists, let

$$C_b = \lim_{h\to 0}\frac{b^h - 1}{h}. \tag{11}$$

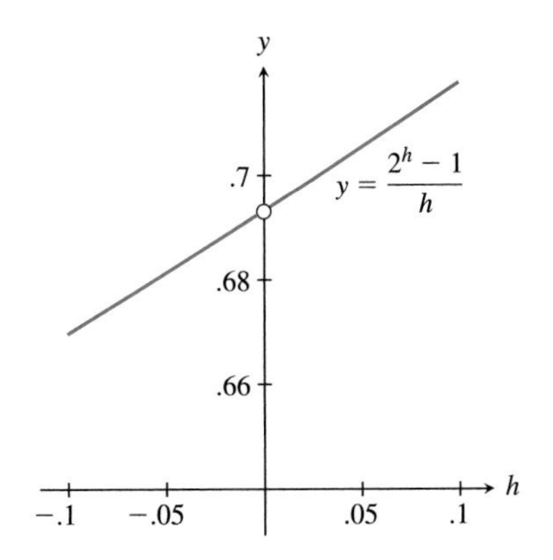

FIGURE 2.26 Graph of $y = (2^h - 1)/h$.

Note that the value of C_b does not depend on x, only on the value of b. Combining (10) and (11), we see that the derivative of b^x is

$$\frac{d}{dx}b^x = C_b b^x.$$

Although the value of C_b does not depend on x, it does depend on the base b of the exponential function. For example, when $b = 2$, we have

$$\frac{d}{dx}2^x = C_2 2^x$$

where

$$C_2 = \lim_{h\to 0}\frac{2^h - 1}{h}. \tag{12}$$

TABLE 2.3

h	$\frac{(2^h - 1)}{h}$
−0.1	0.669670
0.1	0.717735
−0.01	0.690750
0.01	0.695555
−0.001	0.692907
0.001	0.693387
−0.0001	0.693123
0.0001	0.693171

To get an idea of the value of the limit in (12), we can graph $y = (2^h - 1)/h$ for values of h close to 0 (see Fig. 2.26) or calculate the value of this quotient for values of h close to 0 (see Table 2.3). Reading from the table or the graph, we see that $C_2 \approx 0.693$. We could get more accuracy by zooming closer to the graph in Fig. 2.26 or by extending Table 2.3 with smaller values of h. In either case, we find that

$$\frac{d}{dx}2^x \approx (0.693)2^x.$$

Thus, the derivative of the exponential function 2^x is a constant times the function itself, and the constant is approximately 0.693.

We can repeat this procedure for any positive b and find an estimate for C_b. In Exercise 38 you are asked to verify some of the C_b values given in Table 2.4. Using the values in this table, we see that

$$\frac{d}{dx}3^x = C_3 3^x \approx (1.0986)3^x \quad\text{and}\quad \frac{d}{dx}10^x = C_{10}10^x \approx (2.3026)10^x.$$

We summarize the results of the previous discussion in the following statement.

The Derivative of an Exponential Function

Let b be a positive real number with $b \neq 1$. Then

$$\frac{d}{dx}b^x = C_b b^x \tag{13}$$

where the constant C_b is given by

$$C_b = \lim_{h \to 0} \frac{b^h - 1}{h}.$$

Equation (13) reveals one of the most important facts about the exponential function:

The rate of change of b^x with respect to x is proportional to b^x.

This property of the exponential function was illustrated by the function U in the Investigation on page 130. Many familiar quantities change at a rate proportional to the size of the quantity itself. Some examples follow:

Population: The larger a population is, the faster it grows, because more organisms mean more births.

Invested money: Money earns interest at a rate proportional to its amount. A large sum of money earns more interest than a small sum.

Radioactive decay: Radium-227 has a half-life of 6.7 years. This means that a 100-gram sample will decrease to 50 grams in 6.7 years, while a 10-gram sample will decrease to 5 grams over the same time period.

Cooling: A very hot copper wire will lose more heat in 10 minutes than a copper wire just slightly above room temperature.

Exponential functions are used in describing all of these phenomena, and many more.

TABLE 2.4
Some values for C_b.

b	C_b
2.0	0.693147
2.2	0.788457
2.4	0.875469
2.6	0.955511
2.7	0.993252
2.8	1.02962
3.0	1.09861
4.0	1.38629
7.0	1.94591
10.0	2.30259

The Number *e*

The values $C_2 \approx 0.693147$, $C_{2.7} \approx 0.993252$, $C_{2.8} \approx 1.02962$, and $C_3 \approx 1.09861$ in Table 2.4 suggest that C_b increases in value as b increases. Because $C_{2.7}$ is a little less than 1 and $C_{2.8}$ is a little larger than 1, it seems reasonable that there is a value of b between 2.7 and 2.8 with $C_b = 1$. This special value of b is one of the most important numbers in mathematics. We use e to denote this special number. (The letter e honors the great mathematician Leonhard Euler.)

The Number *e*

There is a number e such that $C_e = 1$. Hence

$$\frac{d}{dx}e^x = C_e e^x = e^x.$$

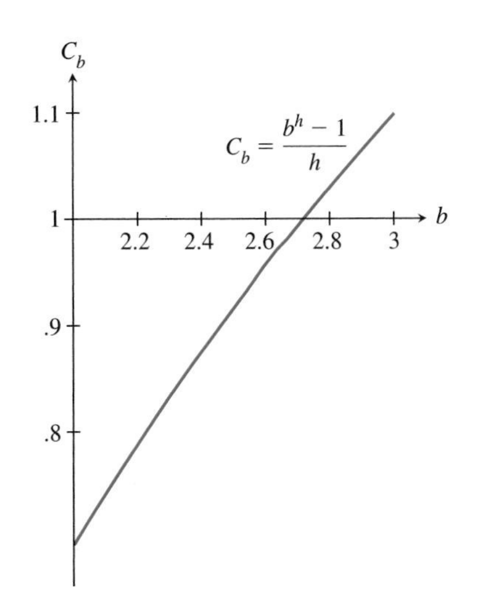

FIGURE 2.27 Graph of $C_b = (b^h - 1)/h$ for $2 \le b \le 3$ and $h = 0.00001$.

Approximating e Because the number e is the special value of b for which

$$C_b = \lim_{h \to 0} \frac{b^h - 1}{h} = 1, \tag{14}$$

we can approximate e by trying different values of b in (14), then numerically or graphically checking the limit to see if it is equal to 1. (See Exercise 22 in Section 1.5 and Exercise 53 at the end of this section.)

Here we will take an alternative approach. Set h equal to a small value (say $h = 0.00001$), then graph

$$C_b = \frac{b^{0.00001} - 1}{0.00001} \tag{15}$$

for b values between 2 and 3. In effect, this allows us to estimate the limit in (14) for all b between 2 and 3. The graph is shown in Fig. 2.27. We are interested in the b value for which (15) is equal to 1. The graph shows that this happens for some value of b between 2.7 and 2.8, so $2.7 < e < 2.8$.

Now take h a little smaller, say $h = 0.000001$, to better approximate the value of the limit in (14). The graph of

$$C_b = \frac{b^{0.000001} - 1}{0.000001}$$

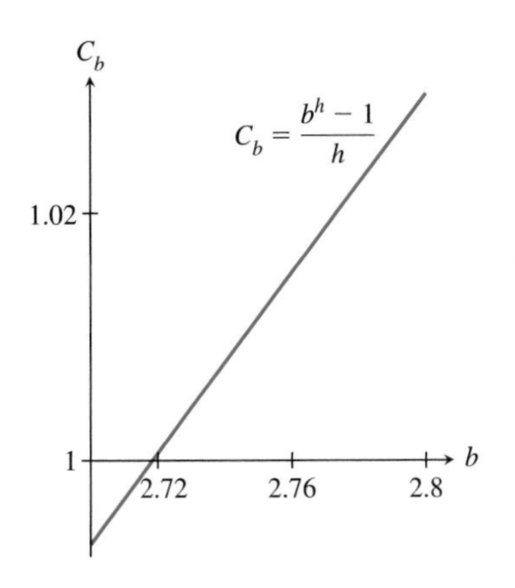

FIGURE 2.28 Graph of $C_b = (b^h - 1)/h$ for $2.7 \le b \le 2.8$ and $h = 0.000001$.

for $2.7 \le b \le 2.8$ is shown in Fig. 2.28. From this graph we see that $C_b = 1$ when b is slightly smaller than 2.72. Hence $e \approx 2.72$ and $2.71 < e < 2.72$.

By continuing this zooming-in process or trying different values of b in (14), we can get better and better approximations to e. With enough patience and computing power, we find that

$$e \approx 2.718281828459045\ldots.$$

Like π, the number e is irrational; that is, neither π nor e can be written as a ratio of integers. This means that the decimal representations for π and e are nonterminating and nonrepeating. Like π, the number e is also transcendental. This means that e is not a zero of any polynomial with integer coefficients. Just as mathematicians and scientists prefer to use radians when working with angles and trigonometric functions, they prefer to use the exponential function with base $b = e$. Calculus with trigonometric functions is "simpler" when done in radians. The calculus of exponential functions is "simplest" when the base is e. Every scientific calculator has a button for evaluating this preferred exponential function. This button is usually labeled e^x or "exp."

The Exponential Function e^x

The exponential function with base e is usually referred to as *the* exponential function.

The Exponential Function

The exponential function, sometimes denoted by "exp," is the function defined for all real x by

$$\exp(x) = e^x.$$

The derivative of the exponential function is

$$\frac{d}{dx}(\exp(x)) = \frac{d}{dx}e^x = e^x. \tag{16}$$

The graph of the exponential function is shown in Fig. 2.29. We end this section with a few examples involving exponential functions.

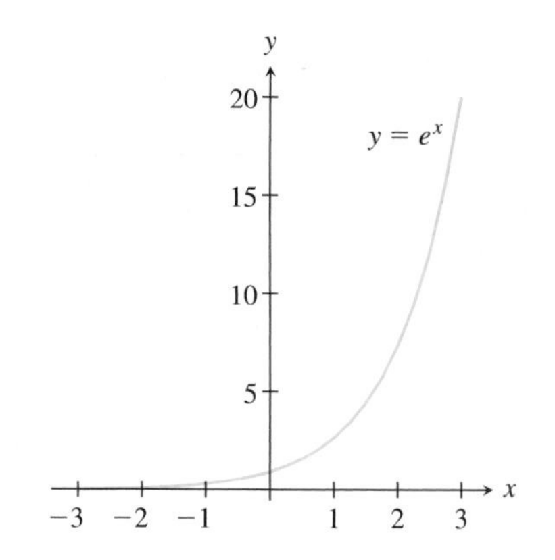

FIGURE 2.29 Graph of $y = e^x$.

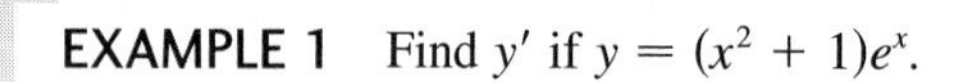

EXAMPLE 1 Find y' if $y = (x^2 + 1)e^x$.

Solution

The function in question is a product of two functions whose derivatives we know. By the product rule,

$$\begin{aligned} y' = ((x^2 + 1)e^x)' &= (x^2 + 1)'e^x + (x^2 + 1)(e^x)' \\ &= 2xe^x + (x^2 + 1)e^x = (x^2 + 2x + 1)e^x. \end{aligned}$$

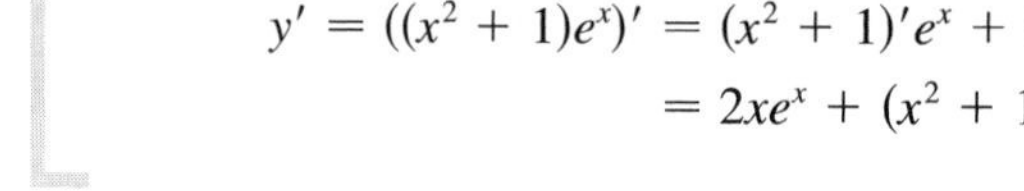

EXAMPLE 2 Let $g(t) = e^{kt}$, where k is a constant. Find $g'(t)$.

Solution

Let $y = g(t) = e^{kt}$. A composition chain for this expression is

$$\begin{aligned} y &= e^u \\ u &= kt. \end{aligned}$$

Then by (16), $\frac{dy}{du} = e^u$, and $\frac{du}{dt} = k$. Hence, by the chain rule,

$$g'(t) = \frac{dy}{dt} = \frac{dy}{du}\frac{du}{dt} = e^u k = ke^{kt}.$$

EXAMPLE 3 Find $h'(x)$ if $h(x) = e^{2x \sin 2x}$.

Solution

The function h is a composition. Let $y = h(x)$ and write a composition chain:

$$\begin{aligned} y &= e^u \\ u &= w \sin w \\ w &= 2x. \end{aligned}$$

We then have $\dfrac{dy}{du} = \dfrac{d}{du}e^u = e^u$ and $\dfrac{dw}{dx} = \dfrac{d}{dx}(2x) = 2$. By the product rule,

$$\frac{du}{dw} = \frac{d}{dw}(w \sin w) = \left(\frac{d}{dw}w\right)\sin w + w\left(\frac{d}{dw}\sin w\right) = \sin w + w\cos w.$$

Now apply the chain rule to get

$$\begin{aligned} h'(x) &= \frac{dy}{dx} = \frac{dy}{du}\frac{du}{dw}\frac{dw}{dx} = e^u(\sin w + w\cos w)\,(2) \\ &= 2(\sin w + w\cos w)e^{w\sin w} = (2\sin 2x + 4x\cos 2x)e^{2x\sin 2x}. \end{aligned}$$

EXAMPLE 4 Bacterial growth

In the early stages of growth, the population P of a colony of bacteria can be described by an exponential function. Assuming time t is in hours, estimate how long it takes a population to double if at time t the population is:

a) $P(t) = 2^t$.
b) $P(t) = e^t$.

Solution

a) At time t_0 the population is $P(t_0) = 2^{t_0}$. Note that one hour later the population is

$$P(t_0 + 1) = 2^{t_0+1} = 2 \cdot 2^{t_0} = 2P(t_0).$$

Hence the population doubles in one hour. Notice that the doubling time is always one hour, regardless of the starting time t_0.

b) Now let $P(t) = e^t$ and suppose that at time t_1 the population is double what it was at time t_0. Then the doubling time is $T = t_1 - t_0$ and

$$P(t_1) = 2P(t_0). \tag{17}$$

This means that

$$e^{t_1} = 2e^{t_0}.$$

Divide both sides of the last equation by e^{t_0} to get

$$2 = \frac{e^{t_1}}{e^{t_0}} = e^{t_1 - t_0} = e^T. \tag{18}$$

The doubling time is T hours, where T is the number for which $e^T = 2$. There are several ways to estimate T. We do this graphically by zooming in on the graph of $y = e^x$ at the point where the graph crosses the line $y = 2$. See Fig. 2.30. From the graph we see that $T \approx 0.693$. Hence the time for the population to double is approximately 0.693 hours, or about 41.6 minutes. Note again that the doubling time is independent of the starting time t_0 of the doubling period.

We could also estimate T by using a calculator to try different possibilities for T in (18) and refining our estimate with each try. Or, you may know that the solution of the equation is $2 = e^T$

$$T = \log_e 2 = \ln 2 \approx 0.693147.$$

We've seen this number earlier in this section. Where?

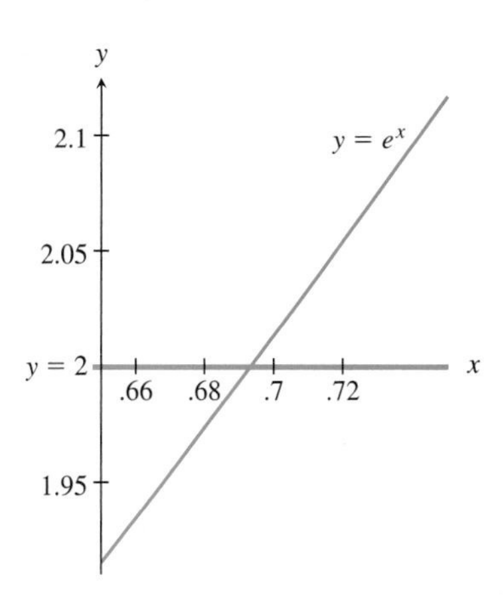

FIGURE 2.30 Graph of $y = e^x$ and $y = 2$ near the intersection point of the two graphs.

EXAMPLE 5 The equation

$$x = e^y \tag{19}$$

defines y as a function of x. Find an expression for $\frac{dy}{dx}$.

Solution

We write $y(x)$ for y to help us remember that y is a function of x. Then (19) becomes

$$x = e^{y(x)}.$$

The composition chain for this expression is

$$x = e^u$$
$$u = y(x).$$

By the chain rule,

$$\frac{dx}{dx} = \frac{dx}{du}\frac{du}{dx}. \tag{20}$$

Because $\frac{dx}{dx} = 1$, $\frac{dx}{du} = e^u$, and $\frac{du}{dx} = y'$, (20) becomes

$$1 = e^u\frac{dy}{dx} = e^y y'.$$

WEB

Solve for y' to get

$$y' = \frac{1}{e^y}.$$

There is nothing wrong with this answer, but from (19) we see that $e^y = x$. Hence we can also write

$$\frac{dy}{dx} = \frac{1}{x}.$$

Exercises 2.6

Exercises 1–7: Find all solutions to the equation. In some cases decimal approximations are appropriate.

1. $2^{-3x} = 16$

2. $5^{2t} - \sqrt{5}\cdot 5^t = 0$

3. $e^\theta = e$

4. $e^r = \frac{1}{2}$

5. $e^{2x} - 3e^x + 2 = 0$

6. $\dfrac{e^{3t} - 1}{e^{3t} + 1} = 4$

7. $\dfrac{e^u + e^{-u}}{2} = 5$

Exercises 8–11: Plot y as a function of x for $-1 \le x \le 1$. Using Table 2.4 as necessary, find $\frac{dy}{dx}$. Find the equation of the line tangent to the curve at the point corresponding to $x = 0.5$.

8. $y = 2^x$

9. $y = 3^x$

10. $y = 4^x$

11. $y = 7^x$

Exercises 12–31: Find the derivative. Use Table 2.4 as necessary.

12. $y = e^{2x}$

13. $y = e^{x/7}$

14. $r = e^{-3t}$

15. $u = e^{-t/2}$

16. $f(x) = \sin(e^x)$

17. $f(x) = \cos(2e^x)$

18. $g(t) = e^{t^2}$

19. $h(t) = e^{-t^2}$

20. $z = e^{e^t}$

21. $w = \sqrt{(e^x)^2 - e^x + 3}$

22. $h(r) = r^4 e^r$

23. $q(s) = 2^s e^{4s}$

24. $s(t) = e^{2t}\cos(5t)$

25. $R(t) = e^{-t}\sin(2t)$

26. $v = \dfrac{e^u + 1}{e^u - 1}$

27. $v = \dfrac{1}{e^u - e^{2u}}$

28. $y = \tan(e^x + 1)$

29. $y = \sec(e^x + 1)$

30. $c(t) = \dfrac{e^t + e^{-t}}{2}$

31. $s(t) = \dfrac{e^t - e^{-t}}{2}$

Exercises 32–35: Find an expression for y′ if the equation defines y as a function of x.

32. $(x + y)^2 = ye^x$

33. $ye^{x+y} = 1$

34. $e^x \tan(xy^2) - 2y = x + y$

35. $e^{y^2} = xy$

36. Suppose you did not have a calculator but needed to find $11^{1/5}$ accurate to within 0.001. Write a short paragraph describing how you would do this "by hand." Write a similar paragraph describing how you would calculate $11^{12/5}$ to an accuracy of 0.001.

37. In defining the exponential function to base b we offhandedly describe it as the function f defined by $f(x) = b^x$. Suppose that $b = 11$ and $x = \sqrt{5}$. Write a short paragraph describing how you would calculate $11^{\sqrt{5}}$ to within 0.001.

T **38.** Verify any three of the C_b values in Table 2.4.

T **39.** Suppose that in investigating the derivative of the exponential function, we decide that we want a base b so that

$$(b^x)' = 2 \cdot b^x.$$

What is the value of b? Find the answer by numerical experimentation, and then by considering the equation $(e^{bx})' = b \cdot e^{bx}$.

T **40.** When the sources of pollution are eliminated, a large polluted body of water eventually "cleanses" itself through natural processes. Observations have shown that during this cleansing period, the concentration of pollutants can be described by an exponential function of the form.

$$C(t) = C_0 e^{kt}.$$

In this expression, $C(t)$ has units of parts per million (ppm) and is the concentration of pollutants at time t, C_0 is the concentration at time $t = 0$, and k is a constant. An environmentalist measures the pollutants in the lake at time $t = 0$ and finds a concentration of 150 ppm. One year later, the concentration is 100 ppm.

a. Find C_0 and k.

b. What will the concentration of pollutants be five years after the initial measurement?

c. When will the concentration of pollutants be less than 10 ppm?

T **41.** When money in a savings account earns interest at a rate of r percent per year, compounded daily, then the balance of the account T years after the money is deposited is well approximated by

$$A(t) = A_0 e^{0.01r \cdot T}$$

where A_0 is the amount of the initial deposit. In the introduction to this section we stated that "a dollar invested at 4 percent interest per year compounded daily will provide your descendants 1000 years later with over ten thousand trillion dollars." Verify this statement.

42. The bacterium *Escherichia coli*, usually called *E. coli*, is found in the human digestive tract. Under ideal conditions each *E. coli* cell divides into two cells $1/3$ hour after its own "birth." Given that the mass of one *E. coli* cell is approximately $5 \cdot 10^{-13}$ grams and assuming no deaths and ideal conditions, what is the mass of an *E. coli* colony after t hours, assuming that it started with one cell? What is the rate of change of the mass of the *E. coli* colony at time $t = 10$ hours?

43. The radioactive substance Carbon-14 (C^{14}) has a half-life of about 5730 years. This means that if we measure the amount of C^{14} in a sample, then measure it again 5730 years later, we will find half as much C^{14} as we started with. A test tube contains 20 g of C^{14}. The tube is sealed and set aside for several thousand years, then the amount of C^{14} is measured again. How much C^{14} remains after 5730 years? After 11,460 years? After 17,190 years? After t years? What is the rate of change of the C^{14} at time $t = 5730$?

44. A metal plate is heated to a temperature of 425°C then allowed to cool in a room where the temperature is kept at 25°C. The plate cools so that the difference between the plate temperature and the room temperature decreases by half of the difference every $1/4$ hour. What is the temperature of the plate after $1/4$ hour? After 1 hour? After t hours? What is the rate of change of the temperature at time $t = 1$ hour?

45. Two thousand years ago Greek astronomers Hipparchus and Ptolemy created a system to quantify the apparent brightness of stars. The 20 brightest visible stars were denoted magnitude 1 stars, and the faintest visible stars were declared to be of magnitude 6. Other stars were assigned magnitudes based on their apparent brightness compared with those of magnitude 1 or 6. In 1850 Pogson refined the system. He noticed that magnitude 1 stars were about 100 times brighter than magnitude 6 stars. Since there are five steps of magnitude from 1 to 6, Pogson proposed that a star of magnitude m should be $\sqrt[5]{100} \approx 2.512$ times brighter than a start of magnitude $m + 1$.

a. Explain how Pogson came up with the number $\sqrt[5]{100}$ and why this number is reasonable.

b. Antares is a magnitude 1.2 star, and Marsik is a magnitude 5.3 star. Let b_1 be the brightness of Antares and b_2 be the brightness of Marsik. Find the numerical value of b_1/b_2.

46. Give a convincing argument that the equation $e^{-2x} = k$ has a solution for every $k > 0$.

47. Give a convincing argument that the equation $e^x + 3e^{3x} = k$ has a solution for every $k > 0$.

48. Consider the equation $e^x + 3e^{-3x} = k$. Find two positive k values for which this equation does *not* have a solution.

49. Suppose that $b > 0$ and $b \neq 1$, and let $r > 0$. Give a convincing argument that the equation

$$b^x = r$$

has exactly one solution. Support your argument with graphs.

50. Let a, b, and k be real numbers. For what values of a and b does the equation

$$e^{ax} + 2e^{bx} = k$$

have a solution for every $k > 0$?

T **51.** Compute e^{C_2}, e^{C_3}, and $e^{C_{10}}$. Make a conjecture about the value of e^{C_b}.

52. a. Differentiate both sides of the equation $(ab)^x = a^x b^x$, and use the result to show that $C_a + C_b = C_{ab}$.

b. Differentiate both sides of the equation $(a/b)^x = a^x/b^x$, and use the result to show that $C_a - C_b = C_{a/b}$.

c. Differentiate both sides of the equation $(a^b)^x = a^{bx}$, and use the result to show that $bC_a = C_{a^b}$.

d. What familiar function satisfies the identities proved in a, b, and c?

T **53.** In finding the value of e, we looked for a number b for which

$$\frac{b^h - 1}{h} \approx 1$$

for small h. Solve this expression for b to obtain

$$b \approx (1 + h)^{1/h}.$$

Substitute small values of h in the expression for b and obtain an approximation to e. Argue that

$$\lim_{h \to 0} (1 + h)^{1/h} = e.$$

54. Find two different functions that have derivative e^x.

55. Find two different functions that have derivative e^{3x}. Check by finding the derivative of your answers.

56. a. Find the equation of the line that passes through the origin and is tangent to the graph of $y = e^x$.

b. Consider the equation

$$e^x = ax.$$

How many solutions does this equation have if $a < 0$? If $a = 0$? If $a > 0$?

T **57.** a. Investigate the behavior of x/e^x, x^2/e^x, x^{10}/e^x, and x^{50}/e^x for very large values of x. (You may do this numerically, graphically, or both.)

b. Make a conjecture about the behavior of x^n/e^x as x grows toward infinity.

58. Simple population models often describe population by an exponential formula

$$P(t) = P_0 e^{kt}$$

where P_0 is the population at some starting time $t = 0$ and k is some positive constant. For such populations, the rate of change is

$$P'(t) = kP_0 e^{kt}.$$

Write a short paragraph explaining the meaning of this rate of change. In particular, explain why larger populations increase faster than smaller ones.

59. If amount A_0 of a radioactive material is present at time $t = 0$, then the amount $A(t)$ present at a later time $t > 0$ is given by

$$A(t) = A_0 e^{-kt}$$

where k is a positive constant. Find the rate of change of the amount of radioactive material with respect to time, and write a short paragraph explaining the meaning of this rate of change. When radioactive material decays, it gives off radiation. Use the rate-of-change equation to explain why it is more dangerous to be near a large amount of radioactive material than near a small amount.

60. The *hyperbolic functions* are defined by the equations

$$\cosh x = \frac{e^x + e^{-x}}{2}, \quad \sinh x = \frac{e^x - e^{-x}}{2}.$$

The function $\cosh x$ is the *hyperbolic cosine* of x, and $\sinh x$ is the *hyperbolic sine* of x. These particular combinations of exponential functions arise often in science and engineering and have been assigned the notations $\cosh x$ and $\sinh x$ for convenience. (See also Exercises 30 and 31.)

a. Show that

$$(\cosh x)' = \sinh x$$

and

$$(\sinh x)' = \cosh x.$$

b. Show that

$$(\cosh x)^2 - (\sinh x)^2 = 1.$$

(This identity is the reason for the word *hyperbolic* in the names of these functions. If $x = \cosh t$ and $y = \sinh t$, then the point (x, y) lies on the hyperbola $x^2 - y^2 = 1$.)

c. Show that

$$\cosh 2x = (\cosh x)^2 + (\sinh x)^2$$

and

$$\sinh 2x = 2 \sinh x \cosh x.$$

d. The *hyperbolic tangent* and *hyperbolic secant* are defined by

$$\tanh x = \frac{\sinh x}{\cosh x}$$

and

$$\operatorname{sech} x = \frac{1}{\cosh x},$$

respectively. Show that

$$(\tanh x)' = (\operatorname{sech} x)^2$$

and

$$(\operatorname{sech} x)' = -\operatorname{sech} x \tanh x.$$

61. When a drug is used in treatment, it is important to know the rate at which the drug is removed from the body. The following data were obtained by monitoring the levels of a drug in a patient's blood. Blood samples were taken every hour, and the concentration of the drug was determined and reported in milligrams per liter.

t hours	Concentration, mg/l
0	10.0
1	7.0
2	5.0
3	3.5
4	2.5
5	2.0
6	1.5
7	1.0
8	0.7
9	0.5

Table for Exercise 61.

Analyze the data as in the Investigation on page 130.

a. Estimate the rate of change of concentration with respect to time.

b. Show that the rate of change is (roughly) proportional to concentration.

c. Come up with an exponential function that seems to describe the data. Plot your function and the data points on the same set of axes.

2.7 Logarithms

In April 1994 Derek Atkins, Michael Graff, Paul Leyland, and Arjen Lenstra announced that the 129-digit number

114381625757888867669235779976146612010218-
296721242362562561842935706935245733897830597123563958705058989075147599290026879543541

can be written as the product of the two numbers

3490529510847650949147849619903898133417764638493387843990820577

and

$$3276913299326670954996198819083446141317764296799294253979828853 3.$$

The 129-digit number, known as RSA 129, appeared in a 1977 paper by Ronald Rivest, Adi Shamir, and Leonard Adelman. In the paper they presented an encoding system now known as "public key encryption." Their encoding system requires a large number, like RSA 129, that is the product of two large prime numbers. Anyone who has the large number and a publicly known encoding number can use these to encode a message. However, these messages can be decoded only by someone who knows the prime factors of the large number. Thus the security of a public key encryption system depends on the feasibility of factoring large numbers. When RSA 129 was presented as an example of a number that could be used for this encryption scheme, the authors estimated that given the current state of knowledge about factoring and the speed of 1977 computers, it would take about 40 quadrillion (4×10^{16}) years to factor the number. Atkins, Graff, Leyland, and Lenstra did it in about eight months.

Of course, computers in 1994 were much faster than the computers of 1977 (and today's computers are even faster!). In addition, the factoring problem was cut into pieces and farmed out to over 600 volunteers worldwide. Each volunteer did a part of the problem and relayed the results to a central location for the final analysis. But the use of faster computers and 600 volunteers is nowhere near enough to cut 40 quadrillion years down to less than 1. The search for factors used a technique called quadratic sieving, developed by Carl Pomerance and announced in a 1982 paper. With this technique, the amount of time required to factor a positive integer N is proportional to

$$L(N) = e^{\sqrt{(\ln N)(\ln \ln N)}} \tag{1}$$

where $\ln N = \log_e N$ is the natural logarithm of N.

Today many businesses and governments use public key encryption to send confidential messages. Hence the speed with which large numbers can be factored is of tremendous importance. As (1) shows, logarithms are essential in assessing the security of public key systems because these codes can be broken only by factoring a large number. This illustrates one way in which logarithms play an important part in today's world. Logarithms are also used in describing the intensity of an earthquake, in measuring the loudness of a Rolling Stones concert, and in establishing a schedule for the administration of medication.

The Logarithm Function

Suppose that we wish to find the number s that solves the equation

$$7^s = 34. \tag{2}$$

Because $7^1 = 7$ and $7^2 = 49$, we know that s is between 1 and 2. With some trial and error and the help of a calculator, we could find s accurate to one or two decimal places. Try it! But no matter how hard we work, we will never be able to write the exact value of s. However, equations similar to (2) show up often, and it is important to be able to work with the solutions to such equations. Although we cannot calculate the exact solution of (2), we can label the solution with the notation

$$s = \log_7 34.$$

We say that "s is the logarithm to the base 7 of 34." This terminology may seem confusing at first, but remember that it is just a convenient means for talking about the solution to an equation such as (2).

DEFINITION Logarithm to the Base b

Let b be a positive real number with $b \neq 1$, and let $r > 0$. The solution s to the equation

$$b^s = r$$

is called the logarithm to the base b of r and is denoted

$$s = \log_b r.$$

In other words, $\log_b r$ is the power to which b must be raised to get an answer of r. That is,

$$b^{\log_b r} = r. \tag{3}$$

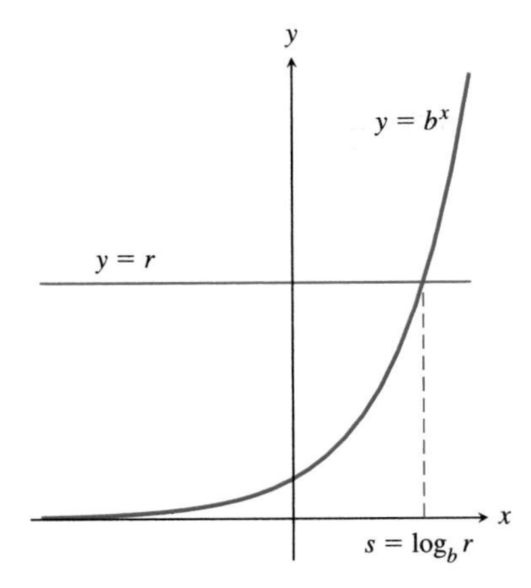

FIGURE 2.31 When $r > 0$, the line $y = r$ meets the graph of $y = b^x$ in exactly one point.

If $b > 0$ and $b \neq 1$, then for each $r > 0$ the equation

$$b^x = r \tag{4}$$

has only one solution $x = s$. To see this, first observe that the horizontal line $y = r$ intersects the graph of $y = b^x$ in exactly one point (see Fig. 2.31). Denote the coordinates of this intersection point by (s, r). Then $b^s = r$, so $s = \log_b r$. Furthermore, there is no other solution to (4). Thus for each $r > 0$, there is one and only one number $\log_b r$ that satisfies (3). This means that the equation

$$f(r) = \log_b r$$

describes a (well-defined) function f whose domain is the set of positive real numbers.

Most scientific or graphing calculators have a button for $\log_{10}$ (usually labeled "log") and a button for $\log_e$ (usually labeled "ln"). The graphs of $y = \log_{10} r$ and $y = \log_e r$ are shown in Fig. 2.32. The graphs remind us of some facts about logarithms to a base $b > 1$. The graphs both contain the point $(1, 0)$. The graphs rise as we move from left to right. As x gets close to 0, the graph falls toward $-\infty$; thus the negative y-axis is an asymptote for each of the graphs. The graphs of $y = \log_e r$ and $y = \log_{10} r$ are unbroken, so these functions are continuous for $r > 0$.

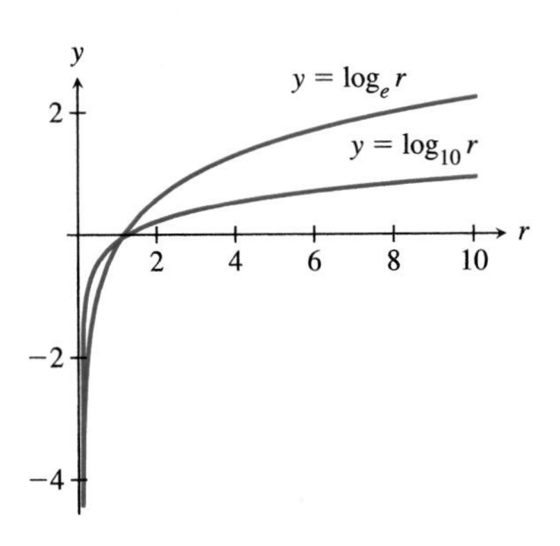

FIGURE 2.32 Graphs of two logarithm functions.

EXAMPLE 1

a) Find the value of:

i) $\log_5 125$.

ii) $\log_5 \dfrac{1}{\sqrt{5}}$.

b) Estimate the value of $\log_{1/2} 5$.

Solution

a) i) Let $c = \log_5 125$. According to the definition of logarithm, we must have

$$5^c = 125 = 5^3.$$

Hence $c = 3$, so

$$\log_5 125 = 3.$$

ii) Now let $d = \log_5 \dfrac{1}{\sqrt{5}}$. Then

$$5^d = \frac{1}{\sqrt{5}} = 5^{-1/2},$$

which means that $d = -1/2$. Hence

$$\log_5 \frac{1}{\sqrt{5}} = -1/2.$$

b) Set $c = \log_{1/2} 5$. Then

$$(1/2)^c = 5.$$

The graph of $y = \left(\frac{1}{2}\right)^x$ is shown in Fig. 2.33. From the graph we see that there is only one value of x for which $(1/2)^x = 5$, and this occurs for some x between -3 and -2. Zoom in on the graph between these values. See Fig. 2.34. We see that $(1/2)^x$ takes the value 5 when x is approximately -2.3. This tells us that $\log_{1/2} 5 \approx -2.3$. Check this answer with your calculator. Is $\left(\frac{1}{2}\right)^{-2.3} \approx 5$?

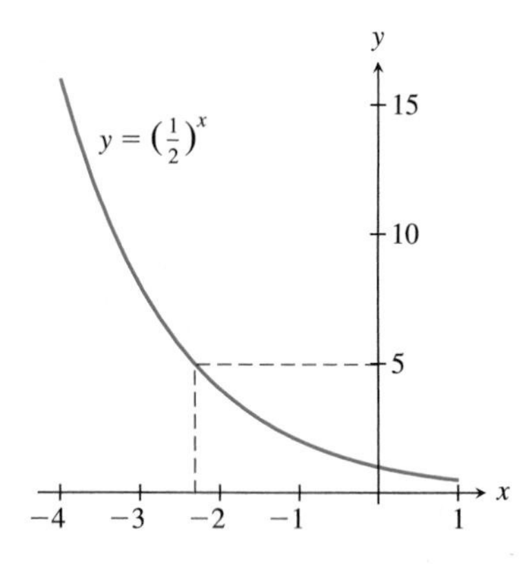

FIGURE 2.33 Graph of $y = (1/2)^x$.

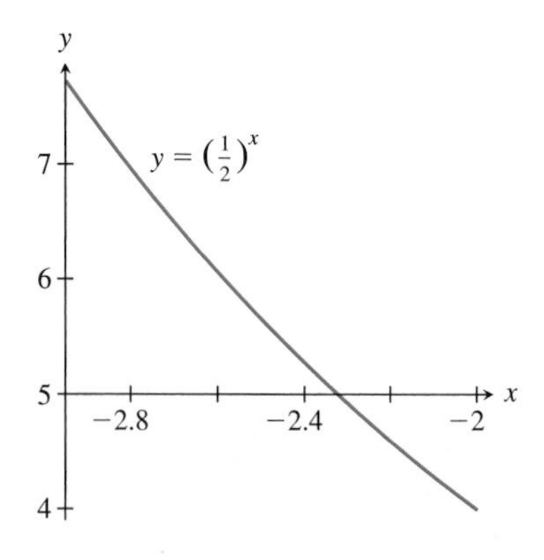

FIGURE 2.34 Graphically estimating the solution to $(1/2)^x = 5$.

Before the invention of hand-held calculators in the 1970s, logarithms were used to speed up certain arithmetic operations. Logarithms enabled people to replace multiplication problems by addition problems, division problems by subtractions, and exponentiations by multiplications. (These techniques are not taught much nowadays, but you may be able to get your teacher to talk about the "old days.") The following identities are the basis for these replacement calculations.

Identities for Logarithms

Let b be positive with $b \neq 1$, let $x > 0$ and $y > 0$, and let r be any real number. Then

a) $\log_b 1 = 0$
b) $\log_b(xy) = \log_b x + \log_b y$
c) $\log_b\left(\dfrac{x}{y}\right) = \log_b x - \log_b y$
d) $\log_b(x^r) = r \log_b x$
e) If $a > 0$ and $a \neq 1$, then $\log_a x = \dfrac{\log_b x}{\log_b a}$.

These identities are still useful in modifying or simplifying complicated expressions involving logarithms. We verify two of these identities here and leave the others for the exercises. See Exercises 29 and 30.

EXAMPLE 2 A logarithm identity

Verify the identity

$$\log_b\left(\frac{x}{y}\right) = \log_b x - \log_b y.$$

Solution

Using the definition of logarithm, we write

$$b^{\log_b(x/y)} = \frac{x}{y} = \frac{b^{\log_b x}}{b^{\log_b y}} = b^{\log_b x - \log_b y}.$$

The first and last expressions in this display are equal, and both are expressed as exponents to base b. Hence the exponents must be equal. (Why?) Equating these exponents, we get

$$\log_b\left(\frac{x}{y}\right) = \log_b x - \log_b y.$$

EXAMPLE 3 A logarithm identity

Verify the identity

$$\log_a x = \frac{\log_b x}{\log_b a}.$$

Solution

Let

$$w = \log_a x. \tag{5}$$

Then $a^w = x$ and

$$\log_b(a^w) = \log_b x. \tag{6}$$

It follows that

$$w \log_b a = \log_b x. \tag{7}$$

Because $a \neq 1$, we know $\log_b a \neq 0$. Divide both sides of (7) by $\log_b a$ to get

$$w = \log_a x = \frac{\log_b x}{\log_b a}.$$

This property of logarithms shows us that there is, in a sense, just one logarithm function. That is, if $a, b > 0$ and $a, b \neq 1$, then $\log_b x$ is a constant multiple of $\log_a x$.

The Derivative of the Logarithm Function

Using the definition of logarithm and results from Section 2.6 about differentiation of the exponential functions, we can find a formula for the derivative of the logarithm function. Let

$$y = \log_b x.$$

By the definition of logarithm,

$$x = b^{\log_b x} = b^y. \tag{8}$$

A composition chain for this statement is

$$\begin{aligned} x &= b^u \\ u &= y. \end{aligned} \tag{9}$$

By the chain rule,

$$\frac{dx}{dx} = \frac{dx}{du}\frac{du}{dx}. \tag{10}$$

The derivative on the left side of (10) is 1. From (9) we have

$$\frac{dx}{du} = C_b b^u \quad \text{and} \quad \frac{du}{dx} = \frac{dy}{dx}, \tag{11}$$

where, as seen in Section 2.6,

$$C_b = \lim_{h \to 0} \frac{b^h - 1}{h}. \tag{12}$$

Substituting (11) into (10), then using (8), we have

$$1 = C_b b^u \frac{dy}{dx} = C_b b^{\log_b x} \frac{dy}{dx} = C_b x \frac{dy}{dx}.$$

Solving for $\frac{dy}{dx}$,

$$\frac{dy}{dx} = \frac{1}{C_b x}.$$

Because $y = \log_b x$ we have

$$\frac{d}{dx} \log_b x = \frac{1}{C_b x}. \tag{13}$$

The Constants C_b The constants C_b defined by (12) arise when we differentiate exponential or logarithm functions. Now that we have defined the logarithm function, we can say more about these constants. Let $y = b^x$. By the definition of logarithm, the statements

$$e^w = b \quad \text{and} \quad \log_e b = w$$

are equivalent. Hence

$$e^{\log_e b} = b.$$

Thus we can write

$$y = b^x = (e^{\log_e b})^x = e^{(\log_e b)x}.$$

We can express this relation as the composition chain

$$\begin{aligned} y &= e^u \\ u &= (\log_e b)x. \end{aligned}$$

Because $\log_e b$ is a constant,

$$\frac{du}{dx} = \frac{d}{dx}((\log_e b)x) = \log_e b.$$

Hence,

$$\frac{dy}{dx} = \frac{dy}{du}\frac{du}{dx} = e^u(\log_e b) = (\log_e b)e^{(\log_e b)x} = (\log_e b)b^x.$$

In the previous section we showed that

$$\frac{dy}{dx} = C_b b^x$$

for some constant C_b. Equating these two expressions for $\frac{dy}{dx}$, we get

$$C_b b^x = (\log_e b)b^x.$$

Dividing both sides of the equation by the nonzero number b^x, we have

$$C_b = \log_e b. \tag{14}$$

We use (14) in restating our formulas for the derivatives of the exponential and logarithm functions.

Derivatives of the Exponential and Logarithm Functions

Let b be a positive real number, with $b \neq 1$. Then

$$\frac{d}{dx} b^x = (\log_e b) b^x$$

and

$$\frac{d}{dx} \log_b x = \frac{1}{(\log_e b) x}.$$

The Natural Logarithm

In the previous section we argued that for calculus the "best base" for the exponential function is the base e. This is because $C_e = 1$ and hence

$$\frac{d}{dx} e^x = e^x.$$

In (13) we saw that

$$\frac{d}{dx} \log_b x = \frac{1}{C_b x}.$$

This derivative is simplest when $C_b = 1$. Hence for calculus, the "logarithm of choice" is the logarithm to base $b = e$. This logarithm is called the *natural logarithm*.

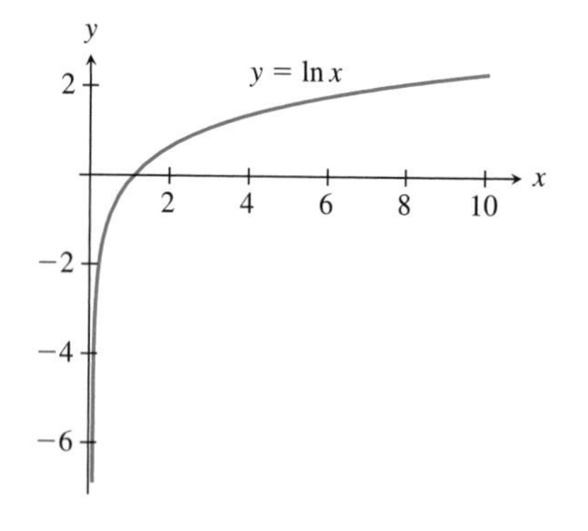

FIGURE 2.35 Graph of the natural logarithm function.

DEFINITION Natural Logarithm

The natural logarithm is the logarithm to the base e. The natural logarithm of x is denoted by

$$\ln x$$

and is defined for all $x > 0$.

Figure 2.35 shows the graph of the natural logarithm.

From the definition of logarithm we see that the natural logarithm and the exponential function are closely related.

The Natural Logarithm and the Exponential Function

If x is a real number, then

$$\ln(e^x) = x. \tag{15}$$

If x is a positive real number, then

$$e^{\ln x} = x. \tag{16}$$

Equations (15) and (16) say that the exponential function is the inverse of the natural logarithm function and that the natural logarithm function is the inverse of the exponential function. We will study inverse functions in the next section.

Engineers, scientists, and mathematicians usually prefer to work with the natural logarithm because its derivative is the "simplest" of the logarithm derivatives.

The Derivative of ln *x*

The derivative of the natural logarithm function is

$$\frac{d}{dx}\ln x = \frac{1}{x}.$$

EXAMPLE 4 Find $\frac{dy}{dt}$ if $y = \ln(2t^2 - 3)$.

Solution

Write the function as the composition chain:

$$y = \ln u$$
$$u = 2t^2 - 3.$$

Then

$$\frac{dy}{du} = \frac{1}{u} \quad \text{and} \quad \frac{du}{dt} = 4t.$$

By the chain rule, we have

$$\frac{dy}{dt} = \frac{dy}{du}\frac{du}{dt} = \frac{1}{u}4t = \frac{4t}{2t^2 - 3}.$$

The preceding example illustrates a general "rule" for differentiating expressions of the form $y = \ln(u(x))$.

EXAMPLE 5 Find a formula for $\frac{d}{dx}\ln(u(x))$.

Solution

Let $y = \ln u(x)$. A composition chain for this expression is

$$y = \ln w$$
$$w = u(x).$$

By the chain rule,

$$\frac{d}{dx}\ln(u(x)) = \frac{dy}{dx} = \frac{dy}{dw}\frac{dw}{dx} = \frac{1}{w}u'(x) = \frac{u'(x)}{u(x)}. \tag{17}$$

EXAMPLE 6 Calculate $f'(x)$ if

$$f(x) = \frac{x^2 \ln x}{x^2 + 1}. \tag{18}$$

Solution

The expression on the right side of (18) is a quotient, so we use the quotient rule,

$$f'(x) = \frac{(x^2 \ln x)'(x^2 + 1) - (x^2 \ln x)(x^2 + 1)'}{(x^2 + 1)^2}.$$

We use the product rule to find the derivative of $x^2 \ln x$. This leads to

$$\begin{aligned} f'(x) &= \frac{((x^2)' \ln x + x^2(\ln x)')(x^2 + 1) - (x^2 \ln x)2x}{(x^2 + 1)^2} \\ &= \frac{\left(2x \ln x + x^2\left(\frac{1}{x}\right)\right)(x^2 + 1) - 2x^3 \ln x}{(x^2 + 1)^2} \\ &= \frac{(2x \ln x + x)(x^2 + 1) - 2x^3 \ln x}{(x^2 + 1)^2}. \end{aligned}$$

The rules of logarithms seen on page 145 can be used to simplify computations by converting products to sums, quotients to differences, and exponentials to products. These rules can also be used to simplify some differentiations that might otherwise involve the product rule, the quotient rule, or the chain rule. The process is sometimes called **logarithmic differentiation.** We illustrate the process with an example.

EXAMPLE 7 Logarithmic differentiation

Find $\frac{dy}{dx}$ if

$$y = \frac{(x^2 + 2)\sqrt{\sin 2x}}{(3x^3 - 2x - 4)^5}. \tag{19}$$

Solution

First note that this derivative could be found using the techniques developed in earlier sections. Because the function is a quotient, we could use the quotient differentiation formula developed in Example 4 of Section 2.2. We would also need the product rule to differentiate the numerator, and the chain rule to differentiate the denominator and the expression $\sqrt{\sin 2x}$ in the numerator.

Instead, start by taking logarithms of both sides, and then use logarithm identities to write the logarithm of the quotient as a difference of logarithms:

$$\ln y = \ln\left(\frac{(x^2+2)\sqrt{\sin 2x}}{(3x^3-2x-4)^5}\right) \tag{20}$$
$$= \ln((x^2+2)(\sin 2x)^{1/2}) - \ln((3x^3-2x-4)^5).$$

Next, converting the logarithm of the product in the last expression to a sum of logs, and then placing the exponents out in front as multipliers, we get

$$\ln y = \ln(x^2+2) + \frac{1}{2}\ln(\sin 2x) - 5\ln(3x^3-2x-4). \tag{21}$$

Now differentiate both sides of (21), remembering that y is a function of x. Using (17), we obtain

$$\frac{y'}{y} = \frac{(x^2+2)'}{x^2+2} + \frac{1}{2}\frac{(\sin 2x)'}{\sin 2x} - 5\frac{(3x^3-2x-4)'}{3x^3-2x-4}$$
$$= \frac{2x}{x^2+2} + \frac{1}{2}\frac{2\cos 2x}{\sin 2x} - 5\frac{9x^2-2}{3x^3-2x-4}.$$

Multiply both sides of this last expression by y and replace y with the expression on the right of (19). We then have an expression for y':

$$y' = \left(\frac{(x^2+2)\sqrt{\sin 2x}}{(3x^3-2x-4)^5}\right)\left(\frac{2x}{x^2+2} + \cot 2x - 5\frac{9x^2-2}{3x^3-2x-4}\right).$$

Exercises 2.7

Exercises 1–6: Find the exact value of the expression.

1. $\log_3 81$
2. $\log_2(1/\sqrt[3]{4})$
3. $\log_{0.23} 0.012167$
4. $\log_e e^{1995}$
5. $\log_{\sqrt{6}}(36\sqrt{6})$
6. $\log_{4.567} 1$

Exercises 7–14: Solve for x.

7. $2^{3x} = 5$
8. $e^x = \frac{1}{2}$
9. $2e^{6x} - 5e^{3x} + 6 = 0$
10. $2e^{6x} + 5e^{3x} - 6 = 0$
11. $\log_2 3x = 5$
12. $\log_4(x+2) + \log_4(x-2) = 1$
13. $(\log_2 x) + (\log_5 x) = \log_{10} x$
14. $(\log_2 x) + (\log_5 x) = (\log_{10} x) + 1$

Exercises 15–24: Find the derivative.

15. $y = \ln(x+2)$
16. $h(t) = t\ln t$
17. $r = \ln(\sin\theta)$
18. $G(u) = \ln(\ln u)$
19. $r(t) = (1 + \ln 2t)^{10}$
20. $f(x) = \dfrac{1+\ln x}{2-\ln x}$
21. $q(z) = \ln\left(\dfrac{1-2z+z^3}{z\sin z}\right)$
22. $y = \dfrac{1}{\sqrt{\ln x}}$
23. $r = e^{(\ln\theta)^2}$
24. $y = e^{x\ln x}$

Exercises 25–28: Find $\frac{dy}{dx}$ by using logarithmic differentiation.

25. $y = \dfrac{x^2}{2x+1}$
26. $y = x^{x+1}$
27. $y = \sqrt{x\tan x}$
28. $y = \left(\dfrac{1+e^x}{1-e^x}\right)^{-10}$
29. Use the definition of logarithm and rules for working with exponentials to prove the identity
$$\log_b xy = \log_b x + \log_b y.$$
30. Use the definition of logarithm and rules for working with exponentials to prove the identity
$$\log_b(x^r) = r\log_b x.$$

31. Why don't we ever talk about logarithms to base 1?

32. In Section 2.3 we saw that if r is a rational number, then

$$\frac{d}{dx}x^r = rx^{r-1}.$$

With the help of logarithms, we can find the derivative of x^a for any real number a. Define $f(x) = x^a$ for $x > 0$, and write

$$f(x) = x^a = e^{a \ln x}.$$

Differentiate this expression and simplify the result to show that

$$f'(x) = ax^{a-1}.$$

Why is the restriction $x > 0$ necessary in defining f?

33. Show that the function $y = \ln|x|$ has derivative

$$\frac{dy}{dx} = \frac{1}{x} \quad (x \neq 0).$$

(*Hint:* First consider the case $x > 0$, and then the case $x < 0$.)

34. Use the chain rule and the result of Exercise 33 to find the derivative of each of the following functions.

a. $y = \ln|\sin x|$

b. $r = \ln|4\theta^2 - 10|$

c. $f(t) = \ln\left|\frac{1-t}{1+t}\right|$

35. Find two different functions with derivative $1/(x-4)$ for $x > 4$.

36. Find two different functions with derivative $2x/(x^2+1)$.

37. Consider the graphs of $y = e^x$ and $y = \ln x$.

a. Tell why if the point (a, b) is on the graph of $y = e^x$, then the point (b, a) is on the graph of $y = \ln x$.

b. Tell why if the point (a, b) is on the graph of $y = \ln x$, then the point (b, a) is on the graph of $y = e^x$.

c. What do parts **a** and **b** imply about the relationship between the two graphs?

T **38.** By looking at the graph of

$$y = \frac{\ln x}{x},$$

determine the behavior of $\frac{\ln x}{x}$ as x gets large and positive.

T **39.** By looking at the graph of

$$y = \frac{\ln x}{\sqrt[4]{x}},$$

determine the behavior of $\frac{\ln x}{\sqrt[4]{x}}$ as x gets large and positive.

T **40.** By looking at the graph of

$$y = x \ln x,$$

determine the behavior of $x \ln x$ as $x \to 0$ through positive values.

41. Let f be a function. What is the domain of

$$\ln(f(x)) + 4\ln(-f(x))?$$

Give reasons for your answer.

42. For what real numbers k does the equation $\ln x = k$ have a solution?

T **43.** Determine (approximately) the positive values of x for which 2^x is larger than x^8.

T **44.** Determine (approximately) the values of x for which $x^{1/10}$ is larger than $\ln x$.

45. Consider the function described by

$$y = \sin 2x - \cos 4x + (2x^3 - 2x^2 + 4)^{1/2}.$$

Explain why logarithmic differentiation would not be very useful in computing dy/dx.

46. A modern-day Rip van Winkle decides to take a very long nap. Having no faith in modern timepieces, he gets a 10-gram sample of radium-226. After t years have passed, the amount of radium-226 that remains will be

$$A = 10e^{k \cdot t}$$

where $k \approx -0.000427869$.

a. Mr. van Winkle solves this equation for t so he can tell the time as a function of the amount A of radium that remains. What is Mr. van Winkle's formula for the time?

b. Mr. van Winkle finally dozes off on January 1, 2000. When he awakens, 1 gram of radium remains. What is the (approximate) date?

47. Given that it took about eight months to factor RSA 129 and that the factoring time is proportional to $L(N)$ as given in (1), estimate the amount of time needed to factor a 150-digit number and a 200-digit number.

48. Suppose that you wanted a public key number that would take one million years to factor using the methods used on RSA 129. Estimate the number of digits needed. What if you wanted your number to be unfactored for one trillion (10^{12}) years?

49. The acidity of a solution can be gauged by measuring the concentration of hydrogen ions in the solution. This ion concentration is usually given in units of moles per liter (mols) and is denoted by $[H^+]$. For distilled water, $[H^+] \approx 10^{-7}$ mols, indicating that each liter of water contains approximately 10^{-7} moles of hydrogen ions. The pH of a solution is defined by

$$\text{pH} = \log_{10}\left(\frac{1}{[H^+]}\right).$$

Thus the pH of distilled water is about 7.

a. If a certain solution has a pH of 5, how does its hydrogen ion concentration compare with that of distilled water? What if the solution has a pH of 10?

b. If solution A has twice the hydrogen ion concentration of solution B, how do the pHs of the two solutions compare?

50. Sound waves transport energy (acoustic energy). The intensity of sound I is the average rate at which the sound wave transmits energy per unit area, so carries units such as watts per square meter. However, when we read about the intensity of various sounds in newspapers and magazines, it is usually described in terms of decibels (dB). Let I_0 be the intensity of sound at the threshold of human hearing. The decibel level of a sound with intensity I is then given by

$$dB = 10 \log_{10}\left(\frac{I}{I_0}\right).$$

a. If the decibel ratings of two sounds differ by 1, how do the intensities of the sounds compare?

b. Suppose that the intensity of a sound doubles. How is the decibel rating affected?

c. A sound of 120 dB is at the human pain threshold. How much more intense is this than a sound at the threshold of human hearing?

51. The magnitude of an earthquake is usually reported by giving the Richter scale measure for the quake. For an earthquake of intensity I, the Richter scale measure is $\log_{10}(I/S)$ where S, a constant, is the intensity of a "standard" quake.

a. The Loma Prieta earthquake of 1989 measured 7.1 on the Richter scale, while the Alaska earthquake of 1964 measured 8.6. How do the intensities of the two earthquakes compare?

b. The 1906 San Francisco earthquake had an intensity 16 times that of the Loma Prieta quake. What was the Richter scale magnitude of the San Francisco quake?

52. The accompanying figure shows the graph of an exponential function $y = e^{kx}$ (solid line) and the graph of $y = \ln x$ (dashed line). Use the method for composing graphs discussed in Section 1.2 to find the graph of $y = \ln(e^{kx})$, and from this graph determine (approximately) the value of k.

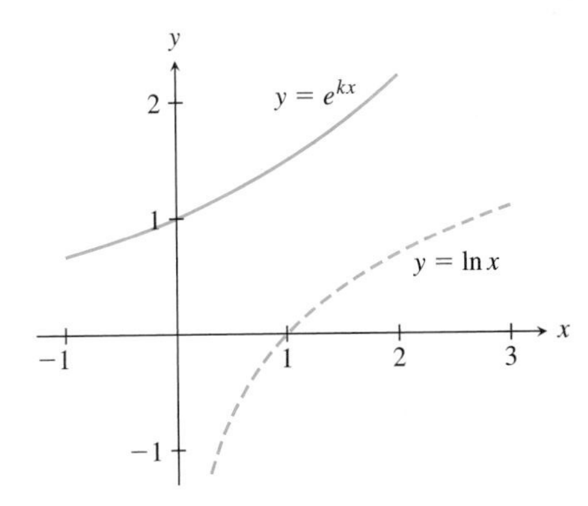

Figure for Exercise 52.

53. The accompanying figure shows the graph of an exponential function $y = e^{kx}$ (solid line) and the graph of $y = \ln x$ (dashed line). Use the method for composing graphs discussed in Section 1.2 to find the graph of $y = \ln(e^{kx})$. From this graph determine (approximately) the value of k.

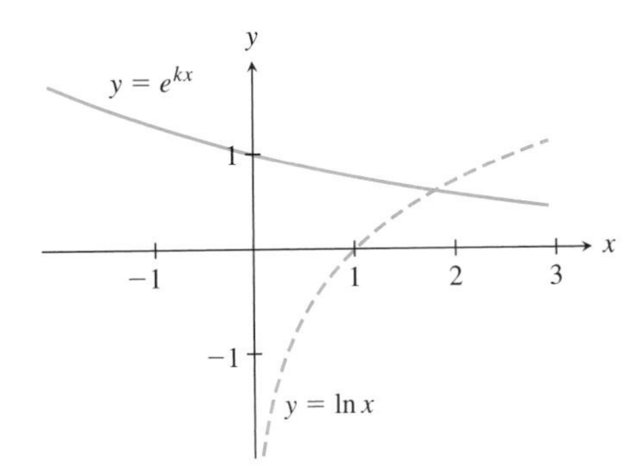

Figure for Exercise 53.

2.8 Inverse Functions

In Bemidji, Minnesota (47.5° north latitude, 95° west longitude), the sun climbs only 10° above the horizon on December 22 but gets as high as 57° above the horizon on June 21. One day between December 22, 1995, and December 22, 1996, you visit Bemidji and notice that the sun is 30° above the horizon at its highest point in the sky. Is this enough to determine the date of your visit? This question is not as strange as it may seem. Pre-twentieth-century navigators determined their latitude using only the date and the angle of the sun above the horizon at noon. Moreover, given any two of these three pieces of information, they could calculate the third.

In spite of this, the answer to our question is, strictly speaking, no. Let $A(t)$ be the angle of the sun above the horizon when the sun is at its highest point in the sky on day t. Between December 22, 1995, and June 21, 1996, $A(t)$ increases from 10° to 57°, so a measurement of 30° is possible on some day before June 21. From June 21, 1996, to December 22, 1996, $A(t)$ decreases from 57° to 10°, so a measurement of 30° is also possible on some day after June 21. The angle $a = A(t)$ (in degrees) on day t is pretty well approximated by

$$a = A(t) = -13.5 + \frac{70.5}{1 + 0.00006(t - 182.5)^2} \quad 0 \le t \le 365, \tag{1}$$

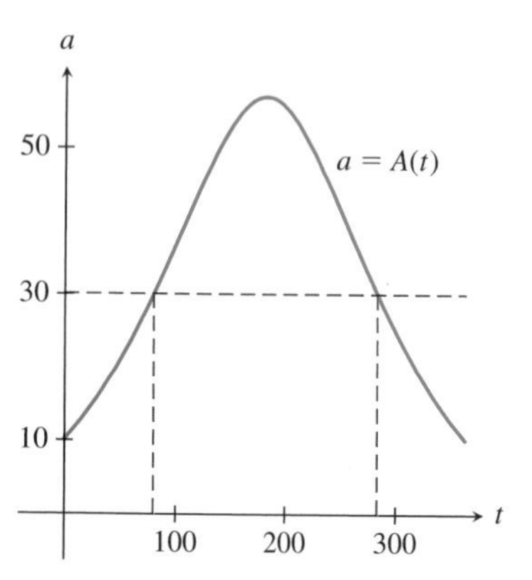

FIGURE 2.36 The angle of the sun in Bemidji, Minnesota.

where $t = 0$ corresponds to December 22. The graph of $a = A(t)$ is shown in Fig. 2.36. Notice that the line $a = 30$ hits the graph twice, indicating that there are two days for which $a = 30°$. So given *only* information about the angle of the sun, it is not possible to determine the date with certainty.

By solving (1) for t, we can find the date(s) that correspond to angle a. Add 13.5 to both sides, and then multiply by $1 + 0.00006(t - 182.5)^2$. We then have

$$(a + 13.5) + 0.00006(a + 13.5)(t - 182.5)^2 = 70.5.$$

Solving for $(t - 182.5)^2$ gives

$$(t - 182.5)^2 = \frac{57 - a}{0.00006(13.5 + a)}.$$

Solving for t we have

$$t = 182.5 - \sqrt{\frac{57 - a}{0.00006(13.5 + a)}} \quad \text{or} \quad t = 182.5 + \sqrt{\frac{57 - a}{0.00006(13.5 + a)}}. \tag{2}$$

These two formulas for t mean that for a value of a between 10° and 57°, there might be two days t when this angle is realized. This can also be seen from the graph in Fig. 2.36.

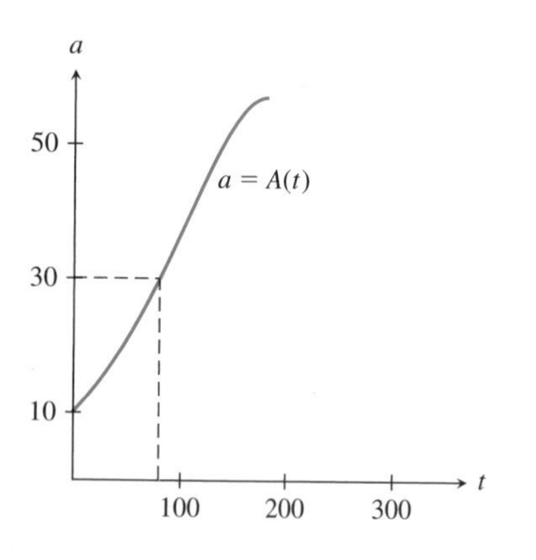

FIGURE 2.37 Each angle corresponds to just one t between 0 and 182.5.

The graph has its high point at $t = 182.5$. (This corresponds to June 21, the summer solstice.) If we know whether the date is before or after June 21, we can decide which value of t gives the correct date. For example, if the date is before June 21, then we are concerned only with the portion of the graph between $t = 0$ and $t = 182.5$. As seen in Fig. 2.37, no horizontal line intersects this portion of the graph more than once. Hence, for a value of a between 10° and 57°, there is only one corresponding value of t. Because $0 \le t \le 182.5$, we know the date is given by the first of the two expressions for t in (2):

$$t = 182.5 - \sqrt{\frac{57 - a}{0.00006(13.5 + a)}}.$$

When we put $a = 30$ into this formula, we get

$$t \approx 80.8,$$

which corresponds to March 13.

The Inverse of a Function

In the Bemidji problem, we were given a function A. The input for A was a number t corresponding to a day of the year, and the output was the angle $a = A(t)$ of the sun above the horizon at its highest point on day t. The angle of 30° was treated as "output" from the function A. We were asked to find the input value that produced this output. This process of determining the input for a function given the output of the function is called *evaluating the inverse of a function*.

In our discussion of this problem, we saw that there may be more than one input that gives a desired output. If we want each output value to correspond to exactly one input value, extra information may be needed to help us choose among candidates for the input. The information that the date was before June 21 told us that $t \leq 182.5$. This allowed us to conclude that the day t was given by

$$t = T(a) = 182.5 - \sqrt{\frac{57 - a}{0.00006(13.5 + a)}}. \tag{3}$$

The function T is the inverse of the function A for $0 \leq t \leq 182.5$. Suppose that some t value with $0 \leq t \leq 182.5$ is used as input in (1), and that the output is a. If we use this output as input in (3), we get back the original t as output. Thus if $0 \leq t \leq 182.5$, then

$$t = T(A(t)) = (T \circ A)(t).$$

We use this relation to define the *inverse of a function.*

DEFINITION Inverse of a Function

Let f be a function with domain $\mathcal{D}$. A function of g is the **inverse of** f if the domain of g is equal to the range of f and

$$(g \circ f)(x) = g(f(x)) = x \tag{4}$$

for every x in $\mathcal{D}$.

FIGURE 2.38 The function $g = f^{-1}$ takes output $f(x)$ from f and returns x.

Equation (4) says that if some value $f(x)$ of f is used as input for g, the output for the function g will be x. Thus g just "undoes" f. We can also illustrate this by thinking of functions as machines. When x is put into the f machine, the output is $f(x)$. This $f(x)$ is then used as input for the g machine, and the resulting output is x. See Fig. 2.38. The definition of the inverse of a function can be expressed in a useful equivalent form. This equivalent form can be stated briefly as follows: g is the inverse of f if

$$y = f(x) \quad \text{exactly when} \quad x = g(y). \tag{5}$$

The symmetry of (5) suggests that g is the inverse of f exactly when f is the inverse of g. See Exercise 38.

The inverse of a function f is usually denoted by f^{-1}. Hence, in the above definition, we have

$$g = f^{-1}.$$

Be careful when you encounter the f^{-1} notation for inverse. In this case, the -1 is not an exponent, but a notational device. In particular, f^{-1} does *not* equal $1/f$ in this context.

EXAMPLE 1 The inverse of the natural logarithm function

Let $f(x) = \ln x$ and $g(x) = e^x$. Show that g is the inverse of f and that f is the inverse of g.

Solution

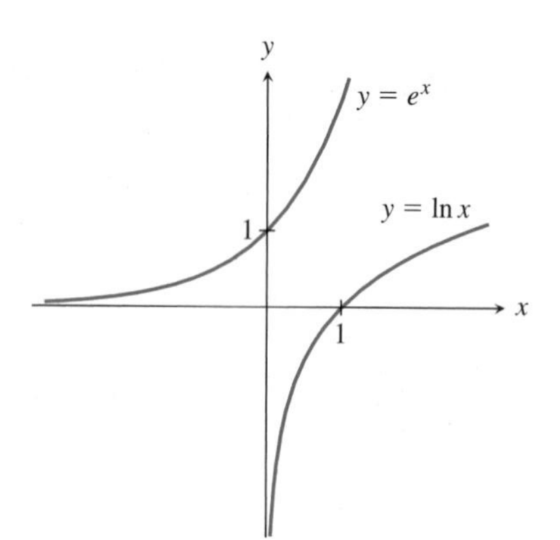

FIGURE 2.39 The graphs of $y = \ln x$ and $y = e^x$.

The graphs of $y = \ln x$ and $y = e^x$ are shown in Fig. 2.39. Let $y = \ln x$. Then by the definition of logarithm we have $x = e^{\ln x} = e^y$. Thus,

$$y = \ln x \quad \text{and} \quad x = e^y.$$

As stated in (5), this implies that the exponential function and the logarithm function are inverses of each other.

We can also verify this relationship by verifying that the logarithm and exponential functions satisfy the definition of inverse function. The range of the natural logarithm is the set of all real numbers. This set is also the domain of the exponential function. Hence the range of g is equal to the domain of f. The natural logarithm function is defined by the relation

$$e^{\ln x} = x \quad (x > 0).$$

This shows that

$$g(f(x)) = x$$

for every x in the domain of the logarithm function. Thus the exponential function g is the inverse of the natural logarithm function f.

Next note that the range of the exponential function g is the set of positive real numbers, and that this set is the domain of the natural logarithm function f. Because

$$(f \circ g)(x) = \ln(e^x) = x(\ln e) = x \cdot 1 = x$$

for all real x, it follows that the natural logarithm function is the inverse of the exponential function.

EXAMPLE 2 Finding an inverse

Let

$$f(x) = \sqrt[3]{x - 1} + 1. \tag{6}$$

Find the function g that is inverse to f and check that the answer is correct.

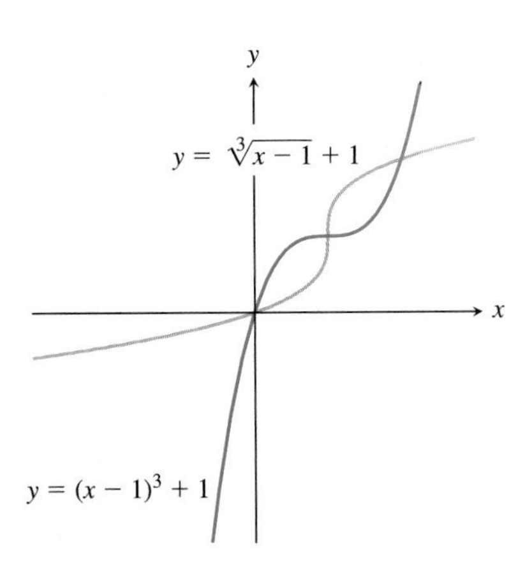

FIGURE 2.40 The graphs of $y = (x-1)^3 + 1$ and $y = (x-1)^{1/3} + 1$.

Solution

Let g be the inverse of f. Then, by (4) and (5),

$$f(g(x)) = x$$

for all x in the domain of g. Combining this with (6), we have

$$x = f(g(x)) = \sqrt[3]{g(x) - 1} + 1.$$

We solve this equation for $g(x)$. Subtract 1 from both sides of the equation to get

$$x - 1 = \sqrt[3]{g(x) - 1}.$$

Next, cube both sides and add 1 to obtain

$$g(x) = (x-1)^3 + 1. \tag{7}$$

Now check that the function g defined by (7) is the inverse of f. Because $f(x) = \sqrt[3]{x-1} + 1$ is defined for all real numbers, we take the domain of f to be the set of all real numbers. The range of f is also the set of all real numbers, and this set is the domain of the polynomial g. For any real number x, we have

$$\begin{aligned}(g \circ f)(x) &= g(f(x)) = (f(x) - 1)^3 + 1 \\ &= \left(\left(\sqrt[3]{x-1} + 1\right) - 1\right)^3 + 1 = x.\end{aligned}$$

Since $(g \circ f)(x) = x$ for all x in the domain of f, we conclude that g is the inverse of f. The graphs of f and its inverse are shown in Fig. 2.40.

EXAMPLE 3 The inverse for the squaring function

a) Let $f(x) = x^2$ for all real numbers x. Show that $g(x) = \sqrt{x}$ is *not* the inverse of f.

b) Let $f(x) = x^2$ for domain $\mathcal{D} = [0, \infty)$. Show that $g(x) = \sqrt{x}$ is the inverse of f.

Solution

a) If $f(x) = x^2$ and $g(x) = \sqrt{x}$, then the range of f is equal to the domain of g. Thus $g(f(x))$ makes sense for all real numbers x. For real x, we have

$$(g \circ f)(x) = g(x^2) = \sqrt{x^2} = |x|.$$

Thus $(g \circ f)(x) \neq x$ if x is negative, so g is not the inverse of f.

b) Now let $f(x) = x^2$ and take the domain of f to be $[0, \infty)$, the set of nonnegative real numbers. As in part **a)**, the range of f is equal to the domain of g. For any nonnegative x, we have

$$(g \circ f)(x) = g(x^2) = \sqrt{x^2} = |x| = x.$$

Hence g is the inverse of f. This example shows that the existence of an inverse function f can depend on the domain of f as well as the definition of f on its domain. The graphs of f and its inverse are shown in Fig. 2.41.

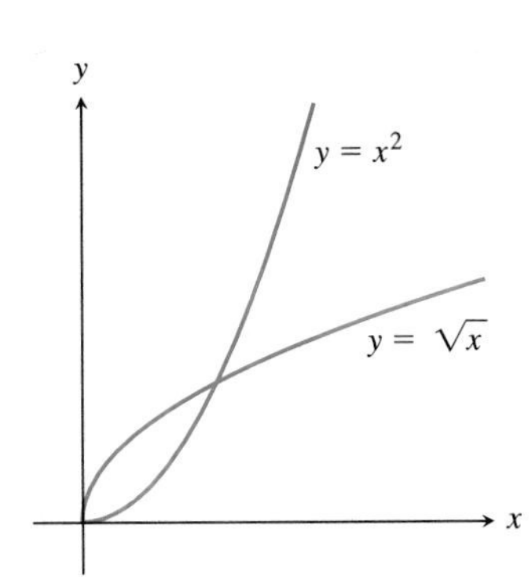

FIGURE 2.41 The graphs of $y = x^2$ and $y = \sqrt{x}$ for $x \geq 0$.

Existence of the Inverse

The Bemidji problem and Example 3**a** show that not every function has an inverse. Suppose that x_1 and x_2 are two different domain (or input) values for a function f and

$$f(x_1) = b = f(x_2).$$

When this happens, f has no inverse function. If the function g were the inverse of f, then we would have

$$g(b) = g(f(x_1)) = x_1 \quad \text{and} \quad g(b) = g(f(x_2)) = x_2.$$

Because g is a function, it cannot give two different values for the input b. Thus f can have no inverse function.

This argument shows that if a function f with domain $\mathcal{D}$ has an inverse, then any two different inputs (domain values) for f must result in different outputs. Functions with this property are said to be *one-to-one* on their domain.

DEFINITION One-to-One Function

Let f be a function with domain $\mathcal{D}$. We say f is **one-to-one** if whenever $x_1, x_2 \in \mathcal{D}$ with

$$x_1 \neq x_2 \quad \text{then} \quad f(x_1) \neq f(x_2).$$

We can quickly check whether or not a function f is one-to-one by looking at its graph.

The Horizontal Line Test

A function f is one-to-one on its domain if and only if each horizontal line intersects the graph of f in at most one point.

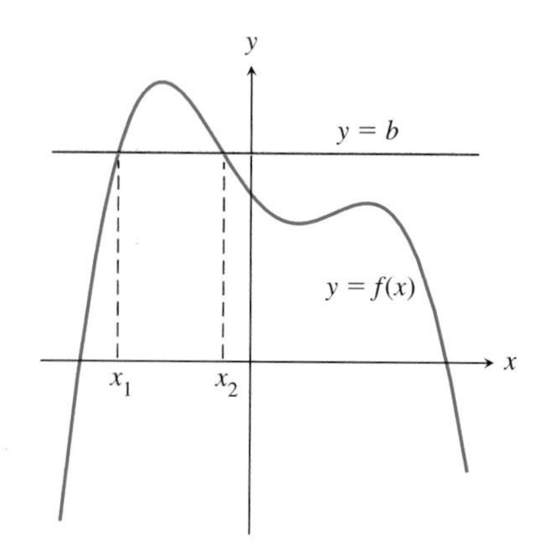

FIGURE 2.42 There is a horizontal line that intersects the graph of f more than once. Hence f is not one-to-one.

Indeed, if a horizontal line $y = b$ intersects the graph of f at two different points (x_1, b) and (x_2, b), then we have

$$f(x_1) = b = f(x_2)$$

although $x_1 \neq x_2$. Hence f is not one-to-one.

On the other hand, if f is not one-to-one, then there are domain values $x_1 \neq x_2$ for which $f(x_1) = f(x_2)$. Call this common value b. Then the horizontal line $y = b$ intersects the graph of f at points (x_1, b) and (x_2, b). See Figs. 2.42 and 2.43.

Checking whether or not a function is one-to-one is equivalent to checking whether or not the function has an inverse. Compare the following result to the vertical line test. See Exercise 29 in Section 1.1.

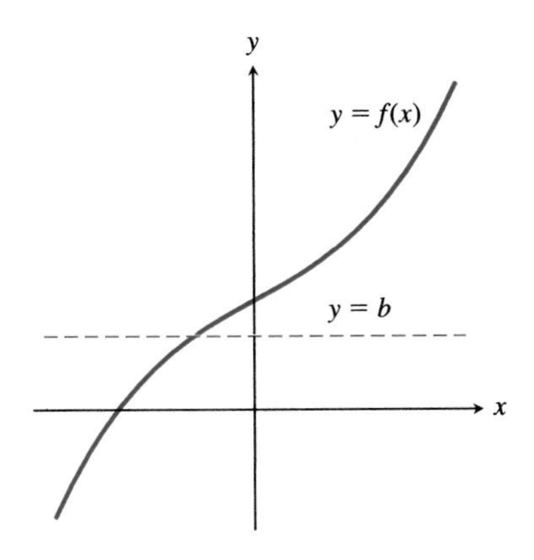

FIGURE 2.43 Each horizontal line intersects the graph of f at most once. Hence f is one-to-one.

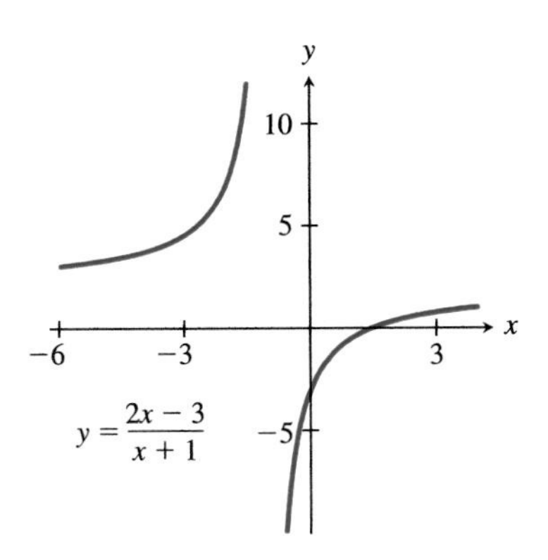

FIGURE 2.44 The graph of f appears to "pass" the horizontal line test.

Existence of an Inverse

A function has an inverse if and only if the function is one-to-one on its domain.

EXAMPLE 4 Let f be defined by

$$f(x) = \frac{2x - 3}{x + 1} \tag{8}$$

for domain $\mathcal{D} = \{x : x \neq -1\}$. Show that f is one-to-one on its domain and find the inverse of f.

Solution

The graph of $y = f(x)$ is shown in Fig. 2.44. We cannot see the graph of f for the whole domain $\mathcal{D}$, but it appears that any horizontal line intersects the graph of f in at most one point. Thus f is one-to-one. We can also check this directly using (8). Suppose that u, v are in $\mathcal{D}$ and that $f(u) = f(v)$. Then

$$\frac{2u - 3}{u + 1} = \frac{2v - 3}{v + 1}.$$

Cross-multiplying leads to

$$(2u - 3)(v + 1) = (2v - 3)(u + 1).$$

Expanding and cancelling like terms on each side leaves

$$5u = 5v,$$

from which we see $u = v$. Hence if $u \neq v$, we must have $f(u) \neq f(v)$, so f is one-to-one.

Because f is one-to-one on its domain, f has an inverse. Let g be this inverse function. Then

$$x = (f \circ g)(x) = f(g(x)) = \frac{2g(x) - 3}{g(x) + 1}. \tag{9}$$

We solve (9) for $g(x)$. To keep the notation simple, replace $g(x)$ by y in (9). This gives

$$x = \frac{2y - 3}{y + 1}.$$

Multiply both sides of this equation by $y + 1$ to get

$$x(y + 1) = 2y - 3.$$

Rearrange this last expression to get

$$xy - 2y = -x - 3.$$

Divide by $x - 2$, and we have

$$y = g(x) = \frac{x + 3}{-x + 2}.$$

How could you check that this g is the inverse of f?

In the Bemidji problem we worked with the function A defined by

$$A(t) = -13.5 + \frac{70.5}{1 + 0.00006(t - 182.5)^2}, \quad 0 \le t \le 365.$$

This function does not have an inverse because it is not one-to-one on its domain. However, we saw that A is one-to-one on the interval $0 \le t \le 182.5$ and that if we restrict A to this part of its domain, then the resulting function does have an inverse.

This idea is an important one. If a function f is not one-to-one on its domain, then perhaps we can find a subset of the domain on which f is one-to-one and restrict f to this subset. We can then find an inverse function for f defined on this subset. We saw this demonstrated in Example 3. The function f described by

$$f(x) = x^2, \quad -\infty < x < \infty$$

does not have an inverse. However, the function F defined by

$$F(x) = x^2, \quad 0 \le x < \infty$$

does have an inverse.

EXAMPLE 5 Restricting a domain

Let f be defined by

$$f(x) = -x^2 + 2x - 7$$

with domain $\mathcal{D}$ the set of all real numbers. Show that f is not one-to-one on $\mathcal{D}$. Find a subset of $\mathcal{D}$ on which f is one-to-one, and find a formula for the inverse of the restricted version of f.

Solution

The graph of $y = f(x)$ is shown in Fig. 2.45. With a quick glance at the graph, we see that there are many horizontal lines that intersect the graph in more than one point. Thus f is not one-to-one on $\mathcal{D}$.

We can also show f is not one-to-one by finding two different real numbers u, v such that $f(u) = f(v)$. For example, let $u = 0$ and $v = 2$. Then

$$f(u) = f(0) = -7 = f(2) = f(v).$$

When a function gives the same output for two different inputs, the function is not one-to-one.

There are many subsets of $\mathcal{D}$ on which f is one-to-one. To find such a subset, erase a portion of the graph of f so that the remaining piece of the graph intersects each horizontal line in at most one point. Then f is one-to-one on the part of the domain corresponding to this piece of the graph. For

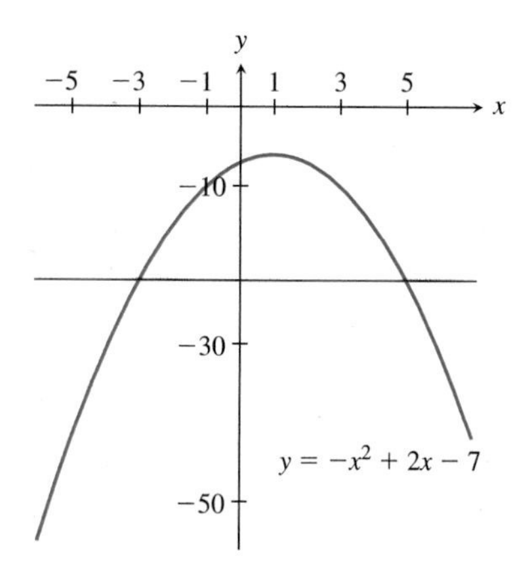

FIGURE 2.45 A horizontal line intersects the graph of f in more than one point.

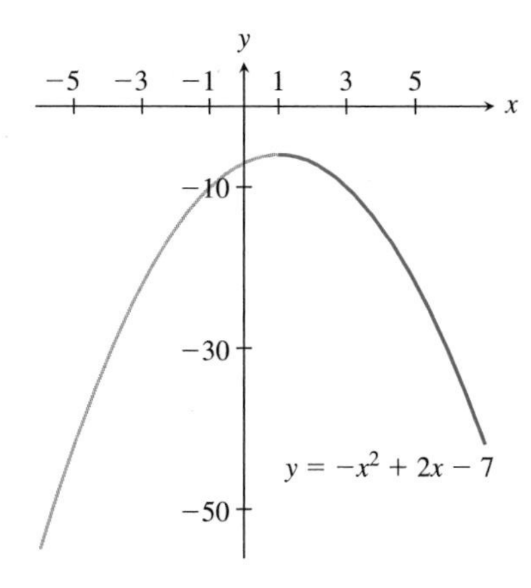

FIGURE 2.46 f is one-to-one on $[1, \infty)$.

example, the portion of the graph for $x \geq 1$ appears to satisfy the horizontal line condition. See Fig. 2.46. Thus f is one-to-one on $[1, \infty) = \{x : x \geq 1\}$.

Now let g be the inverse for the restricted version of f. Note that

$$\text{domain of } g = \text{range of the restricted } f = (-\infty, -6]$$

and

$$\text{range of } g = \text{domain of the restricted } f = [1, \infty).$$

For any x in $(-\infty, -6]$, we have

$$x = f(g(x)) = -(g(x))^2 + 2g(x) - 7.$$

Rewrite this as

$$(g(x))^2 - 2g(x) + (x + 7) = 0.$$

Use the quadratic equation to solve this expression for $g(x)$:

$$g(x) = 1 \pm \sqrt{-x - 6}. \tag{10}$$

The $\pm$ sign in (10) means that there are two possible choices for g. Because we want the range of g to be $[1, \infty)$, we choose the function that corresponds to the $+$ sign. Hence if we restrict f to the interval $[1, \infty)$, this restricted version of f has inverse g with

$$g(x) = 1 + \sqrt{-x - 6}.$$

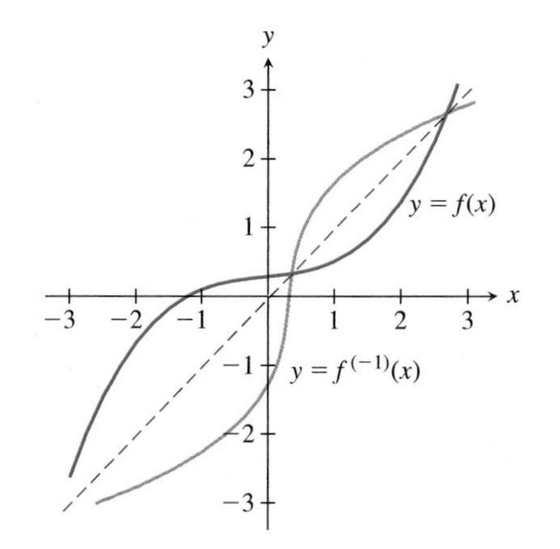

FIGURE 2.47 The graph of the inverse function f^{-1} is the reflection of the graph of f about $y = x$.

The Graph of the Inverse

Let f be a function and $g = f^{-1}$ be its inverse function. If (a, b) is on the graph of $y = f(x)$, then a is in the domain of f and

$$f(a) = b.$$

It follows that

$$a = g(b).$$

In particular, b is in the domain of g and (b, a) is on the graph of $y = g(x)$. Similarly, because f is the inverse of g, we find that if (b, a) is on the graph of $y = g(x)$, then (a, b) is on the graph of $y = f(x)$. Combining these facts, we see that a point (a, b) is on the graph of $y = f(x)$ exactly when the point (b, a) is on the graph of $y = g(x) = f^{-1}(x)$. In the (x, y)-plane, the point (b, a) is the reflection about the line $y = x$ of the point (a, b). This implies the following important relation between the graph of a function and the graph of its inverse. See Fig. 2.47.

The Graph of an Inverse Function

Let g be the inverse of f. The graph of $y = g(x)$ is the reflection about the line $y = x$ of the graph of $y = f(x)$.

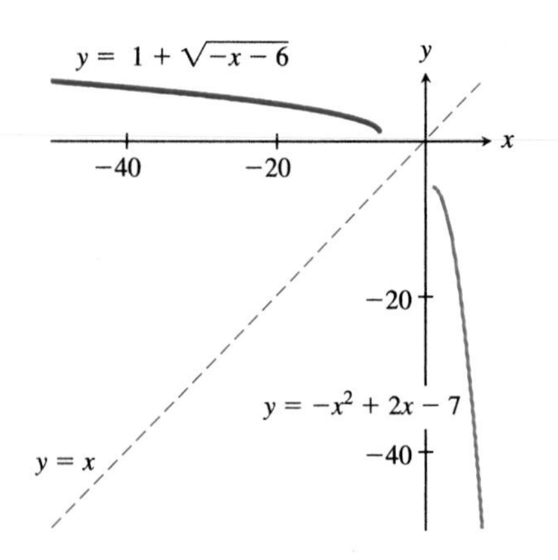

FIGURE 2.48 The graphs of $y = g(x) = f^{-1}(x)$ and $y = f(x)$.

EXAMPLE 6 In Example 5 we found g, the inverse of

$$f(x) = -x^2 + 2x - 7, \quad x \geq 1.$$

Verify that the graph of $y = g(x)$ is the reflection about $y = x$ of the graph of $y = f(x)$.

Solution

The graph of

$$y = g(x) = 1 + \sqrt{-x-6}$$

for $x \leq -6$ is shown in Fig. 2.48. On the graph we have included the graph of $y = f(x)$ for $x \geq 1$. The graph of $y = g(x)$ appears to be the reflection about the line $y = x$ of the graph of $y = f(x)$.

The Derivative of an Inverse

The relationship between the graph of a differentiable function and the graph of its inverse function leads to a relationship between the derivative of a function and the derivative of its inverse. First note that if a line ℓ of slope $m \neq 0$ is reflected about the line $y = x$, the result is a line of slope $1/m$. To see this, assume that (a, b) and (c, d) are points on ℓ. By the slope formula,

$$m = \text{slope of } \ell = \frac{d-b}{c-a}.$$

The reflection of ℓ about $y = x$ is another line. This reflection contains the points (b, a) and (d, c). See Fig. 2.49. The slope of the reflection is

$$\frac{c-a}{d-b} = \frac{1}{m}.$$

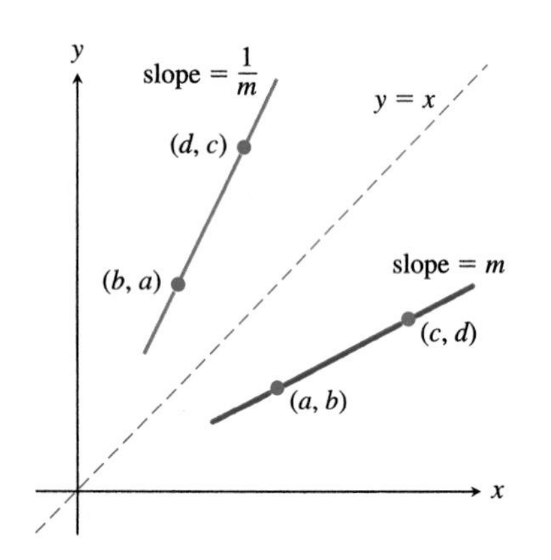

FIGURE 2.49 When a line of slope m is reflected about the line $y = x$, we get a line of slope $1/m$.

Now suppose that a one-to-one function f is differentiable at a point a in its domain and that $f(a) = b$. Then the point (a, b) is on the graph of f. If we zoom in on the graph of $y = f(x)$ near (a, b), we see a "line" of slope $f'(a)$. The point (b, a) is on the graph of $y = g(x) = f^{-1}(x)$ because $g(b) = a$. When we zoom in on the graph of $y = g(x)$ near (b, a), we again see a "line." This line is the reflection about $y = x$ of the "line" of slope $f'(a)$ that we see when zooming in on the graph of $y = f(x)$ near (a, b). Hence near (b, a), the graph of $y = g(x) = f^{-1}(x)$ looks like a line of slope $1/(f'(a))$. (See Fig. 2.50.) This slope is the derivative of g at b. We summarize this discussion in the following result.

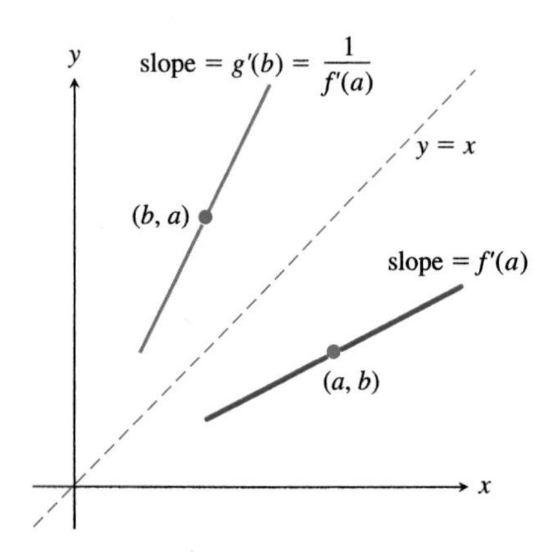

FIGURE 2.50 Zooming in on f and $g = f^{-1}$.

The Derivative of an Inverse

Let $g = f^{-1}$ be the inverse of f. If $f(a) = b$ and $f'(a) \neq 0$, then

$$(f^{-1})'(b) = g'(b) = \frac{1}{f'(a)} = \frac{1}{f'(g(b))}. \tag{11}$$

This result can also be derived using the chain rule. See Exercises 22 and 23. Equation (11) is of interest because it provides geometric insight into the relation between a function and its inverse.

EXAMPLE 7 The derivative of an inverse

Let $f(x) = 4e^{4x-8} - 3$ and observe that $f(2) = 1$. Let $g = f^{-1}$ be the inverse of f. Find $g'(1)$.

Solution

We present two solutions. For the first solution we use (11). Because $f(2) = 1$, we know that $g(1) = 2$. Also, by the chain rule,

$$f'(x) = 4e^{4x-8}(4x - 8)' = 16e^{4x-8}.$$

By (11),

$$g'(1) = \frac{1}{f'(g(1))} = \frac{1}{f'(2)} = \frac{1}{16}.$$

For the second solution, we find $g(x)$, calculate $g'(x)$, then find the value of the derivative when $x = 1$. Because g is the inverse of f, we have

$$x = f(g(x)) = 4e^{4g(x)-8} - 3.$$

Add 3 to both sides of this equation, then divide by 4 to obtain

$$\frac{x+3}{4} = e^{4g(x)-8}.$$

Take the natural logarithm of both sides to obtain

$$\ln\left(\frac{x+3}{4}\right) = \ln(e^{4g(x)-8}) = 4g(x) - 8.$$

It follows that

$$g(x) = 2 + \frac{1}{4}\ln\left(\frac{x+3}{4}\right).$$

By the chain rule,

$$g'(x) = \frac{1}{4}\frac{1}{\left(\frac{x+3}{4}\right)}\left(\frac{x+3}{4}\right)' = \frac{1}{4(x+3)}.$$

Hence $g'(1) = 1/16$.

Exercises 2.8

Exercises 1–8: Assuming that the function f has an inverse, denote the inverse function by g. Find g(x). Check that your answer is correct by verifying that g(f(x)) = x.

1. $f(x) = -3x + 8$

2. $f(x) = \dfrac{2}{3x - 5}$, $x \neq 5/3$

3. $f(x) = 3(x - 2)^3 + 7$

4. $f(x) = \dfrac{-x + 4}{2x + 4}$, $x \neq -2$

5. $f(x) = 2\ln(3x + 4)$, $x > -4/3$

6. $f(x) = -\ln(4 - x) + 5$, $x < 4$

7. $f(x) = 3e^{2x-5} + 6$

8. $f(x) = -2e^{4x+3} - 12$

Exercises 9–12: Let g be the inverse of f. Sketch the graphs of y = f(x) and y = g(x) on the same set of axes.

9. $f(x) = -3x + 2$

10. $f(x) = \dfrac{1}{x}$, $x \neq 0$

11. $f(x) = e^{-x}$

12. $f(x) = \ln(x + 2)$, $x > -2$

13. Show that

$$f(x) = \frac{2 - x}{3 + 2x}$$

is one-to-one on its domain and find a formula for the inverse. What is the domain of the inverse?

14. Show that

$$h(x) = \frac{5x^3}{x^2 + 1}$$

is one-to-one on the real line. What is the domain of the inverse of h?

15. Show that

$$G(t) = \frac{1}{2}\ln(7t)$$

is the inverse of

$$F(t) = \frac{1}{7}e^{2t}.$$

16. Show that the linear function ℓ defined by $\ell(x) = mx + b$ is one-to-one if $m \neq 0$. Write down a formula for the inverse of ℓ.

17. Let $f(x) = x^3 + 2x + 3$, let $g = f^{-1}$ be the inverse function for f, and note that $f(2) = 15$. Find $g'(15)$.

18. Let $F(x) = 2e^{3x-2} + 4$, let $G = F^{-1}$ be the inverse for F, and note that $F(1) = 2e + 4$. Find $G'(2e + 4)$.

19. Let $h(x) = x^2 + 4x - 3$ be defined for $x \geq -2$, let $k = h^{-1}$ be the inverse for h, and note that $h(1) = 2$. Calculate $k'(2)$.

20. Let $f(x) = \sin x$ be defined for $-\pi/2 \leq x \leq \pi/2$, let $g = f^{-1}$ be the inverse function for f, and note that $f(\pi/6) = 1/2$. Find $g'(1/2)$.

21. Let $f(x) = \tan x$ be defined for $-\pi/2 < x < \pi/2$, let $g = f^{-1}$ be the inverse function for f, and note that $f(-\pi/4) = -1$. Find $g'(-1)$.

22. Suppose that g is the inverse of f. Differentiate the expression

$$f(g(x)) = x,$$

and from the result arrive at formula (11) for the derivative of the inverse.

23. Suppose that g is the inverse of f and we write

$$y = f(x) \quad \text{and} \quad x = g(y).$$

Use the chain rule to show that

$$\frac{dy}{dx}\frac{dx}{dy} = 1.$$

24. Sometimes a function f is its own inverse. That is,

$$(f \circ f)(x) = x$$

for all x in the domain of f.

a. Write a short paragraph describing what the graph of such a function f must look like.

b. Draw three graphs for functions that are their own inverses. (No more than one of these graphs should be a line.)

c. Write a formula for a function that is its own inverse. Give two examples, at least one of which is not a line.

25. Estimate the angle of the sun above the horizon as seen from Bemidji, Minnesota, when the sun is at its highest point in the sky on January 31. On October 31. On your birthday. Use (1).

26. What are the possible dates when the sun's angle $A(t)$ in Bemidji measures 20°? When it measures 40°?

Exercises 27–32: Show that the function f is not one-to-one on its domain $\mathcal{D}$. Find a subset of the domain on which f is one-to-one, and then find a formula for the inverse of the restricted version of f.

27. $f(x) = x^2 - 2x + 1, \quad \mathcal{D} = (-\infty, \infty)$

28. $f(x) = -3x^2 + 7x, \quad \mathcal{D} = (-\infty, \infty)$

29. $f(x) = \dfrac{4x}{x^2 + 4}, \quad \mathcal{D} = (-\infty, \infty)$

30. $f(x) = \dfrac{-3}{x^2 - 9}, \quad \mathcal{D} = \{x : x \neq -3, 3\}$

31. $f(x) = \ln(x^2 + 1), \quad \mathcal{D} = (-\infty, \infty)$

32. $f(x) = e^{-x^2/4}, \quad \mathcal{D} = (-\infty, \infty)$

33. Let f be one-to-one on its domain and suppose that g is the inverse of f. Let b be a fixed real number. The function F defined by

$$F(x) = f(x) + b$$

is also one-to-one and so also has an inverse. Find a formula for the inverse of F in terms of g.

34. Let f be one-to-one on its domain and suppose that g is the inverse of f. Let a be a fixed nonzero real number. The function F defined by

$$F(x) = af(x)$$

is also one-to-one and so also has an inverse. Find a formula for the inverse of F in terms of g.

35. Let f be one-to-one on its domain and suppose that g is the inverse of f. Let c be a fixed nonzero real number. The function F defined by

$$F(x) = f(cx)$$

is also one-to-one (but perhaps on a different domain) and so also has an inverse. Find a formula for the inverse of F in terms of g.

36. Let f be one-to-one on its domain. Suppose that we reflect the graph of f about the line $y = -x$.

a. Show that the resulting graph is still the graph of a function; that is, show that this new graph also satisfies the vertical line condition.

b. Let g be the function represented by this new graph. What can you say about $(f \circ g)(x)$?

37. Sometimes the graph of a function and its inverse intersect in one or more points. Explain why these points of intersection either fall on the line $y = x$ or else occur in pairs of the form (a, b), (b, a).

38. Let g and f be functions with

$$\text{domain}(g) = \text{range}(f)$$

and

$$g(f(x)) = x \quad \text{for all } x \text{ in the domain of } f.$$

In other words, g is the inverse of f. Show that f is also the inverse of g.

39. Let f be a function and g be its inverse. Suppose we want to illustrate that the graph of $y = g(x)$ is the reflection about $y = x$ of the graph of $y = f(x)$. Explain why it is important that the scales on the x- and y-axes be the same.

2.9 Inverse Trigonometric Functions

In this section we continue our discussion of inverse functions by studying the inverse trigonometric functions. As expected, the inverse trigonometric functions are useful in analyzing problems and situations involving angles. Every trigonometry student who has "solved a triangle" has worked with the inverse sine function and with the inverses of other trigonometric functions. The inverse trigonometric functions are also useful because they have surprisingly simple derivatives. As a result, these functions can show up in models of many physical situations.

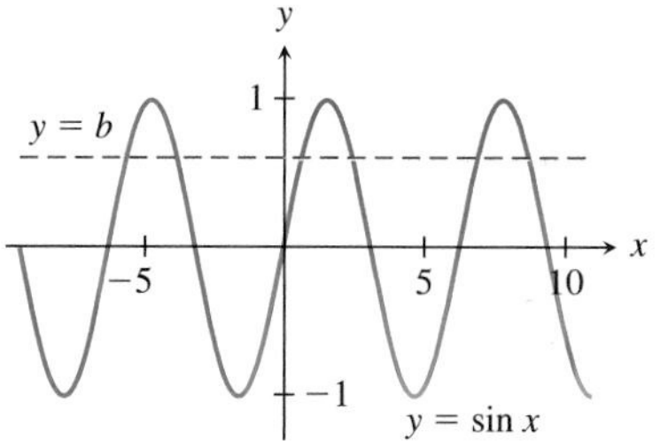

FIGURE 2.51 The sine function is not one-to-one on the real line.

The Inverse Sine Function The sine function has domain $(-\infty, \infty)$ but is not one-to-one on this set. For each real b with $-1 \leq b \leq 1$, the horizontal line $y = b$ intersects the graph of $y = \sin x$ in infinitely many points. See Fig. 2.51. Before we can define an inverse for the sine function, we need to find a subset of the domain on which the function is one-to-one. The domain most often selected is

$$[-\pi/2, \pi/2] = \{x : -\pi/2 \leq x \leq \pi/2\}.$$

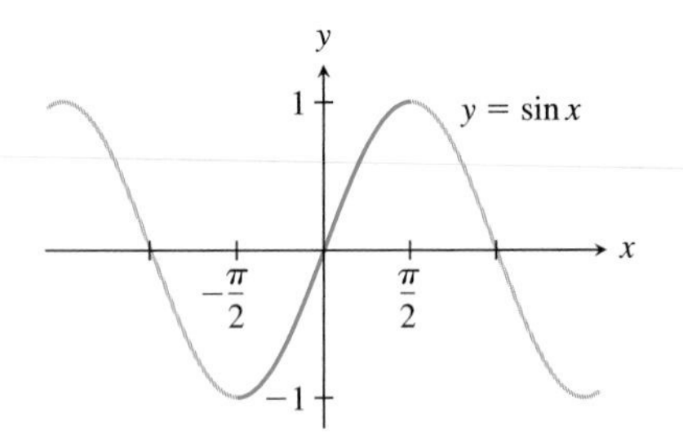

FIGURE 2.52 The sine function is one-to-one on $-\pi/2 \le x \le \pi/2$.

On this domain the sine function is one-to-one and takes on every value from -1 to 1. See Fig. 2.52. The inverse of the sine function on this domain takes as input a number b between -1 and 1 and returns as output the number a between $-\pi/2$ and $\pi/2$ for which $\sin a = b$. This inverse function is called the *arcsine function*.

DEFINITION Inverse Sine Function

The inverse sine function (or arcsine function) has domain

$$[-1,1] = \{x\colon -1 \le x \le 1\}.$$

For x in this domain,

$$\arcsin x = y$$

where y is the number between $-\dfrac{\pi}{2}$ and $\dfrac{\pi}{2}$ for which

$$\sin y = x.$$

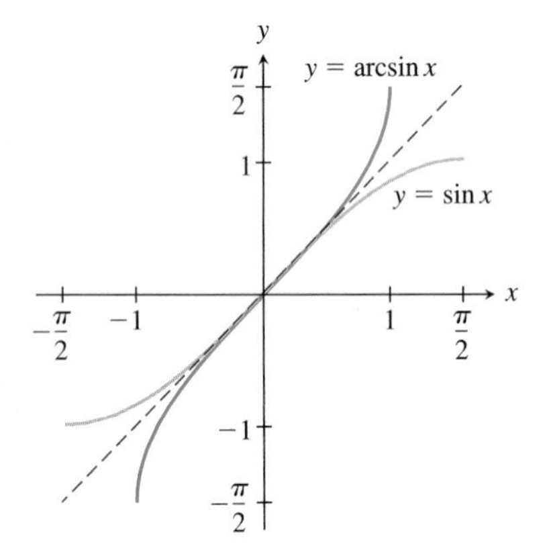

FIGURE 2.53 The graph of $y = \arcsin x$.

Because sine and arcsine are inverse functions, we have

$$\arcsin(\sin x) = x \quad \text{for } -\frac{\pi}{2} \le x \le \frac{\pi}{2} \tag{1}$$

and

$$\sin(\arcsin x) = x \quad \text{for } -1 \le x \le 1. \tag{2}$$

The graph of the inverse sine function is found by reflecting the graph of $y = \sin x$ on $[-\pi/2, \pi/2]$ about the line $y = x$. See Fig. 2.53.

EXAMPLE 1 Find the value of

a) $\arcsin\left(-\frac{\sqrt{2}}{2}\right)$

b) $\sin\left(\arcsin\left(\frac{3}{5}\right)\right)$

c) $\arcsin(\sin 10)$

Solution

a) Because $\sin(-\pi/4) = -\sqrt{2}/2$ and $-\pi/4$ is between $-\pi/2$ and $\pi/2$, we have

$$\arcsin\left(-\frac{\sqrt{2}}{2}\right) = -\frac{\pi}{4}.$$

(Note that $\sin(5\pi/4) = -\sqrt{2}/2$ is also true. However, $\arcsin\left(-\sqrt{2}/2\right) \ne 5\pi/4$ because $5\pi/4 > \pi/2$ and hence is not in the range of the arcsine function.)

b) Because $-1 \leq 3/5 \leq 1$, we can apply (2) to obtain

$$\sin\left(\arcsin\left(\frac{3}{5}\right)\right) = \frac{3}{5}.$$

c) We cannot use (1) because 10 is not between $-\pi/2$ and $\pi/2$. In Fig. 2.54 we show the graph of $y = \sin x$ and the point on the graph with $x = 10$. Note that the number $X = 3\pi - 10$ shown in Fig. 2.54 lies between $-\pi/2$ and $\pi/2$ and that $\sin X = \sin 10$. We use symmetry to calculate the value of X. From the graph, $X = 3\pi - 10$, because X is as far to the left of the origin as 10 is to the right of 3π. Because X is between $-\pi/2$ and $\pi/2$, we can apply (1) to get

$$\arcsin(\sin 10) = \arcsin(\sin X) = X = 3\pi - 10.$$

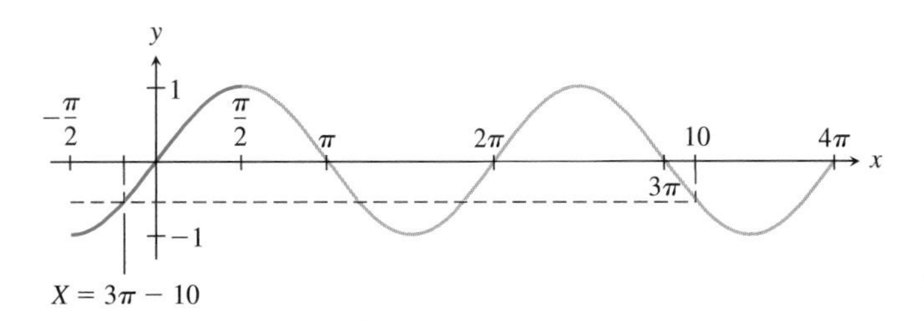

FIGURE 2.54 Example 1**c**: $\sin X = \sin 10$.

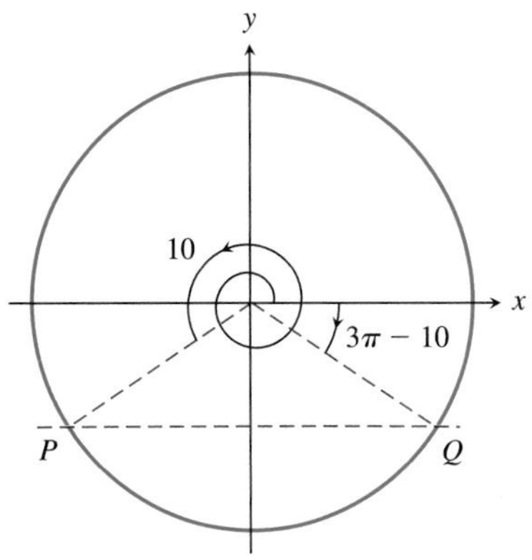

FIGURE 2.55 The third-quadrant angle 10 has the same sine as the fourth-quadrant angle $3\pi - 10$.

For another approach, sketch the angle of measure 10 radians in the unit circle, and let P be at the point at which the terminal side of the angle intersects the unit circle. The y-coordinate of P is $\sin(10)$. We find a second point Q on the unit circle, such that Q and P have the same y-coordinates by drawing a horizontal line through P. See Fig. 2.55. The point Q is in the fourth quadrant and corresponds to the angle of measure $3\pi - 10$. Because $3\pi - 10$ is between $-\pi/2$ and $\pi/2$, we have

$$\arcsin(\sin 10) = \arcsin(\sin(3\pi - 10)) = 3\pi - 10.$$

EXAMPLE 2 Let $y = \arcsin x$. Find $\dfrac{dy}{dx}$.

Solution

We could apply (11) in Section 2.8, but there is really no need to look up a formula. The derivative can quickly be found using (2) and the chain rule. By (2),

$$\sin y = \sin(\arcsin x) = x, \quad -1 \leq x \leq 1. \tag{3}$$

Differentiate the left and right sides of this equation with respect to x, keeping in mind that y is a function of x. By the chain rule,

$$\cos y \frac{dy}{dx} = 1.$$

This leads to

$$\frac{dy}{dx} = \frac{1}{\cos y}. \tag{4}$$

This derivative is defined when $\cos y \neq 0$. Because $-\frac{\pi}{2} \leq y \leq \frac{\pi}{2}$, we see the derivative is defined for

$$-\frac{\pi}{2} < y < \frac{\pi}{2}. \tag{5}$$

There is nothing wrong with the answer in (4), but it can be put into a more useful form. For y satisfying (5), we know $\cos y > 0$. By (3) we see that for such y,

$$\cos y = \sqrt{1 - \sin^2 y} = \sqrt{1 - x^2}.$$

Hence

$$\frac{dy}{dx} = \frac{d}{dx} \arcsin x = \frac{1}{\sqrt{1 - x^2}} \quad \text{for } -1 < x < 1. \tag{6}$$

EXAMPLE 3 Let $y = \arcsin(e^{2x})$. Calculate $\frac{dy}{dx}$.

Solution

A composition chain for the expression is

$$y = \arcsin u$$
$$u = e^w$$
$$w = 2x.$$

By the chain rule and (6)

$$\frac{dy}{dx} = \frac{dy}{du}\frac{du}{dw}\frac{dw}{dx} = \frac{1}{\sqrt{1 - u^2}} \cdot e^w \cdot 2.$$

Replace u by e^w, and then w by $2x$ to get

$$\frac{dy}{dx} = \frac{2e^{2x}}{\sqrt{1 - e^{4x}}}.$$

Most calculators have a built-in arcsine function, though it may be denoted by $\sin^{-1}$ or asin to save space. (Keep in mind that $\sin^{-1} x$ *does not* mean $1/\sin x$.)

The Inverse Tangent Function The tangent function is defined for all real numbers except those for which the cosine is 0. Thus

$$\tan x = \frac{\sin x}{\cos x} \quad \text{for } x \neq \pm\frac{\pi}{2}, \pm\frac{3\pi}{2}, \pm\frac{5\pi}{2}, \ldots.$$

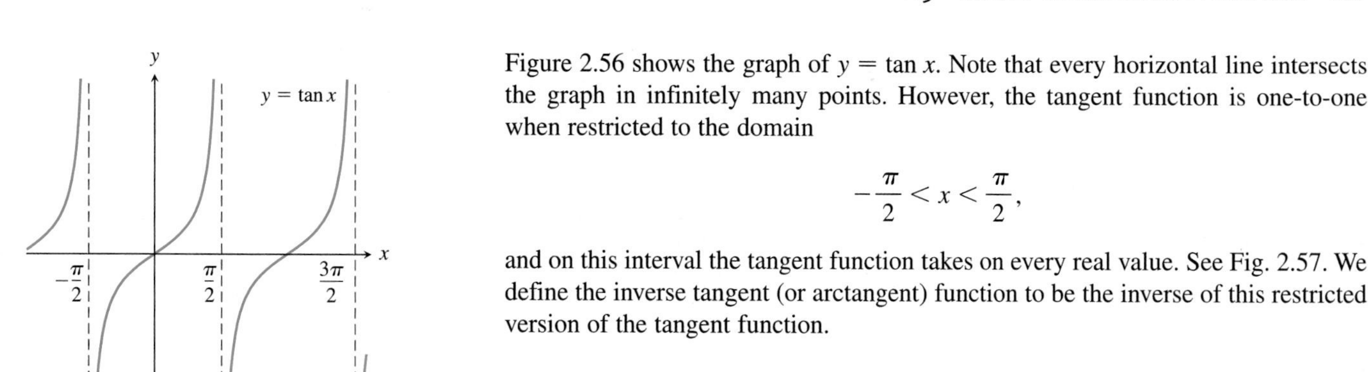

FIGURE 2.56 The tangent function is not one-to-one on the real line.

Figure 2.56 shows the graph of $y = \tan x$. Note that every horizontal line intersects the graph in infinitely many points. However, the tangent function is one-to-one when restricted to the domain

$$-\frac{\pi}{2} < x < \frac{\pi}{2},$$

and on this interval the tangent function takes on every real value. See Fig. 2.57. We define the inverse tangent (or arctangent) function to be the inverse of this restricted version of the tangent function.

DEFINITION Inverse Tangent Function

The inverse tangent function (or arctangent function) has as its domain the set of all real numbers. For real x,

$$\arctan x = y$$

where y is the number between $-\dfrac{\pi}{2}$ and $\dfrac{\pi}{2}$ for which

$$\tan y = x.$$

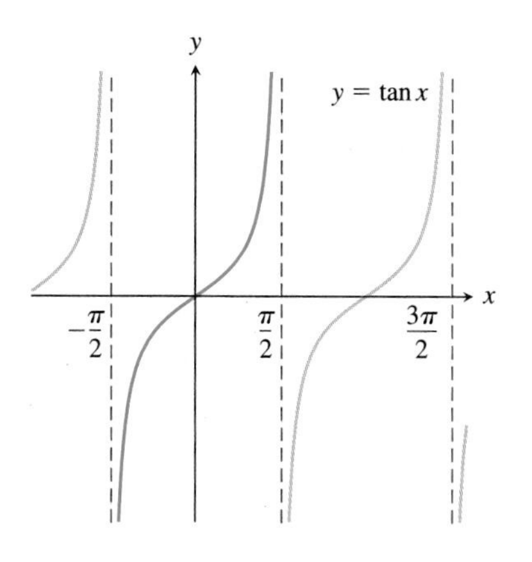

FIGURE 2.57 The tangent function is one-to-one on a suitable domain.

Because tangent and arctangent are inverse functions, we have

$$\arctan(\tan x) = x \quad \text{for } -\frac{\pi}{2} < x < \frac{\pi}{2}$$

and

$$\tan(\arctan x) = x \quad \text{for all real } x. \tag{7}$$

The graph of the inverse tangent function is found by reflecting the graph of $y = \tan x$ $(-\pi/2 < x < \pi/2)$ about the line $y = x$. The graph of $y = \tan x$ has vertical asymptotes at $x = \pm\pi/2$. Under reflection about the line $y = x$, these become horizontal asymptotes $y = \pm\pi/2$ for the graph of $y = \arctan x$. This means that as x gets large and positive, $\arctan x$ approaches $\pi/2$, and as x gets large and negative, $\arctan x$ approaches $-\pi/2$. The graph of the arctangent function is shown in Fig. 2.58.

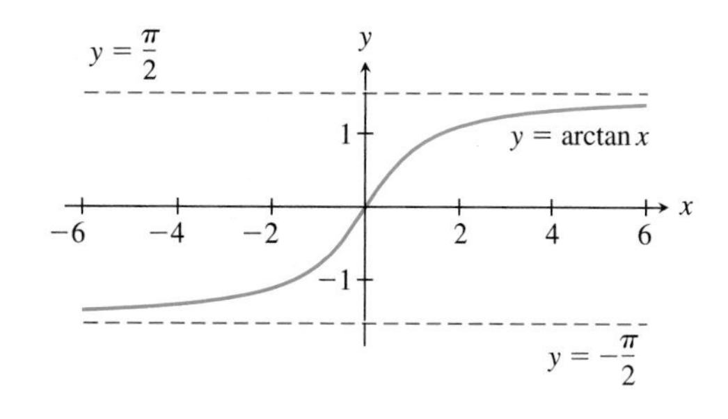

FIGURE 2.58 The graph of $y = \arctan x$ has two horizontal asymptotes.

EXAMPLE 4

a) Find $\arctan\left(-\sqrt{3}\right)$.
b) Simplify $\sin(\arctan x)$.

Solution

a) Because $\tan(-\pi/3) = -\sqrt{3}$ and $-\pi/3$ is between $-\pi/2$ and $\pi/2$, we have

$$\arctan\left(-\sqrt{3}\right) = -\frac{\pi}{3}.$$

b) Simplifications of expressions like this can usually be done with the aid of a reference triangle. Let $\theta = \arctan x$. Then

$$-\frac{\pi}{2} < \theta < \frac{\pi}{2} \quad \text{and} \quad \tan\theta = x.$$

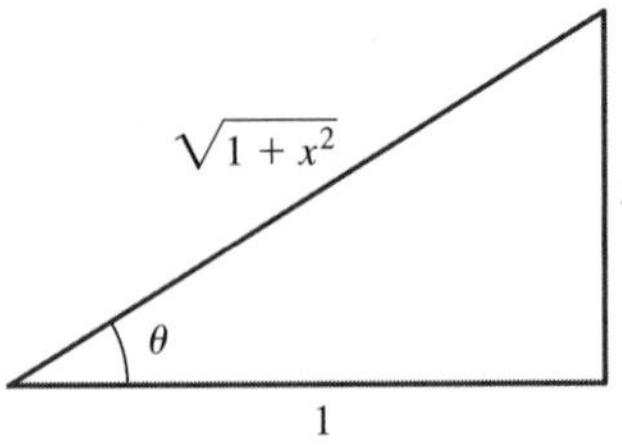

In Fig. 2.59 we have sketched a right triangle illustrating $\tan\theta = x$. In this triangle we have given the side opposite angle θ length x and the side adjacent length 1. By the Pythagorean theorem, the hypotenuse has length $\sqrt{1+x^2}$. We now see $\sin\theta = \dfrac{x}{\sqrt{x^2+1}}$, which leads to

$$\sin(\arctan x) = \sin\theta = \frac{x}{\sqrt{x^2+1}}.$$

See Exercise 37 for the $x < 0$ case.

$\sqrt{1+x^2}$
x
θ
1

FIGURE 2.59 A triangle illustrating $\tan\theta = x$.

EXAMPLE 5 Find the derivative of $\arctan x$.

Solution

Let $y = \arctan x$. We then have

$$\tan y = x. \tag{8}$$

Differentiate both sides of this equation with respect to x to obtain

$$\sec^2 y \frac{dy}{dx} = 1.$$

We solve this equation for dy/dx and use the identity $\sec^2 y = 1 + \tan^2 y$ and (8) to get

$$\frac{dy}{dx} = \frac{1}{\sec^2 y} = \frac{1}{1+\tan^2 y} = \frac{1}{1+x^2}.$$

Hence

$$\frac{d}{dx}(\arctan x) = \frac{1}{1 + x^2}.$$

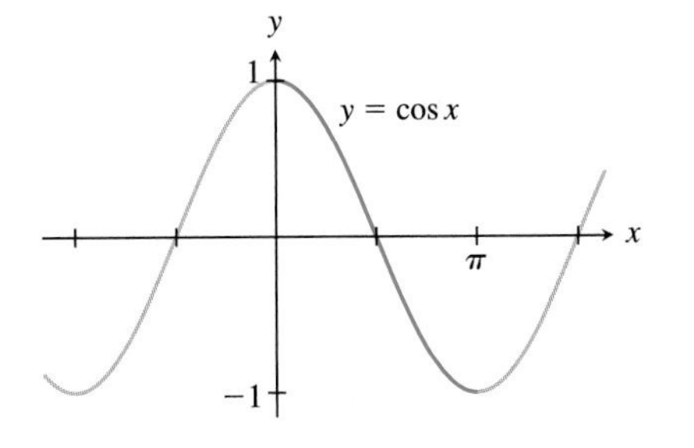

FIGURE 2.60 The cosine function is one-to-one on the interval $0 \leq x \leq \pi$.

Other Inverse Trigonometric Functions Most scientific calculators have arcsine, arctangent, and arccosine as built-in functions. We have already discussed the first two functions in some detail. The inverse for the cosine function can be developed and analyzed by methods similar to those used for the arcsine and arctangent functions. We give a brief discussion of the inverse cosine function, but leave many of the details to the exercises.

The cosine function is one-to-one on the interval $[0, \pi] = \{x : 0 \leq x \leq \pi\}$. For x in this interval, $\cos x$ takes on every value between -1 and 1. See Fig. 2.60. We base the definition of the inverse cosine function on this restricted version of the cosine function.

DEFINITION Inverse Cosine Function

The inverse cosine function (or arccosine function) has domain

$$[-1, 1] = \{x : -1 \leq x \leq 1\}.$$

For x in this domain,

$$\arccos x = y$$

where y is the number between 0 and π for which

$$\cos y = x.$$

Using techniques seen earlier in this section, we can show that

$$\frac{d}{dx}(\arccos x) = -\frac{1}{\sqrt{1 - x^2}}.$$

See Exercise 28.

The inverse secant function can be defined using the same kinds of ideas used in defining the arcsine, arctangent, and arccosine. However, there is an alternative approach. Note that if

$$\sec a = b \quad \text{then} \quad \cos a = \frac{1}{b}.$$

Thus, with proper attention paid to the domain of arcsecant, it would appear that

$$\operatorname{arcsec} b = a = \arccos\left(\frac{1}{b}\right).$$

We formalize this in the following definition.

DEFINITION Inverse Secant Function

The inverse secant function (or arcsecant function) has domain

$$(-\infty, -1] \cup [1, +\infty) = \{x : |x| \geq 1\}.$$

For x in this domain,

$$\operatorname{arcsec} x = y = \arccos\left(\frac{1}{x}\right). \tag{9}$$

Thus y is the number between 0 and π, with $y \neq \pi/2$, and

$$\sec y = x.$$

The arcsecant function also has a surprisingly simple derivative. Using (9) it can be shown that

$$\frac{d}{dx}(\operatorname{arcsec} x) = \frac{d}{dx}\left(\arccos\left(\frac{1}{x}\right)\right) = \frac{1}{|x|\sqrt{x^2 - 1}}.$$

See Exercise 32.

Exercises 2.9

Exercises 1–10: Find the exact value of each of the following expressions.

1. $\arcsin(\sqrt{3}/2)$
2. $\arctan(-1)$
3. $\arcsin(-1)$
4. $\arctan(\tan 6)$
5. $\operatorname{arcsec} 2$
6. $\arccos(-\sqrt{3}/2)$
7. $\sin(\arcsin(-\pi/4))$
8. $\sin(\arctan(3/4))$
9. $\cos(\operatorname{arcsec}(10.6))$
10. $\tan(\operatorname{arcsec} 3)$

Exercises 11–14: Simplify each of the following. Your simplified result should be free of trigonometric and inverse trigonometric functions.

11. $\tan(\arcsin x)$
12. $\sin(\arccos x)$
13. $\sin(\operatorname{arcsec}(2x))$
14. $\cos(\arctan(-5x))$

Exercises 15–26: Find the derivative.

15. $y = \arcsin(2x)$
16. $h(t) = t \arctan t$
17. $r = \arctan\left(\frac{1}{\theta}\right)$
18. $y = \arccos(x^2)$
19. $g(s) = \operatorname{arcsec}(3s - 4)$
20. $z = \frac{1}{\arccos t}$
21. $y = \arcsin\left(\frac{1 + x}{1 - x}\right)$
22. $r(t) = \arcsin t + \arcsin(\sqrt{1 - t^2})$
23. $f(x) = e^{\arctan(x^2)}$
24. $y = \ln(\operatorname{arcsec} x)$
25. $F(w) = \frac{1 + \arccos w}{1 - \arccos w}$
26. $q(r) = \sqrt{1 + \operatorname{arcsec} \sqrt{r}}$
27. Sketch the graph of $y = \arccos x$.

28. Use a technique similar to that used in finding the derivative of the arcsine function to show that

$$\frac{d}{dx}(\arccos x) = -\frac{1}{\sqrt{1-x^2}}$$

for $-1 < x < 1$.

29. The right triangle shown has one side of length x and hypotenuse of length 1.

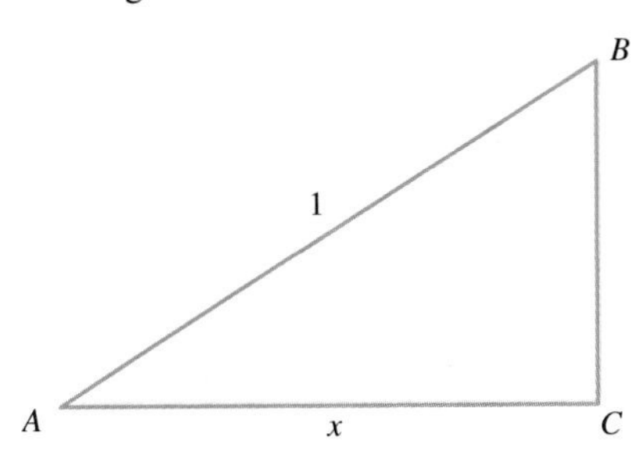

Figure for Exercises 29 and 30.

a. Show that angle B has measure $\arcsin x$ and angle A has measure $\arccos x$.

b. Explain why part **a** suggests the identity

$$\arcsin x + \arccos x = \frac{\pi}{2},$$

and then show that this identity holds for all x with $-1 \le x \le 1$.

c. Use the identity established in **b** to show that

$$\frac{d}{dx}(\arccos x) = -\frac{1}{\sqrt{1-x^2}}$$

for $-1 < x < 1$.

30. Use the right triangle figure to show that

$$\arcsin x = \arccos \sqrt{1-x^2}$$

for $0 \le x \le 1$.

31. Sketch the graph of

$$y = \sec x, \quad 0 \le x \le \pi, x \ne \frac{\pi}{2}.$$

Use this graph as an aid in sketching the graph of

$$y = \operatorname{arcsec} x, \quad |x| \ge 1.$$

32. Use (9) to show that

$$\frac{d}{dx}(\operatorname{arcsec} x) = \frac{1}{|x|\sqrt{x^2-1}}.$$

33. a. Find the equation for the line tangent to the graph of $y = \arctan x$ at the point $(1, \pi/4)$.

b. Sketch the graph of $y = \arctan x$ and the graph of the tangent line.

c. Use the equation for the tangent line to find approximate values for $\arctan(0.95)$ and $\arctan(1.03)$.

34. a. Find the equation for the line tangent to the graph of $y = \arcsin x$ at the point $(-1/2, -\pi/6)$.

b. Sketch the graph of $y = \arcsin x$ and the graph of the tangent line.

c. Use the equation for the tangent line to find approximate values for $\arcsin(-0.48)$ and $\arcsin(-0.56)$.

35. a. Find the equation for the line tangent to the graph of $y = \arccos x$ at the point $(-1/2, 2\pi/3)$.

b. Sketch the graph of $y = \arccos x$ and the graph of the tangent line.

c. Use the equation for the tangent line to find approximate values for $\arccos(-0.48)$ and $\arccos(-0.56)$.

36. a. Find the equation for the line tangent to the graph of $y = \operatorname{arcsec} x$ at the point $(2, \pi/3)$.

b. Sketch the graph of $y = \operatorname{arcsec} x$ and the tangent line.

c. Use the equation for the tangent line to find approximate values for $\operatorname{arcsec}(2.04)$ and $\operatorname{arcsec}(1.93)$.

37. In Example 4 and Fig. 2.59, we used a triangle to illustrate $\tan\theta = x$. The figure is certainly valid if $x > 0$. Is the triangle still useful and the result of Example 4 valid if $x < 0$?

2.10 Modeling: Translating the World into Mathematics

Mathematical models are used in almost every scientific and engineering field. A mathematical model is not a sculpture made of bronze or clay, nor is it made by gluing together pieces of plastic. A mathematical model is an equation or a set of equations. Mathematical models are used to help scientists and engineers better understand physical phenomena, to help economists make predictions about next year's economy, and to help sociologists learn more about human behavior.

A mathematical model usually does not describe exactly the phenomenon it is intended to model. This is because real-world phenomena are usually very complex. There may be factors that play a major role in shaping a situation, as well as many minor factors that have a smaller influence. A mathematical model that attempts to account for all aspects of a situation may require so many equations and pieces of data that it becomes awkward and hard to use. Thus, when mathematicians, scientists, and engineers create models, they try to take into account those factors that have a major influence on a situation but may ignore many of the minor ones. In other words, they make a trade-off between having a model that accounts for everything and having a model that they can work with. Because most mathematical models are not perfect, there is always room for improvement. As new mathematics is developed and faster computers become available, models can be changed to accommodate more of the factors that influence a situation.

In this section we look at some mathematical models that involve derivatives. Such models are not unusual because, perhaps surprisingly, it is often easier to describe the rate of change of a quantity than it is to describe the quantity directly. Thus our models will be equations that involve a function and its derivative. Such an equation is called a *differential equation*. The solution to a differential equation is not a number, but instead is a function or functions. For example, the differential equation

$$x\frac{dy}{dx} + 3y = \frac{e^x}{x^2} \tag{1}$$

describes a function $y = y(x)$ which, when substituted for y on the left side of (1), results in a true equation. You can check that $y = e^x/x^3$ is one such function. See Exercise 1. In this section we do not attempt to solve the differential equations in our models; you will learn to solve such equations in future courses. We have chosen these examples to show the diverse sorts of phenomena that can be modeled mathematically, to illustrate the kinds of assumptions made by mathematical modelers, and to demonstrate how real-world situations are translated into mathematics.

Population Models

Table 2.5 shows U.S. census data for the years 1790–1890. Based on these data, can we predict the population for the year 2000? This is an example of the kind of question that mathematical modelers attempt to answer.

TABLE 2.5

Year	Population (millions)
1790	4
1800	5
1810	7
1820	10
1830	13
1840	17
1850	23
1860	31
1870	40
1880	50
1890	63

A Simple Model The major factors influencing the size of a population are births and deaths. The simplest population model is based on two natural assumptions about the birth rate and the death rate:

1. The number of babies born in a given year is proportional to the population for that year. Hence there is a positive constant β such that
$$B(t) = \beta P(t),$$
where $B(t)$ is the number of births in year t and $P(t)$ is the population in year t.
2. The number of deaths in a given year is proportional to the population for that year. Thus, there is a positive constant δ such that
$$D(t) = \delta P(t),$$
where $D(t)$ is the number of deaths in year t.

The change in the population is the number of births minus the number of deaths. In year t this change is

$$\Delta P = B(t) - D(t) = \beta P(t) - \delta P(t) = kP(t),$$

where $k = \beta - \delta$. Because ΔP is the change in the number of people in a one-year period, the units of ΔP are people per year. Thus ΔP can be interpreted as the rate of change of $P(t)$ with respect to time. Because the rate of change is the derivative, we can describe the change in population by a *differential equation*,

$$\frac{dP}{dt} = kP. \tag{2}$$

Equation (2) is a simple model for population growth. To see how well such a model can describe and predict the U.S. population, we need to solve (2) for P and determine the value of k.

In Section 2.6 we saw that one function that satisfies (2) is the exponential function. We can check that if C is a constant, then

$$P(t) = Ce^{kt}$$

satisfies (2). To determine C and k, we use the data from the table. For convenience, we take $t = 0$ as 1790, so $t = 10$ is 1800, and so on. To determine C, use the 1790 figure ($t = 0$):

$$4 \times 10^6 = P(0) = Ce^{k \cdot 0} = C \cdot 1 = C.$$

Next use, say, the 1840 figure ($t = 50$) to determine k. We have

$$17 \times 10^6 = P(50) = 4 \times 10^6 e^{k \cdot 50}.$$

Divide both sides of this equation by 4×10^6 and then take the logarithm of both sides to get

$$\ln(17/4) = \ln(e^{k \cdot 50}) = 50k(\ln e) = 50k.$$

Thus,

$$k = \frac{\ln(17/4)}{50} \approx 0.0289384,$$

and we have

$$P(t) = 4 \times 10^6 e^{0.0289384t}. \tag{3}$$

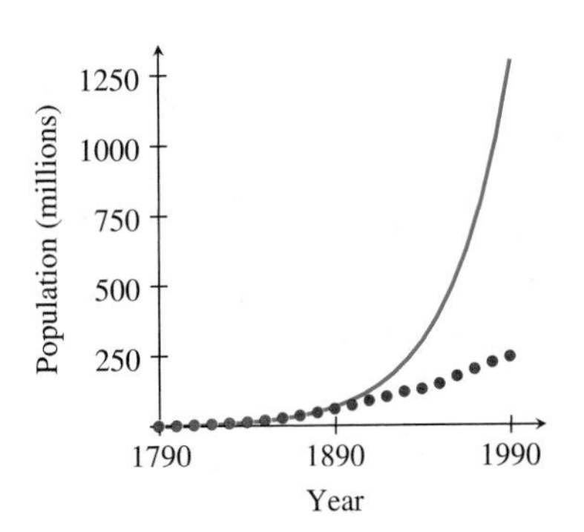

FIGURE 2.61 The simple model does not fit the data well after 1890.

To test this model, we compare it with the U.S. census data. In Fig. 2.61 the points represent the U.S. census data from 1790 through 1990, and the curve is the graph of $P = P(t)$. We see that the curve fits the data very well for the first 100 years or so, but it overestimates the population after 1900. In fact, for $t = 1990$, the P value for the point on the curve is more than five times the actual 1990 population. The year 2000 corresponds to $t = 210$. Putting this value for t into (3), we obtain

$$P(210) = 4 \times 10^6 e^{(0.0289384)(210)} \approx 1.7 \text{ billion}.$$

Judging by Fig. 2.61, this number was not a good prediction of the population for the year 2000.

A Better Model It is doubtful that the U.S. population will ever be 1.7 billion. Long before the population could reach this level the country would face shortages of food, living space, energy, and so on. All of these factors would tend to slow the rate of population growth. With only simple assumptions about birth and death rates, (3) does a good job of predicting population growth under uncrowded conditions (e.g., before 1890). However, nothing in the model accounts for the fact that the United States, or any other nation, cannot accommodate an unlimited number of people. In an effort to improve the predictive power of our model, we modify it to take overcrowding into account.

Because (2) does a good job of describing the change in $P(t)$ when $P(t)$ is small, we keep the kP term in our model. We include a second term to account for the effects of overcrowding. In a population of size P, a particular individual might encounter any of the $P - 1$ other individuals. Because this is true for each of P individuals, there are $P(P - 1)$ different encounters possible. To keep the model simple, we assume that there are P^2 encounters possible. This assumption is reasonable because when P is large, $P(P - 1) \approx P^2$. Of course, not all of these encounters will occur. However, it is reasonable to assume that the number of encounters is proportional to the number of possible encounters. Thus, when P is large, several such encounters take place and people feel crowded. Under such conditions people may be competing for resources such as food, space, and jobs. This will tend to slow the rate of growth of the population. Thus we modify (2) by adding a term $-cP^2$, so that

$$\frac{dP}{dt} = kP - cP^2. \tag{4}$$

In (4) the constant c is much smaller than k because the effects of overcrowding are not felt until P is large.

In Chapter 5 we will learn how to solve (4) using integration. Using the population data for times $t = 0$, 50, and 100 to determine values for the constants in the solution, it can be shown that

$$P(t) = \frac{292 \times 10^6}{1 + 72e^{-0.03t}}. \tag{5}$$

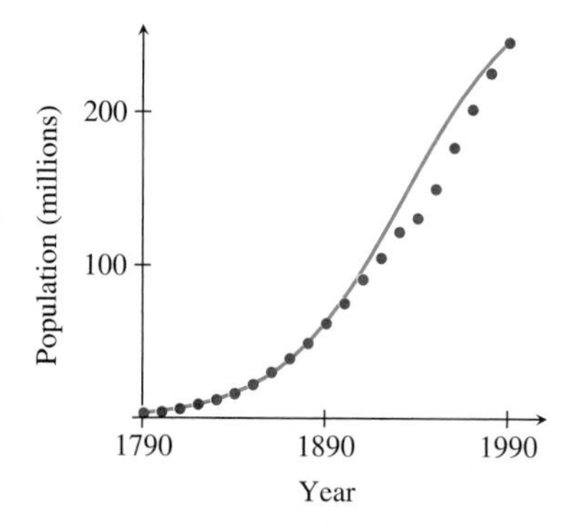

FIGURE 2.62 The model that takes overcrowding into account fits the data pretty well.

In Fig. 2.62, we have plotted the census data and the graph of (5). The figure shows that (5) agrees pretty well with the data and thus might give a good prediction for the population in 2000. The predicted figure is

$$P(210) = \frac{292 \times 10^6}{1 + 72e^{-0.03 \cdot 210}} \approx 258{,}000{,}000.$$

After the 2000 census results are published, we can see how close this prediction is.

In studying animal populations, it is not unusual to run across interrelated species. For example, in a wilderness area populated by rabbits and foxes, the foxes eat the rabbits. This means that the rate of change of the rabbit population depends on the number of foxes (more foxes eat more rabbits). On the other hand, the rate of change of the fox population also depends on the size of the rabbit population (more rabbits mean more food and this leads to more foxes). Thus it is impossible to model the fox population without modeling the rabbit population at the same time. (See Exercise 15.) Models that must account for several things at once usually involve several interrelated equations.

Modeling Red Blood Cell Levels during Surgery

During some surgeries, a patient can lose up to 2.5 liters of blood. During such operations, a patient is given a transfusion of saline solution to maintain blood volume in the body. However, the saline does not contain red blood cells, so cells lost during surgery are not replaced. Normally human blood is about 40% red cells by volume. If the red cell level is too low at the end of surgery, the patient may not survive. To lessen the red cell loss, some of a patient's blood may be removed before surgery and replaced with saline. The blood is then returned to the patient after the surgery. This process, called acute normovolemic hemodilution (ANH), has the effect of decreasing the amount of red cells lost during surgery.

We can model the patient's red cell levels during surgery and use this model to assess the effect of ANH. We make the following assumptions:

1. During surgery 2.5 liters of blood is lost, and the loss occurs at a constant rate.
2. Saline is administered at a rate equal to the rate of blood loss, keeping the combined volume of the blood and saline at 5 liters.
3. The saline mixes immediately with the blood in the body, which means that the red cell concentration is uniform throughout the blood.

Suppose that the surgery lasts T minutes. By assumption 1, the blood is lost at a rate of $(2.5/T)$ liters/minute. Let $R(t)$ be the volume of red cells at time t. Then at time t each liter of blood contains $R(t)/5$ liters of red cells. It follows that the rate of loss of the red cells is

$$\left(\frac{R(t)}{5} \text{ liters cells/liter blood}\right)\left(\frac{2.5}{T} \text{ liters blood/minute}\right)$$
$$= \frac{1}{2T}R(t) \text{ liters cells/minute}$$

Because this rate of change describes a decrease in the cell volume, it makes a negative contribution to the rate of change of the red cell level. Thus we have the differential equation

$$\frac{dR}{dt} = -\frac{1}{2T}R(t). \tag{6}$$

Before surgery or ANH the patient has $0.4 \cdot 5 = 2$ liters of red cells. If L liters of blood are removed and replaced with saline before the operation, then the patient enters surgery with $2 - 0.4 \cdot L$ liters of red cells. Thus if $t = 0$ corresponds to the start of surgery, then we have $R(0) = 2 - 0.4 \cdot L$. This starting (initial) condition together with (6) constitutes our model,

$$\frac{dR}{dt} = -\frac{1}{2T}R(t), \quad R(0) = 2 - 0.4 \cdot L. \tag{7}$$

In Exercise 7 we look at a solution to this model and investigate the effect of the ANH procedure.

This model is based on the article "Calculus in the Operating Room," by Pearl Toy and Stan Wagon, which appeared in the February 1995 issue of *The American Mathematical Monthly*.

Modeling Combat

In a 1916 paper entitled, "Aircraft in Warfare: The Dawn of the Fourth Arm," F. W. Lanchester developed a mathematical model for air combat. Over the years, Lanchester's model has been modified to describe many different kinds of battles, including combat between two ground forces. As an example, we model a simple battle in a field between two armies, the good guys and the bad guys. Let $G(t)$ be the number of good guys at time t and $B(t)$ the number of bad guys at time t. Because battles often last from one to several days, we measure t in hours. During a battle, members of each army are killed by members of the other army. Armies also lose troops through noncombat losses (accidents, desertion, etc.). In addition, the size of an army can increase if reinforcements are available. We make some assumptions to help us describe the rate at which each army gains and loses troops:

1. Both armies move with equal ease about the battlefield. In particular, neither army has a positional advantage over the other (however, see Exercise 12).
2. For each army, combat losses are at a rate proportional to the size of the opposing force. The constant of proportionality reflects the efficiency of the opposing army.
3. An army suffers noncombat losses at a rate proportional to the size of the army.
4. At time t, reinforcements join the good guys at a rate $g(t)$ troops per hour, and join the bad guys at a rate of $b(t)$ troops per hour. (Because reinforcements usually come in sporadically, $b(t)$ and $g(t)$ are usually 0 for most times t.)

According to assumption 2, the good guys suffer combat losses at a rate $-kB(t)$, where the positive constant k is indicative of the efficiency of the bad guys. According to assumption 3, the good guys suffer noncombat losses at a rate $-\ell G(t)$, for some positive constant ℓ. Combining these rates of change with the reinforcement function in assumption 4, we have

$$\frac{dG}{dt} = -kB(t) - \ell G(t) + g(t). \tag{8}$$

Similarly, for the bad guys we have

$$\frac{dB}{dt} = -mG(t) - nB(t) + b(t). \tag{9}$$

To complete the model, we need to know the initial size (size at $t = 0$) of each army. Assume that the good guys start with $G(0) = G_0$ troops and the bad guys with $B(0) = B_0$ troops. Combining these initial conditions with (8) and (9), we have our combat model:

$$\begin{aligned}\frac{dG}{dt} &= -kB(t) - \ell G(t) + g(t)\\ \frac{dB}{dt} &= -mG(t) - nB(t) + b(t)\\ G(0) &= G_0\\ B(0) &= B_0.\end{aligned}$$

Models of this type have been used to analyze ongoing battles and wars. Special adaptations of this model to guerrilla warfare were used to model aspects of the Vietnam War. Other applications have included studies of the Battle of the Bulge, the Battle of Iwo Jima, and the Battle of the Alamo.

Modeling a Bungee Jump

The Dangerous Sports Club was founded in late 1977 by a group of British thrill seekers anxious to add "excitement" to their lives. The club members periodically embark on expeditions in which they participate in unusual, exciting, and frequently

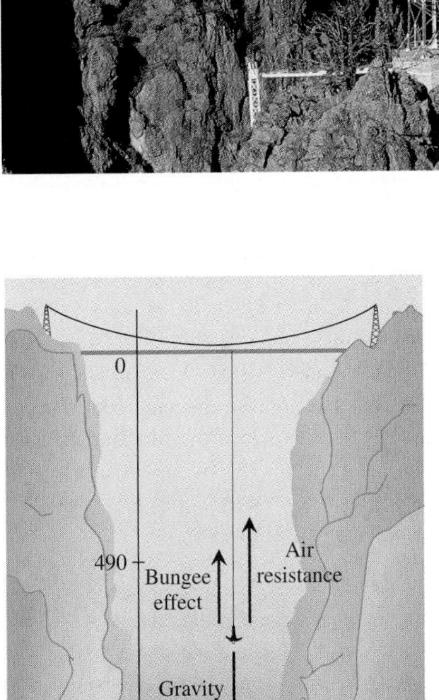

FIGURE 2.63 The breathtaking vista of the Royal Gorge Bridge.

life-threatening activities. Past activities of the club have included hang gliding from the top of Mount Kilimanjaro, skiing an expert alpine slope on everything from a grand piano to a double-decker bus, low-altitude parachute jumping, and formal dining on the lip of an active volcano. In this problem we will work through one of the club's more unusual activities: bridge jumping. In bridge jumping, a participant attaches one end of a bungee cord to himself or herself, attaches the other end to the bridge railing, and then jumps off of the bridge. (Large bungee cords are used to slow jet fighters when they land on the surface of an aircraft carrier.) One of the group's jumps was aired on the television program *That's Incredible*. This jump was made from the Royal Gorge Bridge, a suspension bridge that spans the 1053-foot-deep Royal Gorge in Colorado. One of the jumpers used a 490-foot cord (hoping to touch the bottom of the canyon). By modeling the jump, we might be able to decide whether or not this was a good idea.

Imagine a vertical axis with the origin at the level of the bridge and the positive direction pointing downward. See Fig. 2.63. Label this axis the y-axis, so a positive number y indicates distance y below the bridge. Let $y(t)$ be the position of the jumper at time t. The rate of change of y with respect to t is the speed at time t. We denote the speed by $v(t)$ so

$$\frac{dy}{dt} = v(t).$$

We make the following assumptions:

1. Early in the jump, before the bungee cord starts to retard the jumper's fall, there are two forces acting to affect the jumper's speed. The first is gravity, which acts to increase the downward speed by 32 feet per second every second. The other is air resistance, which acts to reduce the downward speed at a rate proportional to the square of the speed. Thus the effect of air resistance is described by αv^2, where α is a positive constant.
2. Let T_1 be the time that it takes the jumper to fall 490 feet. At this time the bungee cord starts to stretch and acts to decrease the jumper's downward speed. The bungee cord reduces the downward speed at a rate proportional to the amount it is stretched beyond its original length of 490 feet; that is, at a rate proportional to $y - 490$. Hence, the effect of the bungee cord is described by $\beta(y - 490)$, where β is some positive constant. Gravity and air resistance still play a part during this period of the fall.

Assumptions 1 and 2 describe different physical situations, so we need two different sets of differential equations for the model. From the first set of assumptions, we have

$$\begin{aligned} \frac{dy}{dt} &= v(t), \quad y(0) = 0 \\ \frac{dv}{dt} &= 32 - \alpha v^2, \quad v(0) = 0. \end{aligned} \tag{10}$$

This set of equations is valid for $0 \leq t \leq T_1$; that is, until the jumper has fallen 490 feet. The initial conditions $y(0) = 0$ and $v(0) = 0$ indicate that at the start of the jump the bungee jumper is at position 0 and has speed 0.

The second set of assumptions describe the jump from time $t = T_1$ until the time at which the bungee cord starts drawing the jumper back up. If the jumper stops falling at time $t = T_2$, then for $T_1 \leq t \leq T_2$, we have

$$\frac{dy}{dt} = v(t), \quad y(T_1) = 490$$
$$\frac{dv}{dt} = 32 - \alpha v^2 - \beta(y - 490), \quad v(T_1) = V_1, \tag{11}$$

where V_1 is the speed of the jumper at the end of the time period covered by (10).

Equations (10) and (11), together, model the fall of the bungee jumper from time $t = 0$ until time $t = T_2$. If distance is measured in feet and time in seconds, then the value of α can be anywhere from 0.001 to 0.002, depending on whether the jumper is diving head first, falling in spread-eagle position, or holding some other position. The constant β is the "spring constant" of the cord and can be determined by hanging an object of known weight from the cord and measuring the additional distance the cord is stretched under this weight. (By the way, the jumper survived and lived to jump another day.)

Exercises 2.10

1. In (1) we presented

$$x\frac{dy}{dx} + 3y = \frac{e^x}{x^2} \tag{12}$$

as an example of a differential equation.

a. Let

$$y = \frac{e^x}{x^3}. \tag{13}$$

Calculate $\frac{dy}{dx}$, then substitute y and $\frac{dy}{dx}$ into the right side of (12) to verify that (13) is a solution to (12).

b. Let

$$y = \frac{e^x}{x^3} + \frac{c}{x^3}, \tag{14}$$

where c is a constant. Show that (14) is also a solution to (12).

2. Consider the differential equation

$$\frac{dy}{dx} = ky, \tag{15}$$

where k is a constant. Let

$$y = Ce^{kx}, \tag{16}$$

where C is a constant. Show that (16) is a solution to (15).

3. In our first population model, we concluded that the U.S. population at time t might be described by

$$P(t) = Ce^{kt}. \tag{17}$$

We then determined the values of C and k by using the population figures for 1790 and 1840, as given in Table 2.5 on page 174. However, data from other years can be used to determine the values for C and k.

a. Use the population figures for 1790 and 1880 to determine the values of C and k in (17).

b. Plot the data from the table and the function found in part **a** on the same set of axes. Does the function fit the data well?

c. Use the population figures for 1830 and 1890 to determine the values of C and k in (17).

d. Plot the data from the table and the function found in part **c** on the same set of axes. Does the function fit the data well?

4. Money invested by a broker grows at a rate proportional to the value of the investment. Let $A(t)$ be the value of the investment at time t.

a. Write a differential equation describing A.

b. Suppose that the initial investment was \$10,000 and that two years later the investment was worth \$12,500. Find a formula for $A(t)$.

c. Suppose that one year after the money was invested, the investment was worth \$8000 and that three years later it was worth \$11,000. Find a formula for $A(t)$.

5. Suppose that in the population model

$$\frac{dP}{dt} = kP - cP^2$$

we have $k = 0.025$ and $c = 8.33 \times 10^{-11}$.

a. Is the population increasing or decreasing when $P = 250$ million? How about when $P = 400$ million?

b. For what value of P is the population unchanging? Explain what this means in terms of the birth and death rates.

6. In experiments with fruit flies of the species *Drosophila willistoni*, M. E. Gilpin, F. J. Ayala, and J. G. Ehrenfeld found that the number $P(t)$ of flies in a colony at time t is well described by the equation

$$\frac{dP}{dt} = 1.496P - 0.121P^{1.35}.$$

a. Describe how the terms in the equation might represent aspects of the growth of the fruit fly population.

b. For what values of P is the population increasing? Decreasing? Unchanging?

7. In the model of red cell levels during surgery we found that the red cell level, $R(t)$, in a patient is described by

$$\frac{dR}{dt} = -\frac{1}{2T}R(t), \quad R(0) = 2 - 0.4 \cdot L, \tag{18}$$

where L is the volume of blood removed prior to surgery in the ANH procedure.

a. Verify that

$$R(t) = (2 - 0.4 \cdot L)e^{-t/2T} \tag{19}$$

is a solution to (18). Be sure to check that (19) is not only a solution to the differential equation, but also satisfies the initial condition.

b. By setting $L = 0$ we can model the red cell level when ANH is not performed. What is the amount of red cells lost during the T minute surgery?

c. Suppose that $L = 1$, so one liter of the patient's blood is removed and replaced with saline solution before surgery. What is the amount of red cells lost during the T minute surgery? How does this compare with the loss found in part **b**?

8. The radioactive isotope carbon-14 decays at a rate proportional to the amount of C^{14} present. Let $A(t)$ be the amount of C^{14} present in a sample at time t.

a. Write a paragraph explaining why $A(t)$ can be described by the model

$$\frac{dA}{dt} = -kA. \tag{20}$$

Explain the significance of the positive constant k.

b. Verify that $A(t) = Ce^{-kt}$ (with C a constant) is a solution to (20).

c. Suppose that a sample of C^{14} initially has mass 10 grams and that the sample decays to 5 grams in 5568 years. Use these data to find the values of C and k.

9. The concentration of a drug in the bloodstream of an animal decreases at a rate proportional to the concentration of the drug. Let $C(t)$ be the concentration of a drug in the body at time t. Write a differential equation that describes this function.

10. Once the source of pollution is stopped, a polluted body of water will eventually "purify" itself. The concentration of pollutants decreases at a rate proportional to the concentration of pollutants. Let $P(t)$ be the concentration of pollutants in a lake at time t. Write a differential equation that describes P.

11. According to Newton's law of cooling, the temperature of a hot object decreases at a rate proportional to the difference between the temperature of the objects and the temperature of the surroundings. Let $T(t)$ be the temperature of an object at time t and let T_0 be the temperature of the surroundings. Write a differential equation that describes T.

12. In his text *Modeling with Ordinary Differential Equations*, T. P. Dreyer gives a model for the Battle of the Alamo that considers two phases of the battle. At the start of the battle, there were 188 Texans and 3000 Mexicans. In the first phase of the battle, the Texan soldiers were inside the Alamo and were protected by the walls of the fort. The Mexican army was outside in the open, attempting to gain entry. This first phase is modeled by

$$\begin{aligned} \frac{dT}{dt} &= -0.0007TM \\ \frac{dM}{dt} &= -aT \end{aligned} \tag{21}$$

where $T = T(t)$ is the number of Texans alive at time t and $M = M(t)$ is the number of Mexicans alive at time t, with t measured in hours. The positive constant a is a measure of the "efficiency" of the Texan forces. At the end of the first phase of the battle, 100 Texans and 1800 Mexicans were still alive, but the Mexicans had gained entrance to the Alamo. At this time the battle entered its second phase, involving hand-to-hand combat, with neither force having a positional advantage over the other. This second phase is modeled by

$$\begin{aligned} \frac{dT}{dt} &= -bM \\ \frac{dM}{dt} &= -cT. \end{aligned} \tag{22}$$

a. What assumptions do you think were made about the first phase of the battle in coming up with (21) as a model for that phase?

b. How did the assumptions change as the battle entered its second phase? How do these new assumptions lead to (22)?

13. Imagine a pond that contains two species of fish, say A and B, both of which depend on the same food supply. Let $A(t)$ and $B(t)$ be, respectively, the number of fish of species A and B at time t. Because both species eat the same food, a large population for one species has an adverse effect on the other species. Suppose that:

a. The number of offspring produced by species A is proportional to $A(t)$, and a similar statement holds for species B.

b. Competition for scarce resources tends to limit the growth of each population. For each species, this effect is proportional to the number of encounters between members of different species.

   i. Based on these assumptions, write equations for dA/dt and dB/dt. Describe how each term in your equations is reflected in the assumptions.

   ii. Are there any circumstances under which both dA/dt and dB/dt are 0? What do you think happens to the populations in this case?

14. (Continuation of Exercise 13) Every member of a species is in competition for resources with every other member of the species. Thus, too many members of species A will have a detrimental effect on the growth of species A.

a. Alter the model in Exercise 13 to obtain a model that accounts for this intraspecies competition.

b. With this new model, are there any circumstances under which both dA/dt and dB/dt are 0? What do you suppose happens to the populations in this case?

15. Consider an environment with two animal populations, rabbits and foxes. Assume that the rabbits are the only food source for the foxes, but that food for the rabbits is plentiful. Let $F(t)$ and $R(t)$ be, respectively, the number of foxes and rabbits at time t. Assume the following:

a. New rabbits are born at a rate proportional to the rabbit population.

b. Rabbits are killed and eaten by foxes at a rate proportional to the number of encounters between rabbits and foxes.

c. Foxes can reproduce only if they have plenty of food. Thus, new foxes are born at a rate proportional to the number of encounters between rabbits and foxes.

d. The fox population is very vulnerable to population pressures. This effect on the growth rate is proportional to the number of foxes.

   i. Based on these assumptions, write expressions for dR/dt and dF/dt. Tell how each term in your equations is reflected in the assumptions.

   ii. Under what conditions are dR/dt and dF/dt both 0? What happens to the populations in this case?

Review of Key Concepts

In this chapter we derived techniques and formulas for finding derivatives quickly and efficiently. We first developed a collection of "general rules" for finding the derivative of functions built from one or more simple functions. Among these were the product rule, the quotient rule, and the chain rule.

Next we reviewed the elementary functions out of which many other functions are built. These include the power functions, the trigonometric functions, the exponential function, the natural logarithm, and the inverse trigonometric functions. We used the definition of the derivative to find the derivatives of many of these functions and then applied the "general rules" developed earlier for others.

If we remember the derivatives of the elementary functions and know how to apply rules such as the product rule and the chain rule, then we can, with practice, quickly write down the derivatives of many functions. These rules and formulas are valuable because with them we can avoid using the definition of derivative every time we need to perform a differentiation. However, it is important to always keep the definition of derivative in mind. The definition suggests many interpretations of the derivative: as a rate of change, as a slope, and so on. These interpretations are important in helping us understand what the derivative means in the context of a problem.

We summarize the derivatives of elementary functions and the general rules in table form. We also provide a review grid for the elementary functions introduced and reviewed in this chapter.

General Rules for Differentiation

Let f and g be differentiable functions.

The derivative of a constant times a function.

Let c be a constant.

$$(cf(x))' = cf'(x)$$

The derivative of a sum or difference of functions.

$$(f(x) + g(x))' = f'(x) + g'(x)$$
$$(f(x) - g(x))' = f'(x) - g'(x)$$

The product rule.

$$(f(x)g(x))' = f'(x)g(x) + f(x)g'(x)$$

The quotient rule.

$$\left(\frac{f(x)}{g(x)}\right)' = \frac{f'(x)g(x) - f(x)g'(x)}{(g(x))^2}$$

The chain rule.

$$(f \circ g)'(x) = (f(g(x)))' = f'(g(x))g'(x)$$

or

$$\text{If } y = f(u) \text{ and } u = g(x), \text{ then } \frac{dy}{dx} = \frac{dy}{du}\frac{du}{dx}$$

The Derivatives of the Elementary Functions

Function	Derivative	If u is a function of x:	Function	Derivative	If u is a function of x:
c (a constant)	0	—	$\ln x$	$\frac{1}{x}$	$\frac{d}{dx}\ln u = \frac{1}{u}\frac{du}{dx}$
x^n (n constant)	nx^{n-1}	$\frac{d}{dx}u^n = nu^{n-1}\frac{du}{dx}$	a^x $(a > 0, a \neq 1)$	$(\ln a)a^x$	$\frac{d}{dx}a^u = (\ln a)a^u\frac{du}{dx}$
$\sin x$	$\cos x$	$\frac{d}{dx}\sin u = \cos u\frac{du}{dx}$	$\log_a x$ $(a > 0, a \neq 1)$	$\frac{1}{x \ln a}$	$\frac{d}{dx}\log_a u = \frac{1}{u \ln a}\frac{du}{dx}$
$\cos x$	$-\sin x$	$\frac{d}{dx}\cos u = -\sin u\frac{du}{dx}$	$\arcsin x$	$\frac{1}{\sqrt{1-x^2}}$	$\frac{d}{dx}\arcsin u = \frac{1}{\sqrt{1-u^2}}\frac{du}{dx}$
$\tan x$	$\sec^2 x$	$\frac{d}{dx}\tan u = \sec^2 u\frac{du}{dx}$	$\arctan x$	$\frac{1}{1+x^2}$	$\frac{d}{dx}\arctan u = \frac{1}{1+u^2}\frac{du}{dx}$
$\sec x$	$\sec x \tan x$	$\frac{d}{dx}\sec u = \sec u \tan u\frac{du}{dx}$	$\arccos x$	$-\frac{1}{\sqrt{1-x^2}}$	$\frac{d}{dx}\arccos u = -\frac{1}{\sqrt{1-u^2}}\frac{du}{dx}$
e^x	e^x	$\frac{d}{dx}e^u = e^u\frac{du}{dx}$	$\text{arcsec}\, x$	$\frac{1}{\lvert x\rvert\sqrt{x^2-1}}$	$\frac{d}{dx}\text{arcsec}\, u = \frac{1}{\lvert u\rvert\sqrt{u^2-1}}\frac{du}{dx}$

Chapter Summary

Sine and Cosine Functions

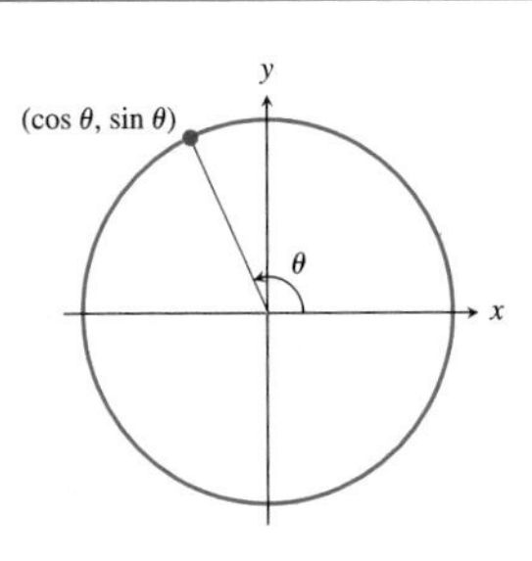

Measure angle θ counterclockwise from the positive x-axis. The terminal side of the angle intersects the circle $x^2 + y^2 = 1$ in the point $(\cos\theta, \sin\theta)$.

Because $(\cos\theta, \sin\theta)$ is a point on the circle $x^2 + y^2 = 1$, many identities can be obtained by using properties of the circle. For example

$$\cos^2\theta + \sin^2\theta = x^2 + y^2 = 1.$$

By reflection about $y = x$,

$$\cos\left(\frac{\pi}{2} - \theta\right) = \sin\theta$$

and

$$\sin\left(\frac{\pi}{2} - \theta\right) = \cos\theta.$$

The Exponential Function

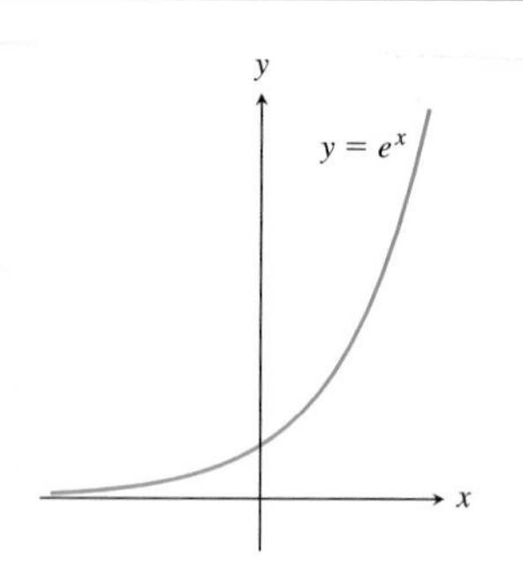

There is a number, denoted by e, such that

$$\lim_{h \to 0} \frac{e^h - 1}{h} = 1.$$

By exploring with a calculator, we find

$$e \approx 2.718281828459045\ldots.$$

The function exp defined by

$$\exp(x) = e^x$$

is called the exponential function.

Let $f(x) = ce^{kx}$. Then

$$f'(x) = kce^{kx} = kf(x).$$

Hence the rate of change of f is proportional to f. Many quantities change at a rate proportional to the quantity itself: population, amount of radioactive material, drug level in a body, and so on. Exponentials are essential in modeling such phenomena.

The Natural Logarithm Function

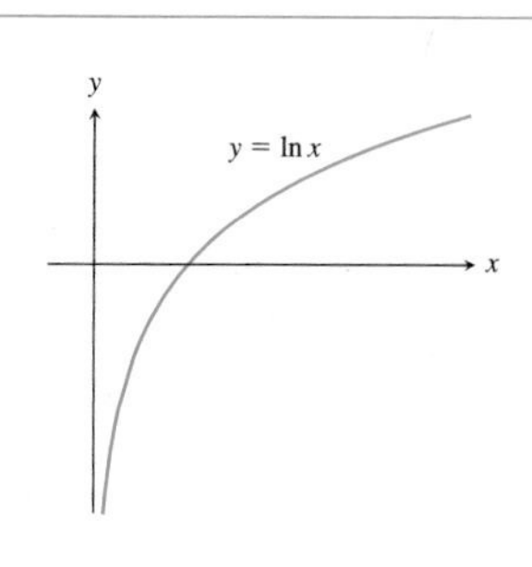

The logarithm of x to the base b is the number s for which $b^s = x$. We use the notation

$$s = \log_b x$$

to describe s. Hence

$$b^{\log_b x} = x.$$

The logarithm to the base e is called the natural logarithm and is denoted ln. Hence for $x > 0$,

$$\ln x = \log_e x.$$

Logarithms satisfy many useful identities. We present some of these for the natural logarithm:

$$\ln(xy) = \ln x + \ln y$$

$$\ln\left(\frac{x}{y}\right) = \ln x - \ln y$$

$$\ln(x^r) = r \ln x.$$

Inverse Function

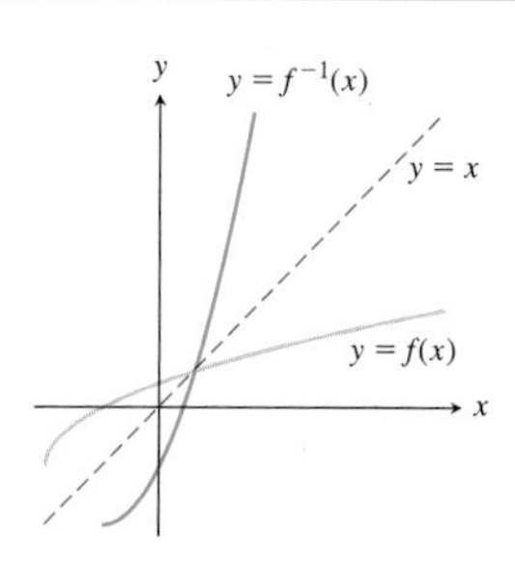

Let f be one-to-one on its domain $\mathcal{D}$. A function g is the inverse of f if the domain of g is equal to the range of f and

$$(g \circ f)(x) = g(f(x)) = x$$

for all x in $\mathcal{D}$. We write $g = f^{-1}$. If g is the inverse of f, then f is also the inverse of g.

If g is the inverse of f, then the graph of $y = g(x)$ is the reflection about the line $y = x$ of the graph of $y = f(x)$. From this it follows that

$$g'(x) = (f^{-1})'(x) = \frac{1}{f'(g(x))}.$$

The Inverse Sine Function		
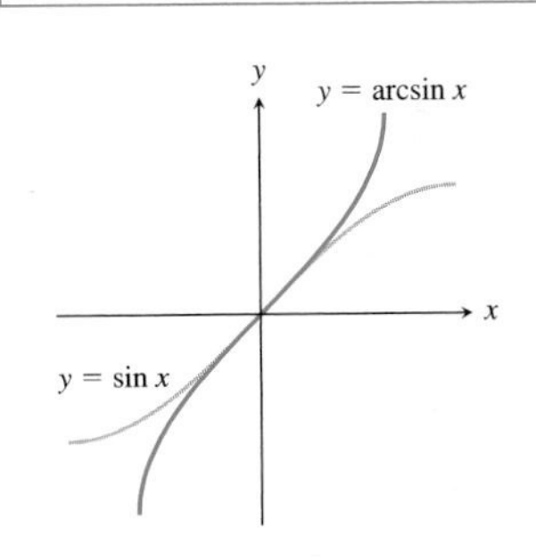	Let $-1 \leq x \leq 1$. The inverse sine of x is the number θ satisfying $$-\frac{\pi}{2} \leq \theta \leq \frac{\pi}{2}$$ and $$\sin \theta = x.$$ We denote the inverse sine function by arcsin and write $$\theta = \arcsin x.$$	If $-1 \leq x \leq 1$, then $$\sin(\arcsin x) = x.$$ If $-\pi/2 \leq \theta \leq \pi/2$, then $$\arcsin(\sin \theta) = \theta.$$

The Inverse Tangent Function		
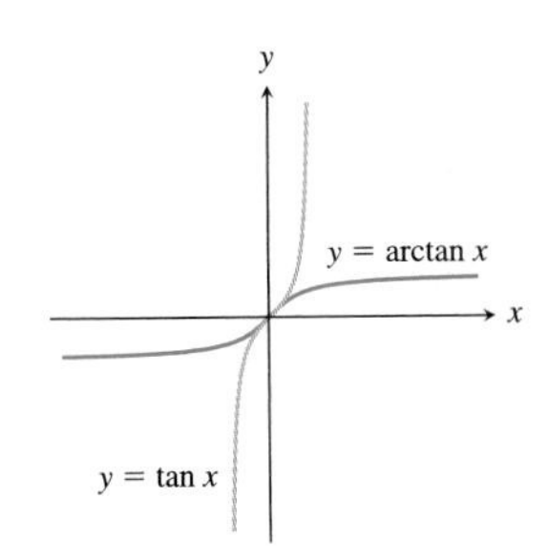	Let x be a real number. The inverse tangent of x is the number θ satisfying $$-\frac{\pi}{2} < \theta < \frac{\pi}{2}$$ and $$\tan \theta = x.$$ We denote the inverse tangent function by arctan and write $$\theta = \arctan x.$$	For any real number x, $$\tan(\arctan x) = x.$$ If $-\pi/2 < \theta < \pi/2$, then $$\arctan(\tan \theta) = \theta.$$

Chapter Review Exercises

Exercises 1–20: Find the derivative.

1. $y = -4x^5 + 7x^4 - \sqrt{2}x^2 + 3x - \pi$
2. $f(x) = (x^4 + x^2 + 1)^4$
3. $r = \left(\theta^2 - \dfrac{3}{\theta}\right) \sin \theta$
4. $Q(t) = \dfrac{3t^2 - 4t + 7}{-t^2 + 4}$
5. $y = \ln(4x^3 - 2x + 3)$
6. $b(s) = se^{s^2}$
7. $k(p) = \dfrac{p}{\sqrt{p^2 + 3}}$
8. $f(x) = \sqrt{\dfrac{2x - 3}{4x + 5}}$
9. $g(x) = \ln\left(\dfrac{2x - 3}{4x + 5}\right)$
10. $r(s) = 2^{2s}3^{3s}$
11. $y = \sqrt{t} \arcsin t^2$
12. $F(x) = (\sqrt{x} + 1)(\sqrt{x} + 2)(\sqrt{x} + 3)(\sqrt{x} + 4)$
13. $s(t) = \ln(\tan 2t)$
14. $z = \sqrt[3]{\dfrac{e^{2w} + 2w}{e^{-2w} - 2w}}$

15. $y = \dfrac{1}{\arctan\left(1 + \dfrac{1}{x}\right)}$

16. $H(z) = \dfrac{(z^2 - z)\sin z}{(1 + \tan z)}$

17. $r = \sin(\cos(1 + \tan 2t))$

18. $y = (ax^2 + bx + c)^d$, a, b, c, d constants

19. $H(s) = e^{as^2+bs+c}$, a, b, c constants

20. $A(t) = \arcsin(at^2 + bt + c)$, a, b, c constants

Exercises 21–24: Use implicit differentiation to find $\dfrac{dy}{dx}$.

21. $y^3 + xy - 3x^2 = 0$

22. $\dfrac{2x - y}{3x + 2y} = xy$

23. $\sin(x + y) = x - y$

24. $e^{x-3y} = 2x + y^2 - 1$

Exercises 25–26: Four graphs are shown on one set of axes. Some of the graphs are mislabeled. Rearrange the labels so the graphs are properly labeled.

25.

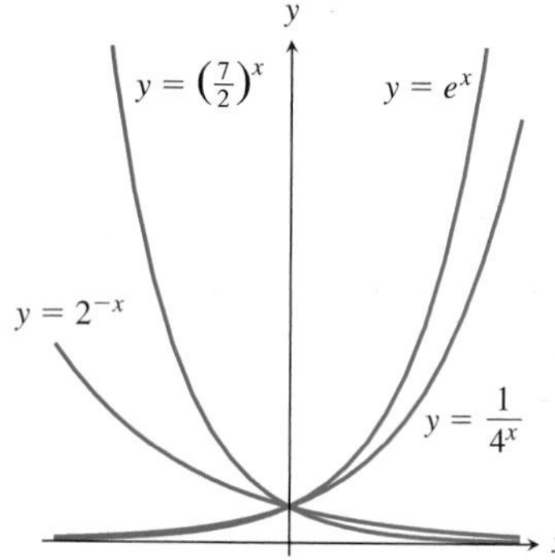

26.

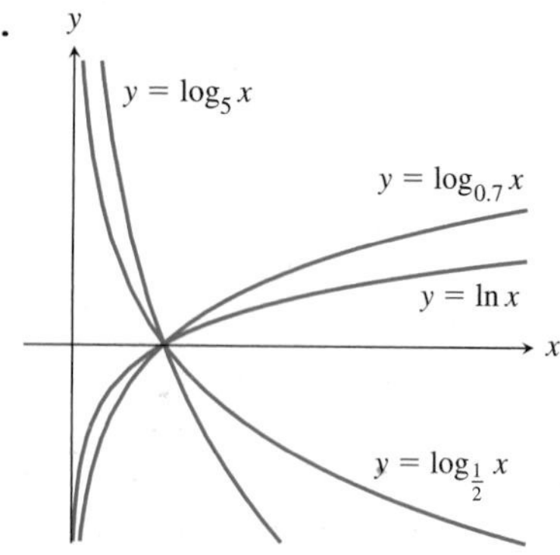

27. The graph of a trigonometric function is shown. Label points where the graph intersects the x-axis and the tick marks on the y-axis if this is the graph of

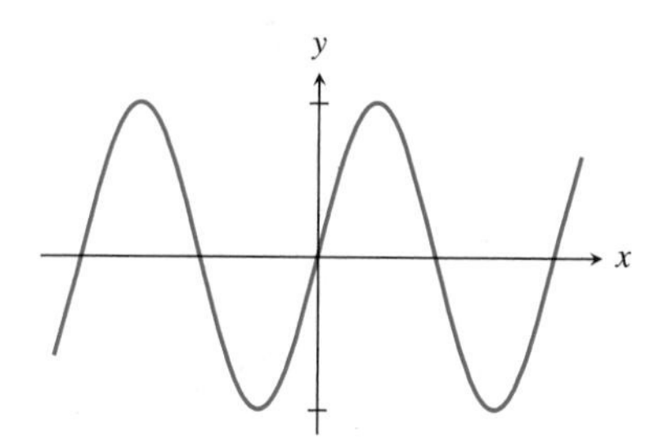

Figure for Exercise 27.

a. $y = \sin 2x$.

b. $y = 4\sin(2\pi x)$.

c. $y = \dfrac{1}{3}\sin\left(\dfrac{4}{\pi}x\right)$.

T **28.** Let $f(x) = e^x$.

a. Find an equation for the line tangent to the graph of $y = f(x)$ at the point $(0, 1)$.

b. Estimate the maximum possible absolute error if we use the tangent line expression to approximate e^x for $-0.25 \le x \le 0.25$.

c. Use the methods of Section 1.6 and 1.7 to find $r > 0$ so that on the interval $-r \le x \le r$, the error in the tangent line approximation is no more than $0.1|x|$ in absolute value.

T **29.** Let $g(x) = \ln x$.

a. Find an equation for the line tangent to the graph of $y = g(x)$ at the point $(1, 0)$.

b. Estimate the maximum possible absolute error if we use the tangent line expression to approximate $\ln x$ for $0.5 \le x \le 1.5$.

c. Use the methods of Section 1.6 and 1.7 to find $r > 0$ so that on the interval $1 - r \le x \le 1 + r$, the error in the tangent line approximation is no more than $0.1|x - 1|$ in absolute value.

T **30.** Let $h(x) = \arctan x$.

a. Find an equation for the line tangent to the graph of $y = h(x)$ at the point $(0, 0)$.

b. Estimate the maximum possible absolute error if we use the tangent line expression to approximate $\arctan x$ for $-0.5 \le x \le 0.5$.

c. Use the methods of Section 1.6 and 1.7 to find $r > 0$ so that on the interval $-r \le x \le r$, the error in the tangent line approximation is no more than $0.1|x|$ in absolute value.

31. A ground camera tracks a 100-foot rocket during a launch. The camera is fixed at a point 300 feet from the launch point and pivots so that it always points at the midpoint of the rocket. Let h be the height of the midpoint of the rocket above the ground.

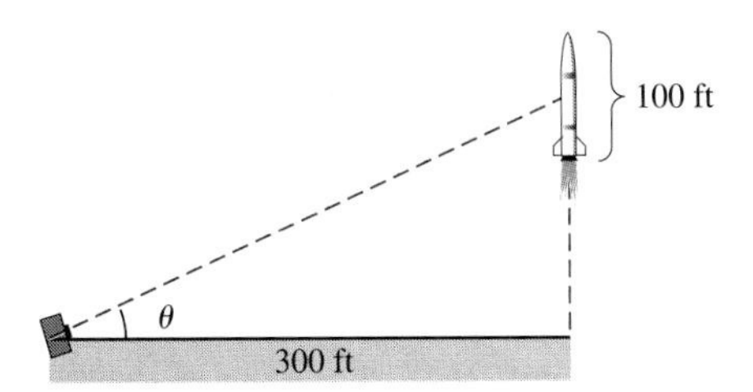

Figure for Exercise 31.

a. Find a formula for the angle θ determined by the ground and the line of sight of the camera. Your answer should be in terms of h and should involve inverse trigonometric functions.

b. For what values of h is your formula valid? What are the possible values for θ?

c. Find $\dfrac{d\theta}{dh}$.

d. What are the units for the derivative? Explain what the derivative means in this situation.

32. A tall cylindrical tower has a base radius of 50 feet. A surveyor stands x feet from the tower, as shown in the accompanying figure.

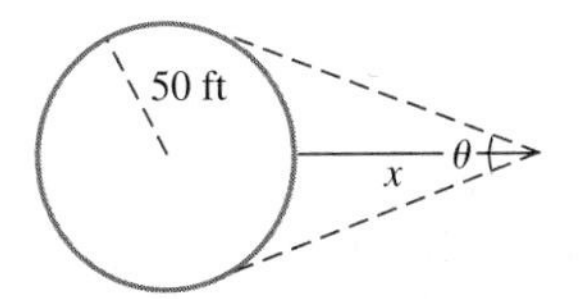

Figure for Exercise 32.

a. Find a formula for the angle θ subtended by the tower. Your answer should be in terms of x and should involve inverse trigonometric functions.

b. For what values of x is your formula valid? What are the possible values for θ?

c. Find $\dfrac{d\theta}{dx}$.

d. What are the units for the derivative? Explain what the derivative means in this situation.

33. An airplane that is 150 feet long flies at a height of 500 feet above the ground. The plane passes directly above a boy on the ground, and the boy watches the plane as it passes over. Let y be the distance from the boy to the tail of the plane.

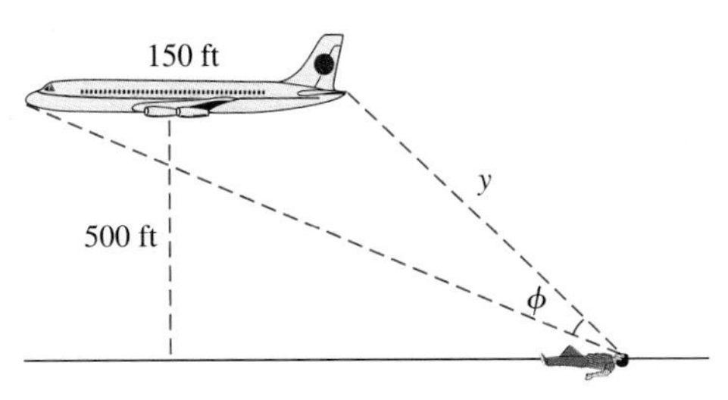

Figure for Exercise 33.

a. Find a formula for the angle of vision ϕ subtended by the plane. Your formula should involve y and inverse trigonometric functions.

b. For what values of y is your formula valid? What are the possible values for ϕ?

c. Find $\dfrac{d\phi}{dy}$.

d. What are the units for the derivative? Explain what the derivative means in this situation.

34. A wheel of radius 1 foot spins counterclockwise at one revolution per minute. A 6-foot rod is attached to the edge of the wheel and pivots as the wheel turns. The free end of the rod moves back and forth in a horizontal grove that is in line with the center of the wheel. Let α be the angle through which the wheel has rotated and let θ be the angle the rod makes with the horizontal groove. See the accompanying diagram.

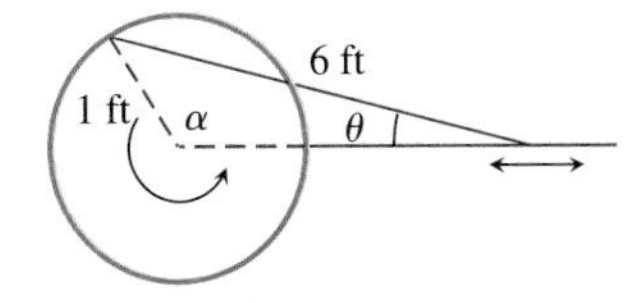

Figure for Exercise 34.

a. Find a formula for θ. Your formula should involve α and inverse trigonometric functions.

b. For what values of α is your formula valid? What are the possible values of θ?

c. Find $\dfrac{d\theta}{d\alpha}$ and $\dfrac{d\theta}{dt}$.

d. What are the units for the derivatives? Explain what the derivatives mean in this situation.

35. In Section 2.8 we gave an approximation for the maximum angle a, of the sun above the horizon on day t in Bemidji, Minnesota. Another good approximation for this angle is given by

$$a = a(t) = -57.3 \arcsin(-0.53 + 0.33 \cos(0.017t))$$

where $t = 0$ corresponds to December 22.

a. Estimate the measure of the angle $a(t)$ on January 31. On October 31.

b. Estimate the date if $a = 30°$ and it is after June 21.

c. Estimate the possible dates if $a = 45°$.

d. Find $\frac{da}{dt}$. What does the derivative mean in this context?

36. In Section 2.7 we remarked that the maximum length of time required to factor a large integer N is proportional to

$$L(N) = e^{\sqrt{(\ln N)(\ln \ln N)}}.$$

Find $\frac{dL}{dN}$ and explain what the derivative means in this context.

37. A Ferris wheel has radius 60 feet and is mounted so the bottom car is 5 feet off the ground. The wheel turns at a rate of one revolution every 40 seconds. You ride the Ferris wheel at a time when the sun is directly overhead.

a. Write a formula that gives the position on the ground of the shadow of your car at any time t. Include all assumptions that you make in coming up with your formula.

b. Find the rate of change of the position of the shadow with respect to time. What does the rate of change mean in this context?

c. When is the rate of change largest? When is it 0? How is the shadow moving at these times?

38. Consider the line $y = \frac{1}{2}t$. For $x \geq 0$, define $A(x)$ to be the area of the triangular region bounded by the t-axis, the line $y = \frac{1}{2}t$, and the vertical line $t = x$. See the accompanying figure.

a. Find a formula for $A(x)$.

b. Find $A'(x)$. What does the derivative mean in this context?

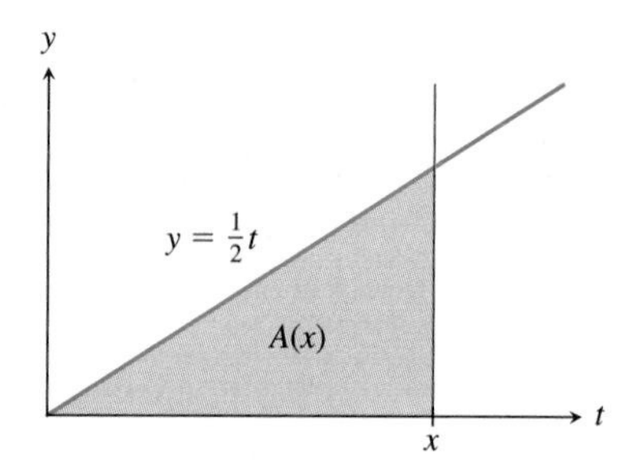

Figure for Exercise 38.

39. Consider the line $y = 4t + 3$. For $x \geq 0$, define $A(x)$ to be the area of the trapezoid bounded by the t-axis, the y-axis, the line $y = 4t + 3$, and the vertical line $t = x$. See the accompanying figure.

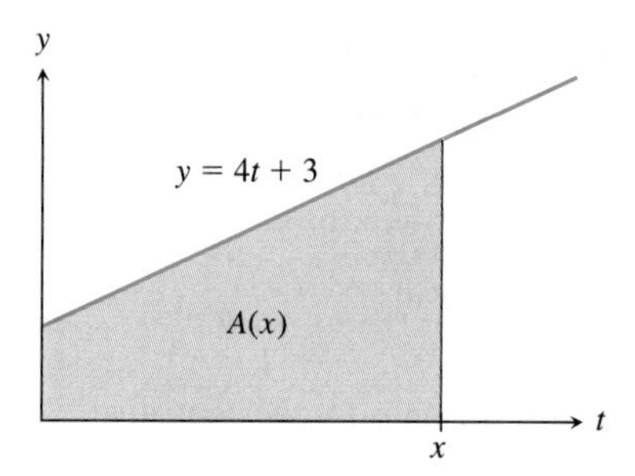

Figure for Exercise 39.

a. Find a formula for $A(x)$.

b. Find $A'(x)$. What does the derivative mean in this context?

40. Let $m \geq 0$ and $b \geq 0$. Consider the line $y = mt + b$. For $x \geq 0$, define $A(x)$ to be the area of the region bounded by the t-axis, the y-axis, the line $y = mt + b$, and the vertical line $t = x$.

a. Find a formula for $A(x)$.

b. Find $A'(x)$. What does the derivative mean in this context?

STUDENT PROJECT

LAMBORGHINI DIABLO VT 6.0 TEST RESULTS

TABLE 2.6

Time (s)	Speed (mph)
0.0	0
2.1	30
2.9	40
3.6	50
4.3	60
5.6	70
6.7	80
7.9	90
8.2	100
9.7	110
11.8	120
13.7	130

Lamborghini Acceleration Data. The graph in Fig. 2.64 represents data from an acceleration test of a 2000 Lamborghini Diablo VT 6.0. From a standing start, the Lamborghini accelerated for several seconds. The times (in seconds) at which the Lamborghini reached certain speeds (in miles per hour) were recorded. The data collected were obtained from *Car and Driver*, July 2000, p. 93, and are summarized in Table 2.6. To obtain Fig. 2.64, the data were graphed and consecutive data points joined by a line segment.

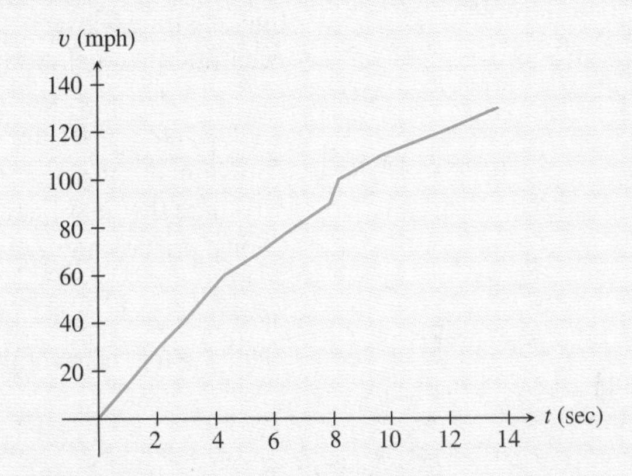

FIGURE 2.64 2000 Lamborghini Diablo VT 6.0 Model Test Results.

In this project, we will first estimate the distance $d(t)$ traveled by the Lamborghini during the first t seconds of the test, for $t = 0, 2.1, 2.9, \ldots, 13.7$. We will do so by estimating the distance traveled by the car in the time between two data points and keeping a running total of the distance traveled. We will also compute the average acceleration for the Lamborghini over the time intervals between consecutive velocity readings.

To get started, let's estimate the distance the Lamborghini traveled during the time interval from 5.6 to 6.7 seconds. During this 1.1-second time interval, the car accelerated from $v(5.6) = 70$ mph to $v(6.7) = 80$ mph. It is reasonable to assume that the average speed during this time was

$$\frac{70 + 80}{2} = 75 \text{ mph}.$$

Now, 1.1 seconds is the same as 1.1/3600 hours. Thus, during this 1.1 seconds of the test from $t = 5.6$ to $t = 6.7$, the Lamborghini traveled approximately

$$(75 \text{ miles/hour})\,(1.1/3600 \text{ hours}) \approx 0.0229 \text{ miles}. \tag{1}$$

In a similar way, we can approximate the distance traveled by the Lamborghini during the time between any two consecutive data points. By adding these distances, we can find the total distance covered by the car during the test.

PROBLEM 1 Using the ideas just outlined, estimate the position $x(t)$ of the Lamborghini at times $0, 2.1, 2.9, 3.6, \ldots, 13.7$. How far did the Lamborghini travel up to the time that its velocity was 130 miles per hour? Using the results of these calculations, prepare a graph of the position $x(t)$ against time t.

PROBLEM 2 Give a geometric interpretation of the computations used in Problem 1. First, take another look at (1). We can rewrite the left side of this equation as

$$\frac{v(5.6) + v(6.7)}{2} \cdot (6.7 - 5.6)/3600.$$

This expression may be interpreted as the area of a certain trapezoid that can be drawn on the velocity graph. (For this it is helpful to think of the units on the t-axis as hours, so all of the t-axis labels should be divided by 3600). Draw the appropriate trapezoid on the graph and carefully explain the connection between the area of this trapezoid and the distance traveled. Finally, relate the numerical value of $x(13.7)$ to an area associated with the velocity graph.

Finding the Acceleration. In Problem 3 we will approximate the Lamborghini's acceleration as a function of time. The average acceleration during a time interval $[t_1, t_2]$ is defined as the

$$\frac{\text{change in speed}}{\text{change in time}} = \frac{v(t_2) - v(t_1)}{t_2 - t_1}.$$

PROBLEM 3 Compute the average acceleration of the Lamborghini on the time intervals $[0, 2.1], [2.1, 2.9], \ldots, [11.8, 13.7]$. Using the velocity graph, give a geometric interpretation of these computed average accelerations. Explain your interpretation carefully.

PROBLEM 4 Find the midpoint of each of the intervals $[0, 2.1], [2.1, 2.9], \ldots, [11.8, 13.7]$. Let t_m denote a typical midpoint. Explain why it is reasonable to use the average accelerations computed in Problem 3 as estimates of the actual acceleration at times $t = t_m$. Using these estimates, sketch a graph of the acceleration $a(t)$ against t. Could you have predicted the rough shape of this acceleration graph by looking at the velocity graph?

3

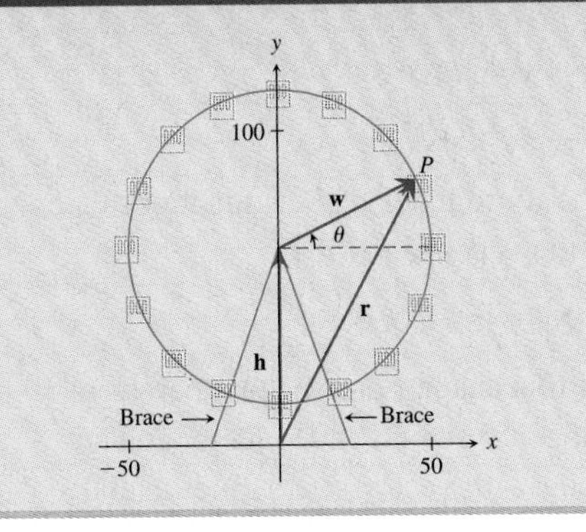

Motion, Vectors, and Parametric Equations

One of the world's tallest Ferris wheels is in Osaka, Japan. If you were riding on this wheel, how could you express your position as a function of time? How could you calculate your velocity?

See Example 3, page 217.

Calculus is used to describe and analyze such motions as that of an electron between charged plates, of a slider-crank mechanism, or of a planet in its orbit about the sun. Calculus is needed because velocity and acceleration, which are key ideas in describing or analyzing motion, are rates of change.

Figure 3.1 shows an aerial photograph of the Tevatron, a particle accelerator at Fermilab in Illinois. The radius of the circular path on which the protons are accelerated is 0.64 miles. After acceleration, the protons follow a line tangent to the circle, moving 0.5 miles toward a fixed target. The two-dimensional motion of a proton can be described and analyzed by a vector giving its position at any time t prior to hitting the target. The velocity and acceleration vectors of the proton can be calculated by differentiation.

Figure 3.2 shows a slider-crank mechanism. The powered crank OP revolves about O and drives the rod PQ, which connects the crank and the slider, causing the slider to move back and forth on the x-axis. This mechanism includes both circular and linear motion. The linear motion of the slider can be described by a single scalar function f of time t. We may express the coordinate x of the slider at time t as $x = f(t)$. The term *scalar* means that $f(t)$ is a real number. To describe the circular motion of the pin P around the crankshaft O, we may use the *parametric equations*

$$x = F(t)$$
$$y = G(t),$$

which give the x- and y-coordinates of P as functions of time t, or we may use the *vector equation*

$$\mathbf{r}(t) = \langle F(t), G(t)\rangle.$$

The main ideas of this chapter are vectors, parametric or vector equations, tangent vectors, and velocity and acceleration vectors. Along with vectors we define several arithmetic operations upon them, including addition, scalar multiplication, and the dot product. The dot product can be used to resolve force, velocity, or acceleration vectors into the sum of two mutually perpendicular vectors. We use all of these ideas to describe the motion of an object without regard to the causes of that motion. This is called *kinematics*. At the end of the chapter we give a brief discussion of Newton's laws, which tell *why* things move. This is called *dynamics*.

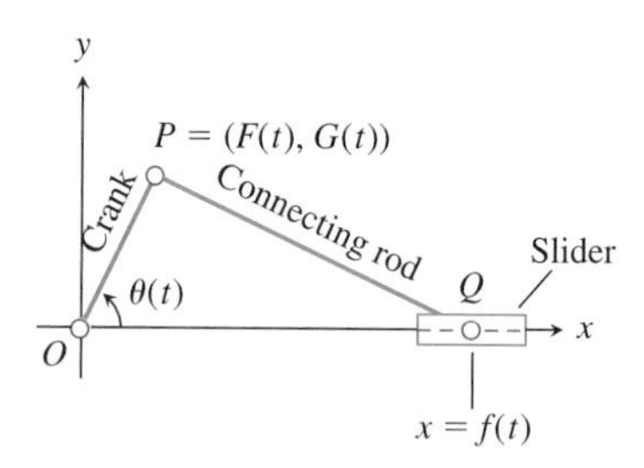

FIGURE 3.2. A slider-crank mechanism for converting rotary motion to straight-line motion.

FIGURE 3.1. The Tevatron at Fermilab, showing the ring on which protons are accelerated.

3.1 Motion along a Line

Before discussing motion in the plane, for which we will use vectors, we discuss motion along a line, for which we can use scalar equations. We introduce the ideas of position, velocity, and acceleration in this simpler context.

Position

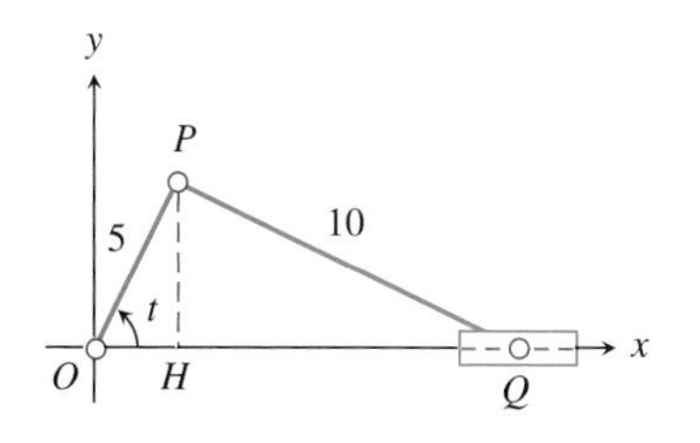

FIGURE 3.3 A simple slider-crank mechanism, with crank OP, connecting rod PQ, and slider Q.

We use the slider-crank mechanism shown in Fig. 3.3 to discuss position and average velocity. We assume that the crank is rotating counterclockwise at 1 radian per second and, in particular, that $\theta(t) = t$ for all time t, and that the lengths of the crank and connecting rod are 5 cm and 10 cm. We assume that $0 \leq t \leq \pi/2$. Because triangle OPH in Fig. 3.3 is a right triangle, $OH = 5 \cos t$ and $PH = 5 \sin t$. Applying the Pythagorean Theorem to right triangle PHQ, $HQ = \sqrt{100 - 25 \sin^2 t}$. Hence, the position of the slider is

$$x = f(t) = OQ = OH + HQ = 5 \cos t + \sqrt{100 - 25 \sin^2 t} \text{ cm.} \tag{1}$$

Although we assumed that $0 \leq t \leq \pi/2$ to simplify the derivation of (1), it is not difficult to show that this equation holds for all t. See Exercise 31.

Velocity

We use the idea of *average velocity on a time interval* (t_1, t_2) to lead up to the idea of the "instantaneous" velocity of the slider at t_1. The average velocity $v(t_1, t_2)$ is the "signed distance" through which the slider moves between t_1 and t_2 divided by the elapsed time $t_2 - t_1$. Thus, using the position function $x = f(t)$ from (1),

$$v(t_1, t_2) = \frac{\text{change in position}}{\text{change in time}} = \frac{\Delta x}{\Delta t} = \frac{f(t_2) - f(t_1)}{t_2 - t_1} \text{ cm/s.}$$

On the time interval $[1, 1.1]$, for example, the average velocity of the slider is

$$v(1,1.1) = \frac{f(1.1) - f(1)}{1.1 - 1} \approx \frac{11.2203 - 11.7733}{1.1 - 1} \approx -5.53061 \text{ cm/s.}$$

TABLE 3.1 The position and average velocity of the slider at several times near $t = 1$.

t	$f(t)$	$v(1,t)$
1.1	11.2203	−5.53061
1.01	11.7187	−5.46875
1.001	11.7679	−5.46113
1.0001	11.7728	−5.46035

In Table 3.1 we give the positions of the slider at $t = 1.1$, 1.01, 1.001, and 1.0001 and the average velocities on the intervals $[1, 1.1]$, $[1, 1.01]$, $[1, 1.001]$, and $[1, 1.0001]$.

From these data, it appears that the average velocity of the slider is approaching a number close to -5.460 as t approaches 1. It is natural to define the "instantaneous velocity" (or, more simply, the velocity) of the slider at $t = 1$ as the limit of the average velocity $v(t,1)$ as $t \to 1$.

You may have noted that the ratios used in calculating the average velocities are numerically equal to the slopes of line segments connecting points of the graph of the function f. Figure 3.4 shows the graph of f and a line segment joining points $(1, f(1))$ and $(t, f(t))$. The slope of this line segment is

$$\frac{f(t) - f(1)}{t - 1} = v(1, t).$$

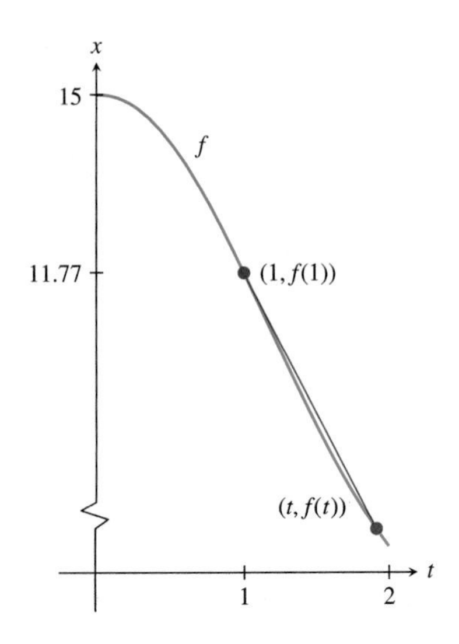

FIGURE 3.4 The graph of the function f and the line segment joining points $(1, f(1))$ and $(t, f(t))$. The slope of this line segment is $(f(t) - f(1))/(t - 1)$, which is the average velocity of the slider between times 1 and t.

You know from Chapter 1 that the limit of this ratio as $t \longrightarrow 1$ is the derivative $f'(1)$ of f at $t = 1$. Applying the chain rule to (1),

$$\begin{aligned} f'(t) &= 5 \sin t + \frac{1}{2\sqrt{100 - 25 \sin^2 t}} \cdot (100 - 25 \sin^2 t)' \\ &= -5 \sin t + \frac{1}{2\sqrt{100 - 25 \sin^2 t}} \cdot (-50) \sin t \cos t. \end{aligned}$$

Evaluating this derivative at $t = 1$,

$$f'(1) = -5 \sin 1 + \frac{1}{2\sqrt{100 - 25 \sin^2 1}} \cdot (-50) \sin 1 \cos 1 \approx -5.460.$$

It appears that the limit of the average velocity $v(1, t)$ as $t \longrightarrow 1$ is the number $f'(1)$.

We summarize this discussion in a definition.

DEFINITION Average Velocity, Velocity, and Speed

Let $x = f(t)$ be the position at time t of an object moving along a line, where f is a function defined on an interval I. The **average velocity** of the object between times t_1 and t_2 in I is

$$v(t_1, t_2) = \frac{f(t_2) - f(t_1)}{t_2 - t_1}. \tag{2}$$

If the function f is differentiable at t_1, we define the **velocity** $v(t_1)$ of the object at time t_1 as the limit of the average velocity $v(t_1, t_2)$ as $t_2 \longrightarrow t_1$. It follows that

$$v(t_1) = \lim_{t_2 \to t_1} \frac{f(t_2) - f(t_1)}{t_2 - t_1} = f'(t_1) = \left.\frac{dx}{dt}\right|_{t=t_1}.$$

The **speed** of the object at time t_1 is the absolute value of its velocity at t_1, that is, $|v(t_1)|$.

WEB The units of average velocity, velocity, and speed are always units of distance divided by units of time. The most common units of velocity are miles per hour (mph), feet per second (ft/s), meters per second (m/s), or kilometers per hour (kph). Usually we'll measure distances in meters, time in seconds, and velocities in meters per second.

EXAMPLE 1 A skydiver jumps from a plane when it is 1000 meters above the Earth. Letting y be her distance above the Earth and t the elapsed time of her jump, her height above the Earth for all t up to the time her chute opens is

$$y = f(t) = 1000 - 4.9t^2, \tag{3}$$

where lengths are measured in meters and time in seconds. (In this model we are ignoring air resistance and any sidewise movement. See Exercise 30 for a more realistic model.) Calculate her average velocity $v(2, t_2)$ for $t_2 = 2.1$, 2.01, 2.001, and 2.0001. From these results, estimate her velocity at $t = 2$.

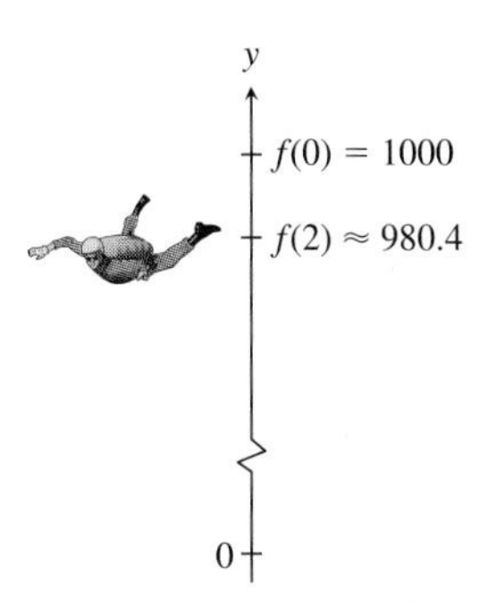

FIGURE 3.5 Positions of a skydiver at times $t = 0$ and $t = 2$.

Compare your estimate with her velocity at $t = 2$. Calculate her speed at $t = 2$.

Solution

Figure 3.5 shows the position of the skydiver at times $t = 0$ and $t = 2$. We use (2), with f given by (3), to calculate the average velocities

$$v(2, t_2) = \frac{f(t) - f(2)}{t - 2}$$

for $t_2 = 2.1, 2.01, 2.001$, and 2.0001. Table 3.2 lists both her height and average velocity at these times.

It appears from the table that for t_2 near $t_1 = 2$, her average velocity $v(2, t_2)$ is near -19.6 m/s. Her velocity at $t = 2$ is

$$v(2) = f'(2) = \frac{d}{dt}(1000 - 4.9t^2)\bigg|_{t=2} = -9.8t\bigg|_{t=2} = -19.6 \text{ m/s}.$$

Her speed is $|v(2)| = |-19.6| = 19.6$ m/s.

TABLE 3.2 The position and average velocity of the skydiver on several intervals.

t	$f(t)$	$v(2,t)$
2	980.400	
2.1	978.391	−20.090
2.01	980.204	−19.649
2.001	980.380	−19.605
2.0001	980.398	−19.600

We give two more examples of motion along a line: the motion of a bullet fired vertically upward and the motion of a railroad "snubber."

EXAMPLE 2 A rifle is fired vertically upward so that at time $t = 0$ the bullet is 1.5 m above ground level and has speed 610 m/s ($\approx$ 2000 ft/s). If we ignore air resistance and choose a coordinate axis with origin at ground level and positive direction upward, the height h of the bullet at time t is

$$h = h(t) = -4.9t^2 + 610t + 1.5, \qquad t \in [0, b],$$

where b is the time when the bullet strikes the Earth. Figure 3.6 shows the graph of the height function h. Calculate b and find the maximum height of the bullet.

Solution

Because the bullet has positive velocity on the way up and negative velocity on the way down, the maximum height of the bullet occurs at the time its speed is zero. We can find this height by differentiating the height equation to find the velocity of the bullet, setting the velocity equation equal to zero, solving this equation for t, and, finally, evaluating the height equation at this time. Differentiating the height equation,

$$v = v(t) = \frac{d}{dt}h(t) = \frac{d}{dt}(-4.9t^2 + 610t + 1.5) = -9.8t + 610. \tag{4}$$

The graph of v is shown at the bottom of Fig. 3.6. To find the time t at which $v(t) = 0$, we solve the equation

$$-9.8t + 610 = 0.$$

The solution is $t = 610/9.8 \approx 62$ s. The maximum height of the bullet is

$$h(610/9.8) = -4.9(610/9.8)^2 + 610(610/9.8) + 1.5 \approx 19{,}000 \text{ m}.$$

This may seem too high, but keep in mind that we are ignoring air resistance.

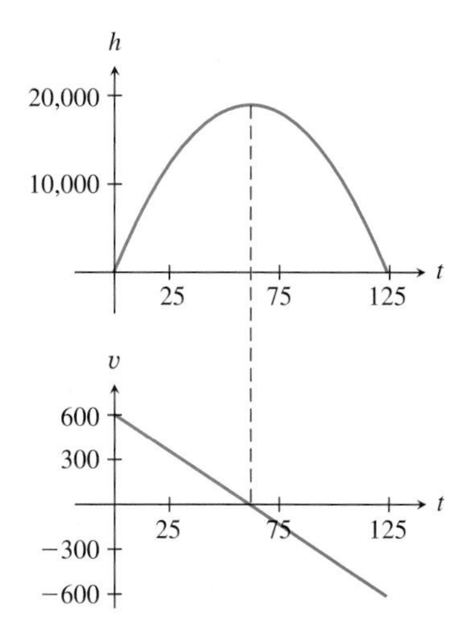

FIGURE 3.6 The bullet's height and velocity as functions of time.

The time b when the bullet hits the ground can be found by solving the equation $h(t) = -4.9t^2 + 610t + 1.5 = 0$ for t. The result is

$$t = \frac{-610 \pm \sqrt{610^2 - 4(-4.9)(1.5)}}{2(-4.9)} \approx -0.0024 \text{ or } 124.$$

Because we took $t = 0$ to be the time at which the bullet was 1.5 meters from the Earth, we ignore $t = -0.0024$ and conclude that the bullet hits the ground after about 124 seconds.

Note: Although the equation $h = h(t)$ describes the motion of the bullet, the bullet does not move on the graph of $h = h(t)$. See Exercise 32.

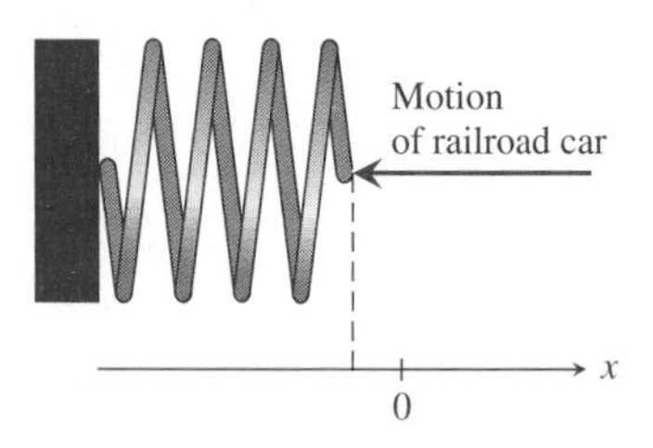

FIGURE 3.7 A model of a snubber.

EXAMPLE 3 Figure 3.7 shows a railroad "snubber." Snubbers are placed at the ends of some railroad tracks so that, when bumped by a railroad car, the direction of motion of the car is reversed and its speed is decreased. Suppose that a freight car with mass 10^4 kilograms and speed 0.5 meters per second hits the snubber. We assume that after first contact between car and snubber, the car and snubber remain in contact during and after the compression of the spring, until the spring returns to its initial length. If we imagine an x-axis located as shown in Fig. 3.7, with the origin aligned with the point of first contact, the position x of the point common to the snubber and car is

$$x = x(t) = -0.398e^{-0.415t}\sin(1.26t), \tag{5}$$

for all times between the time $t = 0$ of first contact and $t \approx 2.5$, the time of last contact. Show that the speed of the car decreases by more than 50 percent between first and last contact with the snubber.

Solution

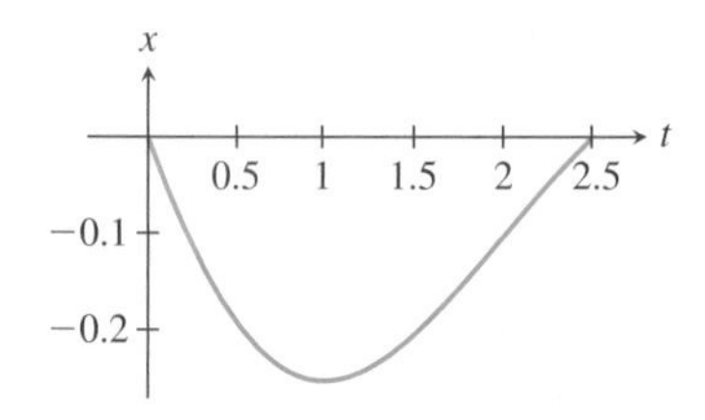

FIGURE 3.8 Position of the contact point between snubber and car as a function of time.

Figure 3.8 shows a graph of x against t. From this graph, the first and last contacts happen at $t = 0$ and $t \approx 2.5$. The time of last contact can be calculated more accurately by setting $x = 0$. From (5), x can be zero only when $\sin(1.26t) = 0$. Hence, $1.26t = 0, \pi, 2\pi, \ldots$. So, the last contact happens at $t = \pi/1.26 \approx 2.49$ seconds.

For any time t, the velocity v of the point of contact is the derivative of its position function. Hence,

$$v = \frac{dx}{dt} = \frac{d}{dt}(-0.398e^{-0.415t}\sin(1.26t)).$$

Applying the product and chain rules,

$$\begin{aligned} v &= -0.398\left(\frac{d}{dt}(e^{-0.415t})\sin(1.26t) + e^{-0.415t}\frac{d}{dt}(\sin(1.26t))\right) \\ &= -0.398(-0.415 \cdot e^{-0.415t}\sin(1.26t) + e^{-0.415t} \cdot 1.26\cos(1.26t)). \end{aligned}$$

Simplifying this expression by factoring out the common exponential,

$$v = -0.398e^{-0.415t}(-0.415\sin(1.26t) + 1.26\cos(1.26t)).$$

To show that the speed of the car decreases by more than 50 percent between first and last contact, we need the speeds of the car at $t = 0$ and $t = 2.49$. We were given that the initial speed is 0.5 m/s. The velocity at $t = 2.49$ is

$$\begin{aligned} v &= -0.398e^{-0.415\cdot 2.49}(-0.415\sin(1.26\cdot 2.49) + 1.26\cos(1.26\cdot 2.49)) \\ &\approx 0.18 \text{ m/s}. \end{aligned}$$

The change from the initial speed of 0.5 m/s to the final speed of $|v(2.49)| \approx 0.18$ m/s is more than a 50 percent decrease in speed.

Acceleration

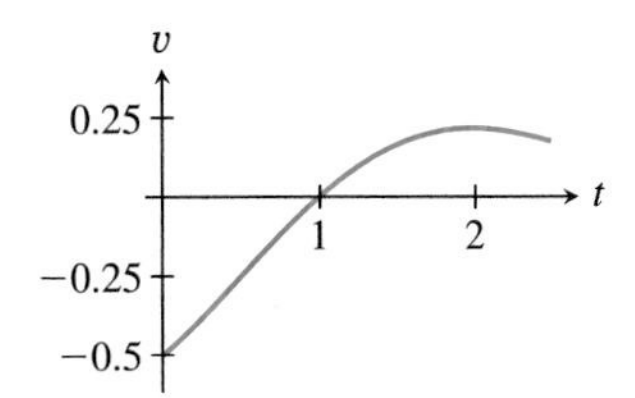

FIGURE 3.9 The velocity of the point of contact between the car and snubber as a function of time t; the rate of change of velocity—or the slope of the velocity graph—is called *acceleration*.

Figure 3.9 shows the graph of velocity $v = v(t)$ of the point of contact between the car and snubber. It is clear from the graph that the velocity changed throughout the time interval from $t = 0$ to $t = 2.49$, the time of last contact. When its velocity changes, the point of contact is said to undergo *acceleration*. We define *acceleration* as the rate of change of velocity with respect to time. If the units of velocity are meters per second (m/s), then the units of acceleration are in meters per second per second (m/s^2).

DEFINITION Acceleration

Let $x = f(t)$ be the position at time t of an object moving along a line during a time interval I. We assume that f and f' are differentiable on I. We define the **acceleration** $a(t)$ of the object at any time t in I by the equation

$$a(t) = v'(t) = \frac{d}{dt}(v) = \frac{d}{dt}\left(\frac{dx}{dt}\right).$$

You may know from high school physics the equations describing the motion of an object undergoing constant acceleration. If we suppose that (i) the motion occurs on an x-axis, (ii) the acceleration is the constant k, and (iii) the initial position (that is, its position at $t = 0$) of the object is x_0 and its initial velocity is v_0, then the position, velocity, and acceleration of the object at any time $t \geq 0$ are given by

$$x = x(t) = \tfrac{1}{2}kt^2 + v_0t + x_0 \tag{6}$$

$$v = v(t) = kt + v_0 \tag{7}$$

$$a = a(t) = k. \tag{8}$$

As you would expect from your own experience, if the acceleration is positive, the velocity will increase for all t; if the acceleration is negative, the velocity will decrease.

Motion in which the acceleration is more or less constant is quite common. If, for example, a speck of moon dust is disturbed, it will settle back to the lunar surface under the essentially constant acceleration due to the gravitational attraction of the moon. Or, provided we ignore the resistance of air, if from a point x_0 meters above the Earth's surface we hurl a baseball vertically upward or downward with velocity v_0, it will fall back to the surface under the very nearly constant acceleration due to the gravitational attraction of the Earth. Equations (6) and (7) may be applied to model these motions. If we choose meters and seconds as units, the constant k will be about ± 1.62 m/s^2 on the moon and ± 9.8 m/s^2 on Earth. The acceleration due to gravitational attraction will be negative if the x-axis is directed upward and positive if it is directed downward. We may also apply these equations to a car undergoing constant acceleration if the car is on a level highway and we ignore the resistance of air and the friction of the tires. We shall often make such simplifying assumptions.

We derive equations (6) and (7) from (8) in Section 3.6. Meanwhile, note that these equations are internally consistent in that

$$\frac{d}{dt}x(t) = \frac{d}{dt}\left(\tfrac{1}{2}kt^2 + v_0 t + x_0\right) = kt + v_0 = v(t)$$

and

$$\frac{d}{dt}v(t) = \frac{d}{dt}(kt + v_0) = k = a(t).$$

EXAMPLE 4 Suppose that a speck of moon dust is stirred up from the lunar surface by an astronaut. It was first noticed when it was 2.0 meters high and moving directly upward at 3.0 m/s. Find its velocity at the time it returns to the surface.

Solution

If as in Fig. 3.10 we put the origin of an x-axis at the surface and take the positive direction upward, the initial position of the speck will be $x_0 = +2.0$ and its initial velocity will be $v_0 = +3.0$. We take the acceleration as $k = -1.62$ because the x-axis is directed upward. We want to find $v(T)$, where T is the time at which the speck returns to the moon's surface, that is, $x(T) = 0$. We use this equation to solve for T and then calculate $v(T)$. From (6),

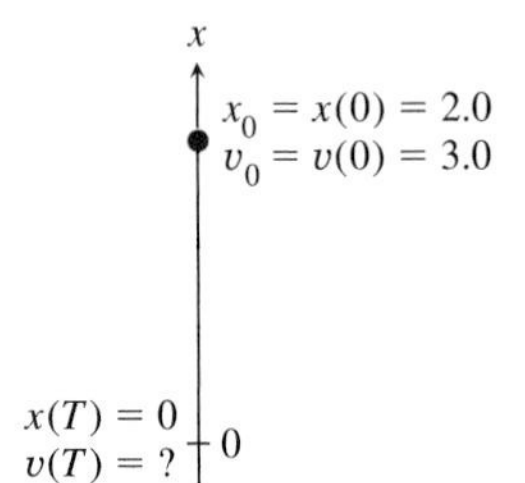

FIGURE 3.10 Motion of a speck of moon dust relative to the chosen x-axis.

$$x = x(t) = \tfrac{1}{2}kt^2 + v_0 t + x_0 = \tfrac{1}{2}(-1.62)t^2 + 3.0t + 2.0.$$

The roots of the quadratic $x(t) = 0$ are

$$t \approx -0.576829, 4.28053.$$

The negative root is the time the speck left the surface. We take T as the other root. Evaluating (7) at this time,

$$v(T) = kT + v_0 \approx -1.62 \cdot 4.28053 + 3.0 \approx -3.93 \text{ m/s}.$$

Hence, the velocity of the speck will be -3.93 m/s when it returns to the surface of the moon.

EXAMPLE 5 Ralph is driving at 40 mph in the fog on a straight road. George is driving at 65 mph and is behind Ralph. George becomes aware of Ralph's car at the time the cars are 180 feet apart. If George instantly applies his brakes, which produce a constant acceleration of -4 ft/s^2, does he avoid hitting Ralph? We took George's acceleration to be negative because it is natural to orient the axis in the direction of motion.

Solution

If we can write expressions $r(t)$ and $g(t)$ for the positions of the two cars at any time t, George will hit Ralph only if the equation $r(t) = g(t)$ has at least one real solution. We use (6) to write expressions for the positions of the two cars, taking $t = 0$ as the time when Ralph applies his brakes. To make the units consistent we use the fact that 60 mph is 88 ft/s. Hence,

$$40 \text{ mph} \longrightarrow \tfrac{176}{3} \text{ ft/s} \quad \text{and} \quad 65 \text{ mph} \longrightarrow \tfrac{286}{3} \text{ ft/s}.$$

We imagine the road as an x-axis, with the origin at George's position at $t = 0$. Letting $r(t)$ and $g(t)$ denote Ralph's and George's positions at any time t,

$$r(t) = \tfrac{176}{3}t + 180$$
$$g(t) = \tfrac{1}{2}(-4)t^2 + \tfrac{286}{3}t = -2t^2 + \tfrac{286}{3}t.$$

Note that at $t = 0$, Ralph's position is $r(0) = 180$ and George's position is $g(0) = 0$.

We set $r(t) = g(t)$ and solve for t. Multiplying both sides of the equation

$$\tfrac{176}{3}t + 180 = -2t^2 + \tfrac{286}{3}t$$

by 3 and simplifying,

$$6t^2 - 110t + 540 = 0.$$

Because the roots of this quadratic are complex (the discriminant $b^2 - 4ac$ is $110^2 - 4 \cdot 6 \cdot 540 < 0$), there is no time t at which $r(t) = g(t)$. Hence, George's car does not hit Ralph's car.

Exercises 3.1

Exercises 1–4: Solve the following. Distances are measured in meters and times in seconds.

1. The crown jewels were accidentally dropped from a bridge 100 m above a roadway, so that their height at time t is $h = h(t) = 100 - 4.9t^2$. Calculate their average velocity $v(1.5, t_2)$ for $t_2 = 1.6$, 1.51, 1.501, and 1.5001. Estimate $v(1.5)$ from your work. Assuming that it will take 5 seconds for the Prince to arrange a safe landing, will the jewels smash into the roadway?

2. The crown jewels were accidentally dropped from a bridge 75 m above a roadway, so that their height at time t is $h = h(t) = 75 - 4.9t^2$. Calculate their average velocity $v(1.25, t_2)$ for $t_2 = 1.35$, 1.26, 1.251, and 1.2501. Estimate $v(1.25)$ from your work. Assuming that it will take 4 seconds for the Princess to arrange a safe landing, will the jewels smash into the roadway?

3. The height h of a rifle bullet is

$$h = h(t) = -4.9t^2 + 700t + 1.3 \text{ m},$$

for $t \in [0, b]$, where b is the time when the bullet strikes the Earth. Find the velocity of the bullet at $t = b$.

4. The height h of a rifle bullet is

$$h = h(t) = -4.9t^2 + 650t + 1.4 \text{ m},$$

for $t \in [0, b]$, where b is the time when the bullet strikes the Earth. Find the velocity of the bullet at $t = b$.

Exercises 5–12: Apply equations (6)–(8) as needed, ignoring, as in the examples, the resistance of air, friction, and so on. Assume that the lengths are measured in meters and times in seconds. Apart from sign, the acceleration due to gravity is $k \approx 9.8$ m/s^2 on Earth, $k \approx 26.5$ m/s^2 on Jupiter, and $k \approx 1.62$ m/s^2 on the Moon.

5. A probe is hovering 100 m above the surface of Jupiter when the engines fail. With what velocity will the probe impact Jupiter?

6. A probe is hovering 100 m above the surface of the moon when the engines fail. With what velocity will the probe impact the moon?

7. A crane is lifting a bucket of cement aggregate straight upward at 1.0 m/s when the cable snaps. If the bucket is 500 m above the ground at the time, how long will it take to hit the ground? What is its greatest speed?

8. A crane is lowering a bucket of cement aggregate straight downward at 1.5 m/s when the cable snaps. If the bucket is 700 m above the ground at the time, how long will it take to hit the ground? What is its greatest speed?

9. A spaceship is traveling in a straight line at 10,000 kph at the time its engines are turned on for 25 s. If the engines provide a steady acceleration of 0.9 m/s^2, how far does the spaceship travel while the engines are on?

10. A spaceship is traveling in a straight line at 10,000 kph when it is reversed and the engines are turned on. If the engines provide a steady acceleration of 0.9 m/s^2, how long does it take for the spaceship to stop?

11. At 2:00 P.M. motorists A and B are traveling in the same direction on a straight road. B is 2 km ahead of A and traveling at 100 kph, and A accelerates at a steady 3 m/s^2 from an initial speed of 50 kph. At what time will A overtake B?

12. At 3:00 P.M. motorists A and B are traveling in the same direction on a straight road. B is 3 km ahead of A and traveling at 120 kph, and A accelerates at 5 m/s^2 from an initial speed of 45 kph. At what time will A overtake B?

13. If the position of the end of a freight car is

$$x = x(t) = -0.5e^{-0.4t}\sin(1.3t),$$

for $t \in [0, \pi/1.3]$, find the velocity of the car at $t = 0$ and $t = \pi/1.3$. Lengths are in meters and time in seconds.

14. If the position of the end of a freight car is

$$x = x(t) = -0.75e^{-0.5t}\sin(1.5t),$$

for $t \in [0, \pi/1.5]$, find the velocity of the car at $t = 0$ and $t = \pi/1.5$. Lengths are in meters and time in seconds.

15. Find the average velocity during ascent of the bullet whose height is $h = h(t) = -4.9t^2 + 610t + 5.4$ m. Time is in seconds. Write a sentence interpreting the average velocity you have found, so that a layperson will understand.

16. Find the average velocity during ascent of the bullet whose height is $h = h(t) = -4.9t^2 + 630t + 3.9$ m. Time is in seconds. Write a sentence interpreting the average velocity you have found, so that a layperson will understand.

17. The accompanying figure shows a weight attached to a spring. The position $x(t)$ of the weight at any time $t \geq 0$ is

$$x(t) = 1.2 - 7.4\cos(0.57t + 1.8).$$

Times are in seconds and distances in meters. Determine the position and velocity of the weight at $t = 0$.

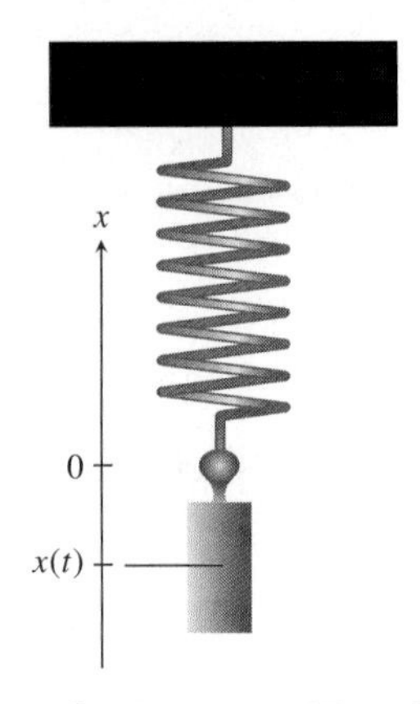

Figure for Exercises 17 and 18.

18. The accompanying figure shows a weight attached to a spring. The position $x(t)$ of the weight at any time $t \geq 0$ is

$$x(t) = 2.25 - 5.6\cos(0.75t + 2.0).$$

Times are in seconds and distances in meters. Determine the position and velocity of the weight at $t = 0$.

19. For $0 \leq t \leq 35.4$ seconds, the position in miles of a Porsche 911RS at time t is

$$x(t) = 10^{-4}(35.8t^{3/2} + 7.9t^2 - 0.9t^{5/2}).$$

Determine its speed in *miles per hour* at $t = 10$ seconds.

20. Find the average velocity during the time interval $[t_1, t_2]$ for the bullet whose height at time t is $h = h(t) = -\frac{1}{2}gt^2 + v_0t + x_0$, where g is the acceleration due to gravity, v_0 is the bullet's initial velocity (we assume $v_0 > 0$), and h_0 is its initial height. Use this result in finding the average velocity of the bullet during its ascent.

21. Find the average velocity during the time interval $[-c/b, -c/b + \pi/(2b)]$ for an object with position $x = x(t) = a\sin(bt + c)$, where a, b, and c are constants.

22. In Exercise 20 the average velocity $v(t_1, t_2)$ was calculated. Calculate $\lim_{t_2 \to t_1} v(t_1, t_2)$ and give a meaning to the result.

23. In Exercise 21 the average velocity $v(t_1, t_2)$ was calculated for specific values of t_1 and t_2. Write out the general case, calculate $\lim_{t_2 \to t_1} v(t_1, t_2)$ and give a meaning to the result. You may wish to use the identity

$$\sin A - \sin B = 2\sin\frac{A - B}{2}\cos\frac{A + B}{2}.$$

24. Your car has a broken speedometer. Plot a velocity graph from the following odometer data. Each data point is of the form (t, x), where t is the time in hours and x is the odometer reading in kilometers.

$(0.0, 11{,}084)$, $(0.08, 11{,}088)$,
$(0.16, 11{,}092)$, $(0.24, 11{,}095)$,
$(0.32, 11{,}097)$, $(0.40, 11{,}099)$,
$(0.48, 11{,}100)$, $(0.56, 11{,}100)$.

25. Your car has a broken speedometer. Plot a velocity graph from the graph of the odometer reading shown in the accompanying figure.

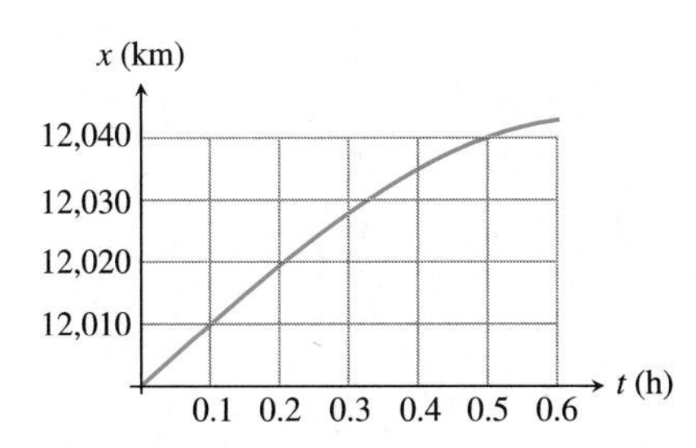

Figure for Exercise 25.

26. An object is moving along a line. A graph of its position versus time is given in the accompanying figure. Sketch a graph of its velocity versus time.

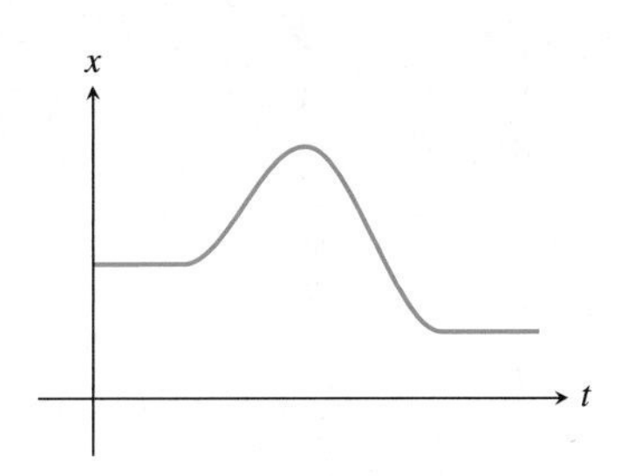

Figure for Exercise 26.

27. An object is moving along a line. A graph of its position versus time is given in the accompanying figure. Sketch a graph of its velocity versus time.

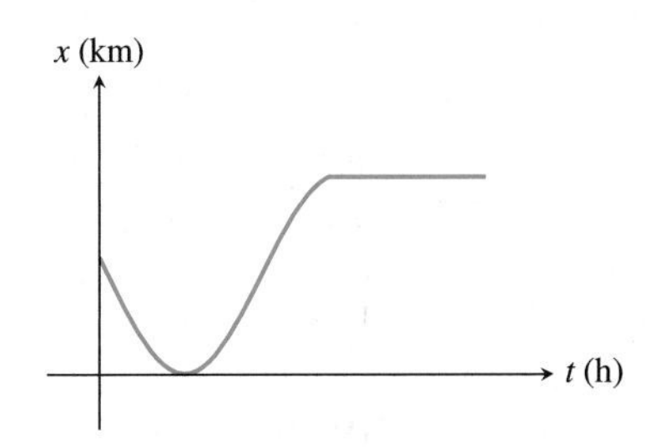

Figure for Exercise 27.

28. In the context of Example 3, show that the velocity

$$v = -0.398e^{-0.415t} \cdot (-0.415 \sin(1.26t) + 1.26 \cos(1.26t))$$

of the point of contact between car and snubber can be written as

$$v = -0.528e^{-0.415t} \sin(1.26t + 1.89).$$

Hint: The expression inside the outer pair of parentheses has the form

$$z = A \sin(Ct) + B \cos(Ct).$$

Show that by writing z in the form

$$\sqrt{A^2 + B^2}\left(\frac{A}{\sqrt{A^2 + B^2}} \sin(Ct) + \frac{B}{\sqrt{A^2 + B^2}} \cos(Ct)\right),$$

an angle $\alpha \in [0, 2\pi)$ can be chosen so that $\cos \alpha = A/\sqrt{A^2 + B^2}$ and $\sin \alpha = B/\sqrt{A^2 + B^2}$.

29. Use the result of the preceding exercise to find the time of maximum compression of the spring in Example 3.

30. The position at time t of a 91.6-kilogram skydiver (this is body mass plus equipment) falling from 1000 meters is

$$x(t) = 1000 - 210 \ln(\cosh(0.22t)), \; t \geq 0,$$

where t is measured in seconds. Calculate the velocity of the skydiver at $t = 2$ seconds. If the skydiver does not deploy a chute, with what velocity will he hit the ground?

You may wish to replace $\cosh w$ with its equal: $\frac{1}{2}(e^w + e^{-w})$. Or you can use the fact that

$$(d/dw) \cosh(w) = \sinh(w) = \frac{e^w - e^{-w}}{2}.$$

Also see Exercise 60, Section 2.6. *Note:* This model assumes that the skydiver falls directly downward in the spread-eagle position (face down). It has been found that under these circumstances, the air resistance is proportional to the square of the skydiver's velocity.

31. Referring to Fig. 3.3, apply the law of cosines to derive the expression

$$x = h(t) = 5 \cos t + \sqrt{100 - 25 \sin^2 t}$$

for the position of the slider at any time t. Include in your discussion a reason for your choice of signs.

32. Write a sentence or two relating the path of the bullet in Example 2 to the graph of the function h.

3.2 Vectors

In Section 3.1 we studied the motion of objects along a line. In this section and the next we prepare to study motion in a plane or in space. Motion in more than one dimension is most easily described and studied using vectors.

While physical or geometric quantities like mass, speed, and length can be described with a single, positive number, quantities like force, velocity, or displacement, which have both magnitude and direction, cannot be so described. To model such quantities we use vectors, which are often represented graphically by arrows. We start with this idea.

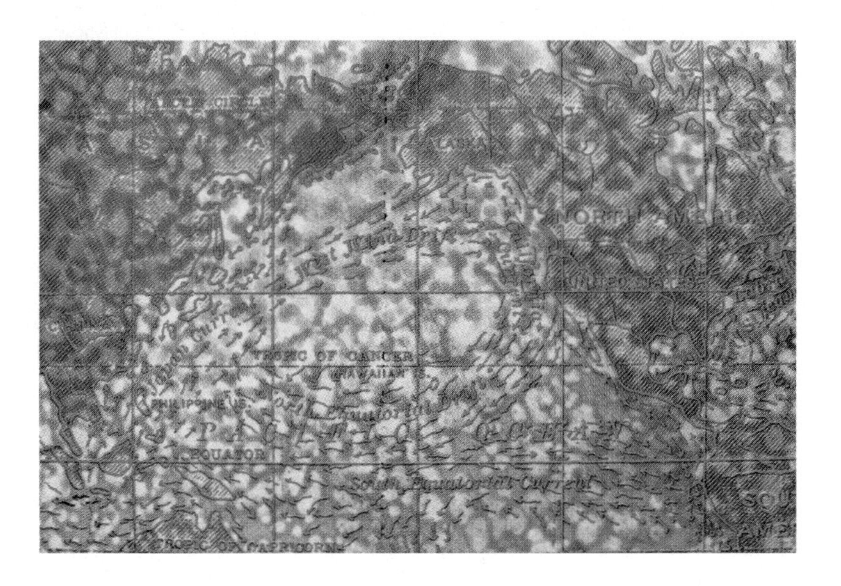

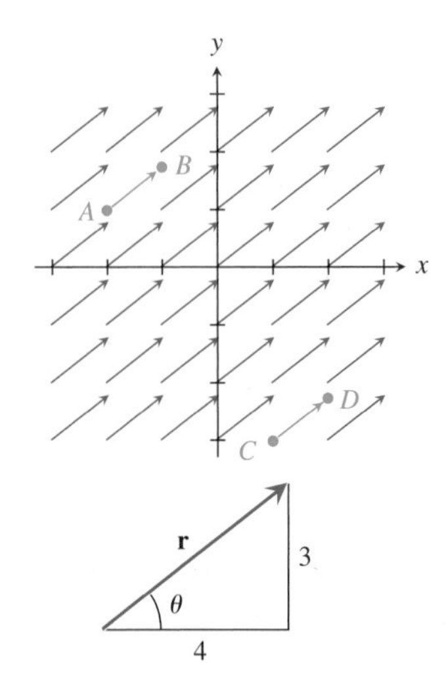

FIGURE 3.11 The flow of a fluid on a plane.

Figure 3.11 is intended to suggest a fluid flowing with constant speed and direction on a horizontal plane during a one-second time interval. The figure both gives a qualitative impression of motion and, if we look more closely, shows a flow in which during the one-second time interval each fluid particle moves the equivalent of four centimeters in the positive x-direction and three centimeters in the positive y-direction. (*Note:* The marks on the axes are four centimeters apart.) The arrows show the *displacements* of several particles of the fluid during the one-second time interval. Because the flow is uniform, all of the arrows have the same length and direction.

We focus on the particle initially at A, with coordinates $(-8, 4)$. After one second, this particle will have been displaced to B, with coordinates $(-8 + 4, 4 + 3) = (-4, 7)$. We describe this displacement with the notation $\overrightarrow{AB}$. Also, shown is the displacement $\overrightarrow{CD}$ of a particle with initial point $(4, -12)$. The terminal point D has coordinates $(4 + 4, -12 + 3) = (8, -9)$. Although $\overrightarrow{AB}$ and $\overrightarrow{CD}$ represent the displacement of different particles, they are the same in that they have the same length and direction. The distance traveled during one second by each fluid particle is $\sqrt{3^2 + 4^2} = 5$, and each moved in the direction $\theta = \arctan(3/4) \approx 37°$. The arrow or vector $\mathbf{r}$ illustrating this common length and direction is shown in the zoom-view at the bottom of the figure.

Equivalent Vectors; Length and Direction of a Vector

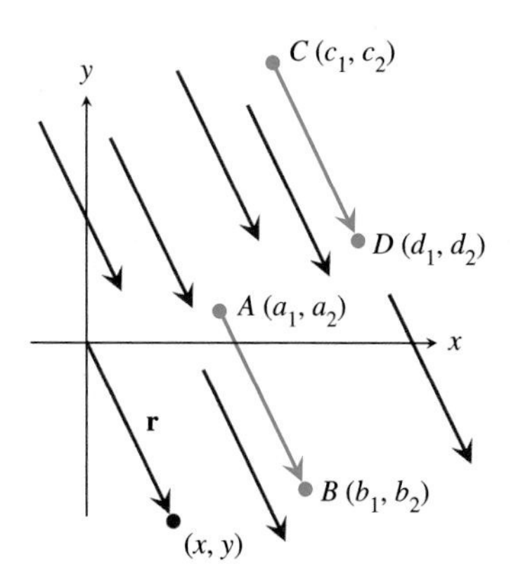

FIGURE 3.12 Equivalent vectors have the same length and the same direction.

Figure 3.12 shows a vector $\overrightarrow{AB}$ from (initial point) A to (terminal point) B and a vector $\overrightarrow{CD}$ from C to D. We say that these vectors are **equivalent** if they have the same length and direction. Using the notation in Fig. 3.12, vectors $\overrightarrow{AB}$ and $\overrightarrow{CD}$ are equivalent if

$$b_1 - a_1 = d_1 - c_1 \quad \text{and} \quad b_2 - a_2 = d_2 - c_2. \tag{1}$$

If these two conditions hold, it follows that the lengths and directions of $\overrightarrow{AB}$ and $\overrightarrow{CD}$ are equal.

Figure 3.12 also shows the vector **r** with initial point at the origin and terminal point (x, y), where $x = b_1 - a_1$ and $y = b_2 - a_2$. This vector is the simplest of the vectors equivalent to $\overrightarrow{AB}$. The vector **r** is determined by the two numbers x and y, which are called **coordinates** of **r**. We denote **r** by

$$\mathbf{r} = \langle x, y \rangle.$$

Because $\mathbf{r} = \langle x, y \rangle$ gives the position of the point (x, y), it is often called a **position vector.**

The length and direction of any vector $\overrightarrow{AB}$ equivalent to $\mathbf{r} = \langle x, y \rangle$, where $x = b_1 - a_1$ and $y = b_2 - a_2$, are the same as the length and direction of **r**. Hence, we may calculate the length and direction of $\overrightarrow{AB}$ in terms of x and y.

Length and Direction of a Vector

Let $\mathbf{r} = \langle x, y \rangle$ be a vector from the origin. The **length** of **r**, and of any vector equivalent to **r**, is

$$\|\mathbf{r}\| = \|\langle x, y \rangle\| = \sqrt{x^2 + y^2}. \tag{2}$$

The **direction** of $\mathbf{r} = \langle x, y \rangle$, and of any vector equivalent to **r**, is the angle θ through which the positive x-axis must be rotated counterclockwise to align with **r**. If $\mathbf{r} = \langle 0, 0 \rangle$, we assign no direction to **r**. See accompanying figure.

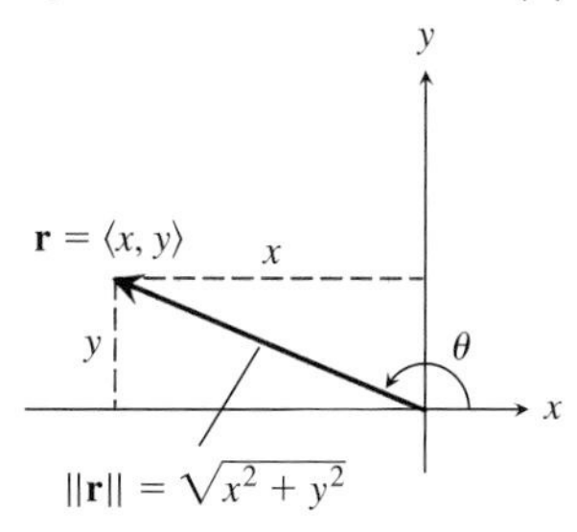

The length $\|\mathbf{r}\|$ and direction θ of a representative vector **r**.

We shall often denote a vector from an initial point P to a terminal point Q by a single boldface letter such as **a** rather than $\overrightarrow{PQ}$. Figure 3.13 shows, for example, a vector **a** from $(-3, 4)$ to $(1, 2)$.

FIGURE 3.13 Equivalent vectors **a**, **b**, and **r**.

EXAMPLE 1 Figure 3.13 shows vectors **a**, **b**, and **r**. Show that these vectors are equivalent. Calculate their common length and direction.

Solution

The vectors **a**, **b**, and **r** are equivalent because, by applying (1),

$$1 - (-3) = 4 - 0 = 4 - 0 = 4 \quad \text{and}$$
$$2 - 4 = -4 - (-2) = -2 - 0 = -2.$$

The common length and direction of these vectors are the length and direction of the position vector $\mathbf{r} = \langle 4, -2 \rangle$. The length of **r** is

$$\|\mathbf{r}\| = \sqrt{4^2 + (-2)^2} = \sqrt{20} \approx 4.47.$$

The direction of **r** is the angle θ shown in Fig. 3.13. Perhaps the easiest way to calculate θ is to calculate the acute angle $2\pi - \theta$ shown in the diagram at the bottom of the figure. The tangent of this angle is $2/4 = 0.5$. Hence

$$\theta = 2\pi - \arctan 0.5 = 2\pi - 0.46364\cdots \approx 5.8.$$

Vector Addition and the Parallelogram Law

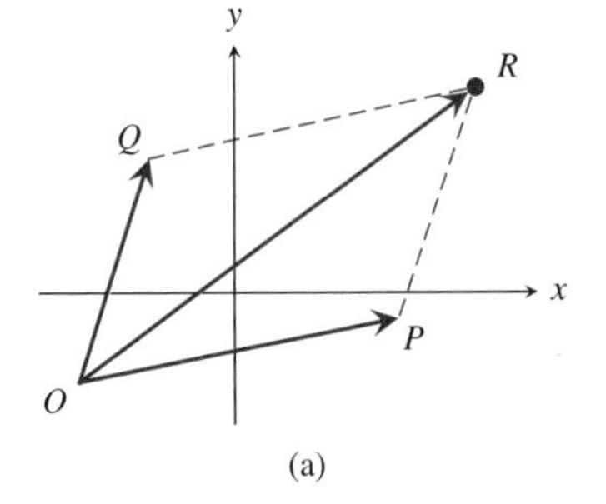

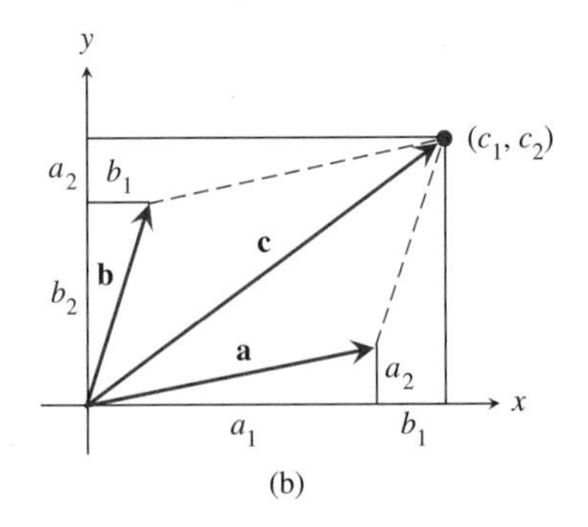

FIGURE 3.14 (a) Adding vectors by applying the parallelogram law; (b) adding position vectors **a** and **b** arithmetically.

Vector addition can be motivated by a physical experiment. Consider two forces acting at a point O of the plane. Figure 3.14(a) shows these forces as the vectors $\overrightarrow{OP}$ and $\overrightarrow{OQ}$. The lengths of the vectors are proportional to the magnitudes of the two forces, and their directions are the same as those of the two forces. It can be shown by experiment that the combined effect of the two forces acting at O is equivalent to the single force corresponding to the vector $\overrightarrow{OR}$, which is the diagonal of the parallelogram with sides $\overrightarrow{OP}$ and $\overrightarrow{OQ}$. This is sometimes called the *parallelogram law*.

The vector addition $\overrightarrow{OP} + \overrightarrow{OQ}$ shown in Fig. 3.14(a) is based on the physical/geometric parallelogram law. For most applications involving vector addition it is more convenient to form the sum $\overrightarrow{OP} + \overrightarrow{OQ}$ arithmetically. We do this by defining the sum $\mathbf{a} + \mathbf{b}$ of the position vectors **a** and **b** equivalent to $\overrightarrow{OP}$ and $\overrightarrow{OQ}$. We show that this definition is consistent with the parallelogram law by showing that $\mathbf{a} + \mathbf{b}$ is equivalent to the vector $\overrightarrow{OP} + \overrightarrow{OQ}$.

Vector Addition

The **sum** of the vectors $\mathbf{a} = \langle a_1, a_2 \rangle$ and $\mathbf{b} = \langle b_1, b_2 \rangle$ is the vector

$$\mathbf{a} + \mathbf{b} = \langle a_1, a_2 \rangle + \langle b_1, b_2 \rangle = \langle a_1 + b_1, a_2 + b_2 \rangle. \tag{3}$$

Figure 3.14(b) gives a geometric interpretation of this definition. Letting **c** be the diagonal vector of the parallelogram formed by **a** and **b**, note that the sum of the segments with lengths a_1 and b_1 is equal to the segment of length c_1, where c_1 is the first coordinate of **c**. Similarly, the sum of the segments with lengths a_2 and b_2 is equal to the segment of length c_2. Thus,

$$\begin{aligned}\mathbf{c} &= \langle c_1, c_2 \rangle \\ &= \langle a_1 + b_1, a_2 + b_2 \rangle \\ &= \langle a_1, b_1 \rangle + \langle a_2, b_2 \rangle = \mathbf{a} + \mathbf{b}.\end{aligned}$$

Hence, adding position vectors by applying (3) is consistent with adding vectors $\overrightarrow{OP}$ and $\overrightarrow{OQ}$ in Fig. 3.14(a) by applying the parallelogram law.

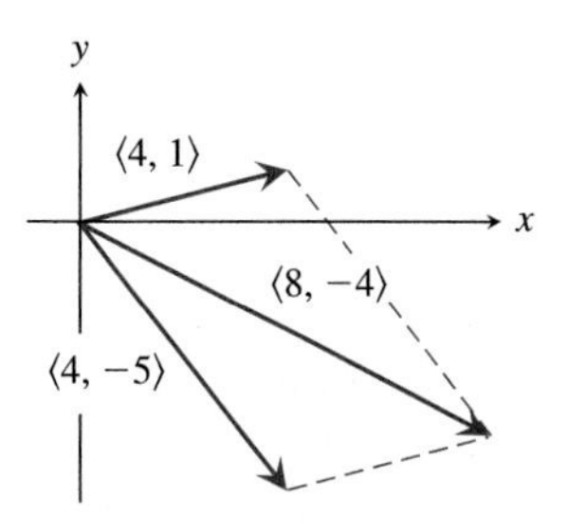

FIGURE 3.15 Adding position vectors.

EXAMPLE 2 Calculate the sum of vectors $\langle 4, 1\rangle$ and $\langle 4, -5\rangle$.

Solution

From the definition of vector addition,

$$\langle 4,1\rangle + \langle 4,-5\rangle = \langle 4+4, 1+(-5)\rangle = \langle 8,-4\rangle.$$

Figure 3.15 shows the given vectors as well as their sum. The vector $\langle 8, -4\rangle$ is a diagonal of the parallelogram determined by $\langle 4, 1\rangle$ and $\langle 4, -5\rangle$.

The consistency mentioned earlier of the physical/geometric sum $\overrightarrow{OP} + \overrightarrow{OQ}$ and the arithmetic sum $\mathbf{a} + \mathbf{b}$ of position vectors $\mathbf{a}$ and $\mathbf{b}$ equivalent to $\overrightarrow{OP}$ and $\overrightarrow{OQ}$ tends to blur the distinction between equivalent vectors and equal vectors. For example, suppose that the points O, P, and Q in Fig. 3.14(a) have coordinates (o_1, o_2), (p_1, p_2), and (q_1, q_2). How can we calculate the coordinates (r_1, r_2) of R in terms of the coordinates of O, P, and Q? If we write $\overrightarrow{OP} = \langle p_1 - o_1, p_2 - o_2\rangle$, $\overrightarrow{OQ} = \langle q_1 - o_1, q_2 - o_2\rangle$, and $\overrightarrow{OR} = \langle r_1 - o_1, r_2 - o_2\rangle$, then because $\overrightarrow{OR} = \overrightarrow{OP} + \overrightarrow{OQ}$,

$$\begin{aligned}\langle r_1 - o_1, r_2 - o_2\rangle &= \langle p_1 - o_1, p_2 - o_2\rangle + \langle q_1 - o_1, q_2 - o_2\rangle \\ &= \langle p_1 + q_1 - 2o_1, p_2 + q_2 - 2o_2\rangle.\end{aligned}$$

Hence, the coordinates r_1 and r_2 of R are

$$\begin{aligned} r_1 &= p_1 + q_1 - o_1 \\ r_2 &= p_2 + q_2 - o_2. \end{aligned}$$

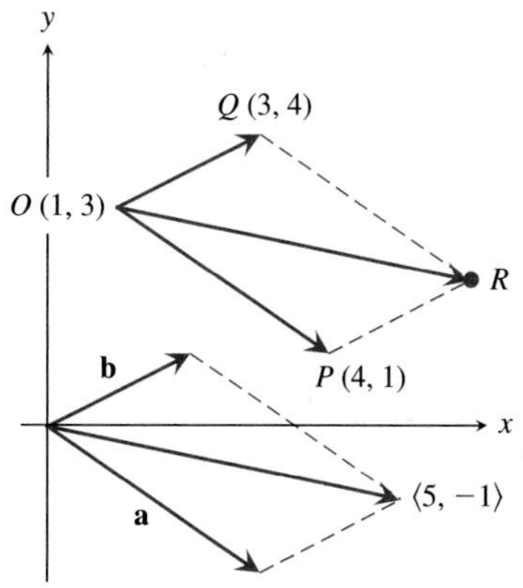

FIGURE 3.16 Adding vectors.

EXAMPLE 3 Let $\overrightarrow{OP}$ and $\overrightarrow{OQ}$ be vectors, where the points O, P, and Q are $(1, 3)$, $(4, 1)$, and $(3, 4)$. See Fig. 3.16. Calculate the length and direction of the sum $\overrightarrow{OR} = \overrightarrow{OP} + \overrightarrow{OQ}$. Also, calculate the coordinates of R.

Solution

The sum of vectors $\overrightarrow{OP}$ and $\overrightarrow{OQ}$ can be found by adding position vectors $\mathbf{a}$ and $\mathbf{b}$ equivalent to $\overrightarrow{OP}$ and $\overrightarrow{OQ}$. The sum $\mathbf{a} + \mathbf{b}$ has the same length and direction as the vector $\overrightarrow{OR}$. From Fig. 3.16,

$$\mathbf{a} = \langle 4-1, 1-3\rangle = \langle 3,-2\rangle \quad \text{and} \quad \mathbf{b} = \langle 3-1, 4-3\rangle = \langle 2,1\rangle.$$

From the definition of vector addition,

$$\mathbf{a} + \mathbf{b} = \langle 3,-2\rangle + \langle 2,1\rangle = \langle 5,-1\rangle.$$

This vector is equivalent to the vector $\overrightarrow{OR}$ and, therefore, has the same length and direction. It follows that

$$\|\overrightarrow{OR}\| = \sqrt{5^2 + (-1)^2} = \sqrt{26}.$$

The direction θ of $\overrightarrow{OR}$ can be found as in Example 1:

$$\theta = 2\pi - \arctan(1/5) \approx 6.09.$$

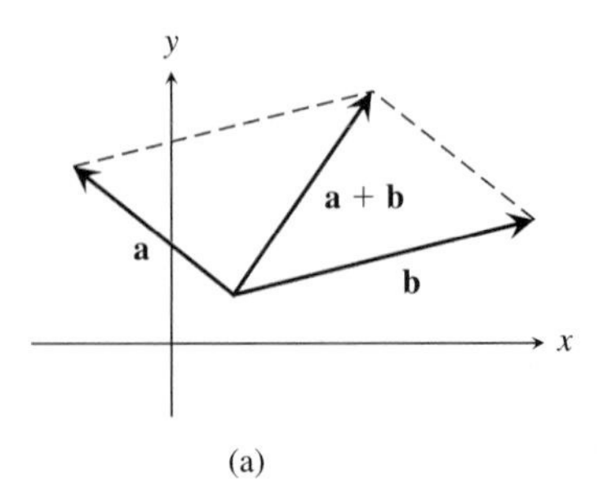

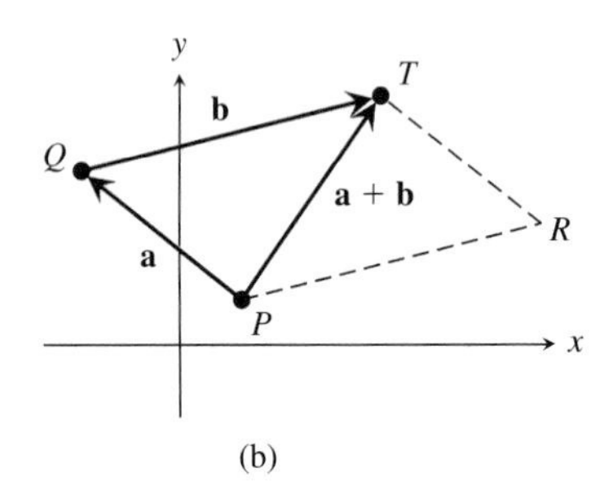

FIGURE 3.17 (a) Vector addition interpreted as, for example, two forces acting at a single point; (b) vector addition interpreted as the sum of successive displacements.

The coordinates r_1 and r_2 of R can be found from the fact that $\overrightarrow{OR}$ is equivalent to $\mathbf{a} + \mathbf{b} = \langle 5, -1 \rangle$, which means that $r_1 - 1 = 5$ and $r_2 - 3 = -1$. Hence, the coordinates of R are $(6, 2)$.

Figure 3.17(a) shows a graphical representation of the sum of **a** and **b**, with the tails of **a**, **b**, and **a** + **b** at the same point. In Fig. 3.17(b), the vectors labeled **a**, **b**, and **a** + **b** are equivalent to the vectors in Fig. 3.17(a), but **a** and **b** are arranged "tip-to-tail." This is useful if, for example, **a** and **b** represent successive displacements of an object. With this interpretation, the sum **a** + **b** represents a single displacement, equivalent to first displacing an object from P to Q and then from Q to T. The sum **a** + **b** is also equivalent to first displacing an object from P to R and then from R to T.

Scalar Multiplication

When we work with both vectors and real numbers in the same context, real numbers are often called **scalars**.

Scalar Multiplication

The **product** of a scalar s and a vector $\mathbf{r} = \langle x, y \rangle$ is

$$s\mathbf{r} = s\langle x, y \rangle = \langle sx, sy \rangle. \qquad (4)$$

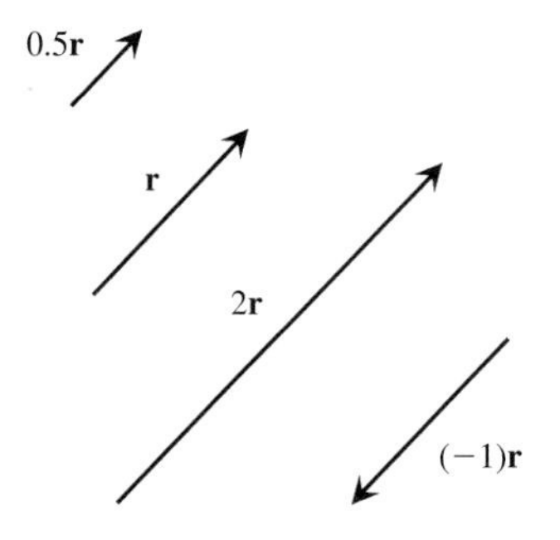

FIGURE 3.18 Multiplying a vector **r** by scalars 0.5, 2, and −1.

Figure 3.18 shows a vector **r** and several scalar multiples of **r**. Because

$$\begin{aligned} \|s\mathbf{r}\| &= \|\langle sx, sy \rangle\| = \sqrt{s^2x^2 + s^2y^2} \\ &= |s|\sqrt{x^2 + y^2} = |s| \cdot \|\mathbf{r}\|, \end{aligned}$$

the length of the vector $s\mathbf{r}$ is $|s|$ times the length of **r**. If $s > 0$, the vector $s\mathbf{r}$ points in the same direction as **r**; if $s < 0$, $s\mathbf{r}$ points in the opposite direction.

Subtracting Vectors

The **difference a − b** of the vectors $\mathbf{a} = \langle a_1, a_2 \rangle$ and $\mathbf{b} = \langle b_1, b_2 \rangle$ is the vector $\mathbf{a} + (-1)\mathbf{b}$. Hence,

$$\begin{aligned} \mathbf{a} - \mathbf{b} &= \langle a_1, a_2 \rangle + (-1)\langle b_1, b_2 \rangle \\ &= \langle a_1, a_2 \rangle + \langle -b_1, -b_2 \rangle = \langle a_1 - b_1, a_2 - b_2 \rangle. \end{aligned} \qquad (5)$$

Figure 3.19(a) shows vectors **a**, **b**, (-1)**b**, and **a** + (-1)**b**. Adding the vectors **a** and (-1)**b** by the parallelogram law gives the vector **a** + (-1)**b** = **a** − **b**. From Fig. 3.19(a), it is clear that the vectors **a** + **b** and **a** − **b** are the two diagonals of the parallelogram formed by the vectors **a** and **b**. Fig. 3.19(b) shows both the sum and difference of the vectors **a** and **b**.

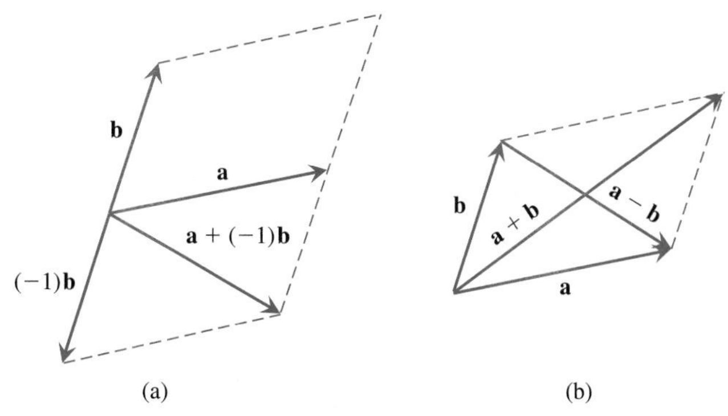

FIGURE 3.19 The sum and difference of vectors **a** and **b** are the two diagonals of the parallelogram formed by **a** and **b**.

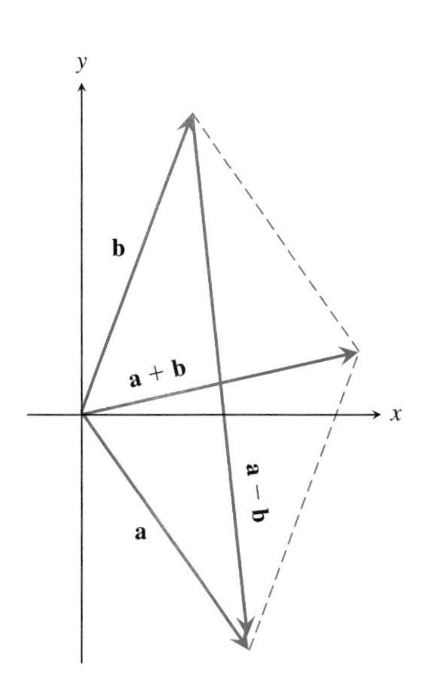

FIGURE 3.20 The sum and difference of vectors **a** and **b** are the two diagonals of the parallelogram formed by **a** and **b**.

EXAMPLE 4 If $\mathbf{a} = \langle 3, -4\rangle$ and $\mathbf{b} = \langle 2, 5\rangle$, calculate $\|\mathbf{a}\|$, $\mathbf{a} + \mathbf{b}$, $\mathbf{a} - \mathbf{b}$, and $-2\mathbf{a} + 7\mathbf{b}$. Sketch the vectors **a**, **b**, $\mathbf{a} + \mathbf{b}$, and $\mathbf{a} - \mathbf{b}$.

Solution

$$\begin{aligned} \|\mathbf{a}\| &= \sqrt{3^2 + (-4)^2} = \sqrt{25} = 5 \\ \mathbf{a} + \mathbf{b} &= \langle 3, -4\rangle + \langle 2, 5\rangle = \langle 5, 1\rangle \\ \mathbf{a} - \mathbf{b} &= \langle 3, -4\rangle - \langle 2, 5\rangle = \langle 1, -9\rangle \\ -2\mathbf{a} + 7\mathbf{b} &= -2\langle 3, -4\rangle + 7\langle 2, 5\rangle = \langle -6, 8\rangle + \langle 14, 35\rangle = \langle 8, 43\rangle. \end{aligned}$$

Figure 3.20 shows the vectors **a**, **b**, $\mathbf{a} + \mathbf{b}$, and $\mathbf{a} - \mathbf{b}$.

The i and j Vectors

The simple vectors $\mathbf{i} = \langle 1, 0\rangle$ and $\mathbf{j} = \langle 0, 1\rangle$ provide an alternative way to express a vector. If $\mathbf{r} = \langle x, y\rangle$, then

$$\mathbf{r} = \langle x, y\rangle = \langle x, 0\rangle + \langle 0, y\rangle = x\langle 1, 0\rangle + y\langle 0, 1\rangle = x\mathbf{i} + y\mathbf{j}. \tag{6}$$

Thus we see that the vectors $\langle x, y\rangle$ and $x\mathbf{i} + y\mathbf{j}$ are different ways of writing the same vector. The $x\mathbf{i} + y\mathbf{j}$ notation is widely used in engineering and physics.

Figure 3.21 shows the vectors **i** and **j**. Also shown are the vectors

$$\langle 2, 3\rangle = 2\langle 1, 0\rangle + 3\langle 0, 1\rangle = 2\mathbf{i} + 3\mathbf{j}$$

and

$$\langle -2, -1\rangle = (-2)\langle 1, 0\rangle - \langle 0, 1\rangle = -2\mathbf{i} - \mathbf{j}.$$

In what follows we use either $\langle a_1, a_2\rangle$ or $a_1\mathbf{i} + a_2\mathbf{j}$ for a vector **a**.

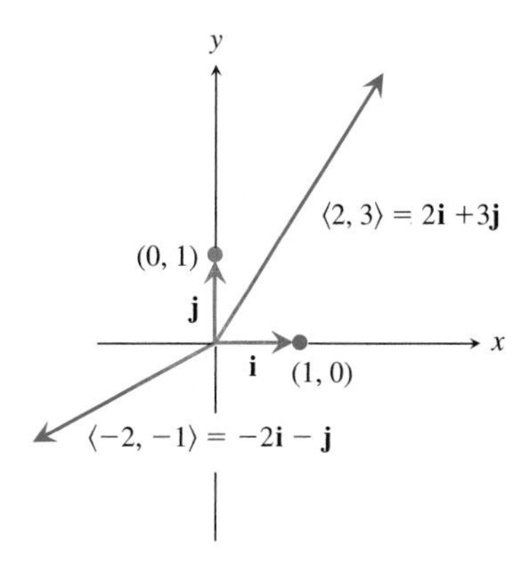

FIGURE 3.21 Writing vectors $\langle 2, 3\rangle$ and $\langle -2, -1\rangle$ in terms of the vectors **i** and **j**.

EXAMPLE 5 Figure 3.22 shows the forces **a** and **b** exerted by two tugboats on a coal barge. The magnitudes of these forces are $\|\mathbf{a}\| = 90$ kilonewtons (approximately 10 tons) and $\|\mathbf{b}\| = 70$ kilonewtons (approximately 8 tons). Find the magnitude and direction of the *resultant force* on the barge, that is, the vector sum of the forces **a** and **b**.

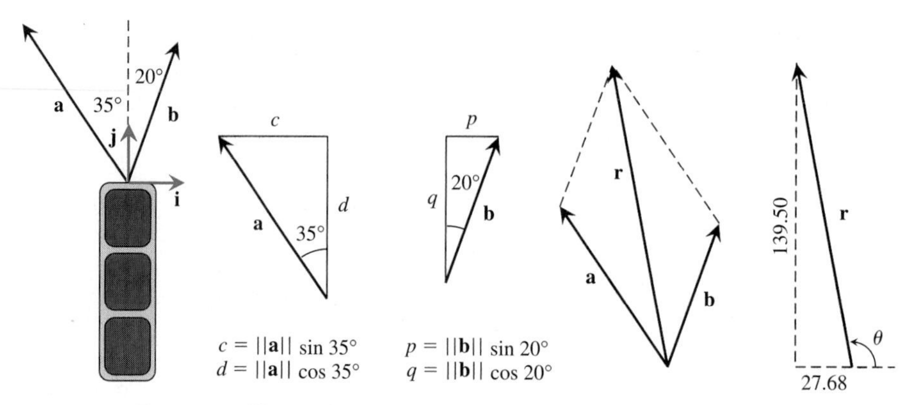

FIGURE 3.22 Forces **a** and **b** on a barge.

Solution

It is natural to think of the barge and the two vectors **a** and **b** in a coordinate system whose origin is at the point on the barge where **a** and **b** act, and the positive x- and y-axes extending to the right and upward, respectively. The **i** and **j** vectors will be as shown at the extreme left of Fig. 3.22. The vectors **a** and **b** are shown separately in the center of Fig. 3.22. The sides c, d, p, and q of the triangles can be calculated using right-triangle trigonometry. Adjusting the signs to fit the directions of **a** and **b**,

$$\mathbf{a} = -c\mathbf{i} + d\mathbf{j}$$
$$\mathbf{b} = p\mathbf{i} + q\mathbf{j}.$$

Hence,

$$\begin{aligned}\mathbf{r} &= \mathbf{a} + \mathbf{b} \\ &= -c\mathbf{i} + d\mathbf{j} + p\mathbf{i} + q\mathbf{j} = (-c + p)\mathbf{i} + (d + q)\mathbf{j} \\ &= (-90 \sin 35° + 70 \sin 20°)\mathbf{i} + (90 \cos 35° + 70 \cos 20°)\mathbf{j} \\ &\approx (-27.68)\mathbf{i} + 139.50\mathbf{j}.\end{aligned}$$

The length or magnitude of the resultant force is

$$\sqrt{(-27.68)^2 + (139.50)^2} = 142.22 \text{ kilonewtons}.$$

Referring to the sketch on the extreme right of the figure, the direction θ of the resultant force is

$$\theta = 180° - \arctan(139.50/27.68) \approx 180° - 78.78° = 101.22°.$$

Unit Vectors

A vector **u** is a unit vector if $\|\mathbf{u}\| = 1$, that is, if the length of **u** is 1. Given a vector $\mathbf{r} \neq \mathbf{0}$, a unit vector **u** in the same direction as **r** is

$$\mathbf{u} = \frac{1}{\|\mathbf{r}\|}\mathbf{r},$$

For example, if $\mathbf{r} = \langle 4, -3 \rangle$, then $\|\mathbf{r}\| = \sqrt{16 + 9} = 5$ and

$$\mathbf{u} = \frac{1}{\|\mathbf{r}\|}\mathbf{r} = \frac{1}{5}\langle 4, -3 \rangle = \langle 4/5, -3/5 \rangle.$$

It is easy to check that $\|\mathbf{u}\| = \sqrt{(4/5)^2 + (-3/5)^2} = 1$.

To find a unit vector in a given direction, we use the standard definition of *sine* and *cosine*. Referring to Fig. 3.23, suppose we want to find a vector in the direction θ. If we place an angle θ in standard position (vertex at the origin and initial side along the positive x-axis), then the coordinates of the point of intersection of the terminal side of the angle with the unit circle are $(\cos\theta, \sin\theta)$. Hence, the position vector $\mathbf{u} = \langle \cos\theta, \sin\theta \rangle$ has unit length and direction θ.

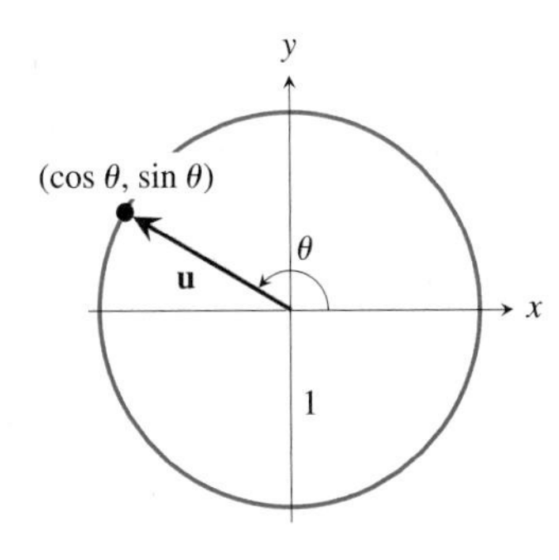

FIGURE 3.23 Corresponding to the angle θ are the point $(\cos\theta, \sin\theta)$ on the unit circle and the unit vector $\mathbf{u} = \langle \cos\theta, \sin\theta \rangle$.

Unit Vector in a Given Direction

A unit vector $\mathbf{u}$ in the direction θ is

$$\mathbf{u} = \langle \cos\theta, \sin\theta \rangle. \tag{7}$$

EXAMPLE 6 Find a unit vector in the direction $\theta = 5\pi/6$ or $150°$. Also, find a vector $\mathbf{F}$ with length 3.5 and direction $\theta = 0.3$.

Solution

Figure 3.23 shows an angle $\theta \approx 150°$. From (7), a unit vector in this direction is

$$\mathbf{u} = \langle \cos 150°, \sin 150° \rangle = \left\langle -\sqrt{3}/2, 1/2 \right\rangle.$$

To find the vector with length 3.5 and direction $\theta = 0.3$, we find a unit vector $\mathbf{u}$ in this direction and then multiply $\mathbf{u}$ by 3.5.

$$\mathbf{u} = \langle \cos 0.3, \sin 0.3 \rangle \approx \langle 0.95534, 0.29552 \rangle.$$

Multiplying this vector by the scalar 3.5,

$$\mathbf{F} = 3.5\mathbf{u} \approx \langle 3.34368, 1.03432 \rangle.$$

As a check, note that $\|\mathbf{F}\| = \sqrt{3.34368^2 + 1.03432^2} \approx \sqrt{12.25001} \approx 3.50000$.

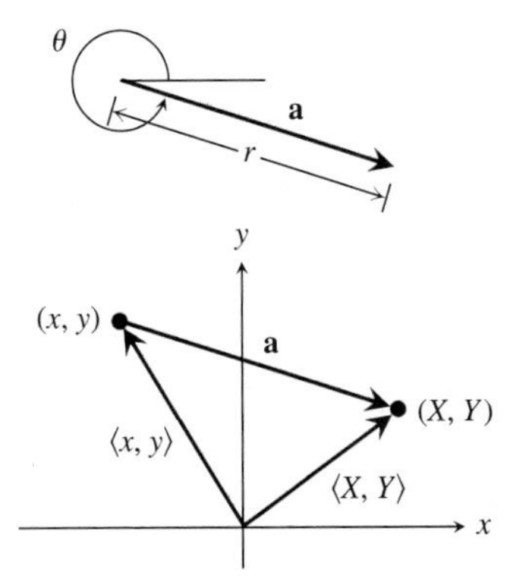

FIGURE 3.24 The displacement $\mathbf{a}$ of an object from (x, y) to (X, Y).

Displacements The displacement of objects along a line or in a plane is often described by vectors. If an object at the point (x, y) is displaced a distance r and in the direction θ, we may describe this displacement by the vector $\mathbf{a} = \langle a_1, a_2 \rangle = r\langle \cos\theta, \sin\theta \rangle$ having magnitude r and direction θ. The coordinates of the object after displacement are $(x + a_1, y + a_2)$. Figure 3.24 shows a representative displacement $\mathbf{a}$ and shows how the "after" coordinates (X, Y) of the object may be calculated from its "before" coordinates (x, y). If $\langle x, y \rangle$ and $\langle X, Y \rangle$ are the position vectors of the object before and after displacement, then directly from the figure we see that

$$\langle X, Y \rangle = \langle x, y \rangle + \mathbf{a}.$$

If an object is subject to successive displacements, the coordinates of its final position can be calculated by adding the several displacement vectors to obtain a single, equivalent displacement.

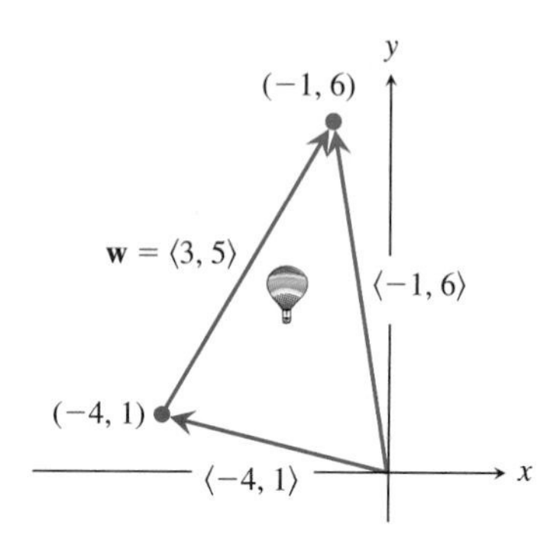

FIGURE 3.25 The displacement $\mathbf{a} = \langle 3, 5\rangle$ of an object initially at $(-4, 1)$.

EXAMPLE 7 A hot-air balloon was first sighted at a point with map coordinates $(-4, 1)$, where distances are measured in kilometers. Assuming that the balloon maintains its altitude and the prevailing breeze displaces it by the vector $\mathbf{w} = \langle 3, 5\rangle$, what is its final position?

Solution

Figure 3.25 shows the balloon at its initial position $(-4, 1)$ and the displacement vector $\mathbf{w} = \langle 3, 5\rangle$. The coordinates (X, Y) of the final position of the balloon satisfy

$$\langle X, Y\rangle = \langle -4, 1\rangle + \mathbf{w} = \langle -4, 1\rangle + \langle 3, 5\rangle = \langle -1, 6\rangle.$$

Hence, $(X, Y) = (-1, 6)$.

We summarize the properties of vector addition and scalar multiplication. Each of these can be proved from the properties of real numbers and the definitions of vector addition and scalar multiplication. See, for example, Exercises 48 and 49.

Properties of Vector Addition and Scalar Multiplication

Let $\mathbf{0} = \langle 0, 0\rangle$ be the zero vector. For vectors $\mathbf{a}$, $\mathbf{b}$, and $\mathbf{c}$ and scalars p and q,

1. $\mathbf{a} + \mathbf{b} = \mathbf{b} + \mathbf{a}$
2. $\mathbf{a} + (\mathbf{b} + \mathbf{c}) = (\mathbf{a} + \mathbf{b}) + \mathbf{c}$
3. $\mathbf{a} + \mathbf{0} = \mathbf{a}$
4. $\mathbf{a} + (-\mathbf{a}) = \mathbf{0}$
5. $1\mathbf{a} = \mathbf{a}$
6. $(p + q)\mathbf{a} = p\mathbf{a} + q\mathbf{a}$
7. $(pq)\mathbf{a} = p(q\mathbf{a})$
8. $p(\mathbf{a} + \mathbf{b}) = p\mathbf{a} + p\mathbf{b}$

Exercises 3.2

Exercises 1–6: Show that $\overrightarrow{PQ}$ and $\overrightarrow{ST}$ are equivalent, find a position vector $\mathbf{r}$ to which both are equivalent, and calculate the common lengths of these vectors. Sketch $\mathbf{r}$, $\overrightarrow{PQ}$ and $\overrightarrow{ST}$.

1. $P\,(1, 1)$, $Q\,(4, -1)$, $S\,(7, 5)$, and $T\,(10, 3)$
2. $P\,(-1, -2)$, $Q\,(-4, -1)$, $S\,(5, -1)$, and $T\,(2, 0)$
3. $P\,(3, -2/3)$, $Q\,(5, -2)$, $S\,(-6, 16/3)$, and $T\,(-4, 4)$
4. $P\,(3, 10)$, $Q\,(1, 4)$, $S\,(-2, -5)$, and $T\,(-4, -11)$
5. $P\,(-2.5, 3.0)$, $Q\,(-0.8, 1.1)$, $S\,(0.0, -1.3)$, and $T\,(1.7, -3.2)$
6. $P\,(2.5, 0.9)$, $Q\,(1.1, -0.3)$, $S\,(2.3, -2.2)$, and $T\,(0.9, -3.4)$

Exercises 7–10: Find $\mathbf{a} + \mathbf{b}$. Use these vectors and their sum to illustrate the parallelogram law.

7. $\mathbf{a} = \langle 2, 5\rangle$ and $\mathbf{b} = \langle 5, 2\rangle$
8. $\mathbf{a} = \langle 2, 6\rangle$ and $\mathbf{b} = \langle 4, 1\rangle$

9. $\mathbf{a} = \langle -3, 3 \rangle$ and $\mathbf{b} = \langle 1, 4 \rangle$

10. $\mathbf{a} = \langle -3, -3 \rangle$ and $\mathbf{b} = \langle 1, -5 \rangle$

Exercises 11–14: Find a position vector equivalent to $\overrightarrow{OR} = \overrightarrow{OP} + \overrightarrow{OQ}$. *Also find the coordinates of R. Sketch* $\overrightarrow{OR}$, $\overrightarrow{OP}$, *and* $\overrightarrow{OQ}$.

11. $O = (1, 1)$, $P = (4, 2)$, $Q = (2, 7)$

12. $O = (-1, -2)$, $P = (4, 1)$, $Q = (5, 5)$

13. $O = (-2.5, 3.0)$, $P = (-0.8, 1.1)$, $Q = (0.0, 3.3)$

14. $O = (2.5, 0.9)$, $P = (1.1, -0.3)$, $Q = (2.3, -2.2)$

Exercises 15–22: Calculate $\|\mathbf{a}\|$, $\mathbf{a} + \mathbf{b}$, $\mathbf{a} - \mathbf{b}$, *and* $h\mathbf{a} + k\mathbf{b}$.

15. $\mathbf{a} = \langle 5, -12 \rangle$, $\mathbf{b} = \langle 1, 4 \rangle$, $h = 1$, and $k = 2$

16. $\mathbf{a} = \langle -24, 7 \rangle$, $\mathbf{b} = \langle 2, -1 \rangle$, $h = 2$, and $k = 1$

17. $\mathbf{a} = -4\mathbf{i} - 5\mathbf{j}$, $\mathbf{b} = 3\mathbf{i} - 2\mathbf{j}$, $h = -2$, $k = 1$

18. $\mathbf{a} = \mathbf{i} + \mathbf{j}$, $\mathbf{b} = 4\mathbf{i} - 7\mathbf{j}$, $h = -2$, $k = 2$

19. $\mathbf{a} = \langle 1, 6 \rangle$, $\mathbf{b} = \langle 1, -4 \rangle$, $h = 0.5$, $k = -0.5$

20. $\mathbf{a} = \langle -1/3, 2/5 \rangle$, $\mathbf{b} = \langle 2/3, -3/4 \rangle$, $h = 1$, $k = 2$

21. $\mathbf{a} = 1.5\mathbf{i} - 2.4\mathbf{j}$, $\mathbf{b} = 0.3\mathbf{i} + 2.2\mathbf{j}$, $h = 1.6$, $k = -3.4$

22. $\mathbf{a} = -11.5\mathbf{i} - 12.4\mathbf{j}$, $\mathbf{b} = 9.6\mathbf{i} + 5.9\mathbf{j}$, $h = 0.5$, $k = -0.7$

Exercises 23–28: Find a unit vector in the direction θ.

23. $\theta = \pi/6$

24. $\theta = \pi/3$

25. $\theta = 210°$

26. $\theta = 240°$

27. $\theta = 5.0$

28. $\theta = 4.5$

29. The mass of the (plastic) stoplight in the figure is approximately 4.5 kilograms. Due to the Earth's attraction, the stoplight exerts a vertical force of 44.1 newtons at the point where the cables are attached. Find the forces the cables exert on the stoplight. *Hint:* The resultant of the cable forces must be equal and opposite (in direction) to the force of 44.1 newtons.

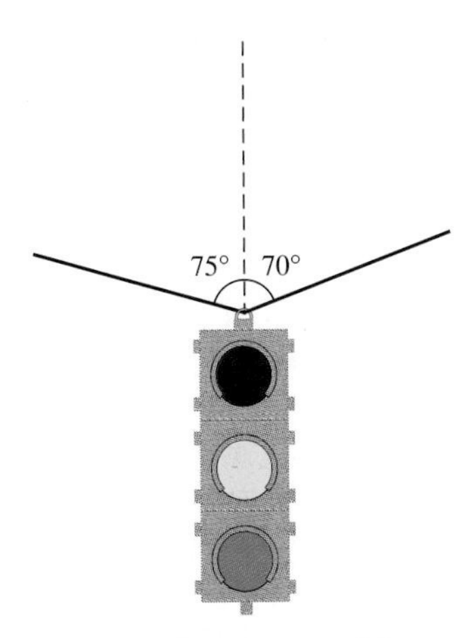

Figure for Exercise 29.

30. As shown in the figure, a tightrope walker has reached the middle of a stretched cable, resulting in a vertical force of 700 newtons at the center point. Find the forces the cables must exert to balance the performer. *Hint:* The resultant of the cable forces must be equal and opposite (in direction) to the force of 700 newtons.

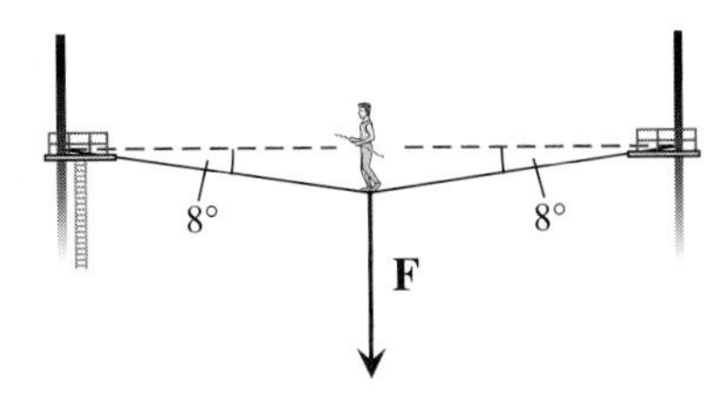

Figure for Exercise 30.

31. Because of bears, two campers have suspended their food pack from ropes attached to two trees, as shown in the figure, resulting in a vertical force of 200 newtons on the rope. Find the forces the ropes must exert to balance the food pack. *Hint:* The resultant of the rope forces must be equal and opposite (in direction) to the vertical force of 200 newtons.

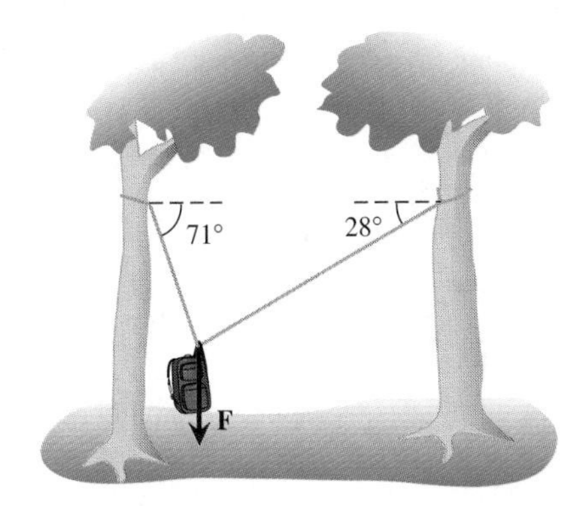

Figure for Exercise 31.

32. What force **F** will just balance the forces **A**, **B**, and **C** shown in the figure, where the magnitudes of these forces are 300 newtons, 350 newtons, and 500 newtons and the angles they make with the dotted line are 30°, 25°, and 20°, respectively?

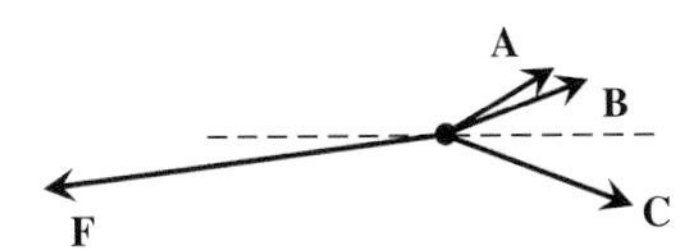

Figure for Exercise 32.

33. Let $\mathbf{r} = \langle 3, 5 \rangle$. Find a vector in the same direction as $\mathbf{r}$ but with length 8.

34. Let $\mathbf{r} = \langle 3, 2\rangle$. Find a vector in the same direction as $\mathbf{r}$ but with length 7.

35. Let A, B, and C be the points $(1, 4)$, $(2, 6)$, and $(-1, 3)$. Find a point D so that $\overrightarrow{CD}$ is twice as long as $\overrightarrow{AB}$ and in the opposite direction.

36. Let A, B, and C be the points $(-3, 1)$, $(3, -4)$, and $(5, 3)$. Find a point D so that $\overrightarrow{CD}$ is twice as long as $\overrightarrow{AB}$ and in the opposite direction.

37. Let $P = (3, 4)$, $Q = (8, 6)$, and $S = (5, 7)$ be points in a plane. Find the coordinates of the fourth vertex of the parallelogram whose sides are $\overrightarrow{PQ}$ and $\overrightarrow{PS}$.

38. Let $P = (1, -3)$, $Q = (-1, -5)$, and $S = (4, -2)$ be points in a plane. Find the coordinates of the fourth vertex of the parallelogram whose sides are $\overrightarrow{PQ}$ and $\overrightarrow{PS}$.

39. Find the coordinates of an object that has been displaced from the point $(-4, 9)$ by the vector $\langle 4, -5\rangle$.

40. Find the coordinates of an object that has been displaced from the point $(0, 4)$ by the vector $\langle -2, -5\rangle$.

41. Find the coordinates of an object that has been displaced from the point $(5, 5)$ a distance of 10 units and in the direction $\theta = 30°$.

42. Find the coordinates of an object that has been displaced from the point $(0, -7)$ a distance of 13 units and in the direction $\theta = 135°$.

43. An object at $(0, 4)$ is displaced to Q by $2\mathbf{i} + 5\mathbf{j}$ and then to T by $-12\mathbf{i} + 13\mathbf{j}$. Calculate the single equivalent displacement and the coordinates of T.

44. An object at $(-4, 9)$ is displaced to Q by $\langle -4, -5\rangle$ and then to T by $\langle 1, 1\rangle$. Calculate the single equivalent displacement and the coordinates of T.

45. Find the coordinates of the object initially at $(1, 1)$ and subjected to successive displacements $\mathbf{a}$, $\mathbf{b}$, and $\mathbf{c}$, where the magnitudes and directions of these displacements are 10, 5, and 7 and 330°, 60°, and 180°, respectively.

46. Find the coordinates of the object initially at $(-10, 15)$ and subjected to successive displacements $\mathbf{a}$, $\mathbf{b}$, and $\mathbf{c}$, where the magnitudes and directions of these displacements are 12, 5, and 15 and 80°, 170°, and 280°, respectively.

47. Find a position vector $\mathbf{r}$ parallel to a line with slope m. What is the slope of a line parallel to the vector $\langle a, b\rangle$, where $a \neq 0$?

48. Show that if $\mathbf{a}$ is a vector and p and q scalars, then $p(q\mathbf{a}) = (pq)\mathbf{a}$. *Hint:* Let $\mathbf{a} = a_1\mathbf{i} + a_2\mathbf{j}$.

49. Show that vector addition is associative: If $\mathbf{a}$, $\mathbf{b}$, and $\mathbf{c}$ are vectors, then $\mathbf{a} + (\mathbf{b} + \mathbf{c}) = (\mathbf{a} + \mathbf{b}) + \mathbf{c}$. *Hint:* Let $\mathbf{a} = \langle a_1, a_2\rangle$, $\mathbf{b} = \langle b_1, b_2\rangle$, and $\mathbf{c} = \langle c_1, c_2\rangle$.

50. The distance and bearing of Site A on the other side of a lake from Base Camp are required. Measuring all bearings counterclockwise from the positive x-axis, a survey team travels 10 km from Base Camp to Station S1 on a bearing of 30°; 15 km from S1 to S2 on a bearing of 110°; and, finally, 12 km on a bearing of 160°. Determine the distance and bearing of Site A from Base Camp.

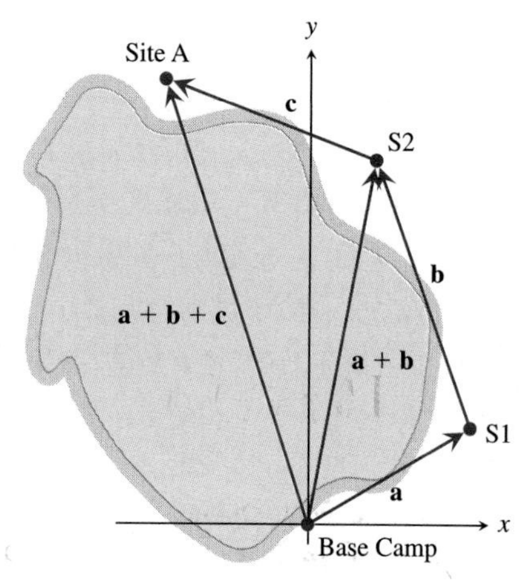

Figure for Exercise 50.

51. Show that for all vectors $\mathbf{a}$ and $\mathbf{b}$,

$$\|\mathbf{a} + \mathbf{b}\| \leq \|\mathbf{a}\| + \|\mathbf{b}\|.$$

This inequality is called the *triangle inequality*. Use a sketch and a few sentences to explain why this name is appropriate.

3.3 Parametric Equations

The orbit of Mars about the sun is an ellipse with one of its two focii at the sun. This ellipse can be described by an equation of the form

$$\frac{(x + c)^2}{a^2} + \frac{y^2}{b^2} = 1,$$

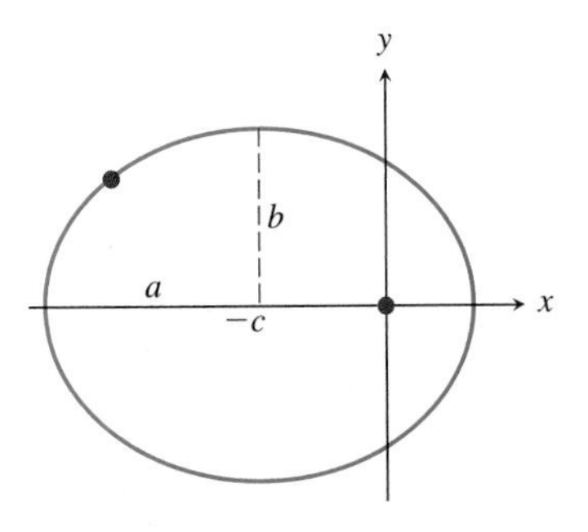

A representative planetary orbit; the sun is at one of the two focii.

where a and b are the lengths of the semimajor and semiminor axes and $c = \sqrt{a^2 - b^2}$ is the distance from the center of the ellipse to either of its two focii. See the accompanying figure. This equation gives the *shape* of the orbit but does not describe Mars' motion in the sense that from it we can determine Mars' coordinate positions at certain times t of interest. For this, we would describe Mars' orbit with *parametric equations* of the form

$$x = f(t)$$

$$y = g(t),$$

where t is a time variable. Such variables as t are called *parameters*. Parametric equations are often written as a position vector

$$\mathbf{r} = \mathbf{r}(t) = \langle f(t), g(t) \rangle.$$

WEB This vector points from the origin to the position of the planet at time t.

We begin our discussion of parametric equations by discussing motion on lines or circles.

Describing Motion along a Line

INVESTIGATION

Describing the Motion of a Dune Buggy

Suppose that in a flat and coordinatized desert a dune buggy is one mile north of the origin at time $t = 0$ and is headed 30° north of east at 40 miles per hour. How can we calculate the coordinates of the dune buggy for times $t > 0$?

Figure 3.26(a) shows the line along which the dune buggy is traveling. We know that this line goes through the point $(0, 1)$ and has slope tan 30°. Using the slope-intercept form, the equation of the line is

$$y = (\tan 30°)x + 1 = \left(1/\sqrt{3}\right)x + 1.$$

This equation tells us the buggy's route but does not tell us its position at any time t. However, see Exercise 26.

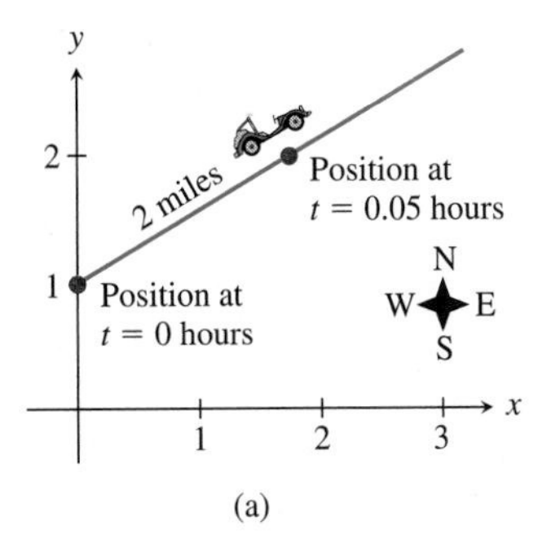

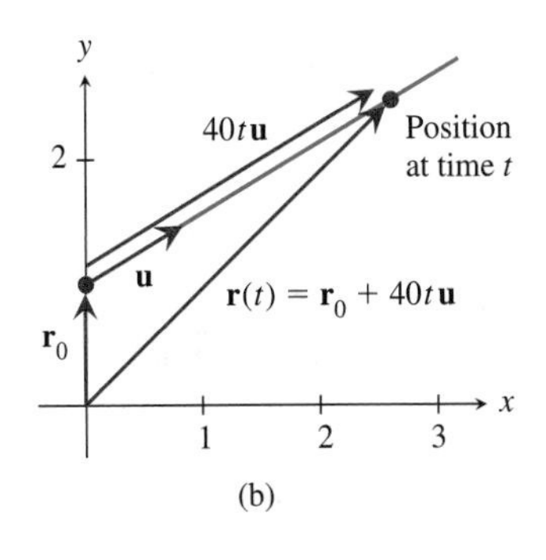

FIGURE 3.26 (a) Path of the dune buggy, showing its positions at $t = 0$ hours and $t = 1/20$ hours (3 minutes); (b) describing the path using vectors.

We use vector addition and scalar multiplication to determine the coordinates $(x(t), y(t))$ of the dune buggy at any time t. Fig. 3.26(b) shows a position vector $\mathbf{r}_0 = \langle 0, 1 \rangle$ for the initial position of the buggy. Because the buggy is moving in the direction $\theta = 30°$, it follows directly from (7) in Section 3.2 that a unit vector in the direction of motion is

$$\mathbf{u} = \langle \cos 30°, \sin 30° \rangle.$$

This vector is shown in Fig. 3.26(b). After t hours, the dune buggy will have traveled $40t$ miles. The vector $40t\mathbf{u}$ gives the displacement at time t of the dune buggy from its starting point. Note that the length $\|40t\mathbf{u}\|$ of this vector is $40t\|\mathbf{u}\| = 40t \cdot 1 = 40t$. Letting $\mathbf{r}(t) = \langle x(t), y(t) \rangle$ be the position vector of the buggy, we see from Fig. 3.26(b) that $\mathbf{r}(t)$ is the sum of the vectors $\mathbf{r}_0$ and $40t\mathbf{u}$, that is,

$$\begin{aligned} \mathbf{r}(t) &= \langle x(t), y(t) \rangle = \mathbf{r}_0 + 40t\mathbf{u} \\ &= \langle 0, 1 \rangle + 40t\langle \cos 30°, \sin 30° \rangle = \langle 20\sqrt{3}t, 1 + 20t \rangle. \end{aligned} \tag{1}$$

For each time $t \geq 0$, the position vector $\mathbf{r}(t)$ points from the origin to the position of the dune buggy. At $t = 1/20 = 0.05$ hours (3 minutes), a position vector of the dune buggy is

$$\mathbf{r}(0.05) = \langle 20\sqrt{3} \cdot 0.05, 1 + 20 \cdot 0.05 \rangle \approx \langle 1.73, 2.00 \rangle.$$

The Cartesian coordinates of the buggy at $t = 0.05$ are $(1.73, 2.00)$, approximately.

Equation (1) is called a **parametric equation.**

EXAMPLE 1 A slow freight train leaves from the point $(50, 40)$ and travels in the direction $\theta = 200°$ at 10 kilometers per hour. Describe its path with a parametric equation. Determine the time at which the freight train crosses the x-axis.

Solution

The vector $\mathbf{r}_0 = \langle 50, 40 \rangle$ is a position vector for the train's initial position. Using (7) from Section 3.2, $\mathbf{u} = \langle \cos 200°, \sin 200° \rangle$ is a unit vector in the direction of travel. After $t \geq 0$ hours, the train will have traveled $10t$ kilometers; hence, the vector $10t\mathbf{u}$ gives the displacement of the train from its starting point. Letting $\mathbf{r}(t) = \langle x(t), y(t) \rangle$ be the position vector of the train, we see from Fig. 3.27 that $\mathbf{r}(t)$ is the sum of the vectors $\mathbf{r}_0$ and $10t\mathbf{u}$, that is,

$$\begin{aligned} \mathbf{r}(t) &= \langle x(t), y(t) \rangle = \mathbf{r}_0 + 10t\mathbf{u} \\ &= \langle 50, 40 \rangle + 10t\langle \cos 200°, \sin 200° \rangle \\ &= \langle 50 + 10t \cos 200°, 40 + 10t \sin 200° \rangle. \end{aligned}$$

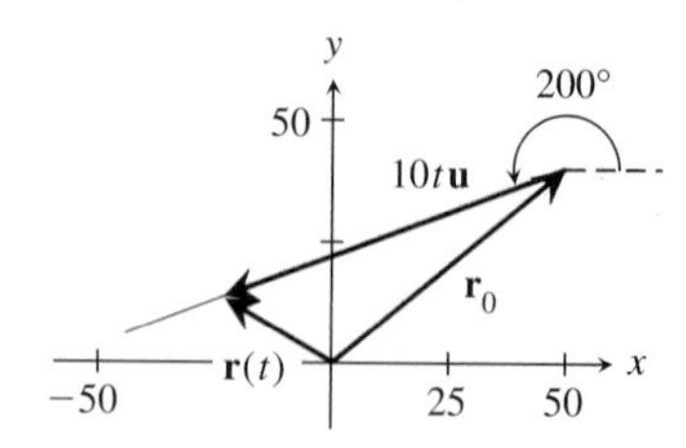

FIGURE 3.27 Path of a train starting from the point with position vector $\mathbf{r}_0 = \langle 50, 40 \rangle$ and heading in the direction $\theta = 200°$ at 10 kilometers per hour.

This equation describes the path of the train in the sense that as t increases from 0, the arrow end of the vector $\mathbf{r}(t)$ traces the path of the train.

To determine the time at which the train crosses the x-axis, we set $y(t)$ equal to 0:

$$40 + 10t \sin 200° = 0.$$

Solving this equation for t, the train crosses the x-axis at

$$t = \frac{-40}{10 \sin 200^\circ} \approx 11.70 \text{ hours.}$$

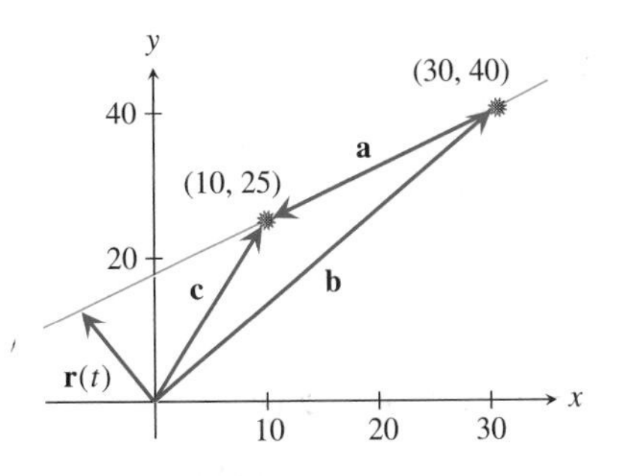

FIGURE 3.28 The path of a meteorite and a vector **a** in the direction of motion.

EXAMPLE 2 This example concerns the motion of a meteorite. For simplicity, we assume that the meteorite moves on a straight line lying in a vertical plane, with the x-axis as the horizon. A meteorite is observed passing through the points $(30, 40)$ and $(10, 25)$ at times $t = 0$ and $t = 0.5$ seconds. Figure 3.28 shows the path of the meteorite. The units on the coordinate axes are in kilometers. When and where will the meteorite hit the Earth?

Solution

This example is much like Example 1. The main difference is that we must calculate the speed and direction from the two observations. Letting $\mathbf{b} = \langle 30, 40 \rangle$ and $\mathbf{c} = \langle 10, 25 \rangle$ be position vectors of the two observations, their difference $\mathbf{a} = \mathbf{c} - \mathbf{b}$ is a vector in the direction of motion. The length of $\mathbf{a}$,

$$\begin{aligned} \|\mathbf{a}\| &= \|\langle 10, 25 \rangle - \langle 30, 40 \rangle\| \\ &= \|\langle -20, -15 \rangle\| = \sqrt{20^2 + 15^2} = 25 \text{ kilometers,} \end{aligned}$$

is equal to the distance the meteorite traveled in 0.5 seconds. Hence, the speed of the meteorite is $25/0.5 = 50$ kilometers per second.

A unit vector $\mathbf{u}$ in the direction of motion is $\mathbf{a}$ divided by $\|\mathbf{a}\| = 25$. Hence, in t seconds the displacement of the meteor is $50t\mathbf{u}$. The position vector $\mathbf{r}(t)$ of the meteor at any time t before it crashes into the x-axis is

$$\begin{aligned} \mathbf{r}(t) &= \mathbf{b} + (50t)\mathbf{u} = \langle 30, 40 \rangle + (50t)\frac{1}{25}\langle -20, -15 \rangle \\ &= \langle 30 - 40t, 40 - 30t \rangle. \end{aligned} \tag{2}$$

To determine the time at which the meteorite crashes into the x-axis, we set the second coordinate of its position vector equal to 0. From (2),

$$40 - 30t = 0.$$

Solving this equation for t, the meteorite hits the x-axis at $t = 40/30 \approx 1.33$ seconds after it was first sighted. Hence, the meteorite hits the x-axis at

$$x = (30 - 40t)\bigg|_{t=40/30} = 30 - 40\left(\frac{4}{3}\right) \approx -23.33 \text{ km.}$$

Eliminating the Parameter For some purposes, a parametric equation like (2) is not necessary. If, for example, we want to describe the path of the meteorite with an equation of the form $y = mx + b$, and we do not need to know where the meteorite is as a function of time t, we can eliminate the parameter t in the parametric equation (2). Letting the coordinates of $\mathbf{r}(t)$ in (2) be

$$x = 30 - 40t$$
$$y = 40 - 30t,$$

we first solve either of these equations for t. From the first equation,

$$t = \frac{30 - x}{40}.$$

Next, we substitute this result into the second equation. This gives

$$y = 40 - 30\left(\frac{30 - x}{40}\right) = \frac{3}{4}x + \frac{70}{4}. \tag{3}$$

Alternatively, recalling that we were given two points $(x_1, y_1) = (30, 40)$ and $(x_2, y_2) = (10, 25)$ on the path of the meteorite, we may use the two-point form

$$y - y_1 = \frac{y_2 - y_1}{x_2 - x_1}(x - x_1)$$

of a line to obtain equation (3).

Describing Motion on a Circle

INVESTIGATION

Describing the Motion of a Slider-Crank Mechanism

Figure 3.29(a) shows the slider-crank mechanism with which we began this chapter. The circle traced by the point P at the end of the crank (length 5 cm) can be described as the graph of the equation $x^2 + y^2 = 5^2$. This describes the circular path well but does not tell where the pin P is at a given time.

Given that at $t = 0$ the pin was at the point $(5, 0)$ and is moving counterclockwise on the circle at 1 radian per second, it follows that after t seconds the angular displacement of the crank will be $\theta(t) = t$ radians. From (7) in Section 3.2, a position vector of the pin P is

$$\mathbf{r}(t) = \langle x(t), y(t)\rangle = \langle 5\cos t, 5\sin t\rangle = 5\langle \cos t, \sin t\rangle. \tag{4}$$

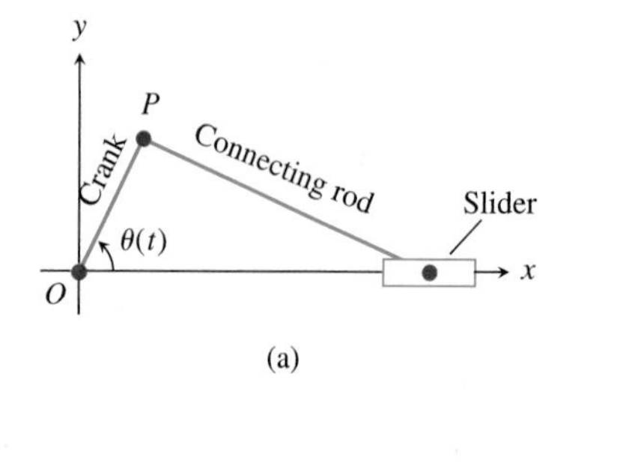

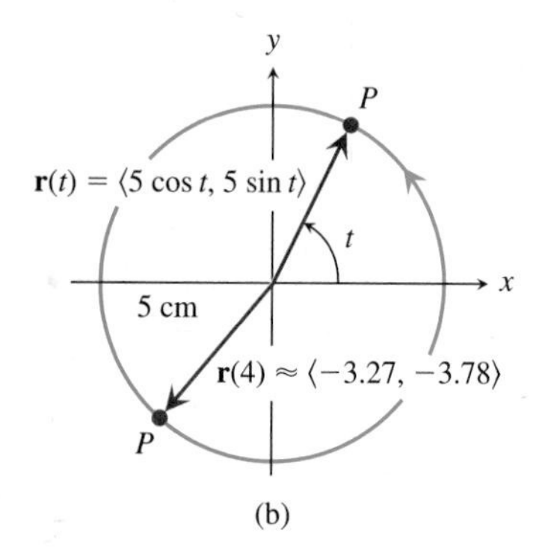

FIGURE 3.29 The location of the pin at P as a function of time t.

From (4), the position of the pin P at any time t can be easily calculated. For example, at $t = 4$ seconds the position of the pin is

$$\mathbf{r}(4) = 5\langle \cos 4, \sin 4 \rangle \approx \langle -3.27, -3.78 \rangle.$$

Figure 3.29(b) shows the position vector $\mathbf{r}(4)$.

Eliminating the Parameter The equation $x^2 + y^2 = 5^2$ describing the circle on which the pin P moves can be recovered from the parametric equation (4) by eliminating the parameter t. Letting $x = 5 \cos t$ and $y = 5 \sin t$, note that

$$x^2 + y^2 = 5^2 \cos^2 t + 5^2 \sin^2 t = 5^2,$$

for all t.

EXAMPLE 3 As of July 1997, the world's tallest Ferris wheel was in Osaka, Japan. See Fig. 3.30. The distance from ground level to its highest point is 112.5 meters and the wheel itself has a diameter of 100 meters. One revolution takes about 15 minutes. Describe the motion of the wheel with a parametric equation and find the maximum speed at which a passenger moves directly downward.

Solution

The center of the ferris wheel will be 50 meters below the high point; hence, a position vector of the center will be $\mathbf{h} = \langle 0, 62.5 \rangle$. We "track" a passenger P on the wheel who, at time $t = 0$, was directly to the right of the center point. Temporarily regarding the center of the ferris wheel as the origin and recalling (7) from Section 3.2, the position vector of P after t minutes will be

$$\mathbf{w} = 50\left\langle \cos\left(\frac{2\pi t}{15}\right), \sin\left(\frac{2\pi t}{15}\right)\right\rangle.$$

The factor $2\pi/15$ comes from the fact that the wheel revolves 2π radians in 15 minutes. We have assumed counterclockwise rotation.

Figure 3.30 shows that the position vector $\mathbf{r} = \mathbf{r}(t)$ of P at any time t is the sum of the constant vector $\mathbf{h}$ and the vector $\mathbf{w}$ with constant length but variable direction, that is,

$$\begin{aligned} \mathbf{r} = \mathbf{r}(t) = \mathbf{h} + \mathbf{w} &= \langle 0, 62.5 \rangle + 50\left\langle \cos\left(\frac{2\pi t}{15}\right), \sin\left(\frac{2\pi t}{15}\right)\right\rangle \\ &= \left\langle 50 \cos\left(\frac{2\pi t}{15}\right), 62.5 + 50 \sin\left(\frac{2\pi t}{15}\right)\right\rangle. \end{aligned} \quad (5)$$

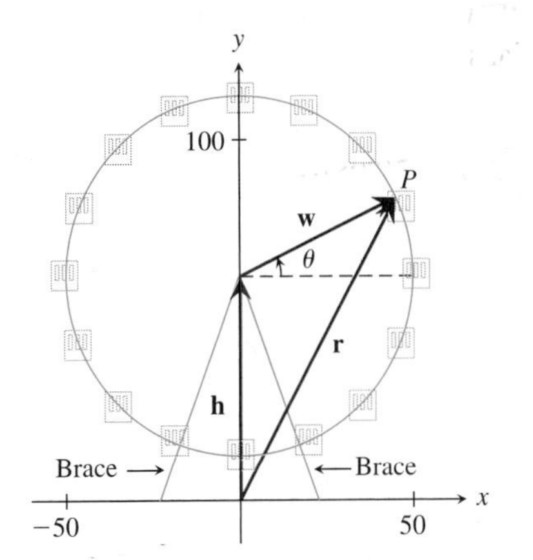

FIGURE 3.30 The position vector **r** of a point on the Ferris wheel, expressed as the sum of a constant vector **h** and a vector **w** that varies with θ.

We can find the vertical speed of the point P by differentiating the y-coordinate of $\mathbf{r} = \mathbf{r}(t) = \langle x(t), y(t)\rangle$. From

$$y(t) = 62.5 + 50 \sin\left(\frac{2\pi t}{15}\right),$$

$$y'(t) = 50\left(\frac{2\pi}{15}\right)\cos\left(\frac{2\pi t}{15}\right) = \left(\frac{20\pi}{3}\right)\cos\left(\frac{2\pi t}{15}\right).$$

Because for all t, $\cos(2\pi t/15)$ is between -1 and 1, the largest value of the speed $|y'(t)|$ is $20\pi/3$ meters per second. This happens at $t = 0$, when cosine is 1 and P is moving upward, and next at $(2\pi/15)t = \pi$, when cosine is -1 and P is moving downward. Hence, the maximum speed directly downward is $20\pi/3 \approx 21$ meters per minute.

Exercises 3.3

Exercises 1–6: A rocket sled is moving in a flat desert. We suppose that a coordinate system has been superimposed on the desert, relative to which the coordinates of the sled at $t = 0$ are (x_0, y_0). We assume that the sled was headed in the given direction at the given speed. Find its position vector $\mathbf{r}(t) = \langle x(t), y(t)\rangle$ at any time t. Calculate its position at time t_1. Sketch the path for $0 \le t \le t_1$. Lengths are in kilometers and time is in hours.

1. $(x_0, y_0) = (0, 1)$; 60° east of north; speed 200 kilometers per hour; $t_1 = 0.001$

2. $(x_0, y_0) = (-1, 0)$; 45° east of north; speed 180 kilometers per hour; $t_1 = 0.002$

3. $(x_0, y_0) = (-5, 3)$; 20° west of north; speed 225 kilometers per hour; $t_1 = 0.001$

4. $(x_0, y_0) = (5, -4)$; 10° north of west; speed 210 kilometers per hour; $t_1 = 0.003$

5. $(x_0, y_0) = (-4.5, 3.2)$; 10° south of east; speed 190 kilometers per hour; $t_1 = 0.001$

6. $(x_0, y_0) = (0.5, -4.6)$; 10° west of south; speed 200 kilometers per hour; $t_1 = 0.002$

Exercises 7–12: A meteorite enters the Earth's atmosphere, moving on a straight line lying in a vertical plane, with the x-axis as the horizon. Its position was recorded at two times. Find its speed and its position vector $\mathbf{r}(t) = \langle x(t), y(t)\rangle$ at any time t before it hits the Earth. Determine where and when it hits Earth. Lengths are in kilometers and time is in seconds.

7. $(50, 40)$ first sighting; $(50 - 20\sqrt{3}, 20)$ second sighting, 5 seconds later

8. $(60, 30)$ first sighting; $(48, 30 - 12\sqrt{3})$ second sighting, 3 seconds later

9. $(35, 30)$ first sighting; $(35 - 16\sqrt{2}, 30 - 16\sqrt{2})$ second sighting, 4 seconds later

10. $(-40, 55)$ first sighting; $(-40 + 24\sqrt{3}, 31)$ second sighting, 6 seconds later

11. $(90.2, 52.7)$ first sighting; $(49.5, 3.3)$ second sighting, 8 seconds later

12. $(-9.9, 65.1)$ first sighting; $(15.9, 34.5)$ second sighting, 5 seconds later

Exercises 13–16: An object is moving on a path described by the parametric equation. Eliminate t to find an equation in x and y describing the path along which the object is moving. Then sketch the path and draw the position vector $\mathbf{r}(t_1)$.

13. $\mathbf{r}(t) = \langle 5t, 2 + 3t\rangle$, $0 \le t \le 1$, $t_1 = 1/2$

14. $\mathbf{r}(t) = (4t - 1)\mathbf{i} - (1 + 3t)\mathbf{j}$, $0 \le t \le 2$, $t_1 = 1$

15. $\mathbf{r}(t) = \langle 3\cos t, 3\sin t\rangle$, $0 \le t \le 2\pi$, $t_1 = \pi/2$

16. $\mathbf{r}(t) = (2\cos t)\mathbf{i} + (2\sin t)\mathbf{j}$, $0 \le t \le 2\pi$, $t_1 = 3\pi/2$

Exercises 17–22: Find the position $\mathbf{r}(t)$ of the object at any time t for the described circular path. Sketch.

17. An object P moves counterclockwise from $(5, 3)$ at 2 radians per second on a circle of radius 2 and center $(3, 3)$.

18. An object P moves counterclockwise from $(14, 5)$ at 3 radians per second in a circle of radius 4 and center $(10, 5)$.

19. An object P moves counterclockwise from $(5, 2)$ at $100°$ per second in a circle of radius 5 and center $(0, 2)$.

20. An object P moves counterclockwise from $(13, 19)$ at $144°$ per second in a circle of radius 7 and center $(6, 19)$.

21. An object P moves clockwise from $(1, 3)$ at 2 radians per second in a circle of radius 2 and center $(3, 3)$.

22. An object P moves clockwise from $(2, 0)$ at 5 radians per second in a circle of radius 1 and center $(2, 1)$.

23. Find a parametric equation giving the position at time t of a passenger on the Osaka Ferris wheel, who, at $t = 0$, boarded the wheel from a platform 12.5 meters from the ground. See Fig. 3.30 and Example 3.

24. The Ferris wheel in Otsu, Japan, has the same diameter as the Osaka wheel, but its overall height is 4 meters less than the 112.5 meters of the Osaka wheel. Find a parametric equation giving the position at time t of a passenger on the Otsu Ferris wheel, who, at $t = 0$, boarded the wheel from a platform 8.5 meters from the ground. Assume that one revolution takes about 14 minutes. See Fig. 3.30 and Example 3.

25. The figure shows a simplified view of the Tevatron, a particle accelerator at Fermilab. The circular path on which protons are accelerated has radius 0.64 miles. At $t = 0$ a proton is injected into the Tevatron at P and moves counterclockwise on the circle at the (for simplicity) constant speed of 18,628 miles per second, which is 10 percent of the speed of light. Write a parametric equation for the path of the proton for $t \geq 0$. Calculate the proton's position at $t = 0.0002$ second. If after one trip around the Tevatron the proton is guided along the fixed-target beam line to a fixed target 0.5 miles from P, what is the total elapsed time from initial injection to target?

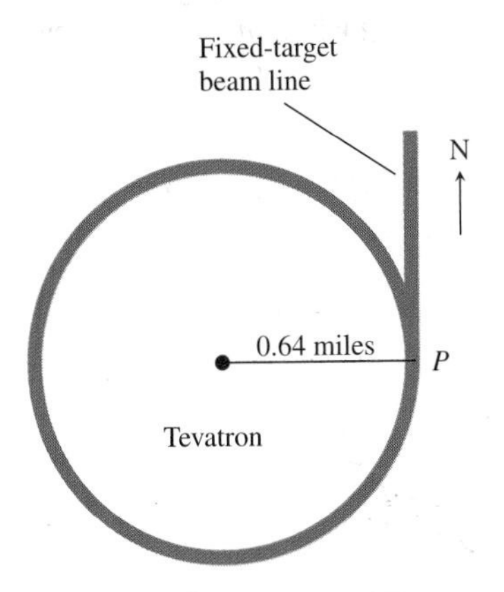

Figure for Exercise 25.

26. Starting from $(0, 1)$ at $t = 0$, a dune buggy is moving at 40 miles per hour into the first quadrant on the graph of the equation

$$y = (\tan 30°)x + 1 = \left(\frac{1}{\sqrt{3}}\right)x + 1.$$

Use non-vector methods to determine the coordinates (x, y) of the dune buggy after t hours. *Hint:* Find (x, y) on the line so that $x \geq 0$ and the distance between $(0, 1)$ and (x, y) is $40t$.

27. A dune buggy moves along a line across a coordinatized flat desert at constant speed v. If the buggy starts at $P = (p_1, p_2)$ and later goes through the point $Q = (q_1, q_2)$, argue that its position $\mathbf{r}(t)$ at any time $t \geq 0$ is

$$\mathbf{r}(t) = \langle p_1, p_2 \rangle + tv\mathbf{u},$$

where $\mathbf{u}$ is a unit vector in the same direction as $\overrightarrow{PQ}$.

28. An object has constant speed and its position at any time $t \geq 0$ is described by $\mathbf{r}(t) = (a + bt)\mathbf{i} + (c + dt)\mathbf{j}$, where a, b, c, and d are constants. What is its speed? Describe the curve on which the object is moving.

29. *Related exercise.* Suppose that an object has constant speed and its position at any time $t \geq 0$ is described by $\mathbf{r}(t) = \mathbf{r}_0 + bt\mathbf{u}$, where $\mathbf{r}_0$ is a constant vector, $\mathbf{u}$ is a constant unit vector, and b is a constant. Show that the speed of the object is $|b|$. If $\mathbf{u}$ were not a unit vector, what would be the speed of the object? Use these ideas to modify the equation $\mathbf{r}(t) = \langle 2, 3 \rangle + 5t\langle 1, 1 \rangle$ so that the speed of the object can be read directly from the equation.

30. An object has constant speed and its position at any time $t \geq 0$ is described by $\mathbf{r}(t) = (a + b\cos(\omega t))\mathbf{i} + (c + b\sin(\omega t))\mathbf{j}$, where a, b, c, and ω are constants. What is its speed?

31. Sketch the path of the object with the position vector $\mathbf{r}(t) = \langle |t - 1|, |t - 2| \rangle$, $0 \leq t \leq 5$. *Hint:* Divide the interval $[0, 5]$ into three parts so that the absolute value signs can be removed.

32. Sketch the path of the object with the position vector $\mathbf{r}(t) = \langle |2t - 5|, 2|-t + 4| \rangle$, $0 \leq t \leq 5$. *Hint:* Divide the interval $[0, 5]$ into three parts so that the absolute value signs can be removed.

3.4 Velocity and Tangent Vectors

In the preceding section, we described the path of an object in motion with a parametric equation of the form $\mathbf{r} = \mathbf{r}(t)$. From such an equation we can infer both the shape of the path and the position of the object at any time t. We use the derivative $\frac{d}{dt}\mathbf{r}(t) = \mathbf{r}'(t)$ to calculate the speed and direction of the object at any time t.

For curves described by a parametric equation $\mathbf{r} = \mathbf{r}(t)$ but having no particular connection with motion, we use the derivative $\frac{d}{dt}\mathbf{r}(t) = \mathbf{r}'(t)$ to calculate the tangent vector or the slope of the curve at any point $\mathbf{r}(t)$.

Paths of Objects in Motion; Velocity Vectors

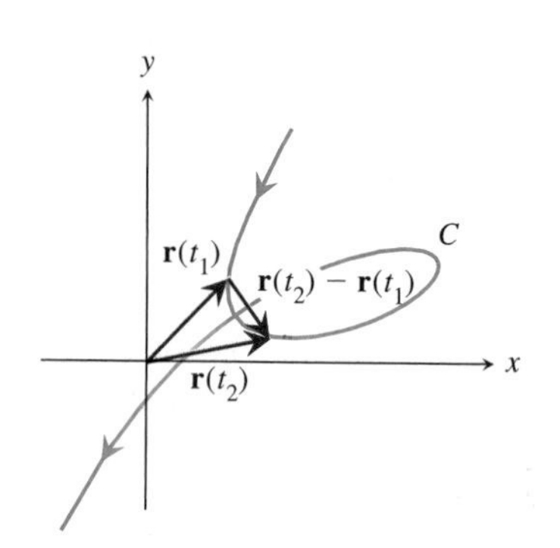

FIGURE 3.31 The vector $\mathbf{r}(t_2) - \mathbf{r}(t_1)$ gives the change in the object's position during the time interval $[t_1, t_2]$.

Suppose that an object is moving on the path C shown in Fig. 3.31 and its position vector at any time t is

$$\mathbf{r}(t) = \langle x(t), y(t) \rangle, \qquad a \le t \le b.$$

As in the one-dimensional case discussed in Section 3.1, we define velocity at t_1 as the limit as $t_2 \to t_1$ of the change in the object's position during the time interval $[t_1, t_2]$ divided by the change $t_2 - t_1$ in time. To "divide" a vector by a number q means to multiply it by the scalar $1/q$.

Figure 3.31 shows the vector $\mathbf{r}(t_2) - \mathbf{r}(t_1)$, which is the change in the object's position during the time interval $[t_1, t_2]$. The product of this vector and the scalar $1/(t_2 - t_1)$ gives the rate at which $\mathbf{r}$ is changing with respect to t on the interval $[t_1, t_2]$. In terms of the coordinate functions $x(t)$ and $y(t)$ of $\mathbf{r}(t)$,

$$\frac{1}{t_2 - t_1}(\mathbf{r}(t_2) - \mathbf{r}(t_1)) = \left\langle \frac{x(t_2) - x(t_1)}{t_2 - t_1}, \frac{y(t_2) - y(t_1)}{t_2 - t_1} \right\rangle.$$

This is the object's **average velocity** over the time interval $[t_1, t_2]$.

DEFINITION Average Velocity, Velocity, and Speed

Let $\mathbf{r} = \mathbf{r}(t)$ be the position at time t of an object moving along a path C, where $\mathbf{r}$ is a function defined on an interval I. The **average velocity** of the object on a subinterval $[t_1, t_2]$ of I is

$$\mathbf{v}(t_1, t_2) = \frac{1}{t_2 - t_1}(\mathbf{r}(t_2) - \mathbf{r}(t_1)) = \left\langle \frac{x(t_2) - x(t_1)}{t_2 - t_1}, \frac{y(t_2) - y(t_1)}{t_2 - t_1} \right\rangle.$$

If the coordinate functions $x = x(t)$ and $y = y(t)$ of $\mathbf{r} = \mathbf{r}(t)$ are differentiable on I, we define the **velocity** $\mathbf{v}(t_1)$ of the object at $\mathbf{r}(t_1)$ as the limit of the average velocity vector on $[t_1, t_2]$ as $t_2 \to t_1$, that is,

$$\begin{aligned}
\mathbf{v}(t_1) &= \lim_{t_2 \to t_1} \frac{1}{t_2 - t_1}(\mathbf{r}(t_2) - \mathbf{r}(t_1)) \\
&= \left\langle \lim_{t_2 \to t_1} \frac{x(t_2) - x(t_1)}{t_2 - t_1}, \lim_{t_2 \to t_1} \frac{y(t_2) - y(t_1)}{t_2 - t_1} \right\rangle \\
&= \langle x'(t_1), y'(t_1) \rangle.
\end{aligned}$$

We write $\mathbf{v}(t_1) = \mathbf{r}'(t_1) = \langle x'(t_1), y'(t_1) \rangle$. The **speed** of the object at time t_1 is the length of its velocity $\mathbf{v}(t_1)$, that is, $\|\mathbf{v}(t_1)\|$.

For motion along a line, both *speed* and *velocity* can be represented by scalars. The speed of an object is always nonnegative; its velocity can be positive, negative, or zero. For motion in a plane or space, the speed of an object remains a scalar; its velocity is a vector. The velocity vector tells both speed and direction.

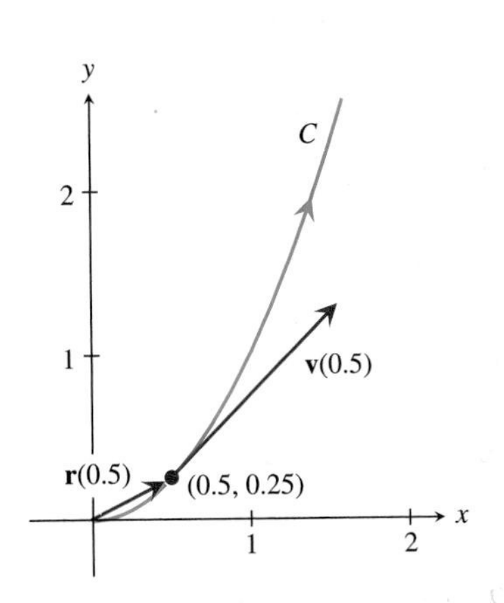

FIGURE 3.32 Path of an object; its velocity vector at $t = 0.5$ s.

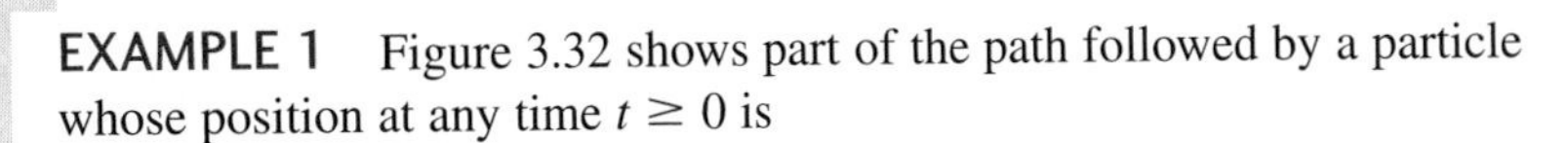

EXAMPLE 1 Figure 3.32 shows part of the path followed by a particle whose position at any time $t \geq 0$ is

$$\mathbf{r}(t) = \langle t, t^2 \rangle.$$

Lengths are measured in meters and time in seconds. Calculate the average velocity of the particle on the intervals $[0.5, 0.6]$, $[0.5, 0.51]$, and $[0.5, 0.501]$. From these results, conjecture the particle's velocity at $t = 0.5$. Then calculate the velocity vector and speed of the particle at $t = 0.5$ s.

Remark: The graph in Fig. 3.32 is the graph of the equation $y = x^2$ for $x \geq 0$. This equation can be derived from the parametric equation $\mathbf{r}(t) = \langle x, y \rangle = \langle t, t^2 \rangle$ by eliminating the parameter t. From the equations $x = t$ and $y = t^2$ we find $y = x^2$.

Solution

The average velocities of the particle on $[0.5, 0.6]$, $[0.5, 0.51]$, and $[0.5, 0.501]$ are

$$\mathbf{v}(0.5, 0.6) = \left\langle \frac{x(0.6) - x(0.5)}{0.6 - 0.5}, \frac{y(0.6) - y(0.5)}{0.6 - 0.5} \right\rangle$$

$$= \left\langle \frac{0.1}{0.1}, \frac{0.6^2 - 0.5^2}{0.1} \right\rangle = \langle 1, 1.1 \rangle \text{ m/s}$$

$$\mathbf{v}(0.5, 0.51) = \left\langle \frac{x(0.51) - x(0.5)}{0.51 - 0.5}, \frac{y(0.51) - y(0.5)}{0.51 - 0.5} \right\rangle$$

$$= \left\langle \frac{0.01}{0.01}, \frac{0.51^2 - 0.5^2}{0.01} \right\rangle = \langle 1, 1.01 \rangle \text{ m/s}$$

$$\mathbf{v}(0.5, 0.501) = \left\langle \frac{x(0.501) - x(0.5)}{0.501 - 0.5}, \frac{y(0.501) - y(0.5)}{0.501 - 0.5} \right\rangle$$

$$= \left\langle \frac{0.001}{0.001}, \frac{0.501^2 - 0.5^2}{0.001} \right\rangle = \langle 1, 1.001 \rangle \text{ m/s}.$$

It appears that as the lengths of subintervals approach 0, the average velocities approach $\langle 1, 1 \rangle$, which means that the velocity of the particle would be $\langle 1, 1 \rangle$.

The particle's velocity vector is

$$\mathbf{v}(t) = \mathbf{r}'(t) = \langle 1, 2t \rangle.$$

Evaluating $\mathbf{v}(t)$ at $t = 0.5$,

$$\mathbf{v}(0.5) = \langle 1, 2 \cdot 0.5 \rangle = \langle 1, 1 \rangle \text{ m/s}.$$

The speed at $t = 0.5$ is the length of $\mathbf{v}(0.5)$. Hence,

$$\|\mathbf{v}(0.5)\| = \sqrt{1^2 + 1^2} = \sqrt{2} \approx 1.4 \text{ m/s}.$$

Fig. 3.32 shows the velocity vector $\mathbf{v}(0.5)$. It is customary to place the tail of the velocity vector at the arrow end of the position vector. The position vector $\mathbf{r}(0.5)$ points to the position of the particle at $t = 0.5$, while the velocity vector shows the direction in which the object is moving at that time.

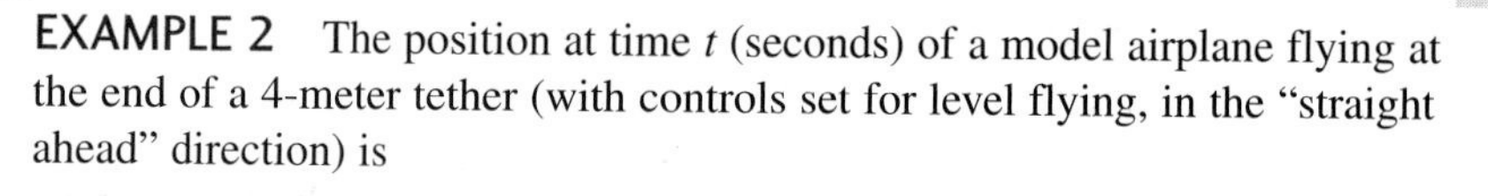

EXAMPLE 2 The position at time t (seconds) of a model airplane flying at the end of a 4-meter tether (with controls set for level flying, in the "straight ahead" direction) is

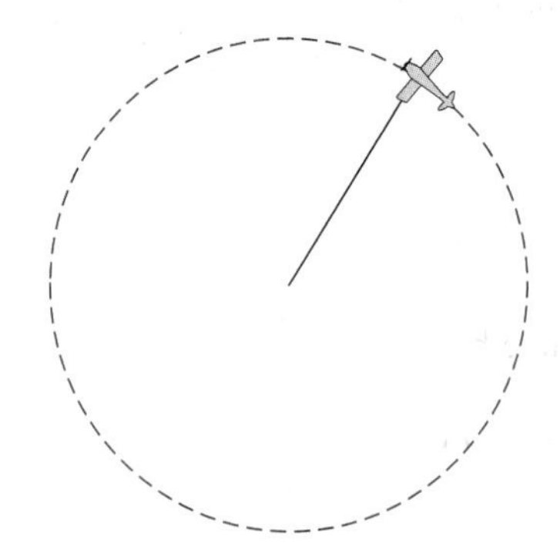

$$\mathbf{r}(t) = 4\cos(1.7t)\mathbf{i} + 4\sin(1.7t)\mathbf{j}.$$

The tether breaks at $t = 3$ seconds. Calculate the speed and direction of the airplane at this time. Sketch the path of the airplane from $t = 0$ to $t = 3$ and draw the vectors $\mathbf{r}(3)$ and $\mathbf{v}(3)$. What happens after $t = 3$?

Solution

The velocity $\mathbf{v}(t)$ of the airplane is the derivative of its position function:

$$\mathbf{v}(t) = \mathbf{r}'(t) = -4 \cdot 1.7\sin(1.7t)\mathbf{i} + 4 \cdot 1.7\cos(1.7t)\mathbf{j}$$
$$= 6.8(-\sin(1.7t)\mathbf{i} + \cos(1.7t)\mathbf{j}).$$

Because the length of the vector $-\sin(1.7t)\mathbf{i} + \cos(1.7t)\mathbf{j}$ is 1, the length of the vector $\mathbf{v}(t)$ is 6.8. Hence, the speed of the airplane is 6.8 meters per second. At $t = 3$,

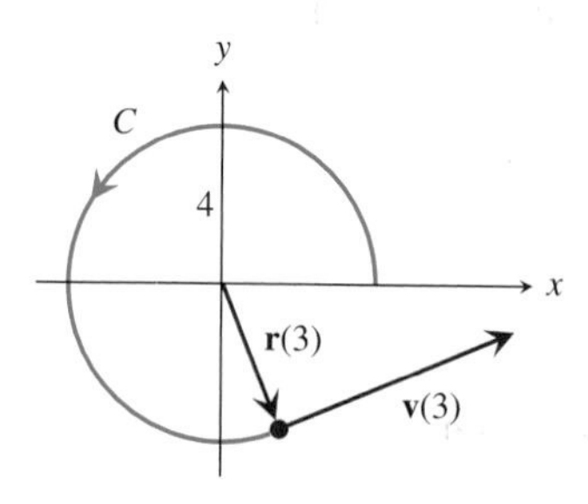

FIGURE 3.33 Model airplane flying on a 4-meter tether.

$$\mathbf{r}(3) = 4\cos(1.7 \cdot 3)\mathbf{i} + 4\sin(1.7 \cdot 3)\mathbf{j} \approx 4(0.38\mathbf{i} - 0.93\mathbf{j})$$
$$\mathbf{v}(3) = 6.8(-\sin(5.1)\mathbf{i} + \cos(5.1)\mathbf{j}) \approx 6.8(0.93\mathbf{i} + 0.38\mathbf{j}).$$

The direction of $\mathbf{v}(3)$ is determined by the vector $-\sin(5.1)\mathbf{i} + \cos(5.1)\mathbf{j}$. Noticing that both coordinates are positive, it follows that the direction of the airplane at $t = 3$ is $\theta = \arctan(-\cos(5.1)/\sin(5.1)) \approx 0.39$ or $22°$.

Figure 3.33 shows the vectors $\mathbf{r}(3)$ and $\mathbf{v}(3)$. If the tether breaks at $t = 3$, the airplane will fly from $\mathbf{r}(3)$ in the direction of the velocity vector $\mathbf{v}(3)$.

Curves and Tangent Vectors

So far in this chapter we have used parametric equations to describe the paths of objects in motion and to calculate their velocities. The *velocity vector* points in the direction of motion and is tangent to the path of the object. For paths/curves not specifically connected with motion, we use *tangent vectors* to indicate the direction of the curve at a point. Before we discuss tangent vectors, we state more carefully what we mean by a *curve*.

DEFINITION Curve

A **curve** C in an (x, y)-plane is the range of a vector function

$$\mathbf{r} = \mathbf{r}(t) = \langle x(t), y(t) \rangle, \qquad t \in I, \tag{1}$$

where I is an interval and the functions $x = x(t)$ and $y = y(t)$ are continuous on I.

We say that the function $\mathbf{r}$ describes C or is a **parametrization** of C. The variable t—or whatever variable is used in (1)—is called a *parameter*.

Describing a curve C with a parametric or vector equation like (1) is different from describing the motion of an object whose path or trajectory is C. This is because a curve can be parametrized in many different ways. One way of thinking about this is to imagine several different particles moving along the same curve C. One particle might move at a constant, slow speed, another at a constant, fast speed, a third with constant acceleration, and a fourth with increasing acceleration. As you would expect, the vector equations describing these different motions would be different. Note, however, that each of these equations would describe the same curve C. Thus, curves may have multiple parametrizations. Would this matter if, for example, we wished to calculate the length of C? Would the curve have the same length whatever parametrization we chose?

The question of multiple parametrizations will come up a few times in the chapters ahead; meanwhile, we discuss several simple curves and some more or less standard parametrizations of them. We give parametrizations for such curves as the trochoid, trisectrix, and tractrix in Exercises 71–78.

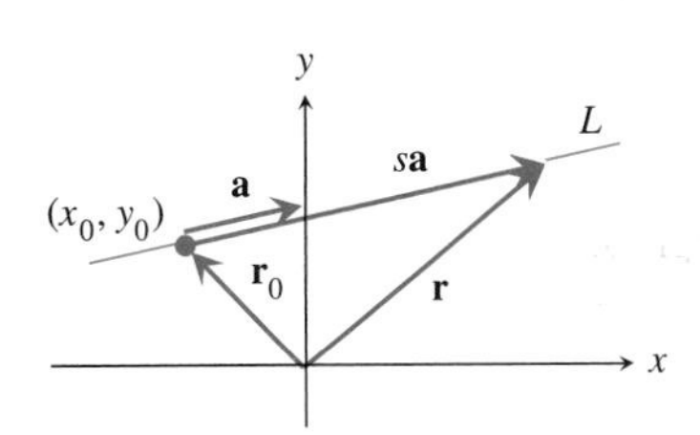

FIGURE 3.34 (Part of) the line through (x_0, y_0) and in the direction of $\mathbf{a}$.

Lines and Circles Are Curves The simplest curves are lines. We give a standard parametric equation for the line L through the point (x_0, y_0) and in the direction of the vector $\mathbf{a} = \langle a_1, a_2 \rangle$. We show that this set of points is a curve.

Figure 3.34 shows the position vector $\mathbf{r}_0 = \langle x_0, y_0 \rangle$ of the point (x_0, y_0) and the vector $\mathbf{a} = \langle a_1, a_2 \rangle$, placed with its tail at (x_0, y_0). If (x, y) is any point on L, the position vector $\mathbf{r} = \langle x, y \rangle$ of this point can be written as the vector sum

$$\mathbf{r} = \mathbf{r}_0 + s\mathbf{a} = \langle x_0, y_0 \rangle + s\langle a_1, a_2 \rangle = \langle x_0 + sa_1, y_0 + sa_2 \rangle, \tag{2}$$

for some value of the scalar s. In the figure, $s \approx 3$. If $\mathbf{r}$ were the position vector of a point on the other side of (x_0, y_0), the scalar s would be negative. To describe the entire line, the parameter s would range over the interval $(-\infty, \infty)$; for the half-line to the right of $\mathbf{r}_0$, s would range over the interval $[0, \infty)$; for the line segment shown in Fig. 3.34, s would range over the interval $[-1, 4]$, approximately.

According to the definition, the line L is a curve because the coordinate functions $x(s) = x_0 + sa_1$ and $y(s) = y_0 + sa_2$ are continuous for all values of the parameter s.

As a specific example, suppose $\mathbf{r}_0 = \langle -1, 1 \rangle$ and $\mathbf{a} = \langle \cos 12°, \sin 12° \rangle$. We recognize $\mathbf{a}$ as a unit vector in the direction 12°. If the parameter s were to vary over the interval $[-1, 4]$, the line segment described by

$$\mathbf{r}(s) = \langle -1, 1 \rangle + s\langle \cos 12°, \sin 12° \rangle \approx \langle -1 + 0.98s, 1 + 0.2s \rangle, \qquad -1 \le s \le 4, \tag{3}$$

would appear as in Fig. 3.34. Figure 3.35 shows this curve as graphed on a calculator in "parametric mode."

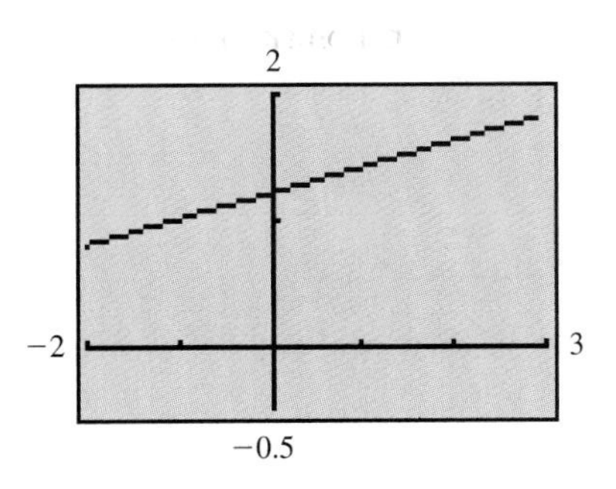

FIGURE 3.35 Graph of $\mathbf{r}(s) = \langle -1 + 0.98s, 1 + 0.2s \rangle$, $-1 \le s \le 4$, in the window $-2 \le x \le 3$ by $-\frac{1}{2} \le y \le 2$.

A Standard Form of a Parametric Equation for a Line

A parametric equation describing the line through the point (x_0, y_0) and parallel to the nonzero vector $\mathbf{a}$ is

$$\mathbf{r} = \mathbf{r}(s) = \mathbf{r}_0 + s\mathbf{a}, \qquad -\infty < s < \infty, \tag{4}$$

where $\mathbf{r}_0 = \langle x_0, y_0 \rangle$. The variable s is a parameter. A parametric equation for a segment of this line is the same as equation (4), except that s would be restricted to a finite interval.

EXAMPLE 3 Give a parametric equation describing the line through the points $(-5, 2)$ and $(-2, 3)$.

Solution

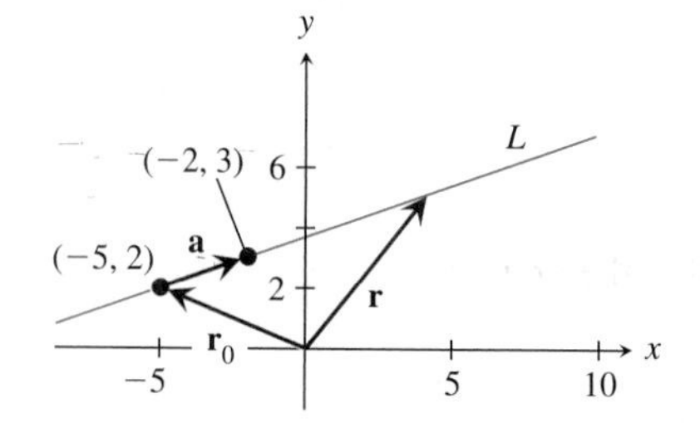

FIGURE 3.36 The line L through the points $(-5, 2)$ and $(-2, 3)$.

Figure 3.36 shows the given points $(-5, 2)$ and $(-2, 3)$. We may take either of these points as (x_0, y_0). If we prefer to think of the line as traced from lower left to upper right, we may take the vector $\mathbf{a} = \langle -2, 3 \rangle - \langle -5, 2 \rangle = \langle -2 - (-5), 3 - 2 \rangle = \langle 3, 1 \rangle$ as parallel to the line. Note that $\mathbf{a}$ is the difference of the position vectors $\langle -2, 3 \rangle$ and $\langle -5, 2 \rangle$ of the given points. Hence, a parametric equation for the line is

$$\mathbf{r} = \mathbf{r}_0 + s\mathbf{a} = \langle -5, 2 \rangle + s\langle 3, 1 \rangle, \qquad -\infty < s < \infty.$$

Although we may use any letter for the parameter, we have used s rather than t here to avoid any association of the curve described by this parametric equation with the path of an object. From now on, we use t, x, s, θ, or other variables as parameters, as suggested by context or convenience. If we intend t to be time, context will make this clear.

The standard form for a parametric equation of a circle follows from our discussion of circular motion in Section 3.3. Figure 3.37 shows a circle with radius a and center at the point with position vector $\mathbf{r}_0 = \langle h, k \rangle$. The position vector $\mathbf{r} = \mathbf{r}(\theta)$ of any point on the circle can be written as the sum of two displacements: The first is $\mathbf{r}_0$ from the origin to the center of the circle; the second is a displacement of length a in the direction θ. Hence,

$$\mathbf{r} = \mathbf{r}(\theta) = \mathbf{r}_0 + a\langle \cos\theta, \sin\theta \rangle.$$

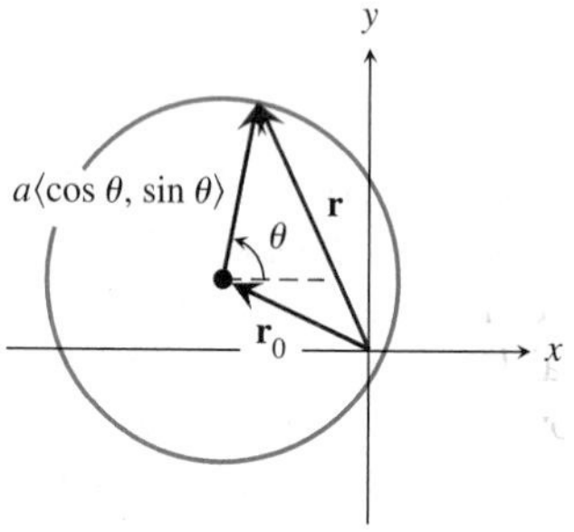

FIGURE 3.37 A circle with radius a and center at the point with position vector $\mathbf{r}_0$.

A Standard Form of a Parametric Equation for a Circle

A parametric equation describing the circle with center at the point (h, k) and with radius a is

$$\mathbf{r} = \mathbf{r}(\theta) = \mathbf{r}_0 + a\langle \cos\theta, \sin\theta \rangle, \qquad 0 \le \theta \le 2\pi, \tag{5}$$

where $\mathbf{r}_0 = \langle h, k \rangle$.

The circle described by (5) is a curve because the coordinate functions

$$x(\theta) = h + a\cos\theta \quad \text{and} \quad y(\theta) = k + a\sin\theta$$

are continuous.

EXAMPLE 4 Give a parametric equation for a circle of radius 3 and with center at the point $(-2, 1)$.

Solution

From (5), with $\mathbf{r}_0 = \langle -2, 1\rangle$ and $a = 3$, a parametric equation for this circle is

$$\mathbf{r} = \mathbf{r}(\theta) = \langle -2, 1\rangle + 3\langle \cos\theta, \sin\theta\rangle, \qquad 0 \le \theta \le 2\pi.$$

Figure 3.37 shows this curve.

Graphs of Continuous Functions Are Curves The graph C of any continuous function f defined on an interval I is a curve, according to our definition. To see that this is so, recall that the graph C of f is the set $\{(x, f(x)) : x \in I\}$. This set is the range of the vector function

$$\mathbf{r} = \mathbf{r}(x) = x\mathbf{i} + f(x)\mathbf{j}, \qquad x \in I.$$

Because x and $f(x)$ are continuous functions of x, the graph of f is a curve.

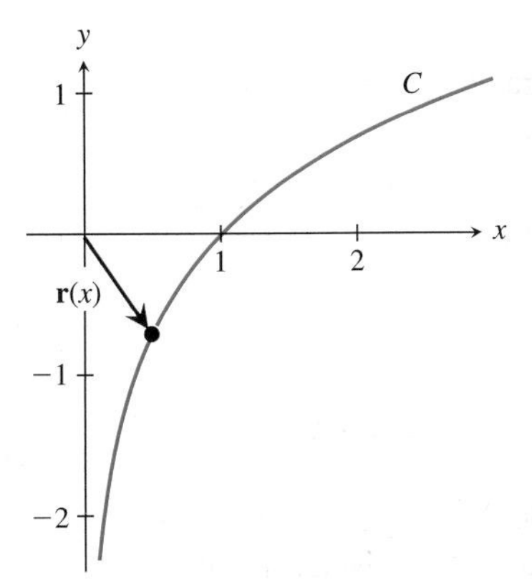

EXAMPLE 5 Show that the graph C of the natural logarithm function defined on $(0, \infty)$ is a curve.

Solution

The tip of the vector $\mathbf{r}(x)$, where

$$\mathbf{r} = \mathbf{r}(x) = x\mathbf{i} + \ln x\mathbf{j}, \qquad 0 < x < \infty,$$

traces the graph of $y = \ln x$ as x varies over $(0, \infty)$. A representative position vector $\mathbf{r}(x)$ of a point on C is shown at the top of Fig. 3.38. Because the coordinate functions x and $\ln x$ are continuous, C is a curve.

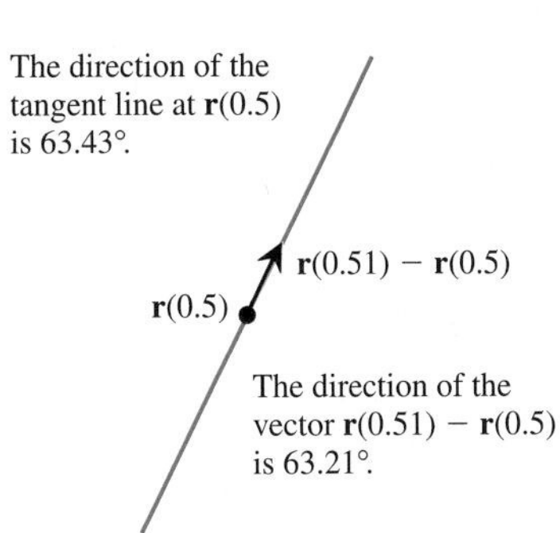

FIGURE 3.38 x as a parameter for the graph of the natural logarithm function.

Tangent Vectors The "representative point" in Example 5 was $\mathbf{r}(0.5) = \langle 0.5, \ln(0.5)\rangle$. Figure 3.38 includes a zoom to the curve C near this point. The figure also shows the tangent line to the graph of the natural logarithm function at this point, as well as the vector $\mathbf{r}(0.51) - \mathbf{r}(0.5)$. We show, as suggested by the figure, that the directions of the tangent line and the vector $\mathbf{r}(0.51) - \mathbf{r}(0.5)$ are very nearly equal.

The slope of the tangent line to the logarithm curve at $(0.5, \ln(0.5))$ is

$$\left.\frac{d}{dx}\ln x\right|_{x=0.5} = \left.\frac{1}{x}\right|_{x=0.5} = 2,$$

from which it follows that the direction of the tangent line is $\arctan(2) \approx 63.43°$. On the other hand, the direction of the vector

$$\mathbf{r}(0.51) - \mathbf{r}(0.5) = \langle 0.51, \ln(0.51)\rangle - \langle 0.5, \ln(0.5)\rangle \approx \langle 0.010000, 0.019803\rangle$$

is approximately arctan(0.019803/0.010000) $\approx 63.21°$. Because these two directions differ by approximately 0.2°, it is not surprising that the tangent line and the vector $\mathbf{r}(0.51) - \mathbf{r}(0.5)$ appear as parallel in Fig. 3.38.

Based on these calculations and our earlier work with velocity vectors for parametrically described paths of objects in motion, we define what is meant by *tangent vector to a curve*.

DEFINITION Tangent Vector to a Curve

Let C be the curve described by the vector function

$$\mathbf{r} = \mathbf{r}(t) = \langle x(t), y(t) \rangle, \qquad t \in I.$$

If the coordinate functions of $\mathbf{r}$ are differentiable at a point $t \in I$, we say that $\mathbf{r}$ is differentiable at t and denote by $\mathbf{r}'(t)$ the vector with coordinate functions $x'(t)$ and $y'(t)$. The vector $\mathbf{r}'(t)$ is a **tangent vector** to the curve C at $\mathbf{r}(t)$.

In the next example we calculate the tangent vector to an ellipse described by a parametric equation. As a check on this calculation we first describe the same ellipse with an equation of the form $x^2/a^2 + y^2/b^2 = 1$ and calculate its slope using implicit differentiation.

The ellipse in Fig. 3.39 is described by the equation

$$\frac{x^2}{5^2} + \frac{y^2}{4^2} = 1. \tag{6}$$

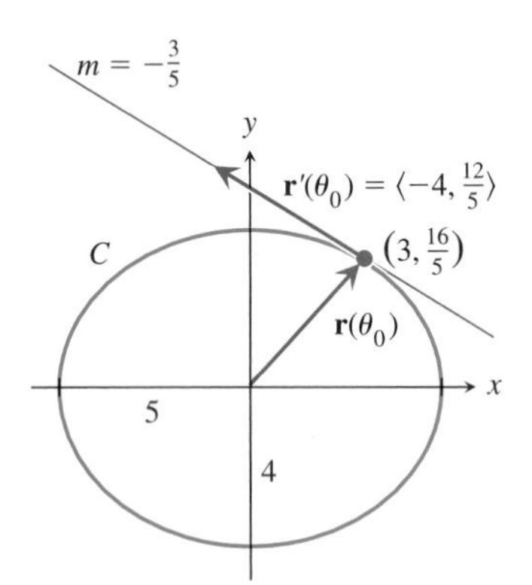

FIGURE 3.39 Relating the slope m of the tangent line and the tangent vector $\mathbf{r}'(\theta_0)$ at the point $(3, 16/5)$.

The point $(3, 16/5)$ is on this ellipse. We calculate the slope of the tangent line to the ellipse at this point by differentiating (6) implicitly:

$$\frac{2x}{5^2} + \frac{2y \cdot y'}{4^2} = 0.$$

Solving for y',

$$y' = \frac{-2x/25}{2y/16} = -\frac{16x}{25y}.$$

At the point $(3, 16/5)$, the slope is

$$y' = -\frac{16 \cdot 3}{25 \cdot (16/5)} = -\frac{3}{5}. \tag{7}$$

EXAMPLE 6 Let C be the curve described by the parametric equation

$$\mathbf{r}(\theta) = \langle 5 \cos \theta, 4 \sin \theta \rangle, \qquad 0 \le \theta \le 2\pi. \tag{8}$$

First set your calculator in "parametric mode" and graph the parametric equation (8). Next, show that

a) every point of the curve C is a point of the graph of equation (6),
b) the point $(3, 16/5)$ has position vector $\mathbf{r}(\theta_0)$, where $\theta_0 = \arccos(3/5)$, and

c) $\mathbf{r}'(\theta_0) = \langle -4, 12/5 \rangle$. Relate this tangent vector to the slope $m = -3/5$ of the ellipse at $(3, 16/5)$.

Solution

Figure 3.40 shows a calculator graph of (8).

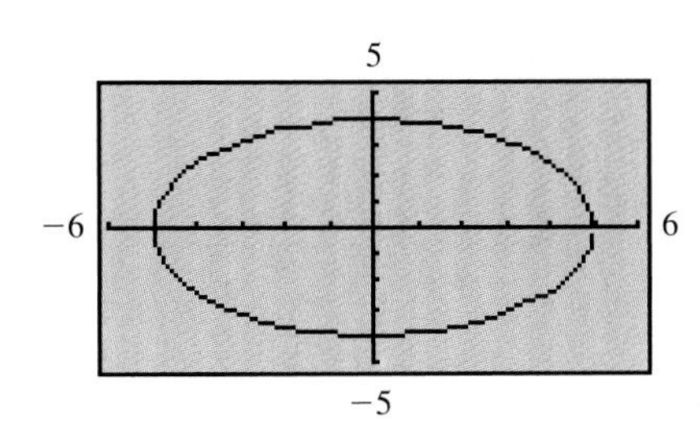

FIGURE 3.40 Graph of $\mathbf{r}(\theta) = \langle 5 \cos \theta, 4 \sin \theta \rangle$, with $0 \le \theta \le 2\pi$, in the window $-6 \le x \le 6$ by $-5 \le y \le 5$.

a) Substituting the coordinates $x = 5 \cos \theta$ and $y = 4 \sin \theta$ of $\mathbf{r}(\theta)$ from (8) into (6),

$$\frac{x^2}{5^2} + \frac{y^2}{4^2} = \frac{5^2 \cos^2\theta}{5^2} + \frac{4^2 \sin^2\theta}{4^2} = 1.$$

Hence, for all θ, the point with position vector $\mathbf{r}(\theta)$ lies on the graph of equation (6). (More is true: Every point of the graph of equation (6) is a point of the curve C. See Exercise 61.)

b) We seek a number θ_0 for which $3 = 5 \cos \theta_0$ and $16/5 = 4 \sin \theta_0$. From the first of these equations, $\cos \theta_0 = 3/5$. One solution of this equation is $\theta_0 = \arccos(3/5)$. Figure 3.41 shows a way of checking that $16/5 = 4 \sin \theta_0$. Hence,

$$\mathbf{r}(\theta_0) = \langle 5 \cos \theta_0, 4 \sin \theta_0 \rangle = \langle 3, 16/5 \rangle.$$

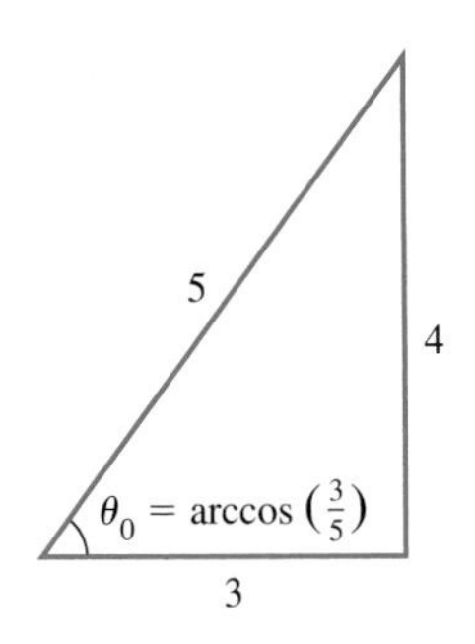

Recall: $\cos \theta_0 = \dfrac{\text{side adjacent}}{\text{hypotenuse}} = \dfrac{3}{5}$

$\sin \theta_0 = \dfrac{\text{side opposite}}{\text{hypotenuse}} = \dfrac{4}{5}$

FIGURE 3.41 The number arccos(3/5) is "the angle whose cosine is 3/5."

c) Differentiating equation (8) gives

$$\mathbf{r}'(\theta) = \langle -5 \sin \theta, 4 \cos \theta \rangle.$$

Evaluating at $\theta_0 = \arccos(3/5)$,

$$\mathbf{r}'(\theta_0) = \langle -5(4/5), 4(3/5) \rangle = \langle -4, 12/5 \rangle.$$

The ratio of rise to run for this vector is $(12/5)/(-4) = -3/5$, which, from (7), is the slope of the tangent line to the ellipse at $(3, 16/5)$. See Fig. 3.39.

The close relation between tangent vector and slope that we noted in this example holds more generally. We state this as the Tangent Vector and Slope Theorem.

THEOREM Tangent Vector and Slope Theorem

Let C be a curve and $\mathbf{r} = \mathbf{r}(t) = \langle x(t), y(t) \rangle$, $t \in I$, a parametrization of C. Assume that the derivative functions $x'(t)$ and $y'(t)$ are continuous for $t \in I$. For each $t \in I$ for which $x'(t) \neq 0$, the slope of C at $\mathbf{r}(t)$ is the ratio

$$\frac{dy}{dx} = \frac{\frac{dy}{dt}}{\frac{dx}{dt}} = \frac{y'(t)}{x'(t)} \tag{9}$$

of the coordinates of the tangent vector $\mathbf{r}'(t) = \langle x'(t), y'(t) \rangle$. See Fig. 3.42.

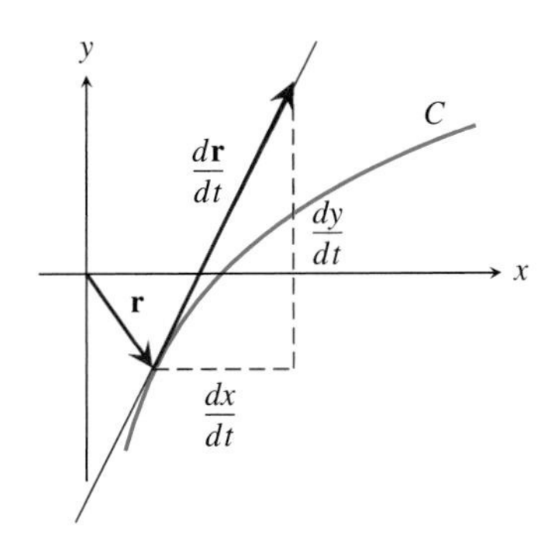

FIGURE 3.42 Slope, as calculated from the tangent vector.

We give an informal proof of this theorem. Assume that t_0 is a point of I for which $x'(t_0) \neq 0$. This assumption means that C is not vertical near $\mathbf{r}(t_0)$ and that C is the graph of an equation $y = f(x)$ near $\mathbf{r}(t_0)$. Hence, for all t near t_0, $y(t) = f(x(t))$. Applying the chain rule to this equation,

$$y'(t) = \frac{dy}{dt} = \frac{df}{dx}\frac{dx}{dt} = \frac{dy}{dx}\frac{dx}{dt} = \frac{dy}{dx} \cdot x'(t).$$

Hence, if $x'(t) \neq 0$, then

$$\frac{dy}{dx} = \frac{y'(t)}{x'(t)},$$

which is (9).

The Tangent Vector and Slope Theorem has a direct geometric interpretation. Figure 3.42 shows the tangent vector $\mathbf{r}'(t) = \dfrac{d\mathbf{r}}{dt}$ at $\mathbf{r}(t)$, with its coordinate functions $x'(t) = \dfrac{dx}{dt}$ and $y'(t) = \dfrac{dy}{dt}$. If the "run" $x'(t)$ is not zero, then the slope dy/dx of the line tangent to C at $\mathbf{r}(t)$ can be found by calculating the ratio

$$\frac{dy}{dx} = \frac{\text{rise}}{\text{run}} = \frac{\dfrac{dy}{dt}}{\dfrac{dx}{dt}} = \frac{y'(t)}{x'(t)}.$$

In the next example we ask you to graph a parabola with a tilted axis. We give a table from which the graph can be sketched if you do not graph the parabola with the help of a CAS or graphing calculator. This curve is discussed further in Exercises 85 and 86, where we relate it to conics whose axes of symmetry are not parallel with the coordinate axes. See also Section 9.1.

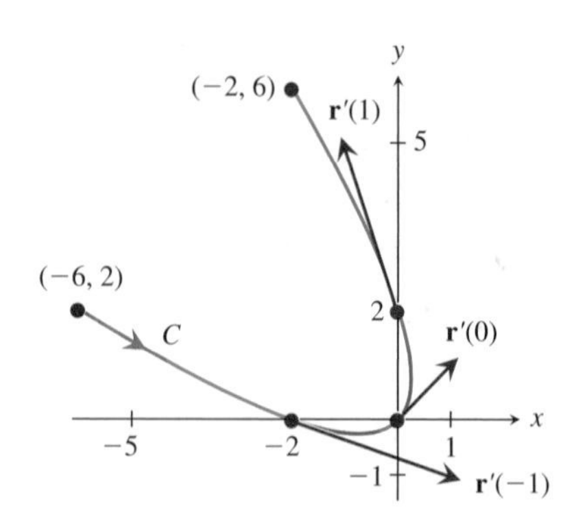

FIGURE 3.43 Parabola with tilted axis.

EXAMPLE 7 Graph the curve C described by

$$\mathbf{r} = \mathbf{r}(t) = x(t)\mathbf{i} + y(t)\mathbf{j} = (t - t^2)\mathbf{i} + (t + t^2)\mathbf{j}, \qquad -2 \le t \le 2. \tag{10}$$

Determine tangent vectors and slopes of C at its x- and y-intercepts.

Solution

Figure 3.43 shows the graph of C. We have shown the endpoints $\mathbf{r}(-2) = \langle -6, 2\rangle$ and $\mathbf{r}(2) = \langle -2, 6\rangle$ of the curve and marked with a small arrowhead the direction in which C is traced as t varies from -2 to 2. The curve may be graphed by calculator, by CAS, or by plotting the points given in Table 3.3 and drawing a smooth curve joining them.

TABLE 3.3 Points on the curve C, corresponding to selected values of the parameter t.

t	$(x(t), y(t))$	t	$(x(t), y(t))$
-2.0	$(-6, 2)$	0.5	$(0.25, 0.75)$
-1.5	$(-3.75, 0.75)$	1.0	$(0, 2)$
-1.0	$(-2, 0)$	1.5	$(-0.75, 3.75)$
-0.5	$(-0.75, -0.25)$	2.0	$(-2, 6)$
0.0	$(0, 0)$		

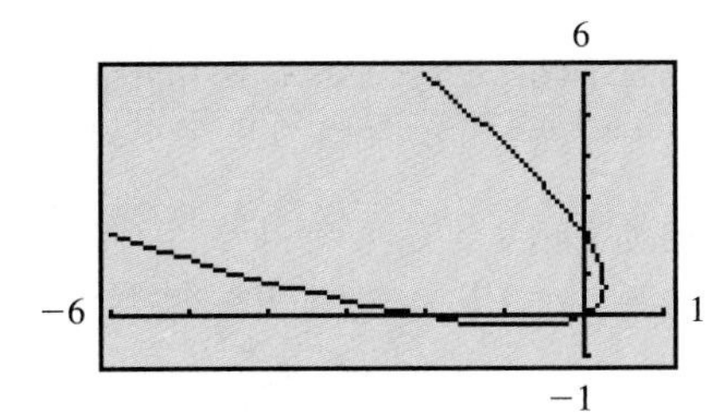

FIGURE 3.44 Graph of $\mathbf{r} = (t - t^2)\mathbf{i} + (t + t^2)\mathbf{j}$, with $-2 \leq t \leq 2$, in the window $-6 \leq x \leq 1$ by $-1 \leq y \leq 6$.

Figure 3.44 shows a calculator graph of (10). By comparing this graph with that shown in Fig. 3.43, we can note the distortion due to the "default scaling" done by the calculator.

At the x-intercepts of C the y-coordinate will be zero; setting $y(t) = t + t^2$ equal to zero gives $t = 0$ or $t = -1$. From $\mathbf{r}(-1) = -2\mathbf{i} + 0\mathbf{j}$ and $\mathbf{r}(0) = 0\mathbf{i} + 0\mathbf{j}$ we see that the x-intercepts are $(-2, 0)$ and $(0, 0)$. At the y-intercepts, the x-coordinate will be zero; setting $x(t) = t - t^2$ equal to zero gives $t = 0$ or $t = 1$. From $\mathbf{r}(0) = 0\mathbf{i} + 0\mathbf{j}$ and $\mathbf{r}(1) = 0\mathbf{i} + 2\mathbf{j}$ we see that the y-intercepts are $(0, 0)$ and $(0, 2)$.

The tangent vector is

$$\mathbf{r}'(t) = (1 - 2t)\mathbf{i} + (1 + 2t)\mathbf{j}.$$

Evaluating $\mathbf{r}'(t)$ at $t = -1$, $t = 0$, and $t = 1$, the tangent vectors at the x- or y-intercepts are

$$\mathbf{r}'(-1) = 3\mathbf{i} - \mathbf{j}$$
$$\mathbf{r}'(0) = \mathbf{i} + \mathbf{j}$$
$$\mathbf{r}'(1) = -\mathbf{i} + 3\mathbf{j}.$$

WEB

By the Tangent Vector and Slope Theorem, the slope of dy/dx at these three points is $dy/dx = (dy/dt)/(dx/dt)$. From the expressions for the three tangent vectors given previously, the slopes at the intercepts $(-2, 0)$, $(0, 0)$, and $(0, 2)$ are $-1/3$, 1, and -3, respectively.

Exercises 3.4

Exercises 1–8: The parametric equation describes the motion of an object. Sketch the path of the object either by using a graphing calculator or CAS, by eliminating the parameter, or by plotting several points and joining them with a smooth curve. Calculate and include in your sketch the object's position and velocity vectors at t_1. Time is measured in seconds and lengths in meters.

1. $\mathbf{r} = \langle 2t, 3t^2 \rangle$, $t \geq 0$; $t_1 = 1$
2. $\mathbf{r} = \langle 4t, t^2 \rangle$, $t \geq 0$; $t_1 = 1$
3. $\mathbf{r} = (1 - t)\mathbf{i} + (1 + 2t^2)\mathbf{j}$, $t \geq 0$; $t_1 = 1$
4. $\mathbf{r} = (2 - 3t)\mathbf{i} + (3t^2 - 2)\mathbf{j}$, $t \geq 0$; $t_1 = 1$
5. $\mathbf{r} = 5\langle \cos 2t, \sin 2t \rangle$, $t \geq 0$; $t_1 = \pi/6$
6. $\mathbf{r} = 2\langle \sin 3t, \cos 3t \rangle$, $t \geq 0$; $t_1 = \pi/2$
7. $\mathbf{r} = \mathbf{i} + 2\mathbf{j} + t\left(\mathbf{i} + \frac{1}{2}\mathbf{j}\right)$, $t \geq 0$; $t_1 = 2$
8. $\mathbf{r} = -2t\mathbf{i} + 3\mathbf{i} + 5\mathbf{j}$, $t \geq 0$; $t_1 = 1$

Exercises 9–12: Sketch the path of the object either by using a graphing calculator or CAS, by eliminating the parameter, or by plotting several points and joining them with a smooth curve. To eliminate the parameter, note that each path is described by a parametric equation of the form $\mathbf{r} = \langle t, f(t)\rangle$ or $\mathbf{r} = t\mathbf{i} + f(t)\mathbf{j}$. Setting $x = t$ and substituting into $y = f(t)$ eliminates the parameter. Calculate the object's average velocity vector in the intervals $[t_1, t_1 + 0.1]$, $[t_1, t_1 + 0.01]$, and $[t_1, t_1 + 0.001]$ and compare the results with $\mathbf{v}(t_1)$. Time is measured in seconds and lengths in meters.

9. $\mathbf{r} = t\mathbf{i} + (\sin t)\mathbf{i}, \quad t \geq 0;\ t_1 = \pi/6$

10. $\mathbf{r} = t\mathbf{i} + (\cos t)\mathbf{j}, \quad t \geq 0;\ t_1 = \pi/3$

11. $\mathbf{r} = \langle t, e^t\rangle, \quad t \geq 0;\ t_1 = 1.0$

12. $\mathbf{r} = \langle t, \ln(t+1)\rangle, \quad t \geq 0;\ t_1 = 1$

Exercises 13–16: The parametric equation describes the motion (in a horizontal plane) of a model airplane, which is being flown by remote control until time t_1, when radio contact with the plane is lost. For $t \geq t_1$, the plane flies horizontally along the tangent to the path at $t = t_1$. Calculate the speed and direction of the airplane at t_1. Sketch the path of the airplane from $t = 0$ to $t = t_1$ and draw the vectors $\mathbf{r}(t_1)$ and $\mathbf{v}(t_1)$. Where is the plane at time $t_1 + 10$ s? Time is measured in seconds and lengths in meters.

13. $\mathbf{r} = 6\langle\cos(2.1t), \sin(2.1t)\rangle, \quad t_1 = 12$

14. $\mathbf{r} = 5\langle\cos(1.5t), \sin(1.5t)\rangle, \quad t_1 = 10$

15. $\mathbf{r} = 2.5t\mathbf{i} + (2.5t^2 + 1)\mathbf{j}, \quad t_1 = 1$

16. $\mathbf{r} = 3t\mathbf{i} + (2t^2 + 1)\mathbf{j}, \quad t_1 = 1$

Exercises 17–20: Write a parametric equation in the standard form $\mathbf{r} = \mathbf{r}_0 + t\mathbf{a}$ for the line with equation $y = mx + b$. Use the given $\mathbf{r}_0$.

17. $y = \frac{2}{3}x + 5, \quad \mathbf{r}_0 = \langle 3, 7\rangle$

18. $y = -2x + 3, \quad \mathbf{r}_0 = \langle 4, -5\rangle$

19. $y = x + 1, \quad \mathbf{r}_0 = \langle 4, 5\rangle$

20. $y = \frac{9}{7}x + 3, \quad \mathbf{r}_0 = \langle 0, 3\rangle$

Exercises 21–24: Write a parametric equation in the standard form $\mathbf{r} = \mathbf{r}_0 + t\mathbf{a}$ for the line determined by the position vectors $\mathbf{p}$ and $\mathbf{q}$. Hint: Their difference determines the direction of the line, and either of them can serve as $\mathbf{r}_0$.

21. $\mathbf{p} = \langle 4, -2\rangle, \mathbf{q} = \langle 0, 3\rangle$

22. $\mathbf{p} = \langle 5, 0\rangle, \mathbf{q} = \langle 0, -2\rangle$

23. $\mathbf{p} = \langle -3, 0\rangle, \mathbf{q} = \langle -3, 5\rangle$

24. $\mathbf{p} = \langle 2, 9\rangle, \mathbf{q} = \langle 2, -13\rangle$

Exercises 25–34: Write the Cartesian equation $y = f(x)$ as a parametric equation $\mathbf{r}(x) = \langle x, f(x)\rangle$ or $\mathbf{r}(x) = x\mathbf{i} + f(x)\mathbf{j}$. Calculate the tangent vector and, using the Tangent Vector and Slope Theorem, the slope at the point with position vector $\mathbf{r}_0$.

25. $y = x^2, -2 \leq x \leq 2, \mathbf{r}_0 = \langle 1, 1\rangle$

26. $y = x^3, -1 \leq x \leq 1.5, \mathbf{r}_0 = \langle 1, 1\rangle$

27. $y = \tan x, 0 \leq x \leq 1.3, \mathbf{r}_0 = \langle \pi/4, 1\rangle$

28. $y = \sec x, -1 \leq x \leq 1, \mathbf{r}_0 = \left\langle \pi/6, 2/\sqrt{3}\right\rangle$

29. $y = x^2 - 6x + 11, 0 \leq x \leq 5, \mathbf{r}_0 = \langle 4, 3\rangle$

30. $y = x^2 + 4x + 3, -4 \leq x \leq 0, \mathbf{r}_0 = \langle -1, 0\rangle$

31. $y = e^{2x}, -1 \leq x \leq 2, \mathbf{r}_0 = \langle 1, e^2\rangle$

32. $y = e^{\sqrt{x}}, 0 \leq x \leq 4, \mathbf{r}_0 = \langle 1, e\rangle$

33. $y = (\ln x)^2, 0.5 \leq x \leq 2, \mathbf{r}_0 = \langle 1, 0\rangle$

34. $y = \ln\sqrt{x}, 0 < x \leq 4, \mathbf{r}_0 = \langle 1, 0\rangle$

Exercises 35–40: Give a parametric equation describing the given motion.

35. An object is moving to the right with speed 10 m/s on the line passing through $(-5, 1)$ and $(6, 7)$. At $t = 0$ the object is at $(-5, 1)$.

36. An object is moving downward with speed 7 m/s on the line passing through $(5, 1)$ and $(6, -10)$. At $t = 0$ the object is at $(5, 1)$.

37. An object is moving counterclockwise with speed 7 m/s on a unit circle centered at $(0, 0)$. At $t = 0$ the object is at $(1, 0)$.

38. An object is moving counterclockwise with speed 10 m/s on a unit circle centered at $(0, 0)$. At $t = 0$ the object is at $(1, 0)$.

39. The velocity of the object for $t \geq 0$ is $\mathbf{v} = \langle 1, 5\rangle$. It was at the point $(0, -2)$ at $t = 0$.

40. The velocity of the object for $t \geq 0$ is $\mathbf{v} = \langle -3, 3\rangle$. It was at the point $(4, 3)$ at $t = 0$.

Exercises 41–44: The trajectory of an object is described by the vector function $\mathbf{r}(t) = \langle x(t), y(t)\rangle$, for $0 \leq t \leq T$. Time is measured in seconds and length in meters. Plot the points with coordinates $\mathbf{r}(0)$, $\mathbf{r}(T/4)$, $\mathbf{r}(T/2)$, $\mathbf{r}(3T/4)$, and $\mathbf{r}(T)$. Use these five points to roughly sketch the trajectory. Calculate its velocity and speed at $t = T/2$. Eliminate the parameter to find an equation in x and y that describes the curve on which the object moves.

41. $\mathbf{r}(t) = \langle t, -t^2 + 16t\rangle; T = 8$

42. $\mathbf{r}(t) = \langle 3t, -2t^2 + 8t\rangle; T = 2$

43. $\mathbf{r}(t) = \langle 4t, -t^2 + 4t\rangle; T = 4$

44. $\mathbf{r}(t) = \langle 2t, -2t^2 + 16t\rangle; T = 8$

Exercises 45–50: Use your calculator to graph the described curve. Calculate the endpoints of the curve. Place an arrowhead on the curve to indicate the tracing order. Eliminate the parameter to find an equation in x and y whose graph includes the parametric curve.

45. $\mathbf{r}(t) = \langle 1 + 2t^2, t + 2\rangle, -2 \leq t \leq 3$

46. $\mathbf{r}(t) = \langle t^2 - 4t + 3, t + 1\rangle, -3 \leq t \leq 3$

47. $\mathbf{r}(t) = \left\langle \frac{1}{2}t^2 + t, t^2 + t\right\rangle, -2 \leq t \leq 2$

48. $\mathbf{r}(t) = \langle t^2 - t, t^2 + 2t \rangle$, $-3 \le t \le 2$

49. $\mathbf{r}(t) = \langle 2t + 1, t - 1 \rangle$, $-2 \le t \le 3$

50. $\mathbf{r}(t) = \langle -t + 3, -2t - 5 \rangle$, $t \ge 0$

Exercises 51–60: Graph the following by using either the parametric form or the equation gotten by eliminating the parameter θ. For curves described by $\mathbf{r} = \langle x, y \rangle = \langle h + a\cos\theta, k + b\sin\theta \rangle$, the parameter may be eliminated by forming the expression $((x - h)/a)^2 + ((y - k)/b)^2$.

51. $\mathbf{r} = \langle 2\cos\theta, 2\sin\theta \rangle$, $0 \le \theta \le 2\pi$

52. $\mathbf{r} = \left\langle \frac{1}{2}\cos\theta, \frac{1}{2}\sin\theta \right\rangle$, $0 \le \theta \le 2\pi$

53. $\mathbf{r} = \langle \cos\theta, \sin\theta \rangle$, $0 \le \theta \le \pi$

54. $\mathbf{r} = \langle 3\cos\theta, 3\sin\theta \rangle$, $0 \le \theta \le 3\pi/2$

55. $\mathbf{r} = \langle 1 + 2\cos\theta, 1 + 2\sin\theta \rangle$, $0 \le \theta \le 2\pi$

56. $\mathbf{r} = \langle -1 + 2\cos\theta, -3 + 2\sin\theta \rangle$, $0 \le \theta \le 2\pi$

57. $\mathbf{r} = \langle 2\cos\theta, 3\sin\theta \rangle$, $0 \le \theta \le 2\pi$

58. $\mathbf{r} = \langle 4\cos\theta, \sin\theta \rangle$, $0 \le \theta \le 2\pi$

59. $\mathbf{r} = \langle 1 + 2\cos\theta, 1 + 3\sin\theta \rangle$, $0 \le \theta \le 2\pi$

60. $\mathbf{r} = \langle -3 + \cos\theta, -5 + 4\sin\theta \rangle$, $0 \le \theta \le 2\pi$

61. Show that corresponding to each point (x, y) satisfying the equation

$$\frac{x^2}{5^2} + \frac{y^2}{4^2} = 1$$

a number θ can be found for which $x = 5\cos\theta$ and $y = 4\sin\theta$.

62. Determine if the point $(-24, -11)$ is on the line with equation $\mathbf{r} = \langle -3, 4 \rangle + t\langle 7, 5 \rangle$.

63. Determine if the point $(1, 15)$ is on the line with equation $\mathbf{r} = \langle -1, 4 \rangle + t\langle 1, 6 \rangle$.

64. Parametrize the graph of $x = y^2$. Can x be used as a parameter? Why or why not?

65. Parametrize the curve with equation $(x + 1)^2 + (y - 2)^2 = 4$.

66. Parametrize the curve with equation $x^2 + y^2 - 2x + 4y = -1$.

67. Parametrize the curve with equation $9x^2 + 4y^2 = 36$.

68. Parametrize the curve with equation $x^2 + 3y^2 - 4x - 30y = -78$.

69. The curve described by $\mathbf{r} = \mathbf{r}(t) = \langle \sin t, \sin(t + \sin t) \rangle$, where $0 \le t \le 2\pi$, has two tangent lines at the point $(0, 0)$. Find the equations of these tangent lines.

70. The ellipse in the figure is described by the equation $x^2/a^2 + y^2/b^2 = 1$ or the parametric equation $\mathbf{r} = \langle a\cos\theta, b\sin\theta \rangle$, $0 \le \theta \le 2\pi$. The lighter curves are circles with radii a and b. Use the figure to interpret the geometric meaning of the parameter θ for the ellipse. *Hint:* Work out the coordinates of the three points shown in the figure.

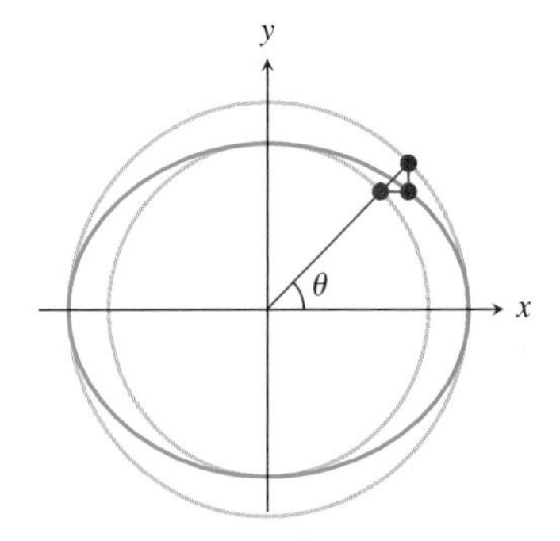

Figure for Exercise 70.

Exercises 71–78: Sketch the "named" curve and calculate a unit tangent vector at **p**.

71. Astroid:

$$\mathbf{r} = \mathbf{r}(t) = \langle a\cos^3 t, \sin^3 t \rangle,$$

where $0 \le t \le 2\pi$. This curve was first discussed by Johann Bernoulli in 1691–1692. It can be formed by rolling a circle of radius $\frac{1}{4}a$ on the inside of a circle of radius a. Take $a = 1$ and $\mathbf{p} = \mathbf{r}(\pi/4)$.

72. Trochoid:

$$\mathbf{r} = \mathbf{r}(t) = \langle at - h\sin t, a - h\cos t \rangle,$$

where $t \ge 0$. Consider a disk of radius a that initially is resting on the x-axis, with its center at $(0, a)$. Choose a point Q on the disk, with coordinates $(0, h)$, where $0 \le h \le a$. If the disk rolls to the right, the curve traced by Q is a trochoid. Take $a = 1$, $h = 0.5$, and $\mathbf{p} = \mathbf{r}(\pi/3)$. Graph on the interval $0 \le t \le 4\pi$.

73. Cycloid:

$$\mathbf{r} = \mathbf{r}(t) = a\langle t - \sin t, 1 - \cos t \rangle,$$

where $t \ge 0$. This curve is a special case of the trochoid discussed in the previous exercise. Take $a = 1$ and $\mathbf{p} = \mathbf{r}(\pi/3)$. Graph on the interval $0 \le t \le 4\pi$.

74. Tractrix:

$$\mathbf{r} = \mathbf{r}(t) = \langle 1/\cosh t, t - \tanh t \rangle,$$

where $t \ge 0$. The hyperbolic tangent function is defined by $\tanh(x) = (\sinh x)/(\cosh x)$, where $\sinh x = (e^x - e^{-x})/2$ and $\cosh x = (e^x + e^{-x})/2$. The tractrix is the path of an object initially at $(1, 0)$ and attached by a string of length 1 to a point moving from $(0, 0)$ upward on the y-axis. Take $\mathbf{p} = \mathbf{r}(1)$. Graph on the interval $0 \le t \le 7$.

75. Witch of Agnesi:

$$\mathbf{r} = \mathbf{r}(t) = \langle 2a \tan t, \cos^2 t \rangle,$$

where $-\pi/2 < t < \pi/2$. This curve, whose name is due to a translation error, was studied by Maria Gaetana Agnesi (1718–1799) in 1748. Take $a = 1$ and $\mathbf{p} = \mathbf{r}(\pi/4)$.

76. Trisectrix of Maclaurin:

$$\mathbf{r} = \mathbf{r}(t) = a\left\langle \frac{\sin 3t}{\sin t}, \frac{\sin 3t}{\cos t} \right\rangle,$$

where $t \neq \pi/2$ and $0 \leq t \leq \pi$. This curve was first studied by Colin Maclaurin in 1742 to trisect angles. Take $a = 1$ and $\mathbf{p} = \mathbf{r}(\pi/3)$.

77. Folium of Descartes:

$$\mathbf{r} = \mathbf{r}(t) = \left\langle \frac{3at}{1+t^3}, \frac{3at^2}{1+t^3} \right\rangle,$$

where $t \neq -1$. This curve was first discussed by René Descartes in 1638. Take $a = 1/3$ and $\mathbf{p} = \mathbf{r}(2)$.

78. Trident of Newton:

$$\mathbf{r} = \mathbf{r}(t) = \langle t, ct^2 + dt + e + f/t \rangle,$$

where $t \neq 0$. This curve appears in a book on cubic curves written by Isaac Newton in 1710. Take $c = d = 1$, $e = 0$, $f = 1/2$ and $\mathbf{p} = \mathbf{r}(-2)$.

Exercises 79–82: Graph the following Lissajous figures. *These curves can be produced on a cathode-ray oscilloscope by applying voltages proportional to x(t) and y(t) on the horizontal and vertical plates between which the electron beam passes. The plates deflect the beam as a function of the applied voltage.*

79. $\mathbf{r} = \sin t\mathbf{i} + \sin 2t\mathbf{j}$, $t \geq 0$

80. $\mathbf{r} = \sin t\mathbf{i} + \sin 3t\mathbf{j}$, $t \geq 0$

81. $\mathbf{r} = \sin 2t\mathbf{i} + \sin 3t\mathbf{j}$, $t \geq 0$

82. $\mathbf{r} = \sin 3t\mathbf{i} + \sin 4t\mathbf{j}$, $t \geq 0$

Exercises 83–84: Sketch the parabola and calculate the slope of the tangent line at the positive x-intercept(s).

83. $\mathbf{r} = \left\langle t + \frac{1}{2}t^2, t - \frac{1}{2}t^2 \right\rangle$, $-5 \leq t \leq 5$

84. $\mathbf{r} = \langle 1 - 2t + t^2, -1 + 2t + t^2 \rangle$, $-5 \leq t \leq 5$

Exercises 85–86: Solve the following Exercises, which concern conics with an xy-term.

85. Eliminating the parameter for the curve C described by (10) in Example 7 gives the equation

$$x - y = -\tfrac{1}{2}(x + y)^2. \tag{11}$$

Show that the set of all points (x, y) equally distant from the line with equation $y = x - \frac{1}{2}$ and the point $\left(-\frac{1}{4}, \frac{1}{4}\right)$ satisfies (11). Also show that any point (x, y) satisfying (11) is equally distant from the given line and point.

86. Eliminating the parameter for the curve C described by (10) in Example 7 gives the equation

$$x - y = -\tfrac{1}{2}(x + y)^2. \tag{12}$$

To show that this equation describes a parabola, we may rotate the axes to eliminate the xy term. Suppose the x- and y-axes are rotated counterclockwise through an angle θ to give a new set of coordinate axes. Label the rotated x-axis as the x'-axis and the rotated y-axis as the y'-axis. If a point has coordinates (x, y) relative to the "old" system and coordinates (x', y') relative to the "new" system, then

$$x = x' \cos\theta - y' \sin\theta$$
$$y = x' \sin\theta + y' \cos\theta.$$

Substitute these rotation equations into the equation of the conic, expand the result, and collect all $x'y'$ terms together. Choose θ so that the coefficient of the resulting single $x'y'$ term is zero. You should find that $\theta = \pi/4$. Show that with this value of θ the transformed equation reduces to $x'^2 = \sqrt{2}y'$, which has the form of a parabola. Check that the graph of this equation matches that of the given equation.

3.5 Dot Product

The *law of cosines* is useful in "solving" certain triangles. For example, if the lengths a, b, and c of the sides of the triangle shown in Fig. 3.45(a) are known, then the angle C can be determined from the law of cosines:

$$c^2 = a^2 + b^2 - 2ab \cos C.$$

Figure 3.45(b) shows triangle ABC in terms of vectors $\mathbf{a}$, $\mathbf{b}$, and $\mathbf{a} - \mathbf{b}$. The sides a, b, and c of the triangle are the lengths of these vectors. Expressed in terms of $\mathbf{a}$ and $\mathbf{b}$, the law of cosines leads to what is called the "dot product" of $\mathbf{a}$ and $\mathbf{b}$.

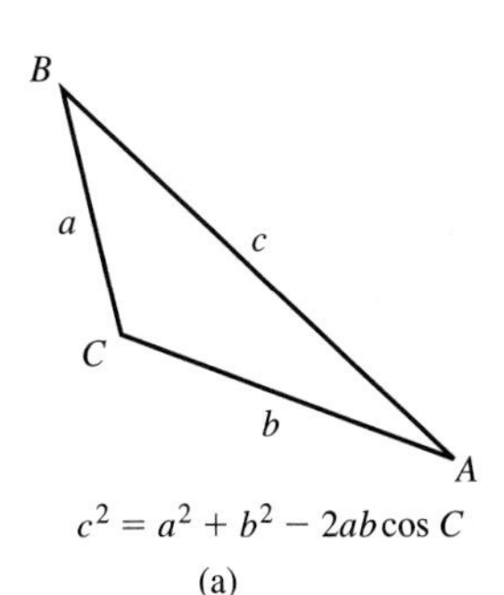

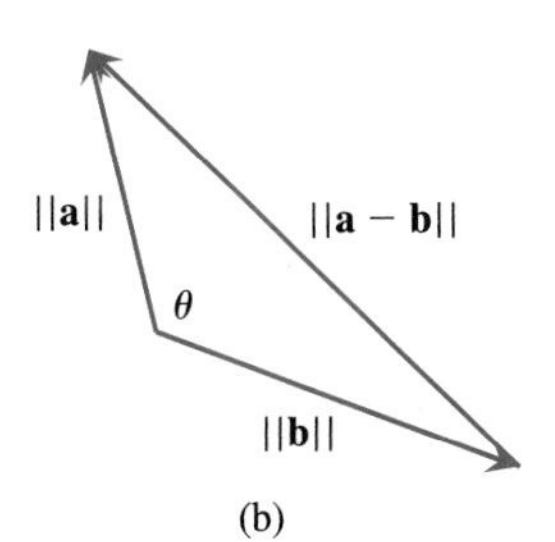

FIGURE 3.45 Determining the angle C with the law of cosines.

Letting θ be the angle between **a** and **b** and applying the law of cosines,

$$\|\mathbf{a} - \mathbf{b}\|^2 = \|\mathbf{a}\|^2 + \|\mathbf{b}\|^2 - 2\|\mathbf{a}\|\,\|\mathbf{b}\|\cos\theta.$$

If $\mathbf{a} = \langle a_1, a_2\rangle$ and $\mathbf{b} = \langle b_1, b_2\rangle$, this expression can be written as

$$(a_1 - b_1)^2 + (a_2 - b_2)^2 = a_1^2 + a_2^2 + b_1^2 + b_2^2 - 2\|\mathbf{a}\|\,\|\mathbf{b}\|\cos\theta.$$

Expanding the left side and simplifying,

$$-2a_1b_1 - 2a_2b_2 = -2\|\mathbf{a}\|\,\|\mathbf{b}\|\cos\theta.$$

Dividing both sides by -2,

$$a_1b_1 + a_2b_2 = \|\mathbf{a}\|\,\|\mathbf{b}\|\cos\theta. \tag{1}$$

The expression $a_1b_1 + a_2b_2$ in this equation is called the *dot product* of the vectors **a** and **b**.

DEFINITION Dot Product of Vectors

The **dot product** of vectors $\mathbf{a} = \langle a_1, a_2\rangle$ and $\mathbf{b} = \langle b_1, b_2\rangle$ is

$$\mathbf{a}\cdot\mathbf{b} = a_1b_1 + a_2b_2.$$

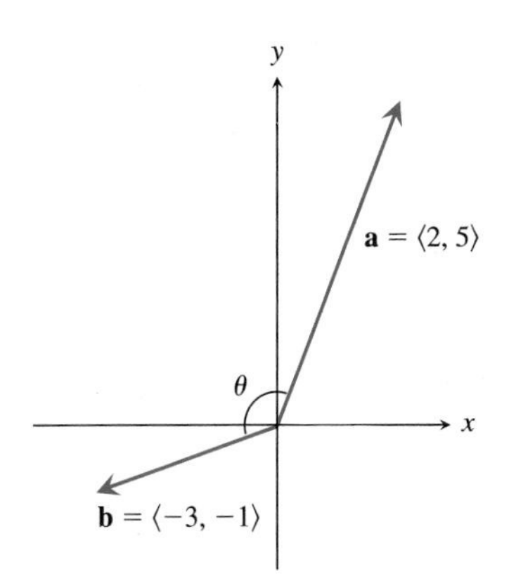

FIGURE 3.46 The vectors $\mathbf{a} = \langle 2, 5\rangle$ and $\mathbf{b} = \langle -3, -1\rangle$ and the angle θ between them.

EXAMPLE 1 Calculate the dot product of vectors $\mathbf{a} = \langle 2, 5\rangle$ and $\mathbf{b} = \langle -3, -1\rangle$ and use equation (1) to calculate the angle between them. See Fig. 3.46.

Solution

The dot product of $\mathbf{a} = \langle 2, 5\rangle$ and $\mathbf{b} = \langle -3, -1\rangle$ is

$$\mathbf{a}\cdot\mathbf{b} = \langle 2, 5\rangle\cdot\langle -3, -1\rangle = (2)(-3) + (5)(-1) = -11.$$

From (1),

$$\cos\theta = \frac{\mathbf{a}\cdot\mathbf{b}}{\|\mathbf{a}\|\,\|\mathbf{b}\|} = \frac{-11}{\sqrt{2^2 + 5^2}\sqrt{(-3)^2 + (-1)^2}}$$
$$= -\frac{11}{\sqrt{29}\sqrt{10}} \approx -0.645942.$$

Recalling that the angles between vectors lie between 0 and π, which is the range of the arccosine function,

$$\theta \approx \arccos(-0.645942) \approx 2.27305 \text{ or about } 130°.$$

Following is a list of the main properties of the dot product. Properties 5–7 show how the dot product relates to the operations of vector sum and scalar product.

Properties of the Dot Product

Let **a** and **b** be vectors and s a scalar.

1. Geometric interpretation of the dot product:
$$\mathbf{a}\cdot\mathbf{b} = \|\mathbf{a}\|\|\mathbf{b}\|\cos\theta, \tag{2}$$
where θ is the angle between **a** and **b**. Note that $0 \le \theta \le \pi$.
2. Assuming that **a** and **b** are nonzero, then **a** and **b** are perpendicular if and only if $\mathbf{a}\cdot\mathbf{b} = 0$.
3. If $\mathbf{a} = \langle 0, 0\rangle$, then $\mathbf{a}\cdot\mathbf{a} = 0$; otherwise, $\mathbf{a}\cdot\mathbf{a} > 0$.
4. The length $\|\mathbf{a}\|$ of a vector **a** is
$$\|\mathbf{a}\| = \sqrt{\mathbf{a}\cdot\mathbf{a}}.$$
5. The dot product is commutative:
$$\mathbf{a}\cdot\mathbf{b} = \mathbf{b}\cdot\mathbf{a}.$$
6. The dot product is distributive over addition and subtraction:
$$\mathbf{a}\cdot(\mathbf{b}\pm\mathbf{c}) = \mathbf{a}\cdot\mathbf{b}\pm\mathbf{a}\cdot\mathbf{c}.$$
7. Scalars can be factored from dot products:
$$(s\mathbf{a})\cdot\mathbf{b} = s(\mathbf{a}\cdot\mathbf{b}).$$

Property (1) is just equation (1).

For property (2), if **a** and **b** are perpendicular, then $\theta = \pi/2$ and, from (2), $\mathbf{a}\cdot\mathbf{b} = 0$. If, on the other hand, $\mathbf{a}\cdot\mathbf{b} = 0$, then, from (2), $\|\mathbf{a}\|\|\mathbf{b}\|\cos\theta = 0$. Because $\|\mathbf{a}\| \neq 0$ and $\|\mathbf{b}\| \neq 0$, it must be true that $\cos\theta = 0$ and, hence, $\theta = \pi/2$.

Properties (3)–(7) are easy to prove and are left to Exercises 48–52.

EXAMPLE 2 Determine which pairs of the vectors
$$\mathbf{a} = \langle 2, 1\rangle, \mathbf{b} = \langle -2, -3\rangle, \mathbf{c} = \langle 3, -2\rangle, \quad \text{and} \quad \mathbf{d} = \langle -1, 2\rangle$$
are perpendicular. Also, find a unit vector perpendicular to $\langle 3, 4\rangle$.

Solution

We use property (2). For this we calculate the dot product of each pair of the four vectors. For example,
$$\mathbf{a}\cdot\mathbf{b} = 2(-2) + 1(-3) = -7 \quad \text{and} \quad \mathbf{b}\cdot\mathbf{c} = (-2)(3) + (-3)(-2) = 0.$$
Because the dot products of **a** with **b**, **c**, and **d** are -7, 4, and 0, the vector **a** is perpendicular to **d** and none of the others; similarly, **b** is perpendicular to **c** and none of the others; and **c** is not perpendicular to **d**. Hence, the only pairs of perpendicular vectors are $\{\mathbf{a}, \mathbf{d}\}$ and $\{\mathbf{b}, \mathbf{c}\}$.

A vector perpendicular to $\langle 3, 4\rangle$ is $\langle -4, 3\rangle$, because the dot product of these two vectors is 0. Dividing $\langle -4, 3\rangle$ by its length will give a unit vector perpendicular to $\langle 3, 4\rangle$, that is,
$$\frac{1}{\sqrt{(-4)^2 + 3^2}}\langle -4, 3\rangle = \frac{1}{5}\langle -4, 3\rangle = \left\langle -\frac{4}{5}, \frac{3}{5}\right\rangle.$$

Resolving Vectors

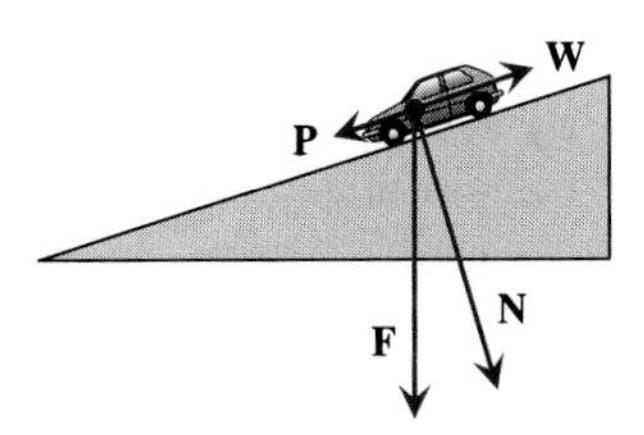

FIGURE 3.47 Resolving the gravitational force **F** on a car to decide whether the car will slide down the hill.

A car is parked on an icy hill with its brakes locked. Will it slide down the hill or remain in place? This question can be answered by analyzing the forces acting on the car. These forces are illustrated in Fig. 3.47. The force **F** is the gravitational force on the car; it acts directly downward. In Fig. 3.47 we have drawn two other vectors, **P** and **N**, one parallel to the hill and the other normal (perpendicular) to the hill. We determined **P** and **N** by the parallelogram law so that $\mathbf{F} = \mathbf{P} + \mathbf{N}$. The force **P** is the component of **F** parallel to the hill and tends to make the car slide down hill. The car is also subject to the force **W** of static friction, which acts in the direction opposite **P**. It is known that $\|\mathbf{W}\| \leq \mu\|\mathbf{N}\|$, where μ is the coefficient of static friction of the tires on ice. If $\|\mathbf{P}\| > \|\mathbf{W}\|$, the car will slide down hill. We use dot products to determine **P** and **N**.

We solve the icy hill problem in the next example, after working out formulas for **P** and **N**. Figure 3.48(a) shows a vector **F** and a unit vector **u**. We show how **F** can be written as a sum of vectors **P** and **N**, where **P** and **N** are parallel and perpendicular to **u**, respectively. The figure shows two cases: In (a) the angle θ between **F** and **u** is less than 90°; in (b) θ is more than 90°. We start with (a).

In the right triangle with sides **F**, **P**, and **N**, the lengths of **P** and **N** are $\|\mathbf{F}\| \cos \theta$ and $\|\mathbf{F}\| \sin \theta$. Hence, $\mathbf{P} = \big(\|\mathbf{F}\| \cos \theta\big)\mathbf{u}$. Because from (2)

$$\mathbf{u} \cdot \mathbf{F} = \|\mathbf{u}\|\,\|\mathbf{F}\| \cos \theta = \|\mathbf{F}\| \cos \theta,$$

it follows that

$$\mathbf{P} = (\mathbf{u} \cdot \mathbf{F})\mathbf{u}.$$

In case (b), $\cos(180° - \theta) = -\cos\theta$ and $\sin(180° - \theta) = \sin\theta$ and the *lengths* of **P** and **N** are $-\|\mathbf{F}\| \cos \theta$ and $\|\mathbf{F}\| \sin \theta$. Hence, $\mathbf{P} = -(-\|\mathbf{F}\| \cos \theta)\mathbf{u}$ and, again from (2),

$$\mathbf{P} = (\mathbf{u} \cdot \mathbf{F})\mathbf{u}.$$

In either case, $\mathbf{N} = \mathbf{F} - \mathbf{P}$. It follows from Fig. 3.48 that **N** is perpendicular to **P**. See Exercise 44 for a nongeometric proof. We summarize these results.

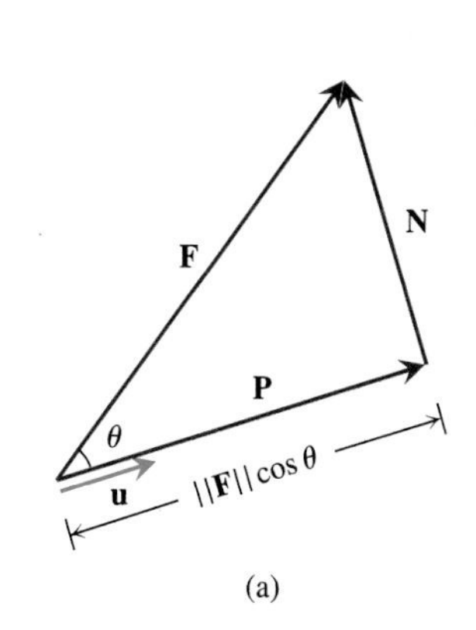

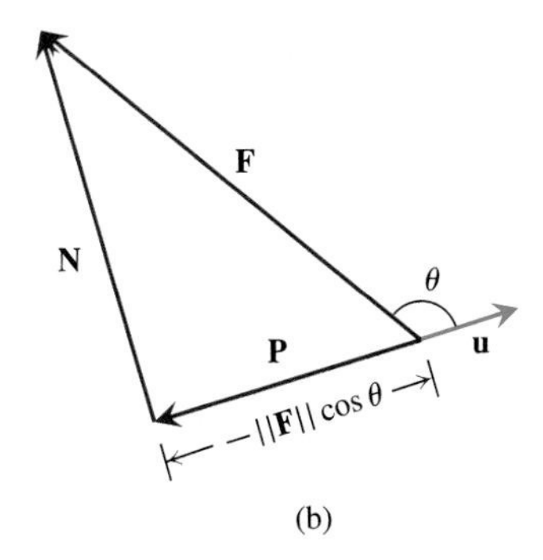

FIGURE 3.48 Resolving **F** into forces parallel and perpendicular to the unit vector **u**.

Resolving a Vector in Directions Parallel and Perpendicular to a Given Unit Vector

Let **u** be a unit vector and **F** any vector.

The vector

$$\mathbf{P} = (\mathbf{F} \cdot \mathbf{u})\mathbf{u} \tag{3}$$

is called the *projection* of **F** on **u**.

The vector $\mathbf{N} = \mathbf{F} - \mathbf{P}$ is perpendicular to the vector **u**. Hence, **F** is the sum of vectors **P** and **N** parallel and perpendicular, respectively, to the vector **u**.

Letting θ be the angle between vectors **u** and **F**, the lengths of the vectors **P** and **N** are

$$\begin{aligned} \|\mathbf{P}\| &= |\mathbf{F} \cdot \mathbf{u}| = \|\mathbf{F}\|\,|\cos \theta| \\ \|\mathbf{N}\| &= \|\mathbf{F}\| \sin \theta. \end{aligned} \tag{4}$$

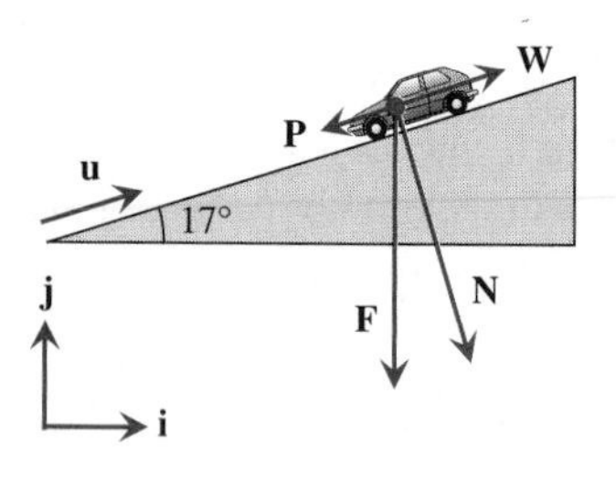

FIGURE 3.49 Resolving the gravitational force **F** on a car to determine if it will slide down the hill.

EXAMPLE 3 Figure 3.49 shows a car on an icy hill. The car is at rest and the brakes are locked. The gravitational force **F** on the car acts directly downward. If the angle of the hill is 17° and the coefficient of static friction μ between tires and ice is 0.14, show that the car will slide down the hill. What is the least angle for which it is certain that the car will slide?

Solution

We resolve **F** into the sum of forces **P** and **N** parallel and perpendicular to the hill. See Fig. 3.49. As stated earlier, the frictional force **W** acts uphill and its magnitude $\|\mathbf{W}\|$ does not exceed $\mu\|\mathbf{N}\|$. Thus, the car will slide down the hill if $\|\mathbf{P}\| > \mu\|\mathbf{N}\| \geq \|\mathbf{W}\|$.

If we orient the unit vectors **i** and **j** as in the figure, then $\mathbf{F} = -c\mathbf{j}$, where c is a positive constant, and $\mathbf{u} = (\cos 17°)\mathbf{i} + (\sin 17°)\mathbf{j}$ is a unit vector in the direction of the hill. Applying (4), where the angle θ between **F** and **u** is $\theta = 17° + 90°$,

$$\|\mathbf{P}\| = \|\mathbf{F}\|\,|\cos\theta| = c|\cos(107°)| \approx 0.29c$$
$$\|\mathbf{N}\| = \|\mathbf{F}\|\sin\theta = c\sin(107°) \approx 0.96c.$$

Because

$$\|\mathbf{P}\| \approx 0.29c$$

and

$$\mu\|\mathbf{N}\| \approx 0.14(0.96c) \approx 0.13c,$$

the frictional force cannot be enough to prevent the car from sliding.

To determine the least angle α for which sliding is certain, we assume that $\|\mathbf{W}\| = \mu\|\mathbf{N}\|$. If the car is to slide, then $\|\mathbf{P}\| > \mu\|\mathbf{N}\|$. By replacing 107° by $\alpha + 90°$ in the previous argument,

$$\|\mathbf{P}\| = c|\cos(\alpha + 90°)| = c\sin\alpha$$
$$\mu\|\mathbf{N}\| = 0.14c\sin(\alpha + 90°) = 0.14c\cos\alpha.$$

Hence, the least angle α for which $\|\mathbf{P}\| > \mu\|\mathbf{N}\|$ satisfies $c\sin\alpha > 0.14c\cos\alpha$. Hence, $\alpha > \arctan(0.14) \approx 8.0°$.

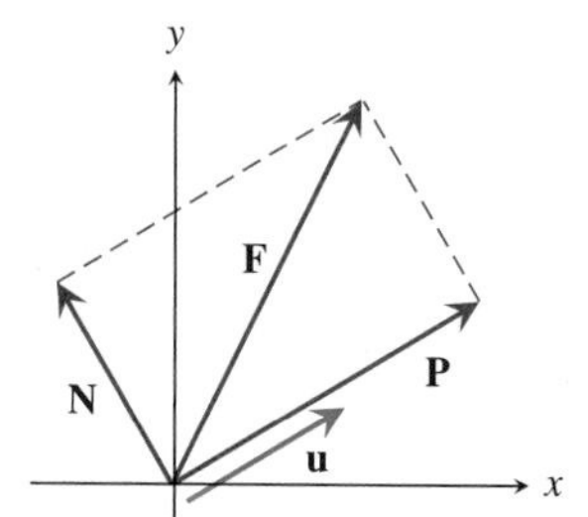

FIGURE 3.50 Writing $\mathbf{F} = \langle 1, 2\rangle$ as the sum of vectors **P** and **N** parallel and perpendicular to the unit vector **u**.

EXAMPLE 4 Write $\mathbf{F} = \langle 1, 2\rangle$ as the sum of vectors parallel and perpendicular to the unit vector $\mathbf{u} = \langle\sqrt{3}/2, 1/2\rangle$. This statement is often shortened to "Resolve **F** into vectors parallel and perpendicular to...."

Solution

Figure 3.50 shows the vectors **F** and **u**. From (3),

$$\mathbf{P} = (\mathbf{F}\cdot\mathbf{u})\mathbf{u} = \left(\langle 1, 2\rangle\cdot\langle\sqrt{3}/2, 1/2\rangle\right)\mathbf{u}$$
$$= \left(\sqrt{3}/2 + 1\right)\mathbf{u} \approx \langle 1.6160, 0.9330\rangle.$$

The vector **N** is

$$\mathbf{N} = \mathbf{F} - \mathbf{P} = \langle 1, 2\rangle - \left(\sqrt{3}/2 + 1\right)\mathbf{u} \approx \langle -0.6160, 1.0670\rangle.$$

EXAMPLE 5 If you have driven a recreational vehicle (RV), you probably know that a tailwind (a wind in the direction of travel) can decrease fuel costs and a strong crosswind (a wind perpendicular to the direction of travel) can overturn the vehicle. The velocity vector **v** of an RV traveling on a curve C is shown in Fig. 3.51(a). At a certain point of C, the RV is heading 40° north of east. There is a steady 40 mph wind blowing in the direction 30° south of east. Calculate the tailwind vector **t**, the crosswind vector **c**, and the speeds of these winds. See the zoom-view in Fig. 3.51(b).

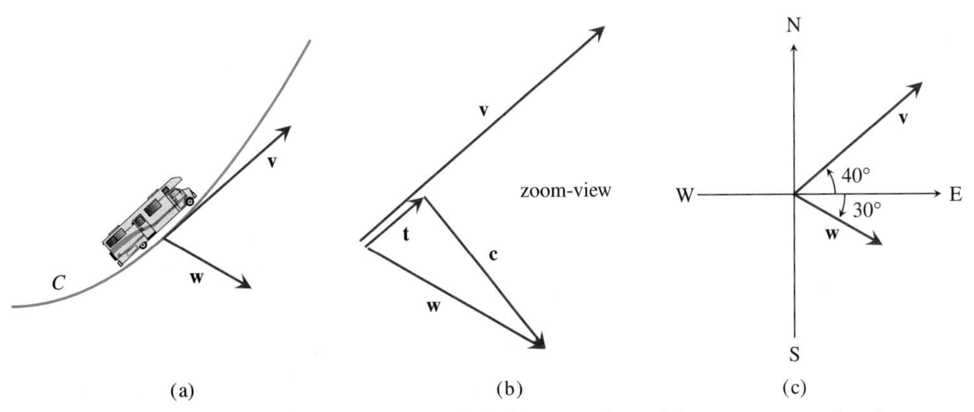

FIGURE 3.51 (a) Resolving a wind blowing on an RV; (b) zoom-view of the vectors **v**, **w**, **t**, and **c**; (c) interpreting "40° north of east" and "30° south of east."

Solution

The tailwind is the projection of the wind vector **w** on a unit vector **u** pointing in the same direction as **v**. Relative to the (x, y)-coordinate system suggested in Fig. 3.51(c),

$$\mathbf{u} = \langle \cos 40°, \sin 40° \rangle$$
$$\mathbf{w} = 40\langle \cos 330°, \sin 330° \rangle.$$

From (3),

$$\begin{aligned} \mathbf{t} &= (\mathbf{w} \cdot \mathbf{u})\mathbf{u} \\ &= (40\langle \cos 330°, \sin 330° \rangle \cdot \langle \cos 40°, \sin 40° \rangle)\mathbf{u} \qquad (5) \\ &\approx 13.7\mathbf{u} \approx \langle 10.5, 8.8 \rangle. \end{aligned}$$

Because $\mathbf{w} = \mathbf{t} + \mathbf{c}$, the crosswind **c** is

$$\begin{aligned} \mathbf{c} = \mathbf{w} - \mathbf{t} &= 40\langle \cos 330°, \sin 330° \rangle - \mathbf{t} \\ &\approx \langle 24.2, -28.8 \rangle. \end{aligned} \qquad (6)$$

The speeds $\|\mathbf{t}\|$ and $\|\mathbf{c}\|$ of the tailwind and crosswind can be obtained from these results. From (5), without calculation,

$$\|\mathbf{t}\| \approx 13.7 \text{ mph}.$$

From (6),

$$\|\mathbf{c}\| \approx 37.6 \text{ mph}.$$

Work

The last topic of this section is the work done by a constant force acting on an object moving along a line. We do more with work in Chapter 4. Figure 3.52(a) shows a person pulling a suitcase up a ramp. If she applies a constant force **F** to the suitcase as it moves a distance s up the ramp, the work W done by the force on the suitcase is the product of the magnitude of the projection of **F** in the direction of motion and the distance s. Hence, the work done is the dot product of the force **F** and the displacement vector $s\mathbf{u}$, where **u** is a unit vector in the direction of motion, that is,

$$W = \big(\|\mathbf{F}\| \cos \theta\big)s = s\big(\|\mathbf{F}\|\|\mathbf{u}\| \cos \theta\big) = s(\mathbf{F}\cdot\mathbf{u}) = \mathbf{F}\cdot(s\mathbf{u}), \tag{7}$$

where θ is the angle between the force and the direction of motion. See Fig. 3.52(b). If the distance is measured in meters and the force in newtons, then work is measured in newton-meters. The work done by a force of 1 N over a distance of 1 meter is 1 newton-meter, also called a joule (J).

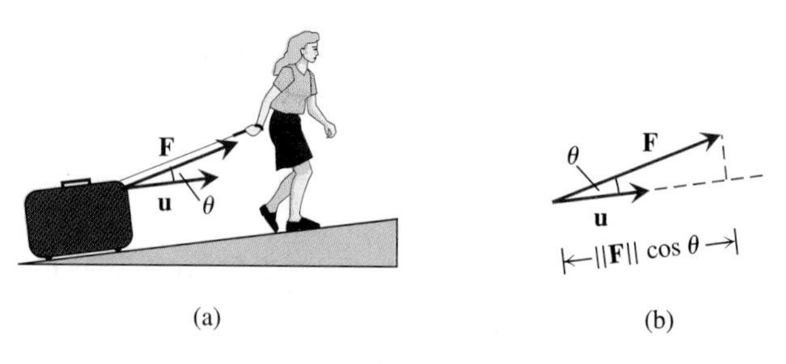

FIGURE 3.52 (a) Force **F** acting on a suitcase; (b) vectors **F** and **u** and the length of the projection of **F** in the direction of motion.

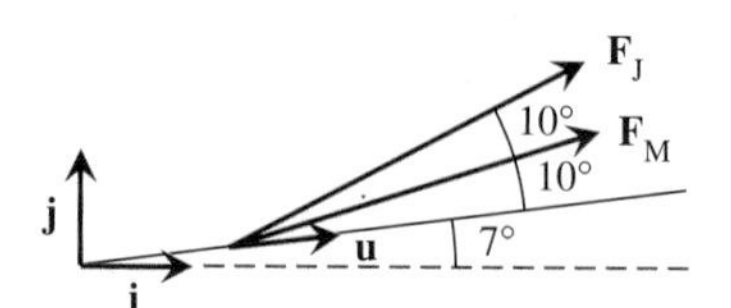

FIGURE 3.53 Forces $\mathbf{F}_J$ and $\mathbf{F}_M$ acting on a suitcase.

EXAMPLE 6 Joe and Mary each pulls a suitcase 10 meters up a 7° ramp. Joe, who is taller, pulls with a force $\mathbf{F}_J$, with magnitude 70 N and at an angle of 20° to the ramp; Mary pulls with a force $\mathbf{F}_M$, with magnitude 68 N and at an angle of 10° to the ramp. See Fig. 3.53. *Note:* The lengths of $\mathbf{F}_J$ and $\mathbf{F}_M$ shown in the figure are not properly scaled relative to the unit vector **u** in the direction of motion. Compare the amounts of work done by Joe and Mary.

Solution

Let **u** be a unit vector pointing in the direction of motion. We use (7) to calculate the work done on the suitcase by Joe and by Mary. Letting W_J and W_M denote the work done by Joe and Mary,

$$W_J = 10\|\mathbf{F}_J\| \cos 20^\circ = 10\cdot 70\cdot \cos 20^\circ \approx 660 \text{ J}$$

$$W_M = 10\|\mathbf{F}_M\| \cos 10^\circ = 10\cdot 68\cdot \cos 10^\circ \approx 670 \text{ J}.$$

Although Mary pulled with less force, she did more work on the suitcase.

Exercises 3.5

Exercises 1–8: Find the dot product of the given vectors **a** *and* **b** *and the angle between them, in radians or degrees, as marked.*

1. $\langle 2,9\rangle, \langle 3,-2\rangle$; degrees
2. $\langle -13,4\rangle, \langle 2,5\rangle$; degrees
3. $\langle 1,1\rangle, \langle 3,-2\rangle$; radians
4. $\langle 2,-8\rangle, \langle 3,-2\rangle$; radians
5. $\langle 1,m\rangle, \langle -m,1\rangle$; radians
6. $\langle p,q\rangle, \langle -q,p\rangle$; radians
7. $3\langle \cos 12°, \sin 12°\rangle, 2.5\langle \cos 87°, \sin 87°\rangle$; degrees
8. $3\langle \cos 110°, \sin 110°\rangle, 2.5\langle \cos 333°, \sin 333°\rangle$; degrees

Exercises 9–12: Determine which pairs of the vectors **a**, **b**, *and* **c** *are perpendicular. Also, find a unit vector perpendicular to the vector* **d**.

9. $\mathbf{a} = \langle 2.3, 4.7\rangle$, $\mathbf{b} = \langle -7.05, 3.45\rangle$, $\mathbf{c} = \langle 4.7, 2.3\rangle$; $\mathbf{d} = \langle 1.5, -2.5\rangle$
10. $\mathbf{a} = \langle -2.88, -1.68\rangle$, $\mathbf{b} = \langle 1.68, 2.88\rangle$, $\mathbf{c} = \langle -2.1, 3.6\rangle$; $\mathbf{d} = \langle -1.5, 3.2\rangle$
11. $\mathbf{a} = 3.08\mathbf{i} + 5.17\mathbf{j}$, $\mathbf{b} = 1.88\mathbf{i} - 1.08\mathbf{j}$, $\mathbf{c} = -0.30\mathbf{i} + 0.19\mathbf{j}$; $\mathbf{d} = 3.2\mathbf{i} + 5.7\mathbf{j}$
12. $\mathbf{a} = -0.75\mathbf{i} + 1.50\mathbf{j}$, $\mathbf{b} = -2.11\mathbf{i} - 1.22\mathbf{j}$, $\mathbf{c} = 1.10\mathbf{i} + 0.55\mathbf{j}$; $\mathbf{d} = -2.55\mathbf{i} - 6.04\mathbf{j}$

Exercises 13–16: Verify the equality of the two sides, given that $\mathbf{a} = -2\mathbf{i} + 5\mathbf{j}$, $\mathbf{b} = 3\mathbf{i} + \mathbf{j}$, $\mathbf{c} = 3\mathbf{i} - 2\mathbf{j}$, *and* $s = -2$.

13. $\mathbf{a}\cdot\mathbf{b} = \mathbf{b}\cdot\mathbf{a}$
14. $\mathbf{a}\cdot(\mathbf{b}+\mathbf{c}) = \mathbf{a}\cdot\mathbf{b} + \mathbf{a}\cdot\mathbf{c}$
15. $(s\mathbf{a})\cdot\mathbf{b} = s(\mathbf{a}\cdot\mathbf{b})$
16. $\|\mathbf{a}\| = \sqrt{\mathbf{a}\cdot\mathbf{a}}$

Exercises 17–24: Find the length of the projection of the vector **a** *on a unit vector in the direction of the vector* **b** *(which may or may not be a unit vector). Sketch.*

17. $\mathbf{a} = \langle 3,2\rangle$, $\mathbf{b} = \langle 1/\sqrt{2}, 1/\sqrt{2}\rangle$
18. $\mathbf{a} = \langle 2,1/2\rangle$, $\mathbf{b} = \langle \sqrt{3}/2, 1/2\rangle$
19. $\mathbf{a} = \mathbf{i} + 5\mathbf{j}$, $\mathbf{b} = 5\mathbf{i} + \mathbf{j}$
20. $\mathbf{a} = 2\mathbf{i} + 7\mathbf{j}$, $\mathbf{b} = 10\mathbf{i} + 3\mathbf{j}$
21. $\mathbf{a} = \langle -4.5, 9.0\rangle$, $\mathbf{b} = \langle -1,-1\rangle$
22. $\mathbf{a} = \langle -10.7, 1.4\rangle$, $\mathbf{b} = \langle -1,1\rangle$
23. $\mathbf{a} = \langle 2,k\rangle$, $\mathbf{b} = \langle -1/\sqrt{2}, 1/\sqrt{2}\rangle$; for the sketch take $k = 3$
24. $\mathbf{a} = \langle h,-2\rangle$, $\mathbf{b} = \langle -\sqrt{3}/2, 1/2\rangle$; for the sketch take $h = 3$
25. An RV is headed directly south and there is a wind from the northwest blowing at 28 mph. Calculate the tailwind and crosswind vectors.
26. An RV is headed 10° east of south, and there is a wind from the northeast blowing at 35 mph. Calculate the tailwind and crosswind vectors.
27. A bicyclist is riding southeast. If the wind is blowing at 20 mph from 5° south of east, what are her headwind and crosswind vectors?
28. A motorcyclist is riding 10° south of due west and encounters winds blowing out of the south at 40 mph. What are his headwind and crosswind vectors?
29. A heavy sledge is resting on a 5° icy hill. The force of the Earth acting on the object is vertical and has magnitude 2000 N. Assuming that the coefficient of static friction between the sledge and the hill is $\mu = 0.20$ and some loggers are attempting to pull the sledge downhill with a force of 250 N, will they move the sledge?
30. A heavy sledge is resting on a 4° icy hill. The force of the Earth acting on the object is vertical and has magnitude 1800 N. Assuming that the coefficient of static friction between the sledge and the hill is $\mu = 0.25$ and some loggers are attempting to pull the sledge downhill with a force of 350 N, will they move the sledge?
31. An object is displaced 15 m up a 10° ramp by a force of 500 N acting at a 45° angle relative to the ramp. Calculate the work done by the force.
32. An object is displaced 7.5 m up a 15° ramp by a force of 750 N acting at a 23° angle relative to the ramp. Calculate the work done by the force.
33. Compare the amounts of work done by forces $\mathbf{F}_1$ and $\mathbf{F}_2$ acting on two objects as they move 50 meters up a 15° hill, where $\mathbf{F}_1$ and $\mathbf{F}_2$ have respective magnitudes 380 N and 550 N and make angles of 24° and 49° with the hill.
34. Compare the amounts of work done by forces $\mathbf{F}_1$ and $\mathbf{F}_2$ acting on two objects as they move 100 meters up a 10° hill, where $\mathbf{F}_1$ and $\mathbf{F}_2$ have respective magnitudes 265 N and 380 N and make angles of 27° and 51° with the hill.
35. Determine a unit vector $\langle x,y\rangle$ perpendicular to $\langle 1,2\rangle$.

36. Determine a unit vector $\langle x, y \rangle$ perpendicular to $\langle -4, 3 \rangle$.

37. A line has slope m. Determine a vector parallel to this line.

38. A line has slope m. Determine a vector perpendicular to this line.

39. Find the angles (in radians) of the triangle with vertices $A = (2, 1)$, $B = (5, 2)$, and $C = (3, 4)$.

40. Find the angles (in degrees) of the triangle with vertices $A = (-1, 1)$, $B = (5, 3)$, and $C = (-3, 4)$.

41. Find t such that the vectors $\langle 9, -3 \rangle$ and $\langle 2t - 3, 4 \rangle$ are perpendicular.

42. Find all values of t for which the vectors $\langle t, -3 \rangle$ and $\langle t + 1, 4 \rangle$ are perpendicular.

43. Use the idea of the length of the projection of one vector in the direction of another vector in calculating the distance from the line $\mathbf{r} = \langle 3, 4 \rangle + t\langle 1, 1 \rangle$ to the point with position vector $\langle 10, 1 \rangle$.

44. Let $\mathbf{F}$ be any nonzero vector and $\mathbf{u}$ a unit vector. If $\mathbf{P}$ and $\mathbf{N}$ are defined by

$$\mathbf{P} = (\mathbf{F} \cdot \mathbf{u})\mathbf{u} \quad \text{and} \quad \mathbf{N} = \mathbf{F} - \mathbf{P},$$

show that (i) $\mathbf{F} = \mathbf{P} + \mathbf{N}$ and (ii) $\mathbf{P}$ and $\mathbf{N}$ are perpendicular. *Hint:* (i) is true by definition of $\mathbf{N}$; for (ii), calculate $\mathbf{P} \cdot \mathbf{N}$.

45. The Earth pulls with a vertical force of 140,000 N on a truck parked on a road inclined at 8° to the horizontal. Ignoring friction, what force must be produced by the brakes to resist the pull of the Earth?

46. At what angle do the sine and cosine curves cross?

47. At what angle does the cosine curve intersect the graph of the equation $y = x$?

48. Show that if $\mathbf{a} = \mathbf{0}$ is the zero vector, then $\mathbf{a} \cdot \mathbf{a} = 0$; otherwise $\mathbf{a} \cdot \mathbf{a} > 0$.

49. Show that for all vectors $\mathbf{a}$,

$$\|\mathbf{a}\| = \sqrt{\mathbf{a} \cdot \mathbf{a}}.$$

50. Show that the dot product is commutative.

51. Show that the dot product is distributive over vector addition.

52. Show that scalars can be factored from dot products.

53. Let $\mathbf{a}$ and $\mathbf{b}$ be unit vectors with directions α and β. By using the dot product to calculate the angle between $\mathbf{a}$ and $\mathbf{b}$, show that

$$\cos(\alpha - \beta) = \cos\alpha \cos\beta + \sin\alpha \sin\beta.$$

54. Assume that $\mathbf{e}_1$ and $\mathbf{e}_2$ are perpendicular unit vectors. The coordinates of a vector $\mathbf{w} = \langle x, y \rangle$ relative to the pair $\mathbf{e}_1$ and $\mathbf{e}_2$ are the numbers x' and y' satisfying the equation

$$\mathbf{w} = x'\mathbf{e}_1 + y'\mathbf{e}_2.$$

Show that $x' = \mathbf{w} \cdot \mathbf{e}_1$ and $y' = \mathbf{w} \cdot \mathbf{e}_2$. Also show that $\|\mathbf{w}\| = \sqrt{x'^2 + y'^2}$. *Hint:* Calculate $\mathbf{w} \cdot \mathbf{e}_1$.

55. Referring to the preceding exercise, let

$$\mathbf{e}_1 = \langle \cos\theta, \sin\theta \rangle$$
$$\mathbf{e}_2 = \langle \cos(\theta + \pi/2), \sin(\theta + \pi/2) \rangle.$$

Show that these vectors are perpendicular unit vectors. Show that the coordinates x' and y' of $\mathbf{w} = \langle x, y \rangle$ relative to the pair $\mathbf{e}_1$ and $\mathbf{e}_2$ are

$$x' = x\cos\theta + y\sin\theta$$
$$y' = -x\sin\theta + y\cos\theta.$$

These two equations are often used to calculate the coordinates (x', y') of a point (x, y) relative to an (x', y')-system obtained from the (x, y)-system by rotating it through an angle θ. Explain this viewpoint.

56. Derive the identity

$$\|\mathbf{a} + \mathbf{b}\|^2 + \|\mathbf{a} - \mathbf{b}\|^2 = 2\|\mathbf{a}\|^2 + 2\|\mathbf{b}\|^2.$$

Interpret this identity geometrically.

57. We showed geometrically that

$$\mathbf{a} \cdot \mathbf{b} = \|\mathbf{a}\|\,\|\mathbf{b}\| \cos\theta$$

for all vectors $\mathbf{a}$ and $\mathbf{b}$. Because $\cos\theta \le 1$ for all θ, it follows that

$$|\mathbf{a} \cdot \mathbf{b}| \le \|\mathbf{a}\|\,\|\mathbf{b}\|,$$

for all vectors $\mathbf{a}$ and $\mathbf{b}$. This result is called the Cauchy-Schwarz inequality. Show that the triangle inequality,

$$\|\mathbf{a} + \mathbf{b}\| \le \|\mathbf{a}\| + \|\mathbf{b}\|,$$

for all vectors $\mathbf{a}$ and $\mathbf{b}$, follows from the Cauchy-Schwarz inequality.

58. The following question was posted on an Internet newsgroup by a graphics software designer. One base of a cylinder is centered at $\mathbf{r}_0 = \langle x_0, y_0, z_0 \rangle$, its height is h, the radius of its base is a, and the axis of the cylinder is in the direction of the unit vector $\mathbf{u}$, drawn from $\mathbf{r}_0$. Given any point $\mathbf{r} = \langle x, y, z \rangle$ in space, how can you determine if $\mathbf{r}$ is on or within the cylinder? If you wish, solve a two-dimensional version of this problem: One side of a rectangle has length a and the center point of this side has position vector $\mathbf{r}_0 = \langle x_0, y_0 \rangle$; the vector from the point with position vector $\mathbf{r}_0$ to the midpoint of the opposite side of the rectangle is $b\mathbf{u}$, where $\mathbf{u}$ is a unit vector. Given the position vector $\mathbf{r}$ of any point (x, y) of the plane, how can you determine if (x, y) is on or within the rectangle?

3.6 Newton's Laws

Mechanics, which is the study of the motions of material bodies, includes both kinematics and dynamics. In kinematics, the path or trajectory of an object is described without regard to the causes of that motion. We discussed kinematics in Sections 3.1–3.4, though much of our effort went toward introducing the ideas from mathematics needed to describe motion. In this section we briefly discuss dynamics.

Isaac Newton (1642–1727) initiated his great works in mechanics, mathematics, and optics just after he graduated from Trinity College of Cambridge University with a Bachelor of Arts degree in 1665. He worked at his family home in Woolsthorpe, to which he had returned because the plague had caused the university to "disperse." Newton later said of his discoveries during this period, "All this was in the two plague years of 1665 and 1666, for in those days I was in the prime of my age for invention, and minded mathematics and [science] more than at any other time since."

In this section we discuss several applications of Newton's second law and his universal gravitation law.

Newton's Laws of Motion and Universal Gravitation

Newton gave his laws of motion in his *Philosophiae Naturalis Principia Mathematica*, first published in 1687.

Newton's Laws of Motion and Universal Gravitation

First law: If a body is at rest or moving at a constant speed along a line, it will remain at rest or keep moving on the line at constant speed unless it is acted upon by a force.

Second law: The rate at which a body's momentum changes is equal to the net force acting on the body.

Third law: The actions of two bodies upon each other are always equal and directly opposite.

Universal gravitation: Any two particles in the universe exert attractive forces on each other proportional to the product of their masses and inversely proportional to the square of the distance between them.

Newton expressed his laws in terms of *mass*, *momentum*, and *force*. We rely upon our intuitive notions of mass and force. The momentum $\mathbf{p}$ of a body is defined as the product of its mass m and velocity $\mathbf{v}$:

$$\mathbf{p} = m\mathbf{v}.$$

If the *net force* acting on a body is $\mathbf{F}$, Newton's second law becomes

$$\mathbf{F} = \frac{d}{dt}\mathbf{p} = \frac{d}{dt}(m\mathbf{v}). \tag{1}$$

If the mass m of the body is constant, Newton's second law can be written as

$$\mathbf{F} = \frac{d}{dt}(m\mathbf{v}) = m\frac{d}{dt}\mathbf{v} = m\mathbf{a}, \tag{2}$$

where the *acceleration* **a** of the body is defined by

$$\mathbf{a} = \frac{d}{dt}\mathbf{v}.$$

This definition extends our earlier definition of *acceleration* in one dimension as the rate of change of velocity. We may think of acceleration as the limit as $\Delta t \to 0$ of the average acceleration

$$\frac{\Delta \mathbf{v}}{\Delta t} = \frac{\mathbf{v}(t + \Delta t) - \mathbf{v}(t)}{\Delta t}$$

on a time interval between t and $t + \Delta t$.

Equation (2) is the starting point for much of dynamics. If we know the net force **F** acting on a body, then we know its acceleration $\mathbf{a} = (1/m)\mathbf{F}$. If the expression for **a** is not too complicated, we can infer from the equation $\mathbf{a} = (1/m)\mathbf{F}$ the velocity vector **v** of the body. And then, because $\mathbf{v} = d\mathbf{r}/dt$, we often can infer the position vector **r** from **v**. After a brief preview later in this section, we discuss this "anti-differentiation" operation in Chapter 5.

Following a brief discussion of units and universal gravitation, we consider three simple examples of forces acting on a body. We consider forces leading to motion along a line, a circle, or a parabola.

Units

We use SI units almost entirely. The abbreviation SI stands for Système International d'Unités (International System of Units). Length is measured in meters (m), mass in kilograms (kg), and time in seconds (s). Units for other quantities are derived from these units. Velocity and speed are measured in meters per second (m/s). Acceleration is measured in meters per second per second ($\mathrm{m/s^2}$). The relation (2) may be used to measure and name a unit of force. The units of force are $\mathrm{kg \cdot m/s^2}$. In honor of Isaac Newton, the force required to accelerate a mass of 1 kg to 1 $\mathrm{m/s^2}$ is called a newton (N). One newton is approximately one-quarter pound.

Universal Gravitation

Newton hypothesized a universal gravitation law to account for the motion of the sun, moon, and planets. Specifically, Newton assumed that the magnitude of the force exerted by either of two particles on the other is proportional to the product of their masses, inversely proportional to the square of the distance between them, and directed along the line joining the particles. Referring to Fig. 3.54, if the particles have masses m and M and are separated by a distance r, the force **F** exerted by the particle of mass M on the particle of mass m is

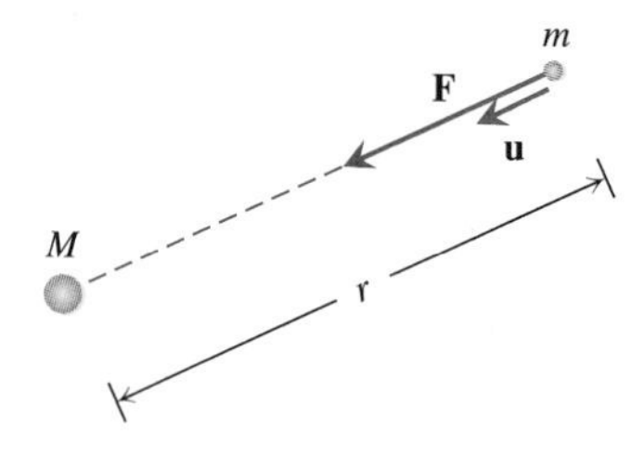

FIGURE 3.54 Universal gravitation.

$$\mathbf{F} = \frac{GmM}{r^2}\mathbf{u}, \tag{3}$$

where $\mathbf{u}$ is a unit vector directed from the particle of mass m toward the particle of mass M. The proportionality constant G is called the *universal gravitational constant*. Its experimentally determined value is 6.67259×10^{-11}. The units of G are $\mathrm{N} \cdot \mathrm{m}^2/\mathrm{kg}^2$.

In applying the universal gravitation law to the sun, moon, or planets, Newton simplified his work by showing that the force exerted on an object by a spherical body S composed of many particles is the same as that exerted on the object by a single particle located at the center of S and having the same mass as S. For example, each atom of the Earth exerts a force on a raindrop. Newton showed that the combined effect of all of these forces is the same as that exerted on the raindrop by a single particle located at the center of the Earth and having the same mass as the Earth. We use this result in deriving a formula for the Earth's gravitational force on a small object near the Earth's surface.

Earth's Gravitational Force

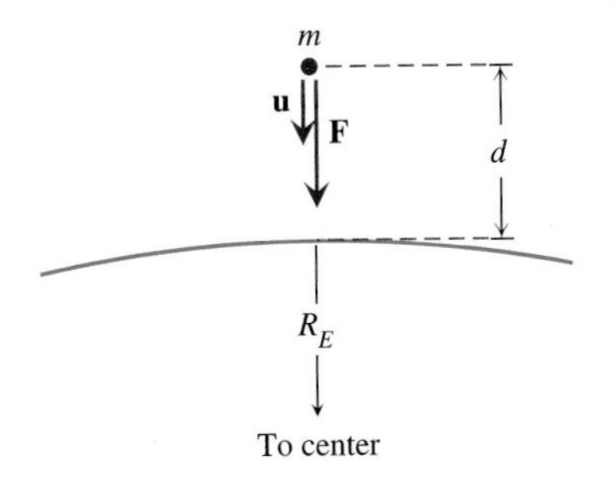

FIGURE 3.55 Force on a small, nearby object.

Figure 3.55 shows an object of mass m at a distance d from the Earth's surface, where m and d are small compared, respectively, to the mass M_E of the Earth and its radius R_E. We assume that the only force acting on this object is the gravitational force. Applying the universal gravitation law to particles of masses m and M_E and separated by a distance $R_E + d$, the force exerted by the Earth on the particle of mass m is

$$\mathbf{F} = \frac{GmM_E}{(R_E + d)^2}\mathbf{u} \approx \frac{GM_E}{R_E^2}m\mathbf{u} = gm\mathbf{u}. \tag{4}$$

In this equation, $\mathbf{u}$ is a unit vector directed from the small object toward the center of the Earth and g denotes the constant GM_E/R_E^2. For small, nearby objects, the replacement of $R_E + d$ by R_E is inconsequential, and we shall use $\mathbf{F} = gm\mathbf{u}$ without further comment.

The accepted value of the constant g, called the *acceleration due to gravity,* is $g = 9.80665 \mathrm{~m/s^2}$. We usually use the value $g = 9.8 \mathrm{~m/s^2}$. Generally accepted values of M_E and R_E are $M_E \approx 5.97 \times 10^{24}$ kg and $R_E \approx 6.378 \times 10^6$ m.

Combining Newton's second law $\mathbf{F} = m\mathbf{a}$ and the equation $\mathbf{F} = gm\mathbf{u}$ for small, nearby objects,

$$m\mathbf{a} = mg\mathbf{u}.$$

Hence

$$\mathbf{a} = g\mathbf{u}. \tag{5}$$

From this result, which is independent of the mass of the object, we note that the object is accelerated toward the center of the Earth, and the magnitude of the acceleration is $g = 9.8 \mathrm{~m/s^2}$.

We use (5) to derive the formula

$$h = h(t) = 1000 - 4.9t^2, \qquad t \geq 0, \tag{6}$$

which was used in the skydiver example in Section 3.1.

INVESTIGATION

Skydiving and the Second Law

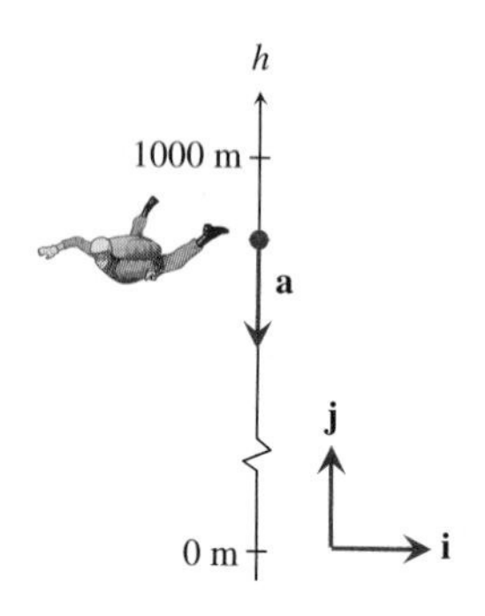

FIGURE 3.56 With no air resistance, the only force acting on the skydiver is the force of gravity; from (5), her acceleration at any time t is $\mathbf{a} = -9.8\mathbf{j}$.

A skydiver falls from a height of 1000 meters above the Earth's surface. If she falls from rest and air resistance is ignored, we show that her height $h = h(t)$ at time t is given by (6).

Figure 3.56 shows the skydiver falling in a coordinate system with positive direction upward and origin at the surface of the Earth. If we assume that the skydiver is a small, nearby object, we can apply (5). Letting $\mathbf{r} = \mathbf{r}(t)$ be the position vector of the skydiver at any time t, her velocity vector is $\mathbf{v} = d\mathbf{r}/dt$ and her acceleration vector is $\mathbf{a} = d\mathbf{v}/dt$. The condition that she falls from rest is

$$\mathbf{v}(0) = \mathbf{0}. \tag{7}$$

With our choice of coordinate system and her given initial position,

$$\mathbf{r}(0) = 1000\mathbf{j}. \tag{8}$$

Conditions (7) and (8) are called *initial conditions*. Because the acceleration is vertically downward, the unit vector $\mathbf{u}$ in (5) is $-\mathbf{j}$. Rewriting (5) by replacing $\mathbf{u}$ by $-\mathbf{j}$ and $\mathbf{a}$ by $d\mathbf{v}/dt$,

$$\frac{d\mathbf{v}}{dt} = 9.8(-\mathbf{j}) = -9.8\mathbf{j}. \tag{9}$$

From this equation we conjecture that the velocity vector $\mathbf{v}$ has the form

$$\mathbf{v}(t) = -9.8t\mathbf{j} + \mathbf{c}_1,$$

where $\mathbf{c}_1$ is a constant vector. Why is this reasonable? The best reason we can give at the moment is that the derivative of $-9.8t\mathbf{j} + \mathbf{c}_1$ is indeed $-9.8\mathbf{j}$, as required by (9). We choose $\mathbf{c}_1$ to fit the initial condition (7). From

$$\mathbf{v}(0) = \mathbf{0} = (-9.8 \cdot 0)\mathbf{j} + \mathbf{c}_1,$$

we find that $\mathbf{c}_1 = \mathbf{0}$ and hence,

$$\frac{d\mathbf{r}}{dt} = \mathbf{v} = -9.8t\mathbf{j}. \tag{10}$$

From (10) it is reasonable to suppose that

$$\mathbf{r}(t) = -\frac{1}{2}9.8t^2\mathbf{j} + \mathbf{c}_2, \tag{11}$$

where $\mathbf{c}_2$ is a constant vector. As before, note that $d\mathbf{r}/dt = \mathbf{v}$, as required by (10). Evaluating equation (11) at $t = 0$,

$$\mathbf{r}(0) = -\frac{1}{2}9.8 \cdot 0^2\mathbf{j} + \mathbf{c}_2 = \mathbf{c}_2.$$

Putting this together with the initial condition (8), we conclude that $\mathbf{c}_2 = 1000\mathbf{j}$. Hence,

$$\mathbf{r}(t) = -4.9t^2\mathbf{j} + 1000\mathbf{j} = (1000 - 4.9t^2)\mathbf{j}.$$

This is equation (6) in vector form. Roughly, we inferred this equation by "undifferentiating" equation (9) to find $\mathbf{v}$ as a function of t; we then undifferentiated equation (10) to find $\mathbf{r}$ as a function of t. In practice, the term *antidifferentiation* is used, not *undifferentiation*.

Most problems in dynamics start with an object on which a known force acts. By applying Newton's second law, we can then write the acceleration of the object as a function of t. From the resulting equation we can then antidifferentiate and apply the initial conditions to infer the velocity and position functions. Problems in which we are asked to determine the velocity and displacement of an object whose acceleration satisfies equations like (5) or (2) and initial conditions like (7) and (8) are called *initial value problems.* In the remainder of this section we solve several initial value problems.

Motion along a Line

Although the motion of real objects takes place in space, it is often possible to model the motion of particular objects in one or two dimensions. If, for example, the only force on a planet is the force exerted by the sun, the motion of the planet can be modeled in two dimensions, in what turns out to be the plane of the orbit. The motion of a subatomic particle in a *linear* accelerator can be modeled in one dimension.

EXAMPLE 1 The mass of a fully loaded passenger jet is 3.600×10^5 kg and its engines provide a force of magnitude 7.700×10^5 N. Assuming that this force is the net force on the jet (thereby ignoring air resistance and runway friction) and the jet starts from rest, what is the minimum length of runway needed for it to reach its takeoff speed of 86 m/s?

Solution

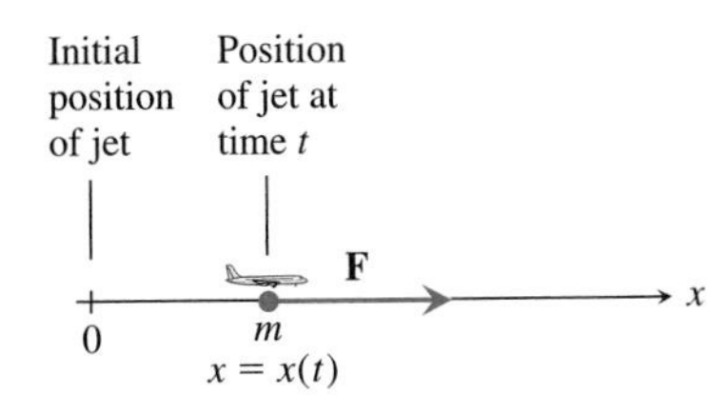

FIGURE 3.57 With given mass and force, what is the minimum runway length required for the jet to reach 86 m/s?

As shown in Fig. 3.57, we choose a coordinate line oriented in the direction of motion and with the origin placed at the jet's initial position. We let $x = x(t)$ be the (unknown) position of the jet at any time t before takeoff, $\mathbf{F}$ the net force acting on the jet, and m its mass. We know that $x(0) = 0$, by our choice of coordinate system, and that $v(0) = x'(0) = 0$, because the jet starts from rest.

Although Newton's second law for constant mass—equation (2)—is not applicable here because jet fuel is being burned, we make the simplifying assumption that the mass of the jet is essentially constant because the mass of the fuel burned during takeoff is relatively small. Applying (2),

$$\mathbf{F} = m\mathbf{a} = m\frac{d\mathbf{v}}{dt}.$$

Because the motion is one dimensional, we may use scalar quantities. Dropping the boldface type,

$$m\frac{dv}{dt} = F$$

or, setting $F/m = k$,

$$\frac{dv}{dt} = k.$$

As we assumed in the Investigation, the function $kt + c_1$, where c_1 is a constant, is the only function whose derivative with respect to t is k. Hence,

$$v = kt + c_1.$$

Because at $t = 0$ the velocity v is 0,

$$0 = k \cdot 0 + c_1.$$

Hence, $c_1 = 0$ and

$$v = \frac{dx}{dt} = kt. \tag{12}$$

Using similar reasoning,

$$x = \frac{1}{2}kt^2 + c_2, \quad \text{where } c_2 \text{ is a constant.}$$

By applying the initial condition $x(0) = 0$, we see that

$$0 = \frac{1}{2}k \cdot 0^2 + c_2. \tag{13}$$

Hence, $c_2 = 0$ and

$$x = \frac{1}{2}kt^2. \tag{14}$$

From (12) we find the time $t = 86/k$ at which $v = 86$ m/s. Substituting this value of t into (14) and replacing k by F/m,

$$x = \frac{1}{2}kt^2 = \frac{1}{2}k\left(\frac{86}{k}\right)^2 = \frac{86^2 m}{2F} = \frac{(86^2)(3.600 \times 10^5)}{(2)(7.700 \times 10^5)} \approx 1729 \text{ meters.}$$

Hence, the runway must be at least 1,729 m long for the jet to reach its takeoff speed of 86 m/s.

Projectile Motion

In the Investigation we used the equation

$$\mathbf{a} = g\mathbf{u} \tag{15}$$

to derive the formula

$$x = x(t) = 1000 - 4.9t^2, \qquad t \geq 0,$$

for the position of a skydiver. We took $\mathbf{u} = -\mathbf{j}$. We used the initial conditions $\mathbf{v}(0) = \mathbf{0}$ and $\mathbf{r}(0) = 1000\mathbf{j}$.

We use (15) in calculating the path, or trajectory, of a rifle bullet. Choosing this model, for either the skydiver or bullet, means that we are ignoring the force due to air resistance in both of these motions. We do this for simplicity.

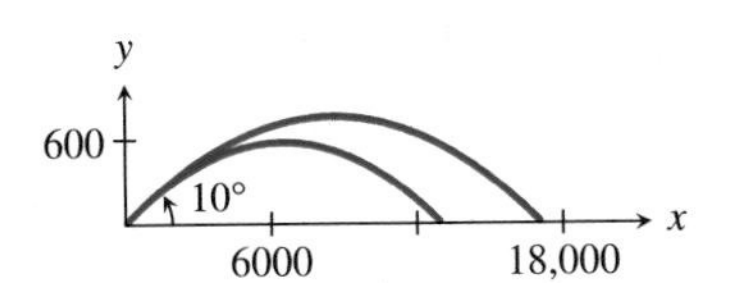

FIGURE 3.58 Trajectories of infantry rifles made in 1900 and 1943.

EXAMPLE 2 In the year 1900, typical infantry rifles fired bullets with muzzle speeds of 610 m/s. By 1943, muzzle speeds had increased to 700 m/s. Calculate the ranges of bullets fired from each of these types of rifles, assuming that both were fired at an angle of 10° from the Earth's surface.

Solution

We put the origin of a rectangular coordinate system at the muzzle of the rifles. The paths of the two test bullets are shown in Fig. 3.58. In this figure, the 10° angle at the origin is exaggerated because of the scaling on the axes.

Letting $\mathbf{r} = \mathbf{r}(t)$, $\mathbf{v} = \mathbf{v}(t)$, and $\mathbf{a} = \mathbf{a}(t)$ denote the position, velocity, and acceleration vectors at any time t of either of the bullets, the initial conditions are

$$\mathbf{r}(0) = \mathbf{0} = 0\mathbf{i} + 0\mathbf{j} \quad \text{and} \quad \mathbf{v}(0) = (b\cos 10°)\mathbf{i} + (b\sin 10°)\mathbf{j}, \tag{16}$$

where $b = 610$ or 700. The velocity vector was formed by multiplying the unit vector $(\cos 10°)\mathbf{i} + (\sin 10°)\mathbf{j}$ in the given direction by a given muzzle speed.

We calculate the position vectors of the two bullets by solving an initial value problem. From (15),

$$\mathbf{a} = \frac{d\mathbf{v}}{dt} = g(-\mathbf{j}) = -g\mathbf{j}.$$

The function $\mathbf{v}$ must have the form

$$\mathbf{v}(t) = -gt\mathbf{j} + \mathbf{c}_1, \tag{17}$$

where $\mathbf{c}_1$ is a constant vector. Applying the initial condition for velocity given in (16),

$$\mathbf{v}(0) = -g \cdot 0\mathbf{j} + \mathbf{c}_1 = (b\cos 10°)\mathbf{i} + (b\sin 10°)\mathbf{j}.$$

Hence,

$$\mathbf{c}_1 = (b\cos 10°)\mathbf{i} + (b\sin 10°)\mathbf{j}.$$

Substituting this value into (17) and replacing $\mathbf{v}$ by $d\mathbf{r}/dt$,

$$\frac{d\mathbf{r}}{dt} = -gt\mathbf{j} + (b\cos 10°)\mathbf{i} + (b\sin 10°)\mathbf{j}.$$

Because $b\cos 10°$ and $b\sin 10°$ are constants, we see from this equation that $\mathbf{r}$ must have the form

$$\mathbf{r}(t) = -\frac{1}{2}gt^2\mathbf{j} + (bt\cos 10°)\mathbf{i} + (bt\sin 10°)\mathbf{j} + \mathbf{c}_2,$$

where $\mathbf{c}_2$ is a constant vector. Because $\mathbf{r}(0) = 0\mathbf{i} + 0\mathbf{j}$ for either bullet,

$$\mathbf{r}(0) = \mathbf{0} = 0\mathbf{i} + 0\mathbf{j} = -\frac{1}{2}g \cdot 0^2\mathbf{j} + (b \cdot 0\cos 10°)\mathbf{i} + (b \cdot 0\sin 10°)\mathbf{j} + \mathbf{c}_2 = \mathbf{c}_2.$$

Hence,

$$\mathbf{c}_2 = 0\mathbf{i} + 0\mathbf{j} = \mathbf{0}.$$

It follows that the position vector of either bullet is

$$\begin{aligned}\mathbf{r} &= -\frac{1}{2}gt^2\mathbf{j} + (bt\cos 10°)\mathbf{i} + (bt\sin 10°)\mathbf{j}\\ &= (bt\cos 10°)\mathbf{i} + \left(-\frac{1}{2}gt^2 + bt\sin 10°\right)\mathbf{j}.\end{aligned}$$

To find the ranges of the two bullets, we calculate the times at which they return to the x-axis by setting their **j**-coordinates equal to zero:

$$-\frac{1}{2}gt^2 + bt\sin 10° = 0.$$

One root of this equation is $t = 0$, which is the time the bullets left the x-axis. The nonzero root is

$$t = \frac{2 \cdot b\sin 10°}{g}.$$

Letting $g = 9.8$, $b = 610$, and then $b = 700$, the times t_1 and t_2 at which the two bullets hit the x-axis are

$$t_1 = \frac{2 \cdot 610\sin 10°}{9.8} = 21.6174\ldots \text{ seconds for the 1900 model}$$

$$t_2 = \frac{2 \cdot 700\sin 10°}{9.8} = 24.8068\ldots \text{ seconds for the 1943 model.}$$

From these values of t_1 and t_2, we may calculate the ranges of the two bullets:

$$\begin{aligned}\mathbf{r}(t_1) &= (610\cos 10°)(21.6174\ldots)\mathbf{i}\\ &\approx 13{,}000\mathbf{i} \text{ meters}\end{aligned}$$

and, similarly,

$$\mathbf{r}(t_2) \approx 17{,}000\mathbf{i} \text{ meters.}$$

We rounded the values of $\mathbf{r}(t_1)$ and $\mathbf{r}(t_2)$ to two significant figures to fit the similar accuracy of g and b. If you doubt that a rifle bullet can have a range of 13 kilometers, or about 8 miles, recall that we ignored air resistance.

Uniform Circular Motion

Many, many objects move on circular paths and with constant angular speed. Such motion is called *uniform circular motion*. Most objects on Earth move with uniform circular motion about the north-south axis of the Earth; many communications satellites move with (nearly) uniform circular motion about the Earth; wheels, gears, and the tips of saw blades usually move with uniform circular motion; and the orbits of Earth, Venus, and Neptune are nearly circular and their angular speeds about the sun are nearly constant.

An object is in uniform circular motion if its motion can be modeled by an equation of the form

$$\mathbf{r} = \mathbf{r}(t) = r\langle\cos\omega t, \sin\omega t\rangle, \qquad t \geq 0, \tag{18}$$

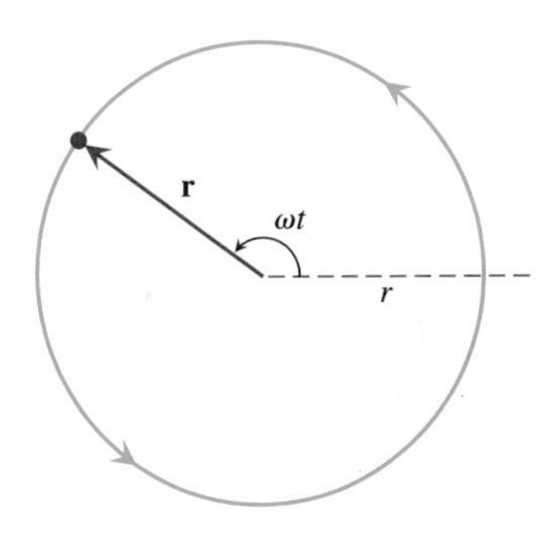

FIGURE 3.59 Object moving counterclockwise with constant angular speed ω on a circle of radius r.

where the constants r and ω are the radius of the circle on which the motion takes place and the angular speed of the object. Figure 3.59 shows a snapshot of an object moving on a circle of radius r. At $t = 0$ the object was at $(r, 0)$ and its angular position was 0 radians. After t seconds its angular position has increased to ωt radians. The rate of change of its angular position is $d(\omega t)/dt = \omega$. The constant ω has units of radians per second.

The velocity $\mathbf{v}$ and acceleration $\mathbf{a}$ of objects in uniform circular motion can be calculated by differentiating their position vectors. From (18),

$$\mathbf{v} = \frac{d\mathbf{r}}{dt} = r\langle -\omega \sin \omega t, \omega \cos \omega t\rangle = r\omega\langle -\sin \omega t, \cos \omega t\rangle \tag{19}$$

and

$$\mathbf{a} = \frac{d\mathbf{v}}{dt} = r\omega\langle -\omega \cos \omega t, -\omega \sin \omega t\rangle = -r\omega^2\langle \cos \omega t, \sin \omega t\rangle. \tag{20}$$

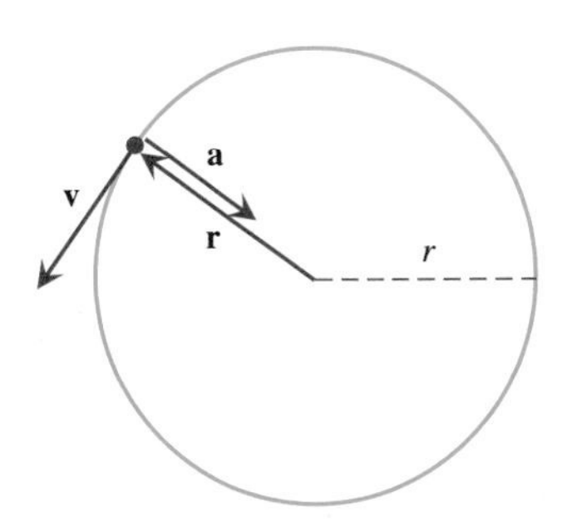

FIGURE 3.60 Velocity and acceleration vectors.

The velocity and acceleration vectors of the object are shown in Fig. 3.60. Note that the direction of $\mathbf{a}$ is opposite that of $\mathbf{r}$. This follows from equations (18) and (20), which show that

$$\mathbf{a} = -\omega^2 \mathbf{r}.$$

A second property of uniform circular motion is that $\mathbf{v}$ is perpendicular to $\mathbf{r}$. Calculating their dot product,

$$\begin{aligned} \mathbf{r} \cdot \mathbf{v} &= r\langle \cos \omega t, \sin \omega t\rangle \cdot r\omega\langle -\sin \omega t, \cos \omega t\rangle \\ &= r^2\omega(-\cos \omega t \sin \omega t + \sin \omega t \cos \omega t) = 0. \end{aligned}$$

Because neither $\mathbf{v}$ nor $\mathbf{r}$ is the zero vector, $\mathbf{v}$ and $\mathbf{r}$ are perpendicular.

From (19) and (20), the speed and the magnitude of the acceleration vector of the object are

$$\begin{aligned} v &= \|\mathbf{v}\| = |r\omega| \cdot 1 = r\omega \\ a &= \|\mathbf{a}\| = |r\omega^2| \cdot 1 = r\omega^2. \end{aligned}$$

From the first of these equations,

$$\omega = \frac{v}{r}. \tag{21}$$

Substituting this into the second equation,

$$a = r\omega^2 = r\frac{v^2}{r^2} = \frac{v^2}{r}. \tag{22}$$

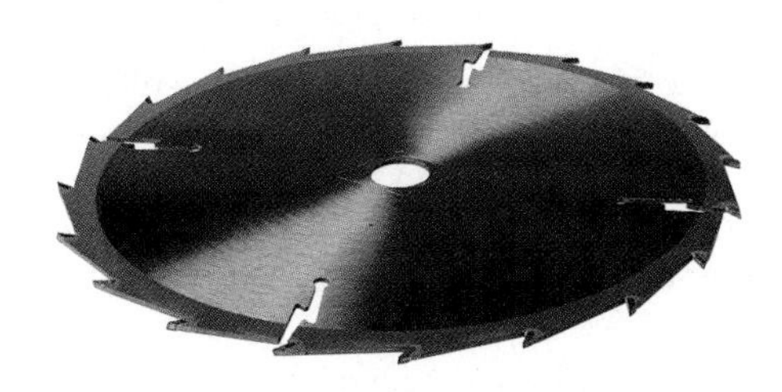

FIGURE 3.61 Carbide cutting tips of a saw blade.

EXAMPLE 3 Figure 3.61 shows part of a circular saw blade with diameter 0.300 m and 20 carbide teeth bonded to the blade. Assume that the bond is safe up to accelerations of 2.00×10^4 m/s^2. If the tips of the blade move at 40 m/s, is this saw blade safe?

Solution

The carbide tips are moving with uniform circular motion. Calculating their acceleration from (22),

$$a = \frac{v^2}{r} = \frac{40^2}{0.150} \approx 10{,}700 \text{ m/s}^2.$$

Because $10{,}700 \text{ m/s}^2 < 2.0 \times 10^4 \text{ m/s}^2 = 20{,}000 \text{ m/s}^2$, the saw blade is safe.

A second application of equation (22) is to the (approximately) uniform circular motion of the moon or other satellite about the Earth. Suppose such a satellite has mass m and the radius of its orbit is r. From Newton's second law and (20), the force $\mathbf{F}$ exerted by the Earth on the satellite is

$$\mathbf{F} = m\mathbf{a} = m(-r\omega^2)\langle \cos \omega t, \sin \omega t \rangle. \tag{23}$$

From this equation and (21), the magnitude F of $\mathbf{F}$ is

$$F = \|\mathbf{F}\| = m(r\omega^2) = m \cdot \frac{v^2}{r} = \frac{mv^2}{r}. \tag{24}$$

We use this equation and the universal gravitation law (3) to relate the speed of the satellite to the radius of its orbit. From (3), $F = GM_E m/r^2$, where M_E is the mass of the Earth. Substituting into (24),

$$\frac{GM_E m}{r^2} = \frac{mv^2}{r}.$$

Hence,

$$v^2 = \frac{GM_E}{r}. \tag{25}$$

Because the satellite travels one circumference per period, its speed v (in meters per second) and period T (in seconds) are related by the equation $v = 2\pi r/T$. Substituting this expression for v into (25),

$$\left(\frac{2\pi r}{T}\right)^2 = \frac{GM_E}{r}.$$

Rearranging this result,

$$r^3 = \frac{GM_E T^2}{4\pi^2}. \tag{26}$$

This equation—which is a version of Kepler's third law (see Exercise 30)—relates the period T and radius r of a satellite's orbit.

EXAMPLE 4 Calculate the altitude of a satellite moving in a circular orbit with a 90-minute period.

Solution

Applying (26), we take $r = R_E + h$, where R_E is the radius of the Earth and h is the altitude of the satellite. Using $M_E \approx 5.97 \times 10^{24}$ kg,

$G \approx 6.67259 \times 10^{-11}\ \text{N}\cdot\text{m}^2/\text{kg}^2$, and $R_E \approx 6.378 \times 10^6$ m in (26),

$$r^3 = \frac{GM_E T^2}{4\pi^2} = \frac{(6.67259 \times 10^{-11})(5.97 \times 10^{24})(90 \cdot 60)^2}{4\pi^2}$$

$$r^3 \approx 2.94237 \times 10^{20}.$$

Replacing r by $(R_E + h)$ and solving for h,

$$h \approx \sqrt[3]{2.94237 \times 10^{20}} - R_E \approx 273 \text{ km}.$$

Hence, if a satellite has a circular orbit about Earth and a 90-minute period, its altitude is about 273 kilometers.

To end the section, we calculate the radius R_M of the moon's orbit about the Earth. Close agreement between this calculation and independently measured values of R_M would be evidence in support of Newton's laws. It may have been this calculation Newton had in mind when he said

> I thereby compared the force requisite to keep the Moon in her Orb with the force of gravity at the surface of the earth, and found them answer pretty nearly.

EXAMPLE 5 Calculate the radius of the moon's orbit, assuming that the moon circles the Earth in $27\frac{1}{3}$ days (this is the *sidereal month*).

Solution

The calculation is essentially the same as in Example 4.

$$R_M{}^3 = \frac{GM_E T^2}{4\pi^2} = \frac{(6.67259 \times 10^{-11})(5.97 \times 10^{24})\left(27\frac{1}{3} \times 24 \times 3600\right)^2}{4\pi^2}$$

$$R_M{}^3 = 5.62758\ldots \times 10^{25}$$

$$R_M = \sqrt[3]{5.62758\ldots \times 10^{25}} \approx 3.83 \times 10^8 \text{ m}.$$

This "answer(s) pretty nearly" to $R_M \approx 3.85 \times 10^8$ m, which is a modern measurement of the mean distance to the moon.

Exercises 3.6

Exercises 1–6: From the given initial height and velocity of a small object near the Earth's surface, determine the time t_1 it hits the Earth and its position $\mathbf{r} = \mathbf{r}(t)$ at any time $t \in [0, t_1]$. Lengths are in meters and time in seconds.

1. $\mathbf{r}(0) = 0\mathbf{j}$ and $\mathbf{v}(0) = 26\mathbf{j}$

2. $\mathbf{r}(0) = 0\mathbf{j}$ and $\mathbf{v}(0) = 24\mathbf{j}$

3. $\mathbf{r}(0) = 50\mathbf{j}$ and $\mathbf{v}(0) = 5\mathbf{j}$

4. $\mathbf{r}(0) = 75\mathbf{j}$ and $\mathbf{v}(0) = 10\mathbf{j}$

5. $\mathbf{r}(0) = 40\mathbf{j}$ and $\mathbf{v}(0) = -5\mathbf{j}$

6. $\mathbf{r}(0) = 60\mathbf{j}$ and $\mathbf{v}(0) = -10\mathbf{j}$

Exercises 7–18: Solve using basic ideas, that is, avoid simply substituting into formulas as much as possible.

7. The mass of a fully loaded passenger jet is 3.600×10^5 kg, and its engines provide a force of 7.700×10^5 N. Assuming that at $t = 0$ the plane has velocity 4 m/s, what is the minimum length of runway needed for it to reach its takeoff speed of 86 m/s?

8. The mass of a fully loaded passenger jet is 3.600×10^5 kg, and its engines provide a force of 6.900×10^5 N. Assuming that at $t = 0$ the plane has velocity 2 m/s, what is the minimum length of runway needed for it to reach its takeoff speed of 86 m/s?

9. Calculate the range of a rifle with muzzle speed 700 m/s, fired with the muzzle 1.5 m above the ground and at an angle of 12° from the horizontal.

10. Calculate the range of a rifle with muzzle speed 610 m/s, fired with the muzzle 1.3 m above the ground and at an angle of 10° from the horizontal.

11. The angular speed of an object moving on a circle with radius 0.5 m is 2000 rpm (revolutions per minute). Find its speed.

12. The speed of an object moving on a circle with radius 0.5 m is 12 m/s. Find its angular speed in radians per second.

13. Assuming a circular orbit, calculate the orbital speed (in km/h) required to maintain a satellite at an altitude of 160 km. What is the period (in minutes) of such a satellite?

14. Assuming a circular orbit, calculate the orbital speed (in km/h) required to maintain a satellite at an altitude of 180 km. What is the period (in minutes) of such a satellite?

15. The speed of the manned USSR spacecraft *Soyuz 10* was 7765 m/s in a nearly circular orbit. What was its altitude in kilometers?

16. Calculate the altitude of a satellite moving in a circular orbit with a period of 89 minutes.

17. A saw blade with diameter 10 inches rotates at 3500 rpm. What is the acceleration (in ft/s^2) on a carbide bit at the cutting edge?

18. A saw blade with diameter 12 inches rotates at 3500 rpm. What is the acceleration (in ft/s^2) on a carbide bit at the cutting edge?

19. The mass of a fully loaded passenger jet is 3.600×10^5 kg, and its engines provide a force of 7.700×10^5 N. Assuming that the plane starts from rest and the runway slopes upward at 2° from the horizontal, what is the minimum length of runway needed for it to reach its takeoff speed of 86 m/s? *Note:* There are two forces acting on the plane.

20. The mass of a fully loaded passenger jet is 3.600×10^5 kg, and its engines provide a force of 7.700×10^5 N. Assuming that the plane starts from rest and the runway slopes downward at 2° from the horizontal, what is the minimum length of runway needed for it to reach its takeoff speed of 86 m/s? *Note:* There are two forces acting on the plane.

21. What muzzle speed would be required for a range of 18,000 m for a rifle fired at an angle of 10° to the horizontal?

22. What muzzle speed would be required for a range of 18,000 m for a rifle fired at an angle of 12° to the horizontal?

23. Calculate the angle of fire for maximum range and calculate this range for a rifle with muzzle speed 700 m/s.

24. Calculate the angle of fire for maximum range and calculate this range for a rifle with muzzle speed 610 m/s.

25. The manned USSR spacecraft *Soyuz 9* moved in an orbit that was approximately circular. Find its altitude if it completed 286 orbits in 424 hours 59 minutes.

26. The manned U.S. spacecraft *Gemini 3* moved in an orbit that was approximately circular. Find its altitude if it completed three orbits in 4 hours 53 minutes.

27. Sketch the position vector, velocity vector, and acceleration vectors at $t = 3$ s for the object whose position vector for any time t is $\mathbf{r} = \mathbf{r}(t) = 2\langle\cos(0.8t), \sin(0.8t)\rangle$. Find the magnitude of the force required to maintain this circular motion if the object has mass 2 kg.

28. Referring to the accompanying figure and the universal gravitation law, the force acting on a small, nearby object of mass m at a distance d from the Earth's surface is

$$\mathbf{F}_1 = \frac{GmM_E}{(R_E + d)^2}\mathbf{u}.$$

This force is commonly approximated by

$$\mathbf{F}_2 = \frac{GmM_E}{R_E^2}\mathbf{u}.$$

Let F_1 and F_2 be the magnitudes of these two forces. If the object has mass 80 kg and is d meters above the Earth's surface, find the least d such that a 1 percent difference between F_1 and F_2 can be observed.

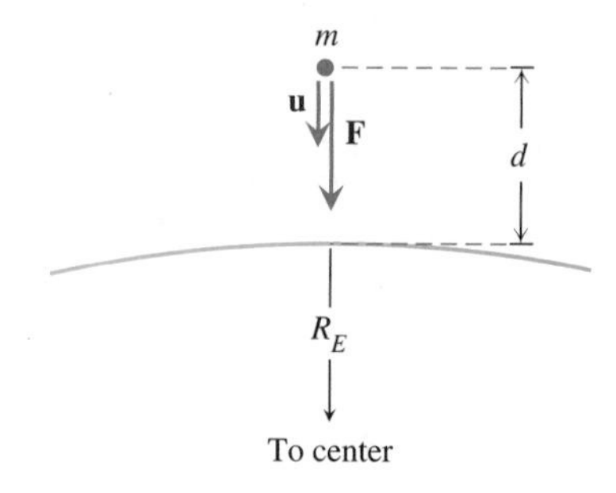

Figure for Exercise 28.

29. A geostationary satellite is one moving so that it stays above a fixed location on the Earth. Assuming that such a satellite has a circular orbit, find its altitude.

30. Kepler discovered three laws of planetary motion. The first is that the planets orbit the sun in ellipses, with the sun at one focus. His third law is that the square of a planet's orbital period T is proportional to the cube of the semimajor axis a of its orbit. Excepting only Mercury and Pluto, the orbits of the planets are very nearly circular. First, verify empirically that Kepler's third law holds. Use the periods and semimajor axes of the orbits of Earth, Venus, Jupiter, and Uranus. Data for the last three are given in the table. The periods are measured in tropical years, the time required for the Earth to complete one orbit about the sun. The semimajor axes are measured in astronomical units (AU), the length of the semimajor axis of Earth's orbit.

	T	a
Venus	0.61521	0.7233316
Jupiter	11.86224	5.202561
Uranus	84.01247	19.21814

Table for Exercise 30.

Does Kepler's third law follow from an equation similar to (26), but with M_E replaced by the mass M_S of the sun? Why or why not?

Review of Key Concepts

This chapter extends the work of Chapter 1 on rates of change into two dimensions. A major goal was to discuss the velocity and acceleration of objects in motion, two important rates of change. To describe the velocity and acceleration of an object moving in two dimensions, we defined vectors and several vector operations. The trajectories or orbits followed by objects in motion were described using parametric equations. We defined the dot product of two vectors and used it to calculate the angle between two vectors, to project a vector on a unit vector, and to calculate the work done by a force acting on an object. We introduced antidifferentiation and initial value problems in connection with Newton's three laws and his universal gravitation law.

Chapter Summary

Equivalent Vectors

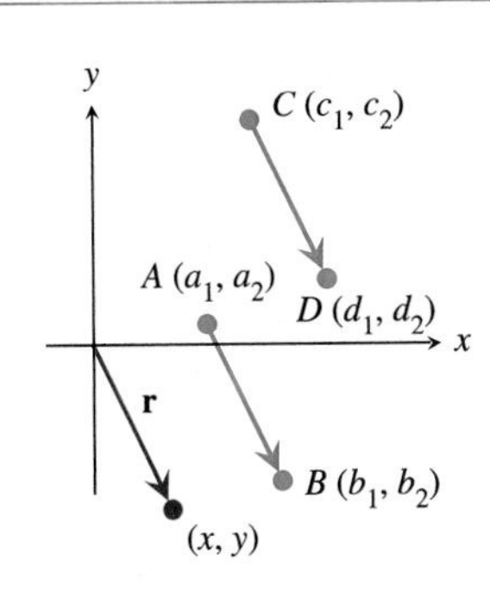

Vectors $\overrightarrow{AB}$, $\overrightarrow{CD}$, and $\mathbf{r}$, where $\mathbf{r} = \langle x, y\rangle$ is a position vector with initial point at the origin, are equivalent if they have the same magnitude and direction. Specifically, vectors $\overrightarrow{AB}$, $\overrightarrow{CD}$, and $\mathbf{r}$ are equivalent if

$$b_1 - a_1 = d_1 - c_1 = x$$

and

$$b_2 - a_2 = d_2 - c_2 = y.$$

If A, B, C, and D are the points

$$(6, 1), (10, -7), (9, 12), \text{and } (13, 4)$$

and $\mathbf{r} = \langle 4, -8\rangle$, then $\overrightarrow{AB}$, $\overrightarrow{CD}$, and $\mathbf{r}$ are equivalent because

$$10 - 6 = 13 - 9 = 4$$

and

$$-7 - 1 = 4 - 12 = -8.$$

Sum and Difference of Vectors

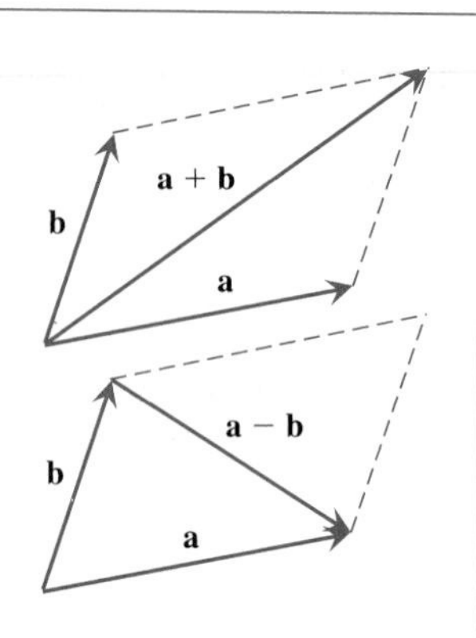

The sum or difference of vectors $\mathbf{a} = \langle a_1, a_2\rangle$ and $\mathbf{b} = \langle b_1, b_2\rangle$ is

$$\mathbf{a} \pm \mathbf{b} = \langle a_1 \pm b_1, a_2 \pm b_2\rangle.$$

For $\mathbf{a} = \langle -5, 2\rangle$ and $\mathbf{b} = \langle 4, 7\rangle$,

$$\mathbf{a} + \mathbf{b} = \langle -1, 9\rangle$$
$$\mathbf{a} - \mathbf{b} = \langle -9, -5\rangle.$$

Product of Scalar and Vector

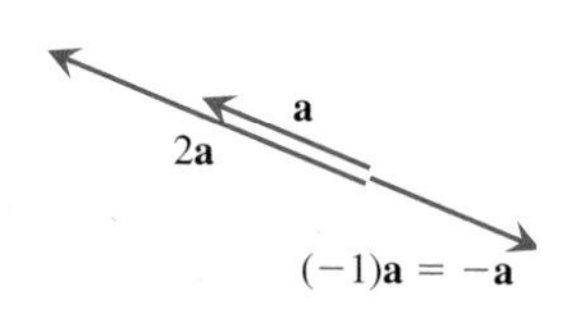

The product of a vector $\mathbf{a} = \langle a_1, a_2\rangle$ and a scalar (real number) s is

$$s\mathbf{a} = s\langle a_1, a_2\rangle = \langle sa_1, sa_2\rangle.$$

For $\mathbf{a} = \langle -5, 2\rangle$ and $s = 2$,

$$s\mathbf{a} = 2\langle -5, 2\rangle = \langle -10, 4\rangle.$$

Length and Direction of a Vector

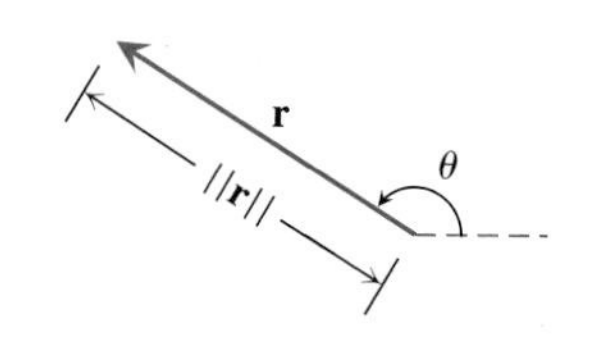

The length of $\mathbf{r} = \langle x, y\rangle$, and of any vector equivalent to $\mathbf{r}$, is

$$\|\mathbf{r}\| = \sqrt{x^2 + y^2}.$$

The direction of $\mathbf{r} = \langle x, y\rangle$, and of any vector equivalent to $\mathbf{r}$, is the angle θ through which the positive x-axis must be rotated counterclockwise to align with $\mathbf{r}$.

For $\mathbf{r} = \langle -9, 5\rangle$,

$$\|\mathbf{r}\| = \sqrt{(-9)^2 + 5^2} = \sqrt{106}.$$

The direction of $\mathbf{r}$ is

$$\theta = \pi - \arctan(5/9) \approx 2.63.$$

Average Velocity and Velocity

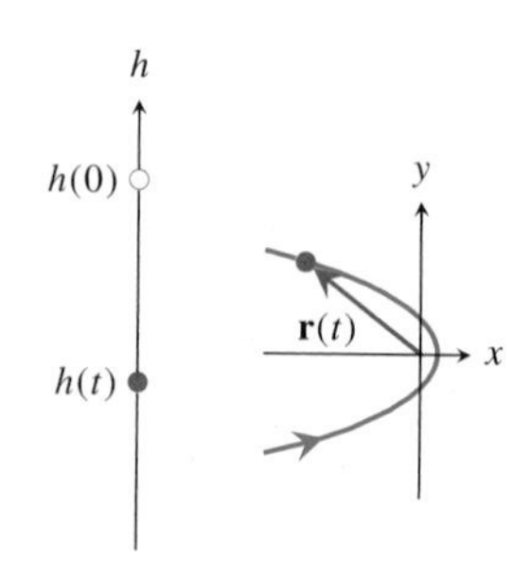

For one-dimensional motion, average velocity on $[t_1, t_2]$ and velocity at t are

$$v(t_1, t_2) = \frac{h(t_2) - h(t_1)}{t_2 - t_1}$$
$$v(t) = \lim_{t_2 \to t} v(t, t_2) = h'(t).$$

For two-dimensional motion, average velocity on $[t_1, t_2]$ and velocity at t are

$$\mathbf{v}(t_1, t_2) = \frac{1}{t_2 - t_1}(\mathbf{r}(t_2) - \mathbf{r}(t_1))$$
$$\mathbf{v}(t) = \lim_{t_2 \to t} \mathbf{v}(t, t_2) = \mathbf{r}'(t).$$

If the position of an object at time t is $\mathbf{r} = (1 - t^2)\mathbf{i} + t\mathbf{j}$, its average velocity on the interval $[0, t_2]$ is

$$\mathbf{v}(0, t_2) = \frac{1}{t_2 - 0}((1 - t_2{}^2)\mathbf{i} + t_2\mathbf{j} - \mathbf{i})$$
$$= -t_2\mathbf{i} + \mathbf{j}.$$

Velocity and Acceleration

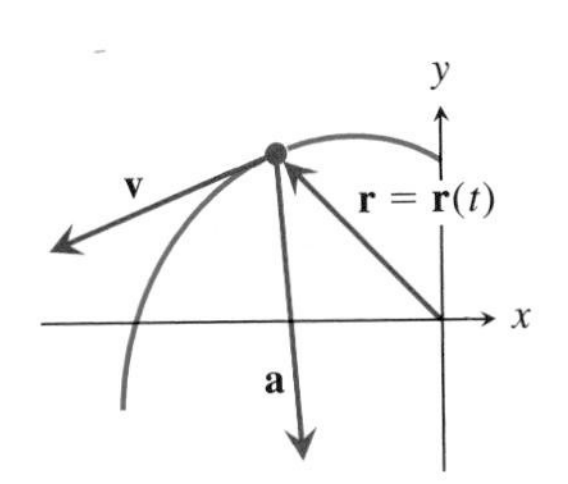

If the position of an object at time t is $\mathbf{r} = \mathbf{r}(t)$, its velocity and acceleration vectors are

$$\mathbf{v} = \frac{d}{dt}\mathbf{r}$$

$$\mathbf{a} = \frac{d}{dt}\mathbf{v}.$$

If the position of an object at time t is $\mathbf{r} = \langle t\cos t, t\sin t\rangle$, its velocity and acceleration vectors at t are

$$\mathbf{v} = \langle \cos t - t\sin t, \sin t + t\cos t\rangle$$

$$\mathbf{a} = \langle -2\sin t - t\cos t, 2\cos t - t\sin t\rangle.$$

Evaluated at $t = 3\pi/4$, $\mathbf{v} \approx \langle -2.4, -1.0\rangle$, and $\mathbf{a} \approx \langle 0.3, -3.1\rangle$..

Parametric Equation of a Line

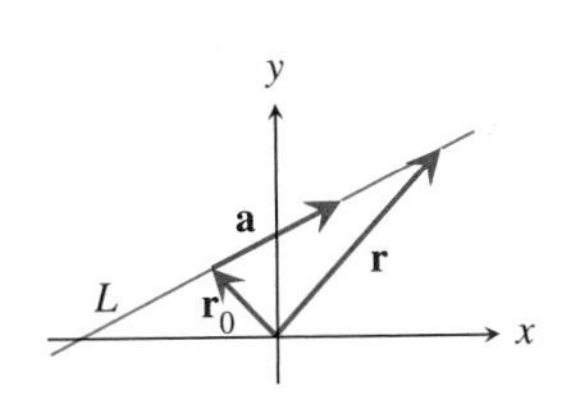

A parametric equation describing the line through a point with position vector $\mathbf{r}_0$ and parallel to the vector $\mathbf{a}$ is

$$\mathbf{r} = \mathbf{r}_0 + s\mathbf{a}, \qquad -\infty < s < \infty.$$

A parametric equation describing the line through the point with position vector $\langle -1, 1\rangle$ and parallel to the vector $\langle \cos 30^\circ, \sin 30^\circ\rangle$ is

$$\mathbf{r} = \langle -1, 1\rangle + s\left\langle \frac{\sqrt{3}}{2}, \frac{1}{2}\right\rangle,$$

$$-\infty < s < \infty.$$

Parametric Equation of a Circle

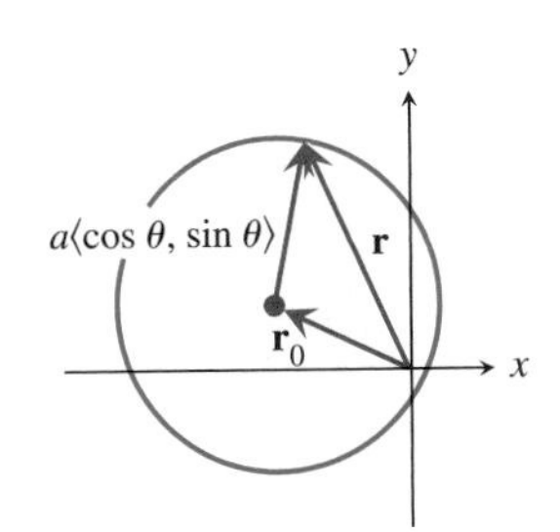

A circle with center $\mathbf{r}_0 = \langle h, k\rangle$ and radius a is described by

$$\mathbf{r} = \mathbf{r}_0 + a\langle \cos\theta, \sin\theta\rangle,$$
$$0 \le \theta \le 2\pi.$$

A parametric equation describing the circle with center $\mathbf{r}_0 = \langle -2, 1\rangle$ and radius 3 is

$$\mathbf{r} = \langle -2, 1\rangle + 3\langle \cos\theta, \sin\theta\rangle,$$
$$0 \le \theta \le 2\pi.$$

Curve, Tangent Vector, Slope of Tangent Line

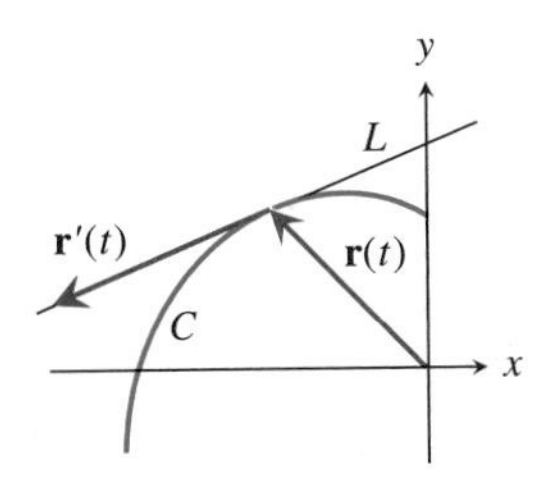

A curve C in an (x, y)-plane is the range of a parametric function

$$\mathbf{r} = \mathbf{r}(t) = \langle x(t), y(t)\rangle, \qquad t \in I,$$

where I is an interval and the functions $x = x(t)$ and $y = y(t)$ are continuous on I. In case the functions x and y are differentiable, a tangent vector to C at $\mathbf{r}(t)$ is

$$\mathbf{r}'(t) = \langle x'(t), y'(t)\rangle.$$

For the curve C described by

$$\mathbf{r}(t) = \langle t\cos t, t\sin t\rangle,$$

the tangent vector and slope at $\mathbf{r}(3\pi/4)$ are

$$\mathbf{r}'(3\pi/4) = \left\langle -\frac{\sqrt{2}}{2} - \frac{3\pi\sqrt{2}}{8}, \frac{\sqrt{2}}{2} - \frac{3\pi\sqrt{2}}{8}\right\rangle$$

Curve, Tangent Vector, Slope of Tangent Line (*continued*)

From the Tangent Vector and Slope Theorem, the slope of the tangent line L to C at $\mathbf{r}(t)$ is

$$\frac{dy}{dx} = \frac{y'(t)}{x'(t)}.$$

$$\frac{dy}{dx} = \left.\frac{dy/dt}{dx/dt}\right|_{t=3\pi/4} = \frac{3\pi - 4}{3\pi + 4} \approx 0.404.$$

Dot Product and Angle between Vectors

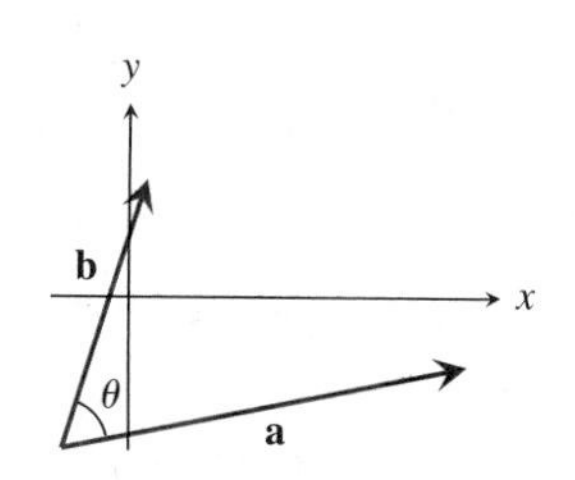

The dot product of vectors $\mathbf{a} = \langle a_1, a_2 \rangle$ and $\mathbf{b} = \langle b_1, b_2 \rangle$ is

$$\mathbf{a} \cdot \mathbf{b} = a_1 b_1 + a_2 b_2.$$

The angle θ between $\mathbf{a}$ and $\mathbf{b}$ can be calculated from

$$\cos \theta = \frac{\mathbf{a} \cdot \mathbf{b}}{\|\mathbf{a}\| \|\mathbf{b}\|}.$$

Nonzero vectors $\mathbf{a}$ and $\mathbf{b}$ are perpendicular if and only if $\mathbf{a} \cdot \mathbf{b} = 0$.

The dot product of the vectors

$$\mathbf{a} = \langle 28.1, 5.4 \rangle \quad \text{and} \quad \mathbf{b} = \langle 6.0, 18.2 \rangle,$$

is $\mathbf{a} \cdot \mathbf{b} = 266.88$. The angle between them is

$$\theta = \arccos\left(\frac{\mathbf{a} \cdot \mathbf{b}}{\|\mathbf{a}\| \|\mathbf{b}\|}\right) \approx 61^\circ.$$

Projection of a Vector on a Unit Vector

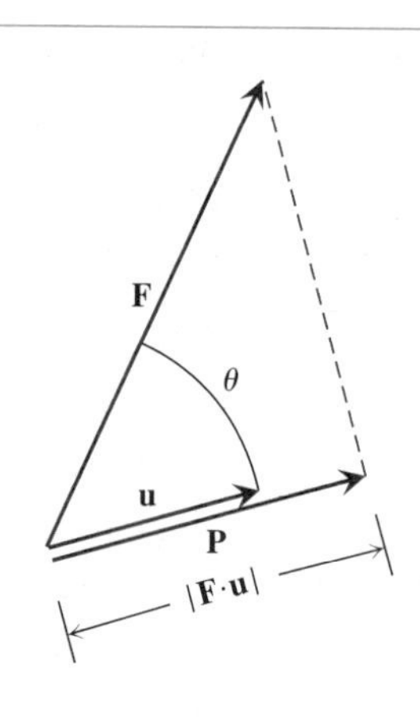

The projection $\mathbf{P}$ of a vector $\mathbf{F}$ onto a unit vector $\mathbf{u}$ is

$$\mathbf{P} = (\mathbf{F} \cdot \mathbf{u})\mathbf{u}.$$

The length of this projection is

$$\|\mathbf{P}\| = |\mathbf{F} \cdot \mathbf{u}| = \|\mathbf{F}\| |\cos \theta|,$$

where θ is the angle between $\mathbf{F}$ and $\mathbf{u}$.

For $\mathbf{F} = \langle 1, 2 \rangle$ and $\mathbf{u} = \langle \cos 15^\circ, \sin 15^\circ \rangle$,

$$\mathbf{P} = (\mathbf{F} \cdot \mathbf{u})\mathbf{u} \approx 1.48\mathbf{u} \approx \langle 1.43, 0.38 \rangle.$$

The length of this projection is

$$\|\mathbf{P}\| \approx 1.48.$$

Chapter Review Exercises

1. Give the definitions of average velocity for objects moving along a line and for objects moving in the plane.

2. An object is moving on the x-axis and its coordinate positions at $t = 36$ s and $t = 39$ s are $x = 35.8$ m and $x = 24.7$ m. What is its average velocity during this 3-second interval?

3. An object is moving in the (x, y)-plane and its position vectors at $t = 5.5$ s and $t = 7.2$ s are $\langle -10, 3 \rangle$ and $\langle 1, 4 \rangle$. What is its average velocity during this 1.7-second interval?

4. The position x of the point of contact between a freight car and a snubber (see Section 3.1, Example 3) is

$$x = x(t) = -0.5e^{-0.25t} \sin(1.5t),$$

for all times between the time $t = 0$ of first contact and the time of last contact. Calculate the velocity of the freight car at $t = 2.0$. What is the time of last contact?

5. A rifle is fired vertically upward so that at $t = 0$ the bullet is 20 m below ground level and has speed 610 m/s. Calculate the position, velocity, and acceleration of the bullet under the assumption of no air resistance. What are the bullet's speeds at 1000 m above ground level on the way up and on the way down?

6. Approximately 2.5 hours ago, your odometer broke. It read 99,999 miles at the time. From a close analysis of available speedometer data, it appears that the true velocity function was very nearly

$$v(t) = -26t^3 + 117t^2 - 156t + 65,$$

for $0 \le t \le 2.5$ h. What would the odometer reading be now had it not broken?

7. An object is moving upward and to the right with speed 10 m/s on the line through $(-20.5, -4.8)$ and $(1.5, 12.3)$. If the object was first noticed 15 s ago as it passed through the point $(-20.5, -4.8)$, where is it now?

8. Two tugboats exert forces

$$\mathbf{F}_1 = 120.5\mathbf{i} + 100.5\mathbf{j} \text{ kN}$$

and

$$\mathbf{F}_2 = 150.3\mathbf{i} - 30.1\mathbf{j} \text{ kN}$$

on a barge. What single force on the tug would just balance the combined force of the two tugs? (A kilonewton (kN) is 1000 newtons.)

9. Calculate the coordinates of the two unit vectors making an angle of 0.5 radians with the vector $\langle 6, 1\rangle$.

10. An object at point P is displaced by the vector $\mathbf{a}_1 = \langle 3, 4\rangle$ and then further displaced by $\langle -10, 7\rangle$, $\langle 2, -8\rangle$, and $\langle 20, 0\rangle$. What is the single equivalent displacement of the object? In what direction and distance from P is the object moved by the four displacements?

11. The hyperbolic functions cosh and sinh are defined by $\cosh(x) = (e^x + e^{-x})/2$ and $\sinh(x) = (e^x - e^{-x})/2$. Let C be the curve described by

$$\mathbf{r} = \mathbf{r}(t) = \langle \cosh t, \sinh t\rangle, \qquad t \ge 0.$$

Sketch C. Note that $\cosh^2 t - \sinh^2 t = 1$ for all t.

12. Sketch the graph of the curve C described by

$$\mathbf{r} = \mathbf{r}(t) = \langle 1 - \cos t, -1 + \sin t\rangle,$$

where $0 \le t \le 2\pi$. Find the coordinates of all points on C at which the slope is $2/3$.

13. Particle A moves on the path described by $\mathbf{r}(s) = \langle 2s, -5 + s\rangle$, $s \ge 0$, while particle B moves on the path described by $\mathbf{r} = \langle t + t^2, t - t^2\rangle$, $t \ge 0$. Do the particles collide if both s and t are read from the same clock?

14. Calculate the angle between the lines with equations

$$\mathbf{r} = \langle 2, 1\rangle + t\langle 3, 4\rangle$$

and

$$\mathbf{r} = \langle -1, 5\rangle + t\langle -3, 1\rangle.$$

15. Show that $\mathbf{a} + \mathbf{b}$ is perpendicular to $\mathbf{a} - \mathbf{b}$ if $\mathbf{a}$ and $\mathbf{b}$ are unit vectors and $\mathbf{a} \ne \pm\mathbf{b}$.

16. Find parametric equations describing the two lines tangent to the graph of the equation $y = x^3$ and passing through the point $(1, 1)$. *Hint:* Choose t so that the line described by

$$\mathbf{r}(s) = \langle t, t^3\rangle + s\langle 1, 3t^2\rangle,$$

where $-\infty < s < \infty$, passes through the point $(1, 1)$.

17. The position vector of an object is

$$\mathbf{r}(t) = \langle \cos 2t, \sin 2t\rangle, \qquad t \ge 0.$$

Lengths are in meters and time in seconds. Calculate its acceleration at $t = 3.1$ s.

18. An object's acceleration is $\mathbf{a} = \langle e^t, e^{-t}\rangle$, $t \ge 0$. If $\mathbf{r}(0) = \langle 0, 0\rangle$ and $\mathbf{v}(0) = \langle 1, 0\rangle$, find its position at any time $t \ge 0$.

19. A 1.2-kg object initially at rest at the origin is acted on by a force $\mathbf{F} = \langle 2.4, 1.7\rangle$ N. What is the object's acceleration? Where is the object and how fast is it moving 3.5 s after the force is first applied?

20. The mass of an electron is 9.11×10^{-31} kg. Calculate the gravitational force exerted by one electron on another if they are 1 mm apart.

21. Write a parametric equation for the circle with center at $(-2, 3)$, radius 5, and traversed in the counterclockwise direction.

22. Calculate the slope of the curve described by

$$\mathbf{r}(t) = \langle t^3 + 1, t^2 + t + 1\rangle, \qquad t \ge 0,$$

at the point $\mathbf{r}(2)$. Give a parametric equation of the tangent line to C at this point.

23. Let $\mathbf{e}_1 = \langle 3, 1\rangle$, $\mathbf{e}_2 = \langle -1, 3\rangle$, and $\mathbf{v} = \langle 5, 7\rangle$. Express $\mathbf{v}$ as a sum of vectors $\mathbf{v}_1$ and $\mathbf{v}_2$, where $\mathbf{v}_1$ and $\mathbf{v}_2$ are parallel to $\mathbf{e}_1$ and $\mathbf{e}_2$, respectively.

24. For what value of t are the vectors $\mathbf{a} = \langle 3, 1\rangle$ and $\mathbf{b} = \langle -2, t\rangle$ parallel? Perpendicular?

25. Use vectors to show that any angle inscribed in a semicircle is a right angle.

26. Calculate $\|\mathbf{a} + \mathbf{b}\|$ and $\|\mathbf{a} - \mathbf{b}\|$, given that $\|\mathbf{a}\| = 5$, $\|\mathbf{b}\| = 8$, and the angle between $\mathbf{a}$ and $\mathbf{b}$ is $2\pi/3$.

T **27.** Sketch the graph of the curve C described by the parametric equation

$$\mathbf{r} = \mathbf{r}(t) = \langle t - t^2, t + 2t^2 \rangle,$$

where $-3 \le t \le 3$. Eliminate the parameter to obtain an equation in x and y. Show that the curve described by this equation includes the curve C. Calculate the slope of C at the point $(-2, 10)$.

28. Let $\mathbf{r} = \mathbf{r}_0 + t\mathbf{a}$ describe a line L, and let $\mathbf{q}$ be the position vector of a point Q not on L. Show that the distance d from L to Q can be calculated as follows. Let $\mathbf{p}$ be the vector projection of $\mathbf{q} - \mathbf{r}_0$ onto $\mathbf{a}$ and $\mathbf{n} = \mathbf{q} - \mathbf{r}_0 - \mathbf{p}$. Then $d = \|\mathbf{n}\|$. Use this procedure to calculate the distance from the line through $(1, 5)$ and $(7, 2)$ to the point $(-5, -3)$.

STUDENT PROJECTS

A. TIMING A RIFLE BULLET

A method for calculating the time taken in seconds for a rifle ball to travel from muzzle to target was published in the July 1893 issue of *Scientific American*. This was reported in the "50 and 100 Years Ago" column of the July 1993 issue.

> It may be of interest to amateur riflemen to know the following simple method for ascertaining the effect of gravity upon a bullet: Sight the rifle upon the target, keeping the sights plumb above the center line of the bore of the rifle. Mark where the ball strikes. Then reverse the rifle, so as to have the sights exactly beneath the line of the bore. In this reversed position sight it on the target as before, and mark where the bullet strikes. Divide the difference in elevation of the two bullet marks by 32 and extract the square root. This will give the time in seconds that it took the ball to travel the distance. The distance divided by the time will give the speed of the bullet per second.
> –J. A. G., Grand Rapids, Michigan

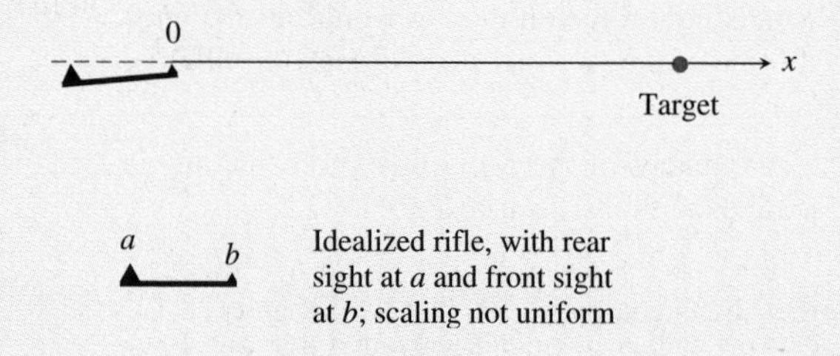

FIGURE 3.62 Diagram for Problem 1.

PROBLEM 1 Verify J. A. G.'s assertions and conclusions, and comment on the assumptions that you and J. A. G. made. See Fig. 3.62.

PROBLEM 2 The word *extract* used by J. A. G. suggests that it was somewhat more difficult to calculate a square root in 1893 than now. Write a short paragraph discussing and contrasting the techniques used by typical students in 1900 and 2000 to calculate, say, $\sqrt{5.73}$.

B. THE QUARTERBACK'S PROBLEM

Figure 3.63 shows a quarterback and receiver at points QB and R on a level playing field. The point R is 15 yards downfield from QB and 10 yards to one side. According to plan A, the receiver will run along the line L at 6 yards per second and receive a pass from the quarterback. The quarterback must pass within 5 seconds after the receiver starts running. The dotted line in the figure is the projection onto the field of the trajectory of the ball in a successful pass. A pass is "successful" if the receiver and ball reach some point P at the same time, that the football leaves the quarterback and arrives at the receiver at the same height, and that the quarterback throws at a speed of 25 yards per second. We use yards and seconds as units.

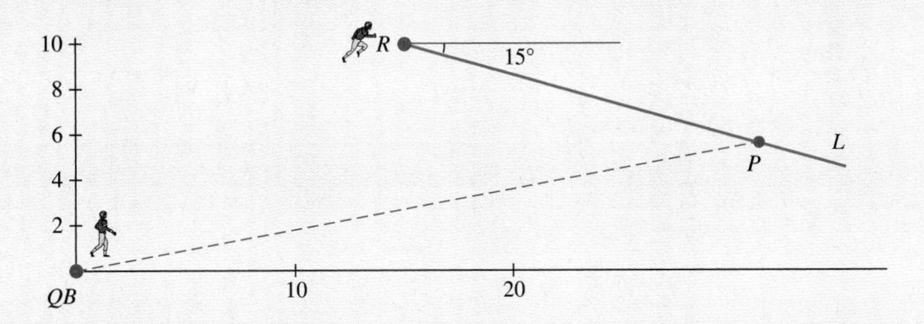

FIGURE 3.63 Plan A.

PROBLEM 1 Assume that the quarterback passes T seconds after the receiver starts running. For each of $T = 1, 3, 5$, determine the quarterback's choices for initial velocity vector. Use $\mathbf{g} = \langle 0, 0, -32/3 \rangle$ as the acceleration due to gravity, and neglect air resistance.

PROBLEM 2 Letting ℓ denote the distance in yards by which the quarterback must "lead" the receiver and θ the angle of elevation (in degrees) of the initial velocity, calculate ℓ and θ for each of the values $T = 1, 3, 5$.

PROBLEM 3 On the basis of your calculations, what is your advice to the quarterback? Why?

4

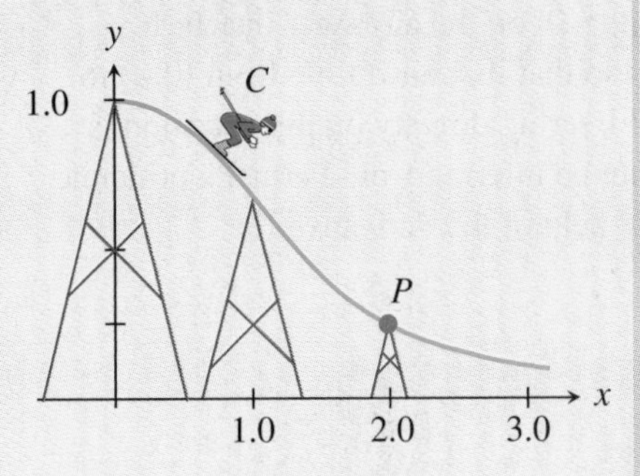

Applications of the Derivative

Joe is skiing down a long ramp to pick up speed before jumping out into the air. Near the bottom of the ramp, Joe's horizontal speed is measured by radar. Can we use this to figure out his vertical speed? This is a problem in related rates.

See Example 2, page 316, for further explanation.

In applying calculus to their own work, engineers and scientists use the idea of derivative in many ways. In earlier chapters, we discussed how the derivative is used to measure the rate of change of a physical or geometric magnitude with respect to time. In this chapter we study applications of the derivative, including the *tangent line approximation,* which may be used to simplify a calculation by replacing a complex expression by a simpler one; *Newton's method,* which is used to solve equations of the form $f(x) = 0$; the analysis of the graph of a function by classifying it as *increasing or decreasing* or *concave up or concave down;* the optimization of a process by locating the maximum or minimum of a function; the calculation of certain limits by applying *l'Hôpital's Rules;* and *Euler's method,* which may be used in solving certain difficult equations describing motion. All of these applications are included in the nine sections listed in the chapter outline, and all are direct applications of the derivative.

We use Fermat's principle of least time to highlight the role of optimization in science. Pierre Fermat (1601–1665) hypothesized that light travels on the path of least time. Consider, for example, the light ray shown in Fig. 4.1. We suppose that this ray emanates from the point $(0, a)$, in air, and is observed later at (b, c), in water. Through which point $(x, 0)$ on the air/water interface should the light pass so that its transit time from $(0, a)$ to (b, c) is a minimum? Bearing directly on this question is Snell's law, which can be inferred from Fermat's principle of least time with the help of the derivative.

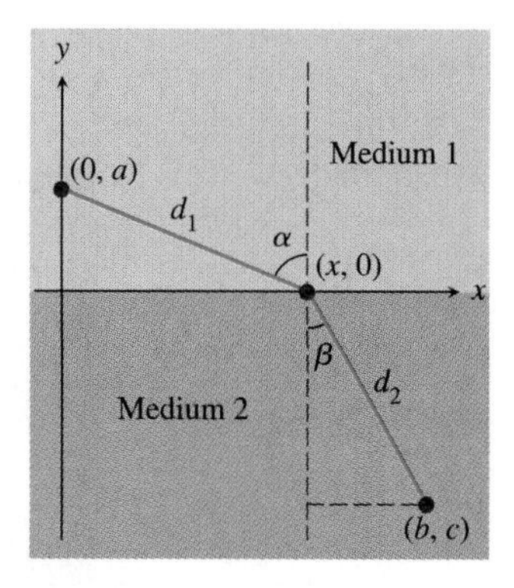

FIGURE 4.1 Possible path of a ray from $(0, a)$ to (b, c).

4.1 The Tangent Line Approximation

WEB Assume that f is a differentiable function and $(a, f(a))$ is a point on its graph. If, as in Section 1.7, we zoom toward $(a, f(a))$, we will notice that the graph of f looks more and more like its tangent line at $(a, f(a))$. The tangent line approximation

$$f(x) \approx f(a) + f'(a)(x - a)$$

is a quantitative statement of this observation.

A simple example of the tangent line approximation is the *small-angle approximation*

$$\sin x \approx x$$

used in physics. Because the sine curve is very nearly coincident with its tangent line when we zoom to $(0, 0)$, the small angle approximation follows by noting that the equation of the tangent line at $(0, 0)$ is

$$y - \sin(0) = \sin'(0) \cdot (x - 0) = \cos(0) \cdot x = 1 \cdot x = x,$$

which simplifies to

$$y = x.$$

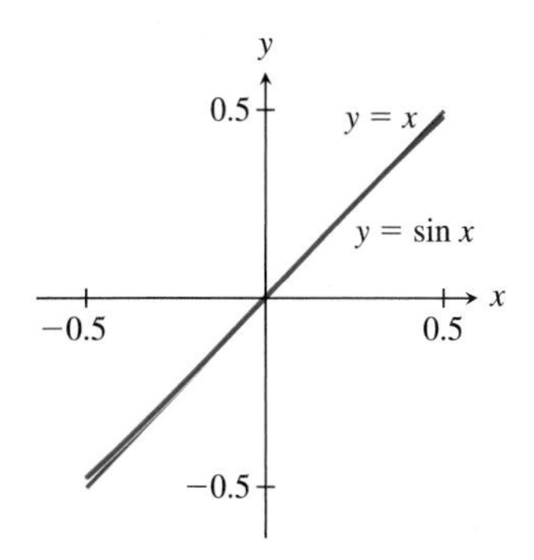

FIGURE 4.2 Graph of the sine function and the line tangent to this graph at $(0,0)$, both graphed on the interval $[-0.5, 0.5]$.

Figure 4.2 shows the graphs of the two functions $y = \sin x$ and $y = x$ on the interval $[-0.5, 0.5]$. The two graphs are nearly indistinguishable for $|x| < 0.4$.

The small-angle approximation is used in several contexts, including a critical simplification of the equation used to model the motion of a simple pendulum. A derivation of the pendulum equation and how this equation can be simplified using the small-angle approximation are outlined in Exercises 36 and 37. The small-angle approximation is also used in deriving the "lensmaker's formula" from Snell's law. Snell's law, which we discuss in Section 4.6, is

$$n_1 \sin \theta_1 = n_2 \sin \theta_2,$$

where n_1 and n_2 are indices of refraction. The lensmaker's formula is

$$n_1\theta_1 = n_2\theta_2,$$

provided that the angles θ_1 and θ_2 are not too large. Both Snell's law and the lensmaker's formula are used in calculating the behavior of light as it moves from one medium to another. In Exercise 19 we compare results inferred from the lensmaker's formula and Snell's law.

The Tangent Line Approximation and the Differential

If f is differentiable at a, then

$$f'(a) = \lim_{x\to a} \frac{f(x) - f(a)}{x - a}.$$

Hence, for x near a,

$$f'(a) \approx \frac{f(x) - f(a)}{x - a}.$$

Solving for $f(x)$,

$$f(x) \approx f(a) + f'(a)(x - a). \tag{1}$$

Because the equation of the tangent line to the graph of f at $(a, f(a))$ is

$$y - f(a) = f'(a)(x - a),$$

or, solving for y,

$$y = f(a) + f'(a)(x - a),$$

the approximation (1) means that the tangent line ℓ is close to the graph of f near $(a, f(a))$. For this reason the approximation (1) is called the *tangent line approximation.* Figure 4.3 shows representative graphs of f and ℓ.

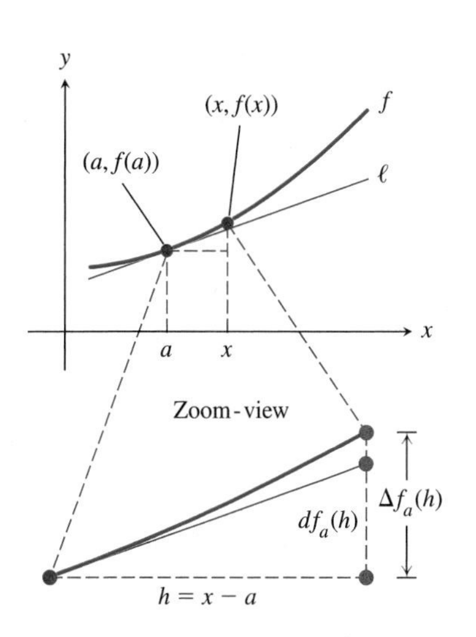

FIGURE 4.3 The tangent line approximation.

Before discussing the zoom-view shown at the bottom of Fig. 4.3, we define the true change in f at a as the difference $f(x) - f(a)$. It is often useful to write this difference in terms of the change $h = x - a$ in x. Using this notation, the true change in f at a is

$$\Delta f_a(h) = f(x) - f(a) = f(a + h) - f(a).$$

If we also define the function $df_a(h) = f'(a)h$, then (1) can be written as

$$\Delta f_a(h) \approx df_a(h).$$

This is an alternative form of the tangent line approximation (1). The function df_a is called the **differential** of f at a.

The zoom-view at the bottom of the figure shows the signed distances $\Delta f_a(h)$ and $df_a(h)$. The figure shows a case in which both $\Delta f_a(h)$ and $df_a(h)$ are positive. The reason why $df_a(h)$ is equal to the distance shown in the figure is that the slope $f'(a)$ of the tangent line is the rise of the tangent line over its run of h, that is,

$$f'(a) = \frac{\text{rise}}{\text{run}} = \frac{\text{rise}}{h}.$$

Hence, rise $= f'(a)h = df_a(h)$.

DEFINITION Differential and Tangent Line Approximation

Let f be a function defined on an interval I and assume that f is differentiable at $a \in I$. The change in f at a is the function Δf_a defined by

$$\Delta f_a(h) = f(a + h) - f(a), \qquad a + h \in I. \tag{2}$$

The **differential** of f at a is the function df_a defined by

$$df_a(h) = f'(a)h, \qquad h \in R. \tag{3}$$

The **tangent line approximation** to f at a is

$$f(x) \approx f(a) + f'(a)(x - a), \qquad \text{for } x \text{ near } a. \tag{4}$$

Letting $h = x - a$, the tangent line approximation to f at a can also be written as

$$f(a + h) \approx f(a) + hf'(a), \qquad \text{for small } |h|, \tag{5}$$

or

$$\Delta f_a(h) \approx df_a(h), \qquad \text{for small } |h|. \tag{6}$$

We often shorten the notations Δf_a and df_a to Δf and df.

We use the tangent line approximation to show that $\sin x \approx x$ for small x.

EXAMPLE 1 Use the tangent line approximation to replace the sine function on $[-0.5, 0.5]$ by a simpler function. Discuss the "worst-case scenario" graphically.

Solution

Because we want to approximate $\sin x$ throughout $[-0.5, 0.5]$, we take $a = 0$, the center point of the interval. From (4), with $a = 0$, $f(0) = \sin 0 = 0$, and $f'(0) = \sin'(0) = \cos 0 = 1$,

$$\sin(x) \approx \sin(0) + \cos(0)(x - 0) = x. \tag{7}$$

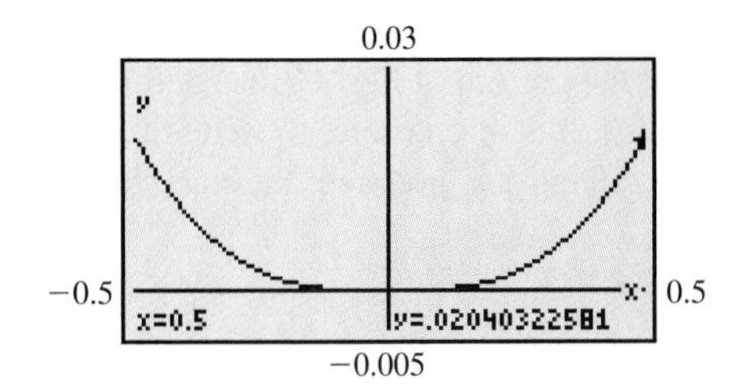

FIGURE 4.4 Graph of $y = |\sin x - x|$, in the window $-0.5 \le x \le 0.5$ by $-0.005 \le y \le 0.03$.

The "worst-case scenario" happens at all points x at which the difference $|\sin x - x|$ is a maximum. Figure 4.4 shows a calculator graph of the difference for $-0.5 \le x \le 0.5$. It is clear that the worst error occurs for $x = \pm 0.5$. The cursor, whose coordinates are shown in the figure, was placed at the point of the graph farthest to the right. From the screen or by a direct calculation, $|\sin 0.5 - 0.5| = 0.020\ldots$.

To summarize, if for $x \in [-0.5, 0.5]$ we replace $\sin x$ by the simpler function x, then the error $|\sin x - x|$ is less than, say, 0.03.

Percentage Error In the next two examples we use the tangent line approximation in calculating the *percentage error* resulting from a measurement. The percentage error in measuring a quantity T is

$$\left|\frac{\text{measured value of } T - \text{true value of } T}{\text{true value of } T}\right| \times 100.$$

Denoting the true value by T and the measured value by $T + \Delta T$, the percentage error is

$$\left|\frac{(T + \Delta T) - T}{T}\right| = \left|\frac{\Delta T}{T}\right| \times 100. \tag{8}$$

This expression for the percentage error in measuring T assumes that the true value of T is known. In practice, the true value is not known. We show in the next example how we can approximate the percentage error made in calculating the volume of a cylinder using a measurement of its radius. This calculation would be of interest in measuring the "displacement" of a gasoline or diesel engine.

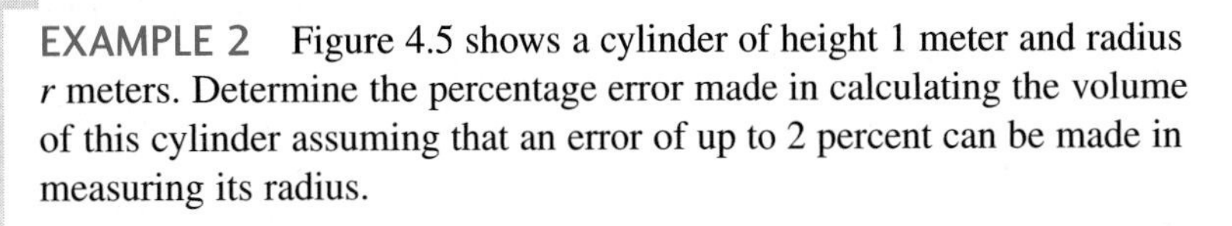

EXAMPLE 2 Figure 4.5 shows a cylinder of height 1 meter and radius r meters. Determine the percentage error made in calculating the volume of this cylinder assuming that an error of up to 2 percent can be made in measuring its radius.

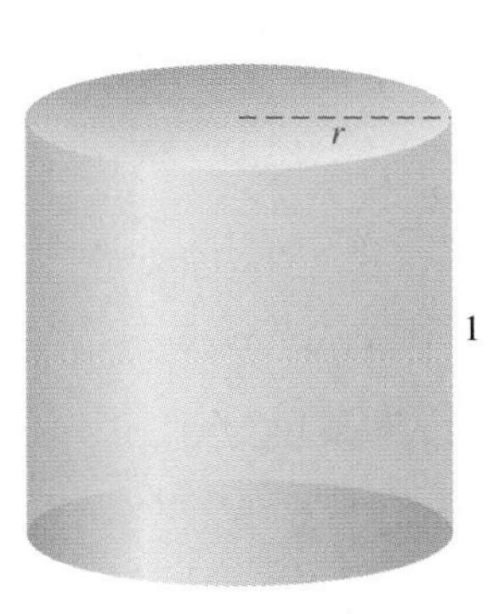

The volume V of a cylinder of height 1 meter and radius r meters is

$V = \pi r^2$

FIGURE 4.5 The volume of a right circular cylinder.

Solution

The volume of a cylinder with height 1 meter and radius r meters is $V = \pi r^2$. We are given that the percentage error in measuring the radius is not more than 2 percent. This means that

$$\left|\frac{(r + \Delta r) - r}{r}\right| \times 100 = \left|\frac{\Delta r}{r}\right| \times 100 \le 2, \tag{9}$$

where $r + \Delta r$ and r are the measured radius and the true radius, respectively.

From (8), the percentage error E in the volume is

$$E = \left|\frac{V + \Delta V - V}{V}\right| \times 100 = \left|\frac{\Delta V}{V}\right| \times 100.$$

Setting $a = r$ and $h = \Delta r$ in the tangent line approximation (6),

$$\Delta V = V(r + \Delta r) - V(r) \approx dV(\Delta r) = V'(r)\Delta r.$$

Hence, the percentage error is

$$E = \left|\frac{\Delta V}{V}\right| \times 100 \approx \left|\frac{V'(r)\Delta r}{V(r)}\right| \times 100.$$

Because $V(r) = \pi r^2$ and $V'(r) = 2\pi r$,

$$E \approx \left|\frac{2\pi r \Delta r}{\pi r^2}\right| \times 100 = 2\left|\frac{\Delta r}{r}\right| \times 100.$$

From (9),

$$E \approx 2\left|\frac{\Delta r}{r}\right| \times 100 \leq 4.$$

Hence, the maximum percentage error in the volume is close to 4%.

EXAMPLE 3 The pH of a solution is defined as $-\log_{10}[H^+]$, where $[H^+]$ is the hydrogen ion concentration. Distilled water has a pH of 7, which corresponds to a hydrogen ion concentration of 10^{-7} moles per liter. Use the tangent line approximation to simplify the calculation of pH of solutions for which $[H^+] = 10^{-7} \pm 4 \times 10^{-8}$. Sketch the pH curve and its tangent line approximation for values of $[H^+]$ in the interval $[10^{-7} - 4 \times 10^{-8}, 10^{-7} + 4 \times 10^{-8}]$. What is the worst possible error in using this approximation?

Solution

Replacing $[H^+]$ by x to simplify the notation, we use the tangent line approximation (4) to approximate the function

$$P(x) = -\log_{10} x = -\frac{\ln x}{\ln 10}$$

near $a = 10^{-7}$. Noting that $P'(x) = -1/(x \ln 10)$, it follows from the tangent line approximation that

$$P(x) \approx P(a) + P'(a)(x - a) = 7 + \left(\frac{-1}{10^{-7} \ln 10}\right)(x - 10^{-7})$$

$$\approx 7 - \frac{x - 10^{-7}}{10^{-7} \ln 10} = 7 - \frac{10^7 x - 1}{\ln 10}.$$

Denoting the tangent line approximation function by p,

$$P(x) \approx p(x) = 7 - \frac{10^7 x - 1}{\ln 10}.$$

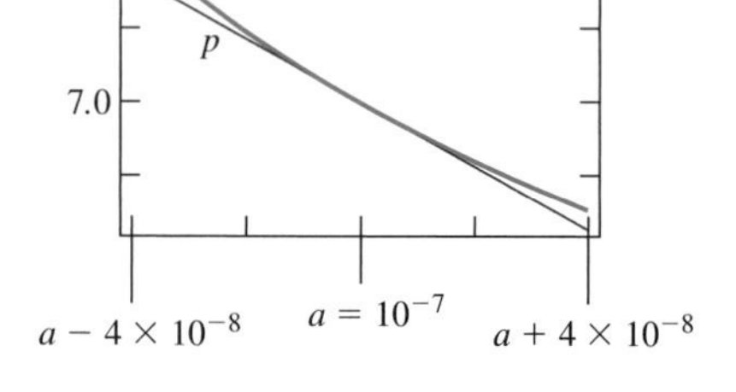

FIGURE 4.6 Graphs of the pH and tangent line approximation functions near $a = 10^{-7}$.

Figure 4.6 shows the graphs of P and the approximating function p.

From the graph, it appears that the worst possible error occurs at $x = a - 4 \times 10^{-8}$. Hence, the worst possible error in approximating $P(x)$ by $p(x)$ is

$$|P(a - 4 \times 10^{-8}) - p(a - 4 \times 10^{-8})| \approx 0.05.$$

In the previous example, the problem was to approximate a function P with a function derived from the tangent line approximation. It was convenient to use the form (4) of the tangent line approximation. In the next example we use form (5).

EXAMPLE 4 The period of a pendulum is the time required for a complete back-and-forth swing. If we measure lengths in meters and time in seconds, the period T of a grandfather's clock with pendulum of length L is

$$T = 2\pi\sqrt{\frac{L}{g}}, \tag{10}$$

where $g \approx 9.8\ \mathrm{m/s^2}$ is the acceleration due to gravity. To what accuracy must the length L be adjusted so that the period of the clock is $2 \pm .0001$ s?

Solution

Let L_2 be the length (in meters) for which the period is exactly 2 s and $L_2 + h$ the actual length of the pendulum, our strategy will be to choose the largest $|h|$ for which

$$|T(L_2 + h) - T(L_2)| \leq 0.001. \tag{11}$$

For this relatively simple function, the largest $|h|$ can be calculated by "exact methods." We outline this solution in Exercise 34. In the solution that follows, we approximate the largest $|h|$ using the tangent line approximation.

The quantity $|T(L_2 + h) - T(L_2)|$ in (11) is the absolute value of the difference between the true period $T(L_2)$ and the period $T(L_2 + h)$ of a pendulum set with an error of h in its length. From (5),

$$|T(L_2 + h) - T(L_2)| \approx |T'(L_2)h| = \frac{2\pi}{\sqrt{g}}\frac{1}{2\sqrt{L_2}}|h| = \frac{\pi}{\sqrt{gL_2}}|h|.$$

Using this approximation, condition (11) can be written as

$$\frac{\pi}{\sqrt{gL_2}}|h| \leq 0.001.$$

Solving for $|h|$,

$$|h| \leq \frac{0.001\sqrt{gL_2}}{\pi}. \tag{12}$$

Because L_2 is the length that gives $T = 2$, it follows from (10) that

$$2 = 2\pi\sqrt{\frac{L_2}{g}}.$$

Hence,

$$L_2 = \frac{g}{\pi^2}.$$

Substituting into (12) and recalling that $g \approx 9.8$,

$$|h| \leq \frac{0.001g}{\pi^2} \approx 0.001\ \mathrm{m}.$$

It follows that the length of L can be off by no more than a millimeter to achieve an accuracy of 0.001 s in the period. Although an accuracy of 0.001 s in the period may appear a bit extreme, note that over a day, which contains $(24 \cdot 3600)/2$ pendulum periods, an error of 0.001 s in the period can cause the clock to be off by as much as 43 seconds. An inaccuracy of 43 seconds per day is about 20 minutes per month.

Exercises 4.1

Exercises 1–8: Calculate the differential $df_a(h)$.

1. $f(x) = (2x + 7)^2$, $a = 3$
2. $f(x) = (x^2 - 1)^{-1}$, $a = 2$
3. $f(x) = e^x$, $a = 1$
4. $f(x) = e^{-3x}$, $a = 1$
5. $f(x) = \dfrac{x}{x^2 + 1}$, $a = -1$
6. $f(x) = \dfrac{\sqrt{x}}{2x - 3}$, $a = 4$
7. $f(x) = \arcsin(\sqrt{x})$, $a = 1/4$
8. $f(x) = \ln\sqrt{x^2 + 1}$, $a = 1$

Exercises 9–16: Write each function value in the form $f(a + h)$, where f and a are chosen so that $f(a)$ and $f'(a)$ can be evaluated without using the "f" key on your calculator. Compare the tangent line approximation to $f(a + h)$ with the value of $f(a + h)$ returned by your calculator when you enter $a + h$ and press the "f" key.

9. $\sin 31°$
10. $\cos 59°$
11. $\sqrt{9.2}$
12. $\sqrt{3.8}$
13. $\ln 2.8$
14. $\ln 2.6$
15. $\arcsin 0.48$
16. $\arctan 1.1$

17. Determine the percentage error made in calculating the volume of a (right) pyramid with square base if its height is known exactly, but a percentage error of up to 3 percent can be made in measuring the length of a side of its base. The volume of a pyramid with height h and square base with side a is $V = \frac{1}{3}a^2h$.

18. Determine the percentage error made in calculating the volume of a (right) prism whose base is an equilateral triangle with side a if its height is known, but a percentage error of up to 3 percent can be made in measuring the length of a side of its base. The volume of such a prism with height h is $V = (\sqrt{3}/4)a^2h$.

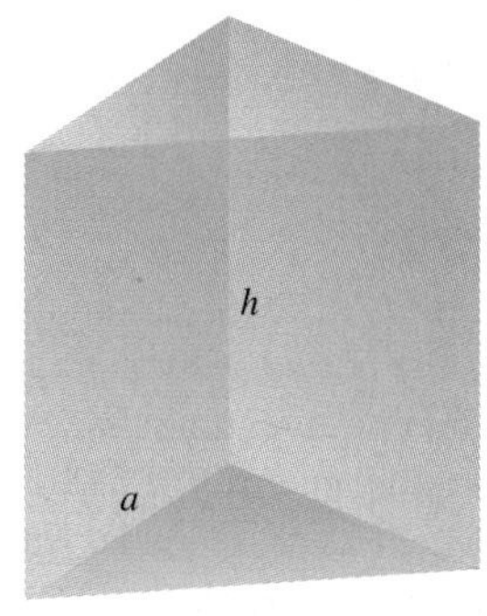

Figure for Exercise 18.

19. The figure shows a ray from point A being refracted at point B of a spherical lens with center at C and refractive index n_2. The ray travels from A to B through air with refractive index n_1 and travels from B to C through the lens. The refractive indices and the angles θ_1 and θ_2 are related by Snell's law:

$$n_1 \sin \theta_1 = n_2 \sin \theta_2.$$

If θ_1 and θ_2 are not too large, apply the small-angle approximation to derive the lensmaker's formula

$$n_1\theta_1 = n_2\theta_2.$$

Snell's law can be used to calculate the refractive index of a lens. Referring to the figure, suppose that the refractive index of air is $n_1 = 1.000293$ and the angles θ_1 and θ_2 were measured as $\theta_1 = 10°$ and $\theta_2 = 6.56°$. Use Snell's law to calculate the refractive index n_2 of the glass in the lens. Compare this result with the value of n_2 obtained by using the lensmaker's formula.

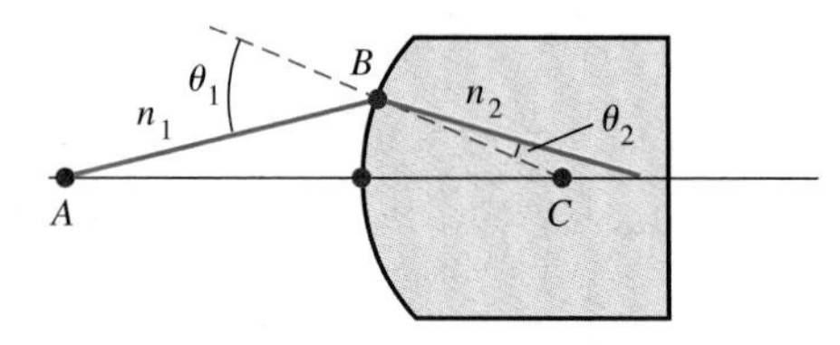

Figure for Exercise 19.

20. The sound intensity in decibels (dB) of a sound of intensity I is

$$\beta = 10 \log_{10} (I/I_0),$$

where $I_0 = 10^{-12}$. The units of intensity are watts per square meter. At 4 meters, a rock band can deliver 120 decibels; eardrums rupture at 160 decibels. Suppose that for $I \geq 1$ the percentage error in measuring I is at most 5 percent. Estimate the worst percentage error in calculating β.

21. Related to the preceding exercise Use the tangent line approximation to find a function that approximates $\beta(I)$ for I near $I = 10$.

22. The period T of a short-swing pendulum of length L is

$$T = 2\pi\sqrt{\frac{L}{g}}.$$

Use the tangent line approximation in deciding how accurately we must set L if we wish the period of this pendulum clock to be within 0.001 seconds of 1 second.

23. The period T of a short-swing pendulum of length L is

$$T = 2\pi\sqrt{\frac{L}{g}}.$$

Use the tangent line approximation in deciding how accurately L must be adjusted if we wish the period of this pendulum clock to be within 0.001 seconds of 1.5 seconds.

24. A coat of paint 0.04 cm thick is applied to a spherical water tower of diameter 30 m. Approximate the amount of paint required for the job. (*Hint:* The volume of a sphere of radius r is $V = \frac{4}{3}\pi r^3$, and there are 264.17 gallons in 1 m^3.)

25. Referring to the figure, assume that a new quarter-dollar coin Q has diameter exactly 2.4 cm and that the moon M moves in a circular orbit with radius 3.85×10^8 m. At EQ $= 2.65$ m from the eye, the quarter covers the moon exactly. Using these data, what is the approximate radius of the moon? What percentage error can be made in determining the distance between eye and coin if the moon's radius is to be determined within 5 percent of the approximate value determined earlier?

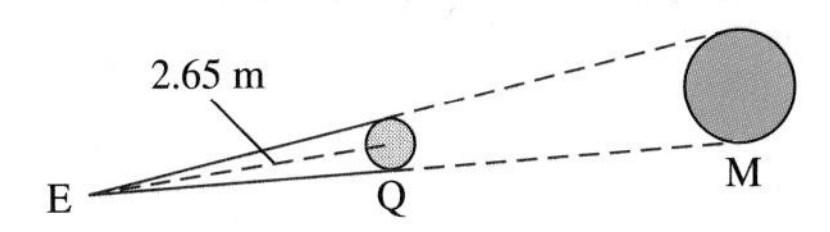

Figure for Exercise 25.

26. What is the tangent line approximation to e^x near $x = 0$? If we were to replace e^x by this tangent line approximation on the interval $[-0.5, 0.5]$, how big does the difference between e^x and this approximation become? Your answer may be based on graphical evidence.

27. What is the tangent line approximation to 2^{-x} near $x = 1$? If we were to replace 2^{-x} by this tangent line approximation on the interval $[0, 2]$, how big does the difference between 2^{-x} and this approximation become? Your answer may be based on graphical evidence.

28. Determine a simple function g that approximates the square root function in the interval $[3, 6]$. How large does $|\sqrt{x} - g(x)|$ become for $x \in [3, 6]$? Give both numerical and graphical evidence of your error analysis.

29. We may use the relation

$$T = 2\pi\sqrt{\frac{L}{g}}$$

connecting the length L and period T of a pendulum to measure the acceleration of gravity. With a pendulum of a fixed length $L = 1.0$ m, how accurately would the period T need to be measured so that g is within 0.05 of 9.8?

30. Determine a polynomial $mt + b$ that approximates $f(t) = 4.5e^{-0.32t}\cos(1.7t)$ near $t = 1$. On the interval $[0.8, 1.2]$ where is the approximation best? Worst?

31. Let f and g be functions defined for all x and let $w = f \circ g$ be their composition, so that $w(x) = f(g(x))$. Show that $dw_a(h) = f'(g(a))dg_a(h)$. If $g(3) = 5$, $g'(3) = 7$, $f'(5) = 9$, and $h = 0.1$, calculate $dw_3(0.1)$.

32. Let $f(x) = x^2$. What relation has the graph of $y = df_2(h)$ to the graph of the tangent line to f at $(2, 4)$?

33. In approximating $f(x)$ by $f(a) + f'(a)(x - a)$, the absolute error is the absolute value of the difference of these two quantities. Express the absolute error in terms of $\Delta f_a(x - a)$ and $df_a(x - a)$. On a figure similar to Fig. 4.3, locate and label the absolute error.

34. Equation (11) in Example 4 is

$$|T(L_2 + h) - T(L_2)| \le E,$$

where $T(L) = 2\pi\sqrt{L/g}$, L_2 satisfies $T(L_2) = 2$, and $E = 0.001$. Show that if $h > 0$, then it follows from this inequality that

$$0 \le h \le \frac{gE(E + 4)}{4\pi^2}$$

and, if $h < 0$, then

$$0 \le -h \le \frac{gE(-E + 4)}{4\pi^2}.$$

What value of h would you use and how does this compare with the h found in Example 4?

35. Sketch a figure like that in Fig. 4.3, but one for which the graph of f lies below all of its tangent lines. On your sketch identify a, x, $(a, f(a))$, $(x, f(x))$, $h = x - a$, $\Delta f_a(h)$, and $df_a(h)$.

36. *Note:* This exercise assumes that you are familiar with vectors. Referring to the accompanying figure, in which the x-axis is directed downward, assume that the pendulum bob has mass m. Let $\mathbf{F} = \langle mg, 0\rangle$ be the gravitational force acting on the pendulum bob and $\mathbf{P}$ the projection of this force in the direction of the unit vector $\langle\cos\theta, \sin\theta\rangle$. Let $\mathbf{N} = \langle mg, 0\rangle - \mathbf{P}$. Show that

$$\mathbf{N} = mg\langle\sin^2\theta, -\sin\theta\cos\theta\rangle.$$

Because $\mathbf{P}$ just balances the tension $\mathbf{T}$ on the bob, the net force on the pendulum bob is $\mathbf{N}$. Because the bob swings in a circle of radius L, its position vector at any time t must be

$$\mathbf{r}(t) = L\langle\cos\theta, \sin\theta\rangle$$

where $\theta = \theta(t)$. We wish to determine the function $\theta = \theta(t)$. To apply Newton's second law $\mathbf{N} = m\mathbf{a}$, calculate the acceleration $\mathbf{a}$ from the expression for $\mathbf{r}$. Specifically, show that

$$\mathbf{a} = L\theta''\langle-\sin\theta, \cos\theta\rangle + L\theta'^2\langle-\cos\theta, -\sin\theta\rangle,$$

where $\theta'' = (\theta')' = d(d\theta/dt)/dt$. By equating the coordinates of both sides of $\mathbf{N} = m\mathbf{a}$, show that

$$g\sin^2\theta = -L\theta''\sin\theta - L\theta'^2\cos\theta$$
$$-g\sin\theta\cos\theta = L\theta''\cos\theta - L\theta'^2\sin\theta.$$

Multiply the first of these two equations by $\sin\theta$ and the second by $-\cos\theta$, add the results, and simplify to obtain

$$L\theta'' + g\sin\theta = 0.$$

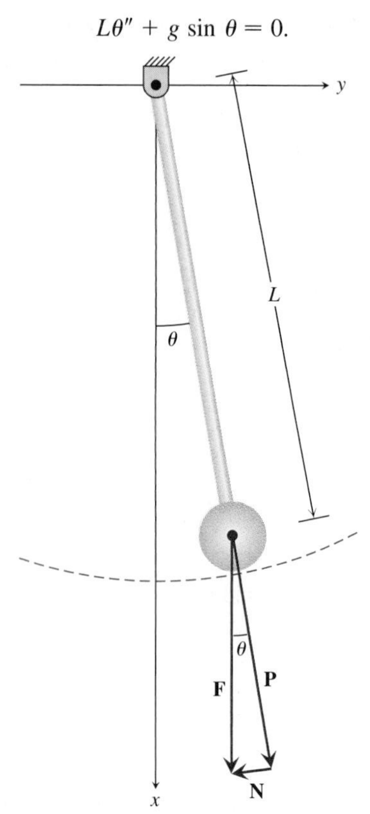

Figure for Exercise 36.

37. Continuation of the preceding pendulum problem If the pendulum does not swing too much from side to side, the equation

$$L\theta'' + g\sin\theta = 0$$

derived in the preceding example may be simplified by using the small-angle approximation. Replacing $\sin\theta$ by θ gives

$$L\theta'' + \frac{g}{L}\theta = 0. \tag{13}$$

If the pendulum is initially displaced to $\theta = \theta_0$ and is released from rest, its subsequent motion is described by

$$\theta = \theta(t) = \theta_0\cos(\omega t), \qquad \text{where } \omega = \sqrt{\frac{g}{L}}.$$

Show that this function satisfies (13) by calculating $\theta''(t)$ and substituting this and $\theta(t)$ into equation (13).

4.2 Newton's Method

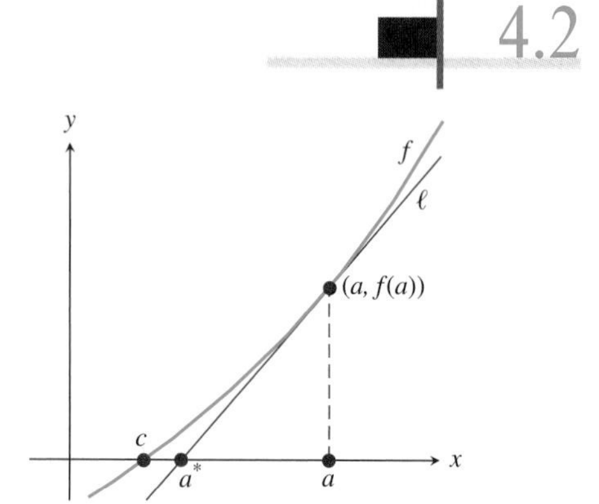

FIGURE 4.7 The geometry of Newton's method: Given an approximation a to c, move vertically from a to $(a, f(a))$, and then slide on the tangent line to its x-intercept a^*, which is (usually) closer to c than a.

Let f be a function defined on an interval of the x-axis. A number c is a **zero** of f if $f(c) = 0$. We also use the terminology "a number c is a root of the equation $f(x) = 0$ if $f(c) = 0$." Newton's method is a procedure or algorithm for approximating the zeros of functions (or, equivalently, the roots of equations). We start with a guess a of a zero c of a function f and use the tangent line approximation to improve this guess.

Figure 4.7 shows the graph of a function f near a zero c of f. The first step in Newton's method is to find an approximation a to the zero c. We may find such an approximation by using numerical or graphical methods. We show an approximation a to c in Fig. 4.7. If we draw the tangent line ℓ to the graph of f at $(a, f(a))$ and a is fairly close to c, then we expect that the x-intercept a^* of the tangent line is closer to c than a. The reason for this is that the x-intercept of the tangent line is close to the x-intercept of the graph of f. The second step in Newton's method is to calculate a^*. The equation of the tangent line is

$$y - f(a) = f'(a)(x - a).$$

To find a^* we set $y = 0$ in this equation and solve for x:

$$0 - f(a) = f'(a)(x - a)$$
$$x = a - \frac{f(a)}{f'(a)}.$$

We denote this value of x by a^*. Usually, the number

$$a^* = a - \frac{f(a)}{f'(a)}$$

is much closer to c than a, as illustrated in Fig. 4.7. Replacing a by a^*, we repeat these steps until a^* is sufficiently close to the zero c of f.

Newton's Method

STEP 1. Determine an approximate value a of a zero c of f.

STEP 2. Calculate

$$a^* = a - \frac{f(a)}{f'(a)}. \tag{1}$$

STEP 3. If a^* is sufficiently close to c, stop; otherwise, replace a by a^* and go back to Step 2.

WEB Perhaps the simplest way of doing Newton's method on your calculator or CAS is to work out symbolically the formula $a - f(a)/f'(a)$ by hand; enter this formula into your calculator; let a be your first guess; evaluate the formula at a and, letting the result be a^*, replace a by a^*; and repeat these steps until a^* is sufficiently close to c. We illustrate this procedure in the first example.

EXAMPLE 1 Approximate $\sqrt{2}$ by applying Newton's method to approximate the positive zero of the function $f(x) = x^2 - 2$. We take $a = 1.5$ as our first guess, basing this on the numerical evidence that $a = 1$ is too small $(1^2 = 1 < 2)$ and $a = 2$ is too large $(2^2 = 4 > 2)$.

Solution

For $f(x) = x^2 - 2$, $f'(x) = 2x$ and equation (1) is

$$a^* = a - \frac{f(a)}{f'(a)} = a - \frac{a^2 - 2}{2a} = \frac{2a^2 - (a^2 - 2)}{2a} = \frac{a^2 + 2}{2a}.$$

Enter this formula into your calculator. Starting with $a = 1.5$, we find

$$a = 1.5: \quad a^* = \frac{a^2 + 2}{2a} = 1.41666666667$$

$$a = 1.41666666667: \quad a^* = \frac{a^2 + 2}{2a} = 1.41421568628$$

$$a = 1.41421568628: \quad a^* = \frac{a^2 + 2}{2a} = 1.41421356237.$$

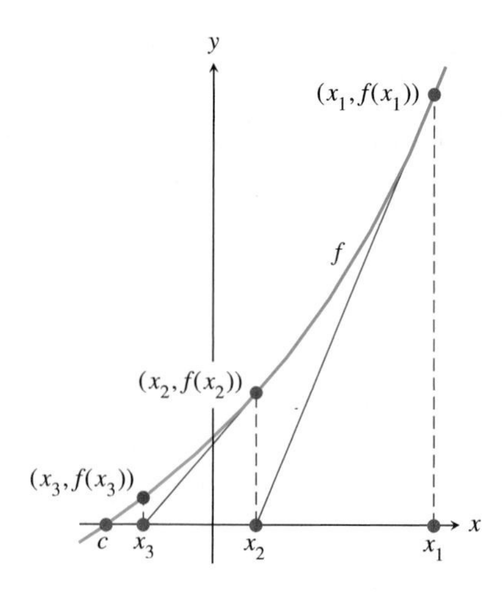

FIGURE 4.8 Three iterations of Newton's method.

The number of correct digits in this calculation approximately doubles with each iteration. This is typical of Newton's method. An accurate approximation of $\sqrt{2}$ is 1.4142135623....

Newton's method is often written in terms of **iterates** $x_1, x_2, x_3, \ldots$ instead of the alternating a and a^* notation. Referring to Fig. 4.8, x_1 is our first guess. The second iterate x_2 is the x-intercept of the tangent line to f at $(x_1, f(x_1))$. Specifically, x_2 is the solution of the equation

$$f(x_1) + f'(x_1)(x - x_1) = 0.$$

After this, the procedure repeats, generating the sequence $x_1, x_2, x_3, \ldots$ of approximations to c, where

$$x_{n+1} = x_n - \frac{f(x_n)}{f'(x_n)}, \qquad n = 1, 2, \ldots. \tag{2}$$

Newton's method usually works very well. Given a reasonable first guess, the sequence $x_1, x_2, x_3, \ldots$ of iterates approaches c rapidly. However, Newton's method can fail even for relatively simple functions. See Exercise 19.

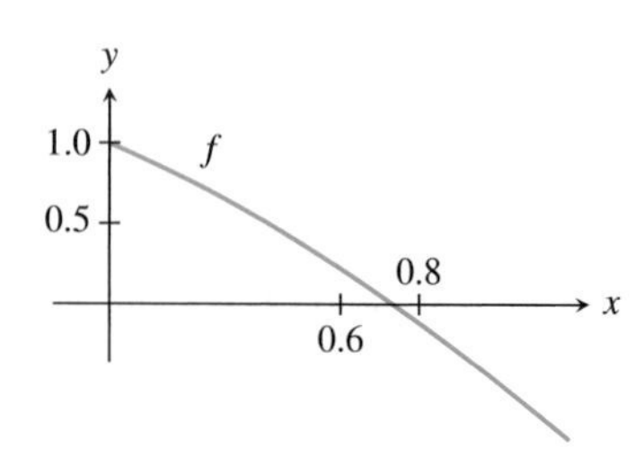

FIGURE 4.9 Locating the one zero of $f(x) = \cos x - x$.

EXAMPLE 2 Use Newton's method to find the one zero of the function $f(x) = \cos x - x$.

Solution

To decide upon a first guess x_1, we graph the function f. Based on Fig. 4.9, we take $x_1 = 0.7$ because f appears to have a zero in the interval $[0.6, 0.8]$.

We calculate the iterates $x_2, x_3, \ldots$ from (2), which for the function $f(x) = \cos x - x$ is

$$x_{n+1} = x_n - \frac{f(x_n)}{f'(x_n)} = x_n - \frac{\cos x_n - x_n}{-\sin x_n - 1} = x_n + \frac{\cos x_n - x_n}{\sin x_n + 1}.$$

We list the iterates in Table 4.1. We continued the algorithm until no further change in the iterates $x_1, x_2, \ldots$ was observed.

Because, to calculator accuracy, $f(x_4) = \cos x_4 - x_4 = 0$, we take x_4 as our approximation to the zero of the function f.

TABLE 4.1 The iterates $x_1, x_2, \ldots, x_5$.

n	x_n
1	0.7
2	0.739436497848
3	0.739085160465
4	0.739085133215
5	0.739085133215

When Is $|c - x_n|$ Sufficiently Small?

Deciding when a Newton's method iterate x_n is sufficiently close to a zero c of a function f can be done in several ways. Often, we simply compare successive iterates x_n and x_{n+1} and, when these agree to, say, five decimals, we assume that x_{n+1}, rounded to four decimals, approximates a zero of f to four decimals. In Example 1 the iterates $x_3 = 1.41421568628$ and $x_4 = 1.41421356237$ agree to four decimals. We may feel confident that $|\sqrt{2} - 1.414| < 0.0005$.

In Example 2 we continued the iteration until $x_4 = x_5$, to calculator accuracy. We then checked that $f(x_4) = f(x_5) = 0$, to calculator accuracy. We may feel confident that the zero c of f satisfies $|c - 0.739085133215| < 0.00000\,00000\,5$.

If we want to be certain that $|c - x_n|$ is sufficiently small, we may use the *Intermediate Value Theorem.* We use the context of Example 1 to discuss this method. The fourth iterate in that example is $x_4 = 1.41421356237$. Suppose that we decide to round x_4 to five decimals, which would be $m = 1.41421$. We show that $|m - \sqrt{2}| < 0.000005$, which means that m is accurate to five decimals—that is, it is less than 5 "off" in the sixth decimal place.

Letting $E = 0.000005$, we calculate

$$f(m - E) = f(1.414205) \approx -2.4 \times 10^{-5}$$

and

$$f(m + E) = f(1.414215) \approx 4.1 \times 10^{-6}.$$

$|m - \sqrt{2}| < \frac{1}{2}(2E) = E$

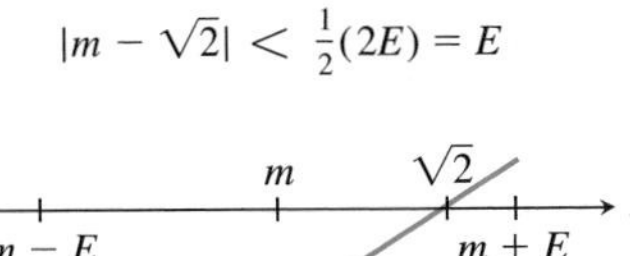

FIGURE 4.10 Changing sign test.

Because, as illustrated in Fig. 4.10, $f(x)$ is negative at $x = m - E$ and positive at $x = m + E$, we expect $f(x) = 0$ for some x between $m - E$ and $m + E$. This point is the positive zero $c = \sqrt{2}$ of f. Because c is between $m - E$ and $m + E$, it cannot be farther from the midpoint m of the interval $[m - E, m + E]$ than one-half the length of this interval, that is

$$|1.41421 - \sqrt{2}| = |m - \sqrt{2}| < \frac{1}{2} \cdot 2E = E = 0.000005.$$

Hence, $m = 1.41421$ is accurate to five decimals.

This argument is based on the Intermediate Value Theorem, which can be stated in several forms. We have chosen a form that is useful in the present context. For a statement of the theorem in which the term "intermediate value" makes sense, see Exercise 21.

THEOREM Intermediate Value Theorem

If f is continuous on an interval $[a, b]$ and $f(a)f(b) < 0$, then f has a zero c in (a, b).

Exercises 4.2

Exercises 1–8: Apply the Intermediate Value Theorem to locate at least one real zero/root of the function/equation between successive "tenths," for example, between 3.7 and 3.8 or between −0.3 and −0.2. Note: *A root of an equation $g(x) = h(x)$ can be found by finding a zero of the function $f(x) = g(x) - h(x)$.*

1. $f(x) = x^3 - 2x - 5$
2. $f(x) = 2^x - 4x$
3. $x = 1 - \dfrac{x^3}{10}$
4. $\sin x = 1 - x$
5. $f(x) = x \tan x - 1, \quad 0 < x < 1.5$
6. $f(x) = x^3 - 4x^2 - x + 3$
7. $\ln x = x - 2, \quad x > 0$
8. $\tan^3\theta - 8\tan^2\theta + 5\tan\theta - 4 = 0, \quad 0 \le \theta < 1.5$

Exercises 9–14: Use Newton's method in finding the zero or root to four decimal places. Use your intuition in deciding the required number of iterations.

9. Find the real zero of the function $f(x) = x^3 - 5$. Let $x_1 = 2.0$.
10. Find the real zero of the function $f(x) = -x + \cos x$. Let $x_1 = 1.0$.

11. Find the real zero of the function $f(x) = \tan x - x$, where $3.3 \leq x \leq 4.7$. Let $x_1 = 4.6$.

12. Find the zero of the function $f(x) = 5x^3 - 7x^2 + 9x - 41$ in the interval $[0, 4]$. Let $x_1 = 2$.

13. Find the zero of the function $f(x) = -6x^5 - 11x^4 + 2x + 2$ in the interval $[-1, 0]$.

14. Find the zeros of the function $f(x) = 2^x - 4x$.

15. Locate between successive tenths the first four positive zeros of the function

$$f(x) = \cos x \cosh x + 1.$$

The function cosh is the hyperbolic cosine function. Its derivative is the hyperbolic sine function, sinh. These functions were defined in Exercise 60 of Section 2.6 as

$$\cosh x = \frac{e^x + e^{-x}}{2} \text{ and } \sinh x = \frac{e^x - e^{-x}}{2}.$$

16. Find the three smallest positive zeros of the equation

$$\cos(68.617\sqrt{x}) \cosh(68.617\sqrt{x}) = -1.$$

See Exercise 15.

17. The equation $x^3 - 2x - 5 = 0$ was used by Wallis in 1685 to illustrate Newton's method. It has been used ever since in works dealing with the numerical solution of equations. Find the real root of this equation.

18. Locate between successive tenths all zeros of the function

$$f(x) = (x - 2)^{1/3} + 2x^2 - 15.$$

19. Find the one zero of $f(x) = \arctan(x)$. Show that for $x_1 = 1.5$ the Newton's method iterates $x_2, x_3, \ldots$ do not tend toward 0. Explain with a graph. What happens if $x_1 = 1.4$ or $x_1 = 1.3$?

20. The function $f(x) = \cos(\ln x)$, $x > 0$, has several zeros between 0 and 1. Explain numerically and graphically what happens when the first guess is $x_1 = 0.5$.

21. The Intermediate Value Theorem is often stated in this way: If f is continuous on an interval I, $u, v \in I$, and W is a number between $f(u)$ and $f(v)$, then there is a number w between u and v for which $f(w) = W$. The name of the theorem arises because W can be reasonably described as being a value of f intermediate between $f(u)$ and $f(v)$. A continuous function defined on an interval takes on all of its intermediate values, that is, skips no intermediate value. Show that these statements follow from the statement we gave earlier, namely, if f is continuous on an interval $[u, v]$ and $f(u)f(v) < 0$, then f has a zero c in (u, v). (*Hint:* Define the function $F(x) = f(x) - W$.)

22. Use either form of the Intermediate Value Theorem (see the preceding exercise) to show that if f and g are continuous functions defined on an interval I and $[f(a) - g(a)][f(b) - g(b)] < 0$ for points a and b of I, then there is a point w between a and b for which $f(w) = g(w)$.

23. By applying Newton's method to the function $f(x) = x^2 - a$, where $a > 0$, justify the ancient divide-and-average algorithm, described next, for approximating $\sqrt{a}$.

STEP 1. Choose a rough approximation g of $\sqrt{a}$.

STEP 2. Divide a by g and then average the quotient with g, that is, calculate $g^* = \dfrac{\frac{a}{g} + g}{2}$.

STEP 3. If g^* is sufficiently accurate, stop. Otherwise, let $g = g^*$ and return to Step 2.

Exercises 24–27: Use the "error bound" E to calculate a number x within E of a zero of f. Recall the discussion in which it was shown that if $f(x - E)f(x + E) < 0$, then x is within E of a zero of f.

24. $f(x) = \ln x - x + 2$, $x \geq 1$; $E = 0.001$

25. $f(x) = \arcsin x - 2x^2$, $0 < x < 1$; $E = 0.001$

26. $f(x) = \sqrt{x} - \tan x$, $0 \leq x \leq 1$; $E = 0.0001$

27. $f(x) = x^5 + x^3 - 1$, $0 \leq x \leq 1$; $E = 0.0001$

Exercises 28–35: Use Newton's method in finding the zero or root to the stated precision. Use your intuition in deciding the required number of iterations.

28. Find to three decimal places the two zeros of $f(x) = (x - 2)^{1/3} + 2x^2 - 15$ in $[-5, 5]$. See Exercise 18.

29. Find to one decimal place the first four positive zeros of the function

$$f(x) = \cos x \cosh x + 1.$$

See Exercise 15.

30. Find to within 0.1° the smallest positive zero of the equation

$$\tan^3\theta - 4\tan^2\theta + \tan\theta - 4 = 0.$$

31. Find to within 0.1° the three smallest positive zeros of the equation

$$\tan^3\theta - 8\tan^2\theta + 17\tan\theta - 8 = 0.$$

32. The volume V (in cubic meters) of 1 mole of gas is related to its temperature T (in kelvins) and pressure P (in pascal's (Pa)) by the ideal gas law $PV = RT$. A more accurate equation is van der Waals' equation:

$$\left(P + \frac{a}{V^2}\right)(V - b) = RT.$$

The constant R is 8.314. For carbon dioxide, $a = 3.592$ and $b = 0.04267$. Find to two decimal places the volume V of 1 mole of carbon dioxide if $P = 20{,}000$ Pa and $T = 320$ K.

33. Find to five decimal places all the zeros of the Chebyshev polynomial

$$128x^8 - 256x^6 + 160x^4 - 32x^2 + 1.$$

You may wish to plot this function to choose initial guesses. You can cut your work in half by an observation. You may check your results by using the fact that the zeros x_k are given by

$$x_k = \cos\left(\frac{(2k+1)\pi}{16}\right),$$

$k = 0, 1, \ldots, 7$.

34. Pulleys of radii R and r are connected by a taut belt of total length L, as shown in the figure. Letting $R = 200$ cm and $r = 100$ cm, and denoting by x the distance between pulley centers, express L in terms of R, r, and x. Find x to within one decimal place for each of the values $L = 2000, 2100, \ldots, 2800$ cm.

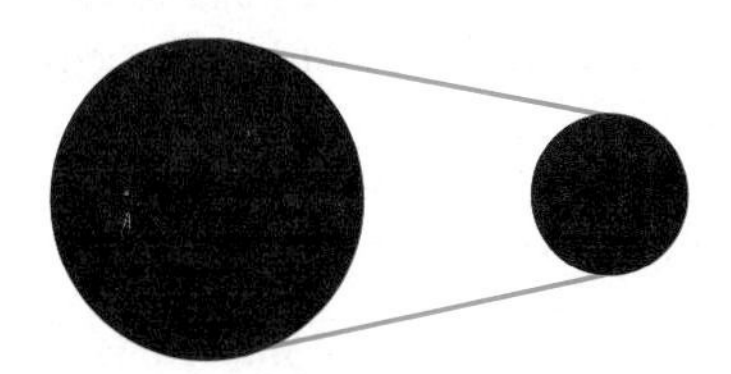

Figure for Exercise 34.

35. Newton's method may be used to find complex zeros. One of the difficulties in finding complex zeros is in locating initial approximations. We illustrate the procedure with the function $f(x) = x^3 - 1$, which we may factor as $f(x) = (x-1)(x^2+x+1)$. The zeros of f are 1 and $-1/2 \pm i\sqrt{3}/2$. We use Newton's method in finding one of the complex zeros. Let $x_1 = -0.4 + i0.8$. Use your calculator to show that $x_2 = -0.516666666667 + i0.866666666667$. Continuing, we find $x_4 = -0.5 + i0.866025403785$. It is clear that the iterates are converging to $-1/2 + i\sqrt{3}/2 \approx -0.5 + i0.866025403785$. Use Newton's method in finding all of the zeros of the polynomial

$$x^4 - 5x^3 + 21x^2 + 13x + 49.$$

There are zeros near $3 + 4i$ and $-0.5 + 1.3i$.

4.3 Increasing/Decreasing Functions; Concavity

A function f is *increasing* if, as x increases, the point $(x, f(x))$ on the graph of f rises; f is *decreasing* if, as x increases, the point $(x, f(x))$ falls. Our first goal in this section is to determine intervals on which a function is either increasing or decreasing. Figure 4.11 shows a function f that is increasing on the intervals I_1 and I_3 and decreasing on I_2.

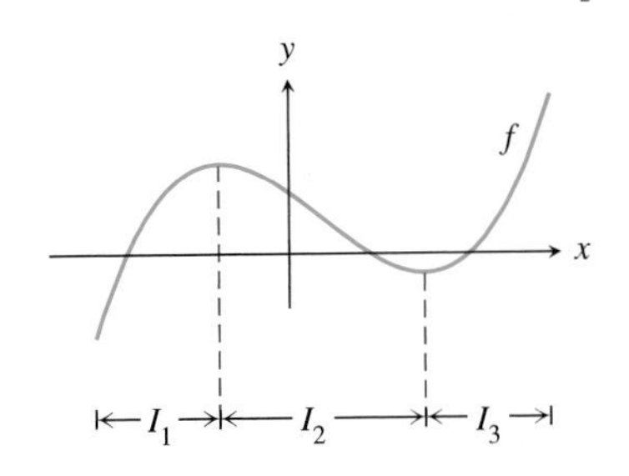

FIGURE 4.11 Intervals on which f is increasing or decreasing.

Increasing or Decreasing Functions

We define what it means for a function to be increasing or decreasing on an interval I. By an "interval" we mean a set of the form (a,b), $(a,b]$, $[a,b)$, $[a,b]$, (a,∞), $[a,\infty)$, $(-\infty,b)$, $(-\infty,b]$, or $(-\infty,\infty)$, where a and b are real numbers and $a < b$.

DEFINITION Increasing Functions; Decreasing Functions

A function f defined on a set including an interval I is **increasing** on I if, for all points u and v of I,

$$f(u) < f(v) \quad \text{whenever} \quad u < v;$$

f is **decreasing** on I if, for all points u and v of I,

$$f(u) > f(v) \quad \text{whenever} \quad u < v.$$

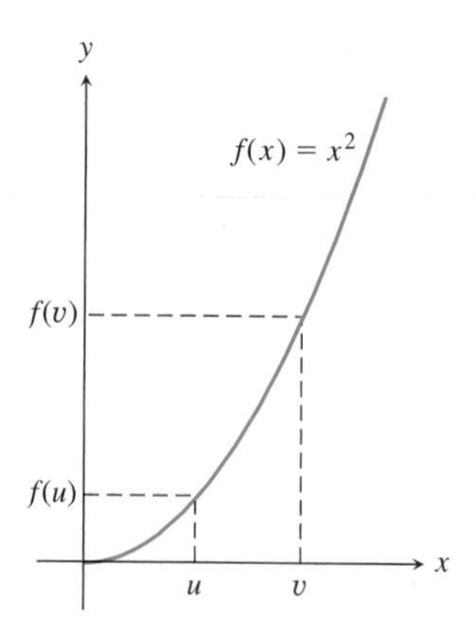

FIGURE 4.12 The squaring function is everywhere increasing on $[0, \infty)$.

EXAMPLE 1 Show that the function $f(x) = x^2$ is increasing on $I = [0, \infty)$.

Solution

Figure 4.12 shows the graph of f. As in the figure, suppose u and v are in I and $0 \le u < v$. We want to show that $f(u) < f(v)$. Because

$$f(v) - f(u) = v^2 - u^2 = (v - u)(v + u), \tag{1}$$

and both $v - u > 0$ and $v + u > 0$, it follows that $f(v) - f(u) > 0$ and $f(v) > f(u)$. Hence, f is increasing on $[0, \infty)$.

In Example 1 we were able to show that the squaring function is increasing on $[0, \infty)$ by applying the definition. For a differentiable function f, we may prefer to apply the I/D Test to determine the intervals on which f is increasing or decreasing.

Increasing/Decreasing Test (I/D Test)

Let f be a function defined on a set including an interval $[a, b]$. If f is continuous on $[a, b]$ and f' is defined on (a, b), then

if $f'(x) > 0$ on (a, b), then f is increasing on $[a, b]$;
if $f'(x) < 0$ on (a, b), then f is decreasing on $[a, b]$.

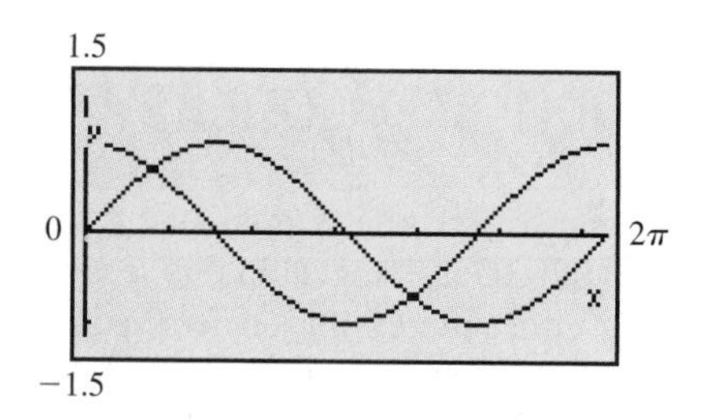

FIGURE 4.13 The graph of the sine function and its derivative, in the window $0 \le x \le 2\pi$ by $-1.5 \le y \le 1.5$.

The I/D Test is a refinement of the observation that if for all x in an interval I the slope of the graph of f at $(x, f(x))$ is positive, then the function must rise with increasing x. We give a proof of the I/D Test at the end of the section.

Figure 4.13 shows a calculator screen of the graph of $f(x) = \sin x$ and its derivative $f'(x) = \cos x$, for $0 \le x \le 2\pi$. According to the I/D Test, when $f'(x) = \cos x > 0$, then f will be increasing; when $f'(x) < 0$, then f will be decreasing. The figure shows that the sine function is increasing/decreasing on intervals in which the cosine function is positive/negative.

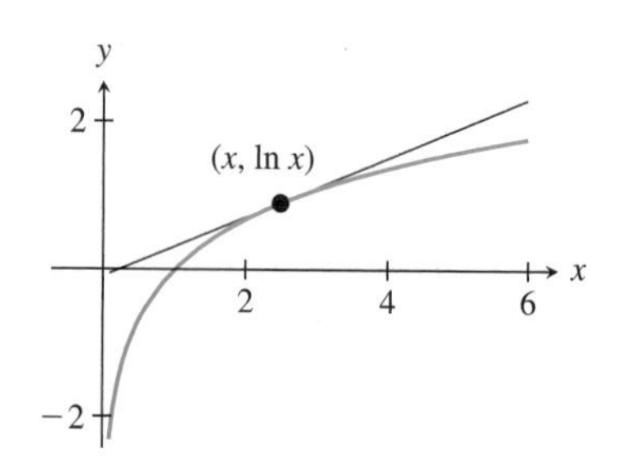

FIGURE 4.14 The natural logarithm function is increasing on its domain.

EXAMPLE 2 Use the I/D Test to show that the function $f(x) = \ln x$, $x > 0$, is increasing on $(0, \infty)$.

Solution

Figure 4.14 shows the graph of $f(x) = \ln x$. It appears from the graph that f is increasing on $(0, \infty)$. To prove this using the I/D Test, let a and b be any two points in the interval $(0, \infty)$, where $a < b$. We show that $f(a) < f(b)$. The natural logarithm function is continuous and differentiable on $[a, b]$. Because

$$f'(x) = \frac{1}{x} > 0, \text{ for all } x \in (a, b),$$

it follows from the I/D Test that f is increasing on $[a, b]$ and, hence, $f(a) < f(b)$. Hence, f is increasing on $(0, \infty)$.

EXAMPLE 3 Use the I/D Test to determine the intervals on which

$$f(x) = 4x^3 - 18x^2 + 15x + 10, \qquad -\infty < x < \infty$$

is increasing or decreasing.

Solution

The derivative of f is

$$\begin{aligned} f'(x) &= 12x^2 - 36x + 15 \\ &= 3(4x^2 - 12x + 5) \\ &= 3(2x - 1)(2x - 5). \end{aligned}$$

From this factorization, we see that f' has the two zeros $x = \frac{1}{2}$ and $x = \frac{5}{2}$. These zeros determine the three intervals $\left(-\infty, \frac{1}{2}\right)$, $\left(\frac{1}{2}, \frac{5}{2}\right)$ and $\left(\frac{5}{2}, \infty\right)$. The sign of f' can be determined by checking the sign of each of the factors $2x - 1$ and $2x - 5$ on each of the three intervals. Table 4.2 summarizes the results.

TABLE 4.2 A "sign table" to determine the sign of f' on several intervals.

	$2x - 1$	$2x - 5$	$f'(x) = 3(2x - 1)(2x - 5)$
$\left(-\infty, \frac{1}{2}\right)$	$-$	$-$	$+$
$\left(\frac{1}{2}, \frac{5}{2}\right)$	$+$	$-$	$-$
$\left(\frac{5}{2}, \infty\right)$	$+$	$+$	$+$

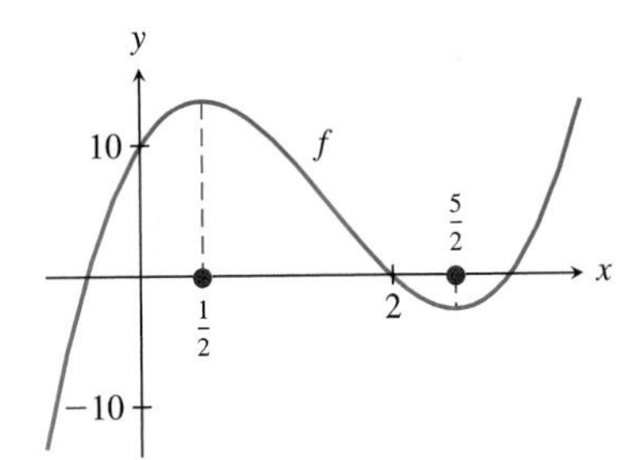

FIGURE 4.15 Use of the I/D Test to determine the intervals on which f is increasing or decreasing.

From the table, for any $x \in \left(-\infty, \frac{1}{2}\right)$, $f'(x) = 3(2x - 1)(2x - 5) > 0$ because both $2x - 1$ and $2x - 5$ are negative; for any $x \in \left(\frac{1}{2}, \frac{5}{2}\right)$, $f'(x) < 0$ because one factor is positive and the other is negative; and for any $x \in \left(\frac{5}{2}, \infty\right)$, $f'(x) > 0$ because both factors are positive.

By applying the I/D Test we conclude that f is increasing on $\left(-\infty, \frac{1}{2}\right]$, decreasing on $\left[\frac{1}{2}, \frac{5}{2}\right]$, and increasing on $\left[\frac{5}{2}, \infty\right)$. Figure 4.15 shows the graph of f and the location of the two zeros of f'.

EXAMPLE 4 In Example 3 of Section 3.1, we discussed the motion of a railroad snubber impacted by a freight car. Figure 4.16 shows the snubber spring and the graph of the position $x(t)$ of the right end of the spring at time t, from first contact at $t = 0$ to last contact at $t = \pi/1.26 \approx 2.5$ seconds. The position function was given by

$$x(t) = -0.398e^{-0.415t} \sin(1.26t).$$

Determine the time intervals in which the spring is decreasing or increasing in length.

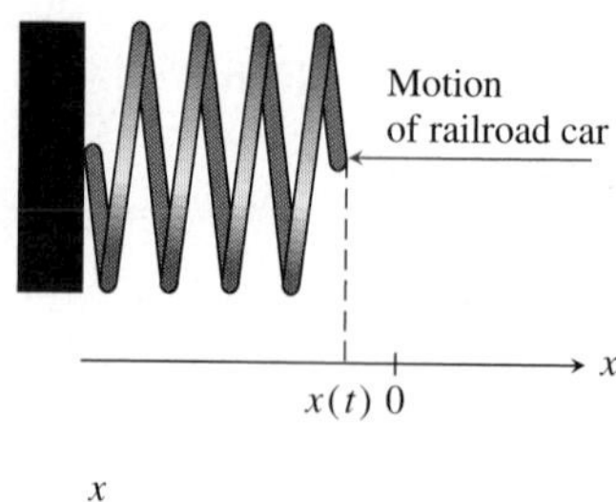

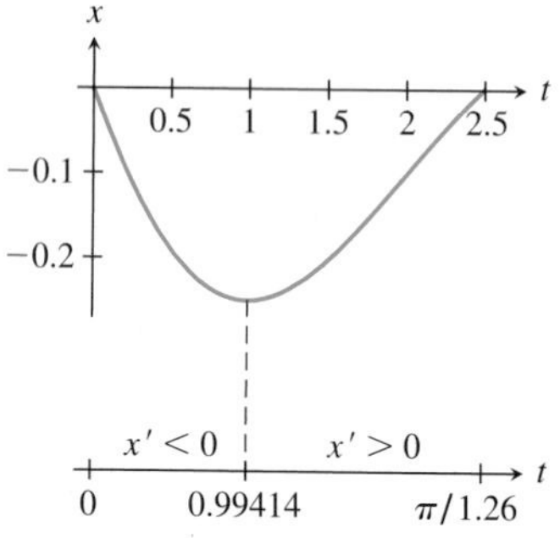

FIGURE 4.16 Use of the I/D Test to determine the time intervals of compression and decompression of the snubber spring.

Solution

From the graph of $x = x(t)$ shown in Fig. 4.16, it appears that the spring compresses until $t \approx 1$ s and then decompresses until $t = \pi/1.26 \approx 2.5$ s. These intervals correspond to the intervals on which the function x is decreasing and increasing. For the I/D Test we need to find the zeros of the derivative x', which is

$$\begin{aligned} x'(t) &= -0.398(-0.415e^{-0.415t}\sin(1.26t) + 1.26 \cdot e^{-0.415t}\cos(1.26t)) \\ &= 0.398e^{-0.415t}(0.415\sin(1.26t) - 1.26\cos(1.26t)). \end{aligned} \tag{2}$$

Because the exponential factor is always positive, we see that $x'(t) = 0$ when

$$0.415\sin(1.26t) - 1.26\cos(1.26t) = 0.$$

This equation can be rearranged as

$$\frac{\sin(1.26t)}{\cos(1.26t)} = \frac{1.26}{0.415}.$$

Hence,

$$\tan(1.26t) = \frac{1.26}{0.415}$$

$$t \approx \frac{1}{1.26}\arctan\left(\frac{1.26}{0.415}\right) \approx 0.99414 \text{ s.}$$

From (2) we see that $x'(0) < 0$; the derivative function x' continues to be negative until it changes sign at $t \approx 0.99414$, after which it is positive (up to $t = \pi/1.26$, at least). This information is summarized at the bottom of Fig. 4.16. From the I/D Test we conclude that the spring decreases in length on the time interval $[0, 0.99414]$ and increases in length on $[0.99414, \pi/1.26]$.

Concave Up or Concave Down Functions

If we hold out one of our hands, with the palm up and the palm and fingers slightly bent, our hand will tend to hold water; if our palm is down, our hand will tend to spill water. Applying these informal criteria to the graph of the function f shown in Fig. 4.17(i), most people would agree that the left side of the graph spills water and the right side holds water. They might further agree that there is a point c between a and b at which the graph changes from spilling water to holding water. Note from Fig. 4.17(i), in which two representative tangent lines have been drawn, that the properties of spilling or holding water appear to be related to whether the graph lies below or above its tangent lines. More formally, this graph is *concave down* on the interval (a, c) and *concave up* on the interval (c, b). The point c at which the graph changes from concave down to concave up is called an *inflection point.*

Figure 4.17(ii) shows the same graph as in (i), but with several tangent lines sketched in. As the point of tangency moves from $(a, f(a))$ to $(c, f(c))$, the tangent lines rotate clockwise. This corresponds to their slopes decreasing from approximately 2 at $(a, f(a))$ to approximately -2 at $(c, f(c))$. This means that the derivative function f' is decreasing on $[a, c]$. As the point of tangency moves from $(c, f(c))$ to $(b, f(b))$, the tangent lines rotate counterclockwise and their slopes increase from ap-

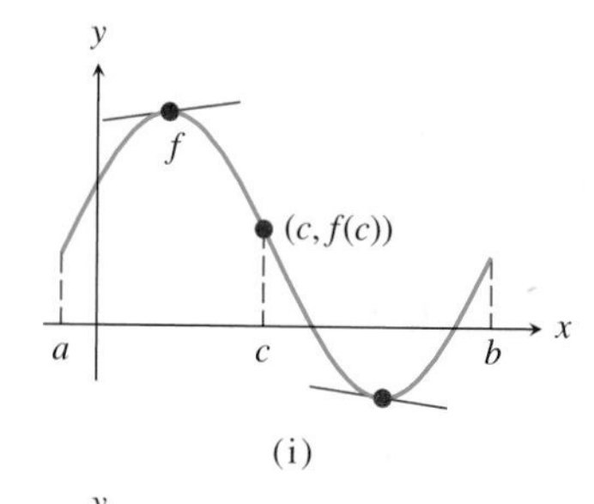

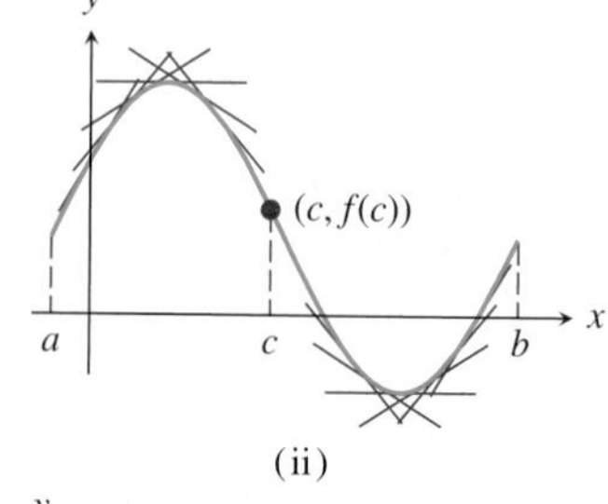

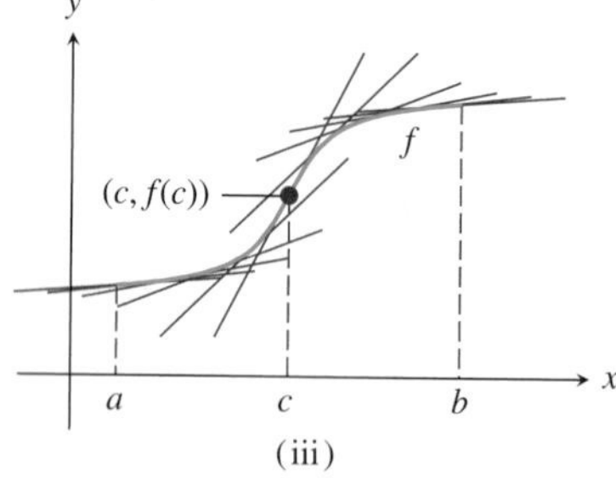

FIGURE 4.17 Tests for concave up and concave down.

proximately -2 at $(c, f(c))$ to approximately 2 at $(b, f(b))$. This means that the derivative function f' is increasing on $[c, b]$.

The graph shown in Fig. 4.17(iii) does not as clearly "hold water" or "spill water" as the graphs in (i) and (ii). However, note that the derivative function f' is increasing on the interval (a, c) and decreasing on (c, b). The point $(c, f(c))$ at which the "concavity" changes is a point of inflection.

DEFINITION Concave Up Functions; Concave Down Functions

A differentiable function f defined on a set $\mathcal{D}$ including an interval (p, q) is **concave up** on (p, q) if f' is increasing on this interval; f is **concave down** on (p, q) if f' is decreasing on this interval.

A point $c \in \mathcal{D}$ is an **inflection point** of f if there are intervals (r, c) and (c, s) in $\mathcal{D}$ for which f is concave up on one of these intervals and concave down on the other.

Because, according to this definition, f is concave up/down on an interval when f' is increasing/decreasing, on this interval, we will often need to calculate the derivative of f'. This is called the *second derivative of f*. If $y = f(x)$, the second derivative is denoted by

$$\frac{d}{dx}\left(\frac{dy}{dx}\right) = \frac{d^2y}{dx^2}, \quad \frac{d}{dx}\left(\frac{df}{dx}\right) = \frac{d^2f}{dx^2}, \quad y'', \quad f'', \quad \text{or} \quad f''(x).$$

Notations for higher order derivatives are similar, although we will usually prefer to denote, say, the fourth derivative of f at a by $f^{(4)}(a)$ instead of $f''''(a)$.

For functions that are twice differentiable (both f' and f'' exist), we can decide where f is concave up or concave down by applying the I/D Test to f'. Thus, if $f''(x) > 0$ on an interval I, then by the I/D Test f' is increasing on I and, by definition, f is concave up on I.

We use the term *open interval* in the Concavity Test. An interval is open provided that it does not include its endpoints. For example, the intervals $(0, 1) = \{x : 0 < x < 1\}$ and $(-\infty, 5) = \{x : x < 5\}$ are open; the intervals $[0, 1] = \{x : 0 \leq x \leq 1\}$, $(0, 1] = \{x : 0 < x \leq 1\}$, and $[0, \infty) = \{x : 0 \leq x\}$ are not.

Concavity Test

Let f be a function defined on a set $\mathcal{D}$ including an open interval I and assume that f is twice differentiable on I.

If $f''(x) > 0$ on I, then f is concave up on I.

If $f''(x) < 0$ on I, then f is concave down on I.

A point $c \in \mathcal{D}$ is an inflection point of f if f'' changes sign at c, which means that there are intervals (r, c) and (c, s) in $\mathcal{D}$ for which f'' is positive on one of these intervals and negative on the other.

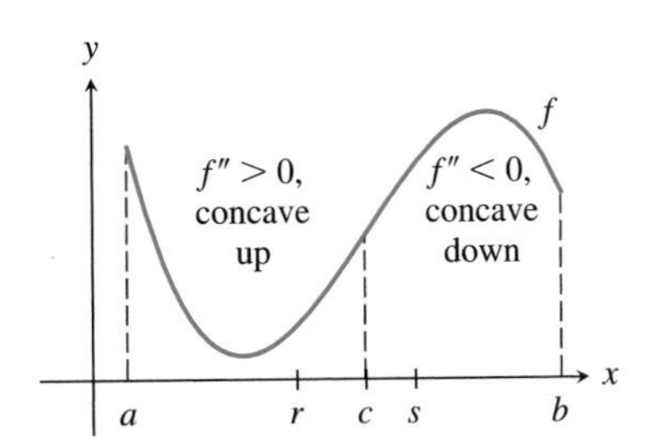

FIGURE 4.18 Concavity test and inflection point test.

Figure 4.18 shows a function f defined on $\mathcal{D} = [a, b]$. According to the Concavity Test, if $f''(x) > 0$ on $I = (a, c)$, then f is concave up on (a, c); if $f''(x) < 0$ on

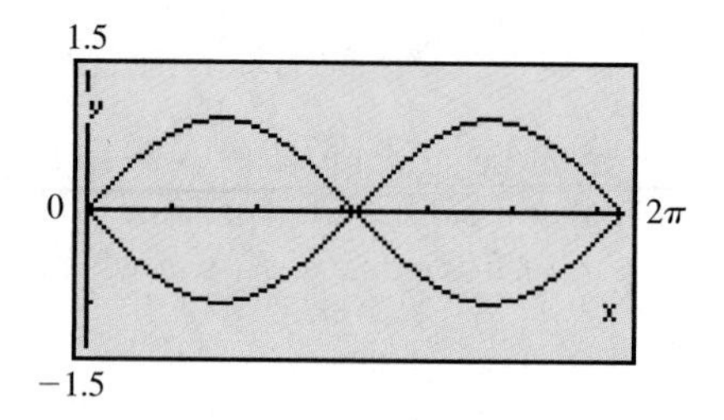

FIGURE 4.19 The graph of the sine function and its second derivative, in the window $0 \le x \le 2\pi$ by $-1.5 \le y \le 1.5$.

$I = (c, b)$, then f is concave down on (c, b); and if $f''(x)$ changes sign at c, then c is an inflection point of f.

Figure 4.19 shows a calculator screen of the graph of $f(x) = \sin x$ and its second derivative $f''(x) = -\sin x$, for $0 \le x \le 2\pi$. According to the Concavity Test, when $f''(x) = -\sin x > 0$, then f will be concave up and when $f''(x) < 0$, then f will be concave down. Hence, as the figure shows, wherever the sine function is positive/negative it will be concave down/concave up.

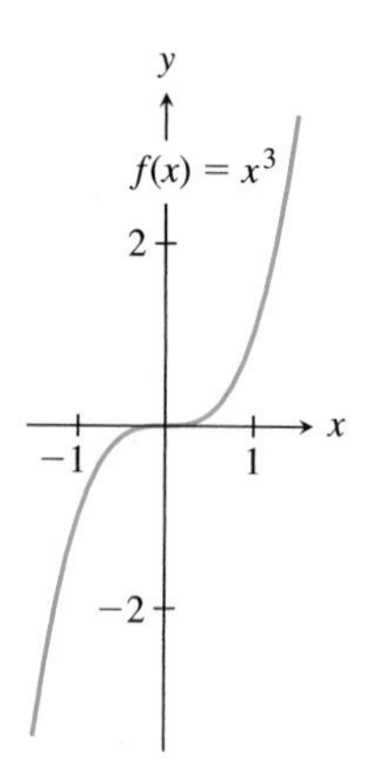

FIGURE 4.20 The function f is concave down on $(-\infty, 0)$, concave up on $(0, \infty)$, and has an inflection point at $x = 0$.

EXAMPLE 5 Determine the intervals on which $f(x) = x^3$ is concave up or concave down and locate any inflection points.

Solution

Because f is a polynomial, it is twice differentiable. For the Concavity Test we calculate

$$f'(x) = 3x^2$$
$$f''(x) = 6x.$$

Because $f''(x) > 0$ on the interval $(0, \infty)$ and $f''(x) < 0$ on $(-\infty, 0)$, f is concave up on $(0, \infty)$ and concave down on $(-\infty, 0)$ by the Concavity Test. The function f has an inflection point at $x = 0$ because f'' changes sign at 0. This is consistent with the graph of f shown in Fig. 4.20.

Note: Often, but not always, inflection points may be found by calculating the zeros of f''. In this example, $x = 0$ is a zero of $f''(x) = 6x$ and is an inflection point. However, for the function $f(x) = x^4$, $f''(0) = 0$ but $x = 0$ is not an inflection point. See Exercise 42.

EXAMPLE 6 Determine the intervals on which

$$f(x) = (x^2 - 1)e^x, \qquad -\infty < x < \infty,$$

is concave up or concave down and locate all inflection points.

Solution

To apply the Concavity Test we calculate f''. Starting with

$$f(x) = (x^2 - 1)e^x$$

and applying the product rule,

$$f'(x) = (2x)e^x + (x^2 - 1)e^x = (x^2 + 2x - 1)e^x$$
$$f''(x) = (2x + 2)e^x + (x^2 + 2x - 1)e^x = (x^2 + 4x + 1)e^x.$$

Because e^x is always positive, $f''(x)$ is positive, negative, or zero when $x^2 + 4x + 1$ is positive, negative, or zero. The zeros of this quadratic are $-2 \pm \sqrt{3} \approx -3.732, -0.268$. Because f'' can change sign only at these zeros, we may determine the sign of f'' by choosing a "test point" x in each of the three intervals

$$I_1 = \left(-\infty, -2 - \sqrt{3}\right), I_2 = \left(-2 - \sqrt{3}, -2 + \sqrt{3}\right), I_3 = \left(-2 + \sqrt{3}, \infty\right).$$

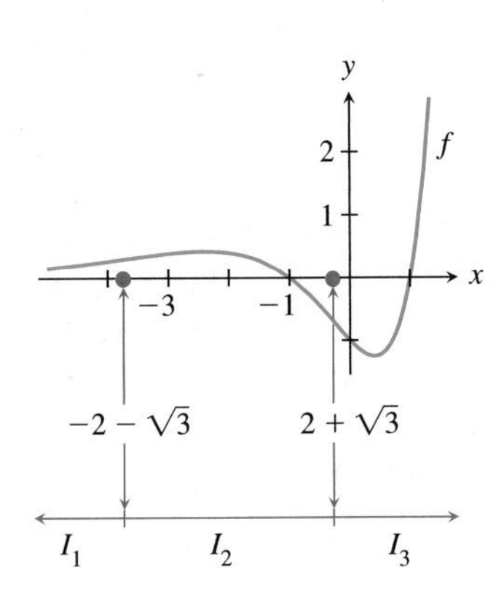

FIGURE 4.21 The graph of $f(x) = (x^2 - 1)e^x$, with intervals of concavity and the two inflection points marked.

TABLE 4.3 Using easy test points x to determine the sign of f'' on several intervals.

	Test point x	$f''(x)$
I_1	−10	+
I_2	−1	−
I_3	10	+

Table 4.3 shows the results of choosing easy test points.

Figure 4.21 shows the location on the x-axis of these zeros and the three intervals I_1, I_2, and I_3. Applying the Concavity Test, we can say that f is concave up on I_1 and I_3 and concave down on I_2 because f'' is positive on I_1 and I_3 and negative on I_2. The points $-2 - \sqrt{3}$ and $-2 + \sqrt{3}$ are inflection points because f'' changes sign at each of these points.

EXAMPLE 7 Find the leftmost inflection point of the function

$$f(x) = \sin x^2, \qquad 0 \le x \le 3.$$

Solution

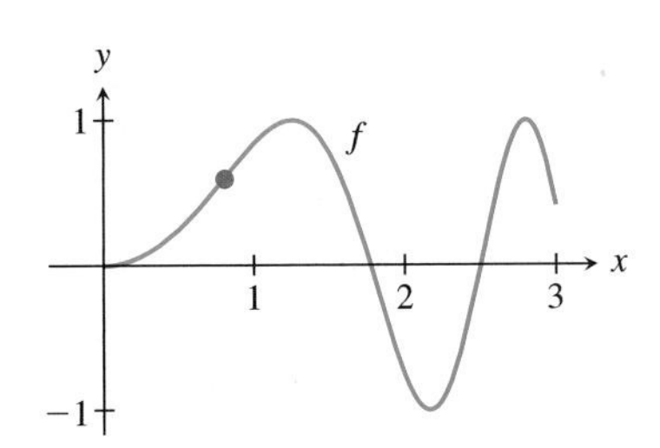

FIGURE 4.22 The graph of f on $[0, 3]$, with an approximate location of the leftmost inflection point.

Figure 4.22 shows the graph of f. Because it appears from this graph that f changes concavity around $x = 0.75$, we expect that the leftmost inflection point will be near this point. We refine this approximation by using the Concavity Test. Calculating f'',

$$f(x) = \sin x^2$$
$$f'(x) = 2x \cos x^2$$
$$f''(x) = 2 \cos x^2 - 4x^2 \sin x^2.$$

Because the expression for f'' is a combination of polynomials and trigonometric functions, we expect that Newton's method will be needed to locate the zeros of f''. Deferring the details to Exercise 41, we find that with $x_1 = 0.84$ as our first guess, the Newton's method iterates are

$$x_1 = 0.84$$
$$x_2 = 0.809658178049$$
$$x_3 = 0.808254919817$$
$$x_4 = 0.808251932950.$$

Hence, $f''(0.808) \approx 0$. Calculating $f''(0.80) \approx 0.075$ and $f''(0.81) \approx -0.016$ as a rough check on whether f'' changes sign at $x \approx 0.808$, we conclude that the leftmost inflection point of f is $x \approx 0.808$.

Proof of the I/D Test

FIGURE 4.23 Geometry of the Mean-value Theorem.

To prove the I/D Test we use the Mean-value Theorem. Let f be defined on $[a, b]$. Referring to Fig. 4.23, the Mean-value Theorem asserts that there is at least one point C on the graph of f at which the slope of the graph is the same as the slope $(f(b) - f(a))/(b - a)$ of the line segment joining the endpoints A and B. For a geometric proof, we may visualize translating the line joining A and B upward until it reaches the point C of "last contact" with the curve. At such a point as C, with coordinates $(c, f(c))$, the slope of the graph of f will be the same as the slope of the line joining A and B, that is, $f'(c) = \dfrac{f(b) - f(a)}{b - a}$. We state the Mean-value Theorem more formally:

THEOREM Mean-value Theorem

If f is continuous on $[a, b]$ and differentiable on (a, b), then there is at least one number $c \in (a, b)$ for which

$$\frac{f(b) - f(a)}{b - a} = f'(c).$$

WEB We use the Mean-value Theorem to prove one part of the I/D Test. Assuming that f satisfies the hypotheses of the Mean-value Theorem, we prove that

$$\text{if } f'(x) > 0 \text{ on } (a, b), \text{ then } f \text{ is increasing on } [a, b].$$

To show that f is increasing on $[a, b]$, we apply the definition: f is increasing on $[a, b]$ if, for all points u and v of $[a, b]$,

$$f(u) < f(v) \qquad \text{whenever} \qquad u < v.$$

If, then, u and v are points of $[a, b]$ for which $a \leq u < v \leq b$, we must show that $f(u) < f(v)$. Applying the Mean-value Theorem to f on the interval $[u, v]$, there is a number $c \in (u, v)$ for which

$$\frac{f(v) - f(u)}{v - u} = f'(c). \tag{3}$$

Because we assumed that $f'(x) > 0$ on (a, b), it must be true that $f'(c) > 0$ because $c \in (a, b)$. It then follows from (3) that

$$f(v) - f(u) = (v - u)f'(c) > 0.$$

Hence, f is increasing on $[a, b]$.

Exercises 4.3

Exercises 1–8: Use the definition of increasing/decreasing functions *to show that the function is increasing or decreasing on its domain I. Do not use the I/D Test.*

1. x^4, $I = (0, \infty)$

2. x^3, $I = (-\infty, \infty)$

3. $\dfrac{1}{x}$, $I = (0, \infty)$

4. $\dfrac{1}{x^2 + 1}$, $I = (0, \infty)$

5. $x^2 + x$, $I = (0, \infty)$

6. $x^2 - x$, $I = \left(\frac{1}{2}, \infty\right)$

7. $\dfrac{x + 1}{x}$, $I = (0, \infty)$

8. $\dfrac{x}{x - 1}$, $I = (1, \infty)$

Exercises 9–16: Use the I/D Test to find the intervals on which the function is increasing or decreasing. The functions are the same as in Exercises 1–8 except that their domains may be different.

9. x^4, $-\infty < x < \infty$

10. x^3, $-\infty < x < \infty$

11. $\dfrac{1}{x}$, $x \neq 0$

12. $\dfrac{1}{x^2 + 1}$, $-\infty < x < \infty$

13. $x^2 + x$, $-\infty < x < \infty$

14. $x^2 - x$, $-\infty < x < \infty$

15. $\dfrac{x + 1}{x}$, $x \neq 0$

16. $\dfrac{x}{x - 1}$, $x \neq 1$

Exercises 17–24: Use the I/D Test to find the intervals on which the function is increasing or decreasing.

17. $\sin x^3$, $0 \leq x \leq 2$

18. $\cos x^3$, $0 \leq x \leq 2$

19. $\sin^2 x$, $0 \leq x \leq \pi$

20. $\dfrac{x}{x^2 + 1}$, $-\infty < x < \infty$

21. $\ln\left(\dfrac{x}{x^2 + 1}\right)$, $x > 0$

22. $\ln\left(\dfrac{x}{x + 1}\right)$, $x > 0$

23. xe^x, $-\infty < x < \infty$

24. xe^{-x}, $-\infty < x < \infty$

Exercises 25–30: Use the Concavity Test to find the open intervals on which the function is concave up or concave down. Also, locate any inflection points.

25. $\sin^2 x$, $0 \leq x \leq \pi$

26. $\cos^2 x$, $0 \leq x \leq \pi$

27. $\dfrac{1}{x^2 + 1}$, $-\infty < x < \infty$

28. $\dfrac{1}{x^3 + 1}$, $x \neq -1$

29. xe^x, $-\infty < x < \infty$

30. xe^{-x}, $-\infty < x < \infty$

Exercises 31–36: Graph the following functions and locate on the graph intervals on which the function is increasing or decreasing, open intervals on which the function is concave up or concave down, and all inflection points.

31. $f(x) = x^2 e^{-x}$, $-\infty < x < \infty$

32. $f(x) = x^3 e^{-x}$, $-\infty < x < \infty$

33. $f(x) = \sin(x^2)$, $-\sqrt{\pi} \leq x \leq \sqrt{\pi}$; it is given that f'' changes sign at $x \approx 0.808252$

34. $f(x) = \cos(x^2)$, $0 \leq x \leq \sqrt{2\pi}$; it is given that f'' changes sign at $x \approx 1.35521$ and $x \approx 2.19450$.

35. $f(x) = \dfrac{x + 1}{x^2 + 1}$, $-5 < x < 5$

36. $f(x) = \dfrac{x^3 + 1}{3x^2 + 1}$, $-2 < x < 2$

37. The *logistic model* of population growth has the form

$$P = P(t) = \frac{a}{1 + be^{-kt}},\ t \geq 0,$$

where a, b, and k are positive constants. The accompanying figure shows a representative graph of the logistic model. Show that the function P is increasing everywhere on its domain. The point of inflection visible on the graph occurs at a time when the rate of increase of the population is beginning to slow down. At what time does the point of inflection occur?

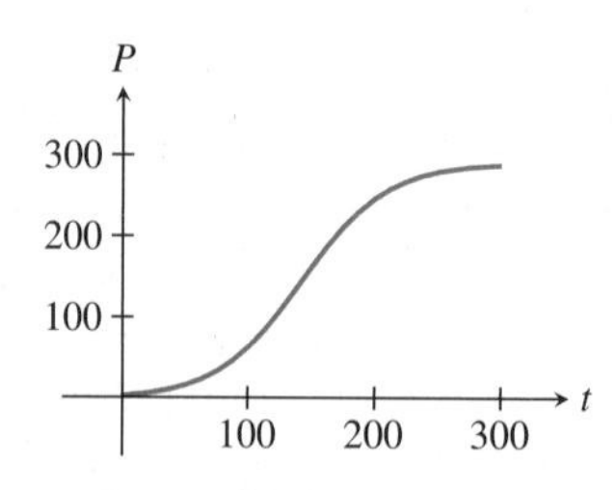

Figure for Exercise 37.

38. The *Gompertz model* of population growth has the form

$$P = P(t) = P_0 e^{k(1-e^{-at})/a},\ t \geq 0,$$

where P_0, k, and a are positive constants. The accompanying figure shows a representative graph of the Gompertz model. Show that the function P is increasing everywhere on its domain. The point of inflection visible on the graph occurs at a time when the rate of increase of the population is beginning to slow down. At what time does the point of inflection occur?

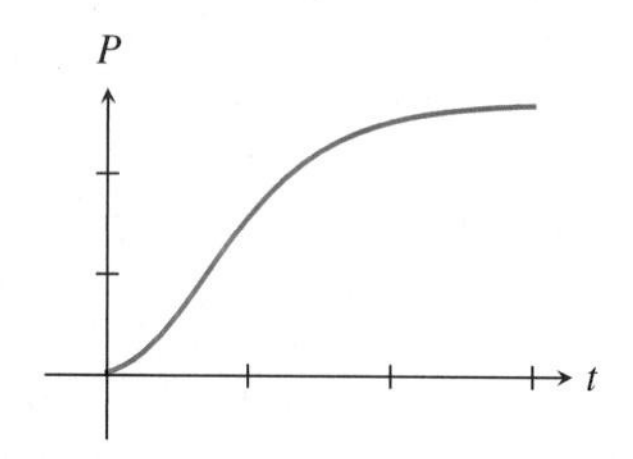

Figure for Exercise 38.

39. In the consecutive chemical reaction

$$A_1 \xrightarrow{k_1} A_2 \xrightarrow{k_2} A_3,$$

where k_1 and k_2 are rate constants, the amount $A_2(t)$ of substance A_2 present at any time t is

$$A_2(t) = -2.0(e^{-0.1t} - e^{-0.05t}),\ t \geq 0.$$

Assuming that time is measured in minutes, determine within the first 60 minutes the intervals in which A_2 is increasing or decreasing.

40. In the consecutive chemical reaction

$$A_1 \xrightarrow{k_1} A_2 \xrightarrow{k_2} A_3,$$

where k_1 and k_2 are rate constants, the amount $A_2(t)$ of substance A_2 present at any time t is

$$A_2(t) = -2.0(e^{-0.08t} - e^{-0.04t}),\ t \geq 0.$$

Assuming that time is measured in minutes, determine within the first 60 minutes the intervals in which A_2 is increasing or decreasing.

T **41.** Find to five significant figures the smallest positive zero of the second derivative of $f(x) = \sin x^2$. For your first guess, solve the equation $f''(x) = 0$ for x, first replacing (using the small-angle approximation) $\cos x^2$ by 1 and $\sin x^2$ by x^2 in the equation $f''(x) = 0$. You should find $x_1 \approx 0.84$.

42. Show that $x = 0$ is not an inflection point of the function $f(x) = x^4$. Is it contrary to the Concavity Test that for $f(x) = x^4$, $f''(0) = 0$ although $x = 0$ is not an inflection point? Explain.

43. Use the definition of an increasing function to determine the intervals on which $f(x) = \sin(\sqrt{x})$, $0 \leq x \leq 10$, is increasing.

44. Use the definition of an increasing function to determine the intervals on which $f(x) = \cos(\sqrt{x})$, $0 \leq x \leq 10$, is increasing.

T **45.** Calculate the rightmost inflection point in Example 7.

T **46.** Calculate the middle inflection point in Example 7.

47. Find the two inflection points of the function $f(x) = (x^2 + 1)e^x$, $-\infty < x < \infty$. Show that f is increasing on $(-\infty, \infty)$.

48. Give formulas for specific functions to illustrate the following pairs of properties:

a. increasing and concave up on $(0, 1)$

b. increasing and concave down on $(0, 1)$

c. decreasing and concave up on $(0, 1)$

d. decreasing and concave down on $(0, 1)$.

49. Discuss the concepts of *increasing, decreasing, concave up, concave down,* and *inflection point* in terms of motion on a line.

50. Do you think it possible that on $(0, \infty)$ a function f can be increasing and concave up? Can you give an example? Do you think it possible that on $(0, \infty)$ a function f can be increasing, concave up, and *bounded above?* "Bounded above" on $(0, \infty)$ means that a constant K can be found so that $f(x) \leq K$ for all $x \in (0, \infty)$.

51. Use the Mean-value Theorem to show that if for all $t > 0$ the derivative of a function f exceeds a fixed positive number q, then $\lim_{t\to\infty} f(t) = \infty$.

52. Show that if f is increasing (decreasing) on an interval I, then $-f$ is decreasing (increasing) on I.

53. Show that if f is concave up (concave down) on an interval (p, q), then $-f$ is concave down (concave up) on (p, q).

54. Let I be an open interval. Show that if f is differentiable on I and $f'(x) = 0$ for all $x = I$, then f is constant on I; that is, there is a constant c for which $f(x) = c$ on I.

55. Continuation of the above problem Use this result to show that if f and g are differentiable on I and $f'(x) = g'(x)$ on I, then f and g "differ by a constant"; that is, there is a constant c for which $f(x) - g(x) = c$ on I.

56. Use the Mean-value Theorem in showing that $\sin x \leq x$ for $x \geq 0$. Next show that $|\sin x| \leq |x|$ for all x.

4.4 Horizontal and Vertical Asymptotes

In this section we extend the definition of

$$\lim_{x \to a} f(x) = L$$

to cases in which either $L = \infty$ or $a = \infty$ and discuss the associated ideas of horizontal or vertical asymptotes. Although these ideas do not explicitly involve derivatives, they, together with the I/D Test and the Concavity Test of Section 4.3, are part of our discussion of how major features of functions and their graphs can be studied with such analytical tools as derivatives and limits. Sections 4.5 (Tools for Optimization) and 4.8 (Indeterminate Forms and l'Hôpital's Rules) will round out this discussion.

In Section 2.10 (and, more generally, in Exercise 37 of Section 4.3), we discussed the logistic equation

$$P(t) = \frac{292}{1 + 72e^{-0.03t}}, \qquad t \geq 0, \tag{1}$$

which describes the U.S. population fairly well up to 1990. Figure 4.24 shows the graph of this equation for $0 \leq t \leq 300$, where t counts the years from 1790. The actual population data for these years is shown as a series of solid dots.

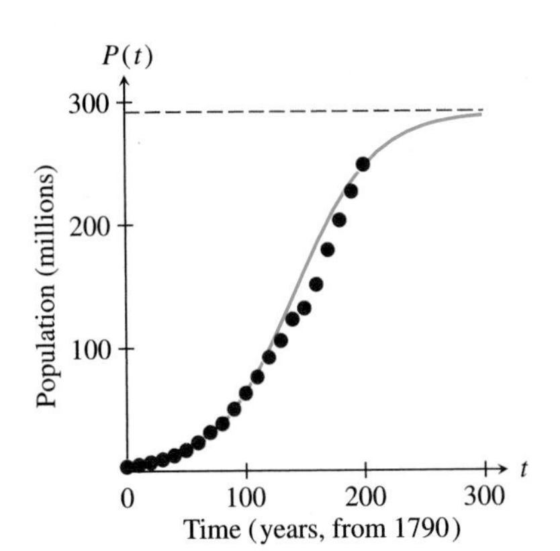

FIGURE 4.24 U.S. population and the logistic equation

The graph of the logistic equation (1) starts to level off after $t \approx 142$, which is roughly the inflection point of P. The graph appears to approach the dotted horizontal line as closely as we wish, provided that t is sufficiently large. If we wish to infer the height of this *horizontal asymptote* from the formula for $P(t)$ given in (1), we might start by calculating $P(300)$, $P(400)$, and $P(500)$. The results are

$$P(300) = 289.43$$
$$P(400) = 291.87$$
$$P(500) = 291.99.$$

We might reasonably infer from these calculations, as well as from Fig. 4.24, that the height of the dotted horizontal line is about 292.

If the graph of a function f approaches a vertical line as x approaches a number a or approaches a line or curve as x grows beyond any preassigned number, that line (or curve) is said to be an *asymptote* of the graph of f. The most common asymptotes are horizontal or vertical lines. We start our study with horizontal asymptotes.

Horizontal Asymptotes

When $\lim_{x \to a} f(x) = L$ was defined in Chapter 1, we said that a number L is the limit of $f(x)$ as x approaches a provided that we can make $f(x)$ as close to L as we like by taking x sufficiently close to a. For horizontal asymptotes, we consider limits in which x is increasing without bound or, more briefly, $x \to \infty$. To define $\lim_{x \to \infty} f(x) = L$, the only change from the definition of $\lim_{x \to a} f(x) = L$ is that we must replace the phrase "by taking x sufficiently close to a" by something that quantifies what it means for x to be "close to infinity."

WEB

We measure how close x is to infinity by specifying how far x is from the origin. We say "by taking x sufficiently large."

DEFINITION $\lim_{x\to\pm\infty} f(x) = L$, $\lim_{x\to\pm\infty} f(x) = \pm\infty$

We say that

$$\lim_{x\to\infty} f(x) = L,$$

where L is a real number, if f is defined on a set including an interval (c, ∞) and we can make $f(x)$ as close to L as we like by taking x sufficiently large.

We say that

$$\lim_{x\to-\infty} f(x) = L,$$

where L is a real number, if f is defined on a set including an interval $(-\infty, b)$ and we can make $f(x)$ as close to L as we like by taking $-x$ sufficiently large. We may prefer saying, "by taking x sufficiently large and negative."

We say that the line with equation $y = L$ is a horizontal asymptote of the graph of f if either $\lim_{x\to\infty} f(x) = L$ or $\lim_{x\to-\infty} f(x) = L$ or both.

We say that

$$\lim_{x\to\infty} f(x) = \infty$$

if f is defined on a set including an interval (c, ∞) and $f(x)$ becomes and remains as large as we like by taking x sufficiently large.

The remaining three possibilities among $\lim_{x\to\pm\infty} f(x) = \pm\infty$ are defined similarly. See Exercise 57.

Limits and Horizontal Asymptotes of Rational Functions A function having the form $p(x)/q(x)$, where $p(x)$ and $q(x)$ are polynomials, is called a *rational function.* Some examples of rational functions are

$$f(x) = \frac{x}{x + \frac{1}{2}}, \qquad g(x) = \frac{x^2 + \sqrt{2}}{x^3 + 3x^2 - x + 1}, \qquad h(x) = \frac{x^3 - \frac{7}{3}x^2 + 1}{\frac{2}{11}x^2 + x + \pi}.$$

The simplest rational functions have the form x^p or $1/x^p$, where p is a positive integer. If we can describe the limiting behavior of these two rational functions, we will have gone a long way toward describing the behavior of any rational function.

Limit of the Power Functions $1/x^p$ and x^p as $x \to \infty$

If p is any positive number, then

$$\lim_{x\to\infty} \frac{1}{x^p} = 0. \tag{2}$$

Hence, the line with equation $y = 0$ is a horizontal asymptote of the graph of the power function $1/x^p$.

If p is any positive number, then

$$\lim_{x\to\infty} x^p = \infty. \tag{3}$$

We show that (2) holds for any $p > 0$. For this we state more precisely the definition of $\lim_{x\to\infty} f(x) = L$ given previously. Taking $L = 0$ and $f(x) = 1/x^p$ in this definition, $\lim_{x\to\infty}(1/x^p) = 0$ provided that we can make $1/x^p$ as close to 0 as we like by taking x sufficiently large. Because the "closeness" of $1/x^p$ to 0 is measured by $|1/x^p - 0|$, we can restate the definition this way: If for each small, positive number E we can take x so large that

$$\left|\frac{1}{x^p} - 0\right| < E, \tag{4}$$

then $\lim_{x\to\infty}(1/x^p) = 0$. The inequality (4) simplifies to

$$\frac{1}{x^p} < E. \tag{5}$$

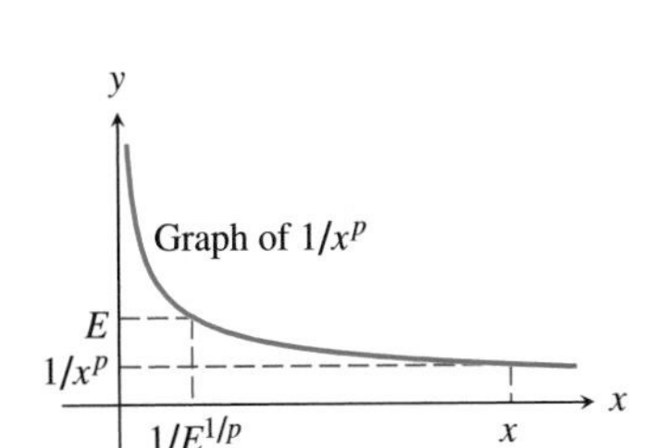

FIGURE 4.25 $1/x^p < E$ when $x > 1/E^{1/p}$.

Let's assume for the moment that this inequality holds. Taking the reciprocal of both sides,

$$x^p > \frac{1}{E}.$$

This inequality is satisfied if

$$x > \frac{1}{E^{1/p}}.$$

Because the steps in this calculation can be reversed, we conclude that (5) holds provided that $x > 1/E^{1/p}$. This proves that $\lim\limits_{x\to\infty} \dfrac{1}{x^p} = 0$. Figure 4.25 shows the geometry of this calculation: Given $E > 0$, we calculate $1/E^{1/p}$; for all $x > 1/E^{1/p}$ we see that $1/x^p < E$. Hence, $\lim_{x\to\infty} 1/x^p = 0$.

We leave the proof of (3) to Exercise 54.

In Fig. 4.25 and some of the examples that follow, the graphs of the functions have both horizontal and vertical asymptotes. We graph these functions in windows that include or suggest all asymptotes, but leave until later discussion of any vertical asymptotes, which are vertical lines near which the function values become unbounded.

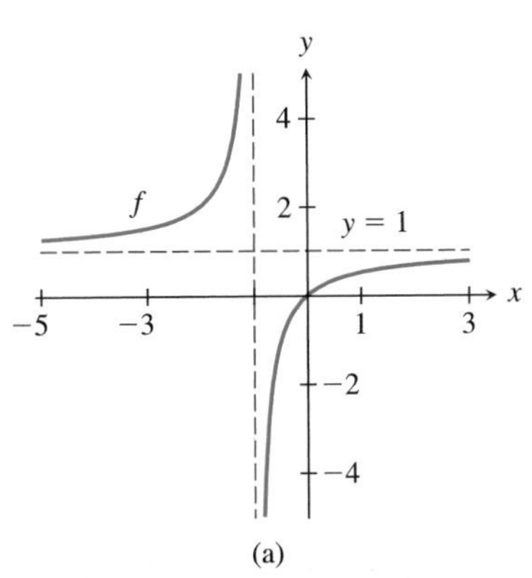

(a)

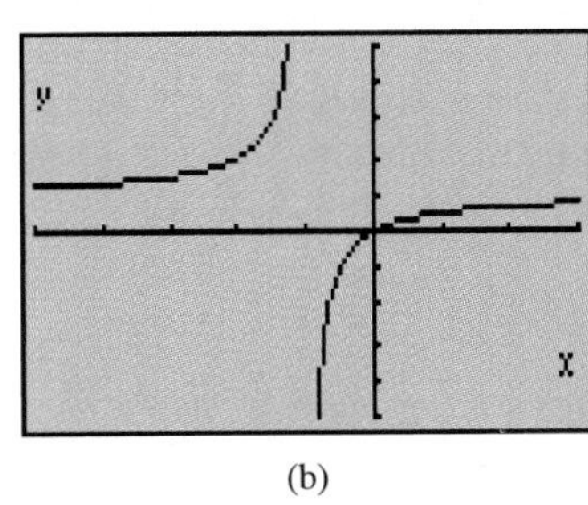

(b)

FIGURE 4.26 The graph and horizontal asymptote of $f(x) = x/(x + 1)$, as drawn by a CAS (a) and a calculator (b). The calculator screen is in the window $-5 \le x \le 3$ by $-5 \le y \le 5$.

EXAMPLE 1 Find the horizontal asymptote of the graph of the rational function

$$f(x) = \frac{x}{x+1}, \qquad x \neq -1.$$

Solution

Figure 4.26(a) shows the graph of f, as drawn by a CAS. Part (b) of the figure shows the same graph as drawn on a calculator screen. It appears from these figures that the graph of f approaches the line with equation $y = 1$ as $x \to \infty$ and as $x \to -\infty$. The data in Table 4.4 give numerical evidence that $\lim_{x\to\infty} x/(x + 1) = 1$ and $\lim_{x\to-\infty} x/(x + 1) = 1$. From these graphical and numerical arguments it appears that the line with equation $y = 1$ is a horizontal asymptote of the graph of f.

TABLE 4.4

x	9	99	999
$f(x)$	0.9	0.99	0.999
x	−11	−101	−1001
$f(x)$	1.1	1.01	1.001

We also give a direct argument that the line with equation $y = 1$ is a horizontal asymptote of f. The first step is to rearrange the expression $x/(x + 1)$ so that we can apply (2). For rational functions, we look for the "dominant term," which is the term with the highest power of x in either numerator or denominator. We then divide numerator and denominator by the dominant term. For the rational function

$$f(x) = \frac{x}{x+1},$$

the dominant term is x. Dividing numerator and denominator by x,

$$f(x) = \frac{\frac{x}{x}}{\frac{x+1}{x}} = \frac{1}{1 + \frac{1}{x}}.$$

Because $\lim_{x\to\infty} 1/x = 0$ by (2),

$$\lim_{x\to\infty} f(x) = \lim_{x\to\infty} \frac{1}{1 + \frac{1}{x}} = \frac{1}{1+0} = 1.$$

A similar argument shows that $\lim_{x\to-\infty} f(x) = 1$. Hence, the line with equation $y = 1$ is a horizontal asymptote of f.

In the direct argument given in this example we assumed that the limit of a sum is the sum of the limits and the limit of a quotient is the quotient of the limits. Limit theorems of this kind were given in Chapter 1 for the case in which $x \to a$, where a is a real number. The same limit theorems hold for limits in which $x \to \infty$ or $x \to -\infty$.

EXAMPLE 2 Determine the horizontal asymptotes of the graph of the function

$$f(x) = \frac{\sqrt{5x^2+3}}{7-x}, \qquad x \neq 7.$$

Solution

We start by calculating the limit of f as $x \to \infty$. For this calculation, we look for the dominant term of $f(x)$. This function is not a rational function because $\sqrt{5x^2 + 3}$ is not a polynomial. However, for large x, $5x^2 + 3 \approx 5x^2$

and, hence, $\sqrt{5x^2+3} \approx \sqrt{5}x$. In the denominator, the term with the highest power is $-x$. Hence, the dominant term of the fraction is x. Dividing numerator and denominator by x,

$$\lim_{x\to\infty} \frac{\sqrt{5x^2+3}}{7-x} = \lim_{x\to\infty} \frac{\dfrac{\sqrt{5x^2+3}}{x}}{\dfrac{7-x}{x}} = \lim_{x\to\infty} \frac{\sqrt{5+\dfrac{3}{x^2}}}{\dfrac{7}{x}-1}$$

$$= \lim_{x\to\infty} \frac{\sqrt{5+0}}{0-1} = -\sqrt{5}.$$

For the limit of f as $x \to -\infty$, we must be careful about taking x under the square root sign. Take a moment and consider how you would explain the following calculation to a friend.

$$\lim_{x\to-\infty} \frac{\sqrt{5x^2+3}}{7-x} = \lim_{x\to-\infty} \frac{\dfrac{\sqrt{5x^2+3}}{x}}{\dfrac{7-x}{x}} = \lim_{x\to-\infty} \frac{\dfrac{\sqrt{5x^2+3}}{-x}}{\dfrac{7-x}{-x}}$$

$$= \lim_{x\to-\infty} \frac{\sqrt{\dfrac{5x^2}{(-x)^2}+\dfrac{3}{(-x)^2}}}{\dfrac{7}{-x}-\dfrac{x}{-x}} = \lim_{x\to-\infty} \frac{\sqrt{5+0}}{0+1} = \sqrt{5}.$$

From these two calculations, we conclude that the lines with equations $y = \sqrt{5}$ and $y = -\sqrt{5}$ are horizontal asymptotes of the graph of f. Figure 4.27 shows the graph of f and these asymptotes. Figure 4.28 shows the same graph as drawn on a calculator screen.

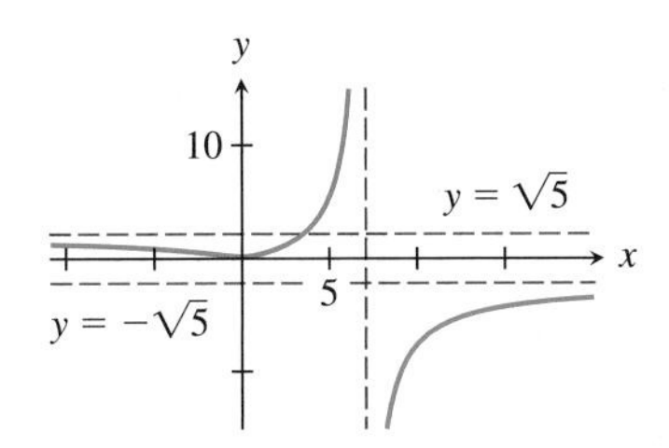

FIGURE 4.27 The graph and the two horizontal asymptotes of $f(x) = \sqrt{5x^2+3}/(7-x)$.

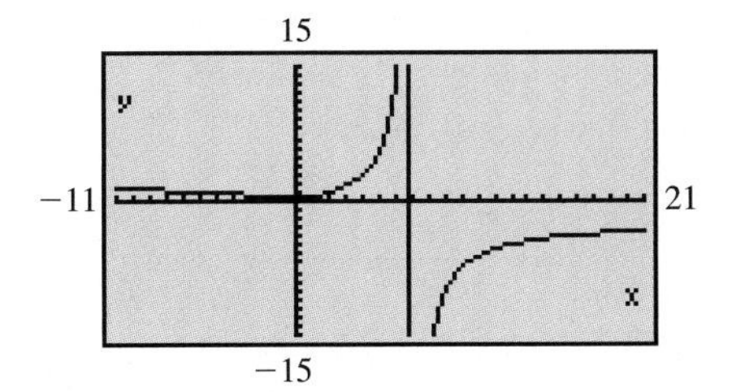

FIGURE 4.28 The graph of $f(x) = \sqrt{5x^2+3}/(7-x)$, in the window $-11 \le x \le 21$ by $-15 \le y \le 15$.

More on Dominant Terms In the preceding examples we used graphing, numerical data, and the idea of a dominant term to work out the limits of functions as $x \to \infty$. To extend the idea of a dominant term to a broader class of functions, we give two results comparing the growth of positive powers of the functions $\ln x$ and e^x to a power of x.

Growth of e^x and $\ln x$ Relative to a Power Function as $x \to \infty$

For any positive constants a and b,

$$\lim_{x\to\infty} \frac{(\ln x)^a}{x^b} = 0 \tag{6}$$

$$\lim_{x\to\infty} \frac{x^a}{e^{bx}} = 0. \tag{7}$$

In words, as $x \to \infty$, any power function x^b dominates any power $(\ln x)^a$ of the logarithm function. And, as $x \to \infty$, any power e^{bx} of the exponential function dominates any power function x^a.

The first of these results follows from the second by a change of variable. If in (7) we set $x = \ln w$,

$$\frac{x^a}{e^{bx}} = \frac{(\ln w)^a}{e^{b \ln w}} = \frac{(\ln w)^a}{e^{\ln(w^b)}} = \frac{(\ln w)^a}{w^b}.$$

As $x \longrightarrow \infty$, $w \longrightarrow \infty$; hence, if (7) is true, then (6) is also true.

We outline a proof of (7) in Exercise 56.

EXAMPLE 3 Does the function

$$f(x) = \frac{\sqrt{x}}{\ln x}, \qquad x > 0 \text{ and } x \neq 1$$

have a horizontal asymptote? What happens to $f(x)$ as $x \longrightarrow 0^+$? Discuss.

Solution

We start with the easier of these two questions. As $x \longrightarrow 0^+$, $\sqrt{x}$ approaches 0 and the absolute value of the denominator becomes large. These "forces" work together to make the fraction approach 0 as $x \longrightarrow 0^+$, that is,

$$\lim_{x \to 0^+} \frac{\sqrt{x}}{\ln x} = 0.$$

This result is consistent with the graph of f shown in Fig. 4.29.

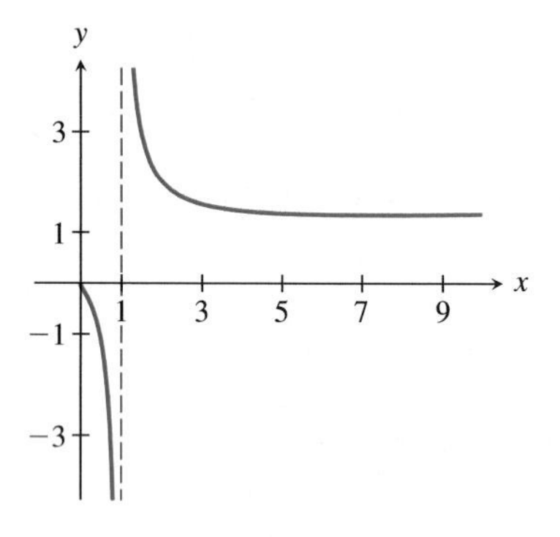

x	10^2	10^4	10^6	10^8
$\frac{\sqrt{x}}{\ln x}$	2.2	10.9	72.4	542.9

FIGURE 4.29 Graph of $f(x) = \sqrt{x}/\ln x$.

Although the graph of f in Fig. 4.29 is shown in a reasonable window, it is not clear whether f has a horizontal asymptote. We note that as $x \longrightarrow \infty$, both the numerator and denominator of $\sqrt{x}/\ln x$ become larger than any fixed number. In what follows, we attempt to decide which of $\sqrt{x}$ or $\ln x$ dominates by evaluating $f(x)$ at a sample of increasing values of x. The data at the bottom of the figure suggest that $f(x)$ increases with x but leave unsettled whether or not

$$\lim_{x \to \infty} \frac{\sqrt{x}}{\ln x} = \infty. \tag{8}$$

We use (6) to show that (8) is true, which means that the graph of f has no horizontal asymptote. From (6), with $a = 1$ and $b = 1/2$,

$$\lim_{x \to \infty} \frac{\ln x}{\sqrt{x}} = 0. \tag{9}$$

Because the ratio $\ln x/\sqrt{x}$ is positive and approaches 0 as $x \longrightarrow \infty$, its reciprocal, $\sqrt{x}/\ln x$, will become greater than any specified number for sufficiently large x, that is,

$$\lim_{x \to \infty} \frac{\ln x}{\sqrt{x}} = \lim_{x \to \infty} \frac{1}{\frac{\ln x}{\sqrt{x}}} = \infty.$$

This result suggests that we should not rely solely on graphs like that in Fig. 4.29 to make claims about asymptotic behavior.

EXAMPLE 4 Show that the graph of

$$f(x) = \frac{(\ln x)^{100}}{x^{0.01}}, \qquad x > 0,$$

has a horizontal asymptote.

Solution

Taking $a = 100$ and $b = 0.01$ in (6),

$$\lim_{x \to \infty} \frac{(\ln x)^{100}}{x^{0.01}} = 0.$$

Hence, the line with equation $y = 0$ is a horizontal asymptote of the graph of f.

Given (6), this argument is correct. However, because it gives little insight into the growth of the logarithm function relative to a power of x, we look at some numerical evidence.

For large values of x, the expression $(\ln x)^{100}$ increases much more rapidly than $\ln x$; at the same time, $x^{0.01}$ increases much less rapidly than x. Let's collect some numerical data (see Table 4.5).

TABLE 4.5 A sample of the values of $f(x)$ for increasing values of x.

x	10	100	1,000
$\frac{(\ln x)^{100}}{x^{0.01}}$	1.6×10^{36}	2.0×10^{66}	8.0×10^{83}

This evidence might tempt us to conclude that $(\ln x)^{100}/x^{0.01} \longrightarrow \infty$ as $x \longrightarrow \infty$. However, we argued from (6) that $\lim_{x \to \infty} f(x) = 0$. So it can't continue to increase. How could you find the point x at which $f(x)$ starts decreasing toward 0? See Exercises 45 and 46 for similar questions.

Vertical Asymptotes

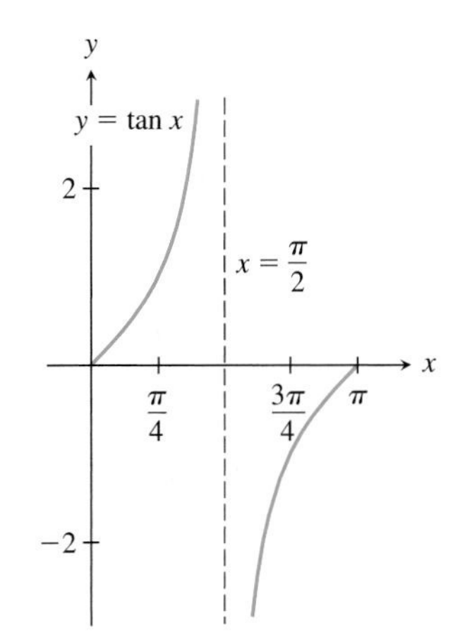

h	0.1	0.01	0.001
$\tan(\pi/2 - h)$	9.97	100.00	1000.00

FIGURE 4.30 One of the vertical asymptotes of the tangent function.

A line with equation $x = c$ is a vertical asymptote of a function f if $f(x)$ increases or decreases without limit as x approaches c from either the left or the right. Figure 4.30 shows the graph of $y = \tan x$ on the set $[0, \pi/2) \cup (\pi/2, \pi]$. Our past experience with the tangent function suggests that as x approaches $\pi/2$ from the left, the values of $\tan x$ become greater than any preassigned number; and, as x approaches $\pi/2$ from the right, the values of $\tan x$ become less (that is, "more negative") than any preassigned number. Hence, we say that the line $x = \pi/2$ is a vertical asymptote of the tangent function. The brief table of values at the bottom of the figure provides some numerical evidence that the line $x = \pi/2$ is a vertical asymptote for the tangent function. More generally, note that by applying the tangent line approximation,

$$\tan(\pi/2 - h) = \frac{\sin(\pi/2 - h)}{\cos(\pi/2 - h)} \approx \frac{\sin(\pi/2) + (-h)(\cos(\pi/2))}{\cos(\pi/2) + (-h)(-\sin(\pi/2))} = \frac{1}{h}.$$

Hence, as in the table, $\tan(\pi/2 - 10^{-n}) \approx 10^n$.

DEFINITION One-Sided, Unbounded Limits

We say that

$$\lim_{x \to c^-} f(x) = \infty$$

if f is defined on a set including an interval (a, c) and we can make $f(x)$ as large as we like by taking $x < c$ and sufficiently close to c.

We say that

$$\lim_{x \to c^+} f(x) = \infty$$

if f is defined on a set including an interval (c, b) and we can make $f(x)$ as large as we like by taking $x > c$ and sufficiently close to c.

The limits

$$\lim_{x \to c^-} f(x) = -\infty \qquad \text{and} \qquad \lim_{x \to c^+} f(x) = -\infty$$

are defined similarly.

We say that the line with equation $x = c$ is a vertical asymptote if at least one of the one-sided limits of f at c is infinite.

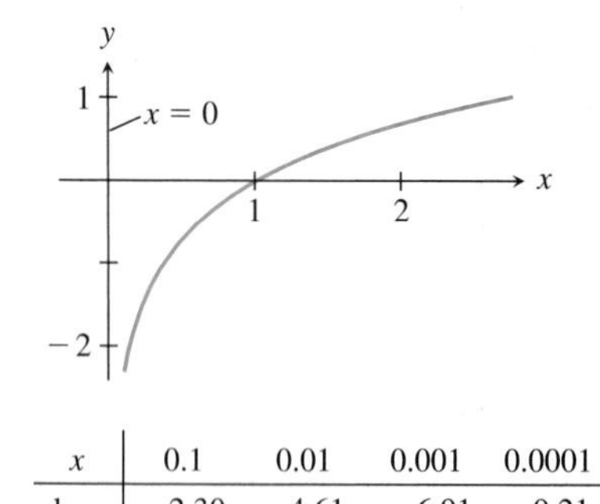

x	0.1	0.01	0.001	0.0001
$\ln x$	−2.30	−4.61	−6.91	−9.21

FIGURE 4.31 The vertical asymptote of the natural logarithm function is the line with equation $x = 0$.

Figure 4.31 shows the graph of the natural logarithm function for $x > 0$. From the figure we expect that

$$\lim_{x \to 0^+} \ln x = -\infty,$$

which would mean that $x = 0$ is a vertical asymptote. Use your calculator to check some of the values shown in the table at the bottom of Fig. 4.31. Also, determine, without recourse to a calculator, a value of $x > 0$ for which $\ln x \leq -100$. *Hint:* Exponentiate both sides of the inequality $\ln x \leq -100$.

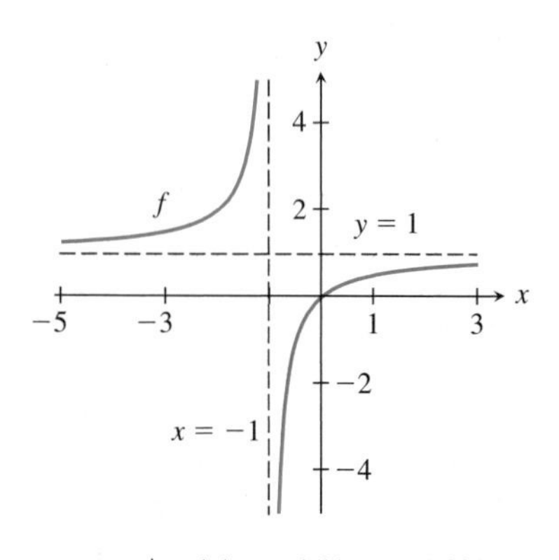

x	−1.1	−1.01	−1.001
$f(x)$	11	101	1001
x	−0.9	−0.99	−0.999
$f(x)$	−9	−99	−999

FIGURE 4.32 The graph and vertical and horizontal asymptotes of $f(x) = x/(x + 1)$.

EXAMPLE 5 Find the vertical asymptote of the function

$$f(x) = \frac{x}{x+1}, \qquad x \neq -1.$$

Solution

We found the horizontal asymptote of the graph of f in Example 1. Figure 4.32 shows the graph of f. It appears from the graph and the table of values shown at the bottom of the figure that the line with equation $x = -1$ is a vertical asymptote of the graph of f and, in particular, that

$$\lim_{x \to -1^+} \frac{x}{x+1} = -\infty \qquad \text{and} \qquad \lim_{x \to -1^-} \frac{x}{x+1} = \infty.$$

The one-sided limits of functions or vertical asymptotes often can be found by doing some mental arithmetic. For example, when $x < -1$ and close to -1, the denominator $x + 1$ is a "small negative" number (that is, $|x + 1|$ is small) and the numerator x is close to -1. Hence, for such x,

$$\frac{x}{x+1} \approx \frac{-1}{\text{"small negative" number}} = \text{large positive number.}$$

As x approaches -1 from the right, the denominator $x + 1$ of $f(x)$ is a small positive number and the numerator x is close to -1. Hence, for such x,

$$\frac{x}{x+1} \approx \frac{-1}{\text{small positive number}} = \text{"large negative" number.}$$

We infer from this mental arithmetic that $x/(x + 1) \to \infty$ as $x \to -1^-$ and $x/(x + 1) \to -\infty$ as $x \to -1^+$. Because at least one of these one-sided limits is infinite, the line with equation $x = -1$ is a vertical asymptote.

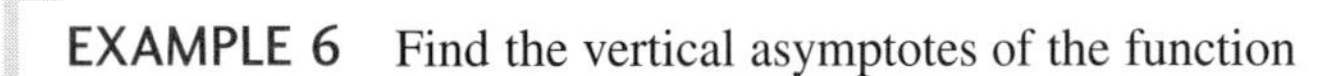

EXAMPLE 6 Find the vertical asymptotes of the function

$$f(x) = \frac{x^2}{(x+2)(x-1)}, \qquad x \neq -2, 1.$$

Solution

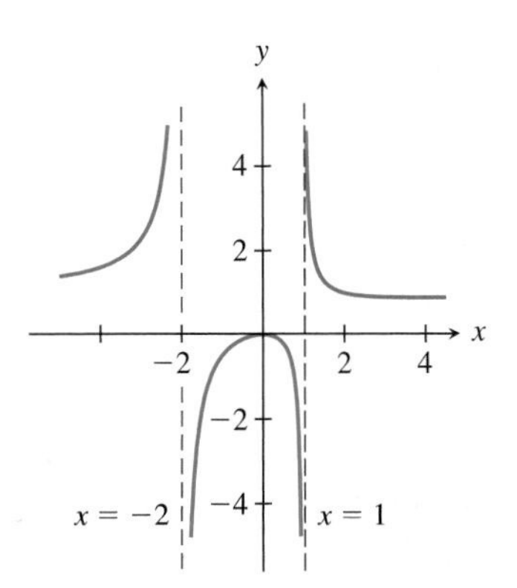

FIGURE 4.33 The graph and vertical asymptotes of $f(x) = x^2/((x + 2)(x - 1))$.

Figures 4.33 and 4.34 show the graph of f. The form of the graph in Fig. 4.33—which was drawn by a CAS—strongly suggests that the vertical asymptotes are the dotted lines shown in the figure. These lines have equations $x = -2$ and $x = 1$.

The solid vertical or near-vertical line segments shown in Fig. 4.34 look like the graphs of the vertical asymptotes with equations $x = -2$ and $x = 1$. These line segments were drawn by our calculator as it plotted the graph of f. To draw the graph of f the calculator chose a sample $x_1, x_2, \ldots, x_n$ of values and plotted the line segments joining $(x_1, f(x_1))$ and $(x_2, f(x_2))$, $(x_2, f(x_2))$ and $(x_3, f(x_3))$, and so on. It happened that, for a certain i, $x_i < -2 < x_{i+1}$. Because both $f(x_i) > 5$ and $f(x_{i+1}) < -5$, the line segment joining the points $(x_i, f(x_i))$ and $(x_{i+1}, f(x_{i+1}))$ was plotted as a near-vertical or vertical line segment. A similar thing happened near $x = 1$.

We may also locate any vertical asymptotes of f by examining its form. Because the denominator of this rational function has two real zeros, $x = -2$ and $x = 1$, we expect the lines with equations $x = -2$ and $x = 1$ to be vertical asymptotes. The one-sided limits of f at the two zeros are

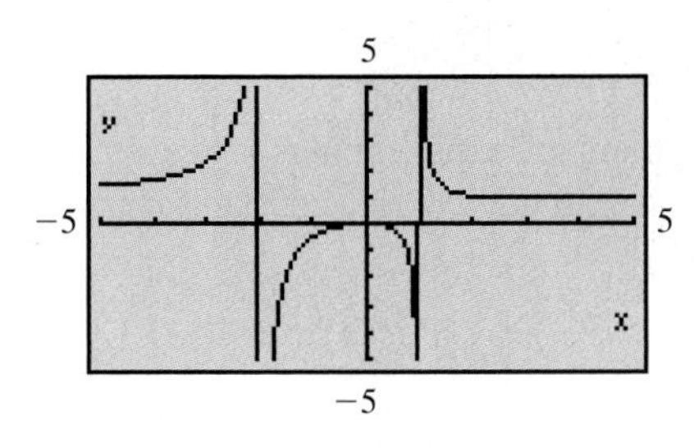

FIGURE 4.34 The graph and vertical asymptotes of $f(x) = x^2/((x + 2)(x - 1))$, in the window $-5 \le x \le 5$ by $-5 \le y \le 5$.

$$\begin{aligned}
\lim_{x\to-2^-} f(x) &= \lim_{x\to-2^-} \frac{x^2}{(x+2)(x-1)} = \infty \\
\lim_{x\to-2^+} f(x) &= \lim_{x\to-2^+} \frac{x^2}{(x+2)(x-1)} = -\infty \\
\lim_{x\to1^-} f(x) &= \lim_{x\to1^-} \frac{x^2}{(x+2)(x-1)} = -\infty \\
\lim_{x\to1^+} f(x) &= \lim_{x\to1^+} \frac{x^2}{(x+2)(x-1)} = \infty.
\end{aligned} \tag{10}$$

These limits can be calculated by some mental arithmetic. For the first limit, we imagine that x is close to -2 but $x < -2$. This means that $x + 2$ is small and negative, $x - 1 \approx -3$, and $x^2 \approx 4$. Hence, for such x,

$$\frac{x^2}{(x+2)(x-1)} \approx \frac{4}{(\text{small negative})(-3)} = \text{large positive}.$$

This suggests that the first limit in (10) is true.

The other one-sided limits can be calculated in a similar way. Hence, the lines with equations $x = -2$ and $x = 1$ are vertical asymptotes of the graph of f.

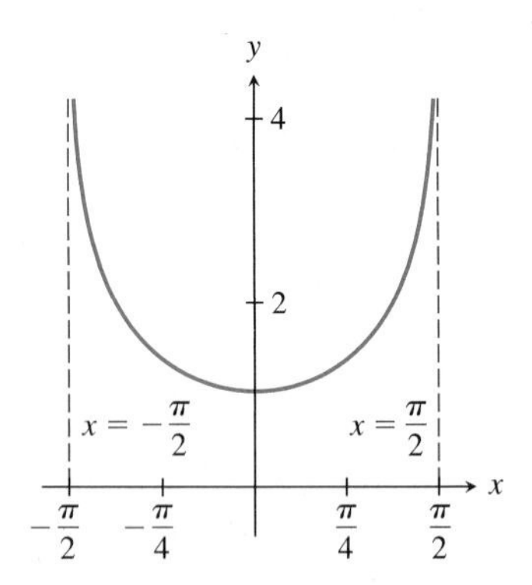

FIGURE 4.35 Graph and vertical asymptotes of $f(x) = 1 - \ln(\cos x)$, $|x| < \pi/2$.

EXAMPLE 7 Show that the lines with equations $x = -\pi/2$ and $x = \pi/2$ are vertical asymptotes of the graph of

$$f(x) = 1 - \ln(\cos x), \qquad |x| < \pi/2.$$

Solution

Because $f(-x) = f(x)$ for all x, the graph of f is symmetric about the y-axis. Using this symmetry, we can infer the shape of the graph of f and its asymptotes on $(-\pi/2, \pi/2)$ by studying f on $[0, \pi/2)$. To show that the line with equation $x = \pi/2$ is a vertical asymptote of f, we consider $\lim_{x\to(\pi/2)^-} f(x)$. As $x \to (\pi/2)^-$, $\cos x$ is positive and is approaching 0. Because $\ln y \to -\infty$ as $y \to 0^+$, we conclude that

$$\lim_{x\to(\pi/2)^-} (1 - \ln(\cos x)) = \infty.$$

Hence, $x = \pi/2$ is a vertical asymptote of the graph of f. Figure 4.35 shows the graph of f and its vertical asymptotes.

Summary of Sections 4.3 and 4.4 In the last two sections, we have discussed several ways of finding and describing the main features of functions or their graphs. With the help of the

- I/D and Concavity Tests

from Section 4.3 we can locate intervals on which f is increasing or decreasing or is concave up or concave down. As discussed in Section 4.4, we can analyze the form of the expression defining f to locate any

- horizontal or vertical asymptotes.

And if the expression defining f contains polynomials, exponentials, or logarithms, we may be able to apply (6) or (7), which state that for any positive constants a and b,

- $\lim_{x\to\infty} \dfrac{(\ln x)^a}{x^b} = 0$
- $\lim_{x\to\infty} \dfrac{x^a}{e^{bx}} = 0.$

We supplement these tools in Section 4.5 with methods for locating any high or low points of the graph of a function and, in Section 4.8, with l'Hôpital's Rules for calculating the limits of certain "indeterminate forms."

Exercises 4.4

Exercises 1–8: Evaluate the limits.

1. $\lim_{x\to\infty} \dfrac{x}{x^2+1}$

2. $\lim_{x\to\infty} \dfrac{3x-1}{x^2+5}$

3. $\lim_{x\to\infty} \dfrac{x^3}{2x+1}$

4. $\lim_{x\to\infty} \dfrac{2x^2+3}{3x-7}$

5. $\lim_{x\to\infty} \dfrac{\sqrt{x^2+1}}{x+1}$

6. $\lim_{x\to\infty} \dfrac{\sqrt{2x^2+9}}{3x-5}$

7. $\lim_{x\to\infty} \dfrac{\sqrt{\ln x}}{\sqrt[3]{x}}$

8. $\lim_{x\to\infty} \dfrac{(\ln x)^3}{\sqrt{x}}$

Exercises 9–16: Evaluate the limits.

9. $\lim_{x\to 3^+} \dfrac{x}{x^2-2x-3}$

10. $\lim_{x\to(-2)^-} \dfrac{2x^3-7}{x^2-3x-10}$

11. $\lim_{x\to 2^+} \dfrac{5}{\sqrt{x^2-x-2}}$

12. $\lim_{x\to(-1)^-} \dfrac{3x-8}{\sqrt{2x^2-3x-5}}$

13. $\lim_{x\to(\pi/2)^-} \ln(\cot x)$

14. $\lim_{x\to 0^+} \ln(\csc x)$

15. $\lim_{x\to 0^+} \dfrac{x^{3/2}}{\sqrt{\sin x}}$

16. $\lim_{x\to 0^+} \dfrac{\sqrt{x}}{\sqrt{\tan x}}$

Exercises 17–42: Determine analytically all horizontal or vertical asymptotes. Graph the function and asymptotes.

17. $f(x) = \dfrac{1}{x-2},\ x \neq 2$

18. $f(x) = \dfrac{1}{x+3},\ x \neq -3$

19. $f(x) = \dfrac{2x-3}{x+1},\ x \neq -1$

20. $f(x) = \dfrac{3x-2}{x-1},\ x \neq 1$

21. $f(x) = \dfrac{x^2+2x-4}{x^2},\ x \neq 0$

22. $f(x) = \dfrac{x^2-x-2}{x^3},\ x \neq 0$

23. $\sec x,\ x \in [0, 2\pi]$ and $x \neq \pi/2, 3\pi/2$

24. $\csc x,\ x \in [0, 2\pi]$ and $x \neq 0, \pi, 2\pi$

25. $f(x) = \arctan x,\ -\infty < x < \infty$

26. $f(x) = e^{-x},\ -\infty < x < \infty$

27. $f(x) = \dfrac{\sqrt{x}+1}{\sqrt{x}-1},\ x \geq 0$ and $x \neq 1$

28. $f(x) = \dfrac{2x}{\sqrt{x^2+1}}$

29. $f(x) = \dfrac{x^3}{(x+1)^2(x-2)},\ x \neq -1, 2$

30. $f(x) = \dfrac{x^2}{2x^2+7x-4},\ x \neq -4, 1/2$

31. $f(x) = \dfrac{\sqrt{2x^2+1}}{3x+1},\ x \neq -1/3$

32. $f(x) = \dfrac{x^2+\sqrt{x+1}}{\sqrt{x^4-1}},\ x > 1$

33. $f(x) = 1 - \ln(\sin x),\ 0 < x \leq \pi/2$

34. $f(x) = 1 + \ln(\tan x),\ 0 < x < \pi/2$

35. $f(x) = \dfrac{x^2}{\ln x},\ x > 0$ and $x \neq 1$

36. $f(x) = \dfrac{\sqrt{x}}{\ln x},\ x > 0$ and $x \neq 1$

37. $f(x) = xe^{-x},\ -\infty < x < \infty$

38. $f(x) = \sqrt{xe^{-x}},\ x > 0$

39. $f(x) = e^{-x/4}\sin x,\ x \geq 0$

40. $f(x) = e^{-x/4}\cos x,\ x \geq 0$

41. $f(x) = \dfrac{x^{1/3}}{x+1},\ x \neq -1$

42. $f(x) = \dfrac{x^{1/5}}{x+1},\ x \neq -1$

43. Determine the inflection point and any horizontal or vertical asymptotes for the logistic equation (1). Write a sentence or two about the meaning of the inflection point in terms of population growth.

44. What happens to the graph of

$$f(x) = \frac{e^{-1/x}}{x^{10}}, \quad x \neq 0$$

as $x \to 0^+$?

45. Use your calculator to figure out how big x must be for $e^{0.01x}$ to catch up to x^{100}.

46. Use your calculator to figure out how big x must be for $e^{0.001x}$ to catch up to x^{1000}.

47. Evaluate

$$\lim_{x\to\infty}\frac{\ln(\ln x)}{\sqrt{x}}.$$

48. Evaluate

$$\lim_{x\to\infty}\frac{\sqrt{1+2^x}}{\sqrt{1+3^x}}.$$

49. Graph $f(x) = x^x$, $0 < x < 2$. You may wish to evaluate f at 0.1, 0.01, and 0.001 to get an idea of its limiting value as $x \to 0^+$. Calculate $\lim_{x\to 0^+} x^x$ by writing $x^x = e^{x\ln x}$ and then making the substitution $x = 1/w$. Does the graph have vertical or horizontal asymptotes?

50. Graph $f(x) = x^{1/x}$, $0 < x < \infty$. Calculate $\lim_{x\to 0^+} x^{1/x}$ by writing $x^{1/x} = e^{(\ln x/x)}$. Does the graph have vertical or horizontal asymptotes?

51. Graph the function $f(x) = x + 1/x$, $x \neq 0$. Give an informal argument as to why the graph of $y = x$ can be regarded as an asymptote to the graph of f.

52. Graph the function $f(x) = (x^3 + 1)/x$, $x \neq 0$. Give an informal argument as to why the graph of $y = x^2$ can be regarded as an asymptote to the graph of f.

53. Show that the graph of $y = (b/a)x$ is an asymptote of the graph of the equation $x^2/a^2 - y^2/b^2 = 1$ in the sense that if the equation for the hyperbola is solved for y in terms of x, then $\lim_{x\to\pm\infty}(y - (b/a)x) = 0$.

54. Show that (3) holds by showing that for any (large) positive number M, a number B can be found for which $x^p > M$ for all $x > B$.

55. Show that $\lim_{x\to 0^+} x \ln x = 0$. *Hint:* Change variables by letting $w = 1/x$. Note that as $x \to 0^+$, $w \to \infty$. The limit can be written as

$$\lim_{x\to 0^+} x\ln x = \lim_{w\to\infty}\frac{-\ln w}{w}.$$

56. Use the following outline to show that for any positive number n,

$$\lim_{x\to\infty}\frac{x^n}{e^x} = 0.$$

Begin by using the I/D Test to show that

$$e^x > 1, \quad x > 0.$$

Next, use the I/D Test and the previous result to show that

$$e^x > 1 + x, \quad x > 0.$$

Continue in this way, showing that for any positive number m,

$$e^x > 1 + x + \frac{x^2}{2} + \cdots + \frac{x^m}{m!}, \quad x > 0.$$

Finally, if $n > 0$, choose a positive integer m for which $m > n$, and use the inequality

$$\frac{x^n}{e^x} < \frac{x^n}{1 + x + x^2/2 + \cdots + x^m/m!}.$$

Use this result to prove (7).

57. Write out definitions of the limits $\lim_{x\to\infty} f(x) = -\infty$, $\lim_{x\to-\infty} f(x) = \infty$, and $\lim_{x\to-\infty} f(x) = -\infty$.

58. Write out definitions of the limits $\lim_{x\to c^-} f(x) = -\infty$ and $\lim_{x\to c^+} f(x) = \infty$.

4.5 Tools for Optimization

This section and the next concern optimization. To *optimize* an object or process generally means to maximize or minimize some aspect of the object or process. For example, a sawyer may wish to cut a rectangular beam of maximum strength (least deflection under uniform loading of the top side) from a circular log. Figure 4.36 shows a cross section of a log of radius a and of a representative beam. If we use the fact (usually stated and proved in courses called "Mechanics of Materials") that the strength S of such a beam is proportional to the product of its width w and the square of its height h, we may solve the sawyer's problem by choosing w so that S is a maximum, where

$$S(w) = kwh^2 = kw(4a^2 - w^2), \qquad 0 \le w \le 2a. \tag{1}$$

In this equation, k is a positive constant (of proportionality), w is the width, and $h^2 = 4a^2 - w^2$ is the square of the height of the log expressed in terms of a and w.

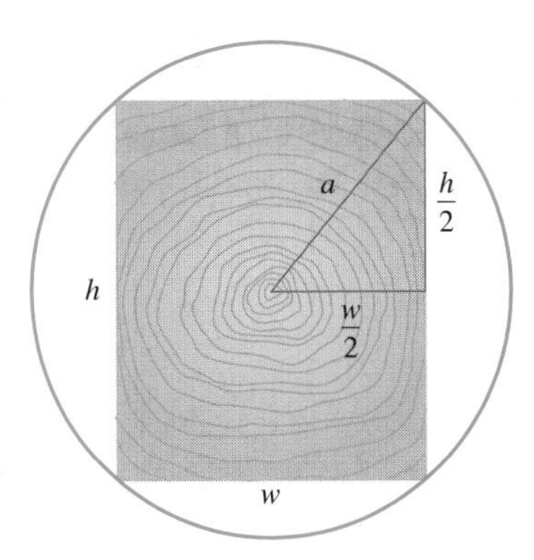

FIGURE 4.36 Rectangular beam cut from a circular log of radius a.

The inequality $0 \le w \le 2a$ is a precise way of stating that the width of the beam must be 0 or more and less than or equal to $2a$, the diameter of the log.

In the present section, we assume that a function such as S in (1) is given and discuss the mathematical tools used to find the values of w at which S takes on its maximum or minimum value. In Section 4.6 we focus more on how to formulate or model optimization problems.

Local Extrema

We begin by defining what it means for a function f to have a local or global maximum or minimum at a point c. The distinction between *local* and *global* agrees with everyday usage. You may be the tallest person locally (for example, in your hometown or neighborhood), but it is not likely that you are the tallest person in the world.

DEFINITION Local or Global Extrema

Let f be a function defined on a set $\mathcal{D}$. We say that f has a local maximum at $c \in \mathcal{D}$ if

$$f(c) \ge f(x) \text{ for all } x \in \mathcal{D} \text{ sufficiently close to } c.$$

We say that f has a local minimum at $c \in \mathcal{D}$ if

$$f(c) \le f(x) \text{ for all } x \in \mathcal{D} \text{ sufficiently close to } c.$$

We say that f has a global maximum at $c \in \mathcal{D}$ if

$$f(c) \ge f(x) \text{ for all } x \in \mathcal{D}.$$

We say that f has a global minimum at $c \in \mathcal{D}$ if

$$f(c) \le f(x) \text{ for all } x \in \mathcal{D}.$$

If f has a local maximum, local minimum, global maximum, or global minimum at $c \in \mathcal{D}$, we say that f has an *extremum* at c and that c is an *extreme point.* The word "global" is often omitted or replaced by the word "absolute."

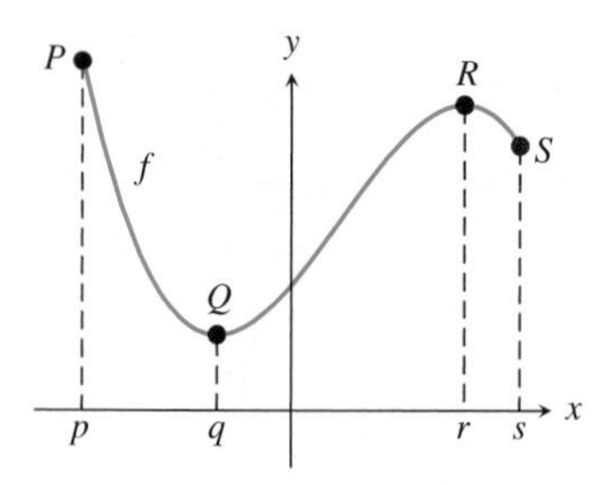

FIGURE 4.37 Highest and lowest points on the graph of f (P and Q); also, points that are relatively high and low on this graph (R and S).

Figure 4.37 shows the graph of a function f defined on the set $\mathcal{D} = [p, s]$. This function has local maxima at p and r, a global maximum at p, local minima at q and s, and a global minimum at q.

The most important tool for locating candidates for extreme points is the Candidate Theorem.

THEOREM Candidate Theorem

Let f be a function defined on an interval I. If f has an extremum at a point c of I, then either

i) c is an endpoint of I,
ii) f is not differentiable at c, or
iii) $f'(c) = 0$.

Points $c \in I$ which are either endpoints of I, points at which f is not differentiable, or points at which f' is zero are called *candidates* or *critical points.*

To prove the Candidate Theorem, we must show that if f has, say, a local maximum at $c \in I$, then c is either an endpoint of I, a point of I at which f is not differentiable, or a point at which f' is zero. If c is an endpoint, then (i) holds. If c is a point of I at which f is not differentiable, then (ii) holds. So suppose that c is not an endpoint and f is differentiable at c. Because f has a local maximum at c,

$$f(c) \geq f(x) \qquad \text{for all } x \text{ sufficiently close to } c. \tag{2}$$

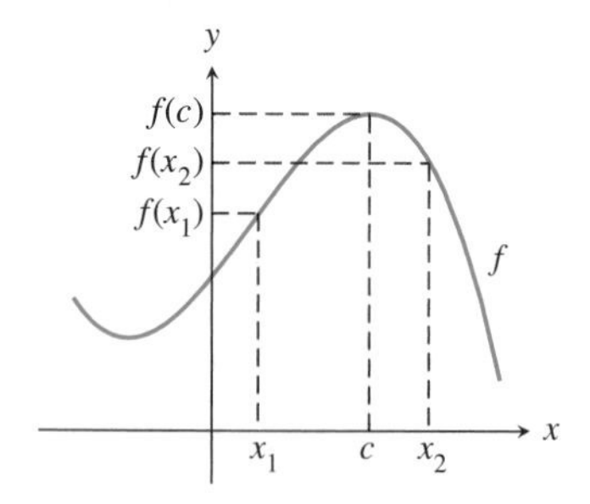

FIGURE 4.38 If f has a local maximum at c, then $(f(x_1) - f(c))/(x_1 - c) \geq 0$ for $x_1 < c$ and $(f(x_2) - f(c))/(x_2 - c) \leq 0$ for $x_2 > c$.

Figure 4.38 shows representative points x_1 and x_2 to the left and right of c. It follows from (2) that for all such points x_1 and x_2 sufficiently close to c,

$$f(x_1) - f(c) \leq 0 \tag{3}$$

$$f(x_2) - f(c) \leq 0. \tag{4}$$

From (3) and the fact that $x_1 < c$,

$$\frac{f(x_1) - f(c)}{x_1 - c} \geq 0. \tag{5}$$

Similarly, from (4) and the fact that $x_2 > c$,

$$\frac{f(x_2) - f(c)}{x_2 - c} \leq 0. \tag{6}$$

It follows from (5) that

$$f'(c) = \lim_{x_1 \to c^-} \frac{f(x_1) - f(c)}{x_1 - c} \geq 0,$$

and from (6) that

$$f'(c) = \lim_{x_2 \to c^+} \frac{f(x_2) - f(c)}{x_2 - c} \leq 0.$$

Hence, $f'(c) = 0$, that is, (iii) holds.

A proof for the case in which f has a local minimum at $c \in I$ is similar.

EXAMPLE 1 Assume that the function

$$f(x) = 1 + 2x - 3\sqrt{x}, \qquad 0 \le x \le 2$$

has both an absolute maximum and an absolute minimum. Use the Candidate Theorem to locate the points at which f takes on its absolute maximum and minimum values.

Solution

Let c be the point at which f has an absolute maximum. By the Candidate Theorem, c must be either (i) an endpoint of $[0, 2]$, (ii) a point where f is not differentiable, or (iii) a point for which $f'(c) = 0$. Because f is not differentiable only at $x = 0$, c must be 0, 2, or a point for which $f'(c) = 0$. Differentiating f,

$$f'(x) = 2 - \frac{3}{2\sqrt{x}}.$$

Solving the equation $f'(x) = 0$ for x,

$$2 - \frac{3}{2\sqrt{x}} = 0$$
$$\frac{3}{2\sqrt{x}} = 2$$
$$x = 9/16.$$

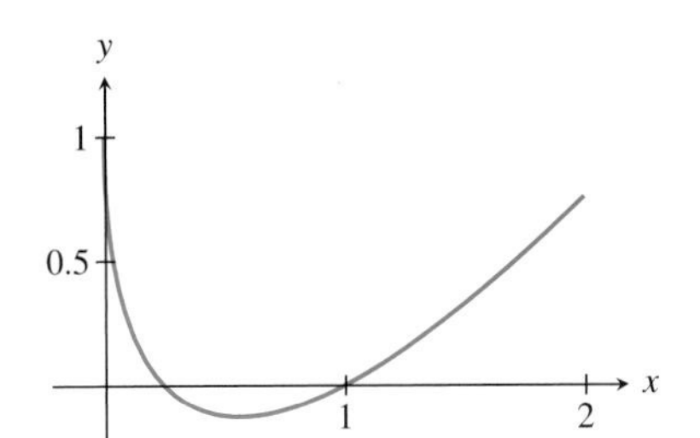

FIGURE 4.39 The graph of $f(x) = 2 + 2x - 3\sqrt{x}$.

The maximum point c must be either 0, 2, or 9/16. Evaluating f at these three candidates,

$$f(0) = 1, \qquad f\left(\frac{9}{16}\right) = 1 + \frac{9}{8} - \frac{9}{4} = -\frac{1}{8},$$
$$f(2) = 1 + 4 - 3\sqrt{2} \approx 0.7574.$$

Hence, the absolute maximum of f occurs at $c = 0$. On the same evidence, the absolute minimum of f occurs at 9/16. The maximum and minimum values of f are 1 and $-1/8 = -0.125$.

Figure 4.39 shows the graph of f.

In Example 1 we said, "Assume that the function . . . has both an absolute maximum and an absolute minimum." We made this assumption so that we could apply the Candidate Theorem, whose second sentence is "If f has an extremum at a point c of I, then. . . ." The Max/Min Theorem gives conditions under which we may be certain that a given function has an absolute maximum and an absolute minimum on an interval and, hence, conditions guaranteeing that the Candidate Theorem may be applied.

THEOREM Max/Min Theorem

If f is continuous on the interval $[a, b]$, then f has a maximum value and a minimum value on $[a, b]$.

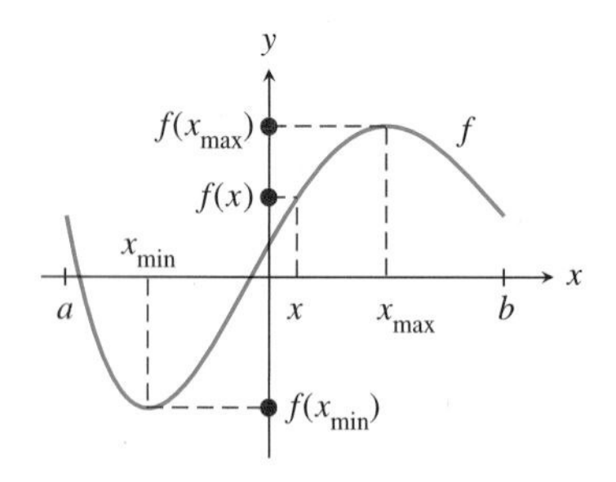

FIGURE 4.40 The high (maximum) and low (minimum) points of a function.

The phrase "has a maximum value and a minimum value on $[a,b]$" means that there are numbers x_{min}, $x_{max} \in [a,b]$ for which

$$f(x_{min}) \le f(x) \le f(x_{max}), \qquad \text{for every } x \in [a,b].$$

See Fig. 4.40. The numbers $f(x_{min})$ and $f(x_{max})$ are the minimum and maximum values of f. The points $(x_{min}, f(x_{min}))$ and $(x_{max}, f(x_{max}))$ are low and high points on the graph of f.

The proof of the Max/Min Theorem is beyond the level of this text.

EXAMPLE 2 Figure 4.41 shows a cross section of a beam of height h and width w, cut from a log of radius a. The strength S of such a beam is proportional to the product of its width w and the square of its height h. Find a "rule of thumb" for Tom, the sawyer, to cut beams of maximum strength.

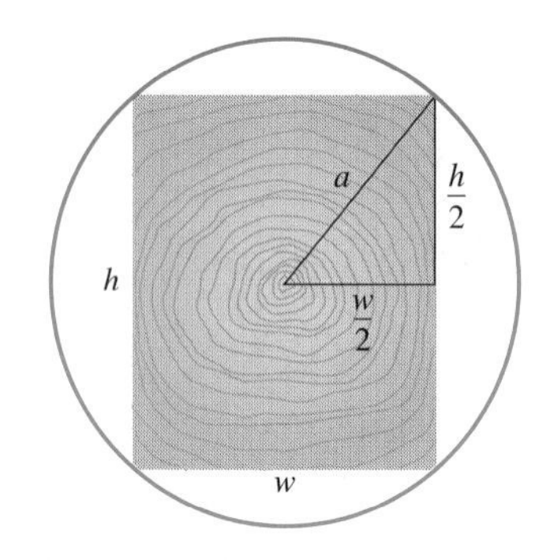

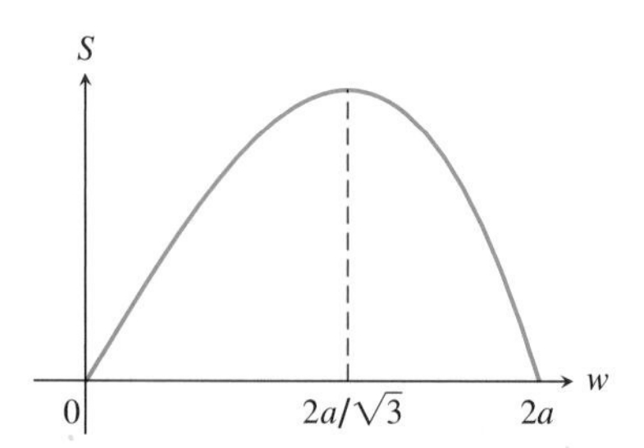

FIGURE 4.41 The graph of the strength S of the beam as a function of its width w.

Solution

The strength S of the beam is $S = kwh^2$, where k is a constant. Because the width and height of the beam are related through the right triangle shown in Fig. 4.41, we can express S in terms of just one of the two variables w and h. From the Pythagorean Theorem,

$$\left(\frac{h}{2}\right)^2 + \left(\frac{w}{2}\right)^2 = a^2. \tag{7}$$

Solving this equation for h^2 and substituting into $S = kwh^2$,

$$S(w) = kwh^2 = kw(4a^2 - w^2) = k(4a^2w - w^3). \tag{8}$$

Given that the diameter of the log is $2a$, we take the domain of S to be the closed interval $[0, 2a]$. Although one might argue that $w = 0$ and $w = 2a$ do not correspond to "real" beams, we use the closed interval as part of our model so that we can apply the Max/Min Theorem.

So, our job now is to maximize the function S defined on the closed interval $[0, 2a]$. Figure 4.41 shows the graph of S.

Because S is continuous on $[0, 2a]$, we know by the Max/Min Theorem that it has both a maximum and a minimum. Because S is differentiable, we know by the Candidate Theorem that the candidates for maximum include only the endpoints 0 and $2a$ and the zeros of S' in $(0, 2a)$. Because

$$S' = k(4a^2 - 3w^2),$$

the only zero of S' in $(0, 2a)$ is $w = 2a/\sqrt{3}$. Evaluating $S(w)$ at the candidates 0, $2a$, and $2a/\sqrt{3}$,

$$S(0) = 0, \qquad S(2a) = 0, \qquad \text{and} \qquad S\left(2a/\sqrt{3}\right) > 0.$$

Hence, the function S has a global maximum at $w = 2a/\sqrt{3}$.

Although we could give Tom the rule "measure a and calculate $w = 2a/\sqrt{3}$," a more practical rule would be useful. For this we calculate the ratio h/w:

$$\frac{h^2}{w^2} = \frac{4a^2 - w^2}{w^2} = \frac{4a^2 - 4a^2/3}{4a^2/3} = 2,$$

from which

$$\frac{h}{w} = \sqrt{2}.$$

A reasonable rule might be "cut it half again as high as it is wide."

The Candidate Theorem gives us a list of candidates for the extrema of a function, but it does not give us any idea whether a candidate is a local minimum or maximum. However, as in Example 2, we can classify the candidates $c_1, c_2, \ldots$ by comparing the values of $f(c_1), f(c_2), \ldots$. And we can use the First Derivative Test, which is a convenient, easily applied test to decide whether a point c at which $f'(c) = 0$ is a local maximum, a local minimum, or neither.

First Derivative Test

Let f be a function defined on an interval I and c an interior point of I. Assume that f is differentiable at and near c.

i) If $f'(c) = 0$ and f' changes from positive to negative at c, then c is a local maximum;
ii) if $f'(c) = 0$ and f' changes from negative to positive at c, then c is a local minimum; and
iii) if f' does not change sign at $x = c$, then c is neither a local maximum nor a local minimum.

Figure 4.42 provides graphical evidence for the First Derivative Test. A proof using the Mean-value Theorem is outlined in Exercise 40.

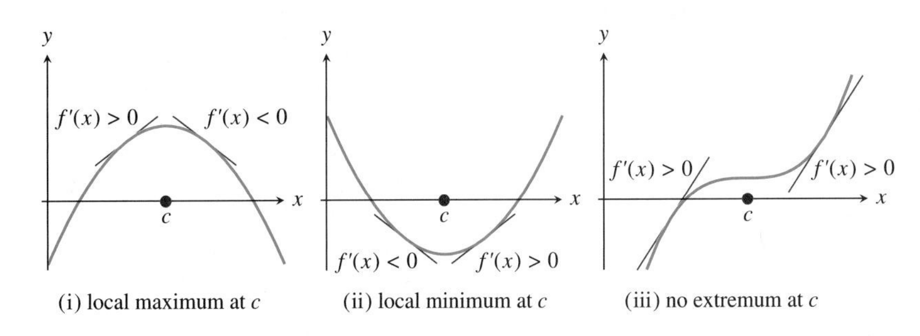

FIGURE 4.42 The three cases of the First Derivative Test.

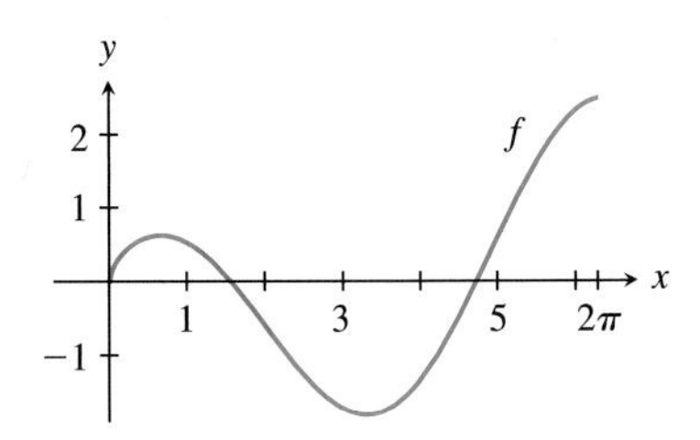

FIGURE 4.43 Graph of $f(x) = \sqrt{x}\cos x$.

EXAMPLE 3 Locate and classify all extrema of the function

$$f(x) = \sqrt{x}\cos x, \qquad \text{where } 0 \le x \le 2\pi.$$

Solution

Figure 4.43 shows a graph of f, from which it appears that f has a local minimum at $x = 0$, a local maximum near $x = 0.5$, a global minimum near $x = \pi$, and a global maximum at $x = 2\pi$.

The function f is differentiable except at $x = 0$. Its derivative is

$$f'(x) = \frac{\cos x}{2\sqrt{x}} - \sqrt{x}\sin x. \tag{9}$$

Setting $f'(x)$ equal to zero and simplifying,

$$2x \sin x - \cos x = 0.$$

We use Newton's method to find the zeros of this equation. From the graph of f, these zeros are near 0.5 and 3.1. Letting $g(x) = 2x \sin x - \cos x$, the Newton's method algorithm is

$$x_{n+1} = x_n - \frac{g(x_n)}{g'(x_n)} = x_n - \frac{2x_n \sin x_n - \cos x_n}{3 \sin x_n + 2x_n \cos x_n}.$$

With $x_1 = 0.5$, we find the zero $c_1 \approx 0.65327$; with $x_1 = 3.1$, we find the zero $c_2 = 3.29231$.

We know from the Candidate Theorem that all extrema are among the points 0, c_1, c_2, and 2π. Evaluating f at these four points,

$$\begin{aligned} f(0) &= 0 \\ f(c_1) &\approx 0.64183 \\ f(c_2) &\approx -1.7939 \\ f(2\pi) &= \sqrt{2\pi} \approx 2.5066, \end{aligned}$$

we conclude that f has a global minimum at c_2 and a global maximum at 2π.

The point $x = 0$ is a local minimum because near $x = 0$ the first term of $f'(x)$ in (9) is much larger than the second term. Hence, $f'(x) > 0$ near $x = 0$, which means that f is increasing. Hence, $x = 0$ is a local minimum.

We show that f has a local maximum at $c_1 \approx 0.65327$ by applying the First Derivative Test. Let's start by calculating f' near c_1, for example,

$$\begin{aligned} f'(0.65) &\approx 0.0058 \\ f'(0.66) &\approx -0.012. \end{aligned}$$

These results strongly suggest that f' changes from positive to negative at c_1 and, hence, that f has a local maximum at c_1.

For candidates like c_1 in Example 3, where except for graphical or numerical evidence we have not proved that f' changes sign at a point like c_1, we may prefer the Second Derivative Test. Instead of checking whether f' changes sign at c_1, we check the sign of the second derivative at c_1.

Second Derivative Test

Let f be a function defined on an interval I and c an interior point of I. Assume that f'' is continuous at and near c and that $f'(c) = 0$.

i) If $f''(c) > 0$, then f has a local minimum at c; and
ii) if $f''(c) < 0$, then f has a local maximum at c.

A proof of the Second Derivative Test is left as Exercise 41. We apply this test to the function f and zero c_1 of f' in Example 3. Because the second derivative of f is

$$f''(x) = -\frac{\sin x}{\sqrt{x}} - \sqrt{x}\cos x - \frac{\cos x}{4x^{3/2}}$$

and the zero $c_1 \approx 0.65327$ is in the first quadrant (that is, $c \in [0, \pi/2]$), it is clear that $f''(c_1) < 0$. It follows from the Second Derivative Test that f has a local maximum at c_1.

Global Extrema

For functions defined and continuous on *closed and bounded* intervals like $[a, b]$, we have applied the Max/Min Theorem to be certain that such functions have a maximum and a minimum. For functions defined and continuous on intervals like (a, b) or (a, ∞), we give an alternative to the Max/Min Theorem.

THEOREM Global Extremum Theorem

Let f be a twice differentiable function on an interval I and c an interior point of I.

i) If $f''(x) > 0$ on I and $f'(c) = 0$, then f has a global minimum at c; and
ii) if $f''(x) < 0$ on I and $f'(c) = 0$, then f has a global maximum at c.

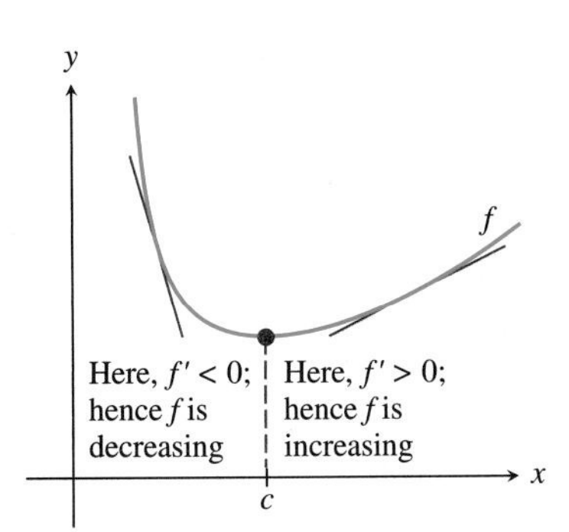

FIGURE 4.44 If $f'(c) = 0$ and the graph of f is concave up everywhere, then c is a global minimum.

Figure 4.44 shows a representative function fitting (i) of the Global Extremum Theorem. At c the first derivative is 0 and for all $x \in (0, \infty)$ the second derivative is positive, which means that the graph is concave up. It appears from the graph that f must have a global minimum at c.

Here's a proof of the case in which I has the form (a, b), $(-\infty, b)$, (a, ∞), or $(-\infty, \infty)$ and (i) is satisfied. Because $f''(x) > 0$ for all $x \in I$, f' is increasing on I by the I/D Test. Because f' is increasing on I and is 0 at $c \in I$, $f'(x) > f'(c) = 0$ for $x > c$. This means that f is increasing on the part of I to the right of c, again using the I/D Test. It follows in the same way that f is decreasing on the part of I to the left of c. Thus, f has a global minimum at c.

EXAMPLE 4 Find all global extrema of the function

$$f(x) = 2\pi x^2 + \frac{2}{x}, \qquad 0 < x < \infty.$$

Solution

This function, whose graph is shown in Fig. 4.45, has no global maximum because $\lim_{x\to 0^+} f(x) = \infty$. We use the Global Extremum Theorem to show that f has a global minimum. For this we calculate f' and f''.

$$f'(x) = 4\pi x - 2x^{-2}$$
$$f''(x) = 4\pi + 4x^{-3}$$

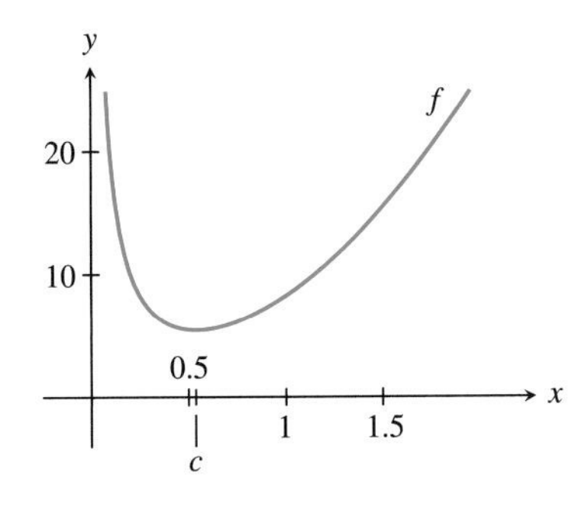

FIGURE 4.45 The graph of $f(x) = 2\pi x^2 + 2/x$ on $(0, \infty)$.

To locate candidates for local extrema, we solve the equation $f'(x) = 0$.

$$4\pi x - \frac{2}{x^2} = 0$$

$$4\pi x = \frac{2}{x^2}$$

$$x^3 = \frac{1}{2\pi}$$

$$x = \sqrt[3]{\frac{1}{2\pi}} \approx 0.54193.$$

Letting $c = \sqrt[3]{1/(2\pi)}$, because $f''(x) > 0$ on $(0, \infty)$ and $f'(c) = 0$, it follows from the Global Extremum Theorem that f has a global minimum at c.

Exercises 4.5

Exercises 1–6: Sketch the graph of a continuous function defined on $[0, 1]$ and having the given characteristics.

1. Global maximum at 0; global minimum at 1
2. Local maximum at 0; local minimum at $1/2$; global maximum at 1
3. Local minimum at 0; global maximum at $1/2$; local minimum at 1
4. Local minima at $1/4$ and 1; local maxima at 0 and $3/4$
5. Local maxima at 0, $1/2$, and 1; local minima at $1/4$ and $3/4$
6. Local maxima at 0 and $1/2$; local minima at $1/4$ and $3/4$; global maximum at 1

Exercises 7–18: Find all candidates for extrema of the function on the given interval. Evaluate the function at any global extrema.

7. $2x^3 - 3x^2 - 12x + 1, \quad [-2, 3]$
8. $4x^3 - 3x^2 - 6x - 2, \quad [-1, 2]$
9. $\dfrac{x}{x^2 + 1}, \quad [-3, 3]$
10. $\dfrac{x - 1}{x^2 + 1}, \quad [-3, 1]$
11. $\dfrac{x^2}{2} + \dfrac{8}{x}, \quad [1/2, 3]$
12. $x^2 + \dfrac{5}{x} + 1, \quad [1, 2]$
13. $\arctan(x - \sqrt{x}), \quad [0, 5]$
14. $\arctan(x^2 - x), \quad [0, 4]$
15. $2x^3 + 9x^2 + 12x + 1, \quad [-1, 1]$
16. $x^4 - 4x + 5, \quad [-1, 2]$
17. $\sin^2 x, \quad [0, 2\pi]$
18. $x^{5/3} - 5x^{1/3}, \quad [0, 2]$

Exercises 19–30: Find all global extrema of the function on the given interval. Evaluate the function at any global extrema.

19. $\dfrac{x^2 + 1}{x^3 + 1}, \quad 0 \le x \le 3$
20. $\dfrac{x^2 + 1}{x^4 + 1}, \quad 0 \le x \le 3$
21. $\dfrac{x}{1 + 2x}, \quad 0 \le x \le 3$
22. $\dfrac{x^2}{1 + x}, \quad 0 \le x \le 2$
23. $x^2\sqrt[3]{x - 1}, \quad -2 \le x \le 2$
24. $-2x^2 + \tan x, \quad 0 \le x \le 1.4$
25. $2xe^{-x}, \quad 0 \le x \le 2$
26. $\cot x + x^2, \quad (0, 3)$
27. $f(x) = x^x$ for $0 < x \le 2$ and $f(0) = 1$
28. $x^2 + (\sin x - 1)^2, \quad 0 \le x \le \pi/2$
29. $2^x - x - 1, \quad [-1, 2]$
30. $\dfrac{x - \ln x}{x}, \quad [1, 5]$

31. Referring to Example 2, justify the rule "cut it half again as high as it is wide."

32. As a variation on the sawyer's problem, suppose the logs have roughly elliptical cross sections and that the greatest diameter of the logs is about 10 percent larger than the least diameter. What rule would you recommend to the sawyer in this case?

T **33.** In Example 3 we defined the function

$$g(x) = 2x \sin x - \cos x.$$

Use Newton's method to find the zeros of g near 0.5 and 3.1 to four significant figures.

34. Locate the global extrema of the function $g(x) = \sqrt{x} \sin x$, $0 \le x \le \pi$.

35. Everyone knows that $-\pi/2 < \arctan x < \pi/2$ for all $x \in R$. Write a clear, succinct sentence arguing that the arctan function has no maximum.

36. Why can't we conclude from the Global Extremum Theorem that the function e^x defined on $(-\infty, \infty)$ has a global minimum? After all, its second derivative is always positive.

37. The Max/Min Theorem has—apart from the assumption that f is continuous—two assumptions on the domain of f, namely, the domain is an interval that is both closed and bounded. Give simple examples showing that if either of these assumptions is not satisfied, f may not have a maximum value.

38. Find the global extrema of the continuous function

$$f(x) = 2x + 1 + 3|2x - 3| - |3x + 1|, \quad -2 \le x \le 3.$$

39. Find the global extrema of the continuous function

$$f(x) = |2x - 5| + |x + 2| - 2|x - 4|, \quad -3 \le x \le 5.$$

40. Use the Mean-value Theorem to prove the First Derivative Test. You may use the following outline. For (i), we show that for x near c, $f(x) \le f(c)$ or $f(x) - f(c) \ge 0$; by the Mean-value Theorem, there is a point w between x and c for which $f(x) - f(c) = f'(w)(x - c)$; consider $x \ge c$ and $x \le c$ separately. The proof of (ii) is similar. For (iii), assume that, say, $f'(x) > 0$ for x near but not equal to c. Use the Mean-value Theorem to show that c is neither a local maximum nor a local minimum.

41. Use the I/D Test in proving the Second Derivative Test.

4.6 Modeling Optimization Problems

Optimization is a major theme in the physical and biological sciences, in economics and business, and in applied mathematics. Here are three examples of optimization.

- Fermat's principle of least time: Light, in moving from one point to another, takes the path of least time.
- Soap bubbles on a wire frame assume shapes that minimize surface tension.
- The fitting of experimental data by least squares or linear regression is based on choosing a line so that the sum of squares of the deviations of the data from the line is a minimum. You may have used least squares in courses in which a straight line is fitted to experimental data. Your calculator probably has a least squares package in its statistics menu.

In this section we discuss a few of the many kinds of optimization problems. Although there is no single method for solving optimization problems, we give some suggestions our students have found effective in formulating and solving such problems.

- *Energy.* Passive reading of the problem and patient waiting for a flash of inspiration are not recommended. Aggressive reading and active exploration strongly improve your chances for flashes of inspiration or glimmers of understanding. False starts and mistakes are normal. The important thing is to try something and then to modify it as necessary.
- *Objectives and variables.* What, exactly, is the problem? Are we, for example, to minimize a distance or maximize a volume? After assigning a variable to the distance, volume, cost, or whatever is to be optimized, relate that variable to the quantities it depends on. If, for example, the problem is to maximize the volume of a cylinder with a given surface area, you will need formulas for the volume and surface area of a cylinder. Recall or otherwise obtain all needed formulas, sketch appropriate diagrams, and relate given and unknown variables as necessary.
- *The mathematical model.* Using these variables, formulas, and relations, formulate a specific extremum problem, including the function to be maximized or minimized and its domain. Usually, common sense or the problem statement provide help in deciding on the domain in which to seek candidates for the extremum.
- *Solve the mathematical model.* Use the Candidate Theorem, Max/Min Theorem, First Derivative Test, Second Derivative Test, or Global Extremum Theorem as necessary in solving the mathematical model. For some problems, a graphical solution may be appropriate.
- *Interpretation.* Solve the original problem by interpreting or otherwise adapting the solution of the mathematical model.

We work through four examples. We calculate the minimum distance from a point to a curve, design the cheapest mayonnaise jar, determine the maximum speed of a piston in a cylinder, and derive Snell's law from Fermat's principle of least time.

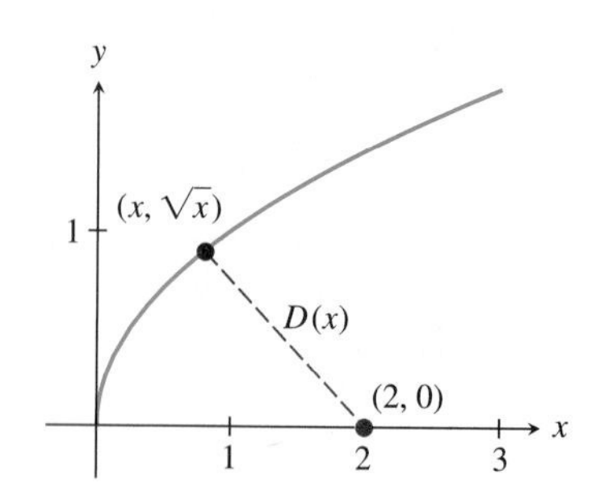

FIGURE 4.46 Finding the point on the graph of $y = \sqrt{x}$ closest to $(2, 0)$.

EXAMPLE 1 Find the point on the graph of the equation $y = \sqrt{x}$, $x \geq 0$, closest to the point $(2, 0)$.

Solution

Figure 4.46 shows the graph of $y = \sqrt{x}$ and the point $(2, 0)$. The distance $D(x)$ between an arbitrary point $\left(x, \sqrt{x}\right)$ on the curve and the point $(2, 0)$ is

$$\sqrt{(x-2)^2 + \left(\sqrt{x} - 0\right)^2}.$$

Hence, the minimum distance problem is solved if we can determine $x \geq 0$ that minimizes the function

$$D(x) = \sqrt{(x-2)^2 + x}, \qquad x \geq 0.$$

In problems involving a distance function $D(x)$, we usually minimize the square $S(x) = D(x)^2$ of the distance function rather than the distance function itself. This gives the value of x that minimizes D because the minimum points of the functions D and S are the same (see Exercise 25). Moreover, we simplify our work because S is simpler than D. Thus, to minimize D we minimize

$$S(x) = (x - 2)^2 + x = x^2 - 3x + 4, \quad x \geq 0.$$

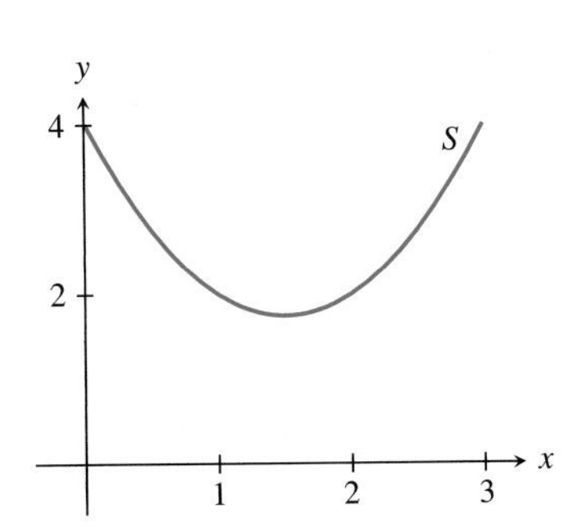

FIGURE 4.47 The graph of the square of the distance function.

Figure 4.47 shows the graph of S. It is clear from the figure that S has a minimum. For proof we use the Global Extremum Theorem. The derivatives of S are

$$S'(x) = 2x - 3$$

$$S''(x) = 2.$$

Because $S'(3/2) = 0$ and $S''(x) > 0$ on $[0, \infty)$, is follows from the Global Extremum Theorem that S has a global minimum at $c = 3/2$. The point $\left(1.5, \sqrt{1.5}\right)$ is the point on the graph of D closest to the point $(2, 0)$.

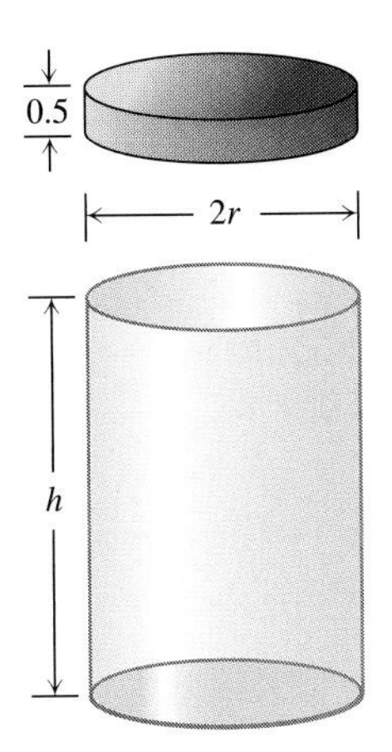

FIGURE 4.48 Exploded view of a cylindrical glass jar with a metal lid.

EXAMPLE 2 A cylindrical glass jar with a metal lid is to contain 1 quart (57.75 cubic inches) of mayonnaise. The diameter of the jar is to be between 2 and 4 inches and the height of the metal lid is 0.5 inch. Given that the lid costs three times as much per square inch as the glass sides and bottom, find the dimensions of the least expensive jar. Figure 4.48 shows an exploded, simplified view of the jar.

Solution

We model the lid and jar with two cylinders, each with one end capped by a disk of radius r. The jar has height h and the lid has height 0.5 inches. The volume of the jar is

$$V = \pi r^2 h. \tag{1}$$

The area of the lid is the sum of the area of a circular disk of radius r and the area of a cylinder of radius r and height 0.5 inches. The area of the jar is the sum of the area of a circular disk of radius r and the area of a cylinder of radius r and height h inches. Given that the metal lid costs three times as much per square inch as the glass sides and bottom, the total cost of the materials for the lid and jar is

$$C = 3k(\pi r^2 + 2\pi r \cdot 0.5) + k(\pi r^2 + 2\pi r \cdot h), \tag{2}$$

where k is a positive constant. In writing this equation we used the fact that the surface area of a cylinder is the circumference of its circular cross section times its height. Given that the jar is to contain $V = 57.75$ cubic inches of mayonnaise, it follows from (1) that

$$h = \frac{57.75}{\pi r^2}. \tag{3}$$

Substituting into (2),

$$C = C(r) = 3k(\pi r^2 + \pi r) + k\left(\pi r^2 + 2\pi r \cdot \frac{57.75}{\pi r^2}\right)$$
$$= 3k(\pi r^2 + \pi r) + k\left(\pi r^2 + \frac{115.5}{r}\right)$$
$$= k\left(4\pi r^2 + 3\pi r + \frac{115.5}{r}\right), \qquad 1 \le r \le 2.$$

The domain $[1, 2]$ of the cost function comes from the requirement that the diameter of the lid be between 2 and 4 inches.

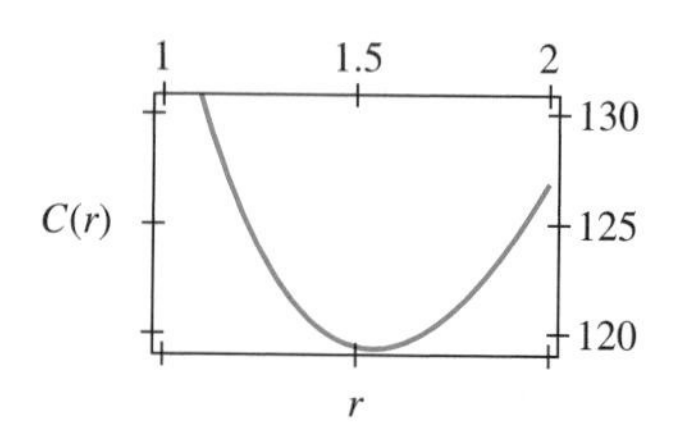

FIGURE 4.49 Graph of the cost function.

Figure 4.49 shows a graph of the cost function. We set $k = 1$ for the purpose of drawing the graph. From the graph we can see that the cost is a minimum for $r \approx 1.5$ inches.

Differentiating the cost function,

$$C'(r) = k\left(8\pi r + 3\pi - \frac{115.5}{r^2}\right).$$

Setting $C'(r) = 0$ and multiplying the resulting equation by r^2/k,

$$8\pi r^3 + 3\pi r^2 - 115.5 = 0.$$

This equation has one real root near $r = 1.5$. By Newton's method, calculator, or CAS, we find that the root is $r \approx 1.55$ inches.

Because

$$C''(r) = k\left(8\pi + \frac{231}{r^3}\right)$$

is positive for all $r \in [1, 2]$, C has a global minimum at $r \approx 1.55$ inches by the Second Derivative Test. From (3) the corresponding value of h is

$$h = \frac{57.75}{\pi r^2} \approx 7.69 \text{ inches.}$$

Hence, the dimensions of the optimum jar and cap are $r \approx 1.55$ inches and $h \approx 7.69$ inches.

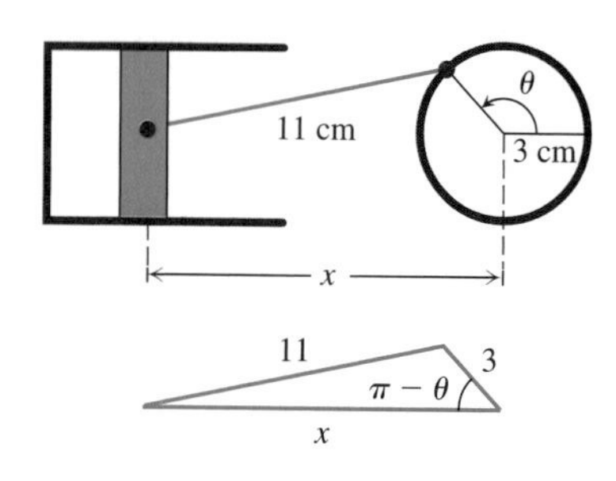

FIGURE 4.50 Relating the distance x and the angular displacement θ for a piston/cylinder mechanism.

EXAMPLE 3 Figure 4.50 shows a piston/cylinder mechanism, with all lengths measured in centimeters. Assuming that the crank is revolving at 250 rpm counterclockwise about a crankshaft whose center is fixed relative to the cylinder in which the piston moves, find the maximum speed of the piston within the cylinder. A graphical solution is sufficient.

Solution

Referring to the figure, both the angle θ and the distance x between the center of the crankshaft and the piston are functions of time t. We wish to find the maximum speed $|dx/dt|$ of the piston. By symmetry, we may restrict our attention to the time interval for which $0 \le \theta \le \pi$. For θ in this range, x is increasing and, hence, $dx/dt \ge 0$; hence, $|dx/dt| = dx/dt$.

Applying the law of cosines to the triangle at the bottom of the figure,

$$11^2 = x^2 + 3^2 - 2 \cdot x \cdot 3 \cdot \cos(\pi - \theta).$$

Replacing $\cos(\pi - \theta)$ by its equal $-\cos\theta$ and solving for x,

$$x = \frac{-6\cos\theta \pm \sqrt{36\cos^2\theta - 4\cdot 1(-112)}}{2}$$

$$x = -3\cos\theta \pm \sqrt{9\cos^2\theta + 112}.$$

Because the length x is positive, we choose the positive root:

$$x = -3\cos\theta + \sqrt{9\cos^2\theta + 112}. \tag{4}$$

Recalling that both x and θ are functions of t, we calculate dx/dt using the chain rule:

$$\frac{dx}{dt} = \frac{dx}{d\theta}\cdot\frac{d\theta}{dt}.$$

Because the angular speed $d\theta/dt$ is a positive constant, we can maximize dx/dt by maximizing $dx/d\theta$.

Differentiating (4) with respect to θ,

$$\frac{dx}{d\theta} = 3\sin\theta + \frac{-18\cos\theta\sin\theta}{2\sqrt{9\cos^2\theta + 112}}.$$

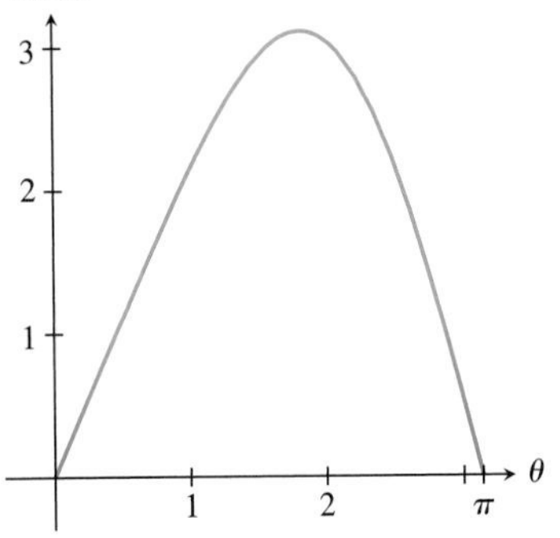

Recalling that "a graphical solution is sufficient," we graph this expression. It appears from the graph in Fig. 4.51 that the maximum of $dx/d\theta$ is about 3.1. From this we can calculate the maximum speed dx/dt by recalling that $dx/dt = (dx/d\theta)(d\theta/dt)$. Because

$$\theta = \frac{250\text{ revolutions}}{\text{minute}}\cdot\frac{2\pi}{1\text{ revolution}}\cdot t\text{ minutes} = 500\pi t\text{ radians},$$

$d\theta/dt = 500\pi$ radians/minute and the maximum speed of the piston is about $3.1\cdot 500\pi \approx 4900$ cm/min.

FIGURE 4.51 The graph of the derivative function $dx/d\theta$, $0 \le \theta \le \pi$.

To explain the refraction of light, Pierre Fermat (1601–1665) assumed that in traveling from a point in one medium to a point in another medium, light follows the path taking the least time. Fermat was able to infer the law of refraction now called Snell's law from these assumptions. Before Fermat's work, Willebrord Snell (1580–1626) had discovered the law of refraction experimentally.

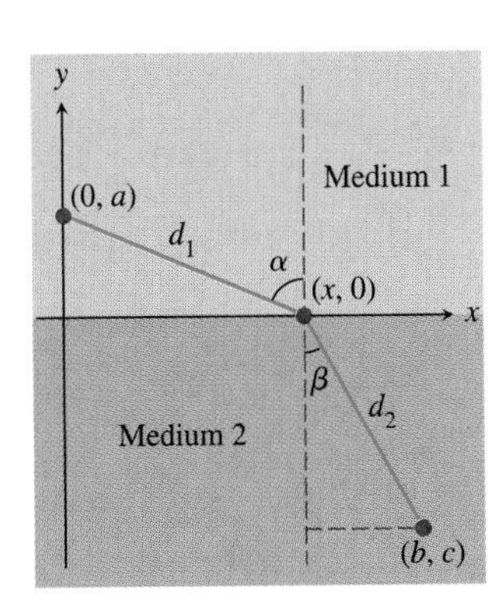

FIGURE 4.52 Possible path of a ray from $(0, a)$ to (b, c).

EXAMPLE 4 Figure 4.52 shows a ray of light moving through two media, which are separated by the x-axis of an (x, y)-coordinate system. A light source at $(0, a)$ in "medium 1" is seen by an observer at (b, c) in "medium 2." Assuming that the velocity of light in medium 1 is v_1 and in medium 2 is v_2, infer Snell's law,

$$\frac{\sin\alpha}{v_1} = \frac{\sin\beta}{v_2}, \tag{5}$$

from Fermat's principle of least time. The angles α and β are shown in the figure. These angles are called, respectively, the angles of incidence and refraction.

Solution

Referring to Fig. 4.52, in traveling from $(0, a)$ to (b, c) the observed ray must cross from medium 1 to medium 2 at some point $(x, 0)$. Let d_1 be the distance from $(0, a)$ to $(x, 0)$ and d_2 the distance from $(x, 0)$ to (b, c). The light ray travels at speed v_1 from $(0, a)$ to $(x, 0)$ and at speed v_2 from $(x, 0)$ to (b, c), taking times $t_1 = d_1/v_1$ and $t_2 = d_2/v_2$. Noting that v_1 and v_2 are constants and that t_1 and t_2 are functions of x, we can find the least-time path by minimizing the function

$$\begin{aligned} T(x) &= \frac{d_1}{v_1} + \frac{d_2}{v_2} \\ &= \frac{\sqrt{x^2 + a^2}}{v_1} + \frac{\sqrt{(x - b)^2 + c^2}}{v_2}, \qquad 0 \le x \le b. \end{aligned} \tag{6}$$

The function T is always positive. On physical or geometric grounds, it is clear that T has a global minimum in $[0, b]$. Or we may use the Global Extremum Theorem to show that T has a global minimum in $[0, b]$. For this we show that T' has a zero and $T''(x) > 0$ for $x \in (-\infty, \infty)$. We leave a calculation of T'' as Exercise 26.

To show that T' has a zero we look at its values near 0 and b. From (6),

$$T'(x) = \frac{2x}{2v_1\sqrt{x^2 + a^2}} + \frac{2(x - b)}{2v_2\sqrt{(x - b)^2 + c^2}}. \tag{7}$$

As $x \to 0^+$, the first term of $T'(x)$ approaches 0 and the second approaches $-b/\left(v_2\sqrt{b^2 + c^2}\right)$, which is negative. So $T'(x)$ is negative for x near 0. As $x \to b$, the second term of $T'(x)$ approaches 0 and the first approaches $b/\left(v_1\sqrt{a^2 + b^2}\right)$, which is positive. So $T'(x)$ is positive for x near b. It follows from the Intermediate Value Theorem that T' has a zero between 0 and b.

The least-time path can be found by solving equation $T'(x) = 0$ for x in terms of a, b, c, v_1, and v_2. However, this requires the solution of a fourth-degree equation. See Exercise 27. Snell's law is a convenient formulation of the condition $T'(x) = 0$ on x that holds at the point x of least time. From (7), the equation $T'(x) = 0$ is

$$\frac{x}{v_1\sqrt{x^2 + a^2}} + \frac{(x - b)}{v_2\sqrt{(x - b)^2 + c^2}} = 0.$$

This equation can be rearranged as

$$\frac{x}{v_1 d_1} = \frac{b - x}{v_2 d_2}.$$

From the right triangles in Fig. 4.52,

$$\frac{\sin \alpha}{v_1} = \frac{\sin \beta}{v_2}.$$

This is Snell's law, which is more often expressed in terms of the indices of refraction, n_1 and n_2, of the two media. These are $n_1 = v_0/v_1$ and $n_2 = v_0/v_2$, where v_0 is the speed of light in a vacuum.

Exercises 4.6

1. Find the point on the graph of the equation $y = \sqrt{2x}$ closest to $(3, 0)$.

2. Find the point on the graph of the equation $y = 3\sqrt{x}$ closest to $(5, 0)$.

3. Find the point on the graph of the equation $y = 2x + 1$ closest to $(1, 1)$.

4. Find the point on the graph of the equation $2x + 3y + 4 = 0$ closest to $(2, 1)$.

5. Find the points on the graph of the equation $y = \sqrt[3]{x}$, $0 \le x \le 2$, farthest from and closest to $(2, 0)$.

6. Find the points on the graph of the equation $y = \sqrt[3]{2x}$, $0 \le x \le 2$, farthest from and closest to $(1, 0)$.

7. If in the mayonnaise jar example (Example 2) the cost of the metal cap (in dollars per square inch) is only twice as much per square inch as the glass sides and bottom, would you expect the jar to become wider or narrower? Justify and quantify your answer.

8. A conical drinking cup has height 3 inches and radius 1 inch. Find the radius of the cylindrical glass rod which, when lowered into the cup along the axis of the cone, displaces the maximum amount of beverage.

9. A soft drink can is to contain 12 fluid ounces. Assume that the (idealized) can is made by forming sheet aluminum into a cylindrical container (open on one end) and then attaching a circular lid. The bottom and sides cost $\$0.02/\text{m}^2$, and the top costs $\$0.05/\text{m}^2$. What is the ratio of height to base diameter of the can that is cheapest to manufacture? If needed, 1 fluid ounce $\approx 2.957 \times 10^{-5}\ \text{m}^3$.

10. A farmer has 800 meters of fence and wishes to enclose a rectangular area for grazing. What are the dimensions of the pasture of maximum area?

11. A farmer has 800 meters of fence and wishes to enclose a rectangular area for grazing. If he has the option of using an existing fence (at least 800 meters long) as one side of the grazing area, what are the dimensions of the pasture of maximum area?

12. What are the dimensions of the rectangle of largest area that can be inscribed in the triangle with vertices $(0, 0)$, $(10, 0)$, and $(0, 5)$? Assume that one vertex of the rectangle is at the origin and the opposite vertex is on the line joining $(10, 0)$ and $(0, 5)$.

13. What are the dimensions of the rectangle of largest area that can be inscribed in the ellipse with equation $x^2/a^2 + y^2/b^2 = 1$? Assume that the center of the rectangle is at the origin and the sides are parallel to the axes.

14. In Example 4, assume that $v_1 = 0.95$, $v_2 = 0.9$, $a = 0.5$, and $(b, c) = (2.0, -1.0)$. Find the point $(x, 0)$ where the observed ray crosses the interface by minimizing T. Assume that $0 \le x \le 2$.

15. A ray of light is moving from air into a diamond. If its angle of incidence is 10°, find its angle of refraction. The indices of refraction of air and diamond are 1.000293 and 2.419, respectively.

16. A ray of light is moving from air into a pool of water. If its angle of incidence is 20°, find its angle of refraction. The indices of refraction of air and water are 1.000293 and 1.333, respectively.

17. A ray of light from a source at $(0, 100)$ is seen at $(200, 50)$ after reflecting from a 200-cm mirror, as in the accompanying figure. Use Fermat's principle of least time to calculate the point on the mirror from which the observed ray reflected. Show that the angles of incidence and reflection are equal.

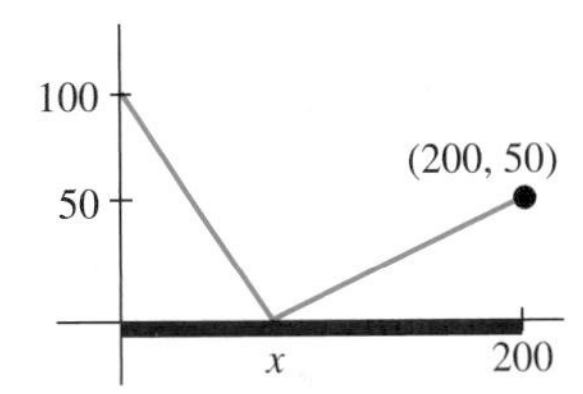

Figure for Exercise 17.

18. In the slider crank linkage shown in the accompanying figure, the crank is rotating counterclockwise at 1 radian per second and the lengths of the crank and connecting rod are 5 cm and 10 cm. Find the maximum speed of the slider for $0 \le \theta \le \dfrac{\pi}{2}$. A graphical solution is sufficient.

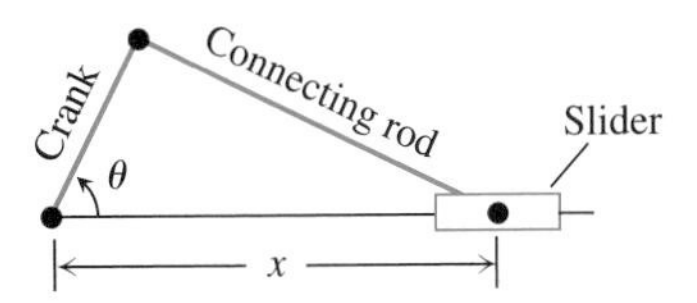

Figure for Exercise 18.

Exercises 19–24: Identify the key variables; write an expression or function to be optimized; use the problem statement or figure to relate the key variables; specify the domain of the function to be optimized; and use the Candidate Theorem, Global Extremum

Theorem, or the Max/Min Theorem to find the maximum or minimum of the function.

19. A box with a square base, rectangular sides, and open top is to contain 1 cubic foot of space. If the material for its base costs $\$3/\text{ft}^2$ and that for its sides costs $\$1/\text{ft}^2$, determine its dimensions so that the cost of the materials is a minimum.

20. A rectangle has perimeter 1 meter. What are the height and width of the rectangle with largest possible area?

21. A rectangle is to be cut from an equilateral triangle with side a so that one side is parallel to a side of the triangle. What are the height and width of the rectangle with the largest possible area?

22. A gutter is made by bending a long piece of sheet metal into thirds along its length, so that a cross section is an open trapezoid, as in the accompanying figure. How should the bending angle θ be chosen so that the area of the cross section is as large as possible?

Figure for Exercise 22.

23. A rectangle is to be cut from a right triangle with sides 3, 5, and $\sqrt{34}$, so that one side is parallel to the side of length 3. What are the dimensions of the rectangle with the largest possible area?

24. Referring to the accompanying figure, an open tray is to be made from a piece of sheet metal 0.6 m by 0.9 m. Cuts of length x are made in the four corners of the piece, each cut parallel to the adjacent side and x units from it. The cuts are shown as dotted lines in the figure. The two rectangles of dimensions x by $0.9 - 2x$ are bent through 90° along the side of length $0.9 - 2x$. Next, the two rectangles of dimensions x by $0.6 - 2x$ are bent through 90° along the side of length $0.6 - 2x$ to form the tray. Finally, the squares of side x are bent through 90° and welded into place. Choose x so that the volume of the tray is as large as possible.

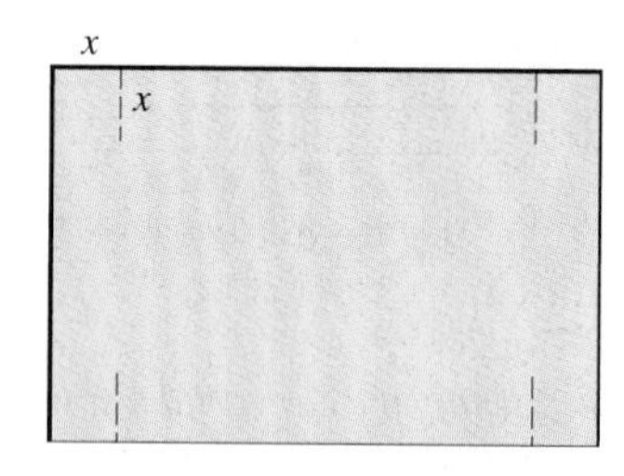

Figure for Exercise 24.

25. Show that a nonnegative function f has a minimum/maximum at c if and only if the function g has a minimum/maximum at c, where $g(x) = f(x)^2$.

26. In Example 4, show that

$$T'' = \frac{c^2}{v_2 d_2^3} + \frac{a^2}{v_1 d_1^3}$$

and, hence, $T''(x) > 0$ for all $x \in (0, b)$.

27. In Example 4, show that the equation

$$Ax^4 + Bx^3 + Cx^2 + Dx + E = 0,$$

where $A = v_1^2 - v_2^2$, $B = -2b(v_1^2 - v_2^2)$, $C = (a^2 + b^2)v_1^2 - (b^2 + c^2)v_2^2$, $D = -2a^2bv_1^2$, and $E = a^2b^2v_1^2$, follows from the equation $T'(x) = 0$. *Hint:* Combine the two terms into a single fraction, set the numerator equal to 0, and then isolate and square the worst-looking radical expression. Because there is an exact procedure for solving any fourth-degree equation (known as Ferrari's method), this result shows that it is possible to explicitly calculate the crossing-point x for the least-time path. However, the result is algebraically complex.

28. The accompanying figure shows a construction for locating the reflecting point of a ray of light moving from $(0, a)$, reflecting from a mirror, and observed at (b, c). The idea is to construct the point $(b, -c)$ and then join $(0, a)$ to $(b, -c)$ by a line segment L. The point where L crosses the mirror is the reflecting point. The dotted ray would be shorter than any other ray and, in particular, the ray shown. The problem of locating the reflecting point is called Heron's problem, after Heron of Alexandria. Show that the construction is a solution to the reflection problem.

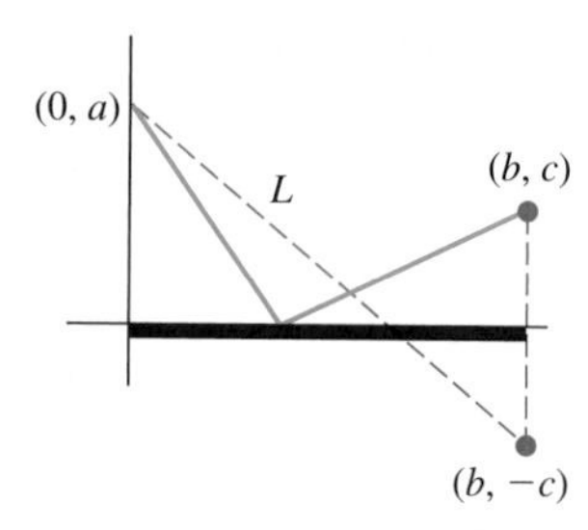

Figure for Exercise 28.

29. A rectangle is inscribed in a semicircle of radius a so that one of its sides lies along the diameter. What are the height and width of the rectangle with largest possible area?

30. Light sources are placed at coordinates 0 and 1 on an x-axis. The illumination at any point x between them from either source is proportional to the "strength" of the source and inversely proportional to the square of the distance to that source. Assume that the source at 0 is L times stronger than that at 1. Show that at the "dimmest" point between 0 and 1, the ratio of the illumination received from the source at 0 to the source at 1 is $\sqrt[3]{L}$.

31. A 10-m-wide east-west canal flows into a 15-m-wide north-south canal (see the figure). Find the length of the longest piece of wood with negligible width that can make the turn.

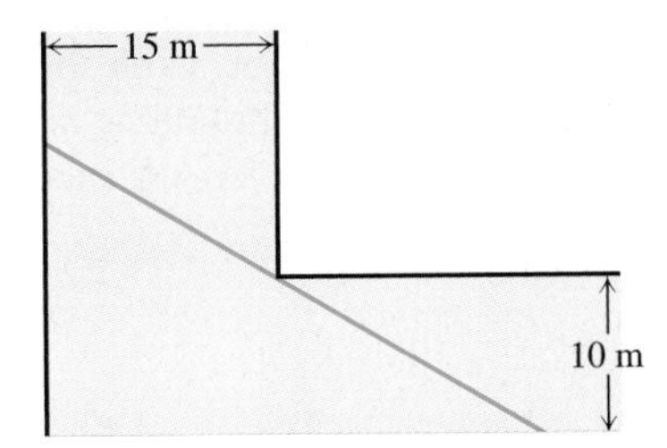

Figure for Exercises 31 and 32.

32. Solve the preceding problem, but replace the wood of negligible width with a barge of width 4 m.

33. Calculate the dimensions of the cylinder of largest volume that can be inscribed in a cone with height 8 m and base radius 5 m. A cross section is shown in the accompanying figure.

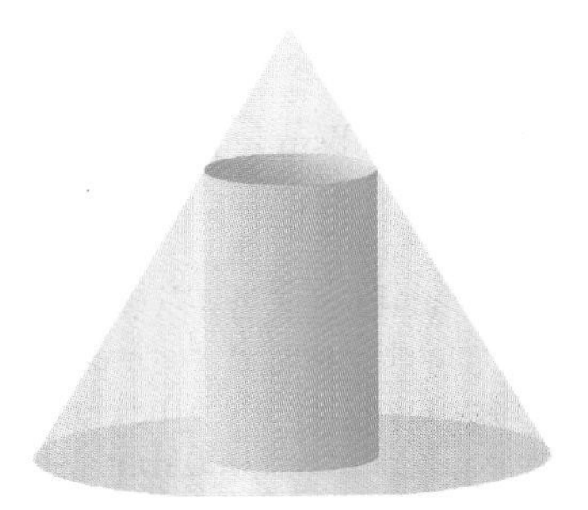

Figure for Exercise 33.

34. Calculate the dimensions of the cylinder of largest volume that can be inscribed in a sphere with radius 5 m. A cross section is shown in the accompanying figure.

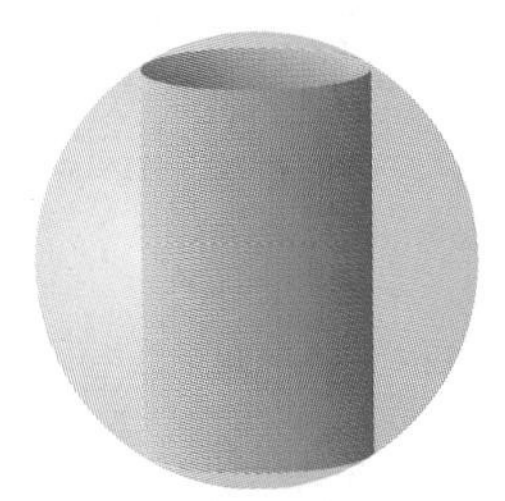

Figure for Exercise 34.

35. A conical cup is made by cutting a sector from a paper circle of radius 15 cm, aligning edges AC and BC, and fastening with tape. See the accompanying figure. How should the cuts be made if the cup is to have the largest possible volume?

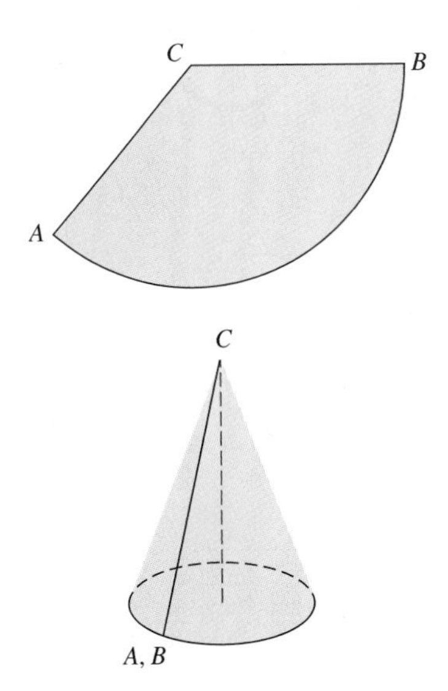

Figure for Exercise 35.

36. The following optimization problem was adapted from *Engineering Economic and Cost Analysis,* second edition, by Courtland A. Collier and William B. Ledbetter (HarperCollins, 1988). One of the key variables in designing electrical transmission lines is the area A of a cross section of the wire. As the area A decreases, the cost of installation decreases; however, the annual cost of the energy lost due to the electrical resistance of the wire increases. The following problem is a simplified design problem for a 200-ft line connecting a transformer to an electrical pump at a municipal water plant. We consider only two costs: those for installing the line and the electrical resistance of the line. The installation cost is a combination of a fixed cost of \$200 for fittings and the cost of the copper per pound. The annual loss of energy due to the electrical resistance of the line is $293.2/A$ kilowatt-hours, where A is measured in square inches. It is given that

- the estimated life of the transmission line is 20 years.
- copper weighs 0.32 lb/in.^3.
- copper costs \$0.80 per pound.
- the cost of energy is \$0.05 per kilowatt-hour.

Calculate the diameter of the wire that minimizes the annual costs. Report both the diameter and the minimum annual cost.

4.7 Related Rates

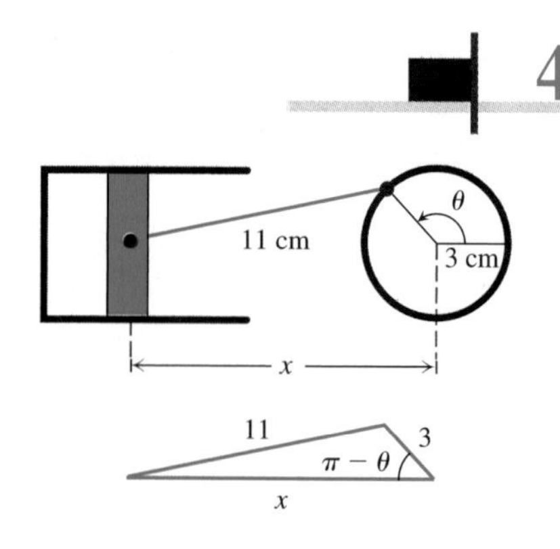

FIGURE 4.53 Relating the rates dx/dt and $d\theta/dt$ in a piston/cylinder mechanism.

Figure 4.53 shows a piston connected to a 3-centimeter crank by an 11-centimeter connecting rod. The crank is revolving about a crankshaft whose center is fixed relative to the cylinder in which the piston moves. In Example 3 of Section 4.6, we assumed that the crank was revolving at 250 rpm counterclockwise and determined the maximum speed of the piston. In the present section, we ask a different question: How can we calculate the velocity dx/dt of the piston from the angular velocity $d\theta/dt$ of the crank at any time t? The answer starts with finding an equation relating these two rates. Problems involving such "related rates" are the subject of this section.

Implicit Differentiation

Related rates problems are often made easier by using implicit differentiation. As part of our discussion of related rates we review implicit differentiation, which was introduced back in Section 2.4. We begin this review in the context of the piston/cylinder mechanism.

INVESTIGATION

Relating Angular and Linear Motion

If the crankshaft of the piston/cylinder mechanism shown in Fig. 4.53 is rotating counterclockwise at 2000 rpm, find the velocity of the piston at the time when $\theta = \pi/2$.

Applying the law of cosines to the triangle at the bottom of the figure,

$$11^2 = x^2 + 3^2 - 2 \cdot x \cdot 3 \cdot \cos(\pi - \theta).$$

Replacing $\cos(\pi - \theta)$ by its equal $-\cos\theta$ and simplifying the resulting equation,

$$x^2 + 6x\cos\theta - 112 = 0. \tag{1}$$

This equation relates the variables x and θ. If either x or θ is given, we can calculate the value of the other. Or, for related rates, if we are given either dx/dt or $d\theta/dt$, we can calculate the value of the other. Suppose, for example, that the crankshaft is turning at an angular velocity equivalent to 2000 rpm. How can we calculate the velocity dx/dt of the piston?

We calculate dx/dt by differentiating (1) implicitly. For this we note that each of x and θ is a function of time t and the left side of equation (1) is a sum of three terms, each of which is a function of t. The sum of these three terms is, according to (1), identically 0 for all t. Hence,

$$\frac{d}{dt}(x^2 + 6x\cos\theta - 112) = \frac{d}{dt}(0) = 0.$$

The derivative of the sum on the left side can be written as the sum of three derivatives, so that

$$\frac{d}{dt}(x^2) + \frac{d}{dt}(6x\cos\theta) - \frac{d}{dt}(112) = 0. \tag{2}$$

Let's look at each of these three terms. The third term is 0 because the derivative of a constant is 0. We must use the chain rule for the first term because its form is $\mathrm{sq}(x(t))$, where sq is the squaring function and $x(t)$ is a function of t. To differentiate this composite function with respect to t we use the chain rule, one form of which is: If $w = \text{sq}(x)$ and $x = x(t)$, then

$$\frac{dw}{dt} = \frac{dw}{dx} \cdot \frac{dx}{dt}.$$

Because $dw/dx = d(x^2)/dx = 2x$,

$$\frac{dw}{dt} = 2x \cdot \frac{dx}{dt}.$$

Dropping the variable w,

$$\frac{d}{dt}(x^2) = 2x \cdot \frac{dx}{dt}.$$

For the second term, we first use the product rule to differentiate the product of the two functions $6x$ and $\cos\theta$ of t.

$$\frac{d}{dt}(6x \cos\theta) = 6\frac{dx}{dt}\cos\theta + 6x\frac{d}{dt}(\cos\theta).$$

To complete the differentiation, we use the chain rule: If $w = \cos\theta$ and $\theta = \theta(t)$, then

$$\frac{dw}{dt} = \frac{dw}{d\theta} \cdot \frac{d\theta}{dt}.$$

Hence,

$$\frac{d}{dt}(6x \cos\theta) = 6\frac{dx}{dt}\cos\theta + 6x(-\sin\theta)\frac{d\theta}{dt}.$$

Having differentiated each of the three terms in (2), the result of differentiating equation (1) with respect to t is

$$2x\frac{dx}{dt} + 6\frac{dx}{dt}\cos\theta - 6x\sin\theta\frac{d\theta}{dt} = 0. \tag{3}$$

We use this equation to find the velocity dx/dt of the piston at times when $\theta = \pi/2$. At time t minutes,

$$\theta = \theta(t) = \frac{2000 \text{ revolutions}}{\text{minute}} \cdot \frac{2\pi}{1 \text{ revolution}} \cdot t \text{ minutes} = 4000\pi t \text{ radians}.$$

Hence, $d\theta/dt = 4000\pi$. From (1), we find that $x = \sqrt{112}$ cm at the time when $\theta = \pi/2$. Substituting $d\theta/dt = 4000\pi$, $\theta = \pi/2$, and $x = \sqrt{112}$ into (3),

$$2 \cdot \sqrt{112}\frac{dx}{dt} + 6\frac{dx}{dt} \cdot \cos(\pi/2) - 6\sqrt{112} \cdot \sin(\pi/2) \cdot 4000\pi = 0.$$

Replacing $\cos(\pi/2)$ by 0 and $\sin(\pi/2)$ by 1 and solving this equation for dx/dt,

$$\frac{dx}{dt} = 12000\pi \approx 38{,}000 \text{ cm/min}.$$

The piston/cylinder problem worked out in the previous Investigation is a representative related rates problem. The main steps in its solution were:

STEP 1. Find an equation relating the primary variables. This equation should hold "in general," not just for a particular time or configuration. Don't get too specific too soon.

STEP 2. Differentiate the equation from Step 1 implicitly with respect to t.

STEP 3. Substitute into the equation from Step 2 any specific rates and other values that are given or can be calculated from the given information. Solve the resulting equation for the unknown rate.

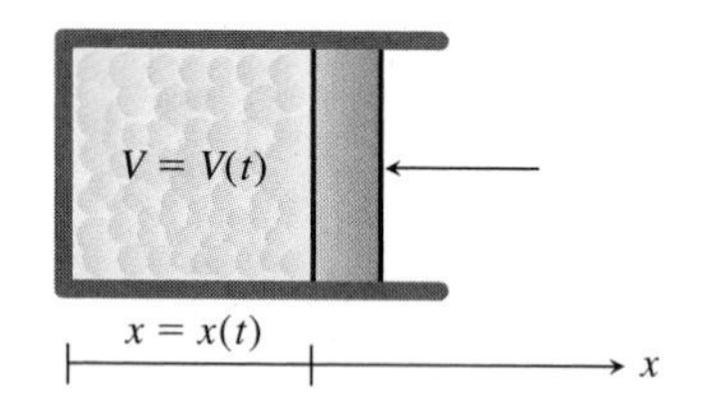

FIGURE 4.54 Relating the rates dV/dt and dx/dt.

EXAMPLE 1 A piston is compressing a gas contained in a cylinder, as illustrated in Fig. 4.54. Given that the piston is moving into the cylinder at 10 cm/s and the diameter of the cylinder is 6 cm, at what rate is the volume of the gas changing?

Solution

The volume V and displacement x of the piston are the primary variables; each is a function of t. The rate of change dx/dt of x with respect to time is given, and we are to calculate dV/dt. Because the volume V of a cylinder of radius r and height h is $V = \pi r^2 h$, the variables V and x are related by

$$V = \pi \cdot 3^2 \cdot x = 9\pi x.$$

Differentiating this equation with respect to t,

$$\frac{dV}{dt} = \frac{d}{dt}(9\pi x) = 9\pi\frac{dx}{dt}. \tag{4}$$

This equation relates the unknown rate dV/dt to the known rate dx/dt. Because the piston is moving into the cylinder, x is decreasing with t and, hence, $dx/dt < 0$. Thus, we take $dx/dt = -10$. Substituting this value into (4),

$$\frac{dV}{dt} = 9\pi(-10) \approx -282.7 \text{ cm}^3/\text{s}. \tag{5}$$

The fact that dV/dt is negative is consistent with the observation that the volume is decreasing with time.

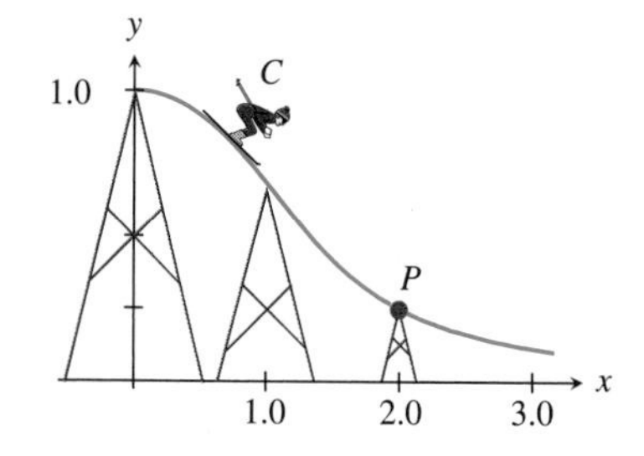

FIGURE 4.55 Joe's path.

EXAMPLE 2 Joe Cool is skiing down the jump shown in Fig. 4.55. As he passes control point P, his velocity in the x-direction is measured as 4.5 meters per second. Given that the curve C is described by the equation

$$x^2y + y^3 = 1 \tag{6}$$

and the x-coordinate of P is $x = 2$, determine Joe's velocity in the y-direction as he passes through the control point.

Solution

Each of the coordinates x and y of Joe's position (x, y) are functions of time t. These functions $x = x(t)$ and $y = y(t)$ satisfy equation (6). The velocities in the x- and y-directions are dx/dt and dy/dt. We know that $dx/dt = 4.5$ at the time Joe passes through P. Differentiating both sides of equation (6) with respect to t will give an equation connecting dy/dt and dx/dt; from this equation we can calculate dy/dt.

Differentiating (6) with respect to t,

$$\frac{d}{dt}(x^2y + y^3) = \frac{d}{dt}1$$

$$2x\frac{dx}{dt}y + x^2\frac{dy}{dt} + 3y^2\frac{dy}{dt} = 0. \tag{7}$$

Note that we used the product rule differentiating the term x^2y. Equation (7) relates dx/dt and dy/dt. At the time Joe passes through the control point, $dx/dt = 4.5$ and $x = 2$. From (6), Joe's y-coordinate at this time satisfies

$$2^2y + y^3 = 1.$$

This equation has exactly one real root, $y \approx 0.246$. Solving (7) for dy/dt and then substituting $x = 2$, $y \approx 0.246$, and $dx/dt = 4.5$,

$$\frac{dy}{dt} = \frac{-2x\frac{dx}{dt}y}{x^2 + 3y^2} \approx \frac{-2 \cdot 2 \cdot 4.5 \cdot 0.246}{2^2 + 3 \cdot 0.246^2} \approx -1.06 \text{ m/s}.$$

Hence, Joe's velocity in the y-direction is about -1.06 m/s. This result is reasonable because we see from the graph in Fig. 4.55 that y is decreasing with t, that is, $dy/dt < 0$.

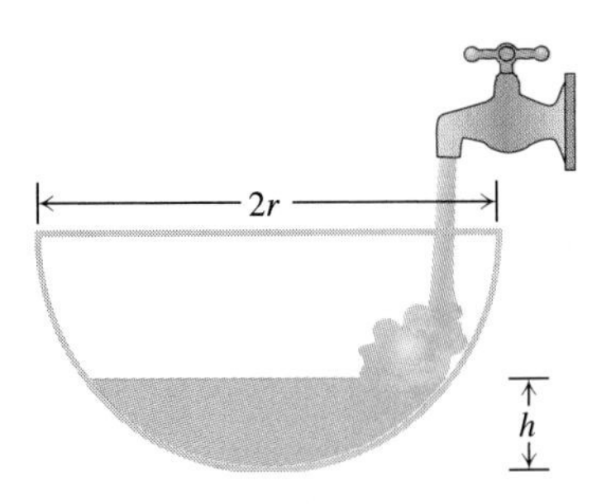

FIGURE 4.56 A hemispherical basin being filled at the rate of 2 gallons per minute.

EXAMPLE 3 Figure 4.56 shows a hemispherical basin of radius 1 foot. A faucet is filling the basin at the rate of 2 gallons per minute (≈ 0.13368 cubic feet per minute). Find the rate at which the level of the water in the basin is rising when it is half full by volume. The volume of a spherical segment of height h (see the shaded cross section in the figure) is $V = \pi\left(rh^2 - \frac{1}{3}h^3\right)$, where r is the radius of the sphere containing the segment.

Solution

Both the volume V of the water in the basin and its height h are functions of time t. The fill rate of 0.13368 cubic feet per minute is the change in the volume of the liquid per minute, that is,

$$\frac{dV}{dt} = 0.13368 \text{ ft}^3/\text{min}.$$

Differentiating the equation for the volume of a spherical segment (first replacing r by 1 and using the brief notations V' and h' for dV/dt and dh/dt),

$$V' = \pi(2hh' - h^2h').$$

Solving this equation for h',

$$h' = \frac{V'}{\pi(2h - h^2)}. \tag{8}$$

We want the value of h' when the basin is half full. Thus, we need the value of h for which the volume of the basin is $\frac{1}{4} \cdot \frac{4}{3}\pi r^3 = \frac{1}{3}\pi$ (recall that $r = 1$). From the equation for the volume of a spherical segment,

$$\frac{1}{3}\pi = \pi\left(rh^2 - \frac{1}{3}h^3\right) = \pi\left(h^2 - \frac{1}{3}h^3\right).$$

This equation simplifies to

$$h^3 - 3h^2 + 1 = 0. \tag{9}$$

The roots of this cubic are $h \approx -0.532089$, 0.652704, and 2.87939. Because the first and third roots are not between 0 and 1 (as h must be), we take $h \approx 0.652704$ ft. From (8), and recalling that $V' \approx 0.13368$,

$$h' \approx \frac{0.13368}{\pi(2 \cdot 0.652704 - 0.652704^2)} \approx 0.048388 \text{ ft/min}.$$

This is the rate at which the water level is rising when the basin is half full.

Exercises 4.7

Exercises 1–10: Functions x and y of t are related by the given equation. Express dy/dt in terms of x, y, and dx/dt. Or, using the notation $y' = dy/dt$ and $x' = dx/dt$, express y' in terms of x, y, and x'.

1. $x = \sqrt{y^2 + 1}$

2. $xy = \sin y$

3. $xy + x^2 - y^2 = 1$

4. $x^3 - xy + y^3 = 1$

5. $x^4 + y^4 = 16$

6. $x^3y + xy^4 = 1$

7. $x \cos y = y^2$

8. $y \sin y = x^2$

9. $e^{xy} = xy^2 + 1$

10. $\arctan(x/y) = x^2y + 1$

11. A hummingbird moves on a circle of radius 5 m. As it passes through $(3, 4)$, the x-coordinate of its velocity is -15 m/s. What is the y-coordinate of its velocity?

12. A hummingbird moves on a circle of radius 7 m. As it passes through $(2, 3\sqrt{5})$, the x-coordinate of its velocity is -20 m/s. What is the y-coordinate of its velocity?

13. Boyle's law says that at a constant temperature, the product of the pressure and volume of a confined ideal gas is constant. If in a constant-temperature environment a cylinder contains 1000 cubic inches of an ideal gas at a pressure of 4 pounds per square inch, find the rate of change in the pressure of the gas if its volume is decreasing by 10 in.^{3}/s.

14. Under "adiabatic conditions," the pressure and volume of a confined gas satisfy the equation $PV^{1.4} = k$, where k is a constant. At a certain time a cylinder contains 1000 cubic inches of natural gas at a pressure of 4 pounds per square inch. Assuming adiabatic conditions, find the rate of change in the pressure of the gas if its volume is decreasing by 10 in.^{3}/s.

15. A particle is moving counterclockwise on the ellipse with equation $x^2/4 + y^2/3 = 1$ so that its angular velocity relative to the origin remains a constant 0.1 radians per second. If lengths are measured in kilometers, find its velocity in the x- and y-directions as it passes through the point $(1, 3/2)$.

16. A particle is moving counterclockwise on the ellipse with equation $x^2/1 + y^2/4 = 1$ so that its angular velocity relative to the origin remains a constant 0.2 radians per second. If lengths are measured in kilometers, find its velocity in the x- and y-directions as it passes through the point $(1/2, \sqrt{3})$.

17. A spherical balloon is filled with water. If the water is leaking out at 2 cm^3/min, find the rate at which the radius is changing when $V = 1000\ \text{cm}^3$.

18. A blood sample is being drawn from your arm into a cylinder of radius 0.5 cm. If a nurse is withdrawing the piston from the cylinder at the rate of 0.25 cm/s, at what rate (in quarts per hour) are you losing blood? (*Fact:* $1\ \text{cm}^3 \approx 1.1 \times 10^{-3}$ qt.)

19. A bucket of cement mixture is placed on point A on level ground, 13.5 m below a pulley. A rope is attached to the bucket, threaded through the pulley, and released. The free end of the rope is 1.0 m above A. A person grasps the end of the rope and, keeping it at height 1.0 m, walks at 1.6 m/s away from A. How fast is the bucket ascending when the free end of the rope is 9 m from its original position? Assume that the pulley and bucket are "points."

20. Because of a bumper crop, field corn is being stored outdoors at the local grain elevator. An overhead boom is adding corn to the top of the pile at 0.3 bushels per second. The pile takes the form of a cone whose diameter is four times its height. Find the rate at which the boom must be raised when the pile contains 100,000 bushels. (One bushel is 0.035 m^3, approximately.)

21. If the temperature is constant in the piston/cylinder mechanism discussed in Example 1, the pressure and volume of the gas are related through Boyle's law. Assume there is 1 mol of an ideal gas in the cylinder and that the temperature is kept at 0°C. In this case Boyle's law is $PV = 2271.0$, where V is measured in cubic meters and P in pascals (1 Pa is 1 N/m^2). If the piston is moving into the cylinder at 10 cm/s, find the rate at which the pressure is changing when the pressure is 1000 Pa.

22. A door 0.8 m wide is swinging shut at the rate of 0.25 radians per second. Calculate the rate at which the free edge of the door is approaching the jamb at the time the door is half open.

T 23. The accompanying figure shows an underground gasoline storage tank. The tank is a cylinder, with length $L = 6$ m and radius of the circular ends $r = 1.5$ m. The tank is being filled at the rate of 0.5 m^3/min. Find the rate at which the level of the liquid in the tank is rising when its depth is 2 m. The area A of a segment of height h in a circle of radius r is

$$A = \tfrac{1}{2}\pi r^2 - r^2 \arcsin(1 - h/r) - (r - h)\sqrt{h(2r - h)}, \quad 0 \le h \le 2r.$$

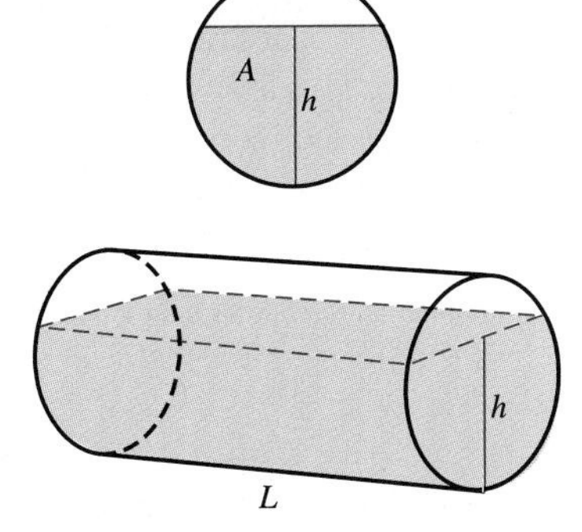

Figure for Exercise 23.

24. A spherical tank of radius 20 m is being filled with liquid propane at the rate of 0.7 m^3/min. Calculate the rate at which the propane level is rising when the tank is 75 percent full by volume. The formula for the volume V of a spherical segment of height h, where a is the radius of the sphere and $0 \le h \le 2a$, is given by $V = \pi h^2(a - h/3)$. See the accompanying figure.

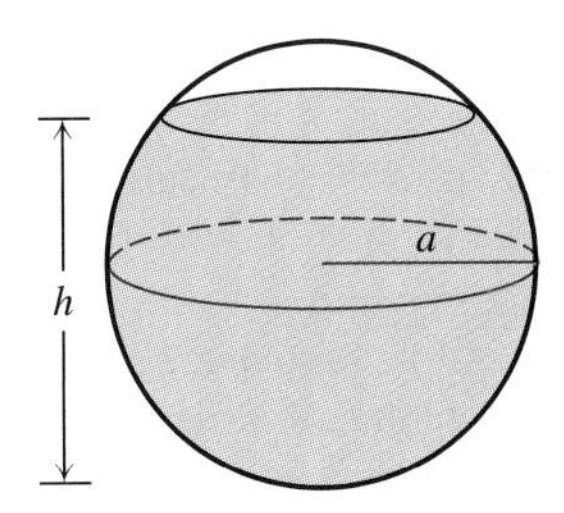

Figure for Exercise 24.

25. You are holding a conical cup (height 2 in. and radius 1 in.) full of milk. A "friend" comes along and pushes his cup (same size) down into your cup, keeping the axes aligned and pushing at the rate of 3 in./s. At what rate is the milk spilling into your lap when the tip of his cup is 1/2 in. from the bottom of your cup?

26. A kitchen drinking cup in the shape of a frustum of a cone measures 4.4 cm across its bottom (diameter) and 6.4 cm across its top. Its depth is 7.3 cm. It is being filled with water at the rate of 30 cm^3/min (which is, approximately, 1 fluid ounce per minute). At what rate is the water level rising when the depth of the water is 2 cm? The formula for the volume of a frustum of a cone is

$$V = \frac{1}{3}\pi H(R^2 + Rr + r^2),$$

where H is the height of the frustum, R the radius of the larger circular base, and r the radius of the smaller circular base.

27. Refer to the figure. A boy starts from A and runs at the rate of 3 m/s toward the center O of a circular room with a single light source at L, where OL is perpendicular to OA. At what rate will his shadow be moving along the wall when he is halfway from A to O?

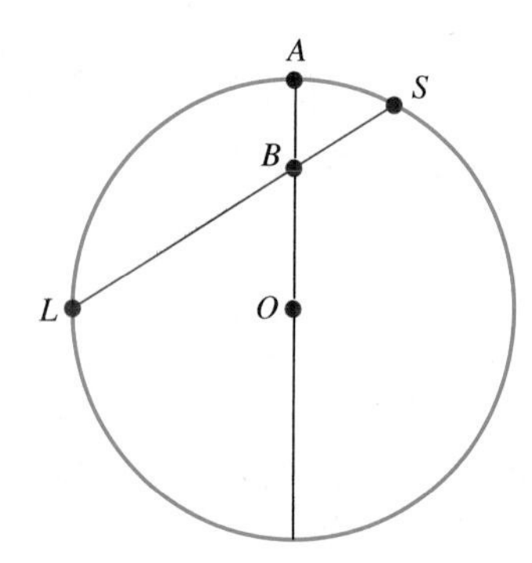

Figure for Exercise 27.

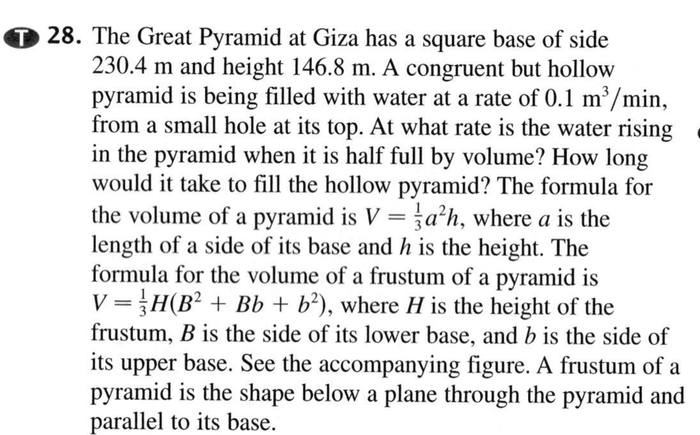

28. The Great Pyramid at Giza has a square base of side 230.4 m and height 146.8 m. A congruent but hollow pyramid is being filled with water at a rate of $0.1\ \text{m}^3/\text{min}$, from a small hole at its top. At what rate is the water rising in the pyramid when it is half full by volume? How long would it take to fill the hollow pyramid? The formula for the volume of a pyramid is $V = \frac{1}{3}a^2h$, where a is the length of a side of its base and h is the height. The formula for the volume of a frustum of a pyramid is $V = \frac{1}{3}H(B^2 + Bb + b^2)$, where H is the height of the frustum, B is the side of its lower base, and b is the side of its upper base. See the accompanying figure. A frustum of a pyramid is the shape below a plane through the pyramid and parallel to its base.

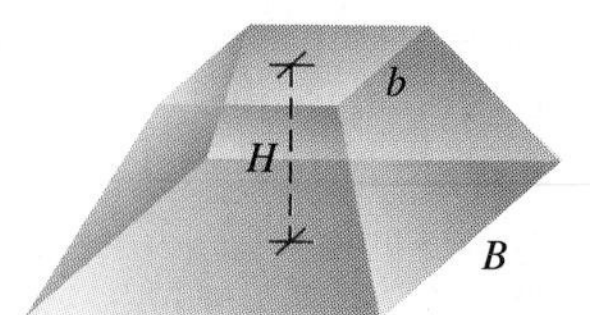

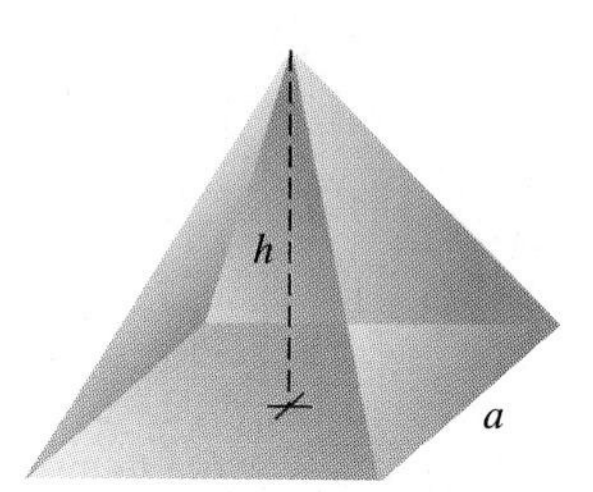

Figure for Exercise 28.

29. According to a story for children, a bird visits the top of a mountain once each century and sharpens its beak on the rock. When the mountain has worn away, one day of eternity will have passed. Mt. McKinley rises approximately 5200 m above its base and is roughly conical, with the diameter of its base about four times its height. Suppose that one "sharpening" removes $0.01\ \text{mm}^3$ of stone and is done so that the proportions of the cone remain fixed. Calculate the rate of decrease in the height of Mt. McKinley when it has been reduced to half its original volume. What is "one day of eternity," measured in centuries?

4.8 Indeterminate Forms and l'Hôpital's Rules

Limits of the form

$$\lim_{x \to a} \frac{f(x)}{g(x)},$$

where both $\lim_{x\to a} f(x) = 0$ and $\lim_{x\to a} g(x) = 0$, are said to have the *indeterminate form* $0/0$. Such limits are called *indeterminate* because they can fail to exist, be infinite, or be finite. This behavior is related to the fact that any real number can be written as the quotient of two "small" numbers.

For example,

$$\frac{10^{-9}}{10^{-9}} = 1, \quad \frac{10^{-9}}{10^{-18}} = 10^9, \quad \frac{10^{-18}}{10^{-9}} = 10^{-9}.$$

The familiar limit

$$\lim_{x \to 0} \frac{\sin x}{x}$$

is a $0/0$ form. As we showed in Section 1.5, this limit is 1. The limits

$$\lim_{x \to 0} \frac{\tan x}{x^2} \tag{1}$$

and

$$\lim_{x \to 0^+} \frac{x}{\sin(x^3)} \tag{2}$$

are also of the form $0/0$. We show that neither has a finite limit.

The limit in (1) does not exist. This follows by writing the quotient as

$$\frac{\tan x}{x^2} = \frac{\sin x}{x} \cdot \frac{1}{x \cos x}.$$

Because $(\sin x)/x \to 1$ and $\cos x \to 1$ as $x \to 0$,

$$\lim_{x \to 0^+} \frac{\sin x}{x} \cdot \frac{1}{x \cos x} = \infty.$$

Similarly,

$$\lim_{x \to 0^-} \frac{\sin x}{x} \cdot \frac{1}{x \cos x} = -\infty$$

Because these two limits do not agree, the limit in (1) does not exist.

For the limit in (2), we write

$$\frac{x}{\sin(x^3)} = \frac{x^3}{\sin(x^3)} \cdot \frac{1}{x^2}.$$

Much as before, as $x \to 0$ the first factor approaches 1 and the second factor approaches ∞. Hence

$$\lim_{x \to 0^+} \frac{x}{\sin(x^3)} = \infty.$$

The limits in (1) and (2) are of the form $0/0$. We also consider the indeterminate forms ∞/∞, $0 \cdot \infty$ (or $\infty \cdot 0$), $\infty - \infty$, 0^0, ∞^0, and 1^∞.

In this section we discuss l'Hôpital's Rules for limits of the form $0/0$ and ∞/∞. The other indeterminate forms can sometimes be reduced to one of these forms by simple algebra or by taking logarithms. l'Hôpital's Rules appeared in the first calculus textbook, written in 1696 by the Marquis l'Hôpital (1661–1704).

l'Hôpital's Rule ($0/0$ form)

Assume that f and g are differentiable on an open interval (a, b) and $g'(x) \neq 0$ for all $x \in (a, b)$. If $f(x)/g(x)$ has the form $0/0$ as $x \to a^+$ and

$$\lim_{x \to a^+} \frac{f'(x)}{g'(x)} = L,$$

where L is finite or ∞ or $-\infty$, then

$$\lim_{x \to a^+} \frac{f(x)}{g(x)} = L.$$

There are corresponding versions of l'Hôpital's Rule (0/0 form) for limits in which $x \to b^-$, $x \to c$ (where $a < c < b$), $x \to \infty$, and $x \to -\infty$. All have similar hypotheses and we use them without further comment. We prove a special case of l'Hôpital's Rule (0/0 form) after several examples.

EXAMPLE 1 Use l'Hôpital's Rule to show that

$$\lim_{x \to 0} \frac{\sin x}{x} = 1.$$

Solution

The ratio $\sin x/x$ has the form 0/0 as $x \to 0$. Because the limit

$$\lim_{x \to 0} \frac{(\sin x)'}{(x)'} = \lim_{x \to 0} \frac{\cos x}{1} = \frac{\cos 0}{1} = 1$$

exists, we may apply l'Hôpital's Rule and conclude that

$$\lim_{x \to 0} \frac{\sin x}{x} = \lim_{x \to 0} \frac{\cos x}{1} = 1.$$

EXAMPLE 2 Use l'Hôpital's Rule to show that

$$\lim_{x \to 1} \frac{\ln x}{x - 1} = 1.$$

Solution

The ratio $\ln x/(x - 1)$ has the form 0/0 as $x \to 1$. We check the limit of the ratio of the derivatives of $f(x) = \ln x$ and $g(x) = x - 1$.

$$\lim_{x \to 1} \frac{f'(x)}{g'(x)} = \lim_{x \to 1} \frac{1/x}{1} = 1. \tag{3}$$

Applying l'Hôpital's Rule, we conclude that

$$\lim_{x \to 1} \frac{\ln x}{x - 1} = \lim_{x \to 1} \frac{1/x}{1} = 1.$$

In the next example the limit of the quotient of derivatives is itself indeterminate, of the form 0/0. In such cases we may wish to apply l'Hôpital's Rule a second time.

EXAMPLE 3 Evaluate the limit

$$\lim_{x \to 0} \frac{x \sin x}{1 - e^{-x^2}}. \tag{4}$$

Solution

The limit in (4) is of the form $0/0$. We can apply l'Hôpital's Rule to evaluate this limit if the limit

$$\lim_{x\to 0}\frac{(x\sin x)'}{(1-e^{-x^2})'}=\lim_{x\to 0}\frac{\sin x + x\cos x}{2xe^{-x^2}} \tag{5}$$

of the quotient of derivatives exists. Because the limit in (5) has the form $0/0$, we may be able to evaluate it by applying l'Hôpital's Rule. Hence, we ask if the limit

$$\lim_{x\to 0}\frac{(\sin x + x\cos x)'}{(2xe^{-x^2})'}=\lim_{x\to 0}\frac{2\cos x - x\sin x}{2e^{-x^2}-4x^2e^{-x^2}}$$

of the quotient of derivatives exists. Because

$$\lim_{x\to 0}\frac{2\cos x - x\sin x}{2e^{-x^2}-4x^2e^{-x^2}}=\frac{2}{2}=1,$$

the limit in (5) exists and is equal to 1, by l'Hôpital's Rule. Because the limit in (5) exists and is equal to 1, the limit in (4) exists and is equal to 1, again by l'Hôpital's Rule.

In practice, this calculation would have been written in one line:

$$\lim_{x\to 0}\frac{x\sin x}{1-e^{-x^2}}=\lim_{x\to 0}\frac{\sin x + x\cos x}{2xe^{-x^2}}=\lim_{x\to 0}\frac{2\cos x - x\sin x}{2e^{-x^2}-4x^2e^{-x^2}}=\frac{2}{2}=1.$$

The string of equalities is conditional until the last step. The third equality is a calculation of an easy limit; the second equality holds because of an application of l'Hôpital's Rule to the functions $\sin x + x\cos x$ and $2xe^{-x^2}$; and the first holds because of an application of l'Hôpital's Rule to the functions $x\sin x$ and $1-e^{-x^2}$. The first application of l'Hôpital's Rule is justified by the outcome of its second application.

In the next example, we show how a limit of the form $\infty - \infty$ can sometimes be evaluated.

EXAMPLE 4 Use l'Hôpital's Rule to show that

$$\lim_{x\to 0^+}\left(\frac{1}{x^2}-\frac{1}{\sin x}\right)=\infty.$$

Solution

This limit has the form $\infty - \infty$ as $x \to 0^+$. It can be written as a fraction having the form $0/0$ as $x \to 0^+$ by combining the two fractions:

$$\frac{1}{x^2}-\frac{1}{\sin x}=\frac{\sin x - x^2}{x^2\sin x}.$$

Applying l'Hôpital's Rule,

$$\lim_{x\to 0^+}\left(\frac{1}{x^2}-\frac{1}{\sin x}\right)=\lim_{x\to 0^+}\frac{\sin x - x^2}{x^2\sin x}$$
$$=\lim_{x\to 0^+}\frac{\cos x - 2x}{2x\sin x + x^2\cos x}=+\infty,$$

because as $x \to 0^+$ the numerator approaches 1 and the denominator approaches 0 through positive values.

This limit can also be evaluated by factoring $1/x$ from both terms and using the fact that $(\sin x)/x \to 1$ as $x \to 0$.

EXAMPLE 5 Evaluate the limit

$$\lim_{x\to\infty} x\arctan(1/x).$$

Solution

This limit has the form $\infty \cdot 0$, not $0/0$. We can convert it to the $0/0$ form by rewriting the limit as

$$\lim_{x\to\infty}\frac{\arctan(1/x)}{1/x}.$$

Applying l'Hôpital's Rule,

$$\lim_{x\to\infty}\frac{\arctan(1/x)}{1/x}=\lim_{x\to\infty}\frac{(\arctan(1/x))'}{(1/x)'}=\lim_{x\to\infty}\frac{\dfrac{-1/x^2}{1+1/x^2}}{-1/x^2}=\lim_{x\to\infty}\frac{1}{1+1/x^2}=1.$$

PROOF OF A SPECIAL CASE OF L'HÔPITAL'S RULE (0/0 FORM) By assuming a bit more about f and g, l'Hôpital's Rule ($0/0$ form) can be proved by a straightforward calculation. Specifically, if we assume that (i) f' and g' are continuous on an interval $[a, b)$, (ii) $f(a) = g(a) = 0$, and (iii) $g'(a) \neq 0$, then

$$\lim_{x\to a^+}\frac{f(x)}{g(x)}=\lim_{x\to a^+}\frac{f(x)-f(a)}{g(x)-g(a)}$$
$$=\lim_{x\to a^+}\frac{\dfrac{f(x)-f(a)}{x-a}}{\dfrac{g(x)-g(a)}{x-a}}=\frac{\displaystyle\lim_{x\to a^+}\frac{f(x)-f(a)}{x-a}}{\displaystyle\lim_{x\to a^+}\frac{g(x)-g(a)}{x-a}}$$
$$=\frac{f'(a)}{g'(a)}=\lim_{x\to a^+}\frac{f'(x)}{g'(x)}.$$

We outline in Exercises 42 and 43 a proof of the more general case of l'Hôpital's Rule stated earlier.

We state l'Hôpital's Rule for the ∞/∞ form. Its proof may be found in more advanced texts.

l'Hôpital's Rule (∞/∞ form)

Assume that f and g are differentiable on an open interval (a, b) and $g'(x) \neq 0$ for all $x \in (a, b)$. If $f(x)/g(x)$ has the form ∞/∞ as $x \longrightarrow a^+$ and

$$\lim_{x \to a^+} \frac{f'(x)}{g'(x)} = L,$$

where L is finite or ∞ or $-\infty$, then

$$\lim_{x \to a^+} \frac{f(x)}{g(x)} = L.$$

There are corresponding versions of l'Hôpital's Rule (∞/∞ form) for limits in which $x \longrightarrow b^-$, $x \longrightarrow c$ (where $a < c < b$), $x \longrightarrow \infty$, and $x \longrightarrow -\infty$. All have similar hypotheses, and we use them without further comment.

EXAMPLE 6 Use l'Hôpital's Rule to evaluate the limit

$$\lim_{x \to \infty} \frac{\ln x}{x}.$$

Solution

We don't need l'Hôpital's Rule for this limit. From equation (6) in Section 4.4,

$$\lim_{x \to \infty} \frac{(\ln x)^a}{x^b} = 0$$

for any a, $b > 0$. Hence, with $a = b = 1$,

$$\lim_{x \to \infty} \frac{\ln x}{x} = 0.$$

On the other hand, applying l'Hôpital's Rule is easy for this limit. Noting that the limit has the ∞/∞ form and

$$\lim_{x \to \infty} \frac{\ln x}{x} = \lim_{x \to \infty} \frac{(\ln x)'}{(x)'} = \lim_{x \to \infty} \frac{1/x}{1} = \lim_{x \to \infty} \frac{1}{x} = 0,$$

it follows that

$$\lim_{x \to \infty} \frac{\ln x}{x} = 0.$$

In the next example we show how limits of the form $0 \cdot \infty$ can be changed to the form ∞/∞.

EXAMPLE 7 Use l'Hôpital's Rule to evaluate the limit

$$\lim_{x \to 0^+} x \ln x.$$

Solution

This limit has the form $0 \cdot (-\infty)$, which we classify as the $0 \cdot \infty$ form. Writing

$$\lim_{x \to 0^+} x \ln x = \lim_{x \to 0^+} \frac{\ln x}{1/x}$$

gives a limit in the ∞/∞ form. Applying l'Hôpital's Rule,

$$\lim_{x \to 0^+} x \ln x = \lim_{x \to 0^+} \frac{\ln x}{1/x} = \lim_{x \to 0^+} \frac{(\ln x)'}{(1/x)'} = \lim_{x \to 0^+} \frac{1/x}{-1/x^2} = \lim_{x \to 0^+} (-x) = 0.$$

The Indeterminate Forms 0^0, ∞^0, and 1^∞ Expressions like $f(x)^{g(x)}$, where $f(x) > 0$, can be written as

$$f(x)^{g(x)} = e^{g(x) \ln(f(x))}.$$

If we can show, perhaps with the use of l'Hôpital's Rule, that $\lim_{x \to a^+} g(x) \ln(f(x)) = L$, where L is finite, then

$$\lim_{x \to a^+} f(x)^{g(x)} = \lim_{x \to a^+} e^{g(x) \ln(f(x))} = e^L.$$

The last equality holds because the exponential function is continuous at L.

A different arrangement of this calculation is to set $y = f(x)^{g(x)}$ and take the logarithm of both sides. This gives

$$\ln y = g(x) \ln(f(x)).$$

If, as earlier, we can show that $\lim_{x \to a^+} g(x) \ln(f(x)) = L$, where L is finite, then

$$\lim_{x \to a^+} f(x)^{g(x)} = \lim_{x \to a^+} y = \lim_{x \to a^+} e^{\ln y} = e^{\lim_{x \to a^+} (\ln y)} = e^L.$$

We illustrate these ideas with two examples.

EXAMPLE 8 Show that $\lim_{x \to 0^+} x^x = 1$.

Solution

The limit has the form 0^0. As shown in Example 7, $\lim_{x \to 0^+} x \ln x = 0$. Hence,

$$\lim_{x \to 0^+} x^x = \lim_{x \to 0^+} e^{x \ln x} = e^{\lim_{x \to 0^+} x \ln x} = e^0 = 1.$$

EXAMPLE 9 Show that for any real number c,

$$\lim_{x\to\infty}\left(1+\frac{c}{x}\right)^x = e^c.$$

Solution

This limit has the form 1^∞. Setting $y = \left(1+\frac{c}{x}\right)^x$ and taking logarithms,

$$\ln y = x\ln\left(1+\frac{c}{x}\right).$$

As $x \to \infty$, the expression on the right has the form $\infty \cdot 0$. Rewriting this so that the indeterminate form becomes $0/0$ and applying l'Hôpital's Rule ($0/0$ form),

$$\lim_{x\to\infty}\ln y = \lim_{x\to\infty}\frac{\ln\left(1+\frac{c}{x}\right)}{\frac{1}{x}} = \lim_{x\to\infty}\frac{\frac{-\frac{c}{x^2}}{1+\frac{c}{x}}}{-\frac{1}{x^2}} = \lim_{x\to\infty}\frac{c}{1+\frac{c}{x}} = c.$$

Hence,

$$\lim_{x\to\infty}\left(1+\frac{c}{x}\right)^x = \lim_{x\to\infty} e^{\ln y} = e^c.$$

Exercises 4.8

Exercises 1–14: Classify the limits by indeterminate form. Do not evaluate. The forms are: $0/0$, ∞/∞, $0\cdot\infty$ *(or* $\infty\cdot 0$*)*, $\infty-\infty$, 0^0, ∞^0, 1^∞.

1. $\displaystyle\lim_{x\to 0^+}\frac{\tan x}{\sqrt{x}}$
2. $\displaystyle\lim_{x\to 0}\frac{1-\cos x}{x^2}$
3. $\displaystyle\lim_{x\to\infty} x\sin(1/x)$
4. $\displaystyle\lim_{x\to\pi^-}(x-\pi)\tan\left(\tfrac{1}{2}x\right)$
5. $\displaystyle\lim_{x\to 0^+}\frac{\ln x}{\ln(e^x-1)}$
6. $\displaystyle\lim_{x\to\infty}\frac{e^x}{x^2}$
7. $\displaystyle\lim_{x\to\infty}\left(\sqrt{x^2-1}-x\right)$
8. $\displaystyle\lim_{x\to(\pi/2)^-}(\tan x-\sec x)$
9. $\displaystyle\lim_{x\to\infty}\left(\sqrt{x}\right)^{1/x}$
10. $\displaystyle\lim_{x\to(\pi/2)^-}(\tan x)^{\cos x}$
11. $\displaystyle\lim_{x\to(\pi/2)^-}(\sin x)^{\tan x}$
12. $\displaystyle\lim_{x\to e^+}(\ln x)^{1/(x-e)}$
13. $\displaystyle\lim_{x\to\infty}\left(\frac{1}{x}\right)^{e^{-x}}$
14. $\displaystyle\lim_{x\to 0^+} x^{\sqrt[3]{x}}$

Exercises 15–40: Evaluate the limits.

15. $\lim_{x\to 0} \frac{\sin^2 x}{x}$

16. $\lim_{x\to 0} \frac{\tan x}{\sqrt{x}}$

17. $\lim_{x\to 0^+} \frac{1 - e^{-2x}}{\sqrt{x}}$

18. $\lim_{x\to \pi} \frac{1 + \cos x}{\pi - x}$

19. $\lim_{x\to \infty} \frac{2^x}{x^2}$

20. $\lim_{x\to \infty} \frac{3^x}{x^3}$

21. $\lim_{x\to 0^+} \left(\frac{1}{\sqrt{x}} - \frac{1}{\tan x}\right)$

22. $\lim_{x\to 0^+} \left(\frac{1}{\sqrt[3]{x}} - \frac{1}{\sin x}\right)$

23. $\lim_{x\to 1^+} \frac{\cos x}{x - 1}$

24. $\lim_{x\to 1^+} \frac{\sin x}{x^2 - 1}$

25. $\lim_{x\to \pi/4} \frac{\sqrt{2}\cos x - 1}{1 - \tan^2 x}$

26. $\lim_{x\to \pi/4} \frac{\pi/4 - x}{\sin(3\pi/4 + x)}$

27. $\lim_{x\to \infty} \left(\sqrt{x}\right)^{1/x}$

28. $\lim_{x\to (\pi/2)^-} (\tan x)^{\cos x}$

29. $\lim_{x\to (\pi/2)^-} (\sin x)^{\tan x}$

30. $\lim_{x\to e^+} (\ln x)^{1/(x-e)}$

31. $\lim_{x\to 0^+} \frac{\tan x}{\sqrt{x}}$

32. $\lim_{x\to 0} \frac{1 - \cos x}{x^2}$

33. $\lim_{x\to \infty} x \sin(1/x)$

34. $\lim_{x\to \pi^-} (x - \pi)\tan\left(\frac{1}{2}x\right)$

35. $\lim_{x\to 0^+} \frac{\ln x}{\ln(e^x - 1)}$

36. $\lim_{x\to \infty} \frac{e^x}{x^2}$

37. $\lim_{x\to \infty} \left(\sqrt{x^2 - 1} - x\right)$

38. $\lim_{x\to (\pi/2)^-} (\tan x - \sec x)$

39. $\lim_{x\to \infty} \left(\frac{1}{x}\right)^{e^{-x}}$

40. $\lim_{x\to 0^+} x^{\sqrt[3]{x}}$

41. Use l'Hôpital's Rule (0/0 form) to prove that

$$\lim_{h\to 0} \frac{f(x + h) - f(x - h)}{2h} = f'(x).$$

Assume that f is continuously differentiable near x.

42. The result stated here is needed in the next exercise Cauchy's Mean-value Theorem: If f and g are continuous on an interval $[a, b]$, differentiable on (a, b), and $g'(x) \neq 0$ for all x in (a, b), then there is at least one number $c \in (a, b)$ for which

$$\frac{f(b) - f(a)}{g(b) - g(a)} = \frac{f'(c)}{g'(c)}.$$

Show that Cauchy's Mean-value Theorem follows by applying the Mean-value Theorem given in Section 4.3 to the function

$$h(x) = (f(b) - f(a))g(x) - (g(b) - g(a))f(x),$$
$$x \in [a, b].$$

43. We outline a proof of l'Hôpital's Rule (0/0 form). Letting $B = \frac{1}{2}(a + b)$, first prove that the functions F and G defined on $[a, B]$ by

$$F(x) = \begin{cases} 0 & x = a \\ f(x) & a < x \le B \end{cases}$$

$$G(x) = \begin{cases} 0 & x = a \\ g(x) & a < x \le B \end{cases}$$

are continuous on $[a, B]$ and differentiable on (a, B). Secondly, applying Cauchy's Mean-value Theorem to F and G on the interval $[a, B]$, show that for any $x \in (a, B)$ there is a $c \in (a, B)$ for which

$$\frac{f(x)}{g(x)} = \frac{F(x) - F(a)}{G(x) - G(a)} = \frac{F'(c)}{G'(c)} = \frac{f'(c)}{g'(c)}.$$

Infer l'Hôpital's Rule (0/0 form) from this result.

4.9 Euler's Method

One of the most important applications of calculus is to the solution of *differential equations,* which are used to model such phenomena as motion, radioactive decay, biological growth, or the formation of new compounds in complex chemical reactions. Because the solutions of many of these differential equations cannot be found by exact methods, approximate solutions are widely used. The simplest method for finding an approximate solution to a differential equation is based on the tangent line approximation. It is called Euler's method.

WEB We discuss Euler's method for initial value problems of the form

$$\frac{dy}{dt} = G(t, y) \qquad (1)$$
$$y(0) = y_0.$$

The number y_0 and the function G are given. By a solution of this initial value problem on a time interval $[0, b]$, we mean a function $y = f(t)$ satisfying

$$f'(t) = G(t, f(t)), \text{ for all } t \in [0, b], \quad \text{and} \quad f(0) = y_0. \qquad (2)$$

A simple example of an initial value problem is

$$\frac{dy}{dt} = G(t, y) = t\sqrt{y} \qquad (3)$$
$$y(0) = 1.$$

The solution of this problem is the function $y = f(t) = \left(\frac{1}{4}t^2 + 1\right)^2$. It's easy to check that this solution satisfies the differential equation and the initial condition. Differentiating y,

$$\frac{dy}{dt} = 2\left(\frac{1}{4}t^2 + 1\right) \cdot \frac{1}{4} \cdot (2t) = \left(\frac{1}{4}t^2 + 1\right)t = t\sqrt{y}.$$

For the initial condition, $f(0) = \left(\frac{1}{4} \cdot 0^2 + 1\right)^2 = 1$.

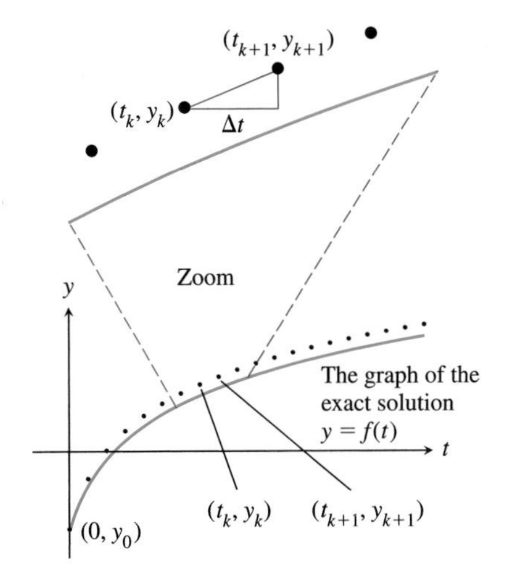

FIGURE 4.57 A geometric view of Euler's method.

Figure 4.57 shows the graph of the exact solution $y = f(t)$ to a representative initial value problem (1) and an approximate solution calculated with Euler's method. As in the figure, an approximate solution is often not a curve, rather it is a sequence of points $(0, y_0), (t_1, y_1), (t_2, y_2), \ldots$ that more or less closely follows the exact solution. The values $0, t_1, t_2, \ldots$ of t are equally spaced, with

$$t_{k+1} - t_k = \Delta t, \qquad k = 0, 1, 2, \ldots.$$

The idea of Euler's method is this: Starting with the known point $(t_0, y_0) = (0, y_0)$ and a fixed time interval of length Δt, we approximate the point $(t_0 + \Delta t, f(t_0 + \Delta t))$ on the exact solution by using the tangent line approximation at (t_0, y_0); denoting this approximation by (t_1, y_1), we then approximate the point $(t_1 + \Delta t, f(t_1 + \Delta t))$ on the exact solution by again using the tangent line approximation, this time at (t_1, y_1). We continue this process until we have generated a sequence of points that approximates the exact solution on the interval $[0, b]$.

We calculate the "new point" (t_{k+1}, y_{k+1}) from the "old point" (t_k, y_k) using the tangent line approximation at (t_k, y_k).

$$f(t_{k+1}) = f(t_k + \Delta t) \approx f(t_k) + \Delta t f'(t_k).$$

A problem with this approximation is that $f(t_k)$ and $f'(t_k)$ are not known, because $(t_k, f(t_k))$ is on the "true" solution curve. However, the value $f(t_k)$ is approximately y_k and the value $f'(t_k)$ can be approximated using the fact that the function f satisfies the differential equation in (2). Hence,

$$f(t_{k+1}) = f(t_k + \Delta t) \approx f(t_k) + \Delta t f'(t_k) \approx y_k + \Delta t G(t_k, y_k).$$

To get started, we use the given point $(0, y_0) = (t_0, y_0)$ and calculate (t_1, y_1); knowing (t_1, y_1), we calculate (t_2, y_2), and so on.

We describe Euler's method as a procedure or algorithm.

Euler's Method

To calculate an approximate solution to the initial value problem

$$\frac{dy}{dt} = G(t, y)$$
$$y(0) = y_0$$

on the time interval $[0, b]$, begin by choosing $\Delta t = b/n$, where n is a positive integer.

STEP 1. The initial point in the approximate solution is $(t_0, y_0) = (0, y(0))$. Set $k = 0$.

STEP 2. Given the "old point" (t_k, y_k), the "new point" of the approximate solution is

$$(t_{k+1}, y_{k+1}) = (t_k + \Delta t, y_k + \Delta t G(t_k, y_k)). \tag{4}$$

STEP 3. If $t_{k+1} + \Delta t = b$, quit; otherwise, increase k by 1 and go to Step 2.

We give two examples to show how Euler's method works. The first is a very simple initial value problem with a known solution; the second is a realistic model of the motion of a paratrooper.

EXAMPLE 1 Use Euler's method to find an approximate solution to the initial value problem

$$\frac{dy}{dt} = G(t, y) = t + y$$
$$y(0) = 1$$

on the interval $[0, 2]$. Take $\Delta t = 0.1$.

Solution

The exact solution to this initial value problem is

$$y = f(t) = 2e^t - t - 1.$$

We check that this function satisfies the initial condition and the differential equation. Evaluating y at $t = 0$, $y(0) = 2 \cdot 1 - 0 - 1 = 1$; hence the initial condition $y(0) = 1$ is satisfied. Differentiating $y = 2e^t - t - 1$,

$$\frac{dy}{dt} = 2e^t - 1;$$

substituting y and dy/dt into the differential equation $dy/dt = t + y$,

$$2e^t - 1 = t + (2e^t - t - 1) = 2e^t - 1.$$

Hence, $y = f(t) = 2e^t - t - 1$ is the solution to the given initial value problem. Figure 4.58 shows the graph of f on the interval $[0, 2]$.

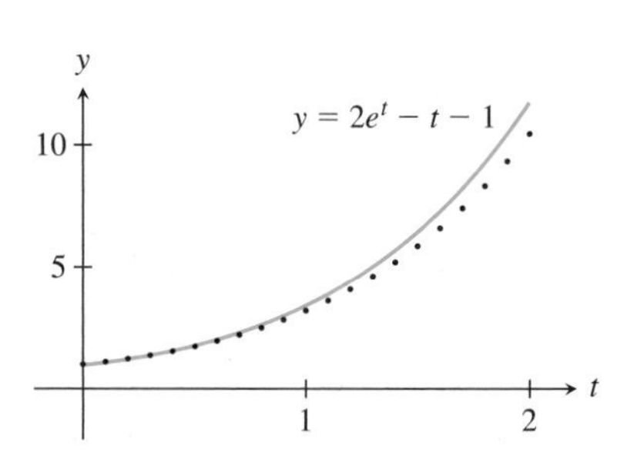

FIGURE 4.58 The graphs of the exact and approximate solutions to the initial value problem $dy/dt = t + y$ and $y(0) = 1$.

We now apply (4) to calculate an approximate solution on the time interval $[0, 2]$. We take $\Delta t = 0.1$. The main step in the algorithm is

$$(t_{k+1}, y_{k+1}) = (t_k + \Delta t, y_k + \Delta t G(t_k, y_k)) = (t_k + 0.1, y_k + 0.1(t_k + y_k)). \quad (5)$$

Starting with $(t_0, y_0) = (0, 1)$, we show the calculation of the first several points of the approximate solution.

$$(t_0, y_0) = (0, 1)$$

$$\begin{aligned}(t_1, y_1) &= (t_0 + \Delta t, y_0 + \Delta t(t_0 + y_0)) \\ &= (0 + 0.1, 1 + 0.1(0 + 1)) = (0.1, 1.1)\end{aligned}$$

$$\begin{aligned}(t_2, y_2) &= (t_1 + \Delta t, y_1 + \Delta t(t_1 + y_1)) \\ &= (0.1 + 0.1, 1.1 + 0.1(0.1 + 1.1)) = (0.2, 1.22)\end{aligned}$$

$$\begin{aligned}(t_3, y_3) &= (t_2 + \Delta t, y_2 + \Delta t(t_2 + y_2)) \\ &= (0.2 + 0.1, 1.22 + 0.1(0.2 + 1.22)) = (0.3, 1.362).\end{aligned}$$

The approximate solution is listed in Table 4.6. These data are plotted as dots in Fig. 4.58.

TABLE 4.6 An approximate solution to the initial value problem $dy/dt = t + y$ and $y(0) = 1$, using Euler's method.

k	(t_k, y_k)	k	(t_k, y_k)
0	(0.0, 1.000)	10	(1.0, 3.187)
1	(0.1, 1.100)	11	(1.1, 3.606)
2	(0.2, 1.220)	12	(1.2, 4.077)
3	(0.3, 1.362)	13	(1.3, 4.605)
4	(0.4, 1.528)	14	(1.4, 5.195)
5	(0.5, 1.721)	15	(1.5, 5.854)
6	(0.6, 1.943)	16	(1.6, 6.590)
7	(0.7, 2.197)	17	(1.7, 7.409)
8	(0.8, 2.487)	18	(1.8, 8.320)
9	(0.9, 2.816)	19	(1.9, 9.332)
		20	(2.0, 10.455)

The "fit" of the approximate solution to the exact solution can be improved by taking Δt smaller. See Exercise 11.

EXAMPLE 2 A paratrooper with mass 91.6 kg (this is 72.6 kg body mass plus 19 kg equipment) is falling at 50 m/s at the time his parachute opens. Use Euler's method to calculate and graph the velocity of the trooper during the first 1.5 seconds in which his parachute was open, assuming that

i) at any time t, the magnitude of the retarding force due to air resistance is proportional to the square of his speed and

ii) after a few seconds, the paratrooper's speed is close to its limiting value of approximately 5 m/s.

Take $\Delta t = 0.01$ seconds.

Solution

We model the paratrooper's motion using Newton's second law.

$$\mathbf{F} = m\mathbf{a} = m\frac{d\mathbf{v}}{dt}, \tag{6}$$

where $\mathbf{F}$ is the net force on the paratrooper and $\mathbf{a} = d\mathbf{v}/dt$ is his acceleration. Because the motion is along a vertical line, we may use scalars instead of vectors. Referring to Fig. 4.59, in which the positive direction of the y-axis is downward, the net force on the trooper is the sum of two forces: the force of the Earth on the trooper, which is mg, and the upward force due to the parachute, which is $-kv^2$, where k is a positive proportionality constant. Adding these two forces and substituting into (6),

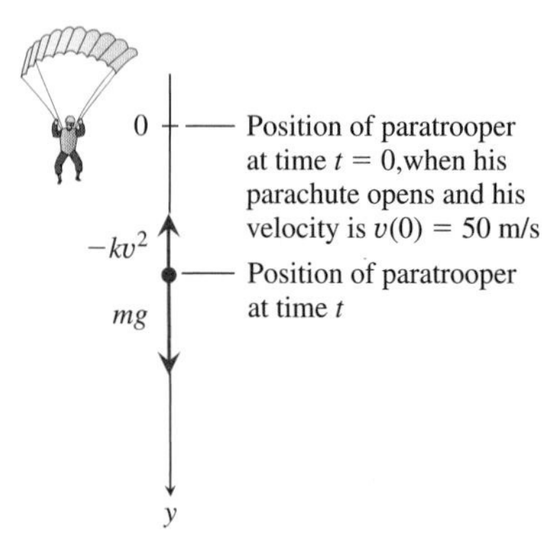

FIGURE 4.59 The gravitational and frictional forces on a paratrooper.

$$91.6 \cdot 9.8 - kv^2 = 91.6\frac{dv}{dt}. \tag{7}$$

The constant k can be evaluated by using the fact that the trooper's limiting velocity is 5 m/s. As stated in (ii), this means that after some time, his velocity is close to its limiting value of 5 m/s. Because v is not changing much as it nears its limiting value, $dv/dt \approx 0$. It follows from (7) that after a few seconds

$$91.6 \cdot 9.8 - k \cdot 5^2 \approx 91.6 \cdot 0.$$

Solving this equation for k,

$$k \approx \frac{91.6 \cdot 9.8}{5^2}.$$

Pairing the initial condition $v(0) = 50$ with the differential equation obtained from (7) by dividing both sides by 91.6 and replacing k by $91.6 \cdot 9.8/5^2$ gives the initial value problem

$$\begin{gathered}\frac{dv}{dt} = G(t, v) = 9.8 - \frac{9.8}{5^2}v^2 \\ v(0) = 50.\end{gathered} \tag{8}$$

We take $\Delta t = 0.01$ and approximate his velocity at $t = 0.01, 0.02, \ldots, 1.50$ using Euler's method. From (4)—but replacing the variable y by v—and (8), the main step in the calculation is

$$v_{k+1} = v_k + \Delta t G(t_k, v_k) = v_k + \Delta t\left(9.8 - \frac{9.8}{5^2}v_k^2\right).$$

How can this calculation be done easily and repeatedly? Perhaps the simplest way is to enter the formula

$$v + 0.01\left(9.8 - \frac{9.8}{5^2}v^2\right)$$

TABLE 4.7 An approximate solution to the initial value problem $dv/dt = 9.8 - (9.8/5^2)v^2$ and $v(0) = 50$, using Euler's method.

k	t_k	v_k
0	0.00	50.0
1	0.01	40.3
2	0.02	34.0
3	0.03	29.6
4	0.04	26.3
5	0.05	23.7
⋮		
146	1.46	5.0
147	1.47	5.0
148	1.48	5.0
149	1.49	5.0
150	1.50	5.0

into a calculator. With $v = 50$ as input, evaluate the formula. This gives an updated value of v; with this as input, again evaluate the formula. Repeat this procedure, recording the results with the corresponding values of t. Table 4.7 gives the first several values, rounded to one decimal place.

Figure 4.60 shows a sketch including both the exact and approximate solutions. To simplify the graph we have plotted only the approximate velocities corresponding to $t = 0.00, 0.04, 0.08, \ldots$. See Exercise 12 for the exact solution and an exercise relating to the size of Δt.

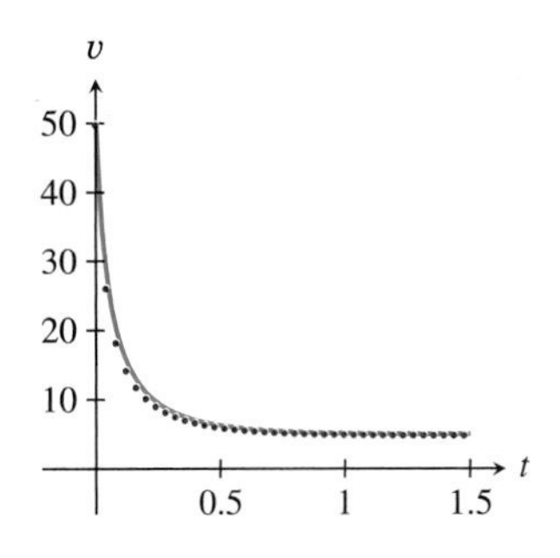

FIGURE 4.60 Approximating the paratrooper's velocity using Euler's method.

Exercises 4.9

Exercises 1–4: Suppose numbers t_{k+1} and y_{k+1} are defined by the formulas $t_{k+1} = t_k + \Delta t$ and $y_{k+1} = y_k + \Delta t f(t_k, y_k)$, where f, Δt, t_0, and y_0 are given. Calculate $(t_0, y_0), \ldots, (t_5, y_5)$.

1. $f(t, y) = t + y$, $\Delta t = 0.5$, $t_0 = 0$, and $y_0 = 1.00$

2. $f(t, y) = 3t - 2y$, $\Delta t = 0.1$, $t_0 = 0$, and $y_0 = 5.0$

3. $f(t, y) = \sqrt{t^2 + y^2}$, $\Delta t = 0.1$, $t_0 = 0$, and $y_0 = 2.00$

4. $f(t, y) = \sqrt{t^2 + 2y^2}$, $\Delta t = 0.2$, $t_0 = 0$, and $y_0 = 10.00$

Exercises 5–10: Use Euler's method to calculate an approximate solution to the initial value problem on the given interval. Take $\Delta t = 0.1$. Graph together the approximate solution and the given exact solution.

T **5.** $\dfrac{dy}{dt} = y$, $y(0) = 1$, $[0, 1]$; $y = f(t) = e^t$

T **6.** $\dfrac{dy}{dt} = \dfrac{1}{2}y$, $y(0) = 0.5$, $[0, 1]$; $y = f(t) = \dfrac{1}{2}e^{t/2}$.

T **7.** $\dfrac{dy}{dt} = -t + y$, $y(0) = 2$, $[0, 1]$; $y = f(t) = e^t + t + 1$.

T **8.** $\dfrac{dy}{dt} = -\dfrac{1}{2}t + \dfrac{1}{3}y$, $y(0) = -1$, $[0, 1]$;
$y = f(t) = -\dfrac{11}{2}e^{t/3} + \dfrac{3}{2}t + \dfrac{9}{2}$.

T **9.** $\dfrac{dy}{dt} = t^2 - y$, $y(0) = 1$, $[0, 1]$;
$y = f(t) = -e^{-t} + t^2 - 2t + 2$.

T **10.** $\dfrac{dy}{dt} = t^2 + y$, $y(0) = -1$, $[0, 1]$;
$y = f(t) = e^t - t^2 - 2t - 2$.

T **11.** In Example 1 take $\Delta t = 0.05$ and calculate an approximate solution using Euler's method. Compare the accuracy of your solution at $t = 2.0$ with that in Example 1 and the exact solution.

T **12.** In Example 2 we took $\Delta t = 0.01$. Using either a combined graph of the approximate and exact solutions or by comparing a few of the values of the approximate and exact solutions, show that taking $\Delta t = 0.05$ does not give a satisfactory approximate solution. The exact solution to the initial value problem is (with, however, constants rounded to three significant figures)

$$v = \frac{5(1 + 1.228e^{3.92t})}{-1 + 1.22e^{3.92t}}.$$

T 13. The U.S. population can be modeled by the *logistic equation*

$$P' = \frac{3}{100}\left(P - \frac{P^2}{292}\right)$$
$$P(0) = 4.$$

where the time is measured in years and the population $P = P(t)$ at time t is measured in millions of persons. The time $t = 0$ corresponds to the year 1790. This equation can be solved exactly. In Chapter 5 we show that the solution is

$$P(t) = \frac{292}{1 + 72e^{-0.03t}}, \qquad t \geq 0.$$

Solve the above initial value problem for the years 1790 to 2000 with Euler's method, with $\Delta t = 10$. Plot the approximate solution and the true solution on the same graph.

T 14. In experiments with fruit flies, M. E. Gilpin, F. J. Ayala, and J. G. Ehrenfeld found that if at $t = 0$ there are P_0 fruit flies in a colony, the number $P(t)$ of fruit flies any subsequent time t, with time measured in days, is well described by the initial value problem

$$\frac{dP}{dt} = 1.496P - 0.121P^{1.35}$$
$$P(0) = P_0.$$

This differential equation was given in Exercise 4, Section 2.10. It is similar to one of the differential equations (also in Section 2.10) used to model the population of the United States. Find an approximate solution to this initial value problem on the time interval $[0, 15]$, taking $P_0 = 100$ and $\Delta t = 1$. Graph this solution.

T 15. Newton determined empirically that if an object is placed into an environment held at a constant temperature A, the rate of change of the temperature of the object is proportional to the difference between the ambient temperature A and the object's temperature. Letting $T(t)$ be the temperature of the object at time t, this means that

$$T'(t) = c(A - T(t)),$$

where c is a constant depending on the physical properties of the object.

Newton's law of cooling can be used to find the approximate time of death of a corpse. In practice, the detective must have found the body before it has cooled to within a few degrees of room temperature, and he or she must be sure that the body has remained since death in a room of constant temperature. If temperatures are measured in degrees Fahrenheit, it is known that $c \approx 0.05$. Assuming that at the time of death the victim's temperature was 98.6°F, the initial value problem is

$$\frac{dT}{dt} = 0.05(A - T)$$
$$T_0 = 98.6.$$

Detective C. Chan found a body at 11 A.M. He immediately measured the temperatures of the room and the body and found them to be 65°F and 80°F, respectively. Acting as Detective Chan's technical assistant, use Euler's method with $\Delta t = 1$ hour to calculate the approximate time of death.

16. The differential equation in the preceding exercise was

$$\frac{dT}{dt} = 0.05(65 - T).$$

Show that the function $T(t) = 65 + ce^{-0.05t}$ satisfies this differential equation for any constant c. Choose this constant so that the initial condition $T(0) = 98.6$ is satisfied. Use your exact solution to determine the time of death. How does it compare with the value obtained by Euler's method?

17. The tangent line approximation can be written as

$$f(a + h) = f(a) + hf'(a) + hE(h),$$

where function E goes to 0 as $h \to 0$. The exact form of E depends upon f, h, and the base point a. We use this equation to describe the error in Euler's method. For Step 1 of Euler's method, we have

$$f(a + h) = f(a) + hf'(a) + hE_1(h).$$

We shorten $E_1(h)$ to E_1. Show that for Step 2, we have

$$f(a + 2h) = f(a) + h(f'(a) + f'(a + h)) + h(E_1 + E_2).$$

Show that for Step n, we have

$$f(a + nh) = f(a) + h(f'(a) + \cdots + f'(a + (n - 1)h)) + h(E_1 + \cdots + E_n).$$

The error in Euler's method is the difference

$$|f(a + nh) - f(a) - h(f'(a) + \cdots + f'(a + (n - 1)h))|,$$

which is equal to

$$|h(E_1 + \cdots + E_n)|.$$

Show that if each of the errors $E_1, \ldots, E_n$ is less than some number E, then for a given step size h, the error in Euler's method is proportional to the number of steps.

Review of Key Concepts

This chapter—titled "Applications of the Derivative"—began with the important idea of the tangent line approximation. We applied it to calculations of percentage error and to the approximation of functions by simpler functions. A second application was to Newton's method, an efficient algorithm for determining the zeros of functions. In the last section of the chapter, we applied the tangent line approximation to solve initial value problems. The resulting algorithm is called Euler's method.

The first and second derivatives were used in the I/D and Concavity Tests for determining whether a function is increasing, decreasing, concave up, or concave down on an interval. Points at which the concavity changes are called *inflection points.* The Mean-value Theorem was brought in to prove the I/D Test.

We discussed the asymptotic behavior of functions, including horizontal and vertical asymptotes and the relative growth of the logarithm and exponential functions in comparison with a power function. l'Hôpital's Rules were discussed in a later section, as a tool to evaluate limits that have one of the indeterminate forms $0/0$, ∞/∞, $0 \cdot \infty$ (or $\infty \cdot 0$), $\infty - \infty$, 0^0, ∞^0, or 1^∞.

In the first of two sections on optimization, we presented the main tools for solving optimization problems: the Candidate Theorem, the Max/Min Theorem, and the Global Extremum Theorem. We also stated the First Derivative and Second Derivative Tests for local extrema. In the second section on optimization, the emphasis was on setting up or modeling optimization problems.

Related-rates problems arise in applications in which two or more time-dependent variables are related by an equation. To determine one rate in terms of another rate, we differentiated implicitly the equation relating the two variables.

Chapter Summary

Tangent Line Approximation

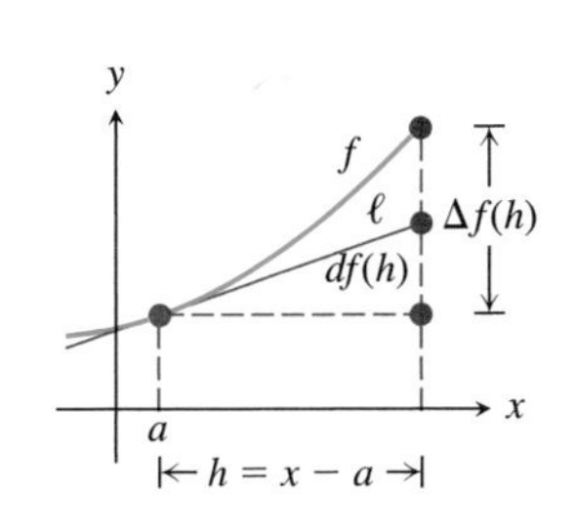

1. The tangent line approximation to $f(a+h)$ at a is
$$f(a+h) \approx f(a) + hf'(a).$$
2. The differential of f at a is
$$df(h) = hf'(a).$$
3. In terms of $\Delta f(h)$ and $df(h)$, the tangent line approximation is
$$\Delta f(h) \approx df(h).$$

If $f(x) = x^2$ and $a = 1$,
$$f(1+h) = (1+h)^2 \approx f(1) + hf'(1) = 1 + 2h$$
$$df(h) = hf'(1) = 2h$$
$$\Delta f(h) = f(1+h) - f(1) = h^2 + 2h \approx hf'(1) = 2h$$

Newton's Method

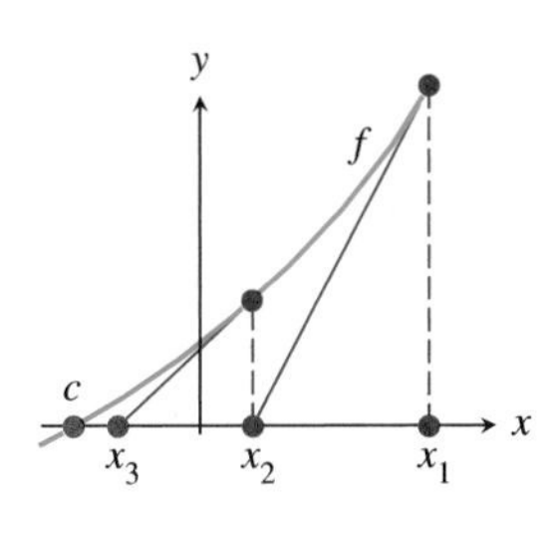

For finding a zero c of a function f, with initial guess x_1: calculate iterates
$$x_{n+1} = x_n - \frac{f(x_n)}{f'(x_n)}, \quad n = 1, 2, \ldots,$$
until x_{n+1} is sufficiently close to c.

With $f(x) = x^3 - 2$ and $x_1 = 1.5$,
$$x_1 = 1.5$$
$$x_2 = 1.29630$$
$$x_3 = 1.26093$$
$$x_4 = 1.25992$$
$$x_5 = 1.25992$$

I/D Test

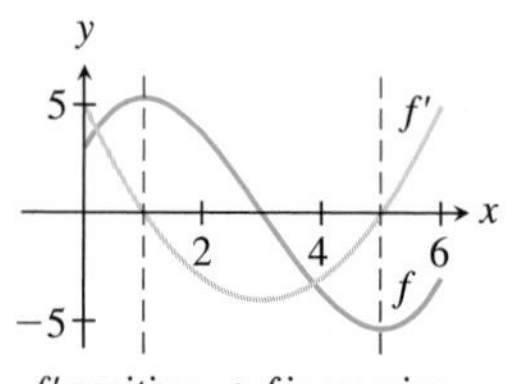

f' positive $\Rightarrow f$ increasing
f' negative $\Rightarrow f$ decreasing

1. If $f'(x) > 0$ on an interval (a, b), then f is increasing on $[a, b]$.
2. If $f'(x) < 0$ on an interval (a, b), then f is decreasing on $[a, b]$.

If $f(x) = \frac{1}{3}x^3 - 3x^2 + 5x + 3$, $0 \le x \le 6$, then $f'(x) = (x - 5)(x - 1)$. Because $f'(x) < 0$ for $1 < x < 5$ and $f'(x) > 0$ for $0 < x < 1$ and $5 < x < 6$, f is decreasing on $[1, 5]$ and increasing on $[0, 1]$ and $[5, 6]$ by the I/D Test.

Concavity Test

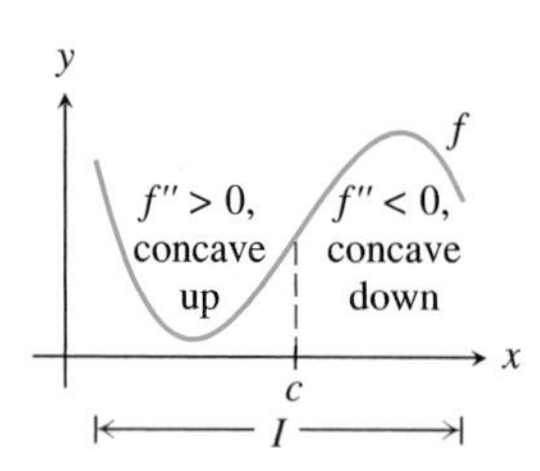

Let f be a function defined on a set including an open interval I and assume that f is twice differentiable on I. Referring to the figure,

if $f''(x) > 0$ on I, then f is concave up on I;

if $f''(x) < 0$ on I, then f is concave down on I.

A point c is an inflection point of f if f'' changes sign at c.

Let $f(x) = \sin x$ on $[0, 2\pi]$. Because $f''(x) = -\sin x$ is negative on $(0, \pi)$, positive on $(\pi, 2\pi)$, and changes sign at π, f is concave down on $(0, \pi)$; concave up on $(\pi, 2\pi)$, and has an inflection point at $x = \pi$.

Mean-value Theorem

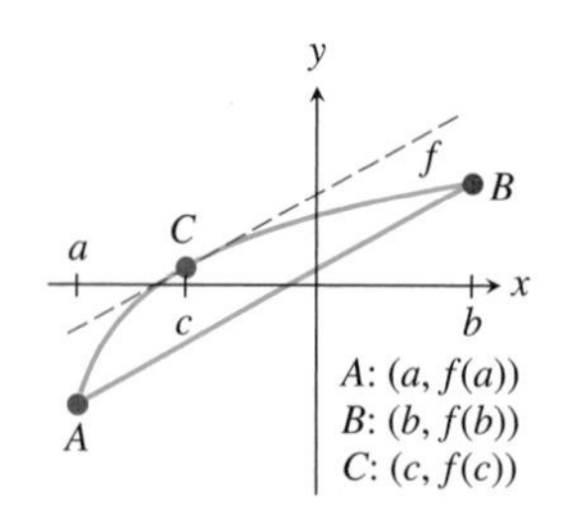

If f is continuous on $[a, b]$ and is differentiable on (a, b), then there is at least one number $c \in (a, b)$ for which

$$\frac{f(b) - f(a)}{b - a} = f'(c).$$

If $f(x) = x^3$ on $[0, 1]$, then the point c for which $(f(1) - f(0))/(1 - 0) = f'(c)$ is $c = 1/\sqrt{3}$.

Asymptotes

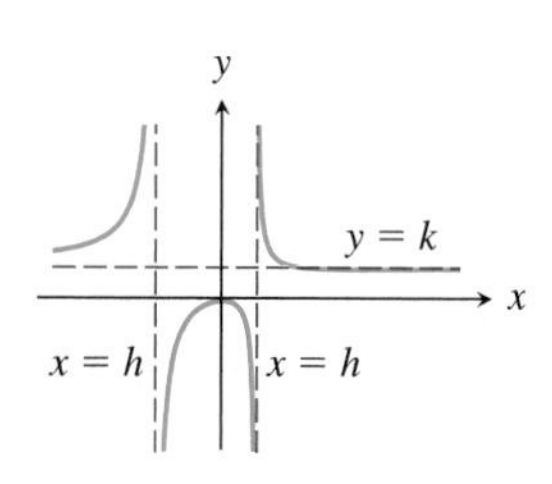

The graph of f has a vertical asymptote at $x = h$ if $\lim_{x \to h^+} f(x) = \pm\infty$ or $\lim_{x \to h^-} f(x) = \pm\infty$. The graph of f has a horizontal asymptote at $y = k$ if $\lim_{x \to \infty} f(x) = k$ or $\lim_{x \to -\infty} f(x) = k$.

Let $a, b > 0$. The exponential function e^{ax} dominates x^b as $x \to \infty$; the logarithm function $(\ln x)^a$ is dominated by x^b as $x \to \infty$.

The function

$$f(x) = \frac{x^2}{(x + 2)(x - 1)}$$

has vertical asymptotes at $x = 1, -2$ and a horizontal asymptote at $y = 1$.

Max/Min Theorem

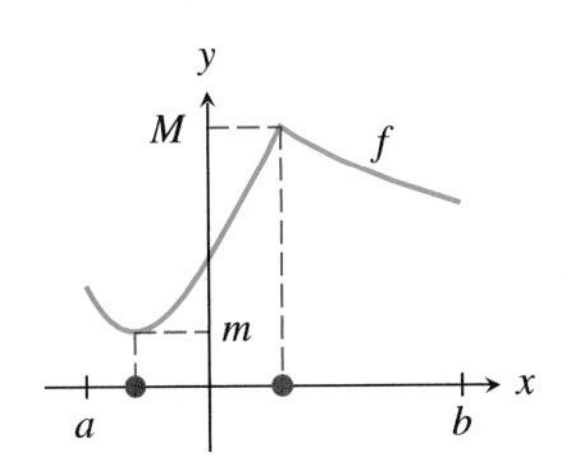

If f is continuous on the interval $[a, b]$, then f has a maximum and a minimum on $[a, b]$.

Applied to the function $f(x) = x^3 - x + 1$ defined on the interval $[0, 1]$, the only possible candidates for the maximum and minimum of f are 0, 1, and $1/\sqrt{3}$. Evaluating f at these three points, $f(0) = 1$, $f(1) = 1$, and $f(1/\sqrt{3}) \approx 0.62$. Hence, the maximum of f is at $x = 0$ (and $x = 1$) and the minimum of f is at $x = 1/\sqrt{3}$.

Candidate Theorem

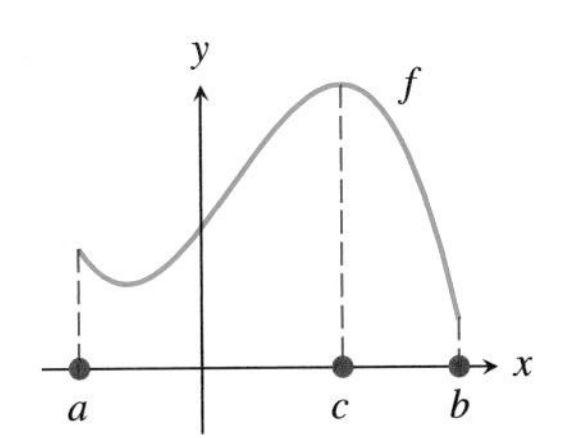

Let f be a function defined on an interval I. If f has an extremum at a point $c \in I$, then either

i) c is an endpoint of I,
ii) f is not differentiable at c, or
iii) $f'(c) = 0$.

Points $c \in I$ which are either endpoints of I, points at which f is not differentiable, or points for which $f'(c) = 0$ are called *candidates.*

Applied to the function $f(x) = x^3 - x + 1$ defined on the interval $[0, 1]$, the only possible candidates for extrema of f are 0, 1, and $1/\sqrt{3}$. The third of these candidates is a zero of $f'(x) = 3x^2 - 1$.

Global Extremum Theorem

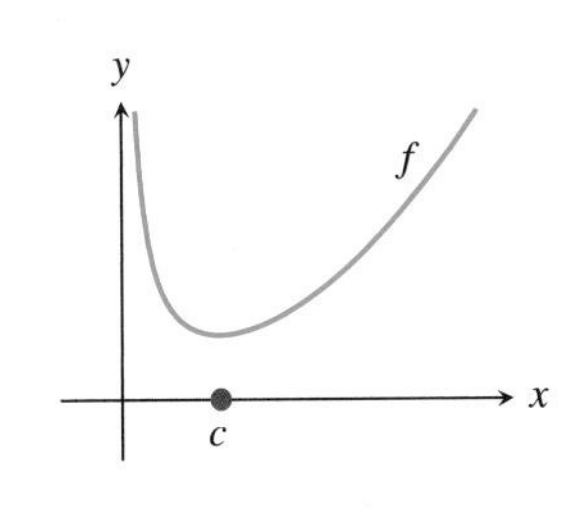

Let f be a twice-differentiable function on an interval I. If c is an interior point of I, then

i) if $f''(x) > 0$ on I and $f'(c) = 0$, then f has a global minimum at c; and
ii) if $f''(x) < 0$ on I and $f'(c) = 0$, then f has a global maximum at c.

Let $f(x) = x^2 + 16/x$, $0 < x < \infty$. Noting that $f'(x) = (2x^3 - 16)/x^2$ and $f''(x) = 2 + 32/x^3$, the point $x = 2$ is a global minimum of f because $f'(2) = 0$ and $f''(x) > 0$ on $(0, \infty)$.

Related Rates

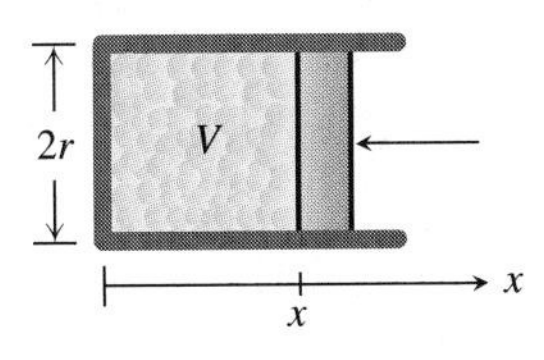

Variables $x = x(t)$ and $V = V(t)$ are related by an equation; either of the rates dx/dt or dV/dt can be calculated in terms of the other by differentiating the equation implicitly.

From the equation

$$V = \pi r^2 x$$

relating V and x (r is constant), we find

$$\frac{dV}{dt} = \pi r^2 \frac{dx}{dt}$$

by differentiating the equation implicitly with respect to t. If, for example, we know dx/dt, we can calculate dV/dt.

l'Hôpital's Rules

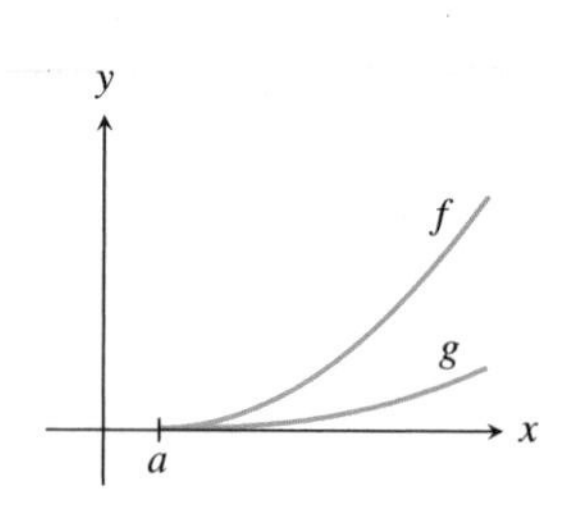

If $f(x)/g(x)$ has the form $0/0$ as $x \to a^+$ and

$$\lim_{x\to a^+} \frac{f'(x)}{g'(x)} = L,$$

then

$$\lim_{x\to a^+} \frac{f(x)}{g(x)} = L.$$

If $f(x)/g(x)$ has the form ∞/∞ as $x \to a^+$ and

$$\lim_{x\to a^+} \frac{f'(x)}{g'(x)} = L,$$

then

$$\lim_{x\to a^+} \frac{f(x)}{g(x)} = L.$$

The limit

$$\lim_{x\to 0^+} \frac{\sin x}{1 - \cos x}$$

is of $0/0$ form. Applying l'Hôpital's Rule,

$$\lim_{x\to 0^+} \frac{\sin x}{1 - \cos x} = \lim_{x\to 0^+} \frac{\cos x}{\sin x} = \infty.$$

The limit

$$\lim_{x\to 0^+} \frac{\ln x}{e^{1/x}}$$

is of ∞/∞ form. Applying l'Hôpital's Rule,

$$\lim_{x\to 0^+} \frac{\ln x}{e^{1/x}} = \lim_{x\to 0^+} \frac{1/x}{e^{1/x}(-1/x^2)}$$

$$= -\lim_{x\to 0^+} \frac{x}{e^{1/x}} = 0.$$

Euler's Method

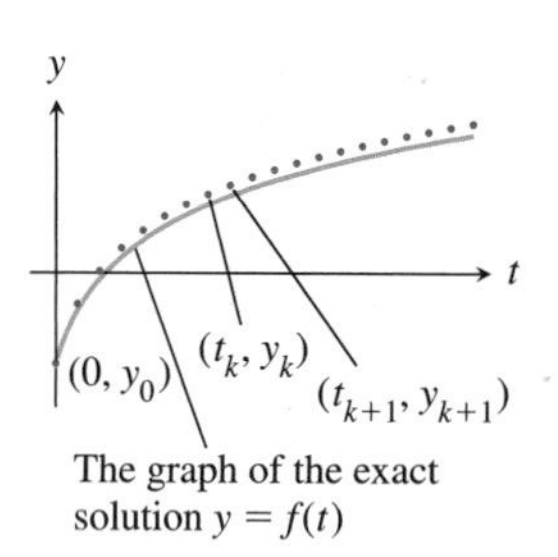

The graph of the exact solution $y = f(t)$

Euler's method for solving the initial value problem $dy/dt = G(t, y)$, $y(0) = y_0$: Given Δt and $(t_0, y_0) = (0, y_0)$,

$$t_{k+1} = (k + 1)\Delta t$$

$$y_{k+1} = y_k + \Delta t G(t_k, y_k),$$

for $k = 0, 1, 2, \ldots$.

With $G(t, y) = y$, $\Delta t = 0.1$, and $y_0 = 1$, the first few steps of Euler's method are

$t_0 = 0$ $\quad y_0 = 1$
$t_1 = 0.1$ $\quad y_1 = 1.1$
$t_2 = 0.2$ $\quad y_2 = 1.21$
$t_3 = 0.3$ $\quad y_3 = 1.331$

Chapter Review Exercises

1. How far apart are $\sqrt[3]{x}$ and its tangent line approximation at $(8, 2)$ on the interval $[6, 10]$?
2. Show that $\sqrt{1 + x} \approx 1 + (1/2)x$ near $x = 0$. How good is this approximation for $-1/2 \le x \le 3/2$?
3. The diameter of a ball bearing is measured as 1.54 cm, with a possible error of 1 percent. Approximate the maximum percentage error in the mass of the bearing. If needed, the density of the bearing is exactly 7500 kg/m^3.
4. A coat of paint 0.04 cm thick is applied to the outside of each of 1000 closed wooden boxes. The bottoms and tops of the boxes are 40 cm by 80 cm, and the height of the boxes is 50 cm. Use the tangent line approximation in calculating the amount of paint needed for the job. Next, calculate the exact amount of paint needed. Using this result, what percentage error would be made in ordering paint if the tangent line approximation were used?
5. Let $p(x) = x^3 + 2x + 1$.
   a. Show that p has an inverse function. *Hint:* Show that p is an increasing function.
   b. Show that p has a zero between -1 and 0.
   c. How many zeros does p have? Why?
   d. Use Newton's method to calculate the zero of p to within 0.001.

6. Use Newton's method to calculate the positive zero of the function

$$f(x) = x^2 - \cos x$$

to three decimal places.

7. Graph the function

$$f(x) = \frac{x^2 - 3x + 1}{x^2 + 1}, \quad x \in R.$$

Label all extreme points, inflection points, and asymptotes.

8. Let $f(x) = 2 \ln(x^2 + 1)$, $x \geq 0$.

a. Sketch the graph of f, showing intervals on which it is increasing, decreasing, concave up, or concave down.

b. What is the range of f?

c. Show that f has an inverse function.

d. Give a formula for f^{-1}.

9. Locate the inflection point of the function

$$f(x) = (x - a)(x - b)(x - c).$$

10. Does the function $f(x) = x^4 - x$ have an inflection point at $x = 0$? Why or why not?

11. Let $f(x) = xe^{-ax}$, $x \geq 0$, where a is a positive constant. In terms of a, locate any local or global extrema and intervals on which f is increasing, decreasing, concave up, or concave down. Sketch a generic graph, including any horizontal or vertical asymptotes.

12. Let

$$f(x) = (x^2 - 3)e^x, \quad x \in R.$$

a. On what intervals is f increasing?

b. Determine the points at which f changes concavity.

c. Determine the minimum value of f.

13. A particle is moving counterclockwise on the unit circle so that its angular speed is 2π radians per minute. At $t = 0$ the particle is at $(1, 0)$. A second particle, also initially located at $(1, 0)$, moves upward on the line $x = 1$ with a speed of 200 m/min. What is the rate at which the particles are separating when $t = 1/12$ minute?

14. Coffee is poured at a uniform rate of 2 cm^3/s into a cup shaped like a truncated cone. If the upper and lower radii of the cup are 4 cm and 2 cm and the height of the cup is 6 cm, how fast will the coffee level be rising when the coffee is halfway up? The volume of a truncated cone of height h and upper and lower radii R and r is

$$V = \frac{1}{3}\pi h(R^2 + Rr + r^2).$$

15. When air expands adiabatically (without heating or cooling), its pressure P and volume V are related by the equation $PV^{1.4} = k$, where k is a constant. Suppose that a quantity of air is undergoing adiabatic expansion. If the pressure of the air is decreasing at 5 kPa/min, at what rate is the volume increasing at the time the volume is 500 cm^3 and the pressure is 100 kPa? (A kilopascal is a pressure of 1000 N/m^2.)

16. A particle is moving on the curve with equation $x^4y + 2y^3 = 3$. Its velocity in the x-direction is a constant 2 m/s. Calculate its velocity in the y-direction at the time when $x = 2$.

17. Gas is pumped into a spherical balloon at 1 ft^3/min. How fast is the diameter of the balloon increasing when the balloon contains 36 ft^3 of gas?

18. A spherical water tank has many coats of old paint and a radius, with paint, of 20 feet. The paint is sandblasted off, with a net decrease of 0.2 in. in the radius. Approximate the net volume of paint removed.

19. The surface area of a spherical hailstone is increasing at the rate of 0.1 cm^2/min at the time the radius of the stone is 1 cm. Find the rate at which the volume is increasing at this time.

20. Variables u and w are functions of time t and are related by the equation

$$u = \sqrt{w^2 + 1}.$$

Express w' in terms of u' and w.

21. Graph the function

$$f(x) = \frac{x}{\sqrt[4]{x^4 + 1}}.$$

Include any horizontal or vertical asymptotes.

22. Graph the function

$$f(x) = \frac{x^2 + 1}{x^2 - 1}.$$

Include any horizontal or vertical asymptotes.

Exercises 23–30: Decide if the limit exists. If it does, determine its value.

23. $\displaystyle\lim_{x \to 0} \frac{\arcsin x}{x}$

24. $\displaystyle\lim_{x \to 0} \frac{\arctan x}{x}$

25. $\displaystyle\lim_{x \to \infty} \frac{e^{2x}}{x^3}$

26. $\displaystyle\lim_{x \to \infty} \frac{(ax + b)^3}{(x + c)^3}$

27. $\displaystyle\lim_{x \to 1^+} \left(\frac{1}{x - 1} - \frac{x}{\sqrt{x - 1}} \right)$

28. $\displaystyle\lim_{x \to \infty} \left(\ln\sqrt{4x + 2} - \ln\sqrt{x + 3} \right)$

29. $\displaystyle\lim_{x \to \infty} (1 + 2/x)^{3x}$

30. $\displaystyle\lim_{x \to \infty} (1 - 3/x)^{2x}$

31. A wire of length 12 inches can be bent into a circle, bent into a square, or cut into two pieces to make both a circle and a square. How much wire should be used for the circle if the total area enclosed by the figure(s) is to be a minimum? A maximum?

32. A wire of length 1 m can be bent into a regular octagon, bent into a regular pentagon, or cut into two pieces to make both a regular octagon and a regular pentagon. How much wire should be used for each if the total area enclosed by the figure(s) is to be a minimum? A maximum?

33. Referring to the accompanying figure, choose P so that θ is a maximum.

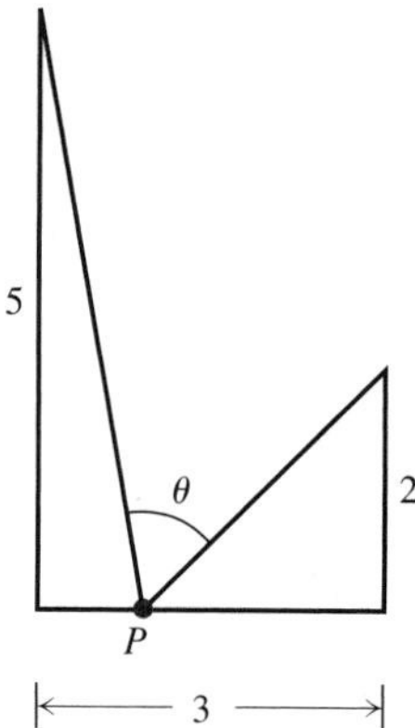

Figure for Exercise 33.

34. Find the maximum and minimum values and the points at which they occur for

$$f(x) = 1 + \frac{3}{x} - \frac{1}{x^2}, \quad \frac{1}{2} \leq x \leq 10.$$

35. Find the maximum and minimum of the function $f(x) = 3^x - 2^x$, $-4 \leq x \leq 2$.

36. Determine the dimensions of the right circular cone of the smallest volume that can be circumscribed about a hemisphere of radius a.

37. Locate the maximum and minimum of the function

$$f(x) = \frac{1}{2}x^3 - \frac{3}{2}x^2 + 5, \ -1.5 \leq x \leq 2.5.$$

38. What are the local and global extrema of the function

$$f(x) = x^{2/3} - x, \ -1 \leq x \leq 8?$$

39. A rectangle has one vertex at $(0, 0)$ and the diagonal vertex in the first quadrant and on the line with equation $2x + y = 1$. Determine the rectangle with the largest area.

40. A rectangle has one vertex at $(0, 0)$ and the diagonal vertex in the first quadrant and on the curve with equation $y = 1/(1 + x^2)$. Determine the rectangle with the largest area.

41. A student hands in a paper with the following statement: "Because $f'(c) = 0$ and c is an interior point of the domain of f, there is a local maximum or minimum at c." If this answer is correct, give a reason why it is correct. If it is false, sketch a figure showing why the answer is false.

T **42.** A student hands in a paper with the following statement: "Because $f''(x) > 0$ everywhere on $(-\infty, \infty)$, it has a minimum somewhere." If this answer is correct, give a reason why it is correct. If it is false, give an example that proves that it is false.

43. Use Euler's method to obtain an approximate solution to the initial value problem

$$y' = t + y^{1.5}, \quad y(0) = 1$$

for $0 \leq x \leq 1$. Use step size $h = 0.1$. A graph of the approximate solution is not necessary.

T **44.** Use Euler's method to obtain an approximate solution to the initial value problem

$$y' = t + y^{1.3}, \quad y(0) = 1$$

for $0 \leq x \leq 1$. Use step size $h = 0.1$. A graph of the approximate solution is not necessary.

STUDENT PROJECTS

A. RETROGRADE MOTION OF MARS

If at each midnight for the next several years you were to observe the position of Mars against the background of the fixed stars, you would find that its angular position along the ecliptic generally increases, but that occasionally it can decrease for several weeks. As seen from Earth, the motion of Mars is said to be either direct or retrograde, depending on whether its angular position is increasing or decreasing. See the assumptions that follow for descriptions of the *ecliptic* and *celestial sphere.*

The retrograde motion of Mars was observed by ancient astronomers. The following careful description of the retrograde motion of Mars was taken from a Babylonian clay tablet found in the ruins of Nineveh.

> Mars at its greatest power becomes very bright and remains so for several weeks; then its motion becomes retrograde for several weeks, after which it resumes its prescribed course.

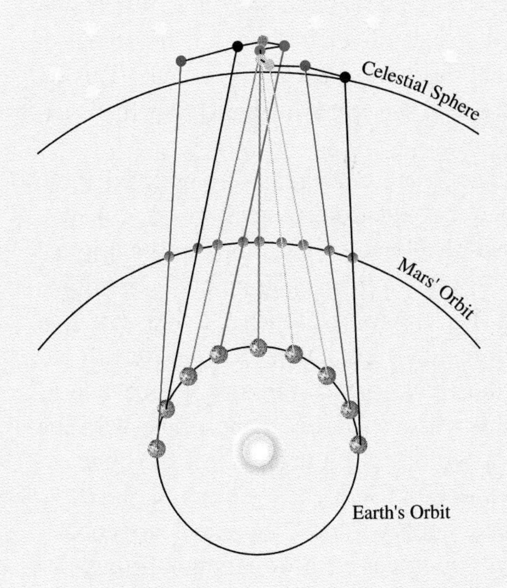

In the discussion and questions that follow we have made several simplifying assumptions about the motions of Earth and Mars. These assumptions do not change in any essential way the phenomenon of retrograde motion.

- The plane of Mars' orbit coincides with the plane of Earth's orbit. (The actual plane is tilted 1.8° from that of Earth.) The intersection of the plane of Earth's orbit with the *celestial sphere* is called the *ecliptic*. The celestial sphere may be regarded as a sun-centered sphere. It is the "background of the fixed stars." Its radius is very, very large relative to solar system distances.
- The orbits of Mars and Earth are sun-centered circles. (The actual orbits are ellipses, with eccentricities 0.093 and 0.017, respectively. The eccentricity of a circle is 0.) We use the astronomical unit (AU) as our unit distance. A distance of 1 AU is defined as the mean distance from Earth to the sun ($\approx 1.496 \times 10^8$ km). The radius of Earth's orbit is 1.0 AU and that of Mars 1.52 AU.
- At some time, which we take as $t = 0$, the sun, Earth, and Mars are aligned, in this order.
- The period of Mars is 1.88 Earth years.

From these assumptions it is clear that the orbits of Earth and Mars have the form

$$\begin{aligned} \mathbf{r}_E &= \mathbf{r}_E(t) = r_E\langle \cos \omega_E t, \sin \omega_E t\rangle, \qquad t \geq 0 \\ \mathbf{r}_M &= \mathbf{r}_M(t) = r_M\langle \cos \omega_M t, \sin \omega_M t\rangle, \qquad t \geq 0. \end{aligned} \tag{1}$$

PROBLEM 1 Assign values to the constants in the vector/parametric equations (1). Graph the two orbits on the same set of axes.

PROBLEM 2 For any fixed $t \geq 0$, let S_t and S_t^* be the intersections of the celestial sphere and the lines with equations

$$\begin{aligned} \mathbf{r} &= \mathbf{r}(u) = \mathbf{r}_E(t) + u(\mathbf{r}_M(t) - \mathbf{r}_E(t)), \qquad u \geq 0, \quad \text{and} \\ \mathbf{r} &= \mathbf{r}(v) = v(\mathbf{r}_M(t) - \mathbf{r}_E(t)), \qquad v \geq 0. \end{aligned}$$

As seen from Earth at time t, the angular displacement $\theta(t)$ of Mars is the angle between the line from $\mathbf{r}_E(t)$ to S_0 and the line from $\mathbf{r}_E(t)$ to S_t. Show that we may instead measure the angular displacement with the angle $\theta^*(t)$ between the line from the origin to S_0 and the line from the origin to S_t^*.

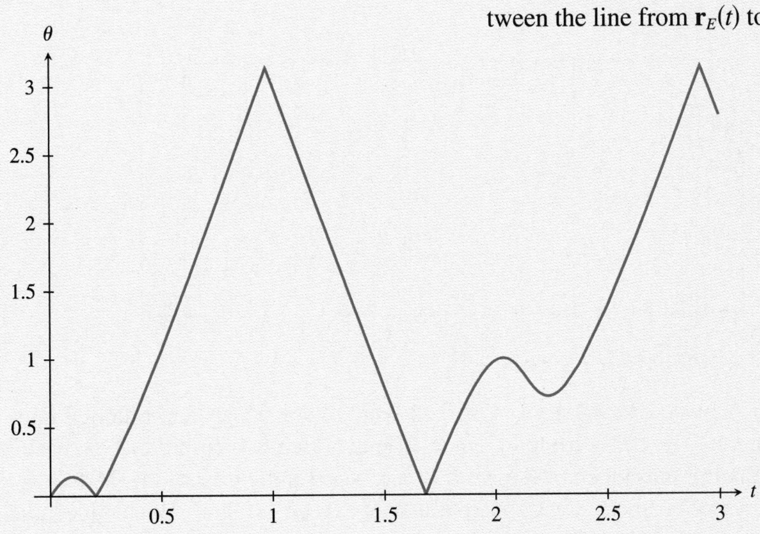

FIGURE 4.61 Angular position of Mars on ecliptic as seen from Earth.

PROBLEM 3 Recall that if **a** and **b** are vectors, the angle $\theta \in [0, \pi]$ between them may be calculated from

$$\theta = \arccos\left(\frac{\mathbf{a} \cdot \mathbf{b}}{\|\mathbf{a}\|\|\mathbf{b}\|}\right).$$

We have plotted in Fig. 4.61 the angle θ between the vectors $\mathbf{r}_M(t) - \mathbf{r}_E(t)$ and $\langle 1, 0\rangle$, $0 \leq t \leq 3$. Explain and reproduce this graph.

PROBLEM 4 For what $t \in [0, 3]$ is Mars brightest? Dimmest?

PROBLEM 5 Explain the comment taken from the Babylonian tablet.

B. FASTER THAN LIGHT?

In the December 1997 issue of *Scientific American,* a reader commented that "I was stunned by your carelessness... that anything can travel faster than the speed of light." The reader was reacting to a statement made in an earlier issue that "This intersection widens so quickly (in fact, faster than the speed of light) that these expanding rings appear as flattened disks."

The editors replied to the stunned reader: "Many readers have wondered about [the statement], but it is correct." To clarify the situation, the editors gave a two-dimensional analog of the three-dimensional phenomenon described in the original article. Figure 4.62 shows a ripple moving outward from a center point O. The ripple is seen when it is $1, 2, \ldots, 7$ meters from O. The ripple first touches a thin wire L at the 4-meter mark. When it is 5 meters from O, the ripple intersects L at B and C. If the ripple has speed v m/s, show that as the ripple passes through A', the rate of change of the distance w between the points of intersection of the ripple with the wire is $\frac{10}{3}v$ m/s. Hence, if v is close to the speed of light, the length of the intersection would grow at a speed greater than the speed of light. The editors of *Scientific American* explained (in the three-dimensional context) that "[this] does not contradict the laws of relativity [because] no physical object or information-carrying signal is moving faster than light."

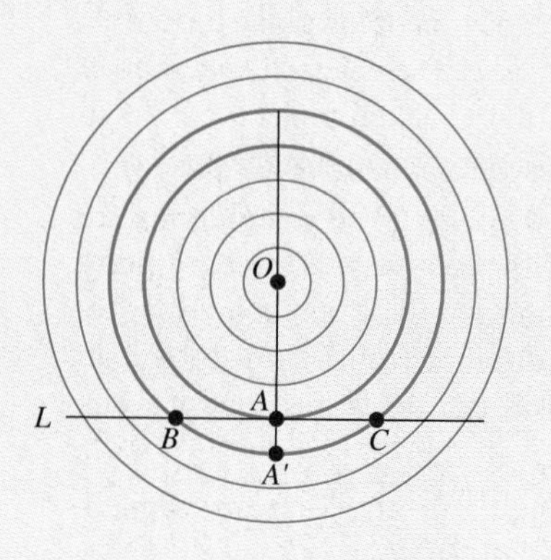

FIGURE 4.62 Ripples intersecting a fixed line.

C. OPTIMAL BOX PROBLEM

Before cutting, folding, and stapling or gluing, cardboard boxes are flat pieces of cardboard on which cutting and folding lines are marked, as in Fig. 4.63. A company wishes to arrange the cutting and folding lines so that with a fixed amount of cardboard, a box of maximum volume results.

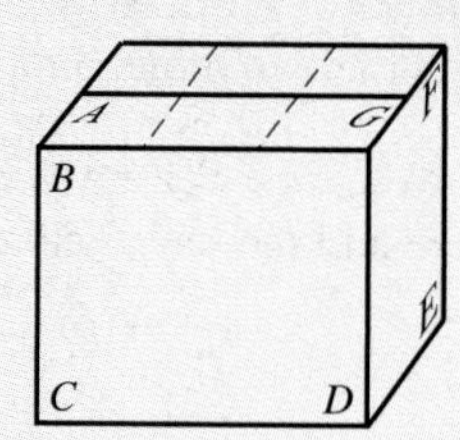

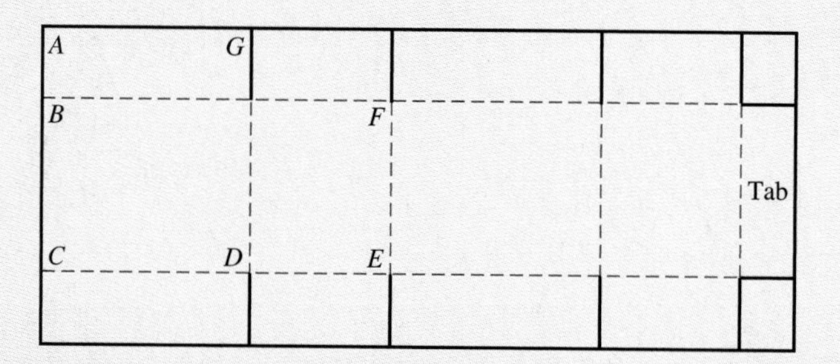

FIGURE 4.63 After and before.

PROBLEM A piece of cardboard is marked with 10 solid lines and 6 dotted lines along which cuts and folds are to be made. The cuts free two corner pieces, which are not used in the subsequent folding of the cardboard and gluing of the tab to form a box. The area of the piece of cardboard must not exceed 1400 in.^{2}. When glued, the tab overlaps BC by 1.25 in. For strength, the dimensions of the eight pieces forming the top and bottom must satisfy $AB = \frac{1}{2}DE = \frac{1}{3}CD$. Choose dimensions AB and BC so that the volume of the box is a maximum.

5

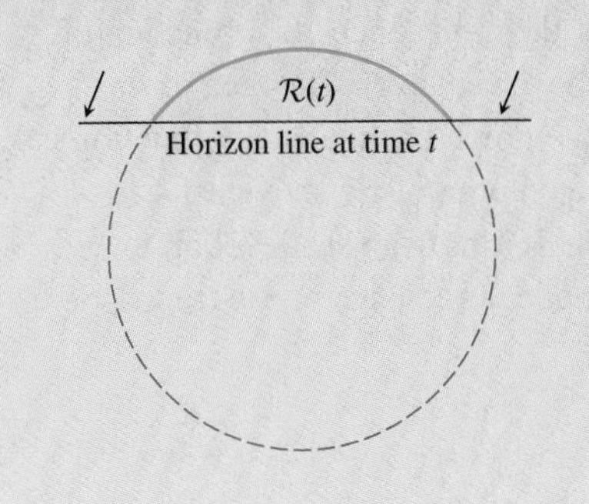

The Integral

As the moon gradually rises above the horizon, the area of the visible segment increases with time. How fast does this area increase? You can calculate this rate by using the chain rule and the Fundamental Theorem of Calculus.

See Example 4, page 368, for further explanation.

Suppose we know the position $x(t)$ of a comet at all times t near a specific time t_0. How can we calculate its velocity at t_0? This is a problem in *differential* calculus, a problem we answered in the first four chapters. In the present chapter we ask the inverse question: How can we calculate the position of a comet at any time $t \in [a, b]$ if we know its position at $t = a$ and its velocity for all $t \in [a, b]$? This is a problem in *integral* calculus.

Applications of integral calculus include the calculation of the positions of objects, lengths of curves, areas of regions in the plane or of surfaces in space, and the volumes of solids. Other applications include the calculation of the mass and center of mass of a solid and the gravitational force exerted by one object on the other, for example, the force exerted by the Earth on the moon.

We motivate and define the definite integral in Section 5.2, using the context of area for much of our discussion. Not only is area an important application, but several other applications can be discussed in terms of area. For example, Fig. 5.1 shows the graph of the (average) density $\delta(x)$ in kg/m^3 of the atmosphere as a function of the height x in meters above sea level. The area of the shaded region beneath the graph of the density function is equal to the mass M of a column of air above a 1×1 square meter patch of the Earth, from sea level up to 11,000 meters. This altitude is the edge of the troposphere.

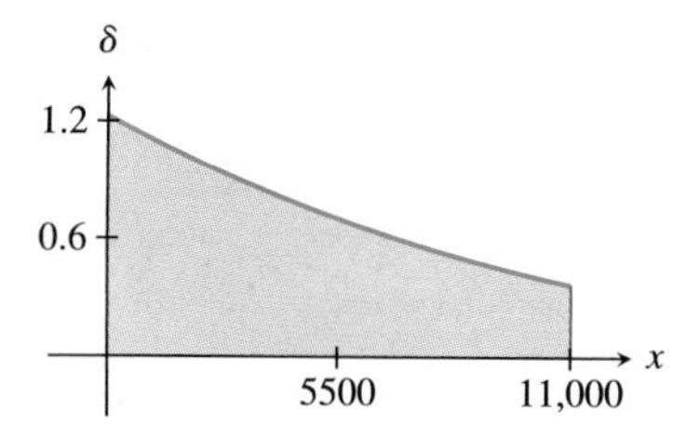

FIGURE 5.1 Graph of the density function of the atmosphere.

5.1 Summation Notation

The sum of the squares of the first 10 integers can be written in full as

$$1^2 + 2^2 + 3^2 + 4^2 + 5^2 + 6^2 + 7^2 + 8^2 + 9^2 + 10^2.$$

We can write this sum more briefly by using the *ellipsis* symbol $\cdots$ to denote terms omitted but understood to follow the pattern suggested by the given terms:

$$1^2 + 2^2 + \cdots + 10^2.$$

Or we may use the summation notation

$$\sum_{i=1}^{10} i^2.$$

This notation contains the summation symbol Σ (which is the uppercase Greek letter sigma), an index of summation i, the general term i^2, the beginning value $i = 1$ of the index, and its ending value $i = 10$. The index of summation advances in steps of 1. More generally, if f is a function defined for integers $i = m, m + 1, \ldots, n$, we shall write the sum

$$f(m) + f(m + 1) + \cdots + f(n)$$

as

$$\sum_{i=m}^{n} f(i).$$

EXAMPLE 1 Evaluate each of the following sums:

a) $\sum_{i=1}^{5} i^2$ b) $\sum_{k=10}^{100} 2.$

Solution

a) It is easy to evaluate this sum by adding five terms:

$$\sum_{i=1}^{5} i^2 = 1^2 + 2^2 + 3^2 + 4^2 + 5^2 = 1 + 4 + 9 + 16 + 25 = 55.$$

b) In this sum the index of summation k takes on values $10, 11, \ldots, 100$ but is not present in the "general term" 2. In this case the general term $f(k)$ is 2 for all k. This means that we are adding $100 - 10 + 1 = 91$ copies of the general term 2. Hence

$$\sum_{k=10}^{100} 2 = \underbrace{2 + 2 + \cdots + 2}_{\text{91 terms}} = 91 \cdot 2 = 182.$$

Another way of looking at the sum (b) is to write it as

$$\sum_{k=10}^{100} 2 \cdot k^0.$$

With k visible in the general term,

$$\sum_{k=10}^{100} 2 = \sum_{k=10}^{100} 2 \cdot k^0 = \underbrace{2 \cdot 10^0 + 2 \cdot 11^0 + \cdots + 2 \cdot 100^0}_{\text{91 terms}} = 91 \cdot 2 = 182.$$

The "summation operator" Σ has several properties.

Properties of Summation

If f and g are functions defined for successive integers $i = m, m + 1, \ldots, n$ and c is a real number, then

$$\sum_{i=m}^{n} (f(i) + g(i)) = \sum_{i=m}^{n} f(i) + \sum_{i=m}^{n} g(i) \tag{1}$$

$$\sum_{i=m}^{n} cf(i) = c \sum_{i=m}^{n} f(i) \tag{2}$$

$$\left| \sum_{i=m}^{n} f(i) \right| \leq \sum_{i=m}^{n} |f(i)| \tag{3}$$

The first of these results is a direct consequence of the commutative and associative laws of addition. Here is an informal argument:

$$\begin{aligned}\sum_{i=m}^{n}(f(i)+g(i)) &= (f(m)+g(m))+(f(m+1)+g(m+1))+\cdots+(f(n)+g(n))\\ &= (f(m)+f(m+1)+\cdots+f(n))+(g(m)+g(m+1)+\cdots+g(n))\\ &= \sum_{i=m}^{n} f(i)+\sum_{i=m}^{n} g(i).\end{aligned}$$

Arguments for Properties (2) and (3) are outlined in Exercises 45 and 46.

In the next section we shall need the formulas

$$\sum_{i=1}^{m} 1 = m, \tag{4}$$

$$\sum_{i=1}^{m} i = \frac{m(m+1)}{2}, \tag{5}$$

$$\sum_{i=1}^{m} i^2 = \frac{m(m+1)(2m+1)}{6}, \tag{6}$$

$$\sum_{i=1}^{m} i^3 = \frac{m^2(m+1)^2}{4}, \tag{7}$$

$$\sum_{i=1}^{m} i^4 = \frac{m(m+1)(2m+1)(3m^2+3m-1)}{30}, \tag{8}$$

and

$$\sum_{i=1}^{m} \sin(ix) = \frac{\sin\left(\frac{1}{2}mx\right)\sin\left(\frac{1}{2}(m+1)x\right)}{\sin\left(\frac{1}{2}x\right)}. \tag{9}$$

In the following examples we show that (4), (5), and (6) hold. The others are left as Exercises 43, 44, and 48.

EXAMPLE 2 Show that (4) holds for $m = 1, 2, \ldots$.

Solution

As in part (b) of Example 1, the result follows because

$$\sum_{i=1}^{m} 1 = \underbrace{1+1+\cdots+1}_{m\ \text{terms}} = m\cdot 1 = m.$$

EXAMPLE 3 Show that (5) holds for $m = 1, 2, \ldots$.

Solution

Letting $S = \sum_{i=1}^{m} i$, we write this sum in two ways:

$$\begin{aligned} S &= 1 + 2 + \cdots + (m-1) + m\\ S &= m + (m-1) + \cdots + 2 + 1\end{aligned}$$

Adding by columns,

$$2S = (m+1) + (m+1) + \cdots + (m+1) + (m+1).$$

Because there are m terms in this sum,

$$2S = m(m+1).$$

Hence,

$$S = \tfrac{1}{2}m(m+1).$$

This is Equation (5).

We use the idea of a "telescoping sum" in the next example. The reference to a telescope is to suggest a telescope made from tubes nested together. If the outermost and innermost tubes are pushed together, most of the inner tubes disappear from view. Figure 5.2 shows the extended and collapsed views.

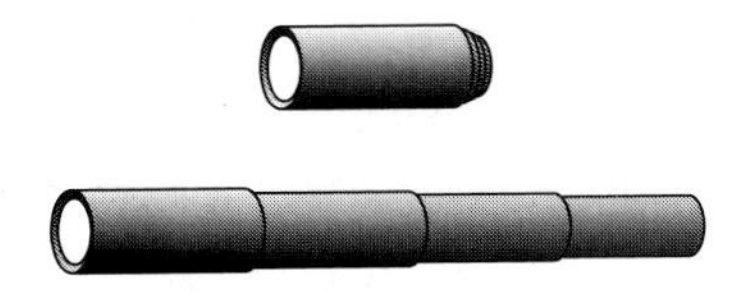

FIGURE 5.2 A telescope, extended and collapsed.

The sum

$$\sum_{i=1}^{4}\left(\frac{1}{j} - \frac{1}{j+1}\right)$$

is a telescoping sum. This becomes clear when we write out the sum as

$$\sum_{i=1}^{4}\left(\frac{1}{j} - \frac{1}{j+1}\right) = \left(\frac{1}{1} - \frac{1}{2}\right) + \left(\frac{1}{2} - \frac{1}{3}\right) + \left(\frac{1}{3} - \frac{1}{4}\right) + \left(\frac{1}{4} - \frac{1}{5}\right).$$

Note that the numbers adjacent to the + signs cancel in pairs, that is,

$$-\frac{1}{2} + \frac{1}{2} = 0, \quad -\frac{1}{3} + \frac{1}{3} = 0, \quad \text{and} \quad -\frac{1}{4} + \frac{1}{4} = 0.$$

Thus, the sum simplifies to the sum of what might be called its first and last terms:

$$\sum_{i=1}^{4}\left(\frac{1}{j} - \frac{1}{j+1}\right) = \frac{1}{1} - \frac{1}{5} = \frac{4}{5}.$$

EXAMPLE 4 Prove that (6) holds for $m = 1, 2, \ldots$.

Solution

We use the fact that $\sum_{i=1}^{m}(i^3 - (i+1)^3)$ is a telescoping sum:

$$\sum_{i=1}^{m}(i^3 - (i+1)^3) = (1^3 - 2^3) + (2^3 - 3^3) + (3^3 - 4^3) + \cdots + (m^3 - (m+1)^3) = 1^3 - (m+1)^3. \tag{10}$$

Because

$$i^3 - (i+1)^3 = i^3 - (i^3 + 3i^2 + 3i + 1) = -3i^2 - 3i - 1,$$

it follows from (10) that

$$1^3 - (m+1)^3 = \sum_{i=1}^{m}(-3i^2 - 3i - 1).$$

"Expanding" this sum by using Properties (1) and (2),

$$1^3 - (m+1)^3 = -3\sum_{i=1}^{m} i^2 - 3\sum_{i=1}^{m} i - \sum_{i=1}^{m} 1.$$

Solving this expression for $\Sigma_{i=1}^{m} i^2$,

$$\sum_{i=1}^{m} i^2 = \frac{1}{3}\left((m+1)^3 - 1 - 3\sum_{i=1}^{m} i - \sum_{i=1}^{m} 1\right).$$

Expanding $(m+1)^3$ and using (4) and (5), which we proved in the first two examples,

$$\sum_{i=1}^{m} i^2 = \frac{1}{3}\left(m^3 + 3m^2 + 3m - \frac{3}{2}m(m+1) - m\right)$$

$$= \frac{1}{6}(2m^3 + 3m^2 + m).$$

Hence,

$$\sum_{i=1}^{m} i^2 = \frac{m(2m^2 + 3m + 1)}{6} = \frac{m(m+1)(2m+1)}{6}.$$

This is (6).

Exercises 5.1

Exercises 1–14: Rewrite using summation notation. Factor out any common factors.

1. $1 + 1/2 + 1/3 + \cdots + 1/10$
2. $1 + 1/2^1 + 1/2^2 + \cdots + 1/2^{10}$
3. $-\ln(1/1) - \ln(1/2) - \cdots - \ln(1/100)$
4. $-e^{1+1/1} - e^{1+1/2} - \cdots - e^{1+1/10000}$
5. $1^2 + 2^2 + \cdots + 10^2$
6. $1^{1/1} + 2^{1/2} + \cdots + 7^{1/7}$
7. $2/3 + 2/4 + 2/5 + \cdots + 2/35$
8. $7/13 + 8/13 + 9/13 + \cdots + 27/13$
9. $\sqrt{h^2(1/19)} + \sqrt{h^2(2/19)} + \cdots + \sqrt{h^2(21/19)}$
10. $\sqrt{3/h^2} + \sqrt{4/h^2} + \cdots + \sqrt{37/h^2}$
11. $\pi^{1+1} + \pi^{1+2} + \pi^{1+3} + \cdots + \pi^{1+21}$
12. $0.01 \cdot \ln(1) + 0.01 \cdot \ln(2) + \cdots + 0.01 \cdot \ln(50)$
13. $1/2 + 2/3 + 3/4 + \cdots + 15/16$
14. $x^1 + x^2 + \cdots + x^n$

Exercises 15–26: Expand each sum and evaluate. For example, $\Sigma_{i=1}^{4} i^3 = 1^3 + 2^3 + 3^3 + 4^3 = 100$.

15. $\sum_{i=1}^{5} i^2$
16. $\sum_{i=1}^{3} i^4$
17. $\sum_{i=1}^{6} (i^2 + 1)$
18. $\sum_{i=1}^{3} (2i^2 - 5i + 1)$
19. $\sum_{j=1}^{2} \sqrt{j+1}$
20. $\sum_{j=1}^{4} \sqrt{3j-2}$
21. $\sum_{k=0}^{5} \sin(k/10)$
22. $\sum_{k=0}^{5} \cos(k/10)$

23. $\sum_{i=0}^{5} 1/(2i + 1)$

24. $\sum_{i=0}^{5} (3i - 5)/3$

25. $\sum_{j=1}^{3} j^0$

26. $\sum_{j=1}^{3} 1$

Exercises 27–40: Evaluate each sum. Use (4)–(8) as needed.

27. $\sum_{j=1}^{25} j^2$

28. $\sum_{j=1}^{25} 3j^3$

29. $\sum_{j=1}^{10} j(j + 1)$

30. $\sum_{j=1}^{10} 2j(j - 1)$

31. $\sum_{j=1}^{m} j(2j + 3)$

32. $\sum_{j=1}^{m} j(3j - 1)$

33. $\sum_{j=1}^{m} 5j(3 - j^2)$

34. $\sum_{j=1}^{m} 3j(2 - j^2)$

35. $\sum_{j=1}^{m} j^2 + j + 1$

36. $\sum_{j=1}^{m} j^2 - j + 1$

37. $\sum_{j=1}^{m} j(7j + 3)^2$

38. $\sum_{j=1}^{m} j(2j - 1)^2$

39. $\sum_{j=3}^{m} (2j + 1)^4$

40. $\sum_{j=3}^{m} (3j + 5)^4$

Exercises 41–44: Use mathematical induction to prove that each formula holds for $m = 1, 2, \ldots$. Proving a statement S_m by mathematical induction has two parts: (1) Show that S_1 is true; (2) show that, for any $k \geq 1$, if statement S_k is true, then so is S_{k+1}. Note that (1) and (2) together show that S_2 is true; applying (2) again shows that S_3 is true; and so on. A brief discussion of mathematical induction is given in the Appendix.

41. $\sum_{i=1}^{m} i = \dfrac{m(m + 1)}{2}$

42. $\sum_{i=1}^{m} i^2 = \dfrac{m(m + 1)(2m + 1)}{6}$

43. $\sum_{i=1}^{m} i^3 = \dfrac{m^2(m + 1)^2}{4}$

44. $\sum_{i=1}^{m} i^4 = \dfrac{m(m + 1)(2m + 1)(3m^2 + 3m - 1)}{30}$

45. Use the distributive property $a(b + c) = ab + ac$ in showing that (2) holds.

46. Prove that (3) holds using the following outline: First show that $|a + b| \leq |a| + |b|$ by considering cases $a, b \geq 0$; $a \geq 0, b < 0$; $a < 0, b \geq 0$; and $a < 0, b < 0$. Next, show that $|a + b + c| \leq |a| + |b| + |c|$, etc. *Hint:* Let $b + c = B$.

47. Show that the identity

$$\sin\alpha \sin\beta = \frac{\cos(\alpha - \beta) - \cos(\alpha + \beta)}{2}$$

holds by subtracting expressions for $\cos(\alpha - \beta)$ and $\cos(\alpha + \beta)$.

48. Use the product-to-sum identity in Exercise 47 in proving (9), which is: for $m = 1, 2, \ldots$ and $1/2x \neq 0, \pm\pi, \pm 2\pi, \ldots,$

$$\sum_{i=1}^{m} \sin(ix) = \frac{\sin\left(\frac{1}{2}mx\right)\sin\left(\frac{1}{2}(m + 1)x\right)}{\sin\left(\frac{1}{2}x\right)}.$$

Multiply both sides of this equation by $\sin\left(\frac{1}{2}x\right)$ and apply the product-to-difference identity to show that

$$\sum_{i=1}^{m} \sin(ix)\sin\left(\tfrac{1}{2}x\right) = \frac{1}{2}\sum_{i=1}^{m}\Big(\cos\left(\tfrac{1}{2}(2i - 1)x\right) - \cos\left(\tfrac{1}{2}(2i + 1)x\right)\Big).$$

Expand this sum and notice that because the terms adjacent to the + signs add to 0, the sum "telescopes" to two terms:

$$\frac{1}{2}\sum_{i=1}^{m}\Big(\cos\left(\tfrac{1}{2}(2i - 1)x\right) - \cos\left(\tfrac{1}{2}(2i + 1)x\right)\Big) = \frac{1}{2}\Big(\cos\left(\tfrac{1}{2}x\right) - \cos\left(\tfrac{1}{2}(2m + 1)x\right)\Big).$$

Apply the product-to-difference identity again.

49. Show that

$$\sum_{i=1}^{m} \cos(ix) = \frac{\sin\left(\frac{1}{2}mx\right)\cos\left(\frac{1}{2}(m + 1)x\right)}{\sin\left(\frac{1}{2}x\right)}.$$

Hint: Use the (other) product-to-sum identity

$$\sin\alpha \cos\beta = \frac{\sin(\alpha + \beta) + \sin(\alpha - \beta)}{2}$$

and adapt the outline in Exercise 48.

5.2 The Definite Integral

We motivate the definite integral by showing how we may approximate the area of the region beneath the graph of a function.

INVESTIGATION 1

Approximating the Area beneath a Parabola

Figure 5.3(a) shows the graph of the function $f(x) = x^2$ on the interval $[0, 1]$. We show how the region $\mathcal{R}$ that is bounded above by the graph of f, below by the x-axis, and on the right by the line with equation $x = 1$ can be approximated by a union of rectangular regions. We approximate the area of $\mathcal{R}$ by summing the areas of the rectangular regions whose union is approximately $\mathcal{R}$.

Figure 5.3(b) shows how we can approximate the area of $\mathcal{R}$ by summing the areas of four rectangles. We formed these rectangles by (i) dividing the interval $[0, 1]$ into four equal subintervals with the points $x_0 = 0/4 = 0$, $x_1 = 1/4$, $x_2 = 2/4$, $x_3 = 3/4$, and $x_4 = 4/4 = 1$; (ii) choosing the midpoints $m_1 = 1/8$, $m_2 = 3/8$, $m_3 = 5/8$, and $m_4 = 7/8$ of each of these four subintervals; and (iii) forming

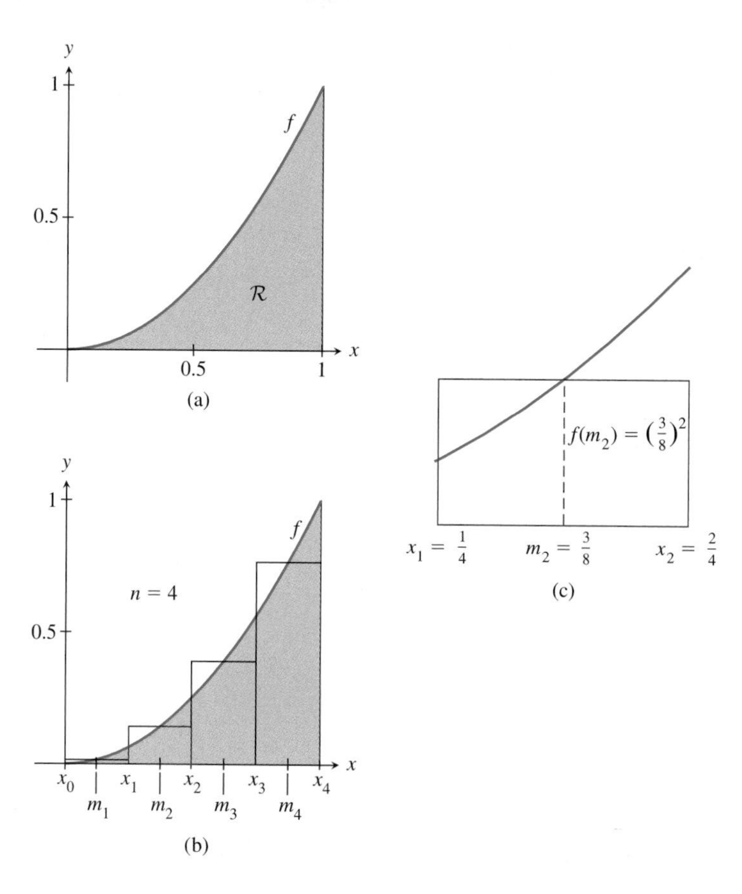

FIGURE 5.3 (a) The region $\mathcal{R}$ beneath the graph of $f(x) = x^2$; (b) approximating the area of $\mathcal{R}$ with $n = 4$ rectangles; (c) zoom-view of the second rectangle.

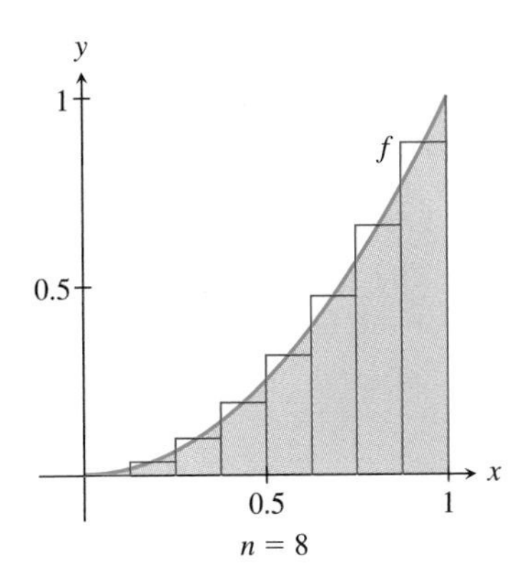

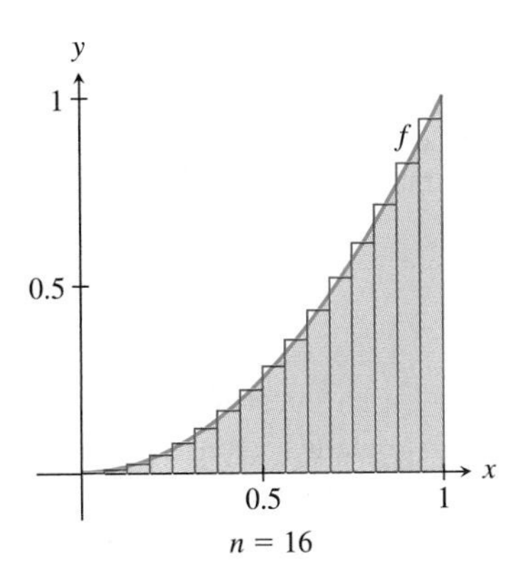

FIGURE 5.4 Approximations to the area of the region beneath the graph of $f(x) = x^2$, using $n = 8$ and $n = 16$ rectangles.

rectangles on the subintervals, where the rectangles have common width $1/4$ and heights $f(m_1)$, $f(m_2)$, $f(m_3)$, and $f(m_4)$.

The first rectangle has height $f(m_1) = f(1/8) = (1/8)^2$ and width $x_1 - x_0 = 1/4$; its area is $(1/8)^2(1/4) = 1/256$ square units. The second rectangle, which is shown in Fig. 5.3(c), has height $f(3/8) = (3/8)^2$ and width $x_2 - x_1 = 1/4$; its area is $(3/8)^2(1/4) = 9/256$ square units. The third rectangle has height $f(5/8) = (5/8)^2$ and width $x_3 - x_2 = 1/4$; its area is $(5/8)^2(1/4) = 25/256$ square units. The fourth rectangle has height $f(7/8) = (7/8)^2$ and width $x_4 - x_3 = 1/4$; its area is $(7/8)^2(1/4) = 49/256$ square units.

The sum M_4 of the areas of the four rectangles is

$$
\begin{aligned}
M_4 &= f(m_1) \cdot \frac{1}{4} + f(m_2) \cdot \frac{1}{4} + f(m_3) \cdot \frac{1}{4} + f(m_4) \cdot \frac{1}{4} \\
&= \frac{1}{4}\left(f\left(\frac{1}{8}\right) + f\left(\frac{3}{8}\right) + f\left(\frac{5}{8}\right) + f\left(\frac{7}{8}\right)\right) \\
&= \frac{1}{4}\left(\left(\frac{1}{8}\right)^2 + \left(\frac{3}{8}\right)^2 + \left(\frac{5}{8}\right)^2 + \left(\frac{7}{8}\right)^2\right) \\
&= \frac{1}{4}\left(\frac{1}{64} + \frac{9}{64} + \frac{25}{64} + \frac{49}{64}\right) = \frac{21}{64} = 0.328125 \text{ square units.}
\end{aligned}
$$

This number is quite close to the exact area of $\mathcal{R}$, which is $1/3$ square units.

Figure 5.4 suggests that if we increase the number of rectangles from 4 to 8, and then to 16, the unions of the rectangular regions more nearly approximate the region $\mathcal{R}$. A numerical calculation, similar to that done for M_4, strengthens the visual evidence. We find

$$M_8 = \frac{85}{256} \approx 0.332031$$

and

$$M_{16} = \frac{341}{1024} \approx 0.333008.$$

The numbers M_4, M_8, and M_{16} appear to be approaching $1/3$ as the interval $[0, 1]$ is more finely divided.

INVESTIGATION 2

Approximating the Area beneath an Exponential Function

In the first investigation we evaluated the function at the midpoints of the subintervals. Although this is a reasonable choice of "evaluation points," other choices are possible. If, for example, we wanted to overestimate the area beneath the graph of the decreasing function $f(x) = e^{-x}$ on the interval $[-1, 2]$—see Fig. 5.5—we could divide the interval $[-1, 2]$ into, say, $n = 10$ subintervals of equal length and

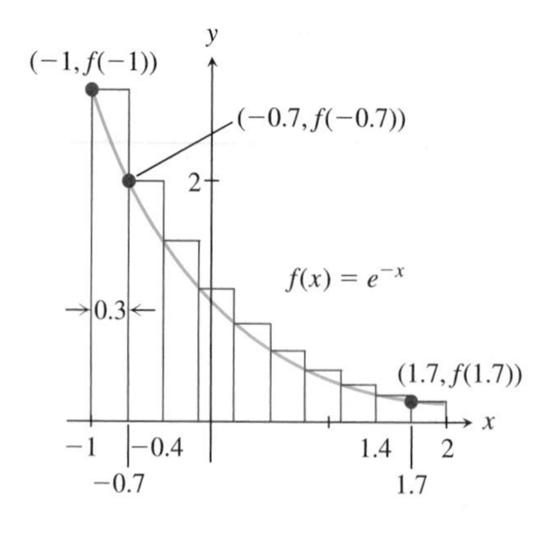

FIGURE 5.5 Approximations to the area of the region beneath the graph of $f(x) = e^{-x}$, using $n = 10$ rectangles.

choose the evaluation points as the left boundaries of the subintervals. This gives, for decreasing functions, *circumscribed* rectangles, whose areas are larger than the regions beneath the graph of f and above the subintervals.

The common length of the subintervals is $(2 - (-1))/10 = 0.3$. The area of the leftmost circumscribed rectangle is $f(-1.0) \cdot 0.3$, the area of the second circumscribed rectangle is $f(-0.7) \cdot 0.3$, and so on. The area of the rightmost circumscribed rectangle is $f(1.7) \cdot 0.3$. The sum C_{10} of the areas of the circumscribed rectangles is

$$\begin{aligned} C_{10} &= f(-1.0) \cdot 0.3 + f(-0.7) \cdot 0.3 + \cdots + f(1.7) \cdot 0.3 \\ &= 0.3(f(-1.0) + f(-0.7) + \cdots + f(1.7)) = 0.3 \sum_{i=1}^{10} f(-1.0 + (i - 1) \cdot 0.3) \\ &= 0.3 \sum_{i=1}^{10} e^{-(-1.0+(i-1)\cdot 0.3)} \\ &\approx 0.3 \cdot 9.96577 = 2.98973 \text{ square units.} \end{aligned}$$

This number overestimates the true area beneath the graph of f on $[-1, 2]$. The true area is $e - 1/e^2 \approx 2.58295$ square units.

Figure 5.6 shows the graph of a function f defined on an interval $[a, b]$. As in Investigations 1 and 2, but with more generality, we show how the region $\mathcal{R}$ bounded above by the graph of f, below by the x-axis, and on the sides by the lines with equations $x = a$ and $x = b$ can be approximated by a union of rectangular regions. We approximate the area of $\mathcal{R}$ by summing the areas of the approximating rectangular regions.

The interval $[a, b]$ has been subdivided into 10 equal subintervals by the points $a = x_0, x_1, x_2, \ldots, x_9, x_{10} = b$. Instead of choosing the midpoints $m_1, m_2, \ldots, m_{10}$ of the subintervals as evaluation points, as in Investigation 1, or as the left boundaries of the subintervals, as in Investigation 2, we choose "arbitrary" evaluation points $c_1, c_2, \ldots, c_{10}$ in the subintervals, with c_1 in the first subinterval $[x_0, x_1]$, c_2 in $[x_1, x_2]$, and so on. We approximate the region $\mathcal{R}$ with a union of rectangular regions and sum the areas of these rectangular regions to approximate the area A of $\mathcal{R}$. The heights of these rectangles are $f(c_1), f(c_2), \ldots, f(c_{10})$, which are values of f at the evaluation points.

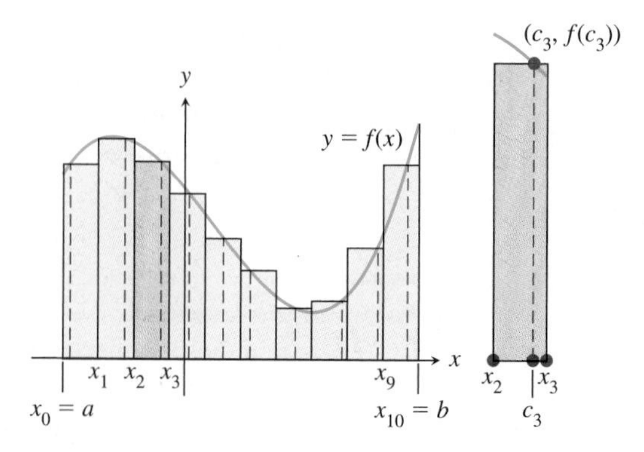

FIGURE 5.6 An approximation to the area of the region $\mathcal{R}$ between the graph of f and the interval $[a, b]$ on the x-axis.

Figure 5.6 shows a zoom-view of a representative rectangle. The area of the rectangle is $f(c_3) \cdot (x_3 - x_2) = f(c_3)\,\Delta x$, where $\Delta x = (b - a)/10$ is the common length of the subintervals. We approximate the area A of $\mathcal{R}$ by summing the areas of the approximating rectangular regions. Using summation notation,

$$A \approx f(c_1)\,\Delta x + f(c_2)\,\Delta x + \cdots + f(c_{10})\,\Delta x = \sum_{i=1}^{10} f(c_i)\,\Delta x.$$

We model the following definition of the definite integral on the approximations we have discussed. We state the definition for any *bounded* function f defined on an interval $[a, b]$. We say that a function f is bounded provided that its graph lies between some pair of horizontal lines. Figure 5.7 shows the graph of a representative function f and region $\mathcal{R}$ and a representative approximating rectangle.

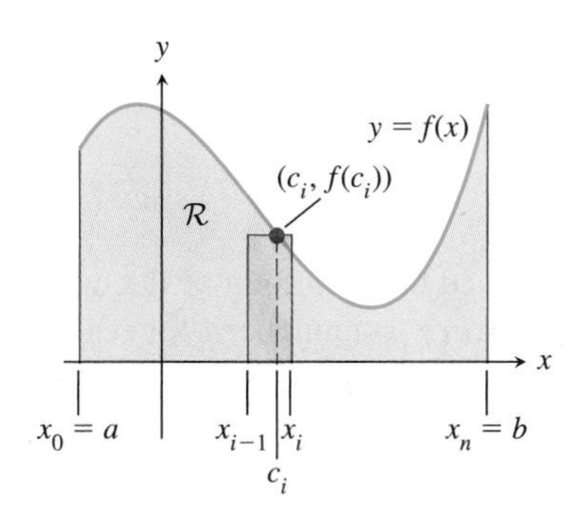

FIGURE 5.7 Some of the ideas and notation used in defining the integral, in graphic form.

DEFINITION The Definite Integral

Let f be a bounded function defined on $[a, b]$. Let n be a positive integer and $\Delta x = (b - a)/n$. The points $\{x_0, x_1, x_2, \ldots, x_n\}$, where

$$x_0 = a, x_1 = a + \Delta x, x_2 = a + 2\,\Delta x, \ldots,$$

$$x_i = a + i\,\Delta x, \ldots, x_n = a + n\,\Delta x = b,$$

divide the interval $[a, b]$ into n subintervals of length Δx. The set $\mathcal{S}_n$ of these subintervals is called a **regular subdivision** of $[a, b]$.

Let $\{c_1, c_2, \ldots, c_n\}$ be any set of numbers for which $x_{i-1} \leq c_i \leq x_i$, $i = 1, 2, \ldots, n$. Note that c_i is in the ith subinterval. We refer to this set as an **evaluation set** for $\mathcal{S}_n$.

For each regular subdivision $\mathcal{S}_n$ and each evaluation set $\mathcal{C} = \{c_1, c_2, \ldots, c_n\}$ for $\mathcal{S}_n$ we define the **Riemann sum**

$$R_n = \sum_{i=1}^{n} f(c_i)\,\Delta x. \quad (1)$$

If there is a number I for which all Riemann sums R_n can be made as close to I as we like by taking n sufficiently large, we say that f is **integrable** on $[a, b]$ and that I is the value of the **definite integral of f on** $[a, b]$. We write this limit as

$$\lim_{n\to\infty} R_n = \lim_{n\to\infty} \sum_{i=1}^{n} f(c_i)\,\Delta x = I. \quad (2)$$

WEB

We also write

$$\int_a^b f(x)\,dx = I. \quad (3)$$

The end points a and b of the **interval of integration** $[a, b]$ are called **lower and upper limits of integration.** The function f is called the **integrand.**

The German mathematician Gottfried Leibniz (1646–1716), who with Newton is credited with the discovery of calculus, was the first to use the notation $\int$ for the integral. This symbol, shaped like an elongated letter S, was chosen for its suggestion of *summation.* Some 200 years later Bernhard Riemann (1826–1866) carefully defined the (Riemann) integral and put it on a sound, mathematical basis.

Must Subdivisions Be Regular? No. We defined the integral using regular subdivisions to avoid unnecessary generality. However, because we will encounter subdivisions which are not regular in Chapter 11, we briefly discuss them now. For the balance of this chapter we consider only regular subdivisions.

Suppose that $[a, b]$ has been divided into n subintervals, not necessarily of equal length, by the points $a = x_0, x_1, \ldots, x_n = b$. Denoting the set of subintervals by $\mathcal{S}_n$, we can choose an evaluation set and form a Riemann sum for $\mathcal{S}_n$ just as for a regular subdivision. See Fig. 5.8. Letting $\{c_1, c_2, \ldots, c_n\}$ be an evaluation set, where $x_{i-1} \leq c_i \leq x_i$, and $\Delta x_i = x_i - x_{i-1}$, the Riemann sum is

$$R_n = \sum_{i=1}^{n} f(c_i)\,\Delta x_i.$$

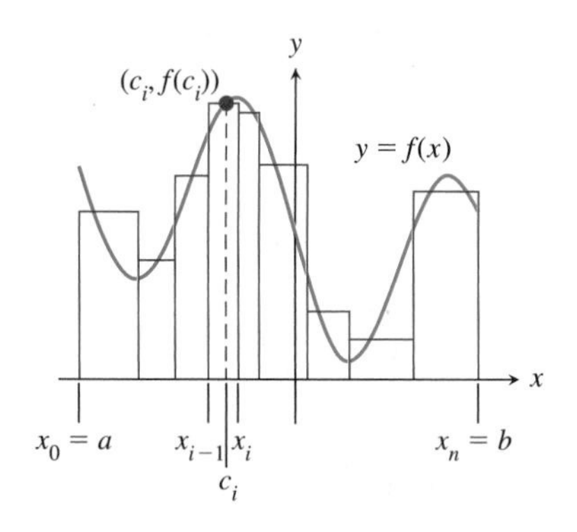

FIGURE 5.8 The elements of a Riemann sum for f on $[a, b]$, including a non-regular subdivision for $[a, b]$ and an arbitrary choice of evaluation points.

If subdivisions need not be regular, we must change the definition of the definite integral: The function f is integrable on $[a, b]$ if there is a number I for which all Riemann sums R_n can be made as close to I as we like by requiring that the longest subinterval in $\mathcal{S}_n$ is sufficiently small.

Are All Functions Integrable?

No. If f fails to be continuous at too many points of $[a, b]$, then the limit in (2) may fail to exist. For continuous functions f the following theorem holds.

Continuous Functions Are Integrable

Each function defined and continuous on an interval $[a, b]$ is integrable on $[a, b]$.

We illustrate the definition of the definite integral with three examples. All functions considered in this chapter are continuous and all subdivisions are regular.

EXAMPLE 1 Calculate the Riemann sum R_n for the integral $\int_0^1 x^2\,dx$, where the evaluation set consists of the coordinates of the left boundaries of the subintervals, that is $\mathcal{C} = \{x_0, x_1, \ldots, x_{n-1}\}$. Use formula (6) from Section 5.1 to simplify R_n. Using this expression for R_n, calculate R_{10}, R_{100}, and R_{1000}. Finally, show that $\lim_{n\to\infty} R_n = 1/3$. Argue that the value of this limit is the area beneath the curve.

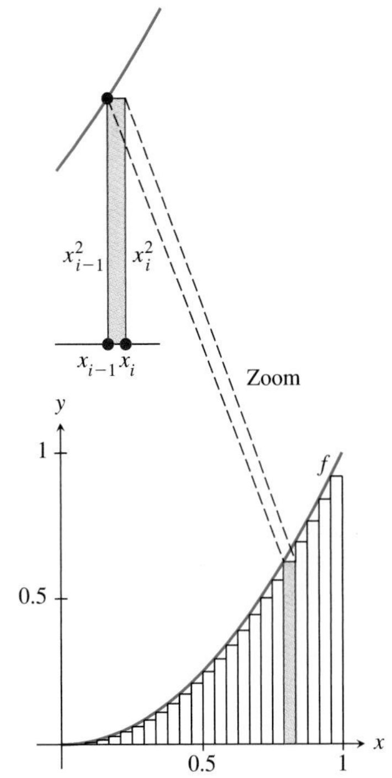

FIGURE 5.9 Approximations to the area of the region beneath the graph of $f(x) = x^2$, $0 \le x \le 1$, using "inscribed" rectangles.

Solution

We opened this section with this integral, using midpoints of subintervals as evaluation sets. We calculated Riemann sums for $n = 4$, 8, and 16, finding $M_4 \approx 0.32815$, $M_8 \approx 0.332031$, and $M_{16} \approx 0.333008$. We noted there that these numbers appeared to be approaching $1/3$.

Figure 5.9 shows the graph of $f(x) = x^2$ on $[0, 1]$ and a subdivision $0 = x_0, x_1, \ldots, x_n = 1$. The common length of the subintervals is

$$\Delta x = (b - a)/n = (1 - 0)/n = 1/n.$$

Note that $x_0 = 0$, $x_1 = 1/n$, $x_2 = 2/n$, and, in general, $x_i = i/n$. In the subinterval $[x_{i-1}, x_i]$, the evaluation point c_i is x_{i-1}. Because the function is increasing on $[0, 1]$, the rectangles are *inscribed* beneath the curve; the height of the rectangle above the subinterval $[x_{i-1}, x_i]$ is $f(x_{i-1}) = x_{i-1}^2$. See Fig. 5.9. The Riemann sum R_n for this choice of evaluation points is

$$R_n = \sum_{i=1}^{n} f(c_i)\,\Delta x = \sum_{i=1}^{n} f(x_{i-1})\,\Delta x.$$

Because $x_{i-1} = (i - 1)/n$ and $f(x) = x^2$,

$$R_n = \sum_{i=1}^{n} \frac{(i-1)^2}{n^2}\,\Delta x.$$

Factoring $\Delta x/n^2$ from each term,

$$R_n = \frac{\Delta x}{n^2} \sum_{i=1}^{n} (i - 1)^2.$$

Because $\Delta x = 1/n$,

$$R_n = \frac{1}{n^3} \sum_{i=1}^{n} (i - 1)^2 = \frac{1}{n^3}(0^2 + 1^2 + \cdots + (n - 1)^2).$$

Dropping the 0^2 from this expression,

$$R_n = \frac{1}{n^3}(1^2 + \cdots + (n - 1)^2).$$

This is a Riemann sum for the given evaluation set.

Recalling formula (6) from Section 5.1, which is

$$\sum_{i=1}^{M} i^2 = 1^2 + 2^2 + \cdots + M^2 = \frac{M(M + 1)(2M + 1)}{6},$$

and taking $M = n - 1$,

$$\begin{aligned} R_n &= \frac{1}{n^3}(1^2 + \cdots + (n - 1)^2) \\ &= \frac{1}{n^3}\frac{(n - 1)(n)(2n - 1)}{6} = \frac{2n^3 - 3n^2 + n}{6n^3}. \end{aligned}$$

Factoring the dominant term n^3 from numerator and denominator,

$$R_n = \frac{n^3}{n^3} \cdot \frac{2 - 3/n + 1/n^2}{6} = \frac{2 - 3/n + 1/n^2}{6}. \tag{4}$$

Using this simplified form of R_n, it is easy to calculate

$$R_{10} = \frac{2 - 3/10 + 1/10^2}{6} = \frac{57}{200} \approx 0.285000$$

$$R_{100} = \frac{2 - 3/100 + 1/100^2}{6} = \frac{6567}{20000} \approx 0.328350$$

$$R_{1000} = \frac{2 - 3/1000 + 1/1000^2}{6} = \frac{665667}{2000000} \approx 0.332834.$$

Finally, from (4),

$$\int_0^1 x^2\,dx = \lim_{n\to\infty} R_n = \lim_{n\to\infty} \frac{2 - 3/n + 1/n^2}{6} = \frac{2 - 0 + 0}{6} = \frac{1}{3}.$$

To argue that this number is the area of the region $\mathcal{R}$ beneath the graph of f we use Fig. 5.9. We note that for each n, R_n is the sum of the areas of n rectangular regions. As n increases, the union of these regions comes closer and closer to filling $\mathcal{R}$ and the sums R_n of the areas of the rectangular regions approach the area of $\mathcal{R}$.

In the next example we evaluate the integral $\int_0^2 x(x - 1)\,dx$ by applying the definition of the definite integral. Figure 5.10(b) shows the graph of this function. Because the graph of f lies both above and below the x-axis it is not clear if $\int_0^2 x(x - 1)\,dx$ can be interpreted as the area of a region $\mathcal{R}$ between the graph and the x-axis. In any case, the definition of the integral as a limit of Riemann sums does not depend upon such an interpretation. We comment on the idea of "net area" after the example.

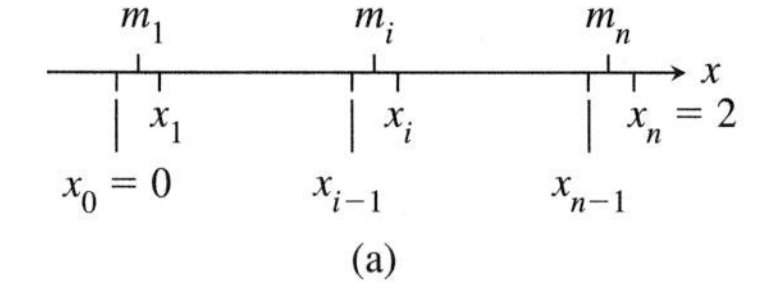

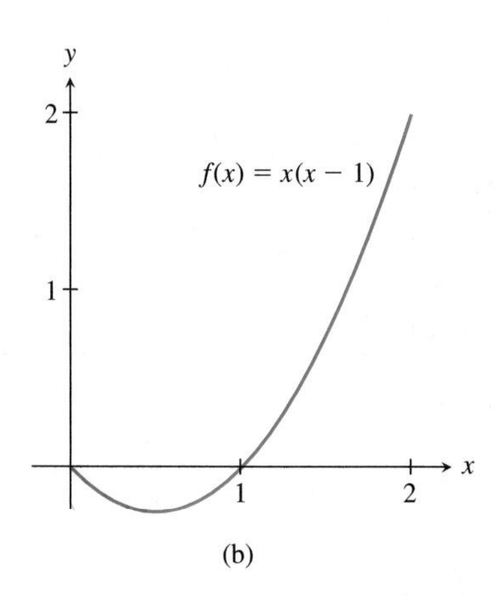

FIGURE 5.10 (a) A subdivision of $[0, 2]$, with midpoints as evaluation points; (b) the graph of $f(x) = x(x - 1)$, $0 \le x \le 2$.

EXAMPLE 2 Apply the definition to evaluate $\int_0^2 x(x - 1)\,dx$. Take the evaluation sets as the midpoints of the subintervals of the regular subdivisions of $[0, 2]$.

Solution

The interval $[0, 2]$ is divided into n equal subintervals by the points $x_0, x_1, \ldots, x_n$, where $x_i = 2i/n$, $i = 0, 1, 2, \ldots, n$. See Fig. 5.10(a). The common length of the subintervals is $\Delta x = 2/n$. The midpoint m_i of the subinterval $[x_{i-1}, x_i]$ is

$$m_i = \frac{1}{2}(x_{i-1} + x_i) = \frac{1}{2}\left(\frac{2(i - 1)}{n} + \frac{2i}{n}\right) = \frac{2i - 1}{n}.$$

The Riemann sum R_n for the given evaluation set is

$$R_n = \sum_{i=1}^{n} f(m_i)\,\Delta x = \frac{2}{n}\sum_{i=1}^{n}\left(\frac{2i-1}{n}\right)\left(\frac{2i-1}{n}-1\right)$$

$$= \frac{2}{n^3}\sum_{i=1}^{n}((2i-1)^2 - n(2i-1))$$

$$= \frac{2}{n^3}\sum_{i=1}^{n}(4i^2 + (-2n-4)i + (n+1))$$

$$= \frac{2}{n^3}\left(4\sum_{i=1}^{n} i^2 + (-2n-4)\sum_{i=1}^{n} i + (n+1)\sum_{i=1}^{n} 1\right).$$

Applying formulas (4), (5), and (6) from Section 5.1,

$$R_n = \frac{2}{n^3}\left(4\frac{n(n+1)(2n+1)}{6} + (-2n-4)\frac{n(n+1)}{2} + (n+1)n\right).$$

Factoring $n/3$ from each term,

$$R_n = \frac{2n}{3n^3}(2(n+1)(2n+1) - 3(n+2)(n+1) + 3(n+1)).$$

Expanding and simplifying,

$$R_n = \frac{2}{3n^2}(n^2-1) = \frac{2}{3}\left(1 - \frac{1}{n^2}\right).$$

The integral $\int_0^2 x(x-1)\,dx$ is the limit of R_n as $n \to \infty$. Hence

$$\int_0^2 x(x-1)\,dx = \lim_{n\to\infty}\frac{2}{3}\left(1 - \frac{1}{n^2}\right) = \frac{2}{3}.$$

Interpreting $\int_a^b f(x)\,dx$ as an Area

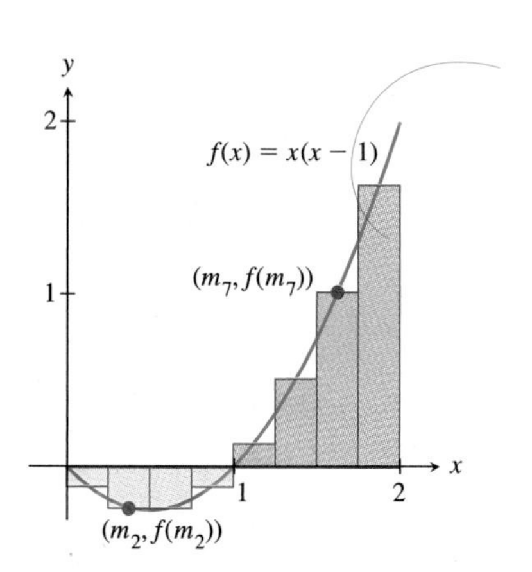

FIGURE 5.11 Interpreting a Riemann sum in terms of "positive areas" and "negative areas," for the function $f(x) = x(x-1)$ on $[0, 2]$.

For nonnegative functions f on $[a, b]$ it is easy to agree that the number $\int_a^b f(x)\,dx$ can be interpreted as the area of the region beneath the graph of f and above the segment $[a, b]$ of the x-axis. We now give an interpretation in terms of area of the number $\int_a^b f(x)\,dx$ for functions f that take on negative values. Figure 5.11 shows the graph of the function $f(x) = x(x-1)$, $0 \le x \le 2$. In Example 2 we showed that $\int_0^2 x(x-1)\,dx = 2/3$. This calculation was done by formula, with no reference to area. We argue here that the number $2/3$ can be interpreted as the *net area* of the region $\mathcal{R}$ between the graph of f and the x-axis. The region $\mathcal{R}$ is the union of two regions: the region above the graph of f and below the x-axis, which is the region with $0 \le x \le 1$, and the region below the graph of f and above the x-axis, which is the region with $1 \le x \le 2$.

We consider the Riemann sum R_8 and take the evaluation points as the midpoints $m_1, m_2, \ldots, m_8$. The values of f at the midpoints $m_1, \ldots, m_4$ of the subintervals $[x_0, x_1], \ldots, [x_3, x_4]$ are negative. See, for example, the rectangle with height $-f(m_2)$ and width $\Delta x = 1/4$. This rectangle corresponds to the term $f(m_2)\,\Delta x$ in the Riemann sum

$$R_8 = \sum_{i=1}^{8} f(m_i)\,\Delta x.$$

Thus, the sum of the first four terms of R_8 corresponds to the negative of the sum of the areas of the yellow rectangles. Similarly, the values of f at the midpoints $m_5, \dots, m_8$ of the subintervals $[x_4, x_5], \dots, [x_7, x_8]$ are positive. See, for example, the rectangle with height $f(m_7)$ and with $\Delta x = 1/4$. This rectangle corresponds to the term $f(m_7)\,\Delta x$ in the Riemann sum R_8. Thus, the sum of the last four terms of R_8 corresponds to the sum of the areas of the light blue rectangles. Evaluating R_8,

$$\begin{aligned} R_8 &= \left(-\frac{7}{64}\right)\Delta x + \left(-\frac{15}{64}\right)\Delta x + \left(-\frac{15}{64}\right)\Delta x + \left(-\frac{7}{64}\right)\Delta x \\ &\quad + \left(\frac{9}{64}\right)\Delta x + \left(\frac{33}{64}\right)\Delta x + \left(\frac{65}{64}\right)\Delta x + \left(\frac{105}{64}\right)\Delta x \\ &= \Delta x\left(-\frac{44}{64}\right) + \Delta x\left(\frac{212}{64}\right) \\ &= \Delta x \cdot \frac{168}{64} = \frac{1}{4} \cdot \frac{168}{64} = \frac{21}{32} \approx 0.65625. \end{aligned}$$

This value of R_8 approximates $\int_0^2 x(x-1)\,dx = 2/3$.

If we say that regions lying wholly beneath the x-axis have "negative area" and regions lying wholly above the x-axis have "positive area," it appears from our discussion that the Riemann sum R_8 approximates the net area of the region between the graph of f and the x-axis and, more generally, that the Riemann sum R_n approximates the net area, with the approximation improving with n. Hence, we expect that the integral $\int_a^b f(x)\,dx$, which is the limit of the Riemann sums, is equal to the net area of this region. We formalize this idea in the next section.

EXAMPLE 3

i) Apply the definition of the integral to evaluate $\int_{-1}^{4} (x-2)\,dx$. For the regular subdivision $\mathcal{S}_n$ of $[a, b]$, take the evaluation set as the set of coordinates of the right boundaries of the subintervals. Use Equations (4) and (5) from Section 5.1 to simplify the approximating Riemann sum.

ii) Calculate the net area of the region between the graph of $f(x) = x - 2$ and the interval $[-1, 4]$ of the x-axis by using the results of part (i). Check this by calculating the area of two triangles.

Solution

i) Figure 5.12(a) shows the graph of f; part (c) of this figure shows an expanded view of the subdivision $\mathcal{S}_n$ of the interval $[-1, 4]$. The interval is divided into n equal subintervals by the points $x_0, x_1, \dots, x_n$, where the common length of the subintervals is $\Delta x = 5/n$. Beginning with $x_0 = -1$, the other points of the subdivision are $x_1 = -1 + 5/n$, $x_2 = -1 + 2 \cdot (5/n)$, and, in general, $x_i = -1 + 5i/n$, $i = 0, 1, 2, \dots, n$. The evaluation set is $\mathcal{C} = \{x_1, x_1, \dots, x_n\}$. The Riemann sum R_n for the given evaluation set is

$$\begin{aligned} R_n &= \sum_{i=1}^{n} f(x_i)\,\Delta x = \Delta x \sum_{i=1}^{n} (x_i - 2) = \frac{5}{n} \sum_{i=1}^{n} \left(-1 + \frac{5}{n}i - 2\right) \\ &= \frac{5}{n^2} \sum_{i=1}^{n} (-3n + 5i) = \frac{5}{n^2}\left(-3n \sum_{i=1}^{n} 1 + 5 \sum_{i=1}^{n} i\right). \end{aligned}$$

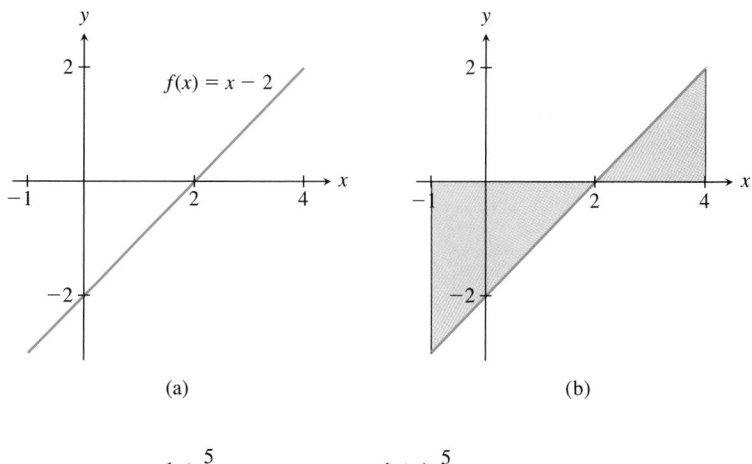

FIGURE 5.12 (a) The graph of $f(x) = x - 2$ on the interval $[-1, 4]$; (b) the two triangular regions; (c) an expanded view of the subdivision $\mathcal{S}_n$.

Applying formulas (4) and (5) from Section 5.1,

$$R_n = \frac{5}{n^2}\left(-3n^2 + 5\frac{n(n+1)}{2}\right) = \frac{5}{2n^2}(-6n^2 + 5n^2 + 5n)$$
$$= \frac{5}{2n^2}(-n^2 + 5n) = \frac{5}{2}\left(-1 + \frac{5}{n}\right).$$

The integral $\int_{-1}^{4} (x - 2)\, dx$ is the limit of R_n as $n \to \infty$. Hence

$$\int_{-1}^{4} (x - 2)\, dx = \lim_{n\to\infty} \frac{5}{2}\left(-1 + \frac{5}{n}\right) = -5/2.$$

ii) The value $-5/2$ of the integral just calculated is the net area of the region between the graph of $f(x) = x - 2$ and the interval $[-1, 4]$ of the x-axis. Referring to Fig. 5.12(b), the area of the triangle between the graph of f and the segment $[-1, 2]$ is $\frac{1}{2} \cdot 3 \cdot 3 = 9/2$. Because the graph of f is below the x-axis on this interval, this triangle contributes $-9/2$ to the net area. The area of the triangle between the graph of f and the segment $[2, 4]$ is $\frac{1}{2} \cdot 2 \cdot 2 = 2$. Because the graph of f is above the x-axis on this interval, this triangle contributes 2 to the net area. Hence, the net area is $-9/2 + 2 = -5/2$. This agrees with the value of the integral calculated in (i).

EXAMPLE 4

a) Apply the definition to evaluate the integral $\int_0^{\pi} \sin x\, dx$. For the regular subdivision $\mathcal{S}_n$ of $[a, b]$, take the evaluation set to be the coordinates of the right boundaries of the subintervals, that is, $\mathcal{C} = \{x_1, x_2, \ldots, x_n\}$. Use formula (9) from Section 5.1 to simplify the approximating Riemann sum.

b) Conjecture the value of the integral $\int_0^{2\pi} \sin x\, dx$.

Solution

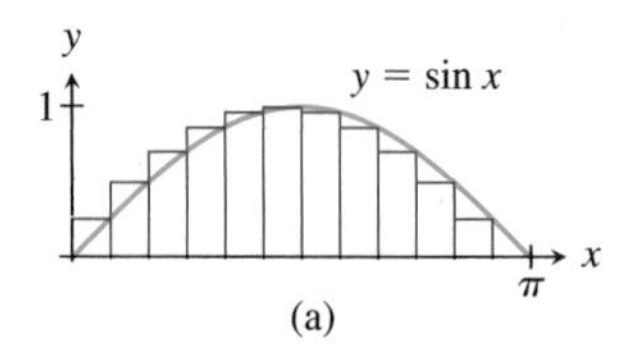

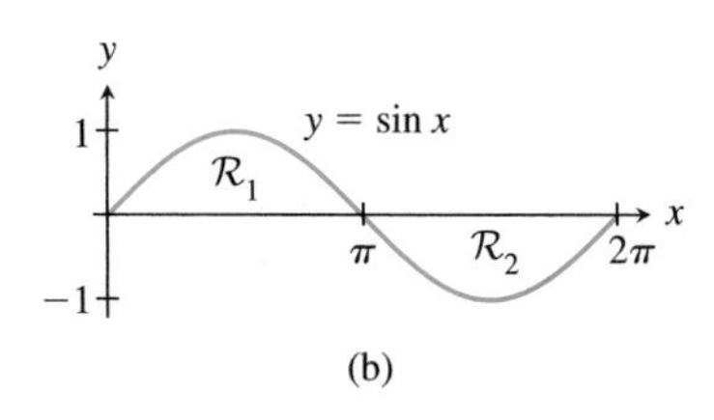

FIGURE 5.13 (a) Graph of $f(x) = \sin x$ on $[0, \pi]$, with a regular subdivision; (b) graph of $f(x) = \sin x$ on $[0, 2\pi]$.

a) Figure 5.13(a) shows the graph of the sine function on $[0, \pi]$. Also shown are the rectangles corresponding to a regular subdivision $\mathcal{S}_n$ of $[0, \pi]$ and an evaluation set $\mathcal{C}$. The interval $[0, \pi]$ is divided into n equal subintervals by the points $x_0, x_1, \ldots, x_n$, where $x_i = \pi i/n$, $i = 0, 1, 2, \ldots, n$. The common length of the subintervals is $\Delta x = \pi/n$. The points $\{c_1, c_2, \ldots, c_n\}$ of the evaluation set are equal to the right boundaries $x_1, x_2, \ldots, x_n$ of the subintervals.

The corresponding Riemann sum R_n is

$$R_n = \sum_{i=1}^{n} f(x_i)\,\Delta x = \sum_{i=1}^{n} \sin(\pi i/n)\pi/n$$
$$= \frac{\pi}{n}\sum_{i=1}^{n} \sin(i(\pi/n)).$$

We evaluate the last sum using (9) from Section 5.1:

$$\sum_{i=1}^{m} \sin(ix) = \frac{\sin\left(\frac{1}{2}mx\right)\sin\left(\frac{1}{2}(m+1)x\right)}{\sin\left(\frac{1}{2}x\right)},$$

where $x = \pi/n$. From the above expression for R_n,

$$R_n = \frac{\pi}{n}\sum_{i=1}^{n} \sin(i(\pi/n)) = \frac{\pi}{n}\left(\frac{\sin\left(\frac{1}{2}n(\pi/n)\right)\cdot\sin\left(\frac{1}{2}(n+1)(\pi/n)\right)}{\sin(\pi/(2n))}\right)$$
$$= \frac{\pi}{n}\left(\frac{1\cdot\sin\left(\frac{1}{2}(n+1)(\pi/n)\right)}{\sin(\pi/(2n))}\right) = \frac{2\sin\left(\frac{1}{2}\pi(1+1/n)\right)}{\dfrac{\sin(\pi/(2n))}{\pi/(2n)}}.$$

Recalling that $\dfrac{\sin\theta}{\theta} \longrightarrow 1$ as $\theta \longrightarrow 0$, we see that with $\theta = \pi/(2n)$, $\theta \longrightarrow 0$ as $n \longrightarrow \infty$. Hence,

$$\lim_{n\to\infty} R_n = \lim_{n\to\infty}\frac{2\sin\left(\frac{1}{2}\pi(1+1/n)\right)}{\dfrac{\sin(\pi/(2n))}{\pi/(2n)}} = \frac{2\cdot 1}{1} = 2.$$

Hence, $\int_0^{\pi} \sin x\, dx = 2$.

b) Figure 5.13(b) shows the graph of the sine function on $[0, 2\pi]$. Because of symmetry, that is, $\sin x = -\sin(x+\pi)$, we expect the regions $\mathcal{R}_1$ and $\mathcal{R}_2$ to have the same area. However, because the Riemann sums for the integral $\int_0^{2\pi} \sin x\, dx$ approximate the *net area,* we conjecture that $\int_0^{2\pi} \sin x\, dx = 0$. See Exercise 24.

Exercises 5.2

Exercises 1–8: Calculate the Riemann sum for the given integral and evaluation set specified by the given point c_i in the subinterval $[x_{i-1}, x_i]$. All subdivisions are regular.

1. R_4 for the integral $\int_0^1 (2x+1)\,dx$, with $c_i = x_{i-1}$.

2. R_4 for the integral $\int_0^1 (3x+2)\,dx$, with $c_i = x_{i-1}$.

3. R_5 for the integral $\int_0^1 (x^2+1)\,dx$, with $c_i = x_i$.

4. R_5 for the integral $\int_0^1 (2x^2+1)\,dx$, with $c_i = x_i$.

5. R_6 for the integral $\int_0^1 \sqrt{x}\,dx$, with $c_i = \frac{1}{2}(x_{i-1} + x_i)$.

6. R_6 for the integral $\int_0^1 \sqrt{x+1}\,dx$, with $c_i = \frac{1}{2}(x_{i-1} + x_i)$.

7. R_8 for the integral $\int_1^5 (2x-1)\,dx$, with $c_i = \frac{1}{2}(x_{i-1} + x_i)$.

8. R_8 for the integral $\int_1^5 (3x-2)\,dx$, with $c_i = \frac{1}{2}(x_{i-1} + x_i)$.

Exercises 9–14: Apply the definition to evaluate the integral $\int_a^b f(x)\,dx$. Take the evaluation sets as the midpoints of the subintervals of the regular subdivisions of $[a, b]$.

9. $\int_1^5 (2x+1)\,dx$

10. $\int_1^5 (3x-2)\,dx$

11. $\int_0^1 (x^2+x)\,dx$

12. $\int_0^1 (x^2+2x)\,dx$

13. $\int_{-1}^1 (2x^2+x+1)\,dx$

14. $\int_{-1}^1 (x^2-2x+3)\,dx$

Exercises 15–18: Apply the definition to evaluate $\int_a^b f(x)\,dx$, using the evaluation sets consisting of the coordinates of the left boundaries of the subintervals of regular subdivisions. Interpret the result in terms of net area. Check your work by calculating the areas of some triangles.

15. $\int_0^3 (-x+2)\,dx$

16. $\int_0^4 (x-3)\,dx$

17. $\int_{-2}^2 x\,dx$

18. $\int_{-1}^2 (1-x)\,dx$

Exercises 19–20: For a given n and associated regular subdivision, the least Riemann sum R_n for the given integral $I = \int_a^b f(x)\,dx$ can be calculated by choosing the evaluation point c_i in the subinterval $[x_{i-1}, x_i]$ to be the point at which f is a minimum in that interval; the greatest Riemann sum R_n can be calculated by choosing c_i to be the point at which f is a maximum in $[x_{i-1}, x_i]$. We denote these two Riemann sums by L_n and G_n. Approximate I with the average $A_{10} = \frac{1}{2}(L_{10} + G_{10})$. State why $|I - A_{10}| \le \frac{1}{2}(G_{10} - L_{10})$.

19. $I = \int_0^1 \sqrt{x}\,dx$

20. $I = \int_0^1 \sin(x^2)\,dx$

21. Calculate the Riemann sum R_n for the integral $\int_0^1 x^3\,dx$, where the evaluation set consists of the coordinates of the left boundaries of the subintervals, that is $\mathcal{C} = \{x_0, x_1, \ldots, x_{n-1}\}$. Use a formula to simplify R_n. Calculate R_8, R_{32}, and R_{100}. Show that $\lim_{n\to\infty} R_n = 1/4$.

22. Calculate the Riemann sum R_n for the integral $\int_0^1 x^4\,dx$, where the evaluation set consists of the coordinates of the left boundaries of the subintervals, that is $\mathcal{C} = \{x_0, x_1, \ldots, x_{n-1}\}$. Use a formula to simplify R_n. Calculate R_8, R_{32}, and R_{100}. Show that $\lim_{n\to\infty} R_n = 1/5$.

23. Show that for any $b > 0$, $\int_0^b x^2\,dx = \frac{1}{3}b^3$. *Hint:* Adapt the calculations used in Example 1.

24. Apply the definition to evaluate $\int_0^{2\pi} \sin x\,dx$. Follow the argument in Example 4, making the necessary adjustments.

T **25.** The density function δ for the graph in Fig. 5.1 is

$$\delta(x) = \frac{353.2(1 - 0.00002256x)^{5.259}}{288.2 - 0.0065x},$$

where $0 \le x \le 11{,}000$. Specifically, at height x meters, the density of the air is $\delta(x)$ kg/m^3. Argue that the area of the shaded region beneath the graph of the density function is numerically (but not dimensionally) equal to the mass M of a column of air above a 1×1 square meter patch of Earth, from sea level up to 11,000 meters, which is the edge of the troposphere. Calculate R_{20} for the integral $\int_0^{11000} \delta(x)\,dx$, using any convenient evaluation set.

5.3 The Fundamental Theorem of Calculus

In Section 5.2 we defined the definite integral $\int_a^b f(x)\,dx$ and interpreted this number as the net area of the region $\mathcal{R}$ between the graph of f and the x-axis. Figure 5.14(a) shows a representative function, interval, and region. We defined the integral as the limit

$$\int_a^b f(x)\,dx = \lim_{n\to\infty} R_n = \lim_{n\to\infty} \sum_{i=1}^{n} f(c_i)\,\Delta x$$

of Riemann sums. We were able to evaluate each of several integrals by first summing the associated Riemann sum R_n with the help of a formula and then taking the limit as $n \to \infty$ of the resulting expression. This process—summing by formula and then taking the limit—is feasible only for relatively simple functions.

In this section we study the Fundamental Theorem of Calculus, which greatly simplifies the evaluation of $\int_a^b f(x)\,dx$ for many functions. We use the context of motion to give a heuristic argument for the Fundamental Theorem of Calculus. For this reason, we use the notation $\int_a^b v(t)\,dt$ instead of $\int_a^b f(x)\,dx$, where $v(t)$ is the velocity of a car moving along a straight highway during the time interval $[a, b]$. Figure 5.14 shows graphs of identical functions $y = f(x)$ and $v = v(t)$ on $[a, b]$. Part (a) of the figure interprets $\int_a^b f(x)\,dx$ in terms of the area of a region $\mathcal{R}$. We use part (b) to interpret $\int_a^b v(t)\,dt$ in terms of motion.

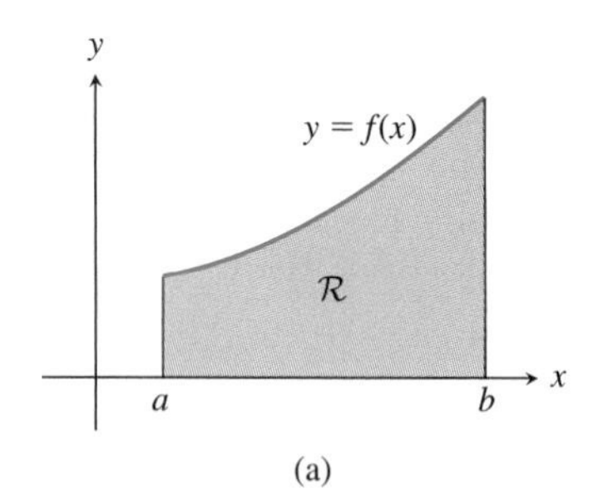

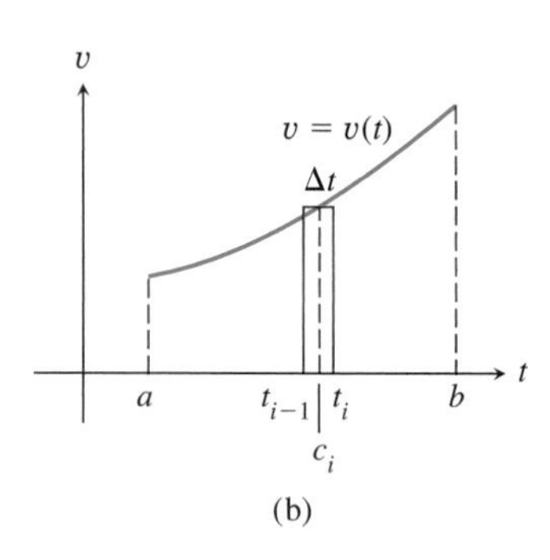

FIGURE 5.14 (a) The area of the shaded region $\mathcal{R}$ is $\int_a^b f(x)\,dx$. (b) Interpreting the integral $\int_a^b v(t)\,dt$ in terms of motion.

We assume that distances are measured in kilometers, time in hours, and velocity in kilometers per hour (kph). We argue that the area of the region beneath the velocity graph is equal to the distance traveled by the object. We lay out the main steps in the argument first and then discuss them in more detail.

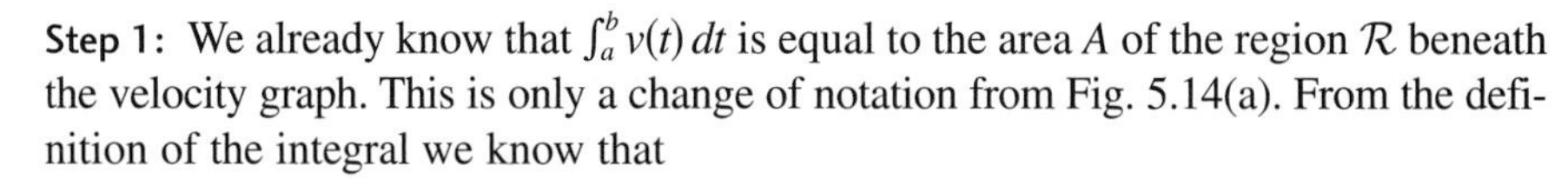

Step 1: We already know that $\int_a^b v(t)\,dt$ is equal to the area A of the region $\mathcal{R}$ beneath the velocity graph. This is only a change of notation from Fig. 5.14(a). From the definition of the integral we know that

$$A = \lim_{n\to\infty} \sum_{i=1}^{n} v(c_i)\,\Delta t = \int_a^b v(t)\,dt, \tag{1}$$

where $\mathcal{C} = \{c_1, c_2, \ldots, c_n\}$ is an evaluation set for the regular subdivision $\mathcal{S}_n$ of $[a, b]$.

Step 2: Referring to Fig. 5.14(b), if for each $i = 1, \ldots, n$ we can show that there is a time c_i in the subinterval $[t_{i-1}, t_i]$ so that the number $v(c_i)\,\Delta t$ is the *exact* distance traveled by the car during this subinterval, then the Riemann sum $R_n = \sum_{i=1}^{n} v(c_i)\,\Delta t$ is *exactly* equal to the distance traveled by the car during the time interval $[a, b]$.

Step 3: Letting $x(t)$ be the position of the car at any time $t \in [a, b]$,

$$x(b) - x(a) = \int_a^b v(t)\,dt. \tag{2}$$

The key to the entire argument is in Step 2, to show that there is a time $c_i \in [t_{i-1}, t_i]$ for which the number $v(c_i)\,\Delta t$ is the exact distance traveled by the car during this subinterval. This follows from the Mean-value Theorem. Letting $x(t)$ be the position of the car for all $t \in [t_{i-1}, t_i]$, the Mean-value Theorem (see Section 4.3, page 282) states that there is a number $c_i \in (t_{i-1}, t_i)$ for which

$$x(t_i) - x(t_{i-1}) = x'(c_i)\,(t_i - t_{i-1}). \tag{3}$$

Because $x'(t) = v(t)$ and $x(t_i) - x(t_{i-1})$ is the distance traveled by the car during the subinterval $[t_{i-1}, t_i]$, Equation (3) states that a time c_i can be chosen in the subinterval

$[t_{i-1}, t_i]$ so that the number $v(c_i)\,\Delta t$ is the exact distance traveled by the car during this subinterval. *Note:* We are assuming that the velocity is never negative during the time interval $[a, b]$, which is consistent with the graph of v shown in Fig. 5.14(b).

Because $v(c_i)\,\Delta t$ is the exact distance traveled by the car during the subinterval $[t_{i-1}, t_i]$, it follows that R_n is the exact distance $x(b) - x(a)$ traveled by the car during the entire interval $[a, b]$. This also follows from

$$\begin{aligned}\sum_{i=1}^{n} v(c_i)\,\Delta t &= (x(t_1) - x(t_0)) + (x(t_2) - x(t_1)) + \cdots + (x(t_n) - x(t_{n-1})) \\ &= x(t_n) - x(t_0) = x(b) - x(a).\end{aligned}$$

With the evaluation sets chosen so that (3) is satisfied, each Riemann sum R_n is equal to $x(b) - x(a)$. It follows that $\lim_{n\to\infty} R_n = x(b) - x(a)$. But it is also true that $\lim_{n\to\infty} R_n = A$. Equation (2) now follows from (1).

How can we use this result to evaluate an integral $\int_a^b v(t)\,dt$? Well, if we can find a function $x = x(t)$ for which $x'(t) = v(t)$, then, according to (2),

$$\int_a^b v(t)\,dt = x(b) - x(a).$$

Here's a quick example: Recall that we showed in Example 1 of Section 5.2 that $\int_0^1 x^2\,dx = 1/3$ or, switching the integration variable to t, $\int_0^1 t^2\,dt = 1/3$. We can now calculate this integral more easily. All we must do is find a function $x = x(t)$ whose derivative is $v = v(t) = t^2$. A moment's thought shows that $\left(\frac{1}{3}t^3\right)' = t^2$. Hence, if we take $x(t) = \frac{1}{3}t^3$,

$$\int_0^1 t^2\,dt = x(1) - x(0) = \left(\frac{1}{3}1^3\right) - \left(\frac{1}{3}0^3\right) = \frac{1}{3}.$$

To discuss the Fundamental Theorem of Calculus much further, we need two properties of the integral. Although we assume these without proof, we show that they are reasonable when interpreted in terms of area. We end the section by listing two additional properties of the integral. These properties follow from the Fundamental Theorem of Calculus and are left as exercises. All functions considered in this chapter are assumed to be continuous.

The Fundamental Theorem of Calculus

Figure 5.15 shows the region $\mathcal{R}$ lying beneath the graph of the nonnegative function f. This region can be described more precisely as

$$\mathcal{R} = \{(x, y) : a \le x \le b, \qquad 0 \le y \le f(x)\}.$$

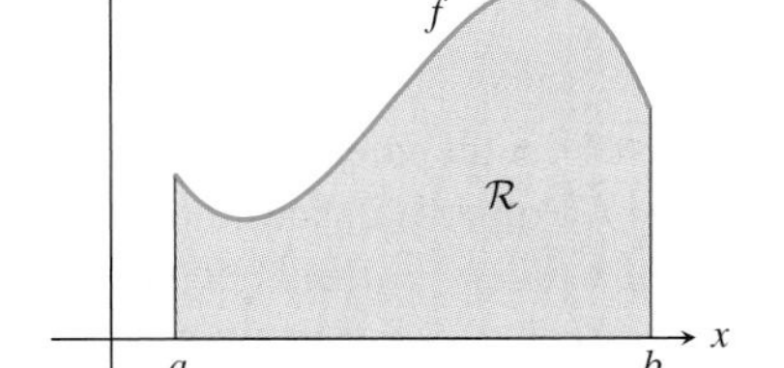

FIGURE 5.15 $\mathcal{R}$ is the region lying beneath the graph of f.

DEFINITION 1 Area and Net Area

Let f be a continuous, nonnegative function on $[a, b]$. The area of the region $\mathcal{R}$ beneath the graph of f is the number $\int_a^b f(x)\,dx$. If we assume only that f is continuous on $[a, b]$, then the net area of the region between f and the segment $[a, b]$ of the x-axis is the number $\int_a^b f(x)\,dx$.

DEFINITION 2 Assigning a Value to the Definite Integral $\int_a^b f(x)\,dx$ When $b = a$

For any continuous function f defined on an interval $[a, b]$, we define

$$\int_a^a f(x)\,dx = 0. \tag{4}$$

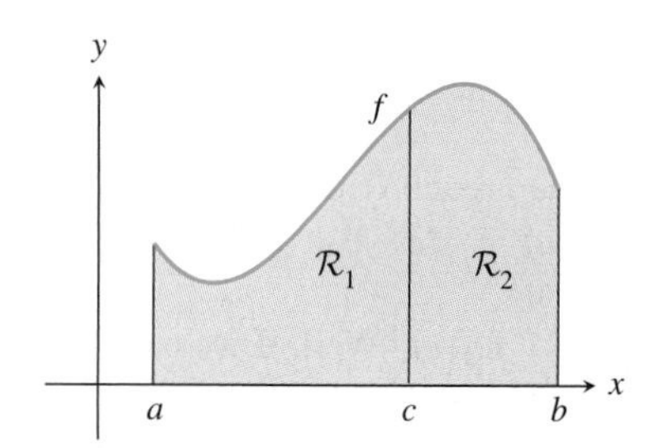

FIGURE 5.16 The area beneath the curve, from a to b, is the sum of the areas of the regions $\mathcal{R}_1$ and $\mathcal{R}_2$.

The intuitive content of Definition 2 is that the net area between the graph of f and the (degenerate) interval $[a, a]$ is 0. Look back at Fig. 5.15 and imagine that the right boundary ($x = b$) is moved towards the left boundary ($x = a$). As $b \to a$, the net area of the region $\mathcal{R}$ approaches 0. For $b = a$, the area of $\mathcal{R}$ is 0.

PROPERTY 1: The Integral Is Additive

For any continuous function f defined on an interval $[a, b]$ and any point c in $[a, b]$,

$$\int_a^c f(x)\,dx + \int_c^b f(x)\,dx = \int_a^b f(x)\,dx. \tag{5}$$

If we assume that f is nonnegative on $[a, b]$ and interpret the integral as an area, this property is easy to understand. Referring to Fig. 5.16, the sum of the areas $\int_a^c f(x)\,dx$ and $\int_c^b f(x)\,dx$ of the regions $\mathcal{R}_1$ and $\mathcal{R}_2$ beneath the graph of f is equal to the area $\int_a^b f(x)\,dx$ of the combined region $\mathcal{R}_1 \cup \mathcal{R}_2$ beneath the graph of f.

PROPERTY 2: Mean-value Theorem for Integrals

For any continuous function f defined on an interval $[a, b]$ there is a point c in $[a, b]$ for which

$$\int_a^b f(x)\,dx = f(c)\,(b - a). \tag{6}$$

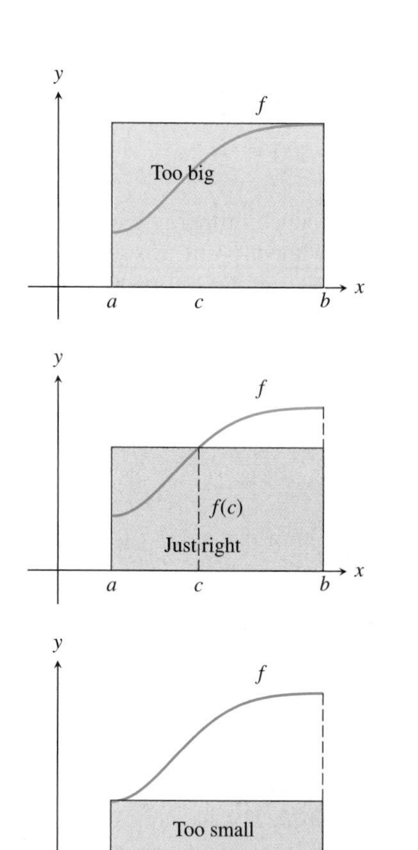

FIGURE 5.17 Mean-value Theorem for Integrals.

If we assume that f is nonnegative on $[a, b]$ and interpret the integral as an area, this property is easy to understand. Referring to Fig. 5.17, Property 2 states that there is an "average value" $f(c)$ of f such that the rectangle of width $b - a$ and height $f(c)$ is equal to the area $\int_a^b f(x)\,dx$ of the region beneath the graph of f. For the increasing function f shown in the figure, think of a "dynamic rectangle," one with constant width $(b - a)$ and heights increasing from $f(a)$ to $f(b)$. The area of the rectangle at the bottom of Fig. 5.17, with height $f(a)$, is too small; the area of the rectangle at the top of the figure, with height $f(b)$, is too big; and, in the middle of the figure, Property 2 says that there is an average value $f(c)$ of f for which the area of the rectangle is "just right," that is, equal to $\int_a^b f(x)\,dx$.

We use additivity and the Mean-value Theorem for Integrals in discussing the Fundamental Theorem of Calculus. We also need to understand integrals of the form

$$\int_a^x f(w)\,dw, \qquad \text{where } a < x < b. \tag{7}$$

Note that in (7) we used w, not x, as the *integration variable.* We did this to avoid confusion with the upper limit of integration. The integration variable plays the same role as an index of summation. Just as we recognize that both of the sums

$$\sum_{i=1}^{3} i^2 \quad \text{and} \quad \sum_{j=1}^{3} j^2$$

are equal to $1^2 + 2^2 + 3^2$, the values of the integrals

$$\int_a^b f(x)\,dx \quad \text{and} \quad \int_a^b f(w)\,dw$$

are equal.

The value of the integral in (7) is a function of x. Denoting this function by F and referring to Fig. 5.18, $F(x) = \int_a^x f(w)\,dw$ is the area of the region beneath the graph of f, between $w = a$ and $w = x$.

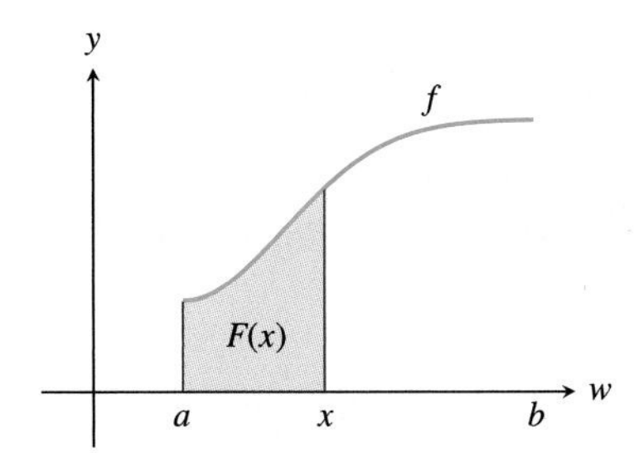

FIGURE 5.18 Area $F(x)$ as a function of x.

For stating the Fundamental Theorem, it is convenient to define the term "antiderivative." We say that a function F is an *antiderivative* of f if $F'(x) = f(x)$ for x in some interval $[a, b]$.

We now have all of the preliminary definitions and properties we need to state and prove the Fundamental Theorem of Calculus.

THEOREM The Fundamental Theorem of Calculus

Assume that f is continuous on $[a, b]$.

Part I: The function F defined on $[a, b]$ by

$$F(x) = \int_a^x f(w)\,dw, \qquad x \in [a, b], \tag{8}$$

is an antiderivative of f; that is,

$$F'(x) = \frac{d}{dx}\int_a^x f(w)\,dw = f(x), \qquad x \in [a, b]. \tag{9}$$

Part II: If G is any antiderivative of f on $[a, b]$, then

$$\int_a^b f(x)\,dx = G(b) - G(a). \tag{10}$$

PROOF OF PART I By definition, the derivative of F at x is

$$F'(x) = \lim_{h\to 0} \frac{F(x + h) - F(x)}{h}. \tag{11}$$

Assuming that $h > 0$, it follows directly from Fig. 5.19 that the numerator of this difference quotient is

$$F(x+h) - F(x) = \int_a^{x+h} f(w)\,dw - \int_a^x f(w)\,dw = \int_x^{x+h} f(w)\,dw, \tag{12}$$

because $F(x + h)$ is the area from $w = a$ up to $w = x + h$ and $F(x)$ is the area from $w = a$ to $w = x$. This also follows from Property 1. See Exercise 47. See Exercise 48 for the $h < 0$ case.

Applying the Mean-value Theorem for Integrals to the interval $[x, x + h]$, there is a $c \in [x, x + h]$ for which

$$F(x+h) - F(x) = \int_x^{x+h} f(w)\,dw = f(c)h. \tag{13}$$

Such a c is shown in Fig. 5.19. The area $f(c)h$ of the yellow rectangle is equal to the area of the region between the graph of f and the interval $[x, x + h]$. Hence, from (11),

$$F'(x) = \lim_{h\to 0^+} \frac{F(x+h) - F(x)}{h} = \lim_{h\to 0^+} f(c) = f(x).$$

The last equality holds because f is continuous at x and the number c is trapped between x and $x + h$. Hence, as $h \to 0$, $c \to x$. A similar argument holds in case $h < 0$.

$F(x)$ = area of □
$F(x + h)$ = area of □ ∪ □
$F(x + h) - F(x)$ = area of □

PROOF OF PART II We gave the idea of this proof in the introduction to this section. For each integer n, we choose the evaluation set $\{c_1, c_2, \ldots, c_n\}$ for the regular subdivision $\mathcal{S}_n$ of $[a, b]$ as follows: applying the Mean-value Theorem to the function G on the subinterval $[x_{i-1}, x_i]$, there is a point $c_i \in (x_{i-1}, x_i)$ for which

$$G(x_i) - G(x_{i-1}) = G'(c_i)(x_i - x_{i-1}) = f(c_i)(x_i - x_{i-1}).$$

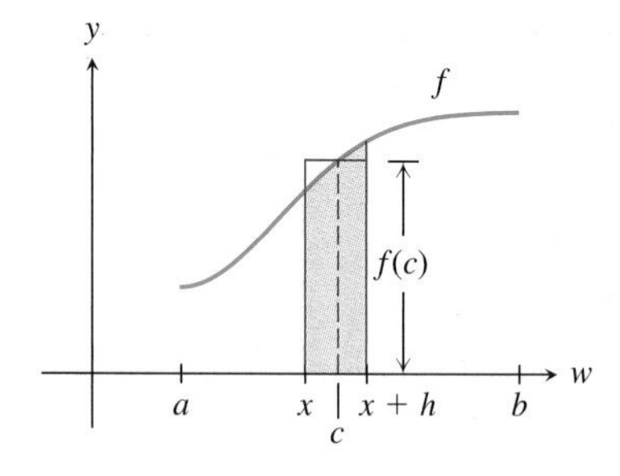

FIGURE 5.19 The regions with areas $F(x)$, $F(x + h)$, and $F(x + h) - F(x)$.

Hence, the Riemann sum R_n is

$$\begin{aligned} R_n &= \sum_{i=1}^{n} f(c_i)\,\Delta x = \sum_{i=1}^{n} f(c_i)(x_i - x_{i-1}) \\ &= \sum_{i=1}^{n} (G(x_i) - G(x_{i-1})). \\ &= (G(x_1) - G(x_0)) + (G(x_2) - G(x_1)) + \cdots + (G(x_n) - G(x_{n-1})). \end{aligned}$$

This sum telescopes to

$$R_n = G(x_n) - G(x_0) = G(b) - G(a).$$

Because f is continuous, we know that $\int_a^b f(x)\,dx$ exists and all Riemann sums approach this number. We also know that, for the particular choices of evaluation points made with the help of the Mean-value Theorem, all Riemann sums are equal to the number $G(b) - G(a)$. Hence,

$$\int_a^b f(x)\,dx = \lim_{x\to\infty} R_n = \lim_{n\to\infty}(G(b) - G(a)) = G(b) - G(a).$$

This is (10), which we were to prove.

We review the Fundamental Theorem in the following examples.

EXAMPLE 1 Find the area of the region beneath the first arch of the sine curve.

Solution

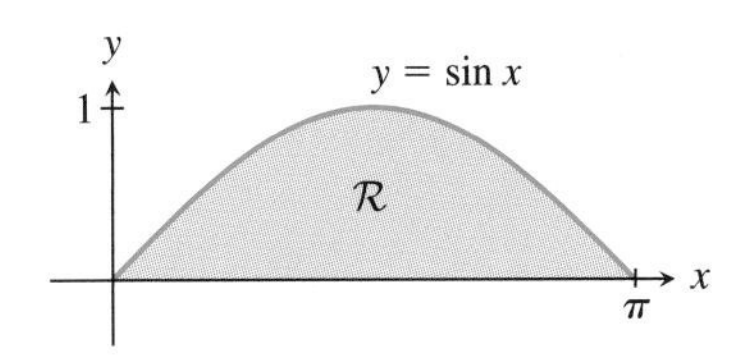

FIGURE 5.20 The region beneath the first arch of the sine curve.

Figure 5.20 shows the region $\mathcal{R}$ whose area we wish to calculate. We calculated the area of this region in Example 4 of Section 5.2. The calculation that follows is *much* easier! The area of $\mathcal{R}$ is the value of the integral $\int_0^\pi \sin x\,dx$. From (10), the value of the integral is $G(\pi) - G(0)$, where G is an antiderivative of $\sin x$. Because $d(-\cos x)/dx = \sin x$, we may take $G(x) = -\cos x$. From (10),

$$\int_0^\pi \sin x\,dx = G(\pi) - G(0) = -\cos \pi - (-\cos 0) = -(-1) - (-1) = 2.$$

The preceding calculation depended on both the idea of the integral and the Fundamental Theorem of Calculus. The first step was our recognition that the integral can be used to calculate an area. The second step was the use of the Fundamental Theorem to evaluate the integral by determining an antiderivative. The integral and the Fundamental Theorem, working together, provide us with a powerful tool for calculating the areas of planar regions, the volumes of solids of revolution, and the length of curves.

In the next example we use the convenient "evaluation notation" $G(x)\big|_a^b$, which, if G is a function defined on an interval $[a, b]$, denotes the number $G(b) - G(a)$, that is,

$$G(x)\Big|_a^b = G(b) - G(a).$$

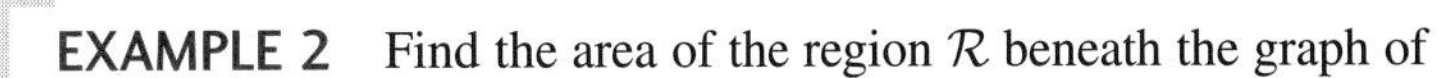

EXAMPLE 2 Find the area of the region $\mathcal{R}$ beneath the graph of

$$f(x) = (x-1)^2, \qquad 0 \le x \le 3.$$

Solution

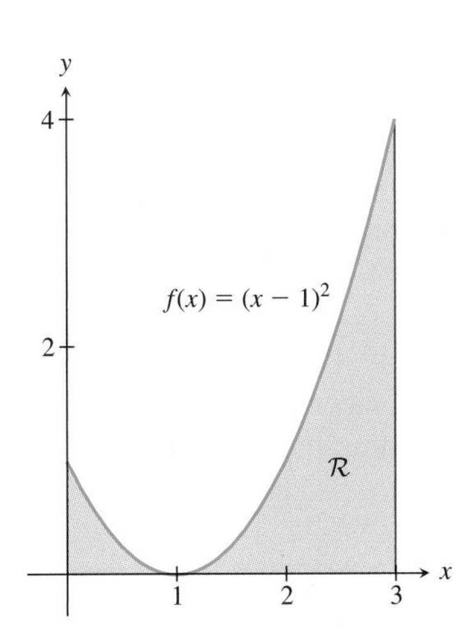

FIGURE 5.21 The region beneath the graph of $f(x) = (x-1)^2$, $0 \le x \le 3$.

A good first step is to sketch the region whose area we wish to calculate. Investing time and effort here often clarifies the problem, exposes subproblems whose solutions we must find, and helps avoid major blunders. Figure 5.21 shows the graph of f. We added the line with equation $x = 3$ and shaded the region $\mathcal{R}$. The area of $\mathcal{R}$ is the value of the integral

$$\int_0^3 (x-1)^2\,dx.$$

An antiderivative of the integrand is $\frac{1}{3}(x-1)^3$. Hence,

$$\int_0^3 (x-1)^2\,dx = \frac{1}{3}(x-1)^3\Big|_0^3 = \frac{1}{3}(3-1)^3 - \frac{1}{3}(0-1)^3 = \frac{8}{3} + \frac{1}{3} = 3.$$

Part II of the Fundamental Theorem of Calculus was used in these two examples. Next we give two examples of Part I.

EXAMPLE 3 Let A be the area function defined by

$$A(x) = \int_0^x \sin w\,dw, \qquad 0 \le x \le \pi. \tag{14}$$

Determine $A'(\pi/4)$.

Solution

We calculated the area of the region beneath the first arch of the sine function in Example 1. Here we are asked to consider the area function A defined in (14) and illustrated in Fig. 5.22. The value of $A(x)$ for a representative x is the area of the region beneath the sine curve and between the lines $w = 0$ and $w = x$. In Example 1 we calculated $A(\pi) = 2$ by noting that $-\cos w$ is an antiderivative of $\sin w$. From Part II of the Fundamental Theorem of Calculus,

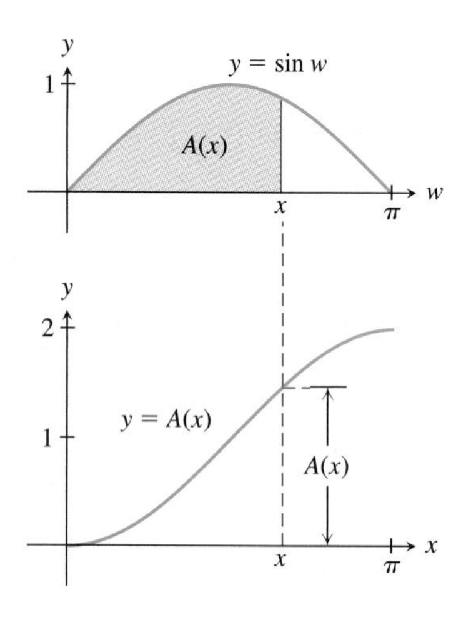

FIGURE 5.22 The region beneath the curve $y = \sin w$, from $w = 0$ to $w = x$.

$$A(x) = \int_0^x \sin w\,dw = -\cos w\Big|_0^x = 1 - \cos x. \tag{15}$$

Hence,

$$\begin{aligned} A'(x) &= 0 - (-\sin x) = \sin x \\ A'\left(\frac{\pi}{4}\right) &= \sin\left(\frac{\pi}{4}\right) = \frac{\sqrt{2}}{2}. \end{aligned} \tag{16}$$

This calculation can be done more easily by using Part I of the Fundamental Theorem of Calculus. From (9),

$$A'(x) = \frac{dA}{dx} = \frac{d}{dx}\left(\int_0^x \sin w\,dw\right) = \sin x. \tag{17}$$

This agrees with (16) and avoids finding an antiderivative needed for (15).

In the next example we show how a rate of change can be found using the chain rule together with Part I of the Fundamental Theorem of Calculus.

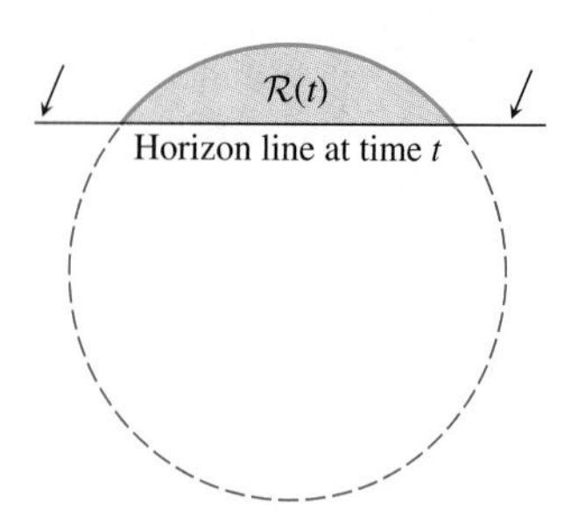

FIGURE 5.23 The region $\mathcal{R}(t)$ of the rising moon at time t after moonrise.

EXAMPLE 4 Figure 5.23 shows a schematic of the moon rising over the eastern horizon. At time t after the start of moonrise, the area $A(t)$ of the region $\mathcal{R}(t)$ of the moon above the horizon is

$$A(t) = \pi a^2 - 2\int_{-a}^{a-kt} \sqrt{a^2 - w^2}\,dw, \qquad 0 \le t \le \frac{2a}{k}, \tag{18}$$

where a is the radius of the moon and k is a positive constant. Find the rate of change $A'(t)$ of $A(t)$ at time $t = a/(2k)$.

Solution

Although it is possible to find an antiderivative for $\sqrt{a^2 - w^2}$ and obtain an explicit expression for $A(t)$ by applying Part II of the Fundamental Theorem, it is far easier to apply Part I. For this we rewrite (18) as

$$A(t) = \pi a^2 - 2\int_{-a}^{x} \sqrt{a^2 - w^2}\,dw, \quad \text{where } x = a - kt. \tag{19}$$

By the chain rule,

$$A'(t) = \frac{dA}{dt} = \frac{dA}{dx}\frac{dx}{dt}. \tag{20}$$

Because $x = a - kt$, $dx/dt = -k$. Calculating the derivative dA/dx by Part I of the Fundamental Theorem, we may write (20) as

$$\begin{aligned} A'(t) &= -k \cdot \frac{d}{dx}\left(\pi a^2 - 2\int_{-a}^{x} \sqrt{a^2 - w^2}\,dw\right) \\ &= -k \cdot \left(0 - 2\sqrt{a^2 - x^2}\right) = 2k\sqrt{a^2 - x^2}. \end{aligned}$$

Replacing x by $a - kt$,

$$A'(t) = 2k\sqrt{a^2 - (a - kt)^2}.$$

At $t = a/(2k)$

$$A'\left(\frac{a}{2k}\right) = 2k\frac{a\sqrt{3}}{2} = ka\sqrt{3} \text{ square units per time unit.}$$

We end the section by listing in the summary that follows several other properties and definitions relating to the integral. For completeness, we have included earlier properties and definitions as well. In Definition 1 we omit the definition of *net area.*

Definitions and Properties of the Integral

Assume that f and g are continuous on $[a, b]$, c is a real number, and r, s, $t \in [a, b]$.

Definition 1: Area. Let f be a nonnegative function on $[a, b]$. The area of the region $\mathcal{R}$ beneath the graph of f is the number $\int_a^b f(x)\,dx$.

Definition 2: $\displaystyle\int_a^a f(w)\,dw = 0.$

Definition 3: $\displaystyle\int_b^a f(x)\,dx = -\int_a^b f(x)\,dx.$

Definition 4: Average value. The average value of f on $[a, b]$ is

$$\frac{1}{b - a}\int_a^b f(x)\,dx.$$

Property 1: Additivity.

$$\int_r^s f(x)\,dx + \int_s^t f(x)\,dx = \int_r^t f(x)\,dx.$$

Property 2: Mean-value Theorem for Integrals. There is a point $c \in [a, b]$ such that

$$\int_a^b f(x)\,dx = f(c)\,(b - a).$$

Property 3: Linearity.

$$\int_a^b (f(x) + g(x))\,dx = \int_a^b f(x)\,dx + \int_a^b g(x)\,dx$$

and

$$\int_a^b cf(x)\,dx = c\int_a^b f(x)\,dx.$$

Property 4: $\left|\int_a^b f(x)\,dx\right| \le \int_a^b |f(x)|\,dx.$

COMMENTS ON DEFINITION 3 This definition, together with Definition 2, makes it possible to state Property 1 for any choices of $r, s, t \in [a, b]$, not just for $a \le r < s < t \le b$.

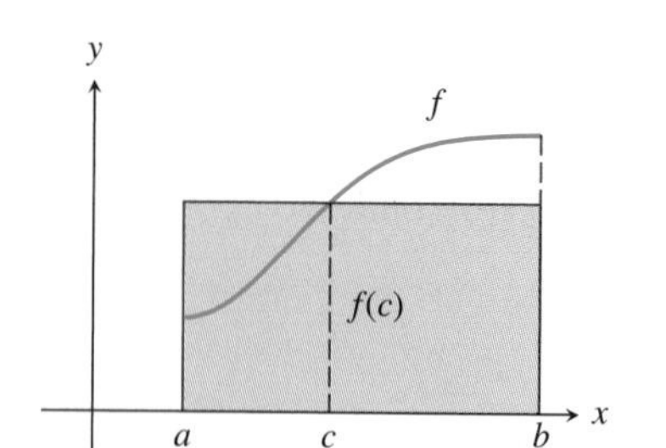

FIGURE 5.24 Mean-value Theorem for Integrals.

COMMENTS ON DEFINITION 4 Note from Property 2 that the average value of f is $f(c)$. Figure 5.24, taken from our discussion of the Mean-value Theorem for Integrals, shows that the function value $f(c)$ is an average value of f in the sense that the rectangle with width $b - a$ and height $f(c)$ has the same area as the region beneath the graph of f.

COMMENTS ON PROPERTY 3 Functions like $L(x) = mx$ are called *linear* because

$$L(x_1 + x_2) = m(x_1 + x_2) = mx_1 + mx_2 = L(x_1) + L(x_2) \tag{21}$$

and

$$L(bx) = m(bx) = b(mx) = bL(x). \tag{22}$$

The derivative is called *linear* because the derivative of a sum of two functions is the sum of their derivatives and the derivative of a scalar times a function is the product of the scalar and the derivative of that function. These properties are analogous to properties (21) and (22).

The definite integral is called *linear* because it satisfies properties analogous to properties (21) and (22). The two parts of Property 3 may be stated in words: The integral of a sum is the sum of the integrals, and the integral of a scalar times a function is the scalar times the integral of the function.

We show in Exercises 49 and 50 that the integral satisfies Property 3.

PROPERTY 4 We show in Exercises 51 and 52 that the integral satisfies Property 4.

Exercises 5.3

Exercises 1–12: Sketch the graph of the function and use Part II of the Fundamental Theorem of Calculus in calculating the area of the region beneath the graph and between the lines described by $x = a$ and $x = b$.

1. $f(x) = \cos x$, $a = 0$, $b = \pi/2$
2. $f(x) = \sin x$, $a = \pi/6$, $b = \pi/2$
3. $f(x) = (x - 1)^3$, $a = 1$, $b = 3$
4. $f(x) = \frac{1}{3}(x - 3)^4$, $a = 2$, $b = 5$
5. $f(x) = e^x$, $a = 0$, $b = 1$
6. $f(x) = 2^x$, $a = 0$, $b = 1$
7. $f(x) = \sec^2 x$, $a = 0$, $b = \pi/4$
8. $f(x) = \sec x \tan x$, $a = 0$, $b = \pi/3$
9. $f(x) = 1/(x^2 + 1)$, $a = 0$, $b = 1$
10. $f(x) = 1/\sqrt{1 - x^2}$, $a = 0$, $b = 1/2$
11. $f(x) = 1/x$, $a = 1$, $b = 2$
12. $f(x) = 1/\sqrt{x}$, $a = 1$, $b = 4$

Exercises 13–22: Use Part I of the Fundamental Theorem of Calculus in differentiating the integral. Use the chain rule as in Example 4 when the upper limit is not simply x.

13. $\int_1^x \ln w\, dw$
14. $\int_0^x \sin w^2\, dw$
15. $\int_1^x \sqrt{1 + t^2}\, dt$
16. $\int_0^x \sqrt{1 - t^2}\, dt$
17. $\int_1^{x^2} \ln w\, dw$
18. $\int_0^{x^3} \cos\sqrt{w}\, dw$
19. $\int_1^{\sqrt{x}} \sqrt{1 + t^2}\, dt$
20. $\int_0^{1/x} \tan t^2\, dt$
21. $\int_0^{e^x} (\arctan t - t^2)\, dt$
22. $\int_0^{\sin x^2} \sin t^2\, dt$

Exercises 23–34: Find an antiderivative of the function f defined on $a \le x \le b$ and then evaluate $\int_a^b f(x)\, dx$.

23. $f(x) = \sqrt{x}$, $0 \le x \le 4$
24. $f(x) = x^{4/3}$, $0 \le x \le 8$
25. $f(x) = x^3 - 7x$, $0 \le x \le 1$
26. $f(x) = x - 2x^4$, $0 \le x \le 1$
27. $f(x) = x(3x - 1)$, $0 \le x \le 4$
28. $f(x) = (x + 1)(x^2 + 1)$, $0 \le x \le 2$
29. $f(x) = (x^2 + 1)/x$, $1 \le x \le 3$
30. $f(x) = (x^3 + 2x + 1)/x^2$, $1 \le x \le 4$
31. $f(x) = x^{1/3} - 2x^{-2/3}$, $1 \le x \le 8$
32. $f(x) = x^{-1/4} - 7\sqrt{x}$, $1 \le x \le 16$
33. $f(x) = 2x \sin x^2$, $0 \le x \le 1$
34. $f(x) = 2xe^{x^2}$, $0 \le x \le 1$

Exercises 35–38: For the given function f, let $F(x) = \int_a^x f(w)\, dw$. Evaluate F at the given points.

35. f as defined in the accompanying graph; $a = 2$; $F(2)$, $F(3)$, and $F(5)$

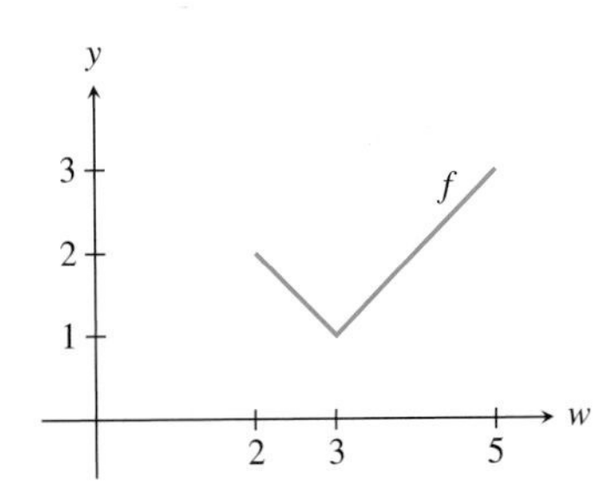

Figure for Exercise 35.

36. f as defined in the accompanying graph; $a = 0$; $F(0)$, $F(1)$, and $F(2)$

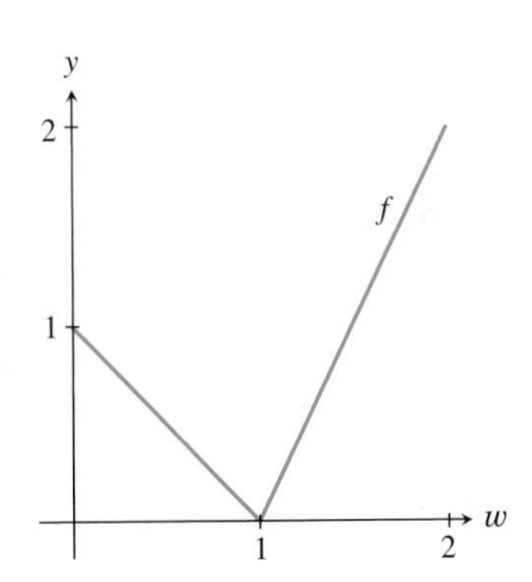

Figure for Exercise 36.

37. $f(w) = \sin w$, $0 \le w \le \pi$; $a = 0$; $F(0)$, $F(\pi/4)$, $F(\pi/2)$, and $F(\pi)$

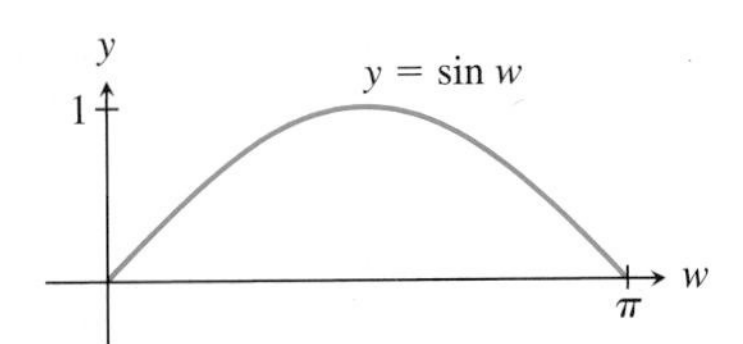

Figure for Exercise 37.

38. $f(w) = \cos w, 0 \leq w \leq \pi; a = 0; F(0), F(\pi/2)$, and $F(\pi)$

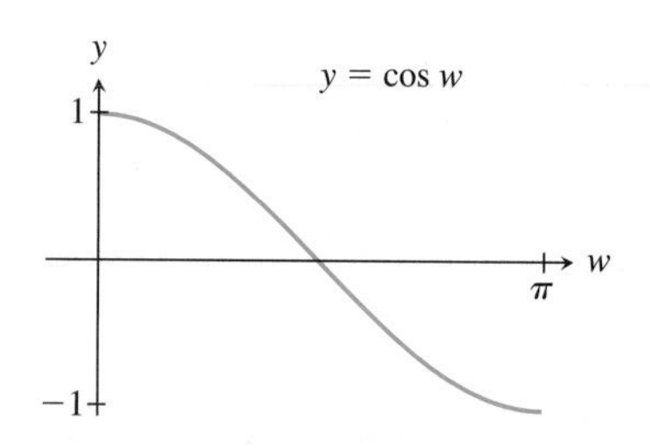

Figure for Exercise 38.

Exercises 39–44: A function f has an antiderivative F. Calculate $\int_a^b f(x)\,dx$.

39. $F(x) = x + 1/x; a = 1, b = 3$

40. $F(x) = 1/\sqrt{x^2+1}; a = 0, b = \sqrt{8}$

41. $F(x) = \sin(x^2); a = 0, b = 1$

42. $F(x) = \tan(\sqrt{x}); a = 0, b = 4$

43. $F(x) = e^{x^2}; a = 0, b = 1$

44. $F(x) = 2^x; a = 1, b = 4$

45. What is the average value of the cosine function on $[0, \pi/2]$? Determine the height of a rectangle with base the interval $[0, \pi/2]$ and area equal to that under the graph of the cosine function on $[0, \pi/2]$. Sketch the graph of the cosine function and the rectangle. Write a simple sentence or two about the meaning of the sketch.

46. What is the average value of the function $f(x) = e^x$ on $[0, 2]$? Determine the height of a rectangle with base the interval $[0, 2]$ and area equal to that under the graph of the exponential function on $[0, 2]$. Sketch the graph of f on $[0, 2]$ and the rectangle. Write a simple sentence or two about the meaning of the sketch.

47. Show that (12) follows from Property 1. *Hint:* Apply Property 1 to justify

$$\begin{aligned} F(x+h) - F(x) &= \int_a^{x+h} f(w)\,dw - \int_a^x f(w)\,dw \\ &= \int_x^a f(w)\,dw + \int_a^{x+h} f(w)\,dw \\ &= \int_x^{x+h} f(w)\,dw. \end{aligned}$$

48. Complete the proof of Part I of the Fundamental Theorem of Calculus by working through the case in which $h < 0$. *Hint:* See the preceding exercise.

49. Use the Fundamental Theorem of Calculus in verifying the first part of Property 3. *Hint:* Let F and G be antiderivatives of f and g, and write an antiderivative of $f + g$ in terms of F and G.

50. Use the Fundamental Theorem of Calculus in verifying the second part of Property 3. *Hint:* Let F be an antiderivative of f, express $c \int_a^b f(x)\,dx$ in terms of F, and, finally, express an antiderivative of cf in terms of F.

51. Make a convincing argument that if f is continuous and nonnegative on $[a, b]$, then $\int_a^b f(x)\,dx \geq 0$.

52. Apply the result stated in the preceding exercise to prove Property 4 of the integral. You may use the following outline: First, observe that $|f(x)| \geq f(x)$ on $[a, b]$; infer that $|f(x)| - f(x) \geq 0$ on $[a, b]$; apply Exercise 51 to show that $\int_a^b |f(x)|\,dx \geq \int_a^b f(x)\,dx$. Next, observe that $|f(x)| \geq -f(x)$ on $[a, b]$; infer that $|f(x)| + f(x) \geq 0$ on $[a, b]$; apply Exercise 51 to show that $\int_a^b |f(x)|\,dx \geq -\int_a^b f(x)\,dx$. Finally, infer Property 4.

53. The statements of Property 1 near Equation (5) and in the summary of the definitions and properties of the integral are not the same. Show that the more general statement follows from Equation (5).

54. Fill in the details of the following proof that for all $a, b > 0$

$$\lim_{x\to\infty} \frac{(\ln x)^a}{x^b} = 0.$$

This limit was given in Section 4.4. Let $p > 0$; then for $w \geq 1$, $w^{-1} \leq w^{-1+p}$. By the result stated in Exercise 51,

$$\int_1^x \frac{1}{w}\,dw \leq \int_1^x w^{-1+p}\,dw.$$

Hence

$$\ln x \leq \frac{x^p}{p} - \frac{1}{p} < \frac{x^p}{p}.$$

Using this result, we have

$$\frac{(\ln x)^a}{x^b} < \frac{x^{pa}}{p^a x^b}.$$

After showing that p can be chosen so that $pa - b < 0$, let $x \to \infty$ and infer the desired result.

55. Let F be defined on $[0, 1.5]$ by

$$F(x) = \int_0^{x^2} \frac{\sin t}{t+1}\,dt.$$

Where does the maximum value of F occur?

T 56. Use your calculator or CAS to generate 100 random numbers in the interval $[0, \pi]$. Calculate the sines of these numbers and average the 100 function values thus generated. This number approximates the average value of the sine function on $[0, \pi]$. Most calculators have a built-in random number generator that returns numbers randomly chosen from $[0, 1]$. A labor-intensive procedure is to press the RAND button and then the SIN button, and add the result to a variable whose initial value is 0. Repeat this 100 times.

Divide the sum by 100. If you can program your calculator, this procedure can be automated. If you repeat the program, say, 20 times and average the outcomes, you should obtain $2/\pi = 0.6366\ldots$, more or less. It may be necessary to arrange for a different "seed" for the random number generator as you repeat the program.

57. The accompanying figure shows a valve opening in a circular pipe of radius 1 meter. At any time $t \in [0, 1]$, the left edge of the valve lies along the line with equation $x = 2t^2$ and water is flowing in the shaded region to the left of this line. At time t, the flow $W(t)$ through the valve, measured in cubic meters per minute, is three times the area of the shaded region. What is the rate of change of the flow at $t = 0.5$ minutes?

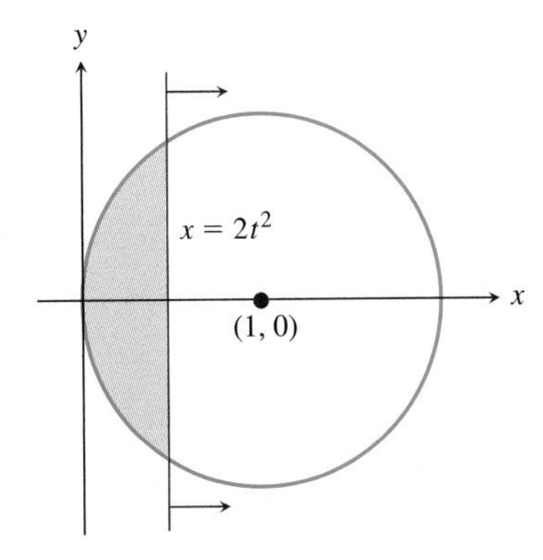

Figure for Exercise 57.

5.4 The Indefinite Integral

To evaluate a definite integral $\int_a^b f(x)\,dx$ it is sufficient to find an antiderivative F of f because, by the Fundamental Theorem of Calculus,

$$\int_a^b f(x)\,dx = F(x)\Big|_a^b = F(b) - F(a).$$

For this reason our main efforts in Sections 5.5–5.8 will be to explore ways of finding an antiderivative F for a given function f. To facilitate this discussion, it is convenient to introduce the idea of the *indefinite integral.*

A function F is an antiderivative of f provided that the derivative of F is f; if we have a variable like x in mind, we may write $F'(x) = f(x)$. This language is descriptive but awkward. In what follows, we use the notation

$$\int f(x)\,dx$$

for an antiderivative of $f(x)$. This notation frees us from choosing a letter like F to denote an antiderivative of f. Though usually not made explicit, when we use this notation we understand that there is an interval I on which f is defined and continuous.

While the definite integral $\int_a^b f(x)\,dx$ is a number, the indefinite integral $\int f(x)\,dx$ denotes a function, one whose derivative is $f(x)$ for all x in some interval I. The term "integral" may refer to either an indefinite or a definite integral.

Here are some examples:

$$\int x^2\,dx = \tfrac{1}{3}x^3,$$

$$\int \cos x\,dx = \sin x,$$

$$\int \frac{1}{x}\,dx = \ln x,$$

and, more generally,

$$\int f(x)\,dx = \int_a^x f(w)\,dw.$$

The first three of these antiderivatives are easily checked by differentiation. For example, the first equation is correct because the derivative of $F(x) = \frac{1}{3}x^3$ is $f(x) = x^2$. The functions f and F are defined on any interval $[a, b]$. The second equation holds because the derivative of $F(x) = \sin x$ is $f(x) = \cos x$. The functions f and F are defined on any interval $[a, b]$. The third formula is correct if we assume that $F(x) = \ln x$ and $f(x) = 1/x$ are defined on an interval $[a, b]$, where $b > a > 0$. On such an interval, $F'(x) = f(x)$. The fourth formula is correct by Part I of the Fundamental Theorem of Calculus.

You may have noticed that in the first and second paragraphs of this section we said "*an* antiderivative," not "*the* antiderivative." The reason for this is that a function f can have more than one antiderivative. For example, if $f(x) = x^2$, then for any choice of the constant C the function

$$\tfrac{1}{3}x^3 + C$$

is an antiderivative of $f(x) = x^2$. All of these antiderivatives of f are the same except for an additive constant, that is, they "differ by a constant." You may ask, "Are there other antiderivatives of x^2?"

We show in Exercises 25 and 26 that, in fact, any two antiderivatives of a function f differ by a constant. This result justifies the common practice of writing

$$\int f(x)\,dx = F(x) + C,$$

where F is a specific antiderivative of f. This is intended to suggest that all antiderivatives of f have the form of any specific antiderivative plus an arbitrary constant C.

The simplest indefinite integrals are those based directly on a differentiation formula. For example, the indefinite integral

$$\int x^n\,dx = \frac{x^{n+1}}{n+1} + C, \qquad n \neq -1, \tag{1}$$

corresponds to the formula

$$\frac{d}{dx}x^{n+1} = (n+1)x^n.$$

We list in Table 5.1 several common indefinite integrals. Each of these is included in the larger table of integrals listed on the inside back cover of this book.

TABLE 5.1 Short table of indefinite integrals.

(A) $\int x^n\,dx = \frac{x^{n+1}}{n+1} + C, n \neq -1$	(B) $\int \frac{1}{x}\,dx = \ln\lvert x\rvert + C$
(C) $\int \sin x\,dx = -\cos x + C$	(D) $\int \cos x\,dx = \sin x + C$
(E) $\int e^x\,dx = e^x + C$	(F) $\int a^x\,dx = \frac{a^x}{\ln a} + C$
(G) $\int \frac{dx}{x^2 + a^2} = \frac{1}{a}\arctan\frac{x}{a} + C$	(H) $\int \frac{dx}{\sqrt{a^2 - x^2}} = \arcsin\frac{x}{a} + C$

Each of these formulas is easily checked. For example, we know that formula (G) holds because

$$\frac{d}{dx}\left(\frac{1}{a}\arctan\frac{x}{a}+C\right)=\frac{1}{a}\left(\frac{1}{1+\left(\frac{x}{a}\right)^2}\cdot\frac{d}{dx}\left(\frac{x}{a}\right)\right)=\frac{1}{x^2+a^2}.$$

Indefinite integrals inherit their properties directly from the properties of derivatives. For example, the antiderivative of a sum is the sum of the antiderivatives, that is,

$$\int (f(x)+g(x))\,dx=\int f(x)\,dx+\int g(x)\,dx. \tag{2}$$

Also, if a is a number, then

$$\int af(x)\,dx=a\int f(x)\,dx. \tag{3}$$

EXAMPLE 1 Apply Properties (2) and (3) and one or more of formulas (A)–(H) to evaluate the indefinite integral

$$\int \left(3x^2+5\sqrt{x}\right)dx.$$

Solution

Applying Properties (2) and (3) of antiderivatives,

$$\int \left(3x^2+5\sqrt{x}\right)dx=\int 3x^2\,dx+\int 5x^{1/2}\,dx=3\int x^2\,dx+5\int x^{1/2}\,dx.$$

Applying formula (A) twice,

$$\int \left(3x^2+5\sqrt{x}\right)dx=3\cdot\frac{1}{2+1}x^3+5\cdot\frac{1}{1+\frac{1}{2}}x^{3/2}+C.$$

We included just one constant in this calculation because the sum of two constants is again a constant. Simplifying,

$$\int \left(3x^2+5\sqrt{x}\right)dx=x^3+\tfrac{10}{3}x^{3/2}+C.$$

How could we check this answer?

EXAMPLE 2 Apply Properties (2) and (3) and formulas (A)–(H) to evaluate the indefinite integral

$$\int e^{x-2}\,dx.$$

Solution

Because $e^{x-2} = e^x \cdot e^{-2}$,

$$\int e^{x-2}\,dx = e^{-2}\int e^x\,dx.$$

Applying formula (E),

$$\int e^{x-2}\,dx = e^{-2}e^x + C = e^{x-2} + C.$$

Note that formula (E) wasn't really necessary; we could have solved the problem by simply observing that an antiderivative of e^{x-2} is e^{x-2}.

Exercises 5.4

Exercises 1–22: Use Properties (2) and (3), formulas (A)–(H), or standard differentiation formulas to evaluate the indefinite integral.

1. $\int (5x^2 - 7)\,dx$

2. $\int (13 - 2x^3)\,dx$

3. $\int (x^3 + 5x^{1/3})\,dx$

4. $\int (x^5 - 7x^{3/4})\,dx$

5. $\int (5x)^3\,dx$

6. $\int (-2x)^4\,dx$

7. $\int \frac{2\,dx}{3x}$

8. $\int \frac{-5\,dx}{9x}$

9. $\int 2^{x+1}\,dx$

10. $\int 3^{x-2}\,dx$

11. $\int \frac{dx}{x^2 + 1}$

12. $\int \frac{dx}{x^2 + 4}$

13. $\int \frac{dx}{\sqrt{4 - x^2}}$

14. $\int \frac{dx}{\sqrt{2 - x^2}}$

15. $\int (3\sqrt{x} + 5\cos x)\,dx$

16. $\int (3x^{-1/2} + 2\sin x)\,dx$

17. $\int (3^x + x^3)\,dx$

18. $\int (10^x + x^{1/10})\,dx$

19. $\int (2x + 3)^2\,dx$

20. $\int (x - 1/x)^2\,dx$

21. $\int \sec x \tan x\,dx$

22. $\int \sec^2 x\,dx$

23. Differentiate the expressions on the right side of formulas (F) and (H) and show that the derivatives simplify to the integrands of the corresponding integrals.

24. Extend Table 1. For example, formulas for

$$\int \sec^2 x\,dx, \quad \int \sec x \tan x\,dx, \quad \text{and} \quad \int \sinh x\,dx$$

could be added. The hyperbolic sine and cosine functions are defined by: $\sinh x = (e^x - e^{-x})/2$ and $\cosh x = (e^x + e^{-x})/2$.

25. Use the Mean-value Theorem to show that if f' is identically zero, then f is a constant. More precisely, show that if f' is defined on $I = [a, b]$ and $f'(x) = 0$ for all $x \in I$, then $f(x) = f(a)$ for all $x \in I$.

26. Use Exercise 25 in showing that if the derivatives of functions f and g are continuous and equal, then they differ by a constant. *Hint:* Consider the function $f - g$.

5.5 Integration by Substitution

Integration by substitution is based on the following observation: If we know the antiderivative formula

$$\int f(x)\,dx = F(x) + C, \tag{1}$$

then we also know the more general formula

$$\int f(g(x))g'(x)\,dx = F(g(x)) + C, \tag{2}$$

where g is any differentiable function. This is true because, by the chain rule,

$$\frac{d}{dx}F(g(x)) = F'(g(x))g'(x) = f(g(x))g'(x).$$

Hence, the composite function $F(g(x))$ is an antiderivative of the function $f(g(x))g'(x)$; hence, Equation (2) holds.

To illustrate, from formula (A) in Table 1 we know that

$$\int x^2\,dx = \tfrac{1}{3}x^3 + C.$$

If we take $g(x) = \sin x$, it then follows from (2) that

$$\int (\sin x)^2 \cos x\,dx = \tfrac{1}{3}(\sin x)^3 + C.$$

The antidifferentiation formula (2) came from formula (1) by the chain rule. Usually we apply this result by going in the opposite direction, that is, by inferring the simpler formula (1) from an integral having the form (2). A method for facilitating this calculation is called *integration by substitution.* We first identify a substitution function $u = g(x)$, next write the differential dg of $u = g(x)$ as $du = dg = g'(x)\,dx$, and then substitute into the left side of (2):

$$\int \underbrace{f(g(x))}_{f(u)}\underbrace{g'(x)\,dx}_{du} = \int f(u)\,du.$$

If our choice of g was a good one, we may be able to find an antiderivative F of f and write

$$\int f(g(x))g'(x)\,dx = \int f(u)\,du = F(u)\Big|_{u=g(x)} + C = F(g(x)) + C. \tag{3}$$

EXAMPLE 1 Evaluate the integral

$$\int (\ln x)^2 \frac{1}{x}\,dx.$$

Solution

The substitution $u = g(x) = \ln x$ is suggested by the observation that $g'(x) = (\ln x)' = 1/x$. Calculating $du = (\ln x)'\,dx = (1/x)\,dx$

and substituting,

$$\int (\ln x)^2 \frac{1}{x}\, dx = \int u^2\, du = \tfrac{1}{3}u^3 \Big|_{u=\ln x} + C = \tfrac{1}{3}(\ln x)^3 + C.$$

We may easily verify that $\frac{1}{3}(\ln x)^3 + C$ is an antiderivative of $(\ln x)^2(1/x)$.

EXAMPLE 2 Evaluate the definite integral

$$\int_0^2 x\sqrt{x^2+1}\, dx \tag{4}$$

by first finding the indefinite integral $\int x\sqrt{x^2+1}\, dx$ and then applying the Fundamental Theorem of Calculus.

Solution

Because the factor x is, except for a factor of 2, the derivative of $g(x) = x^2 + 1$, we try the substitution $u = x^2 + 1$. Differentiating this equation, $du = 2x\, dx$, which we may rewrite as $x\, dx = \frac{1}{2}du$. Substituting,

$$\int x\sqrt{x^2+1}\, dx = \int \sqrt{x^2+1}\,(x\, dx) = \int \sqrt{u}\left(\tfrac{1}{2}du\right).$$

Hence,

$$\int x\sqrt{x^2+1}\, dx = \tfrac{1}{2}\int \sqrt{u}\, du = \tfrac{1}{2}\cdot\tfrac{2}{3}u^{3/2}\Big|_{u=x^2+1} + C$$
$$= \tfrac{1}{3}(x^2+1)^{3/2} + C.$$

We have now found the indefinite integral $\int x\sqrt{x^2+1}\, dx$, that is, we have found the most general antiderivative of $x\sqrt{x^2+1}$.

According to the Fundamental Theorem of Calculus, we may calculate the definite integral $\int_0^2 x\sqrt{x^2+1}\, dx$ by evaluating an antiderivative of $x\sqrt{x^2+1}$ at the two endpoints. Using the antiderivative we found in the last paragraph,

$$\int_0^2 x\sqrt{x^2+1}\, dx = \tfrac{1}{3}(x^2+1)^{3/2} + C\Big|_0^2 \tag{5}$$
$$= \left(\tfrac{1}{3}(2^2+1)^{3/2} + C\right) - \left(\tfrac{1}{3}(0^2+1)^{3/2} + C\right) = \tfrac{1}{3}\left(5\sqrt{5} - 1\right).$$

Note that in Equation (5) the arbitrary constant C was "subtracted out." This means that we did not need to include it in our calculation of the definite integral. We show more generally that if we wish to evaluate a definite integral $\int_a^b f(g(x))g'(x)\, dx$, all we need is *an* antiderivative $F(u)$ of $f(u)$, not the most general antiderivative $F(u) + C$. Moreover, having changed variables from x to u, we need not change back to x at the end of the calculation (recall that (3) includes the step $F(u)\big|_{u=g(x)} + C = F(g(x)) + C$). Specifically, we infer directly from (3) and the

Fundamental Theorem of Calculus that if in the integral $\int_a^b f(g(x))g'(x)\,dx$ we make the substitution $u = g(x)$, then

$$\begin{aligned}\int_a^b f(g(x))g'(x)\,dx &= \int_c^d f(u)\,du \\ &= F(u)\Big|_c^d = F(d) - F(c) = F(g(b)) - F(g(a)),\end{aligned} \tag{6}$$

where $F(x)$ is an antiderivative of $f(x)$, $c = g(a)$, and $d = g(b)$.

Equation (6) need not be memorized. As we show in the next several examples, the steps from the beginning to the end of this equation are more or less self-guided.

EXAMPLE 3 Calculate the area A of the region $\mathcal{R}$ beneath the graph of $f(x) = 1/(2x - 3)$, $2 \le x \le 4$.

Solution

Figure 5.25 shows the graph of f, which although undefined at $x = 3/2$ is positive and continuous on the range of integration $[2, 4]$. The area is given by

$$\int_2^4 \frac{1}{2x - 3}\,dx.$$

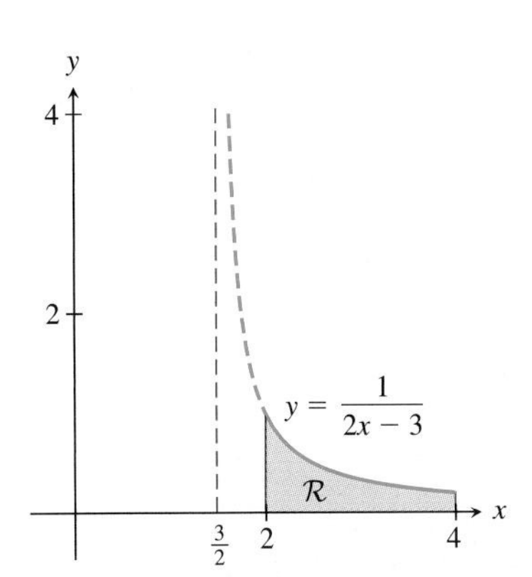

FIGURE 5.25 The region $\mathcal{R}$ beneath the graph of $f(x) = 1/(2x - 3)$, $2 \le x \le 4$.

We try the substitution $u = g(x) = 2x - 3$ in this integral, not so much because we have the functions f and g in (6) well in mind, but because this substitution simplifies the fraction $1/(2x - 3)$. Noting that $du = 2\,dx$ and substituting into the integral for the area A,

$$A = \int_2^4 \frac{1}{2x - 3}\,dx = \int_1^5 \frac{1}{u}\left(\tfrac{1}{2}du\right) = \tfrac{1}{2}\int_1^5 \frac{1}{u}\,du.$$

The limits 1 and 5 in the second integral were calculated as $c = g(2) = 1$ and $d = g(4) = 5$. Noting that $1/u > 0$ on the interval $[1, 5]$, an antiderivative of $1/u$ is $\ln u$; hence,

$$A = \tfrac{1}{2}\int_1^5 \frac{1}{u}\,du = \tfrac{1}{2}\ln u\Big|_1^5 = \tfrac{1}{2}\ln 5 - \tfrac{1}{2}\ln 1 = \tfrac{1}{2}\ln 5.$$

EXAMPLE 4 Evaluate the indefinite integral $\displaystyle\int \frac{1}{2x - 3}\,dx$.

Solution

The integrand of this indefinite integral is the same as the integrand of the definite integral in Example 3. Noting that the integrand is not defined at $x = 3/2$, we assume that x varies over an interval $[a, b]$ entirely to the right of $x = 3/2$ or entirely to the left of $x = 3/2$. As in Example 3, we use the substitution $u = 2x - 3$. As before, $du = 2\,dx$ and

$$\int \frac{1}{2x - 3}\,dx = \int \frac{1}{u}\left(\tfrac{1}{2}du\right) = \tfrac{1}{2}\int \frac{1}{u}\,du.$$

If $2x - 3 > 0$, then $u > 0$ and because

$$\frac{d}{du} \ln u = \frac{1}{u},$$

an antiderivative of $1/u$ is $\ln u$. If $2x - 3 < 0$, then $u < 0$ and because

$$\frac{d}{du} \ln(-u) = \frac{1}{-u} \cdot (-1) = \frac{1}{u},$$

an antiderivative of $1/u$ is $\ln(-u)$.

Recalling that $|u| = u$ when $u > 0$ and $|u| = -u$ when $u < 0$, these results can be combined in one formula:

$$\frac{d}{du} \ln |u| = \frac{1}{u}.$$

Hence,

$$\int \frac{1}{2x-3}\, dx = \tfrac{1}{2} \int \frac{1}{u}\, du = \tfrac{1}{2} \ln |u| \Big|_{u=2x-3} + C = \tfrac{1}{2} \ln |2x - 3| + C.$$

EXAMPLE 5 Evaluate the integral $\int_0^1 x\sqrt{1-x}\, dx$.

Solution

We try the substitution $u = 1 - x$ because it simplifies the factor $\sqrt{1-x}$ of the integrand. Calculating $du = -dx$ and noting from the substitution equation that $u = 1$ when $x = 0$, and $u = 0$ when $x = 1$,

$$\int_0^1 x\sqrt{1-x}\, dx = \int_1^0 x\sqrt{u}\,(-du). \tag{7}$$

To express the factor x in terms of u, we solve the substitution equation $u = 1 - x$ for x. We find $x = 1 - u$. Substituting this into (7) gives

$$\int_0^1 x\sqrt{1-x}\, dx = \int_1^0 (1-u)\sqrt{u}\,(-du). \tag{8}$$

Recalling from Definition 3 in the Definitions and Properties of the Integral, pages 369–370, that $\int_b^a f(x)\, dx = -\int_a^b f(x)\, dx$, it follows from (8) that

$$\int_0^1 x\sqrt{1-x}\, dx = \int_0^1 (1-u)\sqrt{u}\, du.$$

Expanding the integrand of the last integral and integrating the result,

$$\int_0^1 x\sqrt{1-x}\, dx = \int_0^1 (u^{1/2} - u^{3/2})\, du$$

$$= \tfrac{2}{3}u^{3/2} - \tfrac{2}{5}u^{5/2} \Big|_0^1 = \frac{2}{3} - \frac{2}{5} = \frac{4}{15}.$$

If we wish to evaluate an integral using integration by substitution, the substitution must lead to a "known integral," an integral for which an antiderivative can be recognized. We may depend upon our memory, as in the last two examples, when we recalled that an antiderivative of $1/u$ is $\ln u$, or we may use the Short Table of Indefinite Integrals given in Table 1, Section 5.4. The eight indefinite integrals (A)–(H) in this table are included in the larger table of integrals listed on the inside back cover of this book. These eight integrals are among the most basic and most frequently occurring indefinite integrals. We shall use them without comment in the remainder of the text. Many integrals are just one substitution away from these integrals.

EXAMPLE 6 Evaluate the integral

$$\int_0^1 \frac{dx}{2x^2 + 3}.$$

Solution

The integrand most resembles formula (G). There are two ways of changing the form of the integrand so that formula (G) applies, either by using algebra to rearrange the integrand so that it looks like $1/(x^2 + a^2)$ or by substitution. We first rearrange the integrand.

With the form $x^2 + a^2$ in mind,

$$2x^2 + 3 = 2\left(x^2 + \frac{3}{2}\right) = 2\left(x^2 + \left(\sqrt{\frac{3}{2}}\right)^2\right).$$

Using this result,

$$\int_0^1 \frac{dx}{2x^2 + 3} = \frac{1}{2}\int_0^1 \frac{dx}{\left(x^2 + \left(\sqrt{\frac{3}{2}}\right)^2\right)}.$$

From formula (G), with $a = \sqrt{3/2}$,

$$\int_0^1 \frac{dx}{2x^2 + 3} = \frac{1}{2} \cdot \frac{1}{\sqrt{\frac{3}{2}}} \arctan \frac{x}{\sqrt{\frac{3}{2}}}\Bigg|_0^1 = \frac{1}{\sqrt{6}} \arctan \sqrt{\frac{2}{3}} = 0.2795\ldots.$$

For a solution based on a substitution, we write

$$2x^2 + 3 = (\sqrt{2}x)^2 + (\sqrt{3})^2.$$

If we make the substitution $u = \sqrt{2}x$, then $du = \sqrt{2}\,dx$. Hence, $dx = du/\sqrt{2}$ and

$$\int_0^1 \frac{dx}{2x^2 + 3} = \frac{1}{\sqrt{2}}\int_0^{\sqrt{2}} \frac{du}{u^2 + (\sqrt{3})^2}.$$

The limits 0 and $\sqrt{2}$ were obtained by setting $x = 0$ and $x = 1$ in the equation $u = \sqrt{2}x$. From formula (G), with $a = \sqrt{3}$,

$$\int_0^1 \frac{dx}{2x^2 + 3} = \frac{1}{\sqrt{2}} \cdot \frac{1}{\sqrt{3}} \arctan \frac{u}{\sqrt{3}}\Bigg|_0^{\sqrt{2}} = \frac{1}{\sqrt{6}} \arctan \sqrt{\frac{2}{3}} = 0.2795\ldots.$$

In the next example, we use a "trigonometric substitution" to transform an integral to one for which an antiderivative is more easily found.

EXAMPLE 7 Evaluate the integral

$$\int_0^1 \sqrt{1 - x^2}\, dx. \tag{9}$$

Solution

The form of the integrand suggests the unit circle because if we solve the "unit circle equation" $x^2 + y^2 = 1$ for y, one solution is $y = \sqrt{1 - x^2}$. Figure 5.26 shows the graph of the equation $x^2 + y^2 = 1$ and the graph of $y = \sqrt{1 - x^2}$, $0 \leq x \leq 1$. The unit circle equation $x^2 + y^2 = 1$ is closely associated with the trigonometric identity $\sin^2\theta + \cos^2\theta = 1$. If we solve this identity for $\cos\theta$, as we solved the equation $x^2 + y^2 = 1$ for y, we find $\cos\theta = \sqrt{1 - \sin^2\theta}$.

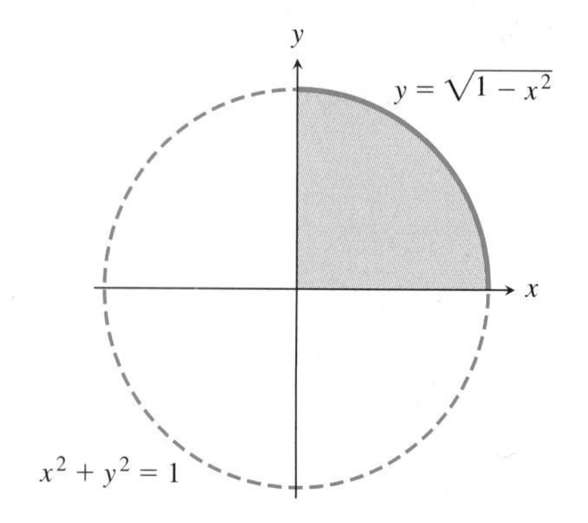

FIGURE 5.26 The unit circle with equation $x^2 + y^2 = 1$; the graph of $y = \sqrt{1 - x^2}$, $0 \leq x \leq 1$; and the region (shaded) whose area is equal to the integral (9).

We are thus led to the substitution $x = \sin\theta$ in (9). Because $0 \leq x \leq 1$, we can restrict θ to the interval $[0, \pi/2]$, so that $\cos\theta \geq 0$. For $x = 0$, $\sin\theta = 0$; hence $\theta = 0$. For $x = 1$, $\sin\theta = 1$; hence, $\theta = \pi/2$. Because $dx = \cos\theta\, d\theta$, the integral (9) becomes

$$\int_0^1 \sqrt{1 - x^2}\, dx = \int_0^{\pi/2} (\cos\theta)(\cos\theta\, d\theta) = \int_0^{\pi/2} \cos^2\theta\, d\theta. \tag{10}$$

This integral can be evaluated by using the trigonometric identity $\cos^2\theta = \frac{1}{2}(1 + \cos 2\theta)$. Making this replacement in (10),

$$\int_0^1 \sqrt{1 - x^2}\, dx = \frac{1}{2}\int_0^{\pi/2} (1 + \cos 2\theta)\, d\theta.$$

An antiderivative of $1 + \cos 2\theta$ is $\theta + \frac{1}{2}\sin 2\theta$. Hence,

$$\int_0^1 \sqrt{1 - x^2}\, dx = \frac{1}{2}\left(\theta + \frac{1}{2}\sin 2\theta\right)\Bigg|_0^{\pi/2}$$

$$= \frac{1}{2}\left(\left(\frac{\pi}{2} + 0\right) - (0 + 0)\right) = \frac{\pi}{4}.$$

We could have anticipated this result because the integral in (9) is an integral for the area of the first quadrant of the unit circle. See the shaded region in Fig. 5.26.

We explore other trigonometric substitutions in Exercises 55–66.

Completing the Square For integrands involving quadratic expressions of the form $ax^2 + bx + c$, a useful change of variable may be suggested by "completing the square." Here's a brief review of how to complete the square:

$$ax^2 + bx + c = a\left(x^2 + \frac{b}{a}x\right) + c = a\left(x^2 + \frac{b}{a}x + \frac{b^2}{4a^2} - \frac{b^2}{4a^2}\right) + c$$

$$= a\left(x + \frac{b}{2a}\right)^2 - \frac{b^2}{4a} + c = a\left(x + \frac{b}{2a}\right)^2 - \frac{b^2 - 4ac}{4a}.$$

In words, to complete the square for the expression $ax^2 + bx + c$, first factor out the coefficient a of the x^2 term from $ax^2 + bx$; next, inside the large parentheses, add and subtract $b^2/(4a^2)$, which is the square of one-half the coefficient of the x term; move the term $-b^2/(4a^2)$ outside of the large parentheses by multiplying it by a; and, finally, arrange the result as the sum or difference of two squares.

EXAMPLE 8 Evaluate

$$\int \frac{dx}{\sqrt{3 + 2x - x^2}}.$$

Solution

Scanning formulas (A)–(H), the only possibility is formula (H), which shows a difference of two squares beneath a radical. We rearrange the expression $3 + 2x - x^2$ as the difference of two squares:

$$\begin{aligned} 3 + 2x - x^2 &= 3 - (x^2 - 2x) \\ &= 3 - (x^2 - 2x + 1) + 1 \\ &= 4 - (x - 1)^2. \end{aligned}$$

Comparing this result to formula (H) suggests the change of variable $u = x - 1$. Noting that $du = dx$ and applying formula (H),

$$\begin{aligned} \int \frac{dx}{\sqrt{3 + 2x - x^2}} &= \int \frac{dx}{\sqrt{4 - (x - 1)^2}} = \int \frac{du}{\sqrt{2^2 - u^2}} \\ &= \arcsin \frac{u}{2} + C = \arcsin\left(\frac{x - 1}{2}\right) + C. \end{aligned}$$

Exercises 67–76 provide some practice in completing the square.

Exercises 5.5

Exercises 1–10: Integrate using the suggested substitution.

1. $\int_0^1 x\sqrt{x^2 + 1}\, dx; \quad u = x^2 + 1$

2. $\int_1^2 x^2(x^3 - 1)^2\, dx; \quad u = x^3 - 1$

3. $\int_0^{\pi/2} \sin^3 x \cos x\, dx; \quad u = \sin x$

4. $\int_0^{\pi/2} \cos^3 x \sin x\, dx; \quad u = \cos x$

5. $\int_1^4 \left(e^{\sqrt{x}}/\sqrt{x}\right) dx; \quad u = \sqrt{x}$

6. $\int_1^4 2^{\sqrt{x}}/\sqrt{x}\, dx; \quad u = \sqrt{x}$

7. $\int_0^1 x\sqrt{3 + 2x}\, dx; \quad u = 3 + 2x$

8. $\int_1^3 (x - 1)\sqrt{x + 1}\, dx; \quad u = x + 1$

9. $\int_1^3 (1/(x + 1))\, dx; \quad u = x + 1$

10. $\int_3^7 (1/(2x - 5))\, dx; \quad u = 2x - 5$

Exercises 11–40: Use a substitution to express each integral in terms of one of the integrals in Table 1.

11. $\int \sqrt{2x-1}\,dx$

12. $\int \sqrt{5+2x}\,dx$

13. $\int \frac{1}{(5-x)^{1/3}}\,dx$

14. $\int \frac{1}{(3x-8)^{7/5}}\,dx$

15. $\int \frac{1}{2x+1}\,dx$

16. $\int \frac{9}{3x+2}\,dx$

17. $\int \sin(1-2x)\,dx$

18. $\int \sin\left(\frac{1}{2}x+3\right)dx$

19. $\int \cos\left(\frac{1}{3}+\frac{2}{7}x\right)dx$

20. $\int \cos(-x+1)\,dx$

21. $\int e^{2x+3}\,dx$

22. $\int e^{-0.1x+0.6}\,dx$

23. $\int 2^{-x+1}\,dx$

24. $\int 5^{7x+3}\,dx$

25. $\int \frac{dx}{2x^2+1}$

26. $\int \frac{dx}{3x^2+1}$

27. $\int \frac{dx}{(2x+1)^2+9}$

28. $\int \frac{dx}{(5x-2)^2+9}$

29. $\int \frac{dx}{\sqrt{1-2x^2}}$

30. $\int \frac{dx}{\sqrt{1-3x^2}}$

31. $\int \frac{dx}{\sqrt{7-2x^2}}$

32. $\int \frac{dx}{\sqrt{5-3x^2}}$

33. $\int \frac{dx}{\sqrt{1-(2x+1)^2}}$

34. $\int \frac{dx}{\sqrt{9-(7+2x)^2}}$

35. $\int \sin^5(3\theta)\cos(3\theta)\,d\theta$

36. $\int \cos^4(5\theta)\sin(5\theta)\,d\theta$

37. $\int \tan^4\theta\sec^2\theta\,d\theta$

38. $\int \tan^3(2\theta)\sec^2(2\theta)\,d\theta$

39. $\int \sec^3\theta(\sec\theta\tan\theta)\,d\theta$

40. $\int \sec^5(2\theta)\sec(2\theta)\tan(2\theta)\,d\theta$

Exercises 41–46: Given that $F'(u)=f(u)$, use substitution to evaluate the integral in terms of F.

41. $\int x^2 f(x^3+1)\,dx$

42. $\int \cos x\, f(\sin x)\,dx$

43. $\int e^x f(e^x)\,dx$

44. $\int \frac{f(\ln x)}{x}\,dx$

45. $\int \frac{x f(\arctan(x^2))}{1+x^4}\,dx$

46. $\int \frac{x f(\arcsin(x^2))}{\sqrt{1-x^4}}\,dx$

47. Assume that $1 \le x \le 2$. Show that

$$\int_0^x \sqrt{2w-w^2}\,dw = \tfrac{1}{4}\left(\pi + 2(x-1)\sqrt{2x-x^2} + 2\arcsin(x-1)\right)$$

by writing $2w-w^2 = 1-(w-1)^2$ and using the substitution $w-1=\sin u$.

48. Calculate $\int_{-3}^{-2} (1/x)\,dx$.

49. Evaluate the integral

$$\int \frac{x\,dx}{\sqrt{1-x^2}}$$

by using a substitution to change it to formula (A) in Table 1.

50. Use formula (B) from Table 1 and a substitution to evaluate $\int_0^1 \tan x\,dx$.

51. Unthinking application of an integral formula similar to those listed in Table 1 or in the table of integrals (see inside back cover) can lead to incorrect results. Criticize the following calculation.

$$\int_{-1}^{1} \frac{1}{x^2}\,dx = \left.\frac{x^{-1}}{-1}\right|_{-1}^{1} = -1-1=-2$$

52. With the preceding exercise in mind, comment on $\int_{-3}^{0} (1/(x+1))\,dx$.

53. Assume that f is continuous on an interval $[-a,a]$. Show that if f is an even function on $[-a,a]$, that is, $f(-x)=f(x)$ for all $x\in[-a,a]$, then

$$\int_{-a}^{a} f(x)\,dx = 2\int_0^a f(x)\,dx.$$

Hint: Write the integral as the sum of integrals on $[-a,0]$ and $[0,a]$ and make the substitution $u=-x$ in one of them.

54. Assume that f is continuous on an interval $[-a,a]$. Show that if f is an odd function on $[-a,a]$, that is, $f(-x)=-f(x)$ for all $x\in[-a,a]$, then

$$\int_{-a}^{a} f(x)\,dx = 0.$$

Hint: Write the integral as the sum of integrals in $[-a,0]$ and $[0,a]$ and make the substitution $u=-x$ in one of them.

Exercises 55–66: Evaluate the integral using the suggested substitution. These problems are similar to the trigonometric substitution discussed in Example 7, which was based on the trigonometric identity $1-\sin^2 u=\cos^2 u$. The substitutions suggested in the exercises are based on this identity or on one of the identities $\tan^2 u+1=\sec^2 u$ or $\sec^2 u - 1 = \tan^2 u$.

55. Try $x=(\sin u)/2$ in

$$\int_0^{1/2} \sqrt{1-4x^2}\,dx.$$

56. Try $x=\left(1/\sqrt{2}\right)\sin u$ in

$$\int_0^{1/\sqrt{2}} \sqrt{1 - 2x^2}\, dx.$$

57. Try $x = \sqrt{3/2} \sin u$ in

$$\int_0^{\sqrt{3/2}} \sqrt{3 - 2x^2}\, dx.$$

58. Try $x = \sqrt{2/3} \sin u$ in

$$\int_0^{\sqrt{2/3}} \sqrt{2 - 3x^2}\, dx.$$

59. Try $x = \tan u$ in

$$\int_0^1 \frac{dx}{(x^2 + 1)^{3/2}}.$$

60. Try $x = 2 \tan u$ in

$$\int_0^1 \frac{dx}{(x^2 + 4)^{3/2}}.$$

One of the new limits will be $\arctan(1/2)$. Note that $\sin(\arctan(1/2)) = 1/\sqrt{5}$.

61. Try $x = \tan u$ in

$$\int_0^1 \sqrt{1 + x^2}\, dx.$$

You will find the integral $\int \sec^3 x\, dx$ in the table of integrals (see inside back cover).

62. Try $x = 3 \sin u$ in

$$\int_1^2 \frac{dx}{x^2\sqrt{9 - x^2}}.$$

What is the derivative of $-\cot u$?

63. Try $x = \tan u$ in

$$\int_{1/\sqrt{3}}^1 \frac{dx}{x(x^2 + 1)}.$$

Write $\cot u$ as $(\cos u)/(\sin u)$ and use formula (B) in the table of integrals.

64. Try $x = 2 \sin u$ in finding the area of one-quarter of the ellipse with equation $x^2/4 + y^2/9 = 1$.

65. Try $x = \sin u$ in finding the area of one-quarter of the ellipse with equation $x^2 + y^2/2 = 1$.

66. Try $x = \sec u$ in

$$\int_{\sqrt{5}}^{\sqrt{10}} \frac{x^2\, dx}{\sqrt{x^2 - 1}}.$$

You will find the integral $\int \sec^3 u\, du$ in the table of integrals (see inside back cover).

Exercises 67–76: Evaluate the integral by completing the square and making a change of variable. The resulting integral will have the form of one of the formulas (A)–(H) or can be done by a trigonometric substitution (see Exercises 55–66).

67. $\displaystyle\int \frac{dx}{2x^2 - 6x + 5}$

68. $\displaystyle\int \frac{dx}{9x^2 - 12x + 5}$

69. $\displaystyle\int \frac{dx}{\sqrt{24 - 10x - 25x^2}}$

70. $\displaystyle\int \frac{dx}{\sqrt{-40 + 28x - 4x^2}}$

71. $\displaystyle\int \frac{dx}{x^2 + x + 1}$

72. $\displaystyle\int \frac{dx}{x^2 - x + 1}$

73. $\displaystyle\int_0^1 \frac{dx}{(2x^2 + 6x + 5)^{3/2}}$

74. $\displaystyle\int_{-1/2}^{1/2} \frac{dx}{\sqrt{x^2 + x + 1}}$

You will find the integral $\int \sec u\, du$ in the table of integrals (see inside back cover). Your final answer should be $\ln(\sqrt{7} + 2) - \ln\sqrt{3}$.

75. $\displaystyle\int_{-1}^1 \frac{x + 1}{\sqrt{x^2 + 2x + 5}}\, dx$

76. $\displaystyle\int_2^3 \frac{\sqrt{3 + 2x - x^2}}{x - 1}\, dx$

5.6 Areas between Curves

In Definition 1 of Section 5.3, we defined the area of the region $\mathcal{R}$ beneath the graph of a nonnegative function f to be the value of the integral $\int_a^b f(x)\, dx$. Although we have calculated the areas of several regions in the preceding sections, our focus has been on how to evaluate integrals, not on regions and their areas. In this section we show how to calculate the areas of such regions as the cross section of the airfoil shown in Fig. 5.27. The upper and lower boundaries of the cross section are the graphs of functions f and g.

WEB

FIGURE 5.27 A representative cross section of an airfoil.

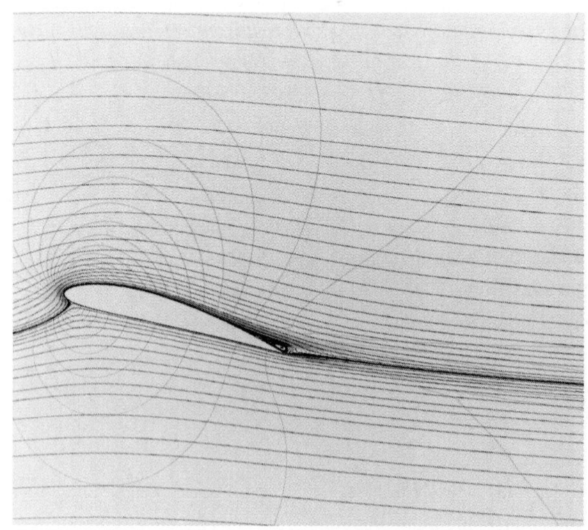

Related applications include the calculation of the mass, center of mass, and moment of inertia of planar areas with variable density. Other applications of the definite integral include finding the volumes of solids, the lengths of curves, and the work done by a variable force acting on an object moving on a curve. We study these applications in Chapter 6.

Area "Elements" Figure 5.28(a) shows the region $\mathcal{R}$ beneath the graph of a non-negative function f. The area of $\mathcal{R}$ is

$$A = \int_a^b f(x)\,dx = \lim_{n\to\infty} \sum_{i=1}^{n} f(c_i)\,\Delta x. \tag{1}$$

The points $c_1, c_2, \ldots, c_n$ in the Riemann sum are the points of an evaluation set for the subdivision $\mathcal{S}_n$. Figure 5.28(b) shows a representative point c_i in the subinterval $[x_{i-1}, x_i]$ and the associated rectangle with height $f(c_i)$ and width $\Delta x = (b - a)/n$.

For calculating the area of the region $\mathcal{R}$, Equation (1) relates the integral $\int_a^b f(x)\,dx$ to the underlying geometry. The reasoning leading to (1) can be broken down into three steps:

Step 1: Draw a representative rectangle with height $f(c_i)$ and width $\Delta x = (b - a)/n$. The area $f(c_i)\,\Delta x$ of this rectangle approximates the area beneath the graph of f and above the subinterval $[x_{i-1}, x_i]$.

Step 2: Form the Riemann sum $\Sigma_{i=1}^{n} f(c_i)\,\Delta x$, which sums the areas of the rectangles to approximate the area beneath the graph of f.

Step 3: Take the limit of these Riemann sums as $n \to \infty$. The value of this limit is $\int_a^b f(x)\,dx$, which is the area A of $\mathcal{R}$.

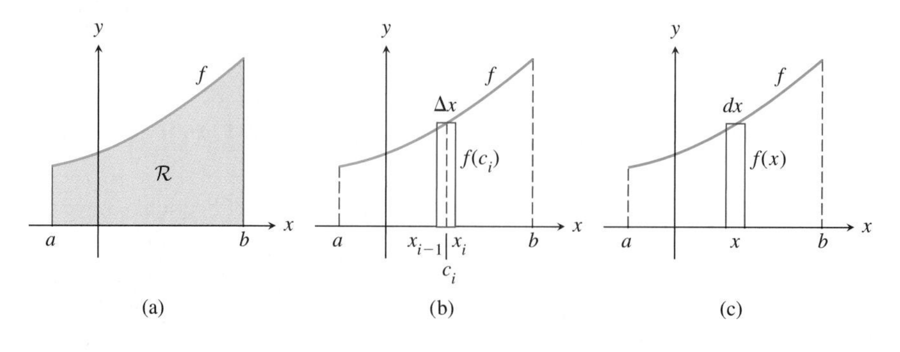

FIGURE 5.28 Viewing a Riemann sum as a sum of area elements.

Figure 5.28(c) shows a simplified representative rectangle with height $f(x)$, width dx, and area $dA = f(x)\,dx$. We shall often use the term *area element* to refer to either this rectangle or its area. We locate the area element with the representative point x instead of c_i. The area $dA = f(x)\,dx$ of the element is the integrand of the area integral $A = \int_a^b f(x)\,dx$.

If we regard the area element $dA = f(x)\,dx$ as part of a Riemann sum corresponding to a very large value of n, so that the width of the rectangle is very small, then the Riemann sum $\Sigma\, dA = \Sigma\, f(x)\,dx$ is very nearly equal to $\int_a^b f(x)\,dx$.

This brief notation can help us connect the geometry to the integral in fewer steps. Once we have identified $f(x)\,dx$ as the area of a representative rectangle, we can write

$$A = \int_a^b f(x)\,dx,$$

which in some sense sums these area elements. In future applications of the definite integral, our practice will be to identify a representative element of area $dA = f(x)\,dx$ and then, without writing a Riemann sum, to sum these elements with a definite integral. This practice compresses Steps 2 and 3 mentioned earlier.

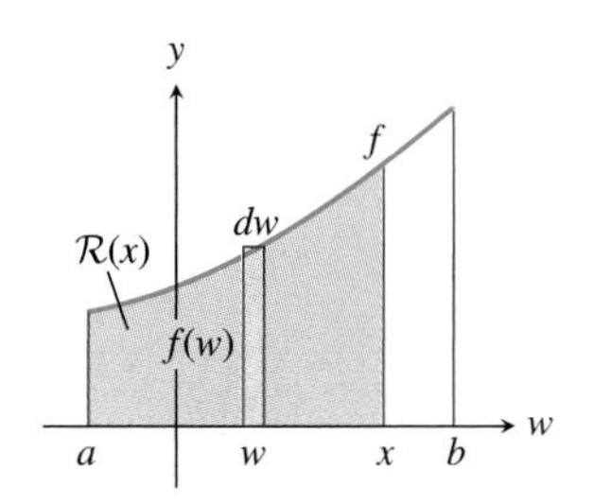

FIGURE 5.29 The "variable" region $\mathcal{R}(x)$ beneath the graph of f.

We comment briefly on the use of the differential notation dA in the area element $dA = f(x)\,dx$. Figure 5.29 shows a shaded region $\mathcal{R}(x)$ beneath the graph of $y = f(w)$. We use w as the integration variable because we want to use x for the right-hand boundary of $\mathcal{R}(x)$. The area $A(x)$ of $\mathcal{R}(x)$ is

$$A(x) = \int_a^x f(w)\,dw. \tag{2}$$

By Part I of the Fundamental Theorem of Calculus,

$$dA = \frac{dA}{dx}\,dx = \left(\frac{d}{dx}\int_a^x f(w)\,dw\right)dx = f(x)\,dx. \tag{3}$$

This calculation helps explain the use of dA for the area element $f(x)\,dx$. We also write $dA = y\,dx$ if we are thinking about the curve as described by the equation $y = f(x)$.

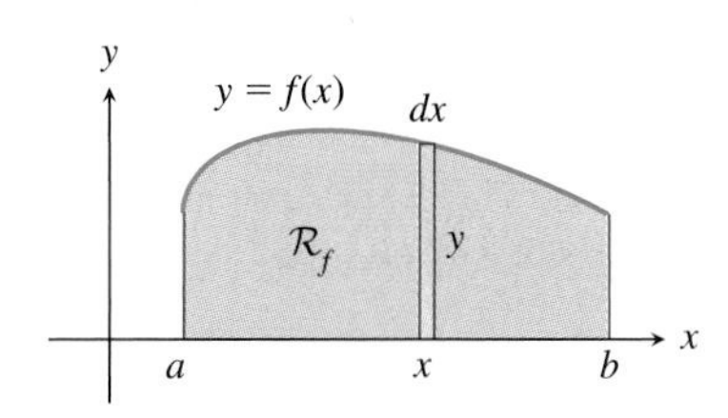

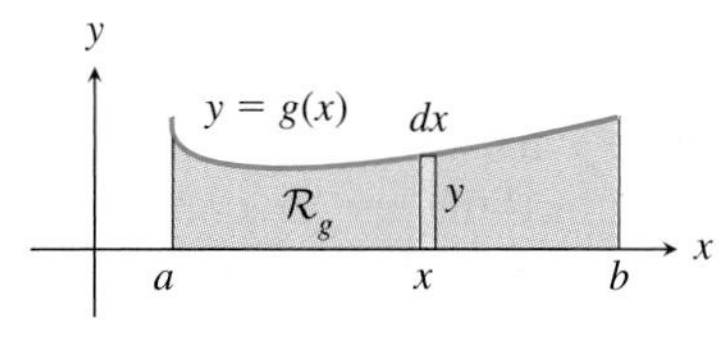

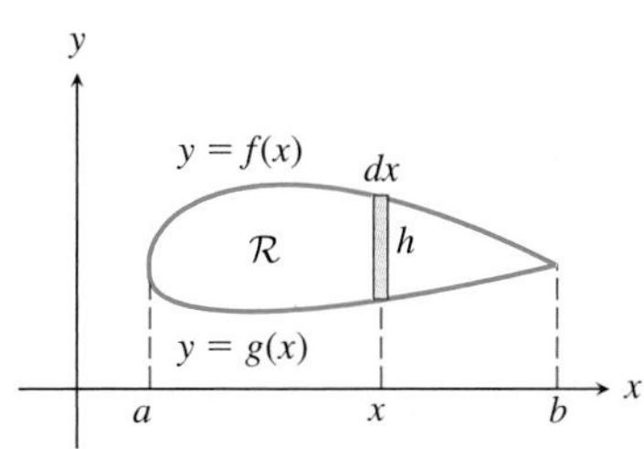

The area element is $dA = h(x)\,dx$, where $h(x) = f(x) - g(x)$.

FIGURE 5.30 Using elements of area in calculating the area of the region $\mathcal{R}$ between two graphs.

The Area of the Region between Two Curves The graph at the bottom of Fig. 5.30 shows a region $\mathcal{R}$ between the graphs of functions f and g. We use the idea of an element of area to calculate the area A of $\mathcal{R}$. Before we do this, we calculate A by subtraction, using what we already know. The area A is the difference between the areas A_f and A_g of the regions $\mathcal{R}_f$ and $\mathcal{R}_g$, where $\mathcal{R}_f$ is the region beneath the graph of f, and $\mathcal{R}_g$ is the region beneath the graph of g. These regions are shown in the Fig. 5.30. Thus

$$A = A_f - A_g = \int_a^b f(x)\,dx - \int_a^b g(x)\,dx. \tag{4}$$

We now calculate A more directly. The element of area shown in the graph at the bottom of Fig. 5.30 is a thin rectangle of height $h = h(x) = f(x) - g(x)$ and width dx. The coordinate variable x locates this element. The height h of the rectangle varies with x. The area of this element is

$$dA = h(x)\,dx = (f(x) - g(x))\,dx. \tag{5}$$

Summing these elements gives the area A of the region $\mathcal{R}$:

$$A = \int_a^b dA = \int_a^b h(x)\,dx = \int_a^b (f(x) - g(x))\,dx. \tag{6}$$

The expressions for A in Equations (4) and (6) are equal, not only because of our intuition-based notion of area, but because of the linearity property of integrals, Property 3 on page 370.

EXAMPLE 1 Calculate the area of the region between the graphs of the functions

$$f(x) = \tfrac{6}{5}x^{1/2} - \tfrac{1}{25}x^{3/2}, \qquad 0 \le x \le 30$$
$$g(x) = \tfrac{1}{30}x^{4/3} - x^{1/3}, \qquad 0 \le x \le 30.$$

Solution

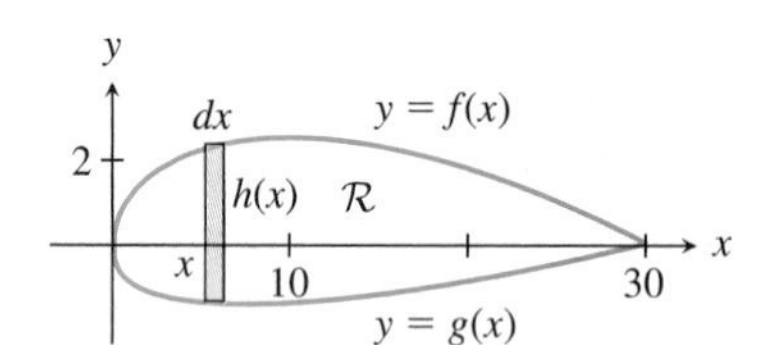

FIGURE 5.31 Using an element of area in calculating the area of the region $\mathcal{R}$ between two graphs.

Figure 5.31 shows the graphs of f and g and an area element $dA = h(x)\,dx$ located at x, where $h(x) = f(x) - g(x)$. Summing these elements gives the area A of the region $\mathcal{R}$:

$$\begin{aligned} A &= \int_a^b dA = \int_a^b h(x)\,dx = \int_a^b (f(x) - g(x))\,dx \\ &= \int_0^{30} \left(\tfrac{6}{5}x^{1/2} - \tfrac{1}{25}x^{3/2} - \tfrac{1}{30}x^{4/3} + x^{1/3}\right) dx \\ &= \tfrac{6}{5}\cdot\tfrac{2}{3}x^{3/2} - \tfrac{1}{25}\cdot\tfrac{2}{5}x^{5/2} - \tfrac{1}{30}\cdot\tfrac{3}{7} + \tfrac{3}{4}x^{4/3}\Big|_0^{30} \\ &= \tfrac{4}{5}30^{3/2} - \tfrac{2}{125}30^{5/2} - \tfrac{1}{70}30^{7/3} + \tfrac{3}{4}30^{4/3} \approx 82.544 \text{ square units.} \end{aligned}$$

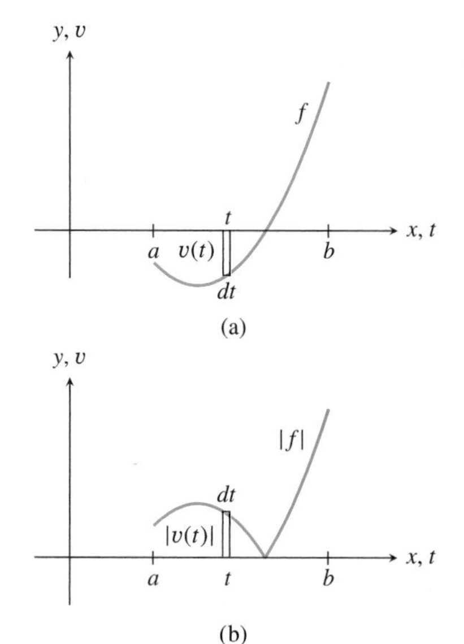

FIGURE 5.32 The graph of the functions f and $|f|$ on the interval $[a, b]$.

Interpreting $\int_a^b |f(x)|\,dx$ or $\int_a^b |v(t)|\,dt$ Figure 5.32(a) shows the graph of a function f defined on an interval $[a, b]$. The axes are labeled with x and y or with t and v. We first work with the labels x and y. In Section 5.2 we defined the *net area* of the region between the graph of f and the interval $[a, b]$ to be the value of the integral $\int_a^b f(x)\,dx$. In some applications we do not want net area, but rather the area of the region between the graph of f and the interval $[a, b]$ of the x-axis, without regard to whether some subregions have "negative area". Figure 5.32(b) shows the graph of the absolute value of the function f. Taking the absolute value converts subregions with "negative area" to regions with positive area and leaves unchanged subregions that have positive area. The total area of the region between the graph of f and the interval $[a, b]$ is $\int_a^b |f(x)|\,dx$.

The integrals $\int_a^b f(x)\,dx$ and $\int_a^b |f(x)|\,dx$ may also be interpreted in the context of the motion of a car moving along a highway during the time interval $[a, b]$. We'll refer to the highway as the w-axis. For this discussion we switch to the labels t and v in Fig. 5.32 and write $v(t) = f(t)$. First look at the velocity graph in Fig. 5.32(a). The element $v(t)\,dt$ is approximately the distance the object moved in time dt in case $v(t) \ge 0$ and is the negative of the distance the object moved in time dt in case $v(t) < 0$. Hence, the element $v(t)\,dt$ is the amount added to or subtracted from the odometer during the time dt (we assume that the odometer is reversible in this interpretation). Letting od (a) and od (b) denote the odometer readings at $t = a$ and $t = b$, the sum of od (a) and the (Riemann) sum $\Sigma\, v(t)\, dt$ approximates od (b). In the limit,

$$\text{od}\,(b) = \text{od}\,(a) + \int_a^b v(t)\, dt.$$

Next, we look at the graph in Fig. 5.32(b) and assume that the odometer is not reversible. This means that we are assuming that it will increase whether we are going in the positive or negative direction on the w-axis. The element $|v(t)|\, dt$ is the (nonnegative) amount added to the odometer during the time dt. Hence, the (Riemann) sum $\Sigma\, |v(t)|\, dt$ approximates the "total" distance D traveled by the car during the time interval $[a, b]$. In the limit,

$$D = \int_a^b |v(t)|\, dt.$$

EXAMPLE 2 The velocity of a car is $v(t) = 10t(t - 1)$ during the time interval $[0, 2]$, where distances are measured in kilometers and time in hours. Find the total distance traveled by the car during the two hours.

Solution

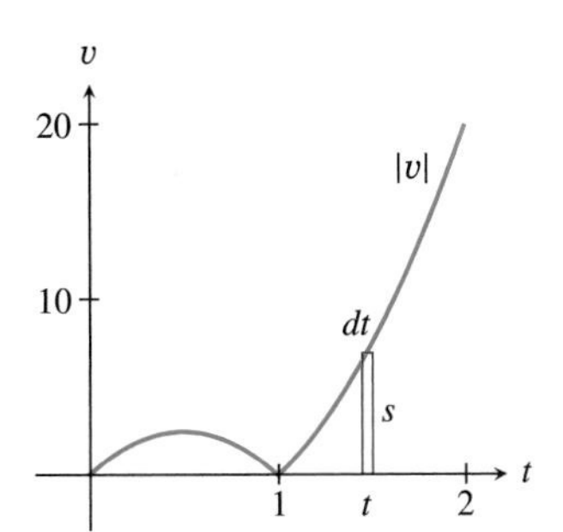

FIGURE 5.33 The graph of the speed $|v(t)|$ of the car for $0 \le t \le 2$.

Figure 5.33 shows the graph of the speed $|v(t)|$ of the car. The total distance traveled by the car during the two hours is $\int_0^2 |v(t)|\, dt$. We may calculate this integral by noticing that $v(t) \le 0$ for $0 \le t \le 1$ and $v(t) \ge 0$ for $1 \le t \le 2$. Hence, the speed $|v(t)|$ is given by

$$|v(t)| = \begin{cases} -10t(t - 1), & 0 \le t \le 1 \\ 10t(t - 1), & 1 \le t \le 2. \end{cases}$$

The total distance D traveled by the car during the two hours is

$$\begin{aligned} D &= \int_0^2 |v(t)|\, dt \\ &= \int_0^1 |v(t)|\, dt + \int_1^2 |v(t)|\, dt \\ &= \int_0^1 (-10t(t - 1))\, dt + \int_1^2 (10t(t - 1))\, dt \\ &= -10\int_0^1 (t^2 - t)\, dt + 10\int_1^2 (t^2 - t)\, dt. \end{aligned}$$

Evaluating these integrals,

$$\begin{aligned} D &= -10\left(\frac{1}{3}t^3 - \frac{1}{2}t^2\right)\Bigg|_0^1 + 10\left(\frac{1}{3}t^3 - \frac{1}{2}t^2\right)\Bigg|_1^2 \\ &= -10\left(\frac{1}{3} - \frac{1}{2}\right) + 10\left(\frac{8}{3} - 2 - \left(\frac{1}{3} - \frac{1}{2}\right)\right) = 10 \text{ kilometers.} \end{aligned}$$

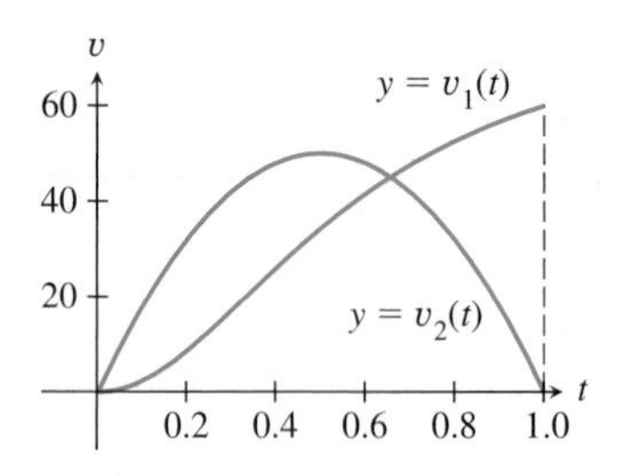

FIGURE 5.34 Velocity profiles of two cars in a race.

EXAMPLE 3 During a one-hour race down a straight road the velocities of two cars were

$$v_1(t) = 80\left(1 - \frac{1}{3t^2 + 1}\right) \text{ mph}, \qquad 0 \le t \le 1$$

$$v_2(t) = 200(t - t^2) \text{ mph}, \qquad 0 \le t \le 1.$$

Figure 5.34 shows the graphs of v_1 and v_2. If both cars were at the mile 0 mark at the beginning of the race, which car won?

Solution

Car 1 started slowly but accelerated throughout the hour; car 2 started fast but had engine trouble halfway through the race. Before you look at the solution, but after you have inspected Fig. 5.34, which car, in your judgment, won? If you looked at the regions between the two graphs and judged that the area of the region in which car 2 is going faster is larger than the area of the region in which car 1 is going faster, you have a good eye.

Because the velocities of the two cars are nonnegative, the distances d_1 and d_2 they have traveled up to $t = 1$ will determine which car won the race. The distances d_1 and d_2 are equal to the areas beneath the graphs of v_1 and v_2, between $t = 0$ and $t = 1$. We calculate $d_1 - d_2$.

$$d_1 - d_2 = \int_0^1 80\left(1 - \frac{1}{3t^2 + 1}\right) dt - \int_0^1 200(t - t^2)\, dt.$$

Using formula (A) and anticipating the use of formula (G),

$$d_1 - d_2 = \left(80t - 100t^2 + \frac{200}{3}t^3\right)\Bigg|_0^1 - \frac{80}{3}\int_0^1 \frac{dt}{t^2 + \left(1/\sqrt{3}\right)^2}.$$

Evaluating and using formula (G),

$$d_1 - d_2 = 80 - 100 + \frac{200}{3} - \left(\frac{80}{3}\sqrt{3}\arctan\left(\sqrt{3}t\right)\right)\Bigg|_0^1$$

$$= -20 + \frac{200}{3} - \frac{80}{3}\sqrt{3}\arctan\sqrt{3} \approx -1.7 \text{ miles.}$$

Hence, $d_1 - d_2 < 0$, which means that $d_1 < d_2$. Car 2 won the race.

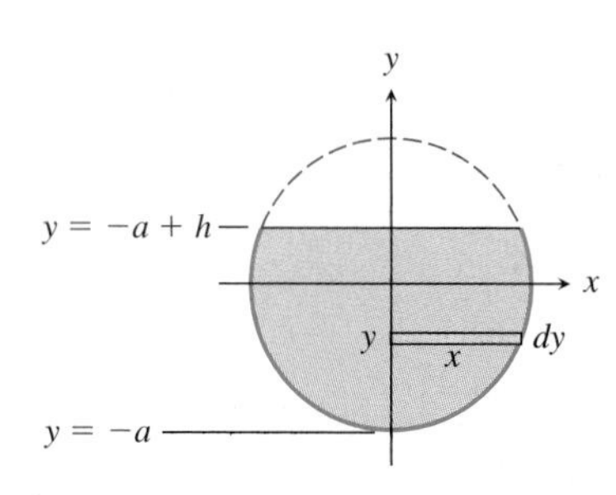

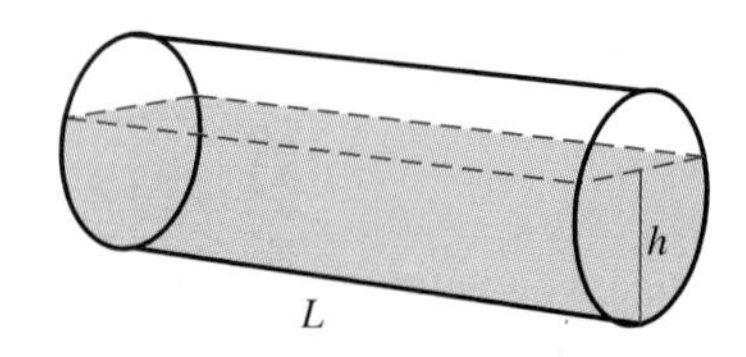

FIGURE 5.35 Finding the volume of gasoline in a cylindrical tank, which reduces to finding the area of a segment of a circle.

Horizontal Area Elements

So far, we have used vertical area elements with width dx in calculating the areas of regions. For some regions it is more natural to use horizontal area elements, with width dy. Suppose, for example, that we are asked to calculate the volume of the gasoline contained in the horizontal tank shown at the bottom of Fig. 5.35. If the depth of the gasoline is h, the volume of the tank will be $V = L \cdot A(h)$, where $A(h)$ is the area of the shaded region (called a *segment*) shown at the top of the figure and L is the length of the tank.

We use the horizontal element shown in the shaded region to calculate the area of this region. Given a representative area element located at y, where h is the depth

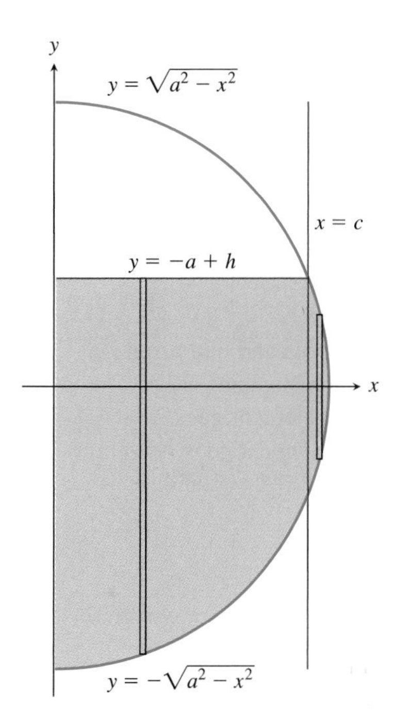

FIGURE 5.36 Zoom-view of part of Fig. 5.35, showing why two integrals would be needed if vertical area elements were used.

of the gasoline and $-a \le y \le -a + h$, the length x of this element satisfies the equation $x^2 + y^2 = a^2$. Solving this equation for x, the area dA of the element is

$$dA = x\,dy = \sqrt{a^2 - y^2}\,dy.$$

Summing these elements, the area of the right half of the shaded region is $\int_{-a}^{-a+h} dA$. Hence, by symmetry the area $A(h)$ of the entire shaded region is

$$A(h) = 2\int_{-a}^{-a+h} dA = 2\int_{-a}^{-a+h} x\,dy = 2\int_{-a}^{-a+h} \sqrt{a^2 - y^2}\,dy.$$

Replacing x by y in formula (7) in the table of integrals (see inside back cover),

$$\begin{aligned} A &= 2\left(\frac{y}{2}\sqrt{a^2 - y^2} + \frac{a^2}{2}\arcsin\frac{y}{a}\right)\Bigg|_{-a}^{-a+h} \\ &= 2\left(\frac{-a+h}{2}\sqrt{a^2 - (-a+h)^2} + \frac{a^2}{2}\arcsin\left(\frac{-a+h}{a}\right)\right) \\ &\quad - 2\left(\frac{-a}{2}\cdot 0 + \frac{a^2}{2}\arcsin(-1)\right) \\ &= (h-a)\sqrt{2ah - h^2} + a^2\arcsin\left(\frac{h-a}{a}\right) - a^2\left(\frac{-\pi}{2}\right) \\ &= \tfrac{1}{2}\pi a^2 + (h-a)\sqrt{2ah - h^2} + a^2\arcsin\left(\frac{h-a}{a}\right). \end{aligned}$$

If we had used a vertical area element, the area A of the shaded region would have been

$$\begin{aligned} A = 2\Bigg(&\int_0^c \left(-a + h - \left(-\sqrt{a^2 - x^2}\right)\right)dx \\ &+ \int_c^a \left(\sqrt{a^2 - x^2} - \left(-\sqrt{a^2 - x^2}\right)\right)dx\Bigg), \end{aligned} \tag{7}$$

where c is the x-coordinate of the point of intersection of the circle and the line $y = -a + h$. Two integrals would be required because, as the zoom-view in Fig. 5.36 shows, to the left of the line $x = c$ the element of area stretches from the bottom half of the circle to the line $y = -a + h$, while to the right of $x = c$ the element stretches from the lower half of the circle to the upper half. Using a horizontal element avoids this complication.

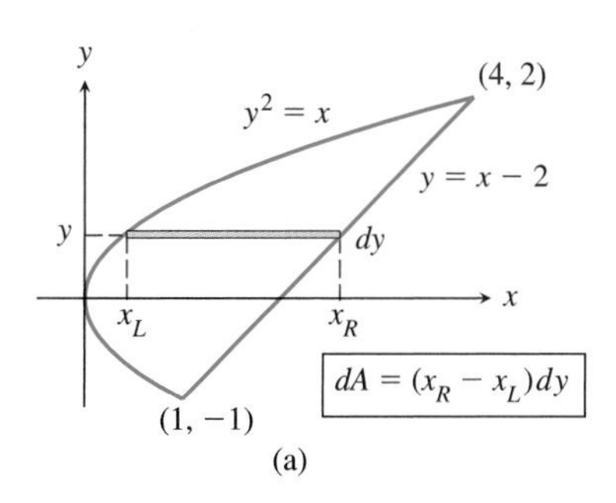

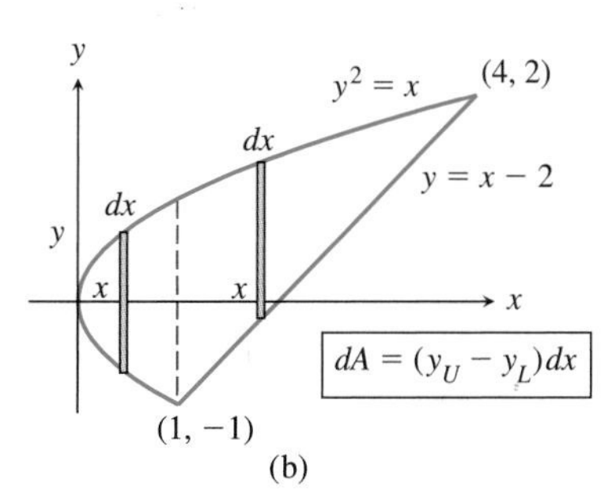

FIGURE 5.37 Calculating an area by using either horizontal or vertical area elements.

EXAMPLE 4 Use both horizontal and vertical area elements to calculate the area of the region bounded by the graphs of the equations $y^2 = x$ and $y = x - 2$. See Fig. 5.37.

Solution

Figure 5.37 shows the coordinates $(4, 2)$ and $(1, -1)$ of the points of intersection of the parabola and line described by the equations $y^2 = x$ and $y = x - 2$. These coordinates were found by solving simultaneously the equations describing the two curves.

We first calculate the area bounded by the parabola and line by using a horizontal area element, as in Fig. 5.37(a). The rectangular element is dy by

$x_R - x_L$, where x_R is "the x of the line" and x_L is "the x of the parabola." The subscripts R and L are intended to suggest "right" and "left." The area of this element is

$$dA = (x_R - x_L)\,dy = ((y + 2) - y^2)\,dy.$$

Using the integral to sum these elements,

$$A = \int_{-1}^{2} dA = \int_{-1}^{2} ((y + 2) - y^2)\,dy = \tfrac{1}{2}y^2 + 2y - \tfrac{1}{3}y^3 \Big|_{-1}^{2} = \frac{9}{2}.$$

Next we calculate the area bounded by the parabola and line using vertical area elements. Figure 5.37(b) shows two such elements. Two elements are needed because to the left of $x = 1$, the element goes from parabola to parabola, while to the right of $x = 1$ the element goes from line to parabola. The formula for dA will be different for $0 \le x \le 1$ and $1 \le x \le 4$. We write

$$dA = (y_U - y_L)\,dx$$

for both cases, where y_U denotes the upper curve and y_L the lower curve. To the left of the line $x = 1$ the upper curve is the parabola with equation $y = \sqrt{x}$ and the lower curve is the parabola with equation $y = -\sqrt{x}$. To the right of the line $x = 1$ the upper curve is the parabola with equation $y = \sqrt{x}$ and the lower curve is the line with equation $y = x - 2$. Thus,

$$\begin{aligned} dA &= (y_U - y_L)\,dx = \left(\sqrt{x} - \left(-\sqrt{x}\right)\right)dx, \quad \text{for } 0 \le x \le 1 \\ dA &= (y_U - y_L)\,dx = \left(\sqrt{x} - (x - 2)\right)dx, \quad \text{for } 1 \le x \le 4. \end{aligned} \tag{8}$$

The area of the entire parabolic segment is the sum of the areas of the two regions separated by the line with equation $x = 1$. From (8),

$$A = \int_0^1 \left(\sqrt{x} - \left(-\sqrt{x}\right)\right)dx + \int_1^4 \left(\sqrt{x} - (x - 2)\right)dx = \frac{9}{2}.$$

Exercises 5.6

Exercises 1–24: Sketch the bounded region between the curves and calculate its area. Use substitution or integral tables as appropriate.

1. $y = (x - 2)^2$, $y = 2x + 4$
2. $y = x^2$, $y = 2x$
3. $y = \sqrt{x}$, $y = \sqrt[3]{x}$, where $x \ge 0$
4. $y = \sqrt[3]{x}$, $y = \sqrt[4]{x}$, where $x \ge 0$
5. $y = x^2 + 1$, $y = 2x + 1$, where $0 \le x \le 2$
6. $y = x^2 + 2x$, $y = x/2$
7. $x + 1 = y^2$, $x = y + 1$
8. $x + 2 = 2y^2$, $x = y + 1$
9. $y = x^5 - x^3$, $y = x - x^3$, where $0 \le x \le 1$
10. $y = x - x^5$, $y = x^2 - x^4$, where $0 \le x \le 1$
11. $x = 3y + 2$, $x = 2y^2$
12. $x = -y^2$, $x = 4 - 2y^2$
13. $x = y^4$, $x = (y - 2)^2$
14. $x = y^4$, $4x = (y - 1)^2$
15. $y = 2x + 1$, $y = e^x$
16. $y = 2x$, $y = \tan x$, where $x \in [0, \pi/2)$
17. $x = y^2$, $x = \frac{1}{2}y^2 + 2$
18. $y = x^2$, $(y - 2)^2 = x - 1$
19. $x^2 + y^2 = 4$, $xy = 1$, where $x > 0$

20. $x^2 + y^2 = 4$, $y = 1/x^2$, where $x > 0$

21. $y = 1/(x^2 + 1)$, $y = x^2$

22. $y = 1/(x^2 + 1)$, $y = x^4$

23. $x^2 + y^2 = 1$, $(x - 1.1)^2 + y^2 = 0.5^2$

24. $x^2 - y^2 = 1$, $x^2 + y^2 = 2^2$, where $x \geq 1$

Exercises 25–28: The velocity of a car on a highway is given. Calculate the final odometer reading at the end of the time interval if it is assumed that the odometer starts at 0 and is reversible.

25. $v(t) = 30(t - 1)(t - 2)$ kilometers per hour on the time interval $[0, 2]$ hours.

26. $v(t) = 30(t - \sqrt{t})$ kilometers per hour on the time interval $[0, 4]$ hours.

27. $v(t) = 75 \cos t$ kilometers per hour on the time interval $[0, 2]$ hours.

28. $v(t) = 100 \sin t - 50$ kilometers per hour on the time interval $[0, 2]$ hours.

Exercises 29–32: The velocity of a car on a highway is given. Calculate the total distance traveled by the car during the time interval.

29. $v(t) = 30(t - 1)(t - 2)$ kilometers per hour on the time interval $[0, 2]$ hours.

30. $v(t) = 30(t - \sqrt{t})$ kilometers per hour on the time interval $[0, 4]$ hours.

31. $v(t) = 75 \cos t$ kilometers per hour on the time interval $[0, 2]$ hours.

32. $v(t) = 100 \sin t - 50$ kilometers per hour on the time interval $[0, 2]$ hours.

33. During a one-hour race, the velocities of two cars are

$$v_1(t) = 75\left(1 - \frac{1}{2t^2 + 1}\right),$$
$$v_2(t) = 185(t - t^2),$$

$0 \leq t \leq 1$. If at the beginning of the race both cars were at the mile 0 mark, which car won?

34. During a one-hour race, the velocities of two cars are

$$v_1(t) = 40(1 - \cos(\pi t)),$$
$$v_2(t) = 75t,$$

$0 \leq t \leq 1$. If at the beginning of the race both cars were at the mile 0 mark, which car won?

T 35. Referring to the accompanying figure, a parabolic region is removed from a copper disk with radius 4 cm and thickness 0.5 cm. Determine the mass of the remaining part of the disk, given that the density of copper is 8920 kg/m^3 and the equation of the parabola is $y = -0.4x^2 + 3.2$.

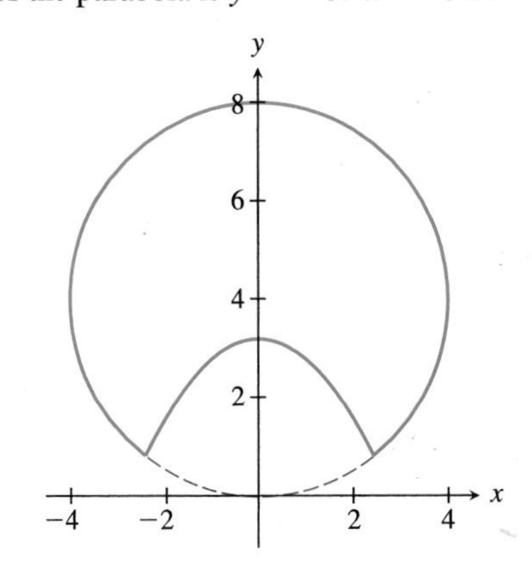

Figure for Exercise 35.

5.7 Integration by Parts

Newton determined empirically that if an object is placed into an environment held at a constant temperature A, the rate of change of the temperature of the object is proportional to the difference between the temperature A and the object's temperature. Letting $T(t)$ be the temperature of the object at time t, Newton's Law of Cooling is

$$T'(t) = c(A - T(t)), \tag{1}$$

where c is a constant depending on the physical properties of the object. The temperature of the environment is often called the *ambient* temperature. As we show in Section 5.9, the differential equation (1) can be solved for the function $T = T(t)$ if we can find the antiderivative

$$\int \frac{dT}{A - T}.$$

FIGURE 5.38 An array of solar collectors.

Making the substitution $u = A - T$,

$$du = -dT \quad \text{and} \quad \int \frac{dT}{A - T} = \int \frac{-du}{u} = -\ln|u| \bigg|_{u=A-T} = -\ln|A - T|.$$

If the temperature A of the environment is a function $A = A(t)$ of time t, Equation (1) holds but finding an antiderivative may not be as easy as for the constant ambient temperature case. If, for example, we were studying the temperature variation of a solar collector we might use a function like

$$A(t) = 65 + 10\cos\left(\tfrac{1}{12}\pi t\right) - 5\cos\left(\tfrac{1}{6}\pi t\right) - 15\sin\left(\tfrac{1}{12}\pi t\right), \qquad 0 \le t \le 24, \quad (2)$$

to model the temperature of the air during a 24-hour period. A graph of A is shown in Fig. 5.39.

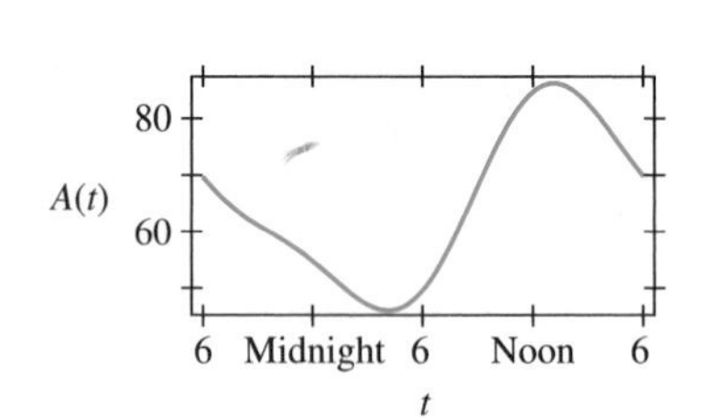

FIGURE 5.39 Air temperature in degrees Fahrenheit during a 24-hour period, from 6 P.M. to 6 P.M.

In the differential equations course that often follows calculus it is shown that the differential equation

$$T'(t) = c(A(t) - T(t)),$$

where $A = A(t)$ is defined in (2), can be solved if we can find the antiderivative of

$$\int e^{kt}A(t)\,dt = \int e^{kt}\left(65 + 10\cos\left(\tfrac{1}{12}\pi t\right) - 5\cos\left(\tfrac{1}{6}\pi t\right) - 15\sin\left(\tfrac{1}{12}\pi t\right)\right)dt. \quad (3)$$

This antiderivative can be found using *integration by parts.* See Examples 4 and 5.

Integration by parts is based on the product rule

$$(uv)' = uv' + vu',$$

where $u = u(x)$ and $v = v(x)$ are differentiable functions of x on an interval $[a, b]$. From the viewpoint of the Fundamental Theorem of Calculus, the product formula states that uv is an antiderivative of $uv' + vu'$. Hence

$$\int_a^b (uv' + vu')\,dx = \int_a^b (uv)'\,dx = uv\bigg|_a^b.$$

Using the linearity of the integral, this result may be rearranged as

$$\int_a^b uv'\,dx = uv\bigg|_a^b - \int_a^b vu'\,dx,$$

or, in terms of the differentials $du = u'\,dx$ and $dv = v'\,dx$,

$$\int_a^b u\,dv = uv\Big|_a^b - \int_a^b v\,du.$$

This result is called the *integration-by-parts formula (for definite integrals).* For convenient reference, we highlight in (4) and (5) this formula together with the corresponding formula for indefinite integrals.

Integration by Parts

If $u = u(x)$ and $v = v(x)$ are differentiable on $[a, b]$, then

$$\int_a^b u\,dv = uv\Big|_a^b - \int_a^b v\,du \tag{4}$$

$$\int u\,dv = uv - \int v\,du. \tag{5}$$

We give Newton's geometric demonstration of formula (4) at the end of the section.

To evaluate an integral $\int_a^b f(x)\,dx$ using integration by parts we express $f(x)\,dx$ as a product of "parts" u and dv, chosen so that $u\,dv = f(x)\,dx$. We may then use (4) to write

$$\int_a^b f(x)\,dx = uv\Big|_a^b - \int_a^b v\,du.$$

This equation gives an alternative to evaluating the integral on the left. Instead of trying to find an antiderivative $F(x)$ of $f(x)$, we can look for an antiderivative of $v(x)u'(x)$, which might be easier.

EXAMPLE 1 The probability $P(t)$ that a certain component of a system will not fail for t time units is governed by the *gamma distribution*

$$P(t) = \int_0^t \frac{1}{(\alpha - 1)!}x^{\alpha-1}e^{-x}\,dx,$$

where α depends upon the characteristics of the part. We consider a component for which $\alpha = 2$. The probability that such a component will not fail for 3 time units is

$$\int_0^3 xe^{-x}\,dx.$$

Evaluate this integral.

Solution

The key step in using integration by parts is to set the given integrand equal to $u\,dv$ in (4) and then decide which part should be u and which part should be dv. It is sometimes useful to set u equal to any polynomial factor of the integrand because in the new integral, $\int_a^b v\,du$, the factor $du = u'\,dx$ is a polynomial of degree 1 less than the degree of u. If, accordingly, we set $u = x$, then dv must be everything else, that is, $dv = e^{-x}dx$. From u and dv we calculate du and v. To find u, we differentiate:

$$\text{From } u = x, \qquad du = dx.$$

To find v, we integrate:

$$\text{From } dv = e^{-x}dx, \qquad v = \int dv = \int e^{-x}dx = -e^{-x}.$$

From (4),

$$\begin{aligned}\int_0^3 xe^{-x}dx &= uv\Big|_0^3 - \int_0^3 v\,du \\ &= x(-e^{-x})\Big|_0^3 - \int_0^3 (-e^{-x})\,dx = -3e^{-3} - e^{-x}\Big|_0^3 \\ &= -3e^{-3} - (e^{-3} - 1) = 1 - 4e^{-3} \approx 0.80.\end{aligned}$$

In Example 1 we set u equal to the polynomial factor. As Example 2 shows, this doesn't always work.

EXAMPLE 2 Evaluate the integral $\int x \ln x\,dx$.

Solution

If we were to set $u = x$ and $dv = \ln x\,dx$, then $du = dx$ and we must evaluate $\int \ln x\,dx$ to find v. This appears to be no easier than evaluating $\int x \ln x\,dx$, our original problem. A different choice of parts is $u = \ln x$ and $dv = x\,dx$. In this case, $v = \int x\,dx = \frac{1}{2}x^2$. The differential du is

$$du = u'\,dx = (\ln x)'\,dx = \frac{1}{x}\,dx.$$

Substituting these results into (5),

$$\begin{aligned}\int x \ln x\,dx &= uv - \int v\,du \\ &= (\ln x)\left(\tfrac{1}{2}x^2\right) - \int \tfrac{1}{2}x^2 \cdot \frac{1}{x}\,dx \\ &= \tfrac{1}{2}x^2 \ln x - \tfrac{1}{4}x^2 + C \\ &= \tfrac{1}{4}x^2(2 \ln x - 1) + C.\end{aligned}$$

The essential feature of the parts assignment in the last example is that the derivative u' of the natural logarithm function is an algebraic function, namely, $1/x$. The product of this algebraic function and v gives an expression whose antiderivative is known. Because the derivatives of the functions arcsin and arctan are also algebraic, we could use a similar choice of parts for the integrals

$$\int x^n \arcsin x\,dx \qquad \text{or} \qquad \int x^n \arctan x\,dx,$$

that is, let $u = \arcsin x$ or $u = \arctan x$ and set dv equal to everything else.

Reduction Formulas

Integration by parts is often used to find reduction formulas for integrals. An example of a reduction formula is

$$\int_0^\pi \sin^n x\,dx = \frac{n-1}{n}\int_0^\pi \sin^{n-2} x\,dx, \qquad n \geq 2. \tag{6}$$

Reduction formulas are used to evaluate integrals by applying the formula as many times as necessary to "reduce" the integral to one with a known antiderivative. Suppose, for example, we wish to evaluate the integral

$$\int_0^\pi \sin^5 x\,dx.$$

Setting $n = 5$ in (6),

$$\int_0^\pi \sin^5 x\,dx = \frac{4}{5}\cdot\int_0^\pi \sin^3 x\,dx.$$

To continue the reduction to lower powers of the sine function, we apply (6) again, this time with $n = 3$. This gives

$$\int_0^\pi \sin^5 x\,dx = \frac{4}{5}\left(\frac{2}{3}\cdot\int_0^\pi \sin x\,dx\right) = \frac{8}{15}\int_0^\pi \sin x\,dx.$$

Hence,

$$\int_0^\pi \sin^5 x\,dx = \frac{8}{15}\cdot(-\cos x)\Big|_0^\pi = \frac{8}{15}\cdot 2 = \frac{16}{15}.$$

Formulas (25)–(37) in the table of integrals (see inside back cover) are reduction formulas. Each may be derived by integration by parts, except for formula (30), which is based on the identity $\tan^2 x + 1 = \sec^2 x$.

EXAMPLE 3 Use integration by parts to derive (6), which is formula (28) in the table of integrals.

Solution

We write the integral as $\int_0^\pi \sin^n x\,dx = \int_0^\pi \sin^{n-1} x \sin x\,dx$ and choose the parts $u = \sin^{n-1} x$ and $dv = \sin x\,dx$. This choice, whatever the value of $n \geq 2$, has the advantage that $\int dv$ can be evaluated. Calculating

$$du = (n-1)\sin^{n-2} x \cos x\,dx \quad \text{and} \quad v = \int \sin x\,dx = -\cos x$$

and substituting into (4),

$$\int_0^\pi \sin^n x\,dx = uv\Big|_0^\pi - \int_0^\pi v\,du$$

$$= -\sin^{n-1}x\cos x\Big|_0^\pi + (n-1)\int_0^\pi \sin^{n-2}x\cos^2 x\,dx.$$

Evaluating the first term on the right and replacing $\cos^2 x$ by $1 - \sin^2 x$ in the second term,

$$\int_0^\pi \sin^n x\,dx = 0 + (n-1)\int_0^\pi (\sin^{n-2}x - \sin^n x)\,dx.$$

This result may be written in the form

$$\int_0^\pi \sin^n x\,dx = (n-1)\int_0^\pi \sin^{n-2}x\,dx - (n-1)\int_0^\pi \sin^n x\,dx.$$

Letting $F_n = \displaystyle\int_0^\pi \sin^n x\,dx$, this equation can be written as

$$F_n = (n-1)\int_0^\pi \sin^{n-2}x\,dx - (n-1)F_n.$$

Solving for F_n,

$$(1 + (n-1))F_n = (n-1)\int_0^\pi \sin^{n-2}x\,dx.$$

Dividing by n,

$$F_n = \int_0^\pi \sin^n x\,dx = \frac{n-1}{n}\int_0^\pi \sin^{n-2}x\,dx, \qquad n \geq 2. \tag{7}$$

This is (6).

Recursion Formulas

We state the reduction formula (6) as a *recursion formula*. This form of (6) is more convenient for calculation. Using the notation $F_n = \int_0^\pi \sin^n x\,dx$ from equation (7) in Example 3 and defining the "starting values"

$$F_0 = \int_0^\pi \sin^0 x\,dx = \int_0^\pi dx = \pi \quad\text{and}\quad F_1 = \int_0^\pi \sin x\,dx = -\cos x\Big|_0^\pi = 2,$$

the reduction formula (6) can be stated as

$$\begin{aligned} F_0 &= \pi \\ F_1 &= 2 \\ F_n &= \frac{n-1}{n} F_{n-2}, \qquad n \geq 2. \end{aligned} \tag{8}$$

The three equations listed in (8) can be entered more or less as given into a calculator or CAS. This makes it possible to evaluate integrals like $\int_0^\pi \sin^n x\, dx$ easily. On our calculator, it took less than 1 second to calculate

$$\int_0^\pi \sin^{20} x\, dx = F_{20} = \frac{19}{20} \cdot \frac{17}{18} \cdot \cdots \cdot \frac{5}{6} \cdot \frac{3}{4} \cdot \frac{1}{2} \cdot \pi \approx 0.55354.$$

More Integration by Parts

In the derivation of the reduction formula (6), integration by parts led to an equation in which it was possible to solve for the unknown integral. This idea is used in the next example, in which we derive formulas needed in solving Newton's Law of Cooling in the case where the ambient temperature A is not a constant.

EXAMPLE 4 Derive the integration formulas

$$\int e^{ax} \sin bx\, dx = \frac{e^{ax}}{a^2 + b^2}(a \sin bx - b \cos bx) \tag{9}$$

and

$$\int e^{ax} \cos bx\, dx = \frac{e^{ax}}{a^2 + b^2}(a \cos bx + b \sin bx), \tag{10}$$

where a and b are nonzero constants. These formulas are listed in the back of this text as formulas (23) and (24).

Solution

For formula (9), we start by choosing $u = e^{ax}$ and $dv = \sin bx\, dx$ in the integration-by-parts formula $\int u\, dv = uv - \int v\, du$. Calculating du and v,

$$du = u'\, dx = ae^{ax}\, dx \quad \text{and} \quad v = \int dv = \int \sin bx\, dx = -\frac{1}{b} \cos bx.$$

Substituting into the integration-by-parts formula,

$$\int e^{ax} \sin bx\, dx = -\frac{1}{b} e^{ax} \cos bx + \frac{a}{b} \int e^{ax} \cos bx\, dx. \tag{11}$$

The integral $\int e^{ax} \cos bx\, dx$ resembles the integral $\int e^{ax} \sin bx\, dx$ in (9). If we apply the integration-by-parts formula again, perhaps we will obtain an equation that can be solved for the original integral. Applying the integration-by-parts formula to $\int e^{ax}\cos bx\,dx$, we choose $u = e^{ax}$ and $dv = \cos bx$. This gives

$$du = ae^{ax}\,dx \quad \text{and} \quad v = \frac{1}{b}\sin bx.$$

From (11) we then have

$$\int e^{ax}\sin bx\,dx = -\frac{1}{b}e^{ax}\cos bx + \frac{a}{b}\left(\frac{1}{b}e^{ax}\sin bx - \frac{a}{b}\int e^{ax}\sin bx\,dx\right)$$

$$\int e^{ax}\sin bx\,dx = -\frac{1}{b}e^{ax}\cos bx + \frac{a}{b^2}e^{ax}\sin bx - \frac{a^2}{b^2}\int e^{ax}\sin bx\,dx.$$

Solving the last equation for the "unknown" $\int e^{ax}\sin bx\,dx$,

$$\left(1 + \frac{a^2}{b^2}\right)\int e^{ax}\sin bx\,dx = \frac{e^{ax}}{b^2}(a\sin bx - b\cos bx).$$

Dividing both sides of this equation by $(1 + a^2/b^2)$ and simplifying,

$$\int e^{ax}\sin bx\,dx = \frac{e^{ax}}{a^2 + b^2}(a\sin bx - b\cos bx).$$

Formula (9) now follows. Formula (10) follows from (9) by using (11). See Exercise 17.

We use formulas (9) and (10) to evaluate the integral

$$\int e^{kt}A(t)\,dt = \int e^{kt}\left(65 + 10\cos\left(\tfrac{1}{12}\pi t\right) - 5\cos\left(\tfrac{1}{6}\pi t\right) - 15\sin\left(\tfrac{1}{12}\pi t\right)\right)dt. \quad (12)$$

This is Equation (3), which we discussed at the beginning of the section in connection with Newton's Law of Cooling. Although the application of formulas (9) and (10) to (12) is straightforward, the calculation is quite lengthy. We use a CAS in the next example.

EXAMPLE 5 Evaluate the integral (12) using a convenient computer algebra system. Take $k = 1$.

Solution

Figure 5.40 shows the results of a CAS calculation. The ambient temperature function $A(t)$ from (12) was entered in input line 1. The antiderivative of $e^tA(t)$ was calculated in input line 2; the result, named "anti," is given in output line 2. From the complex numbers appearing in the denominator of the last three fractions of anti, we conclude that during the integration the CAS used complex numbers and, in reporting the result, did not simplify. If we do a "FullSimplify" on anti, the CAS returns output line 3. The CAS found that $(a + ib)(a - ib) = a^2 + b^2$ in its simplification.

```
In[1]:=A[t_]:=65+10Cos[Pi*t/12]-5Cos[Pi*t/6]-15Sin[Pi*t/12]
In[2]:=anti=Integrate[Exp[t]*A[t],t]
```

$$\text{Out[2]}= 65e^t + \frac{180e^t(8+\pi)\cos\left(\frac{\pi t}{12}\right)}{(-12i+\pi)(12i+\pi)} - \frac{180e^t\cos\left(\frac{\pi t}{6}\right)}{(-6i+\pi)(6i+\pi)} + \frac{120e^t(-18+\pi)\sin\left(\frac{\pi t}{12}\right)}{(-12i+\pi)(12i+\pi)} - \frac{30e^t\pi\sin\left(\frac{\pi t}{6}\right)}{(-6i+\pi)(6i+\pi)}$$

```
In[3]:=FullSimplify[anti]
```

$$\text{Out[3]}= 5e^t\left(13 + \frac{36(8+\pi)\cos\left(\frac{\pi t}{12}\right)}{144+\pi^2} + \frac{24(-18+\pi)\sin\left(\frac{\pi t}{12}\right)}{144+\pi^2} - \frac{6\left(6\cos\left(\frac{\pi t}{6}\right)+\pi\sin\left(\frac{\pi t}{6}\right)\right)}{36+\pi^2}\right)$$

FIGURE 5.40 An antiderivative calculation using Mathematica, a computer algebra system.

For some integrals $\int f(x)\,dx$, the right choice of parts is to take $u = f(x)$ and $dv = dx$. This can be useful when the factor $f'(x)$ in $du = f'(x)\,dx$ is algebraic.

EXAMPLE 6 Evaluate the integral $\displaystyle\int \arcsin x\,dx$.

Solution

If we choose $u = \arcsin x$ and $dv = dx$ in the integration-by-parts formula, then

$$du = \frac{dx}{\sqrt{1-x^2}} \quad \text{and} \quad v = x.$$

Hence,

$$\int \arcsin x\,dx = uv - \int v\,du = x \arcsin x - \int \frac{x\,dx}{\sqrt{1-x^2}}.$$

Making the substitution $w = 1 - x^2$ in the second integral, $dw = -2x\,dx$ and

$$\begin{aligned}\int \arcsin x\,dx &= x \arcsin x - \int \frac{-\frac{1}{2}\,dw}{w^{1/2}}\\ &= x \arcsin x + \tfrac{1}{2}\int w^{-1/2}\,dw\\ &= x \arcsin x + \tfrac{1}{2}\cdot 2w^{1/2}\Big|_{w=1-x^2}\\ &= x \arcsin x + \sqrt{1-x^2} + C.\end{aligned}$$

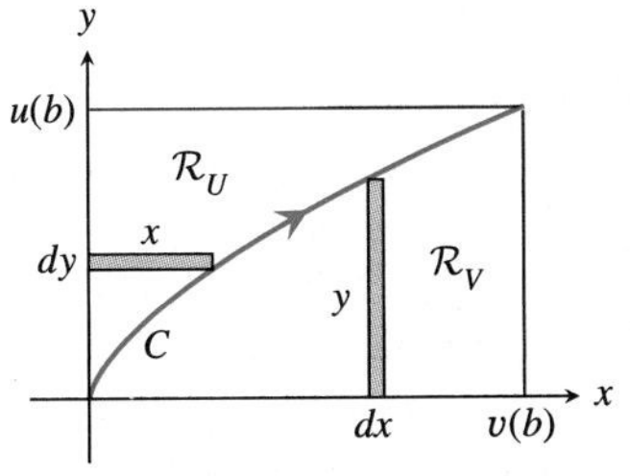

Newton's Geometric Argument

Newton gave a simple, geometric argument for a special case of the integration-by-parts formula, basing his argument on a sketch similar to that shown in Fig. 5.41. As in the figure, suppose that a curve C lies in a rectangular region, starting and ending

FIGURE 5.41 Newton's geometric argument for the integration-by-parts formula.

at opposite corners of the rectangle. We assume, in particular, that C is described parametrically by

$$x = v(t) \qquad a \le t \le b,$$
$$y = u(t),$$

where u and v are increasing on $[a, b]$ and $\mathbf{r}(a) = \langle v(a), u(a) \rangle = \langle 0, 0 \rangle$.

Under these assumptions, the area of the rectangle is $u(b)v(b)$. The area of the rectangle is also equal to the sum of the areas A_L and A_U of the regions $\mathcal{R}_L$ and $\mathcal{R}_U$ beneath and above the curve C. Because

$$dA_L = y\,dx = u(t)v'(t)\,dt$$

and

$$dA_U = x\,dy = v(t)u'(t)\,dt,$$

it follows that

$$u(b)v(b) = A_L + A_U = \int_a^b u(t)v'(t)\,dt + \int_a^b v(t)u'(t)\,dt. \tag{13}$$

This result may be rearranged to give a special case of the integration-by-parts formula (4):

$$\int_a^b u\,dv = u(b)v(b) - \int_a^b v\,du. \tag{14}$$

Exercises 5.7

Exercises 1–16: Evaluate by either of the integration-by-parts formulas (4) or (5).

1. $\int_0^1 xe^{2x}\,dx$
2. $\int_1^2 xe^{-x}\,dx$
3. $\int \ln x\,dx$
4. $\int x^2 \ln 2x\,dx$
5. $\int_0^1 x2^x\,dx$
6. $\int_{-1}^2 x10^x\,dx$
7. $\int \sqrt{x} \ln x\,dx$
8. $\int x^{1/3} \ln x\,dx$
9. $\int x \sin x\,dx$
10. $\int x \cos x\,dx$
11. $\int \arccos x\,dx$. Let $u = \arccos x$ and $dv = dx$.
12. $\int \arctan x\,dx$. Let $u = \arctan x$ and $dv = dx$.
13. $\int e^{\sqrt{x}}\,dx$. Start with the substitution $x = u^2$.
14. $\int \sin\sqrt{x}\,dx$. Start with the substitution $x = u^2$.
15. $\int x \arcsin x\,dx$. After integration by parts use formula (10) in the table of integrals.
16. $\int x \arccos x\,dx$. After integration by parts use formula (10) in the table of integrals.
17. Use (11) in deriving formula (10) from (9).
18. Derive the reduction formula
$$\int x^n e^x\,dx = x^n e^x - n \int x^{n-1} e^x\,dx.$$
This is formula (27) in the table of integrals, with $a = 1$.
19. Use the reduction formula in Exercise 18 to evaluate $\int x^3 e^x\,dx$.
20. Find the area of the region of finite extent that is between the curves described by $y = \ln x$ and $y = (\ln x)^2$.
21. Find the area of the region of finite extent that is between the curves described by $y = e^2 x$ and $y = xe^x$.

22. Find the area of the region of finite extent that is between the curves described by $y = x \ln x$ and $y = \sin x$. *Note:* When you evaluate the expression $x^2 \ln x$ at $x = 0$, think about it as the limit $\lim_{x\to 0^+} x^2 \ln x$.

23. Find the area of the region of finite extent that is between the curves described by $y = x \ln x$ and $y = \sqrt{x}$. *Note:* When you evaluate the expression $x^2 \ln x$ at $x = 0$, think about it as the limit $\lim_{x\to 0^+} x^2 \ln x$.

24. Evaluate $\int x \arctan x\, dx$. In the second integral, note that

$$\frac{x^2}{1 + x^2} = 1 - \frac{1}{1 + x^2}.$$

25. Evaluate $\int 10^{-x} \cos x\, dx$. *Hint:* Express 10^{-x} as e^z.

26. Evaluate $\int 2^x \sin x\, dx$. *Hint:* Express 2^x as e^z.

27. Derive the reduction formula

$$\int (\ln x)^n\, dx = x(\ln x)^n - n \int (\ln x)^{n-1}\, dx.$$

28. Use the reduction formula in the preceding exercise to evaluate $\int_1^e (\ln x)^4\, dx$.

29. Find an antiderivative of $\sin(\ln x)$.

30. Show that the recursion formula given in (8) may be given in the nonrecursive form

$$F_n = \frac{n-1}{n}\frac{n-3}{n-2}\cdots\frac{1}{2}\cdot \pi,$$

if n is even, and

$$F_n = \frac{n-1}{n}\frac{n-3}{n-2}\cdots\frac{2}{3}\cdot 2$$

if n is odd. Use these formulas for calculating $\int_0^\pi \sin^8 x\, dx$ and $\int_0^\pi \sin^9 x\, dx$.

Ⓣ **31.** Show that for $n \geq 2$

$$\int_0^{\pi/2} \cos^n x\, dx = \frac{n-1}{n} \int_0^{\pi/2} \cos^{n-2} x\, dx.$$

Show that this formula may be written as the recursion formula

$$F_0 = \pi/2,$$

$$F_1 = 1,$$

$$F_n = \left(\frac{n-1}{n}\right) F_{n-2}.$$

Use the recursion formula to show that $F_{15} = 2048/6435$.

32. Use integration by parts in deriving the reduction formula

$$\int (k^2 - x^2)^n\, dx = \frac{x(k^2 - x^2)^n}{2n+1} + \frac{2k^2 n}{2n+1} \int (k^2 - x^2)^{n-1}\, dx.$$

33. Does the reduction formula in Exercise 32 hold for $n = 1$? Why?

34. Use the reduction formula in Exercise 32 to evaluate

$$\int_0^1 (1 - x^2)^5\, dx.$$

35. Derive the reduction formula

$$\int \sec^n x\, dx = \frac{\sec^{n-2} x \tan x}{n-1} + \frac{n-2}{n-1} \int \sec^{n-2} x\, dx, \qquad n \geq 2.$$

To use this formula for odd n, see formula (18) in the table of integrals.

36. Use the reduction formula in Exercise 35 to evaluate the integral $\int \sec^3 x\, dx$.

37. Use the reduction formula in Exercise 35 to evaluate the integral $\int_0^{\pi/4} \sec^5 x\, dx$.

38. Let

$$I_n = \int_0^{\pi/3} \sec^n x\, dx.$$

Use the reduction formula in Exercise 35 to derive the reduction formula

$$I_n = \frac{2^{n-2}}{n-1}\sqrt{3} + \frac{n-2}{n-1} I_{n-2}.$$

39. Derive the reduction formula

$$\int \frac{dx}{(x^2 + k^2)^n} = \frac{x}{2k^2(n-1)(x^2 + k^2)^{n-1}} + \frac{2n-3}{2k^2(n-1)} \int \frac{dx}{(x^2 + k^2)^{n-1}}.$$

Hint: Let $dv = dx$.

40. Use the reduction formula in the preceding exercise to evaluate

$$\int_0^1 \frac{dx}{(x^2 + 2)^3}.$$

5.8 Integration by Partial Fractions

Unlike either integration by substitution or by parts, which offer hope but no certainty, integration by partial fractions is guaranteed to work. Indeed, all integrals of the form

$$\int \frac{p(x)}{q(x)}\,dx,$$

where p and q are polynomials, can be evaluated by the method of partial fractions.

Ratios of the form $R(x) = p(x)/q(x)$, where p and q are polynomial functions, are called **rational functions.** The technique of integrating a rational function R by partial fractions is based on an algebraic procedure that decomposes R into a sum of simpler fractions, each of which has a known antiderivative.

INVESTIGATION 1

In algebra you learned how to combine a sum of two or more fractions into one fraction. For example, by finding a "common denominator" you could combine the sum

$$1 + \frac{1}{x} + \frac{1}{x-1} \tag{1}$$

into the fraction

$$\frac{x^2 + x - 1}{x^2 - x}. \tag{2}$$

In this section, we give a procedure for decomposing a single fraction into a sum of simpler fractions. If, for example, we apply the procedure to the fraction (2), we will end up with the sum (1) of simpler fractions. The procedure will make it possible to integrate any *rational function.*

Suppose, for example, that we wish to evaluate the integral

$$\int \frac{x^2 + x - 1}{x^2 - x}\,dx.$$

From (1) we may, instead, integrate each of three simpler fractions:

$$\int \frac{x^2 + x - 1}{x^2 - x}\,dx = \int 1\,dx + \int \frac{1}{x}\,dx + \int \frac{1}{x-1}\,dx. \tag{3}$$

Substituting $u = x - 1$ in the third integral,

$$\begin{aligned}\int \frac{x^2 + x - 1}{x^2 - x}\,dx &= x + \ln|x| + \int \frac{du}{u}\\ &= x + \ln|x| + \ln|u| + C\\ &= x + \ln|x| + \ln|x-1| + C = x + \ln|x(x-1)| + C.\end{aligned}$$

The main goal of this section is to describe a systematic procedure for decomposing rational functions into a sum of simpler fractions, often called *partial fractions.*

The decomposition procedure has three main steps, all algebraic:

1. *Divide* the numerator of the rational function by its denominator to obtain quotient and remainder terms.
2. *Factor* the denominator of the remainder term.
3. *Decompose* the remainder term.

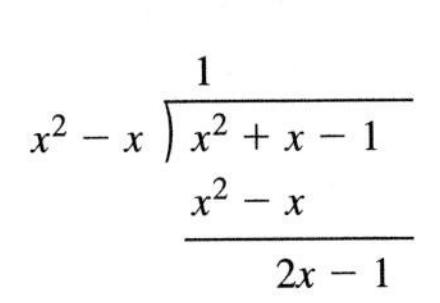

FIGURE 5.42 Long division.

Divide $x^2 + x - 1$ by $x^2 - x$. Figure 5.42 is a reminder of the long-division process, which results in a quotient of 1 and a remainder of $2x - 1$; that is,

$$\frac{x^2 + x - 1}{x^2 - x} = 1 + \frac{2x - 1}{x^2 - x}.$$

Factor the denominator $x^2 - x$ of the remainder term:

$$\frac{x^2 + x - 1}{x^2 - x} = 1 + \frac{2x - 1}{x^2 - x} = 1 + \frac{2x - 1}{x(x - 1)}. \tag{4}$$

Leaving the quotient term for a moment, we *decompose* the remainder term into simpler fractions. The "standard decomposition" of the remainder term has the form

$$\frac{2x - 1}{x(x - 1)} = \frac{A}{x} + \frac{B}{x - 1}, \tag{5}$$

where A and B are constants that we need to find. We discuss the form of the "standard decomposition" later. "Clearing fractions" by multiplying both sides of (5) by the factored denominator,

$$2x - 1 = A(x - 1) + Bx. \tag{6}$$

We can generate a system of two equations in the two unknowns A and B by setting x equal to two different values in this equation. The best values to try are the zeros 0 and 1 of the factored denominator $x(x - 1)$; substituting these into (6),

$$x = 0: \; 2 \cdot 0 - 1 = A(0 - 1) + B \cdot 0 = -A$$
$$x = 1: \; 2 \cdot 1 - 1 = A(1 - 1) + B \cdot 1 = B.$$

Hence, $A = 1$ and $B = 1$. Substituting these values into (5),

$$\frac{2x - 1}{x(x - 1)} = \frac{1}{x} + \frac{1}{x - 1}.$$

From this result and (4),

$$\frac{x^2 + x - 1}{x^2 - x} = 1 + \frac{1}{x} + \frac{1}{x - 1},$$

as shown in (3). We are now ready for the integration following (3).

There are several methods for determining the unknowns A and B from equation (6). We discuss a method based on equating coefficients in Exercise 42.

We work through a second example before giving a more general description of the partial fractions procedure.

INVESTIGATION 2

Calculate the value of the integral

$$\int_1^3 \frac{x+2}{2x^3 - x^2 + 2x - 1}\, dx.$$

Because the degree of the numerator of the rational function

$$\frac{x+2}{2x^3 - x^2 + 2x - 1} \tag{7}$$

is smaller than that of the denominator, we skip the division step (the quotient will be the zero polynomial). The factoring and decomposition steps require more effort than in the first example, although this depends on your algebraic skills or your access to a calculator or CAS. The denominator polynomial factors as

$$2x^3 - x^2 + 2x - 1 = (2x - 1)(x^2 + 1). \tag{8}$$

The factor $2x - 1$ corresponds to the zero $1/2$, and the factor $x^2 + 1$ corresponds to the complex zeros i and $-i$. All polynomials with real coefficients can be factored in this way, as a product of *real* factors. The complex zeros are wrapped up in *irreducible quadratic* factors like $x^2 + 1$. A quadratic factor is said to be irreducible if it has no real zeros. In the standard decomposition, an irreducible quadratic factor leads to a fraction with a numerator of the form $Ax + B$. Thus, the standard decomposition of the fraction (7) is

$$\frac{x+2}{(2x-1)(x^2+1)} = \frac{Ax+B}{x^2+1} + \frac{C}{2x-1}. \tag{9}$$

Clearing fractions by multiplying both sides of (9) by $(x^2 + 1)(2x - 1)$,

$$x + 2 = (Ax + B)(2x - 1) + C(x^2 + 1). \tag{10}$$

We generate three equations in the unknowns A, B, and C by replacing x in (10) by three different numbers. We use the real zero $1/2$ for one equation and $x = 0$ and $x = 1$ for the other two equations (the complex zeros i and $-i$ can also be used; see Exercise 39). These substitutions give

$$\begin{aligned} \text{for } x = \frac{1}{2}: &\quad \frac{5}{2} = C\frac{5}{4} \\ \text{for } x = 0: &\quad 2 = (B)(-1) + C(1) \\ \text{for } x = 1: &\quad 3 = (A + B)(1) + C(2). \end{aligned}$$

From the first equation, $C = 2$; substituting this into the second equation gives $B = 0$; substituting $C = 2$ and $B = 0$ into the third equation gives $A = -1$. Hence

$$\frac{x+2}{(2x-1)(x^2+1)} = \frac{-x}{x^2+1} + \frac{2}{2x-1}.$$

For the integration, the substitutions $u = x^2 + 1$ in the first integral and $w = 2x - 1$ in the second integral give

$$\begin{aligned}\int_1^3 \frac{x + 2}{2x^3 - x^2 + 2x - 1}\,dx &= -\int_1^3 \frac{x\,dx}{x^2 + 1} + \int_1^3 \frac{2\,dx}{2x - 1} \\ &= -\tfrac{1}{2}\int_2^{10} \frac{du}{u} + \int_1^5 \frac{dw}{w} \\ &= -\tfrac{1}{2}(\ln 10 - \ln 2) + \ln 5 \\ &= \tfrac{1}{2}\ln 5.\end{aligned}$$

In these Investigations we illustrated the main steps—divide, factor, and decompose—in integrating a rational function. We now give a more general description of this procedure, followed by examples to illustrate each step.

A rational function $R(x) = p(x)/q(x)$, where $p(x)$ and $q(x)$ are polynomials with real coefficients, can be written as

$$R(x) = \frac{p(x)}{q(x)} = Q(x) + \frac{r(x)}{q(x)},$$

where $Q(x)$ and $r(x)$ are polynomials and the degree of $r(x)$ is less than the degree of the divisor $q(x)$. The polynomials $Q(x)$ and $r(x)$ result from the long division of $p(x)$ by $q(x)$. If the degree of $p(x)$ is less than the degree of $q(x)$, then $Q(x)$ is identically 0. Because

$$\int R(x)\,dx = \int Q(x)\,dx + \int \frac{r(x)}{q(x)}\,dx$$

and the first integral on the right is easy, we focus on the second integral.

We begin by finding all of the real roots of the equation $q(x) = 0$. If c is a root of this equation, then $x - c$ will be a factor. If it happens that c is a multiple root, say of order m, then $(x - c)^m$ will be a factor of $q(x)$. We refer to $(x - c)^m$ as a power of a linear factor or, more briefly, as a linear factor.

Next, we find all of the complex roots of the equation $q(x) = 0$. Because q has real coefficients, the complex roots come in pairs, that is, if $u + iv$ is a complex root of $q(x) = 0$, then so is $u - iv$. These two roots correspond to the factors $x - (u + iv)$ and $x - (u - iv)$. If we multiply these factors together we find

$$(x - (u + iv))\,(x - (u - iv)) = x^2 + Ux + V,$$

where $U = -2u$ and $V = u^2 + v^2$ are real numbers. This quadratic—called an *irreducible quadratic*—is a factor of $q(x)$. If it happens that $u + iv$ is a multiple root, say of order n, then $(x^2 + Ux + V)^n$ will be a factor of $q(x)$. We refer to $(x^2 + Ux + V)^n$ as a power of an irreducible quadratic factor or, more briefly, as an irreducible quadratic factor.

After we have found all of the roots of $q(x) = 0$, real or complex, write $q(x)$ as the product of a real number D, linear factors of the form $(x - c)^m$, and irreducible quadratic factors of the form $(x^2 + Ux + V)^n$.

From advanced algebra it is known that the factored form of $q(x)$ can be decomposed into a sum of terms. Each linear factor $(x - c)^m$ contributes the m terms

$$\frac{A_1}{(x-c)^1} + \cdots + \frac{A_m}{(x-c)^m}. \tag{11}$$

Each irreducible quadratic factor $(x^2 + Ux + V)^n$ contributes the n terms

$$\frac{B_1 x + C_1}{(x^2 + Ux + V)^1} + \cdots + \frac{B_n x + C_n}{(x^2 + Ux + V)^n}. \tag{12}$$

Each of the linear factors or irreducible quadratic factors of $q(x)$ contributes a group of terms of the form (11) or (12); the sum of all such groups is called the **standard decomposition** of $r(x)/q(x)$. The individual terms in such sums are called **partial fractions.** The constants $A_1, B_1, C_1, \ldots$ in the standard decomposition are "unknowns," which must be determined.

Integration by Partial Fractions

To find the integral $\int R(x)\,dx$, where $R(x) = p(x)/q(x)$ is a rational function and $p(x)$ and $q(x)$ have no common factors:

STEP 1. Determine polynomials $Q(x)$ and $r(x)$ for which

$$R(x) = Q(x) + \frac{r(x)}{q(x)},$$

where $Q(x)$ and $r(x)$ are the quotient and remainder polynomials of the division of $p(x)$ by $q(x)$. Recall that the degree of $r(x)$ is less than the degree of $q(x)$. If the degree of $p(x)$ is less than that of $q(x)$, then a division is not necessary; $Q(x)$ is the zero polynomial and $r(x) = p(x)$.

STEP 2. Factor $q(x)$ into powers of linear and irreducible quadratic factors

$$q(x) = D(x - c)^m \cdots (x^2 + Ux + V)^n \cdots.$$

STEP 3. Guided by (11) and (12), write out the standard decomposition of $r(x)/q(x)$, with a group of terms for each of the polynomial factors of $q(x)$. Denote each unknown coefficient by a distinct letter, and equate the standard decomposition to $r(x)/q(x)$.

STEP 4. Clear fractions in the equation formed in Step 3 and evaluate the resulting equation at conveniently chosen values of x. If the decomposition has k unknown coefficients, then k values of x will be required.

STEP 5. Solve the resulting system of k equations in k unknowns.

STEP 6. Integrate $Q(x)$ and all partial fractions.

Any nonzero quotient term in Step 6 is easy to integrate. For the remaining terms, we may try integration by substitution, integration by parts, or apply a formula from the table of integrals. For convenience we have collected in Table 5.2 six integration formulas to use in Step 6 as needed, labeling them to match similar formulas in the main table of integrals, adding a prime (′) or two (″) if the formula has been rearranged or slightly generalized.

TABLE 5.2 Six integrals for partial fractions. Assume that n is a positive integer and, in (A'), $n \neq 1$.

(B′) $$\int \frac{dx}{x+a} = \ln|x+a| + C$$

(A′) $$\int \frac{dx}{(x+a)^n} = \frac{(x+a)^{-n+1}}{-n+1} + C$$

(G) $$\int \frac{dx}{x^2+a^2} = \frac{1}{a}\arctan\frac{x}{a} + C$$

(B″) $$\int \frac{x\,dx}{x^2+a^2} = \tfrac{1}{2}\ln(x^2+a^2) + C$$

(37) $$\int \frac{dx}{(x^2+a^2)^n} = \frac{x}{2a^2(n-1)(x^2+a^2)^{n-1}} + \frac{2n-3}{2a^2(n-1)}\int \frac{dx}{(x^2+a^2)^{n-1}}$$

(A″) $$\int \frac{x\,dx}{(x^2+a^2)^n} = \frac{(x^2+a^2)^{-n+1}}{2(-n+1)} + C$$

EXAMPLE 1 Evaluate

$$\int_5^{10} \frac{dx}{(x-1)(x-2)}.$$

Solution

Because the numerator has degree 0, we can skip Step 1. The denominator has two linear factors, $x-1$ and $x-2$. The standard decomposition of the integrand is

$$\frac{1}{(x-1)(x-2)} = \frac{A}{x-1} + \frac{B}{x-2}, \tag{13}$$

where A and B are unknowns. Clearing fractions by multiplying both sides of this equation by $(x-1)(x-2)$,

$$1 = A(x-2) + B(x-1).$$

Setting $x = 2$ gives $B = 1$ because

$$1 = A(2-2) + B(2-1) = B.$$

Setting $x = 1$ gives $A = -1$. From (13),

$$\int_5^{10} \frac{dx}{(x-1)(x-2)} = \int_5^{10}\left(\frac{-1}{x-1} + \frac{1}{x-2}\right)dx$$

$$= -\int_5^{10}\frac{dx}{x-1} + \int_5^{10}\frac{dx}{x-2}.$$

Each of these integrals has the form of formula (B′) in Table 5.2. Hence,

$$= -\ln|x-1|\Big|_5^{10} + \ln|x-2|\Big|_5^{10}$$

$$= (-\ln 9 + \ln 4) + (\ln 8 - \ln 3) = \ln\frac{32}{27} \approx 0.170.$$

EXAMPLE 2 Evaluate

$$\int_1^2 \frac{2x+1}{x(x^2+1)}\,dx.$$

Solution

Because the degree of the numerator is less than the degree of the denominator, we can skip Step 1. This rational function has a linear factor and an irreducible quadratic factor. The standard decomposition of the integrand is

$$\frac{2x+1}{x(x^2+1)} = \frac{A}{x} + \frac{Bx+C}{x^2+1} \tag{14}$$

where A, B, and C are unknowns. Clearing fractions gives

$$2x + 1 = A(x^2+1) + (Bx + C)x.$$

Setting $x = 0$ in this equation leads to the equation $0 + 1 = A(1) + 0$; hence, $A = 1$. Setting $x = 1$ and $x = -1$ leads to the equations

$$\begin{aligned} 2A + B + C &= 3 \\ 2A + B - C &= -1. \end{aligned} \tag{15}$$

Because $A = 1$, this system reduces to two equations in two unknowns. Adding the two equations gives $4 + 2B = 2$; hence, $B = -1$. By substituting $A = 1$ and $B = -1$ into either equation of (15), we find $C = 2$. With these values of A, B, and C it follows from (14) that

$$\begin{aligned} \int_1^2 \frac{2x+1}{x(x^2+1)}\,dx &= \int_1^2 \left(\frac{1}{x} + \frac{-x+2}{x^2+1}\right)dx \\ &= \int_1^2 \frac{dx}{x} - \int_1^2 \frac{x\,dx}{x^2+1} + 2\int_1^2 \frac{dx}{x^2+1}. \end{aligned}$$

Applying formulas (B′), (B″), and (G) from Table 2,

$$\begin{aligned} \int_1^2 \frac{2x+1}{x(x^2+1)}\,dx &= \ln x\Big|_1^2 - \tfrac{1}{2}\ln(x^2+1)\Big|_1^2 + 2\arctan x\Big|_1^2 \\ &= \ln 2 - \left(\frac{1}{2}\ln(5) - \frac{1}{2}\ln(2)\right) + 2(\arctan(2) - \arctan(1)) \\ &= \frac{3}{2}\ln(2) - \frac{1}{2}\ln(5) + 2\arctan(2) - \frac{\pi}{2} \approx 0.879. \end{aligned}$$

EXAMPLE 3 Evaluate

$$\int_{-5}^{-2} \frac{7x^2+12x+4}{(2x+3)^2(x^2+x+1)}\,dx.$$

Solution

Because the degree of the numerator is less than the degree of the denominator, we can skip Step 1. This rational function has a repeated linear factor and an irreducible quadratic factor (the zeros of $x^2 + x + 1$ are complex because its discriminant $b^2 - 4ac$ is $1 - 4 \cdot 1 \cdot 1 = -3 < 0$). The standard decomposition of the integrand is

$$\frac{7x^2 + 12x + 4}{(2x + 3)^2(x^2 + x + 1)} = \frac{A}{2x + 3} + \frac{B}{(2x + 3)^2} + \frac{Cx + D}{x^2 + x + 1}. \tag{16}$$

Clearing fractions by multiplying by $(2x + 3)^2(x^2 + x + 1)$,

$$\begin{aligned} 7x^2 + 12x + 4 &= A(2x + 3)(x^2 + x + 1) \\ &\quad + B(x^2 + x + 1) + (Cx + D)(2x + 3)^2. \end{aligned} \tag{17}$$

To find A, B, C, and D, set x equal to four different values in (17). Starting with the real zero $x = -3/2$,

$$\begin{aligned} 7\left(-\frac{3}{2}\right)^2 + 12\left(-\frac{3}{2}\right) + 4 = A \cdot 0 + B\left(\left(-\frac{3}{2}\right)^2 + \left(-\frac{3}{2}\right) + 1\right) \\ + \left(C\left(-\frac{3}{2}\right) + D\right) \cdot 0. \end{aligned}$$

Multiplying both sides by 4,

$$\begin{aligned} 7 \cdot 9 - 24 \cdot 3 + 16 &= B(9 - 6 + 4) \\ 7B &= 7. \end{aligned}$$

Hence, $B = 1$. Setting $B = 1$ in (17), we find A, C, and D by setting $x = 0$, 1, -1 in (17). This gives the equations

$$\begin{aligned} 3A \phantom{{}+25C} + 9D &= 3 \\ 15A + 25C + 25D &= 20 \\ A - C + D &= -2. \end{aligned}$$

The solution of this system is $A = -2$, $C = 1$, and $D = 1$. Now that we know A, B, C, and D,

$$\begin{aligned} &\int_{-5}^{-2} \frac{7x^2 + 12x + 4}{(2x + 3)^2(x^2 + x + 1)}\,dx \\ &\quad = \int_{-5}^{-2} \left(\frac{-2}{2x + 3} + \frac{1}{(2x + 3)^2} + \frac{x + 1}{x^2 + x + 1}\right) dx \\ &\quad = \int_{-5}^{-2} \frac{-2\,dx}{2x + 3} + \int_{-5}^{-2} \frac{dx}{(2x + 3)^2} + \int_{-5}^{-2} \frac{x + 1}{x^2 + x + 1}\,dx. \end{aligned} \tag{18}$$

These three integrals may be evaluated using formulas (B′), (A′), (B″), and (G) in Table 2. For the first integral we write $2x + 3 = 2(x + 3/2)$ and use formula (B′).

$$\int_{-5}^{-2} \frac{-2}{2x + 3}\,dx = -\int_{-5}^{-2} \frac{dx}{x + 3/2}\,dx = -\ln\left|x + \frac{3}{2}\right|\Bigg|_{-5}^{-2} = \ln 7. \tag{19}$$

For the second integral from (18),

$$\begin{aligned} \int_{-5}^{-2} \frac{1}{(2x + 3)^2}\,dx &= \frac{1}{4} \cdot \int_{-5}^{-2} \frac{1}{(x + (3/2))^2}\,dx \\ &= \frac{1}{4} \cdot \frac{(x + 3/2)^{-2+1}}{-2 + 1}\Bigg|_{-5}^{-2} = \frac{3}{7}. \end{aligned} \tag{20}$$

For the third integral, we apply formulas (B″) and (G). For this we must complete the square. To complete the square of the quadratic

$$Q(x) = x^2 + x + 1$$

we add and subtract the square of one-half the coefficient of x:

$$Q(x) = x^2 + x + \left(\frac{1}{2}\right)^2 - \left(\frac{1}{2}\right)^2 + 1.$$

We may now write

$$Q(x) = \left(x + \frac{1}{2}\right)^2 + 1 - \frac{1}{4} = \left(x + \frac{1}{2}\right)^2 + \frac{3}{4}.$$

We may now write the third integral as

$$\int_{-5}^{-2} \frac{x+1}{x^2+x+1}\,dx = \int_{-5}^{-2} \frac{x+1}{\left(x+\frac{1}{2}\right)^2 + \frac{3}{4}}\,dx.$$

To more easily apply formulas (B″) and (G), we make the substitution $u = x + \frac{1}{2}$ and set $a = \sqrt{3}/2$. This gives

$$\begin{aligned}\int_{-5}^{-2} \frac{x+1}{x^2+x+1}\,dx &= \int_{-9/2}^{-3/2} \frac{u+\frac{1}{2}}{u^2+a^2}\,du \\ &= \int_{-9/2}^{-3/2} \frac{u}{u^2+a^2}\,du + \frac{1}{2}\int_{-9/2}^{-3/2} \frac{du}{u^2+a^2} \\ &= \frac{1}{2}\ln(u^2+a^2)\Big|_{-9/2}^{-3/2} + \frac{1}{2a}\arctan\left(\frac{u}{a}\right)\Big|_{-9/2}^{-3/2}\end{aligned}$$

$$\int_{-5}^{-2} \frac{x+1}{x^2+x+1}\,dx = -\frac{1}{2}\ln 7 - \frac{\pi}{3\sqrt{3}} + \frac{1}{\sqrt{3}}\arctan 3\sqrt{3}. \tag{21}$$

Putting (19), (20), and (21) together,

$$\int_{-5}^{-2} \frac{7x^2+12x+4}{(2x+3)^2(x^2+x+1)}\,dx = \frac{1}{2}\ln 7 + \frac{3}{7} - \frac{\pi}{3\sqrt{3}} + \frac{1}{\sqrt{3}}\arctan 3\sqrt{3}$$
$$\approx 1.594.$$

Exercises 5.8

Exercises 1–6: Apply Step 1 of the partial fractions procedure to the rational function.

1. $\dfrac{3x^3+5x^2+x-1}{x^2+x+1}$

2. $\dfrac{x^3+2x-1}{(x-1)(2x+1)}$

3. $\dfrac{x^3+3}{(x+1)(x-2)(x+5)}$

4. $\dfrac{x^3+1}{x^3-1}$

5. $\dfrac{2x-1}{(x^2-1)^2}$

6. $\dfrac{x^3}{(x^2+1)^2}$

Exercises 7–12: Factor each polynomial, as in Step 2 of the partial fractions procedure. All zeros are either rational or complex.

7. $x^4 + x^3 - x^2 + x - 2$

8. $2x^5 - 7x^4 - 3x^3 - 3x^2 - 5x + 4$

9. $x^4 - 1$

10. $x^6 - 1$

11. $x^4 + x^3 - x - 1$

12. $x^4 - 3x^3 - x^2 - 3x + 18$

Exercises 13–18: Factor, as in Step 2 of the partial fractions procedure. You may need a numerical algorithm like Newton's method to find some of the zeros.

13. $x^3 - 5x^2 + 19x + 25$

14. $x^4 - 3x^3 - 5x^2 + 29x - 30$

15. $x^4 - 4x^3 - x^2 - 2x + 3$

16. $x^4 - 3x^3 + x^2 - 10x + 6$

17. $x^4 - x^2 - 2$

18. $x^5 + 2x^3 - 35x$

Exercises 19–30: Apply Steps 1–5 of the partial fractions procedure.

19. $\dfrac{-2x + 7}{(x + 4)(x + 1)}$

20. $\dfrac{x + 13}{(x + 5)(x - 3)}$

21. $\dfrac{x^3 - x^2 - 7x - 7}{(x + 2)(x - 3)}$

22. $\dfrac{2x^3 - x^2 - x + 6}{(x^2 - 1)}$

23. $\dfrac{x + 4}{(x + 3)^2}$

24. $\dfrac{3x + 8}{(x + 1)^2}$

25. $\dfrac{3x^2 - 7x + 7}{(x^2 + 1)(x - 2)}$

26. $\dfrac{x^2 + 7x + 4}{(x^2 + x + 1)(2x + 1)}$

27. $\dfrac{x^2 + 18x + 30}{(2x + 3)(x + 3)(x - 2)}$

28. $\dfrac{-5x^2 + 9x + 20}{6(x + 2)(x + 1)(x - 1)}$

29. $\dfrac{1}{(x^2 - 5x + 6)^2(x - 1)}$

30. $\dfrac{1}{(x^2 - 3x + 2)^2(x + 1)}$

Exercises 31–38: Integrate, using Table 2.

31. $\displaystyle\int_0^1 \left(\frac{1}{x + 1} + \frac{1}{x + 4}\right) dx$

32. $\displaystyle\int_2^5 \left(\frac{2}{x} + \frac{1}{2x - 3}\right) dx$

33. $\displaystyle\int \left(\frac{1}{(x + 1)^2} - \frac{3}{x^2 + 4}\right) dx$

34. $\displaystyle\int \left(\frac{1}{x^2 + 25} + \frac{5}{(x + 2)^3}\right) dx$

35. $\displaystyle\int \left(\frac{1}{x^2 + 2x + 3} + \frac{1}{x + 1}\right) dx$

36. $\displaystyle\int \left(\frac{1}{x^2 + 6x + 13} - \frac{2}{x + 9}\right) dx$

37. $\displaystyle\int \left(\frac{x}{(x^2 + 4)^2} + \frac{1}{x^2 + 4}\right) dx$

38. $\displaystyle\int \frac{x + 1}{(x^2 + 2)^2}\, dx$

39. In Example 2 the unknowns A, B, and C were found by setting $x = 0, 1, -1$. Try setting $x = 0$ and $x = i$. For the latter you should obtain the equation $2i + 1 = -B + Ci$. By equating the real and imaginary parts of these two complex numbers, determine B and C.

40. Show that each of the formulas (B′), (A′), (B″), and (A″) can be reduced to formulas (A) or (B) in the table of integrals by a substitution.

41. The integrand of the integral

$$\int_0^4 \frac{dx}{(x - 1)(x - 2)}$$

is the same as that of the integral in Example 1. Comment on the calculation

$$\int_0^4 \frac{dx}{(x - 1)(x - 2)} = \ln \left|\frac{x - 2}{x - 1}\right| \Bigg|_0^4$$
$$= \ln \left|\frac{2}{3}\right| - \ln |2| = -\ln 3.$$

42. A second method of generating a system of equations from the standard decomposition is based on equating coefficients. We explain by example. Starting with the equation (from Investigation 1)

$$2x - 1 = A(x - 1) + Bx,$$

rearrange the right side as

$$2x - 1 = (A + B)x - A. \tag{22}$$

In this form, the coefficients of the polynomial $A(x - 1) + Bx$ are explicit. Because the two polynomials in (22) are to be equal for all values of x, their coefficients must be identical. Hence,

$$\begin{aligned} A + B &= 2 \\ -A &= -1 \end{aligned}$$

Solving this system gives $A = 1$ and $B = 1$.

As a second example, we take the equation (from Investigation 2)

$$x + 2 = (Ax + B)(2x - 1) + C(x^2 + 1).$$

Rearranging the right side,

$$x + 2 = (2A + C)x^2 + (2B - A)x + (C - B).$$

The coefficient of x^2 on the left is 0. Equating coefficients,

$$\begin{aligned} 2A + \quad\quad + C &= 0 \\ -A + 2B \quad\quad &= 1 \\ - \quad B + C &= 2. \end{aligned}$$

Solving this system gives $A = -1$, $B = 0$, and $C = 2$. Determine in a similar way a system of equations for the unknowns A, B, C, and D in Example 3.

Exercises 43–68: Evaluate the integral by partial fractions or a formula from Table 2.

43. $\displaystyle\int \frac{x}{x^2 - 1}\,dx$

44. $\displaystyle\int \frac{x}{x^2 - 5x + 6}\,dx$

45. $\displaystyle\int \frac{dx}{x^4 - 1}$

46. $\displaystyle\int \frac{dx}{x^4 - 16}$

47. $\displaystyle\int \frac{dx}{(x + 1)^2(x - 1)}$

48. $\displaystyle\int \frac{x^2 - 1}{(x + 2)^2(x + 3)}\,dx$

49. $\displaystyle\int \frac{x^2 - 1}{x^2 + 5x + 6}\,dx$

50. $\displaystyle\int \frac{2x^2 + 1}{2x^2 - x - 1}\,dx$

51. $\displaystyle\int \frac{dx}{x^2 + 2x + 2}$

52. $\displaystyle\int \frac{dx}{x^2 - 4x + 13}$

53. $\displaystyle\int \frac{x^3}{x - 1}\,dx$

54. $\displaystyle\int \frac{x^4}{(x - 1)^2}\,dx$

55. $\displaystyle\int \frac{dx}{(x + 1)^4}$

56. $\displaystyle\int \frac{dx}{(2x + 3)^2}$

57. $\displaystyle\int \frac{x}{x^2 + x + 1}\,dx$

58. $\displaystyle\int \frac{3x - 2}{2x^2 - x + 2}\,dx$

59. $\displaystyle\int \frac{dx}{x^4 - 2x^3 + x^2 - 4x + 4}$

60. $\displaystyle\int \frac{x}{x^3 + 1}\,dx$

61. $\displaystyle\int \frac{x - 1}{x^2 - 4x + 13}\,dx$

62. $\displaystyle\int \left(\frac{x + 2}{x - 1}\right)^2 dx$

63. $\displaystyle\int_0^1 \frac{x + 2}{(x^2 + 1)^4}\,dx$

64. $\displaystyle\int_0^1 \frac{x^3 - 1}{x^3 + 1}\,dx$

65. $\displaystyle\int \frac{x^2 - 3x + 2}{x^3 + 2x^2 + x}\,dx$

66. $\displaystyle\int \frac{x^4}{x^6 - 1}\,dx$

67. $\displaystyle\int \frac{x^2(2x^2 - x + 3)}{(x^2 + 1)^2(x - 1)}\,dx$

68. $\displaystyle\int \frac{x^4 + 1}{(x - 1)^2(x + 1)^2}\,dx$

5.9 Solving Simple Differential Equations

As discussed in Section 2.10, a differential equation is an equation relating a function and its derivative(s). For example,

- the function $y(t)$ giving the current at time t in an electrical circuit with a resistance R, an inductance L, and an applied voltage $E\sin(\omega t)$, all connected in series, satisfies the differential equation

$$Ly' + Ry = E\sin(\omega t); \tag{1}$$

- the function $P(t)$ giving the U.S. population at time t can be modeled by the logistic equation

$$\frac{dP}{dt} = \frac{3}{100}\left(P - \frac{P^2}{292}\right), \tag{2}$$

where t is measured in years and P is measured in millions of persons. We assume that $P(0) = 4$, where the time $t = 0$ corresponds to the year 1790; and

- the velocity $v(t)$ of a paratrooper at time t can be modeled by the equation

$$91.6 \cdot 9.8 - kv^2 = 91.6\frac{dv}{dt}, \tag{3}$$

where k is a known constant, the paratrooper's mass is 91.6 kg (this is 72.6 kg body mass plus 19 kg equipment), and he is falling at 50 m/s at the time his parachute opens.

Each of these differential equations can be put into the form

$$y' = f(t, y),$$

where f is a function of the *independent variable* t and the *dependent variable* y. For the electrical circuit,

$$y' = f(t, y) = \frac{E \sin(\omega t) - Ry}{L};$$

for the population example,

$$P' = f(t, P) = \frac{3}{100}\left(P - \frac{P^2}{292}\right),$$

where we have used P instead of y; and for the paratrooper example,

$$v' = f(t, v) = \frac{91.6 \cdot 9.8 - kv^2}{91.6},$$

where we have used v instead of y.

The simplest kind of differential equation $y' = f(t, y)$ is one in which the variables can be "separated." By this we mean that the function $f(t, y)$ can be written as a product $f(t, y) = g(t)h(y)$ of two functions, one a function of t alone and the other a function of y alone. Of the differential equations (1)–(3), the second and third are separable and the first is not.

By replacing y' by dy/dt we may write a separable differential equation

$$y' = f(t, y) = g(t)h(y) \tag{4}$$

as

$$\frac{dy}{dt} = g(t)h(y).$$

We separate the variables y and t by multiplying both sides of this equation by $dt/h(y)$:

$$\frac{dy}{h(y)} = g(t)\, dt. \tag{5}$$

The form of this equation suggests that we try "to integrate both sides."

This means that we seek antiderivatives $H(y)$ and $G(t)$ for $1/h(y)$ and $g(t)$. Assuming that such antiderivatives can be found, we show that the equation

$$H(y) = G(t) + c, \tag{6}$$

where c is an "arbitrary constant," defines a solution of (4). Here's the argument: Suppose that we can, at least theoretically, solve the equation $H(y) = G(t) + c$ for y in terms of t. We show that the resulting function $y = y(t)$ satisfies (4). Regarding y as a function of t and differentiating the equation $H(y) = G(t) + c$,

$$\frac{d}{dt}H(y) = \frac{d}{dt}(G(t) + c).$$

Using the chain rule on the left and recalling that $G'(t) = g(t)$ and $H'(y) = 1/h(y)$,

$$\frac{d}{dy}H(y) \cdot \frac{dy}{dt} = g(t)$$

$$H'(y) \cdot y'(t) = g(t)$$

$$\frac{1}{h(y)} \cdot y'(t) = g(t)$$

$$y' = h(y)g(t) = f(t, y).$$

Hence, $y = y(t)$ satisfies (4).

This solution procedure can be written symbolically as

$$\frac{dy}{dt} = g(t)h(y)$$

$$\frac{dy}{h(y)} = g(t)\,dt$$

$$\int \frac{dy}{h(y)} = \int g(t)\,dt$$

$$H(y) = G(t) + c.$$

WEB The expression $H(y) = G(t) + c$ is sometimes called the *general solution* of the differential equation (4). Although a solution $y = y(t)$ of (4) may not be given explicitly in this general solution, such a solution y is implicitly defined by this equation.

We give three examples of this procedure.

EXAMPLE 1 In some chemical reactions the rate of conversion of a substance at any time t is proportional to the quantity of the substance remaining. Hence, if $y = y(t)$ is the amount of the substance remaining at any time t,

$$\frac{dy}{dt} = ky, \tag{7}$$

where k is a constant. Because y is a decreasing function of t, $k < 0$. Find the general solution of this differential equation.

Solution

Separating the variables,

$$\frac{dy}{dt} = ky$$

$$\frac{dy}{y} = k\,dt.$$

$$\int \frac{dy}{y} = \int k\,dt$$

$$\ln|y| = kt + c.$$

Because $y > 0$,

$$y = e^{kt+c} = e^c \cdot e^{kt}.$$

It is not difficult to show that $y = e^c \cdot e^{kt}$ is a solution to the differential equation (7). Differentiating y with respect to t,

$$y' = e^c \cdot ke^{kt}.$$

Substituting y' and y into the differential equation (7),

$$e^c \cdot ke^{kt} = y' = ky = k(e^c \cdot e^{kt}).$$

This shows that (7) is satisfied for all t by $y = e^c e^{kt}$.

In our second example, we solve an initial value problem, which is a differential equation together with an initial condition.

EXAMPLE 2 A paratrooper falls from rest. If he experiences air resistance proportional to the square of his velocity, his motion is described by

$$\frac{dv}{dt} = b(a^2 - v^2), \tag{8}$$

$$v(0) = 0, \tag{9}$$

where b and a are positive constants. Taking the downward direction as positive, the velocity v of the paratrooper is nonnegative and his acceleration dv/dt is positive for all $t \geq 0$. Find his velocity at any time $t \geq 0$.

Solution

Separating the variables in (8),

$$\frac{dv}{a^2 - v^2} = b\,dt. \tag{10}$$

We may integrate the left side of this equation by applying formula (7) from the table of integrals (see inside back cover). Or we can apply the partial fractions procedures to rewrite the fraction $1/(a^2 - v^2)$ as

$$\frac{1}{a^2 - v^2} = \frac{1}{2a}\left(\frac{1}{a - v} + \frac{1}{a + v}\right).$$

We may now rewrite (10) as

$$\frac{1}{2a}\left(\frac{dv}{a - v} + \frac{dv}{a + v}\right) = b\,dt.$$

Integrating both sides of this equation,

$$\frac{1}{2a}\big(-\ln|a - v| + \ln|a + v|\big) = bt + c. \tag{11}$$

Applying the initial condition (9), that $v = 0$ when $t = 0$,

$$\frac{1}{2a}\big(-\ln|a| + \ln|a|\big) = c. \tag{12}$$

Hence, $c = 0$ and (11) becomes

$$\ln\left|\frac{a+v}{a-v}\right| = 2abt. \tag{13}$$

Recalling that $dv/dt > 0$ for all $t \geq 0$, it follows from (8) that $v < a$. Hence, (13) can be written as

$$\ln\left(\frac{a+v}{a-v}\right) = 2abt.$$

Exponentiating both sides,

$$\frac{a+v}{a-v} = e^{2abt}.$$

Solving this equation for v,

$$\begin{aligned} a + v &= ae^{2abt} - ve^{2abt} \\ v(1 + e^{2abt}) &= ae^{2abt} - a \\ v &= a\frac{e^{2abt} - 1}{e^{2abt} + 1}. \end{aligned}$$

For our third example, we solve the U.S. population problem described earlier, in Equation (2).

EXAMPLE 3 Solve the initial value problem

$$\frac{dP}{dt} = \frac{3}{100}\left(P - \frac{P^2}{292}\right) \tag{14}$$

$$P(0) = 4. \tag{15}$$

The time variable is measured in years and the population P is measured in millions of persons. Recall that $t = 0$ corresponds to the year 1790.

Solution

Separating the variables in (14),

$$\frac{dP}{P\left(1 - \frac{1}{292}P\right)} = \frac{3}{100}dt. \tag{16}$$

The standard decomposition of the left side of this equation is

$$\frac{1}{P\left(1 - \frac{1}{292}P\right)} = \frac{A}{P} + \frac{B}{1 - \frac{1}{292}P}.$$

Clearing fractions,

$$1 = A\left(1 - \tfrac{1}{292}P\right) + BP.$$

Setting $P = 0$ and $P = 292$, we find $A = 1$ and $B = 1/292$. Hence, from (16),

$$\int \left(\frac{1}{P} + \frac{1}{292 - P} \right) dP = \int \frac{3}{100}\, dt = 0.03 \int dt$$
$$\ln |P| - \ln |292 - P| = 0.03t + c$$
$$\ln \left| \frac{P}{292 - P} \right| = 0.03t + c.$$

We find the constant c by applying the initial condition (15), that $P(0) = 4$. Assuming that $P < 292$ (see Exercise 30) and substituting $t = 0$ and $P = 4$ in the last equation,

$$c = -\ln 72.$$

Hence,

$$\ln \left(\frac{P}{292 - P} \right) = 0.03t - \ln 72$$
$$\ln \left(\frac{72P}{292 - P} \right) = 0.03t.$$

Exponentiating both sides,

$$\frac{72P}{292 - P} = e^{0.03t}.$$

Solving this equation for P,

$$P = \frac{292e^{0.03t}}{72 + e^{0.03t}} = \frac{292}{1 + 72e^{-0.03t}}.$$

Exercises 5.9

Exercises 1–6: Show that the function $y = y(t)$ is a solution of the differential equation $y' = f(t, y)$.

1. $y = ce^t$; $y' = y$

2. $y = \frac{1}{2}t^2 + c$; $y' = t$

3. $y = c \cos t$; $y' = -y \tan t$

4. $y = c\sqrt{1 + t^2}$; $y' = \dfrac{ty}{1 + t^2}$

5. $y = \tan(c - 1/t)$; $y' = \dfrac{1 + y^2}{t^2}$

6. $y = \tan(\ln|t| + c)$; $y' = \dfrac{1 + y^2}{t}$

Exercises 7–14: Find the general solution of the differential equation.

7. $y' = \dfrac{t}{y}$

8. $y' = \dfrac{t^2}{y^3}$

9. $y' = \dfrac{ty}{1 + t^2}$. Assume that $y > 0$.

10. $y' = -\dfrac{1 + y^2}{1 + t^2}$

11. $y' = \dfrac{y^2 - 5y + 6}{1 + t}$. Assume that $2 < y < 3$.

12. $y' = \dfrac{y^2 - 5y + 4}{1 + t}$. Assume that $y > 4$.

13. $y' = e^{t-y}$

14. $y' = e^{t+y}$

Exercises 15–22: Solve the initial value problem.

15. $y' = \dfrac{t}{y^2}$; $y(0) = 3$

16. $y' = \dfrac{t^2}{\sqrt{y}}$; assume that $y > 0$ and $y(1) = 1$.

17. $y' = \dfrac{y}{1 + t}$; assume that $y, t > 0$ and $y(1) = 1$.

18. $y' = \dfrac{\sqrt{1 - y^2}}{\sqrt{1 - t^2}}$; assume that $0 \le t, y < 1$ and $y(0) = 1$.

19. $y' = \dfrac{t \sin t}{y}$; $y\left(\frac{1}{2}\pi\right) = 4$

20. $y' = \dfrac{t \cos t}{y^2}$; $y\left(\frac{1}{2}\pi\right) = 3$

21. $y' = \dfrac{e^{t-y}}{y}$; $y(0) = 0$

22. $y' = \dfrac{e^{t+y}}{y}$; $y(0) = 0$

23. The rate of decay of radium 226 at any time t is proportional to the amount of radium remaining at that time. If at $t = 0$ we have 1 kilogram of radium 226 and at $t = 1599$ years 0.5 kilogram will remain, how much will remain after 1000 years?

24. The rate of decay of uranium 232 at any time t is proportional to the amount of uranium remaining at that time. If at $t = 0$ we have 1 kilogram of uranium 232 and after 68.9 years 0.5 kilogram will remain, how much will remain after 25 years?

25. A cup of liquid initially at 150°F is cooling in a room with constant temperature 72°F. The rate of change of its temperature $T(t)$ at any time t is proportional to $72 - T(t)$. If the liquid has cooled to 110°F in 10 minutes, how many more minutes will it be until the liquid has cooled to 100°F?

26. A cup of liquid initially at 200°F is cooling in a room with constant temperature 72°F. The rate of change of its temperature $T(t)$ at any time t is proportional to $72 - T(t)$. If the liquid has cooled to 190°F in 5 minutes, how many more minutes will it be until the liquid has cooled to 110°F?

27. Solve the population problem

$$P' = \frac{3}{100}\left(P - \frac{1}{290}P^2\right)$$
$$P(0) = 3.$$

The time variable is measured in years and the population P is measured in millions of persons. If the initial population is 3 million persons, how many years will it take for the population to reach 10 million persons?

28. Solve the population problem

$$P' = \frac{3}{100}\left(P - \frac{1}{294}P^2\right)$$
$$P(0) = 2.$$

The time variable is measured in years and the population P is measured in millions of persons. If the initial population is 2 million persons, how many years will it take for the population to reach 16 million persons?

29. The velocity $v(t)$ of a paratrooper at time t can be modeled by the equation

$$\frac{dv}{dt} = 9.8 - \frac{9.8}{25}v^2.$$

If the paratrooper is falling at 50 m/s at the time his parachute opens, find his velocity 1 second later.

30. In Example 3 we assumed that $P < 292$ in simplifying the solution to a differential equation. Show that the assumption that $P > 292$ is not compatible with choosing c so that the initial condition $P(0) = 4$ is satisfied.

5.10 Numerical Integration

The integrals used to calculate the length of Mars' orbit or the angular position of a simple pendulum are among the many integrals that cannot be evaluated by substitution, integration by parts, partial fractions, or, indeed, by any known integration technique.

Suppose that we wish to evaluate an integral $\int_a^b f(x)\,dx$ but cannot determine an antiderivative F of f. We may, of course, apply the definition of the definite integral and calculate one or more Riemann sums to approximate $\int_a^b f(x)\,dx$. Using Riemann sums is, however, often inefficient relative to using one of several "numerical integration" algorithms to approximate $\int_a^b f(x)\,dx$. These algorithms are often referred to as "rules." We study three of these rules: the trapezoid rule, the midpoint rule, and Simpson's rule.

Another application of these numerical integration algorithms is to situations in which we don't have a formula for f, but rather a set of data

$$\{(x_1, f(x_1)), \ldots, (x_n, f(x_n))\}.$$

For example, we can approximate the area of the cross section of an airfoil from a data set defining its profile, or we can approximate the position of an incoming missile by "integrating" a data set giving its velocity at times $t_0, t_1, \ldots, t_n$.

Trapezoid Rule

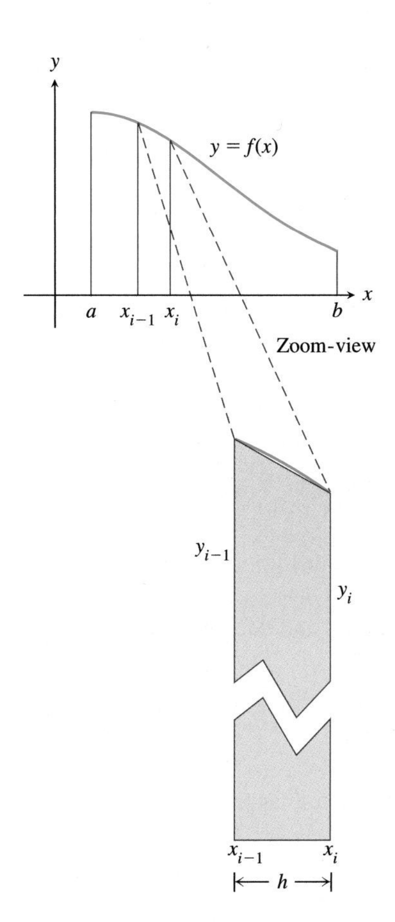

FIGURE 5.43 The trapezoid approximation to the integral $\int_a^b f(x)\,dx$.

Figure 5.43 shows the simple idea leading to the trapezoid rule. On a subinterval $[x_{i-1}, x_i]$ of a regular subdivision $\mathcal{S}_n$ of $[a, b]$, we approximate the area beneath f and above $[x_{i-1}, x_i]$ with the trapezoid with height h and bases $y_{i-1} = f(x_{i-1})$ and $y_i = f(x_i)$. The area of this trapezoid is $\frac{1}{2}(y_{i-1} + y_i)h$. The trapezoid is shown more clearly in the zoom-view at the bottom of the figure.

Adding the trapezoid approximations on the n subintervals gives an approximation to the area beneath the graph of f or to the integral $\int_a^b f(x)\,dx$. Denoting this sum by T_n,

$$\int_a^b f(x)\,dx \approx T_n = \sum_{i=1}^{n} h\frac{f(x_{i-1}) + f(x_i)}{2} = \frac{h}{2}\sum_{i=1}^{n}(y_{i-1} + y)$$
$$= \tfrac{1}{2}h((y_0 + y_1) + (y_1 + y_2) + \cdots + (y_{n-1} + y_n)).$$

For more efficient calculation, we remove the inner parentheses, add adjacent identical terms, and factor the 2 from $2y_1 + 2y_2 + \cdots + 2y_{n-1}$. The result is the Trapezoid Rule:

Trapezoid Rule

Let $a = x_0 < x_1 < \cdots < x_n = b$ be a regular subdivision of $[a, b]$, $h = (b - a)/n$, and let

$$y_0 = f(x_0), y_1 = f(x_1), \ldots, y_n = f(x_n).$$

The trapezoid rule is

$$\int_a^b f(x)\,dx \approx T_n = \tfrac{1}{2}h(y_0 + 2(y_1 + y_2 + \cdots + y_{n-1}) + y_n). \qquad (1)$$

As the number n of subdivisions increases we expect that the difference between T_n and $\int_a^b f(x)\,dx$ will decrease. We explore the question of how to choose n so that T_n is, say, within 0.01 of $\int_a^b f(x)\,dx$ in our first example.

The formula for T_n in (1) does not depend upon f being nonnegative. Although we argued in terms of area, the idea of the trapezoid rule may be expressed in other terms. The approximation is based on replacing the function f on a representative subinterval $[x_{i-1}, x_i]$ by a function whose graph is the line joining the points $(x_{i-1}, f(x_{i-1}))$ and $(x_i, f(x_i))$. If f is nonnegative, this line is the top edge of the trapezoid. See Exercise 15.

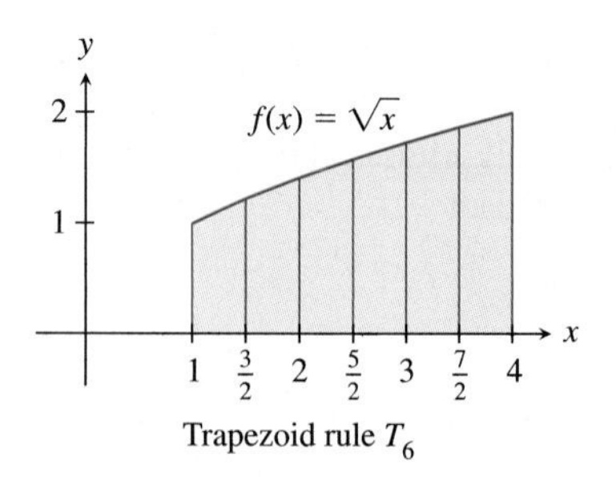

FIGURE 5.44 The approximation T_6 to the integral $\int_1^4 \sqrt{x}\,dx$.

For the first example, we use the trapezoid rule to approximate the value of an integral whose value we know. This will tell us something about the accuracy of the trapezoid rule.

EXAMPLE 1 Use the trapezoid rule with $n = 6$ subdivisions to approximate the integral

$$\int_1^4 \sqrt{x}\,dx = \tfrac{2}{3}x^{3/2}\Big|_1^4 = \frac{14}{3} = 4.666\ldots.$$

Calculate the exact error $|14/3 - T_6|$ of this approximation.

Solution

Figure 5.44 shows the graph of $f(x) = \sqrt{x}$ on $[1,4]$, as well as the trapezoids whose areas sum to T_6. As the figure makes clear, although $n = 6$ is not large, the trapezoids fit quite well. Their tops are barely distinguishable from the graph of $f(x) = \sqrt{x}$.

For $n = 6$, $h = (4 - 1)/6 = 1/2$. The subdivision of $[1,4]$ is

$$x_0 = 1, x_1 = \tfrac{3}{2}, x_2 = \tfrac{4}{2}, x_3 = \tfrac{5}{2}, x_4 = \tfrac{6}{2}, x_5 = \tfrac{7}{2}, x_6 = 4.$$

From (1),

$$\begin{aligned}\int_1^4 \sqrt{x}\,dx \approx T_6 &= \tfrac{1}{2}h\left(\sqrt{x_0} + 2\left(\sqrt{x_1} + \cdots + \sqrt{x_5}\right) + \sqrt{x_6}\right)\\ &\approx \frac{1}{2}\cdot\frac{1}{2}\left(\sqrt{1} + 2\left(\sqrt{\frac{3}{2}} + \sqrt{\frac{4}{2}} + \sqrt{\frac{5}{2}} + \sqrt{\frac{6}{2}}\right.\right.\\ &\qquad \left.\left. + \sqrt{\frac{7}{2}}\right) + \sqrt{4}\right) = 4.661488\ldots.\end{aligned}$$

The difference between the value of the integral and the approximation T_6 is $|14/3 - T_6| = |4.666\ldots - 4.661488\ldots| < 0.0052$.

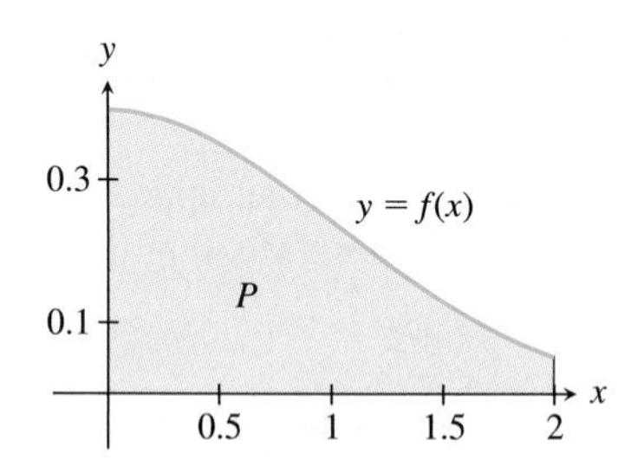

FIGURE 5.45 Part of the graph of $f(x) = \left(1/\sqrt{2\pi}\right)e^{-x^2/2}$.

EXAMPLE 2 Use the trapezoid rule (1) with $n = 5$ to approximate the integral

$$P = \int_0^2 f(x)\,dx, \quad \text{where } f(x) = \frac{1}{\sqrt{2\pi}}e^{-x^2/2}. \tag{2}$$

The graph of f is part of the well-known "bell-shaped curve." Integrals similar to (2) are very important in the fields of probability and statistics. Referring to Fig. 5.45, this number can be interpreted as the area of the region beneath the graph of f on the interval $[0,2]$.

Solution

We calculate T_5 in some detail. From (1), with $n = 5$,

$$T_5 = \tfrac{1}{2}h(y_0 + 2(y_1 + y_2 + y_3 + y_4) + y_5). \tag{3}$$

A geometric interpretation of the trapezoid rule for $n = 5$ is shown at the top of Fig. 5.46. Note that except for the leftmost trapezoid, the areas of

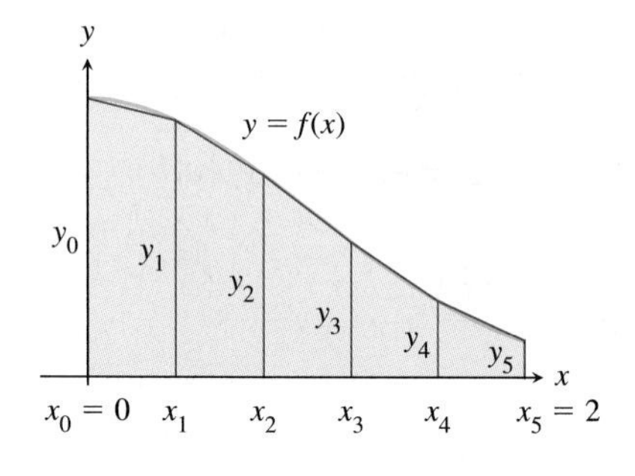

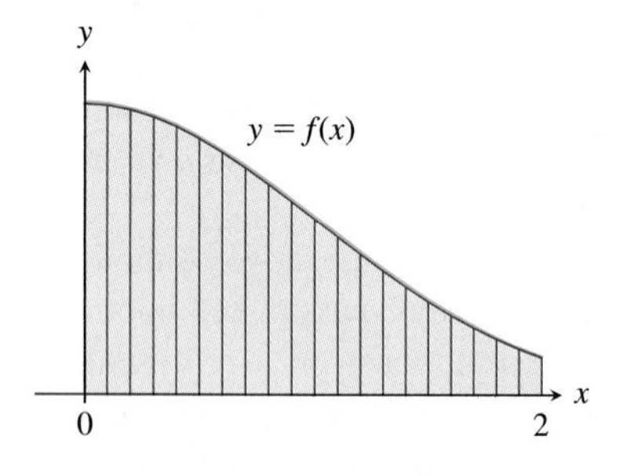

FIGURE 5.46 The trapezoid approximations T_5 and T_{20} to $\int_0^2 \left(1/\sqrt{2\pi}\right)e^{-x^2/2}\,dx$.

TABLE 5.3 Data for T_5.

i	x_i	$y_i = f(x_i)$
0	0.0	0.3989
1	0.4	0.3682
2	0.8	0.2896
3	1.2	0.1941
4	1.6	0.1109
5	2.0	0.0539

the approximating trapezoids appear to be close to the areas beneath the graph of f and above the subintervals $[x_1, x_2,], \ldots, [x_4, x_5]$. For $n = 5$, $h = (b - a)/n = (2 - 0)/5 = 0.4$, $x_0 = 0$, $x_1 = h$, $x_2 = 2h, \ldots, x_5 = 5h = 2$. Data for calculating T_5 are given in Table 5.3. Although only four decimals are displayed in this table, we used the full internal accuracy of our calculator for all calculations.

From (3),

$$T_5 \approx \frac{0.4}{2}(0.3989 + 2(0.3682 + 0.2896 + 0.1941 + 0.1109) + 0.0539)$$
$$\approx 0.4758.$$

To gain some idea of the accuracy of T_5, we calculated $T_{10} \approx 0.4769$ and $T_{20} \approx 0.4772$. The trapezoids for T_{20} are shown at the bottom of Fig. 5.46. Although proof is lacking, it appears that P is within 0.01 of the number 0.48. The change from T_5 to T_{10} is 0.0011, and the change from T_{10} to T_{20} is 0.0003. It appears that further increases in n will not affect the second decimal place.

The Midpoint Rule

The midpoint rule, like the trapezoid rule, is a procedure for approximating the integral $\int_a^b f(x)\,dx$ of an integrable function f. We again divide the interval $[a, b]$ into n equal subintervals with the points $a = x_0, x_1, \ldots, x_n = b$. Each subinterval has length $h = (b - a)/n$.

For convenience, we temporarily assume that $f(x) \geq 0$ on $[a, b]$, as in Fig. 5.47. The midpoint rule is based on the idea that it may be better to approximate the area beneath the graph of f on $[x_{i-1}, x_i]$ by evaluating f at the average of x_{i-1} and x_i rather than to average the values $y_{i-1} = f(x_{i-1})$ and $y_i = f(x_i)$. (The latter idea is equivalent to the trapezoid rule.) For the midpoint rule, we evaluate f at the midpoint $m_i = \frac{1}{2}(x_{i-1} + x_i)$ of the subinterval $[x_{i-1}, x_i]$ and approximate the area beneath f and above this subinterval with the number $hf(m_i)$. For many functions the midpoint rule is indeed more accurate than the trapezoid rule. We give some numerical evidence for this in Examples 3 and 4. In Exercise 16 we outline a proof that for functions that are concave up or down on $[a, b]$ the midpoint rule is better than the trapezoid rule.

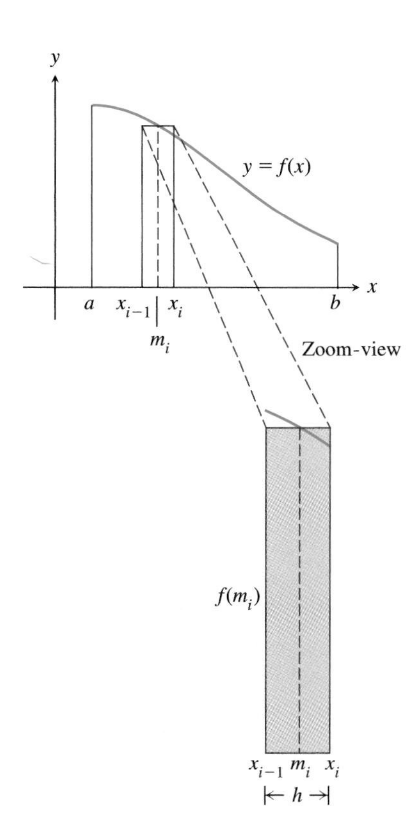

FIGURE 5.47 The midpoint approximation to the integral $\int_a^b f(x)\,dx$.

Midpoint Rule

Let $a = x_0 < x_1 < \cdots < x_n = b$ be a regular subdivision of $[a, b]$, $h = (b - a)/n$, and denote the midpoints of the n subintervals by

$$m_1 = \tfrac{1}{2}(x_0 + x_1), \ldots, m_n = \tfrac{1}{2}(x_{n-1} + x_n).$$

The midpoint rule is

$$\int_a^b f(x)\,dx \approx M_n = h(f(m_1) + f(m_2) + \cdots + f(m_n)). \tag{4}$$

This formula for M_n does not depend upon f being nonnegative. The approximation is based on replacing the function f on a representative subinterval $[x_{i-1}, x_i]$

by the constant function $g_i(x) = f(m_i)$. The graph of g_i is the horizontal line with equation $y = f(m_i)$, above or below the subinterval $[x_{i-1}, x_i]$. See Fig. 5.47.

EXAMPLE 3 Calculate the midpoint rule approximation M_6 and M_{12} for the integral

$$\int_1^4 \sqrt{x}\,dx = \tfrac{2}{3}x^{3/2}\Big|_1^4 = \frac{14}{3} = 4.666\ldots.$$

Given the trapezoid rule approximations $T_6 = 4.661488\ldots$ and $T_{12} = 4.665366\ldots$, compare the accuracy of the trapezoid and midpoint rules.

Solution

Letting $n = 6$, $h = (4-1)/6 = 1/2$. The subdivision of $[1,4]$ is $x_0 = 1, x_1 = \frac{3}{2}, x_2 = \frac{4}{2}, x_3 = \frac{5}{2}, x_4 = \frac{6}{2}, x_5 = \frac{7}{2}, x_6 = 4$. From (4),

$$\int_1^4 \sqrt{x}\,dx \approx M_6 = h\left(\sqrt{\tfrac{1}{2}(x_0 + x_1)} + \sqrt{\tfrac{1}{2}(x_1 + x_2)} + \cdots + \sqrt{\tfrac{1}{2}(x_5 + x_6)}\right)$$
$$= \tfrac{1}{2}\left(\sqrt{1.25} + \sqrt{1.75} + \cdots + \sqrt{3.75}\right) = 4.669244\ldots.$$

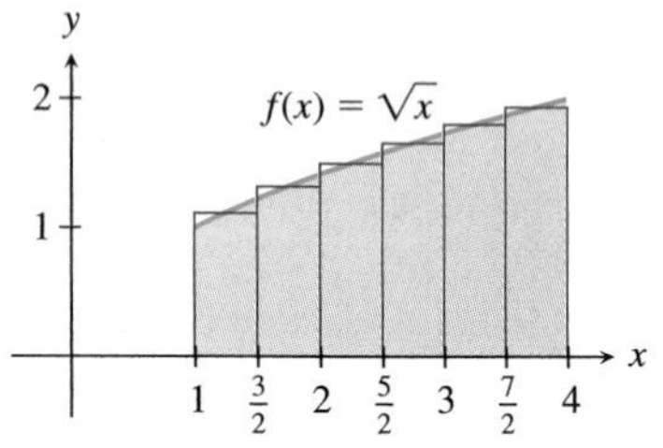

Figure 5.48 shows the graph of $f(x) = \sqrt{x}$ on $[1,4]$, as well as the rectangles associated with M_6.

To compare T_6 and M_6, we calculate the errors $|T_6 - 14/3|$ and $|M_6 - 14/3|$:

$$\left|T_6 - \frac{14}{3}\right| = 0.00517\ldots$$
$$\left|M_6 - \frac{14}{3}\right| = 0.00257\ldots.$$

For $n = 6$, the error in the trapezoid rule approximation is roughly twice that in the midpoint rule. For $n = 12$, $M_{12} = 4.667316\ldots$. The errors associated with T_{12} and M_{12} are

$$\left|T_{12} - \frac{14}{3}\right| = 0.00130\ldots$$
$$\left|M_{12} - \frac{14}{3}\right| = 0.00064\ldots.$$

Again, for $n = 12$, the error in the trapezoid rule is roughly twice that in the midpoint rule.

FIGURE 5.48 The approximation M_6 to the integral $\int_1^4 \sqrt{x}\,dx$.

EXAMPLE 4 Use the midpoint rule with $n = 10$ to approximate the integral

$$P = \int_0^2 \frac{1}{\sqrt{2\pi}} e^{-x^2/2}\,dx. \tag{5}$$

Use the fact that, correct to eight decimals, $P = 0.47724987$ to calculate the magnitude of the error $|P - M_{10}|$.

Solution

Let

$$f(x) = \frac{1}{\sqrt{2\pi}} e^{-x^2/2}.$$

For the midpoint rule M_{10}, $h = (2 - 0)/10 = 1/5 = 0.2$ and the subdivision of $[0, 2]$ is $x_0 = 1, x_1 = 0.2, x_2 = 0.4, \ldots, x_9 = 1.8, x_{10} = 2.0$. From (4),

$$\begin{aligned} M_{10} &= h\big(f\big(\tfrac{1}{2}(x_0 + x_1)\big) + f\big(\tfrac{1}{2}(x_1 + x_2)\big) + \cdots + f\big(\tfrac{1}{2}(x_9 + x_{10})\big)\big) \\ &= \tfrac{1}{5}(f(0.1) + f(0.3) + \cdots + f(0.9)) \approx 0.47742962. \end{aligned}$$

With this value of M_{10} and the given value of P,

$$|P - M_{10}| \approx |0.47724987 - 0.4774296| \approx 0.0002.$$

Simpson's Rule

The trapezoid and midpoint formulas are based on approximating the function f on the subintervals $[x_{i-1}, x_i]$ of a subdivision of $[a, b]$ by functions whose graphs are lines. Simpson's rule is based on approximating f on the union of the intervals $[x_{i-1}, x_i]$ and $[x_i, x_{i+1}]$ by a function of the form $y = Ax^2 + Bx + C$, whose graph is a parabola with a vertical axis. Because for most functions we can more closely fit the graph of f with parabolas than with lines, we expect Simpson's rule to be more accurate.

To determine the three unknown constants A, B, and C, we shall need the values $y_{i-1} = f(x_{i-1})$, $y_i = f(x_i)$, and $y_{i+1} = f(x_{i+1})$ of f at the three points x_{i-1}, x_i, and x_{i+1}. This is the reason why we take the subintervals two at a time in Simpson's rule and, hence, must assume that the number n of subintervals is even.

Simpson's rule will result from fitting a parabola $Ax^2 + Bx + C$ to the three points (x_{i-1}, y_{i-1}), (x_i, y_i), and (x_{i+1}, y_{i+1}), calculating $\int_{x_{i-1}}^{x_{i+1}} (Ax^2 + Bx + C)\,dx$ as an approximation to $\int_{x_{i-1}}^{x_{i+1}} f(x)\,dx$, and then adding together these $n/2$ approximations.

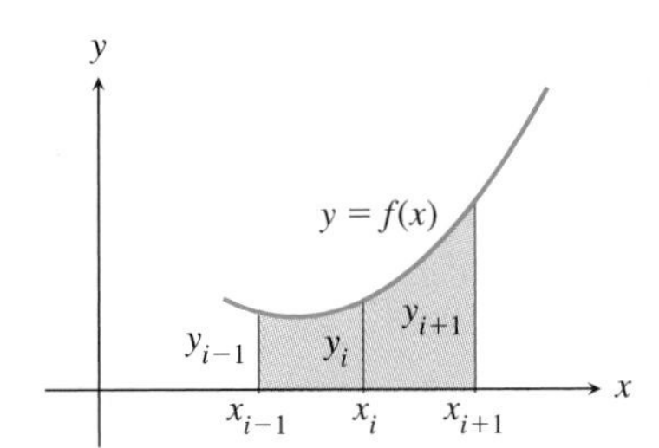

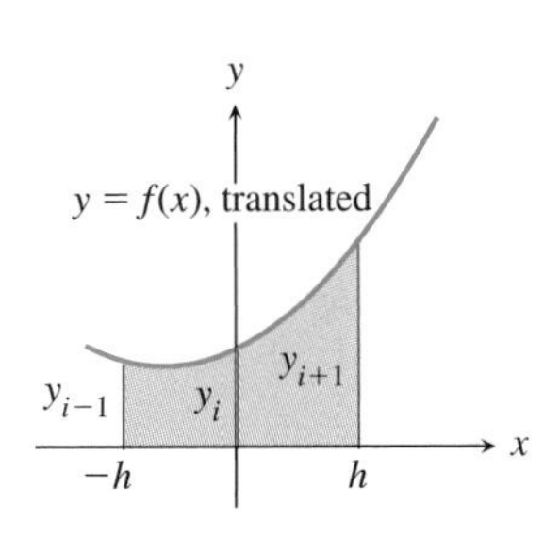

FIGURE 5.49 Translating a graph, without changing the area beneath a curve.

Figure 5.49 shows the graph of f on the subintervals $[x_{i-1}, x_i]$ and $[x_i, x_{i+1}]$. We can greatly simplify our work by translating f horizontally, so that the two subintervals, each of length $h = x_i - x_{i-1} = x_{i+1} - x_i = (b - a)/n$, are centered about the origin. This strategy works because the area of the shaded region beneath the graph of f and above $[x_{i-1}, x_{i+1}]$ is equal to the area of the shaded region beneath the translated f and above the interval $[-h, h]$. We fit a parabola in this new setting and calculate the area beneath it.

Substituting the points $(-h, y_{i-1})$, $(0, y_i)$, and (h, y_{i+1}) one at a time into the equation $y = Ax^2 + Bx + C$ gives the equations

$$\begin{aligned} y_{i-1} &= Ah^2 - Bh + C \\ y_i &= C \\ y_{i+1} &= Ah^2 + Bh + C. \end{aligned}$$

The second equation gives $C = y_i$; adding the first and third equations gives A, and subtracting the first from the third equation gives B. We find

$$A = \frac{y_{i-1} - 2y_i + y_{i+1}}{2h^2}, \quad B = \frac{y_{i+1} - y_{i-1}}{2h}, \quad \text{and} \quad C = y_i.$$

With these values of A, B, and C, the area of the region beneath the quadratic function $Ax^2 + Bx + C$ on $[-h, h]$ is approximately the area of the region beneath the graph of $f(x)$ on $[x_{i-1}, x_{i+1}]$, that is,

$$\int_{-h}^{h} (Ax^2 + Bx + C)\,dx \approx \int_{x_{i-1}}^{x_{i+1}} f(x)\,dx. \tag{6}$$

Evaluating the integral on the left, but not yet replacing A, B, and C by their calculated values,

$$\begin{aligned}\int_{-h}^{h} (Ax^2 + Bx + C)\,dx &= \left(\tfrac{1}{3}Ax^3 + \tfrac{1}{2}Bx^2 + Cx\right)\Big|_{-h}^{h}\\ &= \left(\tfrac{1}{3}Ah^3 + \tfrac{1}{2}Bh^2 + Ch\right) - \left(-\tfrac{1}{3}Ah^3 + \tfrac{1}{2}Bh^2 - Ch\right)\\ &= \tfrac{2}{3}Ah^3 + 2Ch.\end{aligned}$$

Substituting the values for A, B, and C found earlier,

$$\begin{aligned}\int_{-h}^{h} (Ax^2 + Bx + C)\,dx &= \tfrac{2}{3}h^3\frac{y_{i-1} - 2y_i + y_{i+1}}{2h^2} + 2hy_i\\ &= \tfrac{1}{3}h(y_{i-1} + 4y_i + y_{i+1}).\end{aligned}$$

With this result, we may write (6) as

$$\int_{x_{i-1}}^{x_{i+1}} f(x)\,dx \approx \tfrac{1}{3}h(y_{i-1} + 4y_i + y_{i+1}).$$

Adding the approximations for all n pairs of subintervals gives

$$\begin{aligned}\int_a^b f(x)\,dx &= \int_{x_0}^{x_2} f(x)\,dx + \int_{x_2}^{x_4} f(x)\,dx + \cdots + \int_{x_{n-2}}^{x_n} f(x)\,dx\\ &\approx \tfrac{1}{3}h(y_0 + 4y_1 + y_2) + \tfrac{1}{3}h(y_2 + 4y_3 + y_4)\\ &\quad + \cdots + \tfrac{1}{3}h(y_{n-2} + 4y_{n-1} + y_n)\\ &\approx \tfrac{1}{3}h((y_0 + 4y_1 + y_2) + (y_2 + 4y_3 + y_4)\\ &\qquad + \cdots + (y_{n-2} + 4y_{n-1} + y_n)).\end{aligned}$$

Simpson's rule is an efficient rearrangement of this expression.

WEB

Simpson's Rule

For even n, let $a = x_0 < x_1 < \cdots < x_n = b$ be a regular subdivision of $[a, b]$, $h = (b - a)/n$, and

$$y_0 = f(x_0), y_1 = f(x_1), \ldots, y_n = f(x_n).$$

Simpson's rule is

$$\begin{aligned}\int_a^b f(x)\,dx \approx S_n = \tfrac{1}{3}h(y_0 &+ 4(y_1 + y_3 + \cdots + y_{n-1})\\ &+ 2(y_2 + y_4 + \cdots + y_{n-2}) + y_n).\end{aligned} \tag{7}$$

EXAMPLE 5 Use Simpson's rule with $n = 6$ to approximate the integral

$$\int_0^{\sqrt{2\pi}} \sin x^2\,dx. \tag{8}$$

Solution

For Simpson's rule with $n = 6$ (which, as required by Simpson's rule, is even), $h = (\sqrt{2\pi} - 0)/6$ and the subdivision of $[0, \sqrt{2\pi}]$ is

$$x_0 = 0, x_1 = \tfrac{1}{6}\sqrt{2\pi}, x_2 = \tfrac{2}{6}\sqrt{2\pi}, \ldots, x_6 = \tfrac{6}{6}\sqrt{2\pi}.$$

From (7),

$$\begin{aligned} S_6 &= \tfrac{1}{3}h[y_0 + 4(y_1 + y_3 + y_5) + 2(y_2 + y_4) + y_6] \\ &= \tfrac{1}{18}\sqrt{2\pi}\Big[\sin 0^2 + 4\big(\sin(\tfrac{1}{6}\sqrt{2\pi})^2 + \sin(\tfrac{3}{6}\sqrt{2\pi})^2 + \sin(\tfrac{5}{6}\sqrt{2\pi})^2\big) \\ &\qquad + 2\big(\sin(\tfrac{2}{6}\sqrt{2\pi})^2 + \sin(\tfrac{4}{6}\sqrt{2\pi})^2\big) + \sin(\tfrac{6}{6}\sqrt{2\pi})^2\Big] \\ &= 0.404602\ldots. \end{aligned}$$

An accurate value of the integral (8)—calculated with the built-in algorithms on your calculator or CAS—is 0.430408. Hence, the error of approximation in this case is about 0.03.

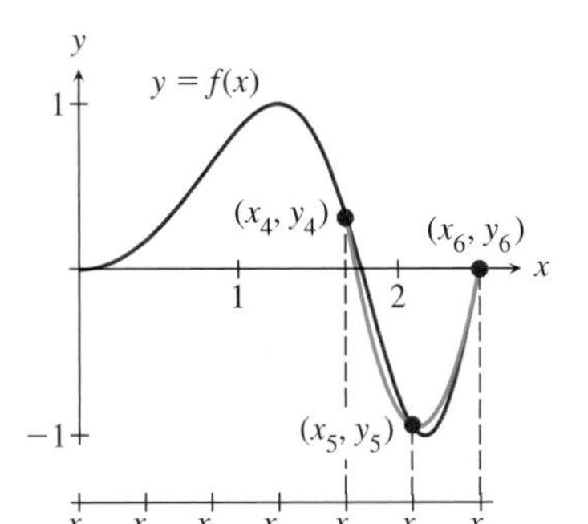

FIGURE 5.50 Fitting a parabola to the points (x_4, y_4), (x_5, y_5), (x_6, y_6) on the graph of $f(x) = \sin x^2$; the parabola is shown in blue.

One reason the error is relatively large in Example 5 may be seen in Fig. 5.50. Because of the position of the subintervals $[x_4, x_5]$ and $[x_5, x_6]$ relative to the minimum of f, the parabola fits relatively poorly. The equation of the parabola and its contribution to S_6 are explored in Exercises 19 and 20.

In Example 4 we used the midpoint rule with $n = 10$ to approximate the integral

$$P = \int_0^2 \frac{1}{\sqrt{2\pi}} e^{-x^2/2}\,dx. \tag{9}$$

We use Simpson's rule with $n = 10$ to again approximate this integral.

EXAMPLE 6 Use Simpson's rule with $n = 10$ to approximate the integral (9). Compare $|P - S_{10}|$ and $|P - M_{10}|$. Recall from Example 4 that $P \approx 0.47724987$ and $|P - M_{10}| \approx 0.0002$.

Solution

Let

$$f(x) = \frac{1}{\sqrt{2\pi}} e^{-x^2/2}.$$

For Simpson's rule with $n = 10$, $h = (2 - 0)/10 = 1/5 = 0.2$ and the subdivision of $[0, 2]$ is $x_0 = 1, x_1 = 0.2, x_2 = 0.4, \ldots, x_9 = 1.8, x_{10} = 2.0$. From (7),

$$\begin{aligned} S_{10} &= \tfrac{1}{3}h(y_0 + 4(y_1 + y_3 + y_5 + y_7 + y_9) \\ &\qquad + 2(y_2 + y_4 + y_6 + y_8) + y_{10}) \approx 0.47724887. \end{aligned}$$

We note that $|P - S_{10}| \approx |0.47724987 - 0.47724887| \approx 0.00001$, which is about 1/200th of the error made with the midpoint rule M_{10}.

Error Analysis for the Trapezoid Rule

We used the trapezoid rule to approximate

$$P = \int_0^2 \frac{1}{\sqrt{2\pi}} e^{-x^2/2}\, dx \tag{10}$$

in Example 2, but said little about choosing the number n of subdivisions so that T_n is within some error tolerance E of P. We end the section by showing how to choose n so that

$$|P - T_n| < 0.001.$$

Our discussion is based on the fact that the error $|\int_a^b f(x)\, dx - T_n|$ in the trapezoid approximation T_n to the integral $\int_a^b f(x)\, dx$ satisfies the inequality

$$\left| \int_a^b f(x)\, dx - T_n \right| \le \frac{b-a}{12} h^2 M, \tag{11}$$

where $h = (b - a)/n$ and M is a number for which $-M \le f''(x) \le M$, that is, $|f''(x)| \le M$, for all $a \le x \le b$. Proofs of this inequality—and similar inequalities for the midpoint rule and Simpson's rule—may be found in books on numerical analysis.

The number M is chosen so that the graph of f'' lies entirely between the lines with equations $y = -M$ and $y = M$. To determine a value of M, we calculate f'' and then estimate M, perhaps by graphing f''. The second derivative of f is

$$f''(x) = \frac{x^2 - 1}{\sqrt{2\pi}} e^{-x^2/2}.$$

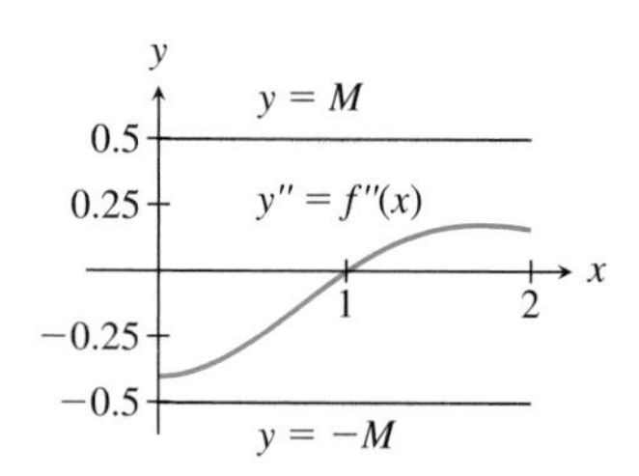

FIGURE 5.51 The graph of f'' on $[0, 2]$; the graph lies between the lines with equations $y = -0.5$ and $y = 0.5$.

A graph of this function is shown in Fig. 5.51. From the figure it appears that the graph of f'' lies between the lines with equations $y = -0.5$ and $y = 0.5$, which means that we can take $M = 0.5$. We could do a little better, but to be on the safe side (and not spend too much time or energy on choosing M), we take $M = 0.5$.

Next, we choose n so that

$$\left| \int_0^2 \frac{1}{\sqrt{2\pi}} e^{-x^2/2}\, dx - T_n \right| \le \frac{b-a}{12} h^2 M \le 0.001. \tag{12}$$

The first inequality in (12) comes from (11), where $h = (b - a)/n = 2/n$ and $M = 0.5$. We force the second inequality to hold by choosing n sufficiently large that

$$\left| \int_0^2 \frac{1}{\sqrt{2\pi}} e^{-x^2/2}\, dx - T_n \right| \le \frac{2-0}{12} h^2 \cdot 0.5 = \frac{1}{3n^2} < 0.001. \tag{13}$$

How do we choose n so that the last inequality holds? We can solve the inequality $1/(3n^2) < 0.001$ for n or we can determine n experimentally, by trying several values. For example, if we calculate $1/(3n^2)$ for $n = 5$, 10, 15 and 20, we find, rounding to two significant figures, 0.013, 0.0033, 0.0015, and 0.00083. Hence, $n = 20$ is large enough. Solving the inequality is not difficult:

$$\frac{1}{3n^2} < 0.001 = \frac{1}{1000}$$

$$\frac{1000}{3} < n^2.$$

Hence,

$$\sqrt{\frac{1000}{3}} < n.$$

This inequality can be written as

$$n > \sqrt{\frac{1000}{3}} \approx 18.26.$$

Taking $n = 19$, we find that $T_{19} = 0.47715\ldots$. Recalling that $P = 0.47724987$,

$$|P - T_{19}| \approx |0.47724987 - 0.47715| = 0.00009987.$$

Hence, T_{19} is well within 0.001 of P.

Exercises 5.10

1. Use the natural logarithm key on your calculator to obtain a good approximation to ln 3. Because $\ln 3 = \int_1^3 dx/x$, we may approximate ln 3 by numerical integration. Calculate the trapezoid rule approximations T_5 and T_{10} and the differences between them and the calculator value of ln 3. Retain six decimals.

2. Repeat Exercise 1 but use the midpoint rule approximations M_5 and M_{10}.

3. Repeat Exercise 1 but use the Simpson's rule approximations S_4 and S_8.

4. Approximate the value of

$$P = \frac{1}{\sqrt{2\pi}} \int_0^1 e^{-x^2/2}\, dx$$

with T_4, T_8, and T_{16}. Using this evidence, try to choose p and d so that $P = p \pm d$. The number d should be positive and as small as possible.

5. Repeat Exercise 4 but use the midpoint rule.

6. Repeat Exercise 4 but use Simpson's rule.

7. Approximate to within 0.001 the integral

$$\int_0^{\pi/2} \ln\left(1 + \sqrt{1 - 0.25 \sin^2 x}\right) dx.$$

8. Use Simpson's rule to calculate arcsin 0.5. Determine n empirically so that S_n is within 0.0001 of arcsin 0.5. Recall that

$$\arcsin 0.5 = \int_0^{0.5} \frac{dx}{\sqrt{1 - x^2}}.$$

9. The length L of the orbit of a planet moving on the ellipse described by $x^2/a^2 + y^2/b^2 = 1$ is given by the integral

$$L = 4a \int_0^{\pi/2} \sqrt{1 - e^2 \sin^2 x}\, dx,$$

where the eccentricity e of the ellipse is related to the lengths a and b of the semimajor and semiminor axes by $e^2 = 1 - b^2/a^2$. Use the midpoint rule to calculate L to four significant figures. Use Pluto's orbit data: $a = 39.78$ AU and $e = 0.2539$ (1 AU $= 149.6 \times 10^6$ km).

10. Calculate the mass M of a column of air above a 1×1 square meter patch of Earth, from sea level up to 11,000 meters, which is the edge of the troposphere. The density (in kg/m^3) of the atmosphere at height x meters is

$$\delta(x) = \frac{353.2(1 - 0.00002256x)^{5.259}}{288.2 - 0.0065x}.$$

11. In the error analysis for the trapezoid rule, we calculated f'' for the function $f(x) = \left(1/\sqrt{2\pi}\right)e^{-x^2/2}$. On the interval $[0, 2]$, we estimated that $|f''(x)| \le 0.5$. Show, in fact, that $|f''(x)| \le 0.4$ on $[0, 2]$. Does this affect the calculation of n in Equation (13)?

12. An airfoil is shown in the accompanying figure. The upper and lower surfaces are the graphs of the functions f and g, respectively. Data for the upper and lower surfaces are listed in the accompanying table. Calculate the cross-sectional area of the airfoil from this data by summing the areas of the trapezoids on the subintervals of the x-axis. Note that the data do not come from a regular subdivision of $[0, 30]$.

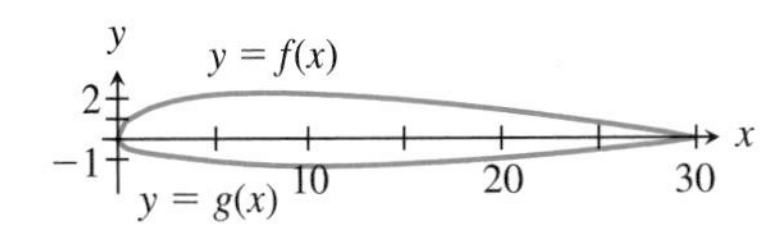

Figure for Exercise 12.

x	$f(x)$	$g(x)$
0	0	0
0.375	0.801	−0.369
0.75	1.083	−0.513
1.50	1.473	−0.678
2.25	1.74	−0.783
3.00	1.929	−0.876
4.50	2.157	−1.050
6.00	2.25	−1.191
7.50	2.280	−1.284
9.00	2.265	−1.338
12.00	2.142	−1.344
15.00	1.923	−1.251
18.00	1.641	−1.101
21.00	1.308	−0.900
24.00	0.924	−0.648
27.00	0.504	−0.369
28.50	0.276	−0.210
30.00	0	0

Table for Exercise 12.

13. The range of hills shown in the accompanying figure has uniform cross sections in the planes perpendicular to the horizontal line joining B and A. The figure is not drawn to scale. A cross section was surveyed, with elevations made every 100 feet along CB. Starting at C, with all measurements given in feet, the survey data were recorded as (x_i, y_i) pairs. These data are given in the accompanying table. A cut for a road is to be made through the hills, perpendicular to line BA. The base of the cut is to lie in the plane determined by A, B, and C. The width of the cut is to be 75 feet. Use the trapezoid rule in determining the volume of the cut.

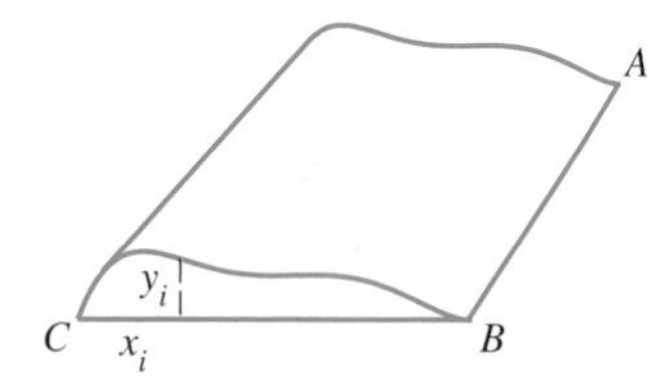

Figure for Exercise 13.

i	x_i	y_i	i	x_i	y_i
0	0	0.0	9	900	20.1
1	100	20.5	10	1000	20.3
2	200	29.2	11	1100	20.0
3	300	30.8	12	1200	18.8
4	400	28.9	13	1300	16.1
5	500	25.7	14	1400	11.9
6	600	22.8	15	1500	6.9
7	700	20.9	16	1600	2.2
8	800	20.1	17	1700	0.0

Table for Exercise 13.

14. The function

$$\mathrm{Si}(x) = \int_0^x \frac{\sin t}{t}\,dt, \ -\infty < x < \infty$$

is called the **sine integral.** Based on the fact that $\lim_{\theta\to 0}(\sin\theta/\theta) = 1$, the integrand is defined to be 1 at $t = 0$. With this understanding, the integrand is continuous everywhere. Calculate Si(1) to five significant figures.

15. Show that the trapezoid rule results from replacing the function f on each subinterval by an approximating function, integrating the approximating function, and summing the results. On the representative subinterval $[x_{i-1}, x_i]$, take the approximating function as the function whose graph is the line joining the points $(x_{i-1}, f(x_{i-1}))$ and $(x_i, f(x_i))$.

16. In the text we gave some numerical evidence that the midpoint rule is better than the trapezoid rule for some functions. Fill in the details of the following argument that for a continuous function f that is concave down on $[a, b]$,

$$0 \le M_n - \int_a^b f(x)\,dx \le \int_a^b f(x)\,dx - T_n. \qquad (14)$$

In the accompanying figure, let Q_1 denote the area of triangle APB, Q_2 the area of triangle PCD, T the area of trapezoid $GFCA$, Q_3 the area of triangle PCE, and Q_4 the area of triangle ACP. Let $I = \int_a^b f(x)\,dx$. Step 1: Show that $Q_1 + Q_2 = Q_3 = Q_4$. This is easy geometry. Step 2: Show that the midpoint approximation M on $[x_i, x_{i+1}]$ is equal to the area beneath any line passing through P. (This is also geometry.) Step 3: Observe that $I \ge T + Q_4$ and, hence,

$$0 \le M - I \le M - T - Q_4 = Q_1 + Q_2.$$

Step 4: Verify the inequality

$$0 \le M - I \le Q_1 + Q_2 = Q_3 = Q_4 \le I - T.$$

Because this inequality holds on each subinterval, (14) follows. A similar argument holds for functions that are concave up.

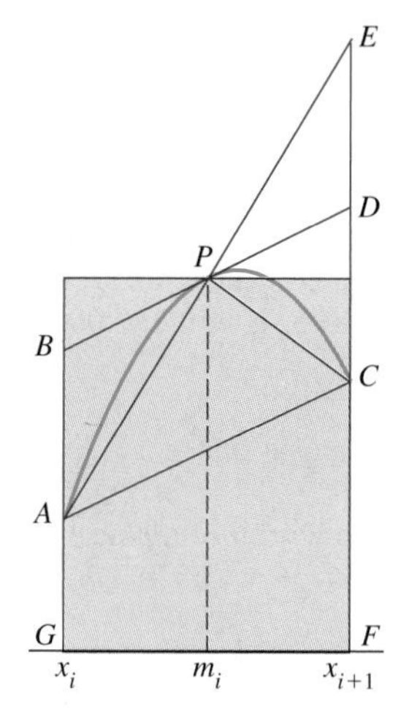

Figure for Exercise 16.

17. Argue that Simpson's rule is exact for polynomials of degree 2 or less. Show experimentally that Simpson's rule is exact for polynomials of degree 3 or less.

18. Continuation of Exercise 17 Use a CAS to prove that Simpson's rule is exact for polynomials of degree 3 or less. *Hint:* For a general cubic f, calculate the coefficients a, b, and c of a quadratic passing through the points $(p-h, f(p-h))$, $(p, f(p))$, and $(p+h, f(p+h))$. Show that the integral of this quadratic on the interval $[p-h, p+h]$ is equal to the integral of the general cubic on the same interval.

19. Show that the parabola through the points (x_4, y_4), (x_5, y_5), (x_6, y_6) in Example 5 is $y = 6.36386x^2 - 26.9957x + 27.6829$, approximately.

20. Continuation of Exercise 19 An accurate value of the integral $\int_{x_4}^{x_6} \sin x^2\,dx$ is -0.446740. Compare this with the integral on the same interval of the approximating parabola given in the preceding exercise.

Review of Key Concepts

We began this chapter by defining the definite integral $\int_a^b f(x)\,dx$ as a limit of Riemann sums $R_n = \Sigma_{i=1}^n f(c_i)\,\Delta x$. When values of f are nonnegative on $[a, b]$ we often interpreted R_n as the sum of the areas of n rectangular regions more or less beneath the graph of f.

The Fundamental Theorem of Calculus, which we discussed first in the context of motion, is the most important tool used to evaluate $\int_a^b f(x)\,dx$.

Much of the remainder of the chapter was spent on techniques for evaluating $\int_a^b f(x)\,dx$. For functions f for which an antiderivative F can be more or less easily found, we discussed the following "integration techniques" for finding F: searching in a table of integrals, substitution, integration by parts, and partial fractions. For functions f for which it is not possible or practical to determine F, we discussed the trapezoid rule, midpoint rule, and Simpson's rule.

One application of the indefinite integral $\int f(x)\,dx$, which is an antiderivative of f, was to the solution of separable differential equations.

Chapter Summary

Definition of the Integral

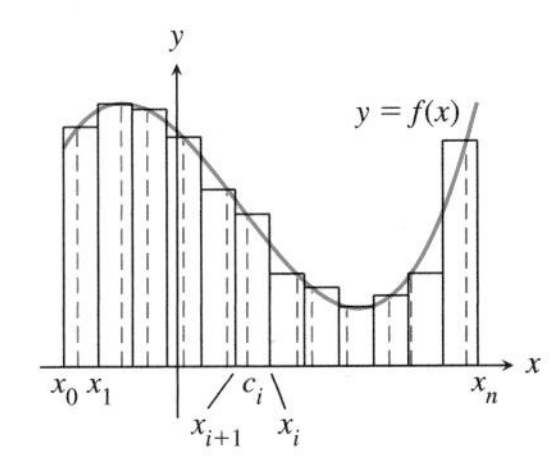

Let f be a bounded function defined on $[a, b]$. If there is a number I for which

$$\lim_{n\to\infty} R_n = \lim_{n\to\infty} \sum_{i=1}^{n} f(c_i)\,\Delta x = I,$$

where $\Delta x = (b-a)/n$, $\{x_0, x_1, \ldots, x_n\}$ is a regular subdivision of $[a, b]$, and $\{c_1, c_2, \ldots, c_n\}$ is any evaluation set for this subdivision, then we say that f is **integrable** on $[a, b]$ and write

$$\int_a^b f(x)\,dx = I.$$

For $f(x) = \ln x$, $1 \le x \le 2$, we form R_4, where the evaluation set is the set of midpoints of the subintervals. With $\Delta x = (2-1)/4 = 1/4$, $c_1 = 9/8$, $c_2 = 11/8$, $c_3 = 13/8$, and $c_4 = 15/8$,

$$R_4 = \Delta x\left(\ln\left(\frac{9}{8}\right) + \ln\left(\frac{11}{8}\right) + \ln\left(\frac{13}{8}\right) + \ln\left(\frac{15}{8}\right)\right) = 0.691219\ldots.$$

This number approximates

$$\int_1^2 \frac{1}{x}\,dx = \ln 2 - \ln 1 = 0.693147\ldots.$$

Definitions and Properties of the Integral

Definition 1: Area. Let f be a nonnegative function on $[a,b]$. The area of the region $\mathcal{R}$ beneath the graph of f is the number $\int_a^b f(x)\,dx$.

Definition 2: $\displaystyle\int_a^a f(w)\,dw = 0.$

Definition 3: $\displaystyle\int_b^a f(x)\,dx = -\int_a^b f(x)\,dx.$

Definition 4: Average value. The average value of f on $[a,b]$ is

$$\frac{1}{b-a}\int_a^b f(x)\,dx.$$

Property 1: Additivity.

$$\int_r^s f(x)\,dx + \int_s^t f(x)\,dx = \int_r^t f(x)\,dx.$$

Property 2: Mean-value Theorem for Integrals. There is a point $c \in [a,b]$ such that

$$\int_a^b f(x)\,dx = f(c)\,(b-a).$$

Property 3: Linearity.

$$\int_a^b (f(x)+g(x))\,dx = \int_a^b f(x)\,dx + \int_a^b g(x)\,dx$$

and

$$\int_a^b cf(x)\,dx = c\int_a^b f(x)\,dx.$$

Property 4: $\displaystyle\left|\int_a^b f(x)\,dx\right| \le \int_a^b |f(x)|\,dx.$

Fundamental Theorem of Calculus

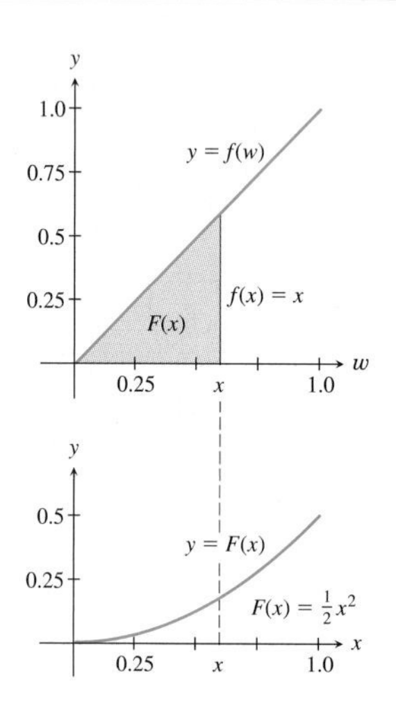

Part I: If f is continuous on $[a,b]$ and the function F is defined by

$$F(x) = \int_a^x f(w)\,dw, \qquad x \in [a,b],$$

then F is an antiderivative of f; that is, for all $x \in [a,b]$,

$$F'(x) = \frac{d}{dx}\int_a^x f(w)\,dw = f(x).$$

Part II: If f is defined on $[a,b]$ and G is any antiderivative of f on $[a,b]$, then

$$\int_a^b f(x)\,dx = G(b) - G(a).$$

If $f(x) = x$, then

$$F(x) = \int_0^x f(w)\,dw = \int_0^x w\,dw$$
$$= \tfrac{1}{2}x^2 - \tfrac{1}{2}0^2 = \tfrac{1}{2}x^2.$$

Note that $F'(x) = x = f(x)$.

Areas of Regions between Curves

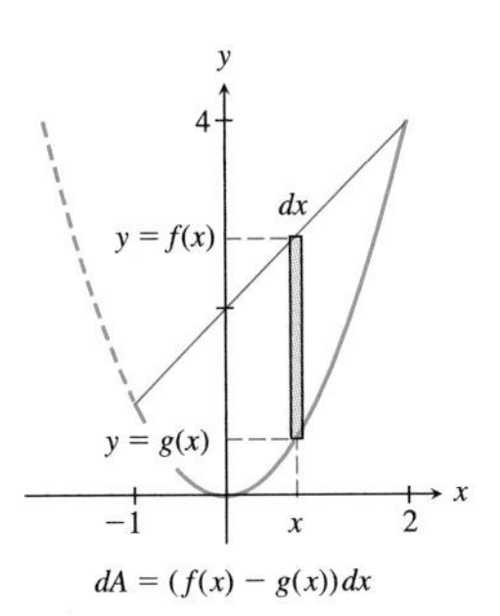

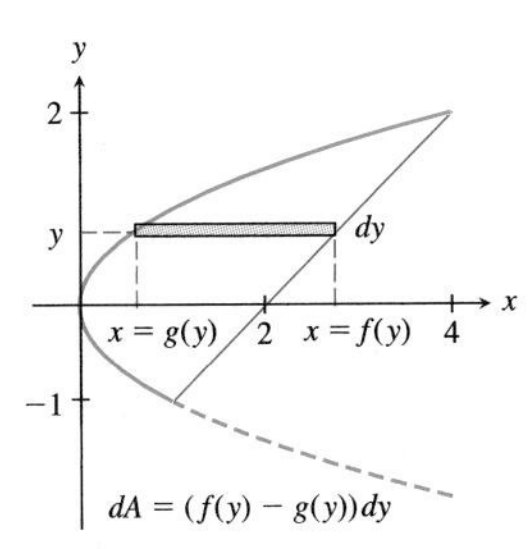

The vertical element of area for calculating the area of the region between graphs of $y = g(x)$ and $y = f(x)$ is

$$dA = (f(x) - g(x))\,dx.$$

The horizontal element of area for calculating the area of the region between graphs of $x = g(y)$ and $x = f(y)$ is

$$dA = (f(y) - g(y))\,dy.$$

The area of the region between the graphs of $y = g(x) = x^2$ and $y = f(x) = x + 2$—see top figure—is $\int_{-1}^{2}((x+2) - x^2)\,dx = 9/2$. The area of the region—see bottom figure—between the graphs of $x = g(y) = y^2$ and $x = f(y) = y + 2$ is $\int_{-1}^{2}(y + 2 - y^2)\,dy = 9/2$.

Integration Techniques

1. Tables of integrals
2. Substitution
3. Integration by parts
4. Partial fractions

1. See the table of integrals on the inside back cover.
2. With the change of variable $u = g(x)$.

$$\int_c^d f(u)\,du = \int_a^b f(g(x))g'(x)\,dx,$$

where $c = g(a)$ and $d = g(b)$.

3. $\displaystyle\int_a^b u\,dv = uv\Big|_a^b - \int_a^b v\,du$
4. Given a rational function $p(x)/q(x)$, divide, factor, decompose, and integrate the partial fractions.

1. Formula (G):

$$\int \frac{dx}{x^2 + a^2} = \frac{1}{a}\arctan\frac{x}{a} + C$$

2. With the substitution $x = u - 1$,

$$\int_0^1 \sqrt{x+1}\,dx = \int_1^2 u^{1/2}\,du.$$

3. With $u = x$ and $dv = e^x\,dx$,

$$\int_0^1 xe^x\,dx = xe^x\Big|_0^1 - \int_0^1 e^x\,dx.$$

4. For the integrand

$$\frac{x^2}{(x^2 - 5x + 6)(x^2 + 1)},$$

the standard decomposition is

$$\frac{A}{x-2} + \frac{B}{x-3} + \frac{Cx + D}{x^2 + 1}.$$

Numerical Integration

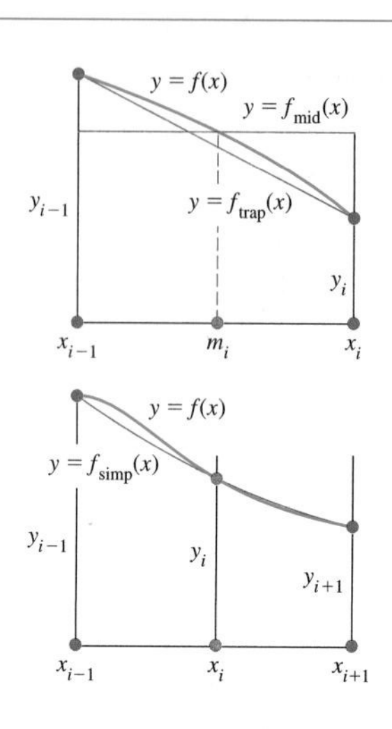

Three approximations to

$$\int_a^b f(x)\,dx$$

- Trapezoid rule: T_n
- Midpoint rule: M_n
- Simpson's rule: S_n (n must be even)

In the trapezoid rule, f is replaced on a representative subinterval $[x_{i-1}, x_i]$ by f_{trap}, whose graph is the line segment joining (x_{i-1}, y_{i-1}) and (x_i, y_i). See top figure. For the midpoint rule, f is replaced by f_{mid}, whose graph is the line segment joining $(x_{i-1}, f(m_i))$ and $(x_i, f(m_i))$. See top figure. For Simpson's rule, f is replaced by f_{simp}, whose graph is the parabola through (x_{i-1}, y_{i-1}), (x_i, y_i), and (x_{i+1}, y_{i+1}). See bottom figure.

$$T_n = \tfrac{1}{2}h(y_0 + 2(y_1 + y_2 + \cdots + y_{n-1}) + y_n)$$

$$M_n = h(f(m_1) + f(m_2) + \cdots + f(m_n))$$

$$S_n = \tfrac{1}{3}h(y_0 + 4(y_1 + y_3 + \cdots + y_{n-1}) + 2(y_2 + y_4 + \cdots + y_{n-2}) + y_n)$$

Chapter Review Exercises

1. Express using summation notation:

$$\frac{3}{2} + \frac{5}{4} + \frac{7}{8} + \cdots + \frac{21}{1024}.$$

2. Calculate $\Sigma_{j=1}^{6} (j - 2)^2$.

3. Let f be a continuous function on an interval $[a, b]$. For a given regular subdivision $\mathcal{S}_n$ of $[a, b]$, the least Riemann sum is $L_n = \Sigma_{i=1}^{n} f(p_i)\,\Delta x$, where $f(p_i)$ is the minimum value of f on $[x_{i-1}, x_i]$ and the greatest Riemann sum is $G_n = \Sigma_{i=1}^{n} f(q_i)\,\Delta x$, where $f(q_i)$ is the maximum value of f on $[x_{i-1}, x_i]$. In addition to the trapezoid rule, midpoint rule, and Simpson's rule, we may use the sums L_n, G_n, or, perhaps, $A_n = \frac{1}{2}(L_n + G_n)$ as approximations to $\int_a^b f(x)\,dx$.

Approximate $\int_0^1 x^4\,dx$ with A_5. Compare A_5 to the exact value of this integral. How is A_5 related, if at all, to the trapezoid approximation T_5?

4. (See previous exercise.) Calculate L_6, G_6, and $A_6 = \frac{1}{2}(L_6 + G_6)$ for the integral $\int_0^1 \sqrt[3]{x}\,dx$. Compare A_6 with the actual value of this integral. Sketch the graph of $\sqrt[3]{x}$ and the approximating rectangles from L_6 and G_6 on the third subinterval.

5. Determine the area of the region bounded by the graphs of $y = 3x$ and $y = 3(x - 2)^2$.

6. Determine the area of the region between the curves with equations $y = 4 - (x - 2)^2$ and $y = x$.

7. Determine the area of the region between the x-axis and the graph of $f(x) = x - x^2$.

8. The velocity of an object is $v(t) = t/(t + 1)$ for $t \geq 0$ and its initial position is $x(0) = 1$. Graph its position for $0 \leq t \leq 5$.

9. The velocity of the end E of a spring is $v(t) = e^{-t} \sin t$ for $t \geq 0$. The initial position of E is $x(0) = 0$. Plot the position of E for $0 \leq t \leq 5$.

10. If $F(x) = \int_0^x \sin(w^2)\,dw$, $x \geq 0$, write the equation of the tangent line to the graph of F at $(1, F(1))$. It is given that $F(1) \approx 0.31$.

11. The graph of a function f defined on the interval $[0, 6]$ is shown in the accompanying figure. A new function F is defined on $[0, 6]$ by

$$F(x) = \int_0^x f(w)\,dw.$$

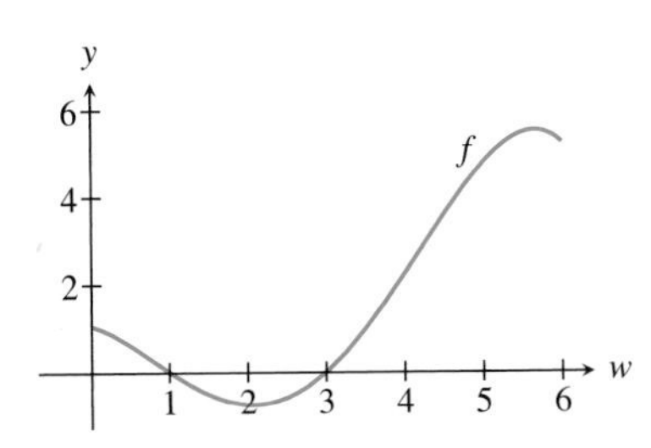

Figure for Exercise 11.

a. Sketch a graph of F and give a justification for the general shape of your sketch. Answers to the next two parts may help you with the graph.

b. On what intervals is F increasing? Decreasing? Why?

c. What are the x-coordinates of the high points of the graph of F? How do you know?

d. Sketch the graph of F'.

12. Let

$$G(x) = \int_0^{1/x} e^{w^2}\,dw \quad 1 \le x \le 10.$$

Sketch a rough graph of G and calculate the slope of the graph at $x = 2$.

13. Find the form of the standard decomposition of the function. Do not solve for the unknowns.

a. $\dfrac{2x - 1}{x^2 - 5x + 6}$

b. $\dfrac{x(x^3 + 1)}{x^4 - 1}$

c. $\dfrac{1}{(5x - 3)^2(x^2 + 1)^3}$

d. $\dfrac{x^2 + x + 1}{(x^2 + 1)^2(x + 1)(x^2 - 2)^2}$

e. $\dfrac{5x^2 + 4x + 2}{(2x^2 + 2x + 1)(x + 1)}$

14. Show that the *average velocity* of an object on an interval $[a, b]$ is equal to the *average value* of the velocity function on $[a, b]$.

15. Evaluate the integral.

a. $\displaystyle\int_0^1 (2x + 1)^5\,dx$

b. $\displaystyle\int \frac{dx}{2x + 3}$

c. $\displaystyle\int \frac{dx}{2x^2 + 1}$

d. $\displaystyle\int \frac{dx}{\sqrt{7 - x^2}}$

e. $\displaystyle\int_1^{\pi^2/4} \frac{\sin\sqrt{x}}{\sqrt{x}}\,dx$

f. $\displaystyle\int_1^e x^{-2}\ln x\,dx$

g. $\displaystyle\int_3^5 \frac{dx}{x^2 - 2x}$

h. $\displaystyle\int_2^3 (x^2 - 2x + 1)^{-1}\,dx$

i. $\displaystyle\int_0^{\pi/4} \frac{dx}{\cos^2 x}$

j. $\displaystyle\int \frac{dx}{\sqrt{x}(1 + x)}$

k. $\displaystyle\int_0^{\pi/2} x\sin x\,dx$

l. $\displaystyle\int \frac{1 + e^x}{1 - e^x}\,dx$

m. Use the trigonometric substitution $x = \sin u$ for the integral

$$\int_0^{1/2} \frac{x^2}{\sqrt{1 - x^2}}\,dx.$$

n. For positive constants a and b, show that

$$\int_{-b}^{b} \sqrt{1 + \sinh^2\left(\frac{x}{a}\right)}\,dx = 2a\sinh\left(\frac{b}{a}\right).$$

Recall the identity $\cosh^2 x - \sinh^2 x = 1$.

16. The integral

$$\int_{-1}^{(\sqrt{5}-2)/2} \frac{dx}{\sqrt{5 - (x + 1)^2}}$$

is transformed to $\int_c^d f(u)\,du$ by the substitution $x + 1 = \sqrt{5}\sin u$. Express $f(u)$, c, and d in simplest form.

17. Using either the tangent line approximation or numerical integration, approximate $s(3.1)$ if $s(3) \approx 9.74709$ and

$$s(x) = \int_0^x \sqrt{1 + 4w^2}\, dw.$$

18. Find a recursion formula for the integral

$$I_n = \int_1^e x(\ln x)^n\, dx$$

and then use it to evaluate I_3.

19. If a stream of water were running down the graph of

$$f(x) = 2x^3 - 15x^2 + 35x - 21,$$

a region would fill and overflow. Calculate the area of the filled region.

20. Calculate and simplify the *exact value* of the integral

$$\int_0^1 \frac{x^4(1 - x)^4}{1 + x^2}\, dx.$$

Plot the integrand on $[0, 1]$. Do the results amuse you? Does the first result have any significance?

21. Establish the identity

$$\arctan x + \arctan \frac{1}{x} = \frac{\pi}{2}, \qquad x > 0$$

by first showing that

$$\int_x^1 \frac{dw}{1 + w^2} = \int_1^{1/x} \frac{dw}{1 + w^2}.$$

(*Hint:* Let $w = 1/u$ in one of these integrals, and then evaluate each integral.)

22. A reduction formula for evaluating integrals containing powers of the logarithm function is

$$\int x^b(\ln ax)^n\, dx = \frac{x^{b+1}}{b + 1}(\ln ax)^n - \frac{n}{b + 1}\int x^b(\ln ax)^{n-1}\, dx,$$

where n is a positive integer, $a > 0$, and $b \neq -1$.

a. Use integration by parts to derive this formula.

b. Use the reduction formula to evaluate $\int \sqrt{x}(\ln x)^2\, dx$.

23. Let

$$I_n = \int_0^x w^n(w^2 + a^2)^{-1/2}\, dw.$$

Use integration by parts to show that

$$nI_n = x^{n-1}\sqrt{x^2 + a^2} - (n - 1)a^2 I_{n-2}(x), \qquad n \geq 2.$$

24. Solve the initial value problem

$$y' = -\frac{t}{y}$$

$$y(0) = 1.$$

25. Solve the initial value problem

$$y' = \frac{y^2 + 1}{t^3 + t}$$

$$y\left(\frac{1}{\sqrt{e^2 - 1}}\right) = 1.$$

26. Use the trapezoid rule with $n = 10$ to approximate $\pi/4$. Note that

$$\frac{\pi}{4} = \int_0^1 \frac{dx}{1 + x^2}.$$

How accurate is the approximation?

27. Use the midpoint rule with $n = 10$ to approximate

$$\int_0^{0.5} x^2 e^{-x^2}\, dx.$$

28. Use Simpson's rule with $n = 10$ to approximate ln 3. Note that

$$\ln 3 = \int_1^3 \frac{1}{x}\, dx.$$

How accurate is the approximation?

29. Use the inequality

$$\left|\int_a^b f(x)\, dx - T_n\right| \leq \frac{b - a}{12} h^2 M$$

in approximating $\int_0^{\pi/2} \cos(x^2)\, dx$ to within 0.001. Show all calculations.

STUDENT PROJECTS

A. SMALL-ROCKET DATA

Acceleration data (ft/s^2) for a small rocket with burnout time of 40 s are given in the accompanying table. Ignoring air resistance, calculate the maximum height reached and the total flight time. This problem comes from *Applied Methods for Digital Computation,* third edition, 1985, by M. L. James *et al.*, HarperCollins Publishers.

t	$a(t)$	t	$a(t)$
0.0	38.4	22.0	11.3
2.0	39.3	24.0	9.2
4.0	39.3	26.0	7.5
6.0	38.1	28.0	6.2
8.0	35.7	30.0	5.2
10.0	35.2	32.0	4.5
12.0	28.7	34.0	4.0
14.0	24.7	36.0	3.7
16.0	20.7	38.0	3.4
18.0	17.1	40.0	3.2
20.0	14.0		

B. IMPROVED CALCULATION OF "HYPERBOLA-AREAS"

In the autumn of 1665, Isaac Newton, newly graduated from Cambridge, discovered how to calculate the area beneath a hyperbola. He recorded his discovery in a notebook. We reproduce part of this notebook below, with only minor changes in his notation. We note that $\|$ means parallel, $\perp$ means perpendicular, "valor" means "value" and "&c." means "et cetera." We have made use of *The Mathematical Papers of Isaac Newton,* D. T. Whiteside (ed.), vol. I, Cambridge University Press, 1967, pp. 134–141.

Improved Calculation of Hyperbola-Areas

If $ea \| vb \| dc \perp ac \| ev = vb = a.$ & $bc = x.$ & $dc = y = \dfrac{aa}{a + x}$. Then is vd a Hyperbola &c:

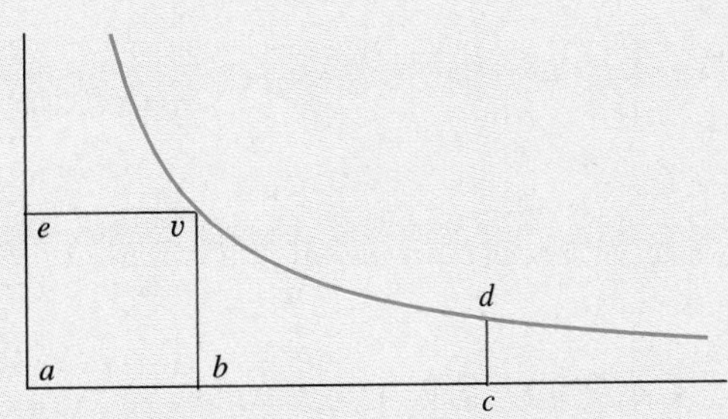

And if $\dfrac{aa}{a+x}$ bee divided as in decimall fractions ye product is

$$\frac{aa}{a+x} = y = a - x + \frac{xx}{a} - \frac{x^3}{aa} + \frac{x^4}{a^3} - \frac{x^5}{a^4} + \frac{x^6}{a^5} - \frac{x^7}{a^6} + \frac{x^8}{a^7}$$

$$- \frac{x^9}{a^8} + \frac{x^{10}}{a^9} - \frac{x^{11}}{a^{10}} + \frac{x^{12}}{a^{11}} - \frac{x^{13}}{a^{12} + a^{11}x} \text{\&c.}$$

Which valor of y being each terme thereof multiplyed by x & divided by ye number of its dimensions: The product will bee ye area *vbcd*. viz:

$$vbcd = ax - \frac{xx}{2} + \frac{x^3}{3a} - \frac{x^4}{4a^2} + \frac{x^5}{5a^3} - \frac{x^6}{6a^4} + \frac{x^7}{7a^5} - \frac{x^8}{8a^6} + \frac{x^9}{9a^7} - \frac{x^{10}}{10a^8} + \frac{x^{11}}{11a^9} - \frac{x^{12}}{12a^{10}} \&c.$$

As for example. If $a = 1.$ & $x = 0.1$. The calculation is as Followeth. [We show a few of the calculations Newton did by hand, copied out by hand, and summed by hand.]

$$0.10000,00000,00000,00000,00000,00000,00000,00000,00000,00000,00000 = ax$$

$$0.00033,33333,33333,33333,33333,33333,33333,33333,33333,33333,33333 = \frac{x^3}{3a}$$

$$0.00000,20000,00000,00000,00000,00000,00000,00000,00000,00000,00000 = \frac{x^5}{5a^3}$$

$$0.00000,00142,85714,28571,42857,14285,71428,57142,85714,28571,42857 = \frac{x^7}{7a^5}$$

$$\vdots$$

$$0.00000,00000,00000,00000,00000,00000,00000,00000,00000,00000,00196 = \frac{x^{51}}{51a^{49}}$$

$$0.00000,00000,00000,00000,00000,00000,00000,00000,00000,00000,00001 = \frac{x^{53}}{53a^{51}}$$

$$\vdots$$

$$0.00500,00000,00000,00000,00000,00000,00000,00000,00000,00000,00000 = \frac{xx}{2}$$

$$0.00002,50000,00000,00000,00000,00000,00000,00000,00000,00000,00000 = \frac{x^4}{3a^2}$$

$$0.00000,01666,66666,66666,66666,66666,66666,66666,66666,66666,66666 = \frac{x^6}{5a^4}$$

$$0.00000,00012,50000,00000,00000,00000,00000,00000,00000,00000,00000 = \frac{x^8}{7a^6}$$

$$\vdots$$

$$0.00000,00000,00000,00000,00000,00000,00000,00000,00000,00000,02000 = \frac{x^{50}}{50a^{48}}$$

$$0.00000,00000,00000,00000,00000,00000,00000,00000,00000,00000,00019 = \frac{x^{52}}{52a^{50}}$$

The [Difference of these two summes] is equall to ye area *bcdv*, viz:

$$bcdv = 0.09531,01798,04324,86004,39521,23280,76509,22206,05365,30864,41992$$

This ends Newton's calculation. Note that Newton knew, in 1665, that the area beneath the graph of $y = x^n$, $0 \le x \le b$, is $b^{n+1}/(n+1)$.

PROBLEM 1 Explain and verify the details of Newton's work.

PROBLEM 2 Calculate Newton's result using a "modern" function, one Newton made no mention of in his notebook. Explain your work.

PROBLEM 3 Did Newton include sufficiently many terms to obtain 52 decimal places of accuracy, as he claimed? Explain your answer.

C. HOW TO DISCOVER AN INTEGRATION ALGORITHM

The trapezoid rule for approximating $\int_a^b f(x)\,dx$ is given by

$$\int_a^b f(x)\,dx \approx T_n = \tfrac{1}{2}h\left(f(a) + f(b) + 2\sum_{i=1}^{n-1} f(a+ih)\right),$$

where $h = (b-a)/n$. In applying the trapezoid rule, we attempt to control the error $E_n = T_n - \int_a^b f(x)\,dx$ by choosing n. By studying how E_n depends on n, Werner Romberg discovered in 1955 a way to improve the efficiency of the trapezoid rule. Romberg showed that several applications of the trapezoid rule can be combined to obtain a strong increase in accuracy at the cost of relatively little additional calculation.

For a function f and interval $[a,b]$, Romberg looked at the approximation sequence $T_2, T_4, T_8, \ldots$. This sequence is efficient because in calculating T_4 roughly half of necessary function evaluations are included in T_2. In the interval $[0,1]$, for example, T_4 requires that we evaluate $f(0)$, $f(1/4)$, $f(2/4)$, $f(3/4)$, and $f(1)$. Of these, three are included in T_2. Similarly, for T_8 we may use T_4 and the additional values $f(1/8)$, $f(3/8)$, $f(5/8)$, and $f(7/8)$.

In general, the calculation of T_4 from the function data needed for T_2, the calculation of T_8 from T_4, and so on, is given by the formula

$$R(m+1,1) = \tfrac{1}{2}R(m,1) + \frac{b-a}{2^{m+1}}\sum_{i=1}^{2^m} f\left(a + (2i-1)\frac{b-a}{2^{m+1}}\right), \tag{1}$$

where $R(m,1)$ denotes T_{2^m}, $m = 1,2,3,\ldots$. We ask you to verify (1) in Problem 1.

For Problems 2–5, we shall need the value I of the integral

$$I = \int_0^\pi \frac{\sin x}{x}\,dx = 1.85193705198\ldots \tag{2}$$

PROBLEM 1 Verify (1) for the special cases $m = 1,2,3$. Do this for arbitrary f, a, and b. This problem is to be done "by hand."

PROBLEM 2 Use (1) in calculating the data for Table 5.4.

TABLE 5.4 Romberg approximations and error analysis.

m	2^m	$R(m,1)$	$E_m = R(m,1) - I$	E_m/E_{m+1}
1	2	1.78539816340	−0.0665389	4.05011
2	4	1.83550812328	−0.0164289	4.01220
3	8	1.84784230644	−0.00409475	4.00303
4	16	1.85091414037	−0.00102291	4.00076
5	32	1.85168137241	−0.00025568	4.00019
6	64	1.85187313511	−0.0000639169	4.00005
7	128	1.85192107295	−0.000015979	4.00001
8	256	1.85193305724	−0.00000399475	4.00000
9	512	1.85193605330	−0.000000998686	4.00000
10	1024	1.85193680231	−0.000000249671	

PROBLEM 3 Show that the results in Table 5.4 lead to the conjecture

$$I \approx \frac{4R(m+1,1) - R(m,1)}{4-1}.$$

PROBLEM 4 Calculate the data for Table 5.5, where

$$R(m+1,2) = \frac{4R(m+1,1) - R(m,1)}{4-1}.$$

TABLE 5.5 A Romberg table.

m	2^m	$R(m,1)$	$R(m,2)$	$R(m,2) - I$
1	2	1.78539816340		
2	4	1.83550812328	1.85221144324	2.7×10^{-4}
3	8	1.84784230644	1.85195370083	1.7×10^{-5}
4	16	1.85091414037	1.85193808501	1.0×10^{-6}
5	32	1.85168137241	1.85193711643	6.4×10^{-8}
6	64	1.85187313511	1.85193705601	4.0×10^{-9}
7	128	1.85192107295	1.85193705223	2.5×10^{-10}
8	256	1.85193305724	1.85193705200	1.6×10^{-11}
9	512	1.85193605330	1.85193705198	9.8×10^{-13}
10	1024	1.85193680231	1.85193705198	6.1×10^{-14}

PROBLEM 5 These results suggest forcibly that the arithmetic combination

$$R(m+1,2) = \frac{4R(m+1,1) - R(m,1)}{4-1}$$

of $R(m+1,1)$ and $R(m,1)$ better approximates the integral (2) than either $R(m,1)$ or $R(m+1,1)$ alone. Table 5.5 gives evidence that the sequence $R(2,2), R(3,2), \ldots$ of numbers converges to I faster than $R(1,1), R(2,1), \ldots$. Provide evidence that the *convergence factor* of the former sequence is $4^2 = 16$ while, as shown earlier, the trapezoid sequence has convergence factor 4. This reasoning can be repeated to obtain sequences having even larger convergence factors. The general case is

$$R(m+1,k+1) = \frac{4^k R(m+1,k) - R(m,k)}{4^k - 1}, \qquad m = k, k+1, \ldots.$$

The calculations can be displayed in tabular form as follows:

$R(1,1)$				
$R(2,1)$	$R(2,2)$			
$R(3,1)$	$R(3,2)$	$R(3,3)$		
$R(4,1)$	$R(4,2)$	$R(4,3)$	$R(4,4)$	
$\vdots$	$\vdots$	$\vdots$	$\vdots$	$\ddots$

Such a table is called a *Romberg table.* The numbers in the first column are the trapezoid rule estimates corresponding to $2^1, 2^2, \ldots$ subintervals. For the integral (2), a partial Romberg table is

1.78539816340			
1.83550812328	1.85221144324		
1.84784230644	1.85195370083	1.85193651801	
1.85091414037	1.85193808501	1.85193704395	1.85193705230
1.85168137241	1.85193711643	1.85193705186	1.85193705198

The accuracy in the fourth column is striking. These estimates arise from the trapezoid estimates $T_2, T_4, \ldots, T_{32}$. Most calculations using the Romberg algorithm do not go beyond three or four columns. This provides sufficient accuracy for most purposes and avoids errors due to rounding that may occur in larger Romberg tables.

PROBLEM 6 Use the Romberg algorithm to calculate to nine decimal places the value of the elliptic integral

$$\int_0^1 \frac{1}{\sqrt{1 - 0.5 \sin^2 x}}\, dx.$$

Use your judgement as to when you have attained nine places of accuracy.

PROBLEM 7 Give an empirical argument that the Romberg algorithm is more efficient than the trapezoid rule. Measure efficiency by the number of function evaluations required to achieve a given accuracy.

PROBLEM 8 Show algebraically that $R(m + 1, 2)$ is Simpson's rule with 2^{m+1} intervals.

6

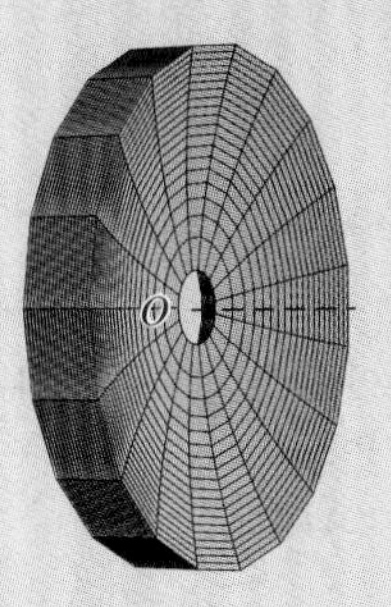

Applications of the Integral

Radio telescopes are in the shape of parabolic reflectors that focus all the incoming radio waves at one point on the axis of symmetry of the dish. Parabolic mirrors work the same way. We can determine the mass of such a mirror by treating it as a solid of revolution.

See Example 5, p. 451, for further explanation.

Newton's and Liebniz's work on the integral, done shortly after 1650, greatly extended the work of such Greek geometers as Archimedes (287–212 B.C.), who knew how to calculate the areas of regions bounded by rectangles, circles, parabolas, and spirals; the volumes of regions bounded by parallelepipeds, spheres, and pyramids; and the lengths of lines and circles.

Through Newton's and Leibniz's work we can calculate the areas, volumes, or lengths of a large variety of geometric objects and can model many mathematical or physical phenomena. A modern area in which integration plays a key role is computerized tomography (CT) in medicine. In CT scanning, a source of X-rays is directed towards the patient from many directions. After passing through the body, the intensity of the rays is recorded by several hundred X-ray photon detectors. These data are integrated to construct an accurate image of the patient's interior organs.

In this chapter we use the integral to model and calculate

- the volumes of solids,
- the lengths of curves,
- the work done by force fields on objects,
- the masses and the centers of mass of objects,
- the curvature of a curve and the tangential and normal acceleration vectors of moving objects; and we also
- classify and evaluate "improper integrals."

A major theme in this chapter is the use of "elements" of area, volume, arc length, work, and mass. Such elements increase understanding and decrease the need for memorized formulas by making direct use of underlying geometrical concepts or physical principles.

6.1 Volumes by Cross Section

Figure 6.1 includes the Great Pyramid. Although the Great Pyramid is a three-dimensional object, we can calculate its volume by integrating a function of one variable. We use the fact that, relative to the line from its vertex to the center of its base, horizontal cross sections of the Great Pyramid are squares.

FIGURE 6.1 The Pyramids of Giza.

INVESTIGATION

Volume of the Great Pyramid

Figure 6.2(a) shows a view of a pyramid with the same dimensions as the Great Pyramid. The side of its square base is $a = 230.4$ m and its height, which is the length of OX, is $h = 146.8$ m. The line OX goes from the vertex of the pyramid to the center of its base and is perpendicular to the base. We show how the volume of this pyramid may be calculated by using a definite integral to sum volume elements dV. This is similar to the procedure we used in Chapter 5 to calculate the area beneath a curve by summing area elements $dA = y\,dx$.

Referring to Fig. 6.2(a), imagine slicing the pyramid into thin cross sections, called *volume elements,* perpendicular to the line OX. A representative volume element is shown in Fig. 6.2(a). This volume element is a "thick square," located x meters from the vertex O. If its dimensions are s by s by dx, then its volume is $dV = s^2\,dx$. Our immediate goal is to express s in terms of x.

For this, draw the triangle OXY, where Y is the midpoint of a side of the base of the pyramid. Figures 6.2(b) and (c) show this triangle. Using similar triangles,

$$\frac{x}{y} = \frac{h}{a/2}.$$

Solving this equation for y,

$$y = \frac{ax}{2h}.$$

Because $s = 2y$, the side s of the volume element is

$$s = \frac{ax}{h} = \frac{a}{h}x.$$

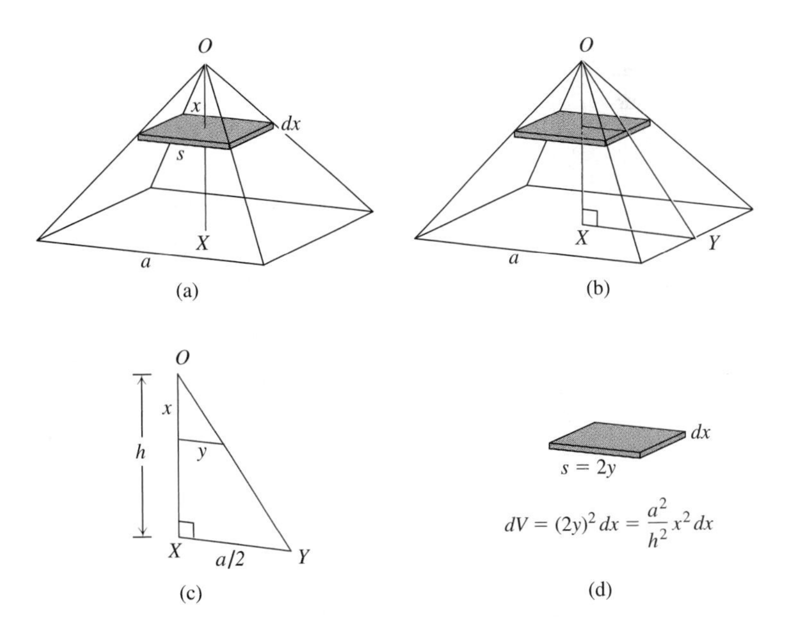

FIGURE 6.2 (a) The pyramid and volume element; (b) the triangle OXY; (c) using similar triangles to express the volume element dV in terms of x; (d) the volume element dv.

Hence, $dV = s^2\,dx = (a^2/h^2)x^2\,dx$. Summing the volumes of these elements gives the approximation

$$V \approx \sum dv = \sum \frac{a^2}{h^2}x^2\,dx$$

to the volume of the pyramid. The limit of this Riemann sum gives the exact volume

$$V = \int_0^h \frac{a^2}{h^2}x^2\,dx = \frac{a^2}{h^2}\cdot\frac{1}{3}x^3\Big|_0^h = \tfrac{1}{3}ha^2. \tag{1}$$

Substituting $a = 230.4$ m and $h = 146.8$ m, the volume of the Great Pyramid is

$$V = \tfrac{1}{3}146.8 \cdot 230.4^2 = 2.59758 \times 10^6 \text{ cubic meters.}$$

If this volume of stone were evenly stacked on a football field (300 feet by 160 feet), the stack would be about 2,000 feet high.

The geometry and calculations in this investigation are an outline of the methods discussed in this section. If we can write a formula for the volumes of thin, parallel cross sections of a solid—called *elements of volume*—as a function of their position along a line, then we can determine the volume of the solid by using the integral to sum these elements.

Calculating Volumes by Cross Section

The area of the cross section of S at w is $g(w)$; the volume of this cross section is $dV = g(w)dw$.

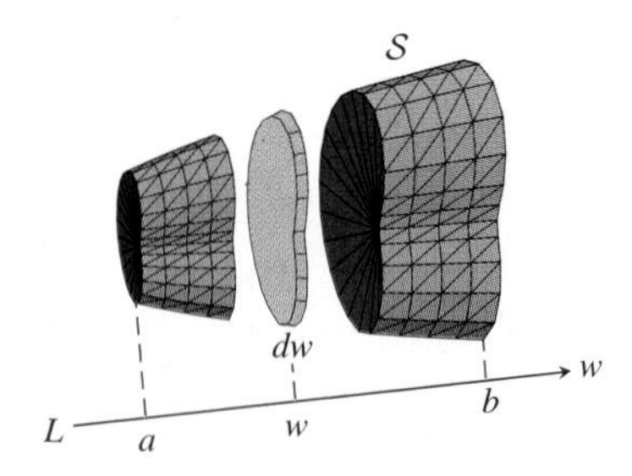

FIGURE 6.3 Volumes by cross section.

Figure 6.3 shows an exploded view of a solid S and coordinate line L. Assume that L is a "w-axis" and that S lies between planes perpendicular to L and passing through the points $w = a$ and $w = b$. If the area of the cross section at a representative point w of L is $g(w)$, then the volume dV of a slice of thickness dw is approximately $g(w)\,dw$.

If we regard the point w as an evaluation point c_i of a regular subdivision $a = w_0, w_1, \ldots, w_n = b$ of $[a, b]$ and set $dw = (b - a)/n$, then

$$dV \approx g(c_i)\,dw.$$

Hence, the volume V of the entire solid is approximately equal to the Riemann sum

$$\sum_{i=1}^{n} g(c_i)\,dw.$$

As the number n of subdivisions of $[a, b]$ increases, these sums will approach the integral $\int_a^b g(w)\,dw$. Hence,

$$V = \lim_{n\to\infty}\sum_{i=1}^{n} g(c_i)\,dw = \int_a^b g(w)\,dw. \tag{2}$$

We give several examples of calculating the volume of a solid with known cross sections.

EXAMPLE 1 Figure 6.4(a) shows a solid with a parabolic base and square cross sections. The base is the parabolic curve shown in part (b) of the figure and described by $x = 1 - y^2$, $0 \le x \le 1$. A representative cross section of

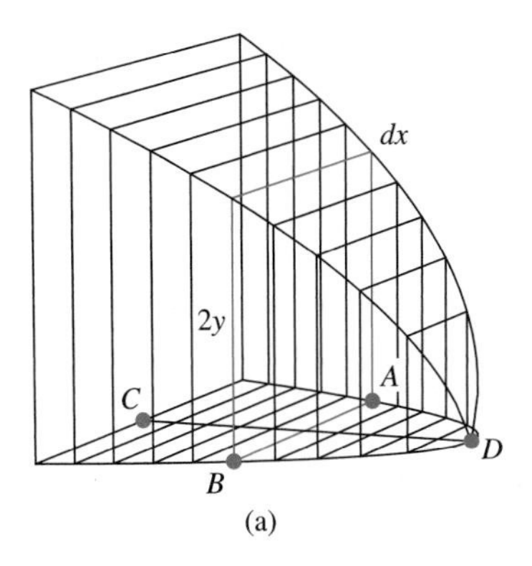

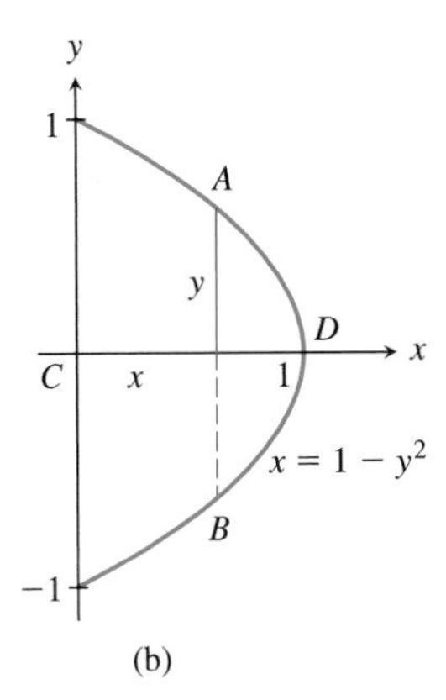

FIGURE 6.4 (a) A solid with a parabolic base and square cross sections. (b) The parabolic base.

the solid, in a plane perpendicular to the line segment CD, is shown in color. Calculate the volume of the solid.

Solution

Comparing the solid here with the general solid shown in Fig. 6.3, we may take the line through C and D as the line L, which we shall take as the x-axis. We are given that cross sections perpendicular to the x-axis are squares. A graph of the parabola described by $x = 1 - y^2$ is shown in Fig. 6.4(b). The side AB has length $2y$, where x and y satisfy the equation $x = 1 - y^2$. Solving this equation for y, the length of AB is $2y = 2\sqrt{1 - x}$. Hence, the area of the square cross section shown in part (a) of the figure is $(2y)^2$ and the volume of the cross section of thickness dx is approximately $dV = (2y)^2\,dx = 4(1 - x)\,dx$. Summing these cross sections as x varies from 0 to 1, the volume of the solid is

$$V = \int_0^1 dV = \int_0^1 4(1 - x)\,dx = 4\left(x - \tfrac{1}{2}x^2\right)\Big|_0^1 = 2 \text{ cubic units.}$$

EXAMPLE 2 Figure 6.5(a) shows a solid cut from a steel rod of radius 1 cm. The bottom base of the solid is the circle with diameter AA' and the top base is the ellipse with major axis EE'. The line segments AA' and EE' lie in the same plane and, if extended, meet at an angle of 30° at the point F. The length of the segment EA is $2 - 1/\sqrt{3}$. Calculate the volume of the solid.

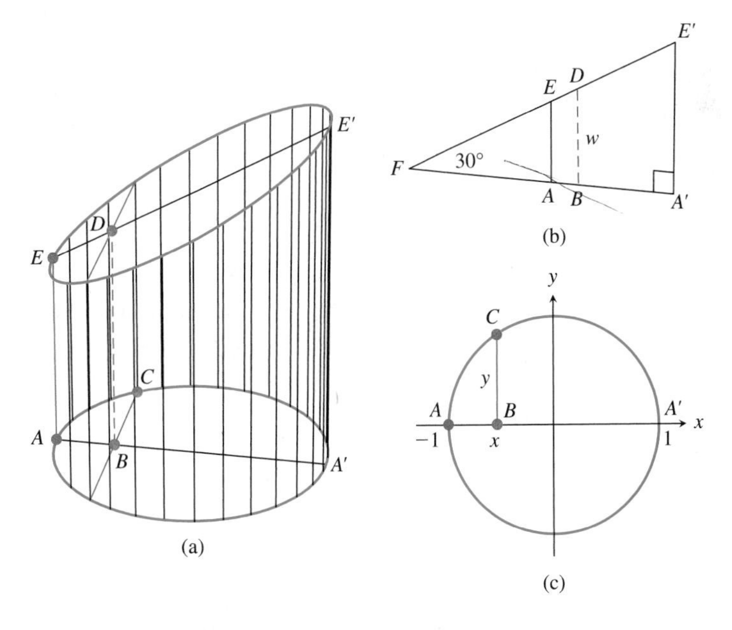

FIGURE 6.5 (a) A cylindrical shape cut from a rod by two planes; (b) using the tangent function to express w in terms of x; (c) the coordinatized circular base.

Solution

We view the solid in Fig. 6.5(a) relative to the diameter AA'. Cross sections perpendicular to this line are rectangles; a representative rectangle is shown in color, with centerline BD. If we view the circular base of the solid as the region bounded by the circle with equation $x^2 + y^2 = 1$, as seen in part (c) of the figure, the cross section at position x, where x varies from -1 to 1, has area $2BC \cdot DB = 2y \cdot DB$. From the equation of the circle, $y = BC = \sqrt{1 - x^2}$. From the triangle in Fig. 6.5(b),

$$\tan 30° = \frac{EA}{FA} = \frac{2 - 1/\sqrt{3}}{FA}.$$

Hence,

$$FA = \frac{2 - 1/\sqrt{3}}{\tan 30°} = \frac{2 - 1/\sqrt{3}}{1/\sqrt{3}} = 2\sqrt{3} - 1.$$

Noting that the length of AB is $x - (-1) = x + 1$,

$$\tan 30° = \frac{DB}{FA + x + 1} = \frac{w}{x + 2\sqrt{3}}.$$

Solving this equation for w,

$$w = \tan 30°\left(x + 2\sqrt{3}\right) = \frac{1}{\sqrt{3}}\left(x + 2\sqrt{3}\right) = 2 + \frac{x}{\sqrt{3}}.$$

Hence, the volume of a representative cross section of thickness dx is

$$dV = 2y \cdot DB\,dx = 2yw\,dx = 2\sqrt{1 - x^2}\left(2 + \frac{x}{\sqrt{3}}\right)dx.$$

Summing these volume elements,

$$V = \int_{-1}^{1} dV = 2\int_{-1}^{1} \sqrt{1 - x^2}\left(2 + \frac{x}{\sqrt{3}}\right)dx$$
$$= 4\int_{-1}^{1} \sqrt{1 - x^2}\,dx + \frac{2}{\sqrt{3}}\int_{-1}^{1} x\sqrt{1 - x^2}\,dx.$$

The first integral can be done with formula (7) from the table of integrals in the inside rear cover or, more simply, by noting that, apart from the factor of 4, the value of the integral is just one-half of the area of a circle of radius 1. The second integral is 0 because the graph of $x\sqrt{1 - x^2}$ is symmetric about the origin. Hence,

$$V = 4 \cdot \tfrac{1}{2}\pi \cdot 1^2 = 2\pi \text{ cm}^3.$$

TABLE 6.1 Areas of cross sections of a construction site.

x	$A(x)$
0	200
10	225
20	230
30	200
40	300
50	290
60	250
70	250
80	300
90	300
100	250

EXAMPLE 3 At the construction site shown in Fig. 6.6 a contractor wishes to level the ground to a given horizontal reference plane. All of the earth at the site lies above a rectangular region of the reference plane. Table 6.1 lists

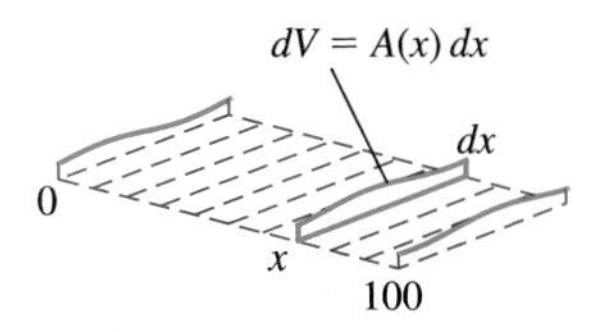

FIGURE 6.6 Cross sections of a construction site lying above a rectangular region.

the areas of 11 cross sections of the site. The cross sections are in planes perpendicular to one side of the rectangular region. Figure 6.6 shows a rough sketch of the construction site and three cross sections. The areas $A(x)$ (in square meters) of the cross sections are listed, where x is the distance (in meters) from one end of the given side. Estimate the volume of earth that must be moved.

Solution

If for all x between 0 and 100 we knew the area $A(x)$ of the cross section at x, the volume dV of a thin cross section would be $dV = A(x)\,dx$ and the volume of earth to be moved would be

$$V = \int_0^{100} A(x)\,dx.$$

However, we know $A(x)$ only at 11 values of x.

We approximate the integral $V = \int_0^{100} A(x)\,dx$ using the trapezoid rule with $n = 10$. Letting $y_0 = A(0), y_1 = A(10), \ldots, y_{10} = A(100)$,

$$V = \int_0^{100} A(x)\,dx \approx T_{10} = \tfrac{1}{2}h(y_0 + 2(y_1 + y_2 + \cdots + y_9) + y_{10}),$$

where $h = (b - a)/10 = 100/10 = 10$. Substituting from Table 6.1,

$$\begin{aligned} V &\approx 5(200 + 2(225 + 230 + 200 + 300 + 290 + 250 \\ &\qquad + 250 + 300 + 300) + 250) \\ &\approx 5(200 + 2(2345) + 250) = 25{,}700 \text{ cubic meters.} \end{aligned}$$

Given that the survey data is of limited accuracy and, moreover, the contractor needs only a rough estimate, this procedure is probably sufficient.

Disks and Washers

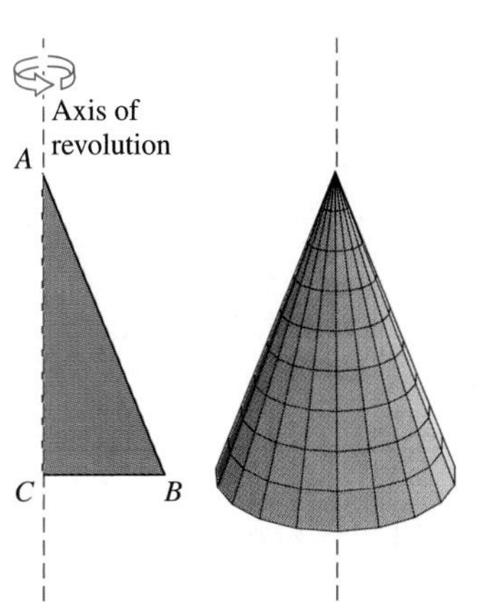

FIGURE 6.7 A cone generated by revolving a triangular region about a line.

If we revolve a two-dimensional region about a line in the plane of the region, we say that the region "generates" a solid called a *solid of revolution*. The line about which the rotation takes place is called the *axis of revolution*. Figure 6.7 shows the solid cone generated by revolving the triangular region ABC about an axis of revolution. We sometimes say that the region ABC "sweeps out" the solid cone.

Figure 6.8(a) shows a solid of revolution S generated by revolving the region beneath the curve C in Fig. 6.8(b) about the x-axis, which is, in this case, the axis of revolution. This generating curve is the graph of a function f defined on an interval $[a, b]$.

If we sketch in the region beneath C an inscribed rectangle of width dx, height $y = f(x)$, and located at x on the x-axis, this rectangle will generate a disk as it and the curve C are revolved about the x-axis. See Fig. 6.8(c). This disk is a right circular cylinder with height dx. Recalling that the volume of a right circular cylinder is the area of its circular cross section times the height of the cylinder, the volume of the disk is $dV = \pi y^2\,dx = \pi(f(x))^2\,dx$.

(a)

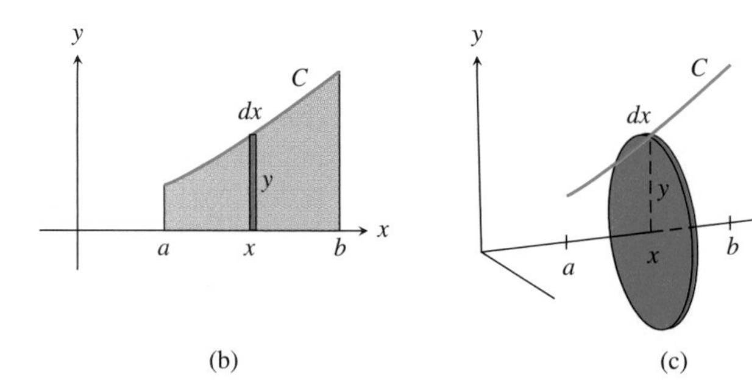

FIGURE 6.8 (a) The solid $\mathcal{S}$ of revolution generated by revolving a region about the x-axis; (b) the generating region; (c) a disk element of volume.

Using the integral to sum these volume elements as x varies from a to b, the volume of the solid $\mathcal{S}$ is

$$V = \int_a^b \pi y^2\,dx = \int_a^b \pi f(x)^2\,dx.$$

This procedure for calculating the volume of a solid of revolution is a special case of the procedure we described for calculating the volumes of solids sectioned by planes perpendicular to a given line. For solids of revolution, the sections are circular and the line L is the axis of revolution.

We give three examples in which we calculate the volume of a solid of revolution.

EXAMPLE 4 Calculate the volume of a right circular cone of height h and base radius r.

Solution

Figure 6.9(a) shows a right circular cone, which we suppose to have height h and base radius r. This cone is a solid of revolution. Figure 6.9(b) and (c) illustrate how the cone can be generated and suggest a coordinate system. The cone can be generated by revolving line OY about the axis of revolution OX. We slice the cone into thin cross sections or volume elements, with the sections perpendicular to the axis of revolution. The solid disk

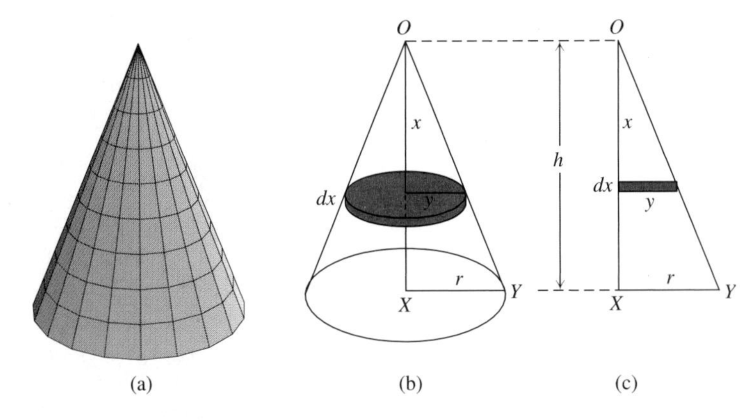

FIGURE 6.9 (a) Right circular cone; (b) a thin, circular cross section of the cone; (c) the area element generating the thin, circular cross section.

shown in Fig. 6.9(b) is generated by revolving the area element $y\,dx$ shown in Fig. 6.9(c) about OX. The volume dV of the representative disk is $dV = \pi y^2\,dx$, which is the area πy^2 of the base of the disk times its height dx.

To express y in terms of x, we use the similar triangles shown in the figure. Because $y/x = r/h$, the volume element is

$$dV = \pi y^2\,dx = \pi\frac{r^2}{h^2}x^2\,dx.$$

We use the integral to sum the volume elements as x varies from 0 to h. Thus,

$$V = \int_0^h dV = \int_0^h \pi y^2\,dx = \int_0^h \pi\frac{r^2}{h^2}x^2\,dx$$

$$= \pi\frac{r^2}{h^2}\frac{x^3}{3}\bigg|_0^h = \pi\frac{r^2}{h^2}\frac{h^3}{3} = \tfrac{1}{3}\pi r^2 h \text{ cubic units}$$

If a parabola is revolved around its axis of symmetry, the resulting shape is called a *paraboloid.* Paraboloids are used to receive and focus television signals, to reflect and focus light in flashlights and car headlights, and to collect and focus starlight in telescopes. In the next example, we calculate the mass of a parabolic mirror. In a telescope, such a mirror might receive starlight and reflect it to another mirror, which in turn reflects the light back through a hole in the parabolic mirror to film or to a light-sensitive instrument.

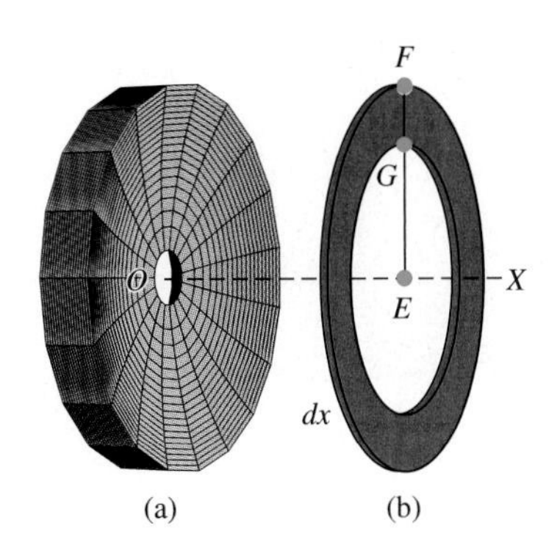

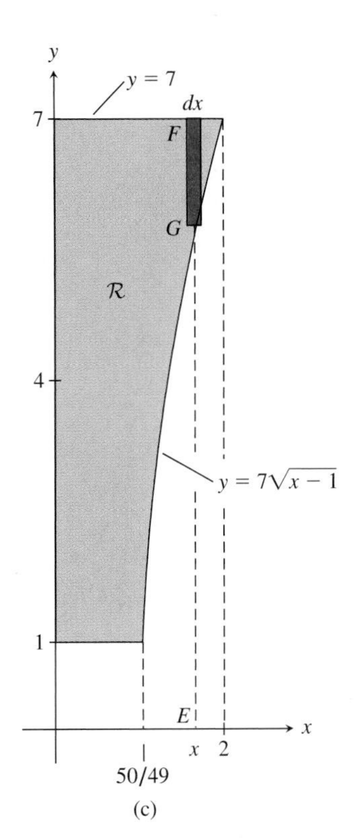

FIGURE 6.10 (a) Parabolic mirror in three dimensions; (b) a cross section of the mirror in a plane perpendicular to the OX axis; (c) the two-dimensional region $\mathcal{R}$ that, when rotated, generates the mirror.

EXAMPLE 5 Figure 6.10(a) shows a parabolic mirror. This mirror can be generated by revolving about the x-axis the shaded region $\mathcal{R}$ shown in Fig. 6.10(c). The region $\mathcal{R}$ has been enlarged relative to the mirror it generates. The region $\mathcal{R}$ is bounded on the right by the curve with equation $y = 7\sqrt{x-1}$, on the left by the y-axis, and lies between the lines with equations $y = 1$ and $y = 7$. All lengths are in centimeters. The density of the glass from which the mirror was cast is $2.590\ \text{g/cm}^3$. Determine the mass of the mirror.

Solution

The x-axis is the axis of revolution. As the region $\mathcal{R}$ revolves about the x-axis, it generates the parabolic mirror. As the area element shown in $\mathcal{R}$ revolves about the x-axis it generates the "washer" shown in part (b) of the figure. The volume of this washer can be calculated by taking the difference between the volumes of two disks with a common axis and common height dx. The radius of the larger disk is the length of EF, which for all x in $[0, 2]$ is 7 cm. The radius of the smaller cylinder is the length of EG, which for all x in $[0, 50/49]$ is 1 cm and for all x in $[50/49, 2]$ is $7\sqrt{x-1}$. Hence, the volume of the washer is

$$dV = \begin{cases} \pi \cdot 7^2\,dx - \pi \cdot 1^2\,dx = 48\pi\,dx, & 0 \le x \le 50/49 \\ \pi \cdot 7^2\,dx - \pi\left(7\sqrt{x-1}\right)^2 dx = 49\pi(1 - x + 1)\,dx, & 50/49 \le x \le 2. \end{cases}$$

Summing these volume elements as x varies from 0 to 2,

$$V = \int_0^2 dV = \int_0^{50/49} 48\pi\,dx + \int_{50/49}^2 49\pi(2 - x)\,dx$$

$$= 48\pi \cdot \frac{50}{49} - \tfrac{1}{2}49\pi(2 - x)^2\Big|_{50/49}^{2} = \frac{3552\pi}{49} \approx 227.733 \text{ cm}^3.$$

Multiplying this volume by 2.590 g/cm^3, which is the density of the glass, the mass of the mirror is about 589.830 grams.

(a)

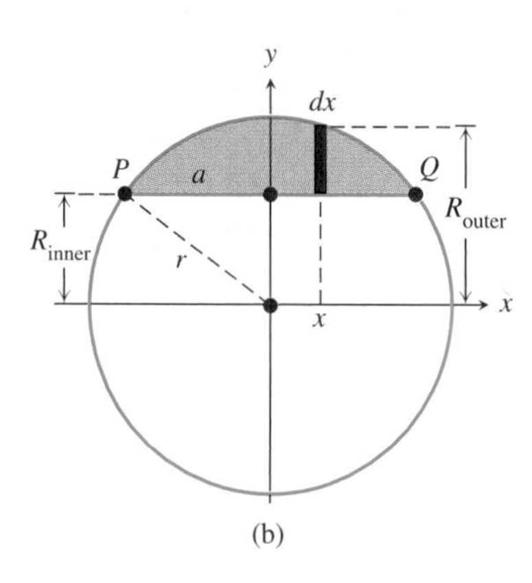

(b)

FIGURE 6.11 (a) Drilled sphere/bead; (b) the region (shaded) that, when revolved about the x-axis, generates this bead.

EXAMPLE 6 A hole of length $2a$ is bored through a solid sphere of radius r, along a diameter. What is the volume of the remaining solid? Figure 6.11(a) shows the "bead" resulting from this operation and Fig. 6.11(b) shows the region (shaded) that, when revolved about the x-axis, generates the bead. The length $2a$, where $0 < 2a < 2r$, is shown in the cross section as the line segment PQ.

Solution

The bead can be generated by revolving about the x-axis the shaded region shown in Fig. 6.11(b). Revolving the shaded element—which is located at x and has width dx—about the x-axis generates a washer. This washer is a cross section of the bead by a plane perpendicular to the x-axis. The outer radius of the washer is $R_{\text{outer}} = \sqrt{r^2 - x^2}$, which is the distance from the x-axis to the circle with equation $x^2 + y^2 = r^2$. The inner radius of the washer is the fixed length $R_{\text{inner}} = \sqrt{r^2 - a^2}$ of the vertical leg of the right triangle shown in Fig. 6.11(b). Hence, the volume of the washer is

$$dV = \pi R_{\text{outer}}^2\,dx - \pi R_{\text{inner}}^2\,dx = \pi(r^2 - x^2 - r^2 + a^2)\,dx = \pi(a^2 - x^2).$$

Summing these volume elements,

$$\begin{aligned} V &= \pi \int_{-a}^{a} (a^2 - x^2)\,dx = \pi\left(a^2x - \tfrac{1}{3}x^3\right)\Big|_{-a}^{a} \\ &= \pi\left(a^3 - \tfrac{1}{3}a^3 - \left(-a^3 + \tfrac{1}{3}a^3\right)\right) = \tfrac{4}{3}\pi a^3. \end{aligned}$$

You may have noticed in this example that the volume of the bead depends only on the length $2a$ of the drilled hole, which we may call the *height* of the bead, not on the radius of the sphere. So if we want a bead of height 1 inch, we may drill either a basketball or a softball. The volume will be $\frac{4}{3}\pi(1/2)^3$ in either case.

Exercises 6.1

Exercises 1–12: Use the disk method in calculating the volume of the solid of revolution generated by revolving a region about the x-axis. The region is in the (x, y)-plane and is bounded above by a curve and below by the x-axis. Make a simple sketch of the solid and a representative area element which, when revolved about the axis of revolution, generates a disk.

1. $y = \sqrt{a^2 - x^2}$, $-a \le x \le a$

2. $y = b\sqrt{1 - x^2/a^2}$, $-a \le x \le a$

3. $y = x^2$, $0 \le x \le 2$

4. $y = x^3$, $0 \le x \le 3/2$

5. $y = \sqrt{x}$, $0 \le x \le 2$

6. $y = \sqrt[3]{x}$, $0 \le x \le 2$

7. $y = \sin x$, $0 \le x \le \pi$

8. $y = \cos x$, $0 \le x \le \pi/2$

9. $y = \sec x$, $0 \le x \le \pi/4$

10. $y = \tan x$, $0 \le x \le \pi/4$

11. $y = e^x$, $-1 \le x \le 1$

12. $y = e^{-x}$, $0 \le x \le 2$

Exercises 13–20: Use washers in calculating the volume of the solid of revolution generated by revolving a region about the x-axis. The region is in the (x, y)-plane and is bounded by two curves. Make a simple sketch of the solid and a representative area element which, when revolved about the axis of revolution, generates a washer. Note: A calculator may be needed to find the intersection points of some of the curves.

13. $y = x^2$, $y = x^3$, $0 \le x \le 1$

14. $y = \sqrt{x}$, $y = \sqrt[3]{x}$, $0 \le x \le 1$

15. $y = -x(x - 5)$, $y = x$

16. $y = x + 1$, $y = (x - 1)^2$

17. $y = \sin x$, $y = (2/\pi)x$, $0 \le x \le \pi/2$. You may need the identity $\sin^2 x = (1 - \cos 2x)/2$.

18. $y = \sec x$, $\pi\sqrt{3}(y - 2) = 4\left(\sqrt{3} - 1\right)(x - \pi/3)$, $|x| \le \pi/2$

19. $y = \ln x$, $y = -(x - 1)(x - 3)$, $x \ge 1$

20. $y = e^x$, $y = -5x(x - 3)$

21. Show that the volume of the oblique cone shown in the figure is $V = \frac{1}{3}\pi r^2 h$.

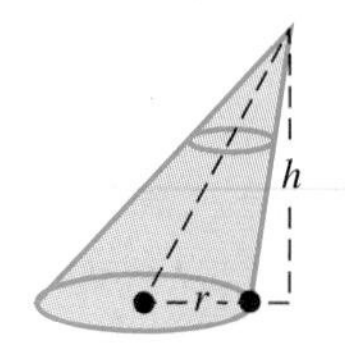

Figure for Exercise 21.

22. Show that the volume of the oblique cylinder shown in the figure is $V = \pi r^2 h$.

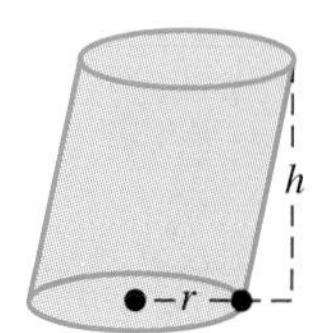

Figure for Exercise 22.

23. Calculate the volume of the solid formed by revolving the line joining $(0, b)$ and (H, B), where $0 < b < B$ and $H > 0$, about the x-axis.

24. The accompanying sketch shows a frustum of a pyramid. Use the method of cross sections to calculate the volume of the frustum with square bases having sides of lengths b and B, where $0 < b < B$, and height H, which is the distance between the bases.

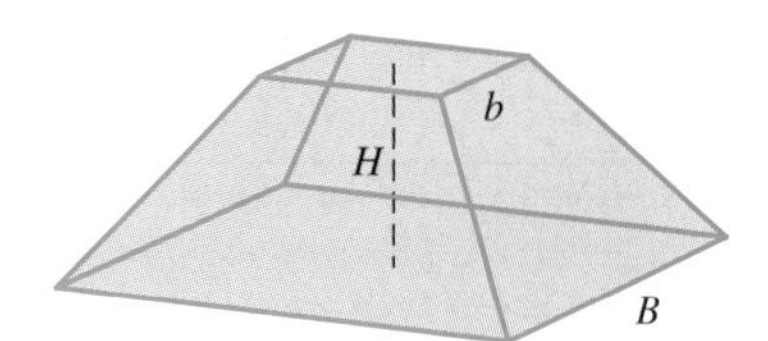

Figure for Exercise 24.

25. We stated that the builders of the Great Pyramid at Giza used the formula $V = \frac{1}{3}a^2h$. Although this is quite probable, all that is known for sure is that they used the formula $V = \frac{1}{3}H(a^2 + ab + b^2)$ for the volume of a frustum of a pyramid, where H is the height of the frustum and a and b are the lengths of the sides of its square bases. Derive the formula for the volume of a frustum from that for the volume of a pyramid.

26. The region bounded on the left by the graph of the equation $(y - 3)^2 = x - 4$ and on the right by the line joining the points $(5, 2)$ and $(8, 5)$ is revolved about the x-axis. Calculate the volume of the resulting solid.

27. The larger of the two regions bounded by the circle with equation $(x - 2)^2 + (y - 2)^2 = 1$ and the line through $(1, 2)$ and $(2, 3)$ is revolved about the x-axis. Calculate its volume.

28. The base of a solid is the right triangle OBA, with sides OB and OA of lengths b and a, respectively. Sections of the solid which are perpendicular to the side OB are right triangles with the vertex of the right angle on OB. For a representative cross section, the ratio of the length of the vertical side to the horizontal side is c/a. Sketch the solid and calculate its volume.

29. Express the volume $V(h)$ of punch in a 16-in.-diameter hemispherical bowl as a function of the depth h of the punch. See the accompanying figure.

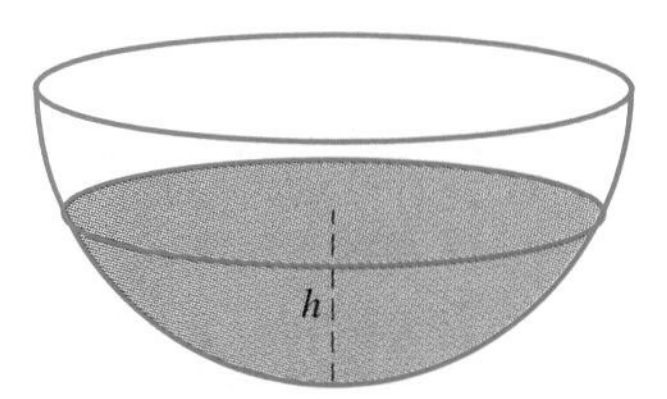

Figure for Exercises 29 – 31.

30. Continuation of Exercise 29 Given that 1 quart is 57.75 in.^{3}, at what depth does the bowl contain 8 quarts of punch?

T 31. Continuation of Exercise 29 "Calibrate" the punch bowl by engraving marks on a great (semi)circle. If the marks are to be located by measuring from the bottom of the bowl along this circle, at what distances should the 4-, 8-, 12-, and 16-quart marks be placed?

32. Find the volume of the torus generated by revolving about the x-axis the region inside a circle of radius a and centered at $(0, b)$, where $0 < a < b$.

33. A solid has a circular base of radius r. Sections of the solid perpendicular to the diameter AB of this base are squares. Calculate the volume of the solid. Two cross sections of this solid are shown in the accompanying figure. Also shown are the curves along which the top corners of the square sections fall.

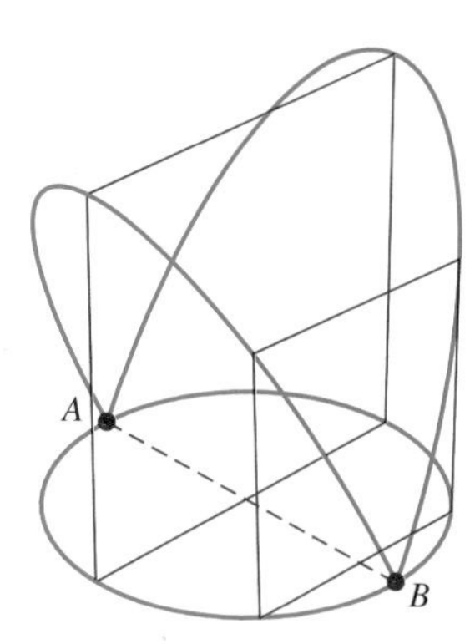

Figure for Exercise 33.

34. A 30° wedge is cut from a brass rod, as illustrated in the accompanying figure. One plane of the wedge is perpendicular to the axis of the rod. The planes intersect along a diameter OX of the circular cross section of the rod. Calculate the volume of the wedge if OX has length $2a$.

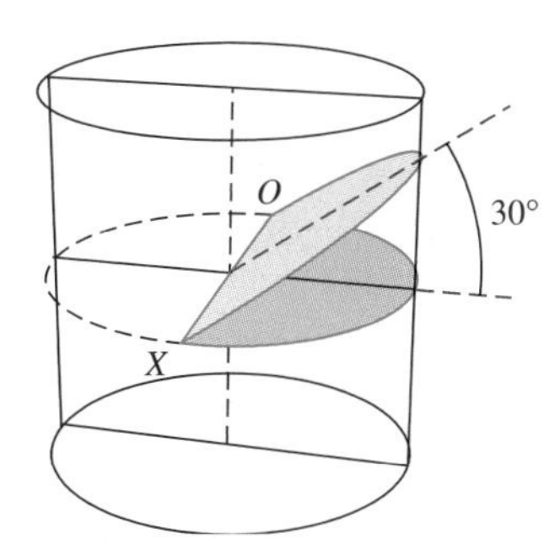

Figure for Exercise 34.

35. Calculate the volume of the wedge in the preceding exercise by using rectangular cross sections.

36. Generalize Exercise 34 by considering a wedge with angle $\theta \in (0, \pi/2)$.

37. A hole of diameter 10 cm is to be drilled through a steel rod of the same diameter. The axis of the cylinder in which the drill moves intersects the axis of the rod at a right angle. The figure shows the cylindrical rod and the cylinder of metal removed by the drill. Calculate the mass of the metal removed by the mill. The density of steel is 7850 kg/m^3.

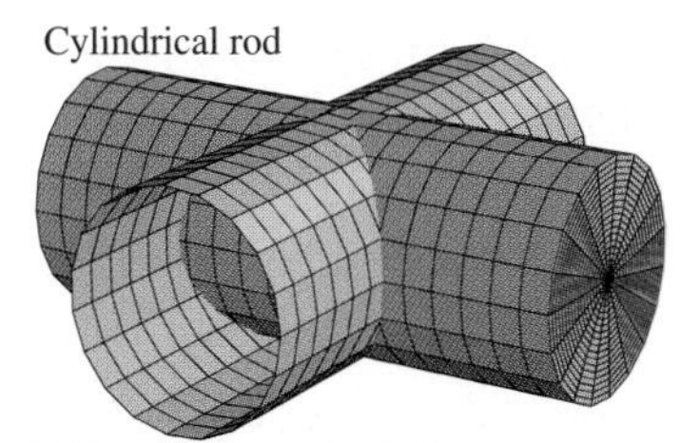

Figure for Exercise 37.

38. Referring back to Example 6, but not using its results, calculate the volume of the drilled hole. Note that part of the drilled hole is a right circular cylinder, whose volume can be calculated by formula.

39. Show that the volume of an elliptical bead of height $h < 2a$ is $\frac{1}{6}\pi h^3(b^2/a^2)$, where the ellipse has equation $x^2/a^2 + y^2/b^2 = 1$. The bead is drilled along the x-axis (with a circular drill). See Example 6.

40. The primary mirror for a telescope can be described as the region generated by revolving a certain region $\mathcal{R}$ described about the x-axis. The region $\mathcal{R}$ is bounded on the right by the graph of the function $f(x) = \sqrt{14.4x}$, $0.005 \le x \le 0.1$, on the left by the line with equation $x = -0.27$, and between the lines with equations $y = f(0.005)$ and $y = f(0.1)$. All lengths are in meters. Calculate the mass of the mirror if the density of the glass from which the mirror was cast is 2590 kg/m^3.

41. The volume of the washer element in Example 6 was given as

$$dV = \pi R_{\text{outer}}^2\,dx - \pi R_{\text{inner}}^2\,dx.$$

A common mistake in using washers is to write

$$dV = \pi(R_{\text{outer}} - R_{\text{inner}})^2\,dx.$$

Explain this mistake.

42. Bonaventura Cavalieri (1598–1647) conjectured that if two solids have cross sections of equal area when cut by planes perpendicular to a common line segment, then the two solids have equal volumes. This is sometimes called *Cavalieri's Principle*. Prove Cavalieri's Principle for the right circular cone and oblique circular cone shown in the accompanying figure. Note that the cones have equal heights and equal base radii.

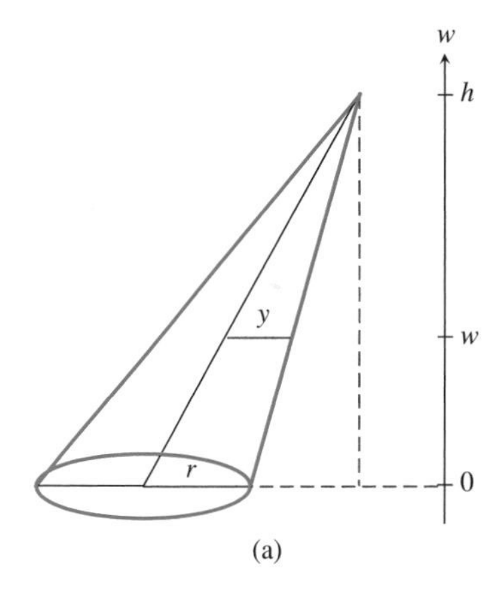

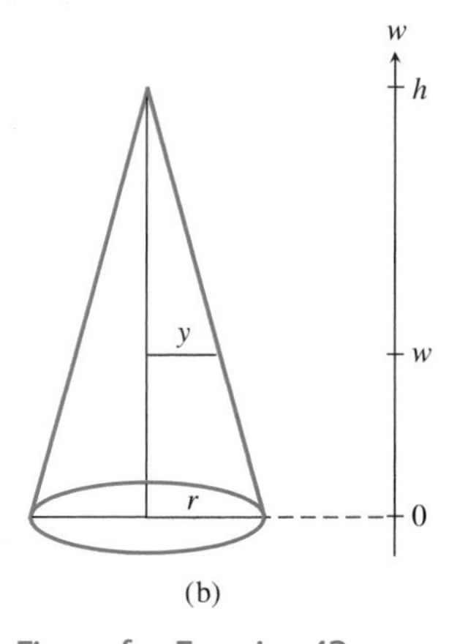

Figure for Exercise 42.

Exercises 43–45: These problems concern the prismoidal formula and its use in such applications as using survey data to approximate the amount of earth to be moved in constructing a highway.

43. The prismoidal formula is: If $p(x) = Ax^3 + Bx^2 + Cx + D$ is a polynomial of degree 3 or less, then

$$\int_a^b p(x)\,dx = \frac{b-a}{6}(p(a) + 4p(m) + p(b)),$$

where $m = (a + b)/2$.

a. Verify the prismoidal formula for the special cases $p(x) = x^n$, $n = 0, 1, 2, 3$.

b. Show that the formula fails for $p(x) = x^4$.

c. Show that the formula holds for $p(x) = Ax^3 + Bx^2 + Cx + D$.

44. A **prismoid** is a solid $\mathcal{S}$ lying between two parallel planes P and P'. Assume that P and P' lie along a line L perpendicular to both planes, with P at $x = a$ and P' at $x = b$. The cross sections of $\mathcal{S}$ at $x = a$ and $x = b$ are called *bases* of $\mathcal{S}$. Assume that for all $a \le x \le b$, the area $A(x)$ of the cross section of $\mathcal{S}$ at coordinate position x is numerically equal to $p(x)$, where p is a polynomial of degree 3 or less. Show that the volume V of such a prismoid is given by

$$V = \tfrac{1}{6}h(B + 4M + B'), \tag{3}$$

where B and B' are the areas of the bases of $\mathcal{S}$, and M is the area of the cross section at $x = (a + b)/2$.

45. Show that the solids $\mathcal{S}1$–$\mathcal{S}3$ are prismoids and calculate their volumes by the prismoidal formula.

a. $\mathcal{S}1$ is bounded by (one nappe of) a cone and two parallel planes, each perpendicular to the axis of the cone.

b. $\mathcal{S}2$ is bounded by a sphere and two parallel planes.

c. $\mathcal{S}3$ is a pyramid.

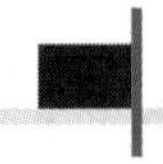

6.2 Volumes by Shells

For solids of revolution, the method of *cylindrical shells* is an alternative to either the disk or washer methods. Figure 6.12(a) shows a solid of revolution. We can think of this solid as having been generated by revolving the shaded region $\mathcal{R}$—see (b) of the figure—about the y-axis. The cylindrical shell shown in Fig. 6.12(c) was generated by revolving the area element shown in (b) about the y-axis. The area element is located by the coordinate x, its height is y, and its thickness is dx. As x varies from a to b the area elements sweep out the region $\mathcal{R}$ and the corresponding cylindrical shells fill out the solid of revolution.

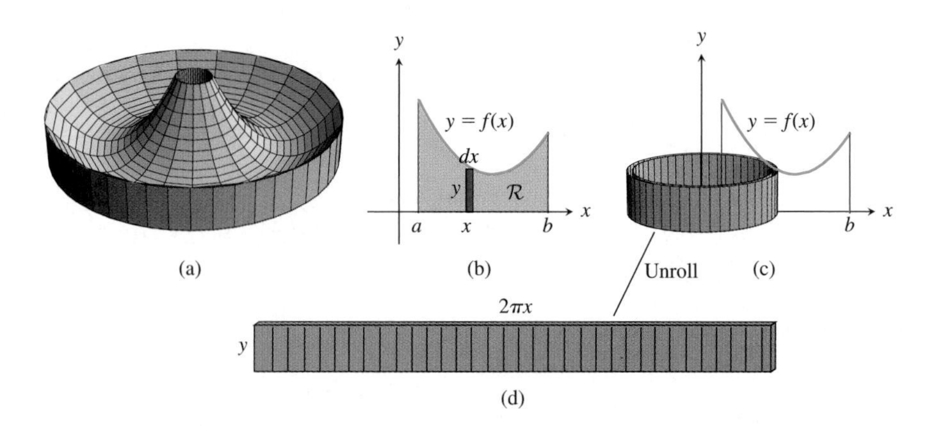

FIGURE 6.12 The solid of revolution shown in (a) can be generated by revolving the region $\mathcal{R}$ in (b) about the y-axis; the shaded area element shown in (b) generates the cylindrical shell shown in (c); (d) shows the cylindrical shell cut along a vertical line and unrolled.

The approximate volume of the representative cylindrical shell in Fig. 6.12(c) can be calculated by cutting the shell along a vertical line and flattening it into a "thick rectangle," as shown in Fig. 6.12(d). The length of one side of the rectangle is the circumference $2\pi x$ of the cylinder, the length of the adjacent side is y, and the thickness of the rectangle is dx. Hence,

$$dV \approx 2\pi x \cdot y\,dx = 2\pi x f(x)\,dx. \tag{1}$$

Using the integral to sum these volume elements, the volume of the solid of revolution is

$$V = \int_a^b 2\pi x f(x)\,dx = 2\pi \int_a^b x f(x)\,dx. \tag{2}$$

EXAMPLE 1 The function f whose graph was shown in Fig. 6.12(b) and (c) is

$$f(x) = \tfrac{1}{4}(x-5)^2 + 2, \qquad 1 \le x \le 8.$$

Calculate the volume of the solid generated by revolving the region $\mathcal{R}$ about the y-axis.

Solution

Noting that $f(x) = \frac{1}{4}(x^2 - 10x + 33)$ and applying Equation (2),

$$\begin{aligned} V = 2\pi \int_a^b x f(x)\,dx &= \frac{\pi}{2}\int_1^8 (x^3 - 10x^2 + 33x)\,dx \\ &= \frac{\pi}{2}\left(\frac{x^4}{4} - 10\frac{x^3}{3} + 33\frac{x^2}{2}\right)\Bigg|_1^8 \\ &= \frac{4319\pi}{24} \approx 565.356 \text{ cubic units.} \end{aligned}$$

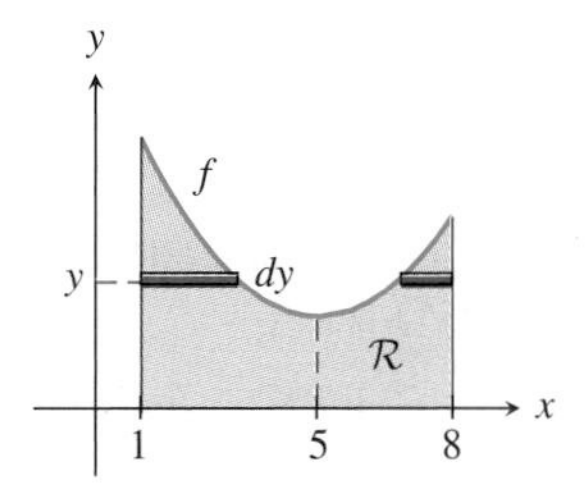

FIGURE 6.13 Using the washer method to calculate the volume of the solid of revolution generated by revolving the region $\mathcal{R}$ about the y-axis.

For the solid of revolution considered in Example 1, the shell method is simpler than the washer method. Notice in Fig. 6.13 that if we attempt to draw area elements in $\mathcal{R}$ that when revolved about the y-axis give washers, the shape of the region $\mathcal{R}$ forces us to divide the region into two parts. The graph of f is (part of) a parabola with vertex on the line $x = 5$. If we divide the region $\mathcal{R}$ along this line, then we can apply the washer method in each portion. However, this sometimes doubles the work.

Rather than simply memorize formula (2), which, as stated, applies only to solids generated by revolving a region about the y-axis, it is preferable to have in mind a "coordinate-free" version of this formula. We can, for example, note that the factor $2\pi x$ is the circumference of the shell, which is 2π times the radius of the shell; the factor $f(x)$ is the height of the shell, measured along its axis; and the factor dx is the thickness of the shell. Thus, we can remember the volume element for the shell method as

$$dV = 2\pi \cdot \text{radius of the shell} \cdot \text{height of the shell} \cdot \text{thickness of the shell}.$$

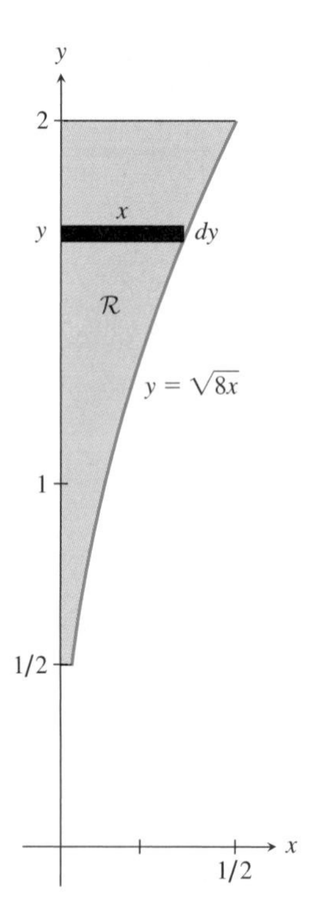

FIGURE 6.14 Parabolic mirror generated by revolving region $\mathcal{R}$ about the x-axis.

EXAMPLE 2 Use the shell method in calculating the mass of the parabolic mirror generated by revolving about the x-axis the shaded region $\mathcal{R}$ shown in Fig. 6.14. The region $\mathcal{R}$ is bounded on the right by the curve with equation $y = \sqrt{8x}$, on the left by the y-axis, and lies between the lines with equations $y = 1/2$ and $y = 2$. All lengths are in centimeters. The density of the glass from which the mirror was cast is 2.590 g/cm^3.

Solution

We can generate cylindrical shells by revolving area elements like that shown in Fig. 6.14 about the x-axis. The volume of this shell is

$$\begin{aligned} dV &= 2\pi \cdot \text{radius of the shell} \cdot \text{height of the shell} \cdot \text{thickness of the shell} \\ &= 2\pi \cdot y \cdot x \cdot dy. \end{aligned}$$

The height of the shell is x, where x and y are related by the equation $y = \sqrt{8x}$. Solving this equation for x gives $x = y^2/8$. Hence,

$$dV = 2\pi y \cdot \left(\frac{y^2}{8}\right) \cdot dy = \tfrac{1}{4}\pi y^3\, dy.$$

Summing these elements as y varies from $1/2$ to 2, the volume occupied by the mirror is

$$V = \frac{\pi}{4}\int_{1/2}^{2} y^3\, dy = \frac{\pi}{16}y^4 \bigg|_{1/2}^{2} = \frac{255\pi}{256} \text{ cm}^3.$$

Multiplying this volume by the density of the glass, the mass of the mirror is about 8.1 grams.

In the next example, we discuss a way of calculating the volume of an unbounded solid of revolution.

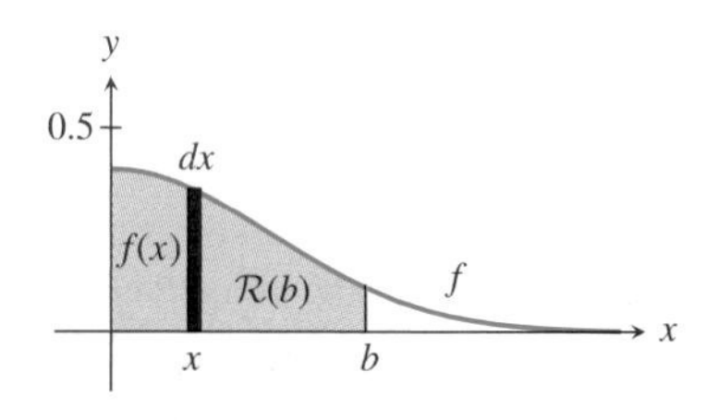

FIGURE 6.15 The region $\mathcal{R}(b)$ beneath the graph of the function $f(x) = \left(1/\sqrt{2\pi}\right)e^{-x^2/2}$.

EXAMPLE 3 Calculate the volume of the solid of revolution formed by revolving the region $\mathcal{R}(b)$ shown in Fig. 6.15 about the y-axis. The region $\mathcal{R}(b)$ is beneath the graph of the function

$$f(x) = \frac{1}{\sqrt{2\pi}}e^{-x^2/2}, \qquad 0 \le x \le b. \tag{3}$$

If the right-hand boundary of region $\mathcal{R}(b)$ generating the solid were extended indefinitely to the right, what could you say about the volume of the resulting unbounded solid of revolution?

Solution

For this solid of revolution, for which the axis of revolution is the y-axis, the shell method is simpler than the disk method. Can you see why? For the shell method, we draw an element of area which, when revolved about the axis of revolution, generates a cylindrical shell. The volume of the cylindrical shell is

$$dV = 2\pi \cdot x \cdot f(x)\, dx.$$

Summing these elements as x varies from 0 to b, the volume $V(b)$ of the solid generated by the region $\mathcal{R}(b)$ is

$$V(b) = 2\pi \int_0^b \frac{1}{\sqrt{2\pi}} x e^{-x^2/2}\, dx$$
$$= \sqrt{2\pi} \int_0^b x e^{-x^2/2}\, dx.$$

If we make the change of variable $u = x^2/2$, then $du = x\,dx$ and the integral becomes

$$V(b) = \sqrt{2\pi} \int_0^{b^2/2} e^{-u}\, du = -\sqrt{2\pi} e^{-u}\Big|_0^{b^2/2} = \sqrt{2\pi}(1 - e^{-b^2/2}) \text{ cubic units.}$$

This is the volume of the solid of revolution generated by region $\mathcal{R}(b)$.

To answer the question concerning the volume V of the unbounded solid of revolution obtained by extending the region $\mathcal{R}(b)$ indefinitely to the right, it is reasonable to take this volume as the limiting value of $V(b)$ as $b \to \infty$. Hence, $V = \lim_{b\to\infty} V(b) = \sqrt{2\pi}$.

Exercises 6.2

Exercises 1–8: Use the shell method in calculating the volume of the solid of revolution generated by revolving the given region about the given axis. Make a simple sketch of the region and, in the region, a representative area element which, when revolved about the axis of revolution, generates a cylindrical shell.

1. The region bounded by the parabola with equation $(y-3)^2 = x - 1$ and the line with equation $x + y = 6$; x-axis.

2. The region bounded by the parabola with equation $(y-3)^2 = 2(x-1)$ and the line with equation $y = 2x - 1$; x-axis.

3. The region between the curves described by $y = \sqrt{x}$ and $y = x$; x-axis.

4. The region between the curves described by $y = x^2$ and $y = x^3$; x-axis.

5. The region bounded below by the segment $[1, 2]$ of the x-axis, on the left by $y = \ln x$, $1 \le x \le e$, and on the right by the curve with equation $y^2 = (x-2)/(e-2)$, $2 \le x \le e$; x-axis.

6. The region bounded below by the segment $[1, 2]$ of the x-axis, on the left by $y = (x-1)/x$, $1 \le x \le 3$, and on the right by the curve with equation $y = (2x-4)/x$, $2 \le x \le 3$; x-axis.

7. The region beneath the graph of $y = \sin x$ and above $[0, \pi/2]$; x-axis.

8. The region beneath the graph of $y = \tan x$ and above $[0, \pi/4]$; x-axis.

Exercises 9–26: Use the shell method in calculating the volume of the solid.

9. Calculate the capacity in liters of a circular reservoir whose depth profile in meters from center to edge is shown in the accompanying figure. *Note:* One cubic meter is 1000 liters.

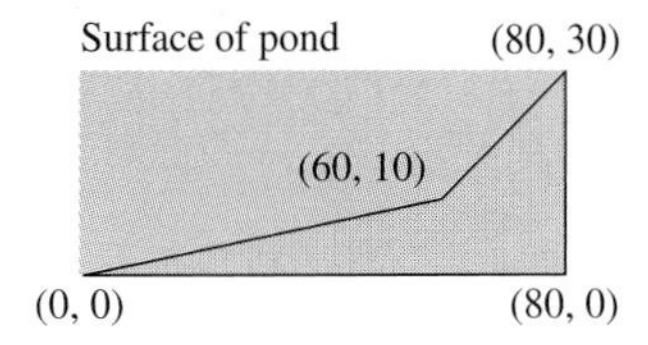

Figure for Exercise 9.

10. Calculate the capacity in liters of a circular reservoir whose depth profile in meters from center to edge is shown in the accompanying figure. *Note:* One cubic meter is 1000 liters.

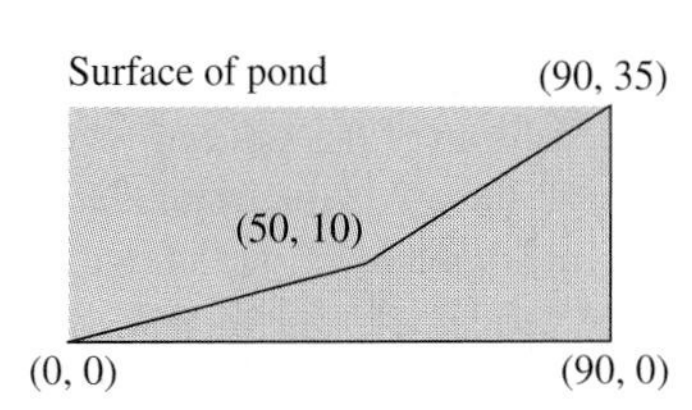

Figure for Exercise 10.

11. Calculate the mass of Earth. Assume that Earth is a solid of revolution, generated by revolving an elliptical region about the y-axis (the polar axis). The mean density of Earth is $5515\ \mathrm{kg/m^3}$, its equatorial radius is 6378.1 km, and its polar radius is 6356.8 km. *Hint:* Consider doing the general case $x^2/a^2 + y^2/b^2 = 1$ first.

12. Calculate the mass of Jupiter. Assume that Jupiter is a solid of revolution, generated by revolving an elliptical region about the y-axis (the polar axis). The mean density of Jupiter is $1326\ \mathrm{kg/m^3}$, its equatorial radius is 71,492 km, and its polar radius is 66,854 km. *Hint:* Consider doing the general case $x^2/a^2 + y^2/b^2 = 1$ first.

13. Calculate the mass of the thrust bearing shown in the accompanying figure. The density of steel is $7850\ \mathrm{kg/m^3}$.

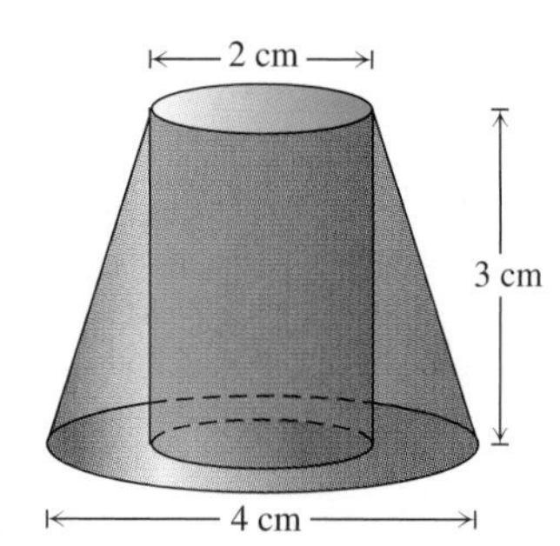

Figure for Exercise 13.

14. Calculate the mass of the thrust bearing shown in the accompanying figure. The density of steel is $7850\ \mathrm{kg/m^3}$.

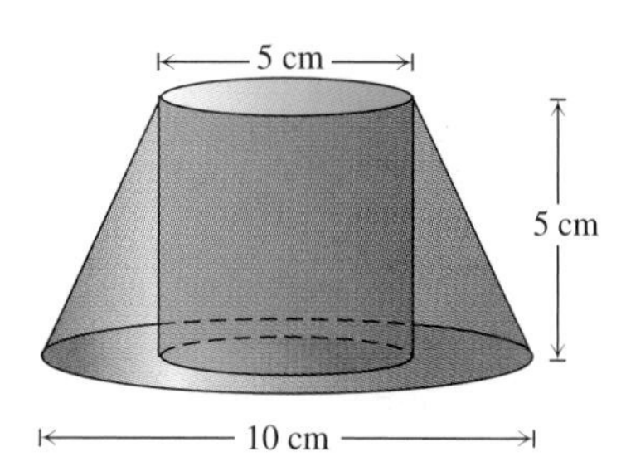

Figure for Exercise 14.

15. Calculate the volume of a right circular cone with height h and base radius r.

16. Calculate the volume of a sphere of radius a.

17. Calculate the volume of the solid of revolution generated by revolving about the x-axis the triangular region with vertices $(2, 1)$, $(1, 4)$, and $(5, 4)$.

18. Calculate the volume of the solid generated by revolving about the x-axis the region beneath the graph of $y = \arcsin x$ and above the interval $[0, 1]$.

19. Calculate the volume of the solid generated by revolving about the x-axis the region beneath the graph of $y = \arctan x$ and above the interval $[0, 1]$. Evaluate the integral for the volume by using the trapezoid rule with $n = 10$; T_{10} will be within 0.01 of the integral.

20. Let A be the region bounded by the graphs of the equations $y = x/2$ and $y = \sqrt{x}$.

a. Calculate the volume of the solid generated by revolving A about the line with equation $x = 0$.

b. Calculate the volume of the solid generated by revolving A about the line with equation $x = 4$.

21. Let A be the region bounded by the graphs of the equations $y = x$ and $y = (x - 2)^2$.

a. Calculate the volume of the solid generated by revolving A about the line with equation $x = 1$.

b. Calculate the volume of the solid generated by revolving A about the line with equation $x = 4$.

22. Calculate the volume of the solid of revolution generated by rotating about the y-axis the region between the graph of $y = x^4 - 2x^2 + 1$ and the x-axis, $0 \le x \le 2$.

23. Calculate the volume of the solid of revolution generated by rotating about the y-axis the region beneath the graph of $f(x) = e^{-x}$ and above the interval $[0, b]$, where $b > 0$. If the region generating the solid were extended indefinitely to the right, what could you say about the volume of the resulting unbounded solid of revolution?

24. Calculate the volume of the solid of revolution generated by rotating about the y-axis the region beneath the graph of $f(x) = 1/x^2$ and above the interval $[1, b]$, where $b > 1$. If the region generating the solid were extended indefinitely to the right, what could you say about the volume of the resulting unbounded solid of revolution?

25. Calculate the volume of the solid of revolution generated by rotating about the x-axis the region between the graphs with equations $y = 1/\sqrt[3]{x}$ and $y = 1$, for $a \le x \le 1$, where $0 < a < 1$. If the region generating the solid were extended by letting $a \to 0^+$, what could you say about the volume of the resulting unbounded solid of revolution?

26. Calculate the volume of the solid of revolution generated by rotating about the x-axis the region between the graphs with equations $y = 1/x$ and $y = 1$, for $a \le x \le 1$, where $0 < a < 1$. If the region generating the solid were extended by letting $a \to 0^+$, what could you say about the volume of the resulting unbounded solid of revolution?

6.3 Polar Coordinates and Parametric Equations

The applications of the integral we take up in Sections 6.4 (how to calculate the length of a curve), 6.5 (how to calculate the area of a region described by a polar equation), and 6.8 (curvature) will depend in part upon some familiarity with polar coordinates, the parametric description of a polar curve, and the description of motion using vector or parametric functions. For this reason we review polar coordinates, polar equations, and parametric equations in this section.

The polar coordinates of a point are intimately related to the position vector of that point. Figure 6.16 shows the position vector $\mathbf{r} = \langle x, y \rangle$ of the point (x, y). This vector has length $r = \sqrt{x^2 + y^2}$ and direction θ, but neither of these is explicit in the notation $\langle x, y \rangle$. To make them explicit, note that $x = r \cos \theta$ and $y = r \sin \theta$ and, hence,

$$\mathbf{r} = \langle x, y \rangle = \langle r \cos \theta, r \sin \theta \rangle = r\langle \cos \theta, \sin \theta \rangle.$$

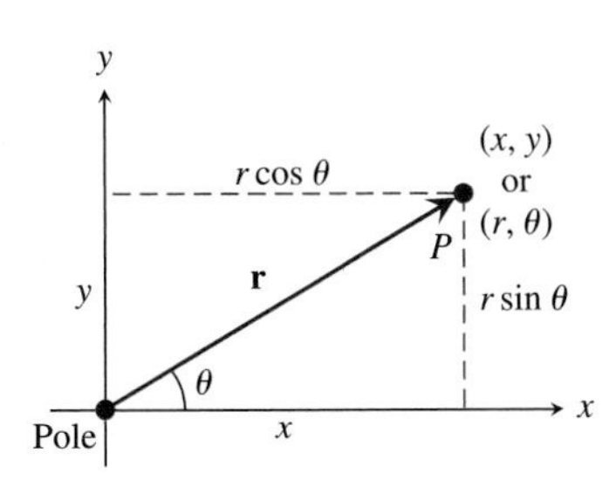

FIGURE 6.16 Polar and Cartesian coordinates.

The pair (r, θ) are the *polar coordinates* of the point with *Cartesian or rectangular coordinates* (x, y).

Cartesian coordinates for a plane are based on two intersecting lines. A point P is located by the ordered pair (x, y), where x and y are the **Cartesian coordinates** of P measured on these lines. For some purposes it is simpler to locate P with the ordered pair (r, θ) of **polar coordinates,** where r is the distance to P from a fixed point called the **pole** and θ is an angle measured from a fixed direction. Figure 6.16 shows both the Cartesian and polar coordinates of a point P. If the pole of the polar coordinate system is placed at the origin of the Cartesian system and the angle is measured counterclockwise from the positive x-axis, the relationship between the Cartesian coordinates (x, y) and the polar coordinates (r, θ) of P can be summarized as follows.

WEB

Relating the Polar and Cartesian Coordinates of a Point P

The Cartesian coordinates (x, y) and polar coordinates (r, θ) of P are related by the equations

$$\begin{aligned} x &= r \cos \theta \\ y &= r \sin \theta. \end{aligned} \tag{1}$$

If we are given polar coordinates r and θ of P, we may use (1) to calculate the Cartesian coordinates x and y of P. If we are given the Cartesian coordinates x and y, how do we calculate the polar coordinates r and θ? For P in the first quadrant, as in Fig. 6.16,

$$\begin{aligned} r &= \sqrt{x^2 + y^2} \\ \theta &= \arctan\left(\frac{y}{x}\right). \end{aligned} \tag{2}$$

For P in other quadrants we calculate r in the same way and adjust θ for the quadrant. We illustrate these calculations in an example.

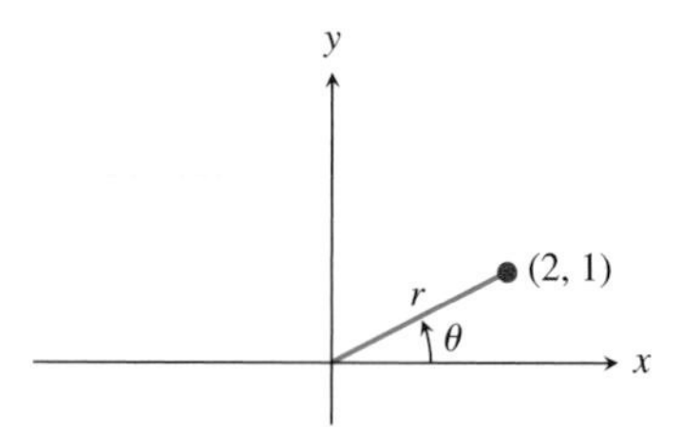

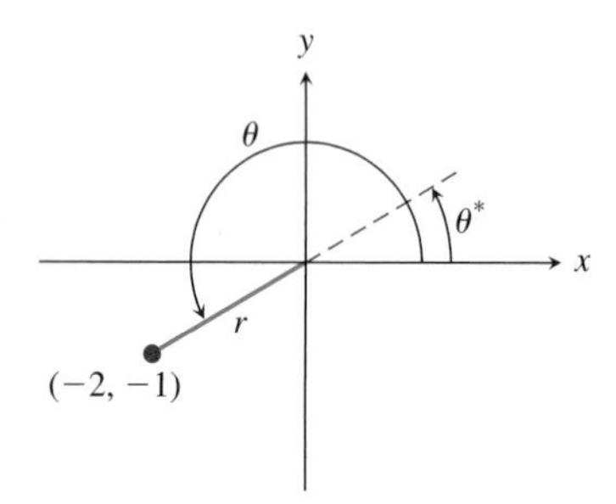

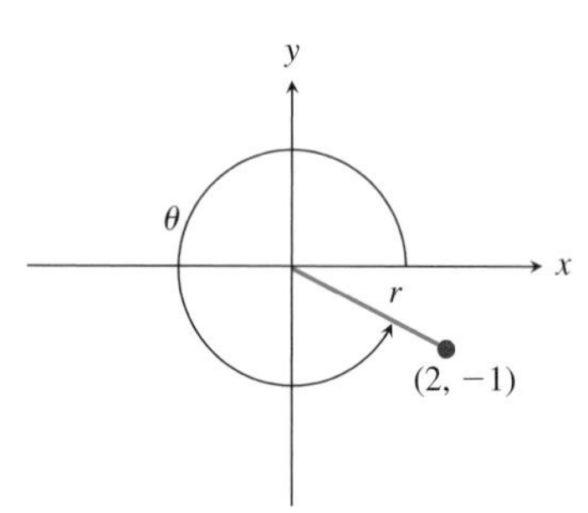

FIGURE 6.17 Converting to rectangular coordinates.

EXAMPLE 1 Figure 6.17 shows points with rectangular coordinates $(2, 1)$, $(-2, -1)$, and $(2, -1)$. Determine the polar coordinates of these points.

Solution

The point $(2, 1)$ is in the first quadrant. Applying (2), the polar coordinates of this point are

$$r = \sqrt{2^2 + 1^2} = \sqrt{5}$$
$$\theta = \arctan(1/2) \approx 0.4636.$$

The point $(-2, -1)$ is in the third quadrant. Applying (2)

$$r = \sqrt{(-2)^2 + (-1)^2} = \sqrt{5}$$

and

$$\theta^* = \arctan((-1)/(-2)) = \arctan(1/2) \approx 0.4636.$$

The relation between the first-quadrant angle θ^* and the third-quadrant angle θ is

$$\theta = \pi + \theta^* = \pi + \arctan(1/2) \approx 3.6052.$$

The point $(2, -1)$ is in the fourth quadrant. Applying (2),

$$r = \sqrt{2^2 + (-1)^2} = \sqrt{5}$$

and

$$\theta = \arctan((-1)/2) = \arctan(-1/2) = -0.4636\ldots.$$

The calculator has used a standard convention and, in effect, decided that $\theta = -0.4636\ldots$. Substituting these values of r and θ into (1), $x = 2$ and $y = -1$. We may prefer the polar coordinates $r = \sqrt{5}$ and $\theta = 2\pi - \arctan(1/2) \approx 5.8195$. These polar coordinates are shown in the figure.

We noticed in Example 1 that circumstances can lead us to accept several sets of polar coordinates for a point. In fact, we shall agree that for any values of r and θ whatsoever, (r, θ) will be a *proper* set of polar coordinates for a point P with Cartesian or rectangular coordinates (x, y) provided that Equations (1) are satisfied. We illustrate this in the next example.

FIGURE 6.18 Three sets of polar coordinates for the point P.

EXAMPLE 2 Show that the point P with rectangular coordinates $\left(\sqrt{3}, 1\right)$ has polar coordinates

$$\left(2, \frac{\pi}{6}\right), \left(2, \frac{\pi}{6} + 2\pi\right), \text{ and } \left(-2, \frac{\pi}{6} + \pi\right).$$

Solution

Figure 6.18 shows the point P with these polar coordinates. These polar coordinate pairs are proper because

$$x = r\cos\theta = 2\cos(\pi/6) = \sqrt{3} \qquad y = r\sin\theta = 2\sin(\pi/6) = 1$$
$$x = r\cos\theta = 2\cos(\pi/6 + 2\pi) = \sqrt{3} \qquad y = r\sin\theta = 2\sin(\pi/6 + 2\pi) = 1$$
$$x = r\cos\theta = -2\cos(\pi/6 + \pi) = \sqrt{3} \qquad y = r\sin\theta = -2\sin(\pi/6 + \pi) = 1.$$

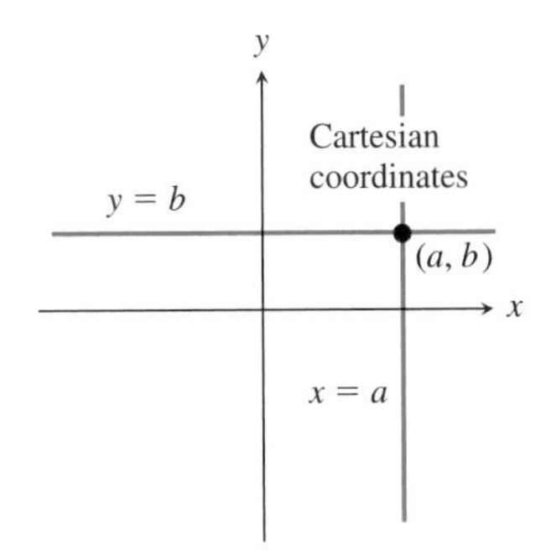

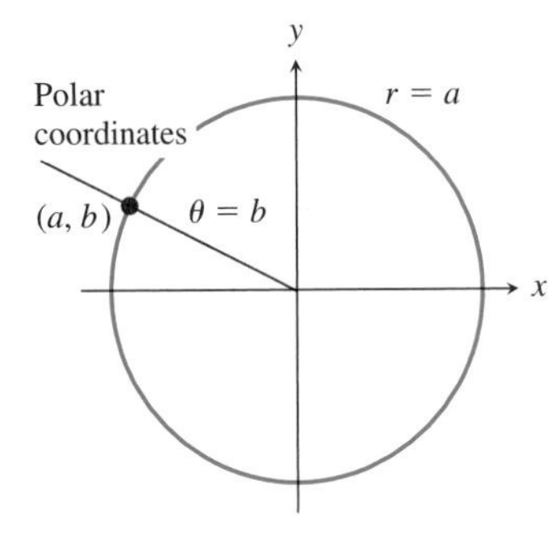

FIGURE 6.19 Cartesian coordinate curves and polar coordinate curves.

In the second pair of polar coordinates, θ has been increased by 2π. For the third pair, in which $r < 0$, we first located the ray determined by θ and then plotted P on its backward extension through the origin.

In Cartesian coordinates the simplest graphs of equations are $x = a$ and $y = b$, where a and b are constants. These are a vertical line through $(a, 0)$ and a horizontal line through $(0, b)$. These "coordinate curves" are shown at the top of Fig. 6.19. The coordinate curves in polar coordinates are shown at the bottom of the figure. These have equations $r = a$ and $\theta = b$. The graph of the equation $r = 1$, for example, is a circle of radius 1 centered at the pole. This is the set of all points (r, θ) in the plane for which $r = 1$. The graph of the equation $\theta = \pi/4$, for example, is the ray from the origin making an angle of $\pi/4$ with the positive x-axis. If r can be negative, the graph of $\theta = \pi/4$ is the entire graph of the equation $y = x$.

We give several examples of plotting polar equations $r = g(\theta)$. The first is similar to graphing $y = x$ in Cartesian coordinates. The last two examples will relate graphs of polar equations to parametric equations.

EXAMPLE 3 Graph the polar equation $r = \theta$ for $0 \leq \theta \leq 2\pi$.

Solution

We must sketch all points (r, θ) satisfying the equation $r = \theta$. For $\theta = 0$, $\pi/2$, 4, 5, and 2π, the corresponding r values are $r = 0$, $\pi/2$, 4, 5, and 2π. Hence, the points with polar coordinates $(0, 0)$, $(\pi/2, \pi/2)$, $(4, 4)$, $(5, 5)$, and $(2\pi, 2\pi)$ are on the graph. Figure 6.20 shows these five points.

As shown in the figure, as θ increases from 0 to 2π, the point $(r, \theta) = (\theta, \theta)$ traces a curve through these five points. This curve is called a *spiral*.

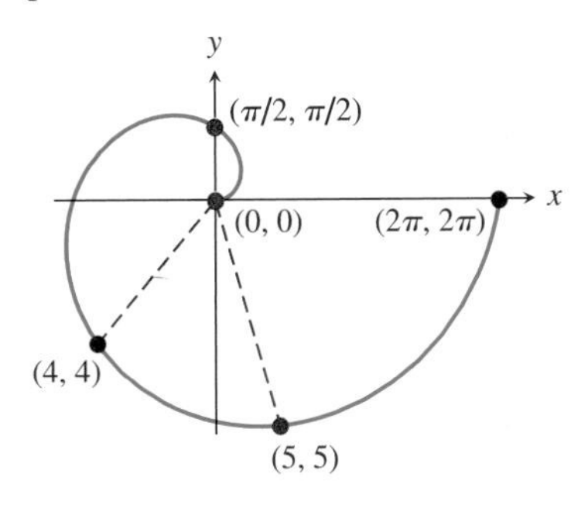

FIGURE 6.20 The spiral $r = \theta$, $0 \leq \theta \leq 2\pi$.

EXAMPLE 4 Graph the polar equation $r = \cos\theta$.

Solution

The graph of this equation is the curve C shown in Fig. 6.21. We show that the curve is a circle at the end of this example.

Let's start by letting θ vary over the "first quadrant," that is, let θ increase from 0 to $\pi/2$. As θ increases, $r = \cos\theta$ decreases from 1 to 0. The curve C is thus traced counterclockwise from $(1, 0)$ to $(0, \pi/2)$, where the pairs are polar coordinates. This is the top half of C. The tracing of the polar curve may be correlated with the graph of the cosine curve with equation $y = \cos x$ shown at the bottom of Fig. 6.21. As x varies over the first quadrant, y decreases from 1 to 0.

As θ varies over the "second quadrant," from $\pi/2$ to π, $r = \cos\theta$ continues to decrease, going from 0 to -1. Because $r \leq 0$, we plot the point (r, θ) on the backwards extension through the origin of the ray defined by θ. The angle θ shown in Fig. 6.21 is approximately 0.7 (radians). The point $(\cos(0.7), 0.7)$ of C corresponding to $\theta = 0.7$ is in the first quadrant. The point $(\cos(\pi - 0.7), \pi - 0.7) = (-\cos(0.7), \pi - 0.7)$ of C, which

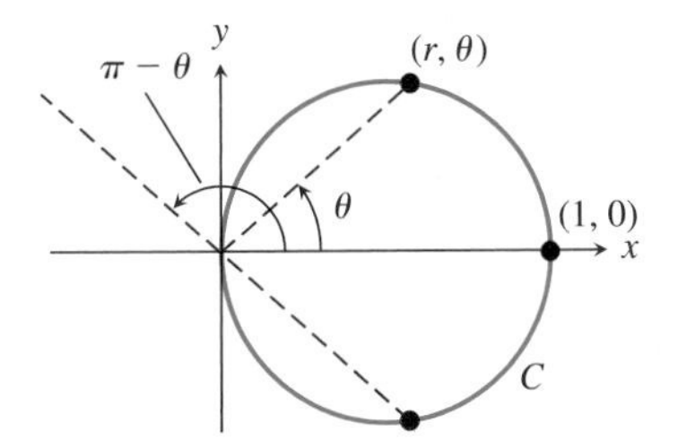

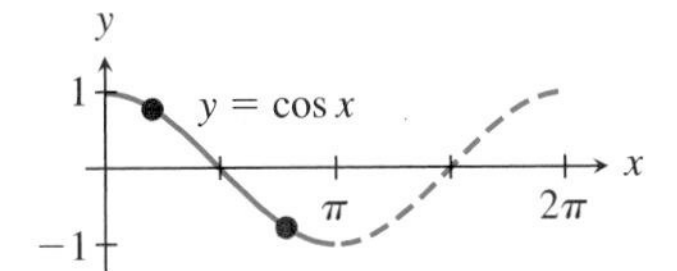

FIGURE 6.21 The polar circle described by $r = \cos\theta$.

corresponds to the second-quadrant angle $\theta = \pi - 0.7$, is in the fourth quadrant.

As θ varies from π to 2π, the circle is retraced.

To show that the curve C with equation $r = \cos\theta$ is a circle, we transform this equation to Cartesian coordinates. Multiplying both sides of this equation by r,

$$r^2 = r\cos\theta.$$

Because $r^2 = x^2 + y^2$ and $x = r\cos\theta$ (these come from (2) and (1)),

$$x^2 + y^2 = x.$$

Moving the x term to the left and completing the square on $x^2 - x$ gives

$$x^2 - x + (-1/2)^2 + y^2 = (-1/2)^2$$
$$(x - 1/2)^2 + y^2 = (1/2)^2.$$

We recognize the last equation as that of a circle with center $(1/2, 0)$ and radius $1/2$. This is the circle shown in Fig. 6.21. *Note:* We have not quite shown that the graph of $r = \cos\theta$ is a circle, only that this graph is part of a circle. See Exercise 40.

In Sections 3.3 and 3.4 we discussed parametric equations for lines and circles and showed how curves described by equations $y = f(x)$ can be described by parametric equations. We now show how curves described by polar equations $r = g(\theta)$ can be described by parametric equations. For any point $(r, \theta) = (g(\theta), \theta)$ on the graph of the equation $r = g(\theta)$ it follows from Equations (1) that the rectangular coordinates (x, y) of this point are

$$x = r\cos\theta = g(\theta)\cos\theta$$
$$y = r\sin\theta = g(\theta)\sin\theta.$$

The parametric equation corresponding to these equations is

$$\mathbf{r} = \mathbf{r}(\theta) = \langle g(\theta)\cos\theta, g(\theta)\sin\theta\rangle = g(\theta)\langle\cos\theta, \sin\theta\rangle. \tag{3}$$

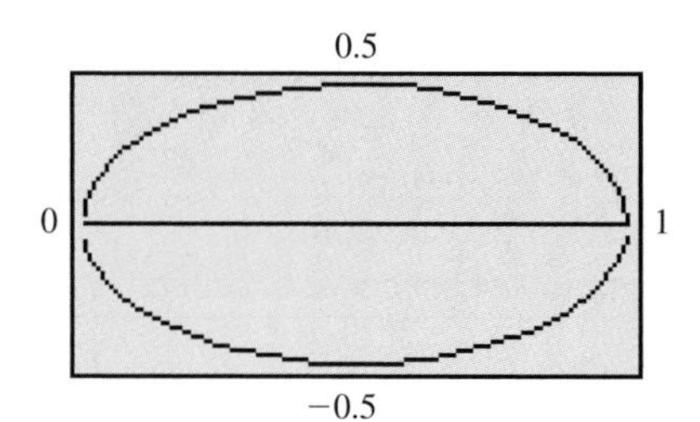

FIGURE 6.22 Graph of $\mathbf{r} = \langle\cos^2\theta, \cos\theta\sin\theta\rangle$, with $0 \le \theta \le \pi$, in the window $0 \le x \le 1$ by $-0.5 \le y \le 0.5$.

EXAMPLE 5 Find a parametric equation for the polar equation $r = \cos\theta$ and plot the parametric equation with a CAS or calculator.

Solution

We found in Example 4 that the graph of the equation $r = \cos\theta$ is a circle. Applying (3) to this equation,

$$\mathbf{r} = \cos\theta\langle\cos\theta, \sin\theta\rangle = \langle\cos^2\theta, \cos\theta\sin\theta\rangle, \qquad 0 \le \theta \le 2\pi. \tag{4}$$

Figure 6.22 shows a calculator graph of this parametric equation. The chosen window, which is the smallest window showing the complete graph, gives a distorted view of the graph.

Figure 6.23 shows a calculator graph of this parametric equation, with a window chosen to make the scales on the two axes nearly equal.

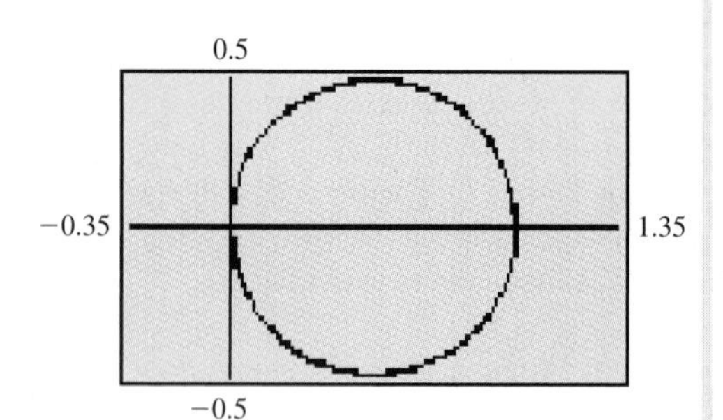

FIGURE 6.23 Graph of $\mathbf{r} = \langle\cos^2\theta, \cos\theta\sin\theta\rangle$, with $0 \le \theta \le \pi$, in the window $-0.35 \le x \le 1.35$ by $-0.5 \le y \le 0.5$.

Next we graph a *limaçon*, one of a class of curves having polar equations of the form $r = a + b \sin \theta$, where a and b are constants.

EXAMPLE 6 Graph the polar curve $r = 1 - \sqrt{2} \sin \theta$, $0 \leq \theta \leq 2\pi$, using both the given polar equation and the parametric Equation (3). Calculate a tangent vector to this curve at the origin.

Solution

To graph polar curves for which r can be negative, it is often useful to find the values of θ at which r is zero. These are the values of θ at which r may change sign, that is, where the graph may pass through the origin. Setting $1 - \sqrt{2} \sin \theta$ equal to zero,

$$\sin \theta = \frac{1}{\sqrt{2}}$$

$$\theta = \frac{\pi}{4}, \frac{3\pi}{4}.$$

These are the angles in $[0, 2\pi)$ at which r may change sign.

If you are graphing by hand, the graph of $r = 1 - \sqrt{2} \sin \theta$ is more quickly understood if you first sketch the graph of

$$y = 1 - \sqrt{2} \sin x$$

in rectangular coordinates.

Starting with the graph of $y = \sin x$, which is shown in Figure 6.24(a), the factor $\sqrt{2}$ stretches the sine curve in the y direction; the factor -1 reflects the result about the x-axis; and adding 1 shifts the reflection upward by one unit. The final result is shown in Fig. 6.24(b). Recalling from above that $1 - \sqrt{2} \sin x$ is zero for $x = \pi/4, 3\pi/4$, we see that for $\pi/4 \leq x \leq 3\pi/4$, the graph lies below the x-axis. Switching back to r and θ, note that r is negative for θ between $\pi/4$ and $3\pi/4$. For $\theta = \pi/2$, for example, $r = 1 - \sqrt{2} \approx -0.414$. Because $r < 0$ we plot the point with polar coordinates $\left(-1 + \sqrt{2}, \pi/2 + \pi\right)$.

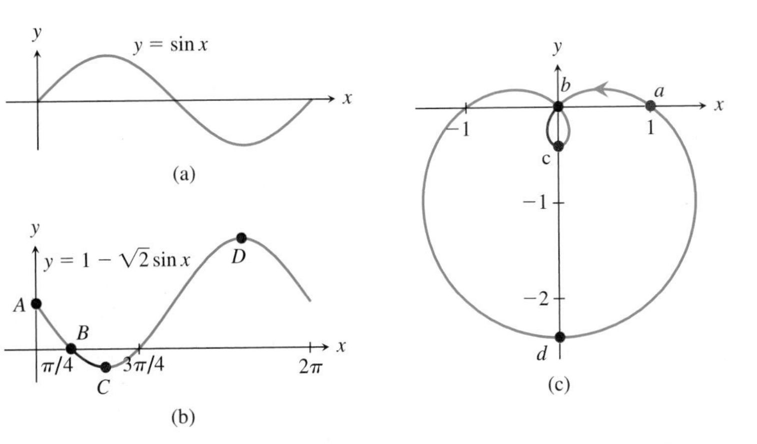

FIGURE 6.24 (a) Graph of $y = \sin x$; (b) graph of $y = 1 - \sqrt{2} \sin x$; (c) graph of the limaçon with equation $r = 1 - \sqrt{2} \sin \theta$.

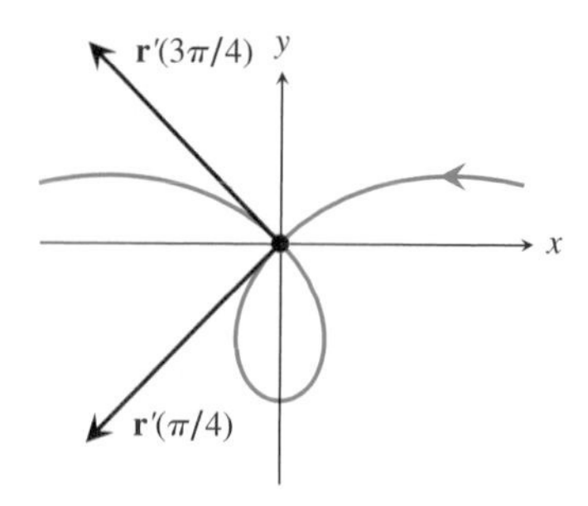

FIGURE 6.25 At $r(\pi/4)$ and $r(3\pi/4)$ the tangent vectors are
$\mathbf{r}'(\pi/4) = -(1/\sqrt{2})\mathbf{i} - (1/\sqrt{2})\mathbf{j}$
$\mathbf{r}'(3\pi/4) = -(1/\sqrt{2})\mathbf{i} + (1/\sqrt{2})\mathbf{j}$.

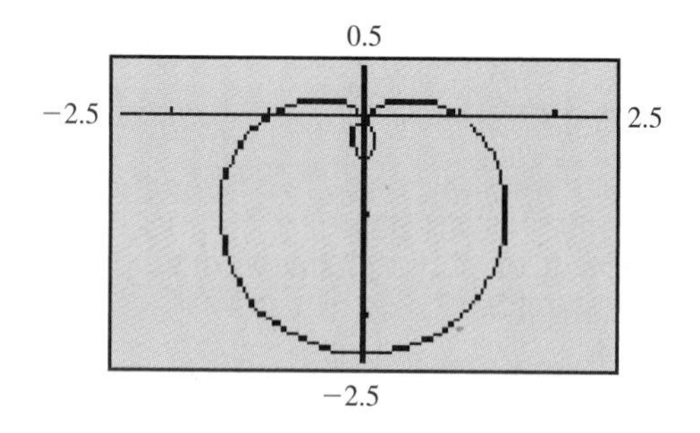

FIGURE 6.26 Graph of $\mathbf{r}(\theta) = \left(1 - \sqrt{2}\sin\theta\right)(\cos\theta\mathbf{i} + \sin\theta\mathbf{j})$, with $0 \le \theta \le 2\pi$, in the window $-2.5 \le x \le 2.5$ by $-2.5 \le y \le 0.5$.

We may draw the limaçon by mentally tracing the graph in Fig. 6.24(b), in which x (think θ) varies over the three intervals $[0, \pi/4]$, $[\pi/4, 3\pi/4]$, and $[3\pi/4, 2\pi]$. The points A, B, C, and D on the middle graph correspond to the points a, b, c, and d on the graph of the limaçon. For $\theta \in [0, \pi/4]$, r decreases from 1 to 0. The curve starts at $(1, 0)$ and, as $\theta \to \pi/4$, the curve approaches the origin. As θ moves through $\pi/4$, r becomes negative and the trace moves into the third quadrant. The arc bc of the curve in Fig. 6.24(c) corresponds to arc BC in (b). At $\theta = \pi/2$, $r = 1 - \sqrt{2}$. This corresponds in Fig. 6.24(b) to the bottom of the dip between $\pi/4$ and $3\pi/4$. As θ moves through $3\pi/4$, r becomes positive and the trace moves into the second quadrant. We continue in this way. At $\theta = 3\pi/2$, $r = 1 + \sqrt{2} \approx 2.414$, which is the maximum value of r.

Figure 6.25 shows a zoom-view of the limaçon near the origin. The limaçon passes through the pole as θ passes through $\pi/4$ and $3\pi/4$. Tangent vectors for these two values of θ can be found by first describing the limaçon with the parametric equation

$$\mathbf{r}(\theta) = r(\cos\theta\mathbf{i} + \sin\theta\mathbf{j}) = \left(1 - \sqrt{2}\sin\theta\right)(\cos\theta\mathbf{i} + \sin\theta\mathbf{j}). \qquad (5)$$

Differentiating this equation,

$$\begin{aligned}\mathbf{r}'(\theta) &= \left(-\sqrt{2}\cos\theta\right)(\cos\theta\mathbf{i} + \sin\theta\mathbf{j}) \\ &\quad + \left(1 - \sqrt{2}\sin\theta\right)(-\sin\theta\mathbf{i} + \cos\theta\mathbf{j}).\end{aligned}$$

Evaluating $\mathbf{r}'(\theta)$ for $\theta = \pi/4$ and $\theta = 3\pi/4$,

$$\begin{aligned}\mathbf{r}'(\pi/4) &= -\left(1/\sqrt{2}\right)\mathbf{i} - \left(1/\sqrt{2}\right)\mathbf{j} \\ \mathbf{r}'(3\pi/4) &= -\left(1/\sqrt{2}\right)\mathbf{i} + \left(1/\sqrt{2}\right)\mathbf{j}.\end{aligned}$$

These tangent vectors are shown in Fig. 6.25.

Figure 6.26 shows a calculator graph of the parametric equation (5), with the calculator set in parametric mode.

The Cycloid

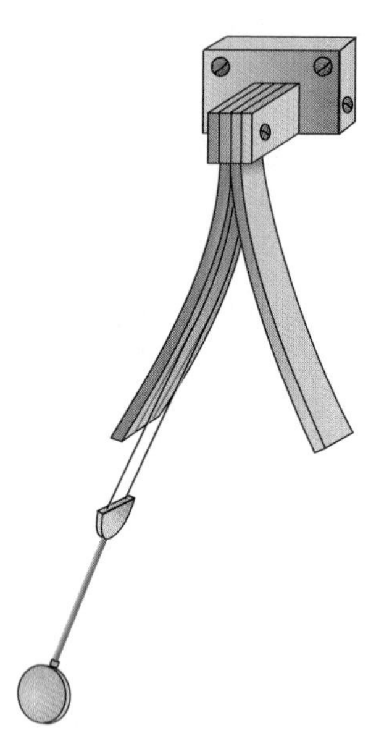

FIGURE 6.27 Isochronal pendulum.

In the seventeenth century, the search for a device or method to accurately determine longitude at sea led Christian Huygens (1629–1695) to the invention of a pendulum clock. Huygens's clock was not the common pendulum clock, but one based on an isochronal pendulum, which has a period independent of the amplitude of the swing. In a common pendulum clock the pendulum is driven by a spring. As the spring winds down, the force exerted on the pendulum decreases; this causes the pendulum to swing through shorter arcs and the clock to speed up.

A part of Huygens's clock is sketched in Fig. 6.27. The cord supporting the bob was constrained to move between two plates, each bent into the shape of the curve called an *isochrone*. The two curved plates are shown in the figure. As the bob swings to the left, the double cord gradually comes into contact with the left plate. At the moment shown in the figure, the cord is in contact with only the upper part of the left plate. As the cord gradually comes in contact with the plate, the bob swings through a curve identical in form to the plate! Huygens's work on pendulum clocks was published as *Horologium Oscillatorium* (Paris, 1673).

The isochrone is also called a *cycloid*. This name is used because the curve can be generated by a point on a rolling circle. We use this idea in deriving a parametric equation for a cycloid/isochrone.

A circle of radius 1 is placed on the x-axis, just touching the origin. If the circle rolls to the right, the curve traced by the point of the circle initially at the origin is called a cycloid. The curve is shown at the top of Fig. 6.28. A zoom-view of the first half of the first arch of the cycloid is shown at the bottom of the figure. The generating circle is shown both in its initial position and after it has turned about its own center through the angle ϕ. We describe C parametrically, using ϕ as a parameter.

Referring to Fig. 6.28, we wish to express the position vector $\mathbf{r}$ of the generating point P in terms of the angle ϕ through which the circle has turned about its own center. In the figure we show the generating point at the point P corresponding to $\phi \approx 1.2$ (about 70°). Letting $\mathbf{u}$ be a vector giving the horizontal displacement of the center of the circle, $\mathbf{v}$ a vector giving the vertical displacement of the center, and $\mathbf{w}$ a vector from the center of the circle to the point P, we see from the figure that

$$\mathbf{r} = \mathbf{u} + \mathbf{v} + \mathbf{w}$$

for any value of $\phi \geq 0$. As the circle rolls and ϕ increases, the vector $\mathbf{v} = \langle 0, 1 \rangle$ is constant and the vectors $\mathbf{u}$ and $\mathbf{w}$ change with ϕ.

Now visualize the circle rolling back to its initial position, noting in particular that the arc subtended by ϕ would roll over the segment of length $\|\mathbf{u}\|$. Because in a unit circle the measure of an arc is the same as the (radian) measure of the central angle subtending that arc, $\mathbf{u} = \langle \phi, 0 \rangle$.

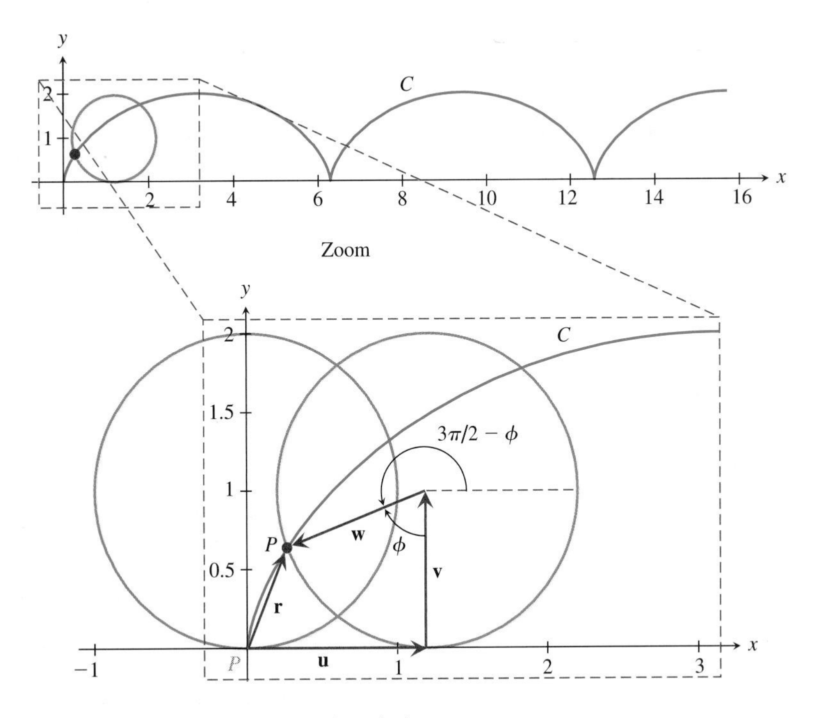

FIGURE 6.28 Cycloid and generating circle.

To write the vector $\mathbf{w}$ in terms of ϕ, first note that $3\pi/2 - \phi$ is the measure of an angle in standard position, that is, measured from the dotted line shown in the figure. Hence,

$$\mathbf{w} = \langle \cos(3\pi/2 - \phi), \sin(3\pi/2 - \phi) \rangle = \langle -\sin \phi, -\cos \phi \rangle.$$

Putting these results together,

$$\begin{aligned} \mathbf{r} = \mathbf{r}(\phi) &= \mathbf{u} + \mathbf{v} + \mathbf{w} \\ &= \langle \phi, 0 \rangle + \langle 0, 1 \rangle + \langle -\sin \phi, -\cos \phi \rangle. \end{aligned}$$

Adding these vectors,

$$\mathbf{r}(\phi) = \langle \phi - \sin \phi, 1 - \cos \phi \rangle, \qquad \phi \geq 0. \tag{6}$$

This equation describes the curve C parametrically using ϕ as a parameter.

EXAMPLE 7 Wherever possible, calculate the slope of the cycloid described by (6).

Solution

From (6), the tangent vector at the point $\mathbf{r}(\phi)$ of the cycloid is

$$\mathbf{r}'(\phi) = \langle 1 - \cos \phi, \sin \phi \rangle.$$

Using the Tangent Vector and Slope Theorem from Section 3.4, we can calculate the slope wherever $x'(\phi) \neq 0$. Because $x'(\phi) = 1 - \cos \phi$, we avoid $\phi = 0, 2\pi, 4\pi, \ldots$. These values of ϕ correspond to the "cusps" of the cycloid, the points of C where the generating point P returns to the x-axis. At these points the tangent vector is the zero vector and slope is not defined. The slope is zero at the tops of the arches, where $y' = 0$. Except for points with $\phi = 0, 2\pi, 4\pi, \ldots$, the slope of the cycloid at $\mathbf{r}(\phi)$ is

$$\frac{dy}{dx} = \frac{\dfrac{dy}{d\phi}}{\dfrac{dx}{d\phi}} = \frac{\sin \phi}{1 - \cos \phi}.$$

The Involute of a Circle

The curve most commonly used for the profile of a gear tooth is the involute of a circle. Figure 6.29(a) shows a pair of gears and (b) shows a zoom-view of the point of contact between the teeth of two gears. The involute is a curve for which the friction at the point of contact is a minimum.

A simple way of describing the involute of a circle is to imagine the unwinding of a string wound around a fixed circle. If the string is held taut in the plane of the circle, its end traces an *involute* of the circle.

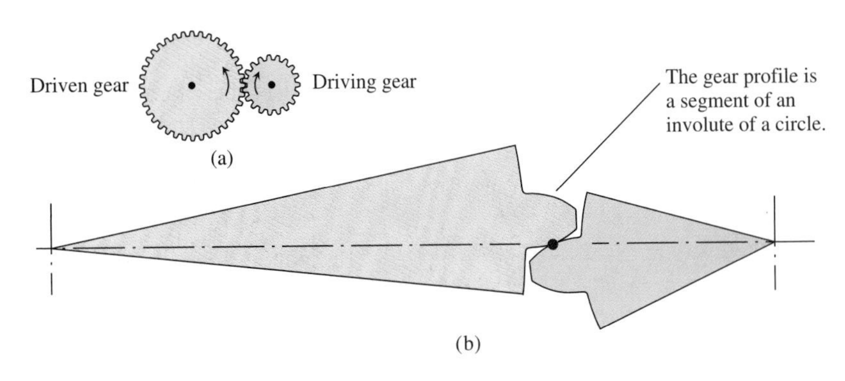

FIGURE 6.29 (a) A pair of gears; (b) a zoom-view of a pair of gear teeth.

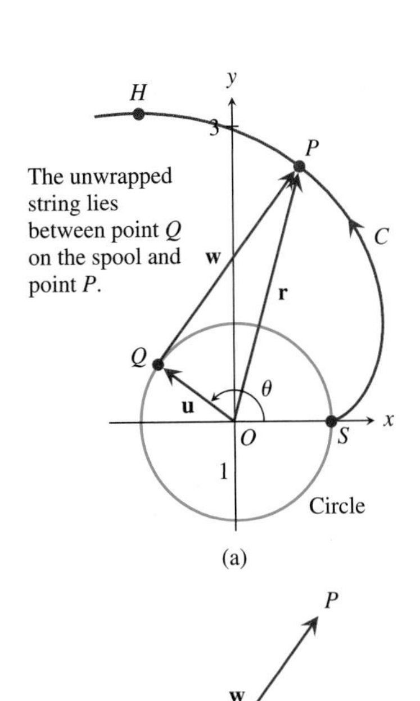

FIGURE 6.30 (a) String unwrapping from a circle; (b) the vectors **u** and **w**.

Figure 6.30(a) shows a circle of radius 1 meter. Suppose the end of the string is at the point P, having started at S. The taut string lies along $\overrightarrow{QP}$. Let θ be the angle from the positive x-axis counterclockwise around to $\overrightarrow{OQ}$, $\mathbf{u}$ the position vector of Q, $\mathbf{r}$ the position vector of P, and $\mathbf{w} = \overrightarrow{QP}$. If we can write $\mathbf{r}$ in terms of θ we will have a parametric equation for the involute C. We first note from Fig. 6.30 that $\mathbf{r} = \mathbf{u} + \mathbf{w}$.

Because the circle has radius 1 meter,

$$\mathbf{u} = \langle \cos\theta, \sin\theta \rangle.$$

Referring to Fig. 6.30(b), note that because $\overrightarrow{OQ}$ is perpendicular to $\overrightarrow{QP}$, the angle between $\mathbf{w}$ and the positive x-axis is $\theta - \pi/2$. Thus, a unit vector in the direction of $\mathbf{w}$ is

$$\mathbf{v} = \langle \cos(\theta - \pi/2), \sin(\theta - \pi/2) \rangle = \langle \sin\theta, -\cos\theta \rangle,$$

The length of $\mathbf{w}$, which is the length of the part of the string that has been unwound, is $1 \cdot \theta$. This comes from the formula $s = a\theta$ for the length of the arc of the sector of a circle with radius a and central angle θ. If we multiply the unit vector $\mathbf{v}$ by θ, we obtain a vector of length θ. Hence,

$$\mathbf{w} = \theta\langle \sin\theta, -\cos\theta \rangle.$$

Putting everything together, the involute C is described by

$$\mathbf{r} = \mathbf{r}(\theta) = \mathbf{u} + \mathbf{w} = \langle \cos\theta + \theta\sin\theta, \sin\theta - \theta\cos\theta \rangle, \qquad \theta \geq 0.$$

EXAMPLE 8 Figure 6.30(a) shows the curve C traced by the end of a string unwrapping from a circle of radius 1 meter. This curve is described by

$$\mathbf{r} = \mathbf{r}(\theta) = \langle \cos\theta + \theta\sin\theta, \sin\theta - \theta\cos\theta \rangle, \qquad \theta \geq 0.$$

Use this equation and the Tangent Vector and Slope Theorem (Section 3.4) to calculate the slope of C at the point P in the figure. The position vector of P is $\mathbf{r}(2.44)$. Also, locate the point H in the second quadrant at which the curve has a horizontal tangent.

Solution

To answer these questions, we calculate the tangent vector and then use the Tangent Vector and Slope Theorem. The tangent vector is

$$\begin{aligned}\mathbf{r}' &= \langle -\sin\theta + \sin\theta + \theta\cos\theta, \cos\theta - \cos\theta + \theta\sin\theta\rangle \\ &= \langle \theta\cos\theta, \theta\sin\theta\rangle.\end{aligned} \tag{7}$$

To find the slope at $\mathbf{r}(2.44)$, we first evaluate $\mathbf{r}'(\theta)$ at $\theta = 2.44$. From (7),

$$\mathbf{r}'(2.44) \approx \langle -1.86, 1.57\rangle.$$

Applying the Tangent Vector and Slope Theorem, the slope at $\mathbf{r}(2.44)$ is

$$m = \frac{y'}{x'} \approx \frac{1.57}{-1.86} \approx -0.845.$$

To locate the point at which the tangent line is horizontal, we use (7) to write

$$\frac{dy}{dx} = \frac{y'(\theta)}{x'(\theta)} = \frac{\theta\sin\theta}{\theta\cos\theta} = \tan\theta.$$

The tangent line will be horizontal when $dy/dx = 0$, that is, when $\tan\theta = 0$. Hence, $\theta = 0, \pi, 2\pi, \ldots$. The first of these values gives the point $\mathbf{r}(0) = (1, 0)$. The tangent line is horizontal at this point but it is not a second-quadrant point. The second value of θ for which $dy/dx = 0$ is $\theta = \pi$. This gives the point H in Fig. 6.30(a), whose position vector is $\mathbf{r}(\pi) = \langle -1, \pi\rangle$.

Exercises 6.3

Exercises 1–12: If the Cartesian coordinates of P are given, find polar coordinates (r, θ) of P, where $r \geq 0$ and $0 \leq \theta \leq 2\pi$. If the polar coordinates of P are given, find the Cartesian coordinates (x, y) of P.

1. $(r, \theta) = \left(\sqrt{2}, \pi/4\right)$
2. $(r, \theta) = \left(\sqrt{2}, 3\pi/4\right)$
3. $(r, \theta) = (2, 1)$
4. $(r, \theta) = (3, 1.5)$
5. $(r, \theta) = (1, 3.5)$
6. $(r, \theta) = (2, 5.2)$
7. $(x, y) = (2, 1)$
8. $(x, y) = (1, 5)$
9. $(x, y) = (-2, 0)$
10. $(x, y) = (0, -5)$
11. $(x, y) = (-1.5, 2.5)$
12. $(x, y) = (-2.6, -4.3)$

Exercises 13–24: Graph the polar equation. If θ is not restricted, plot all (r, θ) satisfying the equation.

13. $r = \theta^2, 0 \leq \theta \leq \pi$
14. $r = \sqrt{\theta}, 0 \leq \theta \leq 2\pi$
15. $r = \theta(2 - \theta), 0 \leq \theta \leq \pi$
16. $r = \theta(1 - \theta), 0 \leq \theta \leq \pi/2$
17. $r = 2\cos\theta$
18. $r = -\cos\theta$
19. $r = 2\cos(\pi/2 - \theta)$
20. $r = 2\sin(\pi/2 - \theta)$
21. The limaçon described by $r = -1 + \sqrt{3}\sin\theta$.
22. The limaçon described by $r = \sqrt{2}\cos\theta - 1$.
23. The four-petaled polar curve $r = \cos 2\theta$. It is useful to first sketch the Cartesian equation $y = \cos 2x$.
24. The four-petaled polar curve $r = \sin 2\theta$. It is useful to first sketch the Cartesian equation $y = \sin 2x$.

25. Calculate the slope of the limaçon described by $r = 1 - 2\sin\theta$ at the point corresponding to $\theta = 5\pi/4$.

26. Calculate the slope of the limaçon described by $r = 1 + 2\sin\theta$ at the point corresponding to $\theta = \pi/2$.

T 27. Determine a parametric equation for the involute of a circle of radius 2 meters. Sketch the involute. Calculate the slope of the curve at the point corresponding to the end of the string when 0.6 meters have been unwound.

T 28. Determine a parametric equation for the involute of a circle of radius 3 meters. Sketch the involute. Calculate the slope of the curve at the point corresponding to the end of the string when 1.0 meters have been unwound.

29. Determine a parametric equation for the cycloid curve C generated by a circle of radius 2 rolling along a line. Sketch one arch of C. Calculate a tangent vector to C at the point $\left(4\pi/3 - \sqrt{3}, 3\right)$.

30. Determine a parametric equation for the cycloid curve C generated by a circle of radius 1.5 rolling along a line. Sketch one arch of C. Calculate a tangent vector to C at the point $(3\pi/4 - 3/2, 3/2)$.

T 31. Ellipses may be described by polar equations of the form $r = p/(1 - e\cos\theta)$, where e is the eccentricity of the ellipse. For ellipses, $0 \le e < 1$. With units chosen so that the length of the semimajor axis of the Earth's orbit is one unit, the values of p and e for the orbits of Halley's comet, Mars, and Earth are

i. Halley's comet: $p = 1.15432$ and $e = 0.9673$;

ii. Mars: $p = 1.5104$ and $e = 0.0933865$;

iii. Earth: $p = 0.999721$ and $e = 0.016718$. (These data were adapted from Peter Duffett-Smith's *Practical Astronomy with Your Calculator,* 2nd edition, Cambridge University Press, 1981.)

Graph these three ellipses on the same set of axes.

T 32. See the preceding problem. The values of p and e for the orbits of the Pons-Brooks comet, Jupiter, and Earth are

i. Pons-Brooks comet: $p = 1.51317$ and $e = 0.955$;

ii. Jupiter: $p = 5.19034$ and $e = 0.0484658$;

iii. Earth: $p = 0.999721$ and $e = 0.016718$. (These data were adapted from Peter Duffett-Smith's *Practical Astronomy with Your Calculator,* 2nd edition, Cambridge University Press, 1981.)

Graph these three ellipses on the same set of axes.

33. Suppose that the wheel generating the cycloid discussed in Example 8 is rolling to the right at 2 meters per second. Describe the motion of a particle that started at $(0, 0)$ and is sticking to the wheel. What is its velocity and speed after $t = 3$ seconds?

34. Suppose that the wheel generating the cycloid discussed in Example 8 is rolling to the right at 1.5 meters per second. Describe the motion of a particle that started at $(0, 0)$ and is sticking to the wheel. What is its velocity and speed after $t = 5$ seconds?

T 35. The *folium of Descartes* is described by

$$\mathbf{r}(t) = \frac{1}{t^3 + 1}\langle 3t, 3t^2\rangle, \qquad t \ne -1.$$

Graph this curve. Find an equation in x and y for the folium by eliminating the parameter. Describe the tracing order and how various subintervals of the t-axis correspond to the three quadrants in which the folium occurs. Locate all points on the folium at which the tangent line is either horizontal or vertical. How many t values correspond to these points?

36. Referring to Fig. 6.30, calculate the slope of the tangent line to the curve of the string unwrapping at $\mathbf{r}(2.6)$. Locate the point $\mathbf{r}(\theta)$ at which the tangent line is vertical.

37. A cycloid is described by the equation

$$\mathbf{r}(\phi) = \langle \phi - \sin\phi, 1 - \cos\phi\rangle,$$

where $\phi \ge 0$. Show that the first half of the first period of this cycloid can be described by the equation $x = \arccos(1 - y) - \sin(\arccos(1 - y))$.

38. Related to the preceding problem Use the equation $x = \arccos(1 - y) - \sin(\arccos(1 - y))$ in calculating the slope of the cycloid at the point $(\pi/2 - 1, 1)$. Check this in a second way.

T 39. Graph the curve described by $r = \ln\theta$, $0 < \theta \le 2\pi$. Find polar coordinates of the point where the curve crosses itself.

40. Show that every point (r, θ) satisfying $r = \cos\theta$ lies on the graph of the equation $(x - 1/2)^2 + y^2 = (1/2)^2$. *Hint:* Multiply both sides of the polar equation by r and then change from polar to rectangular coordinates. Also show that every point satisfying $(x - 1/2)^2 + y^2 = (1/2)^2$ lies on the graph of $r = \cos\theta$.

6.4 Arc Length and Unit Tangent Vectors

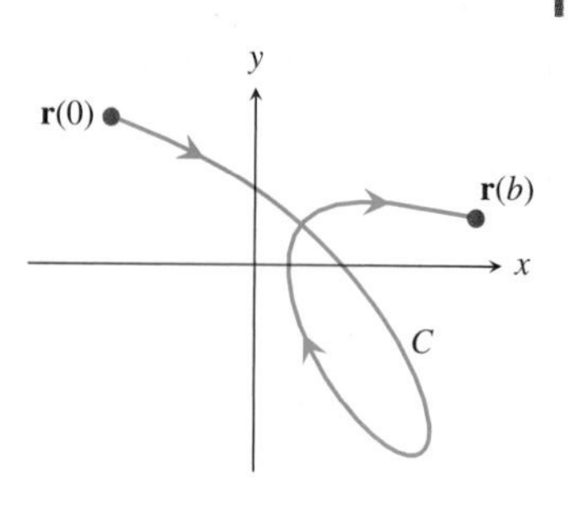

FIGURE 6.31 How can we calculate the distance traveled by a particle during a given time interval $[0, b]$ if we know its position at all times t in this interval?

Figure 6.31 shows the path C of a moving particle. If we know the position vector $\mathbf{r}(t)$ of the particle for all times t in the time interval $[0, b]$, how can we calculate the total distance $s(b)$ traveled by the particle?

To answer this question, recall that at any time t the length $\|\mathbf{r}'(t)\|$ of the velocity vector $\mathbf{r}'(t)$ of the particle is the particle's speed. If we divide the interval $[0, b]$ into n equal subintervals and choose an evaluation point c_i in the subinterval $[t_{i-1}, t_i]$, then the speed of the particle throughout this subinterval is approximately $\|\mathbf{r}'(c_i)\|$ and the distance it travels in time $\Delta t = (b - 0)/n$ is approximately $\|\mathbf{r}'(c_i)\| \Delta t$. Summing over the n subintervals, the total distance $s(b)$ traveled by the particle is approximately

$$R_n = \sum_{i=1}^{n} \|\mathbf{r}'(c_i)\| \Delta t.$$

Because R_n is a Riemann sum for the function $f(t) = \|\mathbf{r}'(t)\|$, it is plausible that

$$s(b) = \lim_{n \to \infty} R_n = \int_0^b \|\mathbf{r}'(t)\| \, dt. \tag{1}$$

Arc Length of a Curve

The formula for the length of a path given in (1) was based on our understanding of motion. More generally, we shall think of length as a number associated with a one-dimensional geometric object such as a line segment or curve, just as we regard area as a number associated with a region of a plane.

Suppose that C is a curve described by a vector equation

$$\mathbf{r} = \mathbf{r}(t) = \langle x(t), y(t) \rangle, \qquad a \le t \le b.$$

In earlier chapters we assumed that the functions x and y were continuous on $[a, b]$. For calculating arc length we assume that these functions have continuous derivatives on $[a, b]$. Curves described by such functions are said to be **smooth curves.** A representative smooth curve C is shown in Fig. 6.32(a).

As we have done with area, mass, or volume, we subdivide the geometric object. For arc length, we can do this by subdividing the domain $[a, b]$ of the parameter t with equally spaced points $a = t_0, t_1, \ldots, t_n = b$. Then the points with position vectors $\mathbf{r}(t_0), \mathbf{r}(t_1), \ldots, \mathbf{r}(t_n)$ subdivide C as shown in Fig. 6.32(a). We let $\Delta t = (b - a)/n$ denote the common length of the subintervals of $[a, b]$.

Figure 6.32(b) shows a zoom-view of the segment of C corresponding to a representative subinterval $[t_{i-1}, t_i]$. The arc length Δs_i of the segment of C between the points with position vector $\mathbf{r}(t_{i-1})$ and $\mathbf{r}(t_i)$ is approximated by the length of the vector $\mathbf{r}(t_i) - \mathbf{r}(t_{i-1})$, which is the same as the length of the line segment joining the points $\mathbf{r}(t_{i-1})$ and $\mathbf{r}(t_i)$. Hence,

$$\Delta s_i \approx \|\mathbf{r}(t_i) - \mathbf{r}(t_{i-1})\| = \sqrt{(x(t_i) - x(t_{i-1}))^2 + (y(t_i) - y(t_{i-1}))^2}. \tag{2}$$

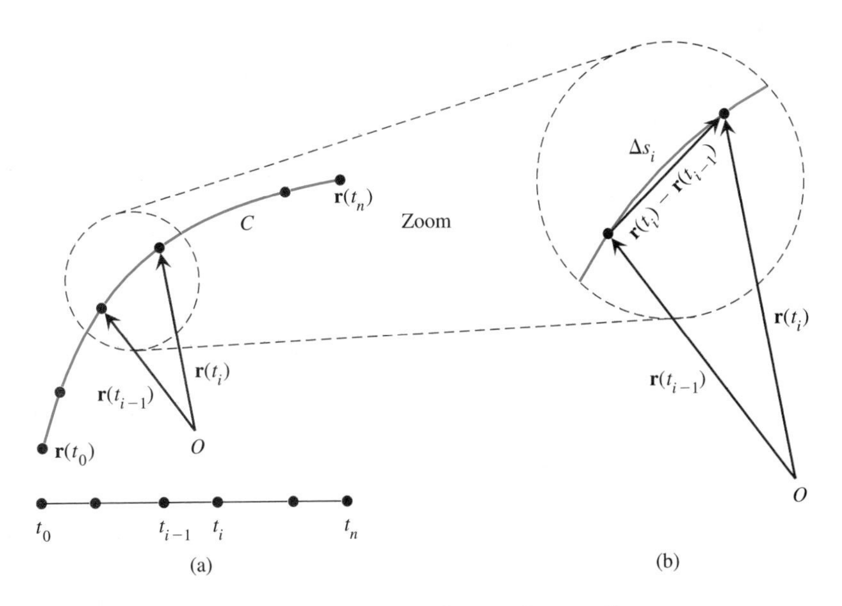

FIGURE 6.32 Approximating the arc length of a smooth curve C.

Before adding these approximations together to approximate the arc length s of C, we simplify them with the tangent line approximation. Recall that the tangent line approximation is: If f is a differentiable function on a short interval $[p, q]$, then

$$f(q) - f(p) \approx (q - p)f'(p).$$

Because $x = x(t)$ and $y = y(t)$ are differentiable on $[t_{i-1}, t_i]$,

$$x(t_i) - x(t_{i-1}) \approx (t_i - t_{i-1})x'(t_{i-1}) = x'(t_{i-1})\,\Delta t$$
$$y(t_i) - y(t_{i-1}) \approx (t_i - t_{i-1})y'(t_{i-1}) = y'(t_{i-1})\,\Delta t.$$

It then follows from (2) that

$$\Delta s_i \approx \sqrt{(x'(t_{i-1})\,\Delta t)^2 + (y'(t_{i-1})\,\Delta t)^2} = \sqrt{x'(t_{i-1})^2 + y'(t_{i-1})^2}\,\Delta t.$$

Hence,

$$s = \sum_{i=1}^{n} \Delta s_i \approx \sum_{i=1}^{n} \sqrt{(x'(t_{i-1}))^2 + (y'(t_{i-1}))^2}\,\Delta t. \tag{3}$$

You may have noticed that (3) depends upon two approximations. The first was needed because the length of the vector $\mathbf{r}(t_i) - \mathbf{r}(t_{i-1})$ is not, in general, equal to the arc length Δs_i of C between the points $\mathbf{r}(t_{i-1})$ and $\mathbf{r}(t_i)$. The second was needed because the tangent line approximations we used are not, in general, exact.

The sum on the right side of (3) is a Riemann sum R_n for the function

$$f(t) = \sqrt{x'(t)^2 + y'(t)^2},$$

where the evaluation points $c_1, c_2, \ldots, c_n$ are the subdivision points $t_0, t_1, \ldots, t_{n-1}$. It therefore appears that the arc length s of C is given by

$$s = \int_a^b \|\mathbf{r}'(t)\|\,dt = \int_a^b \sqrt{x'(t)^2 + y'(t)^2}\,dt.$$

This expression for the arc length of C is consistent with our earlier calculation of the distance $s(b)$ traveled during the time interval $[0, b]$ by a particle with position vector $\mathbf{r} = \mathbf{r}(t)$. Recall from (1) that

$$s(t) = \int_0^b \|\mathbf{v}(t)\|\, dt = \int_0^b \|\mathbf{r}'(t)\|\, dt. \tag{4}$$

DEFINITION Arc Length of a Smooth Curve

Let C be a smooth curve described by

$$\mathbf{r} = \mathbf{r}(t) = \langle x(t), y(t)\rangle, \qquad a \le t \le b.$$

The **arc length** s of C is defined as

$$s = \int_a^b \|\mathbf{r}'(t)\|\, dt = \int_a^b \sqrt{x'(t)^2 + y'(t)^2}\, dt. \tag{5}$$

The arc length function $s = s(t)$ is defined to be the arc length of C from $\mathbf{r}(a)$ to $\mathbf{r}(t)$, so that

$$s(t) = \int_a^t \|\mathbf{r}'(u)\|\, du, \qquad a \le t \le b \tag{6}$$

EXAMPLE 1 Calculate the arc length of the cycloid described by

$$\mathbf{r}(t) = \langle t - \sin t, 1 - \cos t\rangle, \qquad 0 \le t \le 2\pi. \tag{7}$$

Solution

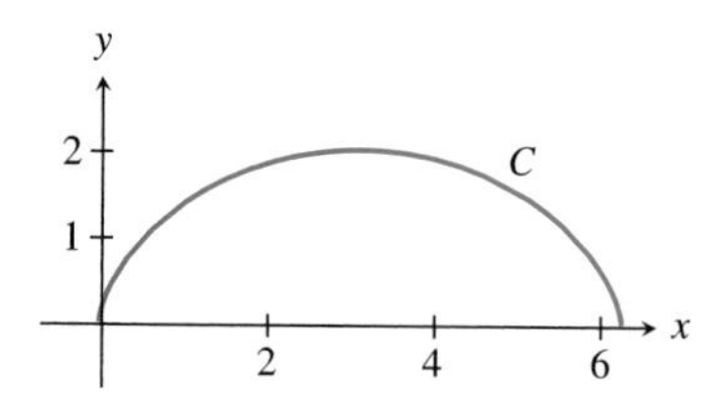

FIGURE 6.33 One arch of a cycloid.

Figure 6.33 shows the graph of the parametric Equation (7). Before substituting into (5), we calculate and simplify $x'(t)^2 + y'(t)^2$. We use the trigonometric identity

$$\sin^2\theta = \frac{1 - \cos 2\theta}{2} \tag{8}$$

in the simplification. Calculating $x'(t)$ and $y'(t)$,

$$\begin{aligned} x'(t)^2 + y'(t)^2 &= (1 - \cos t)^2 + (\sin t)^2 = 1 - 2\cos t + \cos^2 t + \sin^2 t \\ &= 2(1 - \cos t). \end{aligned}$$

Letting $t = 2\theta$ in (8),

$$x'(t)^2 + y'(t)^2 = 2 \cdot 2\sin^2 \tfrac{1}{2}t.$$

Hence,

$$\begin{aligned} s &= \int_0^{2\pi} \sqrt{x'(t)^2 + y'(t)^2}\, dt = \int_0^{2\pi} \sqrt{4\sin^2 \tfrac{1}{2}t}\, dt \\ &= 2\int_0^{2\pi} \left|\sin \tfrac{1}{2}t\right| dt = 2\int_0^{2\pi} \sin \tfrac{1}{2}t\, dt = 8 \text{ units}. \end{aligned}$$

In the last line we were able to replace $\left|\sin \frac{1}{2}t\right|$ by $\sin \frac{1}{2}t$ because for $0 \le t \le 2\pi$, $\sin \frac{1}{2}t \ge 0$. A glance at Fig. 6.33 shows that the length of the arch is about 8 units.

Although arc length calculations are usually quite straightforward, there are a few points we should mention. First, while we were able to calculate the areas of regions associated with many familiar functions by finding an antiderivative and applying the Fundamental Theorem of Calculus, the integrals for the arc lengths of many simple curves must be done by numerical integration.

The second point is that if we wish to calculate the length of a curve C, we must be careful to choose a parametrization of C for which the curve is traced only once. We recall from Example 4 in Section 6.3 the circle C with polar equation $r = \cos\theta$. The circle is traced once for $0 \le \theta \le \pi$ and again for $\pi \le \theta \le 2\pi$. If we were to parametrize C by

$$\mathbf{r} = \mathbf{r}(\theta) = \langle x(\theta), y(\theta)\rangle = \cos\theta\langle\cos\theta, \sin\theta\rangle, \qquad 0 \le \theta \le 2\pi,$$

the integral

$$\int_0^{2\pi} \sqrt{x'(\theta)^2 + y'(\theta)^2}\, d\theta$$

would be equal to 2π, which is twice the circumference $2\pi \cdot (1/2) = \pi$ of a circle of radius $1/2$.

The third point is that the arc length of a curve is the same for any parametrization for which the curve is traced just once. Suppose, for example, that a hare and a tortoise are racing on a circular track of radius 10 meters. The position of the hare at any time t (seconds) might be

$$\mathbf{r}_H(t) = 10\langle\cos(0.9t), \sin(0.9t)\rangle, \qquad 0 \le t \le 2\pi/0.9$$

while the position of the tortoise might be

$$\mathbf{r}_T(t) = 10\langle\cos(0.1t), \sin(0.1t)\rangle, \qquad 0 \le t \le 2\pi/0.1.$$

These parametrizations are different but they describe the same path. If we were to calculate the arc length of the racecourse using either parametrization we would expect—and would find—that the length is the same, $2\pi \cdot 10$ m.

Arc Length Formula for Cartesian Curves The arc length of the graph of a Cartesian equation $y = f(x)$, $a \le x \le b$, or of the graph of a polar equation $r = g(\theta)$, $\alpha \le \theta \le \beta$, can be calculated with the arc length formula (5) if we can describe these graphs parametrically. We discuss the graph of the equation $y = f(x)$ first.

Recall from Section 3.4 that the graph C of the Cartesian equation $y = f(x)$ can be described by the vector equation

$$\mathbf{r}(x) = \langle x, f(x)\rangle, \qquad a \le x \le b. \tag{9}$$

Because for an arc length calculation we require that C be a smooth curve, we assume that f and f' are continuous. The derivative $\mathbf{r}'(x)$ and its length $\|\mathbf{r}'(x)\|$ are

$$\mathbf{r}'(x) = \langle 1, f'(x)\rangle$$
$$\|\mathbf{r}'(x)\| = \sqrt{1 + (f'(x))^2}.$$

Hence, from (5),

Arc Length Formula for Cartesian Curves

The arc length s of the smooth curve C described by $y = f(x)$, $a \leq x \leq b$, where f and f' are continuous on $[a, b]$, is

$$s = \int_a^b \sqrt{1 + (f'(x))^2}\, dx = \int_a^b \sqrt{1 + y'^2}\, dx. \tag{10}$$

EXAMPLE 2 Calculate the arc length of the parabolic arc described by $y = x^2$, for $0 \leq x \leq 1$.

Solution

Figure 6.34 shows the graph of $y = x^2$ on $[0, 1]$. From (10),

$$s = \int_0^1 \sqrt{1 + y'^2}\, dx = \int_0^1 \sqrt{1 + 4x^2}\, dx. \tag{11}$$

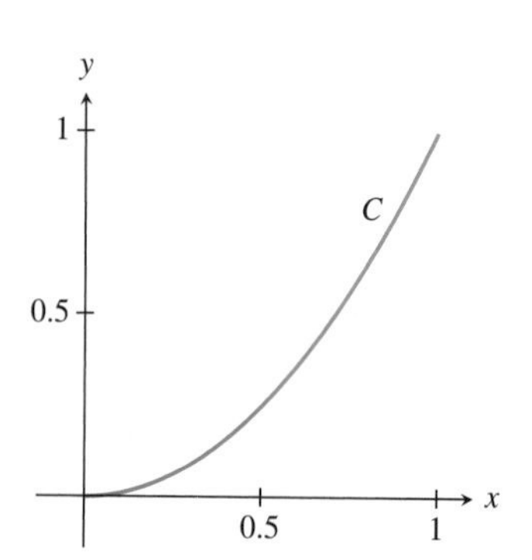

FIGURE 6.34 Parabolic arc.

We make the change of variable $2x = u$ so that we can use formula (4) from the table of integrals to evaluate this integral. From $2x = u$ we have $2\,dx = du$ and

$$\begin{aligned} s = \int_0^1 \sqrt{1 + 4x^2}\, dx &= \frac{1}{2}\int_0^2 \sqrt{1 + u^2}\, du \\ &= \frac{1}{2}\left(\frac{u}{2}\sqrt{u^2 + 1} + \frac{1}{2}\ln\left|u + \sqrt{u^2 + 1}\right|\right)\Bigg|_0^2 \\ &= \frac{\sqrt{5}}{2} + \frac{1}{4}\ln\left(2 + \sqrt{5}\right) \approx 1.48 \text{ units.} \end{aligned}$$

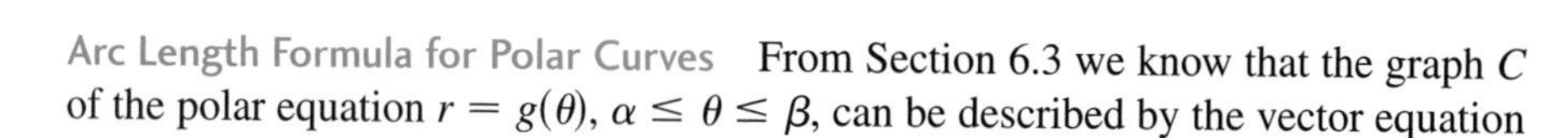

Arc Length Formula for Polar Curves From Section 6.3 we know that the graph C of the polar equation $r = g(\theta)$, $\alpha \leq \theta \leq \beta$, can be described by the vector equation

$$\mathbf{r}(\theta) = \langle g(\theta) \cos \theta, g(\theta) \sin \theta \rangle, \qquad \alpha \leq \theta \leq \beta. \tag{12}$$

Figure 6.35 shows a representative graph and point P on this graph.

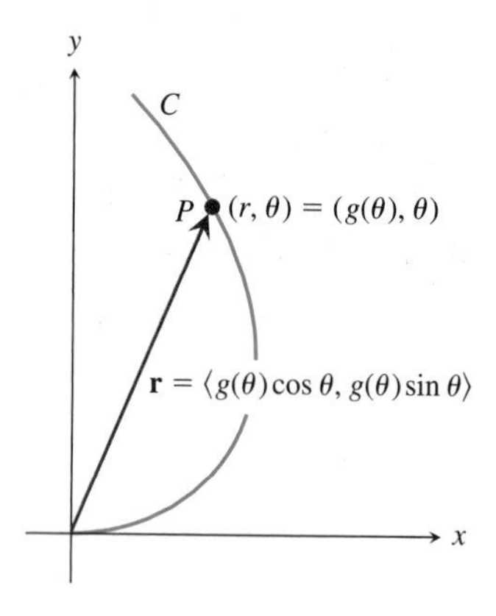

FIGURE 6.35 A point P on the graph of $r = g(\theta)$.

For calculating arc length, we want C to be a smooth curve. Hence, we assume that g and g' are continuous. We need to calculate the derivative $\mathbf{r}'(\theta)$ and its length $\|\mathbf{r}'(\theta)\|$. From (12), but using the shorter notations r and r' instead of $g(\theta)$ and $g'(\theta)$,

$$\begin{aligned} \mathbf{r}'(\theta) &= \langle g'(\theta) \cos \theta + g(\theta)(-\sin \theta), g'(\theta) \sin \theta + g(\theta) \cos \theta \rangle \\ &= \langle r' \cos \theta + r(-\sin \theta), r' \sin \theta + r \cos \theta \rangle. \end{aligned}$$

The square of the length of this vector is

$$\begin{aligned} \|\mathbf{r}'(\theta)\|^2 &= (r' \cos \theta + r(-\sin \theta))^2 + (r' \sin \theta + r \cos \theta)^2 \\ &= r'^2 \cos^2\theta - 2r'r \cos \theta \sin \theta + r^2 \sin^2\theta \\ &\quad + r'^2 \sin^2\theta + 2r'r \sin \theta \cos \theta + r^2 \cos^2\theta \\ &= r'^2(\cos^2\theta + \sin^2\theta) + r^2(\sin^2\theta + \cos^2\theta) \\ &= r'^2 + r^2 = g'(\theta)^2 + g(\theta)^2. \end{aligned}$$

Hence, from (5),

Arc Length Formula for Polar Curves

The arc length of the curve C described by the polar equation $r = g(\theta)$, $\alpha \leq \theta \leq \beta$, where g and g' are continuous on $[\alpha, \beta]$, is

$$s = \int_{\alpha}^{\beta} \sqrt{g(\theta)^2 + g'(\theta)^2}\, d\theta = \int_{\alpha}^{\beta} \sqrt{r^2 + r'^2}\, d\theta. \tag{13}$$

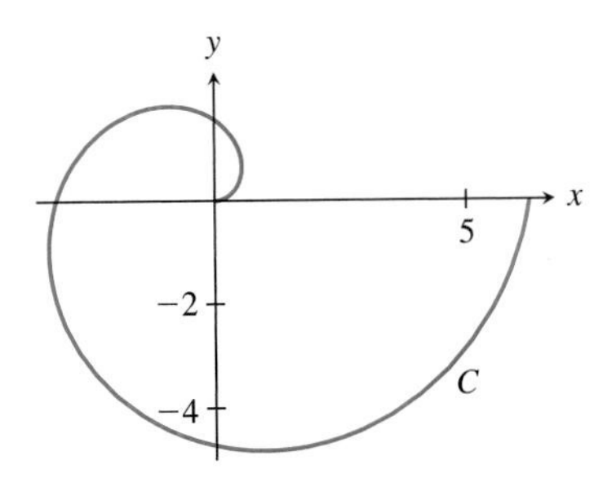

FIGURE 6.36 The Archimedean spiral described by $r = \theta$.

EXAMPLE 3 Calculate the arc length of the Archimedean spiral described by the polar equation $r = \theta$, for $0 \leq \theta \leq 2\pi$. See Fig. 6.36.

Solution

Taking $r = g(\theta)$ and applying (13), the arc length of the spiral is

$$s = \int_{0}^{2\pi} \sqrt{r^2 + r'^2}\, d\theta = \int_{0}^{2\pi} \sqrt{\theta^2 + 1^2}\, d\theta.$$

This integral has the same form as the integral in Equation (11) of Example 2. We may evaluate it with the help of formula (4) from the table of integrals.

$$s = \int_{0}^{2\pi} \sqrt{\theta^2 + 1^2}\, d\theta = \left(\frac{\theta}{2}\sqrt{\theta^2 + 1} + \frac{1}{2}\ln\left|\theta + \sqrt{\theta^2 + 1}\right|\right)\Bigg|_{0}^{2\pi}$$

$$= \pi\sqrt{4\pi^2 + 1} + \frac{1}{2}\ln\left(2\pi + \sqrt{4\pi^2 + 1}\right) \approx 21.26 \text{ units.}$$

Unit Tangent Vectors

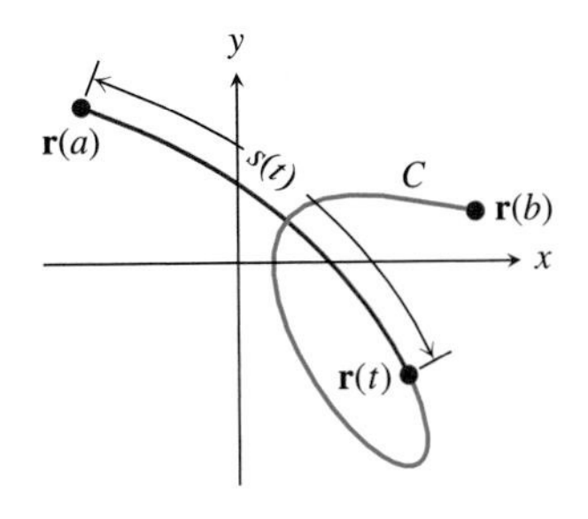

FIGURE 6.37 The arc length $s(t)$ as a function of t.

In modeling and analyzing the motion of a particle, the length of its velocity vector—which is the speed of the particle—is an important aspect of its motion. On the other hand, if we are studying a certain curve as a geometric object, the direction of the tangent vector often is of more importance than its length. In this case we often calculate the *unit tangent vector*. This calculation is related to arc length.

Figure 6.37 shows a smooth curve C described by an equation $\mathbf{r} = \mathbf{r}(t)$, $a \leq t \leq b$, and, measured from $\mathbf{r}(a)$, the arc from $\mathbf{r}(a)$ to $\mathbf{r}(t)$. The length of this arc is

$$s(t) = \int_{a}^{t} \|\mathbf{r}'(w)\|\, dw.$$

From Part I of the Fundamental Theorem of Calculus,

$$\frac{ds}{dt} = \frac{d}{dt}\int_{a}^{t} \|\mathbf{r}'(w)\|\, dw = \|\mathbf{r}'(t)\|. \tag{14}$$

The differential ds of the arc length function is

$$ds = s'(t)\, dt = \|\mathbf{r}'(t)\|\, dt. \tag{15}$$

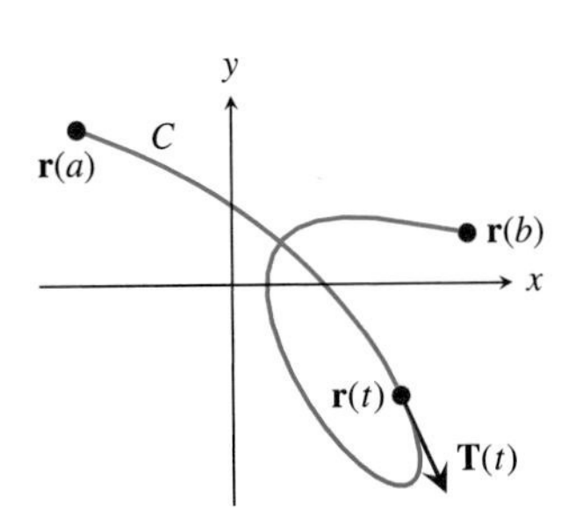

FIGURE 6.38 The unit tangent vector $\mathbf{T}(t)$ at the point $\mathbf{r}(t)$ of C.

Given the curve C shown in Fig. 6.37 and again in Fig. 6.38, the unit tangent vector $\mathbf{T}(t)$ at a representative point $\mathbf{r}(t)$ is the tangent vector $\mathbf{r}'(t)$ divided by its length, that is,

$$\mathbf{T}(t) = \frac{1}{\|\mathbf{r}'(t)\|}\mathbf{r}'(t).$$

From (14) we can also write

$$\mathbf{T}(t) = \frac{1}{ds/dt}\mathbf{r}'(t).$$

Unit Tangent Vector

Let C be a smooth curve described by

$$\mathbf{r} = \mathbf{r}(t) = \langle x(t), y(t)\rangle, \qquad a \le t \le b. \tag{16}$$

At all points $\mathbf{r}(t)$ of C at which $\|\mathbf{r}'(t)\| \neq 0$, the unit tangent vector $\mathbf{T}$ is

$$\mathbf{T} = \mathbf{T}(t) = \frac{1}{\|\mathbf{r}'(t)\|}\mathbf{r}'(t) = \frac{1}{ds/dt}\mathbf{r}'(t). \tag{17}$$

EXAMPLE 4 Find the unit tangent $\mathbf{T}(4\pi/3)$ for the cycloid described in Example 1 by the equation

$$\mathbf{r}(t) = \langle t - \sin t, 1 - \cos t\rangle, \qquad 0 \le t \le 2\pi.$$

Figure 6.39 shows the graph of this cycloid.

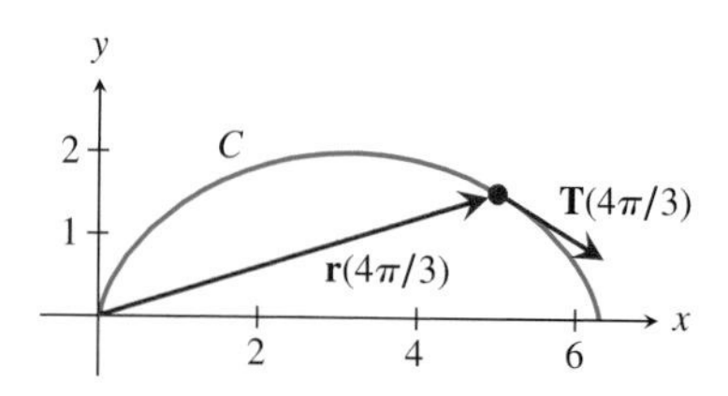

FIGURE 6.39 The unit tangent vector at a point of the cycloid.

Solution

To find the unit tangent $\mathbf{T}(t)$, we calculate $\mathbf{r}'(t)$ and its length $\|\mathbf{r}'(t)\|$:

$$\begin{aligned}\mathbf{r}' &= \langle 1 - \cos t, \sin t\rangle \\ \|\mathbf{r}'\| &= \sqrt{(1 - \cos t)^2 + (\sin t)^2} \\ &= \sqrt{2(1 - \cos t)}.\end{aligned}$$

Substituting into (17),

$$\mathbf{T} = \frac{1}{\|\mathbf{r}'\|}\mathbf{r}' = \left\langle \frac{1 - \cos t}{\sqrt{2(1 - \cos t)}}, \frac{\sin t}{\sqrt{2(1 - \cos t)}}\right\rangle.$$

As might be expected, the unit tangent is not defined at the "cusp" points $\mathbf{r}(0)$ and $\mathbf{r}(2\pi)$, where $\mathbf{r}' = \mathbf{0} = \langle 0, 0\rangle$. The expression for $\mathbf{T}$ can be simplified by noting that $1 - \cos t = 2\sin^2\left(\frac{1}{2}t\right)$ and $\sin t = 2\sin\frac{1}{2}t\cos\frac{1}{2}t$. Leaving the details to Exercise 34,

$$\mathbf{T}(t) = \left\langle \sin\tfrac{1}{2}t, \cos\tfrac{1}{2}t\right\rangle, \ t \neq 0, 2\pi.$$

The unit tangent at $\mathbf{r}(4\pi/3)$ is

$$\begin{aligned}\mathbf{T}(4\pi/3) &= \langle \sin(2\pi/3), \cos(2\pi/3)\rangle \\ &= \left\langle \sqrt{3}/2, -1/2\right\rangle \approx \langle 0.86603, -0.50000\rangle.\end{aligned}$$

Figure 6.39 shows the position vector $\mathbf{r}(4\pi/3)$ and the unit tangent vector $\mathbf{T}(4\pi/3)$ at this point. We drew the unit tangent somewhat longer than one unit so that it could be seen easily.

Exercises 6.4

Exercises 1–6: An object is traveling on a path C during a time interval. Sketch the path and calculate the distance traveled by the object. Assume units are meters and seconds.

1. $\mathbf{r} = \langle 3, -4\rangle + t\langle 5, 6\rangle,\ 0 \le t \le 25$

2. $\mathbf{r} = \langle 1, 1\rangle - 3t\langle 5, 6\rangle,\ 0 \le t \le 25$

3. $\mathbf{r} = \langle 2, -1\rangle + 3\langle \cos 0.05t, \sin 0.05t\rangle,\ 0 \le t \le 50$

4. $\mathbf{r} = \langle -2, 1\rangle + 5\langle \sin 0.25t, \cos 0.25t\rangle,\ 0 \le t \le 20$

5. $\mathbf{r} = 3\langle 2t - \sin 2t, 1 - \cos 2t\rangle,\ 0 \le t \le \pi$

6. $\mathbf{r} = -2\langle t - \sin t, 1 - \cos t\rangle,\ 0 \le t \le 2\pi$

Exercises 7–18: Sketch the described curve and calculate its arc length. Assume units are meters.

7. $y = x^2,\ 0 \le x \le 2$

8. $y = \frac{1}{2}x^2,\ 0 \le x \le 5$

9. $y = x^{3/2},\ 0 \le x \le 5$

10. $y = 1 + \frac{1}{2}x^{3/2},\ 0 \le x \le 1$

11. $r = 1 + \cos\theta,\ 0 \le \theta \le \pi$

12. $r = 1 - \sin\theta,\ -\pi/2 \le \theta \le \pi/2$

13. $r = e^{\theta},\ 0 \le \theta \le \pi$

14. $r = 2^{\theta},\ 0 \le \theta \le \pi$

15. $\mathbf{r} = \langle \cos\theta + \theta\sin\theta, \sin\theta - \theta\cos\theta\rangle,\ 0 \le \theta \le \pi$

16. $\mathbf{r} = \langle \sin\theta + \theta\cos\theta, \cos\theta - \theta\sin\theta\rangle,\ 0 \le \theta \le \pi$

17. $\mathbf{r} = \langle t\cos t, t\sin t\rangle,\ 0 \le t \le 1$

18. $\mathbf{r} = \langle t\sin t, t\cos t\rangle,\ 0 \le t \le 1$

Exercises 19–22: The arc length integrals of many curves must be evaluated numerically, using either, say, the midpoint rule discussed in Section 5.7 or the numerical integration algorithm on your calculator or CAS. The given value of n is sufficient for M_n to be within 0.01 of the arc length. Sketch the curve and approximate its arc length to within 0.01. Assume units are meters.

19. $\mathbf{r} = \langle 3\cos t, 2\sin t\rangle,\ 0 \le t \le \pi/2;\ n = 5$

20. $\mathbf{r} = \langle \cos t, 2\sin t\rangle,\ 0 \le t \le \pi/2;\ n = 7$

21. $y = x^3,\ 0 \le x \le 1;\ n = 7$

22. $y = 1/x,\ 1 \le x \le 2;\ n = 5$

*Exercises 23–26: Calculate the unit tangent **T** at the given point of the curve. Sketch the curve and unit tangent at the given point.*

23. $\mathbf{r} = \langle 3\cos t, 2\sin t\rangle,\ 0 \le t \le 2\pi;\ \mathbf{r}(\pi/4)$

24. $\mathbf{r} = \langle \cos t, 2\sin t\rangle,\ 0 \le t \le \pi/2;\ \mathbf{r}(\pi/6)$

25. $y = x^3,\ 0 \le x \le 2;\ (1, 1)$

26. $y = 1/x,\ 1/2 \le x \le 2;\ (1, 1)$

T **27.** Show that the length of the ellipse described by

$$\mathbf{r} = \mathbf{r}(\theta) = \langle a\cos\theta, b\sin\theta\rangle,$$

where $0 \le \theta \le 2\pi$, is

$$s = 4\int_0^{\pi/2} \sqrt{c^2\sin^2\theta + b^2}\,d\theta,$$

where $c^2 = a^2 - b^2$. We assume that $a \ge b$. Apply this formula to calculate the length of the orbit of Halley's comet in astronomical units (AU). The constants a and b for this comet are $a = 17.9435$ AU and $b = 4.55110$ AU. *Note:* The integral for the length of an ellipse is an *elliptic integral* and cannot be evaluated in "closed form." If you evaluate the integral

$$I = \int_0^{\pi/2} \sqrt{c^2\sin^2\theta + b^2}\,d\theta$$

with the trapezoid rule, it is given that $|I - T_{47}| < 0.01$ AU.

T **28.** Show that the length of the ellipse described by

$$\mathbf{r} = \mathbf{r}(\theta) = \langle a\cos\theta, b\sin\theta\rangle,$$

where $0 \le \theta \le 2\pi$, is

$$s = 4\int_0^{\pi/2} \sqrt{c^2\sin^2\theta + b^2}\,d\theta,$$

where $c^2 = a^2 - b^2$. We assume that $a \ge b$. Apply this formula to calculate the length of the orbit of the Pons-Brook comet in astronomical units (AU). The constants a and b for this comet are $a = 17.200$ AU and $b = 5.10162$ AU. *Note:* The integral for the length of an ellipse is an *elliptic integral* and cannot be evaluated in "closed form." If you evaluate the integral

$$I = \int_0^{\pi/2} \sqrt{c^2\sin^2\theta + b^2}\,d\theta$$

with the trapezoid rule, it is given that $|I - T_{42}| < 0.01$ AU.

29. Use (3) to approximate the arc length of the curve C described by $\mathbf{r} = \langle t, \ln t\rangle,\ 1/2 \le t \le 2$. Divide the interval $[1/2, 2]$ into five parts. Is the approximation too big or too small? Would the approximation improve if the interval were replaced by $[3/2, 3]$? Why?

30. Use (3) to approximate the arc length of the curve C described by $\mathbf{r} = \langle t, e^{-t}\rangle,\ 1 \le t \le 2$. Divide the interval $[1, 2]$ into five parts. Is the approximation too big or too small? Would the approximation improve if the interval were replaced by $[2, 3]$? Why?

31. Evaluate the integral in Example 2 by first making the substitution $2x = \tan\theta$ and then applying formula (19) from the table of integrals.

32. The equations

$$\mathbf{r} = \langle \cos(2\pi t), \sin(2\pi t) \rangle,$$

where $0 \le t \le 1$, and

$$\mathbf{r} = \langle \cos(p(t)), \sin(p(t)) \rangle$$
$$\mathbf{r} = \langle \cos(2\pi(1-t)^2)), \sin(2\pi(1-t)^2) \rangle,$$

where $0 \le t \le 1$ and $p(t) = 2\pi t(4t^2 - 6t + 3)$, describe the same curve C. Use each parametric equation to calculate the arc length of C. Explain your results.

33. In the preceding exercise, a curve was described by two different vector equations. Assume that these equations describe the motions of a hare and tortoise. Which equation describes the motion of the hare? Why? Are the speeds of the two objects the same or different at $t = 0.2$? At $t = 0.5$?

34. Fill in the details of the simplification of $\mathbf{T}$ in Example 4.

35. In Example 4 we found $\mathbf{T}(t) = \left\langle \sin\frac{1}{2}t, \cos\frac{1}{2}t \right\rangle$ for all $t \in [0, 2\pi]$ except $t = 0$ and $t = 2\pi$. Given that the expression $\left\langle \sin\frac{1}{2}t, \cos\frac{1}{2}t \right\rangle$ exists and is continuous for all t, can we in fact say that $\mathbf{T}(t) = \left\langle \sin\frac{1}{2}t, \cos\frac{1}{2}t \right\rangle$ for all t?

6.5 Areas of Regions Described by Polar Equations

Cams are used in many kinds of mechanical devices to convert rotary motion into a motion along a line. Cams are used, for example, in engines to control the opening and closing of valves. Regions like the planar face $\mathcal{R}$ of the cam shown in Fig. 6.40 are often described by polar equations of the form $r = g(\theta)$. When we wish to calculate the mass, center of mass, or moment of inertia (about its center of rotation) of a cam, it is convenient to make such calculations in polar coordinates.

Orbits of planets orbiting the sun are very nearly ellipses with the sun at one focus. Such orbits are usually described by a polar equation of the form

$$r = \frac{p}{1 - e\cos\theta}, \qquad 0 \le \theta \le 2\pi. \tag{1}$$

A representative orbit is shown in Fig. 6.41. Kepler's second law states that the line from the sun to a planet sweeps out equal areas in equal times. For example, if a planetary year were divided into 12 equal months, then, according to Kepler's second law, a line drawn from sun to planet would sweep out 1/12 of the area of the entire ellipse each month. Figure 6.41 shows the area swept out in one month. For our

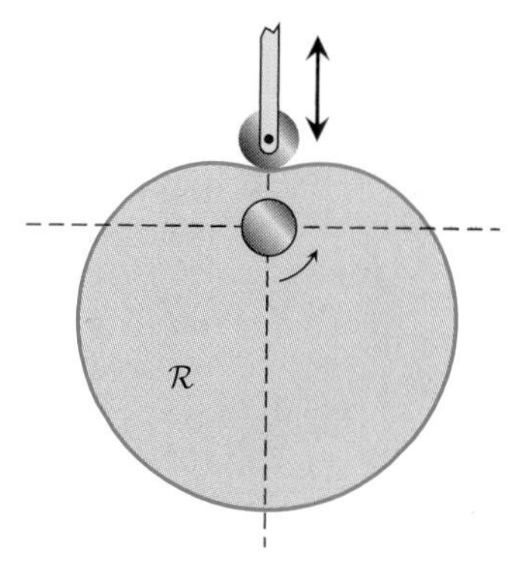

FIGURE 6.40 Planar face of a rotating cam, with follower.

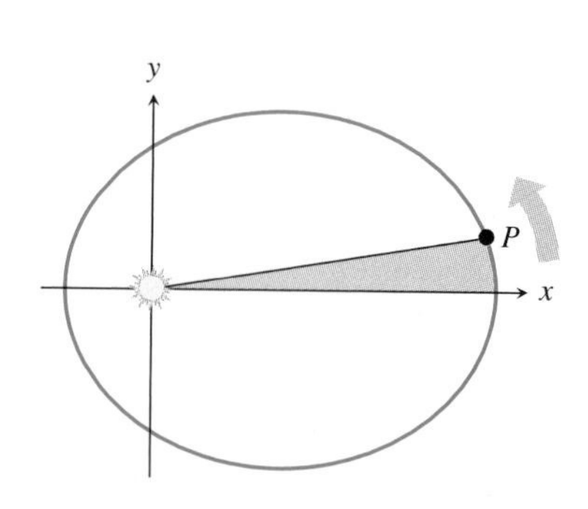

FIGURE 6.41 Orbit of a planet about the sun, with the area "swept out" each month.

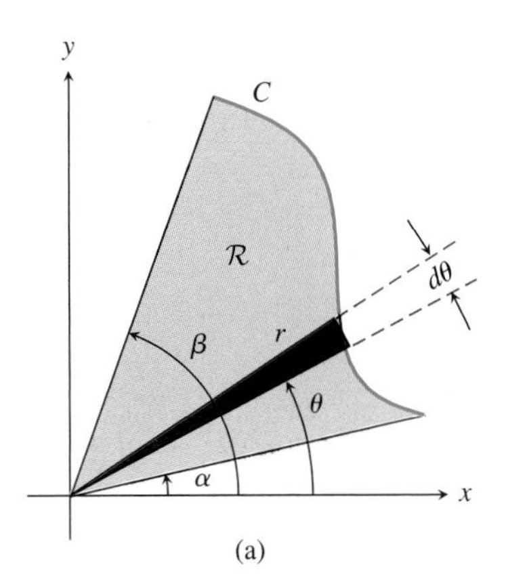

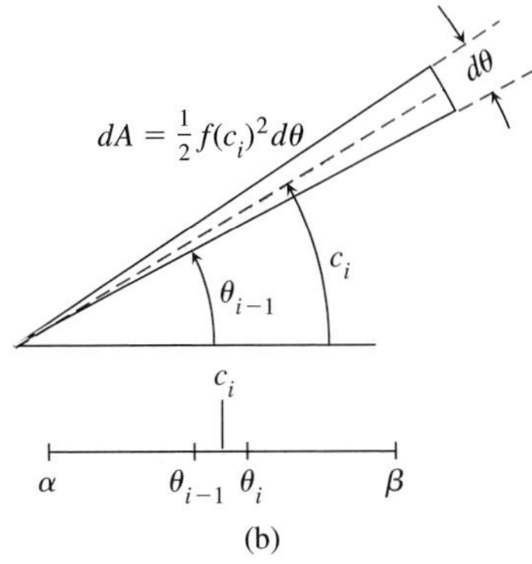

FIGURE 6.42 (a) A representative polar region; (b) an area element in the form of a sector of a circle.

derivation in Section 6.8 of Kepler's second law we need to be able to calculate areas of regions described by polar coordinates.

Figure 6.42(a) shows a representative polar region $\mathcal{R}$, bounded by the graph C of a polar equation $r = g(\theta)$, $\alpha \leq \theta \leq \beta$, and segments of the rays $\theta = \alpha$ and $\theta = \beta$. We approximate the region $\mathcal{R}$ with a union of area elements in the form of narrow circular sectors. A representative area element is shown in Fig. 6.42(a). This area element is a circular sector with central angle $d\theta$ and radius r.

The area dA of this sector is $dA \approx \frac{1}{2}r^2\,d\theta$. This comes from the formula $A = \frac{1}{2}a^2\omega$ for the area of a circular sector with central angle ω and radius a. Figure 6.43 gives a derivation of this formula; the derivation is based on the assumption that the ratio of the area of the sector to the area of the circle is equal to the ratio of the central angle ω of the sector to the central angle 2π of the circle.

To approximate the area of $\mathcal{R}$, we subdivide the interval $[\alpha, \beta]$ into n equal subintervals with the points $\alpha = \theta_0, \theta_1, \ldots, \theta_{i-1}, \theta_i, \ldots, \theta_n = \beta$, denote the common length of the subintervals by $\Delta\theta = d\theta = (\beta - \alpha)/n$, and choose an evaluation set $\{c_1, \ldots, c_n\}$, where $\theta_{i-1} \leq c_i \leq \theta_i$. The circular sector corresponding to the subinterval $[\theta_{i-1}, \theta_i]$ and evaluation point c_i are shown in Fig. 6.42(b). The area of this sector is $dA = \frac{1}{2}g(c_i)^2\,d\theta$. The union of the sectors corresponding to the given subdivision of $[\alpha, \beta]$ approximates the region $\mathcal{R}$, and the sum

$$R_n = \sum_{i=1}^{n} \tfrac{1}{2}g(c_i)^2\,d\theta$$

approximates the area of $\mathcal{R}$. Assuming that g is continuous on the interval $[\alpha, \beta]$, we expect that the area A of $\mathcal{R}$ is

$$A = \lim_{n\to\infty} R_n = \int_\alpha^\beta \tfrac{1}{2}g(\theta)^2\,d\theta.$$

Area of a Region Described by a Polar Equation

Let g be a continuous, nonnegative function defined on the interval $[\alpha, \beta]$. The area A of the region $\mathcal{R}$ bounded by the graphs of the polar equation $r = g(\theta)$ and the rays $\theta = \alpha$ and $\theta = \beta$ is

$$A = \int_\alpha^\beta \tfrac{1}{2}r^2\,d\theta = \int_\alpha^\beta \tfrac{1}{2}g(\theta)^2\,d\theta. \qquad (2)$$

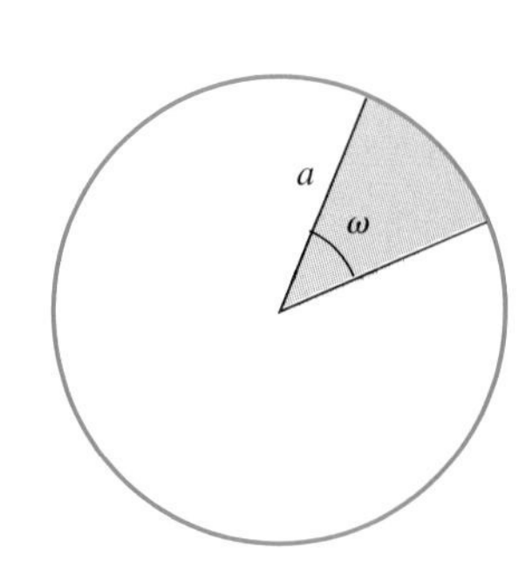

Solving the proportion

$$\frac{A_{\text{seg}}}{A_{\text{cir}}} = \frac{\omega}{2\pi}$$

for A_{seg},

$$A_{\text{seg}} = A_{\text{cir}} \cdot \frac{\omega}{2\pi} = \pi a^2 \cdot \frac{\omega}{2\pi} = \frac{1}{2}a^2\omega$$

$$A_{\text{seg}} = \frac{1}{2}a^2\omega$$

FIGURE 6.43 A simple derivation of the formula for the area of a sector of a circle.

EXAMPLE 1 Calculate the area A of the region $\mathcal{R}$ bounded by the polar circle C with equation $r = \sin\theta$.

Solution

Figure 6.44 shows the region $\mathcal{R}$ and its boundary C. It is important to note that the curve C is traced once as θ varies over the interval $[0, \pi]$, twice if θ varies over $[0, 2\pi]$. Also shown in the figure is a representative element of

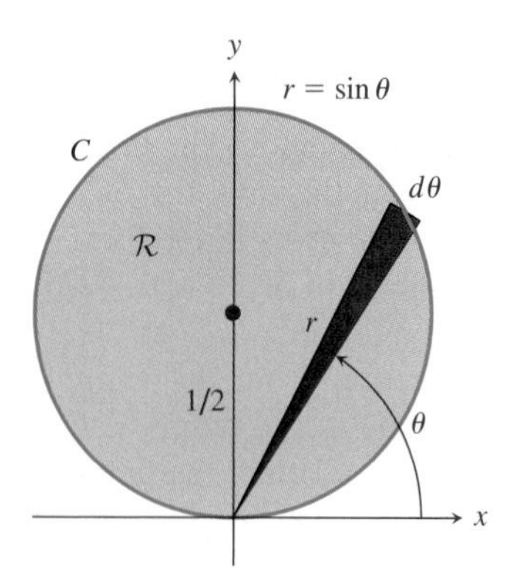

FIGURE 6.44 The polar circle described by $r = \sin\theta$.

area $dA = \frac{1}{2}r^2\,d\theta$. This element "sweeps out" the region $\mathcal{R}$ as θ varies from 0 to π. From (2),

$$A = \int_0^{\pi} \tfrac{1}{2}r^2\,d\theta = \tfrac{1}{2}\int_0^{\pi} \sin^2\theta\,d\theta$$

$$= \frac{1}{2}\int_0^{\pi} \frac{1-\cos(2\theta)}{2}\,d\theta = \frac{1}{4}\left(\theta - \frac{1}{2}\sin(2\theta)\right)\Bigg|_0^{\pi} = \frac{\pi}{4}.$$

This is the area $\pi(1/2)^2$ of a circle of radius $1/2$.

Next we calculate the area of a region lying between two polar curves. We use an area element analogous to the element used for the region between two Cartesian curves.

EXAMPLE 2 Calculate the area of the region in the first quadrant and between the polar curves described by $r = \theta$ and $r = \sin\theta$.

Solution

Figure 6.45 shows the two curves. The region between them is bounded by the graphs of the circle, the spiral, and the ray segment $\theta = \beta = \pi/2$, between $r = 1$ and $r = \pi/2$. We take the element of area dA to be the difference of two sector elements, that is,

$$dA = \tfrac{1}{2}\theta^2\,d\theta - \tfrac{1}{2}\sin^2\theta\,d\theta = \tfrac{1}{2}(\theta^2 - \sin^2\theta)\,d\theta.$$

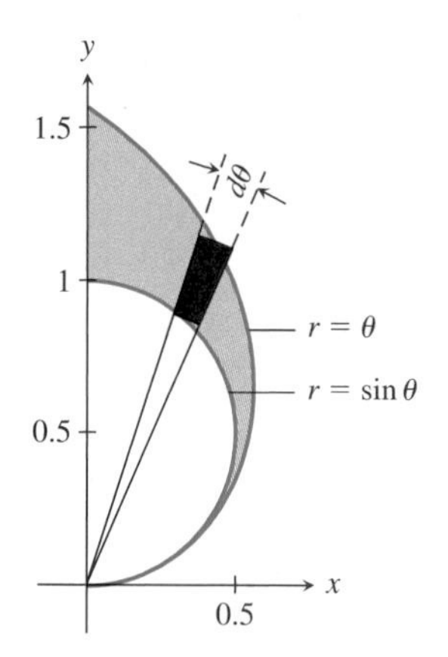

FIGURE 6.45 Calculating the area of the region between a spiral and a circle.

The region with area dA is shaded in the figure. As θ varies from 0 to $\pi/2$, this area element sweeps out the region between the two curves. Summing these elements,

$$A = \int_{\alpha}^{\beta} \tfrac{1}{2}(\theta^2 - \sin^2\theta)\,d\theta$$

$$= \tfrac{1}{2}\int_0^{\pi/2} (\theta^2 - \sin^2\theta)\,d\theta$$

$$= \tfrac{1}{2}\int_0^{\pi/2} \left(\theta^2 - \frac{1-\cos(2\theta)}{2}\right)d\theta$$

$$= \tfrac{1}{2}\left(\tfrac{1}{3}\theta^3 - \tfrac{1}{2}\theta + \tfrac{1}{4}\sin 2\theta\right)\Bigg|_0^{\pi/2}$$

$$A = \frac{\pi}{48}(\pi^2 - 6) \approx 0.253.$$

Kepler's first law states that each planet moves about the sun along an ellipse, with the sun at one focus. Such orbits are usually described in polar coordinates, with the sun at the origin. The orbit shown in Fig. 6.46 is described by the polar equation

$$r = \frac{1}{1 - \frac{3}{5}\cos\theta}, \qquad 0 \leq \theta \leq 2\pi. \tag{3}$$

This equation gives the form of the orbit but does not describe the position of a planet as a function of time.

Kepler's second law states that the line from the sun to a planet sweeps out equal areas in equal times. Figure 6.46 shows a region swept out by the line from the sun to the planet during one month, which we suppose corresponds to $1/12$ of the area of the ellipse. For the ellipse described by (3), the lengths of the semimajor and semiminor axes are $a = 25/16$ and $b = 5/4$. The area of this ellipse is $A = \pi ab = 125\pi/64$.

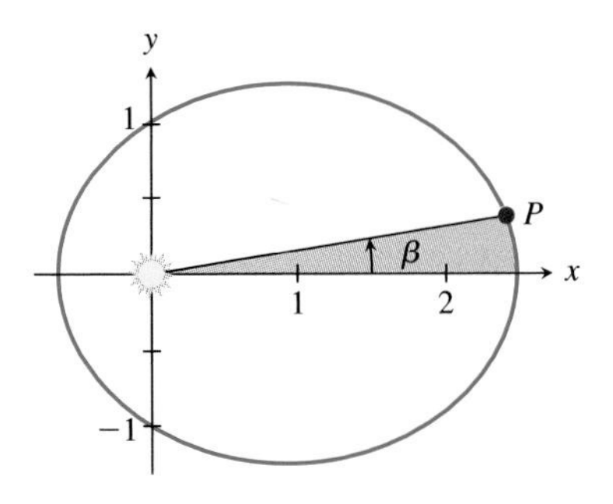

FIGURE 6.46 Planetary orbit and area swept out by the line from the sun to the planet during one month.

EXAMPLE 3 Suppose that a planet is moving counterclockwise on the ellipse described by (3) and initially was at $(r, \theta) = (5/2, 0)$. Find the point P on the ellipse for which the (shaded) area swept by the line from the sun to the planet is $(1/12)A = 125\pi/(12 \cdot 64)$. See Fig. 6.46.

Solution

From the area formula for polar curves and (3), we want to find a number β for which

$$\frac{1}{12}\left(\frac{125\pi}{64}\right) = \frac{1}{2}\int_0^\beta r^2\,d\theta = \frac{1}{2}\int_0^\beta \frac{d\theta}{\left(1 - \frac{3}{5}\cos\theta\right)^2}. \tag{4}$$

This equation can be solved for β in several ways. For example, we can use Newton's method and a numerical integration algorithm to determine β (from which we can calculate P). For this approach, see Exercise 34.

Or we can integrate (4) and solve the resulting equation. Leaving the details of the integration to Exercise 31,

$$\frac{1}{2}\int_0^\beta \frac{d\theta}{\left(1 - \frac{3}{5}\cos\theta\right)^2} = \frac{125\arctan(2\tan(\beta/2))}{64} - \frac{75\sin\beta}{32(3\cos\beta - 5)}. \tag{5}$$

With this result, we may rewrite Equation (4) in the form $g(\beta) = 0$, where

$$g(\beta) = \frac{125\arctan(2\tan(\beta/2))}{64} - \frac{75\sin\beta}{32(3\cos\beta - 5)} - \frac{125\pi}{12 \cdot 64}. \tag{6}$$

To solve the equation $g(\beta) = 0$, we may

i) graph g and estimate β from the screen of our calculator or CAS,
ii) use the equation-solver of our calculator or CAS, or
iii) use Newton's method.

We leave as Exercises 32 and 33 solutions using (ii) or (iii). Figure 6.47 shows a graphical solution. It appears from the graph that $\beta \approx 0.16$. Hence, with the planet initially at $(r, \theta) = (5/2, 0)$, the line joining sun and planet will have swept out $1/12$ of the area of the ellipse when $\theta = \beta \approx 0.16$.

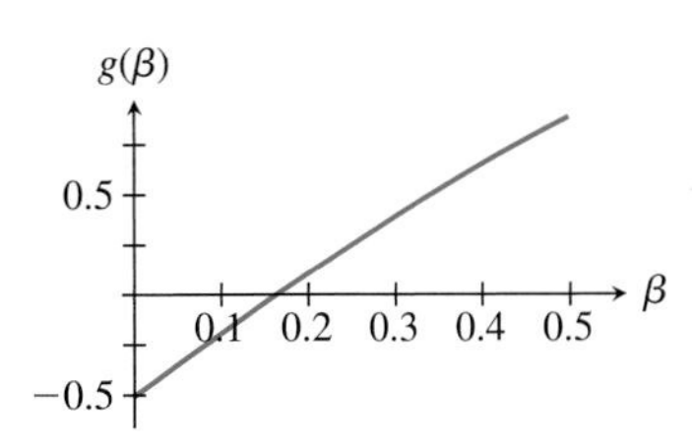

FIGURE 6.47 Graph of the function g and the approximate value of β such that $g(\beta) = 0$.

Exercises 6.5

Exercises 1–22: Sketch and find the area of the polar region $\mathcal{R}$.

1. $\mathcal{R}$ is bounded by the spiral $r = \theta$ and the rays $\theta = \pi/3$ and $\theta = \pi/2$.

2. $\mathcal{R}$ is bounded by the spiral $r = \frac{1}{2}\theta$ and the rays $\theta = \pi/2$ and $\theta = \pi$.

3. $\mathcal{R}$ is bounded by the spiral $r = e^{\theta}$ and the rays $\theta = 0$ and $\theta = \pi/2$.

4. $\mathcal{R}$ is bounded by the spiral $r = 2^{\theta}$ and the rays $\theta = \pi/4$ and $\theta = \pi$.

5. $\mathcal{R}$ is bounded by the circle $r = \cos\theta$ and the rays $\theta = 0$ and $\theta = \pi/6$.

6. $\mathcal{R}$ is bounded by the circle $r = 3\sin\theta$ and the rays $\theta = \pi/3$ and $\theta = \pi/2$.

7. $\mathcal{R}$ is bounded by the cardioid $r = 1 + \cos\theta$.

8. $\mathcal{R}$ is bounded by the cardioid $r = 1 - \sin\theta$.

9. $\mathcal{R}$ is outside the cardioid $r = 1 + \cos\theta$ and inside the circle $r = 3\cos\theta$.

10. $\mathcal{R}$ is outside the cardioid $r = 1 + \sin\theta$ and inside the circle $r = 3\sin\theta$.

11. $\mathcal{R}$ is inside the cardioid $r = 1 + \cos\theta$ and outside the circle $r = 3\cos\theta$.

12. $\mathcal{R}$ is inside the cardioid $r = 1 + \sin\theta$ and outside the circle $r = 3\sin\theta$.

13. $\mathcal{R}$ is inside both circles $r = \cos\theta$ and $r = \sin\theta$.

14. $\mathcal{R}$ is inside both circles $r = a$ and $r = 2a\cos\theta$.

15. $\mathcal{R}$ is inside the smaller loop of the limaçon $r = 1 + 2\cos\theta$.

16. $\mathcal{R}$ is inside the smaller loop of the limaçon $r = 1 + 2\sin\theta$.

17. $\mathcal{R}$ is inside $r = 2$ and above $r = \csc\theta$, $0 < \theta < \pi$.

18. $\mathcal{R}$ is inside $r = 2$ and to the right of $r = \sec\theta$, $-\pi/2 < \theta < \pi/2$.

19. $\mathcal{R}$ is inside the lemniscate with equation

$$(x^2 + y^2)^2 = x^2 - y^2.$$

Hint: Transform the equation to polar coordinates.

20. $\mathcal{R}$ is inside the lemniscate with equation

$$(x^2 + y^2)^2 = 4(x^2 - y^2).$$

Hint: Transform the equation to polar coordinates.

21. $\mathcal{R}$ is inside both $r = \cos\theta$ and $r = \cos(\theta - \pi/4)$.

22. $\mathcal{R}$ is inside both $r = 1 + \cos\theta$ and $r = 1 - \cos\theta$.

23. The profile of a steel cam is described by $r = 6.0 - 4.0\cos\theta$, with units in millimeters. Sketch the cam's profile. Calculate the mass of the cam if it is 10 mm thick. The density of steel is 7.850 g/cm^3.

24. The profile of a steel cam is described by $r = 7.0 - 4.5\cos\theta$, with units in millimeters. Sketch the cam's profile. Calculate the mass of the cam if it is 10 mm thick. The density of steel is 7.850 g/cm^3.

25. With $0 \le \theta \le \pi/2$, determine the area of the region between the graphs of $r = \cos\theta$ and $r = \theta$.

26. With $0 \le \theta \le \pi/2$, determine the area of the region between the graphs of $r = \cos\theta$ and $r = \frac{1}{2}\theta$.

27. The polar curve with equation $r = \csc\theta - \sqrt{2}$, $\pi/6 < \theta < 5\pi/6$, has a loop. Find the area of the region inside the loop.

28. The polar curve with equation $r = \sec\theta - \sqrt{2}$, $5\pi/3 < \theta < 7\pi/3$, has a loop. Find the area of the region inside the loop.

29. Find the area of the region

$$\mathcal{R} = \{(r, \theta) : 0 \le \theta \le 0.5 \text{ and } \theta/2 \le r \le \arcsin\theta\}.$$

Hint: The integral for the area can be done using integration by parts. Alternatively, evaluate it numerically, either using a built-in numerical integrator or, say, the midpoint rule.

T **30.** Calculate the area of the region inside the loop of the curve $r = g(\theta) = \ln\theta$, $0 < \theta < 3\pi/2$. *Hint:* Note that for $0 < \theta < 1$, $r = g(\theta)$ is negative and, hence, the curve will be traced in the third quadrant. The value of θ for which $-g(\theta) = g(\theta + \pi)$ will locate the point at which the curve self-intersects.

31. Use the substitution $u = \tan(\theta/2)$ in verifying (5). Using a few trigonometric identities, it follows from the substitution equation that $d\theta = 2du/(1 + u^2)$, $\sin\theta = 2u/(1 + u^2)$, and $\cos\theta = (1 - u^2)/(1 + u^2)$. Show that

$$\frac{1}{2}\int \frac{d\theta}{\left(1 - \frac{3}{5}\cos\theta\right)^2} = \frac{25}{4}\int \frac{(1 + u^2)\,du}{(1 + 4u^2)^2}.$$

This is a partial fractions problem. Show that (5) follows.

T **32.** Use the equation-solver on your calculator or CAS to show that the zero of

$$g(\beta) = \frac{125\arctan(2\tan(\beta/2))}{64} - \frac{75\sin\beta}{32(3\cos\beta - 5)} - \frac{125\pi}{12 \cdot 64}$$

near 0.2 is $\beta \approx 0.165861$.

33. Use Newton's method to show that the zero of

$$g(\beta) = \frac{125 \arctan(2 \tan(\beta/2))}{64} - \frac{75 \sin \beta}{32(3 \cos \beta - 5)} - \frac{125\pi}{12 \cdot 64}$$

near 0.2 is $\beta \approx 0.165861$.

34. Use Newton's method to calculate to three decimal places the zero of the function

$$f(\beta) = \frac{125\pi}{12 \cdot 64} - \frac{1}{2} \int_0^\beta \frac{d\theta}{\left(1 - \frac{3}{5} \cos \theta\right)^2},$$

where $0 \le \beta \le \pi/2$. Take $\beta_1 = 0.2$ as a first approximation to the zero. *Outline:* If Newton's method is applied to the function f, the iterates $\beta_2, \beta_3, \ldots$ satisfy the equation

$$\beta_{n+1} = \beta_n - \frac{f(\beta_n)}{f'(\beta_n)},$$

$n = 1, 2, \ldots$. Use Part I of the Fundamental Theorem of Calculus to calculate $f'(\beta_n)$ and use the trapezoid rule with $n = 22$ to calculate $f(\beta_n)$. This value of n guarantees that $|T_{22} - f(\beta_n)| < 0.00005$. You should find $\beta_2 \approx 0.165544$.

35. For the ellipse described by (1), with $0 < e < 1$, show that the lengths a and b of the semimajor and semiminor axes are $a = p/(1 - e^2)$ and $b = p/\sqrt{1 - e^2}$.

6.6 Work

In Section 3.5 we briefly discussed the concept of *work* and calculated the work W done by a constant force $\mathbf{F}$ acting on an object moving a distance s along a line. We used the formula

$$W = (\mathbf{F} \cdot \mathbf{u})s, \tag{1}$$

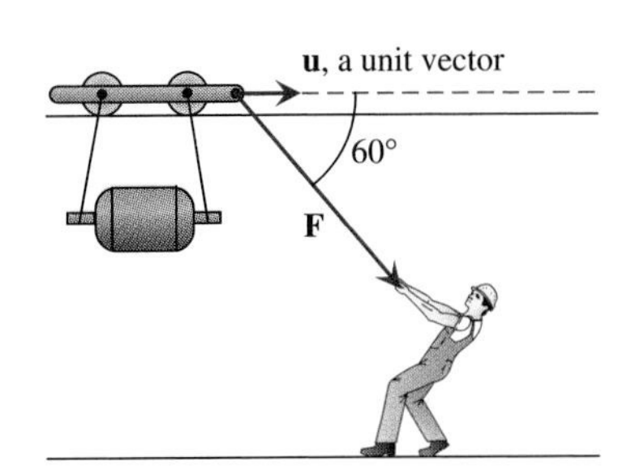

FIGURE 6.48 Constant force acting over a distance.

where $\mathbf{u}$ is a unit vector in the direction of motion. The work done by a force of 1 newton over a distance of 1 meter is 1 newton-meter, also called a joule (J).

Figure 6.48 shows a large electric motor suspended from an overhead trolley. If a force $\mathbf{F}$ is applied to the trolley as in the figure, the trolley will move in the direction of the unit vector $\mathbf{u}$. Assume that the force $\mathbf{F}$ has a constant magnitude of 200 N and makes an angle of 60° with $\mathbf{u}$. If the person walks a distance of $s = 10$ meters while applying the force $\mathbf{F}$, the work done by $\mathbf{F}$ on the motor would be

$$W = (\mathbf{F} \cdot \mathbf{u})s = (\|\mathbf{F}\| \|\mathbf{u}\| \cos 60°)10 = 200 \cdot 1 \cdot 0.5 \cdot 10 = 1000 \text{ J}.$$

In this section we calculate the work done by a variable force acting on an object moving on a smooth or piecewise smooth curve.

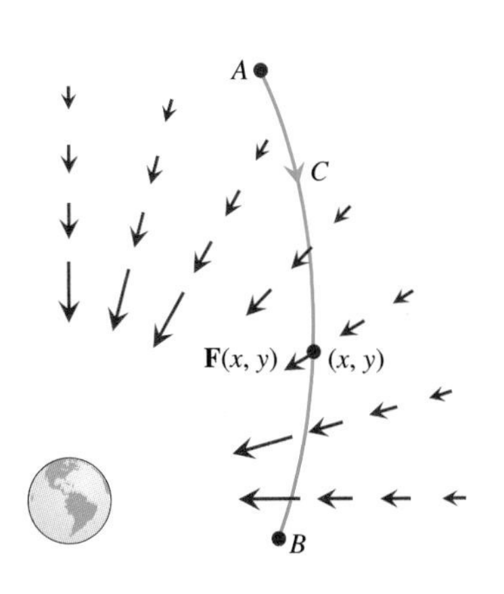

FIGURE 6.49 Earth's gravitational field.

Figure 6.49 shows the curved path C of, say, an asteroid moving in the Earth's gravitational field $\mathbf{F}$. We assume that the path of the asteroid lies in a plane containing the center of the Earth. The gravitational force $\mathbf{F}(x, y)$ acting on the asteroid as it passes through the point (x, y) lies in the plane of C and is directed toward the center of the Earth. The magnitude of this force is inversely proportional to the square of the distance from (x, y) to the center of the Earth.

To approximate the work done by the gravitational field acting on an asteroid of unit mass as it moves on C from A to B, we divide C into short segments, use Equation (1) on each segment to calculate the work done by the "nearly constant" field as the asteroid moves over the "nearly straight" segment, and then sum all of these work elements. The limit of these sums, as we subdivide C more and more finely, is the work done by the gravitational field on the asteroid as it moves from A to B.

Suppose, then, that C is described by the parametric equation

$$\mathbf{r} = \mathbf{r}(t) = \langle x, y \rangle = \langle x(t), y(t) \rangle, \qquad a \le t \le b, \tag{2}$$

where $\mathbf{r}(a)$ and $\mathbf{r}(b)$ are position vectors of A and B. At the point $(x, y) = (x(t), y(t))$ of C, with position vector $\mathbf{r}(t) = \langle x(t), y(t) \rangle$, the force acting on the asteroid is $\mathbf{F}(x(t), y(t))$, or, more briefly, $\mathbf{F}(\mathbf{r}(t))$.

We subdivide C by subdividing the interval $[a, b]$, as shown at the bottom of Fig. 6.50. Corresponding to the representative subinterval $[t, t + dt]$ of length dt is the segment of C from $\mathbf{r}(t)$ to $\mathbf{r}(t + dt)$. These two points may be seen in the zoom-view at the right of the figure. During the short time dt, the speed of the asteroid is approximately $\|\mathbf{r}'(t)\|$; hence, the distance Δs it moved during this time is $\Delta s \approx ds = \|\mathbf{r}'(t)\| \, dt$.

Letting $\mathbf{T}(t)$ be the unit tangent vector to C at $\mathbf{r}(t)$, the work ΔW done by the gravitational field on the asteroid during the time interval $[t, t + dt]$ is

$$\Delta W \approx dW = (\mathbf{F}(\mathbf{r}(t)) \cdot \mathbf{T}(t)) \, ds = (\mathbf{F} \cdot \mathbf{T}) \, ds. \tag{3}$$

This comes directly from (1). The factor $\mathbf{F} \cdot \mathbf{T}$ is approximately the magnitude of the force in the direction of motion and ds is approximately the distance the asteroid moved.

The total work W done by the force field $\mathbf{F}$ as the asteroid moves on C from $A = \mathbf{r}(a)$ to $B = \mathbf{r}(b)$ is approximated by a sum of such work elements. As the number of subdivisions of $[a, b]$ grows without bound, we are led to an integral for W. We give this integral after agreeing on notation and rewriting dW in a convenient form.

To rewrite dW in (3), we recall from Section 6.4 that at the point $\mathbf{r}(t)$ the unit tangent vector is

$$\mathbf{T} = \mathbf{T}(t) = \frac{1}{\|\mathbf{r}'(t)\|} \mathbf{r}'(t).$$

Because $ds/dt = \|\mathbf{r}'(t)\|$, we may write $\mathbf{T}$ as

$$\mathbf{T} = \frac{1}{ds/dt} \mathbf{r}'(t).$$

Substituting this result into (3),

$$dW = \mathbf{F}(\mathbf{r}(t)) \cdot \mathbf{T} \, ds = \mathbf{F}(\mathbf{r}(t)) \cdot \frac{1}{ds/dt} \mathbf{r}'(t) \, ds.$$

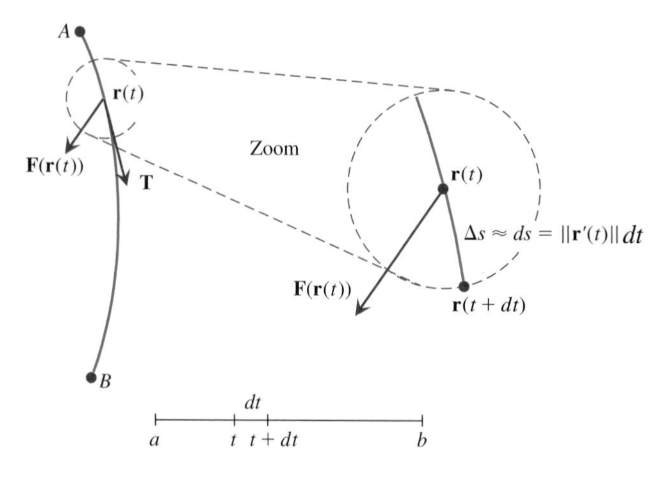

FIGURE 6.50 A variable force acting along a curve.

Because $ds = (ds/dt)\,dt$,

$$dW = \mathbf{F}(\mathbf{r}(t)) \cdot \mathbf{r}'(t)\,dt = \mathbf{F}(\mathbf{r}(t)) \cdot \mathbf{dr}, \tag{4}$$

where the differential $\mathbf{dr}$ of $\mathbf{r}$ is defined by $\mathbf{dr} = \mathbf{r}'(t)\,dt$.

DEFINITION Work

The **work** W done by a force $\mathbf{F}$ acting on an object moving from $\mathbf{r}(a)$ to $\mathbf{r}(b)$ on a smooth curve C is

$$W = \int_C dW = \int_C \mathbf{F} \cdot \mathbf{dr} = \int_a^b \mathbf{F}(\mathbf{r}(t)) \cdot \mathbf{r}'(t)\,dt, \tag{5}$$

where C is described by the equation $\mathbf{r} = \mathbf{r}(t)$, $a \le t \le b$.

The notation $\int_C \mathbf{F} \cdot \mathbf{dr}$ for the work integral reminds us that the work done by $\mathbf{F}$ is the limit of a sum of work increments dW along the curve C. We sometimes say that we are "integrating along the curve C." To evaluate this work integral we choose a vector function $\mathbf{r} = \mathbf{r}(t)$, $a \le t \le b$, describing C and evaluate a Riemann integral on the interval $[a, b]$.

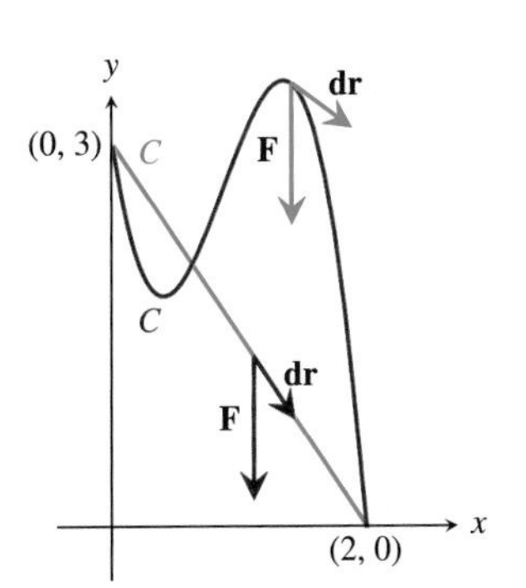

FIGURE 6.51 Work done by a field on an object that moves from $(0, 3)$ to $(2, 0)$ on two different paths.

EXAMPLE 1 Figure 6.51 shows a vertical plane through the point $(0, 0)$ on the Earth's surface. At any point (x, y) near $(0, 0)$, the Earth's gravitational force field $\mathbf{F}$ is very nearly constant. On an object of mass m at (x, y), this field exerts a force of $\mathbf{F} = \langle 0, -mg\rangle$. Calculate the work done by this constant force field as an object moves from $(0, 3)$ to $(2, 0)$ on the line C connecting these points. Also, calculate the work done as the object moves from $(0, 3)$ to $(2, 0)$ on any smooth curve C described by $\mathbf{r} = \mathbf{r}(t) = \langle x(t), y(t)\rangle$, $a \le t \le b$.

Solution

We use (5) in calculating the work done by the gravitational force. We first assume that the object moves from $(0, 3)$ to $(2, 0)$ on the straight line C. We may describe C by $\mathbf{r} = \mathbf{r}(t) = \langle 0, 3\rangle + t\langle 2, -3\rangle$, where $0 \le t \le 1$. The vector $\mathbf{r}'(t) = \langle 2, -3\rangle$ points in the direction of motion. The work done by the field as the object moves on C is

$$W = \int_C \mathbf{F} \cdot \mathbf{dr} = \int_a^b \mathbf{F}(\mathbf{r}(t)) \cdot \mathbf{r}'(t)\,dt = \int_0^1 \langle 0, -mg\rangle \cdot \langle 2, -3\rangle\,dt$$

$$= \int_0^1 3mg\,dt = 3mgt\Big|_0^1 = 3mg \text{ joules.}$$

Next we assume that the object is moving from $(0, 3)$ to $(2, 0)$ on a smooth curve C described by $\mathbf{r} = \mathbf{r}(t) = \langle x(t), y(t)\rangle$, $a \le t \le b$. We assume that $\mathbf{r}(a) = \langle 0, 3\rangle$ and $\mathbf{r}(b) = \langle 2, 0\rangle$. In this case, the work done by the field is

$$W = \int_a^b \mathbf{F}(\mathbf{r}(t)) \cdot \mathbf{r}'(t)\,dt = \int_a^b \langle 0, -mg\rangle \cdot \langle x'(t), y'(t)\rangle\,dt$$

$$= -mg\int_a^b y'(t)\,dt = -mgy(t)\Big|_a^b = -mg(y(b) - y(a))$$

$$= -mg(0 - 3) = 3mg \text{ joules.}$$

The constant force field $\mathbf{F} = \langle 0, -mg\rangle$ is an example of a **conservative force field.** A conservative field is one for which the work done by the field is *independent of path.* In Example 1, we showed that the work done by the field $\mathbf{F} = \langle 0, -mg\rangle$ on an object moving on each of several paths joining $(0, 3)$ and $(2, 0)$ is $3mg$ joules. We discuss the nonconstant case of the Earth's field in Example 5, showing that it also is conservative.

In the next example, we calculate the work done in stretching an ideal spring. This is a variable-force problem because such a spring resists stretching or compression with a force whose magnitude is proportional to the distance it is stretched or compressed from its relaxed state. This is Hooke's law.

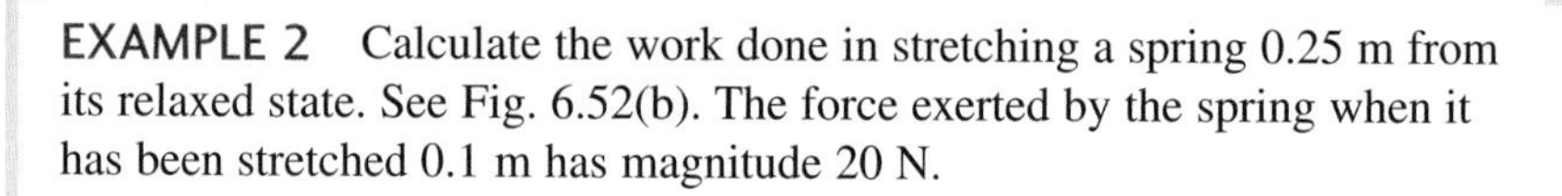

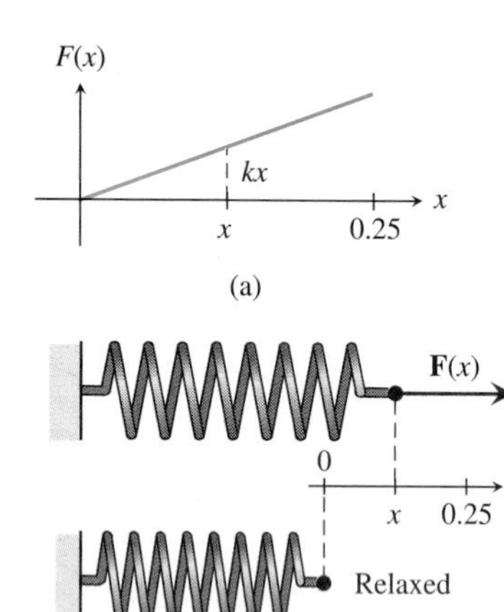

FIGURE 6.52 Work and Hooke's law.

EXAMPLE 2 Calculate the work done in stretching a spring 0.25 m from its relaxed state. See Fig. 6.52(b). The force exerted by the spring when it has been stretched 0.1 m has magnitude 20 N.

Solution

The relaxed spring is shown at the bottom of Fig. 6.52. Also shown is an x-axis, with origin aligned with the free end of the relaxed spring. The axis may be oriented either way; we have chosen the positive orientation in the direction of stretching. The force $\mathbf{F}(x)$ required to stretch the spring when it has been extended x meters from its relaxed state is the negative of the force exerted by the spring. By Hooke's law, the magnitude $F(x)$ of $\mathbf{F}(x)$ satisfies $F(x) = kx$, where k is the spring constant. Because the spring exerts a force of 20 N when it has been stretched 0.1 m, $20 = k(0.1)$. Hence, $k = 200$ N/m and $F(x) = 200x$. A plot of $F(x)$ against x is shown in Fig. 6.52(a).

The idea of an object moving on a curve is less explicit in this example. We may, however, imagine a particle at the point of application of the stretching force. This object moves along the x-axis from $x = 0$ to $x = 0.25$.

The amount of work done by the stretching force as the object moves from x to $x + dx$ is approximately $dW = F(x)\,dx = 200x\,dx$ joules. Hence, the work done in stretching the spring 0.25 m is approximately $W = \Sigma\,dW = \Sigma\,200x\,dx$ joules. In the limit,

$$W = \int_0^{0.25} dW = \int_0^{0.25} 200x\,dx = (200)\tfrac{1}{2}x^2\Big|_0^{0.25} = 6.25 \text{ J.}$$

We may choose to use formula (5) for the calculation. For this we take $F(x) = 200x$ and describe the curve by $r(x) = x$, $0 \le x \le 0.25$. The element of work is $dW = \mathbf{F} \cdot \mathbf{r}'\,dx = 200x\,dx$, as before.

EXAMPLE 3 An object moves counterclockwise on a semicircle C of radius a, from $(a, 0)$ to $(-a, 0)$, acted on by a force that attracts the particle toward the point $(-a, 0)$ and has magnitude proportional to the distance between $(-a, 0)$ and the object. Calculate the work done by this force.

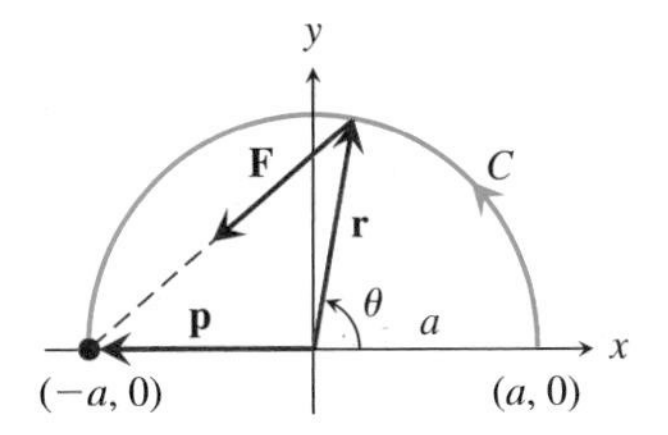

FIGURE 6.53 The work done on a particle by an attractive force.

Solution

Figure 6.53 shows a representative position of the object. Let $\mathbf{p} = \langle -a, 0\rangle$ be the position vector of the point $(-a, 0)$ and describe C by

$$\mathbf{r} = a\langle \cos t, \sin t\rangle, \qquad 0 \le t \le \pi.$$

At $\mathbf{r}$ the force $\mathbf{F}$ is in the direction of $\mathbf{p} - \mathbf{r}$. Its magnitude is $k\|\mathbf{p} - \mathbf{r}\|$, where k is a proportionality constant. To find $\mathbf{F}$ we divide the vector $\mathbf{p} - \mathbf{r}$ by its own length (to find a unit vector in the direction of the force) and then multiply by the magnitude of $\mathbf{F}$. Hence,

$$\begin{aligned}\mathbf{F} &= k\|\mathbf{p} - \mathbf{r}\|\frac{1}{\|\mathbf{p} - \mathbf{r}\|}(\mathbf{p} - \mathbf{r})\\ &= k(\mathbf{p} - \mathbf{r}) = k(\langle -a, 0\rangle - a\langle \cos t, \sin t\rangle)\\ &= -ka\langle 1 + \cos t, \sin t\rangle.\end{aligned}$$

The element of work $dW = \mathbf{F}\cdot d\mathbf{r} = \mathbf{F}\cdot \mathbf{r}'\,dt$ is

$$\begin{aligned}dW = \mathbf{F}\cdot\mathbf{r}'\,dt &= -ka\langle 1 + \cos t, \sin t\rangle \cdot a\langle -\sin t, \cos t\rangle\,dt\\ &= -ka^2(-\sin t - \sin t\cos t + \sin t\cos t)\,dt\\ &= ka^2 \sin t\,dt.\end{aligned}$$

Summing these work elements,

$$\begin{aligned}W &= \int_C \mathbf{F}\cdot d\mathbf{r} = \int_0^{\pi} \mathbf{F}\cdot\mathbf{r}'\,dt = ka^2\int_0^{\pi}\sin t\,dt\\ &= -ka^2\cos t\Big|_0^{\pi} = 2ka^2 \text{ work units}\end{aligned}$$

Pumping Fluid from a Tank

The energy needed to pump liquid from a tank is equal to the work required to lift all of the particles of water to the top of the tank. To lift a particle we must exert a force equal and opposite to the gravitational force on the particle and displace it to the top of the tank. We give an example of pumping water from a reservoir with trapezoidal cross section.

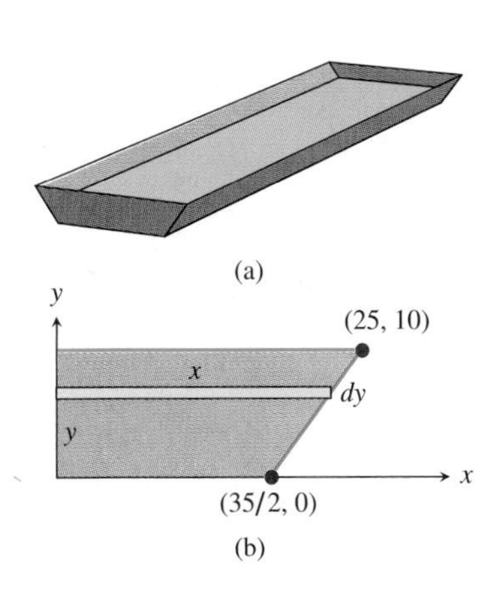

FIGURE 6.54 (a) A trapezoidal reservoir and a thin, horizontal layer of water; (b) the right half of the near end of the reservoir, where the thin, shaded strip is half of the end of the blue, horizontal layer of water shown in (a).

EXAMPLE 4 Figure 6.54 shows a reservoir with trapezoidal cross section. A cross section of the reservoir has height 10 m and its bases have lengths 50 m and 35 m. The length of the reservoir is 200 m. Assuming that the reservoir is filled with water, calculate the work required to pump all of the water to the top of the reservoir. The density of water is $\delta = 1000\ \text{kg/m}^3$.

Solution

Figure 6.54(a) shows the reservoir; also shown is a thin, horizontal layer of water. We calculate the work dW required to move this volume of water to the top of the reservoir and then sum these work elements with the integral. The reason for considering a horizontal layer of water is that every molecule in the layer must be displaced by approximately the same amount to reach the top of the reservoir.

Figure 6.54(b) shows a coordinate system that allows us to express simply the geometry of a representative cross section and the volume of a thin, horizontal layer of water. This layer is like a flattened box, with length 200 m, width $2x$ m, and height dy m.

The mass dm of the layer is $dm = \delta(2x)(200)\,dy$. From Newton's second law, to lift this mass of water requires a force $dF = g\,dm$, which just balances the gravitational force. Because dF acts over a distance of $(10 - y)$ m, the work required to pump out this layer is equal to $dW = (10 - y)\,dF$ joules. Hence, the total amount of work required to empty the reservoir is given by

$$W = \int_0^{10} dW = \int_0^{10} (10 - y)\,dF = \int_0^{10} (10 - y)g\,dm.$$

Substituting $dm = \delta(2x)(200)\,dy$,

$$W = g\int_0^{10} (10 - y)\delta(2x)(200)\,dy.$$

To express x in terms of y, we write out the equation of the line joining $(35/2, 0)$ and $(25, 10)$. Using the point-slope form of a line,

$$y - 0 = \frac{10 - 0}{25 - 35/2}(x - 35/2) = \frac{10 - 0}{50 - 35}(2x - 35).$$

Solving this equation for x,

$$x = \tfrac{1}{4}(3y + 70).$$

Hence

$$W = 400g\delta\int_0^{10} (10 - y)x\,dy = 100g\delta\int_0^{10} (10 - y)(3y + 70)\,dy$$

$$W = (4.00 \times 10^8)g \approx 3.92 \times 10^9 \text{ J}.$$

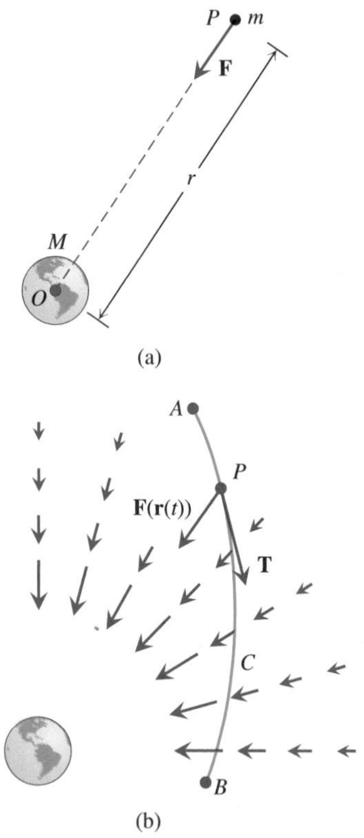

In Example 1, we showed that the constant force field $\mathbf{F} = \langle 0, -mg\rangle$, which approximates the force of attraction of the Earth on an object of mass m and near the surface of the Earth, is conservative. In our final example, we show that the Earth's gravitational field is conservative.

Figure 6.55(a) shows a particle of mass m at a point P in the Earth's gravitational field. According to Newton's universal law of gravitation, the magnitude of the force $\mathbf{F}$ exerted by the Earth on the particle is GMm/r^2, where r is the distance between P and the center O of the Earth, M is the mass of the Earth, and G is a con-

P m F r M O (a)

A P F(r(t)) T C B (b)

FIGURE 6.55 A particle moving in the Earth's force field.

stant. The direction of $\mathbf{F}$ is given by the unit vector $-(1/r)\mathbf{r}$, where $\mathbf{r}$ is the position vector $\overrightarrow{OP}$ of P, relative to the Earth's center at O. The force $\mathbf{F}$ may be expressed as

$$\mathbf{F}(\mathbf{r}) = -\frac{GMm}{r^2}\frac{1}{r}\mathbf{r} = -\frac{GMm}{r^3}\mathbf{r}. \tag{6}$$

Figure 6.55(b) gives an idea of the Earth's gravitational field. The figure shows the force that would be exerted by the Earth on a particle of mass m at a representative point P.

What do we have to do to show that this field is conservative? As discussed in Example 1, we must show that the work done by the field on a particle of mass m moving on a curve C is "independent of path." This means that if we choose any two points A and B and a smooth curve C connecting them, the work done depends only on the endpoints of C, not its shape. We consider only curves lying in a plane containing the center of the Earth.

EXAMPLE 5 Show that the Earth's gravitational field is conservative.

Solution

Let A and B be arbitrary points and C a smooth curve described by

$$\mathbf{r} = \mathbf{r}(t) = \langle x(t), y(t)\rangle, \qquad a \le t \le b,$$

where $\mathbf{r}(a)$ and $\mathbf{r}(b)$ are position vectors of A and B. Suppose a particle of mass m is at P on this curve, where the position vector of P is $\mathbf{r}(t)$. From (6), the field $\mathbf{F}(\mathbf{r}(t))$ at P is

$$\mathbf{F}(\mathbf{r}(t)) = -\frac{GMm}{r^2}\frac{1}{r}\mathbf{r}(t) = -\frac{GMm}{r^3}\mathbf{r}(t),$$

where $r = r(t) = \|\mathbf{r}(t)\|$.

From (5), the work done by the gravitational field on the object is

$$W = \int_C \mathbf{F}\cdot d\mathbf{r} = -GMm\int_a^b \frac{1}{r^3}\mathbf{r}(t)\cdot\mathbf{r}'(t)\,dt.$$

Replacing $\mathbf{r}(t)$ by $\langle x, y\rangle$, $\mathbf{r}'(t)$ by $\langle x', y'\rangle$, and r by $(x^2+y^2)^{1/2}$,

$$\begin{aligned} W &= -GMm\int_a^b \frac{1}{(x^2+y^2)^{3/2}}\langle x, y\rangle\cdot\langle x', y'\rangle\,dt \\ &= -GMm\int_a^b ((x^2+y^2)^{-3/2}(xx'+yy'))\,dt. \end{aligned} \tag{7}$$

This looks difficult, but a remarkable thing happens, which is characteristic of conservative fields. Noting that

$$\frac{d}{dt}(x^2+y^2)^{-1/2} = -\frac{1}{2}(x^2+y^2)^{-3/2}(2xx'+2yy'),$$

we can write

$$\frac{d}{dt}(x^2+y^2)^{-1/2} = -(x^2+y^2)^{-3/2}(xx'+yy'). \tag{8}$$

Substituting (8) into (7),

$$W = GMm\int_a^b \frac{d}{dt}(x^2 + y^2)^{-1/2}\,dt.$$

Hence, by the Fundamental Theorem of Calculus,

$$W = GMm(x(t) + y(t))^{-1/2}\Big|_a^b = \frac{GMm}{\|\mathbf{r}(t)\|}\Big|_a^b = GMm\left(\frac{1}{\|\mathbf{r}(b)\|} - \frac{1}{\|\mathbf{r}(a)\|}\right).$$

This shows that the work W depends only on the endpoints of C, not on the curve itself. Hence, the Earth's gravitational field is conservative.

Kinetic Energy

The definition of *kinetic energy* and a relation between work and kinetic energy follow from an application of Newton's second law to the integral for the work done on an object by a force field. Suppose that we have an object of (constant) mass m moving in a force field $\mathbf{F}$. If the object moves from A to B on a curve C, the work done by the field is

$$W = \int_C \mathbf{F}\cdot d\mathbf{r}.$$

If at any point on C the net force on the object is $\mathbf{F}$ and $\mathbf{a}$ is the acceleration of the object, then, by Newton's second law, $\mathbf{F} = m\mathbf{a}$. If we write the acceleration $\mathbf{a}$ in terms of velocity, then $\mathbf{F} = m\mathbf{v}'$. With this replacement we may rewrite the work integral as

$$W = \int_C \mathbf{F}\cdot d\mathbf{r} = \int_a^b m\mathbf{v}'\cdot\frac{d\mathbf{r}}{dt}\,dt = \int_a^b m\mathbf{v}'\cdot\mathbf{v}\,dt,$$

where C is described by

$$\mathbf{r} = \mathbf{r}(t) = \langle x(t), y(t)\rangle, \qquad a \le t \le b.$$

The vectors $\mathbf{r}(a)$ and $\mathbf{r}(b)$ are position vectors of A and B. Writing $\mathbf{v} = \langle x', y'\rangle$ and $\mathbf{v}' = \langle x'', y''\rangle$,

$$W = m\int_a^b (x'x'' + y'y'')\,dt. \tag{9}$$

The form of the integrand $x'x'' + y'y''$ suggests the chain rule. Specifically,

$$\frac{d}{dt}(x'^2 + y'^2) = 2x'x'' + 2y'y'' = 2(x'x'' + y'y'').$$

It now follows from (9) that

$$W = \tfrac{1}{2}m\int_a^b \frac{d}{dt}(x'^2 + y'^2)\,dt = \tfrac{1}{2}m(x'^2 + y'^2)\Big|_a^b$$
$$= \tfrac{1}{2}m\left(\|\mathbf{r}'(b)\|^2 - \|\mathbf{r}'(a)\|^2\right)$$

If we define the kinetic energy at time t of an object with mass m and speed $v(t) = \|\mathbf{r}'(t)\|$ as $T(t) = \frac{1}{2}mv(t)^2$, then W may be written as

$$W = T(b) - T(a). \tag{10}$$

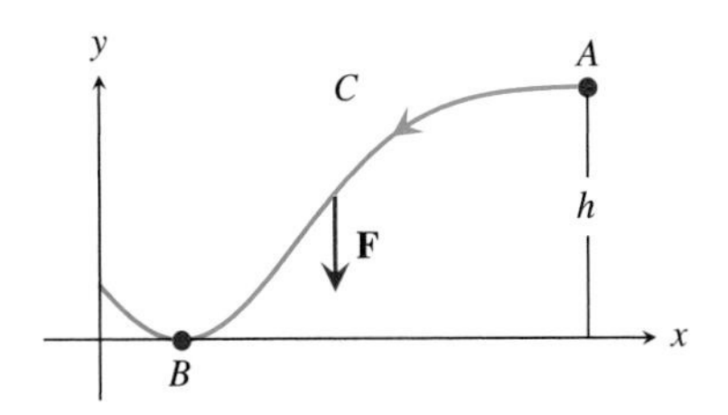

FIGURE 6.56 Death Demon 2.

In words, this equation states that the change in the object's kinetic energy as it moves from A to B is equal to the work done by the force field.

To see how this result can simplify some kinds of calculations, we use the physics lesson Jason gave to Paige in a 1994 *Fox Trot* comic strip. Jason, with calculator in hand, accompanies Paige on the Death Demon 2 roller coaster. Referring to Fig. 6.56, suppose a roller coaster car moves from A to B, a vertical drop of h meters. Jason teaches Paige how to calculate their speed $v(b)$ at point B, assuming that their speed at A was "essentially nothing." From (10),

$$W = T(b) - T(a) = \tfrac{1}{2}mv(b)^2 \tag{11}$$

If we knew W, we could solve this equation for v_b. Jason assumes that the gravitational force is $\langle 0, -mg\rangle$. If this were the net force, we could calculate W by using the fact that the gravitational force is conservative. On real coasters, however, the net force includes friction, and the resulting force field is not conservative. Jason goes

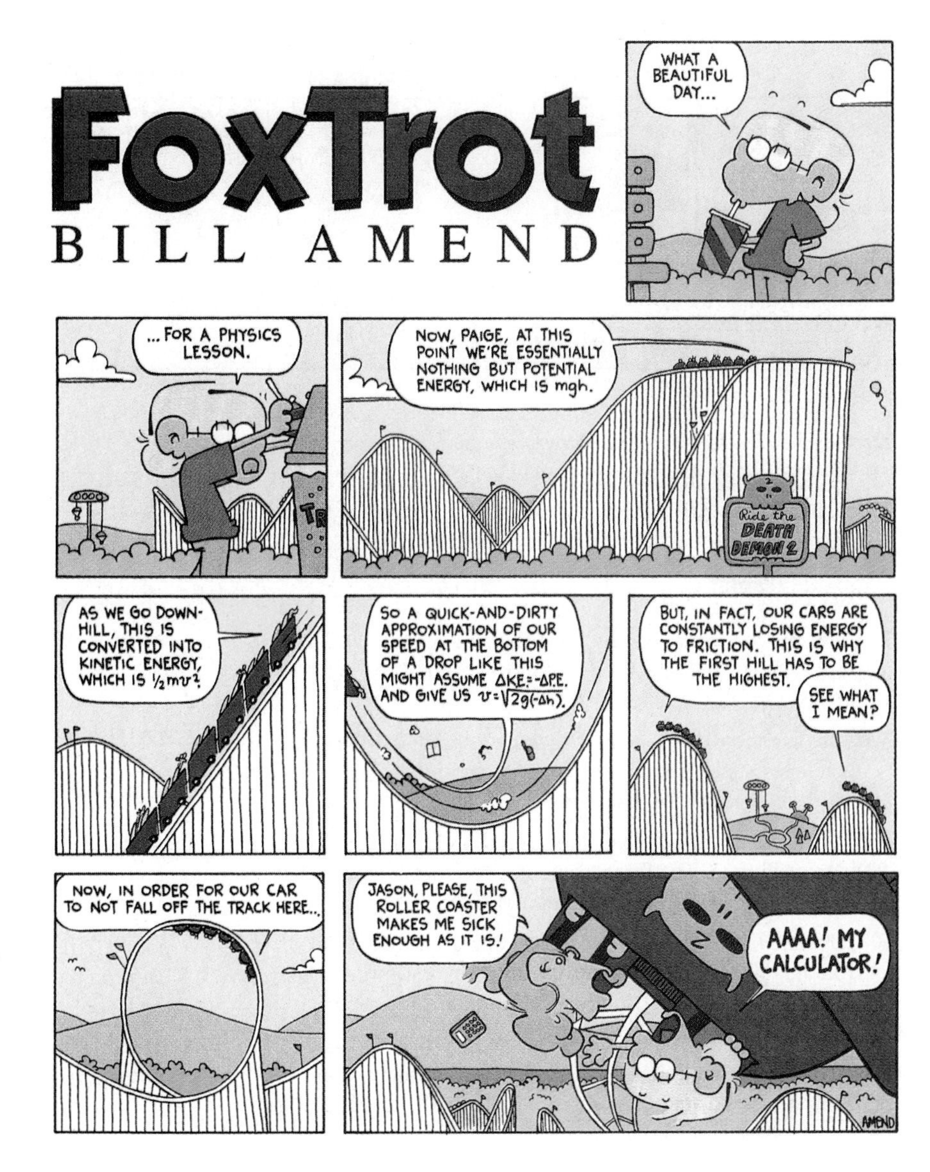

for a "quick-and-dirty" solution by assuming that the net force on the coaster is the conservative gravitational force. Because of this assumption, we may calculate the work done by the gravitational force as the coaster moves from A to B on any smooth curve C. Hence, we do not need to know the actual path of the coaster. The work done by the gravitational force as the coaster moves from A to B is

$$W = \int_C \mathbf{F} \cdot \mathbf{dr} = \int_a^b \langle 0, -mg \rangle \cdot \langle x'(t), y'(t) \rangle, dt$$

$$= -mg \int_a^b y'(t)\, dt = -mgy(t)\Big|_a^b = -mg(y(b) - y(a)).$$

Because $y(b) = 0$ and $y(a) = h$,

$$W = mgh.$$

From (11),

$$mgh = \tfrac{1}{2}mv(b)^2$$

Solving this equation for $v(b)^2$,

$$v(b) = \sqrt{2gh}.$$

For $h = 25$ m (estimated from the comic strip), $v(b) \approx 22$ m/s, which is about 50 mph.

Exercises 6.6

Exercises 1–32: Masses are in kilograms and lengths are in meters. Work will be measured in joules (J).

1. Calculate the work done by the constant gravitational force on an object of mass 5.0 kg as it moves from $(0, 3)$ to $(2, 0)$ on the line described by $\mathbf{r}(x) = \langle x, -3x/2 + 3 \rangle$, where $0 \le x \le 2$.

2. Calculate the work done by the constant gravitational force on an object of mass 2.0 kg as it moves from $(0, 3)$ to $(2, 0)$ on the graph of the function $f(x) = -3x^2/4 + 3, 0 \le x \le 2$.

3. Calculate the work done in stretching a spring 0.30 m from its relaxed state. The force exerted by the spring when it has been stretched 0.05 m has magnitude 25 N.

4. Calculate the work done in stretching a spring 0.20 m from its relaxed state. The force exerted by the spring when it has been stretched 0.05 m has magnitude 30 N.

5. Calculate the work done in compressing a spring 0.30 m from its relaxed state. The force exerted by the spring when it has been stretched 0.05 m has magnitude 25 N. See the accompanying figure.

6. Calculate the work done in compressing a spring 0.20 m from its relaxed state. The force exerted by the spring when it has been stretched 0.05 m has magnitude 30 N. See the accompanying figure.

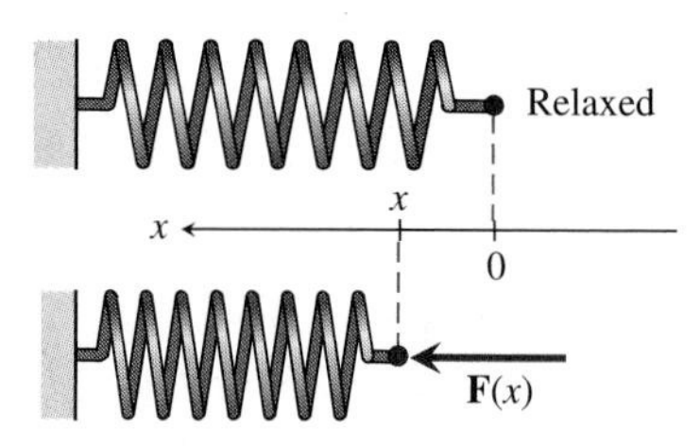

Figure for Exercises 5 and 6.

7. An object moves from $(-1, 0)$ to $(1, 0)$ on the graph of $y = 1 - x^2$, acted on by a force that attracts it toward the point $(1, 0)$ and has magnitude proportional to the distance between the object and $(1, 0)$. Calculate the work done by the force.

8. An object moves from $(-1, 0)$ to $(1, 0)$ on the graph of $y = x^3 - x$, acted on by a force that attracts it toward the point $(1, 0)$ and has magnitude proportional to the distance between the object and $(1, 0)$. Calculate the work done by the force.

9. The top of a reservoir is a rectangle 50 m by 200 m. Vertical cross sections of the reservoir perpendicular to the long side of the rectangle are semicircles. Assuming that the reservoir is filled with water, calculate the work required to pump it dry. The density of water is $\delta = 1000$ kg/m^3.

10. The top of a reservoir is a rectangle 40 m by 180 m. Vertical cross sections of the reservoir perpendicular to the long side of the rectangle are congruent to the parabolic arc described by $y = \frac{1}{20}x^2$, $-20 \le x \le 20$. Assuming that the reservoir is filled with water, calculate the work required to pump it dry. The density of water is $\delta = 1000\ \text{kg/m}^3$.

11. A car of mass m drives down the hill shown in the accompanying figure. All lengths are measured in meters. Calculate the work done by the force of gravity as the car moves from the top to the bottom of the hill.

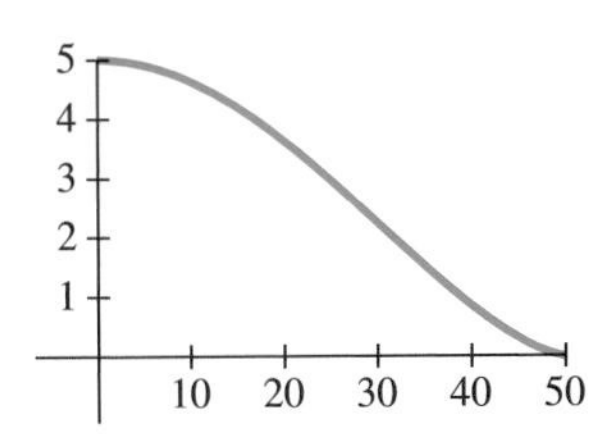

Figure for Exercise 11.

12. A car of mass m drives down the hill shown in the accompanying figure. All lengths are measured in meters. Calculate the work done by the force of gravity as the car moves from the top to the bottom of the hill.

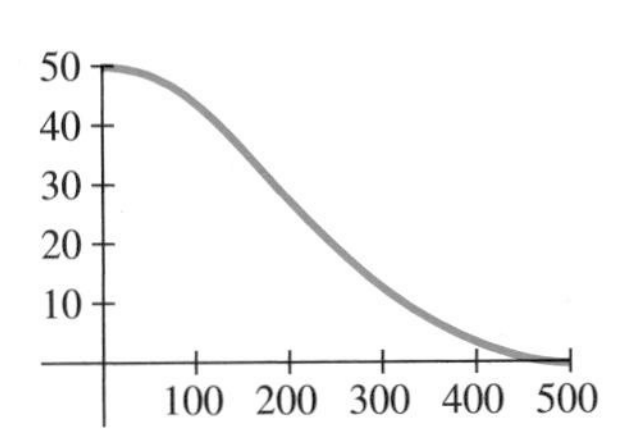

Figure for Exercise 12.

13. A skid is pulled 20 meters along a rough horizontal floor. The force required to just overcome the frictional force varies linearly from 3250 N at the beginning of the motion to 3000 N at the end. Calculate the work done by the force.

14. A skid is pulled 30 meters along a rough horizontal floor. The force required to just overcome the frictional force is $f(x) = 2500/(x + 1)$ N, where x is the distance the skid has moved from its starting point. Calculate the work done by the force.

T **15.** A railroad car bumps into a snubber (the spring/shock absorber mechanism at the end of some tracks). The position $x = x(t)$ of a railroad car during the time the car is in contact with the snubber is given by

$$x = x(t) = -1.7e^{-0.5t}\sin(1.2t).$$

(See the spring and coordinate system in the accompanying sketch.) Lengths are measured in meters and time in seconds. The snubber is in its relaxed state at $t = 0$, compresses during a time interval $[0, b]$, and returns to its initial (relaxed) position during an interval $[b, c]$. If the spring constant is 1.6×10^4, how much work is done by the spring during the time interval $[b, c]$ when it is pushing the car to the right? *Hint:* If you use the coordinate system in the sketch, the force will be $-1.6 \times 10^4 x$ during $[b, c]$.

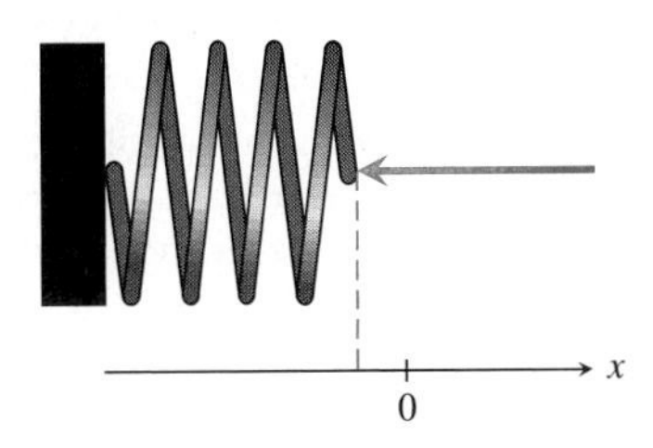

Figure for Exercises 15 and 16.

T **16.** A railroad car bumps into a snubber (the spring/shock absorber mechanism at the end of some tracks). The position $x = x(t)$ of a railroad car during the time the car is in contact with the snubber is given by

$$x = x(t) = -1.7e^{-0.375t}\sin(1.2t).$$

See the spring and coordinate system in the accompanying sketch. Lengths are measured in meters and time in seconds. The snubber is in its relaxed state at $t = 0$, compresses during a time interval $[0, b]$, and returns to its initial (relaxed) position during an interval $[b, c]$. If the spring constant is 1.6×10^4, how much work is done by the spring during the time interval $[b, c]$ when it is pushing the car to the right? *Hint:* If you use the coordinate system in the sketch, the force will be $-1.6 \times 10^4 x$ during $[b, c]$.

17. Calculate the work needed to overcome the Earth's gravitational force $\mathbf{F} = -GM_E m/x^2$ on an object of mass $m = 50{,}000$ kg as it is boosted radially from Earth's surface to an altitude of 160 km. The coordinate x is measured on a line with origin at the center of the Earth and positive direction outward. The value of G, the universal gravitational constant, is $G = 6.672659 \times 10^{-11}$, the mass M_E of the Earth is $M_E = 5.97 \times 10^{24}$ kg, and the radius R_E of the Earth is $R_E \approx 6.378 \times 10^6$ m.

18. Calculate the work needed to overcome the Earth's gravitational force $\mathbf{F} = -GM_E m/x^2$ on an object of mass $m = 60{,}000$ kg as it is boosted radially from Earth's surface to an altitude of 175 km. The coordinate x is measured on a line with origin at the center of the Earth and positive direction outward. The value of G, the universal gravitational constant, is $G = 6.672659 \times 10^{-11}$, the mass M_E of the Earth is $M_E = 5.97 \times 10^{24}$ kg, and the radius R_E of the Earth is $R_E \approx 6.378 \times 10^6$ m.

19. A 500-m steel cable with density 15 kg/m is coiled at the bottom of a vertical shaft. Calculate the work required to lift the cable so that it hangs straight.

20. A 450-m steel cable with density 16 kg/m is coiled at the bottom of a vertical shaft. Calculate the work required to lift the cable so that it hangs straight.

21. An elevator of mass 1500 kg hangs at the end of a 500-m steel cable with density 15 kg/m. Calculate the work required to raise the elevator 490 m.

22. An elevator of mass 2000 kg hangs at the end of a 600-m steel cable with density 15 kg/m. Calculate the work required to raise the elevator 590 m.

23. A bucket with mass 2 kg and a rope of negligible weight are used to draw water from a well that is 20 m deep. The bucket starts with 16 kg of water and is pulled up at a rate of 1 m/s. Water leaks out of a hole in the bucket at a rate of 0.075 kg/s. Find the work done in pulling the bucket to the top of the well. *Hint:* Show that the mass of the bucket as a function of the distance x from the bottom of the well is $m(x) = 18 - 0.075x$.

24. A bucket with mass 2 kg and a rope of negligible weight are used to draw water from a well that is 25 m deep. The bucket starts with 15 kg of water and is pulled up at a rate of 1 m/s. Water leaks out of a hole in the bucket at a rate of 0.075 kg/s. Find the work done in pulling the bucket to the top of the well. *Hint:* Show that the mass of the bucket as a function of the distance x from the bottom of the well is $m(x) = 17 - 0.075x$.

25. A tank in the shape of a sphere with radius 10 m is full of gasoline, with density 680 kg/m^3. Calculate the work required to empty the tank through a hole at the top.

26. A tank in the shape of a sphere with radius 11 m is full of gasoline, with density 680 kg/m^3. Calculate the work required to empty the tank through a hole at the top.

27. A rectangular swimming pool has dimensions $w = 10$ meters by $\ell = 50$ meters. Its depth is constant along lines parallel to the short sides and varies linearly from $h = 1$ meter to $H = 3$ meters along the long side. See the accompanying figure. Assuming the pool is full, calculate the work required to empty the pool. The density of water is $\delta = 1000$ kg/m^3.

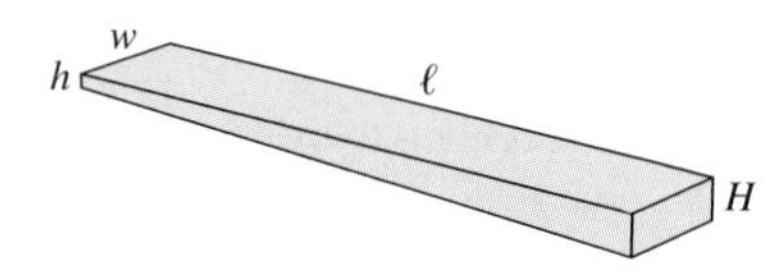

Figure for Exercises 27 and 28.

28. A rectangular swimming pool has dimensions $w = 10$ meters by $\ell = 60$ meters. Its depth is constant along lines parallel to the short sides and varies linearly from $h = 1$ meter to $H = 4$ meters along the long side. See the accompanying figure. Assuming the pool is full, calculate the work required to empty the pool. The density of water is $\delta = 1000$ kg/m^3.

29. Grain is piled in a conical pile to a depth of 5 meters, and the radius of the pile is 10 meters. The grain is to be transferred to a nearby silo of height 40 meters. Find the ratio of the work required to move the grain to the top of the silo assuming that the grain is taken from the bottom of the pile to the work required assuming that the grain is removed in thin, horizontal layers from the top of the (remaining) pile.

30. Grain is piled in a conical pile to a depth of 6 meters, and the radius of the pile is 12 meters. The grain is to be transferred to a nearby silo of height 45 meters. Find the ratio of the work required to move the grain to the top of the silo assuming that the grain is taken from the bottom of the pile to the work required assuming that the grain is removed in thin, horizontal layers from the top of the (remaining) pile.

31. A tank in the shape of a right circular cylinder with radius 3 m and height 10 m is placed so that its axis is horizontal. If the tank is full of gasoline, with density 680 kg/m^3, calculate the work required to empty the tank through a hole at the top of one of the circular ends.

32. A tank in the shape of a right circular cylinder with radius 4 m and height 15 m is placed so that its axis is horizontal. If the tank is full of gasoline, with density 680 kg/m^3, calculate the work required to empty the tank through a hole at the top of one of the circular ends.

33. Assume that the Great Pyramid at Giza is solid granite, with density $\delta = 2500$ kg/m^3. Calculate the work required to put this granite in place, starting from ground level. The side of the square base of the pyramid is $a = 230.4$ m and its height is $h = 146.8$ m.

34. Let O be the origin of an (x, y)-coordinate system and assume that at any point (x, y) other than the origin, a force $\mathbf{F}(x, y)$ has the form

$$\mathbf{F}(x, y) = \frac{k}{x^2 + y^2}\langle x, y\rangle,$$

where k is a constant. Show that this field is conservative.

35. Referring to the accompanying figure and Jason Fox's calculations reported in the text, what is the speed of the roller-coaster car when it reaches B? Why?

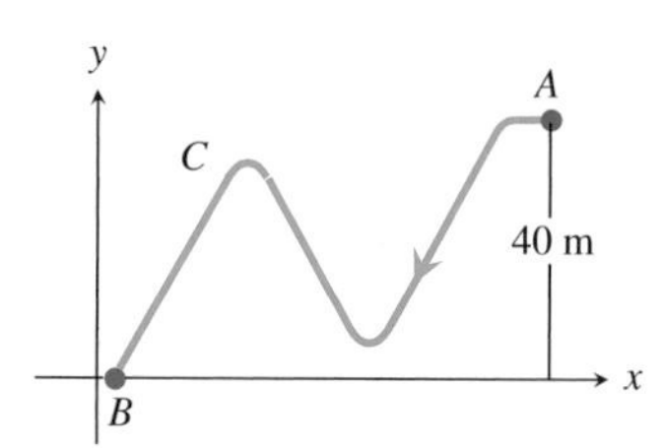

Figure for Exercise 35.

36. We argued that $\Delta s \approx ds = \|\mathbf{r}'(t)\|\, dt$ a few lines above (3). Use the tangent line approximation to derive this result, starting with the observation that during the time dt, the asteroid has moved a distance of

$$\Delta s = \int_a^{t+dt} \|\mathbf{r}'(w)\|\, dw - \int_a^t \|\mathbf{r}'(w)\|\, dw$$

along the curve. *Hint:* Letting $f(t) = \int_a^t \|\mathbf{r}'(w)\|\, dw$, use the tangent line approximation in the form $f(t + dt) - f(t) \approx f'(t)\, dt$.

37. This problem was adapted from *Engineering Mechanics,* by Andrew Pytel and Jaan Kiusalaas (HarperCollins Publishers, 1994). Referring to the accompanying figure, a force $\mathbf{F}$ acts on the end of a spring as the end moves on a curve C. At point A, the spring is relaxed. Show that the work done by the force on the spring is independent of the curve C joining A and B. Let the spring constant be k and assume that C can be described by

$$\mathbf{r} = \mathbf{r}(h), \qquad c \le h \le d,$$

where h is the distance from O to the point of application of $\mathbf{F}$, $\|\mathbf{r}(c)\| = a = \|\overrightarrow{OA}\|$, and $\|\mathbf{r}(d)\| = b = \|\overrightarrow{OB}\|$. Fill in the following outline.

i. Show that $\|\mathbf{r}(h)\| = h$.

ii. Show that the force at $\mathbf{r}(h)$ is

$$\mathbf{F}(\mathbf{r}(h)) = \left(\frac{k(h-c)}{h}\right)\mathbf{r}(h).$$

iii. Show that

$$dW = \left(\frac{k(h-c)}{h}\right)\mathbf{r} \cdot \mathbf{r}'\, dh.$$

iv. Show that $\mathbf{r} \cdot \mathbf{r}' = h$. From parts iii and iv, show that $dW = k(h - c)\, dh$. Now complete the problem.

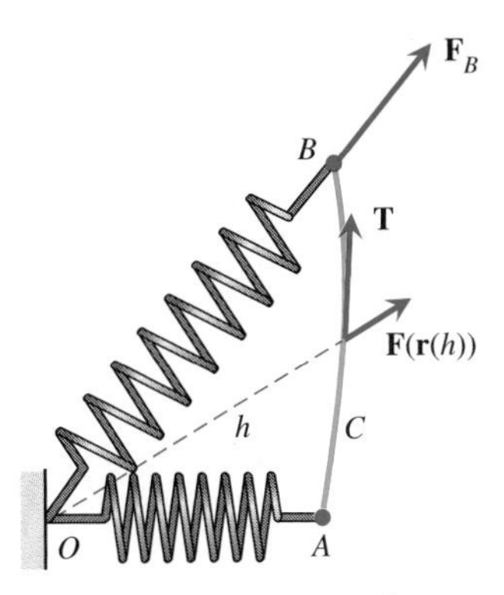

Figure for Exercise 37.

6.7 Center of Mass

If a particle is thrown and the resistance of the air is ignored, it will follow a (possibly degenerate) parabolic path. If a hammer is thrown and the resistance of the air is ignored, most of its constituent atoms will follow complex paths because the hammer rotates during its flight. A relatively simple throw, one in which the spin is confined to a vertical plane, is shown in Fig. 6.57. Both experimental evidence and theoretical analysis show that there is one point of the hammer that follows a parabolic path. That point is called the **center of mass.**

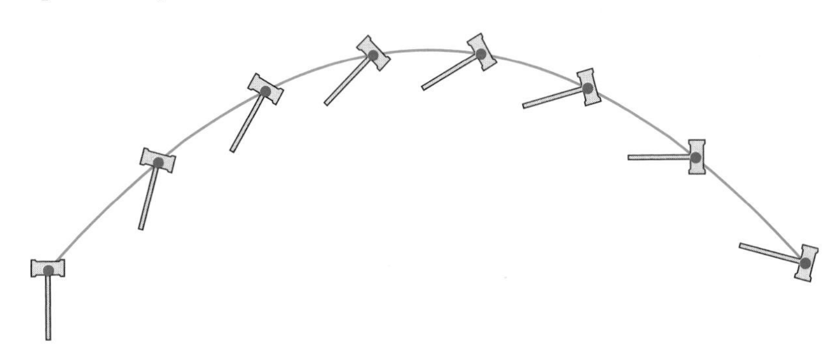

FIGURE 6.57 Tracking the center of mass of a thrown hammer.

Newton used the idea of center of mass in studying the motion of the Earth in the solar system. He replaced each of the systems of atoms constituting the Earth and the sun by "point masses," that is, objects whose masses can be regarded as concentrated at one point. The point mass replacing the Earth, for example, has the same mass as the Earth and is located at the Earth's center. The use of point masses greatly simplified Newton's study of the motion of the Earth.

In this section we study both finite and *continuous* systems. Although all real objects, for example, a hammer or the Earth, are composed of a finite number of atoms, it is often easier to model such objects with line segments, segments of curves, regions of a plane, or solid regions. These regions are called *continuous* because they can be subdivided indefinitely.

Finite Systems

By a system S of **point masses** we mean a set of physical objects located at points on a line, in a plane, or in space. Such systems can model the planets in the solar system, the pellets of shot emerging from a shotgun, or the fragments of a subatomic collision. In this section we study point masses located on a line or in a plane.

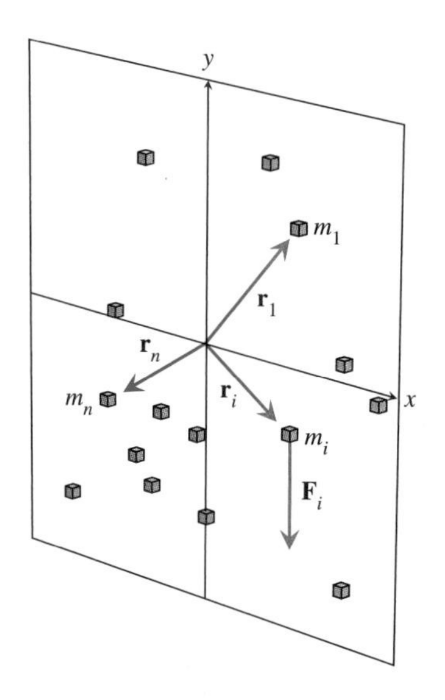

FIGURE 6.58 A two-dimensional system of several point masses.

Figure 6.58 shows a system of n point masses $m_1, m_2, \ldots, m_n$, located in a plane by vectors $\mathbf{r}_1, \mathbf{r}_2, \ldots, \mathbf{r}_n$. We have oriented the plane as if it were vertical so that we may, for example, regard these points as the constituent atoms or molecules of an object like the hammer on which the gravitational force acts.

We suppose that the position vector $\mathbf{r}_i = \langle x_i, y_i \rangle$ of a representative point mass is a function of time t, so that $\mathbf{r}_i = \mathbf{r}_i(t)$, $i = 1, 2, \ldots, n$. If $\mathbf{F}_i$ is the net force acting on the ith particle, then from Newton's second law,

$$\mathbf{F}_i = m_i \mathbf{a}_i = m_i \frac{d^2\mathbf{r}_i}{dt^2}.$$

Letting $\mathbf{F} = \sum_{i=1}^{n} \mathbf{F}_i$ be the net force on the hammer,

$$\mathbf{F} = \sum_{i=1}^{n} \mathbf{F}_i = \sum_{i=1}^{n} m_i \frac{d^2\mathbf{r}_i}{dt^2}.$$

Because the second derivative of a sum is the sum of the second derivatives, the net force on the hammer can be written as

$$\mathbf{F} = \frac{d^2}{dt^2} \sum_{i=1}^{n} m_i \mathbf{r}_i$$

or, letting $m = \sum_{i=1}^{n} m_i$ denote the mass of the hammer, as

$$\mathbf{F} = m \frac{d^2}{dt^2} \left(\frac{1}{m} \sum_{i=1}^{n} m_i \mathbf{r}_i \right).$$

Hence, the point with position vector

$$\mathbf{R} = \frac{1}{m} \sum_{i=1}^{n} m_i \mathbf{r}_i$$

and mass m satisfies the equation

$$\mathbf{F} = m \frac{d^2\mathbf{R}}{dt^2}. \tag{1}$$

This equation has the form of Newton's second law, applied to a particle of mass m and position vector $\mathbf{R} = \mathbf{R}(t)$. If we call the point with position vector $\mathbf{R}$ the *center of mass* of the hammer, we infer from Equation (1) that the center of mass of the hammer moves as if the net force $\mathbf{F}$ were acting on the total mass of the hammer, concentrated at the center of mass. The center of mass of the hammer traces the parabola shown in Fig. 6.57.

Although the net force $\mathbf{F}_i$ on the ith particle of the hammer includes both the gravitational force and the (internal) forces exerted by the other particles, the internal forces cancel in pairs when we form $\mathbf{F} = \Sigma_{i=1}^{n} \mathbf{F}_i$. This follows from Newton's third law. The force on the jth particle due to particle i is exactly balanced by the force on the ith particle due to particle j. It follows that the force $\mathbf{F}$ is the total gravitational force acting on the hammer.

DEFINITION Center of Mass of a System of Point Masses

The center of mass of a system S of n point masses, with masses $m_1, m_2, \ldots, m_n$ and positions $\mathbf{r}_1 = \langle x_1, y_1 \rangle, \mathbf{r}_2 = \langle x_2, y_2 \rangle, \ldots, \mathbf{r}_n = \langle x_n, y_n \rangle$, is the point with position

$$\mathbf{R} = \langle X, Y \rangle = \frac{1}{m} \sum_{i=1}^{n} m_i \mathbf{r}_i, \tag{2}$$

where $m = \Sigma_{i=1}^{n} m_i$ is the total mass of the system.

If the point masses of a system S lie on a line, we drop the vector notation, replace $\mathbf{r}_i$ by x_i, replace $\mathbf{R}$ by X, and write (2) as

$$X = \frac{1}{m} \sum_{i=1}^{n} m_i x_i. \tag{3}$$

In a plane we may choose to write (2) in the form

$$X = \frac{1}{m} \sum_{i=1}^{n} m_i x_i, \qquad Y = \frac{1}{m} \sum_{i=1}^{n} m_i y_i. \tag{4}$$

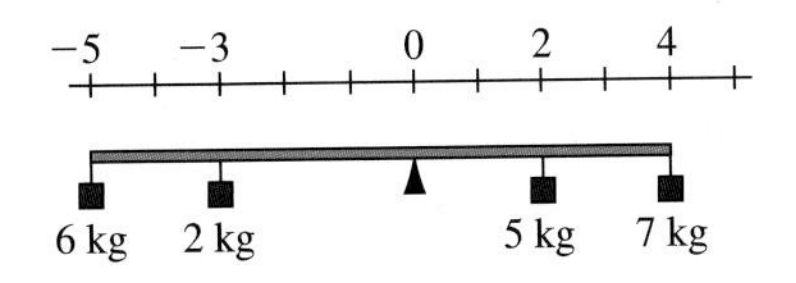

FIGURE 6.59 A one-dimensional system of four point masses.

EXAMPLE 1 Find the center of mass of the four point masses shown in Fig. 6.59. From left to right, their masses are $m_1 = 6$ kg, $m_2 = 2$ kg, $m_3 = 5$ kg, and $m_4 = 7$ kg, with corresponding positions $x_1 = -5$, $x_2 = -3$, $x_3 = 2$, and $x_4 = 4$. The units are kilograms and meters. This system can be thought of as a model of four masses on a (massless) seesaw.

Solution

The center of mass may be calculated using (3), which is the one-dimensional version of (2). The mass of the system is $m = 6 + 2 + 5 + 7 = 20$ kg and the location of the center of mass is

$$X = \frac{1}{m} \sum_{i=1}^{n} m_i x_i = \frac{1}{20}((6)(-5) + (2)(-3) + (5)(2) + (7)(4)) = 0.1 \text{ m}.$$

The center of mass of this system is the point with coordinate $X = 0.1$ m. The seesaw would be in equilibrium if the fulcrum were placed at this point.

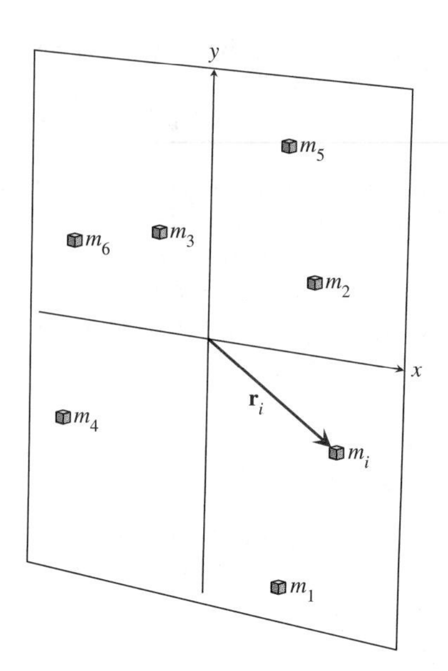

FIGURE 6.60 A two-dimensional system of six point masses.

EXAMPLE 2 The masses and position vectors of the six point masses shown in Fig. 6.60 are $m_1 = 5.5$ kg, $m_2 = 2.5$ kg, $m_3 = 5.8$ kg, $m_4 = 10.0$ kg, $m_5 = 7.7$ kg, and $m_6 = 1.8$ kg, with corresponding positions $\mathbf{r}_1 = \langle 0.40, -0.87\rangle$, $\mathbf{r}_2 = \langle 0.55, 0.27\rangle$, $\mathbf{r}_3 = \langle -0.27, 0.37\rangle$, $\mathbf{r}_4 = \langle -0.80, -0.40\rangle$, $\mathbf{r}_5 = \langle 0.40, 0.75\rangle$, and $\mathbf{r}_6 = \langle -0.75, 0.29\rangle$. Calculate the center of mass of the system. The units are kilograms and meters.

Solution

We use (4) to calculate $\mathbf{R} = \langle X, Y\rangle$. Noting that

$$m = 5.5 + 2.5 + 5.8 + 10.0 + 7.7 + 1.8 = 33.3 \text{ kg},$$

$$X = \frac{1}{33.3}(5.5(0.40) + 2.5(0.55) + 5.8(-0.27) + 10.0(-0.80) + 7.7(0.40) + 1.8(-0.75)) \approx -0.13 \text{ m}$$

$$Y = \frac{1}{33.3}(5.5(-0.87) + 2.5(0.27) + 5.8(0.37) + 10.0(-0.40) + 7.7(0.75) + 1.8(0.29)) = 0.01 \text{ m}.$$

The center of mass of this system is the point with position vector

$$\mathbf{R} = \langle X, Y\rangle \approx \langle -0.13, 0.01\rangle \text{ m}.$$

Continuous Mass Systems

Although it is possible to imagine calculating the center of mass of a solid object by regarding it as a finite system of its constituent atoms and then using (2), it is usually simpler to model solid objects such as rods, curved rods, or thin plates or laminas as "continuous masses."

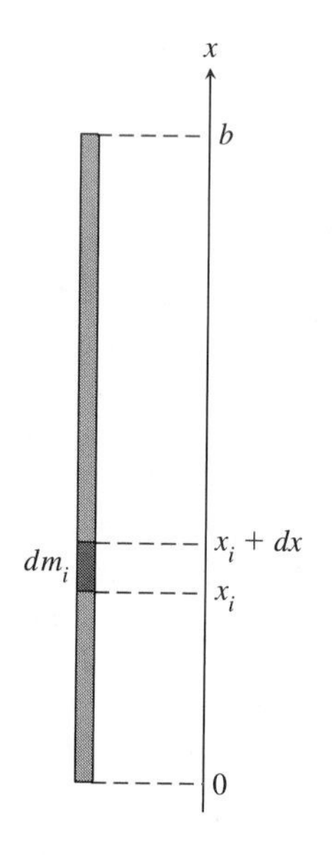

FIGURE 6.61 Metal rod with constant density.

One-Dimensional Continuous Mass Suppose that we wish to calculate the center of mass of the metal rod of length b shown in Fig. 6.61. For a continuous rod, the mass of the rod per unit length is used instead of the mass at a point. Mass per unit length is called *linear density* and is measured in kilograms per meter. We assume at first that the density is constant along the rod and equal to δ kg/m. We divide the rod into n mass elements corresponding to a subdivision of $[0, b]$, view the mass elements as a finite system, and use (3) to calculate the center of mass X_n of the system. The element of mass between x_i and $x_i + dx$ has mass $dm_i = \delta\, dx$. From (3),

$$X_n = \frac{\sum_{i=1}^n dm_i x_i}{\sum_{i=1}^n dm_i} = \frac{\sum_{i=1}^n x_i(\delta\, dx)}{\sum_{i=1}^n \delta\, dx} = \frac{\sum_{i=1}^n x_i\, dx}{\sum_{i=1}^n dx}.$$

In the second and third steps we replaced dm_i by $\delta\, dx$ and removed the common constant factor of δ. The number X_n is the center of mass of the system of n mass elements. We expect that as the number of subdivisions increases, X_n will approach the center of mass X of the rod. Thus

$$X = \frac{\int_0^b x\, dx}{\int_0^b dx} = \frac{\frac{1}{2}b^2}{b} = \frac{1}{2}b.$$

This is hardly surprising—we expect that for a rod of constant density the midpoint of the rod will be the center of mass.

This calculation may be used with very little change to find the center of mass of a rod with variable density. Assuming the density of the rod at position x is $\delta(x)$, the mass dm_i of the element of mass between x_i and $x_i + dx$ is $dm_i \approx \delta(x_i)\,dx$. From (3),

$$X_n \approx \frac{\sum_{i=1}^n dm_i x_i}{\sum_{i=1}^n dm_i} = \frac{\sum_{i=1}^n x_i\,\delta(x_i)\,dx}{\sum_{i=1}^n \delta(x_i)\,dx}.$$

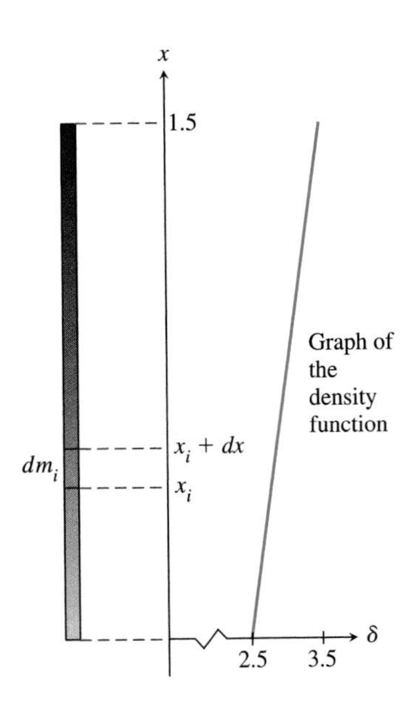

FIGURE 6.62 A 1.5-m rod with variable density.

The number X_n is the center of mass of the system of n mass elements. We expect that as the number of subdivisions increases, X_n will approach the center of mass X of the rod. Thus

$$X = \lim_{n\to\infty} X_n = \frac{\lim_{n\to\infty}\sum_{i=1}^n x_i\,\delta(x_i)\,dx}{\lim_{n\to\infty}\sum_{i=1}^n \delta(x_i)\,dx} = \frac{\int_0^b x\,\delta(x)\,dx}{\int_0^b \delta(x)\,dx}. \tag{5}$$

EXAMPLE 3 Find the mass and center of mass of a 1.5-m rod whose density varies linearly from 2.5 kg/m to 3.5 kg/m from end to end.

Solution

To say that the density "varies linearly" means that if we plot density δ against position x on the rod, the graph will be a line; that is, the density function will have the form $\delta(x) = Ax + B$. Referring to Fig. 6.62, if we place the origin of a coordinate system at the less-dense end, then because $\delta(0) = 2.5$ and $\delta(1.5) = 3.5$,

$$\delta(x) = 2.5 + (1/1.5)x, \qquad 0 \le x \le 1.5.$$

From (5),

$$X = \frac{\int_0^{1.5}(2.5 + (1/1.5)x)x\,dx}{\int_0^{1.5}(2.5 + (1/1.5)x)\,dx} = \frac{3.5625}{4.5} \approx 0.79.$$

The mass of the rod is 4.5 kg, and the center of mass is approximately 0.79 m from the origin.

Next we calculate the center of mass of a curved rod with constant density δ. We may represent such a rod by the curve C shown in Fig. 6.63 and described by $\mathbf{r} = \mathbf{r}(t) = \langle x(t), y(t)\rangle$, $a \le t \le b$. To calculate the center of mass of the rod, we divide it into small point masses and apply (2). To divide C into small point masses, we subdivide the interval $[a, b]$, choose a representative subinterval $[t, t + dt]$, and examine the corresponding mass element dm between $\mathbf{r}(t)$ and $\mathbf{r}(t + dt)$.

Figure 6.63 shows a representative mass element at P. The position of this element is $\mathbf{r}(t)$ and its mass is $dm = \delta\,ds$, where δ is the density of the rod at P and ds the arc length element at P. Using (2) to approximate the center of mass $\mathbf{R}$ of the rod,

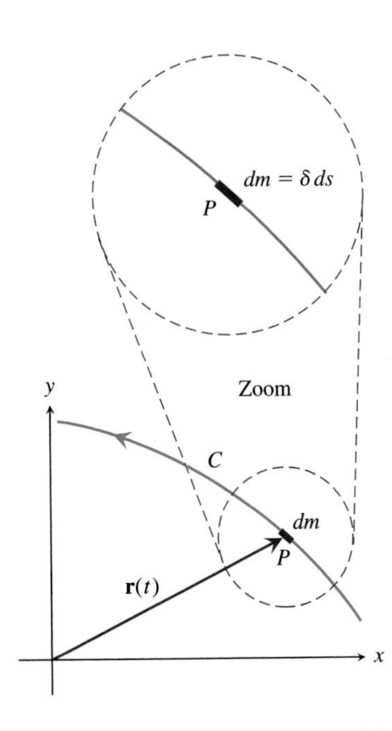

FIGURE 6.63 Curved rod with variable density.

$$\mathbf{R} \approx \frac{1}{\sum dm}\sum dm\,\mathbf{r}(t) = \frac{1}{\delta\sum ds}\delta\sum \mathbf{r}(t)\,ds = \frac{1}{\sum ds}\sum \mathbf{r}(t)\,ds. \tag{6}$$

We used a bare minimum of notation in writing this approximation, omitting the subscripts associated with a subdivision of $[a, b]$ and using the representative mass and arc length elements. We have tried to use enough notation to keep track of what's going on, but not so much that the physical concepts get lost in a flurry of subscripts.

To determine $\mathbf{R}$, we take the limit of the right side of (6) as the number of subdivisions of $[a, b]$ becomes infinite. Recalling that $ds = (ds/dt)\,dt = \|\mathbf{r}'(t)\|\,dt$, it follows that

$$\mathbf{R} = \lim_{n\to\infty}\left(\frac{1}{\sum ds}\sum \mathbf{r}(t)\,ds\right) = \frac{1}{\lim\limits_{n\to\infty}\sum\|\mathbf{r}'(t)\|\,dt}\cdot\lim_{n\to\infty}\sum\|\mathbf{r}'(t)\|\mathbf{r}(t)\,dt.$$

Hence,

$$\mathbf{R} = \frac{1}{\int_a^b\|\mathbf{r}'(t)\|\,dt}\int_a^b\|\mathbf{r}'(t)\|\mathbf{r}(t)\,dt = \frac{1}{L}\int_a^b\|\mathbf{r}'(t)\|\mathbf{r}(t)\,dt,$$

where L is the arc length of C. Recalling that $\mathbf{r}(t) = \langle x(t), y(t)\rangle$,

$$\begin{aligned}\langle X, Y\rangle = \mathbf{R} &= \frac{1}{L}\int_a^b\|\mathbf{r}'(t)\|\,\langle x(t), y(t)\rangle\,dt\\ &= \left\langle\frac{1}{L}\int_a^b\|\mathbf{r}'(t)\|x(t)\,dt, \frac{1}{L}\int_a^b\|\mathbf{r}'(t)\|\,y(t)\,dt\right\rangle.\end{aligned}$$

Hence,

$$\begin{aligned}X &= \frac{1}{L}\int_a^b\|\mathbf{r}'(t)\|x(t)\,dt\\ Y &= \frac{1}{L}\int_a^b\|\mathbf{r}'(t)\|\,y(t)\,dt.\end{aligned}\tag{7}$$

The density of the rod does not appear in these equations. When the density is constant, the center of mass is sometimes called the *centroid*.

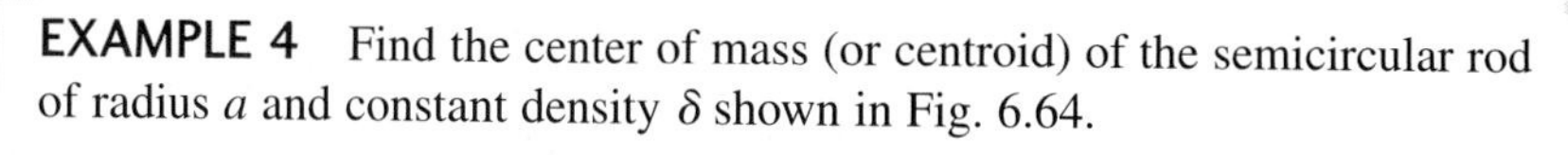

EXAMPLE 4 Find the center of mass (or centroid) of the semicircular rod of radius a and constant density δ shown in Fig. 6.64.

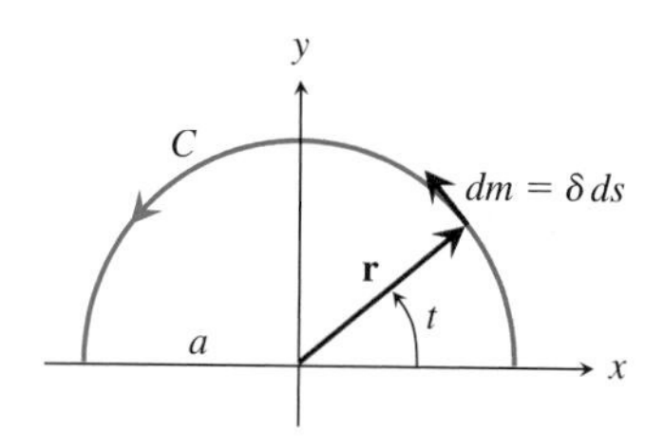

FIGURE 6.64 Semicircular rod with constant density.

Solution

The equation

$$\mathbf{r} = a\langle\cos t, \sin t\rangle, \qquad 0 \le t \le \pi$$

describes the rod. For a semicircle, the arc length part of the calculation is easy. Because

$$\|\mathbf{r}'(t)\| = \|a\langle-\sin t, \cos t\rangle\| = a\cdot 1 = a,$$

the length L of the rod is $\int_0^\pi\|\mathbf{r}'(t)\|\,dt = \pi a$. Applying (7),

$$\begin{aligned}X &= \frac{1}{L}\int_0^\pi\|\mathbf{r}'(t)\|x(t)\,dt = \frac{1}{\pi a}\int_0^\pi a^2\cos t\,dt = 0\\ Y &= \frac{1}{L}\int_0^\pi\|\mathbf{r}'(t)\|\,y(t)\,dt = \frac{1}{\pi a}\int_0^\pi a^2\sin t\,dt = \frac{2a}{\pi}.\end{aligned}$$

For a semicircular rod of radius 1, the center of mass (or centroid) would be $R = \langle 0, 2/\pi\rangle$, a point not on the rod. If the rod were embedded in a rigid piece of plastic with negligible mass and held in a horizontal plane, the balance point would be at the centroid. In future calculations, we take advantage of symmetries like that of the semicircle about the y-axis. We may, for example, infer $X = 0$ directly from the observed symmetry and thereby skip the calculation for X.

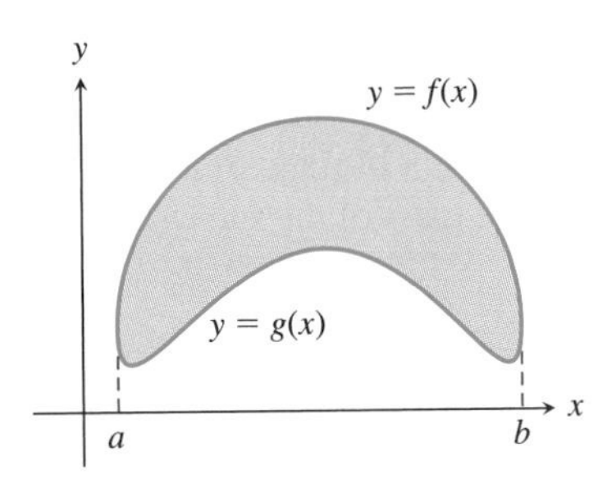

FIGURE 6.65 A representative lamina.

Two-Dimensional Continuous Mass Objects cut from flat sheets of plastic or metal, where dimensions in the plane of the sheet are large relative to the thickness of the sheet, are often called *laminas*. We consider only laminas similar to the lamina shown in Fig. 6.65, which is bounded on the top and bottom by the graphs of functions f and g. The mass of a lamina per unit area is called *planar density* and is measured in kilograms per square meter. We consider only laminas whose planar density is constant.

We start by giving the Mass Subdivision Theorem, which is needed for calculating the mass and center of mass of a lamina.

THEOREM Mass Subdivision Theorem

If a finite system S of point masses is divided into n subsystems $S_1, S_2, \ldots, S_n$, then the center of mass $\mathbf{R}$ of S is equal to the center of mass of the system

$$\{m_1, \mathbf{r}_1\}, \{m_2, \mathbf{r}_2\}, \ldots, \{m_n, \mathbf{r}_n\}, \tag{8}$$

where $m_1, m_2, \ldots, m_n$ and $\mathbf{r}_1, \mathbf{r}_2, \ldots, \mathbf{r}_n$ are the masses and centers of mass of the subsystems $S_1, S_2, \ldots, S_n$.

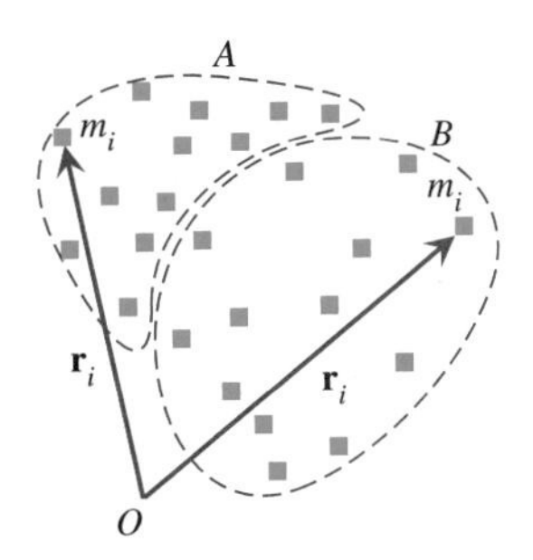

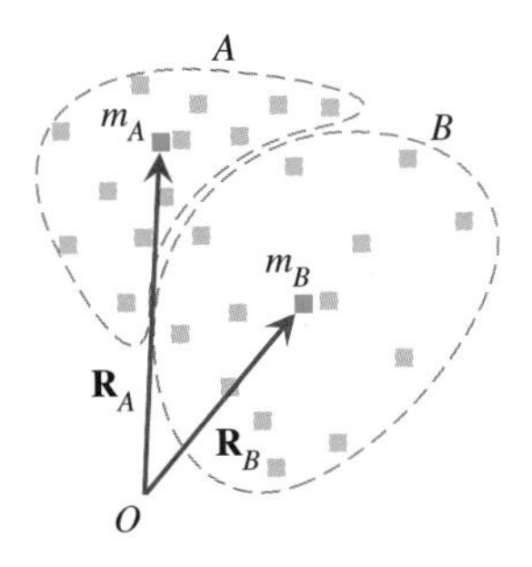

FIGURE 6.66 Mass Subdivision Theorem.

We illustrate the Mass Subdivision Theorem with a system divided into two subsystems. Figure 6.66 shows such a system S of point masses divided into subsystems A and B. If we calculate the centers of mass $\mathbf{R}_A$ and $\mathbf{R}_B$ of the two subsystems and put at these points the total masses m_A and m_B of the subsystems, we have a new system of two point masses. This is shown in the lower sketch in the figure. The Mass Subdivision Theorem states that the center of mass of this two-point mass system is the same as the center of mass of the original system.

Continuing with the case $n = 2$, the Mass Subdivision Theorem can be proved by rearranging the sums $m = \Sigma_S m_i$ and $\Sigma_S m_i \mathbf{R}_i$ for system S into two sums, one for system A and one for system B. Letting m_A and m_B be the masses of the subsystems and $\mathbf{R}$ the center of mass of the system S, and using letters S, A, or B beneath the summation symbols Σ to indicate the system over which the summation is done,

$$\begin{aligned}\mathbf{R} &= \frac{1}{m}\sum_S m_i \mathbf{R}_i = \frac{1}{m_A + m_B}\left(\sum_A m_i \mathbf{R}_i + \sum_B m_i \mathbf{R}_i\right)\\ &= \frac{1}{m_A + m_B}(m_A \mathbf{R}_A + m_B \mathbf{R}_B).\end{aligned}$$

If we divide a complex system into simpler subsystems, this formula gives a way of calculating the center of mass of the complex system from those of the subsystems. For example, the center of mass of a hammer similar to that shown in Fig. 6.67 can be calculated by replacing the hammer by the two-particle system

$$\{m_1, \mathbf{r}_1\}, \{m_2, \mathbf{r}_2\},$$

where m_1 and $\mathbf{r}_1$ are the mass and center of mass of the head, and m_2 and $\mathbf{r}_2$ are the mass and center of mass of the handle. See Exercise 35.

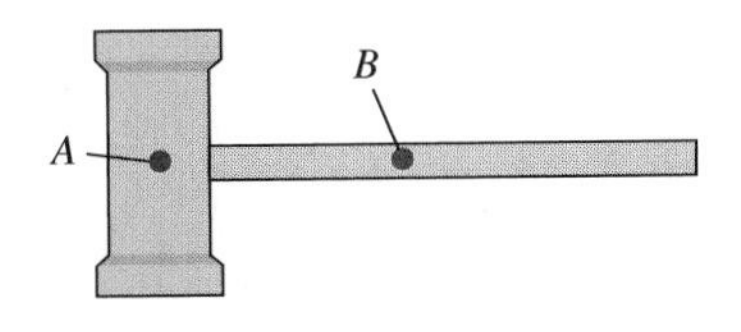

FIGURE 6.67 A hammer divided into two subsystems, head and handle.

We return to our planar lamina of constant planar density bounded on the top and bottom by the graphs of $y = f(x)$ and $y = g(x)$, $a \le x \le b$. To calculate the mass and center of mass (or centroid) of such a lamina, we divide the lamina into

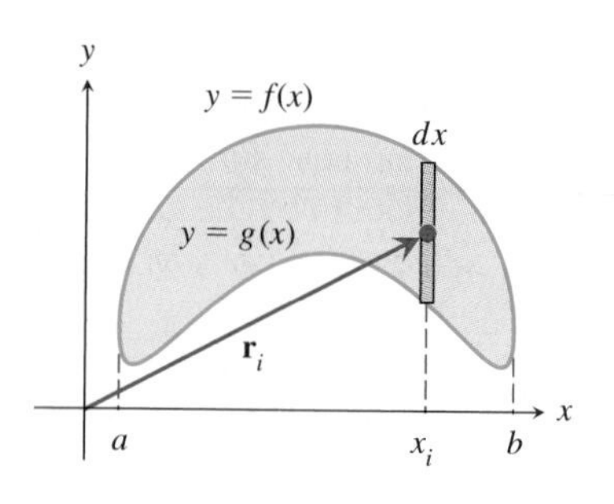

FIGURE 6.68 A representative subsystem of a lamina, with the position vector $\mathbf{r}_i$ marking its center of mass.

n subsystems corresponding to a subdivision of the interval $[a, b]$ into n equal pieces. A representative subsystem is shown in Fig. 6.68. If we regard this subsystem as a rod of length $f(x_i) - g(x_i)$ and use our earlier result that the center of mass of a rod with constant density is at its center, then the mass and center of mass of the subsystem are

$$dm_i = \delta(f(x_i) - g(x_i))\,dx \quad \text{and} \quad \mathbf{r}_i = \left\langle x_i, \tfrac{1}{2}(f(x_i) + g(x_i))\right\rangle.$$

The mass of lamina can be approximated by summing the mass elements $dm_1, \ldots, dm_n$:

$$m \approx \sum_{i=1}^{n} \delta(f(x_i) - g(x_i))\,dx.$$

Taking the limit as $n \longrightarrow \infty$,

$$m = \delta \int_a^b (f(x) - g(x))\,dx. \tag{9}$$

From the Mass Subdivision Theorem and (4), the center of mass $\mathbf{R} = \langle X, Y\rangle$ of the lamina is given by

$$X \approx \frac{1}{m} \sum_{i=1}^{n} \delta x_i (f(x_i) - g(x_i))\,dx$$

$$Y \approx \frac{1}{m} \sum_{i=1}^{n} \delta\left(\tfrac{1}{2}(f(x_i) + g(x_i))\,(f(x_i) - g(x_i))\,dx\right).$$

Taking the limit as $n \longrightarrow \infty$, recalling (9), and removing the common factor δ,

$$X = \frac{\int_a^b x(f(x) - g(x))\,dx}{\int_a^b (f(x) - g(x))\,dx} \tag{10}$$

$$Y = \frac{\frac{1}{2}\int_a^b (f(x)^2 - g(x)^2)\,dx}{\int_a^b (f(x) - g(x))\,dx}. \tag{11}$$

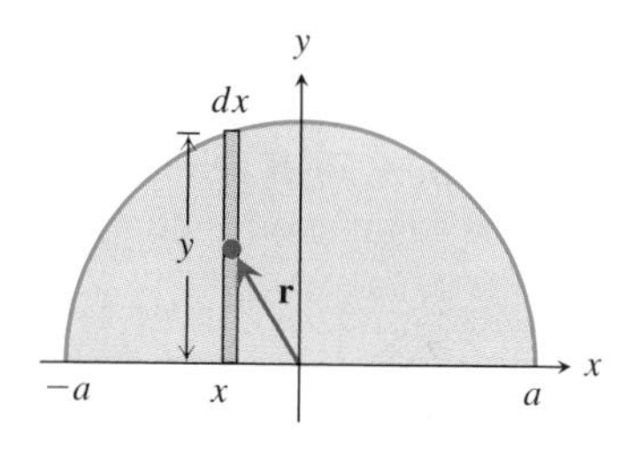

FIGURE 6.69 Semicircular lamina.

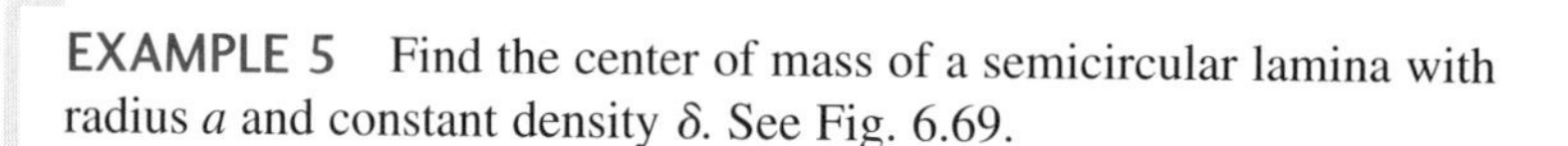

EXAMPLE 5 Find the center of mass of a semicircular lamina with radius a and constant density δ. See Fig. 6.69.

Solution

Rather than simply substitute into formulas (10) and (11), we go through a stripped-down version of the reasoning leading to these results. We apply the Mass Subdivision Theorem to the semicircular lamina after dividing it into strips corresponding to a subdivision of the interval $[-a, a]$. The point x of the subdivision locates a representative subsystem or element. The width and height of this element are dx and y, and its center of mass is at its center, located by the position vector $\mathbf{r} = \left\langle x, \frac{1}{2}y\right\rangle$. The mass dm of a representative element is $\delta y\,dx$. This is the area $y\,dx$ of the element times its density δ. Applying the Mass Subdivision Theorem and then taking a limit as the number of subdivisions becomes infinite,

$$X \approx \frac{\sum dm \cdot x}{\sum dm} = \frac{\delta \sum xy\,dx}{\delta \sum y\,dx} \longrightarrow \frac{\int_{-a}^{a} xy\,dx}{\int_{-a}^{a} y\,dx}$$

$$Y \approx \frac{\sum dm \cdot \frac{1}{2}y}{\sum dm} = \frac{\delta \sum \left(\frac{1}{2}y\right) y\,dx}{\delta \sum y\,dx} \longrightarrow \frac{\int_{-a}^{a} \frac{1}{2}y^2\,dx}{\int_{-a}^{a} y\,dx}.$$

We used the symbol $\longrightarrow$ to indicate the limiting values of these sums as the number of subdivisions becomes infinite. Substituting $y = \sqrt{a^2 - x^2}$ and noting that the integral in the denominators of X and Y is equal to the area $\pi a^2/2$ of a semicircle,

$$X = \frac{\int_{-a}^{a} x\sqrt{a^2 - x^2}\,dx}{\pi a^2/2} = 0$$

$$Y = \frac{\int_{-a}^{a} \frac{1}{2}(a^2 - x^2)\,dx}{\pi a^2/2} = \frac{4a}{3\pi} \approx 0.42a.$$

We have shown that the center of mass of a semicircle is on its line of symmetry, approximately 42 percent of the distance from the center to the curved edge. In practice, rather than calculate $X = 0$, we would infer this from the symmetry of the semicircle about the line $x = 0$.

Exercises 6.7

1. Find the center of mass of the system of point masses with masses 5, 4, 6, and 3 kilograms, with positions $x = -3, -1$, 2, and 7 meters, respectively, on an x-axis.
2. Find the center of mass of the system of point masses with masses 9, 11, 10, and 7 kilograms, with positions $x = -5$, -3, 2, and 10 meters, respectively, on an x-axis.
3. Five objects are arranged on a 10-m board with negligible mass. Starting from the left end, a 5-kg mass is 1 m, a 15-kg mass is 3 m, a 5-kg mass is 4 m, a 20-kg mass is 8 m, and a 10-kg mass is 9 m from the left end. Find the center of mass of this system.
4. Five objects are arranged on a 10-m board with negligible mass. Starting from the left end, a 12-kg mass is 1 m, an 11-kg mass is 4 m, a 7-kg mass is 5 m, a 20-kg mass is 7 m, and a 13-kg mass is 9 m from the left end. Find the center of mass of this system.
5. Find the center of mass **R** of the system of point masses with masses 4, 6, 7, 8, and 1 kg, located by position vectors $\mathbf{r} = \langle -3, 5\rangle, \langle -5, -6\rangle, \langle 5, -1\rangle, \langle 1, 1\rangle$, and $\langle 6, 6\rangle$, respectively. Lengths are in meters.
6. Find the center of mass **R** of the system of point masses with masses 4.5, 3.7, 5.5, 7.7, and 1.5 kg, located by position vectors $\mathbf{r} = \langle -2.4, 5.1\rangle, \langle -4.9, -5.8\rangle, \langle 4.2, -1.0\rangle$, $\langle 1.3, 1.4\rangle$, and $\langle 7.2, 6.9\rangle$, respectively. Lengths are in meters.
7. Find the mass and center of mass of a 0.5-m rod whose density varies linearly from 3.0 kg/m to 3.7 kg/m.
8. Find the mass and center of mass of a 0.7-m rod whose density varies linearly from 2.6 kg/m to 2.9 kg/m.
9. Assuming that the center of mass of a square lamina with uniform density is at its center, use the Mass Subdivision Theorem to locate the center of mass of the lamina shown in the accompanying figure. The edge of the larger square is twice that of the smaller square. Place the origin at the lower left corner of the larger square and assume that the larger square has side $2a$.

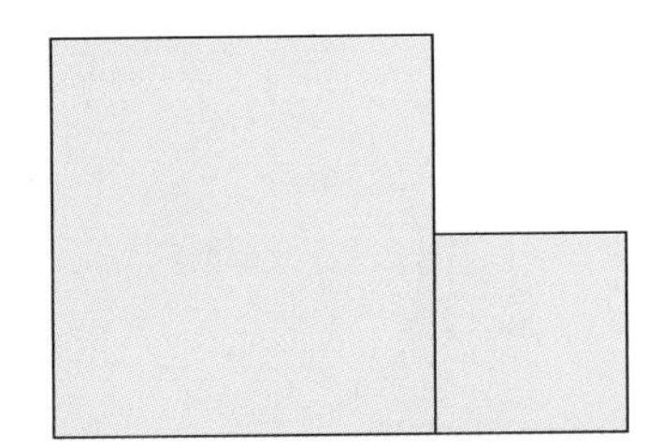

Figure for Exercise 9.

10. Assuming that the center of mass of a triangle with constant density is at the intersection of its medians (lines drawn from a vertex to the midpoint of the opposite side) and that of a square with constant density is at its center, use the Mass Subdivision Theorem to locate the center of mass of the lamina shown in the accompanying figure. Place the origin at the right angle of the isoceles triangle and assume the square has side a.

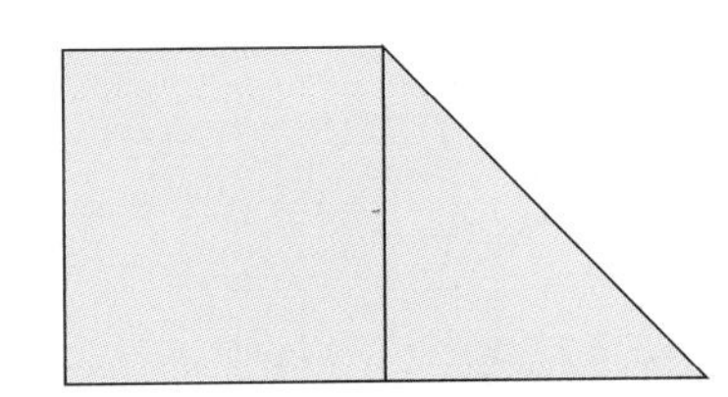

Figure for Exercise 10.

11. Calculate the mass and center of mass of the lamina of constant density δ and bounded by $x = 0$, $x = 1$, $y = 0$, and $y = e^x$.

12. Calculate the mass and center of mass of the lamina of constant density δ and bounded by $x = 0$, $x = \pi$, $y = 0$, and $y = \sin x$.

13. Show that the mass and center of mass of the "second-degree parabolic spandrel" (a shape used to reinforce a joint between beams; it is welded into place) bounded by $x = 0$, $x = b$, $y = h - hx^2/b^2$, and $y = h$ are $m = \delta bh/3$ and $\langle 3b/4, 7h/10\rangle$. Assume that the planar density is δ kg/m^2.

14. Show that the mass and center of mass of the "third-degree parabolic spandrel" (a shape used to reinforce a joint between beams; it is welded into place) bounded by $x = 0$, $x = b$, $y = h - hx^3/b^3$, and $y = h$ are $m = \delta bh/4$ and $\langle 4b/5, 5h/7\rangle$. Assume that the planar density is δ kg/m^2.

15. The accompanying figure shows a sector with central angle θ in a circle of radius a. Show that the centroid of the sector is

$$\mathbf{R} = \frac{2a}{3\theta}\langle \sin\theta, 1 - \cos\theta\rangle.$$

Use this result to show that in a sector of radius r and small central angle $d\theta$, the centroid is $\frac{2}{3}r$ from the vertex of the central angle.

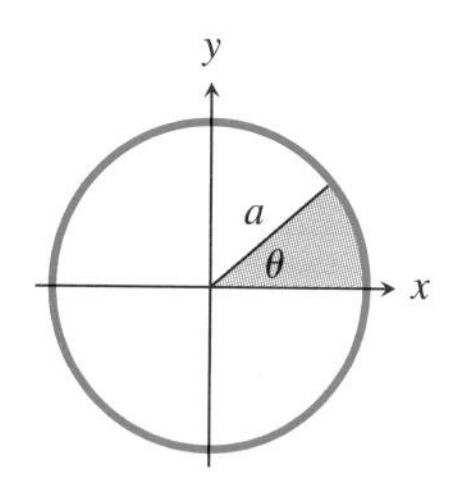

Figure for Exercise 15.

T 16. The accompanying figure shows a segment of height h in a circle of radius a. *Note:* $0 \le h \le 2a$. The area of the shaded region is

$$A = \tfrac{1}{2}\pi a^2 - a^2\arcsin(1 - h/a) - (a - h)\sqrt{h(2a - h)},\ 0 \le h \le 2a.$$

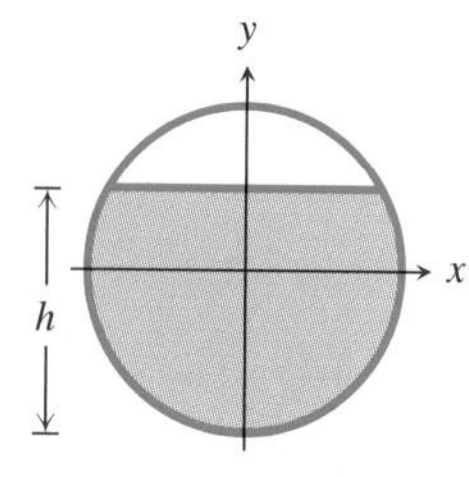

Figure for Exercise 16.

Show that the centroid of the segment is

$$\left\langle 0, -\frac{2}{3A}(h(2a - h))^{3/2}\right\rangle.$$

If $a = 1$ and $h = 1$, the centroid is $\mathbf{R} = \langle 0, -4/(3\pi)\rangle$.

Exercises 17–20: It follows from Exercise 15 that the centroid of the polar element of area $dA = \frac{1}{2}r^2\,d\theta$ has position vector $\frac{2}{3}r\langle\cos\theta, \sin\theta\rangle$. Use this to calculate the centroid. Express your final answer in polar coordinates.

17. The outline of a cam is described by $r = 6.0 - 4.0\cos\theta$, $0 \le \theta \le 2\pi$. Find the coordinates of its centroid.

18. The outline of a cam is described by $r = 7.0 - 4.5\cos\theta$, $0 \le \theta \le 2\pi$. Find the coordinates of its centroid.

19. A machine part is bounded by the graph of $r = \theta$, $0 \le \theta \le \pi$, and a segment of the ray $\theta = \pi$. Find the coordinates of its centroid.

20. A machine part is bounded by the graph of $r = \frac{3}{4}\theta$, $0 \le \theta \le \pi$, and a segment of the ray $\theta = \pi$. Find the coordinates of its centroid.

21. Determine the center of mass of the lamina shown in the figure. The parabola on the right is described by the equation

$$y = a - b(x - 9)^2, \qquad 0 \le x \le 18,$$

where $a = 26$ and $b = 26/81$. The parabola on the left is congruent to the one on the right. Assume that the units are kilograms and meters and that the lamina has constant planar density δ.

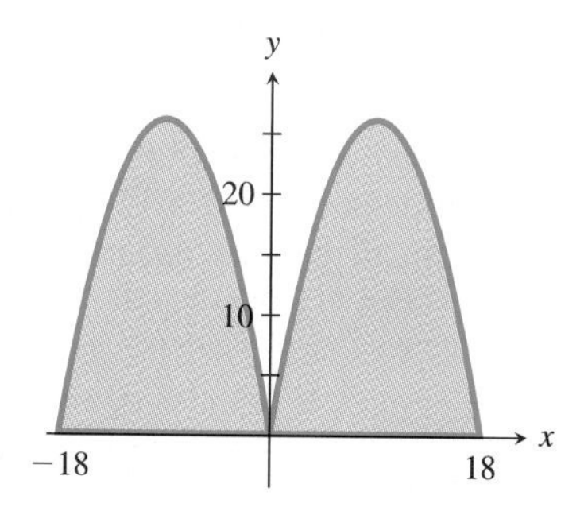

Figure for Exercise 21.

22. Find the centroid of the lamina bounded by the graphs of $y = x$ and $y = x^2 - 3x$.

23. Find the centroid of the lamina bounded by $x = 0$, $x = \pi/2$, $y = 1$, and $y = \sin x$.

24. Find the centroid of the lamina bounded by $x = 0$, $y = 0$, and the first-quadrant part of the ellipse with equation $x^2/a^2 + y^2/b^2 = 1$.

25. Find the centroid of the infinite lamina bounded by $x = 0$, $y = 0$, and $y = e^{-x}$.

26. Find the mass and centroid of the infinite lamina bounded by $x = 1$, $y = 0$, and $y = x^{-3}$.

27. Show that the centroid of a triangle with vertices $(0,0)$, $(0,b)$ and (a,c), where $b, a > 0$, is at the point where the medians of the triangle meet. Recall that a median is a line joining a vertex and the midpoint of the opposite side.

28. A circular lamina with diameter 3 cm has a hole of diameter 1 cm stamped from its center. It is then cut along a diameter to form a half-washer. Use the result of Example 5 in calculating the centroid of the half-washer.

29. Describe in detail how the Mass Subdivision Theorem may be used in determining the center of mass of a lamina formed by punching a circular hole through a circular lamina.

30. How would you go about locating the center of mass of a rod 1 meter in length, the first third of which has density 1 kg/m and the remainder density 2 kg/m?

31. The idea of a rod of variable density can be used to calculate the center of mass of a tapered flagpole made from a material of constant density. Find the mass and center of mass of a 10-m aluminum flagpole, which tapers uniformly from a diameter of 15 cm to 10 cm. Assume the pole is a tube whose walls are 3 cm thick. The density of aluminum is 2702 kg/m^3.

32. Determine the centroid of a rod in the shape of a parabola described by $y = x^2$, $-1 \leq x \leq 1$. If you use the parametric description $\mathbf{r} = \langle x, x^2 \rangle$, you will encounter an integral of the form $\int x^2\sqrt{1 + 4x^2}\,dx$. Use the substitution $2x = \tan u$ and then formula (31) from the table of integrals. Numerical integration is also a possibility.

Ⓣ **33.** Determine the centroid of the cycloid curve described by

$$\mathbf{r} = \langle t - \sin t, 1 - \cos t \rangle, \qquad 0 \leq t \leq 2\pi.$$

The identities $1 - \cos t = 2\sin^2\frac{1}{2}t$, $\sin t = 2\sin\frac{1}{2}t\cos\frac{1}{2}t$, and $\cos t = 1 - 2\sin^2\frac{1}{2}t$ will be useful, together with formula (28) from the table of integrals. Numerical integration is also a possibility.

34. Determine the center of mass of the curved rod forming the rightmost arch of the lamina described in Exercise 21. Assume that the units are kilograms and meters and that the rods have constant linear density δ. See Exercise 32 for help with the integration.

35. Determine the center of mass of a hammer. The steel head is a rectangular parallelepiped, with length 14 cm and a 7-cm-by-7-cm cross section; its density is $\delta = 7800$ kg/m^3. The wooden handle is a cylinder of diameter 2.5 cm and height 35 cm; its density is $\delta = 600$ kg/m^3. Assume that the handle passes through the steel head, and one of its ends is flush with one of the 7-cm-by-14-cm faces.

6.8 Curvature, Acceleration, and Kepler's Second Law

The title of this chapter is "Applications of the Integral." This section, however, is only indirectly an application of integration. We derive Kepler's second law, that the line from the sun to a planet sweeps out equal areas in equal times. The derivation depends upon knowing how to calculate the area of regions described in terms of polar coordinates, which we discussed in Section 6.3. The idea of curvature, a measure of how much a curve bends per unit of arc length, depends upon a knowledge of arc length. We discussed arc length in Section 6.4. Finally, we show how the acceleration vector of an object in motion may be resolved into tangential and normal acceleration vectors. These formulas are useful in the study of motion and the design of machinery. The formulas we derive depend upon the curvature of the path and, hence, depend upon arc length.

Curvature

Figure 6.70 shows a smooth curve C with unit tangent vectors $\mathbf{T}$ drawn at equally spaced points along C. "Equally spaced points along C" means that the length of the curve C between successive points is Δs, a constant. Starting from the point at the

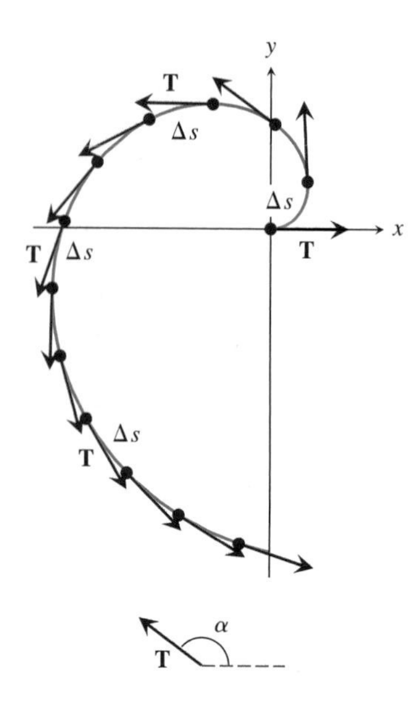

FIGURE 6.70 A curve whose curvature varies smoothly from relatively large near the origin to relatively small at its far end.

origin, it is clear that the directions of the unit tangent vectors are increasing more between successive points near the origin than at the far end of the curve. For example, the directions of the first two unit tangents differ by a little more than $\pi/2$, while near the end of the curve the directions of successive unit tangents differ by much less than $\pi/2$. At any point of C, the absolute value of the rate of change of the direction of the unit tangent vector with respect to arc length is called the *curvature*.

To calculate the curvature at a point of C we recall that the direction of the unit tangent vector $\mathbf{T}$ is the angle α between $\mathbf{T}$ and the positive x-axis, measured counterclockwise from the positive x-axis. See the diagram at the bottom of Fig. 6.70. Because $\mathbf{T}$ is a unit vector,

$$\mathbf{T} = \langle \cos \alpha, \sin \alpha \rangle. \tag{1}$$

To find $|d\alpha/ds|$, which is the curvature, we start with the change $(1/\Delta s)\,\Delta\mathbf{T}$ in the unit tangent per unit change in the arc length and write this in terms of the change $\Delta\alpha$ in α. Thus,

$$\frac{1}{\Delta s}\Delta\mathbf{T} = \frac{\Delta\alpha}{\Delta s}\frac{1}{\Delta\alpha}\Delta\mathbf{T}.$$

The limit of $(1/\Delta s)\Delta\mathbf{T}$ as $\Delta s \to 0$ is

$$\frac{d\mathbf{T}}{ds} = \frac{d\alpha}{ds}\frac{d\mathbf{T}}{d\alpha}.$$

Noting from (1) that

$$\frac{d\mathbf{T}}{d\alpha} = \frac{d}{d\alpha}\langle \cos \alpha, \sin \alpha \rangle = \langle -\sin \alpha, \cos \alpha \rangle, \tag{2}$$

we find

$$\frac{d\mathbf{T}}{ds} = \frac{d\alpha}{ds}\frac{d\mathbf{T}}{d\alpha} = \frac{d\alpha}{ds}\langle -\sin \alpha, \cos \alpha \rangle. \tag{3}$$

Taking the length of both sides of this equation,

$$\left\|\frac{d\mathbf{T}}{ds}\right\| = \left|\frac{d\alpha}{ds}\right|.$$

This result gives us a way of calculating the curvature $|d\alpha/ds|$ at points of C.

Before formally defining curvature, we show that the vector $d\mathbf{T}/ds$ is perpendicular to the unit tangent vector $\mathbf{T}$. This follows by dotting both sides of (3) by $\mathbf{T}$:

$$\begin{aligned}\mathbf{T} \cdot \frac{d\mathbf{T}}{ds} &= \mathbf{T} \cdot \left(\frac{d\alpha}{ds}\frac{d\mathbf{T}}{d\alpha}\right) = \frac{d\alpha}{ds}\left(\mathbf{T} \cdot \frac{d\mathbf{T}}{d\alpha}\right) \\ &= \frac{d\alpha}{ds}(\langle \cos \alpha, \sin \alpha \rangle \cdot \langle -\sin \alpha, \cos \alpha \rangle) = 0.\end{aligned} \tag{4}$$

A vector perpendicular to a tangent vector to a curve is often called a *normal* vector. In this context, the word "normal" is a synonym for "perpendicular."

DEFINITION Curvature Vector, Curvature, and Unit Normal

Let C be a smooth curve described by

$$\mathbf{r} = \mathbf{r}(t) = \langle x(t), y(t)\rangle, \qquad a \le t \le b.$$

Assume that the functions $x = x(t)$ and $y = y(t)$ have continuous second derivatives on $[a, b]$. If $\mathbf{r}'(t) \ne \mathbf{0}$ at the point $\mathbf{r}(t)$ of C and $\mathbf{T}$ is the unit tangent vector at this point, the curvature vector to C at $\mathbf{r}(t)$ is

$$\frac{d\mathbf{T}}{ds}. \tag{5}$$

The curvature $\kappa(t)$ and unit normal $\mathbf{N}(t)$ at $\mathbf{r}(t)$ are

$$\kappa(t) = \left\|\frac{d\mathbf{T}}{ds}\right\| \quad \text{and} \quad \mathbf{N}(t) = \frac{1}{\kappa(t)}\frac{d\mathbf{T}}{ds}. \tag{6}$$

The reciprocal of $\kappa(t)$ is called the **radius of curvature** of C at $\mathbf{r}(t)$.

EXAMPLE 1 Calculate the curvature vector, the curvature, the unit normal, and the radius of curvature at a representative point of the circle of radius a described by

$$\mathbf{r}(\theta) = a\langle\cos\theta, \sin\theta\rangle, \qquad 0 \le \theta \le 2\pi.$$

Solution

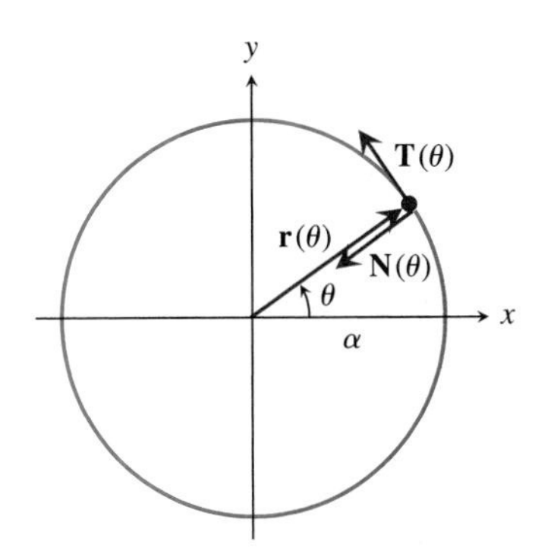

FIGURE 6.71 The unit tangent and unit normal vectors for a circle of radius a.

Figure 6.71 shows the graph of the circle and a representative point $\mathbf{r}(\theta)$. The curvature of a circle is easily found because we know (or can easily calculate) that the arc length s and the parameter θ are related by the equation $s = a\theta$. We'll use this fact after a few preliminary calculations. First, we calculate $\mathbf{r}'(\theta)$.

$$\mathbf{r}'(\theta) = a\langle -\sin\theta, \cos\theta\rangle.$$

Dividing $\mathbf{r}'(\theta)$ by its length a gives the unit tangent vector:

$$\mathbf{T}(\theta) = \langle -\sin\theta, \cos\theta\rangle.$$

Next we calculate the curvature vector $d\mathbf{T}/ds$ with the help of the chain rule and the equation $s = a\theta$ mentioned earlier.

$$\frac{d\mathbf{T}}{ds} = \frac{d\theta}{ds}\frac{d\mathbf{T}}{d\theta} = \frac{1}{ds/d\theta}\frac{d\mathbf{T}}{d\theta} = \frac{1}{a}\frac{d}{d\theta}\langle -\sin\theta, \cos\theta\rangle.$$

Hence,

$$\frac{d\mathbf{T}}{ds} = \left(\frac{1}{a}\right)\langle -\cos\theta, -\sin\theta\rangle.$$

The length of $d\mathbf{T}/ds$ is the curvature:

$$\kappa(\theta) = \left\|\frac{d\mathbf{T}}{ds}\right\| = \frac{1}{a}.$$

As we would expect, the curvature of a circle is the same everywhere. The radius of curvature, which is the reciprocal of the curvature, is a, the radius of the circle. The normal vector is

$$\mathbf{N}(\theta) = \frac{1}{\kappa(\theta)}\frac{d\mathbf{T}}{ds} = \frac{1}{1/a}(1/a)\langle -\cos\theta, -\sin\theta\rangle.$$

The vectors $\mathbf{r}(\theta)$, $\mathbf{T}(\theta)$, and $\mathbf{N}(\theta)$ are shown in the figure. As we proved in (4), the vectors $\mathbf{T}(\theta)$ and $\mathbf{N}(\theta)$ are perpendicular.

In the next example you will notice that it takes some effort to calculate the curvature or unit normal for even a relatively simple curve. Alternative methods are discussed after the example.

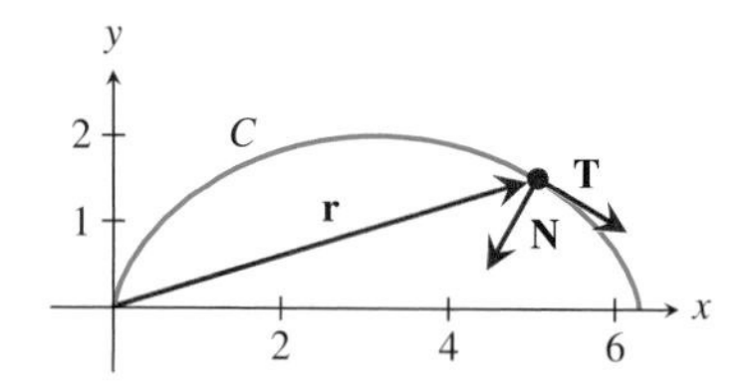

FIGURE 6.72 The unit tangent and unit normal vectors at the point $\mathbf{r}(4\pi/3)$ of a cycloid.

EXAMPLE 2 Figure 6.72 shows the cycloid described by

$$\mathbf{r}(t) = \langle t - \sin t, 1 - \cos t\rangle, \qquad 0 \le t \le 2\pi.$$

Calculate the curvature and unit normal of this cycloid at $\mathbf{r}(4\pi/3)$.

Solution

From Examples 1 and 4 in Section 6.4, where we calculated the arc length and unit tangent vector of this cycloid,

$$\|\mathbf{r}'(t)\| = \frac{ds}{dt} = \sqrt{x'(t)^2 + y'(t)^2} = 2\sin\tfrac{1}{2}t \tag{7}$$

$$\mathbf{T} = \left\langle \sin\tfrac{1}{2}t, \cos\tfrac{1}{2}t\right\rangle, \qquad t \ne 0, 2\pi. \tag{8}$$

From (5) and (6), the curvature vector and curvature are $d\mathbf{T}/ds$ and $\kappa = \|d\mathbf{T}/ds\|$. We calculate $d\mathbf{T}/ds$ first, using (7) and (8). The chain rule is needed because the curvature vector is $d\mathbf{T}/ds$ and $\mathbf{T}$ is given in terms of t.

$$\begin{aligned}\frac{d\mathbf{T}}{ds} &= \frac{dt}{ds}\frac{d\mathbf{T}}{dt} = \frac{1}{ds/dt}\frac{d}{dt}\left\langle \sin\tfrac{1}{2}t, \cos\tfrac{1}{2}t\right\rangle \\ \frac{d\mathbf{T}}{ds} &= \frac{1}{2\sin\frac{1}{2}t}\left\langle \tfrac{1}{2}\cos\tfrac{1}{2}t, -\tfrac{1}{2}\sin\tfrac{1}{2}t\right\rangle.\end{aligned} \tag{9}$$

Because the length of $d\mathbf{T}/ds$ is the curvature κ,

$$\kappa = \frac{1}{4\sin\frac{1}{2}t}\sqrt{\cos^2\tfrac{1}{2}t + \sin^2\tfrac{1}{2}t} = \frac{1}{4\sin\frac{1}{2}t}. \tag{10}$$

Dividing $d\mathbf{T}/ds$ by its length κ gives the unit normal $\mathbf{N}(t)$. From (9) and (10),

$$\mathbf{N}(t) = \left\langle \cos\tfrac{1}{2}t, -\sin\tfrac{1}{2}t\right\rangle. \tag{11}$$

At the point $\mathbf{r}(\frac{4}{3}\pi)$, the curvature and unit normal are

$$\kappa = \frac{1}{4\sin\left(\frac{1}{2}\left(\frac{4}{3}\pi\right)\right)} = \frac{1}{4\left(\sqrt{3}/2\right)} \approx 0.29$$

and

$$\mathbf{N}\left(\tfrac{4}{3}\pi\right) = \left\langle \cos\left(\tfrac{1}{2}\left(\tfrac{4}{3}\pi\right)\right), -\sin\left(\tfrac{1}{2}\left(\tfrac{4}{3}\pi\right)\right)\right\rangle = \left\langle -1/2, -\sqrt{3}/2\right\rangle \approx \langle -0.5, -0.87\rangle.$$

Figure 6.72 shows the unit tangent vector $\mathbf{T}(4\pi/3)$ and the unit normal vector $\mathbf{N}$ to the cycloid at the point $\mathbf{r}(4\pi/3)$. Note that $\mathbf{N}$ points to the "inside" of the curve, the direction in which the curve C turns away from its tangent vector.

Before giving a third example, we give two formulas for curvature.

Formulas for Curvature

Let C be a smooth curve described by

$$\mathbf{r} = \mathbf{r}(t) = \langle x(t), y(t)\rangle, \qquad a \le t \le b.$$

If $\mathbf{r}'(t) \neq \mathbf{0} = \langle 0, 0\rangle$ at $\mathbf{r}(t)$, then the curvature at $\mathbf{r}(t)$ is

$$\kappa(t) = \frac{|x''(t)y'(t) - x'(t)y''(t)|}{(x'(t)^2 + y'(t)^2)^{3/2}}. \tag{12}$$

If C is the graph of a function f defined on $[a, b]$, so that C can be described by

$$\mathbf{r}(x) = \langle x, f(x)\rangle, \qquad a \le x \le b,$$

then the curvature at $\mathbf{r}(x)$ is

$$\kappa(x) = \frac{|f''(x)|}{(1 + f'(x)^2)^{3/2}} = \frac{|y''|}{(1 + y'^2)^{3/2}}. \tag{13}$$

Equation (13) is a special case of the formula (12) for curvature. If C can be described as the graph of a function f, or an equation $y = f(x)$, defined for $a \le x \le b$, then a vector description of C is

$$\mathbf{r} = \mathbf{r}(t) = \langle t, f(t)\rangle, \qquad a \le t \le b.$$

Applying (12),

$$\kappa(t) = \frac{|x''(t)y'(t) - x'(t)y''(t)|}{(x'(t)^2 + y'(t)^2)^{3/2}} = \frac{|0 \cdot f'(t) - 1 \cdot f''(t)|}{(1^2 + f'(t)^2)^{3/2}} = \frac{|f''(t)|}{(1 + f'(t)^2)^{3/2}}.$$

If we replace t by x and simplify, Equation (13) follows. The second form given in (13) uses the dependent variable y instead of function notation.

We outline a proof of (12) in Exercise 32.

EXAMPLE 3 Calculate the unit tangent at the point $(1/2, 1/4)$ on the graph C of the equation $y = x^2$. Calculate the radius of curvature at the points $(1/2, 1/4)$ and $(1, 1)$ of C. Sketch the curve and, at $(1/2, 1/4)$, the tangent vector to the curve.

Solution

We describe C by the vector equation $\mathbf{r}(x) = \langle x, x^2 \rangle$. From $\mathbf{r}'(x) = \langle 1, 2x \rangle$, the unit tangent vector is

$$\mathbf{T} = \frac{1}{\|\mathbf{r}'(x)\|}\mathbf{r}'(x) = \frac{1}{\sqrt{1 + 4x^2}}\langle 1, 2x \rangle.$$

At the point $(1/2, 1/4)$, where $x = 1/2$, the unit tangent vector is

$$\mathbf{T} = \frac{1}{\sqrt{1 + 1}}\langle 1, 1 \rangle = \left\langle \frac{1}{\sqrt{2}}, \frac{1}{\sqrt{2}} \right\rangle.$$

We use (13) to calculate the curvature at the points $(1/2, 1/4)$ and $(1, 1)$ of C. From the equation $y = x^2$, $y' = 2x$ and $y'' = 2$; hence,

$$\kappa = \frac{|y''|}{(1 + y'^2)^{3/2}} = \frac{2}{(1 + 4x^2)^{3/2}}.$$

At $x = 1/2$ and $x = 1$,

$$\kappa(1/2) = \frac{2}{2^{3/2}} = \frac{1}{\sqrt{2}}$$

$$\kappa(1) = \frac{2}{5^{3/2}} = \frac{2}{5\sqrt{5}}.$$

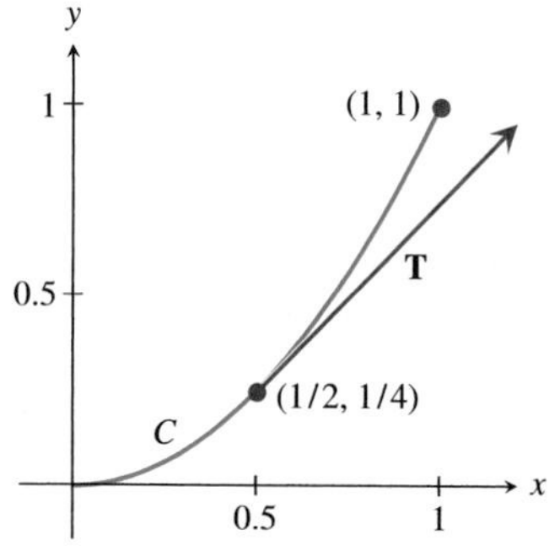

At the point $(1/2, 1/4)$, the radius of curvature is $1/\kappa(1/2) = \sqrt{2} \approx 1.4$; at $(1, 1)$ the radius of curvature is $1/\kappa(1) = 5\sqrt{5}/2 \approx 5.6$. Because the curvature is becoming smaller (or the radius of curvature is becoming larger) as the point moves away from the origin, the curve is becoming flatter.

Figure 6.73 shows the curve C and the unit tangent vector $\mathbf{T}(1/2)$.

FIGURE 6.73 Unit tangent and unit normal vectors.

Acceleration Vectors

The excitement of a roller coaster depends on changes in either the tangential or normal acceleration vectors. Going down a straight piece of track, the acceleration vector points in the tangential direction; on circular turns, it points in the normal direction, towards the center of the circle. On most wooden coasters riders feel these alternately, on different sections of the track. Steel coasters include sections on which riders feel both accelerations simultaneously. Coaster engineers make the turns tighter toward the end of the ride. This increases the magnitude of the normal acceleration, which helps maintain excitement as the coaster loses speed.

In analyzing motion, it is often useful to separate the net force acting on an object (or, through Newton's second law $\mathbf{F} = m\mathbf{a}$, the acceleration) into forces in the tangential and normal directions. The calculation of these acceleration vectors becomes easier if we give two differentiation formulas for vector functions. These formulas are easy to remember because they are very much like the formula for differentiating a product of scalar functions, that is $(fg)' = f'g + fg'$.

Differentiation Formulas

Suppose that $\mathbf{f}$ and $\mathbf{g}$ are vector functions and h a scalar function, all defined and differentiable on $[a, b]$. The derivative of the vector function $h\mathbf{f}$ is

$$(h\mathbf{f})' = h'\mathbf{f} + h\mathbf{f}'. \tag{14}$$

The derivative of the scalar function $\mathbf{f} \cdot \mathbf{g}$ is

$$(\mathbf{f} \cdot \mathbf{g})' = \mathbf{f}' \cdot \mathbf{g} + \mathbf{f} \cdot \mathbf{g}'. \tag{15}$$

We showed earlier that the unit tangent and unit normal vectors are perpendicular. A different proof can be based on formula (15). See Exercise 30.

Tangential and Normal Acceleration The motion of an object on a curve C is often described by a vector equation of the form

$$\mathbf{r} = \mathbf{r}(t) = \langle x(t), y(t) \rangle, \qquad a \le t \le b.$$

The vector $\mathbf{r}(t)$ gives the position of the object at time t. The velocity, speed, and acceleration of the object are

$$\mathbf{v} = \mathbf{v}(t) = \mathbf{r}'(t), \qquad v = v(t) = \|\mathbf{r}'(t)\|, \qquad \mathbf{a} = \mathbf{a}(t) = \mathbf{v}'(t).$$

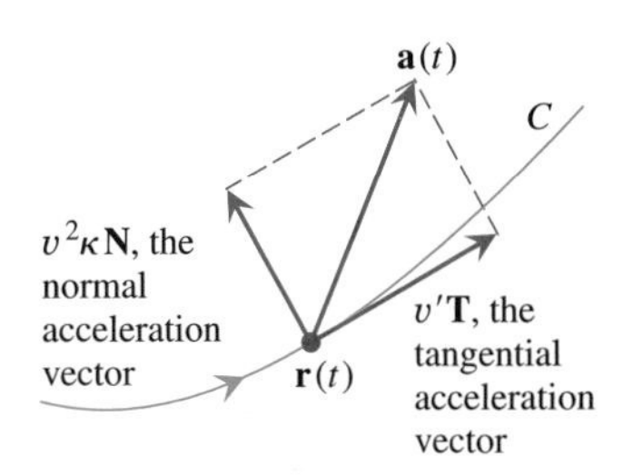

FIGURE 6.74 The projections of the acceleration vector $\mathbf{a}$ on the unit tangent vector $\mathbf{T}$ and unit normal vector $\mathbf{N}$.

Figure 6.74 shows the projections $v'\mathbf{T}$ and $v^2\kappa\mathbf{N}$ of the acceleration vector $\mathbf{a}$ on the unit tangent vector $\mathbf{T}$ and unit normal vector $\mathbf{N}$. These projections, called the *tangential* and *normal acceleration vectors,* are easily derived using (14) and the fact that the velocity vector is the speed times the unit tangent vector.

$$\mathbf{a} = \mathbf{v}' = (v\mathbf{T})' = v'\mathbf{T} + v\mathbf{T}' = v'\mathbf{T} + v\frac{d\mathbf{T}}{ds}\frac{ds}{dt}.$$

Noting that $v = \|\mathbf{r}'(t)\| = ds/dt$ and $\mathbf{N} = (1/\kappa)\, d\mathbf{T}/ds$,

$$\mathbf{a} = v'\mathbf{T} + v^2\kappa\mathbf{N}. \tag{16}$$

This equation expresses the acceleration vector $\mathbf{a}$ as a sum of the tangential and normal acceleration vectors $v'\mathbf{T}$ and $v^2\kappa\mathbf{N}$.

EXAMPLE 4 Calculate the tangential and normal acceleration vectors at time $t = 4\pi/3$ of a molecule of rubber on the circumference of a wheel rolling on the x-axis. Specifically, assume that the molecule's position $\mathbf{r}(t)$ at time t is

$$\mathbf{r}(t) = \langle t - \sin t, 1 - \cos t \rangle, \qquad 0 \le t \le 2\pi.$$

Solution

Most of the work needed to calculate the tangential and normal acceleration vectors was done in Example 2, where we studied the cycloid C on which

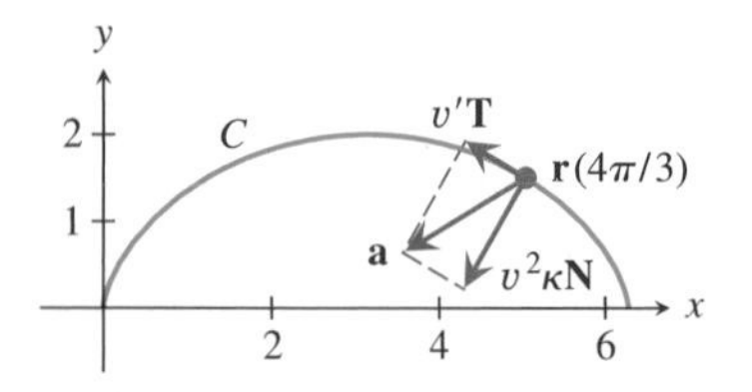

FIGURE 6.75 The acceleration, tangential acceleration, and normal acceleration vectors of a cycloidal motion, all at time $t = 4\pi/3 \approx 4.2$.

the particle is moving. Figure 6.75 shows the curve C. From (7), (8), (10), and (11),

$$v = 2\sin\tfrac{1}{2}t, \tag{17}$$

$$\mathbf{T} = \left\langle \sin\tfrac{1}{2}t, \cos\tfrac{1}{2}t \right\rangle, \tag{18}$$

$$\kappa = \frac{1}{4\sin\frac{1}{2}t}, \tag{19}$$

$$\mathbf{N} = \left\langle \cos\tfrac{1}{2}t, -\sin\tfrac{1}{2}t \right\rangle. \tag{20}$$

From (17) and (18), the tangential acceleration vector is

$$v'\mathbf{T} = \cos\tfrac{1}{2}t\left\langle \sin\tfrac{1}{2}t, \cos\tfrac{1}{2}t \right\rangle.$$

From (19) and (20), the normal acceleration vector is

$$v^2\kappa\mathbf{N} = 4\sin^2\tfrac{1}{2}t\frac{1}{4\sin\frac{1}{2}t}\mathbf{N} = \sin\tfrac{1}{2}t\left\langle \cos\tfrac{1}{2}t, -\sin\tfrac{1}{2}t \right\rangle.$$

Evaluating these vectors at $t = 4\pi/3$, the tangential acceleration vector at $\mathbf{r}(4\pi/3)$ is

$$\begin{aligned} v'\mathbf{T} &= \cos(2\pi/3)\langle \sin(2\pi/3), \cos(2\pi/3)\rangle \\ &= -\frac{1}{2}\left\langle \frac{\sqrt{3}}{2}, -\frac{1}{2} \right\rangle = \left\langle -\frac{\sqrt{3}}{4}, \frac{1}{4} \right\rangle \end{aligned}$$

and the normal acceleration vector is

$$\begin{aligned} v^2\kappa\mathbf{N} &= \sin(2\pi/3)\langle \cos(2\pi/3), -\sin(2\pi/3)\rangle \\ &= \frac{\sqrt{3}}{2}\left\langle -\frac{1}{2}, -\frac{\sqrt{3}}{2} \right\rangle = \left\langle -\frac{\sqrt{3}}{4}, -\frac{3}{4} \right\rangle. \end{aligned}$$

Figure 6.75 shows these vectors at the point $\mathbf{r}(4\pi/3)$.

Radial and Transverse Acceleration The motions of objects like planets or satellites, which are subject to a *central force field,* are usually described in polar coordinates, with the polar origin at the sun or Earth. The central force acting on the planet or, through Newton's second law, the acceleration experienced by the planet, is in the direction of the line connecting the sun and planet, which is called the *radial direction.* If we wish to adjust the orbit of a satellite by firing a jet, we may wish to resolve the satellite's acceleration into radial and transverse acceleration vectors.

We assume that the motion of an object on a curve C is described by an equation in the polar form

$$\mathbf{r} = \mathbf{r}(t) = r(t)\langle \cos\theta(t), \sin\theta(t)\rangle, \qquad a \le t \le b, \tag{21}$$

where the magnitude $r(t)$ and the direction $\theta(t)$ of $\mathbf{r}(t)$ are functions of t.

The radial and transverse directions are the directions of the unit vectors

$$\mathbf{e}_r = \langle \cos\theta(t), \sin\theta(t)\rangle \quad \text{and} \quad \mathbf{e}_\theta = \langle -\sin\theta(t), \cos\theta(t)\rangle, \tag{22}$$

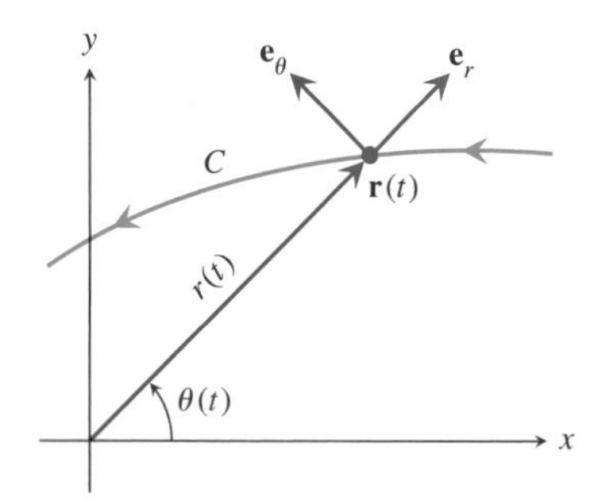

FIGURE 6.76 The vectors $\mathbf{e}_r$ and $\mathbf{e}_\theta$, based at $\mathbf{r}(t)$.

respectively. These vectors are based at $\mathbf{r}(t)$. See Fig. 6.76. The vector $\mathbf{e}_r$ points in the direction of $\mathbf{r}(t)$, which is the **radial direction;** $\mathbf{e}_\theta$ points in the direction of increasing θ, which is the **transverse direction.** The vectors $\mathbf{e}_r$ and $\mathbf{e}_\theta$ are perpendicular.

This follows from

$$\begin{aligned}\mathbf{e}_r \cdot \mathbf{e}_\theta &= \langle \cos\theta, \sin\theta\rangle \cdot \langle -\sin\theta, \cos\theta\rangle \\ &= (\cos\theta)(-\sin\theta) + (\sin\theta)(\cos\theta) = 0.\end{aligned}$$

We will need the derivatives of the vectors $\mathbf{e}_r$ and $\mathbf{e}_\theta$ in our calculations. We shorten $\theta(t)$ to θ, but keep in mind that θ is a function of t. The derivatives $\mathbf{e}_r'$ and $\mathbf{e}_\theta'$ are

$$\begin{aligned}\mathbf{e}_r' &= \langle \cos\theta, \sin\theta\rangle' = \langle(-\sin\theta)\theta', (\cos\theta)\theta'\rangle = \theta'\mathbf{e}_\theta \\ \mathbf{e}_\theta' &= \langle -\sin\theta, \cos\theta\rangle' = \langle(-\cos\theta)\theta', (-\sin\theta)\theta'\rangle = -\theta'\mathbf{e}_r.\end{aligned} \tag{23}$$

With these results, expressing $\mathbf{a}$ in terms of the radial and transverse accelerations takes just a few steps. We first note that

$$\mathbf{r} = \mathbf{r}(t) = r(t)\mathbf{e}_r.$$

Differentiating $\mathbf{r} = r\mathbf{e}_r$ twice with respect to t to get acceleration, using (22), (23), and the differentiation formula (14) as necessary,

$$\begin{aligned}\mathbf{r} &= r\mathbf{e}_r \\ \mathbf{v} = \mathbf{r}' &= r'\mathbf{e}_r + r\mathbf{e}_r' = r'\mathbf{e}_r + r\theta'\mathbf{e}_\theta \\ \mathbf{a} = \mathbf{v}' &= r''\mathbf{e}_r + r'\theta'\mathbf{e}_\theta + (r\theta'' + r'\theta')\mathbf{e}_\theta + r\theta'(-\theta'\mathbf{e}_r) \\ \mathbf{a} &= (r'' - r\theta'^2)\mathbf{e}_r + (2r'\theta' + r\theta'')\mathbf{e}_\theta.\end{aligned} \tag{24}$$

Equation (24) expresses the acceleration vector $\mathbf{a}$ as a sum of the radial and transverse acceleration vectors $(r'' - r\theta'^2)\mathbf{e}_r$ and $(2r'\theta' + r\theta'')\mathbf{e}_\theta$.

We calculate the radial and transverse acceleration vectors for an object moving on a spiral.

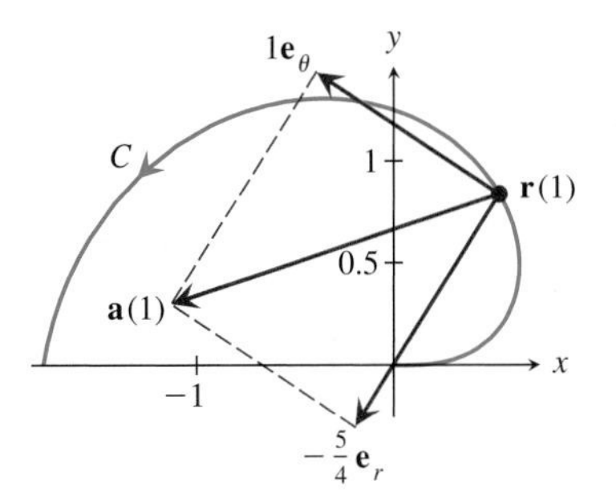

FIGURE 6.77 The acceleration vector and the radial and transverse acceleration vectors at $\mathbf{r}(1)$.

EXAMPLE 5 A coast guard vessel is moving on the search spiral described by

$$\mathbf{r} = \mathbf{r}(t) = \left\langle \sqrt{t}\cos t, \sqrt{t}\sin t\right\rangle, \qquad t \geq 0. \tag{25}$$

See Fig. 6.77. Calculate and sketch its acceleration vector and radial and transverse acceleration vectors at $t = 1$.

Solution

To apply (24), we express the curve in polar form. Comparing (21) and (25), we see that $r(t) = \sqrt{t}$ and $\theta(t) = t$. We calculate the first and second derivatives of $r(t)$ and $\theta(t)$.

$$\begin{aligned} r(t) &= \sqrt{t} = t^{1/2} & \theta(t) &= t \\ r'(t) &= \tfrac{1}{2}t^{-1/2} & \theta'(t) &= 1 \\ r''(t) &= -\tfrac{1}{4}t^{-3/2} & \theta''(t) &= 0.\end{aligned}$$

It now follows from (24) that

$$\begin{aligned}\mathbf{a} &= (r'' - r\theta'^2)\mathbf{e}_r + (2r'\theta' + r\theta'')\mathbf{e}_\theta \\ &= \left(-\tfrac{1}{4}t^{-3/2} - t^{1/2}\right)\mathbf{e}_r + 2\left(\tfrac{1}{2}t^{-1/2}\right)\mathbf{e}_\theta.\end{aligned}$$

For $t = 1$,

$$\mathbf{a} = -\tfrac{5}{4}\mathbf{e}_r + 1\mathbf{e}_\theta.$$

For $t = 1$ and $\theta = 1$ the vectors $\mathbf{e}_r$ and $\mathbf{e}_\theta$ are

$$\mathbf{e}_r = \langle \cos\theta, \sin\theta \rangle = \langle \cos 1, \sin 1 \rangle \approx \langle 0.540, 0.841 \rangle$$
$$\mathbf{e}_\theta = \langle -\sin\theta, \cos\theta \rangle = \langle -\sin 1, \cos 1 \rangle \approx \langle -0.841, 0.540 \rangle.$$

Hence, the radial and transverse acceleration vectors are

$$-\tfrac{5}{4}\mathbf{e}_r \approx \langle -0.675, -1.051 \rangle \quad \text{and} \quad 1\mathbf{e}_\theta = \langle -0.841, 0.540 \rangle.$$

Figure 6.77 shows the acceleration, radial acceleration, and transverse acceleration vectors at $t = 1$.

Kepler's Second Law

Johannes Kepler (1571–1630) inferred his three laws of planetary motion from astronomical data. Kepler's laws are

1. The orbit of each planet is an ellipse with the sun at one focus of the ellipse.
2. The line from the sun to a planet sweeps out equal areas in equal times.
3. The square of the orbital period of a planet is proportional to the cube of its semimajor axis.

Kepler's laws suggest that polar coordinates may be a natural choice for describing planetary motion. Indeed, if we place the sun at the pole or origin of a coordinate system, the elliptical orbits of the planets can be described by (relatively) simple equations. We use our earlier work with the polar area element and our discussion of radial and transverse acceleration vectors in inferring Kepler's second law.

Figure 6.78 shows a representative elliptical orbit, with the sun at one focus. Actual planetary orbits are much more circular than the one shown. Indeed, this ellipse has eccentricity 0.6, which is larger even than that of Pluto, which is the most eccentric of the known planets. The eccentricity of Earth's orbit is 0.017, which means that the orbit is very nearly circular.

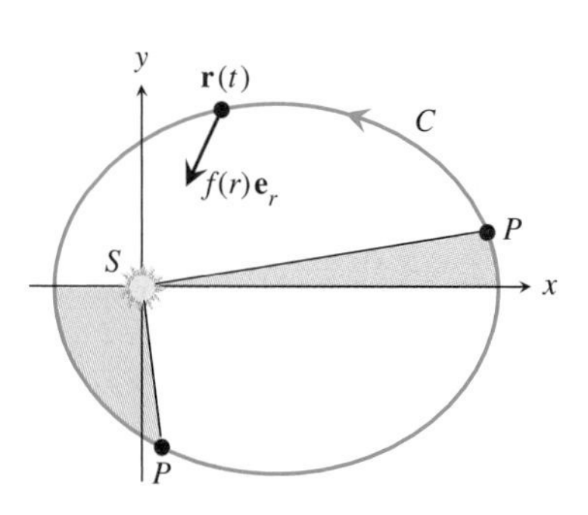

FIGURE 6.78 Kepler's second law.

If a planetary year were divided into, say, 12 equal months, then, according to Kepler's second law, a line drawn from sun to planet would sweep out $1/12$ of the area of the entire ellipse each month. Figure 6.78 shows the regions corresponding to two months. The shaded region on the right side of the figure is swept out during the month just after aphelion (the point on the orbit at which the planet is farthest from the sun and is moving with the least speed). The shaded area on the left is swept out during the month just after perihelion (the point on the orbit at which the planet is closest to the sun and is moving with the greatest speed). The points P in Fig. 6.78 mark the positions of the planet at the ends of these months. The areas of the two regions are equal.

We show that Kepler's second law follows from the assumption that the sole force on a planet is the gravitational force of the sun. We assume that the motion of the planet is described by the parametric equation

$$\mathbf{r} = \mathbf{r}(t) = r(t)\langle \cos\theta(t), \sin\theta(t) \rangle, \qquad t \geq 0,$$

and the force $\mathbf{F}$ exerted by the sun at $\mathbf{r}(t)$ is entirely in the radial direction, that is,

$$\mathbf{F} = \mathbf{F}(\mathbf{r}(t)) = f(r)\mathbf{e}_r.$$

The force $\mathbf{F}$ is related to the acceleration $\mathbf{a}$ by Newton's second law, $\mathbf{F} = m\mathbf{a}$. Replacing $\mathbf{F}$ by $f(r)\mathbf{e}_r$ and writing the acceleration as the sum of the radial and transverse accelerations,

$$f(r)\mathbf{e}_r = m\mathbf{a} = m(r'' - r\theta'^2)\mathbf{e}_r + m(2r'\theta' + r\theta'')\mathbf{e}_\theta. \tag{26}$$

Because the left side of this equation has no $\mathbf{e}_\theta$ term it follows that

$$m(2r'\theta' + r\theta'') = 0 \tag{27}$$

for all t. (Also see Exercise 19.) We rewrite this expression as a derivative by dividing by $2m$ and multiplying by r. This gives

$$0 = rr'\theta' + \tfrac{1}{2}r^2\theta'' = \frac{d}{dt}\left(\tfrac{1}{2}r^2\frac{d\theta}{dt}\right). \tag{28}$$

Recalling that the area element in polar coordinates is $dA = \frac{1}{2}r^2\,d\theta$, which may be written as

$$\frac{dA}{dt} = \tfrac{1}{2}r^2\frac{d\theta}{dt},$$

it follows from (28) that $dA/dt = C$, a constant, because from (28) the derivative of dA/dt is 0 for all t. Hence, between any times t_0 and $t_0 + \Delta t$ the area swept out by the line joining the sun and any planet is

$$\int_{t_0}^{t_0+\Delta t} dA = \int_{t_0}^{t_0+\Delta t} \frac{dA}{dt}\,dt = \int_{t_0}^{t_0+\Delta t} C\,dt = C\,\Delta t,$$

a constant. Note that the area $C\,\Delta t$ swept out between times t_0 and $t_0 + \Delta t$ is dependent only upon the duration Δt of the sweep, not the time t_0 at which it began. Hence, equal areas are swept out in equal times, which is Kepler's second law.

Exercises 6.8

Exercises 1–12: For the described curve, calculate the unit tangent **T**, *unit normal* **N**, *and curvature* κ *at the given point. Sketch the curve and, at the given point, its unit tangent and unit normal. Approximately where on the curve is the curvature the greatest? The least?*

1. $\mathbf{r} = \langle 2t, t^2\rangle$, $-1 \le t \le 2$; $\mathbf{r}(0)$

2. $\mathbf{r} = \left\langle \frac{1}{2}t^2, t\right\rangle$, $-1 \le t \le 2$; $\mathbf{r}(0)$

3. $\mathbf{r} = \langle 3\cos t, 2\sin t\rangle$, $0 \le t \le 2\pi$; $\mathbf{r}(\pi/4)$

4. $\mathbf{r} = \langle \cos t, 2\sin t\rangle$, $0 \le t \le 2\pi$; $\mathbf{r}(\pi/6)$

5. $y = x^3$, $0 \le x \le 2$; $(1, 1)$

6. $y = 1/x$, $1/2 \le x \le 2$; $(1, 1)$

7. The polar curve described by $r = 4\cos\theta$, $0 \le \theta \le \pi/2$; $\left(2\sqrt{2}, \pi/4\right)$

8. The polar curve described by $r = e^\theta$, $0 \le \theta \le \pi/2$; $(3, \ln 3)$

9. The graph of $y = \sin x$, $0 \le x \le \pi/2$; $(\pi/6, 1/2)$

10. The graph of $y = \cos x$, $0 \le x \le \pi/2$; $\left(\pi/4, 1/\sqrt{2}\right)$

11. The graph of $y = e^x$, $0 \le x \le 1.5$; $(1, e)$

12. The graph of $y = \ln x$, $1 \le x \le e^2$; $(e, 1)$

Ⓣ **13.** The motion of an object is described by

$$\mathbf{r} = \mathbf{r}(t) = \langle t - 1, 2t - t^2\rangle, \qquad t \ge 0.$$

Show that the tangential acceleration and normal acceleration vectors are

$$\frac{4(t-1)}{\sqrt{4(1-t)^2 + 1}}\mathbf{T}.$$

and

$$\frac{2}{\sqrt{4(1-t)^2 + 1}}\mathbf{N}.$$

Ⓣ **14.** The motion of an object is given by

$$\mathbf{r} = \mathbf{r}(t) = \langle \cos e^t, \sin e^t\rangle, \qquad t \ge 0.$$

Show that the tangential acceleration and normal acceleration vectors are $e^t\mathbf{T}$ and $e^{2t}\mathbf{N}$.

15. Show that the radial and transverse acceleration vectors at $\mathbf{r}(\pi/18)$ for the motion described by

$$\mathbf{r}(t) = \sin(3t)\langle\cos(3t), \sin(3t)\rangle, \qquad 0 \le t \le \pi/3,$$

are $\langle -9\sqrt{3}/2, -9/2\rangle$ and $\langle -9\sqrt{3}/2, 27/2\rangle$. Sketch the curve and these two vectors. What is the acceleration at this point?

16. Show that the radial and transverse acceleration vectors at $\mathbf{r}(\pi/2)$ for the motion described by

$$\mathbf{r}(t) = e^t\langle\cos(2t), \sin(2t)\rangle, \qquad 0 \le t \le \pi,$$

are $\langle 3e^{\pi/2}, 0\rangle$ and $\langle 0, -4e^{\pi/2}\rangle$. Sketch the curve and these two vectors. What is the acceleration at this point?

17. On the curve described by $f(x) = e^x$, where is the curvature a maximum? What is the maximum curvature?

18. On the curve described by $f(x) = \arctan(x)$, where is the curvature a maximum? What is the maximum curvature?

19. Show that (27) follows from (26) by dotting both sides of the latter equation by $\mathbf{e}_\theta$.

20. A polar curve C is described by $r = r(\theta)$. Show that the curvature $\kappa = \kappa(\theta)$ of C at (r, θ) is given by

$$\kappa(\theta) = \frac{|r^2 + 2r'^2 - rr''|}{(r^2 + r'^2)^{3/2}}.$$

Exercises 21–24: Use the result of Exercise 20 to find the curvature of the curve described by the polar equation.

21. $r = 2^\theta$

22. $r = 3^{-2\theta}$

23. $r = \theta^2,\ \theta > 0$

24. $r = \theta^3,\ \theta > 0$

Exercises 25–28: The radius of curvature is used in fitting the circle of "best fit" to a point of a curve. This circle is called the osculating circle *because it fits the curve well (it "kisses" the curve). At a point $\mathbf{r}(t_0)$ of a curve described by*

$$\mathbf{r} = \mathbf{r}(t), \qquad a \le t \le b,$$

the center of the osculating circle is the point

$$\mathbf{c}(t_0) = \mathbf{r}(t_0) + \frac{1}{\kappa(t_0)}\mathbf{N}(t_0).$$

The osculating circle can be described by

$$\mathbf{f}(\theta) = \mathbf{c}(t_0) + \frac{1}{\kappa(t_0)}\langle\cos\theta, \sin\theta\rangle,$$

where $0 \le \theta \le 2\pi$. For the curve described by $\mathbf{r} = \mathbf{r}(t)$, $a \le t \le b$, graph together the curve and the osculating circle at $\mathbf{r}(t_0)$.

25. $\mathbf{r}(t) = \langle t, t^2\rangle,\ -2 \le t \le 2;\ \mathbf{r}(t_0) = \langle 0, 0\rangle$

26. $\mathbf{r}(t) = \left\langle t, \frac{1}{2}t^2\right\rangle,\ -2 \le t \le 2;\ \mathbf{r}(t_0) = \langle 1, 1/2\rangle$

27. $\mathbf{r}(t) = \langle t, \sin t\rangle,\ 0 \le t \le \pi;\ \mathbf{r}(t_0) = \langle \pi/2, 1\rangle$

28. $\mathbf{r}(t) = \langle t, \cos t\rangle,\ -\pi/2 \le t \le \pi/2;\ \mathbf{r}(t_0) = \langle 0, 1\rangle$

29. A template is to be cut from a square piece of sheet metal, two feet on each side. We coordinatize the sheet with a coordinate system with origin at the center of the sheet and axes perpendicular to the sides. We draw the curve with equation $y = x^2$ on the sheet. We are to remove the metal on the inside of the parabola by making a single pass with a milling machine. The cutting tool of the milling machine is a cylindrical bit with vertical axis and cutting edges on its lateral surface. The vertical axis of the bit will be guided along a curve A lying in the plane of the sheet metal and inside the parabola. We must choose a single bit of radius r for the job. Find r and describe the curve A with a parametric equation.

30. Use formula (15) to prove that the unit tangent and unit normal vectors are perpendicular. *Hint:* First justify and then differentiate the equation $\mathbf{T}(t) \cdot \mathbf{T}(t) = 1$.

31. Use formula (15) in showing that for a differentiable vector function $\mathbf{r} = \mathbf{r}(t)$ defined on an interval $[a, b]$,

$$rr' = \mathbf{r}' \cdot \mathbf{r},$$

where $r^2 = \mathbf{r} \cdot \mathbf{r} = \|\mathbf{r}\|^2$.

32. Starting with

$$\mathbf{r} = \mathbf{r}(t) = \langle x(t), y(t)\rangle, \qquad a \le t \le b,$$

derive the curvature formula (12). The calculation is straightforward but on the long side. You will save time if you simplify and check your calculations as you go.

33. The polar equation of the ellipse in Fig. 6.78 is $r = 1/(1 - 0.6\cos\theta, 0 \le \theta \le 2\pi)$. Determine the rectangular coordinates of the point P furthest the x-axis.

34. Show that formula (14) is true. *Hint:* Two vector functions are equal if their coordinate functions are equal.

35. Show that formula (15) is true. *Hint:* Two vector functions are equal if their coordinate functions are equal.

6.9 Improper Integrals

Improper Integrals of the First Kind

An integral $\int_a^b f(x)\,dx$ is a *proper integral* if f is continuous and the range of integration $[a, b]$ is finite. These are the integrals we have been studying. If the interval of integration has the form $[a, \infty)$, the integral is an *improper integral* of the first kind.

To motivate the definition of this kind of integral, we investigate three instances of the question as to whether a region of infinite extent can have finite area.

INVESTIGATION 1

Can Regions of Infinite Extent Have Finite Area?

A region of the plane has "infinite extent" if it cannot be contained in a circle centered at the origin. We ask: Can a region of infinite extent have finite area and, if so, how can we calculate its area? Figure 6.79 shows region $\mathcal{R}_1$ and $\mathcal{R}_2$, which we suppose to extend indefinitely to the right. Does either region have finite area?

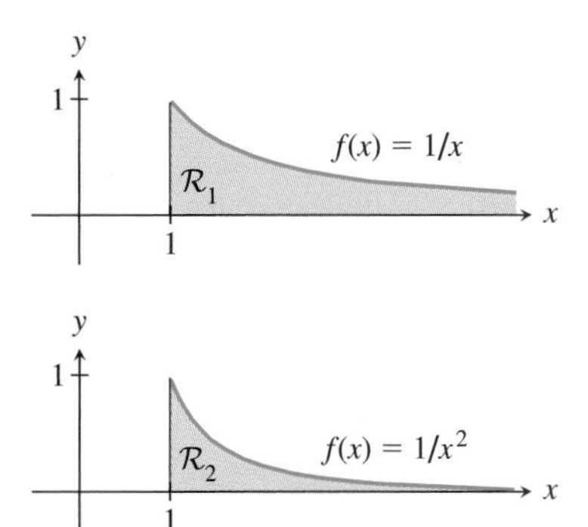

FIGURE 6.79 Two regions of infinite extent.

Figure 6.80 shows regions $\mathcal{R}_1(b)$ and $\mathcal{R}_2(b)$ of finite extent. These regions were obtained by removing from regions $\mathcal{R}_1$ and $\mathcal{R}_2$ everything to the right of $x = b$. The regions $\mathcal{R}_1(b)$ and $\mathcal{R}_2(b)$ come closer to "filling out" the regions $\mathcal{R}_1$ and $\mathcal{R}_2$ as $b \to \infty$. We approach the question of whether the regions $\mathcal{R}_1$ and $\mathcal{R}_2$ have finite area by calculating the areas $A_1(b)$ and $A_2(b)$ of $\mathcal{R}_1(b)$ and $\mathcal{R}_2(b)$ and then letting $b \to \infty$. Because regions $\mathcal{R}_1(b)$ and $\mathcal{R}_2(b)$ have finite extent, we can calculate their areas with the proper integrals

$$A_1(b) = \int_1^b \frac{dx}{x^1} = \ln x \Big|_1^b = \ln b - \ln 1 = \ln b \tag{1}$$

$$A_2(b) = \int_1^b \frac{dx}{x^2} = \frac{x^{-1}}{-1}\Big|_1^b = 1 - \frac{1}{b} \tag{2}$$

Because

$$\lim_{b\to\infty} A_1(b) = \infty \quad \text{and} \quad \lim_{b\to\infty} A_2(b) = 1,$$

we say that region $\mathcal{R}_1$ has infinite area and $\mathcal{R}_2$ has area of 1 square unit. It appears that the graph of $f(x) = 1/x^2$ approaches its asymptote $y = 0$ quickly enough that the area beneath its graph on the interval $[1, \infty)$ is finite.

Figure 6.81 shows a third region $\mathcal{R}$ having infinite extent. Also shown is the region $\mathcal{R}(b)$ of finite extent obtained by removing the part of region $\mathcal{R}$ to the right of $x = b$. These regions are bounded above by the function

$$f(x) = e^{-x^2/2}, \qquad 0 \le x < \infty. \tag{3}$$

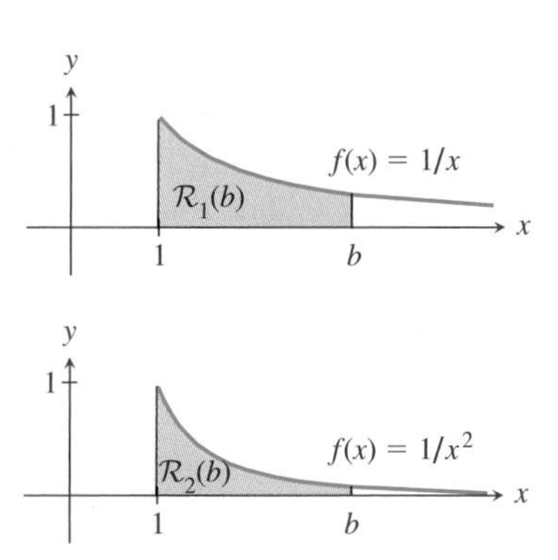

FIGURE 6.80 Two regions of finite extent.

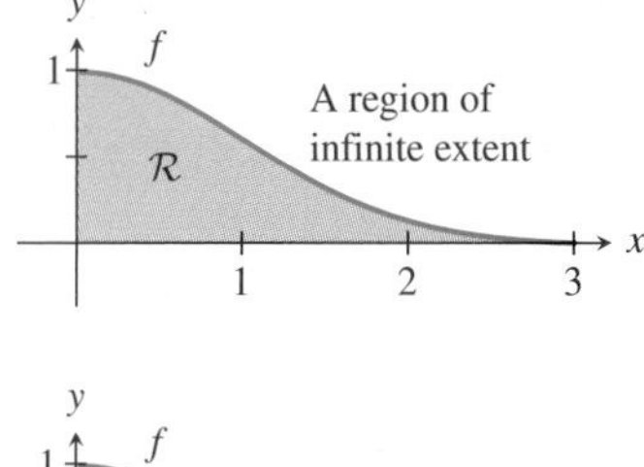

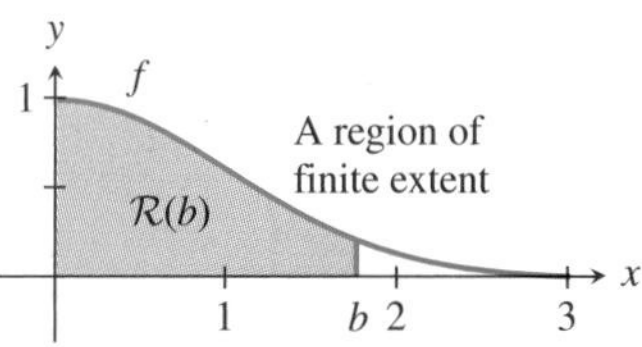

FIGURE 6.81 Regions $\mathcal{R}$ and $\mathcal{R}(b)$ of finite and infinite extent.

Because the function defined in (3) has no antiderivative expressible in terms of elementary functions, we can't calculate $A(b)$ in terms of b and then take a limit as we did for the functions $1/x$ and $1/x^2$. Instead, we compare f to a "test function." For this particular f, we may argue that because

$$\lim_{x\to\infty} \frac{x^2}{e^{x^2/2}} = 0,$$

which comes from Equation (6) in Section 4.4, it must be true that this fraction is eventually less than 1. This means that for large x,

$$\frac{x^2}{e^{x^2/2}} < 1,$$

from which it follows that

$$e^{-x^2/2} < \frac{1}{x^2} \tag{4}$$

for all sufficiently large x. Because we know from (2) that the area of the region beneath the "tail" of the graph of $1/x^2$ is finite, it follows from inequality (4) that the area of the region beneath the tail of the graph of $e^{-x^2/2}$ is also finite.

Figure 6.82 shows the relation between the graphs of the functions $g(x) = 1/x^2$ and $f(x) = e^{-x^2/2}$. It happens that the graph of the exponential lies beneath the graph of the power function $1/x^2$ for all $x > 0$, not just "large x."

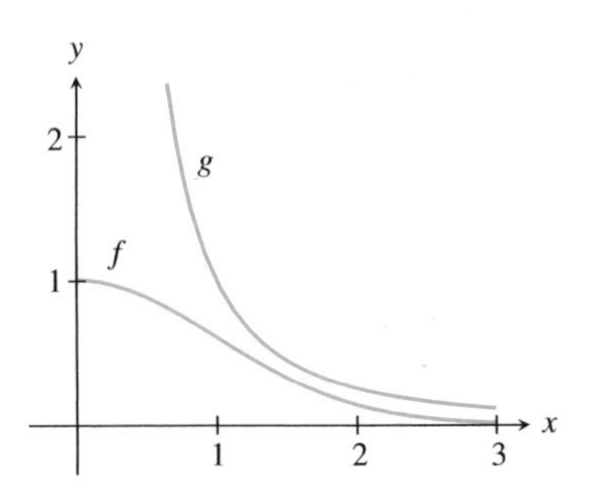

FIGURE 6.82 Estimating the area beneath the graph of f by comparing it to the area beneath the graph of a test function g.

DEFINITION Improper Integrals of the First Kind

If f is continuous on the infinite interval $[a,\infty)$, the improper integral $\int_a^\infty f(x)\,dx$ of the first kind is defined to be

$$\int_a^\infty f(x)\,dx = \lim_{b\to\infty} \int_a^b f(x)\,dx$$

when the limit is a number L. In this case we write $\int_a^\infty f(x)\,dx = L$ and say that the improper integral is *convergent* (or converges to L). If the limit is ∞ (or $-\infty$), we write $\int_a^\infty f(x)\,dx = \infty$ (or $-\infty$) and say that the improper integral is *divergent* (or diverges to infinity or to negative infinity). When the limit is neither finite nor $\pm\infty$, we say that the improper integral is *divergent* (diverges).

We give several examples of improper integrals. Our intent is not only to illustrate the previous definition, but to work informally with the test functions $1/x$ and $1/x^2$ discussed earlier. After the examples, we formalize the use of test functions in Comparison Theorem I.

EXAMPLE 1 Does the improper integral

$$\int_0^\infty \frac{x}{1+x^2}\,dx \tag{5}$$

converge or diverge? If it converges, calculate its value.

Solution

Figure 6.83 shows the graph of the integrand of $f(x) = x/(1+x^2)$. Also shown is the graph of $g(x) = 1/(1+x)$. The function g is of interest because (i) on the interval $[1, \infty)$ the graph of f lies above the graph of g and (ii) the area of the region beneath the graph of g is infinite. If we can show that (i) and (ii) hold, it would follow that $\int_0^\infty f(x)\,dx$ diverges. Here's the (somewhat rough) reasoning: Because $f(x) \geq 0$ on $[0, 1]$ and $f(x) \geq g(x)$ on $[1, \infty)$,

$$\int_0^\infty f(x)\,dx \geq \int_1^\infty f(x)\,dx \geq \int_1^\infty g(x)\,dx = \infty.$$

y
1
0.5
g(x) = 1/(1 + x)
f(x) = x/(1 + x²)
1 2 4 6 x

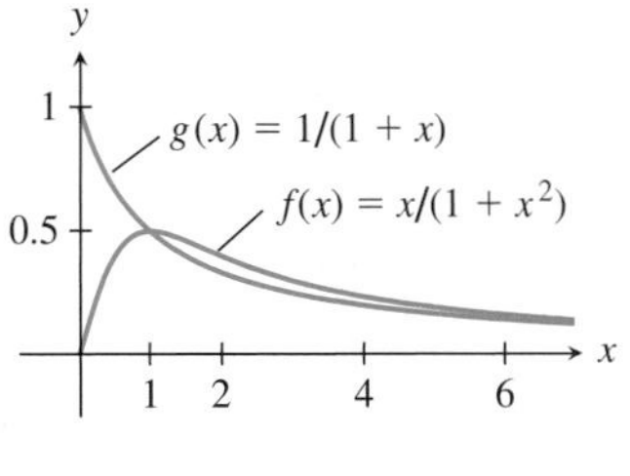

FIGURE 6.83 The graphs of $x/(1+x^2)$ and $1/(1+x)$.

Hence, $\int_0^\infty f(x)\,dx = \infty$ and so diverges. To support this reasoning we must show that (i) and (ii) are true. The first of these follows from

$$f(x) = \frac{x}{1+x^2} = \frac{\frac{x}{x}}{\frac{1+x^2}{x}} = \frac{1}{\frac{1}{x}+x} \geq \frac{1}{1+x} = g(x), \qquad x \geq 1.$$

The truth of statement (ii) follows from

$$\int_1^\infty g(x)\,dx = \int_1^\infty \frac{dx}{1+x}\,dx = \lim_{b\to\infty}\int_1^b \frac{dx}{1+x}\,dx = \lim_{b\to\infty} \ln(1+x)\Big|_1^b = \infty.$$

This argument that the improper integral (5) diverges is somewhat indirect. A more direct argument is based on the substitution $u = 1 + x^2$. Noting that $du = 2x\,dx$,

$$\int_0^b \frac{x}{1+x^2}\,dx = \int_1^{1+b^2} \frac{du}{2u}.$$

Hence,

$$\lim_{b\to\infty}\int_0^b \frac{x}{1+x^2}\,dx = \lim_{b\to\infty}\int_1^{1+b^2} \frac{du}{2u} = \lim_{b\to\infty} \tfrac{1}{2}\ln(1+b^2) = \infty.$$

Hence, the improper integral (5) diverges to infinity.

EXAMPLE 2 Does the improper integral

$$\int_0^\infty \frac{dx}{1+x^2} \tag{6}$$

converge or diverge? If it converges, calculate its value.

Solution

y
2
1.5
1
0.5
g(x) = 1/x²
f(x) = 1/(1 + x²)
2 4 6 x

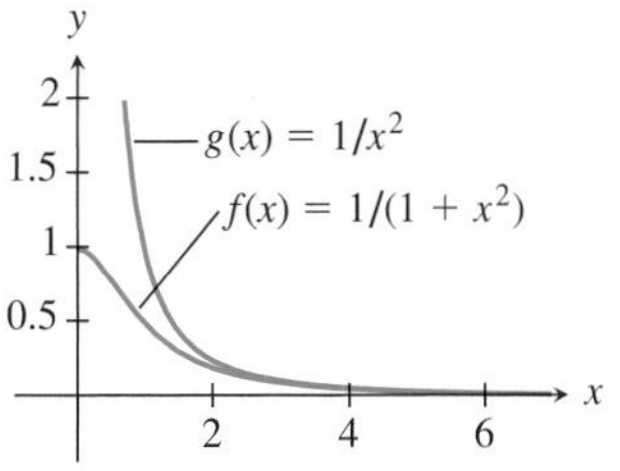

FIGURE 6.84 The graphs of $1/(1+x^2)$ and $1/x^2$.

Figure 6.84 shows the graph of the integrand of $f(x) = 1/(1+x^2)$. Also shown is the graph of $g(x) = 1/x^2$. We argue that $\int_0^\infty f(x)\,dx$ converges because (i) $f(x) \leq g(x)$ on $[1, \infty)$ and (ii) $\int_1^\infty g(x)\,dx = 1$. If we can show that (i) and (ii) hold, it would follow that $\int_0^\infty f(x)\,dx$ converges. Here's the (somewhat rough) reasoning: From Fig. 6.84, $\int_0^1 f(x)\,dx \leq 1$; hence,

$$\int_0^\infty f(x)\,dx = \int_0^1 f(x)\,dx + \int_1^\infty f(x)\,dx \leq 1 + \int_1^\infty g(x)\,dx = 1 + 1 = 2.$$

Hence, $\int_0^\infty f(x)\,dx$ is finite and so (6) converges. To support this reasoning we must show that (i) and (ii) are true. The first of these is true because

$$f(x) = \frac{1}{1+x^2} \leq \frac{1}{x^2} = g(x), \qquad x \geq 1.$$

We showed in Investigation 1 that (ii) holds.

This argument that the improper integral (6) converges is somewhat indirect. A more direct argument—and an argument that gives the value of (6)—is based on formula (G) from the table of integrals. We have

$$\int_0^b \frac{dx}{1+x^2}\,dx = \arctan b,$$

Taking the limit as $b \to \infty$,

$$\int_0^\infty \frac{dx}{1+x^2}\,dx = \lim_{b\to\infty} \int_0^b \frac{dx}{1+x^2}\,dx = \lim_{b\to\infty} \arctan b = \pi/2.$$

Hence, the improper integral (5) converges to $\pi/2$.

The purpose of the next example is to show that an integral may diverge but neither $\lim_{b\to\infty} \int_a^b f(x)\,dx = \infty$ nor $\lim_{b\to\infty} \int_a^b f(x)\,dx = -\infty$.

EXAMPLE 3 Does the improper integral

$$\int_0^\infty \cos x\,dx \tag{7}$$

converge or diverge?

Solution

We consider the integral

$$\int_0^b \cos x\,dx.$$

Evaluating this integral,

$$\int_0^b \cos x\,dx = \sin b.$$

The improper integral (7) diverges because the limit

$$\lim_{b\to\infty} \left(\int_0^\infty \cos x\,dx \right) = \lim_{b\to\infty} \sin b$$

does not exist. Note that the limit approaches neither ∞ nor $-\infty$; it fails to exist because the values of $\sin b$ oscillate between -1 and 1 as b increases.

When attempting to classify a given improper integral as convergent or divergent, we may not be able to find an antiderivative for the integrand. For such integrals we may use a comparison with an improper integral that is known to converge or diverge.

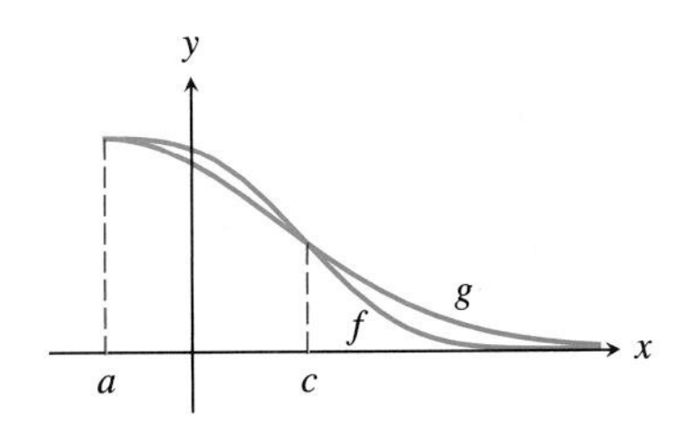

FIGURE 6.85 To use Comparison Theorem I to show that $\int_a^\infty f(x)\,dx$ converges, we look for a function g that eventually becomes and remains larger than f, that is, for some number c, $0 \le f(x) \le g(x)$ for all $x \ge c$, and $\int_a^\infty g(x)\,dx$ converges.

THEOREM Comparison Theorem I

Assume that f and g are continuous on $[a, \infty)$ and, for all sufficiently large x, $0 \le f(x) \le g(x)$. See Fig. 6.85. For improper integrals

$$\int_a^\infty f(x)\,dx \quad \text{and} \quad \int_a^\infty g(x)\,dx$$

of the first kind:

1. If $\displaystyle\int_a^\infty g(x)\,dx$ converges, then so does $\displaystyle\int_a^\infty f(x)\,dx$.
2. If $\displaystyle\int_a^\infty f(x)\,dx$ diverges, then so does $\displaystyle\int_a^\infty g(x)\,dx$.

An argument for Part 2 of Comparison Theorem I is not difficult, particularly if we assume that $0 \le f(x) \le g(x)$ for all $x \ge a$. From Fig. 6.85,

$$G(b) = \int_a^b g(x)\,dx \ge \int_a^b f(x)\,dx = F(b).$$

Because both f and g are nonnegative, the functions F and G increase as $b \to \infty$. Hence, if $\int_a^\infty f(x)\,dx$ diverges, then $F(b) \to \infty$ as $b \to \infty$; but because $G(b) \ge F(b)$, then also $G(b) \to \infty$ as $b \to \infty$. Hence, $\int_a^\infty g(x)\,dx$ diverges.

An argument for Part 1 of Comparison Theorem I is outlined in Exercise 47.

Test Integrals To use the Comparison Theorem in classifying an improper integral $\int_a^\infty f(x)\,dx$ as convergent or divergent, we need an improper integral $\int_a^\infty g(x)\,dx$ of known behavior. There are several families of "test integrals" which are used for this purpose. We have already used the integrals $\int_1^\infty x^{-1}\,dx$ and $\int_1^\infty x^{-2}\,dx$ from one of these families, the power functions.

The improper integral

$$\int_1^\infty \frac{dx}{x^p} \quad \text{is} \quad \begin{cases} \text{convergent if } p > 1 \\ \text{divergent if } p \le 1. \end{cases} \tag{8}$$

We leave a proof of this result to Exercise 21.

In the next example, we show how Comparison Theorem I and a test integral can be teamed up to subdue a relatively difficult integral.

EXAMPLE 4 Classify the following improper integral as convergent or divergent:

$$\int_1^\infty \frac{dx}{\sqrt{x^4 + x^2 + 1}}. \tag{9}$$

Solution

We expect this integral to converge because as x becomes very large, the terms x^2 and 1 are small compared to x^4. Hence the integrand is like $(x^4)^{-1/2} = x^{-2}$. Because the test integral $\int_1^\infty x^{-2}\,dx$ converges, we expect the given integral to converge as well. With these rough preliminaries in mind, we look at the integrand more closely. Because x^{-2} and x^{-4} are positive,

$$\sqrt{x^4 + x^2 + 1} = x^2\sqrt{1 + x^{-2} + x^{-4}} \ge x^2.$$

Taking reciprocals,

$$f(x) = \frac{1}{\sqrt{x^4 + x^2 + 1}} \le \frac{1}{x^2} = g(x), \qquad x \ge 1.$$

Because $\displaystyle\int_1^\infty \frac{1}{x^2}\,dx$ converges, it follows from Comparison Theorem I that the improper integral (9) also converges.

Integrating on $(-\infty, b]$ and $(-\infty, \infty)$ If f is continuous on $(-\infty, b]$, we define

$$\int_{-\infty}^b f(x)\,dx = \lim_{a\to-\infty} \int_a^b f(x)\,dx$$

and say that this improper integral is convergent or divergent depending on whether the limit is finite or not. A corresponding version of Comparison Theorem I holds for these integrals.

If f is continuous on $(-\infty, \infty)$, the improper integral $\int_{-\infty}^\infty f(x)\,dx$ is convergent provided that both improper integrals $\int_{-\infty}^0 f(x)\,dx$ and $\int_0^\infty f(x)\,dx$ are convergent; otherwise it is divergent. If $\int_{-\infty}^\infty f(x)\,dx$ is convergent, we write

$$\int_{-\infty}^\infty f(x)\,dx = \int_{-\infty}^0 f(x)\,dx + \int_0^\infty f(x)\,dx,$$

that is, the value of the integral on the left is the sum of the values of the two integrals on the right.

Improper Integrals of the Second Kind

For improper integrals of the first kind, the interval of integration is not bounded. For improper integrals of the second kind, the interval of integration is bounded but the integrand is not bounded on this interval.

INVESTIGATION 2

Areas of Regions with Vertical Asymptotes

Figure 6.86(a) shows the graph of $f(x) = x^{-3/2}$ for $0 < x \le 1$. In this sketch we have shaded the region $\mathcal{R}$ beneath the graph of f and above the interval $(0, 1]$; in Fig. 6.86(b) we have shaded the region $\mathcal{R}(a)$ beneath the graph of f and above the interval $[a, 1]$, where $0 < a < 1$. The line $x = 0$ is a vertical asymptote to the graph of f. From our experience with improper integrals of the first kind, it makes

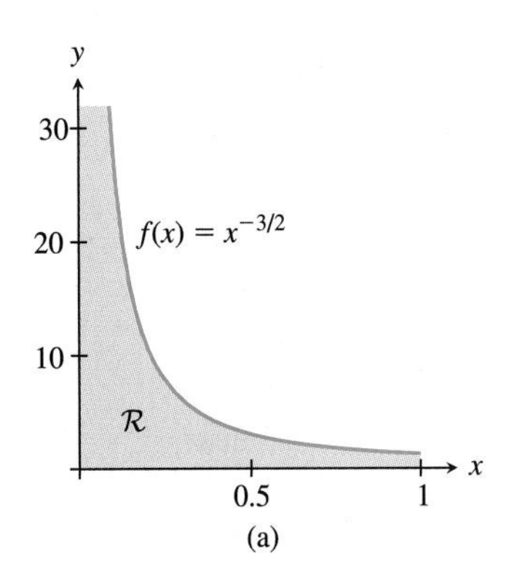

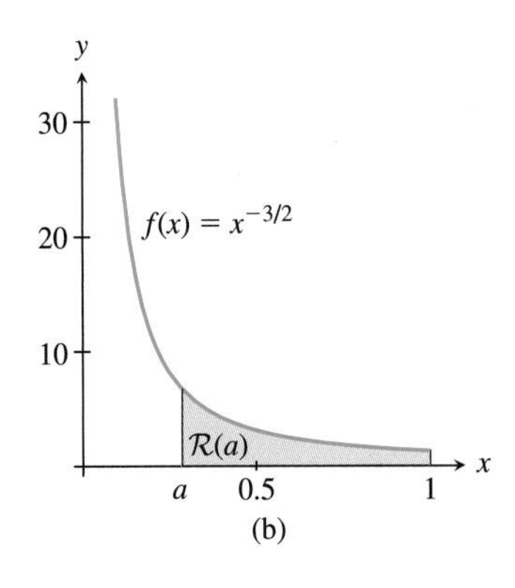

FIGURE 6.86 The area of region $\mathcal{R}$ is finite or infinite depending on the limit of the area of the region $\mathcal{R}(a)$ as $a \to 0^+$.

geometric sense to write

$$\int_0^1 \frac{dx}{x^{3/2}} = \lim_{a\to 0^+} \int_a^1 \frac{dx}{x^{3/2}}. \tag{10}$$

This expresses the idea that the area of the region $\mathcal{R}$ beneath f and above the interval $(0, 1]$ is finite or infinite depending on the limit of the area of the region $\mathcal{R}(a)$ as $a \to 0^+$. Because $-2x^{-1/2}$ is an antiderivative of $x^{-3/2}$,

$$\lim_{a\to 0^+} \int_a^1 \frac{dx}{x^{3/2}} = \lim_{a\to 0^+} (-2x^{-1/2})\Big|_a^1 = \lim_{a\to 0^+} \left(-2 + \frac{2}{\sqrt{a}}\right) = \infty.$$

We say, then, that the region $\mathcal{R}$ has infinite area and the improper integral in (10) is divergent.

Following the pattern for integrals of the first kind, we define improper integrals of the second kind, state a comparison theorem, and give some test integrals.

DEFINITION Convergent and Divergent Improper Integrals of the Second Kind

If f is continuous on the interval $(a, b]$ and $\lim_{x\to a^+} f(x) = \pm\infty$ or the limit does not exist, the improper integral of $\int_a^\infty f(x)\,dx$ of the second kind is defined to be

$$\int_a^b f(x)\,dx = \lim_{w\to a^+} \int_w^b f(x)\,dx$$

when the limit is a number L. In this case we write $\int_a^b f(x)\,dx = L$ and say that the improper integral is *convergent* (or converges to L). If the limit is $\pm\infty$, we write $\int_a^b f(x)\,dx = \pm\infty$ and say that the improper integral is *divergent* (or diverges to $\pm\infty$). When the limit is neither finite nor $\pm\infty$, we say that the improper integral is *divergent* (diverges).

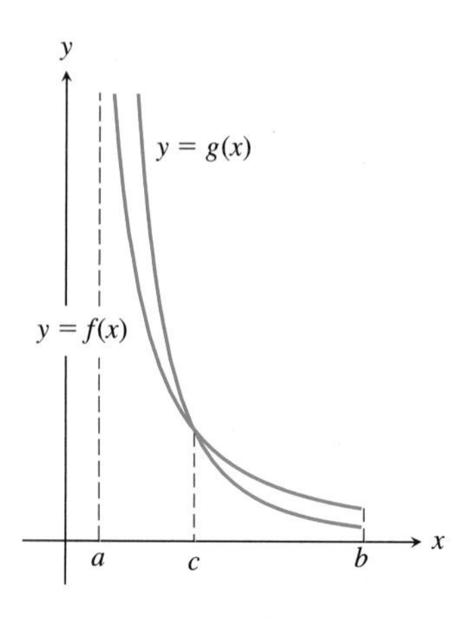

FIGURE 6.87 To use Comparison Theorem II to show that $\int_a^b f(x)\,dx$ converges, we look for a function g that for x near a becomes and remains larger than f, that is, for some number $c > a$, $0 \le f(x) \le g(x)$ for all $x \in (a, c]$ and $\int_a^b g(x)\,dx$ converges.

THEOREM Comparison Theorem II

Assume that f and g are continuous on $(a, b]$ and, for all x near a, $0 \le f(x) \le g(x)$. See Fig. 6.87. For improper integrals

$$\int_a^b f(x)\,dx \quad \text{and} \quad \int_a^b g(x)\,dx$$

of the second kind:

1. If $\int_a^b g(x)\,dx$ converges, then so does $\int_a^b f(x)\,dx$.
2. If $\int_a^b f(x)\,dx$ diverges, then so does $\int_a^b g(x)\,dx$.

Test Integrals

The improper integral

$$\int_0^1 \frac{dx}{x^p} = \lim_{a\to 0^+} \int_a^1 \frac{dx}{x^p} \quad \text{is} \quad \begin{cases} \text{convergent if } p < 1 \\ \text{divergent if } p \ge 1. \end{cases} \tag{11}$$

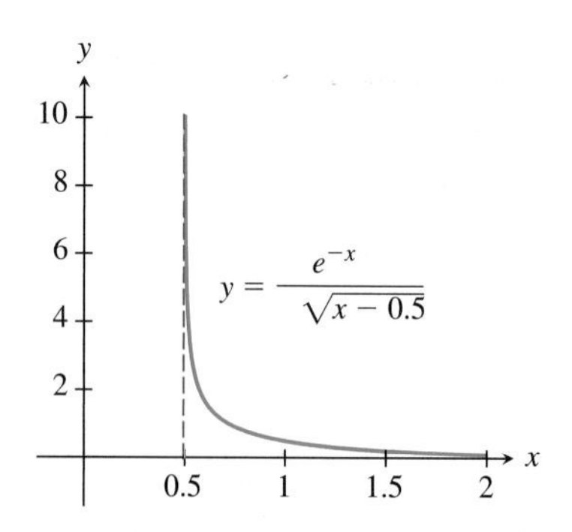

FIGURE 6.88 The region beneath the graph of $y = e^{-x}/\sqrt{x - 0.5}$ and above $(0.5, 2]$ has infinite extent. Is its area finite or infinite?

EXAMPLE 5 Classify the following improper integral as convergent or divergent:

$$\int_{0.5}^2 \frac{e^{-x}\,dx}{\sqrt{x - 0.5}}. \tag{12}$$

Figure 6.88 shows a graph of the integrand.

Solution

To figure out how to apply Comparison Theorem II, we make a rough guess as to whether the integral converges or diverges. We guess that the integral (12) converges because it is like the test integral (11) with $p = 1/2$. We lean in this direction because (i) the behavior of $1/\sqrt{x - 0.5}$ as $x \to 0.5^+$ is the same as $1/\sqrt{x}$ as $x \to 0^+$, (ii) the test integral (11) converges for $p = 1/2$, and (iii) the numerator e^{-x} of this integrand is bounded on $[0.5, 2]$ and hence is not important relative to the question of convergence or divergence.

Next we try to refine our guess. Because e^{-x} is decreasing for all x,

$$e^{-x} \le e^{-0.5} < e^{-0} = 1, \quad \text{for } 0.5 \le x \le 2.$$

Hence,

$$\frac{e^{-x}}{\sqrt{x - 0.5}} < \frac{1}{\sqrt{x - 0.5}}, \qquad 0.5 < x \le 2. \tag{13}$$

If the improper integral

$$\int_{0.5}^{2} \frac{dx}{\sqrt{x - 0.5}} \tag{14}$$

is convergent, then it will follow from (13) and Comparison Theorem II that the improper integral (12) converges.

The improper integral (14) is related to the test integral (11), with $p = 1/2$, through the change of variable $x = w - 0.5$. With this substitution,

$$\int_{0.5}^{2} \frac{dx}{\sqrt{x - 0.5}} = \int_{0}^{1.5} \frac{dw}{\sqrt{w}}.$$

Leaving a few details to Exercise 48, we conclude that the improper integral (14) converges and, hence, the improper integral (12) also converges.

Other Improper Integrals of the Second Kind If f is continuous on $[a, b)$ but its limit as $x \to b^-$ is infinite or does not exist, the definition of the corresponding improper integral and the modification of Comparison Theorem II are similar to the definition and theorem just given. If f is continuous on a finite interval (a, b) but fails to have a finite limit at each of a and b, the improper integral $\int_a^b f(x)\,dx$ is defined as the sum

$$\int_a^b f(x)\,dx = \int_a^c f(x)\,d + \int_c^b f(x)\,dx$$

of two improper integrals, one on $(a, c]$ and the other on $[c, b)$, where c is any point between a and b. The improper integral $\int_a^b f(x)\,dx$ is convergent only when both improper integrals in the sum are convergent. In this case, its value is the sum of the values of the convergent integrals.

Other Improper Integrals Some improper integrals combine the features of improper integrals of both the first and second kinds. For example, the improper integral

$$\int_0^{\infty} x^{-1/2} e^{-x}\,dx \tag{15}$$

has both an infinite interval of integration and an integrand having a vertical asymptote at $x = 0$. Figure 6.89 shows the graph of the integrand. We work further with the improper integral (15) to illustrate the usual definition of convergence for such improper integrals. We write (15) as

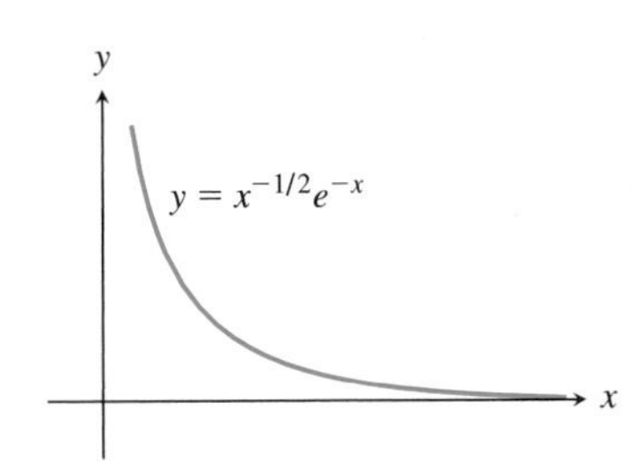

FIGURE 6.89 The integral $\int_0^{\infty} x^{-1/2}e^{-x}\,dx$ is improper for two reasons: It has an infinite interval of integration, and the graph of the integrand has a vertical asymptote.

$$\int_0^{\infty} x^{-1/2} e^{-x}\,dx = \int_0^{c} x^{-1/2} e^{-x}\,dx + \int_c^{\infty} x^{-1/2} e^{-x}\,dx, \tag{16}$$

where c is any positive number, and say that the improper integral (15) converges if both improper integrals in the sum (16) are convergent; otherwise, the improper integral (15) is divergent. Taking $c = 1$ and using an argument similar to that used in Example 5, the integral $\int_0^1 x^{-1/2}e^{-x}\,dx$ is convergent. Because the integral $\int_1^{\infty} x^{-1/2}e^{-x}\,dx$ is also convergent (see Exercise 44) it follows, by definition, that the improper integral (15) is convergent.

Exercises 6.9

Exercises 1–20: Classify each improper integral as convergent or divergent. If it is convergent, calculate its value.

1. $\int_1^\infty \frac{dx}{x+1}$
2. $\int_1^\infty \frac{x}{x^2+1}\,dx$
3. $\int_1^\infty x^{-3/2}\,dx$
4. $\int_1^\infty x^{-5/2}\,dx$
5. $\int_1^\infty \frac{dx}{\sqrt[3]{x^2}}$
6. $\int_1^\infty \frac{dx}{\sqrt[5]{x^6}}$
7. $\int_0^1 \frac{dx}{x^{5/3}}$
8. $\int_1^2 \frac{dx}{(x-1)^2}$
9. $\int_0^1 \frac{dx}{\sqrt{1-x^2}}$
10. $\int_1^2 \frac{dx}{\sqrt{x^2-1}}$
11. $\int_1^2 \frac{dx}{x\sqrt{x^2-1}}$
12. $\int_{1/2}^1 \frac{dx}{x\sqrt{1-x^2}}$
13. $\int_0^{\pi/2} \tan x\,dx$
14. $\int_0^\infty \sin x\,dx$
15. $\int_1^\infty \frac{dx}{x(x+1)}$
16. $\int_1^\infty \frac{dx}{x(2x-1)}$
17. $\int_0^1 \ln x\,dx$
18. $\int_1^\infty \frac{\ln x}{x}\,dx$
19. $\int_{-\infty}^\infty \frac{dx}{x^2+1}$
20. $\int_{-\infty}^0 xe^{-x}\,dx$

21. Show that $\int_1^\infty x^{-p}\,dx$ is convergent for all $p > 1$ and divergent otherwise. This verifies a statement in (8).

22. Show that $\int_0^1 x^{-p}\,dx$ is convergent for all $p < 1$ and divergent otherwise. This verifies a statement in (11).

23. Calculate the area of the region between the graphs of $\cosh x = (e^x + e^{-x})/2$ and $\sinh x = (e^x - e^{-x})/2$ on the interval $[0, \infty)$.

24. Calculate the area of the region between the graphs of $y = x/(x^2+1)$ and $y = 1/x$ on the interval $[1, \infty)$.

25. Classify as convergent or divergent the improper integral
$$\int_0^\infty \frac{dx}{\sqrt{2x^3+x+1}}.$$

26. Classify as convergent or divergent the improper integral
$$\int_0^\infty \frac{dx}{\sqrt{x^4+1}}.$$

27. Classify as convergent or divergent the improper integral
$$\int_0^1 x^{-2/3}e^{-x}\,dx.$$

28. Classify as convergent or divergent the improper integral
$$\int_0^1 x^{-4/3}e^{-x}\,dx.$$

29. Show that the integral $\int_0^\infty e^{-2x}\sin 3x\,dx$ is convergent and calculate its value.

30. Show that the integral $\int_0^\infty e^{-5x}\cos 3x\,dx$ is convergent and calculate its value.

Exercises 31–36: Classify each improper integral as convergent or divergent.

31. $\int_0^\infty 3^{-x^2}\,dx$
32. $\int_1^\infty \frac{\ln x}{x^3}\,dx$
33. $\int_0^{\pi/2} \frac{dx}{\sqrt{x+\sin x}}$
34. $\int_0^1 \frac{\cos x}{x}\,dx$
35. $\int_{-1}^\infty \frac{dx}{\sqrt{1+x^3}}$
36. $\int_0^1 \frac{dx}{\ln(x)\sqrt{1-x}}$

37. How much work does the gravitational field of Earth do on a mass of 1 kg that is moved from the surface of the Earth to the vicinity of Alpha Centauri, a triple star about 4.07×10^{16} meters from Earth? Express your results in terms of the mass of the Earth, its radius, and the universal gravitational constant G. *Hint:* Use an improper integral for the calculation.

38. Show first that for $0 \le x < 3$,
$$\sqrt{\frac{81-5x^2}{9-x^2}} = \sqrt{\frac{81-5x^2}{3+x}} \cdot \frac{1}{\sqrt{3-x}} \le 6 \cdot \frac{1}{\sqrt{3-x}}.$$
Then show that
$$\int_0^3 \sqrt{\frac{81-5x^2}{9-x^2}}\,dx$$
is convergent.

39. Give a divergent improper integral that diverges to neither ∞ nor $-\infty$.

40. Show that if $x \ge 2$, then $x^2 \ge 2x$. Use this in showing that $e^{-x} \ge e^{-x^2/2}$ for $x \ge 2$. Show that $\int_{-\infty}^\infty e^{-x^2}\,dx$ is convergent.

41. Figure 6.81 shows a shaded area $\mathcal{R}(b)$. Because the graph of f is very close to its horizontal asymptote for $x \ge 3$, we may expect that $A(3)$ is a good approximation to $\int_0^\infty f(x)\,dx$. Use the trapezoid rule with $n = 31$ in showing that $A(3) \approx 1.25$. *Note:* This value of n was chosen so that T_n is within 0.001 of $A(3)$.

42. Assume that f is continuous on $[a, \infty)$. Show that if $b > a$, then $\int_a^\infty f(x)\,dx$ is convergent if and only if $\int_b^\infty f(x)\,dx$ convergent. Also show that if $\int_a^\infty f(x)\,dx$ is convergent, then

for any $b > a$ we have

$$\int_a^\infty f(x)\,dx = \int_a^b f(x)\,dx + \int_b^\infty f(x)\,dx.$$

43. Using appropriate definitions, show that

$$\int_{-\infty}^\infty xe^{-x^2}\,dx = 0.$$

From the symmetry of the graph of f about the origin we may be tempted to argue that, obviously, the value of this integral is 0. If, however, we argued similarly that the improper integral

$$\int_{-\infty}^\infty \frac{x}{1+x^2}\,dx$$

is 0, our argument would be more open to question. Why?

44. Show that the improper integral $\int_1^\infty x^{-1/2}e^{-x}\,dx$ is convergent.

45. Referring to the improper integral (15), show that its convergence does not depend on the particular choice of c.

46. Just after Comparison Theorem I we gave an argument for Part 2 of the theorem. In particular, we said that "Because both f and g are nonnegative, the functions F and G increase as $b \to \infty$. . . ." Recalling from Chapter 4 that a function p is increasing provided that $u < v$ implies that $p(u) < p(v)$, criticize our statement.

47. An argument for Part 1 of the Comparison Theorem depends upon the following Theorem: If a function F is defined on an interval of the form $[c, \infty)$ and is both nondecreasing and bounded above, then $\lim_{b\to\infty} F(b)$ exists and is finite. The accompanying figure shows the graph of a function F that is nondecreasing and lies below the graph of the line $y = M$. As x increases, $F(x)$ cannot decrease and yet remains below a horizontal line. It appears that as b increases, $F(b)$ must approach some finite value from below. Using this result, fill in the following outline of an argument for Part 1:

We note that because both f and g are nonnegative, the functions F and G increase as $b \to \infty$. Take $M = \int_a^\infty g(x)\,dx$.

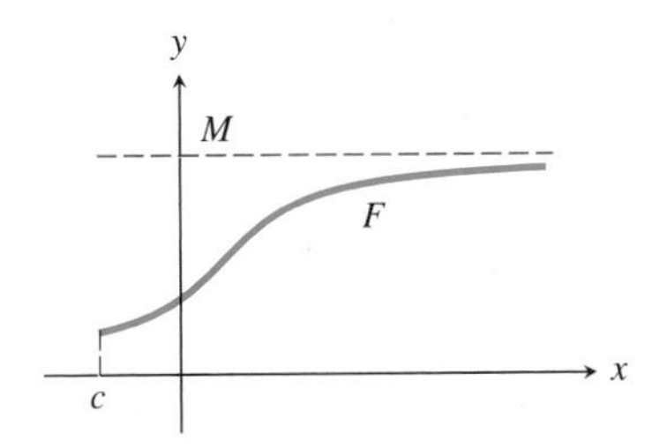

Figure for Exercise 47.

48. Because we have not given a substitution rule for improper integrals, the equation

$$\int_{0.5}^2 \frac{dx}{\sqrt{x-0.5}} = \int_0^{1.5} \frac{dw}{\sqrt{w}}$$

given in Example 5 and based on the substitution $x = w - 0.5$ is questionable. Justify this equation by making a detour through proper integrals, that is, use the fact that

$$\int_{0.5}^2 \frac{dx}{\sqrt{x-0.5}} = \lim_{a\to 0.5^+} \int_a^2 \frac{dx}{\sqrt{x-0.5}}.$$

49. Verify and fill in the details of the following outline for approximating the improper integral

$$\int_0^\infty \sqrt{x}e^{-x^2}\,dx$$

to within 0.01 of its value.

Start by making a change of variable (let $x = u^2$) to simplify the integrand. Split the resulting integral into two parts by finding a number b such that the last integral (which we refer to as the "tail") in

$$\int_0^\infty \sqrt{x}e^{-x^2}\,dx = \int_0^\infty 2u^2e^{-u^4}\,du$$
$$= \int_0^b 2u^2e^{-u^4}\,du + \int_b^\infty 2u^2e^{-u^4}\,du$$

is small. We then use the trapezoid rule to approximate $\int_0^b 2u^2e^{-u^4}\,du$. We must choose n for the trapezoid rule and b for the tail so that

$$\left|\int_0^\infty 2u^2e^{-u^4}\,du - T_n\right| = \left|\int_0^b 2u^2e^{-u^4}\,du - T_n + \int_b^\infty 2u^2e^{-u^4}\,du\right| < 0.01.$$

We choose b so that

$$\int_b^\infty 2u^2e^{-u^4}\,du < 0.005.$$

With b chosen, choose n so that

$$\left|\int_0^b 2u^2e^{-u^4}\,du - T_n\right| < 0.005.$$

To bound the tail, use the fact that $u^2e^{-u^4} < u^2e^{-u^3}$ for $u > 1$. We have

$$\int_b^\infty 2u^2e^{-u^4}\,du < \int_b^\infty 2u^2e^{-u^3}\,du = \frac{2}{3}e^{-b^3}.$$

The inequality $\frac{2}{3}e^{-b^3} < 0.005$ is satisfied for $b = 1.7$. To find n so that

$$\left|\int_0^b 2u^2e^{-u^4}\,du - T_n\right| < 0.005,$$

we recall the trapezoid error formula

$$\left| \int_a^b f(x)\,dx - T_n \right| \le \frac{(b-a)^3}{12n^2} M,$$

where M is at least as large as the maximum value of $|f''|$ on $[a, b]$. Show that we may take $M = 10$, perhaps by graphing the second derivative of $f(u) = 2u^2 e^{-u^4}$. It follows that $n = 29$ and $T_{29} = 0.6126\ldots$

50. See the preceding exercise. The exponential integral function is defined as

$$E_1(x) = \int_x^\infty \frac{e^{-t}}{t}\,dt, \qquad x > 0.$$

Sketch a graph of E_1 by calculating and plotting $E_1(x)$, for $x = 0.2, 0.4, \ldots, 2.0$. For graphing, finding $E_1(x)$ within 0.1 is sufficient. Begin by finding b such that the tail $\int_b^\infty e^{-t}/t\,dt < 0.05$. Next, find a single value of n for the trapezoid rule that is suitable for all of the x values.

Review of Key Concepts

The main goal of this chapter was to show how the definite integral can be used to calculate

- the volumes of certain three-dimensional solids
- the arc lengths of smooth curves
- the areas of polar regions
- the work done by a variable force on an object moving on a curve
- the center of mass of rods with variable density or the center of mass of a lamina
- the curvature of a smooth curve.

The section on improper integrals was less an application of the integral than an extension of it. The discussions of normal, tangential, radial, and transverse acceleration vectors were applications of the integral only in that the idea of arc length was a prerequisite. Kepler's second law followed from the ideas of polar area element and radial and transverse acceleration.

A major theme of the chapter was the use of elements of volume, arc length, work, mass, or polar area. Using an element helps break complex problems into simpler geometric or physical problems. Through the summing and limiting operations implicit in the transition from element to integral, we can calculate the volumes of solids with known cross sections, the volume of a solid of revolution, the length of a curve, the work done by a variable force, or the center of mass of a rod or lamina.

Chapter Summary

Volumes by Cross Section

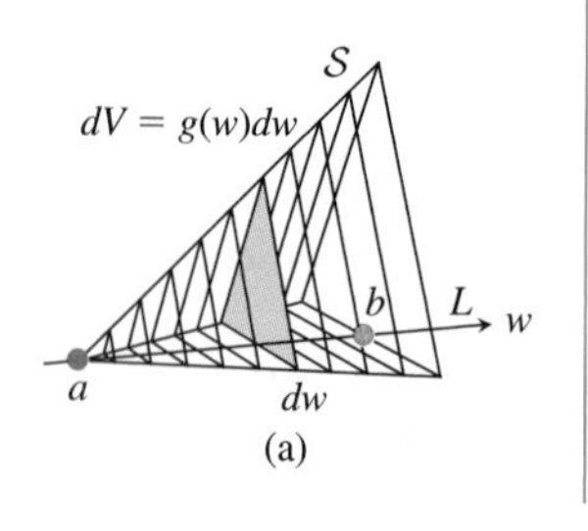

(a)

In part (a) of the figure, the solid S is sectioned by planes perpendicular to a line L. If the area of the cross section at w is $g(w)$, the volume element is

$$dV = g(w)\,dw$$

For part (a), if $a = 0$, $b = 1$, and, for any $w \in [a, b]$, the cross sections are equilateral triangles with side w, then $g(w) = (\sqrt{3}/4)w^2$, $dV = (\sqrt{3}/4)w^2\,dw$, and

$$V = \int_0^1 (\sqrt{3}/4)w^2\,dw = \sqrt{3}/12.$$

Volumes by Cross Section

and, summing the elements with the definite integral, the volume of $\mathcal{S}$ is

$$V = \int_a^b g(w)\,dw.$$

In parts (b) and (c) of the figure, $\mathcal{S}$ is a solid of revolution. If the cross section is a disk—see part (b) of the figure—the volume element is

$$dV = \pi y^2\,dx,$$

where $y = f(x)$ is the radius of the circular cross section at x; the volume of $\mathcal{S}$ is

$$V = \int_a^b \pi f(x)^2\,dx.$$

(b)

For part (b), if $a = 0$, $b = 1$, and $y = f(x) = x^2$, then $dV = \pi(x^2)^2\,dx$, and

$$V = \int_0^1 \pi(x^2)^2\,dx = \pi/5.$$

If the cross section is a washer—see part (c) of the figure—the volume element is

$$dV = (\pi r_{\text{outer}}{}^2 - \pi r_{\text{inner}}{}^2)\,dx,$$

where $r_{\text{outer}} = f(x)$ and $r_{\text{inner}} = g(x)$ are the radii of the washer cross section at x; the volume of $\mathcal{S}$ is

$$V = \int_a^b \pi(f(x)^2 - g(x)^2)\,dx.$$

(c)

For part (c), if $a = 0$, $b = 1$, $y = g(x) = x^3$, and $y = f(x) = x^2$, then $dV = (\pi(x^2)^2 - \pi(x^3)^2)\,dx$, and

$$V = \int_0^1 \pi(x^4 - x^6)\,dx = 2\pi/35.$$

Volumes by Shells

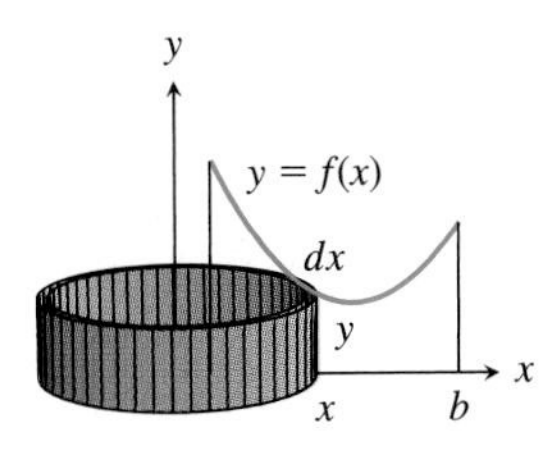

A solid is generated by revolving the region beneath a curve with equation $y = f(x)$ about the y-axis. The volume of the cylindrical shell shown in the figure is

$$dV \approx 2\pi xy\,dx = 2\pi x f(x).$$

Summing these elements, the volume of the solid is

$$V = \int_a^b 2\pi x f(x)\,dx.$$

The figure shows the graph of $f(x) = 2 + \frac{1}{4}(x - 5)^2$, for $1 \le x \le 8$. The volume integral is

$$V = \int_1^8 2\pi x f(x)\,dx$$

$$= 2\pi \int_1^8 x\left(2 + \frac{1}{4}(x - 5)^2\right)dx$$

$$= 4319\pi/24.$$

Polar Area Element

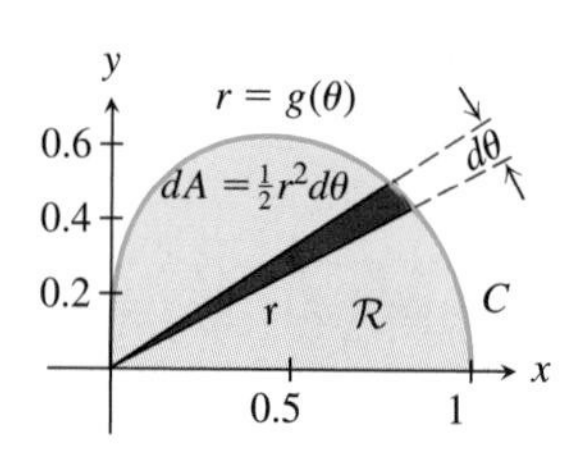

The polar area element for a region $\mathcal{R}$ whose boundary C is described by $r = g(\theta)$, $\alpha \le \theta \le \beta$, and the rays $\theta = \alpha$ and $\theta = \beta$ is

$$dr = \tfrac{1}{2} r^2 \, d\theta.$$

The area A of $\mathcal{R}$ is

$$A = \int_\alpha^\beta \tfrac{1}{2} r^2 \, d\theta = \int_\alpha^\beta \tfrac{1}{2} g(\theta)^2 \, d\theta.$$

The figure shows the graph of $r = g(\theta) = \sqrt{\cos\theta}$, for $0 = \alpha \le \theta \le \beta = \pi/2$. The area integral is

$$A = \int_0^{\pi/2} \tfrac{1}{2} r^2 \, d\theta = 1/2.$$

Arc Length and Unit Tangents

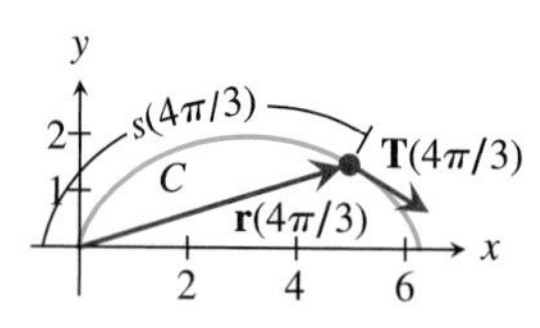

The arc length s of a smooth curve C described by $\mathbf{r} = \mathbf{r}(t) = \langle x(t), y(t) \rangle$, $a \le t \le b$, is

$$s = \int_a^b \|\mathbf{r}'(t)\| \, dt$$

$$= \int_a^b \sqrt{x'(t)^2 + y'(t)^2} \, dt.$$

The arc length s of a smooth Cartesian curve described by $y = f(x)$, $a \le x \le b$, is

$$s = \int_a^b \sqrt{1 + (f'(x))^2} \, dx$$

$$= \int_a^b \sqrt{1 + y'^2} \, dx.$$

The arc length s of a smooth polar curve described by $r = g(\theta)$, $\alpha \le \theta \le \beta$, is

$$s = \int_\alpha^\beta \sqrt{g(\theta)^2 + g'(\theta)^2} \, d\theta$$

$$= \int_\alpha^\beta \sqrt{r^2 + r'^2} \, d\theta.$$

Let C be a smooth curve described by $\mathbf{r} = \mathbf{r}(t) = \langle x(t), y(t) \rangle$, $a \le t \le b$. If $\|\mathbf{r}'(t)\| \ne 0$, the unit tangent vector $\mathbf{T}$ at $\mathbf{r}(t)$ is

$$\mathbf{T} = \mathbf{T}(t) = \frac{1}{\|\mathbf{r}'(t)\|} \mathbf{r}'(t) = \frac{1}{ds/dt} \mathbf{r}'(t).$$

The cycloid shown in the figure is described by $\mathbf{r} = \mathbf{r}(t) = \langle t - \sin t, 1 - \cos t \rangle$, $0 \le t \le 2\pi$. The arc length from $\mathbf{r}(0)$ to $\mathbf{r}(4\pi/3)$ is

$$s(4\pi/3) = \int_0^{4\pi/3} \sqrt{x'(t)^2 + y'(t)^2} \, dt$$

$$= \int_0^{4\pi/3} 2 \sin\left(\tfrac{1}{2} t\right) dt = 6.$$

The unit tangent at $\mathbf{r}(4\pi/3)$ is

$$\mathbf{T}(t) = \frac{1}{\|\mathbf{r}'(t)\|} \mathbf{r}'(t)$$

$$= \left\langle \sin \tfrac{1}{2} t, \cos \tfrac{1}{2} t \right\rangle$$

$$= \left\langle \sqrt{3}/2, -1/2 \right\rangle.$$

Work

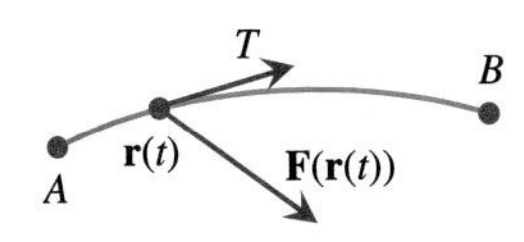

The element of work done by a force field $\mathbf{F}$ as an object moves a distance of ds on a smooth curve C described by $\mathbf{r} = \mathbf{r}(t)$, $a \le t \le b$, is

$$dW = (\mathbf{F} \cdot \mathbf{T})\,ds.$$

The work done by $\mathbf{F}$ is

$$W = \int_a^b \mathbf{F}(\mathbf{r}(t)) \cdot \mathbf{r}'(t)\,dt.$$

If C is described by $\mathbf{r}(t) = \langle t, 2 - t^2 \rangle$, where $-1 \le t \le 1$, and at any point (x, y) the force $\mathbf{F}$ is $\langle x, -y \rangle$, then the work done by the force as the object moves from $\mathbf{r}(-1)$ to $\mathbf{r}(1)$ is

$$W = \int_{-1}^{1} \langle t, -(2 - t^2) \rangle \cdot \langle 1, -2t \rangle\,dt$$
$$= 0.$$

Mass and Center of Mass of a Curved Rod

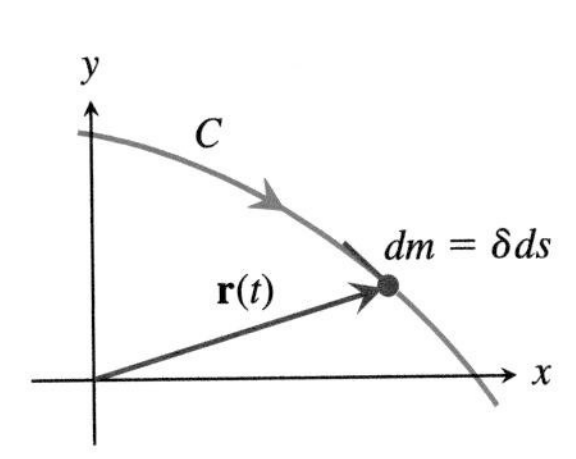

For a curved (or straight) rod C of constant density δ and described by $\mathbf{r} = \mathbf{r}(t)$, $a \le t \le b$, the element of mass is

$$dm = \delta\,ds.$$

The mass of the rod and its center of mass are

$$m = \delta \int_a^b \|\mathbf{r}'(t)\|\,dt$$

$$\mathbf{R} = \frac{\delta}{m} \int_a^b \|\mathbf{r}'(t)\|\mathbf{r}(t)\,dt.$$

The mass and center of mass of the rod described by $\mathbf{r}(t) = \left\langle t, \frac{1}{2}(1 - t^2) \right\rangle$, $0 \le t \le 1$, and having density δ are

$$m = 1.14779\delta$$
$$\mathbf{R} = \langle 0.530998, 0.316971 \rangle.$$

Mass and Center of Mass of a Lamina

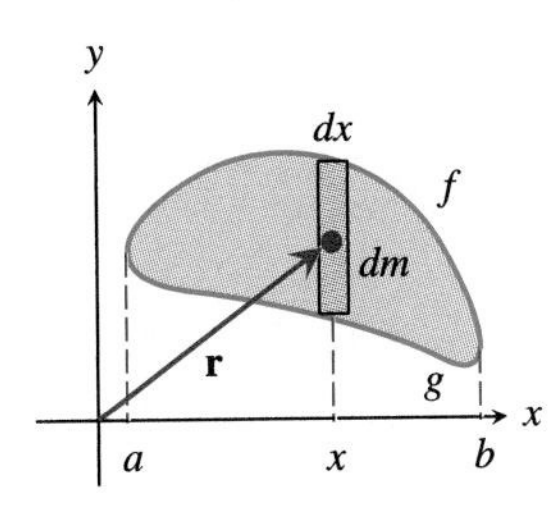

The element of mass of a plane lamina with constant density δ and bounded on the top and bottom by the graphs of $y = f(x)$ and $y = g(x)$ is

$$dm = \delta(f(x) - g(x))\,dx.$$

The mass of the lamina and its center of mass are

$$m = \delta \int_a^b (f(x) - g(x))\,dx$$

$$X = \frac{\delta}{m} \int_a^b x(f(x) - g(x))\,dx$$

$$Y = \frac{\delta}{m} \int_a^b \tfrac{1}{2}(f(x)^2 - g(x)^2)\,dx.$$

The mass and center of mass of the lamina between $y = f(x) = 4x(1 - x)$ and $y = g(x) = \frac{1}{2}x$ are $m = 343\delta/768$ and $\mathbf{R} = \langle X, Y \rangle$, where $X = 7/16$ and $Y = 21/40$.

Curvature and Acceleration

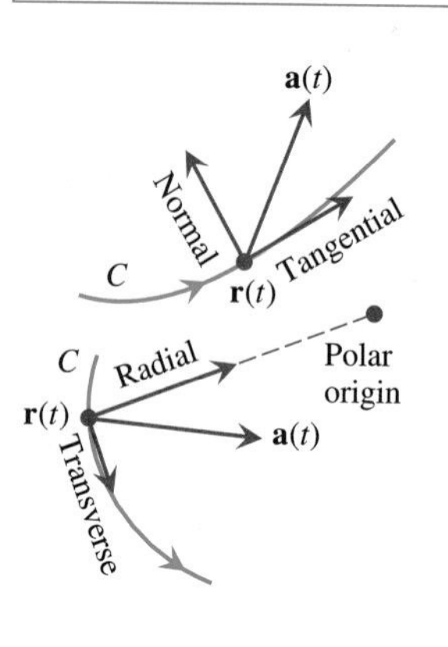

For a smooth curve C, described by $\mathbf{r} = \mathbf{r}(t)$, $a \le t \le b$, the unit tangent $\mathbf{T}$, curvature vector $d\mathbf{T}/ds$, curvature κ, and unit normal $\mathbf{N}$ are

$$\mathbf{T} = \frac{1}{\|d\mathbf{r}/dt\|}\frac{d\mathbf{r}}{dt}$$

$$\frac{d\mathbf{T}}{ds} = \frac{1}{ds/dt}\frac{d\mathbf{T}}{dt} \quad \text{(curvature vector)}$$

$$\kappa = \|d\mathbf{T}/ds\|$$

$$\mathbf{N} = \frac{1}{\kappa}\frac{d\mathbf{T}}{ds}.$$

If $\mathbf{r}(t)$ is the position vector of an object in motion, its acceleration can be written as a sum of tangential and normal accelerations or, in polar coordinates, radial and transverse accelerations:

$$\mathbf{a} = (dv/dt)\mathbf{T} + v^2\kappa\mathbf{N}$$

$$\mathbf{a} = (r'' - r\theta'^2)\mathbf{e}_r + (2r'\theta' + r\theta'')\mathbf{e}_\theta$$

For $\mathbf{r}(t) = e^t\langle\cos t, \sin t\rangle$, at $\mathbf{r}(0)$,

$$\mathbf{T}(0) = \langle 1/\sqrt{2}, 1/\sqrt{2}\rangle$$

$$\frac{d\mathbf{T}}{ds} = \langle -1/2, 1/2\rangle$$

$$\kappa(0) = 1/\sqrt{2}$$

$$\mathbf{N}(0) = \langle -1/\sqrt{2}, 1/\sqrt{2}\rangle.$$

Because we may regard $\mathbf{r}$ as the polar form of an object in motion, with $r(t) = e^t$ and $\theta(t) = t$, and $v = ds/dt = \sqrt{2}e^t$,

$$\mathbf{a} = \sqrt{2}\mathbf{T}(0) + \sqrt{2}\mathbf{N}(0)$$

$$\mathbf{a} = 0\mathbf{e}_r + 2\mathbf{e}_\theta.$$

Improper Integral of the First Kind

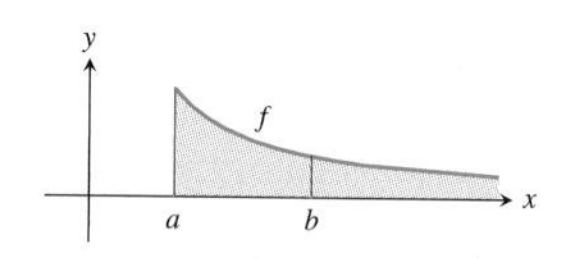

Improper integral of the first kind:

$$\int_a^\infty f(x)\,dx = \lim_{b\to\infty}\int_a^b f(x)\,dx,$$

where f is continuous on $[a, \infty)$. If the limit exists, the improper integral is convergent; otherwise it is divergent.

Test integrals for improper integrals of the first kind:

$$\int_1^\infty \frac{dx}{x^p} \quad \text{is} \quad \begin{cases} \text{convergent if } p > 1 \\ \text{divergent if } p \le 1. \end{cases}$$

Comparison Theorem I

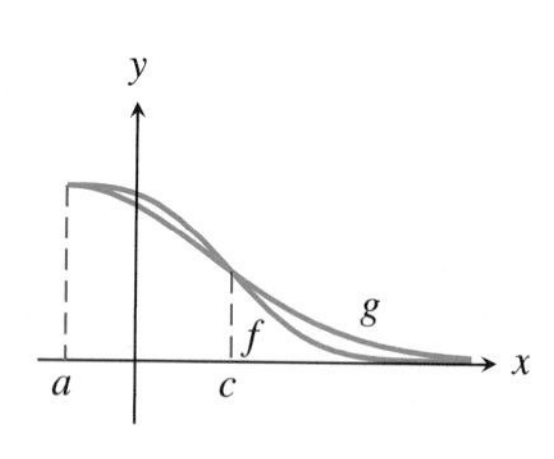

If $0 \le f(x) \le g(x)$ for all sufficiently large x, then

$$\int_a^\infty g(x)\,dx < \infty \Rightarrow \int_a^\infty f(x)\,dx < \infty$$

$$\int_a^\infty f(x)\,dx = \infty \Rightarrow \int_a^\infty g(x)\,dx = \infty$$

Because $0 \le e^{-x^2} \le e^{-x}$ for $x \ge 1$ and $\int_0^\infty e^{-x}dx < \infty$, it follows that $\int_0^\infty e^{-x^2}dx < \infty$.
Because $0 \le 1/x < 1/\ln x$ for $x \ge 2$ and $\int_2^\infty (1/x)\,dx = \infty$, it follows that $\int_2^\infty (1/\ln x)\,dx = \infty$.

Improper Integrals of the Second Kind		
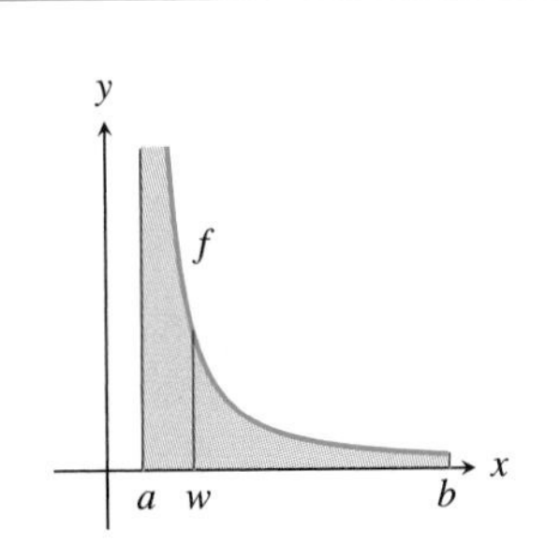	Improper integral of the second kind: $$\int_a^b f(x)\,dx = \lim_{w\to a^+}\int_w^b f(x)\,dx,$$ where f is continuous on $(a,b]$. If the limit exists, the improper integral is convergent; otherwise it is divergent.	Test integrals for improper integrals of the second kind: $$\int_0^1 \frac{dx}{x^p} \text{ is } \begin{cases}\text{convergent if } p<1\\ \text{divergent if } p\geq 1.\end{cases}$$

Comparison Theorem II		
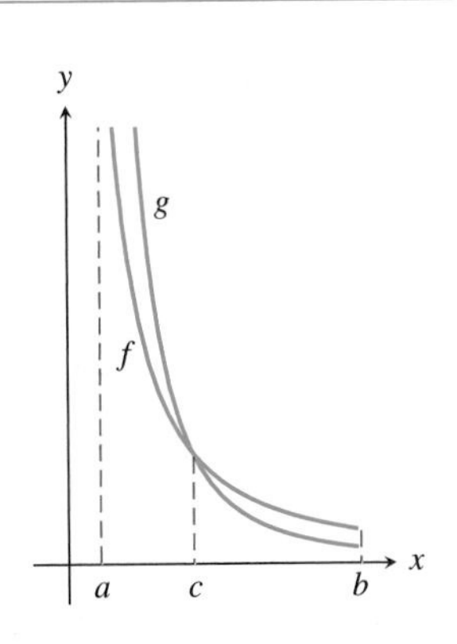	If $0 \leq f(x) \leq g(x)$ for all x near a, then $$\int_a^b g(x)\,dx < \infty \Rightarrow \int_a^b f(x)\,dx < \infty$$ $$\int_a^b f(x)\,dx = \infty \Rightarrow \int_a^b g(x)\,dx = \infty.$$	Because $0 \leq e^{-x}/\sqrt{x} \leq 1/\sqrt{x}$ for $x>0$ and $\int_0^1 (1/\sqrt{x})\,dx < \infty$, it follows that $\int_0^1 (e^{-x}/\sqrt{x})\,dx < \infty$. Because $0 \leq 1/x \leq 1/(x\sin x)$ for $x \in (0,\pi/2)$ and $\int_0^{\pi/2}(1/x)\,dx = \infty$, it follows that $\int_0^{\pi/2}(1/(x\sin x))\,dx = \infty$.

Chapter Review Exercises

1. As seen in the accompanying figure, a solid has a circular base of radius 10 m. Each cross section of the solid by a plane perpendicular to a fixed diameter of the base is an equilateral triangle. Calculate the volume of the solid.

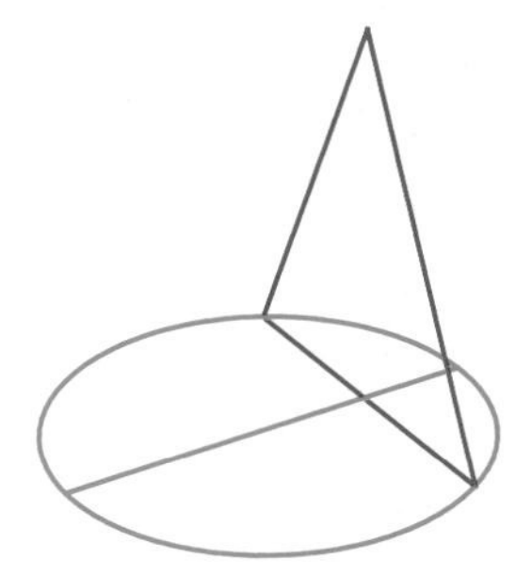

Figure for Exercise 1.

2. As seen in the accompanying figure, a solid has a rectangular base with sides 2 m and 5 m. Cross sections of the solid by planes perpendicular to one of the long sides of the rectangle are trapezoids, with their bases perpendicular to the rectangle. At a distance x from one of the short sides of the rectangle, the bases have lengths $\frac{1}{2}x$ and $2x$. Calculate the volume of the solid.

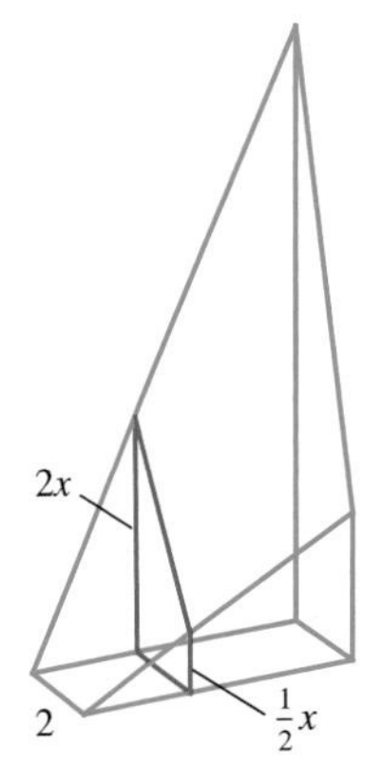

Figure for Exercise 2.

3. Calculate the volume of the solid resulting from revolving a sector of a circle about one of its two bounding radii. Assume that the radius of the circle is a meters and the central angle of the sector is θ, where $0 \le \theta \le \pi$.

4. Show that the area of a right circular cone with height h and base radius r is $\frac{1}{3}\pi r^2 h$ cubic units.

5. Calculate the volume of the solid generated by revolving about the x-axis the region bounded by the curves described by $y = \sqrt{4 - x}$ and $x/4 + y/2 = 1$.

6. The region beneath the graph of $y = \sec x$, $x \in [0, \pi/4]$, is revolved about the x-axis. Calculate its volume.

7. The region beneath the graph of $y = \tan x$, $x \in [0, \pi/4]$, is revolved about the x-axis. Calculate its volume.

8. Find the volume of a doughnut with circular sections and inner radius 1 cm and outer radius 5 cm.

9. Give a simplified but nonevaluated integral whose value is the arc length of the curve described by $x^2 - y^2 = 1$, from $(2, \sqrt{3})$ to $(3, 2\sqrt{2})$.

10. Show that the integral for finding the arc length of the curve described by $y = 3x^{2/3}$, $0 \le x \le 8$, is improper. Rewrite the integrand to remove negative exponents and use the substitution $u = x^{2/3} + 4$ to show that the arc length integral converges to the value $16\sqrt{2} - 8$.

11. What is the arc length of the catenary curve with equation $y = a\cosh(x/a)$, $-b \le x \le b$?

12. Calculate the area of the region bounded by the rays $\theta = -\pi/4$ and $\theta = \pi/4$ and the graph of the polar equation $r = 1 + \sin\theta$.

13. Calculate the area of the region bounded by the lemniscate described by the polar equation $r^2 = \cos 2\theta$.

14. Calculate the area of the region lying inside the curve described by $r = 3\cos\theta$ and outside that described by the polar equation $r = 1 + \cos\theta$.

15. Determine the area of the region in the first quadrant and bounded by the curves with the polar equations $r = 3\theta$ and $r = \sec\theta$.

16. The curve with equation $r = \theta$ divides the first-quadrant portion of the circle with polar equation $r = \cos\theta$ into two regions. Calculate the area of the smaller region.

17. Graph the polar curve with equation $r = 1/(1 - \sin\theta)$ and calculate the area bounded by this curve and the rays $\theta = 0$ and $\theta = \pi/4$. Use numerical integration.

18. A force of 890 N is required to stretch a spring 1 m from its natural length. Calculate the work required to stretch the spring 2 m from its natural length.

19. Before liftoff a launch vehicle includes a 4.3×10^3 kg permanent structure, a 2.6×10^3 kg payload, and 43.1×10^3 kg of fuel. Assuming no air resistance and complete, linear burning of fuel, how much work is required for the vehicle to attain an orbit with height 160 km? It is given that $M_E = 5.97 \times 10^{24}$ kg, $R_E = 6.378 \times 10^6$ m, and $G = 6.67259 \times 10^{-11}$.

20. A small cup in the shape of a frustum of a cone is filled with a soft drink whose density is 1010 kg/m^3. The top and bottom diameters of the cup are 6 cm and 4.6 cm; its height is 7 cm. Calculate the work required to move this liquid to a height 13 cm above the top of the cup.

21. Measuring from one end of a 12-meter beam, there are masses at the 0-m, 4-m, 9-m, and 12-m marks. The masses are 50 kg, 40 kg, 55 kg, and 45 kg, respectively. Assuming the beam has negligible mass, find the balance point.

22. Six masses rest on a plane. Their masses are 18, 10, 15, 17, 19, and 19 kg, and the corresponding position vectors locating them are $\langle -2.2, -4.7\rangle$, $\langle -2.1, 3.1\rangle$, $\langle -4.0, 0\rangle$, $\langle 4.8, -5.4\rangle$, $\langle 6.3, 10.5\rangle$, and $\langle 0.0, 0.0\rangle$. All lengths are in meters. Calculate the center of mass of this system. Recalculate the center of mass by dividing the system into two systems. Suppose system A includes the first three masses and system B includes the last three masses. Calculate the centers of mass of these two systems, and then calculate from these results the center of mass of the two-mass system.

23. Calculate the mass and the center of mass of a 2.0-m rod of bronze, with the density varying linearly from 2.8 kg/m to 3.2 kg/m from end to end.

24. A rod of linear density 2 kg/m has the shape of the graph of $y = \sin x$, $0 \le x \le \pi$. Use numerical integration to find its mass and center of mass.

25. When an oil well burns, sediment is carried into the air by flames and settles to the ground in decreasing amounts, depending on the distance from the well. Experimental evidence indicates that the density (in tons/square mile) at a distance r from the burning oil well is $7/(1 + r^2)$. Find and evaluate an integral that approximates the total amount of sediment deposited within 100 miles of the well.

26. Calculate the center of mass of a lamina in the shape of a rectangle 3 m by 4 m, with a semicircle of diameter 4 m attached along one of the long sides of the rectangle. Assume the density is δ kg/m^2. Avoid integration; use the Mass Subdivision Theorem together with a result worked out in Example 5 in Section 6.7.

27. Calculate the mass and center of mass of a lamina with constant density δ kg/m^2 and in the shape of a sector of a circle. Assume the radius of the circle is a and the central angle of the sector is θ, where $0 \le \theta \le \pi/2$.

28. Calculate the unit tangent, curvature, and unit normal to the curve at the given point. Also calculate $\mathbf{T} \cdot \mathbf{N}$ as a partial check on your work.

a. The curve described by $\mathbf{r} = \langle 2t, t^2 \rangle$; $(-2, 1)$.

b. The curve described by the polar equation $r = \theta^2$; $(1, 1)$.

c. The graph of the function $f(x) = e^x$; $(2, e^2)$.

d. The curve described by $\mathbf{r} = \langle 2 \cos t, \sin t \rangle$, $0 \le t \le 2\pi$; $(2, 0)$.

29. The position vector of a particle is $\mathbf{r} = \langle t \cos t, t \sin t \rangle$ for all $t \ge 0$. Show that the magnitudes of its tangential and normal acceleration vectors are $t/\sqrt{t^2+1}$ and $(t^2+2)/\sqrt{t^2+1}$. Sketch the path.

30. The position vector of a particle is $\mathbf{r} = \sqrt{t}\langle \cos e^t, \sin e^t \rangle$ for all $t \in [0, \ln(2\pi)]$. Show that the magnitudes of its radial and transverse acceleration vectors are $(1 + 4t^2e^{2t})/(4t^{3/2})$ and $e^t\left(\sqrt{t} + 1/\sqrt{t}\right)$. Sketch the path.

31. How would you convince a skeptical, intelligent friend who has not had calculus that a region with infinite extent may have finite area?

32. Give examples of improper integrals of the first and second kinds. Give an example of an improper integral that is not of the first or second kind and explain how questions of convergence or divergence for such integrals are handled.

33. Determine if each of the integrals

$$\int_3^{12} (x-3)^{-3/2}\,dx$$

$$\int_{12}^{\infty} (x-3)^{-3/2}\,dx$$

is convergent or divergent. If it is convergent, determine its value.

34. For each of the integrals

$$\int_0^{\infty} \frac{dx}{1+x^2},$$

$$\int_0^{\infty} \frac{dx}{1+x^3}, \quad \text{and}$$

$$\int_0^{2} \frac{dx}{4x-x^2-3},$$

determine if it is improper. If it is, decide if it is convergent or divergent.

35. Show that the integral

$$\int_{-\infty}^{\infty} \frac{dx}{(x^2+a^2)^{3/2}}$$

is convergent and find its value.

36. For each integral, determine if it is convergent or divergent. If it is convergent, determine its value.

a. $\displaystyle\int_0^{\pi/2} \tan x\,dx$

b. $\displaystyle\int_0^{1} \ln x\,dx$

c. $\displaystyle\int_1^{3} \left(x/\sqrt{x-1}\right)dx$

d. $\displaystyle\int_0^{\infty} (x/(x^2+1))\,dx$

e. $\displaystyle\int_0^{\infty} xe^{-x}\,dx$

STUDENT PROJECTS

A. TSAR-KOLOKOL (TSAR BELL)

Among the Old Believers in Russia are those who believe that on the "Day of the Last Judgment Tsar-Kolokol will rise slowly from its granite pedestal and begin to toll." Tsar-Kolokol is the largest bell ever cast. This massive bell is presently on display inside the Kremlin. It has never tolled. Shortly after it was cast in 1735, the Tsar Bell was badly damaged in a major Moscow fire. Firemen, afraid that the bell might melt in the heat of the fire, sprayed it with cold water. This resulted in a 13-ton section of the bell breaking from its base.

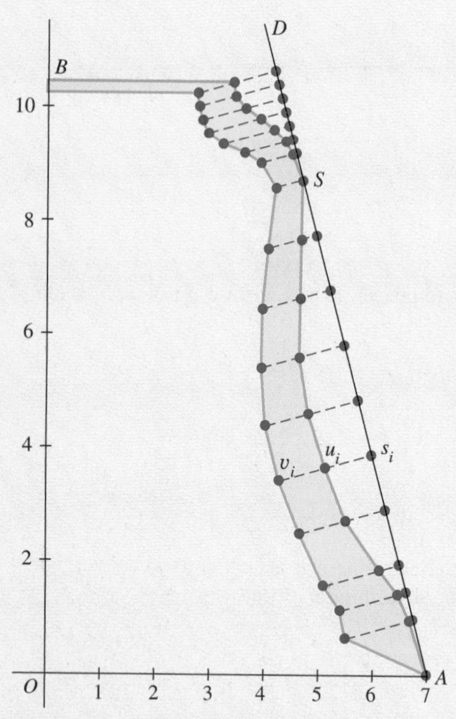

FIGURE 6.90 Profile and layout of the Tsar Bell.

Bell making is a highly complex art. The size, shape, and tone of the bell must be chosen and designed; the molds in which the bell is cast must be designed and built; the alloy must be chosen and metal ordered; furnaces for melting the metal must be planned and built; the bell must be cast without major error; the bell must be separated from its mold, decorated, and tuned; and, especially difficult for very large bells, the bell must be lifted to its final resting place. (Immediately before large bells were cast in Russian bell foundries, icons were placed, candles lighted, prayers offered, and all workmen removed their hats and crossed themselves. This information, the quote at the beginning, and the data for this problem were adapted from the book *The Bells of Russia*, by Edward V. Williams, Princeton University Press, 1985.) In this problem we are concerned with one phase of the making of the Tsar Bell.

The profile of the Tsar Bell is shown in Fig. 6.90. The bottom diameter was divided into 14 equal parts, whose common length was the unit used in the foundry. A line AD was drawn from point A on the bell through S on its shoulder. The bell's profile was given in terms of the lengths of line segments u_i and v_i perpendicular to AD at points s_i on AD, $i = 1, \ldots, 18$, all measured in "foundry units."

These data are given in Table 6.2. The point s_i shown in the figure corresponds to $i = 6$. The point s_6 is four foundry units from A, and the lengths of segments u_6 and v_6 are $7/8$ and $7/8$ foundry units, respectively. It is not clear from Williams's description which other data were specified. In any case, we take OA and angle OAD to be 2.9 meters and 75°, respectively.

TABLE 6.2 Tsar Bell profile data.

i	s_i	u_i	v_i	i	s_i	u_i	v_i
1	0	0	0	10	8	9/32	5/8
2	1	1/16	39/32	11	9	0	1/2
3	1.5	5/32	35/32	12	9.5	1/16	19/32
4	2	3/8	17/16	13	9.75	1/8	25/32
5	3	3/4	7/8	14	10	9/32	31/32
6	4	7/8	7/8	15	10.25	15/32	1
7	5	15/16	13/16	16	10.5	11/16	13/16
8	6	27/32	23/32	17	10.75	13/16	11/16
9	7	9/16	23/32	18	11.0	25/32	11/16

PROBLEM 1 Approximate the volume (in cubic meters) of the Tsar Bell using some numerical integration strategy.

PROBLEM 2 Calculate the mass of the Tsar Bell, given that it is 85 percent copper and 15 percent tin by mass. The densities of copper and tin are 8.96 g/cm^3 and 7.30 g/cm^3, respectively.

PROBLEM 3 Allowing for a 5 percent loss from melting and casting, calculate the amounts of copper and tin that must be ordered.

PROBLEM 4 Determine the height of the Tsar Bell.

B. OPTIMAL SPRAYER

A mobile field irrigation system is shown in Fig. 6.91. Pipe is supported on a framework attached to water-driven wheels w1, w2, w3, ... Water emitters e1, e2, e3, ... are spaced evenly along the pipe. The entire apparatus moves through the field at a uniform speed. We assume that each emitter distributes water uniformly in a circular pattern. Ignoring the special cases of the first and last emitters, determine the spacing of the emitters to optimize the uniformity of coverage. No point in the field is to receive water from more than two emitters.

We may assume that the emitter spray radius is 1 unit. Because no point of the field receives water from more than two emitters, the problem reduces to determining the spacing parameter c. See Fig. 6.92. Note that $1 \le c \le 2$.

PROBLEM 1 Explain why it is useful to find an expression $w(x)$ whose value is proportional to the water received by points along the vertical line through $(x, 0)$, where $0 \le x \le c/2$. Show that $w(x)$ is given by

$$w(x) = \begin{cases} \sqrt{1 - x^2}, & 0 \le x \le c - 1 \\ \sqrt{1 - x^2} + \sqrt{1 - (x - c)^2}, & c - 1 \le x \le c/2. \end{cases}$$

Why may the domain of w be restricted to $[0, c/2]$? Evidently, we have set the proportionality constant equal to 1. Give a justification for this simplification.

PROBLEM 2 The total amount of water received on $[0, c/2]$ is equal to $\int_0^{c/2} w(x)\,dx$. Show by integration that the total amount of water received on $[0, c/2]$ is $\pi/4$. Show also that this result follows directly from Fig. 6.92 by "common sense." The average amount of water received on $[0, c/2]$ is the average value of the function $w(x)$ on this interval.

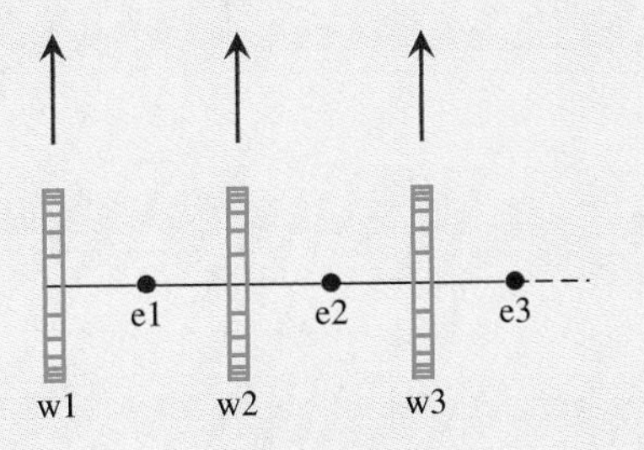

FIGURE 6.91 Sprayer schematic.

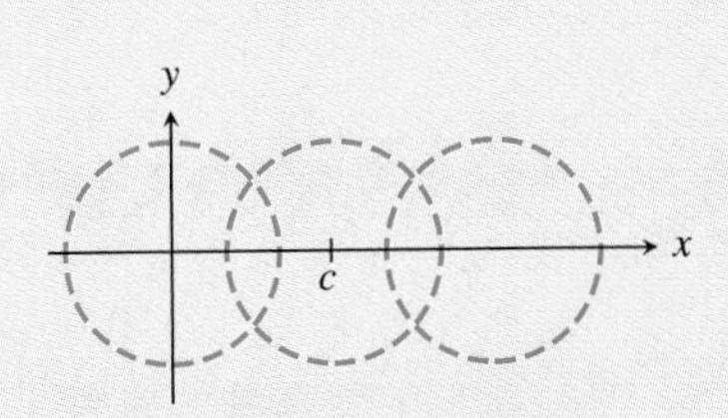

FIGURE 6.92 Emitter coverage.

PROBLEM 3 From the previous calculation of total water, the average value of w on $[0, c/2]$ is $\pi/(2c)$. Give an argument as to why we may optimize the uniformity of coverage by minimizing the "average variation" function

$$g(c) = (2/c)\int_0^{c/2} (w(x) - \pi/(2c))^2\,dx, \qquad 1 \le c \le 2.$$

PROBLEM 4 Show that $g(c)$ may be written in the form

$$g(c) = \frac{k_1}{c} + \frac{k_2}{c^2} + \frac{4}{c}\int_{c-1}^{c/2} \sqrt{1 - x^2}\,\sqrt{1 - (x - c)^2}\,dx, \qquad 1 \le c \le 2,$$

where k_1 and k_2 are constants.

PROBLEM 5 Use a numerical integration algorithm to calculate g at a sufficient number of points so that the point at which g takes its minimum may be found, accurate to one decimal place.

7

Infinite Series, Sequences, and Approximations

A well-known effect of perspective is that each telephone pole in a long line of poles receding into the distance appears to be smaller than the one closer to you. You could describe this effect by a geometric sequence in which each pole is multiplied by a scaling factor less than one.

See Example 3, p. 581, for further explanation.

Set your calculator in radian mode and use it to find the sine of 1.18. You may be struck by two things when you do this: The answer 0.924606012408 appears very quickly, and the 12-decimal-place accuracy is impressive. How does the calculator determine the value to 12 decimal places, and how does it do the calculation so quickly? The procedure that your calculator uses is based on a formula or algorithm that is known to provide a good approximation to the sine function. This formula or algorithm is programmed into the calculator so that it runs very quickly and gives 12-decimal-place accuracy.

Nowadays most calculators use rapidly convergent recursive sequences to calculate values of trigonometric (and other) functions. The algorithms for calculating the terms of the sequences are built into the calculator. In this chapter we study sequences and infinite series. We will see that sequences and infinite series can provide ways to obtain good approximations to functions, definite integrals, and solutions to some equations.

7.1 Taylor Polynomials

We have seen that the equation of a line tangent to the graph of a function can be used to provide a good approximation to that function, at least near the point of tangency. As an example, let $f(x) = e^x$. The line tangent to the graph of $y = e^x$ at the point $(0, 1)$ has equation

$$L(x) = f(0) + f'(0)(x - 0) = e^0 + e^0(x - 0) = 1 + x.$$

If x is close to 0, then $L(x) = 1 + x$ is a good approximation to e^x:

$$e^x \approx 1 + x.$$

This can be seen by looking at the graphs of $y = e^x$ and $y = 1 + x$ on a small interval centered at 0, and noting that the two graphs are very close together. See Fig. 7.1.

When we sketch the graphs of $y = e^x$ and $y = L(x)$ on a larger interval centered at 0, we see that if x is *not* close to 0, then $L(x)$ is not very close to e^x. See Fig. 7.2.

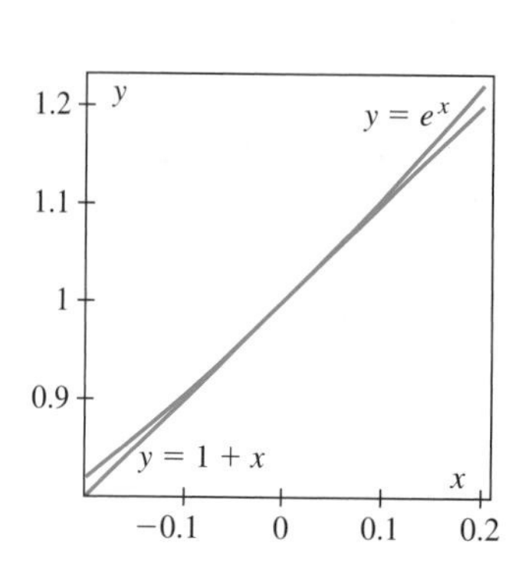

FIGURE 7.1 For $-0.2 \le x \le 0.2$, $1 + x$ is a good approximation to e^x.

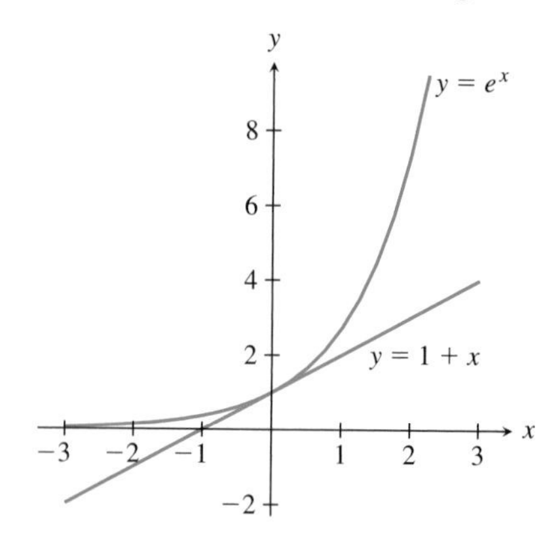

FIGURE 7.2 The line tangent to the graph of $y = e^x$ at $(0, 1)$ is not close to the graph when x is not close to 0.

For example, when $x = 2.1$, the values of e^x and $L(x) = 1 + x$ differ by

$$e^{2.1} - (1 + 2.1) \approx 5.1,$$

that is, by about 62% of the value of $e^{2.1}$. Can we find a better approximation to e^x, one that is closer to e^x throughout the interval $-1 \leq x \leq 1$? One solution is to replace the polynomial approximation $1 + x$ to e^x by a polynomial of degree two or more. We examine this idea in the following Investigation.

INVESTIGATION

Approximating the Exponential Function by a Quadratic

The tangent line approximation $e^x \approx 1 + x$ works well because near the point $(0, 1)$ the graph of $y = e^x$ looks very much like the graph of the tangent line equation $y = L(x) = 1 + x$. This is because

a) $L(x)$ and e^x both take the value 1 at $x = 0$
b) $L(x)$ and e^x have the same derivative at $x = 0$.

However, because the graph of $y = e^x$ is not a straight line, it is not surprising that the graphs of $y = e^x$ and $y = L(x)$ are not close when x is not close to 0. We might get a better approximation if we replace the tangent line equation by a simple equation whose graph "bends" like the graph of $y = e^x$ at $(0, 1)$. To make the bend the same, we choose our approximation so that it has the same second derivative as e^x at $x = 0$. At the same time, we want to retain the features **a)** and **b)** that made the tangent line a good approximation for x near 0. Thus, for our next try at approximating e^x, we seek a quadratic expression

$$Q(x) = a_0 + a_1 x + a_2 x^2 \tag{1}$$

for which

a) $Q(x)$ and e^x take the same value at $x = 0$
b) $Q(x)$ and e^x have the same first derivative at $x = 0$
c) $Q(x)$ and e^x have the same second derivative at $x = 0$.

We use these conditions to find the values of a_0, a_1, and a_2. For notational convenience, we use the alternative notation $\exp(x)$ for e^x. Because

$$a_0 = Q(0) \quad \text{and} \quad \exp(0) = 1,$$

condition **a)** will be satisfied if $a_0 = 1$. Because

$$Q'(0) = a_1 \quad \text{and} \quad \exp'(0) = \exp(0) = 1,$$

condition **b)** will be satisified if $a_1 = 1$. Because

$$Q''(0) = 2a_2 \quad \text{and} \quad \exp''(0) = \exp(0) = 1,$$

condition **c)** will be satisfied if $a_2 = 1/2$. Substituting the values for a_0, a_1, and a_2 into (1), we have

$$Q(x) = 1 + x + \frac{1}{2}x^2.$$

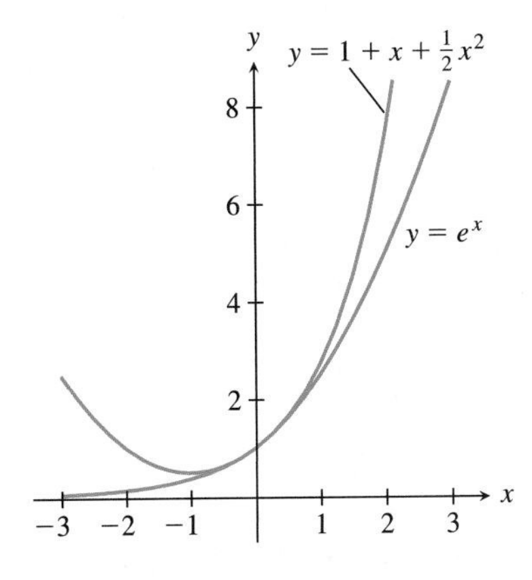

FIGURE 7.3 The quadratic gives a good approximation to e^x for x near 0.

Figure 7.3 shows the graphs of $y = e^x$ and $y = Q(x)$ for $-3 \leq x \leq 3$. Comparing this picture with Fig. 7.2, we see that the quadratic approximation to e^x does a

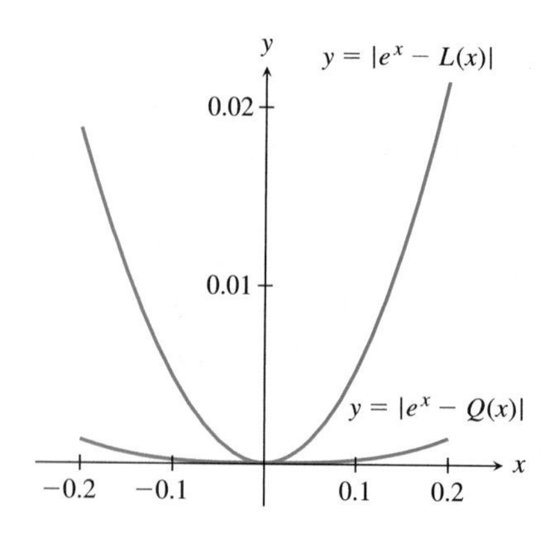

FIGURE 7.4 The quadratic approximation is much better than the tangent line approximation on $-0.2 \le x \le 0.2$.

better job than the tangent line approximation. Neither one is particularly good when $|x| > 1$, but the quadratic $Q(x)$ is close to e^x on a larger interval than was $L(x)$.

Figure 7.4 shows the graphs of

$$y = |e^x - L(x)| \quad \text{and} \quad y = |e^x - Q(x)|$$

for $-0.2 \le x \le 0.2$. These graphs show the absolute value of the errors when $L(x)$ or $Q(x)$ is used to approximate e^x. For x in the interval $-0.2 \le x \le 0.2$, the error made when we use $Q(x)$ to approximate e^x is much less than the error when we use $L(x)$.

As another check, let's calculate the values of e^x, $L(x)$, and $Q(x)$ for $x = 0.15$:

$$e^{0.15} \approx 1.16183424273, \qquad L(0.15) = 1.15, \qquad Q(0.15) = 1.16125.$$

The error in the approximation $e^{0.15} \approx Q(0.15)$ is about $1/20$ of the error in the approximation $e^{0.15} \approx L(0.15)$.

Because of the significant improvement when we approximate e^x with $Q(x)$ it seems natural to try approximating e^x with polynomials of degree three or higher.

EXAMPLE 1 A cubic approximation to e^x

Find a polynomial $C(x)$ of degree 3 that agrees with e^x and its first three derivatives at $x = 0$. Graphically investigate the error when $C(x)$ is used to approximate e^x on $-1 \le x \le 1$. Compare the absolute error functions $|e^x - Q(x)|$ and $|e^x - C(x)|$ on the interval $[-1, 1]$, where $Q(x)$ is the quadratic approximation in the Investigation.

Solution

Let

$$C(x) = a_0 + a_1 x + a_2 x^2 + a_3 x^3. \tag{2}$$

The first three derivatives of $C(x)$ and $e^x = \exp(x)$, all evaluated at $x = 0$, are

$$\begin{aligned} C(0) &= a_0 && \text{and} && \exp(0) = 1, \\ C'(0) &= a_1 && \text{and} && \exp'(0) = 1, \\ C''(0) &= 2a_2 && \text{and} && \exp''(0) = 1, \\ C'''(0) &= 6a_3 && \text{and} && \exp'''(0) = 1. \end{aligned}$$

Equating the expressions in each line, we find $a_0 = 1$, $a_1 = 1$, $a_2 = 1/2$, and $a_3 = 1/6$. Substituting these values in (2) we get

$$C(x) = 1 + x + \frac{1}{2}x^2 + \frac{1}{6}x^3.$$

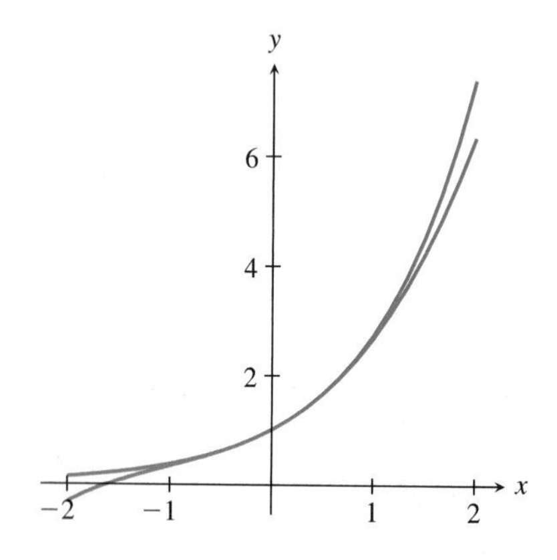

FIGURE 7.5 The graph of $y = e^x$ and $y = C(x) = 1 + x + x^2/2 + x^3/6$ are close for $-2 \le x \le 2$.

Figure 7.5 shows the graphs of $y = e^x$ and $y = C(x)$ on $-2 \le x \le 2$. (We use this larger interval so we can better distinguish between the two graphs.) This figure illustrates how well $C(x)$ approximates e^x. Note that the graphs

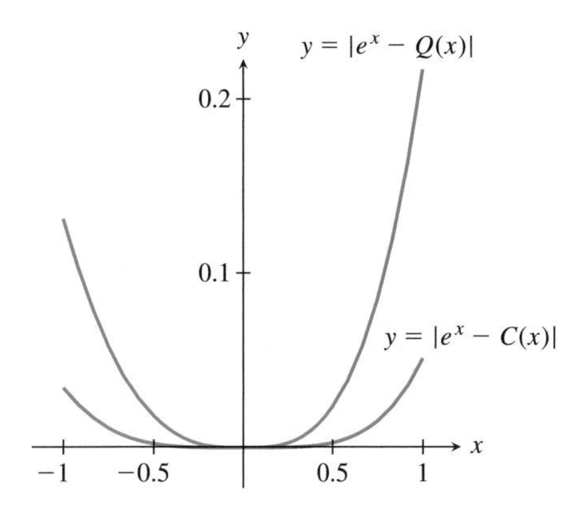

FIGURE 7.6 For $-1 \leq x \leq 1$, we see $|e^x - C(x)|$ is much less than $|e^x - Q(x)|$.

are almost coincident for $-1 \leq x \leq 1$. In Fig. 7.6 we see the graphs of the absolute error functions

$$y = |e^x - C(x)| \quad \text{and} \quad y = |e^x - Q(x)|$$

for $-1 \leq x \leq 1$. These graphs illustrate the errors when $Q(x)$ and $C(x)$ are used to approximate e^x. The figure shows that for $-1 \leq x \leq 1$, $C(x)$ approximates e^x better than $Q(x)$. For example, if we compare the values of $Q(x)$ and $C(x)$ for $x = -0.2$, the approximation $e^x \approx C(x)$ is better than the approximation $e^x \approx Q(x)$:

$$e^{-0.2} \approx 0.818730753078, \qquad Q(-0.2) = 0.82,$$
$$C(-0.2) \approx 0.818667.$$

Taylor Polynomials

We now generalize the idea introduced in the Investigation and in Example 1. Suppose that we wish to use a polynomial of degree n to approximate a function f near a point c. If f has n derivatives at $x = c$, then we can find a polynomial $P(x)$ of degree n that has the same value as f at $x = c$ and such that P and f have the same first derivative, the same second derivative,..., and the same nth derivative at $x = c$.

Because we will be evaluating P and several of its derivatives at $x = c$, it is convenient to express $P(x)$ in powers of $x - c$. Hence, assume that P has the form

$$P(x) = a_0 + a_1(x - c) + a_2(x - c)^2 + \cdots + a_n(x - c)^n. \tag{3}$$

Substituting $x = c$ into (3), we see immediately that

$$P(c) = a_0. \tag{4}$$

It also follows from (3) that

$$P'(x) = a_1 + 2a_2(x - c) + 3a_3(x - c)^2 + \cdots + na_n(x - c)^{n-1}.$$

Substituting $x = c$ into this equation yields

$$P'(c) = a_1. \tag{5}$$

The second derivative of P is

$$P''(x) = (2 \cdot 1)a_2 + (3 \cdot 2)a_3(x - c) + (4 \cdot 3)a_4(x - c)^2 + \cdots$$
$$+ (n(n - 1))a_n(x - c)^{n-2}.$$

Hence,

$$P''(c) = 2a_2. \tag{6}$$

Continuing in this fashion, we see that for $k = 1, 2, \ldots, n$, the kth derivative $P^{(k)}$ of P can be written as

$$P^{(k)}(x) = [k(k - 1)(k - 2) \cdots 1]a_k + [(k + 1)k \cdots 2]a_{k+1}(x - c) + \cdots$$
$$+ [n(n - 2)(n - 2) \cdots (n - k + 1)]a_n(x - c)^{n-k}.$$

Putting $x = c$ in this expression gives

$$P^{(k)}(c) = k(k - 1)(k - 2) \cdots 1a_k = k!\, a_k. \tag{7}$$

Equating the derivatives of P at $x = c$ to the derivatives of f at $x = c$ and taking (4), (5), (6), and (7) into account, we obtain the equations

$$\begin{aligned} f(c) &= P(c) = a_0, \\ f'(c) &= P'(c) = a_1, \\ f''(c) &= P''(c) = 2!\,a_2, \\ &\vdots \\ f^{(k)}(c) &= P^{(k)}(c) = k!\,a_k, \\ &\vdots \\ f^{(n)}(c) &= P^{(n)}(c) = n!\,a_n. \end{aligned}$$

Solving these equations for the coefficients $a_0, a_1, \ldots, a_n$ we find

$$a_0 = f(c), \qquad a_1 = \frac{f'(c)}{1!}, \qquad a_2 = \frac{f''(c)}{2!}, \qquad \ldots,$$
$$a_k = \frac{f^{(k)}(c)}{k!}, \qquad \ldots, \qquad a_n = \frac{f^{(n)}(c)}{n!}.$$

Substituting the values for these coefficients into (3), we find

$$\begin{aligned} P(x) &= f(c) + \frac{f'(c)}{1!}(x - c) + \frac{f''(c)}{2!}(x - c)^2 + \cdots \\ &\quad + \frac{f^{(k)}(c)}{k!}(x - c)^k + \cdots + \frac{f^{(n)}(c)}{n!}(x - c)^n \\ &= \sum_{k=0}^{n} \frac{f^{(k)}(c)}{k!}(x - c)^k. \end{aligned}$$

This polynomial is called a **Taylor polynomial.**

WEB

DEFINITION Taylor Polynomial

Let f be a function, and suppose that the nth derivative of f exists at $x = c$. The nth Taylor polynomial for f at $x = c$ is the polynomial

$$\begin{aligned} T_n(x; c) &= f(c) + \frac{f'(c)}{1!}(x - c) + \frac{f''(c)}{2!}(x - c)^2 + \cdots \\ &\quad + \frac{f^{(k)}(c)}{k!}(x - c)^k + \cdots + \frac{f^{(n)}(c)}{n!}(x - c)^n \qquad (8) \\ &= \sum_{k=0}^{n} \frac{f^{(k)}(c)}{k!}(x - c)^k. \end{aligned}$$

In the last line of (8) we use the convenient conventions $f^{(0)}(c) = f(c)$ and $0! = 1$.

We defined the Taylor polynomial of degree n for f at $x = c$ so that it has the same value and the same first, second, ..., and nth derivatives as f at $x = c$. Thus, we expect that the Taylor polynomial provides a good approximation to $f(x)$ for x close to c.

EXAMPLE 2 Let $f(x) = \tan x$. Find $T_3(x; 0)$ and estimate the maximum error if we use $T_3(x; 0)$ to approximate $\tan x$ on $-0.5 \le x \le 0.5$.

Solution

We begin by finding the first three derivatives of $\tan x$:

$$\begin{aligned}
f^{(0)}(x) &= f(x) = \tan x, \\
f^{(1)}(x) &= f'(x) = (\tan x)' = \sec^2 x, \\
f^{(2)}(x) &= f''(x) = (\sec^2 x)' = 2 \sec x (\sec x)' = 2 \sec^2 x \tan x, \\
f^{(3)}(x) &= f'''(x) = (2 \sec^2 x \tan x)' = 2(\sec^2 x)'(\tan x) + 2(\sec^2 x)(\tan x)' \\
&= 4 \sec^2 x \tan^2 x + 2 \sec^4 x.
\end{aligned}$$

Substitute $x = 0$ in these derivatives to obtain the coefficients of the Taylor polynomial:

$$\frac{f^{(0)}(0)}{0!} = \frac{\tan 0}{1} = 0,$$

$$\frac{f^{(1)}(0)}{1!} = \frac{\sec^2 0}{1} = 1,$$

$$\frac{f^{(2)}(0)}{2!} = \frac{2 \sec^2 0 \tan 0}{2} = 0,$$

$$\frac{f^{(3)}(0)}{3!} = \frac{4 \sec^2 0 \tan^2 0 + 2 \sec^4 0}{6} = \frac{1}{3}.$$

Hence

$$\begin{aligned}
T_3(x; 0) &= \frac{f(0)}{0!} + \frac{f'(0)}{1!}(x - 0) + \frac{f''(0)}{2!}(x - 0)^2 + \frac{f^{(3)}(0)}{3!}(x - 0)^3 \\
&= x + \frac{1}{3}x^3.
\end{aligned}$$

When we use $T_3(x; 0)$ to approximate $\tan x$, the absolute value of the error is

$$|\tan x - T_3(x; 0)|.$$

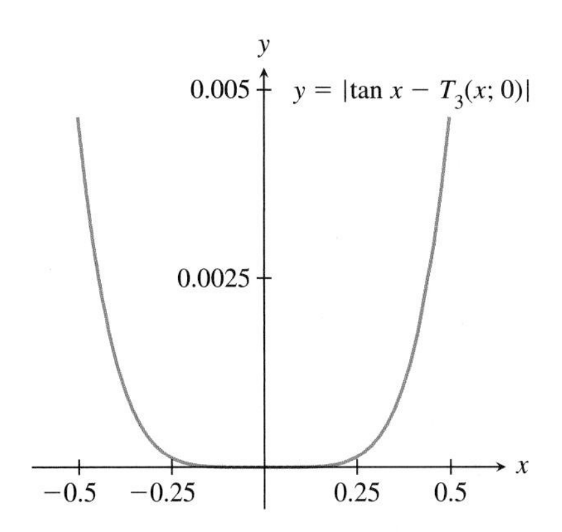

FIGURE 7.7 When $-0.5 \le x \le 0.5$, the difference between $\tan x$ and $T_3(x; 0)$ is no more than 0.005 in absolute value.

Figure 7.7 shows the graph of

$$y = |\tan x - T_3(x; 0)|.$$

This graph of the absolute error made when we approximate $\tan x$ with $T_3(x; 0)$ shows that for $-0.5 \le x \le 0.5$, the error is less than 0.005 in absolute value. Figure 7.8 shows the graph of $y = \tan x$ and $y = T_3(x; 0)$ for $-1 \le x \le 1$. We use this larger interval so we can better distinguish between the two graphs. The two graphs are very close; in fact they are almost coincident on the interval $-0.5 \le x \le 0.5$.

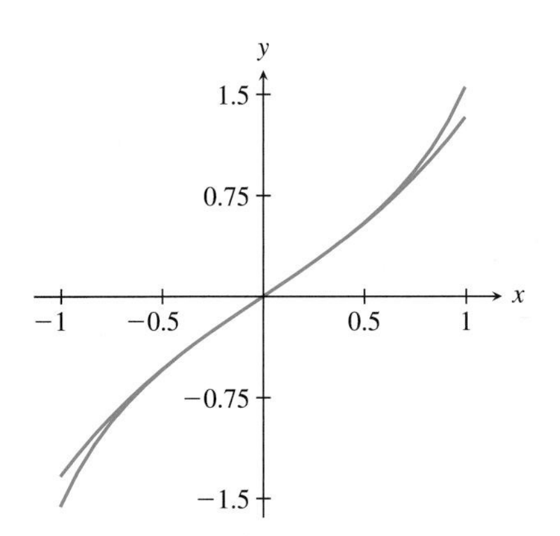

FIGURE 7.8 The graphs of $y = \tan x$ and $y = T_3(x; 0)$. Can you figure out which graph goes with which function?

EXAMPLE 3 Let $p(x) = -2x^3 - 3x^2 + 4x - 5$. Find all of the Taylor polynomials $T_0(x; -2), T_1(x; -2), T_2(x; -2), \dots$ for p.

Solution

We start by finding the values of the derivatives of p at $x = -2$.

$$\begin{aligned}
p(x) &= -2x^3 - 3x^2 + 4x - 5 & p(-2) &= -9,\\
p'(x) &= -6x^2 - 6x + 4 & p'(-2) &= -8,\\
p''(x) &= -12x - 6 & p''(-2) &= 18,\\
p'''(x) &= -12 & p'''(-2) &= -12,\\
p^{(4)}(x) &= p^{(5)}(x) = \cdots = 0 & p^{(4)}(-2) &= p^{(5)}(-2) = \cdots = 0.
\end{aligned}$$

The Taylor polynomials $T_0(x; -2), T_1(x; -2), \dots$ for p are

$$T_0(x; -2) = \frac{p^{(0)}(-2)}{0!} = -9,$$

$$T_1(x; -2) = \frac{p^{(0)}(-2)}{0!} + \frac{p^{(1)}(-2)}{1!}(x - (-2)) = -9 - 8(x + 2),$$

$$\begin{aligned}
T_2(x; -2) &= \frac{p^{(0)}(-2)}{0!} + \frac{p^{(1)}(-2)}{1!}(x - (-2)) + \frac{p^{(2)}(-2)}{2!}(x - (-2))^2\\
&= -9 - 8(x + 2) + 9(x + 2)^2,
\end{aligned}$$

$$\begin{aligned}
T_3(x; -2) &= \frac{p^{(0)}(-2)}{0!} + \frac{p^{(1)}(-2)}{1!}(x - (-2))\\
&\quad + \frac{p^{(2)}(-2)}{2!}(x - (-2))^2 + \frac{p^{(3)}(-2)}{3!}(x - (-2))^3\\
&= -9 - 8(x + 2) + 9(x + 2)^2 - 2(x + 2)^3.
\end{aligned}$$

Because $p^{(k)}(-2) = 0$ for $k = 4, 5, 6, \dots$, we see that in any Taylor polynomial $T_n(x; -2)$ with $n \geq 4$, all terms $(x + 2)^k$ with $k \geq 4$ will have coefficient 0. Hence for $n = 4, 5, 6, \dots$ we have

$$T_n(x; -2) = T_3(x; -2) = -9 - 8(x + 2) + 9(x + 2)^2 - 2(x + 2)^3.$$

In fact, for $n \geq 4$ it turns out that $T_n(x; -2) = T_3(x; -2) = p(x)$. You can verify this by expanding and combining the terms of $T_3(x; -2)$, and then comparing the result with $p(x)$. The graphs of $y = p(x)$ and the first few Taylor polynomials are shown in Fig. 7.9.

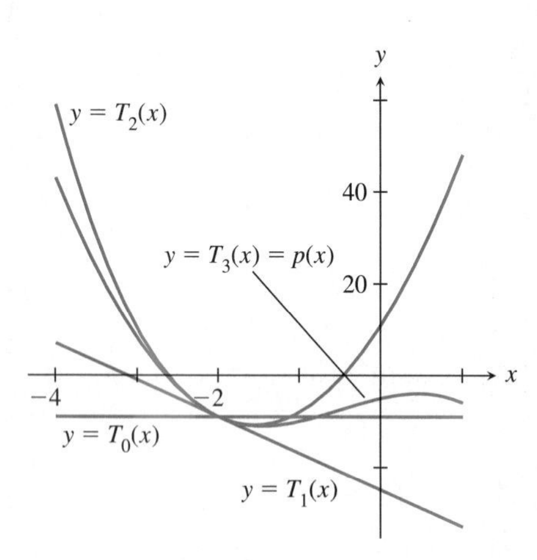

FIGURE 7.9 Graphs of some Taylor polynomials for p. For $n \geq 3$, the graphs of $y = T_n(x; -2)$ and $y = p(x)$ coincide.

EXAMPLE 4 Let $\ell(x) = \ln x$.

a) Let n be a nonnegative integer. Find $T_n(x; 1)$ for ℓ.
b) Investigate the error in the approximation $\ln x \approx T_5(x; 1)$ for $0.8 \leq x \leq 1.2$.

Solution

a) First note that $\ell(1) = \ln 1 = 0$, so the constant term in each Taylor polynomial $T_n(x; 1)$ is 0. In particular, $T_0(x; 1) = 0$. Now let $n \geq 1$. Computing the coefficients for the other $T_n(x; 1)$, we have

$$\ell^{(1)}(x) = (\ln x)' = x^{-1} \qquad a_1 = \frac{\ell^{(1)}(1)}{1!} = 1,$$

$$\ell^{(2)}(x) = (x^{-1})' = -x^{-2} \qquad a_2 = \frac{\ell^{(2)}(1)}{2!} = -\frac{1}{2},$$

$$\ell^{(3)}(x) = (-x^{-2})' = 2x^{-3} \qquad a_3 = \frac{\ell^{(2)}(1)}{3!} = \frac{1}{3},$$

$$\ell^{(4)}(x) = (2x^{-3})' = (-3!)x^{-4} \qquad a_4 = \frac{\ell^{(2)}(1)}{4!} = -\frac{1}{4},$$

$$\vdots \qquad\qquad \vdots$$

$$\ell^{(k)}(x) = ((-1)^k(k-2)!x^{-(k-1)})' \qquad a_k = \frac{\ell^{(k)}(1)}{k!} = (-1)^{k+1}\frac{1}{k},$$

$$= (-1)^{k+1}(k-1)!x^{-k}$$

$$\vdots \qquad\qquad \vdots$$

The coefficients $a_1, a_2, \dots a_n$ for the Taylor polynomial are reciprocals of the positive integers, with alternating signs. Recalling that the constant term of each Taylor polynomial $T_n(x; 1)$ for ℓ is 0, it follows that

$$\begin{aligned} T_n(x; 1) &= \sum_{k=0}^{n} \frac{\ell^{(k)}(1)}{k!}(x-1)^k = 0 + \sum_{k=1}^{n} \frac{\ell^{(k)}(1)}{k!}(x-1)^k \\ &= \sum_{k=1}^{n} \frac{(-1)^{k+1}}{k}(x-1)^k \\ &= (x-1) - \frac{1}{2}(x-1)^2 + \frac{1}{3}(x-1)^3 - \cdots \\ &\quad + \frac{(-1)^{n+1}}{n}(x-1)^n. \end{aligned} \tag{9}$$

b) Setting $n = 5$ in (9), we find

$$T_5(x; 1) = (x-1) - \frac{(x-1)^2}{2} + \frac{(x-1)^3}{3} - \frac{(x-1)^4}{4} + \frac{(x-1)^5}{5}.$$

Figure 7.10 shows the graph of

$$y = |\ln x - T_5(x; 1)|.$$

From this graph we see that when $0.8 \leq x \leq 1.2$, the absolute value of the error in the approximation

$$\ln(x) \approx T_5(x; 1)$$

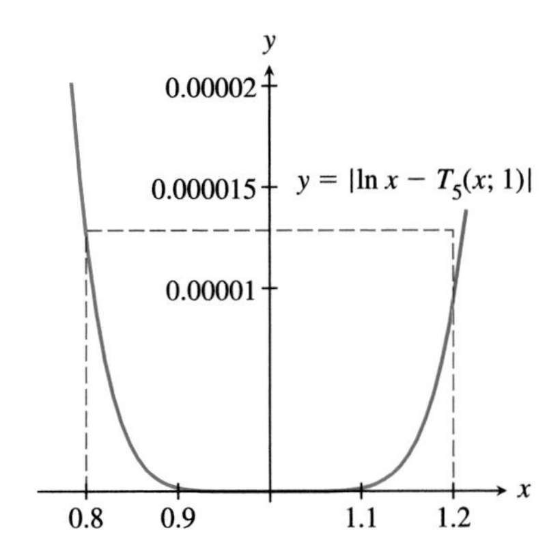

FIGURE 7.10 For $0.8 \leq x \leq 1.2$, the difference between $\ln x$ and $T_5(x; 1)$ is less than 0.000015 in absolute value.

is less than 0.000015. Table 7.1 gives another way of looking at the error. This table shows the values of $\ln x$, $T_5(x; 1)$, and $y = |\ln x - T_5(x; 1)|$ for several values of x between 0.8 and 1.2.

TABLE 7.1

x	$\ln x$	$T_5(x; 1)$	$\|\ln x - T_5(x; 1)\|$
0.80	−0.2231435513	−0.2231306667	1.288×10^{-5}
0.85	−0.1625189295	−0.1625167500	2.179×10^{-6}
0.90	−0.1053605157	−0.1053603333	1.823×10^{-7}
0.95	−0.0512932944	−0.0512932917	2.721×10^{-9}
1.00	0.0	0.0	0.0
1.05	0.0487901642	0.0487901667	2.497×10^{-9}
1.10	0.0953101798	0.0953103333	1.535×10^{-7}
1.15	0.1397619424	0.1397636250	1.683×10^{-6}
1.20	0.1823215568	0.1823306667	9.110×10^{-6}

Exercises 7.1

Exercises 1–6: Find $T_0(x; 0)$, $T_1(x; 0)$, $T_2(x; 0)$, and $T_3(x; 0)$ for the indicated function, and then calculate $f(0.1)$ and $T_3(0.1; 0)$.

1. $f(x) = \dfrac{1}{1 + x}$

2. $f(x) = (3 + x)^{10}$

3. $f(x) = \sin x$

4. $f(x) = \cos x$

5. $f(x) = \arctan x$

6. $f(x) = e^{2x}$

Exercises 7–12: Find $T_0(x; c)$, $T_1(x; c)$, $T_2(x; c)$, and $T_3(x; c)$ for the indicated function and the given value of c. Then calculate $f(a)$ and $T_3(a; c)$ for the given value of a.

7. $f(x) = \dfrac{1}{1 + x}$, $c = -2$, $a = -1.95$

8. $f(x) = \tan x$, $c = \frac{\pi}{4}$, $a = \frac{\pi}{4} - 0.03$

9. $f(x) = x^2 \ln x$, $c = 3$, $a = 2.95$

10. $f(x) = \sin x$, $c = \frac{3\pi}{4}$, $a = \frac{3\pi}{4} + 0.1$

11. $f(x) = \arctan x$, $c = 1$, $a = 1.05$

12. $f(x) = e^x$, $c = \ln 3$, $a = 1$

13. In Fig. 7.5, which graph is the graph of $y = e^x$?

14. In Fig. 7.8, which graph is the graph of $y = \tan x$?

15. In Example 3, verify that $T_n(x; -2) = p(x)$ for $n \geq 3$.

16. Verify at least four of the entries in Table 7.1.

17. For small values of x, scientists sometimes replace the expression $\sqrt{1 + x}$ by $1 + (1/2)x$ to simplify their work. Use a Taylor polynomial to justify this practice. For what range of values of x is the error less than 0.001?

18. The nth Taylor polynomial about 1 for a function f is given by

$$T_n(x; 1) = \sum_{k=0}^{n} \frac{(x - 1)^k}{k!\,4^k}.$$

What is the value of $f^{(8)}(1)$?

19. The nth Taylor polynomial about -2 for a function g is given by

$$T_n(x; -2) = \sum_{k=0}^{n} 3\frac{(x + 2)^k}{k + 1}.$$

What is the value of $g^{(8)}(-2)$?

20. The nth Taylor polynomial about c for a function f is given by

$$T_n(x; c) = \sum_{k=0}^{n} (-1)^k \frac{(x - c)^k}{30}.$$

What is $f^{(k)}(c)$?

21. The nth Taylor polynomial about c for a function g is given by

$$T_n(x; c) = \sum_{k=0}^{n} k^k \frac{(x - c)^k}{(3(k + 1))^{k+1}}.$$

What is $g^{(k)}(c)$?

22. How many different functions f have Taylor polynomial

$$T_2(x; 0) = 1 - x + 2x^2?$$

Explain your answer.

23. How many different functions g have Taylor polynomial

$$T_3(x; -1) = (x + 1) - \frac{(x + 1)}{\pi} + \frac{(x + 1)^2}{e}?$$

Explain your answer.

24. Suppose that we have a Taylor polynomial $T_n(x; c)$ for a function f. Let m be a nonnegative integer with $m < n$. How could we quickly get $T_m(x; c)$ from $T_n(x; c)$?

T **25.** Let $I = [-\pi, \pi]$.

a. Find $T_6(x; 0)$ for $\sin x$.

b. Graph $y = \sin x$ and $y = T_6(x; 0)$ on I.

c. Graph $y = |T_6(x; 0) - \sin x|$ on I.

d. Use the graph to estimate the maximum value of $|T_6(x; 0) - \sin x|$ on I.

e. Calculate $\sin x$ and $T_6(x; 0)$ for $x = -0.2, 0, 0.5$, and 1.

26. Let $I = (-\infty, \infty)$.

a. Find $T_{17}(x; 2)$ for $x^3 - 3x^2 + 4x - 7$.

b. Graph $y = x^3 - 3x^2 + 4x - 7$ and $y = T_{17}(x; 0)$ on I.

c. Graph $y = |T_{17}(x; 0) - (x^3 - 3x^2 + 4x - 7)|$ on I.

T **27.** Let $I = \left[-\frac{3}{4}, \frac{3}{4}\right]$.

a. Find $T_6(x; 0)$ for $\frac{1}{1-x}$.

b. Graph $y = \frac{1}{1-x}$ and $y = T_6(x; 0)$ on I.

c. Graph $y = \left|T_6(x; 0) - \frac{1}{1-x}\right|$ on I.

d. Use the graph to estimate the maximum value of $\left|T_6(x; 0) - \frac{1}{1-x}\right|$ on I.

e. Calculate $\frac{1}{1-x}$ and $T_6(x; 0)$ for $x = -0.9, -0.05, 0.5$, and 0.7.

T **28.** Let $I = [0, 2]$.

a. Find $T_4(x; 1)$ for $\sqrt{x}$.

b. Graph $y = \sqrt{x}$ and $y = T_4(x; 1)$ on I.

c. Graph $y = \left|T_4(x; 1) - \sqrt{x}\right|$ on I.

d. Use the graph to estimate the maximum value of $\left|T_4(x; 1) - \sqrt{x}\right|$ on I.

e. Calculate $\sqrt{x}$ and $T_4(x; 0)$ for $x = 0.4, 0.8, 1, 1.4$, and 1.8.

Exercises 29–33: Let n be a nonnegative integer. Write out the general Taylor polynomial $T_n(x; c)$.

29. $\dfrac{1}{1-x}$, $c = 0$

30. $\sin x$, $c = 0$

31. $\dfrac{2}{3+x}$, $c = -1$

32. $\sqrt{x}$, $c = 1$

33. $\dfrac{1}{x}$, $c = 0$

34. Let $g(x) = \ln x$, let c be a positive real number, and let n be a nonnegative integer. Write out the general Taylor polynomial $T_n(x; c)$ for g.

35. Let $h(x) = \sqrt{x}$, let $c > 0$, and let n be a nonnegative integer number. Write out the general Taylor polynomial $T_n(x; c)$ for h.

36. Let $f(x) = \sin x$, $g(x) = \cos x$, and $h(x) = \sin x + \cos x$. Let $T_4(x; 0)$ be the fourth-degree Taylor polynomial at 0 for f, $P_4(x; 0)$ be the fourth-degree Taylor polynomial at 0 for g, and $Q_4(x; 0)$ be the fourth-degree Taylor polynomial at 0 for h. Is it true that

$$Q_4(x; 0) = T_4(x; 0) + P_4(x; 0)?$$

37. Let f, g, and h be functions with $h(x) = f(x) + g(x)$ for all real x. Let $T_n(x; c)$ be the nth-degree Taylor polynomial at c for f, $P_n(x; c)$ the nth-degree Taylor polynomial at c for g, and $Q_n(x; c)$ the nth-degree Taylor polynomial at c for h. Show that

$$Q_n(x; c) = T_n(x; c) + P_n(x; c).$$

38. Let $T_9(x; 0)$ be the ninth-degree Taylor polynomial for $\sin x$ about 0 and let $Q_8(x; 0)$ be the eighth-degree Taylor polynomial for $\cos x$ about 0. Show that

$$\frac{d}{dx} T_9(x; 0) = Q_8(x; 0).$$

39. Let $T_6(x; 1)$ be the sixth-degree Taylor polynomial for $\ln x$ about 1 and let $Q_5(x; 1)$ be the fifth-degree Taylor polynomial for $1/x$ about 1. Show that

$$\frac{d}{dx} T_6(x; 1) = Q_5(x; 1).$$

40. Let $T_n(x; c)$ be the nth degree Taylor polynomial for a function f and let $Q_{n-1}(x; c)$ be an $(n - 1)$st degree Taylor polynomial for $f'(x)$. Show that

$$\frac{d}{dx} T_n(x; c) = Q_{n-1}(x; c).$$

41. Let $T_6(x; 2)$ be the sixth-degree Taylor polynomial for $1/x$ about 2 and let $Q_5(x; 2)$ be the fifth-degree Taylor polynomial for $-1/x^2$ about 2. Show that

$$T_6(x; 2) = \frac{1}{2} + \int_2^x Q_5(t; 2)\, dt.$$

42. Let $T_9(x; 0)$ be the ninth-degree Taylor polynomial for $\sin x$ about 0 and let $Q_{10}(x; 0)$ be the tenth-degree Taylor polynomial for $\cos x$ about 0. Show that

$$Q_{10}(x; 0) = \cos 0 - \int_0^x T_9(t, 0)\, dt.$$

43. Let $f(x) = x^{14/3}$ and let $T_n(x; 0)$ be the Taylor polynomial of degree n for f at 0. What is the largest n for which T_n exists?

44. Let $g(x) = |x - 1|^3$ and let $T_n(x; 1)$ be the Taylor polynomial of degree n for g at 1. What is the largest n for which T_n exists?

45. Let $f(x) = x^{8/3}$. For what values of c does $T_2(x; c)$ exist? For what values of c does $T_5(x; c)$ exist?

T **46.** It can be shown that for all real x,

$$\lim_{n\to\infty}\left(1 + \frac{x}{n}\right)^n = e^x.$$

Hence when n is large, $P_n(x) = \left(1 + \frac{x}{n}\right)^n$ can be used as an approximation for e^x. Investigate how $P_5(x)$ and $T_5(x; 0)$ each approximate e^x on the interval $[-2, 2]$. Which approximation do you think is better? Why?

T **47.** A quadratic Newton's method Consider the following procedure for approximating a solution to an equation $f(x) = 0$.

a. Make an initial guess r_0 for the root of the equation.

b. Assume that r_k is our kth guess for the root. Find the second-degree Taylor polynomial $T_2(x; r_k)$ for f at r_k. Find a nearby root of $T_2(x; r_k) = 0$ and let this root be r_{k+1}.

c. Repeat Step 2 until you have approximated the desired root of $f(x) = 0$ to the needed accuracy.

Comment on this procedure. What are its advantages over Newton's method? What are its disadvantages? Use the technique described here to approximate the root of $\cos x - 0.6 = 0$ with initial guess $r_0 = 1$. How does the result after three iterations compare with the result for Newton's method after three iterations?

Exercises 48–51: For the indicated function f and value of c, find $T_n(x; c)$ for $n = 0, 1, 2$. For each of these values of n graph $y = |f(x) - T_n(x; c)|$ on the given interval I. Use the graphs to find E_n, the maximum value of $|f(x) - T_n(x; c)|$ on I. Record E_0, E_1, E_2, and then write a short paragraph describing how E_n changes as n gets larger.

T **48.** $f(x) = e^x$, $c = 0$, $I = [-0.5, 0.5]$

T **49.** $f(x) = \dfrac{1}{2 - x}$, $c = 1$, $I = [0.5, 1.5]$

T **50.** $f(x) = \sin x$, $c = \frac{\pi}{4}$, $I = \left[0, \frac{\pi}{2}\right]$

T **51.** $f(x) = \arcsin x$, $c = \frac{1}{2}$, $I = [0, 1]$

7.2 Approximations and Error

Use your calculator to find approximate values for

$$2^{-3.1415}, \quad \tan(16.8), \quad \text{and} \quad \ln(3456.78)$$

In each case, your calculator gives an answer with at least six significant digits. You are probably confident that the answers given by your calculator are correct to six significant digits. It's very easy in this day and age to take the capabilities of calculators and computers for granted. But have you ever wondered how calculators can take such a wide range of inputs for so many different functions and quickly come up with answers that are correct to six or more digits?

When mathematicians and engineers design a calculator algorithm or procedure for approximating $\sin x$, they not only need a fast algorithm, but they must also know in advance the error in the approximation. In particular, they have to know that the calculator will give an answer correct to at least six significant digits. This means that for all allowable inputs x, the absolute value of the difference between, say, $\sin x$ and the calculator value for $\sin x$ is small. Any company selling a calculator whose algorithms fail to be consistently accurate for all allowable inputs runs a high risk of losing market share.

Any time we deal with an approximation to a number or a function, we must be concerned with the error. In Section 7.1 we defined Taylor polynomials and investigated the use of Taylor polynomials to approximate functions. In this section we take a close look at the error for such approximations.

The Remainder

The error in approximating a function f by its Taylor polynomial T_n can be studied by investigating the difference $f(x) - T_n(x; c)$. This difference is often called the *remainder*.

DEFINITION Taylor Remainder

Let f be a function and let $T_n(x; c)$ be the nth Taylor polynomial for f at $x = c$. The remainder associated with $T_n(x; c)$ is

$$f(x) - T_n(x; c).$$

We denote this remainder by $R_n(x; c)$. Hence

$$R_n(x; c) = f(x) - T_n(x; c).$$

The remainder $R_n(x; c)$ is also called the *error* in the approximation of $f(x)$ by the Taylor polynomial of degree n.

The goal of this section is to determine how well $T_n(x; c)$ approximates $f(x)$ on an interval

$$c - r \leq x \leq c + r$$

in the domain of f. See Fig. 7.11. Because we will usually be interested in the magnitude of the error, not its sign, we will study

$$|R_n(x; c)| = |f(x) - T_n(x; c)|.$$

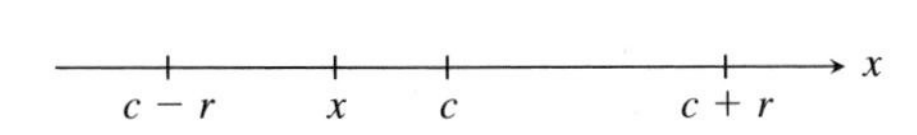

FIGURE 7.11 The interval $c - r \leq x \leq c + r$ on the x-axis.

In many of our applications, we will want to find an answer to one of two problems.

a) Given a Taylor polynomial $T_n(x; c)$, find the maximum value of $|R_n(x; c)|$ on a given interval $c - r \leq x \leq c + r$.

b) Given a number $\epsilon > 0$ and an interval $c - r \leq x \leq c + r$, find n so that

$$|R_n(x; c)| = |f(x) - T_n(x; c)| < \epsilon$$

for $c - r \leq x \leq c + r$.

With the answer to question **a)**, we can judge the accuracy of the approximation $f(x) \approx T_n(x; c)$ for $c - r \leq x \leq c + r$. With the answer to **b)**, we can find out which Taylor polynomial approximates $f(x)$ within a desired tolerance.

EXAMPLE 1 Let $f(x) = \ln x$. Find n so that on the interval $0.5 \leq x \leq 1.5$, the error $R_n(x; 1)$ is no more than 0.0005 in absolute value.

Solution

We need to find n so that

$$|R_n(x; 1)| = |\ln x - T_n(x; 1)| < 0.0005$$

for all x in the interval $0.5 \le x \le 1.5$. In Example 4 of Section 7.1 we showed that

$$T_n(x;1) = \sum_{k=1}^{n} \frac{(-1)^{k+1}}{k}(x-1)^k$$
$$= (x-1) - \frac{(x-1)^2}{2} + \frac{(x-1)^3}{3} - \cdots + (-1)^{n+1}\frac{1}{n}(x-1)^n. \tag{1}$$

We can find an acceptable value of n by graphing

$$y = |\ln x - T_n(x;1)|$$

for different values of n until we find a graph for which the y values stay less than 0.0005 on $0.5 \le x \le 1.5$.

Because we want the error to be quite small, we doubt that $T_1(x;1)$ or $T_2(x;1)$ will be close enough. We try $T_5(x;1)$ first, and check this guess by graphing. Putting $n = 5$ in (1), we get

$$T_5(x;1) = (x-1) - \frac{(x-1)^2}{2} + \frac{(x-1)^3}{3} - \frac{(x-1)^4}{4} + \frac{(x-1)^5}{5}.$$

Figure 7.12 shows the graph of

$$y = |\ln x - T_5(x;1)|.$$

FIGURE 7.12 Near $x = 0.5$ and $x = 1.5$, the error is larger than 0.0005 in absolute value.

We see that near $x = 0.5$ and $x = 1.5$, the absolute value of the error is larger than 0.0005. As supporting numerical evidence, let $x = 1.4$ and use a calculator to see that

$$|\ln 1.4 - T_5(1.4;1)| \approx 0.000509 > 0.0005.$$

We expect that a larger n is needed to get a smaller error, so try $n = 8$ next. Figure 7.13 shows the graph of

$$y = |\ln x - T_8(x;1)|.$$

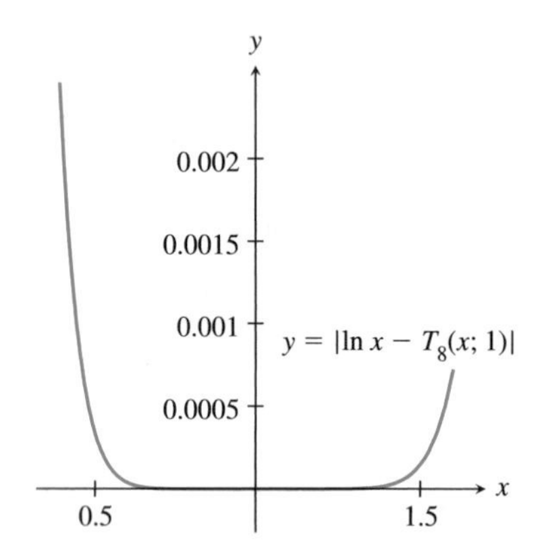

FIGURE 7.13 The graph shows that $|R_8(x;1)| < 0.0005$ for $0.5 \le x \le 1.5$.

Please verify this graph with your own graphing calculator or CAS. We see from the graph that for $0.5 \le x \le 1.5$, the value of

$$|R_8(x;1)| = |\ln x - T_8(x;1)|$$

stays well under 0.0005. Thus for $0.5 \le x \le 1.5$ we have

$$\ln x \approx T_8(x;1)$$

with an error of less than 0.0005 in absolute value. As above we may calculate

$$\ln 1.4 \approx 0.3364722366, \qquad T_8(1.4, 1) \approx 0.3364508038,$$

and

$$|\ln 1.4 - T_8(1.4;1)| \approx 2.143 \times 10^{-5} < 0.0005.$$

What about $T_6(x;1)$ or $T_7(x;1)$? Would either of these result in an absolute error less than 0.0005? What would be the advantages of a value of n smaller than 8?

The Remainder Formula

In Example 1 we worked directly with

$$f(x) - T_n(x; 1)$$

to find information about $R_n(x; 1)$. In our analysis we both evaluated and graphed $R_n(x; 1)$. Note that those calculations required us to evaluate both $f(x)$ and $T_n(x; 1)$. But if we have a need to approximate f, then we may not know how to calculate it conveniently. To better answer questions about approximating functions with their Taylor polynomials, we need another way of working with and estimating $R_n(x; c)$. We will often use the following result.

THEOREM The Taylor Inequality

Let f have a continuous $(n + 1)$st derivative on the interval $c - r \leq x \leq c + r$ and let M_{n+1} be the maximum value of $|f^{(n+1)}(x)|$ for x in this interval. Then for $c - r \leq x \leq c + r$, we have

$$|R_n(x; c)| = |f(x) - T_n(x; c)| \leq \frac{M_{n+1}}{(n + 1)!}|x - c|^{n+1}. \tag{2}$$

INVESTIGATION

Justifying the Taylor Inequality

We justify the Taylor inequality result in the special case $n = 1$. See Exercise 31 for the case $n = 2$. For our argument, we will need the Mean-value Inequality. This inequality follows immediately from the Mean-value Theorem of Section 4.3. For the justification of the result, see Exercise 27.

THEOREM Mean-value Inequality

Let g have a continuous derivative on the interval $I = [a, b]$, and let M be the maximum value of $|g'|$ on I. Then

$$|g(b) - g(a)| \leq M|b - a|. \tag{3}$$

We start by looking at R_1 and its derivative. Consider the expression

$$R_1(t; c) = f(t) - T_1(t; c) = f(t) - f(c) - f'(c)(t - c) \tag{4}$$

for t in the interval $c - r \leq t \leq c + r$. When we compute the derivative of $R_1(t; c)$ with respect to t, we obtain

$$R_1'(t; c) = f'(t) - f'(c).$$

Now apply the Mean-value Inequality (3) to the function f' on $[c - r, c + r]$. If M_2 is the maximum value of $|f''(t)|$ on this interval, then for t in this interval

$$R_1'(t; c) = |f'(t) - f'(c)| \leq M_2|t - c|. \tag{5}$$

We now integrate. For simplicity we assume that $x \geq c$. (For the case $x < c$, see Exercise 30.) Because, by (4), $R_1(c,c) = 0$, we have

$$|R_1(x;c)| = |R_1(x;c) - R_1(c;c)| = \left|\int_c^x R_1'(t;c)\,dt\right|.$$

Now apply Property 4 of the integral (page 370, Section 5.3) to bring the absolute value signs inside the integral. (Note that $c \leq x$ is used here. How?) This gives

$$|R_1(x;c)| = \left|\int_c^x R_1'(t;c)\,dt\right| \leq \int_c^x |R_1'(t;c)|\,dt.$$

If we use (5) to replace the integrand with a larger expression, we have

$$|R_1(x;c)| \leq \int_c^x |R_1'(t;c)|\,dt \leq \int_c^x M_2|t - c|\,dt.$$

Because $c \leq x$ and the variable t of integration runs between c and x, we also have $c \leq t$. Hence, in the last expression, $|t - c| = t - c$. Thus we have

$$|R_1(x;c)| \leq \int_c^x M_2(t - c)\,dt = \frac{1}{2}M_2(x - c)^2 = \frac{M_2}{2!}|x - c|^2.$$

This is the bound on the remainder given by the Taylor inequality (2) for $n = 1$.

Because calculating the exact value of M_{n+1} may be difficult, we sometimes use a number larger than M_{n+1} in place of M_{n+1} when using Taylor's inequality to estimate error. This has the disadvantage of making the error estimate larger than necessary, but it has the advantage of making our calculations easier.

EXAMPLE 2

a) Use the Taylor inequality to estimate the maximum error when $T_7(x;0)$ is used to approximate $\sin x$ for

$$-0.5 \leq x \leq 0.5.$$

Compare this estimate with the graph of

$$y = |\sin x - T_7(x;0)|, \qquad -0.5 \leq x \leq 0.5.$$

b) Suppose we wish to use $T_n(x;0)$ to approximate $\sin x$ on

$$-\pi \leq x \leq \pi$$

with an error of less than 10^{-6}. How large should n be?

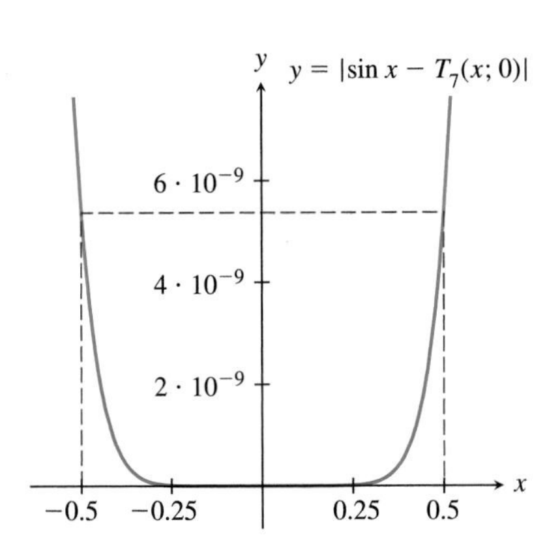

FIGURE 7.14 The graph shows that $|\sin x - T_7(x; 0)| \leq 6 \times 10^{-9}$ on the interval $-0.5 \leq x \leq 0.5$.

TABLE 7.2

n	$\frac{\pi^{n+1}}{(n+1)!}$
0	3.14
1	4.93
2	5.17
3	4.06
4	2.55
5	1.34
6	0.60
7	0.24
8	0.08
9	0.03
10	$7.37 \cdot 10^{-3}$
11	$1.93 \cdot 10^{-3}$
12	$4.66 \cdot 10^{-4}$
13	$1.05 \cdot 10^{-4}$
14	$2.19 \cdot 10^{-5}$
15	$4.30 \cdot 10^{-6}$
16	$7.95 \cdot 10^{-7}$
17	$1.39 \cdot 10^{-7}$

Solution

a) Because

$$\frac{d^8}{dx^8} \sin x = \sin x,$$

we have

$$M_8 = \text{maximum of } |\sin x| \text{ on } [-0.5, 0.5] = \sin 0.5 \leq 0.5.$$

We choose to use the number 0.5 instead of $\sin 0.5 \approx 0.479426$ in our error estimate because 0.5 is easier to work with. By Taylor's inequality, the error when $\sin x$ is approximated by $T_7(x; 0)$ on $-0.5 \leq x \leq 0.5$ must satisfy

$$|\sin x - T_7(x; 0)| = |R_7(x; 0)| \leq \frac{1}{8!} M_8 |x - 0|^8 \leq \frac{1}{8!} 0.5 |x|^8.$$

Because $-0.5 \leq x \leq 0.5$, we have

$$|R_7(x; 0)| \leq \frac{0.5^9}{8!} \approx 5.0 \times 10^{-8}.$$

How does this estimate compare with the actual error? We first compute

$$T_7(x; 0) = x - \frac{1}{3!}x^3 + \frac{1}{5!}x^5 - \frac{1}{7!}x^7.$$

(You should verify this.) Next, graph

$$y = |R_7(x; 0)| = |\sin x - T_7(x; 0)|$$

on the interval $[-0.5, 0.5]$. This graph is shown in Fig. 7.14. From this graph we see that the error is less than 6.0×10^{-9}. This is less than the bound on the error obtained using the Taylor inequality.

b) If we use $T_n(x; 0)$ to approximate $\sin x$ on $-\pi \leq x \leq \pi$, the error $R_n(x; 0)$ satisfies

$$|R_n(x; 0)| \leq \frac{1}{(n+1)!} M_{n+1} |x - 0|^{n+1} \leq \frac{\pi^{n+1}}{(n+1)!} M_{n+1}. \tag{6}$$

Because the $(n+1)$st derivative of $\sin x$ is $\cos x$, $-\sin x$, $-\cos x$, or $\sin x$, we know that $M_{n+1} = \max |f^{(n+1)}(x)| \leq 1$. Hence, from (6),

$$|R_n(x; 0)| \leq \frac{\pi^{n+1}}{(n+1)!}. \tag{7}$$

With a calculator, it is not hard to find a value of n for which this last expression is less than 10^{-6}. The values of the right side of (7) for several values of n are shown in Table 7.2. These data show that if $n \geq 16$, then we can be certain that $T_n(x; 0)$ approximates $\sin x$ on $-\pi \leq x \leq \pi$ with error of less than 10^{-6}.

EXAMPLE 3

a) Use the Taylor inequality to estimate the maximum error when $T_6(x; 100)$ is used to approximate $\sqrt{x}$ for $90 \le x \le 110$. Compare this estimate with the graph of

$$y = \left|\sqrt{x} - T_6(x; 100)\right|, \qquad 90 \le x \le 110.$$

b) Find n so that $T_n(105; 100)$ approximates $\sqrt{105}$ with error of less than 5.0×10^{-11}.

Solution

a) The seventh derivative of $f(x) = \sqrt{x}$ is

$$f^{(7)}(x) = \frac{-11}{2} \cdot \frac{-9}{2} \cdot \frac{-7}{2} \cdot \frac{-5}{2} \cdot \frac{-3}{2} \cdot \frac{-1}{2} \cdot \frac{1}{2} x^{-13/2}$$

$$= \frac{10{,}395}{128x^{13/2}}.$$

Hence

$$M_7 = \text{maximum of } \left|f^{(7)}(x)\right| \text{ on } [90, 110]$$

$$= \frac{10{,}395}{128(90)^{13/2}} \approx 1.611 \times 10^{-11} < 1.7 \times 10^{-11}.$$

By the Taylor inequality, the error when $\sqrt{x}$ is approximated by $T_6(x; 100)$ on $90 \le x \le 110$ satisfies

$$|R_6(x; 100)| \le \frac{1}{7!} M_7 |x - 100|^7 \le 3.4 \times 10^{-15} |x - 100|^7.$$

Because $|x - 100| \le 10$ on $[90, 110]$, we have

$$|R_6(x; 100)| \le 3.4 \times 10^{-15} 10^7 = 3.4 \times 10^{-8}.$$

To see how this bound on the error compares with the actual error, $|R_6(x; 100)| = \left|\sqrt{x} - T_6(x; 100)\right|$ we calculate

$$\sqrt{x} - T_6(x; 100) = \sqrt{x} - \left(10 + \frac{(x-100)}{20} - \frac{(x-100)^2}{8000} + \frac{(x-100)^3}{1{,}600{,}000} - \frac{(x-100)^4}{256{,}000{,}000} + \frac{7(x-100)^5}{256{,}000{,}000{,}000} - \frac{21(x-100)^6}{102{,}400{,}000{,}000{,}000}\right).$$

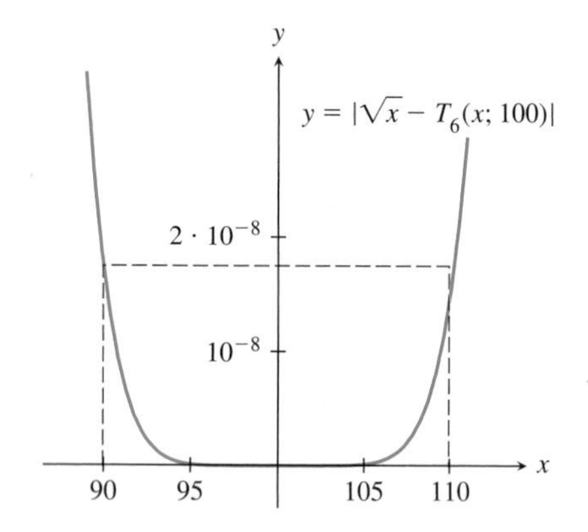

FIGURE 7.15 The graph shows that $|R_6(x; 100)| < 2 \times 10^{-8}$ for $90 \le x \le 110$.

We show the graph of the actual error,

$$y = |R_6(x; 100)| = \left|\sqrt{x} - T_6(x; 100)\right|,$$

in Fig. 7.15.

From the graph we see that the error is less than 2×10^{-8} on the interval $90 \leq x \leq 110$. Thus, in this case the error bound calculated using Taylor's inequality is a reasonably close estimate of the actual error.

b) By looking for a pattern in the first few derivatives of $f(x) = \sqrt{x}$, it is not hard to guess a formula for the $(n + 1)$st derivative:

$$\begin{aligned} f'(x) &= \frac{1}{2}x^{-1/2}, \\ f''(x) &= -\frac{1}{2^2}x^{-3/2}, \\ f'''(x) &= \frac{(3)(1)}{2^3}x^{-5/2}, \\ f^{(4)}(x) &= -\frac{(5)(3)(1)}{2^4}x^{-7/2}, \\ f^{(5)}(x) &= \frac{(7)(5)(3)(1)}{2^5}x^{-9/2}, \\ &\vdots \\ f^{(n+1)}(x) &= (-1)^n\frac{(2n-1)(2n-3)\cdots(5)(3)(1)}{2^{n+1}}x^{-(2n+1)/2}. \end{aligned} \tag{8}$$

We use this result to find M_{n+1} and then estimate the error when $T_n(x; 100)$ is used to approximate $\sqrt{x}$ on $100 \leq x \leq 105$. (We work on the interval $[100, 105]$ instead of $[90, 110]$ because we are concerned with the error in the approximation at $x = 105$.) By (8),

$$\begin{aligned} \left|\frac{d^{n+1}}{dx^{n+1}}\sqrt{x}\right| &= \frac{(2n-1)(2n-3)\cdots(3)(1)}{2^{n+1}}\frac{1}{|x|^{(2n+1)/2}} \\ &\leq \frac{(2n-1)(2n-3)\cdots(3)(1)}{2^{n+1}}\frac{1}{100^{(2n+1)/2}}. \end{aligned}$$

The last step is valid because $\left|\frac{1}{x}\right| \leq \frac{1}{100}$ for $100 \leq x \leq 105$. This shows that

$$M_{n+1} = \frac{(2n-1)(2n-3)\cdots(3)(1)}{2^{n+1}}\frac{1}{100^{(2n+1)/2}}.$$

Next note that $|x - 100| \leq 5$ and apply the Taylor inequality,

$$|R_n(x; 100)| \leq \frac{1}{(n+1)!}M_{n+1}|x - 100|^{n+1} \leq \frac{5^{n+1}}{(n+1)!}M_{n+1}.$$

Hence, on $100 \leq x \leq 105$,

$$|R_n(x; 100)| \leq \frac{5^{n+1}}{(n+1)!} \cdot \frac{(2n-1)(2n-3)\cdots(3)(1)}{2^{n+1}10^{2n+1}}. \tag{9}$$

By experimenting with a calculator or computer, we can find a value of n for which this last expression is less than 5.0×10^{-11}. In Table 7.3 we have calculated the value of the right side of (9) for several values of n. From the table we see that we should take $n \geq 7$ if we want to be assured that $T_n(105; 100)$ approximates $\sqrt{105}$ with error of less than 5.0×10^{-11}.

TABLE 7.3

n	$\dfrac{5^{n+1}(2n-1)(2n-3)\cdots(5)(3)(1)}{(n+1)!\,2^{n+1}10^{2n+1}}$
2	$7.81 \cdot 10^{-5}$
3	$2.44 \cdot 10^{-6}$
4	$8.54 \cdot 10^{-8}$
5	$3.20 \cdot 10^{-9}$
6	$1.26 \cdot 10^{-10}$
7	$5.11 \cdot 10^{-12}$
8	$2.13 \cdot 10^{-13}$

Exercises 7.2

Exercises 1–6: For the indicated function, find $T_2(x; 0)$. Then use the Taylor inequality (2) to find a bound for $R_2(x; 0)$ on the given interval I. Graph $y = |R_2(x; 0)| = |f(x) - T_2(x; 0)|$ and use this graph to find the actual *maximum error on I. How does the error compare to the bound found using the Taylor inequality?*

T **1.** $f(x) = e^{-2x}$ on $I = [-0.3, 0.3]$

T **2.** $f(x) = (3 + x)^{10}$ on $I = [-0.3, 0.3]$

T **3.** $f(x) = \sin x$ on $I = [-\pi/6, \pi/6]$

T **4.** $f(x) = \cos x$ on $I = [-\pi/6, \pi/6]$

T **5.** $f(x) = \dfrac{1}{1 + x}$ on $I = [-1/3, 1/3]$

T **6.** $f(x) = \arctan x$ on $I = [-0.5, 0.5]$

Exercises 7–12: For the given c and function f, use the Taylor inequality to find a value of r for which $T_5(x; c)$ is within 0.01 of $f(x)$ on $[c - r, c + r]$. Check your answer by graphing $y = |R_5(x; c)|$ on $[c - r, c + r]$.

7. $f(x) = \sin x,\ c = 0$

8. $f(x) = \dfrac{1}{x},\ c = 1$

9. $f(x) = x \ln x,\ c = 3$

10. $f(x) = e^x,\ c = \ln 2$

11. $f(x) = 2x^4 - 3x^3 + 6x^2 - 2,\ c = \sqrt{2}$

12. $f(x) = \sqrt{x},\ c = 1$

13. Verify the expression for $T_6(x; 100)$ given in Example 3.

14. Verify the formula given in (8) for the $(n + 1)$st derivative of the square root function.

15. Let $T_n(x; 100)$ be the nth Taylor polynomial about 100 for the square root function $f(x) = \sqrt{x}$. In Example 3 we showed that $T_7(105; 100)$ is within 5.0×10^{-11} of $\sqrt{105}$. Check this by calculating $T_7(105; 100)$ and $\sqrt{105}$ and then computing the difference between the two results.

16. Let $f(x) = \sin 2x$. Find n so that if $T_n(x; 0)$ is used to approximate f on $[-0.5, 0.5]$, then the error is no more than 0.01.

17. Let $H(x) = 1/x$. Find n so that if $T_n(x; 1)$ is used to approximate H on $[2/3, 4/3]$, then the error is no more than 0.01.

18. Let $p(x) = x^4$. Find n so that if $T_n(x; 0)$ is used to approximate p on $[-0.1, 0.1]$, then the error is no more than 0.001.

19. Let $g(t) = \sqrt{t}$. Find n so that if $T_n(t; 4)$ is used to approximate g on $[3, 5]$, then the error is no more than 0.005.

20. Let $r(\theta) = e^{-\theta}$. Find n so that if $T_n(\theta; 0)$ is used to approximate r on $[-3, 3]$, then the error is no more than 0.001.

21. Let $f(x) = \ln(2 - x)$. Find n so that if $T_n(x; 0)$ is used to approximate f on $[-0.5, 0.5]$, then the error is no more than 0.0001.

22. Verify algebraically that

$$\begin{aligned} x^3 &= ((x-2)+2)^3 \\ &= 8 + 12(x-2) + 6(x-2)^2 + (x-2)^3. \end{aligned}$$

Show that the expression on the right is $T_3(x; 2)$ for the function $f(x) = x^3$.

23. Verify algebraically that

$$\begin{aligned} 2x^2 - 7x + 1 &= 2((x+3)-3)^2 - 7((x+3)-3)) + 1 \\ &= 40 - 19(x+3) + 2(x+3)^2. \end{aligned}$$

Show that the expression on the right is $T_2(x; -3)$ for the function $f(x) = 2x^2 - 7x + 1$.

24. Let $p(x)$ be a polynomial of degree n. Based on the previous two exercises, suggest a method for finding $T_n(x; c)$ by expanding the expression $p(x) = p((x - c) + c)$. How could you get Taylor polynomials $T_m(x; c)$ for $m < n$ by this method?

25. Let $p(x) = 3x^3 - 2x + 5$, and let M_n be the maximum value of $|p^{(n)}(x)|$ on the interval $[-100, 100]$. Show that $M_{n+1} = 0$ for $n \geq 3$. Use this fact and the Taylor inequality to show that $T_n(x; 0) = p(x)$ for $n \geq 3$.

26. Let p be a polynomial of degree d. Use the Taylor inequality to show that if $n \geq d$, then $T_n(x; c) = p(x)$ for any c and all real x.

27. The Mean-value Inequality Let f be continuous and differentiable on an interval $I = [a, b]$. Suppose that $|f'(x)| \leq M$ for $a \leq x \leq b$. Show that for any x and t in I,

$$|f(x) - f(t)| \leq M|x - t|.$$

(A good place to start is the Mean-value Theorem of Section 4.3.)

28. Let $f(x)$ be a function defined on an interval $[a, b]$. Suppose that $|f'(x)| \leq 2$ on $[a, b]$. Use the Mean-value Inequality to show that $|f(b) - f(a)| \leq 2|a - b|$.

29. Apply the Mean-value Inequality to show that for any real numbers x and t,

$$|\sin x - \sin t| \leq |x - t|.$$

30. In the Investigation, we derived the Taylor inequality for $|R_1(x; c)|$ in the case $x \geq c$. Complete the argument by considering the case $x < c$.

31. In this problem we outline a proof of the Taylor inequality for the case $n = 2$.

a. Show that

$$\frac{d^2}{dt^2} R_2(t; c) = R_2''(t; c) = f''(t) - f''(c).$$

b. Use the fact that $R_2''(c; c) = 0$ in showing that for t between c and x,

$$|R_2''(t; c)| \leq M_3|t - c|,$$

where M_3 is the maximum value of $f'''(t)$ for t values between c and x.

c. Next assume that $x \geq c$ and show that

$$|R_2'(x; c)| = \left| \int_c^x R_2''(t; c)\, dt \right| \leq \frac{M_3}{2}|x - c|^2.$$

Show that this inequality also holds with $x \leq c$.

d. Note that the inequality

$$|R_2'(x; c)| \leq \frac{M_3}{2}|x - c|^2$$

is also true if x is replaced by any t between c and x. Use this, with one more integration, to show that

$$|R_2(x; c)| \leq \frac{M_3}{3!}|x - c|^3.$$

32. For nonnegative integer n, let $T_n(x; 0)$ be the nth degree Taylor polynomial for the sine function at 0. Is there a value of n such that $T_n(x; 0)$ approximates $\sin x$ to within 0.01 for *all* real numbers x? Justify your answer.

33. For nonnegative integer n, let $T_n(x; 1)$ be the nth degree Taylor polynomial for $\ln x$ at 1. Is there a value of n such that $T_n(x; 1)$ approximates $\ln x$ to within 0.01 for *all* real numbers x in the interval $(0, 2)$? Justify your answer.

34. Show that for positive integer n,

$$(2n-1)(2n-3)\cdots(5)(3)(1) = \frac{(2n)!}{2^n n!}$$

Use this to simplify the bound for $|R_n(x; 100)|$ given in (9).

7.3 Sequences

Set your calculator in radian mode and calculate cos(1.86). Take the answer to this calculation and use it as input for the cosine function. Repeat this process several times, always taking the most recent answer as input for the next cosine calculation. (On TI calculators, this can be done by doing the cos(1.86) calculation, then entering cos(ANS), and then pressing the ENTER button.) Early in the process, the answer displayed by the calculator changes at each step. However, if you continue this process long enough, the answers stop changing.

Let x_n be the number that appears on the screen after the nth calculation. If your calculator is set to display six decimal places, you will see

$$\begin{aligned} x_1 &= \cos 1.86 = -0.285189, \\ x_2 &= \cos x_1 \quad = 0.959608, \\ x_3 &= \cos x_2 \quad = 0.573841, \\ x_4 &= \cos x_3 \quad = 0.839822, \\ &\vdots \qquad\qquad \vdots \\ x_{29} &= \cos x_{28} \quad = 0.739080, \\ x_{30} &= \cos x_{29} \quad = 0.739089, \\ &\vdots \qquad\qquad \vdots \end{aligned} \tag{1}$$

This process—input 1.86, calculate its cosine, and then, step-by-step, calculate the cosine of the preceding result—creates an infinite list of numbers.

An infinite list such as the one described here is called a *sequence*. To describe this sequence, we use either of the notations

$$x_1, x_2, x_3, \ldots \quad \text{or} \quad \{x_n\}_{n=1}^{\infty}$$

and describe how x_n can be computed for any positive integer n. As suggested by (1), we can describe this sequence by giving the first number, x_1, and then describing how subsequent terms are produced:

$$\begin{aligned} x_1 &= \cos 1.86 \\ x_n &= \cos(x_{n-1}), \qquad n \geq 2. \end{aligned}$$

One of the interesting things about this sequence is that it seems to settle down to a fixed number L as n gets large. The value of this fixed number is approximately 0.739085, as you can see by the time you've completed the 40th calculation. We express this fact by writing

$$\lim_{n \to \infty} x_n = L \approx 0.739085.$$

Many important processes in mathematics can be described using sequences. For example, when we use Newton's method to approximate a solution to an equation, we are actually computing the first few terms of an infinite list (or sequence) whose limit is equal to a solution of the equation. A sequence need not be a list of numbers. The Taylor polynomials for a function f form a sequence

$$T_0(x; c),\ T_1(x; c),\ T_2(x; c), \ldots$$

of polynomials. We have seen that when n is large, the Taylor polynomial $T_n(x; c)$ might be a good approximation to f. When this happens, we usually find that the sequence of Taylor polynomials has limit f:

$$\lim_{n\to\infty} T_n(x; c) = f(x).$$

By studying sequences and their limits we will better understand Taylor polynomials and their role in the approximation of functions.

Sequence Notation

In the preceding discussion, we stated that a sequence is an infinite list

$$x_1, x_2, x_3, x_4, \ldots, x_n, \ldots.$$

Each member of the sequence has a definite "position" in the list. The term x_1 is in position 1, term x_2 is in position 2, and so on. Given a position n in the list, there is a definite term x_n in position n. Thus, the term x_n is a *function* of its position n. We use this idea to formally define a sequence.

DEFINITION Sequence

A sequence is a function s with domain equal to the set $\{1, 2, 3, \ldots, n, \ldots\}$ of natural numbers.

If s is a sequence and n is a natural number, we can think of n as referring to a position in a list. The value of the function s at n (i.e., $s(n)$) is simply the nth term in the list. For brevity, we usually write s_n instead of $s(n)$.

Exhibiting a Sequence Sequences can be described in several ways. Sometimes we simply give a formula for the terms of a sequence. For example, the formula

$$s_n = n^2$$

describes the sequence s whose nth term is $s(n) = s_n = n^2$. This sequence might also be denoted by

$$1^2, 2^2, 3^2, \ldots \quad \text{or} \quad \{n^2\}_{n=1}^{\infty}.$$

In the latter case, the subscript 1 and the superscript ∞ indicate that the domain of the function s described here is the set $\{1, 2, 3, \ldots\}$ of natural numbers.

Sometimes a sequence is described by listing the first few terms of the sequence. We assume that the terms listed suggest an obvious pattern and that the reader can continue the list and perhaps guess the formula $s(n)$ that gives the nth term of the list. For example, the list $\frac{1}{2}, \frac{1}{3}, \frac{1}{4}, \ldots$ most likely continues as $\frac{1}{5}, \frac{1}{6}, \frac{1}{7}, \ldots$. The term in position n is

$$s_n = s(n) = \frac{1}{n+1}.$$

Sometimes the value of a term in the sequence may depend on one or more previous terms of the sequence. In these cases, it might be very hard to write down a formula for the nth term. However, if we are told how to generate the nth term from the previous terms, we can find the value of any term of the sequence. As an example, consider the sequence s described by

$$s_1 = 1$$
$$s_n = \frac{1}{2}\left(s_{n-1} + \frac{2}{s_{n-1}}\right), \qquad n \geq 2. \tag{2}$$

We can use (2) to find s_n for any n. For example, to calculate s_2, let $n = 2$ in (2),

$$s_2 = \frac{1}{2}\left(s_1 + \frac{2}{s_1}\right) = \frac{1}{2}\left(1 + \frac{2}{1}\right) = \frac{3}{2} = 1.5.$$

To find s_3, let $n = 3$. This gives

$$s_3 = \frac{1}{2}\left(s_2 + \frac{2}{s_2}\right) = \frac{1}{2}\left(\frac{3}{2} + \frac{2}{(3/2)}\right) = \frac{17}{12} \approx 1.416667,$$

To calculate s_4, let $n = 4$ in (2) to obtain

$$s_4 = \frac{1}{2}\left(s_3 + \frac{2}{s_3}\right) = \frac{1}{2}\left(\frac{17}{12} + \frac{2}{(17/12)}\right) = \frac{577}{408} \approx 1.41422.$$

Although it would be very tedious to find many terms of this sequence by hand, with a programmable calculator or CAS, it is easy to find several terms of the sequence. What do you think happens to this sequence as n gets large? When we have a sequence, such as the one defined by (2), for which the value of a term s_n depends on the value of previous terms, we say that the sequence has been defined *recursively*.

Sometimes it is easier to describe a sequence $s(n)$ by letting n vary over a set different from $\{1, 2, 3, \ldots\}$. For example, it seems natural to describe the sequence

$$\frac{1}{0!}, \frac{1}{1!}, \frac{1}{2!}, \frac{1}{3!}, \ldots$$

by

$$s(n) = \frac{1}{n!} \qquad \text{for } n = 0, 1, 2, 3, \ldots$$

or by $\{1/n!\}_{n=0}^{\infty}$. In practice, we will try to give the simplest formula $s(n)$ to describe a sequence and let the index n vary over an appropriate set of integers. Sometimes we will simply write $\{s_n\}$ to denote a sequence when we are not concerned with the set over which n varies.

EXAMPLE 1 Write out the first five terms of the following sequences.

a) $\left\{\frac{(-1)^n}{n^2 + 1}\right\}_{n=1}^{\infty}$

b) $s_n = \sum_{k=1}^{n} \frac{1}{k}$

c) $F_1(x) = x$
$F_{n+1}(x) = x^{n+1}F_n(x), \quad n \geq 2$

Solution

a) The nth term of the sequence is $s_n = (-1)^n/(n^2+1)$. To find the first five terms of the sequence defined by this expression, evaluate it for $n = 1, 2, 3, 4, 5$.

$$s_1 = \frac{(-1)^1}{1^2+1} = -\frac{1}{2},$$
$$s_2 = \frac{(-1)^2}{2^2+1} = \frac{1}{5},$$
$$s_3 = \frac{(-1)^3}{3^2+1} = -\frac{1}{10},$$
$$s_4 = \frac{(-1)^4}{4^2+1} = \frac{1}{17},$$
$$s_5 = \frac{(-1)^5}{5^2+1} = -\frac{1}{26}.$$

b) Here the sequence elements are found by adding the reciprocals of integers. Evaluating s_n for $n = 1, 2, 3, 4, 5$, we find

$$s_1 = \sum_{k=1}^{1} \frac{1}{k} = 1,$$
$$s_2 = \sum_{k=1}^{2} \frac{1}{k} = 1 + \frac{1}{2} = \frac{3}{2},$$
$$s_3 = \sum_{k=1}^{3} \frac{1}{k} = 1 + \frac{1}{2} + \frac{1}{3} = \frac{11}{6},$$
$$s_4 = \sum_{k=1}^{4} \frac{1}{k} = 1 + \frac{1}{2} + \frac{1}{3} + \frac{1}{4} = \frac{25}{12},$$
$$s_5 = \sum_{k=1}^{5} \frac{1}{k} = 1 + \frac{1}{2} + \frac{1}{3} + \frac{1}{4} + \frac{1}{5} = \frac{137}{60}.$$

c) Sometimes, as in this case, the terms of a sequence are functions. Taking $n = 1, 2, 3, 4, 5$ and using the recursion relation given, we find

$$F_1(x) = x,$$
$$F_2(x) = x^2F_1(x) = x^2x = x^3,$$
$$F_3(x) = x^3F_2(x) = x^3x^3 = x^6,$$
$$F_4(x) = x^4F_3(x) = x^4x^6 = x^{10},$$
$$F_5(x) = x^5F_4(x) = x^5x^{10} = x^{15}.$$

The Limit of a Sequence

Let $\{T_n(x;c)\}_{n=0}^{\infty}$ be the sequence of Taylor polynomials associated with a function $f(x)$. We have seen that $T_n(x;c)$ can be a good approximation to $f(x)$ when n is large, and that the approximation gets better as n gets larger. To better understand the behavior of such a sequence for large values of n, we study the limit of a sequence.

DEFINITION Limit of a Sequence of Real Numbers

Let $\{s_n\}$ be the sequence of real numbers and let L be a real number. We say the sequence has **limit** L if for each $\epsilon > 0$ there is a number $N > 0$ such that

$$|s_n - L| < \epsilon \qquad \text{whenever} \qquad n > N.$$

When this is the case we write

$$\lim_{n\to\infty} s_n = L,$$

and say that the sequence *converges to* L. If there is no finite real number L for which these conditions are satisfied, we say the sequence *diverges* or that $\lim_{n\to\infty} s_n$ does not exist.

WEB Roughly, the definition says that if the sequence $\{s_n\}$ has limit L, then when n gets large, s_n gets close to L and stays close to L.

To illustrate the definition graphically, plot the ordered pairs (n, s_n) in the plane and draw the horizontal lines

$$y = L - \epsilon \quad \text{and} \quad y = L + \epsilon.$$

If

$$\lim_{n\to\infty} s_n = L,$$

there must be a positive integer N for which all of the points

$$(N + 1, s_{N+1}), (N + 2, s_{N+2}), (N + 3, s_{N+3}), \ldots$$

lie between the two lines. See Fig. 7.16. This means that for $n > N$, we have

$$|s_n - L| < \epsilon.$$

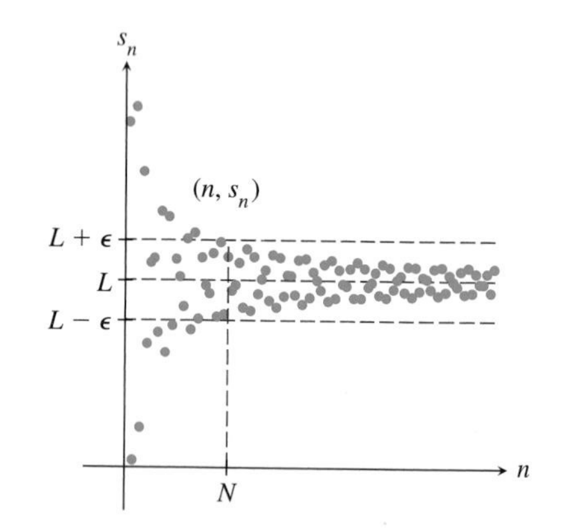

FIGURE 7.16 When $n > N$, s_n is within ϵ of L. In addition, the line $y = L$ is asymptotic to the graph of the points (n, s_n).

Figure 7.16 also illustrates a second way of picturing a convergent sequence: If $\{s_n\}$ converges to L, then the graph of the points (n, s_n), $n = 1, 2, 3, \ldots$, has horizontal asymptote $y = L$.

For another interpretation, suppose that

$$\lim_{n\to\infty} s_n = L$$

and we wish to approximate L with s_n for some n. Applying the definition of limit, the error $|s_n - L|$ can be made less than any small number ϵ by taking n large enough.

EXAMPLE 2 Find the limit of the sequence $\{s_n\}_{n=1}^{\infty} = \left\{\dfrac{3n - 4}{2n + 1}\right\}_{n=1}^{\infty}$.

Solution

For each positive integer n,

$$s_n = \frac{3n - 4}{2n + 1} = \frac{3n - 4}{2n + 1} \cdot \frac{\frac{1}{n}}{\frac{1}{n}} = \frac{3 - \frac{4}{n}}{2 + \frac{1}{n}}.$$

Thus

$$\lim_{n\to\infty} s_n = \lim_{n\to\infty} \frac{3 - \dfrac{4}{n}}{2 + \dfrac{1}{n}} = \frac{3}{2}.$$

EXAMPLE 3 Let $s_n = 1 + (-1)^n$. Investigate $\lim_{n\to\infty} s_n$.

Solution

We list the first few elements of the sequence $\{s_n\}$:

$$0, 2, 0, 2, 0, 2, \ldots$$

We see that the terms of the sequence do not get close and stay close to either 0 or 2 or any other number. Hence the sequence has no limit.

EXAMPLE 4 Let $s_n = \sqrt[n]{n}$. Find the limit of the sequence $\{s_n\}_{n=1}^{\infty}$.

Solution

We investigate the behavior of the sequence in several ways. We first evaluate s_n for a few large values of n and display the results in Table 7.4. From these data it appears that

$$\lim_{n\to\infty} \sqrt[n]{n} = 1. \tag{3}$$

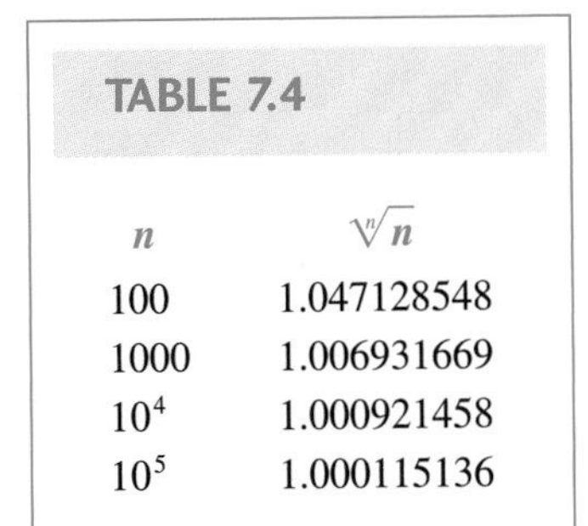

TABLE 7.4

n	$\sqrt[n]{n}$
100	1.047128548
1000	1.006931669
10^4	1.000921458
10^5	1.000115136

For a graphical approach, use a calculator or CAS to graph the ordered pairs (n, s_n) for several values of n. If the sequence $\{s_n\}_{n=1}^{\infty}$ has limit L, then the points plotted should get close to the line $y = L$. It appears from Fig. 7.17 that the points $\left(n, \sqrt[n]{n}\right)$ quickly get close to the line $y = 1$. This suggests that (3) is correct.

Finally, we investigate the limit symbolically. Taking the logarithm of s_n, we have

$$\ln(s_n) = \ln \sqrt[n]{n} = \ln(n^{1/n}) = \frac{\ln n}{n}.$$

Applying l'Hôpital's Rule we find

$$\lim_{n\to\infty} \ln(s_n) = \lim_{n\to\infty} \frac{\ln n}{n} = \lim_{n\to\infty} \frac{\dfrac{1}{n}}{1} = 0.$$

Thus, when n is large, $\ln s_n \approx 0$; and

$$s_n \approx e^0 = 1.$$

This shows that (3) is correct.

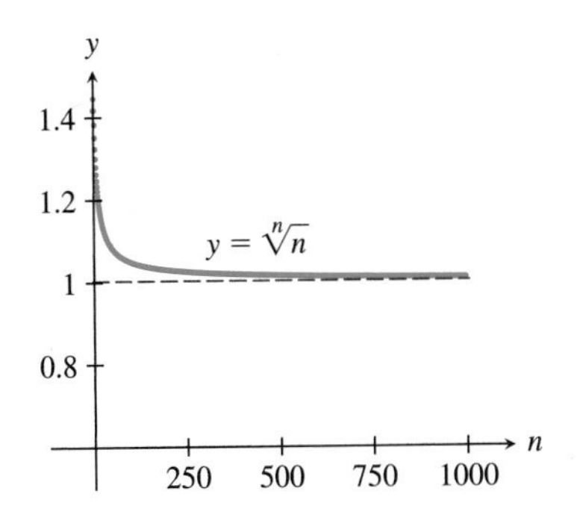

FIGURE 7.17 When n is large, $\sqrt[n]{n}$ is near 1.

EXAMPLE 5 Let $s_n = \sqrt{n^4 + n^3} - n^2$. Investigate $\lim_{n\to\infty} s_n$.

Solution

We first use a calculator to find the approximate value of s_n for a few large values of n. The resulting data are collected in Table 7.5 and suggest that s_n is large when n is large. We verify this symbolically by noting

TABLE 7.5

n	$\sqrt{n^4 + n^3} - n^2$
100	49.8756
1000	499.875
10^4	4999.88
10^5	49999.9

$$s_n = \sqrt{n^4 + n^3} - n^2 = \left(\sqrt{n^4 + n^3} - n^2\right)\frac{\sqrt{n^4 + n^3} + n^2}{\sqrt{n^4 + n^3} + n^2}$$

$$= \frac{n^3}{\sqrt{n^4 + n^3} + n^2} \geq \frac{n^3}{\sqrt{n^4 + n^4} + n^2} \geq \frac{n}{3}.$$

Because s_n always exceeds $n/3$, we can make s_n as large as we like by taking n sufficiently large. In this case, we say that the sequence $\{s_n\}_{n=1}^{\infty}$ diverges to $+\infty$ and write $\lim_{n\to\infty} s_n = +\infty$.

The behavior of the sequence in the last example provides an illustration of the following definition.

DEFINITION Infinite Limits

Let $\{s_n\}$ be a sequence of real numbers. The sequence diverges to $+\infty$ if for each $M > 0$, there is an $N > 0$ so that $s_n > M$ whenever $n > N$. In this case, we write

$$\lim_{n\to\infty} s_n = +\infty.$$

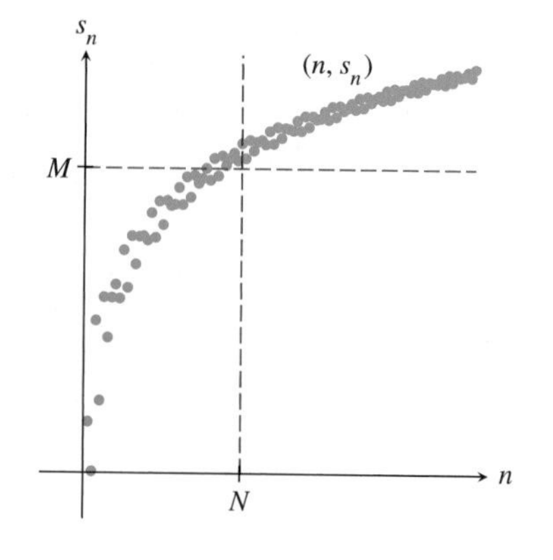

FIGURE 7.18 When $n > N$, s_n is larger than M.

This definition says that if $\lim_{n\to\infty} s_n = +\infty$, then when n gets large enough, the numbers s_n are also large. To illustrate this graphically, choose any number $M > 0$, draw the line $y = M$, and plot the points (n, s_n). If $\lim_{n\to\infty} s_n = +\infty$, then for any such M there is a positive integer N so that the points

$$(N + 1, s_{N+1}), (N + 2, s_{N+2}), (N + 3, s_{N+3}), \ldots$$

all lie above the line $y = M$. See Fig. 7.18.

Monotone and Bounded Sequences

Although it is sometimes difficult to decide whether or not a given sequence has a limit, there are some sequences for which this is immediate.

DEFINITION Monotone Sequence, Bounded Sequence

Let $\{x_n\}_{n=1}^{\infty}$ be a sequence of real numbers.

- The sequence is **monotone increasing** if

$$x_1 \leq x_2 \leq x_3 \leq \cdots \leq x_n \leq x_{n+1} \leq \ldots.$$

The sequence is **monotone decreasing** if

$$x_1 \geq x_2 \geq x_3 \geq \cdots \geq x_n \geq x_{n+1} \geq \ldots.$$

- The sequence is **bounded above** if there is a number A such that

$$x_n \leq A, \quad n = 1, 2, 3, \ldots.$$

The sequence is **bounded below** if there is a number B such that

$$B \leq x_n, \quad n = 1, 2, 3, \ldots.$$

If a sequence is bounded above and bounded below, we say that the sequence is **bounded.** If a sequence is not bounded, we say that it is **unbounded.**

EXAMPLE 6

a) For positive integer n, let $s_n = \sqrt{n^4 + n^3} - n^2$. Show that the sequence $\{s_n\}_{n=1}^{\infty}$ is monotone increasing and unbounded.

b) For positive integer n, let $x_n = \dfrac{(-1)^n}{n}$. Show that the sequence $\{x_n\}_{n=1}^{\infty}$ is bounded but not monotone.

Solution

a) For any positive integer n,

$$\begin{aligned} s_n &= \left(\sqrt{n^4 + n^3} - n^2\right)\frac{\sqrt{n^4 + n^3} + n^2}{\sqrt{n^4 + n^3} + n^2} \\ &= \frac{n^3}{\sqrt{n^4 + n^3} + n^2} = \frac{n}{\sqrt{1 + \frac{1}{n}} + 1}. \end{aligned} \tag{4}$$

The last expression in (4) is an increasing function of n. In particular,

$$s_n = \frac{n}{\sqrt{1 + \frac{1}{n}} + 1} < \frac{n + 1}{\sqrt{1 + \frac{1}{n + 1}} + 1} = s_{n+1}. \tag{5}$$

In (5), substitute $n = 1$, then $n = 2$, then $n = 3$, and so on to see

$$s_1 < s_2 < s_3 < \cdots < s_n < s_{n+1} < \ldots.$$

Therefore, $\{s_n\}_{n=1}^{\infty}$ is monotone increasing. In Example 5 we showed that $\lim_{n\to\infty} s_n = +\infty$. Hence, there is no $M > 0$ such that $s_n \leq M$ for all. This means $\{s_n\}$ is unbounded.

b) The first three terms of the sequence $\{x_n\}_{n=1}^{\infty}$ are

$$x_1 = -1, \qquad x_2 = \frac{1}{2}, \qquad x_3 = -\frac{1}{3}.$$

Thus $x_1 < x_2$, but $x_2 > x_3$. This pattern continues and hence the sequence is neither monotone increasing nor monotone decreasing. The sequence $\{x_n\}_{n=1}^{\infty}$ is bounded because for any n, we have

$$-1 \leq x_n \leq 1.$$

The number -1 is often called a *lower bound* for the sequence and 1 an *upper bound* for the sequence.

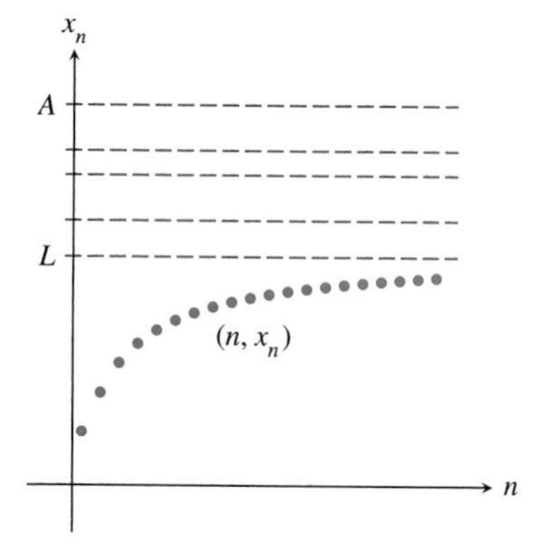

FIGURE 7.19 A sequence that is increasing and bounded above must converge.

When a sequence is both monotone and bounded, it has a limit. We state this result for the case in which the sequence is monotone increasing. See Exercises 43 and 44 for the case in which the sequence is monotone decreasing.

The Monotone Sequence Theorem

Let $\{x_n\}$ be a monotone increasing sequence of real numbers.

a) If $\{x_n\}$ is bounded above, then $\lim_{n\to\infty} x_n$ exists.

b) If $\{x_n\}$ is not bounded above, then

$$\lim_{n\to\infty} x_n = +\infty.$$

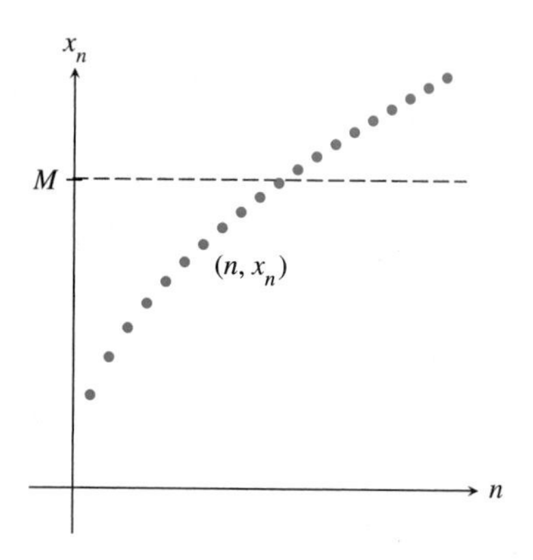

FIGURE 7.20 A sequence that is increasing and not bounded above must diverge to $+\infty$.

We can get some insight into why the Monotone Sequence Theorem is true by looking at two figures. First, assume that $\{x_n\}_{n=1}^{\infty}$ is bounded above. If we graph the ordered pairs (n, x_n), then all of the points of the graph lie below some horizontal line, say $y = A$. Choose the lowest horizontal line $y = L$ that is level with or above all points of the graph. See Fig. 7.19. The sequence $\{s_n\}$ must converge to L.

Now suppose that $\{x_n\}$ is increasing and *not* bounded above. If M is any positive number, then there must be sequence values bigger than M. Because the sequence is increasing, once one sequence value is larger than M, all subsequent sequence values are also larger than M. See Fig. 7.20. This means that $\lim_{n\to\infty} x_n = +\infty$.

The Monotone Sequence Theorem can tell us when a monotone increasing sequence has a limit. However, it does not tell us the value of the limit. Nonetheless, knowing that a sequence has a limit is important information. Once we know that a sequence has a limit, we can attempt to find or approximate that limit.

EXAMPLE 7 Define a sequence $\{a_n\}$ by the recursion relationship

$$a_1 = 1, \qquad a_{n+1} = \sqrt{2a_n}, \qquad n = 1, 2, 3, \ldots.$$

Show that the sequence converges and find its limit.

Solution

We start by calculating a few terms of the sequence,

$$\begin{aligned}
a_1 &= 1,\\
a_2 &= \sqrt{2a_1} = \sqrt{2 \cdot 1} = \sqrt{2} \approx 1.414213,\\
a_3 &= \sqrt{2a_2} = \sqrt{2\sqrt{2}} \approx 1.681793,\\
a_4 &= \sqrt{2a_3} = \sqrt{2\sqrt{2\sqrt{2}}} \approx 1.834008,\\
a_5 &= \sqrt{2a_4} = \sqrt{2\sqrt{2\sqrt{2\sqrt{2}}}} \approx 1.915207.
\end{aligned}$$

Based on these data, it appears that the sequence increases and is bounded above, perhaps by 2. We show that both are true. We first use mathematical induction (see Section A.4 in the Appendix) to show that if any number of the sequence is less than or equal to 2, then so is the next number in the sequence. Suppose that $a_n \le 2$. Then we have

$$a_{n+1} = \sqrt{2a_n} \le \sqrt{2 \cdot 2} = 2, \tag{6}$$

so $a_{n+1} \le 2$ as well. The result is the key step in showing that $a_n \le 2$ for all n. To finish the proof, all we need to do is observe that $a_1 = 1 \le 2$. Then by (6), if $a_1 \le 2$, then $a_2 \le 2$; if $a_2 \le 2$, then $a_3 \le 2$ and so on. We use this result in showing that the sequence is increasing.

Because

$$a_n = \sqrt{a_n \cdot a_n} \le \sqrt{2a_n} = a_{n+1},$$

we see that the sequence is monotone increasing.

Because the sequence is increasing and bounded above (an upper bound is 2, by (6)) we know that the sequence has a finite limit. Let L be the value of the limit. Because $a_{n+1} = \sqrt{2a_n}$,

$$\lim_{n\to\infty} a_{n+1} = \lim_{n\to\infty} \sqrt{2a_n} = \sqrt{2 \lim_{n\to\infty} a_n}.$$

Because the sequence has limit L, this equation can be rewritten as

$$L = \sqrt{2L}.$$

This equation has three solutions: $L = 0$, $L = 2$, and $L = +\infty$. We know $L \ne 0$ because the sequence terms are positive and increasing. We know that $L \ne +\infty$ because the sequence has a finite limit. Thus we must have $L = 2$.

EXAMPLE 8 Find the limit of the sequence defined by $a_k = \left(1 + \frac{1}{k}\right)^k$.

Solution

To get an idea of how the sequence behaves, calculate a_k for several values of k. The results of some of these calculations are shown in Table 7.6. In Fig. 7.21 we show the graph of the points (k, a_k) for $1 \le k \le 1000$. The data in the table and the graph suggest that the sequence $\{a_k\}$ is increasing and bounded above, perhaps by 3. From the table it appears that

TABLE 7.6

k	$\left(1 + \frac{1}{k}\right)^k$
1	2
10	2.593742
100	2.704814
500	2.715569
1000	2.716924
10^4	2.718146
10^5	2.718268
10^6	2.718280
10^7	2.718282
10^8	2.718282

$$\lim_{k\to\infty} a_k = \lim_{k\to\infty}\left(1 + \frac{1}{k}\right)^k \approx 2.71828\ldots.$$

Hence we guess that $\lim_{k\to\infty}\left(1 + \frac{1}{k}\right)^k = e$.

Assuming that the sequence is in fact increasing and bounded above, we show that indeed $\lim_{k\to\infty} a_k = e$. Let L be the value of the limit (we know that there is such an L by the Monotone Sequence Theorem):

$$L = \lim_{k\to\infty}\left(1 + \frac{1}{k}\right)^k. \tag{7}$$

Take the logarithm of both sides of (7) to get

$$\ln L = \lim_{k\to\infty} \ln\left(1 + \frac{1}{k}\right)^k = \lim_{k\to\infty} k \ln\left(1 + \frac{1}{k}\right) = \lim_{k\to\infty} \frac{\ln\left(1 + \frac{1}{k}\right)}{\frac{1}{k}}.$$

In this last form, the limit can be evaluated by using l'Hôpital's Rule:

$$\ln L = \lim_{k\to\infty} \frac{\frac{d}{dk}\ln\left(1 + \frac{1}{k}\right)}{\frac{d}{dk}\frac{1}{k}} = \lim_{k\to\infty} \frac{-\frac{1}{k(k+1)}}{-\frac{1}{k^2}} = \lim_{k\to\infty} \frac{k^2}{k^2 + k} = 1.$$

Thus

$$\lim_{k\to\infty}\left(1 + \frac{1}{k}\right)^k = L = e^{\ln L} = e^1 = e.$$

This verifies the guess suggested by the data.

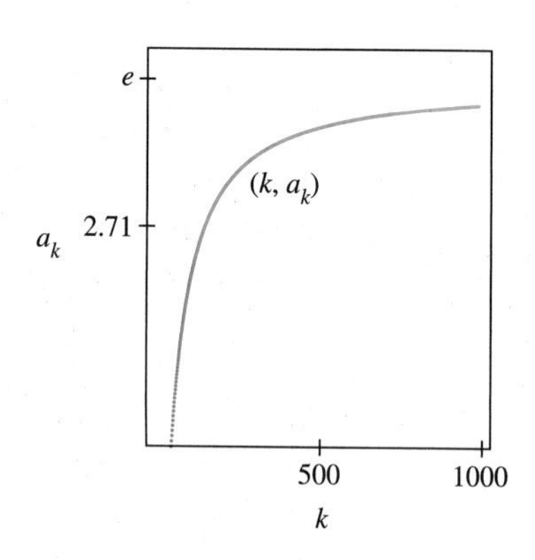

FIGURE 7.21 The sequence $\{a_k\} = \left\{\left(1 + \frac{1}{k}\right)^k\right\}$ is increasing and bounded above.

The Arithmetic of Sequences

Sometimes the terms of a sequence $\{c_k\}$ can be written as the sum, product, or quotient, or other function of terms of other sequences. If we know the behavior of these other sequences, we can often use the following results to say something about the behavior of $\{c_k\}$. The formulas that follow are very similar to the limit theorems discussed in Section 1.5.

Some Limit Theorems

Let $\{a_k\}$ and $\{b_k\}$ be sequences of real numbers with

$$\lim_{k\to\infty} a_k = A \quad \text{and} \quad \lim_{k\to\infty} b_k = B.$$

a) Let r be a real number and let $c_k = ra_k$. Then

$$\lim_{k\to\infty} c_k = \lim_{k\to\infty} (ra_k) = r \lim_{k\to\infty} a_k = rA.$$

b) Let $s_k = a_k + b_k$. Then

$$\lim_{k\to\infty} s_k = \lim_{k\to\infty} (a_k + b_k) = \lim_{k\to\infty} a_k + \lim_{k\to\infty} b_k = A + B.$$

c) Let $p_k = a_k b_k$. Then

$$\lim_{k\to\infty} p_k = \lim_{k\to\infty} (a_k b_k) = \left(\lim_{k\to\infty} a_k\right)\left(\lim_{k\to\infty} b_k\right) = AB.$$

d) Assume that $b_k \neq 0$ for all k and that $B \neq 0$. If $q_k = \dfrac{a_k}{b_k}$, then

$$\lim_{k\to\infty} q_k = \lim_{k\to\infty} \frac{a_k}{b_k} = \frac{\lim_{k\to\infty} a_k}{\lim_{k\to\infty} b_k} = \frac{A}{B}.$$

e) Suppose that the function f is defined for all a_k and that f is continuous at A. If $h_k = f(a_k)$, then

$$\lim_{k\to\infty} h_k = \lim_{k\to\infty} f(a_k) = f\left(\lim_{k\to\infty} a_k\right) = f(A).$$

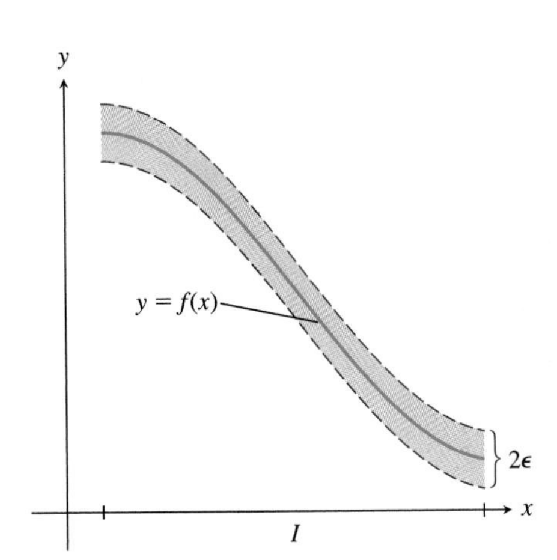

FIGURE 7.22 The ϵ-corridor about f.

Sequences of Functions

Most of the sequence examples we have seen so far have been sequences of numbers. But in Example 1 we saw that the terms of a sequence can be functions. We have also defined and worked with the sequence of Taylor polynomials $\{T_n(x; c)\}$ associated with a function. When we work with a sequence $\{f_n(x)\}_{n=1}^{\infty}$ of functions, we usually want to know about the convergence of the sequence for all values of x in some interval. When the sequence of functions is a sequence of Taylor polynomials, we often deal with uniform convergence. Before defining uniform convergence, we give some convenient terminology.

Let f be a function defined on an interval I and let ϵ be a positive real number. The **ϵ-corridor** about f is the region between the graphs of $y = f(x) - \epsilon$ and $y = f(x) + \epsilon$.

The ϵ-corridor about f is the shaded area in Fig. 7.22. Note that the graph of $y = f(x)$ lies in the ϵ-corridor about f.

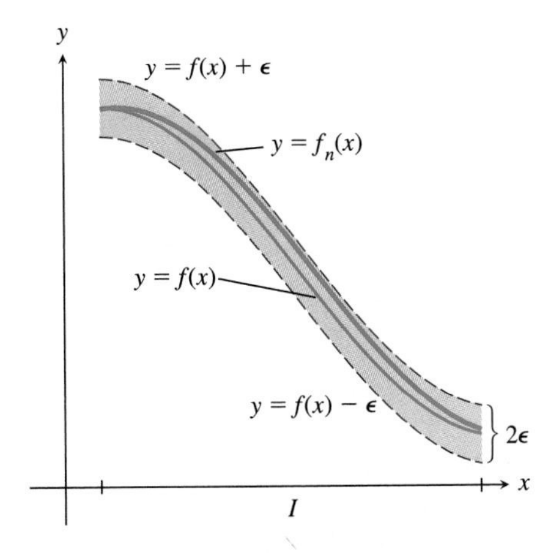

FIGURE 7.23 When n is large, the graph of $y = f_n(x)$ lies in the ϵ-corridor about f.

DEFINITION Uniform Convergence

Let $\{f_n\}_{n=1}^{\infty}$ be a sequence of functions each defined on an interval I and let f also be a function defined on I. We say that the sequence $\{f_n\}$ converges uniformly to f on I if for any $\epsilon > 0$ there is a positive integer N so that for all $n > N$, the graph of $y = f_n(x)$ lies in the ϵ-corridor about f. See Fig. 7.23.

The idea of uniform convergence is closely related to an idea that we worked with in Section 7.2. Note that:

> The graph of $y = f_n(x)$ for x in I lies in the ϵ-corridor about f exactly when $|f(x) - f_n(x)| < \epsilon$ for all x in I. This is the case precisely when the graph of $y = |f(x) - f_n(x)|$ lies below the line $y = \epsilon$. See Fig. 7.24.

In Example 2 of Section 7.2, we showed that if $f(x) = \sin x$ and $T_n(x; 0)$ is the nth Taylor polynomial for f about 0, then

$$|\sin x - T_{16}(x; 0)| \leq 10^{-6} \quad \text{for} \quad -\pi \leq x \leq \pi. \tag{8}$$

Thus, on $-\pi \leq x \leq \pi$, $T_{16}(x; 0)$ lies in the 10^{-6}-corridor about the sine function.

The idea of uniform convergence is very closely related to approximation. When a sequence $\{f_n\}$ converges to a function f uniformly throughout an interval I, the graph of $y = f_n(x)$ can be made close to the graph of $y = f(x)$ by taking n large enough. See Fig. 7.23. Also, as mentioned earlier, if the graph of $y = f_n(x)$ lies in the ϵ-corridor about f, then we have

$$|f(x) - f_n(x)| < \epsilon$$

for each x in I. This is illustrated in Fig. 7.24. Hence when ϵ is a small positive number, f_n will be a good approximation to f for all x in I.

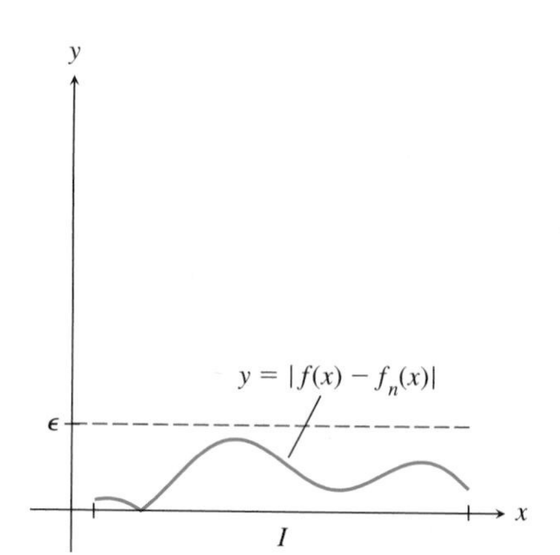

FIGURE 7.24 f_n is in the ϵ-corridor about f exactly when the graph of $y = |f(x) - f_n(x)|$ lies below the line $y = \epsilon$.

In many cases, the sequence of Taylor polynomials $\{T_n(x; c)\}_{n=1}^{\infty}$ for a function f converges uniformly to f on an interval $[c - r, c + r]$ centered at c. This is why Taylor polynomials are often used to approximate functions. For example, from (8) we conclude that for any x in the interval $[-\pi, \pi]$,

$$\sin x \approx T_{16}(x; 0)$$

with an error of at most 10^{-6}.

EXAMPLE 9 Let $f(x) = \ln x$ and let $\{T_n(x; 1)\}_{n=0}^{\infty}$ be the sequence of Taylor polynomials for f about $x = 1$.

a) Show that the sequence of Taylor polynomials converges uniformly to $\ln x$ on $\left[\frac{1}{2}, \frac{3}{2}\right]$.
b) Show that the sequence of Taylor polynomials does not converge uniformly to $\ln x$ on the interval $(0, 2)$.

Solution

a) We will show that the absolute error

$$|\ln x - T_n(x; 1)| = |R_n(x; 1)|$$

can be made as small as we like throughout $\left[\frac{1}{2}, \frac{3}{2}\right]$ by taking n large enough. If we can make $|R_n(x; 1)| < \epsilon$, then $|\ln x - T_n(x; 1)| < \epsilon$. This means that the graph of $y = T_n(x; 1)$ will lie in the ϵ-corridor about f, and hence that $\{T_n\}$ converges uniformly to f on the desired interval. See

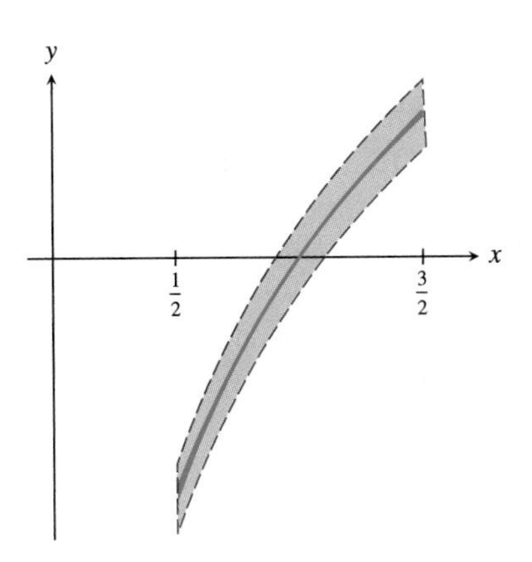

FIGURE 7.25 The 0.1-corridor about ln x, and the graphs of $y = \ln x$ and $y = T_3(x; 1)$.

Fig. 7.25. With $f(x) = \ln x$, we have

$$\begin{aligned} f'(x) &= x^{-1}, \\ f''(x) &= -x^{-2}, \\ f'''(x) &= 2 \cdot 1x^{-3}, \\ f^{(4)}(x) &= -3 \cdot 2 \cdot 1x^{-4}, \\ &\vdots \end{aligned}$$

From this, we see that

$$f^{(n)}(x) = (-1)^{n-1}(n-1)!\,x^{-n},$$

and

$$f^{(n+1)}(x) = (-1)^n n!\,x^{-n-1}. \tag{9}$$

The Taylor estimate for the error $R_n(x; 1)$ on $\left[\frac{1}{2}, \frac{3}{2}\right]$ is

$$|R_n(x; 1)| \leq \frac{M_{n+1}}{(n+1)!}|x-1|^{n+1}, \tag{10}$$

where M_{n+1} is the maximum value of $|f^{(n+1)}(x)|$ on $\left[\frac{1}{2}, \frac{3}{2}\right]$. From (9) we have

$$|f^{(n+1)}(x)| = \left|(-1)^n \frac{n!}{x^{n+1}}\right| = \frac{n!}{|x|^{n+1}}.$$

The maximum value for $\dfrac{n!}{|x|^{n+1}}$ on $\left[\frac{1}{2}, \frac{3}{2}\right]$ occurs when the denominator is smallest, that is, when $x = \frac{1}{2}$. Hence

$$M_{n+1} = \frac{n!}{(1/2)^{n+1}} = n!\,2^{n+1}.$$

For x in $\left[\frac{1}{2}, \frac{3}{2}\right]$ we have $|x-1| \leq \frac{1}{2}$. Using this bound in (10) and our calculation of M_{n+1}, we find

$$\begin{aligned} |R_n(x; 1)| &\leq \frac{M_{n+1}}{(n+1)!}|x-1|^{n+1} \\ &= \frac{n!\,2^{n+1}}{(n+1)!}|x-1|^{n+1} = \frac{2^{n+1}}{n+1}|x-1|^{n+1} \\ &\leq \frac{1}{n+1}2^{n+1}\left(\frac{1}{2}\right)^{n+1} = \frac{1}{n+1}. \end{aligned} \tag{11}$$

Now suppose that we want to make the error

$$|\ln x - T_n(x; 1)| = |R_n(x; 1)| < \epsilon.$$

From (11), we can do this by picking n so that

$$\frac{1}{n+1} < \epsilon;$$

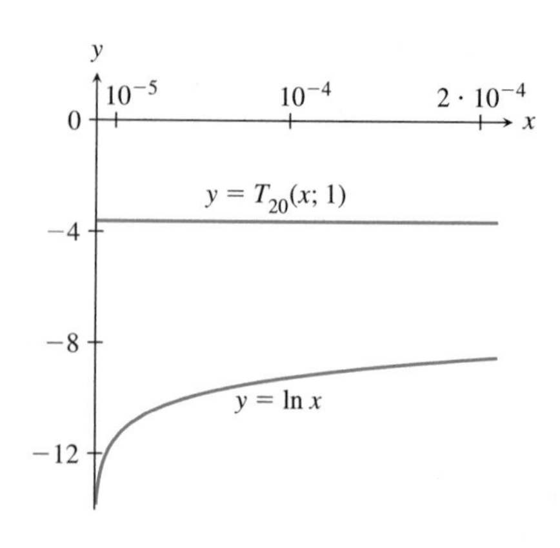

FIGURE 7.26 For x near 0, we see that $T_n(x; 1)$ is not a good approximation to $\ln x$. Why does the graph of $y = T_{20}(x; 1)$ look like a straight line in this picture?

that is, by taking $n > N = \frac{1}{\epsilon} - 1$. This shows that given $\epsilon > 0$, there is an $N(= 1 - \frac{1}{\epsilon})$ so that for $n > N$ we have $|\ln x - T_n(x; 1)| < \epsilon$ for $\frac{1}{2} \le x \le \frac{3}{2}$. This means that the sequence $\{T_n(x; 1)\}_{n=0}^{\infty}$ converges to $\ln x$ uniformly on $\left[\frac{1}{2}, \frac{3}{2}\right]$.

b) The sequence $\{T_n(x; 1)\}_{n=0}^{\infty}$ does not converge uniformly to $\ln x$ on $(0, 2)$ because $T_n(x; 1)$ cannot approximate $\ln x$ near 0. Because $T_n(x; 1)$ is a polynomial, it stays bounded when x is close to 0. On the other hand, $\ln x$ tends to $-\infty$ as x goes to 0 in $(0, 2)$. Hence, no matter what n is,

$$|\ln x - T_n(x; 1)|$$

is very large when x is close to 0. This is illustrated in Fig. 7.26.

Exercises 7.3

Exercises 1–7: Write out the first five terms of each sequence.

1. $\{2^n\}_{n=1}^{\infty}$

2. $\{1 - k + 2k^2\}_{k=0}^{\infty}$

3. $\{3\}_{k=1}^{\infty}$

4. $\left\{\dfrac{j+1}{j-2}\right\}_{j=3}^{\infty}$

5. $\left\{\cos\dfrac{k\pi}{4}\right\}_{k=1}^{\infty}$

6. $\left\{\sum_{k=0}^{n} k\right\}_{n=1}^{\infty}$

7. $\left\{\sum_{k=1}^{n} \dfrac{(-1)^{k-1}}{k^k}\right\}_{n=1}^{\infty}$

Exercises 8–13: The first four terms of a sequence are shown. Write the most likely next three terms and the tenth term. Then write a formula or description for the nth term of the sequence.

8. $-1, 1, -1, 1, \ldots$

9. $3, 3.1, 3.14, 3.141, \ldots$

10. $1, 3, 6, 10, \ldots$

11. $1, 4, 27, 256, \ldots$

12. $\frac{1}{2}, \frac{3}{4}, \frac{7}{8}, \frac{15}{16}, \ldots$

13. $1, 1 + x, 1 + x + x^2, 1 + x + x^2 + x^3, \ldots$

Exercises 14–20: A sequence $\{s_n\}$ is defined recursively. Write out the first five terms of the sequence.

14. $s_1 = 1$
$s_n = n + s_{n-1} \quad (n \ge 2)$

15. $s_0 = 1$
$s_n = ns_{n-1} \quad (n \ge 1)$

16. $s_1 = 2$
$s_n = \dfrac{1}{s_{n-1} + 1} \quad (n \ge 2)$

17. $s_1 = 1$
$s_n = \dfrac{n+1}{n+2} s_{n-1} \quad (n \ge 2)$

18. $s_1 = 1$
$s_2 = 2$
$s_n = s_{n-1} + s_{n-2} \quad (n \ge 3)$

19. $s_1 = 1$
$s_2 = 2$
$s_n = \dfrac{1}{2}(s_{n-1} + s_{n-2}) \quad (n \ge 3)$

20. $s_1 = 1$
$s_2 = 2$
$s_n = \sqrt{s_{n-2}s_{n-1}} \quad (n \ge 3)$

Exercises 21–24: The first few terms of a sequence are listed. Write down a recursive formula for the sequence. (Don't forget to define the first term.)

21. $1, 1 + \frac{1}{2}, 1 + \frac{1}{2} + \frac{1}{3}$

22. $2, \sqrt{1+2}, \sqrt{1+\sqrt{1+2}}, \sqrt{1+\sqrt{1+\sqrt{1+2}}}, \ldots$

23. $2, 2^2, 2^{(2^2)}, 2^{2^{(2^2)}}, \ldots$

24. $1, 1 + x, 1 + x + x^2, 1 + x + x^2 + x^3, \ldots$

Exercises 25–36: Decide if the sequence has a limit. Find or estimate the limit in those cases when it exists. Use analytic, numerical, or graphical means.

25. $\left\{\dfrac{n-1}{n+1}\right\}_{n=1}^{\infty}$ **26.** $\left\{\dfrac{10^k}{k!}\right\}_{k=0}^{\infty}$

27. $\left\{\dfrac{3j^2-2j+1}{100j+2000}\right\}_{j=1}^{\infty}$ **28.** $\{\sqrt{n^2+n}-n\}_{n=0}^{\infty}$

29. $\{\arctan k\}_{k=1}^{\infty}$ **30.** $\left\{n\sin\dfrac{1}{n}\right\}_{n=1}^{\infty}$

31. $\{\cos\sqrt{j}\}_{j=0}^{\infty}$ **32.** $\left\{\sum_{k=1}^{n}\dfrac{1}{2^k}\right\}_{n=1}^{\infty}$

33. $\left\{\sum_{k=1}^{n}\dfrac{1}{\sqrt{k}}\right\}_{n=1}^{\infty}$ **34.** $\left\{\left(1+\dfrac{1}{k}\right)^{2k}\right\}_{k=1}^{\infty}$

35. $\left\{\left(1+\dfrac{2}{k}\right)^{k}\right\}_{k=1}^{\infty}$ **36.** $\left\{\left(1+\dfrac{5}{k}\right)^{k}\right\}_{k=1}^{\infty}$

37. In Fig. 7.25, which graph is the graph of $y=\ln x$ and which is the graph of $y=T_3(x;1)$?

38. Let $s_n=\left(1-\frac{1}{n}\right)^2$. Show that the sequence $\{s_n\}_{n=1}^{\infty}$ is increasing and bounded above. What is the limit of the sequence?

T 39. Let $s_n=\left(1-\frac{1}{n}\right)^n$. Provide graphical and/or numerical evidence that the sequence $\{s_n\}_{n=1}^{\infty}$ is increasing and bounded above. What do you think is the limit of the sequence?

T 40. Let $s_n=(n!)^{1/n}$. Provide graphical and/or numerical evidence that the sequence $\{s_n\}_{n=1}^{\infty}$ is increasing but not bounded above.

T 41. Let $s_n=(n!)^{1/n}$, as in the preceding problem. Plot several points (n,s_n), including some for large values of n. The points seem to lie on a line. Use two points with large n values to find the slope of the line, and relate this to the graph produced in Exercise 40. (When large n values are used to compute the slope, the answer should be $\approx 1/e$.)

T 42. Define a sequence recursively by

$$s_1=\sqrt{2}$$
$$s_n=\sqrt{2+\sqrt{s_{n-1}}}\quad (n\ge 2).$$

Provide graphical and/or numerical evidence that the sequence is increasing and bounded above. Estimate the limit of the sequence.

43. State the decreasing-sequence version of the Monotone Sequence Theorem.

T 44. Define a sequence recursively by

$$s_1=4$$
$$s_n=\frac{1}{2}\left(s_{n-1}+\frac{2}{s_{n-1}}\right)\quad (n\ge 2).$$

Provide graphical and/or numerical evidence that the sequence is decreasing and bounded below. Estimate the limit of the sequence.

T 45. Let c be a positive real number. Define a sequence recursively by

$$s_1=c$$
$$s_n=c^{s_{n-1}}\quad (n\ge 2).$$

With a calculator or computer algebra system, find the first several terms of the sequence for different values of c. For what c does the sequence appear to converge? For what c does it appear to diverge?

46. For integer $n\ge 1$, let

$$s_n=\sum_{k=1}^{n}(-1)^{k-1}\frac{1}{k}=1-\frac{1}{2}+\frac{1}{3}-\cdots+(-1)^{n-1}\frac{1}{n}.$$

We first study the even indexed terms of the sequence, that is, $s_2, s_4, s_6, \ldots$.

a. Show that for $k\ge 2$,

$$s_{2k}=s_{2k-2}+\left(\frac{1}{2k-1}-\frac{1}{2k}\right).$$

Use this to show that the sequence $s_2, s_4, s_6, \ldots$ is monotone increasing.

b. Show that the sequence $s_2, s_4, s_6, \ldots$ is bounded above by 1.

c. Tell why the sequence $s_2, s_4, s_6, \ldots$ converges.

d. Now consider the whole sequence $s_1, s_2, s_3, s_4, \ldots$. Give an argument showing that this sequence converges to the same limit as the sequence of even-indexed terms.

e. The limit of this sequence is ln 2. How close is s_{100} to the limit?

47. For integer $n\ge 1$, let

$$s_n=\sum_{k=1}^{n}(-1)^{k-1}\frac{4}{2k-1}=4-\frac{4}{3}+\frac{4}{5}-\cdots+(-1)^{n-1}\frac{4}{4n-1}.$$

Use the method of the previous problem to show that this sequence converges. What is your guess for the limit of the sequence?

48. Let $a>0$ be constant. Determine

$$\lim_{k\to\infty}\left(1+\tfrac{a}{k}\right)^k.$$

Hint: Use Example 8 and note that

$$\left(1+\tfrac{a}{k}\right)^k = \left(\left(1+\tfrac{a}{k}\right)^{k/a}\right)^a.$$

49. Use the technique illustrated in Example 8 to show that

$$\lim_{k\to\infty}\left(1-\frac{1}{k}\right)^k = \frac{1}{e}.$$

Combine this result with the previous exercise to show that for any real a,

$$\lim_{k\to\infty}\left(1+\frac{a}{k}\right)^k = e^a.$$

Exercises 50–55: For the given function f, real number c, and interval I, show that the sequence $\{T_n(x;c)\}$ of Taylor polynomials converges to f uniformly on I.

50. $f(x) = \cos x$, $c = 0$, $I = [-1, 1]$

51. $f(x) = \sin x$, $c = \frac{\pi}{4}$, $I = \left[0, \frac{\pi}{2}\right]$

52. $f(x) = e^{3x}$, $c = 0$, $I = [-10, 10]$

53. $f(x) = -7x^3 - 3x^2 + 4$, $c = 5$, $I = (-\infty, \infty)$

54. $f(x) = \dfrac{1}{1-x}$, $c = 0$, $I = [-0.25, 0.25]$

55. $f(x) = \ln x$, $c = 6$, $I = [4, 8]$

7.4 Infinite Series

In Example 9 of Section 7.3 we showed that the sequence $\{T_n(x;1)\}_{n=1}^{\infty}$ of Taylor polynomials for $\ln x$ converges uniformly to $\ln x$ on $I = [1/2, 3/2]$. Hence, for all x in this interval,

$$\lim_{n\to\infty} T_n(x;1) = \lim_{n\to\infty}\sum_{k=1}^{n}\frac{(-1)^{k-1}}{k}(x-1)^k = \ln x. \tag{1}$$

In view of (1), it seems natural to write

$$\sum_{k=1}^{\infty}\frac{(-1)^{k-1}}{k}(x-1)^k = \ln x, \quad \frac{1}{2}\le x\le\frac{3}{2}. \tag{2}$$

The sum shown in (2) is an example of an infinite series.

Infinite series show up in the solutions of many problems in science, engineering, and mathematics. We will see how infinite series can be used to approximate functions and definite integrals and to describe some physical phenomena.

Definition of Infinite Series

We will usually think of an infinite series such as (2) as a "sum of infinitely many terms." We use limits to make this idea precise.

DEFINITION Infinite Series

Let $\{a_k\}_{k=1}^{\infty}$ be a sequence of real numbers or of functions. The **infinite series** with terms $\{a_k\}_{k=1}^{\infty}$ is written

$$\sum_{k=1}^{\infty} a_k \quad\text{or}\quad a_1 + a_2 + a_3 + \dots. \tag{3}$$

For each positive integer n,

$$s_n = \sum_{k=1}^{n} a_k$$

is called the nth **partial sum** of the series (3). The sequence

$$s_1, s_2, s_3, \ldots = \{s_n\}_{n=1}^{\infty}$$

WEB

is called the **sequence of partial sums** associated with (3). We say the infinite series (3) converges to S or has sum S if

$$\lim_{n\to\infty} s_n = S.$$

In this case we write

$$\sum_{k=1}^{\infty} a_k = S.$$

If the sequence of partial sums $\{s_n\}_{n=1}^{\infty}$ has no limit, we say that the series (3) diverges. If $\lim_{n\to\infty} s_n = +\infty$, we say that the series (3) diverges to $+\infty$.

It is important to remember that there are two sequences associated with the infinite series

$$\sum_{k=1}^{\infty} a_k.$$

The first is the sequence $\{a_k\}_{k=1}^{\infty}$ of terms of the series. The second is the sequence $\{s_n\}_{n=1}^{\infty}$ of partial sums of the series. Both of these sequences are important in understanding infinite series, but they tell us different things about the series. It is important to distinguish between these two sequences and to understand what each sequence tells us about the behavior of the associated series.

EXAMPLE 1 A telescoping series

Show that the sum of the infinite series

$$\sum_{k=1}^{\infty} \frac{1}{k(k+1)}$$

is 1.

Solution

The sum of an infinite series is defined to be the limit of the sequence of partial sums of the series. For each positive integer n, the nth partial sum of the series is

$$s_n = \sum_{k=1}^{n} \frac{1}{k(k+1)}.$$

The first few terms of the sequence of partial sums are

$$s_1 = \sum_{k=1}^{1} \frac{1}{k(k+1)} = \frac{1}{1\cdot 2} = \frac{1}{2},$$

$$s_2 = \sum_{k=1}^{2} \frac{1}{k(k+1)} = \frac{1}{1\cdot 2} + \frac{1}{2\cdot 3} = \frac{2}{3},$$

$$s_3 = \sum_{k=1}^{3} \frac{1}{k(k+1)} = \frac{1}{1\cdot 2} + \frac{1}{2\cdot 3} + \frac{1}{3\cdot 4} = \frac{3}{4},$$

$$s_4 = \sum_{k=1}^{4} \frac{1}{k(k+1)} = \frac{1}{1\cdot 2} + \frac{1}{2\cdot 3} + \frac{1}{3\cdot 4} + \frac{1}{4\cdot 5} = \frac{4}{5}.$$

From these examples, it seems reasonable to guess that

$$s_n = \frac{n}{n+1}. \tag{4}$$

If this guess is correct, then we conclude that the sum of the series is

$$\sum_{k=1}^{\infty} \frac{1}{k(k+1)} = \lim_{n\to\infty} \sum_{k=1}^{n} \frac{1}{k(k+1)} = \lim_{n\to\infty} s_n = \lim_{n\to\infty} \frac{n}{n+1} = 1. \tag{5}$$

Although guessing and conjecturing are an important part of problem solving, you should always back up your guesses with solid reasoning. We can verify the guess given in (4) by using the partial fractions decomposition

$$\frac{1}{k(k+1)} = \frac{1}{k} - \frac{1}{k+1}.$$

Using this we have

$$\begin{aligned} s_n &= \sum_{k=1}^{n} \frac{1}{k(k+1)} = \sum_{k=1}^{n} \left(\frac{1}{k} - \frac{1}{k+1}\right) \\ &= \left(1 - \frac{1}{2}\right) + \left(\frac{1}{2} - \frac{1}{3}\right) + \left(\frac{1}{3} - \frac{1}{4}\right) + \cdots + \left(\frac{1}{n} - \frac{1}{n+1}\right). \end{aligned}$$

Note that in this last expression, $-1/2 + 1/2 = 0$, $-1/3 + 1/3 = 0$, and so on, so the sum reduces to $1 - \dfrac{1}{n+1}$. (The partial sum is said to telescope. We saw this phenomenon earlier in Section 5.1.) Hence,

$$s_n = 1 - \frac{1}{n+1} = \frac{n}{n+1}.$$

This verifies (4), so the sum of the series is 1, as shown by (5).

EXAMPLE 2 Show that the infinite series $\sum_{k=1}^{\infty} 1$ diverges.

Solution

For each positive integer n, the nth partial sum of the infinite series is

$$s_n = \sum_{k=1}^{n} = \underbrace{1 + 1 + \cdots + 1}_{n\ 1\text{'s}} = n.$$

Because

$$\lim_{n\to\infty} s_n = \lim_{n\to\infty} n = +\infty,$$

the series diverges to $= +\infty$.

EXAMPLE 3 A geometric series

Let r be any real number. The infinite series

$$1 + r + r^2 + r^3 + \cdots = \sum_{k=0}^{\infty} r^k$$

is called a *geometric series*. Show that this series converges if $|r| < 1$ and diverges if $|r| \geq 1$.

Solution

We first consider the special cases $r = 1$ and $r = -1$. For $r = 1$, the geometric series $\Sigma_{k=1}^{\infty} 1^k$ is the same as the divergent series in Example 2. When $r = -1$, the nth partial sum is

$$s_n = \sum_{k=0}^{n-1} (-1)^k = 1 + (-1) + \cdots + (-1)^{n-1} = \begin{cases} 1 & (n \text{ odd}) \\ 0 & (n \text{ even}). \end{cases}$$

Thus, when $r = -1$, the sequence $\{s_n\}_{n=1}^{\infty}$ is alternating 1s and 0s. Hence $\lim_{n\to\infty} s_n$ does not exist, and so $\Sigma_{k=0}^{\infty} r^k$ diverges for $r = -1$.

When $|r| \neq 1$, we write the partial sum s_n in a form that is easier to work with. We have

$$s_n = 1 + r + r^2 + r^3 + \cdots + r^{n-1}. \tag{6}$$

Multiply both sides of (6) by r to obtain

$$rs_n = r + r^2 + r^3 + \cdots + r^{n-1} + r^n. \tag{7}$$

Subtract (7) from (6). After canceling like terms, we see

$$(1 - r)s_n = 1 - r^n.$$

From this expression, we obtain

$$s_n = \frac{1 - r^n}{1 - r}.$$

When $|r| < 1$, we have $\lim_{n\to\infty} r^n = 0$, so

$$\lim_{n\to\infty} s_n = \lim_{n\to\infty} \frac{1 - r^n}{1 - r} = \frac{1}{1 - r}.$$

Hence for $|r| < 1$ the geometric series $\Sigma_{k=0}^{\infty} r^k$ converges and,

$$\sum_{k=0}^{\infty} r^k = \frac{1}{1 - r}.$$

When $r > 1$, we have $\lim_{n\to\infty} r^n = +\infty$. Thus for $r > 1$,

$$\lim_{n\to\infty} s_n = \lim_{n\to\infty} \frac{1 - r^n}{1 - r} = +\infty,$$

so the geometric series $\Sigma_{k=0}^{\infty} r^k$ diverges to $+\infty$.

When $r < -1$, the terms of the sequence $\{r^n\}_{n=1}^{\infty}$ alternate in sign and tend toward ∞ in absolute value. Hence when $r < -1$,

$$\lim_{n\to\infty} s_n = \lim_{n\to\infty} \frac{1 - r^n}{1 - r} \text{ does not exist.}$$

Collecting these results,

$$\sum_{k=0}^{\infty} r^k \begin{cases} = \dfrac{1}{1 - r} & (-1 < r < 1) \\ \text{diverges to } +\infty & (r \geq 1) \\ \text{diverges} & (r \leq -1). \end{cases}$$

We summarize the results of this example.

Geometric Series

Let a and r be real numbers with $a \neq 0$. The infinite series

$$\sum_{k=0}^{\infty} ar^k = a + ar + ar^2 + ar^3 + \cdots$$

is called a **geometric series.** The number r is called the **common ratio** of the series. If $|r| < 1$, the series converges and

$$\sum_{k=0}^{\infty} ar^k = \frac{a}{1 - r}. \qquad (8)$$

If $|r| \geq 1$, then the series diverges.

EXAMPLE 4 Converting a repeating decimal to a fraction

Use geometric series to convert the repeating decimal

$$0.31313131\ldots$$

to a fraction

Solution

We have

$$\begin{aligned} 0.31313131\ldots &= 0.31 + 0.0031 + 0.000031 + 0.00000031 + \cdots \\ &= \frac{31}{100} + \frac{31}{100^2} + \frac{31}{100^3} + \frac{31}{100^4} + \cdots. \end{aligned}$$

This is a geometric series with first term of $a = \frac{31}{100}$ and common ratio $r = \frac{1}{100}$. Because $|r| < 1$, the series converges. By (8) the sum is

$$\frac{\frac{31}{100}}{1 - \frac{1}{100}} = \frac{31}{99}.$$

Hence

$$0.31313131\cdots = \frac{31}{99}.$$

How could you check that this answer is correct?

Some real-life phenomena can be analyzed using geometric series.

EXAMPLE 5 Paying taxes

A worker earns an amount M of money. Assume that some of the money is paid to the government in the form of taxes and that the rest is spent. The portion that is spent becomes income for other people. Suppose that this income is also taxed, and that the remainder is spent to become income for someone else. Assume that the tax rate is r, $0 < r < 1$, (e.g., for a tax rate of 22 percent, $r = 0.22$) and that this "tax-and-spend" pattern continues indefinitely. How much of the original amount M ultimately goes to the government?

Solution

The worker pays amount rM in taxes and spends the remaining $M - rM = (1 - r)M$. This spent sum now is income for other people. When this sum is taxed at rate r, the amount of taxes paid is $r(1 - r)M$. The remaining amount $(1 - r)M - r(1 - r)M = (1 - r)^2M$ is spent and becomes income for other people. The tax on this second spent sum is $r(1 - r)^2M$. This sum is paid to the government and the remaining $(1 - r)^2M - r(1 - r)^2M = (1 - r)^3M$ is spent and becomes new income. The pattern developing here is clear: Evidently in the kth round of the tax-and-spend cycle, $r(1 - r)^{k-1}M$ is paid in taxes. The tax total that goes to the government is found by summing the taxes paid in each round. This total is

$$rM + r(1 - r)M + r(1 - r)^2M + r(1 - r)^3M + \cdots = \sum_{j=0}^{\infty} rM(1 - r)^j.$$

The sum is a geometric series with first term rM and common ratio $1 - r$. Because $0 < r < 1$, it follows that $0 < 1 - r < 1$. Thus the geometric series converges. By (8) the sum is

$$\text{total to government} = \sum_{j=0}^{\infty} rM(1 - r)^j = \frac{rM}{1 - (1 - r)} = M.$$

So under our assumptions, all money eventually goes to the government. Of course, the government then spends its income to put the money back into the economy.

EXAMPLE 6 The harmonic series

The infinite series

$$1 + \frac{1}{2} + \frac{1}{3} + \cdots = \sum_{k=1}^{\infty} \frac{1}{k}$$

is called the *harmonic series*. Determine whether this series converges or diverges.

Solution

The harmonic series and its partial sums occur in many unexpected places in engineering, science, and mathematics. We first give a short argument to show that the series diverges. For a positive integer m, consider the partial sum s_{2^m} of the first 2^m terms of the series. Group the terms as shown,

$$\begin{aligned} s_{2^m} &= 1 + \frac{1}{2} + \frac{1}{3} + \frac{1}{4} + \frac{1}{5} + \frac{1}{6} + \frac{1}{7} + \frac{1}{8} + \cdots + \frac{1}{2^m} \\ &= 1 + \left(\frac{1}{2}\right) + \left(\frac{1}{3} + \frac{1}{4}\right) + \left(\frac{1}{5} + \frac{1}{6} + \frac{1}{7} + \frac{1}{8}\right) \\ &\quad + \cdots + \left(\frac{1}{2^{m-1}+1} + \cdots + \frac{1}{2^m}\right). \end{aligned}$$

The first set of parentheses has one term, and this term is $\frac{1}{2}$. The second set has two terms, the last term being $\frac{1}{4} = \frac{1}{2^2}$. The third set has $4 = 2^2$ terms, the last term being $\frac{1}{8} = \frac{1}{2^3}$. The mth (last) set of parentheses has 2^{m-1} terms, and the last term in this set is $\frac{1}{2^m}$. In each set of parentheses, we replace each term by the last term in the set. This makes the sum smaller, and so

$$\begin{aligned} s_{2^m} &\geq 1 + \left(\frac{1}{2}\right) + \left(\frac{1}{4} + \frac{1}{4}\right) + \left(\frac{1}{8} + \frac{1}{8} + \frac{1}{8} + \frac{1}{8}\right) \\ &\quad + \cdots + \left(\frac{1}{2^m} + \frac{1}{2^m} + \cdots + \frac{1}{2^m}\right) \\ &= 1 + \frac{1}{2} + 2\frac{1}{4} + 4\frac{1}{8} + \cdots + 2^{m-1}\frac{1}{2^m} \\ &= 1 + \underbrace{\frac{1}{2} + \frac{1}{2} + \cdots + \frac{1}{2}}_{m\ \frac{1}{2}\text{'s}} \\ &= 1 + \frac{m}{2}. \end{aligned}$$

This shows that we can make s_{2^m} as large as we like by taking m sufficiently large. Because the sequence of partial sums is increasing,

$$s_1 < s_2 < s_3 < \cdots < s_n < s_{n+1} < \cdots$$

it follows from the Monotone Sequence Theorem that $\lim_{n\to\infty} s_n = +\infty$. Hence, the harmonic series diverges to $+\infty$.

For an alternative way of investigating the partial sums s_n, we compare some areas. In the plane, construct a rectangle with base the interval $[1, 2]$ on the x-axis and height 1. Next to it put a rectangle with base $[2, 3]$ and

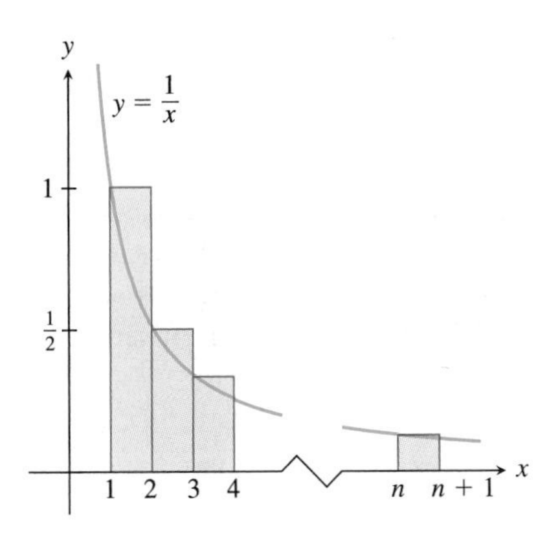

FIGURE 7.27 The area enclosed by the rectangles is larger than the area under the graph for $1 \le x \le n + 1$.

height $1/2$, then a rectangle with base $[3,4]$ and height $1/3$. Continue this process until you have n rectangles, the last with base $[n, n+1]$ and height $1/n$. See Fig. 7.27. The total area enclosed by these rectangles is

$$1 \cdot 1 + 1 \cdot \frac{1}{2} + 1 \cdot \frac{1}{3} + \cdots + 1 \cdot \frac{1}{n} = s_n.$$

The graph of $y = 1/x$ passes through the upper left corner of each rectangle and lies below the top of each rectangle, as shown in Fig. 7.27. Hence the total area enclosed by the n rectangles is greater than the area under the graph of $y = 1/x$ for $1 \le x \le n + 1$. From this observation,

$$s_n \ge \int_1^{n+1} \frac{1}{x}\,dx = \ln x\Big|_1^{n+1} = \ln(n+1). \tag{9}$$

It follows from (9) that

$$\lim_{n\to\infty} s_n \ge \lim_{n\to\infty} \ln(n+1) = +\infty.$$

This again shows that the harmonic series diverges to $+\infty$. In Exercise 46, we ask you to modify this argument to show that $s_n \le 1 + \ln n$.

The Arithmetic of Series

Sometimes we recognize an infinite series $\Sigma_{k=1}^{\infty} a_k$ as a combination of other series. If we know the behavior (or sum) of these other series, we may be able to determine the behavior (or sum) of $\Sigma_{k=1}^{\infty} a_k$. We describe some of the circumstances in which we can do so.

THEOREM Sums and Multiples of Series

Let $\Sigma_{k=1}^{\infty} a_k$ and $\Sigma_{k=1}^{\infty} b_k$ be infinite series.

a) Let c be a real number. If $\Sigma_{k=1}^{\infty} a_k = S$, then

$$\sum_{k=1}^{\infty} ca_k = c\sum_{k=1}^{\infty} a_k = cS.$$

If $c \ne 0$ and $\Sigma_{k=1}^{\infty} a_k$ diverges, then $\Sigma_{k=1}^{\infty} ca_k$ also diverges.

b) Suppose that $\Sigma_{k=1}^{\infty} a_k = S$ and $\Sigma_{k=1}^{\infty} b_k = T$. Then

$$\sum_{k=1}^{\infty} (a_k + b_k) = \sum_{k=1}^{\infty} a_k + \sum_{k=1}^{\infty} b_k = S + T.$$

These results follow from the results on the arithmetic of sequences given in Section 7.3. See Exercises 50 and 51.

A Test for Divergence

To determine whether an infinite series Σa_k converges or diverges, we need to determine the behavior of the sequence $\{s_n\}_{n=1}^{\infty}$ of partial sums. In Examples 1, 2, and

3 we found a formula for s_n and then used this formula to determine $\lim_{n\to\infty} s_n$. In Example 5 we did not find a formula for s_n but were able to find estimates from which we could conclude $\lim_{n\to\infty} s_n = +\infty$. Determining whether a series converges or diverges by working with the sequence of partial sums is difficult for all but a few special series. In the next section we will develop several tests to help us determine the behavior of a series. No such test works all of the time, but for most series that we encounter, we will be able to find a test that applies. We conclude this section with a simple but useful test that helps us quickly identify some divergent series.

Suppose that the infinite series

$$\sum_{k=1}^{\infty} a_k$$

converges to S. Then we know that

$$\lim_{n\to\infty} s_n = \lim_{n\to\infty} \sum_{k=1}^{n} a_k = S. \tag{10}$$

Also note that for $n \geq 2$,

$$a_n = \sum_{k=1}^{n} a_k - \sum_{k=1}^{n-1} a_k = s_n - s_{n-1}. \tag{11}$$

Combining (10) and (11), we have

$$\lim_{n\to\infty} a_n = \lim_{n\to\infty} (s_n - s_{n-1}) = \lim_{n\to\infty} s_n - \lim_{n\to\infty} s_{n-1} = S - S = 0.$$

This shows that if the infinite series $\Sigma_{k=1}^{\infty} a_k$ converges, then the sequence $\{a_k\}_{k=1}^{\infty}$ of terms of the series has limit 0. Note, however, that the information $\lim_{k\to\infty} a_k = 0$ does *not* imply that $\Sigma_{k=1}^{\infty} a_k$ converges. We can only say that when $\lim_{k\to\infty} a_k \neq 0$, the series diverges.

THEOREM The nth Term Divergence Test

If $\lim_{k\to\infty} a_k$ does not exist or if $\lim_{k\to\infty} a_k$ exists but is not equal to 0, then

$$\sum_{k=1}^{\infty} a_k \text{ diverges.}$$

EXAMPLE 7 Use the nth term Divergence Test to show that the geometric series $\Sigma_{k=0}^{\infty} r^k$ diverges when $|r| \geq 1$.

Solution

Observe that

$$\lim_{k\to\infty} r^k = \begin{cases} +\infty & (r > 1) \\ 1 & (r = 1) \\ \text{does not exist} & (r \leq -1). \end{cases} \tag{12}$$

Hence for $|r| \geq 1$, $\lim_{k\to\infty} r^k \neq 0$. Thus, for these values of r, the series $\Sigma_{k=0}^{\infty} r^k$ diverges.

EXAMPLE 8 Show that $\lim_{k\to\infty} a_k = 0$ does not imply that $\Sigma_{k=1}^{\infty} a_k$ converges.

Solution

Let $a_k = \dfrac{1}{k}$. In Example 5 we showed that the harmonic series $\displaystyle\sum_{k=1}^{\infty} \frac{1}{k}$ diverges. However,

$$\lim_{k\to\infty} a_k = \lim_{k\to\infty} \frac{1}{k} = 0.$$

This example demonstrates that $\lim_{k\to\infty} a_k = 0$ does *not* imply that $\Sigma_{k=1}^{\infty} a_k$ converges.

Exercises 7.4

Exercises 1–4: For each infinite series, write out the first five terms and the first five partial sums.

1. $\displaystyle\sum_{k=1}^{\infty} k$

2. $\displaystyle\sum_{k=1}^{\infty} \frac{(-1)^{k-1}}{k}$

3. $\displaystyle\sum_{k=0}^{\infty} \frac{(k+1)x^k}{2^k}$

4. $\displaystyle\sum_{k=2}^{\infty} \frac{1}{(\ln k)^2}$

Exercises 5–13: For each infinite series, find a formula for s_n, the nth partial sum of the series. Then find the limit of the sequence of partial sums.

5. $\displaystyle\sum_{k=1}^{\infty} \pi$

6. $\displaystyle\sum_{k=0}^{\infty} \frac{1}{5^k}$

7. $\displaystyle\sum_{n=0}^{\infty} \left(\sqrt{n+1} - \sqrt{n}\right)$

8. $\displaystyle\sum_{n=1}^{\infty} \left(\frac{1}{\sqrt[3]{n}} - \frac{1}{\sqrt[3]{n+1}}\right)$

9. $\displaystyle\sum_{k=1}^{\infty} \left(\frac{2}{k+3} - \frac{2}{k+4}\right)$

10. $\displaystyle\sum_{k=1}^{\infty} (-4)^{k-1}$

11. $\displaystyle\sum_{k=0}^{\infty} e^{-2k}$

12. $\displaystyle\sum_{k=1}^{\infty} \left(\frac{1}{k+3} - \frac{1}{k+5}\right)$

13. $\displaystyle\sum_{k=2}^{\infty} 5(0.8)^k$

Exercises 14–22: Decide whether the infinite series converges or diverges.

14. $\displaystyle\sum_{k=1}^{\infty} (-2)$

15. $\displaystyle\sum_{k=0}^{\infty} \frac{1}{(3.2)^k}$

16. $\displaystyle\sum_{r=0}^{\infty} (-1)^r \frac{\ln(r+5)}{\ln(r+2)}$

17. $\displaystyle\sum_{k=1}^{\infty} (-1)^k \frac{\ln k + 5}{\ln k + 2}$

18. $\displaystyle\sum_{m=1}^{\infty} \frac{2^m}{m^2}$

19. $\displaystyle\sum_{n=1}^{\infty} \frac{8}{n}$

20. $\displaystyle\sum_{q=1}^{\infty} \ln\left(\frac{q+2}{q+1}\right)$

21. $\displaystyle\sum_{k=1}^{\infty} \left(\frac{1}{2^k} - \frac{5}{3^k}\right)$

22. $\displaystyle\sum_{n=2}^{\infty} \left(\frac{1}{n+2} - \frac{1}{n+5}\right)$

Exercises 23–28: Convert the decimal to a fraction.

23. 0.15151515...

24. 0.346346346...

25. 0.819191919...

26. 0.00233233233...

27. 0.45000000...

28. 0.023444444...

Exercises 29–31: Find r.

29. $\Sigma_{k=0}^{\infty} r^k = 5$

30. $\Sigma_{k=3}^{\infty} r^k = 10$

31. $\Sigma_{k=1}^{\infty} r^k = -e$

Exercises 32–35: A formula for the partial sum s_n of the series $\Sigma_{k=1}^{\infty} a_k$ is given. Find a_1, a_2, a_3, and a_k.

32. $s_n = n + 2$

33. $s_n = \dfrac{4+n}{3+n}$

34. $s_n = (-1)^n$

35. $s_n = \sin n$

Exercises 36–39: Find the values of x for which the geometric series converges.

36. $\displaystyle\sum_{k=0}^{\infty} (2x-3)^k$

37. $\displaystyle\sum_{k=2}^{\infty} (x^2 - 4x - 5)^k$

38. $\displaystyle\sum_{k=0}^{\infty} \left(\frac{1}{x^2-1}\right)^k$

39. $\displaystyle\sum_{k=0}^{\infty} (2\sin x)^k$

40. When a new tennis ball is dropped to a concrete floor from height h, it rebounds to a height of $0.55h$, as illustrated in the accompanying figure. Suppose the ball is dropped from a height of 10 feet and after each bounce rebounds to 55% of the height it attained on the previous bounce. Find the total up-and-down distance the ball travels.

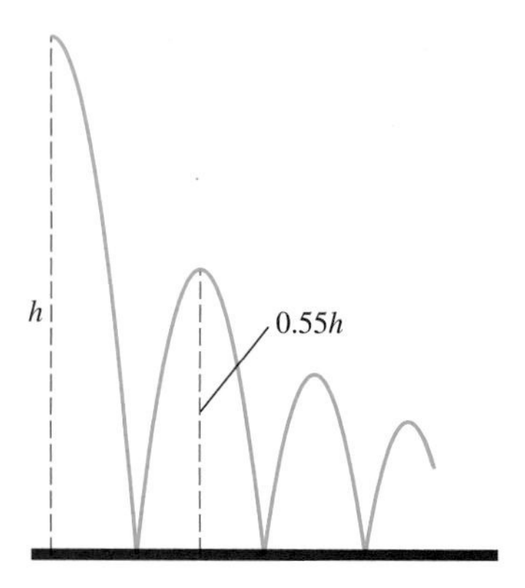

Figure for Exercise 40.

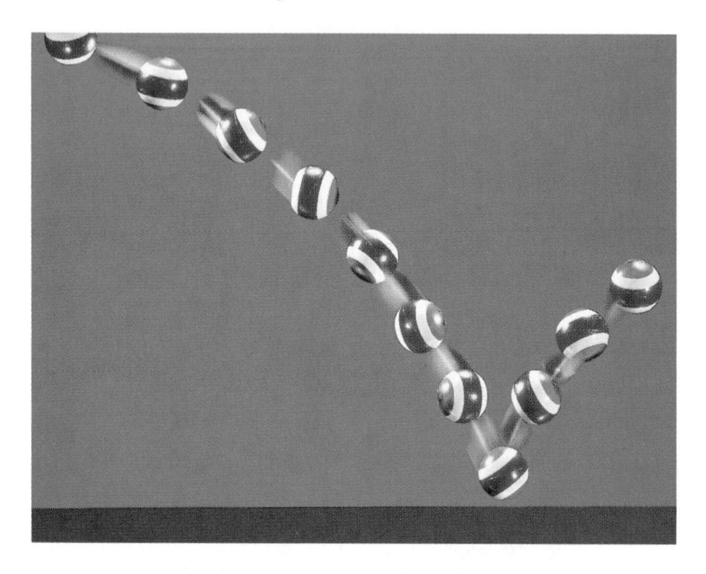

41. Let D be the total up-and-down distance traveled by the tennis ball in the previous problem. If we want the ball to travel distance $2D$, from how high should it be dropped?

42. How long (in seconds) does the tennis ball in Exercise 40 keep bouncing? How about the ball in Exercise 41?

43. A tennis ball dropped from a height of 10 feet actually stops bouncing after about 15 bounces. What is the total up-and-down distance traveled by the tennis ball during these bounces? How many seconds do these 15 bounces take?

44. Economists estimate that, on the average, each person spends 65% of his or her income. When this money is spent, it becomes income for someone else, who in turn spends 65%, and so forth. The effective impact of this money on the economy is the total of the amounts spent as it passes from person to person, each spending 65% of what is received. Suppose Marie earns \$1000. What is the effective impact of this income on the economy?

45. The federal government wishes to stimulate the economy by giving tax rebates to the public so that the effective impact (see Exercise 44) of the rebates on the economy is \$50 billion. How much should the government spend on these rebates?

46. In this problem we produce an upper bound for the nth partial sum of the harmonic series, $\sum_{k=1}^{\infty} \frac{1}{k}$.

a. In the plane, draw a rectangle with base the interval $[1, 2]$ on the x-axis and height $1/2$. Next to it, draw a rectangle with base $[2, 3]$ and height $1/3$, then a rectangle of base $[3, 4]$ and height $1/4$. Continue until you have $n - 1$ rectangles, the last with base $[n - 1, n]$ and height $1/n$. Let A_n be the total area enclosed by these $n - 1$ rectangles. Show that

$$A_n = \frac{1}{2} + \frac{1}{3} + \cdots + \frac{1}{n}.$$

b. Now draw the graph of $y = 1/x$. As in Example 6, compare the area under the graph to the area enclosed by the rectangles. Show that

$$A_n < \ln n.$$

c. Let s_n be the nth partial sum of the harmonic series. Use parts **a** and **b** to find an upper bound for s_n.

47. Consider the infinite series $\sum_{k=1}^{\infty} \frac{1}{k^2}$.

a. Use an argument similar to that given in Example 6 to show that

$$s_n = \sum_{k=1}^{n} \frac{1}{k^2} \geq \int_1^{n+1} \frac{1}{x^2}\,dx.$$

b. Use an argument similar to the one outlined in Exercise 46 to show that

$$s_n = \sum_{k=1}^{n} \frac{1}{k^2} \leq 1 + \int_1^{n} \frac{1}{x^2}\,dx.$$

c. Show that the sequence of partial sums $\{s_n\}_{n=1}^{\infty}$ is increasing and bounded above. What does this imply about the sum of the series?

48. Consider the infinite series $\sum_{k=1}^{\infty} \frac{1}{\sqrt{k}}$.

a. Use an argument similar to that given in Example 6 to show that

$$s_n = \sum_{k=1}^{n} \frac{1}{\sqrt{k}} \geq \int_1^{n+1} \frac{1}{\sqrt{x}}\,dx.$$

b. Use an argument similar to the one outlined in Exercise 46 to show that

$$s_n = \sum_{k=1}^{n} \frac{1}{\sqrt{k}} \le 1 + \int_1^n \frac{1}{\sqrt{x}}\,dx.$$

c. Does the infinite series $\sum_{k=1}^{\infty} \frac{1}{\sqrt{k}}$ converge or diverge?

49. Let $\Sigma_{k=1}^{\infty} a_k$ be an infinite series with $a_k > 0$ for all k. Suppose that $a_{k+1}/a_k > 1$ for every positive integer k. Show that the series $\Sigma_{k=1}^{\infty} a_k$ diverges.

50. Let $\Sigma_{k=1}^{\infty} a_k$ be an infinite series and let c be a real number. Use the results on arithmetic of sequences from Section 7.3 to show the following.

a. If $\Sigma_{k=1}^{\infty} a_k = S$, then

$$\sum_{k=1}^{\infty} ca_k = c\sum_{k=1}^{\infty} a_k = cS.$$

b. If $\Sigma_{k=1}^{\infty} a_k$ diverges and $c \neq 0$, then

$$\sum_{k=1}^{\infty} ca_k \text{ also diverges.}$$

51. Let $\Sigma_{k=1}^{\infty} a_k$ and $\Sigma_{k=1}^{\infty} b_k$ be convergent infinite series with

$$\sum_{k=1}^{\infty} a_k = S \quad \text{and} \quad \sum_{k=1}^{\infty} b_k = T.$$

Show that the series $\Sigma_{k=1}^{\infty} (a_k + b_k)$ also converges and that

$$\sum_{k=1}^{\infty} (a_k + b_k) = S + T.$$

52. Let $\Sigma_{k=1}^{\infty} a_k$ and $\Sigma_{k=1}^{\infty} b_k$ be convergent infinite series with

$$\sum_{k=1}^{\infty} a_k = S \quad \text{and} \quad \sum_{k=1}^{\infty} b_k = T.$$

Is it true that

$$\sum_{k=1}^{\infty} a_k b_k = ST?$$

If yes, give a supporting argument. If no, give one example showing the assertion is not true.

53. Let $\Sigma_{k=1}^{\infty} a_k$ and $\Sigma_{k=1}^{\infty} b_k$ be convergent infinite series with

$$\sum_{k=1}^{\infty} a_k = S \quad \text{and} \quad \sum_{k=1}^{\infty} b_k = T.$$

Suppose that $b_k \neq 0$ for $k = 1, 2, 3, \ldots$ and that $T \neq 0$. Is it true that

$$\sum_{k=1}^{\infty} \frac{a_k}{b_k} = \frac{S}{T}?$$

If yes, give a supporting argument. If no, give one example showing the assertion is not true.

54. The harmonic bridge Imagine that you have access to an unlimited number of identical cards each of length 3 inches. Take n of these cards and stack them on a table in the following way:

The first card is placed so $3 - 1/n$ inches are on the table and $1/n$ inches project beyond the end of the table.

The second card is placed so $3 - 1/(n-1)$ inches of the card are on top of the first card and $1/(n-1)$ inches project beyond the end of the first card.

The third card is placed so $3 - 1/(n-2)$ inches of the card are on top of the second card and $1/(n-2)$ inches project beyond the end of the second card.

⋮

The nth card is placed so $3 - 1/1 = 2$ inches of the card are on the $(n-1)$th card and 1 inch projects beyond the end of the $(n-1)$th card. (See the accompanying figure.)

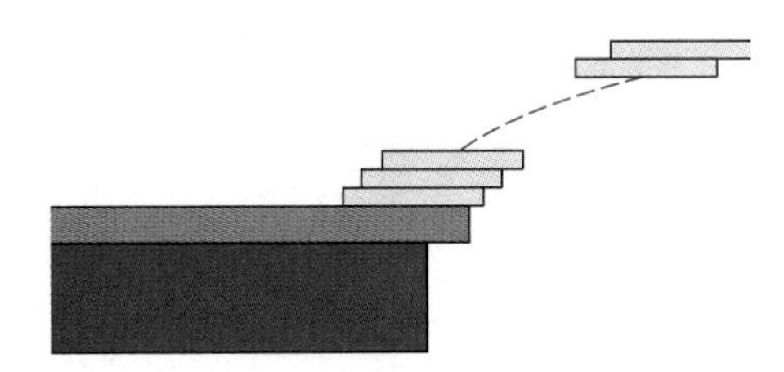

Figure for Exercise 54.

a. Show that the center of gravity of the nth card placed lies over the card below it (that is, over the $(n-1)$st card). Thus the last card will not "tip" off of the card below it.

b. Show that the center of gravity of the nth and $(n-1)$st card lies over the $(n-2)$nd card. Hence the last two cards placed will not fall off the $(n-2)$nd card.

c. Show that the center of gravity of the whole set of n cards lies above the table.

d. Based on **a**, **b**, and **c**, give an argument showing that the card structure is stable.

e. How far beyond the edge of the table is the end of the nth card? Tell why with enough cards you could make this end as far beyond the end of the table as you like.

f. Estimate the number of cards required to build a structure so that the end of the last card is 10 feet beyond the end of the table. Exercise 46 and Example 6 might be helpful.

7.5 Tests for Convergence

When we work with an infinite series $\Sigma_{k=1}^{\infty} a_k$, we are usually interested in answering two questions:

- Does the series converge?
- If the series converges, what is its sum?

The second of these questions may be difficult or impossible to answer unless the series happens to be, say, geometric, telescoping, or associated with a known Taylor series. Often we must be content with an approximation to the sum of the series. Before we start calculating partial sums, however, we need to know whether the series converges or diverges. Otherwise we cannot know whether the partial sums are indeed approximating the sum of the series. Thus it is important to answer the first of the questions listed. In this section we develop several tests to help us determine whether or not a series converges.

Series with Nonnegative Terms

Because the sum of an infinite series is defined to be the limit of its sequence of partial sums, many of the things we learned about sequences can be applied to infinite series. In Section 7.3 we learned that an increasing sequence either converges to a finite number or diverges to infinity. We use this fact to study series of nonnegative terms.

Let $\Sigma_{k=1}^{\infty} a_k$ be an infinite series with $a_k \geq 0$ for all k. (We will refer to such a series as a *series of nonnegative terms.*) Then for each positive integer n,

$$s_{n+1} = \sum_{k=1}^{n+1} a_k = \left(\sum_{k=1}^{n} a_k\right) + a_{n+1} \geq \sum_{k=1}^{n} a_k = s_n.$$

Hence

$$s_1 \leq s_2 \leq s_3 \leq \cdots,$$

that is, the sequence $\{s_n\}_{n=1}^{\infty}$ of partial sums is an increasing sequence. Given that $\{s_n\}_{n=1}^{a}$ is increasing, the Monotone Sequence Theorem says (see Section 7.3):

- If the sequence $\{s_n\}_{n=1}^{\infty}$ is bounded above, then

  $$\lim_{n\to\infty} s_n = S$$

  for some finite real number S. In this case, the series has sum S.
- If the sequence $\{s_n\}_{n=1}^{\infty}$ is not bounded above, then

  $$\lim_{n\to\infty} s_n = +\infty.$$

In this case, the infinite series diverges to (or has sum) $+\infty$. We used these ideas in previous work. In Example 6 of Section 7.4 we worked with the harmonic series

$$\sum_{k=1}^{\infty} \frac{1}{k}. \tag{1}$$

We saw that the sequence of partial sums

$$s_n = \sum_{k=1}^{n} \frac{1}{k}, \qquad n = 1, 2, 3, \ldots,$$

is not bounded above, and hence concluded that (1) diverges to $+\infty$.

If you did Exercise 47 in Section 7.4, you saw that the partial sums of the infinite series

$$\sum_{k=1}^{\infty} \frac{1}{k^2} \tag{2}$$

are bounded above by 2. That is, for each positive integer n,

$$s_n = \sum_{k=1}^{n} \frac{1}{k^2} < 2. \tag{3}$$

Because the partial sums of this series are increasing and bounded above, the limit of the sequence of partial sums exists and is finite. This means that the series converges. Although we can conclude from (3) that the sum of the series (2) is a number less than or equal to 2, we cannot say much more about the sum without extra work. In more advanced texts, it is shown that the series (2) has sum $\pi^2/6$.

The Comparison Tests If two infinite series are similar, knowing the behavior of one of the series might be enough to determine the behavior of the other. The Comparison Tests tell us one way that this can be done.

THEOREM The Comparison Test

Let $\Sigma_{k=1}^{\infty} a_k$ be a series of nonnegative terms.

a) $\Sigma_{k=1}^{\infty} a_k$ converges if there is a *convergent* series $\Sigma_{k=1}^{\infty} c_k$ of nonnegative terms with $a_k \le c_k$ for all sufficiently large k.

b) $\Sigma_{k=1}^{\infty} a_k$ diverges if there is a *divergent* series $\Sigma_{k=1}^{\infty} d_k$ of nonnegative terms with $a_k \ge d_k$ for all sufficiently large k.

To see why part **a)** of the Comparison Test works, let $\Sigma_{k=1}^{\infty} c_k = C$, let $s_n = \Sigma_{k=1}^{n} a_k$, and suppose that $a_k \le c_k$ for $k \ge N$. Then the sequence $\{s_n\}_{n=1}^{\infty}$ of partial sums is monotone increasing and, if $n > N$,

$$\begin{aligned} s_n &= s_N + (a_{N+1} + a_{N+2} + \cdots + a_n) \\ &\le s_N + (c_{N+1} + c_{N+2} + \cdots + c_n) \\ &\le s_N + C. \end{aligned}$$

It follows that

$$s_1 \le s_2 \le \cdots \le s_N \le s_{N+1} \le \cdots \le s_n \le \cdots \le s_N + C.$$

Because N is a fixed positive integer, $s_N + C$ is a finite upper bound for $\{s_n\}_{n=1}^{\infty}$. Therefore, by the Monotone Sequence Theorem, $\lim_{n\to\infty} s_n$ is a finite real number S. This means that $\Sigma_{k=1}^{\infty} a_k$ converges to S.

For an argument in support of part **b)**, see Exercise 32.

EXAMPLE 1 Determine whether the infinite series

$$\sum_{k=1}^{\infty} \frac{1}{\sqrt{2k^2 + 3}} \tag{4}$$

converges or diverges.

Solution

When k is large,

$$\frac{1}{\sqrt{2k^2 + 3}} \approx \frac{1}{\sqrt{2}k}. \tag{5}$$

Thus we expect that the series $\sum_{k=1}^{\infty} \frac{1}{\sqrt{2k^2 + 3}}$ will behave like the series $\sum_{k=1}^{\infty} \frac{1}{\sqrt{2}k}$. The latter series diverges because it is a constant multiple of the harmonic series. To prove the series (4) diverges, we must show that the terms of (4) become and remain larger than those of a divergent series of nonnegative terms. Because of (5), we compare the series (4) with a variation of the harmonic series. For $k \geq 1$,

$$\frac{1}{\sqrt{2k^2 + 3}} \geq \frac{1}{\sqrt{2k^2 + 3k^2}} = \frac{1}{\sqrt{5}k}. \tag{6}$$

Because $\sum_{k=1}^{\infty} \frac{1}{\sqrt{5}k}$ diverges, it follows from the Comparison Tests that $\sum_{k=1}^{\infty} \frac{1}{\sqrt{2k^2 + 3}}$ also diverges.

In the preceding example, we suspected that $\sum_{k=1}^{\infty} \frac{1}{\sqrt{2k^2 + 3}}$ was divergent as soon as we saw estimate (5). The rest of the solution consisted of the algebra aimed at coming up with the formal comparison in (6). Actually, an estimate such as (5) is enough to determine the behavior of a series. This is more precisely stated in the following result.

THEOREM The Limit Comparison Test

Let $\Sigma_{k=1}^{\infty} a_k$ and $\Sigma_{k=1}^{\infty} b_k$ be series of nonnegative terms. Suppose that

$$\lim_{k\to\infty} \frac{a_k}{b_k} = L \quad \text{where } L \neq 0 \text{ and } L \neq \infty. \tag{7}$$

Then $\Sigma_{k=1}^{\infty} a_k$ and $\Sigma_{k=1}^{\infty} b_k$ both converge or both diverge.

Very roughly, (7) says that when k is large, $\frac{a_k}{b_k} \approx L$, that is $a_k \approx Lb_k$. Thus it seems reasonable that for large N,

$$\sum_{k=N}^{\infty} a_k \approx L \sum_{k=N}^{\infty} b_k,$$

and hence that both series converge or both diverge. See Exercise 47 for a more formal argument in support of the Limit Comparison Test. See also Exercises 48 and 49 for a discussion of the cases $L = 0$ and $L = +\infty$.

The Limit Comparison Test is most useful when we know the behavior of one of the two series and wish to determine the behavior of the other.

EXAMPLE 2 Determine whether the infinite series $\sum_{k=1}^{\infty} \frac{3}{k^2 + 3k + 2}$ converges or diverges.

Solution

In Example 1 of Section 7.4 we showed that $\sum_{k=1}^{\infty} \frac{1}{k(k+1)}$ converges. Because

$$\lim_{k\to\infty} \frac{\dfrac{3}{k^2 + 3k + 2}}{\dfrac{1}{k(k+1)}} = \lim_{k\to\infty} \frac{3k^2 + 3k}{k^2 + 3k + 2} = 3,$$

we conclude that $\sum_{k=1}^{\infty} \frac{3}{k^2 + 3k + 2}$ also converges.

To effectively use the Comparison Test, we need to have a collection of series whose behavior we already know. The so-called p-series are used more than any others for this purpose.

The p-series

Let p be a positive real number. The infinite series

$$\sum_{k=1}^{\infty} \frac{1}{k^p}$$

converges if $p > 1$ and diverges if $p \le 1$.

For many values of p the behavior of the p-series can be deduced from the Comparison Test. For example, if $p \le 1$, then

$$\frac{1}{k^p} \ge \frac{1}{k} \quad \text{for } k = 1, 2, 3, \ldots.$$

Because the harmonic series $\sum_{k=1}^{\infty} \frac{1}{k}$ diverges, it follows from the Comparison Tests that $\sum_{k=1}^{\infty} \frac{1}{k^p}$ also diverges.

If $p \geq 2$, then

$$\frac{1}{k^p} \leq \frac{1}{k^2} \quad \text{for } k = 1, 2, 3, \ldots.$$

The series $\sum_{k=1}^{\infty} \frac{1}{k^2}$ converges, so by the Comparison Tests, $\sum_{k=1}^{\infty} \frac{1}{k^p}$ also converges.

For $1 < p < 2$ we can prove that the p-series converges by using the rectangle area techniques used in Example 6 of Section 7.4. See also Exercises 46, 47, and 48 in Section 7.4 and Exercises 35 and 36 in this section.

The Ratio Test In the next section we will work with Taylor series, that is, series whose partial sums are the Taylor polynomials for a function:

$$\sum_{k=0}^{\infty} \frac{f^{(k)}(c)}{k!}(x - c)^k = \lim_{n\to\infty} \sum_{k=0}^{n} \frac{f^{(k)}(c)}{k!}(x - c)^k = \lim_{n\to\infty} T_n(x; c).$$

We will want to know the values of x for which the Taylor series converges. Sometimes this information can be found by working with the remainder $R_n(x; c)$, as in Section 7.2. An alternative is the Ratio Test.

THEOREM The Ratio Test

Let $\Sigma_{k=1}^{\infty} a_k$ be an infinite series of positive terms. Suppose that

$$\lim_{k\to\infty} \frac{a_{k+1}}{a_k} = \rho$$

where $0 \leq \rho \leq \infty$.

a) If $\rho < 1$, then $\Sigma_{k=1}^{\infty} a_k$ converges.
b) If $\rho > 1$, then $\Sigma_{k=1}^{\infty} a_k$ diverges.

If $\rho = 1$, then there is not enough information to decide if the series converges or diverges. Some other method or test must be used.

Because we will use this test often in upcoming sections, we spend some time now showing why the test works.

Suppose

$$\lim_{k\to\infty} \frac{a_{k+1}}{a_k} = \rho < 1. \tag{8}$$

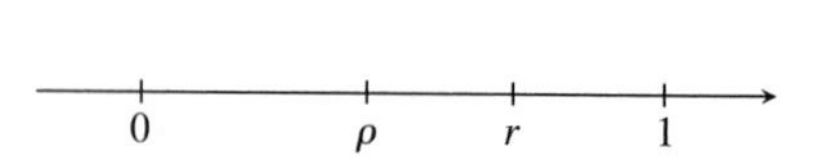

FIGURE 7.28 r is the average of ρ and 1.

Let $r = (1 + \rho)/2$, so $\rho < r < 1$. See Fig. 7.28. According to (8), the numbers a_{k+1}/a_k get close to ρ as $k \to \infty$. This means that eventually these numbers get closer to ρ than they are to r. Hence there is a positive integer N such that

$$\frac{a_{k+1}}{a_k} < r \quad \text{when} \quad k \geq N.$$

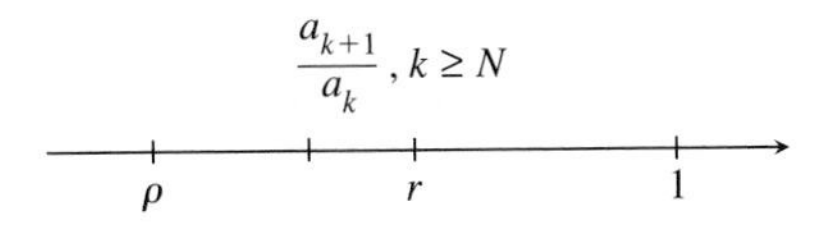

FIGURE 7.29 When k is large enough, $a_{k+1}/a_k < r$.

See Fig. 7.29. Because $a_k > 0$ for all k, we can rewrite the last expression as

$$a_{k+1} < a_k r \quad \text{when} \quad k \geq N. \tag{9}$$

Applying (9) with $k = N$, then $k = N + 1$, then $k = N + 2$, and so on, and using each result in the next line,

$$\begin{aligned} a_{N+1} &< a_N r \\ a_{N+2} &< a_{N+1} r < a_N r^2 \\ a_{N+3} &< a_{N+2} r < a_N r^3 \\ &\vdots \end{aligned}$$

In general

$$a_{N+k} < a_N r^k, \quad \text{for } k > 0.$$

Setting $k = j - N$ in this inequality, we get the equivalent statement

$$a_j < a_N r^{j-N}, \quad j > N.$$

Hence for $n > N$ we have

$$\begin{aligned} s_n &= a_1 + a_2 + \cdots + a_{N-1} + a_N + a_{N+1} + \cdots + a_n \\ &\leq s_{N-1} + a_N + a_N r + a_N r^2 + \cdots + a_N r^{n-N} \\ &\leq s_{n-1} + \sum_{j=0}^{\infty} a_N r^j. \end{aligned}$$

Recall that the sum of a convergent geometric series is the first term over 1 minus the common ratio. Hence,

$$s_n \leq s_{N-1} + \sum_{j=0}^{\infty} a_N r^j = s_{N-1} + \frac{a_N}{1 - r}.$$

Thus the sequence $\{s_n\}_{n=1}^{\infty}$ of partial sums of $\Sigma_{k=1}^{\infty} a_k$ is monotone increasing and bounded above by the number $s_{N-1} + a_N/(1 - r)$. Because it is also monotonic, it follows that $\Sigma_{k=1}^{\infty} a_k$ converges.

Next assume that

$$\lim_{k\to\infty} \frac{a_{k+1}}{a_k} = \rho > 1$$

and let $r = (1 + \rho)/2$. Note that now $1 < r < \rho$. Arguing as before, we can find a positive integer N so that

$$a_j > a_N r^{j-N} \quad \text{when } j \geq N. \tag{10}$$

Because $r > 1$ and $a_N > 0$, we see that

$$\lim_{j\to\infty} a_N r^{j-N} = \infty. \tag{11}$$

Combining (10) and (11), we conclude that

$$\lim_{j\to\infty} a_j \geq \lim_{j\to\infty} a_N r^{j-N} = +\infty. \tag{12}$$

It follows from the nth-term Divergence Test that $\Sigma_{k=1}^{\infty} a_k$ diverges.

EXAMPLE 3 Give examples to show that when $\rho = 1$ in the Ratio Test, the series could either converge or diverge.

Solution

First consider the harmonic series $\sum_{k=1}^{\infty} \frac{1}{k}$. This series has terms $a_k = 1/k$. Applying the Ratio Test, we have

$$\rho = \lim_{k\to\infty} \frac{a_{k+1}}{a_k} = \lim_{k\to\infty} \frac{\frac{1}{k+1}}{\frac{1}{k}} = \lim_{k\to\infty} \frac{k}{k+1} = 1.$$

This example shows that we can have $\rho = 1$ for a divergent series.

In Example 1 of Section 7.4 we showed that the series $\Sigma_{k=1}^{\infty} 1/(k(k+1))$ converges. Applying the Ratio Test to this series, we find

$$\rho = \lim_{k\to\infty} \frac{a_{k+1}}{a_k} = \lim_{k\to\infty} \frac{\frac{1}{(k+1)(k+2)}}{\frac{1}{k(k+1)}} = \lim_{k\to\infty} \frac{k}{k+2} = 1.$$

Thus we can also have $\rho = 1$ for a convergent series.

The Ratio Test is particularly useful for series whose terms a_k involve factorials or powers.

EXAMPLE 4 For each of the following series decide whether the series converges or diverges.

a) $\sum_{k=0}^{\infty} \frac{k^2 - k + 2}{2^k}$

b) $\sum_{k=1}^{\infty} \frac{(k!)^2}{k^{2k+1}}$

Solution

a) With $a_k = (k^2 - k + 2)/2^k$, we have

$$\frac{a_{k+1}}{a_k} = \frac{\frac{(k+1)^2 - (k+1) + 2}{2^{k+1}}}{\frac{k^2 - k + 2}{2^k}} = \frac{1}{2}\frac{k^2 + k + 2}{k^2 - k + 2}.$$

Thus

$$\lim_{k\to\infty} \frac{a_{k+1}}{a_k} = \lim_{k\to\infty} \frac{1}{2}\frac{k^2 + k + 2}{k^2 - k + 2} = \frac{1}{2} < 1.$$

By the Ratio Test, $\Sigma_{k=0}^{\infty} (k^2 - k + 2)/2^k$ converges.

b) Now let $a_k = (k!)^2/k^{2k+1}$. Noting that $(k+1)! = (k+1)k!$,

$$\frac{a_{k+1}}{a_k} = \frac{\dfrac{((k+1)!)^2}{(k+1)^{2(k+1)+1}}}{\dfrac{(k!)^2}{k^{2k+1}}}$$

$$= \frac{(k!)^2(k+1)^2}{(k+1)^{2k+3}} \cdot \frac{k^{2k+1}}{(k!)^2}$$

$$= \frac{k^{2k+1}}{(k+1)^{2k+1}}$$

$$= \left(\frac{k}{k+1}\right)^{2k+1}.$$

Hence

$$\lim_{k\to\infty} \frac{a_{k+1}}{a_k} = \lim_{k\to\infty} \left(\frac{k}{k+1}\right)^{2k+1}. \tag{13}$$

To determine the limit in (13), first note that

$$\left(\frac{k}{k+1}\right)^{2k+1} = \left(\frac{k}{k+1}\right)^{2k}\left(\frac{k}{k+1}\right)$$

$$= \frac{1}{\left(1+\dfrac{1}{k}\right)^{2k}} \cdot \frac{k}{k+1}$$

for all k. In Example 8 of Section 7.3 we showed that

$$\lim_{k\to\infty} \left(1+\frac{1}{k}\right)^k = e.$$

Using this when we evaluate the limit of the expression in (13), we find

$$\lim_{k\to\infty} \frac{a_{k+1}}{a_k} = \lim_{k\to\infty} \left(\frac{1}{\left(1+\dfrac{1}{k}\right)^{2k}} \cdot \frac{k}{k+1}\right) = \frac{1}{e^2} < 1.$$

Hence, by the Ratio Test, the series $\sum_{k=1}^{\infty} \frac{(k!)^2}{k^{2k+1}}$ converges.

Absolute and Conditional Convergence

The Comparison Tests and the Ratio Test are tests to determine the convergence or divergence of series with nonnegative terms. What about series whose terms can be positive or negative? The Comparison and Ratio Tests can be used to determine whether some of these series converge or diverge. For some series these tests lead to no conclusion. We first look at a series for which one of these tests can be successfully used.

EXAMPLE 5 Determine whether the series

$$\sum_{k=1}^{\infty} (-1)^k \frac{k2^k}{k^2 + 1} \tag{14}$$

converges or diverges.

Solution

We cannot use the Comparison or Ratio Tests directly on (14) because the series is not a series of nonnegative terms. However, we can apply the Ratio Test to the series of absolute values

$$\sum_{k=1}^{\infty} \left| (-1)^k \frac{k2^k}{k^2 + 1} \right| = \sum_{k=1}^{\infty} \frac{k2^k}{k^2 + 1}. \tag{15}$$

The kth term of the series of absolute values is $a_k = k2^k/(k^2 + 1)$. Hence

$$\lim_{k\to\infty} \frac{a_{k+1}}{a_k} = \lim_{k\to\infty} \frac{\dfrac{(k + 1)2^{k+1}}{(k + 1)^2 + 1}}{\dfrac{k2^k}{k^2 + 1}} = \lim_{k\to\infty} 2\left(\left(\frac{k + 1}{k}\right)\left(\frac{k^2 + 1}{k^2 + 2k + 2}\right)\right) = 2.$$

Because the limit of the ratio of terms is greater than 1, the series (15) diverges. But we saw in (12) that when a series diverges by the Ratio Test, the terms of the series do not tend to 0. Thus $\lim_{k\to\infty} a_k \neq 0$. Hence,

$$\lim_{k\to\infty} (-1)^k \frac{k2^k}{k^2 + 1} \neq 0.$$

It follows that the series (14) diverges.

Be careful not to draw too many conclusions from this example. There are also examples of series $\Sigma_{k=1}^{\infty} a_k$ for which $\Sigma_{k=1}^{\infty} |a_k|$ diverges, but $\Sigma_{k=1}^{\infty} a_k$ converges. However, the following result tells us that if the series of absolute values converges, then it is always true that the original series also converges.

THEOREM The Absolute Convergence Test

Let $\Sigma_{k=1}^{\infty} a_k$ be an infinite series. If the series $\Sigma_{k=1}^{\infty} |a_k|$ converges, then the series $\Sigma_{k=1}^{\infty} a_k$ also converges.

To justify the Absolute Convergence Test, first observe that for $k \geq 1$,

$$0 \leq (a_k + |a_k|) \leq 2|a_k|. \tag{16}$$

Hence

$$\sum_{k=1}^{\infty} (a_k + |a_k|) \tag{17}$$

is a series of nonnegative terms. Because $\Sigma_{k=1}^{\infty} 2|a_k|$ converges, it follows from (16) and the Comparison Tests that (17) also converges. Thus

$$\sum_{k=1}^{\infty} a_k = \sum_{k=1}^{\infty} (a_k + |a_k| - |a_k|) = \sum_{k=1}^{\infty} (a_k + |a_k|) - \sum_{k=1}^{\infty} |a_k|$$

is the difference of two convergent series. It follows that $\Sigma_{k=1}^{\infty} a_k$ also converges. See the results about sums and multiples of series on page 585 of Section 7.4.

EXAMPLE 6 Show that the series $\Sigma_{k=1}^{\infty} (-1)^{k-1} \frac{1}{k^2}$ converges.

Solution

The series of absolute values of the terms,

$$\sum_{k=1}^{\infty} \left| (-1)^{k-1} \frac{1}{k^2} \right| = \sum_{k=1}^{\infty} \frac{1}{k^2},$$

is the p-series with $p = 2$. Because we know that this series converges, it follows from the Absolute Convergence Test that $\Sigma_{k=1}^{\infty} (-1)^{k-1} \frac{1}{k^2}$ also converges.

DEFINITION Absolute Convergence

If $\Sigma_{k=1}^{\infty} a_k$ is a series for which $\Sigma_{k=1}^{\infty} |a_k|$ converges, then we say the series $\Sigma_{k=1}^{\infty} a_k$ is **absolutely convergent.**

With this new terminology, the Absolute Convergence Test says:

If an infinite series is absolutely convergent, then it is also convergent.

In Example 6 we showed that the series $\Sigma_{k=1}^{\infty} (-1)^{k-1} \frac{1}{k^2}$ is absolutely convergent.

Alternating Series The Absolute Convergence Test says that if $\Sigma_{k=1}^{\infty} |a_k|$ converges, then the series $\Sigma_{k=1}^{\infty} a_k$ also converges. However, if $\Sigma_{k=1}^{\infty} |a_k|$ diverges, then the Absolute Convergence Test does not apply. As we shall see, when $\Sigma_{k=1}^{\infty} |a_k|$ diverges, it may not be possible to conclude anything about the behavior of $\Sigma_{k=1}^{\infty} a_k$. In some cases this series may converge, and in other cases it may diverge (see Example 5). To decide which is the case may take a careful analysis of the series with the roles of positive and negative terms taken into account. Analysis of this sort can be very difficult. However, if the terms of the series alternate in sign and decrease to 0 in absolute value, we can immediately conclude that the series converges.

THEOREM The Alternating Series Test

Let $\{b_k\}_{k=1}^{\infty}$ be a sequence of positive real numbers. If

$$b_1 \geq b_2 \geq b_3 \geq \cdots$$

and

$$\lim_{k\to\infty} b_k = 0,$$

then

$$\sum_{k=1}^{\infty} (-1)^{k-1} b_k = b_1 - b_2 + b_3 - b_4 + - \cdots \tag{18}$$

converges.

To justify the Alternating Series test, we show that the sequence $\{s_n\}$ of partial sums of the series converges. Let $2m$ be an even integer. Then

$$\begin{aligned} s_{2m} &= (b_1 - b_2) + (b_3 - b_4) + \cdots + (b_{2m-3} - b_{2m-2}) + (b_{2m-1} - b_{2m}) \\ &= s_{2m-2} + (b_{2m-1} - b_{2m}). \end{aligned} \tag{19}$$

Because $b_{2m-1} \geq b_{2m}$, it follows from (19) that

$$s_{2m-2} \leq s_{2m}.$$

Thus, the sequence $s_2, s_4, s_6, \ldots, s_{2m}, \ldots$ of even-indexed sums is a monotone increasing sequence. In addition, for any even integer $2m$

$$\begin{aligned} s_{2m} &= (b_1 - b_2) + (b_3 - b_4) + \cdots + (b_{2m-1} - b_{2m}) \\ &= b_1 - (b_2 - b_3) - (b_4 - b_5) - \cdots - (b_{2m-2} - b_{2m-1}) - b_{2m} \\ &\leq b_1. \end{aligned}$$

Thus the sequence of even-indexed partial sums is increasing and bounded above by b_1:

$$s_2 \leq s_4 \leq \cdots \leq s_{2m} \leq \cdots \leq b_1.$$

By the Monotone Sequence Theorem, this sequence has a limit that we call S:

$$\lim_{m\to\infty} s_{2m} = S. \tag{20}$$

The sum of an odd number of terms of the series is of the form

$$s_{2m+1} = s_{2m} + b_{2m+1}.$$

Because $\lim_{m\to\infty} b_{2m+1} = 0$, we have, using (20),

$$\lim_{m\to\infty} s_{2m+1} = \lim_{m\to\infty} s_{2m} + \lim_{m\to\infty} b_{2m+1} = S.$$

Because the sequence of even-indexed partial sums and the sequence of odd-indexed partial sums both converge to S, the sequence $\{s_n\}$ of partial sums also converges to S, that is,

$$\sum_{k=1}^{\infty} (-1)^{k-1} b_k = \lim_{n\to\infty} \sum_{k=1}^{n} (-1)^{k-1} b_k = \lim_{n\to\infty} s_n = S.$$

EXAMPLE 7 Show that the alternating harmonic series

$$\sum_{k=1}^{\infty}(-1)^{k-1}\frac{1}{k} = 1 - \frac{1}{2} + \frac{1}{3} - \frac{1}{4} + \frac{1}{5} - \cdots \tag{21}$$

converges but is not absolutely convergent.

Solution

The terms of (21) alternate in sign. Because

$$1 \geq \frac{1}{2} \geq \frac{1}{3} \geq \cdots$$

and

$$\lim_{k\to\infty} \frac{1}{k} = 0,$$

the series converges by the Alternating Series Test.

The series of absolute values is

$$\sum_{k=1}^{\infty}\left|(-1)^{k-1}\frac{1}{k}\right| = \sum_{k=1}^{\infty}\frac{1}{k}.$$

Because the harmonic series diverges, the alternating harmonic series is not absolutely convergent.

The alternating harmonic series is an example of a *conditionally convergent series*.

DEFINITION Conditionally Convergent Series

If the series $\Sigma_{k=1}^{\infty} a_k$ converges but the series $\Sigma_{k=1}^{\infty} |a_k|$ diverges, then we say that the series $\Sigma_{k=1}^{\infty} a_k$ is **conditionally convergent.**

Estimating the Sum of an Alternating Series Let

$$\sum_{k=1}^{\infty}(-1)^{k-1}b_k = b_1 - b_2 + b_3 - \cdots$$

be a series satisfying the conditions of the Alternating Series Test. That is,

$$b_1 \geq b_2 \geq b_3 \geq \cdots \quad \text{and} \quad \lim_{k\to\infty} b_k = 0.$$

We showed that under these conditions the sequence of even-indexed partial sums is increasing, and converges to the sum S of the series, that is

$$s_2 \leq s_4 \leq \cdots \leq s_{2m} \leq \ldots, \tag{22}$$

and $\lim_{n\to\infty} s_{2n} = S$. On the other hand, the sequence $\{s_{2m+1}\}_{m=0}^{\infty}$ of odd-indexed partial sums is a decreasing sequence that also converges to S:

$$s_1 \geq s_3 \geq \cdots \geq s_{2m+1} \cdots \quad \text{and} \quad \lim_{m\to\infty} s_{2m+1} = S. \tag{23}$$

See Exercise 52. Combining (22) and (23), we have

$$s_2 \le s_4 \le \cdots \le s_{2m} \cdots \le S \cdots \le s_{2m+1} \le \cdots \le s_3 \le s_1.$$

Thus any even-indexed partial sum underestimates S, while any odd-indexed partial sum overestimates S. Therefore, for any positive integer n, we have

$$\begin{aligned} s_n \le S \le s_{n+1} \quad & (n \text{ even}) \\ s_{n+1} \le S \le s_n \quad & (n \text{ odd}). \end{aligned} \tag{24}$$

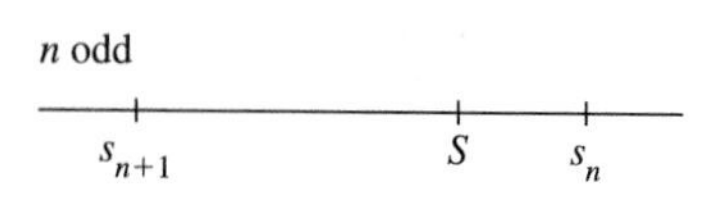

FIGURE 7.30 S is between s_n and s_{n+1}.

This is illustrated in Fig. 7.30. Because $|s_{n+1} - s_n| = b_{n+1}$, it follows from (24) that

$$|s_n - S| \le b_{n+1}.$$

This gives us a simple but effective way to estimate the error when we use the partial sum s_n to estimate S.

Estimating the Sum of an Alternating Series

Let

$$\sum_{k=1}^{\infty} (-1)^{k-1} b_k = b_1 - b_2 + b_3 - b_4 + \cdots$$

be an alternating series whose terms satisfy the hypotheses of the Alternating Series Test, and let S be the sum of the series. Then for any positive integer n,

$$|S - s_n| = \left| S - \sum_{k=1}^{n} (-1)^{k-1} b_k \right| \le b_{n+1}. \tag{25}$$

That is, the partial sum s_n differs from S by no more than the absolute value of the $(n + 1)$st term.

EXAMPLE 8 Approximate the sum of the alternating harmonic series

$$\sum_{k=1}^{\infty} (-1)^{k-1} \frac{1}{k} = 1 - \frac{1}{2} + \frac{1}{3} - \frac{1}{4} + \cdots$$

with an error of less than 0.01.

Solution

In Example 7 we showed that the alternating harmonic series satisfies the hypotheses of the Alternating Series Test and hence converges. Let S be the sum of the series. It follows from (25) that for any positive integer n,

$$|S - s_n| = \left| S - \sum_{k=1}^{n} (-1)^{k-1} \frac{1}{k} \right| \le \frac{1}{n+1}.$$

If we take $n = 100$, then the difference between s_n and S will be less than or equal to $1/101$, which is less than 0.01. Hence (with the aid of a calculator or CAS),

$$S \approx s_{100} = \sum_{k=1}^{100} (-1)^{k-1}\frac{1}{k} \approx 0.688172.$$

The actual sum of the alternating harmonic series is ln 2. Is the estimate 0.688172 within 0.01 of ln 2?

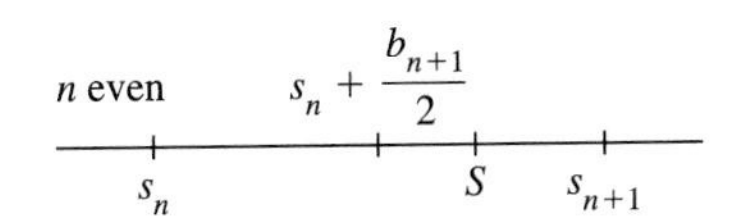

FIGURE 7.31 Both S and $s_n + (-1)^n b_{n+1}/2$ are between s_n and s_{n+1}.

With a little more work (see Exercise 52) it is possible to do better than (25) in estimating the sum of an alternating series. If we use the midpoint of the interval shown in Fig. 7.31 as our estimate for S, we have

$$S \approx \frac{s_n + s_{n+1}}{2} = s_n + \frac{(-1)^n b_{n+1}}{2}.$$

In this case, the error is

$$\left| S - \left(s_n + \frac{(-1)^n b_{n+1}}{2} \right) \right| \le \frac{b_{n+1}}{2}.$$

Exercises 7.5

Exercises 1–15: Decide whether the infinite series converges or diverges.

1. $\sum_{j=1}^{\infty} \frac{j+3}{j^2+2j+100}$

2. $\sum_{t=1}^{\infty} (-1)^{t+1}\frac{\sqrt{t+1}}{t^2+1}$

3. $\sum_{n=0}^{\infty} \frac{n!}{2^n(n+1)}$

4. $\sum_{k=2}^{\infty} \left(\frac{\ln k}{k}\right)^k$

5. $\sum_{l=1}^{\infty} \frac{1}{\sqrt{2l^2+3l+5}}$

6. $\sum_{n=1}^{\infty} \frac{1}{\sqrt{(2n^2+3n+5)^3}}$

7. $\sum_{p=1}^{\infty} pe^{-2p}$

8. $\sum_{k=1}^{\infty} (-1)^{k+1}\frac{2}{2k+3}$

9. $\sum_{j=0}^{\infty} \frac{2^j}{j!}$

10. $\sum_{m=2}^{\infty} \frac{1}{(\ln m)^m}$

11. $\sum_{s=2}^{\infty} (-1)^s\frac{1}{\ln s}$

12. $\sum_{i=1}^{\infty} \sin\left(\frac{\pi i}{4}\right)$

13. $\sum_{k=1}^{\infty} \left(\frac{1}{2^k} + (-1)^{k+1}\frac{1}{k}\right)$

14. $\sum_{z=1}^{\infty} \frac{\sin z}{z^2}$

15. $\sum_{n=2}^{\infty} \frac{\ln n}{n^{1.01}}$

Exercises 16–21: Each of the indicated series has some positive and some negative terms. Decide if the series converges absolutely, converges conditionally, or neither.

16. The series in Exercise 2.

17. The series in Exercise 8.

18. The series in Exercise 11.

19. The series in Exercise 12.

20. The series in Exercise 13.

21. The series in Exercise 14.

Exercises 22–27: Verify that the series satisfies the hypotheses of the Alternating Series Test. Then for the given number E, estimate the sum of the series to within E. Use a computer or calculator as needed.

22. $\sum_{k=1}^{\infty} (-1)^{k+1}\frac{1}{k^2+1}, \quad E = 0.01$

23. $\sum_{r=1}^{\infty} (-1)^{r-1}\frac{1}{\sqrt{r^2+1}}, \quad E = 0.01$

24. $\sum_{j=0}^{\infty} (-1)^j\frac{1}{4^j}, \quad E = 0.0001$

25. $\sum_{k=1}^{\infty} (-1)^{k-1}\frac{1}{2k-1}, \quad E = 0.005$

26. $\sum_{\ell=1}^{\infty} (-1)^{\ell-1}\frac{1}{(\ln(\ell+1))^5}, \quad E = 0.01$

27. $\sum_{j=0}^{\infty} (-1)^j\frac{1}{j!}, \quad E = 10^{-6}$

Exercises 28–31: Let a_k be the kth term of the series. Let $\rho = \lim_{k\to\infty}\left|\frac{a_{k+1}}{a_k}\right|$. For the given value of ρ, find all possible values of x for which,

28. $\sum_{k=0}^{\infty} \frac{x^k}{3^k}, \quad \rho = \frac{1}{2}$

29. $\sum_{k=0}^{\infty} kx^k, \quad \rho = 3$

30. $\sum_{k=0}^{\infty} \frac{(x+1)^k}{5^k}, \quad \rho = e$

31. $\sum_{k=0}^{\infty} \frac{x^k}{k!}, \quad \rho = 0$

32. Write an argument justifying part **b)** of the Comparison Tests. Your argument can be something like the discussion showing why part **a)** of the test is true.

33. Sometimes students use the following *incorrect* version of the Ratio Test:

Let $\Sigma_{k=1}^{\infty} a_k$ be an infinite series with $a_k > 0$ for all k. Suppose that $a_{k+1}/a_k < 1$ for every positive integer k. Then the series $\Sigma_{k=1}^{\infty} a_k$ converges.

Give an example showing that this statement is not always true.

34. Let $\Sigma_{k=1}^{\infty} a_k$ be an infinite series with $a_k > 0$ for all k. Suppose there is a number $\rho < 1$ such that

$$\frac{a_{k+1}}{a_k} < \rho$$

is true for all k. Show that the series $\Sigma_{k=1}^{\infty} a_k$ converges. How does this test differ from the Ratio Test? How is it similar?

35. (In this exercise, you are asked to establish the Integral Test, another test for convergence of series of nonnegative terms.) Let $f(x)$ be a nonnegative, decreasing function defined for $x \geq 1$.

a. Using the accompanying figure as a guide, show that

$$\int_1^{n+1} f(x)\,dx \leq \sum_{k=1}^{n} f(k).$$

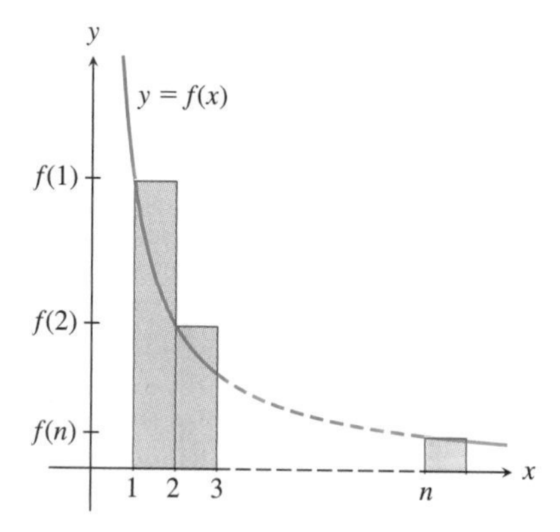

Figure for Exercise 35a.

b. Based on the inequality established in **a**, show that if $\int_1^{\infty} f(x)\,dx$ diverges, then $\Sigma_{k=1}^{\infty} f(k)$ diverges.

c. Next show that

$$\sum_{k=1}^{n} f(k) \leq f(1) + \int_1^n f(x)\,dx.$$

The accompanying figure may be helpful.

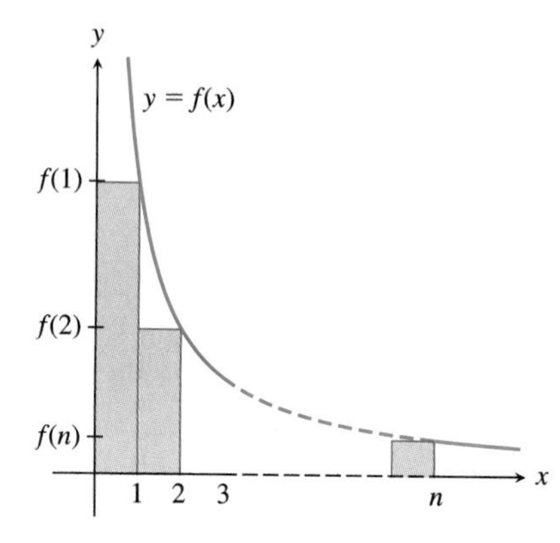

Figure for Exercise 35c.

d. Use the inequality established in **c** to show that if $\int_1^{\infty} f(x)\,dx$ converges, then $\Sigma_{k=1}^{\infty} f(k)$ converges.

e. The Integral Test says that if f is a nonnegative, decreasing function defined for $x \geq 1$, then

$$\sum_{k=1}^{\infty} f(k) \quad \text{and} \quad \int_1^{\infty} f(x)\,dx$$

either both diverge or both converge. Explain why this is true.

36. Use the Integral Test, developed in the previous exercise, to show that the p-series

$$\sum_{k=1}^{\infty} \frac{1}{k^p}$$

converges for $p > 1$ and diverges for $0 < p \leq 1$. (*Hint:* Set $f(x) = 1/x^p$. You will have to consider the case $p = 1$ separately.)

37. Use the Integral Test, developed in Exercise 35, to show that the series

$$\sum_{k=2}^{\infty} \frac{1}{k(\ln k)^p}$$

converges for $p > 1$ and diverges for $0 < p \leq 1$.

38. Use the Integral Test, developed in Exercise 35, to show that the geometric series

$$\sum_{k=1}^{\infty} c^k$$

converges for $0 < c < 1$. The Integral Test cannot be used to show that the series diverges for $c > 1$. Why not?

39. (In this exercise we see how the Integral Test can be used to estimate the error when we approximate the sum of a convergent series by a partial sum.) Let f be a nonnegative, decreasing function defined for $x \geq 1$. Suppose that $\int_1^\infty f(x)\,dx$ converges. Then we know from the Integral Test that $\Sigma_{k=1}^{\infty} f(k)$ also converges. Let $\Sigma_{k=1}^{\infty} f(k) = S$. Show that for positive integer n,

$$\left| S - \sum_{k=1}^{n} f(k) \right| = \left| \sum_{k=n+1}^{\infty} f(k) \right| \leq \int_n^\infty f(x)\,dx.$$

40. Consider the convergent series $\displaystyle\sum_{k=1}^{\infty} \frac{1}{k^{3/2}}$. Let S be the sum of the series.

a. Use the technique developed in the previous problem to show that

$$\left| S - \sum_{k=1}^{100} \frac{1}{k^{3/2}} \right| \leq 0.2.$$

b. What value of n should you take if you want to approximate S by

$$s_n = \sum_{k=1}^{n} \frac{1}{k^{3/2}}$$

with an error of at most 0.005?

41. How many terms of the series $\displaystyle\sum_{k=1}^{\infty} \frac{1}{k^{1.01}}$ are needed to approximate the sum with an error of at most 0.001?

42. How many terms of the series $\displaystyle\sum_{k=2}^{\infty} \frac{1}{k(\ln k)^2}$ are needed to approximate the sum with an error of at most 0.001?

43. How many terms of the series $\displaystyle\sum_{k=2}^{\infty} \frac{1}{k(\ln k)^{1.01}}$ are needed to approximate the sum with an error of at most 0.001?

44. Let $s_n = \displaystyle\sum_{k=1}^{n} \frac{1}{k}$ be a partial sum of the harmonic series. How large (roughly) must n be to make $s_n > 10$? To make $s_n > 100$?

45. Let $s_n = \displaystyle\sum_{k=1}^{n} \frac{1}{\sqrt{k}}$ be a partial sum of the p-series with $p = 1/2$. How large (roughly) must n be to make $s_n > 10$? To make $s_n > 1000$?

46. Let $s_n = \displaystyle\sum_{k=2}^{n} \frac{1}{k \ln k}$. How large (roughly) must n be to make $s_n > 10$? To make $s_n > 100$?

47. Let $\Sigma_{k=1}^{\infty} a_k$ and $\Sigma_{k=1}^{\infty} b_k$ be series of positive terms, and suppose that

$$\lim_{k\to\infty} \frac{a_k}{b_k} = L$$

where $0 < L < \infty$.

a. Explain why there is a positive integer K such that for all $k > K$,

$$\frac{1}{2}L < \frac{a_k}{b_k} < \frac{3}{2}L.$$

b. Show that if the assertion in part **a** is true, then for $k > K$,

$$\frac{1}{2}Lb_k < a_k < \frac{3}{2}Lb_k.$$

c. Explain why it follows from the Comparison Test that $\Sigma_{k=1}^{\infty} a_k$ and $\Sigma_{k=1}^{\infty} b_k$ either both converge or both diverge.

48. Let $\Sigma_{k=1}^{\infty} a_k$ and $\Sigma_{k=1}^{\infty} b_k$ be series of positive terms, and suppose that

$$\lim_{k\to\infty} \frac{a_k}{b_k} = 0.$$

a. Show that if $\Sigma_{k=1}^{\infty} b_k$ converges, then $\Sigma_{k=1}^{\infty} a_k$ also converges.

b. Show by example that the convergence of $\Sigma_{k=1}^{\infty} a_k$ does not imply the convergence of $\Sigma_{k=1}^{\infty} b_k$.

49. Let $\Sigma_{k=1}^{\infty} a_k$ and $\Sigma_{k=1}^{\infty} b_k$ be series of positive terms, and suppose that

$$\lim_{k\to\infty} \frac{a_k}{b_k} = \infty.$$

a. Show that if $\Sigma_{k=1}^{\infty} a_k$ converges, then $\Sigma_{k=1}^{\infty} b_k$ also converges.

b. Show by example that the convergence of $\Sigma_{k=1}^{\infty} b_k$ does not imply the convergence of $\Sigma_{k=1}^{\infty} a_k$.

50. Does the series

$$\sum_{k=1}^{\infty} \frac{1}{k^{1+1/k}}$$

converge or diverge? Justify your answer.

51. Let $\Sigma_{k=1}^{\infty} a_k$ be a convergent series of positive terms. Is it always true that $\Sigma_{k=1}^{\infty} a_k^2$ also converges?

52. Let $\Sigma_{k=1}^{\infty} (-1)^{k-1} b_k$ satisfy the hypotheses of the Alternating Series Test and let $s_n = \Sigma_{k=1}^{n} (-1)^{k-1} b_k$.

a. Show that the odd-indexed partial sums form a decreasing sequence. That is, show

$$s_1 \geq s_3 \geq s_5 \geq \cdots \geq s_{2m-1} \geq \ldots.$$

b. Show that if S is the sum of the series $\Sigma_{k=1}^{\infty} (-1)^{k-1} b_k$ and n is an odd integer, then

$$s_{n+1} \leq S \leq s_n.$$

Establish a similar inequality when n is an even integer.

c. Use the inequalities established in part **b** to show that for positive integer n,

$$\left| S - \left(s_n + \frac{(-1)^n b_{n+1}}{2} \right) \right| \leq \frac{1}{2} b_{n+1}.$$

7.6 Power Series and Taylor Series

Many functions are described by formulas like

$$f(x) = x^2, \qquad g(x) = \frac{1}{(1 + x)^2}, \quad \text{and} \quad h(x) = \sqrt{x}. \tag{1}$$

Such formulas describe a finite set of arithmetic/algebraic operations on x that result in $f(x)$ or $g(x)$ or $h(x)$. For the functions such as those described by

$$s(x) = \sin x, \qquad \ell(x) = \ln x, \quad \text{or} \quad F(x) = \arctan x, \tag{2}$$

there are no finite formulas like those in (1). There are, however, algorithms or formulas with which we can approximate the values of these functions. For example, for any real x,

$$\sin x = x - \frac{x^3}{3!} + \frac{x^5}{5!} - \frac{x^7}{7!} + \cdots. \tag{3}$$

Thus, for any real x, $\sin x$ can be approximated by a partial sum of (3).

There are still other functions that cannot be expressed by simple formulas as in (1) and are not among the well-known elementary functions such as those in (2). Such functions often show up as solutions to problems in wave phenomena, fluid flow, or quantum mechanics. These functions can often be represented by infinite series. For example, the equations that describe the vibrations of a membrane often involve the Bessel function of order zero,

$$J_0(x) = 1 + \sum_{n=1}^{\infty} (-1)^n \frac{x^{2n}}{2^{2n}(n!)^2}. \tag{4}$$

In this section we will study infinite series similar to the ones shown in (3) and (4). Such series are called *power series* or *Taylor series*. We will see how to find the interval of convergence for such a series, how to express some functions as a power series, and how to use the partial sums of these series to approximate the function to a desired accuracy.

Power Series

DEFINITION Power Series

Let c be a real number and $\{a_k\}_{k=0}^{\infty}$ be a sequence of real numbers. An infinite series of the form

$$\sum_{k=0}^{\infty} a_k(x - c)^k = a_0 + a_1(x - c) + a_2(x - c)^2 + \cdots \tag{5}$$

is called a **power series** about c. The numbers $a_0, a_1, a_2, \ldots$ are the coefficients of the power series.

For all x for which (5) converges, we can define a function f by

$$f(x) = \sum_{k=0}^{\infty} a_k(x - c)^k.$$

If we know all of the coefficients a_k for the series and if we can evaluate

$$\lim_{k\to\infty} \frac{a_{k+1}}{a_k},$$

then we can use the Ratio Test or some of the other convergence tests to determine the domain of f.

EXAMPLE 1 Show that the power series

$$\sum_{k=1}^{\infty} \frac{kx^k}{3^k}$$

converges when $|x| < 3$ and diverges when $|x| > 3$.

Solution

The Ratio Test is a test for series with positive terms. Hence we first consider the series

$$\sum_{k=1}^{\infty} \left|\frac{kx^k}{3^k}\right|.$$

To apply the Ratio Test to this series, we look at the ratio of the $(k + 1)$st term to the kth term and take the limit:

$$\rho = \lim_{k\to\infty} \frac{\left|\frac{(k+1)x^{k+1}}{3^{k+1}}\right|}{\left|\frac{kx^k}{3^k}\right|} = \lim_{k\to\infty} \frac{1}{3}\left(\frac{k+1}{k}\right)|x| = \frac{1}{3}|x|. \tag{6}$$

Because

$$\rho = \frac{1}{3}|x| < 1 \quad \text{when } |x| < 3,$$

we see that $\sum_{k=1}^{\infty} \left|\frac{kx^k}{3^k}\right|$ converges when $|x| < 3$. By the Absolute Convergence Test, the series

$$\sum_{k=1}^{\infty} \frac{kx^k}{3^k}$$

also converges with $|x| < 3$.

Next note that

$$\rho = \frac{1}{3}|x| > 1 \quad \text{when } |x| > 3.$$

In our discussion of the Ratio Test in Section 7.5, we showed that when $\rho > 1$, the terms of the series do not tend to 0. Hence if $|x| > 3$, we have

$$\lim_{k\to\infty} \left|\frac{kx^k}{3^k}\right| \neq 0, \quad \text{so} \quad \lim_{k\to\infty} \frac{kx^k}{3^k} \neq 0.$$

Thus, by the nth-term Divergence Test,

$$\sum_{k=1}^{\infty} \frac{kx^k}{3^k}$$

diverges when $|x| > 3$.

When we used the Ratio Test in Example 1, we ignored the special case $x = 0$. The series $\sum_{k=1}^{\infty} \frac{kx^k}{3^k}$ certainly converges when $x = 0$. However, the limit in (6) does not exist in this case because for $k \geq 0$ there is a 0 in the denominator of the ratio. In the future we will not mention this special case. We note here that the series $\Sigma_{k=0}^{\infty} a_k(x - c)^k$ always converges when $x = c$, and we will assume that $x \neq c$ when we apply the Ratio Test to such series.

The process we used in analyzing the series in Example 1 can be streamlined somewhat. Let $\Sigma_{k=0}^{\infty} a_k(x - c)^k$ be a power series and suppose that

$$\lim_{k\to\infty} \left| \frac{a_{k+1}}{a_k} \right| = \rho$$

where $0 \leq \rho \leq \infty$. Define

$$R = \begin{cases} \dfrac{1}{\rho} & (0 < \rho < \infty) \\ \infty & (\rho = 0) \\ 0 & (\rho = \infty). \end{cases} \tag{7}$$

Then for $x \neq c$,

$$\lim_{k\to\infty} \frac{|a_{k+1}(x - c)^{k+1}|}{|a_k(x - c)^k|} = \rho|x - c| = \frac{|x - c|}{R} \text{ is } \begin{cases} <1 \text{ for } |x - c| < R \\ >1 \text{ for } |x - c| > R. \end{cases}$$

Hence $\Sigma_{k=0}^{\infty} a_k(x - c)^k$ converges for $|x - c| < R$ and diverges for $|x - c| > R$. The number R is called the *radius of convergence* of the power series $\Sigma_{k=0}^{\infty} a_k(x - c)^k$.

DEFINITION Radius of Convergence

Let $\Sigma_{k=0}^{\infty} a_k(x - c)^k$ be a power series. There is a number R, $0 \leq R \leq \infty$, such that

$$\sum_{k=0}^{\infty} a_k(x - c)^k \text{ converges when } |x - c| < R.$$

$$\sum_{k=0}^{\infty} a_k(x - c)^k \text{ diverges when } |x - c| > R.$$

The number R is called the **radius of convergence** of $\Sigma_{k=0}^{\infty} a_k(x - c)^k$. Furthermore, if $\lim_{k\to\infty} \left| \frac{a_{k+1}}{a_k} \right| = \rho$ exists, or is $+\infty$, then the radius of convergence is

$$R = \frac{1}{\rho}.$$

The value of the fraction is interpreted as in (7).

EXAMPLE 2 Calculate the radius of convergence of

a) $\sum_{j=0}^{\infty} \frac{x^j}{3^j j!}$

b) $\sum_{k=0}^{\infty} \frac{2^k}{(k+1)^4}(x-\pi)^{3k}$

Solution

a) The series has the form $\Sigma\, a_j(x-c)^j$ with $a_j = 1/(3^j j!)$ and $c = 0$. We first calculate

$$\rho = \lim_{j\to\infty}\left|\frac{a_{j+1}}{a_j}\right| = \lim_{j\to\infty}\frac{\dfrac{1}{3^{j+1}(j+1)!}}{\dfrac{1}{3^j j!}} = \lim_{j\to\infty}\frac{1}{3(j+1)} = 0.$$

The radius of convergence is $R = 1/p = 1/0 = \infty$, so the series converges for all real x.

b) Because the exponents on $x - \pi$ are multiples of three, we cannot use the radius of convergence formula. Why? Instead we apply the Ratio Test to the series of absolute values,

$$\sum_{k=0}^{\infty}\left|\frac{2^k}{(k+1)^4}(x-\pi)^{3k}\right|.$$

We calculate

$$\lim_{k\to\infty}\frac{\left|\dfrac{2^{k+1}}{(k+2)^4}(x-\pi)^{3(k+1)}\right|}{\left|\dfrac{2^k}{(k+1)^4}(x-\pi)^{3k}\right|} = \lim_{k\to\infty}2\left(\frac{k+1}{k+2}\right)^4|x-\pi|^3 = 2|x-\pi|^3.$$

This ratio is less than 1 if and only if $|x-\pi| < 1/\sqrt[3]{2}$ and exceeds 1 if $|x-\pi| > 1/\sqrt[3]{2}$. In the former case, the series converges, while in the latter case, the series diverges. Thus, the radius of convergence of the series is $1/\sqrt[3]{2}$.

Approximating the Sum of a Power Series If $\Sigma_{k=0}^{\infty} a_k(x-c)^k$ is a power series with radius R, then the series converges for all x satisfying $c - R < x < c + R$. Hence for x in this interval we can define a function f by

$$f(x) = \sum_{k=0}^{\infty} a_k(x-c)^k. \tag{8}$$

The series in (8) may or may not converge at either of the endpoints of the interval, that is, at $x = c - R$ or $x = c + R$. We shall usually ignore these endpoints, though the behavior of the series at these points can often be determined by using the Comparison Tests. See Exercises 22–25.

When a function f is defined by a series such as (8), we may want to approximate the value of f at some point in its domain. This can be done using the partial sums of the series. We often use the Ratio Test to find information about the error of such an approximation. Before exploring the error, we discuss the nature of the convergence of power series.

THEOREM Convergence of Power Series

Let $\Sigma_{k=0}^{\infty} a_k(x - c)^k$ be a power series with radius of convergence R. Define

$$f(x) = \sum_{k=0}^{\infty} a_k(x - c)^k, \qquad c - R < x < c + R.$$

If $0 < r < R$, then the series converges uniformly on the interval $[c - r, c + r]$. That is, the sequence

$$\left\{\sum_{k=0}^{n} a_k(x - c)^k\right\}_{n=0}^{\infty}$$

of partial sums of the power series converges to f uniformly on $[c - r, c + r]$. See Fig. 7.32.

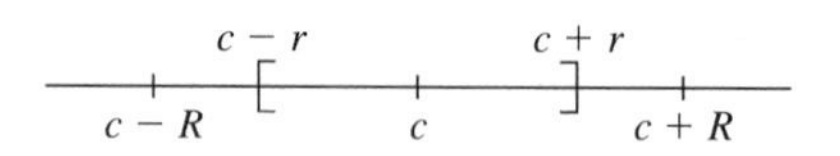

FIGURE 7.32 A power series converges uniformly on an interval $[c - r, c + r] \subset (c - R, c + R)$.

This result can be interpreted graphically. Let ϵ be a small positive number and draw the ϵ-corridor around the graph of $y = f(x)$ for $c - r \leq x \leq c + r$. Then if n is large enough, the graph of $y = \Sigma_{k=0}^{n} a_k(x - c)^k$ lies in the ϵ-corridor about f. This is equivalent to saying that when n is large enough,

$$\left|f(x) - \sum_{k=0}^{n} a_k(x - c)^k\right| < \epsilon$$

for all x satisfying $c - r \leq x \leq c + r$. This means that for $c - r \leq x \leq c + r$,

$$f(x) \approx \sum_{k=0}^{n} a_k(x - c)^k,$$

with error of at most ϵ. See Figs. 7.33 and 7.34.

The following Investigation illustrates how the Ratio Test can be used to show that a power series converges uniformly. We also show how information from the Ratio Test can be used to find an approximation to the sum of the series.

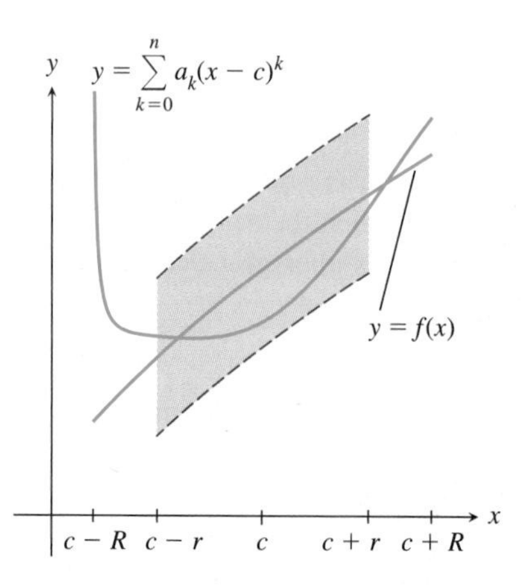

FIGURE 7.33 When n is large, the graph of the nth partial sum lies in the ϵ-corridor about f for $c - r \leq x \leq c + r$. This need not be the case outside this interval.

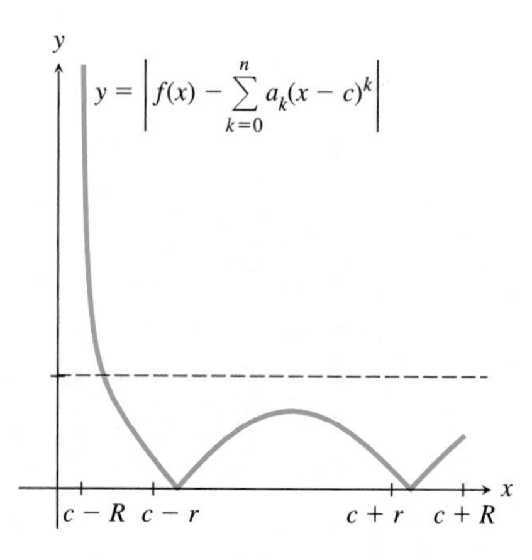

FIGURE 7.34 On $[c - r, c + r]$, the error $|f(x) - \Sigma_{k=0}^{n} a_k(x - c)^k|$ is small when n is large.

INVESTIGATION

Approximating the Sum of a Power Series

In Example 1 we showed that the radius of convergence of the power series

$$\sum_{k=1}^{\infty} \frac{kx^k}{3^k} \tag{9}$$

is 3. For $|x| < 3$, let $f(x)$ be the sum of the series. We illustrate how geometric series can be used to estimate the error when the partial sum of a power series is used to estimate the sum of the series. This approach works because on any closed interval I in the interval of convergence $(c - R, c + R)$ of a power series, the series converges by the Ratio Test. As we saw in Section 7.5 when a series converges by the Ratio Test its terms are eventually dominated by the terms of a convergent geometric series. To illustrate we find a partial sum of (9) that approximates $f(x)$ on $[-2, 2]$ with an error of no more that 0.001. In particular, we find a positive integer N such that

$$\left| f(x) - \sum_{k=1}^{N} \frac{kx^k}{3^k} \right| = \left| \sum_{k=N+1}^{\infty} \frac{kx^k}{3^k} \right| < 0.001, \qquad -2 \le x \le 2 \tag{10}$$

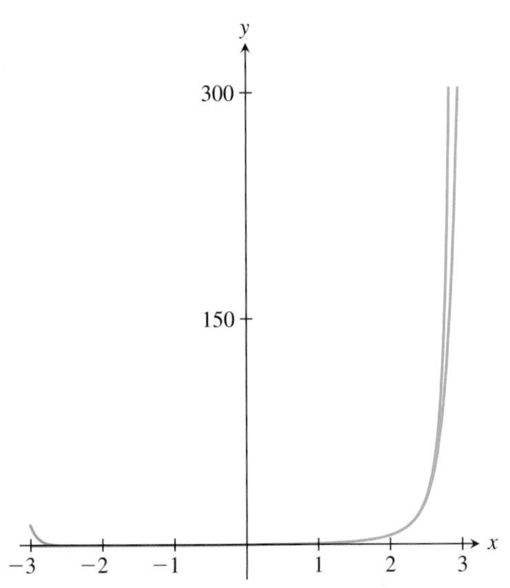

FIGURE 7.35 Graphs $y = f(x)$ and $y = \Sigma_1^{28} kx^k/3^k$.

Our overall plan is to find a convergent geometric series that is independent of x and dominates the series $\Sigma_{k=N+1}^{\infty} \frac{kx^k}{3^k}$ in (10). We use the fact that

$$\lim_{k\to\infty} \frac{k}{a^k} = 0 \qquad \text{for } a = 9/8. \tag{11}$$

Keeping in mind that $|x| \le 2$,

$$\left| \sum_{k=N+1}^{\infty} \frac{kx^k}{3^k} \right| \le \sum_{k=N+1}^{\infty} \left| \frac{kx^k}{3^k} \right| \le \sum_{k=N+1}^{\infty} \frac{k2^k}{3^k} = \sum_{k=N+1}^{\infty} k\left(\frac{2}{3}\right)^k. \tag{12}$$

We claim that we can find N so that $k(2/3)^k < (3/4)^k$ for all $k > N$. To see this, rewrite the inequality as

$$k < \left(\frac{2}{3}\right)^{-k} \left(\frac{3}{4}\right)^k = \left(\frac{9}{8}\right)^k. \tag{13}$$

We know from (11) that we can take k sufficiently large that (13) holds. With a little calculator work we find that (13) holds for $k \ge 29$. Hence, with $N + 1 \ge 29$ in (12),

$$\left| \sum_{k=N+1}^{\infty} \frac{kx^k}{3^k} \right| \le \sum_{k=N+1}^{\infty} k\left(\frac{2}{3}\right)^k \le \sum_{k=N+1}^{\infty} \left(\frac{3}{4}\right)^k = \frac{(3/4)^{N+1}}{1 - \frac{3}{4}} = 3\left(\frac{3}{4}\right)^N. \tag{14}$$

Using a calculator once more, we find that $3(3/4)^N < 0.001$ for all $N \ge 28$. Hence from (14), we see that (10) holds for $N \ge 28$. Thus,

$$f(x) = \sum_{k=1}^{\infty} \frac{kx^k}{3^k} \approx \sum_{k=1}^{28} \frac{kx^k}{3^k}, \qquad -2 \le x \le 2,$$

with an error of at most 0.001. In Fig. 7.35 we have graphed $y = f(x)$ and this approximation on the interval $[-3, 3]$. We can detect no difference in the graphs on the interval $[-2, 2]$. In Fig. 7.36 we have graphed

$$y = \left| f(x) - \sum_{k=1}^{28} \frac{kx^k}{3^k} \right|.$$

We see that the error in the approximation is actually much less than 0.001.

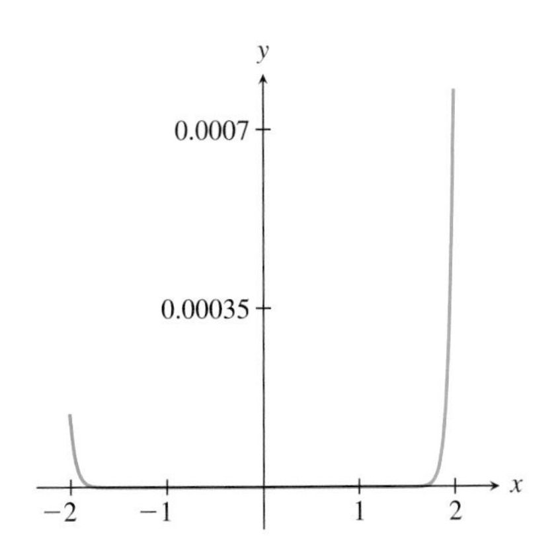

FIGURE 7.36 When we use $\Sigma_1^{28} kx^k/3^k$ to approximate $f(x)$ on $[-2, 2]$, the error is actually much less than 0.001.

In Section 7.1 we introduced the Taylor polynomials. The Taylor polynomials

$$T_n(x; c) = \sum_{k=0}^{n} \frac{f^{(k)}(c)}{k!}(x - c)^k,$$

are the partial sums of a special type of power series called a *Taylor series.*

DEFINITION Taylor Series

Let f be a function, let c be a real number in the domain of f, and suppose that f has derivatives of all orders at c. The **Taylor series** for f at the point c is the power series

$$\sum_{k=0}^{\infty} \frac{f^{(k)}(c)}{k!}(x - c)^k.$$

Given a Taylor series for a function, we will usually be interested in answering two questions:

1. What is the radius of convergence, R, of the series?
2. Does the Taylor series converge to f on the interval $c - R < x < c + R$?

We explore these questions in two examples.

EXAMPLE 3 Find the Taylor series at 0 for $\sin x$. Show that this series converges to $\sin x$ for all real x.

Solution

Let $f(x) = \sin x$. Then

$$\begin{aligned} f^{(0)}(0) &= \sin 0 = 0, \\ f^{(1)}(0) &= \cos 0 = 1, \\ f^{(2)}(0) &= -\sin 0 = 0, \\ f^{(3)}(0) &= -\cos 0 = -1, \\ f^{(4)}(0) &= \sin 0 = 0, \\ &\vdots \end{aligned}$$

If we continue this list, we see that the 0, 1, 0, -1 pattern repeats. Thus, the Taylor series for $\sin x$ is

$$\begin{aligned} \sum_{k=0}^{\infty} \frac{f^{(k)}(c)}{k!}(x - c)^k &= \frac{0}{0!}x^0 + \frac{1}{1!}x^1 + \frac{0}{2!}x^2 + \frac{(-1)}{3!}x^3 + \frac{0}{4!}x^4 + \frac{1}{5!}x^5 + \cdots \\ &= x - \frac{x^3}{3!} + \frac{x^5}{5!} - \frac{x^7}{7!} + \frac{x^9}{9!} - \cdots \\ &= \sum_{j=0}^{\infty} (-1)^j \frac{x^{2j+1}}{(2j + 1)!}. \end{aligned}$$

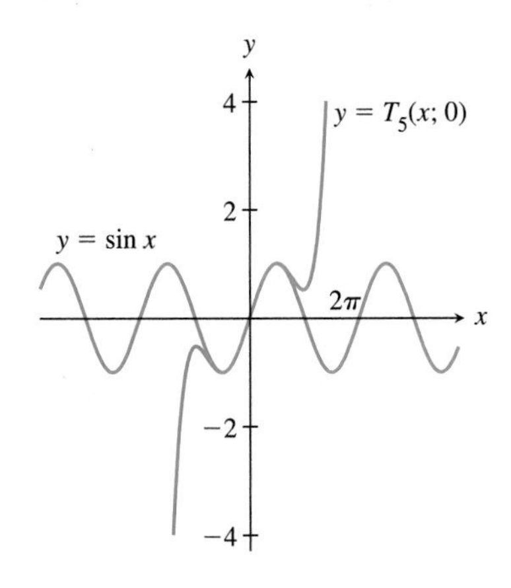

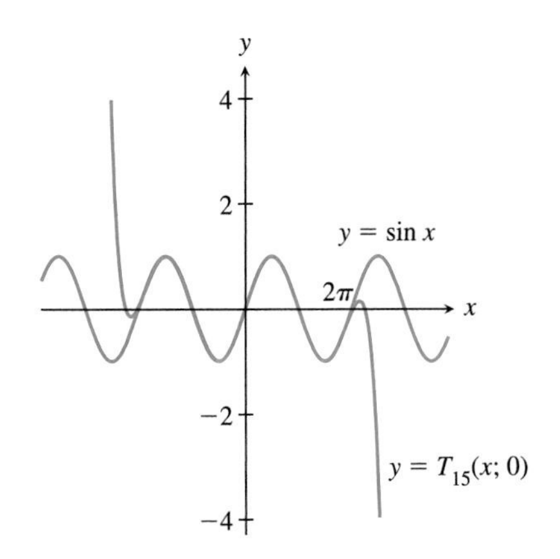

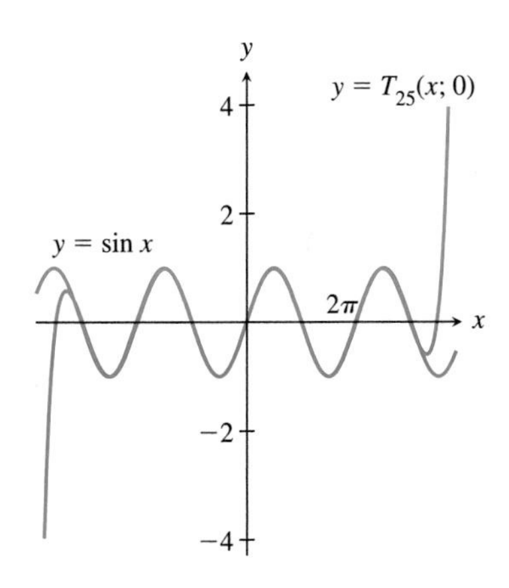

FIGURE 7.37 Graphs $y = \sin x$ and $y = T_n(x; 0)$ for $n = 5, 15, 25$.

To show this series converges for all real x, we use the Ratio Test. For any real number x,

$$\lim_{j\to\infty}\left|\frac{(-1)^{j+1}\dfrac{x^{2j+3}}{(2j+3)!}}{(-1)^j\dfrac{x^{2j+1}}{(2j+1)!}}\right| = \lim_{j\to\infty}\frac{|x|^2}{(2j+3)(2j+2)} = 0 < 1.$$

Hence, for any real number x, the series

$$\sum_{j=0}^{\infty}(-1)^j\frac{x^{2j+1}}{(2j+1)!}$$

converges that is, the series has radius of convergence $R = 1/0 = \infty$.

Next we show that for any x the sum of this Taylor series is indeed $\sin x$. Let $T_n(x; c)$ be the Taylor polynomial of degree n for $\sin x$. Then by the Taylor inequality (Section 7.2),

$$|\sin x - T_n(x; c)| = |R_n(x; c)| \le \frac{M_{n+1}}{(n+1)!}|x|^{n+1}, \tag{15}$$

where

$$\begin{aligned} M_{n+1} &= \text{maximum of } \left|\frac{d^{n+1}}{dt^{n+1}}\sin t\right| \text{ for } t \text{ between 0 and } x \\ &= \text{maximum of } |\sin t| \text{ or } |\cos t| \text{ for } t \text{ between 0 and } x \\ &\le 1. \end{aligned}$$

Putting this upper bound for M_{n+1} into (15), we find

$$|\sin x - T_n(x; c)| \le \frac{|x|^{n+1}}{(n+1)!}.$$

For fixed real x,

$$\lim_{n\to\infty}\frac{|x|^{n+1}}{(n+1)!} = 0.$$

This means that we can make $|\sin x - T_n(x; c)|$ as small as we like by taking n large enough. Hence

$$\sin x = \lim_{n\to\infty}T_n(x; c) = \sum_{k=0}^{\infty}\frac{f^{(k)}(c)}{k!}(x-c)^k = \sum_{j=0}^{\infty}(-1)^j\frac{x^{2j+1}}{(2j+1)!}$$

for all real x. In Fig. 7.37 we show the graphs of $y = \sin x$ and some of the Taylor polynomials for $\sin x$. We see that the Taylor polynomials approximate $\sin x$ better and better on larger and larger intervals as n gets bigger.

In the next example we discuss a Taylor series that can be recognized as a geometric series. As we'll see, this allows us to avoid working with the remainder $R_n(x; c)$.

EXAMPLE 4 Find the Taylor series about 2 for

$$f(x) = \frac{1}{x}.$$

Show that the series converges to $f(x)$ for $0 < x < 4$.

Solution

We begin by finding the coefficients for the Taylor series. Noting that

$$f^{(0)}(x) = \frac{1}{x},$$

$$f^{(1)}(x) = -\frac{1}{x^2},$$

$$f^{(2)}(x) = \frac{2 \cdot 1}{x^3},$$

$$f^{(3)}(x) = -\frac{3 \cdot 2 \cdot 1}{x^4},$$

$$\vdots$$

it appears that

$$f^{(k)}(x) = (-1)^k \frac{k!}{x^{k+1}}.$$

Hence

$$f^{(k)}(2) = (-1)^k \frac{k!}{2^{k+1}},$$

and the Taylor series is

$$\sum_{k=0}^{\infty} \frac{f^{(k)}(2)}{k!}(x-2)^k = \sum_{k=0}^{\infty} (-1)^k \frac{(x-2)^k}{2^{k+1}}. \tag{16}$$

We can rewrite the last series in (16) as

$$\sum_{k=0}^{\infty} \frac{1}{2}\left(\frac{-(x-2)}{2}\right)^k.$$

This shows that the Taylor series is a geometric series with first term $a = 1/2$ and common ratio $r = -(x-2)/2$. Hence the series converges if and only if

$$|r| = \left|\frac{-(x-2)}{2}\right| < 1, \quad \text{that is, if and only if} \quad 0 < x < 4.$$

For such x the series converges to

$$\frac{a}{1-r} = \frac{1/2}{1 - \left(\dfrac{-(x-2)}{2}\right)} = \frac{1}{x}.$$

Thus for $0 < x < 4$, the Taylor series at 2 for $f(x) = 1/x$ converges to $1/x$.

In the next section we will study techniques for calculating power series and Taylor series. For this it will be useful to know some of the more important Taylor series and the intervals on which they converge.

Important Taylor Series

$$\frac{1}{1-x} = 1 + x + x^2 + x^3 + \cdots = \sum_{k=0}^{\infty} x^k, \qquad -1 < x < 1$$

$$\sin x = x - \frac{x^3}{3!} + \frac{x^5}{5!} - \frac{x^7}{7!} + \cdots = \sum_{k=0}^{\infty} (-1)^k \frac{x^{2k+1}}{(2k+1)!}, \qquad -\infty < x < \infty$$

$$\cos x = 1 - \frac{x^2}{2!} + \frac{x^4}{4!} - \frac{x^6}{6!} + \cdots = \sum_{k=0}^{\infty} (-1)^k \frac{x^{2k}}{(2k)!}, \qquad -\infty < x < \infty$$

$$e^x = 1 + x + \frac{x^2}{2!} + \frac{x^3}{3!} + \cdots = \sum_{k=0}^{\infty} \frac{x^k}{k!}, \qquad -\infty < x < \infty$$

$$\sqrt{1+x} = 1 + \frac{1}{2}x - \frac{1}{2^2}\frac{x^2}{2!} + \frac{1 \cdot 3}{2^3}\frac{x^3}{3!} - \frac{1 \cdot 3 \cdot 5}{2^4}\frac{x^4}{4!} + \cdots$$

$$= 1 + \frac{x}{2} + \sum_{k=2}^{\infty} (-1)^{k-1} \frac{1 \cdot 3 \cdots (2k-3)}{2^k k!} x^k, \qquad -1 < x \leq 1$$

In the Exercises we ask you to verify some of the entries in this table.

Exercises 7.6

Exercises 1–6: Find the radius of convergence of each power series.

1. $\sum_{k=1}^{\infty} k^2 x^k$

2. $\sum_{k=1}^{\infty} \frac{(-1)^{k-1}}{2^k}(x+2)^k$

3. $\sum_{j=0}^{\infty} \frac{(j+1)x^j}{j!}$

4. $\sum_{n=2}^{\infty} \frac{x^{2n}}{(\ln n)^2}$

5. $\sum_{j=1}^{\infty} \frac{j!\left(x - \frac{1}{2}\right)^j}{2^j j^j}$

6. $\sum_{k=0}^{\infty} \frac{k(k+1)(k+2)(k+3)}{5^k}(x+\pi)^k$

Exercises 7–15: Find the Taylor series for the function f about the point c. Then find the radius of convergence of the Taylor series.

7. $f(x) = e^x, \quad c = 0$

8. $f(x) = \cos x, \quad c = 0$

9. $f(x) = 2e^{3x}, \quad c = 0$

10. $f(x) = \sqrt{x-2}, \quad c = 3$

11. $f(x) = -2x^3 + 4x^2 - 7x + 4, \quad c = -2$

12. $f(x) = \frac{1}{3+x} + \frac{1}{2-x}, \quad c = 0$

13. $f(x) = \frac{1}{3+x} + \frac{1}{2-x}, \quad c = 5$

14. $f(x) = \frac{1}{3+x} + \frac{1}{2-x}, \quad c = -5$

15. $f(x) = \sqrt{1+x}, \quad c = 0$

Exercises 16–21: Show that the power series converges on the interval I. How many terms of the series are needed to approximate the sum to within 0.001 on I?

16. $\sum_{k=1}^{\infty} x^k, \quad I = \left[-\frac{1}{3}, \frac{1}{3}\right]$

17. $\sum_{k=1}^{\infty} \frac{x^k}{(2k)!}$, $I = [-1, 1]$

18. $\sum_{j=0}^{\infty} \frac{(x-2)^j}{3^{j+1}}$, $I = [1, 3]$

19. $\sum_{m=1}^{\infty} \frac{(-1)^{m-1}}{m} x^m$, $I = \left[-\frac{1}{2}, \frac{1}{2}\right]$

20. $\sum_{j=0}^{\infty} \frac{j(j+1)}{j!}(x+3)^j$, $I = [-5, -1]$

21. $\sum_{k=1}^{\infty} \frac{x^k}{k^k}$, $I = [-3, 3]$

22. The power series $\sum_{k=1}^{\infty} \frac{x^k}{k^2}$ has radius of convergence $r = 1$. Suppose $x = 1$ is substituted into the series. Does the resulting series converge or diverge? What if $x = -1$ is substituted into the series?

23. The power series $\sum_{k=1}^{\infty} \frac{x^k}{k}$ has radius of convergence $r = 1$. Suppose $x = 1$ is substituted into the series. Does the resulting series converge or diverge? What if $x = -1$ is substituted into the series?

24. The power series $\Sigma_{k=1}^{\infty} kx^k$ has radius of convergence $r = 1$. Suppose $x = 1$ is substituted into the series. Does the resulting series converge or diverge? What if $x = -1$ is substituted into the series?

25. Let the series $\Sigma_{k=0}^{\infty} a_k(x-c)^k$ be a power series with radius of convergence R, where $0 < R < \infty$. Substitute the value $x = c + R$ into the series. What can you say about the behavior of the resulting series? What if $x = c - R$ is substituted into the series? (See the previous three exercises before answering!)

26. Find the radius of convergence of the infinite series shown in (4).

27. Write down a power series about $c = 5$ with radius of convergence $r = 2$.

28. Write down a power series about $c = -\pi$ with radius of convergence $r = e$.

29. Let $f(x) = 1/(1-x)$ and let c be a real number with $-1 < c < 1$. When f is expanded in a Taylor series about c, what is the radius of convergence of the resulting series?

30. Let $f(x) = 1/(1-x)$ and let c be a real number with $|c| > 1$. When f is expanded in a Taylor series about c, what is the radius of convergence of the resulting series?

31. Let $g(x) = \cos 3x$ and let c be a real number. When g is expanded in a Taylor series about c, what is the radius of convergence of the resulting series?

32. For what values of x does the series $\sum_{k=0}^{\infty} \frac{k}{x^k}$ converge?

33. For what values of x does the series

$$\sum_{k=0}^{\infty} \frac{(k+1)3^k}{(x-2)^k}$$

converge?

34. For what values of x does the series $\Sigma_{k=0}^{\infty} (2 \sin x)^k$ converge?

35. For what values of x does the series

$$\sum_{k=1}^{\infty} \frac{1}{x^2 + k^2}$$

converge?

36. Verify the third entry in the table on page 615. That is, show that

$$\cos x = \sum_{k=0}^{\infty} (-1)^k \frac{x^{2k}}{(2k)!}$$

for $-\infty < x < \infty$.

37. Verify the fourth entry in the table on page 615. That is, show that

$$e^x = \sum_{k=0}^{\infty} \frac{x^k}{k!}$$

for $-\infty < x < \infty$.

7.7 Working with Power Series

For most of the Taylor series

$$\sum_{k=0}^{\infty} \frac{f^{(k)}(c)}{k!}(x-c)^k$$

we have considered, much of our effort has gone into calculating the derivatives $f^{(k)}(c)$ for $k = 0, 1, 2, \ldots$. In this section we shall see that some Taylor series can be generated in other ways, for example by differentiating, integrating, or algebraically combining one or more of the series for $1/(1-x)$, $\sin x$, $\cos x$, e^x, and $\sqrt{1+x}$ listed at the end of Section 7.6.

Algebraic Operations with Taylor Series

Substitution Many Taylor series can be found by making a substitution in a known series.

EXAMPLE 1 Find the Taylor series about $c = \pi/2$ for $\sin x$.

Solution

Because

$$\sin x = \cos\left(x - \frac{\pi}{2}\right),$$

a series for $\sin x$ can be found by making the substitution $t = x - \pi/2$ in the series

$$\cos t = \sum_{k=0}^{\infty} (-1)^k \frac{t^{2k}}{(2k)!}, \qquad -\infty < t < \infty. \tag{1}$$

With this substitution,

$$\sin x = \cos\left(x - \frac{\pi}{2}\right) = \sum_{k=0}^{\infty} (-1)^k \frac{\left(x - \frac{\pi}{2}\right)^{2k}}{(2k)!}. \tag{2}$$

Since (1) is true for all real t, Equation (2) is true for all real x.

EXAMPLE 2 Find the Taylor series about $c = -2$ for the function

$$f(x) = \frac{2x + 1}{x^2 - 5x + 6}, \qquad x \neq 2, 3.$$

What is the radius of convergence of the series?

Solution

We will obtain the series for f by substitution in two geometric series. First, we use partial fractions to break $f(x)$ into a sum of simpler fractions:

$$\frac{2x + 1}{x^2 - 5x + 6} = \frac{7}{x - 3} - \frac{5}{x - 2}. \tag{3}$$

We will find a Taylor series for each of these fractions and then combine the two series to get a series for f.

Take $7/(x - 3)$ first. Because we want a series about $c = -2$, we look for a series with powers of $(x - c) = (x - (-2)) = (x + 2)$. Hence, we try to express $7/(x - 3)$ in terms of $x + 2$. We can do this algebraically by writing

$$\frac{7}{x - 3} = \frac{7}{x + 2 - 5} = \left(\frac{-\frac{7}{5}}{1 - \frac{x + 2}{5}}\right). \tag{4}$$

The factoring in the last step was done so that we can directly apply the formula

$$\frac{a}{1-r} = a + ar + ar^2 + \cdots = \sum_{k=0}^{\infty} ar^k, \qquad |r| < 1, \tag{5}$$

for the sum of a geometric series. We use this result to find a series for the last expression in (4). Taking $a = -7/5$ and $r = (x+2)/5$ in (5),

$$\frac{-\dfrac{7}{5}}{1 - \dfrac{x+2}{5}} = -\frac{7}{5}\sum_{k=0}^{\infty}\left(\frac{x+2}{5}\right)^k = -7\sum_{k=0}^{\infty}\frac{1}{5^{k+1}}(x+2)^k. \tag{6}$$

Because (5) holds when $|r| < 1$, it follows that (6) is true for $|(x+2)/5| < 1$. Applying a similar procedure to the fraction $5/(x-2)$,

$$\begin{aligned}\frac{5}{x-2} = \frac{5}{x+2-4} &= \left(\frac{-\dfrac{5}{4}}{1 - \dfrac{x+2}{4}}\right)\\ &= -\frac{5}{4}\sum_{k=0}^{\infty}\left(\frac{x+2}{4}\right)^k = -5\sum_{k=0}^{\infty}\frac{1}{4^{k+1}}(x+2)^k.\end{aligned} \tag{7}$$

This result holds for $|(x+2)/4| < 1$. Combining (6) and (7) and (3),

$$\begin{aligned}\frac{2x+1}{x^2-5x+6} &= \frac{7}{x-3} - \frac{5}{x-2}\\ &= -7\sum_{k=0}^{\infty}\frac{1}{5^{k+1}}(x+2)^k + 5\sum_{k=0}^{\infty}\frac{1}{4^{k+1}}(x+2)^k \qquad (8)\\ &= \sum_{k=0}^{\infty}\left(\frac{5}{4^{k+1}} - \frac{7}{5^{k+1}}\right)(x+2)^k.\end{aligned}$$

Because (6) holds for $|x+2| < 5$ and (7) holds for $|x+2| < 4$, we conclude that (8) holds for $|x+2| < 4$. Hence the series has radius of convergence 4.

Multiplication of Series Power or Taylor series about the same point c can be multiplied in much the same way as we multiply polynomials. For series $\Sigma_{k=0}^{\infty} a_k x^k$ and $\Sigma_{k=0}^{\infty} b_k x^k$, we show how to arrange the multiplication so that we can find all terms of the product $\left(\Sigma_{k=0}^{\infty} a_k x^k\right)\left(\Sigma_{k=0}^{\infty} b_k x^k\right)$ through a given degree. If formulas for the general terms a_k and b_k of the two series are known, we give in Exercise 26 a general formula for the product.

EXAMPLE 3 Let $f(x) = e^x \sin x$. Find the terms through degree 5 of the series for f about 0.

Solution

Start by writing out the first few terms of the series for e^x and $\sin x$,

$$e^x = 1 + x + \frac{x^2}{2} + \frac{x^3}{6} + \frac{x^4}{24} + \frac{x^5}{120} + \cdots,$$

$$\sin x = x - \frac{x^3}{6} + \frac{x^5}{120} - \cdots.$$

To find the product of these two series, think of them as "infinitely long polynomials" and proceed as you would if multiplying two polynomials. First, write one series above the other, keeping terms with like powers of x aligned in columns. Then multiply every term in the series on the bottom by every term in the series on the top. Keep the results of these multiplications aligned in columns by powers of x. When we have *all* products that will give a term of degree 5 or less, stop multiplying and add the columns. The result is the first several terms of the series for $e^x \sin x$.

e^x **series**	$1 + x$	$+\frac{x^2}{2}$	$+\frac{x^3}{6}$	$+\frac{x^4}{24}$	$+\frac{x^5}{120}$	$+\cdots$
$\sin x$ **series**	x		$-\frac{x^3}{6}$		$+\frac{x^5}{120}$	$-\cdots$
Multiply top row by x	x	$+x^2$	$+\frac{x^3}{2}$	$+\frac{x^4}{6}$	$+\frac{x^5}{24}$	$+\cdots$
Multiply top row by $-\frac{x^3}{6}$			$-\frac{x^3}{6}$	$-\frac{x^4}{6}$	$-\frac{x^5}{12}$	$-\cdots$
Multiply top row by $\frac{x^5}{120}$					$+\frac{x^5}{120}$	$+\cdots$
Add the columns	x	$+x^2$	$+\frac{x^3}{3}$		$-\frac{x^5}{30}$	$\pm\cdots$

From this result, the terms through degree 5 of the series for f about 0 are

$$x + x^2 + \frac{x^3}{3} - \frac{x^5}{30}.$$

How well does the fifth-degree Taylor polynomial for f approximate f on the interval $[-1, 1]$?

Integration and Differentiation

Integration To show how integration can be used to generate new power series from old ones, suppose that $F'(x) = f(x)$, where

$$f(x) = \sum_{k=0}^{\infty} a_k(x - c)^k, \qquad c - R < x < c + R.$$

It seems reasonable that we can obtain a series for F by integrating the series for f. That is, for $c - R < x < c + R$,

$$F(x) - F(c) = \int_c^x f(t)\,dt = \int_c^x \sum_{k=0}^{\infty} a_k(t-c)^k\,dt$$
$$= \sum_{k=0}^{\infty}\left(\int_c^x a_k(t-c)^k\,dt\right) = \sum_{k=0}^{\infty} \frac{a_k}{k+1}(x-c)^{k+1}.$$

We summarize these results.

THEOREM Integration of Power Series

Suppose that $F'(x) = f(x)$ for $c - R < x < c + R$ and that

$$f(x) = \sum_{k=0}^{\infty} a_k(x-c)^k, \qquad c - R < x < c + R.$$

Then

$$F(x) = F(c) + \int_c^x f(t)\,dt = F(c) + \sum_{k=0}^{\infty} \frac{a_k}{k+1}(x-c)^{k+1},$$

for $c - r < x < c + r$. The series for F about c has the same radius of convergence as the series for f about c.

EXAMPLE 4 Find a power series about 0 that is equal to arctan x in an interval centered at 0.

Solution

Because

$$\arctan x = \arctan x - \arctan 0 = \int_0^x \frac{1}{1+t^2}\,dt, \tag{9}$$

we start by finding a series about 0 for $1/(1 + t^2)$. We do this by substituting $-t^2$ for r in the geometric series

$$\frac{1}{1-r} = \sum_{k=0}^{\infty} r^k, \qquad |r| < 1.$$

Hence, when $|t| < 1$,

$$\frac{1}{1+t^2} = \frac{1}{1-(-t^2)} = \sum_{k=0}^{\infty}(-t^2)^k = \sum_{k=0}^{\infty}(-1)^k t^{2k}.$$

Integrating this last series and using (9),

$$\arctan x = \int_0^x \frac{1}{1+t^2}\,dt = \int_0^x \left(\sum_{k=0}^{\infty}(-1)^k t^{2k}\right)dt$$
$$= \sum_{k=0}^{\infty}\left(\int_0^x (-1)^k t^{2k}\,dt\right) = \sum_{k=0}^{\infty}\frac{(-1)^k}{2k+1}x^{2k+1}.$$

This series converges to arctan x for $-1 < x < 1$. Why?

We have used summation notation in this calculation. This is convenient and powerful but not essential. Here is the calculation done with $\cdots$ to indicate a continuation for the same pattern:

$$\begin{aligned}\arctan x &= \int_0^x \frac{1}{1+t^2}\,dt \\ &= \int_0^x (1 - t^2 + t^4 - \cdots)\,dt = x - \frac{1}{3}x^3 + \frac{1}{5}x^5 - \ldots.\end{aligned}$$

In the next example, we show how an integral can be evaluated by expanding the integrand in a series and then integrating term by term.

EXAMPLE 5 Estimate the value of

$$\int_0^3 e^{-x^2}\,dx \tag{10}$$

to within 0.001.

Solution

We will solve this problem in two ways. First, because

$$e^t = \sum_{k=0}^{\infty} \frac{t^k}{k!}$$

holds for all real t, we can make the substitution $t = -x^2$ to obtain

$$e^{-x^2} = \sum_{k=0}^{\infty} \frac{(-x^2)^k}{k!} = \sum_{k=0}^{\infty} \frac{(-1)^k x^{2k}}{k!}. \tag{11}$$

Hence, for all real w,

$$\int_0^w e^{-x^2}\,dx = \sum_{k=0}^{\infty}\left(\int_0^w (-1)^k \frac{x^{2k}}{k!}\,dx\right) = \sum_{k=0}^{\infty} \frac{(-1)^k}{(2k+1)k!} w^{2k+1}.$$

Recalling (10) and setting $w = 3$,

$$\int_0^3 e^{-x^2}\,dx = \sum_{k=0}^{\infty} (-1)^k \frac{3^{2k+1}}{(2k+1)k!}. \tag{12}$$

We know that the series in (12) converges; we would like to approximate its sum with an error of at most 0.001. To find out how many terms of the series in (12) we need, we use the error estimate from the Alternating Series Test. Letting

$$b_k = \frac{3^{2k+1}}{(2k+1)k!},$$

the series in (12) is $\Sigma_{k=0}^{\infty}(-1)^k b_k$. We know that $\lim_{k\to\infty} b_k = 0$ because the series converges. To check that the sequence $\{b_k\}$ is decreasing, we calculate several values of b_k. The results are shown in Table 7.7.

It appears that the sequence $\{b_k\}$ is decreasing from b_7 onwards. (See Exercise 40 for a proof of this assertion.) Assuming b_k decreases for $k \geq 7$, we can say that (except for the first six terms) (12) satisfies the hypotheses

TABLE 7.7

k	$b_k = \dfrac{3^{2k+1}}{(2k+1)k!}$
0	3.0000
1	9.0000
2	24.3000
3	52.0714
4	91.1250
5	134.2023
6	170.3337
7	189.8004
8	188.4048
9	168.5727
⋮	⋮
25	0.0027
26	0.0009

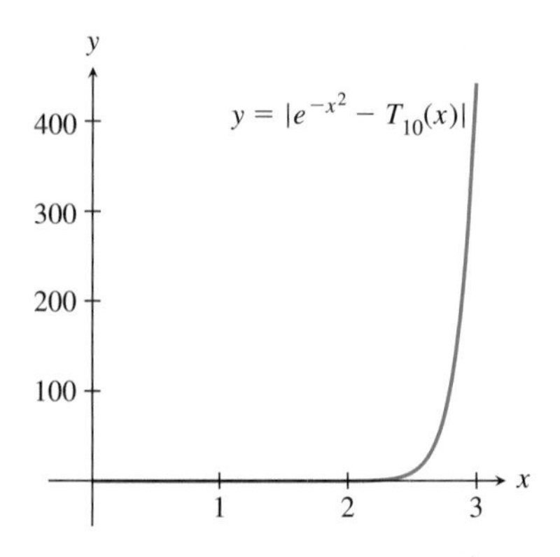

FIGURE 7.38 $T_{10}(x)$ is not a good approximation to e^{-x^2} on $[0, 3]$.

of the Alternating Series Test. Therefore, for $n \geq 7$,

$$\left| \int_0^3 e^{-x^2}\,dx - \sum_{k=0}^{n} (-1)^k \frac{3^{2k+1}}{(2k+1)k!} \right| \leq b_{n+1} = \frac{3^{2n+3}}{(2n+3)(n+1)!}.$$

The difference between the integral and the partial sum will be less than 0.001 if we choose n such that $b_{n+1} < 0.001$. We can find n by trial and error. From Table 7.7, we see that $b_{n+1} < 0.001$ for $n = 25$. Hence,

$$\int_0^3 e^{-x^2}\,dx \approx \sum_{k=0}^{25} (-1)^k \frac{3^{2k+1}}{(2k+1)k!} \approx 0.885522,$$

and the error in this approximation is no more than 0.001.

For a second solution to this problem, we rely more on trial and error and graphs. First, we find a partial sum $T_n(x)$ of the series (11) so that

$$\left| \int_0^3 e^{-x^2}\,dx - \int_0^3 T_n(x)\,dx \right| \leq \int_0^3 |e^{-x^2} - T_n(x)|\,dx < 0.001. \tag{13}$$

Because we are integrating the difference $|e^{-x^2} - T_n(x)|$ over the interval $[0, 3]$, the inequality in (13) is satisfied if $|e^{-x^2} - T_n(x)| < \frac{1}{3}(0.001)$. Hence,

$$\int_0^3 e^{-x^2}\,dx \approx \int_0^3 T_n(x)\,dx$$

with an error of less than 0.001 provided that the graph of $y = |e^{-x^2} - T_n(x)|$ stays below the line $y = \frac{1}{3}(0.001)$.

Figures 7.38–7.40 show these graphs for $n = 10, 20, 29$. We used a CAS to find the partial sums $T_n(x)$ and then graphed the desired difference. By trial and error, we found $n = 29$ was the smallest integer for which $|e^{-x^2} - T_n(x)| < \frac{1}{3}(0.001)$ for $0 \leq x \leq 3$. Hence,

$$\int_0^3 e^{-x^2}\,dx \approx \int_0^3 \left(\sum_{k=0}^{29} (-1)^k \frac{x^{2k}}{k!} \right) dx = \sum_{k=0}^{29} (-1)^k \frac{3^{2k+1}}{(2k+1)k!} \approx 0.886201$$

with an error of no more than 0.001.

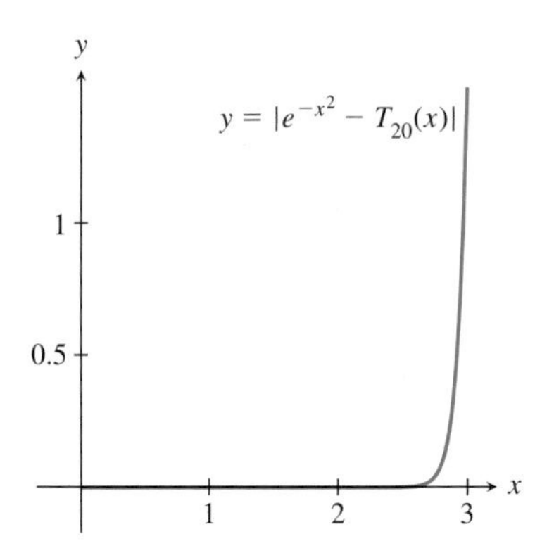

FIGURE 7.39 $|e^{-x^2} - T_{20}(x)|$ is not small enough for all x in the interval $[0, 3]$.

Differentiation We give the principal result used to generate new power series from old ones by differentiation.

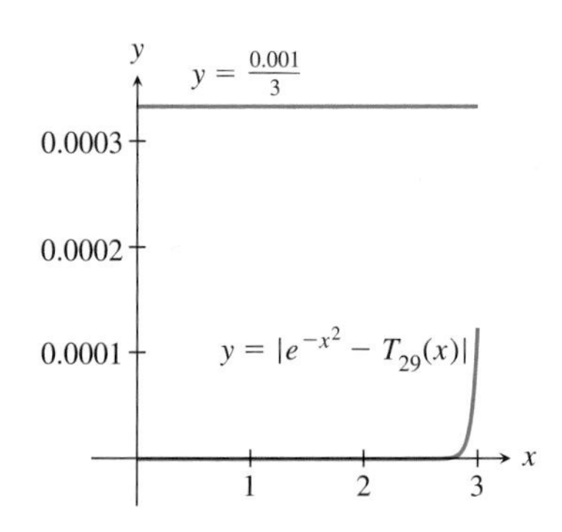

FIGURE 7.40 The graph shows $|e^{-x^2} - T_{29}(x)| < 0.001/3$ for $0 \leq x \leq 3$.

THEOREM Differentiation of Power Series

Suppose that

$$f(x) = \sum_{k=0}^{\infty} a_k (x - c)^k \quad \text{for} \quad c - r < x < c + r.$$

Then

$$f'(x) = \sum_{k=0}^{\infty} k a_k (x - c)^{k-1} \quad \text{for} \quad c - r < x < c + r.$$

The power series for f' has the same radius of convergence as the power series for f.

EXAMPLE 6 Find a power series about 0 for $1/(1-x)^2$.

Solution

Because

$$\frac{1}{(1-x)^2} = \frac{d}{dx}\left(\frac{1}{1-x}\right),$$

we can find a series for $1/(1-x)^2$ by differentiating a power series for $1/(1-x)$. If we view this expression as the sum of a geometric series, then

$$\frac{1}{1-x} = \sum_{k=0}^{\infty} x^k \quad \text{for} \quad -1 < x < 1.$$

Hence,

$$\frac{1}{(1-x)^2} = \frac{d}{dx}\left(\frac{1}{1-x}\right) = \frac{d}{dx}\left(\sum_{k=0}^{\infty} x^k\right) = \sum_{k=0}^{\infty} \frac{d}{dx}(x^k) = \sum_{k=0}^{\infty} kx^{k-1},$$

for $-1 < x < 1$.

How could we have used series multiplication to get this result?

Exercises 7.7

Exercises 1–12: Use the techniques illustrated in this section and, as needed, the five Taylor series given at the end of Section 7.6 to find the Taylor series for f about the point x = c. Give the interval of convergence for each series.

1. $f(x) = e^{-3x}, \quad c = 0$
2. $f(x) = \dfrac{3}{x+4}, \quad c = 0$
3. $f(x) = 2\cos x - 3\sin x, \quad c = 0$
4. $f(x) = \sqrt{1-x^2}, \quad c = 0$
5. $f(x) = \cos^2 x, \quad c = 0$
6. $f(x) = (x^2+1)\sin 2x, \quad c = 0$
7. $f(x) = \cos x, \quad c = \dfrac{\pi}{4}$
8. $f(x) = \dfrac{1}{\sqrt{1+x}}, \quad c = 0$
9. $f(x) = \sqrt{3+2x}, \quad c = -1$
10. $f(x) = \dfrac{x+2}{x^2-3x}, \quad c = 2$
11. $f(x) = \ln(x+2), \quad c = 0$
12. $f(x) = x^2 \sin x, \quad c = 1$

Exercises 13–16: Find the first four nonzero terms of the Taylor series for g about x = c.

13. $g(x) = \dfrac{e^x}{1-x}, \quad c = 0$
14. $g(x) = (\sin x)\sqrt{1+x}, \quad c = 0$
15. $g(x) = \arcsin x, \quad c = 0$
16. $g(x) = \sqrt{\dfrac{1+x}{1-x}}, \quad c = 0$
17. Let $\sum_{k=0}^{\infty} a_k(x-c)^k$ be a power series with
$$\lim_{k\to\infty} \frac{a_k}{a_{k+1}} = R.$$
Show that the series $\sum_{k=0}^{\infty} ka_k(x-c)^{k-1}$ also has radius of convergence R.
18. Let $\sum_{k=0}^{\infty} a_k(x-c)^k$ be a power series with
$$\lim_{k\to\infty} \frac{a_k}{a_{k+1}} = R.$$
Show that the series $\displaystyle\sum_{k=0}^{\infty} \frac{a_k}{k+1}(x-c)^{k+1}$ also has radius of convergence R.

T **19.** Let $f(x) = \sum_{k=1}^{\infty} \frac{\sqrt{k}}{3^k} x^k$ for $-3 < x < 3$. Approximate $\int_0^1 f(x)\,dx$ with an error of less than 0.001.

T **20.** Let $f(x) = \sum_{k=0}^{\infty} (-1)^k \frac{x^k}{(k!)^2}$ for $-\infty < x < \infty$. Approximate $\int_0^1 f(x)\,dx$ with an error of less than 0.001.

T **21.** Approximate $\int_0^{\pi/2} \sin(x^2)\,dx$ with an error of less than 0.001.

Exercises 22–25: Find the sum of the series. Each series was obtained by applying substitution, integration, and/or differentiation to one of the series listed at the end of Section 7.6.

22. $\sum_{k=0}^{\infty} \left(\frac{x-3}{5}\right)^k$

23. $\sum_{k=0}^{\infty} (-1)^{k-1} \frac{(x+2)^{2k}}{k!}$

24. $\sum_{k=0}^{\infty} (k+1)(k+2)x^k$

25. $\sum_{k=1}^{\infty} \frac{1}{k}\left(\frac{x-2}{2}\right)^k$

26. Suppose the product of the two power series $\sum_{k=0}^{\infty} a_k x^k$ and $\sum_{k=0}^{\infty} b_k x^k$ is

$$\left(\sum_{k=0}^{\infty} a_k x^k\right)\left(\sum_{k=0}^{\infty} b_k x^k\right) = \sum_{c=0}^{\infty} c_n x^n.$$

Show that

$$c_0 = a_0 b_0$$
$$c_1 = a_0 b_1 + a_1 b_0$$
$$c_2 = a_0 b_2 + a_1 b_1 + a_2 b_0$$

and that in general

$$\begin{aligned} c_n &= \sum_{k=0}^{n} a_k b_{n-k} \\ &= a_0 b_n + a_1 b_{n-1} + \cdots + a_k b_{n-k} \\ &\quad + \cdots + a_n b_0. \end{aligned}$$

27. By equating the coefficients of x^n on each side of

$$\begin{aligned} \sum_{k=0}^{\infty} \frac{2^k}{k!} x^k &= e^{2x} \\ &= e^x e^x = \left(\sum_{k=0}^{\infty} \frac{1}{k!} x^k\right)\left(\sum_{k=0}^{\infty} \frac{1}{k!} x^k\right), \end{aligned}$$

show that

$$\begin{aligned} \frac{2^n}{n!} &= \frac{1}{0!}\frac{1}{n!} + \frac{1}{1!}\frac{1}{(n-1)!} \\ &\quad + \cdots + \frac{1}{k!}\frac{1}{(n-k)!} + \cdots + \frac{1}{n!}\frac{1}{0!}. \end{aligned}$$

28. Assume that the Taylor series for $\sec x$ at $c = 0$ is $\sum_{k=0}^{\infty} b_k x^k$ and that this series converges to $\sec x$ for x in some interval $-r < x < r$. Use the identity

$$\begin{aligned} 1 &= (\cos x)(\sec x) \\ &= \left(\sum_{k=0}^{\infty} (-1)^k \frac{x^{2k}}{(2k)!}\right)\left(\sum_{k=0}^{\infty} b_k x^k\right) \end{aligned}$$

to find the first three nonzero terms of the series for $\sec x$.

29. With the aid of the identity

$$(\cos x)(\tan x) = \sin x,$$

find the first four nonzero terms of the series for $\tan x$ about $c = 0$.

30. Let $f(x) = \sum_{k=0}^{\infty} a_k (x-c)^k$ for $c - R < x < c + R$. Show that the power series is also the Taylor series for f at $x = c$, that is, show that

$$a_k = \frac{f^{(k)}(c)}{k!}.$$

31. Suppose that $\sum_{k=0}^{\infty} a_k (x-c)^k = 0$ for $c - R < x < c + r$. Show that $a_k = 0$ for $k = 0, 1, 2, \ldots$.

32. Suppose that

$$\sum_{k=0}^{\infty} a_k (x-c)^k = \sum_{k=0}^{\infty} b_k (x-c)^k$$

for $c - R < x < c + R$. Show that $a_k = b_k$ for $k = 0, 1, 2, \ldots$. (*Hint:* Do Exercise 31 first.)

33. Suppose that

$$f(c) = f'(c) = f''(c) = \cdots = f^{(m-1)}(c) = 0$$

and that

$$f(x) = \sum_{k=0}^{\infty} a_k (x-c)^k$$

for $c - R < x < c + R$. Show that for x in this interval,

$$f(x) = (x-c)^m \sum_{k=0}^{\infty} a_{k+m} (x-c)^k.$$

With the help of this result, evaluate

$$\lim_{x \to c} \frac{f(x)}{(x-c)^m}.$$

Exercises 34–39: Using power series to evaluate limits *If* $\lim_{x\to c} f(x) = \lim_{x\to c} g(x) = 0$ *and both f and g are equal to their power series about c, then we can use the power series to determine* $\lim_{x\to c} f(x)/g(x)$. *Use the idea developed in the previous exercise to replace each of* $f(x)$ *and* $g(x)$ *by power series multiplied* *by a factor of the form $(x - c)^m$. Cancel common factors and then determine the limit.*

34. $\lim_{x \to 0} \dfrac{\sin x}{2x}$

35. $\lim_{x \to 0} \dfrac{\cos x - 1}{x \sin x - x}$

36. $\lim_{x \to 0} \dfrac{e^{x^2} - 1 - x^2}{x^2 \cos x}$

37. $\lim_{x \to 1} \dfrac{(\ln x)^2}{2(x - 1)^2}$

38. $\lim_{x \to 0} \dfrac{\ln(1 + ax)}{x}, \quad a > 0$

39. $\lim_{x \to \pi/2} \dfrac{\cos x + (x - \pi/2)}{(x - \pi/2)^3}$

40. Show that the sequence $\{b_k\}$, where $b_k = 3^{2k+1}/((2k + 1)k!)$ is decreasing from $k = 7$ onwards. *Hint:* Simplify the ratio b_{k+1}/b_k and then define a continuous function $f(x)$ by replacing k by x in the ratio. Show that f is decreasing for $x \geq 1$. Also, determine $x \geq 1$ for which $f(x) = 1$.

41. One of the Fresnel integrals (sometimes used in optics) is defined by

$$S(x) = \int_0^x \sin\left(\frac{\pi t^2}{2}\right) dt.$$

By starting with the Taylor series for the sine function about $c = 0$, show that

$$S(x) \approx \frac{\pi x^3}{6} - \frac{\pi^3 x^7}{336} + \frac{\pi^5 x^{11}}{42{,}240} - \frac{\pi^7 x^{15}}{9{,}676{,}800}.$$

Show that $S(0.5) \approx 0.0647324$. How could you check on the accuracy of this value of $S(0.5)$?

Review of Key Concepts

Approximation is important in every area of science, engineering, and mathematics. In this chapter we improved on the tangent line approximation by using Taylor polynomials. The Taylor polynomial $T_n(x; c)$ for a function f takes the value $f(c)$ at $x = c$. In addition, the first n derivatives of T_n and f are the same at $x = c$. Because of this, the Taylor polynomial $T_n(x; c)$ is often a good approximation to $f(x)$ when x is near c. When $n = 1$, the Taylor polynomial is the tangent line approximation. For $n \geq 1$, the Taylor polynomial is usually a better approximation to f than the tangent line expression.

To use an approximation effectively, we need to know something about the error in the approximation. The error in the Taylor polynomial approximation is

$$R_n(x; c) = f(x) - T_n(x; c).$$

We saw how to use this estimate to find a Taylor polynomial for which $|R_n(x; c)| < E$, where E is a given upper bound for the error.

To better understand Taylor polynomials, we next studied sequences and infinite series. We observed that the Taylor polynomials for a function f can be used to give a good approximation to f if the sequence $\{T_n(x; c)\}$ of Taylor polynomials converges to f. Because the sequence of Taylor polynomials for f is the sequence of partial sums of the Taylor series for f, some questions about the behavior of Taylor polynomials were answered by using results from our study of infinite series.

Next we studied power series and Taylor series. Using results developed in earlier sections, we were able to find the radius of convergence for such series. In addition, we saw that if the Taylor series for a function f converges to f on an interval $c - R < x < c + R$, then the corresponding sequence $\{T_n(x; c)\}$ of Taylor polynomials for f converges uniformly to f on any closed interval $[c - r, c + r]$ contained in $(c - R, c + R)$. This means that the Taylor polynomials for f can be used to approximate f as closely as we like on $[c - r, c + r]$.

If Taylor series and Taylor polynomials are to be a useful tool, then it is important that we be able to generate Taylor series and polynomials quickly and accurately. We saw that with knowledge of a few Taylor series (see Section 7.6) we can use substitution, algebra, differentiation, and integration to find Taylor series for many other functions.

Chapter Summary

The Comparison Tests

Let $\Sigma_{k=1}^{\infty} b_k$ be a series of nonnegative terms and let $\Sigma_{k=1}^{\infty} c_k$ be a convergent series, also with nonnegative terms. If there is a positive integer N such that

$$b_k \leq c_k \quad \text{for all } k \geq N,$$

then

$$\sum_{k=1}^{\infty} b_k \quad \text{also converges.}$$

The Comparison Tests (continued)

Let $\Sigma_{k=1}^{\infty} b_k$ be a series of nonnegative terms and let $\Sigma_{k=1}^{\infty} d_k$ be a divergent series of nonnegative terms. If there is a positive integer N such that

$$d_k \leq b_k \quad \text{for all } k \geq N,$$

then

$$\sum_{k=1}^{\infty} b_k \qquad \text{also diverges.}$$

Let $\Sigma_{k=1}^{\infty} a_k$ and $\Sigma_{k=1}^{\infty} b_k$ both be series of positive terms. If

$$0 < \lim_{k\to\infty} \frac{a_k}{b_k} < \infty,$$

then the two series both converge or the two series both diverge.

The Ratio Test

Let $\Sigma_{k=1}^{\infty} a_k$ be a series of positive terms. Suppose that

$$\lim_{k\to\infty} \frac{a_{k+1}}{a_k} = \rho, \quad \text{where } 0 \leq \rho \leq \infty.$$

If

$\rho < 1$, then the series converges

$\rho > 1$, then the series diverges

$\rho = 1$, then the series could either converge or diverge.

Absolute Convergence Test

Let $\Sigma_{k=1}^{\infty} a_k$ be an infinite series. If

$$\sum_{k=1}^{\infty} |a_k|$$

converges, then

$$\sum_{k=1}^{\infty} a_k$$

also converges.

Alternating Series Test

Let $\{b_k\}_{k=1}^{\infty}$ be a sequence of nonnegative real numbers. If

$$b_1 \geq b_2 \geq b_3 \geq \cdots \geq b_k \geq b_{k+1} \geq \cdots$$

and

$$\lim_{k\to\infty} b_k = 0,$$

then

$$\sum_{k=1}^{\infty} (-1)^{k-1} b_k = b_1 - b_2 + b_3 - \cdots$$

converges.

Taylor Polynomials

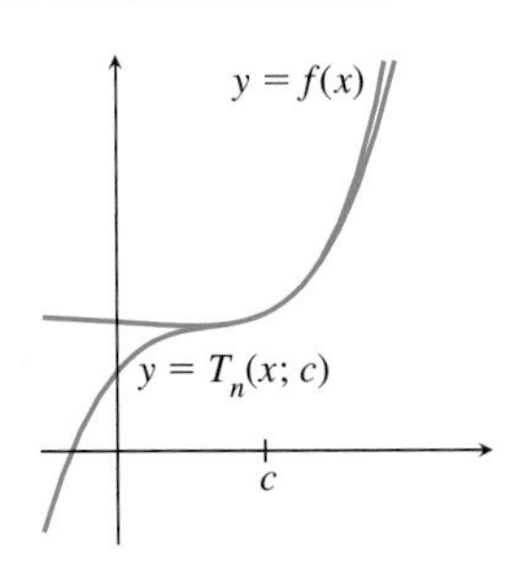

Let f be a function. If $f(c)$, $f'(c)$, $f''(c), \ldots f^{(n)}(c)$ are all defined, then the nth Taylor polynomial for f at c is

$$T_n(x; c) = \sum_{k=0}^{n} \frac{f^{(k)}(c)}{k!}(x - c)^k.$$

When n is large, $T_n(x; c)$ is often a good approximation to $f(x)$ for x values near c.

If $f(x) = \ln x$ and $c = 3$, then

$$f(c) = \ln 3$$

$$f'(c) = \frac{1}{3}$$

$$f''(c) = -\frac{1}{3^2} = -\frac{1}{9}.$$

Hence the second Taylor polynomial about $c = 3$ for f is

$$T_2(x; 3) = \ln 3 + \frac{1}{3}(x - 3) - \frac{1}{18}(x - 3)^2.$$

Error in a Taylor Approximation

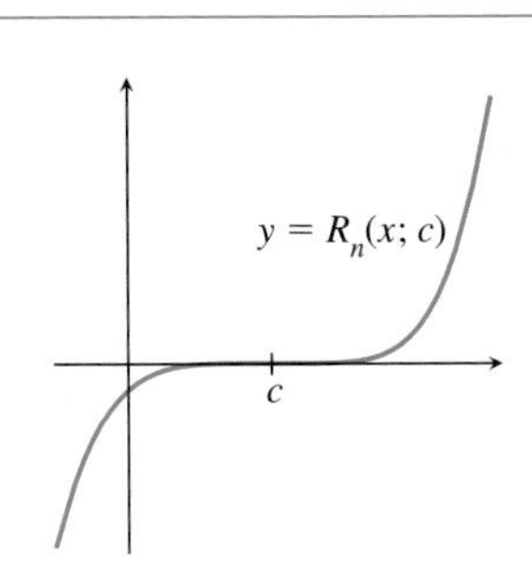

Let f be a function and let $T_n(x; c)$ be the nth Taylor polynomial for f at c. The difference between $f(x)$ and $T_n(x; c)$ is

$$R_n(x; c) = f(x) - T_n(x; c)$$

and is called the error in the approximation by the Taylor polynomial of degree n. If $|R_n(x; c)|$ is small, then $T_n(x; c)$ may be a good approximation to $f(x)$.

Let $f(x) = \ln x$ and $c = 3$. If we use the approximation

$$\ln x \approx T_2(x; 3)$$

then the error is

$$R_n(x; 3) = \ln x - T_2(x; 3) = \ln x - \left(\ln 3 + \frac{(x - 3)}{3} - \frac{(x - 3)^2}{18}\right).$$

Estimating the Error in a Taylor Approximation

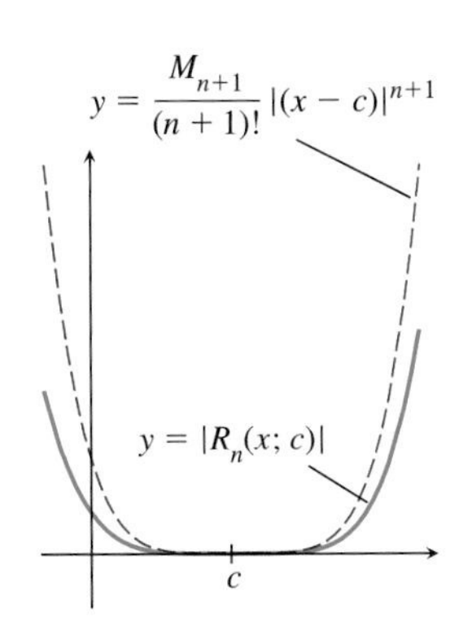

Let f be a function and $T_n(x; c)$ the nth Taylor polynomial for f at c. Suppose that $|f^{(n+1)}(x)| \leq M_{n+1}$ on the interval $c - r \leq x \leq c + r$. Then for x in this interval

$$|R_n(x; c)| \leq \frac{M_{n+1}}{(n + 1)!}|x - c|^{n+1}.$$

This estimate for the error is useful in estimating the error in the approximation $f(x) \approx T_n(x; c)$.

Let $f(x) = \ln x$, $c = 3$, and $n = 2$. Then

$$f^{n+1}(x) = \frac{d^3}{dx^3} \ln x = \frac{6}{x^3}.$$

For $2 \leq x \leq 4$, we have

$$\left|\frac{6}{x^3}\right| \leq \frac{3}{4}.$$

Hence, on this interval,

$$|R_2(x; 3)| \leq \frac{3/4}{3!}|x - 3|^3 \leq \frac{1}{8}.$$

The Limit of a Sequence

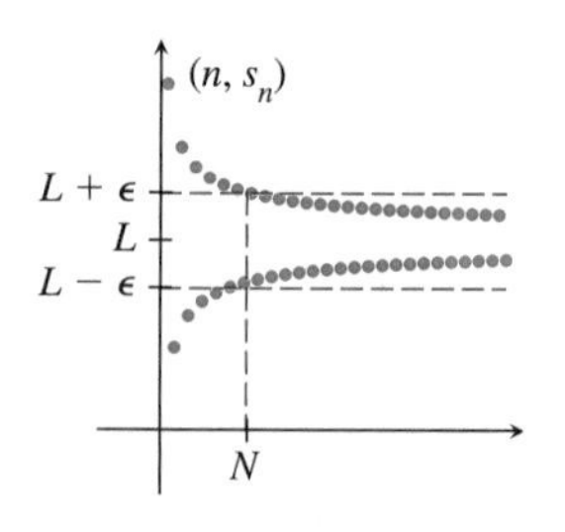

Let $\{s_n\}$ be a sequence of real numbers. The sequence converges to L if for any number $\epsilon > 0$ we can find a number $N > 0$ such that

$$|s_n - L| < \epsilon$$

whenever $n > N$. Roughly, this means that the numbers s_n get close to L as n gets large.

Let $s_n = 1 + (-1)^n \frac{1}{\sqrt{n}}$. When n is very large, $\frac{1}{\sqrt{n}}$ is close to 0. Hence for large n,

$$s_n = 1 + (-1)^n \frac{1}{\sqrt{n}} \approx 1.$$

Thus

$$\lim_{n\to\infty} s_n = \lim_{n\to\infty} \left(1 + (-1)^n \frac{1}{\sqrt{n}}\right) = 1.$$

Increasing, Bounded Above

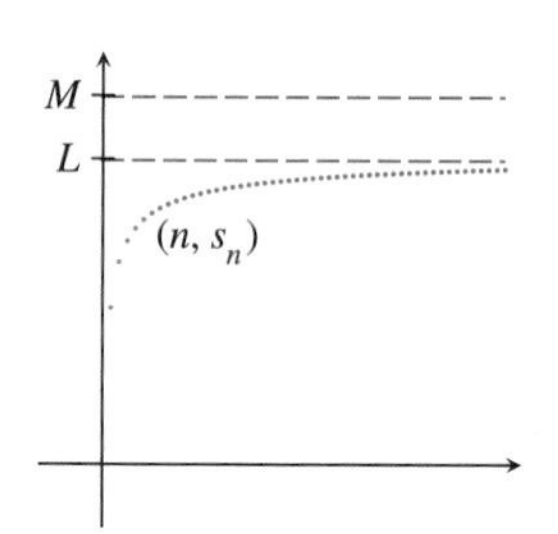

Let $\{s_n\}$ be a sequence of real numbers. The sequence is increasing if

$$s_1 \le s_2 \le \cdots \le s_n \le s_{n+1} \le \ldots.$$

The sequence is bounded above if there is a number A such that

$$s_n \le A$$

for all n. If a sequence of real numbers is increasing and bounded above, then it converges. That is

$$\lim_{n\to\infty} s_n = L$$

for some number L.

Let $s_n = \left(1 + \frac{1}{n}\right)^n$. Experimentation with a calculator or computer suggests that the sequence $\{s_n\}$ is increasing and that $s_n \le 3$ for all n. Hence the sequence has a limit. In Section 7.3, we saw that

$$\lim_{n\to\infty} s_n = \lim_{n\to\infty} \left(1 + \frac{1}{n}\right)^n = e.$$

Uniform Convergence

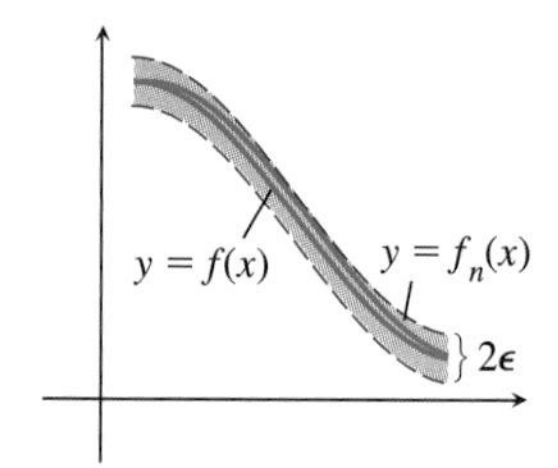

Let $\{f_n\}$ be a sequence of functions defined on an interval I, and let f be another function defined on I. We say that the sequence $\{f_n(x)\}$ converges to $f(x)$ uniformly on I if for each number $\epsilon > 0$ there is a number N so that whenever $n > N$, we have

$$|f(x) - f_n(x)| \le \epsilon$$

for all x in the interval I. This means that for $n > N$, the graph of $y = f_n(x)$ lies in the ϵ-corridor about the graph of $y = f(x)$.

Let

$$T_n(x) = \sum_{k=0}^{n} \frac{x^k}{k!}$$

be the nth Taylor polynomial for f about 0. Then the sequence $\{T_n(x)\}$ converges uniformly to e^x on any interval of the form $[-r, r]$.

Estimating an Alternating Series

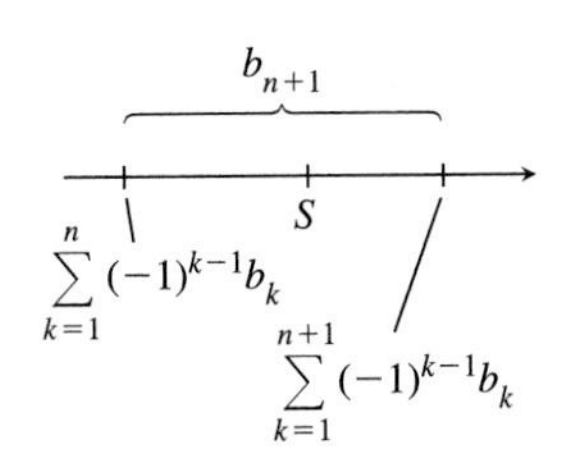

Suppose that $\Sigma_{k=1}^{\infty}(-1)^{k-1}b_k$ converges by the Alternating Series Test and let S be the sum of the series. For positive integer n,

$$\left|S - \sum_{k=0}^{n}(-1)^{k-1}b_k\right| \le |b_{n+1}|.$$

Hence, if we use a partial sum of the series to estimate the sum of the series, the error is no more than the absolute value of the first omitted term.

It can be shown that

$$\sum_{k=1}^{\infty}\frac{(-1)^{k-1}}{2k-1} = 1 - \frac{1}{3} + \frac{1}{5} - \cdots = \frac{\pi}{4}.$$

Hence

$$\left|\frac{\pi}{4} - \sum_{k=1}^{1000}\frac{(-1)^{k-1}}{2k-1}\right| \le \frac{1}{2001}.$$

Radius of Convergence

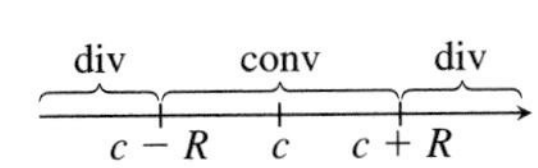

Consider a power series

$$\sum_{k=0}^{\infty} a_k(x-c)^k.$$

There is a number R, $0 \le R \le \infty$, such that

the series converges when $|x - c| < R$

the series diverges when $|x - c| > R$.

The number R is called the *radius of convergence* of the power series. If

$$\lim_{k\to\infty}\left|\frac{a_{k+1}}{a_k}\right| = \rho,$$

where $0 \le \rho \le \infty$, then

$$R = \frac{1}{\rho}.$$

Consider the power series

$$\sum_{k=0}^{\infty}\frac{x^k}{k!}.$$

We have

$$\lim_{k\to\infty}\left|\frac{a_{k+1}}{a_k}\right| = \lim_{k\to\infty}\frac{1/(k+1)!}{1/k!} = 0.$$

The radius of convergence of the power series is

$$R = \frac{1}{\rho} = \frac{1}{0} = \infty.$$

This means that the series converges for every real number x.

Chapter Review Exercises

Exercises 1–9: Find the limit of the sequence.

1. $\left\{\dfrac{10^{2n}}{n!}\right\}$

2. $\left\{\sqrt[n]{100} - 1\right\}$

3. $\{\tan j\}$

4. $\left\{\sum_{j=0}^{n}\dfrac{1}{3^j}\right\}$

5. $\{k^{(-1)^k}\}$

6. $\left\{\left(1 + \dfrac{2}{k}\right)^k\right\}$

7. $\{\sqrt{2n^2 + n + 1} - \sqrt{2}n\}$

8. $\{n^{3/2} - (n+1)^{3/2}\}$

9. $\left\{\sum_{k=0}^{n}\dfrac{1}{k!}\right\}$

Exercises 10–18: Decide whether the series converges or diverges.

10. $\sum_{k=1}^{\infty}\dfrac{2k}{\sqrt{k^2+1}}$

11. $\sum_{k=1}^{\infty}(\sqrt[k]{k} - 1)^k$

12. $\sum_{j=0}^{\infty}6(0.123)^{2j}$

13. $\sum_{m=0}^{\infty}\dfrac{10^m}{m!}$

14. $\sum_{k=1}^{\infty} (-1)^{k-1}\frac{k+1}{2k}$

15. $\sum_{\ell=1}^{\infty} \frac{\ell!}{\ell^{\ell}}$

16. $\sum_{q=1}^{\infty} (e-1)^{q}$

17. $\sum_{r=10}^{\infty} (-1)^{r}\frac{1}{\ln(\ln r)}$

18. $\sum_{j=2}^{\infty} \frac{1}{j^{3/2}-(j-2)^{3/2}}$

Exercises 19–22: Find the radius of convergence of the power series.

19. $\sum_{k=0}^{\infty} k^{10}x^{k}$

20. $\sum_{j=0}^{\infty} \frac{(2.7)^{j}}{j!}(x-2)^{2j}$

21. $\sum_{m=1}^{\infty} \frac{1\cdot 3\cdot 5\cdots(2m-1)}{3\cdot 6\cdot 9\cdots(3m)}x^{m}$

22. $\sum_{r=1}^{\infty} \frac{2\cdot 6\cdot 10\cdots(4r-2)}{3\cdot 7\cdot 11\cdots(4r-1)}(x+3)^{r}$

Exercises 23–28: Find the Taylor series for f about the point c. State the radius of convergence of the series.

23. $f(x)=\cos 2x, \quad c=0$

24. $f(x)=e^{2x}, \quad c=-\ln 2$

25. $f(x)=\frac{x+10}{(x+1)(x+4)}, \quad c=1$

26. $f(x)=x^{10}, \quad c=-2$

27. $f(x)=\int_{0}^{x}\sin(t^{2})\,dt, \quad c=0$

28. $f(x)=\sin^{2}4x, \quad c=\pi/8$

Exercises 29–34: Find the sum of the series. (Some of these series were obtained by substituting a number into a known Taylor series.)

29. $\sum_{k=0}^{\infty} 0.9^{k}$

30. $\sum_{j=0}^{\infty} \frac{2}{(j+1)(j+3)}$

31. $\sum_{k=1}^{\infty} \left(\frac{e-1}{e}\right)^{k}$

32. $\sum_{r=0}^{\infty} \frac{2^{r}}{r!}$ (*Hint:* Recall the series for e^{x}.)

33. $\sum_{k=1}^{\infty} (-1)^{k-1}\frac{1}{k2^{k}}$ (*Hint:* Recall the series for $\ln(1-x)$.)

34. $\sum_{k=0}^{\infty} (-1)^{k}\frac{\pi^{2k+1}}{(2k+1)!}$

35. Find a value of n so that $\sum_{k=1}^{n}\frac{1}{\sqrt[3]{k}}>1000$.

36. Find n so that the nth partial sum of the series $\sum_{k=2}^{\infty}(-1)^{k-1}/\ln k$ is within 0.001 of the sum of the series.

37. Find n so that the nth partial sum of the series $\sum_{k=1}^{\infty}1/k^{4}$ is within 0.0001 of the sum of the series.

38. Find n so that the nth partial sum of the series $\sum_{k=2}^{\infty}1/[k(\ln k)^{4}]$ is within 0.001 of the sum of the series.

39. Find n so that the nth partial sum of the series $\sum_{k=2}^{\infty}(-1)^{k-1}/\sqrt{k}$ is within 10^{-9} of the sum of the series.

40. Let $T_n(x;0)$ be the nth Taylor polynomial for e^{2x} about the point $x=0$. Find n so that $T_n(x;0)$ is within 0.01 of e^{2x} for all x in the interval $-2\le x\le 2$.

41. Let $T_n(x;\pi/4)$ be the nth Taylor polynomial for $\cos x$ about the point $x=\pi/4$. Find n so that $T_n(x;\pi/4)$ is within 0.001 of $\cos x$ for all x in the interval $\pi/4-1\le x\le \pi/4+1$.

42. Let $T_n(x;0)$ be the nth Taylor polynomial for $\arctan x$ about the point $x=0$. Find n so that $T_n(x;0)$ is within 0.01 of $\arctan x$ for all x in the interval $-3/4\le x\le 3/4$.

43. Let $T_n(x;-1)$ be the nth Taylor polynomial for $1/(1-x)$ about the point $x=-1$. Find n so that $T_n(x;-1)$ is within 0.001 of $1/(1-x)$ for all x in the interval $-2\le x\le 0$.

44. Let $i=\sqrt{-1}$, so $i^{2}=-1$, $i^{3}=-i$, etc. Substitute $i\theta$ for x in the series identity

$$e^{x}=\sum_{k=0}^{\infty}\frac{x^{k}}{k!},$$

then do some algebraic manipulations to arrive at the identity

$$e^{i\theta}=\cos\theta+i\sin\theta.$$

Use this identity to evaluate $e^{i\pi}$. Also evaluate $e^{i\pi/4}$.

45. Let $i=\sqrt{-1}$. Use series to prove the identities

$$\cos x=\frac{e^{ix}+e^{-ix}}{2}$$

and

$$\sin x=\frac{e^{ix}-e^{-ix}}{2i}.$$

46. The hyperbolic cosine function, denoted cosh, is defined by

$$\cosh x=\frac{e^{x}+e^{-x}}{2}.$$

Find the Taylor series about $x=0$ for the hyperbolic cosine function.

47. The hyperbolic sine function, denoted sinh, is defined by

$$\sinh x=\frac{e^{x}-e^{-x}}{2}.$$

Find the Taylor series about $x=0$ for the hyperbolic sine function.

48. By combining the results of the previous three problems, show that

$$\cos(ix)=\cosh x$$

and

$$-i\sin(ix)=\sinh x.$$

STUDENT PROJECT

SIGNAL PROCESSING AND FOURIER SERIES

Suppose that the power series $\Sigma_{k=0}^{\infty} a_k x^k$ has radius of convergence 1. Then the series converges for each $x \in (-1, 1)$, so we can define a function f on this interval by $f(x) = \Sigma_{k=0}^{\infty} a_k x^k$. Now suppose that a friend across the country needs to approximate $f(x)$ for several values of x, so she asks us to tell her what f is. We could just write out the first several terms of the series and send them, but there's a shorter way. We could instead just send the first few series coefficients $\{a_0, a_1, a_2, \ldots, a_n\}$ and tell our friend that these are the coefficients of the first $n + 1$ terms of a power series about 0. With this information our friend could reconstruct a partial sum of the series and obtain her approximations. Why does this work? It works because a function that can be expressed as a power series about 0 is completely determined by its power series coefficients.

This idea is used several million times every day in the communications industry, although the series involved are Fourier series instead of power series. This application is based on the following fact.

Fourier Series and Convergence

Let f be a function defined on the interval $0 \leq t \leq 2$. If f is "nice" enough, then there are real numbers $a_0, a_1, a_2, \ldots$ and $b_1, b_2, b_3, \ldots$ such that

$$f(t) = a_0 + \sum_{k=1}^{\infty} (a_k \cos(k\pi t) + b_k \sin(k\pi t)) \qquad (1)$$

for all $0 < t < 2$. Furthermore, the series (1) converges uniformly to $f(t)$ on each interval of the form $[\delta, 2 - \delta]$, where $0 < \delta < 1$.

We will not attempt to define "nice" here. However it can be shown that any function with a continuous first derivative on $[0, 2]$ is "nice." This means that most of the functions we use on a day-to-day basis are "nice" functions.

A series like that in (1) is called a *Fourier series*. The coefficients $\{a_k\}$ and $\{b_k\}$ are called the *Fourier coefficients for f*. In practice, $f(t)$ might be the amplitude or strength of a signal (e.g., a sound) at time t. To communicate information about the signal, we calculate several of the coefficients $a_0, a_1, \ldots b_1, b_2, \ldots$ and transmit them instead of attempting to transmit the signal itself. At the receiving end, the Fourier coefficients are used to obtain a partial sum of the Fourier series for f, and hence an approximation for f. With this information an approximation to the signal can be constructed, perhaps resulting in a voice or a picture from space. In this project we look at some examples to see how the partial sums of a Fourier series approximate a function, then indicate how the approximation process can be implemented in practice.

PROBLEM 1 Calculating Fourier Coefficients

a) Let k and m be positive integers. Verify that

$$\int_0^2 \cos k\pi t \, dt = 0, \qquad \int_0^2 \sin k\pi t \, dt = 0$$

and that

$$\int_0^2 \cos k\pi t \sin m\pi t \, dt = 0.$$

Show also that if $k \neq m$, then

$$\int_0^2 \cos k\pi t \cos m\pi t = 0 \quad \text{and} \quad \int_0^2 \sin k\pi t \sin m\pi t = 0.$$

First, show that for positive integer k,

$$\int_0^2 \cos^2 k\pi t \, dt = 1 = \int_0^2 \sin^2 k\pi t \, dt.$$

b) Assume that f is continuous on $[0, 2]$ and that (1) is true. By integrating both sides of (1), show that

$$a_0 = \frac{1}{2}\int_0^2 f(t)\, dt. \tag{2}$$

c) Assume that f is continuous on $[0, 2]$ and (1) is true. Let m be a positive integer. Multiply both sides of (1) by $\cos m\pi t$, then integrate from 0 to 2. Tell why this leads to

$$a_m = \int_0^2 f(t) \cos m\pi t \, dt. \tag{3}$$

Use a similar argument to show that

$$b_m = \int_0^2 f(t) \sin m\pi t \, dt. \tag{4}$$

PROBLEM 2 An Example of Fourier Approximation Let $f(t) = t$ for all t in $[0, 2]$.

a) Use (2), (3), and (4) to find the Fourier coefficients for f.
b) With the values for the a_k's and b_k's found in (a), graph

$$y = a_0 + \sum_{k=1}^{5} (a_k \cos k\pi t + b_k \sin k\pi t).$$

How well does this partial sum of the Fourier series approximate f on $[0, 2]$?
c) Next, graph

$$y = a_0 + \sum_{k=1}^{20} (a_k \cos k\pi t + b_k \sin k\pi t).$$

How well does this partial sum of the Fourier series approximate f on $[0, 2]$? Where does the partial sum appear to be a good approximation to $f(t) = t$? Where is it a poor approximation?

PROBLEM 3 A Second Example of Fourier Approximation Repeat the previous problem with $f(t) = t^2$.

PROBLEM 4 A Third Example of Fourier Approximation Repeat Problem 2 with the "sawtooth" function.

$$f(t) = \begin{cases} t & (0 \le t \le 1) \\ 2 - t & (1 < t \le 2). \end{cases}$$

The graph of this function is shown in Fig. 7.41.

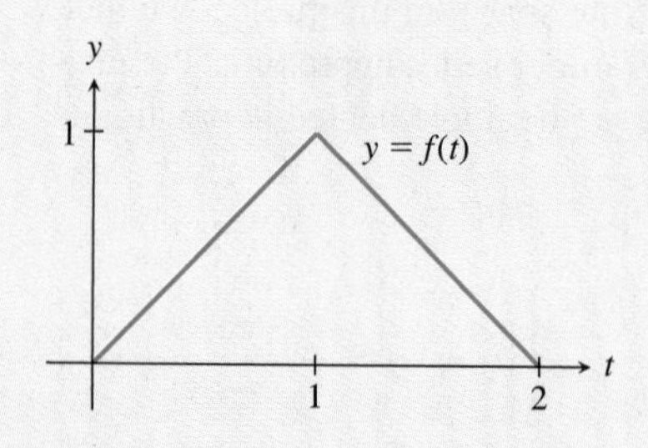

FIGURE 7.41 The "sawtooth" function.

In practice, a formula for the function f that gives the strength of a signal at time t is not known. In such cases the signal is sampled at several different times t and its strength $f(t)$ at each of these times is recorded. With this data the Fourier coefficients (2), (3), and (4) are then approximated using numerical integration.

As an example, suppose a signal is sampled at the ten times 0.2, 0.4, 0.6, ..., 2.0 to give the data shown in Table 7.8. The ten sample times t are equally spaced and divide the interval $[0, 2]$ into ten equal pieces each of length 0.2. Using 0.2 as the width of a rectangle and the $f(t)$ values as the "heights" of rectangles, we can approximate the Fourier coefficients using Riemann sums. For example:

TABLE 7.8

t	$f(t)$
0.2	2.412
0.4	5.138
0.6	−1.334
0.8	−0.0605
1.0	4.0
1.2	−2.412
1.4	−5.138
1.6	1.334
1.8	0.0605
2.0	−4.0

$$
\begin{aligned}
b_1 &= \int_0^2 f(t)\sin(1\cdot\pi t)\,dt \approx \sum_{j=1}^{10} f(0.2j)\,(\sin(1\cdot\pi 0.2j))\,(0.2)\\
&\approx (2.412)\,(\sin(0.2\pi))0.2 + (5.138)\,(\sin(0.4\pi))0.2\\
&\quad + (-1.334)\,(\sin(0.6\pi))0.2 + (-0.0605)\,(\sin(0.8\pi))0.2\\
&\quad + (4.0)\,(\sin(1.0\pi))0.2 + (-2.412)\,(\sin(1.2\pi))0.2\\
&\quad + (-5.138)\,(\sin(1.4\pi))0.2 + (1.334)\,(\sin(1.6\pi))0.2\\
&\quad + (0.0605)\,(\sin(1.8\pi))0.2 + (-4.0)\,(\sin(2.0\pi))0.2\\
&\approx 2.0.
\end{aligned}
$$

Hence, in the Fourier series for the signal f, the coefficient of $\sin(\pi t)$ is approximately 2. The constant term a_0 of the series is approximated by

$$a_0 = \frac{1}{2}\int_0^2 f(t)\,dt \approx \sum_{j=1}^{10} f(0.2j)0.2 \approx 0.$$

PROBLEM 5 Approximation of Fourier Coefficients For the data given in Table 7.8, compute approximations to a_1, a_3, a_3, b_2, and b_3. Plot the data and the graph of

$$y = a_0 + \sum_{k=1}^{3} (a_k \cos(k\pi t) + b_k \sin(k\pi t))$$

on the same set of axes. How well does this partial sum of the Fourier series fit the data?

PROBLEM 6 Approximation of Fourier Coefficients Suppose that the 100 ordered pairs

$$(0.02j, (0.02j)\sin(0.02j + 1)), \qquad j = 1, 2, 3, \ldots, 100$$

are obtained as data points for a signal f defined for $0 \le t \le 2$. Approximate the Fourier coefficients $a_0, a_1, \ldots, a_{10}$ and $b_1, b_2, \ldots, b_{10}$ for f. Plot the data points and the graph of

$$y = a_0 + \sum_{k=1}^{10} (a_k \cos(k\pi t) + b_k \sin(k\pi t))$$

on the same set of axes. How well does this partial sum fit the data?

PROBLEM 7 Approximation of Fourier Coefficients Your instructor will provide 100 ordered pairs of data points for a signal f defined for $0 \le t \le 2$. Use this data to approximate the Fourier coefficients $a_0, a_1, \ldots, a_{10}$ and $b_1, b_2, \ldots, b_{10}$ for f. Plot the data points and the graph of

$$y = a_0 + \sum_{k=1}^{10} (a_k \cos(k\pi t) + b_k \sin(k\pi t))$$

on the same set of axes. How well does this partial sum fit the data?

In practice, the important process of using numerical integration to approximate the Fourier coefficients for a signal is now accomplished by means of the Fast Fourier Transform, or FFT. Mathematicians and engineers have studied the numerical integration process as it is used to approximate Fourier coefficients. They found many inefficiencies in these computations; for example, many multiplications performed in approximating one Fourier coefficient are repeated in approximating another Fourier coefficient. The FFT eliminates many of these redundancies and provides a fast algorithm for approximating Fourier coefficients. With today's high-speed computers, the FFT can be implemented several times every second, resulting in rapid, efficient, and accurate communications.

8

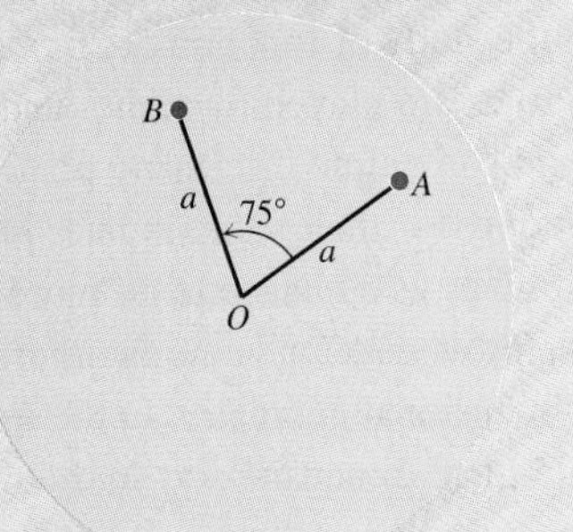

Vectors and Linear Functions

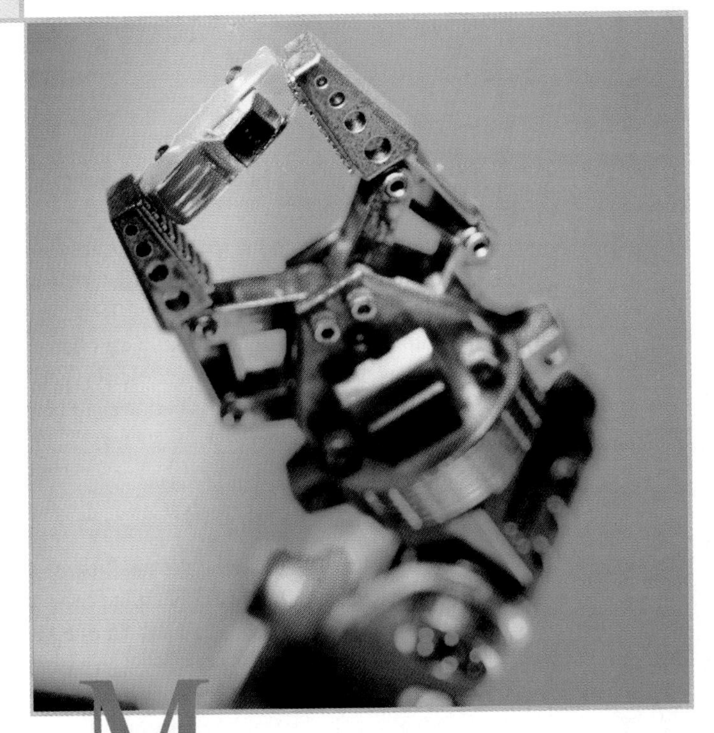

Modern robotic welding and drilling machines are an important part of today's industrial technology. A crucial element in using such robots is controlling their motion so the welds or holes are made at exactly the right locations in the right sequence. To program such motion, we might use a combination of vectors and linear functions.

See p. 678 for further explanation.

In this chapter we lay the groundwork for the study of functions of several variables. Most of the ideas in this chapter are generalizations to three dimensions of ideas developed when we studied functions of one variable and vectors and motion in two dimensions.

We start with a study of three-dimensional vectors. Most of the definitions and results that hold for two-dimensional vectors extend very naturally to the three-dimensional setting. We next study matrices. Matrices will be important in working with vectors and in developing many ideas associated with the calculus of several variables. When we studied the calculus of functions of one variable in Chapters 1–7, lines and equations of lines played a central role. In the study of functions of more than one variable, these roles are played by planes and linear functions. We spend some time discussing planes and linear functions, not only for their value in later sections, but also because it gives us a chance to do more work with vectors and helps us develop some familiarity and intuition for three-dimensional geometry. In the final section, we use vectors in three dimensions to study motion in space.

8.1 Vectors in Three Dimensions

In Chapters 3 and 6 we studied motion in the plane and saw that two-dimensional vectors are a powerful tool in the analysis of such motions. But the world we live in has (at least) three dimensions. To describe the path of a spacecraft to Mars, to program a robotic arm to pick up an object, or to analyze the structure of a crystal lattice, we need to be able to describe and analyze the position of an object in space. As a first step toward this end, we study the algebra and geometry of vectors in three-dimensional space.

FIGURE 8.1 Visualizing the x-, y-, and z-axes.

Cartesian Three-Space

In the plane we describe the position of a point by giving its position relative to two perpendicular lines, usually called the x-axis and the y-axis. To describe the position of a point in space, we add a third line, usually called the z-axis, perpendicular to the plane of the x- and y-axes. We can visualize these three axes by looking at a corner in a classroom. The origin is the point where two walls and the floor meet. The axes run along the line of intersection of the two walls and the lines of intersection of each wall with the floor. To determine which axis is which, stand in the corner with your back to the corner and put your heels together so your feet make a right angle. Your right foot points in the direction of the positive x-axis, your left foot points in the direction of the positive y-axis, and the line from your heels to your head points in the direction of the z-axis. The plane of the floor is determined by the x- and y-axes, so it is referred to as the (x, y)-plane. The two walls lie in the (x, z)- and (y, z)-planes. See Fig. 8.1. In three dimensions, we can select many different viewpoints. We will usually show the view as seen by someone standing in the room looking toward the origin. This is the view shown in Figs. 8.2 and 8.3.

To describe the position of a point in space, we use an ordered triple (x, y, z). The first coordinate is measured along the x-axis, the second along the y-axis, and the third along the z-axis. To locate a point (x, y, z) in three-dimensional space, first locate the point (x, y) in the (x, y)-plane following the usual conventions; we denote this point by $(x, y, 0)$. Next move up or down until you are level with the number z on

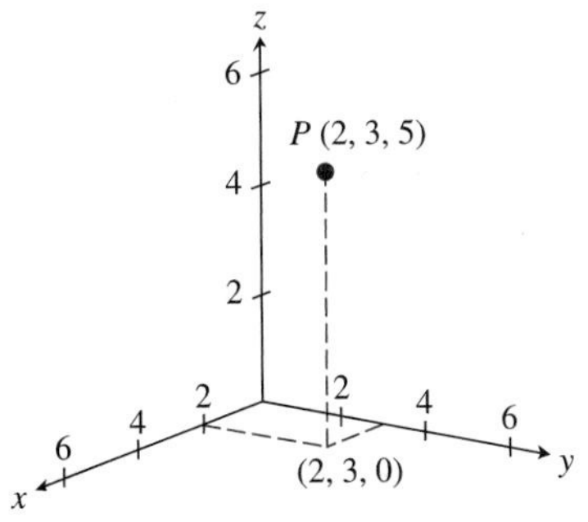

FIGURE 8.2 The point $(2, 3, 5)$.

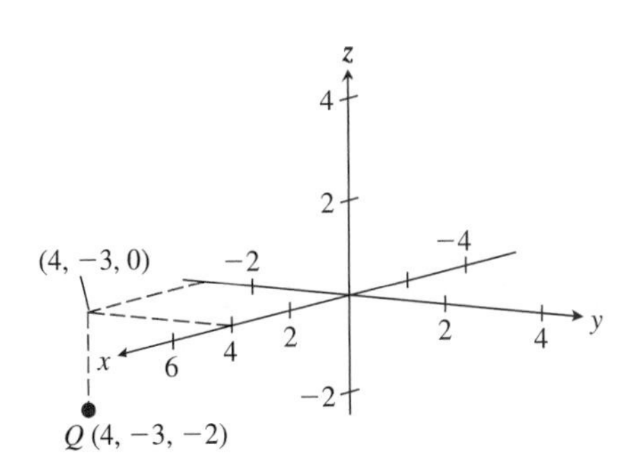

FIGURE 8.3 The point $(4, -3, -2)$.

the z-axis. In Fig. 8.2 we have located the point P with coordinates $(2, 3, 5)$, and in Fig. 8.3 we show the point Q with coordinates $(4, -3, -2)$. We have included dashed lines to help give a sense of the placement of the points in space. In Fig. 8.3 we have shown a portion of each of the negative axes. Usually, as in Fig. 8.2, the negative portions of the axes will not be shown because this makes for a less cluttered picture. Occasionally, it is useful to show a diagram in three-space with a different orientation than that shown in Figs. 8.2 and 8.3. One example of this is shown in Fig. 8.4, where we show a point P in a system in which the x-axis points upward. It is important to pay attention to the axes labels in such diagrams and to clearly label the axes in your own diagrams. The set of all points in space is denoted R^3 and is called *Cartesian three-space* or *Euclidean three-space.*

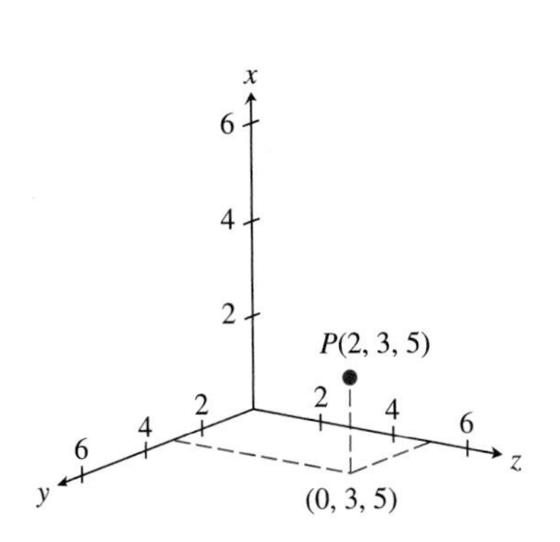

FIGURE 8.4 Another view of the point $(2, 3, 5)$.

We also use ordered triple notation to denote a vector in space. For example, the notation $\mathbf{v} = \langle 2, 3, 5 \rangle$ defines $\mathbf{v}$ as the position vector with initial point the origin $(0, 0, 0)$ and terminal point $(2, 3, 5)$. This vector is shown in Fig. 8.5.

Most of the definitions and formulas for two-dimensional vectors given in Sections 3.2 and 3.5 are meaningful for three-dimensional vectors. These include position vectors, vectors based at points other than the origin, the length of a vector, addition of vectors, the parallelogram figure for vector addition, multiplication of a vector by a scalar, parallel vectors, the dot product, the angle between vectors, and projection. We look at some examples to review these ideas. We use without further comment such definitions as

$$\begin{aligned}
\mathbf{u} + \mathbf{v} &= \langle u_1, u_2, u_3 \rangle + \langle v_1, v_2, v_3 \rangle = \langle u_1 + v_1, u_2 + v_2, u_3 + v_3 \rangle, \\
a\mathbf{v} &= a\langle v_1, v_2, v_3 \rangle = \langle av_1, av_2, av_3 \rangle, \\
\|\mathbf{v}\| &= \|\langle v_1, v_2, v_3 \rangle\| = \sqrt{v_1^2 + v_2^2 + v_3^2}, \\
\mathbf{u} \cdot \mathbf{v} &= \langle u_1, u_2, u_3 \rangle \cdot \langle v_1, v_2, v_3 \rangle = u_1 v_1 + u_2 v_2 + u_3 v_3, \\
&= \|\mathbf{u}\| \, \|\mathbf{v}\| \cos \theta,
\end{aligned}$$

where θ is the angle between $\mathbf{v}$ and $\mathbf{u}$. We discuss the direction of a vector later in this section.

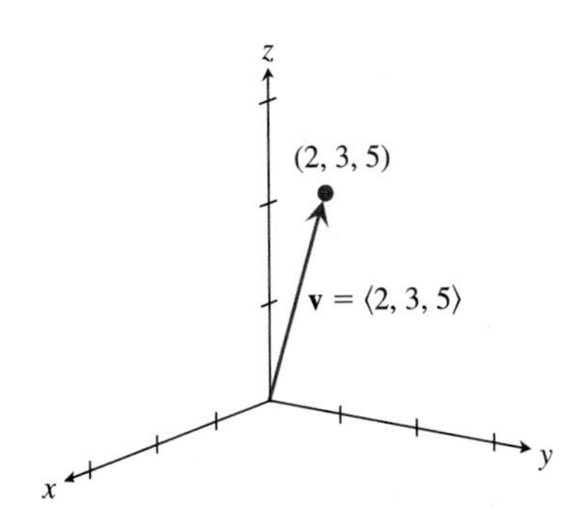

FIGURE 8.5 The vector $\mathbf{v} = \langle 2, 3, 5 \rangle$.

EXAMPLE 1 Let $A = (5, 0, 0)$, $B = (2, 5, 1)$, and $C = (1, 1, 6)$ be three points in space. Find $\overrightarrow{AB}$, $\overrightarrow{AC}$, the vector sum $\overrightarrow{AB} + \overrightarrow{AC}$, and the difference $\overrightarrow{AB} - \overrightarrow{AC}$. Illustrate the sum and difference with a diagram.

Solution

We use the notation $\mathbf{A} = \langle 5, 0, 0\rangle$ for the position vector to the point $(5, 0, 0)$. We first find $\overrightarrow{AB}$ and $\overrightarrow{AC}$. A representation for $\overrightarrow{AB}$ is found by the coordinate-wise subtraction of A from B,

$$\overrightarrow{AB} = \mathbf{B} - \mathbf{A} = \langle 2, 5, 1\rangle - \langle 5, 0, 0\rangle = \langle 2 - 5, 5 - 0, 1 - 0\rangle = \langle -3, 5, 1\rangle.$$

Similarly,

$$\overrightarrow{AC} = \mathbf{C} - \mathbf{A} = \langle 1, 1, 6\rangle - \langle 5, 0, 0\rangle = \langle 1 - 5, 1 - 0, 6 - 0\rangle = \langle -4, 1, 6\rangle.$$

The vector $\overrightarrow{AC} = \langle -4, 1, 6\rangle$ can be interpreted as a displacement vector that gives instructions on how to move from point A to point C. Starting from A, first move -4 in the x-direction, that is, parallel to the x-axis and 4 units in the negative x-direction. From this point, move 1 unit in the y-direction, and then 6 units in the z-direction. Once $\overrightarrow{AB}$ and $\overrightarrow{AC}$ are known, the sum and difference can be calculated:

$$\begin{aligned}\overrightarrow{AB} + \overrightarrow{AC} &= \langle -3, 5, 1\rangle + \langle -4, 1, 6\rangle = \langle -3 + (-4), 5 + 1, 1 + 6\rangle \\ &= \langle -7, 6, 7\rangle\end{aligned}$$

and

$$\begin{aligned}\overrightarrow{AB} - \overrightarrow{AC} &= \langle -3, 5, 1\rangle - \langle -4, 1, 6\rangle = \langle -3 - (-4), 5 - 1, 1 - 6\rangle \\ &= \langle 1, 4, -5\rangle.\end{aligned}$$

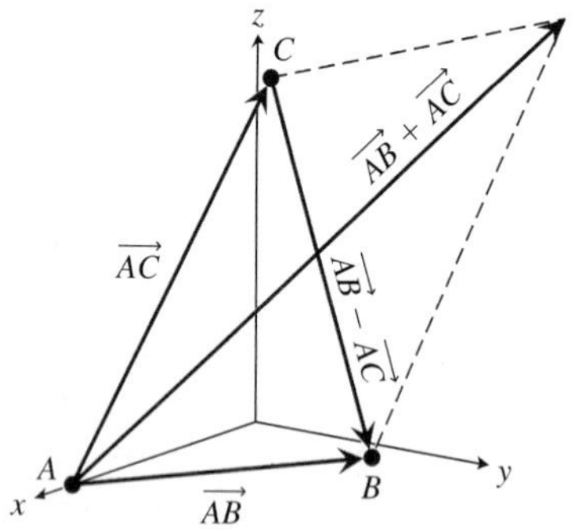

Figure 8.6 shows the parallelogram diagram illustrating the sum and the difference of **AB** and **AC**.

FIGURE 8.6 The sum and difference of vectors are the diagonals of a parallelogram.

EXAMPLE 2 Let $\mathbf{v} = \langle -1, 3, 4\rangle$ and $\mathbf{w} = \langle -2/3, 1/3, 2/3\rangle$.

a) Find the length of $\mathbf{w}$.
b) Find the angle between $\mathbf{v}$ and $\mathbf{w}$.
c) Find the projection of $\mathbf{v}$ on $\mathbf{w}$.

Solution

a) The length of $\mathbf{w}$ is

$$\|\mathbf{w}\| = \sqrt{\left(-\frac{2}{3}\right)^2 + \left(\frac{1}{3}\right)^2 + \left(\frac{2}{3}\right)^2} = 1.$$

Thus $\mathbf{w}$ is a unit vector.

b) Let θ be the angle between $\mathbf{v}$ and $\mathbf{w}$. Then

$$\cos\theta = \frac{\mathbf{v}\cdot\mathbf{w}}{\|\mathbf{v}\|\,\|\mathbf{w}\|}. \tag{1}$$

From

$$\|\mathbf{v}\| = \sqrt{(-1)^2 + 3^2 + 4^2} = \sqrt{26}, \qquad \|\mathbf{w}\| = 1,$$

and

$$\mathbf{v}\cdot\mathbf{w} = \langle -1,3,4\rangle \cdot \left\langle -\frac{2}{3},\frac{1}{3},\frac{2}{3}\right\rangle$$
$$= (-1)\left(-\frac{2}{3}\right) + 3\left(\frac{1}{3}\right) + 4\left(\frac{2}{3}\right) = \frac{13}{3},$$

we find from (1) that

$$\cos\theta = \frac{13/3}{\sqrt{26}} \approx 0.849837. \tag{2}$$

Hence $\theta = \arccos\big(13/(3\sqrt{26})\big) \approx 0.555121$, which is about 31.8°.

c) Because the angle between $\mathbf{v}$ and $\mathbf{w}$ is less than 90°, the projection of $\mathbf{v}$ onto $\mathbf{w}$ is in the same direction as $\mathbf{w}$. See Fig. 8.7. Let l be the length of the projection. From Fig. 8.7 and (2),

$$l = \|\mathbf{v}\| \cos\theta = \sqrt{26}\,\frac{13}{3\sqrt{26}} = \frac{13}{3}. \tag{3}$$

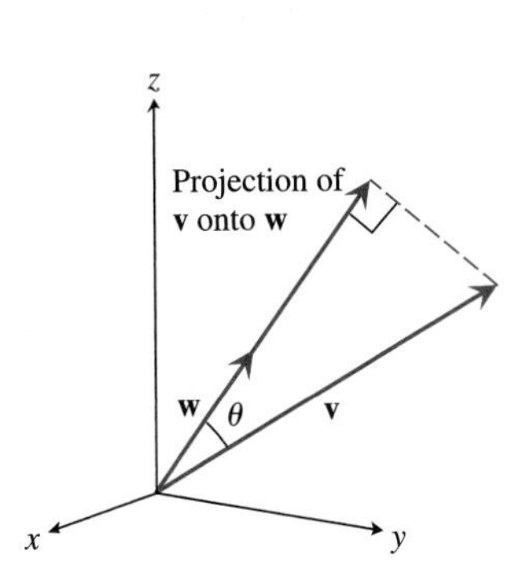

FIGURE 8.7 The projection of $\mathbf{v}$ onto $\mathbf{w}$.

Because $\mathbf{w}$ is a unit vector in the same direction as the projection, the projection of $\mathbf{v}$ onto $\mathbf{w}$ is

$$l\mathbf{w} = \frac{13}{3}\left\langle -\frac{2}{3},\frac{1}{3},\frac{2}{3}\right\rangle = \left\langle -\frac{26}{9},\frac{13}{9},\frac{26}{9}\right\rangle.$$

Combining (1) and (3) we obtain a formula for the projection of $\mathbf{v}$ onto the unit vector $\mathbf{w}$,

$$\text{proj}_{\mathbf{w}}\mathbf{v} = \big(\|\mathbf{v}\| \cos\theta\big)\mathbf{w} = (\mathbf{v}\cdot\mathbf{w})\mathbf{w}.$$

Direction of a Vector In the plane, we can specify the direction of a vector by giving the angle measured from the positive x-axis to the vector. In three dimensions, we describe the direction of a vector $\mathbf{v}$ as a unit vector in the same direction as $\mathbf{v}$. (However, see also Exercise 60.) Two vectors $\mathbf{a}$ and $\mathbf{b}$ are said to be in the same direction if one is a positive scalar multiple of the other, that is, if there is a positive real number k such that $\mathbf{a} = k\mathbf{b}$. This leads to the following definition.

DEFINITION Direction of a Vector

The direction of a nonzero vector $\mathbf{v}$ is the unit vector

$$\mathbf{u} = \frac{1}{\|\mathbf{v}\|}\mathbf{v}. \tag{4}$$

The zero vector $\mathbf{0} = \langle 0,0,0\rangle$ has no direction.

It follows from (4) that if $\mathbf{v}$ is a vector in the direction of the unit vector $\mathbf{u}$, then

$$\mathbf{v} = \|\mathbf{v}\|\mathbf{u}.$$

EXAMPLE 3 Let $\mathbf{v} = \langle -3, 4, -5 \rangle$. Find the direction of $\mathbf{v}$.

Solution

The direction of $\mathbf{v}$ is the unit vector

$$\mathbf{u} = \frac{1}{\|\mathbf{v}\|}\mathbf{v} = \frac{1}{5\sqrt{2}}\langle -3, 4, -5 \rangle = \left\langle -\frac{3}{5\sqrt{2}}, \frac{4}{5\sqrt{2}}, -\frac{1}{\sqrt{2}} \right\rangle.$$

How can you check that $\mathbf{u}$ is indeed a unit vector? Is it true that $\mathbf{v} = \|\mathbf{v}\|\mathbf{u}$?

Notation It is often convenient to have notation for vectors of length 1 in the direction of the positive coordinate axes. Extending the notation for such vectors introduced in Section 3.2, we define

$$\mathbf{i} = \langle 1, 0, 0 \rangle, \qquad \mathbf{j} = \langle 0, 1, 0 \rangle, \qquad \mathbf{k} = \langle 0, 0, 1 \rangle.$$

With this notation, we have

$$\langle x, y, z \rangle = x\mathbf{i} + y\mathbf{j} + z\mathbf{k}.$$

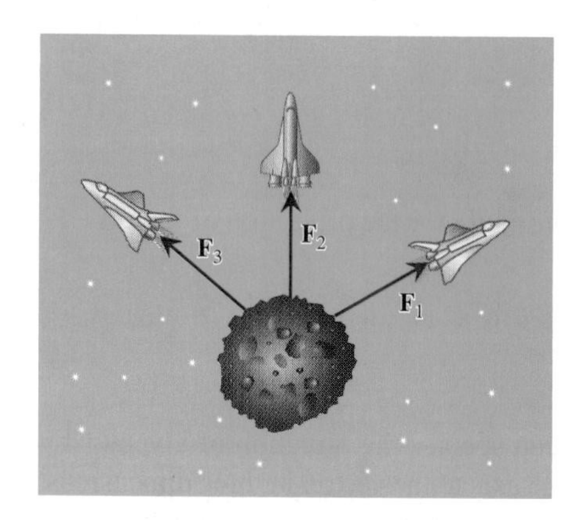

FIGURE 8.8 Towing an asteroid.

EXAMPLE 4 Three space tugs tow a small asteroid. The first tug exerts force $\mathbf{F}_1 = -3\mathbf{i} + \mathbf{j} + 0.3\mathbf{k}$ kilonewtons (kN), and the second one $\mathbf{F}_2 = \mathbf{i} + 2\mathbf{j} + 4\mathbf{k}$ kN. What force $\mathbf{F}_3$ must be exerted by the third tug if we want the resultant force to be in the direction $\mathbf{k}$? See Fig. 8.8.

Solution

The resultant force $\mathbf{F}$ must have the form $\mathbf{F} = a\mathbf{k}$, where $a > 0$. Because $\mathbf{F}_1 + \mathbf{F}_2 + \mathbf{F}_3 = \mathbf{F}$, we have

$$\begin{aligned}\mathbf{F}_3 = \mathbf{F} - \mathbf{F}_1 - \mathbf{F}_2 &= a\mathbf{k} - (-3\mathbf{i} + \mathbf{j} + 0.3\mathbf{k}) - (\mathbf{i} + 2\mathbf{j} + 4\mathbf{k}) \\ &= 2\mathbf{i} - 3\mathbf{j} + (a - 4.3)\mathbf{k}.\end{aligned}$$

Because $a > 0$, $a - 4.3 > -4.3$. Hence, $\mathbf{F}_3$ can be any force of the form $\mathbf{F}_3 = 2\mathbf{i} - 3\mathbf{j} + b\mathbf{k}$ kN, where $b > -4.3$.

Lines in Three Dimensions

Lines in three-space can be described parametrically in much the same way that we describe lines in the plane. Given a point $P_0 = (x_0, y_0, z_0)$ and a vector $\mathbf{v} = \langle a, b, c \rangle$, the line through P_0 and in the direction of $\mathbf{v}$ is the set of all points $P = (x, y, z)$ for which $\overrightarrow{P_0P}$ is parallel to $\mathbf{v}$. Thus, a point P is on the line if and only if there is a real number t with

$$\overrightarrow{P_0P} = t\mathbf{v}.$$

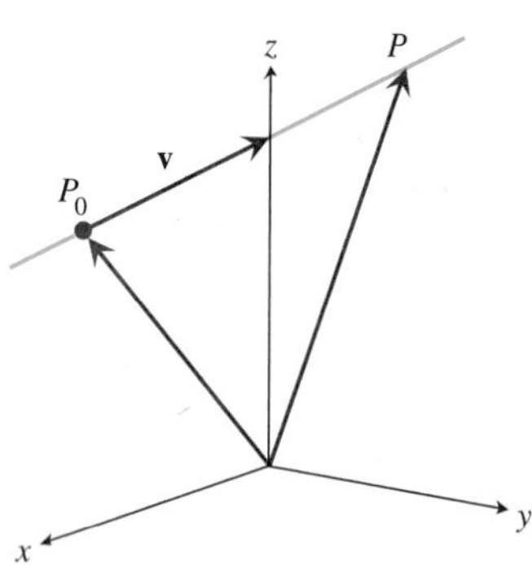

FIGURE 8.9 A line in space is determined by a point P_0 on the line and a vector $\mathbf{v}$ parallel to the line.

This line is illustrated in Fig. 8.9. Recall that $\overrightarrow{P_0P} = \mathbf{P} - \mathbf{P}_0$, where $\mathbf{P}$ and $\mathbf{P}_0$ are position vectors based at the origin. Solving the equation $\overrightarrow{P_0P} = \mathbf{P} - \mathbf{P}_0 = t\mathbf{v}$ for $\mathbf{P}$, we obtain

$$\mathbf{P} = \mathbf{P}(t) = \mathbf{P}_0 + t\mathbf{v}, \; -\infty < t < \infty. \qquad (5)$$

Equation (5) is a parametric or vector equation for a line in space.

Parametric or Vector Equation for a Line

Let $P_0 = (x_0, y_0, z_0)$ be a given point and $\mathbf{v} = \langle a, b, c\rangle$ be a nonzero vector. The line through P_0 and in the direction of $\mathbf{v}$ is described by the vector equation

$$\mathbf{P} = \mathbf{P}(t) = \mathbf{P}_0 + t\mathbf{v}. \tag{6}$$

If we let $\mathbf{P}(t) = \langle x(t), y(t), z(t)\rangle$, then we can rewrite (6) as

$$\mathbf{P}(t) = \langle x(t), y(t), z(t)\rangle = \langle x_0 + at, y_0 + bt, z_0 + ct\rangle. \tag{7}$$

In (6) we have a formula for the position vector $\mathbf{P}(t)$ of a point on the line. When $\mathbf{P}(t)$ is resolved into components, as in (7), we see the coordinates of such a point. Thus, we interpret the ordered triple (7) in two ways: as giving the coordinates $(x_0 + at, y_0 + bt, z_0 + ct)$ of a point on the line or as a position vector from the origin to that point on the line. The first interpretation might be more useful when we work with the line as a geometric object made up of points, but the latter interpretation may be better when the line is the path of a particle moving in space. To remind ourselves of this "motion connection" and to be consistent with notation established in earlier chapters, we will usually use $\mathbf{r}(t)$ in place of $\mathbf{P}(t)$ in (6) and (7).

EXAMPLE 5 Find an equation for the line ℓ through the points $P = (-2, 3, 3)$ and $Q = (4, 0, -1)$. Sketch the line. Is the point $(2, 4, 5)$ on ℓ? How about the point $(-20, 12, 15)$?

Solution

Because P and Q are both on ℓ, the vector

$$\overrightarrow{PQ} = \mathbf{Q} - \mathbf{P} = \langle 4, 0, -1\rangle - \langle -2, 3, 3\rangle = \langle 6, -3, -4\rangle$$

is parallel to ℓ. Hence, we can take $\mathbf{v} = \overrightarrow{PQ}$ in (6). Take P as the given point on the line and let $\mathbf{r}(t)$ denote the position vector of an arbitrary point on ℓ. Then

$$\mathbf{r}(t) = \mathbf{P} + t\left(\overrightarrow{PQ}\right) = \langle -2 + 6t, 3 - 3t, 3 - 4t\rangle. \tag{8}$$

To produce a sketch of the line, simply plot the points P and Q and draw the line determined by these points. See Fig. 8.10.

Now check to see if $(2,4,5)$ is on ℓ. If the first coordinate of a point on the line is 2, then we must have $-2 + 6t = 2$ for some t. This is only possible if $t = 2/3$. However,

$$\mathbf{r}\left(\frac{2}{3}\right) = \left\langle 2, 1, \frac{1}{3}\right\rangle \neq \langle 2, 4, 5\rangle.$$

Thus $(2, 4, 5)$ is not on ℓ. Next we check to see if $(-20, 12, 15)$ is on the line. The first coordinate in (8) is -20 only if $-2 + 6t = -20$, that is, only if $t = -3$. Because

$$\mathbf{r}(-3) = \langle -2 + 6(-3), 3 - 3(-3), 3 - 4(-3)\rangle = \langle -20, 12, 15\rangle,$$

we see that $(-20, 12, 15)$ is on ℓ.

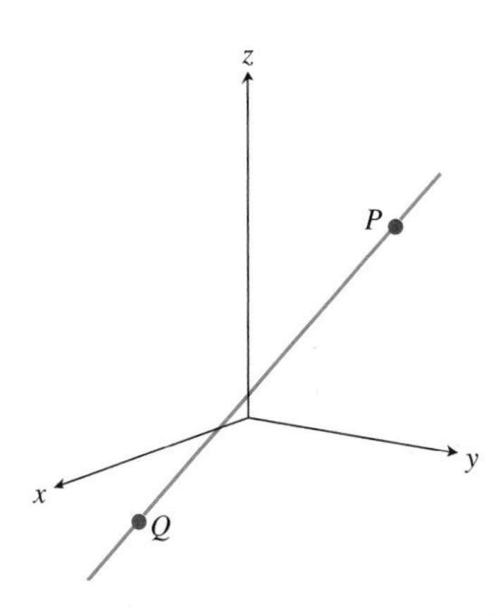

FIGURE 8.10 The line containing P and Q.

EXAMPLE 6 The equations

$$\mathbf{r}_1(t) = \langle 4 + 2t, -3 - t, 5 + 4t \rangle$$

and

$$\mathbf{r}_2(t) = \langle -2 - 6t, 3t, -7 - 12t \rangle$$

are both equations for a line. Show that these equations describe the same line.

Solution

The first equation describes a line ℓ_1 parallel to the vector $\mathbf{v}_1 = \langle 2, -1, 4 \rangle$, while the second equation describes a line ℓ_2 parallel to $\mathbf{v}_2 = \langle -6, 3, -12 \rangle$. Because $\mathbf{v}_2 = -3\mathbf{v}_1$, the vectors $\mathbf{v}_1$ and $\mathbf{v}_2$ are parallel. Hence lines ℓ_1 and ℓ_2 are either distinct parallel lines or are the same line. If the two lines have a point in common, it follows that they are the same line. Setting $t = 0$ in the equation for ℓ_2 gives $\mathbf{r}_2(0) = \langle -2, 0, -7 \rangle$. This point is on line ℓ_2. As in the previous example, it is easy to check that this point is also on ℓ_1; indeed, $\mathbf{r}_1(-3) = \langle -2, 0, -7 \rangle$. Thus ℓ_1 and ℓ_2 are the same line.

The Cartesian Equation for a Line Given the parametric form for a line,

$$\mathbf{r}(t) = \langle x, y, z \rangle = \langle x(t), y(t), z(t) \rangle = \langle x_0 + at, y_0 + bt, z_0 + ct \rangle,$$

we can produce a point on the line by replacing t by any real number. Conversely, if (x, y, z) is a point on the line, then there is a real number t such that

$$\begin{aligned} x &= x_0 + at \\ y &= y_0 + bt \\ z &= z_0 + ct. \end{aligned} \tag{9}$$

Assume for the moment that none of a, b, and c is zero. Then we can solve each of the equations for t to get

$$\frac{x - x_0}{a} = t, \qquad \frac{y - y_0}{b} = t, \qquad \frac{z - z_0}{c} = t. \tag{10}$$

If (x, y, z) is on the line, then the t values produced in (10) are all the same. Hence we can write

$$\frac{x - x_0}{a} = \frac{y - y_0}{b} = \frac{z - z_0}{c}.$$

This is the *symmetric form* for the equation of the line. Any ordered triple (x, y, z) that satisfies this set of equations is a point on the line. If, say, the number $c = 0$ in (9), then we can still solve the first two equations for t. The third equation in (9) is then $z = z_0$ and tells us that every point on the line has third coordinate z_0. In this case, the symmetric form is given by

$$\frac{x - x_0}{a} = \frac{y - y_0}{b}, \qquad z = z_0.$$

EXAMPLE 7 Consider the lines described by the equations

$$\frac{x-3}{2} = \frac{y+1}{-4} = \frac{z}{3} \tag{11}$$

and

$$\frac{x+5}{4} = y + 12 = \frac{z - 9/2}{1/2}. \tag{12}$$

Sketch the lines. Decide whether or not the lines intersect. If they intersect, find the point of intersection.

Solution

The line ℓ_1 described by (11) contains the point $(3, -1, 0)$ (we set each fraction to $t = 0$) and the point $(5, -5, 3)$ (we set each fraction to $t = 1$). We can sketch this line by plotting these points and drawing the line determined by these points. For the line ℓ_2 associated with Equation (12), we set the fractions $t = 1$ and then $t = 2$. In this way we find the ℓ_2 contains the points $(-1, -11, 5)$ and $(3, -10, 11/2)$. The graph is shown in Fig. 8.11. It is impossible to tell from the graph whether or not the lines intersect. Why?

If ℓ_1 and ℓ_2 intersect, then there is one ordered triple (x, y, z) that satisfies both (11) and (12). The x and y values must then satisfy the system of equations

$$\frac{x-3}{2} = \frac{y+1}{-4}$$

$$\frac{x+5}{4} = y + 12.$$

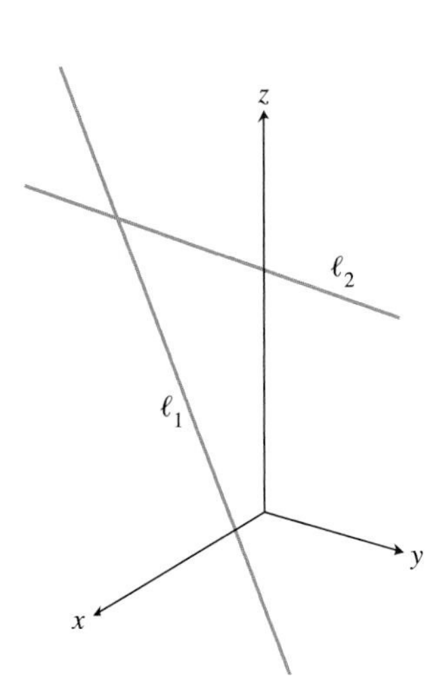

FIGURE 8.11 Do lines ℓ_1 and ℓ_2 intersect?

With some algebra, this system becomes

$$4x + 2y = 10$$

$$x - 4y = 43.$$

The solution to this system is $x = 7$, $y = -9$. We now know that *if* there is an intersection point, then its first coordinate is $x = 7$ and its second coordinate is $y = -9$. Putting $x = 7$ and $y = -9$ into (11), we obtain $z = 6$. When these x and y values are put into (12), we again get $z = 6$. Thus the ordered triple $(7, -9, 6)$ satisfies both (11) and (12). This means that the two lines have a point in common, and this intersection point is $(7, -9, 6)$.

Exercises 8.1

Exercises 1–10: Plot or illustrate the following in R^3.

1. The point $(2, 5, 4)$
2. The point $(2, 5, -4)$
3. The point $(-1, -2, -3)$
4. The point $(3, -1, 4)$
5. The vector $\mathbf{v} = \langle 6, 0, 3\rangle$
6. The vector $\mathbf{v} = \langle -2, -1, 5\rangle$
7. The vector $\overrightarrow{PQ}$ where $P = (-4, 3, 3)$ and $Q = (1, 0, 0)$

8. The vector $\overrightarrow{PQ}$ where $P = (2, 1, 3)$ and $Q = (3, 1, -2)$

9. The line through the points $P = (2, 0, 5)$ and $Q = (0, 4, 4)$

10. The line through the points $P = (-10, 10, -20)$ and $Q = (30, 40, 20)$

Exercises 11–22: Let $\mathbf{u} = \langle 2, -1, 1\rangle$, $\mathbf{v} = \langle -5, 0, 3\rangle$, *and* $\mathbf{w} = \langle 4, 3, 2\rangle$. *Find the following.*

11. The length of $\mathbf{v}$

12. The length of $-\mathbf{w}$

13. $\mathbf{u} - 2\mathbf{v}$

14. $\mathbf{v} + 3\mathbf{w} - 2\mathbf{u}$

15. $\mathbf{w} \cdot \mathbf{u}$

16. $\mathbf{v} \cdot \mathbf{w}$

17. A vector of length 1 in the direction of $\mathbf{v}$

18. A vector of length 1 in the direction opposite to $\mathbf{w}$

19. The angle between $\mathbf{u}$ and $\mathbf{w}$

20. The angle between $\mathbf{u}$ and $\mathbf{v}$

21. The projection of $\mathbf{v}$ onto a unit vector in the direction of $\mathbf{w}$

22. The projection of $\mathbf{w}$ onto a unit vector in the direction of $\mathbf{v}$

Exercises 23–36: Let $\mathbf{u} = \mathbf{i} - 2\mathbf{j} + 2\mathbf{k}$, $\mathbf{v} = -3\mathbf{i} + 8\mathbf{j} - \frac{1}{2}\mathbf{k}$, *and* $\mathbf{w} = \mathbf{k}$. *Find the following.*

23. The length of $\mathbf{v}$

24. The length of $-\mathbf{w}$

25. $\mathbf{u} - 2\mathbf{v}$

26. $\mathbf{v} + 3\mathbf{w} - 2\mathbf{u}$

27. $\mathbf{w} \cdot \mathbf{u}$

28. $\mathbf{v} \cdot \mathbf{w}$

29. A vector of length 1 in the direction of $\mathbf{v}$

30. A vector of length 1 in the direction opposite to $\mathbf{w}$

31. The angle between $\mathbf{u}$ and $\mathbf{w}$

32. The angle between $\mathbf{u}$ and $\mathbf{v}$

33. The projection of $\mathbf{v}$ onto a unit vector in the direction of $\mathbf{w}$

34. The projection of $\mathbf{w}$ onto a unit vector in the direction of $\mathbf{v}$

35. Three space shuttles tow a small asteroid. The first exerts a force $\mathbf{F}_1 = 2000\mathbf{i}$ kN, the second a force of $\mathbf{F}_2 = 400\mathbf{i} - 3000\mathbf{j} + 1000\mathbf{k}$ kN, and the third a force of $\mathbf{F}_3 = -300\mathbf{i} + 2500\mathbf{j} - 2500\mathbf{k}$ kN. What is the net (resultant) force on the asteroid?

36. A tug, a submarine, and a helicopter tow a barge. The submarine pulls with a force of $\mathbf{F}_s = \langle 800, 0, -150\rangle$ kN and the tug with a force of $\mathbf{F}_t = \langle 300, 150, 0\rangle$ kN. If the barge is to be towed in the $\mathbf{i}$ direction, what can be said about the force exerted by the helicopter?

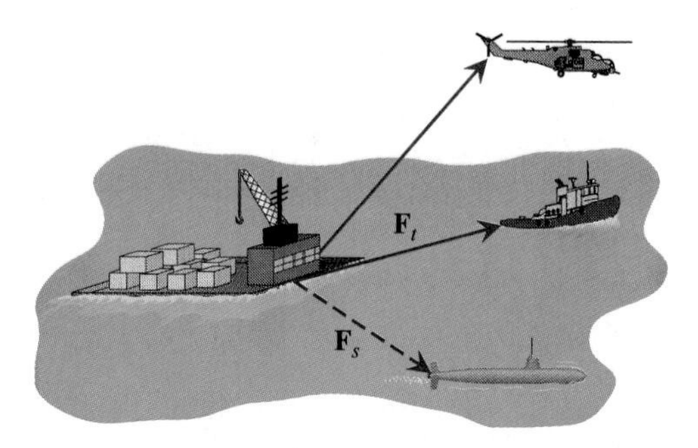

Figure for Exercise 36.

37. A traffic light with a mass of 20 kg hangs from three wires, each of length 15 meters. The wires are at right angles to each other, the light hangs below the point where the three wires come together, and the light is in a position symmetric to the wires. See the accompanying figure. What is the magnitude of the tension (force) on each wire?

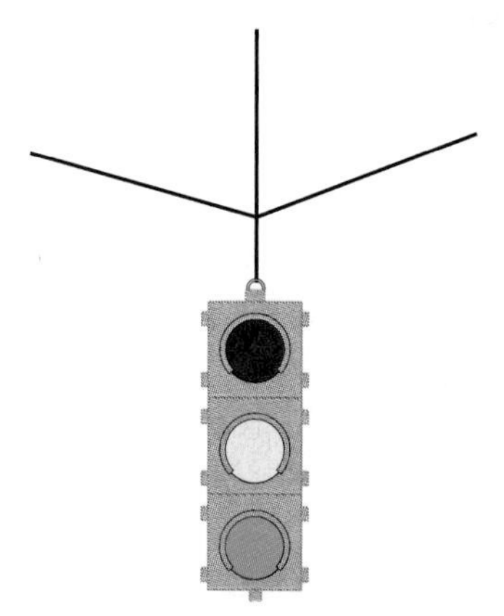

Figure for Exercise 37.

38. Three campers want to cache their food in a tree. They pack the food in a bundle on the ground, attach three ropes, and throw each rope over a tree branch. Then each camper pulls one rope. At the start of the pulling process, the first rope pulls on the bundle with force of magnitude a in direction $\langle -1/\sqrt{3}, 1/\sqrt{3}, 1/\sqrt{3}\rangle$; the second, with force of magnitude b in the direction $\langle 1/\sqrt{2}, 0, 1/\sqrt{2}\rangle$; the third, with force of magnitude c in the direction $\langle -1/\sqrt{14}, -2/\sqrt{14}, 3/\sqrt{14}\rangle$. Find values of a, b, and c (all positive) so that initially the food cache is pulled straight up, that is, in the $\mathbf{k}$ direction.

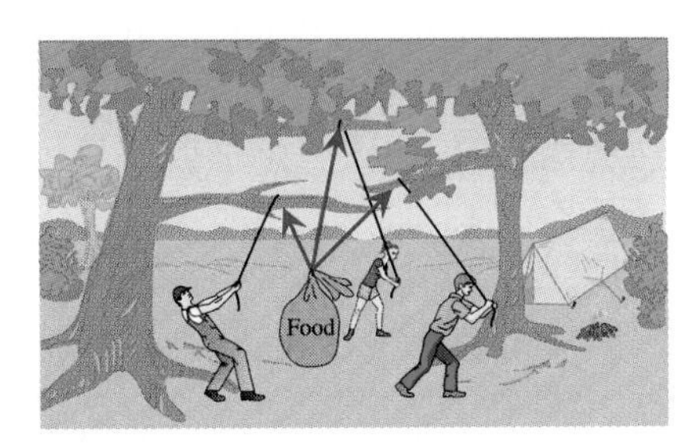

Figure for Exercise 38.

39. Find the parametric and symmetric equations for the line through $P = (-3, 5, 0)$ and in the direction of $\mathbf{v} = \langle -2, 4, 5 \rangle$. Sketch the line.

40. Find the parametric and symmetric equations for the line through $P = (2, 1, -4)$ and in the direction of $\mathbf{v} = \mathbf{k}$. Sketch the line.

41. Find the parametric and symmetric equations for the line through $P = (101, 201, 301)$ and in the direction of $\mathbf{v} = \langle 101, 201, 301 \rangle$. Sketch the line.

42. Find the parametric and symmetric equations for the line through $P = (-3, 5, 9)$ and in the direction of $\mathbf{v} = -2\mathbf{i} + 7\mathbf{j}$. Sketch the line.

43. Find the parametric and symmetric equations for the line through the points $P = (1, -1, 2)$ and $Q = (-3, 5, 5)$. Sketch the line.

44. Find the parametric and symmetric equations for the line through the points $P = (1, 0, 0)$ and $Q = (0, 1, 0)$. Sketch the line.

45. Is the point $A = (-2, 3, 1)$ on the line with equation $\mathbf{r}(t) = \langle -t, 4t - 1, t + 1 \rangle$? Is the point $B = (-5, 19, 6)$ on this line? Provide a sketch.

46. Is the point $A = (-12, -20, -8)$ on the line through the points $P = (-3, 1, 1)$ and $Q = (0, 8, 4)$? Is the point $B = (12, 36, -16)$ on this line? Provide a sketch.

47. Let P and Q be two points in space. Show that the midpoint of $\overline{PQ}$ is on the line through P and Q.

48. Let ℓ_1 be the line described by

$$\mathbf{r}_1(t) = \langle -3 + 2t, 4 + t, 2 - 5t \rangle$$

and ℓ_2 the line described by

$$\mathbf{r}_2(t) = \langle 4t, 5 - 3t, 1 + t \rangle.$$

Are these the same line or different lines? Give reasons for your answer.

49. Let ℓ_1 be the line described by

$$\mathbf{r}_1(t) = \langle 4 - t, t, 1 + 3t \rangle$$

and ℓ_2 the line described by

$$\mathbf{r}_2(t) = \langle 2 + t, 2 - t, 7 - 3t \rangle.$$

Are these the same line or different lines? Give reasons for your answer.

50. Let ℓ_1 be the line described by

$$\mathbf{r}_1(t) = \langle 6 + 3t, 3 + 2t, 2 - 3t \rangle$$

and ℓ_2 the line described by

$$\mathbf{r}_2(t) = \langle 18 + 9t, 9 + 6t, 6 - 9t \rangle.$$

Are these the same line or different lines? Give reasons for your answer.

51. The line ℓ has parametric representation

$$\mathbf{r}(t) = \langle 4 + 2t, -5t, 7 \rangle.$$

Find two other parametric representations for ℓ.

52. The line ℓ has parametric representation

$$\mathbf{r}(t) = \langle 6 - 5t, 3 + 6t, -8 - 9t \rangle.$$

Find two other parametric representations for ℓ.

53. Let ℓ_1 be the line with equation

$$\mathbf{r}(t) = \langle -3 + t, 2t, -5t + 3 \rangle$$

and ℓ_2 be the line with equation

$$\mathbf{r}(t) = \langle 2t, -4t + 1, t \rangle.$$

Do ℓ_1 and ℓ_2 intersect? *Caution:* If the lines intersect, the intersection point may occur for one value of t for ℓ_1 and a different value of t for ℓ_2. How can you allow for this?

54. Let ℓ_1 be the line with equation

$$\mathbf{r}(t) = \langle 1 + t, 2 + t, 3 + t \rangle$$

and ℓ_2 be the line with equation

$$\mathbf{r}(t) = \langle -1 - 3t, 8 + t, 7 \rangle.$$

Do ℓ_1 and ℓ_2 intersect? *Caution:* If the lines intersect, the intersection point may occur for one value of t for ℓ_1 and a different value of t for ℓ_2. How can you allow for this?

55. Let ℓ_1 be the line with equation

$$\mathbf{r}(t) = \langle -3t + 6, 4t, -t + 7 \rangle$$

and ℓ_2 be the line with equation

$$\mathbf{r}(t) = \langle -t - 9, t + 18, -t - 2 \rangle.$$

Do ℓ_1 and ℓ_2 intersect?

56. Let ℓ_1 be the line with equation

$$\frac{x - 3}{2} = \frac{y + 3}{-2} = \frac{z + 1}{5}$$

and ℓ_2 be the line with equation

$$x = y = z.$$

Do ℓ_1 and ℓ_2 intersect?

57. Let ℓ_1 be the line with equation

$$\frac{x-4}{3} = \frac{y+7}{2}, \qquad z = 3$$

and ℓ_2 be the line with equation

$$-x = \frac{y+19}{4} = \frac{z-7}{-2}.$$

Do ℓ_1 and ℓ_2 intersect?

58. Let ℓ_1 be the line with equation

$$\frac{x-3/2}{1/2} = \frac{y-4}{-1/3} = \frac{z+1}{1/5}$$

and ℓ_2 be the line with equation

$$\frac{x-1}{1/4} = \frac{y-1/3}{-2/3} = \frac{z-14}{2}.$$

Do ℓ_1 and ℓ_2 intersect?

59. Let ℓ_1 be the line with equation

$$\frac{x+4}{3} = \frac{z}{-2}, \qquad y = 4$$

and ℓ_2 be the line with equation

$$\frac{x-2}{2/3} = \frac{z+4}{4}, \qquad y = -1.$$

Do ℓ_1 and ℓ_2 intersect?

60. For a nonzero vector $\mathbf{v} = \langle v_1, v_2, v_3 \rangle$, let θ_x be the angle between $\mathbf{v}$ and the positive x-axis, θ_y the angle between $\mathbf{v}$ and the positive y-axis, and θ_z the angle between $\mathbf{v}$ and the positive z-axis. See the accompanying figure. The numbers $\cos\,\theta_x$, $\cos\,\theta_y$, and $\cos\,\theta_z$ are often called the *direction cosines* for $\mathbf{v}$.

a. Use the dot product to show that $v_1 = \|\mathbf{v}\| \cos\,\theta_x$, $v_2 = \|\mathbf{v}\| \cos\,\theta_y$, and $v_3 = \|\mathbf{v}\| \cos\,\theta_z$.

b. Show that $\cos^2\theta_x + \cos^2\theta_y + \cos^2\theta_z = 1$.

c. Show that the unit vector in the direction of $\mathbf{v}$ is $\mathbf{u} = \langle \cos\,\theta_x, \cos\,\theta_y, \cos\,\theta_z \rangle$.

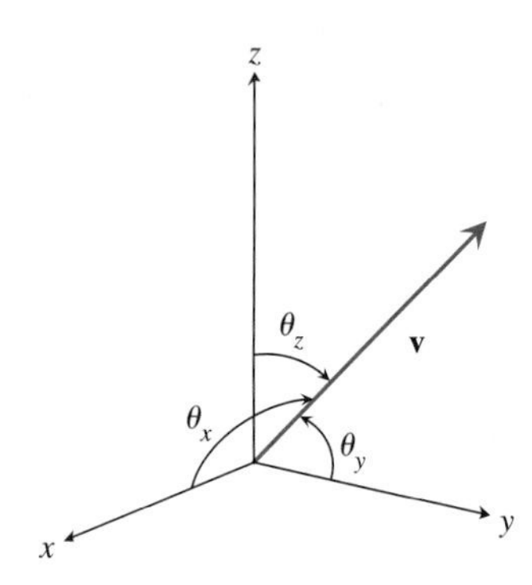

Figure for Exercise 60.

61. Show that the line segment joining points P and Q is described by

$$\mathbf{r}(t) = (1-t)\mathbf{P} + t\mathbf{Q},$$

for $0 \le t \le 1$.

62. Find a parametric equation for the segment joining the points $(-3, 5, 7)$ and $(-2, 0, 10)$.

63. Find a parametric equation for the segment joining the points $(0, 4, -4)$ and $(7, 9, -5)$.

8.2 Matrices and Determinants

According to the *1994 Statistical Abstract of the United States,* the number of school-age (under 20) people in the United States in 1990 was 71,715,000, while the number of people past school age (20 and over) was 176,995,000. The study also provides the following information about changes in the population in the course of a year:

- Ninety-nine percent of the 20-and-over population survive the year.
- Among the under-20 population, 5.5 percent turn 20 each year.
- The 20-and-over population produces children at a rate of 2.3 per 100 individuals; by comparison, the birth rate for the under-20 population is negligible.
- Ninety-four percent of the under-20 population are alive and still under 20 at the end of the year.

We can use this information about population changes to make predictions about the populations in subsequent years. Let x_0 be the 1990 population of the

20-and-over group and y_0 the 1990 population of the under-20 group. Let x_1 and y_1 be the populations of these groups in 1991. According to the information from the *Statistical Abstract,*

$$\begin{aligned} x_1 &= 0.99x_0 + 0.055y_0 \\ y_1 &= 0.023x_0 + 0.94y_0. \end{aligned} \tag{1}$$

The equations (1) can be written in the form

$$\begin{pmatrix} x_1 \\ y_1 \end{pmatrix} = \begin{pmatrix} 0.99x_0 + 0.055y_0 \\ 0.023x_0 + 0.94y_0 \end{pmatrix} = \begin{pmatrix} 0.99 & 0.055 \\ 0.023 & 0.94 \end{pmatrix} \begin{pmatrix} x_0 \\ y_0 \end{pmatrix}. \tag{2}$$

The right-hand expression in (2) has several advantages. It clearly displays both the input information

$$\begin{pmatrix} x_0 \\ y_0 \end{pmatrix}$$

and an array of four constants that tells us how the populations are changing. By interpreting (2) as a product, it is a simple matter to find estimates for the populations one, two, three, or more years in the future.

Matrices

An expression such as

$$U = \begin{pmatrix} x_0 \\ y_0 \end{pmatrix} \quad \text{or} \quad V = \begin{pmatrix} 0.99 & 0.055 \\ 0.023 & 0.94 \end{pmatrix}$$

is called a *matrix.* The matrix U has two rows and one column; V has two rows and two columns. **Rows** are horizontal and **columns** are vertical. Matrices will play an important role in our work with functions of several variables. We will use matrices to describe linear functions, to approximate functions of several variables, and to facilitate change of variables in multiple integrals.

DEFINITION Matrix

A **matrix** is a rectangular array of numbers.

- A matrix with m rows and n columns is called an *m by n matrix,* or an $m \times n$ matrix. We say the matrix is of **dimension** $m \times n$.
- We say that the entry in row i and column j of the matrix is in the i, j-position of the matrix.
- A matrix with just one row is called a **row vector.**
- A matrix with just one column is called a **column vector.**

We use capital letters to denote matrices. If A is a matrix with m rows and n columns, we write $A_{m\times n}$ when it is important to remember the dimension of A. We will denote the entry in the i, j-position of matrix A by a_{ij}. We will work only with matrices containing three or fewer rows and three or fewer columns, but most of what we say will also apply to matrices with larger dimensions.

EXAMPLE 1 Let

$$B = \begin{pmatrix} -1 & \sqrt{2} \\ 0 & 0 \\ 23 & x \end{pmatrix}.$$

a) What is the dimension of B?
b) What is b_{11}? b_{32}? b_{23}?
c) What are the positions of the two entries equal to 0?

Solution

a) Matrix B has 3 rows and 2 columns and is therefore of dimension 3×2. We express this by writing $B_{3\times 2}$.
b) Entry b_{11} is in the 1,1-position, that is, in row 1 and column 1. The entry in this position is -1, so $b_{11} = -1$. The entry in row 3 and column 2 is x, so $b_{32} = x$. Because B has only two columns, there is no 2, 3-position in the matrix. Hence b_{23} is undefined in this matrix.
c) The number 0 appears in two places in the matrix, in row 2, column 1, and in row 2, column 2. Hence $b_{21} = 0$ and $b_{22} = 0$.

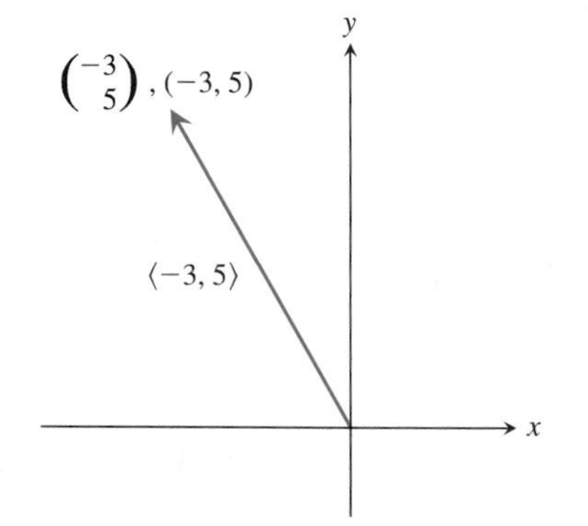

FIGURE 8.12 The row vector **v** and the column vector $\mathbf{v}_c$ are represented graphically by the same vector.

We shall be flexible with vector and matrix notation. You may recall that we have often referred to a point in space using either an ordered triple (x, y, z), or a position vector $\langle x, y, z\rangle$. In what follows, we sometimes interpret the ordered triple (x, y, z) as a row vector. Similarly, we will sometimes interpret the column vector

$$\begin{pmatrix} x \\ y \\ z \end{pmatrix}$$

as an ordered triple giving the position of a point in space, or as a position vector. Similar interpretations hold for row and column vectors of length two. Thus the vectors

$$\mathbf{v} = \langle -3, 5\rangle \quad \text{and} \quad \mathbf{v}_c = \begin{pmatrix} -3 \\ 5 \end{pmatrix}$$

may both represent the point with coordinates $(-3, 5)$ or the position vector to this point. See Fig. 8.12.

Matrix Operations

According to the definition of *matrix,* every vector is also a matrix. Hence it is not surprising that many of the operations defined for vectors extend naturally to operations on matrices.

DEFINITION Matrix Operations

Let A and B be matrices and let k be a real number.

a) Matrices A and B are **equal** if A and B are of the same dimension and $a_{ij} = b_{ij}$ for all i, j.

b) The **sum** of A and B is defined when A and B are of the same dimension. In this case the sum $A + B$ is the matrix C of the same dimension as each of A and B and

$$c_{ij} = a_{ij} + b_{ij} \quad \text{for all } i, j.$$

In other words, simply add elements of A and B that are in like positions. We write $C = A + B$.

c) The **product** of the real number k and a matrix A is the matrix D of the same dimension as A with

$$d_{ij} = ka_{ij} \quad \text{for all } i, j.$$

In other words, simply multiply each element of A by k. We write $D = kA$.

EXAMPLE 2 Let

$$A = \begin{pmatrix} -4 & 3 \\ 2 & 6 \\ 0 & 9 \end{pmatrix}, \quad B = \begin{pmatrix} -2 & 7 & a \\ 4 & 3 & 3 \end{pmatrix}, \quad \text{and} \quad C = \begin{pmatrix} 1 & 1 \\ -2 & 6 \\ 1 & 3 \end{pmatrix}.$$

Find $A + B$, $A + C$, and $-2B$.

Solution

Because $A = A_{3\times2}$ and $B = B_{2\times3}$ are of different dimensions, $A + B$ is not defined.

Because A and C are both of dimension 3×2, the sum $A + C$ is defined. By the definition of *matrix sum,*

$$A + C = \begin{pmatrix} -4 & 3 \\ 2 & 6 \\ 0 & 9 \end{pmatrix} + \begin{pmatrix} 1 & 1 \\ -2 & 6 \\ 1 & 3 \end{pmatrix} = \begin{pmatrix} -4+1 & 3+1 \\ 2+(-2) & 6+6 \\ 0+1 & 9+3 \end{pmatrix} = \begin{pmatrix} -3 & 4 \\ 0 & 12 \\ 1 & 12 \end{pmatrix}.$$

To calculate $(-2)B$, multiply each element of B by -2:

$$(-2)B = -2\begin{pmatrix} -2 & 7 & a \\ 4 & 3 & 3 \end{pmatrix} = \begin{pmatrix} 4 & -14 & -2a \\ -8 & -6 & -6 \end{pmatrix}.$$

Matrix Multiplication In (2), we wrote

$$\begin{pmatrix} x_1 \\ y_1 \end{pmatrix} = \begin{pmatrix} 0.99x_0 + 0.055y_0 \\ 0.023x_0 + 0.94y_0 \end{pmatrix} = \begin{pmatrix} 0.99 & 0.055 \\ 0.023 & 0.94 \end{pmatrix} \begin{pmatrix} x_0 \\ y_0 \end{pmatrix}.$$

The notation on the right was introduced to emphasize that we are given input in the form of a (column) vector

$$\mathbf{v} = \begin{pmatrix} x_0 \\ y_0 \end{pmatrix}$$

and that this input is "acted on" by the 2×2 matrix

$$A = \begin{pmatrix} 0.99 & 0.055 \\ 0.023 & 0.94 \end{pmatrix} \text{ to give } \begin{pmatrix} 0.99x_0 + 0.55y_0 \\ 0.023x_0 + 0.94y_0 \end{pmatrix}.$$

to give the output on the right side of (2). As we shall see, this output is the product of the matrix A and the column vector $\mathbf{v}$. The fundamental operation in calculating the product of two matrices is the product of a row vector and a column vector.

DEFINITION The Product of a Row Vector and a Column Vector

Let $\mathbf{r}$ be a $1 \times n$ row vector and $\mathbf{c}$ be an $n \times 1$ column vector. The product $\mathbf{rc}$ is defined by

$$\begin{aligned} \mathbf{rc} &= (r_1, r_2, \ldots, r_n) \begin{pmatrix} c_1 \\ c_2 \\ \vdots \\ c_n \end{pmatrix} \\ &= \langle r_1, r_2, \ldots, r_n \rangle \cdot \langle c_1, c_2, \ldots, c_n \rangle \\ &= r_1c_1 + r_2c_2 + \cdots + r_nc_n, \end{aligned} \tag{3}$$

where the product in the second line of (3) is a dot product. If the vectors $\mathbf{r}$ and $\mathbf{c}$ have different lengths, then the product $\mathbf{rc}$ is not defined.

For clarity we write a row vector $(r_1\, r_2 \ldots r_n)$ as $(r_1, r_2, \ldots, r_n)$.

EXAMPLE 3 Let

$$\mathbf{r} = (-2, 5, 7) \quad \text{and} \quad \mathbf{c} = \begin{pmatrix} 12 \\ 2 \\ 0 \end{pmatrix}.$$

Calculate the product $\mathbf{rc}$.

Solution

Using (3),

$$\begin{aligned} \mathbf{rc} &= (-2, 5, 7) \begin{pmatrix} 12 \\ 2 \\ 0 \end{pmatrix} \\ &= \langle -2, 5, 7 \rangle \cdot \langle 12, 2, 0 \rangle = (-2)12 + 5 \cdot 2 + 7 \cdot 0 = -14. \end{aligned}$$

To find the product of two matrices, we calculate several row vector/column vector products.

DEFINITION Matrix Product

Let A be an $m \times p$ matrix and B be a $p \times n$ matrix (so the number of columns of A is equal to the number of rows of B). Let $\mathbf{r}_1, \mathbf{r}_2, \ldots, \mathbf{r}_m$ be the rows of A and $\mathbf{c}_1, \mathbf{c}_2, \ldots, \mathbf{c}_n$ be the columns of B. The matrix product AB is the $m \times n$ matrix P defined by

$$P = AB = \begin{pmatrix} \mathbf{r}_1\mathbf{c}_1 & \mathbf{r}_1\mathbf{c}_2 & \cdots & \mathbf{r}_1\mathbf{c}_n \\ \mathbf{r}_2\mathbf{c}_1 & \mathbf{r}_2\mathbf{c}_2 & \cdots & \mathbf{r}_2\mathbf{c}_n \\ \vdots & \vdots & \ddots & \vdots \\ \mathbf{r}_m\mathbf{c}_1 & \mathbf{r}_m\mathbf{c}_2 & \cdots & \mathbf{r}_m\mathbf{c}_n \end{pmatrix}.$$

In particular, for $1 \le i \le m$ and $1 \le j \le n$, p_{ij} is the product of two vectors:

$$p_{ij} = \mathbf{r}_i\mathbf{c}_j = \langle a_{i1}, a_{i2}, \ldots, a_{ip} \rangle \cdot \langle b_{1j}, b_{2j}, \ldots, b_{pj} \rangle,$$

which is the product of the ith row of A and the jth column of B. See Fig. 8.13.

$$\begin{pmatrix} a_{11} & a_{12} & a_{13} \\ a_{21} & a_{22} & a_{23} \\ a_{31} & a_{32} & a_{33} \end{pmatrix} \begin{pmatrix} b_{11} & b_{12} \\ b_{21} & b_{22} \\ b_{31} & b_{32} \end{pmatrix} = \begin{pmatrix} p_{11} & p_{12} \\ p_{21} & p_{22} \\ p_{31} & p_{32} \end{pmatrix}$$

FIGURE 8.13 $p_{32} = \langle a_{31}, a_{32}, a_{33} \rangle \cdot \langle b_{12}, b_{22}, b_{32} \rangle = a_{31}b_{12} + a_{32}b_{22} + a_{33}b_{32}$.

EXAMPLE 4 In (2), we wrote

$$\begin{pmatrix} 0.99x_0 + 0.055y_0 \\ 0.023x_0 + 0.94y_0 \end{pmatrix} = \begin{pmatrix} 0.99 & 0.055 \\ 0.023 & 0.94 \end{pmatrix} \begin{pmatrix} x_0 \\ y_0 \end{pmatrix}.$$

Show that this multiplication is correct.

Solution

Let

$$A = \begin{pmatrix} 0.99 & 0.055 \\ 0.023 & 0.94 \end{pmatrix} \quad \text{and} \quad \mathbf{v} = \begin{pmatrix} x_0 \\ y_0 \end{pmatrix}.$$

Because A is a 2×2 matrix and $\mathbf{v}$ is a 2×1 matrix, the number of columns of A is equal to the number of rows of $\mathbf{v}$. Hence, the product $A\mathbf{v}$ is defined and the result is a 2×1 matrix. By the definition of matrix product, we have

$$A\mathbf{v} = \begin{pmatrix} 0.99 & 0.055 \\ 0.023 & 0.94 \end{pmatrix} \begin{pmatrix} x_0 \\ y_0 \end{pmatrix} = \begin{pmatrix} \langle 0.99, 0.055 \rangle \cdot \langle x_0, y_0 \rangle \\ \langle 0.023, 0.94 \rangle \cdot \langle x_0, y_0 \rangle \end{pmatrix}$$

$$= \begin{pmatrix} 0.99x_0 + 0.055y_0 \\ 0.023x_0 + 0.94y_0 \end{pmatrix}.$$

This shows that the product displayed in (2) is correct.

EXAMPLE 5 Let

$$A = \begin{pmatrix} -3 & 4 \\ 1 & -2 \end{pmatrix} \quad \text{and} \quad B = \begin{pmatrix} 6 & 3 \\ -2 & 0 \\ 1 & -5 \end{pmatrix}.$$

Calculate the matrix products AB and BA.

Solution

First consider AB. The left matrix in the product, A, has two columns, and the right matrix, B, has three rows. Because

$$\text{number of columns of } A \neq \text{number of rows of } B,$$

the product AB is not defined.

Next consider BA. The left matrix in the product, B has two columns and the right matrix, A, has two rows. Because

$$\text{number of columns of } B = \text{number of rows of } A,$$

the product BA is defined. Because B has three rows and A has two columns, the product $P = BA$ will be a 3×2 matrix. The six entries of $P_{3\times 2}$ are

$$\begin{aligned} p_{11} &= \mathbf{r}_1\mathbf{c}_1 = \langle 6, 3 \rangle \cdot \langle -3, 1 \rangle = -15 \\ p_{12} &= \mathbf{r}_1\mathbf{c}_2 = \langle 6, 3 \rangle \cdot \langle 4, -2 \rangle = 18 \\ p_{21} &= \mathbf{r}_2\mathbf{c}_1 = \langle -2, 0 \rangle \cdot \langle -3, 1 \rangle = 6 \\ p_{22} &= \mathbf{r}_2\mathbf{c}_2 = \langle -2, 0 \rangle \cdot \langle 4, -2 \rangle = -8 \\ p_{31} &= \mathbf{r}_3\mathbf{c}_1 = \langle 1, -5 \rangle \cdot \langle -3, 1 \rangle = -8 \\ p_{32} &= \mathbf{r}_3\mathbf{c}_2 = \langle 1, -5 \rangle \cdot \langle 4, -2 \rangle = 14. \end{aligned}$$

Thus

$$P = BA = \begin{pmatrix} 6 & 3 \\ -2 & 0 \\ 1 & -5 \end{pmatrix} \begin{pmatrix} -3 & 4 \\ 1 & -2 \end{pmatrix} = \begin{pmatrix} -15 & 18 \\ 6 & -8 \\ -8 & 14 \end{pmatrix}.$$

This example shows that matrix multiplication is not commutative; that is, AB need not equal BA. If one of these two products is defined, the other may not be. Even if both AB and BA are defined, the two products may not be equal. See Exercises 15 and 16. With this exception, matrix addition and multiplication are very much like addition and multiplication of real numbers. We list some of the properties that they have in common.

Properties of Matrices

Let A, B, and C be matrices and let k be a real number. For each of the following, assume that the operations are defined.

- Matrix addition is commutative and associative:
$$A + B = B + A \quad \text{and} \quad (A + B) + C = A + (B + C).$$
- Multiplication by a real number (scalar) distributes over matrix addition:
$$k(A + B) = kA + kB.$$
- Matrix multiplication is associative:
$$A(BC) = (AB)C.$$
- Scalar multiples may be factored out of matrix products:
$$k(AB) = (kA)B = A(kB).$$
- Matrix multiplication distributes over matrix addition:
$$A(B + C) = AB + AC$$
$$(B + C)A = BA + CA.$$

We will work only with matrices that have at most three rows and columns. For such matrices, any of these properties can be verified by calculating both sides of the given identity. Such calculations can be quickly done with a computer algebra system (CAS). As an example, we verify $A(BC) = (AB)C$ when A and B are both 2×2 matrices and C is a 2×1 matrix (or column vector). Let

$$A = \begin{pmatrix} a_{11} & a_{12} \\ a_{21} & a_{22} \end{pmatrix}, \quad B = \begin{pmatrix} b_{11} & b_{12} \\ b_{21} & b_{22} \end{pmatrix}, \quad \text{and} \quad C = \begin{pmatrix} x \\ y \end{pmatrix}.$$

We then have

$$\begin{aligned}(AB)C &= \left[\begin{pmatrix} a_{11} & a_{12} \\ a_{21} & a_{22} \end{pmatrix}\begin{pmatrix} b_{11} & b_{12} \\ b_{21} & b_{22} \end{pmatrix}\right]\begin{pmatrix} x \\ y \end{pmatrix} \\ &= \begin{pmatrix} a_{11}b_{11} + a_{12}b_{21} & a_{11}b_{12} + a_{12}b_{22} \\ a_{21}b_{11} + a_{22}b_{21} & a_{21}b_{12} + a_{22}b_{22} \end{pmatrix}\begin{pmatrix} x \\ y \end{pmatrix} \\ &= \begin{pmatrix} (a_{11}b_{11} + a_{12}b_{21})x + (a_{11}b_{12} + a_{12}b_{22})y \\ (a_{21}b_{11} + a_{22}b_{21})x + (a_{21}b_{12} + a_{22}b_{22})y \end{pmatrix}.\end{aligned} \tag{4}$$

Performing the multiplication BC first,

$$\begin{aligned}A(BC) &= \begin{pmatrix} a_{11} & a_{12} \\ a_{21} & a_{22} \end{pmatrix}\left[\begin{pmatrix} b_{11} & b_{12} \\ b_{21} & b_{22} \end{pmatrix}\begin{pmatrix} x \\ y \end{pmatrix}\right] \\ &= \begin{pmatrix} a_{11} & a_{12} \\ a_{21} & a_{22} \end{pmatrix}\begin{pmatrix} b_{11}x + b_{12}y \\ b_{21}x + b_{22}y \end{pmatrix} \\ &= \begin{pmatrix} a_{11}(b_{11}x + b_{12}y) + a_{12}(b_{21}x + b_{22}y) \\ a_{21}(b_{11}x + b_{12}y) + a_{22}(b_{21}x + b_{22}y) \end{pmatrix}.\end{aligned} \tag{5}$$

A quick check shows that the last matrices in (4) and (5) are equal. This means that $(AB)C = A(BC)$ for the given matrices A, B, and C.

In Exercises 33 and 34 we ask you to verify some of the other matrix properties.

Determinants

A matrix in which the number of rows is the same as the number of columns is called a *square matrix*. The determinant of a square matrix is a number associated with the matrix. The determinant has many applications to geometry and calculus in two and three dimensions. We will use determinants to calculate areas of triangles and parallelograms and volumes of parallelepipeds, in the calculation of cross products, and when we change variables in multiple integrals. Although the determinant can be defined for square matrices of all sizes, we work only with the determinants of 1×1, 2×2, and 3×3 matrices.

DEFINITION Determinant

Let A be a square matrix. The **determinant** of A is denoted by

$$|A| \quad \text{or} \quad \det(A).$$

The method for computing the determinant of A depends on the dimension of A.

- If $A = (a_{11})$ is a 1×1 matrix, then
$$\det(A) = a_{11}.$$
- If $A = \begin{pmatrix} a_{11} & a_{12} \\ a_{21} & a_{22} \end{pmatrix}$ is a 2×2 matrix, then
$$\det(A) = \det\begin{pmatrix} a_{11} & a_{12} \\ a_{21} & a_{22} \end{pmatrix} = \begin{vmatrix} a_{11} & a_{12} \\ a_{21} & a_{22} \end{vmatrix} = a_{11}a_{22} - a_{12}a_{21}.$$
- If $A = \begin{pmatrix} a_{11} & a_{12} & a_{13} \\ a_{21} & a_{22} & a_{23} \\ a_{31} & a_{32} & a_{33} \end{pmatrix}$ is a 3×3 matrix, then

$$\det(A) = \begin{vmatrix} a_{11} & a_{12} & a_{13} \\ a_{21} & a_{22} & a_{23} \\ a_{31} & a_{32} & a_{33} \end{vmatrix}$$

$$= a_{11}\begin{vmatrix} a_{22} & a_{23} \\ a_{32} & a_{33} \end{vmatrix} - a_{12}\begin{vmatrix} a_{21} & a_{23} \\ a_{31} & a_{33} \end{vmatrix} + a_{13}\begin{vmatrix} a_{21} & a_{22} \\ a_{31} & a_{32} \end{vmatrix} \tag{6}$$

$$= a_{11}(a_{22}a_{33} - a_{23}a_{32}) - a_{12}(a_{21}a_{33} - a_{23}a_{31}) + a_{13}(a_{21}a_{32} - a_{22}a_{31}). \tag{7}$$

When $A = (a_{11})$ is a 1×1 matrix, the notation $|a_{11}|$ for the determinant of the matrix A could easily be confused with the absolute value notation. To avoid this, we will always write $\det(a_{11})$ in the 1×1 case.

The determinant of a 3×3 matrix A is found by calculating the determinants of several 2×2 matrices made up from the entries of A. Looking closely at (6), we

$$\begin{pmatrix} a_{11} & a_{12} & a_{13} \\ a_{21} & a_{22} & a_{23} \\ a_{31} & a_{32} & a_{33} \end{pmatrix}$$

Eliminate the row and column containing a_{11}.

$$\begin{pmatrix} a_{11} & a_{12} & a_{13} \\ a_{21} & a_{22} & a_{23} \\ a_{31} & a_{32} & a_{33} \end{pmatrix}$$

Eliminate the row and column containing a_{12}.

$$\begin{pmatrix} a_{11} & a_{12} & a_{13} \\ a_{21} & a_{22} & a_{23} \\ a_{31} & a_{32} & a_{33} \end{pmatrix}$$

Eliminate the row and column containing a_{13}.

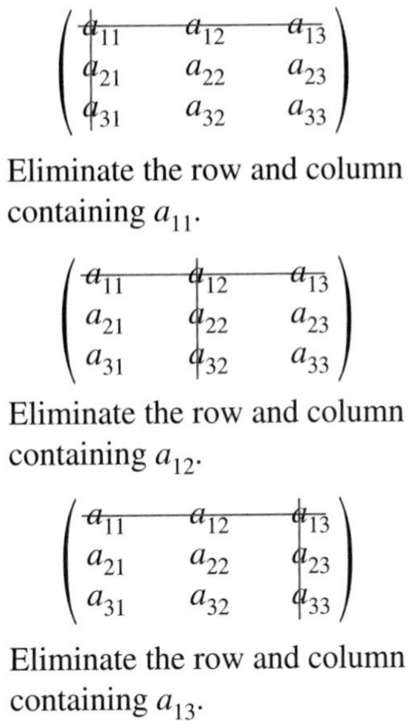

FIGURE 8.14 Evaluating the determinant of a 3×3 matrix.

see that the coefficients a_{11}, a_{12}, and a_{13} are entries in the first row of A and that the sign ($+1$ or -1) in front of a_{1j} is $(-1)^{1+j}$. This accounts for the alternating signs $+$, $-$, $+$ in the sum. The determinant multiplying a_{11} is the determinant of the 2×2 matrix that remains when the row and column containing a_{11} are eliminated from A. Similarly, the determinant multiplying a_{12} is the determinant of the 2×2 matrix that remains when the row and column containing a_{12} are eliminated from A. An analogous statement can be made about the determinant that multiplies a_{13}. See Fig. 8.14.

EXAMPLE 6 Find the determinants of

$$A = \begin{pmatrix} -3 & 5 \\ 4 & -7 \end{pmatrix} \quad \text{and} \quad B = \begin{pmatrix} -1 & 3 & 5 \\ 4 & -3 & 4 \\ 11 & 3 & 0 \end{pmatrix}.$$

Solution

Using the definition of the determinant of a 2×2 matrix, we have

$$\det(A) = |A| = \begin{vmatrix} -3 & 5 \\ 4 & -7 \end{vmatrix} = (-3)(-7) - 5 \cdot 4 = 21 - 20 = 1.$$

Using the preceding discussion and Fig. 8.14 as a guide, we have

$$\begin{aligned} \det(B) &= \begin{vmatrix} -1 & 3 & 5 \\ 4 & -3 & 4 \\ 11 & 3 & 0 \end{vmatrix} \\ &= (-1)^{1+1}(-1)\begin{vmatrix} -3 & 4 \\ 3 & 0 \end{vmatrix} + (-1)^{1+2}(3)\begin{vmatrix} 4 & 4 \\ 11 & 0 \end{vmatrix} \\ &\quad + (-1)^{1+3}(5)\begin{vmatrix} 4 & -3 \\ 11 & 3 \end{vmatrix} \\ &= -((-3)0 - 4 \cdot 3) - 3(4 \cdot 0 - 4 \cdot 11) + 5(4 \cdot 3 - (-3)11) \\ &= 369. \end{aligned}$$

In defining the determinant of a 3×3 matrix, we took the coefficients for the 2×2 determinants from the first row of the matrix. There is nothing special about the first row. The determinant can be evaluated by taking the coefficients from any row or column, multiplying each coefficient by the appropriate factor $(-1)^{i+j}$ and the appropriate 2×2 determinant, then adding the results. When the coefficients come from a particular row or column, we say that the determinant is *expanded* on that row or column. We demonstrate this by expanding B from Example 6 on the third column to evaluate the 3×3 matrix from the previous example.

EXAMPLE 7 Evaluate

$$\det\begin{pmatrix} -1 & 3 & 5 \\ 4 & -3 & 4 \\ 11 & 3 & 0 \end{pmatrix}$$

by expanding on the third column.

$$\begin{pmatrix} -1 & 3 & 5 \\ 4 & -3 & 4 \\ 11 & 3 & 0 \end{pmatrix}$$

Eliminate the row and column containing $a_{13} = 5$.

$$\begin{pmatrix} -1 & 3 & 5 \\ 4 & -3 & 4 \\ 11 & 3 & 0 \end{pmatrix}$$

Eliminate the row and column containing $a_{23} = 4$.

$$\begin{pmatrix} -1 & 3 & 5 \\ 4 & -3 & 4 \\ 11 & 3 & 0 \end{pmatrix}$$

Eliminate the row and column containing $a_{33} = 0$.

FIGURE 8.15 Evaluating the determinant of a 3×3 matrix.

Solution

The third column expansion is illustrated in Fig. 8.15. In this figure, we see the 2×2 matrices formed for each element of the third column, as we remove the row and column containing this element. The expansion is

$$\begin{vmatrix} -1 & 3 & 5 \\ 4 & -3 & 4 \\ 11 & 3 & 0 \end{vmatrix} = (-1)^{1+3}(5)\begin{vmatrix} 4 & -3 \\ 11 & 3 \end{vmatrix} + (-1)^{2+3}(4)\begin{vmatrix} -1 & 3 \\ 11 & 3 \end{vmatrix}$$
$$+ (-1)^{3+3}(0)\begin{vmatrix} -1 & 3 \\ 4 & -3 \end{vmatrix}$$
$$= 5(45) - (4)(-36) + 0 = 369.$$

Several useful properties of determinants can be used to simplify some work with determinants. A few of these properties are listed here.

Properties of Determinants

Let A be a square matrix.

a) Let B be a matrix obtained from A by interchanging any two rows (or any two columns) of A. Then

$$\det(B) = -\det(A).$$

b) Let k be a real number and let C be the matrix obtained by multiplying all of the entries in one row (or column) of A by k. Then

$$\det(C) = k\det(A).$$

c) Let D be the matrix obtained from A by interchanging the rows and columns of A (e.g., the first column of A is the first row of D, the second column of A is the second row of D, and so on). The matrix D is called the **transpose** of A and is denoted A'. Then

$$\det(D) = \det(A') = \det(A).$$

d) Let E be a matrix obtained from A by adding any scalar multiple of one row (column) of A to any other row (column) of A. Then

$$\det(E) = \det(A).$$

e) If one row (column) of A is equal to any other row (column), then

$$\det(A) = 0.$$

f) If one row (column) of A can be obtained by adding scalar multiples of one or more other rows (columns), then

$$\det(A) = 0.$$

g) If A and B are both $m \times m$ matrices, then

$$\det(AB) = \det(A)\det(B).$$

Most of these properties can be proved for 2×2 and 3×3 matrices by direct calculation. Others can be deduced easily from previously established properties. For example, property (e) can be deduced from (a) as follows. Let $\mathbf{r}_1, \mathbf{r}_2, \ldots, \mathbf{r}_m$ be the rows of the $m \times m$ matrix A, and suppose two of these rows are identical, say $\mathbf{r}_i = \mathbf{r}_k$. Let B be the matrix obtained when rows $\mathbf{r}_i$ and $\mathbf{r}_k$ are exchanged. Then $B = A$. However, by property (a) we have

$$\det(A) = -\det(B) = -\det(A),$$

and it follows that $\det(A) = 0$. We leave verification of some of the other properties as exercises.

Exercises 8.2

Exercises 1–12: For the matrices

$$A = \begin{pmatrix} -2 & 3 \\ 4 & 1 \\ 0 & 8 \end{pmatrix}, \quad B = \begin{pmatrix} 2 & 4 \\ 4 & 8 \\ -4 & -2 \end{pmatrix}, \quad C = \begin{pmatrix} -4 & 3 \\ 0 & 2 \end{pmatrix}, \quad D = \begin{pmatrix} 5 \\ 3 \\ 1 \end{pmatrix}, \quad \text{and} \quad E = (2 \quad -2 \quad 3),$$

decide whether the matrix operation is defined. If not, give reasons. If so, perform the operation.

1. $-3A$

2. $B + A$

3. $A - 5C$

4. DE

5. ED

6. AC

7. CA

8. AB

9. BA

10. $(A + B)C$

11. CBA

12. ACC

13. For the matrices A, B, C, D, and E given in the previous exercises, find

a. a_{12}

b. e_{13}

c. b_{32}

d. b_{23}

e. d_{12}

14. For the matrices A, B, C, D, and E given in the previous exercises, find the position of the element(s)

a. 4 in A

b. 4 in B

c. 0 in C

d. 3 in D

e. 3 in E

15. Let

$$A = \begin{pmatrix} 2 & 3 \\ -1 & 6 \end{pmatrix} \quad \text{and} \quad B = \begin{pmatrix} -1 & 5 \\ 0 & 2 \end{pmatrix}.$$

Show that the products AB and BA are both defined, but that $AB \neq BA$.

16. Let

$$C = \begin{pmatrix} c_{11} & c_{12} & c_{13} \\ c_{21} & c_{22} & c_{23} \end{pmatrix} \quad \text{and} \quad D = \begin{pmatrix} d_{11} & d_{12} \\ d_{21} & d_{22} \\ d_{31} & d_{32} \end{pmatrix}.$$

Show that CD and DC are both defined. Is it possible that $CD = DC$? Explain.

17. Find two different 2×2 matrices A and B such that $A \neq B$ and $AB = BA$.

Exercises 18–23: Find the determinant of the matrix.

18. $A = \begin{pmatrix} -3 & 5 \\ 9 & 5 \end{pmatrix}$

19. $B = \begin{pmatrix} a & b \\ c & d \end{pmatrix}$

20. $A = (2001)$

21. $B = \begin{pmatrix} 0 & 0 & 0 \\ 1 & 1 & 1 \end{pmatrix}$

22. $C = \begin{pmatrix} 2 & -2 & 4 \\ 1 & 3 & -3 \\ 2 & -3 & 3 \end{pmatrix}$

23. $A = \begin{pmatrix} 1 & 0 & 1 \\ 1 & 0 & 1 \\ 1 & 0 & 1 \end{pmatrix}$

Exercises 24–28: Verify that $\det(AB) = \det(A)\det(B)$ *for the given matrices.*

24. $A = \begin{pmatrix} 2 & -3 \\ -1 & 4 \end{pmatrix}$, $B = \begin{pmatrix} 0 & -1 \\ 3 & 3 \end{pmatrix}$

25. $A = \begin{pmatrix} 4 & 5 \\ 6 & 7 \end{pmatrix}$, $B = A$

26. $A = \begin{pmatrix} a_{11} & a_{12} \\ a_{21} & a_{22} \end{pmatrix}$, $B = \begin{pmatrix} b_{11} & b_{12} \\ b_{21} & b_{22} \end{pmatrix}$

27. $A = \begin{pmatrix} -2 & 3 & 3 \\ 4 & 0 & 7 \\ -4 & 3 & -2 \end{pmatrix}$, $B = \begin{pmatrix} 1 & 2 & 3 \\ -1 & -2 & -3 \\ 3 & 5 & 7 \end{pmatrix}$

28. $A = \begin{pmatrix} 8 & 1/2 & 4 \\ 1/2 & 0 & -4 \\ 2 & 2 & 2 \end{pmatrix}$, $B = \begin{pmatrix} 2 & 2 & 2 \\ 3 & -12 & 7 \\ 5 & 5 & -2 \end{pmatrix}$

29. Find x and y, if possible, so that

$$\begin{pmatrix} -3 & 4 \\ 1 & 5 \end{pmatrix}\begin{pmatrix} x \\ y \end{pmatrix} = \begin{pmatrix} -10 \\ -3 \end{pmatrix}.$$

30. Find x and y, if possible, so that

$$\begin{pmatrix} -6 & 2 \\ 3 & -1 \end{pmatrix}\begin{pmatrix} x \\ y \end{pmatrix} = \begin{pmatrix} -4 \\ 0 \end{pmatrix}.$$

31. Find x, y, and z, if possible, so that

$$\begin{pmatrix} 1 & 1 & 1 \\ 0 & -1 & 5 \\ 3 & 2 & -2 \end{pmatrix}\begin{pmatrix} x \\ y \\ z \end{pmatrix} = \begin{pmatrix} 1 \\ -15 \\ 18 \end{pmatrix}.$$

32. Find x, y, and z, if possible, so that

$$\begin{pmatrix} 1 & -1 & 2 \\ 3 & 4 & 1 \\ 1 & 6 & -3 \end{pmatrix}\begin{pmatrix} x \\ y \\ z \end{pmatrix} = \begin{pmatrix} 8 \\ 12 \\ 6 \end{pmatrix}.$$

33. Let A, B, and C be 3×2 matrices. Show that

$$A + B = B + A$$

and that

$$A + (B + C) = (A + B) + C.$$

34. Let A and B be 3×2 matrices and let C be a 2×3 matrix. Verify that

$$(A + B)C = AC + BC.$$

T **35.** With the aid of a computer algebra system (CAS), verify that

$$\det(AB) = \det(A)\det(B)$$

is true for 3×3 matrices A and B. If you do not have access to a CAS, verify it by hand for 2×2 matrices.

36. Let A be a 3×3 matrix and let B be a matrix that results when two rows of A are interchanged. Show that $\det(B) = -\det(A)$.

37. Suppose that

$$A = \begin{pmatrix} a_1x + b_1y & a_2x + b_2y & a_3x + b_3y \\ a_1 & a_2 & a_3 \\ b_1 & b_2 & b_3 \end{pmatrix},$$

so that the first row of A is a sum of constant multiples of the other two rows. Show that $\det(A) = 0$.

38. Let

$$A = \begin{pmatrix} 0.99 & 0.055 \\ 0.023 & 0.94 \end{pmatrix} \quad \text{and} \quad \mathbf{v} = \begin{pmatrix} x_0 \\ y_0 \end{pmatrix},$$

where x_0 and y_0 are, respectively, the populations of 20-and-over and under-20 people in the United States in 1990. In (2) we saw that $A\mathbf{v}$ gives the populations of these groups in 1991. Give a possible interpretation for $A^2\mathbf{v} = A(A\mathbf{v})$. Give an interpretation for $A^3\mathbf{v}$.

8.3 The Cross Product

Perpendicularity plays an important role in science, mathematics, and engineering. For example, we describe the position of a point in the plane or space by reference to mutually perpendicular axes. In analyzing the motion of a particle in the plane, we find the tangential and normal components of acceleration, which are perpendicular to each other. The cross product gives us a way to find a vector perpendicular to each of two given vectors. This will be useful when we analyze motion in space, find linear approximations to functions of two variables, and study geometry in three dimensions. We use this geometric idea—that the cross product of vectors **a** and **b** is perpendicular to each of **a** and **b**—to motivate the definition of *cross product.*

INVESTIGATION

Finding a Vector Perpendicular to Two Given Vectors

Let $\mathbf{a} = \langle a_1, a_2, a_3 \rangle$ and $\mathbf{b} = \langle b_1, b_2, b_3 \rangle$ be nonzero vectors in three-space. Find a nonzero vector $\mathbf{v} = \langle v_1, v_2, v_3 \rangle$ perpendicular to both $\mathbf{a}$ and $\mathbf{b}$.

If $\mathbf{v}$ is perpendicular to both $\mathbf{a}$ and $\mathbf{b}$, then $\mathbf{a} \cdot \mathbf{v} = 0$ and $\mathbf{b} \cdot \mathbf{v} = 0$. This means that

$$\begin{aligned} \mathbf{a} \cdot \mathbf{v} &= a_1 v_1 + a_2 v_2 + a_3 v_3 = 0 \\ \mathbf{b} \cdot \mathbf{v} &= b_1 v_1 + b_2 v_2 + b_3 v_3 = 0. \end{aligned} \tag{1}$$

If we regard v_3 as a constant, then we can think of (1) as a system of two equations in the two unknowns v_1 and v_2. Solving this system, we find

$$v_1 = \frac{a_2 b_3 - a_3 b_2}{a_1 b_2 - a_2 b_1} v_3 \quad \text{and} \quad v_2 = \frac{a_3 b_1 - a_1 b_3}{a_1 b_2 - a_2 b_1} v_3, \tag{2}$$

where we assume that $a_1 b_2 - a_2 b_1 \neq 0$. Because any nonzero choice for v_3 leads to a nonzero vector $\mathbf{v}$, we take

$$v_3 = a_1 b_2 - a_2 b_1.$$

This choice simplifies (2) by eliminating the denominators. Hence,

$$v_1 = a_2 b_3 - a_3 b_2 \quad \text{and} \quad v_2 = a_3 b_1 - a_1 b_3.$$

Thus, a vector perpendicular to both $\mathbf{a}$ and $\mathbf{b}$ is

$$\mathbf{v} = \langle v_1, v_2, v_3 \rangle = \langle a_2 b_3 - a_3 b_2, a_3 b_1 - a_1 b_3, a_1 b_2 - a_2 b_1 \rangle. \tag{3}$$

We test this result in an example.

EXAMPLE 1 Let $\mathbf{a} = \langle 0.8, 0.6, 0 \rangle$ and $\mathbf{b} = \langle 0.2, 0.8, 0.5 \rangle$. Find the vector $\mathbf{v}$ given by (3) and verify that this vector is perpendicular to both $\mathbf{a}$ and $\mathbf{b}$.

Solution

Let $\mathbf{v} = \langle v_1, v_2, v_3 \rangle$. Using (3), we have

$$\begin{aligned} v_1 &= a_2 b_3 - a_3 b_2 = (0.6)(0.5) - (0)(0.8) = 0.3, \\ v_2 &= a_3 b_1 - a_1 b_3 = (0)(0.2) - (0.8)(0.5) = -0.4, \\ v_3 &= a_1 b_2 - a_2 b_1 = (0.8)(0.8) - (0.6)(0.2) = 0.52, \end{aligned}$$

so $\mathbf{v} = \langle 0.3, -0.4, 0.52 \rangle$. To check that $\mathbf{v}$ is perpendicular to $\mathbf{a}$ and $\mathbf{b}$, we calculate

$$\begin{aligned} \mathbf{a} \cdot \mathbf{v} &= \langle 0.8, 0.6, 0 \rangle \cdot \langle 0.3, -0.4, 0.52 \rangle = 0 \\ \mathbf{b} \cdot \mathbf{v} &= \langle 0.2, 0.8, 0.5 \rangle \cdot \langle 0.3, -0.4, 0.52 \rangle = 0. \end{aligned}$$

Because $\mathbf{v} \cdot \mathbf{a} = 0$ and $\mathbf{v} \cdot \mathbf{b} = 0$, it follows that $\mathbf{v}$ is perpendicular to both $\mathbf{a}$ and $\mathbf{b}$. These three vectors are shown in Fig. 8.16.

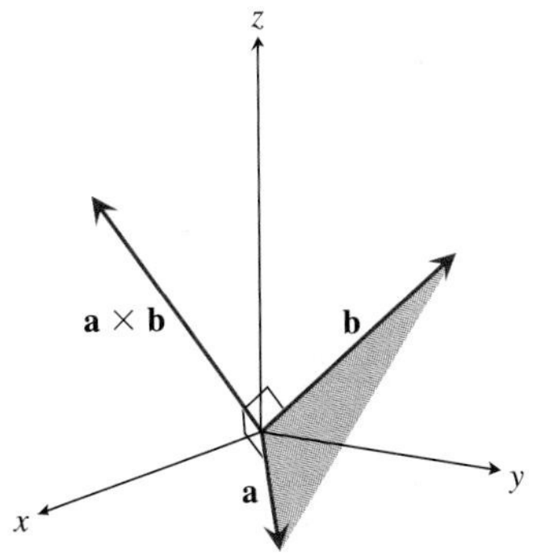

FIGURE 8.16 The vector $\mathbf{a} \times \mathbf{b}$ is perpendicular to the plane containing $\mathbf{a}$ and $\mathbf{b}$.

We take (3) as our definition of the cross product.

DEFINITION The Cross Product

Let $\mathbf{a} = \langle a_1, a_2, a_3 \rangle$ and $\mathbf{b} = \langle b_1, b_2, b_3 \rangle$ be two vectors in three-space. The **cross product** of $\mathbf{a}$ and $\mathbf{b}$ is

$$\mathbf{a} \times \mathbf{b} = \langle a_2b_3 - a_3b_2, a_3b_1 - a_1b_3, a_1b_2 - a_2b_1 \rangle. \tag{4}$$

Note that the cross product of two vectors is another vector.

Writing (4) with the **i**, **j**, **k** notation, we see that the cross product can be calculated using a determinant,

$$\begin{aligned}\mathbf{a} \times \mathbf{b} &= (a_1\mathbf{i} + a_2\mathbf{j} + a_3\mathbf{k}) \times (b_1\mathbf{i} + b_2\mathbf{j} + b_3\mathbf{k})\\ &= (a_2b_3 - a_3b_2)\mathbf{i} - (a_1b_3 - a_3b_1)\mathbf{j} + (a_1b_2 - a_2b_1)\mathbf{k}\\ &= \det\begin{pmatrix}\mathbf{i} & \mathbf{j} & \mathbf{k}\\ a_1 & a_2 & a_3\\ b_1 & b_2 & b_3\end{pmatrix}.\end{aligned} \tag{5}$$

EXAMPLE 2 Let $\mathbf{a} = \langle -3, 1, 5 \rangle$ and $\mathbf{b} = \langle 0, 4, -4 \rangle$. Calculate

$$\mathbf{a} \times \mathbf{b}, \quad \mathbf{b} \times \mathbf{a}, \quad \text{and} \quad \mathbf{a} \times \mathbf{a}.$$

Solution

Using (5) and expansion along the first row of the matrix,

$$\begin{aligned}\mathbf{a} \times \mathbf{b} &= \det\begin{pmatrix}\mathbf{i} & \mathbf{j} & \mathbf{k}\\ -3 & 1 & 5\\ 0 & 4 & -4\end{pmatrix}\\ &= \mathbf{i}(1(-4) - 5 \cdot 4) - \mathbf{j}((-3)(-4) - 5 \cdot 0) + \mathbf{k}((-3)4 - 1 \cdot 0)\\ &= -24\mathbf{i} - 12\mathbf{j} - 12\mathbf{k} = \langle -24, -12, -12 \rangle.\end{aligned}$$

For $\mathbf{b} \times \mathbf{a}$, we have

$$\mathbf{b} \times \mathbf{a} = \det\begin{pmatrix}\mathbf{i} & \mathbf{j} & \mathbf{k}\\ 0 & 4 & -4\\ -3 & 1 & 5\end{pmatrix}.$$

This determinant can be obtained from the one used to calculate $\mathbf{a} \times \mathbf{b}$ by interchanging the last two rows. In Section 8.2 we noted that such a row interchange changes the sign of the determinant. Hence

$$\mathbf{b} \times \mathbf{a} = -(\mathbf{a} \times \mathbf{b}) = -\langle -24, -12, -12 \rangle = \langle 24, 12, 12 \rangle.$$

Finally,

$$\mathbf{a} \times \mathbf{a} = \det\begin{pmatrix}\mathbf{i} & \mathbf{j} & \mathbf{k}\\ -3 & 1 & 5\\ -3 & 1 & 5\end{pmatrix} = 0\mathbf{i} + 0\mathbf{j} + 0\mathbf{k} = \langle 0, 0, 0 \rangle = \mathbf{0}.$$

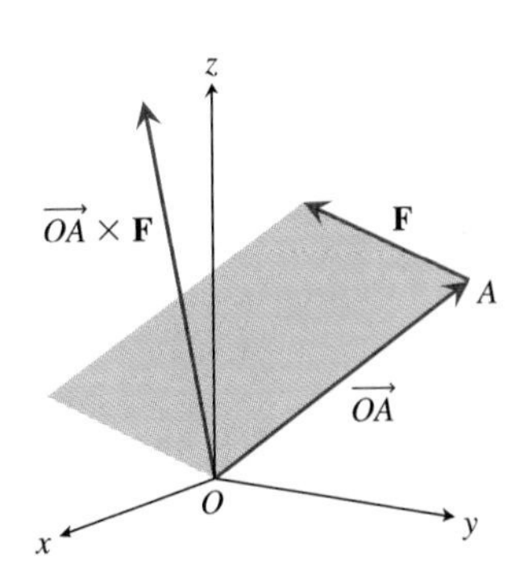

FIGURE 8.17 A force **F** applied at A causes a torque at O.

Torque The cross product is used in physics to calculate torque. Consider an object that is free to rotate about a point O. If a force F acts on the object, and the line of action of F does not pass through O, then the object will rotate about the point O. Suppose that the force **F** is applied at a point A, as shown in Fig. 8.17. We know from experience that the rotational effect is greatest when **F** acts in a direction perpendicular to the *lever arm* OA and that the rotational effect increases as A moves farther from O. For example, when we open a swinging door, we push as far as possible from the hinges and try to keep the direction of our push perpendicular to the door. Physicists use a quantity called **torque,** denoted by $\boldsymbol{\tau}$, to measure this rotational effect at O. The torque about O of a force **F** acting at a point A is defined by

$$\boldsymbol{\tau} = \overrightarrow{OA} \times \mathbf{F}.$$

Thus torque is a vector perpendicular to the plane of **F** and $\overrightarrow{OA}$. If **F** is measured in newtons and distance in meters, then torque has units of newtons · meters = joules.

EXAMPLE 3 A mechanic has two wrenches, one of length 0.3 meters and the other of length 0.5 meters. A torque of magnitude 15 joules is needed to loosen a nut. What force should be applied (at right angles) to each wrench?

Solution

Let the nut be at the origin O with the line of the bolt to which the nut is attached in the positive z-direction. Let the free end of the wrench be at $A = (\ell, 0, 0)$ on the positive x-axis, and the force **F**, of magnitude F, applied in the positive y-direction. See Fig. 8.18. Then $\mathbf{F} = F\mathbf{j}$ and $\overrightarrow{OA} = \ell\mathbf{i}$. The torque is

$$\boldsymbol{\tau} = \overrightarrow{OA} \times \mathbf{F} = \langle \ell, 0, 0 \rangle \times \langle 0, F, 0 \rangle = \langle 0, 0, \ell F \rangle = \ell F\mathbf{k}.$$

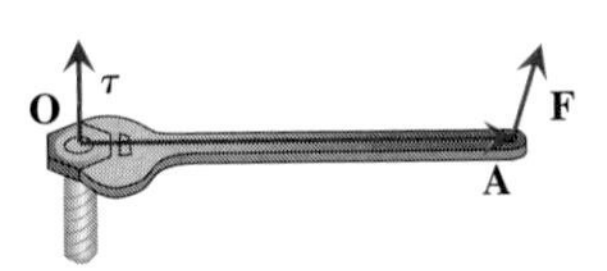

FIGURE 8.18 A force **F** applied at A causes a torque about O.

The magnitude of the torque is $\|\boldsymbol{\tau}\| = \ell F$. For the 0.3 m wrench, $\ell = 0.3$. Because we want $\|\boldsymbol{\tau}\| = 0.3F = 15$, we must have $F = 50$ newtons. For the 0.5 m wrench, we must have $\|\boldsymbol{\tau}\| = 0.5F = 15$, so $F = 30$ newtons.

Some of the arithmetic properties of the cross product are similar to the properties of products of real numbers.

Properties of the Cross Product

Let **a**, **b**, and **c** be vectors in R^3.

a) The cross product is *anticommutative,* that is,

$$\mathbf{a} \times \mathbf{b} = -(\mathbf{b} \times \mathbf{a}).$$

b) The cross product distributes over vector addition:

$$\mathbf{a} \times (\mathbf{b} + \mathbf{c}) = \mathbf{a} \times \mathbf{b} + \mathbf{a} \times \mathbf{c}.$$

c) Scalars can be factored out of a cross product: If k is a real number, then

$$(k\mathbf{a}) \times \mathbf{b} = k(\mathbf{a} \times \mathbf{b}) = \mathbf{a} \times (k\mathbf{b}).$$

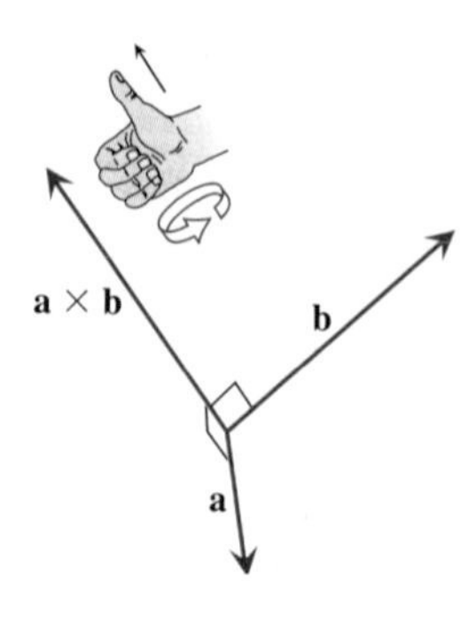

FIGURE 8.19 Using the right-hand rule to determine the direction of **a** × **b**.

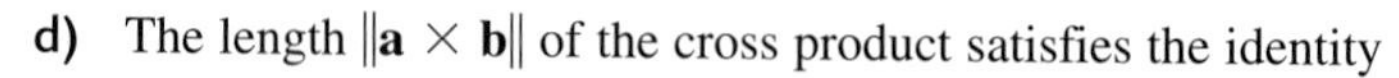

d) The length $\|\mathbf{a} \times \mathbf{b}\|$ of the cross product satisfies the identity

$$\|\mathbf{a} \times \mathbf{b}\|^2 = \|\mathbf{a}\|^2\|\mathbf{b}\|^2 - (\mathbf{a} \cdot \mathbf{b})^2.$$

e) The identity in **d)** can also be written in the form

$$\|\mathbf{a} \times \mathbf{b}\| = \|\mathbf{a}\|\,\|\mathbf{b}\| \sin\theta,$$

where θ, $0 \leq \theta \leq \pi$, is the angle between **a** and **b**.

f) If $\mathbf{a} \times \mathbf{b}$ is not $\mathbf{0} = \langle 0, 0, 0\rangle$, then $\mathbf{a} \times \mathbf{b}$ is perpendicular to both **a** and **b**. The direction of $\mathbf{a} \times \mathbf{b}$ is determined by the right-hand rule. Let θ, $0 < \theta < \pi$, be the angle measured from **a** to **b**. Place your right hand with your fingers pointing in the direction of **a** so that when you bend your fingers, they bend toward **b** through angle θ. Your thumb then points in the direction of $\mathbf{a} \times \mathbf{b}$. See Fig. 8.19.

g) $\mathbf{i} \times \mathbf{j} = \mathbf{k}, \qquad \mathbf{j} \times \mathbf{k} = \mathbf{i}, \qquad \mathbf{k} \times \mathbf{i} = \mathbf{j}.$

Most of these properties can be verified by using (4) or (5) and some algebra. For example, to verify **d)**, we simply expand the right side and then regroup terms and factor to get the left side. We note in advance that the terms of the form $a_1^2b_1^2$, $a_2^2b_2^2$, and $a_3^2b_3^2$ occur in both terms of the difference and add to zero in pairs.

$$\begin{aligned}
\|\mathbf{a}\|^2\|\mathbf{b}\|^2 - (\mathbf{a} \cdot \mathbf{b})^2 &= (a_1^2 + a_2^2 + a_3^2)(b_1^2 + b_2^2 + b_3^2) - (a_1b_1 + a_2b_2 + a_3b_3)^2 \\
&= (a_1^2b_2^2 - 2a_1a_2b_1b_2 + a_2^2b_1^2) + (a_1^2b_3^2 - 2a_1a_3b_1b_3 + a_3^2b_1^2) \\
&\quad + (a_2^2b_3^2 - 2a_2a_3b_2b_3 + a_3^2b_2^2) \\
&= (a_1b_2 - a_2b_1)^2 + (a_1b_3 - a_3b_1)^2 + (a_2b_3 - a_3b_2)^2 \\
&= \|\mathbf{a} \times \mathbf{b}\|^2.
\end{aligned}$$

To deduce property **e)**, we use **d)** and the identity $\mathbf{a} \cdot \mathbf{b} = \|\mathbf{a}\|\,\|\mathbf{b}\| \cos\theta$:

$$\begin{aligned}
\|\mathbf{a} \times \mathbf{b}\|^2 &= \|\mathbf{a}\|^2\|\mathbf{b}\|^2 - (\mathbf{a} \cdot \mathbf{b})^2 \\
&= \|\mathbf{a}\|^2\|\mathbf{b}\|^2 - \|\mathbf{a}\|^2\|\mathbf{b}\|^2 \cos^2\theta \\
&= \|\mathbf{a}\|^2\|\mathbf{b}\|^2(1 - \cos^2\theta) \\
&= \|\mathbf{a}\|^2\|\mathbf{b}\|^2 \sin^2\theta.
\end{aligned}$$

Because $\sin\theta \geq 0$ for $0 \leq \theta \leq \pi$, we can take the square root of both sides to get

$$\|\mathbf{a} \times \mathbf{b}\| = \|\mathbf{a}\|\,\|\mathbf{b}\| \sin\theta.$$

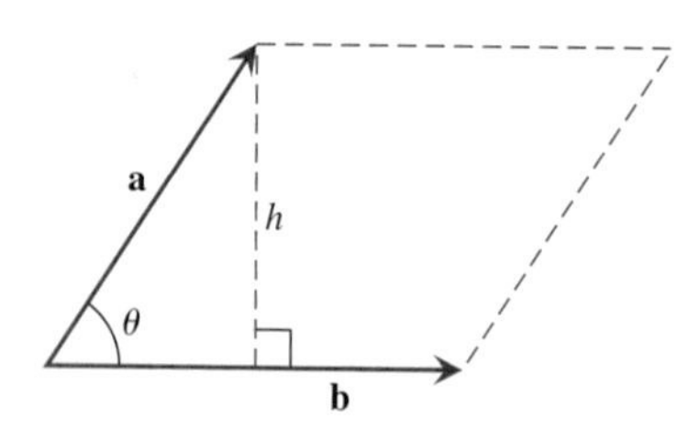

FIGURE 8.20 The parallelogram determined by **a** and **b** has area $\|\mathbf{a} \times \mathbf{b}\|$.

The expression on the right of **e)** might look familiar. In fact, it is a formula for the area of the parallelogram determined by **a** and **b**. Referring to Fig. 8.20, and recalling that the area of a parallelogram is given by the product base times height, we may take $\|\mathbf{b}\|$ as the length of the base and h as its height. Because

$$\frac{h}{\|\mathbf{a}\|} = \sin\theta,$$

the area of the parallelogram is then given by

$$h\|\mathbf{b}\| = \|\mathbf{a}\|\,\|\mathbf{b}\| \sin\theta = \|\mathbf{a} \times \mathbf{b}\|.$$

Hence,

Area of a Parallelogram

The parallelogram determined by vectors **a** and **b** has area $\|\mathbf{a} \times \mathbf{b}\|$.

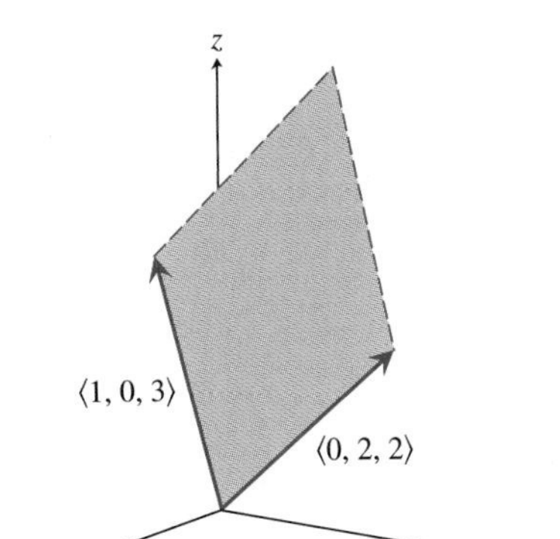

FIGURE 8.21 The parallelogram determined by $\langle 1, 0, 3\rangle$ and $\langle 0, 2, 2\rangle$.

EXAMPLE 4 Find the area of the parallelogram determined by the vectors $\langle 1,0,3\rangle$ and $\langle 0,2,2\rangle$. Sketch the parallelogram. Assume that lengths are in centimeters.

Solution

From (5), we have

$$\langle 1,0,3\rangle \times \langle 0,2,2\rangle = \det\begin{pmatrix} \mathbf{i} & \mathbf{j} & \mathbf{k} \\ 1 & 0 & 3 \\ 0 & 2 & 2 \end{pmatrix} = -6\mathbf{i} - 2\mathbf{j} + 2\mathbf{k}.$$

The area of the parallelogram, shown in Fig. 8.21, is

$$\begin{aligned} \|\langle 1,0,3\rangle \times \langle 0,2,2\rangle\| &= \|-6\mathbf{i} - 2\mathbf{j} + 2\mathbf{k}\| \\ &= \sqrt{(-6)^2 + (-2)^2 + 2^2} = \sqrt{44} \approx 6.63325 \text{ cm}^2. \end{aligned}$$

To calculate the area of the parallelogram determined by two two-dimensional vectors **v** and **w**, append a z-coordinate of 0 to each vector, and then calculate the length of the cross-product of the resulting three-dimensional vectors.

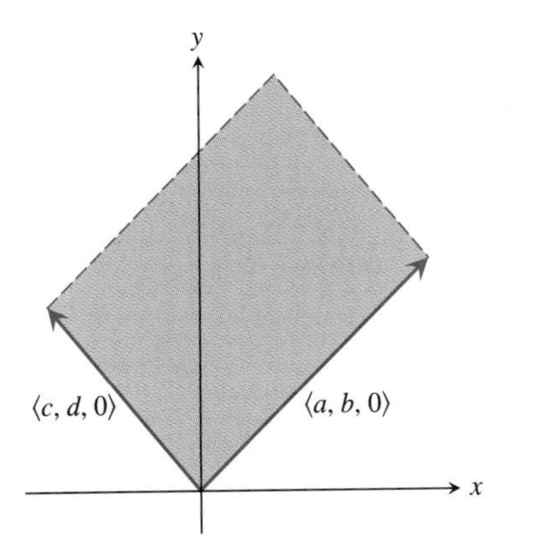

FIGURE 8.22 Append a third coordinate of 0 to treat an R^2 vector like an R^3 vector.

EXAMPLE 5 Find the area of the parallelogram determined by the two-dimensional vectors $\mathbf{v} = \langle a,b\rangle$ and $\mathbf{w} = \langle c,d\rangle$.

Solution

Referring to Fig. 8.22, the area of the parallelogram determined by **v** and **w** is equal to the area of the parallelogram determined by the R^3 vectors $\langle a,b,0\rangle$ and $\langle c,d,0\rangle$. The area is

$$\begin{aligned} \|\langle a,b,0\rangle \times \langle c,d,0\rangle\| &= \|\langle b\cdot 0 - 0\cdot d, 0\cdot c - a\cdot 0, ad - bc\rangle\| \\ &= \|\langle 0,0,ad - bc\rangle\| \\ &= |ad - bc| = \left|\det\begin{pmatrix} a & b \\ c & d \end{pmatrix}\right|. \end{aligned}$$

This area result will be useful in later sections.

Area of a Planar Parallelogram

The area A of the parallelogram determined by vectors $\langle a,b\rangle$ and $\langle c,d\rangle$ is

$$A = \left|\det\begin{pmatrix} a & b \\ c & d \end{pmatrix}\right|.$$

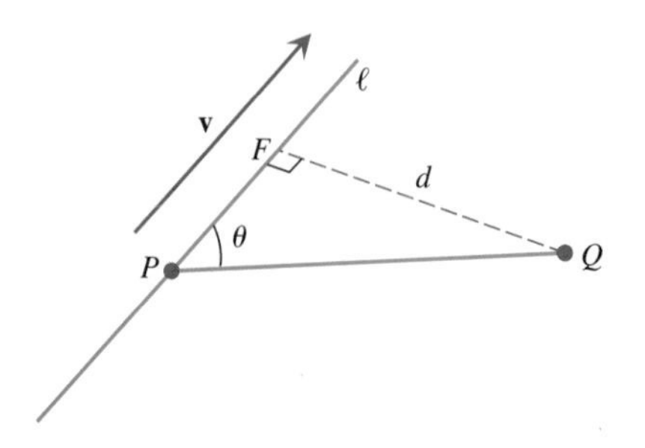

FIGURE 8.23 The distance from a point to a line is the length of the perpendicular from the point to the line.

The dot product provides a connection between two vectors and the cosine of the angle between these two vectors. We saw that in some instances it is convenient to express the cosine of an angle in terms of a dot product. For example, this was the case when we defined the projection of one vector onto another. Because the cross product is linked to the sine of the angle between two vectors—see property **e)**—it may be a useful idea to think about using the cross product when we see the sine function. This is illustrated in the following example, where we find the distance from a point to a line. The distance from a point to a line is defined to be the length of the perpendicular from the point to the line. See Fig. 8.23.

EXAMPLE 6 Let ℓ be the line described by

$$\mathbf{r}(t) = \langle -2, 0, 5\rangle + t\langle 1, 7, 2\rangle.$$

a) Find the distance from the point $Q = (4, 4, 5)$ to ℓ.
b) Find the point on ℓ that is closest to Q.

Solution

a) The line ℓ contains point $P = (-2, 0, 5)$ and is parallel to the vector $\mathbf{v} = \langle 1, 7, 2\rangle$. Let θ be the angle between $\mathbf{v}$ and $\overrightarrow{PQ}$, let F be the point at the foot of the perpendicular from Q to ℓ, and let d be the length of this perpendicular. See Fig. 8.23. Triangle PQF is a right triangle with right angle at F and angle θ at P. Because $\sin\,\theta = d/\|\overrightarrow{PQ}\|$,

$$\text{the distance from } Q \text{ to } \ell = d = \|\overrightarrow{PQ}\| \sin\,\theta. \tag{6}$$

We also know from **e)** in Properties of the Cross Product that

$$\sin\,\theta = \frac{\|\mathbf{v}\times\overrightarrow{PQ}\|}{\|\mathbf{v}\|\,\|\overrightarrow{PQ}\|}. \tag{7}$$

Substituting (7) into (6), we have

$$d = \frac{\|\mathbf{v}\times\overrightarrow{PQ}\|}{\|\mathbf{v}\|}. \tag{8}$$

Calculating $\|\mathbf{v}\| = \sqrt{54}$ and

$$\|\mathbf{v}\times\overrightarrow{PQ}\| = \|\langle 1, 7, 2\rangle \times \langle 6, 4, 0\rangle\| = \|\langle -8, 12, -38\rangle\| = \sqrt{1652}$$

and substituting these numbers in (8), we have

$$d = \text{the distance from } Q \text{ to } \ell = \frac{\sqrt{1652}}{\sqrt{54}} = \frac{\sqrt{826}}{\sqrt{27}} \approx 5.53106.$$

b) We seek the point F on the line for which $\text{dist}(Q, F) = \sqrt{826}/\sqrt{27}$. Because F is on the line, it has the form

$$F = (-2 + t, 7t, 5 + 2t)$$

for some real number t. To find this t we note that, by the distance formula and part **a)**,

$$\text{dist}(Q, F) = \sqrt{(4 - (-2 + t))^2 + (4 - 7t)^2 + (5 - (5 + 2t))^2} = \frac{\sqrt{826}}{\sqrt{27}}.$$

Squaring both sides and simplifying, we have

$$729t^2 - 918t + 289 = 0.$$

This quadratic equation has one solution, $t = 17/27$. When this t value is substituted into the equation for the line, we obtain $F = (-37/27, 119/27, 169/27)$. This is the point on ℓ closest to Q. See also Exercises 42, 43, and 44.

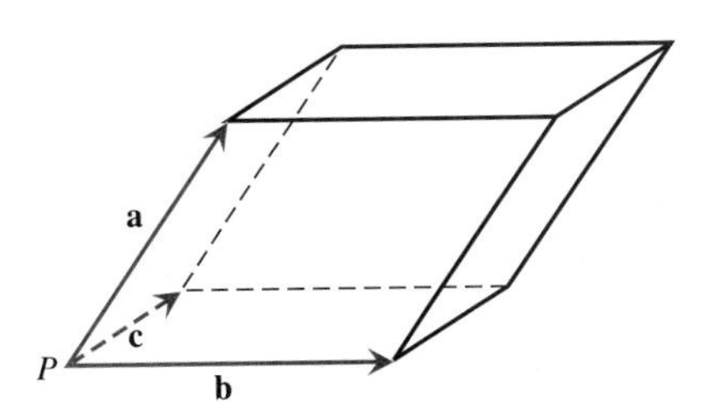

FIGURE 8.24 The parallelepiped determined by **a**, **b**, and **c**.

The Volume of a Parallelepiped The dot product and the cross product can be used together to find the volume of the parallelepiped determined by three vectors. A parallelepiped is a solid with six faces, each of which is a parallelogram. Opposite faces are congruent and parallel. See Fig. 8.24. Let **a**, **b**, and **c** be three non-zero vectors based at a point P, and assume that there is no plane containing all three vectors. Vectors **a** and **b** determine a parallelogram, as do the pairs **a**, **c** and **b**, **c**. These parallelograms are three faces of a parallelepiped. The opposite faces can be filled in to complete the solid as shown in Fig. 8.24.

The volume V of a parallelepiped is given by

$$V = Bh, \tag{9}$$

where B is the area of a chosen base and h is the height of the parallelepiped relative to that base. If we take the base to be the parallelogram determined by **b** and **c**, then

$$B = \|\mathbf{b} \times \mathbf{c}\|. \tag{10}$$

The height h of the parallelepiped relative to this base is the perpendicular distance between this base and the face opposite the base. See Fig. 8.25. Because $\mathbf{b} \times \mathbf{c}$ is a vector perpendicular to both **b** and **c**, it is parallel to any segment used to measure the height of the parallelepiped. It follows that if θ is the angle between **a** and $\mathbf{b} \times \mathbf{c}$, then

$$h = \|\mathbf{a}\|\,|\cos\theta|. \tag{11}$$

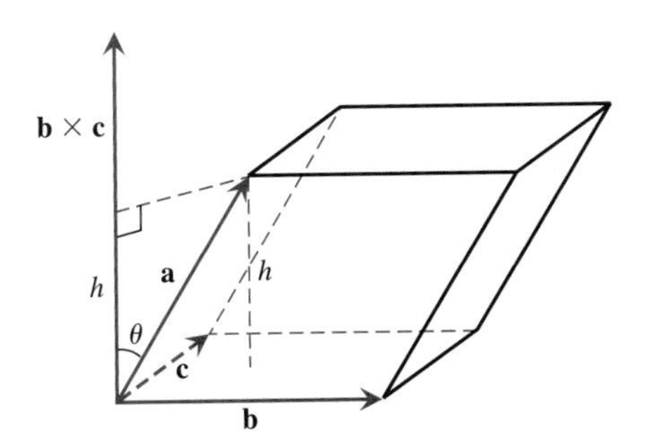

FIGURE 8.25 The height of the parallelepiped can be measured along the vector $\mathbf{a} \times \mathbf{b}$.

See Fig. 8.25. With **a**, **b**, and **c** oriented as in the figure, $\mathbf{b} \times \mathbf{c}$ points from the face determined by **b**, **c** to the opposite face. Hence, $\cos\theta \geq 0$. If $\mathbf{b} \times \mathbf{c}$ points away from the opposite face, then $\cos\theta < 0$ and the absolute values are needed in (11). See also Exercise 54.

Combining (9), (10), and (11),

$$V = \|\mathbf{a}\|\,\|\mathbf{b} \times \mathbf{c}\|\,|\cos\theta| = |\mathbf{a} \cdot (\mathbf{b} \times \mathbf{c})|.$$

Hence,

Volume of a Parallelepiped

The volume of the parallelepiped determined by three noncoplanar vectors **a**, **b**, and **c** is

$$V = |\mathbf{a} \cdot (\mathbf{b} \times \mathbf{c})|. \tag{12}$$

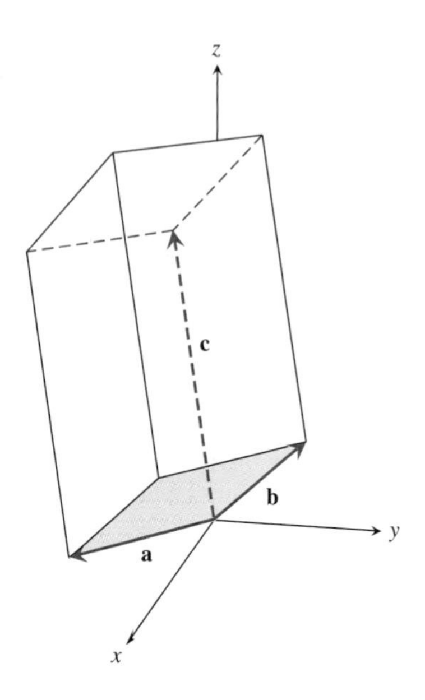

FIGURE 8.26 A parallelepiped determined by three vectors.

EXAMPLE 7 Sketch and find the volume of the parallelepiped determined by the vectors

$$\mathbf{a} = \langle 3, -2, 0 \rangle, \quad \mathbf{b} = \langle 2, 2, 2 \rangle, \quad \text{and} \quad \mathbf{c} = \langle -1, -1, 5 \rangle.$$

Solution

To find the volume V of the parallelpiped, first find the cross product

$$\mathbf{b} \times \mathbf{c} = \langle 2, 2, 2 \rangle \times \langle -1, -1, 5 \rangle = \langle 12, -12, 0 \rangle.$$

By (12),

$$V = |\mathbf{a} \cdot (\mathbf{b} \times \mathbf{c})| = |\langle 3, -2, 0 \rangle \cdot \langle 12, -12, 0 \rangle| = 60.$$

Figure 8.26 shows the parallelepiped determined by **a**, **b**, and **c**. We have shaded the parallelogram determined by **a** and **b** to better see the relative positions of **a**, **b**, and **c**. This figure, produced by a computer algebra system (CAS), was drawn in perspective, so that segments that are actually parallel may not appear so. Get to know how your CAS or calculator draws representations of three-dimensional objects, and learn to be careful in interpreting what you see.

The expression $\mathbf{a} \cdot (\mathbf{b} \times \mathbf{c})$ in (12) is often called a *triple product.* The triple product calculation is easily done with determinants. In Exercise 63 we ask you to verify that

$$\mathbf{a} \cdot (\mathbf{b} \times \mathbf{c}) = \det \begin{pmatrix} a_1 & a_2 & a_3 \\ b_1 & b_2 & b_3 \\ c_1 & c_2 & c_3 \end{pmatrix}. \tag{13}$$

This leads to an alternative formula for the volume of a parallelepiped.

Volume of a Parallelepiped

The volume V of the parallelepiped determined by three noncoplanar vectors $\mathbf{a} = \langle a_1, a_2, a_3 \rangle$, $\mathbf{b} = \langle b_1, b_2, b_3 \rangle$, and $\mathbf{c} = \langle c_1, c_2, c_3 \rangle$ is

$$V = \left| \det \begin{pmatrix} a_1 & a_2 & a_3 \\ b_1 & b_2 & b_3 \\ c_1 & c_2 & c_3 \end{pmatrix} \right|.$$

Exercises 8.3

Exercises 1–10: Let $\mathbf{r} = \langle 3, 4, -2 \rangle$, $\mathbf{u} = \langle 2, -1, 1 \rangle$, $\mathbf{v} = \langle -5, 0, 3 \rangle$, *and* $\mathbf{w} = \langle 4, 3, 2 \rangle$. *Find the following.*

1. $\mathbf{u} \times \mathbf{v}$

2. $\mathbf{r} \times \mathbf{v}$

3. $\mathbf{w} \times \mathbf{r}$

4. $\mathbf{u} \times \mathbf{w}$

5. $\mathbf{u} \cdot (\mathbf{w} \times \mathbf{u})$

6. $\mathbf{r} \cdot (\mathbf{u} \times \mathbf{v})$

7. A unit vector perpendicular to both **r** and **v**

8. A unit vector perpendicular to both **u** and **w**

9. A vector of length 2 perpendicular to both **u** and **w**

10. Two vectors of length 4, each perpendicular to both **v** and **w**

Exercises 11–20: Let $\mathbf{r} = -2\mathbf{i} - 3\mathbf{k}$, $\mathbf{u} = \mathbf{i} - 2\mathbf{j} + 2\mathbf{k}$, $\mathbf{v} = -3\mathbf{i} + 8\mathbf{j} - \frac{1}{2}\mathbf{k}$, *and* $\mathbf{w} = \mathbf{k}$. *Find the following.*

11. $\mathbf{u} \times \mathbf{v}$

12. $\mathbf{r} \times \mathbf{v}$

13. $\mathbf{w} \times \mathbf{r}$

14. $\mathbf{u} \times \mathbf{w}$

15. $\mathbf{u} \cdot (\mathbf{w} \times \mathbf{u})$

16. $\mathbf{r} \cdot (\mathbf{u} \times \mathbf{v})$

17. A unit vector perpendicular to both **r** and **v**

18. A unit vector perpendicular to both **u** and **w**

19. A vector of length 2 perpendicular to both **u** and **w**

20. Two vectors of length 4, each perpendicular to both **v** and **w**

Exercises 21–28: Illustrate the figure in R^2 or R^3. If the figure is a parallelogram, calculate its area. If the figure is a parallelepiped, calculate its volume.

21. The parallelogram determined by $\mathbf{v} = \langle 3, 2 \rangle$ and $\mathbf{w} = \langle 4, -2 \rangle$

22. The parallelogram determined by $\mathbf{a} = \langle 19, 13 \rangle$ and $\mathbf{b} = \langle 4, -6 \rangle$

23. The parallelogram determined by $\mathbf{v} = \langle -1, 0, 1 \rangle$ and $\mathbf{w} = \langle 4, 2, 0 \rangle$

24. The parallelogram determined by $\mathbf{v} = \langle 5, -1, 6 \rangle$ and $\mathbf{w} = \langle -3, 2, 7 \rangle$

25. The parallelogram determined by $\mathbf{v} = \langle 1, 2, 3 \rangle$ and $\mathbf{w} = \langle -2, -2, 5 \rangle$

26. The parallelepiped determined by $\mathbf{a} = \langle 3, 2, 0 \rangle$, $\mathbf{b} = \langle 4, -2, 0 \rangle$, and $\mathbf{c} = \langle 3, 0, 5 \rangle$

27. The parallelepiped determined by $\mathbf{a} = \langle 0, 5, 6 \rangle$, $\mathbf{b} = \langle 2, -4, 0 \rangle$, and $\mathbf{c} = \langle 3, 4, -3 \rangle$

28. The parallelepiped determined by $\mathbf{a} = \langle 1, 1, 1 \rangle$, $\mathbf{b} = \langle 1, 0, -1 \rangle$, and $\mathbf{c} = \langle 0, 0, -3 \rangle$

29. Find an equation for the line through the point $P = (4, -3, 1)$ and perpendicular to the vectors $\mathbf{v} = \langle -1, 3, 4 \rangle$ and $\mathbf{w} = \langle -2, 4, 6 \rangle$. Sketch the line.

30. Find an equation for the line through the point $P = (0, 1, -1)$ and perpendicular to the vectors $\mathbf{v} = \mathbf{i}$ and $\mathbf{w} = \mathbf{i} - 2\mathbf{k}$. Sketch the line.

31. On the same set of axes sketch the vectors $\mathbf{i}$, $\mathbf{i} \times \mathbf{j}$, and $\mathbf{j} \times \mathbf{i}$. Also sketch the parallelogram determined by **i** and **j**.

32. On the same set of axes sketch the vectors $\mathbf{u} = \langle -2, 0, 1 \rangle$, $\mathbf{v} = \langle 3, 1, 1 \rangle$, $\mathbf{u} \times \mathbf{v}$, and $\mathbf{v} \times \mathbf{u}$. Also sketch the parallelogram determined by **u** and **v**.

33. A mechanic has a wrench that is 0.4 m long and uses it to loosen a nut. Calculate the magnitude of the torque if the mechanic applies a force of 50 newtons

a. at a right angle to the handle of the wrench.

b. so that the angle between the line of the wrench and the force vector is 30°.

c. so that the angle between the line of the wrench and the force vector is 135°.

34. A mechanic has a wrench that is 0.6 m long and uses it to loosen a nut. Calculate the magnitude of the torque if the mechanic applies a force of 70 newtons

a. at a right angle to the handle of the wrench.

b. so that the angle between the line of the wrench and the force vector is 45°.

c. so that the angle between the line of the wrench and the force vector is 150°.

35. A biker pushes on the pedal of a bike with a force of magnitude 50 newtons at an angle of 135° to the pedal shaft. The pedal shaft is 20 cm long. Find the magnitude of the torque about the pivot point P of the shaft. See the accompanying figure.

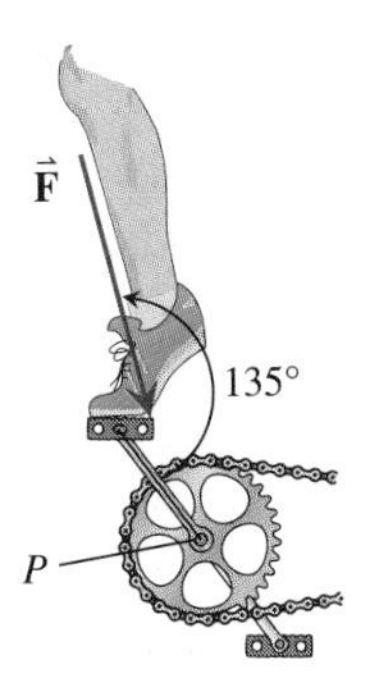

Figure for Exercise 35.

36. A force of 60 newtons is applied to a lever 2 meters from the fulcrum. If the force is applied at an angle of 60° to the lever, what is the magnitude of the torque at the fulcrum? See the accompanying figure.

Figure for Exercise 36.

37. Find the distance from the point $P = (-3, 5, 6)$ to the line with equation

$$\mathbf{r}(t) = \langle -2t, -t + 4, 2t - 1 \rangle.$$

Also find the point on the line that is closest to P.

38. Find the distance from the point $P = (1, 2, 0)$ to the line with equation

$$\mathbf{r}(t) = \langle 6, -t, t + 1 \rangle.$$

Also find the point on the line that is closest to P.

39. Find a formula for the distance from the origin to the line with equation

$$\mathbf{r}(t) = \langle a_1 + b_1 t, a_2 + b_2 t, a_3 + b_3 t \rangle.$$

Also find the point on the line that is closest to the origin.

40. Find the distance from the point $P = (5, 0, 6)$ to the line with equation

$$\frac{x - 3}{2} = y + 1 = \frac{z - 4}{3}.$$

Also find the point on the line that is closest to P.

41. Find the distance from the point $P = (-3, 1, 1)$ to the line with equation

$$\frac{x + 2}{-2} = \frac{z + 1}{3}, \qquad y = 5.$$

Also find the point on the line that is closest to P.

42. Let ℓ be the line with equation

$$\mathbf{r}(t) = \langle -1 + 2t, 3 - t, 5 - 3t \rangle.$$

Let $d(t)$ denote the distance from the origin to the point $\mathbf{r}(t)$.

a. Write a formula for $d(t)$.

b. Using the max/min techniques developed in Chapter 4, find the minimum value of $d(t)$. What is the significance of this minimal value? (*Hint:* Work with $D(t) = d(t)^2$.)

c. Find the point on ℓ closest to the origin.

43. Let ℓ be the line with equation

$$\mathbf{r}(t) = \langle -3t, 5, t + 3 \rangle,$$

let $P = (-2, 4, 1)$, and let $d(t)$ denote the distance from P to the point $\mathbf{r}(t)$.

a. Write a formula for $d(t)$.

b. Using the max/min techniques developed in Chapter 4, find the minimum value of $d(t)$. What is the significance of this minimal value? (*Hint:* Work with $D(t) = d(t)^2$.)

c. Find the point on ℓ closest to P.

44. Let ℓ be the line with equation

$$\mathbf{r}(t) = \left\langle \frac{1}{2}t - 1, -\frac{1}{3}t, t - \frac{2}{3} \right\rangle,$$

let $P = (1, 1, 1)$, and let $d(t)$ denote the distance from P to the point $\mathbf{r}(t)$.

a. Write a formula for $d(t)$.

b. Using the max/min techniques developed in Chapter 4, find the minimum value of $d(t)$. What is the significance of this minimal value? (*Hint:* Work with $D(t) = d(t)^2$.)

c. Find the point on ℓ closest to P.

45. Given a vector $\mathbf{v}$, what is $\mathbf{v} \times \mathbf{v}$? First, justify your answer by using the formula for the cross product. Next, justify your answer by relating the cross product to the area of a parallelogram.

46. Let $\mathbf{v} = \langle -3, 5, 2 \rangle$ and $\mathbf{w} = \langle 1, 3, 3 \rangle$. Is there a vector $\mathbf{u}$ such that $\mathbf{v} \times \mathbf{u} = \mathbf{w}$? How many such vectors $\mathbf{u}$ are there?

47. Let $\mathbf{v} = \langle 2, 1, 1 \rangle$ and $\mathbf{w} = \langle -7, 3, 11 \rangle$. Is there a vector $\mathbf{u}$ such that $\mathbf{v} \times \mathbf{u} = \mathbf{w}$? How many such vectors $\mathbf{u}$ are there?

48. The points $(0, 0, 0)$, $(4, -2, 1)$, and $(5, 5, 5)$ are three of the four vertices of a parallelogram. Find all possibilities for the fourth vertex.

49. The points $(-3, 4, 0)$, $(1, 2, 1)$, and $(-5, 2, 1)$ are three of the four vertices of a parallelogram. Find all possibilities for the fourth vertex.

50. The points $(0, 0, 0)$, $(0, 0, 3)$, $(0, 3, 3)$, and $(3, 0, 0)$ are four of the eight vertices of a cube. What are the other four vertices?

51. The points $(-2, 3, 5)$, $(3, 8, 10)$, and $(3, 8, 5)$ are three of the eight vertices of a cube. What are the other five vertices?

52. The points $(0, 0, 0)$, $\left(\sqrt{2}, -\sqrt{2}, \sqrt{2}\right)$, $(2, 1, -1)$, and $\left(0, \sqrt{3}, \sqrt{3}\right)$ are four of the eight vertices of a cube of side $\sqrt{6}$. Find the other four vertices.

53. Let $\mathbf{a}$ and $\mathbf{b}$ be vectors. Describe the parallelepiped determined by vectors $\mathbf{a}$, $\mathbf{b}$, and $\mathbf{a} \times \mathbf{b}$. What is the volume of this parallelepiped?

54. Consider the parallelepiped determined by $\mathbf{a}$, $\mathbf{b}$, and $\mathbf{c}$. If the base of the parallelogram is determined by $\mathbf{b}$ and $\mathbf{c}$, show that the height is the length of the projection of $\mathbf{a}$ onto $\mathbf{b} \times \mathbf{c}$.

55. Let **u** and **v** be vectors and let $\mathbf{w} = a\mathbf{u} + b\mathbf{v}$, where a and b are real numbers. Describe the parallelepiped determined by these three vectors and find its volume.

56. Verify **a)** in the list of properties of the cross product given in this section.

57. Verify **b)** in the list of properties of the cross product given in this section.

58. Verify **c)** in the list of properties of the cross product given in this section.

59. In **f)** in the list of properties of the cross product, we stated that the direction of the cross product can be determined by the right-hand rule. Verify this statement for vectors **a** and **b** with third component of 0. It may help to use polar coordinates. Let

$$\mathbf{a} = \langle r_1 \cos \alpha, r_1 \sin \alpha, 0\rangle$$

and

$$\mathbf{b} = \langle r_2 \cos(\alpha + \theta), r_2 \sin(\alpha + \theta), 0\rangle,$$

where $0 < \theta < \pi$. Show that

$$\mathbf{a} \times \mathbf{b} = \langle 0, 0, c\rangle$$

for some positive number c. Verify that this direction is the one predicted when we use the right-hand rule to determine the direction of $\mathbf{a} \times \mathbf{b}$.

60. Find an equation for the line that passes through $P = (-2, 3, 3)$ and intersects the line with equation $\mathbf{r}(t) = \langle -t, 2t, -3t\rangle$ at a right angle.

61. Find an equation for the line that passes through $P = (2, -1, 1)$ and intersects the line with equation $\mathbf{r}(t) = \langle 5, -t + 1, -3t + 5\rangle$ at a right angle.

62. In the parallelepiped determined by vectors **a**, **b**, and **c**, what is the significance of $\mathbf{a} + \mathbf{b} + \mathbf{c}$?

63. Prove that the determinant formula for the triple product given in (13) is correct.

8.4 Linear Functions

In Section 8.2 we worked with the equation

$$\begin{pmatrix} x_1 \\ y_1 \end{pmatrix} = \begin{pmatrix} 0.99x_0 + 0.055y_0 \\ 0.023x_0 + 0.94y_0 \end{pmatrix} = \begin{pmatrix} 0.99 & 0.055 \\ 0.023 & 0.94 \end{pmatrix} \begin{pmatrix} x_0 \\ y_0 \end{pmatrix}, \tag{1}$$

where x_0 and x_1 are the numbers of 20-and-over people in the country in one year and the next, and y_0 and y_1 are the analogous numbers for the under-20 group. We can think of (1) as defining a function L that takes as input a (column) vector $\mathbf{v} = \begin{pmatrix} x_0 \\ y_0 \end{pmatrix}$ from R^2 and returns, as output, another vector $L(\mathbf{v}) = \begin{pmatrix} x_1 \\ y_1 \end{pmatrix}$ in R^2. We can write

$$L(\mathbf{v}) = L\left(\begin{pmatrix} x_0 \\ y_0 \end{pmatrix}\right) = \begin{pmatrix} 0.99 & 0.055 \\ 0.023 & 0.94 \end{pmatrix} \begin{pmatrix} x_0 \\ y_0 \end{pmatrix}. \tag{2}$$

A function that takes vector input then multiplies the vector by a matrix of constants to get the output is called a *linear function.*

DEFINITION Linear Function

Let L be a function whose domain is R^n ($n = 1$, 2, or 3) and whose range is a subset of R^m ($m = 1, 2$, or 3). We say that L is a **linear function** if there is an $m \times n$ matrix A of constants such that

$$L(\mathbf{v}) = A\mathbf{v} \tag{3}$$

for each (column) vector $\mathbf{v}$ in R^n. The matrix A is called **the matrix associated with** L.

To avoid the awkward double parentheses in (2), we write

$$L\begin{pmatrix} x \\ y \end{pmatrix} \quad \text{instead of} \quad L\left(\begin{pmatrix} x \\ y \end{pmatrix}\right).$$

EXAMPLE 1 Let L be the function with domain R^3 and range in R^2 defined by

$$L(\mathbf{v}) = L\begin{pmatrix} x \\ y \\ z \end{pmatrix} = \begin{pmatrix} 3x - y + 2z \\ x + 5z \end{pmatrix} \quad \text{for each } \mathbf{v} = \begin{pmatrix} x \\ y \\ z \end{pmatrix} \text{ in } R^3. \tag{4}$$

a) Show that L is a linear function.

b) Let

$$\mathbf{v} = \begin{pmatrix} -1 \\ 3 \\ 2 \end{pmatrix} \quad \text{and} \quad \mathbf{w} = \begin{pmatrix} a \\ b \\ a + b \end{pmatrix}.$$

Evaluate $L(\mathbf{v})$ and $L(\mathbf{w})$.

Solution

a) The domain of L is R^3, and the range is in R^2. Hence if L is a linear function, the matrix A associated with L must be a 2×3 matrix. Such a matrix has the form

$$A = \begin{pmatrix} a_{11} & a_{12} & a_{13} \\ a_{21} & a_{22} & a_{23} \end{pmatrix},$$

where the entries a_{ij} are constants. Furthermore, we must have

$$\begin{aligned} \begin{pmatrix} 3x - y + 2z \\ x + 5z \end{pmatrix} &= L\begin{pmatrix} x \\ y \\ z \end{pmatrix} \\ &= \begin{pmatrix} a_{11} & a_{12} & a_{13} \\ a_{21} & a_{22} & a_{23} \end{pmatrix}\begin{pmatrix} x \\ y \\ z \end{pmatrix} = \begin{pmatrix} a_{11}x + a_{12}y + a_{13}z \\ a_{21}x + a_{22}y + a_{23}z \end{pmatrix}, \end{aligned} \tag{5}$$

for every column vector

$$\mathbf{v} = \begin{pmatrix} x \\ y \\ z \end{pmatrix}.$$

Because the first and last matrices in (5) are equal, it follows that

$$a_{11}x + a_{12}y + a_{13}z = 3x - y + 2z \quad \text{and} \quad a_{21}x + a_{22}y + a_{23}z = x + 5z$$

for every choice of x, y, and z. This will be the case only if we take $a_{11} = 3$, $a_{12} = -1$, $a_{13} = 2$, $a_{21} = 1$, $a_{22} = 0$, and $a_{23} = 5$. Thus with

$$A = \begin{pmatrix} 3 & -1 & 2 \\ 1 & 0 & 5 \end{pmatrix},$$

we have

$$A\mathbf{v} = \begin{pmatrix} 3 & -1 & 2 \\ 1 & 0 & 5 \end{pmatrix}\begin{pmatrix} x \\ y \\ z \end{pmatrix} = \begin{pmatrix} 3x - y + 2z \\ x + 5z \end{pmatrix} = L(\mathbf{v}) \tag{6}$$

for every

$$\mathbf{v} = \begin{pmatrix} x \\ y \\ z \end{pmatrix}.$$

This shows that L is a linear function with associated matrix A. Note that the rows of A are just the coefficients of the entries in the output vector in (4).

b) We can evaluate $L(\mathbf{v})$ by substituting $x = -1$, $y = 3$, and $z = 2$ into (4). The result is

$$L(\mathbf{v}) = L\begin{pmatrix} -1 \\ 3 \\ 2 \end{pmatrix} = \begin{pmatrix} 3(-1) - 3 + 2(2) \\ -1 + 5(2) \end{pmatrix} = \begin{pmatrix} -2 \\ 9 \end{pmatrix}.$$

We use (6) to calculate $L(\mathbf{w})$,

$$\begin{aligned} L(\mathbf{w}) = L\begin{pmatrix} a \\ b \\ a+b \end{pmatrix} &= \begin{pmatrix} 3 & -1 & 2 \\ 1 & 0 & 5 \end{pmatrix}\begin{pmatrix} a \\ b \\ a+b \end{pmatrix} \\ &= \begin{pmatrix} 3a - b + 2(a+b) \\ a + 5(a+b) \end{pmatrix} = \begin{pmatrix} 5a + b \\ 6a + 5b \end{pmatrix}. \end{aligned}$$

In (3), (4), and (5) we see two ways to describe a linear function. We can display the matrix-vector product $A\mathbf{v}$ as in (3) and (5), or we can display the output vector by providing a formula for each component, as in (4). We will use both ways of describing linear functions. In the latter case, the matrix associated with L can be found as in the previous example.

EXAMPLE 2 Let L be the function with domain $R = R^1$ and range in R defined by

$$L(\mathbf{v}) = L(x) = (mx),$$

where m is a real number. Show that L is a linear function.

Solution

Let A be the 1×1 matrix (m). If $\mathbf{v} = (x)$ is a vector in R, then

$$A\mathbf{v} = (m)(x) = (mx) = L(\mathbf{v}).$$

Hence L is a linear function and the associated matrix is the 1×1 matrix (m).

The linear function in Example 2 reminds us of the equation $y = mx$ describing a line in the plane. Indeed, linear functions can be thought of as a generalization of the functions that describe planar lines through the origin. In the expression

$$L(\mathbf{v}) = A\mathbf{v},$$

the $n \times 1$ matrix (column vector) $\mathbf{v}$ is analogous to the input variable x in $y = mx$ and the $m \times n$ matrix A is analogous to the slope m. In future chapters we will study the calculus of functions of several variables. We will see that linear functions play a role in this study very similar to the role played by equations of lines in the study of functions of one variable.

Properties of Linear Functions The word *linear* in the name *linear function* refers to two very important properties of such functions. We describe these properties in the following.

The Linearity Properties

Let L be a linear function with domain R^n, $n = 1, 2$, or 3. Let $\mathbf{v}$ and $\mathbf{w}$ be (column) vectors in R^n, and let c be a real number. Then

$$L(\mathbf{v} + \mathbf{w}) = L(\mathbf{v}) + L(\mathbf{w}) \tag{7}$$

and

$$L(c\mathbf{v}) = cL(\mathbf{v}). \tag{8}$$

Combining (7) and (8), we see that if $\mathbf{v}$ and $\mathbf{w}$ are vectors in R^n and if c and d are real numbers, then

$$L(c\mathbf{v} + d\mathbf{w}) = cL(\mathbf{v}) + dL(\mathbf{w}). \tag{9}$$

These properties follow easily from the arithmetic properties of matrices discussed in Section 8.2. Let A be the matrix associated with L, and let $\mathbf{v}$ and $\mathbf{w}$ be two vectors from the domain of L. Then $L(\mathbf{v} + \mathbf{w})$ is calculated by multiplying the matrix A by the (column) vector $\mathbf{v} + \mathbf{w}$,

$$L(\mathbf{v} + \mathbf{w}) = A(\mathbf{v} + \mathbf{w}).$$

Because matrix multiplication distributes over matrix (or vector) addition, we have

$$L(\mathbf{v} + \mathbf{w}) = A(\mathbf{v} + \mathbf{w}) = A\mathbf{v} + A\mathbf{w} = L(\mathbf{v}) + L(\mathbf{w}).$$

This verifies (7).

If c is a real number, then

$$L(c\mathbf{v}) = A(c\mathbf{v}).$$

Because multiplicative constants can be factored from matrix products,

$$L(c\mathbf{v}) = A(c\mathbf{v}) = c(A\mathbf{v}) = cL(\mathbf{v}).$$

This verifies (8).

Finally, because

$$L(c\mathbf{v} + d\mathbf{w}) = L(c\mathbf{v}) + L(d\mathbf{w}) = cL(\mathbf{v}) + dL(\mathbf{w}),$$

(9) follows from (7) and (8). We further explore linear functions in three examples.

EXAMPLE 3 Let L be a linear function with domain in R^n and range in R^m. Let $\mathbf{0}_n$ denote the zero vector in R^n, that is, the (column) vector in which every component is 0. Show that

$$L(\mathbf{0}_n) = \mathbf{0}_m. \tag{10}$$

Solution

First note that if $\mathbf{v}$ is a vector, then

$$0\mathbf{v} = \mathbf{0}.$$

Thus, by (8),

$$L(\mathbf{0}_n) = L(0\,\mathbf{0}_n) = 0L(\mathbf{0}_n) = \mathbf{0}_m.$$

EXAMPLE 4 Suppose that L is a linear function with domain R^2 and range in R^2 and

$$L\begin{pmatrix}1\\0\end{pmatrix} = \begin{pmatrix}-2\\3\end{pmatrix} \quad \text{and} \quad L\begin{pmatrix}0\\1\end{pmatrix} = \begin{pmatrix}4\\1\end{pmatrix}.$$

Use this information to calculate $L\begin{pmatrix}-6\\4\end{pmatrix}$ and to find a formula for $L\begin{pmatrix}x\\y\end{pmatrix}$.

Solution

Applying (9),

$$\begin{aligned} L\begin{pmatrix}-6\\4\end{pmatrix} &= L\left(-6\begin{pmatrix}1\\0\end{pmatrix} + 4\begin{pmatrix}0\\1\end{pmatrix}\right) \\ &= -6L\begin{pmatrix}1\\0\end{pmatrix} + 4L\begin{pmatrix}0\\1\end{pmatrix} = -6\begin{pmatrix}-2\\3\end{pmatrix} + 4\begin{pmatrix}4\\1\end{pmatrix} = \begin{pmatrix}28\\-14\end{pmatrix}. \end{aligned}$$

Similarly,

$$L\begin{pmatrix}x\\y\end{pmatrix} = xL\begin{pmatrix}1\\0\end{pmatrix} + yL\begin{pmatrix}0\\1\end{pmatrix} = x\begin{pmatrix}-2\\3\end{pmatrix} + y\begin{pmatrix}4\\1\end{pmatrix} = \begin{pmatrix}-2x+4y\\3x+y\end{pmatrix}.$$

What is the matrix associated with L?

EXAMPLE 5 Let L be a linear function with domain R^n and range in R^p, and let M be a linear function with domain R^p and range in R^m. The composition of M and L, denoted by $M \circ L$, is the function defined by

$$(M \circ L)(\mathbf{v}) = M(L(\mathbf{v})), \qquad \mathbf{v} \text{ in } R^n. \tag{11}$$

Show that $M \circ L$ satisfies the linearity properties (7) and (8).

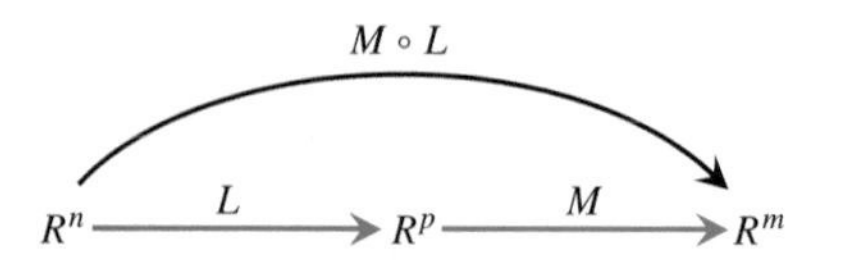

FIGURE 8.27 The function $M \circ L$ is calculated by applying the function M after applying the function L.

Solution

We first show that the definition of $M \circ L$ given in (11) makes sense. Given $\mathbf{v}$ in R^n, we know that $L(\mathbf{v})$ is defined and is a vector in R^p. Because M has domain R^p, the quantity $M(L(\mathbf{v}))$ is also defined, and the result is a vector in R^m. See Fig. 8.27.

Next we show that $M \circ L$ satisfies (7). Because L and M are linear functions, both satisfy (7). Let $\mathbf{v}$ and $\mathbf{w}$ be vectors in R^n. We apply (7) first to L and then to M to see

$$\begin{aligned}(M \circ L)(\mathbf{v} + \mathbf{w}) &= M(L(\mathbf{v} + \mathbf{w})) = M(L(\mathbf{v}) + L(\mathbf{w})) \\ &= M(L(\mathbf{v})) + M(L(\mathbf{w})) = (M \circ L)(\mathbf{v}) + (M \circ L)(\mathbf{w}).\end{aligned}$$

Now let $\mathbf{v}$ be a vector in R^n and let c be a real number. Applying (8) to L and then to M, we find

$$(M \circ L)(c\mathbf{v}) = M(L(c\mathbf{v})) = M(cL(\mathbf{v})) = cM(L(\mathbf{v})) = c(M \circ L)(\mathbf{v}).$$

Thus $M \circ L$ also satisfies (8).

This example can also be done by showing that the composition of linear functions is again a linear function.

Composition of Linear Functions

Let L be a linear function with domain R^n and range in R^p, and let M be a linear function with domain R^p and range in R^m. Let A be the matrix associated with L and B the matrix associated with M. Then $M \circ L$ is a linear function and its associated matrix is BA. Thus, for $\mathbf{v} \in R^n$,

$$(M \circ L)(\mathbf{v}) = (BA)\mathbf{v}.$$

We can verify this result by using the fact that matrix multiplication is associative. If $\mathbf{v}$ is a vector in R^n, we have

$$(M \circ L)(\mathbf{v}) = M(L(\mathbf{v})) = M(A\mathbf{v}) = B(A\mathbf{v}) = (BA)\mathbf{v}.$$

Because $(M \circ L)(\mathbf{v})$ is equal to the matrix product $(BA)\mathbf{v}$, we see that $M \circ L$ is a linear function and has associated matrix BA.

EXAMPLE 6 Let L and M be the linear functions defined by

$$L\begin{pmatrix} x \\ y \end{pmatrix} = \begin{pmatrix} x + y \\ -2y \\ 3x - 4y \end{pmatrix} \quad \text{and} \quad M\begin{pmatrix} x \\ y \\ z \end{pmatrix} = \begin{pmatrix} 2x - z \\ x - y + z \end{pmatrix}.$$

Find a formula for

$$(M \circ L)\begin{pmatrix} x \\ y \end{pmatrix}.$$

Solution

We solve this problem in two ways. For the first solution, we use direct substitution into the formulas for M and L. This gives

$$(M \circ L)\begin{pmatrix} x \\ y \end{pmatrix} = M\left(L\begin{pmatrix} x \\ y \end{pmatrix}\right) = M\begin{pmatrix} x + y \\ -2y \\ 3x - 4y \end{pmatrix}$$
$$= \begin{pmatrix} 2(x + y) - (3x - 4y) \\ (x + y) - (-2y) + (3x - 4y) \end{pmatrix} = \begin{pmatrix} -x + 6y \\ 4x - y \end{pmatrix}.$$

For the second solution, we find the matrices associated with L and M; the product of these matrices is the matrix associated with $M \circ L$. The matrices associated with M and L are, respectively,

$$B = \begin{pmatrix} 2 & 0 & -1 \\ 1 & -1 & 1 \end{pmatrix} \quad \text{and} \quad A = \begin{pmatrix} 1 & 1 \\ 0 & -2 \\ 3 & -4 \end{pmatrix}.$$

The matrix associated with $M \circ L$ is the product of B and A. We have

$$BA = \begin{pmatrix} 2 & 0 & -1 \\ 1 & -1 & 1 \end{pmatrix}\begin{pmatrix} 1 & 1 \\ 0 & -2 \\ 3 & -4 \end{pmatrix} = \begin{pmatrix} -1 & 6 \\ 4 & -1 \end{pmatrix}.$$

Hence

$$(M \circ L)\begin{pmatrix} x \\ y \end{pmatrix} = \begin{pmatrix} -1 & 6 \\ 4 & -1 \end{pmatrix}\begin{pmatrix} x \\ y \end{pmatrix} = \begin{pmatrix} -x + 6y \\ 4x - y \end{pmatrix}.$$

Exercises 8.4

Exercises 1–6: Evaluate $L(\mathbf{v})$ for the given function L and vectors $\mathbf{v}$.

1. $L\begin{pmatrix} x \\ y \end{pmatrix} = \begin{pmatrix} -2x + y \\ x + 4y \end{pmatrix}$

$\mathbf{v} = \begin{pmatrix} 2 \\ -3 \end{pmatrix}, \quad \mathbf{v} = \begin{pmatrix} 0 \\ 1 \end{pmatrix}$

2. $L\begin{pmatrix} x \\ y \end{pmatrix} = \begin{pmatrix} x + 4y \\ -x \\ 2y \end{pmatrix}$

$\mathbf{v} = \begin{pmatrix} -3 \\ 1 \end{pmatrix}, \quad \mathbf{v} = \begin{pmatrix} 4 \\ -4 \end{pmatrix}$

3. $L(x) = \begin{pmatrix} \dfrac{x}{2} \end{pmatrix}$

$\mathbf{v} = (-5), \quad \mathbf{v} = (2002)$

4. $L\begin{pmatrix} x \\ y \end{pmatrix} = \begin{pmatrix} -2 & 4 \\ 1 & 8 \end{pmatrix}\begin{pmatrix} x \\ y \end{pmatrix}$

$\mathbf{v} = \begin{pmatrix} 1 \\ -4 \end{pmatrix}, \quad \mathbf{v} = \begin{pmatrix} 1 \\ 0 \end{pmatrix}$

5. $L\begin{pmatrix} x \\ y \\ z \end{pmatrix} = \begin{pmatrix} 1 & 2 & 3 \\ 1 & 0 & 8 \end{pmatrix}\begin{pmatrix} x \\ y \\ z \end{pmatrix}$

$\mathbf{v} = \begin{pmatrix} 0 \\ 7 \\ -6 \end{pmatrix}, \quad \mathbf{v} = \begin{pmatrix} 1 \\ 0 \\ 1 \end{pmatrix}$

6. $L\begin{pmatrix} x \\ y \\ z \end{pmatrix} = (3 \quad 10 \quad -3)\begin{pmatrix} x \\ y \\ z \end{pmatrix}$

$$\mathbf{v} = \begin{pmatrix} -2 \\ 3 \\ -1 \end{pmatrix}, \quad \mathbf{v} = \begin{pmatrix} 1 \\ -2 \\ 1 \end{pmatrix}$$

Exercises 7–12: Express the function as a matrix-vector product. State the domain R^n and space R^m containing the range, and give the dimension of the matrix.

7. $L\begin{pmatrix} x \\ y \end{pmatrix} = \begin{pmatrix} -3x + y \\ 4x + 7y \end{pmatrix}$

8. $L\begin{pmatrix} x \\ y \\ z \end{pmatrix} = (-4x + z)$

9. $L\begin{pmatrix} x \\ y \\ z \end{pmatrix} = \begin{pmatrix} x \\ y \\ z \end{pmatrix}$

10. $L(x) = (4x)$

11. $L\begin{pmatrix} x \\ y \\ z \end{pmatrix} = (-2x + 3y + 4z)$

12. $L\begin{pmatrix} x \\ y \end{pmatrix} = \begin{pmatrix} x - y \\ x - y \\ \pi y \end{pmatrix}$

Exercises 13–18: For the given matrix A, write out the formula in vector form for the linear function L defined by $L(\mathbf{v}) = A\mathbf{v}$. Give the domain R^n and the space R^m containing the range.

13. $A = \begin{pmatrix} -2 & 3 \\ 6 & 0 \end{pmatrix}$

14. $A = \begin{pmatrix} 2 & 8 & -5 \\ 1 & 2 & -6 \end{pmatrix}$

15. $A = (a \quad b \quad c)$

16. $A = \begin{pmatrix} a \\ b \\ c \end{pmatrix}$

17. $A = (14.92)$

18. $A = \begin{pmatrix} s & e & e \\ t & h & e \\ c & a & t \end{pmatrix}$

19. Let

$$L\begin{pmatrix} x \\ y \end{pmatrix} = \begin{pmatrix} -2x + y \\ 3x + 2y \end{pmatrix}$$

and

$$M\begin{pmatrix} x \\ y \end{pmatrix} = \begin{pmatrix} -x \\ x + 4y \end{pmatrix}.$$

Find the matrix associated with the linear function $M \circ L$. Find the matrix associated with $L \circ M$.

20. Let

$$L\begin{pmatrix} x \\ y \\ z \end{pmatrix} = \begin{pmatrix} x + y + z \\ x - y - z \\ 3y \end{pmatrix}$$

and

$$M\begin{pmatrix} x \\ y \\ z \end{pmatrix} = \begin{pmatrix} -4x + 2y \\ z \\ y - 3z \end{pmatrix}.$$

Find the matrix associated with the linear function $M \circ L$. Find the matrix associated with $L \circ M$.

21. Let

$$L\begin{pmatrix} x \\ y \end{pmatrix} = \begin{pmatrix} 2x \\ 3x - 2y \end{pmatrix}$$

and

$$M\begin{pmatrix} x \\ y \end{pmatrix} = \begin{pmatrix} \frac{1}{2}x \\ \frac{3}{4}x - \frac{1}{2}y \end{pmatrix}.$$

Find the matrix associated with the linear function $M \circ L$ and the matrix associated with $L \circ M$. Describe in words the nature of the composition functions.

22. Let

$$L\begin{pmatrix} x \\ y \\ z \end{pmatrix} = \begin{pmatrix} x + 2y + z \\ -x + 3z \\ 4y + 2z \end{pmatrix}$$

and

$$M\begin{pmatrix} x \\ y \\ z \end{pmatrix} = \begin{pmatrix} x - \frac{1}{2}z \\ -\frac{1}{6}x - \frac{1}{6}y + \frac{1}{3}z \\ \frac{1}{3}x + \frac{1}{3}y - \frac{1}{6}z \end{pmatrix}.$$

Find the matrix associated with the linear function $M \circ L$ and the matrix associated with $L \circ M$. Describe in words the nature of the composition functions.

23. Let

$$A = \begin{pmatrix} a_{11} & a_{12} \\ a_{21} & a_{22} \end{pmatrix}$$

and let L be the linear function on R^2 given by

$$L(\mathbf{v}) = A\mathbf{v}.$$

Show that the columns of A are $L\begin{pmatrix} 1 \\ 0 \end{pmatrix}$ and $L\begin{pmatrix} 0 \\ 1 \end{pmatrix}$.

24. Let

$$A = \begin{pmatrix} a_{11} & a_{12} & a_{13} \\ a_{21} & a_{22} & a_{23} \\ a_{31} & a_{32} & a_{33} \end{pmatrix}$$

and let L be the linear function on R^3 given by

$$L(\mathbf{v}) = A\mathbf{v}.$$

Show that the columns of A are

$$L\begin{pmatrix} 1 \\ 0 \\ 0 \end{pmatrix}, \quad L\begin{pmatrix} 0 \\ 1 \\ 0 \end{pmatrix}, \quad \text{and} \quad L\begin{pmatrix} 0 \\ 0 \\ 1 \end{pmatrix}.$$

25. Show that any function L of the form

$$L\begin{pmatrix} x \\ y \end{pmatrix} = \begin{pmatrix} a_{11}x + a_{12}y \\ a_{21}x + a_{22}y \end{pmatrix}$$

is a linear function.

26. Show that any function L of the form

$$L\begin{pmatrix} x \\ y \\ z \end{pmatrix} = \begin{pmatrix} a_{11}x + a_{12}y + a_{13}z \\ a_{21}x + a_{22}y + a_{23}z \end{pmatrix}$$

is a linear function.

27. Let x be the number of people 20 or over in the United States in a given year, and let y be the number of people under 20 in the same year. Let x_1 and y_1 be these two populations in the following year. In (1), we found that

$$\begin{pmatrix} x_1 \\ y_1 \end{pmatrix} = \begin{pmatrix} 0.99 & 0.055 \\ 0.023 & 0.94 \end{pmatrix} \begin{pmatrix} x \\ y \end{pmatrix}.$$

In 1990 the 20-and-over population was 176,995,000 and the under-20 population was 71,715,000. Use the previous equation to find these two populations in 1991. Once you have the 1991 figures, how could you find the populations in 1992? In 1993?

28. From 1990 to 1991, net immigration/emigration also accounted for some changes in the population. It accounted for an increase of 985,000 in the 20-and-over population and an increase of 400,000 in the under-20 population. How can this information be incorporated into (1)?

29. Find $\begin{pmatrix} x \\ y \end{pmatrix}$ so that

$$\begin{pmatrix} -2 & 3 \\ 1 & 7 \end{pmatrix} \begin{pmatrix} x \\ y \end{pmatrix} = \begin{pmatrix} 0 \\ 2 \end{pmatrix}.$$

30. Find $\begin{pmatrix} x \\ y \end{pmatrix}$ so that $\begin{pmatrix} 4 & -2 \\ 0 & -1 \end{pmatrix} \begin{pmatrix} x \\ y \end{pmatrix} = \begin{pmatrix} 5 \\ -4 \end{pmatrix}$.

31. Find three different vectors $\begin{pmatrix} x \\ y \\ z \end{pmatrix}$ so that

$$\begin{pmatrix} 1 & 1 & 1 \\ 3 & 0 & -2 \end{pmatrix} \begin{pmatrix} x \\ y \\ z \end{pmatrix} = \begin{pmatrix} 3 \\ 1 \end{pmatrix}.$$

32. Find three different vectors $\begin{pmatrix} x \\ y \\ z \end{pmatrix}$ so that

$$\begin{pmatrix} -4 & 5 & 1 \\ 2 & -3 & 1 \end{pmatrix} \begin{pmatrix} x \\ y \\ z \end{pmatrix} = \begin{pmatrix} 0 \\ 0 \end{pmatrix}.$$

33. Let

$$A = \begin{pmatrix} a_{11} & a_{12} & a_{13} \\ a_{21} & a_{22} & a_{23} \end{pmatrix}$$

and let L be the linear function with associated matrix A. Let $\mathbf{r}_1$ and $\mathbf{r}_2$ be the rows of A. Show that any (column) vector of the form

$$\mathbf{v} = k(\mathbf{r}_1 \times \mathbf{r}_2), \qquad k \text{ real},$$

is a solution of the equation

$$L(\mathbf{v}) = \begin{pmatrix} 0 \\ 0 \end{pmatrix}.$$

(*Hint:* First tell why $k(\mathbf{r}_1 \times \mathbf{r}_2)$ is perpendicular to $\mathbf{r}_1$ and $\mathbf{r}_2$.)

34. Use the result established in Exercise 33 to find infinitely many solutions to the equation

$$\begin{pmatrix} -3 & 0 & 6 \\ 2 & -1 & -1 \end{pmatrix} \begin{pmatrix} x \\ y \\ z \end{pmatrix} = \begin{pmatrix} 0 \\ 0 \end{pmatrix}.$$

35. Use the result established in Exercise 33 to find infinitely many solutions to the equation

$$\begin{pmatrix} 1 & 2 & -3 \\ 4 & 1 & -6 \end{pmatrix} \begin{pmatrix} x \\ y \\ z \end{pmatrix} = \begin{pmatrix} 0 \\ 0 \end{pmatrix}.$$

36. Use the result established in Exercise 33 to find infinitely many solutions to the system of equations

$$\begin{aligned} -2x + 4y - 2z &= 0 \\ 8x - y + 4z &= 0. \end{aligned}$$

37. Use the result established in Exercise 33 to find infinitely many solutions to the system of equations

$$\begin{aligned} x - 7y + 2z &= 0 \\ 4x + 3y \quad &= 0. \end{aligned}$$

38. Let L be defined by

$$L\begin{pmatrix} x \\ y \end{pmatrix} = \begin{pmatrix} -2x + 3y + 1 \\ 4x + 3y \end{pmatrix}.$$

Calculate $L(\mathbf{0}_2)$ and tell why the result shows that L is not a linear function.

39. Let L be defined by

$$L\begin{pmatrix} x \\ y \end{pmatrix} = \begin{pmatrix} -x^2 + 3y \\ x - 2y \end{pmatrix}.$$

Let

$$\mathbf{v} = \begin{pmatrix} 2 \\ 2 \end{pmatrix} \quad \text{and} \quad \mathbf{w} = \begin{pmatrix} -3 \\ 1 \end{pmatrix}.$$

Show that

$$L(\mathbf{v} + \mathbf{w}) \neq L(\mathbf{v}) + L(\mathbf{w}).$$

Explain why this result proves that L is not a linear function.

40. Let L be defined by

$$L\begin{pmatrix} x \\ y \\ z \end{pmatrix} = \begin{pmatrix} -x + 3y + z \\ x - 2y + xz \end{pmatrix},$$

and let

$$\mathbf{v} = \begin{pmatrix} 2 \\ 1 \\ -2 \end{pmatrix}.$$

Show that

$$L(3\mathbf{v}) \neq 3L(\mathbf{v}).$$

Explain why this result proves that L is not a linear function.

Exercises 41–44: Show that the given function L is not linear by finding specific vectors for which (7), (8), or (9) does not hold. See Exercises 38, 39, and 40.

41. $L\begin{pmatrix} x \\ y \end{pmatrix} = \begin{pmatrix} -2x + y - 1 \\ x - 4y + 3 \end{pmatrix}$

42. $L\begin{pmatrix} x \\ y \\ z \end{pmatrix} = \begin{pmatrix} x \\ x + y \\ \sin(2z) \end{pmatrix}$

43. $L\begin{pmatrix} x \\ y \end{pmatrix} = (xy + x + y)$

44. $L\begin{pmatrix} x \\ y \\ z \end{pmatrix} = \begin{pmatrix} |x + y| \\ |x| - |z| \\ |x| + |y| + |z| \end{pmatrix}$

8.5 The Geometry of Linear Functions

On CNC (computer numerically controlled) drilling machines, the drilling mechanism can be programmed to drill a series of holes in virtually any pattern in a workpiece. See Fig. 8.28. Because the movement is programmed, the process can be repeated to produce large numbers of nearly identical pieces.

Suppose we wish to drill two holes in each of several circular metal disks. The first hole A is to be drilled at distance a from the center O. The second hole B is to be drilled again at distance a from the center, but so that angle AOB has measure 75°. See Fig. 8.29. The disk is then removed and a new one put in place. Instructions for the motion of the drill can be described in terms of linear functions. On some machines the matrices associated with these functions are the input that tells a computer how to guide the drill mechanism. To find the linear function that describes the location of the second hole, it is helpful to consider the geometry of the situation.

FIGURE 8.28 Programmable drilling machine.

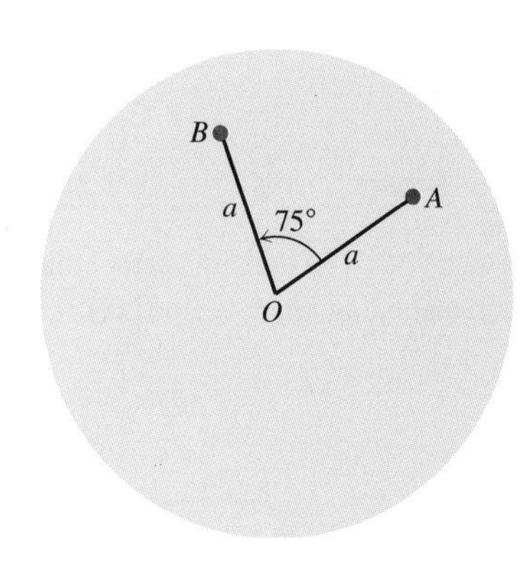

FIGURE 8.29 Drill holes at A and B.

Assume that the disk to be drilled lies in R^2 with the center of the disk at the origin, and that the location of the first hole is given by the position vector $\mathbf{v} = \begin{pmatrix} x \\ y \end{pmatrix}$ (or $\langle x, y \rangle$). Because we are concerned with the angle between the position vectors for the two holes, we express the components of these vectors in polar coordinates. The point (x, y) has polar coordinates (a, θ) for some θ. Hence

$$\mathbf{v} = \begin{pmatrix} x \\ y \end{pmatrix} = \begin{pmatrix} a \cos \theta \\ a \sin \theta \end{pmatrix}.$$

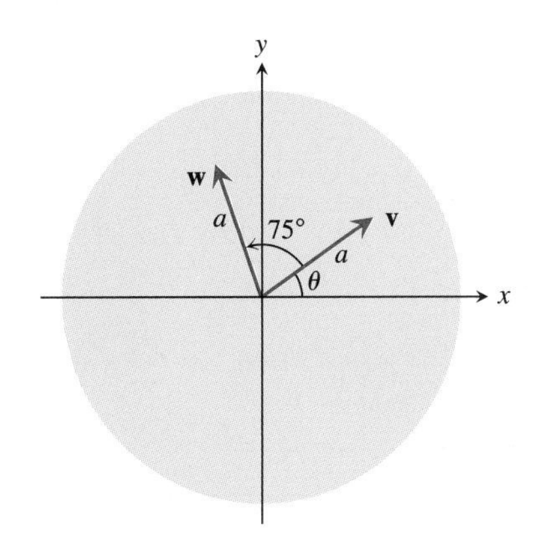

FIGURE 8.30 Position vectors for the holes.

Let $\mathbf{w}$ be the position vector for the second hole. This vector also has length a, and the angle from $\mathbf{v}$ to $\mathbf{w}$ is 75°. See Fig. 8.30. Using a trigonometric identity,

$$\begin{aligned} \mathbf{w} &= \begin{pmatrix} a \cos(\theta + 75^\circ) \\ a \sin(\theta + 75^\circ) \end{pmatrix} = \begin{pmatrix} a \cos \theta \cos 75^\circ - a \sin \theta \sin 75^\circ \\ a \cos \theta \sin 75^\circ + a \sin \theta \cos 75^\circ \end{pmatrix} \\ &= \begin{pmatrix} \cos 75^\circ & -\sin 75^\circ \\ \sin 75^\circ & \cos 75^\circ \end{pmatrix} \begin{pmatrix} a \cos \theta \\ a \sin \theta \end{pmatrix} = \begin{pmatrix} \cos 75^\circ & -\sin 75^\circ \\ \sin 75^\circ & \cos 75^\circ \end{pmatrix} \mathbf{v}. \end{aligned} \tag{1}$$

Thus if $\mathbf{v}$ describes the position of the first hole, the position of the second is found by calculating

$$\mathbf{w} = L(\mathbf{v}), \tag{2}$$

where L is the linear function with associated matrix

$$A = \begin{pmatrix} \cos 75^\circ & -\sin 75^\circ \\ \sin 75^\circ & \cos 75^\circ \end{pmatrix} \approx \begin{pmatrix} 0.258819 & -0.965926 \\ 0.965926 & 0.258819 \end{pmatrix}. \tag{3}$$

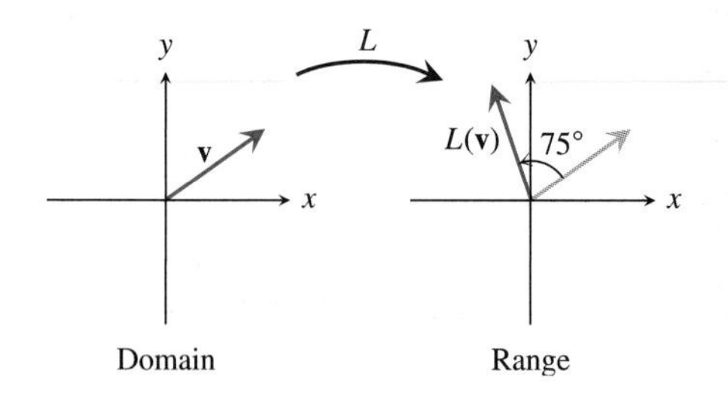

FIGURE 8.31 The domain vector **v** and the corresponding range vector $L(\mathbf{v})$.

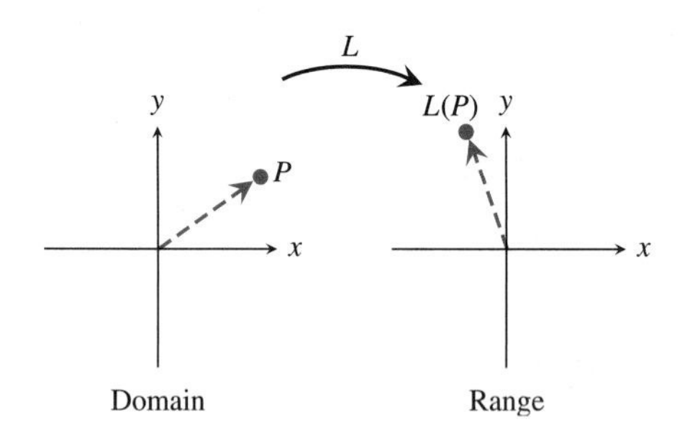

FIGURE 8.32 We can think of a linear function as mapping points.

Graphs of Linear Functions

The linear function L defined by (2) and (3) is completely described by the associated matrix A. But we also see from (1) that L can be described in another way: as taking an input vector **v** and returning a vector **w** of the same length and so that the angle from **v** to **w** has measure 75°. We can display this graphically by exhibiting two planes side by side. The plane on the left represents the domain, R^2, and the plane on the right represents the set R^2 that contains the range. See Fig. 8.31. To "graph" L, we draw an input vector **v** in the domain plane and then draw the corresponding output vector $L(\mathbf{v})$ in the range plane. The resulting picture is not a complete description of L but may give us a way to visualize how the input and output vectors for L are related.

Sometimes it is useful to think of linear functions as mapping points instead of vectors. Let $P = (x, y)$ be a point in the plane, and let $\mathbf{P} = \begin{pmatrix} x \\ y \end{pmatrix}$ or $\langle x, y\rangle$ be the position vector from the origin to P. The corresponding output vector $L(\mathbf{P})$ is the position vector for a point in the range plane. We shall refer to this point as $L(P)$ and say that L *maps* the point P to the point $L(P)$. See Fig. 8.32. The idea of mapping points enables us to investigate how L transforms sets. For example, consider the set S of all points on and inside the square with vertices $(0,0)$, $(1,0)$, $(1,1)$, and $(0,1)$. We shall refer to this square as the **unit square in R^2**. For each point P in the unit square, $L(P)$ is found by rotating P about the origin through 75°. Hence the set of all points $L(P)$ is found by rotating S through 75°. See Fig. 8.33. The set $L(S)$ is called the *image* of S and is defined by

$$L(S) = \{L(P)\colon P \text{ is a point in } S\}.$$

Because the point notation $P = (x, y)$ and the vector notation $\mathbf{P} = \langle x, y\rangle$ convey similar information, we will often use $\mathbf{P}$ to denote both the point $P = (x, y)$ and the vector $\mathbf{P} = \langle x, y\rangle$. Thus, $L(\mathbf{P})$ may be interpreted as the image of the point P or the vector $\mathbf{P}$. When one interpretation is intended instead of the other it will usually be clear from context.

Images of Segments and Polygons Sketches like those in Figs. 8.31, 8.32, and 8.33 can be useful in understanding linear functions. To facilitate this understanding we study the mapping properties of linear functions.

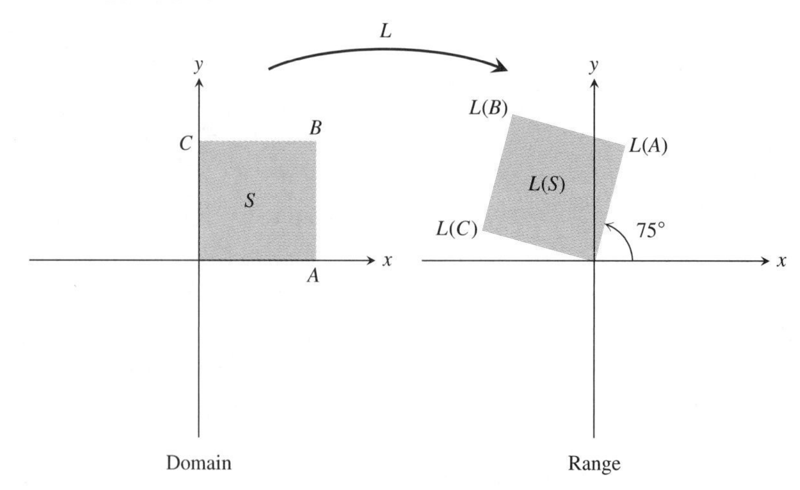

FIGURE 8.33 The image of the unit square under L.

FIGURE 8.34 A linear function maps a segment to a point or a segment.

Linear Functions and Line Segments

Let L be a linear function with domain R^n and range in R^m. Let S be a line segment in R^n, and suppose S is described by

$$\mathbf{r}(t) = t\mathbf{v} + \mathbf{c}, \qquad a \le t \le b,$$

where a and b are real numbers with $a < b$. Then $L(S)$ is a point or a line segment in R^m. In particular, $L(S)$ is a single point if $L(\mathbf{v}) = \mathbf{0}$, and it is a line segment if $L(\mathbf{v}) \neq \mathbf{0}$. See Fig. 8.34.

To prove this assertion we use the linearity properties of linear functions discussed in Section 8.4. Let segment S be the set of points described by the position vector $\mathbf{r}(t)$ as t varies from a to b. Then $L(S)$ is the set of points described by the position vector $L(\mathbf{r}(t))$ as t varies from a to b. Thus the points on $L(S)$ are described by

$$\begin{aligned} \mathbf{w}(t) &= L(\mathbf{r}(t)) = L(t\mathbf{v} + \mathbf{c}) \\ &= L(t\mathbf{v}) + L(\mathbf{c}) = tL(\mathbf{v}) + L(\mathbf{c}). \end{aligned} \tag{4}$$

If $L(\mathbf{v}) = \mathbf{0}$, then (4) reduces to

$$\mathbf{w}(t) = \mathbf{0}t + L(\mathbf{c}) = \mathbf{0} + L(\mathbf{c}) = L(\mathbf{c}), \qquad a \le t \le b.$$

So the image $L(S)$ of S is simply the point in R^m corresponding to the position vector $L(\mathbf{c})$. If $L(\mathbf{v}) \neq \mathbf{0}$, then

$$\mathbf{w}(t) = L(\mathbf{v})t + L(\mathbf{c}), \qquad a \le t \le b,$$

describes the line segment joining the points

$$\mathbf{w}(a) = aL(\mathbf{v}) + L(\mathbf{c}) \quad \text{and} \quad \mathbf{w}(b) = bL(\mathbf{v}) + L(\mathbf{c}) \tag{5}$$

in R^m. This segment is parallel to the vector $L(\mathbf{v})$. This discussion also shows that parallel line segments in the domain map to parallel line segments in the range. See Exercise 54.

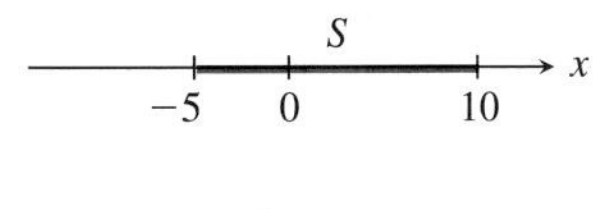

L

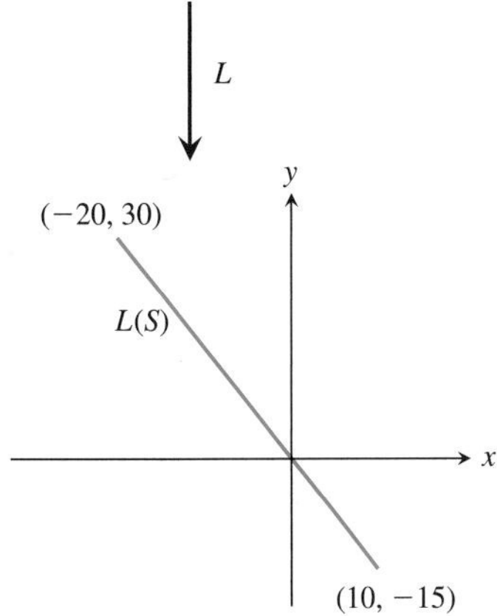

Range

FIGURE 8.35 The linear function L maps the real interval $[-5, 10]$ to a segment in R^2.

EXAMPLE 1 Let L be the linear function with domain $R = R^1$ and range in R^2 given by

$$L(\mathbf{v}) = L(x) = \begin{pmatrix} -2x \\ 3x \end{pmatrix} = \begin{pmatrix} -2 \\ 3 \end{pmatrix}(x) \quad \text{for } \mathbf{v} = (x) \text{ in } R^1.$$

Let S be the R^1 segment described by

$$\mathbf{r}(t) = t(5), \qquad -1 \le t \le 2. \tag{6}$$

Describe the graph $L(S)$ in R^2.

Solution

First note that the position vector $\mathbf{r}(t)$ in (6) describes the points on the x-axis with coordinates $x = 5t$, $-1 \le t \le 2$. Thus S is the segment (or interval) $-5 \le x \le 10$ in R. See Fig. 8.35. The points on $L(S)$ are described by the position vector

$$\mathbf{w}(t) = L(\mathbf{r}(t)) = \begin{pmatrix} -2 \\ 3 \end{pmatrix} t(5) = \begin{pmatrix} -10t \\ 15t \end{pmatrix} = \begin{pmatrix} -10 \\ 15 \end{pmatrix} t \tag{7}$$

for $-1 \le t \le 2$. Thus $L(S)$ is the line segment joining the points

$$\mathbf{w}(-1) = \begin{pmatrix} 10 \\ -15 \end{pmatrix} \quad \text{and} \quad \mathbf{w}(2) = \begin{pmatrix} -20 \\ 30 \end{pmatrix}.$$

In Fig. 8.35 we show S and $L(S)$.

As suggested by the previous example and by (5), if S is the line segment with endpoints A and B, then $L(S)$ is the line segment with endpoints $L(A)$ and $L(B)$. This observation leads to another way to graph and describe the set $L(S)$.

EXAMPLE 2 Let L be a linear function with domain R^3, range in R^2, and with

$$L(\mathbf{v}) = \begin{pmatrix} -1 & 0 & 2 \\ 2 & -2 & 3 \end{pmatrix}\mathbf{v}, \qquad \mathbf{v} \text{ in } R^3. \tag{8}$$

Let S_1 be the R^3 segment described by

$$\mathbf{r}_1(t) = \begin{pmatrix} -4 \\ 0 \\ 1 \end{pmatrix} t + \begin{pmatrix} -1 \\ 1 \\ 2 \end{pmatrix}, \qquad 0 \le t \le 1,$$

and let S_2 be the segment given by

$$\mathbf{r}_2(t) = \begin{pmatrix} 4 \\ 7 \\ 2 \end{pmatrix} t + \begin{pmatrix} -1 \\ 1 \\ 2 \end{pmatrix}, \qquad 0 \le t \le 1.$$

Describe and graph $L(S_1)$ and $L(S_2)$.

Solution

The endpoints of S_1 are

$$\mathbf{A} = \mathbf{r}_1(0) = \begin{pmatrix} -1 \\ 1 \\ 2 \end{pmatrix} \quad \text{and} \quad \mathbf{B} = \mathbf{r}_1(1) = \begin{pmatrix} -5 \\ 1 \\ 3 \end{pmatrix}.$$

Because L maps segments to segments, it follows that $L(S_1)$ is the segment with endpoints

$$L(\mathbf{A}) = L\begin{pmatrix} -1 \\ 1 \\ 2 \end{pmatrix} = \begin{pmatrix} 5 \\ 2 \end{pmatrix} \quad \text{and} \quad L(\mathbf{B}) = L\begin{pmatrix} -5 \\ 1 \\ 3 \end{pmatrix} = \begin{pmatrix} 11 \\ -3 \end{pmatrix}.$$

We can describe $L(S_1)$ as the segment in R^2 with endpoints $\begin{pmatrix} 5 \\ 2 \end{pmatrix}$ and $\begin{pmatrix} 11 \\ -3 \end{pmatrix}$. We graph $L(S_1)$ by plotting these points then drawing the segment that joins them. In Fig. 8.36 we show S_1 and $L(S_1)$. Because S_1 is a segment in R^3, the domain part of this graph is a picture of R^3. The range of L is in R^2, so the range portion shows R^2.

Next note that segment S_2 has endpoints

$$\mathbf{C} = \mathbf{r}_2(0) = \begin{pmatrix} -1 \\ 1 \\ 2 \end{pmatrix} \quad \text{and} \quad \mathbf{D} = \mathbf{r}_2(1) = \begin{pmatrix} 3 \\ 8 \\ 4 \end{pmatrix}.$$

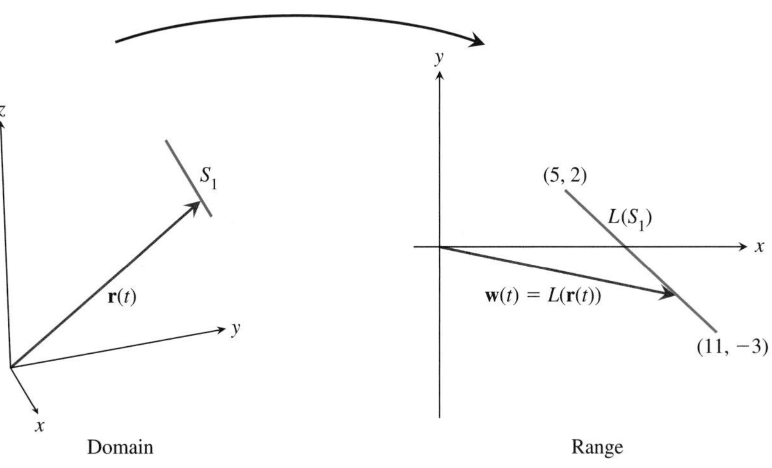

FIGURE 8.36 The linear function L maps the segment S_1 mR^3 to a segment in R^2.

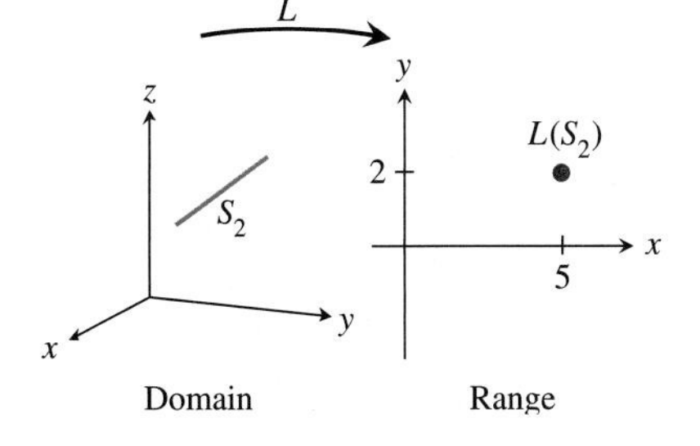

FIGURE 8.37 The linear function L maps the segment S_2 in R^3 to a point in R^2.

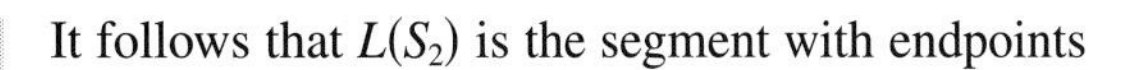

It follows that $L(S_2)$ is the segment with endpoints

$$L(\mathbf{C}) = L\begin{pmatrix} -1 \\ 1 \\ 2 \end{pmatrix} = \begin{pmatrix} 5 \\ 2 \end{pmatrix} \quad \text{and} \quad L(\mathbf{D}) = L\begin{pmatrix} 3 \\ 8 \\ 4 \end{pmatrix} = \begin{pmatrix} 5 \\ 2 \end{pmatrix}.$$

Because these two endpoints are identical, it follows that $L(S_2)$ is just the point with coordinates $\begin{pmatrix} 5 \\ 2 \end{pmatrix}$ (or $(5, 2)$) in R^2. See Fig. 8.37.

Once we know that linear functions map line segments to line segments or points (degenerate line segments), and that parallel segments map to parallel segments, we can say something about images of polygons. Of particular importance is the case in which the polygon is a parallelogram.

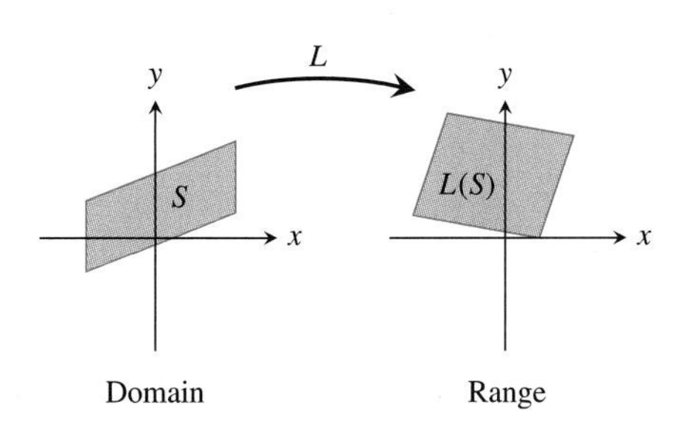

FIGURE 8.38 Linear functions map parallelograms to parallelograms.

Linear Functions and Parallelograms

Let L be a linear function with domain R^n and range in R^m. Let S be the set of points on and inside a parallelogram in R^n. Then $L(S)$ is either a single point, a line segment, or the set of points on and inside a parallelogram. See Fig. 8.38. In particular, let S be determined by the vectors **a** and **b**. Then $L(S)$ is

- a single point if $L(\mathbf{a}) = L(\mathbf{b}) = \mathbf{0}$.
- a nondegenerate line segment if $L(\mathbf{a})$ and $L(\mathbf{b})$ are parallel, nonzero vectors or if one of $L(\mathbf{a})$ and $L(\mathbf{b})$ is equal to $\mathbf{0}$ and the other is not.
- the set of points on and inside of a parallelogram if $L(\mathbf{a})$ and $L(\mathbf{b})$ are not parallel and neither one is $\mathbf{0}$.

We discuss this result for the special case in which $R^n = R^m = R^2$ and S is a rectangle with one vertex at the origin and sides parallel to the coordinate axes. Let

$$\mathbf{a} = \begin{pmatrix} a \\ 0 \end{pmatrix} \quad \text{and} \quad \mathbf{b} = \begin{pmatrix} 0 \\ b \end{pmatrix}$$

be the vectors determined by the sides of S. If $P = (x, y)$ is a point in S, then the position vector

$$\mathbf{P} = \begin{pmatrix} x \\ y \end{pmatrix}$$

for P can be written as

$$\mathbf{P} = \begin{pmatrix} x \\ y \end{pmatrix} = \frac{x}{a}\begin{pmatrix} a \\ 0 \end{pmatrix} + \frac{y}{b}\begin{pmatrix} 0 \\ b \end{pmatrix} = \frac{x}{a}\mathbf{a} + \frac{y}{b}\mathbf{b}.$$

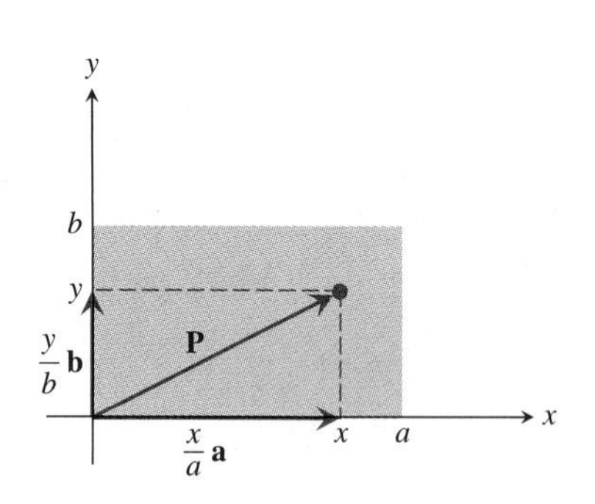

FIGURE 8.39 For P in the rectangle, we express **P** as the sum of a scalar multiple of **a** and a scalar multiple of **b**.

Because $0 \le x \le a$ and $0 \le y \le b$, the numbers x/a and y/b are both between 0 and 1. See Fig. 8.39. By the linearity properties for L, we have

$$L(\mathbf{P}) = L\left(\frac{x}{a}\mathbf{a} + \frac{y}{b}\mathbf{b}\right) = \frac{x}{a}L(\mathbf{a}) + \frac{y}{b}L(\mathbf{b}). \tag{9}$$

We now consider three cases.

Case 1: If $L(\mathbf{a}) = L(\mathbf{b}) = \mathbf{0}$, then it follows from (9) that $L(\mathbf{P}) = \mathbf{0}$ for every point P in S. In this case $L(S)$ is simply the origin in R^2; that is, $L(S)$ is a single point.

Case 2: Suppose that $L(\mathbf{a})$ and $L(\mathbf{b})$ are parallel, nonzero vectors or (as one of two similar cases) $L(\mathbf{a}) \neq \mathbf{0}$ and $L(\mathbf{b}) = \mathbf{0}$. Then there is a constant k such that $L(\mathbf{b}) = kL(\mathbf{a})$. Substituting into (9), we find that

$$L(\mathbf{P}) = \left(\frac{x}{a} + k\frac{y}{b}\right)L(\mathbf{a}). \tag{10}$$

As P moves about in S, x/a and y/b each take on every value in $[0, 1]$, so the points $L(\mathbf{P})$ described by (10) all lie on a segment that contains the origin and is parallel to $L(\mathbf{a})$. See Fig. 8.40.

Case 3: Next assume that $L(\mathbf{a})$ and $L(\mathbf{b})$ are not parallel and that neither is equal to $\mathbf{0}$. Because x/a is in $[0, 1]$, the vector $(x/a)L(\mathbf{a})$ points in the same direction as $L(\mathbf{a})$

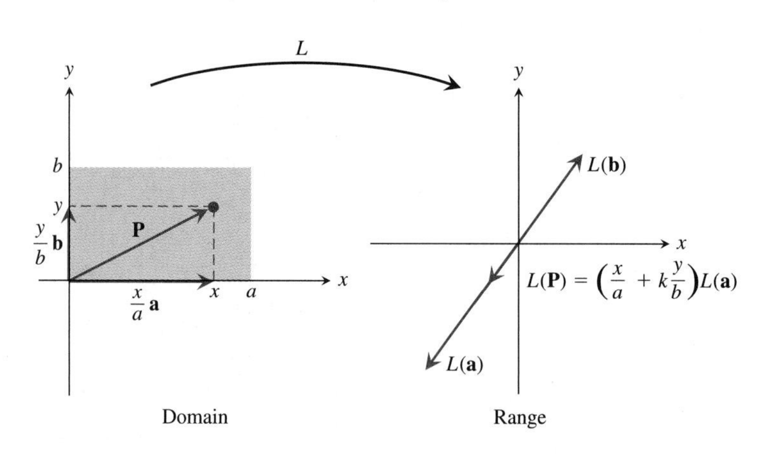

FIGURE 8.40 The image of a parallelogram may be a line segment.

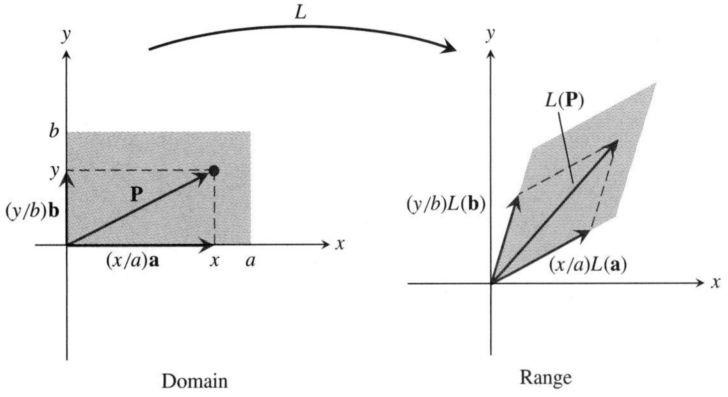

FIGURE 8.41 Because $\mathbf{P} = (x/a)\mathbf{a} + (y/b)\mathbf{b}$, we have $L(\mathbf{P}) = (x/a)L(\mathbf{a}) + (y/b)L(\mathbf{b})$.

and is shorter than $L(\mathbf{a})$. A similar statement holds for $(y/b)L(\mathbf{b})$. When we use the parallelogram law to add these two vectors, we see that the result is a vector on or inside of the parallelogram determined by $L(\mathbf{a})$ and $L(\mathbf{b})$. See Fig. 8.41. It follows from (9) that for each P in S, the point $L(P)$ is on or inside the parallelogram determined by $L(\mathbf{a})$ and $L(\mathbf{b})$. Furthermore, by varying x between 0 and a and y between 0 and b, the point $L(\mathbf{P})$ can be made to coincide with any given point on or inside of the parallelogram determined by $L(\mathbf{a})$ and $L(\mathbf{b})$ (see also Exercise 48). Hence $L(S)$ is the set of points on or inside of this parallelogram.

We have already seen that if L is a linear function and S is a line segment, then $L(S)$ is also a line segment, and that L maps the endpoints of S to the endpoints of $L(S)$. From this it follows that if S_1 is a parallelogram, then the vertices of S_1 map to the vertices of $L(S_1)$.

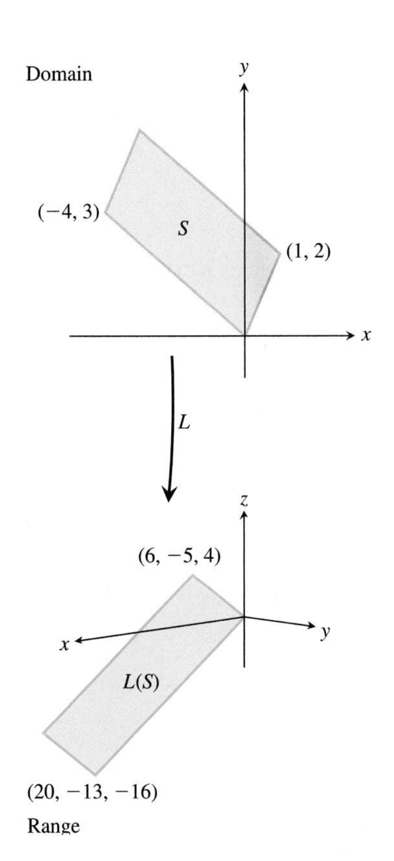

FIGURE 8.42 The parallelogram S in R^2 maps to the parallelogram $L(S)$ in R^3.

EXAMPLE 3 Let L be defined by

$$L\begin{pmatrix} x \\ y \end{pmatrix} = \begin{pmatrix} -2 & 4 \\ 1 & -3 \\ 4 & 0 \end{pmatrix}\begin{pmatrix} x \\ y \end{pmatrix}.$$

Let S be the parallelogram with one vertex at the origin and determined by the position vectors $\mathbf{a} = \begin{pmatrix} -4 \\ 3 \end{pmatrix}$ and $\mathbf{b} = \begin{pmatrix} 1 \\ 2 \end{pmatrix}$. Describe $L(S)$ and sketch both S and $L(S)$.

Solution

A picture of S is shown in the top graph in Fig. 8.42. The vertices of S are $(0,0)$, $(-4,3)$, $(-3,5)$, and $(1,2)$ or, writing them as column vectors,

$$\begin{pmatrix} 0 \\ 0 \end{pmatrix}, \quad \begin{pmatrix} -4 \\ 3 \end{pmatrix}, \quad \begin{pmatrix} -3 \\ 5 \end{pmatrix}, \quad \text{and} \quad \begin{pmatrix} 1 \\ 2 \end{pmatrix}.$$

The vertices of $L(S)$ are

$$L\begin{pmatrix}0\\0\end{pmatrix}=\begin{pmatrix}0\\0\\0\end{pmatrix},\quad L\begin{pmatrix}-4\\3\end{pmatrix}=\begin{pmatrix}20\\-13\\-16\end{pmatrix},\quad L\begin{pmatrix}-3\\5\end{pmatrix}=\begin{pmatrix}26\\-18\\-12\end{pmatrix},$$

$$\text{and}\quad L\begin{pmatrix}1\\2\end{pmatrix}=\begin{pmatrix}6\\-5\\4\end{pmatrix}.$$

The parallelogram with these four vertices (in R^3) is shown in Fig. 8.42. Note also that $L(S)$ is the parallelogram determined by the position vectors

$$L(\mathbf{a})=L\begin{pmatrix}-4\\3\end{pmatrix}=\begin{pmatrix}20\\-13\\-16\end{pmatrix}\quad\text{and}\quad L(\mathbf{b})=L\begin{pmatrix}1\\2\end{pmatrix}=\begin{pmatrix}6\\-5\\4\end{pmatrix}.$$

When L is a linear function with both domain and range in R^2, the areas of a parallelogram and its image are closely related. We investigate this relationship in the next example.

EXAMPLE 4 Let L, given by

$$L\begin{pmatrix}x\\y\end{pmatrix}=\begin{pmatrix}a&b\\c&d\end{pmatrix}\begin{pmatrix}x\\y\end{pmatrix},$$

be a linear function with domain R^2 and range in R^2.

a) Let S be the points on and inside the square with vertices $(0,0)$, $(s,0)$, (s,s), and $(0,s)$. Show that $L(S)$ has area

$$s^2\left|\det\begin{pmatrix}a&b\\c&d\end{pmatrix}\right|.$$

b) Let S' be the points on and inside the square with vertices (k,l), $(k+s,l)$, $(k+s,l+s)$, and $(k,l+s)$. Show that $L(S')$ also has area

$$s^2\left|\det\begin{pmatrix}a&b\\c&d\end{pmatrix}\right|.$$

Solution

a) Because S is determined by the vectors

$$\mathbf{v}=\begin{pmatrix}s\\0\end{pmatrix}\quad\text{and}\quad\mathbf{w}=\begin{pmatrix}0\\s\end{pmatrix},$$

the set $L(S)$ is determined by the vectors

$$L(\mathbf{v})=L\begin{pmatrix}s\\0\end{pmatrix}=\begin{pmatrix}as\\cs\end{pmatrix}\quad\text{and}\quad L(\mathbf{w})=L\begin{pmatrix}0\\s\end{pmatrix}=\begin{pmatrix}bs\\ds\end{pmatrix}.$$

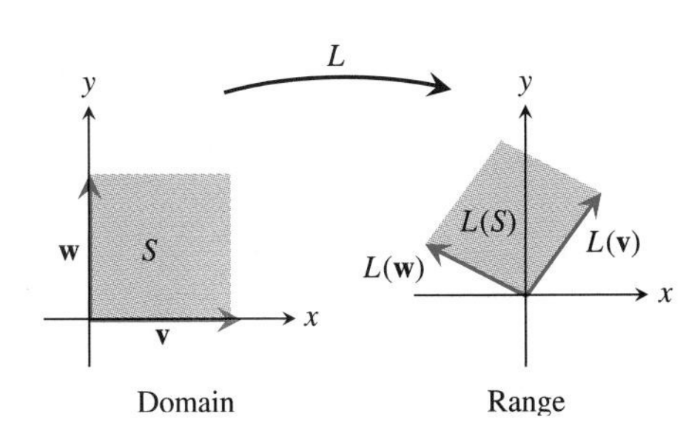

FIGURE 8.43 The square determined by **v** and **w** maps to the parallelogram determined by $L(\mathbf{v})$ and $L(\mathbf{w})$.

Because $L(S)$ is the region on and inside the parallelogram determined by these vectors, the area of $L(S)$ is

$$\text{area}(L(S)) = \left|\det\begin{pmatrix} as & bs \\ cs & ds \end{pmatrix}\right| = s^2\left|\det\begin{pmatrix} a & b \\ c & d \end{pmatrix}\right|,$$

where we have used one of the properties of the determinant seen in Section 8.2 to take a factor of s from each row of the matrix. In Fig. 8.43 we show S and $L(S)$.

b) The vertices of S' are

$$P = \begin{pmatrix} k \\ l \end{pmatrix}, \quad Q = \begin{pmatrix} k + s \\ l \end{pmatrix}, \quad R = \begin{pmatrix} k + s \\ l + s \end{pmatrix}, \quad \text{and} \quad T = \begin{pmatrix} k \\ l + s \end{pmatrix}.$$

Hence, the vertices of the parallelogram $L(S)$ are

$$L(P) = \begin{pmatrix} ak + bl \\ ck + dl \end{pmatrix}, \quad L(Q) = \begin{pmatrix} a(k + s) + bl \\ c(k + s) + dl \end{pmatrix},$$

$$L(R) = \begin{pmatrix} a(k + s) + b(l + s) \\ c(k + s) + d(l + s) \end{pmatrix},$$

$$\text{and} \quad L(T) = \begin{pmatrix} ak + b(l + s) \\ ck + d(l + s) \end{pmatrix}.$$

The sides that share vertex $L(P)$ in this parallelogram are the vectors

$$L(Q) - L(P) = \begin{pmatrix} as \\ cs \end{pmatrix} \quad \text{and} \quad L(T) - L(P) = \begin{pmatrix} bs \\ ds \end{pmatrix}.$$

See Fig. 8.44. As was the case in part (a), the area of this parallelogram is

$$\text{area}(L(S')) = \left|\det\begin{pmatrix} as & bs \\ cs & ds \end{pmatrix}\right| = s^2\left|\det\begin{pmatrix} a & b \\ c & d \end{pmatrix}\right|.$$

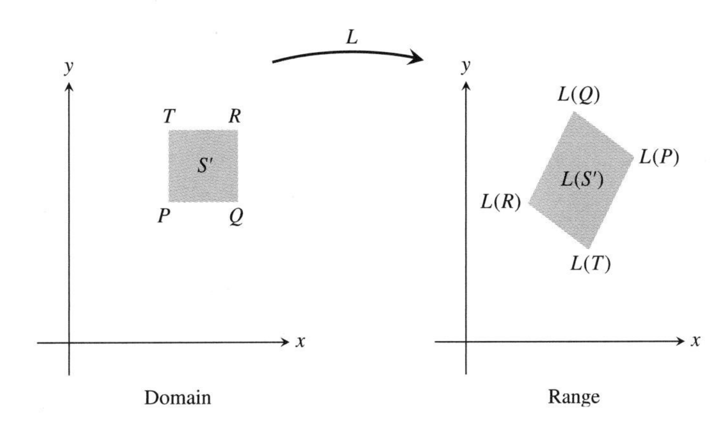

FIGURE 8.44 The square S' and the parallelogram $L(S')$.

This example illustrates the following important result about area and linear functions.

Linear Functions and Area

Let L, defined by

$$L\begin{pmatrix} x \\ y \end{pmatrix} = \begin{pmatrix} a & b \\ c & d \end{pmatrix}\begin{pmatrix} x \\ y \end{pmatrix},$$

be a linear function with domain R^2 and range in R^2. If S is a parallelogram in R^2, then

$$\text{area}(L(S)) = \left|\det\begin{pmatrix} a & b \\ c & d \end{pmatrix}\right| \text{area}(S).$$

Images of Parallelepipeds Because linear functions map parallelograms to parallelograms, it is not surprising that in R^3 these functions also map parallelepipeds in a predictable way. This is especially important when the parallelepiped is a rectangular parallelepiped (or box), that is, a parallelepiped of which every face is a rectangle. Of special significance is the **unit cube** in R^3, that is, the cube with vertices $(0,0,0)$, $(1,0,0)$, $(1,1,0)$, $(0,1,0)$, $(0,0,1)$, $(1,0,1)$, $(1,1,1)$, and $(0,1,1)$.

Linear Functions and Parallelepipeds

Let L, defined by

$$L\begin{pmatrix} x \\ y \\ z \end{pmatrix} = \begin{pmatrix} a_{11} & a_{12} & a_{13} \\ a_{21} & a_{22} & a_{23} \\ a_{31} & a_{32} & a_{33} \end{pmatrix}\begin{pmatrix} x \\ y \\ z \end{pmatrix}$$

be a linear function with domain R^3 and range in R^3, and let S be a parallelepiped in R^3. Then $L(S)$ is either a point, a line segment, a parallelogram, a hexagon, or a parallelepiped. Moreover, if S is determined by the vectors **a**, **b**, and **c**, then $L(S)$ is determined by the vectors $L(\mathbf{a})$, $L(\mathbf{b})$, and $L(\mathbf{c})$. See Fig. 8.45. In addition,

$$\text{volume}(L(S)) = \left|\det\begin{pmatrix} a_{11} & a_{12} & a_{13} \\ a_{21} & a_{22} & a_{23} \\ a_{31} & a_{32} & a_{33} \end{pmatrix}\right| \text{volume}(S). \tag{11}$$

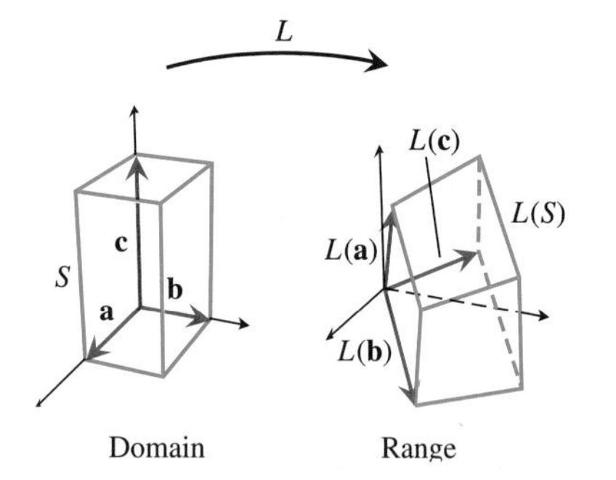

FIGURE 8.45 The image of a rectangular box under a linear function L.

In Exercise 53 we ask you to discuss some aspects of this result.

EXAMPLE 5 Let L, defined by

$$L\begin{pmatrix} x \\ y \\ z \end{pmatrix} = \begin{pmatrix} -2 & 3 & 1 \\ 0 & 6 & 3 \\ 2 & -3 & 1 \end{pmatrix}\begin{pmatrix} x \\ y \\ z \end{pmatrix},$$

be a linear function with domain R^3 and range in R^3. Let C be the unit cube in R^3. Sketch $L(C)$ and find the volume of $L(C)$.

Solution

Because C is determined by the vectors

$$\mathbf{i} = \begin{pmatrix} 1 \\ 0 \\ 0 \end{pmatrix}, \quad \mathbf{j} = \begin{pmatrix} 0 \\ 1 \\ 0 \end{pmatrix}, \quad \text{and} \quad \mathbf{k} = \begin{pmatrix} 0 \\ 0 \\ 1 \end{pmatrix},$$

the parallelepiped $L(C)$ is determined by the vectors

$$L(\mathbf{i}) = \begin{pmatrix} -2 \\ 0 \\ 2 \end{pmatrix}, \quad L(\mathbf{j}) = \begin{pmatrix} 3 \\ 6 \\ -3 \end{pmatrix}, \quad \text{and} \quad L(\mathbf{k}) = \begin{pmatrix} 1 \\ 3 \\ 1 \end{pmatrix}.$$

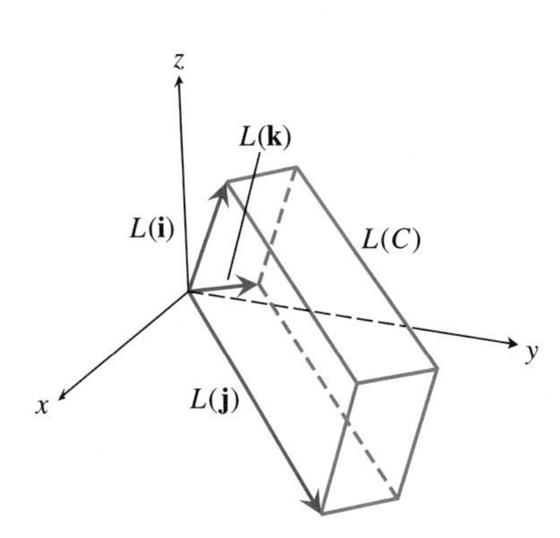

FIGURE 8.46 The image of C is the parallelepiped determined by $L(\mathbf{i})$, $L(\mathbf{j})$, and $L(\mathbf{k})$.

This parallelepiped is shown in Fig. 8.46. The volume of the parallelepiped determined by these vectors is

$$\begin{aligned} \text{volume}(L(C)) &= |L(\mathbf{i}) \cdot (L(\mathbf{j}) \times L(\mathbf{k}))| \\ &= \left| \det \begin{pmatrix} -2 & 0 & 2 \\ 3 & 6 & -3 \\ 1 & 3 & 1 \end{pmatrix} \right| = \left| \det \begin{pmatrix} -2 & 3 & 1 \\ 0 & 6 & 3 \\ 2 & -3 & 1 \end{pmatrix} \right| = 24. \end{aligned}$$

Because the volume of C is 1, this result also illustrates (11).

Translations of Linear Functions In Section 8.3 we stated that linear functions play a role in R^n similar to the role played by

$$f(x) = mx \tag{12}$$

when we deal with real-valued functions of a real variable. The tangent line approximation to a function f near $x = a$ is the line with equation

$$y = f(a) + f'(a)(x - a) = f'(a)x + (f(a) - af'(a)), \qquad x \in R.$$

We have seen that all linear functions L from R to R have the form $L(x) = mx$, where m is a constant. Hence the tangent line approximation is not a linear function unless it happens that $f(a) - af'(a) = 0$. If $f(a) - af'(a) \neq 0$, then the tangent line approximation has the form $y = mx + b$, which is a *translation* of the linear function $y = mx$.

DEFINITION Translation of a Linear Function

Let T be a function with domain R^n and range in R^m. We say that T is a **translation of the linear function** L if there is a constant vector $\mathbf{b}$ in R^m such that

$$T(\mathbf{v}) = L(\mathbf{v}) + \mathbf{b} \tag{13}$$

for every vector $\mathbf{v}$ in R^n. A function with the form seen in (13) is also called an **affine function.**

EXAMPLE 6 Let T, defined by

$$T\begin{pmatrix} x \\ y \end{pmatrix} = \begin{pmatrix} 3x - 2y + 7 \\ -x - 2 \\ x + y + 1 \end{pmatrix},$$

be a function with domain R^2 and range in R^3. Show that T is a translation of a linear function.

Solution

We have

$$T\begin{pmatrix} x \\ y \end{pmatrix} = \begin{pmatrix} 3x - 2y + 7 \\ -x - 2 \\ x + y + 1 \end{pmatrix} = \begin{pmatrix} 3x - 2y \\ -x \\ x + y \end{pmatrix} + \begin{pmatrix} 7 \\ -2 \\ 1 \end{pmatrix}$$

$$= \begin{pmatrix} 3 & -2 \\ -1 & 0 \\ 1 & 1 \end{pmatrix}\begin{pmatrix} x \\ y \end{pmatrix} + \begin{pmatrix} 7 \\ -2 \\ 1 \end{pmatrix}.$$

Thus if $\mathbf{v}$ is a vector in R^2, then

$$T(\mathbf{v}) = L(\mathbf{v}) + \mathbf{b},$$

where L is the linear function defined by

$$L\begin{pmatrix} x \\ y \end{pmatrix} = \begin{pmatrix} 3 & -2 \\ -1 & 0 \\ 1 & 1 \end{pmatrix}\begin{pmatrix} x \\ y \end{pmatrix} \quad \text{and} \quad \mathbf{b} = \begin{pmatrix} 7 \\ -2 \\ 1 \end{pmatrix}.$$

This shows that T is a translation of the linear function L. Note that $\mathbf{b} = T(\mathbf{0})$. Is this always the case?

The translation features of functions of the form (13) is illustrated by the way these functions map subsets of the domain.

EXAMPLE 7 Let T be the function with domain R^2 and range in R^2 defined by

$$T\begin{pmatrix} x \\ y \end{pmatrix} = \begin{pmatrix} 5 & 2 \\ -3 & 2 \end{pmatrix}\begin{pmatrix} x \\ y \end{pmatrix} + \begin{pmatrix} -2 \\ 5 \end{pmatrix}.$$

Let S be the set of points on and inside the unit square. Describe and sketch $T(S)$.

Solution

Let L be the linear function defined by

$$L\begin{pmatrix} x \\ y \end{pmatrix} = \begin{pmatrix} 5 & 2 \\ -3 & 2 \end{pmatrix}\begin{pmatrix} x \\ y \end{pmatrix}$$

and let $\mathbf{b} = \begin{pmatrix} -2 \\ 5 \end{pmatrix}$. Because S is determined by the vectors

$$\mathbf{i} = \begin{pmatrix} 1 \\ 0 \end{pmatrix} \quad \text{and} \quad \mathbf{j} = \begin{pmatrix} 0 \\ 1 \end{pmatrix},$$

the set $L(S)$ is determined by

$$L(\mathbf{i}) = L\begin{pmatrix} 1 \\ 0 \end{pmatrix} = \begin{pmatrix} 5 \\ -3 \end{pmatrix} \quad \text{and} \quad L(\mathbf{j}) = L\begin{pmatrix} 0 \\ 1 \end{pmatrix} = \begin{pmatrix} 2 \\ 2 \end{pmatrix}.$$

The parallelogram $L(S)$ is shown on the right in Fig. 8.47. If P is a point in S, then $T(P)$ is the point designated by the position vector

$$T(\mathbf{P}) = L(\mathbf{P}) + \mathbf{b}.$$

Thus $T(P)$ can be found by translating the point $L(P)$ of $L(S)$ through the vector $\mathbf{b}$. It follows that $T(S)$ is found by translating the set $L(S)$ through $\mathbf{b}$. We can express this by writing $T(S) = L(S) + \mathbf{b}$. See Fig. 8.47.

More directly, the vertices of $T(S)$ are the images of the vertices of S,

$$T\begin{pmatrix} 0 \\ 0 \end{pmatrix} = \begin{pmatrix} -2 \\ 5 \end{pmatrix}, \quad T\begin{pmatrix} 1 \\ 0 \end{pmatrix} = \begin{pmatrix} 3 \\ 2 \end{pmatrix}, \quad T\begin{pmatrix} 1 \\ 1 \end{pmatrix} = \begin{pmatrix} 5 \\ 4 \end{pmatrix}, \quad \text{and} \quad T\begin{pmatrix} 0 \\ 1 \end{pmatrix} = \begin{pmatrix} 0 \\ 7 \end{pmatrix}.$$

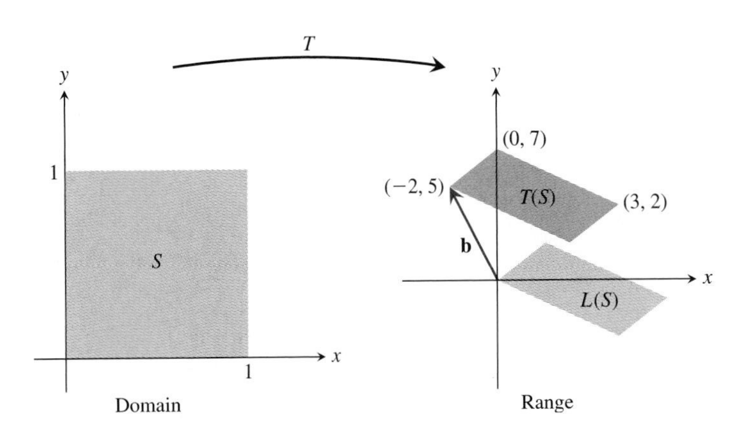

FIGURE 8.47 $T(S)$ can be found by translating $L(S)$ through $\mathbf{b}$.

Exercises 8.5

Exercises 1–16: Sketch S in the domain space and $L(S)$ in the range space. Describe $L(S)$ in words.

1. $L\begin{pmatrix} x \\ y \end{pmatrix} = \begin{pmatrix} x - y \\ 2x - 3y \end{pmatrix}$;
S is described by $\mathbf{r}(t) = \langle 0, 1\rangle + t\langle -4, 1\rangle$, $0 \le t \le 1$.

2. $L\begin{pmatrix} x \\ y \end{pmatrix} = \begin{pmatrix} 0.2x + 0.9y \\ 0.5x + y \end{pmatrix}$;
S is described by $\mathbf{r}(t) = \langle 6, 8\rangle + t\langle 3, 0\rangle$, $0 \le t \le 4$.

3. $L\begin{pmatrix} x \\ y \end{pmatrix} = \begin{pmatrix} -3x + y \\ 6x - 3y \\ 4y \end{pmatrix}$;
S is described by $\mathbf{r}(t) = t\langle 3, 0\rangle$, $-2 \le t \le 3$.

4. $L\begin{pmatrix} x \\ y \end{pmatrix} = \begin{pmatrix} 3x \\ x - y \end{pmatrix}$;
S is the unit square in R^2.

5. $L\begin{pmatrix} x \\ y \end{pmatrix} = \begin{pmatrix} 2 & 1 \\ 2 & 3 \end{pmatrix}\begin{pmatrix} x \\ y \end{pmatrix}$;
S is the unit square in R^2.

6. $L\begin{pmatrix} x \\ y \end{pmatrix} = \begin{pmatrix} 6 & 1 \\ 2 & \frac{1}{3} \\ -1 & 2 \end{pmatrix}\begin{pmatrix} x \\ y \end{pmatrix}$;
S is the square with vertices $(-2, 4)$, $(1, 4)$, $(1, 7)$, and $(-2, 7)$.

7. $L\begin{pmatrix} x \\ y \end{pmatrix} = \begin{pmatrix} 4 & 0 \\ 0 & 3 \end{pmatrix}\begin{pmatrix} x \\ y \end{pmatrix}$;
S is the parallelogram with vertices $(0, 0)$, $(2, 3)$, $(10, -3)$, and $(8, -6)$.

8. $L\begin{pmatrix} x \\ y \end{pmatrix} = \begin{pmatrix} x - 4y \\ 2x + 3y \\ x - y \end{pmatrix}$;
S is the parallelogram with vertices $(2, 3)$, $(9, 9)$, $(0, 1)$, and $(-7, -5)$.

9. $L\begin{pmatrix} x \\ y \end{pmatrix} = (-2x + 7y)$;
S is the parallelogram with vertices $(0, 0)$, $(1, 2)$, $(0, 4)$, and $(-1, 2)$.

10. $L\begin{pmatrix} x \\ y \\ z \end{pmatrix} = \begin{pmatrix} 2x - z \\ x + y + z \\ -2y \end{pmatrix}$;
S is the line segment given by $\mathbf{r}(t) = \langle -2, 3, 5\rangle + t\langle 1, 2, -3\rangle$, $0 \le t \le 5$.

11. $L\begin{pmatrix} x \\ y \\ z \end{pmatrix} = \begin{pmatrix} -8x + y - z \\ x - y + z \\ y - z \end{pmatrix}$;
S is the line segment given by $\mathbf{r}(t) = \langle 4, 0, 5\rangle + t\langle -3, 0, 3\rangle$, $-1 \le t \le 1$.

12. $L\begin{pmatrix} x \\ y \\ z \end{pmatrix} = \begin{pmatrix} x + 2z \\ x + y + z \end{pmatrix}$;
S is the line segment given by $\mathbf{r}(t) = \langle -1, 1, 1\rangle + t\langle 4, -2, -2\rangle$, $7 \le t \le 11$.

13. $L\begin{pmatrix} x \\ y \\ z \end{pmatrix} = \begin{pmatrix} 2x - z \\ x + y + z \\ -2y \end{pmatrix}$;
S is the unit cube in R^3.

14. $L\begin{pmatrix} x \\ y \\ z \end{pmatrix} = \begin{pmatrix} -8 & 1 & -1 \\ 1 & -1 & 1 \\ 0 & 1 & -1 \end{pmatrix}\begin{pmatrix} x \\ y \\ z \end{pmatrix}$;
S is the unit cube in R^3.

15. $L\begin{pmatrix} x \\ y \\ z \end{pmatrix} = \begin{pmatrix} 1 & 0 & 2 \\ 0 & 1 & -1 \end{pmatrix}\begin{pmatrix} x \\ y \\ z \end{pmatrix}$;
S is the unit cube in R^3.

16. $L\begin{pmatrix} x \\ y \\ z \end{pmatrix} = (-x + 2y + 4z)$;
S is the unit cube in R^3.

Exercises 17–20: Find the area of $L(S)$. Sketch S and $L(S)$.

17. $L\begin{pmatrix} x \\ y \end{pmatrix} = \begin{pmatrix} 4x \\ -3y \end{pmatrix}$;
S is the rectangle with vertices $(-1, 2)$, $(-1, 7)$, $(4, 7)$, and $(4, 2)$.

18. $L\begin{pmatrix} x \\ y \end{pmatrix} = \begin{pmatrix} 4x + y \\ x - 3y \end{pmatrix}$;
S is the rectangle with vertices $(4, 4)$, $(5, 4)$, $(5, 9)$, and $(4, 9)$.

19. $L\begin{pmatrix} x \\ y \end{pmatrix} = \begin{pmatrix} 2 & 5 \\ -1 & 3 \end{pmatrix}\begin{pmatrix} x \\ y \end{pmatrix}$;
S is the parallelogram with vertices $(0, 0)$, $(4, 4)$, $(3, 7)$, and $(-1, 3)$.

20. $L\begin{pmatrix} x \\ y \end{pmatrix} = \begin{pmatrix} 0 & 4 \\ -3 & 0 \end{pmatrix}\begin{pmatrix} x \\ y \end{pmatrix}$;
S is the parallelogram with vertices $(-2, 3)$, $(5, 5)$, $(8, 7)$, and $(1, 5)$.

Exercises 21–24: Find the volume of $L(S)$. Sketch S and $L(S)$.

21. $L\begin{pmatrix} x \\ y \\ z \end{pmatrix} = \begin{pmatrix} x \\ -2y \\ 3z \end{pmatrix}$;
S is the rectangular box with vertices $(0, 0, 0)$, $(3, 0, 0)$, $(3, 6, 0)$, $(0, 6, 0)$, $(0, 6, 3)$, $(0, 0, 3)$, $(3, 0, 3)$, and $(3, 6, 3)$.

22. $L\begin{pmatrix} x \\ y \\ z \end{pmatrix} = \begin{pmatrix} x + 2y - z \\ x + y + z \\ -y + 3z \end{pmatrix}$;
S is the rectangular box with vertices $(0, 0, 0)$, $(-5, 0, 0)$, $(-5, 2, 0)$, $(0, 2, 0)$, $(0, 2, -3)$, $(0, 0, -3)$, $(-5, 0, -3)$, and $(-5, 2, -3)$.

23. $L\begin{pmatrix} x \\ y \\ z \end{pmatrix} = \begin{pmatrix} 2 & -1 & 5 \\ -3 & 1 & 0 \\ 2 & 0 & -5 \end{pmatrix}\begin{pmatrix} x \\ y \\ z \end{pmatrix}$;
S is the rectangular box with vertices $(4, 2, 1)$, $(9, 2, 1)$, $(9, 10, 1)$, $(4, 10, 1)$, $(4, 10, 9)$, $(4, 2, 9)$, $(9, 2, 9)$, and $(9, 10, 9)$.

24. $L\begin{pmatrix} x \\ y \\ z \end{pmatrix} = \begin{pmatrix} 3 & -2 & -3 \\ 4 & 0 & 3 \\ 0 & 0 & -3 \end{pmatrix}\begin{pmatrix} x \\ y \\ z \end{pmatrix}$;
S is the rectangular box with vertices $(-4, 5, -1)$, $(0, 5, -1)$, $(0, 10, -1)$, $(-4, 10, -1)$, $(-4, 10, -9)$, $(-4, 5, -9)$, $(0, 5, -9)$, and $(0, 10, -9)$.

Exercises 25–28: Show that the function T is a translation of a linear function. Sketch S and T(S) for the given set S.

25. $T\begin{pmatrix} x \\ y \end{pmatrix} = \begin{pmatrix} -x + 3y - 4 \\ 4x + 1 \end{pmatrix}$;
S is the unit square in R^2.

26. $T\begin{pmatrix} x \\ y \\ z \end{pmatrix} = \begin{pmatrix} x + y + z + 1 \\ 2y - 3z \\ x - y - z - 4 \end{pmatrix}$;
S is the unit cube in R^3.

27. $T\begin{pmatrix} x \\ y \end{pmatrix} = \begin{pmatrix} x + 3y - 8 \\ 2x - y \end{pmatrix}$;
S is the unit square in R^2.

28. $T\begin{pmatrix} x \\ y \\ z \end{pmatrix} = \begin{pmatrix} 2x - 3z \\ x - 2y + z - 3 \\ z \end{pmatrix}$;
S is the unit cube in R^3.

29. Let S be the unit square in R^2 and let S_1 be the parallelogram with vertices $(0, 0)$, $(-2, 5)$, $(4, 4)$, and $(6, -1)$.

a. Find a linear function L with domain R^2 and such that $L(S) = S_1$.

b. How many different linear functions L with domain R^2 are there with $L(S) = S_1$?

c. Find a linear function K with domain R^2 and such that $K(S_1) = S$.

d. How many different linear functions K with domain R^2 are there with $K(S_1) = S$?

30. Let S be the square with vertices $(1, 1)$, $(2, 1)$, $(2, 2)$, and $(1, 2)$. Let S_1 be the parallelogram with vertices $(-1, 6)$, $(4, 3)$, $(-2, 12)$, and $(-7, 15)$.

a. Find a linear function L with domain R^2 and such that $L(S) = S_1$.

b. How many different linear functions L with domain R^2 are there with $L(S) = S_1$?

c. Find a linear function K with domain R^2 and such that $K(S_1) = S$.

d. How many different linear functions K with domain R^2 are there with $K(S_1) = S$?

31. Let S be the square with vertices $(1, 1)$, $(2, 1)$, $(2, 2)$, and $(1, 2)$. Let S_1 be the parallelogram with vertices $(0, 0)$, $(-2, 5)$, $(4, 4)$, and $(6, -1)$.

a. How many different linear functions L with domain R^2 are there with $L(S) = S_1$?

b. How many different linear functions K with domain R^2 are there with $K(S_1) = S$?

32. Let C be the unit cube in R^3 and let S be the rectangular box having one vertex at the origin and determined by the position vectors $\langle 3, 0, 0\rangle$, $\langle 0, -2, 0\rangle$, and $\langle 0, 0, 1\rangle$.

a. Find a linear function L with domain R^3 and such that $L(C) = S$.

b. How many different linear functions L with domain R^3 are there with $L(C) = S$?

c. Find a linear function K with domain R^3 and such that $K(S) = C$.

d. How many different linear functions K with domain R^3 are there with $K(S) = C$?

33. Let C be the unit cube in R^3 and let S be the parallelepiped having one vertex at the origin and determined by the position vectors $\langle 1, 2, 1\rangle$, $\langle -1, 0, 2\rangle$, and $\langle 3, -4, 5\rangle$.

a. Find a linear function L with domain R^3 and such that $L(C) = S$.

b. How many different linear functions L with domain R^3 are there with $L(C) = S$?

c. Find a linear function K with domain R^3 and such that $K(S) = C$.

d. How many different linear functions K with domain R^3 are there with $K(S) = C$?

34. Let θ be a real number and let L be defined by

$$L\begin{pmatrix} x \\ y \end{pmatrix} = \begin{pmatrix} x \cos\theta - y \sin\theta \\ x \sin\theta + y \cos\theta \end{pmatrix}.$$

a. Show that if $\mathbf{v}$ is a vector in R^2, then $\|L(\mathbf{v})\| = \|\mathbf{v}\|$ and $L(\mathbf{v})$ is the rotation of $\mathbf{v}$ through an angle of measure θ. See the accompanying diagram. *Hint:* Reread the discussion at the beginning of this section.

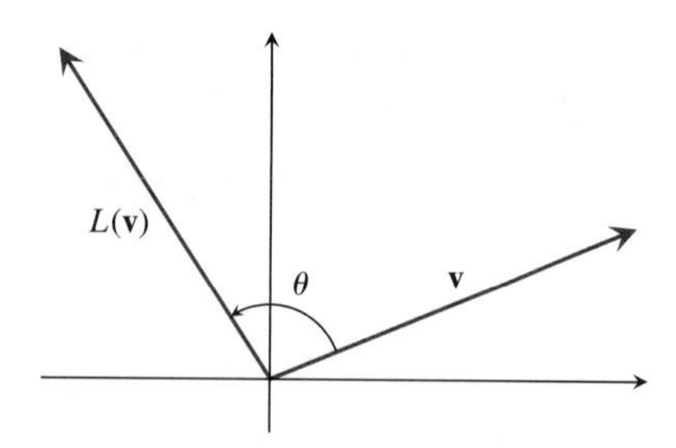

Figure for Exercise 34a.

b. Show that if **v** and **w** are vectors in R^2, the angle between $L(\mathbf{v})$ and $L(\mathbf{w})$ is the same as the angle between **v** and **w**. (In this case we say that L preserves the angle between vectors.)

35. Find a linear function L on R^2 that rotates vectors through an angle of $\pi/3$ and preserves the length of all vectors. That is, $\|\mathbf{v}\| = \|L(\mathbf{v})\|$ for all vectors **v** in R^2. See Exercise 34.

36. Find a linear function L on R^2 that rotates vectors through an angle of $-100°$ and preserves the length of all vectors. That is, $\|\mathbf{v}\| = \|L(\mathbf{v})\|$ for all vectors **v** in R^2. See Exercise 34.

37. Find a linear function L on R^2 that rotates vectors through an angle of $-100°$ and doubles the length of all vectors. That is, $\|L(\mathbf{v})\| = 2\|\mathbf{v}\|$ for all vectors **v** in R^2.

38. For real number θ let L_θ be the linear function with associated matrix

$$\begin{pmatrix} \cos\theta & -\sin\theta \\ \sin\theta & \cos\theta \end{pmatrix}.$$

Show that for any vector $\mathbf{v} = \begin{pmatrix} x \\ y \end{pmatrix}$,

$$L_\theta(L_{-\theta}(\mathbf{v})) = L_{-\theta}(L_\theta(\mathbf{v})) = \mathbf{v}.$$

Explain what this means about the relation between the functions L_θ and $L_{-\theta}$.

39. Find a linear function L with domain R^2 that maps the unit vector

$$\begin{pmatrix} 1/2 \\ \sqrt{3}/2 \end{pmatrix} \quad \text{to} \quad \begin{pmatrix} 1 \\ 0 \end{pmatrix},$$

preserves the angle between any two vectors, and preserves the length of all vectors.

40. Find a linear function L with domain R^2 that maps the unit vector

$$\begin{pmatrix} 3/5 \\ -4/5 \end{pmatrix} \quad \text{to} \quad \begin{pmatrix} 1 \\ 0 \end{pmatrix},$$

preserves the angle between any two vectors, and preserves the length of all vectors.

41. Find a linear function L with domain R^2 that maps the vector

$$\begin{pmatrix} -2 \\ 3 \end{pmatrix} \quad \text{to} \quad \begin{pmatrix} -1 \\ 7 \end{pmatrix}$$

and preserves the angle between any two vectors.

42. Find a linear function L with domain R^2 that maps the vector

$$\begin{pmatrix} -2 \\ -6 \end{pmatrix} \quad \text{to} \quad \begin{pmatrix} 4 \\ 4 \end{pmatrix}$$

and preserves the angle between any two vectors.

43. Let

$$L_1\begin{pmatrix} x \\ y \\ z \end{pmatrix} = \begin{pmatrix} (\sqrt{2}/2)x + (\sqrt{2}/2)y \\ -(\sqrt{2}/2)x + (\sqrt{2}/2)y \\ z \end{pmatrix}.$$

Show that if **v** is a vector in R^3, then $L_1(\mathbf{v})$ can be obtained by rotating **v** through a clockwise angle of $\pi/4$ about the z-axis.

44. Continuation Let

$$L_2\begin{pmatrix} x \\ y \\ z \end{pmatrix} = \begin{pmatrix} x \\ (\sqrt{2}/2)y + (\sqrt{2}/2)z \\ -(\sqrt{2}/2)y + (\sqrt{2}/2)z \end{pmatrix}.$$

Show that if **v** is a vector in R^3, then $L_2(\mathbf{v})$ can be obtained by rotating **v** through a clockwise angle of $\pi/4$ about the x-axis.

45. Continuation Let L_1 and L_2 be the linear functions from the previous two problems, and let L be the linear function defined by

$$L(\mathbf{v}) = L_2(L_1(\mathbf{v})).$$

Describe how $L(\mathbf{v})$ can be found from **v** by using rotations.

46. Let

$$L\begin{pmatrix} x \\ y \\ z \end{pmatrix} = \begin{pmatrix} (1/\sqrt{3})x + (1/\sqrt{2})y - (1/\sqrt{6})z \\ (1/\sqrt{3})x + (2/\sqrt{6})z \\ (1/\sqrt{3})x - (1/\sqrt{2})y - (1/\sqrt{6})z \end{pmatrix}.$$

Show that $\|L(\mathbf{v})\| = \|\mathbf{v}\|$ for every **v** in R^3. Show that if **v** and **w** are two vectors in R^3, then the angle between $L(\mathbf{v})$ and $L(\mathbf{w})$ is equal to the angle between **v** and **w**.

47. Let

$$L\begin{pmatrix} x \\ y \\ z \end{pmatrix} = \begin{pmatrix} -(1/\sqrt{26})x + (1/\sqrt{3})y + (7/\sqrt{78})z \\ (3/\sqrt{26})x - (1/\sqrt{3})y + (5/\sqrt{78})z \\ (4/\sqrt{26})x + (1/\sqrt{3})y - (2/\sqrt{78})z \end{pmatrix}.$$

Show that $\|L(\mathbf{v})\| = \|\mathbf{v}\|$ for every **v** in R^3. Show that if **v** and **w** are two vectors in R^3, then the angle between $L(\mathbf{v})$ and $L(\mathbf{w})$ is equal to the angle between **v** and **w**.

48. Let L be a linear function with domain R^2 and range in R^2 and let S be the set of points on and inside of a rectangle with one vertex at the origin and determined by the vectors **a** and **b**. Let Q be a point on or inside the parallelogram determined by $L(\mathbf{a})$ and $L(\mathbf{b})$. Use a sketch to show that there are real numbers k and m, both between 0 and 1, such that

$$\mathbf{Q} = kL(\mathbf{a}) + mL(\mathbf{b}).$$

Then show that $Q = L(P)$ where

$$\mathbf{P} = k\mathbf{a} + m\mathbf{b}$$

is a point of S. See Fig. 8.41 on page 685.

49. Let L and K be linear functions, both with domain R^2 and range in R^2. The sum of L and K is the function $L + K$ and is defined by

$$(L + K)\begin{pmatrix} x \\ y \end{pmatrix} = L\begin{pmatrix} x \\ y \end{pmatrix} + K\begin{pmatrix} x \\ y \end{pmatrix}.$$

is $L + K$ always a linear function on R^2?

50. Let L and K be linear functions, both with domain R^2 and range in R. The (dot) product of L and K is the function $L \cdot K$ and is defined by

$$(L \cdot K)\begin{pmatrix} x \\ y \end{pmatrix} = L\begin{pmatrix} x \\ y \end{pmatrix} \cdot K\begin{pmatrix} x \\ y \end{pmatrix}.$$

Is $L \cdot K$ always a linear function on R^2?

51. Let L be a linear function with domain R^2 and range in R^2, and let S be the set of points on and inside of the parallelogram with vertices $(0, 0)$, (a, b), (c, d), and $(a + c, b + d)$. Prove that $L(S)$ is a point, a line segment, or the set of points on and inside of the parallelogram determined by the vectors

$$L\begin{pmatrix} a \\ b \end{pmatrix} \quad \text{and} \quad L\begin{pmatrix} c \\ d \end{pmatrix}.$$

Hint: First draw a picture to show that if $P = (x, y)$ is a point in S, then we can write

$$\mathbf{P} = k\begin{pmatrix} a \\ b \end{pmatrix} + m\begin{pmatrix} c \\ d \end{pmatrix}$$

where k and m are real numbers between 0 and 1. Use this expression to find $L(\mathbf{P})$ in terms of

$$L\begin{pmatrix} a \\ b \end{pmatrix} \quad \text{and} \quad L\begin{pmatrix} c \\ d \end{pmatrix}.$$

Based on this expression, discuss the possibilities for S.

52. Let L be given by

$$L\begin{pmatrix} x \\ y \end{pmatrix} = \begin{pmatrix} a & b \\ c & d \end{pmatrix}\begin{pmatrix} x \\ y \end{pmatrix}$$

and let S be the set of points on and inside of a parallelogram in R^2. Show that

$$\text{area}(L(S)) = |ad - bc| \,\text{area}(S).$$

53. Let S be the set of points on and inside of a rectangular box with one vertex at the origin, and let L be a linear function with domain R^3 and range in R^3. Prove that $L(S)$ is either a point, a line segment, the set of points on and inside of a parallelogram, the set of points on and inside of a hexagon, or the set of points on and inside of a parallelepiped. *Hint:* Let S be determined by the vectors **a**, **b**, and **c**. Show that if $P = (x, y, z)$ is a point in S, then

$$\mathbf{P} = k\mathbf{a} + l\mathbf{b} + m\mathbf{c},$$

where k, l, and m are real numbers between 0 and 1. Use this expression to find $L(\mathbf{P})$. Analyze the various possibilities by paralleling the proof for the analogous result about mappings of rectangles in R^2.

54. Let L be a linear function and let S_1 and S_2 be parallel line segments in the domain of L. Show that $L(S_1)$ and $L(S_2)$ are either parallel line segments or both points. Note too that $L(S_1)$ and $L(S_2)$ may be overlapping sets or may be the same set.

55. Let T be a translation of a linear function. Is it always true that

$$T(\mathbf{v} + \mathbf{w}) = T(\mathbf{v}) + T(\mathbf{w})$$

for all vectors **v** and **w** in the domain of T?

56. Let T be a translation of a linear function. Is it always true that

$$T(k\mathbf{v}) = kT(\mathbf{v})$$

for all real numbers k and all vectors **v** in the domain of T?

57. Let T, with domain R^2, be a translation of a linear function and let S be the set of points on and inside of a parallelogram in R^2. What are the possibilities for $T(S)$?

58. Let T, with domain R^3, be a translation of a linear function and let S be the set of points on and inside of a rectangular box in R^3. What are the possibilities for $T(S)$?

59. Let T be a translation of a linear function. Prove that the function L defined by

$$L(\mathbf{v}) = T(\mathbf{v}) - T(\mathbf{0})$$

is a linear function.

60. Let T be a translation of a linear function and suppose that

$$T(\mathbf{v}) = L(\mathbf{v}) + \mathbf{b},$$

where L is a linear function and **b** is a constant vector. Show that $\mathbf{b} = T(\mathbf{0})$.

8.6 Planes

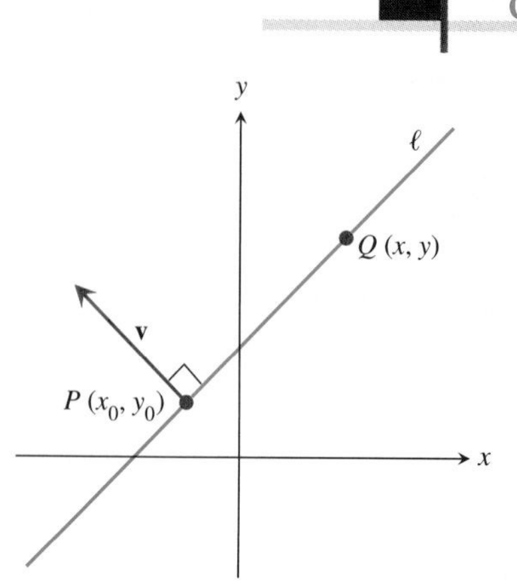

FIGURE 8.48 A line in the plane is determined by a point and a perpendicular vector.

To find parametric equations for a line in two-dimensional space (the plane), we need a point on the line and a vector in the direction of the line. In Section 8.1 we saw that this same idea can be used to describe lines in three-dimensional space.

There is another way to use vectors to describe a line ℓ in the plane. Let $P = (x_0, y_0)$ be a point on ℓ and $\mathbf{v} = \langle a, b\rangle$ be a nonzero vector perpendicular to ℓ. See Fig. 8.48. We can describe ℓ as "the line through P and perpendicular to $\mathbf{v}$." Indeed, given this information we can find an equation for ℓ. Let $Q = (x, y)$ be a point in the plane. Then Q is on ℓ precisely when

$$\overrightarrow{PQ} = (0,0) \qquad \text{or} \qquad \overrightarrow{PQ} \perp \mathbf{v},$$

that is, if and only if

$$\mathbf{v} \cdot \overrightarrow{PQ} = 0. \tag{1}$$

Because $\overrightarrow{PQ} = \langle x - x_0, y - y_0\rangle$, the dot product in (1) can be expanded to give

$$0 = \mathbf{v} \cdot \overrightarrow{PQ} = \langle a, b\rangle \cdot \langle x - x_0, y - y_0\rangle = a(x - x_0) + b(y - y_0).$$

Expanding the last expression and setting $-ax_0 - by_0 = c$, we see that (x, y) is on ℓ if and only if

$$ax + by + c = 0.$$

This is a standard form for the line ℓ. In this form the coefficients of x and y are the components of a vector perpendicular (or **normal**) to ℓ.

We use this "perpendicular vector" idea in three dimensions to find the equation for a plane.

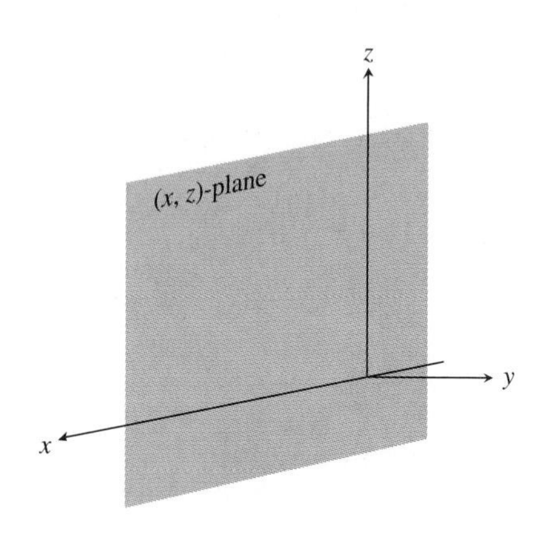

FIGURE 8.49 The (x, z)-plane.

Planes and Normal Vectors

We have already worked with some planes in R^3. For example, the (x, z)-plane is the plane that contains the x- and z-axes. See Fig. 8.49. This plane can be described as the plane containing the origin and perpendicular to the y-axis. This description reminds us of the previous discussion, where we used a point and a perpendicular vector to describe a line. This idea extends readily to planes in R^3.

INVESTIGATION

The Equation for a Plane

Let $P = (x_0, y_0, z_0)$ be a point and $\mathbf{n} = \langle a, b, c\rangle$ be a nonzero vector in R^3. Let Π be the plane that contains P and is perpendicular to the vector $\mathbf{n}$. Find an equation satisfied by the points on Π.

Let $Q = (x, y, z)$ be a point on plane Π. Then the vector

$$\overrightarrow{PQ} = \langle x - x_0, y - y_0, z - z_0\rangle$$

is either the zero vector, $\mathbf{0} = \langle 0, 0, 0\rangle$, or is perpendicular to $\mathbf{n}$. See Fig. 8.50. Hence,

$$\mathbf{n} \cdot \overrightarrow{PQ} = 0.$$

Expanding this dot product, we have

$$\begin{aligned} 0 = \mathbf{n} \cdot \overrightarrow{PQ} &= \langle a, b, c\rangle \cdot \langle x - x_0, y - y_0, z - z_0\rangle \\ &= a(x - x_0) + b(y - y_0) + c(z - z_0). \end{aligned} \tag{2}$$

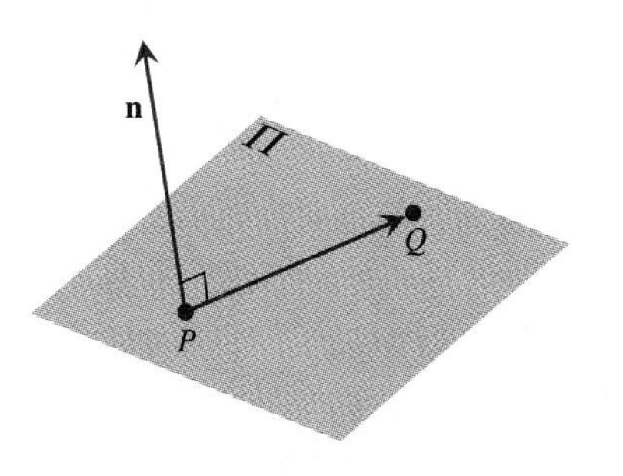

FIGURE 8.50 A plane in space is determined by a point P and a perpendicular vector $\mathbf{n}$.

If (x, y, z) is a point on Π, then x, y, and z satisfy (2).

Conversely, if x, y, and z satisfy (2), then

$$\langle a, b, c \rangle \cdot \langle x - x_0, y - y_0, z - z_0 \rangle = 0.$$

This implies that $\langle x - x_0, y - y_0, z - z_0 \rangle$ is either the zero vector or is perpendicular to $\mathbf{n}$, and hence that (x, y, z) is in the plane containing P and perpendicular to $\mathbf{n}$. Thus a point (x, y, z) satisfies (2) if and only if the point (x, y, z) is on the plane Π.

Expanding the last expression in (2) and setting $-ax_0 - by_0 - cz_0 = d$, we obtain the general form

$$ax + by + cz + d = 0.$$

for the equation of a plane.

We summarize the results of this investigation:

Planes in R^3

Let Π be the plane that contains the point $P = (x_0, y_0, z_0)$ and is perpendicular to the nonzero vector $\mathbf{n} = \langle a, b, c \rangle$. A point $Q = (x, y, z)$ is on Π if and only if

$$a(x - x_0) + b(y - y_0) + c(z - z_0) = 0. \tag{3}$$

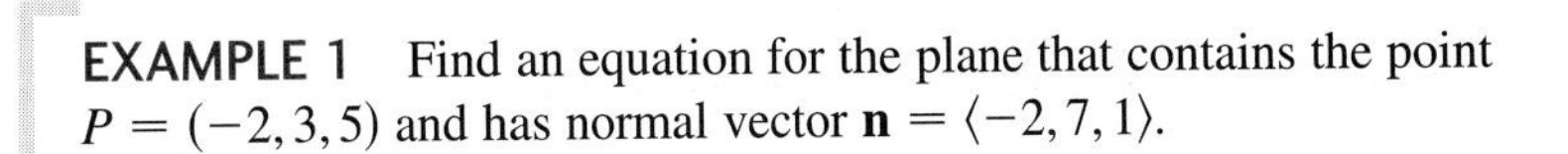

The vector $\mathbf{n}$ is also called a **normal vector** to the plane. Conversely, if a, b, c, and d are constants and $\langle a, b, c \rangle \neq \langle 0, 0, 0 \rangle$, then

$$ax + by + cz + d = 0 \tag{4}$$

is the equation of a plane with normal vector $\mathbf{n} = \langle a, b, c \rangle$; that is, the set of points (x, y, z) that satisfy (4) is a plane.

EXAMPLE 1 Find an equation for the plane that contains the point $P = (-2, 3, 5)$ and has normal vector $\mathbf{n} = \langle -2, 7, 1 \rangle$.

Solution

Let $Q = (x, y, z)$ be a point in R^3. Then Q is on the plane if and only if

$$\begin{aligned} 0 &= \mathbf{n} \cdot \overrightarrow{PQ} \\ &= \langle -2, 7, 1 \rangle \cdot \langle x - (-2), y - 3, z - 5 \rangle \\ &= -2(x + 2) + 7(y - 3) + (z - 5) \\ &= -2x + 7y + z - 30. \end{aligned}$$

Hence an equation for the plane is $-2x + 7y + z - 30 = 0$. The plane is shown in Fig. 8.51.

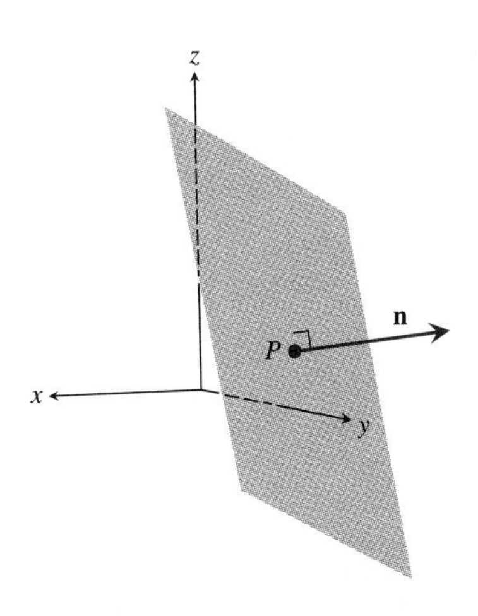

FIGURE 8.51 The plane containing $P = (-2, 3, 5)$ and normal to the vector $\mathbf{n} = \langle -2, 7, 1 \rangle$.

EXAMPLE 2 Describe and sketch the plane Π with equation

$$2x + 3y = 6. \tag{5}$$

Solution

Because (5) has no z-term, we can write the equation as

$$2x + 3y + 0z - 6 = 0. \tag{6}$$

The coefficients of x, y, and z are the components of a vector $\mathbf{n} = \langle 2, 3, 0 \rangle$ normal to Π. To find a point on Π, simply find an ordered triple (x, y, z) that satisfies (5) or (6). We can find such a point by inspection: Set $x = z = 0$ and solve (6) to get $y = 2$. Hence the point $(0, 2, 0)$ is on the plane. We can describe Π as the plane containing $(0, 2, 0)$ and perpendicular to $\mathbf{n} = \langle 2, 3, 0 \rangle$.

Because the third coordinate of $\mathbf{n} = \langle 2, 3, 0 \rangle$ is 0, $\mathbf{n}$ is parallel to the (x, y)-plane, and the plane Π is perpendicular to the (x, y)-plane. The set of points common to Π and the (x, y)-plane (i.e., their intersection) is the set of all points $(x, y, 0)$ such that $2x + 3y - 6 = 0$. Hence the intersection is just a *line,* $2x + 3y - 6 = 0$, in the (x, y)-plane. Thus we can also describe Π as the set of all points that lie directly above, below, or on the line in R^2 described by the equation $2x + 3y = 6$. See Fig. 8.52.

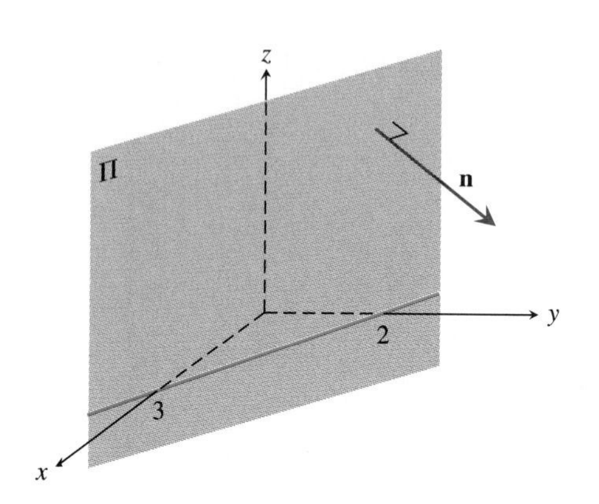

FIGURE 8.52 The plane $2x + 3y = 6$ is perpendicular to the (x, y)-plane.

A plane can be described in many different ways. Given such a description, we can use the information provided to find a point on the plane and a vector normal to the plane. Once these are known, we can write an equation for the plane.

EXAMPLE 3 Find an equation for the plane Π containing the point $(3, 0, -5)$ and parallel to the plane Φ with equation $4x - 3y + 2z = 0$.

Solution

The vector $\mathbf{n} = \langle 4, -3, 2 \rangle$ is normal to Φ. Because Π is parallel to Φ, $\mathbf{n}$ is also normal to Π. See Fig. 8.53. Because the point $(3, 0, -5)$ is on Π, an equation for the plane is

$$4(x - 3) + (-3)(y - 0) + 2(z - (-5)) = 0.$$

This simplifies to

$$4x - 3y + 2z - 2 = 0.$$

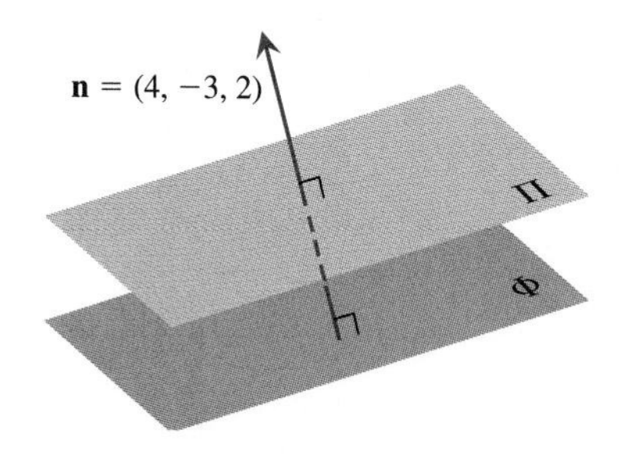

FIGURE 8.53 Parallel planes are perpendicular to the same vector.

EXAMPLE 4 Find an equation for the plane containing the three points $P = (-1, 2, 4)$, $Q = (-4, 2, 1)$, and $R = (1, -2, 0)$.

Solution

Three points determine a unique plane if and only if they are not collinear, so we first check that the three points do not lie on one line. Because the vectors

$$\overrightarrow{PQ} = \langle -3, 0, -3 \rangle \quad \text{and} \quad \overrightarrow{PR} = \langle 2, -4, -4 \rangle$$

are not parallel, the three given points are not collinear.

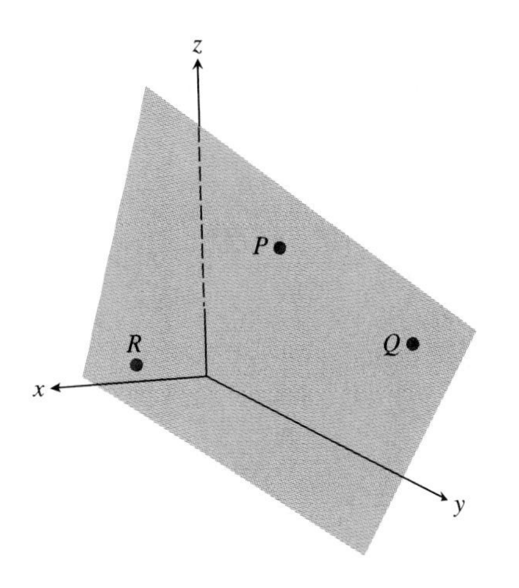

FIGURE 8.54 The plane determined by P, Q, and R.

To find a normal vector, we note that the cross product $\overrightarrow{PQ} \times \overrightarrow{PR}$ is perpendicular to both $\overrightarrow{PQ}$ and $\overrightarrow{PR}$. Because $\overrightarrow{PQ}$ and $\overrightarrow{PR}$ are both parallel to the plane, the cross product is perpendicular to the plane. Hence, we can take

$$\mathbf{n} = \overrightarrow{PQ} \times \overrightarrow{PR} = \det\begin{pmatrix} \mathbf{i} & \mathbf{j} & \mathbf{k} \\ -3 & 0 & -3 \\ 2 & -4 & -4 \end{pmatrix} = \langle -12, -18, 12 \rangle,$$

as a normal vector. Using P as a given point on the plane and $\mathbf{n}$ as the normal vector, an equation for the plane is

$$-12(x - (-1)) - 18(y - 2) + 12(z - 4) = 0. \tag{7}$$

Dividing both sides of (7) by 6 does not affect the triples (x, y, z) that satisfy the equation. Further simplification leads to the equation

$$-2x - 3y + 2z - 4 = 0.$$

The plane is shown in Fig. 8.54. How could we check that this equation is correct?

EXAMPLE 5 Let ℓ_1 be the line described by

$$\mathbf{r}(t) = \langle -2t + 1, 4t, -3t - 8 \rangle$$

and ℓ_2 be the line described by

$$\mathbf{s}(t) = \langle t + 4, -3t - 1, 2t - 6 \rangle.$$

Show that ℓ_1 and ℓ_2 intersect, and then find an equation for the plane containing both lines.

Solution

If ℓ_1 and ℓ_2 have a point in common, then there must be real numbers t_1 and t_2 such that $\mathbf{r}(t_1) = \mathbf{s}(t_2)$; that is,

$$\langle -2t_1 + 1, 4t_1, -3t_1 - 8 \rangle = \langle t_2 + 4, -3t_2 - 1, 2t_2 - 6 \rangle.$$

Equating coordinates of these vectors,

$$\begin{aligned} -2t_1 + 1 &= t_2 + 4 \\ 4t_1 &= -3t_2 - 1 \\ -3t_1 - 8 &= 2t_2 - 6. \end{aligned} \tag{8}$$

To find t_1 and t_2, we first solve the system given by the first two equations in (8). We may write the first two equations as

$$\begin{aligned} 2t_1 + t_2 &= -3 \\ 4t_1 + 3t_2 &= -1. \end{aligned} \tag{9}$$

Multiplying the first of these equations by 2 and subtracting it from the second, we find $t_2 = 5$. Substituting this value for t_2 into either of the equations in (9), we find $t_1 = -4$. We know that $t_1 = -4$ and $t_2 = 5$ satisfy the first two equations in (8). Noting that the values $t_1 = -4$ and $t_2 = 5$ also satisfy the third equation, these values of t_1 and t_2 are a solution to (8). Because

$$\mathbf{r}(-4) = \langle 9, -16, 4\rangle = \mathbf{s}(5),$$

the point $(9, -16, 4)$ is on both ℓ_1 and ℓ_2. Thus, the lines intersect.

Because ℓ_1 and ℓ_2 intersect, the two lines lie in a plane. This plane is illustrated in Fig. 8.55. The point $P = (9, -16, 4)$ is in this plane. Vectors $\mathbf{v}_1 = \langle -2, 4, -3\rangle$ and $\mathbf{v}_2 = \langle 1, -3, 2\rangle$ are parallel to ℓ_1 and ℓ_2, respectively, so these two vectors are also parallel to the plane containing the lines. A vector $\mathbf{n}$ normal to the plane can be found by taking the cross product of $\mathbf{v}_1$ and $\mathbf{v}_2$,

$$\mathbf{n} = \mathbf{v}_1 \times \mathbf{v}_2 = \langle -2, 4, -3\rangle \times \langle 1, -3, 2\rangle = \langle -1, 1, 2\rangle.$$

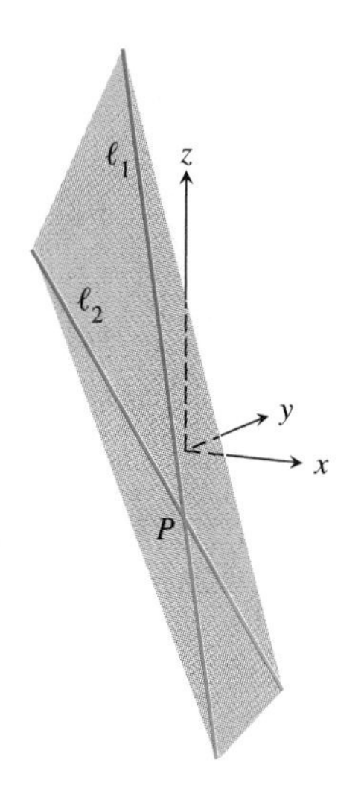

FIGURE 8.55 A plane determined by intersecting lines.

An equation for the plane containing P and with normal vector $\mathbf{n}$ is

$$-1(x - 9) + 1(y - (-16)) + 2(z - 4) = 0.$$

This simplifies to

$$-x + y + 2z + 17 = 0. \tag{10}$$

We can check that this plane does indeed contain both lines. If a point Q is on ℓ_1, then Q has the form (x, y, z) where

$$x = -2t + 1, \quad y = 4t, \quad \text{and} \quad z = -3t - 8.$$

Because these expressions for x, y, and z satisfy (10),

$$-x + y + 2z + 17 = -(-2t + 1) + 4t + 2(-3t - 8) + 17 = 0,$$

the point Q is in the plane. This means that every point of ℓ_1 is in the plane. A similar calculation shows that every point of ℓ_2 is also in the plane.

Exercises 8.6

Exercises 1–10: Write an equation for the plane determined by the given conditions. Sketch the plane.

1. Normal vector $\mathbf{n} = \langle -1, 3, 5\rangle$ and containing the point $P = (2, 0, -5)$
2. Normal vector $\mathbf{n} = \langle 1, 0, 1\rangle$ and containing the point $P = (0, -3, 5)$
3. Normal vector $\mathbf{n} = \mathbf{k}$ and containing the point $P = (0, 0, 0)$
4. Containing the line with equation $\mathbf{r}(t) = \langle -2 + 3t, t, 5 - 7t\rangle$ and the point $P = (7, 8, 2)$
5. Perpendicular to the line with equation $\mathbf{r}(t) = \langle -2 + 3t, t, 5 - 7t\rangle$ and containing the point $P = (7, 8, 2)$
6. Parallel to the plane with equation $-2x + y - z = 6$ and containing the point $P = (0, 1, 2)$
7. Containing the points $P = (-1, 3, 4)$, $Q = (4, 6, -3)$, and $R = (0, 2, 2)$
8. Containing the points $P = (0, 0, 0)$, $Q = (1, 1, 1)$, and $R = (-1, 1, 1)$
9. Perpendicular to the plane with equation $-2x + 3y - 4z = 2\sqrt{2}$ and the plane with equation $4x - 3z = 2$ and containing the point $P = (4, -3, -1)$
10. Perpendicular to the plane with equation $-2x + 3y - 4z = 2\sqrt{2}$, parallel to the line with equation $\mathbf{r}(t) = \langle -3 + t, 4 - 2t, 1 + t\rangle$, and containing the point $P = (2, 0, 3)$

Exercises 11–14: Given two intersecting planes, let θ, $0 \le \theta \le \pi$, be the angle determined by the normals to the planes. The dihedral angle *determined by the planes is defined to be the smaller of θ and*

$\pi - \theta$. (Hence the dihedral angle is an angle between 0 and $\pi/2$.) Find the dihedral angle determined by the two planes.

11. $-3x + 2y + 2z = 4$ and $-x + 3z = 1$

12. $x = 0$ and $x - y + 12z = -3$

13. $-7x + 4y - 10z = e$ and $2x + 2y - 4z = \pi$

14. $x + 3y - 2z = -4$ and $-8x + 3y + 4z = 0$

15. Let Π be the plane with equation

$$-3x + y - z = 4.$$

Find a plane that makes a dihedral angle of $\pi/4$ with Π.

16. Let Π be the plane with equation

$$x - 2y = 0.$$

Find a plane that makes a dihedral angle of $\pi/3$ with Π.

17. Let Π be the plane with equation

$$2x - 4y + z = -8.$$

Find a plane that makes a dihedral angle of 1 radian with Π.

Exercises 18–21: Find the equation of the line in R^2 determined by the given conditions. Sketch the line.

18. Perpendicular to the vector $\mathbf{v} = \langle -6, 5\rangle$ and containing the point $(-4, 2)$

19. Perpendicular to the vector $\mathbf{v} = \langle 0, -1\rangle$ and containing the point $(7, 0)$

20. Perpendicular to the line with equation $2x + 3y - 6 = 0$ containing the point $(1, 1)$

21. Perpendicular to the line with equation $-x + 4y + 3 = 0$ and containing the point $(-2, 3)$

22. Let ℓ be the line with equation

$$\mathbf{r}(t) = \langle -3 + t, 4 + 4t, -7 + 2t\rangle$$

and Π the plane with equation

$$-2x + 3y - 4z = 5.$$

a. By comparing a vector parallel to ℓ with a vector normal to Π, show that ℓ and Π intersect in exactly one point.

b. Find the intersection point.

23. Let ℓ be the line with equation

$$\mathbf{r}(t) = \langle 1 + 2t, 3, 2 - 2t\rangle$$

and Π the plane with equation

$$x + 4y - 4z = 12.$$

a. By comparing a vector parallel to ℓ with a vector normal to Π, show that ℓ and Π intersect in exactly one point.

b. Find the intersection point.

24. Let ℓ be the line with equation

$$\mathbf{r}(t) = \langle -2 + 5t, 2t, -3 - 3t\rangle$$

and Π the plane with equation

$$3x - 6y + z = 12.$$

What is the intersection of ℓ and Π?

25. Consider the lines ℓ_1 and ℓ_2 given by

$$\mathbf{r}_1(t) = \langle -2 + 3t, 4t, -1 - t\rangle$$

and

$$\mathbf{r}_2(t) = \langle -9t, 7 - 12t, 3t\rangle,$$

respectively.

a. Show that ℓ_1 and ℓ_2 are parallel and distinct.

b. Find an equation for the plane containing ℓ_1 and ℓ_2.

c. Show that the plane does indeed contain ℓ_1 and ℓ_2.

26. Consider the lines ℓ_1 and ℓ_2 given by

$$\mathbf{r}_1(t) = \langle t, 1 + t, 2 + t\rangle$$

and

$$\mathbf{r}_2(t) = \langle 2 + 2t, 4 + 2t, 8 + 2t\rangle,$$

respectively.

a. Show that ℓ_1 and ℓ_2 are parallel and distinct.

b. Find an equation for the plane containing ℓ_1 and ℓ_2.

c. Show that the plane does indeed contain ℓ_1 and ℓ_2.

27. Consider the lines ℓ_1 and ℓ_2 given by

$$\mathbf{r}_1(t) = \langle 3, 3 - t, -9 + 3t\rangle$$

and

$$\mathbf{r}_2(t) = \langle -5 + 4t, -16 + 8t, 14 - 7t\rangle,$$

respectively.

a. Show that ℓ_1 and ℓ_2 intersect.

b. Find an equation for the plane containing ℓ_1 and ℓ_2.

c. Show that the plane does indeed contain ℓ_1 and ℓ_2.

28. Consider the lines ℓ_1 and ℓ_2 given by

$$\mathbf{r}_1(t) = \langle -2t, 1 - 3t, 4 + 9t\rangle$$

and

$$\mathbf{r}_2(t) = \langle -3t - 11, 6t + 37, t - 9\rangle,$$

respectively.

a. Show that ℓ_1 and ℓ_2 intersect.

b. Find an equation for the plane containing ℓ_1 and ℓ_2.

c. Show that the plane does indeed contain ℓ_1 and ℓ_2.

29. Consider the lines ℓ_1 and ℓ_2 given by

$$\mathbf{r}_1(t) = \langle -3t, 4 + 4t, 5 - t\rangle$$

and

$$\mathbf{r}_2(t) = \langle -2 - t, 3 + 4t, 7\rangle,$$

respectively. Show that ℓ_1 and ℓ_2 are not parallel and do not intersect. Such lines are called **skew** lines.

30. Consider the lines ℓ_1 and ℓ_2 given by

$$\mathbf{r}_1(t) = \langle 2 + 10t, 4 - 3t, 2 + 5t\rangle$$

and

$$\mathbf{r}_2(t) = \langle -2 + 4t, 7t, -t\rangle,$$

respectively. Show that ℓ_1 and ℓ_2 are not parallel and do not intersect.

31. Let ℓ be the line given by

$$\mathbf{r}(t) = \langle a_0 + a_1 t, b_0 + b_1 t, c_0 + c_1 t\rangle$$

and let Π be the plane with equation

$$a_2 x + b_2 y + c_2 z = d_0.$$

Develop an easy test that tells when ℓ and Π intersect in exactly one point. Explain why your test works.

32. Let $P = (-3, 1, 2)$ and Π be the plane with equation

$$-2x + 4y - z = 2.$$

Find the perpendicular distance from P to Π using each of the following two methods.

a. Find an equation for the line ℓ through P and perpendicular to Π. Find the point Q of intersection of ℓ and Π, and then determine the distance PQ.

b. Let $R \neq P$ be a point on the plane and F the foot of the perpendicular from P to Π. Then $\triangle PFR$ is a right triangle with

$$PF = \text{distance from } P \text{ to } \Pi.$$

Use right-triangle trigonometry and the dot product to determine PF. See the accompanying figure.

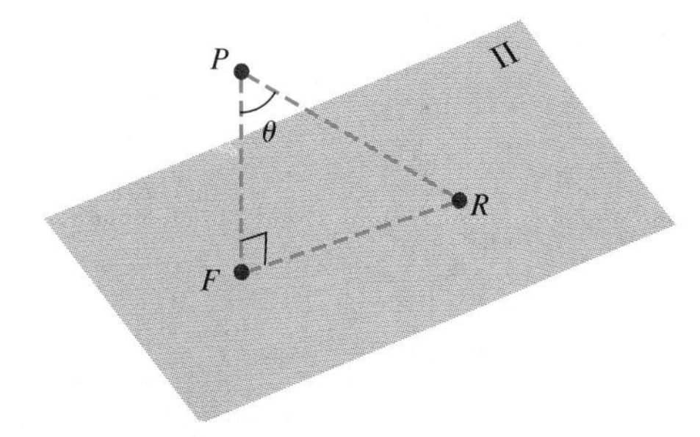

Figure for Exercise 32.

33. Let $P = (0, 1, 4)$ and Π be the plane with equation

$$x - y + z = 0.$$

Use each of the methods mentioned in the previous problem to find the perpendicular distance from P to Π.

34. Let $P = (0, 0, 0)$ and Π be the plane with equation

$$-3x + 4y - z = 4.$$

Use each of the methods described in Exercise 32 to find the perpendicular distance from P to Π.

35. Let $P = (0, 0, 0)$ and Π be the plane with equation

$$ax + by + cz = d.$$

Use either of the methods described in Exercise 32 to find the perpendicular distance from P to Π.

36. Let L be the linear function given by

$$L\begin{pmatrix} x \\ y \\ z \end{pmatrix} = \begin{pmatrix} -2x + y \\ x + y - z \\ 2x - 3z \end{pmatrix},$$

and let S be the set of points on the plane with equation

$$-4x + y + 2z - 4 = 0.$$

Show that $L(S)$ is also a plane.

37. Let L be the linear function given by

$$L\begin{pmatrix} x \\ y \\ z \end{pmatrix} = \begin{pmatrix} z \\ x - z \\ 3x - y - z \end{pmatrix},$$

and let S be the set of points on the plane with equation

$$4y - 3z = 0.$$

Show that $L(S)$ is also a plane.

38. Let L be the linear function with domain R^3 and range in $R = R^1$ and described by

$$L\begin{pmatrix} x \\ y \\ z \end{pmatrix} = (-2x + 3y - 4z).$$

Let S be the set of points in R^3 that satisfy the equation

$$L\begin{pmatrix} x \\ y \\ z \end{pmatrix} = (6).$$

Show that S is a plane in R^3. Give a vector normal to the plane and a point on the plane.

39. Let L be the linear function with domain R^3 and range in $R = R^1$ and described by

$$L\begin{pmatrix} x \\ y \\ z \end{pmatrix} = (4x + 7z).$$

Let S be the set of points in R^3 that satisfy the equation

$$L\begin{pmatrix} x \\ y \\ z \end{pmatrix} = (0).$$

Show that S is a plane in R^3. Give a vector normal to the plane and a point on the plane.

40. Let T be the function with domain R^3 and range in $R = R^1$ and described by

$$T\begin{pmatrix} x \\ y \\ z \end{pmatrix} = (x + y - 3z + 8).$$

Let S be the set of points in R^3 that satisfy the equation

$$T\begin{pmatrix} x \\ y \\ z \end{pmatrix} = (-2).$$

Show that S is a plane in R^3. Give a vector normal to the plane and a point on the plane.

41. Let T be the function with domain R^3 and range in $R = R^1$ and described by

$$T\begin{pmatrix} x \\ y \\ z \end{pmatrix} = (4y - 3).$$

Let S be the set of points in R^3 that satisfy the equation

$$T\begin{pmatrix} x \\ y \\ z \end{pmatrix} = (\sqrt{3}).$$

Show that S is a plane in R^3. Give a vector normal to the plane and a point on the plane.

8.7 Motion in Three Dimensions

The *Viking* spacecraft approaching Mars.

Why do planets orbit the sun in planar orbits? What are the magnitudes of the forces on a roller coaster as it travels around a spiral? What path will water follow as it flows down a hillside? How do scientists land a spacecraft on Mars? To answer these questions, we need to understand motion in three dimensions.

Earlier in this chapter we saw that most of the concepts associated with vectors in two dimensions extend in a natural way to vectors in three dimensions. The same is true for many of the ideas surrounding motion. Most of the definitions and facts we learned about motion in the plane carry over very naturally to motion in three-dimensional space.

Space Curves

When an object moves in R^3, its position changes with time. We denote the position of the object at time t by a position vector

$$\mathbf{r}(t) = \langle x(t), y(t), z(t) \rangle, \tag{1}$$

where $x(t)$, $y(t)$, and $z(t)$ are the x-, y-, and z-coordinates of the position of the object at time t. The function $\mathbf{r}$ defined by (1) is called the *position function* for the motion of the object. If we are interested in the path followed by the particle for times t between t_0 and t_1, we can use (1) to find the position of the object for the times of interest, and then plot the corresponding points in R^3. If $x(t)$, $y(t)$, and $z(t)$ are continuous functions of t, then the result will be a space curve in R^3, perhaps like the one

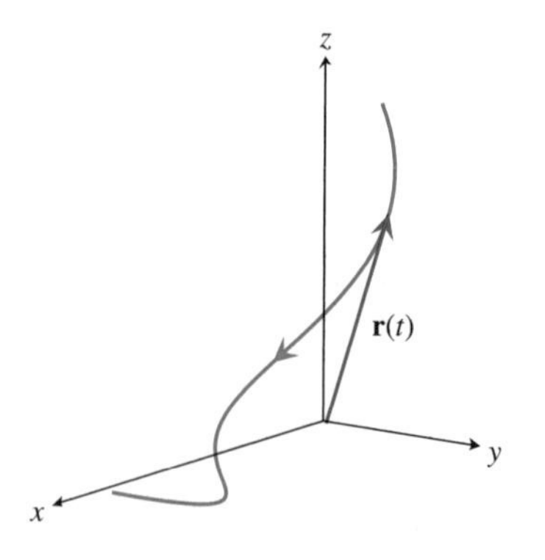

FIGURE 8.56 A space curve.

shown in Fig. 8.56. This curve shows the path of the object as it moves through space. Sometimes we put an arrow on the path to indicate the direction of the motion.

EXAMPLE 1 Describe and sketch the path of an object with position function

$$\mathbf{r}(t) = \langle 3 - 4t, 1 - t, 6t \rangle, \qquad 0 \le t \le 2. \tag{2}$$

Solution

The position function $\mathbf{r}(t)$ in (2) has the form of a vector equation for a line. Hence the path described by (2) is the line segment with endpoints

$$\mathbf{r}(0) = \langle 3, 1, 0 \rangle \quad \text{and} \quad \mathbf{r}(2) = \langle -5, -1, 12 \rangle.$$

As t increases from $t = 0$ to $t = 2$, the object moves along the segment from $\mathbf{r}(0)$ to $\mathbf{r}(2)$. The path is shown in Fig. 8.57.

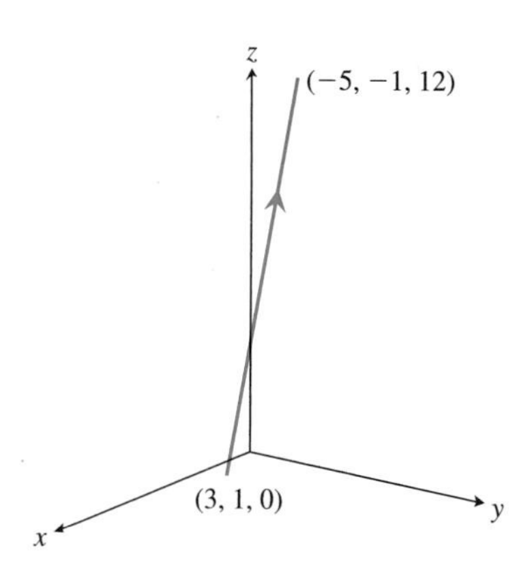

FIGURE 8.57 The path $\mathbf{r}(t) = \langle 3 - 4t, 1 - t, 6t \rangle$ for $0 \le t \le 2$.

Graphing Space Curves Many computer algebra systems have commands for producing graphs of space curves. The graphs that result are, of course, projections onto a two-dimensional screen of three-dimensional objects. After you view some CAS-produced graphs of space curves, you will realize that although you may view the graph on the screen it may be hard to understand what the path really looks like in three dimensions. In addition, we know from experience that objects can look very different when viewed from different positions. Most CASs have commands that enable us to change the direction from which we view a graph in R^3. Which view (or views) best convey the nature of the graph? Before producing a graph, it is important to think about what the graph might look like from different viewpoints. Once we have decided which viewpoint might give the most informative graph, we can use the CAS to produce a representation of the graph as seen from this viewpoint.

One way to get an idea of how a curve looks from different viewpoints is to project the curve into one of the coordinate planes. For example, to project the curve described by (1) into the (x, y)-plane, replace the term $z(t)$ by 0. The resulting position function

$$\mathbf{r}_{xy}(t) = \langle x(t), y(t), 0 \rangle$$

describes a curve in the (x, y)-plane. This curve is found by projecting the space curve vertically onto the (x, y)-plane. The result corresponds to what we would see if we viewed the curve from above. We can project a curve into the other coordinate planes by similar means.

EXAMPLE 2 Describe and sketch the curve defined by

$$\mathbf{r}(t) = \langle 2\cos t, 2\sin t, t \rangle, \qquad 0 \le t \le 4\pi. \tag{3}$$

Solution

Project the path into the (x, y)-plane by setting the third coordinate in $\mathbf{r}(t)$ to 0. We obtain

$$\mathbf{r}_{xy}(t) = \langle 2\cos t, 2\sin t, 0 \rangle, \qquad 0 \le t \le 4\pi.$$

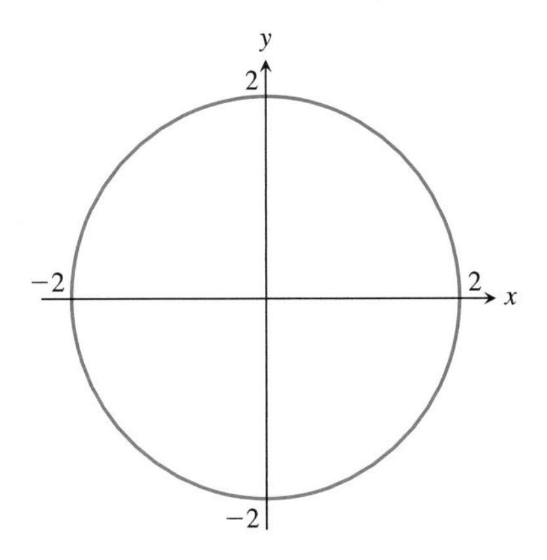

FIGURE 8.58 A top view of the curve $\mathbf{r}(t) = \langle 2\cos t, 2\sin t, t\rangle$.

In earlier chapters we saw that $\mathbf{r}_{xy}(t)$ describes a circle of radius 2 centered at the origin and lying in the (x, y)-plane. As t runs from 0 to 4π, the circle is traced twice. Hence when we look from above down along the z-axis, the curve described by $\mathbf{r}(t)$ will look like a circle of radius 2. See Fig. 8.58.

Next set the second coordinate in $\mathbf{r}(t)$ to 0. We obtain

$$\mathbf{r}_{xz}(t) = \langle 2\cos t, 0, t\rangle, \qquad 0 \le t \le 4\pi.$$

Comparing the first and third coordinates, we see the relationship $x = 2\cos z$. Thus, $\mathbf{r}_{xz}(t)$ describes the curve $x = 2\cos z$ in the (x, z)-plane. The graph of $\mathbf{r}_{xz}(t)$ shows what the curve $\mathbf{r}(t)$ looks like when viewed from the side. See Fig. 8.59. Notice how different the path looks in Figs. 8.58 and 8.59.

Combining these views, we can describe the motion in space. As the point moves around the circle of radius 2 (top view), it rises in the z-direction; this is the influence of the third coordinate, $z = t$. We conclude that the path looks like a spring or helix. This analysis suggests that we might do well to choose a viewpoint that looks down from above so that we can see the circular shape and then move off to the side a little bit so that we can see the rise in the helix. Figure 8.60 is a view of the graph as seen from the point $(2, 2, 15)$.

In Example 2 the circle determined by the first two coordinates was very important in helping us understand the graph of the space curve. Once the circle was identified, we had only to understand that the third coordinate made the graph rise while the circle was being traced. If we can recognize a projection as a familiar two-dimensional curve it may be useful to concentrate on this projection and try to understand how the third coordinate influences the graph. This is demonstrated in the next example.

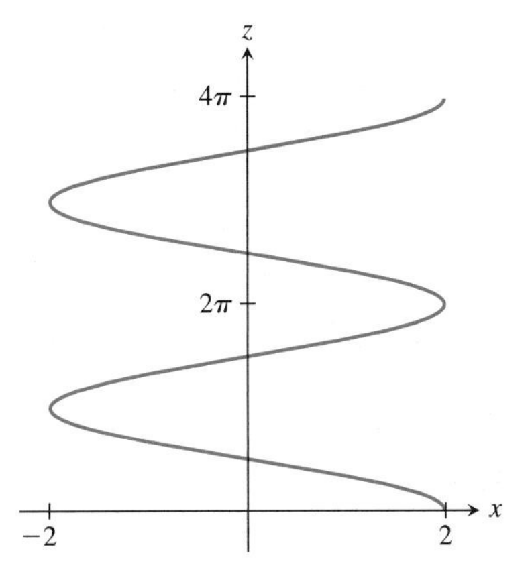

FIGURE 8.59 A side view of the curve $\mathbf{r}(t) = \langle 2\cos t, 2\sin t, t\rangle$.

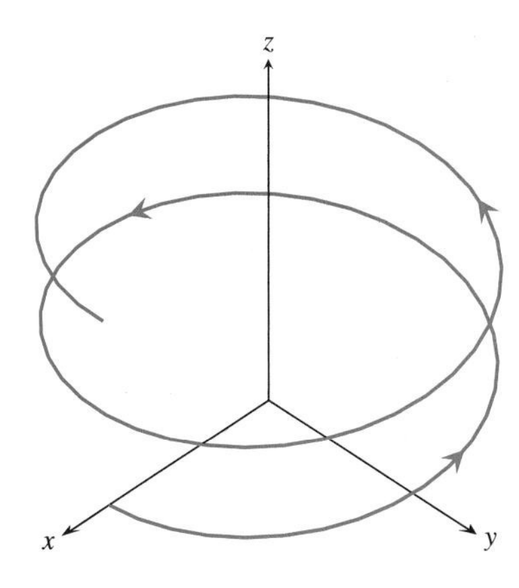

FIGURE 8.60 The path described by $\mathbf{r}(t) = \langle 2\cos t, 2\sin t, t\rangle$ is a helix.

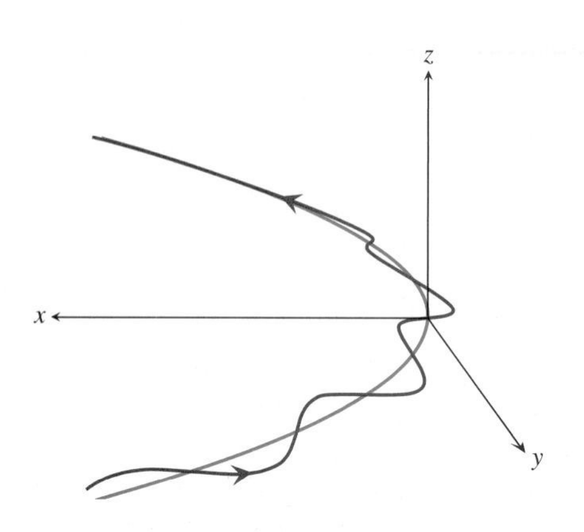

FIGURE 8.61 The path described by $\mathbf{r}(t) = \langle t^2, \sin(10t), t\rangle$ roughly follows the graph of $x = z^2$ in the (x, z)-plane.

EXAMPLE 3 Describe and sketch the curve given by

$$\mathbf{r}(t) = \langle t^2, \sin(10t), t\rangle, \qquad -1.5 \leq t \leq 1.5. \tag{4}$$

Solution

Project the curve into the (x, z)-plane by setting the y-coordinate to 0. The result is

$$\mathbf{r}_{xz}(t) = \langle t^2, 0, t\rangle \qquad -1.5 \leq t \leq 1.5.$$

The path described by this equation is part of the parabola $x = z^2$ in the (x, z)-plane. Hence if we stand far out on the y-axis and look at the graph, we will see a parabola. The $\sin(10t)$ term in (4) causes the y-coordinate of the position function to oscillate rapidly between -1 and 1. Thus as an object moves along the path (4), it roughly follows the parabola $x = z^2$ in the (x, z)-plane but also moves back and forth in the y-direction. In Fig. 8.61 we see the curve as viewed from near the point $(1, 6, 1)$. The graph of the parabola $x = z^2$ is shown for reference.

Motion

In Chapters 3 and 6 we defined *velocity, speed,* the *unit tangent vector,* the *unit normal vector, arc length, acceleration,* and *curvature* for planar motion described by a position function of the form

$$\mathbf{r}(t) = \langle x(t), y(t)\rangle.$$

Many of these ideas extend readily to motion in three-space described by position functions of the form

$$\mathbf{r}(t) = \langle x(t), y(t), z(t)\rangle.$$

We will not formally define all of these concepts for motion in R^3 (however, see Exercises 25, 26, and 27). Instead, we illustrate these ideas in several examples.

EXAMPLE 4 A particle moves along the helix of Example 2 so that at time $t \geq 0$ it is at position

$$\mathbf{r}(t) = \langle 2\cos t, 2\sin t, t\rangle. \tag{5}$$

a) Find the velocity and speed of the particle at any time $t \geq 0$. Sketch the position and velocity vectors for time $t = 2$.
b) Write a position function, $\mathbf{w}(s)$, for the curve using arc length, s, as a parameter. Verify that for this parameterization $\|d\mathbf{w}/ds\| = 1$.

Solution

a) The velocity is given by the vector-valued function

$$\begin{aligned}\mathbf{v}(t) &= \frac{d}{dt}\mathbf{r}(t) = \left\langle \frac{d}{dt}2\cos t, \frac{d}{dt}2\sin t, \frac{d}{dt}t\right\rangle \\ &= \langle -2\sin t, 2\cos t, 1\rangle.\end{aligned}$$

The speed at time t is

$$\|\mathbf{v}(t)\| = \sqrt{(-2\sin t)^2 + (2\cos t)^2 + 1^2} = \sqrt{5}.$$

Hence the particle moves at constant speed along the helix. At time $t = 2$, the position of the particle is

$$\mathbf{r}(2) = \langle 2\cos 2, 2\sin 2, 2\rangle \approx \langle -0.832294, 1.818595, 2\rangle,$$

and the velocity is

$$\mathbf{v}(2) = \langle -2\sin 2, 2\cos 2, 1\rangle \approx \langle -1.818595, -0.832294, 1\rangle.$$

The vectors $\mathbf{r}(2)$ and $\mathbf{v}(2)$ are shown in Fig. 8.62. Note that, as in earlier chapters, we draw the position vector $\mathbf{r}(2)$ from the origin to the curve, but we draw the velocity vector $\mathbf{v}(2)$ with its initial point at position $\mathbf{r}(2)$. The vector $\mathbf{v}(2)$ is tangent to the helix at this point.

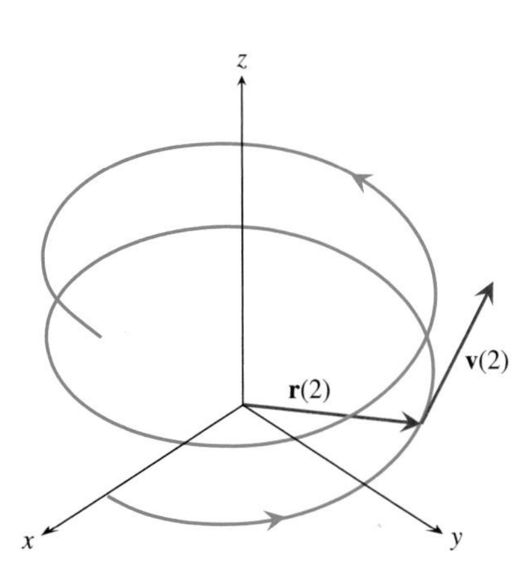

FIGURE 8.62 The position and velocity vectors for the helix at time $t = 2$. The velocity vector $\mathbf{v}(2)$ is tangent to the helix at $\mathbf{r}(2)$.

b) The length of the arc of the helix from $\mathbf{r}(0)$ to $\mathbf{r}(t)$ is just the distance traveled by the particle from time 0 to time t. Because the speed at any time is $\sqrt{5}$, the distance traveled in time t is

$$s = s(t) = (\text{speed})\,(\text{time}) = \sqrt{5}t.$$

It follows that $t = s/\sqrt{5}$, giving us an expression for time t in terms of arc length s. Substituting $s/\sqrt{5}$ for t in (5) gives a parameterization of the helix with arc length as the parameter,

$$\mathbf{w}(s) = \left\langle 2\cos\left(\frac{s}{\sqrt{5}}\right), 2\sin\left(\frac{s}{\sqrt{5}}\right), \frac{s}{\sqrt{5}}\right\rangle.$$

We then have

$$\frac{d\mathbf{w}}{ds} = \left\langle -\frac{2}{\sqrt{5}}\sin\left(\frac{s}{\sqrt{5}}\right), \frac{2}{\sqrt{5}}\cos\left(\frac{s}{\sqrt{5}}\right), \frac{1}{\sqrt{5}}\right\rangle,$$

so

$$\left\|\frac{d\mathbf{w}}{ds}\right\| = \sqrt{\left(-\frac{2}{\sqrt{5}}\sin\left(\frac{s}{\sqrt{5}}\right)\right)^2 + \left(\frac{2}{\sqrt{5}}\cos\left(\frac{s}{\sqrt{5}}\right)\right)^2 + \left(\frac{1}{\sqrt{5}}\right)^2} = 1.$$

Whenever arc length is used as the parameter, the derivative vector has length 1. In particular, $d\mathbf{w}/ds$ is a unit vector tangent to the curve.

EXAMPLE 5 The position vector for an object at time $t \geq 0$ is

$$\mathbf{r}(t) = \left\langle t^2, e^{t/2}, \frac{t}{1+t}\right\rangle,$$

where time is measured in seconds and distance in meters.

a) Find $a_T(1)$, the tangential component of the acceleration at time $t = 1$. Sketch $\mathbf{a}(1)$, $a_T(1)\mathbf{T}$, and $\mathbf{a}(1) - a_T(1)\mathbf{T}(1)$.

b) If $\mathbf{r}(t)$ gives the position of an object of mass 2 kg at time t, what is the magnitude of the force on the object at time $t = 1$?

c) Find t so that the object travels a distance of 10 meters from time 0 to time t.

Solution

a) The velocity and acceleration functions are

$$\mathbf{v}(t) = \frac{d\mathbf{r}}{dt} = \left\langle \frac{d}{dt}t^2, \frac{d}{dt}e^{t/2}, \frac{d}{dt}\left(\frac{t}{1+t}\right)\right\rangle$$
$$= \left\langle 2t, \frac{1}{2}e^{t/2}, \frac{1}{(1+t)^2}\right\rangle \text{ m/s}$$

and

$$\mathbf{a}(t) = \frac{d\mathbf{v}}{dt} = \left\langle \frac{d}{dt}(2t), \frac{d}{dt}\left(\frac{1}{2}e^{t/2}\right), \frac{d}{dt}\left(\frac{1}{(1+t)^2}\right)\right\rangle$$
$$= \left\langle 2, \frac{1}{4}e^{t/2}, -\frac{2}{(1+t)^3}\right\rangle \text{ m/s}^2.$$

At time $t = 1$, the unit tangent vector is

$$\mathbf{T}(1) = \frac{1}{\|\mathbf{v}(1)\|}\mathbf{v}(1) = \frac{1}{\sqrt{2^2 + \left(\sqrt{e}/2\right)^2 + (1/4)^2}}\left\langle 2, \frac{\sqrt{e}}{2}, \frac{1}{4}\right\rangle$$
$$\approx \langle 0.918, 0.379, 0.115\rangle.$$

The acceleration at time $t = 1$ is

$$\mathbf{a}(1) = \left\langle 2, \frac{\sqrt{e}}{4}, -\frac{2}{8}\right\rangle \approx \langle 2, 0.412, -0.25\rangle \text{ m/s}^2. \tag{6}$$

The component of $\mathbf{a}(1)$ in the direction of $\mathbf{T}(1)$ is

$$a_T(1) = \mathbf{a}(1)\cdot\mathbf{T}(1) \approx \langle 2, 0.412, -0.25\rangle \cdot \langle 0.918, 0.379, 0.115\rangle \approx 1.964 \text{ m/s}^2.$$

Hence the tangential acceleration is

$$a_T(1)\mathbf{T}(1) \approx \langle 1.804, 0.744, 0.225\rangle \text{ m/s}^2. \tag{7}$$

Combining (6) and (7), we have

$$\mathbf{a}(1) - a_T(1)\mathbf{T}(1) \approx \langle 0.196, -0.331, -0.475\rangle \text{ m/s}^2.$$

Is $\mathbf{a}(1) - a_T(1)\mathbf{T}(1)$ perpendicular to $\mathbf{T}(1)$? How do you know? The vectors $\mathbf{a}(1)$, $a_T(1)\mathbf{T}(1)$, and $\mathbf{a}(1) - a_T(1)\mathbf{T}(1)$ are shown in Fig. 8.63.

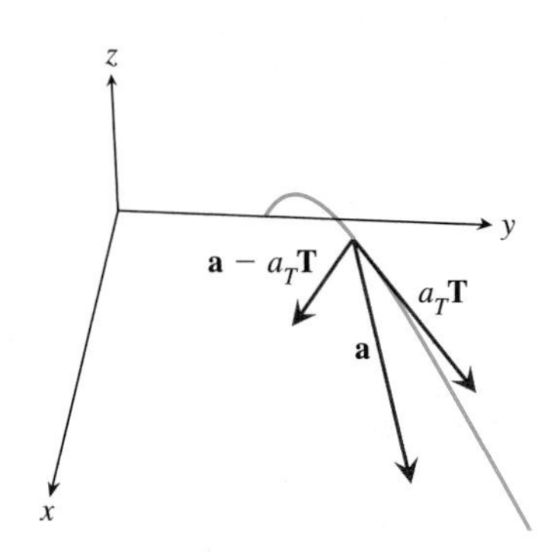

FIGURE 8.63 The acceleration vector $\mathbf{a}(1)$, the tangential acceleration $a_T\mathbf{T}$, and $\mathbf{a} - a_T\mathbf{T}$.

b) By Newton's second law, $\mathbf{F} = m\mathbf{a}$. The magnitude of the force at time $t = 1$ is

$$\|\mathbf{F}(1)\| = \|m\mathbf{a}(1)\| \approx 2\|\langle 2, 0.412, -0.25\rangle\| \approx 4.114 \text{ newtons}.$$

c) The distance $s(t)$ traveled by the object from time 0 to time t is the length of the path between $\mathbf{r}(0)$ and $\mathbf{r}(t)$:

$$s(t) = \int_0^t \|\mathbf{v}(\tau)\|\,d\tau = \int_0^t \sqrt{4\tau^2 + \frac{e^\tau}{4} + \frac{1}{(1+\tau)^4}}\,d\tau. \tag{8}$$

We first use a numerical integration package or CAS to find a rough estimate of a value of t for which $s(t) = 10$. Trying some values of t in the formula for s, we find

$$s(1) \approx 1.38, \quad s(2) \approx 4.57, \quad \text{and} \quad s(3) \approx 9.88.$$

Hence we take $t_1 = 3$ seconds as a rough guess for the solution. We further refine this estimate by applying Newton's method to the equation

$$s(t) - 10 = 0. \tag{9}$$

If t_n is an approximation of a solution to this equation, then an application of Newton's method produces the next iterate,

$$t_{n+1} = t_n - \frac{s(t_n) - 10}{s'(t_n)}.$$

By the Fundamental Theorem of Calculus,

$$s'(t) = \frac{d}{dt}\int_0^t \|\mathbf{v}(\tau)\|\,d\tau = \|\mathbf{v}(t)\|.$$

At each step of Newton's method, we calculate the next iterate numerically. With $t_1 = 3$ as our first guess for the solution to (9),

$$t_2 = t_1 - \frac{s(t_1) - 10}{\|\mathbf{v}(t_1)\|} = 3 - \frac{s(3) - 10}{\|\mathbf{v}(3)\|} \approx 3.01938 \text{ seconds}.$$

Subsequent values of t_n are given in Table 8.1. From the table we conclude that the object has traveled a distance of 10 when $t \approx 3.0193187$ seconds. In Exercise 30, we ask you to verify the entries in the table. In Exercises 31 and 32, we look at other ways to use a calculator or CAS to approximate the solution to (9).

TABLE 8.1

n	t_n
1	3.0
2	3.01938477
3	3.01931873
4	3.01931873
5	3.01931873

In Section 6.8 we proved Kepler's second law ("equal areas in equal times") under the assumption that an object moving under the influence of a central force must move in a plane. Many familiar objects move under the influence of such a force and hence move in planar paths: the planets in the solar system, satellites orbiting the Earth, and the rings of Saturn and Jupiter. In the next example, we show that the planar path assumption made in Section 6.8 is justified. Recall that a moving object with position vector $\mathbf{r}(t)$ is subject to a central force if there is a fixed point O such that at any time t, the force vector is parallel to the direction from position $\mathbf{r}(t)$ to O. We usually take O to be the origin. See Fig. 8.64.

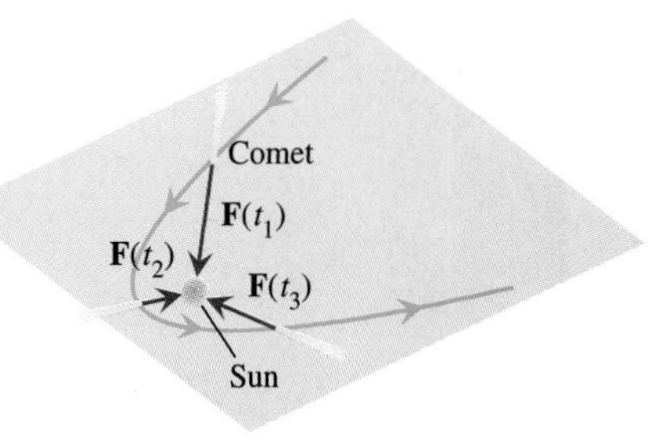

FIGURE 8.64 A moving object subject to a central force.

EXAMPLE 6 Suppose that the only force acting on an object is of the form

$$\mathbf{F}(t) = m\mathbf{a}(t) = f(t)\mathbf{r}(t) \qquad t \geq 0. \tag{10}$$

Show that the path described by $\mathbf{r}(t)$ lies in a plane.

Solution

From (10) we see that the vectors

$$\mathbf{r}''(t) = \mathbf{a}(t) = \frac{f(t)}{m}\mathbf{r}(t) \quad \text{and} \quad \mathbf{r}(t) = \langle x(t), y(t), z(t)\rangle$$

are parallel. Because the cross product of parallel vectors is **0**,

$$\langle 0,0,0\rangle = \mathbf{r} \times \mathbf{r}'' = \langle yz'' - zy'', zx'' - xz'', xy'' - yx''\rangle, \tag{11}$$

where the primes denote differentiation with respect to t and for notational convenience we have written x, x'', $y, \ldots$ instead of $x(t)$, $x''(t)$, $y(t), \ldots$. Equating the first components from the left and right expressions in (11), we find

$$yz'' - zy'' = 0, \qquad t \geq 0 \tag{12}$$

Next, note that

$$\frac{d}{dt}(yz' - zy') = (y'z' + yz'') - (z'y' + zy'') = (yz'' - zy'').$$

Combining this with (12), we conclude that there is a constant c_1 such that

$$yz' - zy' = c_1, \qquad t \geq 0$$

Similar reasoning applies to the other components in (11). Hence there is a constant vector $\mathbf{c} = \langle c_1, c_2, c_3\rangle$ such that

$$\langle yz' - zy', zx' - xz', xy' - yx'\rangle = \mathbf{c}, \qquad t \geq 0$$

We recognize the left side of this expression as $\mathbf{r}(t) \times \mathbf{r}'(t)$, so for all t,

$$\mathbf{r}(t) \times \mathbf{r}'(t) = \mathbf{c}, \qquad t \geq 0$$

If the constant vector **c** is not the zero vector, then $\mathbf{r}(t)$ is always perpendicular to **c**. Hence the path described by $\mathbf{r}(t)$ lies in a plane perpendicular to the constant vector **c**. See Fig. 8.65. This shows that when an object moves under the influence of a central force (and no other force), the path of motion lies in a plane. In particular, if we choose a coordinate system so that **c** is parallel to **k**, then the motion is in the (x, y)-plane. See Exercises 46 and 47 for the case $\mathbf{c} = \langle 0,0,0\rangle$.

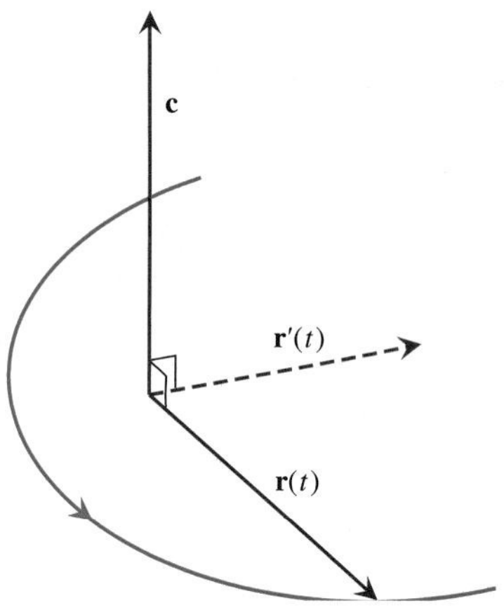

FIGURE 8.65 The object moves in a plane perpendicular to **c**.

The Unit Normal and Curvature Vectors

In Section 6.8 we introduced curvature and the curvature vector to measure the rate at which a planar curve bends. The unit normal **N** was defined to be the unit vector in the direction of the curvature vector. We then saw that acceleration can be decomposed into tangential and normal parts, that is, into a sum of two vectors, one in the direction of **T** and one in the direction of **N**. We might suspect that to analyze acceleration for motion in space we will need to introduce a third vector perpendicular to **T** and **N**. Surprisingly, a third vector is not necessary to discuss acceleration. When the definitions of *curvature* and *unit normal* are naturally extended to include smooth space curves described by

$$\mathbf{r}(t) = \langle x(t), y(t), z(t)\rangle,$$

the two vectors **T** and **N** are still sufficient to express the acceleration vector. See, however, Exercises 53–56.

DEFINITION Curvature Vector, Curvature, and Unit Normal in R^3

Let C be the smooth curve described by

$$\mathbf{r}(t) = \langle x(t), y(t), z(t) \rangle, \qquad a \leq t \leq b, \tag{13}$$

where the coordinate functions x, y, and z have continuous second derivatives on the interval $a \leq t \leq b$. If for a given t, $\mathbf{r}'(t) \neq \langle 0, 0, 0 \rangle$, then the **curvature vector,** $\boldsymbol{\kappa}(t)$, of C at $\mathbf{r}(t)$ is defined to be

$$\boldsymbol{\kappa}(t) = \frac{d\mathbf{T}}{ds}, \tag{14}$$

where s denotes arc length. The **curvature** $\kappa(t)$ and **radius of curvature** $\rho(t)$ at $\mathbf{r}(t)$ are defined by

$$\kappa(t) = \|\boldsymbol{\kappa}(t)\| = \|\boldsymbol{\kappa}(t)\| = \left\|\frac{d\mathbf{T}}{ds}\right\| \quad \text{and} \quad \rho(t) = \frac{1}{\kappa(t)}. \tag{15}$$

When $\kappa(t) \neq 0$, the **unit normal** $\mathbf{N}(t)$ is defined by

$$\mathbf{N}(t) = \frac{1}{\kappa(t)} \frac{d\mathbf{T}}{ds}. \tag{16}$$

The unit normal $\mathbf{N}(t)$ is undefined when $\kappa(t) = 0$.

EXAMPLE 7 Consider again the position function that describes the helix of Example 2,

$$\mathbf{r}(t) = \langle 2 \cos t, 2 \sin t, t \rangle, \qquad t \geq 0.$$

Find the curvature vector, curvature, and unit normal at a point $\mathbf{r}(t)$ on the curve.

Solution

As seen in Example 4, the velocity vector is

$$\mathbf{v}(t) = \frac{d\mathbf{r}}{dt} = \langle -2 \sin t, 2 \cos t, 1 \rangle,$$

and for any value of t, the speed is

$$\frac{ds}{dt} = \|\mathbf{v}(t)\| = \sqrt{5}.$$

Hence at $\mathbf{r}(t)$ the unit tangent vector is

$$\mathbf{T}(t) = \frac{1}{\|\mathbf{v}(t)\|}\mathbf{v}(t) = \left\langle -\frac{2}{\sqrt{5}} \sin t, \frac{2}{\sqrt{5}} \cos t, \frac{1}{\sqrt{5}} \right\rangle.$$

By (14) and the chain rule, the curvature vector is

$$\boldsymbol{\kappa}(t) = \frac{d\mathbf{T}}{ds} = \frac{d\mathbf{T}}{dt}\frac{dt}{ds} = \left\langle -\frac{2}{\sqrt{5}}\cos t, -\frac{2}{\sqrt{5}}\sin t, 0\right\rangle \frac{1}{\sqrt{5}} = \left\langle -\frac{2}{5}\cos t, -\frac{2}{5}\sin t, 0\right\rangle. \tag{17}$$

The vector in (17) has length 2/5. Hence the curvature and radius of curvature at $\mathbf{r}(t)$ are, respectively,

$$\kappa(t) = \left\|\frac{d\mathbf{T}}{ds}\right\| = \frac{2}{5} \quad \text{and} \quad \rho(t) = \frac{1}{\kappa(t)} = \frac{5}{2}.$$

The unit normal is the unit vector in the direction of the curvature vector,

$$\mathbf{N}(t) = \frac{1}{\kappa(t)}\frac{d\mathbf{T}}{ds} = \langle -\cos t, -\sin t, 0\rangle.$$

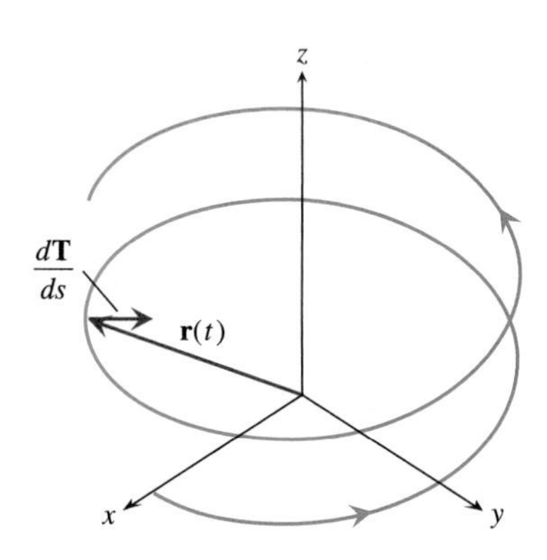

FIGURE 8.66 The curvature vector $d\mathbf{T}/ds$, at $\mathbf{r}(t)$.

The curvature vector is shown in Fig. 8.66. It is interesting to note that, for this curve, the curvature vector (and the unit normal) are parallel to the (x, y)-plane and point toward the center of the helix. It is also of interest to note that even though the helix is somewhat like motion around a circle of radius 2, the curvature is 2/5 rather than 1/2. Evidently the rise in the helix described by the third component of t "decreases" the rate of bending from that of a circle of radius 2.

Earlier in this section we stated that in R^3 the acceleration vector can be expressed in terms of the unit tangent and unit normal vectors. We verify this assertion in the next example.

EXAMPLE 8 Let C be a smooth curve described by the position function

$$\mathbf{r}(t) = \langle x(t), y(t), z(t)\rangle,$$

and assume that the functions x, y, z all have continuous second derivatives. Show that for any t such that $\|\mathbf{r}'(t)\| \neq 0$ and $\kappa(t) \neq 0$,

$$\mathbf{a}(t) = \frac{d^2s}{dt^2}\mathbf{T} + \left(\frac{ds}{dt}\right)^2 \kappa(t)\mathbf{N}. \tag{18}$$

Solution

Because we have assumed that $\|\mathbf{r}'(t)\| \neq 0$, the unit tangent $\mathbf{T}$ exists. We have

$$\mathbf{v}(t) = \|\mathbf{v}(t)\|\mathbf{T} = \frac{ds}{dt}\mathbf{T}.$$

Differentiate both sides of this expression with respect to t. For the derivative on the right, use the product rule for the derivative of a scalar function times a vector function (see Section 6.8). It follows that

$$\mathbf{a}(t) = \frac{d\mathbf{v}}{dt} = \frac{d}{dt}\left(\frac{ds}{dt}\right)\mathbf{T} + \frac{ds}{dt}\frac{d\mathbf{T}}{dt} = \frac{d^2s}{dt^2}\mathbf{T} + \frac{ds}{dt}\left(\frac{d\mathbf{T}}{ds}\frac{ds}{dt}\right).$$

Using (14) and (16), we can express $d\mathbf{T}/ds$ in terms of $\mathbf{N}$ and $\kappa(t)$. Substituting into the previous expression,

$$\mathbf{a}(t) = \frac{d^2s}{dt^2}\mathbf{T} + \left(\frac{ds}{dt}\right)^2 \kappa(t)\mathbf{N}.$$

This verifies (18). See Exercise 48 for the $\kappa(t) = 0$ case.

Equation (18) gives us explicit formulas for the tangential and normal components of acceleration. The tangential and normal components of acceleration are denoted by a_{T} and a_{N}, respectively. By (16), we have

$$a_{\mathrm{T}} = a_{\mathrm{T}}(t) = \frac{d^2s}{dt^2} \quad \text{and} \quad a_{\mathrm{N}} = a_{\mathrm{N}}(t) = \left(\frac{ds}{dt}\right)^2 \kappa(t).$$

Because $a_T = \dfrac{d}{dt}\left(\dfrac{ds}{dt}\right)$ is the rate of change of speed, we see that the tangential component of acceleration tells us whether the object is speeding up or slowing down at a given instant. The normal component of acceleration, a_N, tells us about the change in direction of the object. These quantities are of great importance in controlling the path of an object such as a satellite or the shuttle in space.

Even for simple functions $\mathbf{r}(t)$, the calculations of d^2s/dt^2 and $\kappa(t)$ can be lengthy. As the next example shows, there are alternative ways to calculate these quantities.

EXAMPLE 9 The position of an object at all times $t \geq 0$ is given by

$$\mathbf{r}(t) = \langle \cos^2 t, \sin t \cos t, \sin t \rangle. \tag{19}$$

Find and graph the tangential and normal accelerations at $\mathbf{r}(1.3)$.

Solution

First differentiate (19) to get

$$\mathbf{v}(t) = \frac{d\mathbf{r}}{dt} = \langle -2 \sin t \cos t, \cos^2 t - \sin^2 t, \cos t \rangle.$$

Thus

$$\mathbf{v}(1.3) \approx \langle -0.515501, -0.856889, 0.267499 \rangle.$$

Divide this vector by its length to get

$$\mathbf{T} = \mathbf{T}(1.3) \approx \langle -0.497992, -0.827784, 0.258413 \rangle.$$

The acceleration vector is

$$\begin{aligned} \mathbf{a}(t) &= \frac{d\mathbf{v}}{dt} = \langle 2 \sin^2 t - 2 \cos^2 t, -4 \sin t \cos t, -\sin t \rangle \\ &= \langle -2 \cos 2t, -2 \sin 2t, -\sin t \rangle. \end{aligned}$$

Setting $t = 1.3$, we have

$$\mathbf{a}(1.3) \approx \langle 1.71378, -1.03100, -0.963558 \rangle.$$

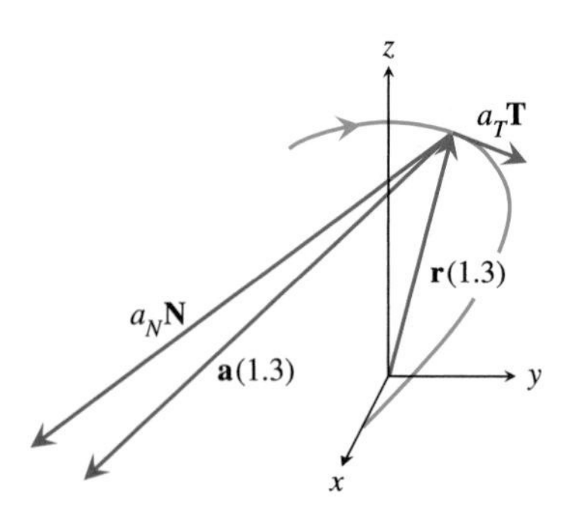

FIGURE 8.67 The tangential and normal components of $\mathbf{a}(1.3)$.

The tangential component of acceleration, a_T, is the component of $\mathbf{a}$ in the direction of $\mathbf{T}$. Hence the tangential acceleration is $a_T\mathbf{T}$ and is given by

$$\begin{aligned} a_T\mathbf{T} &= (\mathbf{a}(1.3) \cdot \mathbf{T})\mathbf{T} \\ &\approx -0.248996\mathbf{T} \approx \langle 0.123998, 0.206115, -0.0643438 \rangle. \end{aligned}$$

The normal acceleration is then given by

$$a_N\mathbf{N} = \mathbf{a}(1.3) - a_T\mathbf{T} \approx \langle 1.58978, -1.23712, -0.899214 \rangle.$$

These vectors are shown in Fig. 8.67.

Another Formula for Curvature Because the result of differentiating the unit tangent vector $\mathbf{T} = \frac{1}{\|\mathbf{r}'(t)\|}\mathbf{r}'(t)$ may be a complicated expression, calculating $\kappa(t)$ by finding the length of $d\mathbf{T}/ds$ may be a lengthy procedure. Another method for calculating curvature is motivated by the fact that $\kappa(t)$ is part of the coefficient of $\mathbf{N}$ in (16). We can isolate this coefficient by using the cross product:

$$\begin{aligned} \mathbf{r}'(t) \times \mathbf{r}''(t) &= \frac{ds}{dt}\mathbf{T} \times \left(\frac{d^2s}{dt^2}\mathbf{T} + \left(\frac{ds}{dt}\right)^2 \kappa(t)\mathbf{N} \right) \\ &= \frac{ds}{dt}\frac{d^2s}{dt^2}(\mathbf{T} \times \mathbf{T}) + \left(\frac{ds}{dt}\right)^3 \kappa(t)\,(\mathbf{T} \times \mathbf{N}). \end{aligned}$$

In this equation, we used the fact that the cross product distributes over vector addition and the fact that scalar multiples can be factored from a cross product. See Section 8.3. Next note that $\mathbf{T} \times \mathbf{T} = \langle 0, 0, 0 \rangle$ and, because $\mathbf{T}$ and $\mathbf{N}$ are perpendicular vectors of length 1, $\|\mathbf{T} \times \mathbf{N}\| = 1$. Using these facts and taking the lengths of the vectors on both sides of the equation for $\mathbf{r}'(t) \times \mathbf{r}''(t)$,

$$\|\mathbf{r}'(t) \times \mathbf{r}''(t)\| = \left\| \left(\frac{ds}{dt}\right)^3 \kappa(t)\,(\mathbf{T} \times \mathbf{N}) \right\| = \left(\frac{ds}{dt}\right)^3 \kappa(t) = \|\mathbf{r}'(t)\|^3 \kappa(t).$$

Solving for $\kappa(t)$ results in the following formula for curvature.

Curvature

Let C be a smooth curve described by

$$\mathbf{r}(t) = \langle x(t), y(t), z(t) \rangle, \qquad a \le t \le b,$$

and assume that the coordinate functions x, y, and z have continuous second derivatives on the interval $a \le t \le b$. Let $\mathbf{r}(t)$ be a point on the curve C, and assume that $\|\mathbf{r}'(t)\| \neq 0$. Then the curvature at $\mathbf{r}(t)$ is

$$\kappa = \kappa(t) = \frac{\|\mathbf{r}'(t) \times \mathbf{r}''(t)\|}{\|\mathbf{r}'(t)\|^3}. \tag{20}$$

Exercises 8.7

Exercises 1–10: Describe and sketch the space curve.

1. $\mathbf{r}(t) = \langle -2t, 4 + 3t, -4 + 2t \rangle, \quad 0 \le t \le 3$
2. $\mathbf{r}(t) = \langle 2t^2, -2 + t^2, 3 \rangle, \quad 0 \le t \le 1$
3. $\mathbf{r}(t) = \langle \cos t, \sin t, 2t \rangle, \quad 0 \le t \le 3\pi$
4. $\mathbf{r}(t) = \langle 4 \sin t, t, 4 \cos t \rangle, \quad 0 \le t \le 6\pi$
5. $\mathbf{r}(t) = \langle 2 \cos 2t, 2 \sin 2t, 3t \rangle, \quad 0 \le t \le 2\pi$
6. $\mathbf{r}(t) = \langle t, 3 \cos t, \sin t \rangle, \quad 0 \le t \le 4\pi$
7. $\mathbf{r}(t) = \langle 4 \sin t, 3t, 2 \cos t \rangle, \quad 0 \le t \le 4\pi$
8. $\mathbf{r}(t) = \langle t \cos t, t \sin t, t \rangle, \quad 0 \le t \le 4\pi$
9. $\mathbf{r}(t) = \langle e^t \cos t, e^t \sin t, t \rangle, \quad 0 \le t \le 4\pi$
10. $\mathbf{r}(t) = \langle e^{-t} \cos t, e^{-t} \sin t, t \rangle, \quad 0 \le t \le 4\pi$

Exercises 11–14: Describe and sketch what you would see if the space curve were projected into each of the coordinate planes. Based on these views, sketch the curve.

11. $\mathbf{r}(t) = \langle \sin t, \cos t, 4 \rangle, \quad 0 \le t \le 2\pi$
12. $\mathbf{r}(t) = \langle t, t, t^3 \rangle, \quad -1 \le t \le 1$
13. $\mathbf{r}(t) = \langle t, e^t, t^2 \rangle, \quad 0 \le t \le 2\pi$
14. $\mathbf{r}(t) = \langle 4 \sin t, t, \cos t \rangle, \quad 0 \le t \le 4\pi$

Exercises 15–18: Find the velocity and acceleration functions for each position function. Sketch $\mathbf{r}(1)$, $\mathbf{v}(1)$, *and* $\mathbf{a}(1)$.

15. The position function of Exercise 1
16. The position function of Exercise 3
17. The position function of Exercise 6
18. The position function of Exercise 10

Exercises 19–24: Find the unit tangent $\mathbf{T}(t)$, *the unit normal* $\mathbf{N}(t)$, *the curvature* $\kappa(t)$, *the tangential component of acceleration* $a_T(t)$, *and the normal component of acceleration* $a_N(t)$ *for the given t.*

19. The position function of Exercise 2, with $t = 2$
20. The position function of Exercise 4, with $t = \pi/2$
21. The position function of Exercise 6, with $t = \pi$
22. The position function of Exercise 10, with $t = 0$
23. The position function of Exercise 12, with $t = 2$
24. The position function of Exercise 13, with $t = 0$
25. Let $\mathbf{r}(t) = \langle x(t), y(t), z(t) \rangle$ be a position function describing motion along a curve C in three dimensions.
    a. Using the definition from Section 3.4 as a model, carefully define the average velocity, $\mathbf{v}[t_0, t_1]$, for the time period $t_0 \le t \le t_1$. Write a careful description of the physical significance of the average velocity.
    b. Write a careful definition of the velocity, $\mathbf{v}(t_0)$, at time t_0. (Your definition should involve limits.)
26. Let $\mathbf{r}(t) = \langle x(t), y(t), z(t) \rangle$ be a position function describing motion along a curve C in three dimensions.
    a. What does it mean for C to be a smooth curve?
    b. Write a definition for the acceleration, $\mathbf{a}(t)$, at time t.
    c. Write a definition for the unit tangent, $\mathbf{T} = \mathbf{T}(t)$, to C at the point $\mathbf{r}(t)$.
27. Let
$$\mathbf{r}(t) = \langle x(t), y(t), z(t) \rangle, \qquad a \le t \le b,$$
describe a smooth curve C in R^3. The arc length s of C is defined to be
$$s = \int_a^b \|\mathbf{v}(t)\|\,dt = \int_a^b \sqrt{x'(t)^2 + y'(t)^2 + z'(t)^2}\,dt.$$
Write a complete discussion, modeled on the one given in Section 6.4, to motivate this definition.
28. Let C be the curve described by $\mathbf{r}(t) = \langle t, t^2, t^3 \rangle$. Find the length of the piece of C between $\mathbf{r}(0)$ and $\mathbf{r}(2)$. Use a numerical integration package on a calculator or CAS if necessary.
29. Let C be the curve described by $\mathbf{r}(t) = \langle 2 \sin t, 3 \cos t, 4t \rangle$. Find the length of the piece of C between $\mathbf{r}(0)$ and $\mathbf{r}(2\pi)$. Use a numerical integration package on a calculator or CAS if necessary.
30. (T) In Example 5 we approximated the solution to (9) using Newton's method with initial guess $t_1 = 3$. The subsequent approximations t_2, t_3, t_4, and t_5 are displayed in Table 8.1 on page 709. Check that these entries are correct.
31. (T) Solve (9) of Example 5 graphically by graphing $s = s(t)$ and $s = 10$ on the same set of axes and then zooming in on the point of intersection until you can read the intersection point to two-decimal-place accuracy.
32. (T) Solve (9) of Example 5 by trial and error. That is, guess a value for t, then use a numerical integration application on your calculator or computer to approximate the value of $s(t)$. Based on the answer (greater than 10 or less than 10?), refine your guess for t. Repeat this process until you have the desired value of t accurate to two decimal places.
33. Let C be the curve described by $\mathbf{r}(t) = \left\langle -\cos t, \sin t, \sqrt{7}t \right\rangle$. The length of the piece of C between $\mathbf{r}(0)$ and $\mathbf{r}(t_1)$, $t_1 > 0$, is 12. Find t_1.
34. (T) Let C be the curve described by $\mathbf{r}(t) = \langle t, t^2, t^3 \rangle$. The length of the piece of C between $\mathbf{r}(0)$ and $\mathbf{r}(t_1)$, $t_1 > 0$, is 12. Find t_1 to two decimal places.

T **35.** Let C be the curve described by $\mathbf{r}(t) = \langle 2\sin t, 3\cos t, 4t\rangle$. The length of the piece of C between $\mathbf{r}(0)$ and $\mathbf{r}(t_1)$, $t_1 > 0$, is $3\sqrt{3}$. Find t_1 to two decimal places.

36. The position of a particle at time t is $\mathbf{r}(t) = \langle b_1 + a_1 t, b_2 + a_2 t, b_3 + a_3 t\rangle$. Describe the motion.

37. At time $t = 0$ an object is at the point $(-2, 4, 3)$. The object moves along a line and passes through $(2, 8, 5)$ at time $t = 1$. Find a position function for the motion.

38. At time $t = 0$ an object is at the point $(-2, 4, 0)$. The object moves with speed of 5 on a line and passes through the point $(0, 3, 2)$. Find a position function for the motion.

39. At time $t = 0$ an object is at the point $(0, 0, 0)$. The object moves at a constant speed of s on a line and passes through the point (a_1, a_2, a_3). Find a position function for the motion.

40. An object moves along a helix. When viewed from above the (x, y)-plane, the object appears to travel around a circle of radius 3 centered at the origin. The object moves at speed 8 and at time $t = 0$ is at $(3, 0, 0)$. Find four different position functions that describe such a motion.

41. An object moves along a helix. When viewed from above the (x, y)-plane, the object appears to travel around a circle of radius 5 centered at the origin. The object moves at speed 20 and at time $t = 0$ is at $(0, 5, 0)$. Find four different position functions that describe such a motion.

42. An object moves along a helix. When viewed from the positive y-axis looking toward the (x, z)-plane, the object appears to travel around a circle of radius 5 centered at the origin. The object moves at speed 19 and at time $t = 0$ is at $(5, 0, 0)$. Find four different position functions that describe such a motion.

43. An airliner of mass 3.6×10^5 kg descends for a landing. For several seconds the plane follows a spiral path described by

$$\mathbf{r}(t) = \left\langle 6000\cos\left(\frac{\pi t}{100}\right), 6000\sin\left(\frac{\pi t}{100}\right), 5000 - 5t\right\rangle,$$

where distance is measured in meters. What is the force on the airliner at time $t = 250$ seconds?

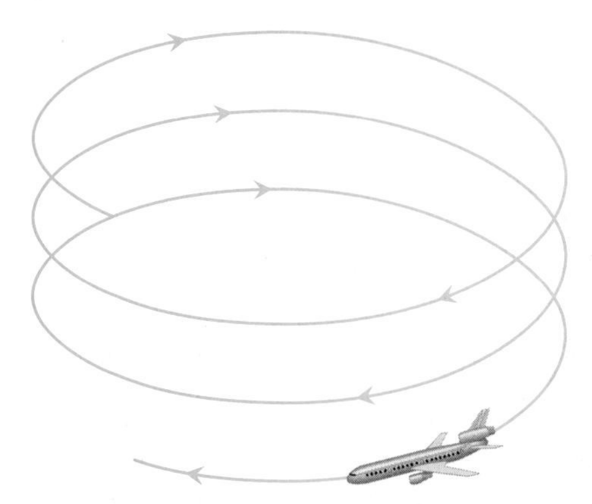

Figure for Exercise 43.

44. A submarine of mass 5.4×10^5 kg ascends from the ocean floor. During its ascent the sub follows a spiral path described by

$$\mathbf{r}(t) = \left\langle 200\cos\left(\frac{\pi t}{20}\right), 200\sin\left(\frac{\pi t}{20}\right), 6t\right\rangle,$$

where distance is measured in meters. What is the force on the submarine at time $t = 200$ seconds?

45. Let C be a smooth curve described by $\mathbf{r}(t) = \langle x(t), y(t), z(t)\rangle$. Assume that x, y, and z have continuous second derivatives. Show that if $\|\mathbf{r}'(t)\| \neq 0$, then the curvature at $\mathbf{r}(t)$ is given by

$$\kappa(t) = \frac{\left(\|\mathbf{r}'(t)\|^2\|\mathbf{r}''(t)\|^2 - (\mathbf{r}'(t)\cdot\mathbf{r}''(t))^2\right)^{1/2}}{\|\mathbf{r}'(t)\|^3}.$$

46. The $\mathbf{c} = \langle 0, 0, 0\rangle$ case of Example 6 With the notation used in Example 6, assume that

$$\mathbf{r}(t) \times \mathbf{r}'(t) = \langle 0, 0, 0\rangle$$

for all t and that neither $\mathbf{r}(t)$ nor $\mathbf{r}'(t)$ is the zero vector.

a. Show that there is a nonzero real valued function k such that

$$x'(t) = k(t)x(t)$$
$$y'(t) = k(t)y(t)$$
$$z'(t) = k(t)z(t).$$

b. Show that there is a constant vector $\mathbf{b} = \langle b_1, b_2, b_3\rangle$ such that

$$\mathbf{r}(t) = e^{K(t)}\mathbf{b} \tag{21}$$

where $K(t)$ is an antiderivative of $k(t)$.

c. Show that the motion described by (21) either is motion along a line or describes a fixed point.

47. Continuation Based on the results of the previous problem, show that if $\mathbf{r}(t) = \langle 0, 0, 0\rangle$ for some time t, then $\mathbf{r}(t) = \langle 0, 0, 0\rangle$ for *all* times t.

48. In Example 8 we showed that when $\|\mathbf{r}'(t)\| \neq 0$ and $\kappa(t) \neq 0$,

$$\mathbf{a}(t) = \frac{d^2 s}{dt^2}\mathbf{T} + \left(\frac{ds}{dt}\right)^2 \kappa(t)\mathbf{N}.$$

Rework Example 8 under the assumption that $\kappa(t) = 0$. What can be concluded about the tangent/normal decomposition of acceleration in this case?

49. Show that the curve described by (19) lies on the sphere of center $(0, 0, 0)$ and radius 1.

50. Let C be the curve described by

$$\mathbf{r}(t) = \langle -3t, 1 + 4t, 1 - t\rangle, \qquad t \geq 0.$$

a. Find a parameterization for C using arc length s as the parameter.

b. Use this parameterization and (14) to find the curvature $\kappa(s)$.

51. Let C be the curve described by

$$\mathbf{r}(t) = \langle -2\cos 3t, 2\sin 3t, 5t\rangle, \qquad t \geq 0.$$

a. Find a parameterization for C using arc length s as the parameter.

b. Use this parameterization and (15) to find the curvature $\kappa(s)$.

52. Let C be the curve described by

$$\mathbf{r}(t) = \langle e^t \cos 4t, e^t \sin 4t, 4e^t \rangle, \qquad t \geq 0.$$

a. Find a parameterization for C using arc length s as the parameter.

b. Use this parameterization and (15) to find the curvature $\kappa(s)$.

Exercises 53–56: The **binormal vector B**, *defined by* $\mathbf{B} = \mathbf{T} \times \mathbf{N}$, *is often used when a third vector perpendicular to both* $\mathbf{T}$ *and* $\mathbf{N}$ *is needed. The binormal arises in problems of spacecraft control. Find the binormal vector to the given path for the given value of t.*

53. $\mathbf{r}(t) = \langle 4 \cos t, -4t, 4 \sin t \rangle$ for $t = \pi/2$

54. $\mathbf{r}(t) = \langle t^2, 0, 2t \rangle$ for $t = 1$

55. $\mathbf{r}(t) = \langle e^t, e^t \sin t, e^t \cos t \rangle$ for $t = 0$

56. $\mathbf{r}(t) = \langle \cos(3t), \sin(3t), 4t \rangle$ for $t = \pi/12$

Exercises 57–60: Let C be a smooth curve described by $\mathbf{r} = \mathbf{r}(t)$. *If* $\|\mathbf{r}'(t_0)\| \neq 0$, *then the line tangent to C at* $\mathbf{r}(t_0)$ *is the line through* $\mathbf{r}(t_0) = \langle x(t_0), y(t_0), z(t_0) \rangle$ *and parallel to* $\mathbf{r}'(t_0)$. *Find an equation for the line tangent to the curve at the given point.*

57. $\mathbf{r}(t) = \langle t^2, 3t, t^3 - 1 \rangle$ at $\mathbf{r}(1)$

58. $\mathbf{r}(t) = \langle 3 \sin t, -3 \cos t, 3t \rangle$ at $\mathbf{r}(\pi/2)$

59. $\mathbf{r}(t) = \langle te^t, t \sin t, t \cos t \rangle$ at $\mathbf{r}(0)$

60. $\mathbf{r}(t) = \langle 1/(t+1), \ln(t+1), 0 \rangle$ at $\mathbf{r}(1)$

61. Let $\mathbf{r}(t) = \langle te^t, t^2 + t, -4 \cos t \rangle$ describe a curve C in R^3.

a. Find an equation for the line tangent to C at $\mathbf{r}(0)$.

b. Use the equation found in **a** to approximate $\mathbf{r}(0.1)$.

c. How could you describe the error in the approximation?

62. Let $\mathbf{r}(t) = \langle \ln t, t \cos \pi t, t \sin \pi t \rangle$ describe a curve C in R^3.

a. Find an equation for the line tangent to C at $\mathbf{r}(1)$.

b. Use the equation found in **a** to approximate $\mathbf{r}(0.97)$.

c. How could you describe the error in the approximation?

63. Verify the formula

$$\kappa(t) = \frac{|x''(t)y'(t) - x'(t)y''(t)|}{(x'(t)^2 + y'(t)^2)^{3/2}}$$

for the curvature of a curve in R^2 described by $\mathbf{r}(t) = \langle x(t), y(t) \rangle$.

Review of Key Concepts

We started this chapter with a discussion of vectors in three dimensions. Many of the definitions and results for planar vectors extended naturally to the three-dimensional setting. These results and definitions included the length of a vector:

$$\|\langle a_1, a_2, a_3 \rangle\| = \sqrt{a_1^2 + a_2^2 + a_3^2};$$

the vector sum:

$$\langle a_1, a_2, a_3 \rangle + \langle b_1, b_2, b_3 \rangle = \langle a_1 + b_1, a_2 + b_2, a_3 + b_3 \rangle;$$

the dot product:

$$\langle a_1, a_2, a_3 \rangle \cdot \langle b_1, b_2, b_3 \rangle = a_1 b_1 + a_2 b_2 + a_3 b_3;$$

the perpendicularity condition: two nonzero vectors **a** and **b** are perpendicular if and only if $\mathbf{a} \cdot \mathbf{b} = 0$; and the vector equation for the line through point P and parallel to the vector **v**:

$$\mathbf{r}(t) = \mathbf{P} + t\mathbf{v}.$$

A linear function L is a function with domain R^n, range in R^m, and defined by

$$L(\mathbf{v}) = A\mathbf{v}, \qquad \mathbf{v} \text{ in } R^n$$

where A is an $m \times n$ matrix and **v** is a (column) vector. Linear functions are generalizations of the real-value functions f of the form

$$f(x) = mx, \qquad x \text{ in } R. \tag{1}$$

When we study functions of several variables in later chapters, linear functions will play a role similar to the role played by functions of form (1) for real-valued functions of one variable.

The determinant of a square matrix is a number associated with the matrix. For 2×2 and 3×3 matrices, the determinant is defined, respectively, by

$$\det\begin{pmatrix} a_{11} & a_{12} \\ a_{21} & a_{22} \end{pmatrix} = a_{11}a_{22} - a_{12}a_{21}$$

and

$$\det\begin{pmatrix} a_{11} & a_{12} & a_{13} \\ a_{21} & a_{22} & a_{23} \\ a_{31} & a_{32} & a_{33} \end{pmatrix}$$

$$= a_{11} \det\begin{pmatrix} a_{22} & a_{23} \\ a_{32} & a_{33} \end{pmatrix} - a_{12} \det\begin{pmatrix} a_{21} & a_{23} \\ a_{31} & a_{33} \end{pmatrix} + a_{13} \det\begin{pmatrix} a_{21} & a_{22} \\ a_{31} & a_{32} \end{pmatrix}.$$

The determinant is used to calculate cross products, areas and volumes of parallelograms and parallelepipeds, and areas and volumes of the images of these shapes under a linear function.

We finished the chapter with a discussion of motion in three dimensions. Again we saw that most of the concepts defined and used to study motion in the plane extend naturally to motion in three dimensions. These ideas are highlighted in the Chapter Summary.

Chapter Summary

Points and Vectors in R^3

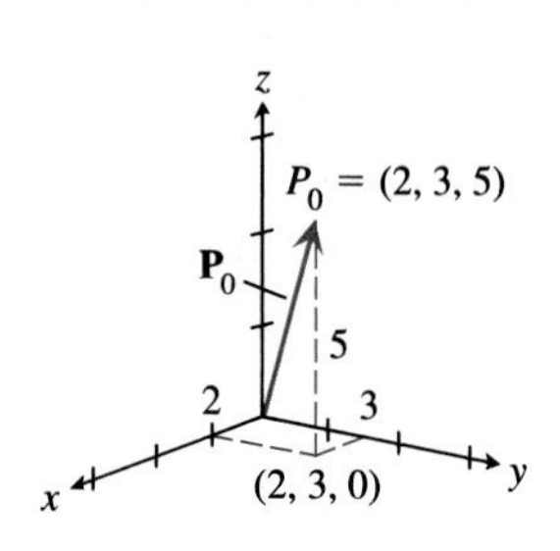

Let P_0 be a point in R^3. To describe the position of P_0 we use an ordered triple (x_0, y_0, z_0). The ordered triple describes the position of P_0 relative to three mutually perpendicular axes, the x-axis, the y-axis, and the z-axis. We write $P_0 = (x_0, y_0, z_0)$.

We also use the ordered triple notation to denote the position vector from the origin to P_0, that is,

$$\mathbf{P}_0 = \langle x_0, y_0, z_0 \rangle.$$

To find the point P with coordinates $(2, 3, 5)$, first find $(2, 3, 0)$ in the (x, y)-plane. The point P is 5 units above this point.

The position vector from the origin to P is

$$\mathbf{P} = \langle 2, 3, 5 \rangle.$$

Sometimes we write the position vector as a column vector,

$$\mathbf{P} = \begin{pmatrix} 2 \\ 3 \\ 5 \end{pmatrix}.$$

Lines in R^3

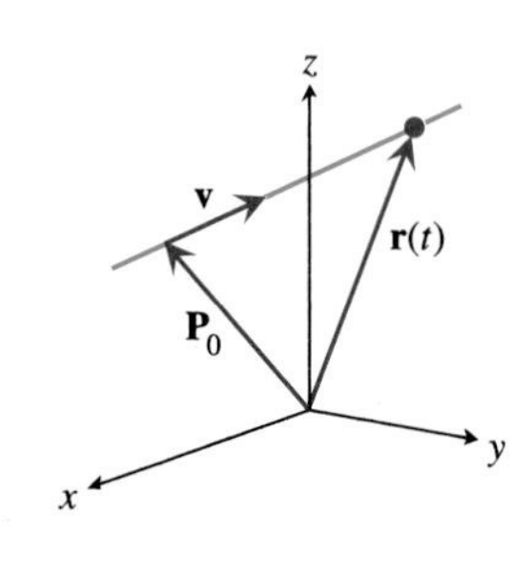

Let P_0 be a point and $\mathbf{v}$ be a vector in R^3. The line through P_0 and parallel to $\mathbf{v}$ consists of all points Q for which

$$\overrightarrow{P_0Q} \text{ is parallel to } \mathbf{v}.$$

The points on the line can also be described by the vector equation

$$\mathbf{r}(t) = \mathbf{P}_0 + t\mathbf{v}.$$

Let $A = (-1, 3, 5)$ and $B = (0, 4, 0)$ be two points in R^3. The line determined by A and B is parallel to

$$\overrightarrow{AB} = \mathbf{B} - \mathbf{A} = \langle 1, 1, -5 \rangle.$$

Because A is on the line, an equation for the line is

$$\begin{aligned} \mathbf{r}(t) &= \mathbf{A} + t(\overrightarrow{AB}) \\ &= \langle -1, 3, 5 \rangle + t\langle 1, 1, -5 \rangle. \end{aligned}$$

The Cross Product

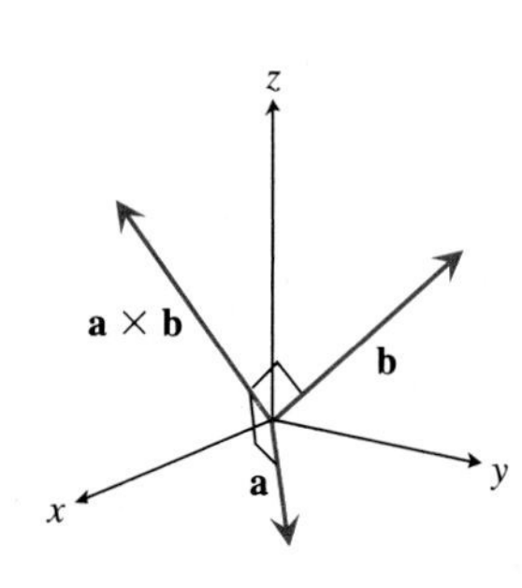

Let $\mathbf{a} = \langle a_1, a_2, a_3 \rangle$ and $\mathbf{b} = \langle b_1, b_2, b_3 \rangle$ be vectors in R^3. The cross product of $\mathbf{a}$ and $\mathbf{b}$ is

$$\begin{aligned} \mathbf{a} \times \mathbf{b} &= \det \begin{pmatrix} \mathbf{i} & \mathbf{j} & \mathbf{k} \\ a_1 & a_2 & a_3 \\ b_1 & b_2 & b_3 \end{pmatrix} \\ &= (a_2 b_3 - a_3 b_2)\mathbf{i} \\ &\quad + (a_3 b_1 - a_1 b_3)\mathbf{j} \\ &\quad + (a_1 b_2 - a_2 b_1)\mathbf{k}. \end{aligned}$$

The vector $\mathbf{a} \times \mathbf{b}$ is perpendicular to each of $\mathbf{a}$ and $\mathbf{b}$. In addition,

$$\|\mathbf{a} \times \mathbf{b}\| = \|\mathbf{a}\| \|\mathbf{b}\| \sin\theta,$$

where θ is the angle between $\mathbf{a}$ and $\mathbf{b}$.

Let $\mathbf{a} = \langle -1, 4, 2 \rangle$ and $\mathbf{b} = \langle 0, 2, 5 \rangle$. Then

$$\mathbf{a} \times \mathbf{b} = \langle 16, 5, -2 \rangle.$$

Because

$$\mathbf{a} \cdot (\mathbf{a} \times \mathbf{b}) = \langle -1, 4, 2 \rangle \cdot \langle 16, 5, -2 \rangle = 0$$

and

$$\mathbf{b} \cdot (\mathbf{a} \times \mathbf{b}) = \langle 0, 2, 5 \rangle \cdot \langle 16, 5, -2 \rangle = 0,$$

we see that $\mathbf{a} \times \mathbf{b}$ is perpendicular to each of $\mathbf{a}$ and $\mathbf{b}$.

The Area of a Parallelogram

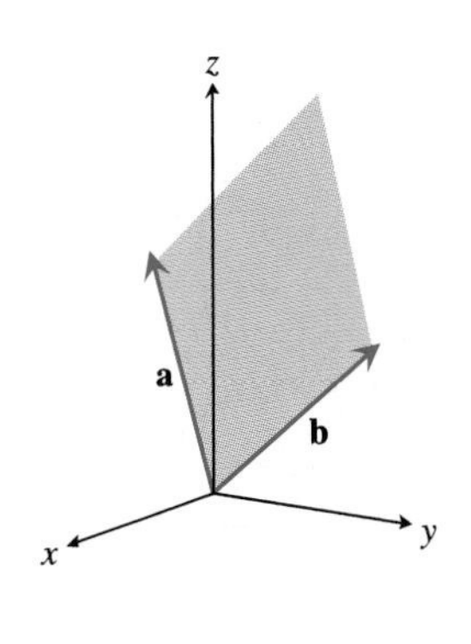

Let **a** and **b** be vectors in R^3. The area of the parallelogram determined by **a** and **b** is equal to the length of the cross-product vector:

$$\text{area} = \|\mathbf{a} \times \mathbf{b}\|.$$

If $\mathbf{a} = \langle a_1, a_2 \rangle$ and $\mathbf{b} = \langle b_1, b_2 \rangle$, then the area of the parallelogram determined by these vectors is

$$\begin{aligned}\text{area} &= \|\langle a_1, a_2, 0\rangle \times \langle b_1, b_2, 0\rangle\| \\ &= \left|\det\begin{pmatrix} a_1 & a_2 \\ b_1 & b_2 \end{pmatrix}\right|.\end{aligned}$$

Let $\mathbf{a} = \langle -1, 4, 2 \rangle$ and $\mathbf{b} = \langle 0, 2, 5 \rangle$. The area of the parallelogram determined by **a** and **b** is

$$\begin{aligned}\|\langle -1,4,2\rangle \times \langle 0,2,5\rangle\| &= \|\langle 16, 5, -2\rangle\| \\ &= \sqrt{285}.\end{aligned}$$

The Volume of a Parallelepiped

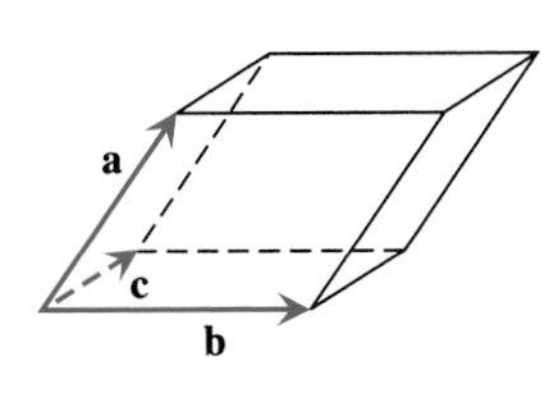

Let $\mathbf{a} = \langle a_1, a_2, a_3 \rangle$, $\mathbf{b} = \langle b_1, b_2, b_3 \rangle$, and $\mathbf{c} = \langle c_1, c_2, c_3 \rangle$ be three vectors in R^3. The volume of the parallelepiped determined by these three vectors is

$$|\mathbf{a} \cdot (\mathbf{b} \times \mathbf{c})| = \left|\det\begin{pmatrix} a_1 & a_2 & a_3 \\ b_1 & b_2 & b_3 \\ c_1 & c_2 & c_3 \end{pmatrix}\right|.$$

Let $\mathbf{a} = \langle 1, 1, 1 \rangle$, $\mathbf{b} = \langle -1, 2, 0 \rangle$, and $\mathbf{c} = \langle 4, 5, -3 \rangle$. The volume of the parallelepiped determined by **a**, **b**, and **c** is

$$\begin{aligned}&|\langle 1,1,1\rangle \cdot (\langle -1,2,0\rangle \times \langle 4,5,-3\rangle)| \\ &\quad = |\langle 1,1,1\rangle \cdot \langle -6,-3,-13\rangle| = 22.\end{aligned}$$

Linear Functions from R^2 to R^2

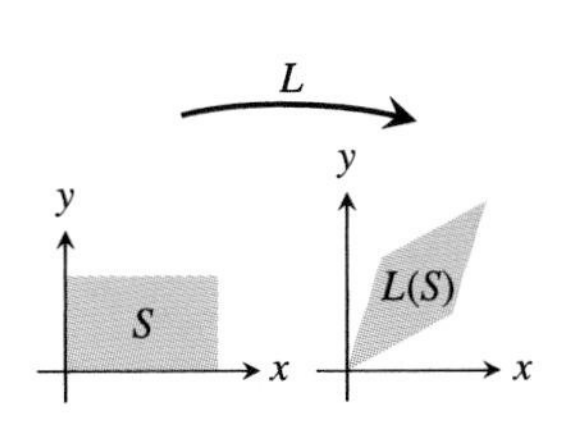

Let L be a linear function with domain R^2, range in R^2, and given by

$$L\begin{pmatrix} x \\ y \end{pmatrix} = \begin{pmatrix} a_{11} & a_{12} \\ a_{21} & a_{22} \end{pmatrix}\begin{pmatrix} x \\ y \end{pmatrix}.$$

If S is a parallelogram in the domain of L, then $L(S)$ is a point, a line segment, or a parallelogram. The areas of S and $L(S)$ are related by

$$\text{area}(L(S)) = \left|\det\begin{pmatrix} a_{11} & a_{12} \\ a_{21} & a_{22} \end{pmatrix}\right| \text{area}(S)$$

Let L be given by

$$L\begin{pmatrix} x \\ y \end{pmatrix} = \begin{pmatrix} -2 & 5 \\ 1 & 7 \end{pmatrix}\begin{pmatrix} x \\ y \end{pmatrix},$$

and let S be the unit square in R^2. Then $L(S)$ is the parallelogram determined by $L\begin{pmatrix} 1 \\ 0 \end{pmatrix} = \begin{pmatrix} -2 \\ 1 \end{pmatrix}$ and $L\begin{pmatrix} 0 \\ 1 \end{pmatrix} = \begin{pmatrix} 5 \\ 7 \end{pmatrix}$. The area of $L(S)$ is

$$\left|\det\begin{pmatrix} -2 & 5 \\ 1 & 7 \end{pmatrix}\right| \text{area}(S) = |-19| \cdot 1 = 19.$$

Linear Functions from R^3 to R^3

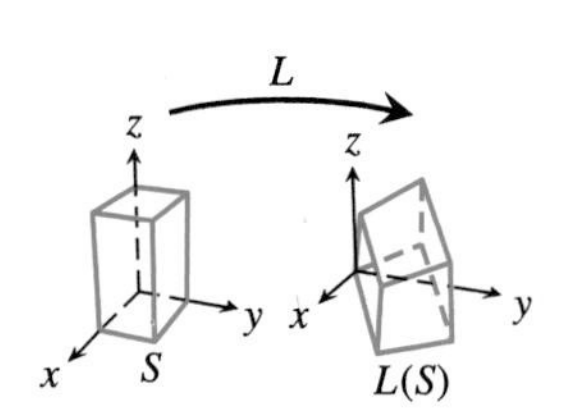

Let L be a linear function with domain R^3, range in R^3, and given by

$$L\begin{pmatrix} x \\ y \\ z \end{pmatrix} = \begin{pmatrix} a_{11} & a_{12} & a_{13} \\ a_{21} & a_{22} & a_{23} \\ a_{31} & a_{32} & a_{33} \end{pmatrix}\begin{pmatrix} x \\ y \\ z \end{pmatrix}.$$

Let L be given by

$$L\begin{pmatrix} x \\ y \\ z \end{pmatrix} = \begin{pmatrix} 0 & -2 & 4 \\ -5 & 2 & 3 \\ 1 & -4 & 5 \end{pmatrix}\begin{pmatrix} x \\ y \\ z \end{pmatrix},$$

Linear Functions from R^3 to R^3 (continued)

If S is a parallelepiped in the domain of L, then $L(S)$ is a point, a line segment, a parallelogram, a hexagon, or a parallelepiped. The volume of $L(S)$ is equal to

$$\left|\det\begin{pmatrix} a_{11} & a_{12} & a_{13} \\ a_{21} & a_{22} & a_{23} \\ a_{31} & a_{32} & a_{33} \end{pmatrix}\right| \text{volume}(S).$$

and let S be the unit cube in R^3. Then $L(S)$ is the parallelepiped determined by

$$L\begin{pmatrix}1\\0\\0\end{pmatrix} = \begin{pmatrix}0\\-5\\1\end{pmatrix}$$

$$L\begin{pmatrix}0\\1\\0\end{pmatrix} = \begin{pmatrix}-2\\2\\-4\end{pmatrix}$$

$$L\begin{pmatrix}0\\0\\1\end{pmatrix} = \begin{pmatrix}4\\3\\5\end{pmatrix}.$$

The volume of $L(S)$ is

$$\left|\det\begin{pmatrix} 0 & -2 & 4 \\ -5 & 2 & 3 \\ 1 & -4 & 5 \end{pmatrix}\right| \text{volume}(S) = 16 \cdot 1$$
$$= 16.$$

Planes

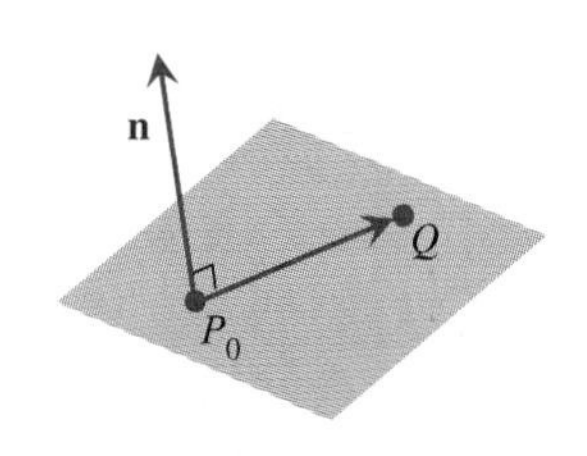

Let $P_0 = (x_0, y_0, z_0)$ be a point in R^3 and $\mathbf{n} = \langle a, b, c\rangle$ a nonzero vector. The plane Π that contains P_0 and has normal vector $\mathbf{n}$ is the set of all points $Q = (x, y, z)$ for which

$$\overrightarrow{P_0Q} \cdot \mathbf{n} = 0.$$

Because $\overrightarrow{P_0Q} = \langle x - x_0, y - y_0, z - z_0\rangle$, we can rewrite this relationship as

$$a(x - x_0) + b(y - y_0) + c(z - z_0) = 0.$$

The plane Π is the set of all points (x, y, z) that satisfy this equation.

Let $A = (-1, 2, 3)$, $B = (0, 2, 1)$, and $C = (3, 4, 1)$. The plane Π containing these three points has normal vector

$$\overrightarrow{AB} \times \overrightarrow{AC} = \langle 1, 0, -2\rangle \times \langle 4, 2, -2\rangle$$
$$= \langle 4, -6, 2\rangle.$$

Because A is a point on Π, an equation for Π is

$$4(x - (-1)) - 6(y - 2) + 2(z - 3) = 0.$$

This can be rewritten as

$$2x - 3y + z = -5.$$

Space Curves

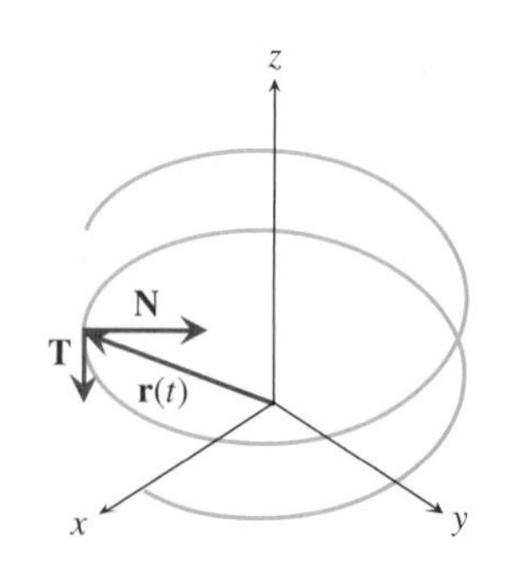

Let

$$\mathbf{r}(t) = \langle x(t), y(t), z(t)\rangle, \qquad a \le t \le b$$

describe a curve C in space. Assuming x, y, and z have the necessary derivatives and $\|\mathbf{r}'(t)\| \neq 0$, the unit tangent, $\mathbf{T}(t)$, to C is given by

$$\mathbf{T} = \mathbf{T}(t) = \frac{\mathbf{r}'(t)}{\|\mathbf{r}'(t)\|}.$$

Let C be the helix described by

$$\mathbf{r}(t) = \langle a \cos t, a \sin t, bt\rangle,$$

where a and b are nonzero constants. Then

$$\mathbf{r}'(t) = \langle -a \sin t, a \cos t, b\rangle,$$

so $\|\mathbf{r}'(t)\| = \sqrt{a^2 + b^2}$. Hence

$$\mathbf{T}(t) = \frac{1}{\sqrt{a^2 + b^2}}\langle -a \sin t, a \cos t, b\rangle.$$

Space Curves

The curvature vector is

$$\frac{d\mathbf{T}}{ds} = \frac{d\mathbf{T}/dt}{ds/dt},$$

where $ds/dt = \|\mathbf{r}'(t)\|$ is the speed of the particle at time t. The curvature is

$$\kappa = \kappa(t) = \left\|\frac{d\mathbf{T}}{ds}\right\|.$$

If $\kappa(t) \neq 0$, then the unit normal $\mathbf{N}(t)$ is

$$\mathbf{N} = \mathbf{N}(t) = \frac{1}{\kappa}\frac{d\mathbf{T}}{ds}.$$

Because

$$\frac{d\mathbf{T}}{dt} = -\frac{a}{\sqrt{a^2 + b^2}}\langle \cos t, \sin t, 0\rangle,$$

the curvature vector is

$$\frac{d\mathbf{T}}{ds} = \frac{d\mathbf{T}/dt}{\|r'(t)\|}$$

$$= -\frac{a}{a^2 + b^2}\langle \cos t, \sin t, 0\rangle.$$

Then $\kappa = \|d\mathbf{T}/ds\| = |a|/(a^2 + b^2)$ and

$$\mathbf{N} = \frac{1}{\kappa}\frac{d\mathbf{T}}{ds} = -\frac{a}{|a|}\langle \cos t, \sin t, 0\rangle.$$

Tangent, Normal Acceleration

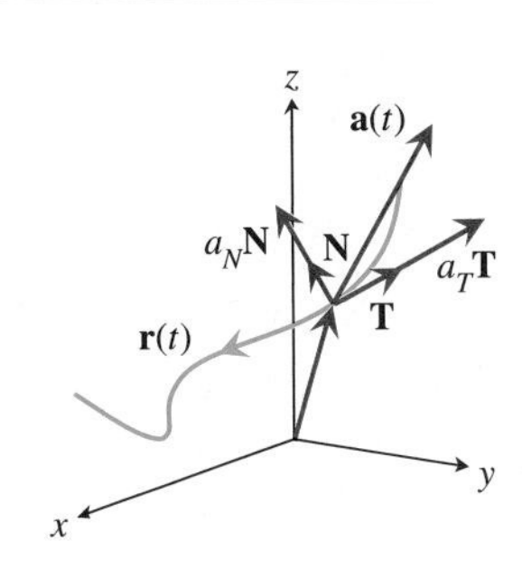

Let

$$\mathbf{r}(t) = \langle x(t), y(t), z(t)\rangle,$$

$a \le t \le b$, describe a curve C in space. If the unit tangent and unit normal vectors $\mathbf{T}$ and $\mathbf{N}$ exist, then the acceleration, $\mathbf{a}(t)$, can be written in the form

$$\mathbf{a}(t) = \frac{d^2s}{dt^2}\mathbf{T} + \left(\frac{ds}{dt}\right)^2 \kappa(t)\mathbf{N}.$$

The tangent component of acceleration,

$$a_T = a_T(t) = \frac{d^2s}{dt^2},$$

tells the rate of change of the speed of the object. The normal component of acceleration,

$$a_N = a_N(t) = \left(\frac{ds}{dt}\right)^2 \kappa(t),$$

describes the rate of change of the direction of the motion.

Let C be the helix described by

$$\mathbf{r}(t) = \langle a \cos t, a \sin t, bt\rangle,$$

where a and b are nonzero constants. Then

$$\frac{ds}{dt} = \|\mathbf{r}'(t)\| = \sqrt{a^2 + b^2}$$

is constant. Hence

$$a_T(t) = \frac{d^2s}{dt^2} = \frac{d}{dt}\left(\frac{ds}{dt}\right) = 0.$$

As seen earlier, $\kappa(t) = \dfrac{|a|}{a^2 + b^2}$, so

$$a_N(t) = \left(\frac{ds}{dt}\right)^2 \kappa(t) = |a|.$$

Thus

$$\mathbf{a}(t) = \langle -a \cos t, -a \sin t, 0\rangle = -|a|\mathbf{N}.$$

In particular,

$$\mathbf{N} = \frac{1}{|a|}\mathbf{a}(t) = -\frac{|a|}{a}\langle \cos t, \sin t, 0\rangle,$$

as was shown previously.

Chapter Review Exercises

Exercises 1–12: Let $\mathbf{a} = \langle -1, 2, 3\rangle$, $\mathbf{b} = \langle 3, 0, 7\rangle$, $\mathbf{c} = \langle 3, 10, -7\rangle$, *and* $\mathbf{d} = \langle 3, -6, 9\rangle$. *Find the following:*

1. $2\mathbf{a} - 5\mathbf{b}$

2. $\mathbf{b} \cdot \mathbf{c}$

3. $\mathbf{a} \times \mathbf{c}$

4. $\mathbf{d} \times \mathbf{a}$

5. The area of the parallelogram determined by **c** and **d**

6. The volume of the parallelepiped determine by **a**, **b**, and **c**

7. The volume of the parallelepiped determined by **a**, **c**, and **d**

8. A vector of length 3 perpendicular to **a**

9. The projection of **c** onto **a**

10. $\|\mathbf{a}\|$

11. $(\mathbf{a} \times \mathbf{b}) \cdot (\mathbf{c} \times \mathbf{d})$

12. $(\mathbf{a} + \mathbf{b}) \cdot (\mathbf{c} + \mathbf{d})$

Exercises 13–24: Let $A = \begin{pmatrix} -4 & 7 & 3 \\ 0 & 2 & -6 \end{pmatrix}$, $B = \begin{pmatrix} 7 & -3 \\ 10 & -4 \end{pmatrix}$,

$C = \begin{pmatrix} 3 & -4 \\ 4 & 0 \\ 5 & 11 \end{pmatrix}$, $D = \begin{pmatrix} 1 & -2 & 3 \\ -4 & 5 & -6 \\ 7 & -8 & 9 \end{pmatrix}$, *and* $E = \begin{pmatrix} 4 & 2 \\ -3 & 7 \end{pmatrix}$.

Find the following:

13. $-2A$

14. $B + 2C$

15. $-B + 5E$

16. $\det A$

17. $\det B$

18. $\det D$

19. AB

20. BA

21. AC

22. CA

23. $\det(BC)$

24. $\det(CB)$

25. Let $A = (2, -3, 6)$, $B = (1, 5, 9)$, and $C = (-2, 0, 6)$ be points in R^3.

a. Show that A, B, and C are not collinear.

b. Find the area of the triangle with vertices A, B, and C.

26. Let $A = (-3, 4, 1)$ and $B = (-3, 0, 2)$ be points in R^3. Find a point C so that triangle ABC is a right triangle of area 10.

27. Let L be defined by

$$L\begin{pmatrix} x \\ y \end{pmatrix} = \begin{pmatrix} -2 & 8 \\ 1 & 3 \end{pmatrix}\begin{pmatrix} x \\ y \end{pmatrix},$$

and let S be the parallelogram with vertices $(2, 1)$, $(1, 4)$, $(4, 8)$, and $(3, 11)$. Describe and sketch $L(S)$. Find the area of $L(S)$.

28. Let T be defined by

$$T\begin{pmatrix} x \\ y \end{pmatrix} = \begin{pmatrix} 4 & -3 \\ 2 & 30 \end{pmatrix}\begin{pmatrix} x \\ y \end{pmatrix} + \begin{pmatrix} 2 \\ -6 \end{pmatrix},$$

and let S be the parallelogram with vertices $(-2, 3)$, $(-4, 6)$, $(2, 7)$, and $(0, 10)$. Describe and sketch $T(S)$. Find the area of $T(S)$.

29. Show that the line described by

$$\mathbf{r}(t) = \langle 0, 14, 6\rangle + t\langle 2, -5, -3\rangle$$

is the intersection of the planes with equations

$$2x - y + 3z = 4$$
$$x + y - z = 8.$$

30. Write down equations for three planes so that the intersection of the three planes is the single point $(-2, 4, 1)$.

31. Write down equations for two planes so that the intersection of the two planes is a line with equation

$$\mathbf{r}(t) = \langle -4, 0, 3\rangle + t\langle 1, 2, 3\rangle.$$

32. Find an equation for the plane containing the point $P_0 = (-2, 4, 0)$ and with normal vector $\mathbf{n} = \langle -5, 8, 1\rangle$.

33. Find an equation for the plane containing the three points $(1, 9, 2)$, $(8, 11, 4)$, and $(0, 0, 0)$.

34. Find an equation for the plane that contains the point $(-6, 1, -1)$ and the line with equation

$$\mathbf{r}(t) = \langle 3, 7, -4\rangle + t\langle 2, 0, 2\rangle.$$

35. Let L be the linear function given by

$$L\begin{pmatrix} x \\ y \\ z \end{pmatrix} = \begin{pmatrix} -2 & 1 & 0 \\ 2 & 1 & 1 \\ 3 & 2 & -1 \end{pmatrix}\begin{pmatrix} x \\ y \\ z \end{pmatrix}$$

and let S be the line described by

$$\mathbf{r}(t) = \langle 5, 0, -2\rangle + t\langle 4, 5, 2\rangle.$$

Show that $L(S)$ is also a line. Find an equation for $L(S)$.

36. Let L be the linear function given by

$$L\begin{pmatrix} x \\ y \\ z \end{pmatrix} = \begin{pmatrix} 1 & 2 & 3 \\ 0 & -1 & 2 \\ 4 & -1 & 2 \end{pmatrix}\begin{pmatrix} x \\ y \\ z \end{pmatrix}$$

and let S be the plane with equation

$$2x - y + 3z = 8.$$

Show that $L(S)$ is also a plane. Find an equation for $L(S)$.

37. Let L and M be the linear functions defined by

$$L\begin{pmatrix} x \\ y \end{pmatrix} = \begin{pmatrix} 2 & -1 \\ 0 & -3 \end{pmatrix}\begin{pmatrix} x \\ y \end{pmatrix}$$

and

$$M\begin{pmatrix} x \\ y \end{pmatrix} = \begin{pmatrix} 4 & 12 \\ -2 & 5 \end{pmatrix}\begin{pmatrix} x \\ y \end{pmatrix}.$$

Let S be the unit square in R^2. Find the area of

a. $(M \circ L)(S)$

b. $(L \circ M)(S)$

c. $(M \circ M)(S)$

d. $(L \circ M \circ L)(S)$.

38. Let L and M be the linear functions defined by

$$L\begin{pmatrix} x \\ y \\ z \end{pmatrix} = \begin{pmatrix} 4 & 0 & -1 \\ 1 & 1 & 1 \\ 2 & -3 & 4 \end{pmatrix}\begin{pmatrix} x \\ y \\ z \end{pmatrix}$$

and

$$M\begin{pmatrix} x \\ y \\ z \end{pmatrix} = \begin{pmatrix} 3 & 0 & 0 \\ 1 & 6 & -2 \\ 5 & 7 & 2 \end{pmatrix}\begin{pmatrix} x \\ y \\ z \end{pmatrix}.$$

Let S be the unit cube in R^3. Find the volume of

a. $(M \circ L)(S)$

b. $(L \circ M)(S)$

c. $(M \circ M)(S)$

d. $(L \circ M \circ L)(S)$.

39. Let M be defined by

$$M\begin{pmatrix} x \\ y \end{pmatrix} = \det\begin{pmatrix} x & y \\ -2 & 4 \end{pmatrix}.$$

Show that M is a linear function. What are the domain and range of M? Write the matrix associated with M.

40. Let K be defined by

$$K\begin{pmatrix} x \\ y \\ z \end{pmatrix} = \det\begin{pmatrix} x & y & z \\ 7 & -2 & 2 \\ 4 & 3 & 0 \end{pmatrix}.$$

Show that K is a linear function. What are the domain and range of K? Write the matrix associated with K.

41. Let L be defined by

$$L\begin{pmatrix} x \\ y \end{pmatrix} = \begin{pmatrix} 3 & -2 \\ 1 & -4 \end{pmatrix}\begin{pmatrix} x \\ y \end{pmatrix}.$$

Show that L is a one-to-one function and that the range of L is all of R^2.

42. Let L be defined by

$$L\begin{pmatrix} x \\ y \\ z \end{pmatrix} = \begin{pmatrix} 1 & -3 & 2 \\ 2 & -6 & 2 \\ -4 & 6 & 0 \end{pmatrix}\begin{pmatrix} x \\ y \\ z \end{pmatrix}.$$

Show that L is a one-to-one function and that the range of L is all of R^3.

43. Let L be the linear function defined by

$$L\begin{pmatrix} x \\ y \end{pmatrix} = \begin{pmatrix} 3 & -2 \\ 3 & -4 \end{pmatrix}\begin{pmatrix} x \\ y \end{pmatrix}.$$

Find all real numbers a for which the equation

$$L(\mathbf{v}) = a\mathbf{v}$$

has a solution $\mathbf{v} \neq \mathbf{0}$.

44. Let L be the linear function defined by

$$L\begin{pmatrix} x \\ y \end{pmatrix} = \begin{pmatrix} 1 & 0 \\ 3 & 4 \end{pmatrix}\begin{pmatrix} x \\ y \end{pmatrix}.$$

Find all real numbers a for which the equation

$$L(\mathbf{v}) = a\mathbf{v}$$

has a solution $\mathbf{v} \neq \mathbf{0}$.

45. An object moves in space so that its position at time t is

$$\mathbf{r}(t) = \langle t \sin t, t \cos t, t \rangle, \qquad t \geq 0.$$

a. Sketch a graph of the path of the motion and describe the path in words.

b. Find the distance traveled by the object from time $t = 0$ until time $t = 2$.

c. Find the unit tangent and unit normal vectors at time $t = 0$.

d. Find the tangent and normal components of acceleration at time $t = 0$.

46. An object moves in space so that its position at time t is

$$\mathbf{r}(t) = \langle t^2, t, -2t \rangle, \qquad t \geq 0.$$

a. Sketch a graph of the path of the motion and describe the path in words.

b. Find the distance traveled by the object from time $t = 0$ until time $t = 5$.

c. Find the unit tangent and unit normal vectors at time $t = 1$.

d. Find the tangent and normal components of acceleration at time $t = 1$.

STUDENT PROJECT

THE CROSS RATIO AND PHOTO ANALYSIS

If you have ever studied different aerial photographs of the same region, you know how hard it can be to match locations on one photo with locations on another. This is especially true if the photos were taken several years apart, and features shown in one photo are not pictured in the other. In this project we study the cross ratio, a simple-to-use yet powerful tool that can be used to pinpoint corresponding locations on different photos, even if the photos were taken years apart and from different angles.

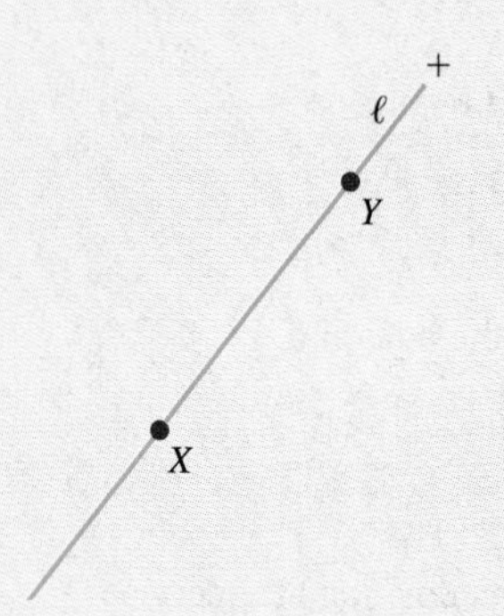

FIGURE 8.68 Distance XY is positive, but distance YX is negative.

The Cross Ratio of Points Let ℓ be a line in the plane and let A, B, C, and D be points on ℓ. Label one direction parallel to ℓ as the positive direction and the opposite direction as the negative direction. Given two points X and Y on ℓ, let XY be the distance from X to Y, where this distance is taken to be positive if the vector $\overrightarrow{XY}$ is in the positive direction and negative if $\overrightarrow{XY}$ is in the negative direction. We will refer to such a distance XY as a *directed distance.* See Fig. 8.68.

The **cross ratio** (AB, CD) of the points A, B, C, D is defined by

$$(AB,CD) = \frac{AC/CB}{AD/DB} = \frac{(AC)(DB)}{(AD)(CB)}, \tag{1}$$

where all distances are directed distances.

PROBLEM 1 Points A, B, C, and D are on the x-axis as shown in Fig. 8.69. Find the value of (AB, CD) and the value of (CA, BD).

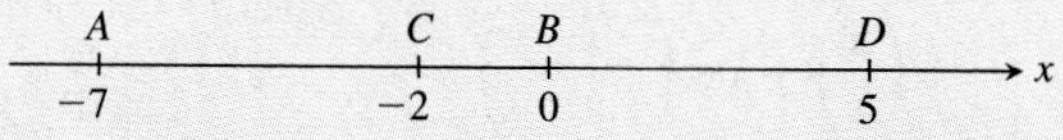

A C B D
−7 −2 0 5 x

FIGURE 8.69

PROBLEM 2 Show that the value of the cross ratio is unchanged if the positive and negative directions on ℓ are reversed.

The cross ratio is useful because it is invariant (unchanged) under projection from a point. These projections are precisely the ones involved in relating two photographic images of the same object. To understand projection from a point and how it affects the cross ratio, see Fig. 8.70. In this figure ℓ and ℓ_1 are two lines, A, B, C, and D are points on ℓ, and F is a point that lies on neither ℓ nor ℓ_1. Now consider the lines FA, FB, FC, and FD, and let these lines intersect ℓ_1 in points A_1, B_1, C_1, and D_1. These points are the projections from F of A, B, C, and D onto ℓ_1. It can be shown that

$$(AB, CD) = (A_1B_1, C_1D_1). \tag{2}$$

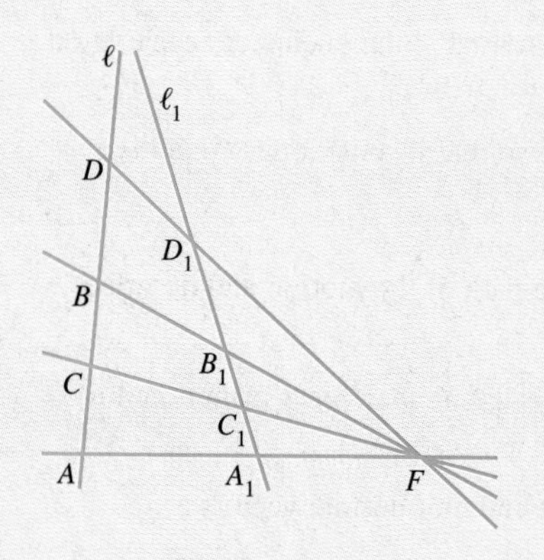

FIGURE 8.70 The cross ratio is invariant under projection from a point.

PROBLEM 3 Show that the cross ratio is invariant under projection from a point by showing that (2) holds.

Note that (2) is valid for any point F and any line ℓ_1. In particular, let G be some other point and let ℓ_2 be some other line. If ℓ_2 intersects lines GA, GB, GC, and GD in points A_2, B_2, C_2, and D_2, then we have

$$(A_1B_1, C_1D_1) = (AB, CD) = (A_2B_2, C_2D_2). \tag{3}$$

This is illustrated in Fig. 8.71.

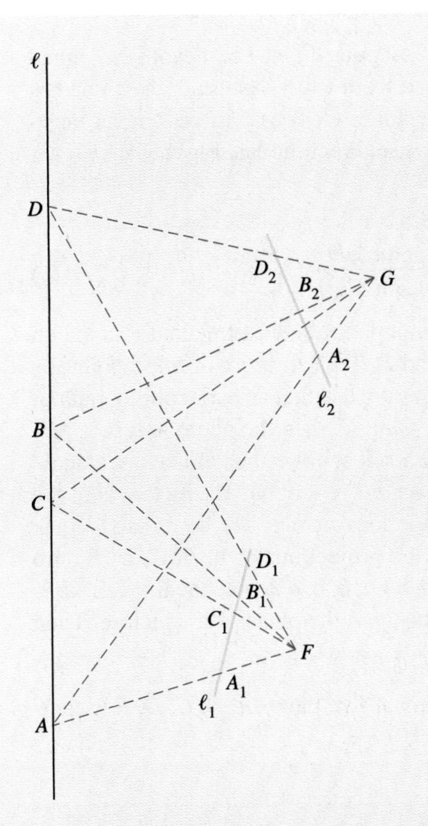

FIGURE 8.71 The cross ratio is invariant under projections from different points.

Now suppose that the points A, B, C, and D represent buildings. A picture of the buildings taken from a point F gives us an image of what we see from point F. In particular, this picture corresponds to the projection of A, B, C, D from F onto a picture-sized rectangle between F and the points A, B, C, D. See Fig. 8.71. If we let A_1, B_1, C_1, D_1 be the points on the picture corresponding to A, B, C, D, then the four points on the picture are also collinear and (2) must hold. Furthermore, if a second picture is taken from a point G, and if A_2, B_2, C_2, D_2 correspond to A, B, C, D on this second picture, then (3) holds. Thus the cross ratios, as calculated from the two photographs, are the same. This gives us a way to match locations on two photos that may have been taken from different angles and at different times.

PROBLEM 4 The pictures in Fig. 8.72 represent photographs taken at three different times. The 1990 photo shows three buildings that lie on a line. In the 1995 photo, an additional building in line with the first three has been added.

a) In the 1990 photo, find the point at which the new building was built.

b) In 1997, a third photo was taken. It was suspected that the new building was actually a mobile missile facility, and that its position was different from that shown in the 1995 photo. Use cross ratios to show that the building had been moved. Note: It might help to make an enlarged photocopy of Fig. 8.72.

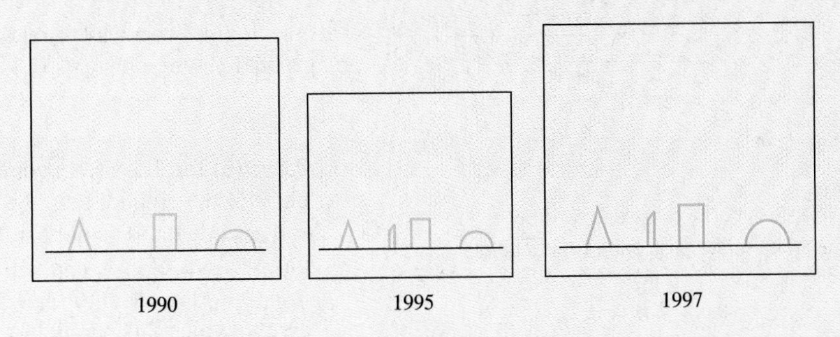

FIGURE 8.72

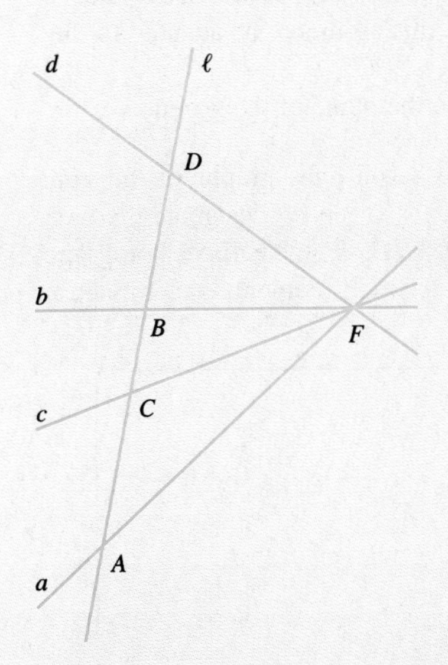

FIGURE 8.73 The cross ratio of four lines can be found by calculating the cross ratio of four points.

The Cross Ratio of Lines Even when points A, B, C, and D are not on a line, the cross ratio can still be used to analyze the positions of the points. This is done by using the cross ratio of lines. Let a, b, c, and d be four lines in a plane with all four lines passing through a point F. Let ℓ be any line in the plane not containing F and intersecting the given lines in points A, B, C, D. See Fig. 8.73. The cross ratio (ab, cd) of lines a, b, c, d is defined to be the cross ratio of the points A, B, C, D; that is

$$(ab, cd) = (AB, CD). \tag{4}$$

PROBLEM 5 At first glance it might seem that the value of (ab, cd) depends on the line ℓ used to obtain points A, B, C, D. Show that this is not the case. That is, show that if two different people use two different lines, say ℓ and ℓ_1, to compute the cross ratio of a, b, c, and d, they will get the same answer. Explain why it is important that the definition of the cross ratio of lines be independent of the line ℓ.

Because (ab, cd) is independent of the line ℓ used in calculating (4), it is not surprising that there is a formula for the cross ratio that makes no use of an auxiliary line. This formula involves the sines of the directed angles between certain pairs of

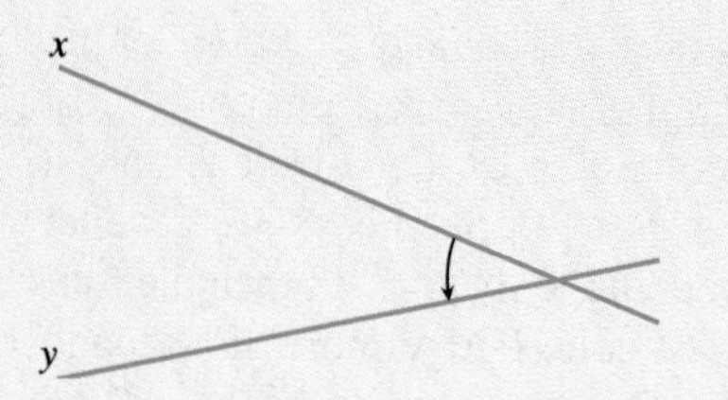

FIGURE 8.74 $\angle xy$ has positive measure, while $\angle yx$ has negative measure.

the lines. Given two lines x and y intersecting at a point F, let $\angle xy$ denote the signed angle of measure between $-\pi$ and π, measured from x to y. As usual, the angle has positive sign if the angle is measured in the counter-clockwise direction and negative sign if the angle is measured in the clockwise direction. See Fig. 8.74.

PROBLEM 6 Show that

$$(ab, cd) = \frac{\sin(\angle ac)\sin(\angle db)}{\sin(\angle cb)\sin(\angle ad)}. \tag{5}$$

Now let A, B, C, D, and F be five points in a plane, and assume that F is not on the line determined by any two of A, B, C, and D. We can think of these points as designating five landmarks on the ground. Now suppose that an aerial photograph of the region is taken. If the photo is taken from point O, then the photo will be a picture of what we see from point O. The picture itself will be shaped like a rectangle and will be the projection from O of some region on the ground. The rectangle of the picture will lie in some plane Π, and this plane may or may not be parallel to the plane of the ground. Let A_1, B_1, C_1, D_1, F_1 be the projections of A, B, C, D, F onto the rectangle of the picture. See Fig. 8.75. Now let a, b, c, d denote the lines AF, BF, CF, DF, and let a_1, b_1, c_1, d_1 be their projections onto the plane of the picture. Then

$$(ab, cd) = (a_1b_1, c_1d_1). \tag{6}$$

This result gives us a way to relate the positions of five points A, B, C, D, F to their images in a photograph.

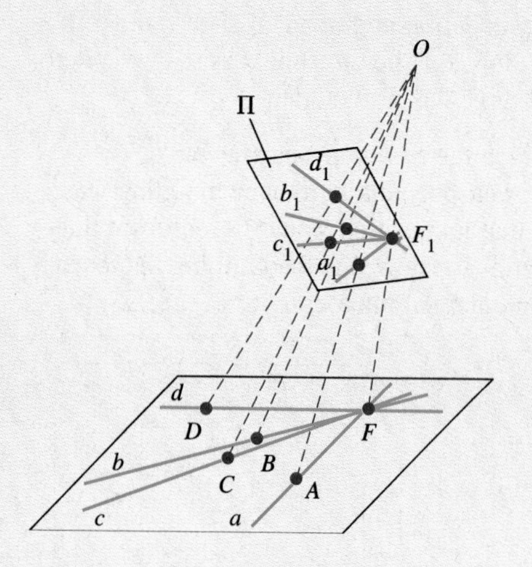

FIGURE 8.75 The cross ratio of lines is invariant under projection from a point.

PROBLEM 7

a) Prove (6) for the case in which the plane Π of the picture is not parallel to the plane of the ground. Let ℓ be the intersection of Π and the plane of the ground, and assume that ℓ is not parallel to any of a, b, c, d. Let A_2, B_2, C_2, D_2 be the intersections of a, b, c, d with ℓ. See Fig. 8.76. Show that both (ab, cd) and (a_1b_1, c_1d_1) are equal to (A_2B_2, C_2D_2). How can this argument be adapted to the case in which ℓ is parallel to, say, a?
b) Prove (6) for the case in which Π is parallel to the plane of the ground.

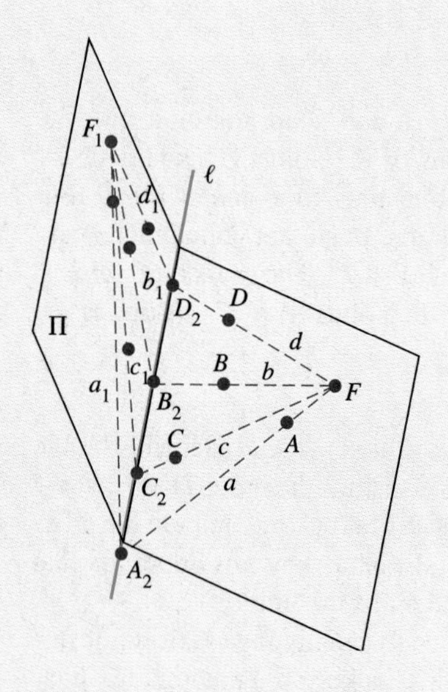

FIGURE 8.76 Diagram for Problem 7.

PROBLEM 8 The pictures in Fig. 8.77 represent aerial photographs of the same region, taken several years apart. In each figure, A, C, D, and F designate the fixed landmarks. The latter picture shows a building marked B. On the earlier photograph, find the (future) location of building B. Note: To make measurements easier make an enlarged copy of Fig. 8.77.

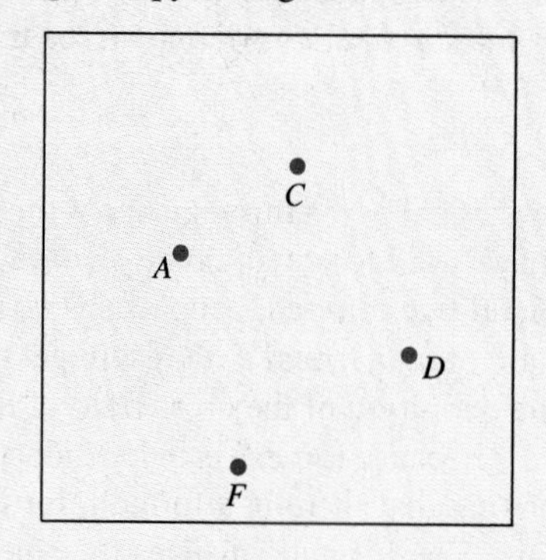

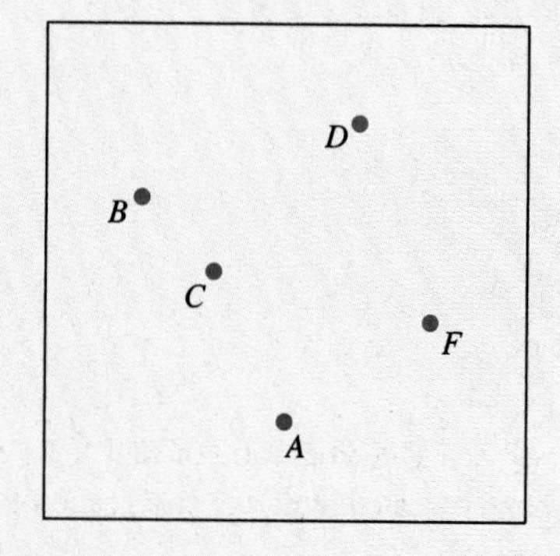

FIGURE 8.77

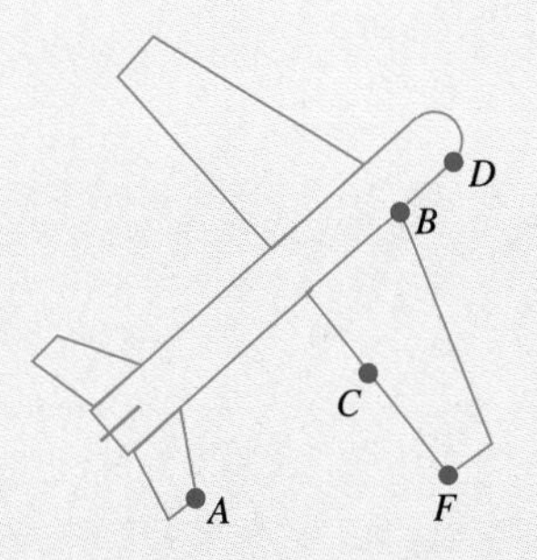

FIGURE 8.78 Five coplanar points on an aircraft.

The cross ratio is used in other identification procedures; for example, in the identification of aircraft from photographs. For each of many aircraft, five coplanar points A, B, C, D, F on the aircraft are identified and recorded, along with the cross ratio of the lines FA, FB, FC, FD. See Fig. 8.78. When the make of an aircraft needs to be determined from a photograph, the five points are found on the picture and the cross ratio is calculated. If the points on different planes are carefully chosen and if the cross ratio can be accurately calculated, then the cross ratio information can be used to identify the model of aircraft, or at least to reduce the number of possible makes of aircraft that need to be considered.

9

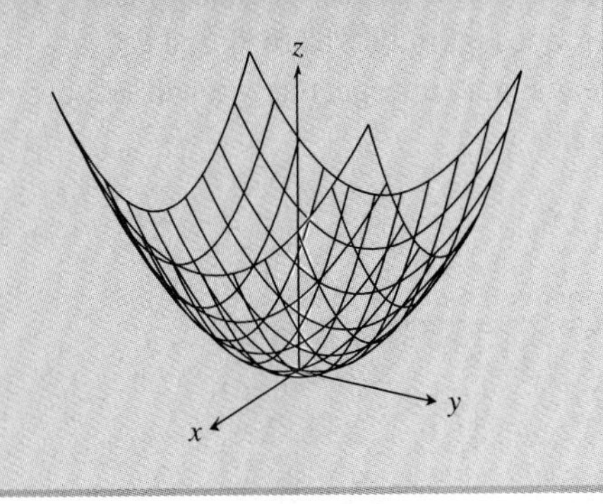

Functions of Several Variables

Most modern radio telescopes, like the Green Bank Telescope shown here, are in the shape of a portion of a paraboloid that focuses all the incoming radio waves at one point on the axis of the paraboloid, above the center of the dish. (Notice the location of the receiver above the reflecting dish.) We can describe such three-dimensional shapes as quadric surfaces.

See Example 4 on p. 733, the Investigation, on p. 760, and p. 769 for further explanation.

We have seen that many physical phenomena can be modeled by functions of one variable. Examples include the motion of a particle in the plane or in space, population growth, and the behavior of a simple harmonic oscillator. On the other hand, many systems seem to depend on more than one variable and so may be difficult or impossible to model accurately with a function of a single variable. Examples include wind velocity (which depends on position and time—see Fig. 9.1), the pressure in a piston chamber (which depends on volume and temperature), and the wind chill on a winter's day (which depends on the air temperature and the wind speed).

In earlier chapters we saw that calculus is a powerful tool for the analysis of situations that can be modeled by functions of one variable. In this chapter we lay the groundwork for the extension of calculus to functions of several variables. When we worked with functions of one variable, we found that graphs contributed significantly to our understanding. Graphs play a similar role in the analysis and understanding of the calculus of functions of several variables. Thus, we discuss in the first four sections of this chapter, several ideas useful to understanding and visualizing such functions. Once these techniques are in place, we define and study rates of change for functions of several variables. We introduce partial derivatives and the gradient vector, two fundamental concepts that we will use in the remainder of the text.

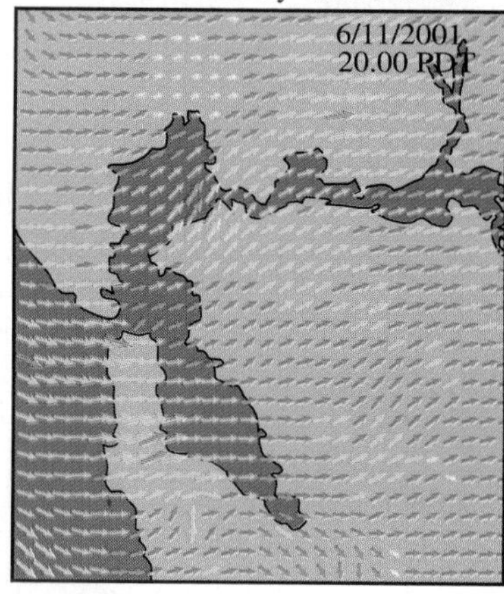

FIGURE 9.1

9.1 Conic Sections

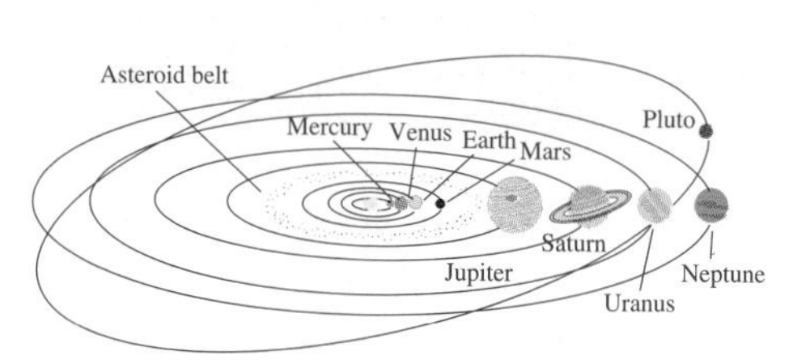

The *conic sections*—parabola, ellipse, and hyperbola—are the key to understanding many diverse phenomena in science and nature. Conic sections are used to describe the orbits of comets and planets, show up in the design of telescopes and whispering galleries, and were used in a WWII navigation system. In this section we define and study the conic sections and investigate some of their applications. Later in this chapter we use conic sections to help analyze several quadric surfaces.

The conic sections are so named because they are the shapes that arise when a plane intersects a cone. As shown in Fig. 9.2, this intersection is an ellipse if the plane intersects only one nappe of the cone, a parabola if the plane is parallel to a line on the cone that passes through the vertex of the cone, and a hyperbola if the plane intersects both nappes of the cone. The conic sections can also be defined as the set of points satisfying certain geometric conditions in the plane.

The Parabola

Let d be a line and F a point not on d, both in the Cartesian plane. The *parabola* with focus F and directrix d is the set of all points that are equidistant from the point and

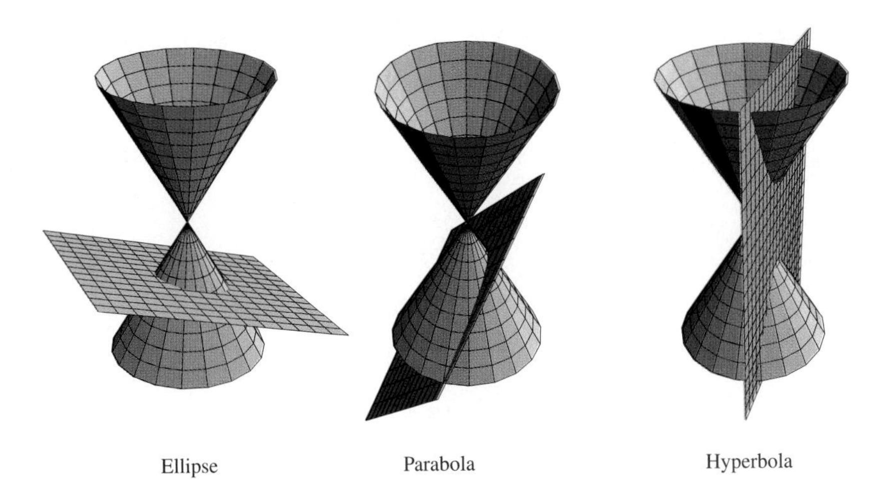

FIGURE 9.2 The intersection of a plane and a cone.

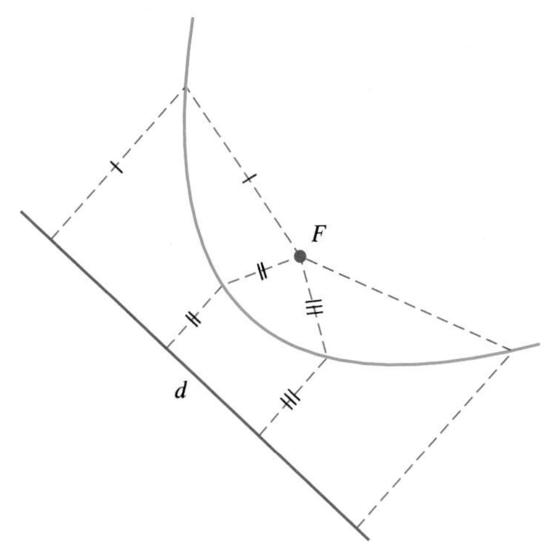

FIGURE 9.3 **A parabola with its directrix and focus. Segments with the same number of marks (|, ||, |||) are of equal length.**

the line. This is illustrated in Fig. 9.3. Given the directrix and the focus, we can find an equation for the parabola.

EXAMPLE 1 Find the equation for the parabola with focus $(p, 0)$ and directrix $x = -p$. Discuss the graph of the parabola.

Solution

In Fig. 9.4 we show the focus, directrix, and parabola in the case $p > 0$. We show the case $p < 0$ in Fig. 9.5. Let $P = (x, y)$ be a point on the parabola. The distance from P to F is

$$\text{dist}(P, F) = \sqrt{(x - p)^2 + y^2}, \tag{1}$$

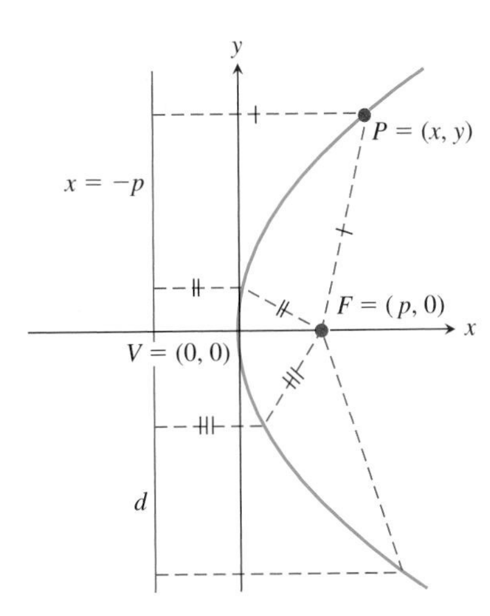

FIGURE 9.4 The parabola with directrix $x = -p$ and focus $F = (p, 0)$, $p > 0$.

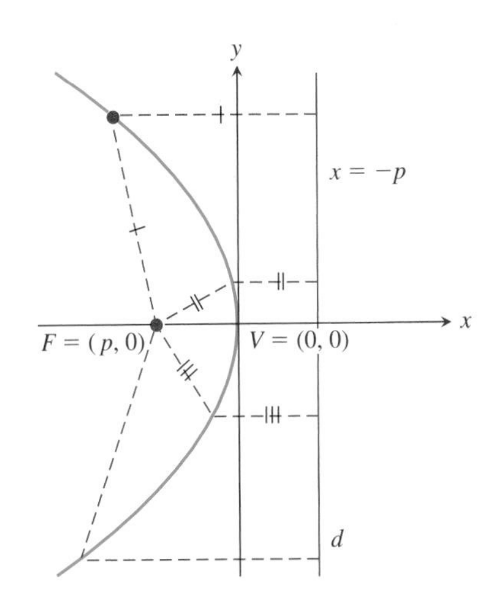

FIGURE 9.5 The parabola with directrix $x = -p$ and focus $F = (p, 0)$, $p < 0$.

and the perpendicular distance from P to d is

$$\text{dist}(P, d) = |x - (-p)|. \tag{2}$$

Because P is on the parabola, the distances in (1) and (2) must be equal,

$$\sqrt{(x-p)^2 + y^2} = |x + p|.$$

Squaring both sides of the equation gives

$$(x-p)^2 + y^2 = (x+p)^2.$$

Expanding and simplifying, we find

$$4px = y^2.$$

This is the equation for the parabola.

As seen in Figs. 9.4 and 9.5, the parabola opens away from the directrix, that is, to the right if $p > 0$ and to the left if $p < 0$. The point $V = (0, 0)$ that is midway between the focus and the directrix is called the *vertex* of the parabola. The parabola is symmetric about the line that is perpendicular to the directrix and passes through the focus. This line of symmetry (the x-axis in Figs. 9.4 and 9.5) is called the *axis* of the parabola.

If we interchange the roles of x and y in Example 1 we obtain

$$4py = x^2. \tag{3}$$

This equation describes a parabola that opens upward ($p > 0$) or downward ($p < 0$).

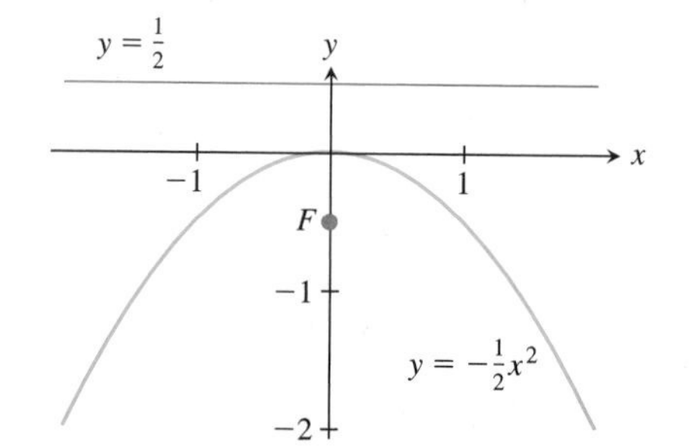

FIGURE 9.6 The parabola $y = -\frac{1}{2}x^2$.

EXAMPLE 2 Find the focus and directrix of the parabola described by

$$y = -\tfrac{1}{2}x^2.$$

Graph the parabola, the focus, and the directrix.

Solution

Rewrite the equation as $-2y = x^2$ so it is in the form seen in (3). We then have $4p = -2$, so $p = -\frac{1}{2}$. The parabola has focus $F = \left(0, -\frac{1}{2}\right)$, directrix $y = \frac{1}{2}$, and opens downward. See Fig. 9.6.

In Chapter 1 we discussed translation of graphs. We can use these ideas to translate parabolas to other positions in the plane. In particular, the equations

$$4p(x-h) = (y-k)^2 \quad \text{and} \quad 4p(y-k) = (x-h)^2 \tag{4}$$

describe parabolas with vertex (h, k). Information about the focus and directrix can be obtained as in Example 1, with appropriate translation.

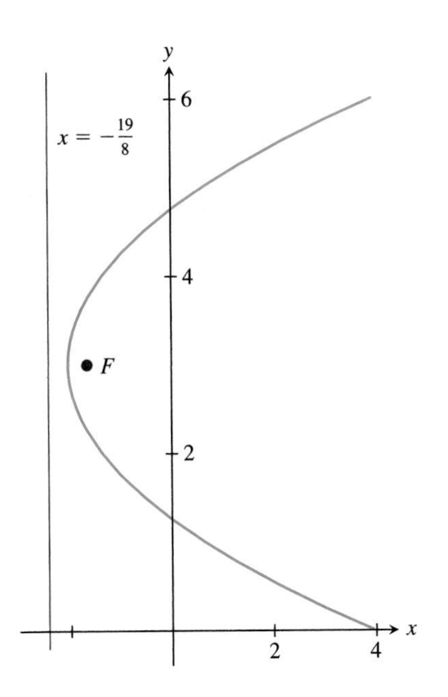

FIGURE 9.7 The parabola $2y^2 - 3x - 12y + 12 = 0$ has vertex $V = (-2, 3)$, directrix $x = -\frac{19}{8}$, and focus $F = \left(-\frac{13}{8}, 3\right)$.

EXAMPLE 3 Find the vertex, focus, and directrix of the parabola described by

$$2y^2 - 3x - 12y + 12 = 0.$$

Graph the parabola and locate the focus and the directrix.

Solution

Group the terms involving y together, then add (and subtract) $2 \cdot 9$ to complete the square in y,

$$2(y^2 - 6y + 9) - 2 \cdot 9 = 3x - 12.$$

With some algebra, this becomes

$$\frac{3}{2}(x + 2) = (y - 3)^2. \tag{5}$$

The graph of (5) is the graph of $\frac{3}{2}x = y^2$, translated so the vertex is at $(-2, 3)$. The parabola opens to the right. Comparing (5) with (4), we find $4p = \frac{3}{2}$, so $p = \frac{3}{8}$. Thus the directrix is $\frac{3}{8}$ to the left of the vertex, so has equation $x = -\frac{19}{8}$, and the focus is $\frac{3}{8}$ to the right of the vertex, so its coordinates are $\left(-\frac{13}{8}, 3\right)$. The graph of the parabola is shown in Fig. 9.7.

The parabola has many practical applications. Parabolas are used in the construction of mirrors in telescopes and headlights and in the design of listening devices. These applications make use of an amazing reflection property of the parabola.

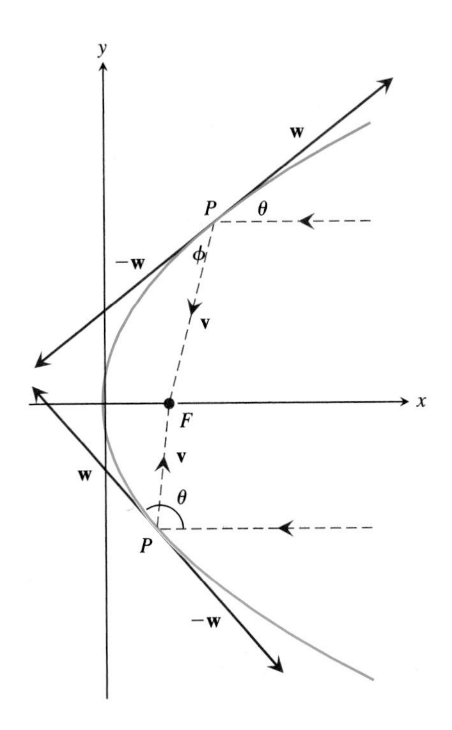

FIGURE 9.8 A ray of light traveling parallel to the axis and into the bowl of the parabola is reflected through the focus.

EXAMPLE 4 A ray of light enters the bowl of a parabolic mirror along a line parallel to the axis of the parabola. Figure 9.8 shows a planar cross section of the three-dimensional mirror, specifically, the plane containing the incident ray and the axis of the parabola. The ray reflects from the parabola so that relative to the tangent line at the point of reflection, the angle of incidence and the angle of reflection are equal. Show that the reflected light ray passes through the focus of the parabola.

Solution

Because we are determining angles, we work with vectors. Assume that the parabola is described by $4px = y^2$, with $p > 0$, so the focus is at $F = (p, 0)$ and the axis of the paraboloid is the x-axis. A parametric representation for the parabola is

$$\mathbf{r}(t) = \left\langle \frac{t^2}{4p}, t \right\rangle.$$

Suppose that the light ray hits the parabolic mirror at $\mathbf{P} = \mathbf{r}(a) = \langle a^2/4p, a \rangle$. A vector tangent to the parabola at this point is

$$\mathbf{w} = \mathbf{r}'(a) = \left\langle \frac{a}{2p}, 1 \right\rangle.$$

Because the incoming light ray travels parallel to the axis, its path is parallel to the vector $\mathbf{u} = \langle 1, 0 \rangle$. The incident angle, θ, is the angle determined by $\mathbf{w}$

and **u**. Now let $\mathbf{v} = \overrightarrow{PF} = \langle p - a^2/(4p), -a\rangle$, and let ϕ be the angle determined by **v** and $-\mathbf{w}$. See Fig. 9.8. We show that $\phi = \theta$, which implies that the reflected light ray travels along PF, and hence passes through the focus. Using dot products,

$$\cos\theta = \frac{\mathbf{w}\cdot\mathbf{v}}{\|\mathbf{w}\|\|\mathbf{v}\|} = \frac{\left\langle \frac{a}{2p}, 1\right\rangle \cdot \langle 1, 0\rangle}{\left\|\left\langle \frac{a}{2p}, 1\right\rangle\right\| \|\langle 1, 0\rangle\|} = \frac{a}{\sqrt{a^2 + 4p^2}} \tag{6}$$

and

$$\cos\phi = \frac{(-\mathbf{w})\cdot\mathbf{v}}{\|-\mathbf{w}\|\|\mathbf{v}\|} = \frac{\left\langle -\frac{a}{2p}, -1\right\rangle \cdot \left\langle \frac{4p^2 - a^2}{4p}, -a\right\rangle}{\left\|\left\langle -\frac{a}{2p}, -1\right\rangle\right\| \left\|\left\langle \frac{4p^2 - a^2}{4p}, a\right\rangle\right\|}$$

$$= \frac{\frac{a}{2} + \frac{a^3}{8p^2}}{\sqrt{\frac{a^2}{4p^2} + 1}\sqrt{\left(p - \frac{a^2}{4p}\right)^2 + a^2}} \tag{7}$$

$$= \frac{\frac{a}{2p}\left(p + \frac{a^2}{4p}\right)}{\sqrt{\frac{a^2}{4p^2} + 1}\left(p + \frac{a^2}{4p}\right)} = \frac{a}{\sqrt{a^2 + 4p^2}}$$

From (6) and (7) we see that $\cos\theta = \cos\phi$. θ and ϕ are either both acute or both obtuse. It follows that $\theta = \phi$.

This reflection property leads to many practical uses for the parabola. Some telescopes and listening devices have parabolic mirrors or reflectors. Light or sound waves enter the device parallel to the axis of the parabola and are all reflected to the focus where they can be seen or heard. In some car headlights, a lightbulb is at the focus of a parabolic mirror. The light from the bulb reflects from the mirror and emerges as a beam of parallel rays.

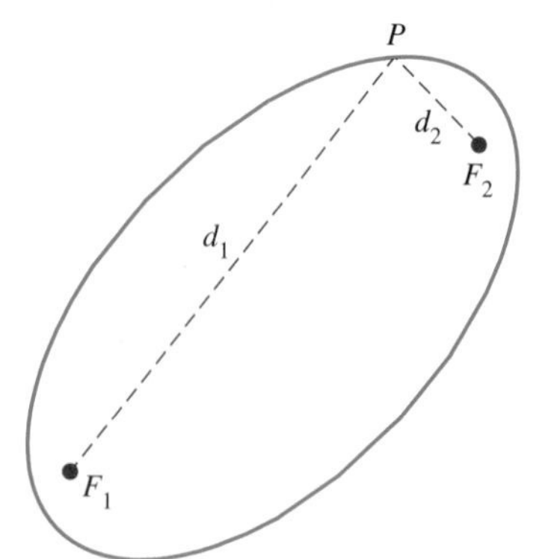

FIGURE 9.9 For every point P on the ellipse, $\text{dist}(P, F_1) + \text{dist}(P, F_2) = d_1 + d_2 = D$.

The Ellipse

Let F_1 and F_2 be two points in the plane, and let D be a positive number with $D > \text{dist}(F_1, F_2)$. The *ellipse* with *foci* F_1 and F_2 and associated distance D is the set of all points P in the plane for which

$$\text{dist}(P, F_1) + \text{dist}(P, F_2) = D.$$

See Fig. 9.9.

This definition suggests an easy way to draw an ellipse: Mark the foci F_1 and F_2 on a blackboard and cut a string of length D. Fasten the ends of the string at the foci, and with the string held taut against a piece of chalk, draw a closed curve. The curve is the ellipse with foci F_1 and F_2 and associated distance D.

Given the foci and the distance D, we can find an equation for the ellipse.

EXAMPLE 5 Let $F_1 = (-c, 0)$, $F_2 = (c, 0)$, and $D = 2a$, with $a > c > 0$. Find the equation for the ellipse with foci F_1 and F_2 and associated distance $2a$.

Solution

Let $P = (x, y)$ be a point on the ellipse. Then

$$2a = \text{dist}(P, F_1) + \text{dist}(P, F_2) = \sqrt{(x+c)^2 + y^2} + \sqrt{(x-c)^2 + y^2}.$$

Rearranging and squaring both sides,

$$\left(2a - \sqrt{(x+c)^2 + y^2}\right)^2 = \left(\sqrt{(x-c)^2 + y^2}\right)^2.$$

Expanding and simplifying,

$$4a^2 - 4a\sqrt{(x+c)^2 + y^2} + 2cx = -2cx,$$

and then

$$\left(a\sqrt{(x+c)^2 + y^2}\right)^2 = (cx + a^2)^2.$$

Expanding and rearranging again,

$$(a^2 - c^2)x^2 + a^2y^2 = a^2(a^2 - c^2).$$

Dividing by $a^2(a^2 - c^2)$, we have

$$\frac{x^2}{a^2} + \frac{y^2}{a^2 - c^2} = 1.$$

For simplicity, we let $b^2 = a^2 - c^2$. The equation for the ellipse is then

$$\frac{x^2}{a^2} + \frac{y^2}{b^2} = 1. \tag{8}$$

The equation is satisfied by the ordered pairs $(a, 0)$, $(-a, 0)$, $(0, b)$, and $(0, -b)$. These are the vertices of the ellipse. The segment joining $(a, 0)$ and $(-a, 0)$ contains the foci and is called the *major axis* for the ellipse. The segment joining $(0, b)$ and $(0, -b)$ is the *minor axis.* Note that $a > b$, so the major axis is always longer than the minor axis. Some of these features are illustrated in Fig. 9.10.

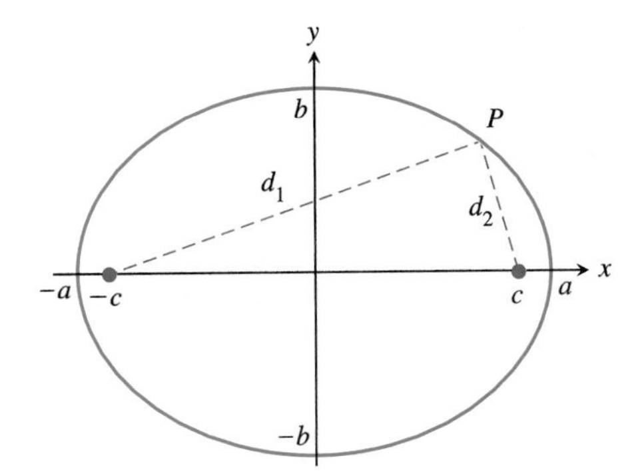

FIGURE 9.10 An ellipse with foci at $F_1 = (-c, 0)$ and $F_2 = (c, 0)$. For each point P on the ellipse, $\text{dist}(P, F_1) + \text{dist}(P, F_2) = d_1 + d_2 = D = 2a$.

If the foci for an ellipse are at $(0, -c)$ and $(0, c)$ and $D = 2a > 2c > 0$, then the equation for the ellipse has form

$$\frac{x^2}{b^2} + \frac{y^2}{a^2} = 1, \tag{9}$$

where $b^2 = a^2 - c^2$. For this ellipse, the major axis is the segment of length $2a$ joining $(0, -a)$ and $(0, a)$, and the minor axis is the segment joining $(-b, 0)$ and $(b, 0)$.

The ellipses described by (8) or (9) are centered at the origin. By shifting the x and y variables, we can translate such an ellipse so its center is at a point (h, k). This leads to the equations

$$\frac{(x-h)^2}{a^2} + \frac{(y-k)^2}{b^2} = 1 \quad \text{and} \quad \frac{(x-h)^2}{b^2} + \frac{(y-k)^2}{a^2} = 1. \tag{10}$$

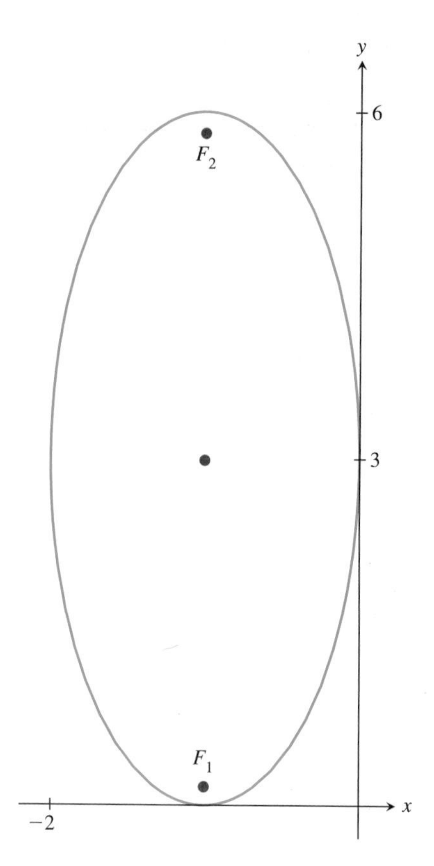

FIGURE 9.11 The ellipse $9x^2 + y^2 + 18x - 6y + 9 = 0$ has center $(-1, 3)$, major axis of length 6, and minor axis of length 2.

EXAMPLE 6 Find the center, foci, major axis, and minor axis for the ellipse described by

$$9x^2 + y^2 + 18x - 6y + 9 = 0.$$

Graph the ellipse.

Solution

Group the x terms together and the y terms together, then complete the square in x and in y. This leads to

$$9(x^2 + 2x + 1 - 1) + (y^2 - 6y + 9 - 9) + 9 = 0,$$

and then to

$$9(x + 1)^2 + (y - 3)^2 = 9.$$

Dividing both sides by 9, we have

$$(x + 1)^2 + \frac{(y - 3)^2}{3^2} = 1.$$

This equation has the form of (9) with the center shifted to the point $(-1, 3)$. For this ellipse, $c^2 = a^2 - b^2 = 9 - 1 = 8$, so the foci are located $\pm\sqrt{8}$ units above and below the center. Thus the foci are at $F_1 = \left(-1, 3 - \sqrt{8}\right)$ and $F_2 = \left(-1, 3 + \sqrt{8}\right)$. The major axis is the segment joining the two vertices $(-1, 6)$ and $(-1, 0)$, and the minor axis is the segment joining the vertices $(-2, 3)$ and $(0, 3)$. The major axis has length $2a = 2 \cdot 3 = 6$ and the minor axis has length $2b = 2 \cdot 1 = 2$. The ellipse is shown in Fig. 9.11.

Like the parabola, the ellipse has an amazing and useful reflection property. Light or sound from one focus of an elliptical reflector is reflected through the other focus of the ellipse. See Fig. 9.12. This property is used in the construction of "whispering galleries." In a room with an elliptical ceiling structure, people talking quietly at one focus of the ellipse can be clearly heard by people listening at the other focus. In Exercise 29, we ask you to prove this reflection property.

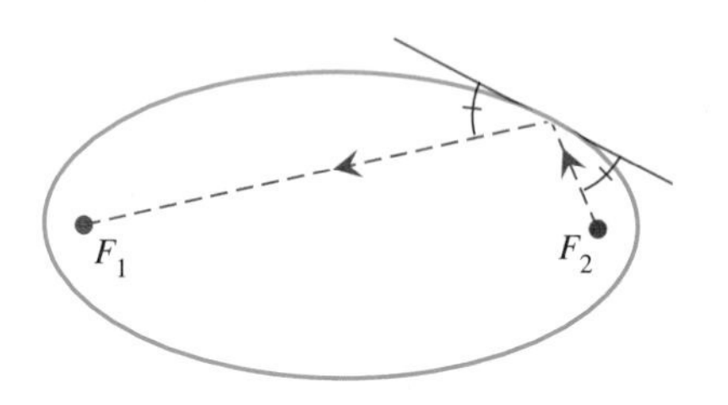

FIGURE 9.12 Light from one focus of an ellipse is reflected through the other focus.

The Hyperbola

Let F_1 and F_2 be two points in the plane, and let D be a positive number with $D < \text{dist}(F_1, F_2)$. The *hyperbola* with *foci* F_1 and F_2 and associated distance D is the set of all points P in the plane for which

$$|\text{dist}(P, F_1) - \text{dist}(P, F_2)| = D. \tag{11}$$

See Fig. 9.13. If we remove the absolute values in (11), we can write (11) using two equations,

$$\text{dist}(P, F_1) - \text{dist}(P, F_2) = D \quad \text{or} \quad \text{dist}(P, F_1) - \text{dist}(P, F_2) = -D.$$

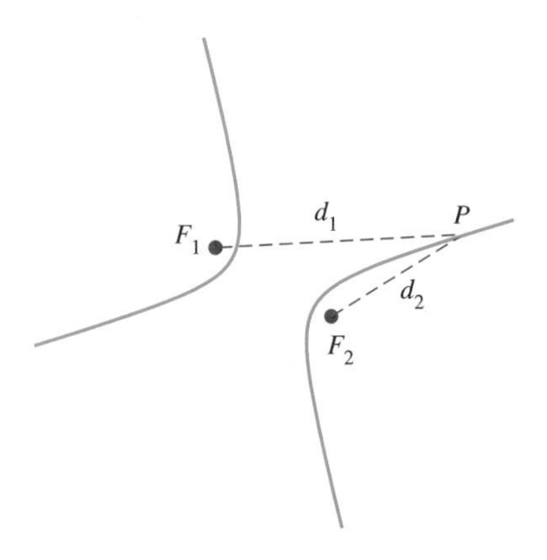

FIGURE 9.13 For every point P on the hyperbola, $|d_1 - d_2| = D$.

The first of these equations describes points on the hyperbola that are closer to F_2 than to F_1, while the second describes points closer to F_1 than to F_2. Each of these equations describes one of the two pieces of the graph seen in Fig. 9.13.

EXAMPLE 7 Let $F_1 = (-c, 0)$, $F_2 = (c, 0)$ and $D = 2a$ with $0 < a < c$. Find the equation for the hyperbola with foci F_1 and F_2 and associated distance $2a$.

Solution

Let $P = (x, y)$ be a point on the hyperbola. Then

$$\pm 2a = \text{dist}(P, F_1) - \text{dist}(P, F_2) = \sqrt{(x+c)^2 + y^2} - \sqrt{(x-c)^2 + y^2}.$$

Rearranging and squaring both sides, we have

$$\left(\pm 2a + \sqrt{(x-c)^2 + y^2}\right)^2 = \left(\sqrt{(x+c)^2 + y^2}\right)^2.$$

Expanding and simplifying leads to

$$4a^2 \pm 4a\sqrt{(x-c)^2 + y^2} - 2cx = 2cx.$$

Adding $2cx - 4a^2$ to both sides, dividing by 4, and squaring again,

$$\left(\pm a\sqrt{(x-c)^2 + y^2}\right)^2 = (cx - a^2)^2.$$

Expanding and rearranging then gives

$$(a^2 - c^2)x^2 + a^2y^2 = a^2(a^2 - c^2).$$

Dividing by $a^2(a^2 - c^2)$ and noting that $a^2 - c^2 < 0$, we obtain

$$\frac{x^2}{a^2} - \frac{y^2}{c^2 - a^2} = 1.$$

If we let $b^2 = c^2 - a^2$, then the equation takes the form

$$\frac{x^2}{a^2} - \frac{y^2}{b^2} = 1. \tag{12}$$

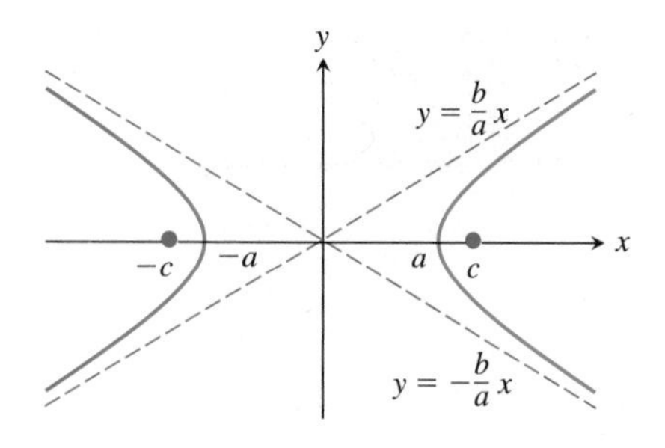

FIGURE 9.14 The hyperbola with foci $(\pm c, 0)$ and associated distance $2a$. The lines $y = \pm \frac{b}{a}x$ are asymptotes to the graph.

This equation is satisfied by the ordered pairs $(a, 0)$ and $(-a, 0)$. These are the vertices of the hyperbola, and they lie on the line through the two foci. Solving (12) for y we get

$$y = \pm b\sqrt{\frac{x^2}{a^2} - 1} = \pm \frac{b}{a}x\sqrt{1 - \frac{a^2}{x^2}}, \qquad |x| \geq a.$$

When $|x|$ is large, the expression under the square root is close to 1 (and a little less than 1), and we have

$$y \approx \pm \frac{b}{a}x.$$

This means that when $|x|$ is large, the graph of the hyperbola is close to the graph of $y = (b/a)x$ or $y = -(b/a)x$. These lines are asymptotes to the graph of the hyperbola. See Fig. 9.14.

If the foci for a hyperbola are at $(0, -c)$ and $(0, c)$ and $D = 2a < 2c$, then the equation takes the form

$$\frac{y^2}{a^2} - \frac{x^2}{b^2} = 1, \tag{13}$$

where $b^2 = c^2 - a^2$. The asymptotes are $y = (a/b)x$ and $y = -(a/b)x$. See Fig. 9.15.

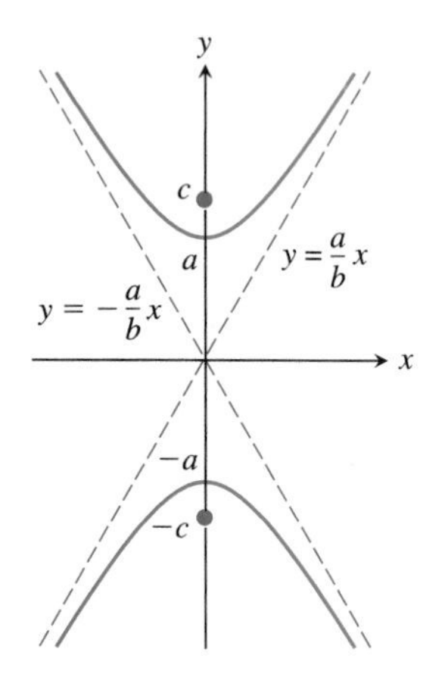

FIGURE 9.15 The hyperbola with foci $(0, \pm c)$ and associated distance $2a$. The lines $y = \pm \frac{a}{b}x$ are asymptotes to the graph.

The hyperbolas in (12) and (13) are centered at the origin. By shifting the x and y variables, we can translate these hyperbolas so their centers are at a point (h, k). The resulting equations are

$$\frac{(x-h)^2}{a^2} - \frac{(y-k)^2}{b^2} = 1 \quad \text{and} \quad \frac{(y-k)^2}{a^2} - \frac{(x-h)^2}{b^2} = 1. \tag{14}$$

The asymptotes for these hyperbolas are similarly translated and take the forms

$$y - k = \pm \frac{b}{a}(x - h) \quad \text{and} \quad y - k = \pm \frac{a}{b}(x - h),$$

respectively.

EXAMPLE 8 Find the center, foci, and asymptotes for the hyperbola with equation

$$-x^2 + 10y^2 - 6x - 20y - 9 = 0. \tag{15}$$

Sketch the graph of the hyperbola, including the center, foci, and asymptotes.

Solution

Completing the square in each of x and y transforms (15) into

$$-(x + 3)^2 + 10(y - 1)^2 = 10,$$

and then

$$(y - 1)^2 - \frac{(x + 3)^2}{10} = 1. \tag{16}$$

This matches the form on the right in (14) and so describes the translation of a hyperbola with foci on the y-axis. The hyperbola has center $(-3, 1)$. In

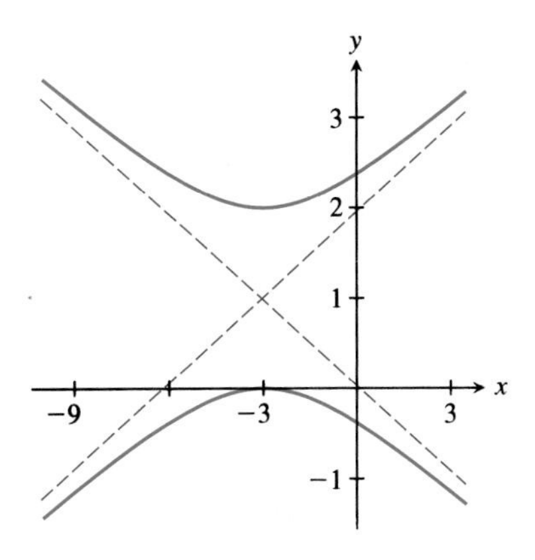

FIGURE 9.16 **The hyperbola $-x^2 + 10y^2 - 6x - 20y - 9 = 0$ has asymptotes $y = 1 \pm \left(1/\sqrt{10}\right)(x + 3)$.**

addition, $c^2 = a^2 + b^2 = 1 + 10 = 11$, so the foci are $\pm\sqrt{11}$ units above and below the center, that is, at the points

$$F_1 = \left(-3, 1 - \sqrt{11}\right) \quad \text{and} \quad F_2 = \left(-3, 1 + \sqrt{11}\right).$$

Solving (16) for $y - 1$ leads to the equations for the asymptotes,

$$(y - 1) = \pm\frac{1}{\sqrt{10}}(x + 3).$$

The graph of the hyperbola is shown in Fig. 9.16.

The LORAN (LOng RAnge Navigation) system, used during World War II (and is still in use today) in both marine and air navigation, is based on the definition (11) of the hyperbola. The LORAN system in a region consists of three transmitting stations M, S_1, and S_2. Each of these stations transmits a regular series of pulses with the pulses from S_1 and S_2 delayed from those of M. Once these time delays are known, the difference $|d - d_1|$ of the distances of a ship from M and S_1 can be calculated, as can the difference $|d - d_2|$ of the distances from M and S_2. With this information, a navigator can determine two hyperbolas on which the ship is located. The intersection of the two hyperbolas—with some dead reckoning to eliminate some multiple intersections—determines the ship's position.

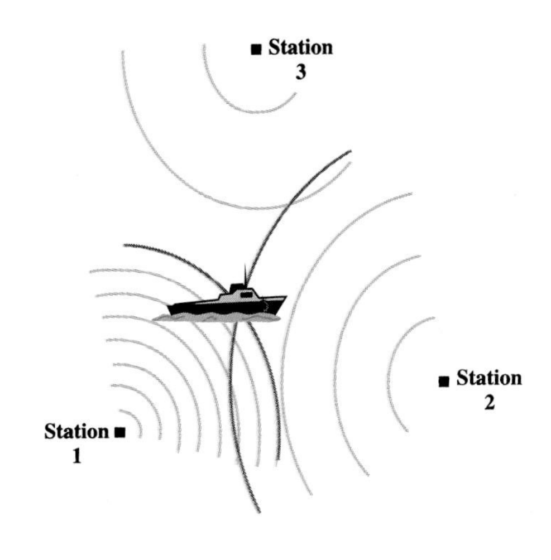

EXAMPLE 9 LORAN navigational transmitters M, S_1, and S_2 are located at $(300, 0)$, $(-300, 0)$, and $(300, 1000)$, respectively. The navigator on a fishing boat listens to pairs S_1, M and S_2, M of transmitters and finds that the difference $|d - d_1|$ of distances from the vessel to M and S_1 is 372 miles, and the difference $|d - d_2|$ of the distances from M and S_2 is 558 miles. Find the possible locations of the boat.

Solution

From the given differences, the boat is on the hyperbola with foci M and S_1 and associated distance $D = 2a = 372$. Taking $c = 300$, $a = 186$ and referring to (12), the equation for the hyperbola is

$$\frac{x^2}{186^2} - \frac{y^2}{300^2 - 186^2} = 1. \tag{17}$$

The boat is also on the hyperbola with foci M and S_2, center $(300, 500)$, and associated distance $2a = 558$. Referring to the right hand equation in (14), the equation for this hyperbola is

$$\frac{(y - 500)^2}{279^2} - \frac{(x - 300)^2}{500^2 - 279^2} = 1. \tag{18}$$

With a calculator or a CAS we can solve equations (17) and (18) simultaneously and obtain four solutions. These are approximately

$$(-193.3, 66.6), \quad (740.3, 906.8), \quad (-1273.3, 1594.1), \quad \text{and} \quad (254.2, 219.3).$$

The boat is at one of these four locations.

A Unified Approach

Though we have treated the three conic sections individually, they can also be studied in a more uniform setting using the focus and directrix approach.

EXAMPLE 10 Let $F = (0, 0)$, let d be the line $x = -p$, $(p > 0)$, and let ϵ be a positive number. Let $\mathcal{C}$ be the set of points P for which

$$\text{dist}(P, F) = \epsilon \text{ dist}(P, d). \tag{19}$$

Show that $\mathcal{C}$ is an ellipse if $0 < \epsilon < 1$, a parabola if $\epsilon = 1$, and a hyperbola if $\epsilon > 1$. The number ϵ is called the *eccentricity* of the conic section.

Solution

Let $P = (x, y)$ be a point on $\mathcal{C}$. Equation (19) then becomes

$$\sqrt{x^2 + y^2} = \epsilon|x + p|.$$

See Fig. 9.17. Squaring both sides and rearranging leads to

$$(1 - \epsilon^2)x^2 - 2p\epsilon^2 x + y^2 = \epsilon^2 p^2. \tag{20}$$

If $\epsilon = 1$, this equation is $y^2 = 2px + p^2$, showing $\mathcal{C}$ is a parabola. If $\epsilon \neq 1$, then in (20) complete the square in x to get

$$(1 - \epsilon^2)\left(x^2 - \frac{2p\epsilon^2}{1 - \epsilon^2}x + \left(\frac{p\epsilon^2}{1 - \epsilon^2}\right)^2\right) + y^2 = \epsilon^2 p^2 + \frac{p^2\epsilon^4}{1 - \epsilon^2}.$$

The right side of this equation simplifies to $\epsilon^2 p^2/(1 - \epsilon^2)$. Dividing both sides of the last equation by this quantity, we obtain

$$\frac{\left(x - \dfrac{p\epsilon^2}{1 - \epsilon^2}\right)^2}{\dfrac{\epsilon^2 p^2}{(1 - \epsilon^2)^2}} + \frac{y^2}{\dfrac{\epsilon^2 p^2}{(1 - \epsilon^2)}} = 1. \tag{21}$$

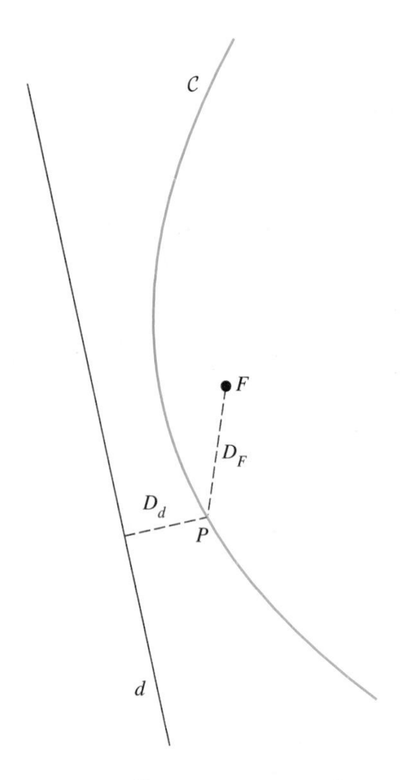

FIGURE 9.17 For every point P on $\mathcal{C}$, $D_F = \epsilon D_d$.

If $0 < \epsilon < 1$, then both terms on the left are positive, so the equation describes an ellipse. Noting that $\sqrt{1 - \epsilon^2} > 1 - \epsilon^2$ and comparing with the forms in (10), we see

$$a = \frac{\epsilon p}{1 - \epsilon^2}, \quad b = \frac{\epsilon p}{\sqrt{1 - \epsilon^2}}, \quad \text{and} \quad c = \sqrt{a^2 - b^2} = \frac{\epsilon^2 p}{1 - \epsilon^2}.$$

Note that $c/a = \epsilon$, giving us a way to calculate the eccentricity from equations of the form in (10).

If $\epsilon > 1$, then the term in (21) involving y^2 is subtracted and the equation describes a hyperbola. Comparing with the right equation in (14), we find

$$a = \frac{\epsilon p}{\epsilon^2 - 1}, \quad b = \frac{\epsilon p}{\sqrt{\epsilon^2 - 1}}, \quad \text{and} \quad c = \sqrt{a^2 + b^2} = \frac{\epsilon^2 p}{\epsilon^2 - 1}.$$

We again see that $c/a = \epsilon$.

In our earlier discussions of ellipses, and hyperbolas, we worked with two foci. The focus used in the *focus-directrix* definition is one of the two foci in the two-focus definition for these conics. See Exercises 51 and 52. Because an ellipse or hyperbola has two foci, there are two focus-directrix pairs associated with each such conic.

The eccentricity of a conic gives an indication of its shape. A conic with eccentricity ϵ near 0 is an ellipse that is almost circular. As ϵ increases to 1 the ellipse stretches, becoming long and skinny. For ϵ close to 1 and a little greater than 1, the conic is a hyperbola that bends sharply near the foci. As ϵ increases, the hyperbola opens up more. For another perspective, refer to Fig. 9.18, where we show a plane cutting one nappe of a cone. Let α be the angle determined by the plane and the axis of the cone, and β the angle measured from the axis of the cone to the side of the cone. It can be shown that the eccentricity is $\epsilon = \cos\alpha/\cos\beta$.

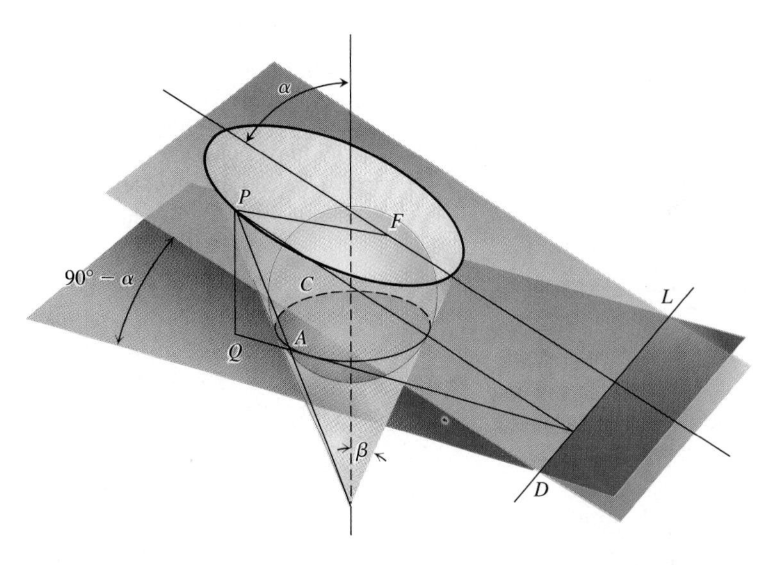

FIGURE 9.18 Let α be the angle determined by the plane and the axis of the cone, and β the angle from the axis to the cone. The eccentricity of the conic is $\cos\alpha/\cos\beta$.

EXAMPLE 11 Find the focus-directrix pairs for the conic with equation

$$9x^2 + y^2 + 18x - 6y + 9 = 0.$$

This conic was discussed in Example 6.

Solution

In Example 6 we saw that the equation has the form

$$(x+1)^2 + \frac{(y-3)^2}{3^2} = 1$$

and describes an ellipse with center $(-1, 3)$ and foci $\left(-1, 3 \pm \sqrt{8}\right)$. Matching this with the equations in (10) we see that $a = 3$, $b = 1$, and $c = \sqrt{a^2 - b^2} = \sqrt{8}$. Hence, the eccentricity is $\epsilon = c/a = \sqrt{8}/3 \approx 0.94 < 1$.

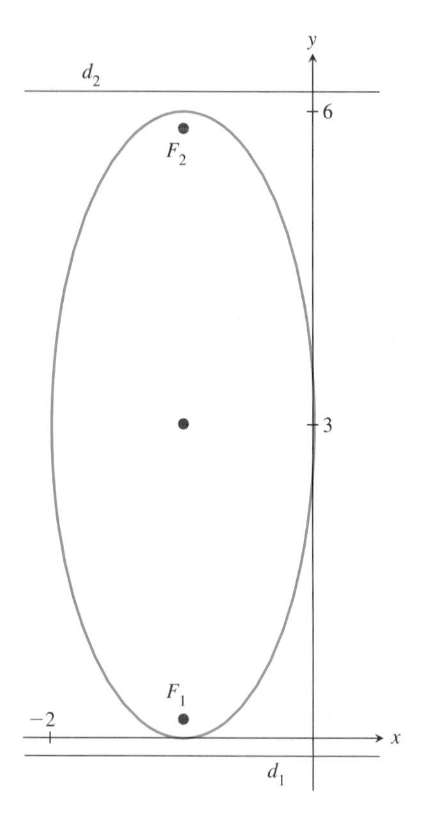

FIGURE 9.19 An ellipse has two focus-directrix pairs. These pairs are F_1, d_1 and F_2, d_2.

Consider the focus $F_1 = \left(-1, 3 - \sqrt{8}\right)$ and the nearer vertex $V_1 = (-1, 0)$. Because $\operatorname{dist}(V_1, F_1) = 3 - \sqrt{8}$, the distance from V_1 to the directrix d_1 associated with F_1 is

$$\operatorname{dist}(V_1, d_1) = \frac{1}{\epsilon}\operatorname{dist}(V_1, F_1) = \frac{3}{\sqrt{8}}\left(3 - \sqrt{8}\right) = \frac{9}{\sqrt{8}} - 3 \approx 0.182.$$

Referring to Fig. 9.19, we see that d_1 is below V_1 so has equation $y = 3 - 9/\sqrt{8}$.

By similar reasoning, the directrix associated with focus $F_2 = \left(-1, 3 + \sqrt{8}\right)$ is $9/\sqrt{8} - 3$ units above vertex $V_2 = (-1, 6)$ so has equation $y = 3 + 9/\sqrt{8}$.

Rotations

In previous examples, we have considered only conic sections with a directrix parallel to one of the coordinate axes, or, equivalently, with foci on a line parallel to one of the coordinate axes. With the rotation ideas discussed in Section 8.5, we can study conics with more general focus and directrix positions.

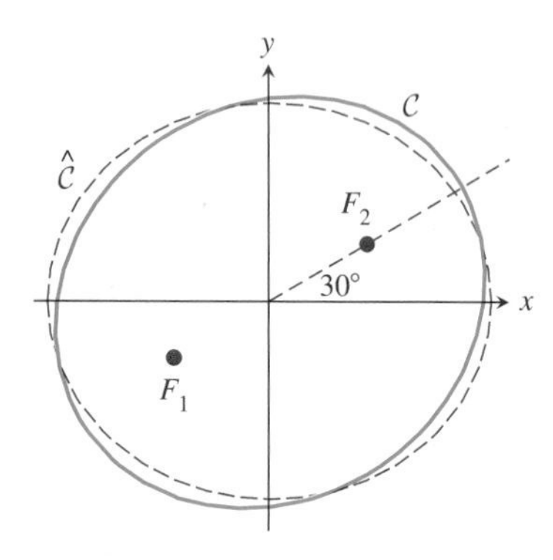

FIGURE 9.20 The ellipse C has foci $F_1 = \left(-\sqrt{3}, -1\right)$ and $F_2 = \left(\sqrt{3}, 1\right)$. The graph of C can be produced by rotating the graph of $\hat{C}$ through 30°.

EXAMPLE 12 Find the equation for the ellipse C with foci $\left(\sqrt{3}, 1\right)$ and $\left(-\sqrt{3}, -1\right)$ and associated distance $D = 8$.

Solution

The ellipse is shown in Fig. 9.20. The foci are each distance 2 from the origin and lie on the line that passes through the origin and makes an angle of 30° with the positive x-axis. Thus, if the ellipse is rotated −30° about the origin, it becomes an ellipse $\hat{C}$ with foci $(\pm 2, 0)$ and associated distance $D = 2a = 8$. Ellipse $\hat{C}$ has equation

$$\frac{x^2}{16} + \frac{y^2}{12} = 1.$$

Now let $P = (x, y)$ be a point on C. If we rotate P through −30°, we obtain a point $\hat{P}$ on $\hat{C}$. This rotation can be accomplished using the linear function L defined by $L(\mathbf{v}) = A\mathbf{v}$, where

$$A = \begin{pmatrix} \cos(-30°) & -\sin(-30°) \\ \sin(-30°) & \cos(-30°) \end{pmatrix} = \begin{pmatrix} \frac{\sqrt{3}}{2} & \frac{1}{2} \\ -\frac{1}{2} & \frac{\sqrt{3}}{2} \end{pmatrix}.$$

Hence

$$\hat{\mathbf{P}} = \begin{pmatrix} \hat{x} \\ \hat{y} \end{pmatrix} = L(\mathbf{P}) = \begin{pmatrix} \frac{\sqrt{3}}{2} & \frac{1}{2} \\ -\frac{1}{2} & \frac{\sqrt{3}}{2} \end{pmatrix}\begin{pmatrix} x \\ y \end{pmatrix} = \begin{pmatrix} \frac{\sqrt{3}}{2}x + \frac{1}{2}y \\ -\frac{1}{2}x + \frac{\sqrt{3}}{2}y \end{pmatrix}.$$

Because $\hat{P}$ is on $\hat{C}$ we must have

$$\frac{\hat{x}}{16}+\frac{\hat{y}}{12}=\frac{\left(\frac{\sqrt{3}}{2}x+\frac{1}{2}y\right)^2}{16}+\frac{\left(-\frac{1}{2}x+\frac{\sqrt{3}}{2}y\right)^2}{12}=1.$$

Expanding and reorganizing this equation, we obtain

$$13x^2-2\sqrt{3}xy+15y^2-192=0.$$

This is the equation for C.

WEB Any conic section in the plane can be described by an equation of the form

$$Ax^2+Bxy+Cy^2+Dx+Ey+F=0. \tag{22}$$

If $B=0$, then the directrix is a line parallel to one of the coordinate axes and the focus or foci lie on a line parallel to the other axis. If $B\neq 0$, then, as illustrated in Example 12, the equation can be analyzed through rotations. Assume that (22) describes a conic C and that when C is rotated about the origin through angle $-\theta$, the result is a conic $\hat{C}$ whose equation has no xy term. If we can find an equation for $\hat{C}$, then we can graph C by first graphing $\hat{C}$, then rotating the graph of $\hat{C}$ through angle θ.

Now let $\hat{P}=(\hat{x},\hat{y})$ be a point on $\hat{C}$. If we rotate $\hat{P}$ through angle θ, we get a point $P=(x,y)$ on C, with

$$\mathbf{P}=\begin{pmatrix}x\\y\end{pmatrix}=\begin{pmatrix}\cos\theta & -\sin\theta\\ \sin\theta & \cos\theta\end{pmatrix}\begin{pmatrix}\hat{x}\\ \hat{y}\end{pmatrix}=\begin{pmatrix}\hat{x}\cos\theta-\hat{y}\sin\theta\\ \hat{x}\sin\theta+\hat{y}\cos\theta\end{pmatrix}.$$

Because this point is on C, it must satisfy (22),

$$\begin{aligned}A(\hat{x}\cos\theta-\hat{y}\sin\theta)^2&+B(\hat{x}\cos\theta-\hat{y}\sin\theta)(\hat{x}\sin\theta+\hat{y}\cos\theta)\\&+C(\hat{x}\sin\theta+\hat{y}\cos\theta)^2+D(\hat{x}\cos\theta-\hat{y}\sin\theta)\\&+E(\hat{x}\sin\theta+\hat{y}\cos\theta)+F=0.\end{aligned}$$

Expanding and rearranging this equation, we have

$$\hat{A}\hat{x}^2+\hat{B}\hat{x}\hat{y}+\hat{C}\hat{y}^2+\hat{D}\hat{x}+\hat{E}\hat{y}+\hat{F}=0, \tag{23}$$

where

$$\begin{aligned}\hat{A}&=A\cos^2\theta+B\cos\theta\sin\theta+C\sin^2\theta\\ \hat{B}&=(C-A)\sin(2\theta)+B\cos(2\theta)\\ \hat{C}&=A\sin^2\theta-B\cos\theta\sin\theta+C\cos^2\theta\\ \hat{D}&=D\cos\theta+E\sin\theta\\ \hat{E}&=-D\sin\theta+E\cos\theta\\ \hat{F}&=F.\end{aligned} \tag{24}$$

Equation (23) is the equation for $\hat{C}$. Because we want this equation to have no $\hat{x}\hat{y}$ term, we must have

$$\hat{B}=(C-A)\sin(2\theta)+B\cos(2\theta)=0.$$

Solving this equation, we have

$$\cot(2\theta) = \frac{A - C}{B} \quad \text{so} \quad \theta = \frac{1}{2}\operatorname{arccot}\left(\frac{A - C}{B}\right). \tag{25}$$

With this result, we have a strategy for graphing equations of the form (22). First use (25) to find the rotation angle θ. Next use (24) to calculate the coefficients $\hat{A}$, $\hat{C}$, $\hat{D}$, $\hat{E}$, and $\hat{F}$, then graph the conic $\hat{\mathcal{C}}$ described by (23). Rotate this graph through angle θ to get the graph of (22).

EXAMPLE 13 Graph the conic section $\mathcal{C}$ described by

$$3x^2 + 10xy + 3y^2 - 4\sqrt{2}x - 12\sqrt{2}y + 8 = 0.$$

Solution

Apply (25) to get the rotation angle

$$\theta = \frac{1}{2}\operatorname{arccot}\left(\frac{3 - 3}{10}\right) = \frac{1}{2}\frac{\pi}{2} = \frac{\pi}{4}.$$

Use this value of θ in (24) to get $\hat{A} = 8$, $\hat{B} = 0$ (as wanted), $\hat{C} = -2$, $\hat{D} = -16$, $\hat{E} = -8$, and $\hat{F} = 8$. Use these coefficients in (23) to obtain the equation

$$8\hat{x}^2 - 2\hat{y}^2 - 16\hat{x} - 8\hat{y} + 8 = 0$$

for a conic $\hat{\mathcal{C}}$. Completing the square in each of x and y, we obtain

$$-(\hat{x} - 1)^2 + \frac{(\hat{y} + 2)^2}{4} = 1.$$

Thus $\hat{\mathcal{C}}$ is a hyperbola with center $(1, -2)$, vertices $(1, -4)$ and $(1, 0)$, and asymptotes $(\hat{y} + 2) = \pm 2(\hat{x} - 1)$. We graph this hyperbola, then rotate it through an angle $\pi/4$ to obtain the graph of $\mathcal{C}$. The graphs of $\hat{\mathcal{C}}$ and $\mathcal{C}$ are shown in Fig. 9.21.

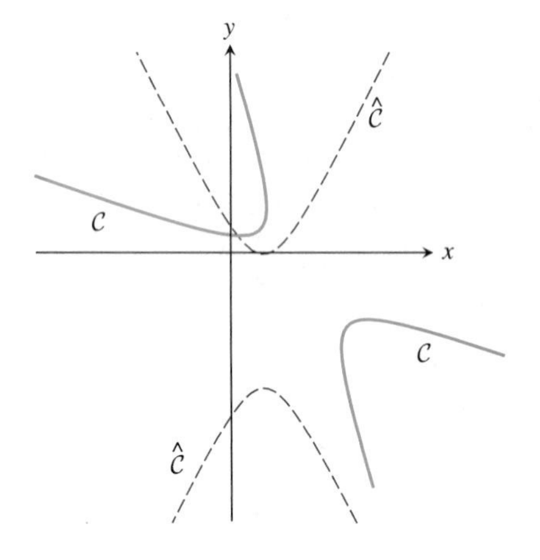

FIGURE 9.21 Hyperbola $\mathcal{C}$ is produced by rotating $\hat{\mathcal{C}}$ through an angle of 45°.

Exercises 9.1

Exercises 1–8: Find the vertex, focus, and directrix for the parabola. Sketch a graph of the parabola.

1. $y = x^2$
2. $y = -8x^2$
3. $y + 2 = 3(x + 1)^2$
4. $9(x - 2) = -9y^2$
5. $y = x^2 + 2x$
6. $y = -2x^2 + 12x - 22$
7. $8x = y^2 + 2y + 25$
8. $8x = -y^2 - 2y + 23$

Exercises 9–16: Identify the conic. If it is an ellipse, find the center, vertices, and foci. If it is a hyperbola, find the center, vertices, foci, and asymptotes. Graph the conic.

9. $\dfrac{x^2}{9} + \dfrac{y^2}{16} = 1$
10. $\dfrac{x^2}{9} - \dfrac{y^2}{16} = 1$
11. $(y - 1)^2 - \dfrac{(x + 1)^2}{2} = 1$

12. $\dfrac{(x+4)^2}{3} + \dfrac{(y+2)^2}{4} = 1$

13. $\left(x - \dfrac{1}{4}\right)^2 + y^2 = 1$

14. $\dfrac{(x-1)^2}{3} - \dfrac{(y+2)^2}{12} = 1$

15. $-(x+3)^2 + \dfrac{(y+2)^2}{4} = 1$

16. $\dfrac{(x-2)^2}{7} + \dfrac{(y+4)^2}{3} = 1$

17. Find the equation of the ellipse with foci $(0, \pm 2)$ and associated distance $D = 8$.

18. Find the equation of the ellipse with foci $(-1, 2)$ and $(7, 2)$ and containing the point $(12, 2)$.

19. Find the equation for the parabola with focus $(2, 2)$ and directrix $y = -6$.

20. Find the equation for the parabola with focus $(0, 0)$ and directrix $x = 4$.

21. Find the equation for the hyperbola with foci $(-2, 1)$ and $(6, 1)$ and associated distance $D = 2$.

22. Find the equation for the hyperbola with foci $(0, 0)$ and $(0, 6)$ and asymptotes $y = \pm 2x + 3$.

Exercises 23–28: Find the equation for the conic section. Identify and graph the conic.

23. focus $(0, 0)$, directrix $x = -3$, eccentricity 2

24. focus $(0, 0)$, directrix $x = -3$, eccentricity $1/2$

25. focus $(0, 0)$, directrix $x = -3$, eccentricity 1

26. focus $(-2, 3)$, directrix $y = 0$, eccentricity 2

27. focus $(-2, 3)$, directrix $y = 0$, eccentricity $1/2$

28. focus $(-2, 3)$, directrix $y = 0$, eccentricity 1

29. Reflection property for the ellipse Let $\mathcal{C}$ be the ellipse with equation

$$\frac{x^2}{a^2} + \frac{y^2}{b^2} = 1$$

and foci F_1 and F_2. Let P be a point on $\mathcal{C}$ and let ℓ be the line tangent to $\mathcal{C}$ at P. Show that the segments $\overline{F_1P}$ and $\overline{F_2P}$ make equal angles with ℓ.

30. Reflection property for the hyperbola Let $\mathcal{C}$ be the hyperbola with equation

$$\frac{x^2}{a^2} - \frac{y^2}{b^2} = 1$$

and foci F_1 and F_2. Let P be a point on $\mathcal{C}$ and let ℓ be the line tangent to $\mathcal{C}$ at P. Show that the segments $\overline{F_1P}$ and $\overline{F_2P}$ make equal angles with ℓ.

31. Consider the general conic equation

$$Ax^2 + Bxy + Cy^2 + Dx + Ey + F = 0$$

for a conic $\mathcal{C}$ and let

$$\hat{A}x^2 + \hat{B}xy + \hat{C}y^2 + \hat{D}x + \hat{E}y + \hat{F} = 0$$

be the equation for the conic $\hat{\mathcal{C}}$ that results from rotating $\mathcal{C}$ through angle $-\theta$. See (24). Show that $A + C = \hat{A} + \hat{C}$. The quantity $A + C$ is called a *rotational invariant* for the conic.

32. Consider the general conic equation

$$Ax^2 + Bxy + Cy^2 + Dx + Ey + F = 0$$

for a conic $\mathcal{C}$ and let

$$\hat{A}x^2 + \hat{B}xy + \hat{C}y^2 + \hat{D}x + \hat{E}y + \hat{F} = 0$$

be the equation for the conic $\hat{\mathcal{C}}$ that results from rotating $\mathcal{C}$ through angle $-\theta$. See (24). Show that $D^2 + E^2 = \hat{D}^2 + \hat{E}^2$. The quantity $D^2 + E^2$ is called a *rotational invariant* for the conic.

33. Consider the general conic equation

$$Ax^2 + Bxy + Cy^2 + Dx + Ey + F = 0$$

for a conic $\mathcal{C}$ and let

$$\hat{A}x^2 + \hat{B}xy + \hat{C}y^2 + \hat{D}x + \hat{E}y + \hat{F} = 0$$

be the equation for the conic $\hat{\mathcal{C}}$ that results from rotating $\mathcal{C}$ through angle $-\theta$. See (24). Show that $B^2 - 4AC = \hat{B}^2 - 4\hat{A}\hat{C}$. The quantity $B^2 - 4AC$ is called a *rotational invariant* for the conic.

34. Consider the general conic equation

$$Ax^2 + Bxy + Cy^2 + Dx + Ey + F = 0$$

for a conic $\mathcal{C}$, and let $R = B^2 - 4AC$. Show that $\mathcal{C}$ is an ellipse if $R < 0$, a parabola if $R = 0$, and a hyperbola if $R > 0$. *Hint:* Let $\hat{A}x^2 + \hat{C}y^2 + \hat{D}x + \hat{E}y + \hat{F} = 0$ be the equation obtained using a rotation that eliminates the xy term. Tell how you can tell the nature of the conic from the sign of $\hat{A}\hat{C}$, and use Exercise 33.

Exercises 35–40: Decide whether the conic is an ellipse, a hyperbola, or a parabola. You may want to use the result in Exercise 34. Calculate an angle of rotation θ that could be used to rotate the conic into one whose equation has no xy term.

35. $-xy + 4x - 3y + 2000 = 0$

36. $4x^2 - 2xy + y^2 - 4x = 0$

37. $-4xy + 3y^2 - 12x + 22y + \pi = 0$

38. $x^2 - 2xy + y^2 + x + y + 1 = 0$

39. $-2x^2 + \sqrt{8}xy - y^2 + 12 = 0$

40. $-x^2 + 2xy - 3y^2 + 4x - 5y + 6 = 0$

41. The parabola with equation $y = x^2 + 1$ is rotated about the origin through an angle of 60°. Find an equation for the resulting parabola.

42. The ellipse with equation $(x^2/4) + y^2 = 1$ is rotated about the origin through an angle of 135°. Find an equation for the resulting ellipse.

43. The hyperbola with equation $y^2 - x^2 = 3$ is rotated about the origin through an angle of $-\pi/4$. Find an equation for the resulting hyperbola.

44. The parabola with equation $x = 3y^2 - 6y + 2$ is rotated about the origin through an angle of $\pi/2$. Find an equation for the resulting parabola.

45. Find an equation for the parabola with focus $(0, 0)$ and directrix the line $x + y = 2$.

46. Find an equation for the ellipse with focus $(0, 0)$, directrix the line $x + y = 2$, and eccentricity $1/2$.

47. Find an equation for the hyperbola with focus $(0, 0)$, directrix the line $x - y = 2$, and eccentricity 2.

48. Find an equation for the parabola with focus $(0, 0)$ and directrix the line $x - y = 4$.

49. Find an equation for the hyperbola with foci $(-1, -\sqrt{3})$ and $(1, \sqrt{3})$ and associated distance $D = 1$.

50. Find an equation for the ellipse with foci $(-1, -\sqrt{3})$ and $(1, \sqrt{3})$ and associated distance $D = 12$.

51. Show that if $0 < \epsilon < 1$, then the equation in (21) does describe an ellipse with one focus at the origin. Where is the other focus?

52. Show that if $\epsilon > 1$, then the equation in (21) does describe a hyperbola with one focus at the origin. Where is the other focus?

Exercises 53–56: Not every equation of the form $Ax^2 + Bxy + Cy^2 + Dx + Ey + F = 0$ describes a nondegenerate conic section. In some cases the set described may be a single line, two intersecting lines, two parallel lines, a point, or the empty set. Determine the set of points described by the following equations.

53. $xy - 3x + 2y - 6 = 0$

54. $x^2 - 2xy + 2y^2 + 1 = 0$

55. $x^2 + 4xy + 4y^2 + x + 2y - 12 = 0$

56. $x^2 - 4xy + 4y^2 = 0$

Exercises 57–60: Calculate the rotation angle and graph the conic.

57. $x^2 - xy + y^2 - 4 = 0$

58. $x^2 + \sqrt{3}xy - 2 = 0$

59. $\sqrt{3}x^2 - 2xy - \sqrt{3}y^2 + 2 = 0$

60. $xy - 4x + 2y + 6 = 0$

9.2 Real-World Functions

In Chapter 8 we studied linear functions with domain R^2 or R^3. Depending on the domain, the input for such a function is a vector $\mathbf{v} = \langle x, y\rangle$ or $\mathbf{w} = \langle x, y, z\rangle$. We think of this input as a single variable that happens to be a vector. However, it can also be useful to think of the input as two real numbers x and y or three real numbers x, y, and z. For example, the linear function L defined by

$$L(\mathbf{v}) = L\begin{pmatrix} x \\ y \\ z \end{pmatrix} = \left\langle -3x + 2y - \frac{1}{2}z \right\rangle \tag{1}$$

can be thought of as a function that requires the three input variables, x, y, and z, and returns a one-dimensional vector (or a real number) as output. When we choose to think of the input as three separate variables, we often write $L(x, y, z)$ instead of the column form in (1).

Many real-world phenomena can be modeled by functions that require more than one real number as input. We refer to such functions as *functions of several variables.* (In our work, *several* means "two or more.") In this section we look at examples of functions of several variables that arise in physics, biology, economics, and engineering. In each case we illustrate some techniques that can help us understand what these functions tell us about the phenomena they are describing or modeling.

Functions of Several Variables

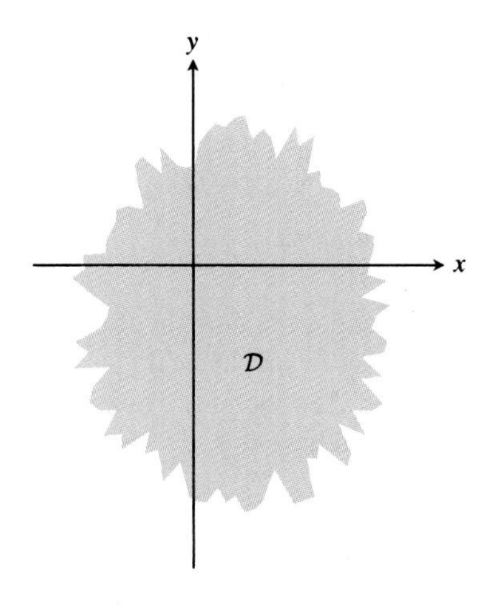

FIGURE 9.22 **The domain of a function f of two variables is a subset $\mathcal{D}$ of the plane.**

Let f be a real-valued function of two variables, say x and y. The domain of such a function is a subset $\mathcal{D}$ of R^2. See Fig. 9.22. We do not always give the domain of f explicitly, often assuming that the domain is all pairs (x, y) for which $f(x, y)$ makes physical or mathematical sense. Thus

$$\mathcal{D}_f = \text{domain of } f = \{(x, y) : f(x, y) \text{ is defined}\}$$

or, if f describes some real-life situation,

$$\mathcal{D}_f = \text{domain of } f = \{(x, y) : f(x, y) \text{ makes real or physical sense}\}.$$

If (x, y) is in the domain of f, then the value of f corresponding to this domain value is denoted by $f(x, y)$. Often we use a letter such as z to denote the value of $f(x, y)$:

$$z = f(x, y).$$

If f is a function of three variables, say x, y, and z, then the domain of f is the set of ordered triples (x, y, z) in R^3 for which $f(x, y, z)$ makes physical or mathematical sense. If f is a real-valued function, we denote the value of $f(x, y, z)$ by a suitable letter such as w, so that

$$w = f(x, y, z).$$

Although most of the functions that we work with will be functions of two or three variables, scientists, engineers, and mathematicians often need to work with functions of four or more variables. In these cases the domain of the function is the set of ordered four-tuples, or five-tuples, and so on, for which the function makes physical or mathematical sense.

Functions of several variables are harder to visualize than functions of one variable. We will find that we must be flexible in choosing a graphical representation for such functions, and may need to choose a representation that emphasizes some features at the expense of others. Sometimes it is helpful to fix all but one of the variables and then study the behavior of the function as that one variable is allowed to change. For example, if f is a function of two variables and

$$z = f(x, y),$$

we can keep x fixed at, say, $x = a$. If y is allowed to change, then we can think of

$$z = f(a, y)$$

as a function of the one variable y. If we understand how z behaves as a function of y, we may get insights into the nature of f. This may be especially true when f describes some physical situation.

EXAMPLE 1 The string on a guitar is 60 cm long and is fastened securely at the ends. When the string is plucked, its motion is very well described (for a short time) by an expression such as

$$s = d(x, t) = \frac{1}{4}\cos(20\pi t)\sin\left(\frac{\pi x}{30}\right) + \frac{1}{16}\cos(40\pi t)\sin\left(\frac{\pi x}{15}\right). \tag{2}$$

In this formula t is time, measured in seconds, with $t = 0$ the time at which the string is plucked, and x is the distance in centimeters from one end of the

FIGURE 9.23 A motionless guitar string is in equilibrium position.

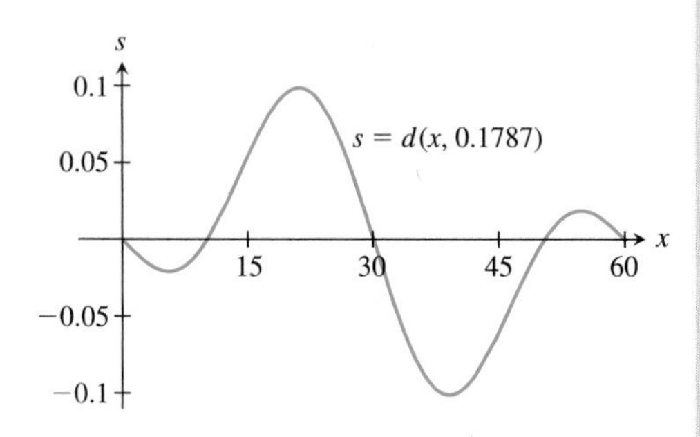

FIGURE 9.24 The graph of $s = d(x, 0.1787)$ shows what the string looks like at time $t = 0.1787$ seconds.

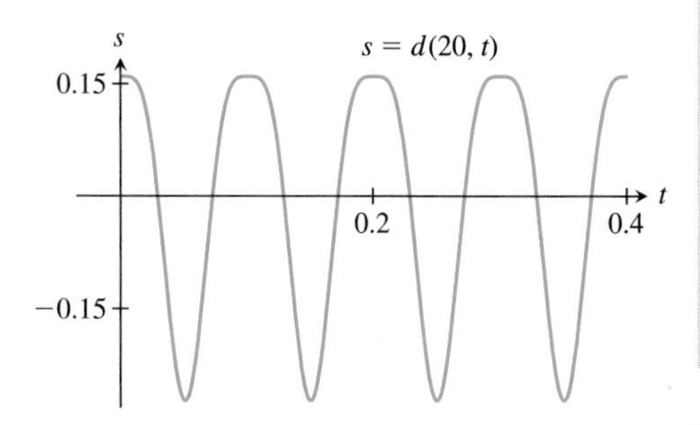

FIGURE 9.25 The graph of $s = d(20, t)$ illustrates the up-and-down motion of a point on the string.

string. The output $s = d(x, t)$ is the displacement in centimeters, from equilibrium position of the point of the string at position x and at time t. (Equilibrium position is taken to be the position of the string when it is in its natural or motionless state. When the string is at rest $s = d(x, t) = 0$ for all x and t. See Fig. 9.23.) Give the domain of d and discuss what (2) tells us about the motion of the guitar string.

Solution

Although (2) makes mathematical sense for all real x and t, the domain of the function is more restricted because the formula describes a physical situation. The guitar string is plucked at time $t = 0$ and then vibrates. Hence we have $t \geq 0$. Because the string is 60 cm long, we also have $0 \leq x \leq 60$. Hence the domain of d is

$$\mathcal{D}_d = \{(x, t) : 0 \leq x \leq 60 \text{ and } t \geq 0\}.$$

To better understand (2) and the motion of the string, let $t = t_1 \geq 0$ be some fixed time. The one-variable expression

$$\begin{aligned} s = f(x) &= d(x, t_1) \\ &= \frac{1}{4}\cos(20\pi t_1)\sin\left(\frac{\pi x}{30}\right) + \frac{1}{16}\cos(40\pi t_1)\sin\left(\frac{\pi x}{15}\right) \end{aligned} \tag{3}$$

describes the shape of the string at time $t = t_1$. If we took a snapshot of the string at time t_1, the picture would look like the graph of (3). For example, at time $t = t_1 = 0.1787$, the string would look like the graph shown in Fig. 9.24. We can get a sense of the motion of the string by graphing (3) for several values of t_1. Many computer algebra systems can animate such a sequence of graphs, thus giving an impression of the guitar string in motion!

Next let $x = x_1$, $0 \leq x_1 \leq 60$, be a fixed position on the string. Then

$$\begin{aligned} s = g(t) &= d(x_1, t) \\ &= \frac{1}{4}\cos(20\pi t)\sin\left(\frac{\pi x_1}{30}\right) + \frac{1}{16}\cos(40\pi t)\sin\left(\frac{\pi x_1}{15}\right) \end{aligned} \tag{4}$$

describes the height of the point of the string at position $x = x_1$ for all times $t \geq 0$. In Fig. 9.25 we show the graph of $s = g(t)$ with $x = x_1 = 20$. This graph shows that the 20-cm point on the string starts at a height of $s = g(0) \approx 0.16238$ cm above equilibrium and then oscillates above and below rest position. Use care in comparing the graphs of Figs. 9.24 and 9.25. Remember that the horizontal axes in the two graphs represent different physical quantities. Also be aware that the scaling on the vertical axes is very different from that on the horizontal axes. Thus, at first glance, the graphs may suggest that there is large up-and-down motion in the string. However, the scaling on the vertical axes indicates that the up-and-down variation is actually very small.

EXAMPLE 2 A ballistic pendulum is a device for measuring the speed of a small, fast-moving object such as a bullet. A wooden block of mass M is suspended by strings at distance H below a ceiling. When a bullet of mass m

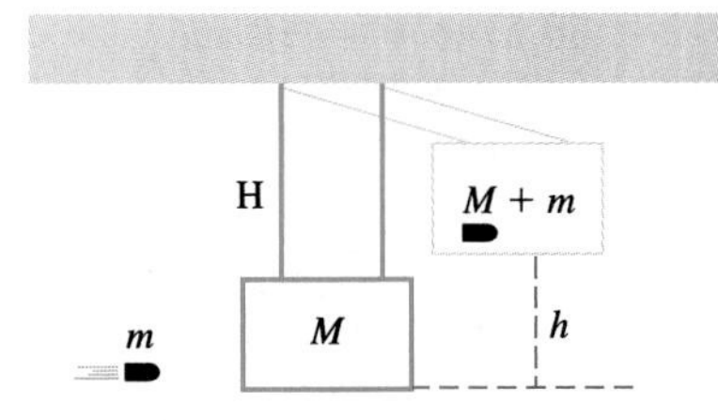

FIGURE 9.26 A ballistic pendulum.

travels horizontally and embeds itself in the block, the block and bullet will swing upward through some height h. See Fig. 9.26. The speed v of the bullet is then given by

$$v = v(m, M, h) = \left(\frac{M + m}{m}\right)\sqrt{2gh}, \tag{5}$$

where g is the acceleration of gravity. State the domain of the function v and discuss what (5) tells us about the speed of the bullet. Assume that distances are measured in centimeters, time in seconds, and masses in grams, so $g = 980\ \text{cm/sec}^2$.

Solution

Formula (5) can be derived using conservation of momentum and conservation of energy. If the speed of the bullet is v, then the momentum p of the bullet in flight is

$$p = (\text{mass})\,(\text{speed}) = mv.$$

When the bullet embeds in the block, the bullet and block together make an object of mass $m + M$. If v_1 is the initial velocity of this combined object, then the momentum of the bullet/block system is $(M + m)v_1$. By conservation of momentum, we have

$$mv = p = (M + m)v_1,$$

so

$$v = \frac{M + m}{n} v_1. \tag{6}$$

To find v_1, we use conservation of energy. At the instant the bullet embeds in the block, the bullet/block system has kinetic energy

$$E = \frac{1}{2}(M + m)v_1^2. \tag{7}$$

As the bullet and block rise to height h, its highest point, this kinetic energy is converted to (gravitational) potential energy. The potential energy of the bullet/block system at height h is

$$E = (M + m)gh. \tag{8}$$

See Exercise 21. By conservation of energy, expressions (7) and (8) must be equal. Equating the expressions and solving for v_1, we find

$$v_1 = \sqrt{2gh}.$$

When this result is substituted into (6), we obtain (5).

Although (5) is defined mathematically for all $m \neq 0$ and all real M and $h \geq 0$, the equation makes physical sense only for $M > 0$, $m > 0$, and $0 < h < H$. Thus the domain of v is

$$\mathcal{D}_v = \{(m, M, h) : m > 0, M > 0, \text{ and } 0 < h < H\}.$$

Let $m = m_0$ and $M = M_0$ be fixed masses. The speed of the bullet is then given by the one-variable expression

$$v = f(h) = v(m_0, M_0, h) = \left(\frac{m_0 + M_0}{m_0}\right)14\sqrt{10h}, \qquad 0 < h < H. \tag{9}$$

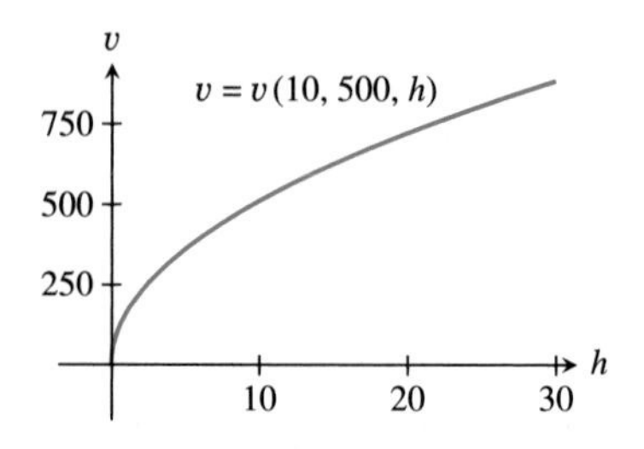

FIGURE 9.27 The speed of the bullet as a function of the height of the pendulum.

In Fig. 9.27 we show the graph of (9) for $0 < h < 30$, with $m_0 = 10$ and $M_0 = 500$. From the graph we see that when h is large, a small change in h does not result in much change in the speed (the graph is relatively flat). In particular, a small error in measuring h when h is large will not have much effect on the accuracy of the speed. On the other hand, when h is small a small error in h can lead to a relatively large error in v (the graph is relatively steep).

Next let $h = h_0$ and $m = m_0$ be fixed, and let M, the mass of the block, vary. The speed of the bullet is then a function of M and is given by

$$v = g(M) = v(m_0, M, h_0) = \left(\frac{m_0 + M}{m_0}\right) 14\sqrt{10h_0}, \qquad M > 0. \tag{10}$$

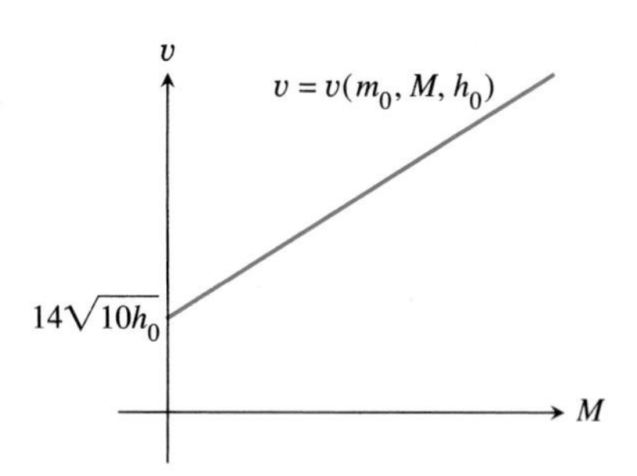

FIGURE 9.28 The velocity is a function of the mass of the block.

The graph of (10) is a portion of the line of slope $14\sqrt{10h_0}/m_0$ and y-intercept $14\sqrt{10h_0}$. See Fig. 9.28. Thus, if an error is made in measuring the mass of the block, there will be a proportional error in the speed of the bullet.

We leave the analysis of the case in which $h = h_0$ and $M = M_0$ to Exercise 11.

EXAMPLE 3 The butterfly species *Papilio glaucus* is brightly colored and is preyed upon by some birds. However, some females of this species are black and look like members of the species *Battus philenor,* a species with an unpleasant taste. If a bird mistakes a black member of the species *Papilio glaucus* for a member of the species *Battus philenor,* it may pass up a tasty meal for fear of getting a bad-tasting butterfly. The effectiveness of mimicry as a survival trait can be modeled mathematically.

Suppose a mimic encounters a predator. Let s be the probability that the predator sees the mimic. Because s is a probability, we have $0 \le s \le 1$. (It is likely that s is near 1, because mimics are often easily seen. They hope to survive by fooling the predator, not by hiding from it.) Assuming that the predator sees the mimic, let f be the probability that the predator is fooled into thinking the mimic is a member of the other species. The probability that the predator is not fooled is $1 - f$. We assume that if the predator is not fooled, it eats the mimic. If the predator is fooled, it still may eat the mimic if it has not had enough encounters with the bad-tasting species to learn that it makes an unpleasant meal. Let ϵ be the probability that the predator eats the mimic even though the predator is fooled. The discussion is summarized in Fig. 9.29. In the figure we see that there are two paths that result in the mimic being eaten. One of these paths has probability labels s and $1 - f$, so the probability that the mimic is seen and is then eaten because it is not mistaken for the other species is $s(1 - f)$. The other path that leads to the mimic being eaten has labels s, f, and ϵ. Hence the probability that the predator sees the mimic, is fooled, but eats the mimic anyway is $sf\epsilon$. Adding these two probabilities gives the probability that the predator eats the mimic,

$$p = p(s, f, \epsilon) = s(1 - f) + sf\epsilon = s[1 - f(1 - \epsilon)]. \tag{11}$$

Find the domain of p and discuss what (11) tells us about mimicry.

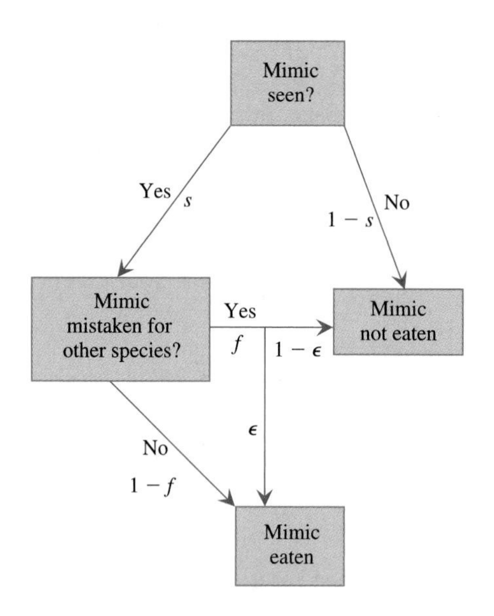

FIGURE 9.29 Will a mimic survive an encounter with a predator?

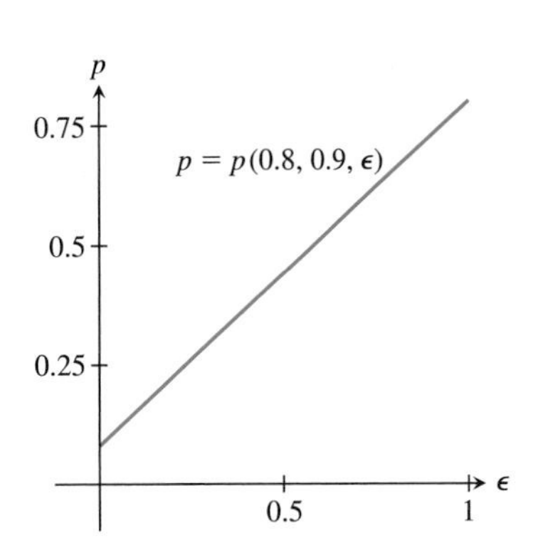

FIGURE 9.30 The probability that a mimic is eaten by a predator is a function of the probability ϵ that the predator eats the mimic even if it is fooled.

Solution

Because $s, f,$ and ϵ are all probabilities, they are all numbers between 0 and 1. Thus the domain of p is

$$\mathcal{D}_p = \{(s, f, \epsilon) : 0 \leq s \leq 1, 0 \leq f \leq 1, \text{ and } 0 \leq \epsilon \leq 1\}.$$

Now fix $s = 0.8$ and $f = 0.9$. Then we have

$$p = f(\epsilon) = p(0.8, 0.9, \epsilon) = 0.8[1 - 0.9(1 - \epsilon)] = 0.08 + 0.72\epsilon, \quad (12)$$

where $0 \leq \epsilon \leq 1$. The graph of (12) is shown in Fig. 9.30. The graph is a line of positive slope and shows, not surprisingly, that the probability that the mimic will be eaten is smallest when ϵ is small. This will be the case if, on mistaking the mimic for the bad-tasting species, the predator usually decides not to eat the mimic. The predator will pass up the mimic if it has had enough encounters with the bad-tasting species to learn that insects colored like the mimic make poor eating. This suggests that for this mimicry to be successful, there should be a lot of bad-tasting species in relation to the mimic.

This example is based on "The Mathematics of Mimicry," by Deborah Charlesworth, in *Maths at Work,* edited by Geoffrey Howson and Ron McLone, Heinemann Educational Books, London, 1983.

Functions Defined by a Table

Not every function is represented by a formula. As is the case with functions of one variable, a function of several variables may be represented by a table of values.

EXAMPLE 4 Table 9.1 provides zero values for U.S. government bonds. The table is taken from *The Basics of Bonds* by Gerald Krefatz, published by Dearborn Financial Publishing, Detroit, 1992. The bonds represented in the table mature to \$1000 in 30 years. The table of zeros is used as follows. Suppose the interest rate is 10 percent when you purchase the bond. You expect to pay \$54, but you can cash the bond in for \$1000 30 years later. Now suppose that 10 years after your purchase the interest rate drops to 8 percent. In a 10 percent market your bond, which is 20 years from maturity, is now worth \$142. However, in the 8 percent market, the bond is worth \$208. This is because at a rate of 8 percent an investor would have to invest \$208 to obtain \$1000 20 years later. (See Exercises 18–20.) Thus an investor might be willing to buy your bond for something between \$142 and \$208. This gives you an extra profit on your investment and allows the investor to invest his or her money at something higher than the current 8 percent rate.

The table of zeros represents a function of two variables. As output, this function gives the current value, V, of a bond that matures to \$1000 in 30 years. The inputs for the function are n, the number of years to maturity, and r, the current interest rate. That is,

$$V = V(n, r).$$

State the domain of V and discuss the behavior of V when one of the variables is held fixed.

TABLE 9.1 Zero values for U.S. government bonds.

Years to Maturity	$ value 8%	$ value 9%	$ value 10%	$ value 11%	$ value 12%	$ value 13%	$ value 14%
30	$95	$71	$54	$40	$30	$23	$17
29	103	78	59	45	34	26	20
28	111	85	85	50	38	29	23
27	120	93	72	56	43	33	26
26	130	101	79	62	48	38	30
25	141	111	87	69	54	43	34
24	152	121	96	77	61	49	39
23	165	132	106	85	69	55	45
22	178	144	117	95	77	63	51
21	193	157	129	106	87	71	58
20	208	172	142	117	97	81	67
19	225	188	157	131	109	91	76
18	244	205	173	146	123	104	88
17	264	224	190	162	138	118	100
16	285	245	210	180	155	133	115
15	308	267	231	201	174	151	131
14	333	292	255	223	196	171	150
13	361	318	281	249	220	195	172
12	390	348	310	277	247	221	197
11	422	380	342	308	278	250	226
10	456	415	377	343	312	284	258
9	494	453	416	381	350	322	296
8	534	494	458	425	394	365	339
7	577	540	505	473	442	414	388
6	625	590	557	526	597	470	444
5	676	644	614	585	558	533	508
4	731	703	677	652	627	604	582
3	790	768	746	725	705	685	666
2	855	839	823	807	792	777	763
1	925	916	907	899	890	882	873
0	1000	1000	1000	1000	1000	1000	1000

Solution

The table of zeros shows only interest rates of $8, 9, \ldots, 14$ percent and whole numbers of years from 30 to 0. However, interest rates may be lower than 8 percent or may be nonintegral, e.g., 8.75 percent. In addition, an investor might buy or sell a bond 15.5 years from maturity and may not be satisfied with rounding to the numbers given in the table. Hence we must decide whether we want the domain of V to reflect only the interest rates and times represented by the table, or all possible interest rates and times. If we are satisfied with the information given by the table, we can take the domain to be exactly those inputs represented by the table:

$$\mathcal{D}_V = \{(n, r) : n = 0, 1, 2, \ldots, 30 \text{ and } r = 8, 9, \ldots, 14\}. \tag{13}$$

If we think that the domain should represent all possible interest rates and all possible times, then

$$\mathcal{D}_V = \{(n, r) : 0 \leq n \leq 30 \text{ and } r > 0\}.$$

Note that in this case the table of zeros is not enough for us to calculate $V(n, r)$ for all possible inputs. We would need more information and perhaps a formula to cover all possible inputs. Which domain should we choose? The answer depends on our needs. For simplicity we will assume that the domain of V is given by (13).

Now fix n at $n = 18$. Then

$$V = V(18, r)$$

gives the value of a 30-year bond 18 years from maturity if the interest rate at that time is r percent. For example,

$$V(18, 8) = 244 \quad \text{and} \quad V(18, 13) = 104.$$

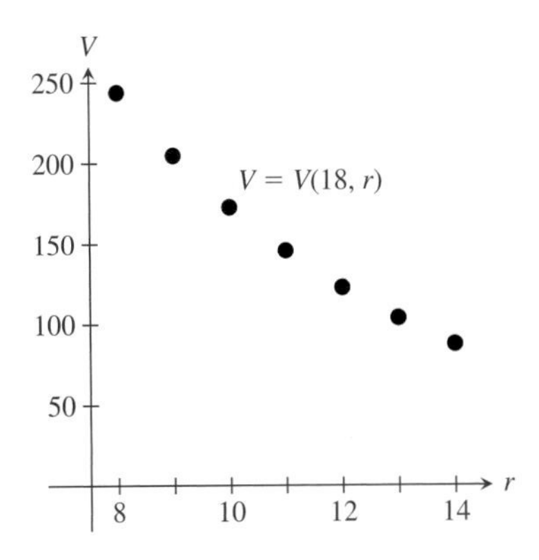

FIGURE 9.31 The lower the interest rate, the higher the value of the bond.

Note that the values of $V(18, r)$ are just the values in the $n = 18$ row of Table 9.1. The graph of $V = V(18, r)$ is shown in Fig. 9.31. From the graph we see that the market value of the bond decreases as the available interest rate increases.

Now let r be fixed at $r = 11$. Then

$$V = V(n, 11)$$

gives the value of the bond n years from maturity when the interest rate is 11 percent. The values of $V(n, 11)$ are found in the $r = 11$ column of Table 9.1. The graph of $V = V(n, 11)$, shown in Fig. 9.32, indicates that the bond grows in value as it nears maturity.

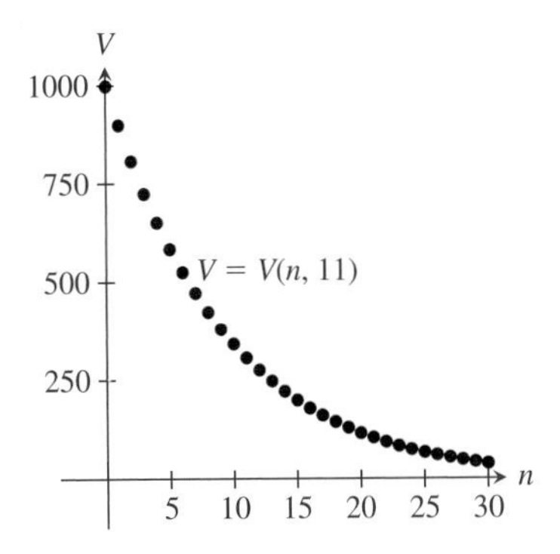

FIGURE 9.32 Bond value as a function of years to maturity.

Sometimes a function of several variables is represented by combining values from a table with a formula.

TABLE 9.2

Age (years) t	Net stumpage value $S(t)$
20	0
30	0
40	43
50	143
60	303
70	497
80	650
90	805
100	913
110	1000
120	1075

EXAMPLE 5 Managing a commercial forest requires long-range planning. Should a stand of trees be cut now and the money invested in, say, the stock market, or will more be earned by letting the trees mature for several years? Suppose a forest is to be cut, sold, and replanted every T years. What should T be to ensure the largest profit?

Table 9.2 gives the net stumpage value $S(t)$ of a stand of fir trees t years after planting. (The *net stumpage* is the difference between the selling price of the trees and the cost of harvesting the trees.) Suppose a crop of trees is to be planted now and harvested in t years. The profit at that time will be $S(t)$. The *present value of the future stand* is the amount of money that must be invested *now* to mature to $S(t)$ in t years. If the interest rate for the investment is r, then this amount is $e^{-rt}S(t)$. See Exercises 18 and 19. When this first crop of trees is harvested, a second crop is immediately planted, with plans to harvest this crop t years later. The present value of this second crop of trees is $e^{-2rt}S(t)$ because if amount $e^{-2rt}S(t)$ is invested now at interest rate r, it will grow to amount $S(t)$ in $2t$ years. Similarly, the present value of the third stand is $e^{-3rt}S(t)$, and so forth. The present value P of all future stands is the sum of the present values of each future stand,

$$P = P(r,t) = e^{-rt}S(t) + e^{-2rt}S(t) + e^{-3rt}S(t) + e^{-4rt}S(t) + \cdots. \quad (14)$$

State the domain of P and discuss the behavior of P as a function of r and of t.

Solution

Because the interest rate r can be any positive number, P is defined for all $r > 0$. The table of net stumpage values gives values of $S(t)$ only for $t = 20, 30, \ldots, 120$. Using the table of values we can probably make a good estimate of $S(t)$ for other values of t between 0 and 120. For example, we can plot the points $(t, S(t))$ as given by the table, connect adjacent points in the graph by a line segment, and then use the graph to define $S(t)$ for other values of t. See Fig. 9.33. Hence we will assume that S and hence P are defined for $0 \leq t \leq 120$, and we take the domain of P to be

$$\mathcal{D}_P = \{(r,t) : r > 0 \text{ and } 0 < t \leq 120\}.$$

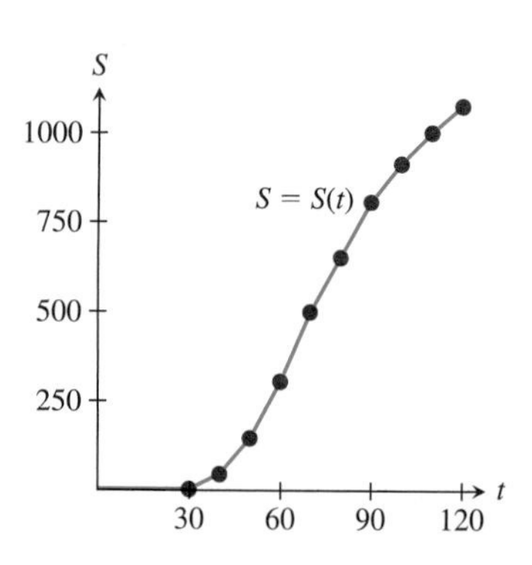

FIGURE 9.33 Net stumpage value.

Before analyzing the behavior of P, note that (14) can be simplified by summing the infinite series on the right side of the equation. The series is a geometric series with first term $e^{-rt}S(t)$ and common ratio e^{-rt}. Because r and t are positive, we have $0 < e^{-rt} < 1$. Hence the series converges. Applying results from Chapter 7 to sum the series, we have

$$P(r,t) = S(t)\frac{e^{-rt}}{1 - e^{-rt}} = \frac{S(t)}{e^{rt} - 1}.$$

Now suppose t is fixed at $t = 60$. This means that the trees are allowed to grow for 60 years; they are then harvested, and a new crop is planted. The expression

$$P = P(r, 60) = \frac{S(60)}{e^{60r} - 1} = \frac{303}{e^{60r} - 1}$$

gives the present value of the future stands of trees as a function of the interest rate r. Note that as r grows, the value of P decreases. (See Fig. 9.34.) This is because when interest rates are high, less money needs to be invested to give some desired future return. Thus in an economy with high interest rates, it might be more profitable to invest money in stocks and bonds instead of trees.

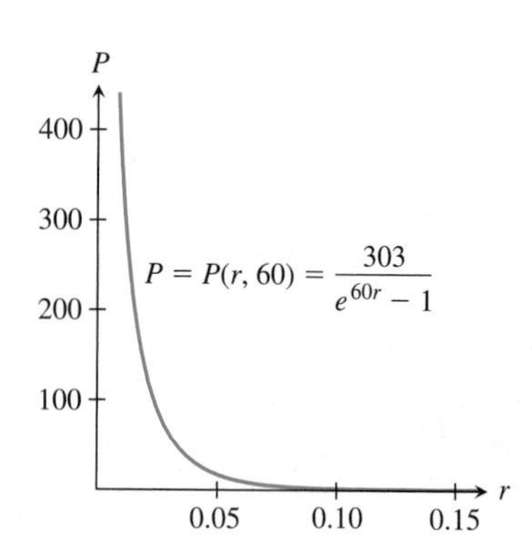

FIGURE 9.34 Present values of future stands as a function of interest rate.

The analysis is more complicated (and interesting) when r is fixed. Suppose the interest rate is 6 percent, so $r = 0.06$. Then

$$P = P(0.06, t) = \frac{S(t)}{e^{0.06t} - 1} \quad (15)$$

gives the present value of all future stands as a function of t, the age of the trees when harvested. To analyze (15), we first find an expression for $S(t)$. Because S is defined by the segments shown in Fig. 9.33, we get different expressions for S on different t intervals. For $50 \leq t \leq 60$, S is defined by

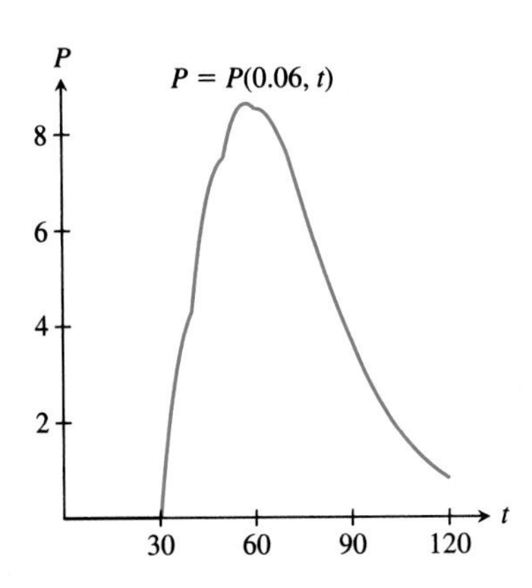

FIGURE 9.35 Present value of future stands as a function of time.

the segment joining $(50, 143)$ and $(60, 303)$. Thus we have

$$S(t) = 16t - 657, \qquad 50 \le t \le 60,$$

so

$$P(0.06, t) = \frac{16t - 657}{e^{0.06t} - 1}, \qquad 50 \le t \le 60. \tag{16}$$

After finding an expression for S on each of the other intervals, we can produce the graph of $P = P(0.06, t)$ shown in Fig. 9.35. From the graph we see that P is maximum when $t \approx 57$. This means that when the interest rate is 6 percent, trees should be planted, allowed to grow for about 57 years, and then harvested and a new crop planted.

This example is based on data and results from the essay "Forestry Management" in *Applying Mathematics,* by D. N. Burghes, I. Huntley, and J. McDonald, Ellis Horwood Limited, Chichester, U.K., 1982.

Exercises 9.2

Exercises 1–4: State the domain of the function f. Sketch the graph of $z = g(x) = f(x, b)$ for the given value of b.

1. $f(x, y) = xy(x - 3y), \qquad b = 2$

2. $f(x, y) = \dfrac{x}{\sqrt{y^2 - 4}}, \qquad b = -3$

3. $f(x, y) = (x + y)\sin(xy), \qquad b = \pi/2$

4. $f(x, y) = \dfrac{e^{xy}}{x + y}, \qquad b = 1$

Exercises 5–10: Describe the domain of the function f. Discuss the behavior of $f(a, y, c)$ for the given values of a and c. Sketch the graph of $w = g(y) = f(a, y, c)$.

5. $f(x, y, z) = \sqrt{x^2 + 2y^2 + 4z^2}, \qquad a = 2, c = 1$

6. $f(x, y, z) = \dfrac{1}{\sqrt{x^2 + 2y^2 + 4z^2}}, \qquad a = 2, c = 1$

7. $f(x, y, z) = \ln(x + y + z), \qquad a = 2, c = -4$

8. $f(x, y, z) = \dfrac{x + y}{y + 2z}, \qquad a = 4, c = 0$

9. $f(x, y, z) = \arcsin(x^2 + y^2 + z^2), \qquad a = -1/4, c = 1/3$

10. $f(x, y, z) = \csc(2x - y + 3z), \qquad a = 0, c = -4$

11. Discuss the behavior of the ballistic pendulum formula (5) of Example 2 when M and h are fixed. What is the physical interpretation of the behavior?

12. In our discussion of the ballistic pendulum (Example 2), we stated that when h is large, a small error in measuring h does not have much effect on the accuracy of the speed, v. On the other hand, when h is small, a small error in measuring h can result in a large error in v. Explain why these statements are true.

13. Discuss the behavior of the mimicry formula (11) of Example 3 when $s = 0.9$ and $\epsilon = 0.3$ are fixed. What does this behavior say about mimicry?

14. Discuss the behavior of the mimicry formula (11) of Example 3 when $f = 0.9$ and $\epsilon = 0.3$ are fixed. What does this behavior say about mimicry?

15. By analyzing (16) in Example 5, find an approximation to the value of t for which $P(0.06, t)$ is maximum.

16. In Example 5, find a formula for $P(0.06, t)$ for $40 \le t \le 50$.

17. In Example 5, find a formula for $P(0.06, t)$ for $100 \le t \le 110$.

18. In Example 5 we stated that if the interest rate is r, where $0 < r < 1$, then an investment of $e^{-rT}P$ dollars now will mature to P dollars in T years. Verify this statement. (Recall that if $A(t)$ is the amount of money in the investment account at time t, then $dA/dt = rA$.)

19. Your first child is born, and you immediately plan to invest money so he or she can attend college at the age of 18. You read that in 18 years, four years at a state university will cost about \$95,000. If you can invest money at 8 percent, how much should you invest now to have the money for your child's education available in 18 years? (See the previous exercise.)

20. In Example 4 we stated that \$208 invested at 8 percent interest will grow to \$1000 in 20 years. Verify this

statement. If your result does not agree with the statement in the example, give a possible reason for the disagreement.

21. When an object is h units above the ground, the gravitational potential energy of the object is equal to the kinetic energy it would have after falling distance h from rest. An object with mass m and traveling at speed v has kinetic energy $mv^2/2$. For an object of mass m at height h, find a formula for the gravitational potential energy in terms of h and m. Use this to verify (5) in Example 2.

22. When pure helium is in a closed container, the temperature T (in Kelvin) as a function of pressure p and volume v is given by the van der Waals law,

$$T = T(p, v) = 12.188\left(p + \frac{0.00007}{v^2}\right)(v - 0.00106).$$

In this equation, p is measured in atmospheres and v is measured relative to the volume of the same body of gas at standard temperature and pressure, with $v = 1$ at standard temperature and pressure.

a. What is the domain of T?

b. Suppose the volume is held fixed at $v = v_0$. How does T behave as a function of p?

c. Suppose the pressure p is held constant at $p = p_0$. How does T behave as a function of v?

d. Let $v = v_0 = 2$ and sketch the graph of $T = T(p, 2)$. According to the graph, what happens to T as p increases? Does this make physical sense?

23. If air resistance and the effect of gravity are ignored, the maximum speed s attained by a rocket is well approximated by

$$s = s(r, m, M) = r \ln\left(\frac{M}{m}\right)$$

where M is the initial mass of the rocket with fuel, m is the mass after the fuel is burned, and r is the speed of the rocket gases relative to the rocket.

a. What is the domain of s?

b. Let $m = m_0$ and $M = M_0$ be fixed. How does s behave as a function of r?

c. Let $m = 1000$ kg and $r = 250$ m/sec. Sketch the graph of $s = s(250, 1000, M)$. What does the graph say about s as a function of M?

d. Let $M = 2500$ kg and $r = 450$ m/sec. Sketch the graph of $s = s(450, m, 2500)$. What does the graph say about s as a function of m?

24. The differential chain block, shown in the accompanying figure, is an ingenious mechanism for lifting heavy loads with very little effort. If force P (for *pull*) is exerted to rotate wheel A through one revolution, the vertical chain labeled C is lowered distance $2\pi r$ while chain D is raised distance $2\pi R$. Hence the mass of weight W is lifted a distance of $\pi(R - r)$. Because the work done by force P is the same as the work done in lifting the weight, we have

$$2\pi RP = \pi(R - r)W.$$

The ratio $M = W/P$ is called the *mechanical advantage* of the mechanism. Hence

$$M = M(r, R) = \frac{2R}{R - r}.$$

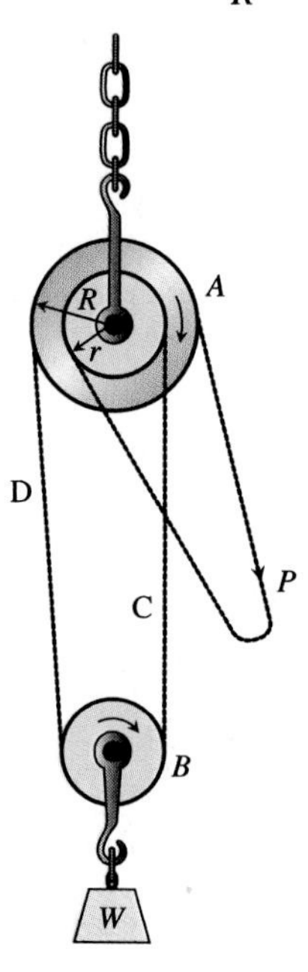

Figure for Exercise 24: The differential chain block.

a. What is the domain of M?

b. Let $R = 50$. Sketch a graph of $M = M(r, 50)$. How does M behave as a function of r?

c. Let $r = 40$. Sketch a graph of $M = M(40, R)$. How does M behave as a function of R? Give a physical interpretation.

T **25.** The accompanying table shows the net stumpage values $S(t)$ for a stand of larch trees of various ages.

Age (years) t	Net Stumpage Value $S(t)$
20	0
30	60
40	132
50	198
60	258
70	319
80	368
90	405
100	429

Table for Exercise 25.

a. Let the domain of S be $\{t : 0 \leq t \leq 100\}$. Plot $S = S(t)$ where the value of $S(t)$ for t between two table values is determined by the line segment connecting adjacent data points. See Example 5.

b. Suppose trees are to be harvested and replanted every t years. As seen in Example 5, the present value of all future stands of trees in an economy with interest rate r is

$$P(r,t) = \frac{S(t)}{e^{rt} - 1}.$$

Assume that the interest rate is $r = 0.06$. Sketch the graph of $P = P(0.06, t)$.

c. When $r = 0.06$, for what value of t is $P(0.06, t)$ maximum?

26. The accompanying table is a record of carbon dioxide concentrations in parts per million by volume (ppmv) for the years 1958 to 1989. These measurements were made on the island of Mauna Loa at sampling sites on the barren lava field of an active volcano. The Mauna Loa site is considered to be one of the best locations for such measurements because possible influences of vegetation and humans are minimal and any influence from volcanic vents can be compensated for. From the way the table is displayed, we can consider CO_2 concentration C to be a function of two variables, the year y and the month m. Hence

$$C = C(y, m).$$

Year	Jan.	Feb.	March	April	May	June	July	Aug.	Sept.	Oct.	Nov.	Dec.	Annual
1958	—	—	316.0	317.6	317.8	—	316.1	315.2	313.4	313.5	—	314.8	—
1959	315.6	316.4	316.8	317.8	318.4	318.2	316.7	315.0	314.0	313.6	315.0	315.8	316.1
1960	316.5	317.1	317.8	319.2	320.1	319.7	318.3	316.0	314.2	314.7	315.1	316.2	317.0
1961	317.9	317.8	318.5	319.5	320.6	319.9	318.7	317.0	315.2	315.5	316.2	317.2	317.7
1962	318.1	318.7	319.8	320.7	321.3	320.9	319.8	317.6	316.5	315.6	316.9	317.9	318.6
1963	318.8	319.3	320.1	321.5	322.4	321.6	319.9	317.9	316.4	316.2	317.1	318.5	319.1
1964	319.6	—	—	—	322.2	321.9	320.4	318.6	316.7	317.2	317.9	318.9	—
1965	319.7	320.8	321.2	322.5	322.6	322.4	321.6	319.2	318.2	317.8	319.4	319.5	320.4
1966	320.4	321.4	322.2	323.5	323.8	323.5	322.2	320.1	318.3	317.7	319.6	320.7	321.1
1967	322.1	322.2	322.8	324.1	324.6	323.8	322.3	320.7	319.0	319.0	320.4	321.7	322.0
1968	322.3	322.9	323.6	324.7	325.3	325.2	323.9	321.8	320.0	319.9	320.9	322.4	322.8
1969	323.6	324.2	325.3	326.3	327.0	326.2	325.4	323.2	321.9	321.3	322.3	323.7	324.2
1970	324.6	325.6	326.6	327.8	327.8	327.5	326.3	324.7	323.1	323.1	324.0	325.1	325.5
1971	326.1	326.6	327.2	327.9	329.2	328.8	327.5	325.7	323.6	323.8	325.1	326.3	326.5
1972	326.9	327.8	328.0	329.9	330.3	329.2	328.1	326.4	325.9	325.3	326.6	327.7	327.6
1973	328.7	329.7	330.5	331.7	332.7	332.2	331.0	329.4	327.6	327.3	328.3	328.8	329.8
1974	329.4	330.9	331.6	332.9	333.3	332.4	331.4	329.6	327.6	327.6	328.6	329.7	330.4
1975	330.5	331.1	331.6	332.9	333.6	333.5	331.9	330.1	328.6	328.3	329.4	330.6	331.0
1976	331.6	332.5	333.4	334.5	334.8	334.3	333.0	330.9	329.2	328.8	330.2	331.5	332.1
1977	332.8	333.2	334.5	335.8	336.5	336.0	334.7	332.4	331.3	330.7	332.1	333.5	333.6
1978	334.7	335.1	336.3	337.4	337.7	337.6	336.2	334.4	332.4	332.2	333.6	334.8	335.2
1979	335.9	336.4	337.6	338.5	339.1	338.9	337.4	335.7	333.6	333.7	335.1	336.5	336.5
1980	337.8	338.2	349.9	340.6	341.2	340.9	339.3	337.3	335.7	335.5	336.7	337.8	338.4
1981	338.8	340.1	340.9	342.0	342.7	341.8	340.0	337.9	336.2	336.3	337.8	339.1	339.5
1982	340.2	341.1	342.2	343.0	343.6	342.9	341.7	339.5	337.8	337.7	339.1	340.4	340.8
1983	341.3	342.5	343.1	344.9	345.8	345.3	344.0	342.4	339.9	340.0	341.2	342.9	342.8
1984	343.7	344.6	345.3	347.0	347.4	346.7	345.4	343.2	341.0	341.2	342.8	344.0	344.3
1985	344.8	345.8	347.2	348.1	348.7	347.9	346.3	344.2	342.9	342.6	344.0	345.3	345.7
1986	346.0	346.7	347.6	349.2	349.9	349.2	347.7	345.5	344.5	343.9	345.3	346.6	346.9
1987	347.7	348.1	349.1	350.6	351.5	351.0	349.2	347.7	346.2	346.2	347.4	348.7	348.6
1988	349.9	351.2	351.9	353.2	353.9	353.3	352.1	350.0	348.5	348.7	349.8	351.1	351.2
1989	352.5	352.8	353.4	355.1	355.4	354.9	353.4	351.3	349.6	349.9			

Table for Exercise 26. Atmospheric concentrations of carbon dioxide* at Mauna Loa.

*Atmospheric CO_2 in parts per million by volume (ppmv). Annual averages based on monthly means. All numbers have been rounded to the nearest tenth.

a. What is the domain of C?

b. Fix $m =$ February. Describe the concentration $C = C(y, \text{Feb.})$ as a function of the year. How have February CO_2 concentrations changed since 1958? Also answer these questions for $m =$ July and $m =$ October.

c. If we compare CO_2 concentrations in different years, why might it be important to compare data from the same month?

d. Fix $y = 1958$. Describe $C = C(1958, m)$ as a function of m. Do the same for $y = 1970$ and $y = 1987$. Do CO_2 concentrations seem to behave the same from year to year?

27. The accompanying table gives data about the shape of the hull of the U.S. Coast Guard cutter *Northland.* Naval architects use these data and numerical integration to find the volume of the portion of the ship below the water. For these purposes, the ship's hull is divided from bow to stern (front to back) into sections called *stations.* At each station the shape of the hull is described by giving the distance d from a vertical line through the center of the station out to the hull. This information is given for various distances measured from the lowest point on the bottom (keel) of the ship. For example, at station 9 and at a height of 8 feet above the lowest point of the ship, the distance from the centerline of the station to the hull is 16.9 feet. See the accompanying figure. For the cutter *Northland,* the spacing between stations is 13.5 feet. (Note, however, that information at half-stations 0.5 and 19.5 is also given). Measurements are made at "waterlines" spaced every 2 feet as measured from the lowest point of the ship. The data in the table give the offset distance d as a function of two variables, the station s and the waterline w. Hence

$$d = d(s, w).$$

a. What is the domain of d?

b. Let $s = 11$. Sketch the graph of $d = d(11, w)$. Explain how this graph relates to the shape of the hull.

c. Sketch the graph of $d = d(1, w)$. How does this graph relate to the shape of the hull? Repeat for $d = d(19, w)$.

d. Let $w = 8$. Sketch the graph of $d = d(s, 8)$. How does this graph relate to the shape of the hull?

e. Team project Suppose the portion of the *Northland* below the 12-foot waterline is underwater. Approximate the volume of the submerged portion of the ship.

	Waterlines									
Station s	2 ft	4 ft	6 ft	8 ft	10 ft	12 ft	14 ft	16 ft	18 ft	20 ft
0	0.0	0.0	0.0	0.0	0.0	0.0	0.17	0.44	0.75	1.08
0.5	0.0	0.25	0.54	0.65	0.93	1.20	1.50	1.96	2.48	2.91
1	0.50	0.90	1.33	1.71	2.14	2.75	3.08	3.55	4.17	4.75
2	1.38	2.40	3.28	4.05	4.78	5.83	6.17	6.87	7.58	8.30
3	2.37	4.00	5.42	6.58	7.61	8.83	9.33	10.10	10.83	11.44
4	3.29	5.64	7.50	9.04	10.32	11.75	12.25	12.94	13.58	14.03
5	4.32	7.26	9.52	11.25	12.65	14.17	14.50	15.21	15.58	15.99
6	5.36	8.81	11.35	13.17	14.58	15.92	16.17	16.71	17.08	17.31
7	6.56	10.32	12.95	14.75	16.03	17.17	17.42	17.81	18.08	18.15
8	7.75	11.61	14.28	15.99	17.08	18.00	18.25	18.48	18.58	18.62
9	8.19	12.64	15.35	16.90	17.86	18.58	18.75	18.83	18.88	18.88
10	8.86	13.22	15.96	17.47	18.26	18.83	18.92	19.00	19.00	19.00
11	8.53	13.21	16.06	17.63	18.43	18.92	19.00	19.00	19.00	19.00
12	7.73	12.67	15.65	17.46	18.36	18.83	19.00	19.00	19.00	19.00
13	5.53	11.36	14.86	16.97	18.05	18.58	18.83	18.90	18.90	18.90
14	0.45	8.70	13.40	16.04	17.51	18.33	18.58	18.70	18.70	18.70
15	0.0	2.12	10.33	14.36	16.58	17.75	18.17	18.29	18.29	18.29
16	0.0	0.0	3.45	11.19	15.02	16.83	17.42	17.59	17.59	17.59
17	0.0	0.0	0.0	5.04	12.38	15.58	16.42	16.74	16.93	16.93
18	0.0	0.0	0.0	0.0	7.21	13.50	15.33	15.75	16.08	16.20
19	0.0	0.0	0.0	0.0	2.99	10.00	13.66	14.36	15.00	15.50
19.5	0.0	0.0	0.0	0.0	0.0	6.50	12.83	14.03	14.50	14.95

Table for Exercise 27. Offsets for 270-ft-class Coast Guard cutter: waterline half-breadths (in feet).*

*Station spacing = 13.5 ft.

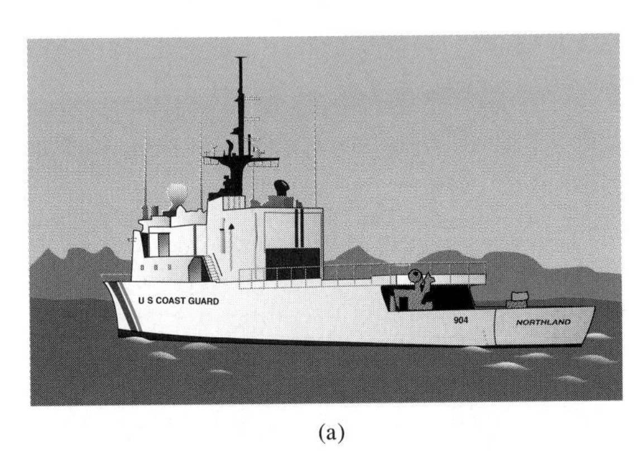

(a)

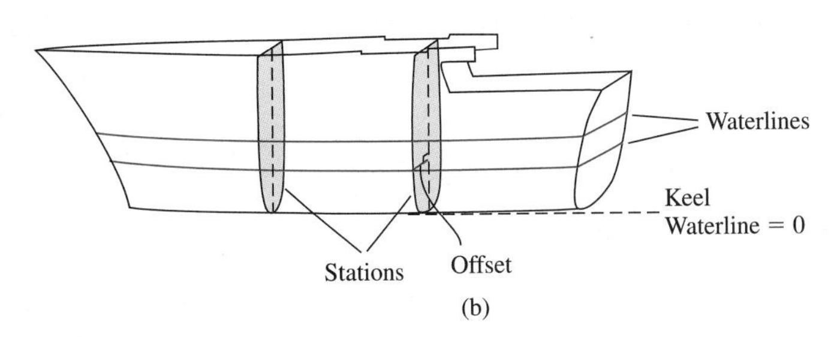

(b)

Figure for Exercise 27.

9.3 Graphing: Surfaces and Level Curves

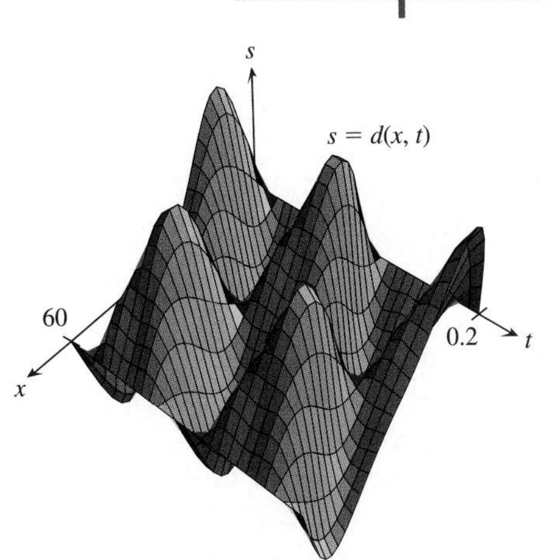

FIGURE 9.36 The motion of a guitar string.

Figure 9.36 shows the graph in R^3 of the function d defined by

$$s = d(x,t) = \frac{1}{4}\cos(20\pi t)\sin\left(\frac{\pi x}{30}\right) + \frac{1}{16}\cos(40\pi t)\sin\left(\frac{\pi x}{15}\right). \tag{1}$$

In the previous section we used this equation to model the motion of a guitar string. Is this graph useful in helping us understand the motion of the guitar string? If so, what does it tell us about the motion? Conversely, what does the previous analysis of (1) tell us about the graph? Graphs like that in Fig. 9.36 can give us a direct, immediate understanding of functions and models in ways that the analysis of the previous section cannot. In this section we look at techniques for producing and understanding graphs of functions of several variables.

Functions of Two Variables

Let f be a function of two variables and let the domain of f be $\mathcal{D}$. To represent $z = f(x, y)$ graphically, we need to provide a picture that conveys the value of $f(x, y)$ for each (x, y) in $\mathcal{D}$. We investigate three methods for doing this.

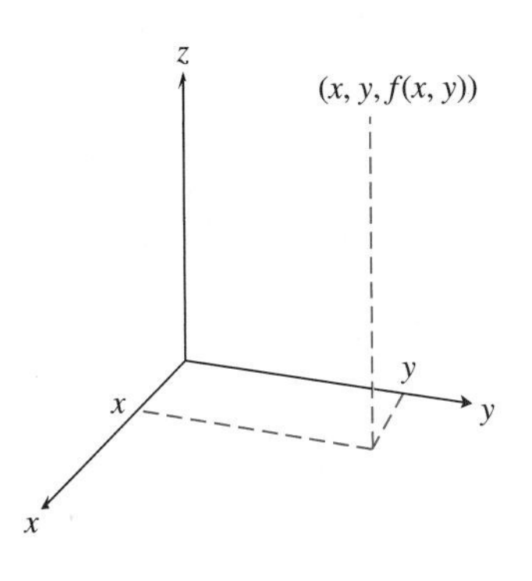

FIGURE 9.37 Plotting the point $(x, y, f(x, y))$.

Graphing $z = f(x, y)$ as a Surface Let f be a function of two variables with domain $\mathcal{D}$. If (x, y) is a point in $\mathcal{D}$ and $z = f(x, y)$, then we use the ordered triple (x, y, z) to associate the input (x, y) with the output z. This information can be represented graphically by plotting the point $P = (x, y, z)$ in R^3. The height z of P above or below the (x, y)-plane represents the value of f at (x, y). See Fig. 9.37. If we plot the points

$$(x, y, z) = (x, y, f(x, y))$$

for all (x, y) in $\mathcal{D}$, we get a *surface* in R^3, as shown in Fig. 9.38. This surface is the graph of $z = f(x, y)$.

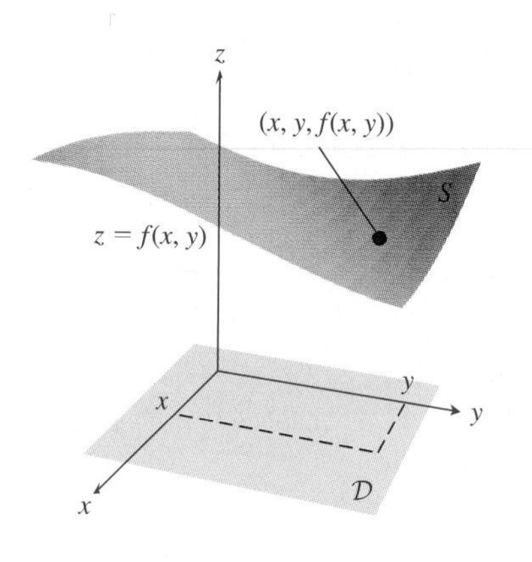

FIGURE 9.38 The graph of $z = f(x, y)$ is a surface in R^3.

DEFINITION Surface Graph

Let f be a function of two variables with domain $\mathcal{D}$ in R^2. The **surface graph** S of f is the set

$$S = \{(x, y, f(x, y)) : \text{where } (x, y) \text{ is a point in } \mathcal{D}\}.$$

See Fig. 9.38.

Drawing a good surface graph for a function is usually more complicated than just plotting a few points. This is because of the difficulties inherent in trying to draw a picture of a three-dimensional object on a two-dimensional piece of paper. Often it is helpful to first do some analysis of the function to be graphed to find out more about the shape of the surface. A computer algebra system that plots the surface graph of a function can be very helpful, but some machine-generated graphs can be hard to interpret without some analysis beforehand. Some of this analysis is demonstrated in the following investigation.

INVESTIGATION

Sketching a Paraboloid

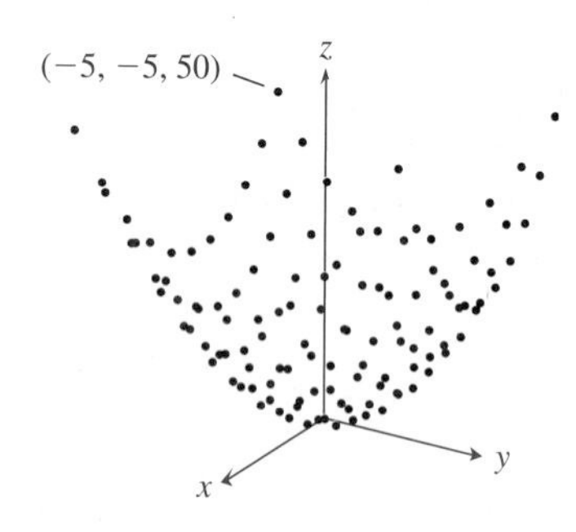

FIGURE 9.39 Several points on the graph of $z = x^2 + y^2$.

Sketch the surface graph of the function f defined by $f(x, y) = x^2 + y^2$.

We start by plotting several points just to see what information might be obtained from such a graph. In Fig. 9.39 we have plotted all the points

$$(x, y, x^2 + y^2)$$

for all pairs (x, y) of integers with x and y between -5 and 5. This gives us a total of $11 \cdot 11 = 121$ points on the graph. Although the picture may be somewhat suggestive of the shape of the surface, it still does not convey much information about the behavior of the function. To get information to help us draw a better graph, we use a technique from the previous section. We fix one of the variables and then see how the function behaves as a function of the other variable.

Fix $x = a$. The graph of $x = a$ is the plane perpendicular to the x-axis and through the point $(a, 0, 0)$. The intersection of this plane and the graph of $z = x^2 + y^2$ is a curve. This curve lies in the plane $x = a$ and is described by the equation

$$z = a^2 + y^2. \tag{2}$$

We call this curve a cross section of the surface, with the plane section perpendicular to the x-axis. From (2) we see that this cross section is a parabola that opens in the positive z-direction. For example, when $a = 3$, (2) becomes $z = 9 + y^2$. This means that the intersection of the plane $x = 3$ with the surface is the parabola with equation $z = 9 + y^2$, drawn in the plane $x = 3$. Setting $a = -1$ shows that the intersection of the surface with the plane $x = -1$ is a parabola with equation $z = y^2 + 1$. See Fig. 9.40. Thus in Fig. 9.39 the points on any plane perpendicular to the x-axis lie on a parabola. In Fig. 9.41 we show many of these parabolic cross sections.

WEB If we set $y = b$ in $z = x^2 + y^2$, we obtain

$$z = x^2 + b^2.$$

This shows that any cross section in a plane perpendicular to the y-axis is also a parabola that opens in the positive z-direction. In Fig. 9.42 these cross sections are included with cross sections of Fig. 9.41. The graph in Fig. 9.42 is often called a "wire frame plot" of the surface.

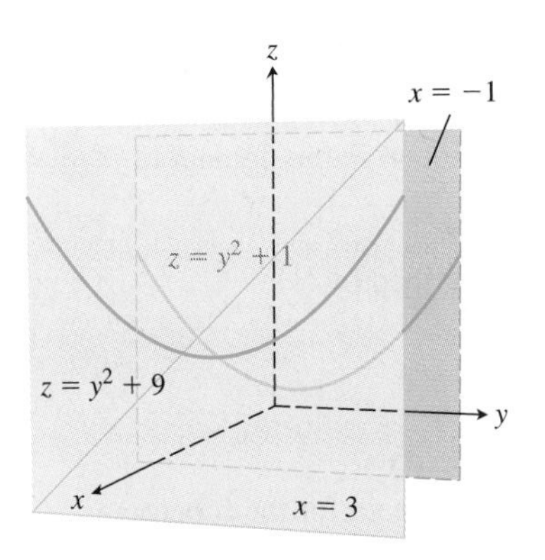

FIGURE 9.40 On the surface $z = x^2 + y^2$, cross sections perpendicular to the x-axis are parabolas.

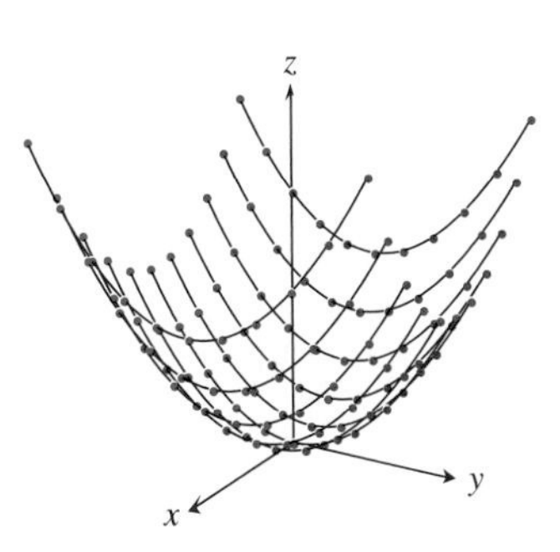

FIGURE 9.41 When we plot several of the cross sections in planes perpendicular to the x-axis, we start to see the shape of the surface.

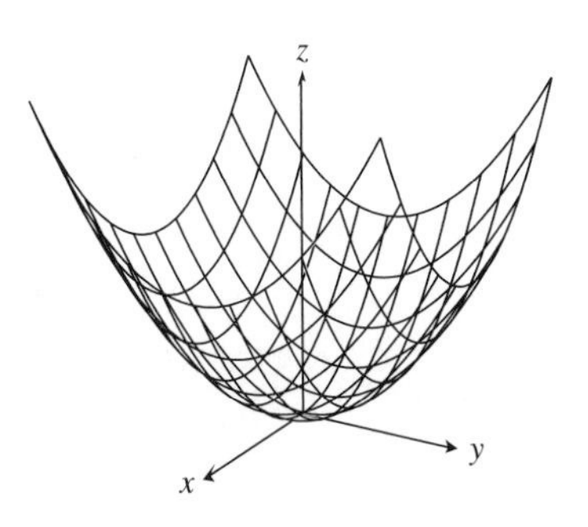

FIGURE 9.42 When we include cross sections perpendicular to the y-axis, we see more of the shape of the surface.

When we fix the z-coordinate, we get information about cross sections taken perpendicular to the z-axis. For $z = c$, this cross section is described by the equation

$$x^2 + y^2 = c. \tag{3}$$

If $c < 0$, there are no values of x and y that satisfy (3). This means that all points on the surface $z = x^2 + y^2$ lie on or above the (x, y)-plane. If $c > 0$, then (3) describes a circle. Thus cross sections to the surface taken perpendicular to the z-axis are circles.

Because the cross sections perpendicular to the x- and y-axes are parabolas opening in the positive z-direction, we know the surface should open upward like a bowl. The cross sections perpendicular to the z-axis tell us that the bowl is circular. When we add to Fig. 9.42 lines suggesting these cross sections, we clearly see the bowl shape, as seen in Fig. 9.43.

Most computer algebra systems and some graphing calculators can graph surfaces in three dimensions. However, the graphs produced in this way can sometimes be hard to interpret. By analyzing cross sections, we can better interpret the graph produced by a computer or calculator, and we can decide on an angle from which to view the graph to best see features of interest. A CAS-produced graph of $z = x^2 + y^2$ is shown in Fig. 9.44. We see the bowl shape in the graph, but it might be hard to recognize if we had not done the previous analysis.

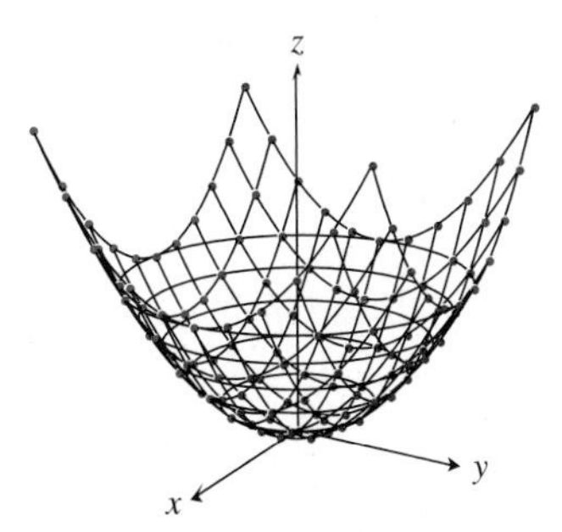

FIGURE 9.43 By sketching cross-section curves, we see that the surface is shaped like a bowl.

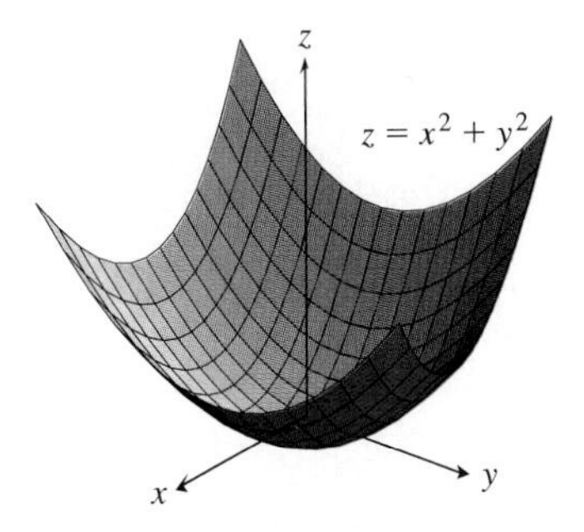

FIGURE 9.44 Computer algebra systems have commands to produce the graphs of surfaces. Such a command was used to produce this graph of $z = x^2 + y^2$.

When the domain of f is not all of R^2, some of the expressions describing the cross sections will not be defined for all real numbers. By paying attention to these restrictions, we obtain the graph of $z = f(x, y)$ for the (x, y) in the domain of f.

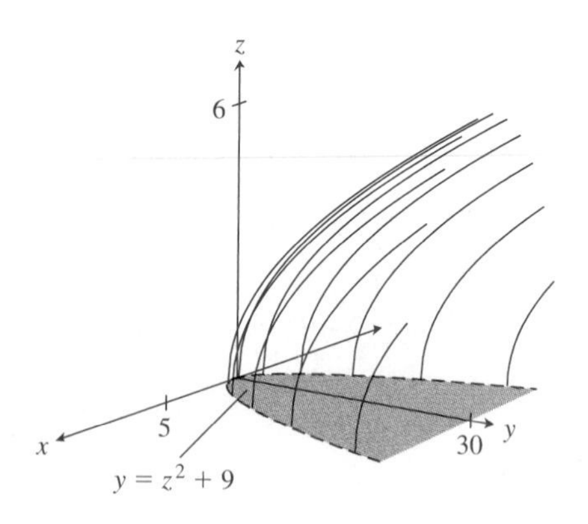

FIGURE 9.45 Cross sections of the graph of $z = \sqrt{y - x^2}$ taken perpendicular to the x-axis.

EXAMPLE 1 Analyze and graph the surface described by

$$z = \sqrt{y - x^2}. \tag{4}$$

Solution

Because (4) is defined only when $y - x^2 \geq 0$, the domain for the function is

$$\mathcal{D} = \{(x, y) : y \geq x^2\}.$$

This domain is the planar region "inside" and on the parabola $y = x^2$ in the (x, y)-plane. The graph of the surface lies above this region. The domain is shaded in Fig. 9.45 and in Fig. 9.46.

Next we look at the cross sections. The cross section in the plane $x = a$ is the graph of

$$z = \sqrt{y - a^2}, \qquad y \geq a^2. \tag{5}$$

Because Equation (5) can be rewritten as

$$y = z^2 + a^2, \qquad z \geq 0,$$

we see that the cross section described by (5) is half of a parabola. For example, setting $x = a = 3$, we find that the cross section in the plane $x = 3$ is the portion of the parabola $y = z^2 + 9$ with $z \geq 0$. This cross section and several others are shown in Fig. 9.45.

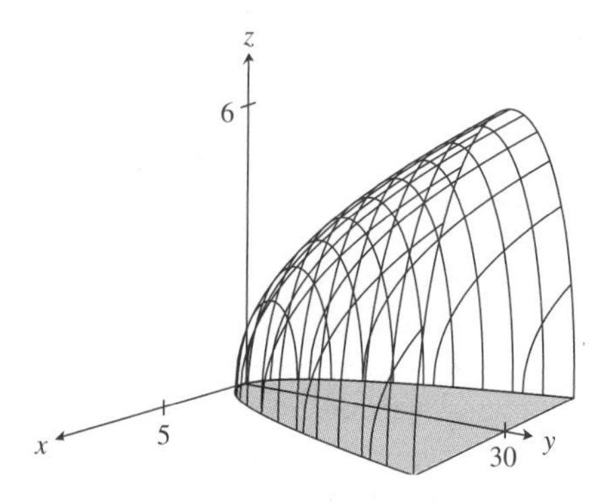

FIGURE 9.46 When we sketch cross sections perpendicular to the x- and y-axes, we start to see the shape of the solid.

The cross section in the plane $y = b$ is the graph of

$$z = \sqrt{b - x^2}. \tag{6}$$

This last expression is meaningful only for $b \geq 0$. Squaring both sides of (6), we have

$$x^2 + z^2 = b, \qquad z \geq 0.$$

Hence (6) describes the upper half of a circle of radius $\sqrt{b}$. In Fig. 9.46 we have included some of these cross sections along with those shown in Fig. 9.45.

Now set $z = c \geq 0$ in (4) to get

$$c = \sqrt{y - x^2}.$$

Square and rearrange to obtain

$$y = x^2 + c^2.$$

Thus the intersection of the surface with the plane $z = c$ is the parabola with equation $y = x^2 + c^2$ in the plane $z = c$. When we include some of these cross sections with those of Fig. 9.46, we get a good picture of the surface. See Fig. 9.47.

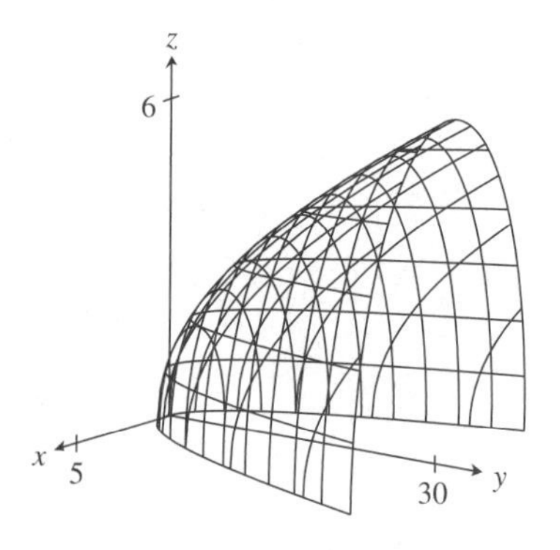

FIGURE 9.47 With cross sections perpendicular to each of the three axes, we see that the surface looks somewhat like an amphitheater.

In Fig. 9.48 we show a graph of (4) produced by a computer algebra system. Because we have analyzed the cross sections, we can recognize the

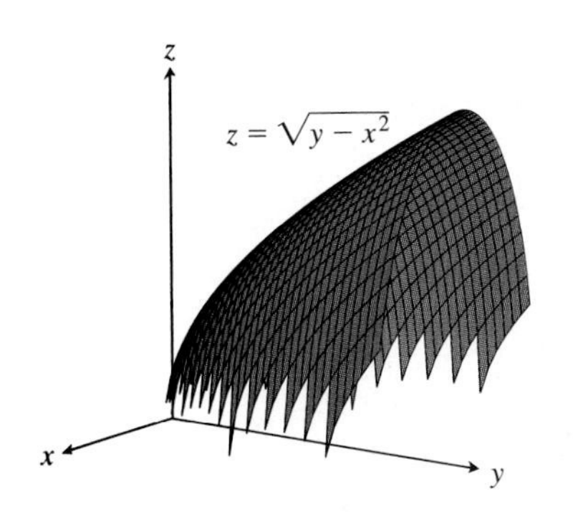

FIGURE 9.48 A graph of $z = \sqrt{y - x^2}$ produced by a computer algebra system.

parabolas and half-circles in the shape of the graph. The bottom edge of the CAS-produced graph is ragged and does not show the smooth edge suggested by Fig. 9.47. This is because when the CAS plots a graph of $z = f(x, y)$, it does so by evaluating $f(x, y)$ for several pairs (x, y) in a rectangular grid. Unfortunately, this rectangular grid does not conform well to the curved boundary of the domain. Because of this, the graph near the boundary of the domain has a jagged appearance. This effect can be somewhat lessened by evaluating $f(x, y)$ at many more points, that is, by including many more points in the grid. Actually, the ragged edge can be completely eliminated by finding a parametric representation of the surface and plotting the surface using this representation. We will look into this in the next section.

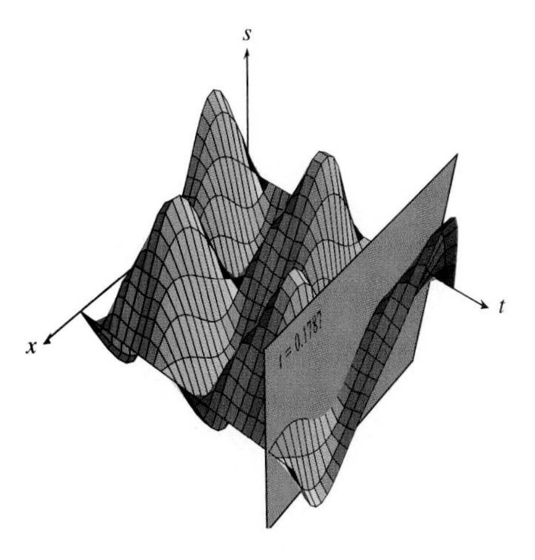

FIGURE 9.49 The intersection of the surface with the plane $t = 0.1787$ shows the state of the string at time $t = 0.1787$.

EXAMPLE 2 In Fig. 9.36 we show the surface graph for $s = d(x, t)$, where

$$s = d(x, t) = \frac{1}{4}\cos(20\pi t)\sin\left(\frac{\pi x}{30}\right) + \frac{1}{16}\cos(40\pi t)\sin\left(\frac{\pi x}{15}\right). \qquad (7)$$

In Example 1 of Section 9.2 we discussed this formula as a model for the motion of a guitar string. In particular, we discussed the significance of $s = d(x, 0.1787)$ and $s = d(20, t)$. Interpret these in terms of cross sections of the surface graph and indicate these cross sections in Fig. 9.36.

Solution

Setting $t = 0.1787$ in (7) gives an equation that describes the intersection of the surface with the plane $t = 0.1787$. The plane and the curve of intersection are shown in Fig. 9.49. As stated in Section 9.2, this curve shows what the string looks like at the instant $t = 0.1787$ seconds. Cross sections in other planes $t = t_1$ show what the string looks like at each instant $t = t_1$. Thus as we move along the t-axis the cross sections perpendicular to this axis illustrate the motion of the string.

Setting $x = 20$ in (7) gives an equation that describes the intersection of the surface with the plane $x = 20$. This plane and the curve of intersection are shown in Fig. 9.50. As discussed in Section 9.2, this curve shows the motion of the point of string 20 centimeters from one end.

Other information about the motion can be discovered by looking at the graph. For example, the cross section perpendicular to the x-axis at $x = 30$ is the line $s = 0$. This tells us that the center point of the string does not move.

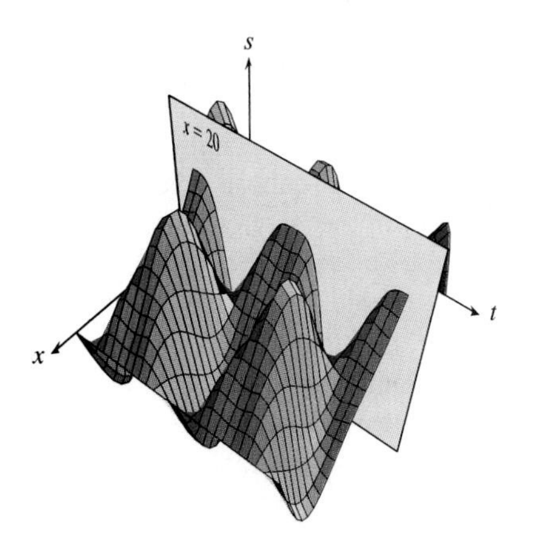

FIGURE 9.50 The intersection of the surface with the plane $x = 20$ shows the motion of the point of the string at the 20-centimeter mark.

Level Curves

If you do much hiking in the wilderness, you have probably used a topographic map. A portion of such a map is shown in Fig. 9.51. On the map we see some curves that indicate altitude of the terrain. For example, one of the curves near Cinder Cone is labeled 6800. This means that all points on this curve are 6800 feet above sea level. Such a line is called a *contour line* or a *level curve.* All of the points on any one of these lines are at the same altitude. Thus, if you were to hike a path

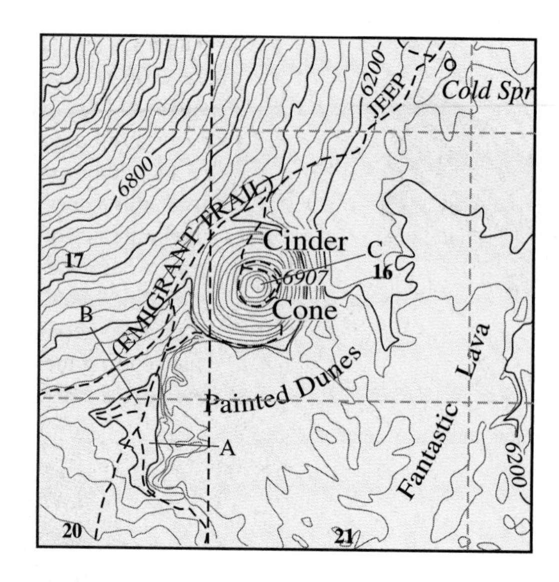

FIGURE 9.51 A topographic map.

following one of these lines, you would never have to go uphill or downhill—your entire hike would be at the same level. By looking at several of these lines and taking the labels into account, we can tell whether the terrain rises, falls, or stays level in a given direction. In particular, when there are many contour lines close together, the terrain is likely to be very steep. This interpretation gives us a glimpse of the derivative: Along any path that crosses several contour lines in a short distance, the rate of change of vertical feet with respect to horizontal feet is large. See Exercise 24 in Section 1.4.

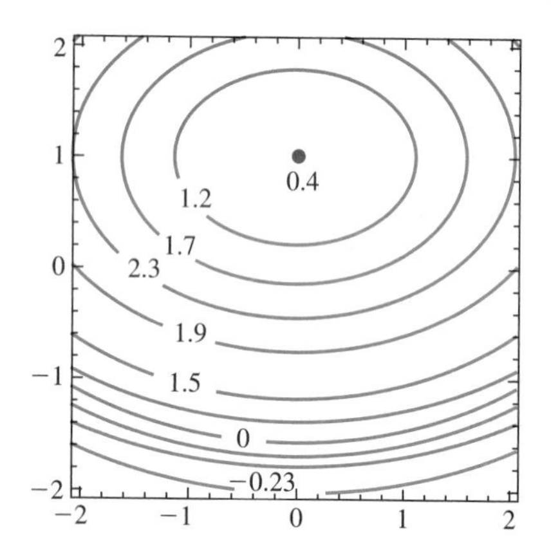

FIGURE 9.52 A level curves graph.

The contour graph, or level curves graph, of a function f of two variables presents information about the function f in the same way that a topographic map presents information about altitude. A contour graph consists of several curves in the (x, y)-plane. Such curves are often labeled with numbers. If a curve in the graph is labeled c, then $f(x, y) = c$ for every point (x, y) on the curve. By noting the labels on different level curves, we can get an idea of how the value of $f(x, y)$ changes as (x, y) changes. An example of such a graph is shown in Fig. 9.52. From this graph we see that function values initially increase as we move away from the point $(0, 1)$, but then start to decrease when we get far enough from this point.

To produce a level curves graph for a function f, we first choose several different values of c. For each value of c we graph the curve in the (x, y)-plane described by the equation $f(x, y) = c$. If this equation describes a familiar curve such as a circle or parabola, we can draw the curve with little trouble. In other cases we can find several solutions (x, y) to the equation $f(x, y) = c$, plot these solutions and then draw a curve through the plotted points. (In fact, this latter method is the way that calculators and CASs produce contour graphs.) After the curve is graphed, we label it c. When we do this for enough values of c, we obtain a contour graph that can help us understand the function f.

It is sometimes useful to note that the solution of the equation

$$c = f(x, y)$$

also describes the intersection of the plane $z = c$ with the surface $z = f(x, y)$. Thus the level curves show the shape of the cross sections perpendicular to the z-axis for the surface graph of f. This can be useful information if we try to visualize the shape of the surface from the level curves graph.

EXAMPLE 3 Produce a level curves graph of $z = f(x, y) = x^2 + y^2$.

Solution

Let c be a real number, and consider the set of solutions (x, y) to

$$x^2 + y^2 = c. \tag{8}$$

If $c < 0$, this equation has no solution. If $c = 0$, the only solution is $(x, y) = (0, 0)$, so the level curve for $c = 0$ is just the origin. For $c > 0$, (8) describes a circle of center $(0, 0)$ and radius $\sqrt{c}$. Thus when $c > 0$, an ordered pair (x, y) satisfies (8) precisely when the point (x, y) lies on the circle of center $(0, 0)$ and radius $\sqrt{c}$. This means that for each $c > 0$, the level curve $f(x, y) = c$ is a circle centered at $(0, 0)$. In Fig. 9.53 we show several of these level curves. Each curve in the figure is labeled with a number corresponding to the value of f on this level curve.

From Fig. 9.53 we see that the value of f increases as we move away from the origin. The level curve labeled c also shows us the shape of the intersection of the plane $z = c$ with the surface graph of $z = f(x, y)$. Because these intersections are circles, we can infer that the surface graph has the bowl shape seen in Figs. 9.43 and 9.44.

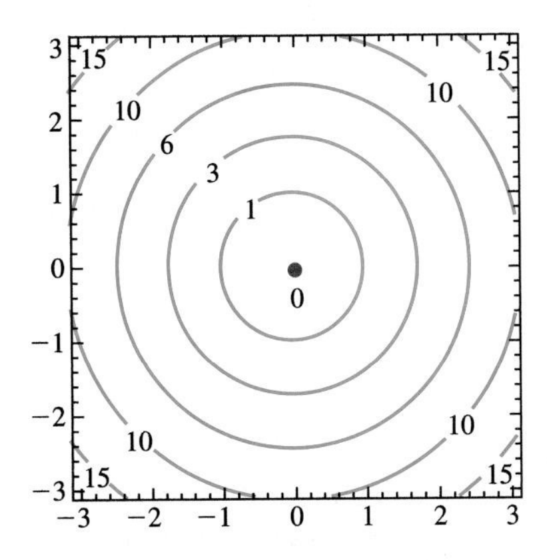

FIGURE 9.53 The level curves for $f(x, y) = x^2 + y^2$ are circles centered at the origin.

We can use a level curves graph to estimate points where a function takes on its maximum or minimum values.

EXAMPLE 4 A school dance is held in a gymnasium that is 200 feet long and 150 feet wide. Loud music is played through three identical speakers each mounted at a height of 20 feet on a wall. One speaker is mounted on the north wall (which is 200 feet long) and is positioned equidistant from the east and west walls. Another speaker is at the junction of the south and west walls, and the third is on the south wall at a point 50 feet from the east wall. The school principal is chaperoning at the dance but finds the music too loud. Where can she stand to get some relative quiet?

Solution

We locate the point on the floor where the intensity of the sound is at a minimum. Put the gym in R^3, with the origin at the point where the south wall, west wall, and floor meet. Let the positive x-axis point east and the positive z-axis point toward the ceiling. Then the speakers are at the points $(0, 0, 20)$, $(150, 0, 20)$, and $(100, 150, 20)$, and are indicated by black rectangles in Fig. 9.54. We treat each speaker as a point source of sound, so at any point, the intensity of the sound from a speaker is inversely proportional to the

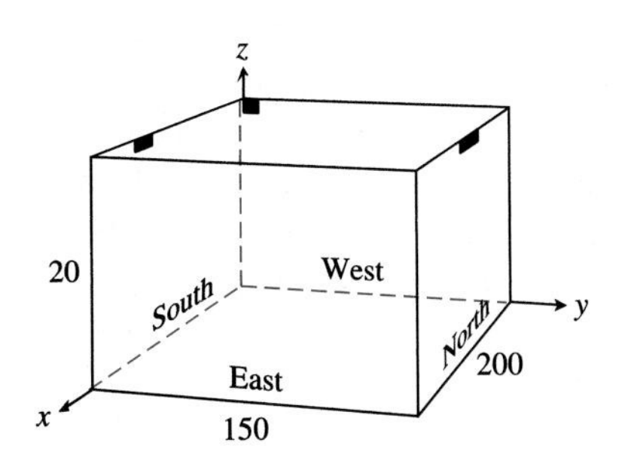

FIGURE 9.54 The gymnasium in R^3.

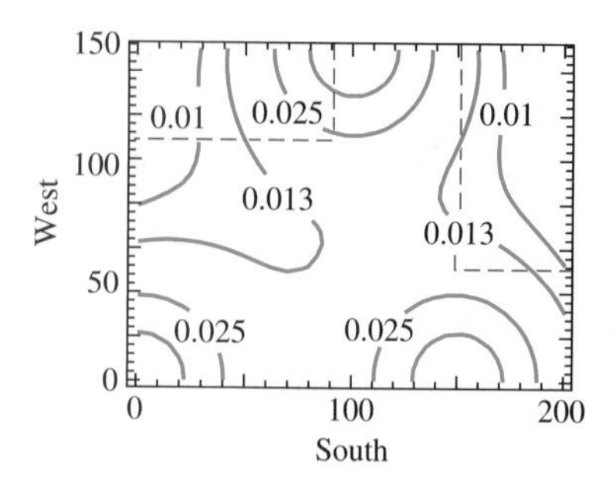

FIGURE 9.55 The intensity $I(x, y)$ of sound on the gym floor. The dashed regions are expanded in Figs. 9.56 and 9.57.

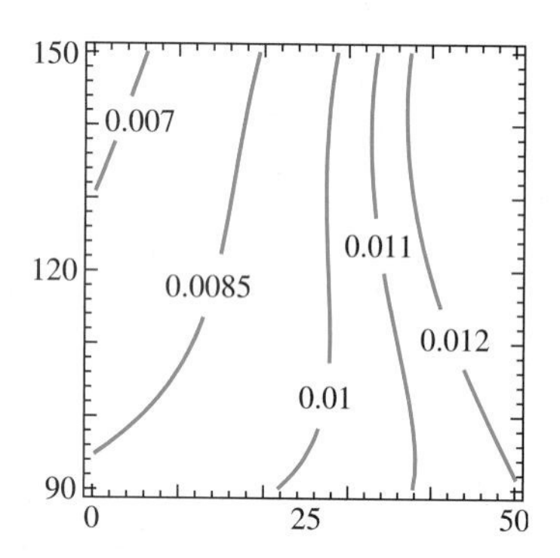

FIGURE 9.56 The intensity of the sound near the northwest corner of the floor.

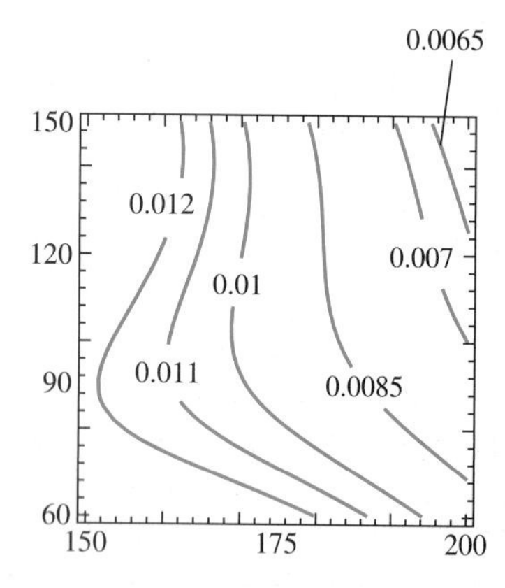

FIGURE 9.57 The intensity of the sound near the northeast corner of the floor.

square of the distance from the point to the speaker. If we ignore sound reflected from the walls and ceiling (or assume that the walls and ceiling absorb sound), then the total intensity of the sound at a point is the sum of the intensities of the sound from the three speakers.

At the point (x, y) (or $(x, y, 0)$) on the gym floor, the distance to the speaker on the north wall is

$$\|\langle x, y, 0\rangle - \langle 100, 150, 20\rangle\| = \sqrt{(x-100)^2 + (y-150)^2 + 20^2}.$$

Recalling that intensity is inversely proportional to the square of distance, the intensity at (x, y) of the sound from the north wall speaker is

$$k\left(\frac{1}{(x-100)^2 + (y-150)^2 + 400}\right),$$

where k is a positive constant of proportionality. Treating the other two speakers similarly, we find that at point (x, y) on the floor the intensity of the sound is

$$I(x, y) = k\left(\frac{1}{(x-100)^2 + (y-150)^2 + 400} + \frac{1}{(x-150)^2 + y^2 + 400} + \frac{1}{x^2 + y^2 + 400}\right).$$

Sound intensity is usually measured in watts/m^2. The intensity of sound from a loud rock band 6 feet away is about 1 W/m^2, which corresponds to a k value of about 40. (We deliberately leave the distance units mixed here to differentiate between different aspects of the problem. Measurements in feet refer to distances in the gym, and those in meters correspond to measurements of sound intensity. This will be helpful when we return to this problem in Section 9.7.) We seek the minimum value of

$$I(x, y) = \frac{40}{(x-100)^2 + (y-150)^2 + 400} + \frac{40}{(x-150)^2 + y^2 + 400} + \frac{40}{x^2 + y^2 + 400}$$

for (x, y) in the gym floor,

$$\{(x, y) : 0 \le x \le 200,\ 0 \le y \le 150\}.$$

A contour plot of $I(x, y)$ for this region is shown in Fig. 9.55. The graph suggests that the sound intensity is high near any of the speakers and is lowest near the northeast or northwest corner of the gym. To get more detailed information about where the principal should stand, we look at a contour graph for the parts of the gym near each of these two corners. These graphs are shown in Figs. 9.56 and 9.57. The graph for the northeast corner shows a level curve labeled 0.0065, but no such curve appears in the graph for the northwest corner. Thus it appears to be a little quieter in the northeast corner than in the northwest corner. We can verify this by evaluating $I(x, y)$ for these corner points:

$$\begin{aligned} &\text{Northwest corner:} && I(0, 150) \approx 0.0064740 \\ &\text{Northeast corner:} && I(200, 150) \approx 0.0060569. \end{aligned}$$

We should look for the principal in the northeast corner of the gym.

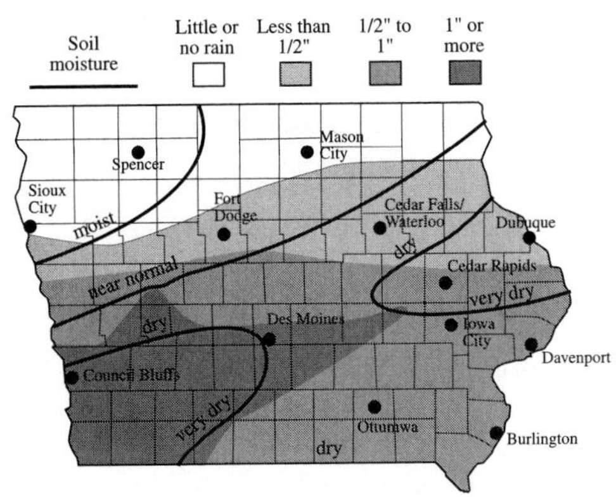

FIGURE 9.58 A map on which different shades of gray indicate different rainfall amounts is an example of a density plot.

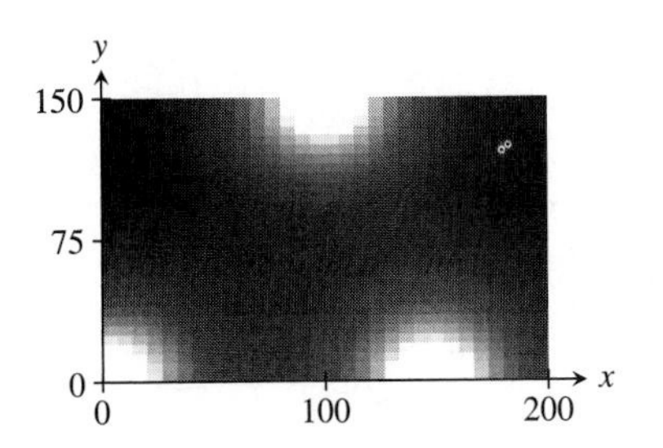

FIGURE 9.59 A density plot of the noise intensity in the gym.

Density Plots

The map shown in Fig. 9.58 shows the rainfall across Iowa from a storm on April 29, 1996. The rainfall amounts for different locations are not recorded directly on the map but are instead denoted by shades of gray. For example, any region that received little or no rain is colored white. This map does not give us precise information about the rainfall amount at any point, but it does give enough information for us to see which regions got a lot of rain, which got a moderate amount, and which got very little. We can think of this map as a representation of the function

$$R = R(x, y),$$

where $R(x, y)$ is the rainfall at point (x, y) in Iowa on April 29, 1996. Like the contour graph, this representation presents information about the graph of $R = R(x, y)$ in a two-dimensional format.

The rainfall map is an example of a *density plot.* A density plot is a representation of a function f of two variables in which the values of f at a point (x, y) is denoted by a color or a shade of gray. Sometimes a density plot is accompanied by a legend that tells us what values of f correspond to different colors or different shades of gray. Density plots can be used to get a quick, initial feel about the behavior of a function. We will not attempt to generate such graphs by hand. However, most computer algebra systems can generate density plots or have commands that shade a contour plot. We encourage you to learn to use such commands if they are available.

In Fig. 9.59 we show a density plot for the function I studied in Example 4. Recall that

$$I(x, y) = \frac{40}{(x - 100)^2 + (y - 150)^2 + 400} + \frac{40}{(x - 150)^2 + y^2 + 400} + \frac{40}{x^2 + y^2 + 400}$$

is the intensity of sound at the point (x, y) on the floor of a gymnasium. In this graph, the larger values of I are indicated by white. The shading grows darker as the values of I decrease. The darkest parts of the graph are in the northwest and northeast corners of the graph. These spots are the quietest in the gym, but the differences in shading are not great enough for us to see that the northeast corner is a little quieter than the northwest. We would need to do more work with the function to find the place where I is minimum, but the density plot does give us an idea of where to look.

Functions of Three Variables

Let f be a function of three variables and set $w = f(x, y, z)$. How can we represent f graphically? When we graph a function of one variable (e.g., $y = g(x)$), we use two mutually perpendicular axes. When we draw the surface graph for a function of two variables (e.g., $z = f(x, y)$), we use three mutually perpendicular axes. Even though we draw the surface graph on a flat piece of paper, we can use perspective and cross-section curves to give the sense of a surface in space. An analogous representation for $w = f(x, y, z)$ would require four mutually perpendicular axes and hence must be represented in four dimensions. However, most of us cannot mentally picture such an object.

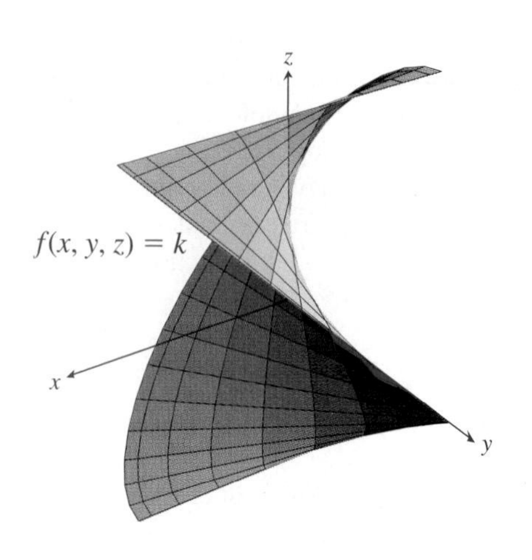

FIGURE 9.60 The function f takes the value k at all points on the level surface $f(x, y, z) = k$.

We have studied other ways to represent a function of two variables. One such technique is the method of level curves. This idea extends readily to functions of three variables with the study of *level surfaces.*

Level Surfaces Let f be a function of three variables and let k be a real number. The set of all solutions to the equation $f(x, y, z) = k$ is the set

$$S_k = \{(x, y, z) : f(x, y, z) = k\}$$

of points in R^3. When all of the points in S_k are plotted in R^3, the resulting graph is called a *level surface* for f. See Fig. 9.60. For any point (x, y, z) on the graph of S_k, we have $f(x, y, z) = k$, so the value of f stays constant (or level) on this surface. By finding the level surfaces S_k for several different values of k, we may get a sense of how the value of $f(x, y, z)$ changes as (x, y, z) changes in R^3.

Because the graph of a level surface is often a surface in R^3, many of the techniques we used to graph functions of two variables are also useful in graphing level surfaces. In addition, some surfaces that cannot be realized as the graph of a single expression of the form $z = f(x, y)$ can be described as level surfaces. This is illustrated in the following example.

EXAMPLE 5 Let f be the function given by

$$f(x, y, z) = 4x^2 + y^2 + \frac{z^2}{9}.$$

Describe the level surfaces for f and graph the surface $f(x, y, z) = 9$.

Solution

First note that $f(x, y, z) \geq 0$ for all (x, y, z). Thus the equation

$$f(x, y, z) = 4x^2 + y^2 + \frac{z^2}{9} = k \tag{9}$$

has solutions only if $k \geq 0$. When $k = 0$, the only solution to (9) is $(0, 0, 0)$. Hence the level surface corresponding to $k = 0$ consists of just one point.

For $k > 0$, we use cross sections to analyze the graph S_k of the solutions to (9). Set $x = a$ in (9) to consider the cross section of S_k in the plane $x = a$. This is described by the equation

$$4a^2 + y^2 + \frac{z^2}{9} = k. \tag{10}$$

Because all terms on the left of (10) are nonnegative, we must have $4a^2 \leq k$. If $4a^2 = k$, then the set described by (10) is the single point $(a, 0, 0)$. If $4a^2 < k$, then (10) can be rearranged to get

$$\frac{y^2}{k - 4a^2} + \frac{z^2}{9(k - 4a^2)} = 1.$$

Thus the cross section in the plane $x = a$ is an ellipse with semi-axes of length $\sqrt{k - 4a^2}$ and $\sqrt{9(k - 4a^2)}$. Note that the semi-axes are longest when $x = a = 0$ and decrease as a^2 grows. Several of these cross sections are shown in Fig. 9.61.

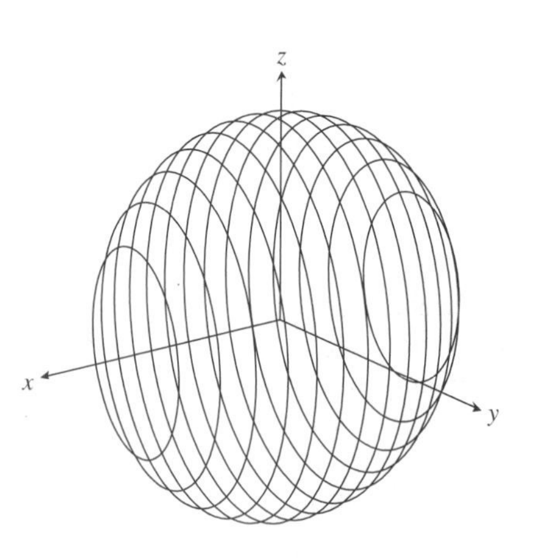

FIGURE 9.61 Some cross sections of the level surface $4x^2 + y^2 + (z^2/9) = k$.

Next set $y = b$ in (9). After some algebra, we have

$$\frac{x^2}{(k - b^2)/4} + \frac{z^2}{9(k - b^2)} = 1.$$

This shows that for $b^2 < k$, the cross section in the plane $y = b$ is an ellipse with semi-axes of length $\sqrt{(k - b^2)/4}$ and $\sqrt{9(k - b^2)}$.

The cross section in the plane $z = c$ is found by setting $z = c$ in (9). This leads to

$$\frac{x^2}{(9k - c^2)/36} + \frac{y^2}{(9k - c^2)/9} = 1.$$

Hence, for $c^2 < 9k$, the cross section in the plane $z = c$ is also an ellipse.

For $k \geq 0$ the level surface S_k is a surface whose cross sections perpendicular to each of the coordinate axes are ellipses. Such a surface is called an *ellipsoid.* When $k = 9$, this ellipsoid is the graph of

$$4x^2 + y^2 + \frac{z^2}{9} = 9.$$

This equation can be rearranged to get

$$\frac{x^2}{(3/2)^2} + \frac{y^2}{3^2} + \frac{z^2}{9^2} = 1.$$

This ellipsoid has vertices at $\pm\frac{3}{2}$ on the x-axis, ± 3 on the y-axis, and ± 9 on the z-axis. The graph of this ellipsoid is shown in Fig. 9.62. Note that the complete ellipsoid cannot be described by a single equation of the form $z = g(x, y)$ but is well described as a level surface of a function of three variables. The graphs in Figs. 9.61 and 9.62 both represent ellipsoids of the form $4x^2 + y^2 + (z^2/9) = k$. The ellipsoids pictured are similar, but they look different because of different scales on the axes.

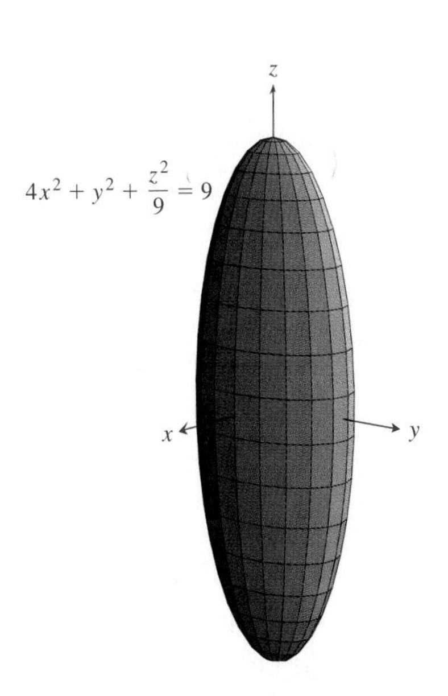

FIGURE 9.62 The level surface $4x^2 + y^2 + (z^2/9) = 9$ is an ellipsoid.

Quadric Surfaces

A surface described by an equation of the form

$$Ax^2 + By^2 + Cz^2 + Dxy + Exz + Fyz + Gx + Hy + Iz + J = 0,$$

with not all of A, B, C equal to 0, is called a *quadric surface.* We have already encountered some quadric surfaces in this section. The paraboloid studied in the Investigation and the ellipsoid seen in Example 5 are both examples of quadric surfaces. Because the equations for these surfaces are quadratic or linear in each variable, cross sections of these surfaces are conic sections. These surfaces often have names that suggest one or more of the cross sections.

EXAMPLE 6 Describe and sketch the graph of the equation

$$\frac{z^2}{4} - x^2 - \frac{y^2}{9} = 1. \tag{11}$$

Solution

The surface described by (11) is called a *hyperboloid of two sheets.* By setting $x = a$ or $y = b$, we see that cross sections perpendicular to the x- or y-axes are hyperbolas. If a point (x, y, z) satisfies (11), then $|z| \geq 2$. For $|c| \geq 2$, the cross section in the plane $z = c$ is an ellipse. The (two-piece) graph of (11) is shown in Fig. 9.63.

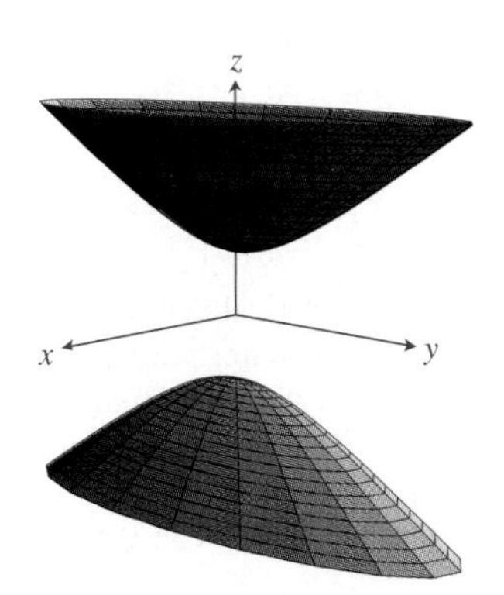

FIGURE 9.63 The hyperboloid of two sheets $\frac{z^2}{4} - x^2 - \frac{y^2}{9} = 1.$

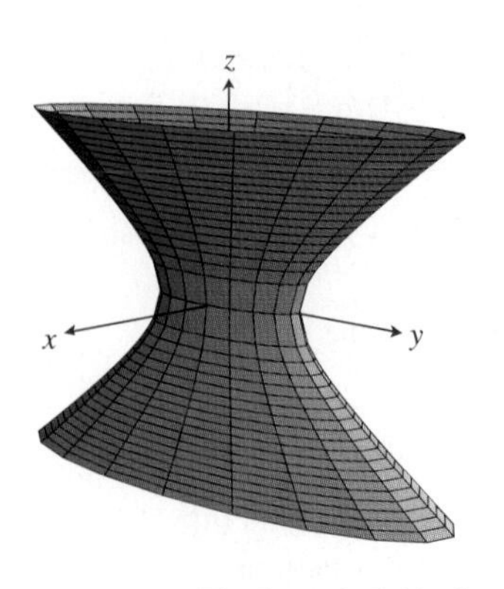

FIGURE 9.64 The hyperboloid of one sheet $x^2 + \frac{y^2}{9} - \frac{z^2}{4} = 1$.

EXAMPLE 7 Describe and sketch the graph of the equation

$$x^2 + \frac{y^2}{9} - \frac{z^2}{4} = 1. \tag{12}$$

Solution

The surface described by (12) is called a *hyperboloid of one sheet.* By setting $x = a$ or $y = b$, we see that cross sections perpendicular to the x- or y-axes are again hyperbolas. For any c, the cross section in the plane $z = c$ is an ellipse. The (one-piece) graph of (12) is shown in Fig. 9.64.

Exercises 9.3

Exercises 1–8: For the surface graph of each function, describe the cross section with respect to each coordinate plane. Sketch some of the cross sections.

1. $z = 2x^2 + y^2$

2. $z = 2x - y^2$

3. $z = 3x - 2y$

4. $z = (x + y)^2$

5. $z = 3x^2 - y^2$

6. $z = \sin(x + 2y)$

7. $z = e^{xy}$

8. $z = \dfrac{1}{(x^2 + y^2)}$

Exercises 9–14: Describe the level curves for each function. Sketch a level curves graph.

9. $f(x, y) = (3x - 3y + 5)^2$

10. $f(x, y) = x^2 + 4y^2$

11. $f(x, y) = \dfrac{2}{\sqrt{x - 2y^2}}$

12. $f(x, y) = xye^{xy}$

13. $f(x, y) = y^2 - x^2$

14. $f(x, y) = \dfrac{1}{x^2 + y^2 + 2}$

Exercises 15–20: Describe and sketch the surface graph for each function.

15. $z = 3x - y + 2$

16. $z = \sqrt{x^2 + y^2}$

17. $z = \sqrt{x} + \sqrt{y}$

18. $z = 2x^2 + 5y^2$

19. $z = \dfrac{1}{1 + x^2 + y^2}$

20. $z = (x + y)^2$

21. When an object of mass m moves at low speed (that is, nonrelativistic speed), the kinetic energy E of the object is given by

$$E = E(m, v) = \frac{1}{2}mv^2.$$

Describe the level curves for the energy function and sketch at least three of the level curves.

22. To study drainage properties of soil, geotechnical engineers use a device called a piezometer to measure water pressure (head) in the soil. In a region to be studied, the pressure at each end is measured and the difference h in these pressures is calculated. If the region has length L, then the flow rate Q of water through the soil (measured in, say, m^3/second) is well approximated by

$$Q = k\frac{h}{L},$$

where k, the permeability of the soil, is a constant. What is the domain for Q? Describe the level curves for the flow rate function and sketch at least three of these level curves.

23. The lens formula relates the position of an object and its image to the focal length of the lens. If an object is at distance s_1 from a lens and the image of the object appears at distance s_2 from the lens, then the focal length f of the lens is given by

$$f = f(s_1, s_2) = \frac{s_1 + s_2}{s_1 s_2}.$$

Describe the level curves for f and sketch at least three of the level curves.

24. The accompanying map of the North Atlantic shows isotherms (lines of constant temperature) in the sea during a summer of the last ice age (18,000 years ago). Carefully describe the function represented by this map. What is the domain of the function? For a given domain element, what determines the value of the function?

25. The accompanying U.S. map shows the annual average for wind power available at various locations. Carefully describe the function represented by this map. What is the domain of the function? For a given domain element, what determines the value of the function?

Exercises 26–31: Describe the level surfaces for each function. Sketch a typical level surface.

26. $f(x, y, z) = -4x - y + 3z + 1$

27. $f(x, y, z) = 4x^2 + 2y^2 + z^2 + 4$

28. $f(x, y, z) = x^2 + z^2$

29. $w = x^2 - y + z^2 + 3$

30. $w = \sqrt{x^2 + y^2 + 4z^2}$

31. $f(x, y, z) = x + y^2 - z^2$

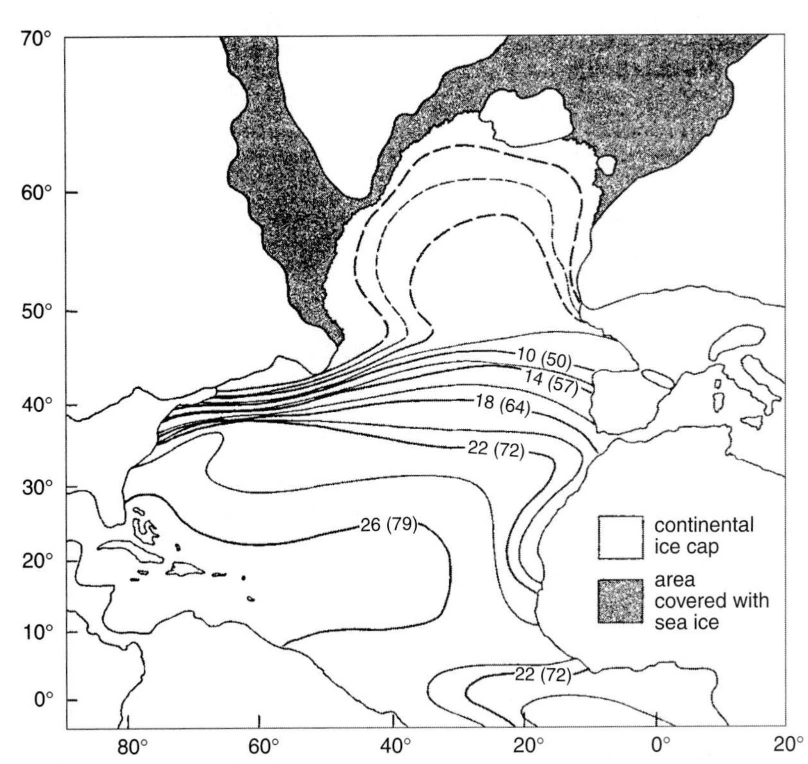

August temperature of sea surface waters [isotherms in C (F)] in the North Atlantic during the height of the last glacial period 18,000 years before present. Broken lines represent extrapolated isotherms. (*After A. McIntyre and N. G. Kipp, Geol. Soc. Amer. Mem. 145, P. 61, 1976*)

Figure for Exercise 24.

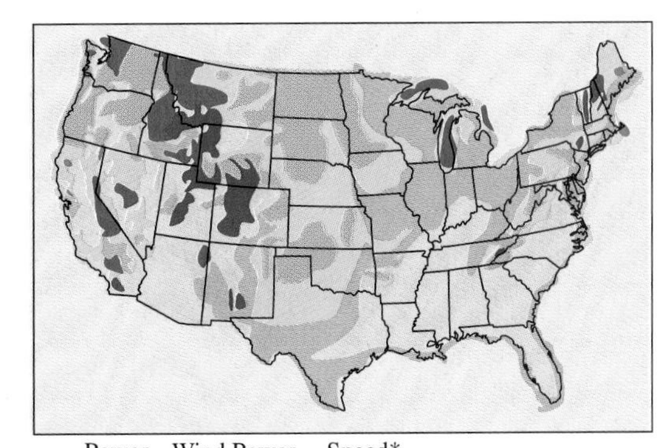

Power Class	Wind Power (W/m²)	Speed* (m/s)
1	<200	<5.6
2	200–300	5.6–6.4
3	300–400	6.4–7.0
4	400–500	7.0–7.5
5	500–600	7.5–8.0
6	600–800	8.0–8.9
7	>800	>8.9

*Equivalent wind speed at sea level for a Rayleigh distribution

Figure for Exercise 25.

Exercises 32–37: Six functions are described. Also shown are the level curves graph and the surface graph for each function. Match each function with its level curves graph and with its surface graph.

32. $f(x, y) = x^2y^2 + 1$

33. $f(x, y) = \sin^2 x + \cos^2 y$

34. $f(x, y) = (2x - 3y)^2$

35. $f(x, y) = \dfrac{1}{(x - 1)^2(y + 2)^2}$

36. $f(x, y) = x^3 + 2xy^2 - 3y + x$

37. $f(x, y) = e^{-x}$

A

B

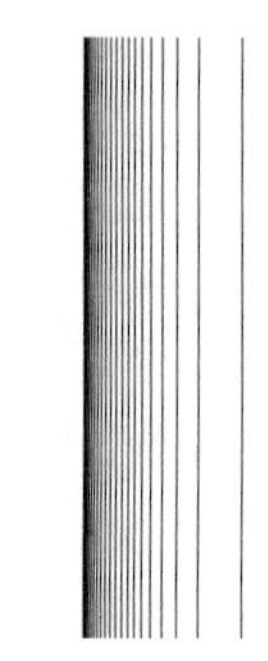

C

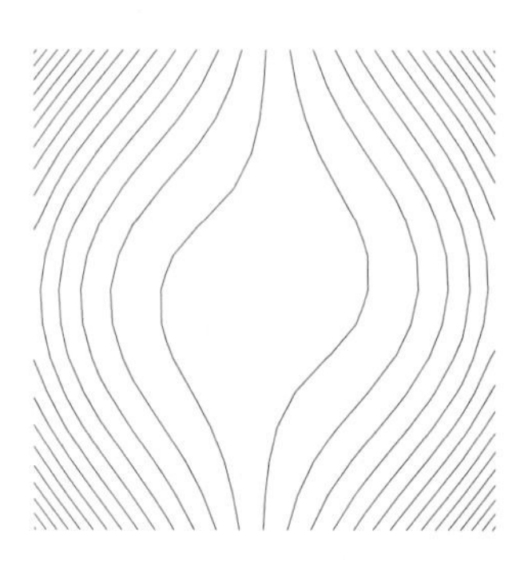

D

E

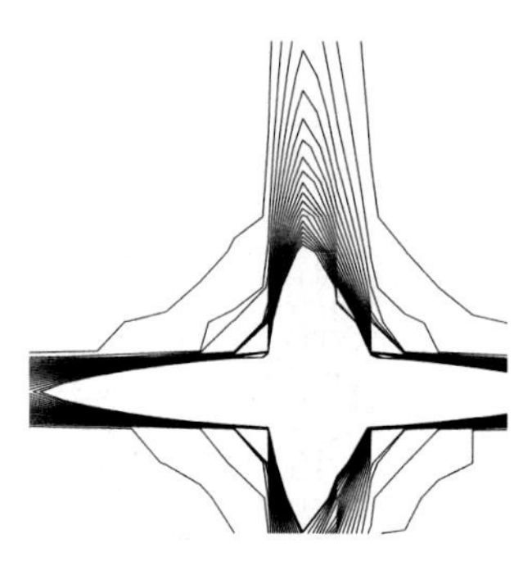

F

i

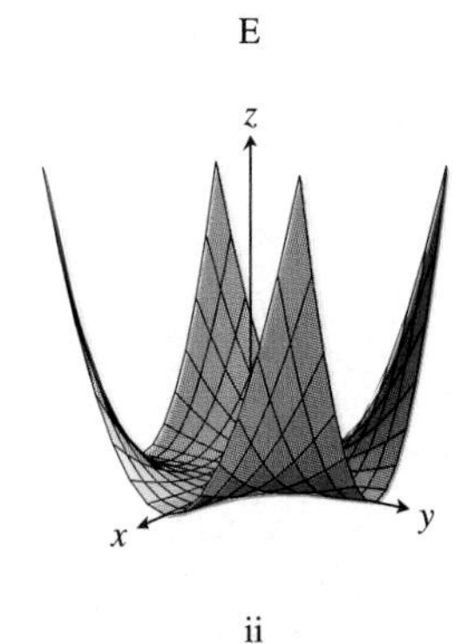

ii

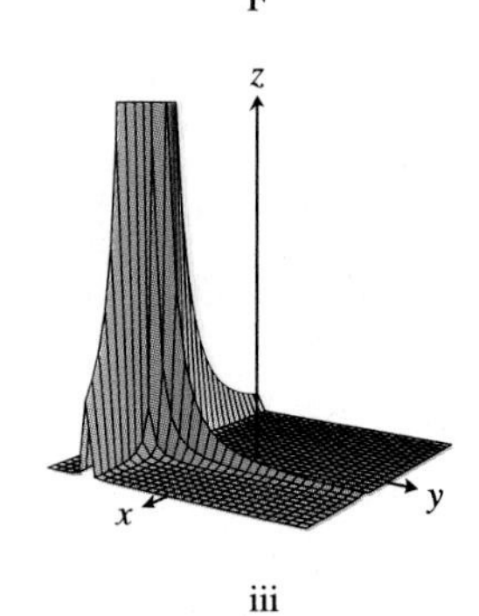

iii

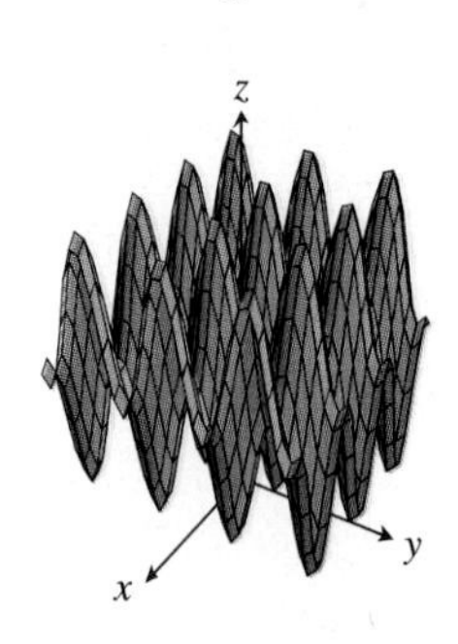

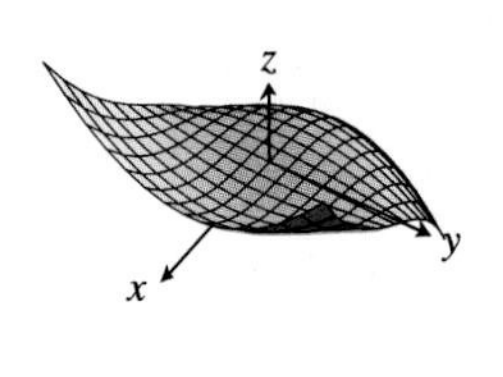

9.4 Graphing: Parametric Representations of Surfaces

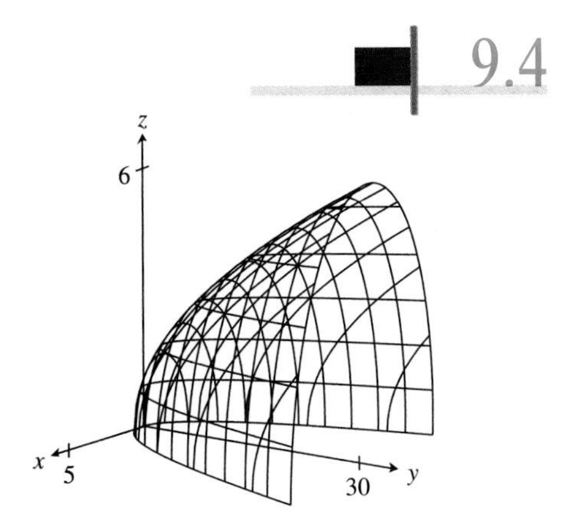

FIGURE 9.65 With cross sections perpendicular to each of the three axes, we see that the surface looks somewhat like an airplane hangar.

In Example 1 of Section 9.3 we produced two graphical representations of the surface described by $z = \sqrt{y - x^2}$. One was a wire frame representation that arose from our investigation of cross sections. The other was the picture produced by a CAS. These figures are reproduced in Figs. 9.65 and 9.66. Both of these figures have some nice features. Figure 9.65 shows the smooth boundary where the surface meets the (x, y)-plane. Figure. 9.66 is nicely shaded and is easy to produce, because it was generated using a command built into the CAS. On the other hand, both figures have some negative aspects. Figure. 9.65 was produced by combining the graphs of several cross sections. This took a lot of work. The graph in Fig. 9.66 has a very irregular edge that badly represents the intersection of the surface with the (x, y)-plane.

In this section we study parametric representations and see how they can be used to produce surface graphs that combine the best features of Figs. 9.65 and 9.66. In later chapters we will see how parametric representations are useful in changing variables in multiple integrals and in evaluating integrals of functions defined on surfaces.

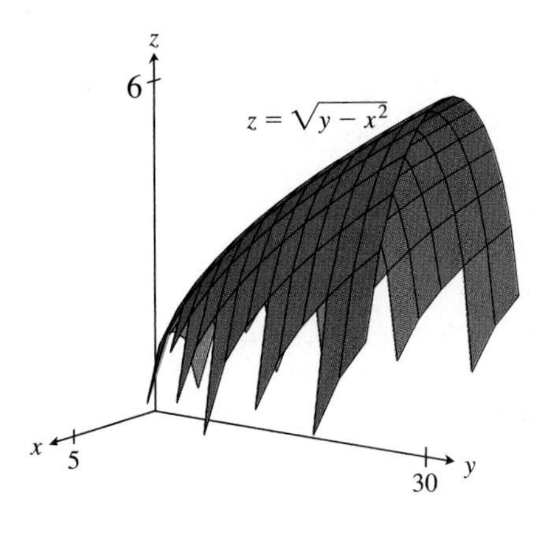

FIGURE 9.66 A graph of $z = \sqrt{y - x^2}$ produced by a computer algebra system.

Representation of Surfaces

We have seen that a curve C in space can be represented parametrically by a vector-valued function of the form

$$\mathbf{r}(t) = \langle x(t), y(t), z(t) \rangle, \qquad a \leq t \leq b.$$

As the variable t ranges from a to b, the points $(x(t), y(t), z(t))$ trace the curve C. Although the curve C "lives" in three-dimensional space, it is a one-dimensional object. This is reflected in the fact that the curve is described by using just one input variable t. A surface in three-space is a two-dimensional object. Thus it is not surprising that we need two input variables to describe a surface parametrically.

DEFINITION Parametric Representation of a Surface

Let

$$\mathbf{r}(u, v) = \langle x(u, v), y(u, v), z(u, v) \rangle \tag{1}$$

be a continuous vector-valued function defined for all points (u, v) in a region $\mathcal{D}$ in the (u, v)-plane. Let S be the set of points (x, y, z) in R^3 for which

$$x = x(u, v), \qquad y = y(u, v), \qquad z = z(u, v),$$

for some ordered pair (u, v) in $\mathcal{D}$. The set S is a surface in R^3, and (1) is called a **parametric representation** of S. See Fig. 9.67.

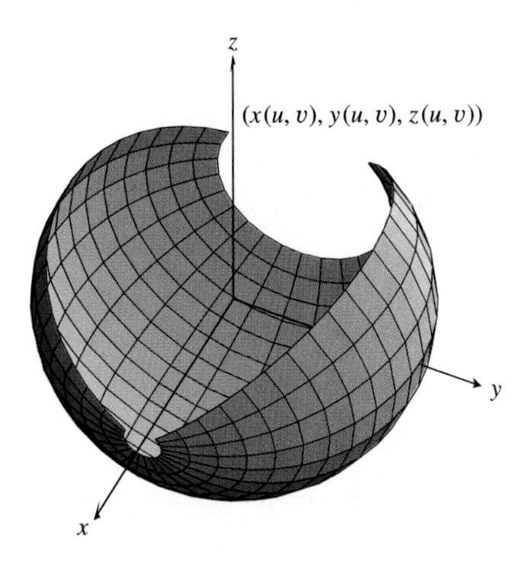

FIGURE 9.67 A parametric surface is a set of points $(x(u, v), y(u, v), z(u, v))$.

Two questions about parametric representations arise immediately:

- Given a parametric representation of a surface S, what can we say about the surface and how can we graph it?
- Given a surface S, how can we find a "good" parametric representation for S? (By "good," we might mean a representation that leads to a good graph of the surface.)

Analyzing Parametric Representations

We use graphical and analytic means to study surfaces that are described parametrically. We can produce graphs of such surfaces directly if we have a calculator or CAS with a parametric plotting command. Although it is true that all surfaces can be represented parametrically, and some can only be represented parametrically, it is also true that some surfaces are best described by a scalar equation $F(x, y, z) = 0$.

EXAMPLE 1 A surface S is described parametrically by

$$\mathbf{r}(u, v) = \langle u - 2v, -u - v, 4u + 5v \rangle, \qquad -\infty < u, v < \infty. \tag{2}$$

Find a scalar equation that represents S. Sketch the graph of S.

Solution

The (x, y, z)-coordinates of a point on S have the form

$$\begin{aligned} x &= x(u, v) = u - 2v \\ y &= y(u, v) = -u - v \\ z &= z(u, v) = 4u + 5v, \end{aligned} \tag{3}$$

for some pair of real numbers u and v. We eliminate the parameters u and v. Solving the first two equations for u and v, we obtain

$$u = \frac{x - 2y}{3} \quad \text{and} \quad v = -\frac{x + y}{3}.$$

Substituting these expressions into the third equation in (3) leads to

$$z = 4u + 5v = 4\left(\frac{x - 2y}{3}\right) + 5\left(\frac{-x - y}{3}\right) = -\frac{1}{3}x - \frac{13}{3}y.$$

Rearranging this expression, we obtain

$$x + 13y + 3z = 0.$$

This shows that S is the plane that passes through the origin and has normal vector $\mathbf{N} = (1, 13, 3)$. The graph of S, produced by using a parametric plot routine to graph (2), is shown in Fig. 9.68.

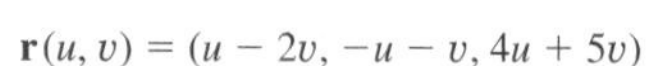

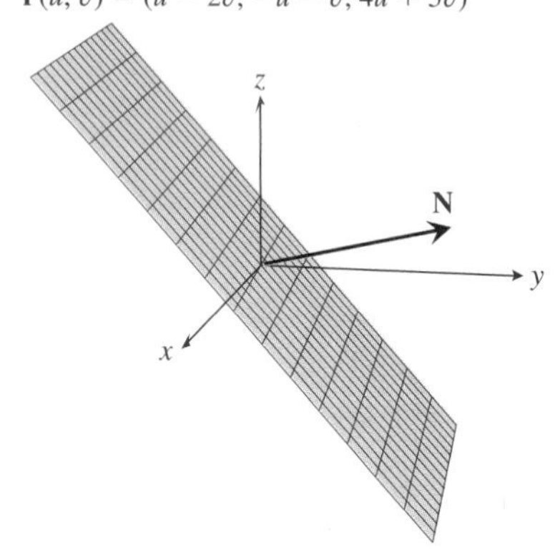

FIGURE 9.68 The parametric surface represented by $\mathbf{r}(u, v) = \langle u - 2v, -u - v, 4u + 5v \rangle$ is a plane.

EXAMPLE 2 Describe and graph the surface S described by

$$\mathbf{F}(r, \theta) = \langle r \cos \theta, r \sin \theta, r^2 \rangle, \qquad 0 \le r \le 5, \qquad 0 \le \theta \le 2\pi. \tag{4}$$

Solution

A point (x, y, z) on the surface S has the form

$$x = x(r, \theta) = r \cos \theta, \qquad y = y(r, \theta) = r \sin \theta, \qquad z = r^2,$$

for some (r, θ) with $0 \le r \le 5$ and $0 \le \theta \le 2\pi$. To get a scalar expression for the surface, note that

$$x^2 + y^2 = (r \cos \theta)^2 + (r \sin \theta)^2 = r^2 = z.$$

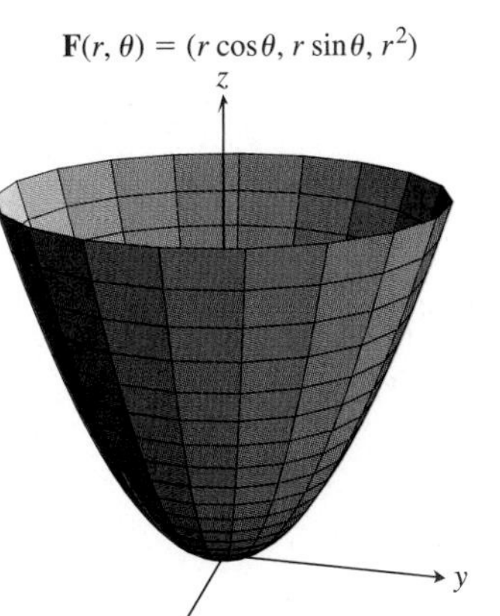

FIGURE 9.69 The parametric surface $\mathbf{F}(r, \theta) = \langle r\cos\theta, r\sin\theta, r^2\rangle$ is a paraboloid.

Hence S is the graph of

$$z = x^2 + y^2. \tag{5}$$

Because there are restrictions on r and θ, there are some restrictions on the x and y inputs in (5). For a fixed r, the ordered pair

$$(x, y) = (r\cos\theta, r\sin\theta), \qquad 0 \le \theta \le 2\pi, \tag{6}$$

traces the circle of radius r and center $(0, 0)$ in the (x, y)-plane. As we let r vary from 0 to 5, these circles fill the disk of center $(0, 0)$ and radius 5. Thus S is the portion of the graph of $z = x^2 + y^2$ that lies above the disk $x^2 + y^2 \le 5$. In Fig. 9.69 we show the graph of S produced by using a parametric plot command to graph (4).

In the Investigation in Section 9.3 we graphed $z = x^2 + y^2$ by analyzing cross sections. We saw that cross sections taken perpendicular to the z-axis are circles. Note that these circular cross sections and the bowl shape of the surface are more clearly shown in Fig. 9.69 than in Fig. 9.44 of the previous section. This is because the parametric representation for S given in (4) conforms to the level curves for the surface. In particular, the level curves are circles centered at the origin, and these circles are traced by (6) when r is fixed and θ varies from 0 to 2π. Finding a parametric representation that conforms to the level curves of a surface is often the key to producing good graphs of surfaces with a CAS.

How Some Plotting Packages Work

Graphing Cartesian Representations The graph in Fig. 9.66 was produced using the Mathematica command for plotting functions of the form $z = f(x, y)$. In particular, the graph resulted from plotting $z = \sqrt{y - x^2}$ for $-5 \le x \le 5$ and $0 \le y \le 30$. For convenience and speed, the all-purpose plot command used to produce Fig. 9.66 responds to all input functions f in a standard way. The result is usually a useful and informative graph, though some important features may be lost or hidden. The jagged edge in Fig. 9.66 resulted because the plot command did not effectively cope with the domain of $\sqrt{y - x^2}$.

The domain of $\sqrt{y - x^2}$ is the set $\mathcal{D} = \{(x, y) : y \ge x^2\}$. The set $\mathcal{D}$ is the region inside the parabola $y = x^2$ and so has a curved boundary. See Fig. 9.70. To graph

$$z = \sqrt{y - x^2} \quad \text{for} \quad -5 \le x \le 5 \quad \text{and} \quad 0 \le y \le 30,$$

Mathematica partitions the interval $-5 \le x \le 5$ into 14 equal intervals using 15 equally spaced points, and partitions the interval $0 \le y \le 30$ in a similar way. These partition points are used as the x- and y-coordinates of points in the plane to get a collection of $15^2 = 225$ points in the rectangle $-5 \le x \le 5$ and $0 \le y \le 30$. See Fig. 9.70. To plot the surface graph, Mathematica attempts to evaluate $z = \sqrt{y - x^2}$ at each of these points. For points outside the domain of the function, the computation results in a complex number. Such results are discarded. For each of the 225 points (x, y) for which $z = \sqrt{y - x^2}$ is real, Mathematica plots the point (x, y, z). Next, the points in each row or column are joined by line segments to produce the cross-section curves seen on the graph (see Figs. 9.71 and 9.72). The graph is then shaded and hidden lines are deleted to produce the final picture.

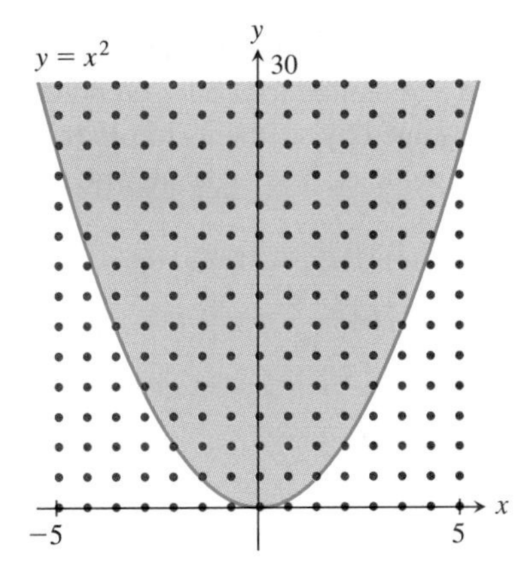

FIGURE 9.70 To plot $z = \sqrt{y - x^2}$ for $-5 \le x \le 5$ and $0 \le y \le 30$, Mathematica attempts to calculate z at 225 points.

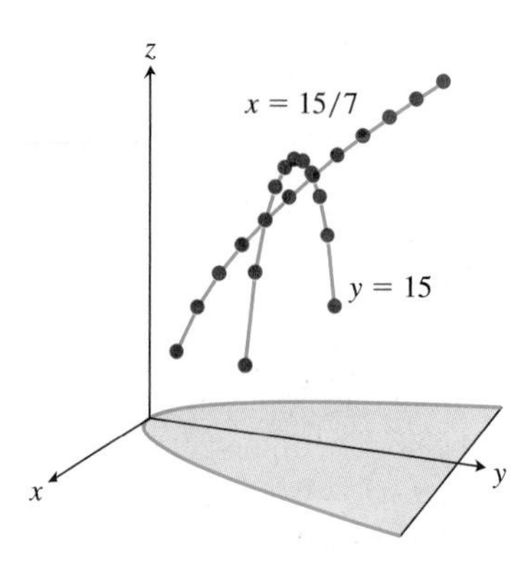

FIGURE 9.71 For a given row or column in the grid, the points (x, y, z) are plotted and connected by line segments.

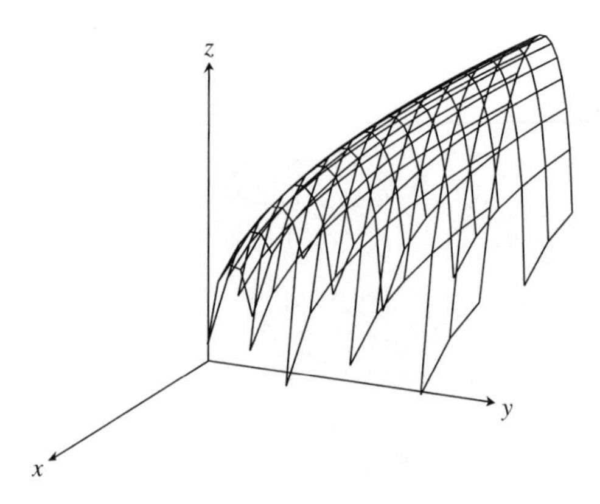

FIGURE 9.72 When the surface $z = \sqrt{y - x^2}$ is graphed with a $z = f(x, y)$ plotting command, we get a ragged edge.

As seen in Fig. 9.70, not every row or column of the grid has a point on the boundary of the domain. When the points (x, y, z) are plotted and joined for such a row or column of the grid, the resulting cross-section curve starts and ends above the z-plane. The low point on one cross section may be at a different height than the low point on another cross section. This leads to the ragged edge seen in Figs. 9.66 and 9.72.

Graphing Parametric Representations In Example 2 we produced the graph of $z = x^2 + y^2$ by using a parametric plotting command to plot

$$\mathbf{F}(r, \theta) = \langle r \cos \theta, r \sin \theta, r^2 \rangle$$

for $0 \leq r \leq 5$ and $0 \leq \theta \leq 2\pi$. The resulting picture, shown in Fig. 9.69, illustrates the bowl-like shape of the surface very well. Mathematica's parametric plotting command works very much like its Cartesian plotting command. In the (r, θ)-plane the intervals $0 \leq r \leq 5$ and $0 \leq \theta \leq 2\pi$ are both cut into 14 equal pieces. The 15 partition points from each interval are used to produce $15^2 = 225$ grid points in the (r, θ)-plane. See Fig. 9.73. Next, $\mathbf{F}(r, \theta)$ is calculated and plotted for each grid point, and adjacent points within each row and column are joined by segments to produce the mesh seen on the surface in Fig. 9.69. For a fixed value of r, say $r = 45/14$, the grid points

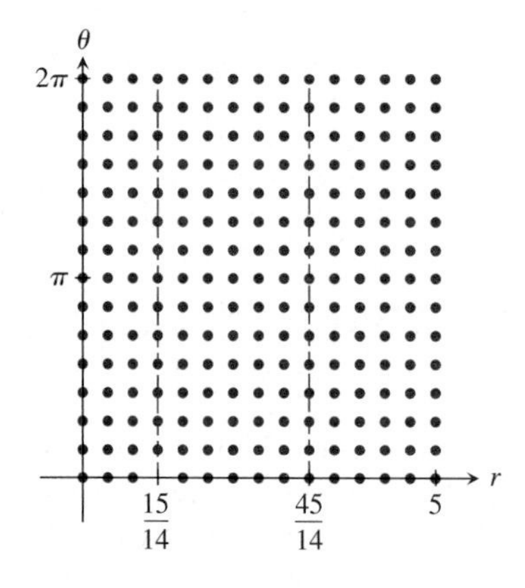

FIGURE 9.73 The grid of 225 points in the (r, θ)-plane.

$$\left(\frac{45}{14}, \theta\right), \qquad \theta = 0, 1 \cdot \frac{2\pi}{14}, 2 \cdot \frac{2\pi}{14}, \ldots, 14 \cdot \frac{2\pi}{14}$$

are mapped by $\mathbf{F}$ to the points

$$\left(\frac{45}{14} \cos \theta, \frac{45}{14} \sin \theta, \left(\frac{45}{14}\right)^2\right), \qquad \theta = 0, 1 \cdot \frac{2\pi}{14}, 2 \cdot \frac{2\pi}{14}, \ldots, 14 \cdot \frac{2\pi}{14}. \tag{7}$$

The first two coordinates of the points (7) are

$$(x, y) = \left(\frac{45}{14} \cos \theta, \frac{45}{14} \sin \theta\right).$$

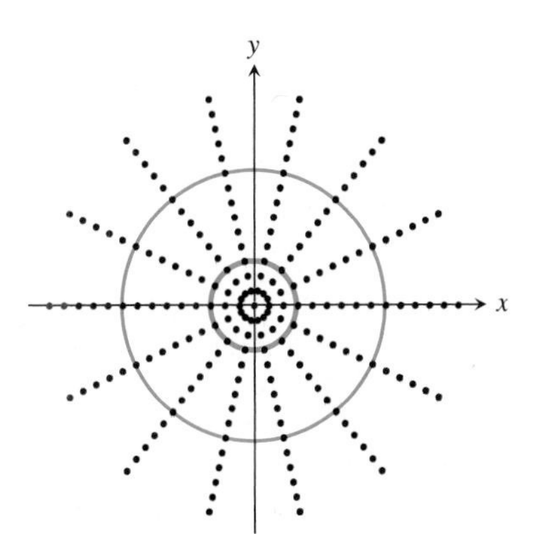

FIGURE 9.74 The lines $r = \frac{15}{14}$ and $r = \frac{45}{15}$ in the (r, θ)-plane correspond to circles of radius $\frac{15}{14}$ and $\frac{45}{15}$ in the (x, y)-plane. These circles are among the 15 level curves for $z = x^2 + y^2$.

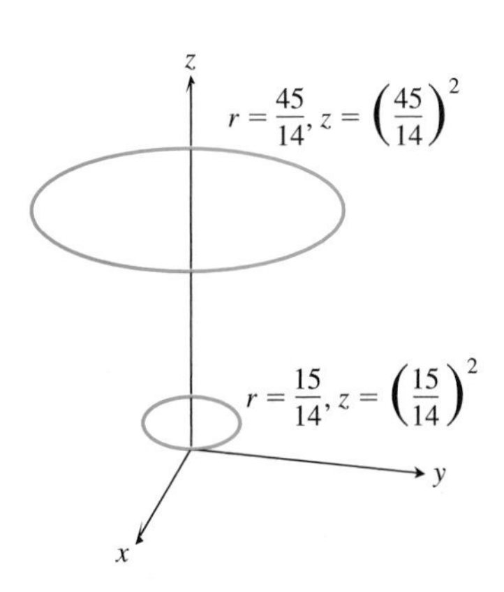

FIGURE 9.75 For each value of r the points $(\cos\theta, r\sin\theta, r^2)$ map to cross sections of the surface.

These points lie on the circle of radius 45/14 centered at the origin, that is, on one of the level curves of the surface $z = x^2 + y^2$. Hence when the points (7) are connected by segments, we see a cross section perpendicular to the z-axis. See Figs. 9.73, 9.74, and 9.75.

Finding Parametric Representations

The use of level curves $z = c$, as illustrated in the previous example, provides a means for producing a parametric representation that might be used to produce a good graph of a surface.

Level Curves and Parametric Representations

Let S be a surface described by $z = f(x, y)$. A parametric representation that might lead to a good graph for S has the form

$$\mathbf{F}(u, v) = \langle x(u, v), y(u, v), f(x(u, v), y(u, v))\rangle,$$
$$a \le u \le b, \qquad c \le v \le d,$$

where for each fixed $v = v_0$ the planar curve

$$(x, y) = (x(u, v_0), y(u, v_0)), \qquad a \le u \le b,$$

traces a level curve of S.

EXAMPLE 3 Let S be the surface described by $z = \sqrt{y - x^2}$. Use the level curves idea to produce a parametric representation of S.

Solution

Because $z = \sqrt{y - x^2} \ge 0$, the level curves for S are characterized by $z = c \ge 0$. These curves are described by

$$\sqrt{y - x^2} = c.$$

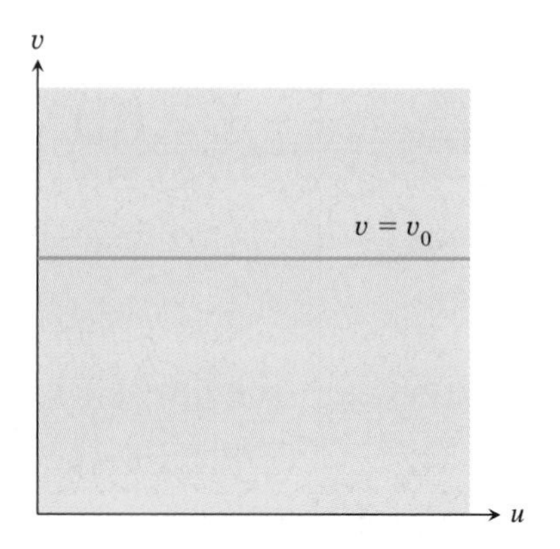

FIGURE 9.76 When $v = v_0$ is fixed, the points (u, v) lie on a horizontal line in the (u, v)-plane.

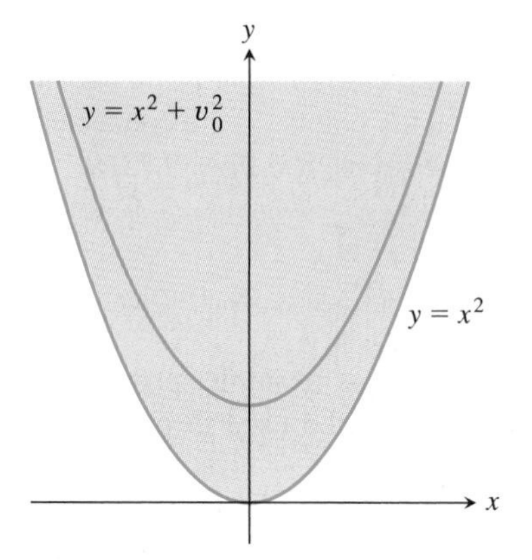

FIGURE 9.77 When $v = v_0$ is fixed and u varies, the points $(x, y) = (u, u^2 + v_0^2)$ trace the parabola $y = x^2 + v_0^2$ in the (x, y)-plane.

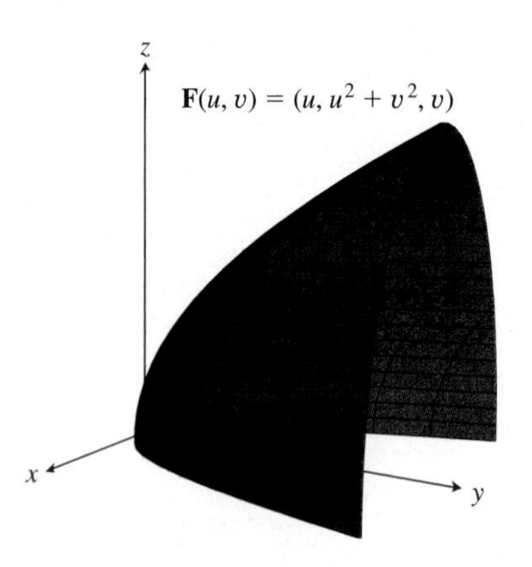

FIGURE 9.78 The graph of $\mathbf{F}(u, v) = \langle u, u^2 + v^2, v\rangle$ produced by a parametric plotting routine.

Solving this for y, we obtain

$$y = x^2 + c^2. \tag{8}$$

Thus the level curves are parabolas. For a fixed value of c, the level curve (8) is described by

$$\langle x, y\rangle = \langle x, x^2 + c^2\rangle, \qquad -\infty < x < \infty. \tag{9}$$

To describe a different level curve, we need only change the value of c. Thus the level curves for S are completely described by (9): We vary x to move along a level curve, and change c to move to a different curve. If we use (9) as the first two coordinates in our parametric representation of S, we will have a representation that conforms to the level curves of S. Thus on the surface a point (x, y, z) with $z = c$ has x- and y-coordinates of the form

$$x = x, \qquad y = x^2 + c^2$$

for some $-\infty < x < \infty$. We use these three coordinate expressions as the coordinate functions for a parametric representation of S:

$$\mathbf{F}(x, c) = \langle x, x^2 + c^2, c\rangle, \qquad -\infty < x < \infty, \qquad c \geq 0. \tag{10}$$

The names of the input variables in (10) do not matter. If we replace x by u and c by v we have

$$\mathbf{F}(u, v) = \langle u, u^2 + v^2, v\rangle, \qquad -\infty < u < \infty, \qquad v \geq 0.$$

This representation is a little better than the one in (10) because it better distinguishes the input variables u, v, from the coordinates x, y, z of a point on S.

For a fixed $v = v_0$ the point $(u, v) = (u, v_0)$ describes a horizontal line in the (u, v)-plane as u varies (see Fig. 9.76). The first two coordinates of $\mathbf{F}(u, v_0)$, that is, $x = u$, $y = u^2 + v_0^2$, then trace a level curve for $z = \sqrt{y - x^2}$ in the (x, y)-plane. (See Fig. 9.77.) In Fig. 9.78 we show the graph of S produced by using this parameterization and a parametric plotting routine. Notice that when $v = 0$, the x and y values trace the parabola $y = x^2$, which is the boundary of the domain of $\sqrt{y - x^2}$. On this boundary $z = 0$. This results in the smooth representation of the bottom edge of the surface shown in Fig. 9.78.

Although we often describe surfaces by equations of the form $z = f(x, y)$, it may sometimes be useful to consider surfaces described by equations such as $y = g(x, z)$ or $x = h(y, z)$.

EXAMPLE 4 Let S be the surface represented by

$$3x^2 - y + z^2 = 4. \tag{11}$$

Find a parametric representation for S, and use it to produce a graph of S.

Solution

Solving (11) for y, we have

$$y = g(x, z) = 3x^2 + z^2 - 4. \tag{12}$$

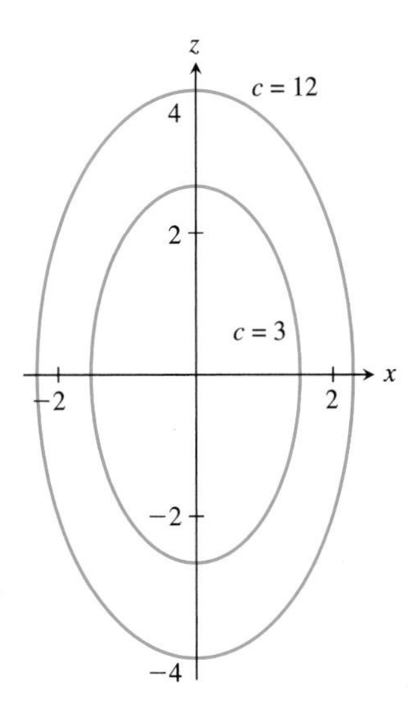

FIGURE 9.79 The level curves $y = c = 3$ and $y = c = 12$ for $y = 3x^2 + z^2 - 4$.

In this equation we think of y as a function of x and z. We investigate the curves in the (x, z)-plane on which y is constant. From (12) we see that $y \geq -4$. For a given $c \geq -4$, the level curve for $y = c$ is described by

$$3x^2 + z^2 - 4 = c.$$

Rewriting this expression as

$$\frac{x^2}{(c+4)/3} + \frac{z^2}{c+4} = 1,$$

we see that the level curve is an ellipse in the (x, z)-plane. See Fig. 9.79. This ellipse has semi-axes of length $\sqrt{(c+4)/3}$ and $\sqrt{c+4}$, so it can be represented parametrically by

$$\langle x, z\rangle = \langle x(t), z(t)\rangle = \left\langle \sqrt{(c+4)/3}\cos t, \sqrt{c+4}\sin t\right\rangle, \qquad 0 \leq t \leq 2\pi. \tag{13}$$

For each fixed $c \geq -4$, Equation (13) describes a level curve for g. When the value of c changes, a different level curve is described. We use (13) to describe the x- and z-coordinates in our parametric representation,

$$x = x(t, c) = \sqrt{\frac{c+4}{3}}\cos t \quad \text{and} \quad z = z(t, c) = \sqrt{c+4}\sin t,$$

where $c \geq -4$ and $0 \leq t \leq 2\pi$. We use these expressions with $y = y(t, c) = c$ to get a parametric representation for the surface,

$$\mathbf{G}(t, c) = \left\langle \sqrt{\frac{(c+4)}{3}}\cos t, c, \sqrt{c+4}\sin t \right\rangle, \qquad c \geq -4, \qquad 0 \leq t \leq 2\pi.$$

When we use this representation as input for a parametric plotting routine, we get the graph shown in Fig. 9.80. The surface is an elliptical paraboloid (or bowl) opening along the positive y-axis.

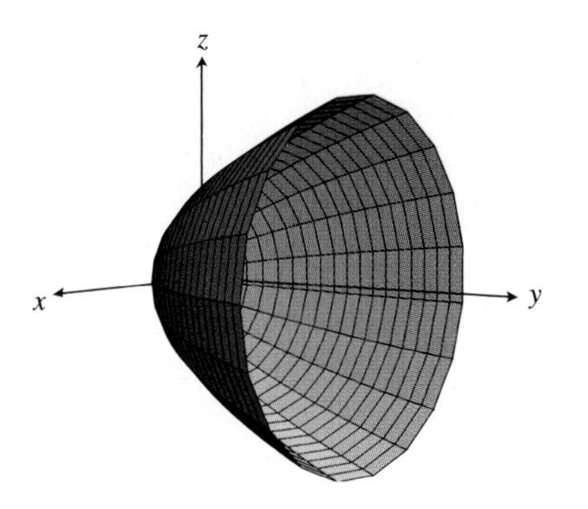

FIGURE 9.80 The graph of $3x^2 - y + z^2 = 4$ is an elliptical paraboloid opening in the positive y-direction.

When we do not have an equation that describes a surface, we may be able to come up with a parametric representation and a graph by working from a written description.

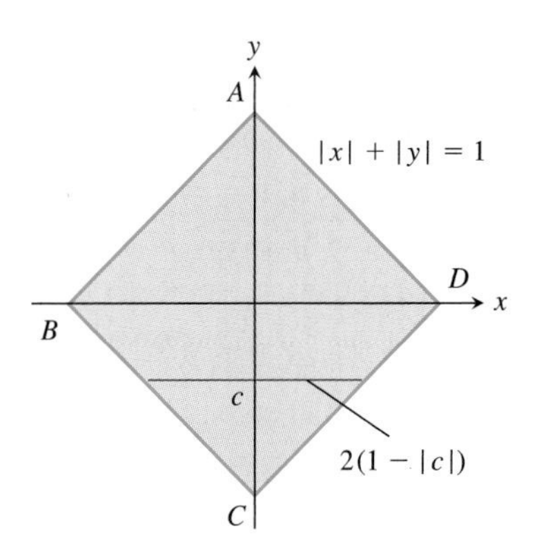

FIGURE 9.81 At position $y = c$ the base has width $2(1 - |c|)$.

EXAMPLE 5 A solid has square base $ABCD$. Any cross section perpendicular to diagonal AC is a semicircle with diameter in the plane of the square. Find a parametric representation for the surface of the solid and produce a graph.

Solution

Let the square be in the (x, y)-plane with vertices $A = (0, 1)$, $B = (-1, 0)$, $C = (0, -1)$, and $D = (1, 0)$. See Fig. 9.81. Note that the edge of the square is the graph of

$$|x| + |y| = 1. \tag{14}$$

Let $-1 \le c \le 1$. By (14), the width of the base at $y = c$ is

$$2|x| = 2\left(1 - |y|\right) = 2\left(1 - |c|\right).$$

Hence the cross section of the solid in the plane $y = c$ is a semicircle with radius $\left(1 - |c|\right)$. See Fig. 9.82.

The x- and z-coordinates of points on this semicircle can be described by

$$x = x(\theta, c) = \left(1 - |c|\right)\cos\theta \quad \text{and} \quad z = z(\theta, c) = \left(1 - |c|\right)\sin\theta,$$

where $0 \le \theta \le \pi$. Combining these equations with the equation $y = y(\theta, c) = c$, we have a parametric representation of the surface of the solid,

$$\mathbf{r}(\theta, c) = \left\langle \left(1 - |c|\right)\cos\theta, c, \left(1 - |c|\right)\sin\theta \right\rangle, \; 0 \le \theta \le \pi, \; -1 \le c \le 1.$$

When this representation is used in a parametric plotting routine, we get the picture of the solid shown in Fig. 9.83.

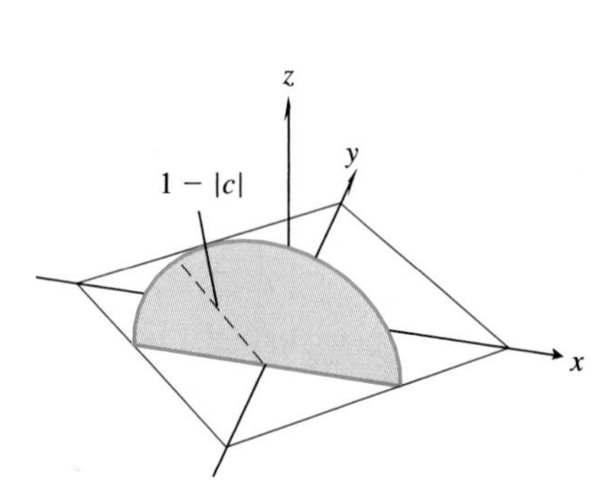

FIGURE 9.82 At position $y = c$, the semicircular cross section has radius $\left(1 - |c|\right)$.

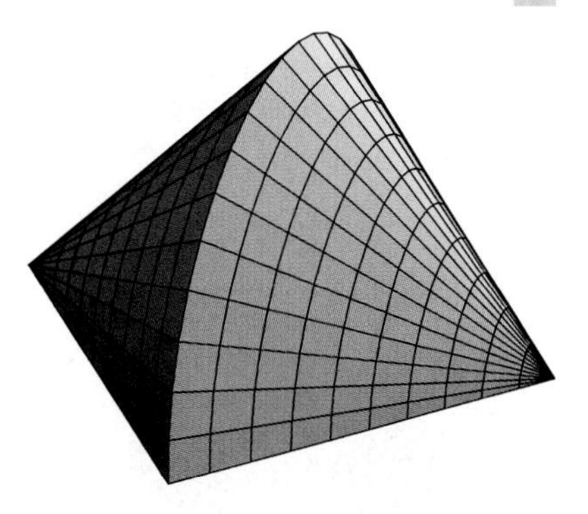

FIGURE 9.83 The solid with square base and semicircular cross sections.

Exercises 9.4

Exercises 1–6: A parametric representation of a surface S and points A and B are given. Decide whether each point lies on S.

1. $\mathbf{r}(u, v) = \langle -2u + 3v, u + 4v, u - v \rangle$, $A = (2, 21, 1)$, $B = (8, 7, 6)$

2. $\mathbf{F}(s, t) = \langle 3s - t, -s - 4t, 5t \rangle$, $A = (-1, 22, 25)$, $B = (-1, 22, -25)$

3. $\mathbf{g}(r, \theta) = \langle r\cos\theta, r^2, r\sin\theta \rangle$, $A = \left(-3/2, 9, 3\sqrt{3}/2\right)$, $B = (-1, 1, 0)$

4. $\mathbf{r}(w, z) = \langle w^4, 3w\sin z, 5w\cos z \rangle$, $A = \left(16, 3\sqrt{3}, 5\sqrt{2}\right)$, $B = (4, -3, -5)$

5. $\mathbf{h}(\theta, \phi) = \langle \cos\theta\sin\phi, \sin\theta\sin\phi, \cos\phi \rangle$, $A = \left(\sqrt{3}/4, 3/4, -1/2\right)$, $B = (0, 0, 1)$

6. $\mathbf{r}(\theta, \phi) = \langle \cos\phi, 2\cos\theta\sin\phi, 3\sin\theta\sin\phi \rangle$, $A = \left(1/\sqrt{2}, 1/\sqrt{2}, 0\right)$, $B = (1/e, e, 3e/2)$

Exercises 7–16: Find a scalar representation (that is, an x, y, z equation) for each of the following parametric representations. Plot the given parametric representation with a parametric plotting routine, and plot your scalar function with a standard plotting package.

7. $\mathbf{r}(u, v) = \langle -3u + v, u, u + 2v \rangle$, $-\infty < u, v < \infty$

8. $\mathbf{F}(u, v) = \langle 2u - 7v + 1, 4u - 3v - 2, u + 3v \rangle$, $-\infty < u, v < \infty$

9. $\mathbf{g}(r, \theta) = \langle r\sin\theta, r\cos\theta, r^2 - 2 \rangle$, $r \ge 0, 0 \le \theta \le 2\pi$

10. $\mathbf{f}(r, \theta) = \langle 3r\cos\theta, 2r\sin\theta, r^2 \rangle$, $r \ge 0, 0 \le \theta \le 2\pi$

11. $\mathbf{F}(r, \theta) = \langle r^2 + 2, 4r\cos\theta, -r\sin\theta \rangle$, $r \ge 0, 0 \le \theta \le 2\pi$

12. $\mathbf{H}(r, \theta) = \langle r\tan\theta, r\sec\theta, r^2 \rangle$, $r \ge 0$, $-\pi/2 < \theta < \pi/2$

13. $\mathbf{f}(r, \theta) = \langle 2r\sec\theta, r^2 + 1, -4r\tan\theta \rangle$, $r \ge 0$, $-\pi/2 < \theta < \pi/2$

14. $\mathbf{g}(r, \theta) = \langle r \cos \theta + r \sin \theta, r^2 \cos 2\theta, r \cos \theta - r \sin \theta \rangle$, $r \geq 0, 0 \leq \theta \leq 2\pi$

15. $\mathbf{r}(u, v) = \langle \ln u, \ln v, uv \rangle$, $u, v > 1$

16. $\mathbf{r}(u, v) = \langle e^{u+v}, 2v, e^{u-v} \rangle$, $-\infty < u, v < \infty$

Exercises 17–26: A surface S is represented by a scalar equation. Find a parametric representation that conforms to the level curves of S. Use the parametric representation in a parametric plotting routine to produce a graph of the surface.

17. $z = 2x^2 + 2y^2$

18. $z = \frac{1}{4}x^2 + \frac{1}{8}y^2$

19. $2x - y^2 - 4z^2 = 2$

20. $z = \sqrt{y - 3x^2}$

21. $z = \sqrt{y + 3x^2}$

22. $x + y + z = 0$

23. $2x - 3y + 5z = 6$

24. $z = e^{xy}$, $x, y > 0$

25. $y = \ln(2x^2 + 4z^2)$

26. $x = \sqrt{4y^2 + 3z^2}$

Exercises 27–30: Just as every curve in space has many different parametric representations, every surface also has many different parametric representations. Find three different parametric representations of the surface represented by the scalar equation.

27. $3x - y + 2z = 5$

28. $z = 2x^2 + 3y^2$

29. $y = e^{x+2z}$

30. $x = \ln(4y + z^2)$

31. Let $z = f(x, y)$ be a Cartesian representation of a surface S. Show that

$$\mathbf{r}(u, v) = \langle u, v, f(u, v) \rangle$$

is a parametric representation of S. Does this parametric representation have any advantages over the Cartesian representation? Why or why not?

32. Let S be the surface represented by

$$\mathbf{r}(\theta, \phi) = \langle \cos \theta \sin \phi, \sin \theta \sin \phi, \cos \phi \rangle$$

for $0 \leq \theta \leq 2\pi$ and $0 \leq \phi \leq \pi$. Show that each point on S is also a point on the sphere of radius 1 centered at the origin.

33. Let S_1 be the surface represented by

$$\mathbf{F}(r, \theta) = (r \cos \theta, 2r \sin \theta, r^2),$$

let S_2 be the surface represented by

$$\mathbf{G}(r, \theta) = \langle r \cos \theta, r^2, 2r \sin \theta \rangle,$$

and let S_3 be the surface represented by

$$\mathbf{H}(r, \theta) = \langle 2r \sin \theta, r \cos \theta, r^2 \rangle.$$

How do the graphs of S_1, S_2, and S_3 compare?

34. Let S_1 be the surface represented by

$$\mathbf{F}(u, v) = \langle u - v, u + 2v, -u + 3v \rangle,$$

let S_2 be the surface represented by

$$\mathbf{G}(u, v) = \langle -u + 3v, u + 2v, u - v \rangle,$$

and let S_3 be the surface represented by

$$\mathbf{H}(u, v) = \langle u + 2v, -u + 3v, u - v \rangle.$$

How do the graphs of S_1, S_2, and S_3 compare?

35. A surface S has a parametric representation

$$\mathbf{r}(u, v) = \langle f(u, v), g(u, v), h(u, v) \rangle.$$

Consider the five functions obtained by rearranging the three coordinate functions:

$$\mathbf{r}_1(u, v) = \langle f(u, v), h(u, v), g(u, v) \rangle,$$
$$\mathbf{r}_2(u, v) = \langle g(u, v), f(u, v), h(u, v) \rangle,$$
$$\mathbf{r}_3(u, v) = \langle g(u, v), h(u, v), f(u, v) \rangle,$$
$$\mathbf{r}_4(u, v) = \langle h(u, v), f(u, v), g(u, v) \rangle,$$
$$\mathbf{r}_5(u, v) = \langle h(u, v), g(u, v), f(u, v) \rangle.$$

Describe carefully how the graph of the surface represented by each of these representations compares to the graph of S. See Exercises 33 and 34.

36. Let S be the surface represented by

$$\mathbf{r}(u, v) = \langle au + bv, cu + dv, fu + gv \rangle,$$

where a, b, c, d, f, and g are constants, and let

$$D = (cg - df)^2 + (ag - bf)^2 + (ad - bc)^2.$$

a. Show that S is a plane if and only if $D \neq 0$. In this case, what is the normal vector for the plane?

b. Show that if $D = 0$, then S is a line or a point.

37. Let S be the surface represented by

$$\mathbf{F}(r, \theta) = \langle r \cos \theta, r \sin \theta, r^2 \rangle,$$

let C_1 be the space curve represented by

$$\mathbf{g}_1(\theta) = \mathbf{F}(3, \theta) = \langle 3 \cos \theta, 3 \sin \theta, 9 \rangle,$$

and let C_2 be the curve represented by

$$\mathbf{g}_2(r) = \mathbf{F}\left(r, \frac{\pi}{4}\right) = \left\langle r \cos\left(\frac{\pi}{4}\right), r \sin\left(\frac{\pi}{4}\right), r^2 \right\rangle.$$

a. Show that C_1 and C_2 lie on the surface S and that both pass through $P = \mathbf{F}(3, \pi/4)$.

b. Graph S and sketch C_1 and C_2 on the graph.

c. Find the vector tangent to C_1 at the point where $\theta = \pi/4$.

d. Find the vector tangent to C_2 at the point where $r = 3$.

e. Find an equation for the plane through $P = \mathbf{F}(3, \pi/4)$ and parallel to the vectors found in parts **c** and **d**. This plane is called the *plane tangent to S at P*.

38. Let S be the surface represented by

$$\mathbf{F}(r, \theta) = \langle 2r \cos \theta, r^2, 3r \sin \theta \rangle,$$

let C_1 be the space curve represented by

$$\mathbf{g}_1(\theta) = \mathbf{F}(1, \theta) = \langle 2 \cos \theta, 1, 3 \sin \theta \rangle,$$

and let C_2 be the curve represented by

$$\mathbf{g}_2(r) = \mathbf{F}\left(r, \frac{2\pi}{3}\right) = \left\langle 2r \cos\left(\frac{2\pi}{3}\right), r^2, 3r \sin\left(\frac{2\pi}{3}\right)\right\rangle.$$

a. Show that C_1 and C_2 lie on the surface S and that both curves pass through $P = \mathbf{F}(1, 2\pi/3)$.

b. Graph S and sketch C_1 and C_2 on the graph.

c. Find the vector tangent to C_1 at the point where $\theta = 2\pi/3$.

d. Find the vector tangent to C_2 at the point where $r = 1$.

e. Find an equation for the plane containing $P = \mathbf{F}(1, 2\pi/3)$ and parallel to the vectors found in parts **c** and **d**. This plane is called the *plane tangent to S at P*.

39. Let S be the surface represented by

$$\mathbf{r}(u, v) = \langle 2u, u^2 + v^2, v + 1 \rangle,$$

let C_1 be the space curve represented by

$$\mathbf{g}_1(u) = \mathbf{r}(u, 4) = \langle 2u, u^2 + 16, 5 \rangle,$$

and let C_2 be the curve represented by

$$\mathbf{g}_2(v) = \mathbf{r}(-2, v) = \langle -4, 4 + v^2, v + 1 \rangle.$$

a. Show that C_1 and C_2 lie on the surface S and that both curves pass through $P = \mathbf{r}(-2, 4)$.

b. Graph S and sketch C_1 and C_2 on the graph.

c. Find the vector tangent to C_1 at the point where $u = -2$.

d. Find the vector tangent to C_2 at the point where $v = 4$.

e. Find an equation for the plane containing $P = \mathbf{r}(-2, 4)$ and parallel to the vectors found in parts **c** and **d**. This plane is called the *plane tangent to S at P*.

40. Let S be the surface represented by

$$\mathbf{r}(u, v) = \langle 2 \cos u \sin v, \sin u \sin v, 2 \cos v \rangle,$$

let C_1 be the space curve represented by

$$\mathbf{g}_1(u) = \mathbf{r}\left(u, \frac{\pi}{4}\right) = \left\langle \sqrt{2} \cos u, \left(\frac{\sqrt{2}}{2}\right) \sin u, \sqrt{2} \right\rangle,$$

and let C_2 be the curve represented by

$$\mathbf{g}_2(v) = \mathbf{r}\left(\frac{\pi}{3}, v\right) = \left\langle \sin v, \left(\frac{\sqrt{3}}{2}\right) \sin v, 2 \cos v \right\rangle.$$

a. Show that C_1 and C_2 lie on the surface S and that both curves pass through $P = \mathbf{r}(\pi/3, \pi/4)$.

b. Graph S and sketch C_1 and C_2 on the graph.

c. Find the vector tangent to C_1 at the point where $u = \pi/3$.

d. Find the vector tangent to C_2 at the point where $v = \pi/4$.

e. Find an equation for the plane containing $P = \mathbf{r}(\pi/3, \pi/4)$ and parallel to the vectors found in parts **c** and **d**. This plane is called the *plane tangent to S at P*.

41. The base of a solid is the square $ABCD$. Each cross section perpendicular to diagonal AC is an isosceles right triangle with the hypotenuse in the base. Find a parametric representation for the surface of the solid. Use your parametric representation to produce a picture of the solid.

42. The base of a solid is a 3-4-5 right triangle. Each cross section perpendicular to the side of length 4 is a half-disk with diameter in the plane of the triangle. Find a parametric representation for the surface of the solid. Use your parametric representation to produce a picture of the solid.

43. The base of a solid is equilateral triangle ABC. Let the altitude from A meet BC at point D. Each cross section perpendicular to AD is a half disk with diameter in the plane of the triangle. Find a parametric representation for the surface of the solid. Use your parametric representation to produce a picture of the solid.

44. The base of a solid is the region that lies inside the parabola $y = 9 - x^2$ and above the x-axis. Each cross section perpendicular to the y-axis is an isosceles right triangle with hypotenuse in the plane of the base. Find a parametric representation for the surface of the solid. Use your parametric representation to produce a picture of the solid.

9.5 Cylindrical and Spherical Coordinates

In the preceding section, we saw that we can use parametric representations to give a useful and sometimes simple description of a surface. We saw in Examples 2 and 5 of that section that polar representations are useful in describing the level curves of some surfaces. The use of polar coordinates to describe the points in one of the coordinate planes is so important and so often used that it deserves special study. In this section we study cylindrical coordinates, a system incorporating both rectangular and polar coordinates. We also study spherical coordinates, another very important coordinate system for three-space. We will see that spherical coordinates can also be used to help us find useful parametric representations for some surfaces.

Cylindrical Coordinates

Let (x, y, z) be the Cartesian coordinates of a point P in space. We can locate P by first locating the point $Q = (x, y, 0)$ in the (x, y)-plane and then moving vertically to the indicated z level. Sometimes it is convenient to describe the position of Q using polar coordinates, say $Q = (r, \theta)$. If we include the r- and θ-coordinates with the z-coordinate information, then we can specify the position of P with the ordered triple

$$(r, \theta, z), \tag{1}$$

where it is now understood that the first two coordinates are to be interpreted as polar coordinates in the (x, y)-plane. Expression (1) gives the **cylindrical coordinates** of P.

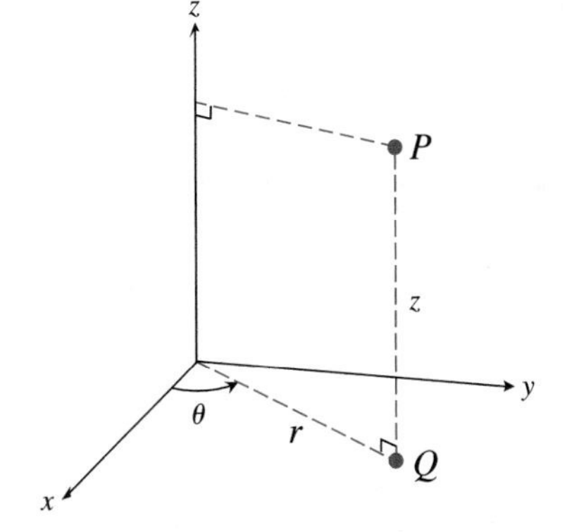

FIGURE 9.84 The point P has cylindrical coordinates (r, θ, z).

Cylindrical Coordinates

Let P be a point in three-space and let Q be the foot of the perpendicular from P to the (x, y)-plane. The cylindrical coordinates of P are

$$(r, \theta, z),$$

where

(r, θ) are the polar coordinates for Q, and

z is the Cartesian z-coordinate for the point P.

See Fig. 9.84.

Because the point Q has many different polar representations, the point P has many different cylindrical representations. Negative r values can also arise. Thus the cylindrical coordinate triples $(2, \pi/3, 5)$, $(2, \pi/3 + 2\pi, 5)$, and $(-2, \pi/3 - \pi, 5)$ all describe the same point in space.

WEB When working in any coordinate system, it is important to know what surfaces are described by holding one of the coordinates fixed and allowing the others to vary. For example, in the Cartesian system we know that the equations $x = a$, $y = b$, and $z = c$ each describe a plane perpendicular to one of the coordinate axes. We have used these planes to study cross sections of surfaces. Knowing the analogous surfaces for cylindrical coordinates can help us recognize when it might be useful to work in cylindrical coordinates.

EXAMPLE 1 Describe and graph the surfaces described by each of the following equations:

a) $r = 4$ **b)** $\theta = \dfrac{2\pi}{3}$ **c)** $z = -2$.

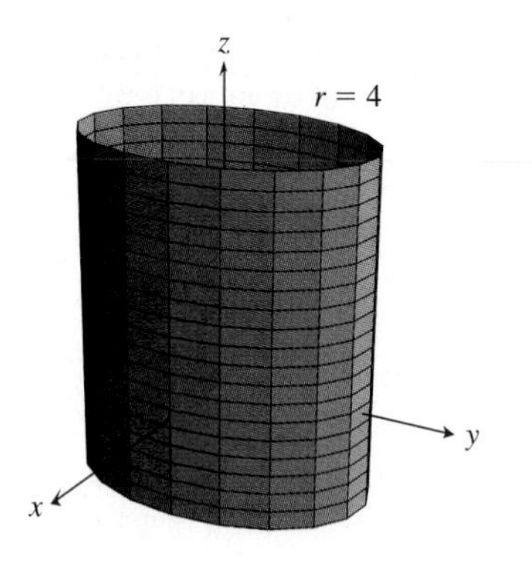

FIGURE 9.85 The graph of $r = c$, $c > 0$, is a cylinder.

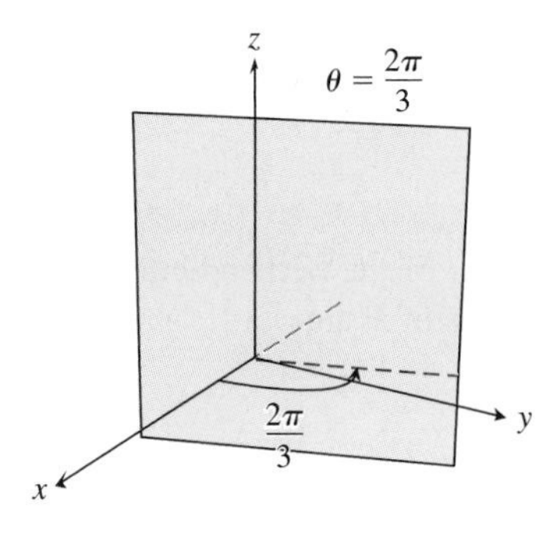

FIGURE 9.86 The graph of $\theta = c$ is a plane.

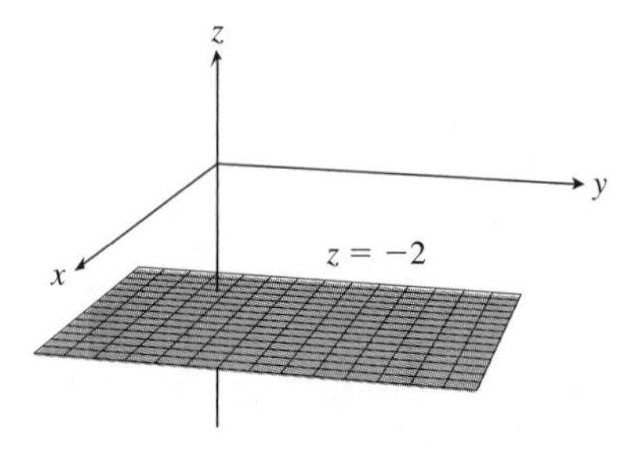

FIGURE 9.87 The graph of $z = c$ is a plane.

Solution

a) The surface S described by $r = 4$ is

$$S = \{(4, \theta, z) : 0 \leq \theta \leq 2\pi,\ -\infty < z < \infty\}.$$

In other words, S consists of all points in space with $r = 4$, but with no restriction on θ or z. Setting $z = 0$, we see that the surface contains the points

$$(4, \theta, 0), \qquad 0 \leq \theta \leq 2\pi, \tag{2}$$

in the (x, y)-plane. These are the points on the circle of radius 4 centered at the origin. If we replace the 0 in (2) by any number z, we still have a point on S. Thus S consists of all points on the circle (2) as well as all points above and below the circle. Hence, S is an infinite cylinder of radius 4 whose axis of symmetry is the z-axis. See Fig. 9.85. This example gives some justification to the name cylindrical coordinates. In this system the surface $r = c$ is a cylinder.

b) The surface S described by $\theta = 2\pi/3$ is

$$S = \left\{\left(r, \frac{2\pi}{3}, z\right) : -\infty < r < \infty,\ -\infty < z < \infty\right\}.$$

Hence S contains all points in three-space with $\theta = 2\pi/3$, but no restriction on the r or z. As in part (a), we start by letting $z = 0$. We then see that S contains the points

$$\left(r, \frac{2\pi}{3}, 0\right), \qquad -\infty < r < \infty, \tag{3}$$

in the (x, y)-plane. These are the points on the line that makes an angle of $2\pi/3$ with the positive x-axis. If we replace the 0 in (3) by any number z, we still have a point on S. Thus S consists of all points above and below the line described by (3). Hence S is the plane that is perpendicular to the (x, y)-plane and intersects the plane in line (3). See Fig. 9.86.

c) The z-coordinate in cylindrical coordinates is the same as the z-coordinate in Cartesian coordinates. Thus the surface described by $z = -2$ is the same in both coordinate systems, that is, a plane perpendicular to the z-axis and intersecting the z-axis at the point $(0, 0, -2)$. See Fig. 9.87.

Scientists, engineers, and mathematicians use cylindrical coordinates because some problems are easier to formulate and solve in cylindrical coordinates than in Cartesian coordinates. As suggested by part (a) of Example 1, cylindrical coordinates are often useful in describing surfaces and other phenomena that have an axis of symmetry along the z-axis.

To study a problem in cylindrical coordinates, we may need to change the problem from Cartesian to cylindrical coordinates and, perhaps, back again. We can do so by using the familiar formulas that allow us to move between polar coordinates and Cartesian coordinates in the plane.

Relationship between Cylindrical and Cartesian Coordinates

Let P have Cartesian coordinates (x, y, z). Then P has cylindrical coordinates (r, θ, z) where

$$r = \sqrt{x^2 + y^2} \quad \text{and} \quad \tan\theta = \frac{y}{x}, \tag{4}$$

where $x \neq 0$ and θ is measured from the positive x-axis and has (x, y) on its terminal side. (If $x = 0$ and $y > 0$, then $\theta = \pi/2$; if $x = 0$ and $y < 0$, then $\theta = -\pi/2$; if $x = y = 0$, then $r = 0$ and θ can have any value.)

Let P have cylindrical coordinates (r, θ, z). Then P has Cartesian coordinates (x, y, z) where

$$x = r\cos\theta \quad \text{and} \quad y = r\sin\theta. \tag{5}$$

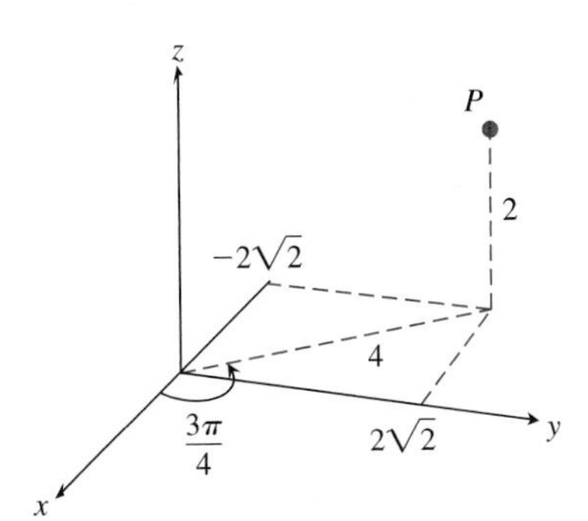

FIGURE 9.88 Point P has cylindrical coordinates $(4, 3\pi/4, 2)$ and Cartesian coordinates $\left(-2\sqrt{2}, 2\sqrt{2}, 2\right)$.

EXAMPLE 2

a) The point P has cylindrical coordinates $(4, 3\pi/4, 2)$. Find the Cartesian coordinates for P.
b) The point R has Cartesian coordinates $(-3, 4, 5)$. Find three different sets of cylindrical coordinates for R.

Solution

a) Using (5), we see that P has Cartesian coordinates (x, y, z) with

$$x = r\cos\theta = 4\cos\left(\frac{3\pi}{4}\right) = -2\sqrt{2}$$

$$y = r\sin\theta = 4\sin\left(\frac{3\pi}{4}\right) = 2\sqrt{2}.$$

Because the z-coordinate is the same in the two coordinate systems, P has Cartesian coordinates $\left(-2\sqrt{2}, 2\sqrt{2}, 2\right)$. See Fig. 9.88.

b) Use (4) to get one set of cylindrical coordinates for R. We have

$$r = \sqrt{x^2 + y^2} = \sqrt{(-3)^2 + (4)^2} = 5.$$

Because $Q = (-3, 4, 0)$, the foot of the perpendicular from R to the (x, y)-plane lies in the second quadrant of the plane, we seek an angle θ so that

$$\tan\theta = \frac{y}{x} = \frac{-4}{3} \quad \text{and} \quad \frac{\pi}{2} < \theta < \pi.$$

As suggested in Fig. 9.89,

$$\theta = \pi - \arctan\frac{4}{3} \approx 2.214297 \approx 126.87°.$$

Hence one cylindrical coordinate representation for R is (approximately) $(5, 2.214297, 5)$. We get other coordinates for R by adding a multiple of 2π to the θ value, or by replacing $r = 5$ by $r = -5$ and adding an odd

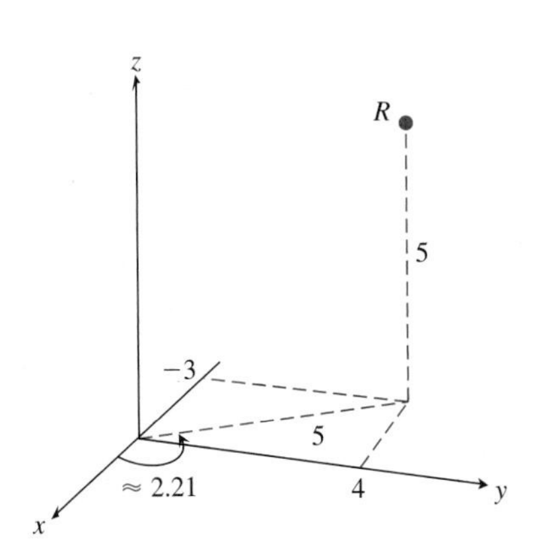

FIGURE 9.89 Point P has Cartesian coordinates $(-3, 4, 5)$ and cylindrical coordinates $\approx (5, 2.214297, 5)$.

multiple of π to θ. Thus R is also described by

$$\left(5, \left(\pi - \arctan\frac{4}{3}\right) + 4\pi, 5\right) \approx (5, 14.780668, 5),$$

and by

$$\left(-5, \left(\pi - \arctan\frac{4}{3}\right) - \pi, 5\right) \approx (-5, -0.927295, 5).$$

EXAMPLE 3 Rewrite the equations in cylindrical coordinates:

a) $z = (x^2 + y^2)^{3/2} - 5$
b) $xyz = 1, \quad x, y, z > 0$

Solution

a) From (4) we have $r^2 = x^2 + y^2$. Substituting this in the expression given in **a)**, we have

$$z = (x^2 + y^2)^{3/2} - 5 = (r^2)^{3/2} - 5 = |r|^3 - 5. \tag{6}$$

b) Solving $xyz = 1$ for z, we have

$$z = \frac{1}{xy}, \qquad x, y > 0. \tag{7}$$

From (5) we have $x = r\cos\theta$ and $y = r\sin\theta$. Because we need $x, y > 0$, we take $0 < \theta < \pi/2$ and $r > 0$. Substituting the expressions for x and y into (7), we have

$$z = \frac{1}{(r\cos\theta)(r\sin\theta)} = \frac{1}{r^2(1/2)\sin 2\theta} = \frac{2}{r^2}\csc 2\theta,$$

with $0 < \theta < \pi/2$ and $r > 0$.

Cylindrical Coordinates and Parametric Representation In Example 2 of Section 9.4 we saw that the parametric representation

$$\mathbf{F}(r, \theta) = \langle r\cos\theta, r\sin\theta, r^2\rangle \tag{8}$$

describes the same surface as the Cartesian expression

$$z = x^2 + y^2. \tag{9}$$

When we apply (4), we see that (9) is equivalent to the cylindrical expression

$$z = r^2.$$

In other words, the cylindrical coordinate representation is found by simply setting z equal to the third coordinate in (8). Indeed, it follows immediately from (4) that a surface described by a parametric expression of the form

$$\mathbf{G}(r, \theta) = \langle r\cos\theta, r\sin\theta, g(r, \theta)\rangle$$

can be described in cylindrical coordinates as

$$z = g(r, \theta).$$

Conversely, given a cylindrical representation

$$z = h(r, \theta)$$

for a surface, we can readily write down a parametric representation,

$$\mathbf{H}(r, \theta) = \langle r \cos \theta, r \sin \theta, h(r, \theta)\rangle.$$

This close relationship between cylindrical and parametric representations suggests two things:

- Cylindrical coordinates can be very useful in constructing a parametric representation of a surface. This can be especially useful when the level curves for the surface are circles, as in Example 2 of Section 9.4. If the cross sections are not circles but are nonetheless curves that enclose the origin, it can still be helpful to look for "almost cylindrical" representations. See Example 4 of Section 9.4, in which the level curves $y = c$ are ellipses in the (x, z)-plane.
- Let

  $$z = h(r, \theta)$$

  be a cylindrical coordinate representation for a surface S. We can graph S by using a parametive plotting package and the parameterization

  $$\mathbf{H}(r, \theta) = \langle r \cos \theta, r \sin \theta, h(r, \theta)\rangle$$

  for S.

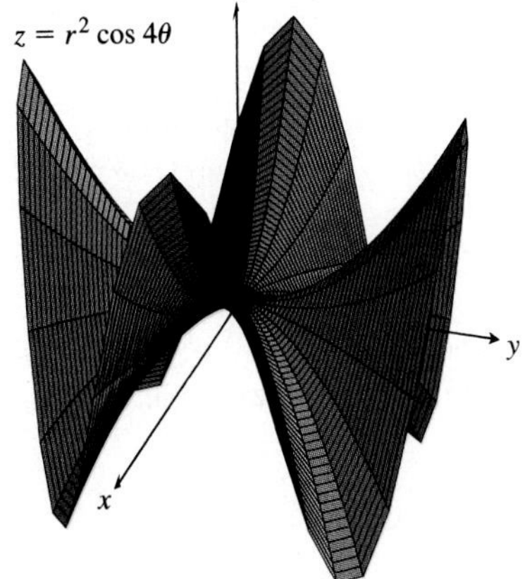

FIGURE 9.90 The graph of $z = r^2 \cos 4\theta$ obtained by using a parametric plotting package.

EXAMPLE 4 Find a parametric representation for the surface described by

$$z = r^2 \cos 4\theta.$$

Use a parametric plotting command to graph the surface.

Solution

By (5) we have $x = r \cos \theta$ and $y = r \sin \theta$. Combining these with the given expression for z, we obtain a parametric representation for S,

$$\mathbf{F}(r, \theta) = \langle r \cos \theta, r \sin \theta, r^2 \cos 4\theta\rangle.$$

Using a parametric plotting command to graph this expression, we obtain Fig. 9.90.

Spherical Coordinates

WEB In Section 9.4 we saw that we can use level curves in finding parametric representations for surfaces that can be represented in the form

$$z = f(x, y) \quad \text{or} \quad y = g(x, z) \quad \text{or} \quad x = h(y, z), \tag{10}$$

where f, g, and h are functions. However, if S is, say, the ellipsoid described by

$$\frac{x^2}{4} + y^2 + 9z^2 = 1,$$

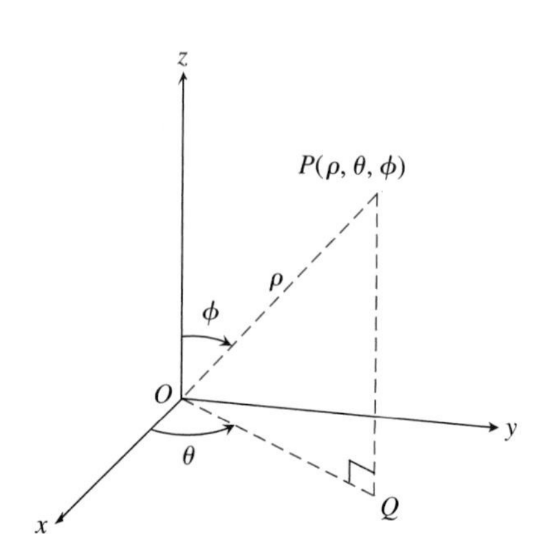

FIGURE 9.91 The point P has spherical coordinates (ρ, θ, ϕ).

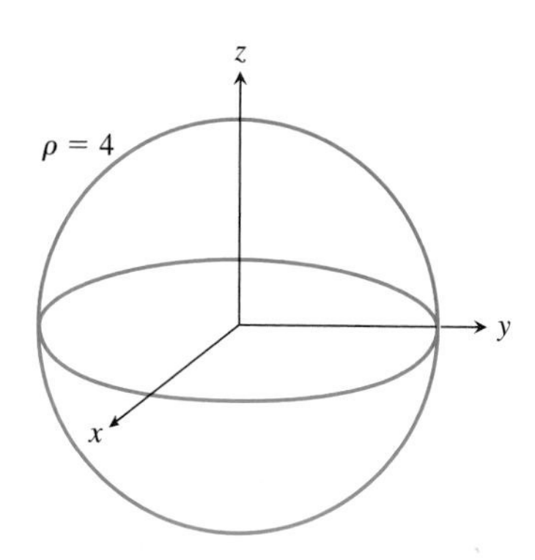

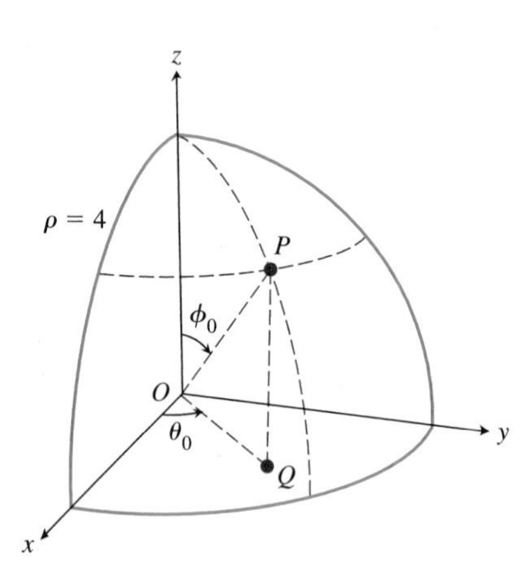

FIGURE 9.92 In spherical coordinates the surface described by $\rho = c$ is a sphere.

then S cannot be described by any of the expressions in (10). For example, if we solve for y, we obtain

$$y = \pm\sqrt{1 - \frac{x^2}{4} - 9z^2},$$

and see that for many input pairs (x, z) we have two possible y values as output. With the techniques studied in Section 9.4, we cannot obtain one parametric representation that accounts for both outputs. Sometimes this difficulty can be overcome by using **spherical coordinates.**

Spherical Coordinates

Let P be a point in three-space, let Q be the foot of the perpendicular from P to the (x, y)-plane, and let $O = (0, 0, 0)$ be the origin. The point P has spherical coordinates

$$(\rho, \theta, \phi), \qquad \rho \geq 0, \qquad 0 \leq \phi \leq \pi,$$

where

$\rho =$ the distance from O to P

$\theta =$ the angle measured from the positive x-axis to vector $\overline{OQ}$

$\phi =$ the angle measured from the positive z-axis to vector $\overline{OP}$.

See Fig. 9.91. Note that θ has the same meaning for spherical coordinates as it has for cylindrical coordinates when $r > 0$. Though θ can take on any real value, we usually need to consider only $0 \leq \theta < 2\pi$. The angle ϕ is usually restricted to $0 \leq \phi \leq \pi$.

Spherical coordinates are so named because the equation $\rho = c$, $c > 0$, describes a sphere in three-space. This is illustrated in the following example.

EXAMPLE 5 Describe the surface given by each of the equations

a) $\rho = 4$, **b)** $\theta = 3\pi/4$, **c)** $\phi = \pi/3$.

Solution

a) The surface S described by $\rho = 4$ is

$$S = \{(4, \theta, \phi) : 0 \leq \theta < 2\pi, 0 \leq \phi \leq \pi\}. \tag{11}$$

Thus all points on S lie on the sphere of radius 4 and centered at $O = (0, 0, 0)$. See Fig. 9.92. We need to show that all points on this sphere are described by (11). Let P be a point on the sphere, let Q be the foot of the perpendicular from P to the (x, y)-plane, and let O be the origin. Note that if $Q \neq O$, then the angle θ_0 from the positive x-axis to OQ is between 0 and 2π and the angle ϕ_0 from the positive z-axis to OP is between 0 and π. In Fig. 9.92 we show a close-up of this situation for $0 < \theta, \phi \leq \pi/2$. Thus P has spherical coordinates $(4, \theta_0, \phi_0)$. As indicated by (11), this point is included on S. Thus S consists of all points on the

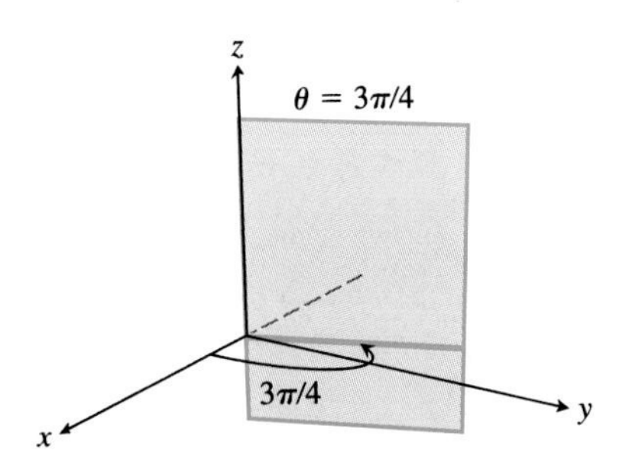

FIGURE 9.93 The surface described by $\theta = c$ is a half-plane.

sphere of radius 4 centered at O. We ask you to consider the case $Q = O$ in Exercise 41.

b) The collection of points described by $\theta = 3\pi/4$ is

$$S = \{(\rho, 3\pi/4, \phi) : \rho \geq 0,\ 0 \leq \phi \leq \pi\}. \tag{12}$$

In cylindrical coordinates the equation $\theta = c$ describes a plane, so we might expect something similar in this case. Let Π be the half-plane that is perpendicular to the (x, y)-plane and intersects the (x, y)-plane in the ray making an angle of $3\pi/4$ with the positive x-axis. Every point on this half-plane has θ-coordinate of $3\pi/4$, and any point not on the half-plane has a different θ-coordinate. Thus in spherical coordinates $\theta = 3\pi/4$ represents the half-plane Π. See Fig. 9.93.

c) The set of points determined by $\phi = \pi/3$ is

$$S = \{(\rho, \theta, \pi/3) : \rho \geq 0,\ 0 \leq \theta \leq 2\pi\}. \tag{13}$$

Let $\theta = 0$. The points

$$(\rho, 0, \pi/3), \qquad \rho \geq 0,$$

lie on the ray that starts from the origin, makes an angle of $\pi/3$ with the positive z-axis, and lies above the positive x-axis. See Fig. 9.94. When we let θ vary from 0 to 2π, this ray is rotated about the z-axis to give the upper nappe of a cone. Thus the surface described by $\phi = \pi/3$ is the upper nappe of the cone with vertex at the origin, axis of symmetry along the positive z-axis, and vertex angle of $\pi/3$. This surface is shown in Fig. 9.94.

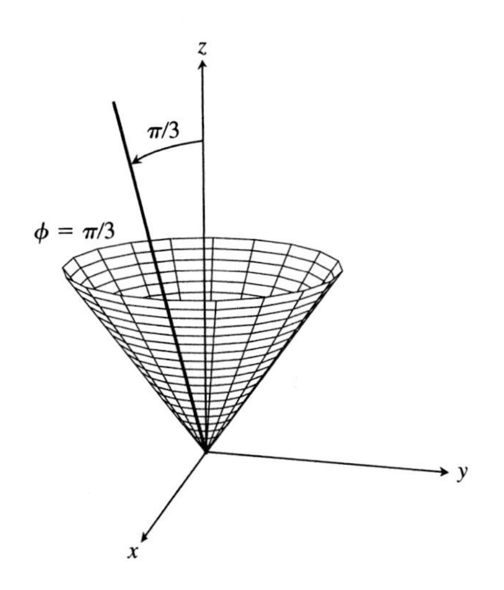

FIGURE 9.94 The surface described by $\phi = c$ is a cone.

To make effective use of spherical coordinates in constructing parametric representations, we need to know how to change from equations in x, y, and z into equations in ρ, θ, and ϕ.

Let P have spherical coordinates (ρ, θ, ϕ) and Cartesian coordinates (x, y, z). Let R be the foot of the perpendicular from P to the z-axis. Note that R has Cartesian coordinates $(0, 0, z)$. Because ϕ is the angle measured from the positive z-axis to OP, it follows that

$$z = \rho \cos \phi. \tag{14}$$

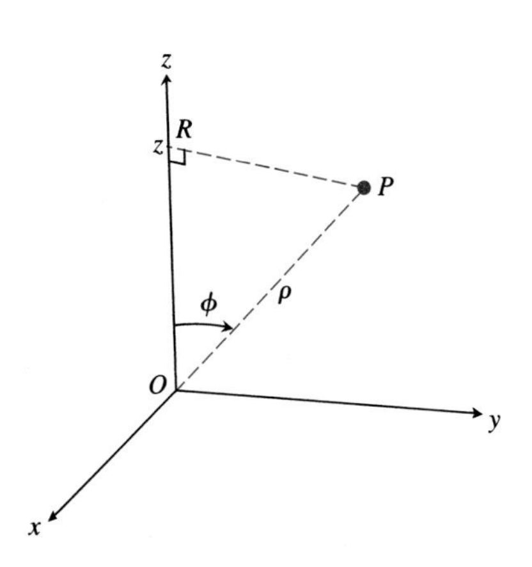

FIGURE 9.95 By trigonometry, $z = \rho \cos \phi$.

See Fig. 9.95. Now let Q be the foot of the perpendicular from P to the (x, y)-plane and let r be the length of $\overline{OQ}$. Then

$$r = \rho \cos\left(\frac{\pi}{2} - \phi\right) = \rho \sin \phi$$

and

$$\begin{aligned} x &= r \cos \theta = \rho \sin \phi \cos \theta \\ y &= r \sin \theta = \rho \sin \phi \sin \theta. \end{aligned} \tag{15}$$

See Fig. 9.96.

It is also useful to have formulas that express spherical coordinates in terms of Cartesian coordinates. Because ρ is the distance from $O = (0, 0, 0)$ to $P = (x, y, z)$, we have

$$\rho = \sqrt{x^2 + y^2 + z^2}. \tag{16}$$

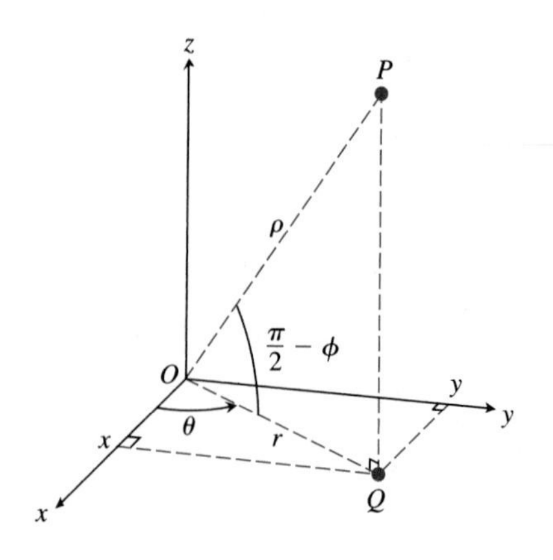

FIGURE 9.96 Expressing x and y in terms of spherical coordinates.

From (14) we see that because ρ is a distance, $\rho \geq 0$ and when $\rho > 0$

$$\cos\phi = \frac{z}{\rho} = \frac{z}{\sqrt{x^2+y^2+z^2}} \quad \text{with} \quad 0 \leq \phi \leq \pi.$$

Hence

$$\phi = \arccos\left(\frac{z}{\sqrt{x^2+y^2+z^2}}\right). \tag{17}$$

From (15) we see that if x and y are not both 0, then θ is the angle that satisfies

$$\cos\theta = \frac{x}{\sqrt{x^2+y^2}} \quad \text{and} \quad \sin\theta = \frac{y}{\sqrt{x^2+y^2}}.$$

Dividing $\sin\theta$ by $\cos\theta$, we can express this last relation as

$$\tan\theta = \frac{y}{x}, \quad \text{where } x \neq 0 \tag{18}$$

and the angle of measure θ measured from the positive x-axis has point (x, y) on its terminal side. We now summarize (14), (15), (16), (17), and (18).

Relationship between Spherical and Cartesian Coordinates

Let P be a point in space with Cartesian coordinates (x, y, z) and spherical coordinates (ρ, θ, ϕ). Then

$$\begin{aligned} x &= \rho\sin\phi\cos\theta \\ y &= \rho\sin\phi\sin\theta \\ z &= \rho\cos\phi. \end{aligned} \tag{19}$$

Also,

$$\begin{aligned} \rho &= \sqrt{x^2+y^2+z^2} \\ \phi &= \arccos\left(\frac{z}{\sqrt{x^2+y^2+z^2}}\right), \quad x, y, z \text{ not all } 0, \\ \tan\theta &= \frac{y}{x} \end{aligned} \tag{20}$$

where $x \neq 0$ and the angle of measure θ measured from the positive x-axis has (x, y) on its terminal side. (If $x = 0$ and $y > 0$ then $\theta = \pi/2$; if $x = 0$ and $y < 0$ then $\theta = 3\pi/2$; if $x = y = 0$ then θ can be any value.)

EXAMPLE 6

a) The point P has spherical coordinates $(5, \pi/3, 3\pi/4)$. Find the Cartesian coordinates for P.

b) The point R has Cartesian coordinates $(-3, 4, 5)$. Find the spherical coordinates for R.

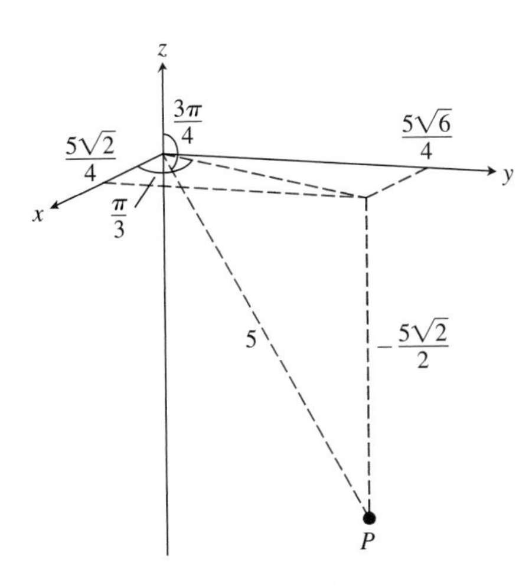

FIGURE 9.97 Point P has spherical coordinates $(5, \pi/3, 3\pi/4)$ and Cartesian coordinates $(5\sqrt{2}/4, 5\sqrt{6}/4, -5\sqrt{2}/2)$.

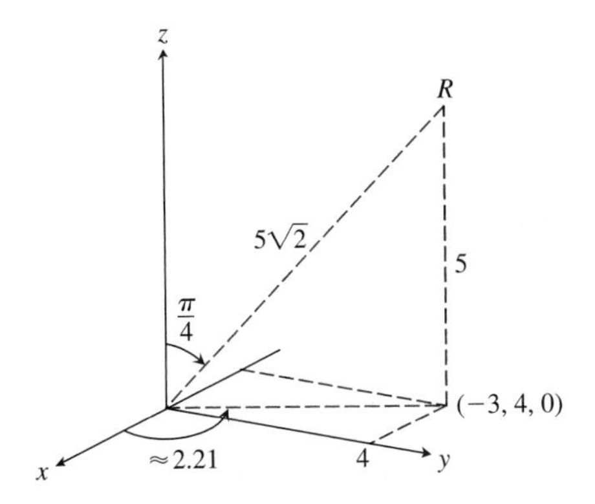

FIGURE 9.98 Point R has Cartesian coordinates $(-3, 4, 5)$ and spherical coordinates $\approx (7.071068, 2.214297, 0.785398)$.

Solution

a) Let P have Cartesian coordinates (x, y, z). According to (19),

$$x = \rho \sin \phi \cos \theta = 5 \sin \frac{3\pi}{4} \cos \frac{\pi}{3} = 5 \cdot \frac{\sqrt{2}}{2} \frac{1}{2} = \frac{5\sqrt{2}}{4},$$

$$y = \rho \sin \phi \sin \theta = 5 \sin \frac{3\pi}{4} \sin \frac{\pi}{3} = 5 \cdot \frac{\sqrt{2}}{2} \frac{\sqrt{3}}{2} = \frac{5\sqrt{6}}{4},$$

$$z = \rho \cos \phi = 5 \cos \frac{3\pi}{4} = 5\left(\frac{-\sqrt{2}}{2}\right) = -\frac{5\sqrt{2}}{2}.$$

Hence the Cartesian coordinates for P are $(5\sqrt{2}/4, 5\sqrt{6}/4, -5\sqrt{2}/2)$. See Fig. 9.97.

b) Let R have spherical coordinates (ρ, θ, ϕ). By (20),

$$\rho = \sqrt{x^2 + y^2 + z^2} = \sqrt{(-3)^2 + 4^2 + 5^2} = 5\sqrt{2},$$

$$\phi = \arccos\left(\frac{z}{\sqrt{x^2 + y^2 + z^2}}\right) = \arccos \frac{z}{\rho} = \arccos \frac{5}{5\sqrt{2}} = \frac{\pi}{4}.$$

Because the point (x, y) is in the second quadrant of the plane, we must have

$$\tan \theta = \frac{y}{x} = -\frac{4}{3} \qquad \text{with} \qquad \frac{\pi}{2} < \theta < \pi.$$

As in part **b)** of Example 2, we find

$$\theta = \pi - \arctan \frac{4}{3} \approx 2.214297.$$

Thus R has spherical coordinates

$$\left(5\sqrt{2}, \pi - \arctan \frac{4}{3}, \pi/4\right) \approx (7.071068, 2.214297, 0.785398).$$

See Fig. 9.98.

EXAMPLE 7 Let S be the sphere of radius $a > 0$ centered at the point $(a, 0, 0)$ on the x-axis. Find a spherical representation for the sphere.

Solution

Let $P = (x, y, z)$ be a point on S. Because the distance from (x, y, z) to the center $(a, 0, 0)$ is a, we have

$$(x - a)^2 + y^2 + z^2 = a^2.$$

Expanding and rearranging gives

$$x^2 + y^2 + z^2 = 2ax.$$

By (20) and (19) we can replace $x^2 + y^2 + z^2$ by ρ^2, and x by $\rho \sin \phi \cos \theta$. We then have

$$\rho^2 = 2a\rho \sin \phi \cos \theta.$$

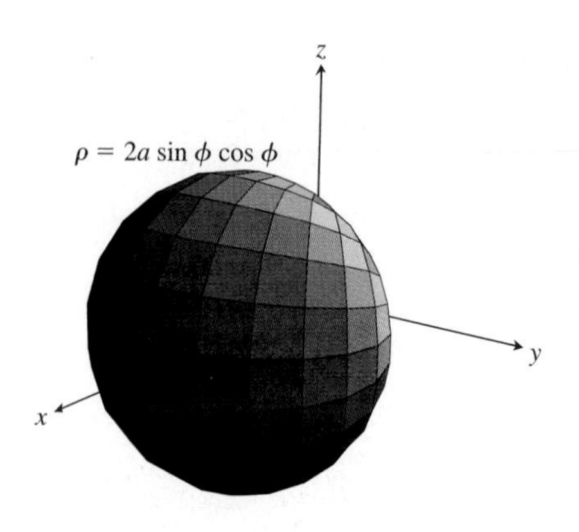

FIGURE 9.99 The surface described by $\rho = 2a \sin\phi \cos\theta$.

If we assume that $\rho > 0$, we can divide both sides of this expression by ρ to get

$$\rho = 2a \sin\phi \cos\theta \tag{21}$$

as our spherical coordinate representation of S. The graph of S is shown in Fig. 9.99.

Sometimes it is useful to describe the same phenomenon in different coordinate systems. Some aspects of a problem may be well described in one system, while other aspects may be better described in another system.

EXAMPLE 8 Two spherical particles, one of mass M and the other of mass m, are in space. According to Newton's law of universal gravitation, each object exerts an attractive force on the other. Either force acts along a line through the centers of the objects and has magnitude proportional to the product of the masses of the objects and inversely proportional to the square of the distance between them. The constant of proportionality is G, the constant of universal gravitation. Find Cartesian and spherical expressions for the force exerted on m. Assume that the object of mass M is centered at the origin O.

Solution

Let the object of mass m be centered at the point $P = (x, y, z)$. The force exerted on this object acts in the direction of the unit vector

$$\mathbf{u} = -\frac{1}{\|\mathbf{P}\|}\mathbf{P} = -\left\langle \frac{x}{\sqrt{x^2+y^2+z^2}}, \frac{y}{\sqrt{x^2+y^2+z^2}}, \frac{z}{\sqrt{x^2+y^2+z^2}} \right\rangle.$$

See Fig. 9.100. The force has magnitude

$$F = \frac{GmM}{\left(\sqrt{x^2+y^2+z^2}\right)^2} = \frac{GmM}{x^2+y^2+z^2}. \tag{22}$$

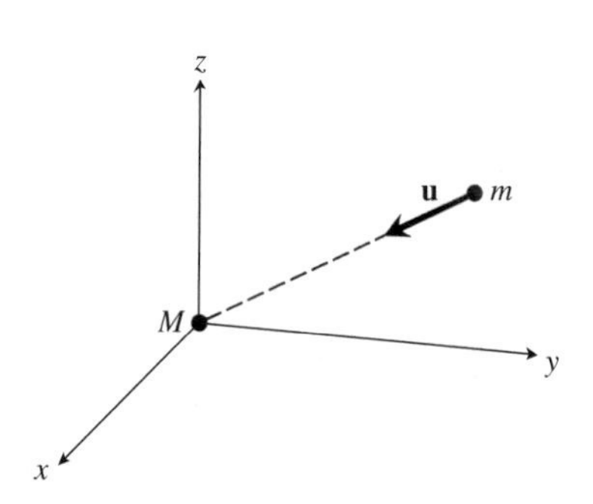

FIGURE 9.100 Two masses in space.

Combining (22) with the direction $\mathbf{u}$ of the force, we see that the force exerted on the object of mass m is

$$\mathbf{F} = F\mathbf{u} = -\frac{GmM}{(x^2+y^2+z^2)^{3/2}}\langle x, y, z\rangle. \tag{23}$$

To express the magnitude of the force in spherical coordinates, substitute ρ^2 for $x^2 + y^2 + z^2$ in (22). We then have

$$F = \frac{GmM}{\rho^2}. \tag{24}$$

The spherical representation for the magnitude of the force is very simple. In particular, (24) clearly illustrates that the magnitude of the force depends only on the distance between the objects and not on the direction of the vector determined by the positions of the objects. On the other hand, the spherical representation for the force $\mathbf{F}$ is not so nice. In (23), replace $x^2 + y^2 + z^2$ by

ρ^2 and x, y, z by the spherical coordinate expressions in (19) to get

$$\begin{aligned}\mathbf{F} &= -\frac{GmM}{\rho^3}\langle \rho \sin\phi \cos\theta, \rho \sin\phi \sin\theta, \rho \cos\phi \rangle \\ &= -\frac{GmM}{\rho^2}\langle \sin\phi \cos\theta, \sin\phi \sin\theta, \cos\phi \rangle.\end{aligned} \tag{25}$$

This expression is not as easy to understand as (23). For those not thoroughly familiar with spherical coordinates, it may not be readily apparent from (25) that the force acts along a vector pointing toward the origin. Thus we see that spherical coordinates do a nice job in describing the magnitude of the force, but Cartesian coordinates are better suited to describing the direction.

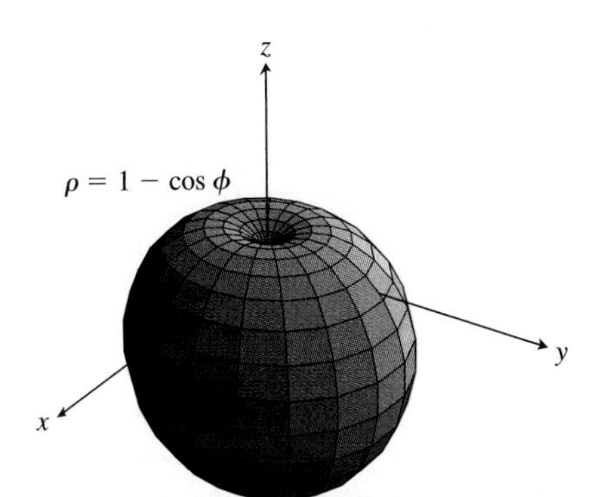

FIGURE 9.101 The surface represented by $\rho = 1 - \cos\phi$ is a "three-dimensional cardioid."

Graphing Spherical Representations Let S be a surface represented by the spherical equation

$$\rho = h(\theta, \phi).$$

Using (19), we can immediately write down a parametric representation of S,

$$\begin{aligned}\mathbf{H}(\theta, \phi) &= \langle x, y, z\rangle = \langle x(\theta, \phi), y(\theta, \phi), z(\theta, \phi)\rangle \\ &= \langle \rho \sin\phi \cos\theta, \rho \sin\phi \sin\theta, \rho \cos\phi \rangle \\ &= \langle h(\theta, \phi) \sin\phi \cos\theta, h(\theta, \phi) \sin\phi \sin\theta, h(\theta, \phi) \cos\phi \rangle.\end{aligned} \tag{26}$$

We can graph this last expression using a parametric plotting command.

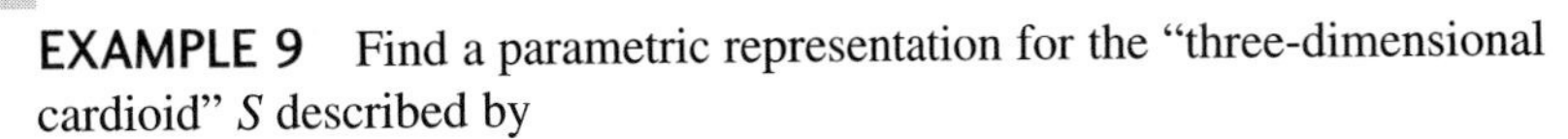

EXAMPLE 9 Find a parametric representation for the "three-dimensional cardioid" S described by

$$\rho = 1 - \cos\phi. \tag{27}$$

Use a parametric plotting routine to produce a graph of S.

Solution

To find a parametric representation, substitute $\rho = h(\theta, \phi) = 1 - \cos\phi$ into (26). This gives

$$\mathbf{H}(\theta, \phi) = \langle (1 - \cos\phi) \sin\phi \cos\theta, (1 - \cos\phi) \sin\phi \sin\theta, (1 - \cos\phi) \cos\phi \rangle.$$

It is interesting to note that even though (27) does not involve θ, this variable still appears in our parametric representation. Using a parametric plotting routine to graph $\mathbf{H}(\theta, \phi)$, we obtain the graph shown in Fig. 9.101. The surface has an indentation at the top; the indentation tapers to a point at the origin. For an alternative viewpoint, the surface can be generated by rotating a planar cardioid about the z-axis. See Fig. 9.102.

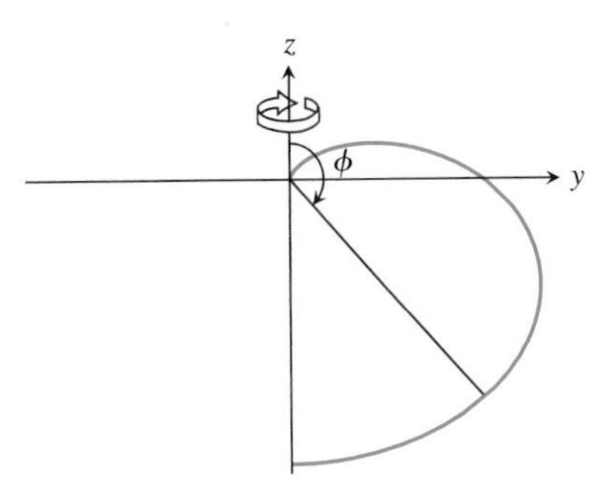

FIGURE 9.102 The surface in Fig. 9.101 can be generated by rotating a cardioid about the z-axis. In the (y, z)-plane, the equation $\rho = 1 - \cos\phi$ may be regarded as a polar equation with ϕ measured from the positive z axis.

In spherical coordinates the sphere of radius 1 centered at the origin is represented by

$$\rho = 1.$$

Using (26) to convert this to a parametric representation, we obtain

$$\mathbf{H}(\theta, \phi) = \langle x, y, z\rangle = \langle x(\theta, \phi), y(\theta, \phi), z(\theta, \phi)\rangle = \langle \sin\phi\cos\theta, \sin\phi\sin\theta, \cos\phi\rangle.$$

Because all of the points represented in this way are at a distance of 1 from the origin, we have

$$\begin{aligned} 1 &= (x(\theta, \phi))^2 + (y(\theta, \phi))^2 + (z(\theta, \phi))^2 \\ &= (\sin\phi\cos\theta)^2 + (\sin\phi\sin\theta)^2 + (\cos\phi)^2. \end{aligned} \tag{28}$$

Indeed, this can be verified directly with a little algebra and trigonometry (see Exercise 32 in Section 9.4). By using (28), we can find parametric representations for some surfaces that can "almost" be easily represented in spherical coordinates.

EXAMPLE 10 Use (28) to find a parametric representation for the ellipsoid S described by the equation

$$\frac{x^2}{4} + y^2 + 9z^2 = 1. \tag{29}$$

Solution

An ellipsoid is similar to a sphere. Furthermore, the representation given in (29) looks similar to the Cartesian equation for a sphere centered at the origin. With the idea of using (28), rewrite (29) as

$$\left(\frac{x}{2}\right)^2 + y^2 + (3z)^2 = 1.$$

Matching the terms in this sum with the terms in (28) suggests letting

$$\frac{x}{2} = \sin\phi\cos\theta, \qquad y = \sin\phi\sin\theta, \qquad 3z = \cos\phi.$$

Solving these equations for x, y, and z leads to a parametric representation for the ellipsoid,

$$\mathbf{H}(\theta, \phi) = \left\langle 2\sin\phi\cos\theta, \sin\phi\sin\theta, \frac{1}{3}\cos\phi\right\rangle$$

for $0 \le \theta < 2\pi$, $0 \le \phi \le \pi$. When we plot the surface represented by this parametric representation, we obtain the graph shown in Fig. 9.103.

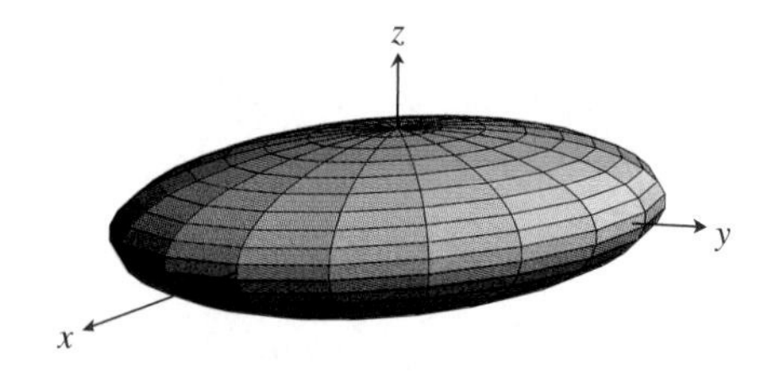

FIGURE 9.103 The ellipsoid $(x^2/4) + y^2 + 9z^2 = 1$.

Exercises 9.5

Exercises 1–4: The Cartesian coordinates for a point in space are given. Find cylindrical and spherical coordinates for the point.

1. $(1, 1, 1)$
2. $(-5, 5\sqrt{3}, 0)$
3. $(1, -1, -\sqrt{2})$
4. $(-5, -12, -13)$

Exercises 5–8: The cylindrical coordinates for a point in space are given. Find the Cartesian and spherical coordinates for the point.

5. $(42, 0, 0)$
6. $(8, 3\pi/4, 6)$
7. $(2, \pi/6, -2)$
8. (π, π, π)

Exercises 9–12: The spherical coordinates (ρ, θ, ϕ) for a point in space are given. Find the Cartesian and cylindrical coordinates for the point.

9. $(42, 0, 0)$

10. $(1, 7\pi/4, 2\pi/3)$

11. $(0, \pi/4, \pi/4)$

12. $(3, 2\pi/3, \pi/3)$

Exercises 13–16: A Cartesian representation for a surface S is given. Find the cylindrical and spherical representations for S.

13. $x^2 + y^2 + z^2 = 16$

14. $z = -\sqrt{x^2 + y^2}$

15. $y = 5$

16. $z = ye^{-x^2-y^2}$

Exercises 17–20: A cylindrical representation for a surface S is given. Find Cartesian and spherical representations for S.

17. $r^2 + z^2 = 9$

18. $z = \sqrt{3}r$

19. $r = 8$

20. $z = r(2\cos\theta - 5\sin\theta)$

Exercises 21–24: A spherical representation for a surface S is given. Find Cartesian and cylindrical representations for S.

21. $\rho = 5$

22. $\phi = 3\pi/4$

23. $\rho = 3\sec\phi$

24. $\rho = 2\cos\phi - 3\sin\phi\sin\theta$

Exercises 25–32: Find a parametric representation for the surface S represented by the Cartesian equation. Use a parametric plotting routine to produce a graph of the surface.

25. $x^2 + y^2 = 5$

26. $z = \sqrt{3x^2 + 3y^2}$

27. $x^2 + y^2 + z^2 - 4y = 0$

28. $x^2 + y^2 + z^2 + 6z = 0$

29. $z = e^{-4x^2-y^2}$

30. $z = \dfrac{x^2}{4} + 9y^2$

31. $4x^2 + 9y^2 + \dfrac{z^2}{4} = 1$

32. $\dfrac{x^2}{2} + \dfrac{y^2}{3} + \dfrac{z^2}{4} = 1$

Exercises 33–38: Given below are the spherical representations for six surfaces (by number), the cylindrical representations for the same surfaces (by letter), and then graphs of the surfaces. Match each surface graph with its spherical and cylindrical representations,

33. $\phi = \dfrac{3\pi}{4}$

34. $\rho = 5\sec\phi$

35. $\rho = 3$

36. $\rho = 3\csc\phi\sin\theta$

37. $\rho = 3\csc\phi$

38. $\rho = \dfrac{3\csc\phi}{\cos\theta + \sin\theta}$

a. $r = 3\sin\theta$

b. $r^2 + z^2 = 9$

c. $r = 3$

d. $z = 5$

e. $r\sin\left(\theta + \dfrac{\pi}{4}\right) = \dfrac{3}{\sqrt{2}}$

f. $z = -r$

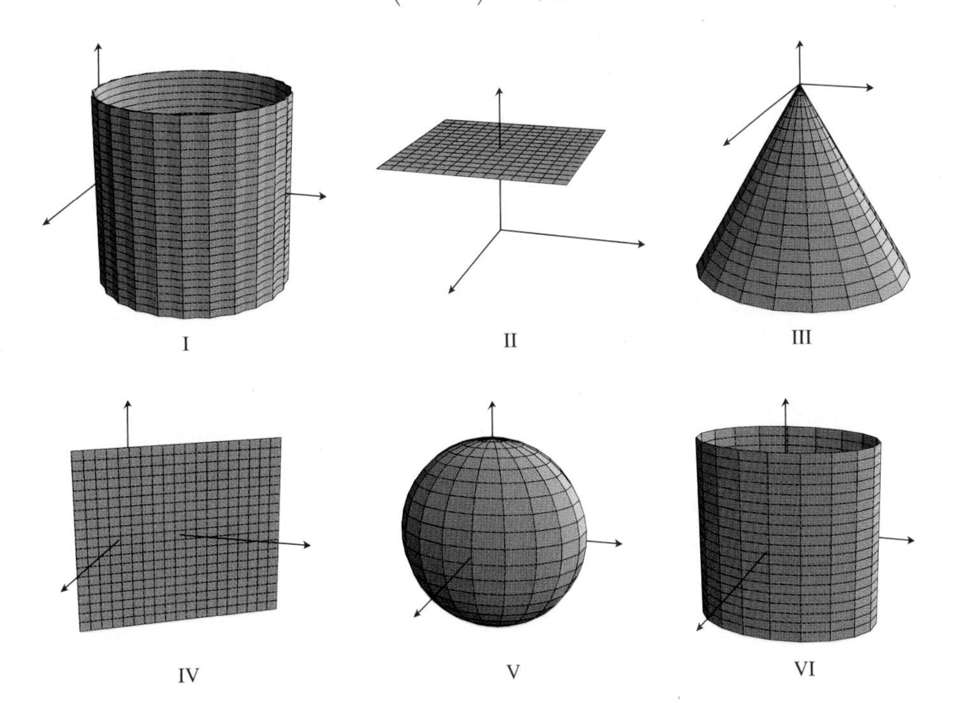

39. In Fig. 9.90 we show the graph of the surface represented by $z = r^2 \cos 4\theta$. Describe the shape of the surface.

40. Let point P have spherical coordinates (ρ, θ, ϕ) and cylindrical coordinates (r, θ, z). Find formulas that express the spherical coordinates in terms of the cylindrical coordinates. Then find formulas that express the cylindrical coordinates in terms of the spherical coordinates.

41. In part **a)** of Example 5 we discussed the surface represented by the equation $\rho = 4$. Complete the example by discussing the case in which $Q = O = (0, 0, 0)$.

42. Let S be the sphere of radius 1 centered at the point with Cartesian coordinates $(1/2, 1/2, \sqrt{2}/2)$. Find an equation that represents S in

a. Cartesian coordinates

b. cylindrical coordinates

c. spherical coordinates.

43. Let S be the sphere of radius 4 centered at the point with Cartesian coordinates $(-\sqrt{3}, 3, -2)$. Find an equation that represents S in

a. Cartesian coordinates

b. cylindrical coordinates

c. spherical coordinates.

44. Let S be the sphere of radius 1 centered at the point with spherical coordinates $(1, \pi/3, 3\pi/4)$. Find an equation that represents S in

a. Cartesian coordinates

b. cylindrical coordinates

c. spherical coordinates.

45. Let S be the sphere of radius 10 centered at the point with spherical coordinates $(10, 3\pi/4, \pi/2)$. Find an equation that represents S in

a. Cartesian coordinates

b. cylindrical coordinates

c. spherical coordinates.

46. Suppose the surface S is represented by the cylindrical equation

$$z = g(r, \theta)$$

and the surface S^* is represented by

$$z = g\left(r, \theta - \frac{\pi}{3}\right).$$

How are S and S^* related? Support your answer with an example and a sketch.

47. Suppose the surface S is represented by the cylindrical-coordinate equation

$$z = g(r, \theta), \qquad r \geq 0,$$

and the surface S^* is represented by

$$z = g(r + 1, \theta), \qquad r \geq 0.$$

How are S and S^* related? Support your answer with an example and a sketch.

48. Suppose the surface S is represented by the spherical-coordinate equation

$$\rho = h(\theta, \phi)$$

and the surface S^* is represented by

$$\rho = 5h(\theta, \phi).$$

How are S and S^* related? Support your answer with an example and a sketch.

49. Suppose the surface S is represented by the spherical-coordinate equation

$$\rho = h(\theta, \phi)$$

and the surface S^* is represented by

$$\rho = h(\theta, \phi) + 1.$$

How are S and S^* related? Support your answer with an example and a sketch.

50. Suppose the surface S is represented by the spherical-coordinate equation

$$\rho = g(\theta, \phi)$$

and the surface S^* is represented by

$$\rho = g\left(\theta - \frac{\pi}{3}, \phi\right).$$

How are S and S^* related? Support your answer with an example and a sketch.

51. Suppose the surface S is represented by the spherical-coordinate equation

$$\rho = g(\theta, \phi), \qquad 0 \leq \phi \leq \pi,$$

and the surface S^* is represented by

$$\rho = g\left(\theta, \phi + \frac{\pi}{6}\right), \qquad 0 \leq \phi \leq \frac{5\pi}{6}.$$

How are S and S^* related? Support your answer with an example and a sketch.

52. Suppose the surface S is represented by the spherical-coordinate equation

$$\rho = g(\theta, \phi), \qquad 0 \leq \phi \leq \pi,$$

and the surface S^* is represented by

$$\rho = g\left(\theta, \phi - \frac{\pi}{6}\right), \qquad \frac{\pi}{6} \leq \phi \leq \pi.$$

How are S and S^* related? Support your answer with an example and a sketch.

53. What problems might arise if there were no restrictions on the ϕ-coordinate in spherical coordinates?

54. Given a point P in the plane, let us say that P has "vector coordinates" $\{a, b\}$ if

$$\mathbf{P} = a\langle 3, -2\rangle + b\langle 4, 1\rangle.$$

a. Let P have Cartesian coordinates $(2, 3)$. What are the vector coordinates for P?

b. Let Q have vector coordinates $\{-5, 1\}$. What are the Cartesian coordinates for Q?

c. Find formulas to express Cartesian coordinates in terms of vector coordinates.

d. Find formulas to express vector coordinates in terms of Cartesian coordinates.

55. Given a point P in space, let us say that P has "vector coordinates" $\{a, b, c\}$ if we have

$$\mathbf{P} = a\langle 1, 1, 1\rangle + b\langle 4, 0, 1\rangle + c\langle 0, 0, 1\rangle.$$

a. Let P have Cartesian coordinates $(4, -4, 2)$. What are the vector coordinates for P?

b. Let Q have vector coordinates $\{-2, 3, 4\}$. What are the Cartesian coordinates for Q?

c. Find formulas to express Cartesian coordinates in terms of vector coordinates.

d. Find formulas to express vector coordinates in terms of Cartesian coordinates.

56. Given a point P in the plane, let us say the point has "oblique coordinates" (a, b) if the point lies on the line $x + y = a$ and is on this line at directed distance b from the point $(a/2, a/2)$. (On the line $x + y = a$, the positive direction is in the direction of the vector $\mathbf{v} = \langle 1, -1\rangle$ and the negative direction in the direction of $-\mathbf{v} = \langle -1, 1\rangle$.) See the accompanying figure.

a. Let P have Cartesian coordinates $(4, -6)$. What are the oblique coordinates for P?

b. Let Q have oblique coordinates $(-5, 8)$. What are the Cartesian coordinates for Q?

c. Find formulas to express Cartesian coordinates in terms of oblique coordinates.

d. Find formulas to express oblique coordinates in terms of Cartesian coordinates.

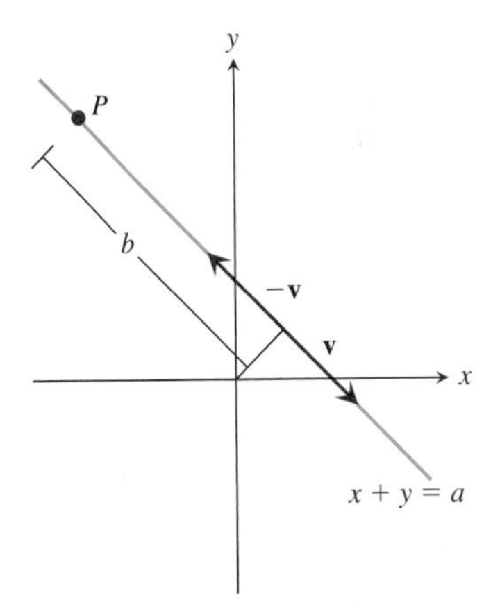

Figure for Exercise 56.

9.6 Limits

Limits played a central role in the study of the calculus of one variable. We used limits in defining the derivative and the integral, in the study of the asymptotic behavior of functions, and in the definition and analysis of sequences and series. Limits also play an important role in the definition and understanding of derivatives and integrals of functions of several variables. In this section we discuss limits of functions of more than one variable.

Interior Points, Boundary Points, and Closed Sets

The domain for any function of two variables is a subset of the plane R^2. The structures of such domains can vary considerably from one function to another. For example, the function f defined by

$$f(x, y) = \sqrt{(4 - x^2 - y^2)(x^2 + y^2 - 1)}$$

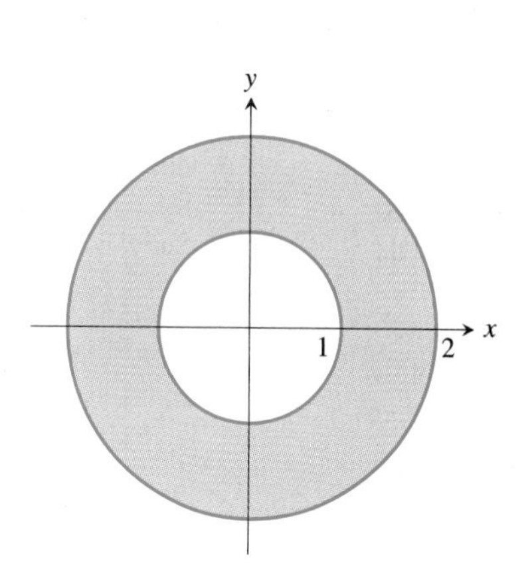

FIGURE 9.104 The domain of $f(x, y) = \sqrt{(4 - x^2 - y^2)(x^2 + y^2 - 1)}$ consists of all points on or between two circles.

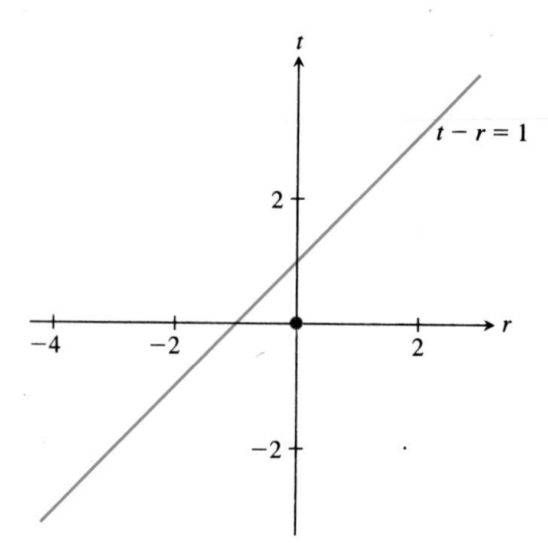

FIGURE 9.105 The domain of $g(r, t) = \sqrt{-(|r| + |t|)|1 + r - t|}$ consists of a point and a line.

has domain

$$\mathcal{D}_f = \{(x, y) : 1 \leq x^2 + y^2 \leq 4\}.$$

This set, shown in Fig. 9.104, consists of all points on or between the circles of radius 1 and 2 centered at the origin. On the other hand, the function g defined by

$$g(r, t) = \sqrt{-(|r| + |t|)|1 + r - t|}$$

is defined only when (r, t) is the origin or on the line with equation $t - r = 1$. This domain is shown in Fig. 9.105.

For our study of functions of two (or more) variables, we will need some terminology about points and sets in the plane.

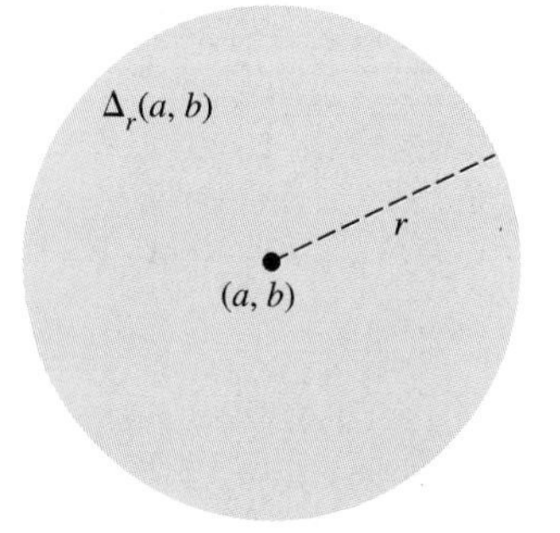

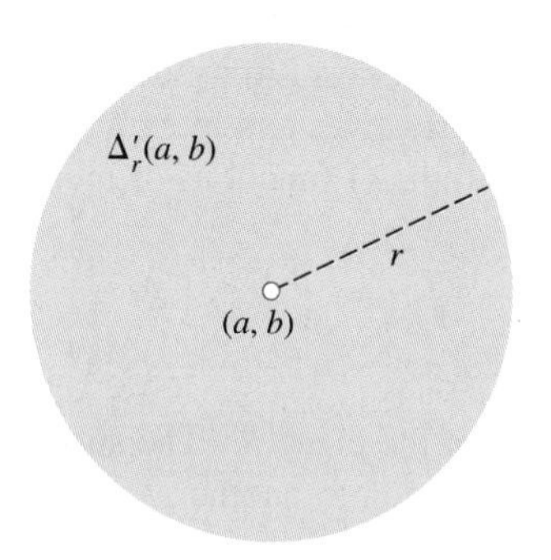

FIGURE 9.106 The disk $\Delta_r(a, b)$ and the punctured disk $\Delta'_r(a, b)$.

DEFINITION Disk

Let $P = (a, b)$ be a point in R^2 and let r be a positive number. The **disk of center P and radius r** is the set of all points inside the circle of center P and radius r. We denote this disk by $\Delta_r(P)$. Thus

$$\Delta_r(P) = \Delta_r(a, b) = \left\{(x, y) : \sqrt{(x - a)^2 + (y - b)^2} < r\right\}.$$

The **punctured disk of center P and radius r** is the disk $\Delta_r(P)$ with the center point, P, removed. We denote the punctured disk by $\Delta'_r(P)$. Hence

$$\Delta'_r(P) = \Delta'_r(a, b) = \left\{(x, y) : 0 < \sqrt{(x - a)^2 + (y - b)^2} < r\right\}.$$

See Fig. 9.106.

We use the definition of *disk* to define interior and boundary points.

DEFINITION Interior Points, Boundary Points, and Closed Sets

Let $\mathcal{D}$ be a subset of the Cartesian plane.

- A point Q in the plane is called an **interior point** of $\mathcal{D}$ if there is a positive number r so that the disk $\Delta_r(Q)$ is contained in $\mathcal{D}$.
- A point P in the plane is called a **boundary point** of $\mathcal{D}$ if every disk $\Delta_r(P)$ centered at P contains points that are in $\mathcal{D}$ and points that are not in $\mathcal{D}$.
- The set $\mathcal{D}$ is called a **closed set** if $\mathcal{D}$ contains all of its boundary points.

We do little with closed sets in the remainder of this chapter. However, closed sets will be important in the next chapter when we discuss the Max/Min Theorem for functions of several variables.

EXAMPLE 1 Let $\mathcal{D}$ be the region in the plane described by

$$\mathcal{D} = \{(x, y): -1 < x \leq 1, -1 \leq y < 1\}.$$

a) What are the boundary points of $\mathcal{D}$?
b) What are the interior points of $\mathcal{D}$?
c) Tell why $\mathcal{D}$ is not a closed set.

Solution

a) The region $\mathcal{D}$ is shown in Fig. 9.107. Every point on the square S with vertices $(1, 1)$, $(1, -1)$, $(-1, -1)$, $(-1, 1)$ is a boundary point of $\mathcal{D}$. Typical disks centered at boundary points P and P_1 are shown in the figure. As seen in Fig. 9.107, each of these disks contains points in $\mathcal{D}$ as well as points not in $\mathcal{D}$. Note that the boundary point P_1 is a point of $\mathcal{D}$, but the boundary point P is not a point of $\mathcal{D}$. Hence a boundary point of a set may or may not be a point of the set.

b) Every point in $\mathcal{D}$ but not on S is an interior point of $\mathcal{D}$. If Q is such a point, then as indicated in Fig. 9.107, we can draw a small disk centered at Q and contained in $\mathcal{D}$. Note that this also shows that an interior point of $\mathcal{D}$ is never a boundary point of $\mathcal{D}$.

c) In part **a)** we noted that the point P in Fig. 9.107 is a boundary point of $\mathcal{D}$ but is not a point of $\mathcal{D}$. Thus $\mathcal{D}$ does not contain all of its boundary points, so it is not a closed set.

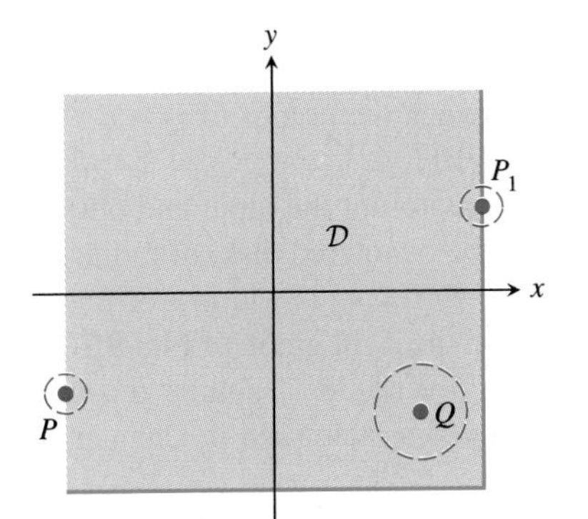

FIGURE 9.107 P and P_1 are boundary points of $\mathcal{D}$ and Q is an interior point of $\mathcal{D}$.

Limits

Limits for functions of two variables can often be interpreted in much the same way that we interpret limits of functions of one variable. For example, the statement

$$\lim_{(x,y)\to(2,-1)} 3x^2y = -12$$

means that the value of $3x^2y$ gets close to -12 as (x, y) gets close to $(2, -1)$. This seems obvious: If (x, y) is close to $(2, -1)$, then x is close to 2, y is close to -1, and the value of $3x^2y$ is close to $3(2^2)(-1) = -12$. However, we will soon see that the value of a limit of a function of two variables is not always clear and that vague phrases like "close to" can be hard to interpret in some circumstances. We can avoid some of these difficulties by defining *limit* carefully.

DEFINITION Limit of a Function of Two Variables

Let f be a real-valued function of two variables with domain $\mathcal{D}$ and let L be a real number. We say that f has limit L as (x, y) approaches (a, b) if

given any number $\epsilon > 0$,

we can find a number $\delta > 0$ such that

when (x, y) is in $\mathcal{D}$ and in the punctured disk $\Delta'_\delta(a, b)$,

then

$$|f(x, y) - L| < \epsilon.$$

In this case we write

$$\lim_{(x,y)\to(a,b)} f(x, y) = L.$$

If there is no such number L then we say that the limit of $f(x, y)$ as (x, y) approaches (a, b) does not exist.

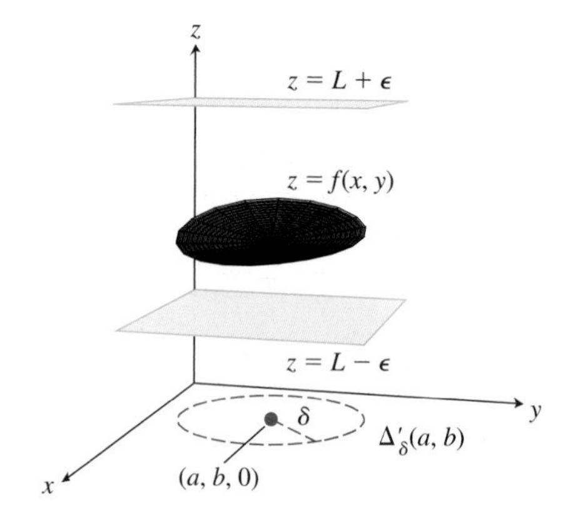

FIGURE 9.108

In plainer, but less precise language, $\lim_{(x,y)\to(a,b)} f(x, y) = L$ means that as (x, y) gets close to (a, b), $f(x, y)$ gets close to L.

We can interpret this definition graphically in several ways. For example, the definition means that the graph of $z = f(x, y)$ lies between the graphs of $z = L - \epsilon$ and $z = L + \epsilon$ whenever $(x, y) \in \Delta_\delta(a, b)$. See Fig. 9.108.

For another way of visualizing the situation, we represent the input and output of f on separate graphs. The input (x, y) comes from the punctured disk, of radius δ, as shown in the left graph in Fig. 9.109. The output, $z = f(x, y)$, is within ϵ of L and so lies in the interval $L - \epsilon < z < L + \epsilon$, as shown in the right graph of Fig. 9.109.

The definition of the limit for a function of several variables is similar in some ways to the definition of the limit for a function of one variable seen in Chapter 1. When we work with

$$\lim_{x\to a} g(x)$$

we understand that the function g may not be defined at $x = a$ and that we are making a statement about the values of $g(x)$ when x is *close to* a. Similarly, when we work with

$$\lim_{(x,y)\to(a,b)} f(x, y)$$

we know that f may not be defined at the point (a, b) but that the limit tells us about the behavior of $f(x, y)$ when (x, y) is *close to* (a, b). The point (a, b) might be an interior point of the domain of f, as suggested in Fig. 9.108, or it might be a boundary point of the domain, as suggested in Fig. 9.109. See Exercise 40 for further discussion about the nature of (a, b).

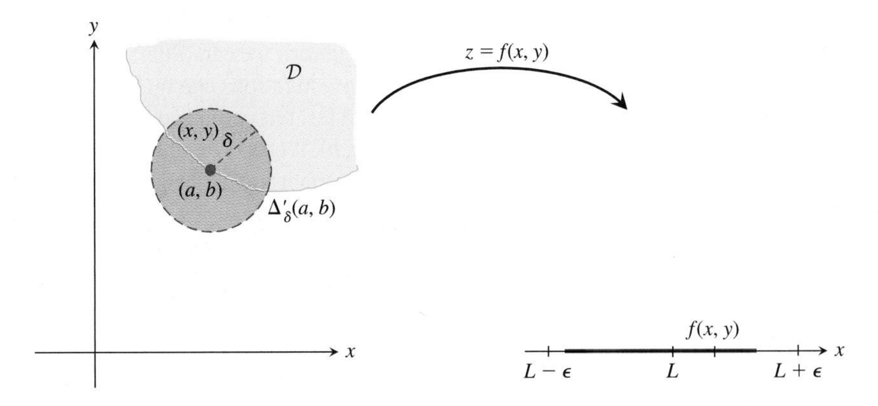

FIGURE 9.109 When (x, y) is in $\mathcal{D}$ and close to (a, b), $z = f(a, b)$ is between $L - \epsilon$ and $L + \epsilon$.

Sometimes the limit of a function of two variables can be evaluated simply by evaluating the function involved. This is true, for example, if the function involved is a polynomial, and for many expressions involving trigonometric, exponential, or logarithmic functions.

EXAMPLE 2 Evaluate the following limits.

a) $\lim_{(x,y)\to(-1,3)} (2(x+1)(y+3)^2 - 3xy)$

b) $\lim_{(r,\theta)\to(2,3\pi/4)} (r+\theta)\cos(2\theta)$

Solution

a) Let $f(x, y) = 2(x+1)(y+3)^2 - 3xy$. If (x, y) is close to $(-1, 3)$, then x is close to -1 and y is close to 3. Hence

$$2(x+1)(y+3)^2 - 3xy \quad \text{is close to} \quad 2(-1+1)(3+3)^2 - 3(-1)(3) = 9,$$

which means that

$$\lim_{(x,y)\to(-1,3)} 2(x+1)(y+3)^2 - 3xy = f(-1, 3) = 9.$$

b) Let $g(r, \theta) = (r+\theta)\cos(2\theta)$. If (r, θ) is close to $(2, 3\pi/4)$, then r is close to 2 and θ is close to $3\pi/4$. It follows that

$$(r+\theta)\cos(2\theta) \quad \text{is close to} \quad \left(2 + \frac{3\pi}{4}\right)\cos\left(2 \cdot \frac{3\pi}{4}\right) = \left(2 + \frac{3\pi}{4}\right) \cdot 0 = 0.$$

Hence,

$$\lim_{(r,\theta)\to(2,3\pi/4)} (r+\theta)\cos(2\theta) = g(2, 3\pi/4) = 0.$$

When $\lim_{(x,y)\to(a,b)} f(x, y)$ cannot be evaluated by simply evaluating $f(a, b)$, we may need to use some care in determining the behavior of $f(x, y)$ as $(x, y) \to (a, b)$. In Chapter 1 we saw that when g is a function of one variable, the limit

$$\lim_{x\to c} g(x)$$

can be investigated graphically or numerically and that from the results of these investigations we can often determine the behavior of the limit. Graphical and numerical investigations can be helpful when we are dealing with limits of functions of two variables, but care is needed. One reason is that when we say the point (x, y) is close to (a, b), it could lie in any direction from (a, b), not just to the left or right. We cannot account for the behavior of a function f in all directions around (a, b) simply by evaluating f at several points. A graph of the surface $z = f(x, y)$ may be deceiving because three-dimensional graphing packages may not give sufficient detail to let us see how a function behaves near a point.

One effective technique for investigating the behavior of

$$\lim_{(x,y)\to(a,b)} f(x, y) \tag{1}$$

is to consider the behavior of $f(x, y)$ as (x, y) approaches (a, b) along a curve C in the (x, y)-plane. In particular, suppose that the curve C is parameterized by

$$\mathbf{r}(t) = \langle x(t), y(t)\rangle \tag{2}$$

and that $\mathbf{r}(t_0) = \langle a, b\rangle$. See Fig. 9.110. Substitute $x = x(t)$ and $y = y(t)$ into $f(x, y)$ and consider the behavior of $f(x(t), y(t))$ as t approaches t_0,

$$\lim_{t\to t_0} f(x(t), y(t)). \tag{3}$$

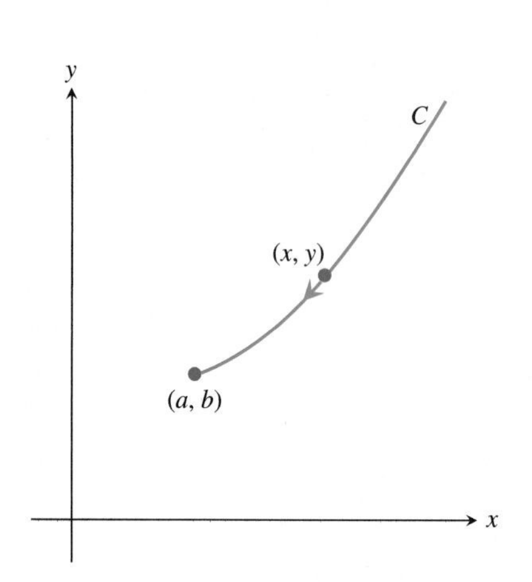

FIGURE 9.110 Investigate a limit by letting (x, y) approach (a, b) along a curve C.

This limit tells us what happens to $f(x, y)$ as (x, y) approaches (a, b) along the curve C. The limit (3) might be easier to evaluate than (1) because it involves only one variable. If we investigate limits such as (3) along several curves C and all of these limits are equal to L, we might suspect that

$$\lim_{(x,y)\to(a,b)} f(x, y) = L.$$

We can then work to establish this fact. On the other hand, if, as suggested by Fig. 9.111, we can find two curves

$$C_1 : \mathbf{r}_1(t) = \langle x_1(t), y_1(t)\rangle \quad \text{and} \quad C_2 : \mathbf{r}_2(t) = \langle x_2(t), y_2(t)\rangle$$

with $\mathbf{r}_1(t_1) = \langle a, b\rangle = \mathbf{r}_2(t_2)$ but

$$\lim_{t\to t_1} f(x_1(t), y_1(t)) = z_1 \neq z_2 = \lim_{t\to t_2} f(x_2(t), y_2(t)),$$

then

$$\lim_{(x,y)\to(a,b)} f(x, y)$$

does not exist. This is because if

$$\lim_{(x,y)\to(a,b)} f(x, y) = L,$$

and (2) describes a curve C with $\mathbf{r}(t_0) = \langle a, b\rangle$, then

$$\lim_{t\to t_0} f(x(t), y(t)) = L.$$

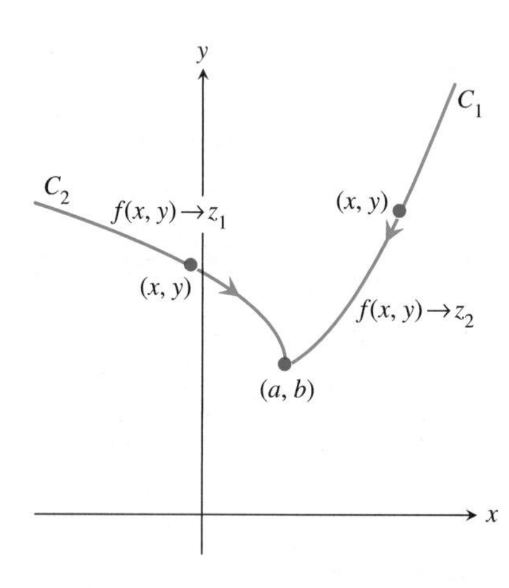

FIGURE 9.111 If we get different limits by approaching (a, b) along different curves, then $\lim_{(x,y)\to(a,b)} f(x, y)$ does not exist.

See Exercise 49.

TABLE 9.3

(x, y)	$f(x,y) = \dfrac{2x^3 + y^2}{x^2 + 2y^2}$
$(0.1, 0)$	0.2
$(0.1, 0.1)$	0.4
$(0, 0.1)$	0.5
$(-0.1, 0.1)$	0.26667
$(-0.1, 0)$	-0.2
$(-0.1, -0.1)$	0.26667
$(0, -0.1)$	0.5
$(0.1, -0.1)$	0.4

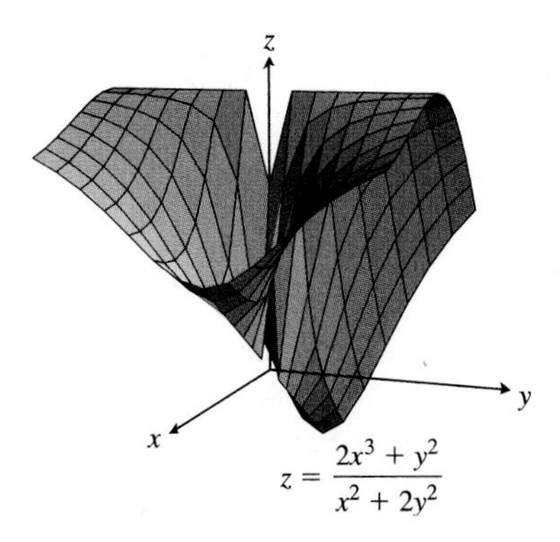

FIGURE 9.112 The graph of $z = f(x, y)$ indicates that we might get different limits if we approach the origin along the x- and y-axes.

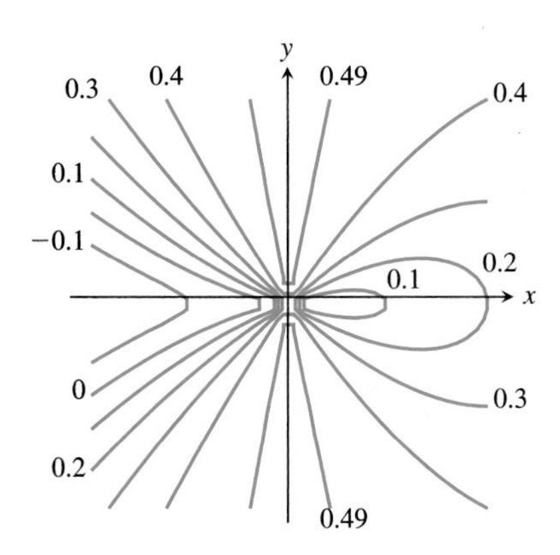

FIGURE 9.113 The level curves graph shows that near the origin, f is small on the x-axis but near $\frac{1}{2}$ on the y-axis.

EXAMPLE 3 Show that

$$\lim_{(x,y)\to(0,0)} \frac{2x^3 + y^2}{x^2 + 2y^2}$$

does not exist by finding two curves along which the function approaches different limits.

Solution

In Table 9.3, we show values of

$$z = f(x, y) = \frac{2x^3 + y^2}{x^2 + 2y^2}$$

for several (x, y) pairs close to $(0, 0)$. Although no conclusions about the limit can be drawn from this data, it does seem that the values of $f(x, y)$ do not settle down to one number as (x, y) approaches $(0, 0)$. Thus it may be the case that the limit does not exist. The graph of $z = f(x, y)$ is shown in Fig. 9.112. In the graph we see a ridge above the y-axis on which the z-coordinate appears to be constant. There is a trough along the x-axis along which the z-coordinate appears to approach 0. Hence we investigate the limit as (x, y) approaches $(0, 0)$ along the x-axis and along the y-axis. On the x-axis, points have the form $(t, 0)$. We let $t \to 0$ to approach the origin along this line. The limit of $f(x, y)$ as (x, y) approaches $(0, 0)$ along the x-axis is

$$\lim_{t\to 0} f(t, 0) = \lim_{t\to 0} \frac{2t^3 + 0^2}{t^2 + 2\cdot 0^2} = 0.$$

To approach the origin along the y-axis, take $(x, y) = (0, t)$ and again let $t \to 0$. We have

$$\lim_{t\to 0} f(0, t) = \lim_{t\to 0} \frac{2\cdot 0^3 + t^2}{0^2 + 2t^2} = \frac{1}{2}.$$

Because we get different results as we approach $(0, 0)$ from different directions, we conclude that

$$\lim_{(x,y)\to(0,0)} \frac{2x^3 + y^2}{x^2 + 2y^2}$$

does not exist.

We could also have inferred possible directions of approach by looking at a level curves plot of f, as shown in Fig. 9.113. The graph shows that level curves for many different values of f appear to meet at the origin. The curves near the x-axis correspond to f values near 0, while those near the y-axis have values near 0.5. This suggests investigating the behavior of f as we approach the origin along the x- and y-axes.

When we talk about a direction in the plane, it is natural to think of a straight-line direction. In the next example we see that such directions do not tell us all we need to know about limits.

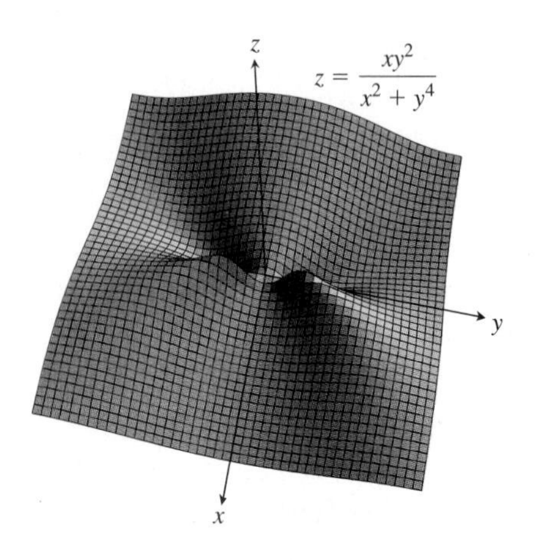

FIGURE 9.114 The ridges and valleys indicate curves along which $h(x, y)$ may not approach 0 as (x, y) approaches the origin.

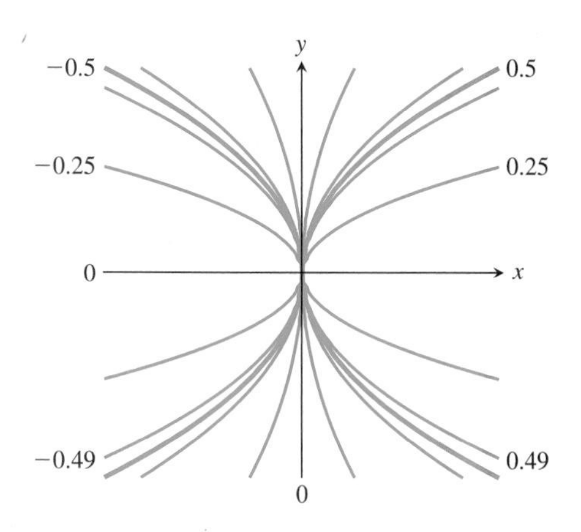

FIGURE 9.115 The level curves graph indicates paths along which the limit is not 0.

EXAMPLE 4 Let

$$h(x, y) = \frac{xy^2}{x^2 + y^4}.$$

Show that the limit of $h(x, y)$ is the same along any line through the origin but that

$$\lim_{(x,y)\to(a,b)} h(x, y) \tag{4}$$

does not exist.

Solution

Any line through the origin can be described in the form

$$\mathbf{r}(t) = \langle x(t), y(t)\rangle = \langle at, bt\rangle, \qquad -\infty < t < \infty,$$

where a and b are not both 0. As (x, y) approaches the origin along this line, $h(x, y)$ approaches

$$\lim_{t\to 0} h(at, bt) = \lim_{t\to 0} \frac{(at)(bt)^2}{(at)^2 + (bt)^4} = \lim_{t\to 0} \frac{ab^2t}{a^2 + b^4t^2} = 0.$$

Because we get the same limit along any line through the origin, it is tempting to say that the limit (4) exists and is 0. But this is not the case!

The graph of $z = h(x, y)$ is shown in Fig. 9.114. From the shading on the surface we see indications of a ridge curving toward the origin along which $h(x, y)$ may not approach 0. This can be seen even better on the level curves graph shown in Fig. 9.115. On this graph we see parabola-like curves on which $h(x, y)$ takes the value 0.5, and a similar curve on which $h(x, y)$ takes the value −0.5. If we let (x, y) approach $(0, 0)$ along the parabola

$$\mathbf{r}(t) = \langle x(t), y(t)\rangle = \langle t^2, t\rangle,$$

then $h(x, y)$ approaches

$$\lim_{t\to 0} h(x(t), y(t)) = \lim_{t\to 0} h(t^2, t) = \lim_{t\to 0} \frac{t^2t^2}{(t^2)^2 + t^4} = \frac{1}{2}.$$

Thus, the limit (4) does not exist because the value of h approaches different numbers as we approach the origin along different curves.

This example shows that we need to be careful when working with limits of functions of two variables. Even if a function has the same limit at a point when we approach the point from many different directions, it is possible that the limit does not exist. Thus if we claim that a limit exists, it is important to justify the claim with some analysis. Sometimes this can be done by relating the limit to a familiar one-variable result.

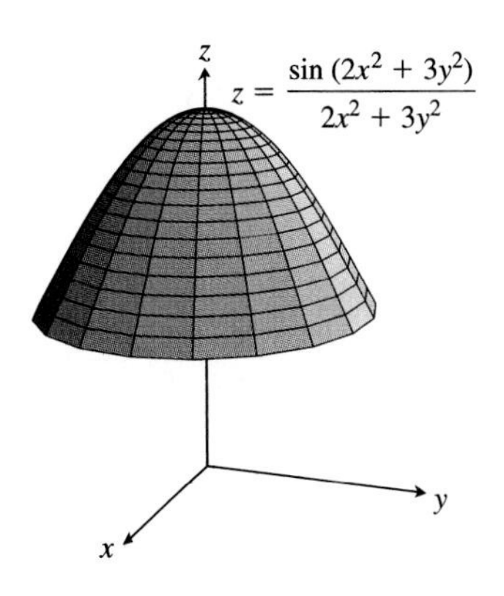

FIGURE 9.116 The graph of $z = s(x, y)$ indicates that $s(x, y)$ approaches a limit as (x, y) approaches $(0, 0)$.

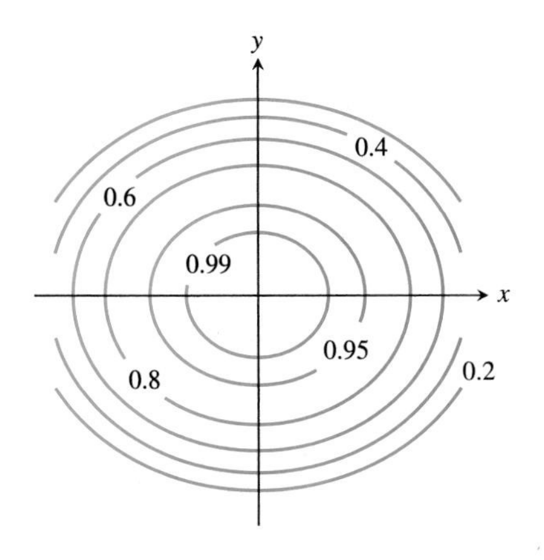

FIGURE 9.117 The level curves graph for $z = s(x, y) = \sin(2x^2 + 3y^2)/(2x^2 + 3y^2)$.

EXAMPLE 5 Let

$$s(x, y) = \frac{\sin(2x^2 + 3y^2)}{2x^2 + 3y^2}. \tag{5}$$

a) Evaluate

$$\lim_{(x,y)\to(0,0)} s(x, y).$$

b) Find a number δ such that if (x, y) is within δ of the origin, then $s(x, y)$ is within 0.01 of the limit found in **a)**. (That is, let $\epsilon = 0.01$ in the definition of limit and find a corresponding value for δ.)

Solution

a) First get a rough idea of how $s(x, y)$ behaves for (x, y) near $(0, 0)$ by looking at the surface graph and a level curves graph. These are shown in Figs. 9.116 and 9.117. Both graphs indicate that $s(x, y)$ seems to approach a single value as (x, y) approaches $(0, 0)$. Alternatively, we can evaluate (5) for a few (x, y) values close to $(0, 0)$. Based on these calculations, shown in Table 9.4, we guess that the limit, if it exists, is 1. To show that the limit is 1, first recall that

$$\lim_{r\to 0} \frac{\sin r}{r} = 1.$$

(See Section 1.5.) In the formula for $s(x, y)$, we let $2x^2 + 3y^2 = r$ and note that as (x, y) approaches $(0, 0)$, r also approaches 0. Hence

$$\lim_{(x,y)\to(0,0)} \frac{\sin(2x^2 + 3y^2)}{2x^2 + 3y^2} = \lim_{r\to 0} \frac{\sin r}{r} = 1.$$

TABLE 9.4

(x, y)	$f(x, y) = \dfrac{\sin(2x^2 + 3y^2)}{2x^2 + 3y^2}$
$(0.1, 0)$	0.999933
$(0.1, 0.1)$	0.999583
$(0, 0.1)$	0.999850
$(-0.1, 0.1)$	0.999583
$(-0.1, 0)$	0.999933
$(-0.1, -0.1)$	0.999583
$(0, -0.1)$	0.999850
$(0.1, -0.1)$	0.999583

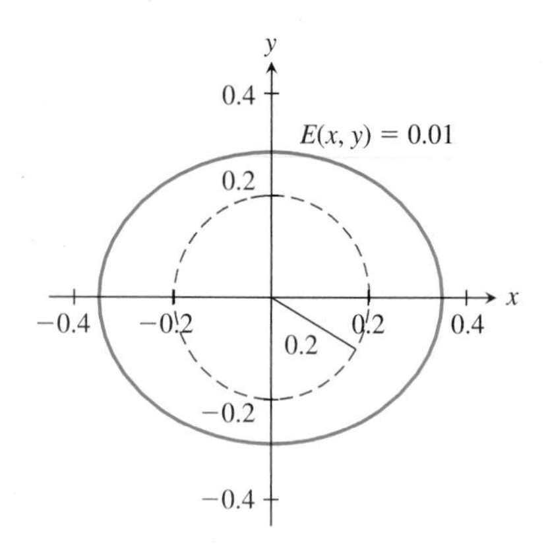

FIGURE 9.118 The disk of center $(0, 0)$ and radius 0.2 lies inside the level curve $E(x, y) = 0.01$.

b) Because the limit found in **a)** is 1, let

$$E(x, y) = |s(x, y) - 1|.$$

We want to make $E(x, y) < 0.01$, so we use a graphics command to generate the level curve $E(x, y) = 0.01$ for (x, y) close to $(0, 0)$. See Fig. 9.118. Assuming that $E(x, y) < 0.01$ for all (x, y) inside of this curve, we can say that this inequality holds inside any disk that lies inside the curve. One such disk is the disk of radius 0.2 centered at $(0, 0)$. Thus, if we take $\delta = 0.2$, we have

$$\left|\frac{\sin(2x^2 + 3y^2)}{2x^2 + 3y^2} - 1\right| < 0.01 \quad \text{when} \quad (x, y) \text{ is in } D_\delta(0, 0).$$

Sometimes when we guess the value of a limit, we can show that our guess is correct by using a "squeezing" argument. This technique is summarized in the following theorem.

A Squeeze Theorem

Let f and g be functions. Suppose that

$$\lim_{(x, y) \to (a, b)} g(x, y) = 0$$

and that for all (x, y) close to (a, b) we have

$$|f(x, y) - L| \le g(x, y).$$

Then

$$\lim_{(x, y) \to (a, b)} f(x, y) = L.$$

EXAMPLE 6 Let

$$f(x, y) = \frac{xy}{\sqrt{x^2 + y^2}}.$$

Find the value of

$$\lim_{(x, y) \to (0, 0)} f(x, y), \tag{6}$$

and use a squeezing argument to show that the answer is correct.

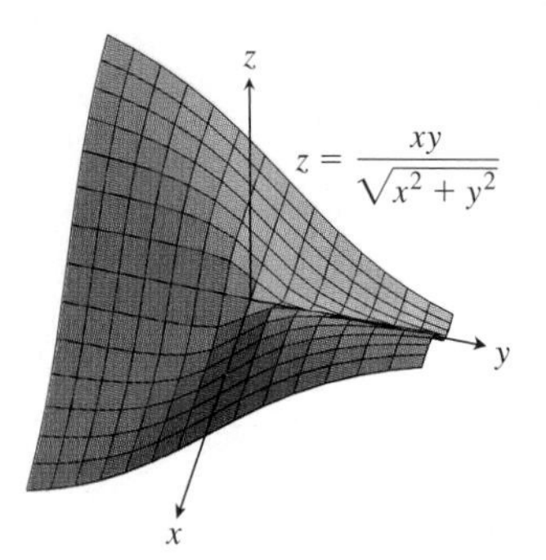

FIGURE 9.119 The graph of $z = xy/\sqrt{x^2 + y^2}$ for (x, y) near $(0, 0)$.

Solution

The graph of $z = f(x, y)$ is shown in Fig. 9.119. The graph suggests that $f(x, y)$ approaches one number as $(x, y) \to (0, 0)$. To determine this number, let $(x, y) \to (0, 0)$ along some line, say the x-axis. A point on the x-axis is represented by $(x, y) = (t, 0)$. We let $t \to 0$ to make this point approach the origin. We then have

$$\lim_{t\to 0} f(t,0) = \lim_{t\to 0} \frac{t\cdot 0}{\sqrt{t^2+0^2}} = 0.$$

Thus if the limit in (6) exists, then the value of the limit must be 0.

To show that the limit is indeed 0, we show that

$$|f(x,y)| \le |y| \quad \text{for} \quad (x,y) \ne (0,0). \tag{7}$$

To establish this fact, first note that if $x = 0$ but $y \ne 0$, then

$$|f(x,y)| = |f(0,y)| = 0 \le |y|. \tag{8}$$

On the other hand, if $x \ne 0$, then

$$|f(x,y)| = \left|\frac{xy}{\sqrt{x^2+y^2}}\right| \le \frac{|xy|}{\sqrt{x^2+0}} = \frac{|xy|}{|x|} = |y|. \tag{9}$$

Combining (8) and (9) gives (7). Because

$$\lim_{(x,y)\to(0,0)} |y| = 0 \quad \text{and} \quad |f(x,y)| \le |y|,$$

it follows from the Squeeze Theorem that

$$\lim_{(x,y)\to(0,0)} f(x,y) = \lim_{(x,y)\to(0,0)} \frac{xy}{\sqrt{x^2+y^2}} = 0.$$

Once a limit is known, we can, in some cases, use the definition of limit to get more information about how the function differs from the limit for various domain values.

EXAMPLE 7 Let $g(x,y) = xy - 3x^2 + 2y$, so

$$\lim_{(x,y)\to(-1,5)} g(x,y) = 2.$$

Show that if (x,y) is in the disk $\Delta_\delta(-1,5)$ of center $(-1,5)$ and radius $\delta < 1$, then

$$|g(x,y) - 2| < 15\delta.$$

In particular, find a δ so that $|g(x,y) - 2| < 0.005$ for (x,y) in $\Delta_\delta(-1,5)$.

Solution

We need to relate

$$g(x,y) - 2 = xy - 3x^2 + 2y - 2 \tag{10}$$

to the distance of (x,y) from $(-1,5)$. We do this by expressing (10) in terms of $x - (-1) = x + 1$ and $y - 5$. We use these expressions because they are related to the distance from $(-1,5)$ to (x,y). We perform the conversion by replacing x by $(x+1) - 1$ and y by $(y-5)+5$ in (10), then expanding the expression while keeping the $(x+1)$ and $(y-5)$ terms intact. For example, the term xy becomes

$$\begin{aligned} xy &= [(x+1)-1][(y-5)+5] \\ &= (x+1)(y+5) + 5(x+1) - (y-5) - 5. \end{aligned}$$

Using this technique on (10), we have

$$\begin{aligned} xy - 3x^2 + 2y - 2 &= [(x+1)-1][(y-5)+5] \\ &\quad - 3[(x+1)-1]^2 + 2[(y-5)-5] - 2 \\ &= (x+1)(y-5) - 3(x+1)^2 \\ &\quad + 11(x+1) + (y-5). \end{aligned} \tag{11}$$

Now suppose that (x, y) is in the disk $\Delta_\delta(x, y)$. Then

$$|x+1| < \delta \quad \text{and} \quad |y-5| < \delta. \tag{12}$$

See Fig. 9.120. Combining (11) and (12), we find

$$\begin{aligned} |g(x,y) - 2| &= |xy - 3x^2 + y - 2| \\ &= |(x+1)(y-5) - 3(x+1)^2 + 11(x+1) + (y-5)|. \end{aligned} \tag{13}$$

Using the triangle inequality (see Appendix Section A1),

$$\begin{aligned} |g(x,y) - 2| &\le |x+1|\,|y-5| + 3|x+1|^2 + 11|x+1| + |y-5| \\ &\le \delta \cdot \delta + 3\delta^2 + 11\delta + \delta = 3\delta^2 + 12\delta. \end{aligned}$$

Because $\delta < 1$, we know that $\delta^2 < \delta$. Using this in (13), we arrive at

$$|g(x,y) - 2| = |xy - 3x^2 + y - 8| < 15\delta. \tag{14}$$

Using (14), we see that $|g(x,y) - 2| < 0.005$ if $15\delta \le 0.005$. In particular, $|g(x,y) - 2| < 0.005$ if (x, y) is within $\delta = 0.005/15 \approx 0.000333$ of $(-1, 5)$.

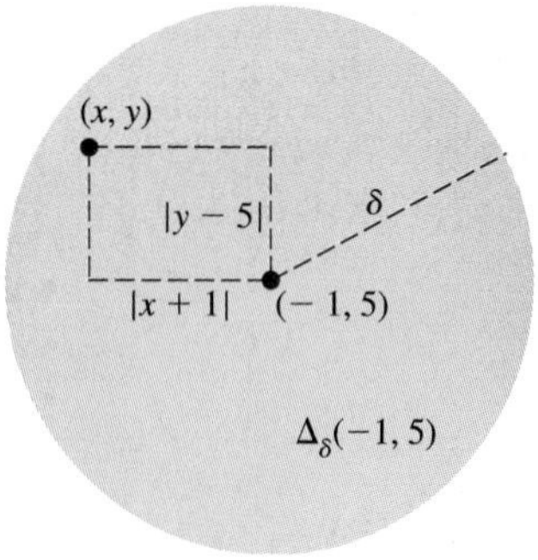

FIGURE 9.120 If (x, y) is in $\Delta_\delta(-1, 5)$, then $|x+1| < \delta$ and $|y-5| < \delta$.

Continuous Functions

A polynomial in two variables x and y is a function that can be represented as a sum of finitely many expressions of the form

$$ax^k y^j,$$

where a is a real number and k and j are nonnegative integers. For example, each of the functions q and r defined by

$$q(x,y) = 3x^3y^2 - 3x^2y + 3x^3 - xy + 12y^4 - 17$$

and

$$r(x,y) = 2(x - 4y)^{15}$$

is a polynomial in x and y. With a little effort we can see that if p is a polynomial and a and b are real numbers, then

$$\lim_{(x,y)\to(a,b)} p(x,y) = p(a,b). \tag{15}$$

More generally, when the limit of a function f at a point (a, b) agrees with the value of f at (a, b), as in (15), we say that f is *continuous* at (a, b).

DEFINITION Continuous Function

Let f be a function of two variables and let (a, b) be in the domain $\mathcal{D}$ of f. We say that f is **continuous** at (a, b) if

$$\lim_{(x,y)\to(a,b)} f(x, y) = f(a, b).$$

If f is not continuous at (a, b), then we say that f is **discontinuous** at (a, b) and that f has a **discontinuity** at (a, b). If f is continuous at every point of $\mathcal{D}$, then we say that f is **continuous on** $\mathcal{D}$.

Just as was the case for functions of one variable, sums and products of continuous functions of several variables are also continuous. The quotient of two continuous functions is continuous at points where the denominator is not zero. Compositions of continuous functions are also continuous.

Properties of Continuous Functions

Let f and g be real-valued functions of two variables, and assume that f and g are continuous on the set $\mathcal{D} \subset R^2$. Then the functions s and m defined by

$$s(x, y) = f(x, y) + g(x, y) \quad \text{and} \quad m(x, y) = f(x, y)g(x, y)$$

are both continuous on $\mathcal{D}$. The function q defined by

$$q(x, y) = \frac{f(x, y)}{g(x, y)}$$

is continuous at all points (x, y) of $\mathcal{D}$ for which $g(x, y) \neq 0$. Let f be a real-valued function of two variables and h a real-valued function of one variable. If f is continuous at (a, b) and h is continuous at $f(a, b)$, then the function $h \circ f$ is continuous at (a, b). In other words,

$$\lim_{(x,y)\to(a,b)} h(f(x, y)) = h(f(a, b)).$$

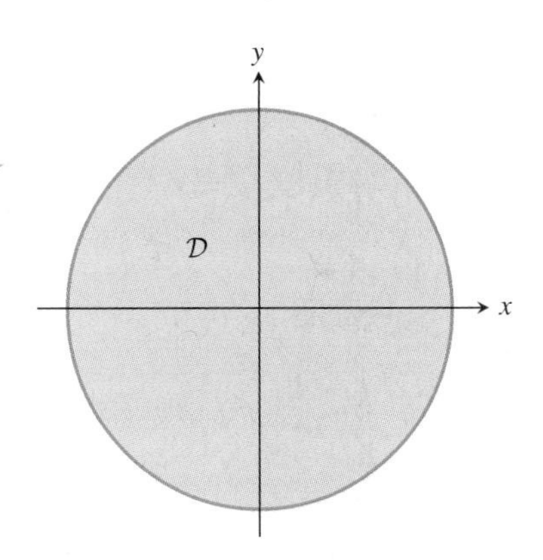

FIGURE 9.121 The function defined by $f(x, y) = \sqrt{1 - x^2 - y^2}$ is continuous on $\mathcal{D}$.

EXAMPLE 8 Let F be the function with domain $\mathcal{D} = \{(x, y) : x^2 + y^2 \leq 1\}$ and defined by

$$F(x, y) = \sqrt{1 - x^2 - y^2}, \qquad (x, y) \text{ in } \mathcal{D}.$$

Is F continuous on $\mathcal{D}$? The set $\mathcal{D}$ is shown in Fig. 9.121.

Solution

Yes. The polynomial $p(x, y) = 1 - x^2 - y^2$ is continuous and nonnegative on $\mathcal{D}$. Because the square root function s is continuous at every positive number and continuous from the right at 0, it follows that the composition

$$(s \circ p)(x, y) = s(p(x, y)) = \sqrt{1 - x^2 - y^2} = F(x, y)$$

is continuous on $\mathcal{D}$.

Other Limit Definitions

Limits of real-valued functions of three (or more) variables are handled much like limits of functions of two variables. We give the definition of such a limit here but leave further discussion to the exercises.

DEFINITION Limit of a Function of Three Variables

Let f be a function of three variables with domain $\mathcal{D}$, let (a, b, c) be a point in three-space, and let L be a real number. We say that f **has limit** L **as** (x, y, z) **approaches** (a, b, c) if

given any number $\epsilon > 0$,

there is a number $\delta > 0$ such that

when (x, y, z) is in $\mathcal{D}$ and $0 < \sqrt{(x-a)^2 + (y-b)^2 + (z-c)^2} < \delta$,

then

$$|f(x, y, z) - L| < \epsilon.$$

When this is the case, we write

$$\lim_{(x,y,z)\to(a,b,c)} f(x, y, z) = L.$$

If there is no such number L, then we say that the limit of $f(x, y, z)$ as (x, y, z) approaches (a, b, c) does not exist.

Sometimes the functions we work with have vector output. We define the limit of a vector-valued function of two variables by using the definition of the limit of a real-valued function.

DEFINITION Limit of a Vector-Valued Function

Let $\mathbf{F}$ be a function with domain and range in R^2 and given by

$$\mathbf{F}(x, y) = \langle f_1(x, y), f_2(x, y) \rangle.$$

We say that $\mathbf{F}(x, y)$ has limit $\mathbf{L} = \langle L_1, L_2 \rangle$ as (x, y) approaches (a, b) if

$$\lim_{(x,y)\to(a,b)} f_1(x, y) = L_1 \quad \text{and} \quad \lim_{(x,y)\to(a,b)} f_2(x, y) = L_2.$$

In this case we write

$$\lim_{(x,y)\to(a,b)} \mathbf{F}(x, y) = \mathbf{L}.$$

Exercises 9.6

Exercises 1–16: Evaluate the limit or show that the limit does not exist.

1. $\lim_{(x,y)\to(-2,4)} (3x^2y^2 - 5x^2y + 6x - 4)$

2. $\lim_{(x,y)\to(1,0)} (x + y)^2 e^{2x-y}$

3. $\lim_{(x,y,z)\to(-2,4,1)} (x^2 - 2y^2 + 3z^2)^5$

4. $\lim_{(x,y,z)\to(1,0,-1)} \dfrac{3xy^2z - 4x + z}{x^2 - 2y + 5z^3}$

5. $\lim_{(x,y)\to(0,0)} \dfrac{x^2 - y^2}{x^2 + y^2}$

6. $\lim_{(x,y)\to(0,0)} \dfrac{x^4 - y^4}{x^2 + y^2}$

7. $\lim_{(x,y)\to(0,0)} \dfrac{3x + 2y}{3|x| + 2|y|}$

8. $\lim_{(x,y)\to(0,0)} \dfrac{3x^2 + 2y^2}{3|x| - 2|y|}$

9. $\lim_{(x,y)\to(1,4)} \dfrac{3x - 2y + 5}{|x - 1| + |y - 4|}$

10. $\lim_{(x,y)\to(0,0)} \dfrac{3xy^2}{2x^2 + 4y^2}$

11. $\lim_{(x,y,z)\to(0,0,0)} \dfrac{\tan(x^2 + 2y^2 + 3z^2)}{x^2 + 2y^2 + 3z^2}$

12. $\lim_{(x,y,z)\to(0,0,0)} \dfrac{x^4 + y^3 + z^2}{x^2 + y^2 + z^2}$

13. $\lim_{(x,y,z)\to(0,0,0)} \ln(2x - 3y + z)$

14. $\lim_{(x,y,z)\to(0,0,0)} e^{-1/(2x^2+3y^2+3z^2)}$

15. $\lim_{(x,y,z)\to(0,0,0)} \left((2x - y + z)\sin\left(\dfrac{1}{4x^2 + y^2 + 8z^2}\right)\right)$

16. $\lim_{(x,y,z)\to(0,0,0)} \dfrac{x^2}{y^2 + (1/z^2)}$

Exercises 17–20: Indicate the boundary points and interior points of the illustrated set.

17.

18.

19.

20.

Exercises 21–24: Sketch the region $\mathcal{D}$. Describe the boundary and interior points of $\mathcal{D}$. Is $\mathcal{D}$ a closed set?

21. $\mathcal{D} = \{(x, y) : x^2 + y^2 \le 1\}$

22. $\mathcal{D} = \{(x, y) : 1 < x^2 + y^2 \le 2\}$

23. $\mathcal{D} = \{(x, y) : 0 \le x \le 2, 0 \le y \le 2 - x\}$

24. $\mathcal{D} = \{(x, y) : |x| + |y| < 1\}$

Exercises 25–26: Let f be a function with $\lim_{(x,y)\to(1,2)} f(x, y) = L$. A level curves graph of $z = |f(x, y) - L|$ is shown. Give a $\delta > 0$ so that $|f(x, y) - L| < 0.01$ when $0 < \sqrt{(x - 1)^2 + (y - 2)^2} < \delta$. Sketch the corresponding disk of radius δ on the illustration.

25.

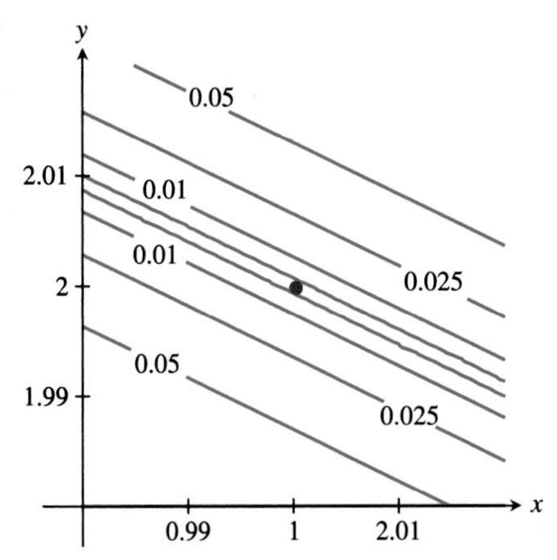

26.

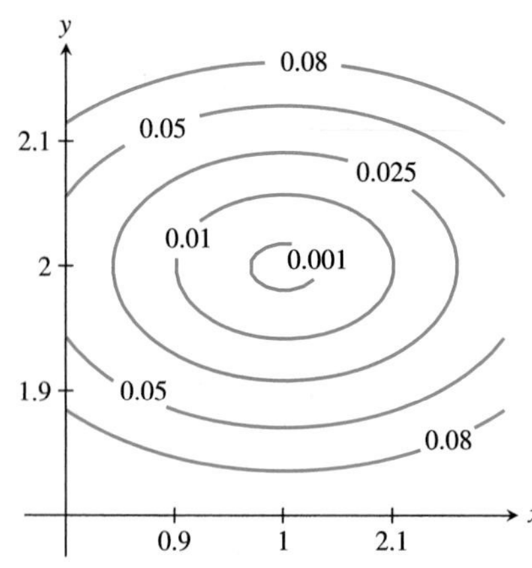

Exercises 27–30: A function f is given and a set $\mathcal{D}$ is described. Sketch $\mathcal{D}$, and list or describe the points of $\mathcal{D}$ at which f is not continuous.

27. $f(x,y) = (3x + y^2 + 4)/(3x - 2y + 1)$,
$\mathcal{D} = \{(x,y): x > 0, y \in R\}$

28. $f(x,y) = \sin(x^2 + y^2)/(x^2 + y^2)$,
$\mathcal{D} = \{(x,y): x^2 + y^2 \leq 1\}$

29. $f(x,y) = \ln(4x - 2y + 1)$,
$\mathcal{D} = \{(x,y): -1 < x, y < 1\}$

30. $f(x,y) = \tan(x^2 + y^2 + \pi/4)$,
$\mathcal{D} = \{(x,y): 1 < x^2 + y^2 < 5\}$

Exercises 31–34: Evaluate $\lim_{(x,y)\to(a,b)} f(x,y)$ for the given function f and (a,b), and denote the value of the limit by L. Then, with the aid of a level curves graph, find a $\delta > 0$ so that $|f(x,y) - L| < 0.005$ when $0 < \sqrt{(x-a)^2 + (y-b)^2} < \delta$.

T **31.** $f(x,y) = 3x^2y - 2xy^2 + 3x$, $(a,b) = (-2,2)$

T **32.** $f(x,y) = \dfrac{2x^2y}{x^2 + y^2}$, $(a,b) = (0,0)$

T **33.** $f(x,y) = \ln(2x^2 + y^3)$, $(a,b) = (1,0)$

T **34.** $f(x,y) = \dfrac{e^{-1/(x^2+y^2)}}{x^2 + y^2}$, $(a,b) = (0,0)$

T **35.** Let $f(x,y) = x^2 + 2xy + 2y^2$, so

$$\lim_{(x,y)\to(1,1)} f(x,y) = 5.$$

a. Let $0 < \delta < 1$. Show that if (x,y) is in $\Delta_\delta(1,1)$, then

$$|f(x,y) - 5| < 15\delta.$$

b. Let $\epsilon = 0.0003$ in the definition of limit. Find a corresponding value of δ.

36. Let $g(x,y) = xy - 2x^2 + y$, so

$$\lim_{(x,y)\to(2,0)} g(x,y) = -8.$$

a. Let $0 < \delta < 1$. Show that if (x,y) is in $\Delta_\delta(2,0)$, then

$$|g(x,y) + 8| < 14\delta.$$

b. Let $\epsilon = 0.001$ in the definition of limit. Find a corresponding value of δ.

37. Let $h(s,t) = s^2t$, so

$$\lim_{(s,t)\to(3,-1)} h(s,t) = -9.$$

a. Let $0 < \delta < 1$. Show that if (s,t) is in $\Delta_\delta(3,-1)$, then

$$|h(s,t) + 9| < 23\delta.$$

b. Let $\epsilon = 0.0001$ in the definition of limit. Find a corresponding value of δ.

38. Let $F(x,y) = (x + y)^3$, so

$$\lim_{(x,y)\to(-2,3)} F(x,y) = 1.$$

a. Let $0 < \delta < 1$. Show that if (x,y) is in $\Delta_\delta(-2,3)$, then

$$|F(x,y) - 1| < 26\delta.$$

b. Let $\epsilon = 0.005$ in the definition of limit. Find a corresponding value of δ.

39. Let p and q be polynomials in two variables with $q(a,b) \neq 0$. Write an argument in support of the assertion

$$\lim_{(x,y)\to(a,b)} \frac{p(x,y)}{q(x,y)} = \frac{p(a,b)}{q(a,b)}.$$

40. Let $\mathcal{D}$ be a set in the plane. A point (a,b) in $\mathcal{D}$ is called an **isolated point** of $\mathcal{D}$ if there is a positive number $r > 0$ such that the punctured disk $\Delta'_r(a,b)$ has no points in common with $\mathcal{D}$. If f is a real-valued function of two variables with domain $\mathcal{D}$ and (a,b) is an isolated point of $\mathcal{D}$, then it is always true that f is continuous at (a,b). Comment on this statement, taking into account the definitions of limit and continuity.

41. In this chapter we defined interior and boundary points for subsets of R^2. By modifying these definitions appropriately, write definitions for interior and boundary points for a subset $\mathcal{D}$ of R^3.

42. Consider the approximation

$$\begin{aligned} f(x,y) &= x^2 + xy + 2y^2 + 3 \\ &\approx -1 + 4x + 2y \end{aligned} \tag{16}$$

for (x,y) close to $(2,0)$.

T a. With the aid of a level curves graph, find $\delta > 0$ so that the error in the approximation is less than 0.001 when (x,y) is within δ of $(2,0)$.

b. Use (16) to estimate the value of $f(2.05, -0.04)$. Compare with the answer given by your calculator, and then use the result of part **a** to estimate the error in the approximation.

43. Consider the approximation

$$f(x, y) = \sqrt{xy} \approx \frac{1}{4}x + y, \tag{17}$$

for (x, y) close to $(4, 1)$.

T **a.** With the aid of a level curves graph, find $\delta > 0$ so that the error in the approximation is less than 0.005 when (x, y) is within δ of $(4, 1)$.

b. Use (17) to estimate the value of $f(3.85, 1.07)$. Compare with the answer given by your calculator, and then use the result of **a** to estimate the error in the approximation.

44. Consider the approximation

$$f(x, y) = \frac{2x - 3y + 6}{x + 4y + 2} \approx 3 - \frac{1}{2}x - \frac{15}{2}y, \tag{18}$$

for (x, y) close to $(0, 0)$.

T **a.** With the aid of a level curves graph, find $\delta > 0$ so that the error in the approximation is less than 0.001 when (x, y) is within δ of $(0, 0)$.

b. Use (18) to estimate the value of $f(-0.04, 0.12)$. Compare with the answer given by your calculator, and then use the result of **a** to estimate the error in the approximation.

45. Consider the square region $\mathcal{D}$ of Example 1.

a. Let Q be a point inside the square; that is, $Q = (x_0, y_0)$ with $-1 < x_0, y_0 < 1$. Describe how you would find a positive number r such that the disk of center Q and radius r is contained in D. What does this prove about the point Q in D?

b. Let R be a point outside of the square; that is, $R = (x_1, y_1)$ with $|x_1| > 1$ or $|y_1| > 1$. Describe how you would find a positive number r such that the disk of center R and radius r does not intersect the region $\mathcal{D}$. What does this imply about R?

46. Let

$$g(x, y) = \frac{x^3y^4}{x^6 + 2y^8}.$$

Show that the limit of $g(x, y)$ as $(x, y) \longrightarrow (0, 0)$ is the same along any line through the origin but that

$$\lim_{(x,y)\to(0,0)} g(x, y)$$

does not exist.

47. Let m and n be positive integers with $m \neq n$ and let

$$F(x, y) = \frac{x^m y^n}{x^{2m} + y^{2n}}.$$

Show that the limit of $F(x, y)$ as $(x, y) \longrightarrow (0, 0)$ is the same along any line through the origin but that

$$\lim_{(x,y)\to(0,0)} F(x, y)$$

does not exist. What is the situation when $m = n$?

48. Let

$$f(x, y) = \begin{cases} \dfrac{e^{-1/x^2}y}{e^{-2/x^2} + y^2} & \text{if } x \neq 0 \\ 0 & \text{if } x = 0. \end{cases}$$

a. Let the curve C be the graph of $y = ax^{m/n}$, for $x \geq 0$, where a is any nonzero constant and m, n are positive integers. Show that as (x, y) approaches $(0, 0)$ along any such curve C, $f(x, y)$ approaches 0.

b. Show that $\lim_{(x,y)\to(0,0)} f(x, y)$ does not exist.

49. Suppose

$$\lim_{(x,y)\to(a,b)} f(x, y) = L.$$

Let C be a curve described by the continuous expression

$$\mathbf{r} = \mathbf{r}(t) = \langle x(t), y(t)\rangle$$

such that $\mathbf{r}(t_0) = \langle a, b\rangle$. Show that

$$\lim_{t\to t_0} f(\mathbf{r}(t)) = \lim_{t\to t_0} f(x(t), y(t)) = L.$$

9.7 Derivatives

Figure 9.122 shows a topographical map for the region around Jackstraw Mountain in Rocky Mountain National Park. Imagine that you are at the point corresponding to P on the map. If you hike in the direction of the arrow marked N, you encounter an upward slope. In the direction opposite N, you move downward at a

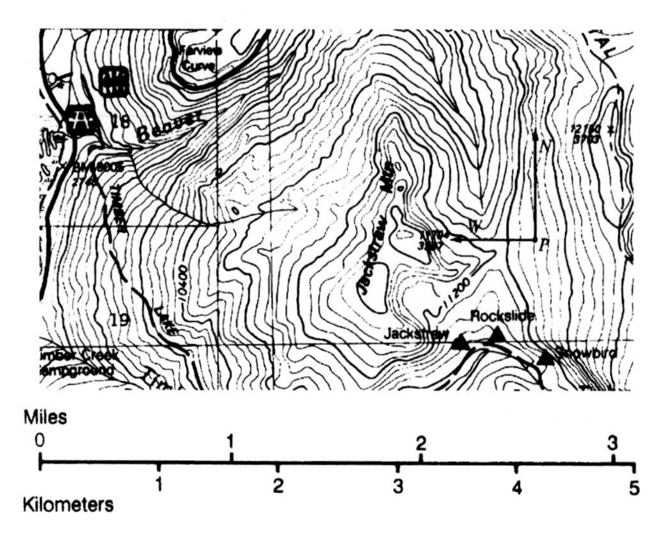

FIGURE 9.122 The region around Jackstraw Mountain in Rocky Mountain National Park.

gradual slope, and in direction W your path is neither rising nor falling because you are moving in the direction of a level curve. The slope of your hiking path, calculated by dividing your change in vertical feet by your change in horizontal feet, tells you the rate of change of vertical feet with respect to horizontal feet. As seen from Fig. 9.122, the slope (or rate of change) is different in different directions. Thus if a friend and you are resting at point P and the friend asks you, "What's the slope here?" you have to respond, "In what direction?"

Recall that a topographical map is a level curves graph of a function H of two variables. Corresponding to an input of a location (x, y) is the altitude $H(x, y)$ above or below sea level at (x, y). As we noted earlier, the slope or rate of change of altitude H with respect to horizontal change at a point may be different depending on the direction of the horizontal change. In this section we study slope as a function of direction for functions of several variables.

Directional Derivatives

INVESTIGATION

Rate of Change of Sound Intensity

In Example 4 of Section 9.3 we discussed the intensity of sound from three speakers in a 200-foot-by-150-foot gymnasium. We imagined the gymnasium as a subset of R^3 with the origin at the lower southwest corner and the positive x-axis running east. We found that the intensity of the sound at the point (x, y) on the floor was proportional to

$$I(x, y) = \frac{40}{(x-100)^2 + (y-150)^2 + 400} + \frac{40}{(x-150)^2 + y^2 + 400} + \frac{40}{x^2 + y^2 + 400},$$

where $I(x, y)$ has units of watts/m^2.

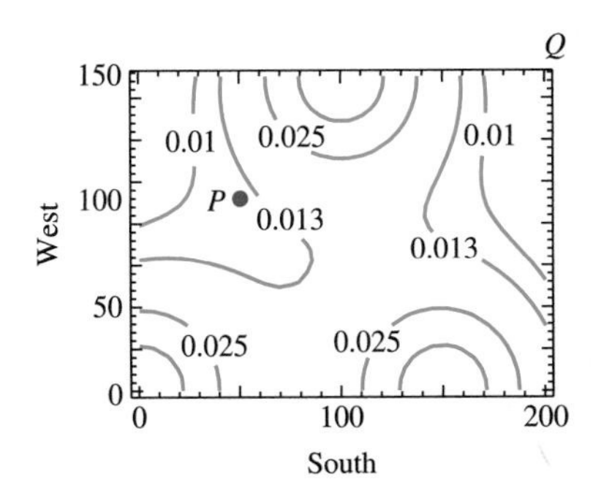

FIGURE 9.123 A level curves graph of sound intensity on the floor of the gymnasium.

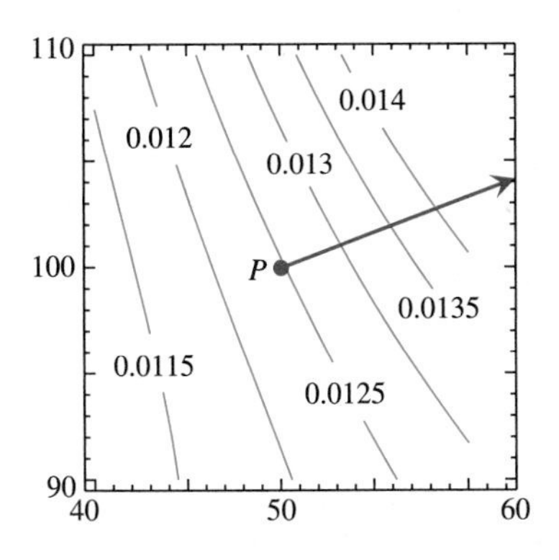

FIGURE 9.124 The principal walks from $P = (50, 100)$ toward the northeast corner of the gym.

By working with the level curves of I, we found that the quietest place in the gym was at the northeast corner, that is, at the point $Q = (200, 150)$. See Fig. 9.123. Now suppose the principal (who values quiet) is at $P = (50, 100)$ on the floor and wants to move to a quieter spot. She heads directly for the northeast corner. As she leaves the point $(50, 100)$, does it appear to get quieter? At what rate? If she wants to move immediately to a quieter area, should she walk in a different direction?

In Fig. 9.124 we have enlarged the region of the floor near the principal's location $P = (50, 100)$. Her direction of travel (directly toward the northeast corner) is indicated by an arrow. As the principal walks in the direction of the arrow, the labels on the level curves increase in value, so the music sounds louder. To get an idea of the rate at which the sound intensity is changing, we calculate the intensity at the starting point and at a nearby point, take the difference, and divide by the distance traveled:

$$\text{Average rate of change of intensity} = \frac{\text{Change in intensity}}{\text{Distance traveled}}.$$

The resulting quotient has units of $(\text{W/m}^2)/\text{ft}$. The direction of travel from $P = (50, 100)$ toward the northeast corner, $Q = (200, 150)$, is in the direction of the vector

$$\mathbf{v} = \mathbf{Q} - \mathbf{P} = \langle 200, 150\rangle - \langle 50, 100\rangle = \langle 150, 50\rangle.$$

The point described by

$$\mathbf{P} + 0.01\mathbf{v} = \langle 50, 100\rangle + 0.01\langle 150, 50\rangle = \langle 51.5, 100.5\rangle$$

is on the path traveled by the principal and is near the starting point. The distance of this point from the starting point P is

$$\|(\mathbf{P} + 0.01\mathbf{v}) - \mathbf{P}\| = \|0.01\mathbf{v}\| = \|\langle 1.5, 0.5\rangle\| = \sqrt{2.5} \approx 1.58114 \text{ ft}.$$

Thus as the principal moves from $(50, 100)$ to $(51.5, 100.5)$, the average rate of change of the intensity of the sound is

$$\begin{aligned}\frac{I(\mathbf{P} + 0.01\mathbf{v}) - I(\mathbf{P})}{\|(\mathbf{P} + 0.01\mathbf{v}) - \mathbf{P}\|} &= \frac{I(51.5, 100.5) - I(50, 100)}{\|0.01\mathbf{v}\|} \\ &\approx \frac{0.0127098 - 0.0124690}{1.58114} \\ &\approx 0.000152295 \ (\text{W/m}^2)/\text{ft}.\end{aligned}$$

This means that as the principal walks from $(50, 150)$ to $(51.5, 100.5)$, the noise intensity increases at an average of $0.000152295 \text{ W/m}^2$ with each foot of progress. Is this figure consistent with Fig. 9.124?

To get a better idea of the rate of change of the sound intensity at the instant that the principal walks from P, we repeat the average-rate-of-change calculation over smaller and smaller distances. As noted in Chapter 8, we will, for convenience, use the boldfaced notation $\mathbf{Q}$ to denote both the point Q and the vector $\mathbf{Q}$. Recalling the definition of the derivative in Chapter 1, we recalculate the average in a way that makes it easy to change the distance traveled. Since the direction the principal walks is given by the vector $\mathbf{v}$, the points on the principal's path are given by

$$\mathbf{P} + h\mathbf{v} = \langle 50, 100\rangle + h\langle 150, 50\rangle = \langle 50 + 150h, 100 + 50h\rangle, \tag{1}$$

where $h > 0$. The average rate of change of sound intensity as the principal walks from **P** to **P** + h**v** is given by

$$\frac{I(\mathbf{P} + h\mathbf{v}) - I(\mathbf{P})}{\|(\mathbf{P} + h\mathbf{v}) - \mathbf{P}\|} = \frac{I(50 + 150h, 100 + 50h) - I(50, 100)}{\|\langle 50 + 150h, 100 + 50h\rangle - \langle 50, 100\rangle\|} = \frac{I(50 + 150h, 100 + 50h) - I(50, 100)}{50\sqrt{10}h}. \tag{2}$$

By substituting various positive values of h into (2), we can find the average rate of change of the sound intensity as the principal walks from P in the direction **a**. To find the instantaneous rate of change of sound at P in the direction of **v**, we take the limit of (2) as $h \to 0^+$. We denote this quantity by $D_{\mathbf{v}}I(P)$ and call it the *directional derivative* of I at P in the direction **v**:

$$D_{\mathbf{v}}I(P) = \lim_{h \to 0^+} \frac{I(\mathbf{P} + h\mathbf{v}) - I(\mathbf{P})}{\|h\mathbf{v}\|} = \lim_{h \to 0^+} \frac{I(50 + 150h, 100 + 50h) - I(50, 100)}{50\sqrt{10}h}.$$

In Exercise 37 we ask you to evaluate this limit. For now, we do so by graphing the expression

$$R = \frac{I(50 + 150h, 100 + 50h) - I(50, 100)}{50\sqrt{10}h}$$

for positive h values near 0. From the graph, shown in Fig. 9.125, we see that the limit is about 1.5×10^{-4} (W/m^2)/ft. Thus, just as the principal moves from P in the direction of **v**, the rate of change of the intensity of the sound is about 1.5×10^{-4} (W/m^2)/ft. Hence, near P, the intensity increases by about 1.5×10^{-4} W/m^2 for each foot she travels.

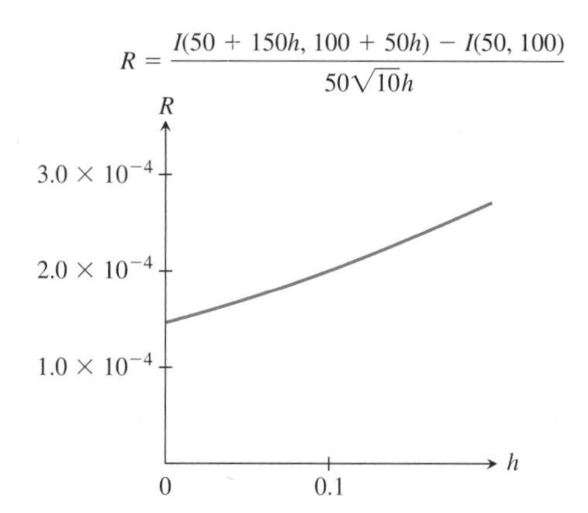

FIGURE 9.125 The graph shows that the limit in (2) is approximately 1.5×10^{-4}.

We summarize the techniques illustrated in the Investigation in the following definition.

DEFINITION Directional Derivative

Let f be a real-valued function defined on a set $\mathcal{D} \subset R^n$ ($n = 2$ or 3). Let P be an interior point of $\mathcal{D}$ and let $\mathbf{v}$ be a nonzero vector in R^n. The **directional derivative of f at P in the direction $\mathbf{v}$** is denoted $D_{\mathbf{v}}f(P)$ and is given by

$$D_{\mathbf{v}}f(P) = \lim_{h\to 0^+} \frac{f(\mathbf{P} + h\mathbf{v}) - f(\mathbf{P})}{\|\mathbf{v}\|h}, \tag{3}$$

provided this limit exists.

When we work with directional derivatives, we will often take $\mathbf{v} = \mathbf{u}$ to be a unit vector. Denoting $\mathbf{v}$ by $\mathbf{u}$, where $\|\mathbf{u}\| = 1$, (3) becomes

$$D_{\mathbf{u}}f(P) = \lim_{h\to 0^+} \frac{f(\mathbf{P} + h\mathbf{u}) - f(\mathbf{P})}{h}. \tag{4}$$

EXAMPLE 1 Let f defined by $f(x,y) = x^2y - 2x + 3$ be a real-valued function defined on R^2 and let $\mathbf{u} = \langle \frac{3}{5}, -\frac{4}{5} \rangle$. Find $D_{\mathbf{u}}f(-1,2)$.

Solution

Taking $\mathbf{P} = \langle -1, 2 \rangle$ and $\mathbf{u} = \langle \frac{3}{5}, -\frac{4}{5} \rangle$ in (4),

$$\begin{aligned} D_{\langle 3/5, -4/5 \rangle}f(-1,2) &= \lim_{h\to 0^+} \frac{f\left(\langle -1,2\rangle + h\langle \frac{3}{5}, -\frac{4}{5}\rangle\right) - f(-1,2)}{h} \\ &= \lim_{h\to 0^+} \frac{f\left(-1 + \frac{3}{5}h, 2 - \frac{4}{5}h\right) - f(-1,2)}{h} \\ &= \lim_{t\to 0^+} \frac{\left[\left(-1 + \frac{3}{5}h\right)^2\left(2 - \frac{4}{5}h\right) - 2\left(-1 + \frac{3}{5}h\right) + 3\right] - 7}{h} \\ &= \lim_{h\to 0^+} \frac{(-36h^3 + 210h^2 - 550h)/125}{h} \\ &= -4.4. \end{aligned}$$

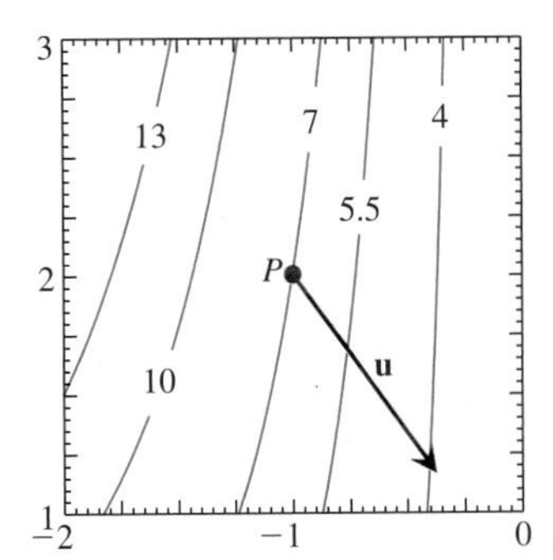

FIGURE 9.126 The level curves graph of $f(x,y) = x^2y - 2x + 3$ near the point $P = (-1,2)$. The arrow indicates the direction $\mathbf{u} = \langle \frac{3}{5}, -\frac{4}{5} \rangle$.

In Fig. 9.126, we show a level curves graph of f near the point $(-1,2)$. We see from the labels of the level curves that f decreases as we leave this point in the direction $\mathbf{u}$. This is consistent with the negative directional derivative in this direction.

EXAMPLE 2 A ball of gas has radius 1000 m and is centered at the origin. The Kelvin temperature at the point (x,y,z) in the sphere is

$$T(x,y,z) = \frac{z^2}{x^2 + 2y^2 + 1}. \tag{5}$$

A probe moves from the point $P = (100, 0, -100)$ toward the origin. At what rate (in Kelvins per meter) does the temperature sensed by the probe change as the probe leaves P?

Solution

The direction of motion is from P toward $O = \langle 0, 0, 0 \rangle$, so is in the direction of the vector $\mathbf{v} = \mathbf{O} - \mathbf{P} = \langle -100, 0, 100 \rangle$. Using (3), the directional derivative of T at P in the direction $\mathbf{v}$ is

$$\begin{aligned}
D_{\mathbf{v}}T(P) &= \lim_{h \to 0^+} \frac{T(\mathbf{P} + h\mathbf{v}) - T(\mathbf{P})}{\|\mathbf{v}\|h} \\
&= \lim_{h \to 0^+} \frac{T(100 - 100h, 0, -100 + 100h) - T(100, 0, -100)}{\|\langle -100, 0, 100 \rangle\|h} \\
&= \lim_{h \to 0^+} \frac{\dfrac{(-100 + 100h)^2}{(100 - 100h)^2 + 2 \cdot 0^2 + 1} - \dfrac{10{,}000}{10{,}001}}{100\sqrt{2}h} \\
&= \lim_{h \to 0^+} \frac{50\sqrt{2}(h - 2)}{10001(10001 - 20000h + 10000h^2)} \\
&= \frac{-100\sqrt{2}}{10001^2} \approx -1.41393 \times 10^{-6} \text{ K/m}.
\end{aligned}$$

Thus as we move from the point $P = (100, 0, -100)$ toward the origin, the temperature decreases. As we leave P, the rate of decrease is about 1.41393×10^{-6} Kelvins per meter.

Partial Derivatives

We next show that for most real-valued functions of two variables, if we know the directional derivatives at (a, b) in two directions, then we can calculate the directional derivative of f at (a, b) in an arbitrary direction. For functions of two variables the two directions usually selected are the **i**- and **j**-directions. The derivatives in these directions are called *partial derivatives.*

DEFINITION Partial Derivatives

Let f be a real-valued function with domain $\mathcal{D} \subset R^2$ and let (a, b) be an interior point of $\mathcal{D}$. The **partial derivative of f with respect to x at** (a, b) is denoted

$$\frac{\partial f}{\partial x}(a, b)$$

and is equal to

$$\frac{\partial f}{\partial x}(a, b) = \lim_{h \to 0} \frac{f(a + h, b) - f(a, b)}{h}, \tag{6}$$

provided that this limit exists. Note that if $(\partial f/\partial x)(a,b)$ exists, then

$$\frac{\partial f}{\partial x}(a,b) = D_{\mathbf{i}} f(a,b).$$

The **partial derivative of f with respect to y at (a,b)** is denoted

$$\frac{\partial f}{\partial y}(a,b)$$

and is defined by

$$\frac{\partial f}{\partial y}(a,b) = \lim_{k \to 0} \frac{f(a,b+k) - f(a,b)}{k}, \tag{7}$$

provided that this limit exists. Note that if $(\partial f/\partial y)(a,b)$ exists, then

$$\frac{\partial f}{\partial y}(a,b) = D_{\mathbf{j}} f(a,b).$$

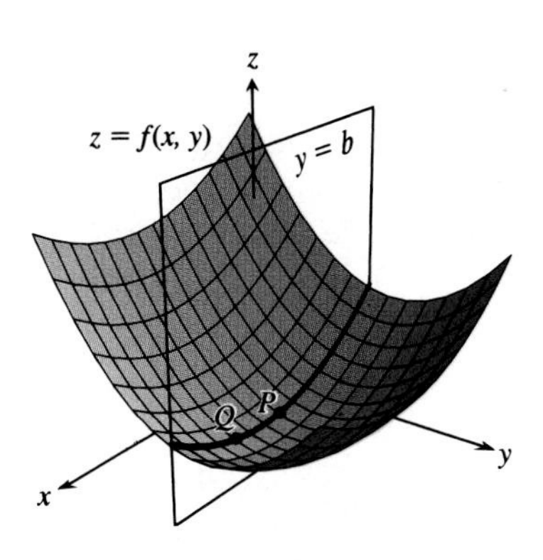

FIGURE 9.127 The cross section $y = b$ of the surface graph of $z = f(x,y)$.

In (6) the y variable is fixed at $y = b$. We can get an intuitive sense of (6) by looking at the $y = b$ cross section of the graph of $z = f(x,y)$. In Fig. 9.127 we show this cross section, and in Fig. 9.128 we have projected it into the (x,z)-plane for closer examination. (In Fig. 1.128, we view the projection from far out on the y-axis.) Figure 9.128 looks very much like the pictures that accompanied our discussions of the derivative in Chapter 1. With this in mind, the quotient

$$\frac{f(a+h,b) - f(a,b)}{h}$$

can be interpreted as the slope, in the plane $y = b$, of the line through points $P = (a,b,f(a,b))$ and $Q = (a+h,b,f(a+h,b))$. As $h \to 0$, we see that $Q \to P$ and the line determined by P and Q approaches the tangent at P to the curve $z = f(x,b)$. Thus we can interpret $(\partial f/\partial x)(a,b)$ as the slope of the line tangent at P to the cross-section curve $z = f(x,b)$. An analogous interpretation can be given to $(\partial f/\partial y)(a,b)$. The interpretation of these partial derivatives as slopes of tangent lines will be important when we discuss planes tangent to the surface described by $z = f(x,y)$.

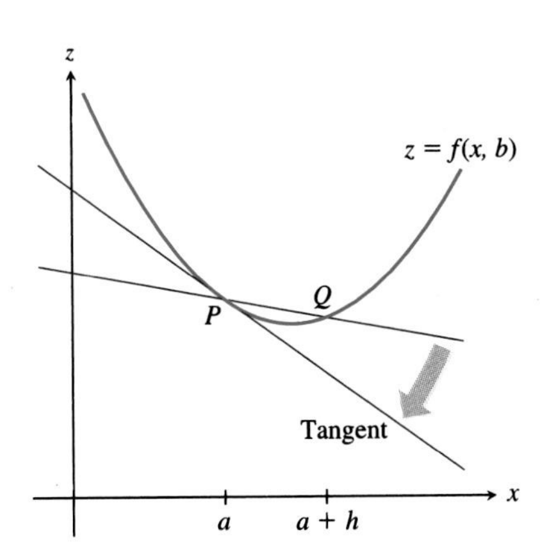

FIGURE 9.128 As Q approaches P, the line determined by P and Q approaches the tangent to the curve at P.

In Chapter 1 we also interpreted the derivative as a rate of change. This interpretation carries over to partial derivatives. If we interpret $z = f(x,b)$ as a function only of the variable x, then the quotient

$$\frac{f(a+h,b) - f(a,b)}{h} = \frac{\text{change in } z}{\text{change in } x}$$

is the average rate of change of $z = f(x,y)$ with respect to x as x changes from a to $a+h$. As $h \to 0$, this quotient approaches $(\partial f/\partial x)(a,b)$, and we interpret the partial derivative as the rate of change of z (or f) with respect to x at (a,b). This means that if we start from (a,b) and let x increase while holding $y = b$ fixed, then at (a,b) the value of z changes at a rate of $(\partial f/\partial x)(a,b)$. This rate of change has units of

$$\frac{z \text{ units}}{x \text{ units}}.$$

A similar interpretation can be given to $(\partial f/\partial y)(a,b)$.

Calculating Partial Derivatives

In (6) we see that we let only the x variable change when we are calculating $(\partial f/\partial x)(a,b)$. The y variable is fixed at $y = b$ and never changes during the limiting process. This means that when we calculate $(\partial f/\partial x)(a,b)$, we can think of f as a function of x only. With this interpretation, (6) looks just like the definition of the derivative of a one-variable function as defined in Chapter 1. Indeed, let y be fixed at $y = b$ and let F be defined by $F(x) = f(x,b)$. Thus F is the one-variable function that arises from f when we set $y = b$. Then

$$\begin{aligned} \frac{\partial f}{\partial x}(a,b) &= \lim_{h\to 0} \frac{f(a+h,b) - f(a,b)}{h} \\ &= \lim_{h\to 0} \frac{F(a+h) - F(a)}{h} = F'(a). \end{aligned} \tag{8}$$

Equation (8) suggests an efficient method for calculating partial derivatives.

Calculating Partial Derivatives

Let f be a function of two variables, say x and y. To find the partial derivative of f with respect to x, regard y as constant and differentiate $f(x,y)$ with respect to x as if it were a function of x alone. To find the partial derivative of f with respect to y, regard x as constant and differentiate $f(x,y)$ with respect to y as if it were a function of y alone.

We illustrate this technique with an example.

EXAMPLE 3 Let f be given by

$$f(x,y) = 2x^2y - 3xy^3 + 2x - y^2 + 3.$$

a) Use the technique described to find

$$\frac{\partial f}{\partial x}(2,-3) \quad \text{and} \quad \frac{\partial f}{\partial y}(2,-3).$$

b) Use the definition of *partial derivative* to calculate $(\partial f/\partial x)(2,-3)$, and check that the answer obtained agrees with that found in **a)**.

Solution

a) To calculate the partial derivative with respect to x at $(2,-3)$, we regard y as constant, in particular $y = -3$. Let F be defined by

$$\begin{aligned} F(x) = f(x,-3) &= 2x^2(-3) - 3x(-3)^3 + 2x - (-3)^2 + 3 \\ &= -6x^2 + 83x - 6. \end{aligned}$$

By (8) we have

$$\frac{\partial f}{\partial x}(2,-3) = \frac{d}{dx}F(x)\bigg|_{x=2} = (-12x + 83)\bigg|_{x=2} = 59.$$

We use a slightly different technique to calculate $(\partial f/\partial y)(2,-3)$. Regard x as a constant, thus thinking of f as a function of y alone. Differentiating with respect to y, we obtain

$$\begin{aligned}\frac{\partial f}{\partial y}(x,y) &= \frac{\partial}{\partial y}(2x^2y - 3xy^3 + 2x - y^2 + 3)\\ &= \frac{\partial}{\partial y}(2x^2y) - \frac{\partial}{\partial y}(3xy^3) + \frac{\partial}{\partial y}(2x) - \frac{\partial}{\partial y}(y^2) + \frac{\partial f}{\partial y}(3) \qquad (9)\\ &= 2x^2\cdot 1 - 3x(3y^2) + 0 - 2y + 0\\ &= 2x^2 - 9xy^2 - 2y.\end{aligned}$$

Note that in these differentiations x was treated like a constant. For example, $\frac{\partial}{\partial y}(2x) = 0$. To find the value of this derivative at $(2,-3)$, just substitute into the last line in (9):

$$\frac{\partial f}{\partial y}(2,-3) = 2(2)^2 - 9\cdot 2\cdot(-3)^2 - 2(-3) = -148.$$

b) By (6),

$$\begin{aligned}\frac{\partial f}{\partial x}(2,-3) &= \lim_{h\to 0}\frac{f(2+h,-3) - f(2,-3)}{h}\\ &= \lim_{h\to 0}\frac{(-6(2+h)^2 + 83(2+h) - 6) - 136}{h}\\ &= \lim_{h\to 0}\frac{-6h^2 + 59h}{h} = 59.\end{aligned}$$

This agrees with the answer obtained in **a)**.

Functions of Three or More Variables Partial derivatives can be defined for real-valued functions of three or more variables. A function of three variables has three partial derivatives, one for each variable. The definitions of these derivatives are similar to the definitions of the partial derivatives of a function of two variables.

DEFINITION Partial Derivatives of a Function of Three Variables

Let f be a real-valued function defined on set $\mathcal{D} \subset R^3$, and let (a,b,c) be an interior point of $\mathcal{D}$ (see Exercise 41 in Section 9.6). The **partial derivative of f with respect to x at** (a,b,c) is denoted

$$\frac{\partial f}{\partial x}(a,b,c)$$

and is given by

$$\frac{\partial f}{\partial x}(a,b,c) = \lim_{h\to 0}\frac{f(a+h,b,c) - f(a,b,c)}{h},$$

provided that this limit exists. The **partial derivative of f with respect to y at (a, b, c)** is denoted

$$\frac{\partial f}{\partial y}(a, b, c)$$

and is defined by

$$\frac{\partial f}{\partial y}(a, b, c) = \lim_{k \to 0} \frac{f(a, b + k, c) - f(a, b, c)}{k},$$

when this limit exists. The **partial derivative of f with respect to z at (a, b, c)** is denoted

$$\frac{\partial f}{\partial z}(a, b, c)$$

and is defined by

$$\frac{\partial f}{\partial z}(a, b, c) = \lim_{m \to 0} \frac{f(a, b, c + m) - f(a, b, c)}{m},$$

if this limit exists.

Note that if $(\partial f/\partial x)(a, b, c)$ exists, then $(\partial f/\partial x)(a, b, c) = D_{\mathbf{i}} f(a, b, c)$, the directional derivative of f at (a, b, c) in the direction $\mathbf{i}$. Similar statements hold for $(\partial f/\partial y)(a, b, c)$ and $(\partial f/\partial z)(a, b, c)$.

The techniques for calculating partial derivatives of functions of two variables extend naturally to functions of three variables.

Notation for Partial Derivatives In Chapter 1 we saw that there are many notations for the derivative. There are also many notations for partial derivatives. For example, if $w = f(x, y, z)$, the partial derivative of f with respect to x is denoted by any of the following:

$$\frac{\partial f}{\partial x}(x, y, z), \qquad \frac{\partial w}{\partial x}(x, y, z), \qquad f_x(x, y, z).$$

We will often drop the (x, y, z) from the notation and simply write $\partial f/\partial x$, $\partial w/\partial x$, or f_x. We use analogous notation for derivatives with respect to any of the other independent variables.

EXAMPLE 4 Let f be a real-valued function of three variables defined by

$$f(x, y, z) = xy \tan(xy^2z^3).$$

Find

$$f_x(x, y, z) \quad \text{and} \quad f_z(x, y, z).$$

Solution

For the partial derivative with respect to x, regard y and z as constant and differentiate with respect to x. Using the product rule and the chain rule, we have

$$
\begin{aligned}
f_x(x, y, z) &= \frac{\partial}{\partial x}(xy \tan(xy^2z^3)) \\
&= \left(\frac{\partial}{\partial x}(xy)\right) \tan(xy^2z^3) + xy\left(\frac{\partial}{\partial x}(\tan(xy^2z^3))\right) \\
&= y \tan(xy^2z^3) + xy \sec^2(xy^2z^3)\left(\frac{\partial}{\partial x}(xy^2z^3)\right) \\
&= y \tan(xy^2z^3) + xy^3z^3 \sec^2(xy^2z^3).
\end{aligned}
$$

Now regard x and y as constants. Then

$$
\begin{aligned}
f_z(x, y, z) &= \frac{\partial}{\partial z}(xy \tan(xy^2z^3)) = xy\left(\frac{\partial}{\partial z}(\tan(xy^2z^3))\right) \\
&= xy \sec^2(xy^2z^3)\left(\frac{\partial}{\partial z}(xy^2z^3)\right) = 3x^2y^3z^2 \sec^2(xy^2z^3).
\end{aligned}
$$

Calculating Directional Derivatives

As the first of many applications of partial derivatives, we show that they can be used in the calculation of directional derivatives. We first use the definition of partial derivative to develop an alternative formula for the change in a function in a given direction. Assume that $f_x(a, b)$ exists and let

$$\frac{f(a + h, b) - f(a, b)}{h} - f_x(a, b) = E_1(h). \tag{10}$$

As $h \to 0$, the left side of this expression has limit 0, so we also have $\lim_{h\to 0} E_1(h) = 0$. With a little algebra, (10) becomes

$$f(a + h, b) - f(a, b) = f_x(a, b)h + E_1(h)h. \tag{11}$$

When h is small, $E_1(h)$ is also small, so $hE_1(h)$ is small compared to h. By a similar argument,

$$f(a, b + k) - f(a, b) = f_y(a, b)k + E_2(k)k \tag{12}$$

where $E_2(k)$ tends to 0 as $k \to 0$.

Now let $\mathbf{u} = \langle \alpha, \beta \rangle$ be a unit vector and suppose we wish to find the directional derivative of f at (a, b) in the direction $\mathbf{u}$. From (4), and adding and subtracting $f(a, b + \beta t)$ to the numerator,

$$
\begin{aligned}
D_{\mathbf{u}} f(a, b) &= \lim_{t\to 0^+} \frac{f(a + \alpha t, b + \beta t) - f(a, b)}{t} \\
&= \lim_{t\to 0^+} \left(\frac{f(a + \alpha t, b + \beta t) - f(a, b + \beta t)}{t} \right. \\
&\qquad \left. + \frac{f(a, b + \beta t) - f(a, b)}{t} \right)
\end{aligned} \tag{13}
$$

From (11), with αt in place of h and $b + \beta t$ in place of b, we have

$$f(a + \alpha t, b + \beta t) - f(a, b + \beta t) = f_x(a, b + \beta t)(\alpha t) + E_1(\alpha t)(\alpha t). \tag{14}$$

From (12), with βt in place of k, we have

$$f(a, b + \beta t) - f(a, b) = f_y(a, b)(\beta t) + E_2(\beta t)(\beta t). \tag{15}$$

Substitute (14) and (15) into (13) to get

$$\begin{aligned} D_{\mathbf{u}} f(a, b) &= \lim_{t \to 0^+} \left(\frac{f_x(a, b + \beta t)(\alpha t) + E_1(\alpha t)(\alpha t)}{t} \right. \\ &\qquad \left. + \frac{f_y(a, b)(\beta t) + E_2(\beta t)(\beta t)}{t} \right) \\ &= \lim_{t \to 0^+} \big(f_x(a, b + \beta t)\alpha + f_y(a, b)\beta + E_1(\alpha t)\alpha + E_2(\beta t)\beta \big). \end{aligned} \tag{16}$$

If we assume that f_x is continuous at (a, b) and recall that $E_1(\alpha t)$ and $E_2(\beta t)$ both tend to 0 as $t \to 0$, it follows from (16) that

$$D_{\mathbf{u}} f(a, b) = f_x(a, b)\alpha + f_y(a, b)\beta.$$

By a similar argument, we can find a formula for the directional derivative of a function of three variables. See Exercise 62. We summarize these results.

Calculating Directional Derivatives

Let f be a function of two variables, say x and y, and let $\mathbf{u} = \langle \alpha, \beta \rangle$ be a unit vector. If the partial derivatives of f are continuous at (a, b), then

$$D_{\mathbf{u}} f(a, b) = f_x(a, b)\alpha + f_y(a, b)\beta. \tag{17}$$

Let g be a function of three variables, say x, y, and z, and let $\mathbf{u} = \langle \alpha, \beta, \gamma \rangle$ be a unit vector. If the partial derivatives of g are continuous at (a, b, c), then

$$D_{\mathbf{u}} g(a, b, c) = g_x(a, b, c)\alpha + g_y(a, b, c)\beta + g_z(a, b, c)\gamma. \tag{18}$$

EXAMPLE 5 Let f defined by $f(x, y) = x^2y - 2x + 3$ be a real-valued function defined on R^2 and let $\mathbf{u} = \langle \frac{3}{5}, -\frac{4}{5} \rangle$. In Example 1 we used the definition of directional derivative to find $D_{\mathbf{u}} f(-1, 2)$. Find this directional derivative using (17).

Solution

We first calculate

$$f_x(x, y) = \frac{\partial}{\partial x}(x^2y - 2x + 3) = 2xy - 2 \quad \text{and} \quad f_y(x, y) = \frac{\partial}{\partial y}(x^2y - 2x + 3) = x^2.$$

Thus, $f_x(-1, 2) = -6$ and $f_y(-1, 2) = 1$. By (15),

$$\begin{aligned} D_{\mathbf{u}} f(-1, 2) &= f_x(-1, 2) \cdot \frac{3}{5} + f_y(-1, 2)\left(-\frac{4}{5}\right) \\ &= -6 \cdot \frac{3}{5} + 1\left(-\frac{4}{5}\right) = -\frac{22}{5}. \end{aligned}$$

The formulas for directional derivatives can be written in a more compact form by introducing some new notation. First observe that the left side of (17) has the form of a dot product:

$$D_{\mathbf{u}}f(a,b) = f_x(a,b)\alpha + f_y(a,b)\beta = \langle f_x(a,b), f_y(a,b)\rangle \cdot \langle \alpha, \beta\rangle. \tag{19}$$

Likewise, we can rewrite (18) as

$$D_{\mathbf{u}}g(a,b,c) = \langle g_x(a,b,c), g_y(a,b,c), g_z(a,b,c)\rangle \cdot \langle \alpha, \beta, \gamma\rangle. \tag{20}$$

In (19) we see a vector with components equal to the partial derivatives of f, and in (20) we find a vector with components equal to the partial derivatives of g. In multivariable calculus this vector, called the *gradient,* often plays the role that the derivative plays in one-variable calculus.

DEFINITION The Gradient

Let f be a function of two variables. Assume that the partial derivatives of f exist at $P = (a,b)$. The **gradient of f at P** is denoted $\nabla f(P)$ or grad $f(P)$ and is given by

$$\text{grad } f(P) = \nabla f(P) = \nabla f(a,b) = \langle f_x(a,b), f_y(a,b)\rangle.$$

If g is a function of three variables and the partial derivatives of g exist at $Q = (a,b,c)$, then the gradient of g at Q is

$$\text{grad } g(Q) = \nabla g(Q) = \nabla g(a,b,c) = \langle g_x(a,b,c), g_y(a,b,c), g_z(a,b,c)\rangle.$$

Combining the gradient notation with (19) and (20), we obtain a simple, easy-to-remember expression for the directional derivative.

Gradient Form for the Directional Derivative

Let f be a function of two or three variables, let f have continuous first partial derivatives at P, and let $\mathbf{u}$ be a unit vector. Then

$$D_{\mathbf{u}}f(P) = \nabla f(P) \cdot \mathbf{u}. \tag{21}$$

EXAMPLE 6 Let $f(x,y,z) = x^2ze^{yz}$ and let $P = (2,0,-1)$.

a) Find $D_{\mathbf{u}}f(P)$ with $\mathbf{u} = \left\langle \frac{1}{3}, -\frac{2}{3}, \frac{2}{3}\right\rangle$.
b) Is there a direction in which the directional derivative of f at P is 0?

Solution

a) We first calculate the gradient of f,

$$\begin{aligned}\nabla f(x,y,z) &= \langle f_x(x,y,z), f_y(x,y,z), f_z(x,y,z)\rangle \\ &= \langle 2xze^{yz}, x^2z^2e^{xy}, (x^2 + x^2yz)e^{yz}\rangle.\end{aligned}$$

Hence $\nabla f(P) = \nabla f(2, 0, -1) = \langle -4, 4, 4 \rangle$. Thus, by (21),

$$D_{\mathbf{u}} f(P) = \nabla f(P) \cdot \mathbf{u} = \langle -4, 4, 4 \rangle \cdot \left\langle \frac{1}{3}, -\frac{2}{3}, \frac{2}{3} \right\rangle = -\frac{4}{3}.$$

b) Let $\mathbf{v}$ be any unit vector and let θ be the angle between $\mathbf{v}$ and $\nabla f(P)$. See Fig. 9.129. We can express the dot product of these two vectors in terms of the length of the vectors and the cosine of the angle between them. Noting that $\|\nabla f(P)\| = \sqrt{3}$,

$$D_{\mathbf{v}} f(P) = \nabla f(P) \cdot \mathbf{v} = \|\nabla f(P)\| \|\mathbf{v}\| \cos \theta = 4\sqrt{3} \cos \theta.$$

It follows that $D_{\mathbf{v}} f(P) = 0$ when $\theta = \pi/2$, that is, when $\mathbf{v}$ is perpendicular to $\nabla f(P)$. Two nonzero vectors are perpendicular if and only if their dot product is 0. Because $\nabla f(P) = \langle -4, 4, 4 \rangle$, we can make the dot product 0 by taking, say, $\mathbf{v} = \langle 1/\sqrt{2}, 0, 1/\sqrt{2} \rangle$. What are some other unit vectors $\mathbf{v}$ for which $D_{\mathbf{v}} f(P) = 0$?

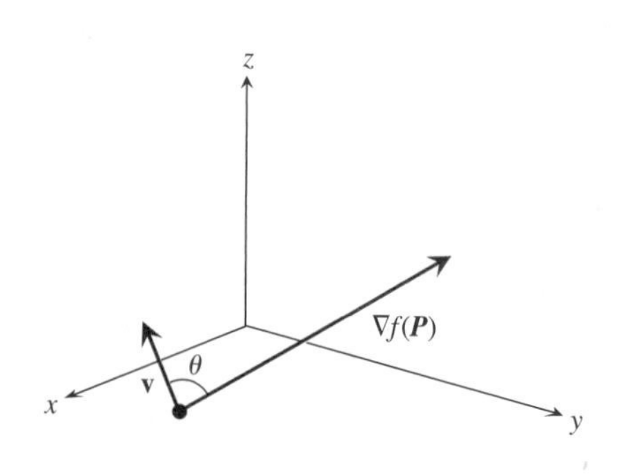

FIGURE 9.129 Let θ be the angle determined by $\nabla f(P)$ and $\mathbf{v}$.

In solving part **b)** of the previous example, we saw that we could control the value of $D_{\mathbf{v}} f(P)$ by controlling the angle θ between $\nabla f(P)$ and $\mathbf{v}$. This is further illustrated in the next example.

EXAMPLE 7 A thin metal plate covers the disk of center $(0, 0)$ and radius 1 in the Cartesian plane. At the point (x, y) on the plane the temperature in degrees centigrade is given by

$$T(x, y) = \frac{x + 2}{(x + 2)^2 + y^2}.$$

From the point $\left(0, \frac{1}{2}\right)$, in what direction is the rate of increase of temperature greatest and what is the rate of increase? Assume that distances in the plane are given in meters.

Solution

The gradient of T is

$$\nabla T(x, y) = \langle T_x(x, y), T_y(x, y) \rangle = \left\langle \frac{-(x + 2)^2 + y^2}{((x + 2)^2 + y^2)^2}, \frac{-2y(x + 2)}{((x + 2)^2 + y^2)^2} \right\rangle.$$

Thus

$$\nabla T\left(0, \frac{1}{2}\right) = \left\langle -\frac{60}{289}, -\frac{32}{289} \right\rangle.$$

Let $\mathbf{u}$ be a unit vector and let θ be the angle between $\nabla T\left(0, \frac{1}{2}\right)$ and $\mathbf{u}$. Then

$$D_{\mathbf{u}} T\left(0, \frac{1}{2}\right) = \nabla T\left(0, \frac{1}{2}\right) \cdot \mathbf{u} = \left\| \nabla T\left(0, \frac{1}{2}\right) \right\| \|\mathbf{u}\| \cos \theta = \frac{4}{17} \cos \theta. \tag{22}$$

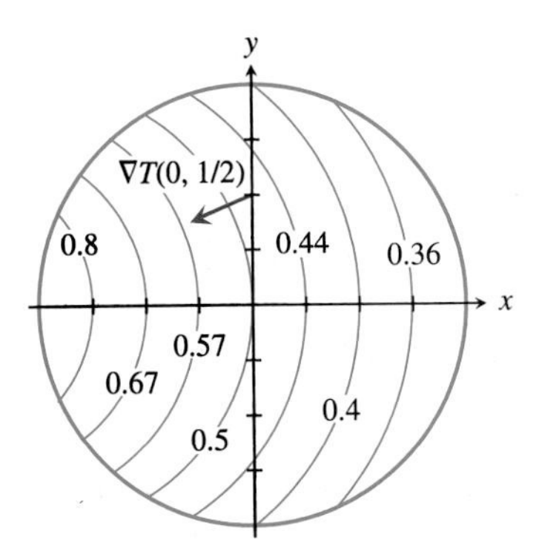

FIGURE 9.130 **The directional derivative $D_{\mathbf{u}} f(P)$ is maximum when $\mathbf{u}$ is in the direction of $\nabla f(P)$.**

The maximum value of this directional derivative occurs when $\cos\theta = 1$, that is, when $\theta = 0$. See Fig. 9.130. This is the case when $\mathbf{u}$ is a unit vector in the same direction as $\nabla T(0, \frac{1}{2})$,

$$\mathbf{u} = \frac{1}{\left\|\nabla T\left(0, \frac{1}{2}\right)\right\|} \nabla T\left(0, \frac{1}{2}\right) = \left\langle -\frac{15}{17}, -\frac{8}{17}\right\rangle.$$

With this choice for $\mathbf{u}$ in (22), we see the directional derivative is

$$D_{\mathbf{u}} T\left(0, \frac{1}{2}\right) = \left\|\nabla T\left(0, \frac{1}{2}\right)\right\| = \frac{4}{17} \,^\circ\text{C/m}.$$

This example illustrates the following beautiful result concerning directional derivatives and the gradient.

Maximum Rate of Change

Let f be a real-valued function of two or three variables, let P be an interior point in the domain of f, and assume that the partial derivatives of f are continuous at P. The directional derivative $D_{\mathbf{u}} f(P)$ is maximum when $\mathbf{u}$ is a unit vector in the direction of $\nabla f(P)$. This is the direction from P in which the rate of increase of f is largest. In this direction the rate of change is

$$D_{\mathbf{u}} f(P) = \|\nabla f(P)\|.$$

The directional derivative of P is minimal in the direction of $-\nabla f(P)$. When $\mathbf{v}$ is a unit vector in this direction, we have

$$D_{\mathbf{v}} f(P) = -\|\nabla f(P)\|.$$

Exercises 9.7

Exercises 1–6: Find f_x and f_y for the given function f.

1. $f(x,y) = -3x + \sqrt{2}y + 4$
2. $f(x,y) = 2x^2y^3 - 4xy^5 + x^2 - 3y + 8$
3. $f(x,y) = \dfrac{x + 2y}{(y - 2x)(y^2 + 1)}$
4. $f(x,y) = \ln(x^2 + 3xy + y^3)$
5. $f(x,y) = \dfrac{2e^x - xe^y}{y^2e^x - e^y}$
6. $f(x,y) = (x - 2)\sec(4xy - 3y^3 + x)$

Exercises 7–12: Find g_x, g_y, and g_z for the given function g.

7. $g(x,y,z) = xy^2z^3 + 2$
8. $g(x,y,z) = e^{2x-5yz+7z^2}$
9. $g(x,y,z) = \dfrac{x^2 + y^2 - z^2}{x^2 - y^2 + z^2}$
10. $g(x,y,z) = x^2\sqrt{y}\arctan(2xy)$
11. $g(x,y,z) = (x + z^2)\cos y \sin 2y$
12. $g(x,y,z) = \ln\left(\dfrac{x + 2y^2 + 3z}{xyz}\right)$.

Exercises 13–20: Find the gradient of the indicated function at the given P.

13. f from Exercise 1, $P = (1, -2)$
14. f from Exercise 2, $P = (0, -1)$
15. f from Exercise 4, $P = (x, y)$
16. f from Exercise 5, $P = (a, b)$
17. g from Exercise 7, $P = (1, 0, -2)$
18. g from Exercise 8, $P = (1, 2, -4)$
19. g from Exercise 10, $P = (t_1, t_2, t_3)$
20. g from Exercise 11, $P = (r, \theta, z)$

Exercises 21–28: For the given function and the given P, find the directional derivative at P in the direction of the unit vector **u**.

21. f from Exercise 1, $P = (1, -2)$, and $\mathbf{u} = \langle 3/5, -4/5 \rangle$

22. f from Exercise 2, $P = (0, -1)$, and $\mathbf{u} = \langle -\sqrt{3}/2, -1/2 \rangle$

23. f from Exercise 4, $P = (a, b)$, and $\mathbf{u} = \langle 1, 0 \rangle$

24. f from Exercise 5, $P = (0, 0)$, and $\mathbf{u} = \langle u_1, u_2 \rangle$

25. g from Exercise 7, $P = (1, 0, -2)$, and $\mathbf{u} = \langle 1/3, 2/3, -2/3 \rangle$

26. g from Exercise 8, $P = (1, 2, -4)$, and $\mathbf{u}\langle -1/2, 1/2, 1/\sqrt{2} \rangle$

27. g from Exercise 10, $P = (0, 1, \pi)$, and $\mathbf{u} = \langle 0, 0, 1 \rangle$

28. g from Exercise 11, $P = (r, \theta, z)$, and $\mathbf{u} = \langle u_1, u_2, u_3 \rangle$

Exercises 29–32: For the given f and P, find the unit vector **u** *for which $D_{\mathbf{u}} f(P)$ is maximum and the unit vector* **v** *for which $D_{\mathbf{v}} f(P)$ is minimum. Give the values of the minimum and maximum directional derivatives.*

29. $f(x, y) = 2x^2 - xy^3 - x + y$, $P = (1, -2)$

30. $f(x, y, z) = (x - 2xy + yz^2 - 1)^2$, $P = (0, 2, 3)$

31. $f(x, y) = xye^{x-y}$, $P = (2, 1)$

32. $f(x, y, z) = \sin(2x - xyz)$, $P = \left(\pi, 0, \frac{1}{2}\right)$

Exercises 33–36: For the given f and P, find all unit vectors **u** *such that $D_{\mathbf{u}} f(P) = 0$.*

33. f and P from Exercise 29

34. f and P from Exercise 30

35. f and P from Exercise 31

36. f and P from Exercise 32

37. Evaluate the limit that appears in (2).

38. Let $f(x, y) = -5xy^2 + 3xy + y^3$ and let $P = (1, -1)$. Find $D_{\mathbf{u}} f(P)$ where **u** is in the direction from P toward the origin.

39. Let $g(x, y) = (x + y)e^{x+2y}$ and let $P = (2, -1)$. Find $D_{\mathbf{u}} g(P)$ where **u** makes an angle of $3\pi/4$ with the positive x -axis.

40. Let $h(x, y) = x \ln(x^2 y + x + y + 2)$ and let $P = (-2, 2)$. Find $D_{\mathbf{u}} h(P)$ where **u** makes an angle of $7\pi/6$ with the positive x-axis.

41. Let $f(x, y, z) = (xyz + 1)/(x^2 + y^2 + z^2 + 1)$ and let $P = (0, -1, 2)$. Find $D_{\mathbf{u}} f(P)$ where **u** is in the direction from P toward the origin.

42. Let $f(x, y, z) = \tan(xyz + \pi/4)$ and let $P = (\pi/6, 2, 1)$. Find $D_{\mathbf{u}} f(P)$ where **u** is in the direction from P toward the point $(\pi/6, -2, 2)$.

43. Let f be a real-valued function of two variables with continuous partial derivatives, let $\mathbf{u} = \langle \sqrt{2}/2, \sqrt{2}/2 \rangle$, and let $\mathbf{v} = \langle -\sqrt{2}/2, \sqrt{2}/2 \rangle$. Given that $D_{\mathbf{u}} f(3, -7) = 8$ and $D_{\mathbf{v}} f(3, -7) = -1$, find $\nabla f(3, -7)$.

44. Let g be a real-valued function of two variables with continuous partial derivatives, let $\mathbf{u} = \left\langle \frac{3}{5}, -\frac{4}{5} \right\rangle$, and let $\mathbf{v} = \left\langle -\frac{5}{13}, -\frac{12}{13} \right\rangle$. Given that $D_{\mathbf{u}} g(3, -7) = 0$ and $D_{\mathbf{v}} g(3, -7) = 1$, find $\nabla g(3, -7)$.

45. Let $f(x, y) = 2x^2 y - xy + 1$ and, for $0 \le \theta \le 2\pi$, let $\mathbf{u}_\theta = \langle \cos\theta, \sin\theta \rangle$ be the unit vector that makes an angle of θ with the positive x-axis. Define the function d on the interval $0 \le \theta \le 2\pi$ by

$$d(\theta) = D_{\mathbf{u}_\theta} f(1, -1).$$

a. Draw the graph of $d = d(\theta)$.

b. Find the θ values where d takes on its maximum and minimum values and the values of θ for which $d(\theta) = 0$.

c. What does the graph of d tell you about the behavior of f near $(1, -1)$?

46. Let $g(x, y) = \sin(2x + y)\cos(x + 2y)$ and, for $0 \le \theta \le 2\pi$, let $\mathbf{u}_\theta = \langle \cos\theta, \sin\theta \rangle$ be the unit vector that makes an angle of θ with the positive x-axis. Define the function d on the interval $0 \le \theta \le 2\pi$ by

$$d(\theta) = D_{\mathbf{u}_\theta} g(\pi/4, -\pi/4).$$

a. Draw the graph of $d = d(\theta)$.

b. Find the θ values where d takes on its maximum and minimum values and the values of θ for which $d(\theta) = 0$.

c. What does the graph of d tell you about the behavior of g near $(\pi/4, -\pi/4)$?

47. Let $f(x, y) = xy + 1$. Let α be the curve formed by the intersection of the surface $z = f(x, y)$ and the plane $y = 2$, and let β be the curve formed by the intersection of the surface and the plane $x = 1$.

a. Find the equation of the line tangent to β at the point $(1, 2, 3)$. (Your answer should be the equation for a line in three-space.)

b. Find the equation of the line tangent to α at the point $(1, 2, 3)$.

c. Find the equation for the plane that contains the lines found in parts **a** and **b**. This plane is the plane tangent to the surface at $(1, 2, 3)$.

48. Let $g(x, y) = 2xe^{x+y}$. Let α be the curve formed by the intersection of the surface $z = g(x, y)$ and the plane $y = -3$, and let β be the curve formed by the intersection of the surface and the plane $x = 4$.

a. Find the equation of the line tangent to β at the point $(4, -3, 8e)$. (Your answer should be the equation for a line in three-space.)

b. Find the equation of the line tangent to α at the point $(4, -3, 8e)$.

c. Find the equation for the plane that contains the lines found in parts **a** and **b**. This plane is the plane tangent to the surface at $(4, -3, 8e)$.

49. Let $h(x, y) = \ln(x^2 y + y^2)$. Let α be the curve formed by the intersection of the surface $z = h(x, y)$ and the plane $y = 1$, and let β be the curve formed by the intersection of the surface and the plane $x = 0$.

a. Find the equation of the line tangent to β at the point $(0, 1, 0)$.

b. Find the equation of the line tangent to α at the point $(0, 1, 0)$.

c. Find the equation for the plane that contains the lines found in parts **a** and **b**. This plane is the plane tangent to the surface at $(0, 1, 0)$.

50. Let f be a real-valued function defined on $S = \{(x, y) : 0 \leq x, y \leq 1\}$. The value of f at several points in S is shown in the accompanying table. Use these data to find an approximation to

a. $D_{\mathbf{u}}f(0.5, 0.5)$ where $\mathbf{u} = \langle 1/\sqrt{2}, 1/\sqrt{2} \rangle$

b. $D_{\mathbf{v}}f(0.5, 0.5)$ where $\mathbf{v} = \langle -2/\sqrt{5}, 1/\sqrt{5} \rangle$

c. $f_x(0.5, 0.5)$ and $f_y(0.5, 0.5)$.

			x		
y	0.3	0.4	0.5	0.6	0.7
0.3	0.62	0.45	0.27	0.071	−0.13
0.4	0.54	0.36	0.17	−0.029	−0.23
0.5	0.45	0.27	0.071	−0.13	−0.32
0.6	0.36	0.17	−0.029	−0.23	−0.42
0.7	0.27	0.071	−0.13	−0.32	−0.5

Table for Exercise 50.

51. Let g be a real-valued function defined on $S = \{(x, y) : 0 \leq x, y \leq 1\}$. The value of g at several points in S is shown in the accompanying table. Use these data to find an approximation to

a. $D_{\mathbf{u}}g(0.5, 0.5)$ where $\mathbf{u}$ makes an angle of $\pi/4$ with the positive x-axis

b. $D_{\mathbf{v}}g(0.5, 0.5)$ where $\mathbf{v}$ makes an angle of $3\pi/4$ with the positive x-axis

c. $g_x(0.5, 0.5)$ and $g_y(0.5, 0.5)$.

			x		
y	0.3	0.4	0.5	0.6	0.7
0.3	0.03	0.036	0.042	0.047	0.052
0.4	0.061	0.071	0.081	0.09	0.099
0.5	0.11	0.13	0.14	0.15	0.17
0.6	0.18	0.2	0.22	0.24	0.26
0.7	0.29	0.31	0.34	0.36	0.38

Table for Exercise 51.

52. The accompanying figure shows the level curves graph for a real-valued function f of two variables. Use the graph to find the approximate value of

a. $D_{\mathbf{u}}f(1, 1)$ where $\mathbf{u} = \langle -1/\sqrt{2}, 1/\sqrt{2} \rangle$

b. $D_{\mathbf{v}}f(0.5, 0.75)$ where $\mathbf{v} = \langle \sqrt{3}/2, -1/2 \rangle$

c. $\nabla f(1, 1)$.

d. Indicate on the graph the direction in which the directional derivative at $(1, 1)$ is maximum and the direction in which the directional derivative at $(1, 1)$ is minimum.

e. Indicate on the graph the directions in which the directional derivative at $(1, 1)$ is 0.

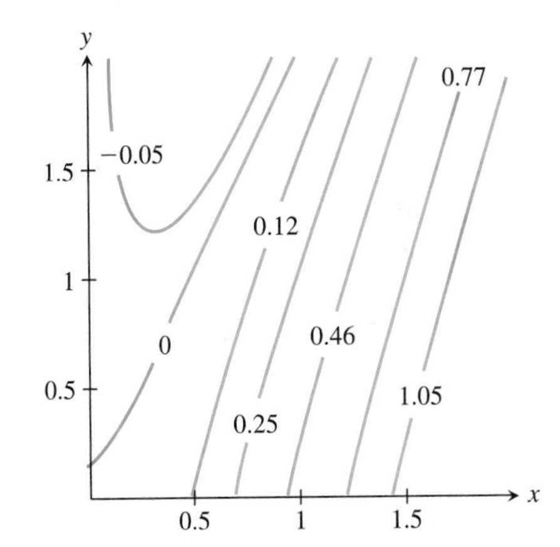

Figure for Exercise 52.

53. The accompanying figure shows the level curves graph for a real-valued function g of two variables. Use the graph to find the approximate value of

a. $D_{\mathbf{u}}g(-1, -0.5)$ where $\mathbf{u} = \langle 1/\sqrt{2}, 1/\sqrt{2} \rangle$

b. $D_{\mathbf{v}}g(-0.5, -0.75)$ where $\mathbf{v}$ makes an angle of $2\pi/3$ with the positive x-axis

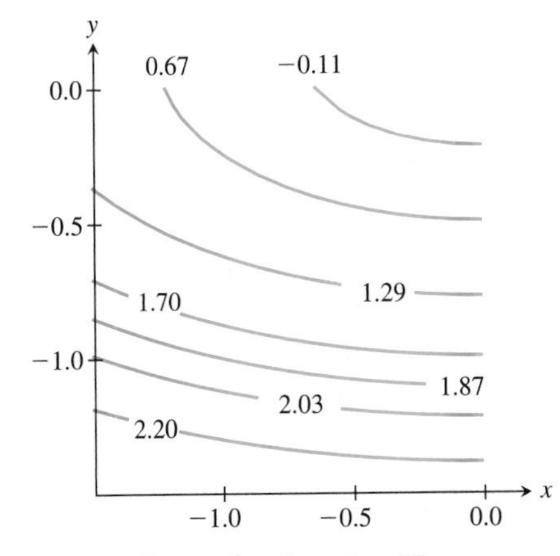

Figure for Exercise 53.

c. $\nabla g(-1,-1)$.

d. Indicate on the graph the direction in which the directional derivative at $(-1,-1)$ is maximum and the direction in which the directional derivative at $(-1,-1)$ in minimum.

e. Indicate on the graph the directions in which the directional derivative at $(-1,-1)$ is 0.

54. Let P_0 be a fixed point in the plane. Given a point $P = (x, y)$ in the plane, let

$$d(x, y) = \text{the distance from } P \text{ to } P_0.$$

If $Q \neq P_0$,

a. Describe the direction $\mathbf{u}$ in which $D_{\mathbf{u}}d(Q)$ is maximum and the direction in which it is minimum.

b. Describe the directions $\mathbf{v}$ in which $D_{\mathbf{v}}d(Q) = 0$.

c. What can be said in answer to these questions when $Q = P_0$?

55. Let P_0 be a fixed point in space. Given a point $P = (x, y, z)$ in R^3, let

$$d(x, y, z) = \text{the distance from } P \text{ to } P_0.$$

If $Q \neq P_0$,

a. Describe the direction $\mathbf{u}$ in which $D_{\mathbf{u}}d(Q)$ is maximum and the direction in which it is minimum.

b. Describe the directions $\mathbf{v}$ in which $D_{\mathbf{v}}d(Q) = 0$.

c. What can be said in answer to these questions when $Q = P_0$?

56. Let ℓ be a line in the plane. Given a point $P = (x, y)$ in the plane, let

$$d(x, y) = \text{the perpendicular distance from } P \text{ to } \ell.$$

If Q is not on ℓ,

a. Describe the direction $\mathbf{u}$ in which $D_{\mathbf{u}}d(Q)$ is maximum and the direction in which it is minimum.

b. Describe the directions $\mathbf{v}$ in which $D_{\mathbf{v}}d(Q) = 0$.

c. What can be said in answer to these questions when Q is on ℓ?

57. Let Π be a plane in space. Given a point $P = (x, y, z)$ in R^3, let

$$d(x, y, z) = \text{the perpendicular distance from } P \text{ to } \Pi.$$

If Q is not on Π,

a. Describe the direction $\mathbf{u}$ in which $D_{\mathbf{u}}d(Q)$ is maximum and the direction in which it is minimum.

b. Describe the directions $\mathbf{v}$ in which $D_{\mathbf{v}}d(Q) = 0$.

c. What can be said in answer to these questions when Q is on Π?

58. Look again at the example about the principal in the noisy gym discussed in the Investigation in this section. From the point $(50, 100)$, in which direction should the principal walk if she wants the sound intensity to decrease as rapidly as possible?

59. A valley is in the shape of the surface

$$z = (x - 2)^2(y + 1)^4 + (x + 2)^2(y - 3)^2.$$

a. A hiker spills water at the point $(1, 0, 82)$. Assuming that water runs downhill in the steepest direction, in which direction will the water flow?

b. If the hiker wishes to hike from $(1, 0, 82)$ in the direction of steepest ascent, in which direction should he go?

60. In Example 3 of Section 9.2 we saw that the probability that a mimic (i.e., a species that looks like another species) is eaten by a predator is

$$p(s, f, \epsilon) = s(1 - f(1 - \epsilon))$$

where s is the probability that the predator sees the mimic, f is the probability that the predator mistakes the mimic for another species, and ϵ is the probability that the mimic is eaten even if the predator misidentifies the mimic. Suppose that for a certain species of butterfly these parameters have values $s = 0.85$, $f = 0.9$, and $\epsilon = 0.3$. Through evolution of the mimic and the predator, all of these factors can change. What should be the "direction" of evolution to most rapidly decrease the mimic's probability of being eaten? Explain the meaning of your answer.

61. When pure helium is in a closed container, the temperature T (in Kelvins) as a function of pressure p and volume v is well approximated by the van der Waals equation

$$T = 12.188\left(p + \frac{0.00007}{v^2}\right)(v - 0.00106).$$

The pressure is measured in atmospheres, and the volume is measured relative to the volume of the same body of gas at standard temperature and pressure. If $p = 0.6$ and $v = 1.5$, in what "direction" should we change (p, v) to make the temperature decrease as rapidly as possible? Explain the meaning of your answer.

62. Give an argument in support of Equation (18). Your argument should be similar to the one used in deriving (17).

63. Let f be a real-valued function of two variables, and assume that f and its partial derivatives are continuous at and near the point (a, b). With the help of the definition of the directional derivative, show that when h and k are small,

$$f(a + h, b + k) \approx f(a, b) + f_x(a, b)h + f_y(a, b)k.$$

Conclude that when (x, y) is close to (a, b), we have

$$f(x, y) \approx f(a, b) + f_x(a, b)(x - a) + f_y(a, b)(y - b).$$

Review of Key Concepts

Many important phenomena can be well modeled only by using functions of several variables. To help us understand such models, we looked at ways to interpret and visualize such functions. For functions that represent physical situations, it may be helpful to hold all but one of the input variables fixed and see what physical phenomenon is described when this one variable is allowed to change. This technique also works well when trying to visualize the graph of the surface described by an equation of the form $z = f(x, y)$. If we fix one of the variables, say $y = b$, then the graph of the one variable function represented by

$$w = w(x) = f(x, b)$$

is a picture of the cross section obtained when the surface is cut by the plane $y = b$. By studying such cross sections, we can better interpret the graph produced by the graphing package on a calculator or CAS. Fixing $z = c$, we get the equation $c = f(x, y)$. The curve described by this equation is called a *level curve* for the function f. When we draw several of these curves in the (x, y)-plane, we get a contour (or level curves) graph of the surface described by $z = f(x, y)$. A *contour graph* gives us an idea of how the values of f rise and fall as (x, y) changes. The level curves also show the shape of the cross sections perpendicular to the z-axis of the surface described by $z = f(x, y)$. When g is a function of three variables, say $w = g(x, y, z)$, the surface described by fixing w is called a *level surface* for g. By studying the level surfaces for such a function, we can gain some insight into how g changes as (x, y, z) changes.

Sometimes more information about a surface can be found by working with a parametric representation. Such representations can be particularly useful in producing good pictures of a surface with a CAS. We saw that such pictures are often generated by using a parametric representation that conforms to the level curves of the surface. We also saw that when a surface can be described very simply in cylindrical or spherical coordinates, we can often use the cylindrical or spherical coordinate expressions to obtain a useful parametric representation for the surface.

In studying the rate of change of a function of several variables, we found it convenient to introduce partial derivatives. If f is a function of two variables, with $z = f(x, y)$, then the partial derivative of f with respect to x at (a, b) is

$$f_x(a, b) = \frac{\partial z}{\partial x}(a, b) = \lim_{h \to 0} \frac{f(a + h, b) - f(a, b)}{h},$$

provided that this limit exists. The partial derivative $f_x(a, b)$ gives the rate of change of f with respect to x when y is held fixed at $y = b$ and x changes from a. We saw that $f_x(a, b)$ is also the slope of the line tangent at (a, b) to the curve of intersection of the surface described by $z = f(x, y)$ and the plane $y = b$. The partial derivative of f with respect to y at (a, b) is given by

$$f_y(a, b) = \frac{\partial z}{\partial y}(a, b) = \lim_{k \to 0} \frac{f(a, b + k) - f(a, b)}{k}.$$

Partial derivatives for functions of three or more variables are defined in an analogous way.

More information about the rate of change of a function can be found by looking at the directional derivative. Let f be a function of several variables and let P be an interior point in the domain of f. If $\mathbf{u}$ is a unit vector, then the directional derivative of f at P in the direction of $\mathbf{u}$ is

$$D_{\mathbf{u}} f(P) = \lim_{h \to 0^+} \frac{f(\mathbf{P} + h\mathbf{u}) - f(\mathbf{P})}{h},$$

provided that this limit exists. The directional derivative is the rate of change of f at P as the input variable changes from P in the direction $\mathbf{u}$.

We noted that the directional derivative can be conveniently expressed in terms of the gradient, $\nabla f(P)$, of f at P and the direction vector $\mathbf{u}$,

$$D_{\mathbf{u}} f(P) = \nabla f(P) \cdot \mathbf{u}.$$

The gradient of f is a vector whose components are the partial derivatives of f. If f is a function of two variables and $z = f(x, y)$, then the gradient of f at $P = (a, b)$ is given by

$$\nabla f(a, b) = \nabla f(P) = \langle f_x(P), f_y(P) \rangle = \langle f_x(a, b), f_y(a, b) \rangle.$$

If g is a function of three variables, with $w = g(x, y, z)$, then the gradient of g at $P = (a, b, c)$ is given by

$$\begin{aligned} \nabla g(a, b, c) &= \nabla g(P) = \langle g_x(P), g_y(P), g_z(P) \rangle \\ &= \langle g_x(a, b, c), g_y(a, b, c), g_z(a, b, c) \rangle. \end{aligned}$$

The direction of the gradient of a function f at P is the direction in which the directional derivative at P is maximum and thus is the direction of maximal rate of increase for f. In this direction, the rate of increase is equal to $\|\nabla f(P)\|$, the length of the gradient of f at P.

Chapter Summary

Cross Sections

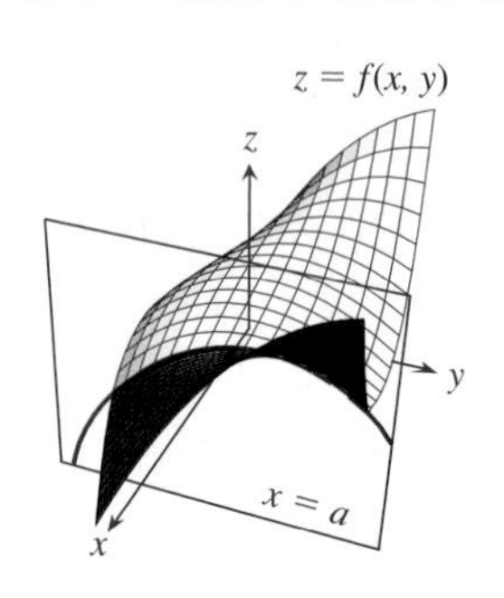

Let f be a function of two variables, with $z = f(x, y)$. Fix $x = a$. The graph of $z = f(a, y)$ in the plane $x = a$ is the curve described by

$$\mathbf{r}(y) = \langle a, y, f(a, y)\rangle.$$

This curve is the intersection of the surface given by $z = f(x, y)$ and the plane $x = a$, and is called the *cross section of the surface in the plane* $x = a$. The cross section in the plane $y = b$ is the curve given by

$$\mathbf{r}(x) = \langle x, b, f(x, b)\rangle.$$

Looking at many such cross sections can help us interpret the graph of the surface.

If $z = f(x, y) = yx^2 - y^2$, then the cross section in the plane $x = 1$ is described by

$$z = f(1, y) = y - y^2$$

and is a parabola in the plane $x = 1$. The cross section in $y = -2$ is given by

$$z = f(x, -2) = -2x^2 - 4$$

and is also a parabola, but in the plane $y = -2$.

Contour Graph

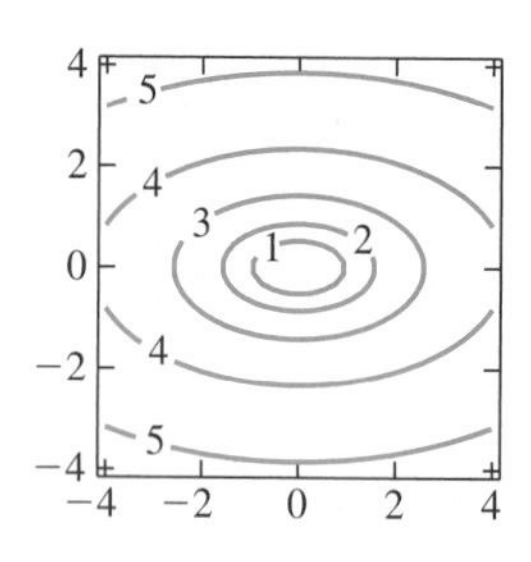

Let f be a function of two variables. Given a real number c, the set of all points (x, y) for which $f(x, y) = c$ is usually a curve in the plane. This curve is called a *level curve for f*. When we plot many of these level curves, we get the *contour* (or level curves) *graph* for f.

Let $f(x, y) = \ln(3x^2 + 10y^2)$ and let c be a number. The set of points (x, y) for which

$$\ln(3x^2 + 10y^2) = c$$

is the set of points (x, y) with

$$3x^2 + 10y^2 = e^c.$$

The points satisfying this equation lie on an ellipse, so the contour graph shows several ellipses.

Density Graph

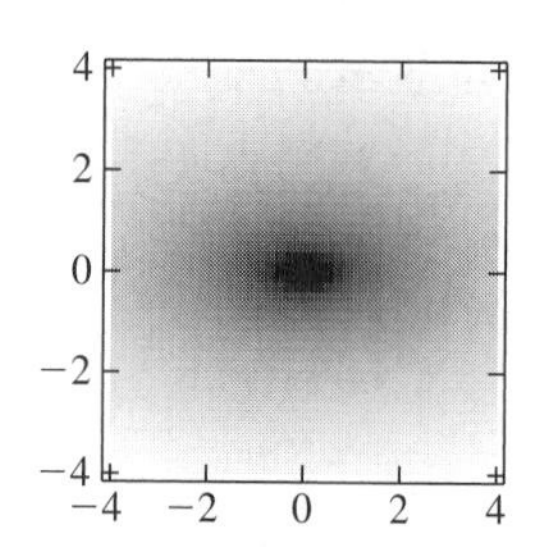

Let f be a function of two variables. If we color or shade points (x, y) in the plane according to the value of $f(x, y)$, we obtain a density graph for $z = f(x, y)$.

Let $f(x, y) = \ln(3x^2 + 10y^2)$. When (x, y) is near the origin, $f(x, y)$ is large and negative. As (x, y) moves from the origin, $f(x, y)$ increases. If we color points (x, y) for which f is large and negative black, points for which f is large and positive white, and use intermediate shades of gray for intermediate values, we obtain the picture to the left.

Parametric Representation

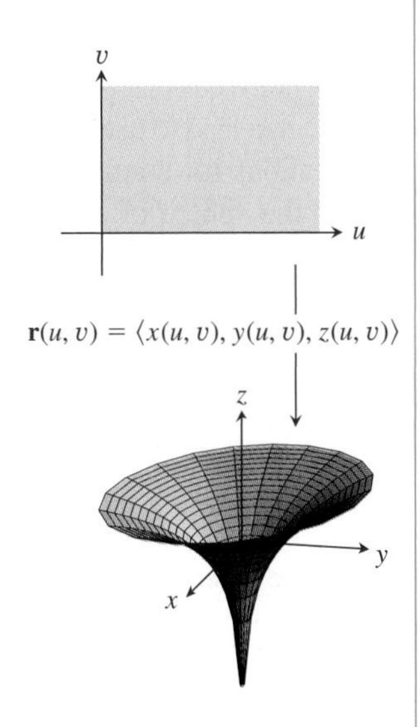

A parametric representation of a surface is a function **r** of the form

$$\mathbf{r}(u,v) = \langle x(u,v), y(u,v), z(u,v)\rangle,$$

where **r** is defined for all points (u, v) in a region $\mathcal{D}$ of the (u, v)-plane. The surface S represented by this function is the set of all (x, y, z) for which $x = x(u, v)$, $y = y(u, v)$, and $z = z(u, v)$ for some (u, v) in $\mathcal{D}$. A parametric representation that conforms to the level curves of a surface S can often be used to produce a good graph of S.

If $f(x, y) = \ln(3x^2 + 10y^2)$, the level curve $f(x, y) = c$ is the ellipse

$$3x^2 + 10y^2 = e^c.$$

These curves can be represented parametrically by

$$\langle x, y\rangle = \left\langle \frac{1}{\sqrt{3}}u \cos v, \frac{1}{\sqrt{10}}u \sin v \right\rangle,$$

where $u^2 = e^c$ and v runs from 0 to 2π. The resulting parametric representation for the surface represented by f is

$$\mathbf{r}(u,v) = \left\langle \frac{1}{\sqrt{3}}u \cos v, \frac{1}{\sqrt{10}}u \sin v, 2 \ln u \right\rangle,$$

where $u > 0$ and $0 \leq v < 2\pi$.

Cylindrical Coordinates

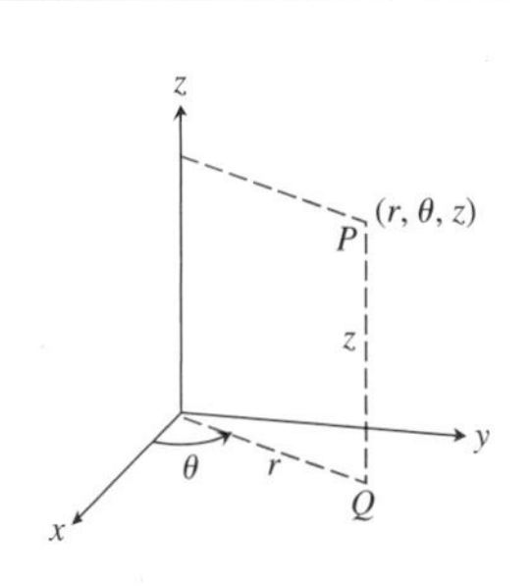

Let P be a point in space and let Q be the foot of the perpendicular from P to the (x, y)-plane. If Q has polar coordinates (r, θ), then P has cylindrical coordinates (r, θ, z), where z is the usual (Cartesian) z-coordinate for P.

If P has cylindrical coordinates (r, θ, z) and Cartesian coordinates (x, y, z), then

$$x = r \cos \theta$$

$$y = r \sin \theta.$$

We also have

$$r = \sqrt{x^2 + y^2}$$

$$\tan \theta = \frac{y}{x}.$$

Some surfaces can be expressed very simply in cylindrical coordinates. For example, consider the half-cone described by

$$z = \sqrt{x^2 + y^2}.$$

Using $r^2 = x^2 + y^2$, we obtain the simple cylindrical equation

$$z = r.$$

Spherical Coordinates

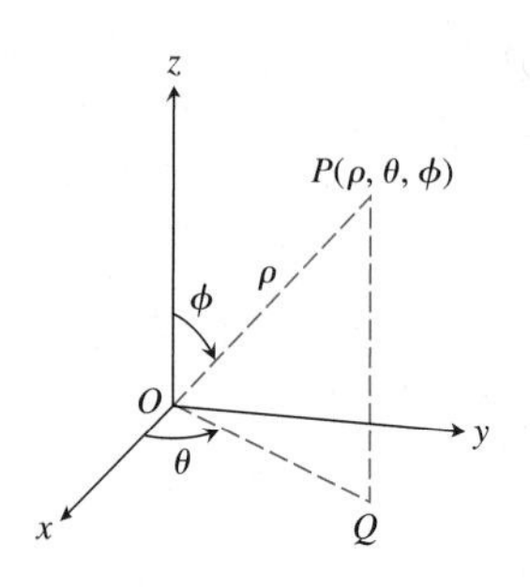

Let P be a point in space and let Q be the foot of the perpendicular from P to the (x, y)-plane. Then P has spherical coordinates (ρ, θ, ϕ) where ρ is the length of OP, θ is the polar angle for the planar point Q, and ϕ $(0 \le \phi \le \pi)$ is the angle measured from the positive z-axis to OP. If P also has Cartesian coordinates (x, y, z), then

$$x = \rho \sin \phi \cos \theta$$

$$y = \rho \sin \phi \sin \theta$$

$$z = \rho \cos \phi,$$

and

$$\rho = \sqrt{x^2 + y^2 + z^2}$$

$$\tan \theta = \frac{y}{x}$$

$$\cos \phi = \frac{z}{\sqrt{x^2 + y^2 + z^2}}.$$

Spherical coordinates can be useful for describing surfaces that enclose the origin. They also lead to useful parametric representations for many surfaces. For example, the sphere of center $(0, 0, 0)$ and radius 1 has spherical equation $\rho = 1$. The Cartesian coordinates (x, y, z) of a point on the sphere have the form

$$x = \sin \phi \cos \theta$$

$$y = \sin \phi \sin \theta$$

$$z = \cos \phi.$$

This leads to the parametric representation

$$\mathbf{r}(\phi, \theta) = \langle \cos \theta \sin \phi, \sin \theta \sin \phi, \cos \phi \rangle,$$

where $0 \le \theta \le 2\pi$ and $0 \le \phi \le \pi$.

Limits

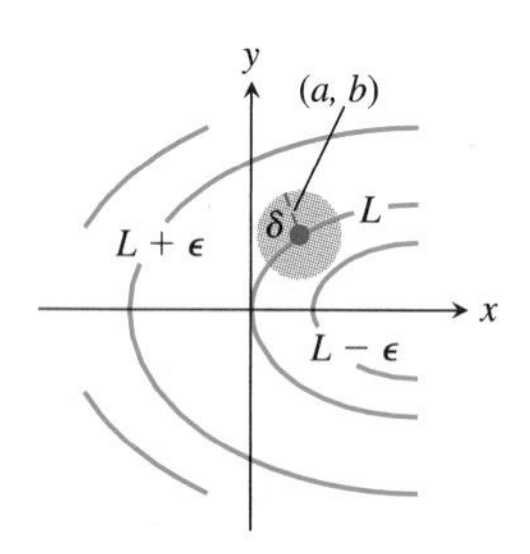

Let f be a function of two variables and let $P = (a, b)$ be a point in the plane. We say that

$$\lim_{(x,y)\to(a,b)} f(x, y) = L$$

if $f(x, y)$ gets close to L as (x, y) gets close to (a, b). More precisely, this means that for a given positive number ϵ, we can find a positive number δ so that

$$|f(x, y) - L| < \epsilon$$

whenever (x, y) is in the domain of f and

$$0 < \sqrt{(x - a)^2 + (y - b)^2} < \delta.$$

Many limits of functions of two (or three) variables can be evaluated by common sense. Sometimes more complicated limits can be analyzed by using the Squeeze Theorem. To evaluate

$$\lim_{(x,y)\to(0,0)} \frac{x^3}{x^2 + y^2},$$

note that for (x, y) close to, but not equal to, $(0, 0)$ we have

$$0 \le \left| \frac{x^3}{x^2 + y^2} \right| \le |x|.$$

Because 0 and $|x|$ both approach 0 as (x, y) approaches $(0, 0)$, we conclude that

$$\lim_{(x,y)\to(0,0)} \frac{x^3}{x^2 + y^2} = 0.$$

Limits That Do Not Exist

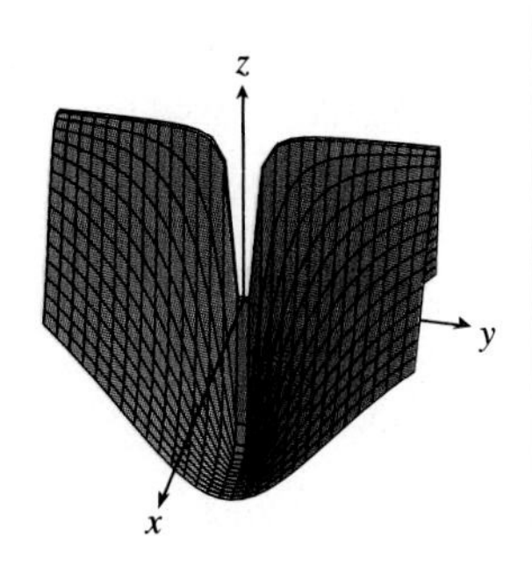

Let f be a function of two variables, let (a, b) be a point in the plane, and suppose that

$$\lim_{(x,y)\to(a,b)} f(x, y) = L.$$

If $\mathbf{r}(t) = \langle x(t), y(t)\rangle$ describes a curve in the plane with $\mathbf{r}(t_0) = \langle a, b\rangle$, then

$$\lim_{t\to t_0} f(x(t), y(t)) = L.$$

Now let $\mathbf{r}_1(t) = \langle x_1(t), y_1(t)\rangle$ and $\mathbf{r}_2(t) = \langle x_2(t), y_2(t)\rangle$ describe two different curves in the plane, and suppose that $\mathbf{r}_1(t_1) = \langle a, b\rangle = \mathbf{r}_2(t_2)$. If

$$\lim_{t\to t_1} f(x_1(t), y_1(t)) \neq \lim_{t\to t_2} f(x_2(t), y_2(t)),$$

or if one or both of these limits fails to exist, then

$$\lim_{(x,y)\to(a,b)} f(x, y)$$

does not exist.

In practice we may be able to show that a limit fails to exist by finding two different curves of approach that yield different limits. Let

$$f(x, y) = \frac{y^2 - x^2}{x^2 + y^2}$$

and consider the limit of f as (x, y) approaches $(0, 0)$ along the x-axis and along the line $y = x$. To approach along the x-axis, set $(x, y) = (t, 0)$ and let $t \to 0$:

$$\lim_{t\to 0} f(t, 0) = \lim_{t\to 0} \frac{0^2 - t^2}{t^2 + 0^2} = -1.$$

To approach along the line $y = x$, let $(x, y) = (t, t)$ and let $t \to 0$:

$$\lim_{t\to 0} f(t, t) = \lim_{t\to 0} \frac{t^2 - t^2}{t^2 + t^2} = 0.$$

Because we obtained different limits by approaching $(0, 0)$ along different curves, $\lim_{(x,y)\to(0,0)} f(x, y)$ does not exist.

Partial Derivatives

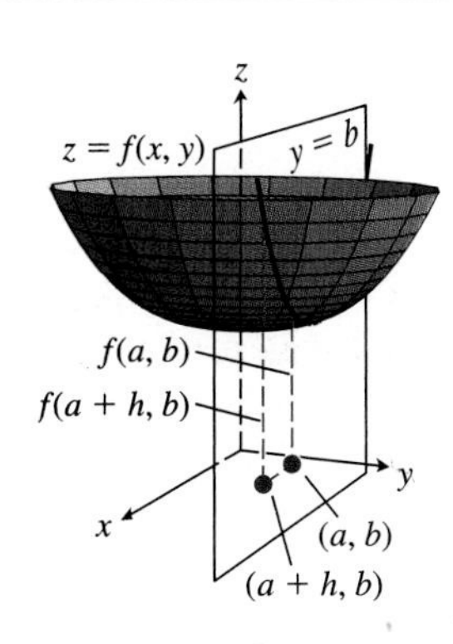

Let f be a function of two variables and let (a, b) be an interior point in the domain of f. The partial derivative of f with respect to x at (a, b) is

$$f_x(a, b) = \lim_{h\to 0} \frac{f(a + h, b) - f(a, b)}{h},$$

provided that this limit exists. The partial derivative of f with respect to y at (a, b) is defined to be

$$f_y(a, b) = \lim_{k\to 0} \frac{f(a, b + k) - f(a, b)}{k}.$$

Partial derivatives of functions of three variables are defined in a similar way.

To compute $f_x(x, y)$, treat y as a constant and perform the differentiation with respect to x using the techniques learned in Chapter 2. If

$$f(x, y) = x^2 \sin(xy + 3),$$

then

$$f_x(x, y) = 2x \sin(xy + 3) + x^2 y \cos(xy + 3),$$

and

$$f_y(x, y) = x^3 \cos(xy + 3).$$

Directional Derivatives

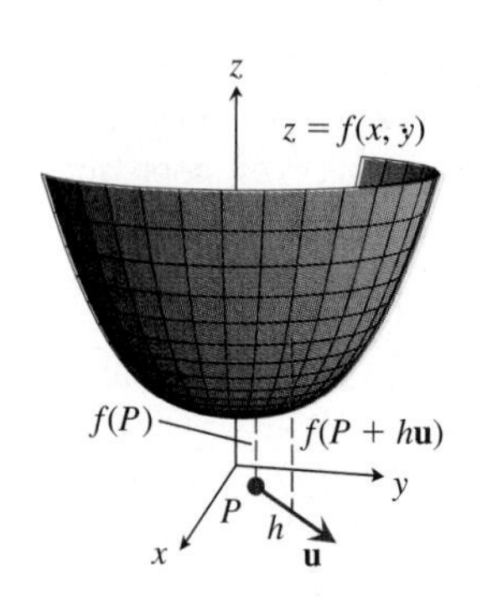

Let f be a function of two variables, let $P = (a, b)$ be an interior point in the domain of f, and let $\mathbf{u} = \langle \alpha, \beta \rangle$ be a unit vector in the plane. The directional derivative of f at (a, b) in the direction of $\mathbf{u}$ gives the rate of change of f as (x, y) changes from (a, b) in the direction of $\mathbf{u}$. This directional derivative is defined by

$$D_{\mathbf{u}}f(a,b) = \lim_{h \to 0} \frac{f(P + h\mathbf{u}) - f(P)}{h}$$
$$= \lim_{h \to 0} \frac{f(a + \alpha h, b + \beta h) - f(a,b)}{h},$$

provided that this limit exists. Directional derivatives of functions of three variables are defined in an analogous way.

The directional derivative of f at P in the direction $\mathbf{u}$ is given by

$$D_{\mathbf{u}}f(P) = \nabla f(P) \cdot \mathbf{u}$$

where $\nabla f(P)$ is the gradient of f at P. If $f(x, y, z) = x^2y + 3z^3$, then the gradient of f is given by

$$\nabla f(x, y, z)$$
$$= \langle f_x(x, y, z), f_y(x, y, z), f_z(x, y, z) \rangle$$
$$= \langle 2xy, x^2, 9z^2 \rangle.$$

The directional derivative of f at $P = (3, 0, -1)$ in the direction of $\mathbf{u} = \langle 0, -3/5, 4/5 \rangle$ is

$$D_{\mathbf{u}}f(P) = \langle 0, 9, 9 \rangle \cdot \left\langle 0, -\frac{3}{5}, \frac{4}{5} \right\rangle$$
$$= \frac{9}{5}.$$

Direction of Maximal Rate of Increase

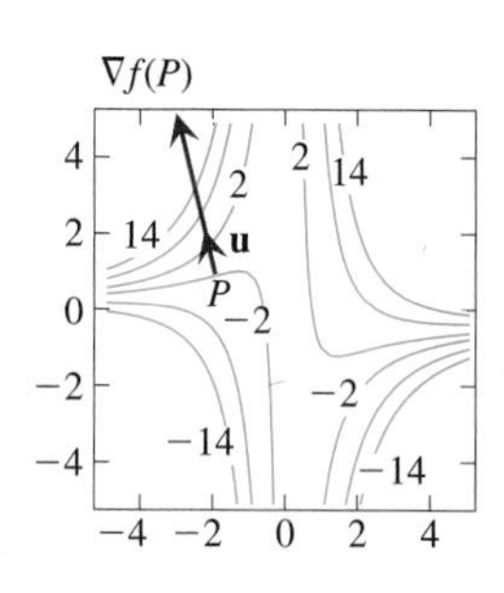

Let f be a function of two or three variables and let $P = (a, b)$ be a point in the interior of the domain of f. If the partial derivatives of f are continuous, then

$$D_{\mathbf{u}}f(P)$$

is maximal when $\mathbf{u}$ is a unit vector in the direction of $\nabla f(P)$. The value of the directional derivative in this direction is

$$\|\nabla f(P)\|.$$

The directional derivative is minimal when $\mathbf{u}$ is a unit vector in the direction of $-\nabla f(P)$. The value of this directional derivative is $-\|\nabla f(P)\|$.

Let $f(x, y) = x^2y + 3x$ and $P = (-2, 1)$. Then $\nabla f(P) = \langle -1, 4 \rangle$ and the unit vector in the direction of this gradient is

$$\mathbf{u} = \left\langle -\frac{1}{\sqrt{17}}, \frac{4}{\sqrt{17}} \right\rangle.$$

This is the direction in which the directional derivative of f at P is maximal. The value of the directional derivative in this direction is

$$\|\nabla f(P)\| = \|\langle -1, 4 \rangle\| = \sqrt{17}.$$

Chapter Review Exercises

Exercises 1–6: Find the equation for each conic section.

1. The parabola with directrix $y = 3$ and focus $(3, 1)$.
2. The ellipse with foci $(2, 0)$ and $(2, 6)$ and associated distance $D = 20$.
3. The hyperbola with foci $(2, 0)$ and $(2, 6)$ and associated distance $D = 1$.
4. The conic section with directrix $x = 4$, focus $(0, 0)$, and eccentricity 4.
5. The conic section with directrix $x = 4$, focus $(0, 0)$, and eccentricity 1.
6. The conic section with directrix $x = 4$, focus $(0, 0)$, and eccentricity $1/4$.

Exercises 7–12: For the indicated function, find the partial derivative with respect to each variable.

7. $f(x, y) = y \tan(xy^2)$
8. $z = (3x^2y - 4xy + y^3)^{5/2}$
9. $\rho = \dfrac{1 + \sin\phi \sin\theta}{1 - \cos\phi \cos\theta}$
10. $w = (x + y)(y + z)(z + x)$
11. $h(u, v, w) = \ln(1 + \sin^2 u + \sin^6 v + \cos^2 2w)$
12. $V(x, y, z) = \dfrac{e^{xyz} - 1}{e^x e^y e^z + 1}$

Exercises 13–18: Find the directional derivative at the given point P and in the indicated direction.

13. $f(x, y) = 3x^2 - 2xy + 4y^2 - y + 1$,
    $P = (-2, 1)$,
    direction: $\mathbf{u} = \langle 1/\sqrt{2}, -1/\sqrt{2} \rangle$
14. $g(u, v) = u \sin(u + v)$,
    $P = (0, \pi/4)$,
    direction: $\mathbf{u} = \langle -0.6, 0.8 \rangle$
15. $\rho(\phi, \theta) = \ln(2\phi^2 + \theta^2 + 1)$,
    $P = (2, -3)$,
    direction: $\mathbf{u} = \langle -1, 0 \rangle$
16. $f(x, y, z) = xze^{y^2+z}$,
    $P = (1, 1, -1)$,
    direction: $\mathbf{u} = \langle 1/3, -2/3, -2/3 \rangle$
17. $g(x, y, z) = \arctan\left(\dfrac{z}{x + y}\right)$,
    $P = (1, 1, 2)$,
    direction: $\mathbf{u} = \langle 1/\sqrt{2}, -1/\sqrt{6}, 1/\sqrt{3} \rangle$
18. $F(r, \theta, z) = r\theta z$,
    $P = (1, \pi, 0)$,
    direction: $\mathbf{u} = \langle \sqrt{3}/2, 0, -1/2 \rangle$

Exercises 19–22: Find the direction in which the rate of change at P is greatest, and calculate the value of this rate of change.

19. f and P of Exercise 13
20. g and P of Exercise 14
21. f and P of Exercise 16
22. g and P of Exercise 17

Exercises 23–26: Describe the cross sections perpendicular to each coordinate axis, then sketch the surface.

23. $z = 4 - x^2 - y^2$
24. $z = \sqrt{25 - x^2 - 2y^2}$
25. $-x^2 + y^2 - z^2 = 1$
26. $y = \sqrt{2x - z^2}$

Exercises 27–30: Find a parametric representation that conforms to the level curves of the surface described. Use the parametric representation and a CAS to generate a picture of the surface.

27. $z = 4 - x^2 - y^2$
28. $z = \sqrt{25 - x^2 - 2y^2}$
29. $y = \sqrt{2x - z^2}$
30. $z = xe^y$

Exercises 31–32: Convert to cylindrical coordinates. Use the cylindrical coordinate expression to obtain a parametric representation for the surface, then use a parametric plotting routine on a CAS to obtain a graph of the surface.

31. $z = 3 + x^2 + y^2$
32. $z = x^2 - 2xy + y^2$

Exercises 33–36: Convert to spherical coordinates. Use the spherical coordinate expression to obtain a parametric representation for the surface, then use a parametric plotting routine on a CAS to obtain a graph of the surface.

33. $x^2 + y^2 + z^2 = 4$
34. $z = \sqrt{2(x^2 + y^2)}$
35. $x = 4$
36. $z^2 - x^2 - y^2 = 4$

Exercises 37–40: Evaluate the limit or show that the limit does not exist.

37. $\displaystyle\lim_{(x,y)\to(1,2)} \frac{3x^2y + 2xy^2 + 3}{4x - 3y + 4}$
38. $\displaystyle\lim_{(x,y)\to(0,0)} \frac{x + y}{|x| + |y|}$

39. $\displaystyle \lim_{(x,y)\to(-1,2)} \frac{(x+1)(y-2)^3}{(x+1)^2+(y-2)^2}$

40. $\displaystyle \lim_{(x,y)\to(0,0)} (x+y)\ln(x^2+y^2)$

T 41. Consider the approximation

$$\sqrt{1+x+2y} \approx 2 + \frac{1}{4}(x-1) + \frac{1}{2}(y-1)$$

for (x, y) near $(1, 1)$. Use a contour plot to find a $\delta > 0$ so that the error in the approximation is less than 0.01 in absolute value when

$$\sqrt{(x-1)^2+(y-1)^2} < \delta.$$

T 42. Consider the approximation

$$\arctan\frac{y}{x} \approx \frac{\pi}{4} - \frac{1}{\sqrt{2}}(x-2) + \frac{1}{\sqrt{2}}(y-2)$$

for (x, y) near $(2, 2)$. Use a contour plot to find a $\delta > 0$ so that the error in the approximation is less than 0.01 in absolute value when

$$\sqrt{(x-2)^2+(y-2)^2} < \delta.$$

43. In Example 1 of Section 9.2 we studied the equation

$$s = d(x,t) = \frac{1}{4}\cos(20\pi t)\sin\left(\frac{\pi x}{30}\right) + \frac{1}{16}\cos(40\pi t)\sin\left(\frac{\pi x}{15}\right)$$

where t is time, x is the distance from one end of a guitar string, and $s = d(x, t)$ is the height above or below equilibrium at time t of the point of string at position x. Calculate $\partial s/\partial t$ and $\partial s/\partial x$. Give the units for each derivative. Discuss what each derivative tells us about the displacement of the guitar string.

44. In Example 2 of Section 9.2 we studied the ballistic pendulum equation

$$v = v(m, M, h) = \left(\frac{M+m}{m}\right)\sqrt{2gh},$$

where M is the mass of the large block, m is the mass of the bullet, h is the maximum height reached by the large block, and v is the velocity of the bullet. Calculate $\partial v/\partial M$, $\partial v/\partial m$, and $\partial v/\partial h$. Give the units for each derivative. Discuss what each derivative tells us about the velocity of the bullet.

45. In Example 8 of Section 9.5 we saw that if a spherical object of mass M is centered at the origin and a spherical object of mass m is centered at (x, y, z), then the magnitude of the gravitational force exerted on the object of mass m is

$$F = F(x, y, z, M, m) = \frac{GmM}{x^2+y^2+z^2}.$$

Calculate the partial derivatives $\partial F/\partial x$, $\partial F/\partial y$, $\partial F/\partial z$, $\partial F/\partial m$, and $\partial F/\partial M$. Give the units for each derivative. Discuss what each derivative tells us about the gravitational force on the object of mass m.

46. In Example 8 of Section 9.5 we saw that if a spherical object of mass M is centered at the origin and a spherical object of mass m is centered at (ρ, θ, ϕ) (in spherical coordinates), then the magnitude of the gravitational force exerted on the object of mass m is

$$F = F(\rho, \theta, \phi, M, m) = \frac{GmM}{\rho^2}.$$

Calculate the partial derivatives $\partial F/\partial \rho$, $\partial F/\partial \theta$, $\partial F/\partial \phi$, $\partial F/\partial m$, and $\partial F/\partial M$. Give the units for each derivative. Discuss what each derivative tells us about the gravitational force on the object of mass m.

47. In Exercise 23 of Section 9.2 we considered the following formula for the maximum speed s attained by a rocket:

$$s = s(r, m, M) = r\ln\left(\frac{M}{m}\right),$$

where M is the initial mass of the rocket with fuel, m is the mass after the fuel is burned, and r is the speed of the rocket gases relative to the rocket. Calculate the partial derivatives $\partial s/\partial r$, $\partial s/\partial M$, and $\partial s/\partial m$. Give the units for each derivative. Discuss what each derivative tells us about the maximum speed attained by the rocket.

48. Common ocean waves and waves that arise during a storm are produced by the wind. For such waves it can be shown that

$$C = C(l, d) = \sqrt{\frac{gl}{2\pi}\left(\frac{e^{2\pi d/l} - e^{-2\pi d/l}}{e^{2\pi d/l} + e^{-2\pi d/l}}\right)},$$

where C is the speed of the wave front moving across the water, l is the length of the wave, d is the depth of the water, and g is the acceleration of gravity. Assume that distances are measured in meters and time in seconds. Calculate each of the partial derivatives $\partial C/\partial l$ and $\partial C/\partial d$. Assign units to each derivative and discuss what each derivative tells us about wave speed.

49. Let f and g be real-valued functions of two variables, each defined on the set $D \subset R^2$. Assume that the partial derivatives of both f and g exist at every point in D. Show that the gradient "operator" ∇ has the following properties:

a. $\nabla(f+g) = \nabla f + \nabla g$

b. $\nabla(cf) = c\nabla f$, (c a real number)

c. $\nabla(fg) = f\nabla g + g\nabla f$

d. $\nabla(f/g) = (g\nabla f - f\nabla g)/g^2$

e. $\nabla f^p = pf^{p-1}\nabla f$

STUDENT PROJECT

RULED SURFACES

A surface is called a *ruled surface* if each of its points lies on a line wholly contained in the surface. A ruled surface is called *developable* if it can be "unrolled" into a plane surface. Among the ruled surfaces are planes, cylinders, and cones. Each of these is developable. Among the quadric surfaces, the elliptic hyperboloid of one sheet and the hyperbolic paraboloid are ruled surfaces.

Ruled surfaces are often described as surfaces generated by the motion of a line. Consider, for example, a plane Π and a line ℓ perpendicular to Π. If ℓ were to move on a circle in the plane Π, it would generate a cylinder. Hence, the cylindrical surface is a ruled surface since each of its points lies on one of the vertical lines and each of these lines lies wholly contained in the surface. In particular, let the (x, y)-plane be Π and let the circle be the unit circle centered at $(0, 0, 0)$, The line ℓ will generate the cylinder shown in Fig. 9.131. On the left, we show the cylinder with shading and surfaces in the back hidden; on the right, we show the line ℓ as it traces the circle. In each sketch, we show only the part of the cylinder between $z = -2$ and $z = 2$.

More generally, suppose that a curve C in R^3 is described by the equation

$$\mathbf{r} = \mathbf{g}(s), \qquad s \in I,$$

where $I \subset R$ is an interval. At each point $\mathbf{g}(s)$ of C, let $\mathbf{u}(s)$ be a unit vector. We show that the surface S described by

$$\mathbf{r} = \mathbf{r}(s,t) = \mathbf{g}(s) + t\mathbf{u}(s), \qquad (s,t) \in I \times R, \tag{1}$$

is a ruled surface. We assume without further comment that functions such as $\mathbf{g}$ and $\mathbf{u}$ used to describe ruled surfaces are continuously differentiable.

To show that S is a ruled surface, let $\mathbf{r}(s_0, t_0)$ be any point of S. This point lies on the line described parametrically by

$$\mathbf{r} = \mathbf{r}(t) = \mathbf{g}(s_0) + t\mathbf{u}(s_0), \qquad t \in R.$$

This line, by definition of S, is wholly contained in S.

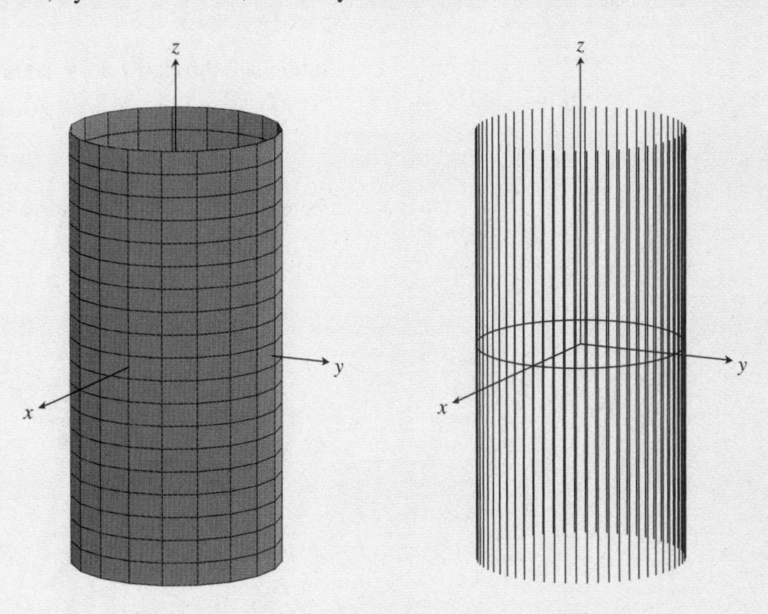

FIGURE 9.131 Part of a cylinder generated by a line.

Some ruled surfaces can be described by equations in which, instead of a curve, the generating line is the organizing principle. All lines in R^3 except for those parallel to the y- or z-axes can be described as the intersection of planes of the form

$$y = ax + b$$
$$z = cx + d,$$

where a, b, c, and d are constants. Thinking of x as a parameter, we may set $x = t$ and describe the line defined by these planes as

$$\mathbf{r} = \mathbf{r}(t) = \langle t, at + b, ct + d \rangle, \qquad t \in R.$$

If we replace the constant a by a variable s and regard the constants b, c, and d as functions of s, where s varies over an interval $I \subset R$, then a surface S is described by

$$\mathbf{r} = \mathbf{r}(s,t) = \langle t, ts + b(s), tc(s) + d(s) \rangle, \qquad (s,t) \in I \times J, \tag{2}$$

where J is an interval in R.

As an example, we take $b(s) = 0$, $c(s) = 1$, $d(s) = s^2$, $I = [0,3]$, and $J = [-5,5]$. The surface S described by

$$\mathbf{r} = \mathbf{r}(s,t) = \langle t, st, t + s^2 \rangle, \qquad (s,t) \in [0,3] \times [-5,5] \tag{3}$$

is shown on the left in Fig. 9.132. The sketch on the right in Fig. 9.132 consists of 31 line segments, which provide snapshots of the generating line.

PROBLEM 1 A cone has the equation $9x^2 + 9y^2 = z^2$. Describe this cone with an equation of the form of (1). Use the parametric equation in graphing the cone between the planes $z = -3$ and $z = 3$. The lines that generate this cone can each be described by

$$\mathbf{h} = \mathbf{h}(t) = \mathbf{r}(s,t), \qquad t \in J,$$

for some fixed $s \in I$. Graph the lines corresponding to approximately 20 equally spaced values of s.

Problems 2 through 7 are related to the surface S described by the equation

$$\frac{x^2}{a^2} + \frac{y^2}{b^2} - \frac{z^2}{c^2} = 1, \tag{4}$$

where $a, b, c > 0$. This quadric surface is called an *elliptic hyperboloid of one sheet.*

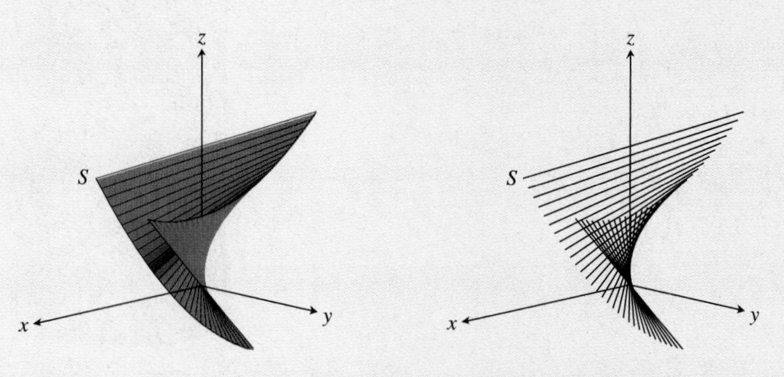

FIGURE 9.132 Two views of a ruled surface.

PROBLEM 2 Equation (4) can be written as

$$\left(\frac{x}{a} + \frac{z}{c}\right)\left(\frac{x}{a} - \frac{z}{c}\right) = \left(1 + \frac{y}{b}\right)\left(1 - \frac{y}{b}\right).$$

The form of this expression suggests the pair of equations

$$\begin{cases} \dfrac{x}{a} + \dfrac{z}{c} = t\left(1 + \dfrac{y}{b}\right) \\ t\left(\dfrac{x}{a} - \dfrac{z}{c}\right) = 1 - \dfrac{y}{b}. \end{cases}$$

Show that, for each value of t, the line defined by these equations can be described by

$$\mathbf{r} = \mathbf{r}(s) = \left\langle \frac{a(-s + 2ct + st^2)}{c(t^2 + 1)}, \frac{b(c + 2st - ct^2)}{c(t^2 + 1)}, s \right\rangle, \qquad s \in R.$$

Show that the elliptic hyperboloid is a ruled surface.

PROBLEM 3 Setting $a = b = c = 1$ in (4) and referring to the preceding problem, graph together the lines corresponding to $t = -2.0, -1.9, \ldots, 1.9, 2.0$. Let s vary over the interval $[-3, 3]$. Explain what you see and, possibly, how "the problem" might be fixed.

PROBLEM 4 With $a = b = c = 1$, an alternative parameterization of S can be based on Equation (1). Letting $\mathbf{g}(s) = \langle \cos s, \sin s, 0 \rangle$ describe the trace of S in the (x, y)-plane, determine, for each $s \in [0, 2\pi]$, a unit vector $\mathbf{u}(s)$ such that the line described by

$$\mathbf{r} = \mathbf{r}(t) = \mathbf{g}(s) + t\mathbf{u}(s), \qquad t \in R,$$

lies in S. Graph together line segments corresponding to $s = j\pi/30$, $j = 0, 1, \ldots, 59$. Choose t so that the endpoints of the lines are in the planes with equations $z = -3$ and $z = 3$.

PROBLEM 5 With $a = b = c = 1$, show that at each point of S the generating line through this point is tangent to S.

PROBLEM 6 With $a = b = c = 1$, show that S is "doubly ruled," that is, there are two sets of generators. You may note either that

$$\begin{cases} \dfrac{x}{a} + \dfrac{z}{c} = t\left(1 - \dfrac{y}{b}\right) \\ t\left(\dfrac{x}{a} - \dfrac{z}{c}\right) = 1 + \dfrac{y}{b} \end{cases}$$

is a second arrangement of (4) or, referring to Problem 4, that the process leading to the unit vector $\mathbf{u}(s)$ has two possible outcomes.

PROBLEM 7 Referring to the preceding problem, the doubly ruled elliptic hyperboloid corresponding to $a = b = c = 1$ can be modeled by two sets of strings, stretched between drilled circular plates in the planes with equations $z = -3$ and

$z = 3$. Fill in the details of the following sparse directions. Clamp two circular plates together and drill a circle of regularly spaced holes through the clamped pair. Keeping the plates parallel and their centers vertically aligned, separate them by 6 units, and rotate the top plate by 2 arctan 3 degrees. Attach strings.

Problems 8 through 11 are related to the surface S described by the equation

$$z = \frac{x^2}{a^2} - \frac{y^2}{b^2}, \tag{5}$$

where $a, b > 0$. This quadric surface is called a *hyperbolic paraboloid.*

PROBLEM 8 Equation (5) can be written as

$$z = \left(\frac{x}{a} + \frac{y}{b}\right)\left(\frac{x}{a} - \frac{y}{b}\right).$$

The form of this expression suggests the pair of equations

$$\begin{cases} z = t\left(\dfrac{x}{a} + \dfrac{y}{b}\right) \\ t = \left(\dfrac{x}{a} - \dfrac{y}{b}\right). \end{cases}$$

Show that for each value of $t \neq 0$, the line defined by these equations can be described by

$$\mathbf{r} = \mathbf{r}(s) = \left\langle \frac{a(s + t)^2}{2t}, \frac{b(s - t^2)}{2t}, s \right\rangle, \qquad s \in R.$$

Discuss the case in which $t = 0$. Show that the hyperbolic paraboloid is a ruled surface.

PROBLEM 9 Sketch the roof shown in Example 2, Section 1, of Chapter 11. First, use the equation given in the example. Second, use equations of the form of (1) to draw the double set of lines used for the figure in the example. See Exercise 21 in Section 1 of Chapter 11.

PROBLEM 10 With $a = b = 1$, describe the surface S by one or more equations of the form of (1). Consider as curves the two branches of the trace of S in the plane $z = 1$. *Hint:* For all s,

$$(\pm\cosh s)^2 - (\sinh s)^2 = 1.$$

PROBLEM 11 Show that S is "doubly ruled," that is, there are two sets of generating lines.

10

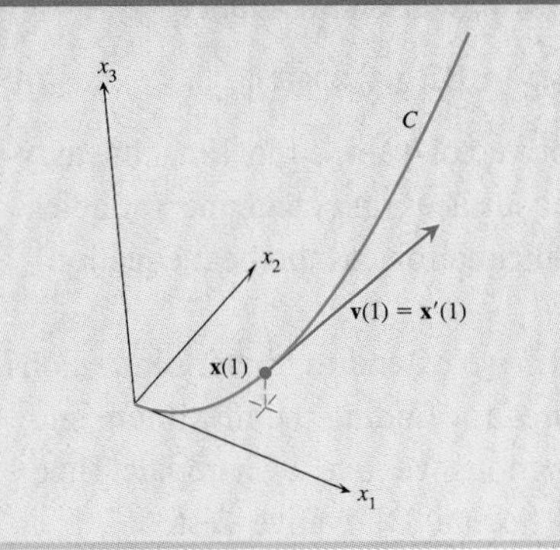

Differentiable Functions of Several Variables

As a jet fighter climbs upward through the air, the air pressure changes dramatically. The rate of change of the pressure depends on the plane's path as well as the changing pressure of the atmosphere with altitude.

See Example 1, p. 864, for further explanation.

An important idea in this book has been the tangent line approximation: If f is differentiable near a, then

$$f(x) \approx f(a) + f'(x)(x - a) \tag{1}$$

for points x near a. For differentiable functions $\mathbf{f}$ of several variables we show in Section 10.1 that

$$\mathbf{f}(\mathbf{x}) \approx \mathbf{f}(\mathbf{a}) + \mathbf{Df}(\mathbf{a})(\mathbf{x} - \mathbf{a}) \tag{2}$$

for points $\mathbf{x}$ near $\mathbf{a}$, where $\mathbf{Df}(\mathbf{a})$ is a matrix of partial derivatives of $\mathbf{f}$. We use this *linearization* of $\mathbf{f}$ to approximate functions and to estimate the percentage error in a variable depending upon several measured variables.

Another important idea has been the chain rule: The derivative of the composition $f \circ g$ of functions $y = f(x)$ and $x = g(t)$ is

$$(f \circ g)'(t) = f'(g(t))g'(t). \tag{3}$$

For the composition of functions $\mathbf{y} = \mathbf{f}(\mathbf{x})$ and $\mathbf{x} = \mathbf{g}(t)$ of several variables, we show in Section 10.2 that

$$\mathbf{D}(\mathbf{f} \circ \mathbf{g})(t) = \mathbf{Df}(\mathbf{g}(t))\mathbf{Dg}(t). \tag{4}$$

We use this chain rule for vector-valued functions in studying level curves and level surfaces and changing variables in such partial differential equations as the heat equation and the wave equation.

In Sections 10.5–10.7 we extend to higher dimensions the one-variable procedures for finding the maximum and minimum values of a function. We give a Candidate Theorem and an analog to the Second Derivative Test.

In extending the ideas of differentiability and the chain rule to higher dimensions, we use the idea of a linear function and its associated matrix, discussed in Chapter 8. This helps to unify our discussion and to facilitate the similar notations used in (1) and (2) and again in (3) and (4).

10.1 Differentiability

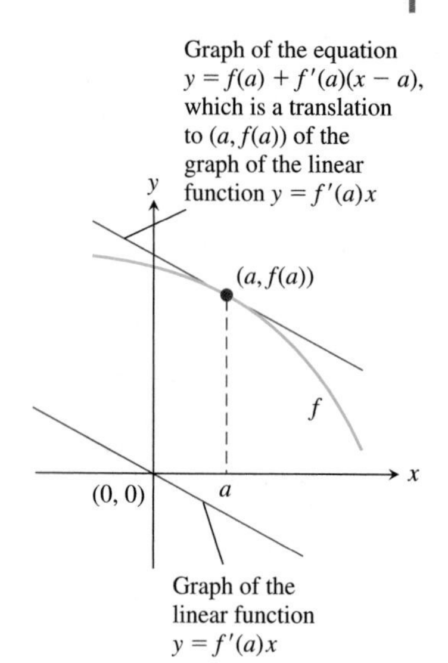

FIGURE 10.1 Graph of f and its tangent line near the point $(a, f(a))$; the tangent line may be regarded as the translation to $(a, f(a))$ of the graph of the *linear* function $y = f'(a)x$.

A real-valued function f of one variable is differentiable at a if, as we zoom to $(a, f(a))$, the graph of f more and more resembles a line through this point. Figure 10.1 shows the graphs of f and its tangent line near the point $(a, f(a))$. Also shown is the graph of the *linear* function defined by the equation $y = f'(a)x$. This function is linear in the sense defined in Chapter 8. The tangent line is a translation to $(a, f(a))$ of the graph of this linear function.

As we show in a moment, f is differentiable at a if there is a linear function $y = mx$ for which

$$\lim_{x \to a} \frac{f(x) - f(a) - m(x - a)}{x - a} = 0. \tag{1}$$

By writing

$$\frac{f(x) - f(a) - m(x - a)}{x - a} = \frac{f(x) - f(a)}{x - a} - m \cdot \frac{x - a}{x - a} = \frac{f(x) - f(a)}{x - a} - m,$$

we see that the existence of the limit in (1) is equivalent to the statement

$$\lim_{x \to a} \frac{f(x) - f(a)}{x - a} = m.$$

We used this form of the limit in Chapter 1 when we first defined the differentiability of f at a.

In the limit (1), we can think of $m(x - a)$ as the value at $h = x - a$ of the linear function $df_a(h) = mh$. This function is defined for all h, and we can think of m as the 1×1 matrix associated with the linear function df_a from R^1 to R^1.

For our discussion of differentiable functions of several variables, we recall from Chapters 8 and 9 that nonvertical planes in R^3 can be described by equations of the form $Ax + By + Cz + D = 0$, where $C \neq 0$. If the plane contains the origin, then $x = y = z = 0$ satisfies the equation and it follows that $D = 0$. This means that we can write the equation of a nonvertical plane through the origin as $z = px + qy$, where $p = -A/C$ and $q = -B/C$.

A real-valued function f of two variables is differentiable at $\mathbf{a} = (a_1, a_2)$ if, as we zoom to $(\mathbf{a}, f(\mathbf{a})) = (a_1, a_2, f(a_1, a_2))$, the graph of f more and more resembles a plane through $(\mathbf{a}, f(\mathbf{a}))$. Figure 10.2 shows the graph of a function f near a point $(\mathbf{a}, f(\mathbf{a}))$. The figure also shows part of the plane tangent to the graph at $(\mathbf{a}, f(\mathbf{a}))$. We regard this tangent plane as the translation to $(\mathbf{a}, f(\mathbf{a}))$ of the graph of a linear function $z = px + qy$. For any point $\mathbf{x} = (x, y)$ in the domain of f, the height of the graph of f above $\mathbf{x}$ is $f(\mathbf{x}) = f(x, y)$, which is the length of the segment from $\mathbf{x}$ to $(x, y, f(x, y))$. For points $\mathbf{x}$ near $\mathbf{a}$, this height is closely approximated by the height of the tangent plane above $\mathbf{x}$, which is the length of the segment from $\mathbf{x}$ to the point Q on the tangent plane. The figure suggests that as we zoom in to $(\mathbf{a}, f(\mathbf{a}))$ the graph of f looks more and more like its tangent plane.

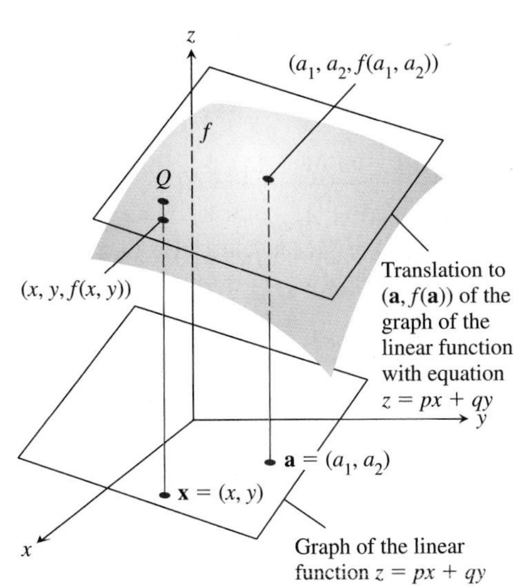

FIGURE 10.2 Graphs of f and the plane tangent to the graph of f for points $\mathbf{x}$ near $\mathbf{a}$; also shown is the graph of the linear function $z = px + qy$, which when translated to $(a, f(a))$ is the tangent plane.

For functions f of one variable, $f'(a)$ exists if and only if as we zoom towards the point $(a, f(a))$ the graph of f more and more resembles its tangent line. It is reasonable to suppose that something like this holds for functions of two variables, that is, if f is a function of two variables, the partial derivatives $f_x(\mathbf{a})$ and $f_y(\mathbf{a})$ of f exist if and only if as we zoom to $(\mathbf{a}, f(\mathbf{a}))$ the graph of f more and more resembles its tangent plane. This equivalence does not, however, hold. It is possible for the partial derivatives of f to exist at $\mathbf{a}$ and, at the same time, the graph of f to never resemble a plane as we zoom to $(\mathbf{a}, f(\mathbf{a}))$. We give an example of such a function at the end of this section.

We use (1) as a guide in defining differentiability for real-valued functions of two variables. In the definition we refer to a linear function $df_{\mathbf{a}}$ from R^2 into R. Recall that if L is a linear function from R^2 into R and M is the matrix associated with L, then for all $\mathbf{h} = \langle h_1, h_2 \rangle$ in R^2,

$$L(\mathbf{h}) = M\mathbf{h} = (m_1 \quad m_2)\begin{pmatrix} h_1 \\ h_2 \end{pmatrix} = m_1 h_1 + m_2 h_2. \tag{2}$$

DEFINITION Differentiability of Real-Valued Functions

Assume that f is a real-valued function with domain $\mathcal{D} \subset R^2$ and that $\mathbf{a} = (a_1, a_2)$ is an interior point of $\mathcal{D}$. The function f is **differentiable** at $\mathbf{a}$ if there is a linear function $df_{\mathbf{a}}$ from R^2 into R for which

$$\lim_{\mathbf{x} \to \mathbf{a}} \frac{f(\mathbf{x}) - f(\mathbf{a}) - df_{\mathbf{a}}(\mathbf{x} - \mathbf{a})}{\|\mathbf{x} - \mathbf{a}\|} = 0. \tag{3}$$

The linear function $df_{\mathbf{a}}$ is called the **differential** of f at $\mathbf{a}$. The matrix associated with $df_{\mathbf{a}}$ is denoted by $Df(\mathbf{a})$ and is called the **derivative matrix** of f at $\mathbf{a}$. We sometimes shorten $df_{\mathbf{a}}$ to df.

We continue to use interchangeably either of the notations $\mathbf{a}$ or (a_1, a_2) for points and, for example, the notations $f(\mathbf{a})$ and $f_x(\mathbf{a})$ for $f(a_1, a_2)$ and $f_x(a_1, a_2)$.

For a function f of several variables, the derivative matrix $Df(\mathbf{a})$ of f at $\mathbf{a}$ plays the same role as $f'(a)$ for a function f of one variable. We show that if f is differentiable at $\mathbf{a}$, then its partial derivatives $f_x(\mathbf{a})$ and $f_y(\mathbf{a})$ exist and the derivative matrix of f at $\mathbf{a}$ is the 1×2 matrix

$$Df(\mathbf{a}) = (f_x(\mathbf{a}) \quad f_y(\mathbf{a})). \tag{4}$$

Because $df_{\mathbf{a}}$ is a linear function with domain R^2 and range in R, its associated matrix is a 1×2 matrix. As in (2), we write this matrix as $(m_1 \quad m_2)$. We may write (3) as

$$\lim_{\mathbf{x}\to\mathbf{a}} \frac{f(x,y) - f(a_1,a_2) - m_1(x-a_1) - m_2(y-a_2)}{\sqrt{(x-a_1)^2 + (y-a_2)^2}} = 0.$$

Assuming that this limit exists, we may choose how $\mathbf{x}$ approaches $\mathbf{a}$. If $\mathbf{x}$ approaches $\mathbf{a}$ from the right, along the line $y = a_2$, then

$$\lim_{x\to a_1^+} \frac{f(x,a_2) - f(a_1,a_2) - m_1(x-a_1)}{x - a_1} = 0.$$

This limit can be rewritten in the form

$$\lim_{x\to a_1^+} \frac{f(x,a_2) - f(a_1,a_2)}{x - a_1} = m_1.$$

From the definition of partial derivatives, the limit on the left of this equation is equal to $f_x(\mathbf{a})$. Hence, if f is differentiable at $\mathbf{a}$, then $m_1 = f_x(\mathbf{a})$. By a similar argument, $f_y(\mathbf{a}) = m_2$. This establishes (4).

The Tangent Plane Approximation

The tangent plane approximation comes directly from (3) and (4). Because the limit of the quotient in (3) is zero and the limit of the denominator is 0, the numerator must go to zero as $\mathbf{x} \to \mathbf{a}$. Hence, for $\mathbf{x}$ close to $\mathbf{a} = (a_1, a_2)$,

$$f(\mathbf{x}) \approx f(\mathbf{a}) + df_{\mathbf{a}}(\mathbf{x} - \mathbf{a}) = f(\mathbf{a}) + f_x(\mathbf{a})(x - a_1) + f_y(\mathbf{a})(y - a_2). \tag{5}$$

This is the "tangent plane approximation" to f near $\mathbf{a}$. From (5) we also see that for points (x, y) near (a_1, a_2) the graph of f is near the plane with equation

$$z = f(a_1,a_2) + f_x(a_1,a_2)(x - a_1) + f_y(a_1,a_2)(y - a_2). \tag{6}$$

From Section 8.6, a normal $\mathbf{n}$ to this plane is $\mathbf{n} = \langle -f_x(a_1,a_2), -f_y(a_1,a_2), 1\rangle$.

Figure 10.3 shows the graph of a differentiable function f for points $\mathbf{x}$ near a point $\mathbf{a} = (a_1, a_2)$. Also shown is an outline of the plane tangent to the graph of f at $(\mathbf{a}, f(\mathbf{a}))$. This plane touches the graph of f at $(\mathbf{a}, f(\mathbf{a}))$ and remains close to the graph of f for points $\mathbf{x}$ not too far from $\mathbf{a}$.

The height $f(\mathbf{x})$ of the function at $\mathbf{x}$ is the segment from $\mathbf{x}$ to P on the graph of f; the height of the tangent plane at $\mathbf{x}$ is the segment from $\mathbf{x}$ to Q. The points P and Q are shown in Fig. 10.3. They lie in the vertical plane through the points $\mathbf{a}$ and $\mathbf{x}$. This plane intersects the graph of f and the tangent plane in the curve and line from $(\mathbf{a}, f(\mathbf{a}))$ to P and Q, respectively. For easy comparison, we have written P and Q in terms of $f(\mathbf{a})$, $\Delta f_{\mathbf{a}}(\mathbf{x} - \mathbf{a}) = f(\mathbf{x}) - f(\mathbf{a})$, and the differential $df_{\mathbf{a}}(\mathbf{x} - \mathbf{a})$. For the point P,

$$(\mathbf{x}, f(\mathbf{x})) = (\mathbf{x}, f(\mathbf{a}) + f(\mathbf{x}) - f(\mathbf{a})) = (\mathbf{x}, f(\mathbf{a}) + \Delta f_{\mathbf{a}}(\mathbf{x} - \mathbf{a})).$$

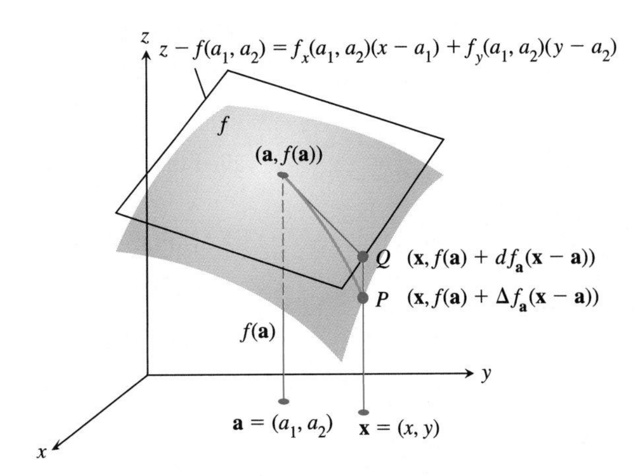

FIGURE 10.3 The tangent plane to the graph of f at $(\mathbf{a}, f(\mathbf{a}))$; a graphical comparison of $\Delta f_{\mathbf{a}}(\mathbf{x} - \mathbf{a})$ and $df_{\mathbf{a}}(\mathbf{x} - \mathbf{a})$.

From (6), the point Q on the tangent plane can be written as

$$(\mathbf{x}, f(\mathbf{a}) + f_x(\mathbf{a})(x - a_1) + f_y(\mathbf{a})(y - a_2)) = (\mathbf{x}, f(\mathbf{a}) + df_{\mathbf{a}}(\mathbf{x} - \mathbf{a})).$$

In this way we see that the true change in f as we move from $\mathbf{a}$ to $\mathbf{x}$ is $\Delta f_{\mathbf{a}}(\mathbf{x} - \mathbf{a})$; this true change is approximated by $df_{\mathbf{a}}(\mathbf{x} - \mathbf{a})$.

We summarize this discussion, work through several examples, comment briefly on nondifferentiable functions, and, at the end of the section, define differentiability for vector-valued functions.

Differentiability

Let f be a real-valued function defined on a region $\mathcal{D} \subset R^2$ and let $\mathbf{a} = (a_1, a_2)$ be an interior point of $\mathcal{D}$. Assume that f is differentiable at $\mathbf{a}$.

Derivative Matrix The derivative matrix of f at $\mathbf{a}$ is

$$Df(\mathbf{a}) = \begin{pmatrix} f_x(\mathbf{a}) & f_y(\mathbf{a}) \end{pmatrix}. \tag{7}$$

This matrix is the matrix associated with the linear function $df_{\mathbf{a}}$.

Tangent Plane The tangent plane to the graph of f at $(\mathbf{a}, f(\mathbf{a}))$ is described by the equation

$$z = f(\mathbf{a}) + f_x(\mathbf{a})(x - a_1) + f_y(\mathbf{a})(y - a_2). \tag{8}$$

The Tangent Plane Approximation For points $\mathbf{x}$ in $\mathcal{D}$ that are near $\mathbf{a}$,

$$f(\mathbf{x}) \approx f(\mathbf{a}) + f_x(\mathbf{a})(x - a_1) + f_y(\mathbf{a})(y - a_2). \tag{9}$$

Or, letting $\Delta f_{\mathbf{a}}(\mathbf{x} - \mathbf{a}) = f(\mathbf{x}) - f(\mathbf{a})$ and $df_{\mathbf{a}}(\mathbf{x} - \mathbf{a}) = f_x(\mathbf{a})(x - a_1) + f_y(\mathbf{a})(y - a_2)$

$$\Delta f_{\mathbf{a}}(\mathbf{x} - \mathbf{a}) \approx df_{\mathbf{a}}(\mathbf{x} - \mathbf{a}). \tag{10}$$

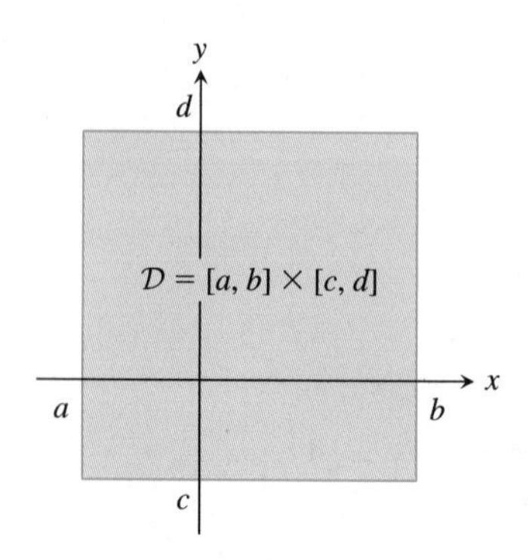

FIGURE 10.4 The Cartesian product $[a, b] \times [c, d]$ of intervals $[a, b]$ and $[c, d]$.

Because we often consider functions defined on rectangular regions similar to the set $\mathcal{D}$ shown in Fig. 10.4, we adopt a brief notation. The set

$$\mathcal{D} = \{(x, y) \in R^2 \,|\, a \leq x \leq b, c \leq y \leq d\}$$

is called the **Cartesian product** of $[a, b]$ and $[c, d]$ and is denoted by $[a, b] \times [c, d]$.

EXAMPLE 1 Figure 10.5 shows the graph of the function f defined by

$$f(x, y) = 3 - x^2 - 2y^2, \qquad (x, y) \in \mathcal{D} = [-1, 1] \times [-1, 1].$$

Assuming that f is differentiable at $\left(\frac{1}{2}, \frac{1}{4}\right)$, write an equation for the plane tangent to the graph of f at $\left(\frac{1}{2}, \frac{1}{4}, f\left(\frac{1}{2}, \frac{1}{4}\right)\right)$ and find a unit normal $\mathbf{N}$ to this plane.

Solution

An equation of the tangent plane is given in (8). For this we need the values of f, f_x, and f_y at $\mathbf{a} = \left(\frac{1}{2}, \frac{1}{4}\right)$.

$$f\left(\tfrac{1}{2}, \tfrac{1}{4}\right) = 3 - \left(\tfrac{1}{2}\right)^2 - 2 \cdot \left(\tfrac{1}{4}\right)^2 = \tfrac{21}{8}.$$

Calculating f_x and f_y at a general point,

$$f_x(x, y) = -2x \quad \text{and} \quad f_y(x, y) = -4y.$$

Evaluating f_x and f_y at $\mathbf{a} = \left(\frac{1}{2}, \frac{1}{4}\right)$,

$$f_x\left(\tfrac{1}{2}, \tfrac{1}{4}\right) = -2\left(\tfrac{1}{2}\right) = -1 \quad \text{and} \quad f_y\left(\tfrac{1}{2}, \tfrac{1}{4}\right) = -4\left(\tfrac{1}{4}\right) = -1.$$

Substituting these values into (8), an equation of the tangent plane at $\left(\frac{1}{2}, \frac{1}{4}, \frac{21}{8}\right)$ is

$$z = \tfrac{21}{8} - 1\left(x - \tfrac{1}{2}\right) - 1\left(y - \tfrac{1}{4}\right).$$

A normal vector to this plane is $\mathbf{n} = \langle 1, 1, 1 \rangle$. A unit normal is

$$\mathbf{N} = \frac{1}{\sqrt{3}} \langle 1, 1, 1 \rangle.$$

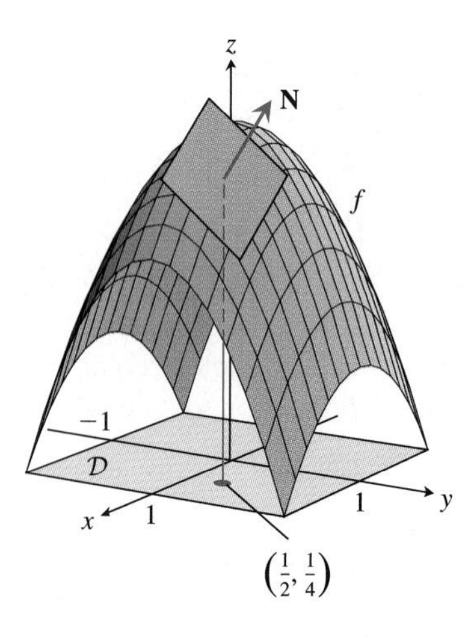

FIGURE 10.5 Graph of f and its tangent plane at $\left(\frac{1}{2}, \frac{1}{4}, f\left(\frac{1}{2}, \frac{1}{4}\right)\right)$.

For now we assume that all functions we consider are differentiable. We give an easy test for differentiability at the end of the section.

EXAMPLE 2 Let f be the function defined by

$$f(x, y) = 1 + \ln(3 - x^2 - y^3), \qquad (x, y) \in \mathcal{D} = [0, 1] \times [0, 1].$$

Figure 10.6 shows a graph of f and its domain $\mathcal{D}$. With $\mathbf{a} = (0.6, 0.4)$ and $\mathbf{x} = (0.8, 0.6)$, use the differential $df_{\mathbf{a}}$ to approximate $f(\mathbf{x}) = f(0.8, 0.6)$. Compare the values of $\Delta f_{\mathbf{a}}(\mathbf{x} - \mathbf{a})$ and $df_{\mathbf{a}}(\mathbf{x} - \mathbf{a})$ numerically and geometrically.

Solution

The purpose of this example is to gain concrete experience with and additional geometric insight into the differential, not to approximate

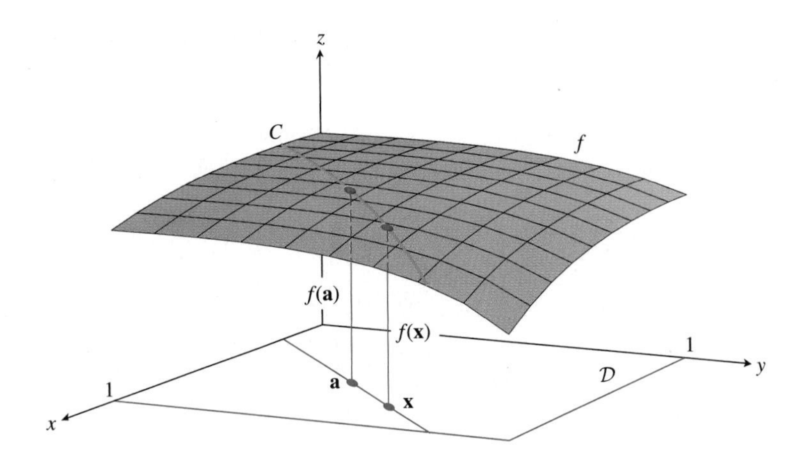

FIGURE 10.6 Approximating $\Delta f(\mathbf{h})$ with $df(\mathbf{h})$.

$f(0.8, 0.6)$, which can be evaluated by calculator or CAS as easily as $f(\mathbf{a})$. To approximate $f(0.8, 0.6)$, we use the tangent plane approximation at $\mathbf{a} = (0.6, 0.4)$. From (9),

$$f(\mathbf{x}) \approx f(\mathbf{a}) + df_{\mathbf{a}}(\mathbf{x} - \mathbf{a}) = f(\mathbf{a}) + f_x(\mathbf{a})\,(x - a_1) + f_y(\mathbf{a})\,(y - a_2). \tag{11}$$

By direct calculation, $f(0.6, 0.4) \approx 1.9462$. We also need the partial derivatives of f. These are:

$$f_x(x, y) = \frac{-2x}{3 - x^2 - y^3} \quad \text{and} \quad f_y(x, y) = \frac{-3y^2}{3 - x^2 - y^3}.$$

Evaluating $f_x(x, y)$ and $f_y(x, y)$ at $\mathbf{a} = (0.6, 0.4)$,

$$f_x(0.6, 0.4) \approx -0.4658 \quad \text{and} \quad f_y(0.6, 0.4) \approx -0.1863.$$

Substituting these values into (11),

$$\begin{aligned} f(0.8, 0.6) &\approx f(\mathbf{a}) + f_x(\mathbf{a})\,(0.8 - a_1) + f_y(\mathbf{a})\,(0.6 - a_2) \\ &\approx 1.9462 + (-0.4658)\,(0.2) + (-0.1863)\,(0.2) \approx 1.8158. \end{aligned} \tag{12}$$

To compare numerically the values of $\Delta f_{\mathbf{a}}(\mathbf{x} - \mathbf{a})$ and $df_{\mathbf{a}}(\mathbf{x} - \mathbf{a})$, we write (11) as

$$\Delta f_{\mathbf{a}}(\mathbf{x} - \mathbf{a}) = f(\mathbf{x}) - f(\mathbf{a}) \approx df_{\mathbf{a}}(\mathbf{x} - \mathbf{a}). \tag{13}$$

From (12),

$$\begin{aligned} df_{\mathbf{a}}(\mathbf{x} - \mathbf{a}) = f_x(\mathbf{a})\,(x - a_1) + f_y(\mathbf{a})\,(y - a_2) &\approx (-0.4658)\,(0.2) \\ &+ (-0.1863)\,(0.2) \approx -0.130. \end{aligned}$$

By calculator,

$$\begin{aligned} \Delta f_{\mathbf{a}}(\mathbf{x} - \mathbf{a}) &= f(\mathbf{x}) - f(\mathbf{a}) = f(0.8, 0.6) - f(0.6, 0.4) \\ &\approx 1.763 - 1.946 = -0.183. \end{aligned}$$

We use Figs. 10.6 and 10.7 in comparing $\Delta f_{\mathbf{a}}(\mathbf{x} - \mathbf{a})$ and $df_{\mathbf{a}}(\mathbf{x} - \mathbf{a})$ geometrically. The points $\mathbf{a} = (0.6, 0.4)$ and $\mathbf{x} = (0.8, 0.6)$ are on a line

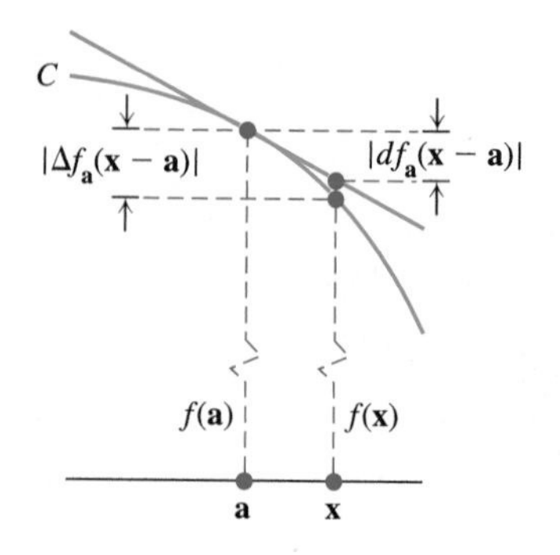

FIGURE 10.7 Using a cross section of the graph of f so that we may graphically compare $\Delta f_{\mathbf{a}}(\mathbf{x}-\mathbf{a})$ and $df_{\mathbf{a}}(\mathbf{x}-\mathbf{a})$.

segment in the domain $\mathcal{D}$ of f. If we section the graph of f by a vertical plane containing $\mathbf{a}$ and $\mathbf{x}$, we obtain a curve C. This curve passes through the points $(\mathbf{a}, f(\mathbf{a}))$ and $(\mathbf{x}, f(\mathbf{x}))$, which are marked as red dots on C. The cross section is shown in Fig. 10.7. The tangent line to C in this two-dimensional cross section is the intersection of the sectioning plane and the tangent plane to the graph of f at $(\mathbf{a}, f(\mathbf{a}))$.

Linearization of Functions *To linearize* a real-valued function f of two variables means to find a first-degree polynomial $A + Bx + Cy$ that approximates $f(x, y)$ near a given point. For differentiable functions, the best linearization is the tangent plane approximation. This is the best linearization at a point $\mathbf{a} = (a_1, a_2)$ in the sense that

$$\lim_{\mathbf{x}\to\mathbf{a}} \frac{f(x,y) - A - B(x-a_1) - C(y-a_2)}{\|\mathbf{x}-\mathbf{a}\|} = 0$$

only when $A = f(a_1, a_2)$, $B = f_x(a_1, a_2)$, and $C = f_y(a_1, a_2)$.

EXAMPLE 3 Linearize the function f defined by

$$f(x,y) = \sqrt{1 - x^2 - y^2}, \qquad x^2 + y^2 \leq 1,$$

near the point $\mathbf{a} = (0.4, 0.6)$.

Solution

The tangent plane approximation to f near $\mathbf{a} = (0.4, 0.6)$ is

$$f(x,y) \approx f(a_1,a_2) + f_x(a_1,a_2)(x - a_1) + f_y(a_1,a_2)(y - a_2). \tag{14}$$

This comes from (9). The linearization of f near $\mathbf{a} = (0.4, 0.6)$ is the polynomial on the right side of (14). To evaluate it, we calculate $f_x(0.4, 0.6)$ and $f_y(0.4, 0.6)$:

$$f_x(0.4, 0.6) = \left.\frac{-2x}{2\sqrt{1 - x^2 - y^2}}\right|_{x=0.4, y=0.6} = \frac{-0.4}{\sqrt{1 - (0.4)^2 - (0.6)^2}} \approx -0.5774$$

$$f_y(0.4, 0.6) = \left.\frac{-2y}{2\sqrt{1 - x^2 - y^2}}\right|_{x=0.4, y=0.6} = \frac{-0.6}{\sqrt{1 - (0.4)^2 - (0.6)^2}} \approx -0.8660.$$

Substituting these values into (14) and calculating $f(0.4, 0.6) \approx 0.6928$,

$$f(x,y) \approx 0.6928 - 0.5774(x - 0.4) - 0.8660(y - 0.6),$$

which simplifies to

$$f(x,y) \approx g(x,y) = 1.4434 - 0.5774x - 0.8660y. \tag{15}$$

The polynomial $g(x, y)$ is what is commonly called the *linearization* of f near $\mathbf{a} = (0.4, 0.6)$.

EXAMPLE 4 Referring to Example 3, analyze the error

$$E(x, y) = f(x, y) - g(x, y) \tag{16}$$

in approximating f by its linearization g near $(0.4, 0.6)$. Use both numerical and graphical analysis.

Solution

From Example 3,

$$f(x, y) = \sqrt{1 - x^2 - y^2}, \qquad x^2 + y^2 \leq 1$$

and, from (15),

$$g(x, y) = 1.4434 - 0.5774x - 0.8660y.$$

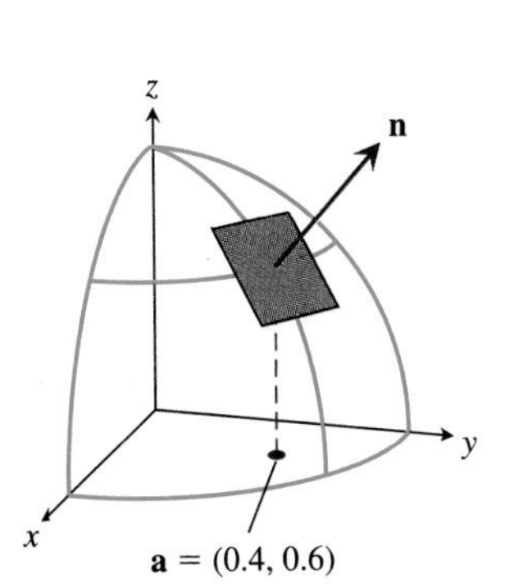

FIGURE 10.8 Spherical surface, tangent plane, and normal vector

Figure 10.8 shows the first-octant portion of the graph of f, which is a hemisphere, the tangent plane at $(0.4, 0.6, f(0.4, 0.6))$, and a normal $\mathbf{n}$. We expect that for points (x, y) near $\mathbf{a} = (0.4, 0.6)$, the error $E(x, y)$ made in approximating f by its linearization g is small. The graph of g is the same as the graph of the tangent plane to the graph of f at $(0.4, 0.6, f(0.4, 0.6))$.

The function $E(x, y) = f(x, y) - g(x, y)$ measures the vertical displacement between the graphs of f and g. We calculate the values of E at nine points, as listed in Table 10.1.

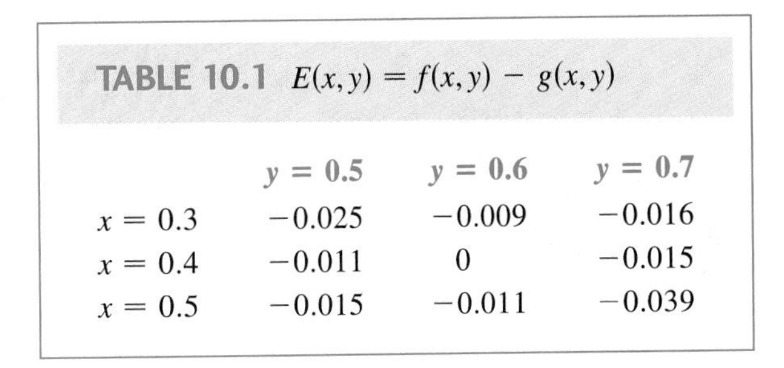

TABLE 10.1 $E(x, y) = f(x, y) - g(x, y)$

	$y = 0.5$	$y = 0.6$	$y = 0.7$
$x = 0.3$	-0.025	-0.009	-0.016
$x = 0.4$	-0.011	0	-0.015
$x = 0.5$	-0.015	-0.011	-0.039

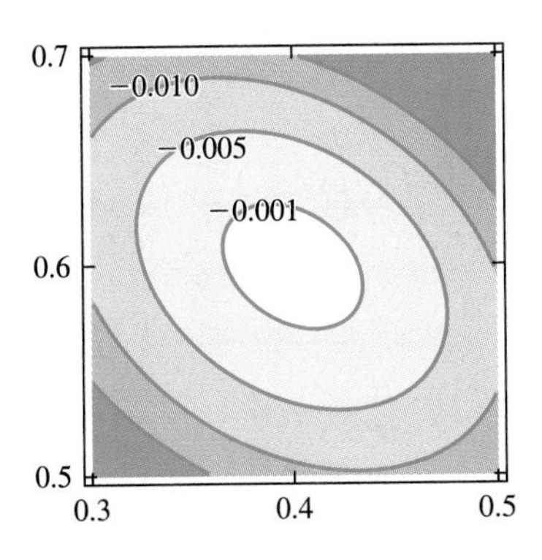

FIGURE 10.9 Contour plot of the linearization error $E(x, y)$ in Example 3.

The error E in the approximation is 0 at the point $\mathbf{a} = (0.4, 0.6)$. This corresponds to the fact that the tangent plane touches the graph of f at $(a_1, a_2, f(a_1, a_2))$. Elsewhere, the tangent plane is a little above the graph of f, so that the values of the error function E are negative.

We now look at the contour plot of E shown in Fig. 10.9. Most CAS programs and some calculators can produce a contour plot similar to that shown in the figure. On the ellipse-like contours, the function E is constant. The values of E on the four contours are -0.001, -0.005, -0.01, and -0.015. As we would expect, the error increases as we move away from the point of tangency.

In the next example we use the ideal gas law. Recall that the law is

$$PV = nRT, \tag{17}$$

where $R = 8.314$ J/mol · K is the universal gas constant, n is the number of moles of gas present (1 mole of any substance is, by definition, 6.022×10^{23} molecules of that substance), P is measured in pascals (Pa, where 1 Pa is a pressure of 1 N/m^2; to give a sense of this unit we note that 1 atmosphere $= 1.01325 \times 10^5$ Pa), V is measured in cubic meters, and T is measured in kelvins (symbol K; temperatures measured in kelvins and on the Celsius scale are related by $K = 273.15 + C$).

EXAMPLE 5 Suppose that we have exactly 1 mole of an ideal gas. The gas is kept at a temperature of 273.15 K (0° C), measured with a percentage error not exceeding 0.5%. The pressure of the gas is adjusted to 1 atmosphere, with a percentage error not exceeding 0.2%. Use the ideal gas law and a differential in approximating the worst possible percentage error in the volume.

Solution

From (17), the volume occupied by 1 mole of an ideal gas at pressure P and temperature T is

$$V = V(P, T) = \frac{nRT}{P}.$$

Letting P and T be the true pressure and temperature of the gas, which are not known, and $P + \Delta P$ and $T + \Delta T$ the measured pressure and temperature, the percentage error in V is

$$P.E. = \frac{V(P + \Delta P, T + \Delta T) - V(P, T)}{V(P, T)} \cdot 100 = \frac{\Delta V(\Delta P, \Delta T)}{V(P, T)} \cdot 100. \tag{18}$$

Using the tangent plane approximation (10), with $\mathbf{a} = (P, T)$ and $\mathbf{x} = (P + \Delta P, T + \Delta T)$,

$$\Delta V(\Delta P, \Delta T) \approx dV(\Delta P, \Delta T) = V_P(P, T)\Delta P + V_T(P, T)\Delta T.$$

From Equation (18),

$$P.E. = \frac{\Delta V(\Delta P, \Delta T)}{V(P, T)} \cdot 100 \approx \frac{V_P(P, T)\Delta P + V_T(P, T)\Delta T}{V(P, T)} \cdot 100. \tag{19}$$

Substituting the partial derivatives

$$V_P(P, T) = \frac{-nRT}{P^2} \quad \text{and} \quad V_T(P, T) = \frac{nR}{P}$$

into (19) and replacing $V(P, T)$ by nRT/P,

$$P.E. = \frac{\Delta V(\Delta P, \Delta T)}{V(T, P)} \cdot 100 \approx \frac{-\dfrac{nRT\Delta P}{P^2} + \dfrac{nR\Delta T}{P}}{\dfrac{nRT}{P}} \cdot 100.$$

Removing the common factors, this expression simplifies to

$$P.E. \approx -\frac{\Delta P}{P} \cdot 100 + \frac{\Delta T}{T} \cdot 100. \tag{20}$$

The terms $(\Delta P/P)\cdot 100$ and $(\Delta T/T)\cdot 100$ are the percentage errors in measuring the pressure and temperature. To maximize the possible percentage error in V, we assume the percentage error in the pressure is negative. From (20), recalling that the percentage errors in the pressure and temperature do not exceed 0.2% and 0.5%,

$$P.E. = \frac{\Delta V}{V}\cdot 100 \approx -(-0.2) + (0.5) = 0.7\%.$$

How to Show That a Function Is Differentiable

Until now we have assumed without proof that the functions we have considered are differentiable. If we want to be certain that a given function f of two variables is differentiable, so that we may, for example, use the tangent plane approximation with confidence, we may use the definition of differentiability, recognize f as a combination of known differentiable functions, or apply a test for differentiability.

In Exercise 40 we outline a proof that the function f in Example 1 is differentiable by using the definition of differentiability. The second method of being certain that a given function f is differentiable is based on such theorems as "the sums and products of differentiable functions are again differentiable." Thus if we know—perhaps by applying the definition of differentiability—that the functions $F(x,y) = c$, $G(x,y) = x$, and $H(x,y) = y$ of two variables are differentiable, where c is a constant, then we would know that all polynomials in x and y are differentiable. Mostly we use the following test for differentiability when we wish to be certain that a given function is differentiable. An argument for this test is outlined in Exercise 49.

THEOREM Test for Differentiability (of Real-Valued Functions)

Let f be a real-valued function and $\mathbf{a}$ an interior point of the domain $\mathcal{D}$ of f. If all of the first-order partial derivatives of f exist and are continuous at $\mathbf{a}$ and all points of $\mathcal{D}$ near $\mathbf{a}$, then f is differentiable at $\mathbf{a}$.

EXAMPLE 6 Show that the function

$$f(x,y,z) = (x+y)\sqrt{z}, \qquad x, y, z \text{ in } R \text{ and } z \geq 0,$$

is differentiable at each point $\mathbf{a} = (a_1, a_2, a_3)$ for which $a_3 > 0$.

Solution

The partials of f at all points (x,y,z) in R for which $z > 0$ are

$$f_x(x,y,z) = \sqrt{z}, \qquad f_y(x,y,z) = \sqrt{z} \quad \text{and} \quad f_z(x,y,z) = \frac{x+y}{2\sqrt{z}}.$$

For any point $\mathbf{a} = (a_1, a_2, a_3)$, where $a_3 > 0$, the first-order partial derivatives of f exist and are continuous at $\mathbf{a}$ and all points of $\mathcal{D}$ near $\mathbf{a}$. Hence, by the Test for Differentiability, f is differentiable at $\mathbf{a}$.

A Nondifferentiable Function

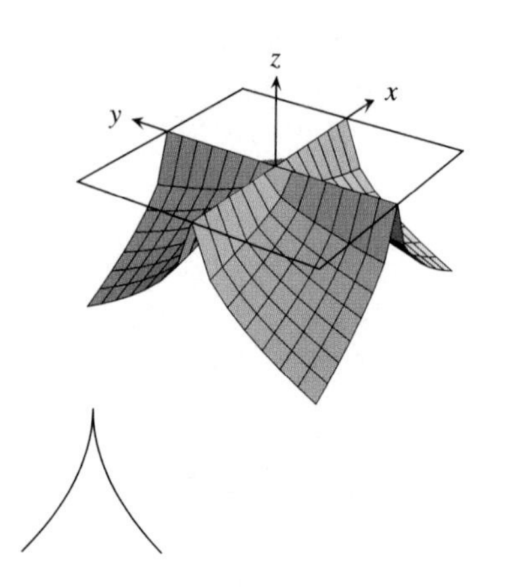

FIGURE 10.10 Graph of the tent function and a representative vertical cross section.

A function of several variables need not be differentiable at a point even if its partial derivatives can be calculated there. Indeed, a function may fail to be continuous at a point where its partials exist! To show how these unexpected phenomena can happen, we consider two "tent functions." The first tent function, whose graph is shown in Fig. 10.10, is not differentiable at $(0, 0)$ even though both partial derivatives $f_x(0, 0)$ and $f_y(0,0)$ exist. We give an informal argument for these assertions.

Imagine two thin, sticky wires stretched along the x- and y-axes. Imagine the wires as supports for a large square of thin, elastic fabric. As the fabric settles onto the wires, it will stick to the wires but sag into the open space between them. Except for the vertical sections containing the x- or y-axis, all other vertical cross sections perpendicular to the x-axis or y-axis resemble the cross section shown in the lower part of the sketch. Because the tent function f is constant along both axes, the partial derivatives $f_x(0, 0)$ and $f_y(0, 0)$ are zero. Because both partials are zero at $(0,0)$, the only candidate for the tangent plane to the graph of the tent function at $(0, 0)$ is described by

$$z - f(0,0) = f_x(0,0)\,(x - 0) + f_y(0,0)\,(y - 0) = 0,$$

that is, the (x, y)-plane itself. The sketches in Fig. 10.10 suggest that this plane (shown in outline, just above the tent) does a poor job of fitting the graph of the tent function near $(0, 0)$. Indeed, the $z = 0$ plane fits so badly that the tent function is not differentiable at $(0, 0)$. A proof for this is outlined in Exercise 44. The function whose graph is shown in Fig. 10.10 is

$$f(x, y) = -\sqrt{|xy|}, \qquad (x, y) \in R^2.$$

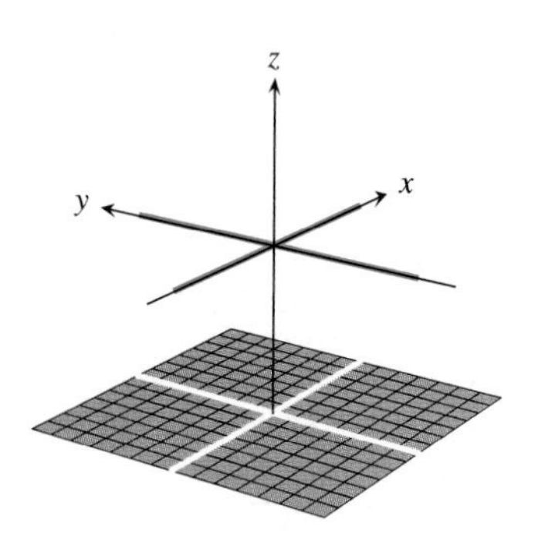

FIGURE 10.11 The "old-tent" function is not continuous at $(0, 0)$.

The tent function is continuous at $(0, 0)$, although it is not differentiable there. Next we describe a function whose partials exist at the origin but which is not continuous there. Again imagine the two thin, sticky wires and let a square of thin, elastic fabric settle over them as before. But this time suppose the fabric is very old, so old that it tears all along the two axes and falls to the plane $z = -1$. The function whose graph is shown in Fig. 10.11 is zero along the axes (where the fabric adhered) and -1 everywhere else. Its partials at $(0, 0)$ are zero. It is not, however, continuous at this point. The old-tent function cannot be differentiable at $(0, 0)$ because, as we show next, differentiability at a point implies continuity there.

Differentiability Implies Continuity

As shown in the discussion of the "old-tent function," the existence of the partial derivatives of a function is not sufficient for continuity. We show that differentiability is sufficient. As a preliminary, recall that a function f is continuous at a point $\mathbf{a}$ of its domain if $f(\mathbf{x})$ becomes and remains as close to $f(\mathbf{a})$ as desired provided that $\mathbf{x}$ is sufficiently close to $\mathbf{a}$. More precisely, f is continuous at $\mathbf{a}$ provided that

$$\lim_{\mathbf{x} \to \mathbf{a}} (f(\mathbf{x}) - f(\mathbf{a})) = 0.$$

From (3) in the definition of differentiability, if f is differentiable at an interior point $\mathbf{a}$ of its domain, then

$$\lim_{\mathbf{x}\to\mathbf{a}} \frac{f(\mathbf{x}) - f(\mathbf{a}) - df(\mathbf{x} - \mathbf{a})}{\|\mathbf{x} - \mathbf{a}\|} = 0.$$

Because the numerator of this expression must approach zero,

$$\begin{aligned}\lim_{\mathbf{x}\to\mathbf{a}} (f(\mathbf{x}) - f(\mathbf{a})) &= \lim_{\mathbf{x}\to\mathbf{a}} df(\mathbf{x} - \mathbf{a}) \\ &= \lim_{\mathbf{x}\to\mathbf{a}} (f_x(a_1, a_2)(x - a_1) + f_y(a_1, a_2)(y - a_2)) = 0.\end{aligned}$$

If follows that f is continuous at $\mathbf{a}$.

Note: from this result it follows that a function that fails to be continuous at a point cannot be differentiable there.

WEB Differentiable Vector-Valued Functions

Vector-valued functions were used in Chapter 9 to describe surfaces. For example, the surface S shown in Fig. 10.12 can be described by a vector or parametric equation of the form

$$\mathbf{r} = \langle x, y, z\rangle = \mathbf{r}(u, v) = \langle r_1(u, v), r_2(u, v), r_3(u, v)\rangle, \qquad (u, v) \in \mathcal{D}. \tag{21}$$

The coordinate functions r_1, r_2, and r_3, as well as $\mathbf{r}$ itself, are functions of two variables. In this subsection, we discuss the differentiability of functions like $\mathbf{r}$ and derive formulas for the normal and the tangent plane to the surface S at a point $\mathbf{r}(u_0, v_0)$.

The definition of a differentiable vector-valued function of several variables has the same form as the definition given earlier (see (3)) for real-valued functions of several variables.

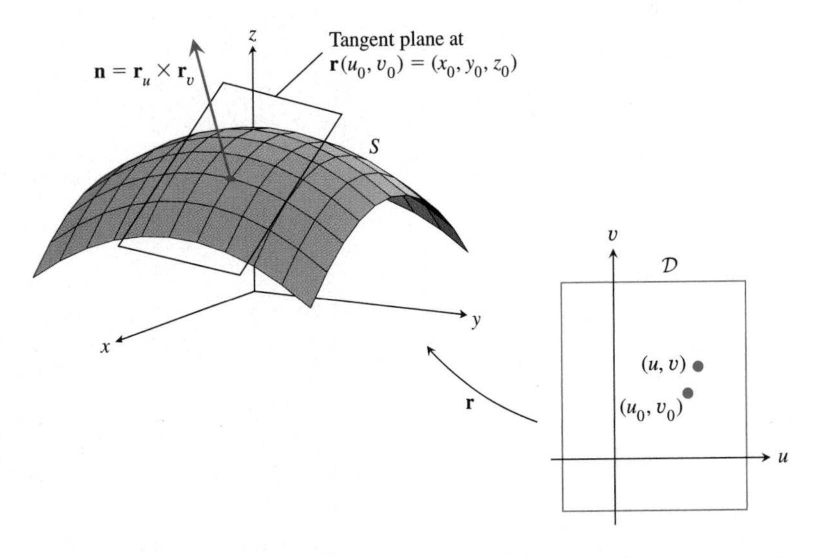

FIGURE 10.12 Surface S described by a parametric function $\mathbf{r}(u, v)$ with domain $\mathcal{D}$.

DEFINITION Differentiability of Vector-Valued Functions

Assume that $\mathbf{f} = \langle f_1, f_2, \ldots, f_m \rangle$ is a vector-valued function with domain $\mathcal{D} \subset R^n$ and that $\mathbf{a}$ is an interior point of $\mathcal{D}$. The function $\mathbf{f}$ is **differentiable** at $\mathbf{a}$ if there is a linear function $\mathbf{df_a}$ from R^n into R^m for which

$$\lim_{\mathbf{x} \to \mathbf{a}} \frac{\mathbf{f}(\mathbf{x}) - \mathbf{f}(\mathbf{a}) - \mathbf{df_a}(\mathbf{x} - \mathbf{a})}{\|\mathbf{x} - \mathbf{a}\|} = \mathbf{0}. \tag{22}$$

The linear function $\mathbf{df_a}$ is called the **differential** of $\mathbf{f}$ at $\mathbf{a}$. The matrix associated with $\mathbf{df_a}$ is denoted by $\mathbf{Df}(\mathbf{a})$ and is called the **derivative matrix** of $\mathbf{f}$ at $\mathbf{a}$. We sometimes shorten $\mathbf{df_a}$ to $\mathbf{df}$. Usually, $m, n \leq 3$.

The form of (22) is the same as that of Equation (3), which was used in defining the differentiability of real-valued functions of two variables, except that boldface type is used in several places. Note that the limit is the zero vector $\mathbf{0}$.

As outlined in Exercise 48, if $\mathbf{f}$ is differentiable at $\mathbf{a}$, then all of the partial derivatives of the coordinate functions $f_1, \ldots, f_m$ exist and the derivative matrix of $\mathbf{f}$ is

$$\mathbf{Df}(\mathbf{a}) = \begin{pmatrix} \dfrac{\partial f_1}{\partial x_1}(\mathbf{a}) & \cdots & \dfrac{\partial f_1}{\partial x_n}(\mathbf{a}) \\ \vdots & & \vdots \\ \dfrac{\partial f_m}{\partial x_1}(\mathbf{a}) & \cdots & \dfrac{\partial f_m}{\partial x_n}(\mathbf{a}) \end{pmatrix}. \tag{23}$$

Describing the Tangent Plane to the Surface Described by $\mathbf{r} = \mathbf{r}(u, v)$

Let S be the surface described by the differentiable function

$$\langle x, y, z \rangle = \mathbf{r} = \mathbf{r}(u, v) = \langle r_1(u, v), r_2(u, v), r_3(u, v) \rangle, \quad (u, v) \in \mathcal{D} \subset R^2, \tag{24}$$

and let $\langle x_0, y_0, z_0 \rangle = \mathbf{r}(u_0, v_0)$ be a point of S. See Figure 10.12. Because $\mathbf{r}$ is differentiable at (u_0, v_0),

$$\mathbf{r}(u, v) = \langle x, y, z \rangle \approx \mathbf{p}(u, v) = \mathbf{r}(u_0, v_0) + \mathbf{Dr}(u_0, v_0) \begin{pmatrix} u - u_0 \\ v - v_0 \end{pmatrix} \tag{25}$$

for points (u, v) near (u_0, v_0). The function $\mathbf{p}$ defined in (25) describes the tangent plane to S at (x_0, y_0, z_0).

To write (25) more briefly, we recall from (24) that $x = r_1(u, v)$, $y = r_2(u, v)$, and $z = r_3(u, v)$. If we further shorten the notation by writing x_u instead of $x_u(u_0, v_0)$,

x_v instead of $x_v(u_0, v_0)$, and so on, then

$$\mathbf{Dr}(u_0, v_0) = \begin{pmatrix} \dfrac{\partial r_1}{\partial u}(u_0, v_0) & \dfrac{\partial r_1}{\partial v}(u_0, v_0) \\ \dfrac{\partial r_2}{\partial u}(u_0, v_0) & \dfrac{\partial r_2}{\partial v}(u_0, v_0) \\ \dfrac{\partial r_3}{\partial u}(u_0, v_0) & \dfrac{\partial r_3}{\partial v}(u_0, v_0) \end{pmatrix} = \begin{pmatrix} x_u & x_v \\ y_u & y_v \\ z_u & z_v \end{pmatrix}.$$

We now write (25) as

$$\mathbf{r}(u, v) = \langle x, y, z \rangle \approx \mathbf{p}(u, v) = \langle x_0, y_0, z_0 \rangle + \begin{pmatrix} x_u & x_v \\ y_u & y_v \\ z_u & z_v \end{pmatrix} \begin{pmatrix} u - u_0 \\ v - v_0 \end{pmatrix}.$$

We may write the function $\mathbf{p}$ in the form of the parametric equations

$$\begin{aligned} x &= x_0 + x_u(u - u_0) + x_v(v - v_0) \\ y &= y_0 + y_u(u - u_0) + y_v(v - v_0) \\ z &= z_0 + z_u(u - u_0) + z_v(v - v_0). \end{aligned} \tag{26}$$

Each of the parameters u and v vary over R. These equations describe the tangent plane to S at (x_0, y_0, z_0).

As outlined in Exercise 43, this tangent plane can also be described by the equation

$$n_1(x - x_0) + n_2(y - y_0) + n_3(z - z_0) = 0, \tag{27}$$

where $\mathbf{n} = \langle n_1, n_2, n_3 \rangle = \mathbf{r}_u(u_0, v_0) \times \mathbf{r}_v(u_0, v_0)$.

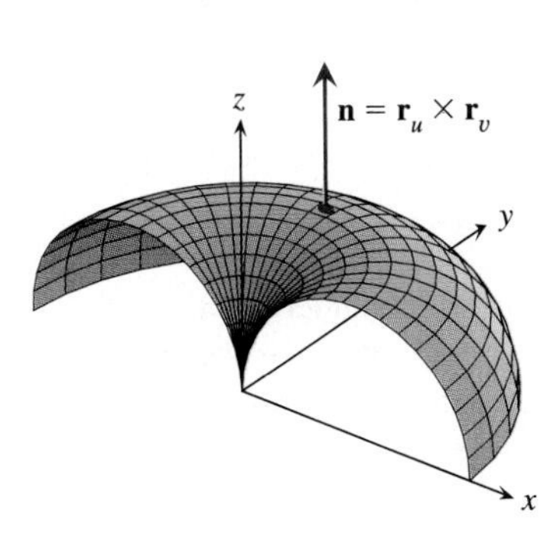

FIGURE 10.13 The surface S described by $\mathbf{r} = \mathbf{r}(u, v)$ and a normal to the tangent plane at the point $\mathbf{r}(\pi/4, \pi/2)$.

EXAMPLE 7 Let S be the surface described by

$$\mathbf{r} = \mathbf{r}(u, v) = \langle \cos v \sin^2 u, \sin v \sin^2 u, \sin u \cos u \rangle, \quad 0 \le u \le \frac{\pi}{2}, 0 \le v \le \pi.$$

Figure 10.13 shows a sketch of S, which is part of the surface that can be described in spherical coordinates by the equation $\rho = \sin \phi$. Use (26) to write parametric equations for the tangent plane to S at the point $\mathbf{r}(\pi/4, \pi/2)$ and use these equations to approximate $\mathbf{r}(0.8, 1.5)$. Compare the result with the true value of $\mathbf{r}(0.8, 1.5)$.

Solution

The partial derivatives of $\mathbf{r}$ with respect to u and v are

$$\begin{aligned} \mathbf{r}_u &= \langle x_u, y_u, z_u \rangle = \langle 2 \sin u \cos u \cos v, 2 \sin u \cos u \sin v, \cos^2 u - \sin^2 u \rangle \\ \mathbf{r}_v &= \langle x_v, y_v, z_v \rangle = \langle -\sin v \sin^2 u, \cos v \sin^2 u, 0 \rangle. \end{aligned}$$

Evaluating these vectors at $(u_0, v_0) = (\pi/4, \pi/2)$,

$$\begin{aligned} \mathbf{r}_u(\pi/4, \pi/2) &= \langle 0, 1, 0 \rangle \\ \mathbf{r}_v(\pi/4, \pi/2) &= \left\langle -\tfrac{1}{2}, 0, 0 \right\rangle. \end{aligned}$$

With these values and $\mathbf{r}(u_0, v_0) = \langle x_0, y_0, z_0 \rangle = \left\langle 0, \frac{1}{2}, \frac{1}{2} \right\rangle$, the parametric equations (26) become

$$\begin{aligned} x &= x_0 + x_u(u - u_0) + x_v(v - v_0) \\ &= 0 + 0 \cdot \left(u - \frac{\pi}{4}\right) + \left(-\frac{1}{2}\right)\left(v - \frac{\pi}{2}\right) \\ y &= y_0 + y_u(u - u_0) + y_v(v - v_0) \\ &= \frac{1}{2} + 1 \cdot \left(u - \frac{\pi}{4}\right) + 0 \cdot \left(v - \frac{\pi}{2}\right) \\ z &= z_0 + z_u(u - u_0) + z_v(v - v_0) \\ &= \frac{1}{2} + 0 \cdot \left(u - \frac{\pi}{4}\right) + 0 \cdot \left(v - \frac{\pi}{2}\right). \end{aligned}$$

These three equations are parametric equations for the plane tangent to S at $\mathbf{r}(\pi/4, \pi/2)$. To approximate $\mathbf{r}(0.8, 1.5)$ we substitute $u = 0.8$ and $v = 1.5$ in these equations. We find

$$\begin{aligned} x &= \left(-\frac{1}{2}\right)\left(1.5 - \frac{\pi}{2}\right) = 0.0353\ldots \\ y &= \frac{1}{2} + 1 \cdot \left(0.8 - \frac{\pi}{4}\right) = 0.514\ldots \\ z &= \frac{1}{2} = 0.5 \end{aligned}$$

These values compare well with $\mathbf{r}(0.8, 1.5) = \langle 0.036\ldots, 0.513\ldots, 0.499\ldots \rangle$.

Test for Differentiability of Vector-Valued Functions

We give a test for differentiability of vector-valued functions; the test is similar to the test for differentiability of real-valued functions.

THEOREM Test for Differentiability (of Vector-valued Functions)

Let $\mathbf{f} = \langle f_1, f_2, \ldots, f_m \rangle$ be a vector-valued function and $\mathbf{a}$ an interior point of the domain $\mathcal{D} \subset R^n$ of $\mathbf{f}$. If all of the first-order partial derivatives of the coordinate functions $f_1, f_2, \ldots, f_m$ exist and are continuous at all points of $\mathcal{D}$ near $\mathbf{a}$, then $\mathbf{f}$ is differentiable at $\mathbf{a}$.

EXAMPLE 8 Show that the vector-valued function

$$\mathbf{r} = \mathbf{r}(u, v) = \langle u^2 v, u^2 + v^2, 2u^2 - 3v^2 \rangle, \qquad (u, v) \in R^2,$$

is differentiable everywhere. Calculate the derivative matrix of $\mathbf{r}$ at $(1, 1)$.

Solution

Using the test for differentiability, it is sufficient to show that all of the first-order partials of the coordinate functions of **r** exist and are continuous everywhere. Because the coordinate functions

$$r_1(u,v) = u^2v, \quad r_2(u,v) = u^2 + v^2, \quad r_3(u,v) = 2u^2 - 3v^2,$$

of **r** are polynomials, their first-order partials exist and are everywhere continuous. Hence, the function **r** is differentiable on R^2.

The derivative matrix of **r** at (u, v) is

$$\mathbf{Dr}(u,v) = \begin{pmatrix} \frac{\partial r_1}{\partial u}(u,v) & \frac{\partial r_1}{\partial v}(u,v) \\ \frac{\partial r_2}{\partial u}(u,v) & \frac{\partial r_2}{\partial v}(u,v) \\ \frac{\partial r_3}{\partial u}(u,v) & \frac{\partial r_3}{\partial v}(u,v) \end{pmatrix} = \begin{pmatrix} 2uv & u^2 \\ 2u & 2v \\ 4u & -6v \end{pmatrix}.$$

Evaluating the derivative matrix at $(1, 1)$,

$$\mathbf{Dr}(1,1) = \begin{pmatrix} 2 & 1 \\ 2 & 2 \\ 4 & -6 \end{pmatrix}.$$

Exercises 10.1

Exercises 1–6: For the given values of **a** *and* $\mathbf{h} = \mathbf{x} - \mathbf{a}$, *calculate* $\Delta f(\mathbf{h}) = f(\mathbf{a} + \mathbf{h}) - f(\mathbf{a})$, $df(\mathbf{h}) = f_x(\mathbf{a})h_1 + f_y(\mathbf{a})h_2$, *and the absolute value of their difference.*

1. $f(x,y) = xy^2$, $\mathbf{a} = (2,3)$, $\mathbf{h} = (0.1, 0.2)$
2. $f(x,y) = 2xy + x^2$, $\mathbf{a} = (1,1)$, $\mathbf{h} = (0.2, -0.3)$
3. $f(x,y) = \ln(xy)$, $\mathbf{a} = (2,1)$, $\mathbf{h} = (0.3, 0.2)$
4. $f(x,y) = e^{xy}$, $\mathbf{a} = (1,-1)$, $\mathbf{h} = (-0.2, 0.1)$
5. $f(x,y,z) = xyz^2$, $\mathbf{a} = (-1,1,1)$, $\mathbf{h} = (0.1, -0.2, 0.1)$
6. $f(x,y,z) = e^{-z}\sin(xy^2)$, $\mathbf{a} = (-1,1,0)$, $\mathbf{h} = (-0.2, 0.05, 0.1)$

Exercises 7–16: Use the test for differentiability to show that f is differentiable at **a**, *find the matrix associated with the linear function df at* **a**, *and calculate* $df(\mathbf{x} - \mathbf{a})$.

7. $f(x,y) = 3x^2y + 5x^3y^2$, $\mathbf{a} = (1,1)$
8. $f(x,y) = xy/(2x - 3y)$, $\mathbf{a} = (4,2)$
9. $f(x,y) = \sin(x/y)$, $\mathbf{a} = (\pi/6, 1)$
10. $f(x,y) = \tan(x/y^2)$, $\mathbf{a} = (\pi/4, 1)$
11. $f(x,y) = e^{-x^2+xy}$, $\mathbf{a} = (1,0)$
12. $f(x,y) = 2^x3^y$, $\mathbf{a} = (1,1)$
13. $f(x,y) = \arcsin(xy^2)$, $\mathbf{a} = \left(1, 1/\sqrt{2}\right)$
14. $f(x,y) = \arctan(1/x + 1/y)$, $\mathbf{a} = (-1,2)$
15. $f(x,y,z) = xy + xz + yz$, $\mathbf{a} = (1,-2,1)$
16. $f(x,y,z) = x/(yz) + y/(xz) + z/(xy)$, $\mathbf{a} = (1,1,-1)$

Exercises 17–22: Find a normal and an equation of the tangent plane to the graph of f at the point $(\mathbf{a}, f(\mathbf{a}))$.

17. $f(x,y) = 2x^2 - 3y^2$, $\mathbf{a} = (2,1)$
18. $f(x,y) = -xy + y^2$, $\mathbf{a} = (1,3)$
19. $f(x,y) = e^{x/y}$, $\mathbf{a} = (2,1)$
20. $f(x,y) = e^{-x^2/y}$, $\mathbf{a} = (2,4)$
21. $f(x,y) = \sin(x^2 + y^2)$, $\mathbf{a} = (-3,2)$
22. $f(x,y) = \tan\sqrt{xy + 1}$, $\mathbf{a} = (3,0)$

Exercises 23–26: The term differential *refers to* $df_{\mathbf{a}}$ *for appropriate f and* **a**.

23. The dimensions of a cardboard box were measured as 90 cm, 50 cm, and 30 cm, with the percentage error in each dimension not exceeding 2%. Use a differential to approximate the worst possible percentage error in the

volume of the box. Assume that the cardboard is of negligible thickness.

24. The base diameter and the height of a cylindrical storage tank were measured as 150 cm and 1600 cm. The percentage errors of these measurements do not exceed 0.3% for base diameter and 0.2% for height. Use a differential in approximating the worst possible percentage error in the storage capacity of the tank. Assume that the walls and ends of the tank are of negligible thickness.

25. A propane storage tank has the form of a cylinder with hemispherical ends. See the accompanying figure. The perimeter of a circular cross section perpendicular to the axis of symmetry of the tank was measured as 12.1 ± 0.05 ft and the perimeter of a cross section lying in a plane containing the axis of symmetry was measured as 37.5 ± 0.1 ft. Use a differential in approximating the worst possible error in the volume of the tank. Assume that the walls and ends of the tank are of negligible thickness.

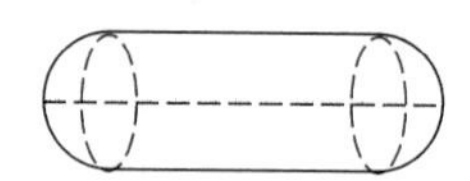

Figure for Exercise 25.

26. The resistance R of a circuit containing two resistors in parallel is

$$\frac{1}{R} = \frac{1}{R_1} + \frac{1}{R_2},$$

where the two resistors have resistances of R_1 and R_2. For resistors with $R_1 = 150 \pm 0.50$ ohm and $R_2 = 200 \pm 0.75$ ohm, use a differential to approximate the error in R.

Exercises 27–32: At the point $\mathbf{r}(u_0, v_0)$ of the surface S described by $\mathbf{r} = \mathbf{r}(u, v)$, find parametric equations of the tangent plane and use them to approximate $\langle x_1, y_1, z_1\rangle = \mathbf{r}(u_1, v_1)$. Compare with the true value of $\mathbf{r}(u_1, v_1)$.

27. $\mathbf{r}(u, v) = \langle u \cos v, u \sin v, u^2\rangle$, $0 \le u \le 2$, $0 \le v \le 2\pi$; $(u_0, v_0) = (1, \pi/4)$; $(u_1, v_1) = (1.1, 0.8)$

28. $\mathbf{r}(u, v) = \langle u, u^2 + v^2, v\rangle$, $u \in R$, $v \ge 0$; $(u_0, v_0) = (1, 2)$; $(u_1, v_1) = (1.1, 1.8)$

29. $\mathbf{r}(u, v) = \langle \ln u, \ln v, uv\rangle$, $u, v \ge 1$; $(u_0, v_0) = (e, e)$; $(u_1, v_1) = (2.6, 2.8)$

30. $\mathbf{r}(u, v) = \langle e^{u+v}, 2v, e^{u-v}\rangle$, $u, v \in R$; $(u_0, v_0) = (-1, 1)$; $(u_1, v_1) = (-1.1, 1.2)$

31. $\mathbf{r}(u, v) = \langle u \tan v, u \sec v, u^2\rangle$, $u \ge 0$, $-\pi/2 < v < \pi/2$; $(u_0, v_0) = (2, \pi/4)$; $(u_1, v_1) = (2.2, 0.8)$

32. $\mathbf{r}(u, v) = \langle \cos v \sin u \cos u, \sin v \sin u \cos u, \cos^2 v\rangle$, $0 \le u \le \pi/2$, $0 \le v \le 2\pi$; $(u_0, v_0) = (\pi/6, \pi/2)$; $(u_1, v_1) = (0.5, 1.5)$

33. Show that the function

$$\mathbf{f}(r, \theta) = \langle r \cos \theta, r \sin \theta\rangle,$$

where $r > 0$ and $0 \le \theta < 2\pi$, is differentiable everywhere, find its derivative matrix at $\mathbf{a} = (1, \pi/6)$, and calculate $\mathbf{df}(\mathbf{x} - \mathbf{a})$, where $\mathbf{x} = (1.3, \pi/6 - 0.2)$.

34. Show that the function

$$\mathbf{f}(u, v) = \langle e^u \cos v, e^u \sin v\rangle,$$

where $(u, v) \in R^2$, is differentiable everywhere, find its derivative matrix at $\mathbf{a} = (0, \pi/2)$, and calculate $\mathbf{df}(\mathbf{x} - \mathbf{a})$, where $\mathbf{x} = (0.1, \pi/2 - 0.1)$.

35. Linearize the function

$$f(x, y) = e^{x/y}$$

near $(1, 1)$. Calculate the absolute value of the difference between f and its linearization at the four points $(1 \pm 0.1, 1 \pm 0.1)$. At which point is the error the largest?

36. Linearize the function

$$f(x, y) = \sqrt{3 + 5x^2 + y^3}$$

near $(1, 2)$. Evaluate f and its linearization g at the four points $(1 \pm 0.25, 2 \pm 0.25)$. At which point is the error the largest?

37. If the initial displacement from equilibrium of a simple pendulum is small, its period T is related to its length L by the equation

$$T = 2\pi\sqrt{\frac{L}{g}}.$$

If $L = 1.000 \pm 0.002$ m and $g = 9.81 \pm 0.02$ m/s^2 are measured with the indicated uncertainties, what is the worst possible error in T?

38. Related problem Show that the percentage error $(dT/T) \cdot 100$ in T is related to the percentage errors $(dL/L) \cdot 100$ and $(dg/g) \cdot 100$ in L and g by the equation

$$\frac{dT}{T} \cdot 100 = \frac{dL}{2L} \cdot 100 - \frac{dg}{2g} \cdot 100.$$

39. Related problem Show that to four significant digits the linearization of the pendulum function T near $L = 1$ and $g = 9.81$ is

$$w(L, g) = 2.006 + 1.003(L - 1) - 0.1022(g - 9.81).$$

What are the errors at the four points $(1 \pm 0.1, 9.81 \pm 0.01)$?

40. Use the definition of differentiability to show that the function f defined by

$$f(x, y) = 3 - x^2 - 2y^2,$$

where $(x, y) \in \mathcal{D} = [-1, 1] \times [-1, 1]$, is differentiable at all interior points $\mathbf{a} = (a_1, a_2)$ of $\mathcal{D}$. Use the following outline if you wish.

a. Show that the numerator of (3) is

$$-(x - a_1)^2 - 2(y - a_2)^2.$$

b. Show that (3) is

$$\lim_{\mathbf{x}\to\mathbf{a}} \frac{-(x-a_1)^2 - 2(y-a_2)^2}{\sqrt{(x-a_1)^2 + (y-a_2)^2}}.$$

c. Show that

$$(x-a_1)^2 + 2(y-a_2)^2 \leq 2((x-a_1)^2 + 2(y-a_2)^2).$$

d. Show that

$$\left|\frac{-(x-a_1)^2 - 2(y-a_2)^2}{\sqrt{(x-a_1)^2 + (y-a_2)^2}}\right| \leq 2\sqrt{(x-a_1)^2 + (y-a_2)^2}.$$

e. Finish the argument to show that f is differentiable.

41. Related to Exercise 40 Apply the definition of differentiability to show that $f(x, y) = x^2 + 3y^3$ is differentiable at $(1, 1)$.

42. Related to Exercise 40 Apply the definition of differentiability to show that $f(x, y) = xy$ is differentiable at $(1, 1)$.

43. Fill in the following outline of how the description (27) of the tangent plane can be inferred from the parametric equations (26). Letting n_1, n_2, and n_3 be the coordinates of the cross product $\mathbf{r}_u(u_0, v_0) \times \mathbf{r}_v(u_0, v_0)$, multiply the first equation of (26) by n_1, the second equation by n_2, the third equation by n_3, and then add the results. As an alternative, show that (27) follows from (26) by eliminating the parameters u and v.

44. Show that the tent function $f(x, y) = -\sqrt{|xy|}$, $(x, y) \in R^2$, is not differentiable at $(0, 0)$. *Hint:* It is sufficient to show that the expression

$$\frac{f(x, y) - f(0, 0)}{\sqrt{x^2 + y^2}}$$

does not go to zero as $\langle x, y\rangle \to \langle 0, 0\rangle$. Consider what happens when $x = y > 0$ and $x \to 0$.

45. Use the test for differentiability to show that the tent function f shown in Fig. 10.10 is differentiable at points (x, y) not on the coordinate axes.

46. Show that the old-tent function g shown in Fig. 10.11 is not continuous at $(0, 0)$. Show also that $g_x(0, 0) = g_y(0, 0) = 0$.

47. Related exercise Show that the old-tent function is not differentiable at $(0, 0)$.

48. If a vector-valued function $\mathbf{f}$ is differentiable at a point $\mathbf{a}$, then the entries of the matrix associated with the linear function $\mathbf{df_a}$ are the partial derivatives of the coordinate functions of $\mathbf{f}$. A proof of this result has the same form as an earlier argument. At the beginning of the section, just after the definition of differentiability, we showed that the entries of the derivative matrix $Df(\mathbf{a})$ are the partial derivatives of f at $\mathbf{a}$. The idea of the proof was to let $\mathbf{x}$ approach $\mathbf{a}$ along specific paths. The same idea works for the more general case. Fill in the details of the following argument.

For simplicity, we take $\mathbf{x} - \mathbf{a} = \langle h_1, 0, \ldots, 0\rangle$, where $h_1 > 0$. The matrix A associated with the linear transformation $\mathbf{df_a}$ is $m \times n$. Show that with $\mathbf{x} - \mathbf{a}$ chosen as above, the ith coordinate of the left side of (22) becomes (leaving the limit aside for a moment)

$$\frac{f_i(\mathbf{x}) - f_i(\mathbf{a}) - a_{i,1}h_1}{h_1}.$$

It follows from this that

$$a_{i,1} = \frac{\partial f_i(\mathbf{a})}{\partial x_1}.$$

Next, consider $\mathbf{x} - \mathbf{a} = \langle 0, h_2, 0, \ldots, 0\rangle$, where $h_2 > 0$, and so on.

49. Fill in the details of the following proof that if f has continuous partial derivatives at all points near an interior point $\mathbf{a}$ of the domain of f, then f is differentiable at $\mathbf{a}$.

i. Let $A = f_x(\mathbf{a})$ and $B = f_y(\mathbf{a})$.

ii. To simplify the notation, let $\mathbf{x} - \mathbf{a} = \mathbf{h}$.

iii. Rewrite the numerator of the fraction

$$\frac{\Delta f(\mathbf{h}) - df(\mathbf{h})}{\|\mathbf{h}\|} \tag{28}$$

from (3) by adding and subtracting the number $q = f(a_1, a_2 + h_2)$. Show that the numerator can be written as

$$\frac{f(a_1 + h_1, a_2 + h_2) - q - Ah_1}{\|\mathbf{h}\|} + \frac{q - f(a_1, a_2) - Bh_2}{\|\mathbf{h}\|}. \tag{29}$$

iv. If we can show that the fraction in (28) approaches zero as $\mathbf{h} \to \mathbf{0}$, or that each of the two terms in (29) approaches zero, it will follow that f is differentiable at $\mathbf{a}$. For the first term, express the fact that

$$\lim_{h_1\to 0} \frac{f(a_1 + h_1, a_2 + h_2) - q}{h_1} = f_x(a_1, a_2 + h_2)$$

in the form

$$f(a_1 + h_1, a_2 + h_2) - q = h_1 f_x(a_1, a_2 + h_2) + h_1 E_1(h_1),$$

where $E_1(h_1) \to 0$ as $h_1 \to 0$. The first term from (29) can then be written as

$$\frac{h_1}{\|\mathbf{h}\|}(f_x(a_1, a_2 + h_2) - A + E_1(h_1)).$$

Because the ratio $h_1/\|\mathbf{h}\|$ lies between -1 and 1 and f_x is continuous at $(a_1, a_2 + h_2)$, the first term in (29) goes to zero as $\|\mathbf{h}\| \to 0$.

10.2 The Chain Rule

If y is a differentiable function of x and x is a differentiable function of t, then y is a differentiable function of t and the derivative of y with respect to t is

$$\frac{dy}{dt} = \frac{dy}{dx}\frac{dx}{dt}. \tag{1}$$

This is the chain rule discussed in Section 2.3 for differentiating the composition $y = (f \circ g)(t) = f(g(t))$ of functions $y = f(x)$ and $x = g(t)$ of one variable. Note that the derivative of the composition of f and g is the *product* of the derivatives of the two functions.

If the vector variable/function $\mathbf{y} = \mathbf{f}(\mathbf{x})$ is a differentiable function of the vector variable/function $\mathbf{x}$ and $\mathbf{x} = \mathbf{g}(\mathbf{t})$ is a differentiable function of the vector variable $\mathbf{t}$, then $\mathbf{y}$ is a differentiable function of $\mathbf{t}$. We give a rough proof of this statement. Our argument will strongly suggest that

$$\mathbf{Dy(t)} = \mathbf{Dy(x)Dx(t)}. \tag{2}$$

This is the chain rule for vector-valued functions. Note that the derivative of the composition of $\mathbf{f}$ and $\mathbf{g}$ is the *product* of the derivative matrices of the two functions. The chain rule (2) for vector functions is a generalization of the chain rule (1) for scalar functions.

The general form (2) of the chain rule makes clear the strong connection between the vector chain rule and the scalar chain rule (1). However, because in many low-dimensional cases the general rule becomes ponderous, we discuss the mechanics of the vector chain rule in stages, with the number of vector variables increasing from one in Case 1 to three in Case 4. Before we start on these four cases, we give a rough proof of (2).

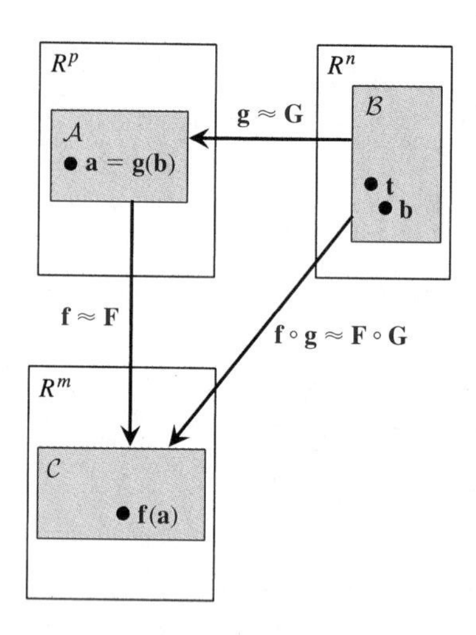

FIGURE 10.14 Because $\mathbf{g(t)} \approx \mathbf{G(t)}$ for $\mathbf{t}$ near $\mathbf{b}$, where $\mathbf{G}$ is the linearization of $\mathbf{g}$ at $\mathbf{b}$, and $\mathbf{f(x)} \approx \mathbf{F(x)}$ for $\mathbf{x}$ near $\mathbf{a}$, where $\mathbf{a} = \mathbf{g(b)}$ and $\mathbf{F}$ is the linearization of $\mathbf{F}$ at $\mathbf{a}$, we expect that $(\mathbf{f} \circ \mathbf{g})(\mathbf{t}) \approx (\mathbf{F} \circ \mathbf{G})(\mathbf{t})$ for $\mathbf{t}$ near $\mathbf{b}$ and that $\mathbf{F} \circ \mathbf{G}$ is the linearization of $\mathbf{f} \circ \mathbf{g}$ at $\mathbf{b}$.

Referring to Fig. 10.14, suppose that $\mathbf{y} = \mathbf{f(x)}$ is a differentiable function of the vector variable $\mathbf{x}$ and $\mathbf{x} = \mathbf{g(t)}$ is a differentiable function of the vector variable $\mathbf{t}$. We assume that the domains of $\mathbf{f}$ and $\mathbf{g}$ are the sets $\mathcal{A}$ and $\mathcal{B}$, respectively, and that the ranges of $\mathbf{f}$ and $\mathbf{g}$ are contained in the sets $\mathcal{C}$ and $\mathcal{A}$. See Fig. 10.14. The composition $\mathbf{f} \circ \mathbf{g}$ is defined on $\mathcal{B}$ and takes its values in $\mathcal{C}$. We argue that $\mathbf{f} \circ \mathbf{g}$ is differentiable at $\mathbf{b} \in \mathcal{B}$ and that its derivative matrix is the product $\mathbf{Df(g(b))Dg(b)}$ of the derivative matrices of $\mathbf{f}$ at $\mathbf{a} = \mathbf{g(b)}$ and $\mathbf{g}$ at $\mathbf{b}$.

Because $\mathbf{f}$ is differentiable at $\mathbf{a} = \mathbf{g(b)}$,

$$\mathbf{f(x)} \approx \mathbf{F(x)} = \mathbf{f(a)} + \mathbf{Df(a)}\,(\mathbf{x} - \mathbf{a}), \tag{3}$$

for $\mathbf{x}$ near $\mathbf{a}$. Because $\mathbf{g}$ is differentiable at $\mathbf{b}$,

$$\mathbf{g(t)} \approx \mathbf{G(t)} = \mathbf{g(b)} + \mathbf{Dg(b)}\,(\mathbf{t} - \mathbf{b}), \tag{4}$$

for $\mathbf{t}$ near $\mathbf{b}$. These approximations come directly from (22) in Section 10.1. The functions $\mathbf{F}$ and $\mathbf{G}$ defined in Equations (3) and (4) are the linearizations of $\mathbf{f}$ at $\mathbf{a} = \mathbf{g(b)}$ and $\mathbf{g}$ at $\mathbf{b}$.

Our argument is based on the reasonable idea that the composition of the linearizations $\mathbf{F}$ and $\mathbf{G}$ is the linearization of the composition $\mathbf{f} \circ \mathbf{g}$ at $\mathbf{b}$. Because $\mathbf{f(x)} \approx \mathbf{F(x)}$ for $\mathbf{x}$ near $\mathbf{a}$ and $\mathbf{g(t)} \approx \mathbf{G(t)}$ for $\mathbf{t}$ near $\mathbf{b}$,

$$(\mathbf{f} \circ \mathbf{g})(\mathbf{t}) = \mathbf{f(g(t))} \approx \mathbf{F(G(t))} = (\mathbf{F} \circ \mathbf{G})(\mathbf{t}).$$

It follows from (3) and (4) that

$$\begin{aligned}(\mathbf{f}\circ\mathbf{g})(\mathbf{t}) &\approx \mathbf{F}(\mathbf{G}(\mathbf{t})) = \mathbf{F}(\mathbf{g}(\mathbf{b}) + \mathbf{Dg}(\mathbf{b})(\mathbf{t}-\mathbf{b}))\\ &\approx \mathbf{f}(\mathbf{a}) + \mathbf{Df}(\mathbf{a})(\mathbf{g}(\mathbf{b}) + \mathbf{Dg}(\mathbf{b})(\mathbf{t}-\mathbf{b}) - \mathbf{a}),\end{aligned}$$

for $\mathbf{t}$ near $\mathbf{b}$. Because $\mathbf{g}(\mathbf{b}) = \mathbf{a}$,

$$(\mathbf{f}\circ\mathbf{g})(\mathbf{t}) \approx \mathbf{f}(\mathbf{g}(\mathbf{b})) + \mathbf{Df}(\mathbf{a})\mathbf{Dg}(\mathbf{b})(\mathbf{t}-\mathbf{b})$$

or

$$(\mathbf{f}\circ\mathbf{g})(\mathbf{t}) \approx (\mathbf{f}\circ\mathbf{g})(\mathbf{b}) + \mathbf{Df}(\mathbf{a})\mathbf{Dg}(\mathbf{b})(\mathbf{t}-\mathbf{b}).$$

This result strongly suggests that $\mathbf{F}\circ\mathbf{G}$ is the linearization of $\mathbf{f}\circ\mathbf{g}$ near $\mathbf{b}$. If this is true, then at $\mathbf{b}$ the derivative matrix $\mathbf{D}(\mathbf{f}\circ\mathbf{g})$ of $\mathbf{f}\circ\mathbf{g}$ is the matrix $\mathbf{Df}(\mathbf{g}(\mathbf{b}))\mathbf{Dg}(\mathbf{b})$, which is the product of the derivative matrices of $\mathbf{f}$ at $\mathbf{a} = \mathbf{g}(\mathbf{b})$ and $\mathbf{g}$ at $\mathbf{b}$.

With the vector chain rule (2) now made plausible, we state it in matrix form.

THEOREM The Chain Rule

If $\mathbf{y} = \mathbf{f}(\mathbf{x})$ and $\mathbf{x} = \mathbf{g}(\mathbf{t})$ are differentiable functions, then $\mathbf{y} = (\mathbf{f}\circ\mathbf{g})(\mathbf{t})$ is differentiable and its derivative matrix $\mathbf{Dy}(\mathbf{t})$ is the product of the derivative matrices of $\mathbf{y} = \mathbf{f}(\mathbf{x})$ and $\mathbf{x} = \mathbf{g}(\mathbf{t})$; that is,

$$\mathbf{Dy}(\mathbf{t}) = \mathbf{Dy}(\mathbf{x})\mathbf{Dx}(\mathbf{t}). \tag{5}$$

If $\mathbf{y} = \langle y_1,\ldots,y_m\rangle$, $\mathbf{x} = \langle x_1,\ldots,x_p\rangle$, and $\mathbf{t} = \langle t_1,\ldots,t_n\rangle$, the chain rule (5) can be stated in matrix form as

$$\begin{pmatrix} \dfrac{\partial y_1}{\partial t_1} & \cdots & \dfrac{\partial y_1}{\partial t_n} \\ \vdots & & \vdots \\ \dfrac{\partial y_m}{\partial t_1} & \cdots & \dfrac{\partial y_m}{\partial t_n} \end{pmatrix} = \begin{pmatrix} \dfrac{\partial y_1}{\partial x_1} & \cdots & \dfrac{\partial y_1}{\partial x_p} \\ \vdots & & \vdots \\ \dfrac{\partial y_m}{\partial x_1} & \cdots & \dfrac{\partial y_m}{\partial x_p} \end{pmatrix} \begin{pmatrix} \dfrac{\partial x_1}{\partial t_1} & \cdots & \dfrac{\partial x_1}{\partial t_n} \\ \vdots & & \vdots \\ \dfrac{\partial x_p}{\partial t_1} & \cdots & \dfrac{\partial x_p}{\partial t_n} \end{pmatrix}. \tag{6}$$

We discuss the chain rule in four cases, in which we gradually increase the number of vector variables.

Case 1: $y = f(\mathbf{x})$, $\mathbf{x} = \mathbf{g}(t)$: **Only x is vector-valued:** The simplest chain rule for functions of several variables is a rule for differentiating the composition of the functions

$$y = f(\mathbf{x}) = f(x_1, x_2, \ldots, x_p) \quad\text{and}\quad \mathbf{x} = \mathbf{g}(t) = \langle x_1(t), x_2(t), \ldots, x_p(t)\rangle,$$

where, of the three variables y, $\mathbf{x}$, and t, only the "chaining variable" $\mathbf{x}$ is a vector variable. As an example, suppose that $y = f(\mathbf{x})$ is the atmospheric pressure in pascals at the point in the atmosphere with position vector $\mathbf{x}$ and that $\mathbf{x} = \mathbf{g}(t)$ describes the position of an aircraft at time t in seconds. The atmospheric pressure on the aircraft as a function of time t is $y = f(\mathbf{g}(t))$, which is the composition of the pressure and position functions. The scalar derivative dy/dt gives the rate of change in air pressure per second on the aircraft at time t.

Assuming that the dimensions of y, $\mathbf{x}$, and t are $m = 1$, p, and $n = 1$, respectively, we see from (6) that the derivative matrix $Dy(t)$ of the composition

$y = f(\mathbf{g}(t))$ is

$$\left(\frac{dy}{dt}\right) = \left(\frac{\partial y}{\partial x_1} \cdots \frac{\partial y}{\partial x_p}\right) \begin{pmatrix} \frac{dx_1}{dt} \\ \vdots \\ \frac{dx_p}{dt} \end{pmatrix}. \tag{7}$$

Hence, Case 1 of the chain rule may be expressed as

$$\frac{dy}{dt} = \frac{\partial y}{\partial x_1}\frac{dx_1}{dt} + \cdots + \frac{\partial y}{\partial x_p}\frac{dx_p}{dt}. \tag{8}$$

In most of our examples and exercises, $p = 2$ or 3.

EXAMPLE 1 Calculate the change in air pressure dy/dt experienced by an aircraft flying along the path C described by

$$\mathbf{x} = \langle x_1, x_2, x_3 \rangle = \langle t, t^2, t^3 \rangle, \qquad t \geq 0,$$

through air whose pressure at $\mathbf{x}$ is

$$y = x_2 x_3 + x_1{}^2.$$

Specifically, calculate dy/dt at $t = 1$.

Solution

Figure 10.15 shows the path C of the aircraft. The pressure is not shown in the figure. Because the functions f and $\mathbf{g}$ are given explicitly, we can express y as a function of t and calculate dy/dt directly. Because $y = x_2x_3 + x_1^2 = t^5 + t^2$, we see that $dy/dt = 5t^4 + 2t$. At $t = 1$, $dy/dt = 7$ pascals per second.

We may also calculate dy/dt by applying (8). With $p = 3$,

$$\begin{aligned}\frac{dy}{dt} &= \frac{\partial y}{\partial x_1}\frac{dx_1}{dt} + \frac{\partial y}{\partial x_2}\frac{dx_2}{dt} + \frac{\partial y}{\partial x_3}\frac{dx_3}{dt} \\ &= (2x_1)\cdot 1 + (x_3)\cdot(2t) + (x_2)\cdot(3t^2).\end{aligned}$$

Because $\langle x_1, x_2, x_3 \rangle = \langle t, t^2, t^3 \rangle$,

$$\frac{dy}{dt} = (2t)\cdot 1 + (t^3)\cdot(2t) + (t^2)\cdot(3t^2) = 2t + 5t^4.$$

Evaluating dy/dt at $t = 1$ gives $dy/dt = 7$ pascals per second, which agrees with our direct calculation.

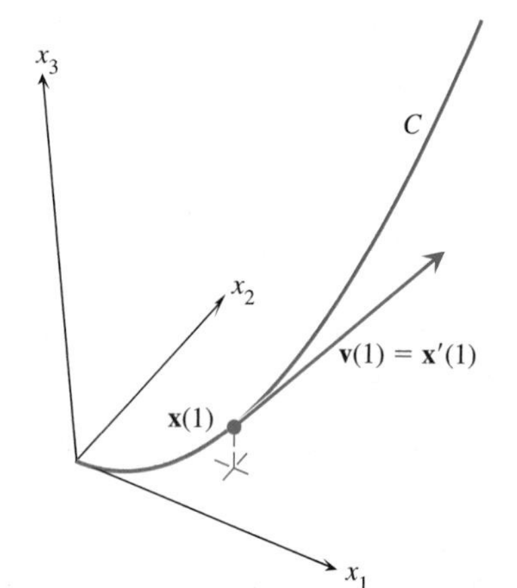

FIGURE 10.15 Aircraft's path through air with variable air pressure; the change in pressure dy/dt experienced by the aircraft is related to its velocity $\mathbf{v}$.

We have chosen to use variables $\mathbf{y}$, $\mathbf{x}$, and $\mathbf{t}$ in discussing Cases 1–4 of the chain rule for functions of several variables so that we can more easily notice the similarities among these cases. However, in many applications it is natural to choose more familiar variables. In Example 1, we may prefer to describe the atmospheric pressure at (x, y, z) by $p = f(x, y, z)$ and the path of the aircraft by $\mathbf{r} = \langle x(t), y(t), z(t)\rangle$. The chain rule (8) then becomes

$$\frac{dp}{dt} = \frac{\partial p}{\partial x}\frac{dx}{dt} + \frac{\partial p}{\partial y}\frac{dy}{dt} + \frac{\partial p}{\partial z}\frac{dz}{dt}.$$

Directional Derivatives Formula (8) for Case 1 of the chain rule is related to the directional derivative of the atmospheric pressure function $y = f(\mathbf{x})$ in the direction of the unit tangent vector $\mathbf{u} = \left(1/\|\mathbf{x}'\|\right)\mathbf{x}'$, where $\mathbf{x}' = \mathbf{v}$ is the velocity vector of the aircraft. Figure 10.15 shows the velocity vector of the aircraft at $t = 1$. The directional derivative

$$D_{\mathbf{u}}y(\mathbf{x}) = \nabla y \cdot \mathbf{u}$$

gives the rate of change of the atmospheric pressure per unit change along the line in the direction $\mathbf{u}$, with units of pascals per meter. The product of the directional derivative and the speed $\|\mathbf{x}'\|$ of the aircraft in meters per second gives the rate of change dy/dt in pascals per second of the pressure function y along the path of the aircraft. In symbols,

$$\begin{aligned}\frac{dy}{dt} &= (D_{\mathbf{u}}y(\mathbf{x})\ \text{Pa/m})\left(\|\mathbf{x}'\|\ \text{m/s}\right) = (\nabla y \cdot \mathbf{u})\|\mathbf{x}'\|\ \text{Pa/s} \\ &= \|\mathbf{x}'\|\left(\nabla y \cdot \left(\frac{1}{\|\mathbf{x}'\|}\mathbf{x}'\right)\right)\ \text{Pa/s}.\end{aligned}$$

Removing the common factor of $\|\mathbf{x}\|$,

$$\begin{aligned}\frac{dy}{dt} &= \nabla y \cdot \mathbf{x}' = \left\langle \frac{\partial y}{\partial x_1}, \frac{\partial y}{\partial x_2}, \frac{\partial y}{\partial x_3}\right\rangle \cdot \left\langle \frac{\partial x_1}{\partial t}, \frac{\partial x_2}{\partial t}, \frac{\partial x_3}{\partial t}\right\rangle \\ &= \frac{\partial y}{\partial x_1}\frac{dx_1}{dt} + \frac{\partial y}{\partial x_2}\frac{dx_2}{dt} + \frac{\partial y}{\partial x_3}\frac{dx_3}{dt},\end{aligned}$$

which is a special case of (8).

Case 2: $y = f(\mathbf{x})$, $\mathbf{x} = \mathbf{g}(\mathbf{t})$: Only x and t are vector-valued: As in Case 1 of the chain rule, the variable $y = f(\mathbf{x})$ often represents a physical variable such as the pressure, temperature, or density at a point $\mathbf{x}$ of a two- or three-dimensional region. The variable $\mathbf{t}$ in Case 2 is usually not time but may describe a change of variables, for instance from rectangular to polar coordinates.

Assuming that the dimensions of y, $\mathbf{x}$, and $\mathbf{t}$ are $m = 1$ and $p = n = 2$, we see from (6) that the derivative matrix $Dy(\mathbf{t})$ of the composition $y = f(\mathbf{g}(\mathbf{t}))$ is

$$\begin{pmatrix} \frac{\partial y}{\partial t_1} & \frac{\partial y}{\partial t_2}\end{pmatrix} = \begin{pmatrix} \frac{\partial y}{\partial x_1} & \frac{\partial y}{\partial x_2}\end{pmatrix}\begin{pmatrix} \frac{\partial x_1}{\partial t_1} & \frac{\partial x_1}{\partial t_2} \\ \frac{\partial x_2}{\partial t_1} & \frac{\partial x_2}{\partial t_2}\end{pmatrix}. \tag{9}$$

Hence, Case 2 of the chain rule may be expressed as

$$\begin{aligned}\frac{\partial y}{\partial t_1} &= \frac{\partial y}{\partial x_1}\frac{\partial x_1}{\partial t_1} + \frac{\partial y}{\partial x_2}\frac{\partial x_2}{\partial t_1} \\ \frac{\partial y}{\partial t_2} &= \frac{\partial y}{\partial x_1}\frac{\partial x_1}{\partial t_2} + \frac{\partial y}{\partial x_2}\frac{\partial x_2}{\partial t_2}.\end{aligned} \tag{10}$$

EXAMPLE 2 The faces of a thin circular plate are insulated—so that heat is neither gained nor lost through the faces—and the circular rim is heated to 100° C. After the temperatures of the plate have stabilized, the temperature u at any point (x, y) is an unknown function $u = f(x, y)$ of x and y. Because the plate is circular we consider changing from rectangular to polar coordinates, using the *change of variable equations*

$$\begin{aligned} x &= r\cos\theta \\ y &= r\sin\theta. \end{aligned} \tag{11}$$

Apply Case 2 of the chain rule to calculate the changes $\partial u/\partial r$ and $\partial u/\partial\theta$ of the temperature in the radial and transverse directions in terms of the changes $\partial u/\partial x$ and $\partial u/\partial y$ of temperature in the x- and y-directions.

Solution

In applying (10) we replace y by u, x_1 by x, x_2 by y, t_1 by r, and t_2 by θ. The composition of $y = f(\mathbf{x})$ and $\mathbf{x} = \mathbf{g}(\mathbf{t})$ then takes the form

$$u = u(x, y), \langle x, y\rangle = \mathbf{g}(r, \theta) = \langle r\cos\theta, r\sin\theta\rangle.$$

The equations (10) allow us to relate the partials of the temperature function with respect to r and θ to the partials of the temperature function with respect to x and y. For this we shall need the partials of x and y with respect to r and θ. From (11),

$$\frac{\partial x}{\partial r} = \cos\theta \qquad \frac{\partial x}{\partial\theta} = -r\sin\theta \qquad \frac{\partial y}{\partial r} = \sin\theta \qquad \frac{\partial y}{\partial\theta} = r\cos\theta.$$

From (10),

$$\begin{aligned} \frac{\partial u}{\partial r} &= \frac{\partial u}{\partial x}\frac{\partial x}{\partial r} + \frac{\partial u}{\partial y}\frac{\partial y}{\partial r} = \frac{\partial u}{\partial x}\cos\theta + \frac{\partial u}{\partial y}\sin\theta \\ \frac{\partial u}{\partial\theta} &= \frac{\partial u}{\partial x}\frac{\partial x}{\partial\theta} + \frac{\partial u}{\partial y}\frac{\partial y}{\partial\theta} = -\frac{\partial u}{\partial x}r\sin\theta + \frac{\partial u}{\partial y}r\cos\theta. \end{aligned}$$

These equations relate the changes $\partial u/\partial r$ and $\partial u/\partial\theta$ of the temperature in the radial and transverse directions to the changes $\partial u/\partial x$ and $\partial u/\partial y$ of temperature in the x- and y-directions.

EXAMPLE 3 The "continuity equation" or "conservation of mass" equation is one of the fundamental equations in fluid mechanics. This equation expresses the condition that the mass of a system of fluid particles remains constant as it moves through a flow field. For a one-dimensional flow, say along an x-axis, with constant velocity c m/s, the continuity equation is

$$\frac{\partial\nu}{\partial t} = -c\frac{\partial\nu}{\partial x}, \tag{12}$$

where $\nu = \nu(x, t)$ is the density (kg/m) of the fluid at coordinate position x and time t. Express the partial differential equation (12) in terms of $\dfrac{\partial\nu}{\partial T}$ and $\dfrac{\partial\nu}{\partial X}$, where the variables T and X are related to the variables t and x by

$$\begin{aligned} x &= \frac{1}{2}X + \frac{1}{2}T \\ t &= -\frac{1}{2c}X + \frac{1}{2c}T. \end{aligned} \tag{13}$$

Solution

We apply (10) to calculate $\partial v/\partial X$ and $\partial v/\partial T$. In terms of the variables X and T, the density function is

$$v = v(x, t) = v\left(\frac{1}{2}X + \frac{1}{2}T, -\frac{1}{2c}X + \frac{1}{2c}T\right).$$

Equations (10) are

$$\begin{aligned} \frac{\partial v}{\partial X} &= \frac{\partial v}{\partial x}\frac{\partial x}{\partial X} + \frac{\partial v}{\partial t}\frac{\partial t}{\partial X} \\ \frac{\partial v}{\partial T} &= \frac{\partial v}{\partial x}\frac{\partial x}{\partial T} + \frac{\partial v}{\partial t}\frac{\partial t}{\partial T}. \end{aligned} \tag{14}$$

From (13)

$$\frac{\partial x}{\partial X} = \frac{1}{2}, \qquad \frac{\partial x}{\partial T} = \frac{1}{2}, \qquad \frac{\partial t}{\partial X} = -\frac{1}{2c}, \quad \text{and} \quad \frac{\partial t}{\partial T} = \frac{1}{2c}.$$

Substituting these results into (14),

$$\begin{aligned} \frac{\partial v}{\partial X} &= \frac{1}{2}\frac{\partial v}{\partial x} - \frac{1}{2c}\frac{\partial v}{\partial t} \\ \frac{\partial v}{\partial T} &= \frac{1}{2}\frac{\partial v}{\partial x} + \frac{1}{2c}\frac{\partial v}{\partial t}. \end{aligned} \tag{15}$$

To express (12) in terms of $\partial v/\partial T$ and $\partial v/\partial X$, we solve the system (15) for $\partial v/\partial x$ and $\partial v/\partial t$ and substitute into (12). Adding and subtracting equations (15),

$$\begin{aligned} \frac{\partial v}{\partial x} &= \frac{\partial v}{\partial X} + \frac{\partial v}{\partial T} \\ \frac{\partial v}{\partial t} &= -c\left(\frac{\partial v}{\partial X} - \frac{\partial v}{\partial T}\right). \end{aligned} \tag{16}$$

Substituting into (12),

$$-c\left(\frac{\partial v}{\partial X} - \frac{\partial v}{\partial T}\right) = -c\left(\frac{\partial v}{\partial X} + \frac{\partial v}{\partial T}\right).$$

This equation simplifies to the simple equation

$$\frac{\partial v}{\partial T} = 0. \tag{17}$$

This is the equation satisfied by $\partial v/\partial T$ and $\partial v/\partial X$. In Exercise 32, we solve this equation and find solutions to the continuity equation (12).

Case 3: $\mathbf{y} = \mathbf{f}(\mathbf{x})$, $\mathbf{x} = \mathbf{g}(t)$: Only $\mathbf{y}$ and $\mathbf{x}$ are vector-valued: Assuming that the dimensions of $\mathbf{y}$, $\mathbf{x}$, and t are $m = 3$, $p = 2$, and $n = 1$, respectively,

$$\mathbf{y} = \langle y_1, y_2, y_3 \rangle = \mathbf{f}(\mathbf{x}) = \mathbf{f}(x_1, x_2) \quad \text{and} \quad \mathbf{x} = \langle x_1, x_2 \rangle = \langle x_1(t), x_2(t) \rangle.$$

We see from (6) that the entries of the derivative matrix $\mathbf{Dy}(t)$ of the composition $\mathbf{y} = \mathbf{f}(\mathbf{g}(t))$ are

$$\begin{aligned} \frac{dy_1}{dt} &= \frac{\partial y_1}{\partial x_1}\frac{dx_1}{dt} + \frac{\partial y_1}{\partial x_2}\frac{dx_2}{dt} \\ \frac{dy_2}{dt} &= \frac{\partial y_2}{\partial x_1}\frac{dx_1}{dt} + \frac{\partial y_2}{\partial x_2}\frac{dx_2}{dt} \\ \frac{dy_3}{dt} &= \frac{\partial y_3}{\partial x_1}\frac{dx_1}{dt} + \frac{\partial y_3}{\partial x_2}\frac{dx_2}{dt}. \end{aligned} \tag{18}$$

Case 3 of the chain rule can be used in the calculation of a tangent vector to a curve lying in a surface. As in the first example, we change to more familiar notations for curves and surfaces by writing $\mathbf{r}$ instead of $\mathbf{y}$ and $\mathbf{g}(t)$ instead of $\mathbf{x}$. Thus,

$$\mathbf{r} = \mathbf{r}(x, y), \quad \text{where } \langle x, y \rangle = \mathbf{g}(t) = \langle x(t), y(t) \rangle.$$

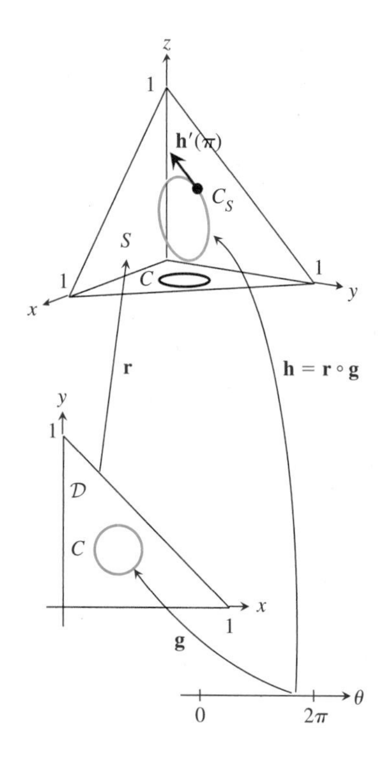

FIGURE 10.16 Describing a curve C_S in a surface S with the composition of two vector-valued functions.

EXAMPLE 4 Figure 10.16 shows an ellipse C_S lying in a surface S. The ellipse can be described by the composition of the functions

$$\mathbf{r} = \mathbf{r}(x, y) = \langle x, y, 1 - x - y \rangle, \qquad (x, y) \in \mathcal{D}, \tag{19}$$

where $\mathcal{D}$ is the region shown in Fig. 10.16, and

$$\langle x, y \rangle = \mathbf{g}(t) = \left\langle \tfrac{1}{3} + \tfrac{1}{7}\cos t,\ \tfrac{1}{3} + \tfrac{1}{7}\sin t \right\rangle, \qquad 0 \le t \le 2\pi. \tag{20}$$

Calculate the position vector of the point on C_S corresponding to $t = \pi$ and use Case 3 of the chain rule in calculating a tangent vector to C_S at this point.

Solution

The surface S is the first-octant part of the plane with equation $x + y + z = 1$. Hence, S is the graph of the equation $z = 1 - x - y$, where (x, y) varies over the triangular region $\mathcal{D}$ in the (x, y)-plane with vertices $(0, 0, 0)$, $(1, 0, 0)$, and $(0, 1, 0)$. If we take x and y as parameters, the surface S can be described by the parametric equation (19). The ellipse C_S is the projection onto S of the circle C in D with center $\left(\frac{1}{3}, \frac{1}{3}\right)$ and radius $\frac{1}{7}$. The circle can be described by (20). The curve C_S can be described by composing $\mathbf{r}$ and $\mathbf{g}$. We write the composition as $\mathbf{h}(t) = \mathbf{r}(\mathbf{g}(t))$.

The position vector of the point on C_S corresponding to $t = \pi$ is

$$\begin{aligned} \mathbf{h}(\pi) &= \mathbf{r}(\mathbf{g}(\pi)) = \mathbf{r}\left(\left\langle \tfrac{1}{3} + \tfrac{1}{7}\cdot(-1), \tfrac{1}{3} + \tfrac{1}{7}\cdot(0) \right\rangle\right) = \mathbf{r}\left(\left\langle \tfrac{4}{21}, \tfrac{1}{3} \right\rangle\right) \\ &= \left\langle \tfrac{4}{21}, \tfrac{1}{3}, 1 - \tfrac{4}{21} - \tfrac{1}{3} \right\rangle = \left\langle \tfrac{4}{21}, \tfrac{1}{3}, \tfrac{10}{21} \right\rangle. \end{aligned}$$

With $\mathbf{h}(t) = \mathbf{r}(\mathbf{g}(t))$, the tangent vector to C_S at $\mathbf{h}(\pi)$ is $\mathbf{h}'(\pi)$. Rewriting Equations (18) from Case 3 of the chain rule in terms of

$\mathbf{r} = \langle r_1, r_2, r_3 \rangle = \langle r_1(x,y), r_2(x,y), r_3(x,y) \rangle$ and $\langle x, y \rangle = \mathbf{g}(t) = \langle x(t), y(t) \rangle$,

$$h_1'(t) = \frac{dr_1}{dt} = \frac{\partial r_1}{\partial x}\frac{dx}{dt} + \frac{\partial r_1}{\partial y}\frac{dy}{dt}$$
$$h_2'(t) = \frac{dr_2}{dt} = \frac{\partial r_2}{\partial x}\frac{dx}{dt} + \frac{\partial r_2}{\partial y}\frac{dy}{dt} \tag{21}$$
$$h_3'(t) = \frac{dr_3}{dt} = \frac{\partial r_3}{\partial x}\frac{dx}{dt} + \frac{\partial r_3}{\partial y}\frac{dy}{dt}.$$

From (19) and (20),

$$h_1'(t) = 1 \cdot \left(-\tfrac{1}{7}\sin t\right) + 0 \cdot \left(\tfrac{1}{7}\cos t\right)$$
$$h_2'(t) = 0 \cdot \left(-\tfrac{1}{7}\sin t\right) + 1 \cdot \left(\tfrac{1}{7}\cos t\right)$$
$$h_3'(t) = (-1) \cdot \left(-\tfrac{1}{7}\sin t\right) + (-1) \cdot \left(\tfrac{1}{7}\cos t\right).$$

Evaluating $\mathbf{h}'(t) = \langle h_1'(t), h_2'(t), h_3'(t) \rangle$ at $t = \pi$,

$$\mathbf{h}'(\pi) = \left\langle 0, -\tfrac{1}{7}, \tfrac{1}{7} \right\rangle.$$

This vector is the tangent vector to C_S at the point with position vector $\mathbf{h}(\pi) = \left\langle \frac{4}{21}, \frac{1}{3}, \frac{10}{21} \right\rangle$.

A **rhumb line** on the Earth's surface is the curve followed by a ship sailing so that it crosses all meridians at a constant angle. A *meridian* is a great circle passing through both north and south poles. Rhumb lines, which simplify the work of helmsmen, are associated with Mercator maps, which simplify the work of navigators. (See the Student Project on Mercator maps at the end of the chapter if you wish to pursue this in more detail.) We calculate the tangent vector to a specific rhumb line in Example 5. We use Fig. 10.17, which is similar to Fig. 10.16.

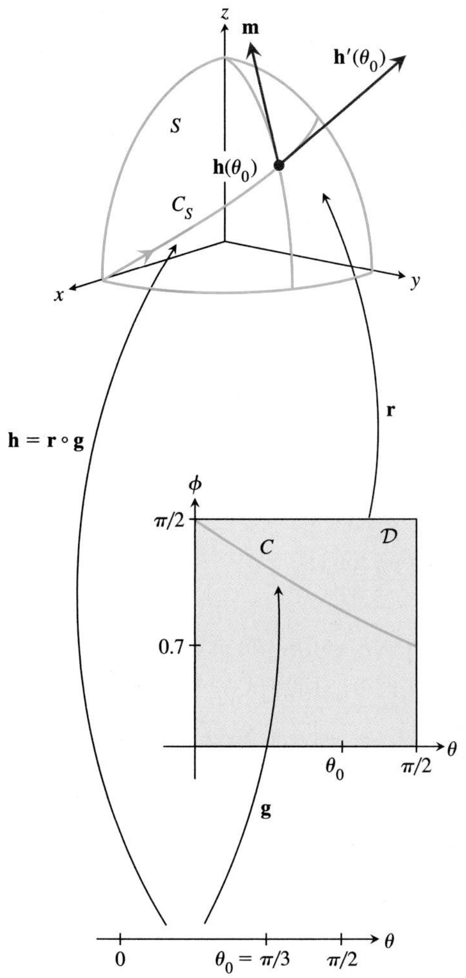

FIGURE 10.17 The plane curve C generating a rhumb line, the rhumb line C_S on the Earth's surface, and a tangent vector $\mathbf{h}'(\theta_0)$ to this rhumb line.

EXAMPLE 5 The first octant portion S of a sphere of radius 1 and centered at the origin is described by

$$\mathbf{r} = \mathbf{r}(\theta, \phi) = \langle x, y, z \rangle = \langle \sin\phi\cos\theta, \sin\phi\sin\theta, \cos\phi \rangle, \tag{22}$$

where $(\theta, \phi) \in \mathcal{D} = [0, \pi/2] \times [0, \pi/2]$. The spherical coordinates of the point $\mathbf{r}(\theta, \phi)$ are $(1, \theta, \phi)$. The surface S is shown in Fig. 10.17. A curve C in the domain $\mathcal{D}$ of $\mathbf{r}$ is described by

$$\mathbf{g}(\theta) = \langle \theta, \phi \rangle = \langle \theta, 2\arctan(e^{-\cot(1)\theta}) \rangle, \qquad 0 \le \theta \le \pi/2. \tag{23}$$

The rhumb line C_S in S is described by the composition $\mathbf{h} = \mathbf{r} \circ \mathbf{g}$, specifically, $\mathbf{h}(\theta) = \mathbf{r}(\mathbf{g}(\theta))$, $0 \le \theta \le \pi/2$. Calculate the tangent vector $\mathbf{h}'(\theta_0)$ to C_S at $\mathbf{h}(\theta_0)$, where $\theta_0 = \pi/3$.

Solution

We calculate $\mathbf{h}'(\theta_0)$ by applying (18). As in the preceding example, we change to a convenient notation. From (22) we write

$$\mathbf{y} = \mathbf{f}(\mathbf{x}) \text{ in the form } \mathbf{r} = \langle x, y, z \rangle = \langle \sin\phi\cos\theta, \sin\phi\sin\theta, \cos\phi \rangle \tag{24}$$

and from (23) we write

$$\mathbf{x} = \mathbf{g}(\mathbf{t}) \text{ in the form } \mathbf{g} = \langle \theta, \phi \rangle = \langle \theta, 2 \arctan(e^{-\cot(1)\theta}) \rangle. \tag{25}$$

In this notation the chain rule for $\mathbf{h}(\theta) = \langle h_1(\theta), h_2(\theta), h_3(\theta) \rangle = \mathbf{r}(\mathbf{g}(\theta))$ is

$$\begin{aligned} h_1'(\theta) &= \frac{\partial x}{\partial \theta}\frac{d\theta}{d\theta} + \frac{\partial x}{\partial \phi}\frac{d\phi}{d\theta} \\ h_2'(\theta) &= \frac{\partial y}{\partial \theta}\frac{d\theta}{d\theta} + \frac{\partial y}{\partial \phi}\frac{d\phi}{d\theta} \\ h_3'(\theta) &= \frac{\partial z}{\partial \theta}\frac{d\theta}{d\theta} + \frac{\partial z}{\partial \phi}\frac{d\phi}{d\theta}. \end{aligned} \tag{26}$$

From (24),

$$\begin{aligned} \frac{\partial x}{\partial \theta} &= -\sin\phi \sin\theta & \frac{\partial x}{\partial \phi} &= \cos\phi\cos\theta \\ \frac{\partial y}{\partial \theta} &= \sin\phi\cos\theta & \frac{\partial y}{\partial \phi} &= \cos\phi\sin\theta \\ \frac{\partial z}{\partial \theta} &= 0 & \frac{\partial z}{\partial \phi} &= -\sin\phi. \end{aligned} \tag{27}$$

From (25),

$$\frac{d\theta}{d\theta} = 1 \qquad \frac{d\phi}{d\theta} = \frac{2e^{-\cot(1)\theta}(-\cot(1))}{1 + e^{-2\cot(1)\theta}}. \tag{28}$$

Substituting these results into (26),

$$\begin{aligned} h_1'(\theta) &= (-\sin\phi\sin\theta)\cdot 1 + (\cos\phi\cos\theta)\cdot\frac{2e^{-\cot(1)\theta}(-\cot(1))}{1 + e^{-2\cot(1)\theta}} \\ h_2'(\theta) &= (\sin\phi\cos\theta)\cdot 1 + (\cos\phi\sin\theta)\cdot\frac{2e^{-\cot(1)\theta}(-\cot(1))}{1 + e^{-2\cot(1)\theta}} \\ h_3'(\theta) &= 0\cdot 1 + (-\sin\phi)\cdot\frac{2e^{-\cot(1)\theta}(-\cot(1))}{1 + e^{-2\cot(1)\theta}}. \end{aligned}$$

Evaluating the tangent vector $\mathbf{h}'(\theta) = \langle h_1'(\theta), h_2'(\theta), h_3'(\theta) \rangle$ at $\theta_0 = \pi/3$,

$$\mathbf{h}'(\theta_0) \approx \langle -0.853917, 0.140791, 0.421183 \rangle.$$

This vector is shown as $\mathbf{h}'(\theta_0)$ in Fig. 10.17. We leave as Exercise 37 an argument that the curve C_S crosses the meridians at an angle of 1 radian.

Case 4: $\mathbf{y} = \mathbf{g}(\mathbf{x})$, $\mathbf{x} = \mathbf{f}(\mathbf{t})$: Any of y, x, and t may be vector-valued: The electric field generated by a system of moving charges has the form $\mathbf{y} = \mathbf{f}(\mathbf{x})$, where $\mathbf{y}$ gives the direction and magnitude of the field and $\mathbf{x}$ the position vector or coordinates of the point at which $\mathbf{y}$ is measured. To find such a field $\mathbf{f}$ we must solve Maxwell's equations, which include the partial derivatives with respect to the variables x_1, x_2, and x_3 of the coordinate functions f_1, f_2, and f_3 of $\mathbf{f}$. The solution procedure can be greatly simplified by choosing a coordinate system fitting the geometry of the system of charges. If we think of $\mathbf{x} = \mathbf{g}(\mathbf{t})$ as the change of variable equations from, say, spherical to rectangular coordinates, Case 4 of the chain rule can be used to find the form taken by Maxwell's equations in spherical coordinates.

For use in the next example, we state the general chain rule (6) for the important case $m = p = n = 3$. If $\mathbf{y} = \mathbf{f}(\mathbf{x})$ and $\mathbf{x} = \mathbf{g}(\mathbf{t})$ are differentiable functions, then $\mathbf{y} = (\mathbf{f} \circ \mathbf{g})(\mathbf{t})$ is differentiable and its derivative matrix is

$$D\mathbf{y}(\mathbf{t}) = D\mathbf{y}(\mathbf{x})D\mathbf{x}(\mathbf{t}).$$

If, in particular, $\mathbf{y} = \langle y_1, y_2, y_2\rangle$, $\mathbf{x} = \langle x_1, x_2, x_3\rangle$, and $\mathbf{t} = \langle t_1, t_2, t_2\rangle$, then

$$\mathbf{Dy}(\mathbf{t}) = \begin{pmatrix} \frac{\partial y_1}{\partial t_1} & \frac{\partial y_1}{\partial t_2} & \frac{\partial y_1}{\partial t_3} \\ \frac{\partial y_2}{\partial t_1} & \frac{\partial y_2}{\partial t_2} & \frac{\partial y_2}{\partial t_3} \\ \frac{\partial y_3}{\partial t_1} & \frac{\partial y_3}{\partial t_2} & \frac{\partial y_3}{\partial t_3} \end{pmatrix} = \begin{pmatrix} \frac{\partial y_1}{\partial x_1} & \frac{\partial y_1}{\partial x_2} & \frac{\partial y_1}{\partial x_3} \\ \frac{\partial y_2}{\partial x_1} & \frac{\partial y_2}{\partial x_2} & \frac{\partial y_2}{\partial x_3} \\ \frac{\partial y_3}{\partial x_1} & \frac{\partial y_3}{\partial x_2} & \frac{\partial y_3}{\partial x_3} \end{pmatrix} \begin{pmatrix} \frac{\partial x_1}{\partial t_1} & \frac{\partial x_1}{\partial t_2} & \frac{\partial x_1}{\partial t_3} \\ \frac{\partial x_2}{\partial t_1} & \frac{\partial x_2}{\partial t_2} & \frac{\partial x_2}{\partial t_3} \\ \frac{\partial x_3}{\partial t_1} & \frac{\partial x_3}{\partial t_2} & \frac{\partial x_3}{\partial t_3} \end{pmatrix}. \tag{29}$$

EXAMPLE 6 At the point with rectangular coordinates $\left(0, \sqrt{2}/2, \sqrt{2}/2\right)$, calculate the partial derivatives $\partial f_1/\partial x$, $\partial f_1/\partial y$, and $\partial f_1/\partial z$ of the electric field

$$\mathbf{f}(x, y, z) = \langle f_1(x, y, z), f_2(x, y, z), f_3(x, y, z)\rangle$$

in terms of the partial derivatives $\partial f_1/\partial \rho$, $\partial f_1/\partial \theta$, and $\partial f_1/\partial \phi$ of f_1 with respect to the spherical coordinates ρ, θ, and ϕ.

Note: The rectangular coordinates $\left(0, \sqrt{2}/2, \sqrt{2}/2\right)$ correspond to the spherical coordinates $(\rho, \theta, \phi) = (1, \pi/2, \pi/4)$.

Solution

Starting with the field equations $\mathbf{y} = \mathbf{f}(x, y, z)$, we change variables with the transformation $(x, y, z) = \mathbf{g}(\rho, \theta, \phi)$ from spherical to rectangular coordinates, specifically,

$$\langle x, y, z\rangle = \mathbf{g}(\rho, \theta, \phi) = \langle \rho \sin\phi \cos\theta, \rho \sin\phi \sin\theta, \rho \cos\phi\rangle. \tag{30}$$

From (29), adjusted to the present notation,

$$\begin{pmatrix} \frac{\partial f_1}{\partial \rho} & \frac{\partial f_1}{\partial \theta} & \frac{\partial f_1}{\partial \phi} \\ \frac{\partial f_2}{\partial \rho} & \frac{\partial f_2}{\partial \theta} & \frac{\partial f_2}{\partial \phi} \\ \frac{\partial f_3}{\partial \rho} & \frac{\partial f_3}{\partial \theta} & \frac{\partial f_3}{\partial \phi} \end{pmatrix} = \begin{pmatrix} \frac{\partial f_1}{\partial x} & \frac{\partial f_1}{\partial y} & \frac{\partial f_1}{\partial z} \\ \frac{\partial f_2}{\partial x} & \frac{\partial f_2}{\partial y} & \frac{\partial f_2}{\partial z} \\ \frac{\partial f_3}{\partial x} & \frac{\partial f_3}{\partial y} & \frac{\partial f_3}{\partial z} \end{pmatrix} \begin{pmatrix} \frac{\partial x}{\partial \rho} & \frac{\partial x}{\partial \theta} & \frac{\partial x}{\partial \phi} \\ \frac{\partial y}{\partial \rho} & \frac{\partial y}{\partial \theta} & \frac{\partial y}{\partial \phi} \\ \frac{\partial z}{\partial \rho} & \frac{\partial z}{\partial \theta} & \frac{\partial z}{\partial \phi} \end{pmatrix}. \tag{31}$$

The partials $\partial f_1/\partial x$, $\partial f_1/\partial y$, and $\partial f_1/\partial z$ of interest appear only in the products of the first row of the derivative matrix $\mathbf{Df}(x, y, z)$ with the three columns of the derivative matrix $\mathbf{Dg}(\rho, \theta, \phi)$. Writing out these products (and equating them to the elements of the first row of the matrix on the left),

$$\begin{aligned} \frac{\partial f_1}{\partial \rho} &= \frac{\partial f_1}{\partial x}\frac{\partial x}{\partial \rho} + \frac{\partial f_1}{\partial y}\frac{\partial y}{\partial \rho} + \frac{\partial f_1}{\partial z}\frac{\partial z}{\partial \rho} \\ \frac{\partial f_1}{\partial \theta} &= \frac{\partial f_1}{\partial x}\frac{\partial x}{\partial \theta} + \frac{\partial f_1}{\partial y}\frac{\partial y}{\partial \theta} + \frac{\partial f_1}{\partial z}\frac{\partial z}{\partial \theta} \\ \frac{\partial f_1}{\partial \phi} &= \frac{\partial f_1}{\partial x}\frac{\partial x}{\partial \phi} + \frac{\partial f_1}{\partial y}\frac{\partial y}{\partial \phi} + \frac{\partial f_1}{\partial z}\frac{\partial z}{\partial \phi}. \end{aligned} \tag{32}$$

From (30) we calculate

$$\mathbf{D}\mathbf{g}(\rho,\theta,\phi) = \begin{pmatrix} \dfrac{\partial x}{\partial \rho} & \dfrac{\partial x}{\partial \theta} & \dfrac{\partial x}{\partial \phi} \\ \dfrac{\partial y}{\partial \rho} & \dfrac{\partial y}{\partial \theta} & \dfrac{\partial y}{\partial \phi} \\ \dfrac{\partial z}{\partial \rho} & \dfrac{\partial z}{\partial \theta} & \dfrac{\partial z}{\partial \phi} \end{pmatrix} = \begin{pmatrix} \sin\phi\cos\theta & -\rho\sin\phi\sin\theta & \rho\cos\phi\cos\theta \\ \sin\phi\sin\theta & \rho\sin\phi\cos\theta & \rho\cos\phi\sin\theta \\ \cos\phi & 0 & -\rho\sin\phi \end{pmatrix}$$

Substituting these results into (32),

$$\begin{aligned} \frac{\partial f_1}{\partial \rho} &= \sin\phi\cos\theta\frac{\partial f_1}{\partial x} + \sin\phi\sin\theta\frac{\partial f_1}{\partial y} + \cos\phi\frac{\partial f_1}{\partial z} \\ \frac{\partial f_1}{\partial \theta} &= -\rho\sin\phi\sin\theta\frac{\partial f_1}{\partial x} + \rho\sin\phi\cos\theta\frac{\partial f_1}{\partial y} + 0\cdot\frac{\partial f_1}{\partial z} \\ \frac{\partial f_1}{\partial \phi} &= \rho\cos\phi\cos\theta\frac{\partial f_1}{\partial x} + \rho\cos\phi\sin\theta\frac{\partial f_1}{\partial y} - \rho\sin\phi\frac{\partial f_1}{\partial z}. \end{aligned}$$

The partial derivatives $\partial f_1/\partial x$, $\partial f_1/\partial y$, and $\partial f_1/\partial z$ can be found by solving this system of equations. Because we want the values of these partials only at the point $(0, \sqrt{2}/2, \sqrt{2}/2)$ we may simplify the system by replacing ρ by 1, θ by $\pi/2$, and ϕ by $\pi/4$. This gives the system

$$\begin{aligned} \frac{\partial f_1}{\partial \rho} &= 0\cdot\frac{\partial f_1}{\partial x} + \frac{\sqrt{2}}{2}\frac{\partial f_1}{\partial y} + \frac{\sqrt{2}}{2}\frac{\partial f_1}{\partial z} \\ \frac{\partial f_1}{\partial \theta} &= -\frac{\sqrt{2}}{2}\frac{\partial f_1}{\partial x} + 0\cdot\frac{\partial f_1}{\partial y} + 0\cdot\frac{\partial f_1}{\partial z} \\ \frac{\partial f_1}{\partial \phi} &= 0\cdot\frac{\partial f_1}{\partial x} + \frac{\sqrt{2}}{2}\frac{\partial f_1}{\partial y} - \frac{\sqrt{2}}{2}\frac{\partial f_1}{\partial z}. \end{aligned}$$

We find $\partial f_1/\partial x$ from the second equation; by adding and subtracting the first and third equations, we find $\partial f_1/\partial y$ and $\partial f_1/\partial z$. The results are

$$\frac{\partial f_1}{\partial x} = -\sqrt{2}\frac{\partial f_1}{\partial \theta}, \quad \frac{\partial f_1}{\partial y} = \frac{1}{\sqrt{2}}\left(\frac{\partial f_1}{\partial \rho} + \frac{\partial f_1}{\partial \phi}\right), \quad \text{and} \quad \frac{\partial f_1}{\partial z} = \frac{1}{\sqrt{2}}\left(\frac{\partial f_1}{\partial \rho} - \frac{\partial f_1}{\partial \phi}\right).$$

Exercises 10.2

Exercises 1–10: Use Case 1 of the chain rule to calculate the derivative of the composite function. Check your result by forming the composite function and differentiating with respect to t.

1. $y = x_1^2 - x_1x_2 + x_2^2$, $\mathbf{x} = \langle x_1, x_2\rangle = \langle e^t, e^{-t}\rangle$
2. $y = x_1^3x_2 - x_1x_2^2$, $\mathbf{x} = \langle x_1, x_2\rangle = \langle \sin t, \cos t\rangle$
3. $f(x,y) = \sin(x/y)$, $\mathbf{g}(t) = \langle x, y\rangle = \langle t, \sqrt{t}\rangle$
4. $f(x,y) = \cos(y/x)$, $\mathbf{g}(t) = \langle x, y\rangle = \langle t, \sqrt{t}\rangle$
5. $y = \arctan(x_1/x_2)$ $\mathbf{x} = \langle x_1, x_2\rangle = \langle t^2, -t^3\rangle$
6. $y = x_1 \arcsin(x_2)$ $\mathbf{x} = \langle x_1, x_2\rangle = \langle t^3, t\rangle$
7. $f(x,y,z) = xyz^2$, $\mathbf{g}(t) = \langle x, y, z\rangle = \langle e^{2t}, e^{-t}, e^t\rangle$
8. $f(x,y,z) = (x+y)/z$, $\mathbf{x} = \langle x, y, z\rangle = \langle 2t-1, 2t+1, t^2\rangle$
9. $y = \sqrt{10 - x_1^2 - x_2^2 - x_3^2}$, $\mathbf{x} = \langle x_1, x_2, x_3\rangle = \langle \cos t, \sin t, 3t\rangle$
10. $y = x_1e^{x_2/x_3}$, $\mathbf{x} = \langle x_1, x_2, x_3\rangle = \langle t^2, t, \sqrt{t}\rangle$

Exercises 11–16: Use Case 2 of the chain rule to calculate the derivative of the composite function.

11. Let $z = y/x$ be a function of two variables and let $x = u\cos v$ and $y = u\sin v$ be change-of-variable equations. Calculate z_u and z_v.

12. Let $z = \arctan(x/y)$ be a function of two variables and let $x = e^u \cos v$ and let $y = e^u \sin v$ be change-of-variable equations. Calculate z_u and z_v.

13. Let $z = f(x, y)$ be a function of two variables and let $x = (u^2 - v^2)/2$ and $y = uv$ be change-of-variable equations. Calculate z_u and z_v in terms of f_x, f_y, u, and v.

14. Let $z = f(x, y)$ be a function of two variables and let $x = \cosh u \cos v$ and $y = \sinh u \sin v$ be change-of-variable equations. Calculate z_u and z_v in terms of f_x, f_y, u, and v.

15. Let $y = f(\mathbf{x}) = \sqrt{1 - x_1^2 - x_2^2}$ and $\mathbf{x} = \mathbf{g}(\mathbf{t}) = \langle t_1 - t_2, t_1 + t_2\rangle$. Calculate $\nabla z(0, 1/2)$, where $z(\mathbf{t}) = f(\mathbf{g}(\mathbf{t}))$.

16. Let $y = f(\mathbf{x}) = \ln(1 + x_1^2 + 3x_2^2)$ and $\mathbf{x} = \mathbf{g}(\mathbf{t}) = \langle t_1 - t_2, t_1 + t_2\rangle$. Calculate $\nabla z(2, -1)$, where $z(\mathbf{t}) = f(\mathbf{g}(\mathbf{t}))$.

Exercises 17–24: Use Case 3 of the chain rule to calculate the derivative matrix of the composite function. Check your result by forming the composite function and differentiating with respect to t.

17. $\mathbf{y} = \langle \sin x_1 \cos x_2, \sin x_1 \sin x_2\rangle$, $\mathbf{x} = \langle x_1, x_2\rangle = \langle t^2, -t^3\rangle$

18. $\mathbf{y} = \langle x_1 \cos x_2, x_1 \sin x_2\rangle$, $\mathbf{x} = \langle x_1, x_2\rangle = \langle t^2, t\rangle$

19. $\mathbf{f}(x, y) = \langle e^{x+y}, e^{x-y}\rangle$, $\langle x, y\rangle = \mathbf{g}(t) = \langle \ln(t+1), \ln t\rangle$

20. $\mathbf{f}(x, y) = \langle x^2 - xy, x^2 - y^2\rangle$, $\langle x, y\rangle = \mathbf{g}(t) = \langle t^2 - t, t^3 - t^2\rangle$

21. $\mathbf{y} = \langle u^2 w, vw, uv\rangle$, $\mathbf{x} = \langle u, v, w\rangle = \langle t, t, \sqrt{t}\rangle$

22. $\mathbf{y} = \langle vw, uw, uv\rangle$, $\mathbf{x} = \langle u, v, w\rangle = \langle \cos t, \sin t, -t\rangle$

23. $\mathbf{f}(x, y, z) = \langle e^z \cos x, e^z \sin x, e^z y\rangle$, $\langle x, y, z\rangle = \mathbf{g}(t) = \langle 2t^2, -2t^2, t\rangle$

24. $\mathbf{f}(x_1, x_2, x_3) = \langle x_1 + x_2 - x_3, x_1 - x_2 + x_3, x_2 + x_3\rangle$, $\langle x_1, x_2, x_3\rangle = \mathbf{g}(t) = \langle 2t - 1, t^3, 2t + 1\rangle$

Exercises 25–28: Show that the curve C described by $\mathbf{r} = \mathbf{r}(t)$ is (or, at least, is part of) a level curve of the function u. Also show that **a** *is on C and use Case 1 of the chain rule in showing that $\nabla u(\mathbf{a})$ is perpendicular to C at* **a**.

25. $u(x, y) = 3x^2 + 2y^2$, $\mathbf{r}(t) = \left\langle \sqrt{2}\cos t, \sqrt{3}\sin t\right\rangle$, $\mathbf{a} = \left(1, \sqrt{3/2}\right)$

26. $u(x, y) = 2x^2 + y^2$, $\mathbf{r}(t) = \left\langle \cos t, \sqrt{2}\sin t\right\rangle$, $\mathbf{a} = \left(1/2, \sqrt{3/2}\right)$

27. $u(x, y) = x^2 - y^2$, $\mathbf{r}(t) = \langle \cosh t, \sinh t\rangle$, $\mathbf{a} = \left(2, \sqrt{3}\right)$

28. $u(x, y) = x^3 - 3xy^2 - 3x$, $\mathbf{r}(t) = \left\langle \sqrt{3}\cosh t, \sinh t\right\rangle$, $\mathbf{a} = \left(5/\sqrt{3}, 4/3\right)$

29. Denote the Fahrenheit temperature F at any point (x, y) of the plane by $F(x, y)$. If the position vector of an object is $\mathbf{r} = \mathbf{r}(t) = e^t\langle \cos t, \sin t\rangle$ at time t, the temperature $T(t)$ of the object at time t is given by $T(t) = F(\mathbf{r}(t))$. Suppose it is found by measurement that at $\mathbf{p} = \mathbf{r}(\pi/2) = e^{\pi/2}\langle 0, 1\rangle$, $F(\mathbf{p}) = 150$, $F_x(\mathbf{p}) = -2.5$, and $F_y(\mathbf{p}) = 1.3$. Calculate $T'(\pi/2)$ and approximate $T(1.5)$.

30. Let S be the upper half of the sphere of radius a centered at the origin, described by

$$\begin{aligned}\mathbf{f} &= \mathbf{f}(x, y) \\ &= \left\langle x, y, \sqrt{a^2 - x^2 - y^2}\right\rangle\end{aligned}$$

where $(x, y) \in \mathcal{D} = \{(x, y) : x^2 + y^2 < a^2\}$. Let C be a curve in $\mathcal{D}$ and described by

$$\mathbf{r} = \mathbf{r}(t) = \langle r_1(t), r_2(t)\rangle,$$

where $c \le t \le d$. Assume that **f** is differentiable. Let the curve C_S be described by

$$\mathbf{h}(t) = \mathbf{f}(\mathbf{r}(t)),$$

where $c \le t \le d$. Show that at any point $\mathbf{h}(t)$, the curve C_S is perpendicular to a normal to S at this point.

31. A circle C in the (x, y)-plane is described by

$$\langle x, y\rangle = \left\langle \frac{1}{3} + \frac{1}{4}\cos\theta, \frac{5}{8} + \frac{1}{4}\sin\theta\right\rangle,$$

where θ varies over the interval $[0, 2\pi]$. This curve is projected vertically upwards onto the plane with equation $x + y/2 + z/3 = 1$, resulting in a curve C_S. Find the tangent vector to the curve C_S at the point on C_S corresponding to the point $(1/3, 7/8)$ on C.

32. With the change of variable equations $x = X/2 + T/2$ and $t = -(1/(2c))X + (1/(2c))T$, the flux-conservative equation

$$\frac{\partial \nu}{\partial t} = -c\frac{\partial \nu}{\partial x} \tag{33}$$

becomes the equation

$$\frac{\partial \nu}{\partial T} = 0, \tag{34}$$

as discussed in Example 3. If f is any differentiable function of X, show that $\nu = f(X)$ is a solution of (34). Infer from this result that $v = f(x - vt)$ is a solution of the flux-conservative equation (33).

33. Given the change of coordinates

$$\begin{aligned}x &= (\cos\psi)X + (\sin\psi)Y \\ y &= -(\sin\psi)X + (\cos\psi)Y \\ z &= Z,\end{aligned}$$

where ψ is a constant, calculate the partial derivatives $\partial f_1/\partial x$, $\partial f_1/\partial y$, and $\partial f_1/\partial z$ of the electric field

$$\mathbf{f}(x, y, z) = \langle f_1(x, y, z), f_2(x, y, z), f_3(x, y, z)\rangle$$

in terms of Z and the partial derivatives $\partial f_1/\partial X$, $\partial f_1/\partial Y$, and $\partial f_1/\partial Z$.

34. Suppose the functions in $\mathbf{y} = \mathbf{f}(\mathbf{x})$ and $\mathbf{x} = \mathbf{g}(\mathbf{t})$ are given by $\mathbf{f}(x_1, x_2) = \langle x_1^2 + x_2^2, x_2/x_1\rangle$ and $\mathbf{g}(t_1, t_2) = \langle t_1 \cos t_2, t_1 \sin t_2\rangle$. Note that $\mathbf{g}(1, \pi/6) = \langle\sqrt{3}/2, 1/2\rangle$. Without forming the composition $\mathbf{f} \circ \mathbf{g}$, calculate $\mathbf{Df}(\sqrt{3}/2, 1/2)$ and $\mathbf{Dg}(1, \pi/6)$. Combine to calculate $\mathbf{Dz}(1, \pi/6)$, where $\mathbf{z}(t) = \mathbf{f}(\mathbf{g}(\mathbf{t}))$. Use this result in approximating $\mathbf{z}(1.05, 0.50)$. What are the matrices associated with the linear transformations $\mathbf{df}(\sqrt{3}/2, 1/2)$, $\mathbf{dg}(1, \pi/6)$, and $\mathbf{dz}(1, \pi/6)$?

35. Suppose the functions in $\mathbf{y} = \mathbf{f}(\mathbf{x})$ and $\mathbf{x} = \mathbf{g}(\mathbf{t})$ are given by $\mathbf{f}(x_1, x_2) = \langle x_1 e^{x_2}, x_2 e^{x_1}\rangle$ and $\mathbf{g}(t_1, t_2) = \langle t_1 + t_2^2, t_1 t_2 + 1\rangle$. Note that $\mathbf{g}(1, 0) = \langle 1, 1\rangle$. Without forming the composition $\mathbf{f} \circ \mathbf{g}$, calculate $\mathbf{Df}(1, 1)$ and $\mathbf{Dg}(1, 0)$ separately. Combine to calculate $\mathbf{Dz}(1, 0)$, where $\mathbf{z}(t) = \mathbf{f}(\mathbf{g}(\mathbf{t}))$. Use this result in approximating $\mathbf{z}(1.05, 0.03)$. What are the matrices associated with the linear transformations $\mathbf{df}(1, 1)$, $\mathbf{dg}(1, 0)$, and $\mathbf{dz}(1, 0)$?

T **36.** Referring to Example 5, show that $\mathbf{h}(\pi/3) \approx \langle 0.404955, 0.701402, 0.586555\rangle$ and $\mathbf{h}(\pi/2) \approx \langle 0, 0.643815, 0.765181\rangle$.

T **37.** Referring to Fig. 10.17, show that the angle between the tangent vector to the rhumb line C_S at $\mathbf{h}(\theta_0)$ and the vector $\mathbf{m}$ tangent to the meridian at $\mathbf{h}(\theta_0)$ is 1 radian.

38. Supposing that f is a differentiable function and $z = f(x/y, y/x)$, express z_x and z_y in terms of the partials of f, x, and y. *Hint:* Think of $f(x/y, y/x)$ as a composition of $f(u, v)$ and two functions u and v of x and y.

39. Fill in the following outline of a proof that for functions $y = f(\mathbf{x})$ and $\mathbf{x} = \mathbf{g}(t)$, where $\mathbf{g}$ is differentiable at b and f is differentiable at $\mathbf{a} = \mathbf{g}(b)$, the composite function $f \circ \mathbf{g}$ is differentiable at b and

$$D(f \circ \mathbf{g})(b) = Df(\mathbf{g}(b))\mathbf{Dg}(b).$$

Begin by showing that because $\mathbf{g}$ is differentiable at b there is a function $\mathbf{E_g}$ for which (i)

$$\lim_{t\to b} \mathbf{E_g}(t) = \mathbf{0}$$

and (ii) for t near b,

$$\mathbf{g}(t) = \mathbf{g}(b) + \mathbf{Dg}(b)(t - b) + \mathbf{E_g}(t)|t - b|.$$

Next, show that because f is differentiable at $\mathbf{a}$, there is a function E_f for which (iii)

$$\lim_{\mathbf{x}\to\mathbf{a}} E_f(\mathbf{x}) = 0$$

and (iv) for $\mathbf{x}$ near $\mathbf{a}$

$$f(\mathbf{x}) = f(\mathbf{a}) + Df(\mathbf{a})(\mathbf{x} - \mathbf{a}) + E_f(\mathbf{x})\|\mathbf{x} - \mathbf{a}\|.$$

Apply these two "error formulas" to show that $f \circ \mathbf{g}$ is differentiable at b.

Start by using the first error formula to write

$$f(\mathbf{g}(t)) = f(\mathbf{g}(b) + \mathbf{Dg}(b)(t - b) + \mathbf{E_g}(t)|t - b|).$$

Next, let $\mathbf{x} = \mathbf{g}(b) + \mathbf{Dg}(b)(t - b) + \mathbf{E_g}(t)|t - b|$ and $\mathbf{a} = \mathbf{g}(b)$ and apply the second error formula.

10.3 Applications of the Chain Rule

In Section 10.2 we gave several examples of how the chain rule is used in contexts in which a change of variables is made. In this section we extend the discussion to include a transformation of the two-dimensional "heat equation" into polar coordinates. We also discuss implicit differentiation of functions and an extension of Taylor's inequality to functions of several variables. We begin the section with a proof that the gradient vector is perpendicular to the level curves or surfaces of functions of several variables.

The Gradient Is Perpendicular to Level Curves or Level Surfaces

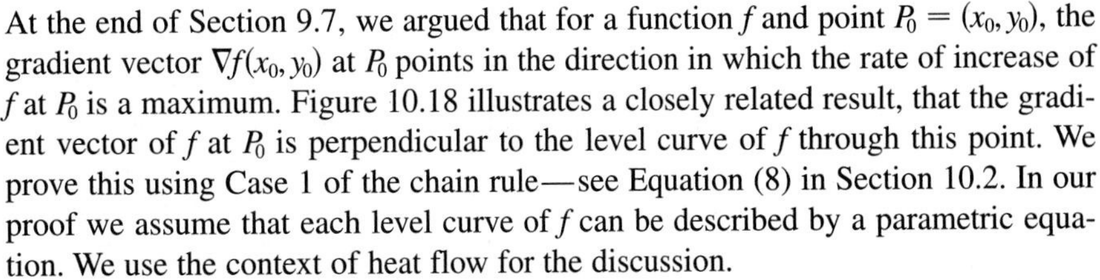

At the end of Section 9.7, we argued that for a function f and point $P_0 = (x_0, y_0)$, the gradient vector $\nabla f(x_0, y_0)$ at P_0 points in the direction in which the rate of increase of f at P_0 is a maximum. Figure 10.18 illustrates a closely related result, that the gradient vector of f at P_0 is perpendicular to the level curve of f through this point. We prove this using Case 1 of the chain rule—see Equation (8) in Section 10.2. In our proof we assume that each level curve of f can be described by a parametric equation. We use the context of heat flow for the discussion.

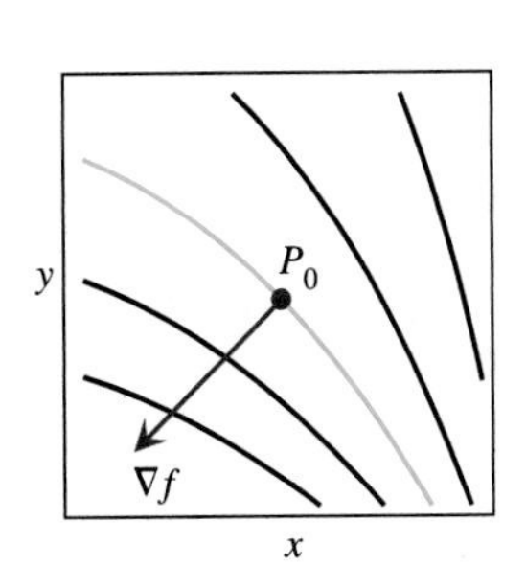

FIGURE 10.18 The gradient vector at P_0 is perpendicular to the level curve of f through this point.

Figure 10.19 shows a thin metal bar π meters wide and extending indefinitely in the y-direction. The bar is insulated on its top and bottom surfaces (so that there is no heat flow through these surfaces), is held at 0° C on the sides $x = 0$ and $x = \pi$, and is held at 1° C on the end $y = 0$. If these *boundary conditions* hold for all $t \geq 0$, the temperature at any point (x, y) in the bar tends over time toward its *steady state;* that is, the temperature no longer changes with time. We denote the steady-state temperature at (x, y) by $u(x, y)$.

Figure 10.20 shows a contour map of the temperature function u in the first 3 meters of the bar. The points in the dark region at the top of the figure are 0.1° C or cooler; points in the "white-hot" region at the bottom are 0.9° C or hotter. The regions are separated by isotherms, or level curves of the temperature function u. We focus on the point with coordinates $(2.5, 1)$, which is (very nearly) on the 0.3° isotherm. Let C be this isotherm and suppose it is described by the parametric equation

$$\mathbf{r} = \mathbf{r}(t) = \langle x(t), y(t) \rangle, \qquad a \leq t \leq b.$$

We assume that $\mathbf{r}(c) = \langle 2.5, 1 \rangle$ for some $c \in (a, b)$.

We show that $\nabla u(2.5, 1)$ is perpendicular to C at $(2.5, 1)$ by showing that the gradient vector $\nabla u(2.5, 1)$ is perpendicular to the tangent vector $\mathbf{r}'(c)$ to C at $(2.5, 1)$. Because for all points (x, y) on C, $u(x, y) = 0.3$, the composite function

$$w = w(t) = u(\mathbf{r}(t)) - 0.3 = u(x(t), y(t)) - 0.3$$

is identically zero for all $t \in [a, b]$. Applying the chain rule (8) from Section 10.2, with $p = 2$,

$$0 = \frac{dw}{dt} = \begin{pmatrix} \dfrac{\partial u}{\partial x} & \dfrac{\partial u}{\partial y} \end{pmatrix} \begin{pmatrix} \dfrac{dx}{dt} \\ \dfrac{dy}{dt} \end{pmatrix} = \nabla u(x, y) \cdot \left\langle \frac{dx}{dt}, \frac{dy}{dt} \right\rangle = \nabla u \cdot \mathbf{r}'.$$

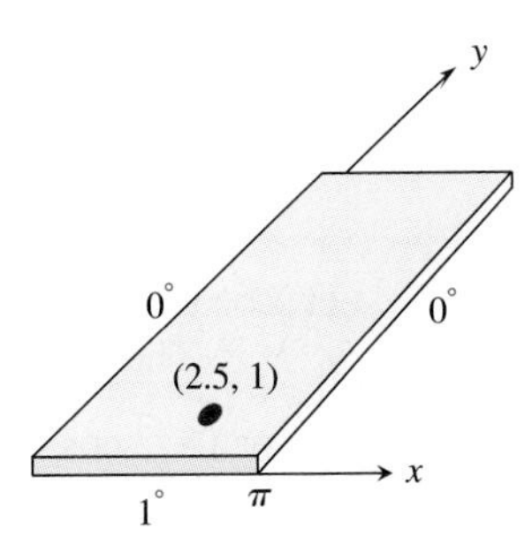

FIGURE 10.19 A thin metal bar subjected to given boundary conditions.

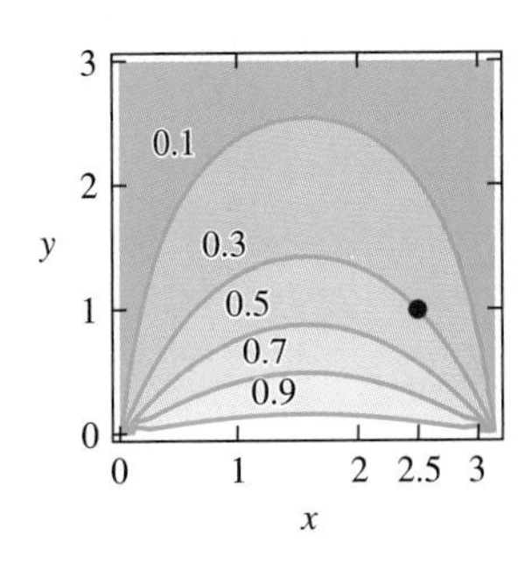

FIGURE 10.20 Contour map of the temperature function u for the thin metal bar.

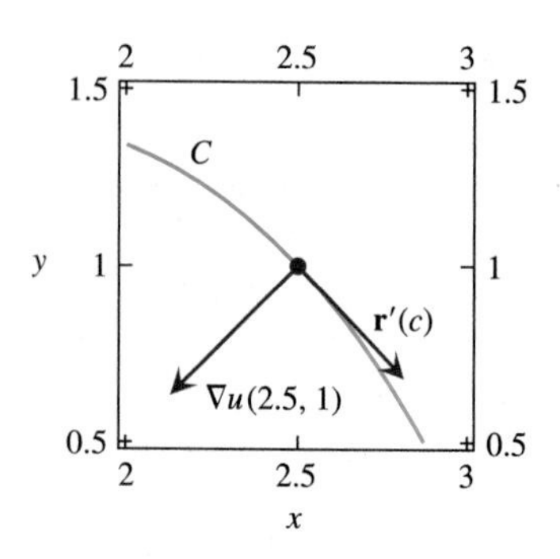

FIGURE 10.21 The curve C (which is the 0.3° isotherm), the tangent vector $\mathbf{r}'(c)$, and the gradient of u at the point (2.5, 1).

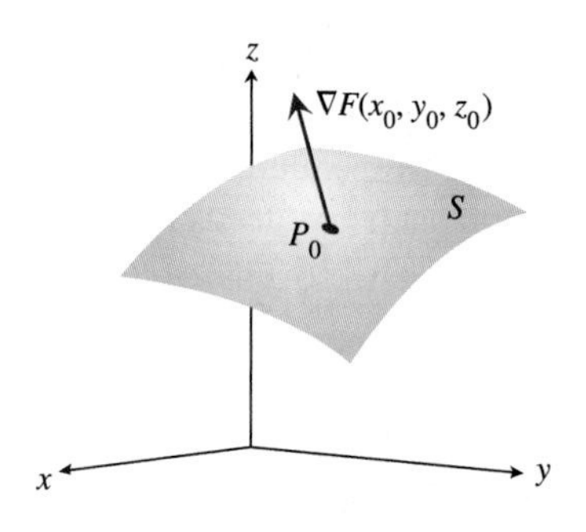

FIGURE 10.22 A surface S described by $F(x, y, z) = c$ and the gradient vector $\nabla F(x_0, y_0, z_0)$ at a representative point $P_0 = (x_0, y_0, z_0)$ on S; the gradient vector is perpendicular to S at P_0.

Because $\nabla u \cdot \mathbf{r}' = 0$, the gradient vector ∇u is perpendicular to the tangent vector $\mathbf{r}'$ to C at all points of C and, in particular, at (2.5, 1).

Figure 10.21 shows the gradient vector $\nabla u(2.5, 1)$, the 0.3° isotherm or level curve C passing through (2.5, 1), and the tangent vector $\mathbf{r}'(c)$ to the isotherm at this point. The fact that the tangent vector $\mathbf{r}'(c)$ to the level curve is perpendicular to the gradient vector $\nabla u(2.5, 1)$ fits with our intuitive sense that if we are standing on a level curve and want to experience the greatest rise in temperature (C° per meter), we should walk in a direction perpendicular to this constant temperature curve.

The above argument—that at a point (x_0, y_0) on a level curve C described by an equation $f(x, y) = k$, where f is differentiable and k is a constant, the gradient vector $\nabla f(x_0, y_0)$ is perpendicular to C—is quite general, although presented in a specific context.

Now let S be a level surface described by an equation $F(x, y, z) = k$, where F is differentiable and k is a constant. We show that at any point $P_0 = (x_0, y_0, z_0)$ on S, the gradient vector $\nabla F(x_0, y_0, z_0)$ is perpendicular to S. Figure 10.22 shows a sketch of a representative surface S and the gradient vector $\nabla F(x_0, y_0, z_0)$ at $P_0 = (x_0, y_0, z_0)$ on S. We show that $\nabla F(x_0, y_0, z_0)$ is perpendicular to S at P_0 by showing that $\nabla F(x_0, y_0, z_0)$ is perpendicular to every smooth curve C lying in S and passing through the point P_0.

Let C be any such smooth curve and assume that C is described by the parametric equation

$$\langle x, y, z\rangle = \mathbf{r}(t) = \langle x(t), y(t), z(t)\rangle, \qquad t \in I,$$

and $\mathbf{r}(t_0) = P_0 = (x_0, y_0, z_0)$ for some interior point t_0 of I. Because C lies in S, the composite function is constant; that is,

$$w(t) = F(x(t), y(t), z(t)) = c \qquad \text{for all } t \in I.$$

Hence $dw/dt = 0$ for all t in the interior of I. Applying the chain rule (8) from Section 10.2, with $p = 3$,

$$\frac{dw}{dt} = \begin{pmatrix} \frac{\partial F}{\partial x} & \frac{\partial F}{\partial y} & \frac{\partial F}{\partial z} \end{pmatrix} \begin{pmatrix} \frac{dx}{dt} \\ \frac{dy}{dt} \\ \frac{dz}{dt} \end{pmatrix} = \nabla F(x, y, z) \cdot \left\langle \frac{dx}{dt}, \frac{dy}{dt}, \frac{dz}{dt} \right\rangle = \nabla F(x, y, z) \cdot \mathbf{r}'(t).$$

Because $dw/dt = 0$, we conclude that the gradient vector $\nabla F(x_0, y_0, z_0)$ is perpendicular to the tangent vector $\mathbf{r}'(t_0)$ to C at $P_0 = (x_0, y_0, z_0)$.

Because C was an arbitrary smooth curve lying in S and passing through P_0, we conclude that the gradient vector $\nabla F(x_0, y_0, z_0)$ is perpendicular to S at P_0.

EXAMPLE 1 Let S be the level surface described by $F(x, y, z) = 6$, where $F(x, y, z) = x^2 + y^2 + z^2$. The smooth curves C_1 and C_2 described by

$$C_1: \quad \mathbf{r}_1(\theta) = \left\langle \sqrt{2}\cos\theta, \sqrt{2}\sin\theta, 2\right\rangle, \qquad 0 \le \theta \le \pi/2,$$

and

$$C_2: \quad \mathbf{r}_2(\phi) = \left\langle \sqrt{3}\sin\phi, \sqrt{3}\sin\phi, \sqrt{6}\cos\phi\right\rangle, \qquad 0 \le \phi \le \pi/2,$$

lie in S. These curves pass through $P_0 = (1, 1, 2)$ for $\theta = \pi/4$ and $\phi = \arccos(2/\sqrt{6})$, respectively. Show by direct calculation that the gradient vector $\nabla F(1, 1, 2)$ at P_0 is perpendicular to each of the smooth curves C_1 and C_2 at P_0.

Solution

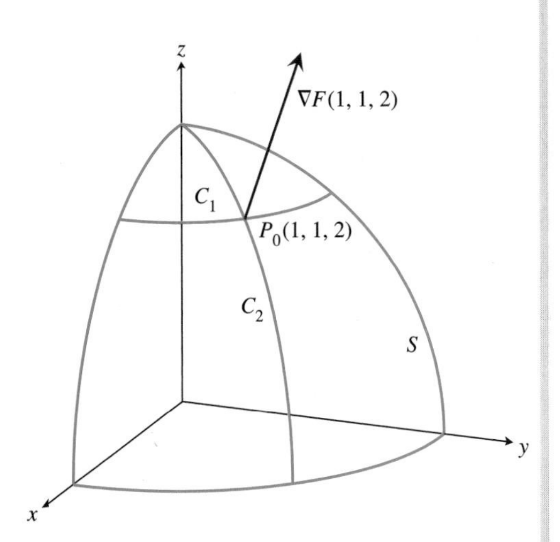

FIGURE 10.23 The level surface S described by $F(x, y, z) = 6$, where $F(x, y, z) = x^2 + y^2 + z^2$; also shown is the point $P_0 = (1, 1, 2)$ on S and the gradient vector $\nabla F(1, 1, 2)$.

Figure 10.23 shows the (circular) traces in the three coordinate planes of the first-octant part of the (spherical) surface S. The curve C_1 is the parallel of latitude though P_0 and C_2 is the meridian through this point.

The gradient of F at $(1, 1, 2)$ is

$$\begin{aligned}\nabla F(x, y, z)\big|_{x=1,y=1,z=2} &= \langle F_x(x, y, z), F_y(x, y, z), F_z(x, y, z)\rangle\big|_{x=1,y=1,z=2} \\ &= \langle 2x, 2y, 2z\rangle\big|_{x=1,y=1,z=2} = \langle 2, 2, 4\rangle.\end{aligned}$$

The tangent vectors to C_1 and C_2 are

$$C_1\colon \mathbf{r}_1'(\theta) = \left\langle -\sqrt{2}\sin\theta, \sqrt{2}\cos\theta, 0\right\rangle$$

and

$$C_2\colon \mathbf{r}_2'(\phi) = \left\langle \sqrt{3}\cos\phi, \sqrt{3}\cos\phi, -\sqrt{6}\sin\phi\right\rangle.$$

Evaluating these tangent vectors at $\theta = \pi/4$ and $\phi = \arccos(2/\sqrt{6})$,

$$C_1\colon \mathbf{r}_1'(\pi/4) = \left\langle -\sqrt{2}/\sqrt{2}, \sqrt{2}/\sqrt{2}, 0\right\rangle = \langle -1, 1, 0\rangle$$

and

$$\begin{aligned}C_2\colon \mathbf{r}_2'\big(\arccos(2\sqrt{6})\big) &= \left\langle \sqrt{3}(2/\sqrt{6}), \sqrt{3}(2/\sqrt{6}), -\sqrt{6}(\sqrt{2}/\sqrt{6})\right\rangle \\ &= \left\langle \sqrt{2}, \sqrt{2}, -\sqrt{2}\right\rangle.\end{aligned}$$

It is easily checked that

$$\nabla F(1, 1, 2)\cdot \mathbf{r}_1'(\pi/4) = 0$$

and

$$\nabla F(1, 1, 2)\cdot \mathbf{r}_1'\big(\arccos(2\sqrt{6})\big) = 0.$$

Hence, the gradient of F is perpendicular to each of the curves C_1 and C_2 at the point $P_0 = (1, 1, 2)$.

Implicit Differentiation

If we know on theoretical grounds that an equation $F(x, y) = 0$ can be solved for y in terms of x, say $y = f(x)$, we say that the function f is implicitly defined by the equation $F(x, y) = 0$. The function f can be found explicitly if the equation $F(x, y) = 0$ is not too complex. For example, the equation $F(x, y) = 3x - 2y - 1 = 0$ can be solved for y in terms of x, obtaining $y = f(x) = \frac{3}{2}x - \frac{1}{2}$. Many equations, however, cannot be solved explicitly. For example, the equation $F(x, y) = \sin(x - y) + x^2y = 0$ cannot be solved explicitly for y in terms of x. However, for pairs (x, y) near $(0, 0)$, it can be shown that there is a function $y = f(x)$ for which $F(x, y) = \sin(x - y) + x^2y = 0$ for all x near 0. The function so determined is said to be defined implicitly.

We discussed implicitly defined functions and a procedure for calculating their derivatives in Section 2.4. With the help of Case 1 of the chain rule, we may apply this procedure to an equation $F(x, y) = 0$ defining a function $y = f(x)$. In this way we can find a formula for y' in terms of the partial derivatives of F. Similarly, we can find formulas for calculating the partial derivatives of a function $z = f(x, y)$ implicitly defined by an equation $F(x, y, z) = 0$.

Functions Defined by an Equation $F(x,y) = 0$ Suppose that $y = f(x)$ is defined implicitly by an equation $F(x, y) = 0$; that is,

$$F(x, f(x)) = 0, \qquad x \in I,$$

where $I \subset R$ is an interval. Assuming that both F and f are differentiable functions, we may differentiate both sides of this equation with respect to x,

$$\frac{d}{dx}F(x, f(x)) = \frac{d}{dx}(0) \qquad x \in I.$$

The derivative on the right is 0. Applying Case 1 of the chain rule to the left side,

$$\frac{d}{dx}F(x, f(x)) = F_x(x, f(x)) \cdot 1 + F_y(x, f(x)) \cdot f'(x) = 0.$$

Assuming that $F_y(x, f(x)) \neq 0$ for $x \in I$ and solving this equation for $f'(x)$,

$$f'(x) = -\frac{F_x(x, f(x))}{F_y(x, f(x))} = -\frac{F_x(x, y)}{F_y(x, y)}.$$

This shows that the derivative of a function $y = f(x)$ implicitly defined by the equation $F(x, y) = 0$ can be calculated in terms of the partials of F whether or not f is explicitly given. In more advanced texts it is shown that the condition $F_y(x, f(x)) \neq 0$ needed for solving for $f'(x)$ is sufficient to guarantee that f exists and is differentiable.

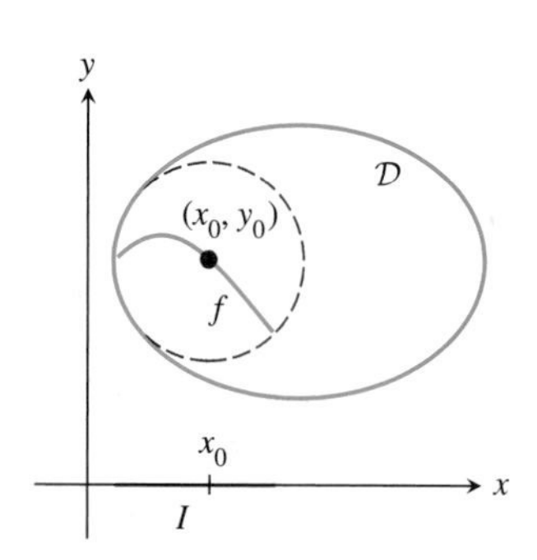

FIGURE 10.24 A function implicitly defined on an interval I.

Calculating f' When f Is Defined Implicitly by $F(x,y) = 0$

Let F be a differentiable real-valued function of two variables defined on a domain $\mathcal{D}$, and let (x_0, y_0) be an interior point of $\mathcal{D}$. Assume that

i) f is a differentiable function on an open interval I containing x_0,
ii) $f(x_0) = y_0$, and
iii) $y = f(x)$ is defined implicitly by the equation $F(x, y) = 0$, that is, $F(x, f(x)) = 0$, $x \in I$.

Figure 10.24 shows a view of the domain $\mathcal{D}$ of F, the graph of f, and the interval I for a representative function F. The derivative of f is

$$f'(x) = -\frac{F_x(x, y)}{F_y(x, y)}, \qquad x \in I, \tag{1}$$

provided that $F_y(x, y) \neq 0$.

EXAMPLE 2 The equation $F(x, y) = x^2 + y^2 - 1 = 0$ defines the differentiable function

$$y = f(x) = \sqrt{1 - x^2}$$

implicitly near $(x_0, y_0) = \left(1/\sqrt{2}, 1/\sqrt{2}\right)$. Calculate $f'(x)$ near x_0 using (1). Also calculate $f'(x)$ by solving the equation $F(x, y) = 0$ explicitly for f and differentiating the result.

Solution

Because F is a polynomial function, it is differentiable on R^2. Hence, (x_0, y_0) is an interior point. To apply (1), we calculate

$$F_x = 2x \quad \text{and} \quad F_y = 2y.$$

At points (x, y) near (x_0, y_0) and at which F_y is not zero,

$$f'(x) = -\frac{F_x(x, y)}{F_y(x, y)} = -\frac{x}{y}. \tag{2}$$

The open interval I containing $x_0 = 1/\sqrt{2}$ must exclude -1 and 1 because if $x = \pm 1$, then $y = 0$ and $F_y(x, y) = 2y = 0$. We may take $I = (-1, 1)$.

Because the function F in this example is relatively simple, we may explicitly solve the equation $F(x, y) = 0$ for y in terms of x, obtaining

$$y = f(x) = \sqrt{1 - x^2}.$$

This function is defined and differentiable for all x near $x_0 = 1/\sqrt{2}$. Figure 10.25 shows the graph of f. The solution $y = -\sqrt{1 - x^2}$, whose graph is the lower half of the unit circle, was excluded because $y_0 = 1/\sqrt{2} \neq -\sqrt{1 - x_0^2}$. Differentiating f near x_0,

$$f'(x) = \frac{1}{2\sqrt{1 - x^2}} \cdot (-2x) = -\frac{x}{y}.$$

This agrees with (2), found by applying formula (1) for calculating the derivative of implicitly defined functions.

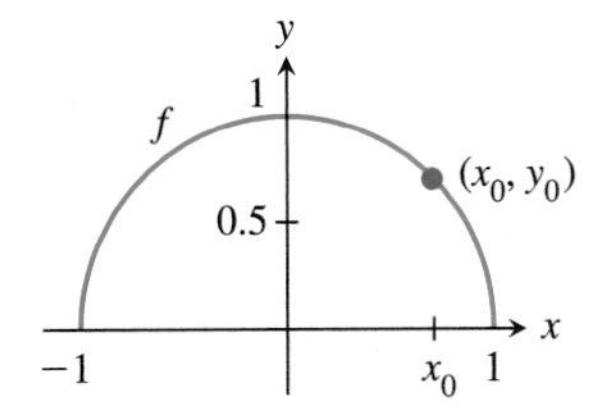

FIGURE 10.25 The graph of the implicitly defined function f.

EXAMPLE 3 Assume that the equation

$$\sin(x - y) + x^2 y = 0$$

defines a differentiable function $y = f(x)$ defined on an interval I containing $x = 0$. Calculate $f'(0)$.

Solution

Letting $F(x, y) = \sin(x - y) + x^2 y$ and $(x_0, y_0) = (0, 0)$, first note that F is differentiable because its first-order partials F_x and F_y are continuous at all points of R^2. Calculating these partials and evaluating them at $(0, 0)$,

$$F_x = \cos(x - y) + 2xy, \qquad F_x(0, 0) = 1,$$
$$F_y = -\cos(x - y) + x^2, \qquad F_y(0, 0) = -1.$$

From (1), we have

$$f'(0) = -\frac{F_x(0,0)}{F_y(0,0)} = -\frac{1}{-1} = 1.$$

Exercises 42 and 43 take up questions concerning the graph of f and the interval I.

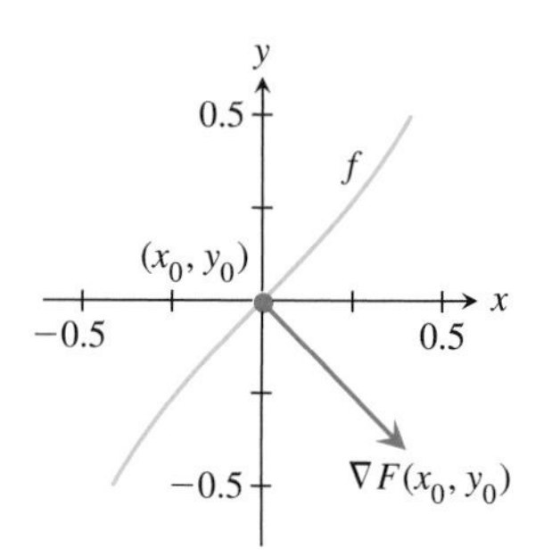

FIGURE 10.26 Graph of f near (x_0, y_0).

The calculations in Example 3 of the derivative of the function f implicitly defined by the equation $F(x, y) = 0$ are related to our work with gradient vectors earlier in this section. Because the graph of f is a subset of the level curve described by $F(x, y) = 0$, we know that $\nabla F(0, 0)$ is perpendicular to the graph of the function f at the point $(0, 0)$. Hence, we can calculate the slope of the graph of f at (x_0, y_0). Thus from $\nabla F(0, 0) = \langle 1, -1 \rangle$ we can find $f'(0) = -(1)/(-1) = 1$. Figure 10.26 shows the graph of f on the interval $(-0.5, 0.5)$ and the gradient vector (not to scale) at $(0, 0)$.

Functions Defined by an Equation $F(x,y,z) = 0$ The methods used to calculate $f'(x)$ for a function defined implicitly by an equation $F(x, y) = 0$ can be extended to the calculation of the partial derivatives of a function $z = f(x, y)$ implicitly defined by an equation of the form $F(x, y, z) = 0$.

Calculating f_x and f_y When f Is Defined Implicitly by $F(x,y,z) = 0$

Let F be a differentiable real-valued function of three variables defined on a domain $\mathcal{D}$, and let (x_0, y_0, z_0) be an interior point of $\mathcal{D}$. Assume that

i) f is a differentiable function defined on a rectangle I containing (x_0, y_0),
ii) $f(x_0, y_0) = z_0$, and
iii) $z = f(x, y)$ is defined implicitly by the equation $F(x, y, z) = 0$, that is, $F(x, y, f(x, y)) = 0$ for $(x, y) \in I$.

Figure 10.27 shows the graph of f and the rectangle I for a representative case. The domain $\mathcal{D}$ of F is not shown. For points $(x, y) \in I$, the partial derivatives of f are given by

$$f_x(x, y) = -\frac{F_x(x, y, z)}{F_z(x, y, z)}$$

$$f_y(x, y) = -\frac{F_y(x, y, z)}{F_z(x, y, z)}, \tag{3}$$

provided that $F_z(x, y, z) \neq 0$. At any point (x, y, z) on the graph of f, the gradient vector $\nabla F(x, y, z)$ is a normal vector to this graph. Near (x_0, y_0, z_0) the graph of f and the level surface described by $F(x, y, z) = 0$ are coincident.

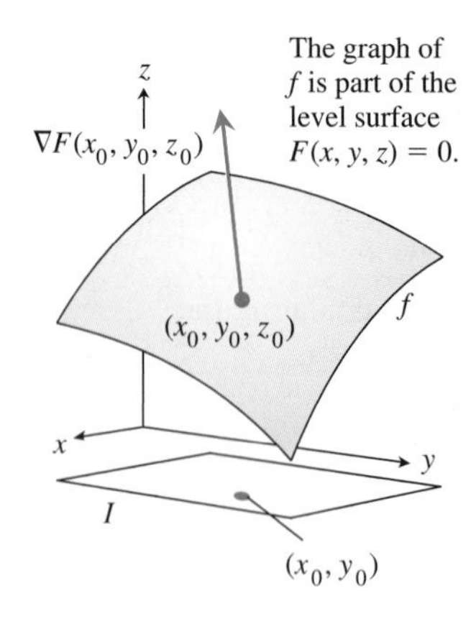

FIGURE 10.27 Graph of f near (x_0, y_0) and the gradient vector $\nabla F(x_0, y_0, z_0)$.

EXAMPLE 4 Assume that the equation $F(x, y, z) = 0$, where

$$F(x, y, z) = x^2 + yz - \tfrac{1}{2}\sqrt{3}\,(y^2 - z^2) - 1, \tag{4}$$

defines a differentiable function $z = f(x, y)$ near the point $(x_0, y_0, z_0) = (0, 1, 1)$. Calculate $f_x(0, 1)$, $f_y(0, 1)$, a unit normal to the graph of f at $(0, 1, 1)$, and an equation of the tangent plane at this point.

Solution

Because F is a polynomial, it is differentiable on R^3. The point $(0, 1, 1)$ is an interior point of R^3; because $F(0, 1, 1) = 0$, the point $(0, 1, 1)$ is on the level surface defined by the equation $F(x, y, z) = 0$; and, given that f is implicitly defined by this equation, $1 = f(0, 1)$. We apply (3) to calculate $f_x(0, 1)$ and $f_y(0, 1)$. For this purpose we first calculate

$$\begin{aligned} F_x(x, y, z) &= 2x, & F_x(0, 1, 1) &= 0 \\ F_y(x, y, z) &= z - \sqrt{3}y, & F_y(0, 1, 1) &= 1 - \sqrt{3} \\ F_z(x, y, z) &= y + \sqrt{3}z, & F_z(0, 1, 1) &= 1 + \sqrt{3}. \end{aligned}$$

Because $F_z(0, 1, 1) \neq 0$,

$$\begin{aligned} f_x(0, 1) &= -\frac{F_x(0, 1, 1)}{F_z(0, 1, 1)} = 0 \\ f_y(0, 1) &= -\frac{F_y(0, 1, 1)}{F_z(0, 1, 1)} = -\frac{1 - \sqrt{3}}{1 + \sqrt{3}}. \end{aligned}$$

We have a choice of methods for determining a normal and an equation for the tangent plane. We know from Section 10.1 that a normal at a point $(0, 1, 1)$ on the graph of f is the vector

$$\langle -f_x(0, 1), -f_y(0, 1), 1\rangle$$

and an equation of the tangent plane is

$$z = f(0, 1) + f_x(0, 1)(x - 0) + f_y(0, 1)(y - 1).$$

From these formulas and the values of $f_x(0, 1)$ and $f_y(0, 1)$ calculated earlier, we can easily find a normal and an equation for the tangent plane to the graph of f at $(0, 1, 1)$.

Or we may recall the statement about the gradient vector $\nabla F(x, y, z)$, just following Equation (3). Denoting the normal $\nabla F(0, 1, 1)$ by $\mathbf{n}$,

$$\mathbf{n} = \langle F_x(0, 1, 1), F_y(0, 1, 1), F_z(0, 1, 1)\rangle = \left\langle 0, 1 - \sqrt{3}, 1 + \sqrt{3}\right\rangle.$$

Dividing this vector by its length $\|\mathbf{n}\| = 2\sqrt{2}$, we obtain the unit normal

$$\mathbf{N} = \left\langle 0, \frac{1 - \sqrt{3}}{2\sqrt{2}}, \frac{1 + \sqrt{3}}{2\sqrt{2}}\right\rangle \approx \langle 0, -0.26, 0.97\rangle.$$

An equation of the tangent plane at $(0, 1, 1)$ is (approximately)

$$-0.26(y - 1) + 0.97(z - 1) = 0.$$

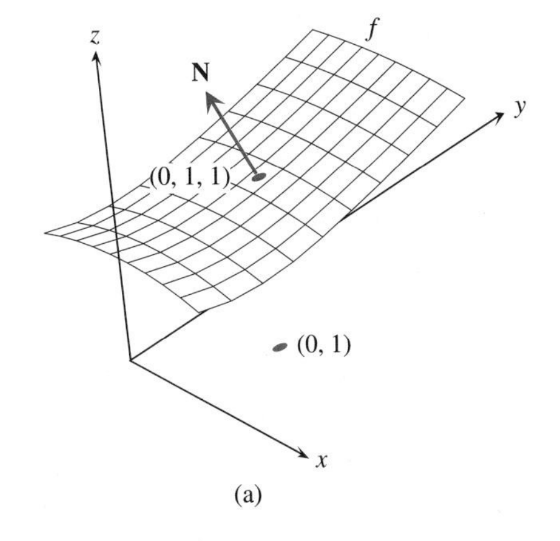

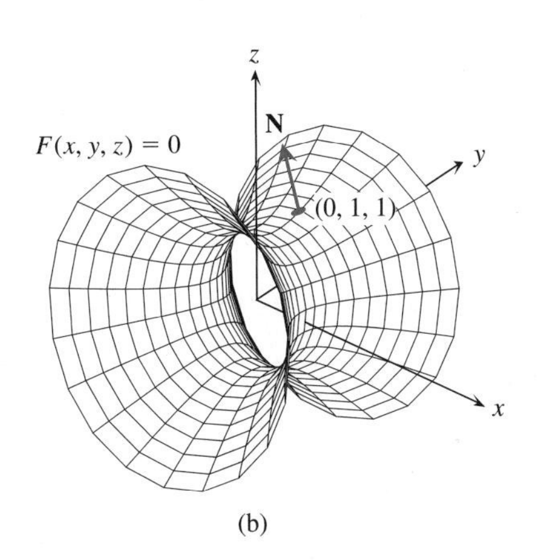

FIGURE 10.28 (a) Unit normal to graph of f at (x_0, y_0, z_0); (b) unit normal to the level surface defined by $F(x, y, z) = 0$; the graph of f in (a) is part of the level surface shown in (b).

Graphing a function $z = f(x, y)$ implicitly defined by an equation $F(x, y, z) = 0$ or graphing the level surface defined by $F(x, y, z) = 0$ can be challenging. Because it happens that the equation $F(x, y, z) = 0$ we considered in Example 4 can be solved explicitly for $z = f(x, y)$, an ordinary "Plot" command can be used. Figure 10.28(a) shows the graph of f. The function f was graphed on the rectangle $\left[-\frac{1}{2}, \frac{1}{2}\right] \times [0, 2]$. Included in this graph are the point $(0, 1)$ and the unit normal $\mathbf{N}$. See Exercise 46.

Figure 10.28(b) shows the level surface described by $F(x, y, z) = 0$. See Exercise 47 for a parametric equation describing this surface. Note that the graph of f does not—and cannot—include all of the graph of the level surface.

The equation $F(x, y, z) = 0$ in Example 4 can be solved for f explicitly (see Exercise 46) by applying the quadratic formula. In the next example, we consider an equation requiring the solution of a cubic equation. Cardan's formula, used for solving cubics, is more complex than the quadratic formula. If you have access to a CAS, try solving for V in terms of P and T. The result is somewhat daunting. It is in fact true that most equations $F(x, y, z) = 0$ cannot be solved algebraically for z in terms of x and y. For such equations we must use (3) to find z_x or z_y.

EXAMPLE 5 For more accurate modeling of the pressure, volume, and temperature of n moles (mol) of a gas, the ideal gas law $PV = nRT$ is replaced by the van der Waals equation

$$\left(P + \frac{n^2a}{V^2}\right)(V - nb) = nRT,$$

where a and b are constants depending upon the specific gas. The constant R occurring in both equations is the universal gas constant $R = 8.314 \text{ J/mol} \cdot \text{K}$. Recall that in SI units, pressure is measured in pascals (where 1 Pa is a pressure of 1 N/m^2), volume V is measured in cubic meters, and temperature T is measured in kelvins (K).

For a 100-mol sample of nitrogen gas ($a = 0.14$ and $b = 3.91 \times 10^{-5}$) at temperature $T_0 = 500$ K (226.85° C) and pressure $P_0 = 5 \times 10^5$ Pa ($\approx$5 atmospheres), what are the volume V_0 of the sample and the rate of change of volume with temperature?

Solution

The two questions suggest that we regard V as a function of T and P. Although we choose not to solve the van der Waals equation for V as an explicit function of P and T, we may solve it numerically for the given values of T_0 and P_0. Substituting $T_0 = 500$ and $P_0 = 5 \times 10^5$ into the van der Waals equation and solving numerically for V, we obtain $V_0 \approx 0.83 \text{ m}^3$. See Exercise 49 for a discussion of this calculation.

To calculate V_T, set

$$F(P, T, V) = \left(P + \frac{n^2a}{V^2}\right)(V - nb) - nRT. \tag{5}$$

Regarding the equation $F(P, T, V) = 0$ as implicitly defining V as a function of P and T and applying (3),

$$V_T = -\frac{F_T}{F_V}.$$

From (5),

$$F_T = -nR$$

$$F_V = \left(P + \frac{n^2a}{V^2}\right) \cdot 1 + (-2n^2aV^{-3})(V - nb).$$

Evaluating each of these partial derivatives at $(P_0, T_0, V_0) = (5 \times 10^5, 500, 0.83)$,

$$F_T \approx -8.3 \times 10^2$$

$$F_V \approx 5.0 \times 10^5.$$

At the given temperature and pressure, the rate of change of volume with respect to temperature is

$$V_T(P_0, V_0) = -\frac{F_T(P_0, T_0, V_0)}{F_V(P_0, T_0, V_0)} \approx 1.7 \times 10^{-3}\ \text{m}^3/\text{K}.$$

Partial Derivatives of Higher Order

In the remaining two applications of the chain rule we will need to calculate, for a given function $z = f(x, y)$, the partial derivatives $(f_x)_x$, $(f_x)_y$, $(f_y)_x$, and $(f_y)_y$ of the first partials $f_x(x, y)$ and $f_y(x, y)$. The standard notations for these "second-order" partial derivatives are similar to the notations y'' and d^2y/dx^2 for the second derivative of a function $y = f(x)$ of a single variable. We write

$$(f_x)_x = f_{xx} = \frac{\partial}{\partial x}\left(\frac{\partial f}{\partial x}\right) = \frac{\partial^2 f}{\partial x^2} = \frac{\partial^2 z}{\partial x^2}$$

$$(f_x)_y = f_{xy} = \frac{\partial}{\partial y}\left(\frac{\partial f}{\partial x}\right) = \frac{\partial^2 f}{\partial y \partial x} = \frac{\partial^2 z}{\partial y \partial x}$$

$$(f_y)_x = f_{yx} = \frac{\partial}{\partial x}\left(\frac{\partial f}{\partial y}\right) = \frac{\partial^2 f}{\partial x \partial y} = \frac{\partial^2 z}{\partial x \partial y}$$

$$(f_y)_y = f_{yy} = \frac{\partial}{\partial y}\left(\frac{\partial f}{\partial y}\right) = \frac{\partial^2 f}{\partial y^2} = \frac{\partial^2 z}{\partial y^2}.$$

Notations for third- and higher-order partials are defined in a similar way.

EXAMPLE 6 Find the second-order partials of

$$f(x, y) = x^2 e^y + y \cos x.$$

Solution

After finding the partials f_x and f_y, we differentiate each of these functions with respect to each of x and y. This will give f_{xx}, f_{xy}, f_{yx}, and f_{yy}.

$$f_x = 2xe^y - y \sin x$$

$$f_y = x^2 e^y + \cos x$$

$$f_{xx} = 2e^y - y \cos x$$

$$f_{xy} = 2xe^y - \sin x$$

$$f_{yx} = 2xe^y - \sin x$$

$$f_{yy} = x^2 e^y$$

We note that for this function, whose second-order partials are continuous, the "mixed partials" f_{xy} and f_{yx} are equal. This result holds generally for functions with continuous second-order partial derivatives. Applications in which we must calculate such higher-order derivatives can be simplified by using the following theorem.

THEOREM Equality of Mixed Partial Derivatives

If $z = f(x, y)$ is defined and has continuous second-order partial derivatives throughout a domain $\mathcal{D}$, then the two functions f_{xy} and f_{yx} are identical in $\mathcal{D}$.

If the third-order partial derivatives of f are also continuous throughout $\mathcal{D}$, then the functions f_{xxy}, f_{xyx}, and f_{yxx} are identical in $\mathcal{D}$ and the functions f_{xyy}, f_{yxy}, and f_{yyx} are identical in $\mathcal{D}$.

A Taylor Inequality for Functions of Two Variables

We begin by recalling from Section 7.2 a special case of the Taylor Inequality for Functions of One Variable. If g is a function defined on the interval $[-1, 1]$ and the third derivative of g is continuous on this interval, then

$$\left| g(t) - \left(g(0) + g'(0)t + \frac{1}{2!}g''(0)t^2 \right) \right| \leq \frac{1}{3!}M_3, \qquad t \in [-1, 1], \tag{6}$$

where M_3 is the maximum value of the function $|g'''|$ on $[-1, 1]$.

We use this result to derive a special case of the Taylor Inequality for Functions of Two Variables, which is needed in Section 10.5.

Let f be a real-valued function of two variables, defined on $\mathcal{D}$ and having continuous partial derivatives through those of third order. Referring to Fig. 10.29, let $\mathbf{a}$ be the center point of a square region S_b lying wholly within $\mathcal{D}$. Let $\mathbf{a} + \mathbf{h} = \langle a_1 + h_1, a_2 + h_2 \rangle$ be any point in S_b and define the function

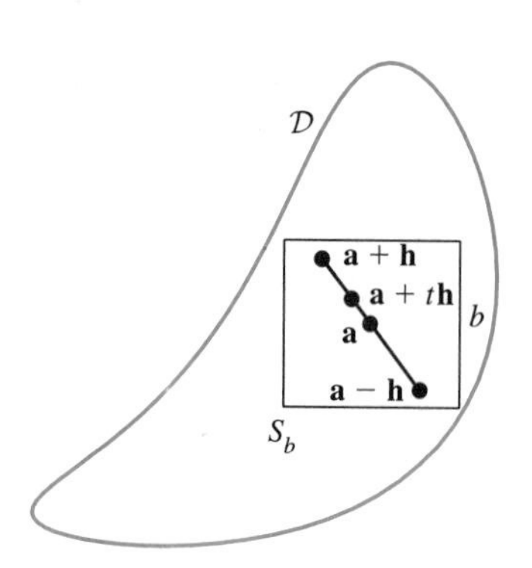

FIGURE 10.29 Domain of f and the square S_b.

$$g(t) = f(\mathbf{a} + t\mathbf{h}), \qquad -1 \leq t \leq 1. \tag{7}$$

We obtain the two-dimensional version of the Taylor Inequality for f by applying the one-dimensional version to g. For this we need the numbers $g'(0)$, $g''(0)$, and $g'''(0)$, which we calculate with the help of the chain rule. The pattern used throughout the calculation is this: If $z = w(x, y)$ is a function of x and y, where $x = a_1 + th_1$ and $y = a_2 + th_2$ are the coordinates of $\mathbf{a} + t\mathbf{h}$, then

$$\frac{dz}{dt} = w_x\frac{dx}{dt} + w_y\frac{dy}{dt} = w_x h_1 + w_y h_2.$$

We set the function w equal to f, f_x, f_y, f_{xx}, f_{xy}, or f_{yy} at different steps in the calculation. Calculating g' first,

$$g'(t) = f_x(\mathbf{a} + t\mathbf{h})h_1 + f_y(\mathbf{a} + t\mathbf{h})h_2. \tag{8}$$

Evaluating this expression at $t = 0$,

$$g'(0) = f_x(\mathbf{a})h_1 + f_y(\mathbf{a})h_2. \tag{9}$$

Next we calculate $g''(t)$ from (8). Temporarily abbreviating $f_{xx}(\mathbf{a} + t\mathbf{h})$ to f_{xx}, $f_{xy}(\mathbf{a} + t\mathbf{h})$ to f_{xy}, and so on,

$$\begin{aligned} g''(t) &= \frac{d}{dt}g'(t) = \frac{d}{dt}(f_x(\mathbf{a} + t\mathbf{h})h_1) + \frac{d}{dt}(f_y(\mathbf{a} + t\mathbf{h})h_2) \\ &= (f_{xx}h_1 + f_{xy}h_2)h_1 + (f_{yx}h_1 + f_{yy}h_2)h_2. \end{aligned}$$

Hence,

$$g''(t) = f_{xx}h_1^2 + f_{xy}h_2h_1 + f_{yx}h_1h_2 + f_{yy}h_2^2.$$

Because f has continuous second-order partial derivatives, we may combine the terms containing f_{xy} or f_{yx}. This gives

$$g''(t) = f_{xx}h_1^2 + 2f_{xy}h_1h_2 + f_{yy}h_2^2. \tag{10}$$

Evaluating this expression at $t = 0$,

$$g''(0) = f_{xx}(\mathbf{a})h_1^2 + 2f_{xy}(\mathbf{a})h_1h_2 + f_{yy}(\mathbf{a})h_2^2. \tag{11}$$

Calculating $g'''(t)$ from (10) in the same way,

$$\begin{aligned} g'''(t) = f_{xxx}h_1^3 &+ f_{xxy}h_1^2h_2 + 2f_{xyx}h_1^2h_2 \\ &+ 2f_{xyy}h_1h_2^2 + f_{yyx}h_1h_2^2 + f_{yyy}h_2^3. \end{aligned} \tag{12}$$

The Taylor Inequality now follows by setting $t = 1$ in (6) and replacing $g(0)$, $g'(0)$, and $g''(0)$ from (7), (9), and (11). Equation (12) is used in finding the error term $Mb^3/3!$ in (13).

THEOREM A Taylor Inequality for Functions of Two Variables

Assume that f is a real-valued function of two variables, defined on a region $\mathcal{D}$ and having continuous partial derivatives through those of third order. Assume that $\mathbf{a}$ is the center point of a square region S_b of side b and lying wholly within $\mathcal{D}$. See Fig. 10.29. For all $\mathbf{a} + \mathbf{h}$ in S_b,

$$\left| f(\mathbf{a}+\mathbf{h}) - \left(f(\mathbf{a}) + f_x(\mathbf{a})h_1 + f_y(\mathbf{a})h_2 + \frac{1}{2!}\left(f_{xx}(\mathbf{a})h_1^2 + 2f_{xy}(\mathbf{a})h_1h_2 + f_{yy}(\mathbf{a})h_2^2\right) \right) \right| \le \frac{1}{3!}Mb^3, \tag{13}$$

where $\mathbf{h} = \langle h_1, h_2 \rangle$ and M is the largest of the maxima of the four continuous functions

$$|f_{xxx}(x,y)|, \quad |f_{xxy}(x,y)|, \quad |f_{xyy}(x,y)|, \quad |f_{yyy}(x,y)|, \quad (x,y) \in S_b.$$

The remainder, or error term, $Mb^3/3!$ is discussed in Exercise 44. The polynomial (in the variables h_1 and h_2) inside the large parentheses in (13) is called the *second-order Taylor polynomial* (for functions of two variables) for f at $\mathbf{a}$.

We give an example to illustrate this Taylor Inequality.

EXAMPLE 7 Approximate $\sqrt{x/y}$ near $(1, 1)$ with the Taylor polynomial of second order. How good is this approximation?

Solution

We take $f(x, y) = \sqrt{x/y}$, $\mathbf{a} = (1, 1)$, and $\mathbf{h} = (x - 1, y - 1)$. If we assume that "near $(1, 1)$" means $0 \le |x - 1|, |y - 1| \le 0.1$, then we may take $b = 0.2$ in (13). See Fig. 10.30. For the Taylor polynomial we evaluate f, f_x,

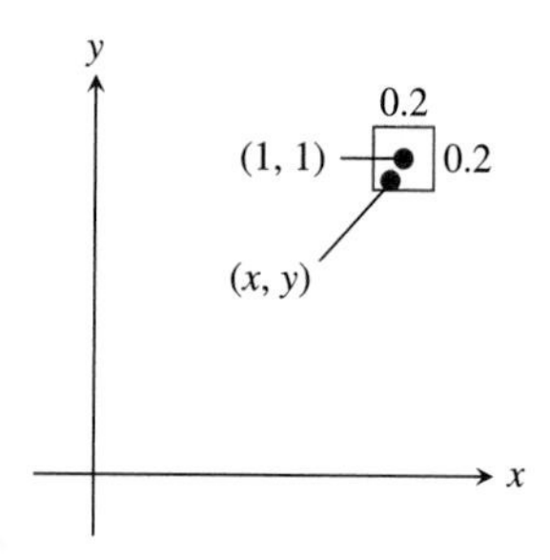

FIGURE 10.30 A representative point (x, y) for which $0 \le |x - 1|, |y - 1| \le 0.1$.

f_y, f_{xx}, f_{xy}, and f_{yy} at $\mathbf{a} = (1, 1)$.

$$f(x,y) = x^{1/2}y^{-1/2} \qquad f(1,1) = 1$$

$$f_x(x,y) = \frac{1}{2}x^{-1/2}y^{-1/2} \qquad f_x(1,1) = \frac{1}{2}$$

$$f_y(x,y) = -\frac{1}{2}x^{1/2}y^{-3/2} \qquad f_y(1,1) = -\frac{1}{2}$$

$$f_{xx}(x,y) = -\frac{1}{4}x^{-3/2}y^{-1/2} \qquad f_{xx}(1,1) = -\frac{1}{4}$$

$$f_{xy}(x,y) = -\frac{1}{4}x^{-1/2}y^{-3/2} \qquad f_{xy}(1,1) = -\frac{1}{4}$$

$$f_{yy}(x,y) = \frac{3}{4}x^{1/2}y^{-5/2} \qquad f_{yy}(1,1) = \frac{3}{4}$$

Substituting these results into (13),

$$\left|\sqrt{\frac{x}{y}} - P(x,y)\right| \le \frac{1}{3!}Mb^3, \tag{14}$$

where $P(x, y)$ is the second-order Taylor polynomial

$$P(x,y) = 1 + \frac{1}{2}(x-1) - \frac{1}{2}(y-1) + \frac{1}{2!}\left(-\frac{1}{4}(x-1)^2 + 2\left(-\frac{1}{4}\right)(x-1)(y-1) + \frac{3}{4}(y-1)^2\right).$$

To find how well $P(x, y)$ fits $f(x, y)$ near $(1, 1)$, we consider the error function

$$E(x,y) = \left|\sqrt{\frac{x}{y}} - P(x,y)\right|, \qquad 0 \le |x-1|, |y-1| \le 0.1. \tag{15}$$

For a graphical look at the error, we may sketch some level curves of the function E. Or we can use (14) after calculating M. This usually overestimates the error.

Figure 10.31 shows some level curves of the error function $E(x, y)$. Evidently, the error is greatest near $(1.1, 0.9)$, where the 0.0005 contour is shown. Near $(1, 1)$, in the center, the error is less than or equal to 0.00001.

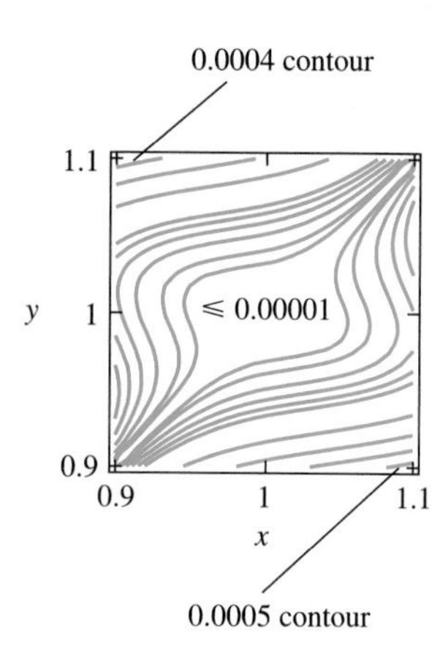

FIGURE 10.31 Level curves of the error function E.

Determining M, from which we can bound the error, can be a lot of work. We reduce the work by overestimating the maximum values of the third derivatives. Here are the calculations. For $0.9 \le x, y \le 1.1$,

$$|f_{xxx}(x,y)| = \left|\frac{3}{8x^{5/2}y^{1/2}}\right| \le \frac{3}{8(0.9)^3} \le 0.6$$

$$|f_{xxy}(x,y)| = \left|\frac{1}{8x^{3/2}y^{3/2}}\right| \le \frac{1}{8(0.9)^3} \le 0.6$$

$$|f_{xyy}(x,y)| = \left|\frac{3}{8x^{1/2}y^{5/2}}\right| \le \frac{3}{8(0.9)^3} \le 0.6$$

$$|f_{yyy}(x,y)| = \left|\frac{-15x^{1/2}}{8y^{7/2}}\right| \le \frac{15(1.1)^{1/2}}{8(0.9)^{7/2}} \le 3.$$

What ideas guided the overestimation shown on the right side of these inequalities? Wanting to overestimate $|f_{xxx}|$ and recalling that $0.9 \le x, y \le 1.1$, we argue that $|f_{xxx}(x, y)|$ is largest when x and y are as small as possible. We therefore set $x = y = 0.9$, which makes it certain that

$$|f_{xxx}(x,y)| \le \frac{3}{8(0.9)^3} = 0.514\cdots \le 0.6, \qquad 0.9 \le x, y \le 1.1.$$

The calculations for $|f_{xxy}|$ and $|f_{xyy}|$ are similar. For $|f_{yyy}|$ we argue in this case that $|f_{yyy}(x,y)|$ is largest when $x = 1.1$ and $y = 0.9$. Hence

$$|f_{yyy}(x,y)| \le \frac{15 \cdot 1.1^{1/2}}{8(0.9)^{7/2}} = 2.84\cdots \le 3, \qquad 0.9 \le x, y \le 1.1.$$

Thus, $M \le 3$ and $E(x,y) \le 3(0.2)^3/6 = 0.004$. From Fig. 10.31 we see that the actual error is approximately $1/10$ of this estimate.

Exercises 10.3

Exercises 1–6: Verify by direct calculation that the gradient vector $\nabla f(x_0, y_0)$ is perpendicular to the level curve $f(x,y) = k$ at (x_0, y_0). Sketch the level curve near (x_0, y_0) and the gradient vector at (x_0, y_0).

1. $f(x,y) = 2x + 3y + 5$; $k = 0$; $(x_0, y_0) = (5, -5)$

2. $f(x,y) = x - 4y + 6$; $k = 0$; $(x_0, y_0) = (2, 2)$

3. $f(x,y) = y - 2x^2$; $k = 1$; $(x_0, y_0) = (1, 3)$

4. $f(x,y) = y + 3x^2$; $k = 2$; $(x_0, y_0) = (1, -1)$

5. $f(x,y) = x^2 + y^2$; $k = 20$; $(x_0, y_0) = (2, 4)$

6. $f(x,y) = x^2/4 + y^2$; $k = 2$; $(x_0, y_0) = (2, 1)$

Exercises 7–10: The equation $F(x,y,z) = 0$ defines a function $z = f(x,y)$ implicitly near (x_0, y_0, z_0). Calculate the vectors $\langle -f_x(x_0, y_0), -f_y(x_0, y_0), 1\rangle$ and $\nabla F(x_0, y_0, z_0)$. Are these vectors parallel? Give an equation for the tangent plane to the graph of f at the point (x_0, y_0, z_0).

7. $x^2 - 3xz + y^2 + 2yz - z^2 = 0$, $(1, 1, 1)$

8. $3x^2yz - 2y^3x - z^2x = 0$, $(1, 1, 1)$

9. $e^z\cos(x + y) + z^2 - 1/2 = 0$, $(\pi/6, \pi/6, 0)$

10. $y\sin(x + z) - x\sin(y + z) - \frac{1}{2}z = 0$, $(0, \pi/4, \pi/2)$

11. For the level surface S described by $F(x,y,z) = 11$, where $F(x,y,z) = x^2 + y^2 + z^2$ and $P_0 = (3, 1, 1)$ is a point on S, show by direct calculation that the gradient vector $\nabla F(3, 1, 1)$ is perpendicular to both the parallel of latitude through P_0 and the meridian through this point. Sketch the first octant of the surface S, the parallel of latitude and the meridian through P_0, and plot the gradient vector at P_0.

12. For the level surface S described by $F(x,y,z) = 9$, where $F(x,y,z) = x^2 + y^2 + z^2$ and $P_0 = (2, 2, 1)$ is a point on S, show by direct calculation that the gradient vector $\nabla F(2, 2, 1)$ is perpendicular to both the parallel of latitude through P_0 and the meridian through this point. Sketch the first octant of the surface S, the parallel of latitude and the meridian through P_0, and plot the gradient vector at P_0.

13. Interpret the surface S described by the equation $x^2 - y^2 = z$ as a level surface for some function $F(x,y,z)$, and use the value of the gradient of F at the point $(2, 1, 3)$ to write the equation of the tangent plane to S at this point.

14. Interpret the surface S described by the equation $xy = z$ as a level surface for some function $F(x,y,z)$, and use the value of the gradient of F at the point $(1, 1, 1)$ to write the equation of the tangent plane to S at this point.

Exercises 15–18: The equation defines a function $y = f(x)$ implicitly near (x_0, y_0). Calculate $f'(x_0)$.

15. $2x^2 + xy - 3y^2 = 0$, $(x_0, y_0) = (1, 1)$

16. $x^2y - 2x + y^3 = 0$, $(x_0, y_0) = (1, -1)$

17. $\arctan(x/y) - y + 1 = 0$, $(x_0, y_0) = (0, 1)$

18. $\ln(x^2 + y) + y^2(1 - x) - 1 = 0$, $(x_0, y_0) = (0, 1)$

Exercise 19–22: The equation defines a function $z = f(x,y)$ implicitly near (x_0, y_0, z_0). Calculate $f_x(x_0, y_0)$, $f_y(x_0, y_0)$, and a normal to the graph of $z = f(x,y)$ at the point (x_0, y_0, z_0).

19. $\sin(xy) + \sin(xz) - z = \sqrt{3} - 1$, $(x_0, y_0, z_0) = (\pi/3, 1, 1)$

20. $xyz - z^4y + 2x^2 = 2$, $(x_0, y_0, z_0) = (1, 1, 1)$

21. $\ln(x^2y + z) + \arctan(z/y) = \pi/4$, $(x_0, y_0, z_0) = (0, 1, 1)$

22. $\dfrac{x}{y^2 + z^2} + \dfrac{y}{x^2 + z^2} = z$, $(x_0, y_0, z_0) = (1, 1, 1)$

Exercises 23–28: Calculate the partial derivative.

23. f_{xx}, where $f(x, y) = x^3y^2 - xy^3$

24. f_{xx}, where $f(x, y) = 5(xy)^3 + x/y$

25. f_{yy}, where $f(x, y) = x^3e^{xy}$

26. f_{yy}, where $f(x, y) = xe^y + y^2$

27. f_{xxy}, where $f(x, y) = x\sin(xy)$

28. f_{yxy}, where $f(x, y) = e^{x/y}$

Exercises 29–32: For second-order partials, verify the hypotheses and the conclusion of the Equality of Mixed Partial Derivatives Theorem.

29. $f(x, y) = x^3y^2 - xy^3$

30. $f(x, y) = 5(xy)^3 + x/y$

31. $f(x, y) = x^3e^{xy}$

32. $f(x, y) = xe^y + y^2$

*Exercises 33–36: Calculate the Taylor polynomial of second order that approximates $f(x, y)$ near **a**.*

33. $f(x, y) = \sqrt{xy}$, $\mathbf{a} = \langle 1, 1\rangle$

34. $f(x, y) = \sqrt[3]{xy}$, $\mathbf{a} = \langle 1, 1\rangle$

35. $f(x, y) = e^x\cos y$, $\mathbf{a} = (0, 0)$

36. $f(x, y) = x^y$, $\mathbf{a} = (1, 0)$

T **37.** Use the error term in the Taylor Inequality to find how well the Taylor polynomial in Exercise 33 approximates $f(x, y) = \sqrt{xy}$ when (x, y) is within 0.25 of $(1, 1)$. Or, if a CAS is available, you may prefer to plot level curves for $E(x, y)$. See Fig. 10.31.

T **38.** Use the error term in the Taylor Inequality to find how well the Taylor polynomial in Exercise 34 approximates $f(x, y) = \sqrt[3]{xy}$ when (x, y) is within 0.25 of $(1, 1)$. Or, if a CAS is available, you may prefer to plot level curves for $E(x, y)$. See Fig. 10.31.

39. Verify the details leading to (14).

40. The equation $f(x, y) = 0$ defines y as a function of x. Express y'' in terms of x, y, and the first and second partial derivatives of f.

41. Let $z = f(u, v)$ and assume that the equation $f(x/y, y/x) = 0$ defines y as a function of x. Express y' in terms of x, y, f_u and f_v.

42. For each fixed $x \in (-0.5, 0.5)$, a value of y satisfying the equation $\sin(x - y) + x^2y = 0$ can be calculated. The data for the graph in Fig. 10.26 were generated in this way, choosing sufficiently many values of x that the graph looks smooth. Explain how this can be done. Use your scheme to verify that $(0.25, 0.266\ldots)$ and $(0.4, 0.476\ldots)$ are points of the graph. If a goal is to graph on $(-0.5, 0.5)$, is it necessary to calculate points for both $x > 0$ and $x < 0$? Why or why not?

43. Figure 10.26 suggests that the interval I mentioned in Example 3 cannot be much larger than $(-1/2, 1/2)$. Attempt to limit I by determining a point (x_1, y_1) such that $F_y(x_1, y_1) = 0$. Show that $(0.703365, 1.756646)$ is such a point by solving the two equations $F_y(x_1, y_1) = 0$ and $F(x_1, y_1) = 0$ simultaneously. Why does this work?

44. Use the triangle inequality in verifying the error term $Mb^3/3!$ in the Taylor Inequality (see Appendix). Start by noting that the left side of (6) is less than or equal to $M_3/3!$, where M_3 is the maximum of $|g'''(t)|$ for $t \in [-1, 1]$. Next, apply the triangle inequality to the expression (12) for $g'''(t)$, noting that $|h_1| \le b/2$ and $|h_2| \le b/2$.

45. Go through the calculations leading from (10) to (12).

T **46.** In Example 4 find f explicitly by solving the equation

$$x^2 + yz - \tfrac{1}{2}\sqrt{3}\,(y^2 - z^2) - 1 = 0$$

for z in terms of x and y. Calculate $f_x(0, 1)$ and $f_y(0, 1)$. If you have convenient access to CAS or graphing calculator, graph f on $[-1/2, 1/2] \times [0, 2]$.

47. Show that the surface described parametrically by

$$\mathbf{r}(w, \theta) = \left\langle \sqrt{1 + w^2}\sin\theta,\ \frac{(\sqrt{3} + 1)w + (\sqrt{3} - 1)\sqrt{1 + w^2}\cos\theta}{2\sqrt{2}},\ \frac{(1 - \sqrt{3})w + (1 + \sqrt{3})\sqrt{1 + w^2}\cos\theta}{2\sqrt{2}} \right\rangle,$$

where $w \in R$ and $0 \le \theta \le 2\pi$, is a subset of the level surface described by $F(x, y, z) = 0$ in Example 4. Figure 10.28 was graphed with $-2 \le w \le 2$.

48. Continuation Show further that this parametric equation can be generated as follows. The graph of the equation $x^2 + z^2 = 1 + y^2$ is a "hyperboloid of one sheet" and can be described parametrically by

$$\langle x, y, z\rangle = \left\langle \sqrt{1 + w^2}\sin\theta, w, \sqrt{1 + w^2}\cos\theta \right\rangle.$$

Now introduce new coordinates by rotating by $-15°$ about the x-axis.

T **49.** In Example 5, the value of V corresponding to $T_0 = 500$ and $P_0 = 5 \times 10^5$ can be obtained in several ways. We explore a solution based on forming the cubic polynomial $V^2F(P, T, V) = 0$ and setting $T = T_0$ and $P = P_0$ after simplification. You should obtain

$$500000V^3 - 417655V^2 + 1400V - 5.474 = 0.$$

(This result was copied directly from a calculator screen. If you do the simplification by hand, the result may be slightly different, either in value or format.) Show that this cubic equation has two complex roots and the real root $V \approx 0.83196$.

50. If you have access to a CAS that will solve cubic equations symbolically, solve the van der Waals equation for V in terms of P and T.

51. Let

$$f(x, y) = \begin{cases} 0, & (x, y) = (0, 0) \\ \dfrac{xy(x^2 - y^2)}{x^2 + y^2}, & (x, y) \neq (0, 0). \end{cases}$$

Show that $f_x(0, 0) = f_y(0, 0) = 0$ and, for $(x, y) \neq (0, 0)$,

$$f_x(x, y) = \frac{y(x^4 + 4x^2y^2 - y^4)}{(x^2 + y^2)^2}$$

and

$$f_y(x, y) = \frac{x(x^4 - 4x^2y^2 - y^4)}{(x^2 + y^2)^2}.$$

By using the definition of partial derivatives, show that $f_{xy}(0, 0) = -1$ and $f_{yx} = 1$. Are all of the second-order partial derivatives of f continuous at $(0, 0)$? Why?

10.4 Further Applications of the Chain Rule

We discuss two equations from mathematical physics, the wave equation and the heat equation. Each of these equations is a *partial differential equation,* which is an equation involving the partial derivatives of an unknown function.

Two-Dimensional Heat Equation

In rectangular coordinates, the two-dimensional heat equation has the form

$$u_{xx} + u_{yy} = 0. \tag{1}$$

At the beginning of the last section, we considered a thin rectangular bar π meters wide and extending indefinitely in the y-direction. The bar is insulated on its top and bottom surfaces (so that there is no heat flow through these surfaces), is held at 0° C on the sides $x = 0$ and $x = \pi$, and is held at 1° C on the end $y = 0$. See Fig. 10.32.

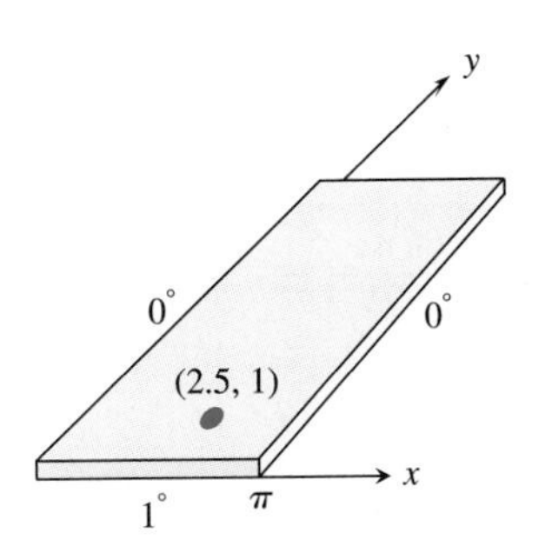

FIGURE 10.32 A thin metal bar subjected to given boundary conditions.

If these *boundary conditions* hold for all $t \geq 0$, the temperature at any point (x, y) in the bar tends over time toward its *steady state;* that is, the temperature no longer changes with time. We denote the steady-state temperature at (x, y) by $u(x, y)$. It is proved in advanced physics courses that the function $u = u(x, y)$ satisfies the heat equation (1).

Because the heat equation (1) is expressed in *rectangular* coordinates and the boundary of the bar can be described by *rectangular* equations (for example, one side of the bar is described by the equation $x = \pi$), standard methods can be used to find the solution

$$u(x, y) = \frac{2}{\pi} \arctan\left(\frac{\sin x}{\sinh y}\right), \qquad 0 \leq x \leq \pi, \qquad y > 0, \tag{2}$$

of this "boundary value problem." While we do not discuss these methods, it is not difficult to show that (2) is a solution. In part, $u(0, y) = 0$ and $u(\pi, y) = 0$ for all $y > 0$. Also, $\lim_{y \to 0^+} u(x, y) = 1$ for all $0 < x < \pi$. We leave as Exercise 26 the calculation that shows that (2) satisfies (1).

INVESTIGATION

The Heat Equation in Polar Coordinates

The "default" coordinate system used to model most physical phenomena is rectangular. If, for example, our job were to study the flow of heat in a two-dimensional region, we would start with the heat equation (1) satisfied by the steady-state temperature $u = u(x, y)$ at any point (x, y) of the region. If the region can be easily described in rectangular coordinates, we would choose a solution procedure based on the fact that the unknown function u satisfies (1). If, however, the region were circular, the solution procedure would be simpler if we were working in polar coordinates. In this case, instead of solving for the function $u = u(x, y)$ satisfying the heat equation (1), we would seek the function $U(r, \theta) = u(r\cos\theta, r\sin\theta)$ satisfying

$$U_{rr} + \frac{1}{r^2}U_{\theta\theta} + \frac{1}{r}U_r = 0. \tag{3}$$

This is the heat equation in polar coordinates. In this section we show how this equation can be derived from (1).

Equation (3) follows from (1) by using the chain rule to express u_{xx} and u_{yy} in terms of r, θ, U, U_r, U_θ, U_{rr}, $U_{r\theta}$, and $U_{\theta\theta}$, and then substituting these expressions into (1). We do these calculations in two stages. First, we express u_x and u_y in terms of r, θ, U, U_r, and U_θ.

For this calculation, instead of thinking about the relation between the functions u and U in the form $U(r, \theta) = u(r\cos\theta, r\sin\theta)$, it is more convenient to consider

$$u(x, y) = U(r, \theta),$$

where r and θ are the functions of x and y implicitly defined by the change-of-variable equations

$$\begin{aligned} x &= r\cos\theta \\ y &= r\sin\theta. \end{aligned} \tag{4}$$

Applying Case 2 of the chain rule to the composition $u(x, y) = U(r, \theta)$,

$$\begin{aligned} u_x &= U_r r_x + U_\theta \theta_x \\ u_y &= U_r r_y + U_\theta \theta_y. \end{aligned} \tag{5}$$

We calculate the partials r_x, r_y, θ_x, and θ_y by thinking of the variables r and θ in (4) as functions of x and y. These equations are then identities in x and y. If we differentiate each of the identities in (4) with respect to x,

$$\begin{aligned} 1 &= r_x\cos\theta - r(\sin\theta)\theta_x \\ 0 &= r_x\sin\theta + r(\cos\theta)\theta_x. \end{aligned} \tag{6}$$

These equations can be regarded as a system of two equations in the two unknowns r_x and θ_x. We give the solution in a moment. If we differentiate each of the identities in (4) with respect to y,

$$\begin{aligned} 0 &= r_y\cos\theta - r(\sin\theta)\theta_y \\ 1 &= r_y\sin\theta + r(\cos\theta)\theta_y. \end{aligned} \tag{7}$$

These equations are a system of two equations in the two unknowns r_y and θ_y. Leaving the details to Exercise 25, the solutions of these systems are

$$r_x = \cos\theta \qquad \theta_x = -\frac{\sin\theta}{r}$$
$$r_y = \sin\theta \qquad \theta_y = \frac{\cos\theta}{r}. \tag{8}$$

Substituting these results into (5),

$$u_x = (\cos\theta)U_r - \left(\frac{\sin\theta}{r}\right)U_\theta$$
$$u_y = (\sin\theta)U_r + \left(\frac{\cos\theta}{r}\right)U_\theta. \tag{9}$$

To complete the calculation leading to (3), we must differentiate each of the equations in (9) to obtain u_{xx} and u_{yy}. For u_{xx}, we differentiate the first equation in (9) with respect to x. Applying the sum and product rules,

$$\frac{\partial}{\partial x}u_x = \frac{\partial}{\partial x}\left((\cos\theta)U_r - \left(\frac{\sin\theta}{r}\right)U_\theta\right)$$
$$u_{xx} = \frac{\partial}{\partial x}((\cos\theta)U_r) - \frac{\partial}{\partial x}\left(\left(\frac{\sin\theta}{r}\right)U_\theta\right) \tag{10}$$
$$u_{xx} = U_r\frac{\partial}{\partial x}\cos\theta + \cos\theta\frac{\partial}{\partial x}U_r - U_\theta\frac{\partial}{\partial x}\left(\frac{\sin\theta}{r}\right) - \frac{\sin\theta}{r}\frac{\partial}{\partial x}U_\theta.$$

Next we calculate the partial derivatives of $\cos\theta$, $(1/r)\sin\theta$, U_r, and U_θ with respect to x. Starting with the first two and keeping (8) in mind,

$$\frac{\partial}{\partial x}(\cos\theta) = -(\sin\theta)\theta_x = \frac{\sin^2\theta}{r},$$

and, by the quotient rule,

$$\frac{\partial}{\partial x}\left(\frac{\sin\theta}{r}\right) = \frac{r(\cos\theta)\theta_x - (\sin\theta)r_x}{r^2} = -\frac{2\sin\theta\cos\theta}{r^2}.$$

To differentiate U_r and U_θ with respect to x in (10), we apply Case 2 of the chain rule (see Section 10.2), thinking of U_r and U_θ as functions of r and θ and, in turn, r and θ as functions of x and y. This gives

$$\frac{\partial}{\partial x}U_r = U_{rr}r_x + U_{r\theta}\theta_x = U_{rr}\cos\theta + U_{r\theta}\left(\frac{-\sin\theta}{r}\right)$$
$$\frac{\partial}{\partial x}U_\theta = U_{\theta r}r_x + U_{\theta\theta}\theta_x = U_{\theta r}\cos\theta + U_{\theta\theta}\left(-\frac{\sin\theta}{r}\right).$$

If we assume that U has continuous second partials, we may replace $U_{\theta r}$ by $U_{r\theta}$ in the last equation.

Substituting all of these results into (10),

$$u_{xx} = \frac{\sin^2\theta}{r}U_r + \cos^2\theta\, U_{rr} - \frac{2\sin\theta\cos\theta}{r}U_{r\theta} + \frac{2\sin\theta\cos\theta}{r^2}U_\theta + \frac{\sin^2\theta}{r^2}U_{\theta\theta}.$$

Forming u_{yy} in the same way,

$$u_{yy} = \frac{\cos^2\theta}{r}U_r + \sin^2\theta\, U_{rr} + \frac{2\sin\theta\cos\theta}{r}U_{r\theta} - \frac{2\sin\theta\cos\theta}{r^2}U_\theta + \frac{\cos^2\theta}{r^2}U_{\theta\theta}.$$

Adding these expressions for u_{xx} and u_{yy},

$$u_{xx} + u_{yy} = \frac{1}{r}U_r + U_{rr} + \frac{1}{r^2}U_{\theta\theta}.$$

Because $u_{xx} + u_{yy} = 0$,

$$U_{rr} + \frac{1}{r^2}U_{\theta\theta} + \frac{1}{r}U_r = 0. \tag{11}$$

This is the steady-state heat equation in polar coordinates.

EXAMPLE 1 Assume that $u = u(x, y)$ satisfies the equation

$$u_{xy} + u_x - u_y = 0. \tag{12}$$

If $U(w, z) = u(x, y)$, where $x = w + z$ and $y = -w + z$, find an equation satisfied by $U = U(w, z)$.

Solution

We apply Case 2 of the chain rule to the equation $u(x, y) = U(w, z)$, where w and z are the functions of x and y implicitly defined by the change-of-variable equations

$$\begin{aligned} x &= w + z \\ y &= -w + z. \end{aligned} \tag{13}$$

From the chain rule,

$$\begin{aligned} u_x &= U_w w_x + U_z z_x \\ u_y &= U_w w_y + U_z z_y. \end{aligned} \tag{14}$$

To calculate w_x, z_x, w_y, and z_y, we differentiate (13), first with respect to x and then with respect to y. We find

$$\begin{aligned} x_x &= 1 = w_x + z_x \\ y_x &= 0 = -w_x + z_x \end{aligned}$$

and

$$\begin{aligned} x_y &= 0 = w_y + z_y \\ y_y &= 1 = -w_y + z_y. \end{aligned}$$

It follows from these two systems of equations that

$$w_x = \frac{1}{2}, \qquad z_x = \frac{1}{2}, \qquad w_y = -\frac{1}{2}, \quad \text{and} \quad z_y = \frac{1}{2}. \tag{15}$$

Substituting these results into (14),

$$\begin{aligned} u_x &= \frac{1}{2}U_w + \frac{1}{2}U_z \\ u_y &= -\frac{1}{2}U_w + \frac{1}{2}U_z. \end{aligned} \tag{16}$$

To differentiate u_x with respect to y in the first equation of (16), we apply Case 2 of the chain rule, thinking of u_x as a function of w and z and, in turn, w and z as functions of x and y. Thus,

$$u_{xy} = \frac{1}{2}(U_{ww}w_y + U_{wz}z_y) + \frac{1}{2}(U_{zw}w_y + U_{zz}z_y)$$
$$= -\frac{1}{4}U_{ww} + \frac{1}{4}U_{wz} - \frac{1}{4}U_{zw} + \frac{1}{4}U_{zz}.$$

If U has continuous second-order partial derivatives, then by the Equality of Mixed Partial Derivatives Theorem $U_{wz} = U_{zw}$ and, hence,

$$u_{xy} = \frac{1}{4}(-U_{ww} + U_{zz}).$$

Using this result together with (16) and (12),

$$0 = u_{xy} + u_x - u_y = \frac{1}{4}(-U_{ww} + U_{zz}) + \left(\frac{1}{2}U_w + \frac{1}{2}U_z\right) - \left(-\frac{1}{2}U_w + \frac{1}{2}U_z\right).$$

Hence, U satisfies the partial differential equation.

$$\frac{1}{4}(-U_{ww} + U_{zz}) + U_w = 0.$$

D'Alembert's Solution of the Wave Equation

We end this section by showing how the French mathematician Jean Le Rond D'Alembert (1717–1783) used a change of variables in solving the heat equation.

If an infinitely long elastic string is stretched along the x-axis and is set in motion by holding it in the shape of the graph of a function f (the graph of f is the curve labeled "$t = 0$" in Fig. 10.33) and then releasing it from rest, the vibration of the string causes the points of the string to move in lines perpendicular to the x-axis provided that the magnitude of the virbration is small. As shown in Exercise 32, it follows from Newton's laws that the vertical displacement $u(x, t)$ at any subsequent time t of a point initially at $(x, f(x))$ satisfies the one-dimensional wave equation

$$u_{tt} = c^2 u_{xx}. \tag{17}$$

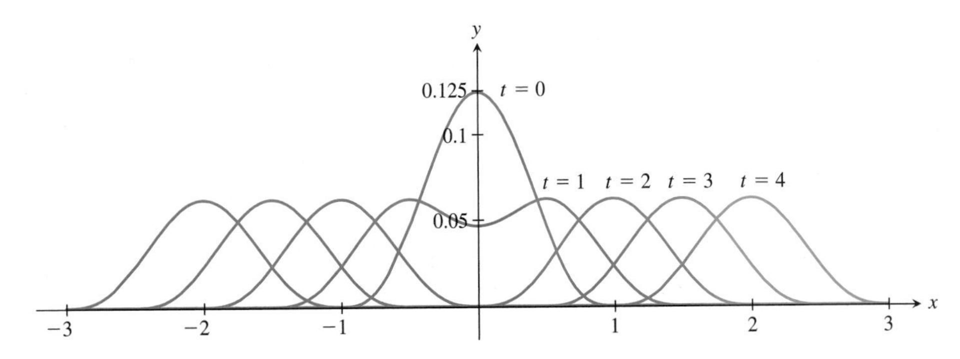

FIGURE 10.33 String at times $t = 0, 1, 2, 3$, and 4 s after release; initial shape ($t = 0$) is described by the function f.

The constant c, which measures the wave speed, is related to the tension T in the string and its density σ by $c^2 = T/\sigma$.

Figure 10.33 shows the shapes of the string at $t = 0$, 1, 2, 3, and 4 seconds after release. Given that the initial shape is symmetric about the y-axis, it is perhaps not surprising that the string separates into two congruent pulses, one traveling to the right and the other to the left. From experiment we expect that each pulse travels at a constant speed, say, c meters per second. Figure 10.33 was drawn with $c = 0.5$ meters per second.

Focusing for a moment on the pulse to the right of the origin and corresponding to $t = 4$, if the equation $y = g(x)$ describes a curve symmetric about the origin and congruent to this pulse, then the equation $y = g(x - 4c)$ describes this pulse. For if we wish to know the height of this pulse at the point x, we calculate the point $x - 4c$, which is, as it were, where the point x was 4 seconds ago, and evaluate g at this point. More generally, after t seconds the pulse moving to the right is described by $y = g(x - ct)$. Similarly, the pulse moving to the left is described by $y = g(x + ct)$.

To describe the motion of the string we seek a function $u = u(x, t)$ satisfying (17), the initial shape condition

$$u(x, 0) = f(x), \qquad -\infty < x < \infty, \tag{18}$$

and the initial velocity condition

$$u_t(x, 0) = 0, \qquad -\infty < x < \infty. \tag{19}$$

This condition states that the string is at rest when it is released at $t = 0$.

D'Alembert's solution of this problem began with the change of variables

$$\begin{aligned} w &= x + ct \\ z &= x - ct \end{aligned} \tag{20}$$

suggested by the equations $y = g(x - ct)$ and $y = g(x + ct)$ discussed earlier.

We define $U(w, z) = u(x, y)$, where x and y are the functions of w and z determined by (20). To find the equation satisfied by U, we calculate u_{tt} and u_{xx} in terms of U and its derivatives and substitute the results into (17). We shall need the partials w_x, w_t, z_x, and z_t. These are

$$\begin{aligned} w_x &= 1 \qquad & w_t &= c \\ z_x &= 1 \qquad & z_t &= -c. \end{aligned}$$

From the chain rule,

$$\begin{aligned} u_x &= U_w w_x + U_z z_x = U_w + U_z \\ u_t &= U_w w_t + U_z z_t = cU_w - cU_z \\ u_{xx} &= U_{ww} w_x + U_{wz} z_x + U_{zw} w_x + U_{zz} z_x = U_{ww} + 2U_{wz} + U_{zz} \\ u_{tt} &= c(U_{ww} w_t + U_{wz} z_t) - c(U_{zw} w_t + U_{zz} z_t) \\ &= c(U_{ww} c + U_{wz}(-c)) - c(U_{zw} c + U_{zz}(-c)) \\ &= c^2 U_{ww} - 2c^2 U_{wz} + c^2 U_{zz}. \end{aligned}$$

In simplifying u_{xx} and u_{tt}, we replaced U_{zw} by U_{wz}. For this we assume that U has continuous second partial derivatives. Using these results in (17),

$$c^2 U_{ww} - 2c^2 U_{wz} + c^2 U_{zz} = c^2(U_{ww} + 2U_{wz} + U_{zz}),$$

which simplifies to

$$U_{wz} = 0. \tag{21}$$

This simpler partial differential equation replaces (17). Rewriting it as

$$\frac{\partial}{\partial z}\left(\frac{\partial U}{\partial w}\right) = 0,$$

we observe that $U_w = \partial U/\partial w$ is a function whose partial derivative with respect to z is 0. From this we conjecture that U_w is a function of w alone, that is,

$$U_w = h(w).$$

We further conjecture that

$$U = U(w, z) = \int h(w)\, dw + k(z) = H(w) + k(z), \tag{22}$$

where H is an antiderivative of h and k is any function of z. The function k plays the role of the arbitrary constant C often added to an antiderivative. For any choice of k, the function U in (22) satisfies (21) because

$$U_w = \frac{\partial}{\partial w}(H(w) + k(z)) = h(w)$$

and, differentiating both sides of this result with respect to z,

$$U_{wz} = 0.$$

Hence, for any twice-differentiable functions h and k, the function U in (22) satisfies (21). In what follows we choose h and k so that the initial conditions (18) and (19) are satisfied.

The initial conditions (18) and (19) on u may be expressed in terms of the (w, z)-coordinate system. Keeping in mind that

$$w = x + ct \quad \text{and} \quad z = x - ct,$$

condition (18) becomes

$$u(x, 0) = f(x) = U(w, z) = U(x, x) = H(x) + k(x). \tag{23}$$

To express condition (19) in the (w, z)-coordinate system, we differentiate the equation $u(x, t) = U(w, z)$ with respect to t and then use (22).

$$u_t(x, t) = U_w w_t + U_z z_t = cU_w - cU_z = cH'(w) - ck'(z). \tag{24}$$

From condition (19),

$$0 = u_t(x, 0) = cH'(x) - ck'(x). \tag{25}$$

As outlined in Exercise 24, it follows from (23) and (25) that

$$H(x) = \frac{1}{2}f(x)$$

$$k(x) = \frac{1}{2}f(x).$$

From these results, d'Alembert's solution of the wave equation (17), with initial conditions (18) and (19), now follows by substituting (20) into (22), that is,

$$u(x, t) = U(w, z) = U(x + ct, x - ct) = \frac{1}{2}(f(x + ct) + f(x - ct)).$$

Exercises 10.4

Exercises 1–12: Show that the function satisfies the heat equation $u_{xx} + u_{yy} = 0$ in rectangular coordinates or the heat equation $U_{rr} + (1/r^2)U_{\theta\theta} + (1/r)U_r = 0$ in polar coordinates.

1. $u(x,y) = x^2 - y^2$
2. $u(x,y) = 2xy$
3. $u(x,y) = x^3 - 3xy^2$
4. $u(x,y) = 3x^2y - y^3$
5. $u(x,y) = \sin x \cosh y$
6. $u(x,y) = \cos x \sinh y$
7. $u(x,y) = \cos x \cosh y$
8. $u(x,y) = \sin x \sinh y$
9. $U(r,\theta) = r^2 \cos 2\theta$
10. $U(r,\theta) = r^2 \sin 2\theta$
11. $U(r,\theta) = r^3 \cos \theta(4\cos^2\theta - 3)$
12. $U(r,\theta) = r^3 \sin \theta(3 - 4\sin^2\theta)$

Exercises 13–16: Find the form of the steady-state heat equation $u_{xx} + u_{yy} = 0$ in the (r,s)-coordinate system.

13. $r = x - y$ and $s = x + y$
14. $r = 2x - 3y$ and $s = -x + 2y$
15. $x = r\cos\theta - s\sin\theta$ and $y = r\sin\theta + s\cos\theta$, where θ is a constant
16. $x = r\cosh\theta + s\sinh\theta$ and $y = r\sinh\theta + s\cosh\theta$, where θ is a constant

Exercises 17–20: Find the form of the partial differential equation $u_{xy} = xy + u_x$ in the (r,s)-coordinate system.

17. $r = x - y$ and $s = x + y$
18. $r = 2x - 3y$ and $s = -x + 2y$
19. $x = r\cos\theta - s\sin\theta$ and $y = r\sin\theta + s\cos\theta$, where θ is a constant
20. $x = r\cosh\theta + s\sinh\theta$ and $y = r\sinh\theta + s\cosh\theta$, where θ is a constant

21. Using the calculations in the text for transforming the heat equation into polar coordinates, show that

$$u_x^2 + u_y^2 = U_r^2 + \frac{1}{r^2}U_\theta^2.$$

22. Verify the correctness of the formula for u_{yy} given a few lines before (11).

23. The initial shape of the string shown in Fig. 10.33 is given by

$$f(x) = 0,\ x \in (-\infty, -1)$$
$$f(x) = (1 + \cos(\pi x))/16,\ x \in [-1, 1]$$
$$f(x) = 0,\ x \in (1, \infty).$$

Does this function have a continuous second derivative? Why might this question be raised?

24. Show that $H(x) = \frac{1}{2}f(x)$ and $k(x) = \frac{1}{2}f(x)$ follow from Equations (23) and (25). *Hint:* Divide by c and integrate (25); then solve a system of equations simultaneously.

25. Solve the two systems (6) and (7). Try to take advantage of the fact that these two systems have the same matrix of coefficients, that is, the unknowns have the same coefficients. Cramer's rule works pretty well here. The solutions are listed in (8).

26. Test your CAS or your own powers of differentiation and simplification in showing that the function u given in (2) satisfies the heat equation (1).

27. It was stated at the beginning of Section 10.3 that the point $(2.5, 1)$ is "very nearly" on the $0.3°$ isotherm. See Fig. 10.20 or 10.21. This can be checked by calculating $u(2.5, 1)$ in (2). Verify that $u(2.5, 1) \approx 0.299861$. *Hints:* If you do not get this answer, perhaps your calculator is not in radian mode. If your calculator does not have a SINH key, then use the fact that the sinh function is defined by

$$\sinh y = \frac{e^y - e^{-y}}{2},\ y \in R.$$

28. The hyperbolic sine function sinh is invertible on its domain $(-\infty, \infty)$. In particular, if $z = \sinh w$, then $w = \operatorname{arcsinh}(z) = \ln\left(z + \sqrt{z^2 + 1}\right)$. Use this to show that the $0.3°$ isotherm discussed in the text and shown in Fig. 10.21 is described by

$$y = \operatorname{arcsinh}\left(\frac{\sin x}{\tan(0.3\pi/2)}\right).$$

Use this equation in reproducing Fig. 10.21.

29. (T) The preceding problem concerned graphing contours, or level curves, for functions f for which the equation $f(x,y) = k$ can be solved for y in terms of x. When this is not possible, a procedure based on Newton's method (or another algorithm for finding the zeros of a function) can be used to generate data from which the contours can be plotted. The accompanying figure shows a contour plot for

$$f(x,t) = e^{-t}\sin x - \frac{1}{9}e^{-9t}\sin 3x.$$

Each curve is the set of points satisfying an equation $f(x,t) = k$. In the figure are shown contours for $k = 0, 0.1, 0.2, \ldots, 0.9, 1.0$. the contour on the left (the vertical line) is the $k = 0$ contour. We outline a procedure for generating points on the 0.2 contour.

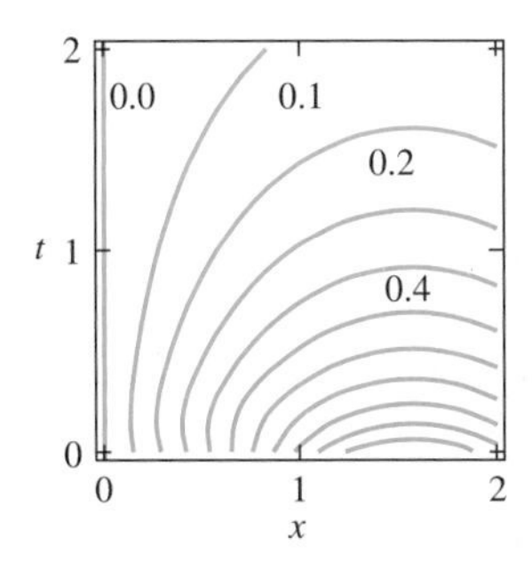

Figure for Exercise 29.

Take $x = 0.3$ and consider the equation $f(0.3, t) = 0.2$. To find the value(s) of t satisfying $f(0.3, t) = 0.2$, we set $g(t) = f(0.3, t) - 0.2$. A rough graph of g shows that it has a zero near 0.4; Newton's method then gives $t \approx 0.38$. Repeat this for, say, $x = 0.6, 0.9, 1.2, 1.5, 1.8$. Use these results to plot the 0.2 contour.

30. Fig. 10.21 shows the 0.3° isotherm from Fig. 10.20 and the gradient vector $\nabla u(2.5, 1)$. Use (2) in showing that this gradient vector is $\nabla u(2.5, 1) \approx \langle -0.344617, -0.338023 \rangle$.

T 31. The initial shape of the string shown in Fig. 10.33 is given by

$$f(x) = 0, x \in (-\infty, -1)$$
$$f(x) = (1 + \cos(\pi x))/16, x \in [-1, 1]$$
$$f(x) = 0, x > 1.$$

Duplicate the graphs shown in this figure on your calculator. If you have access to a CAS, it may be possible to "animate" the motion of the string, by increasing the number of snapshots. In either case, you may wish to enter the function f defined above with the help of an "if...then...else..." command.

32. Fill in the following outline of a derivation of the wave equation from Newton's laws. We make several simplifying assumptions. We assume that the string is stretched horizontally, has uniform lineal density (σ), and is perfectly elastic (it does not resist bending); that the tension in the string is large enough that the gravitational force can be ignored; and that each point of the string moves in a vertical direction. At $t = 0$ we deform the string from its horizontal position along an x-axis and release it. We fix our attention on the point P of the string at time t after release and consider the forces acting on a small segment of the string. See the accompanying figure. Let $U(x, t)$ denote the vertical displacement of P. The only forces $\mathbf{F}_1$ and $\mathbf{F}_2$ acting on the segment are directed along the tangents to the curve of the string (the string does not resist bending) at the points $P = (x, U(x, t))$ and $Q = (x + \Delta x, U(x + \Delta x, t))$. Let T_1 and T_2 be the magnitudes of $\mathbf{F}_1$ and $\mathbf{F}_2$. In terms of the angles θ_1 and θ_2, the forces $\mathbf{F}_1$ and $\mathbf{F}_2$ are given by

$$\mathbf{F}_1 = T_1\langle -\cos\theta_1, -\sin\theta_1 \rangle$$
$$\mathbf{F}_2 = T_2\langle \cos\theta_2, \sin\theta_2 \rangle.$$

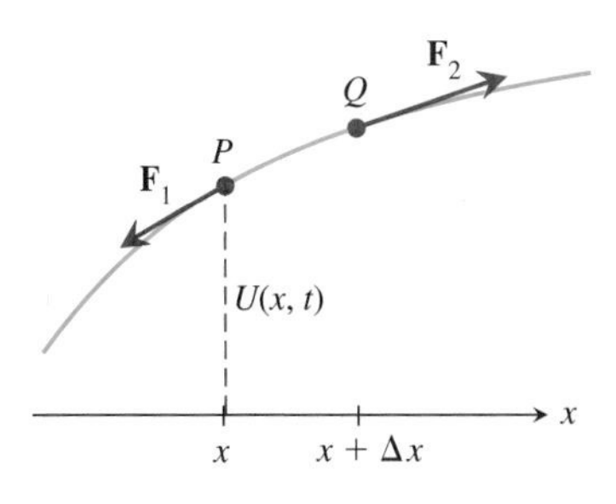

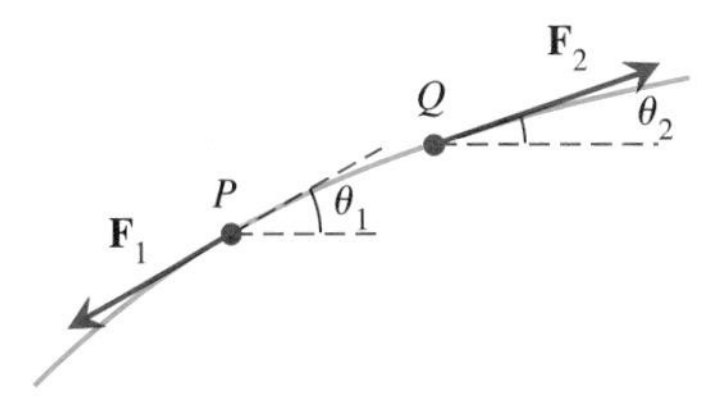

Figure for Exercise 32.

The net force acting on the string segment is

$$\mathbf{F}_1 + \mathbf{F}_2 = \langle -T_1\cos\theta_1 + T_2\cos\theta_2, -T_1\sin\theta_1 + T_2\sin\theta_2 \rangle. \quad (26)$$

Because we assumed that the motion of the points of the string is vertical, the first coordinate of the net force is 0; that is,

$$T = T_1\cos\theta_1 = T_2\cos\theta_2. \quad (27)$$

We refer to T as the tension. For small string motions (think of a violin string vibrating), θ_1 and θ_2 are small and $\|\mathbf{F}_1\| \approx \|\mathbf{F}_2\| \approx T$. The mass of the segment is $\sigma\Delta x$, and its acceleration is $\langle 0, U_{tt} \rangle$. From Newton's second law, we have

$$\mathbf{F}_1 + \mathbf{F}_2 = \langle 0, \sigma\Delta x U_{tt} \rangle.$$

From Equations (26) and (27),

$$\langle 0, \sigma\Delta x U_{tt} \rangle = \langle 0, -T_1\sin\theta_1 + T_2\sin\theta_2 \rangle.$$

Setting the second coordinates equal and dividing by $\Delta x T$,

$$\frac{\sigma}{T}U_{tt} = \frac{\tan\theta_2 - \tan\theta_1}{\Delta x}.$$

Letting $k^2 = \sigma/T$,

$$k^2 U_{tt} \approx \frac{U_x(x + \Delta x, t) - U_x(x, t)}{\Delta x}.$$

The wave equation follows by supposing that $\Delta x \to 0$:

$$k^2 U_{tt} = U_{xx}.$$

10.5 Optimization

The ideas used to locate extrema of functions of several variables are analogous to those used in Section 4.7 for functions of one variable. However, the graphs and calculations are often more complex. For example, instead of solving an equation such as $f'(x) = 0$ to locate a candidate for a point at which a maximum occurs, we must solve a system of equations. Or, instead of calculating the values of f at the two boundary points of its domain $[a, b]$, we must take account of the infinitely many values of f on the boundary of a region $\mathcal{D}$.

We start with definitions of local and global extrema.

DEFINITION Local or Global Extrema

A real-valued function f defined on a set $\mathcal{D} \subset R^n$ has a **local maximum** at $\mathbf{w} \in \mathcal{D}$ if

$$f(\mathbf{w}) \geq f(\mathbf{x}) \qquad \text{for all } \mathbf{x} \in \mathcal{D} \text{ sufficiently close to } \mathbf{w};$$

f has a **local minimum** at $\mathbf{w} \in \mathcal{D}$ if

$$f(\mathbf{w}) \leq f(\mathbf{x}) \qquad \text{for all } \mathbf{x} \in \mathcal{D} \text{ sufficiently close to } \mathbf{w};$$

f has a **global maximum** at $\mathbf{w} \in \mathcal{D}$ if

$$f(\mathbf{w}) \geq f(\mathbf{x}) \qquad \text{for all } \mathbf{x} \in \mathcal{D};$$

f has a **global minimum** at $\mathbf{w} \in \mathcal{D}$ if

$$f(\mathbf{w}) \leq f(\mathbf{x}) \qquad \text{for all } \mathbf{x} \in \mathcal{D}.$$

If f has a local maximum, local minimum, global maximum, or global minimum at $\mathbf{w} \in \mathcal{D}$, we say that f has an *extremum* at $\mathbf{w}$ and that $\mathbf{w}$ is an *extreme point.* The word *global* is often omitted or replaced by the word *absolute*.

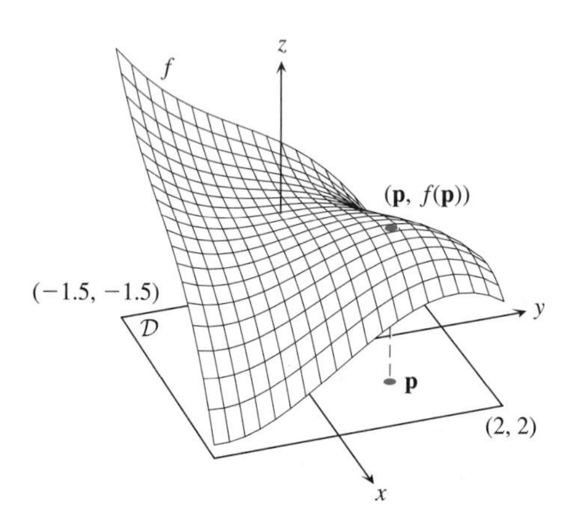

FIGURE 10.34 Local and global extrema of a function f defined on a region $\mathcal{D}$.

Figure 10.34 shows the graph of a function f defined on the rectangular region $\mathcal{D} = [-1.5, 2] \times [-1.5, 2] \subset R^2$. Although judging the relative heights of points on such graphs can be misleading, it appears that f has local extrema at the four corners of $\mathcal{D}$ and a local maximum at $\mathbf{p} \in \mathcal{D}$. A global maximum appears to be located at the vertex $(-1.5, -1.5)$ of $\mathcal{D}$. It is not clear from the figure which of $(-1.5, 2)$ or $(2, -1.5)$ is a global minimum. We note for later use the "mountain pass," or *saddle point,* located above the origin. A saddle point is something like an inflection point.

The most important tool for locating candidates for extreme points is the Candidate Theorem. For functions of one variable, the candidates for extrema were endpoints of the domain, zeros of f', or points at which f was nondifferentiable. The same is true for functions of several variables, except that instead of endpoints, the candidates include *all* boundary points, and instead of solving the equation $f'(x) = 0$, we solve the system of n equations $\nabla f = \mathbf{0}$. Usually, $n = 2$ or $n = 3$ in examples and problems.

Recall that a point $\mathbf{p}$ of a subset $\mathcal{D}$ of R^2 (R^3) is an interior point of $\mathcal{D}$ if all sufficiently small disks (balls) centered at $\mathbf{p}$ are entirely within $\mathcal{D}$. A point $\mathbf{p}$ of R^2 (R^3) is a boundary point of $\mathcal{D}$ if every disk (ball) centered at $\mathbf{p}$ contains points that belong to $\mathcal{D}$ and points that do not belong to $\mathcal{D}$. The boundary of $\mathcal{D}$ is denoted by $\partial\mathcal{D}$.

We state the Candidate Theorem for functions of two variables; the corresponding theorem for functions of three or more variables is entirely analogous.

THEOREM Candidate Theorem

Assume that f is a real-valued continuous function of two variables on its domain $\mathcal{D} \subset R^2$. If f has a local maximum or local minimum at $\mathbf{p} \in \mathcal{D}$, then the only candidates for $\mathbf{p}$ are the boundary points of $\mathcal{D}$, any interior point $\mathbf{w}$ at which $\nabla f(\mathbf{w}) = \mathbf{0}$, or any interior point at which f has no gradient vector.

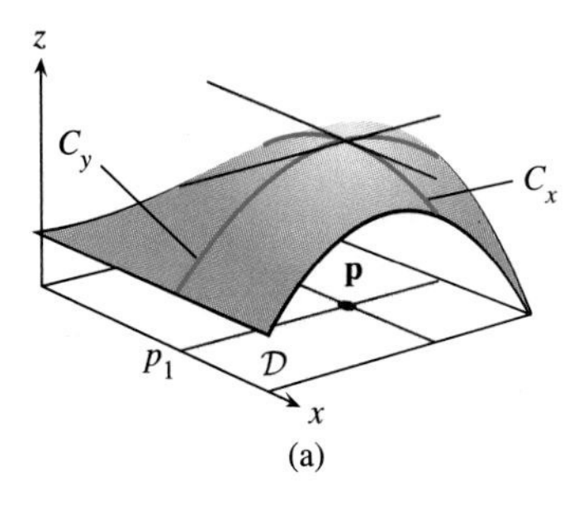

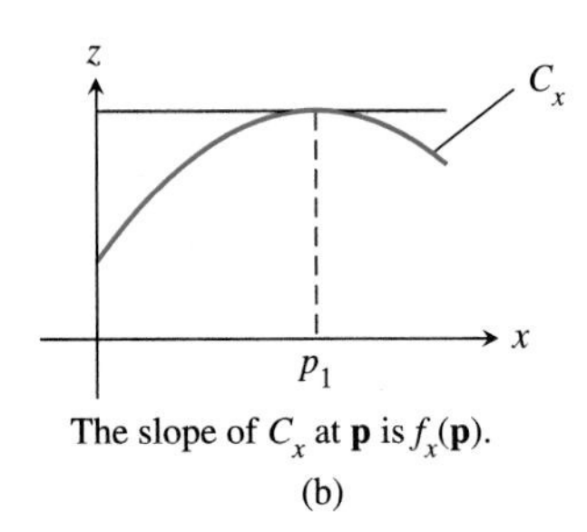

The slope of C_x at $\mathbf{p}$ is $f_x(\mathbf{p})$.

(b)

FIGURE 10.35 Curves C_x and C_y through a high point.

We give a proof based on the Candidate Theorem for functions of one variable, which is stated in Section 4.7. If $\mathbf{p}$ is a point at which f has a local maximum (the proof for the local minimum case is similar) and $\mathbf{p}$ is neither a boundary point of $\mathcal{D}$ nor a point where f has no gradient vector, then $\mathbf{p}$ must be an interior point. It follows from the Candidate Theorem for functions of one variable that $f_x(\mathbf{p}) = f_y(\mathbf{p}) = 0$ and, hence, $\nabla f(\mathbf{p}) = \mathbf{0}$. The idea is illustrated in Fig. 10.35(a), which shows the graph of a function f with a local maximum at $\mathbf{p} = (p_1, p_2)$. The curves C_x and C_y are the intersections of the graph of f and the planes $y = p_2$ and $x = p_1$. The slopes of C_x and C_y at $\mathbf{p}$ are $f_x(\mathbf{p})$ and $f_y(\mathbf{p})$. The projection of the curve C_x onto the (x, z)-plane is shown in Fig. 10.35(b). Because f has a local maximum at $\mathbf{p}$, C_x has a local maximum at p_1. It then follows from the earlier Candidate Theorem that $f_x(\mathbf{p}) = 0$. Similarly, $f_y(\mathbf{p}) = 0$ and, hence, $\nabla f(\mathbf{p}) = \mathbf{0}$. See Exercise 32.

EXAMPLE 1 Find all extrema of the function

$$f(x, y) = xy\left(1 - \frac{1}{3}(x + y)\right), \qquad (x, y) \in \mathcal{D},$$

where $\mathcal{D}$ is the region on and within the triangle shown in Fig. 10.36.

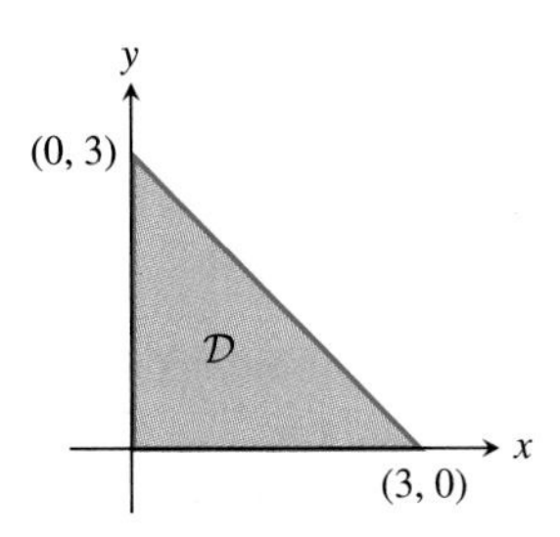

FIGURE 10.36 Domain of f.

Solution

The boundary of $\mathcal{D}$ includes segments on the two axes and a segment of the line described by $x + y = 3$. On the segment along the x-axis, where $y = 0$, $f(x, 0) = x \cdot 0\left(1 - \frac{1}{3}(x + 0)\right) = 0$. On the segment along the y-axis, where $x = 0$, $f(0, y) = 0 \cdot y\left(1 - \frac{1}{3}(0 + y)\right) = 0$. And, along the line joining $(3, 0)$ and $(0, 3)$, which is described by the equation $x + y = 3$, $f(x, y) = xy\left(1 - \frac{1}{3}(3)\right) = 0$. Hence, f is zero at all points on the boundary $\mathcal{D}$.

At all points (x, y) in the interior of $\mathcal{D}$, x, y, and $1 - \frac{1}{3}(x + y)$ are positive; hence, the value of f is positive. Hence, f takes on its absolute minimum value everywhere on the boundary of $\mathcal{D}$. Because f is everywhere differentiable in the interior of $\mathcal{D}$, the only remaining candidates for extrema are interior points $\mathbf{p} = (x, y)$ at which $\nabla f(\mathbf{p}) = \langle f_x(\mathbf{p}), f_y(\mathbf{p})\rangle = \mathbf{0}$. To locate such points, we calculate and simplify f_x and f_y.

$$f_x = y\left(1 - \frac{1}{3}(x + y)\right) + xy\left(-\frac{1}{3}\right) = y\left(1 - \frac{2}{3}x - \frac{1}{3}y\right)$$

$$f_y = x\left(1 - \frac{1}{3}(x + y)\right) + xy\left(-\frac{1}{3}\right) = x\left(1 - \frac{1}{3}x - \frac{2}{3}y\right).$$

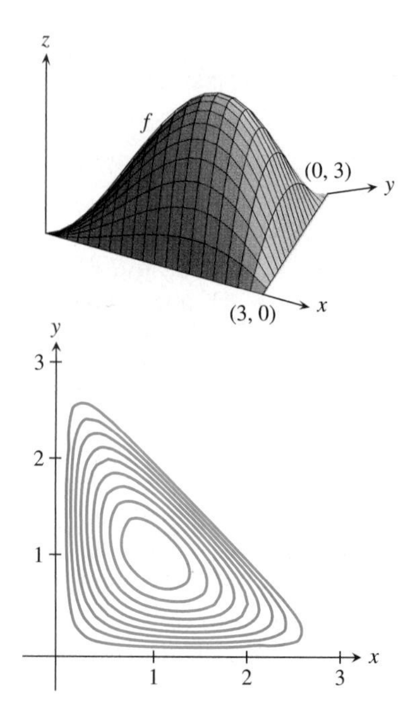

FIGURE 10.37 Graph and contour plot of f.

Setting f_x and f_y equal to zero gives the system of equations

$$y\left(1 - \frac{2}{3}x - \frac{1}{3}y\right) = 0$$

$$x\left(1 - \frac{1}{3}x - \frac{2}{3}y\right) = 0.$$

Because the factors y and x are not zero in the interior of $\mathcal{D}$, the only possible candidate is $(1, 1)$, which is the solution of the system

$$1 - \frac{2}{3}x - \frac{1}{3}y = 0$$

$$1 - \frac{1}{3}x - \frac{2}{3}y = 0.$$

If f has a maximum, then it must be at the point $(1, 1)$. The value of $f(1, 1)$ is $1/3$. Figure 10.37 shows a graph of f and a contour plot. The graph was scaled in the vertical direction. The contour plot provides another view of the variation of f on $\mathcal{D}$. Both views are consistent with our strong impression that the maximum of f is $1/3$ and this occurs at $(1, 1)$. We say "impression" because our argument is not quite conclusive. Why?

The Candidate Theorem is not sufficient to make our argument conclusive because it states only that "If f... has a local maximum...." What we need is Weierstrass' Max/Min Theorem. This theorem and its analog in Chapter 4 are associated with the German mathematician Karl Weierstrass (1815–1897). These theorems are proved in more advanced courses.

THEOREM Max/Min Theorem

If f is a real-valued continuous function on a closed and bounded subset $\mathcal{D}$ of R^2 or R^3, then f has a maximum value and a minimum value on $\mathcal{D}$.

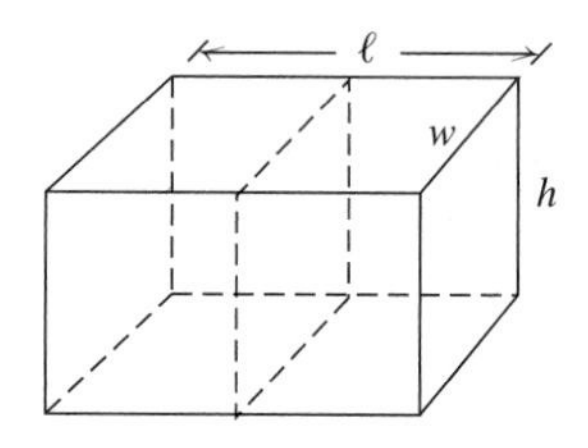

The domain $\mathcal{D}$ of the function f in Example 1 is both closed and bounded ($\mathcal{D}$ is bounded because all of its points are within a fixed distance of the origin; $\mathcal{D}$ is closed because all of its boundary points are in $\mathcal{D}$). Because the function f is continuous, the Max/Min Theorem guarantees that f takes on its minimum and maximum values somewhere in $\mathcal{D}$. The minimum value occurs on the boundary of $\mathcal{D}$. Because f is positive on $\mathcal{D}$, its maximum value occurs at an interior point. With this fact in mind, we used the Candidate Theorem to locate the point $(1, 1)$ at which f takes on its maximum value.

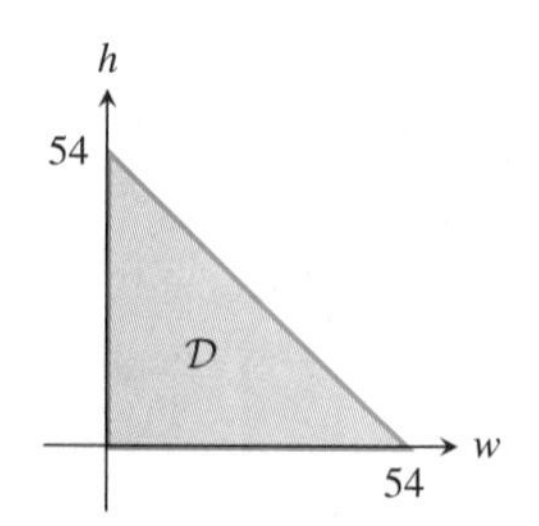

FIGURE 10.38 Representative box with girth shown in color; also, the domain $\mathcal{D}$ of V.

EXAMPLE 2 According to the postal regulations for 2001, a box can be mailed only if the sum of its length and girth does not exceed 108 inches. The length of a box is its longest horizontal dimension, and the girth is twice the sum of the remaining two dimensions. Figure 10.38 shows in color a rectangle whose perimeter is the girth of the box. Within the postal regulations, what are the dimensions of a box with largest volume?

Solution

The volume of a box with length ℓ, width w, and height h is $V = \ell wh$. Because it is clear that the maximum volume occurs when $\ell + 2(w + h) = 108$—see Exercise 33—we replace the postal regulation

$$\ell + 2(w + h) \leq 108$$

by

$$\ell + 2(w + h) = 108.$$

This allows us to express the volume of the box as a function of two variables. Because $\ell = 108 - 2(w + h)$,

$$V = \ell wh = (108 - 2(w + h))wh.$$

Because we want each of $\ell = 108 - 2(w + h)$, w, and h to be nonnegative, the domain of V is the set of all (w, h) for which $w \geq 0$, $h \geq 0$, and $\ell = 108 - 2(w + h) \geq 0$. Simplifying the last inequality, $w + h \leq 54$.

We may now state the problem as follows: Maximize the function

$$V(w, h) = (108 - 2(w + h))wh, \qquad (w, h) \in \mathcal{D}, \tag{1}$$

where $\mathcal{D} = \{(w, h) : w, h \geq 0 \text{ and } w + h \leq 54\}$. Figure 10.39 shows a sketch of $\mathcal{D}$.

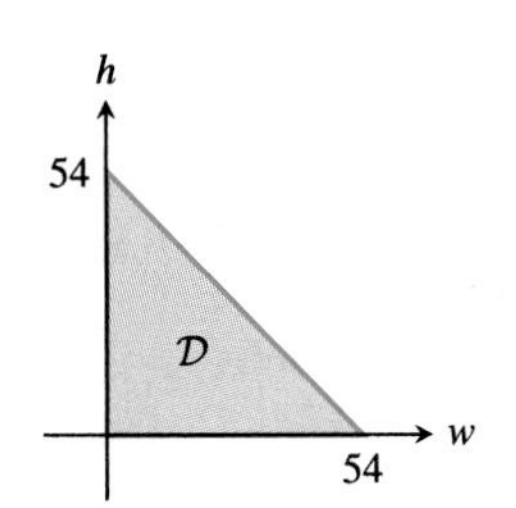

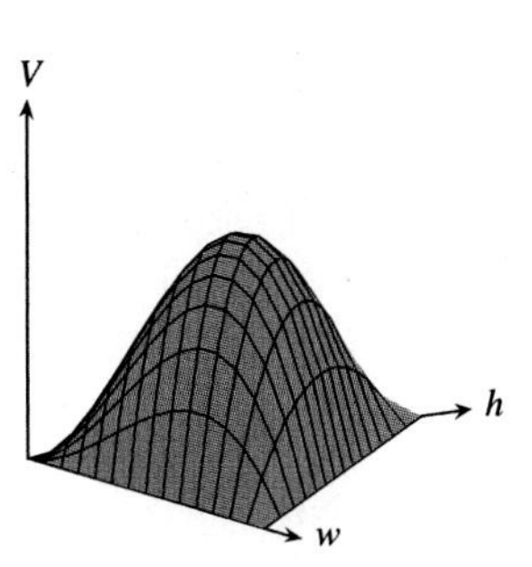

FIGURE 10.39 Graph of the volume function.

Figure 10.39 also shows a graph of V (scaled in the vertical direction because V takes on values larger than 10,000 cubic inches in the interior of $\mathcal{D}$) and its domain $\mathcal{D}$. If you were to insist that boxes cannot have zero width or height, you might argue that the most natural domain for the volume function is

$$\mathcal{D}^* = \{(w, h) : w, h > 0 \text{ and } w + h < 54\}.$$

However, we gain something and lose nothing by considering $\mathcal{D}$ instead of $\mathcal{D}^*$. If f has a maximum in $\mathcal{D}^*$, we will discover it in our search of the interior of $\mathcal{D}$. Moreover, because $\mathcal{D}$ is closed and bounded, we gain the use of the Max/Min Theorem. Because the set $\mathcal{D}^*$ is not closed (it does not contain its boundary points), we cannot apply the Max/Min Theorem, at least not directly.

From both Fig. 10.39 and (1) we see that the function V is 0 on the boundary of $\mathcal{D}$ and positive throughout the interior of this set. From the Max/Min Theorem we know that V has a maximum; because the maximum of V is positive, V takes on its maximum value somewhere in the interior of $\mathcal{D}$. We use the Candidate Theorem to find this point.

The partial derivatives of V are

$$V_w = -2wh + (108 - 2(w + h))h$$
$$V_h = -2wh + (108 - 2(w + h))w.$$

Setting these partial derivatives equal to zero and factoring gives

$$2h(54 - w - h - w) = 0$$
$$2w(54 - w - h - h) = 0.$$

Because any solution (h, w) of this system for which $h = 0$ or $w = 0$ is not an interior point of $\mathcal{D}$, the system may be simplified to

$$\begin{aligned} 2w + h &= 54 \\ w + 2h &= 54. \end{aligned}$$

The solution of this system is $(h, w) = (18, 18)$.

At the point $(18, 18)$ the value of V is 11,664 (11,664 in^3 = 6.75 ft^3). The function V has an absolute minimum on $\mathcal{D}$, namely, 0. By the Candidate Theorem, the absolute maximum of V must occur at $(18, 18)$. Hence, the dimensions of the box of maximum volume are $\ell = 36$ inches and $w = h = 18$ inches.

EXAMPLE 3 Locate the global extrema of the function

$$f(x, y) = 3xy - x^3 - y^3 + 15, \qquad -\frac{3}{2} \le x, y \le 2.$$

Solution

The polynomial function f is continuous and its domain $\mathcal{D}$ is the rectangular region $\left[-\frac{3}{2}, 2\right] \times \left[-\frac{3}{2}, 2\right]$, which is closed and bounded. By the Max/Min Theorem, f has a maximum and a minimum on $\mathcal{D}$. Each of the global extrema of f is an interior point of $\mathcal{D}$ or a boundary point.

By the Candidate Theorem, if $\mathbf{p} = (x, y)$ is an interior point at which f has an extremum, then the only candidates for $\mathbf{p}$ must satisfy $\nabla f(\mathbf{p}) = \langle f_x(\mathbf{p}), f_y(\mathbf{p})\rangle = \mathbf{0}$. We are thus led to the system of equations

$$\begin{aligned} f_x(\mathbf{p}) &= 3y - 3x^2 = 0 \\ f_y(\mathbf{p}) &= 3x - 3y^2 = 0. \end{aligned}$$

From the first equation $y = x^2$. Substituting this into the second equation gives $x - x^4 = x(1 - x^3) = 0$. The real roots of this polynomial are $x = 0$ and $x = 1$. The corresponding y-values are 0 and 1, respectively. Hence, the only interior points at which an extremum may occur are $(0, 0)$ and $(1, 1)$. We evaluate f at these points, obtaining $f(0, 0) = 15$ and $f(1, 1) = 16$. Figure 10.40 shows these data. The candidates are marked by dots, and the corresponding function values are in small, adjacent boxes.

The search for candidates in the interior of $\mathcal{D}$ is complete. Next, we search the boundary, starting with the $x = 2$ boundary.

On the $x = 2$ boundary, points have the form $(2, y)$, where $-3/2 \le y \le 2$. To investigate f on this boundary we consider the function

$$g(y) = f(2, y) = -y^3 + 6y + 7, \qquad -3/2 \le y \le 2.$$

Applying the one-dimensional Candidate Theorem, we evaluate g at the endpoints $y = -3/2$ and $y = 2$ and at any zeros of the equation $g'(y)$ that happen to lie between $-3/2$ and 2. From $g'(y) = -3y^2 + 6$, both of the zeros $y = \pm\sqrt{2}$ of g' lie in this interval. The four values of g are marked along the $x = 2$ boundary in Fig. 10.40. Along this boundary, f has a minimum of approximately 1.3 at $\left(2, -\sqrt{2}\right)$ and a maximum of approximately 12.7 at $\left(2, \sqrt{2}\right)$. A graph of g is shown on the right side of the figure.

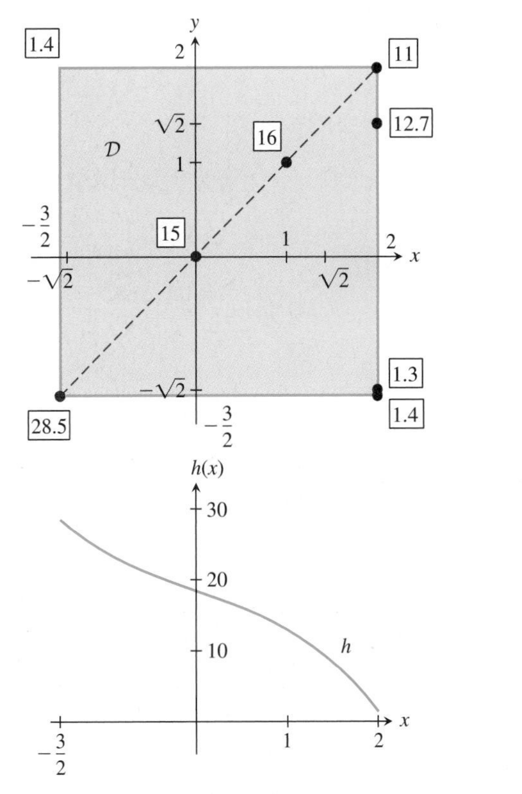

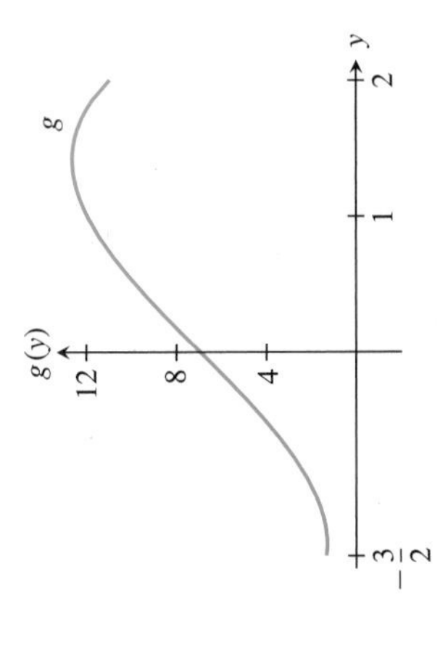

FIGURE 10.40 Summary of search results, including the domain $\mathcal{D}$ of f, the graph of $g(y) = f(2, y)$, and the graph of $h(x) = f(x, -3/2)$.

On the $y = -3/2$ boundary, points have the form $(x, -3/2)$, where $-3/2 \leq x \leq 2$. Hence, we consider the function

$$h(x) = f(x, -3/2) = -x^3 - \frac{9}{2}x + \frac{147}{8}.$$

Because $h'(x) = -3x^2 - \frac{9}{2} < 0$, f decreases on the interval $-3/2 \leq x \leq 2$. We find $f(-3/2, -3/2) = 28.5$ and $f(2, -3/2) \approx 1.4$. A graph of h is shown at the bottom of the figure.

For this particular function we need not examine the other two boundaries in detail. Because $f(x, y) = f(y, x)$, the graph of f is symmetric about the plane $y = x$. Thus, along the $y = 2$ boundary, f behaves as it does on the $x = 2$ boundary. Along the $x = -3/2$ boundary, f behaves as it does on the $y = -3/2$ boundary.

From these data, including both interior and boundary points, we conclude that f takes on its global maximum of approximately 28.5 at $(-3/2, -3/2)$ and its global minimum of approximately 1.3 at $\left(2, -\sqrt{2}\right)$ and $\left(-\sqrt{2}, 2\right)$.

Figure 10.41 shows a graph and a contour plot of f on $\mathcal{D}$. The interior point $(1, 1)$ is shown in the figure. To it corresponds $(1, 1, 16)$, which appears to be on top of a small rise, a local maximum. The point $(0, 0, 15)$ appears to be neither a local minimum nor a local maximum. This inference from the figure is strengthened by inspecting the intersection of the surface with the

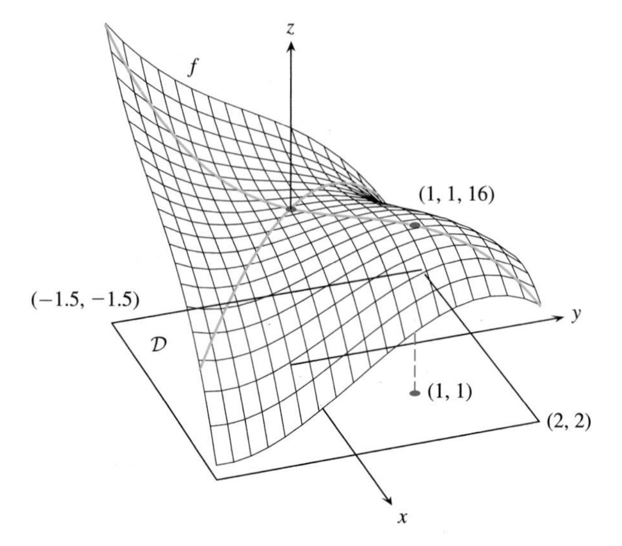

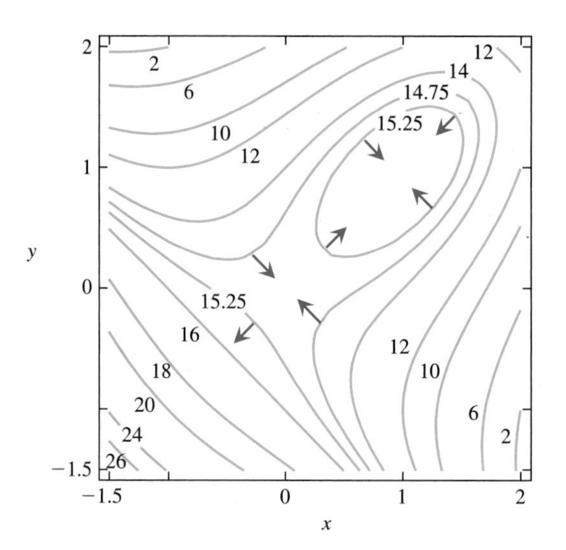

FIGURE 10.41 Graph and contour plot showing local and global extrema of f.

two planes with equations $y = x$ and $y = -x$. These curves of intersection appear in blue. The point $(0, 0, 15)$ appears to be a local minimum on one of these curves and a local maximum on the other.

The contour plot at the bottom of the figure suggests the same conclusions. At four points on the elliptical part of the contour labeled with 15.25, we have plotted four vectors in the directions of the gradient vectors at these points. They show the direction in which f is increasing. Near the origin there are four vectors (one vector is from the first group); they, too, point in the direction of increasing function values. It appears that the point $(0, 0, 15)$ is neither a local maximum nor a local minimum.

Method of Steepest Descent Cauchy's method of steepest descent for locating the minimum of a real-valued function f of two or more variables is based on the fact that, at a point $\mathbf{a}$, the vector $-\nabla f(\mathbf{a})$ points in the direction in which f is decreasing most rapidly. The French mathematician Augustin Cauchy (1789–1857) discovered the method in 1847. Efficient versions of the method of steepest descent are used today to solve complex optimization problems.

We discuss a basic version of the method of steepest descent, one that can be done on a graphing calculator. To understand the ideas and calculations, it is particularly important to follow the example with paper, pencil, and calculator, checking the details as you go. Briefly described, the method of steepest descent goes like this: Somehow locate a base point $\mathbf{a}_1$ close to a point at which the function has a (local) minimum, calculate the vector $-\nabla f(\mathbf{a}_1)$ with the thought in mind that this vector may point in the general direction of a minimum, run a line through $\mathbf{a}_1$ in this direction, and solve a one-dimensional minimization problem along this line. This process often results in a point $\mathbf{a}_2$ closer to the minimum point than $\mathbf{a}_1$. Repeat this procedure, with $\mathbf{a}_2$ as the base point. Continue until the base points are sufficiently close to a local minimum.

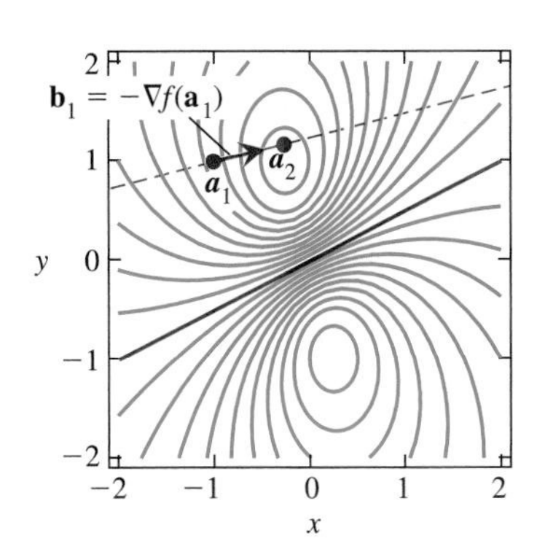

FIGURE 10.42 Steepest descent from $\mathbf{a}_1$.

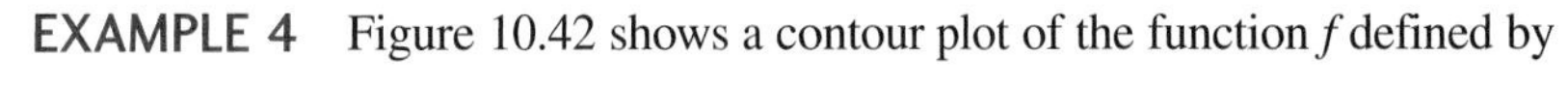

EXAMPLE 4 Figure 10.42 shows a contour plot of the function f defined by

$$f(x, y) = \frac{x - 2y}{1 + 2x^2 + y^2}, \qquad (x, y) \in [-2, 2] \times [-2, 2]. \tag{2}$$

Apply the method of steepest descent to calculate $\mathbf{a}_2$, given $\mathbf{a}_1 = (-1, 1)$.

Solution

Because the denominator in the fraction in (2) is always positive, the sign of $f(x, y)$ depends entirely upon the expression $x - 2y$. The one straight contour shown in Fig. 10.42 is the 0 contour, which is the set of all points (x, y) for which $f(x, y) = 0$, that is $y = x/2$. Above this line the function is always negative; below the line it is positive. The contours are spaced by 0.1, so, starting with the 0 contour, the contours are $\pm 0.1, \pm 0.2, \ldots, \pm 1.0$. By counting contours, it is clear that $-0.8 \le f(\mathbf{a}_1) \le -0.7$ and that the minimum of f is less than -1.0. By actual calculation, $f(\mathbf{a}_1) = f(-1, 1) = -\frac{3}{4} = -0.75$.

The negative gradient of f is given by

$$\begin{aligned}-\nabla f(x, y) &= -\langle f_x(x, y), f_y(x, y)\rangle \\ &= -\left\langle \frac{-2x^2 + 8xy + y^2 + 1}{(1 + 2x^2 + y^2)^2}, \frac{2(-2x^2 - xy + y^2 - 1)}{(1 + 2x^2 + y^2)^2} \right\rangle.\end{aligned}$$

Evaluating this at $\mathbf{a}_1 = \langle -1, 1\rangle$,

$$\mathbf{b}_1 = -\nabla f(\mathbf{a}_1) = \left\langle \frac{1}{2}, \frac{1}{8} \right\rangle = \langle 0.5, 0.125\rangle.$$

This vector, $\mathbf{b}_1$, based at $\mathbf{a}_1$, is shown in Fig. 10.42. It points in the direction in which f is decreasing most rapidly. It is perpendicular to the level curve through $\mathbf{a}_1$. The ray in the domain of f in the direction of $\mathbf{b}_1$ and passing through $\mathbf{a}_1$ is described by

$$\mathbf{r}(t) = \mathbf{a}_1 + t\mathbf{b}_1, \qquad t \in I_1,$$

where I_1 is an interval containing 0. This line, shown in Fig. 10.42 as a dotted line passing through $\mathbf{a}_1$, gives a one-dimensional section through the domain of f. The values of f along this line are given by

$$z = g_1(t) = f(\mathbf{r}(t)) = f(\mathbf{a}_1 + t\mathbf{b}_1) = f\left(-1 + \tfrac{1}{2}t, 1 + \tfrac{1}{8}t\right), \qquad t \in I_1. \tag{3}$$

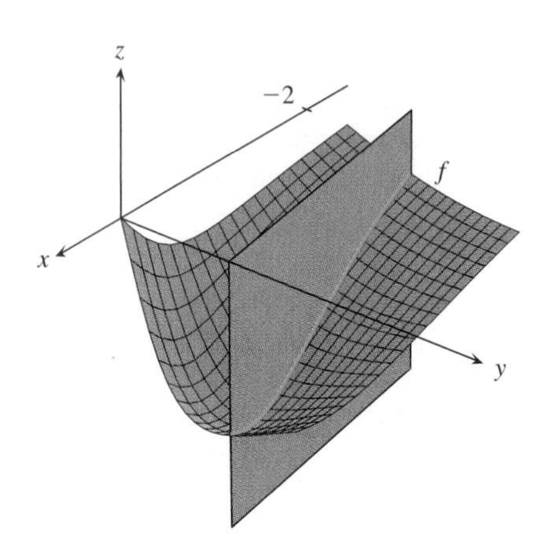

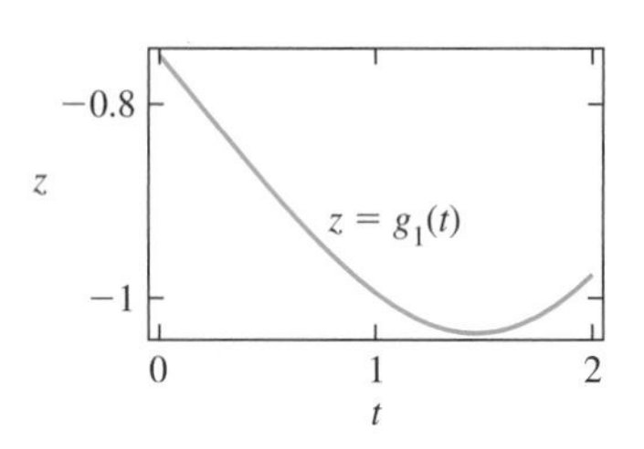

FIGURE 10.43 One-dimensional minimum on the graph of $w = g(t) = f(\mathbf{r}(t))$.

Figure 10.43 shows the graph of f for $(x, y) \in [-2, 0] \times [0, 2]$, which is the part of the graph containing the minimum point. Also shown is the graph of the vertical plane containing the points $\mathbf{a}_1$ and $\mathbf{a}_1 + \mathbf{b}_1$. This plane intersects the graph of f in a curve, whose low point we locate using one-dimensional methods. Figure 10.43 also shows the graph of the function g_1 described in (3). Specifically, the figure shows the graph of g_1 on the interval $I_1 = [0, 2]$. The interval I_1 was found experimentally, by trying a few intervals containing $t = 0$.

Depending upon the form of f, evaluating or plotting the function g_1 can be quite tedious if all calculations must be done by hand. However, with a CAS or a calculator, if f can be defined as a function requiring two inputs, all that is needed to calculate f for the inputs formed from the two coordinates of $\mathbf{a}_1 + t\mathbf{b}_1$.

Use the graph of g_1 to approximate the t-coordinate of the one-dimensional minimum point. We used the cursor to obtain $t = 1.45$, with which we can calculate the next base point $\mathbf{a}_2$:

$$\mathbf{a}_2 = \mathbf{a}_1 + 1.45\mathbf{b}_1 = \langle -0.275, 1.18125 \rangle.$$

The value of f at this point is

$$f(\mathbf{a}_2) = g_1(1.45) = -1.03569.$$

The steepest descent algorithm is: Repeat this procedure, thus obtaining a sequence of base points $\mathbf{a}_1, \mathbf{a}_2, \mathbf{a}_3, \ldots$. Often, these points approach the point $\mathbf{a}$ at which f has a minimum, where we expect the gradient of f to be zero. To decide when the sequence $\mathbf{a}_1, \mathbf{a}_2, \mathbf{a}_3, \ldots$ is sufficiently close to $\mathbf{a}$, we may calculate the lengths of $\nabla f(\mathbf{a}_1), \nabla f(\mathbf{a}_2), \nabla f(\mathbf{a}_3), \ldots$, halting the algorithm when the length of the gradient is sufficiently small. This halting rule is based on our expectation that near a minimum the gradient of f will be close to zero.

See the Student Project, "Cauchy's Method of Steepest Descent," at the end of this chapter for more detail on this algorithm.

Exercises 10.5

Exercises 1–12: Use the Candidate Theorem to locate points at which f may have extrema.

1. $f(x, y) = 2y^2 - x(x - 1)^2$
2. $f(x, y) = y^2 + 2y - x^2(x - 1)$
3. $f(x, y) = x^2 - xy + y^2 + 3x$
4. $f(x, y) = x^2 - 3xy + y^2 + y$
5. $f(x, y) = x \cos y,\ 0 < x, y < 2\pi$
6. $f(x, y) = x \sin y,\ 0 < x, y < 2\pi$
7. $f(x, y) = x^3 - 3xy + y^3$
8. $f(x, y) = y\sqrt{x} - y^2 - x + 6y,\ x \geq 0$
9. $f(x, y) = x^4 + y^4 - 4xy$
10. $f(x, y) = xy/(2 + x^4 + y^4)$
11. $f(x, y, z) = xyz - x^2 - y^2 - z^2$
12. $f(x, y, z) = zy + xz^2 - z^2 - y - x^2$

Exercises 13–22: For the given function, locate its global extrema and corresponding values.

13. $x^3 - y^2 + 6xy, 0 \le x \le 1, 0 \le y \le x$

14. $x^3 - 3x + y^2, 0 \le x \le 2, -1 \le y \le 1$

15. $x^2ye^{-x-y}, 0 \le x, y \le 3$

16. $e^{-x} \sin y, 0 \le x \le 1, 0 \le y \le \pi$

17. $(x - y + 1)^2, -1 \le x \le 1, 0 \le y \le 2$

18. $x^3 + y^3 - 3xy, 0 \le x, y \le 2$

19. $x\sqrt{y} - x^2 + 9x - y, x, y \in [0, 10]$

20. $xy + 20/x + 20/y, x, y \in [1, 3]$

21. $2x^2 + x + y^2 - 2, x^2 + y^2 \le 4$

22. $2x^3 + y^4, x^2 + y^2 \le 1$

Exercises 23–28: Solve and argue why your answer is the absolute minimum or maximum.

23. What is the largest possible volume of an open, rectangular aquarium made from 12 ft^2 of clear glass? Ignore the thickness of the glass.

24. An open, rectangular aquarium is to hold 3.75 ft^3 of water. The bottom is to be made from slate and the sides from glass. If slate costs five times as much as glass, what dimensions should be chosen to minimize the cost? Ignore the thickness of the materials.

25. On the graph described by $x^2 + y^2 + 1/(x^4y^2) = z^2$, find the point closest to the origin.

26. On the graph described by $2x + 3y - z = 10$, find the point closest to the origin.

27. A rectangular box is inscribed in the tetrahedron formed in the first octant by the coordinate planes and the plane with equation $x/a + y/b + z/c = 1$, where $a, b, c > 0$. Assume that the faces of the box are parallel to the coordinate planes. What are the dimensions of the box of largest volume?

28. A rectangular box is inscribed in the ellipsoid described by $x^2/a^2 + y^2/b^2 + z^2/c^2 = 1$ so that its faces are parallel to the coordinate planes. What are the dimensions of the box of largest volume?

29. Find the absolute extrema of the function

$$f(x, y) = \sin x + \sin y + \sin(x + y), 0 \le x, y \le \pi.$$

30. For all $x, y > 0$, the equation $e^z + 3xyz = 0$ determines exactly one value of z and so determines z as a function f of x and y. Find the point on the graph of f closest to the origin.

31. Among all triangles with fixed perimeter p, what are the sides and area of the triangle with maximum area? Is there a triangle with minimum area? *Hint:* The area A of a triangle with sides of length a, b, and c is $A = \sqrt{s(s-a)(s-b)(s-c)}$, where $s = \frac{1}{2}(a + b + c)$.

32. Explain in detail how the Candidate Theorem from Chapter 4 was used to prove the Candidate Theorem for functions of several variables.

33. In Example 2, give a convincing argument as to why "it is clear that the maximum volume occurs when $\ell + 2(w + h) = 108$."

T **34.** Using the notation in Example 4, let

$$f(x, y) = x^2 + 3y^2 - 2xy - x.$$

A contour plot of f near $\mathbf{a}_1 = (1/2, 1/2)$ is shown in the accompanying figure. Calculate the second base point $\mathbf{a}_2$ of the steepest descent algorithm. Plot g_1 on $[0, 0.5]$ and solve the one-dimensional minimum problem symbolically. Show that $\mathbf{a}_2$ is not a minimum point for f by locating a point at which f is smaller than at $\mathbf{a}_2$.

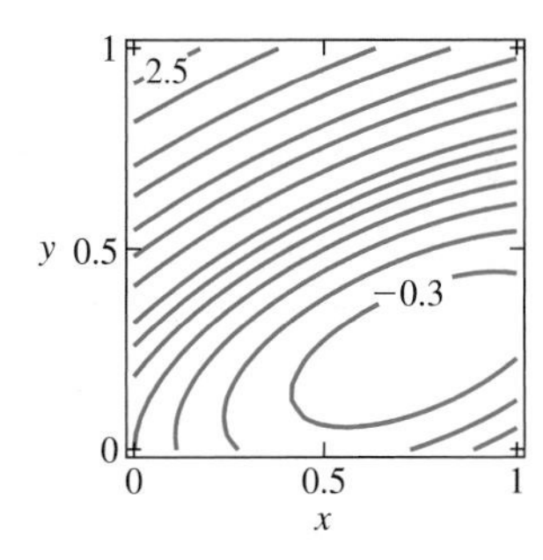

Figure for Exercise 34.

T **35.** Using the notation in Example 4, let

$$f(x, y) = 3x^2 + 2y^2 - xy - 2x.$$

A contour plot of f near $\mathbf{a}_1 = (0, 0)$ is shown in the accompanying figure. Calculate the second base point $\mathbf{a}_2$ of the steepest descent algorithm. Plot g_1 on $[0, 0.5]$ and solve the one-dimensional minimum problem symbolically. Show that $\mathbf{a}_2$ is not a minimum point for f by locating a point at which f is smaller than at $\mathbf{a}_2$.

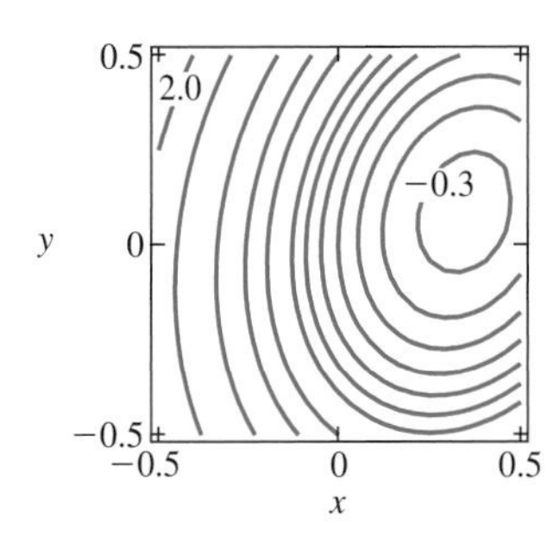

Figure for Exercise 35.

T **36.** Using notation similar to that in Example 4, let

$$f(x, y, z) = x^2 + 2y^2 + z^2 - xz + 2x.$$

Let the first base point be $\mathbf{a}_1 = (0, 0, 0)$. Calculate the second base point $\mathbf{a}_2$ of the steepest descent algorithm. Plot g_1 on $[0, 1]$ and solve the one-dimensional minimum problem symbolically. Show that $\mathbf{a}_2$ is not a minimum point for f by locating a point at which f is smaller than at $\mathbf{a}_2$.

37. Using the notation in Example 4, let

$$f(x, y, z) = 2x^2 + 2y^2 + z^2 - xz + yz + z.$$

Let the first base point be $\mathbf{a}_1 = (0, 0, -1)$. Calculate the second base point $\mathbf{a}_2$ of the steepest descent algorithm. Plot g_1 on $[0, 0.5]$ and solve the one-dimensional minimum problem symbolically. Show that $\mathbf{a}_2$ is not a minimum point for f by locating a point at which f is smaller than at $\mathbf{a}_2$.

38. The accompanying figure shows a cross-section of a proposed tunnel. The lower, middle, and upper rectangles provide for utilities, a roadway, and ventilation. If the tunnel has a circular cross section with radius 1 unit and the air and utility rectangles have equal area, what are the angles a and b giving rectangles whose combined area is as large as possible? *Hints:* It is sufficient to consider one-quarter of the cross section. Use angles a and b to find vertices of the air and roadway rectangles. The combined area of regions A and B is

$$f(a,b) = \cos a \sin a + \cos b(\sin b - \sin a).$$

The domain $\mathcal{D}$ of f is the set of all $(a, b) \in R^2$ for which $0 \le a \le b \le \pi/2$. In the equations $f_x(a, b) = 0$ and $f_y(a, b) = 0$, let $u = \cos a$ and $v = \cos b$.

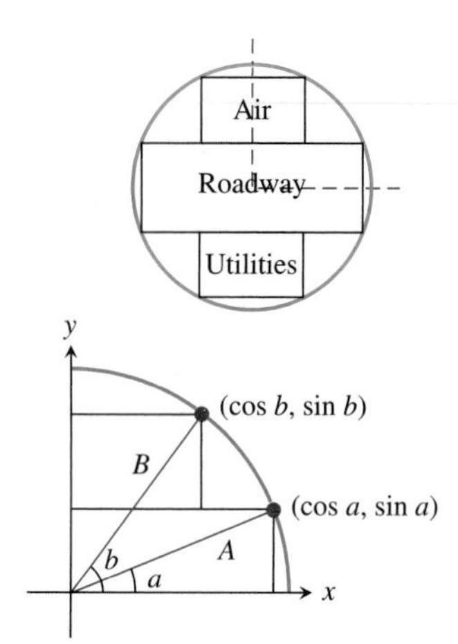

Figure for Exercise 38.

39. Find the absolute maximum of the function defined by

$$f(t, y) = \frac{t}{t^2 + 4}\left(\frac{4a^2}{t^2 + 4} - \frac{2ay}{\sqrt{t^2 + 4}} + y^2\right),$$

where $0 < t < \infty$, $0 \le y \le a/2$, and a is a constant.

40. Referring to Example 3, give a direct argument that $(0, 0)$ is neither a local minimum nor a local maximum. *Hint:* Consider $f(x, x)$ and $f(x, -x)$ for small $|x|$.

10.6 Second Derivatives Test

In Chapter 4 we discussed the Second Derivative Test for local extrema and the Concavity Test for intervals on which a function is concave up or concave down. We also gave a criterion for locating points of inflection. We now give the Second Derivatives Test for functions of several variables. This test can be useful in classifying a point on the graph of a function $z = f(x, y)$ as a local minimum, a local maximum, or a *saddle point.*

Let f be a differentiable function of two variables and S its graph. A point (x, y, z) on S at which the tangent plane is horizontal is called a *stationary point* of S; the corresponding point (x, y) in the domain of f is also called a stationary point. Stationary points are divided into three categories: local minima, local maxima, and saddle points. Examples of such points are flagged in Fig. 10.44. Flag S.P. marks a saddle point (the crest of a pass), flag m marks a local minimum (the deepest point of a mountain lake), and flag M marks a local maximum (the highest point of a peak).

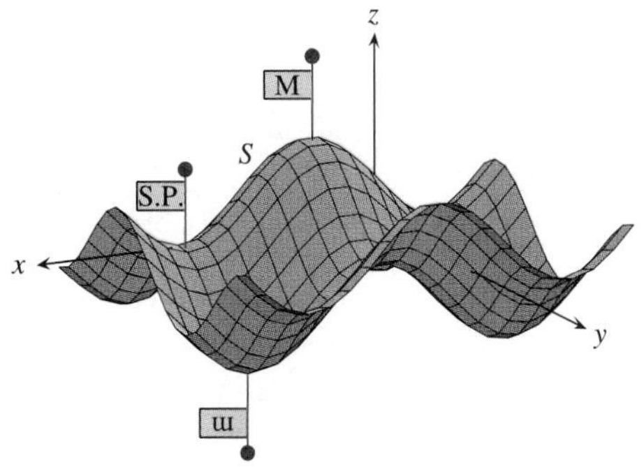

FIGURE 10.44 Saddle point, local minimum, and local maximum on the graph of a function.

DEFINITION Stationary and Saddle Points

Let f be a differentiable real-valued function defined on a set $\mathcal{D} \subset R^2$ and let S be the graph of f. A point $(x, y, z) \in S$ is a **stationary point** of S (or (x, y) is a stationary point of f) if $\nabla f(x, y) = \mathbf{0}$. A stationary point (x, y, z) of S is a **saddle point** of S (or (x, y) is a saddle point of f) if f has neither a local maximum nor a local minimum at (x, y).

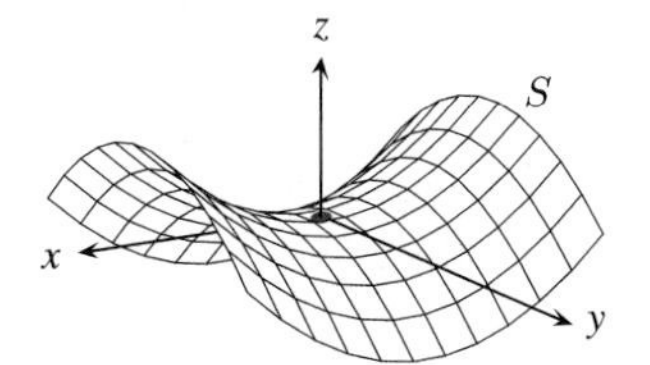

FIGURE 10.45 A saddle point $(0, 0)$ on a hyperbolic paraboloid.

EXAMPLE 1 Show that the point $(0, 0, 0)$ on the graph of the function

$$f(x, y) = x^2 - y^2, \qquad (x, y) \in R^2,$$

is a saddle point of f. The graph of f, shown in Fig. 10.45 as S, is called a *hyperbolic paraboloid.*

Solution

To show that $(0, 0, 0)$ is a saddle point of S, we first show that it is a stationary point. For this, note that

$$\nabla f(x, y) = \langle 2x, -2y \rangle \quad \text{and} \quad \nabla f(0, 0) = \langle 0, 0 \rangle.$$

The point $(0, 0)$ is a saddle point because it is neither a local maximum nor a local minimum. It is not a local maximum because along the x-axis the surface rises from 0 at $(0, 0)$, that is, for any $x \neq 0$,

$$f(x, 0) = x^2 > 0 = f(0, 0).$$

The point $(0, 0)$ is not a local minimum because along the y-axis the surface falls from 0 at $(0, 0)$, that is, for any $y \neq 0$,

$$f(0, y) = -y^2 < 0 = f(0, 0).$$

Hence, the point $(0, 0, 0)$ is a saddle point of f.

In this example we were able to argue directly from the form of the function that $(0, 0)$ was a saddle point. For more complicated functions, the Second Derivatives Test provides a convenient way to categorize many stationary points.

THEOREM Second Derivatives Test

Let f be a real-valued function defined on $\mathcal{D} \subset R^2$ and having continuous third-order partial derivatives. Let $\mathbf{a}$ be an interior stationary point of $\mathcal{D}$. Define the **discriminant** of f at $\mathbf{a}$ to be the number

$$D(\mathbf{a}) = f_{xx}(\mathbf{a})f_{yy}(\mathbf{a}) - (f_{xy}(\mathbf{a}))^2.$$

Case 1: If $D(\mathbf{a}) > 0$ and $f_{xx}(\mathbf{a}) > 0$, then $\mathbf{a}$ is a local minimum point.

Case 2: If $D(\mathbf{a}) > 0$ and $f_{xx}(\mathbf{a}) < 0$, then $\mathbf{a}$ is a local maximum point.

Case 3: If $D(\mathbf{a}) < 0$, then $\mathbf{a}$ is a saddle point.

Case 4: If $D(\mathbf{a}) = 0$, the test fails.

In Cases 1 and 2 for local extrema, the conditions $f_{xx}(\mathbf{a}) > 0$ and $f_{xx}(\mathbf{a}) < 0$ can be replaced by $f_{yy}(\mathbf{a}) > 0$ and $f_{yy}(\mathbf{a}) < 0$.

The Second Derivatives Test can be made plausible with the help of the Taylor Inequality. From (13) in Section 10.3,

$$\left| f(\mathbf{a} + \mathbf{h}) - \left(f(\mathbf{a}) + f_x(\mathbf{a})h_1 + f_y(\mathbf{a})h_2 + \frac{1}{2!}(f_{xx}(\mathbf{a})h_1^2 + 2f_{xy}(\mathbf{a})h_1h_2 + f_{yy}(\mathbf{a})h_2^2) \right) \right| \le \frac{1}{3!}Mb^3, \tag{1}$$

where $h = \langle h_1, h_2 \rangle$.

In determining the nature of f at a stationery point $\mathbf{a}$, we are concerned only with the behavior of f near $\mathbf{a}$. Hence, we choose b so that the difference $Mb^3/3!$ is negligible (see Fig. 10.46). For $\mathbf{a}$ and $\mathbf{a} + \mathbf{h}$ in S_b, it follows from 1 that

$$f(\mathbf{a} + \mathbf{h}) \approx f(\mathbf{a}) + f_x(\mathbf{a})h_1 + f_y(\mathbf{a})h_2 + \frac{1}{2!}(f_{xx}(\mathbf{a})h_1^2 + 2f_{xy}(\mathbf{a})h_1h_2 + f_{yy}(\mathbf{a})h_2^2). \tag{2}$$

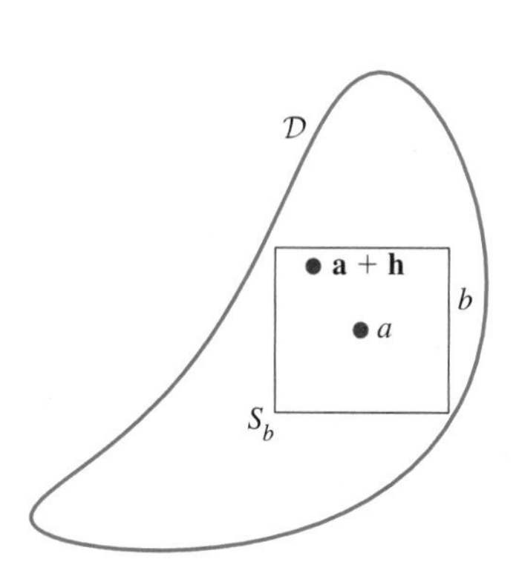

FIGURE 10.46 The domain $\mathcal{D}$ of f, the square region S_b, and points a and $\mathbf{a} + \mathbf{h}$ within S_b.

Because $\mathbf{a}$ is a stationary point, the second and third terms on the right side of (2) are zero. If we write $A = f_{xx}(\mathbf{a})$, $B = f_{xy}(\mathbf{a})$, and $C = f_{yy}(\mathbf{a})$, then (2) becomes

$$f(\mathbf{a} + \mathbf{h}) \approx f(\mathbf{a}) + \frac{1}{2}(Ah_1^2 + 2Bh_1h_2 + Ch_2^2).$$

We assume in what follows that $A \neq 0$. We leave as Exercise 13 the $A = 0$ case. Factor A from all terms except the first:

$$f(\mathbf{a} + \mathbf{h}) \approx f(\mathbf{a}) + \frac{A}{2}\left(h_1^2 + \frac{2B}{A}h_1h_2 + \frac{C}{A}h_2^2 \right).$$

Complete the square by adding and subtracting $B^2h_2^2/A^2$:

$$f(\mathbf{a} + \mathbf{h}) \approx f(\mathbf{a}) + \frac{A}{2}\left(\left(h_1 + \frac{B}{A}h_2 \right)^2 + \frac{AC - B^2}{A^2}h_2^2 \right).$$

Write this expression in terms of the discriminant $D(\mathbf{a}) = AC - B^2$:

$$f(\mathbf{a} + \mathbf{h}) \approx f(\mathbf{a}) + \frac{A}{2}\left(\left(h_1 + \frac{B}{A}h_2\right)^2 + \frac{D(\mathbf{a})}{A^2}h_2^2\right). \tag{3}$$

We are now ready for Cases 1–4 of the Second Derivatives Test.

Case 1: If $D(\mathbf{a}) > 0$ and $A = f_{xx}(\mathbf{a}) > 0$, it follows directly from (3) that $f(\mathbf{a} + \mathbf{h}) \geq f(\mathbf{a})$ for all points $\mathbf{a} + \mathbf{h}$ sufficiently close to $\mathbf{a}$. Hence $\mathbf{a}$ is a local minimum.

Case 2: Minor variation on Case 1. See Exercise 14.

Case 3: Assume that $D(\mathbf{a}) < 0$. We also assume that $A > 0$, leaving for Exercise 15 the subcase in which $A < 0$. We first note that if we take $\mathbf{h} = \langle h_1, 0\rangle \neq \mathbf{0}$ in (3) then $f(\mathbf{a} + \mathbf{h}) > f(\mathbf{a})$. Hence, $f(\mathbf{a})$ is not a local maximum. Next we show that $f(\mathbf{a})$ is not a local minimum. There are two parts to this. First, if $B = 0$ and we take $\mathbf{h} = \langle 0, h_2\rangle \neq \mathbf{0}$ in (3), then (3) becomes

$$f(\mathbf{a} + \mathbf{h}) \approx f(\mathbf{a}) + \frac{A}{2}\cdot\frac{D(\mathbf{a})}{A^2}h_2^2 < f(\mathbf{a}).$$

Hence, $f(\mathbf{a})$ is not a local minimum. Secondly, if $B \neq 0$ and we take $\mathbf{h} = \langle h_1, (-A/B)h_1\rangle \neq \mathbf{0}$ in (3), then

$$f(\mathbf{a} + \mathbf{h}) \approx f(\mathbf{a}) + \frac{A}{2}\cdot\frac{D(\mathbf{a})}{A^2}\left(-\frac{A}{B}h_1\right)^2 < f(\mathbf{a}),$$

again showing that $f(\mathbf{a})$ is not a local minimum. It now follows that $\mathbf{a}$ is a saddle point.

Case 4: This case is discussed in Examples 3 and 4.

EXAMPLE 2 Use the Second Derivatives Test to classify the stationary points of the function

$$f(x, y) = \frac{1}{3}x^3 + xy^2 - x, \qquad (x, y) \in R^2.$$

Solution

A point (x, y) is a stationary point of f if $\nabla f(x, y) = \mathbf{0}$. We therefore calculate the partial derivatives

$$f_x = x^2 + y^2 - 1$$
$$f_y = 2xy,$$

set both f_x and f_y equal to zero, and solve the resulting system of equations. From the equation $2xy = 0$, either $x = 0$ or $y = 0$. If $x = 0$, then from the equation $x^2 + y^2 - 1 = 0$ we see that $y^2 = 1$; hence, $y = \pm 1$. In the same way, if $y = 0$, then $x = \pm 1$. Hence, the stationary points are $(0, -1)$, $(0, 1)$, $(-1, 0)$, and $(1, 0)$. To classify these points, we try the Second Derivatives Test. For this, we calculate the second-order partials

$$f_{xx} = 2x$$
$$f_{xy} = 2y$$
$$f_{yy} = 2x.$$

The discriminant $D(\mathbf{a})$ of f at a general point $\mathbf{a} = (x, y)$ is

$$D(\mathbf{a}) = f_{xx}f_{yy} - f_{xy}^2 = 4x^2 - 4y^2.$$

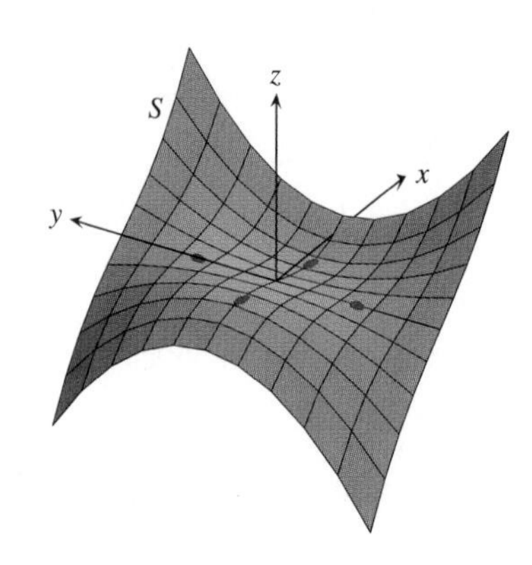

We test the four stationary points of f:

$(0,-1)$: Because $D(0,-1) = -4 < 0$, $(0,-1)$ is a saddle point.

$(0,1)$: Because $D(0,1) = -4 < 0$, $(0,1)$ is a saddle point.

$(-1,0)$: Because $D(-1,0) = 4$ and $f_{xx}(-1,0) = -2 < 0$, $(-1,0)$ is a local maximum.

$(1,0)$: Because $D(1,0) = 4$ and $f_{xx}(1,0) = 2 > 0$, $(1,0)$ is a local minimum.

A graph and contour plot of f are shown in Fig. 10.47, with the stationary points marked with black dots. The local minimum at $(1, 0, -2/3)$ is directly beneath the positive x-axis. If a pool were formed at this point, perhaps from last winter's snowpack on the higher slopes, then the water level would rise to $z = 0$ and then flow over the saddle points.

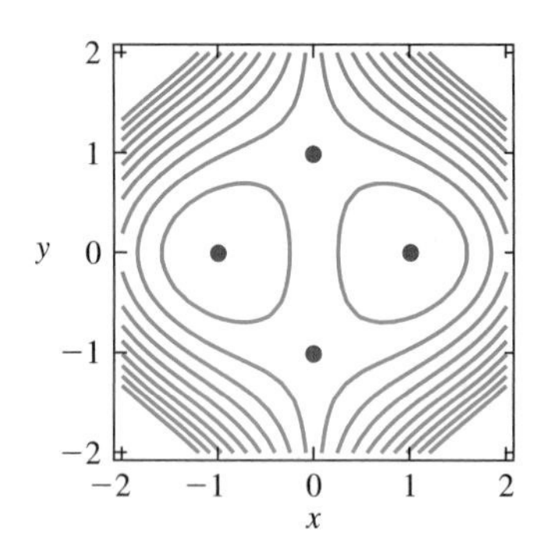

FIGURE 10.47 Contour plot and graph, with two saddle points, a local minimum, and a local maximum, for the function f.

EXAMPLE 3 Classify the stationary points of the function

$$f(x, y) = x^3 - 3xy^2, \qquad (x, y) \in R^2.$$

Solution

Figure 10.48 shows the graph of f, often called a *monkey saddle* because both tail and legs are provided for. To find the stationary points of f, we first calculate the partial derivatives of f.

$$f_x = 3x^2 - 3y^2$$
$$f_y = -6xy.$$

From the equation $-6xy = 0$ we know that $x = 0$ or $y = 0$. Substituting each of these values into the equation $3x^2 - 3y^2 = 0$ shows that the only stationary point is $(0, 0)$. The second-order partial derivatives are

$$f_{xx} = 6x$$
$$f_{xy} = -6y$$
$$f_{yy} = -6x.$$

At $(0,0)$, $D(\mathbf{a}) = (6 \cdot 0)(-6 \cdot 0) - (-6 \cdot 0)^2 = 0$. Hence, the Second Derivatives Test fails.

To determine whether the stationary point $(0, 0)$ is or is not a saddle point, we must either rely on visual evidence or go back to the definition of *saddle point.* It is reasonably clear from Fig. 10.48 that within any small region centered at the origin there are both points of the surface that are higher than $f(0,0) = 0$ and points lower than this value. Hence, the origin is a saddle point. An analytical argument to verify this fact is not difficult. It is enough to notice that $f(x,0) = x^3$ is positive for $x > 0$ and negative for $x < 0$.

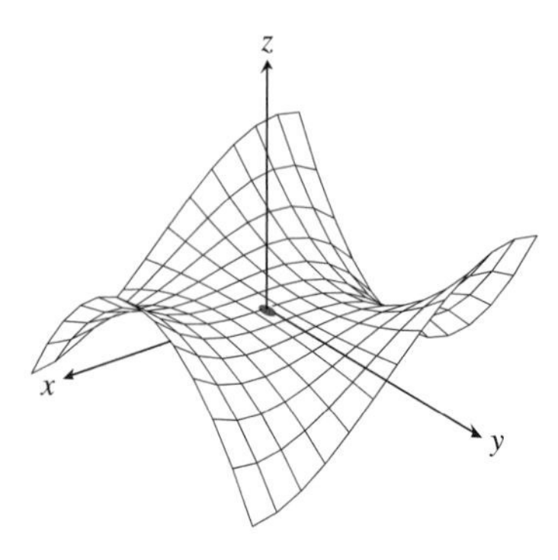

FIGURE 10.48 Monkey saddle.

In the preceding example, $\mathbf{a} = (0, 0)$ is a saddle point and $D(\mathbf{a}) = 0$. It can also happen that a function has a local extremum at a point at which the discriminant is 0.

EXAMPLE 4 Show that the function

$$f(x,y) = x^2, \qquad (x,y) \in R^2,$$

has a local minimum at each point $(0,y)$ of the y-axis and $D(0,y) = 0$.

Solution

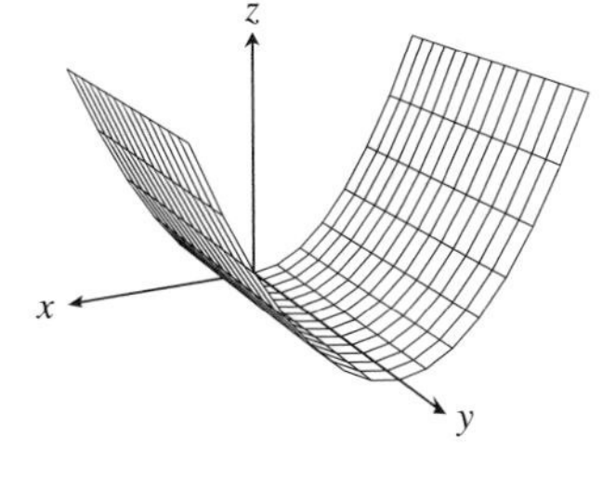

FIGURE 10.49 Parabolic cylinder.

The graph of f is the parabolic cylinder shown in Fig. 10.49. From the graph it appears that every point on the y-axis is a stationary point and a local minimum. We verify this observation by calculating

$$f_x = 2x$$
$$f_y = 0.$$

It follows that each point $(0,y)$ on the y-axis is a stationary point. Because $f(x,y) \geq 0$ everywhere, any points (x,y) at which $f(x,y) = 0$ must be minimum points. Hence, each point $(0,y)$ on the y-axis is a minimum point.

Next we show that the Second Derivatives Test fails at each of these points. The second-order partial derivatives are $f_{xx} = 2$, $f_{xy} = 0$, and $f_{yy} = 0$. At each stationary point $(0,y)$, $D(0,y) = 0$. Hence the Second Derivatives Test fails.

Examples 3 and 4 show that the Second Derivatives Test fails if $D(\mathbf{a}) = 0$, that is, it fails to discriminate among saddle points and local extrema.

Exercises 10.6

1. The quadratic function $f(x,y) = x^2 - y^2$ in Example 1 is one of the simplest functions having a saddle point. As in Example 1, use the definition of saddle point to show that the quadratic function $f(x,y) = xy$ has a saddle point at the origin. Graph this function by hand and erect a flag at the saddle point. *Hint:* The graph resembles the graph shown in Fig. 10.45, but rotated through 45°.

2. For the general quadratic function

$$f(x,y) = Ax^2 + Bxy + Cy^2 + Dx + Ey + F$$

defined on R^2, show that there may be no, one, or infinitely many stationary points. Give specific equations to illustrate these three cases.

Exercises 3–12: Use the Second Derivatives Test to help classify the stationary points of the function. Unless noted, each function is defined on R^2.

3. $f(x,y) = x^2 - y^2 + 3xy$

4. $f(x,y) = x + xy^3 + x^3$

5. $f(x,y) = x^2 - (2y + 3)^2$

6. $f(x,y) = 1 - x^2 - (y - 5)^2$

7. $f(x,y) = x^3 + y^3 - 3xy$

8. $f(x,y) = xye^{-(x^2+y^2)/2}$

9. $f(x,y) = x^4 + y^4 - 4xy$

10. $f(x,y) = x/y + 27/x - y$, $x, y \neq 0$

11. $f(x,y) = (y - 1)\ln(xy)$, $x, y > 0$

12. $f(x,y) = x\cos y$, $0 \leq y \leq \pi$

13. In the proof of the Second Derivatives Test, we assumed that $A \neq 0$. Give an argument in case $A = 0$.

14. Give the argument for Case 2 of the Second Derivatives Test.

15. Give the argument for the subcase $A < 0$ of Case 3 of the Second Derivatives Test.

16. Classify the stationary points of

$$f(x, y) = x^2 - xy^4 + x.$$

17. Classify the stationary points of

$$f(x, y) = x^3 \cos y + xy^2, 0 < x, y < 2.$$

18. Classify the stationary points of

$$f(x, y) = (x^2 - y^3)e^{-(x^2+y^2)/2}, (x, y) \in R^2.$$

19. Use the Second Derivatives Test to show that the point on the plane with equation $x + \frac{1}{2}y + \frac{1}{3}z = 1$ closest to the point $(5, 4, 3)$ corresponds to a local minimum of the square of a distance function.

20. Use the Second Derivatives Test to show that the point on the plane with equation $2x + 3y + z = 1$ closest to the point $(10, 5, 5)$ corresponds to a local minimum of the square of a distance function.

21. Use the Second Derivatives Test to show that the point on the surface with equation $z^2 = xy + 1$ closest to the origin corresponds to a local minimum of the square of a distance function.

22. Use the Second Derivatives Test to show that the point on the surface with equation $z^2 = x^2y + 2$ closest to the origin corresponds to a local minimum of the square of a distance function.

23. On the accompanying contour map of a function f, there are three stationary points. Given that "darker is lower," classify these points as local minima, maxima, or saddle points.

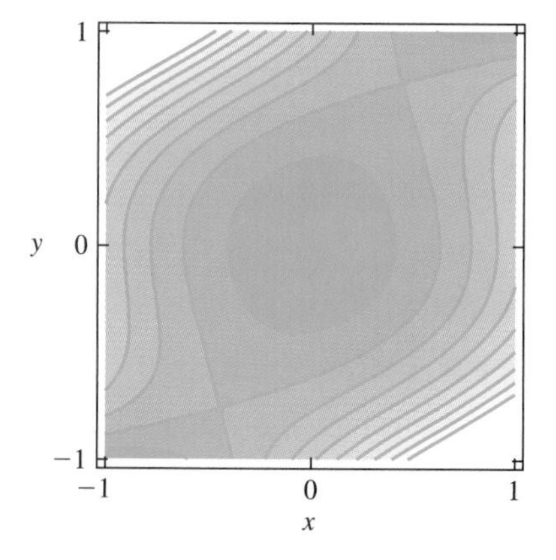

Figure for Exercise 23.

24. On the accompanying contour map of a function f, there are three stationary points. Given that "darker is lower," classify these points as local minima, maxima, or saddle points.

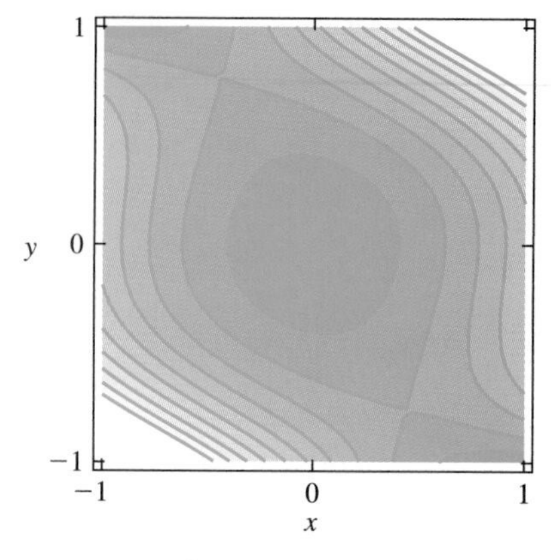

Figure for Exercise 24.

25. Use the Second Derivatives Test to classify the stationary points of the function

$$f(x, y) = x^2 + xy - ky^2 + 2y, \qquad (x, y) \in R^2.$$

Your results will fall into three cases. Describe the most important features of the graph in each case.

Exercises 26–30: These exercises outline the method of least squares, which Victor Katz calls "the most important statistical method of the nineteenth century... a way of collecting numerous observational measurements of a particular event into a single 'best' result." Katz's comments are given in his A History of Mathematics *(New York: HarperCollins, 1993). The method of least squares, most likely invented by both Adrien-Marie Legendre and Carl Friedrich Gauss in the period 1805–1809, became and remains a standard method in astronomy, surveying, statistics, and many other areas in which a "best fit" is wanted to represent empirical data scattered about a line. Look on your calculator for least squares calculation and plotting facilities, probably in the statistics applications. Exercises 29 and 30 were taken from* A First Course in Statistics *(New York: HarperCollins, 1992), written by Sellers, Vardeman, and Hackert.*

26. Suppose we have a set of n data points

$$(x_1, y_1), (x_2, y_2), \ldots, (x_n, y_n).$$

These may be, for example, the one-mile-run records shown in the accompanying figure. The data were given in Chapter 1. In this case the x_i are dates measured in years and the y_i are times measured in minutes. Each data point (x_i, y_i) gives the record time y_i for running the mile and the date on which the record was set. If we believe that the data are somewhat linear and wish to find the line $y = mx + b$ of "best fit," we can use the method of least squares. To attain least squares, we minimize a function of the variables m and b that determine the slope and intercept of the line $y = mx + b$ of best fit. We choose a point (m, b) in (m, b)-space for which the corresponding line best fits the data. How do we define "best"? Among the possible answers, Legendre and Gauss selected the pair (m, b) that minimizes the sum of the squares of the *deviations* d_i. A typical deviation is shown in the second accompanying figure. For each data point (x_i, y_i), the deviation d_i is the number $d_i = y_i - (mx_i + b)$. We wish to choose m and b so that the function

$$f(m, b) = \sum_{i=1}^{n} d_i^2 = \sum_{i=1}^{n} (y_i - (mx_i + b))^2$$

is a minimum. Show that the equation of the least squares line is $y = mx + b$, where m and b satisfy the equations

$$m \sum_{i=1}^{n} x_i + bn = \sum_{i=1}^{n} y_i$$

$$m \sum_{i=1}^{n} x_i^2 + b \sum_{i=1}^{n} x_i = \sum_{i=1}^{n} x_i y_i.$$

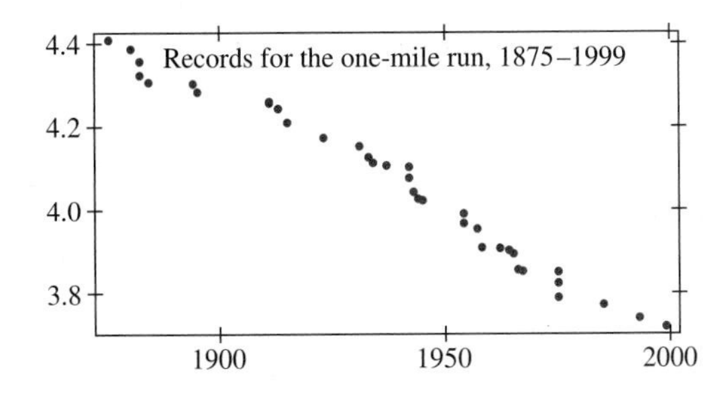

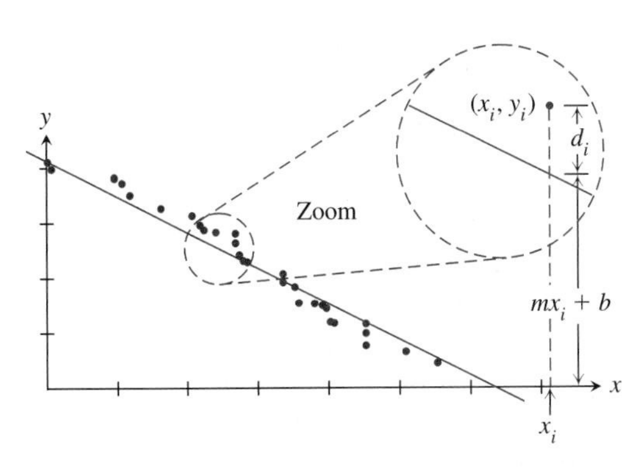

Figures for Exercise 26.

27. Show that the least squares line goes through the centroid $(\overline{x}, \overline{y})$ of the data, where

$$(\overline{x}, \overline{y}) = \left(\frac{1}{n} \sum_{i=1}^{n} x_i, \frac{1}{n} \sum_{i=1}^{n} y_i \right).$$

28. Use the Second Derivatives Test in showing that the least squares solution (m, b) is a local minimum. For this you may need to note that for $n = 3$ (as an example),

$$2(x_1^2 + x_2^2 + x_3^2) + 2(x_1x_2 + x_1x_3 + x_2x_3) \geq 0.$$

This follows from adding three inequalities together, namely, $(x_1 + x_2)^2 \geq 0$ and two similar inequalities.

T 29. For the following data, find the least squares line and plot it together with the data. Reagent 4UD900 has the property that when it is added to water, the temperature of the water increases rapidly. In its development, a series of tests were performed in which various quantities of the reagent were added to a fixed quantity of water at approximately the same initial temperature. The temperatures of the water were measured exactly 60 seconds after the reagent was added, and the differences in temperature readings were found. The amounts of reagent added (in milligrams) and the corresponding changes in temperatures (in degrees Celsius) were recorded as the pairs

$$(27, 57), (45, 64), (41, 80), (19, 46), (35, 62),$$
$$(39, 72), (19, 52), (49, 77), (15, 57), (31, 68).$$

T 30. For the following data, find the least squares line and plot it together with the data. According to a report published by the Federal Energy Administration, proper insulation can increase temperature control by as much as 20 to 30%. A study was conducted to determine the strength of the linear relationship between the thickness x (measured in inches) of insulation in walls and the corresponding increase in temperature control y (measured as a fraction). The following (x, y) pairs are from a laboratory study of the relationship of y to x.

$$(1.0, 0.02), (2.0, 0.05), (3.0, 0.08),$$
$$(4.0, 0.14), (5.0, 0.24), (6.0, 0.28)$$

Exercises 31–34: In these exercises the Second Derivatives Test may be used to study the quadric surface S described by the quadratic function $z = Ax^2 + 2Bxy + Cy^2$, where A, B, and C are constants.

31. Show that $(0, 0)$ is a stationary point of S. Also show that if $AC - B^2 \neq 0$, then $(0, 0)$ is the only stationary point.

32. Assume that $AC - B^2 > 0$. Show that $A \neq 0$ and, hence,

$$z = A\left(x^2 + \frac{2B}{A}xy + \frac{C}{A}y^2 \right).$$

Moreover, show that S has either an absolute minimum or an absolute maximum at $(0, 0, 0)$. Classify these two cases in terms of A.

33. Apply Exercises 31 and 32 to discuss the extrema of the surface S described by $z = x^2 + xy + y^2$.

34. Apply Exercises 31 and 32 to discuss the extrema of the surface S described by $z = -3x^2 + xy - y^2$.

10.7 Optimization with Constraints

We have used the Candidate Theorem, the Max/Min Theorem, and the Second Derivatives Test in locating and classifying the extrema of functions. To these results we add the method of Lagrange multipliers, which is used for problems in which the search for the extreme points of a function is constrained or restricted in some way. Here is a typical optimization problem with a constraint.

INVESTIGATION

Least and Greatest Distances from Points on an Ellipse to Its Center

Figure 10.50 shows the ellipse C described by

$$17x^2 + 12xy + 8y^2 = 100. \tag{1}$$

What are the least and greatest distances between the origin and the ellipse? This equation can be restated in terms of the distance function $d(x, y)$ shown in the figure. What are the minimum and maximum values of d as the point (x, y) varies over the ellipse? Note that if (x, y) were not restricted to the ellipse, the minimum of d would be 0 and it would have no maximum. If (x, y) is restricted to the ellipse, it appears from the graph that the minimum and maximum of d are between 2 and 5.

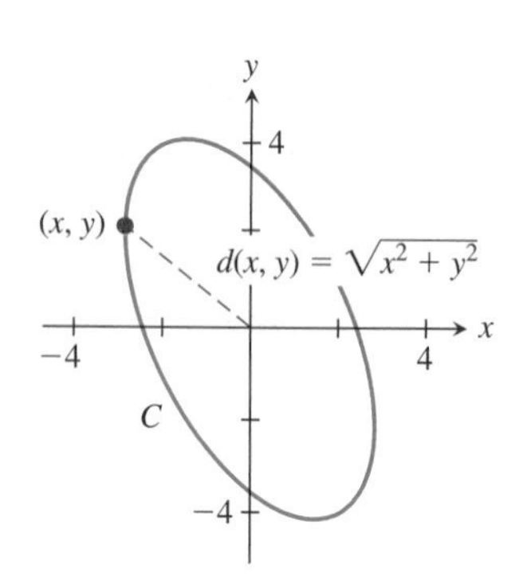

FIGURE 10.50 Determining the least and greatest distances between the origin and the ellipse.

To simplify our discussion, we replace the distance function d by the squared distance function $f(x, y) = x^2 + y^2$. Note that a point on C minimizes/maximizes the distance function d if and only if it minimizes/maximizes the function f. We also note that we may describe the constraint curve C as the 0 contour of the function g defined by

$$g(x, y) = 17x^2 + 12xy + 8y^2 - 100, \qquad (x, y) \in R^2.$$

This constrained optimization problem can be solved in several ways.

Method 1: We may solve the equation $g(x, y) = 0$ for y, obtaining

$$y = \frac{-3x \pm 5\sqrt{8 - x^2}}{4}.$$

If we substitute each of these solutions into $f(x, y) = x^2 + y^2$, we obtain functions of x alone. This reduces the original question to two one-variable optimization problems.

Method 2: As in Method 1, we choose x as the independent variable. However, we sidestep solving the equation $g(x, y) = 0$ for y in terms of x by assuming that this equation defines y implicitly in terms of x. With either solution in mind, we work towards minimizing and maximizing the function $p(x) = f(x, y)$, where y denotes an implicitly defined function of x, by calculating dp/dx and setting it equal to zero. We do this by implicitly differentiating the equation $p(x) = f(x, y) = x^2 + y^2$ with respect to x. We find

$$p'(x) = \frac{d}{dx} f(x, y) = 2x + 2yy'.$$

To obtain candidates for the minimum and maximum we set

$$2x + 2yy' = 0. \tag{2}$$

This is one equation in the unknowns x, y, and y'. We obtain a second equation in x, y, and y' by differentiating both sides of the constraint equation $g(x, y) = 0$ with respect to x. We find

$$34x + 12y + 12xy' + 16yy' = 0. \tag{3}$$

We may eliminate y' from Equations (2) and (3) by solving one of them for y' and substituting into the other. From (3),

$$y' = \frac{-34x - 12y}{12x + 16y}. \tag{4}$$

Substituting this expression for y' into (2) and simplifying gives

$$2x^2 - 3xy - 2y^2 = 0. \tag{5}$$

To find the points (x, y) on the ellipse that are closest and farthest from $(0, 0)$ we solve simultaneously the equations $g(x, y) = 0$ and (5). We find solutions $(2, -4)$, $(2, 1)$, $(-2, 4)$, and $(-2, -1)$. These are the points on the ellipse that are the closest and farthest from $(0, 0)$.

Method 3: A third approach is to note that the tangent lines at the closest and farthest points (x, y) of the ellipse are perpendicular to the lines from the origin to these points. We equate y' from (14), which gives the slope of y' of the ellipse at (x, y), and $-x/y$, which is the slope of the line perpendicular to the line from the origin to (x, y). This gives the equation

$$2x^2 - 3xy - 2y^2 = 0,$$

which is the same as (5). The four extreme points may be found as in Method 2.

Method 4: The method of Lagrange multipliers is related to the second and third methods but is applicable to a much wider range of problems. It is based on the relation between the level curves of f and the constraint curve C described by $g(x, y) = 0$. Shown at the left of Fig. 10.51 are both C and the curves described by

$$f(x, y) = x^2 + y^2 = k^2, \qquad k = 0.375, 0.750, \ldots, 6.75.$$

The points (x, y) on the k-contour correspond to all points $(x, y) \in R^2$ that are k units from the origin. We may find the point or points on the constraint curve C that are closest (farthest) from the origin by imagining that the k contour is moving outwards from the origin. As k increases from 0, we look for the "first contour" ("last

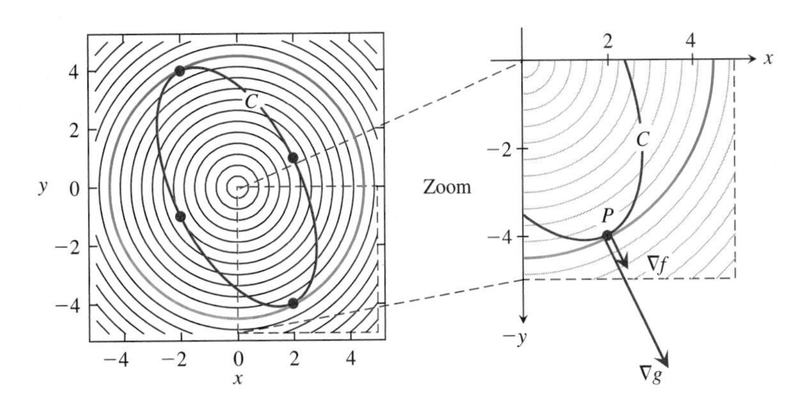

FIGURE 10.51 The "first contour" and the "last contour" are tangent to the constraint curve.

contour") that touches the constraint curve. The first contour touches the ellipse at a point or points closest to the origin. Similarly, the last contour touches the ellipse at a point or points farthest from the origin. The zoom-view of the last contour in the right part of Fig. 10.51 gives an idea of what happens at a point P where the moving contour is just losing contact with the constraint curve C. It appears that at P the contour and constraint curve C are tangent, which means that vectors ∇f (which is perpendicular to the contours) and ∇g (which is perpendicular to the constraint curve C) are parallel. These vectors are also parallel at the point or points on C where the moving contour is first touching the constraint curve C.

It follows that at such minimum and maximum points (x, y)—the points of first and last contact—there is a scalar λ for which

$$\nabla f(x, y) = \lambda \nabla g(x, y). \tag{6}$$

This equation states that $\nabla f(x, y)$ is a scalar multiple of $\nabla g(x, y)$, that is, these two vectors are parallel. The scalar λ is known as a Lagrange multiplier.

We may find the points (x, y) of first and last contact, at which the vectors ∇f and ∇g are parallel, by solving the system of equations

$$g(x, y) = 0 \tag{7}$$

$$\nabla f(x, y) = \lambda \nabla g(x, y) \tag{8}$$

for the unknowns x, y, and λ. The first equation is the constraint equation, which requires (x, y) to be on C. Replacing (8) by the two scalar equations

$$f_x(x, y) = \lambda g_x(x, y) \quad \text{and} \quad f_y(x, y) = \lambda g_y(x, y),$$

the system of Equations (7) and (8) can be written as

$$\begin{aligned} g(x, y) &= 0 \\ f_x(x, y) &= \lambda g_x(x, y) \\ f_y(x, y) &= \lambda g_y(x, y). \end{aligned} \tag{9}$$

This is a system of three equations in the three unknowns x, y, and λ.

Recalling that

$$f(x, y) = x^2 + y^2 \quad \text{and} \quad g(x, y) = 17x^2 + 12xy + 8y^2 - 100,$$

the system (9) is

$$17x^2 + 12xy + 8y^2 - 100 = 0 \tag{10}$$

$$2x = \lambda(34x + 12y) \tag{11}$$

$$2y = \lambda(12x + 16y). \tag{12}$$

For this particular set of equations, we first eliminate λ. From (11) and (12),

$$\lambda = \frac{2x}{34x + 12y}$$

$$\lambda = \frac{2y}{12x + 16y}.$$

Equating the two fractions, cross-multiplying, and simplifying gives

$$2x^2 - 3xy - 2y^2 = 0. \tag{13}$$

Multiplying Equation (13) by 4 and adding the resulting equation to (10) gives $25x^2 - 100 = 0$. Hence, $x = \pm 2$. Substituting $x = 2$ into Equation (13),

$$\begin{aligned} 8 - 6y - 2y^2 &= 0 \\ -2(y^2 + 3y - 4) &= 0 \\ -2(y+4)(y-1) &= 0. \end{aligned}$$

With $x = 2$, $y = -4$, or $y = 1$. Substituting $x = -2$ into Equation (13),

$$\begin{aligned} 8 + 6y - 2y^2 &= 0 \\ -2(y^2 - 3y - 4) &= 0 \\ -2(y-4)(y+1) &= 0. \end{aligned}$$

With $x = -2$, $y = 4$, or $y = -1$. The four points $(2, -4)$, $(2, 1)$, $(-2, 4)$, and $(-2, -1)$ thus determined are the only points (x, y) that lie on the ellipse and at which the gradient vectors ∇f and ∇g are parallel. These are the points on the ellipse that are at the least and greatest distances from the origin. They appear to be at the ends of the minor and major axes of the ellipse. The maximum and minimum distances are $\sqrt{f(2, -4)} = \sqrt{20}$ and $\sqrt{f(2, 1)} = \sqrt{5}$.

The Candidate Theorem for the method of Lagrange multipliers is similar to earlier Candidate Theorems in that it leads to a system of equations whose solutions are candidates for extrema. Once the candidates are found, further work may be required to determine which, if any, correspond to a global maximum or a global minimum.

THEOREM Candidate Theorem for the Method of Lagrange Multipliers

Let f and g be real-valued functions defined on a common domain $\mathcal{D} \subset R^n$. Assume that the first-order partial derivatives of f and g are continuous on $\mathcal{D}$ and let $\mathcal{S}$ be the subset of $\mathcal{D}$ satisfying the constraint equation $g(\mathbf{x}) = 0$. If $\mathbf{p} \in \mathcal{S}$ is a local extremum of f on $\mathcal{S}$ and $\nabla g(\mathbf{p}) \neq \mathbf{0}$, then there is a real number λ for which

$$\nabla f(\mathbf{p}) = \lambda \nabla g(\mathbf{p}).$$

To apply this result to find candidates $\mathbf{p}$ for the extrema of f on $\mathcal{S}$ we solve the system of equations

$$\begin{aligned} \nabla f(\mathbf{p}) &= \lambda \nabla g(\mathbf{p}) \\ g(\mathbf{p}) &= 0. \end{aligned} \tag{14}$$

If $\mathbf{p} = (x_1, \ldots, x_n)$, the equation $\nabla f(\mathbf{p}) = \lambda \nabla g(\mathbf{p})$ may be replaced by the n scalar equations

$$\frac{\partial f}{\partial x_i} = \lambda \frac{\partial g}{\partial x_i}, \qquad i = 1, 2, \ldots, n.$$

Along with the constraint equation $g(x, y) = 0$, this gives a system of $n + 1$ equations in the $n + 1$ unknowns $x_1, \ldots, x_n, \lambda$.

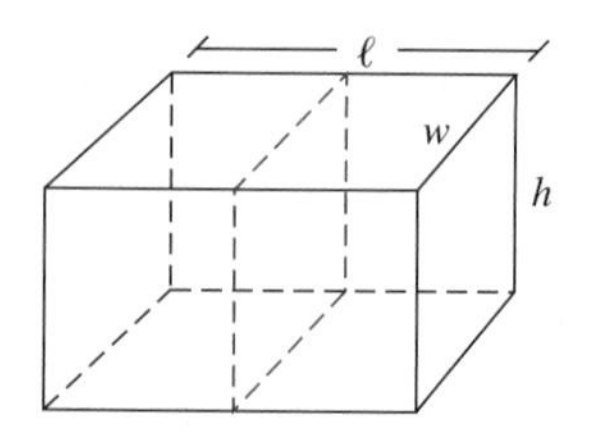

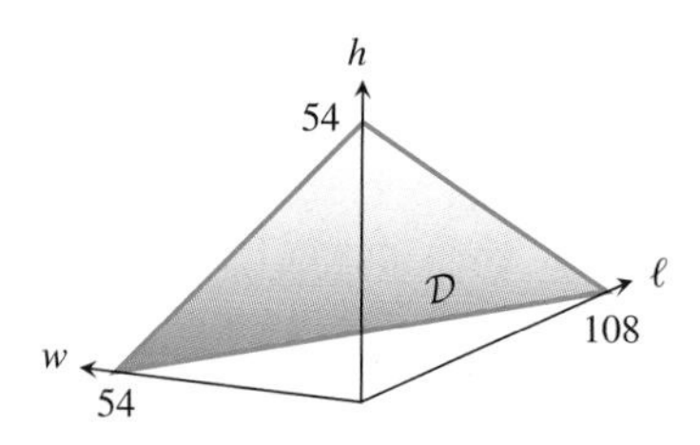

FIGURE 10.52 Representative box with girth shown in color; also, the domain $\mathcal{D}$.

EXAMPLE 1 According to the postal regulations for 2001, a box can be mailed only if the sum of its length and girth does not exceed 108 inches. The length of a box is its longest dimension, and the girth is twice the sum of the remaining two dimensions. Figure 10.52 shows a representative box. Within the postal regulations, what are the dimensions of a box with largest volume? (We saw this problem earlier, in Example 2 from Section 10.5.)

Solution

The volume of a box with length ℓ, width w, and height h is $V = \ell wh$. For maximum volume we replace the postal regulation

$$\ell + 2(w + h) \leq 108$$

by

$$\ell + 2(w + h) = 108.$$

We may now state the problem as follows: Maximize the function

$$V(\ell, w, h) = \ell wh,$$

subject to the constraint equation

$$g(\ell, w, h) = \ell + 2w + 2h - 108 = 0.$$

From (14), the candidate equations are

$$\begin{aligned} \nabla V(\ell, w, h) &= \lambda \nabla g(\ell, w, h) \\ g(\ell, w, h) &= 0. \end{aligned}$$

Calculating the gradients, the system becomes

$$\begin{aligned} \langle wh, \ell h, \ell w \rangle &= \lambda \langle 1, 2, 2 \rangle \\ \ell + 2w + h - 108 &= 0. \end{aligned}$$

Equating coordinates in the vector equation gives the system

$$\begin{aligned} wh &= \lambda \\ \ell h &= 2\lambda \\ \ell w &= 2\lambda \\ \ell + 2w + 2h &= 108 \end{aligned}$$

of four scalar equations in the four unknowns ℓ, w, h, λ. From the first two equations,

$$2wh = \ell h.$$

From the first and third equations,

$$2wh = \ell w.$$

For maximum volume, $\ell, w, h > 0$. Hence, we may remove the nonzero factors h and w from both sides of the last two equations. This gives

$$2w = \ell$$
$$2h = \ell.$$

Substituting these expressions into the constraint equation,

$$\ell + \ell + \ell = 108.$$

Hence, $\ell = 36$ and $w = h = 18$. This is the solution given in Section 10.5. From the discussion in that example, we know this is a maximum point. How can we justify this assertion without appealing to the earlier work?

If the existence of a maximum point is taken as clear from the problem and requiring no proof, then because the system of candidate equations has one and only one solution, this solution must be the maximum point.

In some optimization problems, we can use the Max/Min Theorem to show that a maximum or minimum exists. For this postal package problem, we define the function

$$f(\ell, w, h) = \ell w h,$$

for $(\ell, w, h) \in \mathcal{D} = \{(\ell, w, h) \in R^3 : \ell, w, h \geq 0 \text{ and } \ell + 2w + 2h = 108\}$. Figure 10.52 shows the domain $\mathcal{D}$. Because the graph of the equation $\ell + 2w + 2h = 108$ is a plane passing through the first octant of (ℓ, w, h)-space, we see from the conditions $\ell, w, h \geq 0$ that $\mathcal{D}$ is a triangle together with its interior. Thus, $\mathcal{D}$ is closed and bounded. Because f is continuous, it follows from the Max/Min Theorem that g has a maximum and a minimum. The minimum point is on the boundary, where f is 0, and the maximum point is in the interior of $\mathcal{D}$, where f is always positive. Given the existence of this maximum point, we can calculate its coordinates by applying the Candidate Theorem for Lagrange Multipliers.

EXAMPLE 2 Maximize the function $f(x, y, z) = xyz^2$ on the sphere $x^2 + y^2 + z^2 = 4a^2$.

Solution

The continuous function f with the closed and bounded domain $\mathcal{D} = \{(x, y, z) : x^2 + y^2 + z^2 = 4a^2\}$ has an absolute maximum according to the Max/Min Theorem. If this point is denoted by (x, y, z), then there is a scalar λ for which

$$\nabla f(x, y, z) = \lambda \nabla g(x, y, z),$$

where $g(x, y, z) = x^2 + y^2 + z^2 - 4a^2$. The system (14) is

$$\begin{aligned} yz^2 &= \lambda 2x \\ xz^2 &= \lambda 2y \\ 2xyz &= \lambda 2z \\ x^2 + y^2 + z^2 &= 4a^2. \end{aligned} \tag{15}$$

We ignore all solutions in which any of λ, x, y, or z is zero because if $x = 0$, $y = 0$, or $z = 0$, then $f(x, y, z) = xyz^2$ would be zero and not a maximum. If $\lambda = 0$ it follows that at least one of x, y, or $z = 0$ and again $f(x, y, z) = xyz^2$ would be zero and not a maximum. Assuming then that none of x, y, z, or λ is zero, we solve each of the first three equations for λ, equate the results, and solve for x, y, and z. From the equation

$$2\lambda = \frac{yz^2}{x} = \frac{xz^2}{y} = \frac{2xyz}{z},$$

it follows that $x^2 = y^2 = \frac{1}{2}z^2$ (for example, from $yz^2/x = xz^2/y$, it follows that $x^2 = y^2$). Substituting into the constraint equation,

$$\frac{1}{2}z^2 + \frac{1}{2}z^2 + z^2 = 4a^2, \qquad \text{or, solving for } z, \qquad z = \pm\sqrt{2}a.$$

With each of these two values of z there are two corresponding values of each of x and y, resulting in the eight solutions

$$\left(a, a, \sqrt{2}a\right), \left(a, a, -\sqrt{2}a\right), \left(-a, -a, \sqrt{2}a\right), \left(-a, -a, -\sqrt{2}a\right),$$
$$\left(a, -a, \sqrt{2}a\right), \left(a, -a, -\sqrt{2}a\right), \left(-a, a, \sqrt{2}a\right), \left(-a, a, -\sqrt{2}a\right)$$

of the system (15). The first four of these solutions maximize the function $f(x, y, z) = xyz^2$ on the sphere with equation $x^2 + y^2 + z^2 = 4a^2$. The maximum value of f on the sphere is $2a^4$.

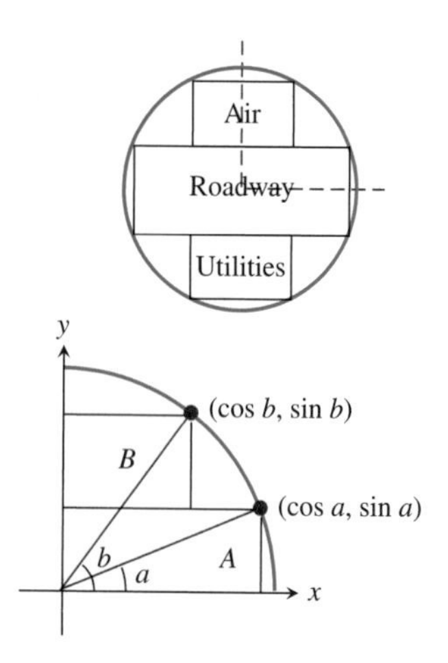

FIGURE 10.53 Cross section of proposed tunnel and one quadrant of a coordinatized cross section.

EXAMPLE 3 The sketch at the top of Fig. 10.53 shows a cross section of a proposed tunnel. The lower, middle, and upper rectangles provide for utilities, a roadway, and ventilation. If the tunnel has a circular cross section with radius 1 unit and the air and utility rectangles have equal area, what are the dimensions of the rectangles so that their combined area is as large as possible? This optimization problem was given earlier, as Exercise 38 in Section 10.5. We now solve this problem under the constraint that the combined areas of the air and utility rectangles be equal to the area of the roadway rectangle.

Solution

The sketch at the bottom of Fig. 10.53 shows one quadrant of the cross section and identifies angles a and b, which determine all rectangles. The constraint that the combined areas of the air and utility rectangles be equal to the area of the roadway rectangle is satisfied if we require that the area of rectangles A and B be equal.

The combined area of these rectangles is

$$f(a, b) = \cos a \sin a + \cos b(\sin b - \sin a). \tag{16}$$

Thus, we must maximize f on the set of all $(a, b) \in R^2$ for which $0 \le a \le b \le \pi/2$, subject to the constraint that the areas of rectangles A and B be equal, that is,

$$\cos a \sin a = \cos b(\sin b - \sin a).$$

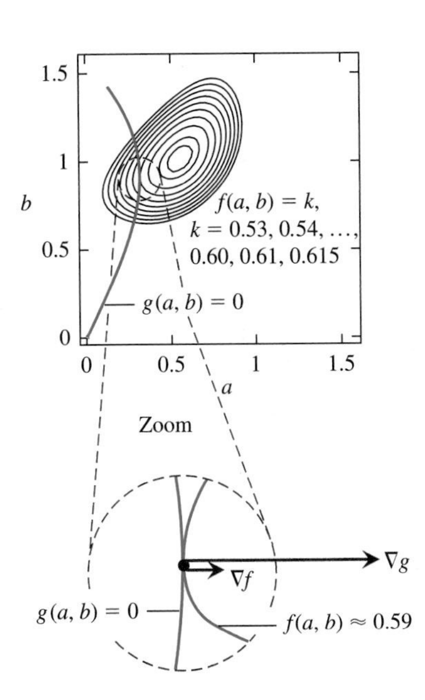

FIGURE 10.54 The last contour is tangent to the constraint curve.

If we define g by

$$g(a,b) = \cos a \sin a - \cos b(\sin b - \sin a),$$

the constraint equation is $g(a,b) = 0$.

Figure 10.54 shows the constraint curve with equation $g(a,b) = 0$ and the level curves described by $f(a,b) = k$ for 10 different values of k. As k increases, the oval contours become smaller. The zoom-view at the bottom of Fig. 10.54 shows a section of the 0.59 contour, which is very nearly tangent to the constraint curve ($k = 0.588\ldots$ gives tangency, as we show shortly).

From the equations $\nabla f = \lambda \nabla g$ and $g(a,b) = 0$,

$$\begin{aligned} -\sin^2 a + \cos^2 a - \cos a \cos b &= \lambda(-\sin^2 a + \cos^2 a + \cos a \cos b) \\ -\sin^2 b + \sin a \sin b + \cos^2 b &= \lambda(\sin^2 b - \sin a \sin b - \cos^2 b) \\ \cos a \sin a - \cos b(\sin b - \sin a) &= 0. \end{aligned} \tag{17}$$

We solve (17), leaving some details to Exercises 30–32. To avoid degenerate cases, we assume that $0 < a \leq b < \pi/2$ and, for a moment, $\lambda \neq 0$. It then follows from the first two equations of (17) that

$$2 \cos 2a(\sin a \sin b + \cos 2b) = 0.$$

The first factor in this equation can be zero only if $a = \pi/4$. If this value is substituted into the constraint equation, we obtain $1 + \sqrt{2} \cos b = \sin(2b)$, which has no solution. Why? Setting the second factor equal to zero,

$$\sin a = -\frac{\cos 2b}{\sin b}. \tag{18}$$

If $\lambda = 0$ in (17), it follows from the second equation that (18) again holds. Using (18) to eliminate a from the constraint equation,

$$w(15w^3 - 27w^2 + 16w - 3) = 0, \tag{19}$$

where $w = \cos^2 b$. Equation (19) has two complex roots, one positive root $w \approx 0.377859$, and $w = 0$.

We ignore the $w = 0$ root because then $b = \pi/2$. Recall that we assumed earlier that $b < \pi/2$. If $w \approx 0.377859$, then $b \approx 0.908788$. From (18), the corresponding value of a is approximately 0.314881. It appears that (0.314881, 0.908788) is a maximum point. From (16), the value of f at this point is 0.59, which is consistent with Fig. 10.54.

Exercises 10.7

Exercises 1–20: Use the method of Lagrange multipliers in solving the optimization problems.

1. Locate the point(s) on the curve described by $x^2y = 16$ closest to the origin.
2. Locate the point(s) on the curve described by $x^3y = 1$ closest to the origin.
3. Locate the point(s) on the curve described by $xy = 2\sqrt{2}$ closest to $(1, 0)$.

4. Locate the point(s) on the curve described by $x^2y = 1$ closest to $(0, 1)$.

5. Locate the point(s) on the surface described by $x^2y^2z = 1$ closest to the origin.

6. Locate the point(s) on the surface with equation $z^2 = 1 + xy$ that are nearest the origin.

7. Locate the point(s) on the sphere $x^2 + y^2 + z^2 = 4$ that are closest to $(3, 1, -1)$.

8. Locate the point(s) on the sphere $x^2 + y^2 + z^2 = 9$ that are closest to $(10, 4, -5)$.

9. Locate the point(s) on the plane $x + y + z = 1$ that minimize $x^4 + y^4 + z^4$.

10. Locate the point(s) on the plane $x + y + z = 3$ that minimize $x^3 + y^3 + z^3$.

11. Assuming that a cylindrical container can be mailed only if the sum of its height and circumference do not exceed 108 inches, what are the dimensions of the cylinder with largest volume that can be mailed?

12. The production level P of a factory during one time period is sometimes modeled by the Cobb-Douglas production function

$$P(x, y) = Kx^{\alpha}y^{1-\alpha},$$

where K is a positive constant, x is the number of units of labor scheduled, and y is the number of units of capital invested. If $\alpha = 1/3$, labor costs \$200/unit, capital costs \$300/unit, and the owner has \$1,500,000 available for one time period, what amounts of labor and capital would maximize production?

13. Related problem Suppose now that the production level is fixed at $2000K$. What amounts x of labor and y of capital minimize the cost function $200x + 300y$?

14. Minimize the function xy on the circle $x^2 + y^2 = 1$.

15. Maximize the function $x/a + y/b + z/c$ on the sphere with equation $x^2 + y^2 + z^2 = 1$, where $a, b, c > 0$ are constants.

16. Maximize the function xyz if it is required that $x^3 + y^3 + z^3 = 1$.

17. Find the dimensions of a box with largest volume if it lies in the first octant, three of its faces lie in the coordinate planes, and the box lies beneath (or touching) the plane with equation $2x + 3y + 4z = 6$.

18. Find the dimensions of a box with largest volume if it lies in the first octant, three of its faces lie in the coordinate planes, and the box lies beneath (or touching) the ellipsoid with equation $x^2 + y^2/4 + z^2/9 = 1$.

19. The topless shipping carton with two dividers shown in the accompanying figure is to be constructed from 25 m^2 of cardboard. Show that, for maximum volume, the length of the box must be four times the height and the width must be two times the height.

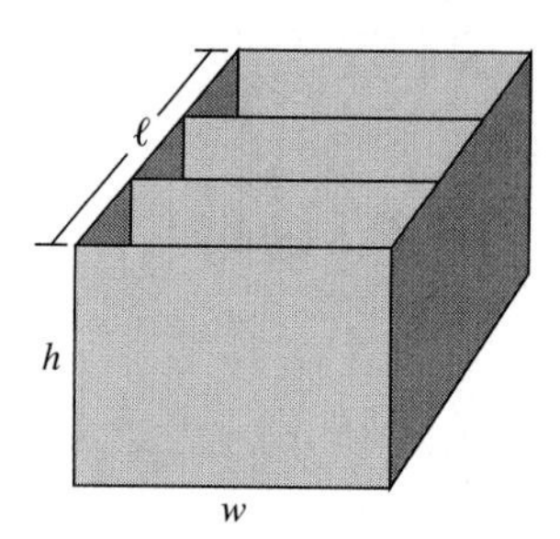

Figure for Exercise 19.

20. Related problem Solve the shipping carton problem if it has a top.

21. The candidate equations in Example 2 have eight distinct solutions, four of which make f a maximum on the sphere. Comment on the other four solutions.

22. Locate the points on the surface described by $x^4 + y^4 + z^4 = 1$ closest to and farthest from the origin.

23. Find the distance from the origin to the plane with equation $ax + by + cz + d = 0$ by the method of Lagrange multipliers. Assume that not all of a, b, and c are zero.

24. Find the dimensions and volume of the largest rectangular box that can be inscribed in the ellipsoid with equation

$$\frac{x^2}{a^2} + \frac{y^2}{b^2} + \frac{z^2}{c^2} = 1.$$

25. The graph of the equation $x^2 - 4xy + 5y^2 = 10$ is an ellipse. Find the greatest and least distances between the ellipse and the origin.

26. The graph of the equation $x^2 + (15/4)xy + 10y^2 = 1$ is an ellipse. Find the greatest and least distances between the ellipse and the origin.

27. Heron's formula for the area A of a triangle with sides a, b, and c is

$$A = \sqrt{s(s-a)(s-b)(s-c)}$$

where $s = (a + b + c)/2$. Show that among all triangles with a given perimeter, the equilateral triangle has the largest area.

28. Show that among all rectangles with a given perimeter, the square has the largest area.

29. Show that the maximum and minimum values of $xy + xz$ on the ellipsoid with equation $x^2 + 2y^2 + 3z^2 = 1$ are $\pm 5/(2\sqrt{30})$.

30. Referring to Example 3, give the details leading to the equation

$$2 \cos 2a(\sin a \sin b + \cos 2b) = 0.$$

Hint: Before solving the first two equations of (17) for λ and equating the results, use the identity

$$\cos 2\theta = \cos^2\theta - \sin^2\theta.$$

31. Related problem Give the details leading to the equation

$$w(15w^3 - 27w^2 + 16w - 3) = 0.$$

Hint: Replace $\sin a$ in the constraint equation, but leave the term $\cos a$ for a moment. After simplification, you should obtain

$$\cos a = \frac{\cos^3 b}{1 - 2\cos^2 b}.$$

Now square and use (18) once more; after simplification, set $w = \cos^2 b$.

32. Related problem Show that the equation

$$15w^3 - 27w^2 + 16w - 3 = 0$$

has two complex roots and one real root $w \approx 0.377859$.

33. Related problem A tunnel problem related to the problem worked out in Example 3 was solved in Exercise 38, Section 10.5. Without reference to the actual values, which of the two maxima would you expect to be larger? Why?

34. Show that for all positive numbers p and q,

$$\sqrt{pq} \leq \frac{p + q}{2}.$$

Hint: Maximize xy subject to the constraint $\frac{1}{2}(x + y) = k$.

35. Show that for all positive numbers u, v, and w,

$$\sqrt[3]{uvw} \leq \frac{u + v + w}{3}.$$

Hint: Maximize xyz subject to the constraint $\frac{1}{3}(x + y + z) = k$.

36. Follow the method given in Exercises 34 and 35 to show that for all positive numbers $p_1, p_2, \ldots, p_n$,

$$\sqrt[n]{p_1 p_2 \cdots p_n} \leq \frac{p_1 + p_2 + \cdots + p_n}{n}.$$

Review of Key Concepts

The central idea of this chapter is that of a differentiable function. A function $\mathbf{f}$ whose domain is contained in R^n and whose range is contained in R^m is differentiable at $\mathbf{a}$ if for $\mathbf{x}$ near $\mathbf{a}$, $\mathbf{f}(\mathbf{x}) - \mathbf{f}(\mathbf{a})$ can be approximated by a linear function $\mathbf{df_a}$ defined on R^n and taking its values in R^m. The matrix $\mathbf{Df}(\mathbf{a})$ of partial derivatives associated with this linear function is the derivative matrix.

The chain rule gives formulas for differentiating the compositions of real- and vector-valued functions of one, two, or more variables. For the composition of functions $\mathbf{y} = \mathbf{f}(\mathbf{x})$ and $\mathbf{x} = \mathbf{g}(\mathbf{t})$, the chain rule is

$$\mathbf{D}(\mathbf{f} \circ \mathbf{g})(\mathbf{t}) = \mathbf{Df}(\mathbf{g}(\mathbf{t}))\mathbf{Dg}(\mathbf{t}).$$

The use of the derivative matrices makes clear that this chain rule is strongly related to the chain rule

$$(f \circ g)'(t) = f'(g(t))g'(t)$$

for the derivative of the composition of the real-valued functions $y = f(x)$ and $x = g(t)$ of real variables x and t. We used the vector chain rule in studying implicit differentiation, calculating the normal to a surface defined implicitly, deriving a Taylor Inequality for real-valued functions of several variables, and transforming the heat equation to polar coordinates.

The chapter ended with several sections on tools for optimization of functions. In Section 10.5 we optimized functions defined on more or less standard domains, using the Candidate Theorem and the Max/Min Theorem. In Section 10.6 we studied the Second Derivatives Test and in Section 10.7 we optimized functions whose domains were constrained by a second function by applying the method of Lagrange multipliers.

Chapter Summary

Differentiable Functions

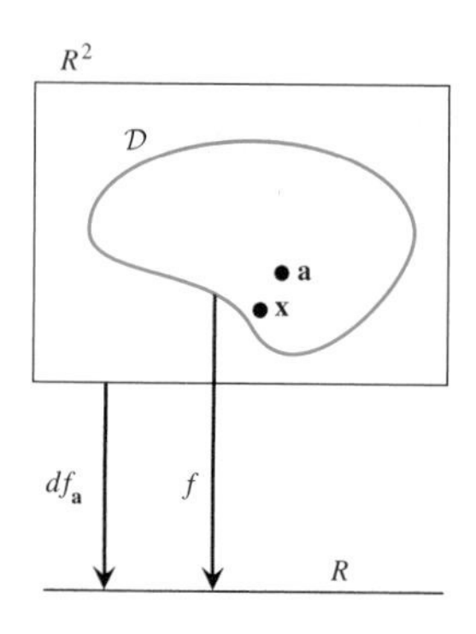

We take $n = 2$ and $m = 1$. A real-valued function f of two variables is differentiable at $\mathbf{a} \in R^2$ if there is a linear function $df_{\mathbf{a}}$ from R^2 into R for which

$$\lim_{\mathbf{x}\to\mathbf{a}} \frac{f(\mathbf{x}) - f(\mathbf{a}) - df_{\mathbf{a}}(\mathbf{x} - \mathbf{a})}{\|\mathbf{x} - \mathbf{a}\|} = 0.$$

The matrix associated with $df_{\mathbf{a}}$ is denoted by $Df(\mathbf{a})$ and is called the **derivative matrix** of f at $\mathbf{a}$. For all $\mathbf{h} \in R^2$,

$$\begin{aligned} df_{\mathbf{a}}(\mathbf{h}) &= Df(\mathbf{a})(\mathbf{h}) \\ &= \left(f_x(\mathbf{a}) \quad f_y(\mathbf{a})\right)\begin{pmatrix} h_1 \\ h_2 \end{pmatrix}. \end{aligned}$$

Let f be the function defined by

$$f(x, y) = x^2 y,$$

with domain R^2 $(n = 2)$ and range in R $(m = 1)$. Letting $\mathbf{a} = (1, 1)$,

$$\begin{aligned} df_{\mathbf{a}}(\mathbf{h}) &= (f_x(\mathbf{a}) \quad f_y(\mathbf{a}))\begin{pmatrix} h_1 \\ h_2 \end{pmatrix} \\ &= (2 \quad 1)\begin{pmatrix} h_1 \\ h_2 \end{pmatrix} \\ &= 2h_1 + h_2. \end{aligned}$$

Tangent Plane and Function Approximation

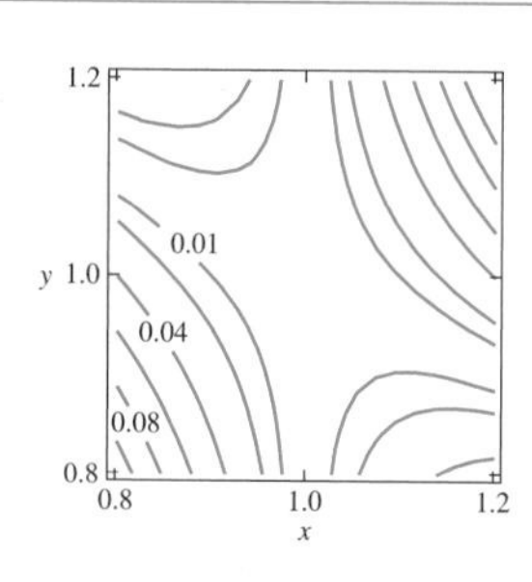

The tangent plane to the graph of a differentiable function f at $\mathbf{p} = (\mathbf{a}, f(\mathbf{a}))$ is the plane with equation

$$\begin{aligned} z = f(\mathbf{a}) &+ f_x(\mathbf{a})(x - a_1) \\ &+ f_y(\mathbf{a})(y - a_2). \end{aligned}$$

The tangent plane approximation to f at $\mathbf{a} = (a_1, a_2)$ is

$$\begin{aligned} f(\mathbf{x}) = f(x, y) &\approx f(\mathbf{a}) \\ &+ f_x(\mathbf{a})(x - a_1) + f_y(\mathbf{a})(y - a_2). \end{aligned}$$

For the function $f(x, y) = x^2 y$ and $\mathbf{a} = (1, 1)$, the tangent plane approximation is

$$f(x, y) \approx 1 + 2(x - 1) + (y - 1).$$

A contour plot of

$$|f(x, y) - (1 + 2(x - 1) + (y - 1))|$$

is shown in the figure, with contours 0.01, 0.02, 0.04, 0.06, 0.08, 0.1.

Chain Rule

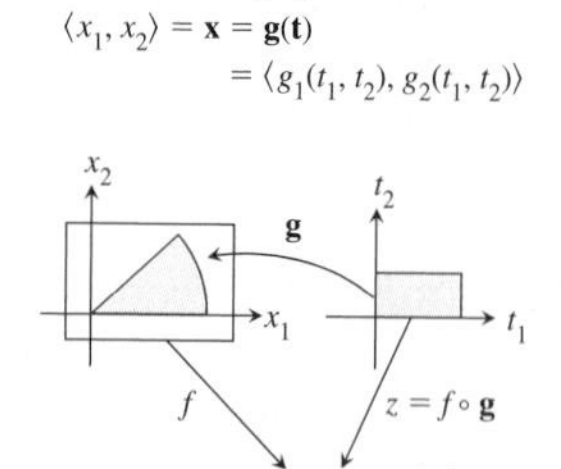

Let $y = f(\mathbf{x})$ and $\mathbf{x} = \mathbf{g}(\mathbf{t})$ be differentiable functions, and let $z = f \circ \mathbf{g}$. Then the composition z is differentiable and its derivative matrix is the product of the derivative matrices of f and $\mathbf{g}$; that is,

$$Dz(\mathbf{t}) = Df(\mathbf{g}(\mathbf{t}))D\mathbf{g}(\mathbf{t}).$$

Let $z = f(x, y) = x^2 - y^2$, $x = ue^v$, and $y = v^2$. Applying the chain rule,

$$\begin{pmatrix} \dfrac{\partial z}{\partial u} & \dfrac{\partial z}{\partial v} \end{pmatrix} = \begin{pmatrix} \dfrac{\partial z}{\partial x} & \dfrac{\partial z}{\partial y} \end{pmatrix} \begin{pmatrix} \dfrac{\partial x}{\partial u} & \dfrac{\partial x}{\partial v} \\ \dfrac{\partial y}{\partial u} & \dfrac{\partial y}{\partial v} \end{pmatrix}.$$

Hence,

$$\begin{aligned} \begin{pmatrix} \dfrac{\partial z}{\partial u} & \dfrac{\partial z}{\partial v} \end{pmatrix} &= (2x \quad -2y)\begin{pmatrix} e^v & ue^v \\ 0 & 2v \end{pmatrix} \\ &= (2xe^v \quad 2xue^v - 4yv) \\ &= (2ue^{2v} \quad 2u^2e^{2v} - 4v^3) \end{aligned}$$

Implicit Functions and Their Partial Derivatives

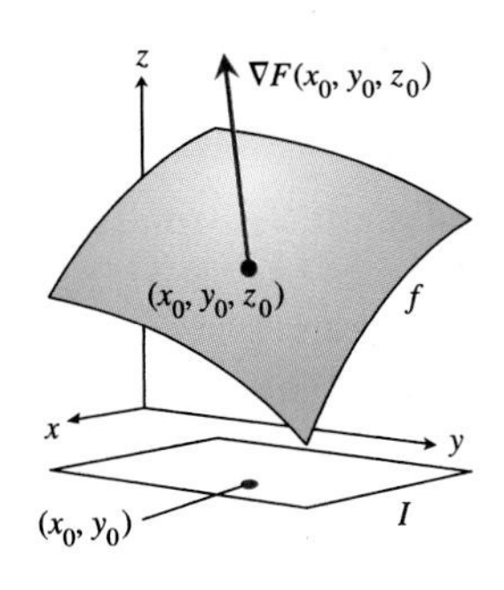

Assume that the equation $F(x, y, z) = 0$ defines $z = f(x, y)$ near (x_0, y_0, z_0). The graph of f coincides with the 0 contour of F near (x_0, y_0, z_0). The partial derivatives f_x and f_y are given by

$$f_x(x, y) = -\frac{F_x(x, y, z)}{F_z(x, y, z)}$$

$$f_y(x, y) = -\frac{F_y(x, y, z)}{F_z(x, y, z)},$$

provided that $F_z(x, y, z) \neq 0$. The gradient vector $\nabla F(x, y, z)$ is normal to the graph of f.

For $F(x, y, z) = x^2 + y^2 + z^2 - 1$, the equation $F(x, y, z) = 0$ defines the function $z = \sqrt{1 - x^2 - y^2}$. For $(x, y) = \left(\frac{1}{2}, \frac{1}{2}\right)$, $z = 1/\sqrt{2}$ and

$$f_x\left(\frac{1}{2}, \frac{1}{2}\right) = -\frac{F_x\left(\frac{1}{2}, \frac{1}{2}, \frac{1}{\sqrt{2}}\right)}{F_z\left(\frac{1}{2}, \frac{1}{2}, \frac{1}{\sqrt{2}}\right)} = -\frac{1}{\sqrt{2}}.$$

Because we can solve for f explicitly in this case, we can also calculate $f_x\left(\frac{1}{2}, \frac{1}{2}\right)$ explicitly:

$$f_x\left(\frac{1}{2}, \frac{1}{2}\right) = \left.\frac{-x}{\sqrt{1 - x^2 - y^2}}\right|_{x=1/2, y=1/2} = -\frac{1}{\sqrt{2}}.$$

A Taylor Inequality

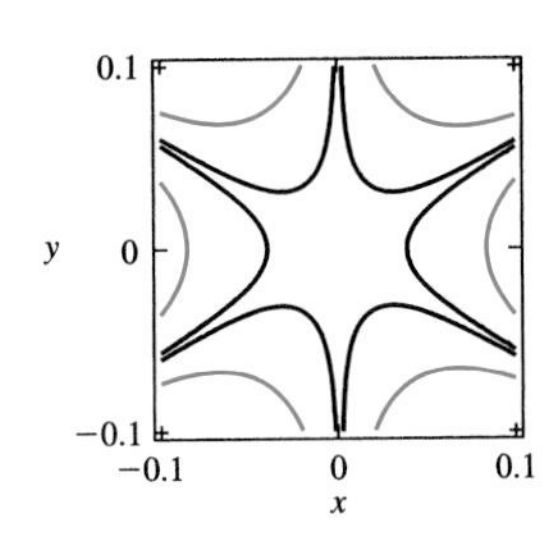

Assume that f is a real-valued function of two variables, defined on a region $\mathcal{D}$ and having continuous partial derivatives through those of third order near a point $\mathbf{a}$ of $\mathcal{D}$. Then

$$\left| f(\mathbf{a} + \mathbf{h}) - \left(f(\mathbf{a}) + f_x(\mathbf{a})h_1 + f_y(\mathbf{a})h_2 + \frac{1}{2!}\left(f_{xx}(\mathbf{a})h_1{}^2 + 2f_{xy}(\mathbf{a})h_1h_2 + f_{yy}(\mathbf{a})h_2{}^2\right)\right)\right| \leq \frac{1}{3!}Mb^3.$$

If $f(x, y) = e^x \cos y$, $\mathbf{a} = \langle 0, 0\rangle$, $\mathbf{h} = \langle x, y\rangle - \mathbf{a}$, and $b = 0.2$, then

$$e^x \cos y \approx 1 + x + \frac{1}{2}(x^2 - y^2),$$

with an error less than 0.0017. This follows because we may take $M = 1.2$ and $Mb^3/3! \leq 0.0017$. The figure shows the 0.00001 contour (in black) and the 0.0001 contour (in blue) for

$$\left| e^x \cos y - 1 - x - \frac{1}{2}(x^2 - y^2)\right|.$$

From the figure, the error estimate 0.0017 appears to be conservative.

Max/Min and Candidate Theorems

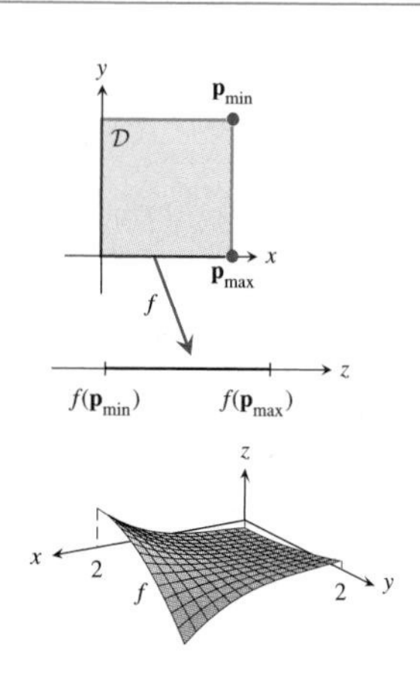

Max/Min Theorem: A continuous function f defined on a closed and bounded subset $\mathcal{D}$ of R^n has a maximum and a minimum.

Candidate Theorem: If f is continuous on a set $\mathcal{D} \subset R^2$ and has a local extremum at $\mathbf{p} \in \mathcal{D}$, then the only candidates for $\mathbf{p}$ are the boundary points of $\mathcal{D}$, any interior point $\mathbf{w}$ at which $\nabla f(\mathbf{w}) = 0$, or any interior point at which f has no gradient vector.

For the function

$$f(x, y) = (y^2 - 2)(x^2 + 1)(1 - x),$$

defined on $[0, 2] \times [0, 2]$, the gradient of f vanishes only at $(1, \sqrt{2})$, where f is 0. On the top boundary, f is decreasing (from 2 to -10); on the bottom boundary, f is increasing (from -2 to 10); on the left boundary, f is increasing (from -2 to 2); and on the right boundary, f is decreasing (from 10 to -10). So, f has a maximum of 10 at $(2, 0)$ and a minimum of -10 at $(2, 2)$. See the figure.

Second Derivatives Test

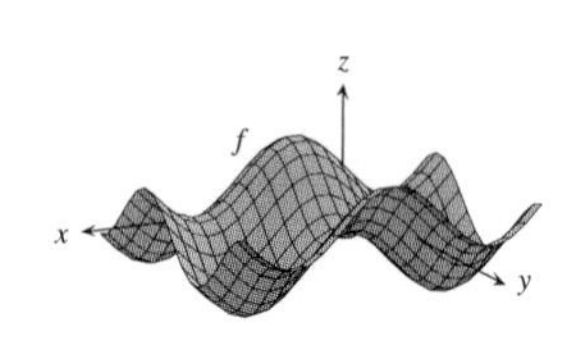

A point (x, y) is a stationary point of a differentiable real-valued function f if $\nabla f(x, y) = \mathbf{0}$. If f does not have a local extremum at (x, y), then (x, y) is a saddle point of f.

Let $\mathbf{a}$ be an interior stationary point of f. Let

$$D(\mathbf{a}) = f_{xx}(\mathbf{a}) f_{yy}(\mathbf{a}) - (f_{xy}(\mathbf{a}))^2.$$

Case 1: If $D(\mathbf{a}) > 0$ and $f_{xx}(\mathbf{a}) > 0$, then $\mathbf{a}$ is a local minimum point.
Case 2: If $D(\mathbf{a}) > 0$ and $f_{xx}(\mathbf{a}) < 0$, then $\mathbf{a}$ is a local maximum point.
Case 3: If $D(\mathbf{a}) < 0$, then $\mathbf{a}$ is a saddle point.
Case 4: If $D(\mathbf{a}) = 0$, the test fails.

The figure (see Fig. 10.44 for a larger figure) shows a graph of a function with $\nabla f(x, y) = 0$ at six interior points: $(\pi/2, 0)$ is a maximum point, $(-\pi/2, \pi)$ and $(3\pi/2, \pi)$ are minimum points, and $(-\pi/2, 0)$, $(\pi/2, \pi)$, and $(3\pi/2, 0)$ are saddle points. The function is $f(x, y) = \sin x + \cos y$, defined on $[-\pi, 2\pi] \times [-\pi, 2\pi]$.

Candidate Theorem for Lagrange Multipliers

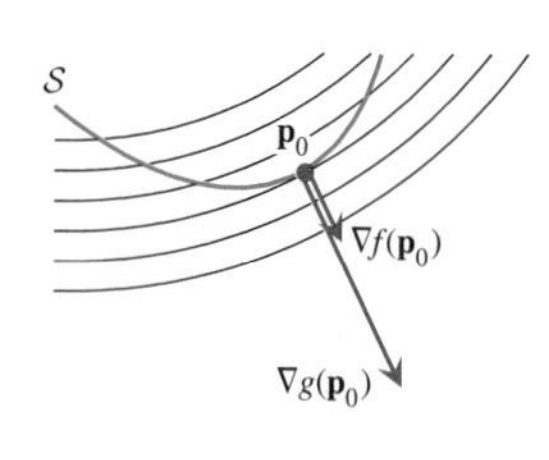

Let f and g be real-valued functions defined on a common domain $\mathcal{D} \subset R^n$ and let $\mathcal{S}$ be the subset of $\mathcal{D}$ satisfying the constraint equation $g(x, y) = 0$. If $\mathbf{p}_0$ is a local extremum of f on $\mathcal{S}$ and $\nabla g(\mathbf{p}_0) \neq \mathbf{0}$, then there is a scalar λ for which

$$\nabla f(\mathbf{p}_0) = \lambda \nabla g(\mathbf{p}_0).$$

The figure shows a family of level curves of f. We seek to maximize or minimize f on the set S. If $\mathbf{p}_0$ is such an extremum, the gradient vectors of f and g at $\mathbf{p}_0$ will be parallel. The extremum satisfies the equations

$$\nabla f(\mathbf{p}_0) = \lambda \nabla g(\mathbf{p}_0)$$
$$g(\mathbf{p}_0) = 0.$$

Chapter Review Exercises

1. Use the definition of differentiability to show that the function f defined by
$$f(x, y) = 5 + x^2 + xy$$
is differentiable at $(1, 1)$.

2. For a function defined by an equation $z = f(x, y)$, explain how the graph of the differential df at $\mathbf{a} = \langle a_1, a_2 \rangle$ is related to the graph of the plane tangent to the graph of f at $(a_1, a_2, f(a_1, a_2))$. If $\mathbf{a} = \langle 1, -1 \rangle$, $\mathbf{x} = \langle 1.2, -1.1 \rangle$, and f is the function defined in the preceding problem, calculate $f(\mathbf{x}) - f(\mathbf{a})$ and $df_{\mathbf{a}}(\mathbf{x} - \mathbf{a})$. Why does the second of these numbers approximate the first?

3. Give an equation describing the tangent plane at the point $(1, 1, 2)$ of the level surface described by $f(x, y, z) = 4$, where $f(x, y, z) = z^3 - xyz^2$.

4. Let $f(x, y) = 5x^2 + y^2$, $\mathbf{a} = (1, 2)$, and $\mathbf{x} = (1.05, 2.1)$. Calculate $f(\mathbf{x}) - f(\mathbf{a})$ and $df_{\mathbf{a}}(\mathbf{x} - \mathbf{a})$. Discuss these two quantities in terms of the graph of f and the tangent plane to the graph of f at $(1, 2, 9)$.

5. Use the Taylor Inequality for Functions of Two Variables in linearizing the function f defined by
$$f(x, y) = \sqrt[3]{1 - x - y^3},$$
where $x + y^3 \leq 1$, near the point $\mathbf{a} = \langle 3/4, 1/2 \rangle$. What percentage error is made if $f(0.7, 0.4)$ is calculated with the linearization formula?

6. Use the notion of the differential in approximating the maximum percentage error in the volume of a rectangular parallelepiped if the maximum percentage errors in the length and width are 2% and the maximum percentage error in the height is 3%.

7. At time $t = 0$ a particle leaves the point $(1, 1, 1)$ at a speed of 10 m/s. Its path lies along the outward normal to the surface with equation $x^2 + 2y^2 + 3z^2 = 6$. At what time does it cross the sphere $x^2 + y^2 + z^2 = 10$? Explain your work and include an informative sketch.

8. The temperature T at the point (x, y) of the plane is given by $T = f(x, y)$. By actual measurement it is known that $\nabla f(\sqrt{3}, 1) = \langle 5, 7 \rangle$. A change to polar coordinates is made. Find $\partial T/\partial r$ and $\partial T/\partial \theta$ at this point.

9. If $H(r, s) = f(r^2 + s^2)$, calculate $H_s(-1, 2)$ if $f'(5) = -3$.

10. Show that the equation of the plane tangent to the surface with equation $x^2/a^2 - y^2/b^2 + z^2/c^2 = 1$ at the point (x_0, y_0, z_0) is given by
$$\frac{xx_0}{a^2} - \frac{yy_0}{b^2} + \frac{zz_0}{c^2} = 1.$$

11. If $f(x, y) = x^2y + y^3$, $(x, y) \in R^2$, what is the matrix associated with the differential of f at $(1, 2)$? What is the linearization of f near this point?

12. If $\mathbf{f}(x, y) = \langle x/y, y/x \rangle$, $x \neq 0$ and $y \neq 0$, what is the matrix associated with the differential of $\mathbf{f}$ at $(1, 2)$? What is the linearization of $\mathbf{f}$ near this point?

13. Let $w = f(ut, t^2)$ be the function of u and t obtained by substituting $x = ut$ and $y = t^2$ in the expression for $f(x, y)$, an unspecified function. Find w_t and w_{tu} at $(t, u) = (2, 1)$, given that $f_x(2, 4) = f_y(2, 4) = 5$ and $f_{xx}(2, 4) = 1$, $f_{xy}(2, 4) = -2$, and $f_{yy}(2, 4) = -1$.

14. If $w = f(x, y, z)$, $x = e^u \cos v$, $y = e^u \sin v$, and $z = e^{2u}$, calculate $\partial w/\partial v$ in terms of u, v, and the partials of f with respect to x, y, and z.

15. For the surface S defined by $z = 4x^2 + y^2$, find an equation for the tangent plane to S at $(1, 1, 5)$ and a unit normal to this tangent plane.

16. Suppose $P = E^2/R$. Use the tangent plane approximation to approximate the change in P as (E, R) changes from $(100, 25)$ to $(100.5, 25.1)$.

17. If $z = f(x, y)$, where $x = 1/t$ and $y = \sqrt{t}$, express dz/dt in terms of f_x, f_y, and t.

18. Show that for small $|x|$ and $|y|$,
$$\sqrt{\frac{1 + x}{1 + y}} \approx 1 + \frac{1}{2}x - \frac{1}{2}y + \frac{1}{2}\left(-\frac{1}{4}x^2 - \frac{1}{2}xy + \frac{3}{4}y^2\right).$$

19. A surface S is the graph of a function $z = f(x, y)$ defined implicitly by the equation $F(x, y, z) = 0$. At any time t a particle is at the point $\mathbf{r} = \langle x, y, z \rangle = \langle 2 \cos t, 2 \sin t, f(2 \cos t, 2 \sin t) \rangle$ of a curve C in surface S. Calculate the velocity of the particle at time $t = \pi$ given that $3 = f(-2, 0)$, $F_x(-2, 0, 3) = 2.5$, $F_y(-2, 0, 3) = -1.75$, and $F_z(-2, 0, 3) = -0.90$.

20. Let $f(x, y) = xy$, for $x, y > 0$. Describe the level curve with equation $f(x, y) = 1$ parametrically, perhaps with $\mathbf{r}(x) = \langle x, 1/x \rangle$, $0 < x < \infty$. At the point $\mathbf{r}(2)$, show that the tangent vector to the level curve is perpendicular to the gradient vector $\nabla f(2, 1/2)$.

21. If the function $u(x, y)$ satisfies the equation $xu_y + yu_x = x + y$, find an equation satisfied by the function $U(r, s) = u(r + s, r - s)$ in the variables r and s, where
$$x = r + s$$
$$y = r - s.$$

22. If the function $u(x, y)$ satisfies the equation $u_x{}^2 + u_y{}^2 = 0$, find an equation satisfied by the function $U(r, s) = u(e^r \cos s, e^r \sin s)$ in the variables r and s, where

$$x = e^r \cos s$$
$$y = e^r \sin s.$$

23. Let $\mathbf{y} = \mathbf{f}(\mathbf{x})$ and $\mathbf{x} = \mathbf{g}(\mathbf{t}) = \langle t_1^2 + t_2^2, t_1 t_2 \rangle$. Express the derivative matrix $\mathbf{D}(\mathbf{f} \circ \mathbf{g})(\mathbf{t})$ in terms of the partial derivatives of f_1 and f_2.

24. If $g(t) = f(e^t, e^{-t})$, calculate $g''(t)$ in terms of t and the partial derivatives of f.

25. If $H(u, v) = f(uv, u/v)$, calculate $\partial^2 H/\partial v^2$ in terms of u, v, and the partials of f.

26. Find the second-order Taylor polynomial for f at $\mathbf{a} = (1, 1)$, where

$$f(x, y) = x \ln(xy).$$

Referring to the Taylor Inequality for Functions of Two Variables, show that if $|x - 1| < 0.2$ and $|y - 1| < 0.2$, then M can be taken as 4.7. Show that in the square $[0.9, 1.1] \times [0.9, 1.1]$, the difference between f and the second-order Taylor polynomial is less than 0.007.

27. Near $(1, 1)$, the equation $F(x, y) = x^3y^3 - 2x^2y + 1 = 0$ determines y as a function f of x. Calculate $f'(1)$ in two ways, first by the method given in Section 2.3 and then by the formula in Section 10.3 for calculating f' when f is defined implicitly.

28. Near $(1, 0, 1)$, the equation $xze^{yz} - 1 = 0$ determines z as a function f of x and y. Calculate $f_x(1, 0)$ and $f_y(1, 0)$. What is an equation of the plane tangent to the graph of f at $(1, 0)$? Approximate the value of $f(0.9, 0.1)$. How could this value be determined accurately with, say, Newton's method?

29. Let $g(x, y) = x^2 - 2xy + y^3 - y$, $(x, y) \in R^2$. Locate and classify each of the stationary points of g as either a local minimum point, a local maximum point, or a saddle point.

30. Let $f(x, y) = x^2y - 3y^3 + 6x$, $(x, y) \in R^2$. Locate and classify each of the stationary points of f as either a local minimum point, a local maximum point, or a saddle point.

31. The function $f(x, y) = x^3 - 3xy + y^2$ is defined on the triangular region T with vertices $(0, 0)$, $(1, 0)$, and $(1, 3)$. Find the points of T at which f takes on its maximum or minimum values. Evaluate f at these points.

32. For

$$f(x, y) = x^3 - 6x^2 - 3y^2,$$

$-1 \le x, y \le 1$, find the absolute maximum and absolute minimum values of f and where they occur.

33. For the function $f(x, y) = xy^2 - 3x^3 + 2y$ defined on R^2, find all points (x, y) at which f has a local extremum or a saddle point. Show that f has neither a maximum nor a minimum on R^2.

34. Find the dimensions and cost of the least expensive box made in the shape of a rectangular parallelepiped with volume 12 ft^3. The box has a reinforced cardboard bottom costing $\$3/\text{ft}^2$, sides costing $\$1/\text{ft}^2$, and no top.

35. Find three positive numbers x, y, and z whose sum is 100 and that are such that $x^a y^b z^c$ is a maximum. Assume that $a, b, c > 0$.

36. Find the distance between the lines with equations

$$\mathbf{f}(t) = \langle 2, 1, 1 \rangle + t\langle -1, 1, 1 \rangle$$

and

$$\mathbf{g}(s) = \langle 2, 0, 3 \rangle + s\langle -1, 4, 1 \rangle,$$

where $-\infty < s, t < \infty$. Do this by minimizing a function F defined on R^2, where $F(s, t)$ is the square of the distance between $\mathbf{f}(t)$ and $\mathbf{g}(s)$.

37. The temperature T in degrees Fahrenheit of a rectangular metal plate with vertices $(0, 0)$, $(1, 0)$, $(0, 1)$, and $(1, 1)$ is given by $T = 20(5 - x^2 - y^2)$. Sketch the 80°, 88.75°, and 95° isotherms in the plate. Where are the hottest and coldest points on the plate?

38. Use Lagrange multipliers in determining the extreme values of the function $f(x, y) = x^2 + y$ on the circle with equation $x^2 + y^2 = 1$.

39. Suppose f is harmonic on the entire plane (this means that $f_{xx} + f_{yy} = 0$ everywhere), has continuous third-order partials everywhere, and has a strict local maximum at $\mathbf{r}_0$ (that is, $f(\mathbf{r}_0 + \mathbf{h}) < f(\mathbf{r}_0)$ for all $\mathbf{h} \neq \mathbf{0}$ and $\|\mathbf{h}\|$ is sufficiently small). Show that $f_{xy}(\mathbf{r}_0) = 0$.

STUDENT PROJECTS

A. CAUCHY'S METHOD OF STEEPEST DESCENT

If you were walking on the surface described by a real-valued function $z = f(x, y)$ of two variables and were searching for a local minimum, you might argue that your best strategy would be that of a drop of water, to flow always in the direction of steepest descent. Usually, the direction of steepest descent will change with your position. It might happen, for example, that from your present position the direction of steepest descent would take you directly down the side of a valley, toward a stream that eventually flows into a lake containing the local minimum. The direction of steepest descent might change rapidly as you waded through the stream.

Although the idea of searching on a surface for a local minimum is useful, it is important to distinguish between the surface or graph of f and the domain of f when reference is made to the direction of steepest descent. As shown in Chapter 9, f increases (decreases) most rapidly at a given point $\mathbf{a}$ of its domain in the direction of the gradient vector $\nabla f(\mathbf{a})\,(-\nabla f(\mathbf{a}))$. The vector $-\nabla f(\mathbf{a})$ specifies the direction of steepest descent in the domain of f; it does not point in your direction of motion on the surface. Said differently, the vector $-\nabla f(\mathbf{a})$ specifies a direction of the planar map of the surface.

In terms of the surface, the method of steepest descent is a search procedure, or algorithm, for locating a local minimum of the surface. At any point $\mathbf{a}$, we may calculate the elevation $f(\mathbf{a})$ of the surface and the direction $-\nabla f(\mathbf{a})$ of steepest descent.

We describe a steepest descent algorithm and then give an example. The algorithm, stated in terms of a differentiable function f, generates a sequence of points $\mathbf{a}_1, \mathbf{a}_2, \ldots$, which may converge to a local minimum of f. We assume that each generated point lies in the domain of f. The algorithm can fail in various ways.

Step 1: Set $n = 1$ and choose a point $\mathbf{a}_1$ near a local minimum of f.

Step 2: Calculate $f(\mathbf{a}_n)$ and the unit negative gradient vector

$$\mathbf{U}_n = \mathbf{U}_n(\mathbf{a}_n) = -\frac{1}{\|-\nabla f(\mathbf{a}_n)\|}\nabla f(\mathbf{a}_n).$$

In the direction of $\mathbf{U}_n$, the values of f will decrease and then increase, eventually rising above the height $f(\mathbf{a}_n)$. Choose a positive number h_n such that

$$f(\mathbf{a}_n + h_n\mathbf{U}_n) > f(\mathbf{a}_n).$$

Let $\mathbf{q}_n = \mathbf{a}_n + h_n\mathbf{U}_n$. As the algorithm cycles through Step 2, the size of the h_n's should gradually decrease.

Step 3: Minimize f along the line segment from $\mathbf{a}_n$ to $\mathbf{q}_n$. For this, let

$$\mathbf{r}(t) = \mathbf{a}_n + t(\mathbf{q}_n - \mathbf{a}_n), \qquad 0 \le t \le 1,$$

and solve the one-dimensional problem of minimizing the function g on $[0, 1]$, where

$$g(t) = f(\mathbf{r}(t)), \qquad 0 \le t \le 1.$$

Denote by $t_{\min}$ the point $[0, 1]$ at which g is a minimum.

Step 4: Denote the point $\mathbf{r}(t_{\min}) = \mathbf{a}_n + t_{\min}(\mathbf{q}_n - \mathbf{a}_n)$ by $\mathbf{a}_{n+1}$ and calculate $\nabla f(\mathbf{a}_{n+1})$. Stop if $\|\nabla f(\mathbf{a}_{n+1})\|$ is sufficiently small. Otherwise, go to Step 2.

Example

A contour plot of the function f defined by

$$f(x,y) = x^4 + 3x^2y + 5y^2 + x + y, \qquad (x,y) \in R^2,$$

is shown in Fig. 10.55. It appears that f has a local minimum near $(-0.9, -0.3)$, where the value of f is less than -0.8. To illustrate the algorithm, we choose $\mathbf{a}_1 = \langle -0.6, 0.1\rangle$. It would be easy to improve upon this particular initial guess.

The gradient of f at any point $\mathbf{a} = \langle x, y\rangle$ is the vector

$$\nabla f(\mathbf{a}) = \langle 4x^3 + 6xy + 1, 3x^2 + 10y + 1\rangle.$$

The unit negative gradient of f at $\mathbf{a}_1$ is

$$\mathbf{U}_1 = \frac{1}{3.088135}\langle 0.224000, -3.080000\rangle = \langle 0.072536, -0.997366\rangle.$$

In what follows, we choose the step size h_1 by comparing $f(\mathbf{a}_1)$ and $f(\mathbf{a}_1 + h_1\mathbf{U}_1)$ for several values of h_1. For this we need the height at $\mathbf{a}_1$, which is $f(\mathbf{a}_1) = -0.212400$. From Fig. 10.55, we guess that a reasonable step size is $h_1 = 0.7$. We calculate

$$\begin{aligned}\mathbf{a}_1 + h_1\mathbf{U}_1 &= \langle -0.6, 0.1\rangle + 0.7\langle 0.072536, -0.997366\rangle \\ &= \langle -0.549225, -0.598156\rangle\end{aligned}$$

and then $f(\mathbf{a}_1 + h_1\mathbf{U}_1) = 0.191266$, which is larger than $f(\mathbf{a}_1) = -0.212400$. We take $\mathbf{q}_1 = \langle -0.549225, -0.598156\rangle$. The point $\mathbf{q}_1$ is across the valley from $\mathbf{a}_1$ and at a higher elevation.

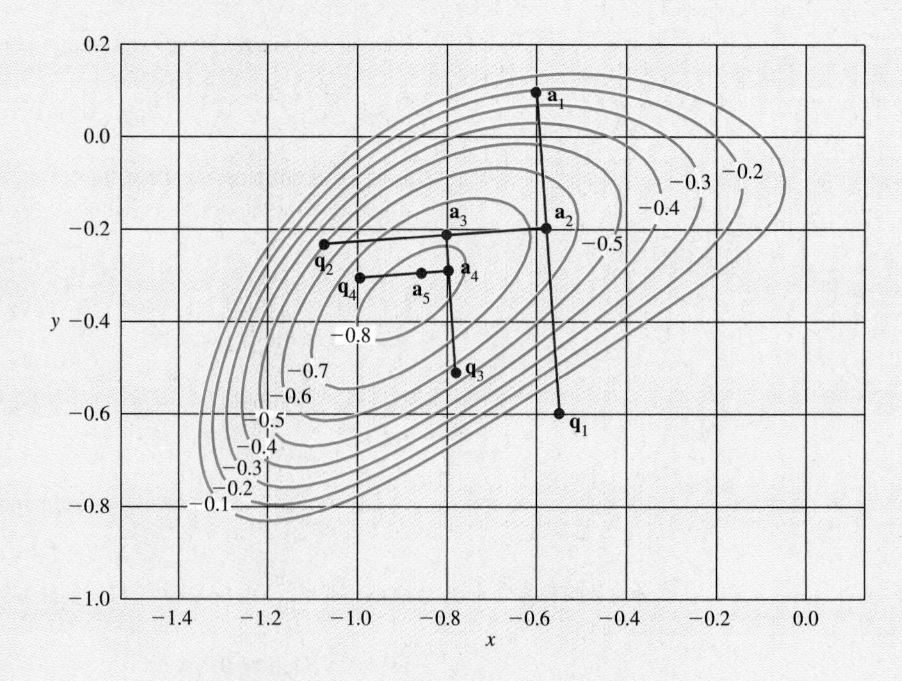

FIGURE 10.55 A sequence $\mathbf{a}_1, \mathbf{a}_2, \mathbf{a}_3, \ldots$ generated by the algorithm.

We define

$$\begin{aligned} g(t) &= f(\mathbf{r}(t)) = f(\mathbf{a}_1 + t(\mathbf{q}_1 - \mathbf{a}_1)) \\ &= -0.212400 - 2.16169t + 2.57107t^2 \\ &\quad -0.00571391t^3 + 6.64660 \cdot 10^{-6}t^4, \qquad 0 \le t \le 1, \end{aligned}$$

and minimize g on $[0, 1]$. We obtain $t_{\min} = 0.420979$, from which

$$\mathbf{a}_2 = \mathbf{a}_1 + t_{\min}(\mathbf{q}_1 - \mathbf{a}_1) = \langle -0.578625, -0.193909 \rangle.$$

We calculate

$$\nabla f(\mathbf{a}_2) = \langle 0.898293, 0.0653307 \rangle \quad \text{and} \quad \|\nabla f(\mathbf{a}_2)\| = 0.900666.$$

Because the length of the gradient vector of f is not "sufficiently small," we go to Step 2 of the algorithm.

Graphical and numerical details of several iterations of the algorithm are given in Fig. 10.55 and Table 10.2. It appears from these calculations and Fig. 10.55 that the points $\mathbf{a}_1, \mathbf{a}_2, \ldots$ are approaching a local minimum point, the values $f(\mathbf{a}_1), f(\mathbf{a}_2) \ldots$ are approaching a local minimum of f, and the length of the gradient is approaching zero, as it should. For this particular function, all candidates for extrema may be found by classical methods. The local minimum point thus found is $\langle -0.886324, -0.335671 \rangle$, approximately. The value of f at this point is -0.832579.

TABLE 10.2 Iterations of steepest descent algorithm.

n	h_n	$\mathbf{q}_n$	$\mathbf{a}_n$	$f(\mathbf{a}_n)$	$\|\nabla f(\mathbf{a}_n)\|$
1	0.7	$\langle -0.55, -0.60 \rangle$	$\langle -0.60, 0.10 \rangle$	-0.21	3.09
2	0.5	$\langle -1.08, -0.23 \rangle$	$\langle -0.58, -0.19 \rangle$	-0.67	0.90
3	0.3	$\langle -0.78, -0.51 \rangle$	$\langle -0.80, -0.21 \rangle$	-0.78	0.84
4	0.2	$\langle -1.00, -0.30 \rangle$	$\langle -0.80, -0.29 \rangle$	-0.82	0.35
5	0.1	$\langle -0.85, -0.39 \rangle$	$\langle -0.86, -0.29 \rangle$	-0.83	0.28
6			$\langle -0.86, -0.32 \rangle$	-0.83	0.13

PROBLEM 1 First, find the low point for the function f in Step 1 of the Example by solving the system of equations $\nabla f(x, y) = 0$. Compare your results with the last line of Table 10.2. Next, taking $\mathbf{p}_1 = \langle -1.1, -0.4 \rangle$ as a starting value, use the steepest descent algorithm to approximate the low point of f. Use Fig. 10.55 in choosing h_1.

PROBLEM 2 Use the steepest descent algorithm in determining the local minimum nearest the origin for the function

$$f(x, y) = \cos(x + y) + \sin x + \cos y.$$

Repeat the steepest descent algorithm until the length of the gradient is less than 0.1. *Extra credit:* Solve this problem using classical methods.

PROBLEM 3 Use the steepest descent algorithm in determining the local minimum for the function

$$f(x, y) = (x - 1)^2 + y^3 - xy.$$

Repeat the steepest descent algorithm until the length of the gradient is less than 0.001.

PROBLEM 4 The circles C_1 and C_2 are described by

$$C_1: \langle x, y, z \rangle = \langle \cos s, 0, \sin s \rangle, \qquad 0 \le s \le 2\pi$$
$$C_2: \langle x, y, z \rangle = \langle 2, 2, 2 \rangle + (\cos t)\mathbf{e}_1 + (\sin t)\mathbf{e}_2, \qquad 0 \le t \le 2\pi,$$

where $\mathbf{e}_1$ and $\mathbf{e}_2$ are the mutually perpendicular unit vectors

$$\mathbf{e}_1 = \left(1/\sqrt{3}\right)\langle -1, 1, 1 \rangle$$
$$\mathbf{e}_2 = \left(1/\sqrt{2}\right)\langle 0, -1, 1 \rangle.$$

Find position vectors $\mathbf{a}_1$ and $\mathbf{a}_2$ of points on C_1 and C_2 for which $\|\mathbf{a}_1 - \mathbf{a}_2\|$ is a minimum. The number $\|\mathbf{a}_1 - \mathbf{a}_2\|$ is the "shortest distance" between the two circles.

PROBLEM 5 The steepest descent algorithm may be used to approximate the solutions of systems of equations. Solve the system of equations

$$x^2 - 10x + y^2 + 8 = 0$$
$$xy^2 + x - 10y + 8 = 0$$

by minimizing the function

$$f(x, y) = (x^2 - 10x + y^2 + 8)^2 + (xy^2 + x - 10y + 8)^2.$$

Look for all (x, y) such that f has a local minimum and $f(x, y) = 0$. One of the two solutions is almost obvious. Repeat the steepest descent algorithm until an appropriate gradient is less than 0.1.

PROBLEM 6 Use the idea outlined in the preceding problem to approximate the solution to the system

$$2\pi \sin(xy) - y - 2\pi x = 0$$
$$(4\pi - 1)(e^{2x-1} - 1) + 4y - 8\pi x = 0.$$

Repeat the steepest descent algorithm until an appropriate gradient is less than 0.1.

PROBLEM 7 Locate the point on the ellipsoid $x^2/2 + y^2/3 + z^2/5 = 1$ closest to the point $(1, 1, 3)$. Repeat the steepest descent algorithm until an appropriate gradient is less than 0.1.

PROBLEM 8 Use the steepest descent algorithm in minimizing the function

$$f(x, y, z) = x^2 + 2y^2 + z^2 - 2xy + 2x - 2.5y - z + 2.$$

Although f is a function of three variables and level surfaces may be plotted, it is doubtful that a graphical approach is useful. Instead, rearrange f by completing the squares as far as possible, noting that if the resulting nonnegative terms are set to zero, we may obtain a promising first guess for $\mathbf{a}_1$.

PROBLEM 9 Suppose we have four objects fixed in space. Their masses and positions are

$$\begin{aligned} m_1 &= 5.5 & \mathbf{r}_1 &= \langle 1.5, 1.3, 1.0\rangle \\ m_2 &= 3.4 & \mathbf{r}_2 &= \langle 2.2, 7.2, 8.6\rangle \\ m_3 &= 4.5 & \mathbf{r}_3 &= \langle 7.9, 1.4, 9.3\rangle \\ m_4 &= 5.3 & \mathbf{r}_4 &= \langle 10.5, 9.7, 2.5\rangle. \end{aligned}$$

Assume that the force exerted on a unit mass by any one of these objects varies directly with the mass of the object and inversely with the square of the distance between the object and the unit mass. Use steepest descent in locating a unit mass at a point $\mathbf{p}_{\text{min}}$ where the magnitude of the combined force on the unit mass is as small as possible. The point $\mathbf{a}_{\text{min}}$ is to be inside the smallest sphere containing the four objects.

PROBLEM 10 In the Example at the beginning of this Project, the function g in the first iteration is very nearly quadratic. It is often assumed in steepest descent algorithms that near a minimum, the function is essentially quadratic. This results in a faster algorithm. At a point $\mathbf{a}_n$, find $\mathbf{U}_n$ and calculate f at three nearby points in this direction. Fit a parabola to these data, and then use a standard formula to calculate its vertex. This gives $\mathbf{a}_{n+1}$. Work out the details of a replacement for Step 3 based on these ideas.

B. WAVE EQUATION

A string is stretched along the x-axis, between the points $(0,0)$ and $(2,0)$. The string is plucked by raising its midpoint to a height h above the x-axis and then releasing it from rest.

Assume that the string vibrates within a vertical plane, that its points move only vertically, and that the amplitude of vibration is small. For any time $t \geq 0$, let $U(x,t)$, $0 \leq x \leq 2$, denote the vertical displacement of point x of the string at time t. By elementary mechanics (see Exercise 32 of Section 10.4), U satisfies the *wave equation*

$$c^2 \frac{\partial^2 U}{\partial t^2} = \frac{\partial^2 U}{\partial x^2}, \tag{1}$$

where c is a constant determined by the tension in the string and its density.

Because we want the ends of the string to be fixed, the displacement U is subject to the **boundary conditions**

$$U(0,t) = 0 \quad \text{and} \quad U(2,t) = 0, \qquad t \geq 0. \tag{2}$$

We assume further that U satisfies two **initial conditions.** Its initial configuration, shown in Fig. 10.56, is the graph of a function f. Thus,

$$U(x,0) = f(x), \qquad 0 \leq x \leq 2. \tag{3}$$

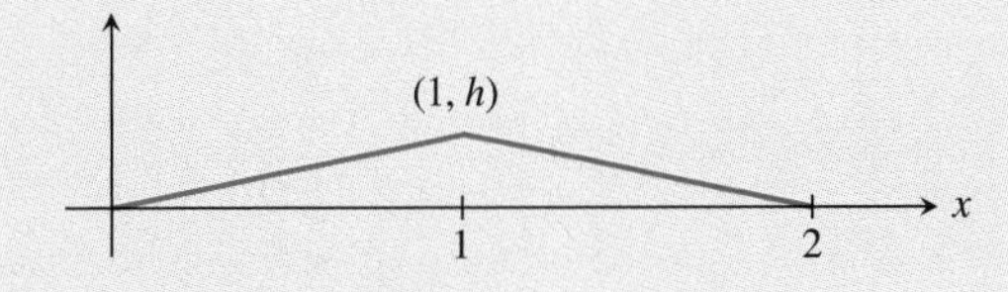

FIGURE 10.56 Plucked string.

The second initial condition is

$$\frac{\partial U(x,0)}{\partial t} = 0, \qquad 0 \le x \le 2, \tag{4}$$

which specifies that the string is released from rest, that is, the velocity of each point of the string is zero at $t = 0$.

Standard techniques for solving the wave equation (1) subject to the given boundary and initial conditions (2)–(4) give the solution as the sum

$$U(x,t) = \sum_{n=1}^{\infty} A_n \sin\frac{n\pi x}{2} \cos\frac{n\pi t}{2c} \tag{5}$$

of *standing waves*. The amplitude A_n of the nth standing wave is

$$A_n = \int_0^2 f(x) \sin\frac{n\pi x}{2}\, dx, \qquad n = 1, 2, \ldots. \tag{6}$$

Except for the suggested animation, the following problems do not require a CAS.

PROBLEM 1 Show that

$$A_n = \frac{8h}{\pi^2 n^2} \sin\frac{n\pi}{2}, \qquad n = 1, 2, \ldots.$$

Note that $A_n = 0$ when n is even.

PROBLEM 2 In practice, U is approximated by truncating the infinite series (5). For example,

$$\begin{aligned} u(x,t) &= \sum_{n=1}^{10} A_n \sin\frac{n\pi x}{2} \cos\frac{n\pi t}{2c} \\ &= \sum_{m=0}^{4} A_{2m+1} \sin\frac{(2m+1)\pi x}{2} \cos\frac{(2m+1)\pi t}{2c}. \end{aligned}$$

Does the function u satisfy (1), (2), and (4)? Support your statements with appropriate calculations.

For the remaining problems, take $h = 0.3$ and $c = 0.05$.

PROBLEM 3 Find how well the function u satisfies (3) by comparing two graphs. You may plot both curves on the same axes or plot their difference.

PROBLEM 4 Show that

$$|U(x,t) - u(x,t)| \le 0.014, \qquad 0 \le x \le 2, \qquad t \ge 0,$$

by filling in the details of the following argument. Try to justify each step.

$$\begin{aligned} |U(x,t) - u(x,t)| &= \left| \sum_{n=11}^{\infty} A_n \sin\frac{n\pi x}{2} \cos\frac{n\pi t}{2c} \right| \le \sum_{n=11}^{\infty} \left| A_n \sin\frac{n\pi x}{2} \cos\frac{n\pi t}{2c} \right| \\ &\le \sum_{n=11}^{\infty} |A_n| = \frac{8h}{\pi^2}\left(\frac{1}{11^2} + \frac{1}{13^2} + \cdots \right) \le \frac{2.4}{\pi^2}\frac{1}{18} \le 0.014. \end{aligned}$$

As part of filling in the details, give a geometric argument for the result that

$$\frac{1}{11^2} + \frac{1}{13^2} + \cdots \le \int_4^{\infty} (2x+1)^{-2}\, dx = \frac{1}{18},$$

which we used earlier.

PROBLEM 5 How would you truncate the series for U if it were required that

$$|U(x,t) - u(x,t)| \leq 0.001, \qquad 0 \leq x \leq 2, \qquad t \geq 0?$$

PROBLEM 6 To simulate the string in motion, run an animation by graphing $u(x,t)$, $0 \leq x \leq 2$, for each of the values $t = 0, 0.03, 0.06, \ldots, 0.21$.

C. RHUMB LINES ON MERCATOR MAPS

Gerardus Mercator (1512–1594), often called the greatest cartographer of the sixteenth century, devised the representation of the Earth's surface on a flat surface known today as "Mercator's projection." Referring to Fig. 10.57, if on a voyage *mpa* on a spherical Earth a ship's helmsman maintains a constant compass heading (each meridian *cpb* is crossed at the same angle, then the ship's path as plotted on Mercator's map is a straight line. The path *mpa* is called a *rhumb line;* it plots on Mercator's map as line *MPA*.

Mercator's maps became popular with both helmsmen and navigators because they simplified parts of their work. However, sailing on a rhumb line usually does not give the shortest path between points. Sailing on a great circle gives shortest paths but usually requires the helmsman to constantly correct the ship's heading. In practice, great circle routes were approximated by sailing on several rhumb lines.

In the problems to follow, we work with the part of the unit sphere described by

$$\mathbf{r} = \mathbf{r}(\theta, \phi) = (\cos\theta \sin\phi, \sin\theta \sin\phi, \cos\phi), \qquad 0 \leq \theta, \phi \leq \pi/2. \tag{7}$$

The point p in Fig. 10.57 has colatitude ϕ. This number measures angle *cop* from *oc* toward the (x, y)-plane. (Geographic latitude is measured from the equator toward the poles.) The longitude θ of p is measured by angle *aob* from *ox* toward *oy*. (Geographic longitude in measured from the Greenwich meridian.) The curve *cpb* described by (7) with θ constant is a meridian, and the curve *npq* described by (7) with ϕ constant is parallel.

Mercator assigned a point (X, Y) of a plane to every point $\mathbf{r}(\theta, \phi)$ of (the first-octant portion of) the sphere, except for the point c. The coordinates X and Y are functions of ϕ and θ, so that

$$(X, Y) = (X(\theta, \phi), Y(\theta, \phi)), \qquad 0 \leq \theta \leq \pi/2, \qquad 0 < \phi \leq \pi/2.$$

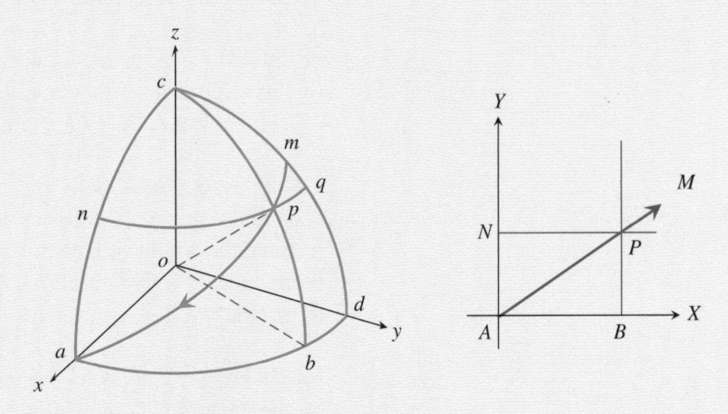

FIGURE 10.57 Earth and Mercator's map.

To find the functions X and Y, Mercator imposed three conditions:

1. C1: Equally spaced meridians must map into vertical, equally spaced lines.
2. C2: The parallels must map into horizontal lines.
3. C3: At all points $p \neq c$ of the sphere, the scale changes in the directions of the meridians and parallels are to be the same.

PROBLEM 1 Explain in simple language how Mercator's maps simplified parts of the work of helmsmen and navigators.

PROBLEM 2 Give heuristic but convincing arguments that conditions C1 and C2 together imply that X and Y have the forms

$$X(\theta, \phi) = k\theta + h \quad \text{and} \quad Y(\theta, \phi) = g(\phi),$$

where k and h are constants and g is a function.

PROBLEM 3 Explain and add detail to the following discussion of the condition C3.

1. For the parallels, the ratio of map to sphere distance is

$$\left| \frac{X(\theta + \Delta\theta, \phi) - X(\theta, \phi)}{\Delta\theta \sin \phi} \right|.$$

For the meridians, the ratio of map to sphere distance is

$$\left| \frac{Y(\theta, \phi + \Delta\phi) - Y(\theta, \phi)}{\Delta\phi} \right|.$$

2. Using limiting arguments and integration as necessary, show that

$$(X, Y) = \left(k\theta + h, -k \ln\left(\tan \frac{1}{2}\phi \right) + q \right),$$

where q is a constant. Use the conditions that a is to map to the origin A of the map and the point d to $(1, 0)$ to evaluate h, k, and q.

PROBLEM 4 To illustrate the distortion on a Mercator map, create a careful sketch of the map image of the "square" of side 0.2 units and centered at the point p with colatitude 30° and longitude 70°. Measure the sides of the square along the meridian and parallel that pass through p. What does this distortion mean from the viewpoint of, say, a young child viewing a Mercator map?

PROBLEM 5 Suppose a ship's course crosses the meridians at a certain angle. Argue directly from Mercator's conditions that on Mercator's map the plotted course crosses all vertical lines at the same angle.

PROBLEM 6 As shown in Fig. 10.57, a ship is sailing in a southwesterly direction, on rhumb line *mpa*. At the point p, the angle between the tangent vector to this rhumb line and the downward-pointing tangent vector to the meridian is 1 radian. Give a parametric equation that describes this rhumb line. Write the equation of the ship's course as plotted on Mercator's map.

PROBLEM 7 Calculate and compare the distances sailed on the rhumb line *mpa* and on the great circle connecting m and a.

11

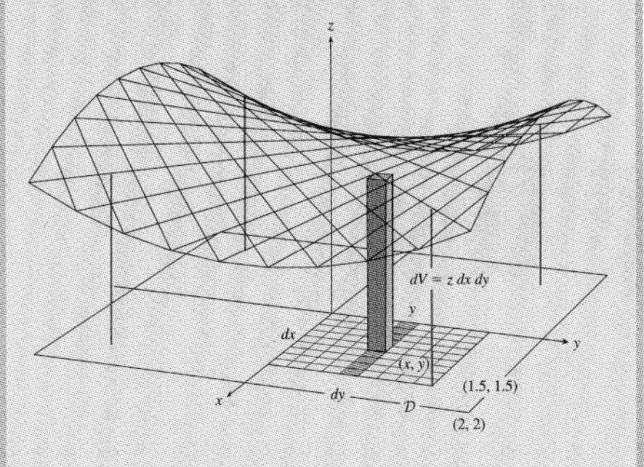

Multiple Integrals

The church of Notre-Dame du Haut in Ronchamp, France, designed by LeCorbusier, was completed in 1955 and is considered one of the great buildings of the 20th century. The volume of space beneath the roof can be calculated by describing the roof with a function and then using a multiple integral.

See Example 2, p. 946, for further explanation.

Just as the concept of a differentiable function of several variables is based on our understanding of the derivative of a function of one variable, the concept of a multiple integral is based on our understanding of one-dimensional integrals of real-valued functions of one variable.

In Chapters 5 and 6 we used one-dimensional integrals to calculate the areas, lengths, masses, and centers of mass of one- or two-dimensional geometric or physical objects. We used a one-dimensional integral to calculate the volumes of regions of R^3 that can be sectioned, by planes perpendicular to a given line, into two-dimensional regions with known areas. Here we use double or triple integrals to calculate volumes, surface areas, masses, and centers of mass of two- or three-dimensional geometric or physical objects. As an example, Fig. 11.1 shows a curved plate with variable density. A double integral can be used to calculate its area, mass, or center of mass.

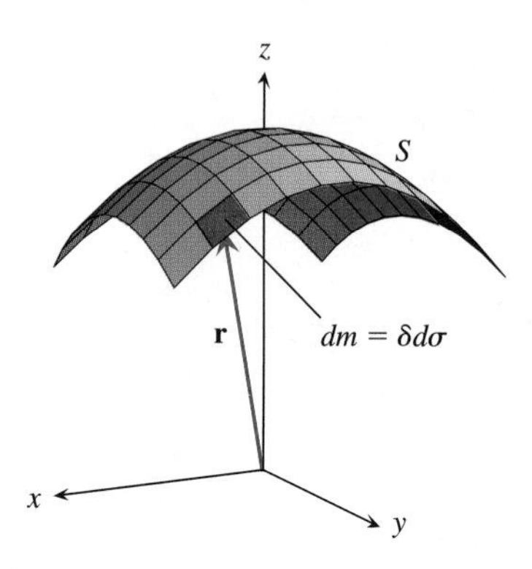

FIGURE 11.1 A thin, nonplanar lamina with variable density.

11.1 The Double Integral on Rectangles

By measuring the total annual rainfall on a watershed, hydrologists can predict the amount of water available for irrigation or driving turbines at a hydroelectric power station. As a guide to defining and evaluating double integrals, we discuss two ways of calculating total annual rainfall. The first method is based on data from 25 subregions of the watershed and the second method on a calculation of the volume of a solid with known cross sections.

INVESTIGATION

Approximating the Rainfall on a Watershed

Figure 11.2 shows rainfall data collected in 25 subregions of a 100-mile-by-100-mile square region $\mathcal{D}$. Listed within each subregion is the total annual rainfall collected in a rain gauge placed at the center of the subregion. To approximate the total annual rainfall received by each subregion, multiply the rainfall collected at the center point by the area of the subregion. For the subregion with center point $c_{2,3}$, for example, the

$$\text{total annual rainfall} \approx 30.77 \cdot 20 \cdot 20 \text{ square miles–inches.}$$

The square mile–inch units of the rainfall received by a region are analogous to the acre-foot or acre-inch units used in reporting the volume of water contained in a lake or reservoir. An acre-inch of water is the volume of water required to cover one acre (about 43,560 square feet) to a depth of one inch.

The annual rainfall V received by the watershed $\mathcal{D}$ is found by summing the rainfall amounts for each of the 20-mile-by-20-mile sections. Thus,

$$\begin{aligned} V &= (43.29 + 39.53 + \cdots + 19.49 + 18.15) \cdot 20 \cdot 20 \\ &\approx 28.04 \times 10^4 \text{ square mile–inches.} \end{aligned} \tag{1}$$

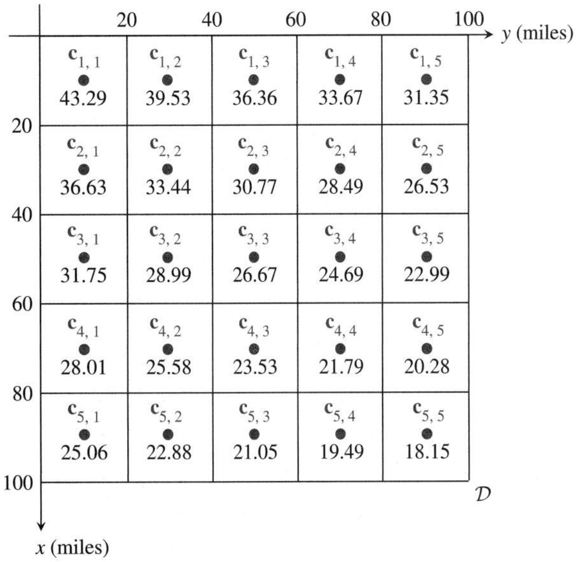

x (miles) \ y (miles)	0–20	20–40	40–60	60–80	80–100
0–20	$\mathbf{c}_{1,1}$ 43.29	$\mathbf{c}_{1,2}$ 39.53	$\mathbf{c}_{1,3}$ 36.36	$\mathbf{c}_{1,4}$ 33.67	$\mathbf{c}_{1,5}$ 31.35
20–40	$\mathbf{c}_{2,1}$ 36.63	$\mathbf{c}_{2,2}$ 33.44	$\mathbf{c}_{2,3}$ 30.77	$\mathbf{c}_{2,4}$ 28.49	$\mathbf{c}_{2,5}$ 26.53
40–60	$\mathbf{c}_{3,1}$ 31.75	$\mathbf{c}_{3,2}$ 28.99	$\mathbf{c}_{3,3}$ 26.67	$\mathbf{c}_{3,4}$ 24.69	$\mathbf{c}_{3,5}$ 22.99
60–80	$\mathbf{c}_{4,1}$ 28.01	$\mathbf{c}_{4,2}$ 25.58	$\mathbf{c}_{4,3}$ 23.53	$\mathbf{c}_{4,4}$ 21.79	$\mathbf{c}_{4,5}$ 20.28
80–100	$\mathbf{c}_{5,1}$ 25.06	$\mathbf{c}_{5,2}$ 22.88	$\mathbf{c}_{5,3}$ 21.05	$\mathbf{c}_{5,4}$ 19.49	$\mathbf{c}_{5,5}$ 18.15

$\mathcal{D}$

FIGURE 11.2 The total annual rainfall received at the center points of 25 subregions, from which the rainfall received on the entire region can be approximated.

To motivate the definition of *multiple integral,* we continue with the rainfall model but now assume a "continuous" rainfall model. Specifically, assume that the total annual rainfall $r(x, y)$ at $(x, y) \in \mathcal{D}$ is given by the function

$$r(x, y) = \frac{100^3}{(x + 100)(y + 200)}, \qquad (x, y) \in \mathcal{D}, \tag{2}$$

where x and y are given in miles and $r(x, y)$ in inches. This function is consistent with the data listed in Fig. 11.2. Figure 11.3 suggests the graph of r by showing its boundary.

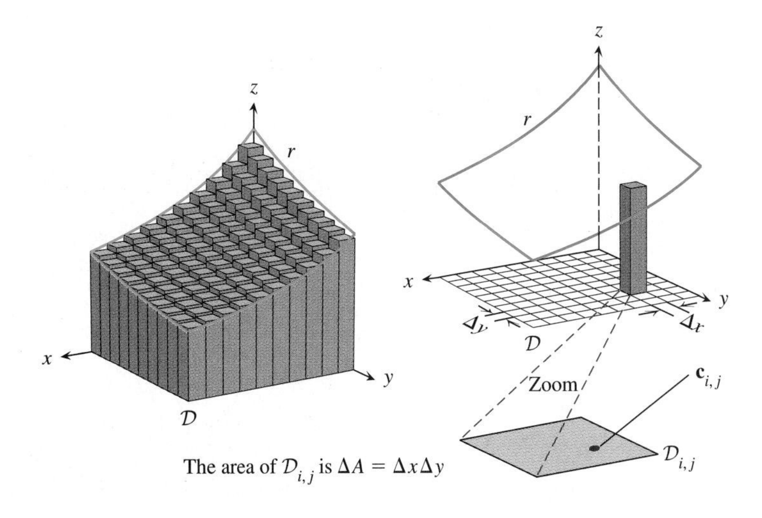

FIGURE 11.3 Collectors or volume elements beneath the graph of the rainfall function r.

We argue that the total annual rainfall on $\mathcal{D}$ is equal to the volume beneath the graph of r. To connect the ideas of total annual rainfall and volume, imagine that $\mathcal{D}$ has been subdivided into $n \times n$ square subregions

$$\mathcal{D}_{1,1}, \mathcal{D}_{1,2}, \ldots, \mathcal{D}_{1,n}, \mathcal{D}_{2,1}, \ldots, \mathcal{D}_{2,n}, \ldots, \mathcal{D}_{n,n},$$

and that collectors with square cross sections stand upon these subregions. Figure 11.3 shows 100 collectors. We assume that each of the square cross sections is sufficiently small that the rainfall on the corresponding subregion of $\mathcal{D}$ is essentially constant. The product of the height of each collector and the area of the subregion on which it stands is approximately equal to the total annual rainfall received on the subregion or to the volume beneath the graph of r and above the subregion. Summing all collectors gives an approximation to the total annual rainfall on $\mathcal{D}$ or the volume beneath the graph of r. We expand these remarks in what follows.

The area of the representative subregion $\mathcal{D}_{i,j}$ shown in Fig. 11.3 is $\Delta A = \Delta x \, \Delta y$. If in each subregion $\mathcal{D}_{i,j}$ we choose an evaluation point $\mathbf{c}_{i,j}$, then the volume of rainfall on $\mathcal{D}_{i,j}$ is approximately $r(\mathbf{c}_{i,j}) \, \Delta x \, \Delta y$. Summing the rainfall amounts over all subregions we see that the total annual rainfall V on $\mathcal{D}$ is

$$V \approx \sum_{i,j} r(\mathbf{c}_{i,j}) \, \Delta A = \sum_{i,j} r(\mathbf{c}_{i,j}) \, \Delta x \, \Delta y. \tag{3}$$

The evaluation point $\mathbf{c}_{i,j}$ in $\mathcal{D}_{i,j}$ may be chosen as the point receiving the minimum rain, the point receiving the maximum rain, or a point at which the received rain is between the minimum and maximum rain. The term $r(\mathbf{c}_{i,j}) \, \Delta A = r(\mathbf{c}_{i,j}) \, \Delta x \, \Delta y$ in the sum (3) is approximately the volume $dV_{i,j}$ between the graph of r and the subregion $\mathcal{D}_{i,j}$.

As the number of subregions becomes large, we expect that the sum in (3) will approach V. This limit will be called the *double integral of r on $\mathcal{D}$*. With the one-dimensional integral in mind, the form of the approximating sum suggests the notations

$$\iint_{\mathcal{D}} r(x, y) \, dA \quad \text{or} \quad \iint_{\mathcal{D}} r(x, y) \, dx \, dy$$

for the double integral.

The brief summation notation used in (3) does not specify an order in which the volume elements $dV_{i,j} = r(\mathbf{c}_{i,j}) \, \Delta x \, \Delta y$ are summed. We may, for example, rearrange (3) as

$$V \approx \sum_{j=1}^{n} \left(\sum_{i=1}^{n} dV_{i,j} \right) = \sum_{j=1}^{n} \left(\sum_{i=1}^{n} r(\mathbf{c}_{i,j}) \, \Delta A \right). \tag{4}$$

Figure 11.4 shows the ordering of the subregions $\mathcal{D}_{1,1}, \ldots, \mathcal{D}_{n,n}$ in (4). The first subscript in $\mathcal{D}_{i,j}$ corresponds to an interval of width Δx along the x-axis while the second subscript corresponds to an interval of width Δy along the y-axis. The arrangement of the x- and y-axes corresponds to the view of the (x, y)-plane shown in Fig. 11.3. With the index j fixed in the outer summation, the inner sum in (4) is

$$\sum_{i=1}^{n} r(\mathbf{c}_{i,j}) \, \Delta A = r(\mathbf{c}_{1,j}) \, \Delta A + \cdots + r(\mathbf{c}_{i,j}) \, \Delta A + \cdots + r(\mathbf{c}_{n,j}) \, \Delta A.$$

From Fig. 11.4 we see that in this inner sum the volume elements are summed in the x-direction first, that is, j is fixed and i increases from 1 to n.

We argue that the summation order in (4) leads to the formula

$$V = \int_0^{100} \left(\int_0^{100} r(x, y) \, dx \right) dy$$

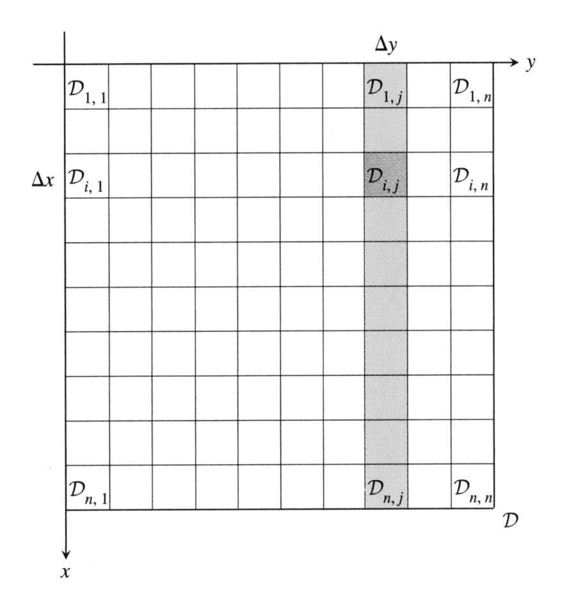

FIGURE 11.4 A view of $\mathcal{D}$ from above, showing the volume elements being summed first in the x-direction.

for the volume V of the region $\mathcal{K}$ beneath the graph of r and above the region $\mathcal{D}$. This expression for V contains two, one-dimensional integrals, each of which can be evaluated with one-variable methods.

Writing the evaluation points $\mathbf{c}_{i,j}$ in (4) as (x_i, y_j), replacing ΔA by $\Delta x\,\Delta y$, and noting that Δx and Δy depend on neither i nor j, we may write (4) as

$$V \approx \sum_{j=1}^{n}\left(\sum_{i=1}^{n} r(x_i, y_j)\,\Delta x\right)\Delta y. \tag{5}$$

If n is sufficiently large, the inner sum in this approximation is very nearly equal to the integral $\int_0^{100} r(x, y_j)\,dx$. Hence,

$$V \approx \sum_{j=1}^{n}\left(\int_0^{100} r(x, y_j)\,dx\right)\Delta y. \tag{6}$$

Applying one-variable methods to evaluate the integral in (6), in which y_j is constant,

$$\begin{aligned}\int_0^{100} r(x, y_j)\,dx &= \int_0^{100} \frac{100^3}{(x+100)(y_j+200)}\,dx \\ &= \frac{100^3}{y_j+200}\int_0^{100}\frac{dx}{x+100} = \frac{100^3}{y_j+200}\ln(x+100)\bigg|_0^{100} \\ &= \frac{100^3}{y_j+200}(\ln 200 - \ln 100) = \frac{100^3}{y_j+200}\ln 2.\end{aligned}$$

If we write this result as

$$\alpha(y_j) = \int_0^{100} r(x, y_j)\,dx = \frac{100^3}{y_j+200}\ln 2, \tag{7}$$

we see from (6) that

$$V \approx \sum_{j=1}^{n} \left(\int_0^{100} r(x, y_j)\, dx \right) \Delta y = \sum_{j=1}^{n} \alpha(y_j)\, \Delta y.$$

Because for large n,

$$\sum_{j=1}^{n} \alpha(y_j)\, \Delta y \approx \int_0^{100} \alpha(y)\, dy,$$

it follows that

$$V \approx \sum_{j=1}^{n} \alpha(y_j)\, \Delta y \approx \int_0^{100} \alpha(y)\, dy = \int_0^{100} \left(\int_0^{100} r(x, y)\, dx \right) dy.$$

As n increases without bound it is plausible that

$$V = \int_0^{100} \alpha(y)\, dy = \int_0^{100} \left(\int_0^{100} r(x, y)\, dx \right) dy. \tag{8}$$

We evaluate the integrals in (8) and compare the result with our earlier approximation of V using 25 subdivisions of $\mathcal{D}$. Recalling (7),

$$\begin{aligned} V &= \int_0^{100} \alpha(y)\, dy = 100^3 \ln 2 \int_0^{100} \frac{1}{y + 200}\, dy = 100^3 \ln 2 \left(\ln(y + 200) \Big|_0^{100} \right) \\ &= 100^3 \ln 2 \ln 1.5 \approx 28.10 \times 10^4 \text{ square mile–inches.} \end{aligned}$$

This value is within about 2 percent of the value $V \approx 28.04 \times 10^4$ square mile–inches from (1).

We use Fig. 11.5 in interpreting (5) and its limiting form (8) geometrically. Figure 11.5(a) shows the row of volume elements corresponding to the inner sum in (5). Recalling that this inner sum is very nearly equal to the integral $\int_0^{100} r(x, y_j)\, dx$, we see that the inner sum in (5) approximates the area of the cross section of $\mathcal{K}$ perpendicular to the y-axis and through $(0, y_j)$. If we multiply this inner sum by Δy, the product approximates the volume of the row of volume elements corresponding to y_j. The outer sum in (5) sums the rows of volume elements and approximates the volume V of $\mathcal{K}$.

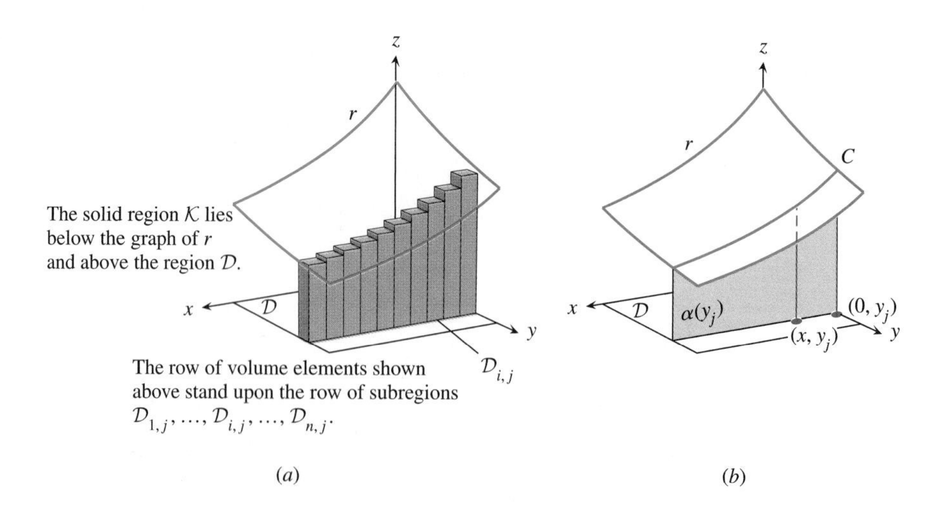

FIGURE 11.5 (a) Calculating the volume of the solid region $\mathcal{K}$ by summing first in the x-direction ($n = 10$); (b) The same as in (a) except n is much larger.

Figure 11.5(b) is similar to Fig. 11.5(a), but we imagine that n is much, much larger. The crude "butter sticks" in Fig. 11.5(a) have been replaced by a line-like volume element $dV = r(x, y)\,dx\,dy$ and the location of a representative volume element is given by (x, y) instead of (x_i, y_j). Note that we have replaced Δx and Δy by dx and dy.

Referring to (8), the value of the inner integral is the area $\alpha(y)$ of the cross section of $\mathcal{K}$ perpendicular to the y-axis and through $(0, y)$. We may imagine that this area is swept out by the volume element as x varies from 0 to 100 and the volume element moves across $\mathcal{D}$ in the vertical cross section. The product of $\alpha(y)$ and dy is the volume of a thin cross section of the region $\mathcal{K}$. The outer integral in (8) sums these cross sections and gives the volume V of $\mathcal{K}$.

Before defining the double integral $\iint_{\mathcal{D}} f(x, y)\,dA$ of a function f defined on a region $\mathcal{D}$ we work through two more examples.

EXAMPLE 1 Calculate the volume V of the prism bounded on top by the graph of the function

$$f(x, y) = 8 - \frac{4}{5}x - \frac{2}{3}y, \qquad (x, y) \in \mathcal{D} = [0, 2] \times [0, 6],$$

on the sides by the vertical planes with equations $y = 6$, $y = 0$, $x = 2$, and $x = 0$, and on the bottom by the plane with equation $z = 0$. Figure 11.6 shows the graph of f and the set $\mathcal{D}$ on which it is defined.

Solution

The region $\mathcal{D}$ has been subdivided into n^2 rectangular subregions by subdividing each of the intervals $[0, 2]$ and $[0, 6]$ into n equal subintervals

$$[x_0, x_1], \ldots, [x_{n-1}, x_n] \quad \text{and} \quad [y_0, y_1], \ldots, [y_{n-1}, y_n].$$

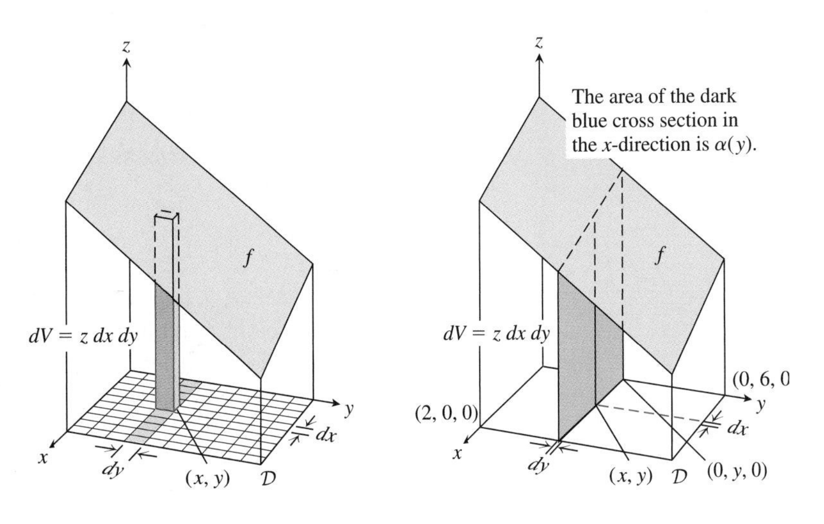

FIGURE 11.6 Two views of a representative volume element $dV = z\,dx\,dy$.

Figure 11.6 shows a representative volume element at a point (x, y) of a subregion of area $dA = dx\,dy$. The height of the volume element is $z = f(x, y)$, which is the distance from the point (x, y) to the graph of f. The volume dV of the volume element is the product of its height z and cross-sectional area $dA = dx\,dy$; that is, $dV = z\,dx\,dy$.

As in the Investigation, we fix y and sum first in the x-direction. In the figure, the darker strip in $\mathcal{D}$ corresponds to a representative, fixed value of y. The volume beneath the graph of f and above the strip is approximately

$$\sum_x dV = \left(\sum_x z\,dx\right) dy.$$

The sum $\Sigma_x z\,dx$ approximates the area $\alpha(y)$ of the vertical cross section through $(0, y, 0)$ and parallel to the (x, z)-plane. The product of this sum and dy is the volume of the thin section with thickness dy. If we sum all such thin sections, we obtain an approximating, repeated sum for the volume, specifically,

$$V \approx \sum_y \left(\sum_x z\,dx\right) dy = \sum_y \left(\sum_x f(x, y)\,dx\right) dy.$$

Based on our geometric intuition and earlier experiences with one-dimensional integrals and volumes by cross sections, we expect that

$$V = \lim_{n\to\infty} \sum_y \left(\sum_x f(x, y)\,dx\right) dy = \int_0^6 \alpha(y)\,dy = \int_0^6 \left(\int_0^2 f(x, y)\,dx\right) dy.$$

This iterated integral (we formally define the term *iterated integral* following Example 2) is not difficult to evaluate. Because y has a constant value in the inner integral,

$$\begin{aligned} V &= \int_0^6 \left(\int_0^2 \left(8 - \frac{4}{5}x - \frac{2}{3}y\right) dx\right) dy \\ &= \int_0^6 \left(8x - \frac{4}{5}\cdot\frac{1}{2}x^2 - \frac{2}{3}y\cdot x\right)\Bigg|_{x=0}^{x=2} dy \\ &= \int_0^6 \left(16 - \frac{8}{5} - \frac{4}{3}y\right) dy \\ &= \frac{72}{5}y - \frac{2}{3}y^2\Bigg|_0^6 = 62.4. \end{aligned}$$

Supposing all lengths were measured in meters, the volume of the prism is 62.4 m^3.

EXAMPLE 2 A building has a thin concrete roof in the shape of a hyperbolic paraboloid. See Fig. 11.7. The contractor plans to use the fact that hyperbolic paraboloids have the property that two sets of cables can be stretched between selected points on the edges of the roof so that each cable lies entirely in the surface, as suggested in the figure. Such surfaces are called "doubly ruled surfaces." (See the Student Project at the end of Chapter 9 for more about ruled surfaces.) In this application cables will support forms in which cement/gravel aggregate will be poured to form the roof. To determine the air conditioning and heat requirements of the building, the contractor wishes to

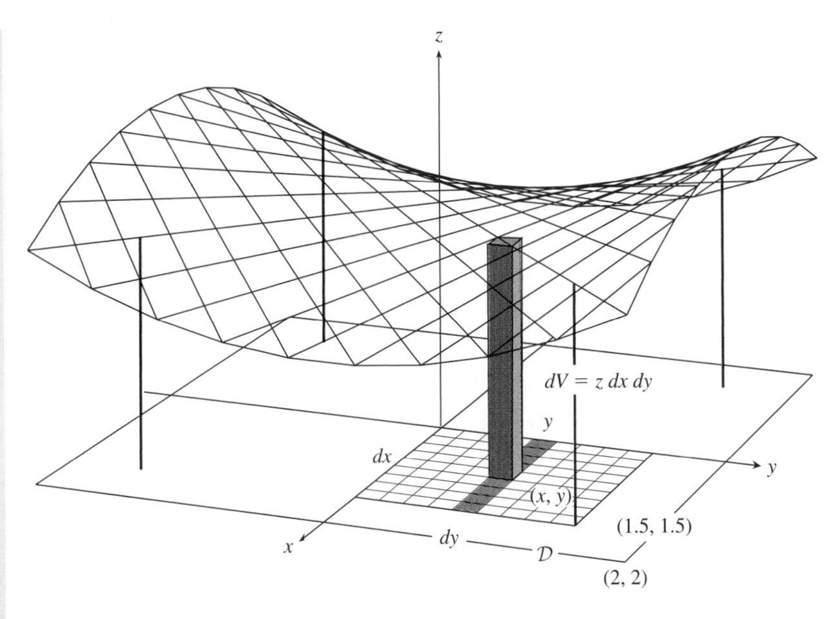

FIGURE 11.7 Roof cast in the form of a hyperbolic paraboloid.

calculate the volume of air in the completed building. It is given that the roof is closely approximated by the surface described by the equation

$$z = \frac{1}{8}(y^2 - x^2) + 1.5, \qquad -2 \leq x, y \leq 2,$$

relative to the coordinate system shown in the figure. The floor of the building is set in 0.5 unit from the edges of the ground-level projection of the roof.

Solution

We take advantage of symmetry by calculating the volume V of the part of the building lying above the region $\mathcal{D} = [0, 1.5] \times [0, 1.5]$ and later multiplying by 4. As in Example 1, we subdivide the set $\mathcal{D}$ into n^2 congruent rectangular subregions. The figure shows a volume element corresponding to a representative subregion. With fixed y, we first sum the volume elements in the x-direction. For each y between 0 and 1.5, x varies from 0 to 1.5. The volume dV of the element is the product of its height z and cross-sectional area $dx\,dy$, so that $dV = z\,dx\,dy$.

The volume beneath the roof and above the darkened strip of width dy is approximately

$$\sum_x dV = \left(\sum_x z\,dx\right) dy.$$

Summing this on y gives

$$V \approx \sum_y \left(\sum_x z\,dx\right) dy.$$

Based on our geometric intuition, we expect that

$$V = \lim_{n\to\infty} \sum_y \left(\sum_x z\,dx\right) dy = \int_0^{1.5} \alpha(y)\,dy = \int_0^{1.5} \left(\int_0^{1.5} \left(\frac{1}{8}(y^2 - x^2) + 1.5\right) dx\right) dy,$$

where $\alpha(y)$ is the area of the cross section perpendicular to the y-axis at $(0, y)$. In the inner integral, y is constant. Factoring $\frac{1}{8}$ from the inner integral and integrating with respect to x,

$$V = \frac{1}{8}\int_0^{1.5}\left(y^2x - \tfrac{1}{3}x^3 + 12x\right)\Big|_{x=0}^{x=1.5} dy = \frac{1}{8}\int_0^{1.5}(1.5y^2 - 1.125 + 18)\,dy$$

$$= \frac{1}{8}\left(1.5\cdot\frac{1}{3}y^3 + 16.875y\right)\Big|_0^{1.5} = \frac{1}{8}\left(\frac{1}{3}\cdot 1.5^4 + 16.875\cdot 1.5\right) = 3.375.$$

The volume of the entire building is $4V = 13.5$ cubic units.

The definition of the *double integral* is similar to that of the *definite integral* defined in Chapter 5.

DEFINITION The Double Integral on Rectangular Regions of the Plane

Let f be a bounded function defined on $\mathcal{D} = [a, b] \times [c, d] \subset R^2$. Let n be a positive integer, $\Delta x = (b - a)/n$, and $\Delta y = (d - c)/n$. The points

$$x_0 < x_1 < \cdots < x_i < \cdots < x_n \quad \text{and} \quad y_0 < y_1 < \cdots < y_j < \cdots < y_n,$$

where $x_i = a + i\Delta x$ and $y_j = c + j\Delta y$, divide the intervals $[a, b]$ and $[c, d]$ into n subintervals of lengths Δx and Δy, respectively. These subintervals generate a subdivision of $\mathcal{D}$ into n^2 congruent subregions

$$\mathcal{D}_{i,j} = [x_{i-1}, x_i] \times [y_{j-1}, y_j], \qquad 1 \le i, j \le n,$$

of area $\Delta A = \Delta x \cdot \Delta y$. A representative subregion $\mathcal{D}_{i,j}$ is shown in Fig. 11.8. For each positive integer n, the set $\mathcal{S}_n$ of these subregions is called a **regular subdivision** of $\mathcal{D}$.

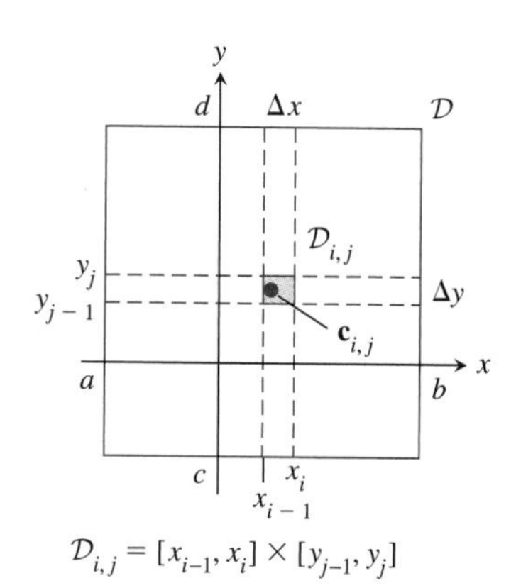

FIGURE 11.8 Representative subregion of $\mathcal{D}$.

Let $\{\mathbf{c}_{i,j}: 1 \le i, j \le n\}$ be any set of n^2 numbers for which $\mathbf{c}_{i,j} \in \mathcal{D}_{i,j}$, $1 \le i, j \le n$. We refer to this set as an *evaluation set* for $\mathcal{S}_n$.

For each regular subdivision $\mathcal{S}_n$ of $\mathcal{D}$ and each evaluation set $\mathcal{C} = \{\mathbf{c}_{i,j}: 1 \le i, j \le n\}$ we define the **Riemann sum**

$$R_n = \sum_{i,j} f(\mathbf{c}_{i,j})\,\Delta A = \sum_{i,j} f(\mathbf{c}_{i,j})\,\Delta x\,\Delta y.$$

If there is a number I for which all Riemann sums R_n can be made as close to I as we like by taking n sufficiently large, we say that f is **integrable** on $\mathcal{D}$ and that I is the value of the **double integral** of f on $\mathcal{D}$. We write this limit in either of the two forms

$$\lim_{n\to\infty} R_n = I = \iint_{\mathcal{D}} f(x, y)\,dA = \iint_{\mathcal{D}} f(x, y)\,dx\,dy. \tag{9}$$

The function f is called the *integrand*.

Are All Functions Integrable?

Not all bounded functions defined on rectangular regions are integrable. Roughly, if the function is not continuous "almost everywhere" on $\mathcal{D}$, the limit of the Riemann sums in (9) will fail to exist. In the present section we consider only functions that are continuous everywhere, and for these the following theorem holds.

Sufficient Condition for Integrability on Rectangular Regions

If f is continuous on the rectangular region $\mathcal{D} = [a, b] \times [c, d]$, then f is integrable on $\mathcal{D}$.

The main tool for evaluating the double integral

$$\iint_{\mathcal{D}} f(x, y)\, dx\, dy$$

is called the Iterated Integral Theorem. This theorem formalizes the calculations of volume in the Investigation and Examples 1 and 2.

THEOREM Iterated Integral Theorem for Rectangular Regions

If f is continuous on the rectangular region $\mathcal{D} = [a, b] \times [c, d]$, then

$$\iint_{\mathcal{D}} f(x, y)\, dx\, dy = \int_c^d \left(\int_a^b f(x, y)\, dx \right) dy \tag{10}$$

and

$$\iint_{\mathcal{D}} f(x, y)\, dx\, dy = \int_a^b \left(\int_c^d f(x, y)\, dy \right) dx. \tag{11}$$

The expressions on the right sides of these equations are called *iterated* (or *repeated*) *integrals.*

EXAMPLE 3 Let f be defined on $\mathcal{D} = [0, 1] \times [0, 1]$ by

$$f(x, y) = x^2 + y^2, \qquad (x, y) \in \mathcal{D}.$$

For $n = 10$, 100, and 1000, calculate the Riemann sums L_n and U_n corresponding to the evaluation sets

$$\{\mathbf{m}_{i,j} : 1 \le i, j \le n\} \quad \text{and} \quad \{\mathbf{M}_{i,j} : 1 \le i, j \le n\},$$

where $\mathbf{m}_{i,j}$ ($\mathbf{M}_{i,j}$) is the point in $\mathcal{D}_{i,j}$ at which f is a minimum (maximum). Also calculate the average $A_n = (L_n + U_n)/2$ of these two Riemann sums. Compare these values with $\iint_{\mathcal{D}} f(x, y)\, dx\, dy$.

Solution

The graph of f is shown in Fig. 11.9(a). Also shown are a subdivision of $\mathcal{D}$ and a volume element with height $f(\mathbf{m}_{i,j})$. Because f increases on $\mathcal{D}$ in all directions outward from the origin into the first quadrant, the Riemann sums L_n and U_n are formed by evaluating f at the point of the subregion closest to and farthest from the origin. The volume element shown in the figure is an inscribed parallelepiped resting on $\mathcal{D}_{8,7}$, in the regular subdivision $\mathcal{S}_{10}$. Its height is $f(\mathbf{m}_{8,7}) = f(0.7, 0.6)$, which is the minimum of f on $\mathcal{D}_{8,7}$. The evaluation point $\mathbf{m}_{8,7}$ is highlighted in Fig. 11.9(b). This volume element contributes $dV = f(0.7, 0.6)\,dx\,dy$ to the sum L_{10}, where $dx = dy = 1/n = 1/10$. The points in $\mathcal{D}$ at which f is evaluated for the sums L_n and U_n are shown in Fig. 11.9(b) and (c).

Following this pattern, for any positive integer n the lower and upper sums are

$$L_n = \sum_{j=1}^{n}\sum_{i=1}^{n}\left(\left(\frac{i-1}{n}\right)^2 + \left(\frac{j-1}{n}\right)^2\right)\frac{1}{n^2}$$
$$U_n = \sum_{j=1}^{n}\sum_{i=1}^{n}\left(\left(\frac{i}{n}\right)^2 + \left(\frac{j}{n}\right)^2\right)\frac{1}{n^2}. \tag{12}$$

These sums can be evaluated by using the formula (see (6) in Section 6.1)

$$1^2 + 2^2 + \cdots + m^2 = \frac{m(m+1)(2m+1)}{6}, \qquad m = 1, 2, \ldots. \tag{13}$$

We leave as Exercise 18 the straightforward calculations leading to

$$L_n = \frac{2n^2 - 3n + 1}{3n^2} \tag{14}$$

$$U_n = \frac{2n^2 + 3n + 1}{3n^2}. \tag{15}$$

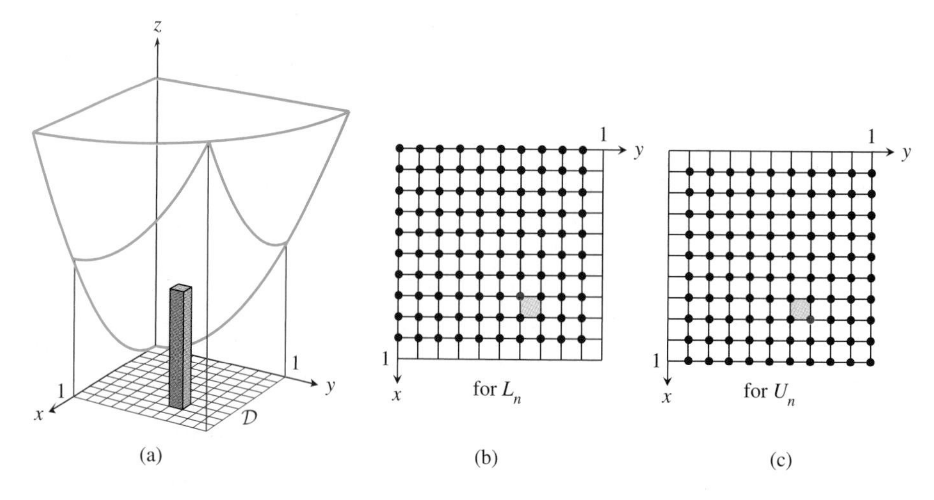

FIGURE 11.9 (a) Area beneath part of a paraboloid; (b) evaluation points for L_n; (c) evaluation points for U_n.

Table 11.1 gives the results of applying these formulas to calculate L_n, U_n, and A_n for $n = 10$, 100, and 1000.

TABLE 11.1 Lower and upper approximations to $\iint_{\mathcal{D}} f(x,y)\,dx\,dy$.

n	L_n	U_n	A_n
10	0.570000	0.770000	0.670000
100	0.656700	0.676700	0.666700
1000	0.665667	0.667667	0.666667

These data are evidence that L_n, U_n, and A_n approach $2/3$ as n increases. This is also evident from (14) and (15). We calculate the double integral of f on $\mathcal{D}$ by applying (8) from the Iterated Integral Theorem.

$$\iint_{\mathcal{D}} f(x,y)\,dx\,dy = \int_0^1 \left(\int_0^1 (x^2 + y^2)\,dx \right) dy = \int_0^1 \left(\frac{1}{3}x^3 + xy^2 \right)\Bigg|_0^1 dy$$

$$= \int_0^1 \left(\frac{1}{3} + y^2 \right) dy = \frac{1}{3}y + \frac{1}{3}y^3 \Bigg|_0^1 = \frac{1}{3} + \frac{1}{3} = \frac{2}{3}.$$

This result agrees with the trend of the numerical data in Table 11.1.

Exercises 11.1

Exercises 1–6: The graph of the function f defined on the set $\mathcal{D}$ is shown in the figure (some figures are scaled in the vertical direction). Calculate the Riemann sum R_n for the given evaluation set $\mathcal{C}$.

1. $f(x,y) = 3(1 - (1/4)x - (1/3)y)$; $\mathcal{D} = [0,2] \times [0,1]$; $n = 2$; the evaluation set $\mathcal{C} = \{\mathbf{c}_{i,j} : 1 \le i, j \le 2\}$ is the set of midpoints of the four rectangles.

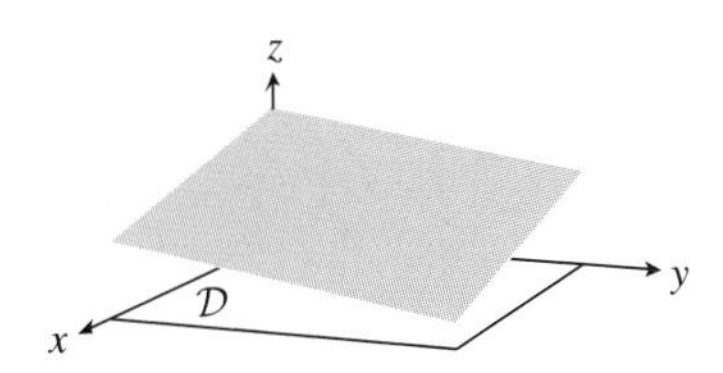

2. $f(x,y) = 4(1 - (1/5)x - (1/3)y)$; $\mathcal{D} = [0,1] \times [0,1]$; $n = 2$; the evaluation set $\mathcal{C} = \{\mathbf{c}_{i,j} : 1 \le i, j \le 2\}$ is the set of midpoints of the four rectangles.

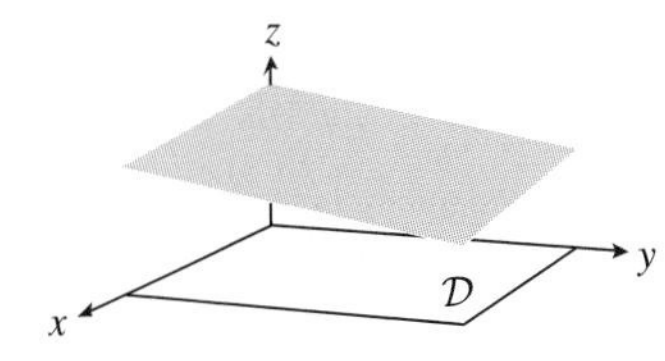

3. $f(x,y) = 16 - x^2 - 2y^2$; $\mathcal{D} = [0,2] \times [0,2]$; $n = 3$; the evaluation set $\mathcal{C} = \{\mathbf{c}_{i,j} : 1 \le i, j \le 3\}$ is the set of points in the nine rectangles at which f takes on its minimum value.

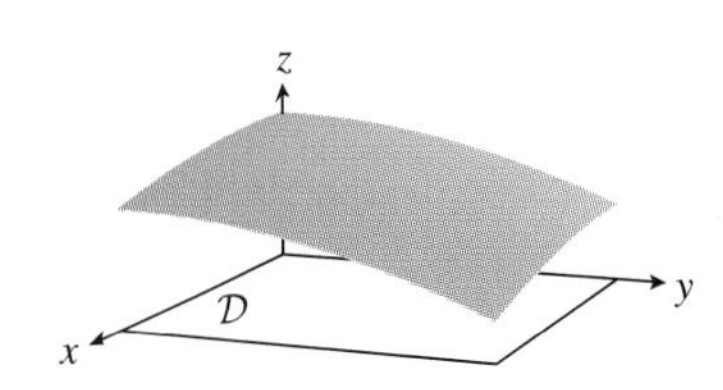

4. $f(x,y) = 1 + x^2 + y^2$; $\mathcal{D} = [-1,1] \times [0,1]$; $n = 3$; the evaluation set $\mathcal{C} = \{\mathbf{c}_{i,j}: 1 \le i,j \le 3\}$ is the set of points in the nine rectangles at which f takes on its maximum value.

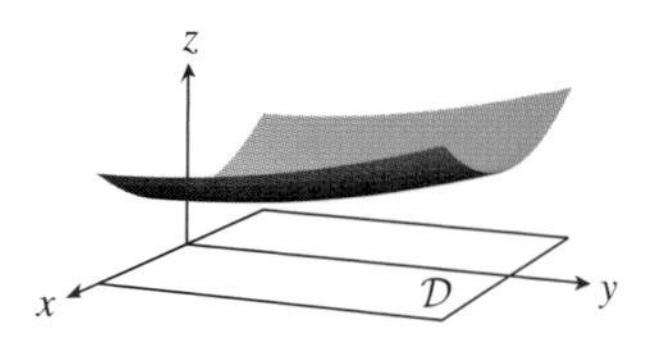

5. $f(x,y) = \sin x \cos y$; $\mathcal{D} = [0, \pi/2] \times [0, \pi/2]$; $n = 4$; the evaluation set $\mathcal{C} = \{\mathbf{c}_{i,j}: 1 \le i,j \le 4\}$ is the set of midpoints of the sixteen rectangles.

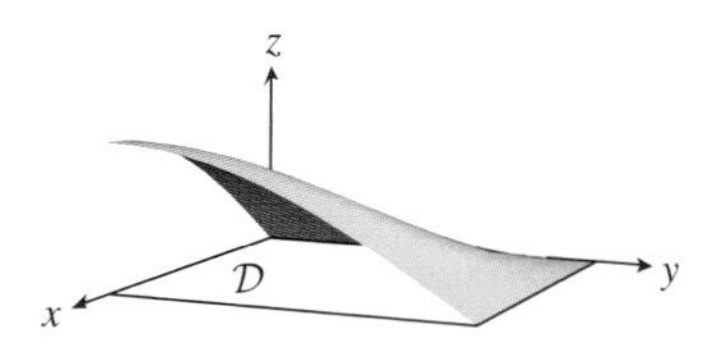

6. $f(x,y) = e^{-x-y}$; $\mathcal{D} = [0, 1.5] \times [0, 1.5]$; $n = 4$; the evaluation set $\mathcal{C} = \{\mathbf{c}_{i,j}: 1 \le i,j \le 4\}$ is the set of midpoints of the sixteen rectangles.

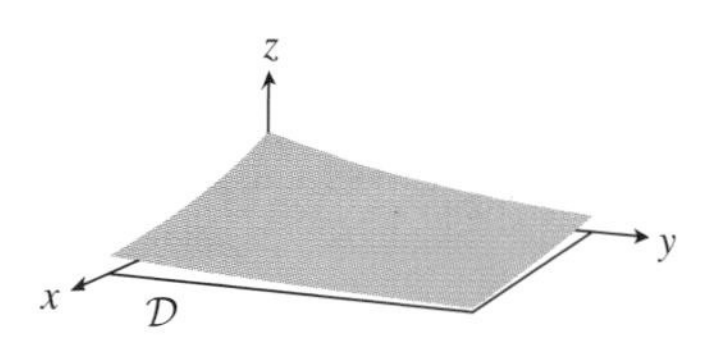

Exercises 7–12: Use the Iterated Integral Theorem in calculating the volume of the region beneath the graph of f and above the set $\mathcal{D}$ on which it is defined. Calculate the percentage error in the Riemann sum.

7. Use the function described in Exercise 1 and the Riemann sum calculated there.

8. Use the function described in Exercise 2 and the Riemann sum calculated there.

9. Use the function described in Exercise 3 and the Riemann sum calculated there.

10. Use the function described in Exercise 4 and the Riemann sum calculated there.

11. Use the function described in Exercise 5 and the Riemann sum calculated there.

12. Use the function described in Exercise 6 and the Riemann sum calculated there.

Exercises 13–16: These problems refer to the discussion of total annual rainfall in the Investigation.

13. Find the total annual rainfall if, with no other changes, the rainfall function is replaced by

$$r(x,y) = \frac{100^2}{2x + y + 200},$$

for $(x,y) \in \mathcal{D}$. Graph r.

14. Find the total annual rainfall if, with no other changes, the rainfall function is replaced by

$$r(x,y) = \frac{100^3}{(x + 75)(y + 250)},$$

for $(x,y) \in \mathcal{D}$. Graph r.

15. Referring to the Investigation, assume that during the year the rain collects upon $\mathcal{D}$ and that it is not absorbed, does not evaporate, and does not run off. The resulting pool will have a horizontal surface and have depth w inches. Give a convincing argument that $100^2 w = V$, where V is the total annual rainfall or the volume of the region $\mathcal{K}$. Can w reasonably be described as the average rainfall on $\mathcal{D}$ or the average value of the rainfall function? Find a point $(x^*, y^*) \in \mathcal{D}$ such that $w = r(x^*, y^*)$.

16. Continuation In Exercise 15, we located a point $(x^*, y^*) \approx (42.3250, 50)$ with average rainfall for the region $\mathcal{D}$. Find the average rainfall on $\mathcal{D}$ if the rainfall function is that given in Exercise 14. Describe the set of all points in $\mathcal{D}$ which receive the average rainfall.

17. The volume of the prism in Example 1 can be calculated directly. Decompose the prism into two parts: a base, which is a rectangular parallelepiped, and the top, which is a bisected rectangular parallelepiped. Calculate the volume using this decomposition.

18. Verify (14) and (15) by applying formula (13) to (12).

19. Continuation Referring to Example 3 but replacing $f(x,y) = x^2 + y^2$ by a general function f defined on $[0,1] \times [0,1]$ and increasing in all directions outward from the origin, use Fig. 11.9(b) in expressing U_n in (12) in terms of L_n. Why might this be a useful result?

20. On the basis of Table 11.1 alone, show that

$$\left| \iint_{\mathcal{D}} (x^2 + y^2)\, dx\, dy - A_{1000} \right| \le 0.001.$$

Explain your reasoning.

21. Show that the hyperbolic paraboloid in Example 2 is doubly ruled. We outline an argument based on the cables shown in Fig. 11.7. If the ends of the cables are projected vertically onto the (x,y)-plane, the resulting points are equally spaced

on a square of side 4. Pair these points by their common cable and write equations for the lines the paired points determine. Use the equations of the lines and that of the surface to show that the cables lie in the surface.

22. To calculate the amount of aggregate required for the roof in Example 2, the contractor notes that the cables divide the surface into regions that are nearly parallelograms, as shown in Fig. 11.7. One of these parallelograms is determined by adjacent vertices $A = (0, 0, 1.5)$, $B = (8/9, -8/9, 1.5)$, and $C = (8/9, 8/9, 1.5)$. Calculate the area of the parallelogram whose sides lie along the vectors $\overrightarrow{AB}$ and $\overrightarrow{AC}$. Is the area of the corresponding roof region larger or smaller than this result? Why?

23. The rectangular region $\mathcal{I} = [0, 3] \times [0, 1]$ can be written as the union of rectangular regions $\mathcal{I}_1$ and $\mathcal{I}_2$, where

$$\mathcal{I}_1 = [0, 1] \times [0, 1], \mathcal{I}_2 = [1, 3] \times [0, 1].$$

Assuming that f is continuous on $\mathcal{I}$, use the Iterated Integral Theorem in showing that

$$\iint_{\mathcal{I}} f(x, y)\, dx\, dy = \iint_{\mathcal{I}_1} f(x, y)\, dx\, dy + \iint_{\mathcal{I}_2} f(x, y)\, dx\, dy.$$

24. The western two-thirds of Massachusetts is shown in the accompanying figure, with numbered stations at which rainfall data were collected. Also shown is an 11-mile-by-11-mile grid. The table below the figure lists the annual rainfall at each of the stations, measured in millimeters. The form of each data point is (r, s), where r is the amount of rain and s the station number. Approximate the total annual volume of rainfall received by western Massachusetts.

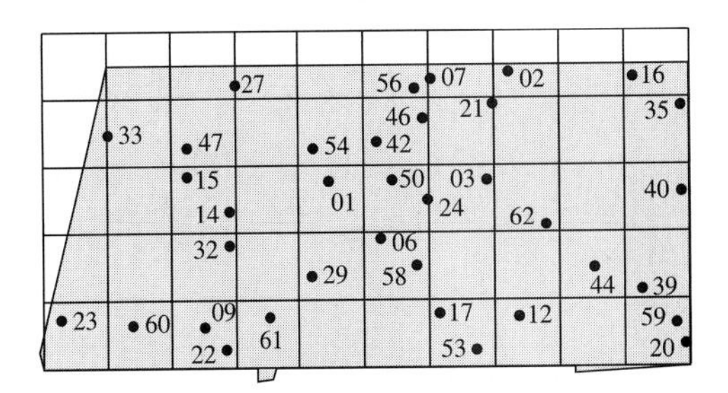

Station rainfall data

(495.10, 1)	(499.20, 21)	(581.80, 44)
(528.50, 2)	(420.20, 22)	(470.10, 46)
(533.50, 3)	(635.90, 23)	(131.40, 47)
(532.80, 6)	(546.10, 24)	(234.10, 50)
(520.90, 7)	(554.20, 27)	(520.60, 53)
(609.20, 9)	(441.60, 29)	(562.00, 54)
(528.50, 12)	(525.90, 32)	(535.00, 56)
(589.30, 14)	(534.40, 33)	(531.40, 58)
(576.50, 15)	(459.00, 35)	(616.50, 59)
(244.50, 16)	(447.30, 39)	(518.60, 60)
(502.30, 17)	(521.40, 40)	(547.30, 61)
(559.80, 20)	(681.80, 42)	(523.90, 62)

Figure for Exercise 24

11.2 Extending the Double Integral and Applications

In Section 11.1 we defined the double integral for bounded functions f defined on rectangular regions $\mathcal{D} = [a, b] \times [c, d]$. We now extend this definition to functions defined on bounded regions. For this purpose we discuss functions that are continuous except on sets having area zero.

Figure 11.10(a) shows the graph of a bounded function defined on the rectangular region $\mathcal{D} = [0, 1] \times [0, 1]$. Figure 11.10(b) shows a zoom-view of the set $\mathcal{D}$,

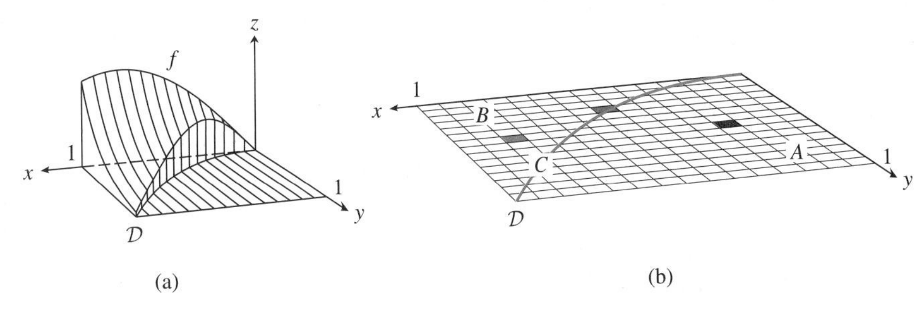

FIGURE 11.10 (a) The graph of a function that is continuous except along the curve C; (b) a regular subdivision of $\mathcal{D} = [0, 1] \times [0, 1]$, with three representative subregions $\mathcal{D}_{i,j}$.

partitioned into three subsets A, B, and C. The flat part of the graph in Fig. 11.10(a), where $f(x, y) = 0$, is coincident with the set A; the "hilly" part of the graph, where $f(x, y) > 0$, lies above the set B; and for the points (x, y) on C, the values $f(x, y)$ cannot be inferred from the figure.

Although it appears from Fig. 11.10 that f may fail to be continuous on C, we might argue that the double integral $\int\int_{\mathcal{D}} f(x, y)\, dA$ not only exists but is equal to the volume of the region beneath the graph of f and above the set B. Let $\mathcal{S}_n$ be a regular subdivision of $\mathcal{D}$ and $\{\mathcal{D}_{i,j} : 1 \leq i, j \leq n\}$ the set of associated subregions. We consider a representative Riemann sum

$$R_n = \sum_{i,j} f(\mathbf{c}_{i,j})\, \Delta x\, \Delta y, \tag{1}$$

where $\mathcal{C} = \{\mathbf{c}_{i,j} : 1 \leq i, j \leq n\}$ is an evaluation set. We sort the sum R_n into three sums: the first, or "red sum," contains all volume elements $dV = f(\mathbf{c}_{i,j})\, \Delta x\, \Delta y$ for which $\mathcal{D}_{i,j} \subset A$; the second, or "blue sum," contains all volume elements $dV = f(\mathbf{c}_{i,j})\, \Delta x\, \Delta y$ for which $\mathcal{D}_{i,j} \subset B$; and the third, or "green sum," contains all volume elements $dV = f(\mathbf{c}_{i,j})\, \Delta x\, \Delta y$ for which $\mathcal{D}_{i,j}$ intersects the curve C. Figure 11.10(b) shows the bases of representative red, blue, and green volume elements.

Each of the red volume elements contributes 0 to the sum R_n because $\mathbf{c}_{i,j} \in A$ and $f(\mathbf{c}_{i,j}) = 0$. Each of the green volume elements, which stand on green subregions $\mathcal{D}_{i,j}$, contributes only a small amount to the sum R_n because the sum of their areas can be made as small as we like by taking n sufficiently large. The blue volume elements are the major contributors to the sum R_n. It follows that the Riemann sum R_n approximates the volume beneath the graph of f and above the set B.

Because the sum of the areas of the set of all subregions (of $\mathcal{S}_n$) that intersect curves like C tends to 0 as $n \to \infty$, we say that the area of C is zero. Using this terminology, the function f considered here is continuous on $\mathcal{D}$, except on a set of area zero.

We now define the double integral on bounded regions. We do this by applying the earlier definition of the double integral on rectangular regions. Using Fig. 11.11 as a general guide to the definition, we start with a function f defined on the bounded region $\mathcal{D}$, enclose $\mathcal{D}$ in a rectangular region $\mathcal{I} = [a, b] \times [c, d]$, define a new function f^* on $\mathcal{I}$, where f^* agrees with f on $\mathcal{D}$ and is zero on the part of $\mathcal{I}$ not in $\mathcal{D}$, and define the double integral of f on $\mathcal{D}$ to be the double integral of f^* on $\mathcal{I}$.

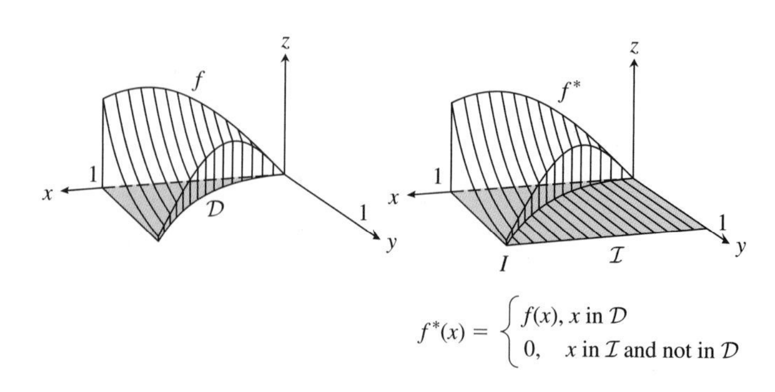

FIGURE 11.11 Extending f on $\mathcal{D}$ to f^* on $\mathcal{I} \supset \mathcal{D}$.

DEFINITION The Double Integral on Bounded Regions

Let f be a bounded function defined on a bounded region $\mathcal{D}$. We say that f is integrable on $\mathcal{D}$ if the function f^* is integrable on a rectangular region $\mathcal{I} = [a,b] \times [c,d]$ containing $\mathcal{D}$, where

$$f^*(x,y) = \begin{cases} f(x,y), & (x,y) \text{ in } \mathcal{D} \\ 0, & (x,y) \text{ in } \mathcal{I} \text{ and not in } \mathcal{D}. \end{cases}$$

In this case we define

$$\iint_{\mathcal{D}} f(x,y)\,dx\,dy = \iint_{\mathcal{I}} f^*(x,y)\,dx\,dy.$$

Before we give a sufficient condition for integrability and an iterated integral theorem for functions defined on bounded regions, we give an example to review the ideas we have discussed so far.

EXAMPLE 1 Calculate the volume beneath the graph of the function

$f(x,y) = \left(1 - \sqrt{y}\right) \sin 2x$, $(x,y) \in \mathcal{D} = \{(x,y) : 0 \le y \le x^2 \text{ and } 0 \le x \le 1\}$.

The region $\mathcal{D}$, shown lightly shaded in Fig. 11.12(a), can also be described as the set of points between C and the x-axis, where C is the curve described by the equation $y = x^2$, $0 \le x \le 1$.

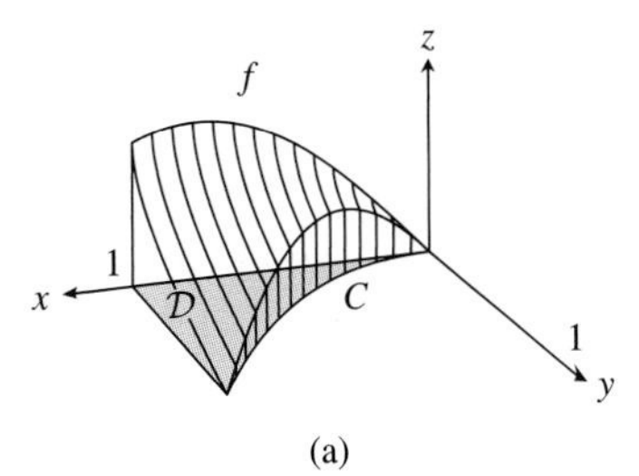

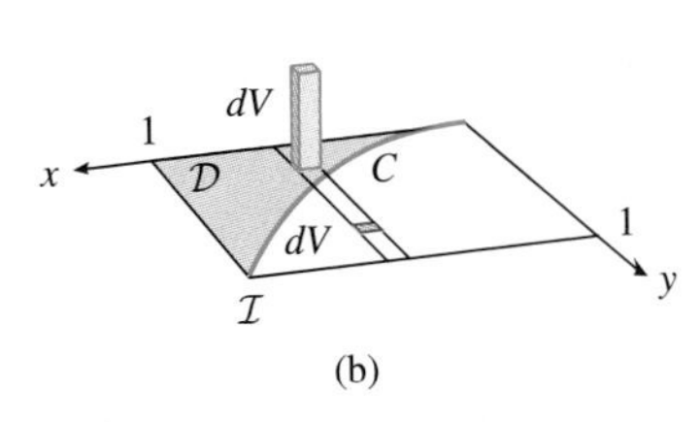

FIGURE 11.12 (a) Graph of a function f defined on a bounded region; the curve C is described by the equation $y = x^2$, $0 \le x \le 1$; (b) the domain of f^* and a volume element.

Solution

Figure 11.12 (a) shows the graph of the function f. We extend f to a function f^* defined on an interval $\mathcal{I}$ containing $\mathcal{D}$. Here we may take $\mathcal{I} = [0,1] \times [0,1]$. See Fig. 11.12(b). The function f^* agrees with f on $\mathcal{D}$ and is zero on the part of $\mathcal{I}$ not in $\mathcal{D}$. From the figure, f^* is continuous on $\mathcal{I}$ except for all points on the curve C other than its two endpoints. While we do not know if the Iterated Integral Theorem can be applied to f^* (because f^* is not everywhere continuous on $\mathcal{I}$), we lean on our earlier discussion about curves with area zero and proceed experimentally. Applying the Iterated Integral Theorem for Rectangular Regions from Section 11.1, the volume beneath the graph of the function is

$$V = \iint_{\mathcal{D}} f^*(x,y)\,dA = \int_0^1 \left(\int_0^1 f^*(x,y)\,dy \right) dx.$$

With this choice of an iterated integral, we fix x and integrate in the y-direction first. Figure 11.12(b) shows the volume element $dv = f(x,y)\,dy\,dx$ in two positions. In the part of $\mathcal{I}$ in which the volume element is blue, the height of the element is $f^*(x, y) = f(x, y)$; in the part of $\mathcal{I}$ in which the volume element is red, its height is 0, because of the definition of f^*. Ignoring the relatively few volume elements whose bases touch the curve C,

$$\int_0^1 f^*(x, y)\, dy = \int_0^{x^2} f^*(x, y)\, dy + \int_{x^2}^1 f^*(x, y)\, dy = \int_0^{x^2} f(x, y)\, dy + 0.$$

Hence,

$$\begin{aligned}\iint_{\mathcal{D}} f^*(x, y)\, dA &= \int_0^1 \left(\int_0^{x^2} \left(1 - \sqrt{y}\right) \sin 2x\, dy \right) dx \\ &= \int_0^1 \left(y - \frac{2}{3} y^{3/2} \right) \Bigg|_0^{x^2} \sin 2x\, dx \\ &= \int_0^1 \left(x^2 - \frac{2}{3} x^3 \right) \sin 2x\, dx.\end{aligned}$$

Applying integration formula (25) from the Table of Integrals or integrating by parts (see Exercise 41),

$$\iint_{\mathcal{D}} f^*(x, y)\, dA = \frac{1}{12}(3 \sin(2) - 5 \cos(2) - 3) \approx 0.150719.$$

Sufficient Condition for Integrability and the Iterated Integral Theorem for Bounded Regions of the Plane

We consider planar regions of two kinds, called Type I and Type II. A region of Type I has the familiar shape of a region bounded below and above by curves with equations

$$y = g_1(x), \quad a \le x \le b, \quad \text{and} \quad y = g_2(x), \quad a \le x \le b,$$

where g_1 and g_2 are continuous on $[a, b]$, and on the sides by lines with equations $x = a$ and $x = b$. A region of Type I is shown in Fig. 11.13(a). Regions of Type II are similar to those of Type I but are rotated through 90°. These regions are bounded on the left and right by curves with equations

$$x = h_1(y), \quad c \le y \le d, \quad \text{and} \quad x = h_2(y), \quad c \le y \le d,$$

where h_1 and h_2 are continuous on $[c, d]$, and on the bottom and top by lines with equations $y = c$ and $y = d$. A region of Type II is shown in Fig. 11.13(b).

The main result for calculating the double integral $\iint_{\mathcal{D}} f(x, y)\, dx\, dy$ where $\mathcal{D}$ is a region of Type I or Type II is an iterated integral theorem.

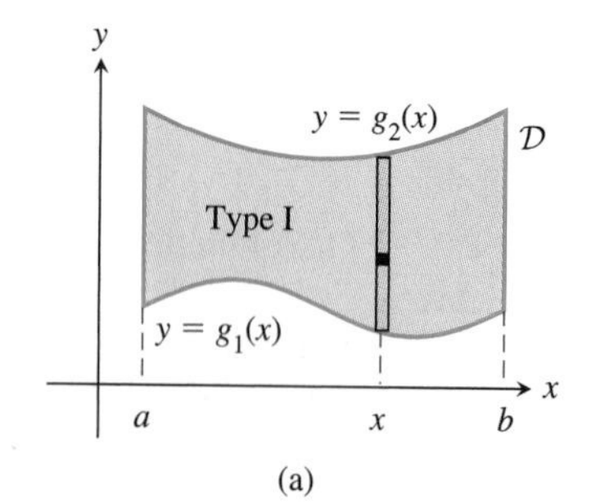

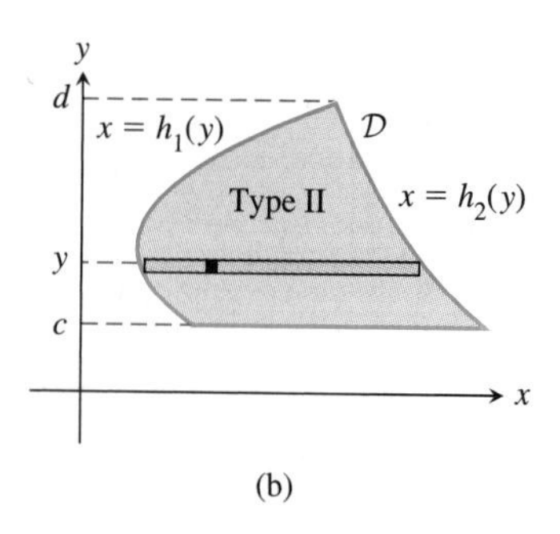

FIGURE 11.13 (a) A region of Type I; (b) a region of Type II.

THEOREM Iterated Integral Theorem for Regions of Type I or Type II

If f is continuous on a region $\mathcal{D}$ of Type I, where $\mathcal{D} = \{(x, y) : a \le x \le b, g_1(x) \le y \le g_2(x)\}$, then f is integrable on $\mathcal{D}$ and

$$\iint_{\mathcal{D}} f(x, y)\, dx\, dy = \int_a^b \left(\int_{g_1(x)}^{g_2(x)} f(x, y)\, dy \right) dx. \tag{2}$$

If f is continuous on a region $\mathcal{D}$ of Type II, where $\mathcal{D} = \{(x, y) : c \leq y \leq d, h_1(y) \leq x \leq h_2(y)\}$, then f is integrable on $\mathcal{D}$ and

$$\iint_{\mathcal{D}} f(x, y)\,dx\,dy = \int_c^d \left(\int_{h_1(x)}^{h_2(x)} f(x, y)\,dx \right) dy. \tag{3}$$

The integrals on the right sides of (2) and (3) are called **iterated integrals.**

We formalize our intuitive understandings of the relationship between the double integral and an area or volume calculation.

DEFINITION Area and Volume

If $\mathcal{D}$ is a region of Type I or Type II, the area of the region $\mathcal{D}$ is the number $\iint_{\mathcal{D}} 1\,dA$.

If f is continuous and nonnegative on a region $\mathcal{D}$ of Type I or Type II, then the volume of the region beneath the graph of f (and above the (x, y)-plane) is the number $\iint_{\mathcal{D}} f(x, y)\,dA$.

EXAMPLE 2 Calculate the volume beneath the graph of the function

$$f(x, y) = \frac{1}{8}(y^2 - x^2) + 1.5, \qquad -2 \leq x, y \leq 2, \tag{4}$$

and above the region $\mathcal{D} = \left\{(x, y) : 0 \leq x \leq 2 \text{ and } 0 \leq y \leq \sqrt{x}\right\}$.

Solution

Figure 11.14 shows the graph of the function. We note from either the figure or Exercise 42 that $f(x, y)$ is nonnegative for $(x, y) \in \mathcal{D}$. Hence, it makes sense to speak of the "volume beneath the graph. . . ." Because the region $\mathcal{D}$ is of Types I and II, we may use either of the iterated integrals from the Iterated Integral Theorem in calculating $V = \iint_{\mathcal{D}} f(x, y)\,dx\,dy$.

Viewing $\mathcal{D}$ as a region of Type I, as seen in Fig. 11.14(a), we take x as fixed and imagine a volume element dV moving in the y-direction, from

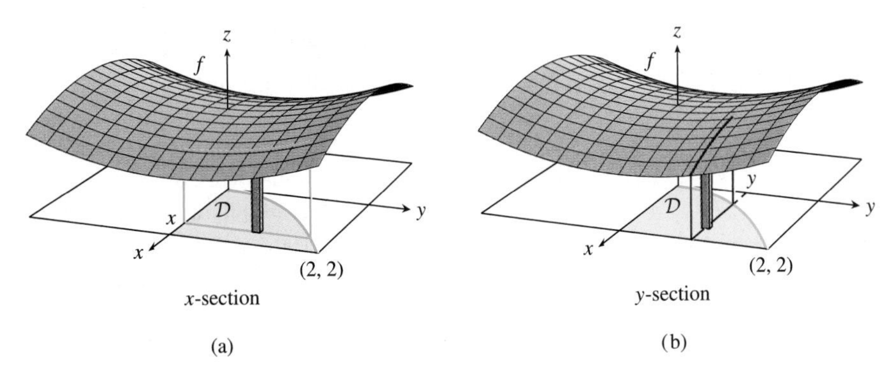

FIGURE 11.14 (a) $\mathcal{D}$ as a region of Type I; (b) $\mathcal{D}$ seen as a region of Type II.

$y = 0$ to $y = \sqrt{x}$. This sweeps out the "blue" x-section shown in the figure. The area of this section is

$$\int_0^{\sqrt{x}} f(x, y)\, dy.$$

Summing the x-sections as x varies from 0 to 2,

$$\begin{aligned}
V &= \int_0^2 \left(\int_0^{\sqrt{x}} \left(\frac{1}{8}(y^2 - x^2) + 1.5 \right) dy \right) dx \\
&= \int_0^2 \left(\frac{1}{24}y^3 - \frac{1}{8}yx^2 + 1.5y \right) \Bigg|_0^{\sqrt{x}} dx \\
&= \int_0^2 \left(\frac{1}{24}x^{3/2} - \frac{1}{8}x^{5/2} + 1.5x^{1/2} \right) dx \\
&= \frac{1}{60}x^{5/2} - \frac{1}{28}x^{7/2} + x^{3/2} \Bigg|_0^2 \\
&= \frac{187\sqrt{2}}{105} \text{ units}^3 \approx 2.51865 \text{ units}^3.
\end{aligned}$$

Next, we view $\mathcal{D}$ as a region of Type II, as seen in Fig. 11.14(b), take y as fixed, and imagine a volume element dV moving in the x-direction, from $x = y^2$ to $x = 2$. This sweeps out the "red" y-section shown in the figure. The area of this section is

$$\int_{y^2}^2 f(x, y)\, dy.$$

Summing the y-sections as y varies from 0 to 2,

$$V = \int_0^2 \left(\int_{y^2}^2 \left(\frac{1}{8}(y^2 - x^2) + 1.5 \right) dx \right) dy = \frac{187\sqrt{2}}{105} \text{ units}^3 \approx 2.51865 \text{ units}^3.$$

EXAMPLE 3 Calculate the volume of the solid bounded above by the plane with equation $x + z = 4$ and below by the set $\mathcal{D}$ of points in the (x, y)-plane that lie between the curves described by $x = y^2$ and $y = x - 2$.

Solution

To determine the region $\mathcal{D}$ of integration, which in this case is the bottom of the solid, we graph the given curves. The first, $x = y^2$, describes a parabola and the second, $y = x - 2$, a line. The intersection points of these curves are $(1, -1)$ and $(4, 2)$. A sketch of $\mathcal{D}$ is shown in Fig. 11.15(a).

It is usually a good idea to attempt some kind of visualization of the solid. The time and effort invested in a sketch, whether made by hand or with the help of a CAS or calculator, often will help you to set up a correct integral for the volume. With some practice, you can learn to make a freehand sketch similar to that shown in Fig. 11.15(b). Start with a graph of $\mathcal{D}$ in the (x, y)-plane; sketch the plane described by $x + z = 4$ by first drawing its traces in the (x, y)- and (y, z)-planes. Use the traces to sketch a rectangle

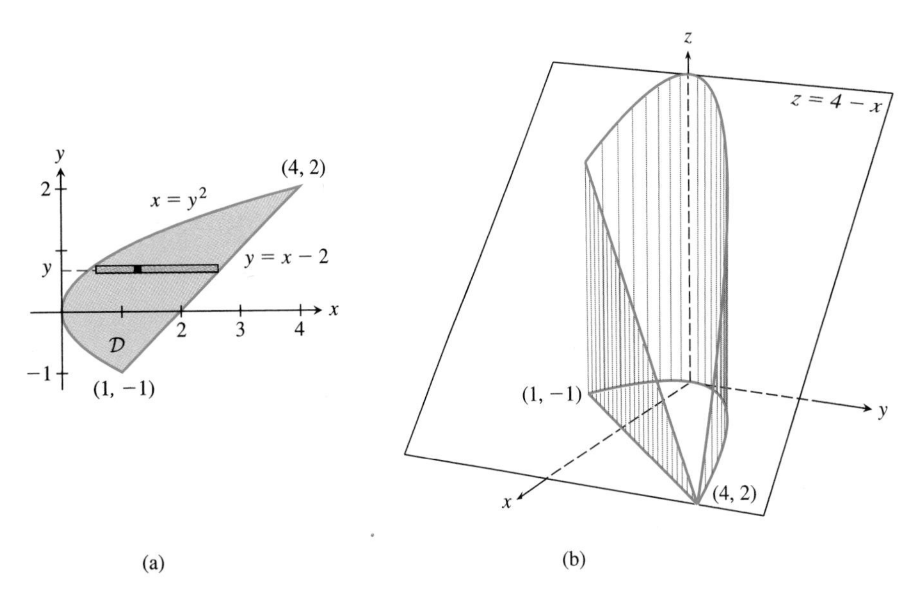

FIGURE 11.15 (a) $\mathcal{D}$ is a region of Type II; (b) the graph of f.

suggesting the plane. Sketch in the projection of the boundary of $\mathcal{D}$ onto the plane; and then add several vertical lines, joining points on the boundary of $\mathcal{D}$ to their projections in the plane.

To calculate the volume $V = \iint_{\mathcal{D}} z\,dx\,dy$, we integrate first in the x-direction because this region is of Type II. For each y between -1 and 2, x varies from $x = y^2$ to $x = 2 + y$. The volume element is $dV = z\,dx\,dy$, where $z = 4 - x$. The base of this element is shown as a small, dark rectangle in Fig. 11.15(a). Its movement in the x-direction is suggested by the horizontal strip. Applying the Iterated Integral Theorem to the volume integral,

$$V = \iint_{\mathcal{D}} (4 - x)\,dx\,dy = \int_{-1}^{2} \left(\int_{y^2}^{y+2} (4 - x)\,dx \right) dy$$

$$= \int_{-1}^{2} \left(4x - \frac{1}{2}x^2 \right) \Bigg|_{y^2}^{y+2} dy = \int_{-1}^{2} \left(\frac{1}{2}y^4 - \frac{9}{2}y^2 + 2y + 6 \right) dy$$

$$= \left(\frac{1}{10}y^5 - \frac{3}{2}y^3 + y^2 + 6y \right) \Bigg|_{-1}^{2} = 10.8 \text{ units}^3.$$

EXAMPLE 4 A bearing has the shape of a solid iron ball of radius 10 cm pierced along a diameter by a square hole of side 8 cm. Given that the density of iron is $\delta = 7.860 \times 10^{-3}$ kg/cm^3, calculate the mass of the bearing.

Solution

Figure 11.16(a) shows a first-octant view of the ball with its center placed at the origin. In drawing this figure and in the subsequent calculations, we take advantage of the symmetry of the bearing. This usually simplifies both the figure and the calculations.

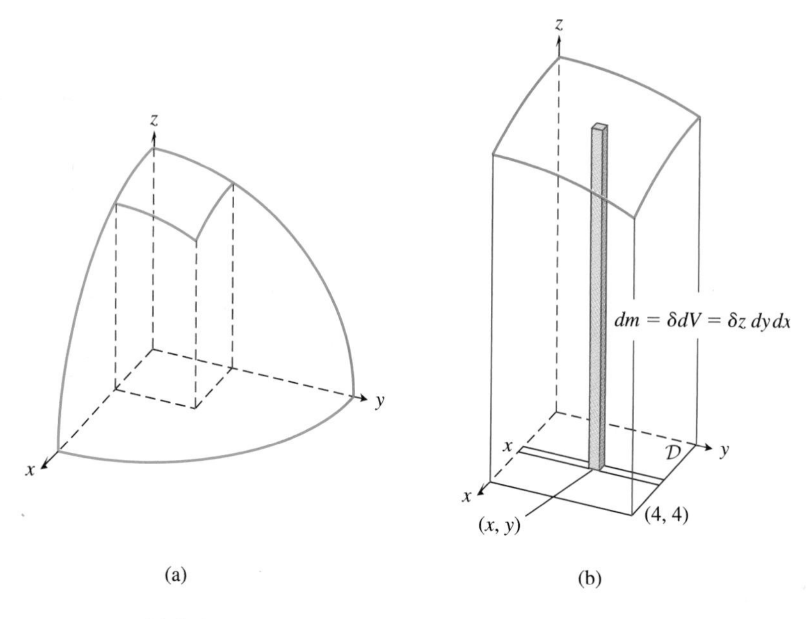

FIGURE 11.16 (a) Sphere with square hole; (b) square hole and mass element $dm = \delta dV$.

The first-octant portion of a sphere of radius 10 and centered at the origin can be described by the equation

$$z = \sqrt{10^2 - x^2 - y^2}, \tag{5}$$

where $\mathcal{D}$ is the first-quadrant portion of the circle described by $x^2 + y^2 \leq 10^2$.

To calculate the mass of the bearing, we first calculate the mass m of the square hole as if it were solid iron and then subtract the result from the mass of the solid ball. For this calculation the region of integration will be a square plus its interior. If we were to calculate the mass of the bearing directly, the region of integration would be the interior of the quarter-circle with a square region removed. See Exercise 35.

Figure 11.16(b) shows a zoom-view of the square hole and a representative mass element $dm = \delta dV = \delta z\, dy\, dx$. We sum in the y-direction first. To help keep this in mind, we noted a fixed x in the figure, drew within $\mathcal{D}$ a strip indicating the initial motion of the element, and wrote dy first in the expression for dm. The units of dm are kilograms because

$$dm = \delta z\, dy\, dx = \frac{\delta \text{ kg}}{\text{cm}^3} \cdot z \text{ cm} \cdot dy \text{ cm} \cdot dx \text{ cm} = \delta z\, dy\, dx \text{ kg}.$$

The mass m of the square hole is given by

$$\frac{1}{8} m = \delta \int_0^4 \left(\int_0^4 \sqrt{100 - x^2 - y^2}\, dy \right) dx. \tag{6}$$

Formula (7) from the Table of Integrals is

$$\int \sqrt{a^2 - x^2}\, dx = \frac{x}{2} \sqrt{a^2 - x^2} + \frac{a^2}{2} \arcsin \frac{x}{a} + C.$$

We can apply this formula to the inner integral in (6), for which the variable of integration is y, by replacing x by y and a^2 by $100 - x^2$. This gives

$$\frac{1}{8}m = \delta\int_0^4 \left(\frac{y}{2}\sqrt{100 - x^2 - y^2} + \frac{100 - x^2}{2}\arcsin\frac{y}{\sqrt{100 - x^2}}\right)\Bigg|_{y=0}^{y=4} dx$$

$$= \delta\int_0^4 \left(2\sqrt{84 - x^2} + \frac{100 - x^2}{2}\arcsin\frac{4}{\sqrt{100 - x^2}}\right) dx.$$

This integral can be evaluated analytically, either by hand or CAS. The first term of the integrand is easy (use formula (7)); the second term takes much more effort. If an exact answer is not needed, we may use numerical integration. For the latter, we may use a CAS, apply the trapezoid rule, or use the numerical integration key on a calculator. We used the built-in numerical integration program on our calculator, obtaining

$$\frac{1}{8}m \approx \delta \cdot 151.1159\cdots.$$

From this result we may calculate the mass B of the bearing, by subtracting m from the mass of a solid ball of iron.

$$B \approx \delta\left(\frac{4}{3}\pi 10^3 - 8 \cdot 151.1159\cdots\right) \approx 23.42 \text{ kg}.$$

Before discussing the calculation of the center of mass of a lamina, we give several properties of the double integral. These properties are analogous to the properties listed in Section 5.3 for one-dimensional integrals.

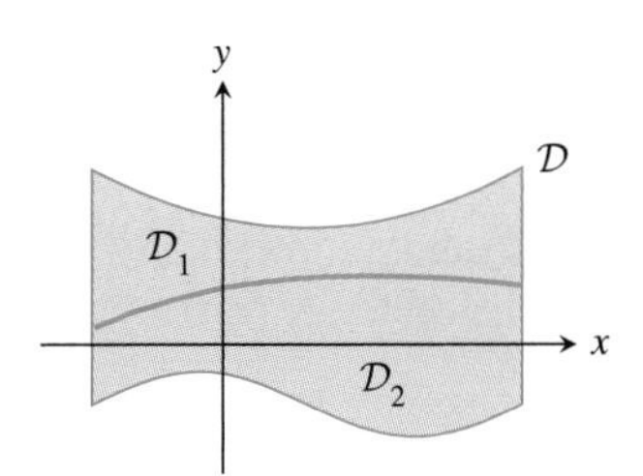

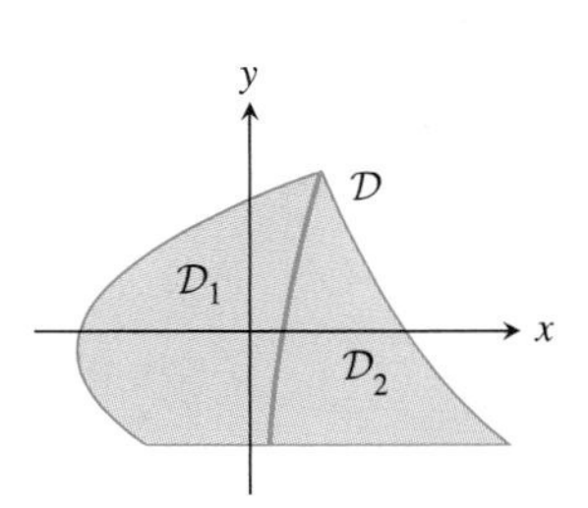

FIGURE 11.17 Two decompositions of $\mathcal{D}$ into subregions $\mathcal{D}_1$ and $\mathcal{D}_2$.

THEOREM Three Properties of the Double Integral

Let f and g be integrable on a region $\mathcal{D}$ of R^2 and suppose that $\mathcal{D}$ has been decomposed into regions $\mathcal{D}_1$ and $\mathcal{D}_2$. We assume that $\mathcal{D}_1$ and $\mathcal{D}_2$ can be described in ways similar to our descriptions of regions of Types I and II. Figure 11.17 shows two relatively simple decompositions of a region $\mathcal{D}$.

PROPERTY 1: ADDITIVITY

The function f is integrable on $\mathcal{D}_1$ and $\mathcal{D}_2$ and

$$\iint_{\mathcal{D}} f(x, y)\,dx\,dy = \iint_{\mathcal{D}_1} f(x, y)\,dx\,dy + \iint_{\mathcal{D}_2} f(x, y)\,dx\,dy.$$

PROPERTY 2: LINEARITY

Let r and s be real numbers. The function h defined on $\mathcal{D}$ by $h(x, y) = rf(x, y) + sg(x, y)$ is integrable on $\mathcal{D}$ and

$$\iint_{\mathcal{D}} (rf(x, y) + sg(x, y))\,dx\,dy = r\iint_{\mathcal{D}} f(x, y)\,dx\,dy + s\iint_{\mathcal{D}} g(x, y)\,dx\,dy.$$

PROPERTY 3: PRESERVATION OF INEQUALITY

If $f(x, y) \le g(x, y)$ on $\mathcal{D}$, then

$$\iint_{\mathcal{D}} f(x, y)\,dx\,dy \le \iint_{\mathcal{D}} g(x, y)\,dx\,dy.$$

Mass and Center of Mass of a Lamina

In the next two examples we calculate the mass or center of mass of a lamina. We use the term *laminas* to describe flat, material objects whose surface dimensions are large relative to the thickness. In Section 6.5 we used one-dimensional integrals to calculate the masses or centers of mass of laminas with constant density. Now that we have defined the double integral, the constant density restriction can be relaxed.

Recall from Section 6.7 that the center of mass of a system S of point masses $m_1, m_2, \ldots, m_n$ located in a plane by position vectors $\mathbf{r}_1, \mathbf{r}_2, \ldots, \mathbf{r}_n$ is the point with position vector

$$\mathbf{R} = \frac{1}{m} \sum_{i=1}^{n} m_i \mathbf{r}_i,$$

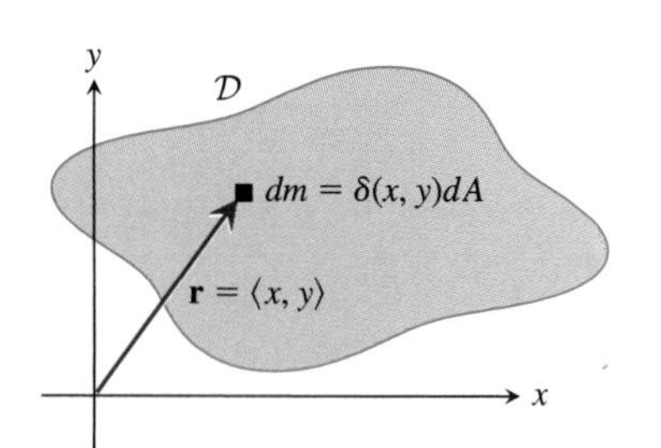

FIGURE 11.18 Lamina with variable density.

where $m = \sum_{i=1}^{n} m_i$ is the mass of the system. Referring to Fig. 11.18, we may approximate the mass and center of mass of the lamina $\mathcal{D}$ by replacing it with a system of point masses associated with a regular subdivision $\mathcal{S}_n$ of $\mathcal{D}$. Shown in the figure is a representative subregion or point mass with position vector $\mathbf{r}$. If at (x, y) the density of the lamina, measured in, say, kilograms per square meter, is $\delta(x, y)$, then the subregion has mass $dm = \delta(x, y)\, dA$. If we apply the above definition of center of mass to the regular subdivision $\mathcal{S}_n$ and evaluation set $\{\mathbf{r}_{i,j} = \langle x_i, y_j\rangle : 1 \le i, j \le n\}$, and recall the definition of the double integral, we are led to the following:

$$\mathbf{R} \approx \frac{1}{m} \sum_{i,j} \mathbf{r}_{i,j} dm_{i,j} = \frac{1}{m} \sum_{i,j} \delta(x_i, y_j) \langle x_i, y_j\rangle\, dA \to \frac{1}{m} \iint_{\mathcal{D}} \delta(x, y) \langle x, y\rangle\, dA.$$

This calculation suggests the following definition.

DEFINITION Mass and Center of Mass of a Lamina

A lamina can be modeled by a bounded region $\mathcal{D}$ of R^2 and an integrable density function δ defined on $\mathcal{D}$. The position vector of its center of mass is

$$\mathbf{R} = \frac{1}{m} \iint_{\mathcal{D}} \mathbf{r}\, dm = \frac{1}{m} \iint_{\mathcal{D}} \delta(x, y) \langle x, y\rangle\, dx\, dy, \qquad (7)$$

where $m = \iint_{\mathcal{D}} dm = \iint_{\mathcal{D}} \delta(x, y)\, dA$ is the mass of the lamina. If the density of the lamina is constant, the center of mass is called the **centroid** of $\mathcal{D}$.

If we write $\mathbf{R} = \langle X, Y\rangle$ then the center of mass can be written as the two scalar equations

$$X = \frac{1}{m} \iint_{\mathcal{D}} x\delta(x, y)\, dx\, dy \quad \text{and} \quad Y = \frac{1}{m} \iint_{\mathcal{D}} y\delta(x, y)\, dx\, dy. \qquad (8)$$

We give two examples to illustrate mass or center of mass calculations.

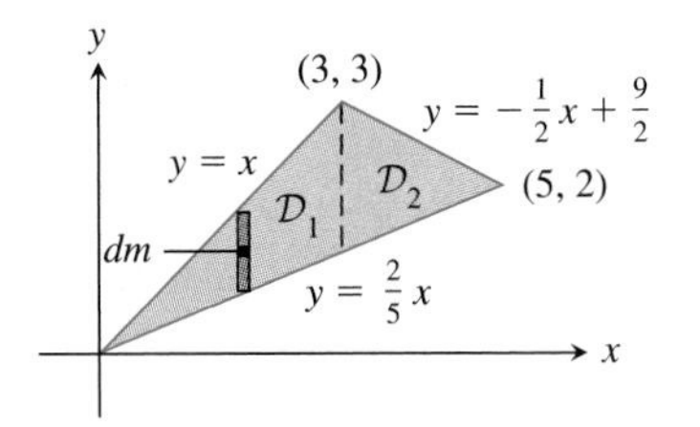

FIGURE 11.19 Triangular lamina with variable density.

EXAMPLE 5 Find the mass of the triangular lamina with vertices $(0, 0)$, $(5, 2)$, and $(3, 3)$. See Fig. 11.19. The density of this lamina at any point (x, y) is proportional to the distance from (x, y) to the y-axis. Units are meters and kilograms.

Solution

The region $\mathcal{D}$ is shown as the union of regions $\mathcal{D}_1$ and $\mathcal{D}_2$, each of Type I. (Note that $\mathcal{D}$ also may be written as the union of two regions of Type II. The solution for this decomposition of $\mathcal{D}$ is left as Exercise 37.)

From the statement that the density $\delta(x, y)$ of the lamina at the point (x, y) is proportional to the distance from (x, y) to the y-axis,

$$\delta(x, y) = kx,$$

where k is a positive constant. The mass of the lamina is

$$m = \iint_{\mathcal{D}} dm = \iint_{\mathcal{D}} \delta(x, y)\, dA = k \iint_{\mathcal{D}} x\, dx\, dy.$$

Applying the additivity property of the double integral,

$$m = \iint_{\mathcal{D}_1} dm + \iint_{\mathcal{D}_2} dm.$$

Applying the Iterated Integral Theorem to the first integral and noting from Fig. 11.19 that for each fixed x between 0 and 3 the mass element dm moves from the line with equation $y = 2x/5$ to the line with equation $y = x$, the mass m_1 of $\mathcal{D}_1$ is

$$m_1 = \iint_{\mathcal{D}_1} dm = k \int_0^3 \left(\int_{2x/5}^{x} x\, dy \right) dx.$$

Similarly, the mass m_2 of $\mathcal{D}_2$ is

$$m_2 = \iint_{\mathcal{D}_2} dm = k \int_3^5 \left(\int_{2x/5}^{-x/2+9/2} x\, dy \right) dx.$$

We work through the evaluation of $\iint_{\mathcal{D}_1} dm$ and leave the evaluation of $\iint_{\mathcal{D}_2} dm$ as Exercise 38. From the expression for m_1,

$$\begin{aligned} m_1 &= k \int_0^3 \left(\int_{2x/5}^{x} x\, dy \right) dx \\ &= k \int_0^3 \left(xy \Big|_{2x/5}^{x} \right) dx = k \int_0^3 \left(\frac{3}{5} x^2 \right) dx \\ &= k \left(\frac{1}{5} x^3 \right) \Big|_0^3 = \frac{27k}{5}. \end{aligned}$$

Similarly,

$$m_2 = \frac{33k}{5}.$$

The total mass of the lamina is $m = m_1 + m_2 = 12k$ kilograms.

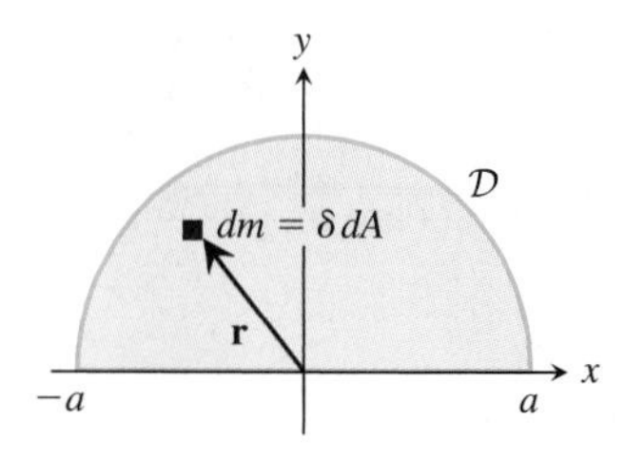

FIGURE 11.20 Semicircular lamina.

EXAMPLE 6 Find the mass and center of mass of the semicircular lamina shown in Fig. 11.20. The density of the lamina at a point (x, y) is proportional to the square of the distance from (x, y) to $(0, 0)$. Units are meters and kilograms.

Solution

The region $\mathcal{D}$ is of Types I and II. We use formula (2) from the Iterated Integral Theorem for regions of Type I. See Exercise 40 for an application of formula (3) for regions of Type II. The density of the lamina at $\mathbf{r} = \langle x, y\rangle$ is $\delta(x, y) = k\|\mathbf{r}\|^2 = k(x^2 + y^2)$, where k is a positive constant. The mass of the lamina is

$$
\begin{aligned}
m &= \iint_{\mathcal{D}} \delta\, dm = \iint_{\mathcal{D}} k(x^2 + y^2)\, dx\, dy \\
&= k \int_{-a}^{a} \int_{0}^{\sqrt{a^2-x^2}} (x^2 + y^2)\, dy\, dx \\
&= k \int_{-a}^{a} \left(x^2 y + \frac{1}{3} y^3 \right) \Bigg|_{0}^{\sqrt{a^2-x^2}} dx = k \int_{-a}^{a} \frac{1}{3}(2x^2 + a^2)\sqrt{a^2 - x^2}\, dx \text{ kg.}
\end{aligned}
$$

Applying formulas (7) and (8) from the Table of Integrals, $m = \frac{1}{4}k\pi a^4$.

It follows from the symmetry of the lamina and the density function that the center of mass is on the y-axis, that is, $\mathbf{R} = (0, Y)$. Thus, we need only calculate Y. From (8),

$$
Y = \frac{1}{m} \iint_{\mathcal{D}} y\delta(x, y)\, dx\, dy = \frac{k}{m} \int_{-a}^{a} \int_{0}^{\sqrt{a^2-x^2}} y(x^2 + y^2)\, dy\, dx.
$$

Leaving the details to Exercise 39,

$$
\mathbf{R} = \langle 0, Y\rangle = \langle 0, 8a/(5\pi)\rangle \text{ m.}
$$

Exercises 11.2

Exercises 1–6: Sketch the region of integration $\mathcal{D}$ and evaluate the iterated integral.

1. $\iint_{\mathcal{D}} y\, dA = \int_0^1 \int_0^x y\, dy\, dx$
2. $\iint_{\mathcal{D}} xy\, dA = \int_1^2 \int_x^2 xy\, dy\, dx$
3. $\iint_{\mathcal{D}} y/x\, dA = \int_1^2 \int_1^y y/x\, dx\, dy$
4. $\iint_{\mathcal{D}} y/x\, dA = \int_1^3 \int_1^{y^2} y/x\, dx\, dy$
5. $\iint_{\mathcal{D}} \sin x^2\, dA = \int_0^1 \int_0^x \sin x^2\, dy\, dx$
6. $\iint_{\mathcal{D}} e^{y/x}\, dA = \int_{0.1}^1 \int_0^{x^3} e^{y/x}\, dy\, dx$

Exercises 7–12: Assume that f is continuous on $\mathcal{D}$. Give an iterated integral equal to $\iint_{\mathcal{D}} f(x, y)\, dx\, dy$.

7. $\mathcal{D}$ is the triangular region with vertices $(1, 0)$, $(2, 0)$, and $(1, 1)$.
8. $\mathcal{D}$ is the triangular region with vertices $(-1, 1)$, $(2, -2)$, and $(2, 1)$.
9. $\mathcal{D}$ is the region bounded by the curves with equations $y^2 = x$ and $y - \sqrt{2} = \sqrt{2}(x - 2)$.
10. $\mathcal{D}$ is the region bounded by the graph of the tangent function from $(0, 0)$ to $(\pi/4, 1)$, the line from $(\pi/4, 1)$ to $(\pi/2, 0)$, and the line from $(0, 0)$ to $(\pi/2, 0)$.

11. $\mathcal{D}$ is the region bounded by the curves with equations $y = x^2$ and $y = 1 - x^2$.

12. $\mathcal{D}$ is the region bounded by the curves with equations $x = 2 - y^2$ and $x = y^2$.

Exercises 13–16: Sketch the region $\mathcal{D}$ of integration, change the order of integration, and evaluate.

13. $\iint_{\mathcal{D}} xy^2\,dA = \int_0^1 \int_{2x}^2 xy^2\,dy\,dx$

14. $\iint_{\mathcal{D}} y/x\,dA = \int_1^4 \int_{\sqrt{y}}^2 y/x\,dx\,dy$

15. $\iint_{\mathcal{D}} y\,dA = \int_0^1 \int_{\arctan y}^{\pi/4} y\,dx\,dy$

16. $\iint_{\mathcal{D}} x\,dA = \int_{1/e}^e \int_{-1}^{\ln x} x\,dy\,dx$

Exercises 17–26: Use a double integral to find the volume of the described region.

17. The region beneath the plane with equation $y + z = 1$ and above the triangle with vertices $(0,0)$, $(1,0)$, and $(0,1)$.

18. The region beneath the plane with equation $x + y + z = 2$ and above the triangle with vertices $(0,0)$, $(1,1)$, and $(0,1)$.

19. The tetrahedron bounded by the planes described by $x + 2y + 3z = 6$, $x = y$, $z = 0$, and $y = 0$.

20. The tetrahedron bounded by the planes described by $x + y + z = 1$, $x = y$, $z = 0$, and $x = 0$.

21. The region beneath the plane with equation $x + y - z = 0$ and above the set of points in the (x, y)-plane and bounded by the curves with equations $y = 0$, $y = x^2$, and $x = 1$.

22. The region beneath the plane with equation $x + y + z = 1$ and above the set of points in the (x, y)-plane and bounded by the curves with equations $x = 0$, $y = \sqrt{x}$, and $x + y = 1$.

23. The region beneath the plane with equation $z = 2x + 1$ and above the disc described by $(x - 1)^2 + y^2 \leq 1$.

24. The region beneath the plane with equation $z = y + 1$ and above the disc described by $x^2 + (y - 1)^2 \leq 1$.

25. The region common to the cylinders $x^2 + z^2 = a^2$ and $y^2 + z^2 = a^2$.

26. The region in the first octant cut from the cylinder $x^2 + z^2 = a^2$ by the planes $x = y$ and $x = 2y$.

Exercises 27–32: Find the mass and center of mass of the lamina occupying region $\mathcal{D}$ and having density δ. Units are meters and kilograms.

27. $\mathcal{D} = \{(x, y) \in R^2 : 0 \leq y \leq 4 - x^2\}$. The density at a point of $\mathcal{D}$ is proportional to the distance from that point to the x-axis.

28. $\mathcal{D} = \{(x, y) \in R^2 : x^2 \leq y \leq 4\}$. The density at a point of $\mathcal{D}$ is proportional to the distance from that point to the x-axis.

29. $\mathcal{D} = \{(x, y) \in R^2 : x^2 \leq y \leq x + 2\}$. The density at a point of $\mathcal{D}$ is proportional to the distance from that point to the y-axis.

30. $\mathcal{D} = \{(x, y) \in R^2 : y^2 \leq x \leq 1 + 3y/2\}$. The density at a point of $\mathcal{D}$ is proportional to the distance from that point to the y-axis.

31. $\mathcal{D} = \left\{(x, y) \in R^2 : 0 \leq y \leq \sqrt{1 - x^2}\right\}$. The density at a point of $\mathcal{D}$ is proportional to the distance from that point to the origin. It is given that the mass is $k\pi/3$.

32. $\mathcal{D} = \left\{(x, y) \in R^2 : x \geq 0, 0 \leq y \leq \sqrt{1 - x^2}\right\}$. The density at a point of $\mathcal{D}$ is proportional to the distance from that point to the origin. It is given that the mass is $k\pi/6$.

33. In Example 3, argue that $\mathcal{D}$ is a Type I region and set up, but do not evaluate, two iterated integrals whose sum is the volume V.

34. Continuation Evaluate the two iterated integrals in the preceding exercise and show that their sum is 10.8.

35. Give a clear, convincing explanation as to why it is easier to calculate the mass of the bearing in Example 4 by working with the square hole instead of the bearing itself.

T **36.** Use your calculator to check the answer to Example 4.

37. Show that the region $\mathcal{D}$ in Example 5 is the union of two regions of Type II. Set up two iterated integrals for the mass of this region.

38. Calculate the mass m_2 of the sublamina $\mathcal{D}_2$ in Example 5.

39. Fill in the details of the integration for the calculation of Y in Example 6.

40. Regard the region $\mathcal{D}$ in Example 6 as a region of Type II and set up, but do not evaluate, iterated integrals for the mass and the center of mass.

41. The integral

$$\int_0^1 \left(x^2 - \frac{2}{3}x^3\right) \sin 2x\,dx$$

was worked out in Example 1 by applying integration formula (25) from the Table of Integrals. Use the integration by parts formula directly in showing that the value of this integral is $\frac{1}{12}(3\sin(2) - 5\cos(2) - 3) \approx 0.150719$.

42. Show that

$$\frac{1}{8}(y^2 - x^2) + 1.5 \geq 0$$

when $0 \leq x \leq 2$ and $0 \leq y \leq \sqrt{x}$.
Hint: $\frac{1}{8}(y^2 - x^2) \geq \frac{1}{8}(0 - 4)$.

43. Calculate the volume V beneath the graph of the function

$$f(x, y) = x^2 + y^2$$

defined on the circular region with boundary described by $x^2 + y^2 = 1$.

44. One of Pappus' theorems If a plane area is revolved about a line that lies in its plane but does not intersect the area, then the volume generated is equal to the product of the area and the distance traveled by its centroid. Fill in the details in the following proof of a special case of this result. Let $\mathcal{D}$ be a region of Type I in the (x, y)-plane, where the defining functions g_1 and g_2 satisfy $0 \leq g_1(x) \leq g_2(x)$ for $x \in [a, b]$. Let $\langle X, Y \rangle$ be the centroid of $\mathcal{D}$, A the area of $\mathcal{D}$, and V the volume of the solid generated by revolving $\mathcal{D}$ about the x-axis. Pappus' result is that $V = 2\pi YA$. *Hint:* Calculate V by the washer method and Y by the methods of this section.

45. Continuation Use Pappus' Theorem in calculating the volume of the torus generated by revolving a circle of radius r about an axis in its plane at a distance $b > r$ from its center.

46. Find the volume of the ellipsoid with equation $x^2/a^2 + y^2/b^2 + z^2/c^2 = 1$. The volume of the ellipsoid is eight times the part of the solid in the first octant. See the accompanying figure. Show that

$$V = 8c \int_0^a \int_0^{b\sqrt{1-x^2/a^2}} z \, dy \, dx,$$

where $z = \sqrt{1 - x^2/a^2 - y^2/b^2}$. Replace $1 - x^2/a^2$ by, say, A^2, and make the substitution $y = bY$ to show that

$$V = 8bc \int_0^a \int_0^A \sqrt{A^2 - Y^2} \, dY \, dx.$$

Either using formula (7) from the Table of Integrals or simply noting that the value of the inner integral is one-quarter of the area of a circle of radius A, show that

$$V = 2\pi bc \int_0^a (1 - x^2/a^2) \, dx = \frac{4}{3} \pi abc.$$

Is the case $a = b = c = r$ significant?

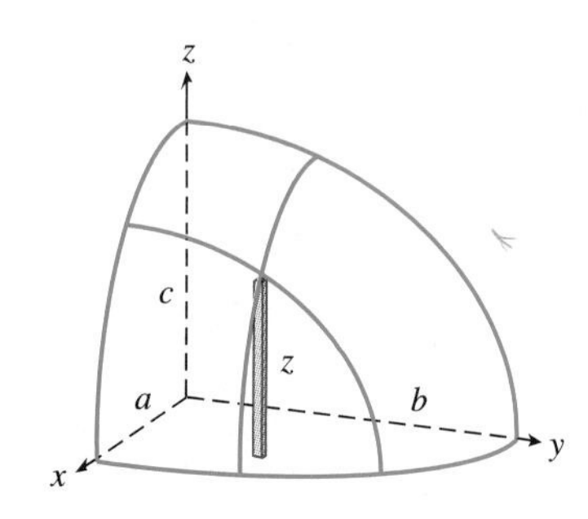

Figure for Exercises 46 and 47.

47. The ellipsoid shown in the accompanying figure is described by the equation $x^2/9 + y^2/25 + z^2/16 = 1$. The volume element is located at $(2, 2)$. What is the height of this element? Give parametric equations of the blue curves passing through the top of the volume element.

11.3 Surface Area

Among the physical variables used to describe or analyze fluid flow is the *flowrate,* which is the amount of fluid passing through a nonmaterial reference surface per unit time. The flowrate depends upon both the geometry of the surface and the velocity field of the fluid. In Chapter 12 these ideas will be used in motivating the idea of a surface integral and in calculating the flowrate of a fluid through a reference surface. In the present section we discuss the geometry of surfaces as it relates to surface area. We also calculate the masses and centers of mass of laminas in the form of surfaces.

Figure 11.21 shows a representative surface, one that can be described by a parametric equation of the form

$$\mathbf{r} = \mathbf{r}(u, v) = \langle r_1(u, v), r_2(u, v), r_3(u, v) \rangle, \qquad (u, v) \in \mathcal{D} = \mathcal{I} \times \mathcal{J}, \tag{1}$$

where $\mathcal{I}$ and $\mathcal{J}$ are intervals on the u- and v-axes. We shall assume in all that follows that $\mathbf{r}$ is one-to-one on $\mathcal{D}$ and that the coordinate functions of $\mathbf{r}$ have continuous partial derivatives. It follows from the Tests for Differentiability in Section 10.1 that the coordinate functions of $\mathbf{r}$ are differentiable and $\mathbf{r}$ itself is differentiable. Surfaces that can be described with coordinate functions having continuous partial derivatives are called **smooth** surfaces; surfaces described by one-to-one functions are called **simple** surfaces.

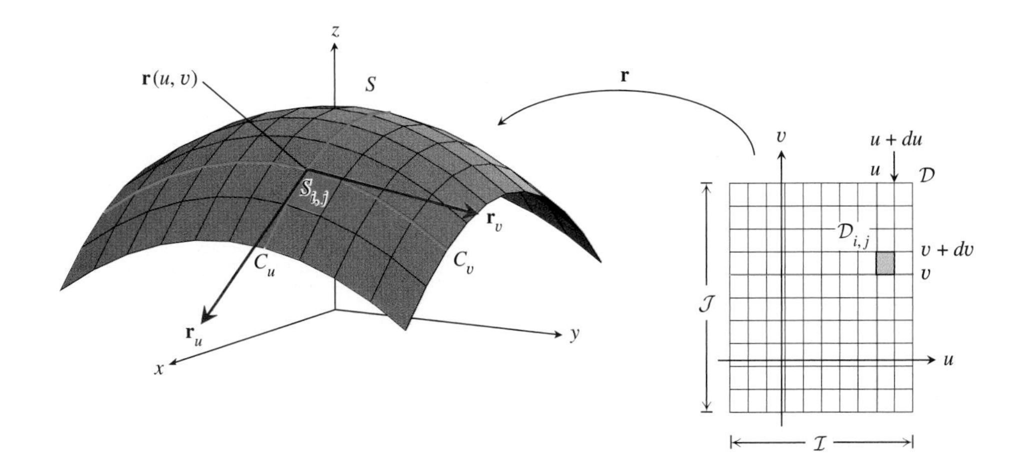

FIGURE 11.21 Tangent vectors and unit normal at a point of a surface S.

We use a geometric argument to motivate the formula

$$A(S) = \iint_{\mathcal{D}} \|\mathbf{r}_u \times \mathbf{r}_v\| \, du \, dv \tag{2}$$

for the area $A(S)$ of the simple, smooth surface S described by (1).

Let $\mathcal{S}_n$ be a regular subdivision of the set $\mathcal{D}$ on which $\mathbf{r}$ is defined. Figure 11.21 shows the representative subdivision $\mathcal{S}_{10}$. The function $\mathbf{r}$ maps the point (u, v), the red lines through (u, v), and the blue subregion $\mathcal{D}_{i,j}$, all in $\mathcal{D}$, to the point $\mathbf{r}(u, v)$, the red curves C_u and C_v, and the blue subregion $S_{i,j}$, all in the surface S. The curves C_u and C_v are described by

$$\begin{aligned} C_u &: \mathbf{r} = \mathbf{r}(u, v), \qquad u \in \mathcal{I} \\ C_v &: \mathbf{r} = \mathbf{r}(u, v), \qquad v \in \mathcal{J}. \end{aligned}$$

Figure 11.21 also shows the tangent vectors $\mathbf{r}_u$ and $\mathbf{r}_v$ to the curves C_u and C_v at the point $\mathbf{r}(u, v)$, where

$$\mathbf{r}_u = \frac{\partial}{\partial u}\mathbf{r}(u, v) = \left\langle \frac{\partial}{\partial u} r_1(u, v), \frac{\partial}{\partial u} r_2(u, v), \frac{\partial}{\partial u} r_3(u, v) \right\rangle$$

and

$$\mathbf{r}_v = \frac{\partial}{\partial v}\mathbf{r}(u, v) = \left\langle \frac{\partial}{\partial v} r_1(u, v), \frac{\partial}{\partial v} r_2(u, v), \frac{\partial}{\partial v} r_3(u, v) \right\rangle.$$

Because $\mathbf{r}$ is differentiable at (u, v),

$$\mathbf{r}(u + h, v + k) \approx \mathbf{r}(u, v) + \begin{pmatrix} \dfrac{\partial r_1}{\partial u} & \dfrac{\partial r_1}{\partial v} \\ \dfrac{\partial r_2}{\partial u} & \dfrac{\partial r_2}{\partial v} \\ \dfrac{\partial r_3}{\partial u} & \dfrac{\partial r_3}{\partial v} \end{pmatrix} \begin{pmatrix} h \\ k \end{pmatrix} \tag{3}$$

for all points $(u+h, v+k)$ near (u, v). With (u, v) fixed for the moment, the expressions on the left and right sides of (3) are functions of h and k. The function on the left, with $0 \leq h \leq du$ and $0 \leq k \leq dv$, maps points of $\mathcal{D}_{i,j}$ onto the surface S; the function on the right side of (3), with $0 \leq h \leq du$ and $0 \leq k \leq dv$, maps points of $\mathcal{D}_{i,j}$ onto a part of the plane tangent to the surface S. Setting (h, k) equal to $(0, 0)$, $(du, 0)$, $(0, dv)$, and (du, dv) in (3), we find the approximations to $\mathbf{r}$ at each of the four corners of $\mathcal{D}_{i,j}$ to be

$$\begin{aligned} \mathbf{r}(u, v) &= \mathbf{r}(u, v) \\ \mathbf{r}(u+du, v) &\approx \mathbf{r}(u, v) + du\,\mathbf{r}_u \\ \mathbf{r}(u, v+dv) &\approx \mathbf{r}(u, v) + dv\,\mathbf{r}_v \\ \mathbf{r}(u+du, v+dv) &\approx \mathbf{r}(u, v) + du\,\mathbf{r}_u + dv\,\mathbf{r}_v. \end{aligned} \tag{4}$$

Figure 11.22 shows a zoom-view of $S_{i,j}$ and of a part of the plane tangent to S at $\mathbf{r}(u, v)$. The points on the left in (4) are points of $S_{i,j}$ corresponding to the vertices of $\mathcal{D}_{i,j}$ and lie on S. Because the right side of (3) is a translation of a linear function, it maps the rectangle $\mathcal{D}_{i,j}$ to a parallelogram $T_{i,j}$. The points on the right in (4) are the vertices of $T_{i,j}$ and lie in the plane tangent to S at $\mathbf{r}(u, v)$. Note that $T_{i,j}$ is determined by the vectors $du\mathbf{r}_u$ and $dv\mathbf{r}_v$, both based at $\mathbf{r}(u, v)$. Because $\mathbf{r}$ is differentiable, we expect the area $A(S_{i,j})$ of $S_{i,j}$ to be approximately equal to the area $A(T_{i,j})$ of the parallelogram, that is,

$$A(S_{i,j}) \approx A(T_{i,j}) = \|(du\,\mathbf{r}_u) \times (dv\,\mathbf{r}_v)\| = \|\mathbf{r}_u \times \mathbf{r}_v\|\,du\,dv. \tag{5}$$

We denote the area $A(T_{i,j})$ of the parallelogram by $d\sigma_{i,j}$.

Because, whatever the number n of subdivisions of $\mathcal{D}$, the sum of the areas $A(S_{i,j})$ is equal to the area $A(S)$ of S, it follows from (5) that

$$A(S) = \sum_{i,j} A(S_{i,j}) \approx \sum_{i,j} d\sigma_{i,j} = \sum_{i,j} \|\mathbf{r}_u \times \mathbf{r}_v\|\,du\,dv.$$

As the number n of subdivisions of $\mathcal{D}$ increases without bound, we expect that

$$A(S) = \lim_{n\to\infty} \sum_{i,j} \|\mathbf{r}_u \times \mathbf{r}_v\|\,du\,dv = \iint_{\mathcal{D}} \|\mathbf{r}_u \times \mathbf{r}_v\|\,du\,dv.$$

With these remarks as motivation, we define the surface area of a simple, smooth surface.

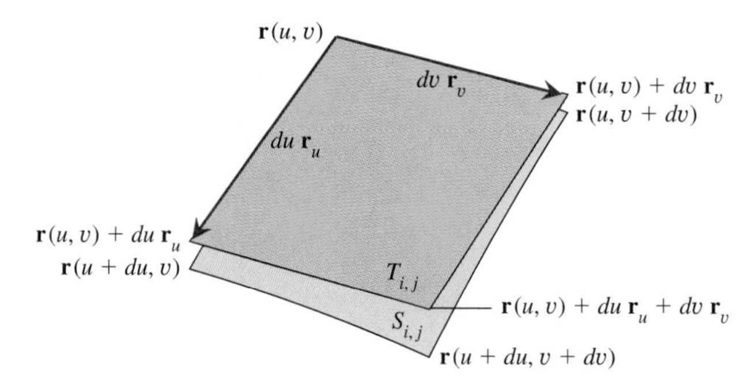

FIGURE 11.22 A zoom-view of the subsurface $S_{i,j}$ and, in the tangent plane, a parallelogram $T_{i,j}$, both corresponding to $\mathcal{D}_{i,j}$.

DEFINITION Surface Area

Let S be a simple, smooth surface defined by the vector function

$$\mathbf{r} = \mathbf{r}(u, v), \qquad (u, v) \in \mathcal{D},$$

where $\mathcal{D}$ is a region of Type I or II. The area $A(S)$ of S is the value of the integral

$$A(S) = \iint_{\mathcal{D}} dv = \iint_{\mathcal{D}} \|\mathbf{r}_u \times \mathbf{r}_v\| \, du \, dv. \tag{6}$$

We will often refer to $\|\mathbf{r}_u \times \mathbf{r}_v\| \, du \, dv$ as an element of surface area and denote it by $d\sigma$. When we include an element of surface area in the graphs of surfaces, we also include the surface normal $\mathbf{n} = \mathbf{r}_u \times \mathbf{r}_v$ to give a sense of the orientation of the element $d\sigma$.

Because a simple, smooth surface S may have more than one parametrization, we note that the area of S is independent of the particular parametrization chosen to describe S. That is, the value of the integral (6) is the same for all parametrizations of S, provided that the parametrizations are one-to-one on their domains and their coordinate functions have continuous partial derivatives. We do not prove this result.

EXAMPLE 1 Find the area of the surface S described by the function

$$\mathbf{r} = \mathbf{r}(u, v) = \langle u \cos v, u \sin v, av \rangle, \qquad (u, v) \in \mathcal{D} = [1/2, 2] \times [0, 2\pi], \tag{7}$$

where a is a positive constant.

Solution

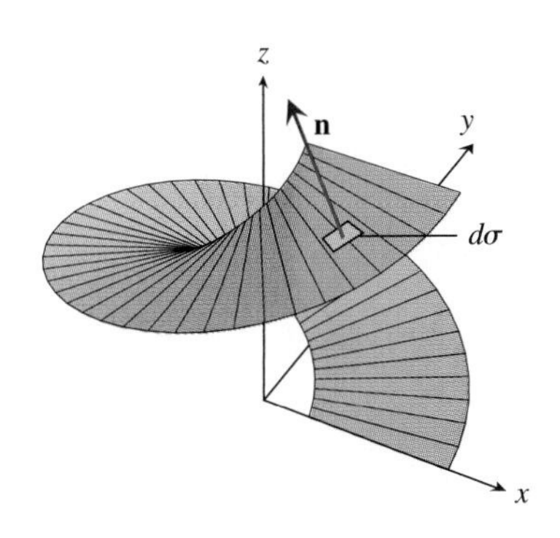

FIGURE 11.23 One turn of a spiral ramp.

Figure 11.23 shows the surface S; for the figure we took $a = 1/2$. The element of surface area $d\sigma$ is shown in the figure, lying in the surface at the base of a normal $\mathbf{n}$ to the surface. From (6), the area of S is

$$A(S) = \iint_{\mathcal{D}} d\sigma = \iint_{\mathcal{D}} \|\mathbf{r}_u \times \mathbf{r}_v\| \, du \, dv.$$

The partial derivatives $\mathbf{r}_u$ and $\mathbf{r}_v$ are

$$\mathbf{r}_u = \langle \cos v, \sin v, 0 \rangle$$
$$\mathbf{r}_v = \langle -u \sin v, u \cos v, a \rangle.$$

It follows that

$$\|\mathbf{r}_u \times \mathbf{r}_v\| = \|\langle a \sin v, -a \cos v, u \rangle\| = \sqrt{u^2 + a^2}.$$

Hence,

$$A(S) = \iint_{\mathcal{D}} \sqrt{u^2 + a^2} \, du \, dv.$$

The region $\mathcal{D} = [1/2, 2] \times [0, 2\pi]$ of integration is of Type I and Type II. Noting that the integrand is independent of v, we integrate on v first, obtaining

$$A(S) = \int_{1/2}^{2} \int_{0}^{2\pi} \sqrt{u^2 + a^2}\, dv\, du = 2\pi \int_{1/2}^{2} \sqrt{u^2 + a^2}\, du.$$

Applying formula (4) from the Table of Integrals,

$$A(S) = \pi\left(u\sqrt{u^2 + a^2} + a^2 \ln\left(u + \sqrt{u^2 + a^2}\right)\right)\Big|_{1/2}^{2}.$$

If $a = 1/2$,

$$A(S) = 12.7953\ldots \text{ square units.}$$

EXAMPLE 2 Determine the area of a sphere of radius a.

Solution

To calculate the area of a surface, we must describe it with an equation or function. How shall we describe the sphere? The equation $x^2 + y^2 + z^2 = a^2$ comes easily to mind. If we solve this equation for z and then regard x and y as parameters, we may describe the upper hemisphere with the vector equation

$$\mathbf{r} = \mathbf{r}(x, y) = \left\langle x, y, \sqrt{a^2 - x^2 - y^2}\right\rangle, \qquad (x, y) \in \mathcal{D}, \tag{8}$$

where $\mathcal{D}$ is the origin-centered disk of radius a in the (x, y)-plane. With this parametrization,

$$\begin{aligned}
\mathbf{r}_x &= \left\langle 1, 0, \frac{-2x}{2\sqrt{a^2 - x^2 - y^2}}\right\rangle \\
\mathbf{r}_y &= \left\langle 0, 1, \frac{-2y}{2\sqrt{a^2 - x^2 - y^2}}\right\rangle \\
\mathbf{r}_x \times \mathbf{r}_y &= \left\langle \frac{x}{\sqrt{a^2 - x^2 - y^2}}, \frac{y}{\sqrt{a^2 - x^2 - y^2}}, 1\right\rangle \\
\|\mathbf{r}_x \times \mathbf{r}_y\| &= \frac{a}{\sqrt{a^2 - x^2 - y^2}}.
\end{aligned}$$

It now follows from (6) that the area of a sphere of radius a is

$$2\iint_{\mathcal{D}} \frac{a}{\sqrt{a^2 - x^2 - y^2}}\, dx\, dy. \tag{9}$$

For a second approach, recall that an origin-centered sphere of radius a is parametrized most naturally with the spherical coordinates θ and ϕ. Moreover, this parametrization leads to a simpler integral for A. Recalling the equations

$$\begin{aligned}
x &= \rho \cos\theta \sin\phi \\
y &= \rho \sin\theta \sin\phi \\
z &= \rho \cos\phi
\end{aligned} \tag{10}$$

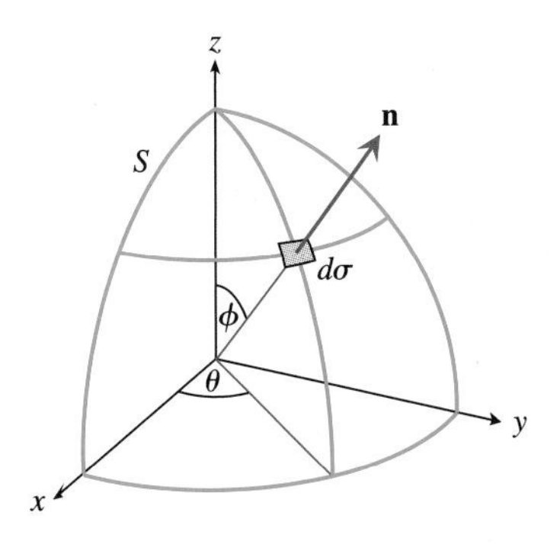

FIGURE 11.24 One-eighth of a sphere, with an element $d\sigma$ of surface area.

relating the rectangular and spherical coordinates of a point in space, the rectangular coordinates of a point on the sphere S of radius a are $(a\cos\theta\sin\phi, a\sin\theta\sin\phi, a\cos\phi)$. Hence, the surface S of the first octant part of a sphere of radius a can be described by

$$\mathbf{r} = \mathbf{r}(\theta,\phi) = \langle a\cos\theta\sin\phi, a\sin\theta\sin\phi, a\cos\phi\rangle, \qquad (\theta,\phi)\in\mathcal{D}, \tag{11}$$

where $\mathcal{D} = [0,\pi/2]\times[0,\pi/2]$. The set $\mathcal{D}$ will be the region of integration. See Fig. 11.24.

Denoting the three coordinate functions of $\mathbf{r}$ by x, y, and z, the cross product $\mathbf{r}_\theta \times \mathbf{r}_\phi$ is

$$\mathbf{r}_\theta\times\mathbf{r}_\phi = \begin{vmatrix} \mathbf{i} & \mathbf{j} & \mathbf{k} \\ x_\theta & y_\theta & z_\theta \\ x_\phi & y_\phi & z_\phi \end{vmatrix} = \begin{vmatrix} \mathbf{i} & \mathbf{j} & \mathbf{k} \\ -a\sin\theta\sin\phi & a\cos\theta\sin\phi & 0 \\ a\cos\theta\cos\phi & a\sin\theta\cos\phi & -a\sin\phi \end{vmatrix}$$
$$= \langle -a^2\cos\theta\sin^2\phi, -a^2\sin\theta\sin^2\phi, -a^2\sin\phi\cos\phi\rangle$$
$$= -a^2\sin\phi\langle\cos\theta\sin\phi, \sin\theta\sin\phi, \cos\phi\rangle.$$

Because the vector $\langle\cos\theta\sin\phi, \sin\theta\sin\phi, \cos\phi\rangle$ is a position vector of a point on a sphere of radius 1, its length is 1 and, hence,

$$\|\mathbf{r}_\theta\times\mathbf{r}_\phi\| = a^2\sin\phi. \tag{12}$$

The area $A(S)$ of S is

$$A(S) = \iint_{\mathcal{D}} \|\mathbf{r}_\phi\times\mathbf{r}_\phi\|\,d\phi\,d\theta = \int_0^{\pi/2}\int_0^{\pi/2} a^2\sin\phi\,d\phi\,d\theta$$
$$= a^2\int_0^{\pi/2} 1\,d\theta = \tfrac{1}{2}\pi a^2.$$

Multiplying by 8, the area of a sphere of radius a is $4\pi a^2$ square units.

You may have noticed that the vector function $\mathbf{r} = \mathbf{r}(\theta,\phi)$ describing S is not one-to-one on $\mathcal{D}$, as required in the definition of the area of a surface. The sets $\mathcal{D}$ and S are shown in Fig. 11.25. All the points along the side of $\mathcal{D}$ for which $\phi = 0$ are transformed by $\mathbf{r}$ to the north pole. For example,

$$\mathbf{r}(0,0) = \mathbf{r}(\pi/6,0) = \mathbf{r}(\pi/2,0) = \langle 0,0,a\rangle.$$

We note that as long as the set of points in $\mathcal{D}$ at which $\mathbf{r}$ is not one-to-one is contained in the boundary of $\mathcal{D}$, the application of (6) gives correct results. A partial justification is discussed in Exercise 21.

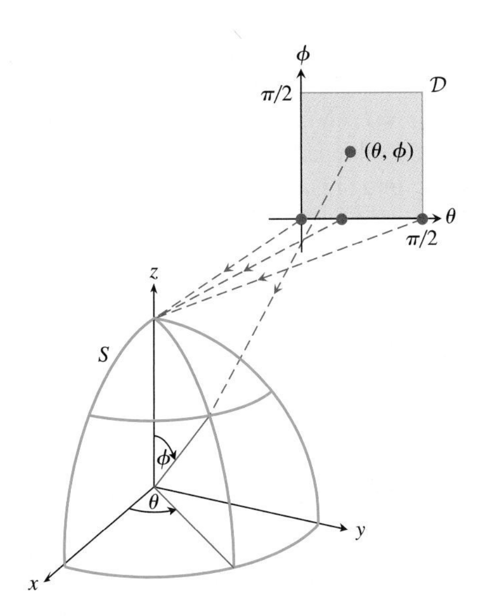

FIGURE 11.25 The parametrization describing S is not everywhere one-to-one.

Area of a Surface Described by the Equation $z = f(x,y)$

Some surfaces are most naturally described as the graph of a function $z = f(x,y)$, $(x,y)\in\mathcal{D}$. If f has continuous partial derivatives, we show that the area of such a surface is $\iint_{\mathcal{D}}\sqrt{1+f_x^2+f_y^2}\,dx\,dy$.

We may apply (6) by describing the graph of f by the parametric equation

$$\mathbf{r} = \mathbf{r}(x,y) = \langle x, y, f(x,y)\rangle, \qquad (x,y)\in\mathcal{D}.$$

The function $\mathbf{r}$ is one-to-one and continuously differentiable. Hence, S is a simple, smooth surface. We calculate $\mathbf{r}_x$ and $\mathbf{r}_y$, their cross product, and the length of the cross product.

$$\mathbf{r}_x \times \mathbf{r}_y = \begin{vmatrix} \mathbf{i} & \mathbf{j} & \mathbf{k} \\ 1 & 0 & f_x \\ 0 & 1 & f_y \end{vmatrix} = \langle -f_x, -f_y, 1 \rangle$$

$$\|\mathbf{r}_x \times \mathbf{r}_y\| = \sqrt{1 + f_x^2 + f_y^2}.$$

From (6), the area $A(S)$ of S is

$$A(S) = \iint_{\mathcal{D}} \sqrt{1 + f_x^2 + f_y^2}\, dx\, dy. \tag{13}$$

EXAMPLE 3 Find the area of the surface S described by

$$z = f(x, y) = \frac{1}{6}(9 - 3x^2 - 2y^2), \qquad (x, y) \in \mathcal{D} = [-1, 1] \times [-1, 1].$$

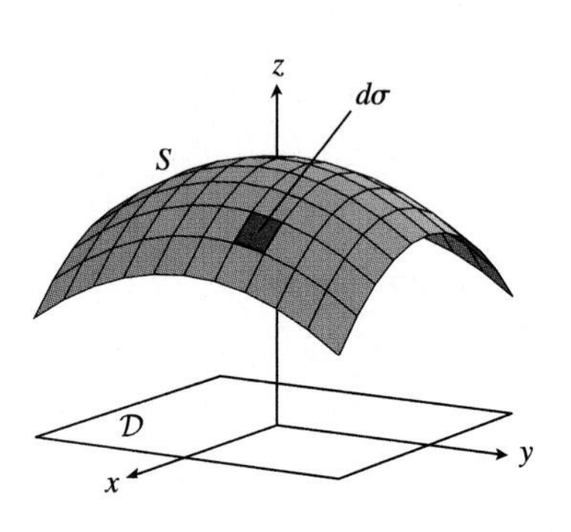

FIGURE 11.26 The surface S and region $\mathcal{D}$.

Solution

The surface S—part of an elliptic paraboloid—is shown in Fig. 11.26. It is a simple, smooth surface because $\mathbf{r} = \langle x, y, f(x, y) \rangle$ is both continuously differentiable and one-to-one. See Exercise 34. The partial derivatives f_x and f_y are $f_x = -x$ and $f_y = -2y/3$. Substituting into (13),

$$A(S) = \iint_{\mathcal{D}} \sqrt{1 + x^2 + \frac{4}{9}y^2}\, dx\, dy.$$

Using symmetry and the Iterated Integral Theorem,

$$A(S) = 4\int_0^1 \int_0^1 \sqrt{x^2 + \frac{4}{9}y^2 + 1}\, dx\, dy.$$

Letting $a^2 = \frac{4}{9}y^2 + 1$ and applying formula (4) from the Table of Integrals,

$$A(S) = 4\int_0^1 \left(\frac{x}{2}\sqrt{x^2 + a^2} + \frac{a^2}{2}\ln\left(x + \sqrt{x^2 + a^2}\right)\right)\Bigg|_0^1 dy$$

$$= 2\int_0^1 \left(\sqrt{1 + a^2} + a^2 \ln \frac{1 + \sqrt{1 + a^2}}{a} \right) dy.$$

Replacing a^2 by $\frac{4}{9}y^2 + 1$,

$$A(S) = 2\int_0^1 \left(\sqrt{2 + \tfrac{4}{9}y^2} + \left(\tfrac{4}{9}y^2 + 1\right) \ln \frac{1 + \sqrt{2 + \frac{4}{9}y^2}}{\sqrt{\frac{4}{9}y^2 + 1}} \right) dy.$$

This integral can be evaluated analytically or numerically. To four significant figures, $A(S) \approx 4.840$ square units.

Area of a Surface of Revolution

We derive a formula for the area of a surface generated by revolving a plane curve about the x-axis. We use the general formula (6) for the area of a surface.

Figure 11.27 shows a smooth curve C lying in the (x, y)-plane and described by an equation

$$\mathbf{r}_C = \mathbf{r}_C(t) = \langle x(t), y(t), 0 \rangle, \qquad a \leq t \leq b.$$

We assume that the curve is not self-intersecting and that $y(t) \geq 0$ for all $t \in [a, b]$. We revolve the curve about the x-axis, thus generating a surface S, one-quarter of which is suggested in the figure.

We use parameters t and θ in describing S. Let $\mathbf{r}$ be the position vector of the point on S obtained by revolving the point P with coordinates $(x(t), y(t), 0)$ through an angle θ, as shown in the figure. We may write $\mathbf{r}$ as the sum of vectors $\mathbf{v}$ and $\mathbf{w}$, where $\mathbf{v} = \langle x(t), 0, 0 \rangle$ lies along the x-axis and $\mathbf{w}$ on the circle of radius $y(t)$ traced by P. Because $\mathbf{w}$ is parallel to the (y, z)-plane, $\mathbf{w} = y(t) \langle 0, \cos\,\theta, \sin\,\theta \rangle$. Hence

$$\begin{aligned} \mathbf{r} = \mathbf{r}(t, \theta)\mathbf{v} + \mathbf{w} &= \langle x(t), 0, 0 \rangle + y(t) \langle 0, \cos\,\theta, \sin\,\theta \rangle \\ &= \langle x(t), y(t)\cos\,\theta, y(t)\sin\,\theta \rangle, \qquad (t, \theta) \in \mathcal{D} = [a, b] \times [0, 2\pi]. \end{aligned}$$

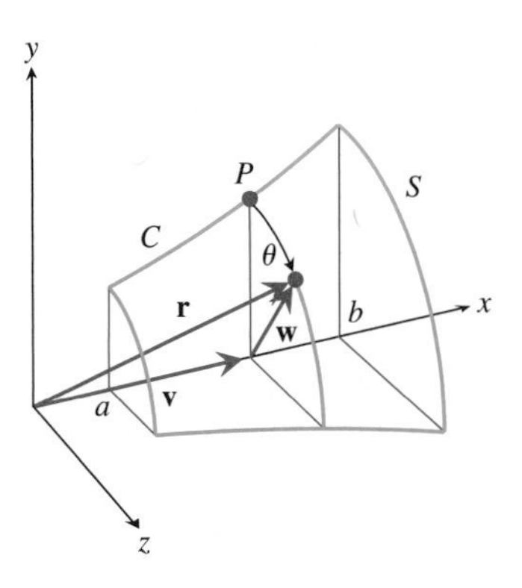

FIGURE 11.27 Surface of revolution generated by a curve C.

We calculate $\mathbf{r}_t$ and $\mathbf{r}_\theta$, their cross product, and the length of the cross product. We abbreviate $y(t)$, $x'(t)$, and $y'(t)$ to y, x', and y'.

$$\mathbf{r}_t \times \mathbf{r}_\theta = \begin{vmatrix} \mathbf{i} & \mathbf{j} & \mathbf{k} \\ x' & y'\cos\,\theta & y'\sin\,\theta \\ 0 & -y\sin\,\theta & y\cos\,\theta \end{vmatrix} = \langle yy', -x'y\cos\,\theta, -x'y\sin\,\theta \rangle$$

$$\|\mathbf{r}_t \times \mathbf{r}_\theta\| = y\sqrt{x'^2 + y'^2}.$$

From (6), the area of the surface of revolution is

$$A = \iint_{\mathcal{D}} y\sqrt{x'^2 + y'^2}\,dt\,d\theta = \int_0^{2\pi} \int_a^b y(t)\sqrt{x'(t)^2 + y'(t)^2}\,dt\,d\theta.$$

Because the limits and integrand of the inner integral are independent of θ, the area of the surface of revolution S is

$$A(S) = 2\pi \int_a^b y(t)\sqrt{x'(t)^2 + y'(t)^2}\,dt. \tag{14}$$

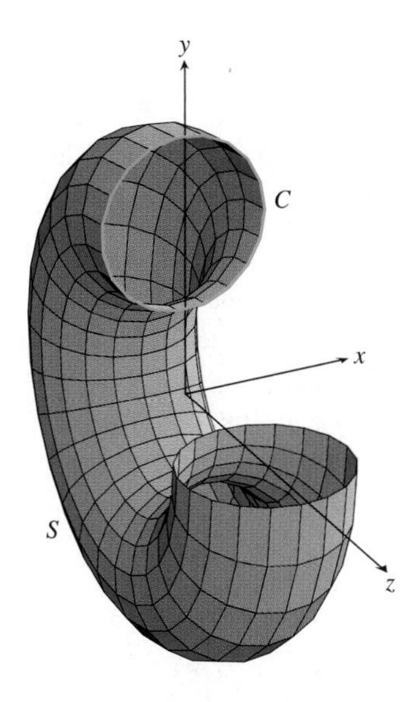

FIGURE 11.28 Three-quarters of the surface of revolution generated by the circle C.

EXAMPLE 4 Find the surface area of the torus or anchor ring generated by revolving the circle C described by

$$\mathbf{r} = \mathbf{r}(t) = \langle 0, b, 0 \rangle + a\langle \cos t, \sin t, 0 \rangle, \qquad 0 \leq t \leq 2\pi, \tag{15}$$

about the x-axis. Assume that $0 < a < b$. The circle C, with center $(0, b, 0)$ and radius a, is shown in the upper part of Fig. 11.28.

Solution

For use in (14), we note from (15) that C is described by

$$x = x(t) = a \cos t, \qquad y = y(t) = b + a \sin t.$$

It follows that the surface area of the torus is

$$\begin{aligned} S(A) &= 2\pi \int_0^{2\pi} (b + a \sin t)\sqrt{(-a \sin t)^2 + (a \cos t)^2}\, dt \\ &= 2\pi \int_0^{2\pi} (b + a \sin t) a\, dt = 4\pi^2 ab \text{ square units.} \end{aligned}$$

Center of Mass of a Curved Lamina

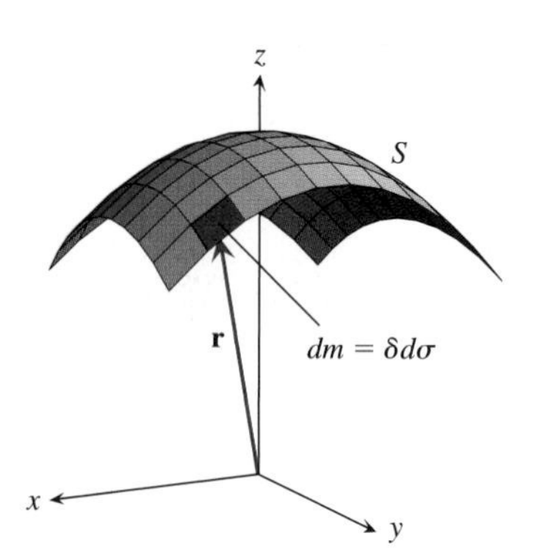

FIGURE 11.29 Thin lamina with variable density.

Figure 11.29 shows a curved lamina, which is something like a curved piece of sheet metal or other material. We derive formulas for calculating the mass and center of mass of a curved lamina.

As in other center of mass calculations, we subdivide the object into point masses dm and then sum the products $dm\,\mathbf{r}$ of the point masses and their position vectors. For a curved lamina with density $\delta = \delta(\mathbf{r}(u, v))$ at $\mathbf{r}(u, v)$, we subdivide S into surface elements of area $d\sigma$. The mass dm of such a representative surface element is $dm = \delta\, d\sigma$. Summing these mass elements to approximate the total mass m of the lamina,

$$m \approx \Sigma\, dm = \Sigma\, \delta\, d\sigma = \Sigma\, \delta(\mathbf{r})\|\mathbf{r}_u \times \mathbf{r}_v\|\, du\, dv.$$

Next, letting $\mathbf{r} = \mathbf{r}(u, v)$ be the position vector of the surface element, we form the products $dm\,\mathbf{r} = \delta\, d\sigma\, \mathbf{r}$, divide by the total mass m, and sum the results to approximate the center of mass, that is,

$$\mathbf{R} = \langle X, Y, Z \rangle \approx \frac{1}{m} \Sigma\, \mathbf{r}\, dm = \frac{1}{m} \Sigma\, \delta(\mathbf{r})\, \|\mathbf{r}_u \times \mathbf{r}_v\|\, \mathbf{r}\, du\, dv.$$

In the limit, as the number of subdivisions increases without bound, the mass and center of mass can be expressed in terms of double integrals.

DEFINITION Center of Mass of a Curved Lamina

Suppose that a curved lamina is modeled by a simple, smooth surface S described by $\mathbf{r} = \mathbf{r}(u, v)$, $(u, v) \in \mathcal{D}$, and a continuous density function δ defined on S. The density of the lamina at $\mathbf{r}(u, v)$ is $\delta(\mathbf{r}(u, v))$. The position vector of the center of mass of the lamina is

$$\mathbf{R} = \langle X, Y, Z \rangle = \frac{1}{m} \iint_{\mathcal{D}} \mathbf{r}\, dm = \frac{1}{m} \iint_{\mathcal{D}} \delta(\mathbf{r}(u, v))\, \|\mathbf{r}_u \times \mathbf{r}_v\|\, \mathbf{r}(u, v)\, du\, dv, \tag{16}$$

where $m = \iint_{\mathcal{D}} dm = \iint_{\mathcal{D}} \delta(\mathbf{r}(u, v))\, d\sigma$ is the mass of the lamina. If the density of the lamina is constant, the center of mass is called the **centroid** of $\mathcal{D}$.

EXAMPLE 5 Find the center of mass of a hemispherical lamina with radius a (meters) and density (kilograms/square meter) proportional to the distance from the plane of the equator.

Solution

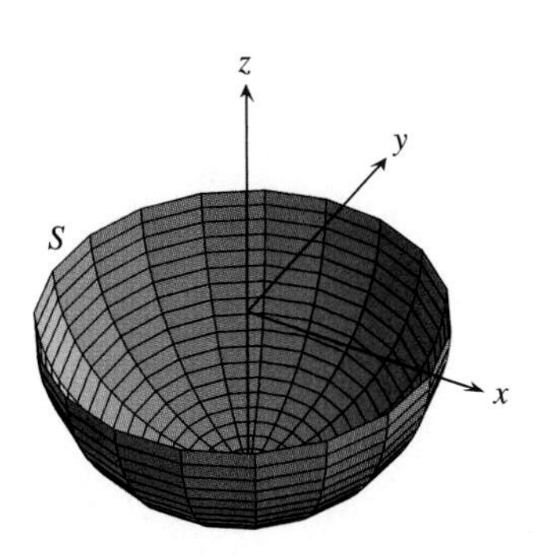

FIGURE 11.30 Center of mass of hemispherical lamina.

We take the bottom half of a sphere as our model, as shown in Fig. 11.30. We may describe S with

$$\mathbf{r} = \mathbf{r}(\theta, \phi) = \langle a \cos\theta \sin\phi, a \sin\theta \sin\phi, a \cos\phi \rangle, \qquad (\theta, \phi) \in \mathcal{D}, \tag{17}$$

where $\mathcal{D} = [0, 2\pi] \times [\pi/2, \pi]$. We used this parametrization in Example 2, but with the domain $\mathcal{D} = [0, 2\pi] \times [0, \pi/2]$. The change of domain corresponds to our choice of the bottom half of the sphere. The density $\delta(x, y, z)$ of the lamina at any point (x, y, z) is $k|z|$. Recalling that $\pi/2 \le \phi \le \pi$ and, from (17), $z = a\cos\phi < 0$, the density function is

$$\delta(\mathbf{r}(\theta, \phi)) = k|a \cos\phi| = -ka \cos\phi. \tag{18}$$

The mass of a surface element $d\sigma$ of the hemisphere is

$$dm = \delta(\mathbf{r}(\phi, \theta))\, d\sigma = -ka \cos\phi \|\mathbf{r}_\theta \times \mathbf{r}_\phi\|\, d\phi\, d\theta.$$

Recalling from (12) that $\|\mathbf{r}_\theta \times \mathbf{r}_\phi\| = a^2 \sin\phi$,

$$\begin{aligned} dm &= -ka \cos\phi (a^2 \sin\phi)\, d\theta\, d\phi = -ka^3 \cos\phi \sin\phi\, d\theta\, d\phi \\ &= -\frac{1}{2} ka^3 \sin(2\phi)\, d\theta\, d\phi. \end{aligned}$$

Hence,

$$m = -\frac{1}{2} ka^3 \iint_{\mathcal{D}} \sin(2\phi)\, d\theta\, d\phi = -\frac{1}{2} ka^3 \int_0^{2\pi} \int_{\pi/2}^{\pi} \sin(2\phi)\, d\theta\, d\phi. \tag{19}$$

Because the limits and integrand are independent of θ,

$$m = -\frac{1}{2} ka^3 (2\pi) \int_{\pi/2}^{\pi} \sin(2\phi)\, d\phi \tag{20}$$

$$= ka^3 \pi \text{ kg}. \tag{21}$$

By symmetry, the center of mass of the hemisphere is on the z-axis. Hence, $X = Y = 0$ and we need to calculate only Z in (16). Noting that $\mathbf{r}(\theta, \phi) = \langle a \sin\phi \cos\theta, a \sin\phi \sin\theta, a \cos\phi \rangle$ in (16),

$$Z = \frac{1}{m} \iint_{\mathcal{D}} \delta(\mathbf{r}(\phi, \theta)) \|\mathbf{r}_\theta \times \mathbf{r}_\phi\| a \cos\phi\, d\theta\, d\phi.$$

From (18),

$$Z = \frac{1}{m} \iint_{\mathcal{D}} (-ka \cos\phi) a^2 \sin\phi (a \cos\phi)\, d\theta\, d\phi.$$

The iterated integral for Z will have the same form as for m in (19). Hence,

$$Z = \frac{1}{m}\int_0^{2\pi}\int_{\pi/2}^{\pi}(-ka\cos\phi)a^2\sin\phi(a\cos\phi)\,d\phi\,d\theta$$
$$= -\frac{ka^4}{ka^3\pi}\cdot 2\pi\int_{\pi/2}^{\pi}\cos^2\phi\sin\phi\,d\phi$$
$$= -2a\int_{\pi/2}^{\pi}\cos^2\phi\sin\phi\,d\phi = -\frac{2}{3}a.$$

Hence, the center of mass of this hemispherical lamina is two-thirds of the way from the plane of the equator toward its south pole.

Exercises 11.3

Exercises 1–8: Sketch the surface and calculate its surface area.

1. The part of the plane with equation $3x + 2y + 6z = 12$ and lying in the first octant. Check your work by using the cross product to calculate the area of a triangle. (The area of triangle ABC is $\frac{1}{2}\|\overrightarrow{AB}\times\overrightarrow{AC}\|$.)

2. The part of the plane with equation $6x + 3y + 2z = 6$ and lying in the first octant. Check your work by using the cross product to calculate the area of a triangle. (The area of triangle ABC is $\frac{1}{2}\|\overrightarrow{AB}\times\overrightarrow{AC}\|$.)

3. The parabolic cylinder with equation $z = \frac{1}{2}y^2$ and lying over the region $[0,3]\times[-1,1]$ in the (x,y)-plane.

4. The parabolic cylinder with equation $z = x^2$ and lying over the region $[0,1]\times[0,4]$ in the (x,y)-plane.

5. The surface described by

$$\mathbf{r} = \mathbf{r}(u,v) = \langle u, u+v, 2u+v\rangle, \qquad 0 \le u, v \le 1.$$

This surface can also be described as the plane with equation $z = x + y$ and lying over the parallelogram with vertices $(0,0)$, $(1,1)$, $(1,2)$, and $(0,1)$.

6. The surface described by

$$\mathbf{r} = \mathbf{r}(u,v) = \langle u-v, u+v, 2-2u\rangle,$$

where $1/2 \le u \le 2$ and $0 \le v \le 1/2$. This surface can also be described as the part of the plane with equation $x + y + z = 2$.

7. The graph of

$$f(x,y) = x^2 + 2y^2, \qquad -1 \le x, y \le 1.$$

After applying the Iterated Integral Theorem, do the first integral by substitution or by using a formula from the Table of Integrals. Use numerical integration to do the outer integral.

8. The graph of

$$f(x,y) = 3x^2 + 2y^2, \qquad -2 \le x, y \le 1.$$

After applying the Iterated Integral Theorem, do the first integral by substitution or by using a formula from the Table of Integrals. Use numerical integration to do the outer integral.

Exercises 9–12: Set up an iterated integral, complete in every detail, for the area of the surface. Do not evaluate.

9. The "screw surface" described by

$$\mathbf{r} = \mathbf{r}(u,v) = \left\langle u\cos v, u\sin v, \tfrac{1}{5}v\right\rangle,$$

where $u \in \left[\frac{1}{2}, 1\right]$, $v \in [0, 2\pi]$.

10. The "screw surface" described by

$$\mathbf{r} = \mathbf{r}(u,v) = \langle 2u\cos v, 2u\sin v, v^2\rangle,$$

where $u \in [0,2]$, $v \in [0,2\pi]$.

11. The part cut from the surface with equation

$$z = x^2 - y^2 + 2, \qquad -5 \le x, y \le 5,$$

by a vertical cylinder of radius 0.1 centered on the point $(1,0)$.

12. The part cut from the surface with equation

$$z = \sqrt{a^2 - x^2 - y^2},$$

where $0 \le x^2 + y^2 \le a^2$, by a vertical cylinder with radius $\frac{1}{2}a$ centered on the point $\left(0, \frac{1}{2}a\right)$.

Exercises 13–16: Find the area of the surface of revolution generated by revolving the curve about the x-axis.

13. The cycloid described by

$$\mathbf{r} = \mathbf{r}(t) = a\langle t - \sin t, 1 - \cos t, 0\rangle, \qquad t \in [0, 2\pi].$$

14. The curve described by

$$\mathbf{r} = \mathbf{r}(t) = e^t\langle \cos t, \sin t, 0\rangle, \qquad t \in [0, \pi/2].$$

15. The curve described by the equation $y = e^x$, $x \in [0, 1]$. *Hint:* In the integral, make the substitution $w = e^t$ and use formula (4) from the Table of Integrals.

16. The curve described by the equation $y = \sin x$, $x \in [0, \pi]$. *Hint:* In the integral, make the substitution $w = \cos t$ and use formula (4) from the Table of Integrals.

Exercises 17–20: Find the mass and center of mass of the lamina with density δ. Assume that all dimensions are measured in meters and the mass in kilograms.

17. The conical lamina with constant density δ and described by

$$\mathbf{r} = \mathbf{r}(u, v) = \langle au \cos v, au \sin v, bu\rangle,$$
$$(u, v) \in [0, 1] \times [0, 2\pi],$$

where $a, b > 0$.

18. A hemispherical lamina with radius a and density at any point proportional to the square of the distance from that point to the plane of the equator.

19. The ceramic surface of a nose cone is described by

$$\mathbf{r} = \mathbf{r}(u, v) = \langle u \cos v, u \sin v, u^2\rangle,$$
$$(u, v) \in [0, 1.5] \times [0, 2\pi].$$

The density δ of the lamina is constant.

20. The ceramic surface of a nose cone is described by

$$\mathbf{r} = \mathbf{r}(u, v) = \left\langle \sqrt{u} \cos v, \sqrt{u} \sin v, u^{1.5}\right\rangle,$$
$$(u, v) \in [0, 4] \times [0, 2\pi].$$

The density δ of the lamina at $\mathbf{r}(u, v)$ is proportional to the distance from $\mathbf{r}(u, v)$ to the plane with equation $z = 8$. *Note:* Use numerical integration as necessary.

21. If in Example 2 we restrict the set on which $\mathbf{r} = \mathbf{r}(\theta, \phi)$ is defined to, say, $\mathcal{D} = [0, \pi/2] \times [\epsilon, \pi/2]$, where ϵ is a small positive number, show that $\mathbf{r}$ is one-to-one. The surface S_ϵ described by $\mathbf{r}$ is identical to S except for a small piece near the north pole. One way of sidestepping the problem mentioned in this example, that $\mathbf{r}$ is not one-to-one on $\mathcal{D}$, is to agree that the area $A(S)$ of S is the limit of the area $A(S_\epsilon)$ of S_ϵ as $\epsilon \to 0$. Show in fact, that $\lim_{\epsilon\to 0} A_\epsilon = \frac{1}{2}\pi a^2$.

22. One of Pappus' theorems If a plane curve is revolved about a line that lies in its plane but does not intersect the curve, then the area of the surface generated is equal to the product of the length of the curve and the distance traveled by its centroid. Fill in the details in the following proof of a special case of this result. Referring to Fig. 11.27, let C be the plane curve and the x-axis the line about which it is revolved. Equation (14) gives the area A of the surface generated by C. Calculate the centroid of C and compare it with S. Only one coordinate of the centroid is needed.

23. Continuation Use Pappus' Theorem to calculate the area of the torus described in Example 4.

24. The half-doughnut lamina shown in the accompanying figure is described by

$$\mathbf{r} = \langle 2 \cos \theta + \cos \psi \cos \theta, 2 \sin \theta + \cos \psi \sin \theta, \sin \psi\rangle,$$

where $0 \le \theta \le \pi$ and $0 \le \psi \le 2\pi$. Find the centroid of this lamina.

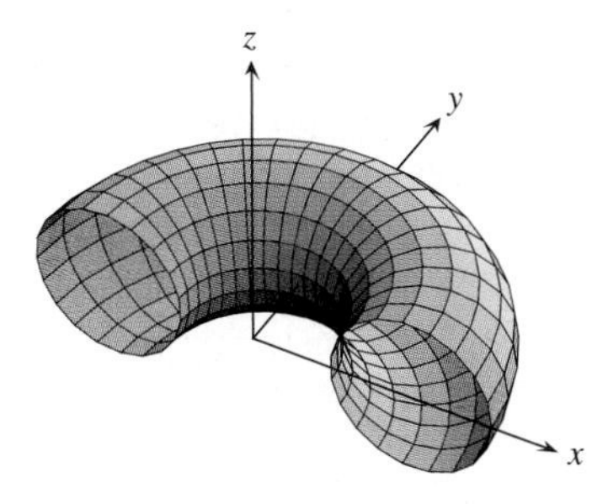

Figure for Exercise 24.

25. Formula (14) for the area of a surface of revolution assumes that the generating curve C in the (x, y)-plane is described by a vector/parametric function. Give a formula for curves described by equations of the form $y = f(x)$, $a \le x \le b$.

26. Show that formula (13) for the area of a surface S described by an equation of the form $z = f(x, y)$, $(x, y) \in \mathcal{D}$, can be written as

$$A = \iint_{\mathcal{D}} \sec \gamma \, dx\, dy,$$

where γ is the angle between $\mathbf{k}$ and the normal $\mathbf{n} = \langle -f_x, -f_x, 1\rangle$ to the surface at the point $(x, y, f(x, y))$. Use this result to express the area of a nonvertical planar surface in terms of the area of its projection onto the (x, y)-plane.

27. What percentage of the Earth's surface would you see from a satellite with altitude 100 kilometers? Assume the Earth is a perfect sphere of radius 6,378 kilometers.

28. The parallels of latitude on the Earth's surface are the intersections of this surface with planes perpendicular to the line joining the north and south poles. Assuming that the Earth's surface is spherical, show that the area of the surface lying between any two parallels of latitude depends upon the distance between the planes of these two circles, not upon their latitudes.

29. Related question A better model of the Earth's surface is an ellipsoid of revolution, with the polar radius shorter than the equatorial radius. Show that the property discussed in the preceding exercise no longer holds true.

30. Related question Calculate the area of the Earth's surface assuming that it is an ellipsoid of revolution with equatorial radius $a = 6{,}378{,}160$ m and polar radius $b = 6{,}356{,}774$ m.

31. A building presently occupied by the Soil & Water Conservation Society, located just north of Des Moines, Iowa, has a roof that resembles a hyperbolic paraboloid. The roof closely conforms to the graph of the function

$$f(x,y) = \frac{x^2}{1725/4} - \frac{y^2}{31{,}050/7} + \frac{566}{69}, \qquad (x,y) \in \mathcal{D},$$

where $\mathcal{D}$ is bounded by the quadrilateral with vertices $(50,0)$, $(0,30)$, $(-65,0)$, and $(0,-30)$. Units are in feet. The roof is 2/3 feet thick and was cast on the site. Find the approximate volume of cement/gravel aggregate required to cast the roof.

32. Using the outline in Example 1, work through the calculation and simplification of $\mathbf{r}_u \times \mathbf{r}_v$ and $\|\mathbf{r}_u \times \mathbf{r}_v\|$. Show that the function $\mathbf{r}$ in Example 1 is one-to-one on $\mathcal{D}$.

33. Show that the function $\mathbf{r}$ in (8) is one-to-one on $\mathcal{D}$. Are $\mathbf{r}_x$ and $\mathbf{r}_y$ continuous on $\mathcal{D}$? The integral in (9) is not well defined as it stands. However, it is a convergent improper double integral. Putting these remarks aside, work through the details leading to (9).

34. Show that the vector function

$$\mathbf{r} = \mathbf{r}(x,y) = \left\langle x, y, \tfrac{1}{6}(9 - 3x^2 - 2y^2)\right\rangle,$$

where $(x,y) \in R^2$, is continuously differentiable and one-to-one. Recall that a function $\mathbf{r}$ is one-to-one if $(x_1,y_1) = (x_2,y_2)$ whenever $\mathbf{r}(x_1,y_1) = \mathbf{r}(x_2,y_2)$.

35. Use (14) in calculating the area of a sphere of radius a. A generating curve is described by

$$\mathbf{r}(\theta) = a\langle \cos\theta, \sin\theta, 0\rangle,$$

where $0 \le \theta \le \pi$.

36. A conical segment with height h and base radius r is generated by revolving the line with equation $y = (r/h)x$, $0 \le x \le h$, about the x-axis. Show that the area of the cone is $\pi r \ell$, where $\ell = \sqrt{r^2 + h^2}$ is the "slant height" of the cone.

37. Related problem Show that the area of the surface generated by revolving the triangle ABC about side AB is $\pi(a+b)p$, where p is the distance from C to side AB and a and b are the lengths of the sides opposite the vertices A and B.

11.4 Change-of-Variables Formula for Double Integrals

For the change of variables $x = g(u)$, where $a \le u \le b$, the one-dimensional Change-of-Variables Theorem is

$$\int_{g(a)}^{g(b)} f(x)\,dx = \int_a^b f(g(u))g'(u)\,du. \tag{1}$$

This formula was given in Section 5.5, where we discussed "integration by substitution." We referred to $x = g(u)$ as a substitution equation. We give a rough proof of (1) as a guide to a proof of the change-of-variables formula for double integrals.

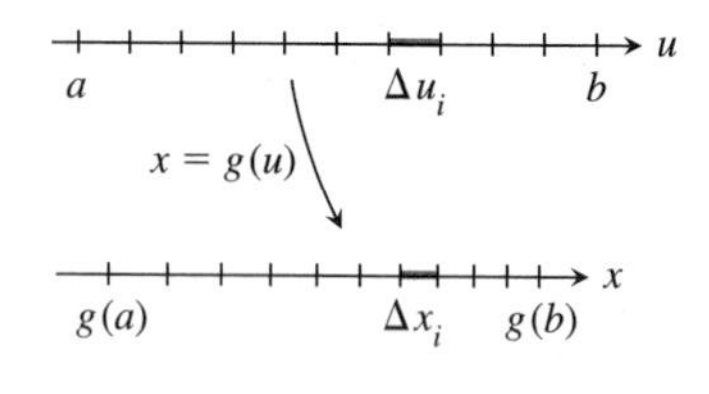

$$\begin{aligned}\Delta x_i = x_i - x_{i-1} &= g(u_i) - g(u_{i-1})\\ &\approx g'(u_{i-1})(u_i - u_{i-1})\\ &\approx g'(u_{i-1})\Delta u_i\end{aligned}$$

FIGURE 11.31 A regular subdivision of $[a,b]$ on the u-axis and the corresponding subdivision of $[g(a), g(b)]$ on the x-axis.

Referring to Fig. 11.31, let $u_0, u_1, \ldots, u_{n-1}, u_n$ be a subdivision of $[a,b]$ and $x_0, x_1, \ldots, x_{n-1}, x_n$, where $x_i = g(u_i)$, the corresponding subdivision of $[g(a), g(b)]$. Assuming that g' is continuous on $[a,b]$,

$$\int_{g(a)}^{g(b)} f(x)\,dx = \sum_{i=1}^{n} \int_{x_{i-1}}^{x_i} f(x)\,dx \approx \sum_{i=1}^{n} f(x_{i-1})\,\Delta x_i.$$

The difference between the second and third terms in this approximation may be made as small as we like by taking n sufficiently large. Because $x_{i-1} = g(u_{i-1})$, $\Delta x_i = g(u_i) - g(u_{i-1})$, and, by the Tangent Line Approximation, $g(u_i) - g(u_{i-1}) \approx g'(u_{i-1})\,\Delta u_i$,

$$\int_{g(a)}^{g(b)} f(x)\,dx \approx \sum_{i=1}^{n} f(g(u_{i-1}))g'(u_{i-1})\,\Delta u_i.$$

Noting that the last term is a Riemann sum for $\int_a^b f(g(u))g'(u)\,du$,

$$\int_{g(a)}^{g(b)} f(x)\,dx \approx \sum_{i=1}^{n} f(g(u_{i-1}))g'(u_{i-1})\,\Delta u_i \approx \int_a^b f(g(u))g'(u)\,du.$$

The change of variable equation (1) follows from this approximation.

A critical step in this proof was

$$\Delta x_i = g(u_i) - g(u_{i-1}) \approx dg_{u_{i-1}}(\Delta u_i) = g'(u_{i-1})\,\Delta u_i. \tag{2}$$

We may read this as: the length Δx_i of the interval $[g(u_{i-1}), g(u_i)]$ is approximately equal to the product of the length Δu_i of the interval $[u_{i-1}, u_i]$ and the determinant $g'(u_{i-1})$ of the 1×1 matrix $(g'(u_{i-1}))$ associated with the linear function $dg_{u_{i-1}}$. This is closely related to the one-dimensional analog of the two-dimensional Linear Functions and Area Theorem: a linear function L from R^2 into R^2 maps rectangular regions to parallelograms, with the area of the parallelogram equal to the product of the absolute value of the determinant of the matrix associated with L and the area of the rectangular region. This theorem was discussed in Section 8.5.

Now suppose that in attempting to evaluate the double integral

$$\iint_{\mathcal{B}} f(x, y)\,dx\,dy$$

the shape of the region $\mathcal{B}$ or the form of the integrand f suggests the change of variables $\langle x, y\rangle = \mathbf{g}(u, v)$, where $\mathbf{g}$ is defined on $\mathcal{D} \subset R^2$ and $\mathbf{g}(\mathcal{D}) = \mathcal{B}$. See Fig. 11.32. We give a rough proof of the change-of-variables formula

$$\iint_{\mathbf{g}(\mathcal{D})} f(x, y)\,dx\,dy = \iint_{\mathcal{D}} f(\mathbf{g}(u, v))\,|\det(\mathbf{Dg}(u, v))|\,du\,dv. \tag{3}$$

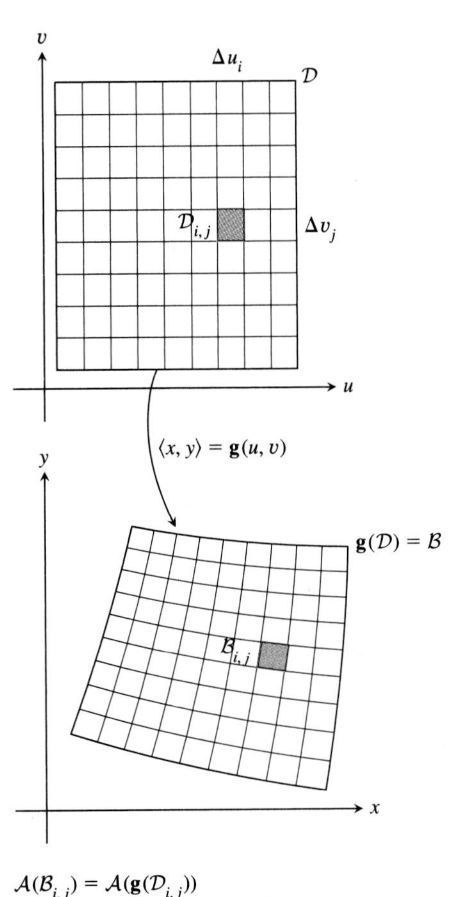

FIGURE 11.32 Regions $\mathcal{D}$ and $\mathbf{g}(\mathcal{D})$.

The factor $|\det(\mathbf{Dg}(u, v))|$, which is the absolute value of the determinant of the derivative matrix of the change of variables $\langle x, y\rangle = \mathbf{g}(u, v)$, is the analog of the factor $g'(u)$ in the one-dimensional formula (1). Our argument for the change-of-variables formula (3) has the same form as our earlier argument for (1).

Let $u_0, u_1, \ldots, u_{n-1}, u_n$, and $v_0, v_1, \ldots, v_{n-1}, v_n$ generate a subdivision of $\mathcal{D}$ and let $\mathcal{D}_{i,j}$ be a representative subregion. Let $\mathcal{B}_{i,j} = \mathbf{g}(\mathcal{D}_{i,j})$ be the corresponding subregion of $\mathbf{g}(\mathcal{D})$ and denote by $\mathcal{A}(\mathcal{B}_{i,j})$ the area of the subregion $\mathcal{B}_{i,j}$. Note the use of the Linear Functions and Area Theorem in moving from line 1 to line 2 of the following argument for (3).

$$\begin{aligned}\iint_{\mathbf{g}(\mathcal{D})} f(x, y)\,dx\,dy &\approx \sum_{i,j} f(x_{i-1}, y_{j-1})\mathcal{A}(\mathcal{B}_{i,j}) = \sum_{i,j} f(\mathbf{g}(u_{i-1}, v_{j-1}))\mathcal{A}(\mathbf{g}(\mathcal{D}_{i,j}))\\ &\approx \sum_{i,j} f(\mathbf{g}(u_{i-1}, v_{j-1}))\,|\det(\mathbf{Dg}(u_{i-1}, v_{j-1}))|\,\Delta u_i\,\Delta v_j\\ &\approx \iint_{\mathcal{D}} f(\mathbf{g}(u,v))\,|\det(\mathbf{Dg}(u, v))|\,du\,dv.\end{aligned}$$

The change-of-variables formula (3) is often stated using a compact notation for the determinant of the derivative matrix of $\mathbf{g}$. Writing $\mathbf{g}$ in terms of its coordinate functions g_1 and g_2, so that

$$\langle x, y\rangle = \mathbf{g}(u, v) = \langle g_1(u, v), g_2(u, v)\rangle,$$

and writing the partial derivatives $\partial g_1/\partial u$, $\partial g_1/\partial v, \ldots$ as $\partial x/\partial u$, $\partial x/\partial v, \ldots$, the determinant of the derivative matrix of $\mathbf{g}$ is often written as

$$\frac{\partial(x, y)}{\partial(u, v)} = \begin{vmatrix} \frac{\partial x}{\partial u} & \frac{\partial x}{\partial v} \\ \frac{\partial y}{\partial u} & \frac{\partial y}{\partial v} \end{vmatrix} = \begin{vmatrix} \frac{\partial g_1}{\partial u} & \frac{\partial g_1}{\partial v} \\ \frac{\partial g_2}{\partial u} & \frac{\partial g_2}{\partial v} \end{vmatrix}. \tag{4}$$

We also write this determinant, often briefly referred to as "the Jacobian" of $\mathbf{g}$, as $\partial(x, y)/\partial(u, v)$.

We can now state the change-of-variables formula.

WEB

THEOREM Change-of-Variables Formula for Double Integrals

Let $\mathbf{g}$ be a function defined on a region $\mathcal{D}$ of R^2. Assume that on $\mathcal{D}$ the coordinate functions $x = g_1(u, v)$ and $y = g_2(u, v)$ of $\mathbf{g}$ are continuously differentiable, $\mathbf{g}$ is one-to-one, and $\partial(x, y)/\partial(u, v) \neq 0$. We refer to $\langle x, y\rangle = \mathbf{g}(u, v)$ as a change-of-variables equation. If f is a continuous function defined on the range $\mathbf{g}(\mathcal{D})$ of $\mathbf{g}$, then

$$\iint_{\mathbf{g}(\mathcal{D})} f(x, y)\,dx\,dy = \iint_{\mathcal{D}} f(\mathbf{g}(u, v)) \left|\frac{\partial(x, y)}{\partial(u, v)}\right| du\,dv. \tag{5}$$

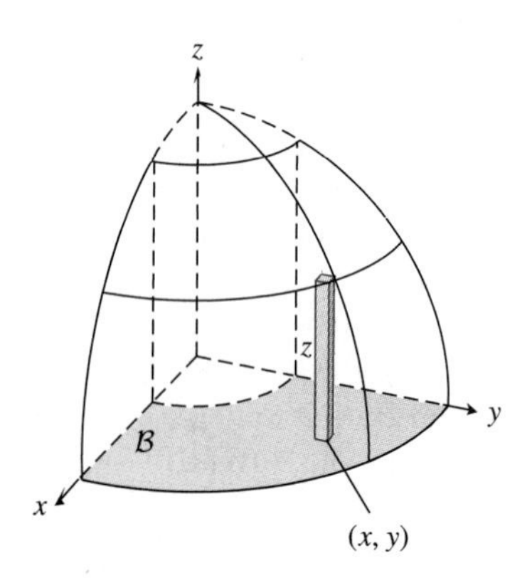

FIGURE 11.33 Sphere with round hole.

EXAMPLE 1 Calculate the mass of a bearing made from a solid iron ball of radius 10 cm by drilling along a diameter a hole of radius 4 cm. One-eighth of this solid is shown in Fig. 11.33. The density of iron is $\delta = 0.00786$ kg/cm^3.

Solution

We calculated the mass of a bearing with a square hole in Example 4 in Section 11.2. We calculate m, the mass in the first octant portion of the bearing. Letting $\mathcal{B}$ denote the integration region shown in Fig. 11.33,

$$m = \iint_{\mathcal{B}} dm = \iint_{\mathcal{B}} \delta z\,dx\,dy$$
$$= \delta \iint_{\mathcal{B}} z\,dx\,dy.$$

Because the height z of the spherical surface above (x, y) is $z = \sqrt{100 - x^2 - y^2}$,

$$m = \delta \iint_{\mathcal{B}} \sqrt{100 - x^2 - y^2}\,dx\,dy. \tag{6}$$

Both the shape of the region and the form of the integrand suggest that we change to polar coordinates. The substitution equation for polar coordinates is

$$\langle x, y\rangle = \mathbf{g}(r, \theta) = \langle r\cos\theta, r\sin\theta\rangle, \qquad (r, \theta) \in \mathcal{D} = [4, 10] \times [0, \pi/2]. \tag{7}$$

The region $\mathcal{D}$ was chosen so that $\mathcal{B} = \mathbf{g}(\mathcal{D})$. Referring to Fig. 11.34, the set $\mathcal{B} = \mathbf{g}(\mathcal{D})$ is the "old" region of integration; the set $\mathcal{D}$ is the "new" region of integration. The Jacobian $\partial(x, y)/\partial(r, \theta)$ is

$$\frac{\partial(x,y)}{\partial(r,\theta)} = \begin{vmatrix} \dfrac{\partial x}{\partial r} & \dfrac{\partial x}{\partial \theta} \\ \dfrac{\partial y}{\partial r} & \dfrac{\partial y}{\partial \theta} \end{vmatrix} = \begin{vmatrix} \cos\theta & -r\sin\theta \\ \sin\theta & r\cos\theta \end{vmatrix} = r\cos^2\theta + r\sin^2\theta = r.$$

Applying (5) to (6),

$$\begin{aligned} m &= \delta \iint_{\mathcal{B}} \sqrt{100 - x^2 - y^2}\, dx\, dy = \delta \iint_{\mathbf{g}(\mathcal{D})} \sqrt{100 - x^2 - y^2}\, dx\, dy \\ &= \delta \iint_{\mathcal{D}} \sqrt{100 - r^2}\, |r|\, dr\, d\theta. \end{aligned} \tag{8}$$

Noting the $r > 0$ in $\mathcal{D}$, it follows from the Iterated Integral Theorem that

$$m = \delta \int_0^{\pi/2} \int_4^{10} \sqrt{100 - r^2}\, r\, dr\, d\theta = (\pi/2)\delta \int_4^{10} \sqrt{100 - r^2}\, r\, dr.$$

Evaluating the last integral, the mass of the bearing is

$$8m = \frac{4}{3}\pi\delta 84^{3/2} \approx 25.35 \text{ kg}.$$

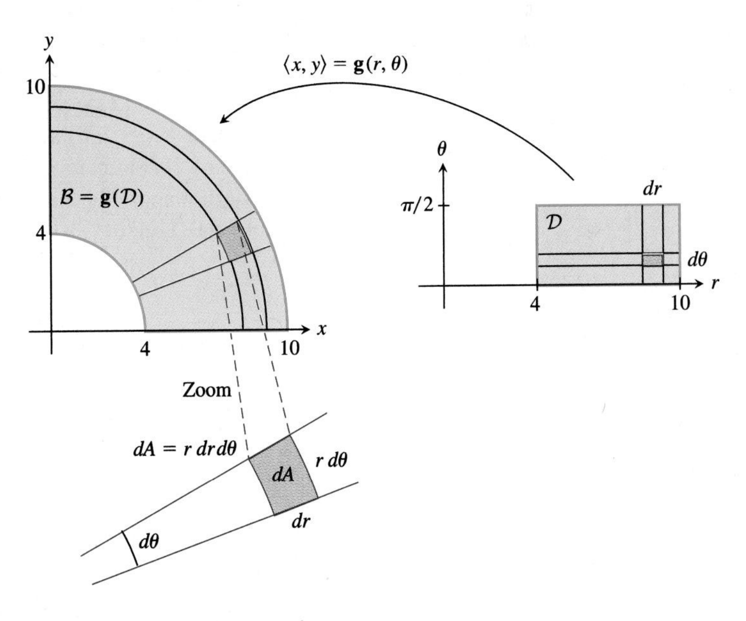

FIGURE 11.34 The polar area element.

When the change of variables is a familiar one, such as the change from rectangular to polar coordinates, the somewhat formal application of the Change-of-Variables Formula used in Example 1 is often replaced by a direct calculation based on elements of area. In the context of Example 1, we consider the mass element $dm = \delta z\,dA$. In rectangular coordinates $z = \sqrt{1 - x^2 - y^2}$ and we must sum dm over the quarter-annulus $\mathcal{B} = \mathbf{g}(\mathcal{D})$ shown in Fig. 11.33 or 11.34. If we choose rectangular coordinates, then $dA = dx\,dy$ and because $\mathbf{g}(\mathcal{D})$ is neither a Type I nor a Type II region, we anticipate that the evaluation by the Iterated Integral Theorem of the double integral $\iint_{\mathcal{B}} z\,dx\,dy$ will be awkward. In polar coordinates, $z = \sqrt{1 - r^2}$, the area element dA takes the shape shown in Fig. 11.34, and the region $\mathcal{D}$ of integration is simpler.

We focus on the quarter-annulus $\mathcal{B}$. We may infer directly from the zoom-view portion of the figure that $dA = r\,dr\,d\theta$. The subregion is very nearly a rectangle with sides dr and $r\,d\theta$, the latter because one side is the arc of a circle with radius r and central angle $d\theta$. The mass element becomes $dm = \delta z\,dA = \delta\sqrt{100 - r^2}\,r\,dr\,d\theta$. We may sum these elements first in the radial direction. For each $\theta \in [0, \pi/2]$, r varies from 4 to 10. In this way we are led directly to write

$$\frac{1}{8}m = \delta\int_0^{\pi/2}\int_4^{10}\sqrt{100 - r^2}\,r\,dr\,d\theta.$$

EXAMPLE 2 Evaluate the integral $\iint_{\mathcal{B}} x^2y^2\,dx\,dy$, where $\mathcal{B}$ is the region bounded by the ellipse with equation $x^2/a^2 + y^2/b^2 = 1$.

Solution

The change of variables

$$(x, y) = \mathbf{g}(X, Y) = \langle aX, bY\rangle, \qquad (X, Y) \in \mathcal{D},$$

where $a, b > 0$ and $\mathcal{D}$ is the unit disk in the (X, Y)-plane, is called a *change of scale.* Referring to Fig. 11.35, the change of variables $\mathbf{g}$ transforms the unit disk into the elliptical region $\mathcal{B}$.

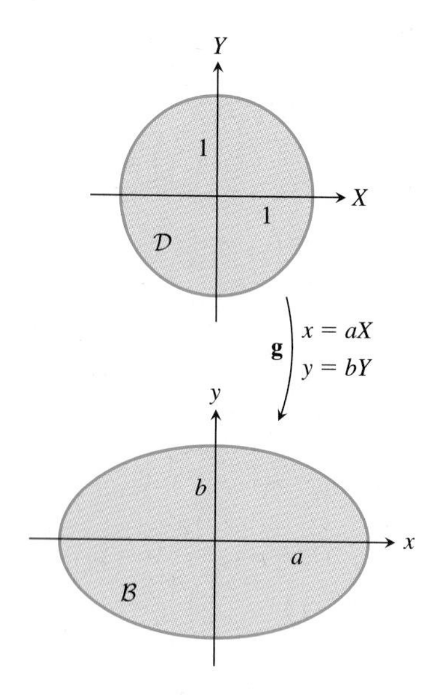

FIGURE 11.35 Transforming a circle into an ellipse.

The change-of-variable equations take a point (X, Y) on the unit circle in the (X, Y)-plane to a point on the ellipse in the (x, y)-plane. For if $X^2 + Y^2 = 1$, then because $x = aX$ and $y = bY$, it follows that $x^2/a^2 + y^2/b^2 = 1$. The interior of $\mathcal{D}$ is transformed to the interior of $\mathcal{B}$.

Before applying the change-of-variables formula (5), we calculate the Jacobian $\partial(x, y)/\partial(X, Y)$:

$$\frac{\partial(x, y)}{\partial(X, Y)} = \begin{vmatrix} \dfrac{\partial x}{\partial X} & \dfrac{\partial x}{\partial Y} \\ \dfrac{\partial y}{\partial X} & \dfrac{\partial y}{\partial Y} \end{vmatrix} = \begin{vmatrix} a & 0 \\ 0 & b \end{vmatrix} = ab.$$

Using this result in the change-of-variables formula gives

$$\begin{aligned}\iint_{\mathcal{B}} x^2y^2\,dx\,dy &= \iint_{\mathbf{g}(\mathcal{D})} x^2y^2\,dx\,dy \\ &= \iint_{\mathcal{D}} (aX)^2(bY)^2\left|\frac{\partial(x, y)}{\partial(X, Y)}\right| dX\,dY \\ &= a^3b^3\iint_{\mathcal{D}} X^2Y^2\,dX\,dY.\end{aligned}$$

Because $\mathcal{D}$ is the unit disk we consider changing variables again; we change to polar coordinates. Replacing X by $r \cos\theta$, Y by $r \sin\theta$, and the area element $dX\,dY$ by $r\,dr\,d\theta$,

$$\iint_{\mathcal{B}} x^2y^2\,dx\,dy = a^3b^3\int_0^{2\pi}\int_0^1 (r^2\cos^2\theta)(r^2\sin^2\theta)r\,dr\,d\theta$$

$$= \frac{a^3b^3}{6}\int_0^{2\pi} \sin^2\theta\cos^2\theta\,d\theta = \frac{1}{24}\pi a^3b^3.$$

The last integral was evaluated by writing

$$\sin^2\theta\cos^2\theta = \frac{1}{4}(2\sin\theta\cos\theta)^2 = \frac{1}{4}\sin^2 2\theta = \frac{1}{8}(1 - \cos 4\theta)$$

or by applying formula (32) from the Table of Integrals.

For some double integrals, a useful change of variables may be suggested by the form of the integrand; for others, it may be the shape of the region of integration. In the first of the remaining three examples, it is the form of the integrand; in the second and third, it is both the integrand and the region.

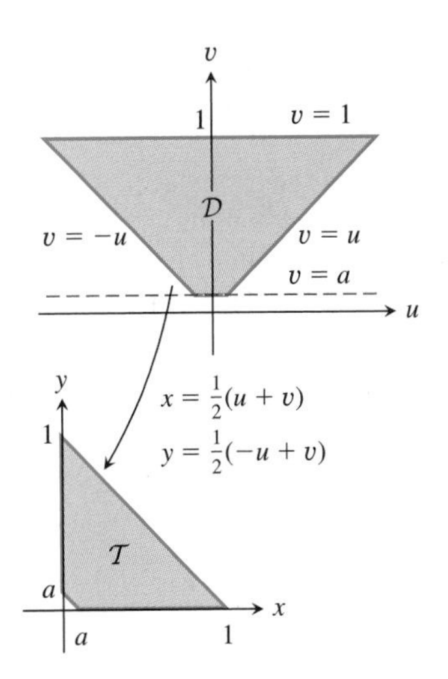

FIGURE 11.36 A change of variables suggested by the integrand.

EXAMPLE 3 Evaluate the integral

$$\iint_{\mathcal{T}} e^{(x-y)/(x+y)}\,dx\,dy,$$

where $\mathcal{T}$ is the trapezoidal region shown in Fig. 11.36. The small number $a \in (0, 1)$ was chosen to exclude the origin from the region of integration because at the origin the integrand is not defined. See Exercise 28.

Solution

It is the form of the integrand more than the geometry of $\mathcal{T}$ that suggests the change of variables

$$x - y = u \tag{9}$$

$$x + y = v. \tag{10}$$

We calculate the Jacobian $\partial(x,y)/\partial(u,v)$ and determine the set $\mathcal{D}$ for which $\mathbf{g}(\mathcal{D}) = \mathcal{T}$. To calculate the Jacobian, it is most convenient to express the change of variables in the form $\langle x, y\rangle = \mathbf{g}\langle u, v\rangle$, that is, to solve (9) and (10) for x and y in terms of u and v. Adding and subtracting these equations gives

$$x = \frac{1}{2}u + \frac{1}{2}v \tag{11}$$

$$y = -\frac{1}{2}u + \frac{1}{2}v. \tag{12}$$

It follows directly that $\partial(x,y)/\partial(u,v) = \frac{1}{2}$. To determine the set $\mathcal{D}$ for which $\mathcal{T} = \mathbf{g}(\mathcal{D})$, we look at the boundary of $\mathcal{T}$, one side at a time. Because $\mathbf{g}$ is a linear function, we know that it transforms line segments to line segments.

1. On the bottom segment of $\mathcal{T}$, where $y = 0$, we note from (12) that $u = v$. This means that the change of variables **g** transforms the line with equation $v = u$ to the line with equation $y = 0$.
2. For the segment of $\mathcal{T}$ on which $x = 0$, it follows from (11) that $v = -u$.
3. For the segment of $\mathcal{T}$ on which $x + y = 1$, it follows from (10) that $v = 1$.
4. For the segment of $\mathcal{T}$ on which $x + y = a$, it follows from (10) that $v = a$.

Referring to Fig. 11.36, we see that **g** transforms the trapezoidal region $\mathcal{D}$ to the trapezoidal region $\mathcal{T}$.

Applying the change-of-variables formula (5),

$$\iint_{\mathcal{T}} e^{(x-y)/(x+y)}\,dx\,dy = \iint_{\mathcal{D}} e^{u/v}\left|\frac{\partial(x,y)}{\partial(u,v)}\right| du\,dv.$$

Recalling that $\partial(x,y)/\partial(u,v) = \frac{1}{2}$ and applying the Iterated Integral Theorem,

$$\begin{aligned}\iint_{\mathcal{T}} e^{(x-y)/(x+y)}\,dx\,dy &= \frac{1}{2}\int_a^1\int_{-v}^{v} e^{u/v}\,du\,dv \\ &= \frac{1}{2}\int_a^1 ve^{u/v}\Big|_{-v}^{v}\,dv = \frac{1}{2}\int_a^1 (ve - ve^{-1})\,dv \\ &= \frac{1}{2}(e - e^{-1})\frac{1}{2}v^2\Big|_a^1 = \frac{1}{4}(e - e^{-1})(1 - a^2).\end{aligned}$$

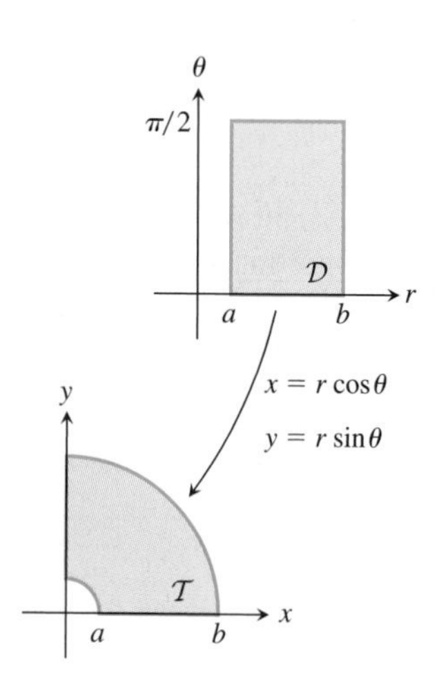

FIGURE 11.37 Transforming a rectangle into a quarter-annulus.

EXAMPLE 4 Evaluate the integral

$$\iint_{\mathcal{T}} e^{-x^2-y^2}\,dx\,dy,$$

where $\mathcal{T}$ is the region shown in Fig. 11.37.

Solution

Both the geometry of the region $\mathcal{T}$ and the expression $x^2 + y^2$ in the integrand strongly suggest that we change to polar coordinates. The origin was excluded from $\mathcal{T}$ because $\partial(x,y)/\partial(r,\theta) = r$ is zero at $r = 0$, which would violate the hypothesis in the Change-of-Variables Formula that $\partial(x,y)/\partial(r,\theta) \neq 0$. From (5),

$$\begin{aligned}\iint_{\mathcal{T}} e^{-x^2-y^2}\,dx\,dy &= \iint_{\mathcal{D}} e^{-r^2} r\,dr\,d\theta \\ &= \int_0^{\pi/2}\int_a^b e^{-r^2} r\,dr\,d\theta.\end{aligned}$$

Because the inner integral is independent of θ,

$$\iint_{\mathcal{T}} e^{-x^2-y^2}\,dx\,dy = \frac{1}{2}\pi\int_a^b e^{-r^2} r\,dr.$$

Making the substitution $w = r^2$,

$$\iint_{\mathcal{T}} e^{-x^2-y^2}\,dx\,dy = \frac{1}{4}\pi\int_{a^2}^{b^2} e^{-w}\,dw = \frac{1}{4}\pi(e^{-a^2} - e^{-b^2}). \tag{13}$$

Formula (40) from the Table of Integrals Formula (40) in the Table of Integrals is

$$\int_0^{\infty} e^{-x^2}\,dx = \frac{\sqrt{\pi}}{2}.$$

We derive this result, which is important in the study of the normal probability distribution, using the Change-of-Variables Formula.

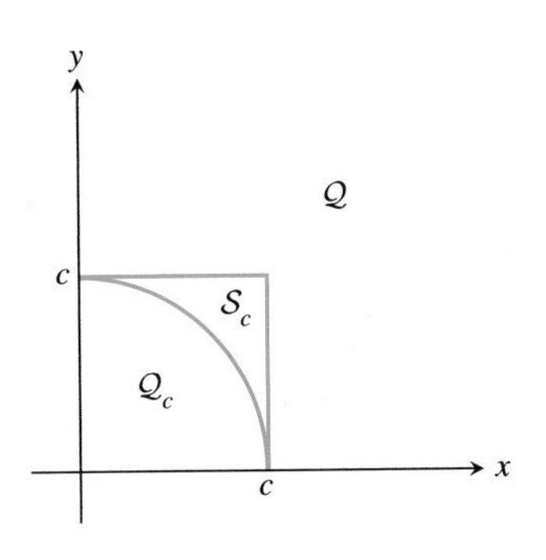

FIGURE 11.38 The sets $\mathcal{Q}$, $\mathcal{Q}_c$, and $\mathcal{S}_c$.

Let $\mathcal{Q}$ denote the first quadrant, that is, $\mathcal{Q} = \{(x,y) \in R^2 : x, y \geq 0\}$; for each $c > 0$, let $\mathcal{Q}_c$ denote the part of $\mathcal{Q}$ within c units of the origin; let $\mathcal{S}_c$ denote the square region with opposite vertices $(0,0)$ and (c,c); and let f be a continuous, bounded function on $\mathcal{Q}$. See Fig. 11.38.

Because our argument for Formula (40) has several steps, we outline the main ideas here. We argue that $\lim_{b\to\infty}\int_0^b e^{-x^2}\,dx = \sqrt{\pi}/2$ by showing that

$$\left(\frac{\sqrt{\pi}}{2}\right)^2 = \lim_{b\to\infty}\left(\int_0^b e^{-y^2}\,dy\right)\left(\int_0^b e^{-y^2}\,dy\right) = \lim_{b\to\infty}\int_0^b\int_0^b e^{-x^2-y^2}\,dx\,dy$$

$$= \lim_{b\to\infty}\iint_{\mathcal{Q}_b} e^{-x^2-y^2}\,dx\,dy.$$

Note that this result relates the value of the single integral we wish to evaluate to the double integral $\iint_{\mathcal{Q}_b} e^{-x^2-y^2}\,dx\,dy$. We evaluate this double integral by changing from rectangular to polar coordinates.

We say that the improper double integral $\iint_{\mathcal{Q}} f(x,y)\,dA$ **converges** if there is a number L for which

$$\lim_{c\to\infty}\iint_{\mathcal{Q}_c} f(x,y)\,dA = L.$$

In this case, we say that L is the value of the improper integral.

Suppose that $0 < a < b$ and define f on $\mathcal{Q}$ by $f(x,y) = e^{-x^2-y^2}$. Because $0 \leq e^{-x^2-y^2} \leq e^0 = 1$ for all $(x,y) \in R^2$, f is bounded on $\mathcal{Q}$. Moreover, it follows from Property 3 on page 961 that

$$0 \leq \iint_{\mathcal{Q}_a} e^{-x^2-y^2}\,dx\,dy \leq \iint_{\mathcal{Q}_a} 1\,dx\,dy = \frac{1}{4}\pi a^2.$$

Taking the limit as $a \to 0^+$ in this inequality,

$$\lim_{a\to 0^+}\iint_{\mathcal{Q}_a} e^{-x^2-y^2}\,dx\,dy = 0. \tag{14}$$

Letting $\mathcal{T}$ be the set of points in $\mathcal{Q}_b$ but not in $\mathcal{Q}_a$—as in Fig. 11.37—it follows from Property 1, page 961, that

$$\iint_{\mathcal{Q}_b} e^{-x^2-y^2}\,dx\,dy = \iint_{\mathcal{Q}_a} e^{-x^2-y^2}\,dx\,dy + \iint_{\mathcal{T}} e^{-x^2-y^2}\,dx\,dy.$$

From (13) in Example 4,

$$\iint_{Q_b} e^{-x^2-y^2}\,dx\,dy = \iint_{Q_a} e^{-x^2-y^2}\,dx\,dy + \frac{1}{4}\pi(e^{-a^2} - e^{-b^2}).$$

With (14) in mind and taking the limit as $a \to 0^+$ of both sides of this equation,

$$\iint_{Q_b} e^{-x^2-y^2}\,dx\,dy = 0 + \frac{1}{4}\pi(1 - e^{-b^2}) = \frac{1}{4}\pi(1 - e^{-b^2}).$$

Because the limit as $b \to \infty$ of this expression is a number, it follows that $\iint_Q e^{-x^2-y^2}\,dx\,dy$ converges and

$$\iint_{Q} e^{-x^2-y^2}\,dx\,dy = \lim_{b\to\infty} \iint_{Q_b} e^{-x^2-y^2}\,dx\,dy = \lim_{b\to\infty} \frac{1}{4}\pi(1 - e^{-b^2}) = \frac{1}{4}\pi. \tag{15}$$

Referring to Fig. 11.38, because $e^{-x^2-y^2} \geq 0$,

$$\iint_{Q_b} e^{-x^2-y^2}\,dx\,dy \leq \iint_{S_b} e^{-x^2-y^2}\,dx\,dy \leq \iint_{Q} e^{-x^2-y^2}\,dx\,dy = \frac{1}{4}\pi.$$

Because the limit as $b \to \infty$ of the left-most integral in this inequality is equal to the right-most integral, it follows from the Squeeze Theorem that

$$\lim_{b\to\infty} \iint_{S_b} e^{-x^2-y^2}\,dx\,dy = \iint_{Q} e^{-x^2-y^2}\,dx\,dy = \frac{1}{4}\pi. \tag{16}$$

From (16) and the Iterated Integral Theorem,

$$\begin{aligned}
\frac{1}{4}\pi = \iint_{Q} e^{-x^2-y^2}\,dx\,dy &= \lim_{b\to\infty} \iint_{S_b} e^{-x^2-y^2}\,dx\,dy \\
&= \lim_{b\to\infty} \int_0^b \int_0^b e^{-x^2} e^{-y^2}\,dx\,dy \\
&= \lim_{b\to\infty} \left(\int_0^b e^{-x^2}\,dx\right)\left(\int_0^b e^{-y^2}\,dy\right).
\end{aligned}$$

These improper integrals converge to the same number. Hence

$$\frac{1}{4}\pi = \left(\int_0^\infty e^{-x^2}\,dx\right)^2.$$

It follows that

$$\int_0^\infty e^{-x^2}\,dx = \frac{\sqrt{\pi}}{2}. \tag{17}$$

This is formula (40) from the Table of Integrals.

Exercises 11.4

Exercises 1–6: Calculate the Jacobian $\partial(x,y)/\partial(u,v)$ of the change of variables.

1. $\langle x,y\rangle = \mathbf{g}(u,v) = \langle u^2 - v^2, uv\rangle$

2. $\langle x,y\rangle = \mathbf{g}(u,v) = \langle au + bv, cu + dv\rangle$

3. $\langle x,y\rangle = \mathbf{g}(u,v) = \langle u^2 \cos v, u^2 \sin v\rangle$

4. $\langle x,y\rangle = \mathbf{g}(u,v) = \langle e^u \cos v, e^u \sin v\rangle$

5. $\langle x,y\rangle = \mathbf{g}(u,v) = \langle (u/v)^{2/3}, (u^4/v)^{1/3}\rangle$

6. $\langle x,y\rangle = \mathbf{g}(u,v) = \langle (v/u)^{2/5}, (v^6/u)^{1/5}\rangle$

Exercises 7–12: Sketch $\mathcal{D}$ and $\mathbf{g}(\mathcal{D})$ from the description of $\mathcal{D}$ and change of variables $\langle x,y\rangle = \mathbf{g}(u,v)$.

7. $\mathcal{D} = [0,1] \times [0,1]$ and $\langle x,y\rangle = \langle 2u, 3v\rangle$

8. $\mathcal{D} = [1,2] \times [0,3]$ and $\langle x,y\rangle = \left\langle \frac{1}{2}v, \frac{3}{4}u\right\rangle$

9. $\mathcal{D} = [1,2] \times [1,3]$ and $\langle u,v\rangle = \langle y/\sqrt{x}, y/x^2\rangle$

10. $\mathcal{D} = [1,2] \times [2,3]$ and $\langle u,v\rangle = \langle y/\sqrt{x}, y/x\rangle$

11. $\mathcal{D} = [0,1] \times [0,1]$ and $\langle x,y\rangle = \langle u - v, u + v\rangle$

12. $D = [-1,1] \times [-2,-1]$ and $\langle x,y\rangle = \langle u + v, u\rangle$

Exercises 13–20: Evaluate the integral using the given change of variables. It may be necessary to express x and y in terms of u and v. Sketch both regions of integration.

13. $\iint_{\mathcal{T}} \sqrt{x+y}\,dx\,dy$, where $\mathcal{T}$ is the parallelogram with sides described by

$$x + y = 0, \quad x + y = 1,$$
$$2x - y = 0, \quad 2x - y = 3.$$

Let $u = x + y$ and $v = 2x - y$.

14. $\iint_{\mathcal{T}} (x - 5y)\,dx\,dy$, where $\mathcal{T}$ is the parallelogram with sides described by

$$y - x = 0, \quad y - x = -2,$$
$$y + 2x = 0, \quad y + 2x = 3.$$

Let $u = x - y$ and $v = 2x + y$.

15. $\iint_{\mathcal{T}} x^2y^2\,dx\,dy$, where $\mathcal{T}$ is the region bounded by the curves with equations

$$xy = 2, \quad xy = 4,$$
$$y^2 = x, \quad y^2 = 3x.$$

Let $xy = u$ and $y^2 = vx$.

16. $\iint_{\mathcal{T}} x^2y^2\,dx\,dy$, where $\mathcal{T}$ is the region bounded by the curves with equations

$$xy = 2, \quad xy = 3,$$
$$y = x, \quad y = 5x.$$

Let $xy = u$ and $y = vx$.

17. $\iint_{\mathcal{T}} xy\,dx\,dy$, where $\mathcal{T}$ is the region bounded by the ellipse with equation $5x^2 + 2xy + 5y^2 = 12$. Let $x = u/2 + v/\sqrt{6}$ and $y = -u/2 + v/\sqrt{6}$.

18. $\iint_{\mathcal{T}} (x^2 + y^2)\,dx\,dy$, where $\mathcal{T}$ is the region bounded by the ellipse with equation $5x^2 + 2\sqrt{3}xy + 7y^2 = 4$. Let $x = \sqrt{3}u + v/\sqrt{2}$ and $y = -u + \sqrt{3}v/\sqrt{2}$.

19. $\iint_{\mathcal{T}} \sqrt{x^2 + y^2}\,dx\,dy$, where $\mathcal{T}$ is a sector of the unit circle, with central angle $\pi/4$. Use polar coordinates.

20. $\iint_{\mathcal{T}} y\,dx\,dy$, where $\mathcal{T}$ is the region bounded by the circle with equation $x^2 + (y-1)^2 = 1$. Use polar coordinates.

21. Use a double integral in calculating the surface area of the "spherical cap" cut from the hemisphere with equation $z = \sqrt{1 - x^2 - y^2}$ by the cylinder $x^2 + y^2 = 0.1^2$. See also Exercise 31.

22. Find the volume beneath the graph of $z = xy$ and above the set of all points (x,y) in R^2 satisfying the inequalities $0 \le x \le 2$ and $0 \le y \le \sqrt{2x - x^2}$.

23. Find the volume of the solid bounded on the sides by the cylinder with equation $x^2 + y^2 - 2y = 0$, on the bottom by the (x,y)-plane, and on the top by the cone with equation $z = 2\sqrt{x^2 + y^2}$.

24. Set up and evaluate the integral in Example 2 without a change of variables. Integrate on y first, make the substitution $x = a \sin\theta$ in the resulting integral, and then use formula (32) in the Table of Integrals.

25. Related problem In Example 2, evaluate the integral

$$\int_0^{2\pi} \sin^2\theta \cos^2\theta\,d\theta$$

with formula (32) from the Table of Integrals.

26. Related problem Verify the evaluation of the integral in Example 2 using the given trigonometric identity.

27. Compose the two changes of variables in Example 2 into one change of variable. How does the Jacobian of the composite function relate to the Jacobians of the two changes of variable?

28. If in Example 3 the region of integration $\mathcal{T}$ were replaced by the triangular region $\mathcal{K}$ with vertices $(0, 0)$, $(1, 0)$, and $(0, 1)$, the integral $\iint_{\mathcal{K}} e^{(x-y)/(x+y)}\,dx\,dy$ would be an improper double integral because the integrand is not continuous at $(0, 0)$. Would this improper double integral be classified as convergent or divergent? What is your evidence? If it is convergent, what is its value?

29. At the beginning of the section, in the informal argument for (1), we assumed that the change-of-variable function g was one-to-one and had a continuous derivative on $[a, b]$. These assumptions are sufficient to guarantee that g must be either strictly increasing or strictly decreasing on $[a, b]$. Use this result to show that a regular subdivision $\mathcal{S}_n$ of the interval $[a, b]$ on the u-axis generates a subdivision $\mathcal{G}_n$ of $[c, d]$. Show by example that the subdivision of $[c, d]$ need not be regular. Use another result to show that the lengths of the subintervals of $\mathcal{G}_n$ go to zero as $n \to \infty$.

30. Evaluate the integral

$$\iint_{R^2} e^{-(x^2+xy+y^2)}\,dx\,dy.$$

Use the change of variables

$$x = \frac{1}{\sqrt{3}}u - v$$

$$y = \frac{1}{\sqrt{3}}u + v$$

and apply (15) to the resulting double integral.

31. A side view of an aspirin tablet is shown in the accompanying figure, where $a = 2$ mm, $h = 4$ mm, and $w = 10$ mm. The curves shown in this view are congruent arcs of a circle. A top view of the tablet would be a circle. Use a double integral in calculating the total surface area of the tablet. Calculate the area of the cylindrical sides by a known formula. See Exercise 21.

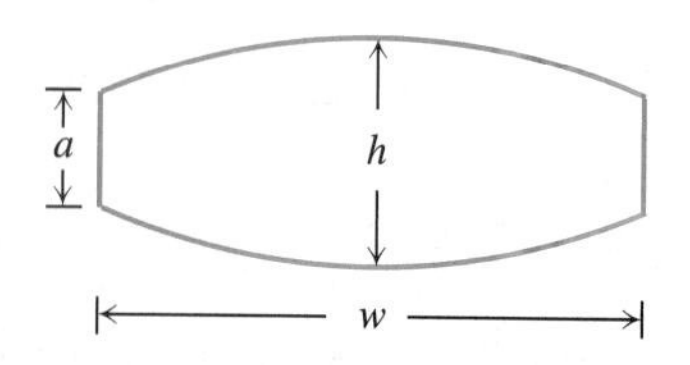

Figure for Exercise 31.

32. Evaluate

$$\iint_{\mathcal{T}} \cos((x - y)/(x + y))\,dx\,dy,$$

where $\mathcal{T}$ is the trapezoidal region with vertices $(a, 0)$, $(b, 0)$, $(0, b)$, and $(0, a)$. Assume that $0 < a < b$.

33. Evaluate $\iint_{\mathcal{T}}(1 + x^2 + y^2)^{-2}\,dx\,dy$, where $\mathcal{T}$ is the triangular region with vertices $(0, 0)$, $(1, 0)$, and $(1, 1)$. It is given that

$$\int_0^{\pi/4} \frac{1}{1 + \sec^2\theta}\,d\theta = \frac{1}{4}\pi - \frac{1}{2}\sqrt{2}\arctan\left(1/\sqrt{2}\right).$$

34. Find the center of mass of the lamina with constant density δ and bounded by the curves with equations $(x - y)^2 + x + y = 0$ and $x + y + 4 = 0$.

35. Evaluate the integral $\int_0^1 \int_{x^2}^{x} (x^2 + y^2)^{-1/2}\,dy\,dx$.

36. Use polar coordinates in calculating the mass of a bearing made in the shape of a solid iron ball of radius 10 cm pierced along a diameter by a square hole of side 8 cm. The density of iron is 7.860×10^{-3} kg/cm^3. See Example 4 in Section 11.2.

37. Find the area of the surface cut from the cone with equation $z = 2\sqrt{x^2 + y^2}$ by the cylinder with equation $x^2 + y^2 - 2y = 0$.

38. Establish Dirichlet's formula

$$\iint_{\mathcal{T}} x^{\alpha}y^{\beta}f(x + y)\,dx\,dy = \frac{\alpha!\,\beta!}{(\alpha + \beta + 1)!}\int_0^1 r^{\alpha+\beta+1}f(r)\,dr,$$

where α and β are nonnegative integers, f is continuous on $[0, 1]$, and $\mathcal{T}$ is the triangular region with vertices $(0, 0)$, $(1, 0)$, and $(0, 1)$. Use the change of variables

$$x = r\cos^2\theta, \qquad y = r\sin^2\theta,$$

where $(r, \theta) \in [0, 1] \times [0, \pi/2]$, and the formula (which follows from formula (32) in the Table of Integrals)

$$2\int_0^{\pi/2} \cos^{2\alpha+1}\theta\,\sin^{2\beta+1}\theta\,d\theta = \frac{\alpha!\,\beta!}{(\alpha + \beta + 1)!}.$$

11.5 Triple Integrals

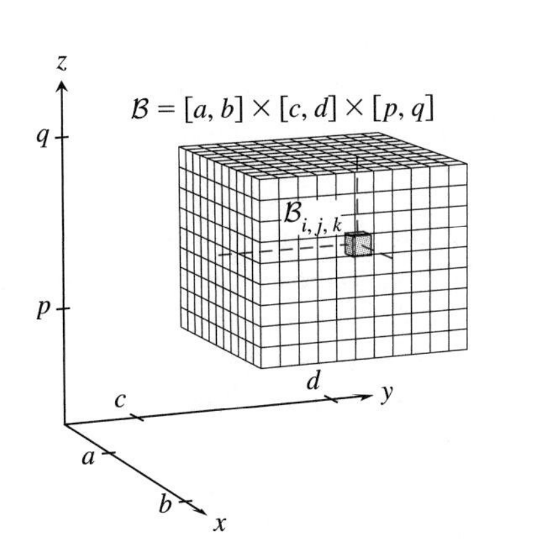

FIGURE 11.39 Cartesian product of three intervals and a representative subregion.

The gravitational and magnetic fields of the Earth are three-dimensional, the air flow around an aircraft is three-dimensional, and the data gathered to model and predict the motion and development of a hurricane are three-dimensional. The modeling and analysis of such three-dimensional systems often require that we sum or "integrate" one or more physical magnitudes that depend upon three space variables. We may have to calculate the net flow of moisture out of a certain region of space, the net force on an asteroid from the combined actions of the planets and sun, or the volume of an underground oil pool. The models and calculations used in studying such phenomena often depend upon the idea of a *triple integral.*

To define and evaluate triple integrals, we extend the ideas and techniques we discussed in Section 11.2 for double integrals. We start with a bounded function f defined on the set

$$\mathcal{B} = [a,b] \times [c,d] \times [p,q] = \{(x,y,z) \in R^3 : a \le x \le b, c \le y \le d, p \le z \le q\}.$$

Figure 11.39 shows the set $\mathcal{B}$, which we shall call a *cuboidal* region. It is the analog of the interval $[a,b]$ for one-dimensional integrals and the rectangular region $[a,b] \times [c,d]$ for double integrals.

DEFINITION The Triple Integral on Cuboidal Regions

Let f be a bounded function defined on $\mathcal{B} = [a,b] \times [c,d] \times [p,q]$. Let n be a positive integer and $\Delta x = (b-a)/n$, $\Delta y = (d-c)/n$, and $\Delta z = (q-p)/n$. The points $x_0 < x_1 < \cdots < x_i < \cdots < x_n$, $y_0 < y_1 < \cdots < y_j < \cdots < y_n$, and $z_0 < z_1 < \cdots < z_k < \cdots < z_n$, where $x_i = a + i\Delta x$, $y_j = c + j\Delta y$, and $z_k = p + k\Delta z$, divide the intervals $[a,b]$, $[c,d]$, and $[p,q]$ into n equal subintervals of lengths Δx, Δy, and Δz, respectively. These subintervals generate a subdivision of $\mathcal{B}$ into n^3 congruent subregions

$$\mathcal{B}_{i,j,k} = [x_{i-1}, x_i] \times [y_{j-1}, y_j] \times [z_{k-1}, z_k], \qquad 1 \le i, j, k \le n,$$

of volume $\Delta V = \Delta x \cdot \Delta y \cdot \Delta z$. A representative subregion $\mathcal{B}_{i,j,k}$ is shown in Fig. 11.39. For each positive integer n, the set $\mathcal{S}_n$ of these subregions is called a **regular subdivision** of $\mathcal{B}$.

Let $\{\mathbf{c}_{i,j,k} : 1 \le i, j, k \le n\}$ be any set of n^3 numbers for which $\mathbf{c}_{i,j,k} \in \mathcal{B}_{i,j,k}$ for all $1 \le i, j, k \le n$. We refer to this set as an *evaluation set* for $\mathcal{S}_n$.

For each regular subdivision $\mathcal{S}_n$ of $\mathcal{B}$ and each evaluation set $\mathcal{C} = \{\mathbf{c}_{i,j,k} : 1 \le i, j, k \le n\}$, we define the **Riemann sum**

$$R_n = \sum_{i,j,k} f(\mathbf{c}_{i,j,k})\,\Delta V = \sum_{i,j,k} f(\mathbf{c}_{i,j,k})\,\Delta x\,\Delta y\,\Delta z. \tag{1}$$

If there is a number I for which all Riemann sums R_n can be made as close to I as we like by taking n sufficiently large, we say that f is **integrable** on $\mathcal{B}$ and that I is the value of the **triple integral** of f on $\mathcal{D}$. We write this limit in either of the two forms

$$\lim_{n\to\infty} R_n = I = \iiint_{\mathcal{B}} f(x,y,z)\,dV = \iiint_{\mathcal{B}} f(x,y,z)\,dx\,dy\,dz. \qquad (2)$$

The function f is called the *integrand.*

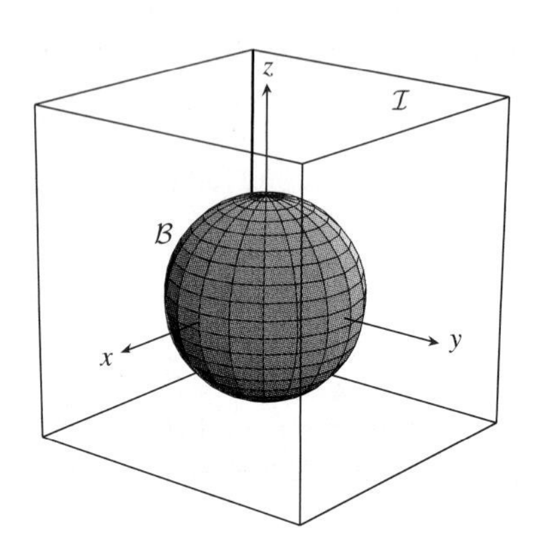

FIGURE 11.40 Extending a function δ defined on $\mathcal{B}$ to a function δ^* defined on $\mathcal{I} \supset \mathcal{B}$.

If $\mathcal{B}$ is a bounded region (this means that $\mathcal{B}$ is contained within some origin-centered spherical region) but not necessarily cuboidal, we define a triple integral on $\mathcal{B}$ in the same way as for double integrals. Figure 11.40 shows a solid region $\mathcal{B}$ whose boundary is a sphere. If $\mathcal{B}$ has variable density and we were asked to use the definition of the triple integral on cuboidal regions in calculating the center of mass of $\mathcal{B}$, we could extend the bounded density function $\delta(x,y,z)$ defined on $\mathcal{B}$ to a density function $\delta^*(x,y,z)$ on a cuboid $\mathcal{I}$ containing $\mathcal{B}$, where

$$\delta^*(x,y,z) = \begin{cases} \delta(x,y,z), & (x,y,z) \text{ in } \mathcal{B} \\ 0, & (x,y,z) \text{ in } \mathcal{I} \text{ but not in } \mathcal{B}. \end{cases}$$

The mass of the ball would be

$$\iiint_{\mathcal{I}} \delta^*(x,y,z)\,dV,$$

provided that the function δ^* is integrable on $\mathcal{I}$.

DEFINITION The Triple Integral on Bounded Regions

Let f be a bounded function defined on a bounded region $\mathcal{B} \subset R^3$. We say that f is integrable on $\mathcal{B}$ if the function f^* is integrable on a cuboidal region $\mathcal{I} = [a,b] \times [c,d] \times [p,q]$ containing $\mathcal{B}$, where

$$f^*(x,y,z) = \begin{cases} f(x,y,z), & (x,y,z) \text{ in } \mathcal{B} \\ 0, & (x,y,z) \text{ in } \mathcal{I} \text{ and not in } \mathcal{B}. \end{cases}$$

In this case we define

$$\iint_{\mathcal{B}} f(x,y,z)\,dV = \iint_{\mathcal{I}} f^*(x,y,z)\,dV.$$

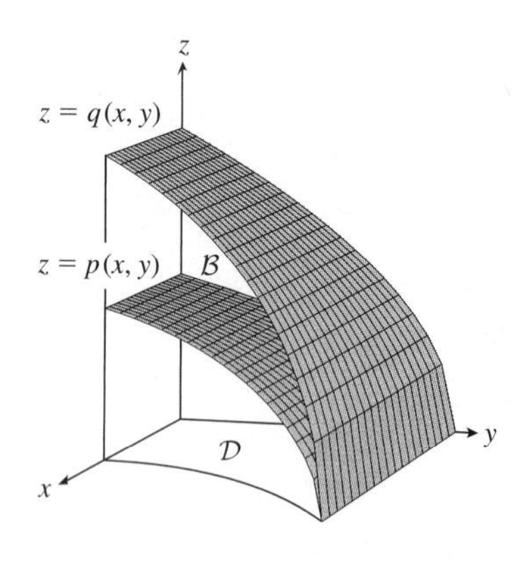

FIGURE 11.41 The region $\mathcal{B}$ lies above the region $\mathcal{D}$ in the (x, y)-plane and between the graphs of p and q.

The main tools for evaluating double integrals are the Iterated Integral Theorem and the Change-of-Variables Theorem. The same is true for triple integrals. For continuous functions f defined on regions $\mathcal{B}$ like that shown in Fig. 11.41, an iterated integral theorem reduces the evaluation of many triple integrals to an iterated integral containing a single and a double integral. The region $\mathcal{B}$ shown in Fig. 11.41 lies between the graphs of bounded, continuous functions p and q, which are defined on a bounded subset $\mathcal{D}$ of the (x, y)-plane. Such regions are said to be of Type I.

Regions of Types II and III can be described by functions defined on subsets of the (x, z)- and (y, z)-planes, respectively. See Exercises 3 and 4. The numbering I, II, and III of the types is in *lexicographic* order—that is, (x, y) precedes (x, z), and (x, z) precedes (y, z).

THEOREM Iterated Integral Theorem for Regions of Type I

Assume that f is continuous on the Type I region

$$\mathcal{B} = \{(x, y, z) \in R^3 : (x, y) \in \mathcal{D}, p(x, y) \leq z \leq q(x, y)\}, \tag{3}$$

where p and q are continuous functions defined on the closed and bounded region $\mathcal{D}$ of the (x, y)-plane. Then f is integrable on $\mathcal{B}$ and

$$\iiint_{\mathcal{B}} f(x, y, z)\, dV = \iint_{\mathcal{D}} \left(\int_{p(x,y)}^{q(x,y)} f(x, y, z)\, dz \right) dx\, dy. \tag{4}$$

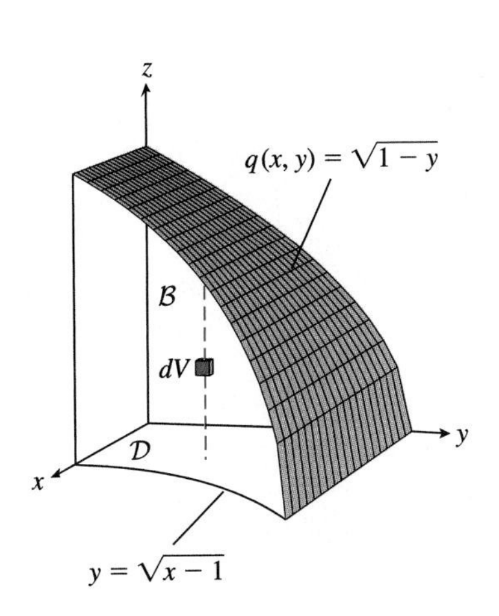

FIGURE 11.42 Region of Type I.

Volume If $\mathcal{B}$ is a region of Type I and $f(x, y, z) = 1$ for all $(x, y, z) \in \mathcal{B}$, then the Riemann sums $\Sigma_{i,j,k} 1\, dx\, dy\, dz$ are sums of volume elements $dv = dx\, dy\, dz$; these sums approximate the volume V of $\mathcal{B}$. We expect that as the number n of subdivisions increases, the Riemann sums approach the volume V of $\mathcal{B}$.

If $\mathcal{B}$ is a region of Type I, Type II, or Type III, the volume of the region $\mathcal{B}$ is the number $\iiint_{\mathcal{B}} 1\, dV$.

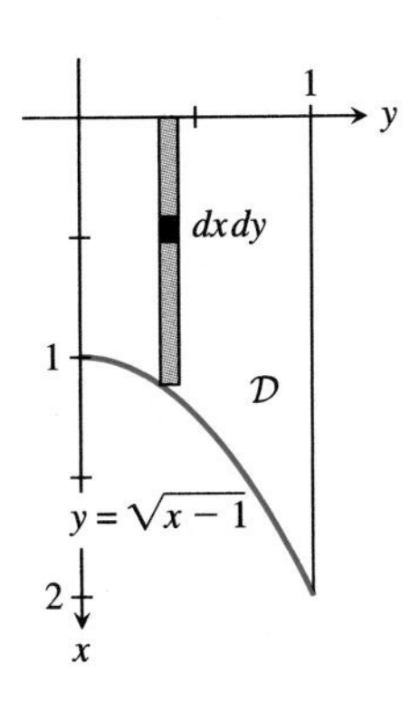

FIGURE 11.43 Region $\mathcal{D}$ for outer double integral when $\mathcal{B}$ is regarded as a region of Type I.

EXAMPLE 1 Calculate the volume V of the region $\mathcal{B}$ between the graphs of $p(x, y) = 0$ and $q(x, y) = \sqrt{1-y}$ and lying above the subset $\mathcal{D}$ shown in Fig. 11.42. The set $\mathcal{D}$ is bounded by the curves with equations $y = \sqrt{x-1}$, $y = 0$, $y = 1$, and $x = 0$.

Solution

The region $\mathcal{B}$ is of Type I, lying between surfaces described by $z = 0$ and $z = \sqrt{1-y}$. From the definition of volume and the Iterated Integral Theorem,

$$V = \iiint_{\mathcal{B}} 1\, dV = \iint_{\mathcal{D}} \left(\int_0^{\sqrt{1-y}} 1\, dz \right) dx\, dy.$$

The integration in the z-direction is suggested in Fig. 11.42 by the dotted vertical line. With fixed (x, y), the first integration sums volume elements $dV = dx\, dy\, dz$ in the z-direction, thus generating a slender column of volume $\sqrt{1-y}\, dx\, dy$. Summing these columns over $\mathcal{D}$ by the double integral gives

$$V = \iint_{\mathcal{D}} \sqrt{1-y}\, dx\, dy.$$

The region $\mathcal{D}$ for this double integral is shown in Figs. 11.42 and 11.43; it is of Type II in the plane. It is generally a good idea to sketch $\mathcal{D}$ separately, as an aid to evaluating the double integral. If, in $\mathcal{D}$, we choose to integrate in the x-direction first, we see from Fig. 11.43 that

$$V = \int_0^1 \int_0^{1+y^2} \sqrt{1-y}\,dx\,dy = \int_0^1 (1 + y^2)\sqrt{1-y}\,dy.$$

In the last integral, we try the substitution $1 - y = w$. Because $-dy = dw$,

$$V = -\int_1^0 (1 + (1-w)^2)\sqrt{w}\,dw = \int_0^1 (2w^{1/2} - 2w^{3/2} + w^{5/2})\,dw$$

$$= 2\frac{1}{3/2} - 2\frac{1}{5/2} + \frac{1}{7/2} = \frac{86}{105} \text{ cubic units.}$$

Referring to Fig. 11.42, the cylindrical surface with equation $z = \sqrt{1-y}$ and the region $\mathcal{D}$ play the major roles in visualizing the region $\mathcal{B}$ in Example 1. Although a CAS was used to draw the cylindrical surface in the figure, a much simpler sketch would have been sufficient for setting up the triple integral for the volume. Figure 11.44(a) shows that a few traces in the coordinate planes and a few lines parallel to the axis of the cylinder give a good sense of the region $\mathcal{B}$. Because the top-bounding surface is a cylinder in the x-direction, we started by drawing its trace in the (y, z)-plane $\left(\text{set } x = 0 \text{ in the equation } z = \sqrt{1-y}\right)$; from this trace we drew several lines parallel to the x-axis. Because $\mathcal{B}$ is to lie above $\mathcal{D}$, we imagined a vertical line moving along the $y = \sqrt{x-1}$ boundary of $\mathcal{D}$. We sketched in the trace made by this line on the cylinder.

Some regions are of more than one type. For example, the region $\mathcal{B}$ in Example 1 is also of Types II and III. We show how to set up an iterated integral for the volume of $\mathcal{B}$ when this region is regarded as Type III. We leave the Type II case as Exercise 35.

We use $\mathcal{E}$ to denote the projection of $\mathcal{B}$ onto the (y, z)-plane. As a region of Type III, $\mathcal{B} = \{(x, y, z) : 0 \le x \le y^2 + 1, (y, z) \in \mathcal{E}\}$. See Figs. 11.44(b) and 11.45. If we wish to integrate in the x-direction first, then with fixed $(y, z) \in \mathcal{E}$, the volume

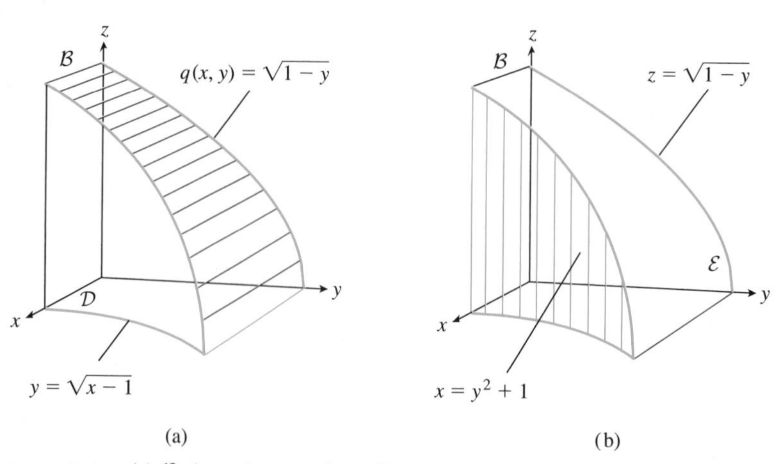

FIGURE 11.44 (a) $\mathcal{B}$ viewed as a region of Type I; (b) $\mathcal{B}$ viewed as a region of Type III.

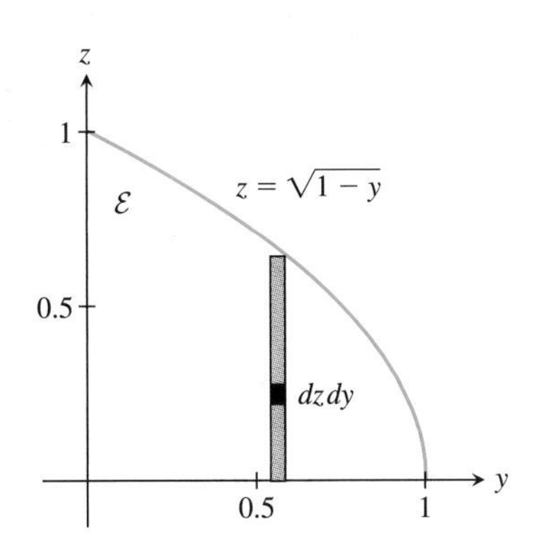

FIGURE 11.45 Region $\mathcal{E}$ for outer double integral when $\mathcal{B}$ is regarded as a region of Type III.

element $dV = dx\,dy\,dz$ moves from the point $(0, y, z)$ on the surface with equation $x = 0$ to the point $(y^2 + 1, y, z)$ on the surface with equation $x = y^2 + 1$.

At the time we sketch $\mathcal{E}$, we can decide on the order of integration for the double integral. Referring to Fig. 11.45, if we decide to integrate in the z-direction first, then we say that the integration order for the triple integral is x, z, and y and write

$$V = \iiint_{\mathcal{B}} 1\,dV = \iint_{\mathcal{E}} \int_0^{y^2+1} dx\,dz\,dy.$$

Note the order of the integration variables in the last expression. Inserting the limits for the double integral,

$$V = \int_0^1 \int_0^{\sqrt{1-y}} \int_0^{y^2+1} dx\,dz\,dy$$

$$= \int_0^1 \int_0^{\sqrt{1-y}} (y^2 + 1)\,dz\,dy = \int_0^1 (y^2 + 1)z \Big|_0^{\sqrt{1-y}} dy.$$

From here on, the calculation is the same as at the end of Example 1.

$$V = \int_0^1 (y^2 + 1)\sqrt{1 - y}\,dy = \frac{86}{105} \text{ cubic units.}$$

In the next example we calculate the mass and center of mass of a solid object. The definitions of *mass* and *center of mass* for three-dimensional objects have the same form as the definitions for one- and two-dimensional objects.

DEFINITION Mass and Center of Mass of a Solid

A solid is modeled by a bounded region $\mathcal{B}$ of R^3 and an integrable density function δ defined on $\mathcal{B}$. The position vector of its center of mass is

$$\mathbf{R} = \frac{1}{m} \iiint_{\mathcal{B}} \mathbf{r}\,dm = \frac{1}{m} \iiint_{\mathcal{B}} \delta(x, y, z)\,\langle x, y, z\rangle\,dx\,dy\,dz, \qquad (5)$$

where $m = \iiint_{\mathcal{B}} dm = \iiint_{\mathcal{B}} \delta(x, y, z)\,dV$ is the mass of the solid. If the density of the solid is constant, the center of mass often is called the **centroid** of $\mathcal{B}$.

If we set $\mathbf{R} = \langle X, Y, Z\rangle$, we may write (5) as the three scalar equations

$$X = \frac{1}{m} \iiint_{\mathcal{B}} x\delta(x, y, z)\,dx\,dy\,dz$$

$$Y = \frac{1}{m} \iiint_{\mathcal{B}} y\delta(x, y, z)\,dx\,dy\,dz$$

$$Z = \frac{1}{m} \iiint_{\mathcal{B}} z\delta(x, y, z)\,dx\,dy\,dz.$$

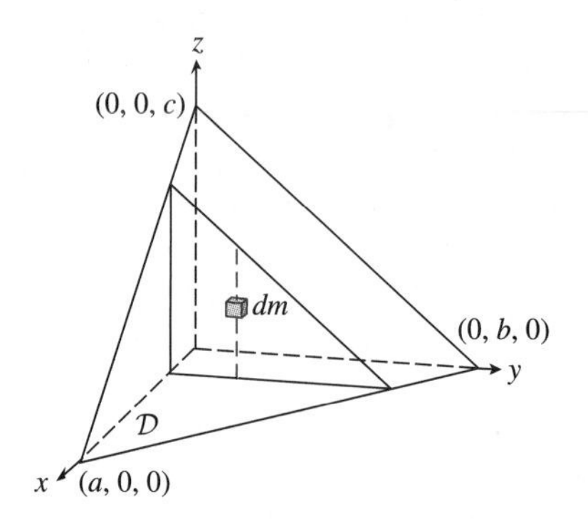

FIGURE 11.46 Right tetrahedron.

EXAMPLE 2 Figure 11.46 shows an outline of a solid right tetrahedron with constant density δ kg/m^3 and volume $\frac{1}{6}abc$ m^3. Show that the center of mass of this tetrahedron is $\mathbf{R} = \frac{1}{4}\langle a, b, c\rangle$ meters.

Solution

The mass element shown in the figure has mass $dm = \delta\, dV$, where $dV = dx\,dy\,dz$ is the volume element. The mass m of the tetrahedron is $m = \delta \cdot V = \frac{1}{6}\delta abc$. We show that the third coordinate Z of $\mathbf{R}$ is $\frac{1}{4}c$. Because

$$\mathbf{R} = \frac{1}{m}\iiint_{\mathcal{B}} \mathbf{R}\, dm = \frac{\delta}{m}\iiint_{\mathcal{B}} \langle x, y, z\rangle\, dx\,dy\,dz,$$

the third coordinate Z of $\mathbf{R}$ is

$$Z = \frac{\delta}{m}\iiint_{\mathcal{B}} z\, dx\,dy\,dz.$$

The tetrahedron is of Type I (and II and III). Because the plane forming the top of the tetrahedron is described by the equation $x/a + y/b + z/c = 1$, the bottom and top surfaces of the tetrahedron are described by the equations

$$z = p(x, y) = 0 \quad \text{and} \quad z = q(x, y) = c(1 - x/a - y/b), \qquad (x, y) \in \mathcal{D},$$

where $\mathcal{D}$ is the triangle with vertices $(0, 0, 0)$, $(a, 0, 0)$, and $(0, b, 0)$. Using the integration order z, y, and x,

$$\begin{aligned}
Z &= \frac{\delta}{m}\int_0^a \int_0^{b(1-x/a)} \int_0^{c(1-x/a-y/b)} z\, dz\,dy\,dx \\
&= \frac{\delta c^2}{2m}\int_0^a \int_0^{b(1-x/a)} \left(1 - \frac{x}{a} - \frac{y}{b}\right)^2 dy\,dx \\
&= \frac{-b\delta c^2}{6m}\int_0^a \left(1 - \frac{x}{a} - \frac{y}{b}\right)^3\Bigg|_0^{b(1-x/a)} dx \\
&= \frac{b\delta c^2}{6m}\int_0^a \left(1 - \frac{x}{a}\right)^3 dx = \frac{-ab\delta c^2}{24m}\left(1 - \frac{x}{a}\right)^4\Bigg|_0^a = \frac{1}{4}c.
\end{aligned}$$

From symmetry, the first and second coordinates of $\mathbf{R}$ are $\frac{1}{4}a$ meters and $\frac{1}{4}b$ meters. Hence $\mathbf{R} = \frac{1}{4}\langle a, b, c\rangle$ meters.

EXAMPLE 3 Calculate the volume of the region $\mathcal{B}$ beneath the graph with equation $z = \frac{1}{2}x^2 + \frac{1}{3}y^2$ and above the triangular region $\mathcal{D}$ with vertices $(0, 0)$, $(1, 1)$, and $(0, 2)$.

Solution

To sketch $\mathcal{B}$ so that we can set up an integral for the volume, we draw a few traces of the graph of the equation. Setting $y = 0$, the trace in the (x, z)-plane has equation $z = \frac{1}{2}x^2$. Sketch this parabola. Next, setting $x = 0$, sketch the parabola with equation $z = \frac{1}{3}y^2$. This trace lies in the (y, z)-plane. These two curves suggest the general shape of the paraboloidal surface. Adding a trace in a plane parallel to the (x, y)-plane is usually helpful. The question is, how high? For this figure, the triangular region $\mathcal{D}$ in the (x, y)-plane helps us choose the height. The heights of the paraboloid above the vertices $(0,0)$, $(1,1)$, and $(0,2)$ are 0, $\frac{5}{6}$, and $\frac{4}{3}$, respectively. In the plane with equation $z = \frac{4}{3}$, we sketch an ellipse. Its equation is $\frac{4}{3} = \frac{1}{2}x^2 + \frac{1}{3}y^2$, but we need not spend much effort on drawing a careful ellipse. The solid region $\mathcal{B}$ lies beneath the paraboloid and above the triangular region $\mathcal{D}$. Add vertical lines through $(1,1)$ and $(0,2)$. The second of these lines rises to $\left(0, 2, \frac{4}{3}\right)$. The other rises to $\left(1, 1, \frac{5}{6}\right)$. The part of the paraboloid lying above the triangle can be suggested by sketching curves on the surface of the paraboloid, as shown in the figure.

An element dV of volume is shown in Fig. 11.47. The vertical line drawn through the element signals that we are viewing $\mathcal{B}$ as a region of Type I and plan to integrate in the z-direction first. The horizontal line through the base of the vertical line signals that we have chosen the integration order z, y, and x.

Applying the Iterated Integral Theorem,

$$V = \iiint_{\mathcal{B}} 1\, dV = \iint_{\mathcal{D}} \int_0^{x^2/2+y^2/3} 1\, dz\, dy\, dx.$$

As seen in Fig. 11.47, the equations of two of the lines forming the boundary of $\mathcal{D}$ are $y = x$ and $y = 2 - x$. Hence,

$$\begin{aligned} V &= \int_0^1 \int_x^{2-x} \left(\frac{1}{2}x^2 + \frac{1}{3}y^2\right) dy\, dx \\ &= \frac{1}{9}\int_0^1 (-11x^3 + 15x^2 - 12x + 8)\, dx. \end{aligned}$$

After integrating with respect to x and simplifying the result,

$$V = \frac{17}{36} \text{ cubic units.}$$

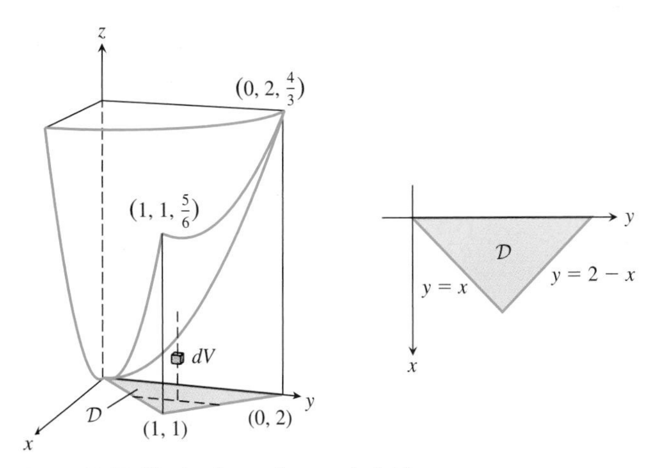

FIGURE 11.47 Region beneath a paraboloid.

EXAMPLE 4 Figure 11.48 shows a pyramid with height h meters and a square base having sides of length $2a$ meters. Assuming that the pyramid is hollow and filled with a liquid of density δ kg/m^3, calculate the work needed to pump the liquid out the top vertex of the pyramid.

Solution

Figure 11.48 shows a mass element $dm = \delta\, dV$ at the point (x, y, z). The force needed on this element to just overcome the gravitational attraction of the Earth is $dF = g\, dm = g\delta\, dV$. The work required to move this mass element to the top vertex of the pyramid is $dw = (h - z)g\delta\, dV = (h - z)g\delta\, dx\, dy\, dz$. By symmetry, it is enough to calculate the work needed to pump out one-eighth of the pyramid, the part above the region $\mathcal{D}$ shown in the figure. The vertices of $\mathcal{D}$ are $(0, 0)$, (a, a), and $(0, a)$. If, as suggested in the figure, we sum first in the z-direction, next in the x-direction, and, finally, in the y-direction, the total work needed to pump out the pyramid is

$$W = 8 \iint_{\mathcal{D}} \int_0^{h(1-y/a)} dw.$$

The face of the pyramid above $\mathcal{D}$ is the plane with equation $y/a + z/h = 1$. This is the plane containing the points $(0, a, 0)$, $(a, a, 0)$, and $(0, 0, h)$. Noting that $\mathcal{D}$ is bounded on the top by the line with equation $x = 0$ and on the bottom by the line with equation $x = y$,

$$\begin{aligned}
W &= 8 \iint_{\mathcal{D}} \int_0^{h(1-y/a)} dw = 8g\delta \int_0^a \int_0^y \int_0^{h(1-y/a)} (h - z)\, dz\, dx\, dy \\
&= 8g\delta \int_0^a \int_0^y \left(hz - \frac{1}{2} z^2 \right) \Bigg|_0^{h(1-y/a)} dx\, dy \\
&= 8g\delta \int_0^a \int_0^y \left(h^2\left(1 - \frac{y}{a}\right) - \frac{1}{2} h^2 \left(1 - \frac{y}{a}\right)^2 \right) dx\, dy \\
&= 8g\delta \cdot \frac{h^2}{2a^2} \int_0^a \int_0^y (a^2 - y^2)\, dx\, dy = 8g\delta \cdot \frac{h^2}{2a^2} \int_0^a y(a^2 - y^2)\, dy \\
&= \frac{4g\delta h^2}{a^2} \left(\frac{1}{2} a^2 y^2 - \frac{1}{4} y^4 \right) \Bigg|_0^a = g\delta h^2 a^2 \text{ joules.}
\end{aligned}$$

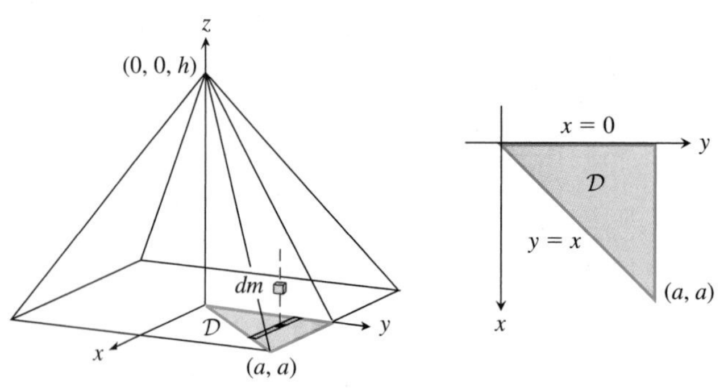

FIGURE 11.48 Element of mass, with first integration in the z-direction, second in the x-direction, and third in the y-direction.

Exercises 11.5

1. From the definition of the triple integral, find an estimate for the mass of the solid region $\mathcal{B} = [1,2] \times [1,3] \times [1,2]$ with density function

$$\delta(x, y, z) = xy^2z, \qquad (x, y, z) \in \mathcal{B}.$$

Use the regular subdivision $\mathcal{S}_2$ of $\mathcal{B}$ obtained by dividing each of the intervals $[1,2]$, $[1,3]$, and $[1,2]$ into two equal subintervals. Use the evaluation set $\mathcal{C}_2$ consisting of the center points of the eight subregions. Compare the average of your estimates with the true mass of $\mathcal{B}$.

2. From the definition of the triple integral, find an estimate for the mass of a solid $\mathcal{B} = [1,3] \times [1,2] \times [1,2]$ with density function

$$\delta(x, y, z) = x^2yz, \qquad (x, y, z) \in \mathcal{B}.$$

Use the regular subdivision $\mathcal{S}_2$ of $\mathcal{B}$ obtained by dividing each of the subintervals $[1,3]$, $[1,2]$ and $[1,2]$ into two equal subintervals. Use the evaluation set $\mathcal{C}_2$ consisting of the center points of the eight subregions. Compare the average of your estimates with the true mass of $\mathcal{B}$.

3. Write out a definition for a Region of Type II and state an iterated integral theorem for functions defined on regions of Type II.

4. Write out a definition for a Region of Type III and state an iterated integral theorem for functions defined on regions of Type III.

5. Sketch and calculate the volume of the solid $\mathcal{B}$ consisting of all points $(x, y, z) \in R^3$ satisfying

$$(x, y) \in \mathcal{D}, \qquad \tfrac{1}{2}\sqrt{1-y} \le z \le \sqrt{1-y},$$

where the region $\mathcal{D}$ is as described in Example 1. Regard $\mathcal{B}$ as a region of Type I.

6. Sketch and calculate the volume of the solid $\mathcal{B}$ consisting of all points $(x, y, z) \in R^3$ satisfying

$$(x, y) \in \mathcal{D}, \qquad y^2 \le z \le \tfrac{1}{2}(y^2 + 1),$$

where the region $\mathcal{D}$ is described in Example 1. Regard $\mathcal{B}$ as a region of Type I.

7. Do Exercise 5, but regard $\mathcal{B}$ as a region of Type III.

8. Do Exercise 6, but regard $\mathcal{B}$ as a region of Type III.

9. A right tetrahedron is bounded by the coordinate planes and the graph of $z = 2(1 - x - y/2)$. Use a triple integral to calculate its volume. See Example 2 if you need help with the limits.

10. A right tetrahedron is bounded by the coordinate planes and the graph of $z = 3(1 - x/2 - y/3)$. Use a triple integral to calculate its volume. See Example 2 if you need help with the limits.

11. Use a triple integral to find the volume of a region $\mathcal{B}$ similar to that shown in Fig. 11.42. The only change is in the set $\mathcal{D}$, which in this exercise is bounded by the curves with equations $y = x - 1$, $y = 0$, $y = 1$, and $x = 0$.

12. Use a triple integral to find the volume of a region $\mathcal{B}$ similar to that shown in Fig. 11.42. The only change is in the set $\mathcal{D}$, which in this exercise is bounded by the curves with equations $x + y = 1$, $y = 0$, and $x = 0$.

13. Use a triple integral to find the volume of a region $\mathcal{B}$ similar to that shown in Fig. 11.47. For this exercise, however, the upper surface has equation $z = x^2 + 2y^2$ and the triangle has vertices $(0,0)$, $(0,1)$, and $(1,0)$.

14. Use a triple integral to find the volume of a region $\mathcal{B}$ similar to that shown in Fig. 11.47. For this exercise, however, the upper surface has equation $z = \frac{1}{2}x^2 + y^2$ and the triangle has vertices $(0,0)$, $(0,1)$, and $(2,0)$.

15. Referring to the accompanying figure, use a triple integral to find the volume of the tetrahedron with vertices $(0,0,0)$, $(2,1,0)$, $(0,2,0)$, and $(0,0,3)$.

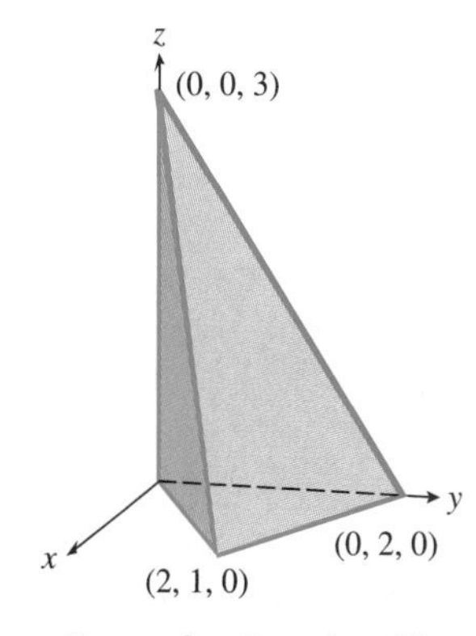

Figure for Exercise 15.

16. Referring to the accompanying figure, use a triple integral to find the volume of the tetrahedron with vertices $(0,0,0)$, $(2,2,0)$, $(0,3,0)$, and $(0,0,3)$.

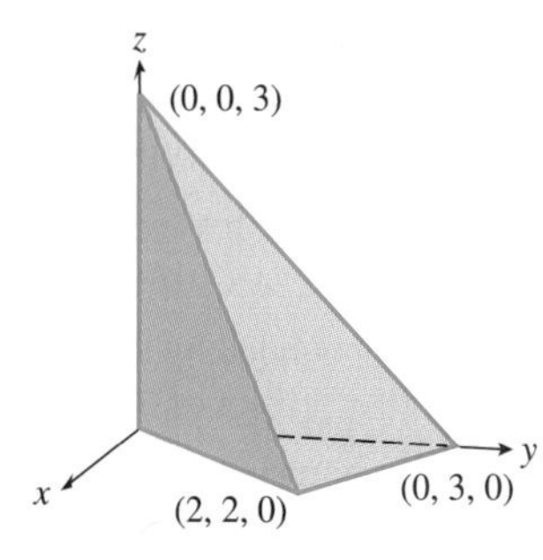

Figure for Exercise 16.

17. Use a triple integral to find the volume of the solid bounded by the cylindrical surfaces with equations $z = \frac{1}{2}y^2$, $x = 0$, and $z = 2 - x$. See the accompanying figure, which shows the trace of the first cylinder in the third. You may wish to integrate in the x-direction first.

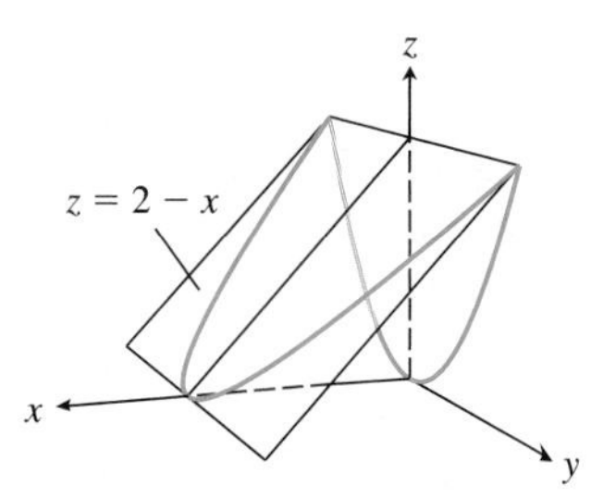

Figure for Exercise 17.

18. Use a triple integral to find the volume of the solid bounded by the cylindrical surfaces with equations $z = y^2$, $x = 0$, and $z = \sqrt{1 - x}$. See the accompanying figure, which shows the trace of the first cylinder in the third. You may wish to integrate in the x-direction first.

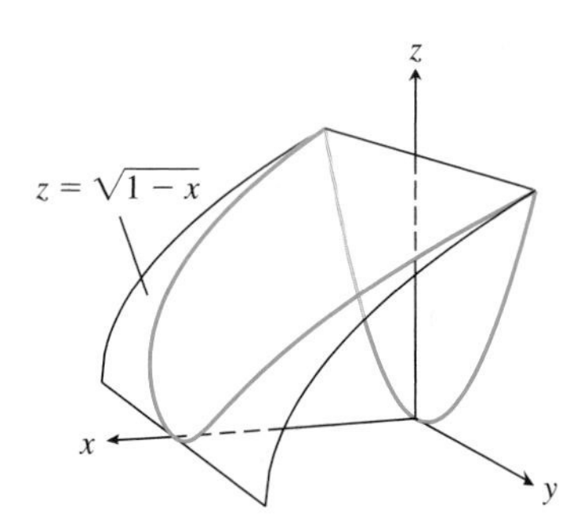

Figure for Exercise 18.

19. Calculate the mass and the center of mass of the tetrahedron shown in Example 2, but assume that the density is proportional to the distance to the plane with equation $z = c$.

20. Suppose the tetrahedron given in Example 2 is filled with material that packs down as the tetrahedron is filled, so that the density of the material is proportional to its depth. Locate the center of mass of the packed tetrahedron.

Exercises 21–26: Evaluate the triple integral.

21. $\iiint_{\mathcal{K}} xy\,dV$, where $\mathcal{K}$ is the solid tetrahedron with vertices $(0,0,0)$, $(1,0,0)$, $(0,1,0)$, and $(0,0,1)$.

22. $\iiint_{\mathcal{K}} yz\,dV$, where $\mathcal{K}$ is the solid tetrahedron with vertices $(0,0,0)$, $(2,0,0)$, $(0,1,0)$, and $(0,0,3)$.

23. $\iiint_{\mathcal{K}} x\,dV$, where $\mathcal{K}$ is bounded by the planes with equations $y + z/3 = 1$, $z = 0$, $x = 0$, $y = 0$, and $x = 2$.

24. $\iiint_{\mathcal{K}} y\,dV$, where $\mathcal{K}$ is bounded by the planes with equations $x + z = 2$, $z = 0$, $y = 0$, $x = 0$, and $y = 3$.

25. $\iiint_{\mathcal{K}} \sin x\,dV$, where $\mathcal{K}$ is the first-octant region bounded above by the graph of $z = xy$ and below by the triangular region with vertices $(0,0,0)$, $(\pi,0,0)$, and $(0,\pi,0)$.

26. $\iiint_{\mathcal{K}} \cos y\,dV$, where $\mathcal{K}$ is the first-octant region bounded above by the graph of $z = x^2 + y^2$ and below by the triangular region with vertices $(0,0,0)$, $(\pi/2,0,0)$, and $(0,\pi/2,0)$.

27. Find the center of mass of a solid right circular cone $\mathcal{B}$ of height h and base radius a if the density at each point of $\mathcal{B}$ is proportional to the distance between that point and the base of the cone.

28. With $a = 1$, $b = 2$, and $c = 3$, find the centroid of the solid below the upper half of the ellipsoid with equation $x^2/a^2 + y^2/b^2 + z^2/c^2 = 1$ and above the (x,y)-plane. It is given that the volume of this solid is 4π. Use symmetry to find the coordinates X and Y of the centroid $\mathbf{R}$.

29. Use a triple integral to find the volume of the region bounded by the parabolic cylinders with equations $z = 4 - x^2$ and $z = 2 + x^2$ and the plane with equations $y = -1$ and $y = 3$.

30. Use a triple integral to find the volume of the region bounded above and below by the surfaces with equations $z = 9 - x^2 - 3y^2$ and $z = 1 + x^2 + y^2$.

31. A right circular cylinder with height h, and base radius a is filled with liquid of density δ. Calculate the work needed to pump the liquid to the top of the cylinder. Assume units of meters and kilograms.

32. A tetrahedron has vertices $(0,0,0)$, $(a,0,0)$, $(0,b,0)$, and $(0,0,c)$. It is filled with material whose density at (x,y,z) is proportional to xyz^2. Assuming units of meters and kilograms, calculate the work required to move the material to the top vertex.

33. A tank of volume V has the shape of a bounded region $\mathcal{B}$ and is filled with liquid with constant density δ. Let P be the highest point of $\mathcal{B}$. Show that the work required to pump the liquid from $\mathcal{B}$ to P is equal to the work required to lift a particle of mass δV from the center of mass of $\mathcal{B}$ to P.

34. Suppose that a body of constant density has been generated by revolving about the z-axis a lamina lying in the (z,y)-plane and of constant density. Is it true that the height of the center of mass of the body above the (x,y)-plane is equal to the height of the center of mass of the lamina above the y-axis?

35. Regard the region $\mathcal{B}$ of Example 1 as Type II and set up an iterated integral (or the sum of two iterated integrals) for its volume.

36. Solid $\mathcal{B}_1$ is bounded on the top by the cylinder with equation $z = 10x - x^2$, $0 \le x \le 10$, on the bottom by the (x,y)-plane, and on the sides by the planes with equations $y = 0$ and $y = 10$. Solid $\mathcal{B}_2$ is the set of all points (x,y,z) for which $0 \le x \le 10$, $4 \le y \le 6$, and $15 \le z \le 20$. Calculate the volume common to the solids $\mathcal{B}_1$ and $\mathcal{B}_2$.

11.6 Change-of-Variables Formula for Triple Integrals

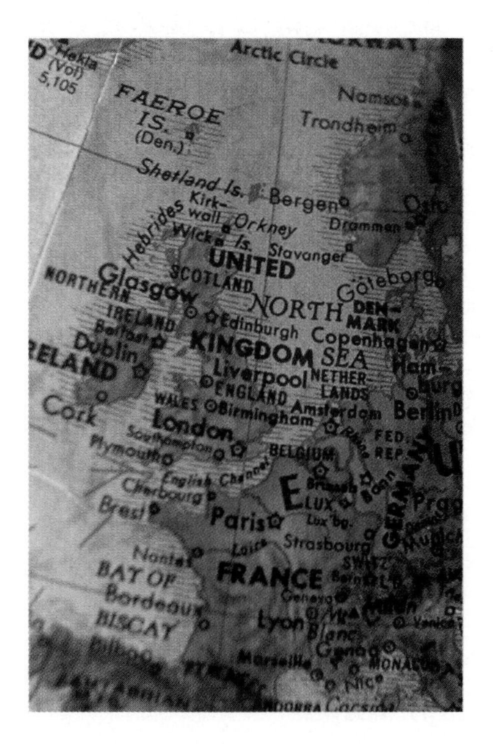

We have seen that in some applications of double integrals we may need to evaluate an integral $\iint_{\mathcal{B}} f(x, y)\,dA$ on regions $\mathcal{B}$ that are difficult to describe using rectangular coordinates. The same is true for triple integrals. Figure 11.49(a) shows, for example, how a cube of material may appear after an elastic deformation into a parallelepiped. We may be interested in the percentage change in volume from cube to parallelepiped, or we may wish to calculate the mass and center of mass of the parallelepiped itself. While it would be possible to write out in rectangular coordinates a sum of several triple integrals for, say, the volume of the parallelepiped, figuring out the limits would take some effort, both in visualization and in the detailed calculations required to find equations for the several planes and lines defining its shape. As an alternative, we may use the Change-of-Variables Formula for Triple Integrals. In this case we may use the function defining the elastic deformation as the change of variables.

Figure 11.49(b) shows a sketch of one octant of a spherical model of the Earth, with an outlined solid $\mathcal{B}$. This solid is the set of all points of the Earth lying between the Earth's center and the surface bounded by the 20° W and 30° W meridians and the 50° N and 60° N parallels. If we were asked to evaluate a triple integral $\iiint_{\mathcal{B}} f(x, y, z)\,dV$ on the region $\mathcal{B}$, we would not willingly choose rectangular coordinates for the job. Rather, we would change from rectangular coordinates to spherical coordinates and use the Change-of-Variables Formula for Triple Integrals. As we show in Example 4, we can write the transformed integral by visualizing the region $\mathcal{B}$ in Fig. 11.49(b) directly in terms of spherical coordinates and using a spherical volume element.

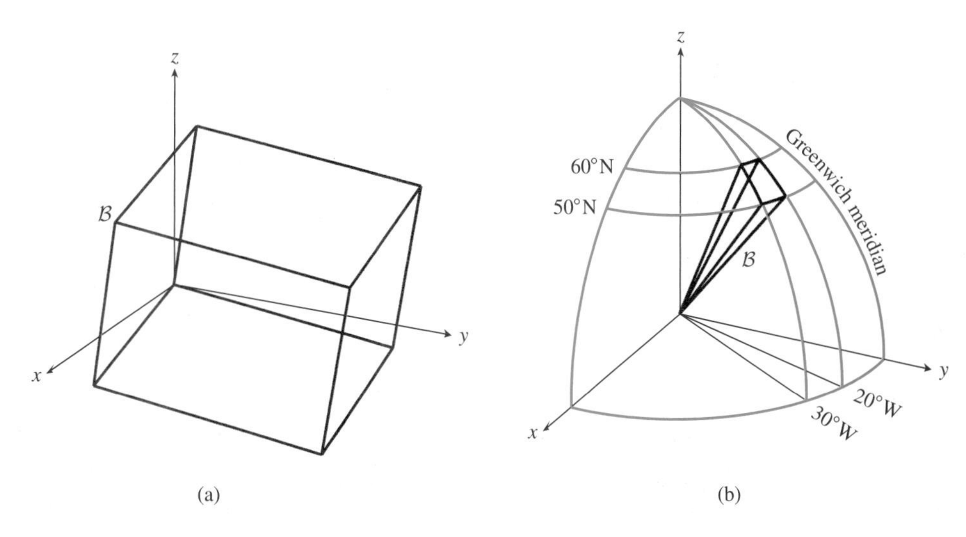

FIGURE 11.49 Evaluating the integral $\iiint_{\mathcal{B}} f(x, y, z)\,dV$ on regions $\mathcal{B}$ not easily describable in terms of rectangular coordinates.

We give the Change-of-Variables Formula for Triple Integrals for the general case. However, apart from one example and a few exercises, we emphasize the change of variables from rectangular to cylindrical coordinates or from rectangular to spherical coordinates.

Suppose that we wish to evaluate the triple integral $\iiint_{\mathcal{B}} f(x, y, z)\, dx\, dy\, dz$, where $\mathcal{B}$ is a bounded subset of R^3. Suppose further that the shape of the region $\mathcal{B}$ or the form of the function f suggests the change of variables $\langle x, y, z\rangle = \mathbf{g}(u, v, w)$, where $\mathbf{g}$ is defined on $\mathcal{D} \subset R^3$ and $\mathcal{B} = \mathbf{g}(\mathcal{D})$. See Fig. 11.50. The change-of-variables formula is

$$\iiint\limits_{\mathbf{g}(\mathcal{D})} f(x, y, z)\, dx\, dy\, dz = \iiint\limits_{\mathcal{D}} f(\mathbf{g}(u, v, w))\left|\det(\mathbf{Dg}(u, v, w))\right| du\, dv\, dw. \tag{1}$$

The factor $\left|\det(\mathbf{Dg}(u, v, w))\right|$, which is the absolute value of the determinant of the Jacobian matrix of the change-of-variable equation $\langle x, y, z\rangle = \mathbf{g}(u, v, w)$, is the analog of the factor $\left|\det(\mathbf{Dg}(u, v))\right|$ in the Change-of-Variables Formula for Double Integrals.

We state the Change-of-Variables Formula for Triple Integrals in terms of the coordinate functions g_1, g_2, and g_3 of $\mathbf{g}$. If we think of the change-of-variables equation $\langle x, y, z\rangle = \mathbf{g}(u, v, w)$ in the form

$$\langle x, y, z\rangle = \mathbf{g}(u, v, w) = \langle g_1(u, v, w), g_2(u, v, w), g_3(u, v, w)\rangle,$$

then we may write such partial derivatives as $\partial g_1/\partial u$ in the form $\partial x/\partial u$. In this notation, the determinant of the Jacobian matrix $\mathbf{Dg}(u, v)$ is often written as

$$\frac{\partial(x, y, z)}{\partial(u, v, w)} = \begin{vmatrix} \dfrac{\partial x}{\partial u} & \dfrac{\partial x}{\partial v} & \dfrac{\partial x}{\partial w} \\ \dfrac{\partial y}{\partial u} & \dfrac{\partial y}{\partial v} & \dfrac{\partial y}{\partial w} \\ \dfrac{\partial z}{\partial u} & \dfrac{\partial z}{\partial v} & \dfrac{\partial z}{\partial w} \end{vmatrix} = \begin{vmatrix} \dfrac{\partial g_1}{\partial u} & \dfrac{\partial g_1}{\partial v} & \dfrac{\partial g_1}{\partial w} \\ \dfrac{\partial g_2}{\partial u} & \dfrac{\partial g_2}{\partial v} & \dfrac{\partial g_2}{\partial w} \\ \dfrac{\partial g_3}{\partial u} & \dfrac{\partial g_3}{\partial v} & \dfrac{\partial g_3}{\partial w} \end{vmatrix}. \tag{2}$$

We also write this determinant, often briefly referred to as "the Jacobian" of $\mathbf{g}$, as $\partial(x, y, z)/\partial(u, v, w)$.

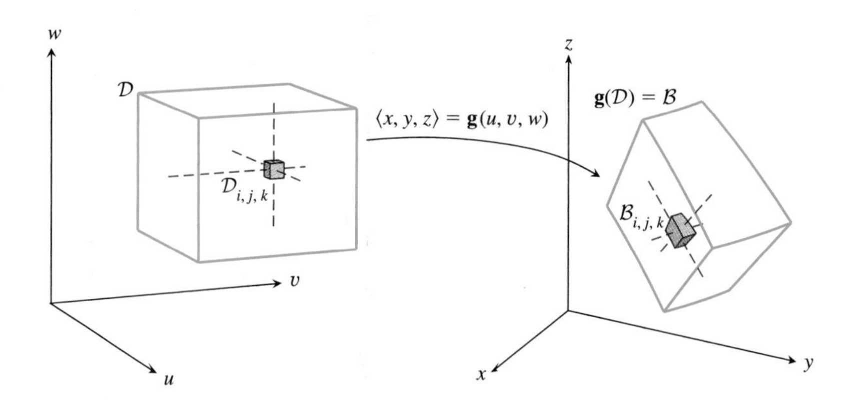

FIGURE 11.50 Domain and range of the change of variables $\mathbf{g}$; also shown are subregions $\mathcal{D}_{i,j,k}$ and $\mathcal{B}_{i,j,k}$ corresponding to a subdivision of $\mathcal{D}$.

Using this notation, the Change-of-Variables Formula for Triple Integrals is

Change-of-Variables Formula for Triple Integrals
Let $\mathbf{g}$ be a function defined on a region $\mathcal{D}$ of R^3. Assume that on $\mathcal{D}$ the coordinate functions $x = g_1(u, v, w)$, $y = g_2(u, v, w)$, and $z = g_3(u, v, w)$ of $\mathbf{g}$ are continuously differentiable, $\mathbf{g}$ is one-to-one, and $\partial(x, y, z)/\partial(u, v, w) \neq 0$. We refer to $\langle x, y, z\rangle = \mathbf{g}(u, v, w)$ as the change-of-variables equation. If f is a continuous function defined on the range $\mathcal{B} = \mathbf{g}(\mathcal{D})$ of $\mathbf{g}$, then

$$\iiint\limits_{\mathbf{g}(\mathcal{D})} f(x, y, z)\, dx\, dy\, dz = \iiint\limits_{\mathcal{D}} f(\mathbf{g}(u, v, w)) \left|\frac{\partial(x, y, z)}{\partial(u, v, w)}\right| du\, dv\, dw. \tag{3}$$

A proof of formula (3) can be made along the lines of the argument we gave for the Change-of-Variables Formula for Double Integrals. The proof depends upon the Linear Functions and Area Theorem from Section 8.5, just as in proofs of the one- and two-dimensional change-of-variables formulas.

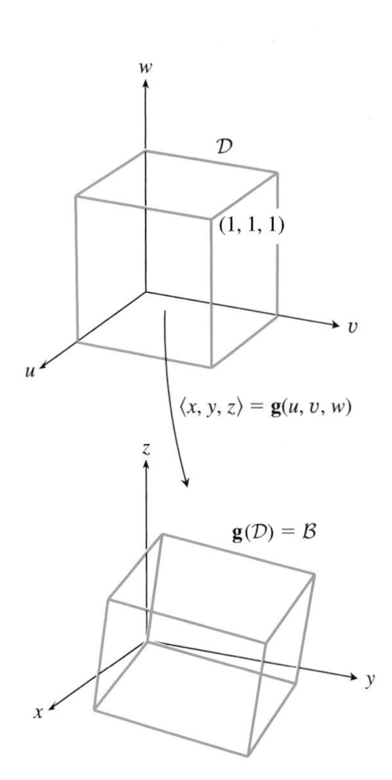

FIGURE 11.51 Domain and range of the change of variables $\mathbf{g}$.

EXAMPLE 1 Use the change of variables

$$\langle x, y, z\rangle = \mathbf{g}(u, v, w) = \left\langle \frac{5}{4}u + \frac{1}{4}v + \frac{1}{4}w, \frac{1}{4}u + \frac{5}{4}v + \frac{1}{4}w, \frac{7}{8}w \right\rangle, \tag{4}$$

where $(u, v, w) \in \mathcal{D}$, to evaluate the integral

$$\iiint\limits_{\mathcal{B}} xyz^2\, dx\, dy\, dz,$$

where $\mathcal{B}$ is the image $\mathbf{g}(\mathcal{D})$ of the unit cube $\mathcal{D}$ shown in Fig. 11.51.

Solution

This example relates to the elastic deformation of a cube mentioned earlier. The change of variables (4) comes from the elastic deformation described by

$$L(u, v, w) = \begin{pmatrix} x \\ y \\ z \end{pmatrix} = \begin{pmatrix} \frac{5}{4} & \frac{1}{4} & \frac{1}{4} \\ \frac{1}{4} & \frac{5}{4} & \frac{1}{4} \\ 0 & 0 & \frac{7}{8} \end{pmatrix} \begin{pmatrix} u \\ v \\ w \end{pmatrix}.$$

To evaluate the integral $\iiint_{\mathcal{B}} xyz^2\, dx\, dy\, dz$, we calculate the Jacobian of $\mathbf{g}$:

$$\frac{\partial(x, y, z)}{\partial(u, v, w)} = \begin{vmatrix} \frac{\partial x}{\partial u} & \frac{\partial x}{\partial v} & \frac{\partial x}{\partial w} \\ \frac{\partial y}{\partial u} & \frac{\partial y}{\partial v} & \frac{\partial y}{\partial w} \\ \frac{\partial z}{\partial u} & \frac{\partial z}{\partial v} & \frac{\partial z}{\partial w} \end{vmatrix} = \begin{vmatrix} \frac{5}{4} & \frac{1}{4} & \frac{1}{4} \\ \frac{1}{4} & \frac{5}{4} & \frac{1}{4} \\ 0 & 0 & \frac{7}{8} \end{vmatrix} = \frac{21}{16}.$$

From (4) and the Change-of-Variables Formula (3),

$$\iiint_{\mathcal{B}} xyz^2\,dx\,dy\,dz = \iiint_{\mathcal{D}} \left(\frac{5}{4}u + \frac{1}{4}v + \frac{1}{4}w\right)\left(\frac{1}{4}u + \frac{5}{4}v + \frac{1}{4}w\right) \cdot \left(\frac{7}{8}w\right)^2 \cdot \frac{21}{16}\,du\,dv\,dw$$

$$= \frac{2401}{7680} \approx 0.31263.$$

See Exercise 28 for a discussion of how, given the vertices of the parallelepiped $\mathcal{B}$, the change of variables $\mathbf{g}$ can be found.

Cylindrical Coordinates

WEB In this and the next subsection, we use the equations relating rectangular coordinates to cylindrical or spherical coordinates to simplify the calculation of certain triple integrals $\iiint_{\mathcal{B}} f(x, y, z)\,dx\,dy\,dz$. We use the Change-of-Variables Formula for Triple Integrals but interpret the Jacobians for these changes of variable in terms of cylindrical or spherical volume elements and read the limits on the cylindrical or spherical variables directly from a figure in (x, y, z)-space. These practices often shorten the calculations.

Cylindrical coordinates extend polar coordinates for the plane into space. The equations for changing from rectangular to cylindrical coordinates are

$$x = r\cos\theta$$
$$y = r\sin\theta$$
$$z = z.$$

The Jacobian $\partial(x, y, z)/\partial(r, \theta, z)$ is

$$\frac{\partial(x, y, z)}{\partial(r, \theta, z)} = \begin{vmatrix} \dfrac{\partial x}{\partial r} & \dfrac{\partial x}{\partial \theta} & \dfrac{\partial x}{\partial z} \\ \dfrac{\partial y}{\partial r} & \dfrac{\partial y}{\partial \theta} & \dfrac{\partial y}{\partial z} \\ \dfrac{\partial z}{\partial r} & \dfrac{\partial z}{\partial \theta} & \dfrac{\partial z}{\partial z} \end{vmatrix} = \begin{vmatrix} \cos\theta & -r\sin\theta & 0 \\ \sin\theta & r\cos\theta & 0 \\ 0 & 0 & 1 \end{vmatrix} = r.$$

We use this result in calculating the volume element in cylindrical coordinates. If in the Change-of-Variable Formula (3) we take f to be identically 1, the volume V of the region $\mathbf{g}(\mathcal{D})$ is

$$V = \iiint_{\mathbf{g}(\mathcal{D})} dx\,dy\,dz = \iiint_{\mathcal{D}} \left|\frac{\partial(x, y, z)}{\partial(r, \theta, z)}\right| du\,dv\,dw.$$

Hence, the volume element in cylindrical coordinates is $dV = r\,dr\,d\theta\,dz$.

This expression for the volume element may also be verified geometrically. In rectangular coordinates, as shown in Fig. 11.52(a), the volume element $dV = dx\,dy\,dz$ at (x, y, z) is the rectangular parallelepiped defined by the points (x, y, z) and $(x + dx, y + dy, z + dz)$. Note that in the second triple, each coordinate

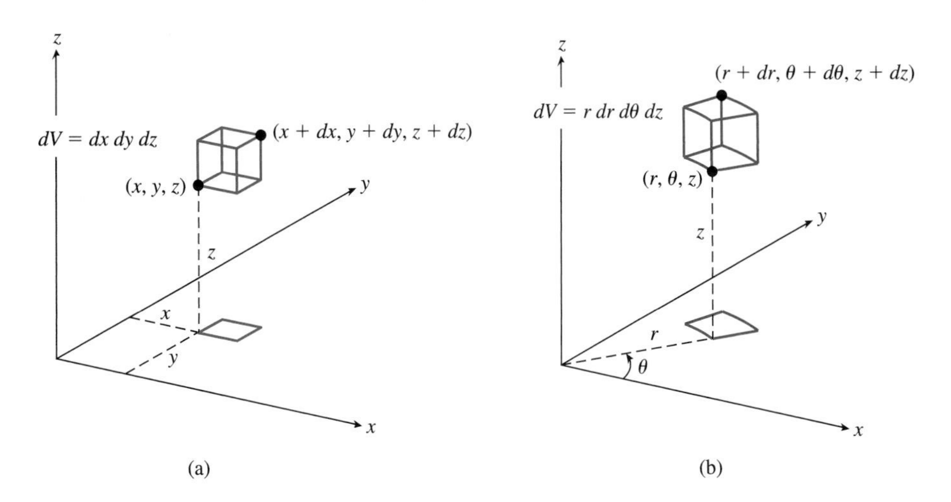

FIGURE 11.52 (a) Volume element in rectangular coordinates; (b) volume element in cylindrical coordinates.

from the first triple has been incremented. In cylindrical coordinates—interpreted directly in (x, y, z)-space—the volume element may be formed in a similar way: Choose a point (r, θ, z) and increment each variable. We show the resulting volume element in Fig. 11.52(b). Because the coordinate surfaces $r = a$, $\theta = b$, and $z = c$ (which are a cylinder, a vertical plane, and a horizontal plane) meet at right angles, the volume element is approximately a rectangular parallelepiped. The edge closest to the origin is the arc of a circle and has length $r\,d\theta$; the straight, horizontal edges have length dr; and the vertical edges have length dz. It follows that the volume of this element is approximately $dV = r\,d\theta\,dr\,dz$.

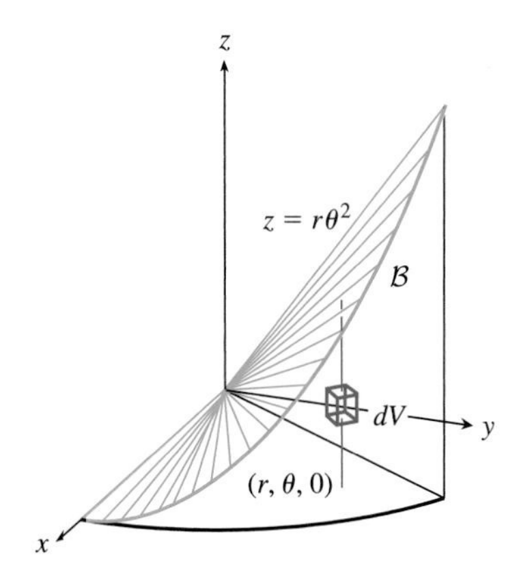

FIGURE 11.53 The region $\mathcal{B}$ beneath a ramp.

EXAMPLE 2 Referring to Fig. 11.53, calculate the volume of the region $\mathcal{B}$ between the (x, y)-plane and the surface described in polar coordinates by $z = r\theta^2$, where $0 \le r \le 1$ and $0 \le \theta \le \pi/3$.

Solution

In cylindrical coordinates, a triple integral for the volume of $\mathcal{B}$ is

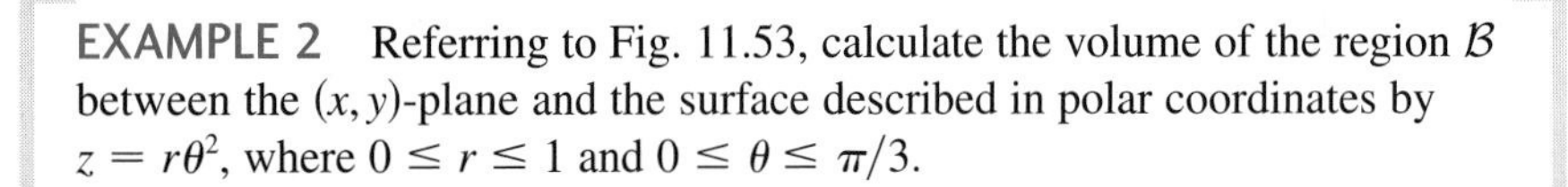

$$V = \iiint_{\mathcal{B}} 1\,dx\,dy\,dz = \iiint_{\mathcal{D}} r\,dr\,d\theta\,dz,$$

where $\mathcal{D}$ is the region in (r, θ, z)-space corresponding to the region $\mathcal{B}$. We use the Iterated Integral Theorem to evaluate this integral but determine directly from Fig. 11.53 the limits on the three iterated integrals. You may find it useful to regard the limits as a kind of program for the movement of the volume element throughout the region $\mathcal{B}$. For this particular region, we begin by holding r and θ fixed and varying z between 0 and $z = r\theta^2$. This programs the element to move vertically between the (x, y)-plane and the ramplike surface, sweeping out a vertical column. Next, with fixed θ, we vary r from 0 to 1; the column, whose height varies with r, sweeps out a thin, vertical wedge of $\mathcal{B}$. Finally, by varying θ from 0 to $\pi/3$, the wedge, whose dimensions vary with θ, moves through $\pi/3$ radians and sweeps out all of $\mathcal{B}$. The iterated integral is

$$V = \iiint_{\mathcal{D}} r\,dr\,d\theta\,dz = \int_0^{\pi/3}\int_0^1\int_0^{r\theta^2} r\,dz\,dr\,d\theta$$

$$= \int_0^{\pi/3}\int_0^1 r^2\theta^2\,dr\,d\theta = \int_0^{\pi/3} \frac{1}{3}r^3\theta^2\Big|_0^1\,d\theta$$

$$= \frac{1}{9}\theta^3\Big|_0^{\pi/3} = \frac{\pi^3}{243} \text{ cubic units} \approx 0.127598 \text{ cubic units.}$$

EXAMPLE 3 Find the mass and the center of mass of the tapered roller bearing shown from the side in Fig. 11.54(a). The height of the bearing is 65 mm, the diameters of the ends are 44 mm and 30 mm, and the bearing is made from high-carbon steel with density $\delta = 7860$ kg/m^3.

Solution

The circular symmetry of the bearing about its vertical axis suggests that cylindrical coordinates may simplify the integrals for its mass and center of mass. The set $\mathcal{B}$ shown in Fig. 11.54(c) represents one-quarter of the bearing. The volume element $dV = r\,dr\,d\theta\,dz$ in cylindrical coordinates is located at the point (r, θ, z). The form of the bearing leads us to integrate in the θ-direction first, from $\theta = 0$ to $\theta = 2\pi$, with r and z fixed. This is suggested by the dotted quarter-circle in the figure. Thus summed, the volume element sweeps out a thin ring about the z-axis. Next, with fixed z, we integrate in the r-direction, from $r = 0$ to $r = q(z)$, where q is a function to be determined. Thus summed, the thin rings generate a thin plate, which is a horizontal cross section of the bearing. In the last integration, z varies from 0 to 65. This sums the cross sections.

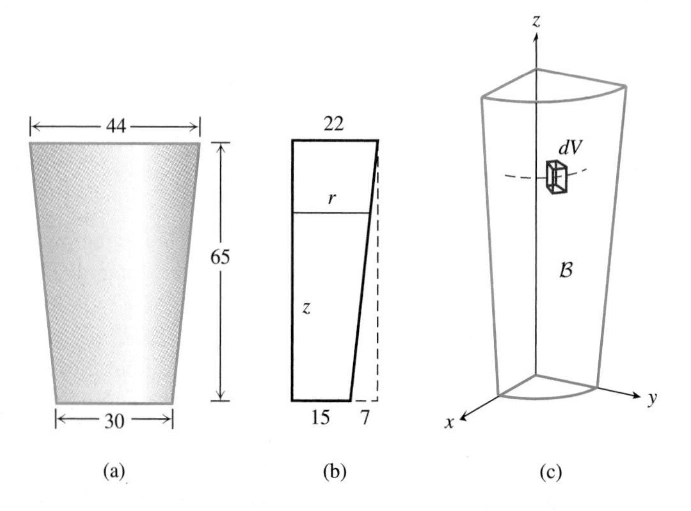

FIGURE 11.54 (a) A side view of the bearing; (b) a vertical section of the bearing; (c) and a placement of the bearing in a coordinate system.

We use the sketch in Fig. 11.54(b) to determine the function $r = q(z)$. The equation of the slanted line in this sketch has the form

$$z = \frac{65}{7} r + c, \tag{5}$$

for some constant c. Because $r = 15$ when $z = 0$,

$$0 = \frac{65}{7} 15 + c.$$

Substituting the value $c = -\frac{975}{7}$ into (5) and solving for r,

$$r = q(z) = \frac{7}{65} z + 15. \tag{6}$$

This is the polar equation for the slanted surface of the bearing. It tells us that for fixed z, r varies from 0 to $\frac{7}{65}z + 15$. Hence, the region

$$\left\{ (r, \theta, z) : 0 \le \theta \le 2\pi, 0 \le r \le \frac{7}{65} z + 15, 0 \le z \le 65 \right\}. \tag{7}$$

We can now set up an iterated integral for the mass of the bearing. The mass element is $dm = \delta\, dV = \delta r\, d\theta\, dr\, dz$. From Fig. 11.54(c) and (6),

$$m = \iiint_{\mathcal{B}} dm = \int_0^{65} \int_0^{7z/65+15} \int_0^{2\pi} \delta r\, d\theta\, dr\, dz \tag{8}$$

$$= 2\pi\delta \int_0^{65} \int_0^{7z/65+15} r\, dr\, dz = 2\pi\delta \int_0^{65} \frac{1}{2} r^2 \Big|_0^{7z/65+15} dz$$

$$= \pi\delta \int_0^{65} \left(\frac{7z}{65} + 15 \right)^2 dz = \frac{67535\pi\delta}{3}.$$

Taking $\delta = 7860 \times 10^{-9}\ \text{kg/mm}^3$ to make the units consistent,

$$m = 0.555879\ \text{kg}.$$

By symmetry, each of the first two coordinates of the center of mass $\mathbf{R} = \langle X, Y, Z \rangle$ of $\mathcal{B}$ are 0. The remaining coordinate is

$$Z = \frac{1}{m} \iiint_{\mathcal{B}} z\, dm = \frac{\delta}{m} \iiint_{\mathcal{B}} z\, dx\, dy\, dz.$$

Evaluating this integral by changing to cylindrical coordinates,

$$Z = \frac{\delta}{m} \int_0^{65} \int_0^{7z/65+15} \int_0^{2\pi} rz\, d\theta\, dr\, dz \tag{9}$$

$$= \frac{2\pi\delta}{m} \int_0^{65} \int_0^{7z/65+15} rz\, dr\, dz = \frac{2\pi\delta}{m} \int_0^{65} \frac{1}{2} z r^2 \Big|_0^{7z/65+15} dz$$

$$= \frac{\pi\delta}{m} \int_0^{65} z \left(\frac{7z}{65} + 15 \right)^2 dz \approx 36.5508\ \text{mm}.$$

The limits in these integrals can also be determined by sketching the region $\mathcal{D}$ in (r, θ, z)-space, where, referring to the Change-of-Variables Formula, $\mathbf{g}(\mathcal{D}) = \mathcal{B}$. Figure 11.55 shows the region $\mathcal{D}$. See (7). This region is of Types I, II, and III. By viewing $\mathcal{D}$ as a region of Type II and applying the Iterated Integral Theorem, we obtain the iterated integrals (8) and (9) used above to calculate m and Z.

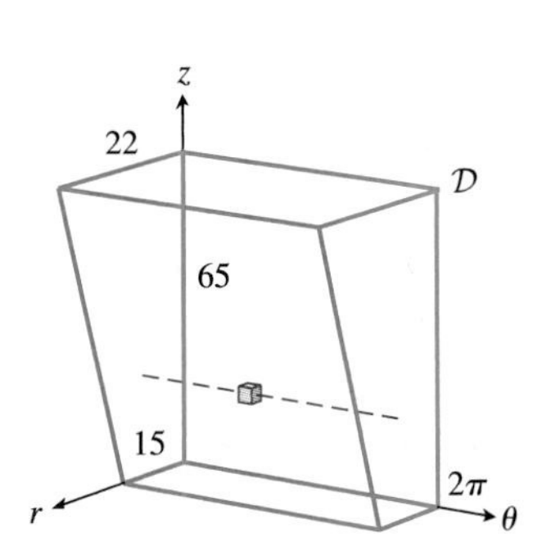

FIGURE 11.55 A view of the bearing in (r, θ, z)-space.

Spherical Coordinates

WEB The equations for changing between the rectangular and spherical coordinate systems are

$$x = \rho \cos \theta \sin \phi$$
$$y = \rho \sin \theta \sin \phi$$
$$z = \rho \cos \phi.$$

The Jacobian of this change of variables is

$$\frac{\partial(x, y, z)}{\partial(\rho, \theta, \phi)} = \begin{vmatrix} \frac{\partial x}{\partial \rho} & \frac{\partial x}{\partial \theta} & \frac{\partial x}{\partial \phi} \\ \frac{\partial y}{\partial \rho} & \frac{\partial y}{\partial \theta} & \frac{\partial y}{\partial \phi} \\ \frac{\partial z}{\partial \rho} & \frac{\partial z}{\partial \theta} & \frac{\partial z}{\partial \phi} \end{vmatrix}$$

$$= \begin{vmatrix} \cos \theta \sin \phi & -\rho \sin \theta \sin \phi & \rho \cos \theta \cos \phi \\ \sin \theta \sin \phi & \rho \cos \theta \sin \phi & \rho \sin \theta \cos \phi \\ \cos \phi & 0 & -\rho \sin \phi \end{vmatrix} = \rho^2 \sin \phi.$$

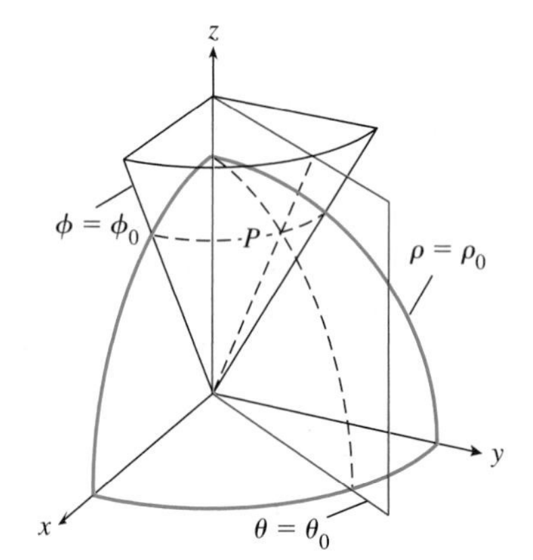

FIGURE 11.56 The point $(\rho_0, \theta_0, \phi_0)$ at the intersection of the mutually perpendicular coordinate surfaces with equations $\rho = \rho_0$, $\theta = \theta_0$, and $\phi = \phi_0$.

Hence, the volume element in spherical coordinates is

$$dV = \rho^2 \sin \phi \, d\rho \, d\theta \, d\phi. \tag{10}$$

We interpret this geometrically.

Figure 11.56 shows a point P with spherical coordinates $(\rho_0, \theta_0, \phi_0)$. This point is common to three coordinate surfaces, an origin-centered sphere with equation $\rho = \rho_0$, a half-plane with equation $\theta = \theta_0$, and a cone with equation $\phi = \phi_0$. These surfaces intersect at right angles. If we increment each variable, we obtain the volume element in spherical coordinates. See Fig. 11.57(a). Diagonal corners of this element are (ρ, θ, ϕ) and $(\rho_0 + d\rho, \theta + d\theta, \phi + d\phi)$.

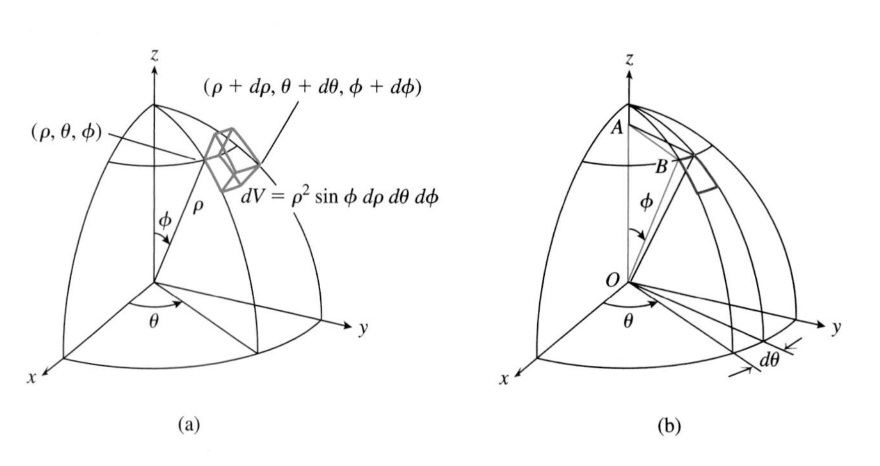

FIGURE 11.57 (a) Volume element in spherical coordinates; (b) calculating the lengths of the edges of the volume element.

One edge of the volume element lies on a ray from the origin; this edge has length $d\rho$. A second edge is an arc of the circle of radius ρ lying in the vertical plane containing the origin and the point (ρ, θ, ϕ). This edge has length $\rho d\phi$. The third edge is also an arc of a circle, but the circle is horizontal and has radius $AB = \rho \sin \phi$. From Fig. 11.57(b) we see that in the triangle ABO, with right angle at A, $\sin \phi = AB/OB = AB/\rho$ and, hence, $AB = \rho \sin \phi$. It now follows that the length of the red arc of the horizontal circle through B is $\rho \sin \phi d\theta$. The product of the lengths of these three edges is

$$dV = d\rho \cdot \rho d\phi \cdot \rho \sin \phi d\theta = \rho^2 \sin \phi d\rho d\theta d\phi. \tag{11}$$

This result agrees with (10).

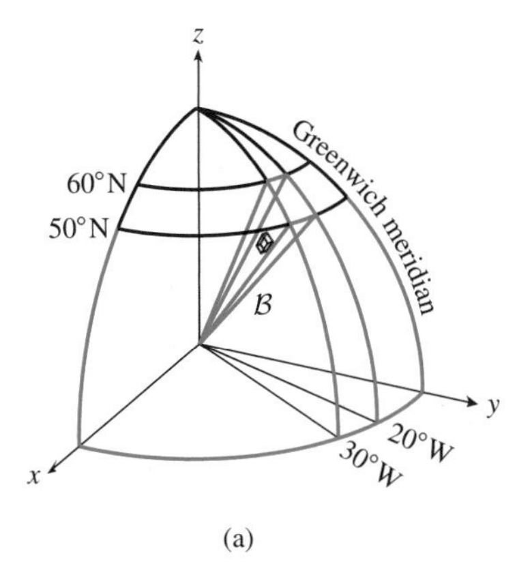

(a)

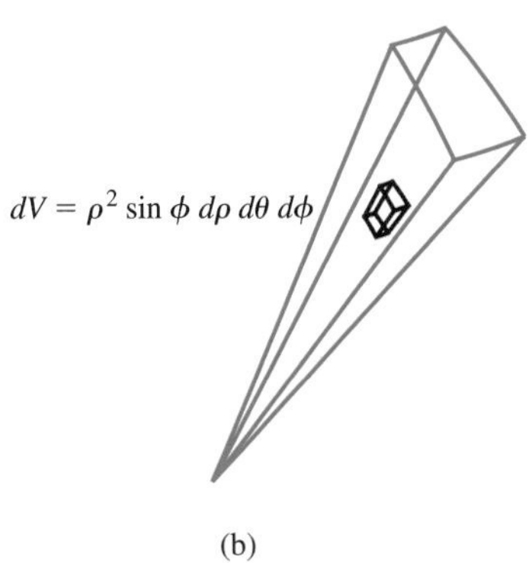

(b)

FIGURE 11.58 (a) The region $\mathcal{B}$ between the center of the Earth and the part of the surface bounded by the meridians 20° W and 30° W and parallels 50° N and 60° N; (b) a zoom-view of the region $\mathcal{B}$ and a spherical volume element within it.

EXAMPLE 4 Find the volume of the wedge-shaped region of the Earth between the center of the Earth and the part of the surface bounded by the meridians 20° W and 30° W and parallels 50° N and 60° N. Assume that the radius of the Earth is about 6378 km.

Solution

Figure 11.58(a) shows the region $\mathcal{B}$ whose volume we wish to calculate. Before setting up the integral for the volume, we switch from geographical longitude and latitude—measured in degrees and established conventions as to origins and positive directions—to spherical coordinates θ and ϕ, measured in radians. Because θ is measured counterclockwise from the positive x-axis (as viewed from high on the positive z-axis), we see that the region lies between the planes $\theta = \pi/3$ and $\theta = \pi/3 + \pi/18$. Note that 10° is equivalent to $10 \cdot \pi/180 = \pi/18$ radians. Because ϕ is measured downwards from the North Pole on great circles through the Pole, we see that the region lies between the cones $\phi = \pi/6$ and $\phi = \pi/6 + \pi/18$.

Figure 11.58(b) is a zoom-view of the region $\mathcal{B}$ and the spherical volume element, which, from (8), has volume $dV = \rho^2 \sin \phi d\rho d\theta d\phi$. With fixed θ and ϕ (fixed longitude and latitude), we sum volume elements in the ρ-direction, from $\rho = 0$ at the center of the Earth to $\rho = 6378$ kilometers at the surface. Next, we sum in the θ-direction, with fixed ϕ. With a given ϕ, θ will vary from $\theta = \pi/3$ to $\theta = \pi/3 + \pi/18 = 7\pi/18$. Finally, we sum in the ϕ-direction, from $\phi = \pi/6$ to $\phi = \pi/6 + \pi/18 = 2\pi/9$. The iterated integral for the volume of $\mathcal{B}$ is

$$V = \int_{\pi/6}^{2\pi/9} \int_{\pi/3}^{7\pi/18} \int_{0}^{6378} \rho^2 \sin \phi d\rho d\theta d\phi.$$

The calculation is straightforward.

$$\begin{aligned} V &= \frac{1}{3} \cdot 6378^3 \int_{\pi/6}^{2\pi/9} \int_{\pi/3}^{7\pi/18} \sin \phi d\theta d\phi \\ &= \frac{1}{3} \cdot 6378^3 \cdot \left(\frac{7\pi}{18} - \frac{\pi}{3} \right) \int_{\pi/6}^{2\pi/9} \sin \phi d\phi \\ &= \frac{1}{3} \cdot 6378^3 \cdot \left(\frac{\pi}{18} \right) (-\cos \phi) \Big|_{\pi/6}^{2\pi/9} \\ &\approx 1.51 \times 10^9 \text{ km}^3. \end{aligned}$$

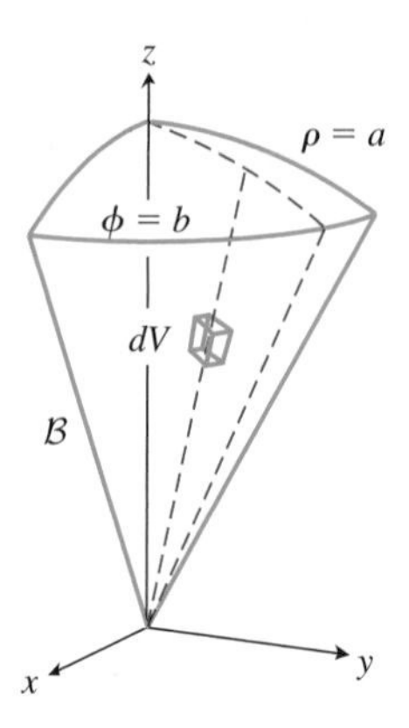

FIGURE 11.59 In spherical coordinates, the region $\mathcal{B}$ is the set of points within both the sphere with equation $\rho = a$ and the cone with equation $\phi = b$.

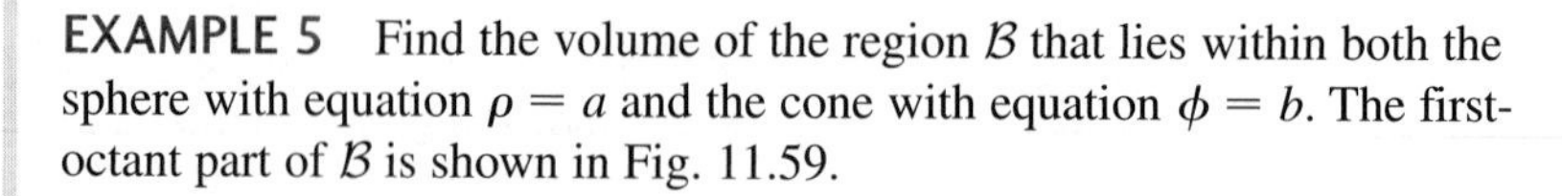

EXAMPLE 5 Find the volume of the region $\mathcal{B}$ that lies within both the sphere with equation $\rho = a$ and the cone with equation $\phi = b$. The first-octant part of $\mathcal{B}$ is shown in Fig. 11.59.

Solution

Using the spherical volume element $dV = \rho^2 \sin\phi\, d\rho\, d\theta\, d\phi$ shown in Fig. 11.59, a natural integration order is ρ, ϕ, and θ. As suggested by the dotted line in the figure, we sum first in the ρ-direction, with ϕ and θ fixed. This generates a thin ray from the origin out to the point (a, θ, ϕ) on the sphere. Next, with θ fixed, we sum the thin rays in the ϕ-direction. This generates the sector of a circle partially outlined with dotted red lines. Finally, summing in the θ-direction sweeps out the region $\mathcal{B}$.

Working directly from the figure,

$$V = \int_0^{2\pi} \int_0^b \int_0^a \rho^2 \sin\phi\, d\rho\, d\phi\, d\theta.$$

Evaluating this triple integral,

$$\begin{aligned} V &= \int_0^{2\pi} \int_0^b \frac{1}{3}\rho^3 \sin\phi \Big|_0^a d\phi\, d\theta = \int_0^{2\pi} \int_0^b \frac{1}{3} a^3 \sin\phi\, d\phi\, d\theta \\ &= \int_0^{2\pi} -\frac{1}{3} a^3 \cos\phi \Big|_0^b d\theta = \frac{1}{3} a^3 (1 - \cos b) \int_0^{2\pi} d\theta \\ &= \frac{2}{3}\pi a^3 (1 - \cos b). \end{aligned}$$

A partial check on this result may be made by setting $b = \frac{1}{2}\pi$. We should obtain $V = \frac{2}{3}\pi a^3$. Why?

EXAMPLE 6 Evaluate the integral

$$I = \iiint_{\mathcal{B}} z^2 e^{-(x^2+y^2+z^2)}\, dx\, dy\, dz$$

where $\mathcal{B}$ is the unit ball centered at the origin.

Solution

Both the region of integration and the form of the integrand suggest that a change to spherical coordinates would simplify our work. In spherical coordinates, the region $\mathcal{B}$ of integration is the set $\{(\rho, \theta, \phi) : 0 \le \rho \le 1, 0 \le \theta \le 2\pi, 0 \le \phi \le \pi\}$. With

$$\mathbf{g}(\rho, \theta, \phi) = \langle \rho \sin\phi \cos\theta, \rho \sin\phi \sin\theta, \rho \cos\phi \rangle,$$

the sphere $\mathcal{B}$ is the set $\mathbf{g}(\mathcal{D})$, where $\mathcal{D}$ is the rectangular parallelepiped in (ρ, θ, ϕ)-space shown in Fig. 11.60. Applying (3) to this integral and recalling that $z = \rho\cos\phi$ and $x^2 + y^2 + z^2 = \rho^2$,

$$\iiint_{\mathbf{g}(\mathcal{D})} z^2 e^{-(x^2+y^2+z^2)}\, dx\, dy\, dz = \iiint_{\mathcal{D}} (\rho^2 \cos^2\phi) e^{-\rho^2} \rho^2 \sin\phi\, d\rho\, d\theta\, d\phi.$$

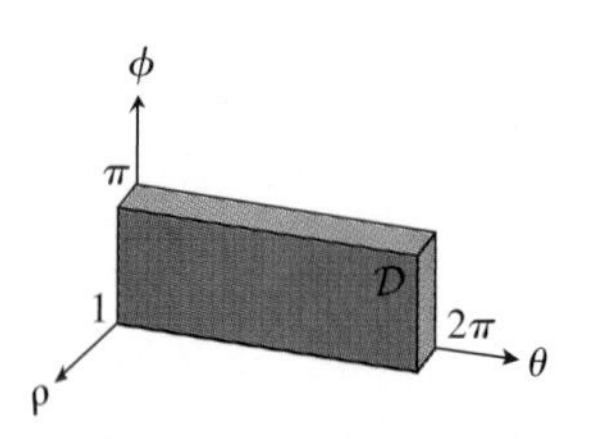

FIGURE 11.60 The set $\mathcal{D}$ in (ρ, θ, ϕ)-space.

Given the shape and the orientation of $\mathcal{D}$, we have a choice of six possible integration orders. We choose an integration order in which the easier integrations come first.

$$
\begin{aligned}
I &= \int_0^1 \int_0^{\pi} \int_0^{2\pi} \rho^4 e^{-\rho^2} \cos^2\phi \sin\phi \, d\theta \, d\phi \, d\rho \\
&= 2\pi \int_0^1 \int_0^{\pi} \rho^4 e^{-\rho^2} \cos^2\phi \sin\phi \, d\phi \, d\rho \\
&= -2\pi \int_0^1 \rho^2 e^{-\rho^4} \left(\frac{1}{3} \cos^3\phi \right) \Big|_0^{\pi} d\rho = \frac{4}{3}\pi \int_0^1 \rho^4 e^{-\rho^2} d\rho.
\end{aligned}
$$

This single integral can be evaluated numerically. Using the built-in numerical integration algorithm on a calculator, we obtained the value 0.10026 87981. Using this result, $I \approx 0.42000\,49593$. For an alternative way of expressing the answer see Exercise 25.

Potential Functions

During the first century after the invention of calculus many physicists and mathematicians worked on the problem of calculating the gravitational attraction of one mass on another; for example, the attraction of the sun on a planet or of the Earth on its moon. Making a direct calculation of, say, the attraction of the Earth on its moon by using a triple integral to sum the contributions to the total force on the moon by each particle of Earth is quite difficult, particularly if the assumption that the Earth is spherical is replaced by the assumption that it is an oblate spheroid (an ellipsoid generated by revolving an ellipse about its minor axis). During the period 1738–1762, such calculations were greatly simplified by the work of Daniel Bernoulli, Leonard Euler, and Joseph Lagrange, all of whom contributed to the idea of a *potential function.*

In his early studies of gravitation, Newton inferred the inverse square law for planets from Kepler's laws. He viewed the planets as point masses subject to the attraction of a sun-centered force. Some twenty years later, Newton stated in the *Principia* his law of universal gravitation, that the force $\mathbf{F}$ exerted on a particle of mass m_1 located at $\mathbf{r}_1$ by a particle of mass m_2 located at $\mathbf{r}_2$ has magnitude proportional to the product of the masses and inversely proportional to the square of the distance between m_1 and m_2. Newton assumed also that $\mathbf{F}$ is directed from $\mathbf{r}_1$ towards $\mathbf{r}_2$. Thus,

$$
\mathbf{F} = \frac{Gm_1m_2}{\|\mathbf{r}_2 - \mathbf{r}_1\|^2} \frac{1}{\|\mathbf{r}_2 - \mathbf{r}_1\|}(\mathbf{r}_2 - \mathbf{r}_1) = \frac{Gm_1m_2}{\|\mathbf{r}_2 - \mathbf{r}_1\|^3}(\mathbf{r}_2 - \mathbf{r}_1). \tag{12}
$$

If this law were used to calculate the total force on the Earth by the sun by summing the forces exerted on each particle of the Earth by each particle of the sun, would it turn out that the total force on the Earth can be calculated directly from (12) by regarding the Earth and sun as point masses? Newton showed in his *Principia* that this is indeed true. We discuss a related problem here and solve it with the help of a potential function.

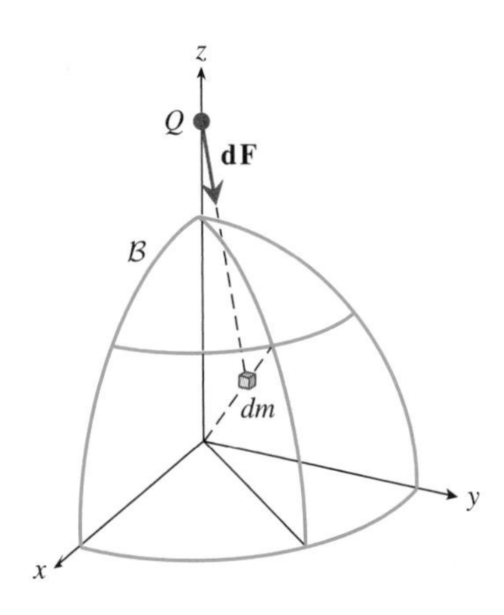

FIGURE 11.61 Attraction on a particle at Q by the solid sphere $\mathcal{B}$.

Figure 11.61 shows part of a solid sphere $\mathcal{B}$ (or ball), a unit mass at Q, and the force $\mathbf{dF}$ on the unit mass due to a particle of mass dm at a representative point

(x, y, z) of $\mathcal{B}$. We show that the total force $\mathbf{F}_Q$ exerted by $\mathcal{B}$ on the particle at Q can be calculated by summing the forces $\mathbf{dF}$ as dm varies over $\mathcal{B}$. We also show that $\mathbf{F}_Q$ can be calculated as if the mass of the ball were concentrated at its center.

We set up an integral for the total force $\mathbf{F}_Q$. Using Newton's law of universal gravitation (12), the element $\mathbf{dF}$ of force exerted on the unit mass at Q by an element dm of mass at (x, y, z) is

$$\mathbf{dF} = \frac{G \cdot 1 \cdot dm}{\|\mathbf{r} - \mathbf{r}_Q\|^3}(\mathbf{r} - \mathbf{r}_Q),$$

where $\mathbf{r}$ and $\mathbf{r}_Q$ are position vectors of the points (x, y, z) and $(0, 0, q)$. If we write the element of mass in terms of a density function $\delta(x, y, z)$ for $\mathcal{B}$ and the volume element $dV = dx\,dy\,dz$, the total force $\mathbf{F}_Q$ at Q is

$$\mathbf{F}_Q = G \iiint_{\mathcal{B}} \frac{\delta(x, y, z)}{(x^2 + y^2 + (z - q)^2)^{3/2}} \langle x, y, z - q \rangle \, dx\,dy\,dz. \tag{13}$$

Even if the density of $\mathcal{B}$ were constant, this integral would look difficult. See, however, Exercise 26. We evaluate the integral (13) indirectly, by defining a potential function.

To discuss the potential function, we put aside for a moment the particle of unit mass at Q and the solid ball $\mathcal{B}$. Figure 11.62 shows a unit mass at $\mathbf{r} = \langle x, y, z \rangle$ and a particle of mass m at $\mathbf{P} = \langle X, Y, Z \rangle$. The force $\mathbf{F}(\mathbf{r})$ exerted on the unit mass by the particle of mass m is

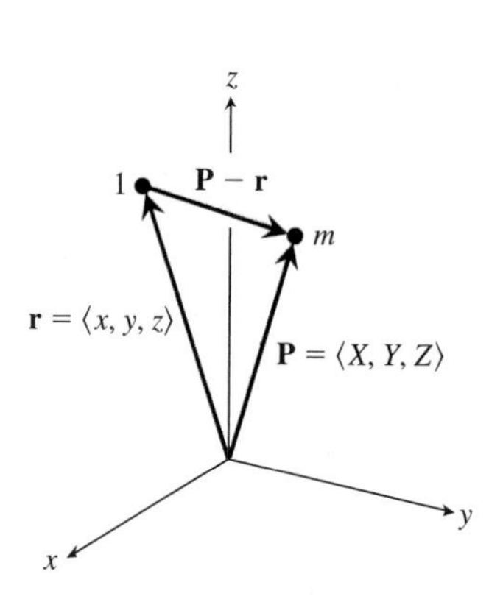

FIGURE 11.62 Calculating the total force at Q.

$$\mathbf{F}(\mathbf{r}) = \frac{Gm}{\|\mathbf{P} - \mathbf{r}\|^2} \frac{1}{\|\mathbf{P} - \mathbf{r}\|}(\mathbf{P} - \mathbf{r}) = \frac{Gm}{\|\mathbf{P} - \mathbf{r}\|^3}(\mathbf{P} - \mathbf{r})$$
$$= \frac{Gm}{\|\mathbf{P} - \mathbf{r}\|^3}\langle X - x, Y - y, Z - z \rangle.$$

We observe that the force $\mathbf{F}(\mathbf{r})$ is the gradient of a scalar function f; specifically, that $\mathbf{F}(\mathbf{r}) = \nabla f(\mathbf{r})$, where

$$f(\mathbf{r}) = \frac{Gm}{\|\mathbf{P} - \mathbf{r}\|} = \frac{Gm}{((X - x)^2 + (Y - y)^2 + (Z - z)^2)^{1/2}}. \tag{14}$$

We verify that the first coordinates of $\nabla f(\mathbf{r})$ and $\mathbf{F}(\mathbf{r})$ are equal. Differentiating f with respect to x,

$$f_x = \frac{\partial}{\partial x} Gm((X - x)^2 + (Y - y)^2 + (Z - z)^2)^{-1/2}$$
$$= Gm\left(-\frac{1}{2}\right)((X - x)^2 + (Y - y)^2 + (Z - z)^2)^{-3/2} \cdot 2 \cdot (X - x) \cdot (-1)$$
$$= \frac{Gm}{\|\mathbf{P} - \mathbf{r}\|^3}(X - x).$$

Note that f_x is the first coordinate of $\mathbf{F}(\mathbf{r})$. The calculations for the other coordinates are similar.

If, instead of a single mass m at P, we have point masses $m_1, \ldots, m_n$ at points with position vectors $\mathbf{P}_1, \ldots, \mathbf{P}_n$, then the total force $\mathbf{F}(\mathbf{r})$ on a unit mass at $\mathbf{r}$ is

$$\mathbf{F}(\mathbf{r}) = \sum_{i=1}^{n} \frac{Gm_i}{\|\mathbf{P}_i - \mathbf{r}\|^3}(\mathbf{P}_i - \mathbf{r}). \tag{15}$$

Applying the observation leading to (14) to each term of this sum,

$$\frac{Gm_i}{\|\mathbf{P}_i - \mathbf{r}\|^3}(\mathbf{P}_i - \mathbf{r}) = \nabla\left(\frac{Gm_i}{\|\mathbf{P}_i - \mathbf{r}\|}\right), \quad i = 1, 2, \ldots, n.$$

It now follows from (15) that

$$\mathbf{F}(\mathbf{r}) = \sum_{i=1}^{n} \frac{Gm_i}{\|\mathbf{P}_i - \mathbf{r}\|^3}(\mathbf{P}_i - \mathbf{r}) = \sum_{i=1}^{n} \nabla\left(\frac{Gm_i}{\|\mathbf{P}_i - \mathbf{r}\|}\right) = \nabla\left(\sum_{i=1}^{n} \frac{Gm_i}{\|\mathbf{P}_i - \mathbf{r}\|}\right). \tag{16}$$

Hence, the total force $\mathbf{F}(\mathbf{r})$ at $\mathbf{r}$ is again the gradient of a scalar function.

Now suppose that, instead of n point masses, we have a "continuous" mass distributed throughout a region $\mathcal{B}$. We approximate the total force $\mathbf{F}(\mathbf{r})$ exerted by the mass $\mathcal{B}$ on a unit mass at $\mathbf{r}$ by subdividing $\mathcal{B}$. Consider a representative subregion located at $\mathbf{P}_{i,j,k} = \langle X_i, Y_j, Z_k\rangle$ and with volume $dX\,dY\,dZ$. The mass of this subregion is $\delta(\mathbf{P}_{i,j,k})\,dX\,dY\,dZ$, approximately. As above,

$$\mathbf{F}(\mathbf{r}) \approx \sum_{i,j,k} \frac{G\delta(\mathbf{P}_{i,j,k})\,dX\,dY\,dZ}{\|\mathbf{P}_{i,j,k} - \mathbf{r}\|^3}(\mathbf{P}_{i,j,k} - \mathbf{r}) = \nabla\left(\sum_{i,j,k} \frac{G\delta(\mathbf{P}_{i,j,k})}{\|\mathbf{P}_{i,j,k} - \mathbf{r}\|}\,dX\,dY\,dZ\right).$$

As the number of subdivisions of $\mathcal{B}$ increases without bound, we expect that

$$\mathbf{F}(\mathbf{r}) = \iiint_{\mathcal{B}} \frac{G\delta(\mathbf{P})}{\|\mathbf{P} - \mathbf{r}\|^3}(\mathbf{P} - \mathbf{r})\,dX\,dY\,dZ = \nabla \iiint_{\mathcal{B}} \frac{G\delta(\mathbf{P})}{\|\mathbf{P} - \mathbf{r}\|}\,dX\,dY\,dZ. \tag{17}$$

Hence, as in (16), the total force $\mathbf{F}(\mathbf{r})$ at $\mathbf{r}$ is the gradient of a scalar function. Specifically, $\mathbf{F}(\mathbf{r}) = \nabla f(\mathbf{r})$, where

$$f(\mathbf{r}) = \iiint_{\mathcal{B}} \frac{G\delta(\mathbf{P})}{\|\mathbf{P} - \mathbf{r}\|}\,dX\,dY\,dZ. \tag{18}$$

We use this result to evaluate the integral (13). Because

$$\mathbf{F}_Q = \mathbf{F}(0, 0, q) = \nabla f(0, 0, q),$$

we set $x = y = 0$ in (18). Moreover, by symmetry, only the third coordinate of $\mathbf{F}_Q(0, 0, q)$ is nonzero. Hence, in calculating the gradient of $f(0, 0, z)$ we need only to differentiate $f(0, 0, z)$ with respect to z.

Focusing for a moment on the denominator $\|\mathbf{P} - \mathbf{r}\|$ in (16), we change from (X, Y, Z)-coordinates to spherical coordinates. This gives

$$\begin{aligned}
\|\mathbf{P} - \mathbf{r}\| &= \sqrt{(X - x)^2 + (Y - y)^2 + (Z - z)^2} = \sqrt{X^2 + Y^2 + (Z - z)^2} \\
&= \sqrt{\rho^2 \sin^2\phi \cos^2\theta + \rho^2 \sin^2\phi \sin^2\theta + (\rho \cos \phi - z)^2} \\
&= \sqrt{\rho^2 \sin^2\phi + \rho^2 \cos^2\phi - 2\rho z \cos \phi + z^2} \\
&= \sqrt{\rho^2 - 2\rho z \cos \phi + z^2}.
\end{aligned}$$

Substituting this result into (18) and applying the Change-of-Variables Formula (3)

$$\begin{aligned}
f(0, 0, z) &= G\delta \int_0^a \int_0^\pi \int_0^{2\pi} \frac{\rho^2 \sin \phi}{\sqrt{\rho^2 - 2\rho z \cos \phi + z^2}}\,d\theta\,d\phi\,d\rho \\
&= 2\pi G\delta \int_0^a \int_0^\pi \frac{\rho^2 \sin \phi}{\sqrt{\rho^2 - 2\rho z \cos \phi + z^2}}\,d\phi\,d\rho.
\end{aligned}$$

Making the substitution $w = \rho^2 - 2\rho z \cos\phi + z^2$ in the inner integral,

$$f(0,0,z) = \frac{2\pi G\delta}{z}\int_0^a \rho(\rho^2 - 2\rho z\cos\phi + z^2)^{1/2}\Big|_0^\pi \, d\rho$$

$$= \frac{2\pi G\delta}{z}\int_0^a \rho\big(|\rho + z| - |\rho - z|\big)\, d\rho.$$

Because we eventually take $z = q > a$, it follows that $z > \rho \geq 0$ and $|\rho + z| - |\rho - z| = 2\rho$. Thus, the potential function f is

$$f(0,0,z) = \frac{4\pi G\delta}{z}\int_0^a \rho^2\, d\rho = \frac{4\pi G\delta a^3}{3z} = \frac{Gm}{z},$$

where $m = \frac{4}{3}\pi\delta a^3$ is the mass of the solid ball $\mathcal{B}$. We can now evaluate the integral (13) by calculating f_z. We have

$$\frac{\partial}{\partial z} f(0,0,z) = -\frac{Gm}{z^2}.$$

$$\mathbf{F}(0,0,z) = \langle 0, 0, f_z(0,0,z)\rangle = \left\langle 0, 0, -\frac{Gm}{z^2}\right\rangle.$$

At the point $(0,0,q)$ the magnitude of the force $\mathbf{F}_Q = \mathbf{F}(0,0,q)$ is

$$\|\mathbf{F}_Q\| = \frac{Gm}{q^2}.$$

We see, then, that the force exerted on a unit mass at $(0,0,q)$ by a ball can be calculated as if the entire mass of the ball were concentrated at its center.

Exercises 11.6

1. The change of variables $(x, y, z) = \mathbf{g}(u, v, w)$ defined by the equations

$$x = u + v + w$$
$$y = u + v$$
$$z = v + w$$

transforms the unit cube $\mathcal{D}$ in (u, v, w)-space to a parallelepiped $\mathcal{B} = \mathbf{g}(\mathcal{D})$ in (x, y, z)-space. Calculate the volume of this parallelepiped in two ways, first from

$$|(\mathbf{g}(1,0,0) \times \mathbf{g}(0,1,0)) \cdot \mathbf{g}(0,0,1)|;$$

next from $|\partial(x, y, z)/\partial(u, v, w)|$. Explain.

2. The change of variables $(x, y, z) = \mathbf{g}(u, v, w)$ defined by the equations

$$x = u + 2v + w$$
$$y = u + v + w$$
$$z = 2u + v + w$$

transforms the unit cube $\mathcal{D}$ in (u, v, w)-space to a parallelepiped $\mathcal{B} = \mathbf{g}(\mathcal{D})$ in (x, y, z)-space. Calculate the volume of this parallelepiped in two ways, first from

$$|(\mathbf{g}(1,0,0) \times \mathbf{g}(0,1,0)) \cdot \mathbf{g}(0,0,1)|;$$

next from $|\partial(x, y, z)/\partial(u, v, w)|$. Explain.

3. In the accompanying figure, the set $\mathcal{D}$ is the region below the graph of the equation $z = r\theta^2$ and above the rectangle $[0, 1] \times [0, \pi/3]$ in the (r, θ)-plane. Fig. 11.53 shows the set $\mathcal{B} = \mathbf{g}(\mathcal{D})$ in (x, y, z)-space. The solution of Example 2 used the integration order z, r, and θ. Calculate the volume of $\mathcal{B}$ using the integration order θ, z and r.

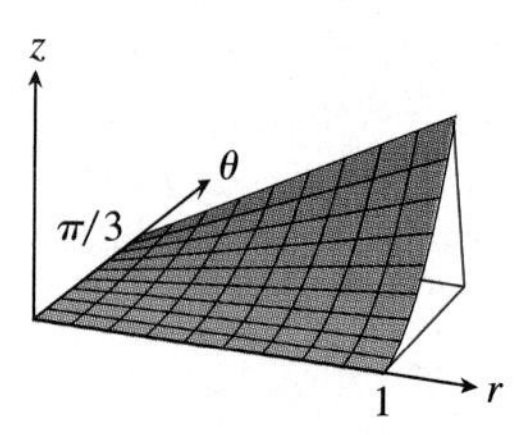

Figure for Exercise 3.

4. In the accompanying figure, the set $\mathcal{D}$ is the region below the graph of the equation $z = r\theta^3$ and above the rectangle $[0, 1] \times [0, \pi/4]$ in the (r, θ)-plane. Fig. 11.53 shows what is very nearly the set $\mathcal{B} = \mathbf{g}(\mathcal{D})$ in (x, y, z)-space. The solution of Example 2 used the integration order z, r, and θ. Calculate the volume of $\mathcal{B}$ using the integration order θ, z, and r.

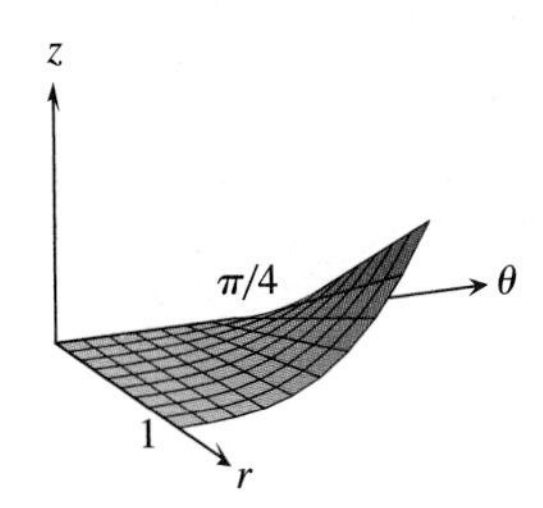

Figure for Exercise 4.

5. Find the volume of the region that is inside both the origin-centered ball of radius a and the cylinder $r = a\cos\theta$.

6. Find the volume of the region that is inside both the origin-centered ball of radius a and the cylinder $r = a\sin\theta$.

7. Find the volume of the region between the paraboloids with equations $z = r^2$ and $z = 8 - r^2$.

8. Find the volume of the region between the paraboloids with equations $z = r^2$ and $z = 18 - r^2$.

9. Find the mass of a solid cylinder of radius a and height h if its density at any point is proportional to the distance from that point to the axis of the cylinder.

10. Find the mass of a solid cylinder of radius a and height h if its density at a point is proportional to the distance from that point to one base of the cylinder.

11. Find the centroid of the region $\mathcal{B}$ described in Example 5.

12. Find the center of mass of a solid occupying the region $\mathcal{B}$ described in Example 5. The density of the solid at any point is proportional to the distance of that point from the tip of the cone.

13. Evaluate $\iiint_{\mathcal{B}} e^{-(x^2+y^2+z^2)^{3/2}}\,dx\,dy\,dz$, where $\mathcal{B}$ is the unit ball centered at the origin.

14. Evaluate $\iiint_{\mathcal{B}} z e^{-(x^2+y^2+z^2)^2}\,dx\,dy\,dz$, where $\mathcal{B}$ is the upper half of the unit ball centered at the origin.

15. Use spherical coordinates in evaluating the integral $\iiint_{\mathcal{B}} x\,dx\,dy\,dz$, where $\mathcal{B}$ is the first-octant portion of the unit ball centered at the origin.

16. Use spherical coordinates in evaluating the integral $\iiint_{\mathcal{B}} y\,dx\,dy\,dz$, where $\mathcal{B}$ is the first-octant portion of the unit ball centered at the origin.

17. Find the volume of the first-octant region lying inside the ellipsoid with equation $x^2/a^2 + y^2/b^2 + z^2/c^2 = 1$, where a, b, and c are positive constants, and between the planes with equations $x = 0$ and $x = y$. Use the change of variables $x = au$, $y = bv$, and $z = cw$.

18. Variation on the preceding problem Find the centroid of the region lying inside the ellipsoid with equation $x^2/a^2 + y^2/b^2 + z^2/c^2 = 1$, where a, b, and c are positive constants, and above the plane with equation $z = 0$. Use the change of variables $x = au$, $y = bv$, and $z = cw$.

19. Use cylindrical coordinates in determining the volume of the region bounded above by the plane with equation $z = y$ and below by the paraboloid $z = x^2 + y^2$.

20. Continuation Find the centroid of the region in Exercise 19.

21. Find the volume of the solid that lies above the cone with equation $\phi = \pi/3$ and inside the torus with equation $\rho = 4\sin\phi$.

22. Find the volume of the solid that lies above the cone with equation $\phi = \pi/3$ and below the sphere with equation $\rho = 4\cos\phi$.

23. Use cylindrical coordinates to find the centroid of a hemisphere of radius a.

24. Use cylindrical or spherical coordinates to find the center of mass of a hemisphere of radius a if its density at any point P is proportional to the distance from P to the center of the base.

25. The error function is defined by

$$\operatorname{erf}(x) = \frac{2}{\sqrt{\pi}}\int_0^x e^{-t^2}\,dt,$$

for $-\infty < x < \infty$. Show that

$$\int_0^1 \rho^4 e^{-\rho^2}\,d\rho = -\frac{5}{4}e^{-1} + \frac{3}{8}\sqrt{\pi}\operatorname{erf}(1)$$

by integrating by parts twice. In each case, take $dv = \rho e^{-\rho^2}\,d\rho$. Use this result in Example 6 to show that

$$I = \frac{\pi}{6e}\left(-10 + 3e\sqrt{\pi}\operatorname{erf}(1)\right).$$

Given that $\operatorname{erf}(1) \approx 0.84270\,07929$, check on the numerical integration used in Example 6.

26. Complete the following outline of a direct calculation of the total force exerted on a unit mass at Q by a solid ball $\mathcal{B}$ of radius a and constant density δ. From (13), argue that the first two coordinates of $\mathbf{F}_Q$ are 0 and, changing to spherical coordinates, the third coordinate F_3 of $\mathbf{F}_Q$ is

$$F_3 = G\delta\int_0^a\int_0^\pi\int_0^{2\pi} g(\rho, \phi)\,dV,$$

where $dV = \rho^2\sin\phi\,d\theta\,d\phi\,d\rho$ and

$$g(\rho, \phi) = \frac{\rho\cos\phi - q}{(\rho^2 - 2\rho q + q^2)^{3/2}}.$$

The integration on θ is easy. For the integration with respect to ϕ, make the substitution $w = \rho^2 - 2\rho q \cos\phi + q^2$. Show that $dw = 2\rho q \sin\phi\, d\phi$ and

$$\rho\cos\phi - q = \frac{\rho^2 - q^2 - w}{2q}.$$

Show that F_3 becomes

$$F_3 = \frac{G\delta\pi}{2q^2}\int_0^a \left(\int_{(\rho-q)^2}^{(\rho+q)^2} ((\rho^2 - q^2)w^{-3/2} - w^{-1/2})\, dw\right)\rho\, d\rho.$$

Show that the inner integral simplifies to -8ρ (keep in mind that $q > \rho$) and

$$F_3 = -\frac{G}{q^2}M.$$

27. Variation on the preceding problem Work through the preceding problem, but assume that the density of $\mathcal{B}$ is a function of the distance from its center. Does the main conclusion—that the total force exerted by $\mathcal{B}$ on the particle at Q can be calculated as if the mass of the sphere were concentrated at its center—remain true?

28. The change of variables

$$(x, y, z) = \mathbf{g}(u, v, w)$$
$$= \left(\frac{5}{4}u + \frac{1}{4}v + \frac{1}{4}w, \frac{1}{4}u + \frac{5}{4}v + \frac{1}{4}w, \frac{7}{8}w\right),$$

used in Example 1 was given without a discussion of how it might have arisen. Given that $\mathcal{B}$ in Fig. 11.51 is the parallelepiped with vertices $(0,0,0)$, $(5/4,1/4,0)$, $(3/2,3/2,0)$, $(1/4,5/4,0)$, $(1/4,1/4,7/8)$, $(3/2,1/2,7/8)$, $(7/4,7/4,7/8)$, $(1/2,3/2,7/8)$, show that the function $\mathbf{g}$ arises by imagining the unit cube in (u, v, w)-space is linearly deformed into $\mathcal{B}$ in (x, y, z)-space. Specifically, think of $\mathbf{g}$ in the form

$$\begin{pmatrix} x \\ y \\ z \end{pmatrix} = \begin{pmatrix} g_{1,1} & g_{1,2} & g_{1,3} \\ g_{2,1} & g_{2,2} & g_{2,3} \\ g_{3,1} & g_{3,2} & g_{3,3} \end{pmatrix}\begin{pmatrix} u \\ v \\ w \end{pmatrix}.$$

The entries $g_{i,j}$ of the matrix associated with $\mathbf{g}$ can be found by using just a few of the given vertices of $\mathcal{B}$.

T 29. The upper portion of the accompanying figure shows a space station formed from a ball of radius 1 by removing the part common to a ball of radius 0.75 and positioned so that its center is 1.4 units directly above the center of the ball of radius 1. A vertical cross section of the two balls is shown in the lower portion of the figure. If the density at a point (x, y, z) of the space station is proportional to the distance from the center of the sphere from which it was formed, find its mass.

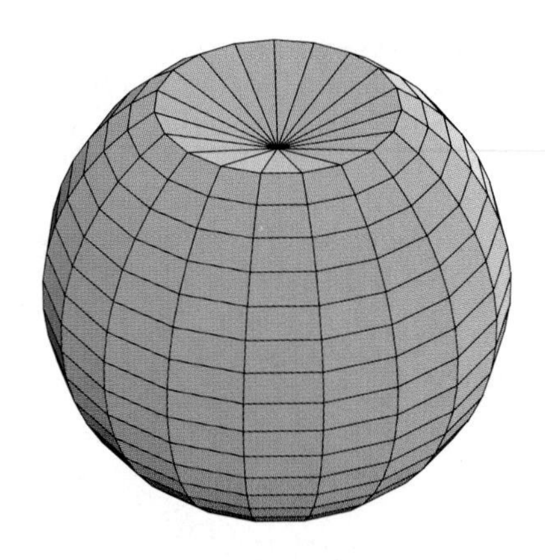

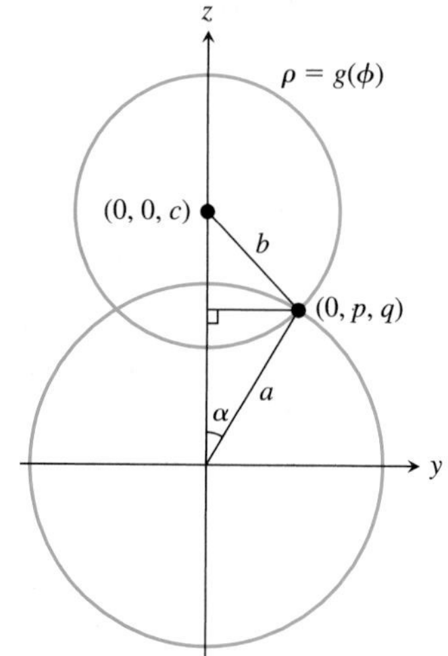

Figure for Exercise 29.

30. A bubble of gas of radius 0.1 cm is introduced into an eyeball, which is assumed to be a sphere of radius 1 cm. The bubble rises to the top of the eyeball, and its bottom more or less flattens to a horizontal plane. What percentage of the inside surface area of the eyeball is covered up by the flattened bubble?

31. Express the integral

$$I = \iiint_{\mathcal{B}} e^{-(x^2+13y^2+2z^2-4xy-6yz)}\, dV$$

as a product of integrals of the form $\int_a^b e^{-t^2}\, dt$. The region of integration $\mathcal{B}$ is bounded by the parallelepiped with vertices $(0,0,0)$, $(1,0,0)$, $(5/3,1/3,0)$, $(2/3,1/3,0)$, $(2/3,1/3,1)$, $(5/3,1/3,1)$, $(7/3,2/3,1)$, $(4/3,2/3,1)$. Note that the expression $x^2 + 13y^2 + 2z^2 - 4xy - 6yz$ from the integrand can be written in the form

$$(x - 2y)^2 + (3y - z)^2 + z^2$$

by completing the square. The form of this result suggests the substitutions

$$u = x - 2y$$
$$v = 3y - z$$
$$w = z.$$

Show that $\mathcal{B} = \mathbf{g}(\mathcal{D})$, where

$$\mathcal{D} = [0,1] \times [0,1] \times [0,1],$$

and

$$I = \frac{1}{3}\int_0^1 e^{-u^2}\,du \int_0^1 e^{-v^2}\,dv \int_0^1 e^{-w^2}\,dw$$
$$= \frac{1}{3}\left(\int_0^1 e^{-t^2}\,dt\right)^3.$$

Review of Key Concepts

In this chapter we defined double and triple integrals, evaluated these integrals by applying an iterated integral theorem or making a change of variables, and applied the idea of multiple integrals to measure several geometric and physical quantities. The definitions of double and triple integrals are patterned after that of the single integral. The change-of-variable formulas for the double and triple integrals are analogous to that for the single integral. The proofs of these formulas, which depend upon the Linear Functions and Area Theorem from Section 8.5, are quite similar to the proof for the change-of-variables formula for the single integral. Elements of volume, surface area, and mass were used throughout the chapter to connect geometric or physical ideas with the formal definitions. We calculated the volumes of the elements of volume for the cylindrical and spherical coordinate systems.

We summarize these ideas in table form.

Chapter Summary

Definition of the Double Integral

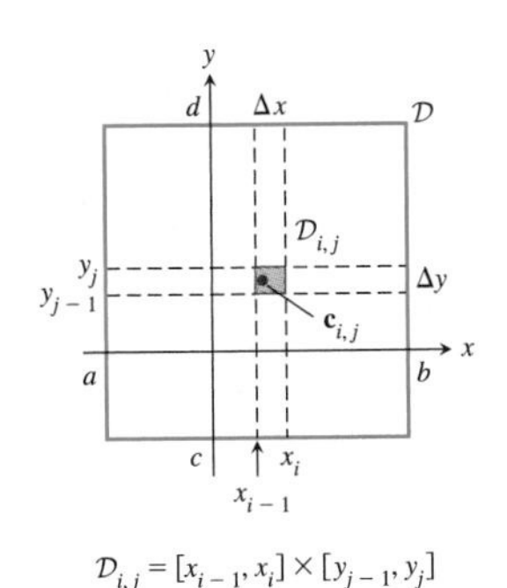

$\mathcal{D}_{i,j} = [x_{i-1}, x_i] \times [y_{j-1}, y_j]$

Let f be a bounded function defined on $\mathcal{D} = [a,b] \times [c,d]$. If there is a number I for which

$$\lim_{n\to\infty} R_n = \lim_{n\to\infty} \sum_{i,j} f(\mathbf{c}_{i,j})\,\Delta x\,\Delta y = I,$$

where $\Delta x = (b-a)/n$ and $\Delta y = (d-c)/n$, $\mathcal{D}_{i,j}$ is a subregion of a regular subdivision $\mathcal{S}_n$ of $\mathcal{D}$, and $\{\mathbf{c}_{i,j}: 1 \le i,j \le n\}$ is an evaluation set for $\mathcal{S}_n$, then we say that f is **integrable** on $\mathcal{D}$, that I is the value of the **double integral,** and write

$$\iint_{\mathcal{D}} f(x,y)\,dA = \iint_{\mathcal{D}} f(x,y)\,dx\,dy = I.$$

For $f(x,y) = xy^2$, $0 \le x,\ y \le 1$, we form R_2, where the evaluation set is the set of midpoints of the four squares making up $\mathcal{S}_2$. With $\Delta x = \frac{1}{2}$, $\Delta y = \frac{1}{2}$, $c_{1,1} = \left(\frac{1}{4}, \frac{1}{4}\right)$, and so on,

$$R_2 = \Delta x\,\Delta y\left(\frac{1}{64} + \frac{3}{64} + \frac{9}{64} + \frac{27}{64}\right)$$
$$= \frac{5}{32} = 0.15625.$$

This number approximates

$$\iint_{\mathcal{D}} f(x,y)\,dx\,dy = \frac{1}{6} \approx 0.16667.$$

Iterated Integral Theorem for Rectangular Regions

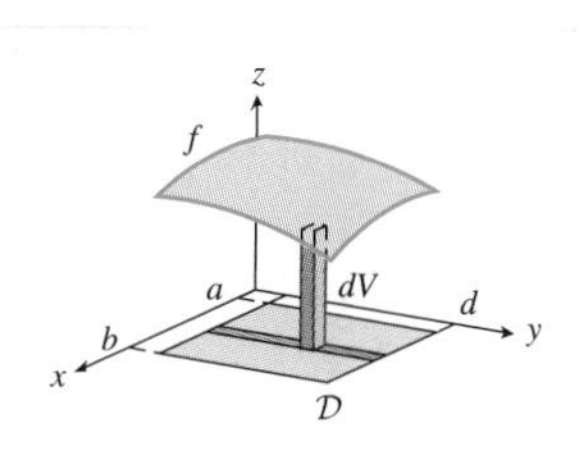

If f is continuous on $\mathcal{D} = [a, b] \times [c, d]$, then $\iint_{\mathcal{D}} f(x, y)\, dA$ is equal to each of the iterated integrals

$$\int_c^d \left(\int_a^b f(x, y)\, dx \right) dy$$

and

$$\int_a^b \left(\int_c^d f(x, y)\, dy \right) dx.$$

For $f(x, y) = 100 - x^2 - y^2$, $a = 1$, $b = 4$, $c = 2$, and $d = 5$, each of the iterated integrals

$$\int_2^5 \left(\int_1^4 (100 - x^2 - y^2)\, dx \right) dy$$

and

$$\int_1^4 \left(\int_2^5 (100 - x^2 - y^2)\, dy \right) dx$$

is equal to the integral $\iint_{\mathcal{D}} f(x, y)\, dA$. The value of all three integrals is 720.

Extending the Double Integral

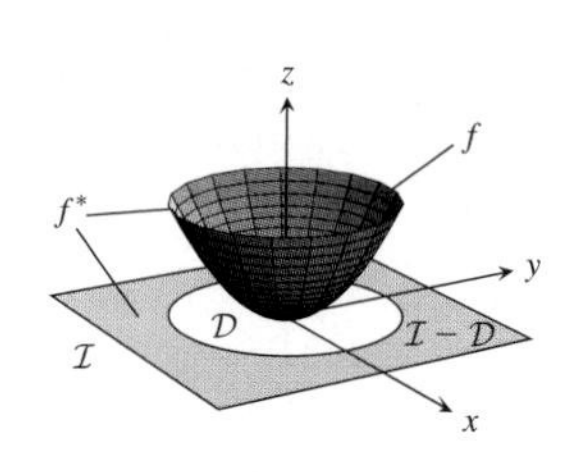

Let f be a bounded function defined on a bounded region $\mathcal{D}$. We say that f is *integrable* on $\mathcal{D}$ if the function f^* is integrable on a rectangular region $\mathcal{I} = [a, b] \times [c, d]$ containing $\mathcal{D}$, where

$$f^*(x, y) = \begin{cases} f(x, y), & (x, y) \in \mathcal{D} \\ 0, & (x, y) \in \mathcal{I} - \mathcal{D}. \end{cases}$$

We define

$$\iint_{\mathcal{D}} f(x, y)\, dx\, dy = \iint_{\mathcal{I}} f^*(x, y)\, dx\, dy.$$

Letting $\mathcal{D}$ be the unit disk centered at the origin and $f(x, y) = x^2 + y^2$ for $(x, y) \in \mathcal{D}$, and taking $\mathcal{I} = [0, 1.5] \times [0, 1.5]$,

$$\iint_{\mathcal{D}} (x^2 + y^2)\, dx\, dy = \iint_{\mathcal{I}} f^*(x, y)\, dx\, dy.$$

Iterated Integral Theorem for Regions

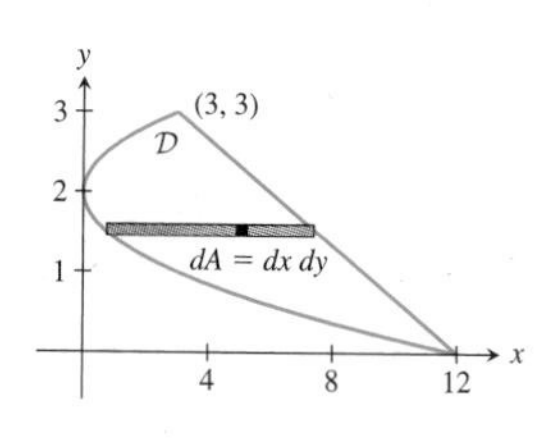

If f is continuous on a region $\mathcal{D}$ of Type I, then

$$\iint_{\mathcal{D}} f\, dA = \int_a^b \left(\int_{g_1(x)}^{g_2(x)} f(x, y)\, dy \right) dx$$

If f is continuous on a region $\mathcal{D}$ of Type II, then

$$\iint_{\mathcal{D}} f\, dA = \int_c^d \left(\int_{h_1(y)}^{h_2(y)} f(x, y)\, dx \right) dy$$

Shown in the figure are curves with equations $3(y - 2)^2 = x$ and $x + 3y = 12$. The region $\mathcal{D}$ they bound is of Type II. The element of area is $dA = dx\, dy$ and

$$\iint_{\mathcal{D}} f\, dA = \int_0^3 \int_{3(y-2)^2}^{12-3y} f(x, y)\, dx\, dy.$$

Definition of Volume and Properties of the Double Integral

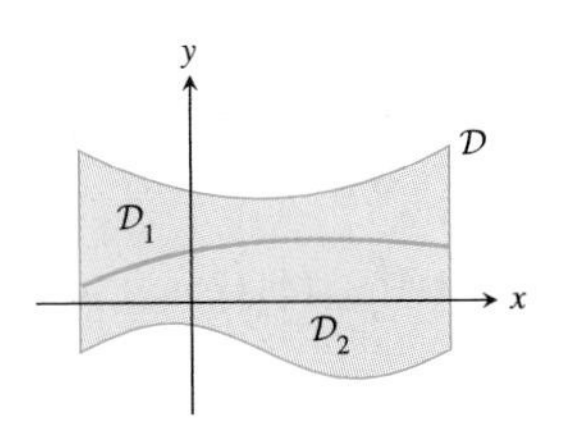

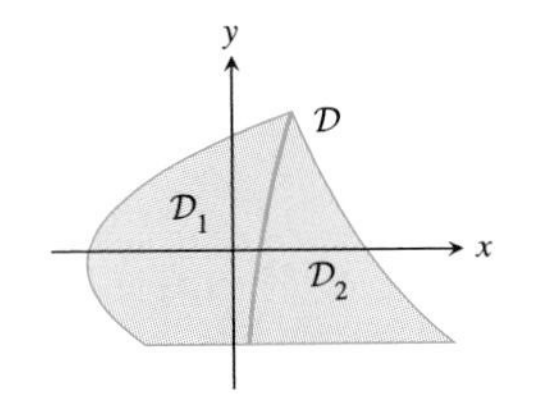

Let f and g be integrable on the region $\mathcal{D}$ and suppose that $\mathcal{D}$ has been decomposed into regions $\mathcal{D}_1$ and $\mathcal{D}_2$.

Definition: Area If f is nonnegative on a region $\mathcal{D}$ of Type I or Type II, the volume of the region beneath the graph of f (and above the (x, y)-plane) is the number $\iint_{\mathcal{D}} f(x, y)\, dA$.

Property 1: Additivity The function f is integrable on $\mathcal{D}_1$ and $\mathcal{D}_2$ and

$$\iint_{\mathcal{D}} f(x, y)\, dx\, dy = \iint_{\mathcal{D}_1} f(x, y)\, dx\, dy + \iint_{\mathcal{D}_2} f(x, y)\, dx\, dy.$$

Property 2: Linearity Let r and s be real numbers. The function h defined on $\mathcal{D}$ by $h(x) = rf(x) + sg(x)$ is integrable on $\mathcal{D}$ and

$$\iint_{\mathcal{D}} (rf(x, y) + sg(x, y))\, dx\, dy = r \iint_{\mathcal{D}} f(x, y)\, dx\, dy + s \iint_{\mathcal{D}} g(x, y)\, dx\, dy.$$

Mass and Center of Mass of a Lamina

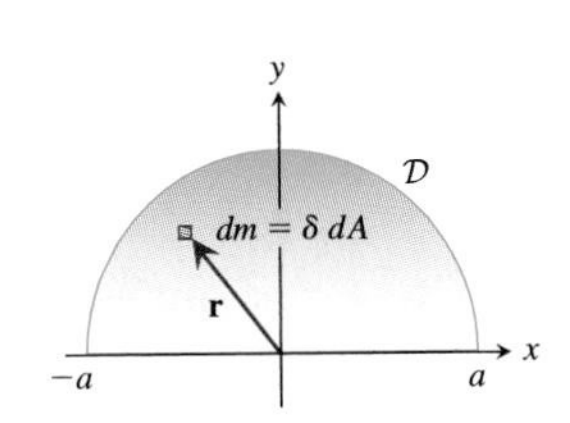

The mass m of a lamina $\mathcal{D}$ with density function δ is

$$m = \iint_{\mathcal{D}} dm = \iint_{\mathcal{D}} \delta(x, y)\, dA.$$

The center of mass of the lamina is located at the point $\mathbf{R}$, where

$$\mathbf{R} = \frac{1}{m} \iint_{\mathcal{D}} \mathbf{r}\, dm$$

$$= \frac{1}{m} \iint_{\mathcal{D}} \delta(x, y) \langle x, y \rangle\, dx\, dy.$$

Integrals for the mass and center of mass of a semicircular lamina with radius a and density proportional to the distance from the diameter are

$$m = \int_{-a}^{a} \int_{0}^{\sqrt{a^2 - x^2}} ky\, dy\, dx = \tfrac{2}{3} ka^3$$

$$\mathbf{R} = \frac{1}{m} \int_{-a}^{a} \int_{0}^{\sqrt{a^2 - x^2}} ky \langle x, y \rangle\, dy\, dx$$

$$= \left(0, \frac{3\pi a}{16}\right).$$

Surface Area

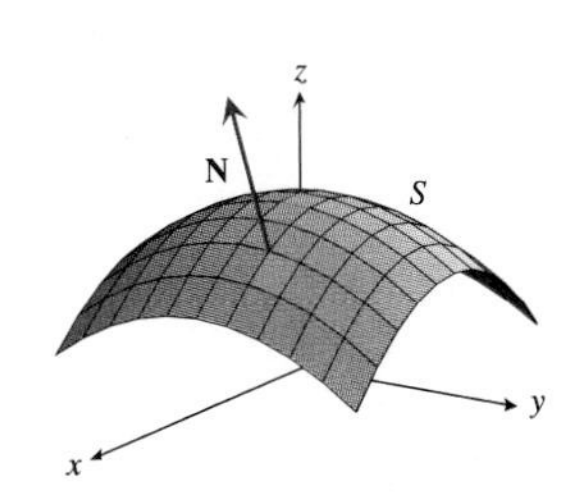

Let S be a simple, smooth surface described by the vector function

$$\mathbf{r} = \mathbf{r}(u, v), \qquad (u, v) \in \mathcal{D},$$

If S is described by an equation of the form $z = f(x, y)$, $(x, y) \in \mathcal{D}$, then

$$A(S) = \iint_{\mathcal{D}} \sqrt{1 + f_x^2 + f_y^2}\, dx\, dy.$$

Surface Area (continued)

where $\mathcal{D}$ is a region of Type I or II. The area A of S is

$$A(S) = \iint_{\mathcal{D}} \|\mathbf{r}_u \times \mathbf{r}_v\| \, du \, dv.$$

If S is generated by revolving the curve C in the (x, y)-plane about the x-axis, where C is described by

$$\mathbf{r}_C = \langle x(t), y(t), 0 \rangle, \qquad a \leq t \leq b,$$

then

$$A(S) = 2\pi \int_a^b y(t)\sqrt{x'(t)^2 + y'(t)^2} \, dt.$$

Change-of-Variables Formula

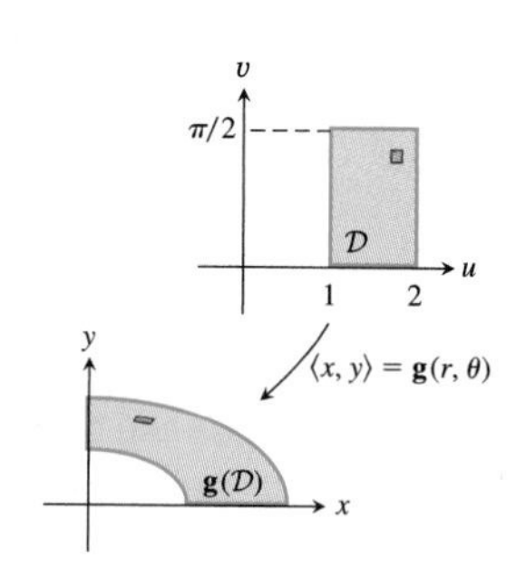

Let $\mathbf{g}$ be a function defined on a region $\mathcal{D}$ of R^2. Assume that on $\mathcal{D}$ the coordinate functions of $\mathbf{g}$ are continuously differentiable, $\mathbf{g}$ is one-to-one, and $\partial(x, y)/\partial(u, v) \neq 0$. If f is a continuous function defined on the range $\mathbf{g}(\mathcal{D})$ of $\mathbf{g}$, then

$$\iint_{\mathbf{g}(\mathcal{D})} f(x, y) \, dx \, dy$$

$$= \iint_{\mathcal{D}} f(\mathbf{g}(u, v)) \left| \frac{\partial(x, y)}{\partial(u, v)} \right| du \, dv.$$

For the integral $\iint_{\mathbf{g}(\mathcal{D})} xy \, dx \, dy$ and the change of variables

$$\langle x, y \rangle = \mathbf{g}(u, v) = \langle 2u \cos v, u \sin v \rangle,$$

where $(u, v) \in [1, 2] \times [0, \pi/2]$, as in the figure, the Jacobian is $2u$ and

$$\iint_{\mathbf{g}(\mathcal{D})} xy \, dx \, dy$$

$$= \int_0^{\pi/2} \int_1^2 2u^2 \sin(2v) \, du \, dv = \frac{14}{3}.$$

Change of Variables from Rectangular to Cylindrical Coordinates

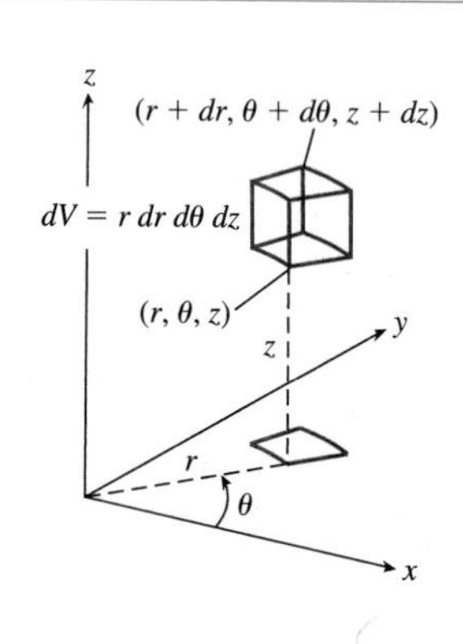

The change-of-variables formula for rectangular to cylindrical coordinates is

$$\iiint_{\mathbf{g}(\mathcal{D})} f(x, y, z) \, dx \, dy \, dz$$

$$= \iiint_{\mathcal{D}} f(r \cos \theta, r \sin \theta, z) r \, dr \, d\theta \, dz.$$

If $\mathbf{g}(\mathcal{D})$ is the cylinder $x^2 + y^2 = 1$, between $z = 0$ and $z = 2$, then

$$\iiint_{\mathbf{g}(\mathcal{D})} x^2 y^2 z \, dx \, dy \, dz$$

$$= \int_0^{2\pi} \int_0^1 \int_0^2 z r^5 \cos^2\theta \sin^2\theta \, dz \, dr \, d\theta$$

$$= \frac{1}{12}\pi.$$

Change of Variable from Rectangular to Spherical Coordinates

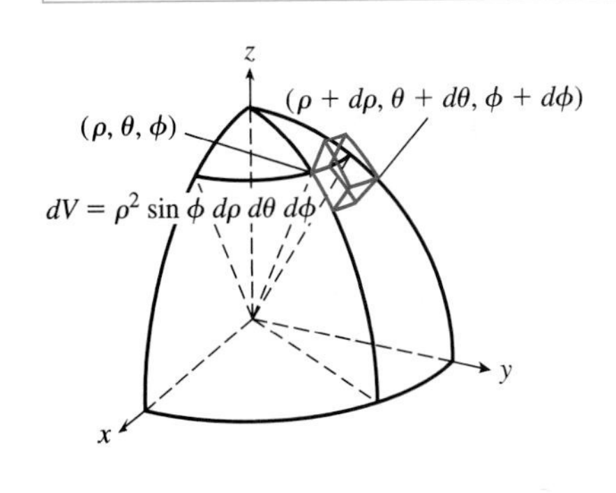

The change-of-variables formula for rectangular to spherical coordinates is

$$\iiint_{\mathbf{g}(\mathcal{B})} f(x, y, z)\, dx\, dy\, dz$$

$$= \iiint_{\mathcal{B}} f(\rho \sin\phi \cos\theta, \rho \sin\phi \sin\theta, \rho \cos\phi)\rho^2 \sin\phi \, d\rho \, d\theta \, d\phi.$$

If $\mathbf{g}(\mathcal{D})$ is the sphere of radius 1 and center $(0, 0, 1)$, then

$$\iiint_{\mathbf{g}(\mathcal{D})} x^2 y^2\, dx\, dy\, dz$$

$$= \int_0^{2\pi}\int_0^{\pi/2}\int_0^{2\cos\phi} \rho^6 \sin^5\phi \sin^2\theta \cdot \cos^2\theta \, d\rho \, d\phi \, d\theta = \frac{4}{15}\pi.$$

Chapter Review Exercises

1. Let f be defined on $\mathcal{D} = [0, 1] \times [0, 1]$ by $f(x, y) = x + 2y^2$. For the regular subdivision $\mathcal{S}_2$ of $\mathcal{D}$, calculate the Riemann sum for the evaluation set consisting of the midpoints of the four subregions of $\mathcal{S}_4$. Calculate $\iint_{\mathcal{D}} f(x, y)\, dx\, dy$ by the Iterated Integral Theorem and the percentage error associated with the Riemann sum.

2. Let $\mathcal{D} = [0, 1] \times [0, 1]$ and define f on $\mathcal{D}$ by

$$f(x, y) = \begin{cases} xy, & x + y \le 1 \\ 0, & \text{otherwise.} \end{cases}$$

Calculate $\iint_{\mathcal{D}} f(x, y)\, dx\, dy$. Justify your calculation.

3. Sketch the region of integration and evaluate the integral:

$$\int_0^1 \int_x^1 e^{x/y}\, dy\, dx.$$

4. Rewrite the integral

$$\int_0^4 \int_{y/2}^{\sqrt{y}} e^x \cos y \, dx\, dy$$

by reversing the order of integration.

5. A solid $\mathcal{K}$ consists of all points below the triangle with vertices $(0, 0, 6)$, $(2, 2, 2)$, and $(0, 2, 4)$ and above the projection of this triangle on the (x, y)-plane. Use a double integral to find the volume of $\mathcal{K}$.

6. Sketch a region $\mathcal{D}$ of Type II for which

$$\iint_{\mathcal{D}} f(x, y)\, dA = \int_0^1 \int_y^{\sqrt{y}} (x^2 + y^2)\, dx\, dy.$$

Noting that $\mathcal{D}$ is also of Type I, set up the corresponding iterated integral.

7. Set up an iterated integral for the mass of the solid bounded by the planes with equations

$$z = 0,\ x = 0,\ x = 2,\ y = 0, \quad \text{and} \quad y = 1$$

and the paraboloid with equation $z = 1 + x^2 + 2y^2$. Assume that the density of the solid at (x, y, z) is $\delta = 2 + z$ gm/cm^3.

8. Find the mass and x-coordinate of the center of mass of a lamina in the shape of the cardioid with equation $r = 2(1 + \cos\theta)$. Assume the density of the lamina is 2 kg/m^2.

9. A plane lamina is bounded by the spiral with equation $r = \theta$, $0 \le \theta \le \pi/2$, and the y-axis. Its density at any point is proportional to the distance from that point to the origin. Completely set up, but do not integrate, an iterated integral for the mass and the center of mass of the lamina.

10. Let $\mathcal{D}$ be the region of the (x, y)-plane between the curves with equations $y = x$ and $(x - 1)^2 = 4y$. Completely set up, but do not integrate, an iterated integral for $\iint_{\mathcal{D}} (x + y)\, dx\, dy$.

11. Let $\mathcal{D}$ be the region of the (x, y)-plane between the curves with equations $y = x$ and $(x - 1/2)^2 = 4y$. Let $\mathcal{K}$ be the solid above $\mathcal{D}$ and below the cylinder with equation $x + z = 8$. Find the mass of $\mathcal{K}$ if $\delta(x, y, z) = 2z$ kg/m^3 is its density at $(x, y, z) \in \mathcal{K}$.

12. Evaluate the integral

$$\int_0^1 \int_{\sqrt{x}}^1 e^{y^3}\, dy\, dx$$

by reversing the order of integration. Include a sketch of the region of integration.

13. Evaluate the integral

$$\int_0^1 \int_x^1 \sin(y^2)\, dy\, dx.$$

Sketch the region of integration. Specifically, what feature(s) of this integral led to your choice of strategy?

14. Find the volume of the region beneath the plane defined by the points $(1, 0, 0)$, $(0, 3, 0)$, and $(0, 0, 1)$ and above the triangular region $\mathcal{D}$ of the (x, y)-plane, where $\mathcal{D}$ has vertices $(1, 0, 0)$ $(0, 3, 0)$, and $(0, 1, 0)$. See the accompanying figure.

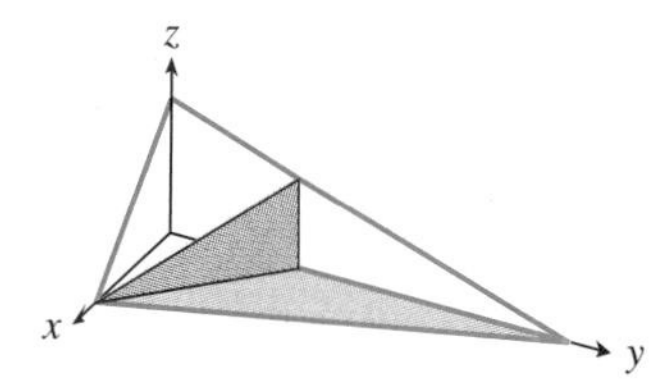

Figure for Exercise 14.

15. Evaluate a triple integral whose value is the volume in the first octant that is bounded by the planes with equations $x = 0$, $z = 0$, $x + y + z = 3$, $y = 1$, $y = 2$, and $y = 2x$. A partial sketch is shown in the accompanying line drawing.

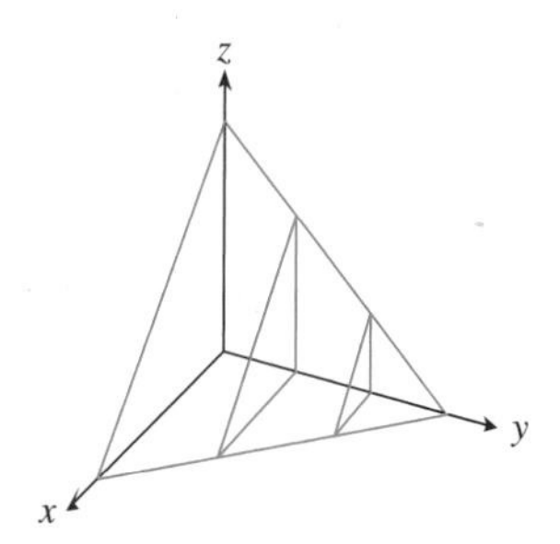

Figure for Exercise 15.

16. Assuming that the surface of the Earth is a perfect sphere of radius 6,378,145 m, calculate the area of the patch of the Earth's surface lying between the meridians with longitudes 0° 00′ 00″ W and 0° 00′ 01″ W and the parallels with latitudes 0° 00′ 00″ N and 0° 00′ 01″ N. Is the piece of the surface the size of a postage stamp, a garden, or a football field?

17. A surface S is described by

$$\mathbf{r} = \mathbf{r}(u, v) = \langle u^2, v^2, u + 2v \rangle,$$

where $0 \le u, v \le 4$. Calculate a unit normal to S at the point $(4, 4, 6)$. Completely set up a double integral whose value is the area of S.

18. Find the area of the surface cut from the graph with equation $x + y + z = 1$ by the solid cylinder described by $9x^2 + 4y^2 \le 1$.

19. Calculate the surface area cut from the parabolic surface with equation $z = (1/2)y^2$ by the planes with equations $y = x$, $y = 2\sqrt{2}$, and $x = 0$.

20. Calculate the mass of a thin spherical lamina S of radius a if its density at any point is proportional to the distance from a fixed point of S. *Hint:* Use spherical coordinates, place the lamina so that the origin is its "South Pole," and take the origin as the fixed point.

21. Set up a double integral for the area of the part of the surface with equation $z = x^2 - y^2$ that lies within the cylinder with equation $x^2 + y^2 = 4$.

22. Set up an iterated integral in polar coordinates for the surface area of the portion of the paraboloid with equation $z = 1 + x^2 + 2y^2$ lying above the right-hand loop of the lemniscate with equation $r = \sqrt{\cos 2\theta}$.

23. Set up an iterated integral for the surface area of that portion of the surface with equation $z = y^2 - x^2$ that, when projected along the y-axis, gives the quarter-circle with equation $z = \sqrt{a^2 - x^2}$, $0 \le x \le a$.

24. Let S be the lamina in the shape of the paraboloid described by

$$\mathbf{r}(u, v) = \langle u \cos v, 1 - u^2, u \sin v \rangle,$$

where $0 \le u \le 1/\sqrt{2}$ and $0 \le v \le 2\pi$ and lengths are measured in meters. Calculate the mass of the lamina if its density is 2 kg/m^2. Set up an integral for the centroid of the lamina.

25. The equation

$$\mathbf{r}(u, v) = \langle (R + \cos u) \cos v, (R + \cos u) \sin v, \sin u \rangle,$$

where $R > 1$, $0 \le u \le 2\pi$, and $0 \le v \le 2\pi$, describes a torus S. Use this parametrization in determining the area of S.

26. A lamina has the shape of the region inside the curve with equation $x^2 + y^2 = 2y$ and outside the curve with equation $x^2 + y^2 = 1$. Find its center of mass if its density at any point is proportional to the distance of that point from the origin.

27. Use the change-of-variables formula in evaluating

$$\int_0^{\sqrt{2}} \int_y^{\sqrt{4-y^2}} \frac{1}{1 + x^2 + y^2}\, dx\, dy.$$

28. Find the volume of the solid lying beneath the plane with equation $z = x + 5$ and above the circle with equation $x^2 + (y - 1)^2 = 1$.

29. Set up an iterated integral for the volume of grain in a conical pile having base radius 20 ft and a height of 12 ft. A vertical vent pipe of radius 1 ft passes through the pile, placed so that the vertical center line of the pile is tangent to the vent pipe and parallel to its axis.

30. Use the change of variables

$$x = 3u + v$$
$$y = u + v$$

in evaluating the integral

$$\iint_{\mathcal{D}} xy\,dx\,dy,$$

where $\mathcal{D}$ is the parallelogram with vertices $(0,0)$, $(3,1)$, $(5,3)$, and $(2,2)$.

31. Evaluate $\iint_{\mathcal{D}}(x+y)^2\,dx\,dy$, where $\mathcal{D}$ is the parallelogram bounded by the lines

$$x+y=0, \qquad x+y=1,$$
$$2x-y=0, \qquad 2x-y=3.$$

32. Use a triple integral in determining the volume of the solid bounded by the elliptic cylinder with equation $4x^2+z^2=4$ and the planes with equations $y=0$ and $y=z+2$.

33. Set up a triple integral for the volume of the solid lying in the first octant and above the plane with equation $x/3+y/3+z/3=1$ and below the plane with equation $x/3+y/2+z/4=1$.

34. A solid $\mathcal{K}$ is bounded below by the surface with equation $z=x^2+y^2+1$ and above by the surface with equation $z=2(-x-y+1)$. Completely set up a triple iterated integral for the volume of $\mathcal{K}$.

35. Set up a triple iterated integral for the volume of the region in the first octant bounded by the surfaces with equations $x+y+z=1$, $y=x^2$, $x=0$, and $z=0$.

36. Find the z-coordinate of the centroid of the solid bounded above by the sphere with equation $\rho=3$ and below by the cone $\phi=\pi/3$.

37. Transform the triple integral

$$\int_0^{2\pi}\int_0^R\int_0^H \frac{r^2}{1+r^2}\,dz\,dr\,d\theta,$$

where (r,θ,z) are cylindrical coordinates and $R, H>0$, to a triple integral in rectangular coordinates.

38. Set up a triple iterated integral in cylindrical coordinates for the volume of the first-octant solid bounded by the plane with equation $z=0$, the sphere with equation $r^2+z^2=4$, the cylinder with equation $r=(4/\pi)\theta$, and the plane with equation $\theta=\pi/2$.

39. Use spherical coordinates in calculating the volume of the solid between the spheres with equations $x^2+y^2+z^2=a^2$ and $x^2+y^2+z^2=b^2$, where $b>a$, and inside the half-cone with equation $z=\sqrt{x^2+y^2}$.

40. Evaluate $\iint_{\mathcal{D}}(x+y)\,dA$, where $\mathcal{D}$ is the triangular region with vertices $(0,0)$, $(-1,1)$, and $(1,1)$.

41. Let $\mathcal{B}$ be the portion of the ball described by the inequality $x^2+y^2+z^2\le 4$ and lying above the cone with equation $z=\sqrt{3(x^2+y^2)}$. Use the change-of-variables formula to express $\iiint_{\mathcal{B}} z\,dV$ as an iterated integral in

i. cylindrical coordinates and

ii. spherical coordinates.

42. Set up a triple iterated integral in spherical coordinates whose value is the mass of the solid in the first octant lying inside the sphere with equation $x^2+y^2+z^2=4$, outside the sphere with equation $x^2+y^2+z^2=2z$, above the cone with equation $3z^2=x^2+y^2$, and below the cone with equation $z^2=x^2+y^2$. The density of the solid at (x,y,z) is $1/(x^2+y^2+z^2+1)\ \text{kg/m}^3$. *Hint:* Sketch a vertical section of the solid.

43. Set up a triple iterated integral for the volume of the solid lying beneath the half-cone with equation $\sqrt{x^2+y^2}+(1/3)z=1$ and above the half-cone with equation $z=\sqrt{x^2+y^2}$.

44. Find the center of mass of the solid bounded above by the surface with equation $\rho=\phi$, $0\le\phi\le\pi/2$, and below by the plane with equation $\phi=\pi/2$. The density of the solid at any point is proportional to the distance from that point to the origin.

STUDENT PROJECT

ON THE ATTRACTION OF SPHEROIDS OF REVOLUTION

In Section 10.6 we discussed Newton's proof that the total force exerted on the Earth by the sun can be calculated as if both bodies were particles, with their total masses concentrated at their geometric centers. In this, Newton assumed that both bodies were spherical. In 1742 Colin Maclaurin (1698–1746), professor at the University of Edinburgh and disciple of Newton, extended Newton's result to ellipsoids of revolution. Adrien Marie Legendre (1752–1833) generalized Newton's result to certain "spheroids of revolution." In this problem we ask you to work through some of Legendre's results. This discussion of Legendre's original work was adapted from *A Source Book in Classical Analysis,* edited by Garrett Birkhoff, Harvard University Press, Cambridge, Mass., 1973.

Legendre calculated the force exerted on a unit mass by the points of a solid of revolution. He was able to express the force in terms of what are now called *Legendre polynomials.* We follow part of Legendre's argument closely, but use modern notation.

In Fig. 11.63 we show a unit mass at Q, which is external to a solid $\mathcal{K}$ of revolution, and a representative point P of the solid. The axis of revolution is the z-axis. We assume that $\mathcal{K}$ has constant density δ. The spherical coordinates of Q and P are $(\rho', 0, \phi')$ and (ρ, θ, ϕ), respectively. The angle between the vectors $\overrightarrow{OP}$ and $\overrightarrow{OQ}$ is α.

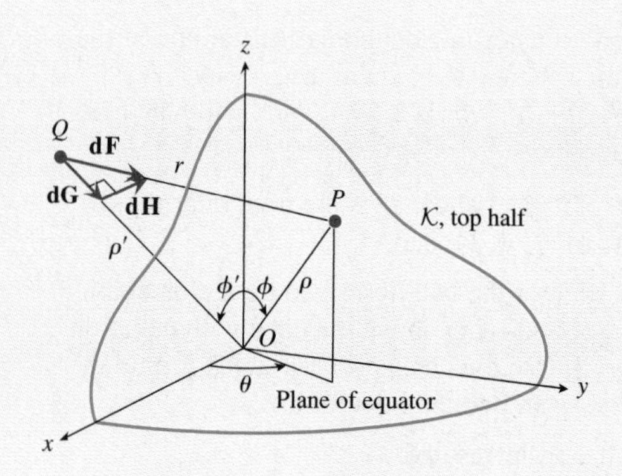

FIGURE 11.63 Force exerted on Q by a mass element at P.

Step 1

Show that

$$\overrightarrow{QP} = \langle \rho \sin \phi \cos \theta - \rho' \sin \phi', \rho \sin \phi \sin \theta, \rho \cos \phi - \rho' \cos \phi' \rangle$$

and

$$\|\overrightarrow{QP}\|^2 = r^2 = \rho^2 + \rho'^2 - 2\rho\rho'(\sin \phi \sin \phi' \cos \theta + \cos \phi \cos \phi').$$

Show also that

$$\|\overrightarrow{QP}\|^2 = r^2 = \rho^2 + p'^2 - 2\rho\rho' \cos \alpha.$$

Step 2

Show that the force $d\mathbf{F}(P)$ exerted on the unit mass at Q by the spherical volume element of mass $\delta\rho^2 \sin \phi d\rho d\theta d\phi$ at P is

$$\mathbf{dF}(P) = \frac{G \cdot 1 \cdot \delta\rho^2 \sin \phi d\rho d\theta d\phi}{r^3} \overrightarrow{QP}.$$

Step 3

Legendre wrote the force $\mathbf{dF}$ as the sum of the projection $\mathbf{dG}$ of $\mathbf{dF}$ in the radial direction $\overrightarrow{QO}$ and a vector $\mathbf{dH}$ perpendicular to $\mathbf{dG}$. Legendre divided his analysis into two parts, corresponding to these two vectors. We discuss the part of his argument related to the forces in the radial direction $\overrightarrow{QO}$. Show that $\mathbf{dG}$ is given by

$$\begin{aligned}\mathbf{dG} &= (\mathbf{dF} \cdot \langle \sin \phi', 0, \cos \phi' \rangle)\langle \sin \phi', 0, \cos \phi' \rangle \\ &= -\frac{G\delta(\rho' - \rho \cos \alpha)}{r^3}\rho^2 \sin \phi d\rho d\theta d\phi \langle \sin \phi', 0, \cos \phi' \rangle.\end{aligned}$$

Step 4

Legendre assumed that the "shape of the meridian" is given by $\rho = S(\phi)$ and that the solid $\mathcal{K}$ is symmetric about the plane of its equator. Show that the magnitude F_R of the radial component of the force exerted on $\mathcal{K}$ by Q is given by

$$F_R = G\delta \int_0^{\pi} \int_0^{\pi} \int_{-S(\phi)}^{S(\phi)} \frac{(\rho' - \rho \cos \alpha)\rho^2}{(\rho^2 - 2\rho\rho' \cos \alpha + \rho'^2)^{3/2}} \sin \phi d\rho d\phi d\theta.$$

Legendre remarks that the innermost integral can be done "exactly, by known methods." He chose, however, to expand the integrand in an infinite series in order to simplify the remaining integrations. Letting

$$v = \rho/\rho', \qquad w = 2(\rho/\rho') \cos \alpha - \rho^2/\rho'^2 = 2v \cos \alpha - v^2,$$

and assuming that $|w| < 1$ (so that the Maclaurin series of $(1 - w)^{-3/2}$ converges), show that the fraction in the integral for F_R can be written as

$$\frac{\rho^2}{\rho'^2}(1 - v\cos\alpha)(1 - w)^{-3/2} = \frac{\rho^2}{\rho'^2}(1 - v\cos\alpha)\left(1 + \frac{1\cdot 3}{1!\cdot 2}w + \frac{1\cdot 3\cdot 5}{2!\cdot 2^2}w^2 + \frac{1\cdot 3\cdot 5\cdot 7}{3!\,2^3}w^3 + \cdots\right).$$

Step 5

Show that the integral for F_R can be written as

$$F_R = G\delta\int_0^{\pi}\int_0^{\pi}\int_{-S(\phi)}^{S(\phi)}\frac{\rho^2}{\rho'^2}\left[1 + (2\cos\alpha)\frac{\rho}{\rho'} + \left(-\frac{3}{2} + \frac{9\cos^2\alpha}{2}\right)\frac{\rho^2}{\rho'^2} + (-6\cos\alpha + 10\cos^3\alpha)\frac{\rho^3}{\rho'^3} + \left(\frac{15}{8} - \frac{75\cos^2\alpha}{4} + \frac{175\cos^4\alpha}{8}\right)\frac{\rho^4}{\rho'^4} + \cdots\right]\sin\phi\, d\rho\, d\theta\, d\phi.$$

Step 6

If this result is integrated with respect to ρ and rearranged, show that F_R can be written as

$$F_R = \frac{2G\delta}{\rho'^2}\int_0^{\pi} S^3\int_0^{\pi}\left[\frac{1}{3}(1) + \frac{3}{5}\left(\frac{3}{2}\cos^2\alpha - \frac{1}{2}\right)\frac{S^2}{\rho'^2} + \frac{5}{7}\left(\frac{5\cdot 7}{2\cdot 4}\cos^4\alpha - \frac{3\cdot 5}{2\cdot 4}\cdot 2\cos^2\alpha + \frac{3\cdot 1}{2\cdot 4}\right)\frac{S^4}{\rho'^4} + \frac{7}{9}\left(\frac{7\cdot 9\cdot 11}{2\cdot 4\cdot 6}\cos^6\alpha - \frac{5\cdot 7\cdot 9}{2\cdot 4\cdot 6}\cdot 3\cos^4\alpha + \frac{3\cdot 5\cdot 7}{2\cdot 4\cdot 6}\cdot 3\cos^2\alpha - \frac{1\cdot 3\cdot 5}{2\cdot 4\cdot 6}\right)\frac{S^6}{\rho'^6} + \cdots\right]\sin\phi\, d\theta\, d\phi.$$

Step 7

The expressions within the large parentheses in Step 6 are Legendre polynomials. The Legendre polynomials P_0, P_2, and P_4 are

$$P_0(x) = 1$$

$$P_2(x) = \frac{1}{2}(3x^2 - 1)$$

$$P_4(x) = \frac{1}{8}(35x^4 - 30x^2 + 3).$$

Step 8

Legendre showed that these polynomials have the property

$$\frac{1}{\pi}\int_0^{\pi} P_{2k}(\cos\alpha)\, d\theta = P_{2k}(\cos\phi)P_{2k}(\cos\phi'), \qquad k = 0, 1, 2, \ldots.$$

Use this result to show that

$$F_R = \frac{3Gm}{\rho'^2} \sum_{k=0}^{\infty} \frac{2k+1}{2k+3} P_{2k}(\cos \phi') \frac{\beta_k}{\rho'^{2k}},$$

where m is the mass of the spheroid and

$$\beta_k = \frac{4\pi\delta}{3m} \int_0^{\pi/2} P_{2k}(\cos \phi) S^{2k+3} \sin \phi \, d\phi, \qquad k = 0, 1, 2, \ldots.$$

Step 9

Explain what Legendre meant when he stated that the β_k depend only upon the "shape of the meridian."

From these results, Legendre went on to obtain his final result on the attraction exerted by $\mathcal{K}$ on a unit mass at Q. For this he used the idea of a potential function, which had just been invented by Pierre Simon Laplace (1749–1827).

12

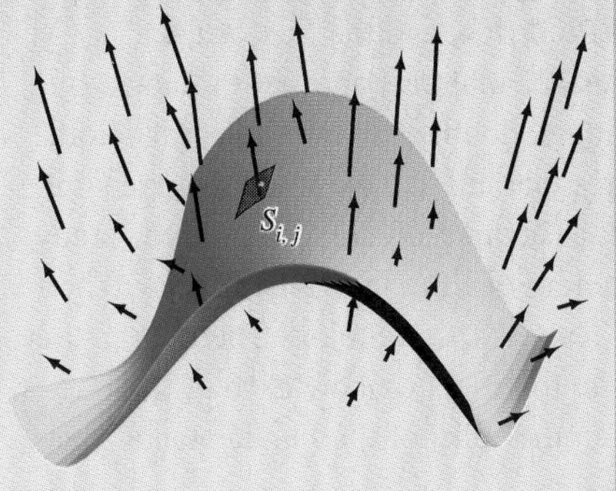

Line and Surface Integrals

Scientists and engineers use wind tunnels to study the flow of air around airplanes, buildings, sculptures, and other structures. Vector calculus plays a major role in the analysis of such air flows.

See Investigation 3 on p. 1094 for an analysis of some aspects of fluid flow.

In Chapter 11 we defined double and triple integrals of functions defined on two- or three-dimensional regions. Our definitions for these integrals were a natural extension of the definition of the definite integral for a function defined on an interval.

In this chapter we study functions defined on curves or surfaces in space. We start by extending the definition of *definite integral* to real- and vector-valued functions defined on curves. We use these integrals—called line integrals—to calculate the mass of a wire and the work done by a force, and to study the flow of a fluid in the plane. In the second half of the chapter, we define the integral of real- and vector-valued functions on surfaces. We use these integrals—called surface integrals—to study such phenomena as three-dimensional fluid flow and the flux of heat.

Many line and surface integrals are best evaluated by converting them to single, double, or triple integrals. Some of the several theorems that tell us how to do this are similar to the Fundamental Theorem of Calculus. As you study the Fundamental Theorem of Line Integrals, Green's Theorem, the Divergence Theorem, and Stokes' Theorem in this chapter, take a moment to reflect on how each of these results says something in the spirit of the Fundamental Theorem of Calculus.

12.1 The Line Integral

The definite integral

$$\int_a^b f(t)\,dt \quad \text{or} \quad \int_a^b f(x)\,dx \tag{1}$$

of a function f defined on an interval $[a, b] \subset R$ is used to model many physical phenomena. For example, if $f(t)$ is the speed of an object at time t, then the first integral in (1) gives the total distance traveled by the object; if $f(x)$ is the density of a wire at position x, then the second integral in (1) is the mass of the wire; if $f(x) \geq 0$ on $a \leq x \leq b$, then the second integral in (1) is the area of the region under the graph of f and above the x-axis.

It is sometimes useful to extend the integral (1) to functions defined on domains other than intervals in R. If, for example, we wish to find the mass of a curved wire C in space, or the distance traveled by an object sliding on C, then it makes sense to consider a function f defined on $C \subset R^3$, where f is either the density of the wire or the speed of the object. To find the mass of the wire or the distance traveled by the object, we need to know how to integrate a function defined at points of the curve C. In Chapter 6 we worked briefly with such integrals when we calculated the work done by a force acting on an object moving along a planar curve. In this section we lay the groundwork for a better understanding of these integrals by studying the line integral with respect to arc length.

The Integral with Respect to Arc Length

INVESTIGATION

Determining the Mass of a Curved Wire

A thin wire in space is in the shape of the helical curve C described by

$$\mathbf{r} = \mathbf{r}(t) = \langle x(t), y(t), z(t)\rangle = \langle 2\cos t, 2\sin t, t\rangle, \qquad \pi \leq t \leq 2\pi.$$

Assume that the density (in grams per centimeter) of the wire at point (x, y, z) is

$$\delta(x, y, z) = x^2 + y^2 + z + 1.$$

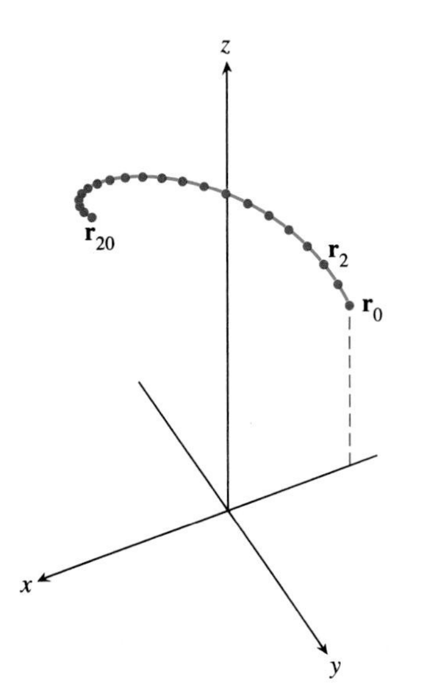

FIGURE 12.1 Subdividing the wire into 20 pieces

Estimate the mass of the wire.

In Section 6.7 we approximated the mass of a curved planar rod by cutting the rod into small pieces, approximating the mass of each piece, then adding the masses to get an approximation of the mass of the wire. We use a similar procedure to estimate the mass of a wire in space.

We first cut the wire into, say, 20 small pieces. We do this by first cutting the interval $\pi \leq t \leq 2\pi$ into 20 equal pieces. The endpoints of the intervals in this regular subdivision of $[\pi, 2\pi]$ are

$$\pi = t_0 < t_1 < t_2 < \cdots < t_{20} = 2\pi,$$

where

$$t_k = \pi + k\Delta t \quad \text{and} \quad \Delta t = \frac{\pi}{20}.$$

The points

$$\mathbf{r}(t_0) = \mathbf{r}_0, \mathbf{r}(t_1) = \mathbf{r}_1, \ldots, \mathbf{r}(t_{20}) = \mathbf{r}_{20}$$

then determine a subdivision of C. We shall call such a subdivision a **regular subdivision of** C. See Fig. 12.1. The kth piece of the wire lies between $\mathbf{r}_{k-1}$ and $\mathbf{r}_k$. Because

$$\mathbf{r}'(t) = \langle -2 \sin t, 2 \cos t, 1 \rangle$$

and

$$\|\mathbf{r}'(t)\| = \sqrt{(-2 \sin t)^2 + (2 \cos t)^2 + 1^2} = \sqrt{5},$$

the length of the kth piece is

$$\Delta s_k = \int_{t_{k-1}}^{t_k} \|\mathbf{r}'(t)\|\, dt = \int_{t_{k-1}}^{t_k} \sqrt{5}\, dt = \sqrt{5}\,(t_k - t_{k-1}) = \sqrt{5}\,\Delta t = \frac{\pi\sqrt{5}}{20}, \quad k = 1, 2, \ldots, 20.$$

To approximate the mass of the kth piece, we find the density of the wire at some point in this piece and multiply this density by the length. If, for example, we choose to find the density at the point $\mathbf{r}(t_k)$, then the density at this point is

$$\begin{aligned} \delta(\mathbf{r}(t_k)) &= \delta(2 \cos t_k, 2 \sin t_k, t_k) = (2 \cos t_k)^2 + (2 \sin t_k)^2 + t_k + 1 \\ &= 5 + t_k = 5 + \pi + k\,\Delta t. \end{aligned}$$

Denoting the mass of the kth piece of wire by Δm_k

$$\Delta m_k \approx \delta(\mathbf{r}(t_k))\,\Delta s_k = \left(5 + \pi + \frac{\pi}{20}k\right)\frac{\pi\sqrt{5}}{20}\ \text{g}, \qquad 1 \leq k \leq 20.$$

Summing these approximations gives an approximation of the mass M of the wire,

$$\begin{aligned} M = \sum_{k=1}^{20} \Delta m_k &\approx \sum_{k=1}^{20} \left(5 + \pi + \frac{\pi}{20}k\right)\frac{\pi\sqrt{5}}{20} \\ &= \frac{\pi\sqrt{5}}{20}\left(20(5 + \pi) + \frac{\pi}{20}\sum_{k=1}^{20} k\right) \\ &= 5\sqrt{5}\,\pi + \sqrt{5}\,\pi^2 + \frac{\sqrt{5}\,\pi^2}{20^2}\cdot 210 \\ &\approx 68.8\ \text{g}. \end{aligned}$$

This approximation to the mass of the wire was based on 20 subdivisions. In a similar manner, given a positive integer n, we can form a regular subdivision of the interval $\pi \leq t \leq 2\pi$ into n pieces and use this to generate a regular subdivision of (the wire) C. It seems reasonable that as the number of subdivisions grows, the approximation to the mass of the wire gets closer and closer to its actual mass M. In other words, the approximating sums approach M as the number of subdivisions tends to infinity. This limit of the sums is called the *integral along C of δ with respect to arc length.* We use the integral symbol to denote this number:

$$M = \text{mass of wire} = \int_C \delta\, ds.$$

We use these ideas to motivate the definition of the integral of a function defined on a curve.

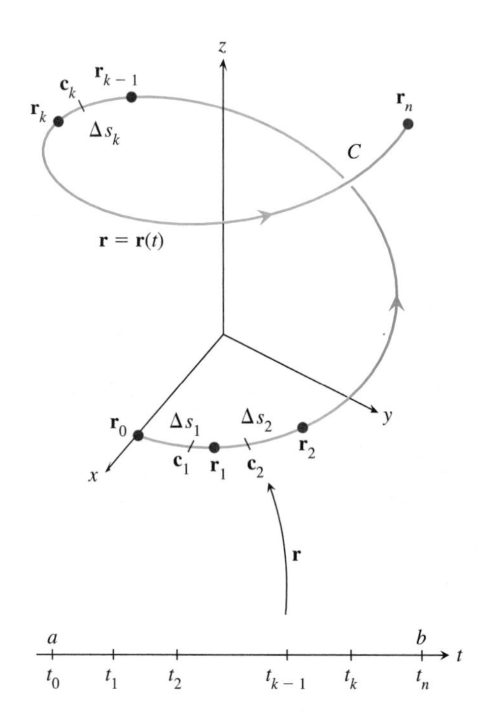

FIGURE 12.2 A regular subdivision of $[a, b]$ leads to a subdivision of C.

DEFINITION The Line Integral with Respect to Arc Length

Let C be a smooth curve in the plane or in space and let f be a real-valued function defined on C. For a regular subdivision $\mathbf{r}_0, \mathbf{r}_1, \ldots, \mathbf{r}_n$ of C, let Δs_k be the length of the kth piece of C, that is, the piece between $\mathbf{r}_{k-1}$ and $\mathbf{r}_k$. See Fig. 12.2. Let $\{\mathbf{c}_1, \mathbf{c}_2, \ldots, \mathbf{c}_n\}$ be any set of points on C for which $\mathbf{c}_k$ is on the kth piece of C. We refer to this set as an **evaluation set.**

For each regular subdivision of C and each associated evaluation set $\{\mathbf{c}_1, \mathbf{c}_2, \ldots, \mathbf{c}_n\}$, we define the Riemann sum

$$R_n = \sum_{k=1}^{n} f(\mathbf{c}_k)\, \Delta s_k.$$

If there is a number I for which all Riemann sums R_n can be made as close to I as we like by taking n sufficiently large, then we say that f is **integrable with respect to arc length on C.** We write

$$I = \lim_{n\to\infty} R_n = \lim_{n\to\infty} \sum_{k=1}^{n} f(\mathbf{c}_k)\, \Delta s_k = \int_C f\, ds.$$

The number I is called **the line integral with respect to arc length of f along C.**

Evaluation of Line Integrals

In Chapter 6 we evaluated "work integrals" in the plane by converting them to definite integrals, then evaluating the resulting integrals. The same technique works for line integrals along curves.

Evaluation of $\int_C f\,ds$

Let C be a smooth curve parameterized by $\mathbf{r} = \mathbf{r}(t)$, $a \le t \le b$, and let f be a function defined on C. If f is continuous, then the line integral with respect to arc length of f along C exists and

$$\int_C f\,ds = \int_a^b f(\mathbf{r}(t))\,\|\mathbf{r}'(t)\|\,dt. \tag{2}$$

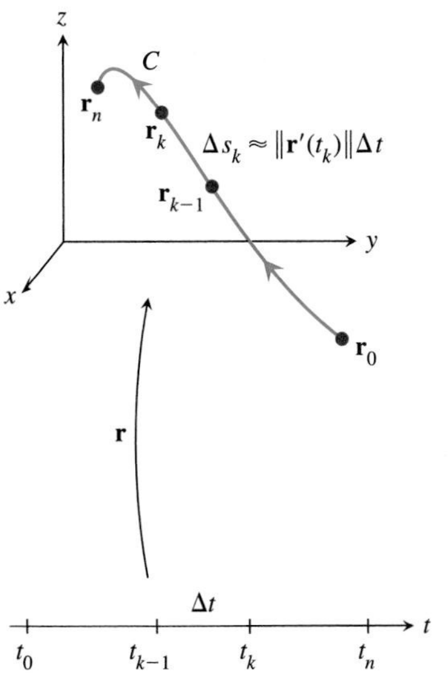

FIGURE 12.3 When Δt is small, $\Delta s_k \approx \|\mathbf{r}'(t_k)\|\,\Delta t$.

We give a brief argument in support of (2). Consider first the integral on the right side of (2). By the definition of *definite integral* given in Section 5.2, this integral can be approximated by a sum associated with a regular subdivision $t_0 < t_1 < \cdots < t_n$ of $[a, b]$. That is,

$$\int_a^b f(\mathbf{r}(t))\,\|\mathbf{r}'(t)\|\,dt \approx \sum_{k=1}^{n} f(\mathbf{r}(t_k))\,\|\mathbf{r}'(t_k)\|\,\Delta t, \tag{3}$$

where $\Delta t = (b - a)/n$ is the length of each interval in the regular subdivision and we have taken the right endpoints $\{t_1, t_2, \ldots, t_n\}$ as the evaluation set for the sum. Now let $\mathbf{r}(t_k) = \mathbf{r}_k$, $k = 0, 1, 2, \ldots, n$. The length of the piece of C between $\mathbf{r}_{k-1}$ and $\mathbf{r}_k$ is

$$\Delta s_k = \int_{t_{k-1}}^{t_k} \|\mathbf{r}'(t)\|\,dt \approx \|\mathbf{r}'(t_k)\|\,\Delta t. \tag{4}$$

This approximation is a good one if $t_k - t_{k-1} = \Delta t$ is small, that is, if n is large. See Fig. 12.3. Combining (3) and (4), we have

$$\int_a^b f(\mathbf{r}(t))\,\|\mathbf{r}'(t)\|\,dt \approx \sum_{k=1}^{n} f(\mathbf{r}_k)\,\Delta s_k. \tag{5}$$

The sum on the right side of this expression is a Riemann sum using the regular subdivision $\{\mathbf{r}_0, \mathbf{r}_1, \ldots, \mathbf{r}_n\}$ of C and the associated evaluation set $\{\mathbf{r}_1, \ldots, \mathbf{r}_n\}$. It follows that for large n,

$$\sum_{k=1}^{n} f(\mathbf{r}_k)\,\Delta s_k \approx \int_C f\,ds. \tag{6}$$

The approximations in (3), (4), (5), and (6) all become better and better as n tends to infinity. This suggests that

$$\int_C f\,ds = \int_a^b f(\mathbf{r}(t))\,\|\mathbf{r}'(t)\|\,dt,$$

as desired.

For our first example, we find the mass of the wire discussed in the Investigation.

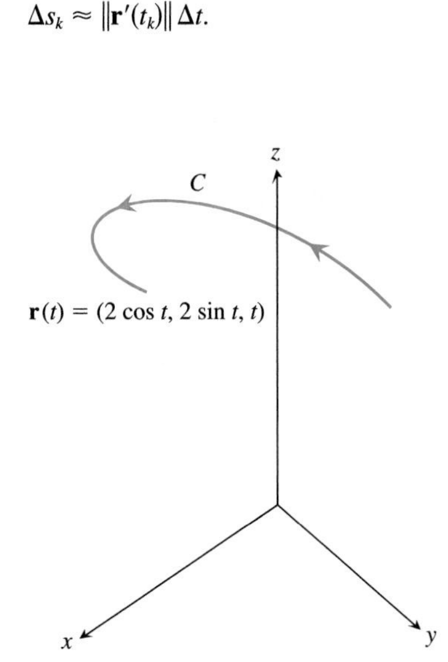

FIGURE 12.4 A wire in the shape of part of a helical curve.

EXAMPLE 1 A wire is in the shape of the helical curve C described by

$$\mathbf{r}(t) = \langle 2\cos t, 2\sin t, t\rangle, \qquad \pi \le t \le 2\pi.$$

See Fig. 12.4. At point (x, y, z) on the wire, the density is

$$\delta = \delta(x, y, z) = x^2 + y^2 + z + 1 \text{ g/cm}.$$

Find the mass of the wire.

Solution

As seen in the Investigation,

$$M = \text{mass of wire} = \int_C \delta\, ds.$$

Next note that

$$\|\mathbf{r}'(t)\| = \|\langle -2 \sin t, 2 \cos t, 1\rangle\| = ((2 \sin t)^2 + (2 \cos t)^2 + 1)^{1/2} = \sqrt{5}$$

and

$$\delta(\mathbf{r}(t)) = \delta(2 \cos t, 2 \sin t, t) = 5 + t.$$

Because the density function δ is continuous, we can apply (2) to get

$$M = \int_C \delta(\mathbf{r})\, ds = \int_\pi^{2\pi} \delta(\mathbf{r}(t))\, \|\mathbf{r}'(t)\|\, dt$$

$$= \int_\pi^{2\pi} (5 + t)\sqrt{5}\, dt = 5\pi\sqrt{5} + \frac{3\pi^2\sqrt{5}}{2} \approx 68.2277 \text{ g.}$$

How does this compare with the estimate for the mass found in the Investigation?

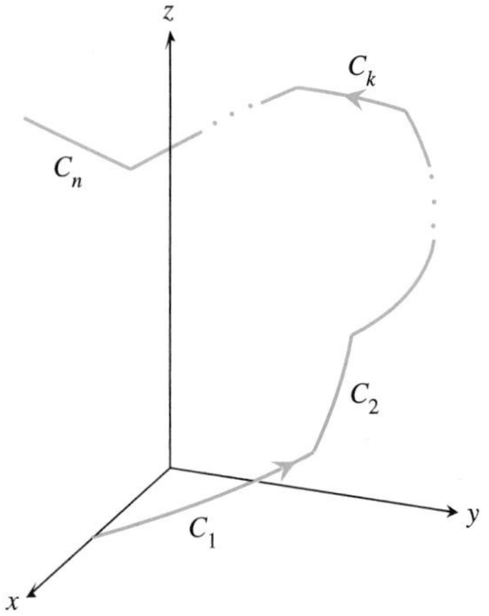

FIGURE 12.5 A piecewise smooth curve is made up of smooth pieces.

Piecewise Smooth Curves A curve C described by a function $\mathbf{r} = \mathbf{r}(t)$ is called *piecewise smooth* if there are points

$$a = T_0 < T_1 < T_2 < \cdots < T_{n-1} < T_n = b$$

such that $\mathbf{r}$ has a continuous derivative on each interval $T_{k-1} \leq t \leq T_k$. This means that for $k = 1, 2, \ldots, n$, the curve C_k described by

$$\mathbf{r} = \mathbf{r}(t), \qquad T_{k-1} \leq t \leq T_k, \qquad k = 1, 2, \ldots, n.$$

is smooth. We write the curve C as $C = C_1 + C_2 + \cdots + C_n$. See Fig. 12.5. If f is continuous on C, then f is continuous on each C_k and the line integral of f along C is defined to be the sum of the line integrals of f along each of the curves C_k,

$$\int_C f ds = \int_{C_1} f ds + \int_{C_2} f ds + \cdots + \int_{C_n} f ds. \tag{7}$$

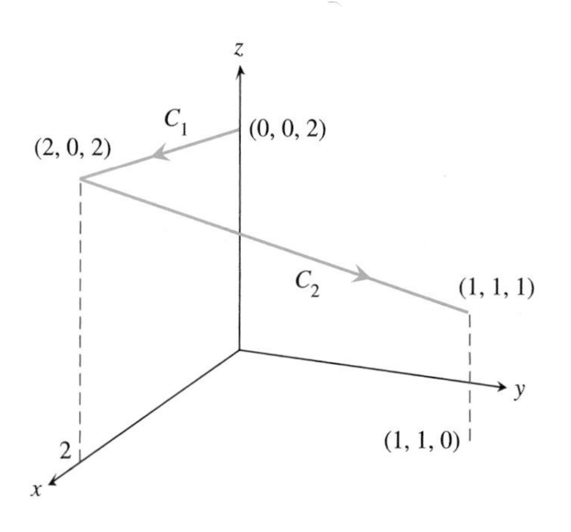

FIGURE 12.6 Curve C consists of two segments.

EXAMPLE 2 Let $f(x, y, z) = x + yz$. Evaluate

$$\int_C f ds,$$

for $C = C_1 + C_2$ where C_1 is the line segment from $(0, 0, 2)$ to $(2, 0, 2)$, and C_2 is the line segment from $(2, 0, 2)$ to $(1, 1, 1)$. See Fig. 12.6.

Solution

We can describe C_1 parametrically by

$$\mathbf{r} = \mathbf{r}_1(t) = \langle t, 0, 2\rangle, \qquad 0 \leq t \leq 2,$$

and C_2 by

$$\mathbf{r} = \mathbf{r}_2(t) = \langle 2, 0, 2\rangle + t\langle 1-2, 1-0, 1-2\rangle = \langle 2-t, t, 2-t\rangle, \quad 0 \le t \le 1.$$

We then have

$$\|\mathbf{r}_1'(t)\| = \|\langle 1, 0, 0\rangle\| = 1 \quad \text{and} \quad \|\mathbf{r}_2'(t)\| = \|\langle -1, 1, -1\rangle\| = \sqrt{3}. \tag{8}$$

Also,

$$f(\mathbf{r}_1(t)) = f(t, 0, 2) = t + 0 \cdot 2 = t \tag{9}$$

and

$$f(\mathbf{r}_2(t)) = f(2-t, t, 2-t) = (2-t) + t(2-t) = -t^2 + t + 2. \tag{10}$$

Applying (7) and using (8), (9), and (10), we have

$$\begin{aligned}
\int_C f\,ds &= \int_{C_1} f\,ds + \int_{C_2} f\,ds \\
&= \int_0^2 f(\mathbf{r}_1(t))\,\|\mathbf{r}_1'(t)\|\,dt + \int_0^1 f(\mathbf{r}_2(t))\,\|\mathbf{r}_2'(t)\|\,dt \\
&= \int_0^2 t\,dt + \int_0^1 (-t^2 + t + 2)\sqrt{3}\,dt \\
&= 2 + \frac{13\sqrt{3}}{6} \approx 5.75278.
\end{aligned}$$

Reparameterization In the previous example, we were given a geometric description of a piecewise smooth curve C. In evaluating $\int_C f\,ds$ we found a parameterization for each of the smooth pieces of C and then used these parameterizations when we applied (2). Because, however, we have seen that any curve has many different parameterizations, a natural question to ask is: When we calculate the value of $\int_C f\,ds$ using (2), can we get different values by using a different parameterization for C? Fortunately, the answer to this question is no, provided we take a little care in defining what we mean by "different parameterization."

DEFINITION Reparameterization of a Smooth Curve

Let C be a smooth curve described parametrically by

$$\mathbf{r} = \mathbf{r}(t), \qquad a \le t \le b.$$

Let t $= \phi(\tau)$ be an increasing function with a continuous, nonzero derivative, defined on an interval $c \le \tau \le d$ and with $a = \phi(c)$ and $b = \phi(d)$. (So ϕ is a one-to-one function and maps the interval $c \le \tau \le d$ onto the interval $a \le t \le b$.) The function $\hat{\mathbf{r}}$ defined by

$$\hat{\mathbf{r}}(\tau) = \mathbf{r}(\phi(\tau)), \qquad c \le \tau \le d, \tag{11}$$

is called a **reparameterization** of C.

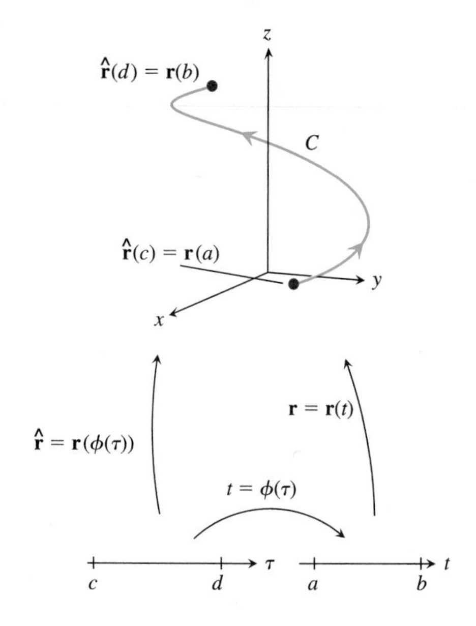

FIGURE 12.7 The function $\hat{\mathbf{r}}$ is a reparameterization of $\mathbf{r}$. Both functions describe the points of C.

The functions $\mathbf{r}$ and $\hat{\mathbf{r}}$ have the same range, so the output of each function describes the points of C. Furthermore, as τ runs from c to d, $\hat{\mathbf{r}}(\tau)$ gives the points of C from $\mathbf{r}(a)$ to $\mathbf{r}(b)$. Thus $\hat{\mathbf{r}}$ and $\mathbf{r}$ are both parameterizations for C and describe C in the same "direction"; that is, the two parameterizations have the same initial and terminal points. See Fig. 12.7.

Now let f be a continuous function defined on the points of C, let $\mathbf{r}$ be a parameterization of C, and let $\hat{\mathbf{r}} = \mathbf{r} \circ \phi$ be a reparameterization of C. Then the value of $\int_C f\,ds$ obtained by using the parameterization $\mathbf{r}$ is the same as that obtained by using the parameterization $\hat{\mathbf{r}}$. To verify this, consider the integral

$$\int_c^d f(\hat{\mathbf{r}}(\tau))\|\hat{\mathbf{r}}'(\tau)\|\,d\tau \tag{12}$$

and make the substitution $t = \phi(\tau)$. Because $a = \phi(c)$, $b = \phi(d)$, $dt = \phi'(\tau)\,d\tau$, and $\phi'(\tau)$ is positive, we have

$$\begin{aligned}\int_c^d f(\hat{\mathbf{r}}(\tau))\,\|\hat{\mathbf{r}}'(\tau)\|\,d\tau &= \int_c^d f(\mathbf{r}(\phi(\tau)))\,\|\mathbf{r}'(\phi(\tau))\phi'(\tau)\|\,d\tau \\ &= \int_a^b f(\mathbf{r}(t))\,\|\mathbf{r}'(t)\|\,dt.\end{aligned} \tag{13}$$

The integral in (12) is the integral that arises when $\hat{\mathbf{r}}$ is used to parameterize C. The last integral in (13) is the one that arises when $\mathbf{r}$ is used. Because these two integrals are equal, we conclude that the value of $\int_C f\,ds$ found by using (2) does not depend on the parameterization of C. We now summarize this result.

Independence of Parameterization

Let C be a smooth curve described by $\mathbf{r} = \mathbf{r}(t)$, $a \le t \le b$, and let $\hat{\mathbf{r}}(\tau)$, $c \le \tau \le d$, be a reparameterization of C. If the real-valued function f is defined and continuous on the points of C, then

$$\int_a^b f(\mathbf{r}(t))\,\|\mathbf{r}'(t)\|\,dt = \int_C f\,ds = \int_c^d f(\hat{\mathbf{r}}(\tau))\,\|\hat{\mathbf{r}}'(\tau)\|\,d\tau.$$

This result can be applied to make some line integral calculations easier.

EXAMPLE 3 Let C be the curve described by

$$\mathbf{r} = \mathbf{r}(t) = \left\langle \frac{t}{\sqrt{1+t^2}}, \frac{1}{1+t^2}, \frac{t^2}{1+t^2} \right\rangle, \qquad 0 \le t \le 2, \tag{14}$$

and let $f(x, y, z) = x + y + z - 1$. Evaluate

$$\int_C f\,ds. \tag{15}$$

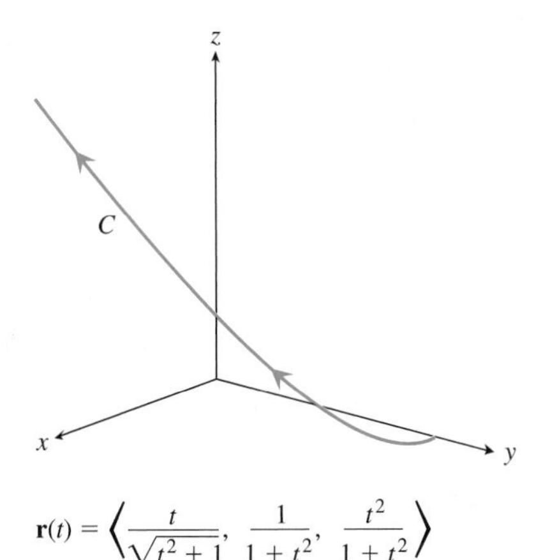

$$\mathbf{r}(t) = \left\langle \frac{t}{\sqrt{t^2+1}}, \frac{1}{1+t^2}, \frac{t^2}{1+t^2} \right\rangle$$

FIGURE 12.8 The curve C can also be described by the simpler function $\mathbf{r}_1(\tau) = \langle \tau, 1 - \tau^2, \tau^2 \rangle$.

Solution

Because the parameterization in (14) is somewhat complicated, we first look for an alternative parameterization. Letting

$$\tau = \frac{t}{\sqrt{1 + t^2}}$$

be the first coordinate in (14), we see that the third coordinate is τ^2 and the second coordinate is $1 - \tau^2$. As t increases from 0 to 2, τ increases from 0 to $2/\sqrt{5}$. Hence C is also described by the parameterization

$$\mathbf{r}_1(\tau) = \langle \tau, 1 - \tau^2, \tau^2 \rangle, \qquad 0 \le \tau \le \frac{2}{\sqrt{5}}.$$

See Fig. 12.8. Furthermore, $\mathbf{r}(t) = \mathbf{r}_1(\phi(t))$ where $\phi(t) = t/\sqrt{t^2 + 1}$ is increasing on $0 \le t \le 2$. If we think of C as a curve parameterized by $\mathbf{r}_1$, then $\mathbf{r}$ is a reparameterization of C. Thus, the integral in (15) can be evaluated using either parameterization. We use the parameterization $\mathbf{r}_1$ because it is simpler. Because

$$f(\mathbf{r}_1(\tau)) = f(\tau, 1 - \tau^2, \tau^2) = \tau + (1 - \tau^2) + \tau^2 - 1 = \tau$$

and

$$\|\mathbf{r}_1'(\tau)\| = \|\langle 1, -2\tau, 2\tau \rangle\| = \sqrt{1 + 8\tau^2},$$

we have

$$\begin{aligned}\int_C f\,ds &= \int_0^{2/\sqrt{5}} f(\mathbf{r}_1(\tau))\,\|\mathbf{r}_1'(\tau)\|\,d\tau \\ &= \int_0^{2/\sqrt{5}} \tau\sqrt{1 + 8\tau^2}\,d\tau = \frac{1}{24}(1 + 8\tau^2)^{3/2}\Big|_0^{2/\sqrt{5}} \\ &= \frac{1}{24}\left(\frac{37}{5}\right)^{3/2} - \frac{1}{24} \approx 0.797091.\end{aligned}$$

Exercises 12.1

Exercises 1–8: Sketch the graph of C and evaluate $\int_C f\,ds$. You may have to evaluate some of the integrals numerically.

1. C is described by

$$\mathbf{r}(t) = \langle t, 2t, 3t \rangle, \qquad 0 \le t \le 2;$$
$$f(x, y, z) = x + 2y + 3z.$$

2. C is described by

$$\mathbf{r}(t) = \langle -t + 1, t + 1, 2t \rangle, \qquad -2 \le t \le 1;$$
$$f(x, y, z) = x^2 - y^2 + z^2.$$

3. C is described by

$$\mathbf{r}(t) = \langle \sin t, \cos t, 2t \rangle, \qquad 0 \le t \le 2\pi;$$
$$f(x, y, z) = xy + z.$$

4. C is described by

$$\mathbf{r}(t) = \langle t, t \sin t, t \cos t \rangle, \qquad 0 \le t \le 2\pi;$$
$$f(x, y, z) = x.$$

5. C is the curve whose graph is the line segment from $(1, -1, 2)$ to $(4, -2, 1)$;

$$f(x, y, z) = 3x - 2y + 4z.$$

6. C is the curve whose graph is the line segment from $(3, 5, 2)$ to $(0, 0, 0)$ followed by the segment from $(0, 0, 0)$ to $(-1, -3, 2)$;

$$f(x, y, z) = xyz.$$

7. C is the curve whose graph is the portion of the circle of center $(0, 0, 0)$ and radius 2 that lies in the (x, z)-plane, has nonnegative z-coordinate, and has initial point $(2, 0, 0)$ and terminal point $(-2, 0, 0)$;

$$f(x, y, z) = x + y - 2.$$

8. C is the triangle with vertices $(0, 0, 0)$, $(2, 2, 2)$, and $(0, 0, 2)$. Assume that $(0, 0, 0)$ is the initial and terminal point of the path and that the path goes from $(0, 0, 0)$ to $(2, 2, 2)$ to $(0, 0, 2)$, then back to $(0, 0, 0)$;

$$f(x, y, z) = xy + yz + zx.$$

Exercises 9–10: Two parameterizations, $\mathbf{r}$ *and* $\hat{\mathbf{r}}$, *of a curve* C *are given. Verify that one of these parameterizations is a reparameterization of the other. Then check that* $\int_C f\,ds$ *gives the same value for both parameterizations. You may need to evaluate some of the integrals numerically.*

9.

$$\mathbf{r}(t) = \langle \cos t, \sin t, t \rangle, \qquad \frac{\pi}{6} \le t \le \frac{\pi}{3},$$

$$\hat{\mathbf{r}}(\tau) = \left\langle \sqrt{1 - \tau^2}, \tau, \arcsin \tau \right\rangle, \qquad \frac{1}{2} \le t \le \frac{\sqrt{3}}{2},$$

and

$$f(x, y, z) = x^2 + y^2 + z^2$$

10.

$$\mathbf{r}(t) = \langle 2e^t, e^{2t}, t \rangle, \qquad 0 \le t \le \frac{1}{2},$$

$$\hat{\mathbf{r}}(\tau) = \langle 2\tau, \tau^2, \ln \tau \rangle, \qquad 1 \le \tau \le \sqrt{e},$$

and

$$f(x, y, z) = 1 + xz$$

Exercises 11–12: Use a regular subdivision with 10 pieces to find an estimate for the mass of the wire. Choose a convenient evaluation set.

11. The wire is in the shape of the helix described by

$$\mathbf{r}(t) = \langle t, 3 \sin t, 3 \cos t \rangle, \qquad 0 \le t \le \frac{\pi}{2}.$$

The density of the wire at point (x, y, z) is

$$\delta(x, y, z) = z^2 + 1.$$

12. The wire is in the shape of the curve described by

$$\mathbf{r}(t) = \left\langle t^2, \sqrt{t}, t \right\rangle, \qquad 1 \le t \le 2.$$

The density of the wire at point (x, y, z) is

$$\delta(x, y, z) = x + z.$$

13. A thin wire is in the shape of the curve described by

$$\mathbf{r}(t) = \langle t, t^2, t^3 \rangle, \qquad 0 \le t \le 1.$$

The density of the wire at point (x, y, z) is

$$\delta(x, y, z) = (x^2 + z^2 + 1) \text{ g/cm}.$$

Set up an integral that gives the mass of the wire; then find or approximate the value of the integral.

14. A thin wire is in the shape of the curve described by

$$\mathbf{r}(t) = \langle t, 2e^t, e^{2t} \rangle, \qquad 0 \le t \le 1.$$

The density of the wire at point (x, y, z) is

$$\delta(x, y, z) = yz \text{ g/cm}.$$

Find the mass of the wire.

15. A thin wire is in the shape of the square with vertices $(0, 0, 0)$, $(1, 1, 0)$, $\left(1, 1, \sqrt{2}\right)$, and $\left(0, 0, \sqrt{2}\right)$. The density of the wire at point (x, y, z) is

$$\delta(x, y, z) = (1 + xyz) \text{ g/cm}.$$

Find the mass of the wire.

16. A thin wire is in the shape of the circle of center $(0, 0, 0)$, radius 1 in the (y, z)-plane. The density of the wire at point (x, y, z) is

$$\delta(x, y, z) = \left(|y| + |z|\right) \text{ g/cm}.$$

Find the mass of the wire.

17. The outside edge of a curved, level road follows the planar curve C described by

$$\mathbf{r} = \mathbf{r}(t) = \langle x(t), y(t) \rangle, \qquad a \le t \le b.$$

A vertical wall is to be erected along this edge to keep careless drivers from veering off the road. At point (x, y) on the curve, the height of the wall, measured perpendicular to the road surface, is $f(x, y)$ feet. (See the accompanying figure.) Assuming that all distances are measured in feet, show that the area of the wall is

$$\int_C f\,ds.$$

Hint: Subdivide the outer curve into small parts and let Δs_k be the length of the kth piece. Argue that the area of the portion of the wall above the kth piece is approximately

$$f(\mathbf{c}_k)\,\Delta s_k$$

for some choice of $\mathbf{c}_k$. Sum these area contributions and show that the sum leads to a line integral.

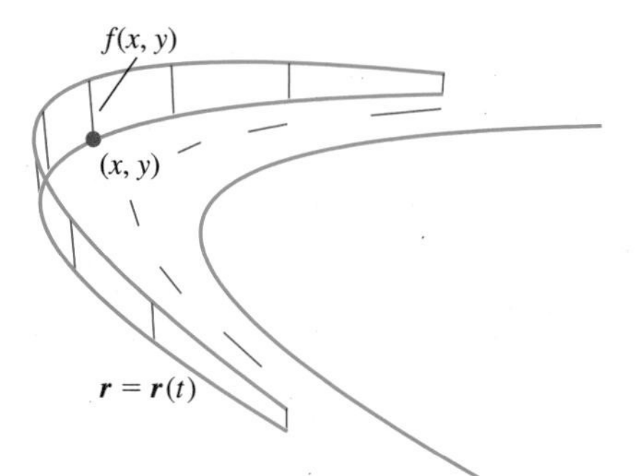

Figure for Exercise 17.

18. Assume that the outer edge of the road in Exercise 17 is the curve given by

$$\mathbf{r}(t) = \langle 150 \cos t, 150 \sin t \rangle, \qquad 0 \le t \le \frac{\pi}{2},$$

and that the height of the safety wall is given by

$$f(x, y) = 0.0005xy.$$

Find the area of the wall.

19. Suppose that the outer edge of the road in Exercise 17 is described by

$$\mathbf{r}(t) = \left\langle \frac{1}{50} t^2, t \right\rangle, \qquad -50 \le t \le 50,$$

and that the safety wall (which is also decorative) has height given by

$$f(x, y) = 2 + \sin\left(\frac{x + y}{4}\right).$$

Find an approximation to the area of the wall.

20. Let C be the curve described by

$$\mathbf{r}(t), \qquad a \le t \le b,$$

and let C_1 be the curve given by

$$\mathbf{r}_1(t) = \mathbf{r}(a + b - t), \qquad a \le t \le b.$$

a. Show that C and C_1 consist of the same set of points, but that as t runs from a to b, the points $\mathbf{r}(t)$ and $\mathbf{r}_1(t)$ move along the curves in opposite directions.

b. Show that if $\mathbf{r}$ is continuously differentiable on $a \le t \le b$ and f is continuous on the points of the curve described by $\mathbf{r}$, then

$$\int_C f\,ds = \int_{C_1} f\,ds.$$

21. In Section 6.7 we found the center of gravity of a curved piece of wire in the plane. In this exercise we adapt that argument to R^3 to get an expression for the center of gravity of a curved wire in space. Assume that a curved wire is in the shape of the curve C described by

$$\mathbf{r} = \mathbf{r}(t) = \langle x(t), y(t), z(t) \rangle,$$

for $a \le t \le b$, and let the density of the wire at point (x, y, z) be $\delta(x, y, z)$.

a. Cut the wire into n small pieces, and argue that the mass of the kth piece is

$$\Delta m_k \approx \delta(\mathbf{r}_k)\,\Delta s_k$$

for an appropriate choice of point $\mathbf{r}_k$.

b. Argue as in Section 6.7 that the position of the center of mass is given approximately by

$$\frac{1}{\sum_{k=1}^{n} \Delta m_k} \sum_{k=1}^{n} \Delta m_k \mathbf{r}_k.$$

See the accompanying figure. Show too that when this expression is resolved into coordinates, we obtain an ordered triple (X, Y, Z) with

$$X_n = \frac{\sum_{k=1}^{n} x(t_k)\,\Delta m_k}{\sum_{k=1}^{n} \Delta m_k}$$

$$Y_n = \frac{\sum_{k=1}^{n} y(t_k)\,\Delta m_k}{\sum_{k=1}^{n} \Delta m_k}$$

$$Z_n = \frac{\sum_{k=1}^{n} z(t_k)\,\Delta m_k}{\sum_{k=1}^{n} \Delta m_k}.$$

c. Finally, argue that as the wire is cut into smaller and smaller pieces, the ordered triple (X_n, Y_n, Z_n) tends to

$$(X, Y, Z) \tag{16}$$

with

$$X = \frac{1}{M} \int_C x(t)\delta(\mathbf{r}(t))\,ds,$$

where M is the mass of the wire and Y, Z are given by similar expressions. The center of mass of the wire is given by (16).

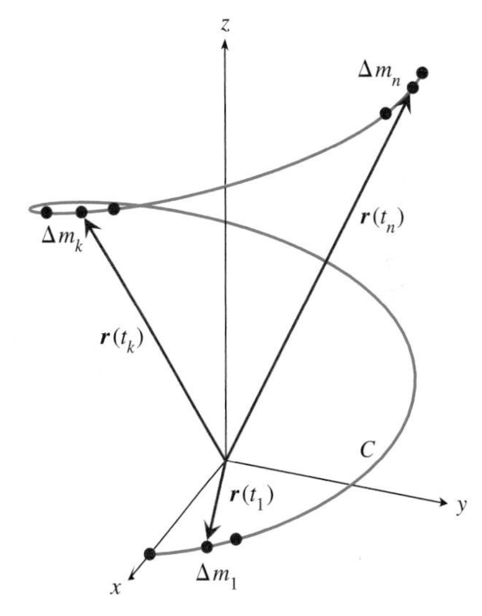

Figure for Exercise 21.

Exercises 22–25: A wire in space is in the shape of the curve C and has density function δ. Use the result of Exercise 21 to find the coordinates of the center of mass of the wire.

22. C is described by

$$\mathbf{r}(t) = \langle t, 2t, 3t \rangle, \qquad 0 \le t \le 2,$$

and the density by

$$\delta(x, y, z) = x + y + z + 1.$$

23. C is given by

$$\mathbf{r}(t) = \langle t, 2e^t, e^{2t} \rangle, \qquad 0 \le t \le 1,$$

and the density by

$$\delta(x, y, z) = yz.$$

See Exercise 14.

24. C is the square with vertices $(0, 0, 0)$, $(1, 1, 0)$, $(1, 1, \sqrt{2})$, and $(0, 0, \sqrt{2})$ and density described by

$$\delta(x, y, z) = 1 + xyz.$$

See Exercise 15.

25. C is given by

$$\mathbf{r}(t) = \langle t, 2 \sin t, 2 \cos t \rangle, \qquad 0 \le t \le 2\pi,$$

and the density by

$$\delta(x, y, z) = 4 + yz.$$

12.2 Vector Fields, Work, and Flows

Many physical phenomena in space can be modeled by associating a vector with each point in space. Examples include electric, magnetic, and gravitational fields and the velocity field for a fluid. A function that associates a vector with a point is called a *vector field.*

Vector Fields

DEFINITION Vector Field

A function **F** with domain a set $\mathcal{D}$ in R^n and range in R^n is called a *vector field.* (In our applications, n will be 2 or 3.)

The functions **F**, **G**, and **H** defined by

$$\begin{aligned} \mathbf{F}(x, y) &= \langle x^2 y, 2xye^x \rangle \\ \mathbf{G}(x, y, z) &= \langle -x, -y, -z \rangle \\ \mathbf{H}(\rho, \theta, \phi) &= \langle \rho \cos \theta, \rho \sin \theta, 0 \rangle \end{aligned} \tag{1}$$

are examples of vector fields. The function **F** has domain and range in R^2, while **G** and **H** have domain and range in R^3.

As suggested by (1), a vector field **F** with domain and range in R^2 has the form

$$\mathbf{F}(x, y) = \langle f_1(x, y), f_2(x, y) \rangle,$$

where f_1 and f_2 are real-valued functions of two variables. We say that **F** is continuous if the functions f_1 and f_2 are continuous on the domain of **F**. Similarly, a vector field

$$\mathbf{G}(x, y, z) = \langle g_1(x, y, z), g_2(x, y, z), g_3(x, y, z) \rangle$$

is continuous if each of the real-valued functions g_1, g_2, and g_3 is continuous on the domain of **G**.

The Graph of a Vector Field

To represent a vector field $\mathbf{F}$ graphically, we consider several points $\mathbf{P}$ in the domain of $\mathbf{F}$. For each such $\mathbf{P}$ we sketch the vector $\mathbf{F}(\mathbf{P})$, with the initial point of the vector at $\mathbf{P}$. The resulting graph consists of several arrows. The arrow from a point $\mathbf{P}$ represents the vector $\mathbf{F}(\mathbf{P})$, so the length of the arrow corresponds to (or is proportional to) the magnitude of $\mathbf{F}(\mathbf{P})$, and the direction of the arrow indicates the direction of $\mathbf{F}(\mathbf{P})$. See Figs. 12.9 and 12.10.

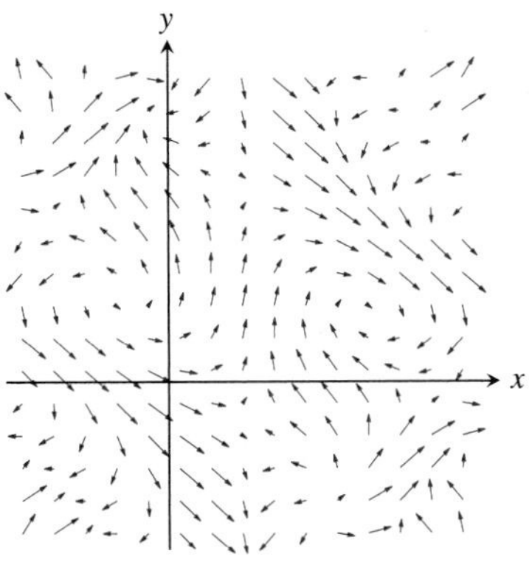

FIGURE 12.9 A vector field in the plane.

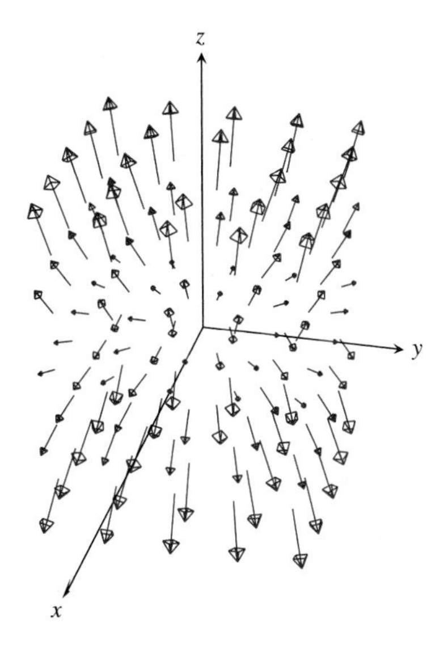

FIGURE 12.10 A vector field in space.

EXAMPLE 1 Fluid flow

A thin fluid (gas or liquid) flows across the plane. At position (x, y) in the plane, the velocity of the fluid is given by

$$\mathbf{v} = \mathbf{v}(x, y) = \langle x^2 - y^2, -2xy \rangle. \tag{2}$$

Sketch a graph of the vector field $\mathbf{v}$, then discuss the physical significance of (2).

Solution

To graph $\mathbf{v}$, we select a point $\mathbf{P}_0 = \langle x_0, y_0 \rangle$ in the plane, calculate $\mathbf{v}(x_0, y_0)$, and then draw the vector $\mathbf{v}(x_0, y_0)$ with initial point at the point $\mathbf{P}_0$. We repeat this process for several different points in the plane. When we are done, we have a figure containing many arrows in the plane. These arrows indicate the velocity of the fluid at several positions (x, y). As examples, consider the points $\mathbf{P}_1 = \langle 1, 0.5 \rangle$ and $\mathbf{P}_2 = \langle -1.5, -1 \rangle$. We have

$$\mathbf{v}(\mathbf{P}_1) = \mathbf{v}(1, 0.5) = \langle 1^2 - 0.5^2, -2 \cdot 1 \cdot 0.5 \rangle = \langle 0.75, -1 \rangle$$

and

$$\mathbf{v}(\mathbf{P}_2) = \mathbf{v}(-1.5, -1) = \langle (-1.5)^2 - (-1)^2, -2(-1.5)(-1) \rangle = \langle 1.25, -3 \rangle.$$

From the point $\mathbf{P}_1$ draw the vector $\langle 0.75, -1 \rangle$, and from $\mathbf{P}_2$ draw the vector $\langle 1.25, -3 \rangle$. See Fig. 12.11. With a computer or calculator we can draw $\mathbf{v}(\mathbf{P})$ for several points $\mathbf{P}$, obtaining a picture like that in Fig. 12.12.

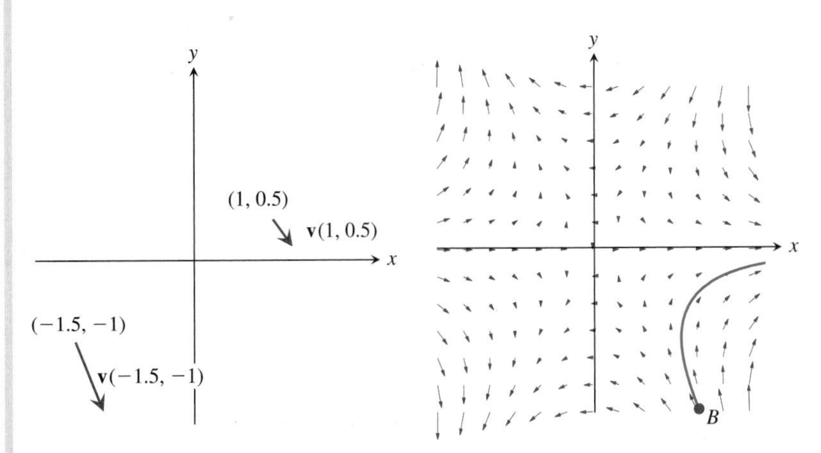

FIGURE 12.11 Two vectors of the vector field $\mathbf{v}(x, y) = \langle x^2 - y^2, -2xy \rangle$.

FIGURE 12.12 The path of a bubble in the flow described by $\mathbf{v}(x, y) = \langle x^2 - y^2, -2xy \rangle$.

With a little imagination, we can even get a sense of the flow of the fluid by looking at the graph. Imagine a small bubble in the fluid at the point marked B in Fig. 12.12. The velocity of the bubble, when it is at point B, is represented by the arrow with initial point at B. Thus, at the instant the bubble is at B, it is moving in the direction of the arrow and at a speed proportional to the length of the arrow. But it may not move at this velocity for long. As soon as the bubble leaves B, it is at a new point where the fluid may have a different velocity, as represented by the arrow at this new point. By following the arrows and adjusting for how they change as the bubble moves from B, we can get a rough idea of the path taken by the bubble. See Fig. 12.12.

The study of fluid flow is of great importance in engineering and physics. In designing aircraft, a significant amount of time and money goes into analyzing the flow of air (a fluid!) around an aircraft. Fluid mechanics is also important in studying the propagation of contaminants in groundwater, in designing effective means for sewage treatment, and in predicting the weather. When describing the flow of a fluid, it is often convenient to think of the fluid as consisting of many small, discrete particles. As the fluid flows, these particles move from one position to another; the speeds and the directions of the particles may change as well. If the velocity of a particle of fluid depends only on its position, and not on time, then a vector field such as (2) can be used to describe the velocity of the fluid at every point.

The graph of a vector field in three dimensions is harder to visualize. To avoid some of the confusion that can result from the presence of too many arrows (as in Fig. 12.12), it is sometimes better to display only a small portion of the vector field.

EXAMPLE 2 Sketch two different graphs of part of the gravitational field generated by the Earth.

Solution

Assume that Earth is centered at $(0, 0, 0)$ and let $R = 6.37 \times 10^6$ meters denote its radius. The gravitational field at a point $\mathbf{P} = \langle x, y, z \rangle$ at distance greater than R from Earth's center is the gravitational force exerted on an object of mass 1 kg at $\mathbf{P}$. The magnitude of the force is inversely proportional to the square of the distance to the center of Earth, and the direction of the force is toward the center of Earth. Thus, the gravitational field $\mathbf{F}$ is given by

$$\mathbf{F}(\mathbf{P}) = \mathbf{F}(x, y, z) = \frac{-GM \cdot 1}{(x^2 + y^2 + z^2)^{3/2}} \langle x, y, z \rangle, \tag{3}$$

where $G \approx 6.67 \times 10^{-11}\ \mathrm{N{\cdot}m^2/kg^2}$ is the universal constant of gravitation and $M \approx 5.97 \times 10^{24}$ kg is the mass of Earth. See Exercise 19 and Section 9.5. The length of the vector on the right-hand side of (3) depends only on the distance of P from the center of Earth. Furthermore, the direction of $\mathbf{F}(\mathbf{P})$ is the direction from $\mathbf{P}$ to the center of Earth. These remarks are well illustrated by graphing $\mathbf{F}(\mathbf{P})$ for all points on some sphere of radius greater than R and centered at the origin. In Fig. 12.13 we show the graph of the vector field for points on a sphere of radius $10R$. The graph was produced by using a vector field plotting package from a CAS.

To illustrate that the magnitude of $\mathbf{F}(\mathbf{P})$ varies with the distance of $\mathbf{P}$ from the origin, we graph the vector field for points $\mathbf{P}$ in the plane $x = 10R$. See Fig. 12.14. In this figure we can clearly see that the length of $\mathbf{F}(\mathbf{P})$ decreases as we move farther from the origin.

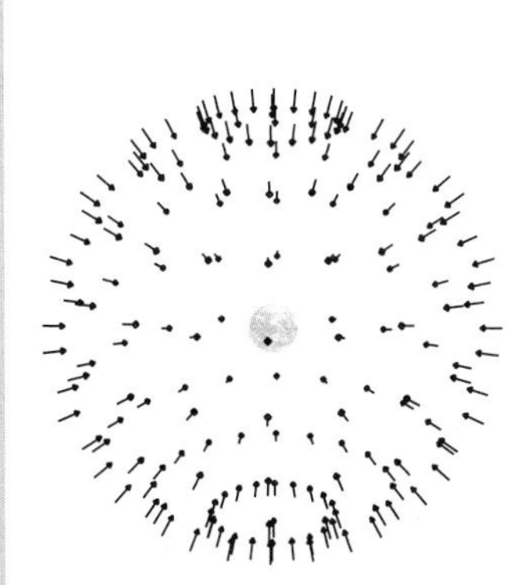

FIGURE 12.13 Earth's gravitational field on a sphere of radius $10R$.

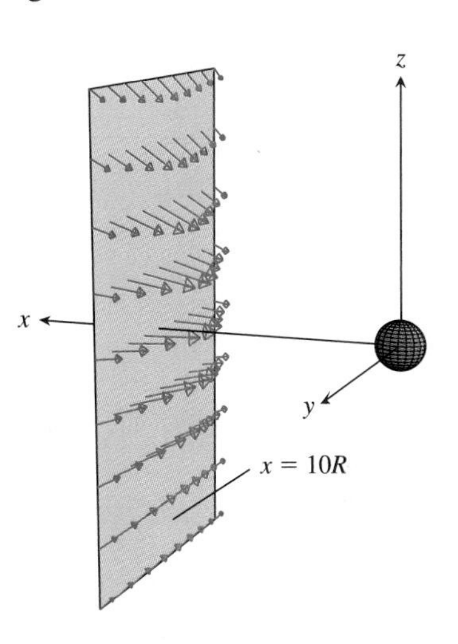

FIGURE 12.14 Earth's gravitational field on the plane described by $x = 10R$.

Vector Fields and Line Integrals

Fluid Flow and Flux When a vector field models a physical phenomenon, line integrals can often be used to help analyze the field. When the vector field is a force field, as in Example 2, we can use line integrals to calculate the work done by the field when an object moves from one point to another in the field. When the vector field models the flow of a fluid, as in Example 1, line integrals can be used to calculate the rate at which fluid flows across a curved boundary.

Flux for a Fluid Flow

Let $\mathbf{v} = \mathbf{v}(x, y)$ describe the velocity field for a fluid flowing in the plane. Assume that $\mathbf{v}$ is continuous, that the units for $\mathbf{v}$ are cm/s, and that the fluid is of constant density δ g/cm^2. Let C be a smooth planar curve described by

$$\mathbf{r}(\tau) = \langle x(\tau), y(\tau) \rangle, \qquad a \leq \tau \leq b. \tag{4}$$

At what rate does fluid mass flow across C? (*Note:* Such curves as C are not actual barriers to the flow; rather they are imaginary and provide means to measure the flow in regions of interest.)

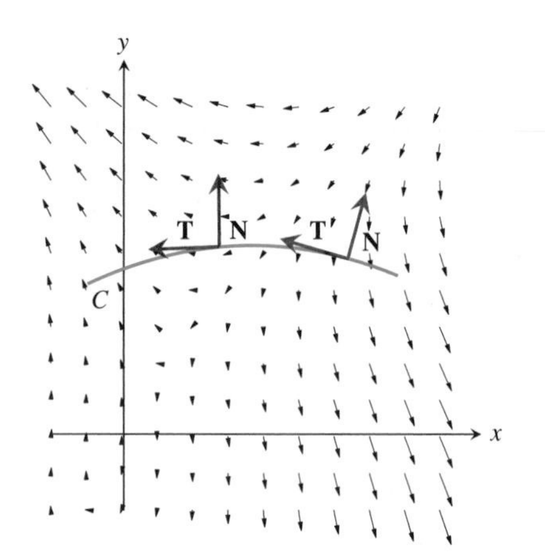

FIGURE 12.15 The normal vector **N** indicates the positive direction across C.

In Fig. 12.15 we show the curve C and the vector field **v** for some points near C. As the figure shows, there are some points on C where fluid flows across C and (roughly) toward the origin, and other points where fluid flows across C and away from the origin. To calculate the net flow across C, we designate one of these directions of flow as "positive" and the other direction as "negative." At a given point of C, flow across C in the positive direction makes a positive contribution to the net flow across C, and flow in the negative direction makes a negative contribution. We will use unit vectors normal to C to define these positive and negative directions. In particular, if **P** is a point on C and $\mathbf{T} = \langle u, v\rangle$ is the unit tangent to C at **P**, let **N** be the unit vector normal to **T** and pointing to the right as we face in the direction of **T**. Hence, $\mathbf{N} = \langle v, -u\rangle$. We take **N** to be the *positive direction across C at* **P**. See Fig. 12.15.

Now consider the subdivision of C arising from a regular subdivision of $[a, b]$ into n segments. Use Δs to denote the length of a representative piece of C arising from this subdivision. Then

$$\Delta s \approx \|\mathbf{r}'(\tau)\|\, d\tau = ds,$$

where τ is in the subinterval of $[a, b]$ corresponding to the piece of C and $d\tau = (b - a)/n$ is the length of the subinterval. See Fig. 12.16. Let **v** represent the velocity of the fluid at a point on this small piece, let **N** be the positive direction normal, and let **T** be the unit tangent at this point. Now imagine the fluid that flows across this small piece of C in a short time Δt. At the beginning of this time interval, we consider the fluid that has just started to flow across the length Δs along C. During the time interval Δt, the fluid that initially was along the length Δs sweeps out, approximately, a parallelogram with sides

$$(\mathbf{v}\text{ cm/s})(\Delta t\text{ s}) = \Delta t\,\mathbf{v}\text{ cm} \quad\text{and}\quad ds\,\mathbf{T}\text{ cm}.$$

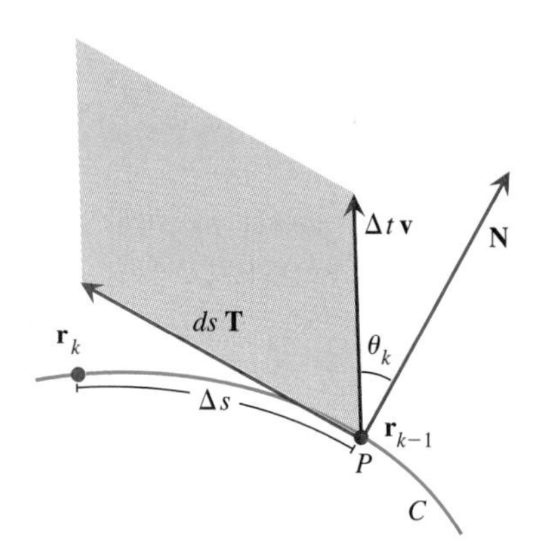

FIGURE 12.16 When $0 \le \theta_k < \pi/2$, the net flow across the segment is positive.

See Figs. 12.16 and 12.17. The area of this parallelogram is

$$\|(\Delta t\mathbf{v}) \times (ds\,\mathbf{T})\| = \Delta t\|\mathbf{v} \times \mathbf{T}\|\, ds = \Delta t|\mathbf{v}\cdot\mathbf{N}|\, ds\text{ cm}^2. \tag{5}$$

We ask you to verify this result in Exercise 24. When we multiply the area of this parallelogram by the fluid density δ and divide by the elapsed time Δt, we obtain dm, the mass-per-second rate of flow across the piece of C,

$$dm = \delta|\mathbf{v}\cdot\mathbf{N}|\, ds\text{ g/s}. \tag{6}$$

If we delete the absolute value signs from (6), we obtain a "signed" rate of flow,

$$d\Phi = \delta(\mathbf{v}\cdot\mathbf{N})\, ds\text{ g/s}. \tag{7}$$

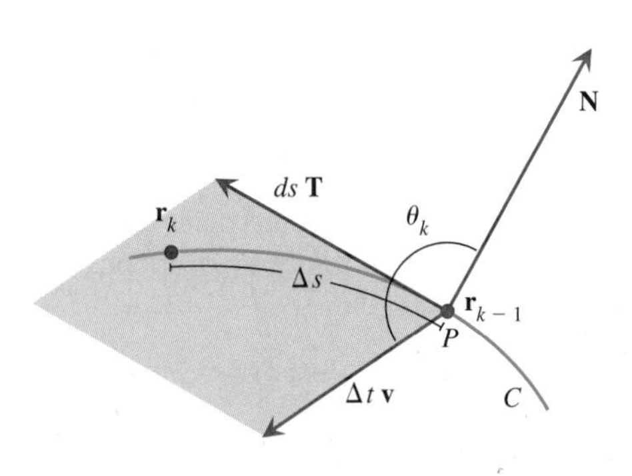

FIGURE 12.17 When $\pi/2 \le \theta_k < \pi$, the net flow across the segment is negative.

This quantity is positive when $\mathbf{v}\cdot\mathbf{N} > 0$, that is, when the angle between **v** and **N** is less than 90°. This is the case when **v** and **N** are on the same side of C, so the net flow of fluid is toward the **N** side of C. If **v** and **N** are on opposite sides of C, then (7) is negative and the net flow of fluid is away from the **N** side of C. See Figs. 12.16 and 12.17.

Adding the contributions (7) for each piece of the subdivision gives an approximation of the net rate of flow of fluid mass across C. This quantity is called the **flux** of the flow across C. Thus, we have

$$\Phi = \text{flux across } C \approx \sum d\Phi = \delta\sum(\mathbf{v}\cdot\mathbf{N})\, ds.$$

As we let $n \to \infty$, these sums become integrals and we find

$$\Phi = \text{flux across } C = \delta \int_C (\mathbf{v} \cdot \mathbf{N})\, ds. \tag{8}$$

Because δ is constant, we factored it from the sum and the integral. See Exercise 32 for the case in which δ is not constant.

The word *flux* does not always indicate a mass rate of flow. Flux can be calculated for any vector field. When the field is an electric field, the flux measures the number of lines of force through a given surface. Flux can also be calculated for gravitational fields and magnetic fields and is associated with heat transfer.

We apply the result of Investigation 1 to the flow given in Example 1.

EXAMPLE 3 Let $\mathbf{v}(x, y) = \langle x^2 - y^2, -2xy\rangle$ be the velocity field for a fluid in the plane, as discussed in Example 1. Assume that the fluid has density of $\delta = 2$ g/cm^2 and let C be the half circle described by

$$\mathbf{r}(t) = \langle \cos t, \sin t\rangle, \qquad \frac{-\pi}{2} \le t \le \frac{\pi}{2}.$$

Find the flux across C.

Solution

For the given description of C, the unit tangent to C at $\mathbf{r}(t)$ is

$$\mathbf{T}\langle\mathbf{r}(t)\rangle = \frac{1}{\|\mathbf{r}'(t)\|}\mathbf{r}'(t) = \frac{1}{1}\langle -\sin t, \cos t\rangle = \langle -\sin t, \cos t\rangle.$$

Recall that if the unit tangent is $\mathbf{T} = \langle a_1, a_2\rangle$, then we have agreed to take the unit normal as $\mathbf{N} = \langle a_2, -a_1\rangle$. Thus the unit normal to C at $\mathbf{r}(t)$ is

$$\mathbf{N}(\mathbf{r}(t)) = \langle \cos t, \sin t\rangle.$$

This normal defines (or indicates) the positive direction across C. In addition,

$$\mathbf{v}(\mathbf{r}(t)) = \mathbf{v}(\cos t, \sin t) = \langle \cos^2 t - \sin^2 t, -2\cos t \sin t\rangle.$$

Hence

$$\begin{aligned}\mathbf{v}(\mathbf{r}(t)) \cdot \mathbf{N}(\mathbf{r}(t)) &= \langle \cos^2 t - \sin^2 t, -2\cos t \sin t\rangle \cdot \langle \cos t, \sin t\rangle \\ &= \cos^3 t - 3\cos t \sin^2 t.\end{aligned}$$

By (8), the flux Φ across C is

$$\Phi = \int_C \delta(\mathbf{v} \cdot \mathbf{N})\, ds = \delta \int_{-\pi/2}^{\pi/2} \mathbf{v}(\mathbf{r}(t)) \cdot \mathbf{N}(\mathbf{r}(t))\, \|\mathbf{r}'(t)\|\, dt.$$

$$= 2\int_{-\pi/2}^{\pi/2} (\cos^3 t - 3\cos t \sin^2 t)\, dt.$$

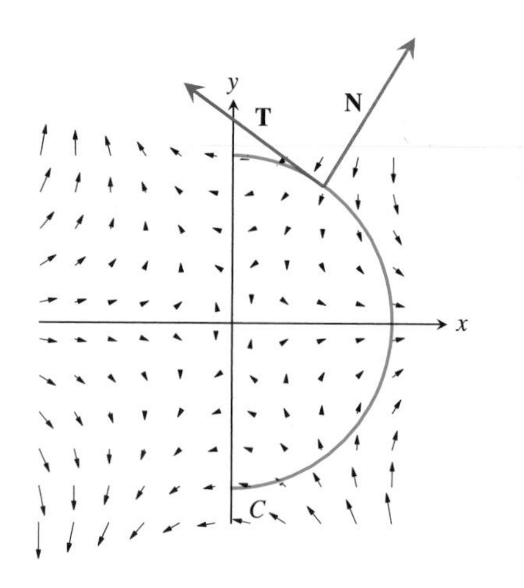

FIGURE 12.18 The flow $\mathbf{v}(x, y) = \langle x^2 - y^2, -2xy \rangle$ past the curve C.

Replace $\cos^3 t$ by $\cos t(1 - \sin^2 t)$ and simplify to get

$$= 2\int_{-\pi/2}^{\pi/2} (\cos t - 4\cos t \sin^2 t)\, dt$$

$$= 2\sin t - \frac{8}{3}\sin^3 t \Big|_{-\pi/2}^{\pi/2} = -\frac{4}{3}.$$

Thus, the flux (or net rate of flow) past C is $-\frac{4}{3}$ g/s. Roughly, this means that in one second, $\frac{4}{3}$ g more fluid flows through C in the negative direction than in the positive direction. The vector field $\mathbf{v}$ and curve C are shown in Fig. 12.18. Does it appear from the figure that there is more flow in the negative direction through C than in the positive direction?

When C is a smooth, simple closed curve in the plane of a flow, the flux across C measures the net rate of flow of fluid out of (or into) the region enclosed by C. If C is parameterized so that C is traced in the counterclockwise direction, then according to our conventions, the positive normals point to the outside of C.

EXAMPLE 4 A fluid flowing in the plane has a constant density of $\delta = 1$ g/cm². The flow of the fluid is described by

$$\mathbf{v}(x, y) = \langle ye^x, y \rangle \text{ cm/s}.$$

Find the flux across the ellipse C described by

$$\mathbf{r}(t) = \langle 2\cos t, \sin t \rangle, \qquad 0 \le t \le 2\pi. \tag{9}$$

Solution

A picture of the vector field $\mathbf{v}$ and the curve C is shown in Fig. 12.19. From this figure, can you tell whether the net rate of flow across C is positive, negative, or zero?

To answer the question more precisely, we consider the flux integral

$$\Phi = \int_C \delta(\mathbf{v} \cdot \mathbf{N})\, ds = \int_0^{2\pi} \delta(\mathbf{v}(\mathbf{r}(t)) \cdot \mathbf{N}(\mathbf{r}(t)))\, \|\mathbf{r}'(t)\|\, dt, \tag{10}$$

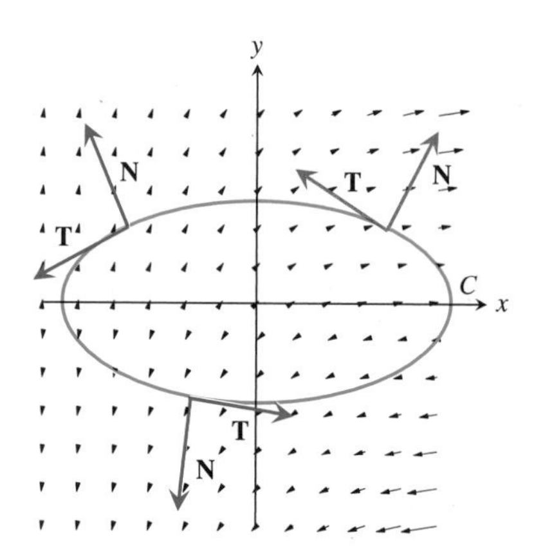

FIGURE 12.19 The flow through an ellipse.

where $\mathbf{N}(\mathbf{r}(t))$ is the positive unit normal to C at $\mathbf{r}(t)$. Because the unit tangent to C at $\mathbf{r}(t)$ is

$$\mathbf{T}(\mathbf{r}(t)) = \frac{1}{\|\mathbf{r}'(t)\|}\mathbf{r}'(t) = \frac{1}{\|\mathbf{r}'(t)\|}\langle -2\sin t, \cos t \rangle,$$

we have

$$\mathbf{N}(\mathbf{r}(t)) = \frac{1}{\|\mathbf{r}'(t)\|}\langle \cos t, 2\sin t \rangle. \tag{11}$$

In addition,

$$\mathbf{v}(r(t)) = \mathbf{v}(2\cos t, \sin t) = \langle e^{2\cos t}\sin t, \sin t \rangle. \tag{12}$$

Setting $\delta = 1$, substituting (11) and (12) into (10), and using (9), we have

$$\begin{aligned}\Phi &= \int_0^{2\pi} \langle e^{2\cos t}\sin t, \sin t\rangle \cdot \langle \cos t, 2\sin t\rangle\, dt \\ &= \int_0^{2\pi} (e^{2\cos t}\sin t\cos t + 2\sin^2 t)\, dt \\ &= \int_0^{2\pi} e^{2\cos t}\sin t\cos t\, dt + 2\int_0^{2\pi} \sin^2 t\, dt = 0 + 2\pi = 2\pi \text{ g/s}.\end{aligned}$$

This means that more fluid is flowing out of the closed curve C than is flowing into it. When this is the case and the fluid has constant density, we say that there is a **source** for the fluid inside of C. If the integral had been negative, then we would have said that there is a **sink** for the fluid inside of C.

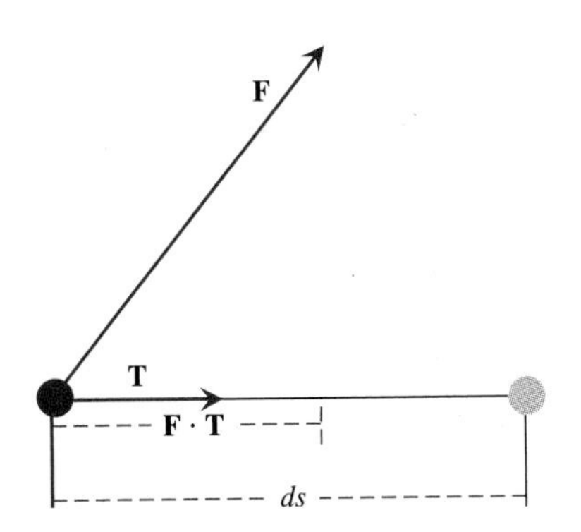

FIGURE 12.20 A force **F** moves an object in a direction parallel to **T**.

Force Fields and Work When a constant force **F** acts on an object through distance ds in a direction parallel to the unit vector **T**, then the work done by **F** is

$$(\mathbf{F}\cdot\mathbf{T})\, ds. \tag{13}$$

See Fig. 12.20. In Section 6.6 we used this basic result to derive an expression for the work done by a nonconstant force as an object moves along a curved path in the plane. The argument used to develop the planar case extends readily to the R^3 situation.

INVESTIGATION 2

Determining the Work Done by a Force

A smooth curve C in R^3 is described by

$$\mathbf{r}(t) = \langle x(t), y(t), z(t)\rangle, \qquad a \le t \le b.$$

A force acts on an object as it moves along C from $\mathbf{r}(a)$ to $\mathbf{r}(b)$. Assume that at the point (x, y, z) on C the force is $\mathbf{F}(x, y, z)$ newtons, where $\mathbf{F}$ is a continuous vector field defined on C. Find the work done by this force as the object moves on C from $\mathbf{r}(a)$ to $\mathbf{r}(b)$.

Use a regular subdivision of $[a, b]$ to generate a subdivision of C into n small pieces. Let Δs be the length of a typical piece of this subdivision. Then

$$\Delta s \approx \|\mathbf{r}'(t)\|\, dt = ds,$$

where t is in the subinterval corresponding to the piece of C and $dt = (b - a)/n$. Let $\mathbf{T}$ be the unit tangent to C and $\mathbf{F}$ the force at a point on this piece of C. By (13), the work ΔW done by the force as the object moves a distance Δs along the small piece of C can be approximated by

$$\Delta W \approx dW = (\mathbf{F}\cdot\mathbf{T})\, ds.$$

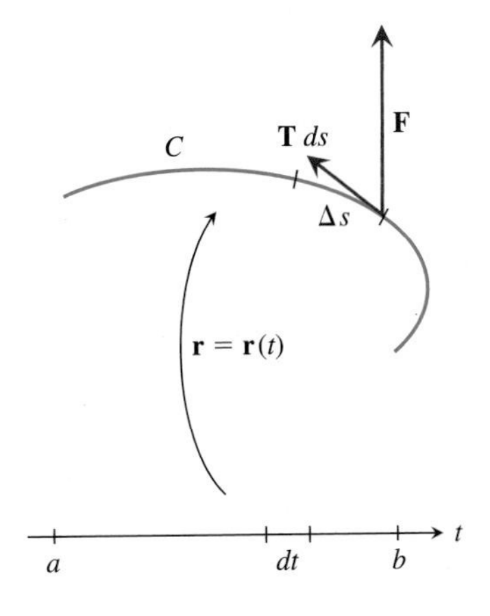

FIGURE 12.21 The direction of the motion is roughly in the direction of **T**.

See Fig. 12.21. The total work W done is approximated by summing these contributions,

$$W = \sum \Delta W \approx \sum dW = \sum (\mathbf{F}\cdot\mathbf{T})\, ds.$$

As n approaches infinity, the last sum becomes a line integral along C. Thus we find

$$W = \int_C (\mathbf{F}\cdot\mathbf{T})\, ds. \tag{14}$$

EXAMPLE 5 A force **F**, given by

$$\mathbf{F}(x, y, z) = \langle y, z, x \rangle \text{ newtons,}$$

acts on an object as it moves along the helical curve described by

$$\mathbf{r}(t) = \langle t, \cos t, \sin t \rangle, \qquad 0 \le t \le 2\pi.$$

Calculate the work done by **F**.

Solution

In Fig. 12.22 we show the curve C, the force field vectors and unit tangent vectors for a few points on C. The unit tangent to C is

$$\mathbf{T} = \mathbf{T}(\mathbf{r}(t)) = \frac{1}{\|\mathbf{r}'(t)\|}\mathbf{r}'(t) = \frac{1}{\|\mathbf{r}'(t)\|}\langle 1, -\sin t, \cos t \rangle \tag{15}$$

and

$$\mathbf{F}(\mathbf{r}(t)) = \mathbf{F}(t, \cos t, \sin t) = \langle \cos t, \sin t, t \rangle. \tag{16}$$

Combining (14), (15), and (16), we have

$$\begin{aligned}
\text{work} &= \int_C (\mathbf{F} \cdot \mathbf{T})\, ds = \int_0^{2\pi} (\mathbf{F}(\mathbf{r}(t)) \cdot \mathbf{T}(\mathbf{r}(t)))\, \|\mathbf{r}'(t)\|\, dt \\
&= \int_0^{2\pi} \langle \cos t, \sin t, t \rangle \cdot \langle 1, -\sin t, \cos t \rangle\, dt \\
&= \int_0^{2\pi} (\cos t - \sin^2 t + t \cos t)\, dt \\
&= \left(\sin t - \frac{t}{2} + \frac{1}{4} \sin 2t + t \sin t + \cos t \right)\Bigg|_0^{2\pi} = -\pi.
\end{aligned}$$

If distance is measured in meters, mass in kilograms, and time in seconds, then the work done is $-\pi$ joules.

In Fig. 12.22 it appears that at each point on C, the angle between **F** and **T** is greater than $\pi/2$. Hence, the dot product $\mathbf{F} \cdot \mathbf{T}$ is negative at each point of C, and the negative answer for the work makes sense. Because $(\mathbf{F} \cdot \mathbf{T})\mathbf{T}$, the tangent component of **F**, points opposite the direction of the motion along C, the force is working against the direction of motion. This could mean, for example, that the force is slowing the object as it moves along C.

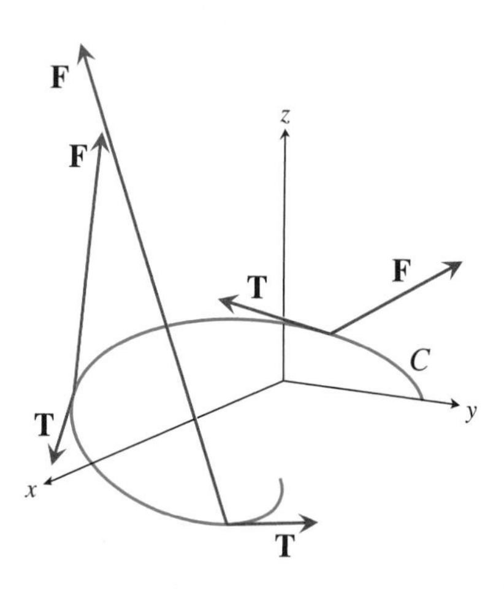

FIGURE 12.22 Some force field vectors and some unit tangent vectors at points on C.

With some force fields, the work done by the field as an object moves on a curve from one point to another depends only on the starting and ending points of the curve. We will study such fields in more detail in the next section, but for now we show that the force field generated by Earth's gravitational field has this property.

EXAMPLE 6 Let **A** and **B** be two points on or above the surface of Earth. An object of mass m kg moves from **A** to **B** along a smooth curve that stays on or above Earth's surface. Show that the work done by Earth's gravitational field is independent of the path from **A** to **B**. As seen in

Example 2, the force $\mathbf{F}$ at point (x, y, z) is

$$\mathbf{F}(x, y, z) = \frac{-GMm}{(x^2 + y^2 + z^2)^{3/2}} \langle x, y, z \rangle. \tag{17}$$

Solution

Let C be a smooth curve described by

$$\mathbf{r}(t) = \langle x(t), y(t), z(t) \rangle, \qquad a \le t \le b,$$

with $\mathbf{r}(a) = \mathbf{A}$ and $\mathbf{r}(b) = \mathbf{B}$. By (14), the work done as the object moves along C is

$$\begin{aligned}
\int_C (\mathbf{F} \cdot \mathbf{T})\, ds &= \int_a^b \left(\mathbf{F}(\mathbf{r}(t)) \cdot \left(\frac{1}{\|\mathbf{r}'(t)\|} \mathbf{r}'(t) \right) \right) \|\mathbf{r}'(t)\|\, dt \\
&= \int_a^b \frac{-GmM}{(x(t)^2 + y(t)^2 + z(t)^2)^{3/2}} \langle x(t), y(t), z(t) \rangle \cdot \langle x'(t), y'(t), z'(t) \rangle\, dt \\
&= -GmM \int_a^b \frac{x(t)x'(t) + y(t)y'(t) + z(t)z'(t)}{(x(t)^2 + y(t)^2 + z(t)^2)^{3/2}}\, dt.
\end{aligned}$$

The form of the integrand leads to the observation that

$$-\frac{x(t)x'(t) + y(t)y'(t) + z(t)z'(t)}{(x(t)^2 + y(t)^2 + z(t)^2)^{3/2}} = \frac{d}{dt}\left(\frac{1}{\sqrt{x(t)^2 + y(t)^2 + z(t)^2}} \right).$$

This greatly simplifies the rest of this calculation. We now have

$$\begin{aligned}
\int_C (\mathbf{F} \cdot \mathbf{T})\, ds &= GmM \int_a^b \frac{d}{dt}\left(\frac{1}{\sqrt{x(t)^2 + y(t)^2 + z(t)^2}} \right) dt \\
&= GmM \left(\frac{1}{\sqrt{x(b)^2 + y(b)^2 + z(b)^2}} - \frac{1}{\sqrt{x(a)^2 + y(a)^2 + z(a)^2}} \right) \\
&= GmM \left(\frac{1}{\|\mathbf{B}\|} - \frac{1}{\|\mathbf{A}\|} \right).
\end{aligned}$$

As this last expression shows, the work depends on the endpoints A and B of the path of integration, but is independent of the path C linking A to B. In fact, because the last expression involves only $\|\mathbf{A}\|$ and $\|\mathbf{B}\|$, we see that the work done by the gravitational field depends only on the initial and final heights of the object above Earth's surface.

Reversing the Direction of Integration

When we studied the Riemann integral, we saw that if the function f is continuous on $[a, b]$, then

$$\int_a^b f(x)\, dx = -\int_b^a f(x)\, dx.$$

In other words, reversing the direction of integration (right to left instead of left to right) changes the sign of the integral.

A similar result holds for line integrals. Let C be a piecewise smooth curve parameterized by $\mathbf{r} = \mathbf{r}(t)$, $a \le t \le b$. Let $-C$ denote the curve described by

$$\mathbf{r}_1(t) = \mathbf{r}(a + b - t), \qquad a \le t \le b.$$

Note that $\mathbf{r}_1(t)$ is a point on C for any $t \in [a, b]$ and that $\mathbf{r}_1(a) = \mathbf{r}(b)$ and $\mathbf{r}_1(b) = \mathbf{r}(a)$. Thus, if $\mathbf{r}(t)$ traces the curve C from $\mathbf{A} = \mathbf{r}(a)$ to $\mathbf{B} = \mathbf{r}(b)$, then $\mathbf{r}_1$ traces C from $\mathbf{B}$ to $\mathbf{A}$. In Exercise 27 we ask you to establish the following important result.

> Let C be a piecewise smooth curve in the plane or in space and let $\mathbf{F}$ be a continuous vector field defined on C. Then
>
> $$\int_{-C} (\mathbf{F} \cdot \mathbf{T})\, ds = -\int_{C} (\mathbf{F} \cdot \mathbf{T})\, ds.$$

Line Integral Notation

The form of the integrals

$$\Phi = \int_C (\mathbf{v} \cdot \mathbf{N})\, ds \quad \text{and} \quad W = \int_C (\mathbf{F} \cdot \mathbf{T})\, ds, \tag{18}$$

for flux Φ and work W are closely related to the underlying physical phenomena. Although this feature of the integrals in (18) is useful, it is often convenient to write these line integrals in a different form.

Let C be a smooth curve in R^3 parameterized by

$$\mathbf{r}(t) = \langle x(t), y(t), z(t) \rangle, \qquad a \le t \le b,$$

and let

$$\mathbf{F}(x, y, z) = \langle L(x, y, z), M(x, y, z), N(x, y, z) \rangle$$

describe a vector field defined in a region including C. The unit tangent to C is given by

$$\mathbf{T} = \mathbf{T}(t) = \frac{1}{\|\mathbf{r}'(t)\|}\mathbf{r}'(t) = \frac{1}{\|\mathbf{r}'(t)\|}\langle x'(t), y'(t), z'(t) \rangle.$$

Thus,

$$\begin{aligned}
\int_C (\mathbf{F} \cdot \mathbf{T})\, ds &= \int_a^b (\mathbf{F}(\mathbf{r}(t)) \cdot \mathbf{T}(t))\, \|\mathbf{r}'(t)\|\, dt \\
&= \int_a^b (\mathbf{F}(\mathbf{r}(t)) \cdot \mathbf{r}'(t))\, dt \\
&= \int_a^b \langle L(\mathbf{r}(t)), M(\mathbf{r}(t)), N(\mathbf{r}(t)) \rangle \cdot \langle x'(t), y'(t), z'(t) \rangle\, dt \qquad (19) \\
&= \int_a^b (L(x(t), y(t), z(t))x'(t) + M(x(t), y(t), z(t))y'(t) \\
&\qquad + N(x(t), y(t), z(t))z'(t))\, dt.
\end{aligned}$$

If we use the differentials

$$dx = x'(t)\,dt, \qquad dy = y'(t)\,dt, \qquad dz = z'(t)\,dt$$

in the last expression of (19) and write $\int_C$ instead of $\int_a^b$, then we are led to the expression

$$\int_C (\mathbf{F}\cdot\mathbf{T})\,ds = \int_C L\,dx + M\,dy + N\,dz. \tag{20}$$

The flux integral can be written in a similar form. In particular, let C be a smooth curve in the plane and let

$$\mathbf{v} = \langle P(x,y), Q(x,y)\rangle$$

describe a vector field defined on C. We then have

$$\int_C (\mathbf{v}\cdot\mathbf{N})\,ds = \int_C -Q\,dx + P\,dy. \tag{21}$$

The left-hand sides of (20) and (21) carry more physically relevant information than do the right sides. The latter, however, are often more convenient for evaluation.

EXAMPLE 7 Evaluate

$$\int_C xyz\,dx + (y - z)\,dy + x^2\,dz$$

where C is the segment from $\mathbf{A} = \langle -1,2,0\rangle$ to $\mathbf{B} = \langle 2,0,2\rangle$.

Solution

The segment C is parallel to the vector

$$\overrightarrow{AB} = \mathbf{B} - \mathbf{A} = \langle 3,-2,2\rangle$$

and contains the point $\mathbf{A}$. A parametric representation for the segment is

$$\mathbf{r}(t) = \mathbf{A} + t\overrightarrow{AB} = \langle -1,2,0\rangle + t\langle 3,-2,2\rangle = \langle -1+3t, 2-2t, 2t\rangle,$$

where $0 \le t \le 1$. With

$$x = x(t) = -1 + 3t, \quad y = y(t) = 2 - 2t, \quad \text{and} \quad z = z(t) = 2t,$$

we have

$$\begin{aligned}
&\int_C xyz\,dx + (y - z)\,dy + x^2\,dz \\
&\quad = \int_0^1 [x(t)y(t)z(t)x'(t) + (y(t) - z(t))y'(t) + x(t)^2 z'(t)]\,dt \\
&\quad = \int_0^1 [(-1+3t)(2-2t)(2t)3 + (2 - 2t - 2t)(-2) + (-1+3t)^2 2]\,dt \\
&\quad = \int_0^1 (-36t^3 + 66t^2 - 16t - 2)\,dt = 3.
\end{aligned}$$

The line integral $\int_C (\mathbf{F} \cdot \mathbf{T})\, ds$ is occasionally encountered in yet a third form. Because

$$\frac{d\mathbf{r}}{dt} = \mathbf{r}'(t) = \langle x'(t), y'(t), z'(t)\rangle,$$

we write

$$d\mathbf{r} = \mathbf{r}'(t)\, dt. \tag{22}$$

But then,

$$d\mathbf{r} = \mathbf{r}'(t)\, dt = \left(\frac{1}{\|\mathbf{r}'(t)\|}\mathbf{r}'(t)\right)\|\mathbf{r}'(t)\|\, dt = \mathbf{T}\|\mathbf{r}'(t)\|\, dt = \mathbf{T}\, ds. \tag{23}$$

Combining (22) and (23), it is reasonable to write

$$\int_C (\mathbf{F} \cdot \mathbf{T})\, ds = \int_C \mathbf{F} \cdot d\mathbf{r}.$$

This gives another form for the line integral on the left. Furthermore, if $\mathbf{F} = \mathbf{F}(x, y) = \langle L(x, y), M(x, y)\rangle$ and we let $d\mathbf{r} = \mathbf{d}\langle x, y\rangle = \langle dx, dy\rangle$, then we can write

$$\int_C \mathbf{F} \cdot d\mathbf{r} = \int_C \langle L(x, y), M(x, y)\rangle \cdot \langle dx, dy\rangle = \int_C L\, dx + M\, dy,$$

which is the two-dimensional version of (20).

Exercises 12.2

Exercises 1–4: For the planar vector field **F**, *draw the vectors* **F(P)** *based at the given* **P**. *If you have access to a vector field graphing routine, use it to produce a graph showing several vectors in the field.*

1. $\mathbf{F}(x, y) = \left\langle \frac{1}{2}x, -\frac{1}{2}y\right\rangle$,
   $\mathbf{P} = \langle 1, 0\rangle, \langle 1, 1\rangle, \left\langle -1, \frac{1}{2}\right\rangle$, and $\langle -1, -1\rangle$
2. $\mathbf{F}(x, y) = \langle -x, 0\rangle$,
   $\mathbf{P} = \langle 1, 0\rangle, \langle 1, 1\rangle, \langle 0, -2\rangle$, and $\langle 0, 0\rangle$
3. $\mathbf{F}(x, y) = \left\langle \sqrt{x^2 + y^2}, \sqrt{x^2 + y^2}\right\rangle$,
   $\mathbf{P} = \langle 1, 1\rangle, \langle -2, 4\rangle, \langle 0, -2\rangle$, and $\left\langle -\sqrt{2}/2, \sqrt{2}/2\right\rangle$
4. $\mathbf{F}(x, y) = \langle y, -x\rangle$,
   $\mathbf{P} = \langle 1, 0\rangle, \langle 2, 1\rangle, \langle 0, -2\rangle$, and $\langle -2, -1\rangle$

Exercises 5–8: The vector field **F** *is defined on* R^3. *Draw the vectors* **F(P)** *based at the given* **P**. *If you have access to a vector field graphing routine, use it to produce a graph showing several vectors on the sphere of center* $(0, 0, 0)$ *and radius* 1 *and also several vectors on the plane* $z = 2$.

5. $\mathbf{F}(x, y, z) = \langle yz, zx, xy\rangle$,
   $\mathbf{P} = \langle 1, 1, 1\rangle, \langle -2, 0, 1\rangle$, and $\langle 0, -3, 0\rangle$
6. $\mathbf{F}(x, y, z) = \left\langle \frac{1}{3}x, \frac{1}{2}y, z + 1\right\rangle$,
   $\mathbf{P} = \langle 0, 0, 0\rangle, \langle 1, 1, 1\rangle$, and $\langle -2, 0, -1\rangle$
7. $\mathbf{F}(x, y, z) = \frac{1}{\sqrt{x^2 + y^2 + z^2}}\langle x, y, z\rangle$,
   $\mathbf{P} = \langle 1, 2, 3\rangle, \langle -1, 4, 0\rangle$, and $\left\langle 1/\sqrt{3}, 1/\sqrt{3}, 1/\sqrt{3}\right\rangle$
8. $\mathbf{F}(x, y, z) = \langle e^{z+y}, 0, e^{x+y}\rangle$,
   $\mathbf{P} = \langle 0, 0, 0\rangle, \langle 2, -1, 1\rangle$, and $\langle 1, 0, 0\rangle$

Exercises 9–12: Evaluate $\int_C (\mathbf{F} \cdot \mathbf{T})\, ds$ *for the given vector field* **F** *and curve C.*

9. $\mathbf{F}(x, y) = \langle x + y, x - y\rangle$,
   C the line segment from $(0, 0)$ to $(3, 4)$

10. $\mathbf{F}(x, y) = \langle -2xy, y + 1 \rangle$,
C given by

$$\mathbf{r}(t) = \langle -2 \cos t, 2 \sin t \rangle$$

for $0 \leq t \leq \pi/2$

11. $\mathbf{F}(x, y, z) = \langle e^{x+y}, xz, y \rangle$,
C the segment from $(1, 2, 3)$ to $(-1, -2, -3)$

12. $\mathbf{F}(x, y, z) = \langle z \cos(x + y), z \cos(x + y), \sin(x + y) \rangle$,
C described by

$$\mathbf{r}(t) = \langle t, 1, 2t \rangle$$

for $0 \leq t \leq 3$

Exercises 13–18: Evaluate the following line integrals.

13. $\int_C (\mathbf{v} \cdot \mathbf{N})\, ds$, where

$$\mathbf{v}(x, y) = \langle x + 2y, 2x - y \rangle$$

and C is the line segment from $(0, 0)$ to $(-3, 4)$

14. $\int_C (\mathbf{v} \cdot \mathbf{N})\, ds$, where

$$\mathbf{v}(x, y) = \left\langle \frac{x}{\sqrt{x^2 + y^2}}, \frac{-y}{\sqrt{x^2 + y^2}} \right\rangle$$

and C is the circle of center $(0, 0)$ and radius 2

15. $\int_C 2xyz\, dx + x^2 z\, dy + x^2 y\, dz$, where C is described by

$$\mathbf{r}(t) = \langle t, t^2, t^3 \rangle, \qquad 0 \leq t \leq 2$$

16. $\int_C 3\, dx + e^{2x+y}\, dy + e^{-2x+z}\, dz$, where C is the line segment from $(0, 0, 1)$ to $(1, 1, 1)$

17. $\int_C \mathbf{F} \cdot d\mathbf{r}$, where

$$\mathbf{F}(x, y) = \left\langle x\sqrt{x^2 + y^2}, y\sqrt{x^2 + y^2} \right\rangle$$

and C is the piece of the parabola $y = x^2$ from $(0, 0)$ to $(-2, 4)$

18. $\int_C \mathbf{F} \cdot d\mathbf{r}$, where

$$\mathbf{F}(x, y, z) = \langle -x + y + z, x - y + z, x + y - z \rangle$$

and C is the helical curve described by

$$\mathbf{r}(t) = \langle \cos t, 2 \sin t, t \rangle, \qquad 0 \leq t \leq 2\pi$$

19. Let $\mathbf{F}(x, y, z)$ be the gravitational force exerted by Earth on an object of mass 1 kg at a point (x, y, z) above Earth's surface. The magnitude of $\mathbf{F}$ is inversely proportional to the square of the distance of the object from the center of Earth, and the constant of proportionality is GM. The vector $\mathbf{F}$ points from (x, y, z) towards the center of Earth. Show that under these assumptions, $\mathbf{F}(x, y, z)$ is given by (3).

20. The vector field $\mathbf{v}$ given by

$$\mathbf{v}(x, y) = \langle e^x \cos y, e^x \sin y \rangle$$

describes the velocity of a fluid flowing in the plane. Find the flux (net mass rate of flow) across the line segment with initial point $(1, 0)$ and terminal point $(0, 1)$. Assume that distance is measured in centimeters, time in seconds, and mass in grams and that the fluid has density of $1\ \text{g/cm}^2$.

21. The vector field $\mathbf{v}$ defined by

$$\mathbf{v}(x, y) = \langle 2x^2 y - 1, 2x - y \rangle$$

describes the velocity of a fluid flowing in the plane. Find the flux (net mass rate of flow) across the portion of the parabola $y = x^2$ from $(-2, 4)$ to $(1, 1)$. Assume that distance is measured in centimeters, time in seconds, and mass in grams and that the fluid has areal density of $1\ \text{g/cm}^2$.

22. The vector field $\mathbf{F}$ given by

$$\mathbf{F}(x, y, z) = \langle x^2, x \cos y, x \sin z \rangle$$

gives the force in newtons exerted on an object at the point (x, y, z). This force acts on an object as it moves from the point $(0, 0, 0)$ to the point $(-2, 1, 4)$ along the segment joining these points. Find the work done by $\mathbf{F}$.

23. The vector field $\mathbf{F}$ given by

$$\mathbf{F}(x, y, z) = \left\langle y \ln z, x \ln z, \frac{xy}{z} \right\rangle$$

gives the force in newtons exerted on an object at the point (x, y, z) with $z > 0$. This force acts on an object as it moves from the point $(1, 0, 1)$ to the point $(-1, 0, 2)$ along the helix described by

$$\mathbf{r}(t) = \langle \cos \pi t, \sin \pi t, t + 1 \rangle, \qquad 0 \leq t \leq 1.$$

Find the work done by $\mathbf{F}$.

24. In (5) of Investigation 1, we used the identity

$$\|\mathbf{v} \times \mathbf{T}\| = |\mathbf{v} \cdot \mathbf{N}|$$

in finding the area of the parallelograms in Figs. 12.16 and 12.17. Prove this identity for the vectors $\mathbf{T}$, $\mathbf{N}$, and $\mathbf{v}$ of Investigation 1.

25. A graph of the velocity field

$$\mathbf{v}(x, y) = \langle x + y, x - y \rangle$$

for a fluid of constant density is shown in the accompanying figure. In the figure we also show the triangle with vertices $(0, 0)$, $(3, 0)$, and $(0, 4)$.

a. Does there appear to be a sink, a source, or neither inside the triangle?

b. Check your answer to part **a** by calculating the flux through the triangle. (See Section 12.1 for a reminder about integration along piecewise smooth curves.)

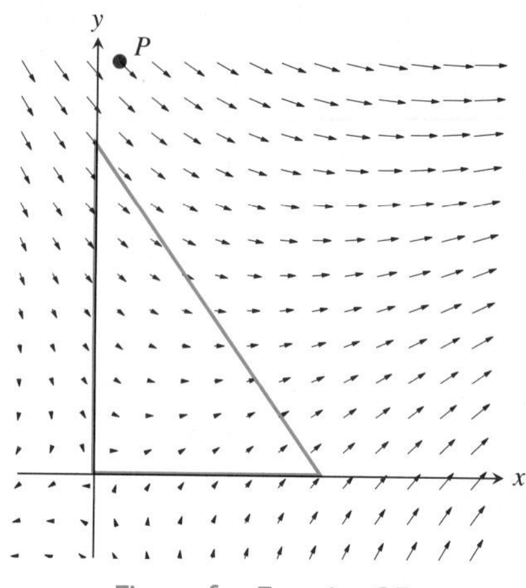

Figure for Exercise 25.

c. Imagine that the point marked P (see figure) is a bubble in the fluid. Trace the path of the bubble as it flows with the fluid.

26. A graph of the velocity field

$$\mathbf{v}(x, y) = \langle x \cos(x + y), -2y \sin(x - y) \rangle$$

for a fluid of constant density is shown in the accompanying figure. In the figure we also show the triangle with vertices $(0, 0)$, $(3, 0)$, and $(3, 4)$.

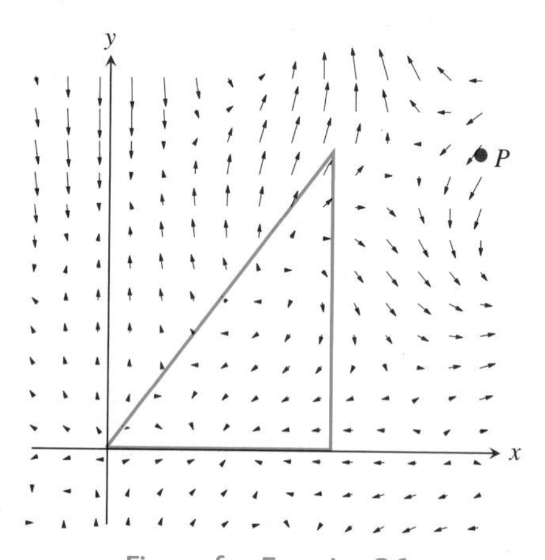

Figure for Exercise 26.

a. Does there appear to be a sink, a source, or neither inside the triangle?

b. Check your answer to part **a** by calculating the flux through the triangle. (See Section 12.1 for a reminder about integration along piecewise smooth curves.)

c. Imagine that the point marked P (see figure) is a bubble in the fluid. Trace the path of the bubble as it flows with the fluid.

27. Let $\mathbf{F}$ be a vector field in space and let C be a smooth curve in space with parametric representation

$$\mathbf{r}(t) = \langle x(t), y(t), z(t) \rangle, \qquad a \le t \le b.$$

Note that as t goes from $t = a$ to $t = b$, the curve C is traced from $\mathbf{A} = \mathbf{r}(a)$ to $\mathbf{B} = \mathbf{r}(b)$. Define $-C$ to be the curve parameterized by

$$\mathbf{r}_1(t) = \mathbf{r}(a + b - t), \qquad a \le t \le b.$$

a. Show that as t goes from $t = a$ to $t = b$, $\mathbf{r}_1$ traces the points of C from $\mathbf{B}$ to $\mathbf{A}$.

b. Show that

$$\int_{-C} (\mathbf{F} \cdot \mathbf{T})\, ds = -\int_{C} (\mathbf{F} \cdot \mathbf{T})\, ds.$$

28. Consider the vector field $\mathbf{F}$ defined by

$$\mathbf{F}(x, y, z) = \langle z \cos(x + y), z \cos(x + y), \sin(x + y) \rangle.$$

(See Exercise 12.) Let $\mathbf{A}$ and $\mathbf{B}$ be two points in R^3 and let C be a smooth curve from $\mathbf{A}$ to $\mathbf{B}$. Prove that

$$\int_C \mathbf{F} \cdot d\mathbf{r}$$

depends only on $\mathbf{A}$ and $\mathbf{B}$; that is, that the value of the integral is the same for *all* smooth curves C from $\mathbf{A}$ to $\mathbf{B}$.

29. Consider the vector field $\mathbf{F}$ defined by

$$\mathbf{F}(x, y, z) = \left\langle y \ln z, x \ln z, \frac{xy}{z} \right\rangle$$

for (x, y, z) in the set

$$\mathcal{D} = \{(x, y, z) : z > 0\}.$$

See Exercise 23. Let $\mathbf{A}$ and $\mathbf{B}$ be two points in $\mathcal{D}$ and let C be a smooth curve from $\mathbf{A}$ to $\mathbf{B}$ in $\mathcal{D}$. Prove that

$$\int_C \mathbf{F} \cdot d\mathbf{r}$$

depends only on $\mathbf{A}$ and $\mathbf{B}$; that is, that the value of the integral is the same for *all* smooth curves C from $\mathbf{A}$ to $\mathbf{B}$ in $\mathcal{D}$.

30. Let f be defined on R^3 by

$$f(x, y, z) = x^2 y - 2xyz + 3z$$

and define $\mathbf{F}$ to be the gradient of f,

$$\mathbf{F}(x, y, z) = \nabla f(x, y, z).$$

Let $\mathbf{A}$ and $\mathbf{B}$ be two points in R^3. Prove that

$$\int_C (\mathbf{F} \cdot \mathbf{T})\, ds$$

has the same value for every smooth curve C from $\mathbf{A}$ to $\mathbf{B}$. What is the value of the integral if C is a closed curve, that is, if $\mathbf{A} = \mathbf{B}$?

31. Let f be defined on R^3 by

$$f(x, y, z) = x^2 \sin(xy + yz)$$

and define $\mathbf{F}$ to be the gradient of f,

$$\mathbf{F}(x, y, z) = \nabla f(x, y, z).$$

Let $\mathbf{A}$ and $\mathbf{B}$ be two points in R^3. Prove that

$$\int_C (\mathbf{F} \cdot \mathbf{T})\, ds$$

has the same value for every smooth curve C from $\mathbf{A}$ to $\mathbf{B}$. Show in addition that if C is any simple closed smooth curve, then

$$\int_C (\mathbf{F} \cdot \mathbf{T})\, ds = 0.$$

32. In Investigation 1 we established the formula

$$\Phi = \text{flux across } C = \delta \int_C (\mathbf{v} \cdot \mathbf{N})\, ds$$

for the case in which the fluid is of constant areal density δ. Find a line integral formula for flux when the density $\delta = \delta(x, y)$ is a continuous function of position.

12.3 The Fundamental Theorem of Line Integrals

In the previous section we learned to evaluate the line integrals

$$\int_C (\mathbf{F} \cdot \mathbf{T})\, ds, \quad \int_C L\,dx + M\,dy + N\,dz, \quad \text{and} \quad \int_C \mathbf{F} \cdot d\mathbf{r} \tag{1}$$

by finding a parameterization for the curve C and then using the parameterization to convert the integrals in (1) to Riemann integrals. When the curve C has a complicated parameterization or is piecewise smooth with a lot of smooth pieces, this evaluation process can take a good deal of effort. In this section we see that this process can be shortened considerably if the function $\mathbf{F}$ has an "anti-gradient," that is, if there is a function f for which grad $f = \mathbf{F}$. This is made more precise in the following result.

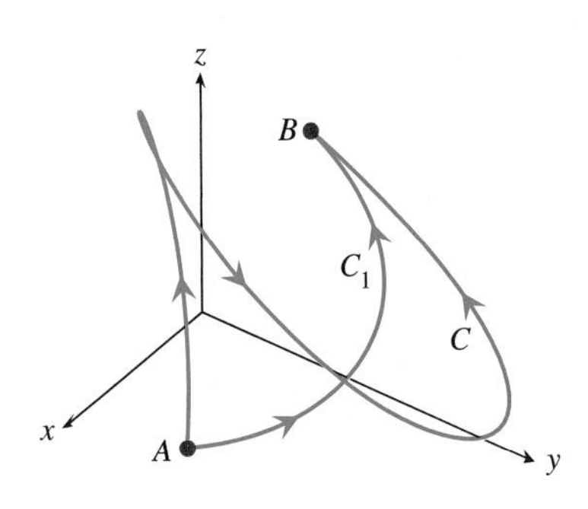

FIGURE 12.123 If $\mathbf{F} = \nabla f$ and curves C and C_1 have the same initial points and the same terminal points, $\mathbf{A}$ and $\mathbf{B}$ then $\int_C (\mathbf{F} \cdot \mathbf{T})\, ds = f(\mathbf{B}) - f(\mathbf{A}) = \int_{C_1} (\mathbf{F} \cdot \mathbf{T})\, ds$.

The Fundamental Theorem of Line Integrals

Let $\mathbf{F}$ be a continuous vector field with domain $\mathcal{D}$ in R^n and range in R^n ($n = 2$ or $n = 3$) and let f be a real-valued function defined on $\mathcal{D}$ with

$$\text{grad } f = \nabla f = \mathbf{F}.$$

If C is any piecewise smooth curve in $\mathcal{D}$ with initial point $\mathbf{A}$ and terminal point $\mathbf{B}$, then

$$\int_C \mathbf{F} \cdot d\mathbf{r} = \int_C (\mathbf{F} \cdot \mathbf{T})\, ds = f(\mathbf{B}) - f(\mathbf{A}). \tag{2}$$

Note that the integral in (2) depends only on the endpoints of C, but not on the piecewise smooth curve C joining $\mathbf{A}$ to $\mathbf{B}$. See Fig. 12.23.

To see why this theorem is true, we assume that the domain $\mathcal{D}$ of $\mathbf{F}$ is a subset of R^3 and that C is a smooth curve in $\mathcal{D}$ described by

$$\mathbf{r}(t) = \langle x(t), y(t), z(t)\rangle, \qquad a \le t \le b.$$

Suppose that $\mathbf{F}$ is given by

$$\mathbf{F}(x, y, z) = \langle L(x, y, z), M(x, y, z), N(x, y, z)\rangle.$$

Because $\mathbf{F}$ is the gradient of f, we have

$$\frac{\partial f}{\partial x} = L(x, y, z), \qquad \frac{\partial f}{\partial y} = M(x, y, z), \qquad \frac{\partial f}{\partial z} = N(x, y, z).$$

For real $t \in [a, b]$,

$$f(\mathbf{r}(t)) = f(x(t), y(t), z(t))$$

defines a real-valued function of the real variable t. Because f and $\mathbf{r}$ are both differentiable, so is $f \circ \mathbf{r}$. To calculate $(f \circ \mathbf{r})'(t)$ we apply the chain rule (see Section 10.2), obtaining

$$\begin{aligned} \frac{d}{dt} f(\mathbf{r}(t)) &= \nabla f \cdot \mathbf{r}'(t) = \left\langle \frac{\partial f}{\partial x}, \frac{\partial f}{\partial y}, \frac{\partial f}{\partial z} \right\rangle \cdot \langle x'(t), y'(t), z'(t)\rangle \\ &= L(x(t), y(t), z(t))x'(t) + M(x(t), y(t), z(t))y'(t) \\ &\quad + N(x(t), y(t), z(t))z'(t). \end{aligned} \tag{3}$$

Using the results on notation for line integrals discussed at the end of Section 12.2,

$$\begin{aligned} \int_C (\mathbf{F} \cdot \mathbf{T})\, ds &= \int_C L\, dx + M\, dy + N\, dz \\ &= \int_a^b (L(x(t), y(t), z(t))x'(t) + M(x(t), y(t), z(t))y'(t) \\ &\quad + N(x(t), y(t), z(t))z'(t))\, dt. \end{aligned}$$

Substituting (3) into this last expression, we find

$$\begin{aligned} \int_C (\mathbf{F} \cdot \mathbf{T})\, ds &= \int_a^b \frac{d}{dt} f(\mathbf{r}(t))\, dt \\ &= f(\mathbf{r}(t))\Big|_a^b = f(\mathbf{r}(b)) - f(\mathbf{r}(a)) \\ &= f(\mathbf{B}) - f(\mathbf{A}), \end{aligned}$$

where $\mathbf{r}(a) = \mathbf{A}$ and $\mathbf{r}(b) = \mathbf{B}$ are the initial and terminal points of C.

Once we have the Fundamental Theorem of Line Integrals for smooth curves, we can get the result for piecewise smooth curves C by applying the smooth curve result on each smooth piece of C. See Exercise 34.

If we recognize a vector field $\mathbf{F}$ as the gradient of a function f, the evaluation of the line integral is almost immediate.

EXAMPLE 1 Let

$$\mathbf{F}(x, y, z) = \langle yz, xz, xy \rangle.$$

a) Show that $\mathbf{F} = \nabla f$, where $f(x, y, z) = xyz$.
b) Let C be the piecewise smooth curve consisting of the line segment from $(-1, 2, 3)$ to $(-2, 3, 1)$, followed by the line segment from $(-2, 3, 1)$ to $(-2, 3, 0)$, and then by the line segment from $(-2, 3, 0)$ to $(0, 1, -4)$. See Fig. 12.24. Evaluate

$$\int_C yz\,dx + xz\,dy + xy\,dz.$$

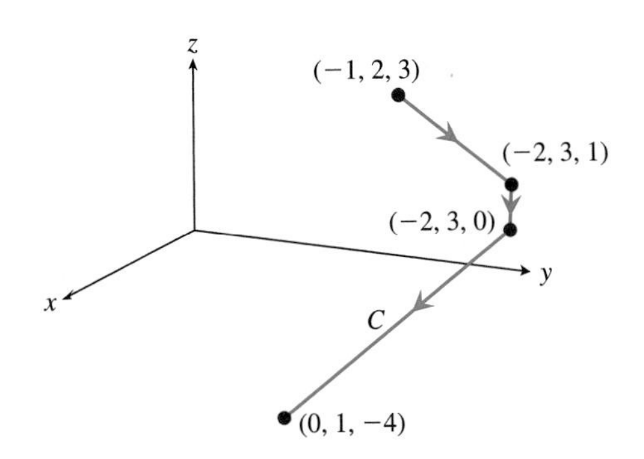

FIGURE 12.24 The piecewise linear curve C.

Solution

a) The gradient of $f(x, y, z) = xyz$ is

$$\begin{aligned}\nabla f(x, y, z) &= \langle f_x(x, y, z), f_y(x, y, z), f_z(x, y, z) \rangle \\ &= \langle yz, xz, xy \rangle = \mathbf{F}(x, y, z).\end{aligned}$$

b) The initial point of C is $\mathbf{A} = \langle -1, 2, 3 \rangle$ and the terminal point is $\mathbf{B} = \langle 0, 1, -4 \rangle$. Because $\nabla f = \mathbf{F}$, we may apply the Fundamental Theorem of Line Integrals:

$$\begin{aligned}\int_C yz\,dx + xz\,dy + xy\,dz &= \int_C (\mathbf{F} \cdot \mathbf{T})\,ds = f(x, y, z)\Big|_{\mathbf{A}}^{\mathbf{B}} \\ &= xyz\Big|_{\langle -1, 2, 3 \rangle}^{\langle 0, 1, -4 \rangle} = 0 - (-6) = 6.\end{aligned}$$

It is interesting to note the similarities between the Fundamental Theorem of Calculus, seen in Chapter 5, and the Fundamental Theorem of Line Integrals. The Fundamental Theorem of Calculus tells us that if g is a continuous function of one variable defined on an interval $[a, b]$ and $G' = g$, then

$$\int_a^b g(x)\,dx = G(b) - G(a).$$

The Fundamental Theorem of Line Integrals says that if $\nabla f = \mathbf{F}$ and C is a piecewise smooth curve from $\mathbf{A}$ to $\mathbf{B}$, then

$$\int_C \mathbf{F} \cdot \mathbf{dr} = \int_C (\mathbf{F} \cdot \mathbf{T})\,ds = f(\mathbf{B}) - f(\mathbf{A}).$$

The function f is called a **potential function** for $\mathbf{F}$, and plays a role much like the role of the antiderivative in the Fundamental Theorem of Calculus. When a vector field $\mathbf{F}$ has a potential function, we say that $\mathbf{F}$ is a **conservative vector field.**

Recognizing Gradients

To effectively use the Fundamental Theorem of Line Integrals, we must be able to answer two questions:

a) Given a vector field **F**, how can we tell whether or not **F** is the gradient of a function f?
b) Once we know **F** is the gradient of some function f, how do we find f?

The following theorem gives a partial answer to the first question. This theorem gives us conditions on **F** that must be true if **F** is the gradient of a function f. We shall see that for most cases we encounter, when these conditions are satisfied **F** is the gradient of a function f.

> **The Gradient Test**
>
> TWO VARIABLES
>
> Let **F**, given by
>
> $$\mathbf{F}(x, y) = \langle M(x, y), N(x, y)\rangle,$$
>
> be a vector field defined on a subset of R^2. If **F** is the gradient of a function f and the coordinate functions M and N of **F** have continuous first partial derivatives, then
>
> $$\frac{\partial N}{\partial x} = \frac{\partial M}{\partial y} \tag{4}$$
>
> at all points in the domain of **F**.
>
> THREE VARIABLES
>
> Let **G**, given by
>
> $$\mathbf{G}(x, y, z) = \langle P(x, y, z), Q(x, y, z), R(x, y, z)\rangle,$$
>
> be a vector field defined on a subset of R^3. If **G** is the gradient of a function g and the coordinate functions P, Q, and R of **G** have continuous first partial derivatives, then
>
> $$\frac{\partial Q}{\partial x} = \frac{\partial P}{\partial y}, \qquad \frac{\partial R}{\partial x} = \frac{\partial P}{\partial z}, \qquad \frac{\partial R}{\partial y} = \frac{\partial Q}{\partial z} \tag{5}$$
>
> at all points in the domain of **G**.

We prove the Gradient Test for the two-variable case. The argument given extends readily to the three-variable case. See Exercise 27. If $\nabla f = \mathbf{F}$, then

$$\nabla f(x, y) = \langle f_x(x, y), f_y(x, y)\rangle = \langle M(x, y), N(x, y)\rangle = \mathbf{F}(x, y).$$

Hence,

$$f_x(x, y) = M(x, y) \quad \text{and} \quad f_y(x, y) = N(x, y).$$

It follows that

$$\frac{\partial N}{\partial x} = f_{yx} \quad \text{and} \quad \frac{\partial M}{\partial y} = f_{xy}.$$

Because f_{xy} and f_{yx} are both continuous, they are equal and it follows that

$$\frac{\partial N}{\partial x} = \frac{\partial M}{\partial y}.$$

Finding a Potential Function

Suppose that we wish to evaluate the line integral

$$\int_C (\mathbf{F} \cdot \mathbf{T})\, ds$$

and have verified that the vector field $\mathbf{F}$ satisfies (4) or (5). If we can find a potential function for $\mathbf{F}$, then we can evaluate the integral using the Fundamental Theorem of Line Integrals. We now investigate a method for finding a potential function, assuming that such a function exists.

Two-Variable Functions We first consider the case in which $\mathbf{F}$ is a vector field defined on a subset of R^2.

INVESTIGATION

Finding a Potential Function for a Vector Field

Find a potential function f for the vector field $\mathbf{F}$ given by

$$\mathbf{F}(x, y) = \langle M(x, y), N(x, y) \rangle = \langle 2xy - 1, x^2 + 4y \rangle.$$

To better understand the process for finding f, first note that the function f defined by

$$f(x, y) = x^2y - x + 2y^2 \tag{6}$$

is a potential function for $\mathbf{F}$. Indeed, we have

$$\frac{\partial f}{\partial x} = \frac{\partial}{\partial x}(x^2y) - \frac{\partial}{\partial x}(x) + \frac{\partial}{\partial x}(2y^2) = 2xy - 1 + 0 = M(x, y) \tag{7}$$

and

$$\frac{\partial f}{\partial y} = \frac{\partial}{\partial y}(x^2y) - \frac{\partial}{\partial y}(x) + \frac{\partial}{\partial y}(2y^2) = x^2 - 0 + 4y = N(x, y),$$

so

$$\nabla f(x, y) = \langle M(x, y), N(x, y) \rangle = \mathbf{F}(x, y).$$

In calculating $\partial f/\partial x$ we treated y as a constant, so the $2y^2$ term differentiated to 0. Similarly, in computing $\partial f/\partial y$, the x term differentiated to 0. In finding a potential function, it is important to remember that terms involving only y differentiate to 0 when differentiated with respect to x, and that terms involving only x differentiate to 0 when differentiated with respect to y.

Now assume that we do not yet know f, but wish to recover it from $\mathbf{F}$. We know that if f is a potential function for $\mathbf{F}$, then

$$\left\langle \frac{\partial f}{\partial x}, \frac{\partial f}{\partial y} \right\rangle = \nabla f(x, y) = \mathbf{F}(x, y) = \langle 2xy - 1, x^2 + 4y \rangle.$$

It follows that

$$\frac{\partial f}{\partial x} = 2xy - 1 \quad \text{and} \quad \frac{\partial f}{\partial y} = x^2 + 4y. \tag{8}$$

We can recover f by integrating the first of these expressions with respect to x or by integrating the second expression with respect to y. If we choose to integrate the first expression with respect to x, then we consider this operation as reversing the differentiation in (7). Thus, we think of y as constant. Integration then leads to

$$f(x, y) = \int \frac{\partial f}{\partial x}\,dx = \int (2xy - 1)\,dx = x^2y - x + C(y), \tag{9}$$

where $C(y)$ is the "constant" of integration. We denote this constant by $C(y)$ because it may be a function of y. Indeed, we see that the $2y^2$ term in (6) differentiated to 0 because we thought of it as constant when differentiating with respect to x. If the integration (9) is to recover $f(x, y)$, then we must account for all possible terms that differentiate to 0 with respect to x. We do so by writing $C(y)$ for the "constant" of integration. To determine f completely, we need to determine $C(y)$. We do this by considering the partial derivative of f with respect to y. From (8) and (9) we have

$$x^2 + 4y = \frac{\partial f}{\partial y} = \frac{\partial}{\partial y}(x^2y - x + C(y)) = x^2 + C'(y).$$

It follows that

$$C'(y) = 4y$$

and hence that

$$C(y) = 2y^2 + c, \tag{10}$$

where c is a numerical constant. Because $C = C(y)$ is a function of y only, c does not depend on x or y, that is, it is a "true" constant. Combining (9) and (10), we conclude that

$$f(x, y) = x^2y - x + 2y^2 + c. \tag{11}$$

It is easy to check that (11) defines a potential function f for $\mathbf{F}$ for any constant c. In practice we usually take $c = 0$, because *any* potential function for $\mathbf{F}$ can be used in applying the Fundamental Theorem of Line Integrals.

In the previous discussion we could have found f by integrating either of the expressions in (8). Sometimes one of these integrals is easier to evaluate than the other. Selecting the easier integral may result in an easier calculation.

EXAMPLE 2 A vector field $\mathbf{F}$ is defined by

$$\mathbf{F}(x, y) = \langle e^x(\sin(x + 3y) + \cos(x + 3y)), 3e^x \cos(x + 3y)\rangle.$$

Verify that $\mathbf{F}$ satisfies the conditions of the Gradient Test. Find a potential function for $\mathbf{F}$.

Solution

Let

$$M(x, y) = e^x(\sin(x + 3y) + \cos(x + 3y)) \tag{12}$$

and

$$N(x, y) = 3e^x \cos(x + 3y)$$

be the coordinate functions of $\mathbf{F}$, so that $\mathbf{F}(x, y) = \langle M(x, y), N(x, y)\rangle$. Note that

$$\frac{\partial N}{\partial x} = 3e^x(\cos(x + 3y) - \sin(x + 3y)) = \frac{\partial M}{\partial y}.$$

Hence, $\mathbf{F}$ satisfies the requirements of the Gradient Test.

Assume f is a potential function for $\mathbf{F}$. Because

$$\frac{\partial f}{\partial x} = M(x, y) \quad \text{and} \quad \frac{\partial f}{\partial y} = N(x, y), \tag{13}$$

we have

$$f(x, y) = \int M(x, y)\, dx \quad \text{and} \quad f(x, y) = \int N(x, y)\, dy.$$

A glance at M and N suggests that the second of these integrals is a little easier to evaluate. Thus we have

$$f(x, y) = \int 3e^x \cos(x + 3y)\, dy = e^x \sin(x + 3y) + C(x), \tag{14}$$

where $C(x)$ is a function of x only. To find $C(x)$, differentiate (14) with respect to x. Using (12), (13), and (14),

$$\begin{aligned} e^x(\sin(x + 3y) + \cos(x + 3y)) = M(x, y) &= \frac{\partial f}{\partial x} \\ &= e^x(\sin(x + 3y) + \cos(x + 3y)) + C'(x). \end{aligned}$$

It follows that $C'(x) = 0$. Hence $C(x) = c$ is a constant and

$$f(x, y) = e^x \sin(x + 3y) + c$$

is a potential function for $\mathbf{F}$. How can we check that this answer is correct?

Three-Variable Functions The process for finding a potential function for a conservative vector field defined on a subset of R^3 is similar to that used for vector fields defined on a subset of R^2. However, because the functions involved are functions of three variables, the "constants" of integration that occur may be functions of two variables. We illustrate with an example.

EXAMPLE 3 Let $\mathbf{F}$ be the vector field defined by

$$\mathbf{F}(x, y, z) = \langle 2xyz + 3y^2, x^2z + 6xy - 2z^3, x^2y - 6yz^2\rangle.$$

a) Check that $\mathbf{F}$ satisfies the conditions of the Gradient Test and then show that $\mathbf{F}$ has a potential function.

b) Evaluate $\int_C (\mathbf{F} \cdot \mathbf{T})\, ds$ where C is the curve described by

$$\mathbf{r}(t) = \left\langle t^2e^t, t + \sqrt{t}, e^t \cos \pi t\right\rangle, \qquad 0 \le t \le 4.$$

Solution

Let

$$P(x, y, z) = 2xyz + 3y^2,$$
$$Q(x, y, z) = x^2z + 6xy - 2z^3,$$
$$R(x, y, z) = x^2y - 6yz^2$$

be the coordinate functions of **F**, so that

$$\mathbf{F}(x, y, z) = \langle P(x, y, z), Q(x, y, z), R(x, y, z)\rangle.$$

a) First note that

$$\frac{\partial Q}{\partial x} = 2xz + 6y = \frac{\partial P}{\partial y},$$
$$\frac{\partial R}{\partial x} = 2xy = \frac{\partial P}{\partial z},$$
$$\frac{\partial R}{\partial y} = x^2 - 6z^2 = \frac{\partial Q}{\partial z}.$$

Hence, **F** satisfies the conditions of the Gradient Test. Now assume that f is a potential function for **F**. We then have

$$\frac{\partial f}{\partial x} = P(x, y, z) = 2xyz + 3y^2,$$
$$\frac{\partial f}{\partial y} = Q(x, y, z) = x^2z + 6xy - 2z^3, \tag{15}$$
$$\frac{\partial f}{\partial z} = R(x, y, z) = x^2y - 6yz^2.$$

Integrating the first expression in (15) with respect to x, we find

$$\begin{aligned} f(x, y, z) &= \int \frac{\partial f}{\partial x}\,dx = \int (2xyz + 3y^2)\,dx \\ &= x^2yz + 3xy^2 + C_1(y, z). \end{aligned} \tag{16}$$

In performing this integration, we have treated y and z as constants. In particular, the "constant" of integration, $C_1(y, z)$, may be an expression involving the variables y and z, but not x. (We can verify that this integration is correct by differentiating with respect to x the right-hand expression in (16). The $C_1(y, z)$ term differentiates to 0 and the derivatives of the other two terms combine to give $P(x, y, z)$.) We find $C_1(y, z)$ by differentiating the expression for f in (16) with respect to y or z and equating the result to the appropriate expression in (15). For example, if we differentiate with respect to z and use the third expression in (15), we obtain

$$x^2y - 6yz^2 = R(x, y, z) = \frac{\partial f}{\partial z} = x^2y + \frac{\partial C_1}{\partial z}(y, z).$$

It follows that

$$\frac{\partial C_1}{\partial z}(y, z) = -6yz^2.$$

Integrating this expression with respect to z, we find

$$C_1(y,z) = \int \frac{\partial C_1}{\partial z}(y,z)\,dz = \int (-6yz^2)\,dz = -2yz^3 + C_2(y),$$

where the "constant" of integration $C_2(y)$ is a function only of y. The constant must have this form because any expression depending only on y will differentiate to 0 when differentiated with respect to z. Substituting this expression for $C_1(y,z)$ into (16), we find that

$$f(x,y,z) = x^2yz + 3xy^2 - 2yz^3 + C_2(y). \tag{17}$$

To determine $C_2(y)$, we differentiate (17) with respect to y and use the second expression in (15). We then find that

$$x^2z + 6xy - 2z^3 = Q(x,y,z) = \frac{\partial f}{\partial y} = x^2z + 6xy - 2z^3 + C_2'(y).$$

Thus $C_2'(y) = 0$ and it follows that $C_2(y)$, which can depend only on y, must be a constant function of y. Hence $C_2(y) = c$ for some real number c, and

$$f(x,y,z) = x^2yz + 3xy^2 - 2yz^3 + c \tag{18}$$

is a potential function for $\mathbf{F}$. We take $c = 0$ for convenience in evaluating the line integral.

b) The initial and terminal points of C are

$$\mathbf{A} = \mathbf{r}(0) = \langle 0,0,1\rangle \quad\text{and}\quad \mathbf{B} = \mathbf{r}(4) = \langle 16e^4, 6, e^4\rangle.$$

Because $\mathbf{F}$ is a conservative vector field and f is a potential function for $\mathbf{F}$, we can evaluate the line integral using the Fundamental Theorem of Line Integrals. Using the expression for f in (18) with $c = 0$, we find

$$\int_C (\mathbf{F}\cdot\mathbf{T})\,ds = f(\mathbf{B}) - f(\mathbf{A}) = f(16e^4, 6, e^4) - f(0,0,1)$$

$$= (1524e^{12} + 1728e^4) - 0 \approx 2.48133 \times 10^8.$$

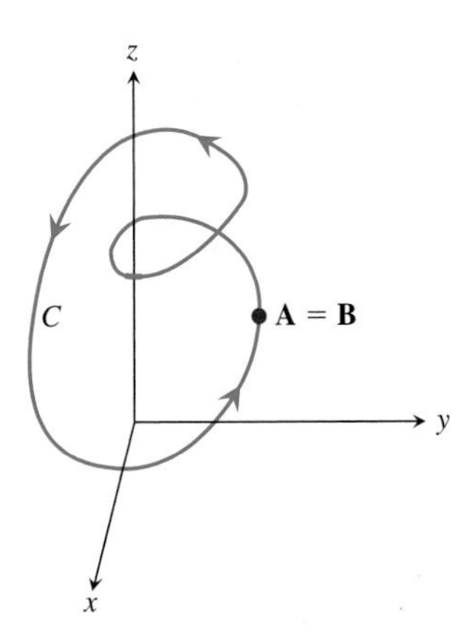

FIGURE 12.25 If $\mathbf{F} = \nabla f$ and C is a closed curve, then $\int_C (\mathbf{F}\cdot\mathbf{T})\,ds = 0$.

Closed Curves

Let $\mathbf{F}$ be a conservative vector field and let C, described by $\mathbf{r}(t)$, $a \le t \le b$, be a closed, piecewise smooth curve in the domain of $\mathbf{F}$. Because C is closed, the initial and terminal points of C are the same; that is,

$$\mathbf{A} = \mathbf{r}(a) = \mathbf{r}(b) = \mathbf{B}.$$

See Fig. 12.25. If f is a potential function for $\mathbf{F}$, then it follows from the Fundamental Theorem of Line Integrals that

$$\int_C (\mathbf{F}\cdot\mathbf{T})\,ds = f(\mathbf{B}) - f(\mathbf{A}) = 0.$$

Thus, we have proved the following important result.

Conservative Vector Fields and Closed Curves

Let $\mathbf{F}$ be a conservative vector field and let C be a closed, piecewise smooth curve in the domain of $\mathbf{F}$. Then

$$\int_C \mathbf{F}\cdot d\mathbf{r} = \int_C (\mathbf{F}\cdot\mathbf{T})\,ds = 0. \tag{19}$$

In physical problems, conservative vector fields often have special significance. The fact that the integral in (19) is 0 for such vector fields may say important things about the physical situation being studied. For example, let $\mathbf{F}$ be a conservative force field in the plane or in space and let f be a potential function for $\mathbf{F}$. If the force acts on an object moving from point $\mathbf{A}$ to point $\mathbf{B}$ along a piecewise smooth curve C, then the work done by $\mathbf{F}$ is

$$\text{work} = \int_C (\mathbf{F}\cdot\mathbf{T})\,ds = f(\mathbf{B}) - f(\mathbf{A}).$$

In particular, if $\mathbf{A} = \mathbf{B}$, then C is a closed curve and the net work done is 0. As a practical example, the gravitational field for Earth is a conservative vector field. See Exercise 23. Thus, if an object moves around in space under the influence of Earth's gravitational field and ends up where it started, the net work done by the gravitational field is 0.

When the path C of integration is a closed path, the symbol $\oint_c$ is sometimes used in place of $\int$. Thus (19) can be written as

$$\oint_C (\mathbf{F}\cdot\mathbf{T})\,ds = 0.$$

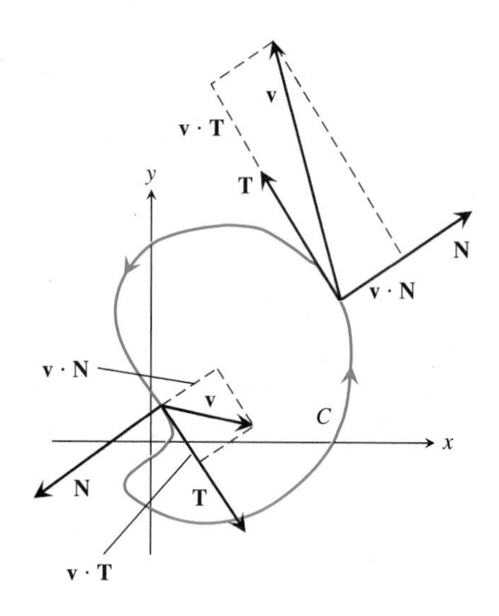

FIGURE 12.26 If $\mathbf{v}$ is a flow and C is a closed curve, then $\int_C (\mathbf{v}\cdot\mathbf{N})\,ds$ gives the flux across C while $\int_C (\mathbf{v}\cdot\mathbf{T})\,ds$ gives the circulation about C.

Flows Let $\mathbf{v}$ be the velocity field for the planar flow of a fluid of constant density δ. Assume that

$$\mathbf{v}(x, y) = \langle P(x, y), Q(x, y)\rangle, \tag{20}$$

and let C be a closed, piecewise smooth planar curve described by

$$\mathbf{r}(t) = \langle x(t), y(t)\rangle, \qquad a \le t \le b.$$

In Section 12.2 we saw that the flux across the closed curve C is given by

$$\begin{aligned}\text{flux} &= \delta\int_C (\mathbf{v}\cdot\mathbf{N})\,ds = \delta\int_a^b \langle P(\mathbf{r}(t)), Q(\mathbf{r}(t))\rangle\cdot\langle y'(t), -x'(t)\rangle\,dt \\ &= \delta\int_C -Q\,dx + P\,dy = \delta\int_C (\tilde{\mathbf{v}}\cdot\mathbf{T})\,ds,\end{aligned}$$

where $\tilde{\mathbf{v}}(x, y) = \langle -Q(x, y), P(x, y)\rangle$. See Fig. 12.26. If the vector field $\tilde{\mathbf{v}}$ is conservative, then the integral in the last expression is 0; that is, the flux across any closed curve is 0. If $\tilde{\mathbf{v}}$ has a potential function, then by the Gradient Test, we must have

$$\frac{\partial P}{\partial x} = -\frac{\partial Q}{\partial y}. \tag{21}$$

Thus, if (21) is true, we can try to find a potential function for $\tilde{\mathbf{v}}$. If such a potential function exists, then the flux integral is 0 and (20) describes a fluid flow for which the flux (net mass rate of flow) across any closed curve is 0.

When (20) describes the velocity field for a planar flow and C is a closed curve, the line integral

$$\int_C (\mathbf{v} \cdot \mathbf{T})\, ds \tag{22}$$

also has important physical significance. We can interpret the integral in (22) as "summing the tangential components of velocity along C." This is illustrated in Fig. 12.26; see also Exercise 35. The value of the integral in (22) is called the **circulation** of the fluid around C and gives information about vortices in the flow. When the circulation integral (22) is 0 for all closed curves C in the domain of $\mathbf{v}$, we say the fluid flow is **irrotational.**

EXAMPLE 4 The velocity field for a fluid flowing in the plane is given by

$$\mathbf{v}(x, y) = \left\langle \frac{y}{x^2 + y^2 + 1}, -\frac{x}{x^2 + y^2 + 1} \right\rangle \text{ cm/s}.$$

Assume the fluid has denisty 1 g/cm^2 and let C be the circle of center $(0, 0)$ and radius 1 in the plane. Find the circulation around C and the flux through C.

Solution

Figure 12.27 shows a graph of the vector field $\mathbf{v}$, the unit circle C, and a representative tangent and normal to C. If

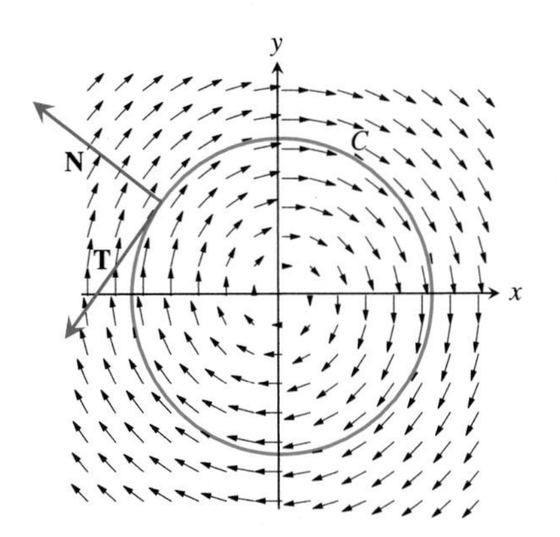

FIGURE 12.27 The vector field $\mathbf{v} = \left\langle \frac{y}{x^2 + y^2 + 1}, -\frac{x}{x^2 + y^2 + 1} \right\rangle$ and the unit circle.

$$P(x, y) = \frac{y}{x^2 + y^2 + 1} \quad \text{and} \quad Q(x, y) = -\frac{x}{x^2 + y^2 + 1}$$

are the coordinate functions for $\mathbf{v}$, then

$$\mathbf{v}(x, y) = \langle P(x, y), Q(x, y) \rangle.$$

The circulation around C is given by

$$\int_C (\mathbf{v} \cdot \mathbf{T})\, ds = \int_C P(x, y)\, dx + Q(x, y)\, dy. \tag{23}$$

Because

$$\frac{\partial Q}{\partial x} = \frac{x^2 - y^2 - 1}{(x^2 + y^2 + 1)^2} \quad \text{and} \quad \frac{\partial P}{\partial y} = \frac{x^2 - y^2 + 1}{(x^2 + y^2 + 1)^2}$$

are not equal, the vector field $\mathbf{v}$ is not conservative. In this case we may evaluate the circulation integral by converting it to a Riemann integral. The path C is parameterized by

$$\mathbf{r}(t) = \langle x(t), y(t) \rangle = \langle \cos t, \sin t \rangle, \qquad 0 \le t \le 2\pi.$$

With $x = \cos t$ and $y = \sin t$ in the second integral in (23), we have

$$\begin{aligned}
\text{circulation} &= \int_C P(x, y)\,dx + Q(x, y)\,dy \\
&= \int_0^{2\pi} P(\cos t, \sin t)\,d(\cos t) + Q(\cos t, \sin t)\,d(\sin t) \\
&= \int_0^{2\pi} \left(\frac{\sin t}{\cos^2 t + \sin^2 t + 1}(-\sin t) + \frac{-\cos t}{\cos^2 t + \sin^2 t + 1}(\cos t) \right) dt \\
&= -\frac{1}{2} \int_0^{2\pi} dt = -\pi.
\end{aligned}$$

This negative result is consistent with the fact that on C the vectors $\mathbf{T}$ and $\mathbf{v}$ are in opposite directions.

The flux Φ across C is given by

$$\Phi = \delta \int_C (\mathbf{v} \cdot \mathbf{N})\,ds = \delta \int_C (\tilde{\mathbf{v}} \cdot \mathbf{T})\,ds \tag{24}$$

where $\tilde{\mathbf{v}}(x, y) = \langle -Q(x, y), P(x, y) \rangle$. Because

$$\frac{\partial P}{\partial x} = \frac{-2xy}{(x^2 + y^2 + 1)^2} = \frac{\partial(-Q)}{\partial y},$$

we see that $\tilde{\mathbf{v}}$ satisfies the conditions of the Gradient Test. It is easy to check that

$$f(x, y) = \frac{1}{2} \ln(x^2 + y^2 + 1)$$

is a potential function for $\tilde{\mathbf{v}}$, that is, $\tilde{\mathbf{v}}$ is a conservative vector field. Because C is a closed curve, it follows that the flux integral (24) is 0 g/s. The 0 flux is consistent with the fact that on C, the vectors $\mathbf{N}$ and $\mathbf{v}$ are perpendicular, that is, the flow is tangent to C, so no fluid flows across C.

Existence of a Potential Function

Let $\mathbf{F}$ be a vector field having the component functions with continuous first partial derivatives. According to the Gradient Test, if $\mathbf{F}$ has a potential function, then (4) or (5) must be true. In particular, if the conditions described in (4) or (5) are not true, then there is no point in looking for a potential function for $\mathbf{F}$. But what if (4) or (5) is true? Is it always the case that we can find a potential function for $\mathbf{F}$? In general, the answer is no. However, if the domain of $\mathbf{F}$ is "nice" enough, then we can be sure that $\mathbf{F}$ has a potential function whenever (4) or (5) is satisfied. We limit our description of "nice" domains to the planar case.

Potential Functions in Planar Domains We first provide an example showing that (4) can be true even when $\mathbf{F}$ does not have a potential function.

EXAMPLE 5 Let $\mathbf{F}$ be the vector field given by

$$\mathbf{F}(x,y) = \langle M(x,y), N(x,y)\rangle = \left\langle \frac{-y}{x^2+y^2}, \frac{x}{x^2+y^2} \right\rangle, \qquad (x,y) \neq (0,0).$$

Show that $\mathbf{F}$ satisfies the conditions of the Gradient Test, but that $\mathbf{F}$ has no potential function on its domain.

Solution

If $(x,y) \neq (0,0)$, then

$$\frac{\partial N}{\partial x} = \frac{y^2 - x^2}{(x^2+y^2)^2} = \frac{\partial M}{\partial y}.$$

Hence $\mathbf{F}$ satisfies the conditions of the Gradient Test on its domain.

Now assume that $\mathbf{F}$ has a potential function f and let C be the circle of radius 1 centered at the origin. If $\mathbf{F}$ has a potential function f, then by (19) we must have

$$\int_C (\mathbf{F} \cdot \mathbf{T})\, ds = 0. \tag{25}$$

However, we can calculate this integral directly using the parameterization

$$\mathbf{r}(t) = \langle x(t), y(t)\rangle = \langle \cos t, \sin t\rangle, \qquad 0 \le t \le 2\pi,$$

for C. With this parameterization, we have

$$\int_C (\mathbf{F} \cdot \mathbf{T})\, ds = \int_0^{2\pi} (M(x(t), y(t))x'(t) + N(x(t), y(t))y'(t))\, dt$$

$$= \int_0^{2\pi} \left(\frac{-\sin t}{\cos^2 t + \sin^2 t}(-\sin t) + \frac{\cos t}{\cos^2 t + \sin^2 t}\cos t \right) dt = 2\pi.$$

Because this result is nonzero, we see that (25) is not true, so $\mathbf{F}$ cannot have a potential function. For more about this example, see Exercise 26.

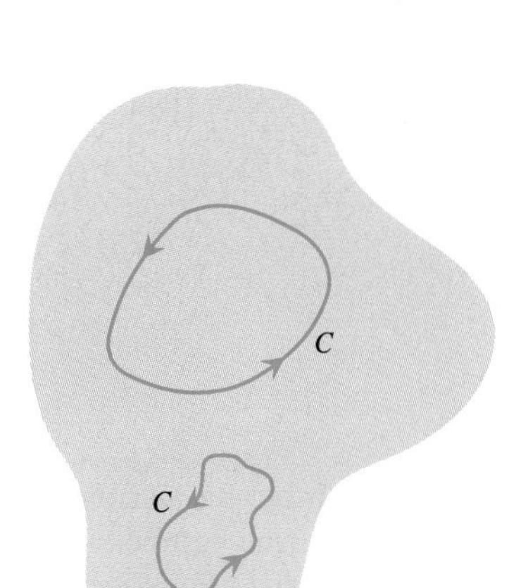

FIGURE 12.28 The set $\mathcal{D}$ is simply connected because for any closed curve C in $\mathcal{D}$, the inside of C is also contained in $\mathcal{D}$.

In Example 5, the domain for $\mathbf{F}$ was the entire plane except for the "hole" at the point $(0,0)$. The existence or nonexistence of holes in the domain of $\mathbf{F}$ is of critical importance in deciding whether or not a function satisfying (4) *must* have a potential function. Domains without holes are called **simply connected.**

DEFINITION Simply Connected Set

Let $\mathcal{D}$ be a subset of the plane such that every point of $\mathcal{D}$ is an interior point of $\mathcal{D}$ and such that $\mathcal{D}$ is connected (that is, consists of only one piece). If for every simple (i.e., non-self-intersecting) closed curve C in $\mathcal{D}$ the region inside of C is also contained in $\mathcal{D}$ (that is, $\mathcal{D}$ has no holes), then we say that the set $\mathcal{D}$ is simply connected. See Figs. 12.28 and 12.29.

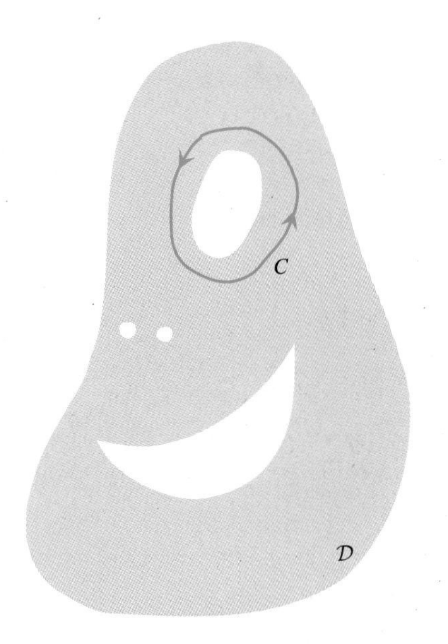

FIGURE 12.29 The set $\mathcal{D}$ is not simply connected because there is a closed curve C in $\mathcal{D}$ such that the inside of C is not in $\mathcal{D}$.

The relationship between simply connected sets and the existence of potential functions is given in the following result.

> **THEOREM** Potential Functions and Simply Connected Sets
>
> Let $\mathcal{D}$ be a simply connected subset of the plane and let
>
> $$\mathbf{F}(x,y) = \langle M(x,y), N(x,y)\rangle$$
>
> describe a continuously differentiable vector field on $\mathcal{D}$. The vector field $\mathbf{F}$ has a potential function f on $\mathcal{D}$ if and only if
>
> $$\frac{\partial N}{\partial x} = \frac{\partial M}{\partial y} \tag{26}$$
>
> at all points of $\mathcal{D}$.

We will not prove this result, but see Exercise 36 for a further discussion on how to find a potential function. In a sense, the previous result is the best possible. In particular, if $\mathcal{D}$ is not simply connected, then there is a vector field $\mathbf{F}$ defined on $\mathcal{D}$ for which (26) holds and such that $\mathbf{F}$ does not have a potential function on $\mathcal{D}$. See Exercise 28.

Exercises 12.3

Exercises 1–4: Show that $\nabla f = \mathbf{F}$, then evaluate $\int_C (\mathbf{F}\cdot\mathbf{T})\,ds$.

1. $f(x,y) = \sqrt{xy} + x^2 - 1,$

$$\mathbf{F}(x,y) = \left\langle \frac{1}{2}\sqrt{\frac{y}{x}} + 2x, \frac{1}{2}\sqrt{\frac{x}{y}} \right\rangle,$$

C described by $\mathbf{r}(t) = \langle t^3\cos^2\pi t + 1, \sqrt{t+1}\rangle$, $0 \le t \le 3$

2. $f(x,y) = \arctan(xy^2),$

$$\mathbf{F}(x,y) = \left\langle \frac{y^2}{1+x^2y^4}, \frac{2xy}{1+x^2y^4} \right\rangle,$$

C the portion of the graph of $y = \tan x$ with $0 \le x \le \pi/4$

3. $f(x,y,z) = xe^{yz} - xyz,$

$$\mathbf{F}(x,y,z) = \langle e^{yz} - yz, xze^{yz} - xz, xye^{yz} - xy\rangle,$$

C the rectangle with vertices $(0,0,0)$, $(1,1,0)$, $(1,1,1)$, and $(0,0,1)$

4. $f(x,y,z) = \ln(x^2 + 2z^2 + 1),$

$$\mathbf{F}(x,y) = \left\langle \frac{2x}{x^2+2z^2+1}, 0, \frac{4z}{x^2+2z^2+1} \right\rangle,$$

C described by $\mathbf{r}(t) = \langle 4\cos t, t^4, 4\sin t\rangle$, $0 \le t \le 2\pi$

Exercises 5–12: Use the Gradient Test to check whether or not $\mathbf{F}$ might have a potential function. If $\mathbf{F}$ satisfies the conditions of the Gradient Test, find a potential function for $\mathbf{F}$.

5. $\mathbf{F}(x,y) = \langle x + y + 2, x - 3y - 4\rangle$

6. $\mathbf{F}(x,y) = \langle (x+2y)e^{-x}, 2xye^{-x}\rangle$

7. $\mathbf{F}(x,y) = \langle y^2\sec^2(xy), xy\sec^2(xy) + \tan(xy)\rangle$

8. $\mathbf{F}(x,y,z) = \langle x\sin y, y\sin z, z\sin x\rangle$

9. $\mathbf{F}(x,y,z) = \langle 0, 2yz - z^2, y^2 - 2yz\rangle$

10. $\mathbf{F}(x,y,z) = \langle x - 2y + 3z, -2x + 4y - 6z, 3x - 6y + 9z\rangle$

11. $\mathbf{F}(x,y,z) = \langle e^{yz}, e^{xz}, e^{xy}\rangle$

12. $\mathbf{F}(x,y,z) = \langle by + ayz + cz, bx + axz + dz, cx + axy + dy\rangle$, a, b, c, d constants

Exercises 13–16: Evaluate the line integral.

13. $\displaystyle\int_C \frac{x}{\sqrt{x^2+y^2}}\,dx + \frac{y}{\sqrt{x^2+y^2}}\,dy,$

where C is the portion of the graph of $y = \cos \pi x$ from $(0,1)$ to $(2,1)$

14. $\displaystyle\int_C x\,dx + z\,dy + y\,dz,$

where C is the circle of intersection of the plane with equation $3x - 2y + z = 0$ and the sphere with equation $x^2 + y^2 + z^2 = 1$

15. $\displaystyle\int_C y\,dx + z\,dy + x\,dz,$

where C is the piecewise linear curve consisting of the segment from $(0, 0, 0)$ to $(1, 1, 0)$, followed by the segment from $(1, 1, 0)$ to $(0, 0, 1)$

16. $\displaystyle\int_C \frac{x}{x^2 + y^2 + z^2}\,dx + \frac{y}{x^2 + y^2 + z^2}\,dy + \frac{z}{x^2 + y^2 + z^2}\,dz,$

where C is described by

$$\mathbf{r}(t) = \langle e^t \sin \pi t, e^t, 2e^t \cos 2\pi t\rangle, \qquad 0 \le t \le 2$$

Exercises 17–20: In Examples 2 and 3, we found formulas for potential functions. Each of these formulas involved an arbitrary constant c. Sometimes we wish our potential function to take on a given value at a particular point. This can be achieved by assigning the proper numerical value to c. For the given vector field **F**, *find the potential function f that satisfies the given condition.*

17. $\mathbf{F}(x, y) = \langle y - 2, x + 1\rangle,$
$f(1, 1) = 0$

18. $\mathbf{F}(x, y) = \left\langle \ln(x + y) + \dfrac{x}{x + y}, \dfrac{x}{x + y}\right\rangle,$
$f(e, 0) = 2e$

19. $\mathbf{F}(x, y, z) = \langle \sin(yz), xz \cos(yz) - y, xy \cos(yz)\rangle,$
$f(2, 1, \pi/2) = -3$

20. $\mathbf{F}(x, y, z) = \langle 0, 2yz^2 - 3z + 1, 2y^2z - 3y + 4z - 2\rangle,$
$f(4, -2, 0) = 0$

21. Let

$$\mathbf{v}(x, y) = \langle x^3 - 3xy^2, 3x^2y - y^3\rangle$$

describe the velocity field for the flow of a fluid in the plane, and let C be the circle of center $(0, 0)$ and radius 1.

a. Show that the flux across C is 0.

b. Show that the circulation about C is 0.

22. The velocity field **v** for a fluid of density 1 g/cm^2 flowing in the plane is described by

$$\mathbf{v}(x, y) = \langle x^3 + 3xy^2, 3x^2y + y^3\rangle \text{ cm/s}.$$

Let C be the circle of center $(0, 0)$ and radius 1.

a. Find the flux across C. Your answer should include units.

b. Find the circulation about C. Your answer should include units.

23. At a point (x, y, z) above the surface of Earth, the gravitational potential is given by the vector field

$$\mathbf{F}(x, y, z) = \frac{-GM}{(x^2 + y^2 + z^2)^{3/2}}\langle x, y, z\rangle,$$

where G is the universal constant of gravitation and M is the mass of Earth. Show that **F** is a conservative vector field by finding a potential function for **F**. What is the domain of the potential function?

24. A force **F** acting on an object in space is given by

$$\mathbf{F}(x, y, z) = \left\langle \frac{x}{(x^2 + y^2 + z^2 + 3)^{3/2}}, \frac{y}{(x^2 + y^2 + z^2 + 3)^{3/2}}, \frac{z}{(x^2 + y^2 + z^2 + 3)^{3/2}}\right\rangle.$$

Find the work done by **F** as the object moves from the point $(-1, 3, 4)$ to the point $(2, 0, 3)$ along any piecewise smooth curve C joining the two points.

25. A force **F** acting on an object in space is given by

$$\mathbf{F}(x, y, z) = \left\langle \frac{y}{z}, \frac{z}{x}, \frac{x}{y}\right\rangle.$$

Find the work done by **F** as the object moves from the point $(1, 1, 1)$ to the point $(3, 3, 3)$ along the line segment joining the two points.

26. Let

$$f(x, y) = \arctan\left(\frac{y}{x}\right), \qquad x \ne 0.$$

a. Show that

$$\nabla f(x, y) = \left\langle \frac{-y}{x^2 + y^2}, \frac{x}{x^2 + y^2}\right\rangle. \tag{27}$$

b. In Example 5 we showed that the vector field (27) does not have a potential on the set

$$\mathcal{D} = \{(x, y) : (x, y) \ne (0, 0)\}.$$

Explain why Example 5 and part **a** are not in contradiction.

27. Prove the Gradient Test in the three-variable case. That is, let $\mathbf{G} = \langle P, Q, R\rangle$ be a vector field defined on a subset of R^3. Show that if P, Q, and R have continuous first partial derivatives and **G** is the gradient of a function g, then

$$Q_x = P_y, \qquad R_x = P_z, \qquad R_y = Q_z.$$

28. Let $\mathcal{D} = \{(x, y) : (x, y) \ne (a, b)\}$ and let **F** be defined on $\mathcal{D}$ by

$$\mathbf{F}(x, y) = \left\langle \frac{-(y - b)}{(x - a)^2 + (y - b)^2}, \frac{x - a}{(x - a)^2 + (y - b)^2}\right\rangle.$$

a. Show that $\mathbf{F}$ satisfies the hypotheses of the Gradient Test on $\mathcal{D}$.

b. Let C be the circle of radius 1 centered at (a, b). Show that

$$\int_C (\mathbf{F} \cdot \mathbf{T})\, ds \neq 0.$$

c. Does $\mathbf{F}$ have a potential function on $\mathcal{D}$?

d. Let $\mathcal{D}_1 = \{(x, y) : x > a\}$ and define f on $\mathcal{D}_1$ by

$$f(x, y) = \arctan\left(\frac{y - b}{x - a}\right).$$

Show that f is a potential function for $\mathbf{F}$ on $\mathcal{D}_1$.

e. Relate your answers to **c** and **d** to the theorem on potential functions and simply connected sets on page 1064.

29. Let $\mathcal{D} = \{(x, y) : (x, y) \neq (0, 0)\}$ and let $\mathbf{F}$ be defined on $\mathcal{D}$ by

$$\mathbf{F} = \left\langle \frac{-x}{(x^2 + y^2)^2}, \frac{-y}{(x^2 + y^2)^2} \right\rangle.$$

a. Show that $\mathbf{F}$ satisfies the conditions of the Gradient Test and that $\mathbf{F}$ has a potential function on $\mathcal{D}$.

b. Note that $\mathcal{D}$ is not simply connected. Why doesn't the existence of a potential for $\mathbf{F}$ contradict the theorem on potential functions and simply connected sets on page 1064?

Exercises 30–33: Sometimes a vector field $\mathbf{F}$ is "almost conservative" in that adding a simple function to one or more components of $\mathbf{F}$ results in a conservative vector field $\mathbf{G}$. In such a case, we can write

$$\int_C (\mathbf{F} \cdot \mathbf{T})\, ds = \int_C (\mathbf{G} \cdot \mathbf{T})\, ds + \int_C ((\mathbf{F} - \mathbf{G}) \cdot \mathbf{T})\, ds.$$

The first of the integrals on the right can be evaluated by finding a potential function for $\mathbf{G}$, and the second integral on the right might be easier than the integral on the left if $\mathbf{F} - \mathbf{G}$ is simple enough. In the following problems, find a function u so that $\mathbf{G}$ is a conservative vector field, and then evaluate $\int_C (\mathbf{F} \cdot \mathbf{T})\, ds$.

30. $\mathbf{F}(x, y) = \langle 3x^2y, x^3 - 4xy \rangle$,
$\mathbf{G}(x, y) = \mathbf{F}(x, y) + \langle u(x, y), 0 \rangle$,
C the line segment from $(1, -1)$ to $(-2, 0)$

31. $\mathbf{F}(x, y) = \langle (2xy + y^2 + 2)e^{xy}, (2x^2 + xy)e^{xy} \rangle$,
$\mathbf{G}(x, y) = \mathbf{F}(x, y) + \langle 0, u(x, y) \rangle$,
C the portion of the graph of $y = 1/x$ that runs from the point $(1, 1)$ to the point $(2, 1/2)$

32. $\mathbf{F}(x, y, z) = \langle 2xy, 2yz, y^2 \rangle$,
$\mathbf{G}(x, y, z) = \mathbf{F}(x, y, z) + \langle 0, u(x, y, z), 0 \rangle$,
C the piecewise linear curve consisting of the segment from $(1, 0, 0)$ to $(4, 0, 0)$, followed by the segment from $(4, 0, 0)$ to $(4, 0, -3)$

33. $\mathbf{F}(x, y, z) = \langle 2xy + z^2, x^2 + 2yz, y^2 \rangle$,
$\mathbf{G}(x, y, z) = \mathbf{F}(x, y, z) + \langle 0, 0, u(x, y, z) \rangle$,
C described by $\mathbf{r}(t) = \langle \cos t, \sin t, 2t \rangle$, $0 \leq t \leq 2\pi$

34. This exercise completes the proof of the Fundamental Theorem of Line Integrals. Let C be a piecewise smooth curve with initial point $\mathbf{A}$ and terminal point $\mathbf{B}$. Let $\mathbf{F}$ be a conservative vector field with potential function f, and assume that f and $\mathbf{F}$ are defined on C. By applying the Fundamental Theorem of Line Integrals to the smooth pieces of C, show that

$$\int_C (\mathbf{F} \cdot \mathbf{T})\, ds = f(\mathbf{B}) - f(\mathbf{A}).$$

35. Let C be a closed, planar curve and let $\mathbf{v}$ describe the flow of a fluid in the plane. Subdivide C into n short pieces of equal length and let Δs denote the length of each of these pieces. For $1 \leq k \leq n$, let $\mathbf{c}_k$ be a point on the kth piece of the curve, and let $\mathbf{T}_k$ be the unit tangent to C at $\mathbf{c}_k$. Consider the sum

$$\sum_{k=1}^{n} (\mathbf{F}(\mathbf{c}_k) \cdot \mathbf{T}_k)\, \Delta s.$$

a. Write an argument justifying the statement that the sum gives an approximation of the "flow tangent to C."

b. Argue that as the number of pieces in the subdivision approaches infinity, the sum approaches the circulation integral,

$$\int_C (\mathbf{F} \cdot \mathbf{T})\, ds.$$

c. What are the units for the previous integral? Is "circulation" a good name for the physical quantity measured by the integral? Why or why not?

36. A planar set $\mathcal{D}$ is convex if whenever A and B are two points in $\mathcal{D}$, the segment $[A, B]$ joining A to B is also contained in $\mathcal{D}$. Let $\mathbf{F}$ be the vector field given by

$$\mathbf{F}(x, y) = \langle M(x, y), N(x, y) \rangle$$

and assume that

$$\int_C M\, dx + N\, dy = 0$$

for every piecewise, smooth, closed curve in $\mathcal{D}$. Now let $P_0 = (a, b)$ be a fixed point in $\mathcal{D}$. For $P = (x, y)$ in $\mathcal{D}$, define

$$f(x, y) = f(P) + \int_{[P_0, P]} M(x, y)\, dx + N(x, y)\, dy.$$

Use the following outline in showing that $\nabla f = \mathbf{F}$.

a. Show that

$$\begin{aligned}\frac{f(c+h,d)-f(c,d)}{h} &= \frac{1}{h}\int_{[P,Q]} M(x,y)\,dx + N(x,y)\,dy \\ &= \frac{1}{h}\int_{[P,Q]} M(x,d)\,dx,\end{aligned}$$

where $h > 0$ and $P = (c,d)$ and $Q = (c+h,d)$ are both points in $\mathcal{D}$.

b. Next show that

$$\begin{aligned}\left|\frac{f(c+h,d)-f(c,d)}{h} - M(c,d)\right| &= \frac{1}{h}\left|\int_c^{c+h} (M(x,d) - M(c,d)\,dx\right| \\ &\le \frac{1}{h}\int_c^{c+h} |M(x,d) - M(c,d)|\,dx.\end{aligned}$$

c. Show that

$$\lim_{h\to 0}\left|\frac{f(c+h,d)-f(c,d)}{h} - M(c,d)\right| = 0,$$

and from this conclude that

$$\frac{\partial f}{\partial x}(c,d) = M(c,d).$$

d. Use a process similar to that just outlined to show that

$$\frac{\partial f}{\partial y}(c,d) = N(c,d).$$

12.4 Green's Theorem

According to the Fundamental Theorem of Line Integrals, if P and Q are defined and have continuous partial derivatives on a simply connected set $\mathcal{D}$, and

$$\frac{\partial P}{\partial y} = \frac{\partial Q}{\partial x} \tag{1}$$

on $\mathcal{D}$, then

$$\oint_C P\,dx + Q\,dy = 0$$

for any piecewise, smooth, simple closed curve C in $\mathcal{D}$. In particular, if (1) holds, then we can evaluate $\int_C P\,dx + Q\,dy$ without finding a parameterization for C. In this section we see that even when (1) does not hold, we can still evaluate $\int_C P\,dx + Q\,dy$ without finding a parameterization for C. The theorem that tells us how to do this is Green's Theorem.

Green's Theorem

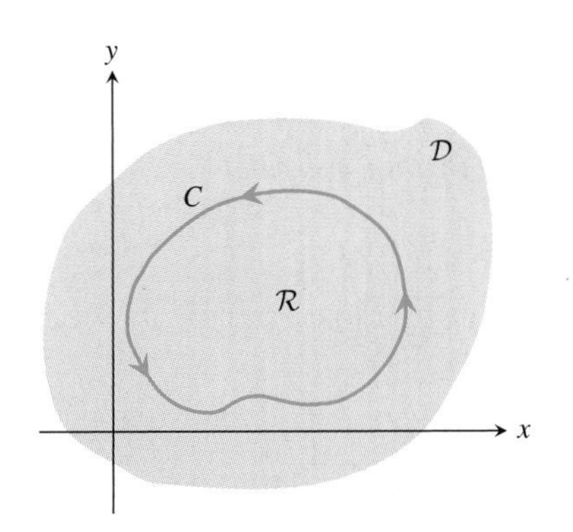

FIGURE 12.30 The curve C and the region $\mathcal{R}$ inside of C are contained in $\mathcal{D}$.

THEOREM Green's Theorem

Let $\mathcal{D}$ be a simply connected set in the plane and let P and Q be defined on $\mathcal{D}$ and have continuous first partial derivatives on $\mathcal{D}$. Let C be a piecewise, smooth, simple closed curve in $\mathcal{D}$ and let $\mathcal{R}$ be the region inside of C. See Fig. 12.30. Then

$$\oint_C P\,dx + Q\,dy = \int_C P\,dx + Q\,dy = \iint_{\mathcal{R}} \left(\frac{\partial Q}{\partial x} - \frac{\partial P}{\partial y}\right) dA. \tag{2}$$

Equation (2) will follow if we can show that

$$\int_C P\,dx = \iint_{\mathcal{R}} \left(-\frac{\partial P}{\partial y}\right) dA \quad \text{and} \quad \int_C Q\,dy = \iint_{\mathcal{R}} \frac{\partial Q}{\partial x}\,dA. \tag{3}$$

If each of these equations holds, then (2) will follow by adding the two equations. To get an idea of why Green's Theorem is true, we prove the first of the equations in (3) for the case in which C is a simple polygon, and then indicate how the polygon case leads to the result for piecewise smooth curves. In Exercises 20, 21, and 22, we provide an outline for a proof of the second equation in (3).

We first investigate the case in which C is a quadrilateral with two sides parallel to the y-axis. The vertical sides of the quadrilateral are described by $x = c$ and $x = d$, and the bottom and top sides, respectively, by the equations

$$y = \ell_1(x) = m_1x + b_1 \quad \text{and} \quad y = \ell_2(x) = m_2x + b_2.$$

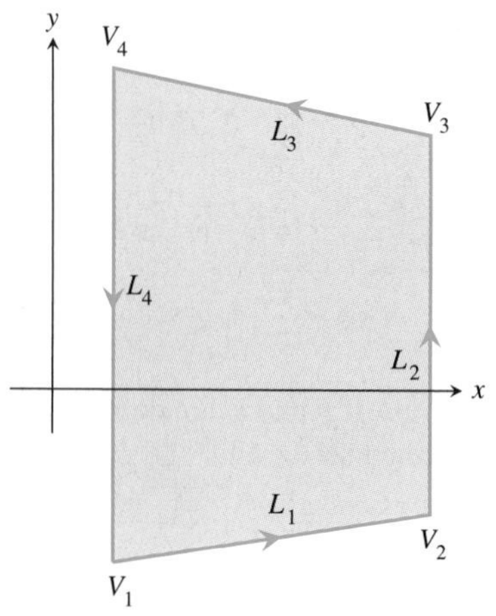

See Fig. 12.31. The vertices of the quadrilateral in counterclockwise order are $V_1 = (c, m_1c + b_1)$, $V_2 = (d, m_1d + b_1)$, $V_3 = (d, m_2d + b_2)$, and $V_4 = (c, m_2c + b_2)$. Let L_1 be the segment from V_1 to V_2, let L_2 be the segment from V_2 to V_3, let L_3 be the segment from V_3 to V_4, and let L_4 be the segment from V_4 to V_1. Then $C = L_1 + L_2 + L_3 + L_4$ is the piecewise smooth curve obtained by joining these four segments.

FIGURE 12.31 The closed curve C is a quadrilateral with two sides parallel to the y-axis.

To verify that

$$\int_C P\,dx = \iint_{\mathcal{R}} \left(-\frac{\partial P}{\partial y}\right) dA \tag{4}$$

holds for C, we change the form of each of the integrals in (4) and then note that the resulting forms are identical. To work with the line integral, we need a parameterization of each of the segments L_j, $j = 1, 2, 3, 4$. These curves are described by

$$\begin{aligned}
&L_1 : \mathbf{r}_1(t) = \langle x_1(t), y_1(t)\rangle = \langle t, m_1t + b_1\rangle, \qquad c \le t \le d,\\
&L_2 : \mathbf{r}_2(t) = \langle x_2(t), y_2(t)\rangle = \langle d, t\rangle, \qquad m_1d + b_1 \le t \le m_2d + b_2,\\
&L_3 : \mathbf{r}_3(t) = \langle x_3(t), y_3(t)\rangle = \langle c + d - t, m_2(c + d - t) + b_2\rangle, \qquad c \le t \le d,\\
&L_4 : \mathbf{r}_4(t) = \langle x_4(t), y_4(t)\rangle = \langle c, (m_1 + m_2)c + (b_1 + b_2) - t\rangle,\\
&\qquad\qquad\qquad\qquad\qquad\qquad m_1c + b_1 \le t \le m_2c + b_2.
\end{aligned}$$

We have

$$\int_C P\,dx = \int_{L_1} P\,dx + \int_{L_2} P\,dx + \int_{L_3} P\,dx + \int_{L_4} P\,dx. \tag{5}$$

Now

$$\int_{L_1} P\,dx = \int_c^d P(x_1(t), y_1(t))x_1'(t)\,dt = \int_c^d P(t, m_1t + b_1)\,dt \tag{6}$$

and

$$\begin{aligned}
\int_{L_3} P\,dx &= \int_c^d P(x_3(t), y_3(t))x_3'(t)\,dt\\
&= \int_c^d P(c + d - t, m_2(c + d - t) + b_2)(-1)\,dt \qquad (7)\\
&= -\int_c^d P(\tau, m_2\tau + b_2)\,d\tau,
\end{aligned}$$

where we have made the substitution $\tau = c + d - t$ in obtaining the last integral. Because $x_2'(t) = 0 = x_4'(t)$, we also have

$$\int_{L_2} P\,dx = \int_{m_1 d+b_1}^{m_2 d+b_2} P(\mathbf{r}_2(t))x_2'(t)\,dt = 0 = \int_{L_4} P\,dx. \tag{8}$$

Combining (5), (6), (7), and (8), we have

$$\begin{aligned}\int_C P\,dx &= \int_c^d P(t, m_1 t + b_1)\,dt - \int_c^d P(\tau, m_2\tau + b_2)\,d\tau \\ &= -\int_c^d (P(t, m_2 t + b_2) - P(t, m_1 t + b_1))\,dt.\end{aligned} \tag{9}$$

Next we work with the integral on the right side of (4). Note that $\mathcal{R}$ is the region between the graphs of $y = m_1 x + b_1$ and $y = m_2 x + b_2$, for $c \le x \le d$. We express the double integral as an iterated integral, with the y integration performed first. This gives

$$\begin{aligned}\iint_{\mathcal{R}} \left(-\frac{\partial P}{\partial y}(x, y)\right) dA &= -\int_c^d \int_{m_1 x+b_1}^{m_2 x+b_2} \frac{\partial P}{\partial y}(x, y)\,dy\,dx \\ &= -\int_c^d P(x, y)\Big|_{y=m_1 x+b_1}^{y=m_2 x+b_2} dx \\ &= -\int_c^d (P(x, m_2 x + b_2) - P(x, m_1 x + b_1))\,dx.\end{aligned} \tag{10}$$

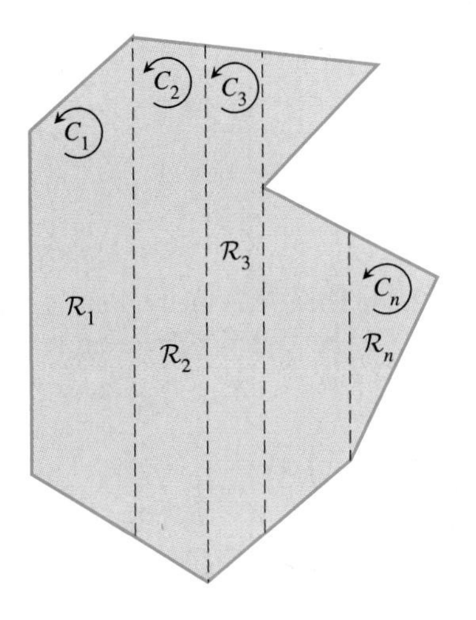

FIGURE 12.32 A polygon subdivided into quadrilaterals.

Because (9) and (10) have the same value, the two integrals in (4) are equal. This shows that (4) is true when C is a quadrilateral with two sides parallel to the y-axis.

Next let C be an arbitrary simple polygon. From each vertex of C, draw a vertical segment. This cuts $\mathcal{R}$ into quadrilaterals with sides parallel to the y-axis, as shown in Fig. 12.32. Let $C_1, C_2, \ldots, C_n$ denote the paths that follow the edges of the quadrilaterals, each with counterclockwise orientation, and let $\mathcal{R}_1, \mathcal{R}_2, \ldots, \mathcal{R}_n$, respectively, be the regions inside of these quadrilaterals. As seen in (8), $\int_L P\,dx$ is 0 for any vertical line L. It follows that

$$\int_C P\,dx = \int_{C_1} P\,dx + \int_{C_2} P\,dx + \cdots + \int_{C_n} P\,dx.$$

Because each of the curves appearing on the right side of this equation is a quadrilateral with sides parallel to the y-axis, this is equal to

$$\iint_{\mathcal{R}_1} \left(-\frac{\partial P}{\partial y}\right) dA + \iint_{\mathcal{R}_2} \left(-\frac{\partial P}{\partial y}\right) dA + \cdots + \iint_{\mathcal{R}_n} \left(-\frac{\partial P}{\partial y}\right) dA = \iint_{\mathcal{R}} \left(-\frac{\partial P}{\partial y}\right) dA.$$

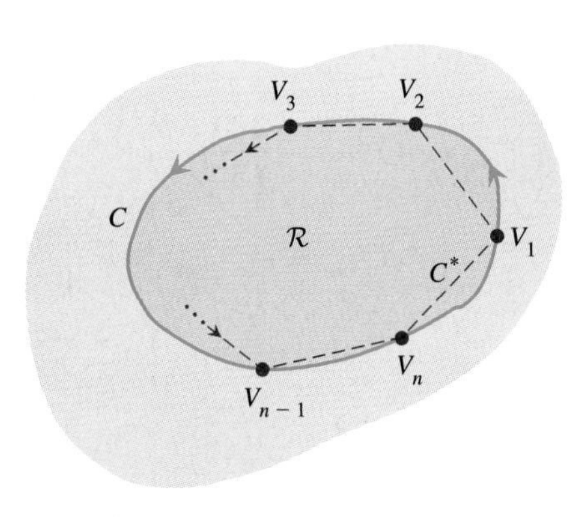

FIGURE 12.33 The curve C is approximated by the polygonal curve C^*.

This verifies that (4) is true when C is a simple polygon.

Finally, if C is a smooth, simple closed curve, let $\mathbf{V}_1, \mathbf{V}_2, \ldots, \mathbf{V}_n$ be points on C, let C^* be the polygon with these points as vertices, and let $\mathcal{R}^*$ be the region inside of C^*. See Fig. 12.33. Because C is a smooth, simple closed curve, we can choose the points so that C^* is a simple polygon. In addition, because P and $\partial P/\partial x$ are continuous, the vertex points can be selected so that

$$\int_C P\,dx \approx \int_{C^*} P\,dx \quad \text{and} \quad \iint_{\mathcal{R}} \left(-\frac{\partial P}{\partial y}\right) dA \approx \iint_{\mathcal{R}^*} \left(-\frac{\partial P}{\partial y}\right) dA.$$

It can be shown that these calculations can be made as accurate as we like by taking enough vertex points and making the distance between consecutive points sufficiently small. Because C^* is a simple polygon,

$$\int_{C^*} P\,dx = \iint_{R^*} \left(-\frac{\partial P}{\partial y}\right) dA,$$

and it follows that

$$\int_{C} P\,dx = \iint_{\mathcal{R}} \left(-\frac{\partial P}{\partial y}\right) dA.$$

This shows that (4) is true whenever C is a simple, closed smooth curve.

We illustrate Green's Theorem with several examples.

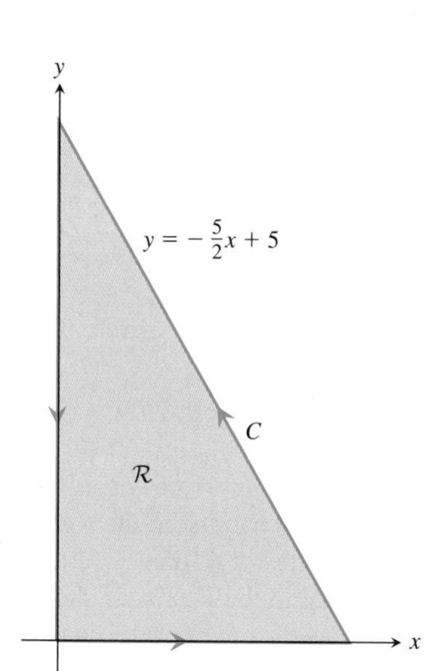

FIGURE 12.34 The triangular curve C and the region $\mathcal{R}$.

EXAMPLE 1 Let C be the triangle with vertices $(0,0)$, $(2,0)$, and $(0,5)$. Evaluate

$$\int_C (x^2 - y^2)\,dx + 2xy\,dy.$$

Solution

Let $\mathcal{R}$ be the region inside of C, as seen in Fig. 12.34. Then $\mathcal{R}$ is the region bounded by the lines $x = 0$, $y = 0$, and $y = -\frac{5}{2}x + 5$. By Green's Theorem, we have

$$\int_C (x^2 - y^2)\,dx + 2xy\,dy = \iint_{\mathcal{R}} \left(\frac{\partial}{\partial x}2xy - \frac{\partial}{\partial y}(x^2 - y^2)\right) dA = \iint_{\mathcal{R}} 4y\,dA.$$

The region $\mathcal{R}$ is bounded below by the line $y = 0$ and above by the line $y = -\frac{5}{2}x + 5$ for $0 \le x \le 2$. Thus we have

$$\begin{aligned}\int_C (x^2 - y^2)\,dx + 2xy\,dy &= \iint_{\mathcal{R}} 4y\,dA \\ &= \int_0^2 \int_0^{-(5/2)x+5} 4y\,dy\,dx \\ &= \int_0^2 2\left(-\frac{5}{2}x + 5\right)^2 dx = \frac{100}{3}.\end{aligned}$$

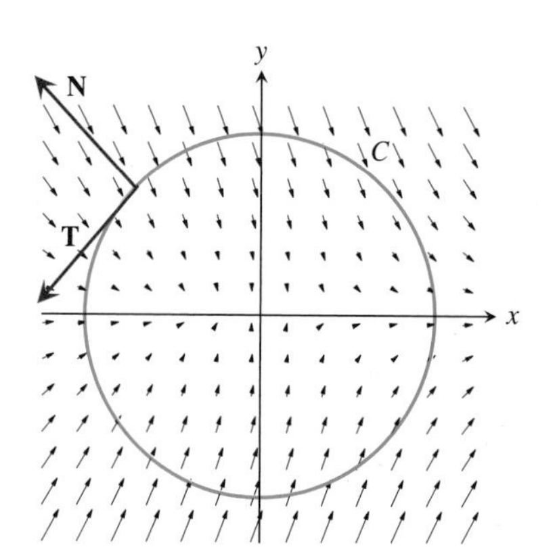

FIGURE 12.35 Planar flow across a circle.

EXAMPLE 2 Let

$$\mathbf{v}(x,y) = \langle \ln(x^2 + y^2 + 1), -2y \rangle \text{ cm/s}$$

be the velocity field for the planar flow of a fluid of constant density δ g/cm^2. Find the flux across the circle of center $(0,0)$ and radius 1. See Fig. 12.35.

Solution

Let C be the unit circle and let $\mathcal{R}$ be the region inside of C. The flux Φ across C is given by

$$\Phi = \delta \int_C (\mathbf{v} \cdot \mathbf{N})\, ds, \tag{11}$$

where $\mathbf{N}$ is the outward normal to C. As seen in Section 12.3,

$$\int_C (\mathbf{v} \cdot \mathbf{N})\, ds = \int_C (\tilde{\mathbf{v}} \cdot \mathbf{T})\, ds, \tag{12}$$

where $\tilde{\mathbf{v}}(x, y) = \langle 2y, \ln(x^2 + y^2 + 1)\rangle$. Combining this with (11) and (12), the flux across C is

$$\Phi = \delta \int_C 2y\, dx + \ln(x^2 + y^2 + 1)\, dy.$$

By Green's Theorem, the latter integral is equal to

$$\iint_{\mathcal{R}} \left(\frac{\partial}{\partial x} \ln(x^2 + y^2 + 1) - \frac{\partial}{\partial y} 2y \right) dA = \iint_{\mathcal{R}} \left(\frac{2x}{x^2 + y^2 + 1} - 2 \right) dA.$$

Because $\mathcal{R}$ is a disk centered at the origin, we use polar coordinates to evaluate the last integral, obtaining

$$\Phi = \delta \int_0^1 \int_0^{2\pi} \left(\frac{2r \cos\theta}{r^2 + 1} - 2 \right) r\, d\theta\, dr = \delta \int_0^1 (-4\pi r)\, dr = -2\pi\delta \text{ g/s}.$$

The negative answer means that more fluid flows into the circle than flows out, that is, there is a sink for the flow inside of C.

EXAMPLE 3 Let C be a smooth, simple closed curve in the plane and let $\mathcal{R}$ be the region inside of C. Show that

$$\text{area of } R = \frac{1}{2} \int_C -y\, dx + x\, dy. \tag{13}$$

Solution

By Green's Theorem,

$$\frac{1}{2} \int_C -y\, dx + x\, dy = \frac{1}{2} \iint_{\mathcal{R}} \left(\frac{\partial}{\partial x} x - \frac{\partial}{\partial y}(-y) \right) dA = \iint_{\mathcal{R}} 1\, dA.$$

Because the last integral is the area of $\mathcal{R}$, (13) follows. This formula for area is the idea behind the *planimeter,* an ingenious mechanical device for finding the area inside of planar curves. See the Student Project titled "The Polar Planimeter" at the end of this chapter.

Regions with Holes

In our statement of Green's Theorem, we required that P and Q be defined and have continuous partial derivatives on the region $\mathcal{R}$, where $\mathcal{R}$ is the region inside of C. A modified version of Green's Theorem can be applied to the case in which $\mathcal{R}$ has holes.

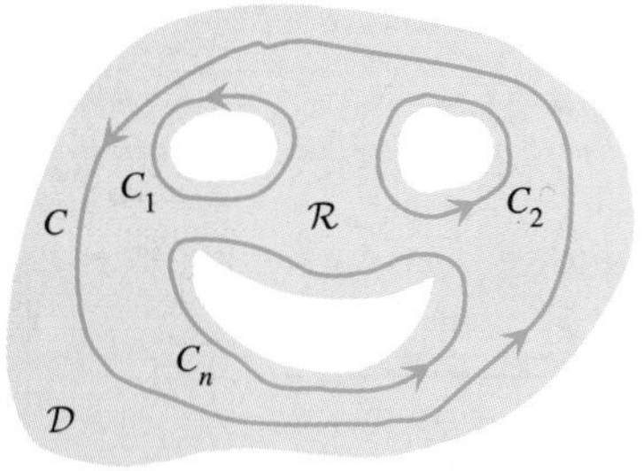

THEOREM A Generalized Green's Theorem

Let $\mathcal{D}$ be a subset of the plane and let P and Q be defined and have continuous first partial derivatives on $\mathcal{D}$. Let C be a piecewise smooth, simple closed curve with counterclockwise orientation in $\mathcal{D}$. Suppose that each of the curves $C_1, C_2, \ldots, C_n$ is a piecewise smooth, simple closed curve and that

- each curve C_j, $1 \leq j \leq n$, lies inside of C;
- for each pair (C_i, C_j) of curves, C_j lies outside curve C_i;
- each curve C_j has counterclockwise orientation.

See Fig. 12.36. Let $\mathcal{R}$ denote the region that is inside of C but is outside of all the curves $C_1, C_2, \ldots, C_n$, and assume that this region is contained in $\mathcal{D}$. Then

$$\int_C P\,dx + Q\,dy - \int_{C_1} P\,dx + Q\,dy - \int_{C_2} P\,dx + Q\,dy - \cdots$$
$$\cdots - \int_{C_n} P\,dx + Q\,dy = \iint_{\mathcal{R}} \left(\frac{\partial Q}{\partial x} - \frac{\partial P}{\partial y}\right) dA. \tag{14}$$

FIGURE 12.36 Closed curves in a region with holes.

In Exercise 28 we ask you to give an argument indicating how the generalized version of Green's Theorem follows from the earlier version of Green's Theorem. To demonstrate the generalized Green's Theorem, we find the flux across the inner and outer boundaries of a ring-shaped region.

EXAMPLE 4 The velocity field $\mathbf{v}$ for a planar fluid flow is given by

$$\mathbf{v}(x, y) = \langle x, -2y \rangle \text{ cm/s}.$$

Let $\mathcal{R}$ be the ring-shaped region between the circle of center $(0,0)$ and radius 4 and the circle of center $(1,0)$ and radius 2. See Fig. 12.37. Find the flux out of $\mathcal{R}$. Note that the boundary of $\mathcal{R}$ includes both the inner and outer circles. Assume that distances are measured in centimeters, time in seconds, and that the fluid has constant density of δ g/cm^2.

Solution

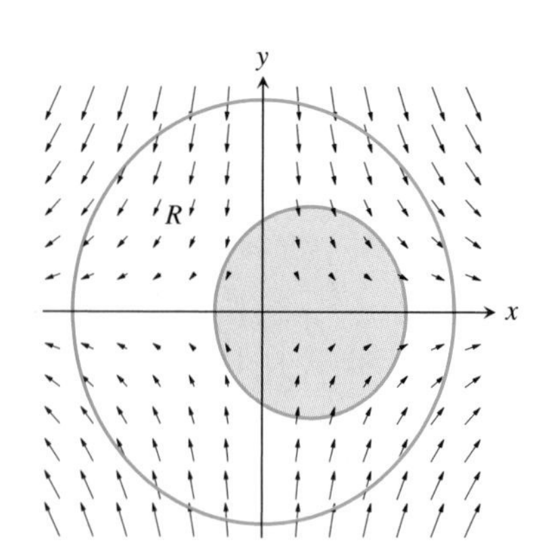

FIGURE 12.37 Measuring the flux across the boundary of a ring-shaped region.

Let C denote the circle of center $(0,0)$ and radius 4, and let C_1 be the circle of center $(1,0)$ and radius 2, both with counterclockwise orientation. When measuring the flux across the boundary of a region, we take the unit normals pointing toward the outside of the region as the "positive" direction, because these normals indicate the direction of flow out of the region. Because the unit normal to a curve points to the right relative to the unit tangent, the normal will point toward the outside of the ring-shaped region if we take a

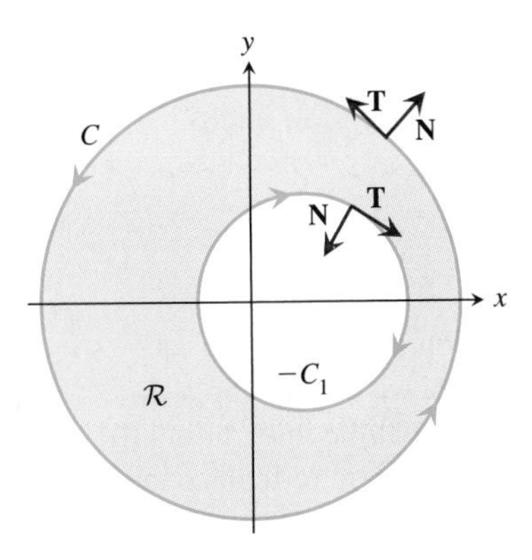

FIGURE 12.38 The positive direction normal points toward the outside of the region.

counterclockwise orientation on the outer boundary curve and a clockwise orientation on the inner boundary curve. Because the clockwise direction on C_1 traces the curve in the opposite direction to the counterclockwise direction, we use $-C_1$ to denote the clockwise tracing. See Fig. 12.38. The flux across the boundary of $\mathcal{R}$ is found by summing the fluxes across the individual boundary circles. Taking

$$\tilde{\mathbf{v}}(x, y) = \langle 2y, x \rangle,$$

the flux Φ is given by

$$\begin{aligned}
\Phi &= \delta \int_C (\mathbf{v} \cdot \mathbf{N})\, ds + \delta \int_{-C_1} (\mathbf{v} \cdot \mathbf{N})\, ds \\
&= \delta \int_C (\tilde{\mathbf{v}} \cdot \mathbf{T})\, ds + \delta \int_{-C_1} (\tilde{\mathbf{v}} \cdot \mathbf{T})\, ds \\
&= \delta \int_C (\tilde{\mathbf{v}} \cdot \mathbf{T})\, ds - \delta \int_{C_1} (\tilde{\mathbf{v}} \cdot \mathbf{T})\, ds \\
&= \delta\left(\int_C 2y\, dx + x\, dy - \int_{C_1} 2y\, dx + x\, dy \right).
\end{aligned} \tag{15}$$

In the third line of this calculation, we used the fact that $\int_{-C_1} = -\int_{C_1}$. See Section 12.2. Applying the generalized Green's Theorem to the last expression in (15), we have

$$\begin{aligned}
\Phi &= \delta \iint_{\mathcal{R}} \left(\frac{\partial}{\partial x}(x) - \frac{\partial}{\partial y}(2y) \right) dA \\
&= \delta \iint_{\mathcal{R}} -1\, dA = -\delta \text{ area}(R) = -12\pi\delta \text{ cm}^2/\text{s}.
\end{aligned}$$

What does this answer mean in relation to the net flow out of (or into) the region $\mathcal{R}$?

In cases when $\partial Q/\partial x - \partial P/\partial y = 0$, the generalized Green's Theorem can be used to simplify some line integrals along closed curves. This is even true in cases where the vector field defined by $\langle P(x, y), Q(x, y) \rangle$ does not have a potential function.

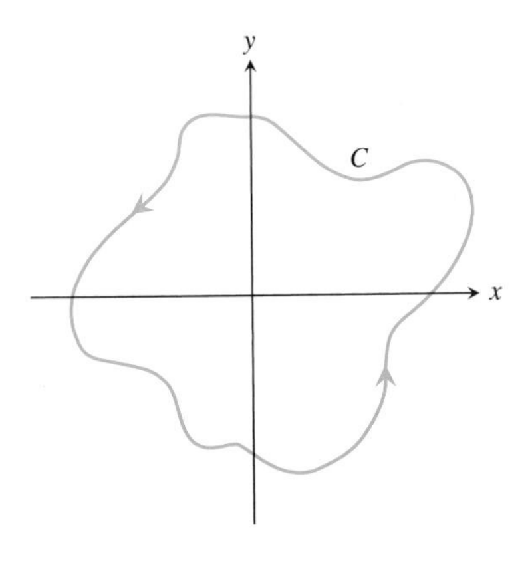

FIGURE 12.39 A closed curve about the origin.

EXAMPLE 5 Let C be a piecewise smooth, simple closed planar curve such that the origin is inside of C. See Fig. 12.39. Let $\mathbf{F}$ be the vector field given by

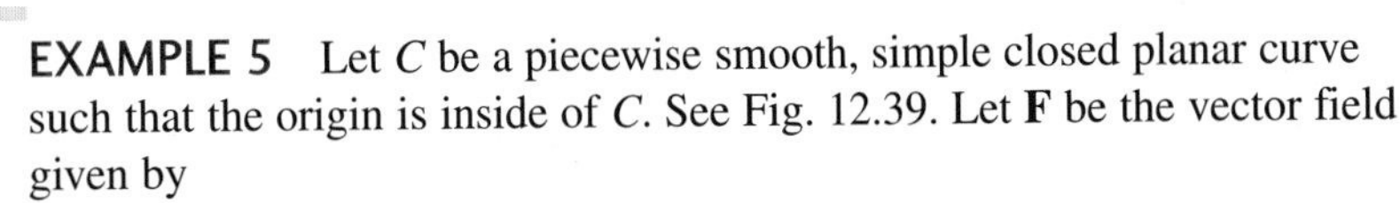

$$\mathbf{F}(x, y) = \langle M(x, y), N(x, y) \rangle = \left\langle \frac{-y}{x^2 + y^2}, \frac{x}{x^2 + y^2} \right\rangle.$$

Evaluate

$$\int_C M(x, y)\, dx + N(x, y)\, dy. \tag{16}$$

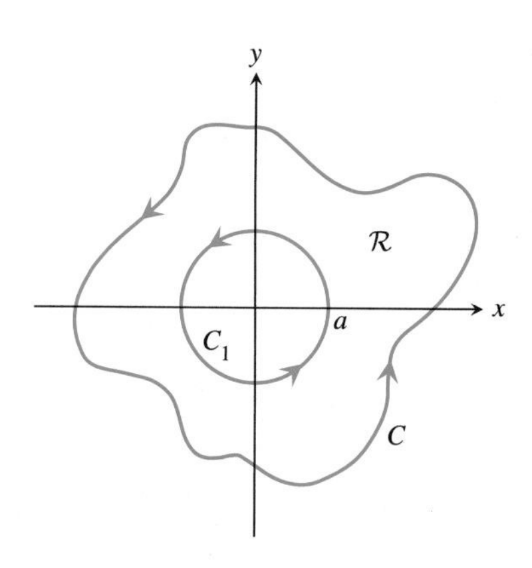

FIGURE 12.40 Curve C_1 is a small circle about the origin and is inside of C.

Solution

In Example 4 of Section 12.3, we saw that for this choice of M and N we have

$$\frac{\partial N}{\partial x}(x,y) = \frac{\partial M}{\partial y}(x,y) \tag{17}$$

for $(x,y) \neq (0,0)$, but that $\mathbf{F}$ has no potential function defined on the set $\{(x,y):(x,y) \neq (0,0)\}$. Thus, even though (17) is true, we cannot apply the Fundamental Theorem of Line Integrals to conclude that the integral in (16) is 0. However, we can use the generalized Green's Theorem to replace the integral by an integral along a different closed curve.

Let C_1 be the circle of radius a centered at the origin, and assume that a is small enough that C_1 is inside of C. See Fig. 12.40. If $\mathcal{R}$ is the region between C and C_1, then by the generalized Green's Theorem,

$$\int_C M\,dx + N\,dy - \int_{C_1} M\,dx + N\,dy = \iint_{\mathcal{R}} \left(\frac{\partial N}{\partial x} - \frac{\partial M}{\partial y}\right) dA = \iint_{\mathcal{R}} 0\,dA = 0.$$

If follows that

$$\int_C M\,dx + N\,dy = \int_{C_1} M\,dx + N\,dy, \tag{18}$$

so we can calculate $\int_C M\,dx + N\,dy$ by instead calculating $\int_{C_1} M\,dx + N\,dy$. The curve C_1 is parameterized by

$$\mathbf{r}(t) = \langle a\cos t, a\sin t\rangle, \qquad 0 \le t \le 2\pi.$$

Thus,

$$\begin{aligned}
\int_{C_1} M\,dx + N\,dy
&= \int_0^{2\pi} [M(x(t),y(t))x'(t) + N(x(t),y(t))y'(t)]\,dt \\
&= \int_0^{2\pi} \left(\frac{-a\sin t}{a^2\sin^2 t + a^2\cos^2 t}(-a\sin t) + \frac{a\cos t}{a^2\sin^2 t + a^2\cos^2 t}(a\cos t)\right) dt \\
&= \int_0^{2\pi} 1\,dt = 2\pi.
\end{aligned}$$

Combining this result with (18), we conclude that

$$\int_C M(x,y)\,dx + N(x,y)\,dy = 2\pi.$$

In this example, we were able to change the closed path of integration from the curve C to the curve C_1 without affecting the value of the integral. This example illustrates the following path independence result for integrals along closed curves.

Path Independence for Integrals along Closed Curves

Let $\mathcal{D}$ be a subset of the plane. Assume that the functions P and Q are defined and have continuous first partial derivatives on $\mathcal{D}$ and that

$$\frac{\partial Q}{\partial x}(x, y) = \frac{\partial P}{\partial y}(x, y)$$

for all (x, y) in $\mathcal{D}$. If C and C_1 are piecewise smooth, simple closed planar curves in $\mathcal{D}$, both with counterclockwise orientation, with C_1 inside of C, and if the region between C and C_1 is contained in $\mathcal{D}$, then

$$\int_C P\,dx + Q\,dy = \int_{C_1} P\,dx + Q\,dy.$$

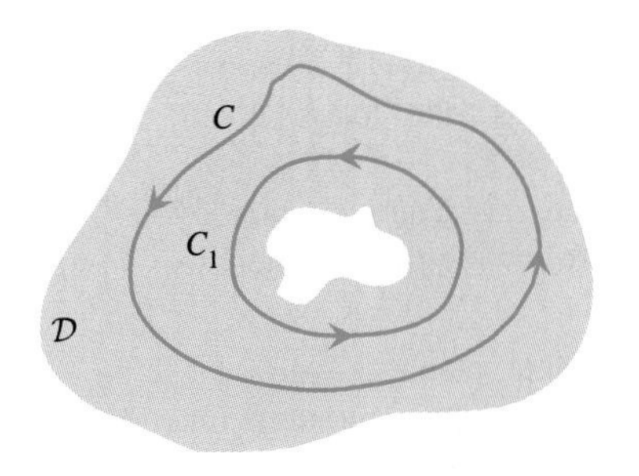

FIGURE 12.41 Two curves, one inside of the other.

The relative positions of curves C and C_1 are shown in Fig. 12.41. In Exercise 29, we ask the reader to show how this result follows from the generalized Green's Theorem. In Exercise 30, we generalize this result to the case in which the curves C and C_1 intersect.

Exercises 12.4

Exercises 1–6: Evaluate each of the integrals in two ways: first by finding a parameterization for C and converting the line integral to a definite integral; second, by applying Green's Theorem. Assume that each closed curve has counterclockwise orientation.

1. $\int_C 2y\,dx + 3\,dy$,
C the circle with center $(0, 0)$ and radius 1

2. $\int_C xy\,dx + (x + y)\,dy$,
C the triangle with vertices $(0, 0)$, $(3, 0)$, and $(0, 3)$

3. $\int_C (\mathbf{F} \cdot \mathbf{T})\,ds$,
$\mathbf{F}(x, y) = \left\langle \frac{x}{x^2 + y^2 + 2}, \frac{y}{x^2 + y^2 + 2} \right\rangle$,
C the circle with center $(0, 0)$ and radius $\sqrt{2}$

4. $\int_C x\,dy$,
C the ellipse given by $\mathbf{r}(t) = \langle a \cos t, b \sin t \rangle$, $a, b > 0$

5. $\int_C (\mathbf{F} \cdot \mathbf{T})\,ds$,
$\mathbf{F}(x, y) = \langle x^2 y - 2xy, xy^2 + 3xy \rangle$,
C the square with vertices $(0, 0)$, $(2, 0)$, $(2, 2)$, and $(0, 2)$

6. $\int_C (\mathbf{F} \cdot \mathbf{T})\,ds$,
$\mathbf{F}(x, y) = \langle ye^x, e^y \rangle$,
C the boundary of the region above the graph of $y = x^2$ and below the graph of $y = 9$

Exercises 7–10: In Example 3 we showed that the area of the region inside of the piecewise, smooth, simple closed curve C is given by $(1/2)\int_C -y\,dx + x\,dy$. By evaluating this integral, find the area inside C.

7. C the circle of center $(0, 0)$ and radius a

8. C the ellipse described by $x^2/4 + y^2/9 = 1$

9. C the cardioid with polar equation $r = 1 + \cos\theta$. *Hint:* In obtaining a parameterization for C, remember that $x = r\cos\theta$ and $y = r\sin\theta$.

10. C described by $\mathbf{r}(t) = \langle \cos^3 t, \sin t \rangle$, $0 \le t \le 2\pi$

Exercises 11–14: A flow in the plane has velocity field **v**. *Use Green's Theorem or the generalized Green's Theorem to calculate the flux across the given curve or curves. Assume the fluid has density* 1 g/cm^2.

11. $\mathbf{v}(x, y) = \langle x + y, x - y \rangle$.
Calculate the flux across the circle with center $(0, 0)$ and radius 1.

12. $\mathbf{v}(x, y) = \langle x^2 - y^2, 2xy \rangle$.
Calculate the flux across the triangle with vertices $(0, 0)$, $(2, 2)$, and $(0, 2)$.

13. $\mathbf{v}(x, y) = \langle \sin x \cos y, -\cos x \sin y \rangle$.
Calculate the flux across the boundary of the region between the circle with center $(0, 0)$ and radius 4 and the circle with center $(0, 2)$ and radius 1.

14. $\mathbf{v}(x, y) = \left\langle \frac{x}{\sqrt{x^2 + y^2}}, \frac{y}{\sqrt{x^2 + y^2}} \right\rangle$.
Calculate the flux across the boundary of the region between the circle with center $(0, 0)$ and radius 2 and the circle with center $(0, 0)$ and radius 1.

Exercises 15–18: Check that the theorem about path independence for integrals along closed paths holds, then use the result in evaluating the integral.

15. $\int_C \frac{x}{\sqrt{x^2 + y^2}}\, dx + \frac{y}{\sqrt{x^2 + y^2}}\, dy,$
C the ellipse with equation $x^2/9 + y^2/25 = 1$

16. $\int_C \frac{y - 2}{(y - 2)^2 + (x + 3)^2}\, dx - \frac{x + 3}{(y - 2)^2 + (x + 3)^2}\, dy,$
C the rectangle with vertices $(0, 0)$, $(0, 3)$, $(-5, 3)$, and $(-5, 0)$

17. $\int_C \left(4xy - \frac{y}{x^2 + y^2}\right) dx + \left(2x^2 + \frac{x}{x^2 + y^2}\right) dy,$
C the circle with center $\left(\sqrt{5}, \sqrt{7}\right)$ and radius 5

18. $\int_C \frac{x}{x^2 + y^2 - 2}\, dx + \frac{y}{x^2 + y^2 - 2}\, dy,$
C a piecewise smooth, simple closed curve with the circle of center $(0, 0)$ and radius $\sqrt{2}$ inside of C

19. Let C be a piecewise smooth, simple closed curve in the plane, let $\mathcal{R}$ be the region inside of C, and let P and Q be functions of two variables. Assume that P, Q, $\mathcal{R}$, and C satisfy the requirements of Green's Theorem, and let $\mathbf{F} = \langle P, Q \rangle$.

a. Show that

$$\int_C (\mathbf{F} \cdot \mathbf{T})\, ds = \iint_{\mathcal{R}} \left(\frac{\partial Q}{\partial x} - \frac{\partial P}{\partial y}\right) dA.$$

b. Show that

$$\int_C (\mathbf{F} \cdot \mathbf{N})\, ds = \iint_{\mathcal{R}} \left(\frac{\partial P}{\partial x} + \frac{\partial Q}{\partial y}\right) dA.$$

20. Let C be a simple, planar quadrilateral with two sides parallel to the x-axis, let $\mathcal{R}$ be the region inside of C, and let Q be a continuous function with continuous partial derivatives on C and $\mathcal{R}$. Show that

$$\int_C Q\, dy = \iint_{\mathcal{R}} \frac{\partial Q}{\partial x}\, dA$$

by evaluating both of the integrals. You may model your solution on the discussion following the statement of Green's Theorem.

21. Continuation Let C be a simple, planar polygon, let $\mathcal{R}$ be the region inside of C, and let Q be a continuous function with continuous partial derivatives on C and $\mathcal{R}$. Show that

$$\int_C Q\, dy = \iint_{\mathcal{R}} \frac{\partial Q}{\partial x}\, dA$$

by subdividing the polygon into quadrilaterals and then applying the result of Exercise 20.

22. Continuation Let C be a simple, piecewise smooth planar curve, let $\mathcal{R}$ be the region inside of C, and let Q be a continuous function with continuous partial derivatives on C and $\mathcal{R}$. Given an argument in support of the equation

$$\int_C Q\, dy = \iint_{\mathcal{R}} \frac{\partial Q}{\partial x}\, dA.$$

You may model your solution on the discussion following the statement of Green's Theorem.

23. Let C be a piecewise smooth, simple closed curve and let $\mathcal{R}$ be the region inside of C. Show that

$$\text{area}(\mathcal{R}) = -\int_C y\, dx = \int_C x\, dy.$$

24. Let

$$P(x, y) = xy^2 + \frac{1}{3}y^3 + y + 2$$

and

$$Q(x, y) = x^2y + xy^2 + 2x - 3.$$

Suppose that C is a piecewise smooth, simple closed curve in the plane and $\mathcal{R}$ is the region inside of C. Show that

$$\text{area}(\mathcal{R}) = \int_C P\, dx + Q\, dy.$$

25. Find functions P and Q of two variables such that the area of the region inside any piecewise smooth, simple closed curve C is given by

$$\int_C P\,dx + Q\,dy.$$

Your P and Q should be different from those in Exercises 23 and 24 and from those in Example 3.

26. Let C be the circle with center $(0, 0)$ and radius 1. Why can't Green's Theorem be used to evaluate

$$\int_C \frac{y}{x^2+y^2}\,dx - \frac{x}{x^2+y^2}\,dy?$$

27. Let C be a piecewise smooth, simple closed curve, and let f and g be continuously differentiable functions of one variable. What can be said about the value of

$$\int_C f(x)\,dx + g(y)\,dy?$$

28. The accompanying figure illustrates a generalized Green's Theorem configuration with two curves, C_1 and C_2, inside of curve C. The dotted segments L_1, L_2, L_3 are introduced to make two simple closed curves, Γ_1 and Γ_2. Show that if P, Q, C, C_1, and C_2 satisfy the hypotheses of the generalized Green's Theorem, then

$$\int_C P\,dx + Q\,dy - \int_{C_1} P\,dx + Q\,dy - \int_{C_2} P\,dx + Q\,dy$$
$$= \int_{\Gamma_1} P\,dx + Q\,dy + \int_{\Gamma_2} P\,dx + Q\,dy.$$

Next, apply Green's Theorem to show that

$$\int_{\Gamma_1} P\,dx + Q\,dy + \int_{\Gamma_2} P\,dx + Q\,dy = \iint_{\mathcal{R}} \left(\frac{\partial Q}{\partial x} - \frac{\partial P}{\partial y}\right) dA.$$

Combine these results to give an argument in support of the generalized Green's Theorem in this special case.

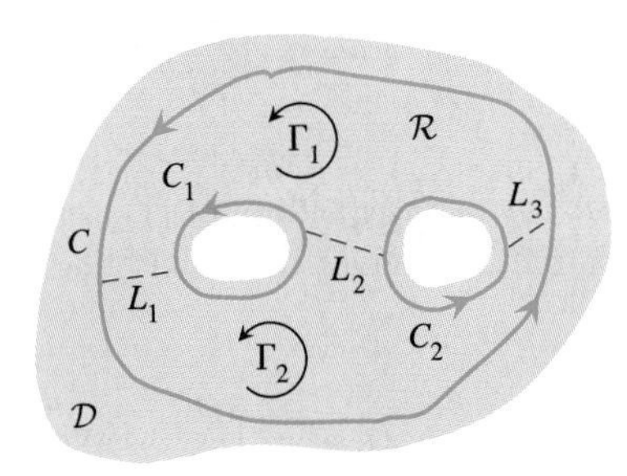

Figure for Exercise 28.

29. In the accompanying figure, we have introduced two segments, L_1 and L_2, into Fig. 12.41 to make two simple closed curves Γ_1 and Γ_2. Show that if P, Q, C, and C_1 satisfy the hypothesis of the theorem about path independence for integrals along closed curves, then

$$\int_{\Gamma_1} P\,dx + Q\,dy = 0 = \int_{\Gamma_2} P\,dx + Q\,dy.$$

Use this result to show that

$$\int_C P\,dx + Q\,dy = \int_{C_1} P\,dx + Q\,dy.$$

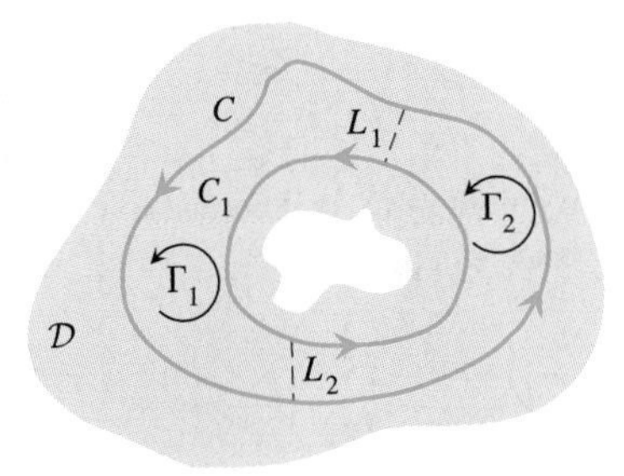

Figure for Exercise 29.

30. Let $\mathcal{D}$ be a subset of the plane. Assume that the functions P and Q are defined and have continuous first partial derivatives on $\mathcal{D}$ and that

$$\frac{\partial Q}{\partial x}(x, y) = \frac{\partial P}{\partial y}(x, y)$$

for all (x, y) in $\mathcal{D}$. Let C and C_1 be piecewise smooth, possibly intersecting, simple closed planar curves in $\mathcal{D}$, and assume that all points between C and C_1 are contained in $\mathcal{D}$. Show that

$$\int_C P\,dx + Q\,dy = \int_{C_1} P\,dx + Q\,dy.$$

Hint: Introduce an appropriate closed curve C_2 that intersects neither C nor C_1, then apply the result on path independence for integrals along closed curves.

12.5 Divergence and Curl

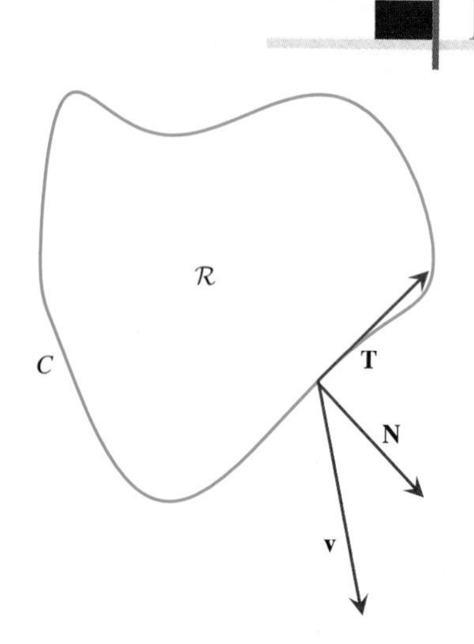

FIGURE 12.42 The unit tangent, unit normal, and a velocity vector at a point on C.

Let $\mathbf{v} = \langle P, Q\rangle$ be the vector field for a flow in the plane, and let C be a piecewise smooth, simple closed planar curve. See Fig. 12.42. In Section 12.2 we learned how to calculate the flux across C; in Section 12.3 we saw how to find the circulation around C. When we apply Green's Theorem to the flux and circulation expressions, we obtain formulas that involve double integrals over the region $\mathcal{R}$ inside of C:

$$\text{flux} = \int_C (\mathbf{v}\cdot\mathbf{N})\,ds = \int_C -Q\,dx + P\,dy = \iint_{\mathcal{R}} \left(\frac{\partial P}{\partial x} + \frac{\partial Q}{\partial y}\right) dA;$$

$$\text{circulation} = \int_C (\mathbf{v}\cdot\mathbf{T})\,ds = \int_C P\,dx + Q\,dy = \iint_{\mathcal{R}} \left(\frac{\partial Q}{\partial x} - \frac{\partial P}{\partial y}\right) dA.$$

The expressions give some insight into the physical behavior of the flow on or inside of C, but they tell us nothing about the behavior of the fluid at a particular point in the plane. To obtain such information, we apply a limit process on the flux and circulation integrals.

Flux and the Divergence

Assume that $\mathbf{v}$, given by

$$\mathbf{v}(x, y) = \langle P(x, y), Q(x, y)\rangle,$$

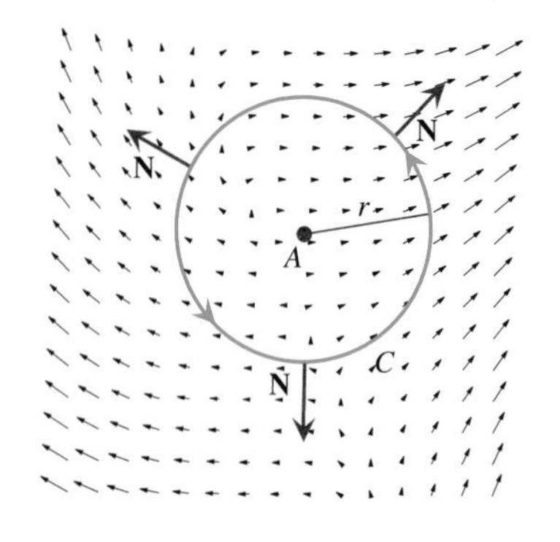

FIGURE 12.43 Calculating the flux across the circle $C = C(A, r)$.

is the velocity field for a planar flow in some domain $\mathcal{D}$, and that P and Q are continuous with continuous first partial derivatives. Let $A = (x_0, y_0)$ be a point in $\mathcal{D}$. Given $r > 0$, we denote the circle with center A and radius r by $C(A, r)$. We denote by $D(A, r)$ the set consisting of the circle $C(A, r)$ and the region inside of this circle. Assume that r is small enough that $D(A, r)$ is contained in $\mathcal{D}$. See Fig. 12.43. We define the average flux across $C(A, r)$ as

$$\begin{aligned}\text{average flux} &= \frac{\text{flux across } C(A, r)}{\text{area}(D(A, r))} \\ &= \frac{1}{\pi r^2}\int_{C(A,r)} (\mathbf{v}\cdot\mathbf{N})\,ds \qquad (1) \\ &= \frac{1}{\pi r^2}\iint_{D(A,r)} \left(\frac{\partial P}{\partial x} + \frac{\partial Q}{\partial y}\right) dA.\end{aligned}$$

We measure the "flux at the point A" by taking the limit of (1) as the radius r of the disk approaches 0. Because the partial derivatives of P and Q are continuous, we know that if r is small, then

$$\frac{\partial P}{\partial x}(x, y) + \frac{\partial Q}{\partial y}(x, y) \approx \frac{\partial P}{\partial x}(x_0, y_0) + \frac{\partial Q}{\partial y}(x_0, y_0)$$

for all (x, y) in $D(A, r)$. Furthermore, we can make the error in the approximation as small as we like by taking r sufficiently small. It follows that for small r,

$$\begin{aligned}\text{average flux} &= \frac{1}{\pi r^2}\iint_{D(A,r)}\left(\frac{\partial P}{\partial x}(x, y) + \frac{\partial Q}{\partial y}(x, y)\right) dA \\ &\approx \frac{1}{\pi r^2}\iint_{D(A,r)}\left(\frac{\partial P}{\partial x}(x_0, y_0) + \frac{\partial Q}{\partial y}(x_0, y_0)\right) dA.\end{aligned}$$

Because the integrand in the last expression is constant, we can factor it out of the integral. This gives

$$\begin{aligned}\text{average flux} &\approx \left(\frac{\partial P}{\partial x}(x_0, y_0) + \frac{\partial Q}{\partial y}(x_0, y_0)\right)\left(\frac{1}{\pi r^2}\iint_{D(A,r)} 1\, dA\right) \\ &= \frac{\partial P}{\partial x}(x_0, y_0) + \frac{\partial Q}{\partial y}(x_0, y_0).\end{aligned}$$

As r approaches 0 through positive values, this approximation becomes exact because the error in the approximation tends to 0 (see Exercise 27). Thus the flux at $A = (x_0, y_0)$ is

$$\lim_{r \to 0+} \frac{\text{flux across } C(A, r)}{\text{area}(D(A, r))} = \frac{\partial P}{\partial x}(x_0, y_0) + \frac{\partial Q}{\partial y}(x_0, y_0). \tag{2}$$

The last expression in (2) is called the *divergence of* $\mathbf{v}$ *at* (x_0, y_0). If we consider the physical units in (1), we note that average flux has units of flux per unit area. Thus we can interpret the divergence as representing the "flux per unit area at a point."

DEFINITION The Divergence

Two dimensions: Let $\mathbf{F}$ be a vector field defined on a set $D \subset R^2$ by

$$\mathbf{F}(x, y) = \langle M(x, y), N(x, y)\rangle.$$

The **divergence** of $\mathbf{F}$ is defined as

$$\text{div } \mathbf{F} = \frac{\partial M}{\partial x} + \frac{\partial N}{\partial y}$$

at every point (x, y) of $\mathcal{D}$ where both partial derivatives exist.

WEB

Three dimensions: Let $\mathbf{G}$ be the vector field defined on a set $D \subset R^3$ by

$$\mathbf{G}(x, y, z) = \langle P(x, y, z), Q(x, y, z), R(x, y, z)\rangle.$$

The divergence of $\mathbf{G}$ is defined as

$$\text{div } \mathbf{G} = \frac{\partial P}{\partial x} + \frac{\partial Q}{\partial y} + \frac{\partial R}{\partial z}$$

at every point (x, y, z) of $\mathcal{D}$ where all three partial derivatives exist.

Divergence and Flow The divergence of the velocity field $\mathbf{v}$ for a planar flow gives information about the flux throughout a region. Suppose that $\mathbf{v}(x, y) = \langle P(x, y), Q(x, y)\rangle$ and that P and Q have continuous partial derivatives in a disk centered at $A = (x_0, y_0)$. If $\operatorname{div} \mathbf{v}(x_0, y_0) > 0$, then the divergence is also positive at all points in a disk $D(A, r)$ of sufficiently small radius r centered at (x_0, y_0). If $C(A, r)$ is the circle that bounds this disk, then by Green's Theorem,

$$\begin{aligned}\text{flux across } C(A, r) &= \int_{C(A,r)} -Q\,dx + P\,dy \\ &= \iint_{D(A,r)} \left(\frac{\partial P}{\partial x} + \frac{\partial Q}{\partial y}\right) dA = \iint_{D(A,r)} \operatorname{div} \mathbf{v}\,dA > 0.\end{aligned}$$

The last integral is positive because the integrand is positive. This means that more fluid flows out than flows in through the circle $C(A, r)$. We may say that fluid flows away from or "diverges" from the point (x_0, y_0).

The ∇ Operator

The divergence of a vector field is often expressed using the "nabla" symbol

$$\nabla = \left\langle \frac{\partial}{\partial x}, \frac{\partial}{\partial y}, \frac{\partial}{\partial z}\right\rangle = \frac{\partial}{\partial x}\mathbf{i} + \frac{\partial}{\partial y}\mathbf{j} + \frac{\partial}{\partial z}\mathbf{k}.$$

The nabla symbol is often treated like a vector; it can be multiplied by scalars, and can be used in cross and dot products. Some care is needed, however, in interpreting operations involving the components of ∇. The most important operation is multiplication. When $\partial/\partial x$ is multiplied on the left by a number or a function, we interpret the result as a number or function multiplying a derivative. Hence

$$3\frac{\partial}{\partial x} \text{ is read "3 times the derivative with respect to } x \text{ of\ldots"}$$

and

$$(x^2 + 2xz)\frac{\partial}{\partial x} \text{ is "}(x^2 + 2xz)\text{ times the derivative with respect to } x \text{ of\ldots."}$$

When $\partial/\partial x$ is multiplied on the right by a number or function, we interpret the result as the derivative with respect to x of the number or function. Hence

$$\frac{\partial}{\partial x}(x^2yz - 3y + 2) \text{ is "the derivative with respect to } x \text{ of } (x^2yz - 3y + 2)\text{"} = 2xyz.$$

With these interpretations, the divergence of a vector field $\mathbf{F} = \langle P, Q, R\rangle$ can be expressed as the dot product of ∇ and $\mathbf{F}$,

$$\begin{aligned}\operatorname{div} \mathbf{F} &= \frac{\partial P}{\partial x} + \frac{\partial Q}{\partial y} + \frac{\partial R}{\partial z} = \frac{\partial}{\partial x}P + \frac{\partial}{\partial y}Q + \frac{\partial}{\partial z}R \\ &= \left\langle \frac{\partial}{\partial x}, \frac{\partial}{\partial y}, \frac{\partial}{\partial z}\right\rangle \cdot \langle P, Q, R\rangle = \nabla \cdot \mathbf{F}.\end{aligned}$$

This notation for the divergence can also be used for the divergence of planar fields if we think of such vector fields as having a third component of 0. That is, if we think of $\mathbf{F} = \langle M, N\rangle$ as $\langle M, N, 0\rangle$, then the notation

$$\operatorname{div}\mathbf{F} = \nabla\cdot\mathbf{F}$$

is still valid.

We saw the ∇ symbol when we defined the gradient in Chapter 9. The ∇ notation for the gradient is consistent with the idea of ∇ as a vector. Indeed, we obtain the gradient of f by multiplying ∇ on the right by the scalar function f,

$$\begin{aligned}\nabla f &= \left\langle \frac{\partial}{\partial x}, \frac{\partial}{\partial y}, \frac{\partial}{\partial z}\right\rangle f = \left\langle \frac{\partial}{\partial x}f, \frac{\partial}{\partial y}f, \frac{\partial}{\partial z}f\right\rangle \\ &= \left\langle \frac{\partial f}{\partial x}, \frac{\partial f}{\partial y}, \frac{\partial f}{\partial z}\right\rangle = \operatorname{grad} f.\end{aligned}$$

When we use the ∇ symbol, the notation ∇f for the gradient and $\nabla\cdot\mathbf{F}$ for the divergence look very much alike. However, it is important to keep in mind that these are very different operations. The gradient takes a scalar function as input and returns a vector field as output; the divergence takes a vector field as input and returns a scalar function as output.

EXAMPLE 1 Let

$$f(x, y, z) = 2x^2yz\tan(xz) \quad\text{and}\quad \mathbf{F}(x, y, z) = \langle x^2z - 2y, ze^{2y}, ye^{z-x}\rangle.$$

Decide which of the following operations are defined, and evaluate those that are defined:

$$\nabla f, \qquad \nabla\cdot f, \qquad \nabla\mathbf{F}, \qquad \nabla\cdot\mathbf{F}.$$

Solution

Because f is a scalar function, ∇f makes sense. We have

$$\begin{aligned}\nabla f &= \left\langle \frac{\partial}{\partial x}, \frac{\partial}{\partial y}, \frac{\partial}{\partial z}\right\rangle 2x^2yz\tan(xz) \\ &= \left\langle \frac{\partial}{\partial x}(2x^2yz\tan(xz)), \frac{\partial}{\partial y}(2x^2yz\tan(xz)), \frac{\partial}{\partial z}(2x^2yz\tan(xz))\right\rangle \\ &= \langle 4xyz\tan(xz) + 2x^2yz^2\sec^2(xz), 2x^2z\tan(xz), \\ &\qquad 2x^2y\tan(xz) + 2x^3yz\sec^2(xz)\rangle.\end{aligned}$$

Because f is a scalar function and the divergence takes a vector field as input, the expression $\nabla\cdot f = \operatorname{div} f$ is undefined.

The gradient operator acts on scalar functions. Because $\mathbf{F}$ is a vector field, $\nabla\mathbf{F} = \operatorname{grad}\mathbf{F}$ is undefined.

Because the divergence takes a vector field as input, $\nabla\cdot\mathbf{F} = \operatorname{div}\mathbf{F}$ is defined. We have

$$\begin{aligned}\nabla\cdot\mathbf{F} &= \left\langle \frac{\partial}{\partial x}, \frac{\partial}{\partial y}, \frac{\partial}{\partial z}\right\rangle\cdot\langle x^2z - 2y, ze^{2y}, ye^{z-x}\rangle \\ &= \frac{\partial}{\partial x}(x^2z - 2y) + \frac{\partial}{\partial y}ze^{2y} + \frac{\partial}{\partial z}ye^{z-x} = 2xz + 2ze^{2y} + ye^{z-x}.\end{aligned}$$

EXAMPLE 2 Let f be a function with domain $\mathcal{D} \subset R^3$ and assume that the second partial derivatives of f exist in $\mathcal{D}$. Evaluate div (grad f).

Solution

First find the gradient of f,

$$\operatorname{grad} f = \nabla f = \left\langle \frac{\partial f}{\partial x}, \frac{\partial f}{\partial y}, \frac{\partial f}{\partial z} \right\rangle.$$

We then have

$$\begin{aligned} \operatorname{div}(\operatorname{grad} f) &= \nabla \cdot (\nabla f) \\ &= \left\langle \frac{\partial}{\partial x}, \frac{\partial}{\partial y}, \frac{\partial}{\partial z} \right\rangle \cdot \left\langle \frac{\partial f}{\partial x}, \frac{\partial f}{\partial y}, \frac{\partial f}{\partial z} \right\rangle = \frac{\partial^2 f}{\partial x^2} + \frac{\partial^2 f}{\partial y^2} + \frac{\partial^2 f}{\partial z^2}. \end{aligned}$$

Because div (grad f) $= \nabla \cdot (\nabla f)$, this expression is often written as $\nabla^2 f$. The operator

$$\nabla^2 = \nabla \cdot \nabla = \frac{\partial^2}{\partial x^2} + \frac{\partial^2}{\partial y^2} + \frac{\partial^2}{\partial z^2}$$

is called the *Laplacian*. The Laplacian often shows up in discussions of fluid mechanics, wave phenomena, or diffusion.

The Curl

We have seen that some operations on vector fields can be described as the scalar product of ∇ with a scalar function and as the dot product of ∇ with a vector field. When we consider the cross product of ∇ with a vector field, we obtain the operation called curl.

DEFINITION The Curl

Let

$$\mathbf{F}(x, y, z) = \langle P(x, y, z), Q(x, y, z), R(x, y, z) \rangle$$

be a vector field defined on some subset $\mathcal{D}$ of R^3. The **curl** of $\mathbf{F}$ is defined as the vector field $\nabla \times \mathbf{F}$. That is,

WEB

$$\begin{aligned} \operatorname{curl} \mathbf{F} = \nabla \times \mathbf{F} &= \det \begin{pmatrix} \mathbf{i} & \mathbf{j} & \mathbf{k} \\ \frac{\partial}{\partial x} & \frac{\partial}{\partial y} & \frac{\partial}{\partial z} \\ P & Q & R \end{pmatrix} \\ &= \left(\frac{\partial R}{\partial y} - \frac{\partial Q}{\partial z} \right)\mathbf{i} + \left(\frac{\partial P}{\partial z} - \frac{\partial R}{\partial x} \right)\mathbf{j} + \left(\frac{\partial Q}{\partial x} - \frac{\partial P}{\partial y} \right)\mathbf{k} \\ &= \left\langle \frac{\partial R}{\partial y} - \frac{\partial Q}{\partial z}, \frac{\partial P}{\partial z} - \frac{\partial R}{\partial x}, \frac{\partial Q}{\partial x} - \frac{\partial P}{\partial y} \right\rangle. \end{aligned} \tag{3}$$

Sometimes we may wish to calculate the curl of a planar vector field. We do so by appending a third component of 0 to the vector field and then using (3) to calculate the curl. Thus, if $\mathbf{F} = \langle M(x,y), N(x,y)\rangle$, then

$$\begin{aligned}\operatorname{curl}\mathbf{F} &= \nabla \times \langle M(x,y), N(x,y), 0\rangle \\ &= \left\langle \frac{\partial}{\partial y}0 - \frac{\partial}{\partial z}N(x,y), \frac{\partial}{\partial z}M(x,y) - \frac{\partial}{\partial x}0, \frac{\partial}{\partial x}N(x,y) - \frac{\partial}{\partial y}M(x,y)\right\rangle \\ &= \left\langle 0, 0, \frac{\partial N}{\partial x} - \frac{\partial M}{\partial y}\right\rangle = \left(\frac{\partial N}{\partial x} - \frac{\partial M}{\partial y}\right)\mathbf{k}.\end{aligned} \tag{4}$$

One of the marks of good mathematical notation is that it suggests or reminds us of useful relationships. The ∇ symbol is a good example of such notation. Thinking of ∇ as a vector and applying known results about scalar, dot, and cross products leads to some useful identities.

EXAMPLE 3 Let

$$\mathbf{F}(x,y,z) = \langle -2xy + yz, -3x^3 + y^2, xyz\rangle.$$

Show that div (curl $\mathbf{F}$) = 0.

Solution

First observe that with the ∇ notation,

$$\operatorname{div}(\operatorname{curl}\mathbf{F}) = \nabla\cdot(\nabla\times\mathbf{F}). \tag{5}$$

Because we know that $\mathbf{a}\cdot(\mathbf{a}\times\mathbf{b}) = 0$ for all vector $\mathbf{a}$ and $\mathbf{b}$, we might conjecture that div (curl $\mathbf{F}$) = 0. However, because ∇ is not a vector in the usual sense we need to do the calculations to verify that the result is indeed 0. We first calculate the curl of $\mathbf{F}$,

$$\begin{aligned}\nabla\times\mathbf{F} &= \det\begin{pmatrix}\mathbf{i} & \mathbf{j} & \mathbf{k} \\ \frac{\partial}{\partial x} & \frac{\partial}{\partial y} & \frac{\partial}{\partial z} \\ -2xy + yz & -3x^3 + y^2 & xyz\end{pmatrix} \\ &= \left\langle \frac{\partial}{\partial y}xyz - \frac{\partial}{\partial z}(-3x^3 + y^2), \frac{\partial}{\partial z}(-2xy + yz) - \frac{\partial}{\partial x}xyz, \right. \\ &\qquad \left. \frac{\partial}{\partial x}(-3x^3 + y^2) - \frac{\partial}{\partial y}(-2xy + yz)\right\rangle \\ &= \langle xz, y - yz, -9x^2 + 2x - z\rangle.\end{aligned}$$

We then have

$$\begin{aligned}\nabla\cdot(\nabla\times\mathbf{F}) &= \left\langle \frac{\partial}{\partial x}, \frac{\partial}{\partial y}, \frac{\partial}{\partial z}\right\rangle \cdot \langle xz, y - yz, -9x^2 + 2x - z\rangle \\ &= \frac{\partial}{\partial x}xz + \frac{\partial}{\partial y}(y - yz) + \frac{\partial}{\partial z}(-9x^2 + 2x - z) \\ &= z + (1 - z) + (-1) = 0.\end{aligned} \tag{6}$$

In Exercise 28 we ask you to prove that div (curl $\mathbf{F}$) = 0 for all vector fields $\mathbf{F}$ whose component functions have continuous second partial derivatives.

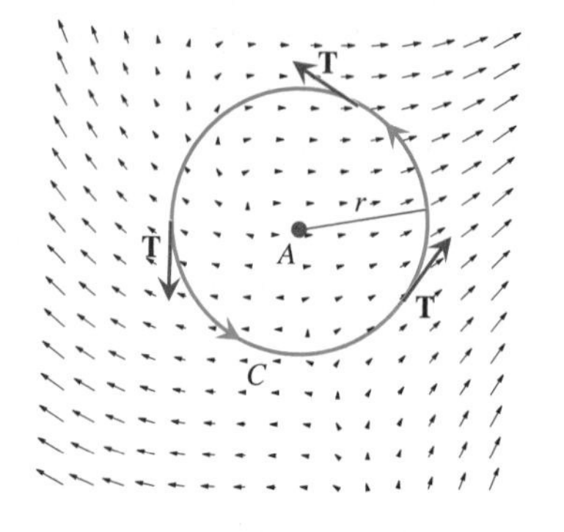

FIGURE 12.44 Calculating the circulation around the circle $C = C(A, r)$.

Circulation and the Curl In (1) we defined the average flux across a closed curve. Using a limit argument, we found that the divergence of the velocity field for a flow gives information about the flux across small curves surrounding a given point. Applying a similar argument to the circulation integral shows that the curl of the velocity field carries information about the circulation around small circles centered at a given point. See Fig. 12.44. As in the flux/divergence discussion, let $\mathbf{v} = \langle P, Q \rangle$ be the velocity field for a planar flow, and let $A = (x_0, y_0)$ be a point in the domain $\mathcal{D}$ of $\mathbf{v}$. Let $C(A, r)$ be the circle of radius r centered at A, and let $D(A, r)$ be the solid disk that has this circle as boundary. The average circulation around $C(A, r)$ is defined as

$$\text{average circulation} = \frac{\text{circulation around } C(A, r)}{\text{area } (D(A, r))}$$
$$= \frac{1}{\pi r^2} \int_{C(A,r)} (\mathbf{v} \cdot \mathbf{T})\, ds.$$

Using Green's Theorem we may express this average circulation as

$$= \frac{1}{\pi r^2} \iint_{D(A,r)} \left(\frac{\partial Q}{\partial x} - \frac{\partial P}{\partial y} \right) dA. \tag{7}$$

We define the circulation at A as the limit of the average circulation as $r \to 0^+$. In Exercise 32 we ask you to show that the result of the limiting process is

$$\text{circulation at } A = \frac{\partial Q}{\partial x}(x_0, y_0) - \frac{\partial P}{\partial y}(x_0, y_0). \tag{8}$$

If we consider the physical units in (7), we see that this quantity carries units of circulation per unit area. In addition, the expression on the right in (8) is the third component of

$$\text{curl } \mathbf{v} = \text{curl } \langle P, Q, 0 \rangle = \left\langle 0, 0, \frac{\partial Q}{\partial x} - \frac{\partial P}{\partial y} \right\rangle$$

evaluated at (x_0, y_0). Thus, for planar flows, the curl carries information about the "circulation per unit area at a point."

Green's Theorem Again

At the beginning of this section, we applied Green's Theorem to the flux and circulation integrals. We saw that if $\mathbf{v} = \langle P, Q \rangle$ is the velocity field for a planar flow and C is a simple closed curve such that C and the region $\mathcal{R}$ inside of C are both in the domain $\mathcal{D}$ of $\mathbf{v}$, then

$$\text{flux across } C = \int_C (\mathbf{v} \cdot \mathbf{N})\, ds = \iint_{\mathcal{R}} \left(\frac{\partial P}{\partial x} + \frac{\partial Q}{\partial y} \right) dA \tag{9}$$

and

$$\text{circulation around } C = \int_C (\mathbf{v} \cdot \mathbf{T})\, ds = \iint_{\mathcal{R}} \left(\frac{\partial Q}{\partial x} - \frac{\partial P}{\partial y} \right) dA. \tag{10}$$

The integrand in (9) is the divergence of $\mathbf{v}$. Making this substitution in (9), we obtain

$$\int_C (\mathbf{v} \cdot \mathbf{N})\, ds = \iint_{\mathcal{R}} (\operatorname{div} \mathbf{v})\, dA = \iint_{\mathcal{R}} (\nabla \cdot \mathbf{v})\, dA. \tag{11}$$

When Green's Theorem is written in this form, it is also called the Divergence Theorem for the plane.

The integrand in (10) is the third component of curl $\mathbf{v}$. We can obtain this third component as a dot product,

$$\frac{\partial Q}{\partial x} - \frac{\partial P}{\partial y} = \left\langle 0, 0, \frac{\partial Q}{\partial x} - \frac{\partial P}{\partial y} \right\rangle \cdot \langle 0, 0, 1 \rangle = (\operatorname{curl} \mathbf{v}) \cdot \mathbf{k}.$$

For reasons that will become clear in subsequent sections, we will let $\mathbf{N} = \mathbf{k}$ and think of $\mathbf{N}$ as the unit vector normal to the plane of the flow. With this convention, (10) can be written as

$$\int_C (\mathbf{v} \cdot \mathbf{T})\, ds = \iint_{\mathcal{R}} (\operatorname{curl} \mathbf{v}) \cdot \mathbf{N}\, dA = \iint_{\mathcal{R}} (\nabla \times \mathbf{v}) \cdot \mathbf{N}\, dA. \tag{12}$$

This form of Green's Theorem is called Stokes' Theorem for the plane.

In the remaining three sections of this chapter, we will study surface integrals. We will see that Green's Theorem can be naturally extended to three dimensions in two ways. These two ways are reflected in (11) and (12).

Exercises 12.5

Exercises 1–6: Calculate the divergence and curl of the vector field.

1. $\mathbf{F}(x, y, z) = \langle yz, zx, xy \rangle$
2. $\mathbf{G}(x, y, z) = \langle x \ln(yz), yze^x, \pi \rangle$
3. $\mathbf{v}(x, y) = \langle x^2 - y^2, 2xy \rangle$
4. $\mathbf{v}(x, y) = \langle x \tan(x^2 + y^2), y \cot(x^2 + y^2) \rangle$
5. $\mathbf{H}(x, y, z) = \operatorname{grad}((x + 2y + 3z)^2)$
6. $\mathbf{f}(x, y, z) = \operatorname{grad}(e^x \sin(e^{xy}))$

Exercises 7–16: Let f be a scalar function and $\mathbf{F}$ a vector field. Write each of the following expressions using the ∇ symbol and then state whether or not the expression is defined.

7. grad **F**
8. curl f
9. curl (div **F**)
10. div (curl **F**)
11. grad (div f)
12. curl (curl **F**)
13. grad (div **F**)
14. div (grad f)
15. div (curl (grad f))
16. curl (grad f)
17. Let **F** and **G** be vector fields with domain and range in R^3 and assume that each of the component functions of **F** and **G** is differentiable. Prove the identity

$$\operatorname{div}(\mathbf{F} + \mathbf{G}) = \operatorname{div} \mathbf{F} + \operatorname{div} \mathbf{G}.$$

When this identity is written with the ∇ symbol, what vector identity is suggested?

18. Let **F** and **G** be vector fields with domain and range in R^3 and assume that each of the component functions of **F** and **G** has partial derivatives. Prove the identity

$$\operatorname{curl}(\mathbf{F} + \mathbf{G}) = \operatorname{curl} \mathbf{F} + \operatorname{curl} \mathbf{G}.$$

When this identity is written with the ∇ symbol, what vector identity is suggested?

19. Let f be a scalar function with domain in R^3, let **F** be a vector field with domain and range in R^3, and assume that

both functions have first partial derivatives. The vector field $f\mathbf{F}$ is defined by

$$(f\mathbf{F})(x,y,z) = f(x,y,z)\mathbf{F}(x,y,z).$$

Show that

$$\text{div}\,(f\mathbf{F}) = (\text{grad}\, f)\cdot\mathbf{F} + f(\text{div}\,\mathbf{F}).$$

20. Let $\mathbf{F}$ and $\mathbf{G}$ be vector fields with domain and range in R^3 and assume that each of the component functions of $\mathbf{F}$ and $\mathbf{G}$ has first partial derivatives. The vector field $\mathbf{F}\times\mathbf{G}$ is defined by

$$(\mathbf{F}\times\mathbf{G})(x,y,z) = \mathbf{F}(x,y,z)\times\mathbf{G}(x,y,z).$$

Show that

$$\text{div}\,(\mathbf{F}\times\mathbf{G}) = \mathbf{G}\cdot\text{curl}\,\mathbf{F} - \mathbf{F}\cdot\text{curl}\,\mathbf{G}.$$

Exercises 21–26: Let $\mathbf{r} = \langle x,y,z\rangle$. *Verify the following identities.*

21. $\nabla\|\mathbf{r}\| = \dfrac{1}{\|\mathbf{r}\|}\mathbf{r}$

22. $\nabla\cdot\mathbf{r} = 3$

23. $\nabla\times\mathbf{r} = \mathbf{0}$

24. $\nabla\|\mathbf{r}\|^2 = 2\mathbf{r}$

25. $\nabla\dfrac{1}{\|\mathbf{r}\|} = -\dfrac{1}{\|\mathbf{r}\|^3}\mathbf{r}$

26. $\nabla\times\left(\dfrac{1}{\|\mathbf{r}\|}\mathbf{r}\right) = \mathbf{0}$

27. In this exercise we fill in some of the details in the argument to obtain (2). Assume that P and Q have continuous first partial derivatives in some domain $\mathcal{D}$ and that $A = (x_0, y_0)$ is a point in $\mathcal{D}$. Let $C(A,r)$ be the circle of center A and radius r, let $D(A,r)$ be the solid disk determined by this circle, and assume that both the disk and the circle are contained in $\mathcal{D}$.

a. For (x,y) close to (x_0,y_0), we can write

$$\frac{\partial P}{\partial x}(x,y) + \frac{\partial Q}{\partial y}(x,y) = \frac{\partial P}{\partial x}(x_0,y_0) + \frac{\partial Q}{\partial y}(x_0,y_0) + E(x,y).$$

Explain why the error $E(x,y)\longrightarrow 0$ as $(x,y)\longrightarrow(x_0,y_0)$.

b. Use this result to explain why

$$\frac{1}{\pi r^2}\iint_{D(A,r)} \frac{\partial P}{\partial x}(x,y) + \frac{\partial Q}{\partial y}(x,y)\,dA$$
$$= \frac{\partial P}{\partial x}(x_0,y_0) + \frac{\partial Q}{\partial y}(x_0,y_0) + \frac{1}{\pi r^2}\iint_{D(A,r)} E(x,y)\,dA.$$

c. Now tell why

$$\lim_{r\to 0^+}\frac{1}{\pi r^2}\iint_{D(A,r)} \frac{\partial P}{\partial x}(x,y) + \frac{\partial Q}{\partial y}(x,y)\,dA$$
$$= \frac{\partial P}{\partial x}(x_0,y_0) + \frac{\partial Q}{\partial y}(x_0,y_0)$$

and explain why this leads to (2).

28. Let $\mathbf{F}$, given by

$$\mathbf{F}(x,y,z) = \langle P(x,y,z), Q(x,y,z), R(x,y,z)\rangle,$$

be a vector field. Show that if the second partial derivatives of P, Q, and R are continuous, then

$$\text{div}\,(\text{curl}\,\mathbf{F}) = 0.$$

29. Let f and g be real-valued functions with domain R^3. Show that if f and g have continuous second partial derivatives, then

$$\text{div}\,(\text{grad}\, f\times\text{grad}\, g) = 0.$$

30. Let $\mathbf{F}$ and $\mathbf{G}$ be vector fields with domain and range in R^3, and assume that the components of $\mathbf{F}$ and $\mathbf{G}$ have continuous second partial derivatives. Show that

$$\text{curl}\,(\mathbf{F}\times\mathbf{G}) = \mathbf{F}\,\text{div}\,\mathbf{G} - \mathbf{G}\,\text{div}\,\mathbf{F} + (\mathbf{G}\cdot\nabla)\mathbf{F} - (\mathbf{F}\cdot\nabla)\mathbf{G}.$$

In the course of your work, state clearly what is meant by $\mathbf{F}\cdot\nabla$, and describe how this operator acts on the vector field $\mathbf{G}$.

31. Let $\mathbf{F}$, defined by

$$\mathbf{F}(x,y) = \langle M(x,y), N(x,y)\rangle,$$

be a vector field and assume that M and N have continuous first partial derivatives. Show that if $\mathbf{F}$ is conservative, then $\text{curl}\,\mathbf{F} = \mathbf{0}$. State and prove an analogous result for vector fields with domain and range in R^3.

32. Complete the argument that for a planar flow $\mathbf{v}$, the "circulation at a point" is $(\text{curl}\,\mathbf{v})\cdot\mathbf{k}$. That is, with the notation used in (7), show that

$$\lim_{r\to 0^+}\frac{\text{circulation around } C(A,r)}{\text{area}(D(A,r))} = \frac{\partial Q}{\partial x}(x_0,y_0) - \frac{\partial P}{\partial y}(x_0,y_0).$$

33. Let f and g be real-valued functions each with domain $\mathcal{D}$ in the plane, and assume that all second partial derivatives of f and g exist and are continuous in $\mathcal{D}$. Let C be a piecewise smooth, simple closed curve in $\mathcal{D}$, and assume that the region $\mathcal{R}$ inside of C is also contained in $\mathcal{D}$. Use the "divergence form" of Green's Theorem, i.e., (11), to prove the First Green's Identity,

$$\iint_{\mathcal{R}} f\nabla^2 g\,dA = \int_C f(\nabla g)\cdot\mathbf{N}\,ds - \iint_{\mathcal{R}} \nabla f\cdot\nabla g\,dA.$$

(Here $\nabla^2 = \dfrac{\partial^2}{\partial x^2} + \dfrac{\partial^2}{\partial y^2}$ is the Laplacian.)

34. Let f and g be real-valued functions each with domain $\mathcal{D}$ in the plane, and assume that all second partial derivatives of f and g exist and are continuous in $\mathcal{D}$. Let C be a piecewise smooth, simple closed curve in $\mathcal{D}$ and assume that the region $\mathcal{R}$ inside of C is also contained in $\mathcal{D}$. Use the "divergence form" of Green's Theorem, i.e., (11), to prove the Second Green's Identity,

$$\iint_{\mathcal{R}} (f\nabla^2 g - g\nabla^2 f)\, dA = \int_C (f\nabla g - g\nabla f)\cdot \mathbf{N}\, ds.$$

(Here $\nabla^2 = \dfrac{\partial^2}{\partial x^2} + \dfrac{\partial^2}{\partial y^2}$ is the Laplacian.)

35. A point in the plane $z = z_0$ moves at constant speed v on a circle of radius $\mathcal{R}$ with center $(0, 0, z_0)$. See the accompanying figure. Let $\mathbf{v} = \mathbf{v}(x, y, z_0)$ be the velocity of the particle when it is at position $\mathbf{r} = \langle x, y, z_0\rangle$. The angular speed of the particle is then $\omega = v/R$ and the angular velocity is defined to be $\boldsymbol{\omega} = (v/R)\mathbf{k}$.

a. Let γ be the angle between $\mathbf{r}$ and $\boldsymbol{\omega}$. Show that

$$v = \omega\|\mathbf{r}\| \sin\gamma,$$

where $v = \|\mathbf{v}\|$ and $\omega = \|\boldsymbol{\omega}\|$. Use this in showing that

$$\mathbf{v} = \boldsymbol{\omega} \times \mathbf{r}.$$

b. Show that

$$\text{curl}\ \mathbf{v} = 2\boldsymbol{\omega}.$$

You may find it helpful to use Exercises 22 and 30.

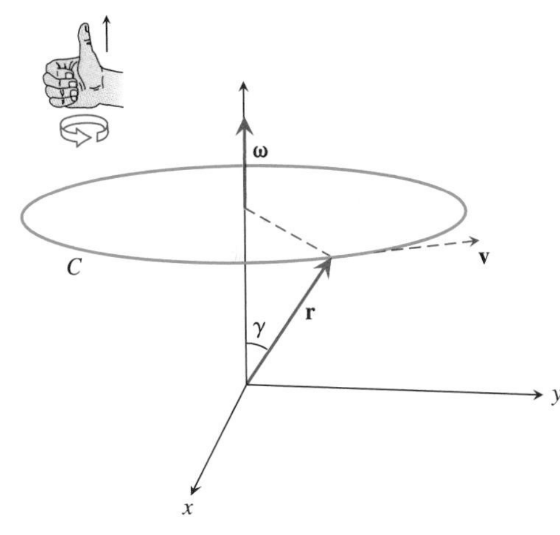

Figure for Exercise 35.

12.6 Surface Integrals

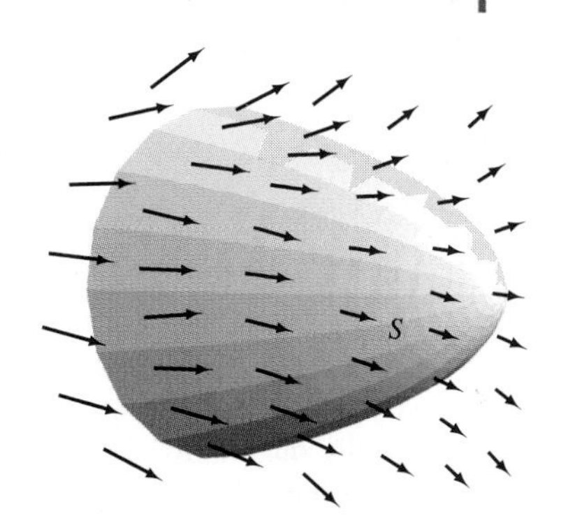

FIGURE 12.45 The velocity field for a fluid flow and a surface S.

Let $\mathbf{v}$ be the velocity field for a fluid flowing through a three-dimensional region, and let S be an imaginary surface in the region of the flow. See Fig. 12.45. How much mass of fluid flows through S in one second? The mass rate of flow, or flux, through S can be calculated in a way paralleling the calculation of flux across a curve. We start by cutting the surface S into small pieces, then estimating the mass rate of flow across each piece. The sum of these estimates gives us an estimate of the flux across S. To make this estimate exact, we observe that as the number of pieces grows and the pieces get small, the sum approaches a definite number. It is not surprising that the result of this process is an integral. But it is the integral of a function defined on a surface S, not on an interval, or a curve, or a region in the plane. This new integral is called a **surface integral.** In this section we define the surface integral and show that the value of such integrals can often be found by evaluating a related double integral. At the end of the section, we return to the problem of finding the flux through a surface.

Defining the Surface Integral

INVESTIGATION 1

The Mass of a Spherical Metal Shell

A thin, spherical metal shell of radius 1 is centered at the origin. At the point $\mathbf{P} = (x, y, z)$ on the sphere, the metal has density of

$$\delta(\mathbf{P}) = \delta(x, y, z)\ \text{g/cm}^2.$$

Find the mass of the metal shell.

Our strategy in solving this problem is, by now, a familiar one: We subdivide the sphere into small pieces, find an estimate for the mass of each piece, then sum these estimates. The result is an approximation for the mass of the sphere. As we take finer and finer subdivisions of the sphere, these approximations get closer and closer to a specific number. We identify this number as the mass of the sphere.

Let S be the sphere described by the equation

$$x^2 + y^2 + z^2 = 1,$$

so the metal sphere is coincident with S. We first subdivide S into small pieces. Although we could probably describe an adequate subdivision of S by using lines of latitude and longitude, our aim here is to develop a method of subdivision that will also work for other, less familiar surfaces. We base this method on parametric representations. In Chapter 9 we saw that S can be described parametrically by

$$\begin{aligned} \mathbf{r} = \mathbf{r}(u, v) &= \langle x(u, v), y(u, v), z(u, v)\rangle \\ &= \langle \cos u \sin v, \sin u \sin v, \cos v\rangle, \end{aligned} \tag{1}$$

where $0 \leq u \leq 2\pi$ and $0 \leq v \leq \pi$. Thus the domain for $\mathbf{r}$ is the rectangle

$$\mathcal{R} = [0, 2\pi] \times [0, \pi] = \{(u, v) : 0 \leq u \leq 2\pi, 0 \leq v \leq \pi\}$$

in the (u, v)-plane. We subdivide S in the same way that we subdivided a surface when defining *surface area* in Section 11.3. We first form a regular subdivision of $\mathcal{R}$ into n^2 rectangles $R_{i,j}$, $1 \leq i, j \leq n$. Each rectangle measures $\Delta u \times \Delta v$, where $\Delta u = 2\pi/n$ and $\Delta v = \pi/n$. When we apply the function $\mathbf{r}$, each rectangle $R_{i,j}$ maps to a small region $S_{i,j}$ on S. The result is a subdivision of S into n^2 pieces. We call this subdivision, arising from a regular subdivision of $\mathcal{R}$, a **regular subdivision of** S. See Fig. 12.46. For each pair i, j, $1 \leq i, j \leq n$, let $\mathbf{c}_{i,j}$ be a point in $S_{i,j}$. The set $\{\mathbf{c}_{i,j} : 1 \leq i, j \leq n\}$ is called an **evaluation set associated with the subdivision of** S.

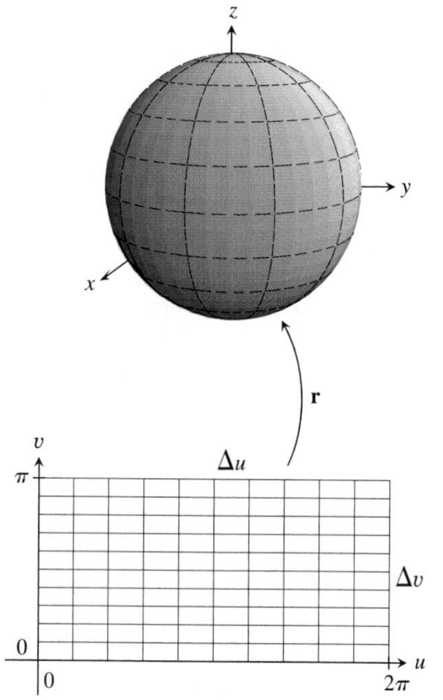

FIGURE 12.46 Mapping an $n \times n$ grid onto the sphere.

Assume now that δ is continuous on S. If piece $S_{i,j}$ is small and has area $\Delta\sigma_{i,j}$, then the mass $\Delta m_{i,j}$ of this piece can be approximated as

$$\Delta m_{i,j} \approx \delta(\mathbf{c}_{i,j})\Delta\sigma_{i,j}.$$

Summing these n^2 approximations gives an approximation of the mass of the sphere

$$\text{mass of metal sphere} = \sum_{i=1}^{n}\sum_{j=1}^{n} \Delta m_{i,j} \approx \sum_{i,j} \delta(\mathbf{c}_{i,j})\Delta\sigma_{i,j}. \tag{2}$$

It seems reasonable that as we take more and more rectangles in our subdivision of $\mathcal{R}$, the last sum in (2) gets closer and closer to M, the mass of the sphere. In other words, the approximating sums approach M as $n \to \infty$. This limit of the sum is called the *integral over S of δ* with respect to surface area. We use a double integral to denote the result of this limiting process,

$$M = \iint_S \delta(x, y, z)\, d\sigma. \tag{3}$$

We summarize this discussion in a definition, but first extend some of the ideas introduced to the case in which the domain of $\mathbf{r}$ is not a rectangle. Let S be a bounded surface in space and assume that S is represented parametrically by a function $\mathbf{r}$ de-

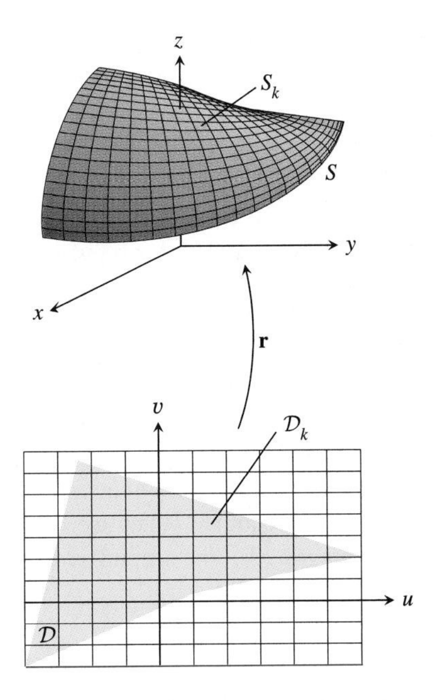

FIGURE 12.47 A rectangle $\mathcal{R}_{i,j}$ from the domain grid maps into a piece $S_{i,j}$ on S.

fined on a bounded set $\mathcal{D}$ in the (u, v)-plane. Then $\mathcal{D}$ is contained in a rectangle $\mathcal{R}$ with sides parallel to the u- and v-axes. For each positive integer n, there is a regular subdivision of $\mathcal{R}$ into n^2 rectangles. This subdivision of $\mathcal{R}$ divides $\mathcal{D}$ into subregions $\mathcal{D}_{11}, \ldots, \mathcal{D}_{pq}$ where $1 \le p, q \le n$. When we apply the function $\mathbf{r}$, the subregions $\mathcal{D}_{11}, \ldots, \mathcal{D}_{pq}$ generate a subdivision of the surface S into subregions $S_{11}, \ldots, S_{pq}$. We call this subdivision a **regular subdivision of** S and call n the **positive integer associated with the subdivision.** For each subregion $S_{i,j}$, let $\mathbf{c}_{i,j}$ be a point of $S_{i,j}$. The set $\{\mathbf{c}_{i,j}\}$ of these points is called an **evaluation set associated with the subdivision of** S. See Fig. 12.47.

DEFINITION The Integral with Respect to Surface Area

Let S be a bounded surface in space and let f be a bounded function defined on S. Assume that S is represented parametrically by a function $\mathbf{r}$ defined on a bounded set $\mathcal{D}$ in the (u, v)-plane. For each regular subdivision $\{S_{i,j}\}$ of S and each associated evaluation set $\{\mathbf{c}_{i,j}\}$, consider the Riemann sum

$$\sum_{i,j} f(\mathbf{c}_{i,j})\Delta\sigma_{i,j},$$

where $\Delta\sigma_{ij}$ is the area of subregion S_{ij}. If there is a number I such that all Riemann sums can be made as close to I as we like by taking n sufficiently large, then we say that f is **integrable with respect to surface area on** S. We write

$$I = \lim_{n\to\infty} \sum_{k=1}^{n} f(\mathbf{c}_{i,j})\Delta\sigma_{i,j} = \iint_S f\, d\sigma.$$

The number I is called the **integral with respect to surface area of** f **on** S.

INVESTIGATION 2

The Mass of a Spherical Metal Shell, continued

Consider the metal sphere discussed in the previous investigation with density function

$$\delta(x, y, z) = (2 - z)\ \mathrm{g/cm^2}.$$

Find and evaluate a double integral that gives the mass of the sphere.

When n is large, the rightmost sum in (2) gives an approximation of the mass. The approximation gets better and better as n grows. To evaluate this sum exactly, we need to know $\Delta\sigma_{i,j}$ and $\delta(\mathbf{c}_{i,j})$. Rather than find these numbers exactly, we use the estimate for $\Delta\sigma_{i,j}$ developed in Section 11.3. When we use this estimate in the rightmost sum in (2), we obtain an expression that we can handle using previously developed ideas about integration.

Let $\mathcal{R}_{i,j}$ be a rectangle in the regular partition of $\mathcal{R}$ and let $S_{i,j}$ be the corresponding piece of S. In Section 11.3 we noted that the area $\Delta\sigma_{i,j}$ of $S_{i,j}$ is close to the area of the parallelogram whose sides are given by the vectors

$$\mathbf{r}_u \Delta u = \left\langle \frac{\partial x}{\partial u}, \frac{\partial y}{\partial u}, \frac{\partial z}{\partial u} \right\rangle \Delta u \quad \text{and} \quad \mathbf{r}_v \Delta v = \left\langle \frac{\partial x}{\partial v}, \frac{\partial y}{\partial v}, \frac{\partial z}{\partial v} \right\rangle \Delta v,$$

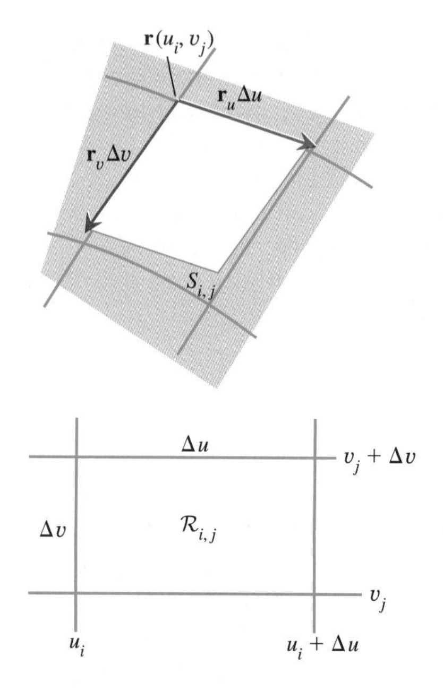

FIGURE 12.48 A piece of $S_{i,j}$ is approximated by a parallelogram with sides $\mathbf{r}_u \Delta u$ and $\mathbf{r}_v \Delta v$.

where the partial derivatives are all evaluated at some point (u_i, v_j) in $\mathcal{R}_{i,j}$. See Fig. 12.48. The area $d\sigma_{i,j}$ of this parallelogram is the length of the cross product of these vectors. Hence we have

$$\Delta\sigma_{i,j} \approx d\sigma_{i,j} = \text{area of parallelogram} = \|\mathbf{r}_u \times \mathbf{r}_v\| \Delta u \, \Delta v. \tag{4}$$

Because (u_i, v_j) is a point in $\mathcal{R}_{i,j}$, the point $\mathbf{c}_{i,j} = \mathbf{r}(u_i, v_j)$ is a point in $S_{i,j}$. Thus we can use $\{\mathbf{c}_{i,j} : i \leq i,j \leq n\}$ as an evaluation set associated with the subdivision. Using this evaluation set and the approximation (4) in (2), we have

$$\begin{aligned} \text{mass of sphere} &\approx \sum_{i,j} \delta(\mathbf{c}_{i,j}) \, \Delta\sigma_{i,j} \\ &\approx \sum_{i,j} \delta(\mathbf{r}(u_i, v_j)) \, \|\mathbf{r}_u(u_i, v_j) \times \mathbf{r}_v(u_i, v_j)\| \, \Delta u \, \Delta v. \end{aligned} \tag{5}$$

Because the functions and partial derivatives involved in (5) are continuous, the two sums can be made as close as we like by taking n large enough. When we let n grow, the first of the sums in (5) tends to the surface integral (3). The second sum is a Riemann sum associated with the double integral

$$\iint_{\mathcal{R}} \delta(\mathbf{r}(u, v)) \, \|\mathbf{r}_u(u, v) \times \mathbf{r}_v(u, v)\| \, dA. \tag{6}$$

See Section 11.1. Hence as n grows, the second sum in (5) approaches the integral in (6). Because both of the sums in (5) also tend to the mass of the sphere as n approaches infinity, it follows that the associated double and surface integrals must be equal. That is,

$$\text{mass} = \iint_S \delta(x, y, z) \, d\sigma = \iint_{\mathcal{D}} \delta(\mathbf{r}(u, v)) \, \|\mathbf{r}_u(u, v) \times \mathbf{r}_v(u, v)\| \, dA.$$

For the sphere in Investigation 1, $\mathcal{D} = [0, 2\pi] \times [0, \pi]$,

$$\delta(x, y, z) = 2 - z, \quad \text{and} \quad \mathbf{r}(u, v) = \langle \cos u \sin v, \sin u \sin v, \cos v \rangle.$$

Hence

$$\begin{aligned} \iint_{\mathcal{D}} \delta(\mathbf{r}(u, v)) \, \|\mathbf{r}_u \times \mathbf{r}_v\| \, dA &= \int_0^{\pi} \int_0^{2\pi} (2 - \cos v) \, \|\langle -\sin u \sin v, \cos u \sin v, 0 \rangle \\ &\qquad \times \langle \cos u \cos v, \sin u \cos v, -\sin v \rangle\| \, du \, dv \\ &= \int_0^{\pi} \int_0^{2\pi} (2 - \cos v) \, |\sin v| \, du \, dv \\ &= 2\pi \int_0^{\pi} (2 - \cos v) \sin v \, dv = 8\pi. \end{aligned}$$

Hence the mass of the sphere is 8π grams.

This discussion can also serve as an outline of a proof of the following result on evaluation of surface integrals.

Evaluation of $\iint_S f\,d\sigma$

Let S be a bounded surface in space and let f be a bounded and continuous real-valued function defined on S. Assume that S is represented parametrically by

$$\mathbf{r} = \mathbf{r}(u, v) = \langle x(u, v), y(u, v), z(u, v)\rangle$$

for (u, v) in a bounded set $\mathcal{D}$ in the (u, v)-plane. If $\mathbf{r}$ has continuous first partial derivatives, then

$$\iint_S f(x, y, z)\,d\sigma = \iint_{\mathcal{D}} f(\mathbf{r}(u, v))\,\|\mathbf{r}_u \times \mathbf{r}_v\|\,du\,dv. \tag{7}$$

We learned in Section 12.1 that when we evaluate a line integral $\int_c f ds$ by choosing a parametrization of C, we may choose—with certain restrictions—any convenient parametrication. The same is true for the surface integral $\iint_s f d\sigma$. We may choose—with certain natural restrictions—any convenient parametrication.

EXAMPLE 1 Let S be the portion of the paraboloid with equation $z = x^2 + y^2$ that lies below the plane $z = 9$, and let f be defined on S by $f(x, y, z) = xz + 1$. Evaluate

$$\iint_S f(x, y, z)\,d\sigma.$$

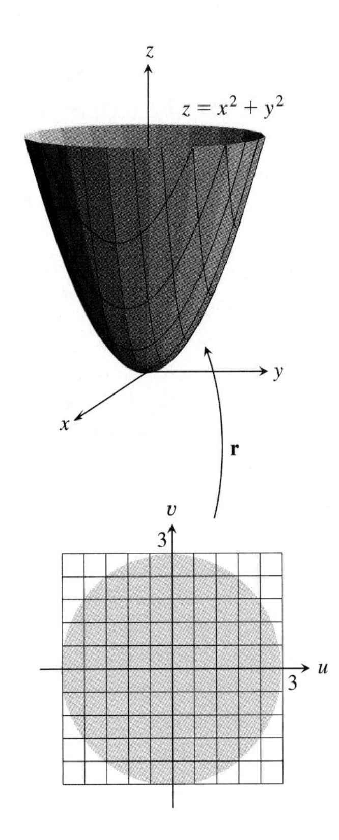

FIGURE 12.49 The surface S represented parametrically by a function $\mathbf{r}$ defined on the disk $\{(u, v) : u^2 + v^2 \le 9\}$.

Solution

The intersection of the plane $z = 9$ and the paraboloid described by $z = x^2 + y^2$ is the circle of center $(0, 0, 9)$ and radius 3 in the plane $z = 9$. Thus S is the portion of the paraboloid that lies on or above the planar disk

$$\mathcal{D} = \{(x, y) : x^2 + y^2 \le 9\}.$$

See Fig. 12.49. Because S is represented by $z = g(x, y) = x^2 + y^2$ for $(x, y) \in \mathcal{D}$, we can represent S parametrically by

$$\mathbf{r}(u, v) = \langle u, v, g(u, v)\rangle = \langle u, v, u^2 + v^2\rangle, \qquad u^2 + v^2 \le 9.$$

Hence by (7)

$$\begin{aligned}
\iint_S f(x, y, z)\,d\sigma &= \iint_{\mathcal{D}} f(\mathbf{r}(u, v))\,\|\mathbf{r}_u \times \mathbf{r}_v\|\,du\,dv \\
&= \iint_{\mathcal{D}} f(u, v, u^2 + v^2)\,\|(1, 0, 2u) \times (0, 1, 2v)\|\,du\,dv \\
&= \iint_{\mathcal{D}} (u(u^2 + v^2) + 1)\,\|\langle -2u, -2v, 1\rangle\|\,du\,dv \\
&= \iint_{\mathcal{D}} (u(u^2 + v^2) + 1)\sqrt{4u^2 + 4v^2 + 1}\,du\,dv.
\end{aligned} \tag{8}$$

Because D is a disk centered at the origin, we use polar coordinates to evaluate the last integral. With $u = r\cos\theta$ and $v = r\sin\theta$, the last integral in (8) becomes

$$\int_0^3 \int_0^{2\pi} \left[(r^3\cos\theta + 1)\sqrt{4r^2+1}\right] r\,d\theta\,dr$$

$$= 2\pi \int_0^3 r\sqrt{4r^2+1}\,dr = \frac{1}{6}\pi(4r^2+1)^{3/2}\Big|_0^3 = \frac{1}{6}\pi(37^{3/2} - 1) \approx 117.319.$$

Flux

Orientable Surfaces Let $\mathbf{v}$ be the vector field defined by

$$\mathbf{v}(x, y, z) = \langle P(x, y, z), Q(x, y, z), R(x, y, z)\rangle, \qquad (x, y, z) \in \mathcal{D}.$$

Suppose that $\mathbf{v}$ is the velocity field for a fluid flowing in $\mathcal{D}$ and that S is a surface in $\mathcal{D}$. The flux of the fluid through S gives a measure of the net rate of flow of fluid across S. When we found the flux across a planar curve C, we chose at each point of C a positive direction. This direction was defined in terms of a unit normal to the curve at the point of interest. We follow an analogous procedure when calculating the flux across a surface. At each point on S we choose a unit normal and use it to define a positive direction through S. Thus, we only work with surfaces (or unions of surfaces) for which we can define a unit normal at each point. We also require that we can define the unit normals so they vary continuously over S, with no abrupt changes to different sides of S. This leads us to the idea of **orientable surface.**

DEFINITION Orientable Surface

Let S be a surface in R^3. If, at each point (x, y, z) of S we can assign a unit normal $\mathbf{N} = \mathbf{N}(x, y, z)$ so that $\mathbf{N}$ is continuous on S, then we say that S is an **orientable surface** and that the function $\mathbf{N}$ defines an **orientation on** S.

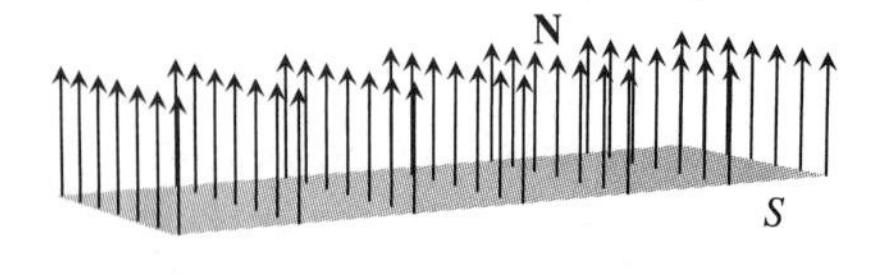

FIGURE 12.50 If S is an orientable surface, then the unit normal $\mathbf{N}$ can be defined so it varies continuously on S.

This definition is illustrated in Fig. 12.50. The arrows in the figure are unit normals to the surface. If a unit normal $\mathbf{N}(x, y, z)$ exists at each point of the surface shown, and if these normals all point to the same "side" of the surface, then it seems reasonable that the function $\mathbf{N}$ should be continuous on S. This just means that as we move around on the surface, the arrow or vector $\mathbf{N}$ should not make any abrupt changes in direction. In particular, the normal shouldn't suddenly point to the "other side" of the surface, as illustrated in Fig. 12.51.

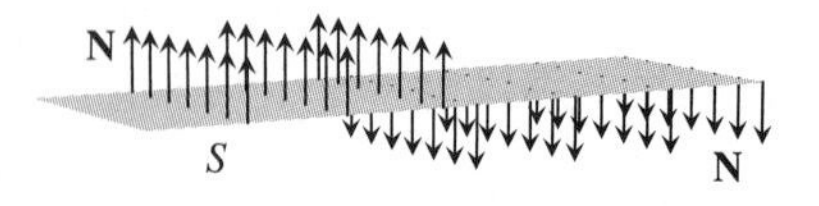

FIGURE 12.51 This choice for unit normals $\mathbf{N}$ is not a continuous function on S.

Most or all of the surfaces that we have already encountered are orientable surfaces. In fact, for many such surfaces an orientation can be defined by using a parametric representation for the surface. This fact is made more precise in the following statement.

WEB

Orientation and Parametric Representation

Let S be a surface in space and suppose that S is described by parametric representation

$$\mathbf{r} = \mathbf{r}(u, v) = \langle x(u, v), y(u, v), z(u, v)\rangle, \qquad (u, v) \in \mathcal{D}.$$

Suppose also that $\mathbf{r}$ has continuous first partial derivatives on $\mathcal{D}$ and that

$$\mathbf{r}_u(u, v) \times \mathbf{r}_v(u, v) \neq \mathbf{0}$$

for all $(u, v) \in \mathcal{D}$. For each point (x, y, z) of S, choose $(u, v) \in \mathcal{D}$ so that $\mathbf{r}(u, v) = \langle x, y, z\rangle$ and define

$$\mathbf{N}(x, y, z) = \frac{\mathbf{r}_u(u, v) \times \mathbf{r}_v(u, v)}{\|\mathbf{r}_u(u, v) \times \mathbf{r}_v(u, v)\|}. \tag{9}$$

Then $\mathbf{N}$ defines an orientation on S, that is, $\mathbf{N}$ is continuous on S.

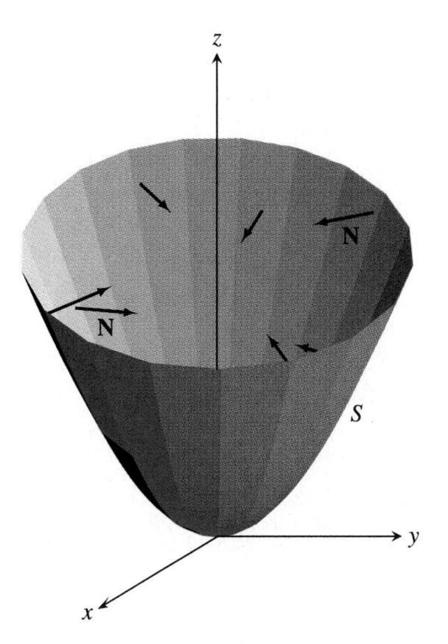

FIGURE 12.52 Some unit normals to the paraboloid with equation $z = x^2 + y^2$.

EXAMPLE 2 Let S be the surface described by

$$z = x^2 + y^2, \qquad 0 \leq x^2 + y^2 \leq 9.$$

Show that S is orientable by finding a continuous unit normal function on S.

Solution

As seen in Example 1, the surface S is described parametrically by

$$\mathbf{r}(u, v) = \langle u, v, u^2 + v^2\rangle, \qquad 0 \leq u^2 + v^2 \leq 9.$$

For any such (u, v), we have

$$\mathbf{r}_u(u, v) \times \mathbf{r}_v(u, v) = \langle 1, 0, 2u\rangle \times \langle 0, 1, 2v\rangle = \langle -2u, -2v, 1\rangle \neq \mathbf{0}.$$

Because $\mathbf{r}_u \times \mathbf{r}_v \neq \mathbf{0}$, we can use (9) to define an orientation on S. If (x, y, z) is a point on S, then $\langle x, y, z\rangle = \langle x, y, x^2 + y^2\rangle = \mathbf{r}(x, y)$. Hence

$$\mathbf{N}(x, y, z) = \left.\frac{\mathbf{r}_u(u, v) \times \mathbf{r}_v(u, v)}{\|\mathbf{r}_u(u, v) \times \mathbf{r}_v(u, v)\|}\right|_{(u,v)=(x,y)} = \frac{\langle -2x, -2y, 1\rangle}{\sqrt{4x^2 + 4y^2 + 1}}$$

defines a continuous unit normal function on S and shows that S is orientable. Note that at a point (x, y, z) on S, $\mathbf{N}(x, y, z)$ is the unit normal with positive third component. Some of these vectors are shown in Fig. 12.52. They are "inside" the paraboloid.

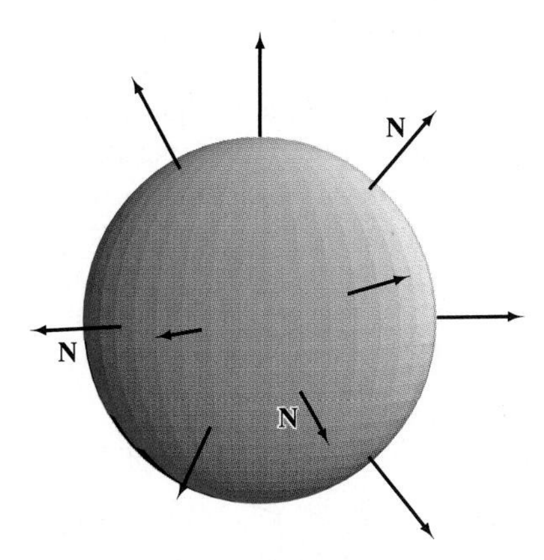

FIGURE 12.53 Outward unit normals on a sphere.

When S is a simple closed surface in space (that is, a surface that completely encloses a bounded region of space and does not intersect itself) we usually define our unit normals $\mathbf{N}(x, y, z)$ to point toward the "outside" of S. For example, if S is the sphere centered at $(0, 0, 0)$ and of radius a, then for (x, y, z) on S, we may define

$$\mathbf{N}(x, y, z) = \frac{1}{\|\langle x, y, z\rangle\|}\langle x, y, z\rangle = \frac{1}{a}\langle x, y, z\rangle.$$

See Fig. 12.53.

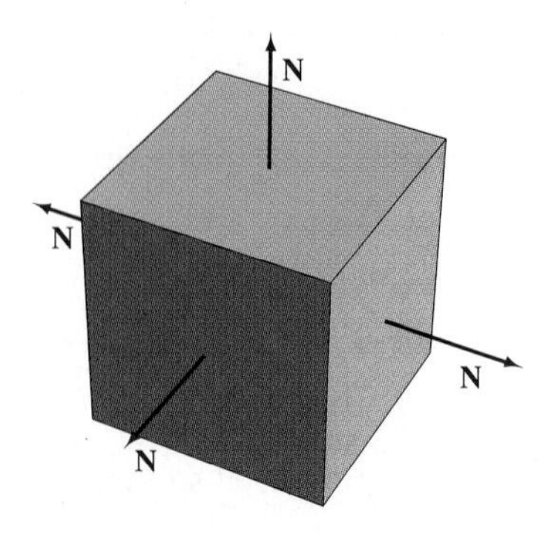

FIGURE 12.54 Some outward unit normals on the cube. There are no normals defined at the edges of the cube.

For some applications we will need to work with a surface S for which we cannot define a continuous unit normal function. If S can be written as a union of finitely many surfaces

$$S = S_1 + S_2 + \cdots + S_n$$

where each of $S_1, S_2, \ldots, S_n$ is orientable, we can define a continuous unit normal function on each of the pieces, but leave the normal function undefined on the boundaries where two pieces meet. For example, if S is a cube and $A = (x, y, z)$ is a corner or edge point of S, then there is no vector normal to S at A. However, we can find a unit normal at any nonedge point of S. Because S is a simple closed surface, we take all of these normals to point toward the outside of S. See Fig. 12.54. Flux calculations on such surfaces are not affected by edges or corners because these parts of the surface have area zero. It is not surprising that there are points on the cube where the unit normal is undefined; after all, the cube is not "smooth." For an example of a "smooth" surface without a continuous unit normal function, see Exercise 30.

Calculating Flux Once we have defined a positive direction on an orientable surface, we can talk about the net mass rate of flow of a fluid through a surface.

INVESTIGATION 3

The Flux for a Fluid Flow in Space

Let $\mathbf{v}$, given by

$$\mathbf{v}(x, y, z) = \langle P(x, y, z), Q(x, y, z), R(x, y, z)\rangle,$$

be a continuous velocity field for the flow of a fluid through a region of space. Assume that the density of the fluid at point (x, y, z) is $\delta(x, y, z)$ and that the function δ is continuous. Let S be an orientable surface in the flow, and assume that S is represented parametrically by

$$\mathbf{r} = \mathbf{r}(u, v), \qquad (u, v) \in \mathcal{D},$$

where $\mathcal{D}$ is a bounded set in R^2. Let $\mathbf{N} = \mathbf{N}(x, y, z)$ define a continuous unit normal on S. Find the flux through S. Assume distance is measured in centimeters, mass in grams, and time in seconds.

We start with a regular subdivision of $\mathcal{D}$. When we apply the mapping $\mathbf{r}$, we obtain a subdivision of S into small pieces $S_{i,j}$. We estimate the mass of fluid that passes through a typical representative piece $S_{i,j}$ in the short time interval Δt. See Fig. 12.55.

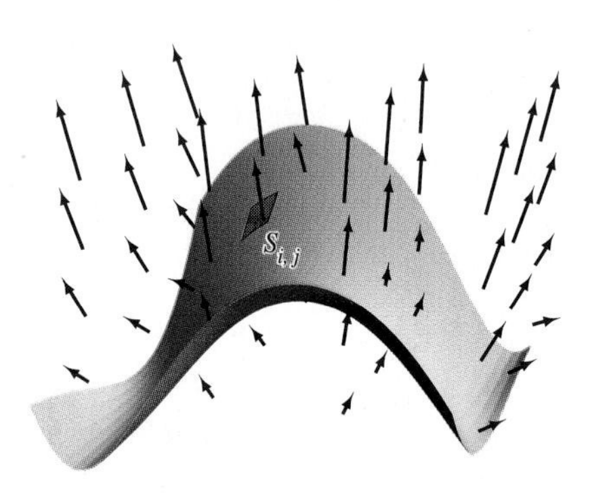

FIGURE 12.55 A flow through a surface and a small region $S_{i,j}$ on the surface. This region is magnified in the next two figures.

Denote the area of $S_{i,j}$ by $\Delta\sigma_{i,j}$, let $\mathbf{N}_{i,j}$ be the unit normal at a representative point on this piece, and let $\mathbf{v}_{i,j}$ be the fluid velocity and $\delta_{i,j}$ the fluid density at this point. Imagine the fluid that flows through $S_{i,j}$ in Δt seconds. We can visualize this fluid as a small patch of area $\Delta\sigma_{i,j}$ moving with approximate velocity $\mathbf{v}_{i,j}$. In Δt seconds this fluid patch travels through

$$(\mathbf{v}_{i,j}\ \text{cm/s})\,(\Delta t\ \text{s}) = \Delta t\,\mathbf{v}_{i,j}\ \text{cm}.$$

During this time the fluid sweeps out a "prism" with base $S_{i,j}$ and side $\Delta t\mathbf{v}_{i,j}$. The volume of this prism is the area $\Delta\sigma_{i,j}$ of its base times its height h. The height is the length of the projection of $\Delta t\mathbf{v}_{i,j}$ onto the normal $\mathbf{N}_{i,j}$. Thus

$$h = |(\Delta t\mathbf{v}_{i,j}) \cdot \mathbf{N}_{i,j}| = \Delta t|\mathbf{v}_{i,j} \cdot \mathbf{N}_{i,j}|\ \text{cm}.$$

See Figs. 12.56 and 12.57.

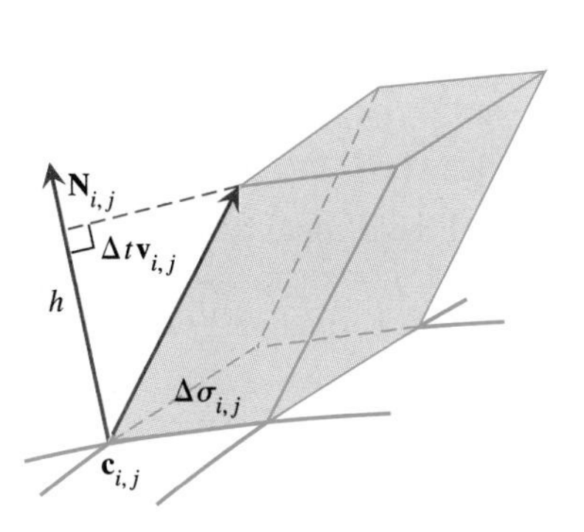

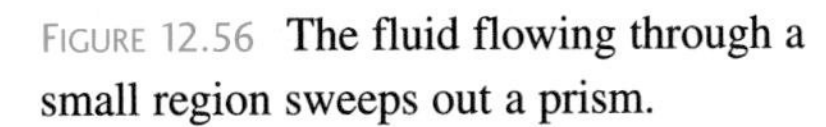

FIGURE 12.56 The fluid flowing through a small region sweeps out a prism.

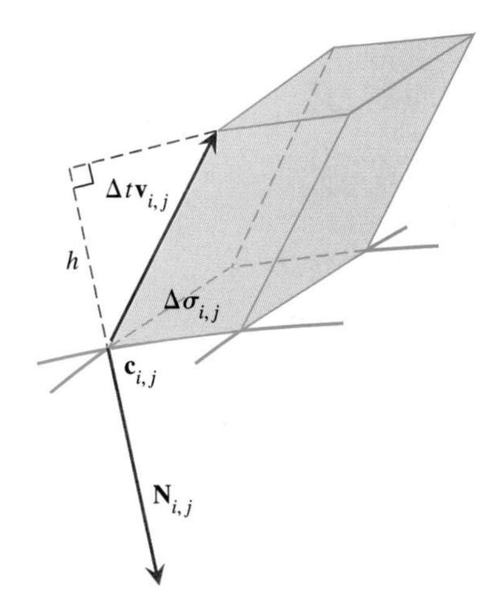

FIGURE 12.57 The fluid flowing through a small region sweeps out a prism-shaped region.

The mass of the fluid in this prism is

$$\text{mass} = (\text{volume})\,(\text{density}) = (h\Delta\sigma_{i,j})\delta_{i,j} = \Delta t\big(|\mathbf{v}_{i,j}\cdot\mathbf{N}_{i,j}|\big)\delta_{i,j}\Delta\sigma_{i,j}\text{ g}.$$

Dividing this quantity by the elapsed time Δt gives the approximate mass rate of flow of fluid through $S_{i,j}$,

$$|\mathbf{v}_{i,j}\cdot\mathbf{N}_{i,j}|\delta_{i,j}\Delta\sigma_{i,j}\text{ g/s}. \tag{10}$$

When we drop the absolute value signs from the dot product in (10), we obtain a "signed" rate of flow

$$\Delta\Phi_{i,j} = (\mathbf{v}_{i,j}\cdot\mathbf{N}_{i,j})\delta_{i,j}\Delta\sigma_{i,j}\text{ g/s}. \tag{11}$$

This expression is positive when $\mathbf{v}_{i,j}\cdot\mathbf{N}_{i,j} > 0$, that is, when the angle between $\mathbf{v}_{i,j}$ and $\mathbf{N}_{i,j}$ is less than 90°. When this is the case, $\mathbf{N}_{i,j}$ and $\mathbf{v}_{i,j}$ are on the same side of S and fluid flows through S towards the $\mathbf{N}_{i,j}$ side of S. When $\mathbf{v}_{i,j}$ and $\mathbf{N}_{i,j}$ are on opposite sides of S, fluid flows away from the $\mathbf{N}_{i,j}$ side of S. Note that in this case, (11) is negative. Hence the direction of flow (toward or away from the $\mathbf{N}_{i,j}$ side of $S_{i,j}$) is reflected in the sign of (11). This is illustrated in Figs. 12.56 and 12.57.

When the contributions (11) for each piece of the subdivision are summed, we obtain an estimate for the net mass rate of flow, or flux, through S,

$$\text{flux} \approx \sum \Delta\Phi_{i,j} = \sum (\mathbf{v}_{i,j}\cdot\mathbf{N}_{i,j})\delta_{i,j}\Delta\sigma_{i,j}.$$

As we let the number of pieces in the regular subdivision approach infinity, the last sum becomes a surface integral and we find

$$\text{flux through } S = \iint_S (\mathbf{v}\cdot\mathbf{N})\,\delta\,d\sigma. \tag{12}$$

The integral has units of g/s.

Integrals similar to the one in (12) arise in many physical problems. Even when **v** is not a velocity field, the value of the integral is referred to as a *flux*.

DEFINITION Flux

Let **F** be a continuous vector field with domain and range in R^3, let S be a bounded orientable surface in the domain of **F**, and let **N** define a continuous unit normal on S. The **flux Φ of F through** S is given by

$$\Phi = \iint_S (\mathbf{F} \cdot \mathbf{N})\, d\sigma. \tag{13}$$

If $\mathbf{F}(x, y, z) = \delta(x, y, z)\mathbf{v}(x, y, z)$, where **v** is a vector field describing the flow of a fluid in a region of space, δ is the density function for the fluid, and both **v** and δ are continuous, then (14) gives the net mass-per-unit time rate of flow of the fluid through S.

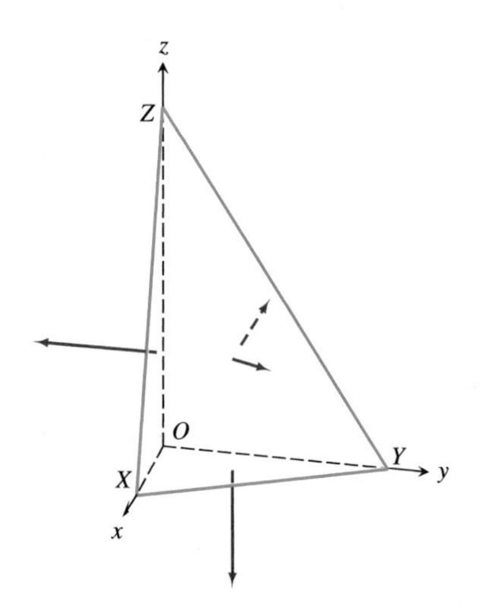

FIGURE 12.58 The tetrahedron S and its outward normals.

EXAMPLE 3 Let **v**, given by

$$\mathbf{v}(x, y, z) = \langle xy, yz, xz \rangle$$

be the velocity field for a fluid of density 1 g/cm^3 flowing in space. Let S be the right tetrahedron with vertices $(0,0,0)$, $(1,0,0)$, $(0,2,0)$, and $(0,0,3)$, as shown in Fig. 12.58. Find the flux Φ through S. Assume that lengths are measured in cm and velocities in cm/s.

Solution

Let $X = (1,0,0)$, $Y = (0,2,0)$, $Z = (0,0,3)$, $O = (0,0,0)$, and consider the four triangular faces $S_{xy} = OXY$, $S_{xz} = OXZ$, $S_{yz} = OYZ$, and $S_{xyz} = XYZ$. Because S is the union of these four triangular faces we write

$$S = S_{xy} + S_{xz} + S_{yz} + S_{xyz}.$$

Each of the four triangular surfaces is orientable, and because each is part of a plane, the normal vector is constant on each surface. If we define the unit normal **N** to be

$$\begin{array}{lll} \langle 0,0,-1 \rangle & \text{on} & S_{xy} \\ \langle 0,-1,0 \rangle & \text{on} & S_{xz} \\ \langle -1,0,0 \rangle & \text{on} & S_{yz} \\ \frac{1}{7}\langle 6,3,2 \rangle & \text{on} & S_{xyz}, \end{array}$$

then we have an orientation on S with outward normals. Because $S = S_{xy} + S_{xz} + S_{yz} + S_{xyz}$, and $\delta = 1$ we have

$$\begin{aligned}\Phi &= \iint_S \delta(\mathbf{v}\cdot\mathbf{N})\,dS \\ &= \iint_{S_{xy}} (\mathbf{v}\cdot\mathbf{N})\,dS + \iint_{S_{xz}} (\mathbf{v}\cdot\mathbf{N})\,dS + \iint_{S_{yz}} (\mathbf{v}\cdot\mathbf{N})\,dS + \iint_{S_{xyz}} (\mathbf{v}\cdot\mathbf{N})\,dS,\end{aligned} \tag{14}$$

Recall that the density function δ is identically 1. The surface S_{xy} is represented parametrically by

$$\mathbf{r}(u,v) = \langle u, v, 0\rangle, \qquad (u,v) \in \mathcal{D}_{xy} = \{(u,v): u, v \geq 0, 2u + v \leq 2\}.$$

Hence

$$\begin{aligned}\iint_{S_{xy}} (\mathbf{v}\cdot\mathbf{N})\,d\sigma &= \iint_{\mathcal{D}_{xy}} \langle \mathbf{v}(\mathbf{r}(u,v))\cdot\langle 0,0,-1\rangle\rangle \|\mathbf{r}_u \times \mathbf{r}_v\|\,dA \\ &= \iint_{\mathcal{D}_{xy}} (\langle uv, 0, 0\rangle \cdot \langle 0,0,-1\rangle)\|\langle 1,0,0\rangle \times \langle 0,1,0\rangle\|\,dA = 0.\end{aligned} \tag{15}$$

Similar calculations show that

$$\iint_{S_{xz}} (\mathbf{v}\cdot\mathbf{N})\,d\sigma = 0 = \iint_{S_{yz}} (\mathbf{v}\cdot\mathbf{N})\,d\sigma. \tag{16}$$

The surface S_{xyz} is represented parametrically by

$$\mathbf{r}(u,v) = \left\langle u, v, 3 - 3u - \frac{3}{2}v\right\rangle,$$
$$(u,v) \in \mathcal{D}_{xyz} = \{(u,v): u, v \geq 0, 2u + v \leq 2\}.$$

Hence,

$$\begin{aligned}\iint_{S_{xyz}} (\mathbf{v}\cdot\mathbf{N})\,d\sigma &= \iint_{\mathcal{D}_{xyz}} \left(\mathbf{v}(\mathbf{r}(u,v))\cdot\frac{1}{7}\langle 6,3,2\rangle\right)\|\mathbf{r}_u \times \mathbf{r}_v\|\,dA \\ &= \frac{1}{7}\iint_{\mathcal{D}_{xyz}} \left(\left\langle uv, 3v - 3uv - \frac{3}{2}v^2, 3u - 3u^2 - \frac{3}{2}uv\right\rangle \cdot \langle 6,3,2\rangle\right) \\ &\quad \times \|\langle 1, 0, -3\rangle \times \langle 0, 1, -3/2\rangle\|\,dA \\ &= \frac{1}{2}\int_0^1 \int_0^{2-2u} \left(-6u^2 - 6uv - \frac{9}{2}v^2 + 6u + 9v\right)dv\,du = \frac{3}{2}.\end{aligned} \tag{17}$$

Combining (14), (15), (16), and (17), we find that $\Phi = \frac{3}{2}$ g/s. Because the flux is positive, more fluid is flowing out of S than is flowing into S.

Notation for Flux Integrals

As defined in (13), we will usually write the flux integral as

$$\iint_S (\mathbf{F}\cdot\mathbf{N})\,d\sigma. \tag{18}$$

However, because this notation is not universal, we briefly look at another common notation.

This second notation is suggested by looking at a representation for **N**. Suppose that S is represented parametrically by

$$\mathbf{r} = \mathbf{r}(u,v) = \langle x(u,v), y(u,v), z(u,v)\rangle, \qquad (u,v)\in\mathcal{D},$$

and that

$$\mathbf{N}(x,y,z) = \mathbf{N}(\mathbf{r}(u,v)) = \frac{\mathbf{r}_u(u,v)\times\mathbf{r}_v(u,v)}{\|\mathbf{r}_u(u,v)\times\mathbf{r}_v(u,v)\|}$$

defines an orientation on S. We then have

$$\begin{aligned}\iint_S (\mathbf{F}\cdot\mathbf{N})\,d\sigma &= \iint_S \left(\mathbf{F}(\mathbf{r})\cdot\frac{\mathbf{r}_u\times\mathbf{r}_v}{\|\mathbf{r}_u\times\mathbf{r}_v\|}\right)\|\mathbf{r}_u\times\mathbf{r}_v\|\,dA\\ &= \iint_{\mathcal{D}} (\mathbf{F}(\mathbf{r})\cdot(\mathbf{r}_u\times\mathbf{r}_v))\,dA.\end{aligned} \tag{19}$$

Now

$$\begin{aligned}\mathbf{r}_u\times\mathbf{r}_v &= \left\langle\frac{\partial x}{\partial u},\frac{\partial y}{\partial u},\frac{\partial z}{\partial u}\right\rangle\times\left\langle\frac{\partial x}{\partial v},\frac{\partial y}{\partial v},\frac{\partial z}{\partial v}\right\rangle\\ &= \left\langle\frac{\partial y}{\partial u}\frac{\partial z}{\partial v}-\frac{\partial z}{\partial u}\frac{\partial y}{\partial v},\frac{\partial z}{\partial u}\frac{\partial x}{\partial v}-\frac{\partial x}{\partial u}\frac{\partial z}{\partial v},\frac{\partial x}{\partial u}\frac{\partial y}{\partial v}-\frac{\partial y}{\partial u}\frac{\partial x}{\partial v}\right\rangle\\ &= \left\langle\frac{\partial(y,z)}{\partial(u,v)},\frac{\partial(z,x)}{\partial(u,v)},\frac{\partial(x,y)}{\partial(u,v)}\right\rangle.\end{aligned} \tag{20}$$

Thus, the coordinates of $\mathbf{r}_u\times\mathbf{r}_v$ can be written in terms of Jacobians. If $\mathbf{F} = \langle P,Q,R\rangle$, then combining (20) with (19) leads to

$$\iint_S (\mathbf{F}\cdot\mathbf{N})\,d\sigma = \iint_{\mathcal{D}}\left(P\frac{\partial(y,z)}{\partial(u,v)}+Q\frac{\partial(z,x)}{\partial(u,v)}+R\frac{\partial(x,y)}{\partial(u,v)}\right)du\,dv. \tag{21}$$

This suggests the notations

$$\frac{\partial(y,z)}{\partial(u,v)}du\,dv = dy\,dz,\qquad \frac{\partial(z,x)}{\partial(u,v)}du\,dv = dz\,dx,\quad\text{and}\quad \frac{\partial(x,y)}{\partial(u,v)}du\,dv = dx\,dy.$$

When this notation is introduced into (21), we arrive at another notation for the flux integral in (18),

$$\iint_S (\mathbf{F}\cdot\mathbf{N})\,d\sigma = \iint_S P\,dy\,dz + Q\,dx\,dz + R\,dx\,dy.$$

Exercises 12.6

Exercises 1–6: Evaluate $\iint_S f(x,y,z)\,d\sigma$ for the given S and f.

1. $f(x,y,z) = z^2$, S the sphere with center $(0,0,0)$ and radius 2
2. $f(x,y,z) = x + y + z$, S the first-octant portion of the plane with equation $2x + 3y + 6z = 12$
3. $f(x,y,z) = \sqrt{x^2 + z^2 + 4}$, S described by $y = \ln(x^2 + z^2)$, $1 \le x^2 + z^2 \le 4$
4. $f(x,y,z) = 8$, S a cube of side a with one vertex at the origin and one at (a,a,a)
5. $f(x,y,z) = x + z$, S described by $\mathbf{r}(u,v) = \langle u - v, u + v, u - 2v + 2\rangle$, $0 \le u \le 2$, $1 \le v \le 4$
6. $f(x,y,z) = x/\sqrt{x^2 + y^2}$, S the first-octant portion of the surface given by $z = 1 - x^2 - y^2$

Exercises 7–10: A piece of thin metal is in the shape of the surface S. At point (x,y,z) on the surface, the material has density $\delta(x,y,z)$. Find the mass of the metal. Assume that length is measured in centimeters, mass in grams, and density in g/cm^2.

7. S the cone with cylindrical equation $z = 2r$, $0 \le r \le 2$, $0 \le \theta \le 2\pi$, $\delta(x,y,z) = xy + 1$
8. S the portion of the cylinder $x^2 + y^2 = 9$ between the planes $z = 0$ and $z = 3$, $\delta(x,y,z) = xy^2 + z + 9$
9. S the portion of the cylinder $x^2 + y^2 = 9$ between the planes $z = 0$ and $z = 3$ along with the disks making up the top and bottom of the cylinder, $\delta(x,y,z) = x^2 + 2y^2 + z^2$
10. S the tetrahedron with vertices $(0,0,0)$, $(6,0,0)$, $(0,3,0)$, and $(0,0,2)$, $\delta(x,y,z) = e^{x+2y+3z}$

Exercises 11–18: Evaluate $\iint_S (\mathbf{F}\cdot\mathbf{N})\,d\sigma$ for the given S and **F**. *Whenever S is a closed surface, take* **N** *to be the outward unit normal.*

11. $\mathbf{F}(x,y,z) = \langle x,y,z\rangle$, S the first-octant portion of the plane with equation $3x + 4y + z = 4$. (Choose the **N** with positive third component.)
12. $\mathbf{F}(x,y,z) = \langle x,z,y\rangle$, S the sphere with equation $x^2 + y^2 + z^2 = 1$
13. $\mathbf{F}(x,y,z) = \langle x,y^2,-z\rangle$, S the portion of the sphere $x^2 + y^2 + z^2 = 4$ with $y \ge 0$. (Choose the **N** with nonnegative second component.)
14. $\mathbf{F}(x,y,z) = \langle 0,-z^2,y\rangle$, S the portion of the graph of $z = \sqrt{x^2 + y^2}$ with $1 \le z \le 4$. (Choose the **N** with positive third component.)
15. $\mathbf{F}(x,y,z) = \langle -z,2,x\rangle$, S the portion of the cylinder $x^2 + y^2 = 9$ between the planes $z = 0$ and $z = 3$. (Choose the **N** pointing towards the outside of the cylinder.)
16. $\mathbf{F}(x,y,z) = \langle yz,zx,xy\rangle$, S the tetrahedron with vertices $(0,0,0)$, $(1,0,0)$, $(0,1,0)$, and $(0,0,3)$
17. $\mathbf{F}(x,y,z) = \langle x,y,z^2\rangle$, S described by $\mathbf{r}(u,v) = \langle u\sin v, u\cos v, u\rangle$, $0 \le u \le 1$, $0 \le v \le 2\pi$. (Choose the **N** with nonnegative third component.)
18. $\mathbf{F}(x,y,z) = \langle a,b,c\rangle$, S the tetrahedron with vertices $(0,0,0)$, $(a,0,0)$, $(0,b,0)$, and $(0,0,c)$. Assume that a, b, c are positive constants.
19. Let S be a surface in space. What can be said about
$$\iint_S 1\,d\sigma?$$
20. Let S be the surface given by
$$z = x^2e^{x+2y}.$$
Find an orientation **N** on S for which the third component of N is always positive.
21. Let S be the surface given by
$$z = x\arctan y.$$
Find an orientation **N** on S for which the third component of **N** is always negative.
22. Let S be the ellipsoid with equation
$$\frac{x^2}{4} + y^2 + \frac{z^2}{16} = 1.$$
Find the "outward normal" orientation **N** on S. Sketch the ellipsoid and a few of the normals.
23. Let S be the surface with equation
$$x^2 - y^2 + z^2 = 1.$$
Find an orientation **N** on S. Sketch S and a few of the normals on S.
24. Let S be a surface in R^3 and suppose that **N** gives an orientation on S.
    a. Is it true that $\tilde{\mathbf{N}} = -\mathbf{N}$ also gives an orientation on S?
    b. Let **F** be a continuous vector field defined on S. How are the values of
    $$\iint_S (\mathbf{F}\cdot\mathbf{N})\,d\sigma \quad \text{and} \quad \iint_S (\mathbf{F}\cdot\tilde{\mathbf{N}})\,d\sigma$$
    related?
25. Let D be a solid object in space and assume that D is bounded and has no holes. (So D could be a solid ball, a cone, a rectangular block, etc.) Let S be the surface of D and let **N** be the outward normal on S. Suppose D is cut into two objects, say D_1 and D_2 with surfaces S_1 and S_2, as shown in the accompanying figure. (In the figure we have separated D_1 and D_2 slightly for a clearer view, but in reality these two objects abut along the surface of the cut.) Let $\mathbf{N}_1$ be the outward unit normal for S_1 and $\mathbf{N}_2$ the outward unit normal to S_2. Let **F** be a continuous vector field defined on a set

containing D and S. Show that

$$\iint_{S_1} (\mathbf{F} \cdot \mathbf{N}_1)\, d\sigma + \iint_{S_2} (\mathbf{F} \cdot \mathbf{N}_2)\, d\sigma = \iint_{S} (\mathbf{F} \cdot \mathbf{N})\, d\sigma.$$

See Exercise 24.

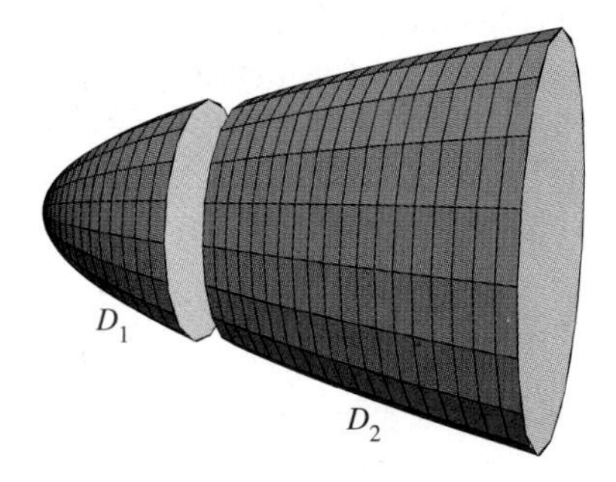

Figure for Exercise 25.

26. Let S be a surface and assume that at the point (x, y, z), the density of the surface is $\delta(x, y, z)$. Assume that δ is a continuous function. Show that the center of gravity of the surface is at the point (x_c, y_c, z_c) where

$$x_c = \frac{1}{M} \iint_S x\delta(x, y, z)\, d\sigma,$$

$$y_c = \frac{1}{M} \iint_S y\delta(x, y, z)\, d\sigma,$$

$$z_c = \frac{1}{M} \iint_S z\delta(x, y, z)\, d\sigma,$$

and M is the mass of S. *Hint:* Subdivide S into small pieces. Find the mass dm of a typical piece, then calculate the weighted average of the positions of these pieces. (See also Section 6.7 and Exercise 21 in Section 12.1.)

27. Let S be the upper half of the sphere of center $(0, 0, 0)$ and radius 1. Assume that the surface is made of a homogeneous material, so the density function is constant. Find the center of mass of S.

28. A tetrahedron shell with vertices $(0, 0, 0)$, $(a, 0, 0)$, $(0, b, 0)$, and $(0, 0, c)$ is made of a material of constant density. Find the center of mass of the tetrahedron.

29. The surface S consists of the portion of the cylinder $y^2 + z^2 = 4$ between the planes $x = 0$ and $x = 4$, and the disks determined by the ends of the cylinder. At point (x, y, z), the density of S is $(2x + 1)$ g/cm^2. Find the center of mass of S.

30. A nonorientable surface The Möbius strip is an example of a "smooth" but nonorientable surface. To construct a Möbius strip, take a long, thin strip of paper, lay it flat on the table, then give one end a half-twist. See the accompanying figure. Bring the two ends together and fasten them with glue or tape. To see that this object models a nonorientable surface,

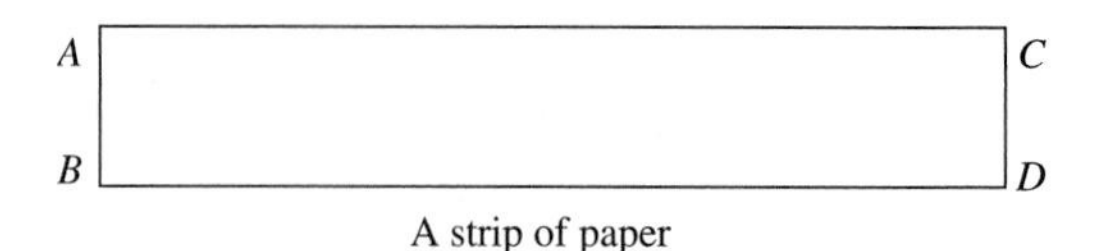

A strip of paper

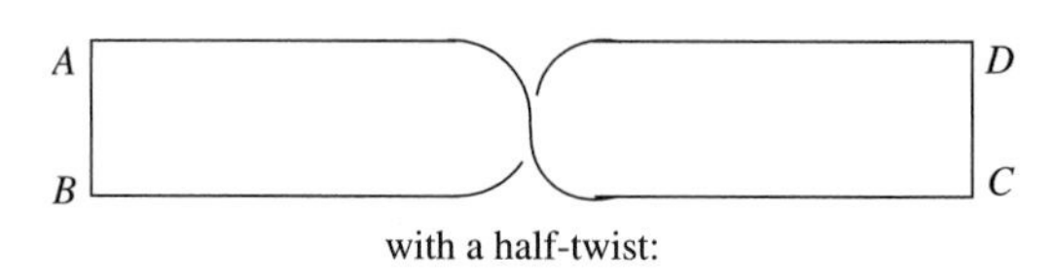

with a half-twist:

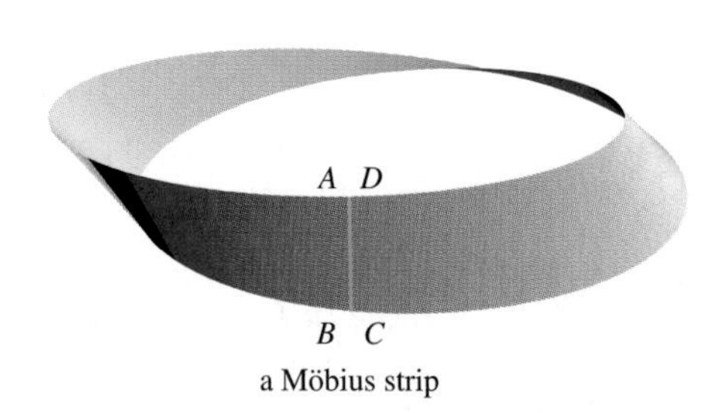

a Möbius strip

Figure for Exercise 30.

mark a point P midway between the edges of the strip. At point P place an arrow (or vector) perpendicular to the surface. Now move this arrow along the centerline of the surface, keeping it perpendicular to the surface at all times. Eventually you are back at the point P but on the "other side" of the strip. (Remember, a surface is just one point thick!) Compare the position of the arrow at the start of the procedure with its present position. Why does this experiment show that the Möbius strip is nonorientable?

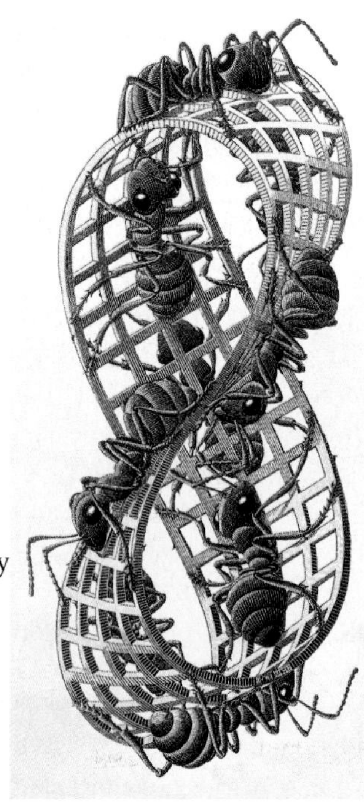

31. More fun with a Möbius strip Construct a Möbius strip as described in the previous exercise. Cut the Möbius strip along the centerline midway between the two edges. What happens? Now make another Möbius strip. Cut the strip lengthwise along a line one-third of the way between the two edges. What happens this time?

12.7 The Divergence Theorem

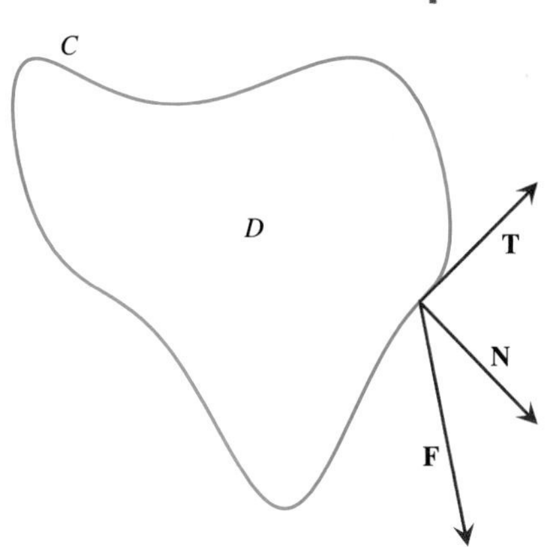

FIGURE 12.59 The closed curve C encloses the region $\mathcal{D}$.

Let C be a simple closed curve, let $\mathcal{D}$ be the region inside of C, and let $\mathbf{F}$ be a continuously differentiable vector field defined on $C \cup \mathcal{D}$ and with range in R^2. See Fig. 12.59. According to Green's Theorem,

$$\int_C (\mathbf{F} \cdot \mathbf{N})\, ds = \iint_{\mathcal{D}} \operatorname{div} \mathbf{F}\, dA \tag{1}$$

and

$$\int_C (\mathbf{F} \cdot \mathbf{T})\, ds = \iint_{\mathcal{D}} (\operatorname{curl} \mathbf{F}) \cdot \mathbf{k}\, dA, \tag{2}$$

where $\mathbf{k} = \langle 0, 0, 1 \rangle$ is perpendicular to the plane of C. In form (1), Green's Theorem gives insight into the flux across C. When expressed as (2), it gives information about the circulation around C. It is natural to ask: Are there corresponding results for fluid flow and other phenomena in three dimensions? The answer is yes, both (1) and (2) generalize to three dimensions. The three-dimensional version of (2) is Stokes' Theorem which, like Green's Theorem, relates a line integral along a closed curve to a surface integral over a region bounded by the curve. We will study Stokes' Theorem in Section 12.8. In the present section we study the three-dimensional version of (1). This result, called the Divergence Theorem, relates a flux integral over a closed surface to a triple integral over the region inside of the surface.

The Divergence Theorem In Section 12.4 we discussed Green's Theorem first for planar regions without holes, then went on to discuss a version of the theorem for regions with holes. Versions of the Divergence Theorem can also be stated and proved for many kinds of regions in space. However, we will limit our discussion to regions that are simultaneously of Types I, II, and III, as defined in Section 11.5. We refer to such regions as **simple regions.** The region bounded by a sphere is an example of a simple region.

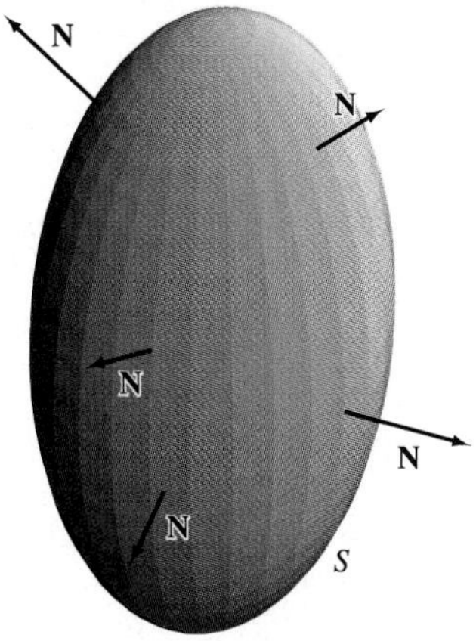

FIGURE 12.60 A simple region $\mathcal{B}$ bounded by the surface S.

THEOREM The Divergence Theorem

Let $\mathcal{B}$ be a simple region in space and assume that S, the boundary surface of $\mathcal{B}$, is orientable. See Fig. 12.60. Let $\mathbf{F}$, given by

$$\mathbf{F}(x, y, z) = \langle P(x, y, z), Q(x, y, z), R(x, y, z) \rangle, \tag{3}$$

be a vector field defined on a region containing $\mathcal{B}$ and S and assume that P, Q, and R have continuous partial derivatives on this domain. Then

$$\iint_S (\mathbf{F} \cdot \mathbf{N})\, d\sigma = \iiint_{\mathcal{B}} \operatorname{div} \mathbf{F}\, dV. \tag{4}$$

Continuing an earlier convention, $\mathbf{N}$ denotes the outward unit normal on S.

As a first step in proving the Divergence Theorem, note that (3) can be rewritten as

$$\mathbf{F} = \langle P, 0, 0 \rangle + \langle 0, Q, 0 \rangle + \langle 0, 0, R \rangle.$$

Incorporating this into (4), we obtain

$$\iint_S (\langle P,0,0\rangle \cdot \mathbf{N} + \langle 0,Q,0\rangle \cdot \mathbf{N} + \langle 0,0,R\rangle \cdot \mathbf{N})\, d\sigma = \iiint_{\mathcal{B}} \left(\frac{\partial P}{\partial x} + \frac{\partial Q}{\partial y} + \frac{\partial R}{\partial z}\right) dV.$$

The Divergence Theorem will be proved if we show that

$$\begin{aligned} \iint_S \langle P,0,0\rangle \cdot \mathbf{N}\, d\sigma &= \iiint_{\mathcal{B}} \frac{\partial P}{\partial x} dV \\ \iint_S \langle 0,Q,0\rangle \cdot \mathbf{N}\, d\sigma &= \iiint_{\mathcal{B}} \frac{\partial Q}{\partial y} dV \\ \iint_S \langle 0,0,R\rangle \cdot \mathbf{N}\, d\sigma &= \iiint_{\mathcal{B}} \frac{\partial R}{\partial z} dV. \end{aligned} \tag{5}$$

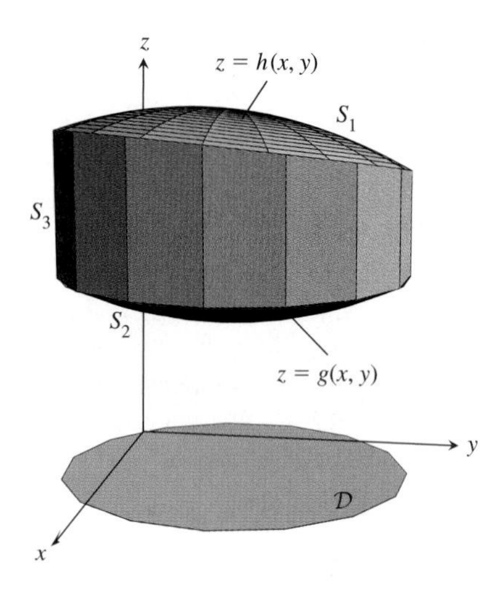

FIGURE 12.61 AU: Add Figure caption

We show that the last of these identities is true for regions $\mathcal{B}$ of Type I. A similar argument can be used to show that the first and second equations in (5) are true for regions $\mathcal{B}$ of Type III and Type II, respectively.

If $\mathcal{B}$ is of Type I, then there are functions g and h and a set $\mathcal{D}$ in the plane such that

$$\mathcal{B} = \{(x,y,z) : (x,y) \in \mathcal{D} \text{ and } g(x,y) \le z \le h(x,y)\}.$$

See Fig. 12.61. Hence,

$$\begin{aligned} \iiint_{\mathcal{B}} \frac{\partial R}{\partial z} dV &= \iint_{\mathcal{D}} \left(\int_{g(x,y)}^{h(x,y)} \frac{\partial R}{\partial z} dz \right) dA \\ &= \iint_{\mathcal{D}} (R(x,y,h(x,y)) - R(x,y,g(x,y)))\, dA. \end{aligned} \tag{6}$$

Next we consider the surface integral $\iint_S \langle 0,0,R\rangle \cdot \mathbf{N}\, d\sigma$. Let S_1 be the top piece of S—that is, the graph of $z = h(x,y)$—let S_2 be the bottom piece, and let S_3 be the surface making up the sides of S, so $S = S_1 + S_2 + S_3$. See Fig. 12.61. Then

$$\begin{aligned} \iint_S \langle 0,0,R\rangle \cdot \mathbf{N}\, d\sigma &= \iint_{S_1} \langle 0,0,R\rangle \cdot \mathbf{N}\, d\sigma \\ &\quad + \iint_{S_2} \langle 0,0,R\rangle \cdot \mathbf{N}\, d\sigma + \iint_{S_3} \langle 0,0,R\rangle \cdot \mathbf{N}\, d\sigma. \end{aligned} \tag{7}$$

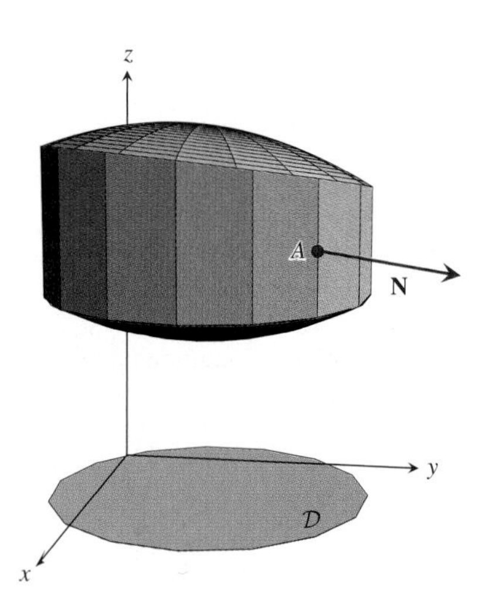

FIGURE 12.62 The normal at a point on the side of a Type I region is parallel to the (x, y)-plane.

If $A = (x_0, y_0, z_0)$ is a point on S_3, then S_3 contains the segment joining the points $(x_0, y_0, g(x_0, y_0))$ and $(x_0, y_0, h(x_0, y_0))$. The normal to S_3 at A is perpendicular to every curve in S_3 through A, so it must be perpendicular to this segment. See Fig. 12.62. Hence the normal to S_3 at A must have a third component of 0, and so has the form

$$\mathbf{N} = \mathbf{N}(x_0, y_0, z_0) = \langle u_0, v_0, 0\rangle.$$

It follows that at any such point A on S_3,

$$\langle 0,0,R\rangle \cdot \mathbf{N} = \langle 0,0,R\rangle \cdot \langle u_0, v_0, 0\rangle = 0,$$

and hence that

$$\iint_{S_3} \langle 0, 0, R\rangle \cdot \mathbf{N}\, d\sigma = \iint_{S_3} 0\, d\sigma = 0. \tag{8}$$

Because S_1 is described by $\mathbf{r}(x, y) = \langle x, y, h(x, y)\rangle$, $(x, y) \in \mathcal{D}$, the outward (upward) unit normal at a point $\mathbf{r}(x, y)$ on S_1 is given by

$$\mathbf{N} = \frac{\mathbf{r}_x \times \mathbf{r}_y}{\|\mathbf{r}_x \times \mathbf{r}_y\|} = \frac{\langle h_x, h_y, 1\rangle}{\sqrt{(h_x)^2 + (h_y)^2 + 1}}. \tag{9}$$

At a point $(x, y, g(x, y))$ of S_2, the outward normal points downward, so has a negative third component. This normal is given by

$$\mathbf{N} = \frac{\langle -g_x, -g_y, -1\rangle}{\sqrt{(g_x)^2 + (g_y)^2 + 1}}. \tag{10}$$

Using (9), we see that

$$\begin{aligned} \iint_{S_1} \langle 0, 0, R\rangle \cdot \mathbf{N}\, d\sigma &= \iint_{\mathcal{D}} \frac{R(x, y, h(x, y))}{\sqrt{(h_x)^2 + (h_y)^2 + 1}} \sqrt{(h_x)^2 + (h_y)^2 + 1}\, dA \\ &= \iint_{\mathcal{D}} R(x, y, h(x, y))\, dA. \end{aligned} \tag{11}$$

Using (10) and a similar process, we find that

$$\iint_{S_2} \langle 0, 0, R\rangle \cdot \mathbf{N}\, d\sigma = -\iint_{\mathcal{D}} R(x, y, g(x, y))\, dA. \tag{12}$$

When we combine (7), (8), (11), and (12), we obtain

$$\iint_{S} \langle 0, 0, R\rangle \cdot \mathbf{N}\, d\sigma = \iint_{\mathcal{D}} (R(x, y, h(x, y)) - R(x, y, g(x, y)))\, dA.$$

Comparing this equality with (6), we conclude that

$$\iint_{S} \langle 0, 0, R\rangle \cdot \mathbf{N}\, d\sigma = \iiint_{\mathcal{B}} \frac{\partial R}{\partial z}\, dV.$$

This is the third part of (5), so we have verified one "part" of the Divergence Theorem. In Exercise 22 we ask you to use a similar argument to verify one of the other equations in (5). We assumed that $\mathcal{B}$ is of Type I in the preceding argument, and assume that it is of Type II or III in Exercise 22. Thus we have an argument in support of the Divergence Theorem for regions $\mathcal{B}$ that are simultaneously of Types I, II, and III. In more advanced texts it is shown that the theorem is true for more general regions.

EXAMPLE 1 Evaluate

$$\iint_{S} \langle x^3 - y, y^3 - z, z^3 - x\rangle \cdot \mathbf{N}\, d\sigma,$$

where S is the sphere of center $(0, 0, 0)$ and radius 1.

Solution

Although we could evaluate this surface integral directly, it is simpler to use the Divergence Theorem and a triple integral. Let $\mathcal{B}$ be the solid ball bounded by S. By the Divergence Theorem,

$$\iint_S \langle x^3 - y, y^3 - z, z^3 - x\rangle \cdot \mathbf{N}\, d\sigma = \iiint_{\mathcal{B}} \operatorname{div}\langle x^3 - y, y^3 - z, z^3 - x\rangle\, dV$$

$$= \iiint_{\mathcal{B}} (3x^2 + 3y^2 + 3z^2)\, dV.$$

We use spherical coordinates to evaluate the last integral. Noting that $3x^2 + 3y^2 + 3z^2 = 3\rho^2$ and replacing dV by $\rho^2 \sin\phi\, d\rho\, d\theta\, d\phi$, we have

$$\iiint_{\mathcal{B}} (3x^2 + 3y^2 + 3z^2)\, dV = \int_0^{\pi}\int_0^{2\pi}\int_0^1 3\rho^4 \sin\phi\, d\rho\, d\theta\, d\phi = \frac{12}{5}\pi.$$

As our second example, we use the Divergence Theorem to evaluate the surface integral from Example 3 of Section 12.6.

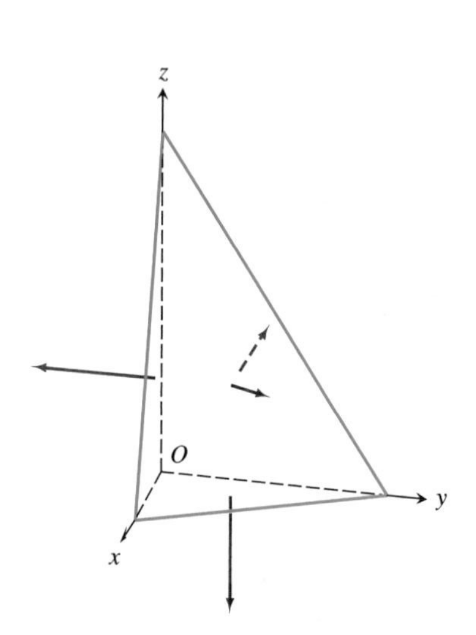

FIGURE 12.63 The surface S is a tetrahedron.

EXAMPLE 2 Evaluate

$$\iint_S \langle xy, yz, xz\rangle \cdot \mathbf{N}\, d\sigma,$$

where S is the tetrahedron with vertices $(0,0,0)$, $(1,0,0)$, $(0,2,0)$, and $(0,0,3)$. See Fig. 12.63.

Solution

Let $\mathcal{B}$ be the bounded solid region determined by the tetrahedron. By the Divergence Theorem,

$$\iint_S \langle xy, yz, xz\rangle \cdot \mathbf{N}\, d\sigma = \iiint_{\mathcal{B}} \operatorname{div}\langle xy, yz, xz\rangle\, dV = \iiint_{\mathcal{B}} (y + z + x)\, dV.$$

Note that $\mathcal{D}$ is the set of points (x, y, z) with

$$0 \le z \le 3 - 3x - \frac{3}{2}y$$

where (x, y) is in the triangular region

$$\{(x, y) : 0 \le y \le 2 - 2x,\ 0 \le x \le 1\}.$$

Thus

$$\iiint_{\mathcal{B}} (y + z + x)\, dV = \int_0^1\int_0^{2-2x}\int_0^{3-3x-\frac{3}{2}y} (x + y + z)\, dz\, dy\, dx = \frac{3}{2}.$$

More about Flux

Flux at a Point Let $\mathbf{v}$ be the velocity field for a planar flow. In Section 12.5 we saw that the divergence of $\mathbf{v}$ at a point can be interpreted as a measure of the flux at that point. This interpretation carries over to flows in three dimensions and is made plausible by applying the Divergence Theorem.

Let

$$\mathbf{v}(x, y, z) = \langle P(x, y, z), Q(x, y, z), R(x, y, z) \rangle$$

be the velocity field for a fluid of density 1 flowing in space, and assume that P, Q, and R have continuous first partial derivatives. Let $A = (x_0, y_0, z_0)$ be a point in the domain of the flow. For $r > 0$, let $S(A, r)$ be the sphere of center A and radius r, and let $B(A, r)$ be the solid ball bounded by this sphere. Assume that when r is small enough, the solid ball is in the region $\mathcal{D}$ of the flow. We define the flux at A by

$$\text{flux at } A = \lim_{r \to 0} \frac{1}{\text{vol}(B(A, r))} \iint_{S(A,r)} (\mathbf{v} \cdot \mathbf{N})\, d\sigma, \tag{13}$$

provided the limit exists. Applying the Divergence Theorem to the surface integral in (13), we find

$$\text{flux at } A = \lim_{r \to 0} \frac{1}{\text{vol}(B(A, r))} \iiint_{B(A,r)} \text{div}\, \mathbf{v}\, dV. \tag{14}$$

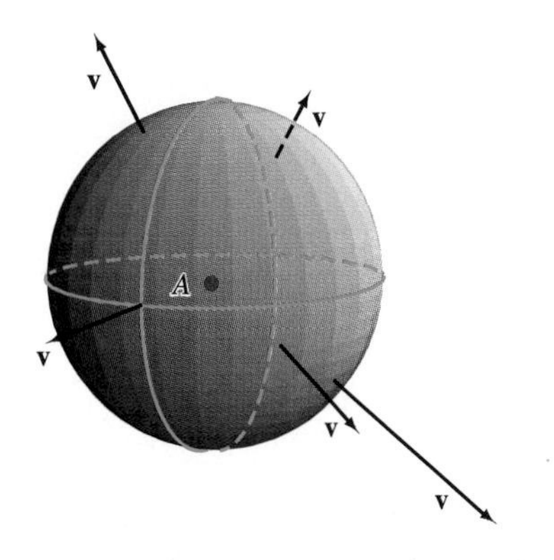

FIGURE 12.64 The flow near a source A.

Because the first partial derivatives of P, Q, and R are continuous at A, the divergence of $\mathbf{v}$ is also continuous at A. Thus if r is small and (x, y, z) is in the solid sphere $B(A, r)$, we have

$$\text{div}\, \mathbf{v}(x, y, z) \approx \text{div}\, \mathbf{v}(x_0, y_0, z_0).$$

The error in this approximation is small when r is small, and approaches 0 as r tends to 0. Thus, for small r, we have

$$\begin{aligned} \frac{1}{\text{vol}(B(A, r))} \iiint_{B(A,r)} \text{div}\, \mathbf{v}(x, y, z)\, dV &\approx \frac{1}{\text{vol}(B(A, r))} \iiint_{B(A,r)} \text{div}\, \mathbf{v}(x_0, y_0, z_0)\, dV \\ &= \frac{\text{div}\, \mathbf{v}(x_0, y_0, z_0)}{\text{vol}(B(A, r))} \iiint_{B(A,r)} 1\, dV \\ &= \text{div}\, \mathbf{v}(x_0, y_0, z_0). \end{aligned} \tag{15}$$

As r approaches 0, this approximation becomes exact. When we combine (13), (14), and (15), we find that

$$\text{flux at } (x_0, y_0, z_0) = \lim_{r \to 0} \frac{1}{\text{vol}(B(A, r))} \iint_{S(A,r)} (\mathbf{v} \cdot \mathbf{N})\, d\sigma = \text{div}\, \mathbf{v}(x_0, y_0, z_0).$$

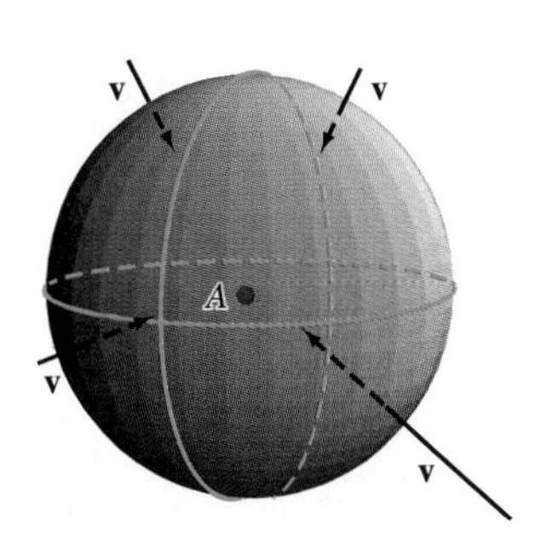

FIGURE 12.65 The flow near a sink A.

This quantity carries units of flux per unit volume.

If $\text{div}\, \mathbf{v}(x_0, y_0, z_0) > 0$, then the flux through a small sphere centered at A is positive. This means that more fluid flows out of the sphere than flows into the sphere. In this case we say that there is a **source** for the flow at A. If $\text{div}\, \mathbf{v}(x_0, y_0, z_0) < 0$, then the flow has a **sink** at A. See Figs. 12.64 and 12.65.

Heat Flux We have noted the physical significance of the flux integral $\iint_S (\mathbf{F} \cdot \mathbf{N})\, d\sigma$ when $\mathbf{F}$ is the velocity field for a fluid flow. As another example, we consider heat flux.

A solid metal object in space occupies a region $\mathcal{B}$. Assume that at point (x, y, z) in $\mathcal{B}$ the temperature of the object is $T(x, y, z)$ and that this temperature does not change with time. (In this case we say the temperature of the object is in *steady state*. The temperature may be different at different points in $\mathcal{B}$, but at a given point the temperature does not change with time.) The **heat flow** associated with T is defined to be the vector field

$$\mathbf{F} = -k \operatorname{grad} T = -k\langle T_x, T_y, T_z\rangle, \tag{16}$$

where the positive constant k is the thermal conductivity of the metal. (The value of k is determined experimentally and has units of calories per centimeter per second per °C.) Recall that $-\operatorname{grad} T(x_0, y_0, z_0)$ gives the direction from (x_0, y_0, z_0) in which the function T decreases most rapidly. Given that heat flows from regions of high temperature to regions of lower temperature, it seems reasonable that the rate of flow should be proportional to the length of the gradient and in the direction opposite the gradient. Thus the term *heat flow* associated with $-\operatorname{grad} T$ seems reasonable.

Now let S be an (imaginary) simple closed surface in $\mathcal{B}$. The **heat flux** across S is defined to be

$$\iint_S (\mathbf{F} \cdot \mathbf{N}) d\sigma = -k \iint_S (\operatorname{grad} T) \cdot \mathbf{N}\, d\sigma. \tag{17}$$

This integral gives a measure of the net amount of heat flowing through the surface S. See also the definition of flux given in Section 12.6.

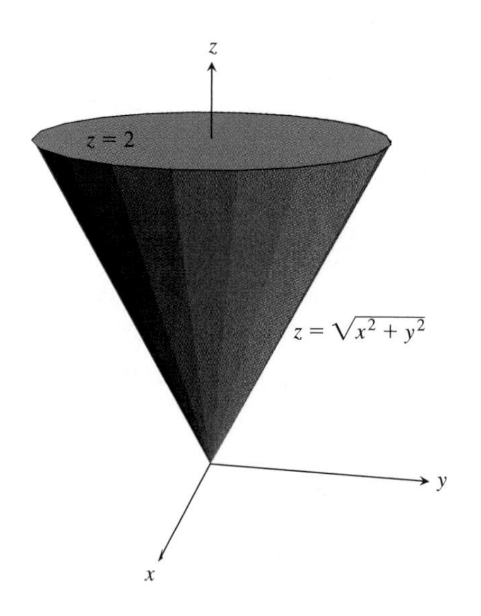

FIGURE 12.66 A conical object made of copper.

EXAMPLE 3 A solid copper cone is modeled by the region

$$B = \{(x, y, z) : \sqrt{x^2 + y^2} \le z \le 2\}.$$

Let S be the surface of R and let $T(x, y, z) = (x^2y^2 + 1)$°C be the steady-state temperature of the cone at the point $(x, y, z) \in B$. See Fig. 12.66. Find the heat flux through S. The thermal conductivity of copper is approximately 0.918 cal/s/cm/°C.

Solution

The heat flow $\mathbf{F}$ is given by

$$\mathbf{F}(x, y, z) = -k \operatorname{grad} T(x, y, z) = -0.918\langle 2xy^2, 2x^2y, 0\rangle.$$

The heat flux across S is

$$\iint_S (\mathbf{F} \cdot \mathbf{N})\, d\sigma = -0.918 \iint_S \langle 2xy^2, 2x^2y, 0\rangle \cdot \mathbf{N}\, d\sigma.$$

Applying the Divergence Theorem to this integral, we find

$$\begin{aligned} \text{heat flux across } S &= -0.918 \iiint_B \operatorname{div}\langle 2xy^2, 2x^2y, 0\rangle \, dV \\ &= -0.918 \iiint_B (2y^2 + 2x^2) dV. \end{aligned} \tag{18}$$

In cylindrical coordinates, the solid region $\mathcal{B}$ is described as

$$\{(r, \theta, z) : r \le z \le 2, 0 \le r \le 2, 0 \le \theta \le 2\pi\}.$$

Using this description to evaluate the last integral in (18), we have

$$\begin{aligned} \text{heat flux across } S &= -0.918 \int_0^{2\pi} \int_0^2 \int_r^2 (2r^2) r \, dz \, dr \, d\theta \\ &= -\frac{32\pi}{5}(0.918) \approx -18.457. \end{aligned}$$

The answer has units of calories/second. Can these units be deduced from the units of the terms in the integral?

Exercises 12.7

Exercises 1–4: Use the Divergence Theorem to evaluate $\iint_S (\mathbf{F} \cdot \mathbf{N})\, d\sigma$ for the given S and **F**. *Assume the unit normal* **N** *is the outward normal.*

1. $\mathbf{F}(x, y, z) = \langle x, y, z\rangle$, S the sphere with radius 1 and centered at the origin
2. $\mathbf{F}(x, y, z) = \langle x^2yz, 2, x + 2z\rangle$, S the cube with sides of length 1, two of whose vertices are $(0, 0, 0)$ and $(1, 1, 1)$
3. $\mathbf{F}(x, y, z) = \langle 1 - x - xz, yz, xy\rangle$, S consists of the boundary of the solid bounded by the paraboloid with equation $x = 1 - y^2 - z^2$ and the (y, z)-plane
4. $\mathbf{F}(x, y, z) = \langle 3x + y^2, x - y + z, 4z\rangle$, S the surface of the ellipse with equation
$$\frac{x^2}{3} + y^2 + 3z^2 = 1$$

Exercises 5–8: Verify the Divergence Theorem by evaluating $\iint_S (\mathbf{F} \cdot \mathbf{N})\, d\sigma$ for the given S and **F**, *and then evaluating $\iiint_{\mathcal{B}} \operatorname{div} \mathbf{F}\, dV$ where $\mathcal{B}$ is the region bounded by S.*

5. $\mathbf{F}(x, y, z) = \langle yz^2, x^2 + 2y, e^{x+y}\rangle$, S the sphere with equation $x^2 + y^2 + z^2 = 1$
6. $\mathbf{F}(x, y, z) = \langle -4xy, 3y^2, x - y\rangle$, S the tetrahedron with vertices $(0, 0, 0)$, $(1, 0, 0)$, $(0, 1, 0)$, and $(0, 0, 1)$
7. $\mathbf{F}(x, y, z) = \langle 2xy^2, yz, 2zx^2\rangle$, S the surface of the region bounded by the cylinder with equation $x^2 + y^2 = 2$ and the planes $z = 1$ and $z = 4$
8. $\mathbf{F}(x, y, z) = \langle x^3, -xz, 3y^2z\rangle$, S the surface of the region bounded by the paraboloid with equation $z = 9 - x^2 - y^2$ and the plane $z = 0$

Exercises 9–12: Let **v** *be the velocity field for a fluid of density δ flowing through a region of space. Calculate the flux across the surface S. Your answer should include units. Assume that the units for length, time, and mass are cm, s, and g, respectively.*

9. $\mathbf{v}(x, y, z) = \langle 2x, -y, z\rangle$, $\delta(x, y, z) = 1$, S the sphere with radius 1 and centered at the origin
10. $\mathbf{v}(x, y, z) = \langle x^3, y^3, z^3\rangle$, $\delta(x, y, z) = 2$, S the cube with sides of length 1, two of whose vertices are $(0, 0, 0)$ and $(1, 1, 1)$
11. $\mathbf{v}(x, y, z) = \langle 2, xz^2, ye^x\rangle$, $\delta(x, y, z) = x^2 + 1$, S the tetrahedron with vertices $(0, 0, 0)$, $(3, 0, 0)$, $(0, 2, 0)$, and $(0, 0, 4)$
12. $\mathbf{v}(x, y, z) = \langle x^3, y^3, z^3\rangle$, $\delta(x, y, z) = z + 1$, S consists of the upper half of the sphere of center $(0, 0, 0)$ and radius 2 and the disk of center $(0, 0)$ and radius 2 in the (x, y)-plane

Exercises 13–16: Describe the set of points in space at which the flow has a source, a sink, or neither a source nor a sink.

13. For the $\mathbf{v}$ of Exercise 10

14. For the $\mathbf{v}$ of Exercise 11

15. For the $\mathbf{v}$ of Exercise 12

16. $\mathbf{v}(x, y, z) = \langle 2x^2 - y, -y^2 + xz, 3z^2 - 8x \rangle$

Exercises 17–20: The temperature T is given for each point of a solid object occupying a region $\mathcal{B}$. Assume that the object is made of a homogeneous material with thermal conductivity k. Let S be an (imaginary) simple closed surface for which S and the region inside of S is contained in $\mathcal{B}$. Find the heat flow and the heat flux across S. Give units with your answer. Assume that length, time, heat, and temperature are measured in cm, s, cal, and °C, respectively.

17. $T(x, y, z) = x + y + z$, $k = 0.408$ (aluminum), S the surface of the region bounded by the paraboloid with equation $z = 2x^2 + 4y^2$ and the plane with equation $4x + y + 3z = 12$

18. $T(x, y, z) = 100x^3$, $k = 0.0001$ (cork), S the rectangular solid bounded by the planes $x = \pm 1/2$, $y = \pm 2$, and $z = \pm 3$

19. $T(x, y, z) = x^2 + y^2 + z^2$, $k = 0.005$ (ice), S the sphere with equation $x^2 + y^2 + z^2 = 4$

20. $T(x, y, z) = x^2z^2 + y$, $k = 1.006$ (silver), S the surface of the region bounded by the cylinder with equation $x^2 + z^2 = 4$ and the planes $y = -1$ and $y = 3$

21. Let $\mathcal{B}$ be a region that is simultaneously of Types I, II, and III and let S be the surface of $\mathcal{B}$. Show that

$$\text{volume of } \mathcal{B} = \frac{1}{3} \iint_S \langle x, y, z \rangle \cdot \mathbf{N}\, d\sigma.$$

Find two different vector fields $\mathbf{F}$ with $\mathbf{F} \neq \frac{1}{3}\langle x, y, z \rangle$ and such that

$$\text{volume of } \mathcal{D} = \iint_S \mathbf{F} \cdot \mathbf{N}\, d\sigma.$$

22. Let $\mathcal{B}$ be a region of Type II; that is, there are functions g and h and a set $\mathcal{D}$ in the (x, z)-plane such that

$$\mathcal{B} = \{(x, y, z) : g(x, z) \leq y \leq h(x, z), (x, z) \in \mathcal{D}\}.$$

Assume that the surface S of $\mathcal{B}$ is orientable and let Q be defined on and have continuous first partial derivatives at all points of $\mathcal{B}$ and S. By paralleling the argument for regions of Type I given in this section, prove that

$$\iint_S \langle 0, Q, 0 \rangle \cdot \mathbf{N}\, d\sigma = \iiint_{\mathcal{B}} \frac{\partial Q}{\partial y}\, dV.$$

23. A solid object occupies a region $\mathcal{B}$ in space. Assume that $\mathcal{B}$ is a simple region and that S is a simple closed surface contained in $\mathcal{B}$. If the object is made of a substance with thermal conductivity k and T is the temperature function for the object, show that the heat flux across S is given by

$$-k \iiint_{\mathcal{B}} \nabla^2 T\, dV.$$

(Recall that $\Delta = \nabla^2 = \nabla \cdot \nabla = \partial^2/\partial x^2 + \partial^2/\partial y^2 + \partial^2/\partial z^2$ is the Laplacian.)

24. A fluid is called *incompressible* if the flux across any orientable, smooth, simple closed surface in the region of the flow is 0. Let $\mathbf{v}$ be the velocity field for the flow of an incompressible fluid in space, and assume that at point (x, y, z) the fluid has density $\delta(x, y, z)$. Assume that $\mathbf{v}$ and δ have continuous first partial derivatives. Show that if $B(A, r)$, the ball of center A and radius r, lies in the region of the flow, then

$$\iiint_{B(A,r)} \operatorname{div}(\delta \mathbf{v})\, dV = 0.$$

Use this result to show that

$$(\operatorname{grad} \delta) \cdot \mathbf{v} + \delta(\operatorname{div} \mathbf{v}) = 0$$

at all points in the region of the flow. (This equation is a simple version of the *equation of continuity,* one of the most important equations in the study of fluid dynamics.)

25. Let S be the upper half of the sphere with equation $x^2 + y^2 + z^2 = 1$, oriented by normals $\mathbf{N}$ that point away from the origin. Consider the surface integral

$$\iint_S \langle x + yz, z^2 \sin x, 4z \rangle \cdot \mathbf{N}\, d\sigma. \tag{19}$$

Even though S is not the surface of a bounded region in space, we can still use the Divergence Theorem to evaluate the integral.

a. Let S_d be the disk of center $(0, 0, 0)$ and radius 1 in the (x, y)-plane and let $S' = S + S_d$ be the surface obtained by joining the surfaces S and S_d. Use the Divergence Theorem to evaluate

$$\iint_{S'} \langle x + yz, z^2 \sin x, 4z \rangle \cdot \mathbf{N}\, d\sigma,$$

where $\mathbf{N}$ is the outward normal to S'.

b. Orient S_d with the downward unit normal $\mathbf{N} = -\langle 0, 0, 1 \rangle$. Evaluate

$$\iint_{S_d} \langle x + yz, z^2 \sin x, 4z \rangle \cdot \mathbf{N}\, d\sigma.$$

c. Combine the results of parts **a** and **b** to find the value of the integral in (19).

26. Let S consist of the sides and the bottom (but not the top) of the cube bounded by the planes $x = 0$, $x = 1$, $y = 0$, $y = 1$, $z = 0$, $z = 1$. Assume that the unit normals $\mathbf{N}$ to S point towards the outside of the cube. Use the idea outlined in Exercise 25 to evaluate

$$\iint_S \langle x^2y, xy^2, xyz\rangle \cdot \mathbf{N}\, d\sigma.$$

27. Let S be the portion of the paraboloid with equation $z = 9 - x^2 - y^2$ that lies above the plane $z = 1$, oriented with "upward" unit normals. Use the idea outlined in Exercise 25 to evaluate

$$\iint_S \langle yze^{yz}, x^3 \ln(2xz + z^3), z\rangle \cdot \mathbf{N}\, d\sigma.$$

28. Assume that the solid region $\mathcal{B}$ and its surface S satisfy the hypotheses of the Divergence Theorem. Let the vector field $\mathbf{F}$ be defined on a set containing $\mathcal{B}$ and S and assume that the component functions for $\mathbf{F}$ have continuous second partial derivatives. Show that

$$\iint_S (\text{curl }\mathbf{F}) \cdot \mathbf{N}\, d\sigma = 0.$$

29. Assume that the solid region $\mathcal{B}$ and its surface S satisfy the hypotheses of the Divergence Theorem. Let f and g be real-valued functions defined on a set containing $\mathcal{B}$ and S and assume f and g have continuous second partial derivatives. Show that

$$\iint_S (f\nabla g) \cdot \mathbf{N}\, d\sigma = \iiint_{\mathcal{B}} (f\nabla^2 g + \nabla f \cdot \nabla g)\, dV.$$

12.8 Stokes' Theorem

The Fundamental Theorem of Calculus for Line Integrals, Green's Theorem, the Divergence Theorem (also called Gauss' Theorem or Ostrogradsky's Theorem), and Stokes' Theorem are relatively recent discoveries in the long history of mathematics. These generalizations of the Fundamental Theorem of Calculus emerged in the nineteenth century from studies of fluid dynamics, electricity and magnetism, and heat flow. The vector notation used to state these theorems was introduced at the beginning of the twentieth century.

Green's Theorem is due to George Green (1793–1841), an English mathematical physicist. It first appeared in 1828, in *An Essay on the Application of Mathematical Analysis to the Theories of Electricity and Magnetism.* A special case of the Divergence Theorem was given by Gauss in 1813. The Ukrainian mathematician Mikhail Ostrogradsky (1801–1861) proved the general case in 1826. Stokes' Theorem was first stated by Lord Kelvin in a letter to George Stokes (1819–1903). It first appeared in print in 1854 when Stokes included it among the questions for the Smith Prize at Cambridge.

James Clerk Maxwell (1831–1879), in his famous *Treatise on Electricity and Magnetism,* which appeared in 1873, made extensive use of the Divergence Theorem and Stokes' Theorem. Near the end of the century, these theorems were stated in vector form by Josiah Willard Gibbs (1839–1903). Gibbs was a professor of mathematical physics at Yale University. He lectured on vector analysis at Yale for many years and published his book *Vector Analysis* in 1901.

In Section 12.5 we remarked that Green's Theorem,

$$\iint_R \left(\frac{\partial Q}{\partial x} - \frac{\partial P}{\partial y}\right) dA = \int_C P\, dx + Q\, dy,$$

can be generalized in at least two ways. If we replace the two-dimensional region R and its one-dimensional boundary C by a three-dimensional solid $\mathcal{B}$ together with its two-dimensional boundary surface S, Green's Theorem generalizes to the Divergence Theorem. If we replace the two-dimensional region R and its one-dimensional boundary C by a two-dimensional surface S in R^3 and its one-dimensional boundary curve C, Green's Theorem generalizes to Stokes' Theorem.

We briefly state the Fundamental Theorem and three of its generalizations in Fig. 12.67. We note two striking similarities among the results listed in the fourth column. First, in each equation both a function and its "derivative" appear. There are F and F', F and its gradient ∇F, $\mathbf{F}$ and its divergence $\nabla \cdot \mathbf{F}$, and $\mathbf{F}$ and its curl $\nabla \times \mathbf{F}$. Second, on the left of these equations the "derivative" function is integrated over a domain, while on the right the function is evaluated on the boundary of that domain.

Stokes' Theorem

For Stokes' Theorem, we consider a vector-valued function $\mathbf{F}$ defined on a domain in R^3. This function may describe the velocity field $\mathbf{v}$ of a fluid, a magnetic field $\mathbf{H}$ due to a current flowing, or a gravitational field. We may wish, for example, to calculate the circulation $\int_C (\mathbf{v} \cdot \mathbf{T})\, ds$ of the fluid around a simple closed curve C. In some cases, such a calculation can be simplified greatly by the use of Stokes' Theorem.

$f : [a, b] \to R$	 a b	$F' = f$	Fundamental Theorem of Calculus $\int_a^b F'(x)\,dx = F(b) - F(a)$
$\mathbf{f} : C \to R^2$ or $\mathbf{f} : C \to R^3$	 A B $C \subset R^2$ or R^3	$\nabla F = \mathbf{f}$	Fundamental Theorem of Line Integrals $\int_C (\nabla F \cdot \mathbf{T})\,ds = F(B) - F(A)$
$f : \mathcal{B} \to R$	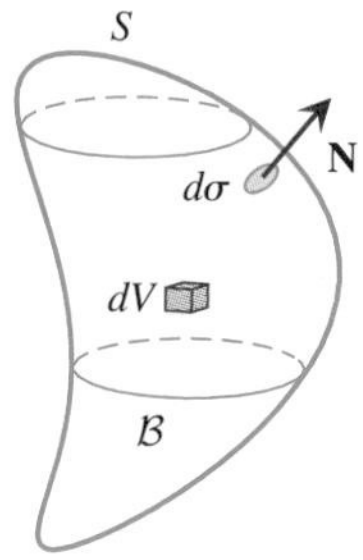 S $d\sigma$ $\mathbf{N}$ dV $\mathcal{B}$	$\nabla \cdot \mathbf{F} = f$	Divergence Theorem $\iiint_{\mathcal{B}} \nabla \cdot \mathbf{F}\,dV = \iint_S \mathbf{F} \cdot \mathbf{N}\,d\sigma$
$\mathbf{f} : S \to R^3$	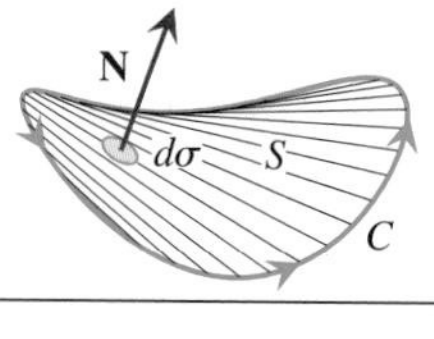 $\mathbf{N}$ $d\sigma$ S C	$\nabla \times \mathbf{F} = \mathbf{f}$	Stokes' Theorem $\iint_S (\nabla \times \mathbf{F}) \cdot \mathbf{N}\,d\sigma = \int_C \mathbf{F} \cdot \mathbf{T}\,ds$

FIGURE 12.67 The Fundamental Theorem of Calculus and its generalizations.

We start by giving Stokes' Theorem for a simple smooth surface S with a smooth boundary curve C. A representative surface S and boundary curve C are shown in Fig. 12.68. We assume that both S and C are contained in the domain of the vector-valued function $\mathbf{F}$. We assume that surface S is described by

$$\mathbf{r} = \mathbf{r}(u, v), \qquad (u, v) \in \mathcal{D}, \tag{1}$$

where $\mathcal{D} \subset R^2$ is a region of Type I or II. Let C^* be the boundary of $\mathcal{D}$. Taking (as usual) the counterclockwise direction on C^* as positive, the positive direction on the boundary curve C of S is that inherited from C^* through Equation 1. We then assign a positive normal direction to S so that if a positive normal vector is grasped by the fingers on the right hand, with the thumb pointing in the direction of this vector, then the fingers point in the positive direction on C.

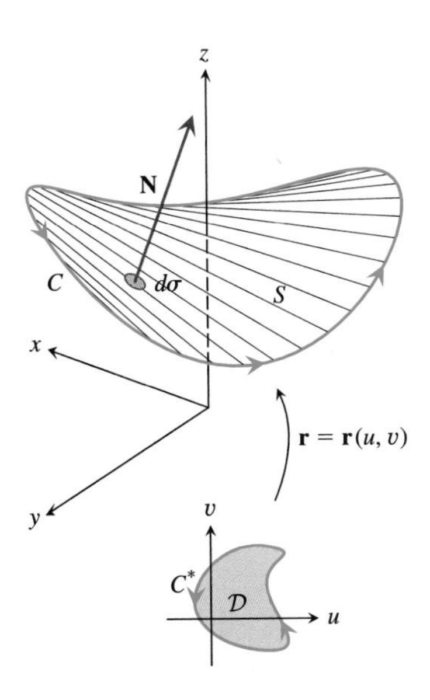

FIGURE 12.68 A simple smooth surface, its boundary curve, and their parameterizations.

THEOREM Stokes' Theorem

Let $\mathbf{F}$ be a continuously differentiable vector-valued function defined on the simple smooth surface S described by (1). Assume that S and its boundary curve C can be oriented as described in the preceding paragraph, $\mathbf{N}$ is the positive unit normal to S, and $\mathbf{T}$ is the positive unit tangent vector to C. Then

$$\iint_S (\nabla \times \mathbf{F}) \cdot \mathbf{N}\, d\sigma = \int_C \mathbf{F} \cdot \mathbf{T}\, ds. \tag{2}$$

We give the first few steps of an argument for Stokes' Theorem at the end of the section and complete it in Exercise 29. Meanwhile, we give examples illustrating and extending this important result.

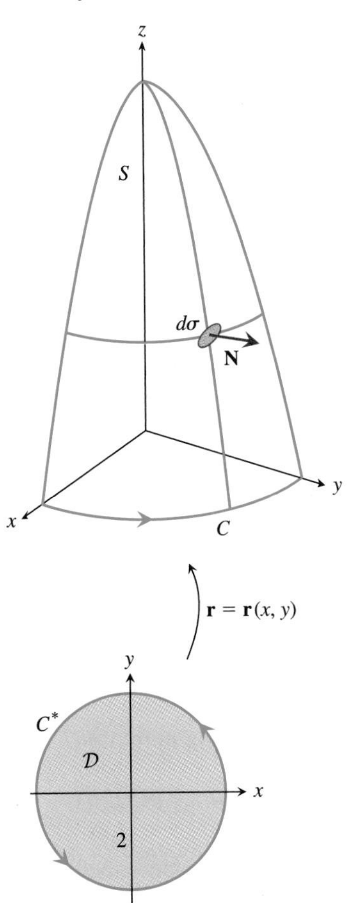

FIGURE 12.69 One-quarter of the paraboloidal surface.

EXAMPLE 1 Verify that the surface integral and line integral in (2) are equal when the vector field F is

$$\mathbf{F}(x, y, z) = \langle -y, x, 1 \rangle, \qquad (x, y, z) \in R^3, \tag{3}$$

and S is the paraboloidal surface described by the equation

$$z = 4 - x^2 - y^2, \qquad (x, y) \in \mathcal{D} = \{(x, y) \mid x^2 + y^2 \leq 4\}.$$

Solution

We leave some details to Exercises 1–3. Part of the surface S is shown in Fig. 12.69. A parameterization of S is

$$\mathbf{r} = \mathbf{r}(x, y) = \langle x, y, 4 - x^2 - y^2 \rangle, \qquad (x, y) \in \mathcal{D}. \tag{4}$$

It is straightforward to check that

i) S is a simple smooth surface,
ii) as the boundary C^* of $\mathcal{D}$ is traced counterclockwise, the boundary C of S is traced in the direction indicated in the figure,
iii) $\mathbf{r}_x \times \mathbf{r}_y = \langle 2x, 2y, 1 \rangle$,

iv) the positive unit normal on S is

$$\mathbf{N} = \frac{1}{\|\mathbf{r}_x \times \mathbf{r}_y\|}(\mathbf{r}_x \times \mathbf{r}_y) = \frac{1}{\sqrt{1 + 4(x^2 + y^2)}}\langle 2x, 2y, 1\rangle, \text{ and}$$

v) the curl of the vector field $\mathbf{F}$ is $\nabla \times \mathbf{F} = \langle 0, 0, 2\rangle$.

We begin by calculating the line integral $\int_C (\mathbf{F} \cdot \mathbf{T})\, ds$. If we describe the curve C by

$$\mathbf{r} = \mathbf{r}(\theta) = \langle 2 \cos \theta, 2 \sin \theta, 0\rangle, \qquad 0 \le \theta \le 2\pi,$$

then the value of the line integral is

$$\begin{aligned} \int_C \mathbf{F} \cdot \mathbf{T}\, ds &= \int_0^{2\pi} \mathbf{F}(2 \cos \theta, 2 \sin \theta, 0) \cdot \langle -2 \sin \theta, 2 \cos \theta, 0\rangle\, d\theta \\ &= \int_0^{2\pi} \langle -2 \sin \theta, 2 \cos \theta, 1\rangle \cdot \langle -2 \sin \theta, 2 \cos \theta, 0\rangle\, d\theta \qquad (5) \\ &= \int_0^{2\pi} (4 \sin^2\theta + 4 \cos^2\theta)\, d\theta = 8\pi. \end{aligned}$$

Next we calculate the value of the surface integral. Substituting $\nabla \times \mathbf{F}$ and $\mathbf{N}$ from (iv) and (v), and using (7) from Section 12.6, we find

$$\begin{aligned} \iint_S (\nabla \times \mathbf{F}) \cdot \mathbf{N}\, d\sigma &= \iint_{\mathcal{D}} \langle 0, 0, 2\rangle \cdot \left(\frac{1}{\|\mathbf{r}_x \times \mathbf{r}_y\|}\langle 2x, 2y, 1\rangle\right) \|\mathbf{r}_x \times \mathbf{r}_y\|\, dx\, dy \\ &= \iint_{\mathcal{D}} \frac{2}{\|\mathbf{r}_x \times \mathbf{r}_y\|}\|\mathbf{r}_x \times \mathbf{r}_y\|\, dx\, dy. \end{aligned} \qquad (6)$$

Hence

$$\iint_S (\nabla \times \mathbf{F}) \cdot \mathbf{N}\, d\sigma = \iint_{\mathcal{D}} 2\, dx\, dy = 2(\pi 2^2) = 8\pi. \qquad (7)$$

This completes the verification that Stokes' Theorem holds for the given vector field, surface, and boundary curve.

In our second example, we use the same vector field as in Example 1 and again verify the equality of the two integrals in Stokes' Theorem. The surface S in Example 2 is different from that in Example 1 in two ways. First, the function $\mathbf{r}$ chosen to describe the surface S is not one-to-one on $\mathcal{D}$. The other difference is that the boundary C of $\mathcal{D}$ is only piecewise smooth.

EXAMPLE 2 Verify that the surface integral and line integral in (2) are equal when the vector field $\mathbf{F}$ is

$$\mathbf{F}(x, y, z) = \langle -y, x, 0\rangle, \qquad (x, y, z) \in R^3,$$

and S is the portion of the sphere of center $(0, 0, 0)$ and radius 1 described by

$$\mathbf{r} = \mathbf{r}(\phi, \theta) = \langle \cos \theta \sin \phi, \sin \theta \sin \phi, \cos \phi\rangle, \qquad 0 \le \phi, \theta \le \frac{\pi}{2}.$$

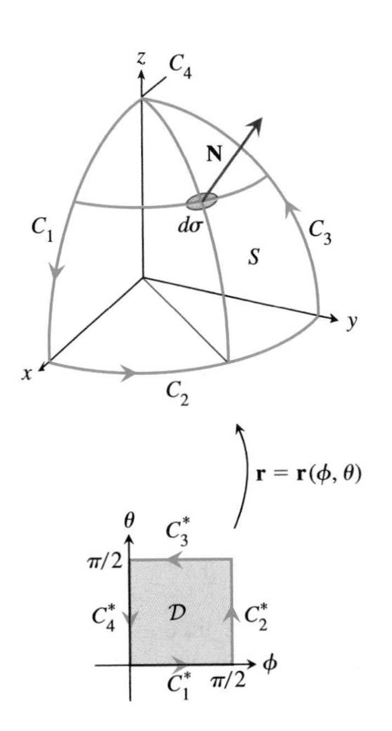

FIGURE 12.70 The surface S and boundary curves.

Solution

Some details are left to Exercises 4–6. The surface S is shown in Fig. 12.70. With few variations, the calculations follow those in Example 1. We start with several comments and calculations.

i) The boundary C^* of $\mathcal{D}$ can be written as a sum of smooth curves C_1^*, C_2^*, C_3^*, and C_4^*. We write $C^* = C_1^* + C_2^* + C_3^* + C_4^*$. These parts of the boundary of C^* are transformed by $\mathbf{r}$ to the subsets C_1, C_2, C_3, and C_4 of C. We write the boundary of S as $C = C_1 + C_2 + C_3 + C_4$. It is on C_4^* that $\mathbf{r}$ is not one-to-one. The corresponding curve C_4 consists of only the point $(0, 0, 1)$.

ii) $\mathbf{r}_\phi \times \mathbf{r}_\theta = (\sin\phi)\mathbf{r}$.

iii) Because S is part of the sphere of center $(0, 0, 0)$ and radius 1, the positive unit normal on S is $\mathbf{N} = \mathbf{r}$. This can also be shown by direct calculation,

$$\mathbf{N} = \frac{1}{\|\mathbf{r}_\phi \times \mathbf{r}_\theta\|}(\mathbf{r}_\phi \times \mathbf{r}_\theta) = \frac{1}{\sin\phi}(\sin\phi)\mathbf{r} = \mathbf{r}.$$

iv) From Example 1, the curl of the vector field $\mathbf{F}$ is $\nabla \times \mathbf{F} = \langle 0, 0, 2\rangle$.

We first calculate the line integral

$$\int_C (\mathbf{F}\cdot\mathbf{T})\,ds = \int_{C_1} (\mathbf{F}\cdot\mathbf{T})\,ds + \int_{C_2} (\mathbf{F}\cdot\mathbf{T})\,ds + \int_{C_3} (\mathbf{F}\cdot\mathbf{T})\,ds + \int_{C_4} (\mathbf{F}\cdot\mathbf{T})\,ds.$$

The line integral on C_1 is 0 because C_1 lies in the (x, z)-plane. Indeed, on this curve $y = 0$ and the second component of $\mathbf{T}$ is 0, so $\mathbf{F}\cdot\mathbf{T} = 0$. A similar observation shows that the line integral on C_3 is 0. Because the curve C_4 is a single point, the fourth line integral is 0. The line integral on C_2 can be evaluated using the parameterization

$$\mathbf{r} = \mathbf{r}(t) = \langle \cos t, \sin t, 0\rangle, \qquad 0 \le t \le \frac{\pi}{2}.$$

It follows that $\int_{C_2} (\mathbf{F}\cdot\mathbf{T})\,ds = \pi/2$.

Before we choose a parameterization of S, we do a preliminary calculation. For this we recall that $\mathbf{N} = \mathbf{r} = \langle x, y, z\rangle$. Hence

$$\iint_S (\nabla \times \mathbf{F})\cdot\mathbf{N}\,d\sigma = \iint_S \langle 0, 0, 2\rangle\cdot\langle x, y, z\rangle\,d\sigma = \iint_S 2z\,d\sigma.$$

Using either of the parameterizations

$$\mathbf{r} = \mathbf{r}(x, y) = \left\langle x, y, \sqrt{1 - x^2 - y^2}\right\rangle$$

or

$$\mathbf{r} = \mathbf{r}(\phi, \theta) = \langle \cos\theta\sin\phi, \sin\theta\sin\phi, \cos\phi\rangle,$$

it follows that $\iint_S (\nabla \times \mathbf{F})\cdot\mathbf{N}\,d\sigma = \pi/2$. This agrees with the value of the line integral calculated earlier and so verifies Stokes' Theorem in this case.

In fluid dynamics, Stokes' Theorem provides a second way of viewing the circulation of a fluid on a curve. For this, we replace the vector field $\mathbf{F}$ with the velocity field $\mathbf{v}$ of the fluid and write the line integral in Stokes' Theorem as $\int_C (\mathbf{v} \cdot \mathbf{T})\, ds$. The value of this integral is a measure of the circulation of the liquid around C. According to Stokes' Theorem, we may instead evaluate the surface integral $\iint_S (\nabla \times \mathbf{v}) \cdot \mathbf{N}\, d\sigma$, which sums the product of the normal component of the "circulation density" $\nabla \times \mathbf{v}$ and the element of area $d\sigma$ over the surface S. As an example, we calculate the circulation of the boiling oil around the boundary of a potato chip.

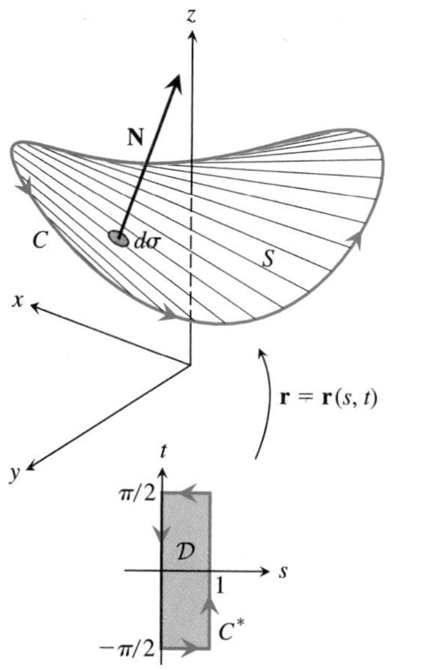

FIGURE 12.71 The surface S and boundary curve C.

EXAMPLE 3 A potato chip has surface S and boundary C, as shown in Fig. 12.71. The curve C is described by

$$\mathbf{r} = \mathbf{g}(t) = \left\langle \cos t, \sin t, 1 + \frac{1}{4}\sin 2t \right\rangle, \qquad 0 \le t \le 2\pi, \tag{8}$$

and the surface S by

$$\mathbf{r} = \mathbf{r}(s,t) = s\mathbf{g}(t) + (1 - s)\mathbf{g}(\pi - t), \qquad (s,t) \in [0,1] \times \left[\frac{-\pi}{2}, \frac{\pi}{2}\right]. \tag{9}$$

Calculate the circulation around the chip of the oil with velocity field given by

$$\mathbf{v}(x,y,z) = \langle yz, xz, xy \rangle, \qquad (x,y,z) \in R^3.$$

Solution

As noted earlier, we may calculate either of the integrals

$$\int_C (\mathbf{v} \cdot \mathbf{T})\, ds \quad \text{or} \quad \iint_S (\nabla \times \mathbf{v}) \cdot \mathbf{N}\, d\sigma.$$

To evaluate the surface integral, we must calculate $\nabla \times \mathbf{v}$. The result is

$$\nabla \times \mathbf{v} = \det \begin{pmatrix} \mathbf{i} & \mathbf{j} & \mathbf{k} \\ \dfrac{\partial}{\partial x} & \dfrac{\partial}{\partial y} & \dfrac{\partial}{\partial z} \\ yz & xz & xy \end{pmatrix} = \langle 0,0,0 \rangle = \mathbf{0}.$$

Hence $\iint_S (\nabla \times \mathbf{v}) \cdot \mathbf{N}\, d\sigma = 0$. By Stokes' Theorem, it follows that $\iint_C (\mathbf{v} \cdot \mathbf{T})\, dt = 0$, so the circulation of the oil around the boundary of the chip is 0.

Because the circulation density $\nabla \times \mathbf{v}$ is easily calculated, it makes sense to calculate it first, before attempting to visualize the velocity field or attempting a direct calculation of the circulation. We show one visualization of the movement of the oil. In using a CAS to graph the velocity vectors, it is often wise to avoid clutter by limiting the number of vectors drawn. If we note from (8) that the boundary of the chip lies above the square $[-1,1] \times [-1,1]$ and between the planes with equations $z = \frac{3}{4}$ and $z = \frac{5}{4}$, it may be sufficient to plot the velocity field $\mathbf{v}$ in just these two planes. We show the resulting graph in Fig. 12.72.

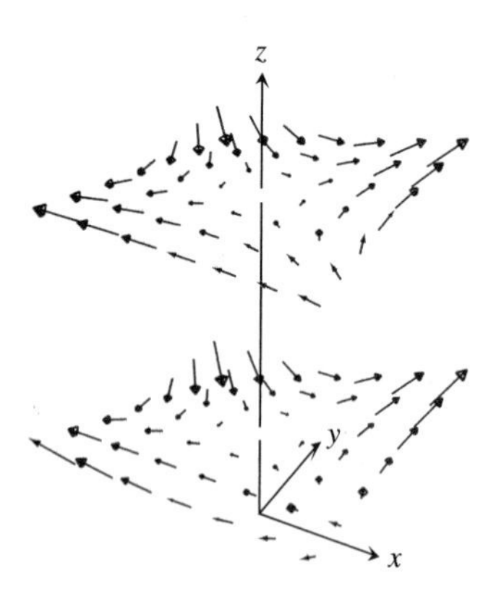

FIGURE 12.72 The velocity field $\mathbf{v}$ in planes with equations $z = 3/4$ and $z = 5/4$.

This figure suggests that the oil does not tend to circulate about the boundary of the chip. Indeed, above the first and third quadrants, the liquid is flowing outward. A velocity field whose curl is $\mathbf{0}$ is called *irrotational.*

Extensions of Stokes' Theorem

We stated Stokes' Theorem for simple smooth surfaces. In Example 2 we gave evidence that Stokes' Theorem remains true if the boundary curve is only piecewise smooth. We illustrate by example other surfaces for which Stokes' Theorem is true and mention one surface for which Stokes' Theorem fails.

EXAMPLE 4 Verify that Stokes' Theorem is true for any continuously differentiable vector field $\mathbf{F}$ defined on the cylindrical surface S described by

$$\mathbf{r} = \mathbf{r}(t, \theta) = \langle \cos\theta, t, \sin\theta \rangle, \qquad (t, \theta) \in \mathcal{D} = [0, 1] \times [0, 2\pi].$$

This surface is shown in Fig. 12.73.

Solution

This surface differs from surfaces considered so far in that its boundary curve is not a single closed curve and, moreover, the points of $\mathcal{D}$ at which $\mathbf{r}$ is not one-to-one are transformed to points that are in some sense interior points of S. Note that the points on the top and bottom boundaries of $\mathcal{D}$ are transformed to the line AB shown in the figure. In effect, the function $\mathbf{r}$ takes $\mathcal{D}$ and bends it into a cylinder, with the ends of length one glued together along AB.

The boundary of S is the union of the curves C_1 and C_2. It is possible to choose a positive normal to S and positive directions for C_1 and C_2 so that the orientations of the normal and curves are consistent. If we were to walk

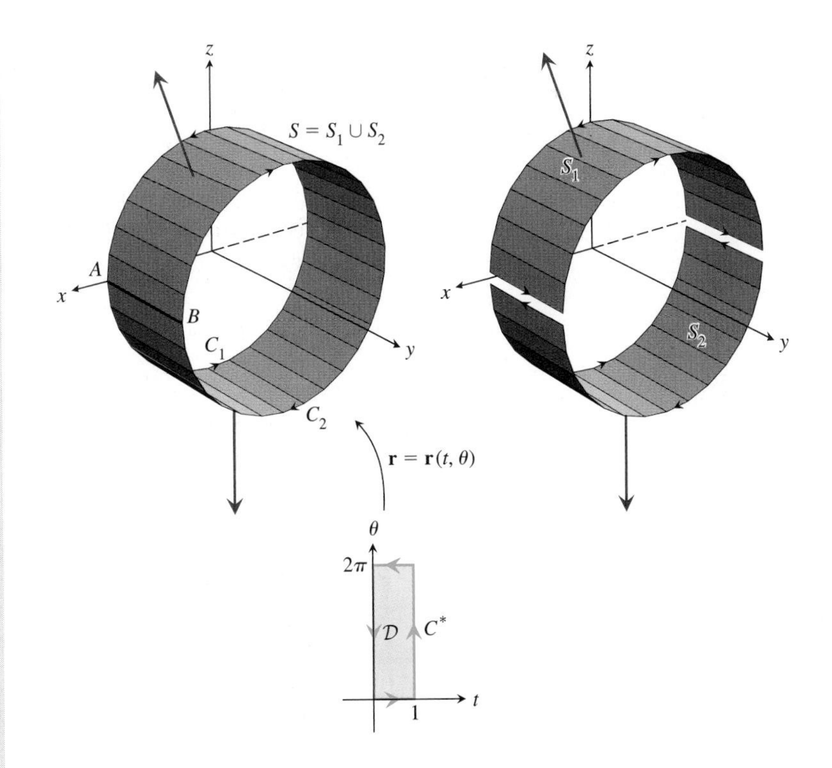

FIGURE 12.73 Putting two simple smooth surfaces together.

on the boundary of S with our heads in the direction of the positive normal, then we would walk in the positive direction of the boundary curves. Put differently, if we were to grasp the positive normal vector with our right hand, thumb pointing in the direction of this normal, then our fingers would curl in the positive directions of the two curves.

We may apply Stokes' Theorem to each of the surfaces S_1 and S_2 shown on the right of Fig. 12.73. These half-cylinders are shown slightly separated in the figure so that our intent is clearer. We may imagine that the cylindrical pipe on the left has been fabricated from the two half-cylinders shown on the right by welding them along the two horizontal seams. Before welding, each of the half-cylinders is a surface to which Stokes' Theorem applies. Hence,

$$\iint_{S_1} (\nabla \times \mathbf{F}) \cdot \mathbf{N}\, d\sigma = \int_{\partial S_1} \mathbf{F} \cdot \mathbf{T}\, ds \tag{10}$$

$$\iint_{S_2} (\nabla \times \mathbf{F}) \cdot \mathbf{N}\, d\sigma = \int_{\partial S_2} \mathbf{F} \cdot \mathbf{T}\, ds, \tag{11}$$

where ∂S_1 and ∂S_2 denote the boundaries of S_1 and S_2. Each of these line integrals can be written as a sum of four line integrals, two on half-circles and two on line segments. The positive normals to the two surfaces and the positive directions of the eight curves are marked. If we add Equations (10) and (11) together, the sum of the line integrals along the welds will be zero. This should be clear from the figure, which shows that the welds are traversed in different directions. The remaining four line integrals can be combined into two line integrals, one on C_1 and the other on C_2.

For the two surface integrals,

$$\iint_{S_1} (\nabla \times \mathbf{F}) \cdot \mathbf{N}\, d\sigma + \iint_{S_2} (\nabla \times \mathbf{F}) \cdot \mathbf{N}\, d\sigma = \iint_{S} (\nabla \times \mathbf{F}) \cdot \mathbf{N}\, d\sigma.$$

This follows by evaluating the first integral on $\mathcal{D}_1$ (the lower half of $\mathcal{D}$), and the second integral on $\mathcal{D}_2$ (the upper half of $\mathcal{D}$), combining the two double integrals, and rewriting as a surface integral on $\mathcal{D}$. It now follows that

$$\iint_{S} (\nabla \times \mathbf{F}) \cdot \mathbf{N}\, d\sigma = \int_{\partial S} \mathbf{F} \cdot \mathbf{T}\, ds = \int_{C_1} \mathbf{F} \cdot \mathbf{T}\, ds + \int_{C_2} \mathbf{F} \cdot \mathbf{T}\, ds. \tag{12}$$

Thus, Stokes' Theorem has been verified for the surface S and any continuously differentiable field $\mathbf{F}$ defined on S.

EXAMPLE 5 Let $\mathbf{F}$ be a continuously differentiable vector field. Show that if S is the surface of a doughnut (torus) and $\mathbf{F}$ is defined on S, then

$$\iint_{S} (\nabla \times \mathbf{F}) \cdot \mathbf{N}\, d\sigma = 0. \tag{13}$$

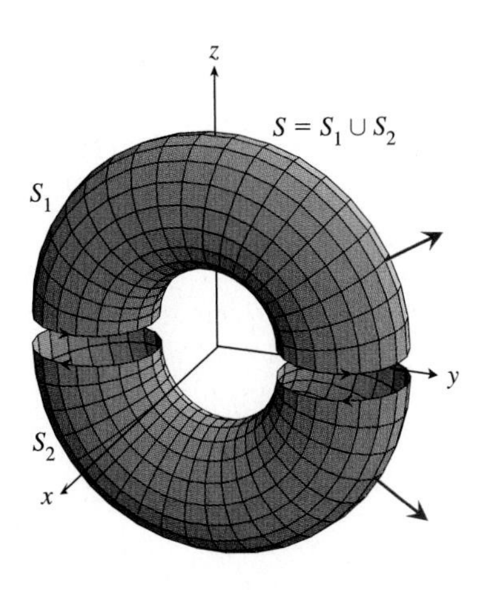

FIGURE 12.74 A closed surface, seen as the union of two cylinderlike surfaces.

Solution

We might expect that Equation (13) is true on the grounds that because a doughnut has no boundary curve, the line integral in Stokes' Theorem is 0. A rigorous argument can be based on Example 4, in which we argued that Stokes' Theorem holds true on a cylinder. We may divide the torus S into two surfaces S_1 and S_2, as shown in Fig. 12.74. We have separated the two surfaces a little so that they may be seen more clearly. Each of the surfaces is like a cylinder in the sense that if it were made from an elastic material, it could be straightened out into a cylinder. Positive normals and positive directions can be assigned on S_1 and S_2 so that if we grasp a positive normal vector of either surface with our right hand, thumb pointing in the positive direction, then our fingers will curl in the positive directions of the boundary curves of the surface. If we apply Stokes' Theorem to the two surfaces and add the results, the four line integrals will add to 0. Equation (13) follows.

In the preceding two examples, we extended Stokes' Theorem to surfaces S that can be written as the union of simple smooth surfaces S_1 and S_2. For each of these surfaces, it was possible to assign positive normal directions to one side and positive directions to the boundary curve so that when fitted together, the contiguous boundary curves were traced in opposite directions and the positive normal vectors for the entire surface varied continuously. For some surfaces this is not possible. A simple example of such a surface is the Möbius strip, formed from a strip of paper by giving one end a half-twist and then smoothly joining the ends together. In Exercise 30 of Section 12.6 we saw that the Möbius strip is an example of a nonorientable surface.

When Is a Vector Field F Conservative?

You may recall from Section 12.3 that for a continuously differentible vector field $\mathbf{F} = \langle P, Q \rangle$ defined on a simply connected domain $\mathcal{D} \subset R^2$, $\mathbf{F} = \nabla f$ if and only if $Q_x = P_y$.

To avoid the more complex notion of simple connectivity in R^3, we give a generalization of this result only for the case in which the domain $\mathcal{B}$ of $\mathbf{F} = \langle P, Q, R \rangle$ is *convex* in R^3. A set is convex if for any two points of the set the line segment joining these points lies wholly within the set. Cubical or spherical solids are convex, as are all "ovoid" solids. Stokes' Theorem is used in proving the following result.

Potential Functions on Convex Sets in R^3

Let $\mathcal{B} \subset R^3$ be a convex set and $\mathbf{F} = \langle P, Q, R \rangle$ a continuously differentiable vector field defined on $\mathcal{B}$. The vector field $\mathbf{F}$ is conservative on $\mathcal{B}$ if and only if $\nabla \times \mathbf{F} = \mathbf{0}$ on $\mathcal{B}$.

The Easy Part of an Argument for Stokes' Theorem

We remarked earlier that Stokes' Theorem is a generalization of Green's Theorem. Moreover, a proof of Stokes' Theorem can be based directly on Green's Theorem. Let the surface S be represented by

$$\langle x, y, z\rangle = \mathbf{r}(u, v) = \langle r_1(u, v), r_2(u, v), r_3(u, v)\rangle, \qquad (u, v) \in \mathcal{D}.$$

Let C^* be the boundary of $\mathcal{D}$, and let

$$\langle u, v\rangle = \mathbf{g}(t) = \langle g_1(t), g_2(t)\rangle, \qquad a \le t \le b,$$

be a parameterization of C^*. Referring to Fig. 12.75, the idea is to prove Stokes' formula

$$\iint_S (\nabla \times \mathbf{F}) \cdot \mathbf{N}\, d\sigma = \int_C \mathbf{F} \cdot \mathbf{T}\, ds \tag{14}$$

by using the functions $\mathbf{r}$ and $\mathbf{g}$ to move the argument "down" to the (u, v)-plane and there apply Green's Theorem.

Let $\mathbf{F} = \langle F_1, F_2, F_3\rangle$ and let $\mathbf{N} = \langle n_1, n_2, n_3\rangle$ be a continuous unit normal defined on S. Then

$$\begin{aligned}(\nabla \times \mathbf{F}) \cdot \mathbf{N} &= \left\langle \frac{\partial F_3}{\partial y} - \frac{\partial F_2}{\partial z}, \frac{\partial F_1}{\partial z} - \frac{\partial F_3}{\partial x}, \frac{\partial F_2}{\partial x} - \frac{\partial F_1}{\partial y} \right\rangle \cdot \langle n_1, n_2, n_3\rangle \\ &= \left(\frac{\partial F_3}{\partial y} - \frac{\partial F_2}{\partial z}\right) n_1 + \left(\frac{\partial F_1}{\partial z} - \frac{\partial F_3}{\partial x}\right) n_2 + \left(\frac{\partial F_2}{\partial x} - \frac{\partial F_1}{\partial y}\right) n_3 \\ &= \left(\frac{\partial F_1}{\partial z} n_2 - \frac{\partial F_1}{\partial y} n_3\right) + \left(\frac{\partial F_2}{\partial x} n_3 - \frac{\partial F_2}{\partial z} n_1\right) + \left(\frac{\partial F_3}{\partial y} n_1 - \frac{\partial F_3}{\partial x} n_2\right).\end{aligned}$$

Hence,

$$\begin{aligned}&\iint_S (\nabla \times \mathbf{F}) \cdot \mathbf{N}\, d\sigma \\ &\quad = \iint_S \left(\left(\frac{\partial F_1}{\partial z} n_2 - \frac{\partial F_1}{\partial y} n_3\right) + \left(\frac{\partial F_2}{\partial x} n_3 - \frac{\partial F_2}{\partial z} n_1\right) + \left(\frac{\partial F_3}{\partial y} n_1 - \frac{\partial F_3}{\partial x} n_2\right) \right) d\sigma.\end{aligned}$$

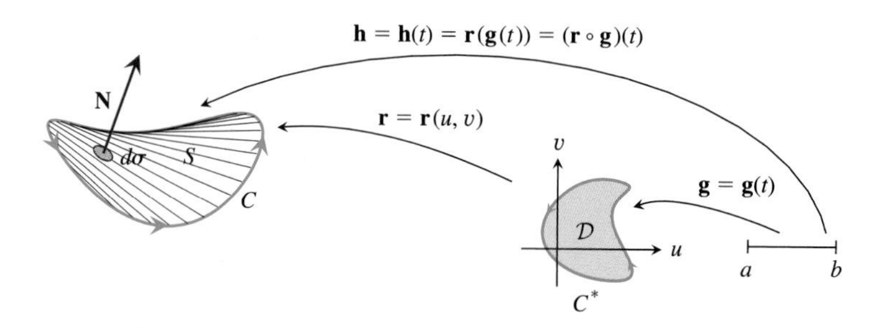

FIGURE 12.75 The functions $\mathbf{g}$ and $\mathbf{h} = \mathbf{r} \circ \mathbf{g}$ that describe the boundaries of $\mathcal{D}$ and S.

Comparing this with

$$\int_C (\mathbf{F}\cdot\mathbf{T})\,ds = \int_C F_1\,dx + F_2\,dy + F_3\,dz,$$

we see that formula (14) will follow if we can show that

$$\int_C F_1\,dx = \iint_S \left(\frac{\partial F_1}{\partial z}n_2 - \frac{\partial F_1}{\partial z}n_3\right)d\sigma, \tag{15}$$

$$\int_C F_2\,dy = \iint_S \left(\frac{\partial F_2}{\partial x}n_3 - \frac{\partial F_2}{\partial z}n_1\right)d\sigma, \quad \text{and} \tag{16}$$

$$\int_C F_3\,dz = \iint_S \left(\frac{\partial F_3}{\partial y}n_1 - \frac{\partial F_3}{\partial x}n_2\right)d\sigma. \tag{17}$$

We do part of the argument to establish (15). Equations (16) and (17) can be established in a similar way. The function $\mathbf{h}$ defined by

$$\mathbf{h}(t) = \langle h_1(t), h_2(t), h_3(t)\rangle = \mathbf{r}(u,v) = \mathbf{r}(\mathbf{g}(t)), \qquad a \le t \le b, \tag{18}$$

is a parameterization of the boundary C of S. Hence,

$$\int_C F_1\,dx = \int_a^b F_1(\mathbf{h}(t))h_1'(t)\,dt.$$

Because

$$h_1(t) = r_1(\mathbf{g}(t)) = r_1(g_1(t), g_2(t)),$$

an application of the chain rule shows that

$$h_1'(t) = \frac{\partial r_1}{\partial u}g_1'(t) + \frac{\partial r_1}{\partial v}g_2'(t).$$

Thus

$$\int_C F_1\,dx = \int_a^b F_1(\mathbf{h}(t))\left(\frac{\partial r_1}{\partial u}g_1'(t) + \frac{\partial r_1}{\partial v}g_2'(t)\right)dt.$$

We express the last integral as a line integral on C^*. Using (18) and noting that $du = g_1'(t)\,dt$ and $dv = g_2'(t)\,dt$, we have,

$$\int_C F_1\,dx = \int_{C^*} (F_1\circ\mathbf{r})(u,v)\frac{\partial r_1}{\partial u}du + (F_1\circ\mathbf{r})(u,v)\frac{\partial r_1}{\partial v}dv.$$

Applying Green's Theorem to the line integral on the right, we obtain

$$\int_C F_1\,dx = \iint_{\mathcal{D}} \left[\frac{\partial}{\partial u}\left((F_1\circ\mathbf{r})\frac{\partial r_1}{\partial v}\right) - \frac{\partial}{\partial v}\left((F_1\circ\mathbf{r})\frac{\partial r_1}{\partial u}\right)\right]du\,dv. \tag{19}$$

The rest of the proof is a calculation to show that the double integral in (19) can be transformed into the surface integral in Equation (15). We leave this for Exercise 29.

Exercises 12.8

1. In Example 1, verify the statements (i)–(v).

2. In Example 1, fill in the details leading to (5).

3. In Example 1, fill in the details leading to (7).

4. In Example 2, verify the statements (i)–(iv).

5. Show that the line integrals on C_1 and C_3 in Example 2 are zero.

6. Choose one of the two parameterizations of the surface S in Example 2 and show that $\int\int_S (\nabla \times \mathbf{F}) \cdot \mathbf{N}\, d\sigma = \pi/2$.

7. In Example 3, show that

$$\mathbf{r}(s,t) = \left\langle (-1+2s)\cos t, \sin t, 1 - \frac{1}{4}\sin 2t + \frac{1}{2}s \sin 2t \right\rangle.$$

8. In Example 3, show that the line integral simplifies to the integral

$$\int_0^{2\pi} \left(\frac{1}{4}\sin 4t + \cos 2t \right) dt.$$

Show by inspection of the integrand and region of integration that the value of this integral is 0.

Exercises 9–13: For the given vector field **F** *surface S, and boundary C, verify that the surface integral and line integral in Stokes' Theorem are equal.*

9. $\mathbf{F}(x,y,z) = \langle 2z, x+y, -yz\rangle$; S is the part of the plane with equation $x+y+z=1$ above the triangular region with vertices $(0,0,0)$, $(1,0,0)$, and $(0,1,0)$.

10. $\mathbf{F}(x,y,z) = \langle x, x+y, z\rangle$; S is the part of the paraboloid with equation $z = 1 - x^2 - y^2$ above the circular region in the (x,y)-plane and bounded by the circle with equation $x^2+y^2=1$.

11. $\mathbf{F}(x,y,z) = \langle z, xz, x\rangle$; S is the surface described by

$$\mathbf{r} = \mathbf{r}(u,v) = \langle v, u, v\rangle, \qquad (u,v) \in \mathcal{D},$$

where $\mathcal{D} = [0,1]\times[0,1]$.

12. $\mathbf{F}(x,y,z) = \langle x-y, y-z, z-x\rangle$; S is the surface cut from the plane with equation $x+y+z=3$ by the cylinder with equation $x^2+y^2=1$. Assume that the boundary of S is oriented counterclockwise when seen from above.

13. $\mathbf{F}(x,y,z) = \langle xy, y^2, z^2\rangle$; S is the frustum of a cone described by $\mathbf{r} = \mathbf{r}(u,v) = \langle u\cos v, u\sin v, 1-u\rangle$, where $1/2 \le u \le 1$ and $0 \le v \le 2\pi$. Assume that the positive normal for the frustum is the outward normal.

Exercises 14–17: For the vector field $\mathbf{F}\langle x,y,z\rangle = \langle -y, x, -z\rangle$ *defined on* R^3, *use Stokes' Theorem to evaluate the integral* $\int_C \mathbf{F}\cdot\mathbf{T}\,ds$ *for the given curve C.*

14. C is the triangle with vertices $(1,1,5)$, $(2,0,1)$, and $(0,3,2)$, traced in this order.

15. C is the circle in which the plane with equation $y = z$ intersects the unit sphere. The positive direction on C is counterclockwise when seen from above.

16. C is the boundary of the surface described by

$$\mathbf{r} = \mathbf{r}(u,v) = \langle u, \cos v, \sin v\rangle,$$

where $0 \le u \le 1$ and $-\pi/2 \le v \le \pi/2$. The positive normal to the surface points into the first octant.

17. C is the ellipse cut from the plane with equation $x+2y+3z=1$ by the cylinder with equation $x^2+y^2=1$. The positive direction on C is counterclockwise when seen from above.

18. As in Example 5, argue that Equation (13) holds on a doughnut or torus by dividing it into four pieces. Using Fig. 12.74, bisect the torus with a plane perpendicular to its axis of symmetry (the x-axis), and then divide the resulting two surfaces (each could be flattened into the region between concentric circles) with the (x,z)-plane.

19. Calculate the circulation of the fluid with velocity field $\mathbf{v} = \langle y^3, xy+3xy^2, z^4\rangle$ on the intersection of the upper half of the ellipsoid with equation $(1/2)x^2 + (1/2)y^2 + z^2 = 1$ with the cylinder with equation $x^2+y^2-y=0$. *Hint:* Use cylindrical coordinates in describing the surface.

20. Let k be an unspecified constant. Is the vector field

$$\mathbf{F}(x,y,z) = \left\langle \frac{-y}{(x^2+y^2)}, \frac{x}{(x^2+y^2)}, k \right\rangle,$$

defined on the set $\mathcal{B} = \{(x,y,z)\,|\,x,y,z>0\}$, conservative? How could this question be answered if the domain were $\mathcal{B} = \{(x,y,z)\,|\,(x,y,z) \ne (0,0,0)\}$? Is the result labeled "Potential Functions on Convex Sets in R^3" applicable in this case? Why or why not?

21. Consider the vector field

$$\mathbf{F}(x,y,z) = \left\langle \frac{1}{y}, -\frac{x}{y^2}, 2z-1 \right\rangle,$$

defined on the set $\mathcal{B} = \{(x,y,z)\,|\,y>0\}$. Is this vector field conservative?

22. In Example 1, note that the curve C is the boundary of many surfaces, not just the paraboloid. For example, C is the boundary of the disk of radius 2 in the (x,y)-plane. If S denotes any simple smooth surface for which $\partial S = C$, then, according to Stokes' Theorem,

$$\iint_S (\nabla \times \mathbf{F}) \cdot \mathbf{N}\, d\sigma = \int_C \mathbf{F} \cdot d\mathbf{r}.$$

If we wish to calculate any of these integrals, we may choose the simplest. In this case, the simplest is the surface integral over the disk of radius 2 in the (x, y)-plane. Show by direct evaluation of this surface integral that its value is 8π.

23. Given the idea expressed in Exercise 22, suppose there is a simple smooth surface S hanging from a circular hoop of radius 2. Assume that the hoop lies in the (x, y)-plane and its center is at $(0, 0, 0)$. We may imagine the surface to be that of a rubber balloon, with its neck terminating in the hoop. Using the vector field $\mathbf{F}$ from Example 1, calculate the circulation on the boundary of S by evaluating a surface integral.

24. Using the ideas in Example 5, show that Stokes' Theorem on the sphere S with equation $(x - h)^2 + (y - k)^2 + (z - \ell)^2 = a^2$ implies

$$\iint_S (\nabla \times \mathbf{F}) \cdot \mathbf{N}\, d\sigma = 0.$$

25. The surface S in Example 3 may be regarded as filling in the curve C by bridging with line segments pairs of "opposite points" of C. Explain. If you have access to a CAS, compare the graphs of the surface S produced by

i. the parameterization in Equation (9) (an alternate form of this parameterization is given in Exercise 7) or

ii. the graph consisting of the union of the graph of the curve C and the bridging line segments mentioned previously.

Which, in your opinion, gives a clearer view?

26. Evaluate $\int_C xy\, dx + z\, dy + y\, dz$ on the curve described by

$$\mathbf{r} = \mathbf{r}(t) = \left\langle \cos t, \sin t, 1 + \frac{1}{4}\sin 2t \right\rangle, 0 \le t \le 2\pi.$$

This curve is the boundary of the chip discussed in Example 3. It may be useful to notice that the surface S of that example can be described by $z = 1 + (1/2)xy$ on a suitable domain.

27. Let $\mathcal{B} \subset R^3$ be a convex set and $\mathbf{F} = \langle F_1, F_2, F_3 \rangle$ a continuously differentiable vector field defined on $\mathcal{B}$. Show that if the vector field $\mathbf{F}$ is conservative on $\mathcal{B}$, then $\nabla \times \mathbf{F} = \mathbf{0}$ on $\mathcal{B}$.

28. Under the conditions stated in Stokes' Theorem and assuming that the real-valued functions f and g have continuous second partial derivatives on S and its boundary, that $\mathbf{a}$ is a constant vector, and that $\mathbf{r} = \langle x, y, z \rangle$, show that

a. $\displaystyle \frac{1}{2}\int_C (\mathbf{a} \times \mathbf{r}) \cdot \mathbf{T}\, ds = \iint_S \mathbf{a} \cdot \mathbf{N}\, d\sigma$

b. $\displaystyle \int_C (f\nabla g) \cdot \mathbf{T}\, ds = \iint_S (\nabla f \times \nabla g) \cdot \mathbf{N}\, d\sigma$

c. $\displaystyle \int_C (f\nabla f) \cdot \mathbf{T}\, ds = 0$ d. $\displaystyle \int_C (f\nabla g + g\nabla f) \cdot \mathbf{T}\, ds = 0.$

29. Complete the proof of Stokes' Theorem by filling in the details of the following outline of the main steps.

a. It follows from the equality of the mixed partial derivatives of the coordinate functions of $\mathbf{r}$ that the integrand of Equation (19) can be written as

$$\frac{\partial}{\partial u}(F_1 \circ \mathbf{r})\frac{\partial r_1}{\partial v} - \frac{\partial}{\partial v}(F_1 \circ \mathbf{r})\frac{\partial r_1}{\partial u}.$$

b. By the chain rule,

$$\frac{\partial}{\partial u}(F_1 \circ \mathbf{r}) = \frac{\partial F_1}{\partial x}\frac{\partial r_1}{\partial u} + \frac{\partial F_1}{\partial y}\frac{\partial r_2}{\partial u} + \frac{\partial F_1}{\partial z}\frac{\partial r_3}{\partial u}$$

$$\frac{\partial}{\partial v}(F_1 \circ \mathbf{r}) = \frac{\partial F_1}{\partial x}\frac{\partial r_1}{\partial v} + \frac{\partial F_1}{\partial y}\frac{\partial r_2}{\partial v} + \frac{\partial F_1}{\partial z}\frac{\partial r_3}{\partial v}.$$

Substitution of these results into the integrand from **a** gives

$$\frac{\partial F_1}{\partial z}\left(\frac{\partial r_3}{\partial u}\frac{\partial r_1}{\partial v} - \frac{\partial r_3}{\partial v}\frac{\partial r_1}{\partial u}\right) - \frac{\partial F_1}{\partial y}\left(\frac{\partial r_1}{\partial u}\frac{\partial r_2}{\partial v} - \frac{\partial r_1}{\partial v}\frac{\partial r_2}{\partial u}\right).$$

c. From this result we note that, apart from a factor of $1/\|\mathbf{r}_u \times \mathbf{r}_v\|$, the expressions in parentheses are the second and third coordinates of the unit normal $\mathbf{N}$ to S. Equation (15) now follows.

Review of Key Concepts

In this chapter we defined line and surface integrals, looked at several techniques for evaluating these integrals, and saw that these integrals are useful in studying fluid mechanics and heat flow. In defining the line and surface integrals we paralleled the definitions of the definite integral and multiple integrals presented in previous chapters. Thus $\int_C f\, ds$ and $\iint_S F\, d\sigma$ were each defined as the limit of certain Riemann sums. We gave examples showing that line or surface integrals can be used to calculate the mass of a curved rod or surface, the circulation and flux for a flowing fluid, and the heat flux through the surface of a solid.

For line and surface integrals, we noted that the Fundamental Theorem of Line Integrals, various forms of Green's Theorem, the

Divergence Theorem, and Stokes' Theorem are generalizations of the Fundamental Theorem of Calculus for definite integrals. These results relate integrals involving "derivatives" of real-valued functions or vector fields to other expressions involving the functions. The "derivatives" involved in these generalizations of the Fundamental Theorem are not the derivatives studied in earlier sections, but instead are the gradient, divergence, or curl, whose definitions follow.

The ∇ symbol. The "nabla" operator ∇ is given by

$$\nabla = \left\langle \frac{\partial}{\partial x}, \frac{\partial}{\partial y}, \frac{\partial}{\partial z} \right\rangle.$$

Although the components of ∇ are not numbers or functions, the nabla operator behaves like a vector in many ways.

The gradient. If f is a real-valued function of three variables, then the gradient of f is

$$\text{grad } f = \nabla f = \left\langle \frac{\partial}{\partial x}, \frac{\partial}{\partial y}, \frac{\partial}{\partial z} \right\rangle f = \left\langle \frac{\partial f}{\partial x}, \frac{\partial f}{\partial y}, \frac{\partial f}{\partial z} \right\rangle.$$

The divergence. If $\mathbf{F} = \langle P, Q, R \rangle$ is a vector field, then the divergence of $\mathbf{F}$ is

$$\begin{aligned} \text{div } f = \nabla \cdot \mathbf{F} &= \left\langle \frac{\partial}{\partial x}, \frac{\partial}{\partial y}, \frac{\partial}{\partial z} \right\rangle \cdot \langle P, Q, R \rangle \\ &= \frac{\partial P}{\partial x} + \frac{\partial Q}{\partial y} + \frac{\partial R}{\partial z}. \end{aligned}$$

The curl. If $\mathbf{F} = \langle P, Q, R \rangle$ is a vector field, then the curl of $\mathbf{F}$ is

$$\begin{aligned} \text{curl } \mathbf{F} = \nabla \times \mathbf{F} &= \left\langle \frac{\partial}{\partial x}, \frac{\partial}{\partial y}, \frac{\partial}{\partial z} \right\rangle \times \langle P, Q, R \rangle \\ &= \det \begin{pmatrix} \mathbf{i} & \mathbf{j} & \mathbf{k} \\ \frac{\partial}{\partial x} & \frac{\partial}{\partial y} & \frac{\partial}{\partial z} \\ P & Q & R \end{pmatrix} \\ &= \left\langle \frac{\partial R}{\partial y} - \frac{\partial Q}{\partial z}, \frac{\partial P}{\partial z} - \frac{\partial R}{\partial x}, \frac{\partial Q}{\partial x} - \frac{\partial P}{\partial y} \right\rangle. \end{aligned}$$

In using the Fundamental Theorem of Line Integrals, it is important to recognize when a vector field may be the gradient of a real-valued function. The Gradient Test provides a quick way to recognize vector fields that are *not* gradients.

The Gradient Test. **Two dimensions:** Let $\mathbf{F} = \langle P, Q \rangle$ be a continuously differentiable vector field. If there is a function f for which grad $f = \mathbf{F}$, then

$$\frac{\partial Q}{\partial x} = \frac{\partial P}{\partial y}. \tag{1}$$

Three dimensions: Let $\mathbf{F} = \langle P, Q, R \rangle$ be a continuously differentiable vector field. If there is a function f for which grad $f = \mathbf{F}$, then

$$\frac{\partial Q}{\partial x} = \frac{\partial P}{\partial y}, \quad \frac{\partial R}{\partial x} = \frac{\partial P}{\partial z}, \quad \text{and} \quad \frac{\partial R}{\partial y} = \frac{\partial Q}{\partial z}. \tag{2}$$

If (1) or (2) do not hold, then $\mathbf{F}$ is not a gradient. If (1) or (2) does hold, then $\mathbf{F}$ may be the gradient of a function f and we may attempt to find f by integration.

Chapter Summary

The Line Integral

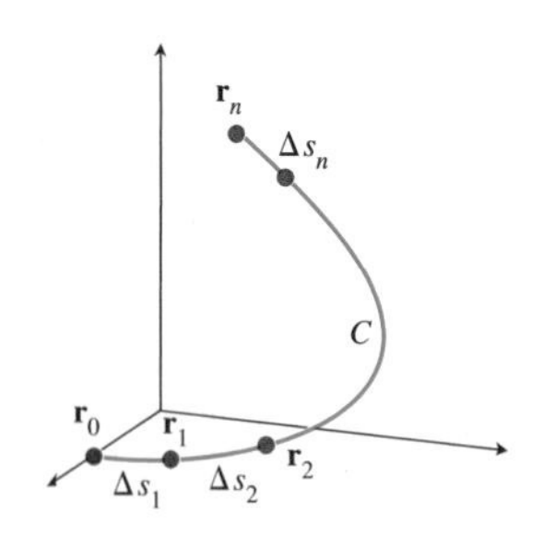

Let C be a smooth curve and f a continuous real-valued function defined on C. The line integral

$$\int_C f\,ds$$

Let $\mathbf{r}_0, \mathbf{r}_1, \ldots, \mathbf{r}_n$ be a subdivision of C and let $\{\mathbf{c}_1, \mathbf{c}_2, \ldots, \mathbf{c}_n\}$ be an evaluation set associated with the subdivision. The Riemann sum associated with this subdivision and evaluation set is

$$\sum_{k=1}^{n} f(\mathbf{c}_k)\,\Delta s_k,$$

The Line Integral

is defined to be the limit of Riemann sums associated with regular subdivisions of C and associated evaluation sets.

where Δs_k is the length of the kth piece of the subdivision. The line integral of f with respect to arc length on C is given by

$$\int_C f\,ds = \lim_{n\to\infty} \sum_{k=1}^{n} f(\mathbf{c}_k)\,\Delta s_k.$$

Evaluation of $\int_C f\,ds$

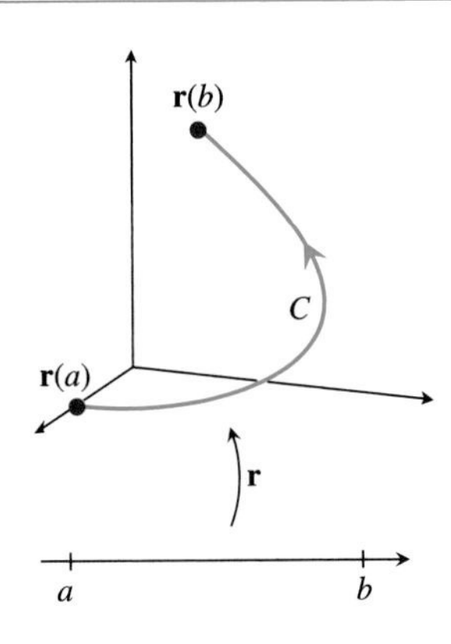

If the smooth curve C is parameterized by

$$\mathbf{r} = \mathbf{r}(t), \qquad a \le t \le b,$$

and f is a continuous real-valued function defined on C, then

$$\int_C f\,ds = \int_a^b f(\mathbf{r}(t))\,\|\mathbf{r}'(t)\|\,dt.$$

Let C be the line segment from $(0,0,0)$ to $(-1,5,2)$ and let

$$f(x,y,z) = xy^2z.$$

Then C is described by

$$\mathbf{r} = \mathbf{r}(t) = \langle -t, 5t, 2t\rangle, \qquad 0 \le t \le 1.$$

We then have

$$f(\mathbf{r}(t)) = (-t)(5t)^2(2t) = -50t^4$$

and

$$\|\mathbf{r}'(t)\| = \|\langle -1,5,2\rangle\| = \sqrt{30},$$

so

$$\int_C f\,ds = \int_0^1 (-50t^4)\sqrt{30}\,dt$$

$$= -10\sqrt{30}.$$

Vector Field

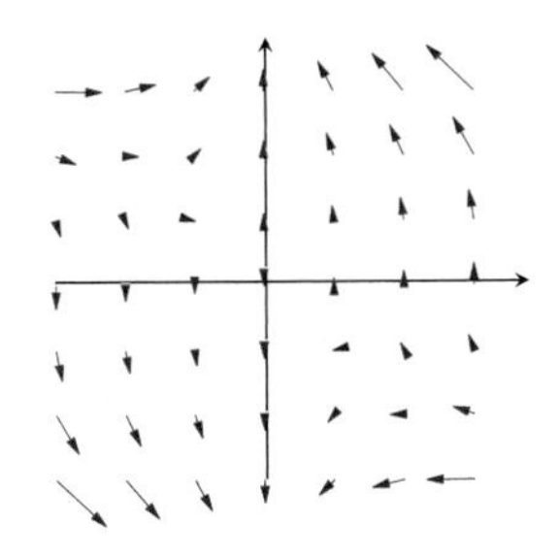

A function $\mathbf{F}$ with domain and range in R^2 or R^3 is called a *vector field.* To represent a vector field graphically, draw the vector $\mathbf{F}(\mathbf{P})$ with initial point at $\mathbf{P}$ for several points $\mathbf{P}$.

Let $\mathbf{F}(x,y) = \langle -2xy^2, x+y\rangle$. If $\mathbf{P} = \langle -2, 1\rangle$, then

$$\mathbf{F}(\mathbf{P}) = \mathbf{F}(-2,1) = \langle 4, -1\rangle.$$

To graphically represent the value of $\mathbf{F}$ at $\mathbf{P}$ place the vector $\langle 4,-1\rangle$ in the plane with its initial point at $\mathbf{P} = \langle -2, 1\rangle$.

The Fundamental Theorem of Line Integrals

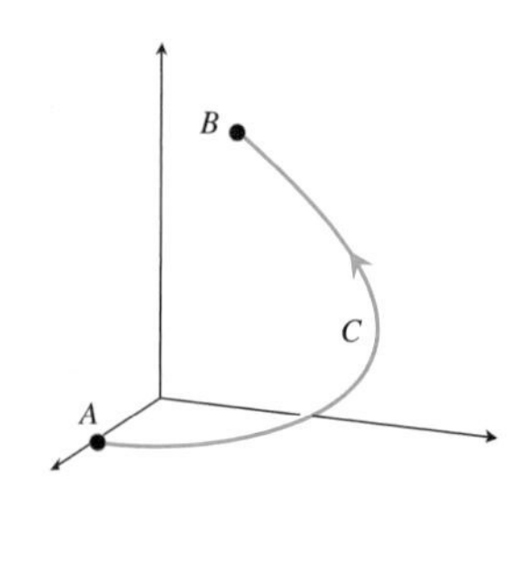

Let C be a smooth curve with initial point $\mathbf{A}$ and terminal point $\mathbf{B}$, and let $\mathbf{F} = \langle P, Q, R\rangle$ be a continuous vector field defined on a region containing C. If

$$\mathbf{F} = \operatorname{grad} f,$$

then

$$\int_C \mathbf{F}\cdot\mathbf{T}\,ds = \int_C P\,dx + Q\,dy + R\,dz$$
$$= f(\mathbf{B}) - f(\mathbf{A}).$$

Let C be a smooth curve with initial point $(-2, 0, 1)$ and terminal point $(1, -1, 2)$. If

$$\mathbf{F}(x, y, z) = \langle 2xe^{yz}, x^2ze^{yz}, x^2ye^{yz}\rangle$$

and

$$f(x, y, z) = x^2e^{yz},$$

then $\mathbf{F} = \operatorname{grad} f$. Hence

$$\int_C \mathbf{F}\cdot\mathbf{T}\,ds$$
$$= \int_C 2xe^{yz}\,dx + x^2ze^{yz}\,dy + x^2ye^{yz}\,dz$$
$$= f(1, -1, 2) - f(-2, 0, 1) = \frac{1}{e^2} - 4$$

Simply Connected Set

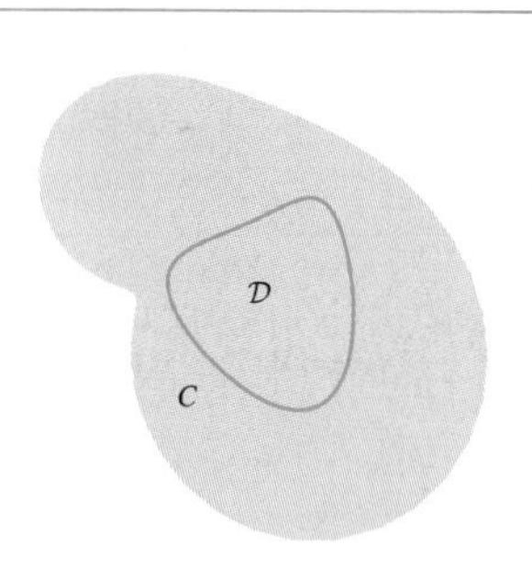

A set $\mathcal{D}$ in the plane is simply connected if for every simple closed curve C in $\mathcal{D}$, the region inside of C is also contained in $\mathcal{D}$.

Let $\mathbf{F} = \langle P, Q\rangle$ be a continuously differentiable vector field defined on a simply connected planar set $\mathcal{D}$. Then $\mathbf{F}$ is the gradient of a function f if and only if

$$\frac{\partial Q}{\partial x} = \frac{\partial P}{\partial y}.$$

Let

$$\mathbf{F}(x, y) = \langle P, Q\rangle = \left\langle \frac{x}{y}, -\frac{x^2}{2y^2}\right\rangle.$$

On the simply connected set $\mathcal{D} = \{(x, y) : y > 0\}$,

$$\frac{\partial Q}{\partial x} = -\frac{x}{y^2} = \frac{\partial P}{\partial y}.$$

Therefore, $\mathbf{F}$ is the gradient of a function defined on $\mathcal{D}$. Indeed, $\mathbf{F} = \operatorname{grad} f$, where

$$f(x, y) = \frac{x^2}{2y}.$$

Conservative Vector Fields

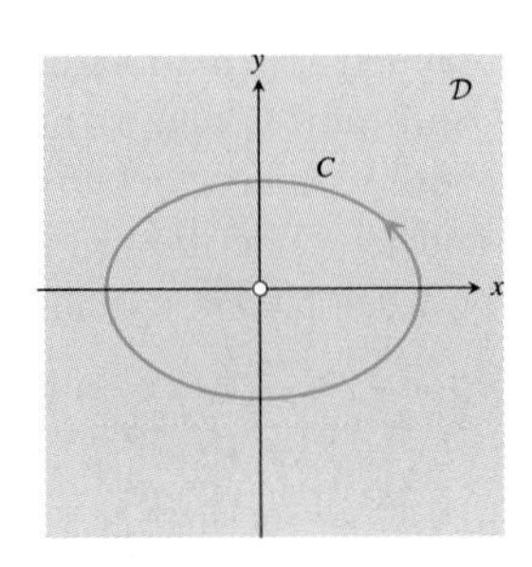

A vector field $\mathbf{F}$ defined on a set $\mathcal{D}$ is conservative if there is a real-valued function f defined on $\mathcal{D}$ with $\operatorname{grad} f = \mathbf{F}$.

If $\mathbf{F}$ is a conservative vector field and C is a closed curve in $\mathcal{D}$, then

$$\int_C \mathbf{F}\cdot\mathbf{T}\,ds = 0.$$

Conversely, if $\mathcal{D}$ is simply connected and $\int_C \mathbf{F}\cdot\mathbf{T}\,ds = 0$ for every closed curve in $\mathcal{D}$, then $\mathbf{F}$ is a conservative vector field on $\mathcal{D}$. (See Exercise 36 in Section 12.3.)

Let

$$\mathbf{F}(x,y) = \left\langle \frac{x}{(x^2+y^2)^2}, \frac{y}{(x^2+y^2)^2} \right\rangle$$

for (x,y) in the set

$$\mathcal{D} = \{(x,y) : (x,y) \neq (0,0)\}.$$

Because

$$\mathbf{F}(x,y) = \operatorname{grad}\left(\frac{-1}{2(x^2+y^2)}\right)$$

on $\mathcal{D}$, it follows that

$$\int_C \mathbf{F}\cdot\mathbf{T}\,ds = \int_C \frac{x}{(x^2+y^2)^2}\,dx + \frac{y}{(x^2+y^2)^2}\,dy = 0$$

for any closed curve C in $\mathcal{D}$.

Green's Theorem

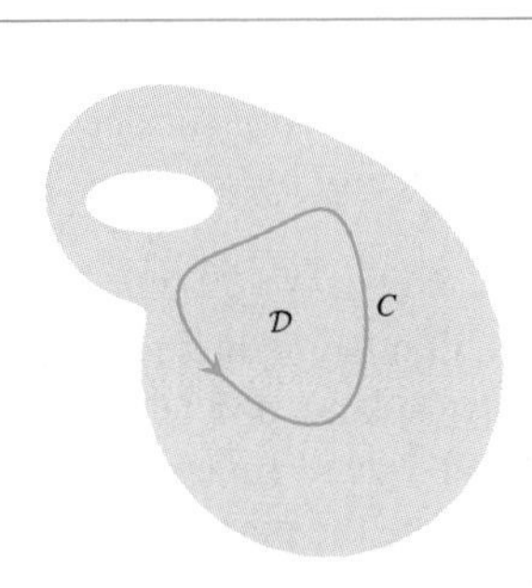

Let C be a simple, closed, piecewise smooth planar curve. Let $\mathbf{F} = \langle P, Q\rangle$ be a continuously differentiable vector field whose domain contains C and the region $\mathcal{D}$ inside of C. Then

$$\int_C (\mathbf{F}\cdot\mathbf{T})\,ds = \int_C P\,dx + Q\,dy = \iint_{\mathcal{D}} \left(\frac{\partial Q}{\partial x} - \frac{\partial P}{\partial y}\right) dA = \iint_{\mathcal{D}} (\operatorname{curl}\mathbf{F})\cdot\mathbf{k}\,dA.$$

An equivalent statement is

$$\int_C (\mathbf{F}\cdot\mathbf{N})\,ds = \int_C -Q\,dx + P\,dy = \iint_{\mathcal{D}} \left(\frac{\partial P}{\partial x} + \frac{\partial Q}{\partial y}\right) dA = \iint_{\mathcal{D}} \operatorname{div}\mathbf{F}\,dA.$$

Let C be the square with vertices $(-1,-1)$, $(1,-1)$, $(1,1)$, $(-1,1)$ and counterclockwise orientation. Then

$$\int_C (x^2+y)\,dx + 2xy\,dy = \int_{-1}^{1}\int_{-1}^{1} (2y-1)\,dy\,dx = -4.$$

Green's Theorem for Regions with Holes

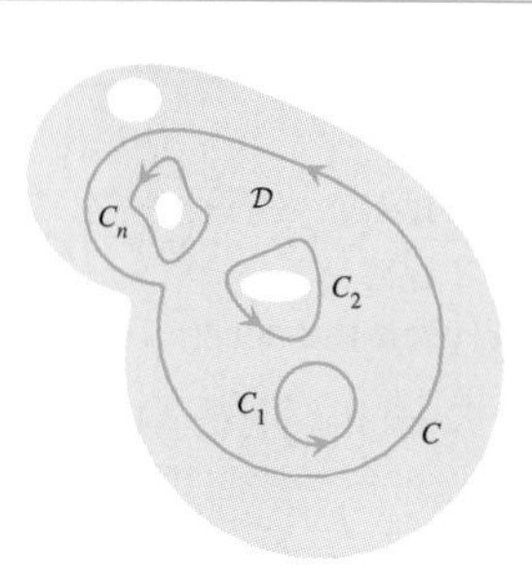

Let $C, C_1, C_2, \ldots, C_n$ be simple, closed, piecewise smooth planar curves such that each C_k lies inside of C, and so that C_j lies outside of C_k for every choice of j and k with $j \neq k$. Assume each of these curves has counterclockwise orientation. Let $\mathbf{F} = \langle P, Q \rangle$ be a continuously differentiable vector field whose domain contains all of these curves and the region $\mathcal{D}$ inside of C but outside each C_k. Then

$$\int_C P\,dx + Q\,dy - \int_{C_1} P\,dx + Q\,dy - \cdots - \int_{C_n} P\,dx + Q\,dy = \iint_{\mathcal{D}} \left(\frac{\partial Q}{\partial x} - \frac{\partial P}{\partial y} \right) dA.$$

Let C and C_1 be the circles of center $(0,0)$ and radii 2 and 1, respectively, both with counterclockwise orientation, and let $\mathcal{D}$ be the ring-shaped region between the two circles. Then

$$\int_C -y\,dx + x\,dy - \int_{C_1} -y\,dx + x\,dy = \iint_{\mathcal{D}} 2\,dA = 2(\text{area of } \mathcal{D}) = 6\pi.$$

The Surface Integral

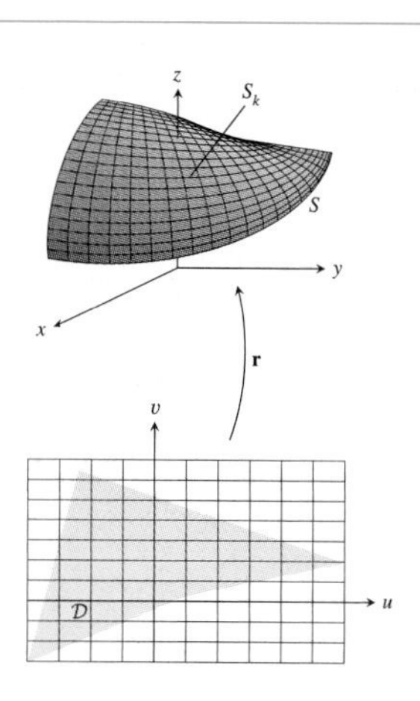

Let S be a smooth surface and let f be a continuous real-valued function defined on S. The surface integral

$$\iint_S f\,d\sigma$$

is defined to be the limit of Riemann sums associated with regular subdivisions of S and associated evaluation sets.

Let S be a smooth surface represented parametrically by

$$\mathbf{r} = \mathbf{r}(u, v)$$

for (u, v) in some bounded region $\mathcal{D}$ of the (u, v)-plane, and let f be a continuous real-valued function defined on S. Let $\{S_{i,j}\}$ be the pieces of S that arise by forming a regular subdivision of $\mathcal{D}$ into n^2 pieces and then using $\mathbf{r}$ to map the subdivision onto S. Let $\{\mathbf{c}_{i,j}\}$ be an evaluation set associated with the subdivision. The Riemann sum associated with this subdivision and evaluation set is

$$\sum f(\mathbf{c}_{i,j})\,\Delta\sigma_{i,j},$$

where $\Delta\sigma_{i,j}$ is the area of $S_{i,j}$. The surface integral of f with respect to surface area on S is given by

$$\iint_S f\,d\sigma = \lim_{n\to\infty} \sum f(\mathbf{c}_{i,j})\,\Delta\sigma_{i,j},$$

provided that this limit exists.

Evaluation of $\iint_S f\,d\sigma$

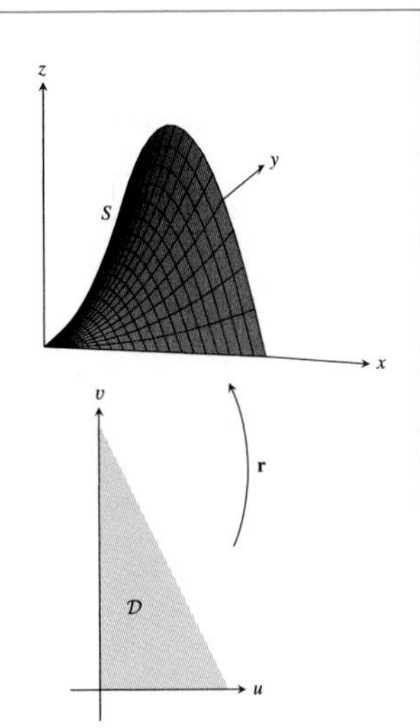

Let S be a smooth surface described by

$$\mathbf{r} = \mathbf{r}(u, v), \qquad (u, v) \in \mathcal{D},$$

for some bounded planar set $\mathcal{D}$. If f is a continuous real-valued function defined on S, then

$$\iint_S f\,d\sigma = \iint_{\mathcal{D}} f(\mathbf{r}(u, v))\,\|\mathbf{r}_u \times \mathbf{r}_v\|\,du\,dv.$$

Let S be the graph of $z = xy$ for (x, y) in the region $\mathcal{D}$ bounded by the triangle with vertices $(0, 0)$, $(2, 0)$ and $(0, 4)$, and let $f(x, y, z) = z$. Then S is represented parametrically by

$$\mathbf{r} = \mathbf{r}(u, v) = \langle u, v, uv \rangle$$

for (u, v) in $\mathcal{D}$. Then

$$\|\mathbf{r}_u \times \mathbf{r}_v\| = \sqrt{u^2 + v^2 + 1}$$

and

$$\iint_S f\,d\sigma = \iint_{\mathcal{D}} uv\sqrt{u^2 + v^2 + 1}\,du\,dv.$$

Orientable Surface

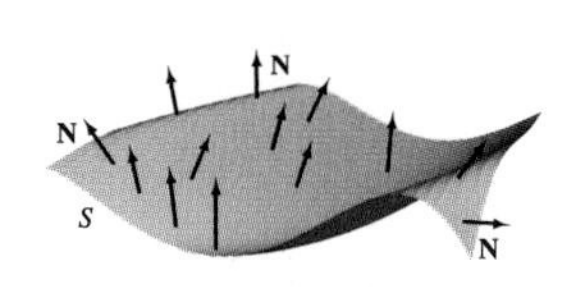

A smooth surface S is orientable if there is a continuous function $\mathbf{N}$ defined on S such that at each point (x, y, z) of S, the vector $\mathbf{N}(x, y, z)$ is a unit vector normal to S at (x, y, z). The function $\mathbf{N}$ is called an *orientation* on S. Orientation on a surface is important when the idea of direction through a surface or direction around a surface is needed.

Let S be a surface described parametrically by

$$\mathbf{r} = \mathbf{r}(u, v), \qquad (u, v) \in \mathcal{D}.$$

Suppose that $\mathbf{r}$ is continuously differentiable and that $\mathbf{r}_u \times \mathbf{r}_v$ is never $\mathbf{0}$. Given (x, y, z) on S, find (u, v) with $\mathbf{r}(u, v) = \langle x, y, z \rangle$. Then

$$\mathbf{N} = \mathbf{N}(x, y, z) = \frac{\mathbf{r}_u \times \mathbf{r}_v}{\|\mathbf{r}_u \times \mathbf{r}_v\|}$$

defines an orientation on S.

The Divergence Theorem

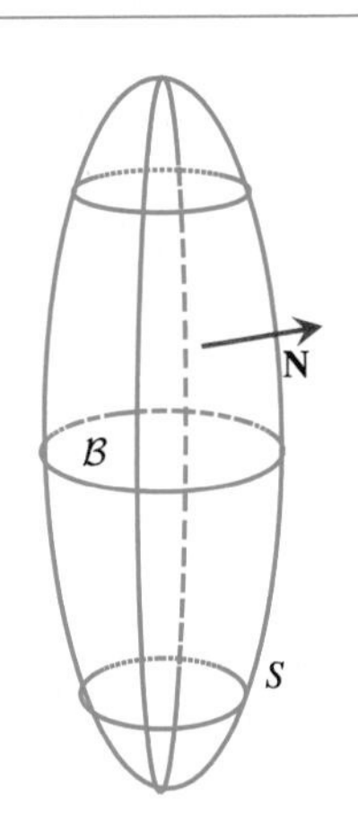

Let S be an orientable surface in space bounding a bounded simple solid region $\mathcal{B}$. Let $\mathbf{N}$ denote the outward normal on S. If $\mathbf{F}$ is a continuously differentiable vector field defined on a set containing S and $\mathcal{B}$, then

$$\iint_S (\mathbf{F} \cdot \mathbf{N})\,d\sigma = \iiint_{\mathcal{B}} \operatorname{div} \mathbf{F}\,dV.$$

Let S be the ellipsoid with equation

$$\frac{x^2}{4} + y^2 + \frac{z^2}{25} = 1$$

and let $\mathcal{B}$ be the region inside of S. Because $\operatorname{div}\langle x, y, z \rangle = 3$, we have

$$\begin{aligned} \iint_S \langle x, y, z \rangle \cdot \mathbf{N}\,d\sigma &= \iiint_{\mathcal{B}} 3\,dV \\ &= 3(\text{vol. of ellipsoid}) \\ &= 3\left(\frac{4}{3}\pi \cdot 2 \cdot 1 \cdot 5\right) = 40\pi. \end{aligned}$$

Stokes' Theorem

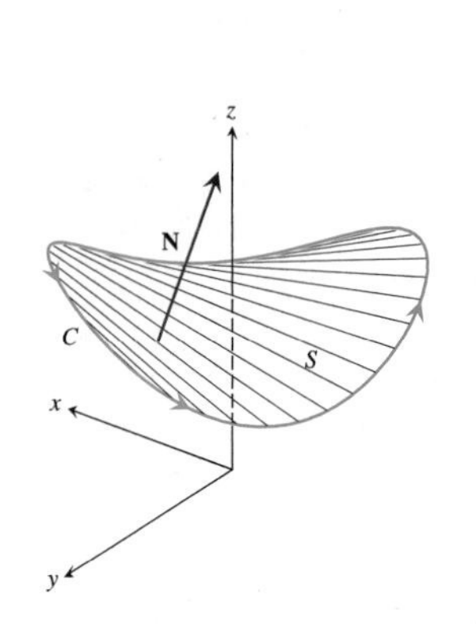

Let S be a simple smooth surface and let C be the boundary of S. Assume that $\mathbf{N}$ is an orientation for S and that with respect to this orientation C is described in the positive (counterclockwise) direction. If $\mathbf{F}$ is a continuously differentiable vector field defined on S and C, then

$$\iint_S (\nabla \times \mathbf{F}) \cdot \mathbf{N}\, d\sigma = \int_C (\mathbf{F} \cdot \mathbf{T})\, ds,$$

where $\mathbf{T}$ is the unit tangent to C.

Let S be the upper half of the sphere with center $(0,0,0)$ and radius 1 and let $\mathbf{F}(x,y,z) = \langle -y, x, z\rangle$. Because C is the circle of center $(0,0,0)$ and radius 1 in the (x,y)-plane, the vector tangent to C at a point $(x,y,0)$ is $\mathbf{T} = \langle -y, x, 0\rangle$. Hence $\mathbf{F} \cdot \mathbf{T} = y^2 + x^2 = 1$ and

$$\iint_S (\nabla \times \mathbf{F}) \cdot \mathbf{N}\, d\sigma = \int_C 1\, ds = 2\pi.$$

Chapter Review Exercises

Exercises 1–16: Evaluate the integral.

1. $\displaystyle\int_C xy^2\, ds,$

C the line segment from $(-2,3)$ to $(1,-4)$

2. $\displaystyle\int_C (\mathbf{F} \cdot \mathbf{T})\, ds,\ \mathbf{F}(x,y,z) = \langle x + y, y + z, z + x\rangle,$

C the circle of center $(0,0,0)$ and radius 1 in the (x,z)-plane. Assume that C is oriented in the direction from $(1,0,0)$ to $(0,0,1)$ to $(-1,0,0)$.

3. $\displaystyle\int_C (2 + xz)xye^{xz}\, dx + x^2e^{xz}\, dy + x^3ye^{xz}\, dz,$

C given by $\mathbf{r}(t) = \langle t^2 + 1, te^{t^2}, \cos \pi t\rangle,\ 0 \le t \le 1$

4. $\displaystyle\int_C x^2y\, dx + (xy^3 + 1)\, dy,$

C the rectangle with vertices $(1,1)$, $(5,1)$, $(5,3)$, and $(1,3)$, oriented in the counterclockwise direction

5. $\displaystyle\int_C (\mathbf{F} \cdot \mathbf{N})\, ds,\ \mathbf{F}(x,y) = \langle 2x^2 - xy, 3xy + y^2 + 1\rangle,$

C the triangle with vertices $(0,0)$, $(3,0)$, and $(0,4)$, oriented in the counterclockwise direction

6. $\displaystyle\int_C \mathbf{F} \cdot \mathbf{N}\, ds,\ \mathbf{F}(x,y) = \langle x^2 - y^2, 2xy\rangle,$

C described by $\mathbf{r}(t) = \langle 2 \cos 2t, 2 \sin 2t\rangle,\ \pi/4 \le t \le \pi/2$

7. $\displaystyle\iint_S (x^2 + y^2)\, d\sigma,$

S the sphere with center $(0,0,0)$ and radius 1

8. $\displaystyle\iint_S (\mathbf{F} \cdot \mathbf{N})\, d\sigma,\ \mathbf{F}(x,y,z) = \langle 2y, 3x, z\rangle,$

S the sphere with equation $x^2 + y^2 + z^2 = 2$ and $\mathbf{N}$ the outward unit normal to S

9. $\displaystyle\iint_S (\nabla \times \mathbf{F}) \cdot \mathbf{N}\, d\sigma,\ \mathbf{F}(x,y,z) = \langle 2y, 3x, z\rangle,$

S the half-sphere given by $x^2 + y^2 + z^2 = 2,\ z \ge 0$, and $\mathbf{N}$ the unit normal pointing away from the origin

10. $\displaystyle\int_C \langle 2x, -y, 3z\rangle \cdot \mathbf{T}\, ds,$

C the curve described by $y^2 + z^2 = 4,\ x = 3$, with counterclockwise orientation relative to the positive x-axis

11. $\displaystyle\int_C \langle 5, 3, \sqrt{2}\rangle \cdot \mathbf{T}\, ds,$

C the curve determined by the intersection of the paraboloid with equation $z = 2x^2 + 5y^2$ and the plane with equation $x + 3y + 7z = 44$. Assume that C is oriented with counterclockwise orientation relative to the positive z-axis.

12. $\int_C \mathbf{F}\cdot\mathbf{T}\,ds$, $\mathbf{F}(x,y,z) = \langle x^2 + y, z, z^4 + 2x\rangle$,
C described by $\mathbf{r}(t) = \langle \cos t, \sin t, \cos t + \sin t\rangle$, $0 \le t \le 2\pi$

13. $\iint_S \mathbf{F}\cdot\mathbf{N}\,d\sigma$, $\mathbf{F}(x,y,z) = \langle x^2 + y, x + y^2, -2yz\rangle$,
S the surface of the tetrahedron with vertices $(3,0,0)$, $(3,4,0)$, and $(0,4,0)$, and $(3,4,-5)$, and $\mathbf{N}$ the outward unit normal to S

14. $\iint_S \mathbf{F}\cdot\mathbf{N}\,d\sigma$, $\mathbf{F}(x,y,z) = \langle x^3 + yz, xz + y^3, -2z\rangle$,
S the surface of the cylinder bounded by the surfaces described by $x^2 + y^2 = 9$, $z = -1$, and $z = 5$, and $\mathbf{N}$ the outward unit normal to S

15. $\iint_S (\nabla \times \mathbf{F})\cdot\mathbf{N}\,d\sigma$, $\mathbf{F}(x,y,z) = \langle 2y, -x, 4z^2\rangle$,
S given by $x^2 + y^2 + z^2 = 1$, $x \ge 0$, $y \ge 0$, $z \ge 0$, and $\mathbf{N}$ the unit normal pointing away from the origin

16. $\int_C \mathbf{F}\cdot\mathbf{T}\,ds$, $\mathbf{F}(x,y,z) = \langle y, x, -6z\rangle$,
C given by $\mathbf{r}(t) = \langle \cos t, \cos 2t, t \sin t\rangle$, $0 \le t \le 2\pi$

Exercises 17–22: Use the Gradient Test to decide whether or not the vector field $\mathbf{F}$ *may have a potential function. Find a potential function if one exists.*

17. $\mathbf{F}(x,y) = \langle y \sin(x^2y) + 2x^2y^2\cos(x^2y), x^3\cos(x^2y)\rangle$

18. $\mathbf{F}(x,y) = \left\langle \frac{y^2 - x^2}{(x^2+y^2)^2}, \frac{-2xy}{(x^2+y^2)^2}\right\rangle$

19. $\mathbf{F}(x,y) = \left\langle \frac{x}{\sqrt{x^2+2y}}, \frac{1}{\sqrt{x^2+2y}}\right\rangle$

20. $\mathbf{F}(x,y,z) = \langle x + y + z, x + y + z, x + y + z\rangle$

21. $\mathbf{F}(x,y,z) = \langle x^2e^{y+z}, y^2e^{x+z}, z^2e^{x+y}\rangle$

22. $\mathbf{F}(x,y,z) = \left\langle \frac{2xy}{x^2y + y^2z}, \frac{x^2 + 2yz}{x^2y + y^2z}, \frac{y^2}{x^2y + y^2z}\right\rangle$

23. Let $\mathbf{v}(x,y) = \langle x^2 + 2xy, -3xy - y^2\rangle$ describe the velocity field for a fluid flowing in the plane. Assume that at point (x,y) the fluid has density $(2 - x)\ \mathrm{g/cm^2}$, and let C be the circle of center $(0,0)$ and radius 1 with counterclockwise orientation.

a. Find the net mass rate of flow (flux) across C.

b. Find the circulation around C.

24. Let $\mathbf{v}(x,y,z) = \langle xyz, z + 1, x - y\rangle$ describe the velocity field for a fluid flowing in space. Assume that the fluid has constant density $1\ \mathrm{g/cm^2}$, and let S be the surface described by $z = 9 - x^2 - y^2$, $z \ge 0$. Orient S using the unit normal with positive third component.

a. Find the net mass rate of flow (flux) across S.

b. Find the circulation around the boundary of S.

25. Let S and S_1 be the boundaries of two simple closed surfaces and assume that S_1 lies inside of S. Let $\mathcal{B}$ be the region between the two surfaces and let $\mathbf{F}$ be a continuously differentiable vector field defined on the set containing S, S_1, and $\mathcal{B}$. Show that

$$\iint_S \mathbf{F}\cdot\mathbf{N}\,d\sigma - \iint_{S_1} \mathbf{F}\cdot\mathbf{N}\,d\sigma = \iiint_{\mathcal{B}} \operatorname{div}\mathbf{F}\,dV,$$

where $\mathbf{N}$ denotes the outward unit normal on each surface. (Assume that $\mathcal{B}$ can be written as a union of a finite number of disjoint simple solids.)

26. Let S and S_1 be the boundaries of two simple surfaces and assume that S_1 lies inside of S. Let $\mathcal{B}$ be the region between the two surfaces and let $\mathbf{F}$ be a continuously differentiable vector field defined on the set containing S, S_1, and $\mathcal{B}$. Show that if $\operatorname{div}\mathbf{F} \equiv 0$ on $\mathcal{B}$, then

$$\iint_S \mathbf{F}\cdot\mathbf{N}\,d\sigma = \iint_{S_1} \mathbf{F}\cdot\mathbf{N}\,d\sigma.$$

Here $\mathbf{N}$ denotes the outward unit normal on each surface. See Exercise 25.

27. An electrically charged particle is positioned at the origin. The electromagnetic field $\mathbf{E}$ generated by the particle is defined at all points $\mathbf{r} = \langle x,y,z\rangle \ne \langle 0,0,0\rangle$. For such $\mathbf{r}$, the magnitude of $\mathbf{E}(\mathbf{r})$ is equal to the magnitude of the force exerted on a particle of unit charge positioned at $\mathbf{r}$. The magnitude of this force is directly proportional to the charge of the particle at the origin, and inversely proportional to the square of the distance of $\mathbf{r}$ from the origin. The force is

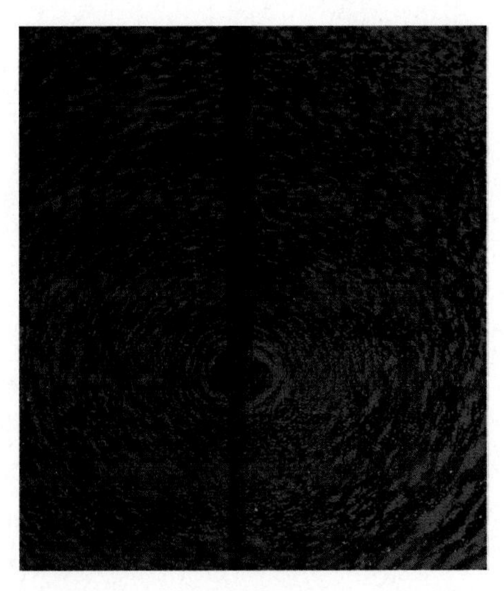

directed towards the origin if the force is attractive, and

away from the origin if it is repulsive. The vector field $\mathbf{E}$ carries units of force per unit of charge.

a. Write a formula for $\mathbf{E}(\mathbf{r})$.

b. Sketch a graph $\mathbf{E}(\mathbf{r})$ for points $\mathbf{r}$ on a sphere centered at the origin.

c. Sketch a graph $\mathbf{E}(\mathbf{r})$ for points $\mathbf{r}$ on a plane that does not pass through the origin.

d. Let S be the sphere of radius R centered at the origin. Show that the flux of $\mathbf{E}$ through S is independent of R.

e. Let S be the boundary surface of any simple solid. Show that if the origin lies outside of S, then the flux of $\mathbf{E}$ through S is 0 and that if the origin lies inside of S, then the flux is the same as the flux through a sphere centered at the origin.

28. An infinitely long wire has constant charge density of c coulombs/cm. In this exercise we ask you to find and work with the electromagnetic field generated by the wire. At a point $\mathbf{r}$ not on the wire, the electromagnetic field $\mathbf{E}(\mathbf{r})$ induced by the charged wire is found by "summing" the electromagnetic field contributions at $\mathbf{r}$ for each small piece of the wire. In particular, assume that the wire runs along the z-axis and that $\mathbf{r} = \langle x_0, y_0, z_0 \rangle$. Consider a piece of wire of length dz at position $\langle 0, 0, z \rangle$ on the wire.

a. If the piece is small enough, we can think of it as a point with charge cdz coulombs. Use the formula for the electromagnetic field for a point charge found in the previous exercise to find the electromagnetic field at $\mathbf{r}$ induced by the small piece of wire.

b. Use integration to "sum" all of the electromagnetic field contributions from each piece of wire. Evaluate the resulting improper integral to obtain a formula for $\mathbf{E}(\mathbf{r})$.

c. How is the magnitude of $\mathbf{E}(\mathbf{r})$ related to the distance of $\mathbf{r}$ from the wire? What is the direction of $\mathbf{E}(\mathbf{r})$?

d. A cylinder of height h and radius R is positioned so its axis lies on the z-axis. Let S be the surface of this cylinder, including the top and bottom. What is the flux of $\mathbf{E}$ through S?

29. Faraday's Law In 1831 Michael Faraday and Joseph Henry, working independently, discovered that when magnetic flux through a loop of conducting wire is varied, an electric current is produced in the wire. Experiments and observations led to the following relation:

$$\int_C \mathbf{E} \cdot \mathbf{T}\, ds = -\iint_S \frac{\partial \mathbf{B}}{\partial t} \cdot \mathbf{N}\, d\sigma, \qquad (1)$$

where $\mathbf{E}$ is the electromagnetic field (see Exercise 27), $\mathbf{B}(x, y, z, t)$ is the magnetic field at point (x, y, x) at time t, C is the closed curve defined by the loop of wire, and S is a smooth, orientable surface bounded by C. Equation (1) is known as Faraday's Law. Use Stokes' Theorem—on the left side of (1)—to derive the alternative form of Faraday's Law,

$$\nabla \times \mathbf{E} + \frac{\partial \mathbf{B}}{\partial t} = \mathbf{0}.$$

STUDENT PROJECTS

A. SOLID ANGLES

Figure 12.76 shows the planar angle determined by a point P and two rays or half-lines H_1 and H_2 emanating from P. The measure of this angle is the length θ of the arc cut from the unit circle with center at P by the rays H_1 and H_2. Solid angles are a natural extension of this idea. If a point source of light at a point P in space illuminates a surface S, the solid angle defined by P and S is, roughly, the bundle of light rays falling on S. A surface S (suggested by its boundary) and some of the rays from P to the boundary of S are shown in Fig. 12.77. To attach a numerical measure to this solid angle, we imagine a unit sphere centered at P. The measure of the solid angle defined by P and S is the area of the part of the unit sphere illuminated by the bundle of rays.

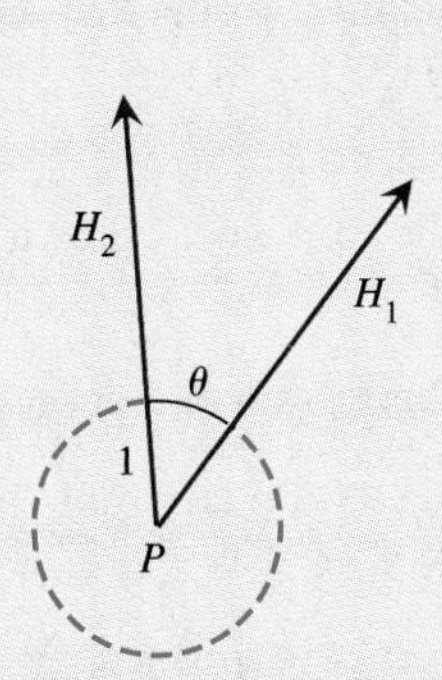

FIGURE 12.76 A planar angle.

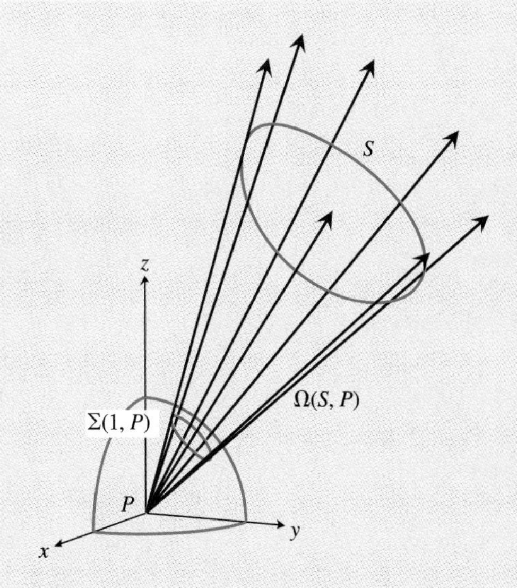

FIGURE 12.77 A solid angle.

DEFINITION Solid Angle

Let P be a given point and S a simple smooth surface not containing P. Assume that all rays from P intersect S at most once. The solid angle subtended by S at P, written as $\Omega(S, P)$, is the collection of all rays from P that intersect S. The measure of $\Omega(S, P)$, written $|\Omega(S, P)|$, is defined to be the area of $\Sigma(1, P)$, where $\Sigma(a, P)$ is the intersection of $\Omega(S, P)$ with the sphere of radius a centered at P.

PROBLEM 1 Let P be a given point and S a simple smooth surface not containing P. Assume that all rays from P intersect S at most once. Show that for any $a > 0$,

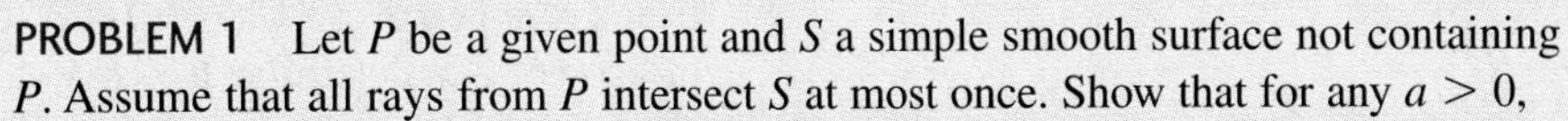

$$|\Omega(S, P)| = \frac{\text{area of } \Sigma(a, P)}{a^2} = \iint_S \frac{\mathbf{r} \cdot \mathbf{N}}{r^3}\, d\sigma. \tag{1}$$

Hint: Apply the Divergence Theorem to the solid V defined to be that part of $\Omega(S, P)$ between S and $\Sigma(a, P)$.

PROBLEM 2 The points $A = (4, 4, 4)$, $B = (2, 4, 5)$, and $C = (1, 1, 6)$ define a plane, triangular surface S. A generic view of S and these points is shown in Fig. 12.78. Calculate the solid angle subtended by S at $P = (0, 0, 0)$ in two ways:

a) Evaluate the surface integral given in (1). For this you may use the parametric description of S given by

$$\mathbf{r} = \mathbf{r}(s, t) = \langle 4, 4, 4 \rangle + s\langle -2, 0, 1 \rangle + t\langle -3, -3, 2 \rangle,$$

where $0 \le s \le 1$ and $0 \le t \le 1 - s$. With this parameterization, the "inner" iterated integral can be done by most CASs. The remaining integral can be done numerically.

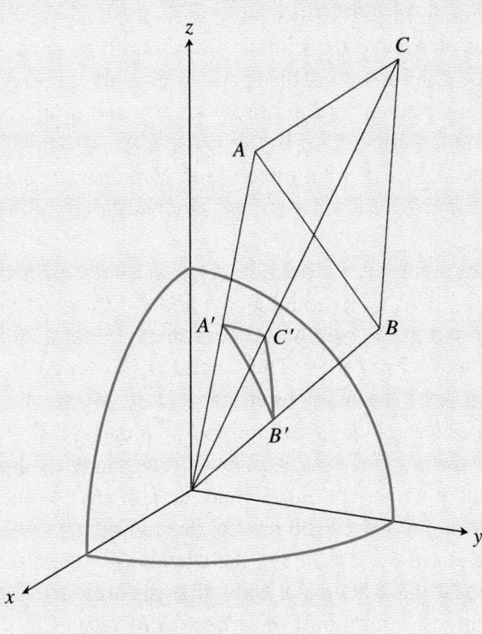

FIGURE 12.78 Figure for Problem 2.

b) From spherical trigonometry, the area of the spherical triangle $\Sigma(1, P)$ formed by intersecting $\Omega(S, P)$ with the unit sphere about P is $\alpha' + \beta' + \gamma' - \pi$, where α', β', and γ' are the radian measures of the angles at the points A', B', and C' on the unit sphere corresponding to A, B, and C. The points A', B', and C' may be found by noting that the point A', for example, is the intersection with the unit sphere of the line through the origin and the point A. It is given that $\alpha' \approx 0.7137243789$ and $\beta' \approx 2.063337179$.

PROBLEM 3 The surface S shown in Fig. 12.77 is part of the ellipsoid with equation

$$\mathbf{r} = \mathbf{r}(\phi, \theta) = \langle 2\cos\theta\sin\phi, 5\sin\theta\sin\phi, 4\cos\phi\rangle, \qquad (\phi, \theta) \in [0, \pi] \times [0, 2\pi].$$

The boundary of S is the curve of the ellipsoid traced by a point moving on the circle with equation

$$\left(\phi - \frac{3}{4}\right)^2 + \left(\theta - \frac{3}{4}\right)^2 = \frac{1}{4}$$

in the (ϕ, θ)-plane. Calculate $|\Omega(S, P)|$, where P is the origin.

B. THE POLAR PLANIMETER

In the physical, biological, and engineering sciences, it is not uncommon to be faced with the problem of finding the area of a region enclosed by a curve. When the curve arises from a series of observations or is traced by a mechanical recording device, an equation for the curve may not be available. The need to repeatedly find areas of such regions led to the invention of a device called the **planimeter**, which can be used to find such areas quickly and to a considerable degree of accuracy. The first planimeter was designed by J. M. Hermann in 1814. In ensuing years the design was improved, and many different kinds of planimeters evolved. According to an advertisement appearing in 1856 or 1857, the planimeter enables persons entirely ignorant of geometry to determine the area of any planar region more correctly and in much shorter time than the most experienced mathematician! In this project we investigate the polar planimeter. For more on other kinds of planimeters, see *Handbook of the Napier Tercentenary Celebration, Volume III,* edited by E. M. Horsburgh and published in 1982 by Tomash Publishers.

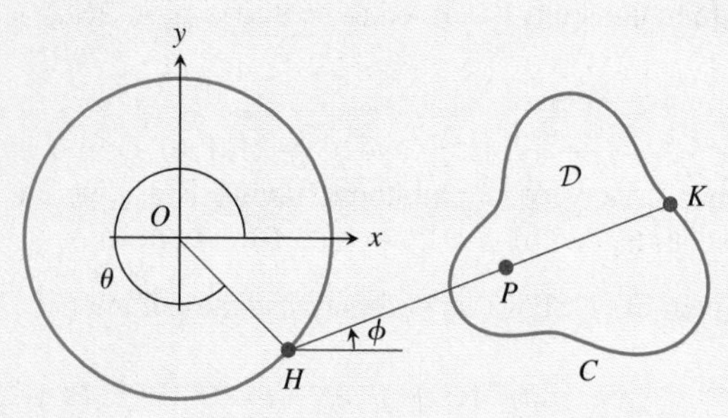

FIGURE 12.79 End H traces a circle, end K traces curve C.

Figure 12.79 shows a planar region $\mathcal{D}$, its boundary C, and an idealized polar planimeter with arms OH and HK of fixed lengths h and k, respectively. The point O of the first arm is fixed at the origin of a coordinate system, and its free end H moves on a circle with O as center. The second arm is hinged to the first arm at H. In the operation of the planimeter, an operator moves the free end K of the second arm counterclockwise around the boundary of $\mathcal{D}$. A recording wheel is mounted on the second arm at point P. The plane of the wheel is perpendicular both to the plane containing $\mathcal{D}$ and the arm HK. As the point K traces the curve C, the wheel at P will roll or slip, depending upon the directions of the line HK and the curve C. Attached to the wheel is a meter that measures the net angular motion of the wheel. We show that the area of $\mathcal{D}$ is directly proportional to the net angular motion of the wheel as K traces C.

We assume that the curve C is simple, smooth, and is described by

$$\mathbf{r} = \mathbf{f}(t), \qquad a \le t \le b. \tag{2}$$

PROBLEM 1 As the point K traces C, we assume that the angles θ and ϕ at O and H can be determined as continuously differentiable functions of t. Show that in terms of $\theta = \theta(t)$ and $\phi = \phi(t)$, the curve C_P generated by the point P may be described by

$$\mathbf{r} = \mathbf{r}(t) = h\langle \cos\theta, \sin\theta\rangle + pk\langle \cos\phi, \sin\phi\rangle, \qquad a \le t \le b, \tag{3}$$

where p is a fixed number between 0 and 1.

A unit tangent vector to C_P is $\left(\dfrac{1}{\|\mathbf{r}'\|}\right)\mathbf{r}'$, and the differential of arc length is $ds = \|\mathbf{r}'\|\,dt$. It follows that the vector $dt\,\mathbf{r}'$ is tangent to C_P and has length equal to ds. The wheel at P measures the length of the projection of this vector on the vector $\langle -\sin\phi, \cos\phi\rangle$, which is perpendicular to the arm HK. As P moves a distance ds along C_P, the wheel will record the displacement

$$d\sigma = dt\,\mathbf{r}' \cdot \langle -\sin\phi, \cos\phi\rangle.$$

Substituting $\mathbf{r}'$ from (3) into the expression for $d\sigma$ and simplifying,

$$d\sigma = pk\phi'\,dt + h\theta' \cos(\theta - \phi)\,dt. \tag{4}$$

PROBLEM 2 Show that the net wheel displacement σ is given by

$$\sigma = h\int_a^b \theta' \cos(\theta - \phi)\,dt. \tag{5}$$

Tell why this shows that σ is independent of the position of the wheel on HK.

We show that $A = k\sigma$, that is, the area $\mathcal{D}$ is directly proportional to the net angular displacement of the wheel. Our argument uses the formula

$$A = \frac{1}{2}\int_C -y\,dx + x\,dy \tag{6}$$

for the area A enclosed by the curve C. This result comes from Green's Theorem. See Example 3 in Section 12.4.

PROBLEM 3 Describe C by

$$\mathbf{r} = \mathbf{f}(t) = \langle x(t), y(t)\rangle = h\langle \cos\theta, \sin\theta\rangle + k\langle \cos\phi, \sin\phi\rangle, \qquad a \le t \le b.$$

Show that the integrand of the line integral in (6) can be written as

$$-yx' + xy' = h^2\theta' + k^2\phi' + 2k\frac{d\sigma}{dt} - 2pk^2\phi' - hk(\theta' - \phi')\cos(\theta - \phi).$$

PROBLEM 4 Using (6) and the results of Problem 2 and Problem 3, show that

$$A = k\sigma,$$

that is, that the area of $\mathcal{D}$ is proportional to the net angular displacement of the wheel.

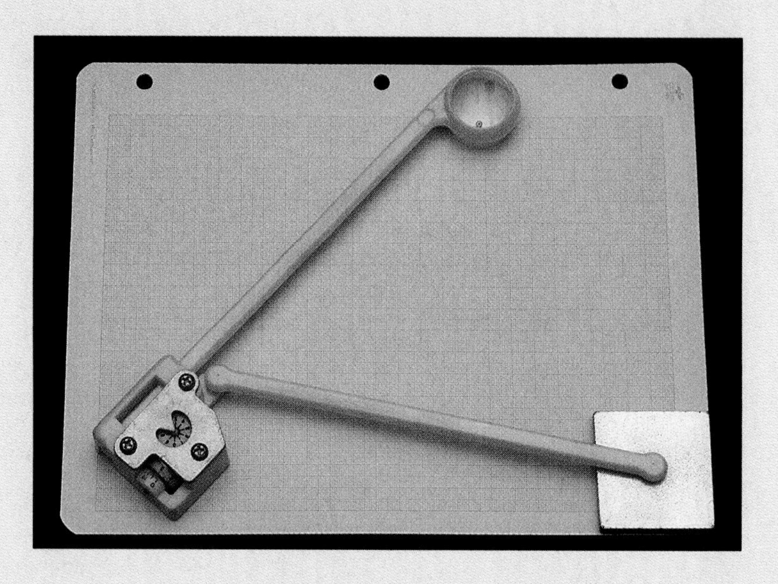

A polar planimeter.

Appendix

A.1 Algebra

We review the following topics from algebra:

- inequalities and the absolute value function
- factoring
- roots of polynomials
- the Binomial Theorem.

Inequalities and the Absolute Value Function

Several operations can be done on the inequalities $a \le b$, or $a < b$ to simplify them or to change their form. We state these for the $\le$ case but note that they hold for the $<$ case as well. Because $a > b$ and $a \ge b$ are equivalent to $b < a$ and $b \le a$, the operations also apply to the $\ge$ and $>$ cases.

Operations on Inequalities

If $a \le b$ and c is any number, then $a + c \le b + c$ (1)

If $a \le b$ and c is any number > 0, then $a \cdot c \le b \cdot c$ (2)

If $a \le b$ and c is any number < 0, then $a \cdot c \ge b \cdot c$ (3)

In words, (1) says that if any number is added to both sides of an inequality, then the inequality is preserved; (2) says that if both sides of an inequality are multiplied by the same positive constant, then the inequality is preserved, and (3) says that if both sides of an equality are multiplied by the same negative constant, then the inequality is "reversed."

The absolute value function $|x|$ is defined as

$$|x| = \begin{cases} x, & \text{if } x \ge 0 \\ -x, & \text{if } x < 0. \end{cases} \tag{4}$$

One application of the absolute value function is in calculating the distance between two points a and b on the x-axis. The distance between a and b is $a - b$ or $b - a$, depending upon which of these is nonnegative. Hence, the distance between a and b is $|a - b|$. We understand that a "distance" is a nonnegative number.

If a is a point on the x-axis and a is less than a distance d from a second point b, then $|a - b| < d$. We show that the condition $|a - b| < d$ is equivalent to the inequality $-d < a - b < d$, that is,

$$|a - b| < d \quad \text{if and only if} \quad -d < a - b < d. \tag{5}$$

We assume that $d > 0$ in the argument, which we split into two cases.

Case I: $a \geq b$. If $a \geq b$, then $|a - b| = a - b \geq 0$. Hence, the condition $|a - b| < d$ is equivalent to $a - b < d$. Because $-d < 0$, it follows directly that $-d < a - b < d$.

Case II: $a < b$. If $a < b$, then $|a - b| = b - a > 0$. Hence, the condition $|a - b| < d$ is equivalent to $b - a < d$ or, multiplying by -1, $-d < a - b$. Because $a - b < 0$ and $d > 0$, it follows that $-d < a - b < d$.

It now follows that if $|a - b| < d$ holds, then $-d < a - b < d$. The converse, that is, if $-d < a - b < d$, then $|a - b| < d$, follows by the definition of the absolute value function. This completes the proof of the equivalence (5).

The Triangle Inequality

The absolute value of a sum of real numbers is less than or equal to the sum of their absolute values, that is, if n is a positive integer and $a_1, a_2, \ldots, a_n$ are real numbers, then

$$|a_1 + a_2 + \cdots + a_n| \leq |a_1| + |a_2| + \cdots + |a_n|. \tag{6}$$

Factoring

We list several factoring identities useful in calculus.

$$a^2 \pm 2ab + b^2 = (a \pm b)^2 \tag{7}$$

$$a^2 - b^2 = (a - b)(a + b) \tag{8}$$

$$a^3 + b^3 = (a + b)(a^2 - ab + b^2) \tag{9}$$

$$a^3 - b^3 = (a - b)(a^2 + ab + b^2) \tag{10}$$

Identities (8) and (10) are special cases of the identity

$$a^n - b^n = (a - b)(a^{n-1} + a^{n-2}b^1 + \cdots + a^1b^{n-2} + b^{n-1}). \tag{11}$$

Roots of Polynomials

We review several results useful in finding the roots of polynomials with real coefficients. We state these results as theorems about the polynomial

$$p(x) = a_nx^n + a_{n-1}x^{n-1} + \cdots + a_1x + a_0.$$

We assume that $a_n \neq 0$ so that $p(x)$ is of degree n. We use the terms *roots* and *factors*. A real or complex number r is a root of the polynomial equation $p(x) = 0$ if $p(r) = 0$. The polynomial $x - r$ is a factor of $p(x)$ if there is a polynomial $q(x)$ for which $p(x) = (x - r)q(x)$.

Roots and Factors Theorem A real or complex number r is a root of the polynomial equation $p(x) = 0$ if and only if $x - r$ is a factor of the polynomial $p(x)$.

Rational Roots Theorem If the coefficients $a_n, a_{n-1}, \ldots, a_1, a_0$ of the polynomial $p(x)$ are integers, then a rational number u/v (where u and v have no common factors) can be a root of $p(x)$ only if v is a factor of a_n and u is a factor of a_0.

Fundamental Theorem of Algebra The polynomial equation $p(x) = 0$ has at least one root.

We mention briefly three other results.

1. The polynomial equation $p(x) = 0$ has n roots, though they are not necessarily distinct.
2. If the polynomial $p(x)$ is of odd degree, then the equation $p(x) = 0$ has at least one real root.
3. If the complex number $a + ib$ is a root of the polynomial equation $p(x) = 0$, then $a - ib$ is also a root, that is, complex roots come in conjugate pairs $a \pm ib$.

We illustrate some of these theorems by solving the polynomial equation

$$p(x) = 3x^5 + 13x^4 + 20x^3 + 16x^2 - 16.$$

Because this polynomial has degree 5, it has five roots, at least one of which must be real. The real roots may be either irrational or rational, but IF they are rational, (say, u/v, where u and v are integers), then v must be a factor of $a_5 = 3$ and u must be a factor of $a_0 = -16$. The factors of 3 are ± 1 and ± 3; the factors of -16 are ± 1, ± 2, ± 4, ± 8, and ± 16. If we go through these possibilities, testing as we go, we'll find that -2 and $2/3$ are roots.

Hence, $x - (-2) = x + 2$ and $x - 2/3$ are factors and can be "divided out." By long division or two synthetic divisions,

$$\frac{3x^5 + 13x^4 + 20x^3 + 16x^2 - 16}{(x + 2)(x - 2/3)} = 3x^3 + 9x^2 + 12x + 12.$$

The remaining roots of $p(x) = 0$ are the roots of the polynomial equation

$$3x^3 + 9x^2 + 12x + 12 = 3(x^3 + 3x^2 + 4x + 4) = 0.$$

We may apply all of the theorems mentioned earlier to this cubic polynomial. We find, for example, using the Rational Roots Theorem, that -2 is a root. This means that -2 is a double root of the polynomial equation $p(x) = 0$. Dividing the cubic by $x + 2$ gives

$$\frac{x^3 + 3x^2 + 4x + 4}{x + 2} = x^2 + x + 2.$$

The roots of $x^2 + x + 2 = 0$ are $-1/2 \pm i\sqrt{7}/2$. Hence, the roots of the polynomial equation $p(x) = 0$ are $-2, -2, 2/3$, and the conjugate pair $-1/2 \pm i\sqrt{7}/2$. Thus we have five roots, three of which are distinct and one of which is repeated.

The Binomial Theorem

Although the Binomial Theorem can be regarded as a factoring identity, as in

$$(x + y)^3 = x^3 + 3x^2y + 3xy^2 + y^3,$$

it is usually discussed separately because it has applications unrelated to factoring.

THEOREM The Binomial Theorem

If a and b are numbers and n is a positive integer, then

$$(a + b)^n = \binom{n}{0}a^n + \binom{n}{1}a^{n-1}b^1 + \cdots + \binom{n}{n-1}a^1b^{n-1} + \binom{n}{n}b^n, \tag{12}$$

where

$$\binom{n}{k} = \frac{n!}{k!\,(n-k)!}, \qquad k = 0, 1, \ldots, n.$$

The binomial coefficient $\binom{n}{k}$ is the number of ways of choosing k things from n distinct objects. There are several widely used methods for generating the binomial coefficients. One way is to use Pascal's triangle. Another is to generate each coefficient from the one just preceding it. We illustrate this method by expanding $(a + b)^4$. Begin by writing the first term, a^4, and regard its coefficient as 1. Then apply the rule: The coefficient of the $(k + 1)$st term is the coefficient of the kth term times the exponent of a in the kth term divided by the "number" of the kth term, which is just k. Hence,

$$\begin{aligned}(a + b)^4 &= a^4 + \frac{1\cdot 4}{1}a^3b^1 + \frac{4\cdot 3}{2}a^2b^2 + \frac{6\cdot 2}{3}a^1b^3 + \frac{4\cdot 1}{4}b^4 \\ &= a^4 + 4a^3b + 6a^2b^2 + 4ab^3 + b^4.\end{aligned}$$

A.2 Trigonometry

Angles

In most work in mathematics, angles are measured in degrees or radians. Because one complete revolution measures 360° or 2π radians, we have the relationship

$$2\pi \text{ radians} = 360°. \tag{13}$$

Using (13) and proportions, we can convert angle measures from degrees to radians and from radians to degrees.

EXAMPLE 1

a) Find the radian measure of an angle of measure 225°.
b) Find the degree measure for an angle of measure 2 radians.

Solution

a) Let r be the radian measure for the angle. Then

$$\frac{r}{2\pi} = \frac{225}{360}, \quad \text{so} \quad r = \frac{450\pi}{360} = \frac{5\pi}{4} \approx 3.927 \text{ radians.}$$

b) Let d be the degree measure for the angle. Then

$$\frac{d}{360} = \frac{2}{2\pi}, \quad \text{so} \quad d = \frac{360}{\pi} \approx 114.6°.$$

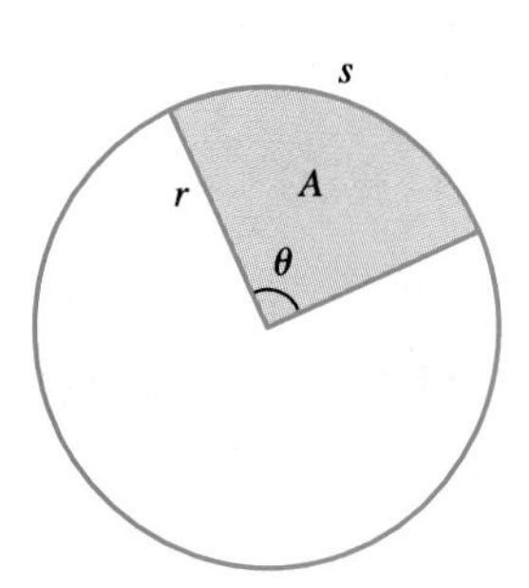

FIGURE A.1 Central angle θ subtends an arc of length s and a sector of area A.

The same work with ratios leads to formulas for the length of circular arcs and the area of sectors of circles. In Fig. A.1 we show a central angle of θ radians in a circle of radius r. The length s of the arc subtended by the angle is the fraction $\theta/2\pi$ of a full circumference. Hence

$$\frac{\theta}{2\pi} = \frac{s}{2\pi r}.$$

It follows that

$$s = r\theta.$$

The area A of the sector is the fraction $\theta/2\pi$ of the area of the circle, so

$$\frac{\theta}{2\pi} = \frac{A}{\pi r^2}.$$

Solving for A,

$$A = \frac{1}{2}r^2\theta.$$

We summarize these results.

> A central angle of measure θ radians $(0 \leq \theta \leq 2\pi)$ in a circle of radius r subtends an arc of length
>
> $$s = r\theta.$$
>
> The area of the sector determined by this angle is
>
> $$A = \frac{1}{2}r^2\theta.$$

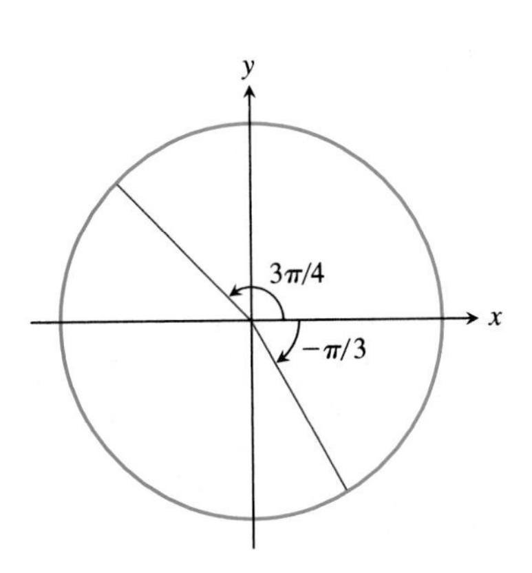

FIGURE A.2 Angles of measure $3\pi/4$ and $-\pi/3$.

When we work with angles in the Cartesian plane, we often place them in standard position with vertex at the origin and initial side on the positive x-axis. Angles of positive measure are measured in the counterclockwise direction, and angles of negative measure are measured in the clockwise direction. In Fig. A.2 we show angles of measure $3\pi/4$ radians and $-\pi/3$ radians. We also work with angles of measure greater than 2π or less than -2π radians. These angles can be illustrated by indicating an appropriate number of complete revolutions before ending at the terminal side of the angle.

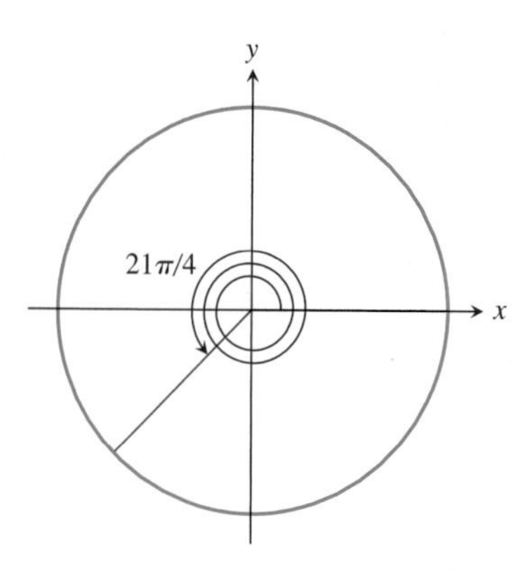

FIGURE A.3 An angle of measure $21\pi/4$.

EXAMPLE 2 Sketch the angle of measure $21\pi/4$ radians.

Solution

Note that

$$\frac{21\pi}{4} = 4\pi + \frac{5\pi}{4} = 2(2\pi) + \frac{5\pi}{4}.$$

This angle is co-terminal with an angle of $5\pi/4$ (225°), so has its terminal side in the third quadrant. The $2(2\pi)$ term means that two complete revolutions are completed before stopping at the terminal side. See Fig. A.3.

The Trigonometric Functions

The trigonometric functions are usually first seen in the study of right triangle trigonometry. Given a right triangle with one acute angle of measure θ, let "opp" denote the length of the side opposite angle θ, "adj" denote the length of the side adjacent to the angle, and "hyp" the length of the hypotenuse. See Fig. A.4. We define

$$\sin\theta = \frac{\text{opp}}{\text{hyp}} \qquad \cos\theta = \frac{\text{adj}}{\text{hyp}} \qquad \tan\theta = \frac{\text{opp}}{\text{adj}}.$$

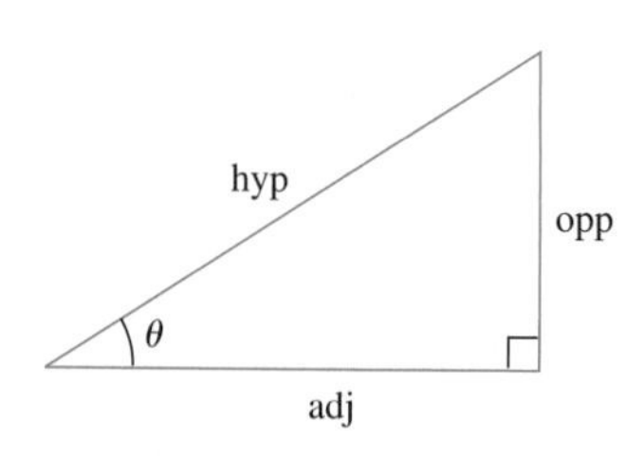

FIGURE A.4 The side opposite angle θ is labeled "opp." The side adjacent to the angle is labeled "adj," and the hypotenuse is labeled "hyp."

These ratios are the same for any right triangle with an acute angle θ because any two such triangles are similar. It is because the values of these functions depend only on the angles in the triangle and not on the size of the triangle that trigonometry is such a useful and powerful method for analyzing geometric figures.

The trigonometric functions can be defined for angles that are not acute angles by referring to the circle of center $O = (0, 0)$ and radius 1. This circle is referred to as the **unit circle.** Starting from the positive x-axis, measure an acute angle of measure θ and let $P = (x_\theta, y_\theta)$ be the point in the first quadrant where the terminal side of the angle intersects the circle. Drop a perpendicular from P to meet the x-axis at Q. Then triangle OPQ is a right triangle with acute angle θ. (See Fig. A.5.) Side PQ, which is opposite θ, has length y_θ; side OQ has length x_θ; and the hypotenuse has length 1. Thus

$$\cos\theta = \frac{\text{adj}}{\text{hyp}} = x_\theta \quad \text{and} \quad \sin\theta = \frac{\text{opp}}{\text{hyp}} = y_\theta.$$

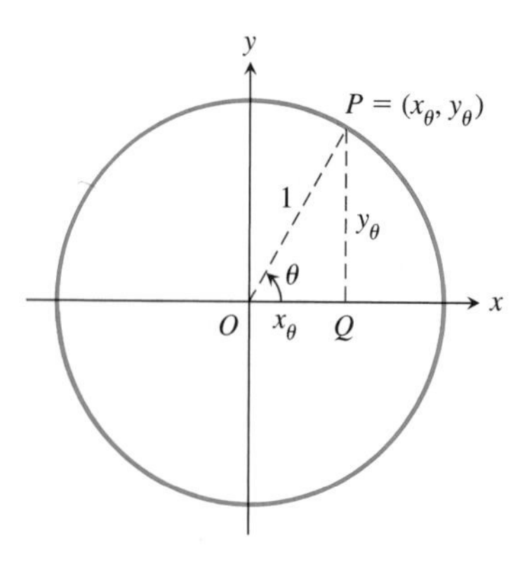

FIGURE A.5 The terminal side of the angle of measure θ meets the unit circle in the point with coordinates $(x_\theta, y_\theta) = (\cos\theta, \sin\theta)$.

Motivated by this matching of the values of $\cos\theta$ and $\sin\theta$ with the x- and y-coordinates of the point on the circle, we extend the domain of the sine and cosine functions from positive acute angles to all angles.

DEFINITION Sine and Cosine

Let θ be any real number. From the positive x-axis, measure an angle of θ radians. Let (x_θ, y_θ) be the coordinates of the point where the terminal side of the angle meets the unit circle. See Fig. A.6. We define

$$\cos\theta = x_\theta \quad \text{and} \quad \sin\theta = y_\theta.$$

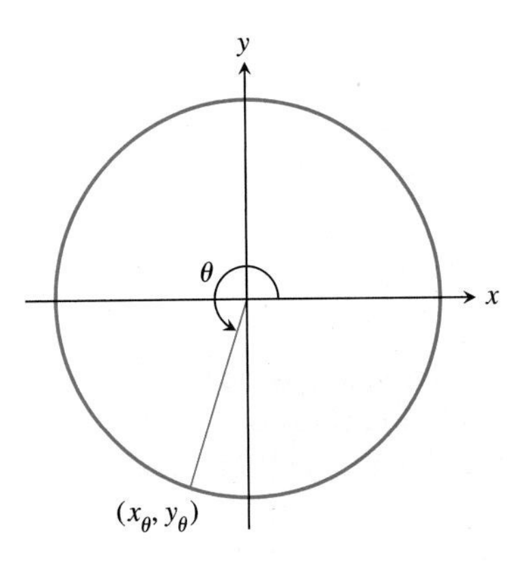

FIGURE A.6 With the unit circle, we can define sine and cosine for all angles.

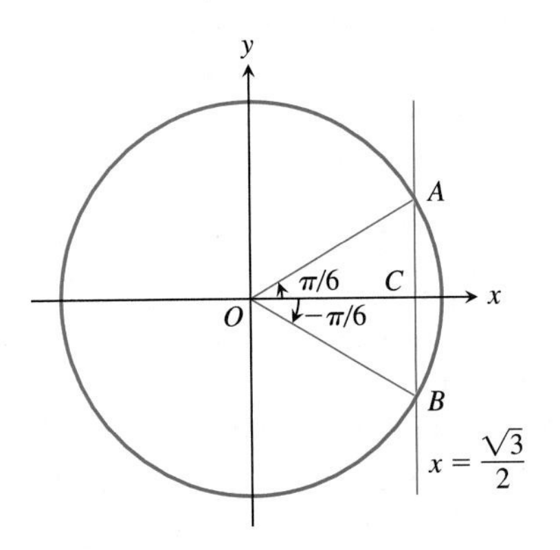

FIGURE A.7 The line $x = \sqrt{3}/2$ intersects the unit circle points that correspond to angles θ with $\cos\,\theta = \sqrt{3}/2$.

We will assume, unless told otherwise, that all angle measures are given in radians. The reason for this is simply because the calculus of trigonometric functions works out better in radians than it does in degrees. We say more about this in Section 2.5.

Many of the facts and identities that you might have learned (or memorized) in high school can be quickly found in the geometry of the unit circle. Get in the habit of using the circle to work with easy trigonometric identities. This will not only increase your understanding of the trigonometric functions, but also shorten the list of facts you might feel you need to memorize. For example, because the angles θ and $\theta + 2\pi$ intersect the unit circle in the same point, we immediately see that

$$\cos(\theta + 2\pi) = \cos(\theta) \quad \text{and} \quad \sin(\theta + 2\pi) = \sin\,\theta. \tag{14}$$

Thus the cosine and sine functions are both periodic with period 2π.

We present more examples of how the unit circle can be used to work with trigonometric functions and prove trigonometric identities.

EXAMPLE 3 Find all angles θ such that $\cos\,\theta = \sqrt{3}/2$.

Solution

In Fig. A.7 we show the unit circle and the line $x = \sqrt{3}/2$. This line intersects the circle in two points, A and B. Any angle with terminal side through A or B is an angle for which the cosine has value $\sqrt{3}/2$. Triangle AOC, shown in Fig. A.7, is a 30-60-90 right triangle with $30° = \pi/6$ radian angle at O. Hence $\cos(\pi/6) = \sqrt{3}/2$. By symmetry, we also have $\cos(\angle COB) = \cos(-\pi/6) = \sqrt{3}/2$. Once we have these two solutions, we can use (14) to write down all solutions,

$$\theta = \frac{\pi}{6} + 2\pi k \quad \text{or} \quad \theta = -\frac{\pi}{6} + 2\pi k,$$

where $k = 0, \pm 1, \pm 2, \pm 3, \ldots$.

EXAMPLE 4 Use the unit circle to show that

$$\cos^2\theta + \sin^2\theta = 1.$$

Solution

When an angle of measure θ is drawn in standard position, the terminal side intersects the unit circle in the point

$$(x_\theta, y_\theta) = (\cos\,\theta, \sin\,\theta).$$

Any point (x_θ, y_θ) on the unit circle satisfies the equation

$$1 = x_\theta^2 + y_\theta^2 = \cos^2\theta + \sin^2\theta.$$

EXAMPLE 5 Use the unit circle to show that

$$\cos\,\theta = \sin\left(\frac{\pi}{2} - \theta\right) \quad \text{and} \quad \sin\,\theta = \cos\left(\frac{\pi}{2} - \theta\right).$$

Solution

First observe that the average of the angles θ and $\pi/2 - \theta$ is

$$\frac{\theta + \left(\frac{\pi}{2} - \theta\right)}{2} = \frac{\pi}{4}.$$

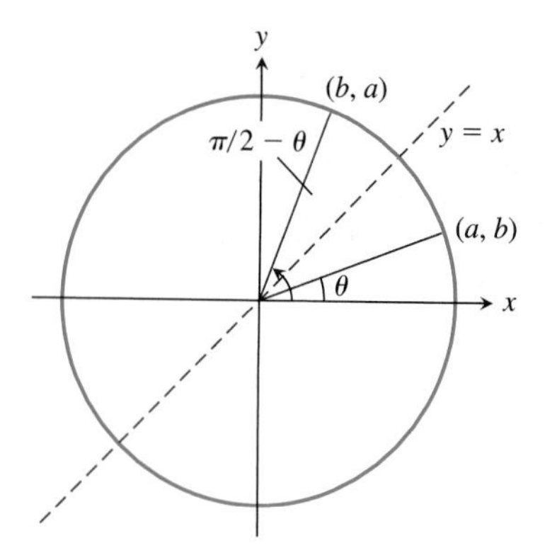

FIGURE A.8 The angles of measure θ and $\pi/2 - \theta$ are symmetric with respect to the line $y = x$.

Thus the terminal sides of θ and $(\pi/2) - \theta$ are symmetrically located with respect to the angle $\pi/4$. Hence one angle is the reflection of the other across the line $y = x$. See Fig. A.8. Thus, if the terminal side of the angle of measure θ intersects the circle at a point

$$P = (a, b) = (\cos\theta, \sin\theta),$$

then the terminal side of the angle of measure $(\pi/2) - \theta$ intersects the circle at point

$$Q = (b, a) = \left(\cos\left(\frac{\pi}{2} - \theta\right), \sin\left(\frac{\pi}{2} - \theta\right)\right).$$

We can then read off

$$\cos\left(\frac{\pi}{2} - \theta\right) = b = \sin\theta \quad \text{and} \quad \sin\left(\frac{\pi}{2} - \theta\right) = a = \cos\theta.$$

EXAMPLE 6 Use the unit circle to prove the double angle formula for the cosine,

$$\cos(2\theta) = \cos^2\theta - \sin^2\theta.$$

Solution

The circle on the left in Fig. A.9 shows the points $A = (1, 0)$ and $B = (\cos(2\theta), \sin(2\theta))$. The circle on the right shows the points $C = (\cos\theta, \sin\theta)$ and $D = (\cos(-\theta), \sin(-\theta))$. Because

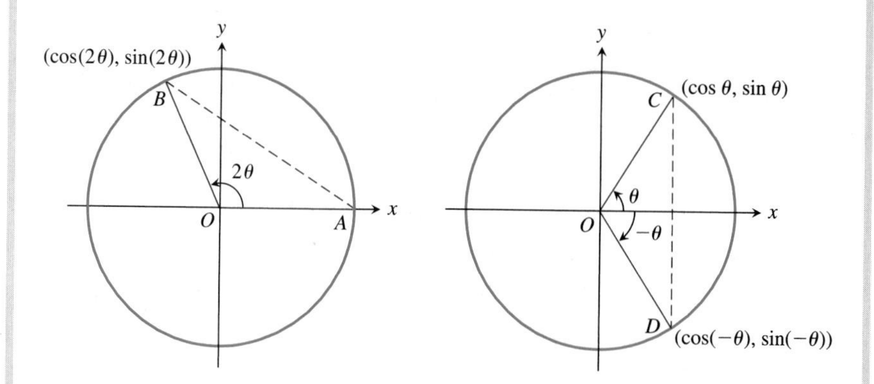

FIGURE A.9 Because $\angle AOB = \angle COD$, we have $AB = CD$.

$\angle AOB = \angle COD = 2\theta$, it follows that $AB = CD$. By the distance formula,

$$AB^2 = (1 - \cos(2\theta))^2 + (0 - \sin(2\theta))^2 \\ = 1 - 2\cos(2\theta) + \cos^2(2\theta) + \sin^2(2\theta) = 2 - 2\cos(2\theta). \tag{15}$$

Again using the distance formula,

$$CD^2 = (\cos\theta - \cos(-\theta))^2 + (\sin\theta - \sin(-\theta))^2 = 4\sin^2\theta.$$

In this last calculation, we have used the identities $\cos(-\theta) = \cos\theta$ and $\sin(-\theta) = -\sin\theta$. (You can see why these formulas are true by looking at the circle on the right in Fig. A.9.) Equating the expressions for AB^2 and CD^2, we have

$$2 - 2\cos(2\theta) = 4\sin^2\theta.$$

Rearranging, we find

$$\cos(2\theta) = 1 - 2\sin^2\theta = (1 - \sin^2\theta) - \sin^2\theta = \cos^2\theta - \sin^2\theta.$$

Once the sine and cosine are found, it is not hard to evaluate the other four trigonometric functions.

DEFINITION Tangent, Cotangent, Secant, and Cosecant

If θ is an angle for which $\cos\theta \neq 0$, (that is, $\theta \neq \pi/2 + \pi k$), then the tangent of θ is defined by

$$\tan\theta = \frac{\sin\theta}{\cos\theta}$$

and the secant of θ by

$$\sec\theta = \frac{1}{\cos\theta}.$$

If θ is an angle for which $\sin\theta \neq 0$, (that is, $\theta \neq \pi k$), then the cotangent of θ is defined by

$$\cot\theta = \frac{\cos\theta}{\sin\theta}$$

and the cosecant of θ by

$$\csc\theta = \frac{1}{\sin\theta}.$$

We list some of the more frequently encountered trigonometric identities. In Example 4 we proved the important identity

$$\cos^2\theta + \sin^2\theta = 1. \tag{16}$$

If we divide both sides of (16) by $\cos^2\theta$, we obtain

$$1 + \tan^2\theta = \sec^2\theta.$$

Dividing both sides of (16) by $\sin^2\theta$ gives

$$\cot^2\theta + 1 = \csc^2\theta.$$

As illustrated in Fig. A.9, the cosine function is an even function and the sine function is an odd function, that is,

$$\cos(-\theta) = \cos\theta \quad \text{and} \quad \sin(-\theta) = -\sin\theta. \tag{17}$$

In Example 5 we used the unit circle to establish the symmetry formulas,

$$\cos\left(\frac{\pi}{2} - \theta\right) = \sin\theta \quad \text{and} \quad \sin\left(\frac{\pi}{2} - \theta\right) = \cos\theta. \tag{18}$$

The angle addition formula for the cosine can be proved in a manner similar to that used in Example 6 to prove the double angle formula. The formula for the sine function follows from the formula for the cosine by using (18):

$$\begin{aligned} \cos(\alpha + \beta) &= \cos\alpha\cos\beta - \sin\alpha\sin\beta \\ \sin(\alpha + \beta) &= \sin\alpha\cos\beta + \cos\alpha\sin\beta. \end{aligned} \tag{19}$$

Replacing β by $-\beta$ in (19) and using (17), we obtain

$$\begin{aligned} \cos(\alpha - \beta) &= \cos\alpha\cos\beta + \sin\alpha\sin\beta \\ \sin(\alpha - \beta) &= \sin\alpha\cos\beta - \cos\alpha\sin\beta. \end{aligned} \tag{20}$$

Setting $\beta = \alpha$ in (19) gives the double-angle formulas,

$$\cos(2\alpha) = \cos^2\alpha - \sin^2\alpha \quad \text{and} \quad \sin(2\alpha) = 2\sin\alpha\cos\alpha. \tag{21}$$

Dividing the second formula in (19) by the first and doing a little algebra gives the angle addition formula for the tangent,

$$\tan(\alpha + \beta) = \frac{\tan\alpha + \tan\beta}{1 - \tan\alpha\tan\beta}.$$

To obtain the double-angle formula, set $\beta = \alpha$,

$$\tan(2\alpha) = \frac{2\tan\alpha}{1 - \tan^2\alpha}.$$

If we start with the basic identity $1 = \cos^2\alpha + \sin^2\alpha$, then add or subtract the double-angle formula $\cos 2\alpha = \cos^2\alpha - \sin^2\alpha$, we obtain the half-angle formulas,

$$\cos^2\alpha = \frac{1 + \cos(2\alpha)}{2} \quad \text{and} \quad \sin^2\alpha = \frac{1 - \cos(2\alpha)}{2}.$$

These formulas are used in integrating some trigonometric functions. By adding or subtracting various combinations of the formulas in (19) and (20), we obtain formulas for resolving products of trigonometric functions into sums:

$$\cos\alpha\cos\beta = \frac{1}{2}\cos(\alpha-\beta) + \frac{1}{2}\cos(\alpha+\beta)$$

$$\sin\alpha\sin\beta = \frac{1}{2}\cos(\alpha-\beta) - \frac{1}{2}\cos(\alpha+\beta)$$

$$\sin\alpha\cos\beta = \frac{1}{2}\sin(\alpha+\beta) + \frac{1}{2}\sin(\alpha-\beta).$$

Graphs of the Trigonometric Functions

The graph of $w = \sin\theta$ is not hard to sketch if we take note of the behavior of the y-coordinate of points on the unit circle as we move around the circle. When $\theta = 0$, this y-coordinate has value $\sin 0 = 0$. As angle θ increases from 0 to $\pi/2$, this y-coordinate increases from 0 to $\sin(\pi/2) = 1$. See Fig. A.10. The value of $\sin\theta$ then decreases from 1 to 0 as θ grows from $\pi/2$ to π, then continues to decrease to -1 as θ increases to $3\pi/2$. As θ increases from $3\pi/2$ to 2π, the value of $\sin\theta$ increases from -1 to 0. We have now moved once around the circle, and as θ grows from 2π to 4π, the $\sin\theta$ values will repeat the behavior just discussed. This behavior is reflected in the familiar graph shown in Fig. A.11.

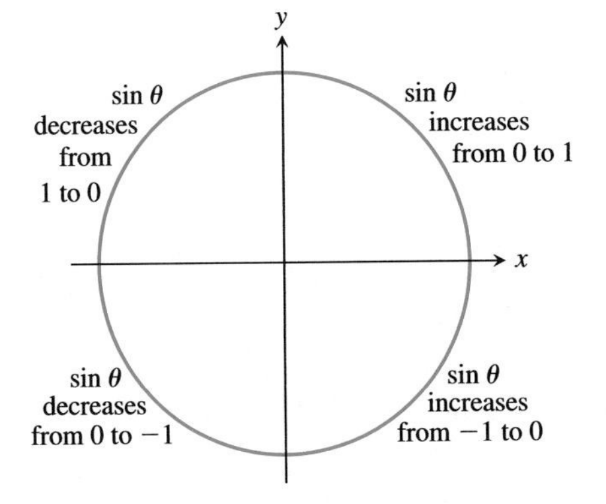

FIGURE A.10 The rise and fall of the sine function.

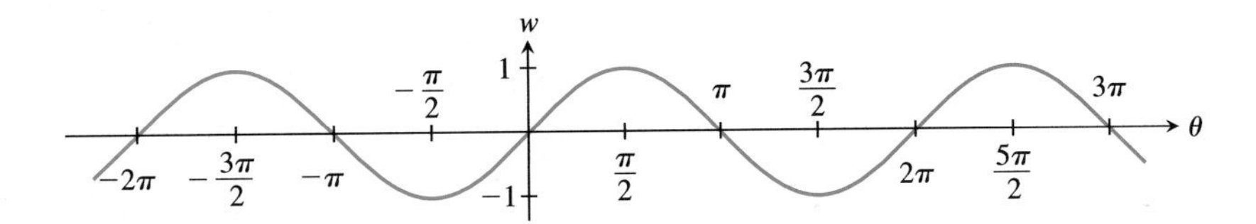

FIGURE A.11 Graph of $w = \sin\theta$.

A similar analysis leads to the graph of $w = \cos\theta$, shown in Fig. A.12.

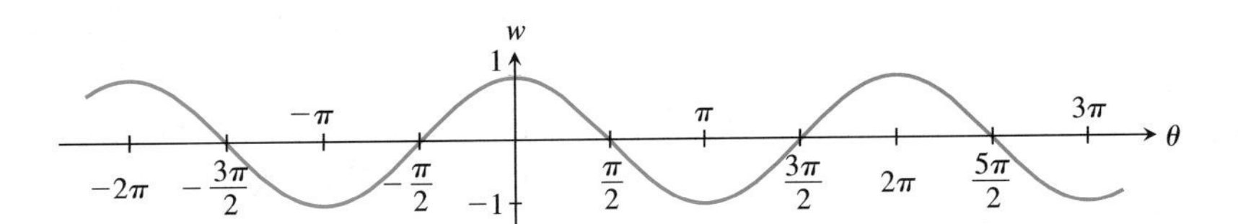

FIGURE A.12 Graph of $w = \cos\theta$.

Note that the graph of the cosine appears to be the graph of the sine function moved $\pi/2$ units to the left. This is reflected in the identity

$$\cos\theta = \sin\left(\theta + \frac{\pi}{2}\right).$$

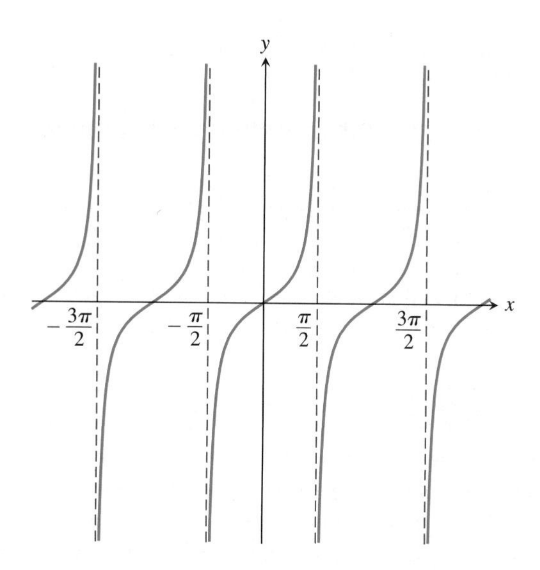

FIGURE A.13 The graph of $y = \tan x$.

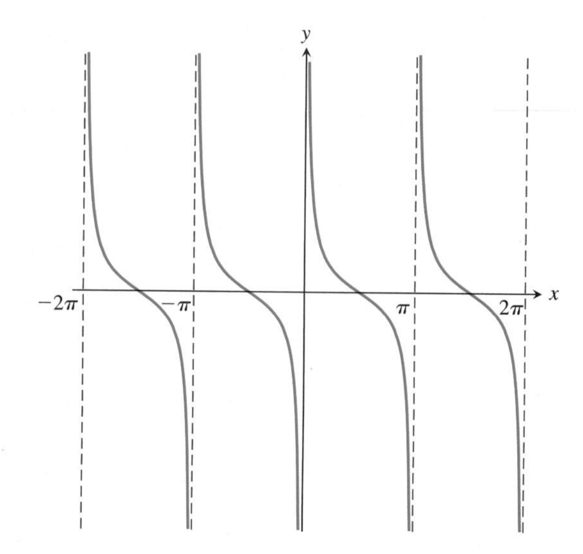

FIGURE A.14 The graph of $y = \cot x$.

The graph of $y = \tan x$ is shown in Fig. A.13. Recall that $\tan x = \sin x/\cos x$ is not defined when $\cos x = 0$, that is, when $x = \pi/2 + \pi k$, $k = 0, \pm 1, \pm 2, \ldots$. The graph has vertical asymptotes at these x-values. Similarly, the graph of $y = \cot x$, shown in Fig. A.14, has vertical asymptotes at the points $x = \pi k$, $k = 0, \pm 1, \pm 2, \ldots$.

Because $\sec x = 1/\cos x$, we can use the graph of $y = \cos x$ to aid us in drawing the graph of $y = \sec x$. Noting that $-1 \leq \cos x \leq 1$, we see that $|\sec x| \geq 1$. Also observe that when $\cos x$ is small and positive, $\sec x$ is large and positive, and when $\cos x$ is small and negative, $\sec x$ is large and negative. The graph of $y = \sec x$ has a vertical asymptote at each x value for which $\cos x = 0$. The graph of $y = \sec x$ is shown in Fig. A.15. Because $\csc x = 1/\sin x$, we can use the graph of $y = \sin x$ to help us draw the graph of $y = \csc x$. See Fig. A.16.

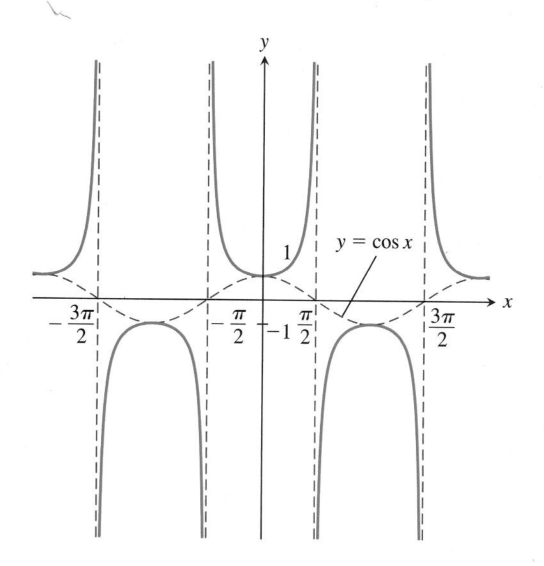

FIGURE A.15 The graph of $y = \sec x$.

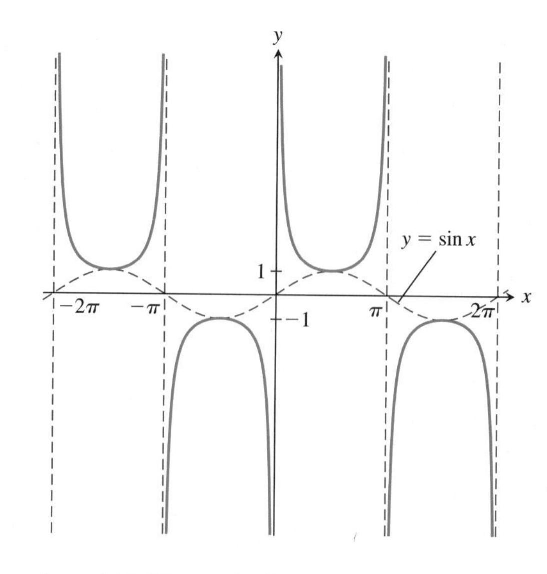

FIGURE A.16 The graph of $y = \csc x$.

A.3 Polar Coordinates

The default coordinate system for the plane, which we may imagine as a set of geometric points, is the *rectangular coordinate system.* It is based on two mutually perpendicular number lines (called *axes* in this context), intersecting at the points corresponding to zero on the two lines. The point of intersection of the two lines is called the *origin.* Figure A.17 shows a representative arrangement, with the positive halves of the lines extending to the right and upwards from the origin. A representative point P is located by two signed numbers, x and y. These numbers are usually listed together as the ordered pair (x, y).

We shall describe a polar coordinate system for the plane relative to a given rectangular system. Referring to Fig. A.17(b), the polar coordinates of P are r, which is the distance from the origin/pole O to P, and the angle θ between the line segment OP and the positive x-axis. The polar angle θ is measured from the positive x-axis counterclockwise to OP—this, by common convention, is the positive direction. In calculus the polar angle is usually measured in radians. Figure A.17(b) shows the polar coordinates (r, θ) of P and, using the definitions of the trigonometric functions, the relations

$$\begin{aligned} x &= r \cos \theta \\ y &= r \sin \theta \end{aligned} \tag{22}$$

between the rectangular coordinates (x, y) and polar coordinates (r, θ) of P.

If we are given polar coordinates r and θ of P, we may calculate the rectangular coordinates x and y of P directly from (22). If, on the other hand, we are given the Cartesian coordinates x and y of P, Equations (22) do not unambiguously determine r and θ unless we establish certain conventions. We may, for example, decide that r must be nonnegative and θ must lie in the interval $[0, 2\pi)$. If we adopt this convention, then, except for $x = y = 0$, Equations (22) do in fact determine r and θ. Because, however, there are contexts in which it is useful to allow r to be negative and θ to be outside of $[0, 2\pi)$, the following convention is often adopted: If in a rectangular coordinate system we have a point P with coordinates (x, y) and r and θ are any values whatsoever satisfying Equations (22), then we say that (r, θ) is a set of polar coordinates for P.

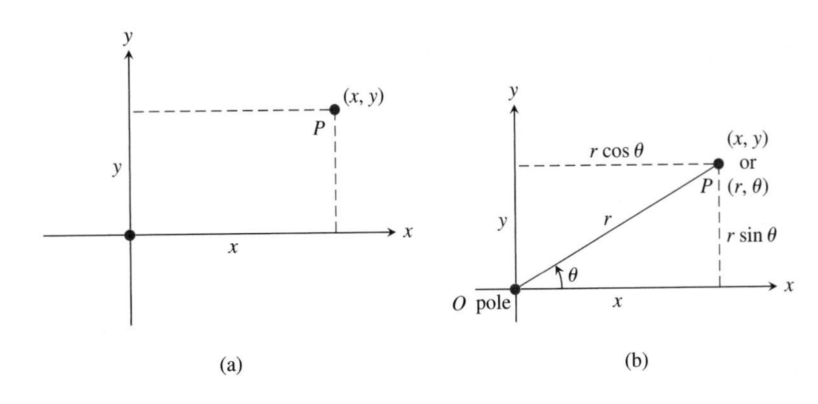

FIGURE A.17 (a) The rectangular coordinates x and y of point P; (b) the polar coordinates r and θ of point P.

We illustrate this convention by showing that the point P with rectangular coordinates $(\sqrt{3}, 1)$ has polar coordinates

$$(2, \pi/6), \quad (2, \pi/6 + 2\pi), \quad \text{and} \quad (-2, \pi/6 + \pi).$$

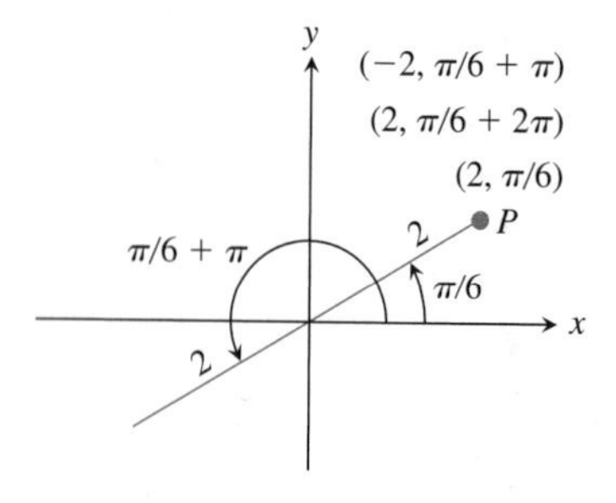

FIGURE A.18 Several polar coordinates for a point P.

Figure A.18 shows the point P with these polar coordinates. These ordered pairs are polar coordinates of P because

$$x = r\cos\theta = 2\cos(\pi/6) = \sqrt{3} \qquad y = r\sin\theta = 2\sin(\pi/6) = 1$$

$$x = r\cos\theta = 2\cos(\pi/6 + 2\pi) = \sqrt{3} \qquad y = r\sin\theta = 2\sin(\pi/6 + 2\pi) = 1$$

$$x = r\cos\theta = -2\cos(\pi/6 + \pi) = \sqrt{3} \qquad y = r\sin\theta = -2\sin(\pi/6 + \pi) = 1.$$

In the second pair of polar coordinates, θ has been increased by 2π. For the third pair, in which $r < 0$, we first located the ray determined by θ and then plotted P on its backward extension through the origin.

Graphing in Polar Coordinates

In rectangular coordinates, the simplest graphs are those of equations $x = a$ and $y = b$, where a and b are constants. These are a vertical line through $(a, 0)$ and a horizontal line through $(0, b)$. These "coordinate curves" are shown in Fig. A.19(a). The coordinate curves $r = a$ and $\theta = b$ in polar coordinates are shown in Fig. A.19(b). The graph of the equation $r = 1$ is a circle of radius 1 centered at the pole. This is the set of all points (r, θ) in the plane for which $r = 1$. The graph of the equation $\theta = \pi/4$ is the ray from the origin making an angle of $\pi/4$ with the positive x-axis. If r can be negative, the graph of $\theta = \pi/4$ is the entire graph of the equation $y = x$.

We graph several polar equations $r = g(\theta)$, beginning with the polar equation $r = \theta$. Our job is to sketch, or at least to suggest, all points (r, θ) satisfying the equation $r = \theta$. For $\theta = 0, \pi/2, 4, 5,$ and 2π, the corresponding r values are $r = 0, \pi/2, 4, 5,$ and 2π. Hence, the points $(0, 0)$, $(\pi/2, \pi/2)$, $(4, 4)$, $(5, 5)$, and $(2\pi, 2\pi)$ are on the graph. Figure A.20 shows these five points. As shown in the figure, as θ increases from 0 to 2π, the point $(r, \theta) = (\theta, \theta)$ traces a curve through these five points. This curve is called a *spiral*. As θ increases beyond 2π, the spiral will continue to wrap around the origin and recede from it.

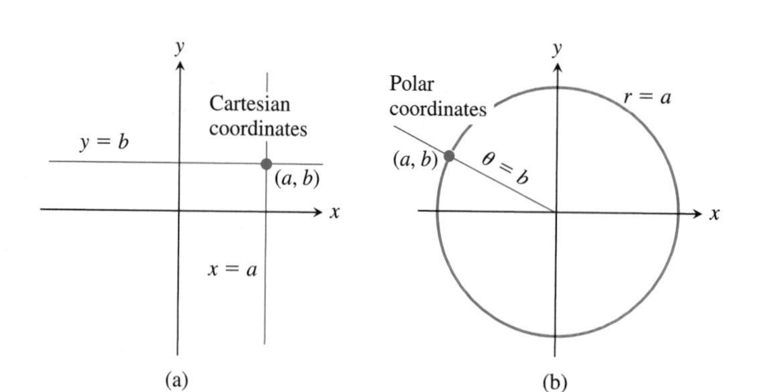

FIGURE A.19 (a) Rectangular coordinate curves; (b) polar coordinate curves.

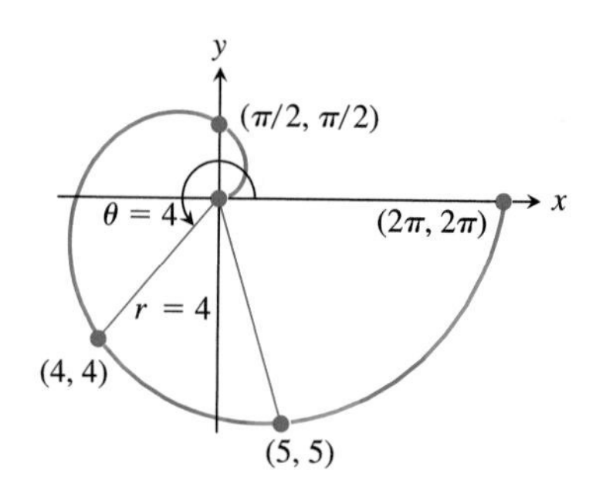

FIGURE A.20 The spiral $r = \theta$.

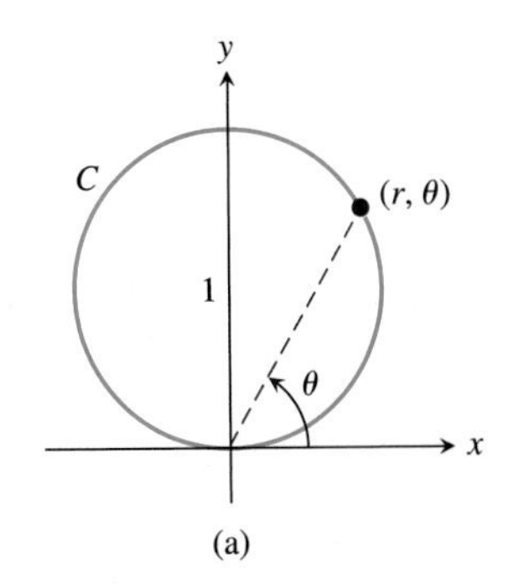

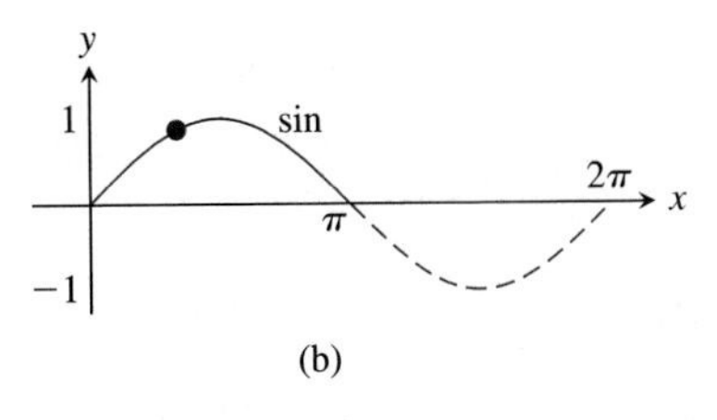

FIGURE A.21 (a) The polar circle described by $r = \sin\,\theta$; (b) the graph of $y = \sin\,x$.

Next we graph the polar equation described by $r = \sin\,\theta$. The graph of this equation is the curve C shown in Fig. A.21(a). To understand this graph, let's start by letting θ vary over the "first quadrant." As θ increases from 0 to $\pi/2$, $r = \sin\,\theta$ increases from 0 to 1. The curve C is thus traced counterclockwise from $(0, 0)$ to $(1, \pi/2)$ (these pairs are polar coordinates). This is the right half of C. The tracing of some polar curves $r = g(\theta)$ is simplified by correlating it with the graph of the curve $y = g(x)$ in rectangular coordinates. In this case we consider the sine curve with equation $y = \sin\,x$, which is shown in Fig. A21(b). As x varies over the first quadrant, y increases from 0 to 1. Continuing, as x varies over the second quadrant, y decreases from 1 to 0. This corresponds to r decreasing from 1 to 0 as θ varies from $\pi/2$ to π. Because the sine curve is symmetric about the $x = \pi/2$ line, we expect the curve C for θ in the second quadrant to mirror the curve C for θ in the first quadrant. As θ varies from π to 2π, $r \le 0$ and the circle is retraced.

To show that the curve C with equation $r = \sin\,\theta$ is a circle, we transform this equation to rectangular coordinates. Multiplying both sides of this equation by r,

$$r^2 = r\sin\,\theta.$$

Because $r^2 = x^2 + y^2$ and $y = r\sin\,\theta$ (these come from (22)),

$$x^2 + y^2 = y.$$

Moving the y term to the left and completing the square on $y^2 - y$ gives

$$\begin{aligned} x^2 + y^2 - y + (-1/2)^2 &= (-1/2)^2 \\ x^2 + (y - 1/2)^2 &= (1/2)^2. \end{aligned}$$

We recognize the last equation as that of a circle with center $(0, 1/2)$ and radius $1/2$.

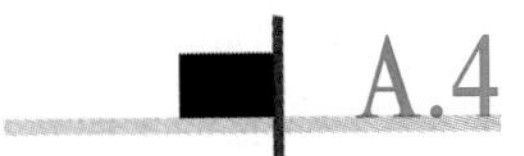

A.4 Mathematical Induction

Statements like

$$1 + 2 + \cdots + n = \frac{n(n+1)}{2}, \qquad n = 1, 2, \ldots, \tag{23}$$

and

$$\frac{d}{dx}x^n = nx^{n-1}, \qquad n = 1, 2, \ldots, \tag{24}$$

are sometimes proved by *mathematical induction,* which is based on one of the axioms for the set $N = \{1, 2, 3, \ldots\}$ of *natural numbers.*

Axiom of Induction

If M is a subset of N and

i) 1 is an element of M
ii) $k + 1$ is an element of M whenever k is an element of M,

then $M = N$.

We illustrate how "proofs by induction" go by proving (23) and (24). To prove that

$$1 + 2 + \cdots + n = \frac{n(n+1)}{2} \tag{25}$$

holds for $n = 1, 2, \ldots$, let M be the set of all numbers n in N for which (25) is true. Our goal is to show that $M = N$, so that (25) will have been shown to be true for $n = 1, 2, \ldots$. Our first step is to show that 1 is in M. This is true because

$$1 = \frac{1(1+1)}{2} = 1.$$

Next, we *assume* that (25) is true for $n = k$; that is, we are assuming that

$$1 + 2 + \cdots + k = \frac{k(k+1)}{2}.$$

Adding $k + 1$ to both sides of this (assumed) equality,

$$1 + 2 + \cdots + k + (k+1) = \frac{k(k+1)}{2} + (k+1)$$

$$\begin{aligned} 1 + 2 + \cdots + (k+1) &= (k+1)\left(\frac{k}{2} + 1\right) \\ &= (k+1)\frac{k+2}{2} = \frac{(k+1)(k+2)}{2}. \end{aligned}$$

Hence

$$1 + 2 + \cdots + (k+1) = \frac{(k+1)(k+2)}{2}.$$

This shows that $k + 1$ is in M. It now follows by the Axiom of Induction that $M = N$, that is, (25) holds for $n = 1, 2, \ldots$.

To prove that

$$\frac{d}{dx}x^n = nx^{n-1} \tag{26}$$

holds for $n = 1, 2, \ldots$, we first verify that $d(x)/dx = 1$ directly from the definition of the derivative. Hence, (26) holds for $n = 1$. Now assume that (26) holds for $n = k$. Because we may write

$$\frac{d}{dx}x^{k+1} = \frac{d}{dx}(x \cdot x^k),$$

we may apply the product rule. As part of the product rule, we use our assumption that (26) holds for $n = k$.

$$\begin{aligned} \frac{d}{dx}x^{k+1} &= \frac{d}{dx}(x \cdot x^k) = 1 \cdot x^k + x \cdot kx^{k-1} \\ &= x^k + kx \cdot x^{k-1} = x^k(1 + k) = (k+1)x^k. \end{aligned}$$

Hence, (26) holds for $n = k + 1$. It now follows that (26) holds for all $n \in N$.

Answers

CHAPTER 1 — RATES OF CHANGE, LIMITS, AND THE DERIVATIVE

1.1 Functions

1. -9 **3.** $12 + 5\sqrt{3}$
5. $8x^3 + 12x^2(2h + 1) + 6x(4h^2 + 4h + 1) + (8h^3 + 12h^2 + 6h + 1)$
7. $\{x : x \text{ a real number}\}$ **9.** $\{x : x \leq -1 \text{ or } x \geq 0\}$
11. $\{r : r > b\}$ **13.** domain: all real numbers; range: $\{4\}$
15. domain: all real numbers; range: $\{y : 0 < y \leq 1\}$
17. domain: $\{x : x \neq 0\}$; range: $\{-1, 1\}$
19. $f(x) = \sqrt{-(x + 1)(x - 3)}$ (Other answers are possible.)
21.

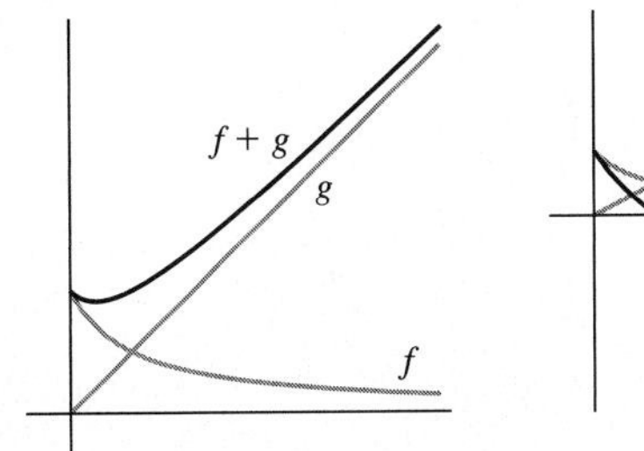

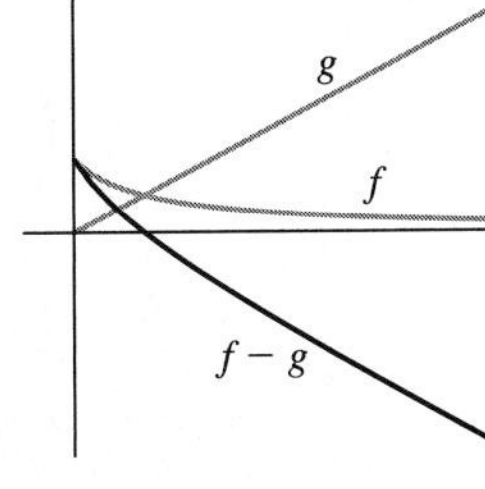

23. $2x + h$

25. $h(t) = \dfrac{7}{2}, t > 3; \quad 2 + \dfrac{1}{2}t, 0 \leq t \leq 3$

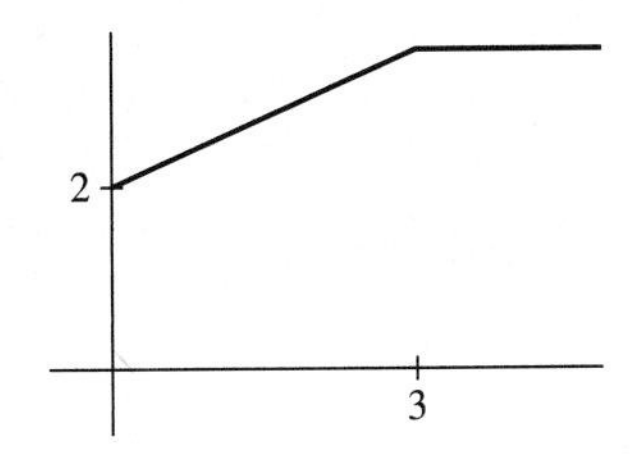

27. 13.5 km/hr
29. If a vertical line is drawn through the point x on the x-axis, the y-coordinates of the points in which the line intersects the graph are the output values that correspond to input value x. If there is no intersection point, the x is not in the domain of the function defined by the graph. If the vertical line intersects the graph in exactly one point, then the y-coordinate is the unique output that is assigned to x. This assignment of a well-defined y-coordinate to an x-coordinate defines y as a function of x. If a vertical line intersects the graph in two or more points, then there are at least two different y values assigned to x. This means the output associated with x is not well defined, so the correspondence does not define a function.
31.

Degree	Normal	Factored
1	2	2
2	6	4
3	10	6
4	14	8
5	18	10

For polynomials of degree exceeding 1, the factored method costs roughly $1/2$ as much as the normal method.
33. Function 1: domain = team; range = $\{0, 1, 2, 3\}$. The function takes a team name as input and returns the number of conference wins as output. Function 2: domain = team; range = $\{0, 1, 2, 3\}$. The function takes a team name as input and returns the number of conference losses as output.
35. domain = places in continential United States; range = deciles of temperatures: -10s, -0s, 0s, 10s, 20s, ... The function takes a location in the continental United States as input and returns the decile for temperature at that location.
37. The graph is not correct. The function f is not defined at $x = -3$, so there should be an open dot at the point $(-3, -6)$ on the line.

1.2 Compositions of Functions

1. $(f \circ g)(x) = \dfrac{1}{x^2 - 6x + 8}$, domain: $\{x : x \neq 2, x \neq 4\}$

3. $(f \circ g)(t) = \dfrac{|\sin t|}{\sin t}$, domain: $\{t : t \neq 0, \pm\pi, \pm 2\pi, \ldots\}$

5. $(h \circ k)(x) = \sqrt{(x - 1)(x - 2)(x - 3)^2(x - 4)}$, domain: $\{x : 1 \leq x \leq 2 \text{ or } x \geq 4\} \cup \{3\}$

7. $G = f \circ g$, $g(x) = x^3 - 3x + 7$, $f(x) = 5x^{10}$

9. $H = f \circ g \circ h$, $h(x) = x + 5$, $g(x) = \sqrt{x}$, $f(x) = \sin x$

11. $r = f \circ g \circ h$, $h(x) = \sin x$, $g(x) = \dfrac{1 - 2x}{1 + 2x}$, $f(x) = x^{1/2}$

13. The graph is wrong. The composition is only defined for $x > 3$.

15.

x	$(s \circ r)(x)$
0	3
1	2

17.

x	$r \circ (r \circ s)(x)$
−2	3
2	3

19. $\{x: -2 \le x \le 2\}$ **21.** $\{x: -\sqrt[3]{11} \le x \le -\sqrt[3]{9}\}$

23. **a.** domain $f \circ g = \{-1/2, 0, 1/2, 1, 3/2, 2\}$, range of $f \circ g = \{-6, -5, 0, 1, 3\}$, domain $g \circ f =$ domain of f range of $g \circ f = \{-13, -11, -1, 1, 5\}$

b. domain $f \circ h = \{-1, 0, 3, 8\}$, range of $f \circ h$ is $\{0, -6, -5, 3\}$, domain $h \circ f = \{-2, -1, 0, 3\}$, range of $h \circ f$ is $\{1, \sqrt{2}, 2\}$

25. **a.** The graph of $y = f(-x)$ is found by reflecting the graph of $y = f(x)$ over the y-axis.

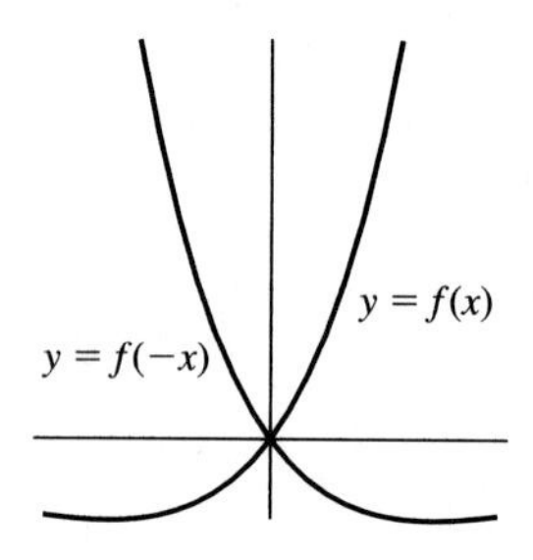

b. The graph of $y = -f(x)$ is found by reflecting the graph of $y = f(x)$ over the x-axis.

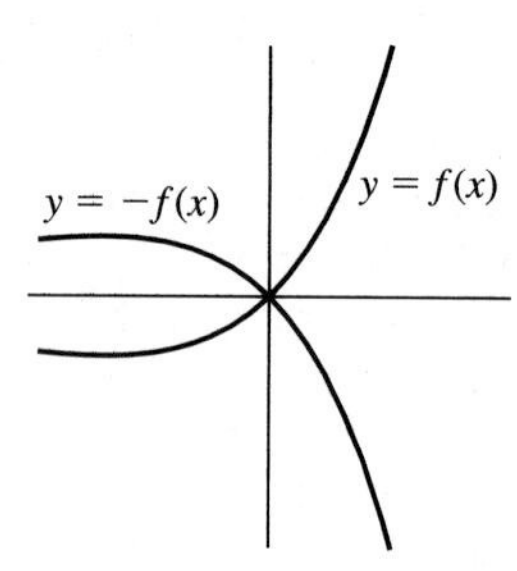

27. **a.** The graph of $y = f(\ell(x))$ is found by shifting the graph of $y = f(x)$ π units to the right, then compressing the resulting graph horizontally by a factor of 2, then reflecting the result in the y-axis.

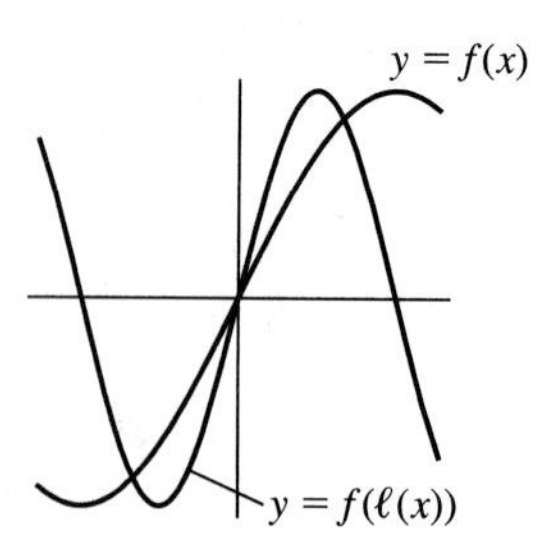

b. The graph of $y = \ell(f(x))$ is found by magnifying the graph of $y = f(x)$ vertically by a factor of 2, then reflecting the result in the x-axis, then shifting that result upward a distance of π units.

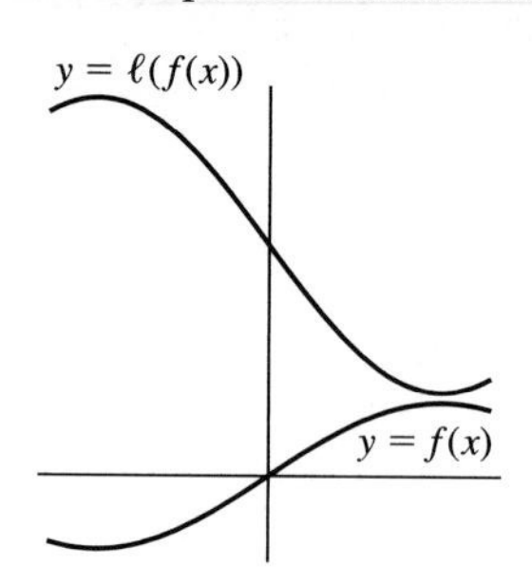

29.

31.

33.

35.

37.

39. As the composition deepens, the graph becomes more and more compressed vertically, getting more like the line $y = 0$.

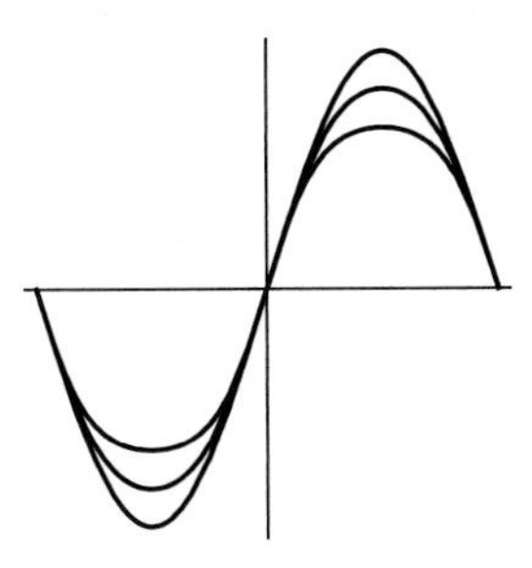

1.3 Slope as a Rate of Change

1. slope = 3, y-intercept = -4

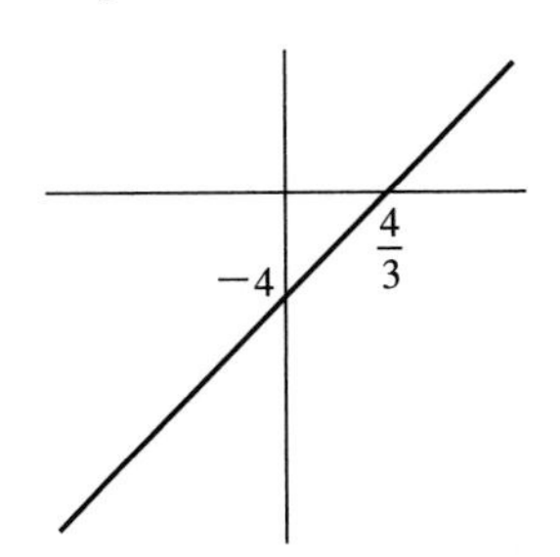

3. $y = -\frac{2}{3}x + \frac{11}{3}$

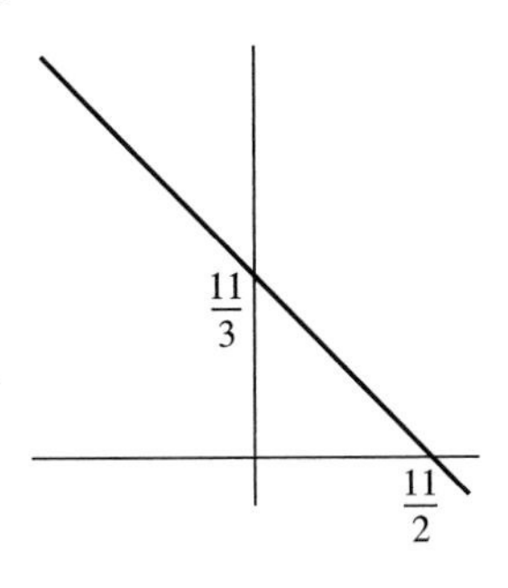

5. $y = -2x + 5$

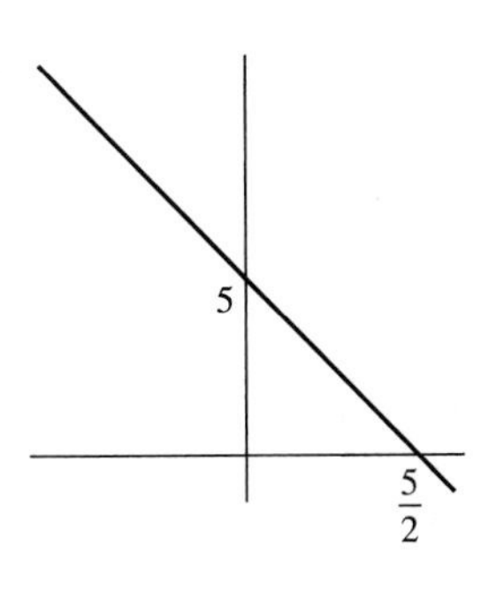

7. $y = 10x + 110$

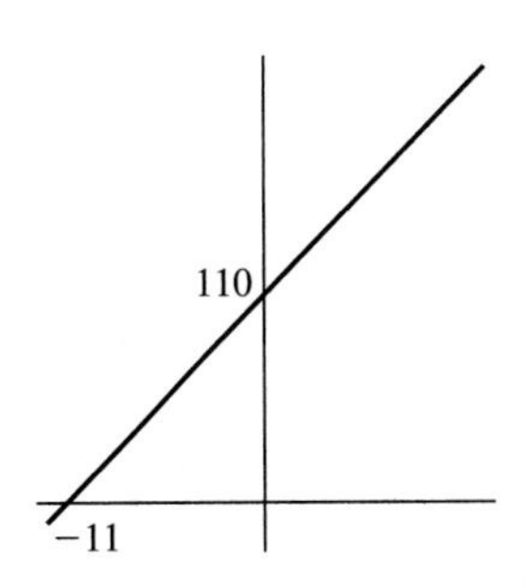

9. angle of 63.4 degrees

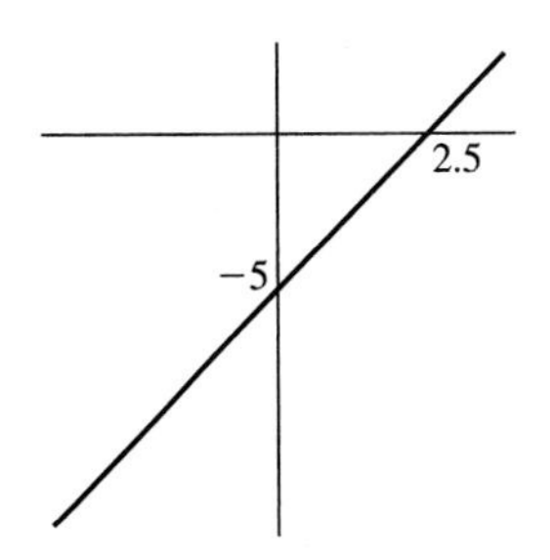

11. She will reach ground after 1000 seconds; $h(t) = 10{,}000 - 10t$, $0 \le t \le 1000$; rate of change $= -10$ ft/s.

13. 2/15 in/min, $d = \frac{2}{15}t + 136$

15. **a.** $h(t) = \frac{10t}{64\pi}$ **b.** The rate of change is $\frac{10}{64\pi}$ cm/s.
c. The can will be full after 402 seconds.

17. ≈ 5556°C/meter. This means we can use the beam as a "thermometer." A one-millimeter increase in length would correspond to a temperature rise of about 5.556°C.

19. No, we will always see a right angle in the graph.

21. distance, time

23. **a.** The tick marks should read from top to bottom as 1, 2/3, 1/3.

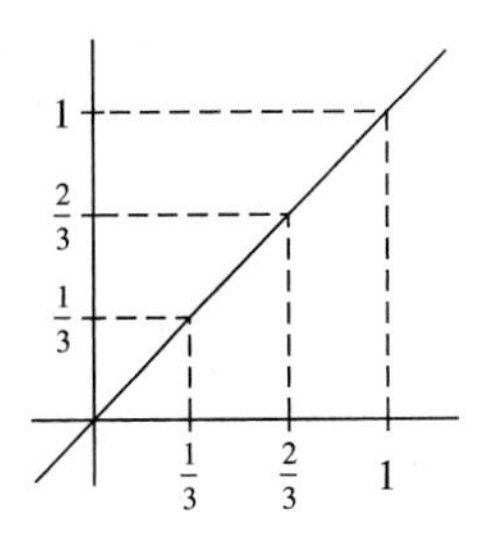

b. The tick marks should read from top to bottom as 3, 2, 1.

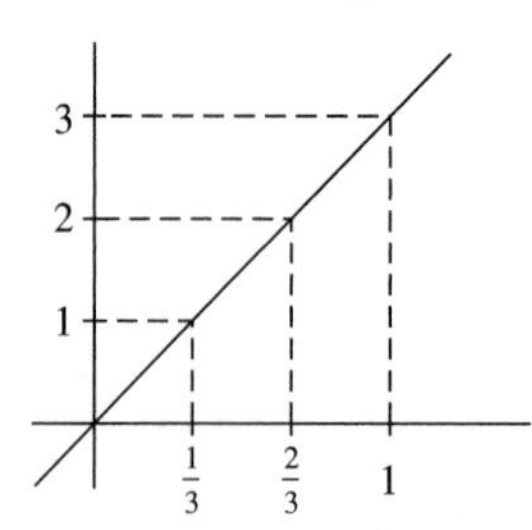

c. The tick marks should read from top to bottom as 1/4, 1/6, 1/12.

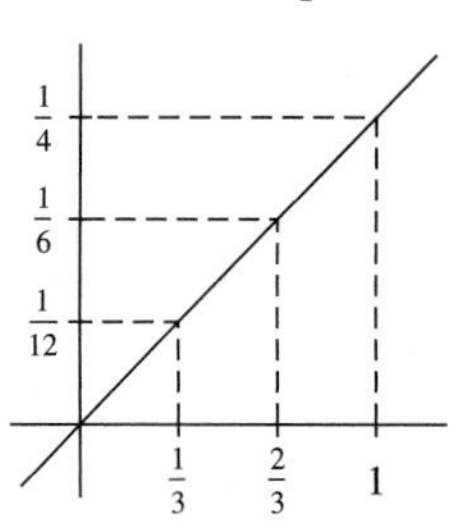

25.

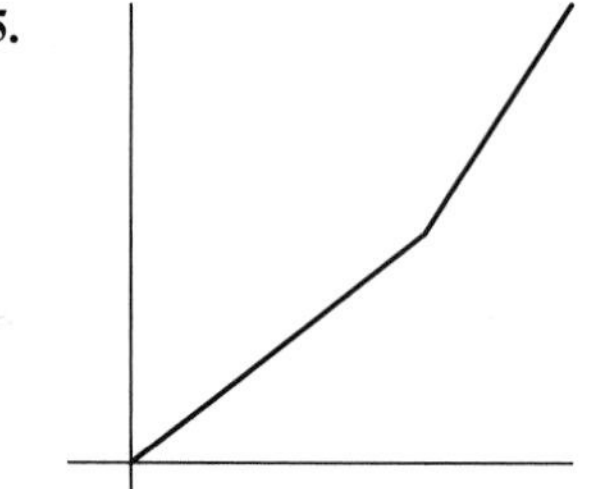

27.

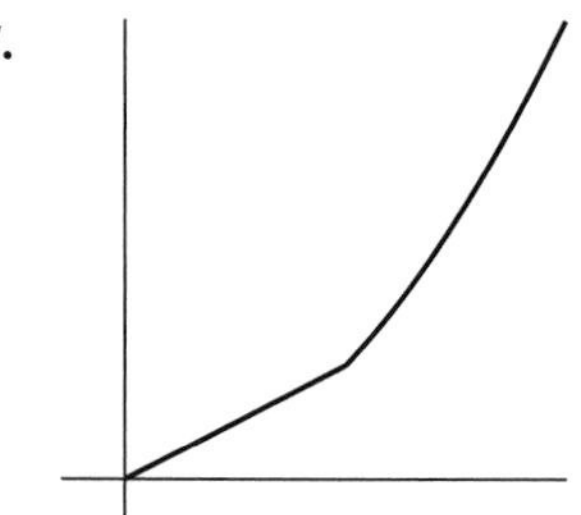

29.

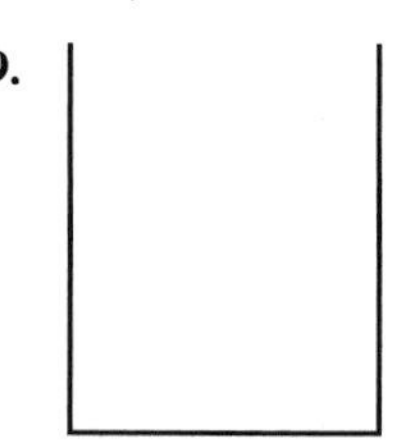

31.

1.4 Calculating Rates of Change

1. 4 **3.** 5 **5.** 12 **7.** $-\frac{1}{2}$ **9.** $\frac{1}{6}$

13. At $x = 0$, the rate of change is ≈ 0.693147. At $x = 1$, the rate of change is ≈ 1.386294. At $x = 2$, the rate of change is ≈ 2.772589. At $x = 4$, the rate of change is ≈ 11.090355.

15. The rate of change is undefined at $x = -3$ and $x = \frac{4}{3}$,

$$\text{r.o.c} = \begin{cases} 5 & x < -3 \\ 7 & -3 < x < 4/3 \\ -5 & x > 4/3 \end{cases}$$

17. $-\frac{3}{2}$ millibars/hr; pressure is falling. The most rapid rise was approximately 5 A.M. Friday. The most rapid drop was at about 2 P.M. Thursday.

19. The graph will look like a horizontal line through the point $(-2, 4)$.

21. At any given $x = a$, the two graphs will have the same rates of change.

23. 1800: ≈ 0.2 million people/year;
1900: ≈ 1.6 million people/year; 1980: ≈ 2 million people/year

25. One hour: estimate 60 miles
One minute: estimate 1 mile
Two seconds: estimate 176 feet
0.01 seconds: estimate 0.88 feet
The last is likely most accurate; the first is likely least accurate. The first is likely least accurate because we do not know if the driver will maintain the speed of 60 miles per hour for the next hour.

1.5 Limits

1. $-\frac{2}{3}$ **3.** 1 **5.** does not exist **7.** 45 **9.** $\frac{1}{2}$

11. does not exist **13.** 1 **15.** 1 **17.** 2

19. **a.** $c = -4, c = 2$ **b.** does not exist **c.** 9
d. does not exist

21. **a.** ≈ 0.693 **b.** ≈ 1.099 **c.** ≈ -0.693 **d.** ≈ -1.099

23. **a.** $a = 0.003, b = 0.001$ **b.** $a = -0.005, b = 0.00001$
c. $a = 10^{-2}, b = 10^{-11}$ **d.** $a = -10^{-11}, b = 10^{-2}$

25. **a.** $f(x) = 5x - 15, g(x) = x - 3$
b. $f(x) = 3\sqrt{3} - \sqrt{3}x, g(x) = x - 3$
c. $f(x) = x - 3, g(x) = x^2 - 6x + 9$

27.

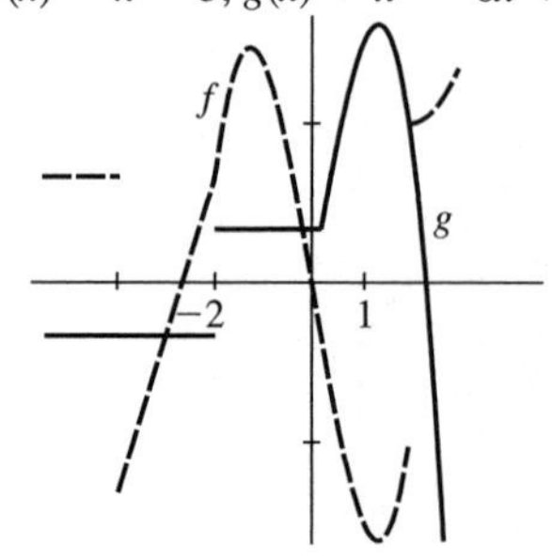

29. -2 **31.** $-1/4$ **33.** 0 **35.** $-1/2$ **37.** M
39. **b.** ≈ 0.203 **c.** any positive $r \le 0.02$
41. **b.** ≈ 0.0133 **c.** any positive $r \le 0.15$

1.6 More Work with Limits

1. **a.** 6 **b.** any positive $\delta < 0.01$
3. **a.** 6 **b.** Any $\delta > 0$ will work.
5. **a.** -3 **b.** Any positive $\delta < 0.1$ will work.
7. **a.** -13 **b.** $1.9995 < x < 2.0005$
9. **a.** 1 **b.** $-0.06 < x < 0.06$
11. **a.** $2\sqrt{5}$ **b.** $4.96 < x < 5.04$
13. **a.** $-\dfrac{1}{4}$ **b.** $0.93 < x < 1.07$
15. **a.** $-\dfrac{4}{3}$ **b.** $\sqrt{2r^2+1} \approx 3 - \dfrac{4}{3}(r+2)$
c. $-2.15 < r < -1.85$ **d.** error of about 1.495×10^{-5}
17. **a.** 2 **b.** $\tan x \approx 1 + 2\left(x - \dfrac{\pi}{4}\right)$ **c.** $0.77 < x < 0.80$
d. error of about 0.00239
19. **a.** $\lim_{t\to 0^+} h(t) = 1$, $\lim_{t\to 0^-} h(t) = -1$ **b.** does not exist
c. not continuous at 0
21. **a.** 1, 1 **b.** 1, 1 **c.** 2, 0 **d.** -1 and 1
23. If $\lim_{x\to a} g(x) = L$ and h is continuous at L, then $\lim_{x\to a} h(g(x)) = h(L)$.
33. $r(b) = 2$ when $b < -4$ or $b > 4$, $r(b) = 1$ when $b = -4$ or $b = 4$, $r(b) = 0$ when $-4 < b < 4$. The graph of $r(b)$ is discontinuous at $b = -4$ and $b = 4$.

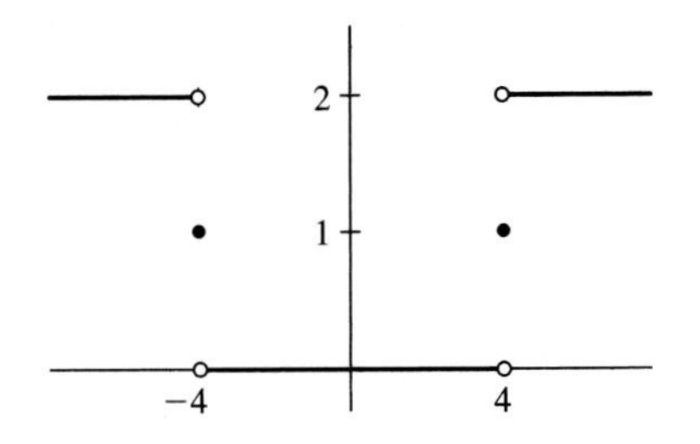

1.7 The Derivative

1. $2x$ **3.** $-6t + 4$ **5.** $\dfrac{4}{(2-x)^2}$ **7.** $2 + \dfrac{3}{2\sqrt{t+9}}$

9. $-\dfrac{a}{(ax+b)^2}$ **11.** $y = -x - 1$

13. $y = 4(r-1) + 6 = 4r + 2$ **15.** $y = -\dfrac{8}{9}(u-2) + \dfrac{2}{3}$

17. $y = 2(x-2) + 4 = 2x$

19. $\dfrac{1}{3.15} \approx -\dfrac{1}{9}(3.15 - 3) + \dfrac{1}{3} \approx 0.316667$, error ≈ 0.000794

$\dfrac{1}{2.85} \approx -\dfrac{1}{9}(2.85 - 3) + \dfrac{1}{3} = 0.35$, error ≈ 0.000877

21. **a.** $y = 2\left(x - \dfrac{\pi}{4}\right) + 1$

b. $\tan\left(\dfrac{\pi}{4} - 0.05\right) \approx 2(-0.05) + 1 = 0.9$

23. -6.051×10^{-7} N/m
25. **a.** $y = x + 2$ **b.** $f(0.05) \approx 2.05$, $f(-0.1) \approx 1.9$
27. $g'(-1) = 5$, $g(-0.88) \approx -8.4$
29. **a.** $y = 12x + 16$ **b.** $r < 0.001$
31. **a.** $h(-4) = f(-(-4)) = f(4) = 3$ **b.** $h'(-4) = -7$
33. **a.** $h(4) = 8f(4) = 8 \cdot 3 = 24$ **b.** $h'(4) = 56$
35. $f(x) = x^9 - 4x^7 + 3x - 2$, $a = 2$, $f'(a) = 515$
37.

Chapter Review Exercises

1. domain f: $\{0, 1, 2, 3, 4, 5\}$; domain g: $\{-3, -0.5, 0, 2, 3\}$
3. domain $f + g$: $\{0, 2, 3\}$

x	$(f+g)(x)$
0	1
2	$-\pi + \sin 1$
3	3

5. domain $\dfrac{f}{g}$: $\{0, 2, 3\}$

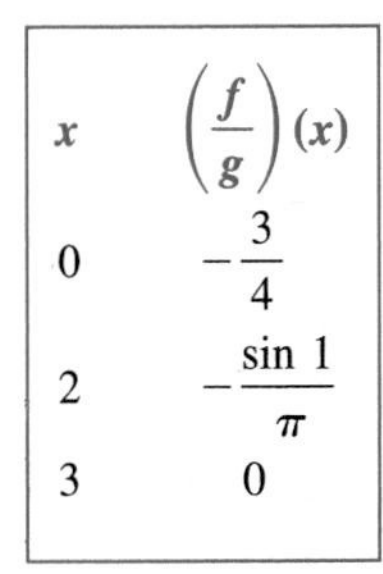

x	$\left(\dfrac{f}{g}\right)(x)$
0	$-\dfrac{3}{4}$
2	$-\dfrac{\sin 1}{\pi}$
3	0

7. Taking the graphs as indicating the full domains, domain f: $0 < x < 5$; domain g: $0 \le x \le 6$.
9. domain $(f + g)$: $0 < x < 5$

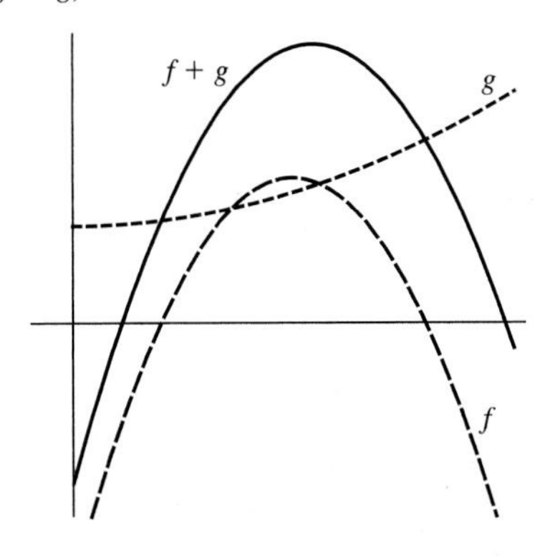

11. Domain $\left(\dfrac{g}{f}\right)$: $\{x : 0 < x < 5 \text{ with } x \neq 1, 4\}$

13. ≈ 4 **15.** m/(cm/s) **19.** -2 **21.** 3

23. $1000 \ln 10 \approx 2302.59$ **25.** $3/5$ **27.** any $\delta \leq 0.035$

29. 2 **31.** $1 - \dfrac{1}{x^2}$

33. $y = 3 + \dfrac{1}{6}(x - 7)$; any positive $r \leq 0.45$ works

CHAPTER 2—FINDING THE DERIVATIVE

2.1 Derivatives of Polynomials

1. $f'(x) = 4$ **3.** $y' = x^3 + x^2 + x + 1$ **5.** $H'(s) = 0$

7. $f'(x) = 4x + 3$ **9. a.** $f'(x) = x^2$ **b.** $g(x) = \dfrac{1}{3}x^3 + 10$

11. a. $y' = 4x^3$ **b.** $h(x) = \dfrac{1}{12}x^4$

13. $-\dfrac{51}{2}$ **15.** -4 **17.** $y = 5(x - 2) + 6$

19. $y = (2ad + b)(x - d) + (ad^2 + bd + c)$

27. $r'(\theta) = 1 - \dfrac{1}{\theta^2}$ **29.** $-\dfrac{8}{r^3} - \dfrac{16}{r^5}$

31. 380 ft

2.2 Derivatives of Products and Quotients

1. $(2x - 3)(2x + 4) + (x^2 - 3x + 1)2$

3. $H'(t) = 2(-9t^5 + 6t^4 - 21t^2 + 8t - 1)(-45t^4 + 24t^3 - 42t + 8)$

5. $f'(x) = \dfrac{(4x - 3)(x^2 - 7) - 2x(2x^2 - 3x + 10)}{(x^2 - 7)^2}$

7. $s'(r) = 2\left(1 - \dfrac{1}{r} + \dfrac{2}{r^2}\right)\left(\dfrac{1}{r^2} - \dfrac{4}{r^3}\right)$

9. $r' = 2\theta(\theta^3 - 3)(\theta^4 - 4) + (\theta^2 - 2)(3\theta^2)(\theta^4 - 4)$
$+ (\theta^2 - 2)(\theta^3 - 3)(4\theta^3)$

11. $p'(r) = \dfrac{1}{2}\dfrac{2r + 2}{\sqrt{r^2 + 2r + 7}}$

13. $F'(s) = \dfrac{1}{2}\dfrac{2}{\sqrt{2s - 1}}\sqrt{6s + 7} + \sqrt{2s - 1}\left(\dfrac{1}{2}\dfrac{6}{\sqrt{6s + 7}}\right)$

15. 13 **17.** 0

19. $\left(-\dfrac{7}{9} + \dfrac{5\sqrt{10}}{18}, \dfrac{238}{243} - \dfrac{625\sqrt{10}}{243}\right) \approx (0.100633, -7.15401)$

$\left(-\dfrac{7}{9} - \dfrac{5\sqrt{10}}{18}, \dfrac{238}{243} + \dfrac{625\sqrt{10}}{243}\right) \approx (-1.65619, 9.11285)$

21. $y = \dfrac{113}{225}(x - 2) + \dfrac{1}{15}$

23. $w(x) = (x^2 - 3x + 8)(4x^3 + 3x - 1)$

25. a. $2u(x)u'(x)$ **b.** $3(u(x))^2u'(x)$ **c.** $4(u(x))^3u'(x)$
d. $5(u(x))^4u'(x)$, $n(u(x))^{n-1}u'(x)$

27. $\dfrac{d}{dx}|x| = \dfrac{x}{|x|}$. This derivative has value 1 if $x > 0$ and value -1 if $x < 0$. The derivative is not defined for $x = 0$.

31. 0.60001 kg/min **33.** $40{,}000\pi$ m²/hr, 200π m³/hr

39. $h(t) = (u \circ v \circ w)(t) = u(v(w(t)))$, where $u(t) = \sqrt[3]{1 + t}$, $v(t) = \sqrt[4]{t}$, $w(t) = t^2 - 1$

2.3 Differentiating Compositions

1. $f'(x) = \dfrac{26}{3}x^{10/3}$ **3.** $y = 23(2x^2 - 3x + 1)^{22}(4x - 3)$

5. $g'(x) = \dfrac{5}{4}\left(x^3 - 5x + \dfrac{1}{2}\right)^{1/4}(3x^2 - 5)$

7. $f'(t) = \dfrac{1}{\sqrt{2t - 7}}$

9. $r' = 12(2\theta - 1)^5(4\theta + \pi)^8 + 32(2\theta - 1)^6(4\theta + \pi)^7$

11. $y' = \dfrac{1}{8}\dfrac{1}{\sqrt{1 + \sqrt{1 + \sqrt{x + 1}}}}\dfrac{1}{\sqrt{1 + \sqrt{x + 1}}}\dfrac{1}{\sqrt{x + 1}}$

13. $h'(u) = \dfrac{1}{3}\left(\dfrac{4u - 1}{3u + 7}\right)^{-2/3}\left(\dfrac{4(3u + 7) - 3(4u - 1)}{(3u + 7)^2}\right)$

15. $H'(z) = 24z(3z^2 - 1)^3(4z^{-3} + 2z^{-2} + 4)\sqrt{1 - \dfrac{1}{z}}$
$+ (3z^2 - 1)^4(-12z^{-4} - 4z^{-3})\sqrt{1 - \dfrac{1}{z}}$
$+ (3z^2 - 1)^4(4z^{-3} + 2z^{-2} + 4)\left(1 - \dfrac{1}{z}\right)^{-1/2}\dfrac{1}{2z^2}$

17. $y' = (2x^2 - 6x - 1)(2x - 3)$ **19.** $-3x^{-5/2}(x^{-3/2} + 1)$

21. $-(2x - 1)^{-3/2}$ **23.** $\dfrac{18x - 6}{9x^2 - 6x + 2}$

25. $\dfrac{(x^2 - x - 2)(2x - 1)}{|x^2 - x - 2|}$

27. $3(|x^3 - x| + 2)^2\left(\dfrac{(x^3 - x)(3x^2 - 1)}{|x^3 - x|}\right)$

29.

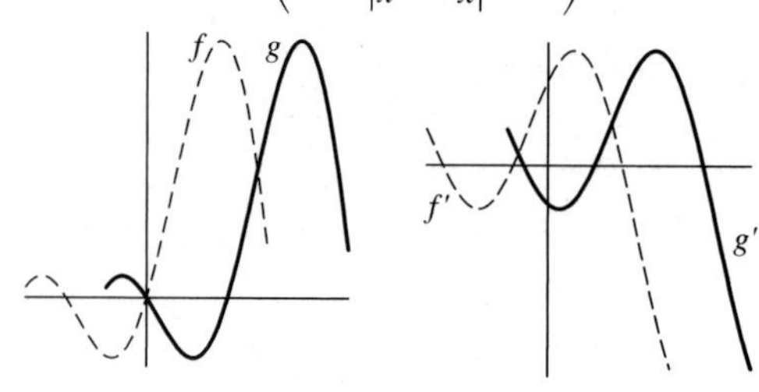

31.

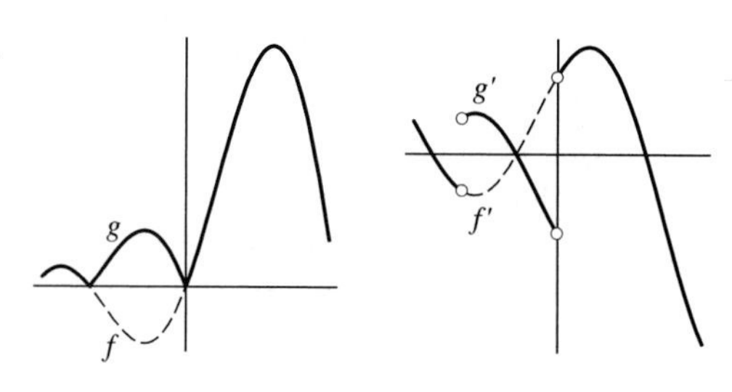

33. $y = \frac{1}{\sqrt{2}} - \frac{1}{2\sqrt{2}}(x - 1)$

35. $y = -\frac{1}{32} + \frac{3}{64}(x - 2)$ **37.** $-8.075°\text{F/s}$

2.4 Implicit Differentiation

1. $\frac{3x}{4y}$ **3.** $\frac{3x^2 - y}{2x^2 - x}$ **5.** $\frac{4x - 3y}{3x - 2y}$ **7.** $\frac{2x - 2y - y^2}{2(x - 4y + xy)}$

9. $\frac{-x^2 + 4y + 6xy^2 + 18y^4}{4x - 2x^2 + 18x^2y + 12y^2 + 12xy^2 - 18y^4}$, or $\frac{3y^2 + x - y}{x - 9y^2 - 6xy - 2}$ **11.** $-\frac{1}{3}$

13. -1 **15.** $y = 1$ **17.** $y = -x$

19. $y = -2 - \sqrt{8 + 2x - x^2}$

21. $y = \frac{5}{4}x + \frac{1}{4}\sqrt{12 + 17x^2}$ **25.** $\frac{1 - 8x}{8y}$

27.

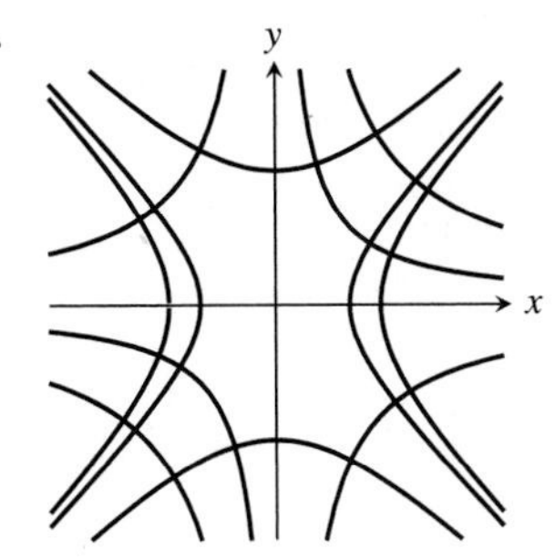

29. **a.** $\frac{1 + 2xy^2 + 2y^3}{3 - 2x^2y - 6xy^2}$ **b.** $\frac{3 - 2x^2y - 6xy^2}{1 + 2xy^2 + 2y^3}$

31. **a.** x-intercepts: $(\pm\sqrt{7}, 0)$, y-intercepts: $(0, \pm\sqrt{28/3})$
b. $(\sqrt{7/2}, \sqrt{14})$ and $(-\sqrt{7/2}, -\sqrt{14})$
c. $(\sqrt{21/2}, \sqrt{14/3})$ and $(-\sqrt{21/2}, -\sqrt{14/3})$

2.5 Trigonometric Functions

1. $y \leq 2\cos(2x + 3)$ **3.** $f'(t) = (6t - 4)\sec^2(3t^2 - 4t + 1)$

5. $f'(x) = \frac{x \sin x + \cos x}{x^2}$ **7.** $T'(\theta) = \frac{2}{(1 - \theta)^2}\sec^2\left(\frac{1 + \theta}{1 - \theta}\right)$

9. $f'(x) = 4x\cos(x^2 + 1)$

11. $g'(t) = \frac{-3\csc^2(3t)(1 + \csc^2(3t)) + 6\cot^2(3t)\csc(3t)}{(1 + \csc^2(3t))^2}$

13. $z' = \cos(2r)$ **15.** $y' = (\cos x - x\sin x)\cos(x\cos x)$

17. $f'(x) = 4\tan(2x)\sec^2(2x)$

19. $y' = -\csc^2 t(\sin(\cot t) + (\cot t)\cos(\cot t))$

21. $-\sqrt{2}$ **23.** -1 **25.** 1

27. $y = 1 + 2\left(x - \frac{\pi}{4}\right)$

$\tan\left(\frac{\pi}{4} + 0.03\right) \approx 1.06$; error of about 0.0019

29. $y = \frac{1}{\sqrt{2}} - \frac{\pi}{\sqrt{2}}\left(x - \frac{1}{4}\right)$

$f(0.265) \approx 0.673785$; error of about 8×10^{-4}

31. 0.8 or -0.8 **33.** $\sqrt{3}$ or $-\sqrt{3}$ **35.** 10

37. $y = -\cos x$ **41.** $2\arctan(1/\sqrt{2})$

45. **a.** t = time is in minutes. At $t = 0$ the cork is at a high point, labeled as height 5. Height $h = 0$ is midway between the high and low points of the cork. (Other answers are possible.)

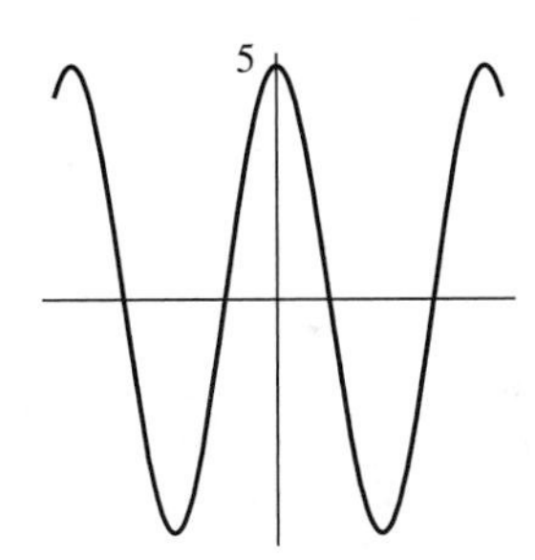

b. $h = 5\cos(12\pi t)$ **c.** $\frac{dh}{dt} = -60\pi\sin(12\pi t)$

47. **a.** $\frac{\sin b - \sin a}{\cos b - \cos a}$ **b.** $y = \left(\frac{\sin b - \sin a}{\cos b - \cos a}\right)(x - 1)$

c. $\left(-\frac{\cos 2b + \cos 2a - 2\cos(b + a)}{2 - 2\cos(b - a)}, -\left(\frac{\sin b - \sin a}{\cos b - \cos a}\right)\left(\frac{\cos 2b + \cos 2a - 2\cos(b + a)}{2 - 2\cos(b - a)} + 1\right)\right)$

e. $\sin(a + b) = -\left(\frac{\sin b - \sin a}{\cos b - \cos a}\right)\left(\frac{\cos 2b + \cos 2a - 2\cos(b + a)}{2 - 2\cos(b - a)} + 1\right)$

$\cos(a + b) = -\frac{\cos 2b + \cos 2a - 2\cos(a + b)}{2 - 2\cos(a - b)}$

2.6 Exponential Functions

1. $-\frac{4}{3}$ **3.** 1 **5.** $x = 0$ or $x = \ln 2 \approx 0.693147$

7. $u = \ln(5 + \sqrt{24}) \approx 2.29243$ and $\ln(5 - \sqrt{24}) \approx -2.29243$

9. $\frac{dy}{dx} = C_3 3^x \approx 1.09861 \cdot 3^x$

$y = \sqrt{3} + C_3\sqrt{3}\left(x - \frac{1}{2}\right) \approx 1.73205 + 1.90285(x - 0.5)$

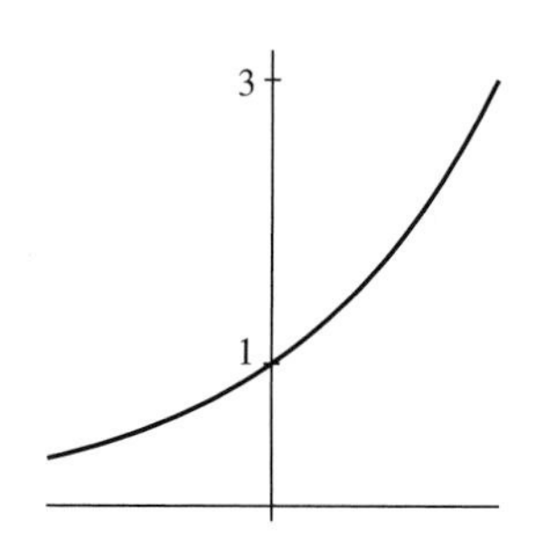

11. $\frac{dy}{dx} = C_7 7^x \approx 1.94591 \cdot 7^x$

$y = \sqrt{7} + C_7\sqrt{7}\left(x - \frac{1}{2}\right) \approx 2.64575 + 5.14839(x - 0.5)$

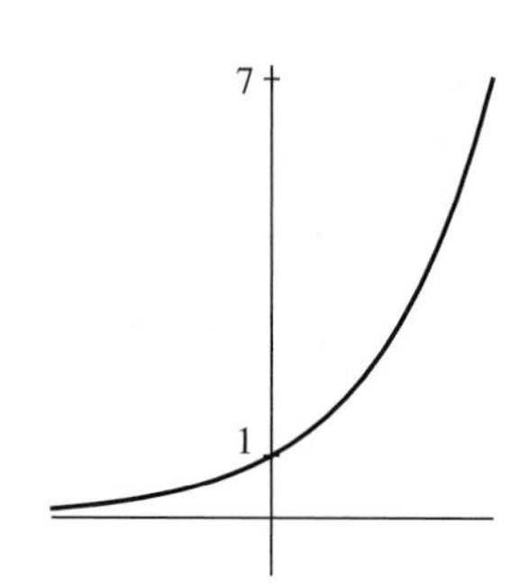

13. $\frac{1}{7}e^{x/7}$ **15.** $-\frac{1}{2}e^{-t/2}$ **17.** $-2e^x \sin(2e^x)$

19. $-2te^{-t^2}$ **21.** $\frac{2e^{2x} - e^x}{2\sqrt{e^{2x} - e^x + 3}}$ **23.** $C_2 2^s e^{4s} + 2^s(4e^{4s})$

25. $e^{-t}(2\cos(2t) - \sin(2t))$ **27.** $\frac{2e^{2u} - e^u}{(e^u - e^{2u})^2}$

29. $e^x \sec(e^x + 1)\tan(e^x + 1)$ **31.** $\frac{e^t + e^{-t}}{2}$

33. $-\frac{y}{y+1}$ **35.** $\frac{y}{2ye^{y^2} - x}$ **39.** $b = e^2$

43. After 5730 years: 10 g
After 11,460 years: 5 g
After 17,190 years: 2.5g
After t years: $20 \cdot 2^{-t/5730}$ g
At time $t = 5730$, rate of change is $(-C_2/573)$ g/yr

45. **b.** $100^{4.1/5} \approx 43.65$

51. $e^{C_2} = 2$, $e^{C_3} = 3$, $e^{C_{10}} = 10$, $e^{C_b} = b$

55. $\frac{1}{3}e^{3x}$ and $\frac{1}{3}e^{3x} + 2001$

59. $A'(t) = -kA_0e^{-kt}$

2.7 Logarithms

1. 4 **3.** 3 **5.** 5 **7.** $\frac{1}{3}\log_2 5 = \frac{1}{3}\frac{\ln 5}{\ln 2} \approx 0.773976$

9. no solution **11.** $\frac{32}{3}$ **13.** 1

15. $\frac{1}{x+2}$ **17.** $\frac{\cos\theta}{\sin\theta}$ $(= \cot\theta)$ **19.** $\frac{10}{t}(1 + \ln(2t))^9$

21. $\frac{-2 + 3z^2}{1 - 2z + z^3} - \frac{\sin z + z\cos z}{z \sin z}$ **23.** $\frac{2\ln\theta}{\theta}e^{(\ln\theta)^2}$

25. $\frac{x^2}{2x+1}\left(\frac{2}{x} - \frac{2}{2x+1}\right)$ **27.** $\sqrt{x}\tan x\left(\frac{1}{2x} + \frac{\sec^2 x}{2\tan x}\right)$

31. $\log_1 x = y$ would mean that $1^y = x$. If $x \neq 1$, there is no possible y value.

35. $\ln(x - 4)$ and $\ln(2(x - 4))$

37. **c.** One graph is the reflection of the other in the line $y = x$.

39. The graph appears to approach 0. **41.** the empty set

43. for approximately $0 < x < 1.1$ and $x > 43.6$

47. For a 150-digit number: $\approx$ 30 years
For a 200-digit number: $\approx$ 110,000 years

49. **a.** pH of 5: about 100 times the number of H^+ ions as water; pH of 10: about 1/1000 times the number of H^+ ions as water
b. The pHs differ by $\log_{10} 2$

51. **a.** The Alaska quake was about $10^{1.5} \approx 31.6$ times as intense as the Loma Prieta quake. **b.** 8.3

53. $k \approx -0.25$

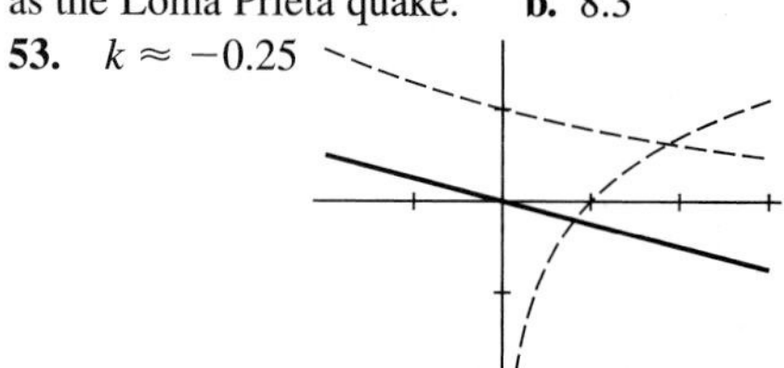

2.8 Inverse Functions

1. $g(x) = -\frac{1}{3}x + \frac{8}{3}$ **3.** $g(x) = 2 + \left(\frac{x-7}{3}\right)^{1/3}$

5. $g(x) = -\frac{4}{3} + \frac{1}{3}e^{x/2}$

7. $g(x) = \frac{5}{2} + \frac{1}{2}\ln\left(\frac{x-6}{3}\right)$, $x > 6$

9.

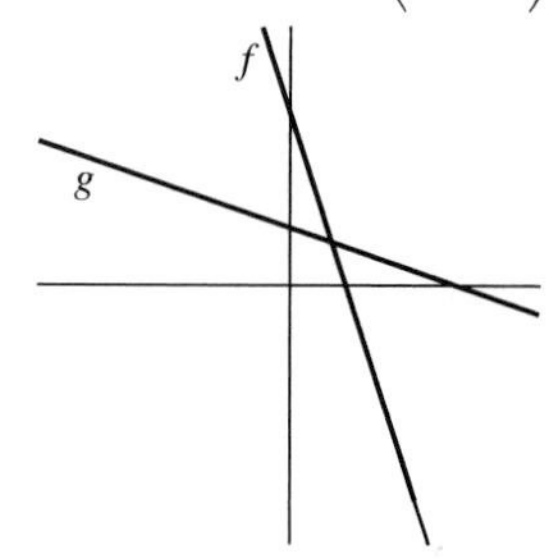

11.

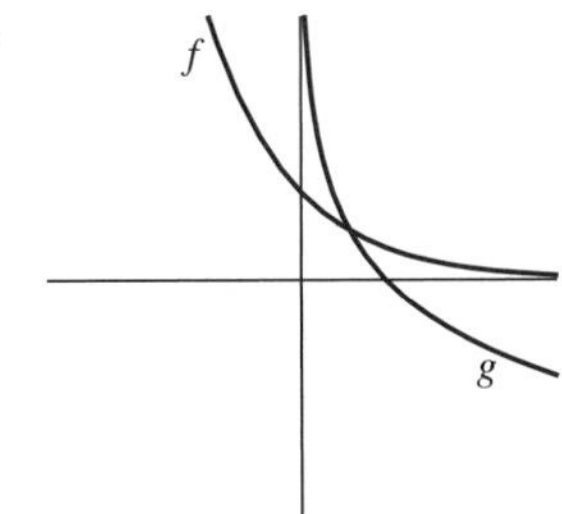

13. $f^{-1}(x) = \dfrac{2-3x}{1+2x}, \{x : x \neq -1/2\}$ 17. $1/14$ 19. $1/6$

21. $1/2$ 25. January 31: $\approx 18.3°$; October 31: $\approx 21.4°$

27. f is one-to-one on $[1, \infty)$. For this restriction, $f^{-1}(x) = \sqrt{x} + 1$.

29. f is one-to-one on $[2, \infty)$.

For this restriction, $f^{-1}(x) = \dfrac{2}{x}\left(1 + \sqrt{1 - x^2}\right)$.

31. f is one-to-one on $(-\infty, 0]$.
For this restriction, $f^{-1}(x) = -\sqrt{e^x - 1}$.

33. $F^{-1}(x) = g(x - b)$ 35. $F^{-1}(x) = \dfrac{1}{c}g(x)$

2.9 Inverse Trigonometric Functions

1. $\dfrac{\pi}{3}$ 3. $-\dfrac{\pi}{2}$ 5. $\dfrac{\pi}{3}$ 7. $-\dfrac{\pi}{4}$ 9. $\dfrac{1}{10.6}$

11. $\dfrac{x}{\sqrt{1-x^2}}$ 13. $\dfrac{\sqrt{4x^2-1}}{2x}$ 15. $\dfrac{2}{\sqrt{1-4x^2}}$

17. $-\dfrac{1}{1+\theta^2}$ 19. $\dfrac{3}{|3s-4|\sqrt{(3s-4)^2-1}}$

21. $\dfrac{2}{(1-x)^2}\dfrac{1}{\sqrt{1-\left(\dfrac{1+x}{1-x}\right)^2}}$ 23. $\dfrac{2x}{1+x^4}e^{\arctan x^2}$

25. $-\dfrac{2}{\sqrt{1-w^2}\,(1-\arccos w)^2}$

27.

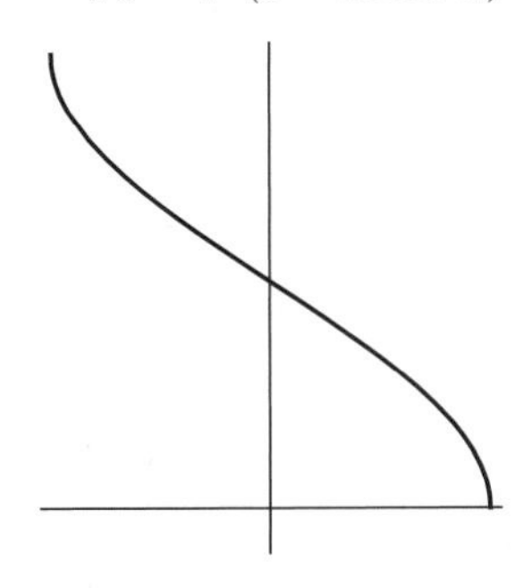

31.

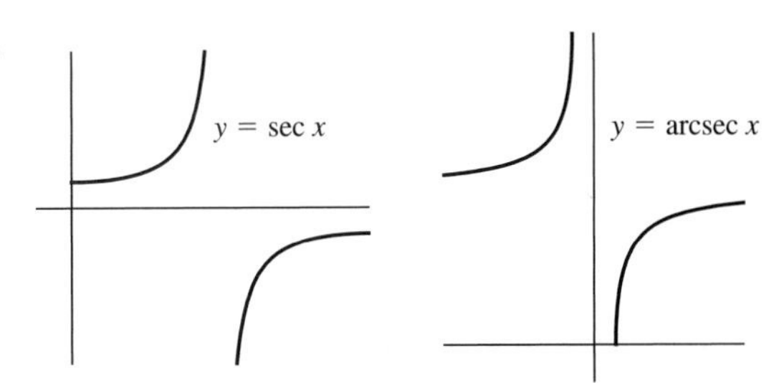

33. a. $y = \dfrac{\pi}{4} + \dfrac{1}{2}(x - 1)$

b.

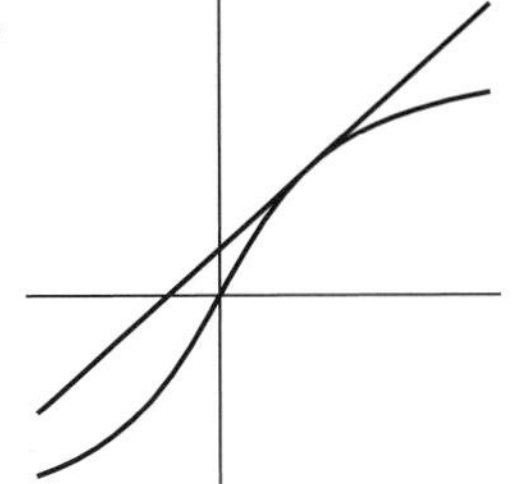

c. $\arctan(0.95) \approx \dfrac{\pi}{4} - 0.025 \approx 0.760398$;

$\arctan(1.03) \approx \dfrac{\pi}{4} + 0.015 \approx 0.800398$

35. a. $y = \dfrac{2\pi}{3} - \dfrac{2}{\sqrt{3}}\left(x + \dfrac{1}{2}\right)$

b.

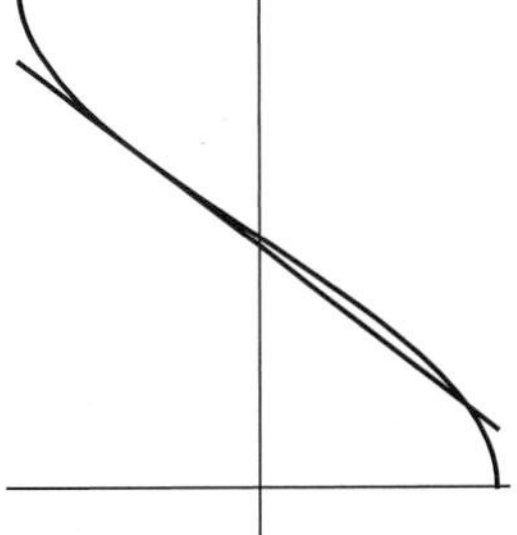

c. $\arccos(-0.48) \approx \dfrac{2\pi}{3} - \dfrac{2}{\sqrt{3}}(0.02) \approx 2.071$;

$\arccos(-0.56) \approx \dfrac{2\pi}{3} + \dfrac{2}{\sqrt{3}}(0.06) \approx 2.16368$

2.10 Modeling: Translating the World into Mathematics

1. a. $\dfrac{dy}{dx} = \dfrac{xe^x - 3e^x}{x^4}$

3. a. $C = 4.0 \cdot 10^6$, $k \approx 0.0280637$, $P(t) \approx 4.0 \cdot 10^6 e^{0.0280637t}$

b.

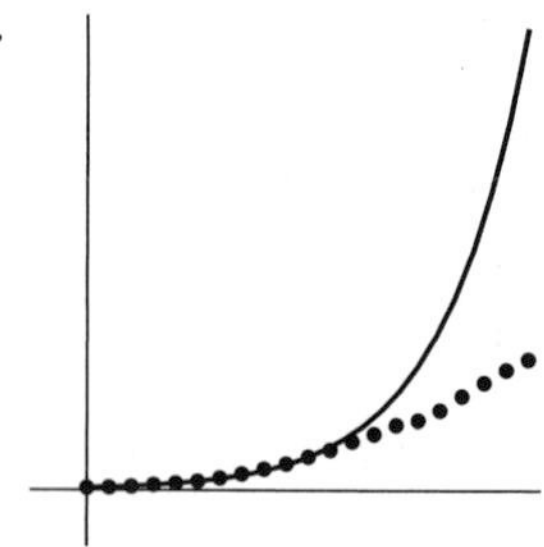

c. $C \approx 4.54 \cdot 10^6$, $k \approx 0.0263031$, $P(t) \approx 4.54 \cdot 10^6 e^{0.0263031t}$, where $t = 0$ corresponds to 1790.

d.

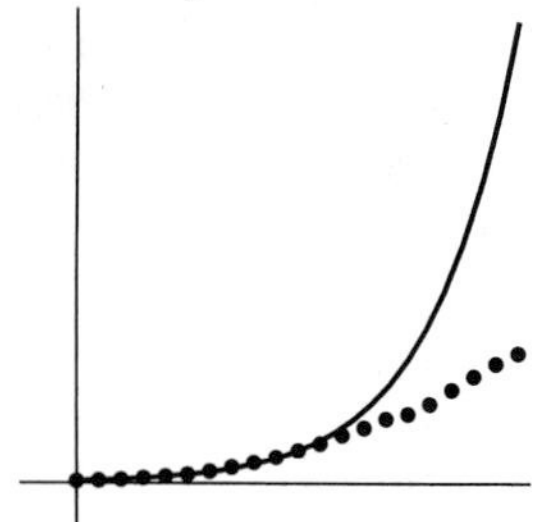

5. **a.** increasing when $P = 250$ million, decreasing when $P = 400$ million
b. $P = k/c \approx 3.00 \cdot 10^8$. If the population ever reaches k/c, then it stays at this figure.

7. **b.** ≈ 0.79 liters **c.** ≈ 0.63 liters

9. $\dfrac{dC}{dt} = -kC$, k a positive constant

11. $\dfrac{dT}{dt} = -k(T - T_0)$, k a positive constant

13. **i.** $\dfrac{dA}{dt} = c_1A - k_1AB$, $\dfrac{dB}{dt} = c_2B - k_2AB$, where c_1, c_2, k_1, k_2 are positive constants
ii. Yes, for example, when $A = c_2/k_2$ and $B = c_1/k_1$

15. **i.** $\dfrac{dR}{dt} = c_1R - k_1RF$, $\dfrac{dF}{dt} = c_2RF - k_2F$, where c_1, c_2, k_1, k_2 are positive constants
ii. when $R = F = 0$ or when $R = k_2/c_2$ and $F = c_1/k_1$

Chapter Review Exercises

1. $-20x^4 + 28x^3 - 2\sqrt{2}x + 3$

3. $\left(2\theta + \dfrac{3}{\theta^2}\right) \sin\theta + \left(\theta^2 - \dfrac{3}{\theta}\right)\cos\theta$ **5.** $\dfrac{12x^2 - 2}{4x^3 - 2x + 3}$

7. $\dfrac{3}{(p^2+3)^{3/2}}$ **9.** $\dfrac{22}{8x^2 - 2x - 15}$

11. $\dfrac{2t^{3/2}}{\sqrt{1-t^4}} + \dfrac{\arcsin(t^2)}{2\sqrt{t}}$

13. $2\csc(2t)\sec(2t)$

15. $\dfrac{1}{(2x^2 + 2x + 1)(\arctan(1 + 1/x))^2}$

17. $-2\sec^2(2t)\sin(1 + \tan(2t))\cos(\cos(1 + \tan(2t)))$

19. $(2as + b)e^{as^2+bs+c}$ **21.** $\dfrac{6x - y}{x + 3y^2}$ **23.** $\dfrac{1 - \cos(x+y)}{1 + \cos(x+y)}$

25.

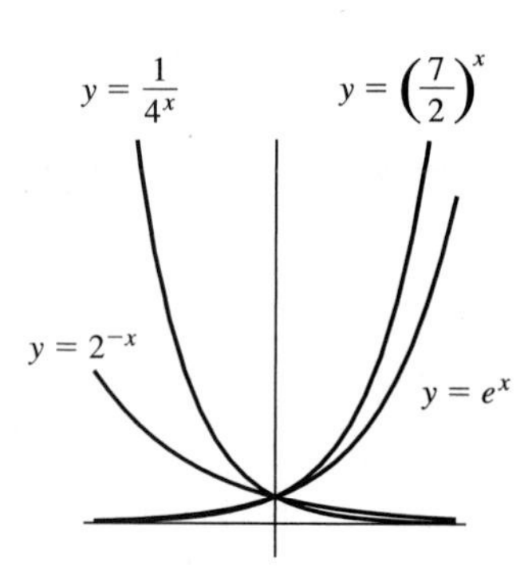

27. **a.**

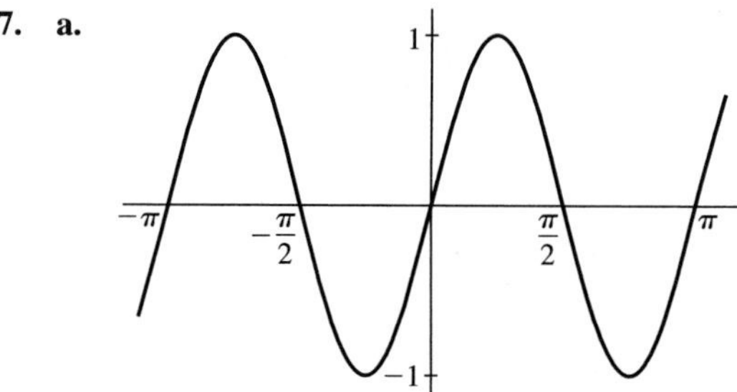

b.

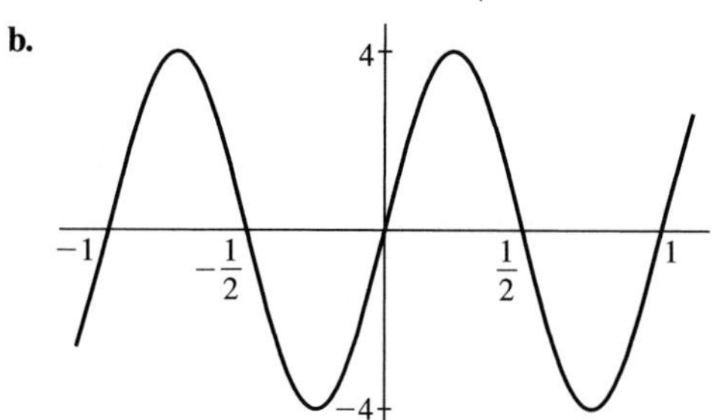

c.

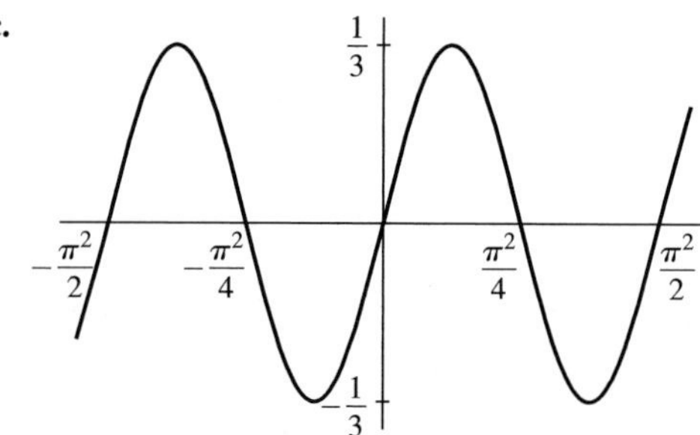

29. **a.** $y = x - 1$ **b.** ≈ 1.93148 **c.** $r \leq 0.17$ works

31. **a.** $\theta = \arctan(h/300)$ **b.** $50 \leq h < \infty$; $\arctan(1/6 \leq \theta < \pi/2)$ **c.** $\dfrac{300}{90{,}000 + h^2}$ **d.** radians/ft

33. **a.** $\phi = \arcsin\left(\dfrac{500}{y}\right) - \arctan\left(\dfrac{500}{150 + \sqrt{y^2 - 500^2}}\right)$

b. $y \geq 500$, $0 \leq \phi < \arctan(3/20)$

c. $\dfrac{d\phi}{dy} = \dfrac{500y}{\sqrt{y^2 - 500^2}\left(22{,}500 + y^2 + 300\sqrt{y^2 - 500^2}\right)} - \dfrac{500}{\sqrt{y^4 - 500^2y^2}}$

d. radians/ft

35. **a.** January 31: $\approx 15.9°$; October 31: $\approx 20°$
b. close to October 1 **c.** close to April 27 or August 23
d. $\approx \dfrac{0.32\sin(0.017t)}{\sqrt{1-(0.53-0.33\cos(0.017t))^2}}$

37. **a.** $x = 60\cos\left(\dfrac{2\pi t}{40} - \dfrac{\pi}{2}\right)$, where $t = 0$ is a time when the car is at its low point and $x = 0$ corresponds to the point on the ground directly beneath the center of the wheel. We also assume that the wheel is turning counterclockwise.
b. $\dfrac{dx}{dt} = -3\pi\sin\left(\dfrac{2\pi t}{40} - \dfrac{\pi}{2}\right)$
c. Largest rate of change: $t = 0, 20, 40, 60, \ldots$. At these times the rate of change is $\pm 3\pi$ and the car is at its highest or lowest point. Smallest rate of change: $t = 10, 30, 50, 70, \ldots$. At these times the rate of change is 0 and the car shadow is at $x = \pm 60$.

39. **a.** $2x^2 + 3x$ **b.** $A'(x) = 4x + 3$

CHAPTER 3—MOTION, VECTORS, AND PARAMETRIC EQUATIONS

3.1 Motion along a Line

1. $v(1.5, 1.6) = -15.19$ m/s, $v(1.5, 1.51) = -14.749$ m/s, $v(1.5, 1.501) = -14.7049$ m/s, $v(1.5, 1.5001) \approx -14.7005$ m/s; $v(1.5) \approx -14.7$ m/s; because $h(5) < 0$ m, the jewels will not be saved.

3. $b \approx 142.859$ s, $v(b) \approx -700.018$ m/s

5. $v(2.74721) \approx -72.8011$ m/s

7. It will take 10.2 s, approximately. Its top speed is 99 m/s.

9. About 70,000 m **11.** 41.4 seconds past 2:00 P.M.

13. $v(0) = -0.65$ m/s; $v(\pi/1.3) \approx 0.247232$ m/s

15. $v(0, 62.2449) = 305$ m/s. During the 62.2449 seconds the bullet took to reach its highest point, the velocity of the bullet decreased from 610 m/s to 0 m/s. If it had gone at its average velocity (305 m/s) throughout its ascent the bullet would have risen to the same level.

17. $x(0) = 2.88130$ m; $v(0) = 4.10769$ m/s

19. $v(10)\cdot 3600 \approx 92.4$ mph

21. $v(-c/b, -c/b + \pi/(2b)) = \dfrac{2ab}{\pi}$

23. Using the given identity,

$$v(t_1, t_2) = ab\frac{\sin\left(\frac{1}{2}b(t_2 - t_1)\right)}{\frac{1}{2}b(t_2 - t_1)}\cos\left(\tfrac{1}{2}b(t_1 + t_2) + c\right).$$

Using the fact that $\lim\limits_{\theta\to 0}\dfrac{\sin\theta}{\theta} = 1$, it follows that

$\lim\limits_{t_2\to t_1} v(t_1, t_2) = ab\cos(bt_1 + c)$, which is the velocity at t_1.

25. From the odometer readings, we find (0.0, 12000), (0.1, 12010), (0.2, 12019), (0.3, 12027), (0.4, 12035), (0.5, 12040), and (0.6, 12042). Readings may vary a bit. The figure gives the average velocities on the intervals $[0, 0.1], \ldots, [0.5, 0.6]$.

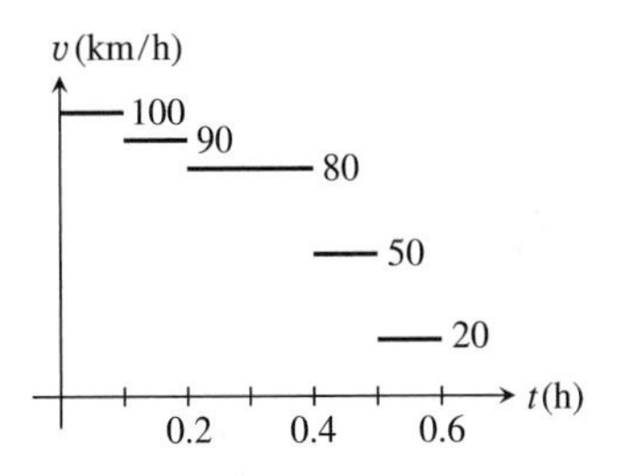

27. The figure shows a velocity graph including the main features.

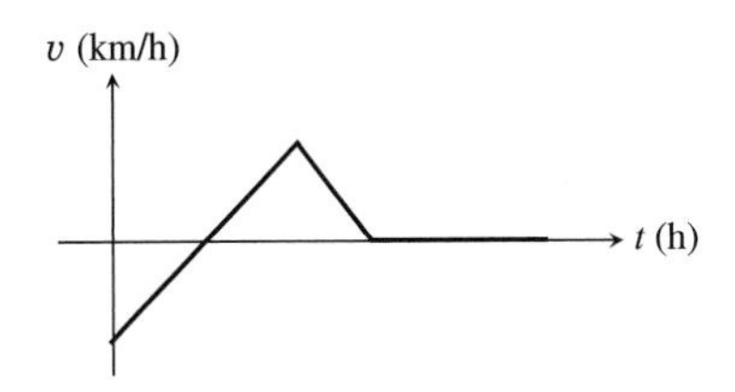

29. The maximum compression will occur at the first time $t \ge 0$ for which $v = v(t) = 0$. From

$$v = -0.528e^{-0.415t}\sin(1.26t + 1.89),$$

we set $1.26t + 1.89 = \pi$, from which it follows that maximum compression occurs at $t \approx 0.993$ seconds.

31. In solving the equation

$$10^2 = 5^2 + x^2 - 2\cdot 5\cdot x\cdot\cos t,$$

choose the + sign because point Q will always be to the right of point P.

3.2 Vectors

1. $\overrightarrow{PQ}$ and $\overrightarrow{ST}$ are equivalent because $4 - 1 = 10 - 7 = 3$ and $-1 - 1 = 3 - 5 = -2$; $\mathbf{r} = \langle 3, -2\rangle$; $\|\mathbf{r}\| = \sqrt{13}$.

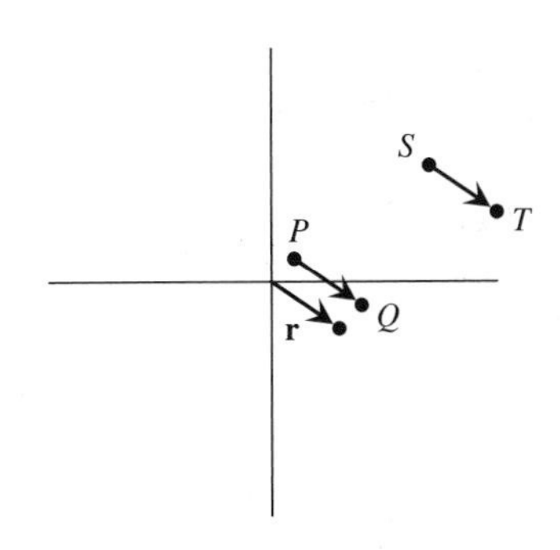

3. $\overrightarrow{PQ}$ and $\overrightarrow{ST}$ are equivalent because $5 - 3 = -4 - (-6) = 2$ and $-2 - (-2/3) = 4 - (16/3) = -4/3$; $\mathbf{r} = \langle 2, -4/3 \rangle$; $\|\mathbf{r}\| = \sqrt{52/9}$.

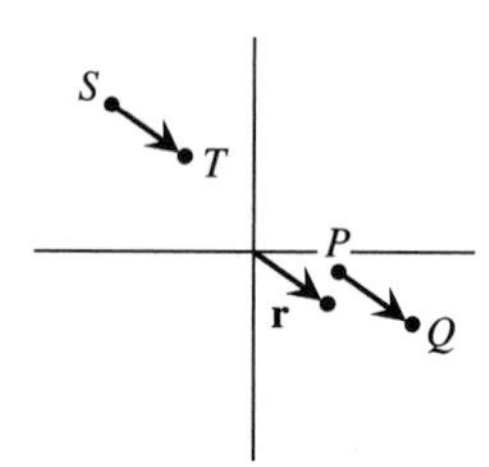

5. $\overrightarrow{PQ}$ and $\overrightarrow{ST}$ are equivalent because $(-0.8) - (-2.5) = 1.7 - 0.0 = 1.7$ and $1.1 - 3.0 = -3.2 - (-1.3) = -1.9$; $\mathbf{r} = \langle 1.7, -1.9 \rangle$; $\|\mathbf{r}\| = \sqrt{6.5}$.

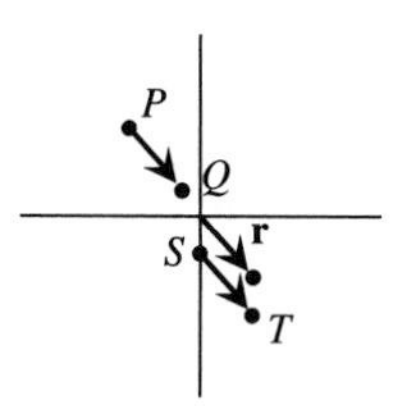

7. $\mathbf{a} + \mathbf{b} = \langle 7, 7 \rangle$

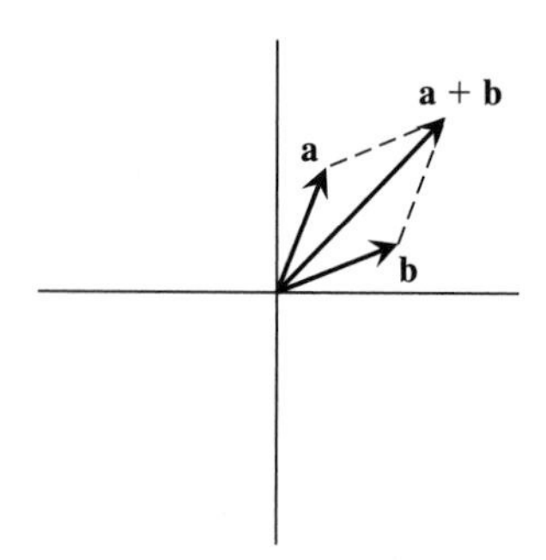

9. $\mathbf{a} + \mathbf{b} = \langle -2, 7 \rangle$

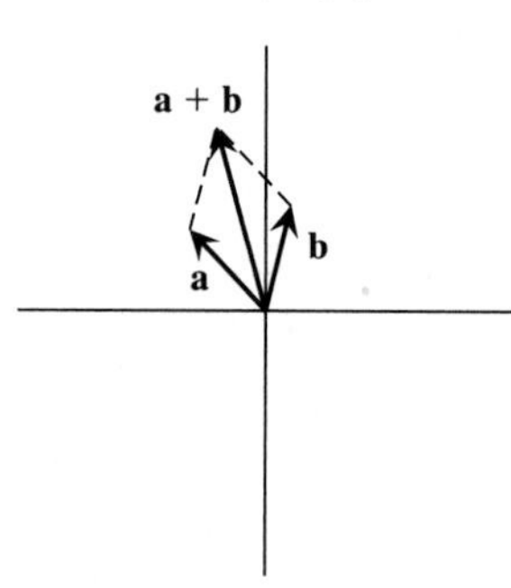

11. $\overrightarrow{OR} = \langle 4, 7 \rangle$, $R = (5, 8)$

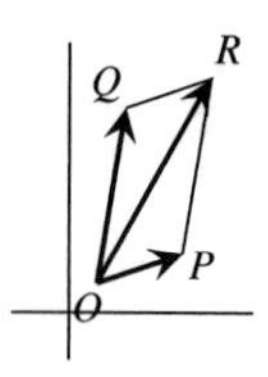

13. $\overrightarrow{OR} = \langle 4.2, -1.6 \rangle$, $R = (1.7, 1.4)$

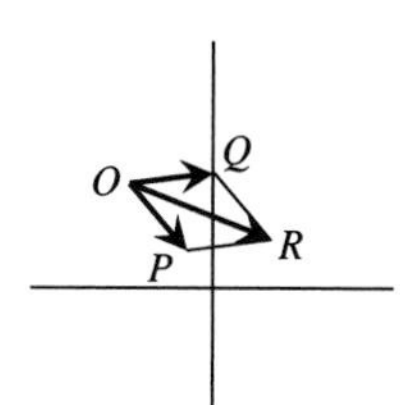

15. $\|\mathbf{a}\| = 13$, $\mathbf{a} + \mathbf{b} = \langle 6, -8 \rangle$, $\mathbf{a} - \mathbf{b} = \langle 4, -16 \rangle$, $h\mathbf{a} + k\mathbf{b} = \langle 7, -4 \rangle$

17. $\|\mathbf{a}\| = \sqrt{41}$, $\mathbf{a} + \mathbf{b} = -\mathbf{i} - 7\mathbf{j}$, $\mathbf{a} - \mathbf{b} = -7\mathbf{i} - 3\mathbf{j}$, $h\mathbf{a} + k\mathbf{b} = 11\mathbf{i} + 8\mathbf{j}$

19. $\|\mathbf{a}\| = \sqrt{37}$, $\mathbf{a} + \mathbf{b} = \langle 2, 2 \rangle$, $\mathbf{a} - \mathbf{b} = \langle 0, 10 \rangle$, $h\mathbf{a} + k\mathbf{b} = \langle 0, 5 \rangle$

21. $\|\mathbf{a}\| \approx 2.83019$, $\mathbf{a} + \mathbf{b} = 1.8\mathbf{i} - 0.2\mathbf{j}$, $\mathbf{a} - \mathbf{b} = 1.2\mathbf{i} - 4.6\mathbf{j}$, $h\mathbf{a} + k\mathbf{b} = 1.38\mathbf{i} - 11.32\mathbf{j}$

23. $\left\langle \sqrt{3}/2, 1/2 \right\rangle$ **25.** $\left\langle -\sqrt{3}/2, -1/2 \right\rangle$

27. $\langle 0.283662, -0.958924 \rangle$

29. The magnitude of the force on the left is 72.2492 newtons and the magnitude of the force on the right is 74.2662 newtons.

31. The magnitude of the force on the left is 178.791 newtons and the magnitude of the force on the right is 65.9253 newtons.

33. $\left(8/\sqrt{34}\right)\langle 3, 5 \rangle$

35. The coordinates of D are $(-3, -1)$.

37. The coordinates of T are $(10, 9)$.

39. $(0, 4)$ **41.** $\left(5 + 5\sqrt{3}, 10\right)$

43. The single equivalent displacement is $\langle -10\mathbf{i} + 18\mathbf{j} \rangle$; the coordinates of T are $(-10, 22)$.

45. $\left(-7/2 + 5\sqrt{3}, -4 + 5\sqrt{3}/2\right)$

47. $\mathbf{r} = \langle 1, m \rangle$; $m = b/a$

51. Referring to the figure, the sum of the distances $\|\mathbf{a}\|$ and $\|\mathbf{b}\|$ is at least as large as the distance $\|\mathbf{a}\| + \|\mathbf{b}\|$.

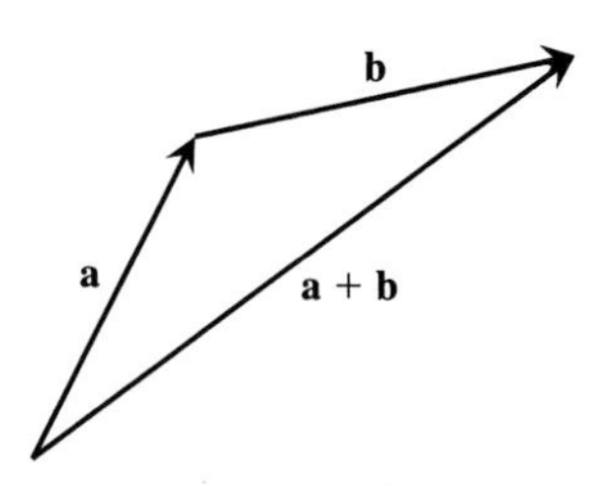

3.3 Parametric Equations

1. $\mathbf{r}(t) = \langle 200t\cos(30°), 1 + 200t\sin(30°) \rangle$, $\mathbf{r}(t_1) \approx \langle 0.173205, 1.1 \rangle$

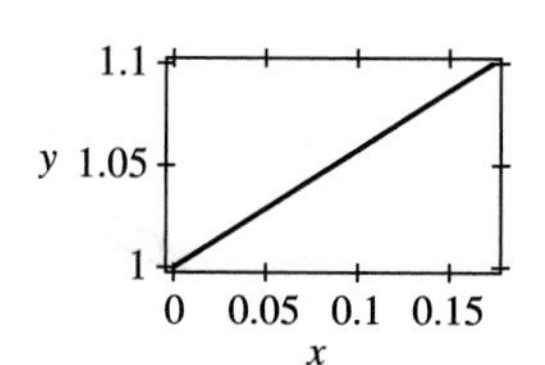

3. $\mathbf{r}(t) = \langle -5 + 225t\cos(110°), 3 + 225t\sin(110°) \rangle$, $\mathbf{r}(t_1) \approx \langle -5.07695, 3.21143 \rangle$

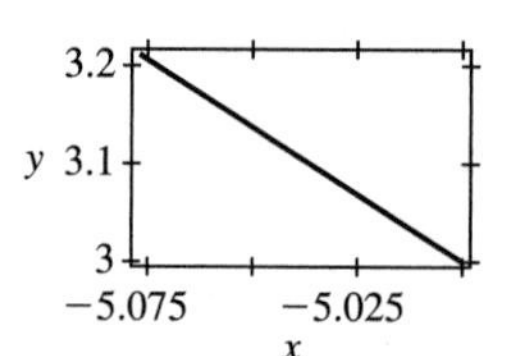

5. $\mathbf{r}(t) = \langle -4.5 + 190t\cos(350°), 3.2 + 190t\sin(350°) \rangle$, $\mathbf{r}(t_1) \approx \langle -4.31289, 3.16701 \rangle$

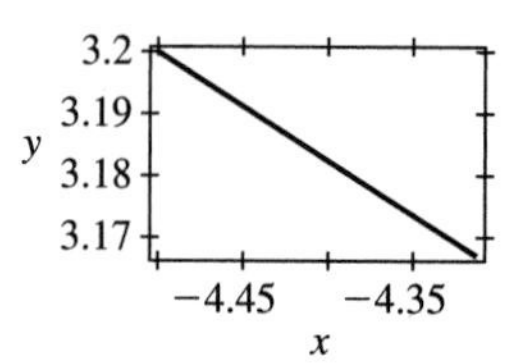

7. Speed is 8 km/s; $\mathbf{r}(t) = \langle 50 - 4\sqrt{3}t, 40 - 4t\rangle$; the meteorite hits the Earth at $(50 - 40\sqrt{3}, 0)$ at $t = 10$ s.
9. Speed is 8 km/s; $\mathbf{r}(t) = \langle 35 - 4\sqrt{2}t, 30 - 4\sqrt{2}t\rangle$; the meteorite hits the Earth at $(5, 0)$ at $t = 15\sqrt{2}/4$ s.
11. Speed is about 8.0 km/s; $\mathbf{r}(t) \approx \langle 90.2 - 5.1t, 52.7 - 6.2t\rangle$; the meteorite hits the Earth near $(46.8, 0)$ at $t \approx 8.5$ s.
13. $x/5 = (y - 2)/3$; $\mathbf{r}(t_1) = \langle 5/2, 7/2\rangle$

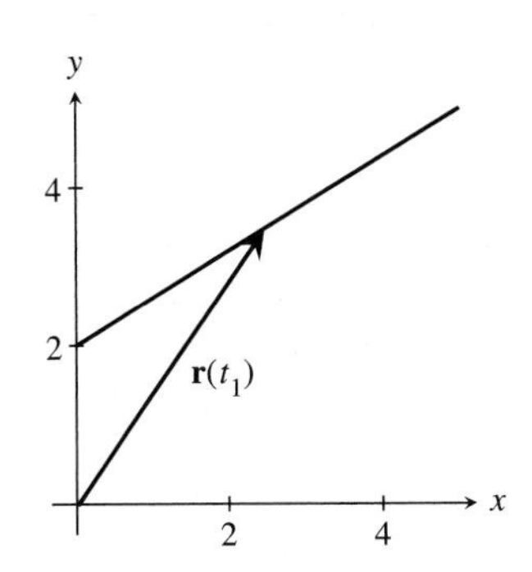

15. $x^2 + y^2 = 9$; $\mathbf{r}(t_1) = \langle 0, 3\rangle$

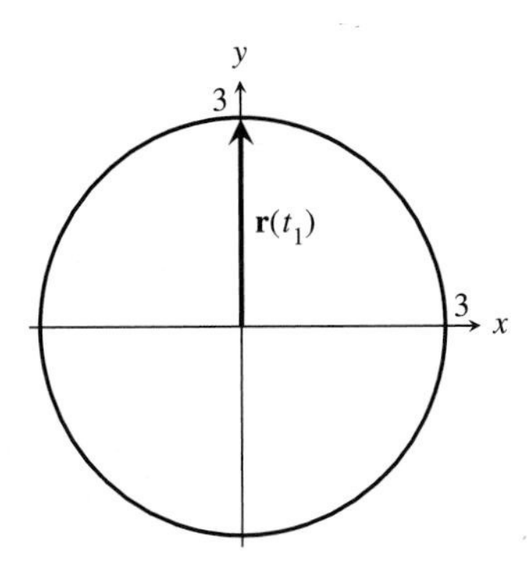

17. $\mathbf{r}(t) = \langle 3, 3\rangle + 2\langle \cos(2t), \sin(2t)\rangle$

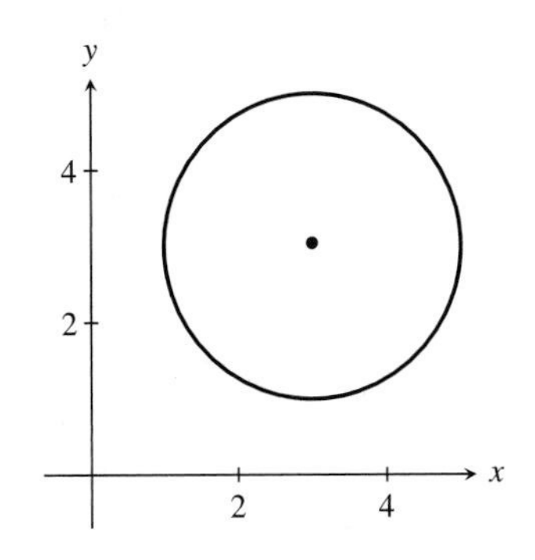

19. $\mathbf{r}(t) = \langle 0, 2\rangle + 5\langle \cos(5\pi t/9), \sin(5\pi t/9)\rangle$

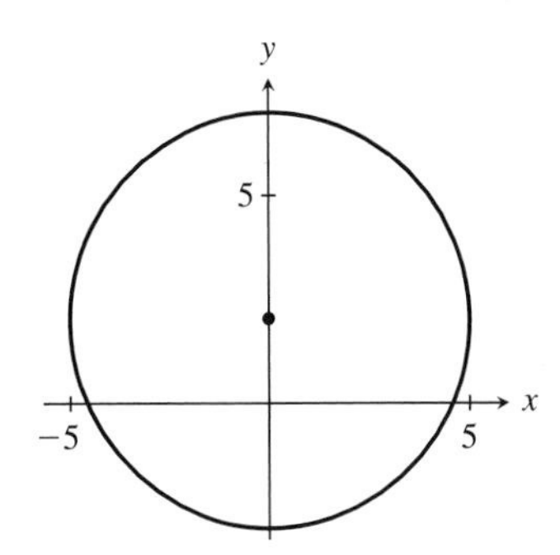

21. $\mathbf{r}(t) = \langle 3, 3\rangle + 2\langle \cos(\pi - 2t), \sin(\pi - 2t)\rangle$, $0 \le t \le \pi$

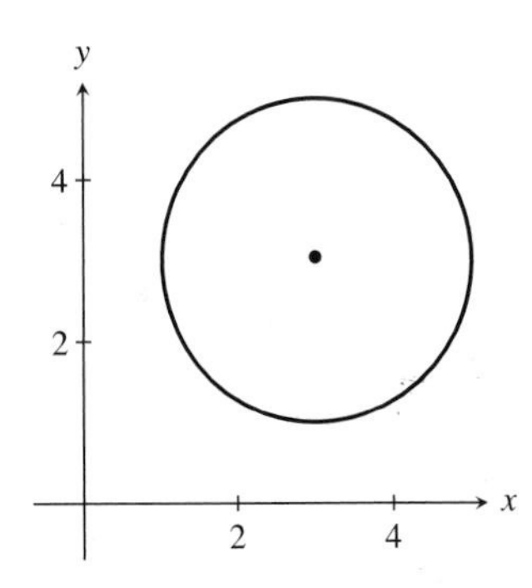

23. $\mathbf{r}(t) =$
$\langle 0, 62.5\rangle + 50\langle \cos(2\pi t/15 + 3\pi/2), \sin(2\pi t/15 + 3\pi/2)\rangle$
25. $\mathbf{r}(t) = 0.64\langle \cos(18628t/0.64), \sin(18628t/0.64)\rangle$; $\mathbf{r}(0.0002) \approx \langle 0.572923, -0.285236\rangle$; total elapsed time is about 0.000242712 seconds.
29. The object is moving along a line. It travels a distance of $\|\mathbf{r}(t+1) - \mathbf{r}(t)\| = |b|$ in one time unit. Because its speed is constant, its speed is $|b|$ units per time unit. If $\mathbf{u}$ were not a unit vector, it follows in the same way that the speed of the object is $|b| \cdot \|u\|$ units per time unit. The equation $\mathbf{r}(t) = \langle 2, 3\rangle + 5t\langle 1, 1\rangle$ can be written as $\mathbf{r}(t) = \langle 2, 3\rangle + (5\sqrt{2})t\mathbf{u}$, where $\mathbf{u}$ is the unit vector $\langle 1/\sqrt{2}, 1/\sqrt{2}\rangle$. It follows that the speed of the object is $5\sqrt{2}$ units per time unit.
31.

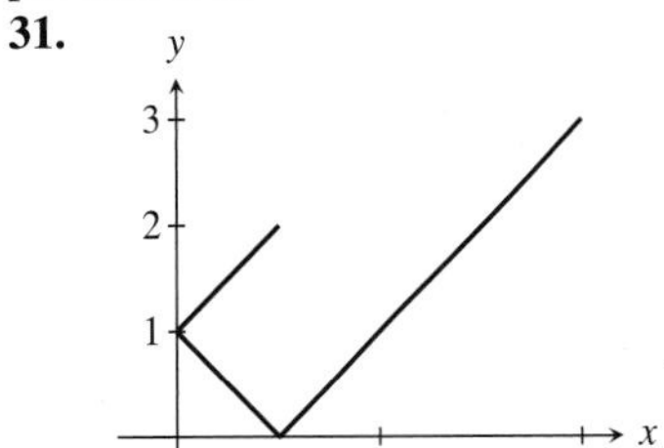

3.4 Velocity and Tangent Vectors

1. $\mathbf{r}(1) = \langle 2, 3\rangle$, $\mathbf{v}(1) = \langle 2, 6\rangle$

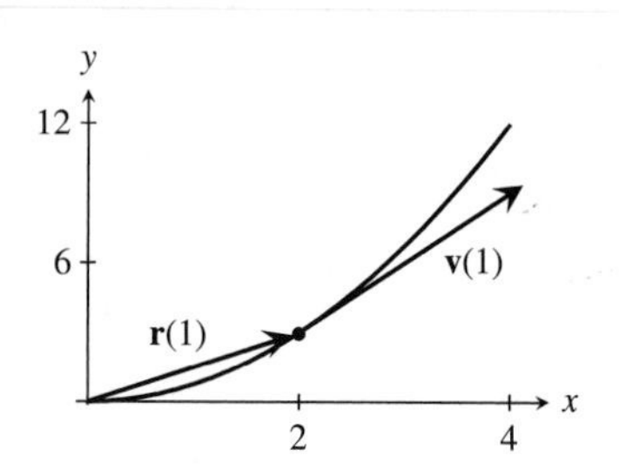

3. $\mathbf{r}(1) = \langle 0, 3\rangle$, $\mathbf{v}(1) = \langle -1, 4\rangle$

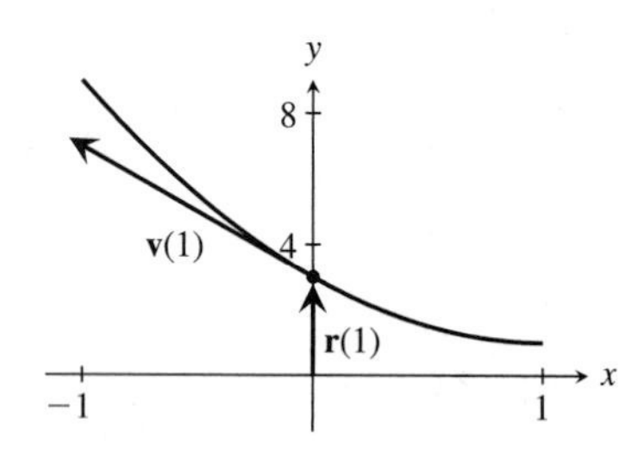

5. $\mathbf{r}(\pi/6) = \left\langle 5/2, 5\sqrt{3}/2\right\rangle$, $\mathbf{v}(\pi/6) = \left\langle -5\sqrt{3}, 5\right\rangle$

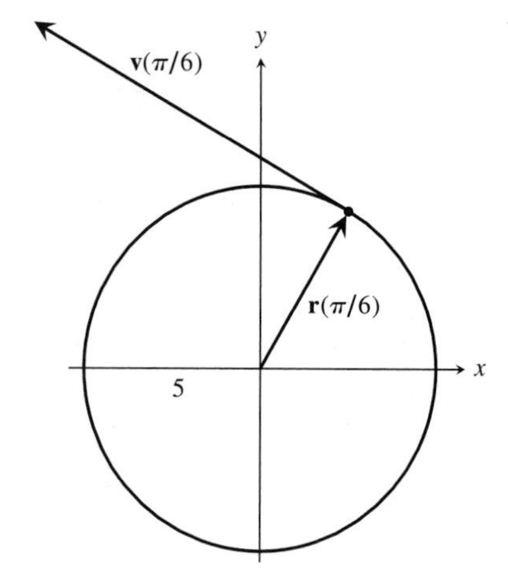

7. $\mathbf{r}(2) = 3\mathbf{i} + 3\mathbf{j}$, $\mathbf{v}(2) = \mathbf{i} + (1/2)\mathbf{j}$

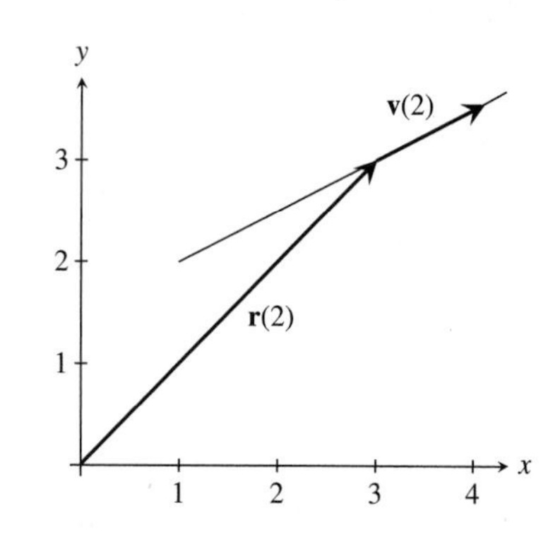

9. $\mathbf{v}(t_1, t_1 + 0.1) = \langle 1, 0.839604\rangle$, $\mathbf{v}(t_1, t_1 + 0.01) = \langle 1, 0.863511\rangle$, $\mathbf{v}(t_1, t_1 + 0.001) = \langle 1, 0.865775\rangle$; $\mathbf{v}(t_1) = \langle 1, 0.866025\rangle$; the differences between the velocity at t_1 and the three average velocities are $\langle 0, 0.03\rangle$, $\langle 0, 0.003\rangle$, and $\langle 0, 0.0003\rangle$, approximately.

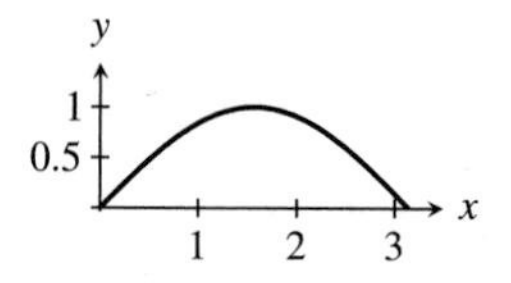

11. $\mathbf{v}(t_1, t_1 + 0.1) = \langle 1, 2.85884\rangle$, $\mathbf{v}(t_1, t_1 + 0.01) = \langle 1, 2.73192\rangle$, $\mathbf{v}(t_1, t_1 + 0.001) = \langle 1, 2.71964\rangle$; $\mathbf{v}(t_1) = \langle 1, 2.71828\rangle$; the differences between the velocity at t_1 and the three average velocities are $\langle 0, 0.14\rangle$, $\langle 0, 0.014\rangle$, and $\langle 0, 0.0014\rangle$, approximately.

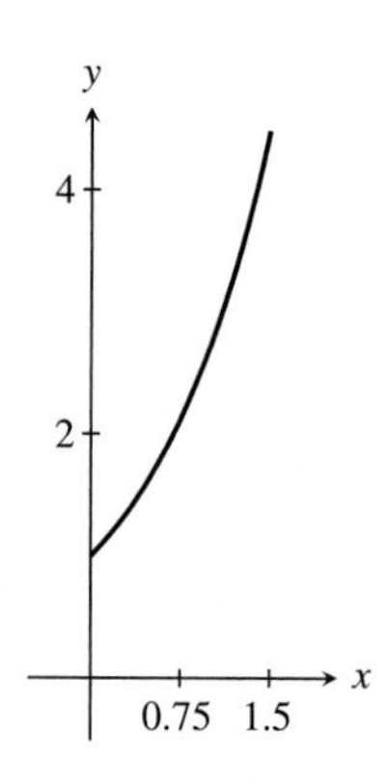

13. Speed and direction at t_1 are 12.6 and 1.63806; the position of the plane 10 seconds after t_1 is $\langle -2.48178, 126.118\rangle$.

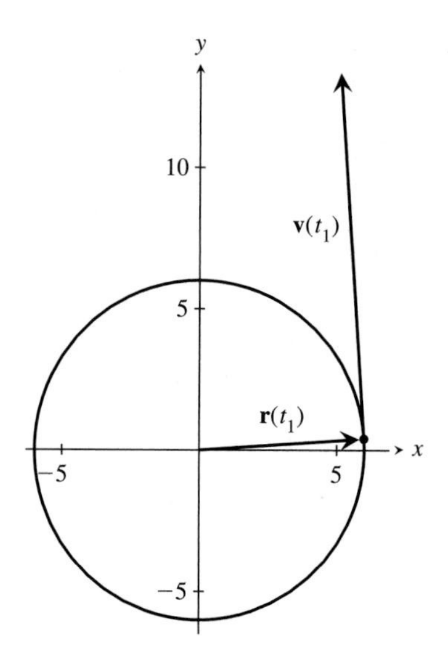

15. Speed and direction at t_1 are 5.59017 and 1.10715; the position of the plane 10 seconds after t_1 is $\langle 27.5, 53.5\rangle$.

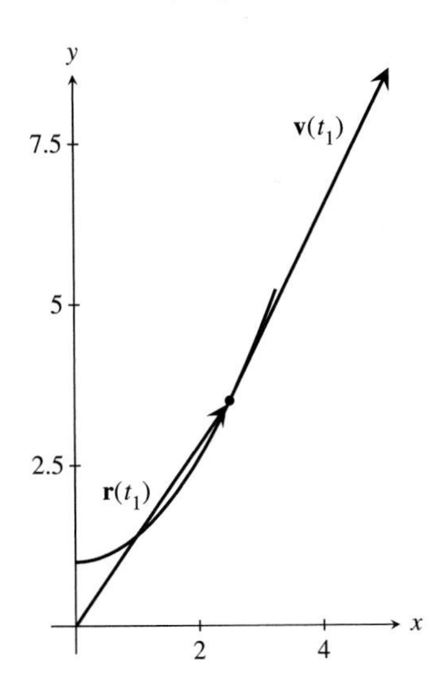

17. $\mathbf{r}(t) = \langle 3, 7\rangle + t\langle 3, 2\rangle, -\infty < t < \infty$
19. $\mathbf{r}(t) = \langle 4, 5\rangle + t\langle 1,1\rangle, -\infty < t < \infty$
21. $\mathbf{r}(t) = \langle 4, -2\rangle + t\langle -4, 5\rangle, -\infty < t < \infty$
23. $\mathbf{r}(t) = \langle -3, 0\rangle + t\langle 0, 5\rangle, -\infty < t < \infty$
25. $\mathbf{r}(x) = x\mathbf{i} + x^2\mathbf{j}$; $\mathbf{r}'(1) = \mathbf{i} + 2\mathbf{j}$; slope is 2
27. $\mathbf{r}(x) = \langle x, \tan x\rangle$; $\mathbf{r}'(\pi/4) = \langle 1, 2\rangle$; slope is 2
29. $\mathbf{r}(x) = x\mathbf{i} + (x^2 - 6x + 11)\mathbf{j}$; $\mathbf{r}'(1) = \mathbf{i} + 2\mathbf{j}$; slope is 2
31. $\mathbf{r}(x) = \langle x, e^{2x}\rangle$; $\mathbf{r}'(1) = \langle 1, 2e^2\rangle$; slope is $2e^2$
33. $\mathbf{r}(x) = x\mathbf{i} + (\ln x)^2\mathbf{j}$; $\mathbf{r}'(1) = \mathbf{i}$; slope is 0
35. $\mathbf{r} = \langle -5, 1\rangle + \frac{10}{\sqrt{157}}t\langle 11, 6\rangle$ **37.** $\mathbf{r} = \langle \cos(7t), \sin(7t)\rangle$
39. $\mathbf{r} = t\langle 1, 5\rangle + \langle 0, -2\rangle$
41. $\mathbf{v}(T/2) = \langle 1, 8\rangle$ m/s; speed at $T/2$ is $\sqrt{65}$ m/s; eliminating the parameter t gives $y = 16x - x^2, 0 \le x \le 8$.

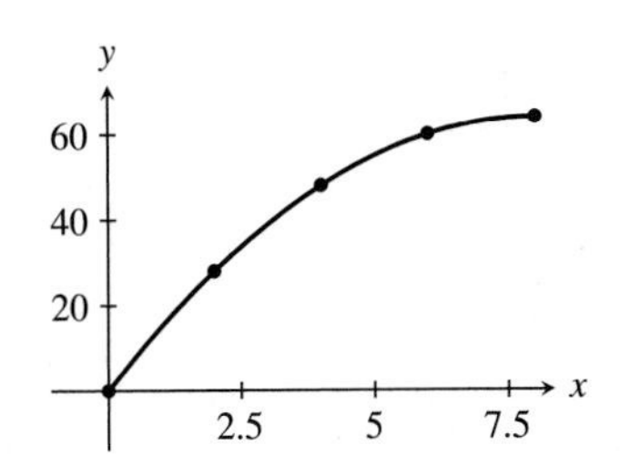

43. $\mathbf{v}(T/2) = \langle 4, 0\rangle$ m/s; speed is 4 m/s; eliminating the parameter t gives $16y = 16x - x^2, 0 \le x \le 16$.

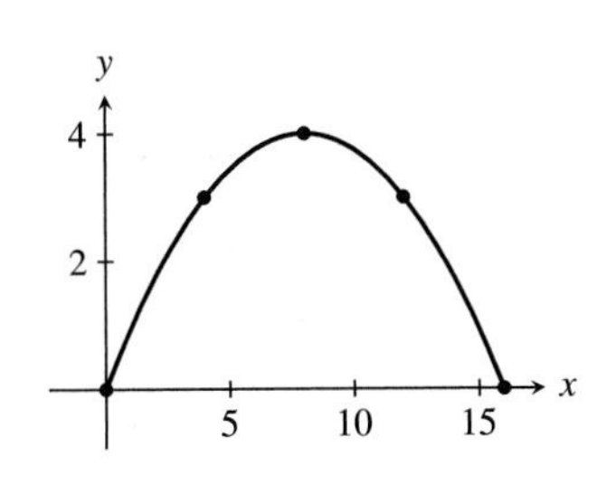

45. $\mathbf{r}(-2) = \langle 9, 0\rangle$; $\mathbf{r}(3) = \langle 19, 5\rangle$; $x = 2y^2 - 8y + 9$

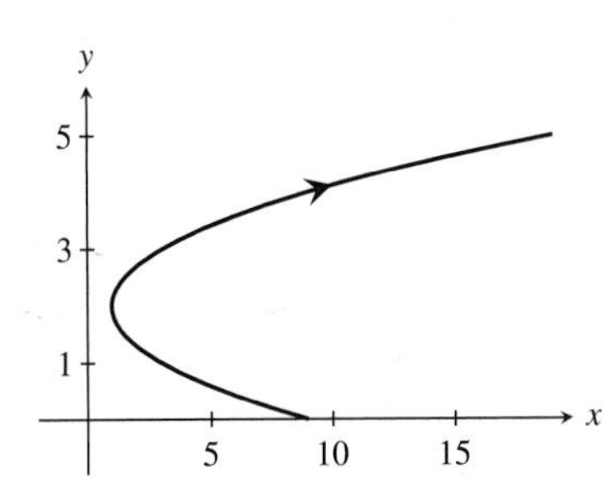

47. $\mathbf{r}(-2) = \langle 0, 2\rangle$; $\mathbf{r}(2) = \langle 4, 6\rangle$; $-y(y - 2) = 4x^2 - 2x(2y - 1)$

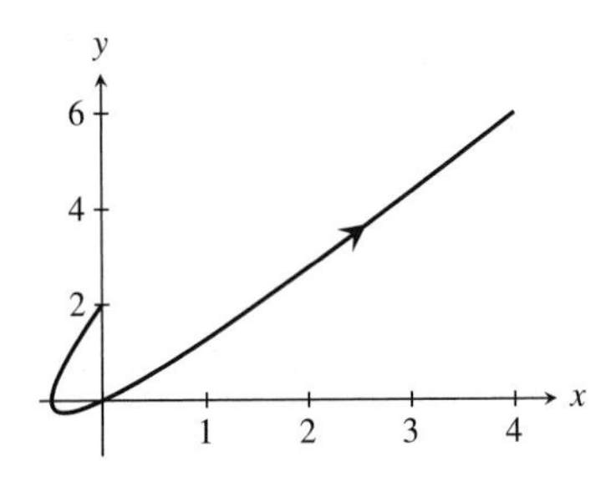

49. $\mathbf{r}(-2) = \langle -3, -3\rangle$; $\mathbf{r}(3) = \langle 7, 2\rangle$; $x = 2y + 3$

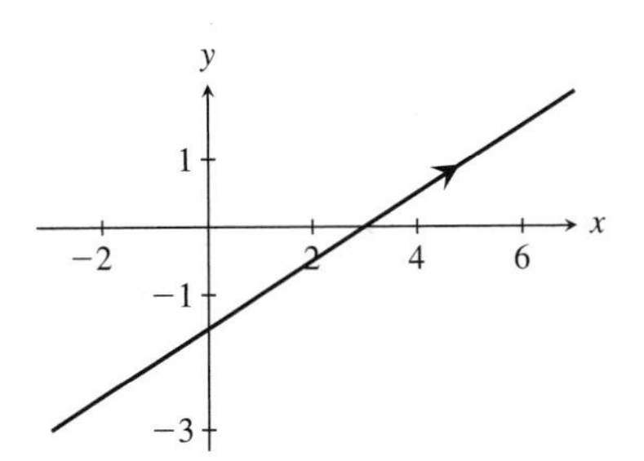

51.

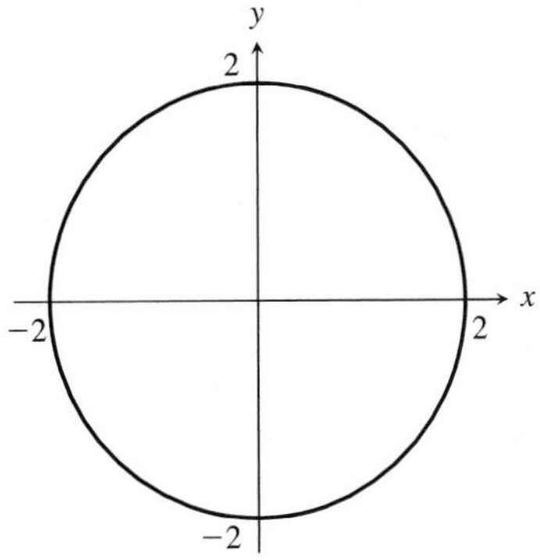

53.

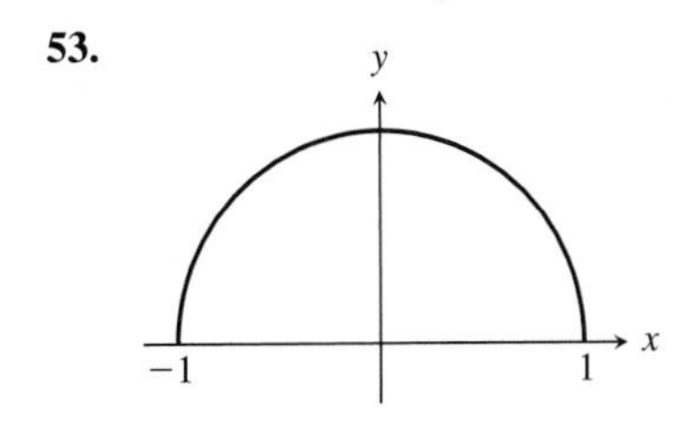

55.

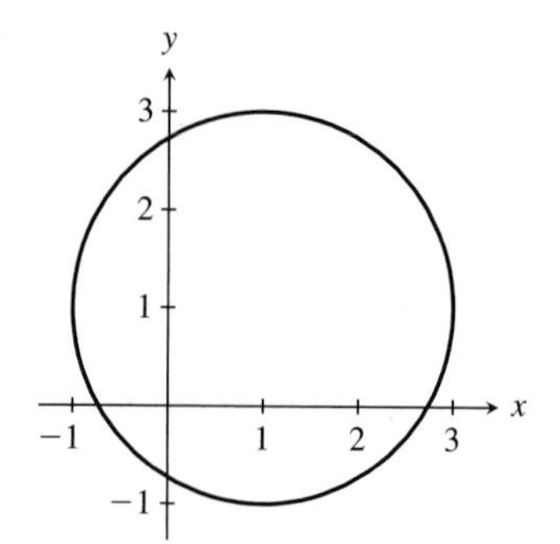

57.

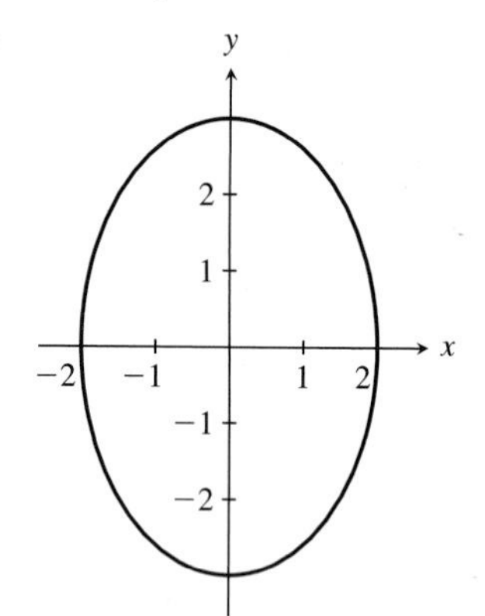

59.

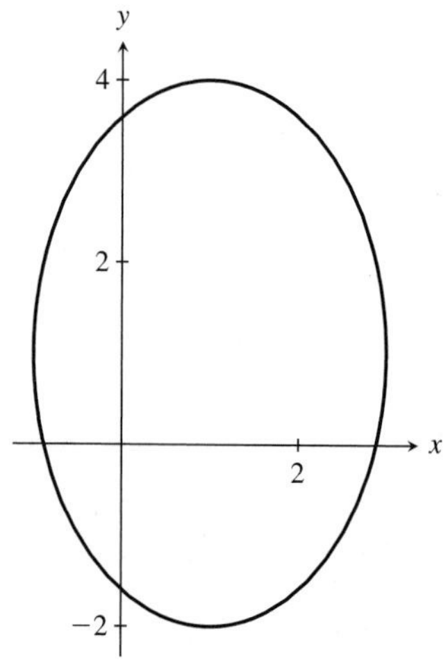

61. Because (x, y) satisfies the equation $x^2/5^2 + y^2/4^2 = 1$, the point $(x/5, y/4)$ satisfies the equation $X^2 + Y^2 = 1$. Hence, there is a number $\theta \in [0, 2\pi)$ for which $x/5 = \cos\theta$ and $y/4 = \sin\theta$.

63. No; if it were there would be a t for which $t - 1 = 1$ and $6t + 4 = 15$. These equations are not consistent.

65. A parametrization is $x + 1 = 2\sin\theta$ and $y - 2 = 2\sin\theta$, $0 \le \theta \le 2\pi$.

67. One parametrization is $x = 2\cos\theta$ and $y = 3\sin\theta$, $0 \le \theta \le 2\pi$.

69. By inspection, $\mathbf{r}(0) = \mathbf{r}(\pi) = \langle 0, 0\rangle$; we find $\mathbf{r}'(0) = \langle 1, 2\rangle$ and $\mathbf{r}'(\pi) = \langle -1, 0\rangle$. Hence equations of the tangent lines are $y = 0$ and $y = 2x$.

71. $\mathbf{T} = \langle -1/\sqrt{2}, 1/\sqrt{2}\rangle$

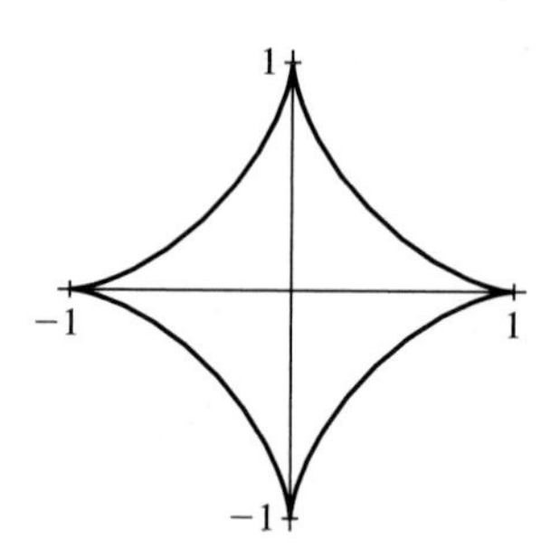

73. $\mathbf{T} = \langle 1/2, \sqrt{3}/2\rangle$

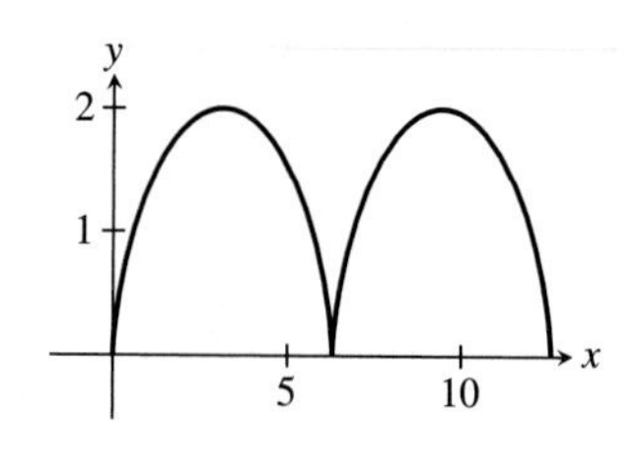

75. $\mathbf{T} = \langle 4/\sqrt{17}, -1/\sqrt{17}\rangle$

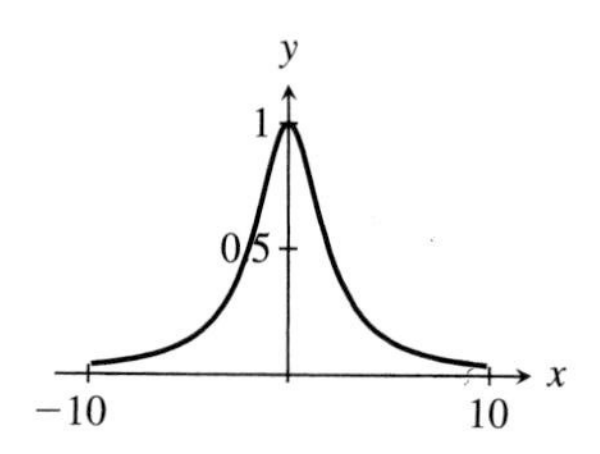

77. $\mathbf{T} = \langle -5/\sqrt{41}, -4/\sqrt{41}\rangle$

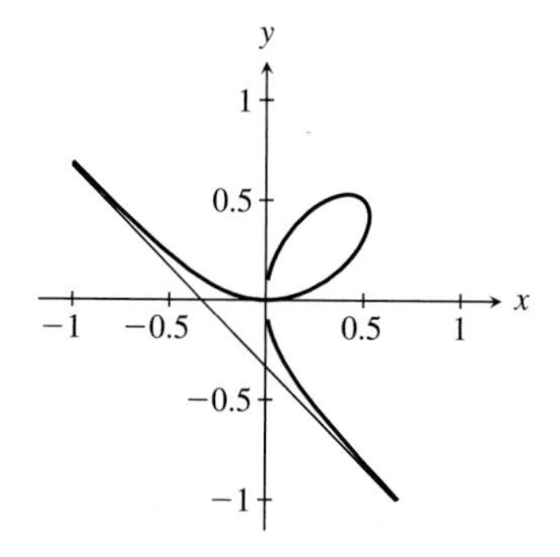

79.

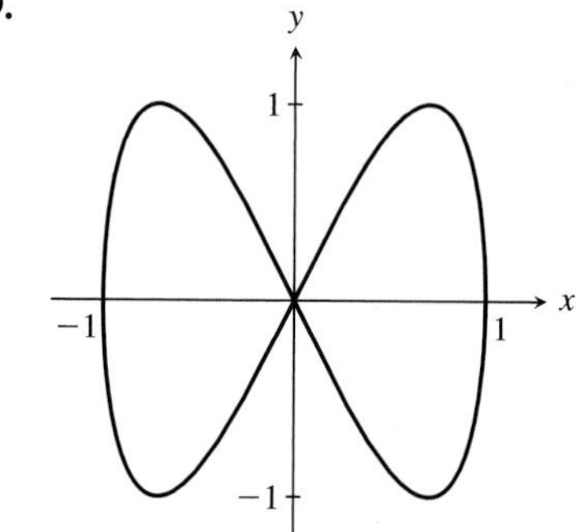

81.

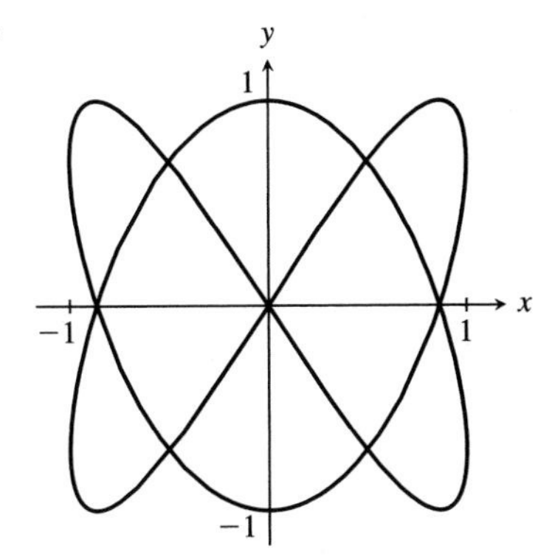

83. For $t = 0$, $\mathbf{r}(0) = \langle 0, 0\rangle$; also, $\mathbf{r}'(0) = \langle 1, 1\rangle$. The slope at this x-intercept is 1. For $t = 2$, $\mathbf{r}(2) = \langle 4, 0\rangle$; also $\mathbf{r}'(2) = \langle 3, -1\rangle$. Hence, the slope at the positive x-intercept is $-1/3$.

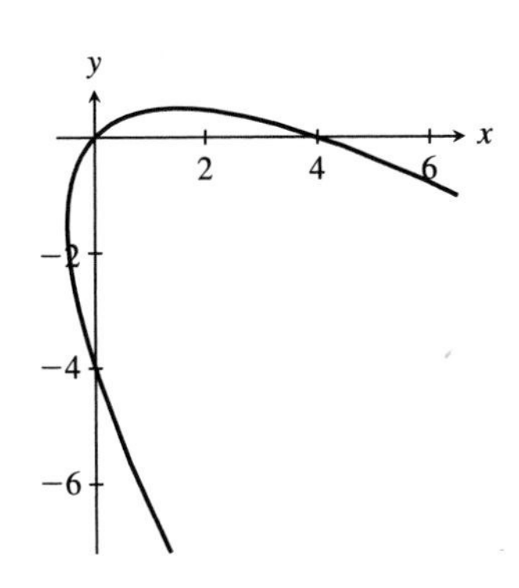

3.5 Dot Product

1. $\mathbf{a} \cdot \mathbf{b} = -12$; the angle between **a** and **b** is 111.161°, approximately.
3. $\mathbf{a} \cdot \mathbf{b} = 1$; the angle between **a** and **b** is 1.37340, approximately.
5. $\mathbf{a} \cdot \mathbf{b} = 0$; the angle between **a** and **b** is $\pi/2$.
7. $\mathbf{a} \cdot \mathbf{b} = 1.94114$; by inspection or calculation, the angle between **a** and **b** is 75°.
9. **a** and **b** are perpendicular; $\langle 0.857493, 0.514496\rangle$
11. none of the three pairs is perpendicular; $\langle -0.871984\mathbf{i} + 0.489535\mathbf{j}\rangle$
17. The length of the projection of **a** on a unit vector in the direction of **b** is $5/\sqrt{2}$.

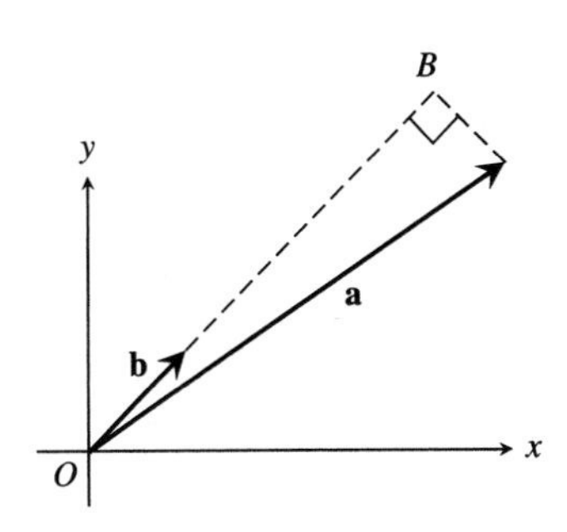

19. The length of the projection of **a** on a unit vector in the direction of **b** is $10/\sqrt{26}$.

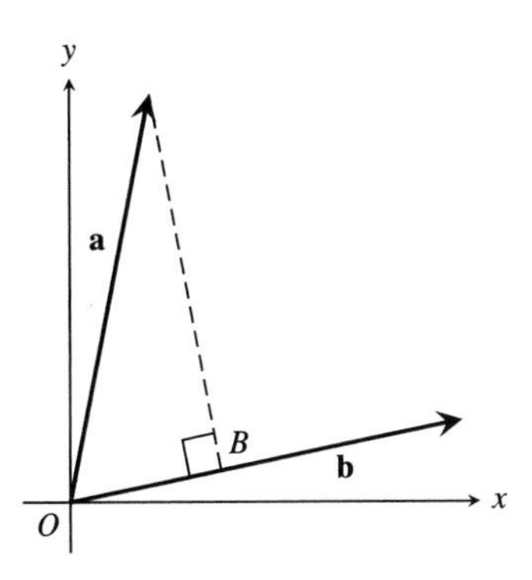

21. The length of the projection of **a** on a unit vector in the direction of **b** is about 3.18198.

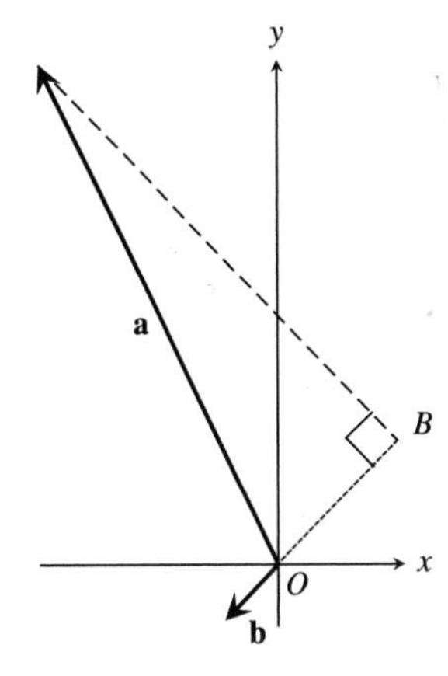

23. The length of the projection of **a** on a unit vector in the direction of **b** is $|-2 + k|/\sqrt{2}$.

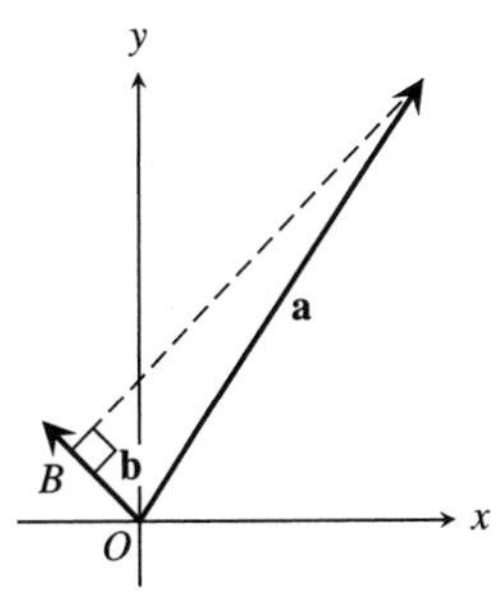

25. $\mathbf{t} = \langle 0, -14\sqrt{2}\rangle$; $\mathbf{c} = \langle 14\sqrt{2}, 0\rangle$
27. $\mathbf{h} \approx \langle -10.8335, 10.8335\rangle$; $\mathbf{c} \approx \langle -9.09039, -9.09039\rangle$
29. Resolving $\mathbf{F} = \langle 0, -2000\rangle$ into the sum $\mathbf{P} + \mathbf{N}$, where **P** is parallel to the hill and **N** is perpendicular to the hill, the magnitude of the frictional force cannot exceed $0.20\|\mathbf{N}\| \approx 398$ N; because $\|\mathbf{P}\| \approx 174$ and $174 + 250 > 398$, the sledge will move.
31. Approximately 5303 J

33. The work done by the force $\mathbf{F}_1$ is approximately 17,357 J; the work done by $\mathbf{F}_2$ is approximately 18,042 J.
35. $\langle x, y\rangle = (1/\sqrt{5})\langle -2, 1\rangle$ **37.** $\langle 1, m\rangle$
39. The angles at A, B, and C are approximately 0.927295, 1.10715, and 1.10715.
41. $t = 13/6$ **43.** $5\sqrt{2}$
45. Approximately $19484\langle \cos 8°, \sin 8°\rangle$ N **47.** 78.96°

3.6 Newton's Laws

1. $t_1 \approx 5.30612$ s; $\mathbf{r}(t) = (-4.9t^2 + 26t)\mathbf{j}$ m
3. $t_1 \approx 3.74507$ s; $\mathbf{r}(t) = (-4.9t^2 + 5t + 50)\mathbf{j}$ m
5. $t_1 \approx 2.39214$ s; $\mathbf{r}(t) = (-4.9t^2 - 5t + 40)\mathbf{j}$ m
7. Approximately 1725 m **9.** Approximately 20,000 m
11. $100\pi/3$ m/s
13. The orbital speed required is approximately 28,100 km/h; its period would be approximately 87.7 minutes.
15. 229 km **17.** Approximately 56,000 ft/s^2
19. Approximately 2058 m **21.** Approximately 718 m/s
23. The angle of fire should be 45° for maximum range. With muzzle speed 700 m/s, the maximum range is approximately 50,000 m.
25. Approximately 232 km
27. The magnitude of the required force is 2.56 N.

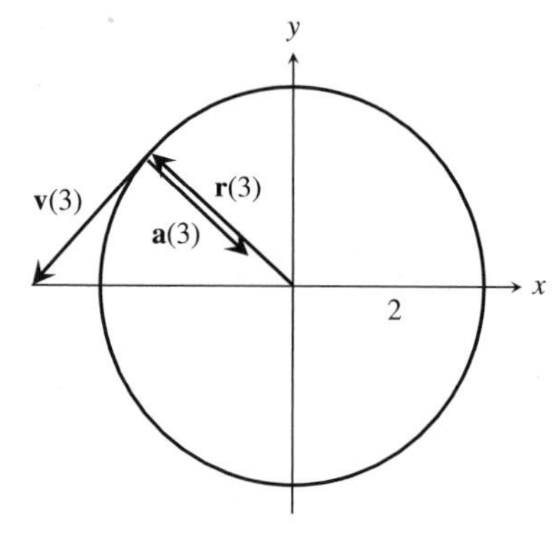

29. Approximately 36,000 km

Chapter Review Exercises

1. If at any time t the position of an object is $x(t)$, then the average velocity of the object on the time interval $[t_1, t_2]$ is $(x(t_2) - x(t_1))/(t_2 - t_1)$. If at any time t the position of an object is $\mathbf{r}(t)$, then the average velocity of the object on the time interval $[t_1, t_2]$ is $(1/(t_2 - t_1))(\mathbf{r}(t_2) - \mathbf{r}(t_1))$.
3. Average velocity is approximately $\langle 6.47059, 0.588235\rangle$
5. Assuming that the y-axis is vertical, with the positive direction upward and the origin at ground level, the position, velocity, and acceleration of the bullet are $y(t) = -4.9t^2 + 610t - 20$, $y'(t) = -9.8t + 610$, and $y''(t) = -9.8$. The bullet's speeds at 1000 m above ground level are approximately 593 m/s, whether going up or down.
7. Approximately (97.9, 87.3) m
9. $0.944459\mathbf{i} - 0.328629\mathbf{j}$ and $0.786825\mathbf{i} + 0.617176\mathbf{j}$
11.

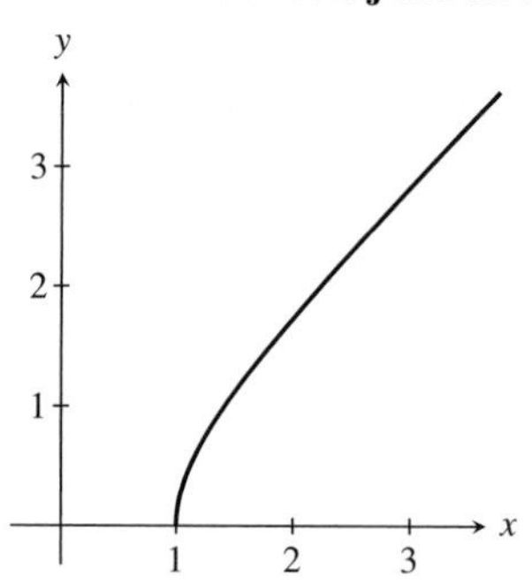

13. The paths intersect at the point $(6, -2)$. The objects do not collide.
17. $\mathbf{a}(3.1) \approx \langle -3.98617, 0.332358\rangle$ m/s^2
19. $\mathbf{a} = (1/1.2)\langle 2.4, 1.7\rangle$; when $t = 3.5$, the object is at the point (12.25, 8.67708) and its speed is 8.57817 m/s, approximately.
21. $\mathbf{r}(t) = (-2 + 5\cos t)\mathbf{i} + (3 + 5\sin t)\mathbf{j}$, $0 \le t \le 2\pi$
23. $\mathbf{v}_1 = \langle 33/5, 11/5\rangle$; $\mathbf{v}_2 = \langle -8/5, 24/5\rangle$
27.

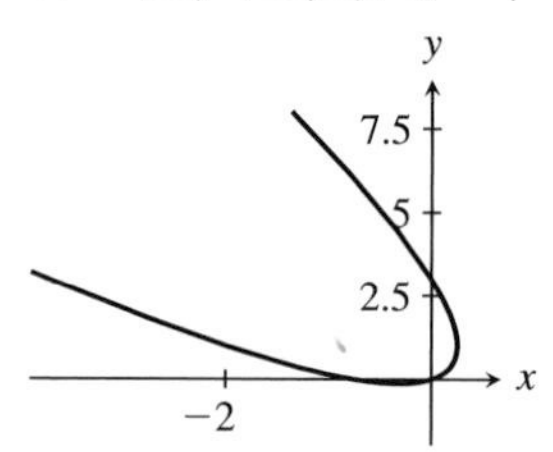

Eliminating t gives $4x^2 + 4xy + y^2 + 3x - 3y = 0$. Slope at $(-2, 10)$ is -3.

CHAPTER 4—APPLICATIONS OF THE DERIVATIVE

4.1 The Tangent Line Approximation

1. $df_a(h) = 52h$ **3.** $df_a(h) = he$ **5.** $df_a(h) = 0$ (for all h)
7. $df_a(h) = 2h/\sqrt{3}$
9. By calculator, $\sin 31° \approx 0.515038$; by tangent line approximation, $\sin 31° \approx 0.515115$.
11. By calculator, $\sqrt{9.2} \approx 3.03315$; by tangent line approximation, $\sqrt{9.2} \approx 3.03333$.
13. By calculator, $\ln(2.8) \approx 1.02962$; by tangent line approximation, $\ln(2.8) \approx 1.03006$.
15. By calculator, $\arcsin(0.48) \approx 0.500655$; by tangent line approximation, $\arcsin(0.48) \approx 0.500505$.
17. The percentage error would be less than 6 percent.
19. Using Snell's law, $n_2 = 1.52043$; using the lensmaker's formula, $n_2 = 1.52484$. These values differ significantly in the third decimal place.
21. $\beta(I) \approx 130 + \dfrac{I - 10}{\ln 10}$ **23.** About 0.74 mm.
25. Approximate radius of the moon is 1740 kilometers. No more than a 5 percent error can be made in determining the distance between eye and coin.

27. Near $x = 1$, $2^{-x} \approx \frac{1}{2}(1 - \ln 2(x - 1))$. The figure shows the graph of $y = |2^{-x} - \frac{1}{2}(1 - \ln 2(x - 1))|$. From the figure, $y \le 0.15$.

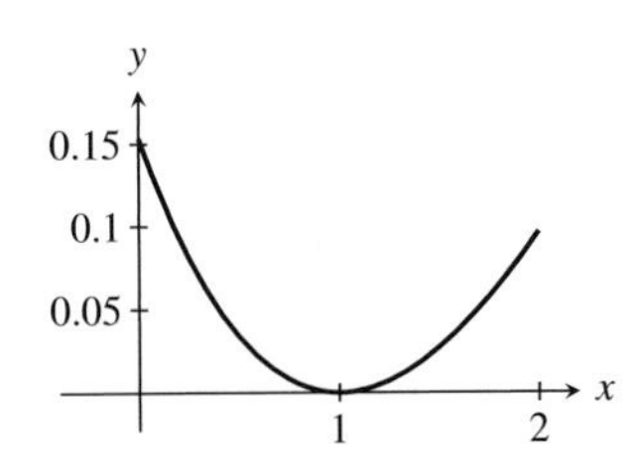

29. Using the tangent line approximation,

$$|f(T_g + h) - f(T_g)| \approx |df(T_g)h| = 8\pi^2 h/T_g^{\,3} \le 0.05,$$

where $9.8T_g^{\,2} = 4\pi^2$. It follows that T must be measured within 0.00512 seconds.

31. $dw_3(0.1) = 9 \cdot 7 \cdot 0.1 = 6.3$

33.

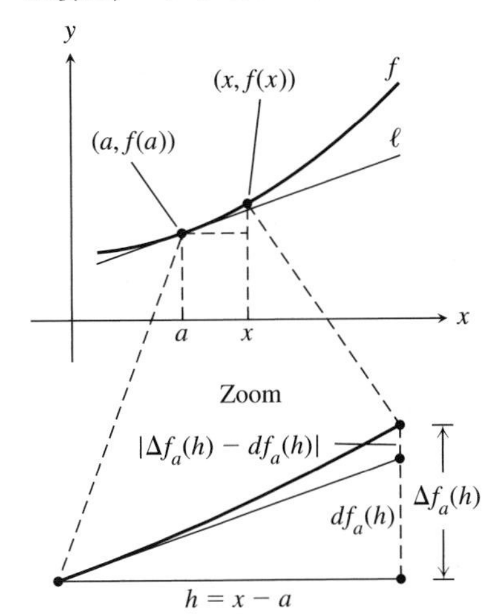

35.

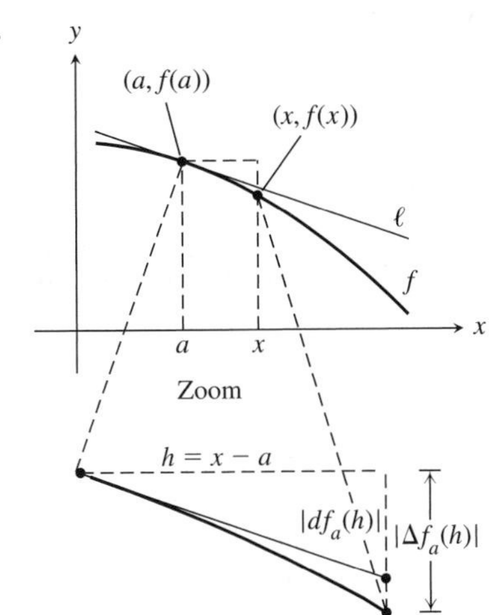

37. Differentiating $\theta(t)$,

$$\theta'(t) = \theta_0(-\sin(\omega t)) \cdot \omega = -\theta_0\omega \sin(\omega t).$$

Differentiating again,

$$\theta''(t) = -\theta_0\omega^2 \cos(\omega t).$$

Substituting into (13) and recalling that $\omega = \sqrt{g/L}$,

$$L\frac{d^2\theta}{dt^2} + g\theta = L(-\theta_0\omega^2 \cos(\omega t)) + g\theta_0 \cos(\omega t)$$
$$= -g\theta_0 \cos(\omega t) + g\theta_0 \cos(\omega t) = 0.$$

Hence, $\theta = \theta(t)$ satisfies (13) for all t.

4.2 Newton's Method

1. A zero lies between 2.0 and 2.1
3. A zero lies between 0.9 and 1.0.
5. A zero lies between 0.8 and 0.9.
7. A zero lies between 3.1 and 3.2.
9. $x_4 = 1.70998$; $x_5 = 1.70998$; we guess that 1.7100 is, to four decimal places, a real zero of f.
11. $x_5 = 4.49341$; $x_6 = 4.49341$; we guess that 4.4934 is, to four decimal places, a real zero of f.
13. With $x_1 = -0.5$ we find that $x_4 = -0.578467$; $x_5 = -0.578467$; we guess that -0.5785 is, to four decimal places, a real zero of f.
15. The first zero lies between 1.8 and 1.9; the second between 4.6 and 4.7; and third between 7.8 and 7.9; and the fourth between 10.9 and 11.0.
17. With $x_1 = 2$ we find that $x_4 = 2.09455$; $x_5 = 2.09455$; we guess that 2.0946 is, to four decimal places, a real zero of f.
19. For $x_1 = 1.5$, the iterates move away from the origin. Specifically, $x_2 \approx -1.69408$ and $x_3 \approx 2.32113$

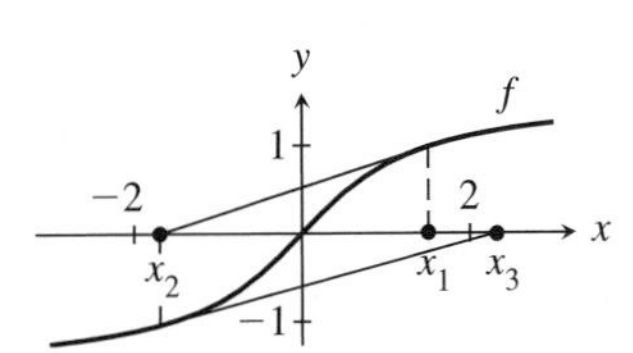

For $x_1 = 1.4$, the iterates also move away from the origin. For $x_1 = 1.3$, the iterates $x_1, x_2, \ldots$ approach 0.
25. With $x_1 = 0.5$, $x_3 = 0.526583$ and $f(x_3 - E) * f(x_3 + E) < 0$.
27. With $x_1 = 0.5$, $x_6 = 0.837669$ and $f(x_6 - E) * f(x_6 + E) < 0$.
29. The first four positive zeros are 1.9, 4.7, 7.9, and 11.0.
31. 33.3°, 68.4°, 78.3°
35. With $x_1 = 3 + 4i$, $x_2 = 2.99556 + 4.00032i$, $x_3 = 2.99555 + 4.00033i$, and $x_4 = 2.99555 + 4.00033i$. With $x_1 = -0.5 + 1.3i$, $x_2 = -0.495499 + 1.31009i$, $x_3 = -0.495554 + 1.31008i$, and $x_4 = -0.495554 + 1.31008i$. The other zeros are $2.99555 - 4.00033i$ and $-0.495554 - 1.31008i$.

4.3 Increasing/Decreasing Functions; Concavity

1. Increasing **3.** Decreasing **5.** Increasing
7. Decreasing
9. Decreasing on $(-\infty, 0]$ and increasing on $[0, \infty)$.
11. Decreasing on $(-\infty, 0) \cup (0, \infty)$.
13. Decreasing on $\left(-\infty, -\frac{1}{2}\right]$ and increasing on $\left[-\frac{1}{2}, \infty\right)$.

15. Decreasing on $(-\infty, 0)$ and decreasing on $(0, \infty)$.
17. Increasing on $\left[0, \sqrt[3]{\pi/2}\right]$ and $\left[\sqrt[3]{3\pi/2}, \sqrt[3]{5\pi/2}\right]$ and decreasing on $\left[\sqrt[3]{\pi/2}, \sqrt[3]{3\pi/2}\right]$ and $\left[\sqrt[3]{5\pi/2}, 2\right]$.
19. Increasing on $[0, \pi/2]$ and decreasing on $[\pi/2, \pi]$.
21. Increasing on $(0, 1]$ and decreasing on $[1, \infty)$.
23. Increasing on $[-1, \infty)$ and decreasing on $(-\infty, -1]$.
25. Concave up on $(0, \pi/4)$ and $(3\pi/4, \pi)$; concave down on $(\pi/4, 3\pi/4)$; inflection points at $\pi/4$ and $3\pi/4$.
27. Concave up on $\left(-\infty, -1/\sqrt{3}\right)$ and $\left(1/\sqrt{3}, \infty\right)$; concave down on $\left(-1/\sqrt{3}, 1/\sqrt{3}\right)$; inflection points at $-1/\sqrt{3}$ and $1/\sqrt{3}$.
29. Concave up on $(-2, \infty)$; concave down on $(-\infty, -2)$; inflection point at -2.
31.

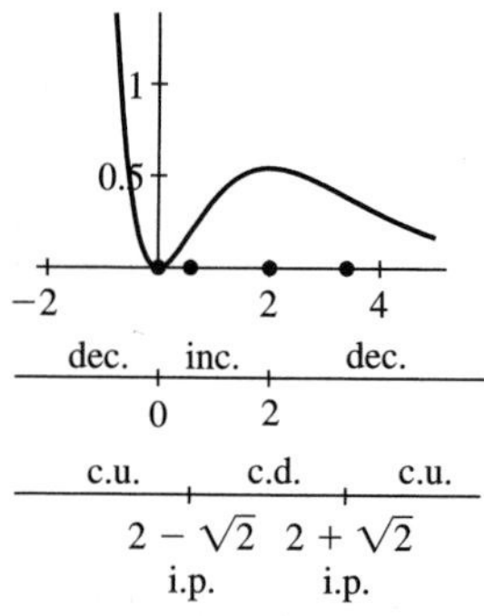

33. Note that the graph is symmetric about the y-axis. We graph only to the right of the origin.

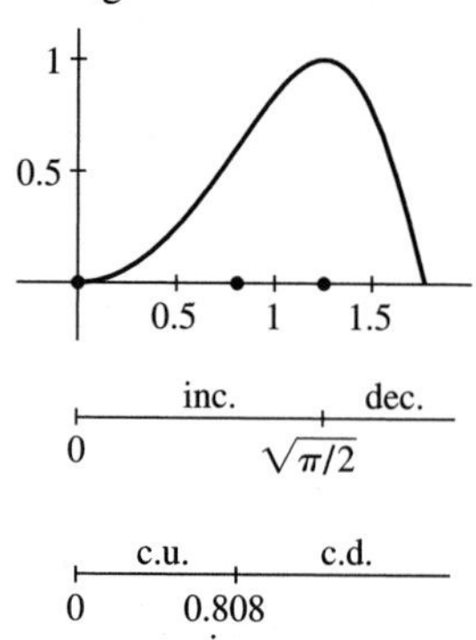

35.

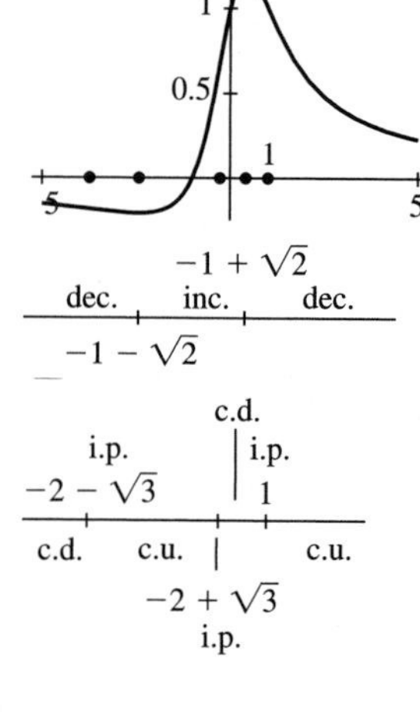

37. Because $P'(t) = abke^{kt}/(b + e^{kt})^2 > 0$, P is increasing on its domain. The point of inflection occurs at $t = \ln(b)/k$.
39. Increasing in $[0, 13.8629]$ and decreasing in $[13.8629, 60]$.
41. $x_5 = 0.80825$
43. Increasing on $[0, (\pi/2)^2]$. **45.** $x \approx 2.52223$
47. Inflection points at $(-3, f(-3))$ and $(-1, f(-1))$; because $f'(x) = (x + 1)^2 e^x \geq 0$ for all x, f is increasing on $(-\infty, \infty)$.
49. Increasing goes with positive velocity; decreasing goes with negative velocity; concave up goes with positive acceleration; concave down goes with negative acceleration; and inflection point goes with a time t at which the acceleration goes from positive to negative or from negative to positive.

4.4 Horizontal and Vertical Asymptotes

1. 0 **3.** ∞ **5.** 1 **7.** 0 **9.** ∞ **11.** ∞ **13.** $-\infty$
15. 0
17. Horizontal asymptote: $y = 0$; vertical asymptote: $x = 2$

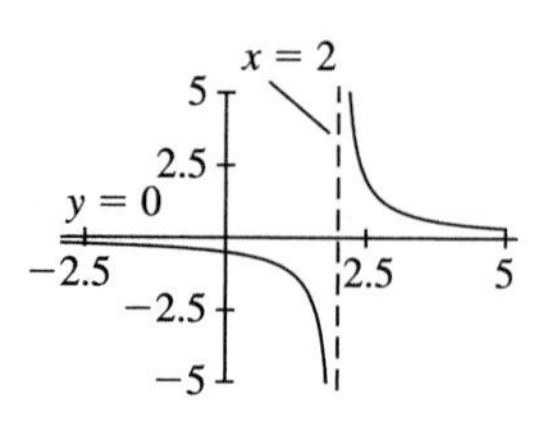

19. Horizontal asymptote: $y = 2$; vertical asymptote: $x = -1$

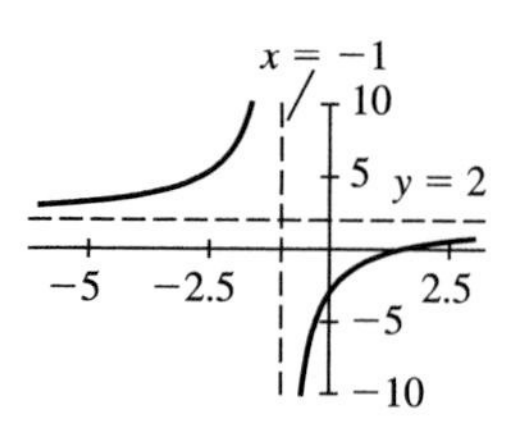

21. Horizontal asymptote: $y = 1$; vertical asymptote: $x = 0$

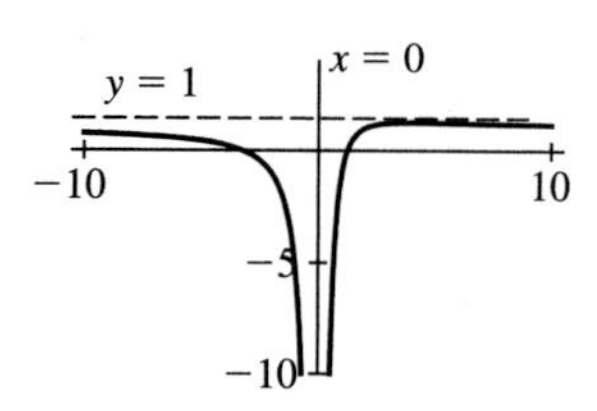

23. Vertical asymptotes: $x = \pi/2$ and $x = 3\pi/2$

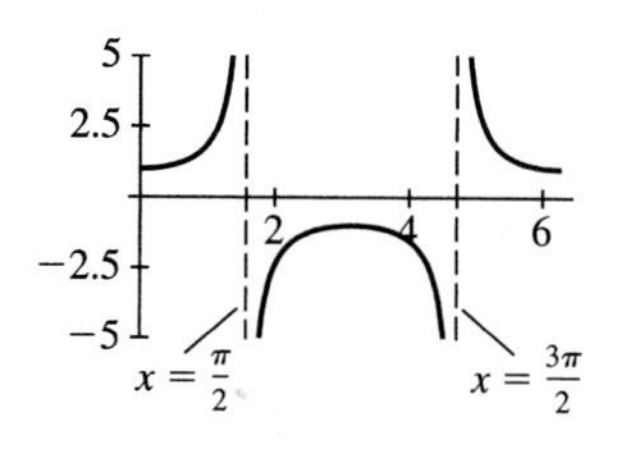

25. Horizontal asymptotes: $y = -\pi/2$ and $y = \pi/2$

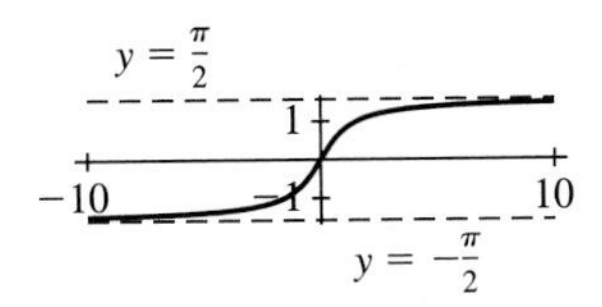

27. Horizontal asymptote: $y = 1$; vertical asymptote: $x = 1$

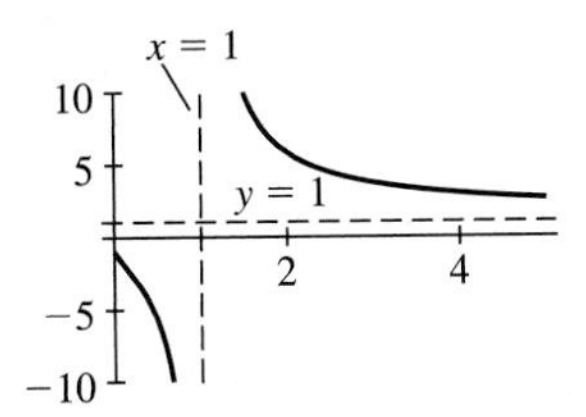

29. Horizontal asymptotes: $y = 1$; vertical asymptotes: $x = -1$ and $x = 2$

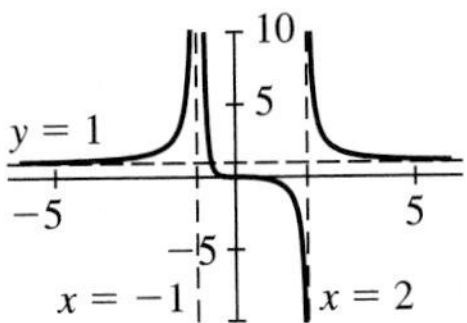

31. Horizontal asymptotes: $y = \sqrt{2}/3$ and $y = -\sqrt{2}/3$; vertical asymptote: $x = -1/3$

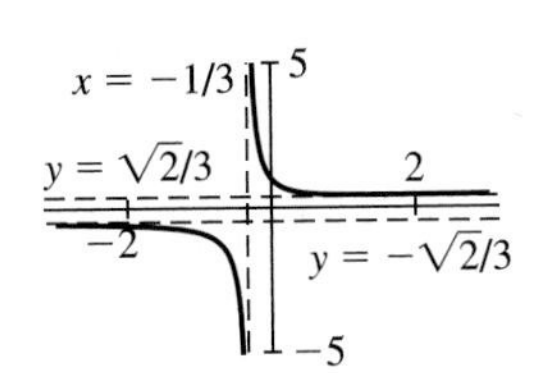

33. Vertical asymptote: $x = 0$

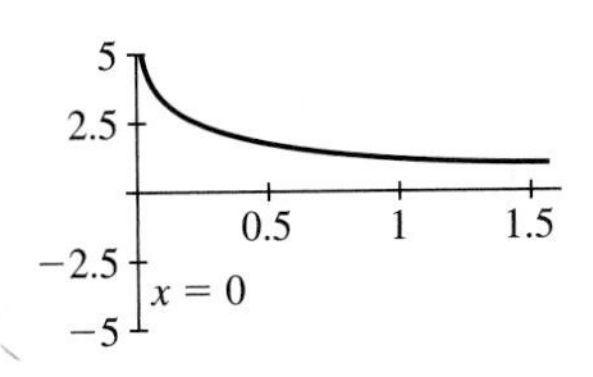

35. Vertical asymptote: $x = 1$

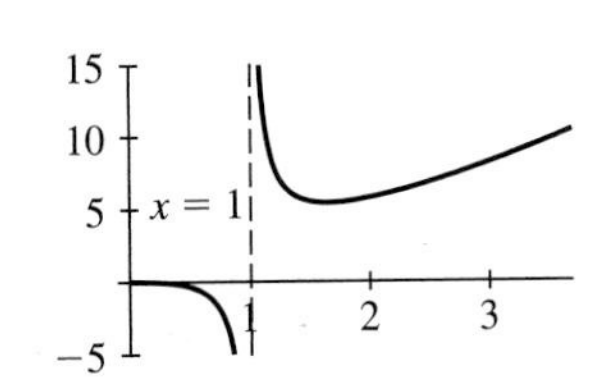

37. Horizontal asymptote: $y = 0$

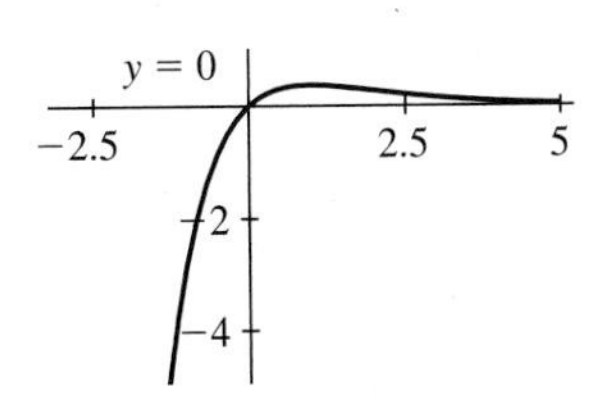

39. Horizontal asymptote: $y = 0$

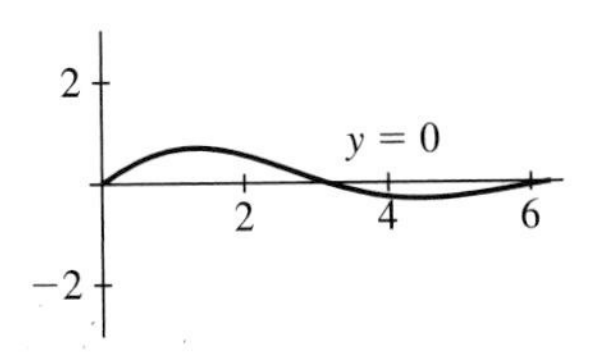

41. Horizontal asymptote: $y = 0$; vertical asymptote: $x = -1$

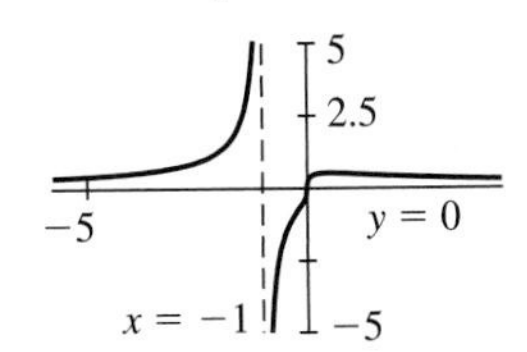

43. No vertical asymptotes; one horizontal asymptote: $y = 292$. Inflection point at $(t, f(t))$, where $t \approx 142.556$.

45. It catches up somewhere between 116,600 and 116,700.

47. 0

49. 1; the graph has neither vertical nor horizontal asymptotes.

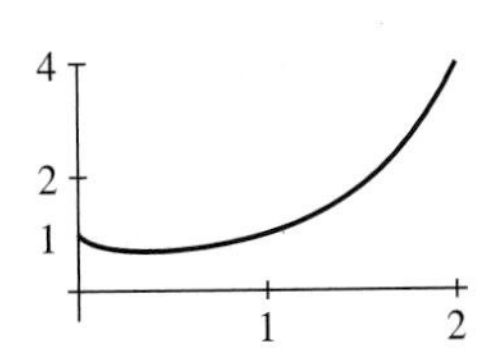

51. As x approaches 0, $f(x) = x + 1/x \approx 1/x$; hence, we expect the graph of f to become as close to the graph of $g(x) = 1/x$ as we wish provided that x is sufficiently close to 0. Similarly, as $|x|$ grows without bound, $f(x) = x + 1/x \approx x$; hence, we expect the graph of f to become as close to the graph of $g(x) = x$ as we wish provided that $|x|$ is sufficiently large.

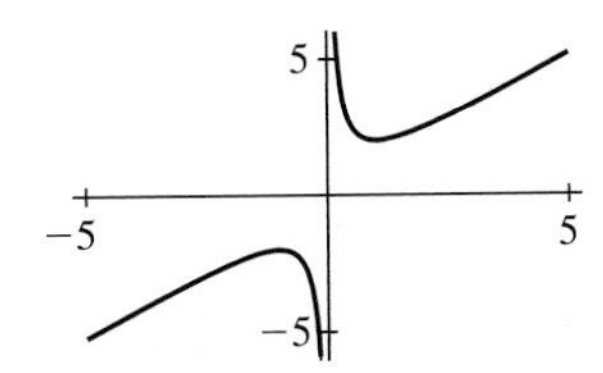

57. $\lim_{x\to\infty} f(x) = -\infty$ if f is defined on an interval (a, ∞) and for each $E < 0$ there is $N > 0$ for which $x > N$ implies $f(x) < E$.

$\lim_{x\to-\infty} f(x) = \infty$ if f is defined on an interval $(-\infty, b)$ and for each $E > 0$ there is $N < 0$ for which $x < N$ implies $f(x) > E$.

$\lim_{x\to-\infty} f(x) = -\infty$ if f is defined on an interval $(-\infty, b)$ and for each $E < 0$ there is $N < 0$ for which $x < N$ implies $f(x) < E$.

4.5 Tools for Optimization

1. The figure shows the graph of the function $f(x) = 1 - x$, $0 \le x \le 1$, which has a global maximum at $x = 0$ and a global minimum at $x = 1$.

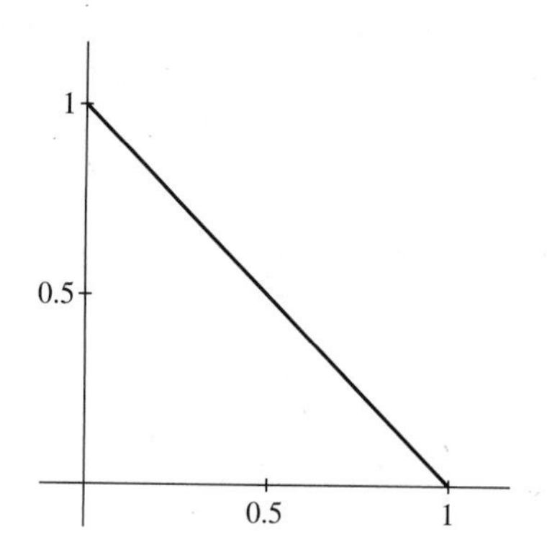

3. The figure shows the graph of the function $f(x) = x(1 - x)$, $0 \le x \le 1$, which has a global maximum at $x = 1/2$ and local minima at $x = 0, 1$.

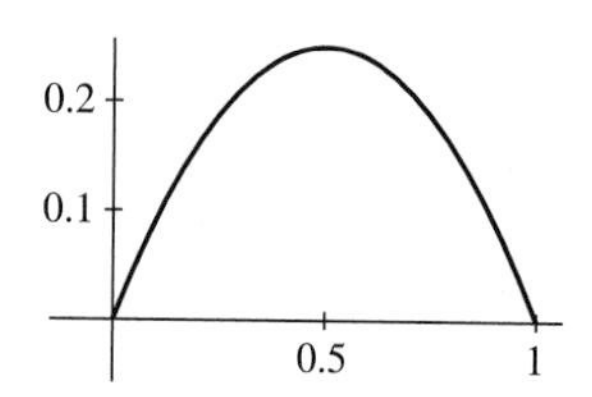

5. The figure shows the graph of the function $f(x) = \cos(4\pi x)$, $0 \le x \le 1$, which has local maxima at $x = 0, 1/2, 1$ and local minima at $x = 1/4, 3/4$.

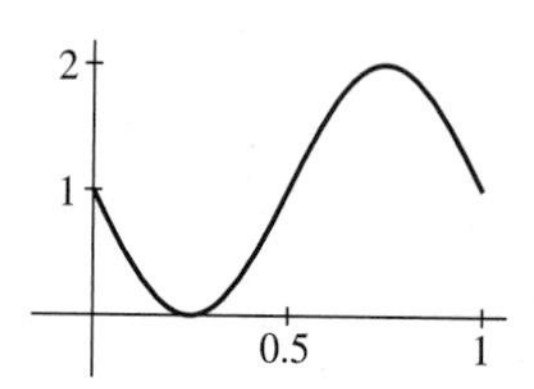

7. The candidates are $-2, -1, 2, 3$; global minimum at $x = 2$ $(f(2) = -19)$; global maximum at $x = -1$ $(f(-1) = 8)$.
9. The candidates are $-3, -1, 1, 3$; global minimum at $x = -1$ $(f(-1) = -1/2)$; global maximum at $x = 1$ $(f(1) = 1/2)$.
11. The candidates are $1/2, 2, 3$; global minimum at $x = 2$ $(f(2) = 6)$; global maximum at $x = 1/2$ $(f(1/2) = 129/8)$.
13. The candidates are $0, 1/4, 5$; global minimum at $x = 1/4$ $(f(1/4) \approx -0.244979)$; global maximum at $x = 5$ $(f(5) \approx 1.22365)$.
15. The candidates are $-1, 1$; global minimum at $x = -1$ $(f(-1) = -4)$; global maximum at $x = 1$ $(f(1) = 24)$.
17. The candidates are $0, \pi/2, \pi, 3\pi/2, 2\pi$; global minimum at $x = 0, \pi, 2\pi$ (f is 0 at all of these points); global maximum at $x = \pi/2, 3\pi/2$ (f is 1 at all of these points).
19. The candidates are $0, 0.596072, 3$; global minimum at $x = 3$ $(f(3) = 5/14)$; global maximum at $x = 0.596072$ $(f(0.596072) = 1.11843)$.
21. The candidates are $0, 3$; global minimum at $x = 0$ $(f(0) = 0)$; global maximum at $x = 3$ $(f(3) = 3/7)$.
23. The candidates are $-2, 0, 6/7, 1, 2$; global minimum at $x = -2$ $(f(-2) = -5.76900)$; global maximum at $x = 2$ $(f(2) = 4)$.
25. The candidates are $0, 1, 2$; global minimum at $x = 0$ $(f(0) = 0)$; global maximum at $x = 1$ $(f(1) = 2/e)$.
27. The candidates are $0, 1/e, 2$; global minimum at $x = 1/e$ $(f(1/e) = e^{-1/e})$; global maximum at $x = 2$ $(f(2) = 4)$.
29. The candidates are $-1, 0.528767, 2$; global minimum at $x = 0.528767$ $(f(0.528767) = -0.0860713)$; global maximum at $x = 2$ $(f(2) = 1)$.
31. We showed there that $h/w = \sqrt{2}$. Because $\sqrt{2} = 1.414\ldots \approx 1.5$, $h \approx 1.5w$. Hence, "cut it half again as high as it is wide" is a reasonable, simple rule.
33. 0.6533 and 3.2923
35. If there were a point $w \in (-\pi/2, \pi/2)$ for which $\arctan(w) \ge \arctan(x)$ for all $x \in R$, then $w \ge x$ for all $x \in R$. This is impossible because $w \in (-\pi/2, \pi/2)$.
37. The function $f(x) = 1/x$ defined on the nonclosed interval $(0, 1)$ is continuous but has no maximum. The function $f(x) = x$ defined on the interval $(0, \infty)$ is continuous but has no maximum.
39. The set of candidates is $\{-3, -2, 5/2, 4, 5\}$. The global maximum is at $x = 5$ $(f(5) = 10)$; the global minimum is at $x = -2$ $(f(-2) = -3)$.

4.6 Modeling Optimization Problems

1. $(2, 2)$ **3.** $(1/5, 7/5)$
5. $(0.00466216, 0.167056)$ is farthest; $(1.72189, 1.19858)$ is the closest.
7. The jar should become wider because the top is cheaper. The minimum-cost jar has dimensions $r = 1.72526$ inch and $h = 6.17580$ inch.
9. $7/4$
11. The farmer should build the enclosure along the existing fence. Two of the sides would be 200 m and one side 400 m.
13. Horizontal dimension is $\sqrt{2}a$ and the vertical dimension is $\sqrt{2}b$.
15. $\beta \approx 4.11773°$ **17.** $x = 400/3$
19. The dimensions of the box of minimum cost are $x = \sqrt[3]{2/3}$ and $y = (3/2)^{2/3}$.
21. The dimensions of the rectangle of maximum area are $x = \frac{1}{2}a$ and $y = (\sqrt{3}/4)a$.
23. The dimensions of the largest rectangle are $x = 3/2$ and $y = 5/2$.
29. Width is $\sqrt{2}a$ and height is $a/\sqrt{2}$.

31. The length of the longest piece of wood is approximately 35.1174 meters.
33. Height is $8/3$ m and base radius is $10/3$ m.
35. Cut so the angle ACB has radian measure $2\pi\sqrt{2/3}$.

4.7 Related Rates

1. $y' = \dfrac{x'\sqrt{1+y^2}}{y}$ **3.** $y' = \dfrac{2xx' + yx'}{2y - x}$

5. $y' = \dfrac{-x^3x'}{y^3}$ **7.** $y' = \dfrac{x' \cos(y)}{2y + x \sin(y)}$

9. $y' = -\dfrac{(e^{xy} - y)yx'}{x(e^{xy} - 2y)}$ **11.** $45/4$ m/s

13. 0.04 pounds per square inch per second
15. $x' = -0.1625$ kilometers/s and $y' = 0.08125$ kilometers/s.
17. $-1/(50 \cdot 6^{2/3}\pi^{1/3}) \approx -0.00413567$ cm/s.
19. The bucket is ascending at approximately 0.93 m/s.
21. Approximately 0.1245 Pa/s.
23. Approximately 0.029 m/min.
25. Milk is spilling at the rate of $-27\pi/16$ cubic inches per second, which is approximately -5.3 cubic inches per second.
27. The shadow will be moving at $24/5$ m/s.
29. The rate of decrease in the height of Mt. McKinley at half-volume is approximately -4.7×10^{-20} m/century. One day of eternity is 5.9×10^{22} centuries.

4.8 Indeterminate Forms and l'Hôpital's Rules

1. $0/0$ **3.** $\infty \cdot 0$ **5.** ∞/∞ **7.** $\infty - \infty$ **9.** ∞^0
11. 1^∞ **13.** 0^0 **15.** 0 **17.** 0 **19.** ∞ **21.** $-\infty$
23. ∞ **25.** $1/4$ **27.** 1 **29.** 1 **31.** 0 **33.** 1
35. 1 **37.** 0 **39.** 1

4.9 Euler's Method

1. (0.0, 1), (0.5, 1.5), (1.0, 2.5), (1.5, 4.25), (2.0, 7.125), (2.5, 11.6875)
3. (0.0, 2), (0.1, 2.2), (0.2, 2.42023), (0.3, 2.66307), (0.4, 2.93107), (0.5, 3.22689)
5. (0, 1.0), (0.1, 1.1), (0.2, 1.21), (0.3, 1.331), (0.4, 1.4641), (0.5, 1.61051), (0.6, 1.77156), (0.7, 1.94872), (0.8, 2.14359), (0.9, 2.35795), (1., 2.59374)

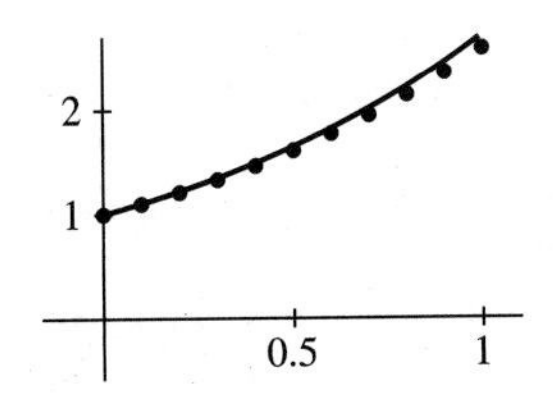

7. (0, 2.0), (0.1, 2.2), (0.2, 2.41), (0.3, 2.631), (0.4, 2.8641), (0.5, 3.11051), (0.6, 3.37156), (0.7, 3.64872), (0.8, 3.94359), (0.9, 4.25795), (1., 4.59374)

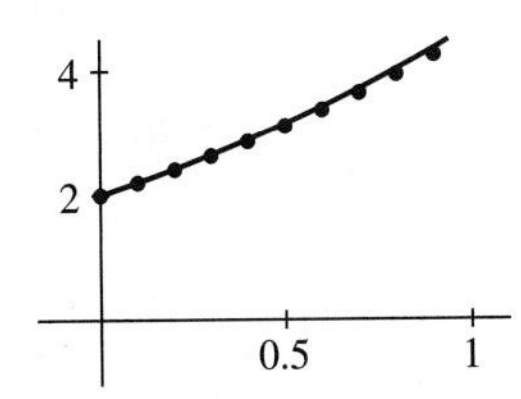

9. (0, 1.), (0.1, 0.9), (0.2, 0.811), (0.3, 0.7339), (0.4, 0.66951), (0.5, 0.618559), (0.6, 0.581703), (0.7, 0.559533), (0.8, 0.55258), (0.9, 0.561322), (1., 0.586189)

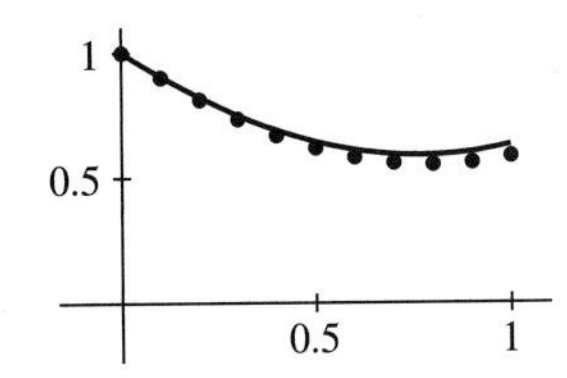

11. With $t = 0.05$, $y_{40} \approx 11.080$; with $t = 0.1$, $y_{20} \approx 10.455$; the exact solution at $t = 2$ is $y(2) \approx 11.778$.
13. (0, 4.0), (10, 5.18356), (20, 6.71102), (30, 8.67806), (40, 11.2041), (50, 14.4364), (60, 18.5532), (70, 23.7655), (80, 30.3148), (90, 38.4651), (100, 48.4845), (110, 60.6147), (120, 75.0244), (130, 91.7488), (140, 110.625), (150, 131.239), (160, 152.915), (170, 174.766), (180, 195.816), (190, 215.166), (200, 232.151), (210, 246.426)

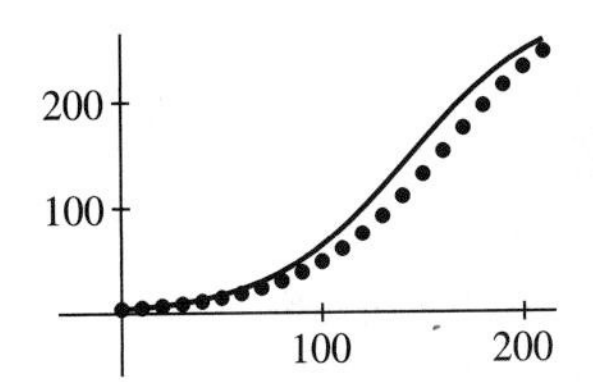

15. From the approximate solution (0, 98.6), (1., 96.92), (2., 95.324), (3., 93.8078), (4., 92.3674), (5., 90.999), (6., 89.6991), (7., 88.4641), (8., 87.2909), (9., 86.1764), (10., 85.1176), (11., 84.1117), (12., 83.1561), (13., 82.2483), (14., 81.3859), (15., 80.5666), (16., 79.7883), (17., 79.0488), (18., 78.3464), (19., 77.6791), (20., 77.0451) we infer that the corpse reached a temperature of 80°F at $t = 16$. This means that the death happened at approximately 7 P.M. of the previous day.

Chapter Review Exercises

1. About 0.016 at $x = 6$.
3. Three percent.
5. **(a)** Because $f'(x) = 3x^2 + 2 > 0$, f is everywhere increasing and, hence, is 1-1, that is, has an inverse function. **(b)** Because $f(-1) < 0$ and $f(0) > 0$, f has a zero between -1 and 0.

(c) Because f is increasing on $(-\infty, \infty)$, it can have at most one zero.
(d) The one zero is -0.453398, approximately.
7.

Minimum at $x = 1$
Maximum at $x = -1$
Inflection points at
$x = -\sqrt{3}, 0, \sqrt{3}$

9. Inflection point at $x = \frac{1}{3}(a + b + c)$.
11. Increasing on $[0, 1/a]$; decreasing on $[1/a, \infty)$; concave down on $(0, 2/a)$; concave up on $(2/a, \infty)$; horizontal asymptote: $y = 0$. Global minimum at $x = 0$; global maximum at $x = 1/a$. Figure plotted with $a = 2$.

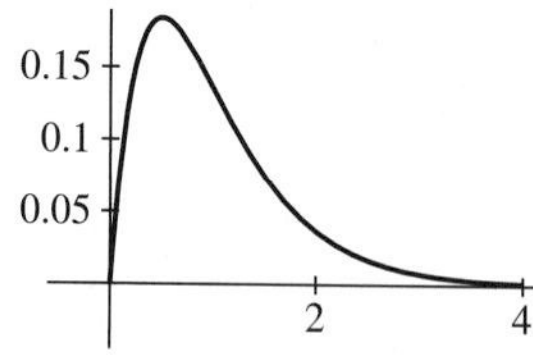

13. The particles are separating at 194.6 m/min.
15. 17.86 cubic centimeters per minute
17. $1/(18\sqrt[3]{\pi}) \approx 0.038$ feet per minute
19. $0.05\ \text{cm}^3/\text{min}$
21.

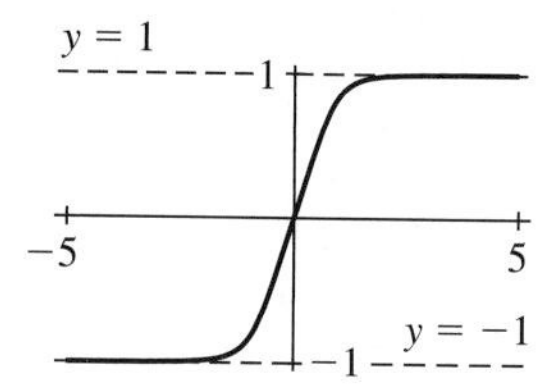

23. The limit exists and is 1. **25.** The limit exists and is ∞.
27. The limit exists and is ∞. **29.** The limit exists and is e^6.
31. Use all of the wire for the circle to have maximum area enclosed. Use $(12\pi)/(\pi + 4)$ to have minimum area enclosed.
33. Maximum at $P = 5 - 2\sqrt{5}$.
35. Maximum of 5 at $x = 2$; minimum of approximately -0.167948 at $x = \ln((\ln 2)/(\ln 3))/\ln(3/2) \approx -1.13588$.
37. Maximum at $x = 0$ and minimum at $x = -1.5$.
39. The rectangle with width $1/4$ and height $1/2$.
41. The figure is of $f(x) = x^3$, $-1 \le x \le 1$; $f'(0) = 0$ but there is no minimum or maximum at $x = 0$.

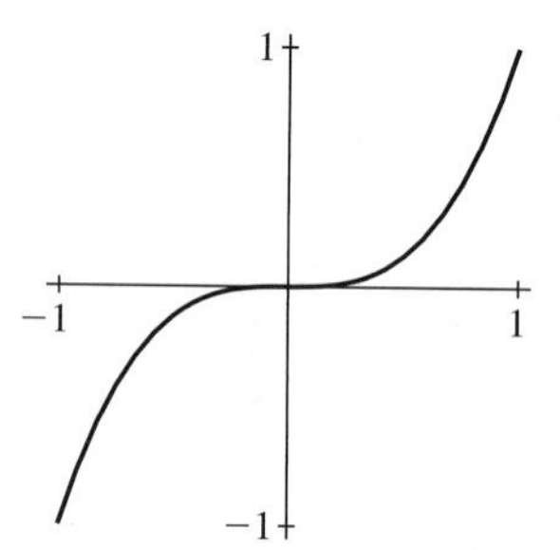

43. (0, 1), (0.1, 1.1), (0.2, 1.22537), (0.3, 1.38101), (0.4, 1.5733), (0.5, 1.81065), (0.6, 2.10429), (0.7, 2.46954), (0.8, 2.92762), (0.9, 3.50855), (1.0, 4.25574)

CHAPTER 5 — THE INTEGRAL

5.1 Summation Notation

1. $\sum_{i=1}^{10} 1/i$ **3.** $-\sum_{i=1}^{100} \ln(1/i)$ **5.** $\sum_{i=1}^{10} i^2$ **7.** $2\sum_{i=3}^{35} 1/i$
9. $(|h|/\sqrt{19})\sum_{i=1}^{21} \sqrt{i}$ **11.** $\pi\sum_{i=1}^{21} \pi^i$ **13.** $\sum_{i=1}^{15} i/(i + 1)$
15. $1^2 + 2^2 + 3^2 + 4^2 + 5^2 = 55$
17. $(1^2 + 1) + (2^2 + 1) + (3^2 + 1) + (4^2 + 1) + (5^2 + 1) + (6^2 + 1) = 97$
19. $\sqrt{2} + \sqrt{3} \approx 3.14626$
21. $\sin(0/10) + \sin(1/10) + \sin(2/10) + \sin(3/10) + \sin(4/10) + \sin(5/10) \approx 1.46287$
23. $1 + 1/3 + 1/5 + 1/7 + 1/9 + 1/11 = 6508/3465 \approx 1.87821$
25. $1^\circ + 2^\circ + 3^\circ = 1 + 1 + 1 = 3$ **27.** 5525 **29.** 440
31. $\dfrac{m(m + 1)(4m + 11)}{6}$ **33.** $-\dfrac{5m(m - 2)(m + 1)(m + 3)}{4}$
35. $\dfrac{m(m^2 + 3m + 5)}{3}$ **37.** $\dfrac{m(m + 1)(49m^2 + 105m + 46)}{4}$
39.
$$\frac{(m - 2)(b_4 m^4 + b_3 m^3 + b_2 m^2 + b_1 m + b_0)}{15},$$
where $b_4 = 48$, $b_3 = 336$, $b_2 = 1112$, $b_1 = 2584$, and $b_0 = 5295$.

5.2 The Definite Integral

1. $7/4$ **3.** $36/25$ **5.** 0.670231 **7.** 20 **9.** 28
11. $5/6$ **13.** $10/3$ **15.** $3/2$; net area of $3/2$ square units
17. 0; net area of 0 square units
19. $L_{10} = 0.610509$; $G_{10} = 0.710509$; $A_{10} = 0.660509$; $|I - A_{10}| \le \frac{1}{2}(G_{10} - L_{10})$ holds because I lies between L_{10} and G_{10} and, hence, cannot be more than $\frac{1}{2}(G_{10} - L_{10})$ away from the midpoint of the interval $[L_{10}, G_{10}]$.
21. $R_n = (n - 1)^2/(4n^2)$; $R_8 \approx 0.191406$; $R_{32} \approx 0.234619$; $R_{100} = 0.245025$.
25. Consider a column of air above a 1×1 square meter patch of earth and, from this column, a section from height i meters to $i + 1$ meters. The mass of the section will be approximately $\delta(i) \cdot 1 \cdot 1 \cdot 1$ kg. The sum of 11,000 sections like this approximates the total mass of air in the column; this sum is also a Riemann sum for the integral $\int_0^{11000} \delta(x)\,dx$. Letting the evaluation points be the midpoints of the subintervals, $R_{20} \approx 8023.7$.

5.3 The Fundamental Theorem of Calculus

1. Area is 1 square unit.

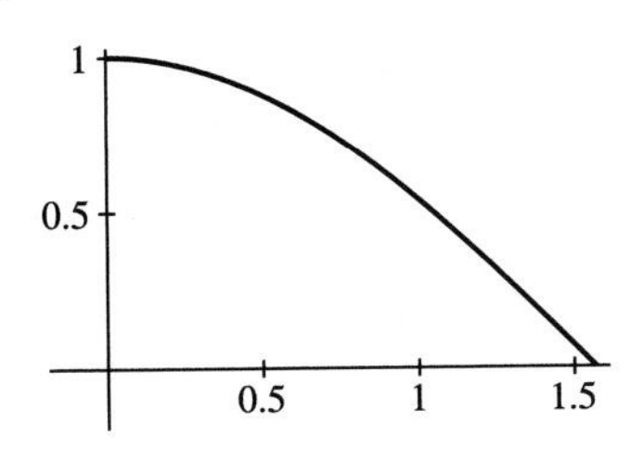

3. Area is 4 square units.

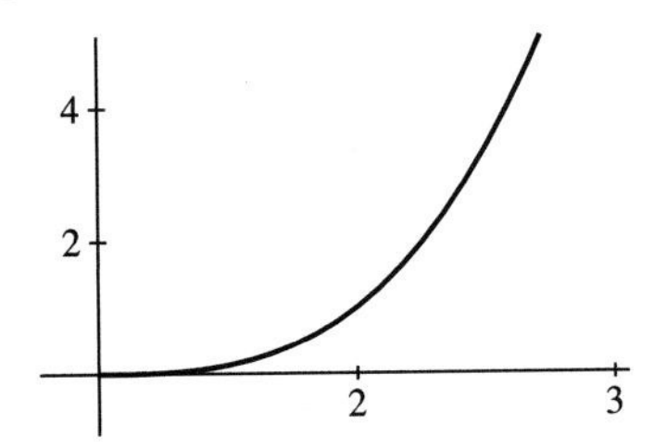

5. Area is $e - 1$ square units.

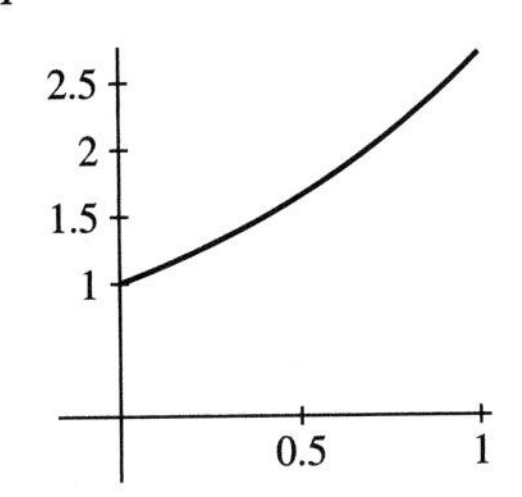

7. Area is 1 square unit.

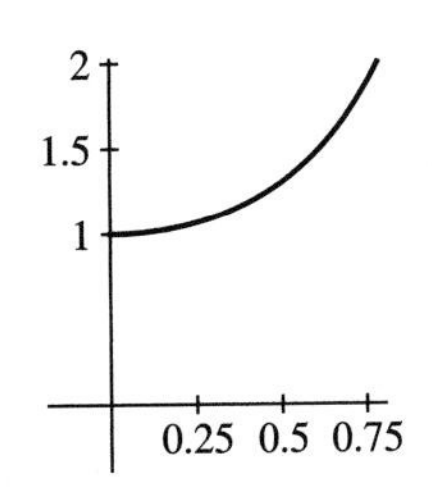

9. Area is $\pi/4$ square units.

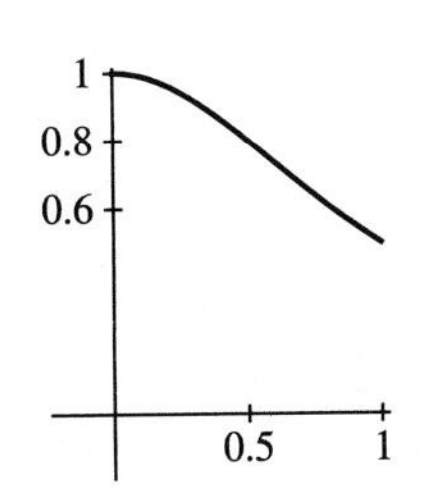

11. Area is $\ln 2 \approx 0.693147$ square units.

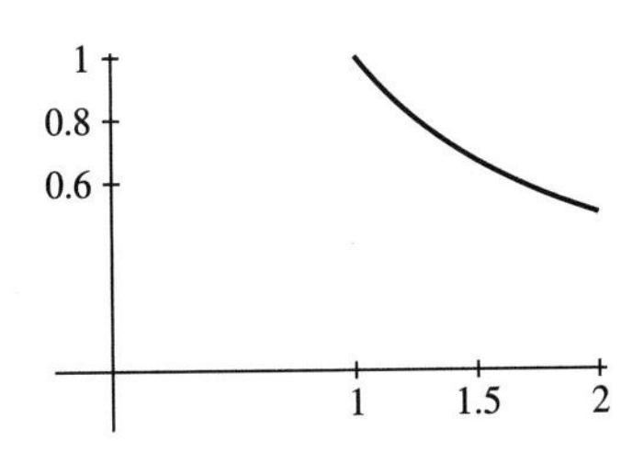

13. $\ln x$ **15.** $\sqrt{1 + x^2}$ **17.** $2x \ln(x^2)$
19. $\sqrt{1 + x}/(2\sqrt{x})$ **21.** $(\arctan(e^x) - e^{2x})e^x$
23. An antiderivative of f is $2x^{3/2}/3$; $16/3$
25. An antiderivative of f is $-\frac{7}{2}x^2 + \frac{1}{4}x^4$; $-13/4$
27. An antiderivative of f is $-\frac{1}{2}x^2 + x^3$; 56
29. An antiderivative of f is $\frac{1}{2}x^2 + \ln x$; $4 + \ln 3$
31. An antiderivative of f is $-6x^{1/3} + \frac{3}{4}x^{4/3}$; $21/4$
33. An antiderivative of f is $-\cos(x^2)$; $1 - \cos 1 \approx 0.459698$
35. $F(2) = 0$; $F(3) = 3/2$; $F(5) = 11/2$
37. $F(0) = 0$, $F(\pi/4) = 1 - \sqrt{2}/2$, $F(\pi/2) = 1$, $F(\pi) = 2$
39. $4/3$ **41.** $\sin 1 \approx 0.841471$ **43.** $e - 1$
45. Average value of f on $[0, \pi/2]$ is $2/\pi$, which is the height of the rectangle of equal area. The area of the rectangle is equal to the area beneath the graph of the cosine function on the interval $[0, \pi/2]$.

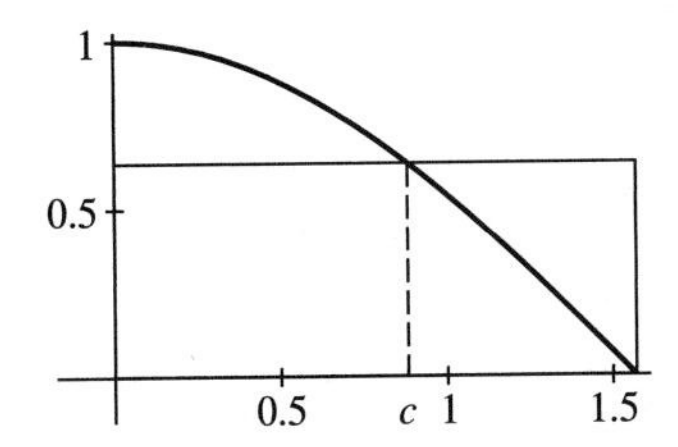

55. The maximum occurs at $x = 1.5$.
57. $6\sqrt{3}$ cubic meters per minute per minute

5.4 The Indefinite Integral

1. $\frac{5}{3}x^3 - 7x + C$ **3.** $\frac{1}{4}x^4 + \frac{15}{4}x^{4/3} + C$ **5.** $\frac{125}{4}x^4 + C$
7. $\frac{2}{3}\ln|x| + C$ **9.** $\dfrac{2^{x+1}}{\ln 2} + C$ **11.** $\arctan x + C$
13. $\arcsin(x/2) + C$ **15.** $2x^{3/2} + 5 \sin x + C$
17. $3^x/\ln(3) + \frac{1}{4}x^4 + C$ **19.** $\frac{1}{6}(2x + 3)^3 + C$
21. $\sec x + C$

5.5 Integration by Substitution

1. $2\sqrt{2}/3 - 1/3$ **3.** $1/4$ **5.** $2e^2 - 2e$ **7.** $3\sqrt{3}/5$
9. $\ln 4 - \ln 2 = \ln 2$
11. $u = 2x - 1$; formula (A);

$$\int \sqrt{2x - 1}\,dx = \tfrac{1}{2}\int u^{1/2}\,du$$

13. $u = 5 - x$; formula (A);

$$\int \frac{1}{(5-x)^{1/3}}\,dx = -\int u^{-1/3}\,du$$

15. $u = 2x + 1$; formula (B);

$$\int \frac{1}{2x+1}\,dx = \tfrac{1}{2}\int \frac{du}{u}$$

17. $u = 1 - 2x$; formula (C);

$$\int \sin(1-2x)\,dx = -\tfrac{1}{2}\int \sin u\,du$$

19. $u = \frac{1}{3} + \frac{2}{7}x$; formula (D);

$$\int \cos\left(\tfrac{1}{3} + \tfrac{2}{7}x\right)dx = \tfrac{7}{2}\int \cos u\,du$$

21. $u = 2x + 3$; formula (E);

$$\int e^{2x+3}\,dx = \tfrac{1}{2}\int e^u\,du$$

23. $u = -x + 1$; formula (F);

$$\int 2^{-x+1}\,dx = -\int 2^u\,du$$

25. $u = \sqrt{2}x$; formula (G);

$$\int \frac{dx}{2x^2+1} = \frac{1}{\sqrt{2}}\int \frac{du}{u^2+1}$$

27. $u = 2x + 1$; formula (G);

$$\int \frac{dx}{(2x+1)^2+9} = \tfrac{1}{2}\int \frac{du}{u^2+9}$$

29. $u = \sqrt{2}x$; formula (H);

$$\int \frac{dx}{\sqrt{1-2x^2}} = \frac{1}{\sqrt{2}}\int \frac{du}{\sqrt{1-u^2}}$$

31. $u = \sqrt{2}x$; formula (H);

$$\int \frac{dx}{\sqrt{7-2x^2}} = \frac{1}{\sqrt{2}}\int \frac{du}{\sqrt{7-u^2}}$$

33. $u = 2x + 1$; formula (H);

$$\int \frac{dx}{\sqrt{1-(2x+1)^2}} = \tfrac{1}{2}\int \frac{du}{\sqrt{1-u^2}}$$

35. $u = \sin 3\theta$; formula (A);

$$\int \sin^5(3\theta)\cos(3\theta)\,d\theta = \tfrac{1}{3}\int u^5\,du$$

37. $u = \tan\theta$; formula (A);

$$\int \tan^4\theta\sec^2\theta\,d\theta = \int u^4\,du$$

39. $u = \sec\theta$; formula (A);

$$\int \sec^3\theta(\sec\theta\tan\theta)\,d\theta = \int u^3\,du$$

41. $\frac{1}{3}F(x^3+1) + C$ **43.** $F(e^x) + C$
45. $\frac{1}{2}F(\arctan(x^2)) + C$ **49.** $-\sqrt{1-x^2} + C$

51. The integrand is not continuous in $[-1, 1]$.
55. $\frac{1}{8}\pi$ **57.** $\frac{1}{8}3\sqrt{2}\pi$ **59.** $\sqrt{2}/2$
61. $\sqrt{2}/2 + \frac{1}{2}\ln(\sqrt{2}+1)$ **63.** $\frac{1}{2}\ln 2$ **65.** $\pi\sqrt{2}/4$
67. $\arctan(2x-3) + C$ **69.** $\frac{1}{5}\arcsin(x + 1/5) + C$
71. $\dfrac{2\sqrt{3}}{3}\arctan\left((2x+1)/\sqrt{3}\right) + C$
73. $-\dfrac{3\sqrt{5}}{5} + \dfrac{5\sqrt{13}}{13}$ **75.** $2\sqrt{2} - 2$

5.6 Areas between Curves

1. 36 square units

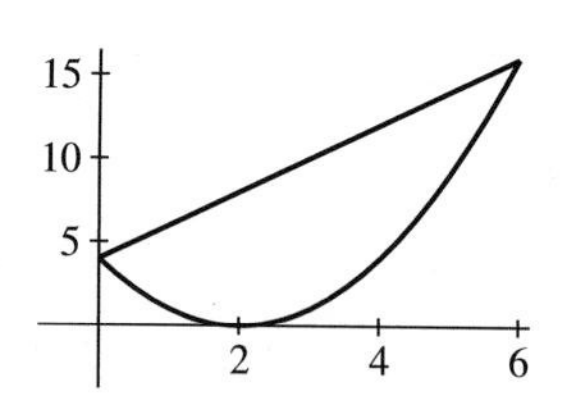

3. 1/12 square units

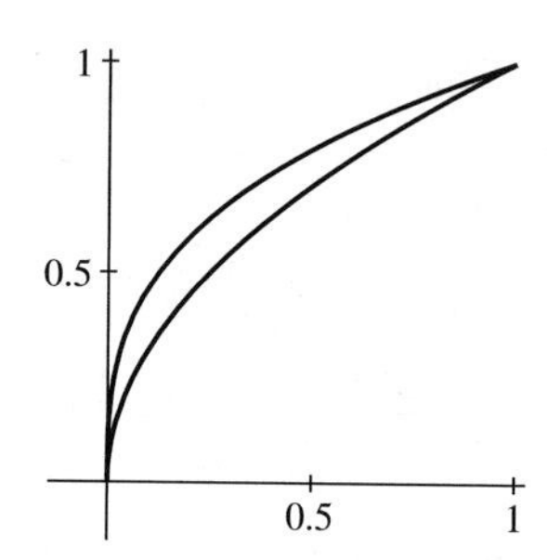

5. 4/3 square units

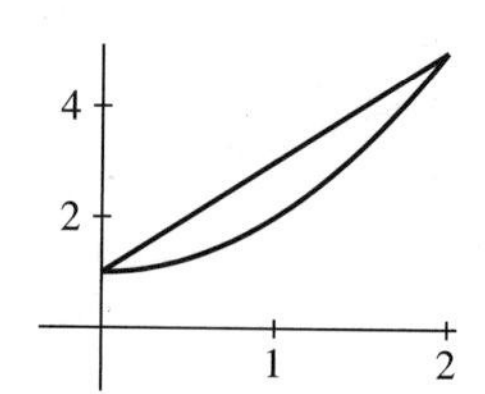

7. 9/2 square units

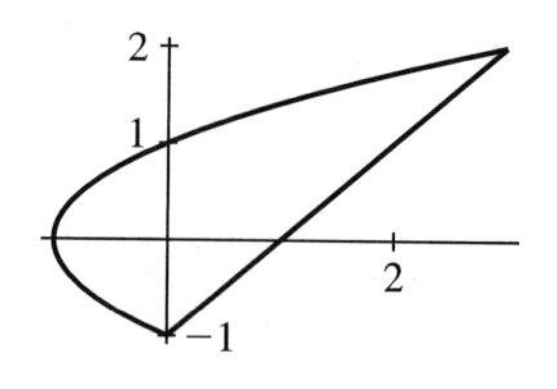

9. 1/3 square units

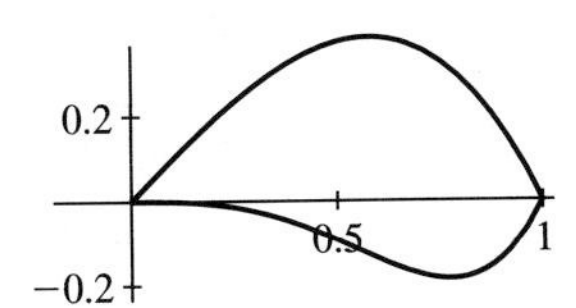

11. 125/24 square units

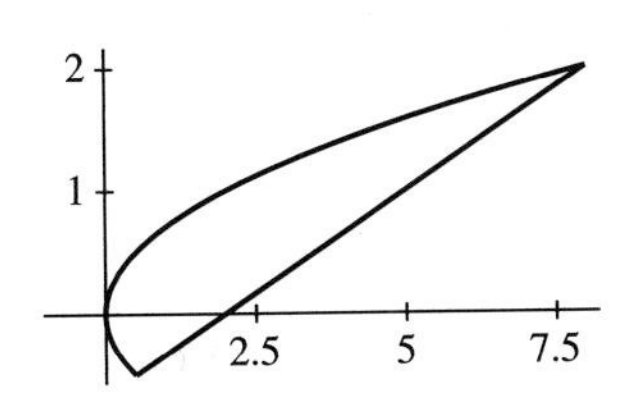

13. 72/5 square units

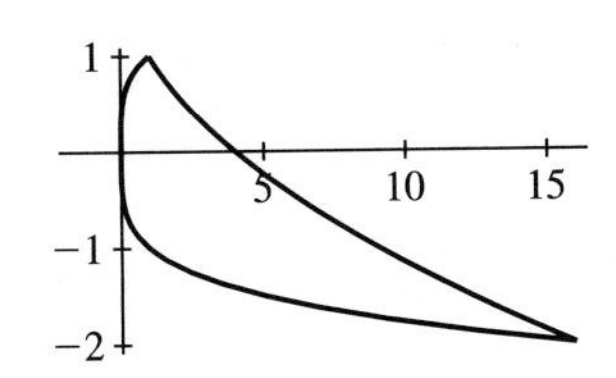

15. 0.322188 square units

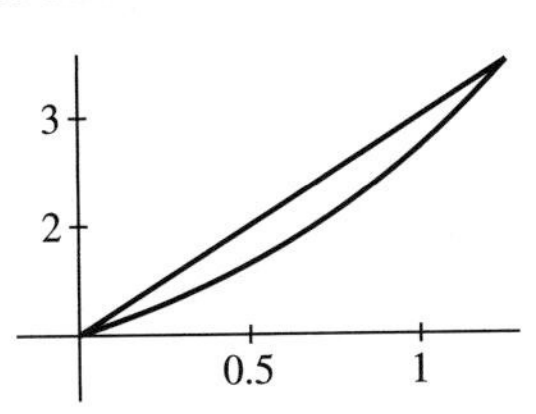

17. 16/3 square units

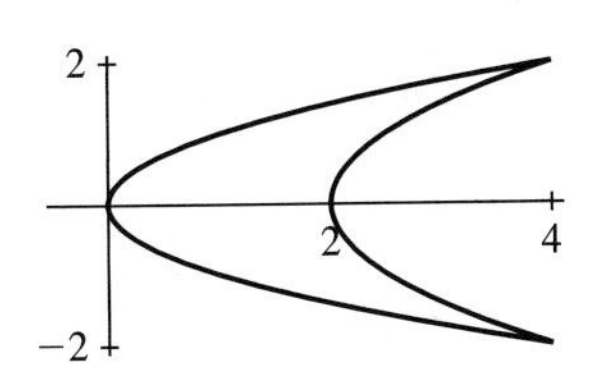

19. 0.777437 square units

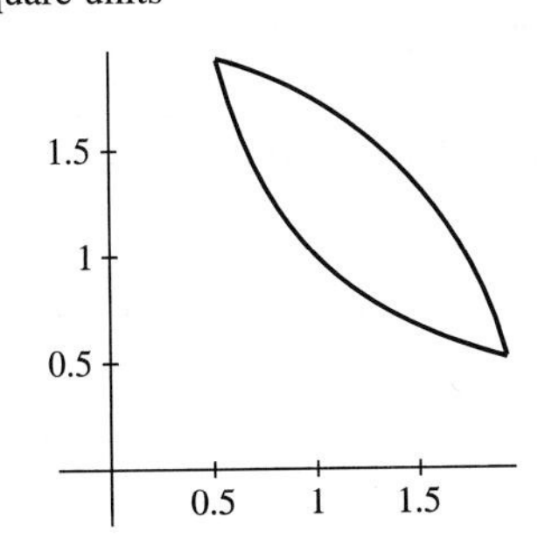

21. 1.00857 square units

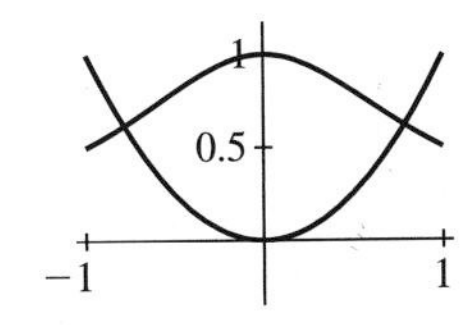

23. 0.256692 square units

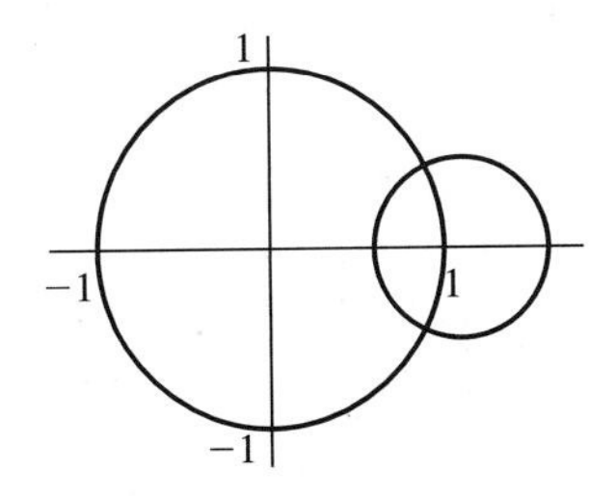

25. 20 kilometers **27.** $75\sin(2) \approx 68.1973$ kilometers
29. 30 kilometers
31. $150 - 75\sin(2) \approx 81.8027$ kilometers
33. Car 2 **35.** 0.177563 kilograms

5.7 Integration by Parts

1. $\frac{1}{4}(e^2 + 1)$ **3.** $x(\ln x - 1) + C$ **5.** $2/\ln(2) - 1/(\ln(2))^2$
7. $-\frac{4}{9}x^{3/2} + \frac{2}{3}x^{3/2}\ln x + C$ **9.** $-x\cos x + \sin x + C$
11. $x \arccos x - \sqrt{1 - x^2} + C$ **13.** $2e^{\sqrt{x}}\left(\sqrt{x} - 1\right) + C$
15. $\frac{1}{2}x^2 \arcsin x + \frac{1}{4}x\sqrt{1 - x^2} - \frac{1}{4}\arcsin x + C$
19. $e^x(x^3 - 3x^2 + 6x - 6) + C$ **21.** $e^2 - 1$ square units
23. 1.49961 square units
25. $-\dfrac{10^{-x}}{1 + (\ln(10))^2}(\cos x + \ln(10)\sin x) + C$
29. $\frac{1}{2}(x\sin(\ln x) - x\cos(\ln x))$
33. Yes. It checks by a direct calculation.
37. $\frac{1}{8}\left(7\sqrt{2} + 3\ln(1 + \sqrt{2})\right)$

5.8 Integration by Partial Fractions

1. $Q(x) = 3x + 2;\ r(x) = -4x - 3$
3. $Q(x) = 1;\ r(x) = -4x^2 + 7x + 13$
5. $Q(x) = 0;\ r(x) = 2x - 1$
7. $(x - 1)(x + 2)(x^2 + 1)$ **9.** $(x - 1)(x + 1)(x^2 + 1)$
11. $(x - 1)(x + 1)(x^2 + x + 1)$ **13.** $(x + 1)(x^2 - 6x + 25)$
15. $\approx (x - 0.697224)(x - 4.30278)(x^2 + x + 1.00002)$
17. $\left(x - \sqrt{2}\right)\left(x + \sqrt{2}\right)(x^2 + 1)$
19. $\dfrac{3}{x + 1} - \dfrac{5}{x + 4}$ **21.** $x - \dfrac{2}{x - 3} + \dfrac{1}{x + 2}$
23. $\dfrac{1}{x + 3} + \dfrac{1}{(x + 3)^2}$ **25.** $\dfrac{2x - 3}{x^2 + 1} + \dfrac{1}{x - 2}$
27. $-\dfrac{1}{x + 3} - \dfrac{1}{2x + 3} + \dfrac{2}{x - 2}$

29. $\dfrac{1}{x-2}+\dfrac{1}{4(x-1)}+\dfrac{1}{(x-2)^2}-\dfrac{5}{4(x-3)}+\dfrac{1}{2(x-3)^2}$
31. $\ln 2 - \ln 4 + \ln 5 = \ln(5/2)$
33. $-1/(x+1) - \frac{3}{2}\arctan(x/2) + C$
35. $\ln|x+1| + \arctan\big((x+1)/\sqrt{2}\big)/\sqrt{2} + C$
37. $-1/(2(x^2+4)) + \frac{1}{2}\arctan(x/2) + C$
43. $\frac{1}{2}\ln|x+1| + \frac{1}{2}\ln|x-1| + C$
45. $\frac{1}{4}\ln|x-1| - \frac{1}{4}\ln|1+x| - \frac{1}{2}\arctan x + C$
47. $\frac{1}{4}\ln|x-1| - \frac{1}{4}\ln|x+1| + 1/(2(x+1)) + C$
49. $x + 3\ln|x+2| - 8\ln|x+3| + C$
51. $\arctan(x+1) + C$
53. $x + \frac{1}{2}x^2 + \frac{1}{3}x^3 + \ln|x-1| + C$
55. $-1/(3(x+1)^3) + C$
57. $\frac{1}{2}\ln|x^2+x+1| - \arctan\big((2x+1)/\sqrt{3}\big)/\sqrt{3} + C$
59. $\frac{1}{8}\ln|x-2| - \frac{1}{4}\ln|x-1| + \frac{1}{16}\ln|x^2+x+2| + \arctan\big((2x+1)/\sqrt{7}\big)/\big(8\sqrt{7}\big) + C$
61. $\frac{1}{3}\arctan((x-2)/3) + \frac{1}{2}\ln|x^2-4x+13| + C$
63. $(15\pi+58)/96$
65. $6/(x+1) + 2\ln|x| - \ln|x+1| + C$
67. $\frac{1}{2}\arctan(x) + \frac{1}{2}x/(x^2+1) + \ln|x-1| + \frac{1}{2}\ln(x^2+1) + C$

5.9 Solving Simple Differential Equations

7. $\frac{1}{2}y^2 = \frac{1}{2}t^2 + c$ **9.** $\ln y = \frac{1}{2}\ln(1+t^2) + c$
11. $\ln(3-y) - \ln(y-2) = \ln(t+1) + c$
13. $e^y = e^t + c$ **15.** $y^3 = \frac{3}{2}t^2 + 27$ **17.** $2y = t + 1$
19. $y = \sqrt{2}\sqrt{7 - t\cos t + \sin t}$
21. $e^y(y-1) = -2 + e^t$ **23.** 0.648244 kilograms
25. 14.2 − 10 minutes = 4.2 minutes
27. 41.0 years **29.** 5.2 meters per second

5.10 Numerical Integration

1. $T_5 \approx 1.110268$; $T_{10} \approx 1.101562$; $|T_5 - \ln 3| \approx 0.011655$; $|T_{10} - \ln 3| = 0.002950$
3. $S_4 = 1.100000$; $S_8 = 1.098725$; $|S_4 - \ln 3| = 0.001388$; $|S_8 - \ln 3| = 0.000113$
5. $M_4 \approx 0.341977$; $M_8 \approx 0.341502$; $M_{16} \approx 0.341384$; $P = 0.3414 \pm 0.0001$
7. Because $S_4 \approx 1.035787$ and $S_6 \approx 1.035786$ we take the integral to be 1.036.
9. Because $M_4 \approx 245.9$ and $M_6 \approx 245.9$ we take the integral to be 245.9.
11. The maximum value of f'' occurs at $x = 0$. Hence, $|f''(0)| \approx 0.3989 \le 0.4$. If we take $M = 0.4$, we find that $n = 17$ is sufficient to obtain the desired accuracy.
13. $31520 \cdot 75 = 2{,}364{,}000$ cubic feet

Chapter Review Exercises

1. $\sum_{j=1}^{10} (2j+1)/2^j$
3. $A_5 = 0.21328$; the difference between A_5 and the exact value 0.2 of the integral is 0.01328; A_5 is the same as the trapezoid approximation T_5
5. 27/2 square units **7.** 1/6 square units
9.

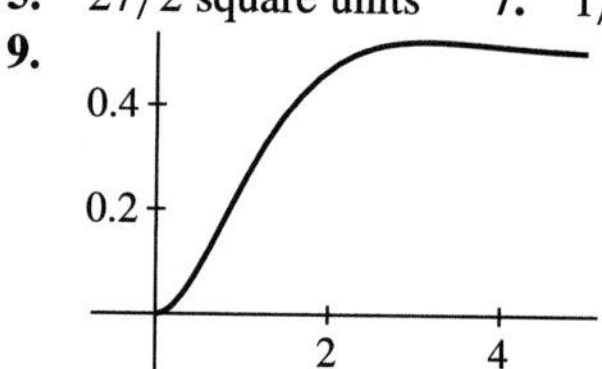

11.

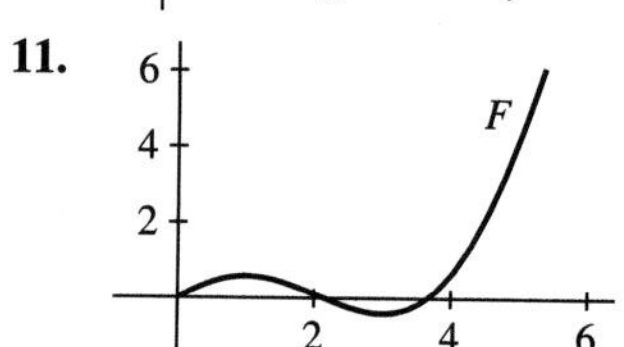

(a) Because f is positive from 0 to 1, F will increase; F decreases on $[1,2]$ and goes to zero near 2 because f looks as if the net area on $[0,2]$ is 0; F continues to decrease until 3; after 3 it increases. **(b)** It is clear that F is increasing on $[0,1]$ and $[3,6]$ and decreasing on $[1,3]$ because $f = F'$ and we have the graph of f. **(c)** $x = 6$ is one high point because F is increasing on $[3,6]$; another is $x = 1$ because F is increasing on $[0,1]$ and decreasing on $[1,3]$. **(d)** The graph of F' is the graph of f, given in the statement of this exercise.
13. **(a)**
$$\frac{A}{x-2}+\frac{B}{x-3}$$
(b)
$$1+\frac{A}{x-1}+\frac{B}{x+1}+\frac{Cx+D}{x^2+1}$$
(c)
$$\frac{A}{5x-3}+\frac{B}{(5x-3)^2}+\frac{Cx+D}{x^2+1}+\frac{Ex+F}{(x^2+1)^2}+\frac{Gx+H}{(x^2+1)^3}$$
(d)
$$\frac{Ax+B}{x^2+1}+\frac{Cx+D}{(x^2+1)^2}+\frac{E}{x+1}+\frac{F}{x-\sqrt{2}}+\frac{F}{\big(x-\sqrt{2}\big)^2}+\frac{G}{x+\sqrt{2}}+\frac{H}{\big(x+\sqrt{2}\big)^2}$$
(e)
$$\frac{Ax+B}{2x^2+2x+1}+\frac{C}{x+1}$$
15. **(a)** 182/3 **(b)** $\frac{1}{2}\ln|2x+3| + C$ **(c)** $\arctan\big(\sqrt{2}x\big)/\sqrt{2} + C$ **(d)** $\arcsin\big(x/\sqrt{7}\big) + C$ **(e)** $2\cos 1$ **(f)** $(e-2)/e$ **(g)** $\ln 3 - \frac{1}{2}\ln 5$ **(h)** 1/2 **(i)** 1 **(j)** $2\arctan\big(\sqrt{x}\big) + C$ **(k)** 1 **(l)** $x - 2\ln|e^x - 1| + C$ **(m)** $\frac{1}{2}\big(\pi/6 - \sqrt{3}/4\big)$
17. Using the tangent line approximation, $s(3.1) \approx 10.36$
19. $75/32 \approx 2.34375$ square units
25. $\arctan y = \ln t - \frac{1}{2}\ln(1+t^2) + 2$
27. $M_{10} \approx 0.035879$
29. $n \ge 48$; $T_{48} \approx 0.848964$ and so the integral is 0.849 to within 0.001

CHAPTER 6—APPLICATIONS OF THE INTEGRAL

6.1 Volumes by Cross Section

1. $\frac{4}{3}\pi a^3$ cubic units

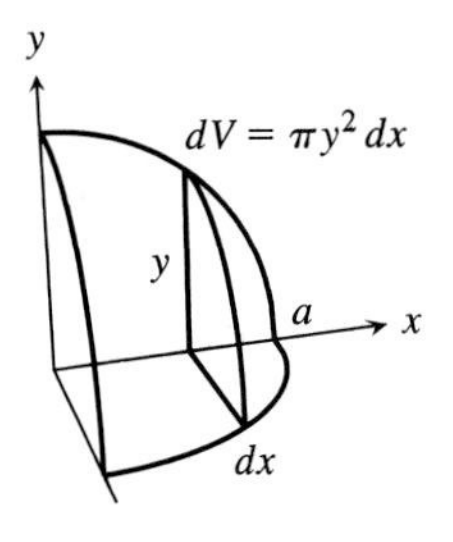

3. $32\pi/5$ cubic units

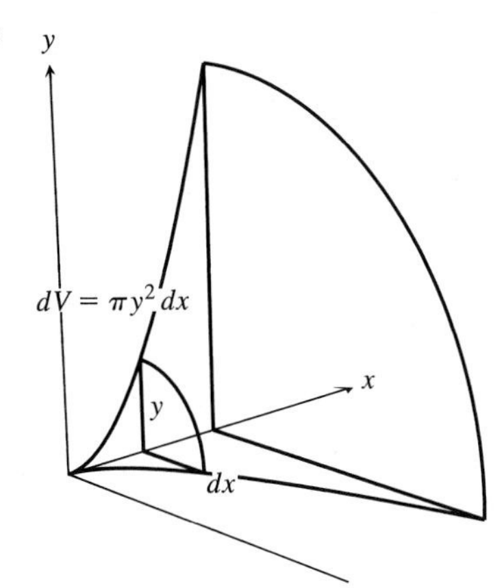

5. 2π cubic units

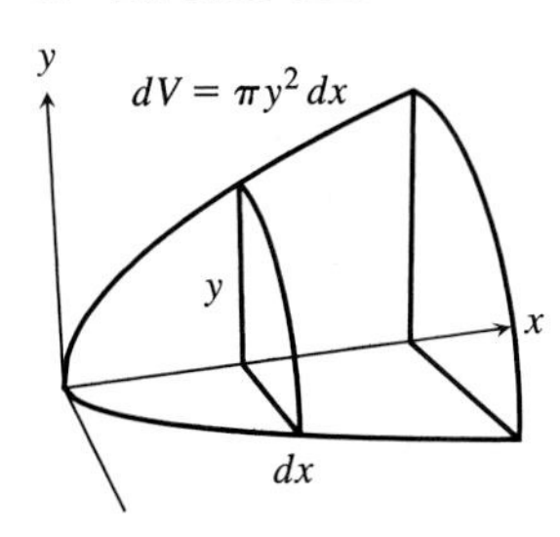

7. $\pi^2/2$ cubic units

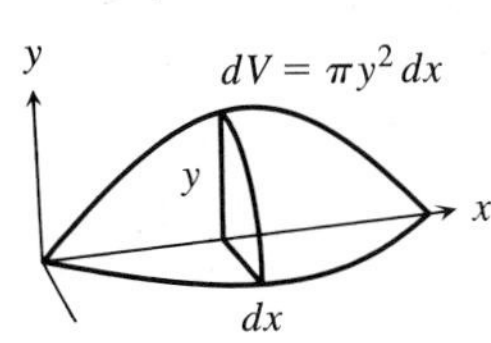

9. π cubic units

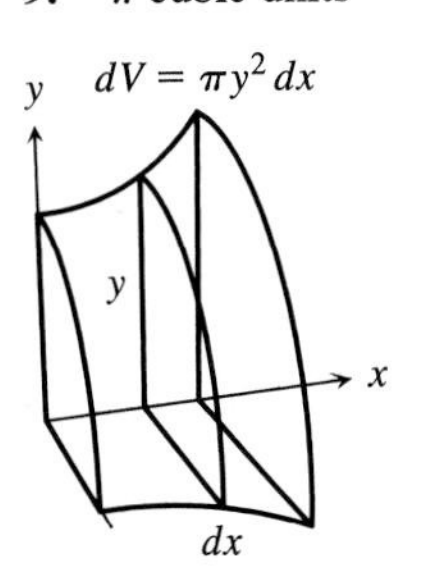

11. $\pi(e^4 - 1)/(2e^2)$ cubic units

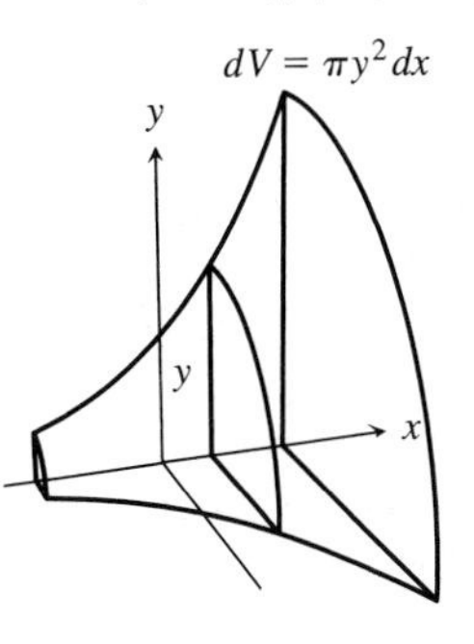

13. $2\pi/35$ cubic units

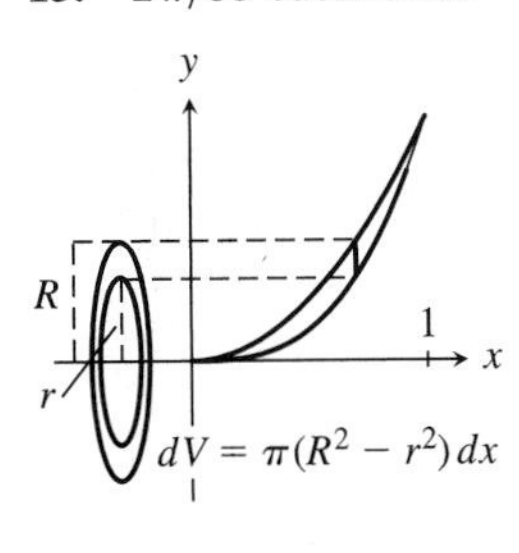

15. $384\pi/5$ cubic units

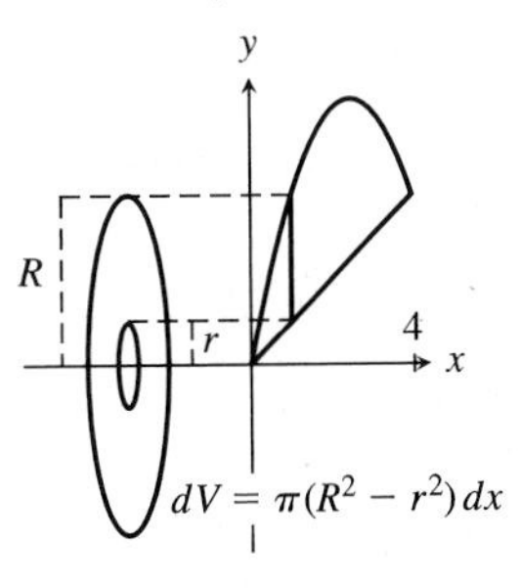

17. $\pi^2/12$ cubic units

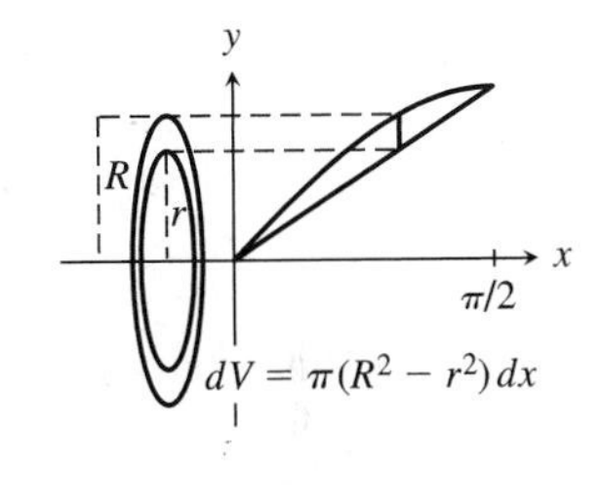

19. 1.43382 cubic units

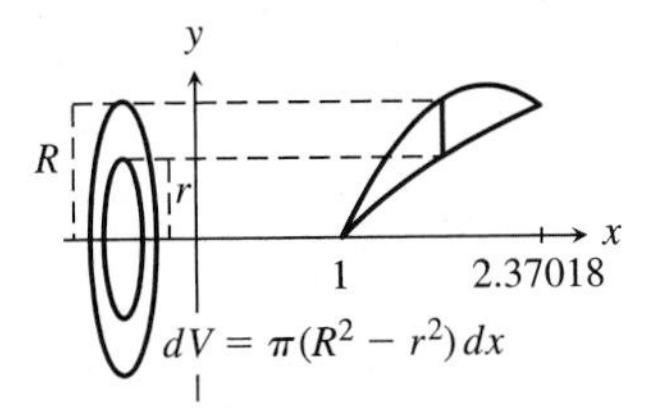

23. $\frac{1}{3}\pi H(B^2 + bB + b^2)$ **27.** $\frac{1}{3}\pi(9\pi + 5)$ cubic units

29. $V(h) = \pi h^2\left(8 - \dfrac{1}{3}h\right)$

31. To the nearest 0.1 inches, measure 7.5 inches, 9.3 inches, 10.6 inches, and 11.8 inches from the bottom.

33. $\frac{16}{3}r^3$ cubic units **35.** $2a^3/(3\sqrt{3})$ cubic units

37. $m = \dfrac{2000}{3}\,\text{cm}^3 \cdot \dfrac{1\ \text{m}^3}{100^3\ \text{cm}^3} \cdot \dfrac{7850\ \text{kg}}{\text{m}^3} \approx 5.23\ \text{kg}$

41. The square of a difference $a - b$ is not the same as the difference $a^2 - b^2$ of the squares.

6.2 Volumes by Shells

1. $45\pi/2 \approx 70.69$ cubic units

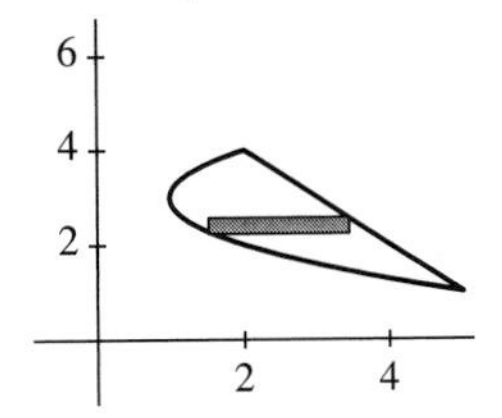

3. $\pi/6 \approx 0.5236$ cubic units

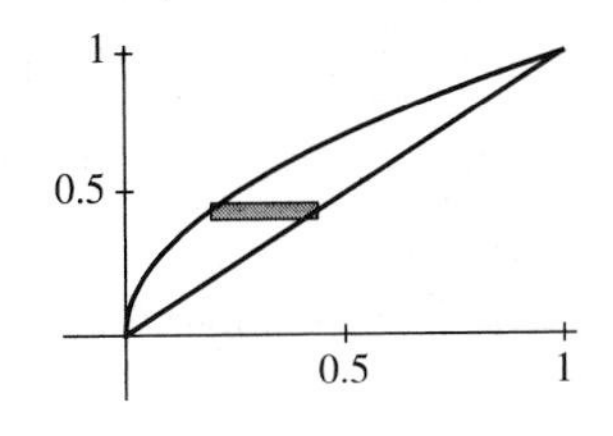

5. $\frac{1}{2}\pi(e-2) \approx 1.128$ cubic units

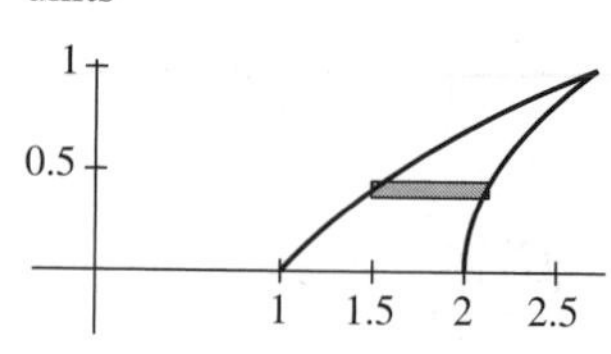

7. $\frac{1}{4}\pi^2 \approx 2.4674$ cubic units

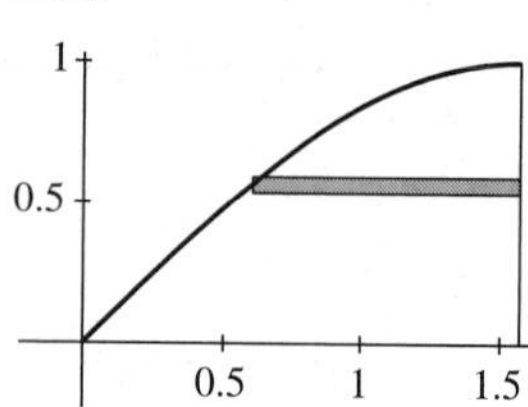

9. 3.5×10^8 liters **11.** 5.97×10^{24} kg **13.** 0.098646 kg
15. $\frac{1}{3}\pi r^2 h$ cubic units **17.** 36π cubic units
19. 0.762278... cubic units
21. **(a)** $27\pi/2$ cubic units; **(b)** $27\pi/2$ cubic units
23. $2\pi(1-(b+1)e^{-b})$ cubic units; the volume of the unbounded solid of revolution is 2π cubic units.
25. $\pi\left(a+2-3\sqrt[3]{a}\right)$ cubic units; the volume of the unbounded solid of revolution is 2π cubic units.

6.3 Polar Coordinates and Parametric Equations

1. $(x, y) = (1, 1)$
3. $(x, y) = (2\cos 1, 2\sin 1) \approx (1.08060, 1.68294)$
5. $(x, y) = (\cos 3.5, \sin 3.5) \approx (-0.936457, -0.350783)$
7. $(r, \theta) = \left(\sqrt{5}, \arctan(1/2)\right) \approx (2.23607, 0.463648)$
9. $(r, \theta) = (2, \pi)$ **11.** $(r, \theta) \approx (2.91548, 2.11122)$

13.

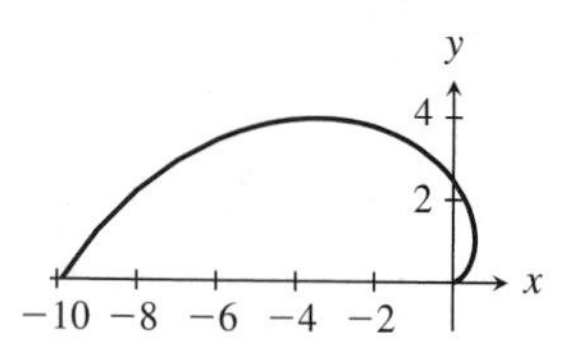

15.

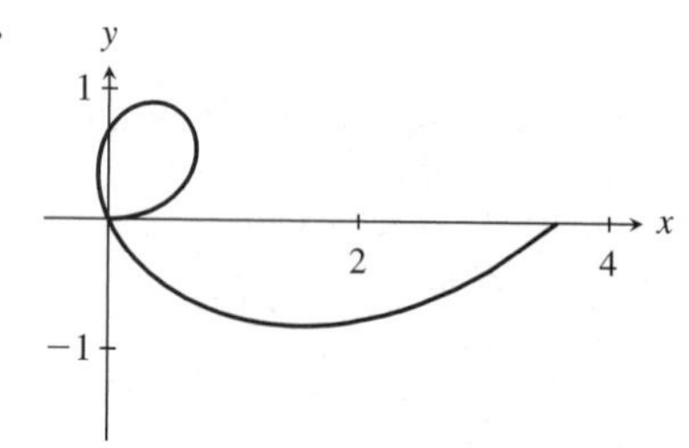

17.

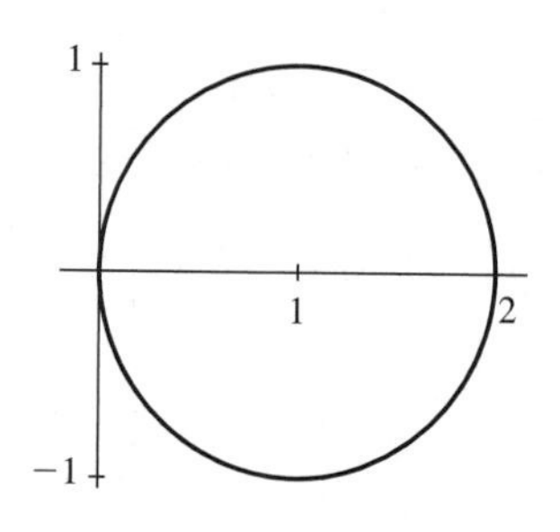

19.

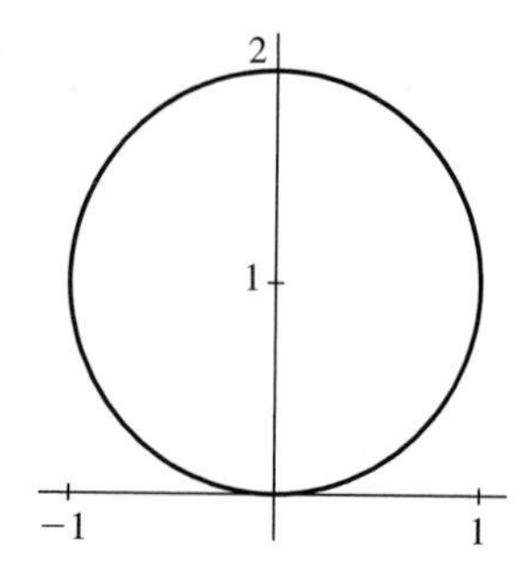

21.

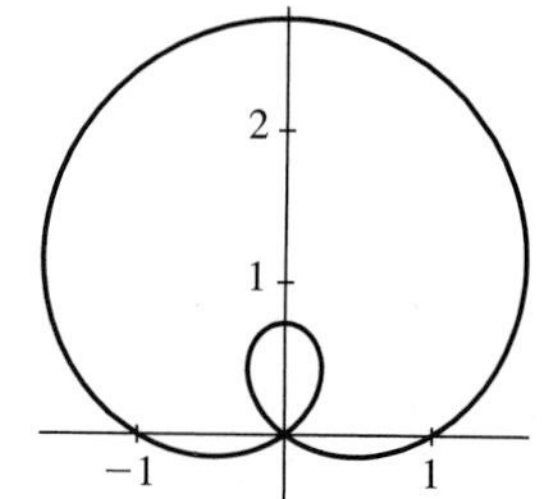

23.

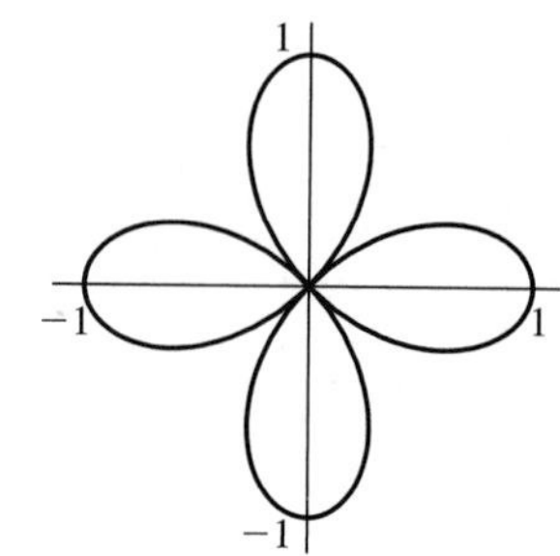

25. $dy/dx \approx -3.82843$
27. Assume that the end of the cable is initially at $(2, 0)$ and is unwrapping in the counterclockwise direction. Then

$$\mathbf{r}(\theta) = \langle 2\cos\theta + 2\theta\sin\theta, 2\sin\theta - 2\theta\cos\theta \rangle.$$

The slope at $\mathbf{r}(0.3)$ is 0.309336.

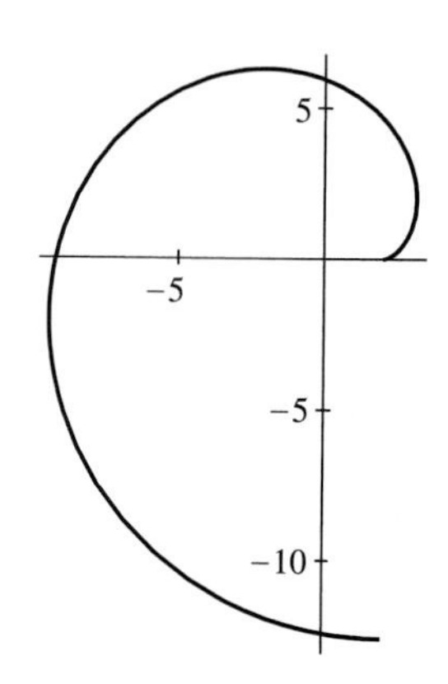

29. $\mathbf{r} = \mathbf{r}(\phi) = 2\langle \phi - \sin\phi, 1 - \cos\phi \rangle$
$\mathbf{r}'(2\pi/3) \approx \langle 3, \sqrt{3} \rangle$

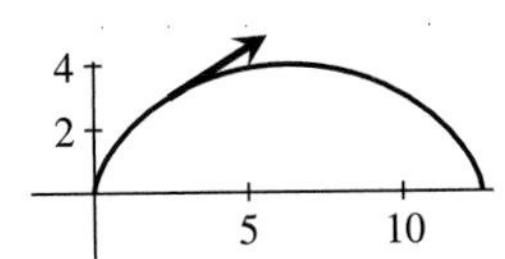

31.

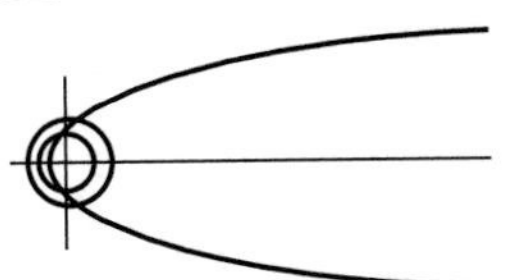

33. The vector $\mathbf{u} = \langle \phi, 0 \rangle$ would be $\langle 2t, 0 \rangle$ m after t seconds. Taking $\phi = 2t$, the motion is described by $\mathbf{r}(t) = \langle 2t - \sin(2t), 1 - \cos(2t) \rangle$; $\mathbf{v}(3) = 2\langle 1 - \cos(6), \sin(6) \rangle$ m/s; $\|\mathbf{v}(3)\| \approx 0.564480$ m/s.

35. $x^3 - 3xy + y^3 = 0$; the fourth-quadrant part of the folium is traced as t varies on the interval $(-\infty, -1)$, with the trace moving downward as t increases toward -1; the second-quadrant part is traced as t varies on $(-1, 0]$, with the trace moving toward the origin as t increases; the first-quadrant part is traced as t varies on $[0, \infty)$, with the trace moving counterclockwise; the points on the folium at which the tangent line is horizontal are $\mathbf{r}(0)$ and $\mathbf{r}(\sqrt[3]{2})$; the point on the folium at which the tangent line is vertical is $\mathbf{r}(1/\sqrt[3]{2})$.

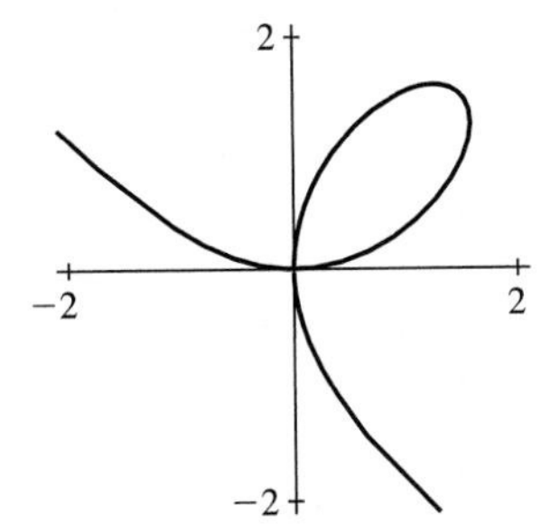

39. $(r, \theta) \approx (1.23340, 3.43829)$

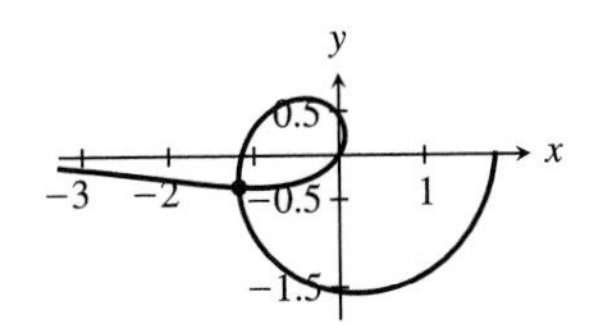

6.4 Arc Length and Unit Tangent

1. $25\sqrt{61}$ meters

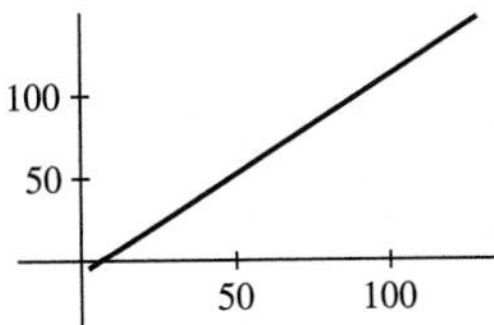

3. 7.5 meters

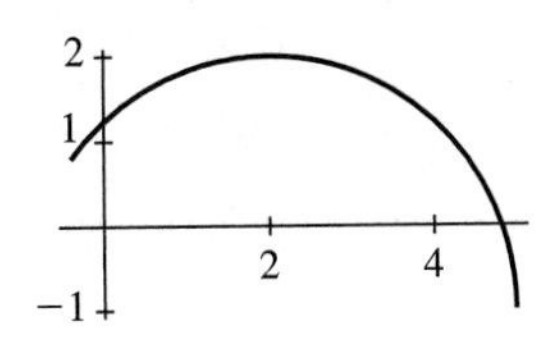

5. 24 meters

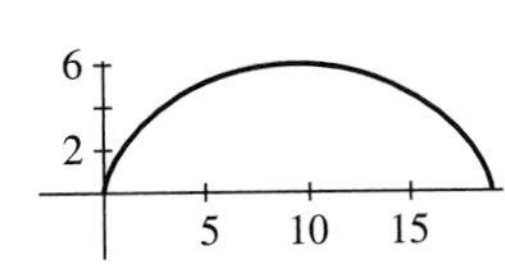

7. 4.64678 meters

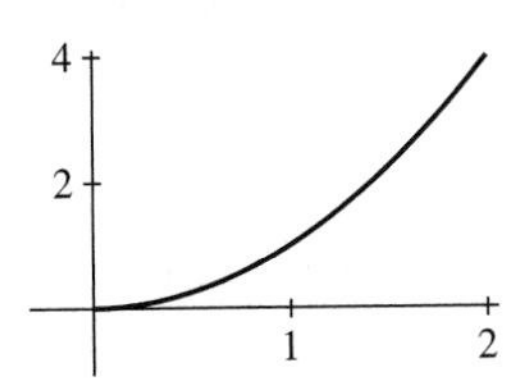

9. 12.4074 meters

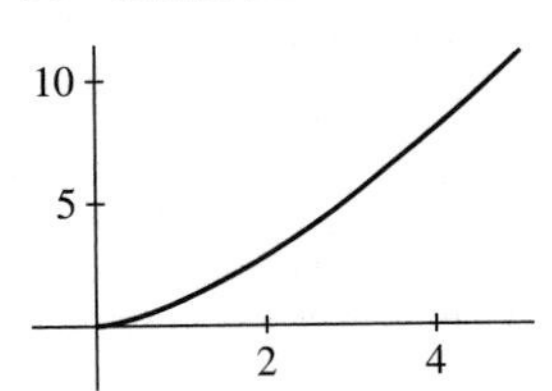

11. 4 meters

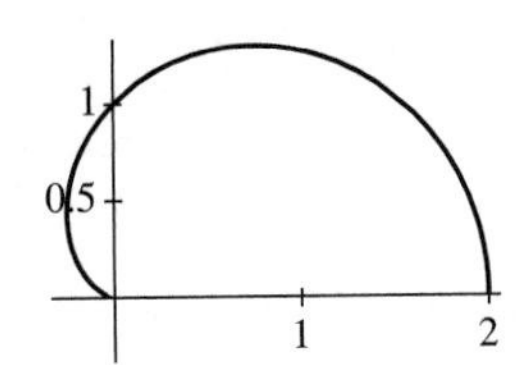

13. 31.3117 meters

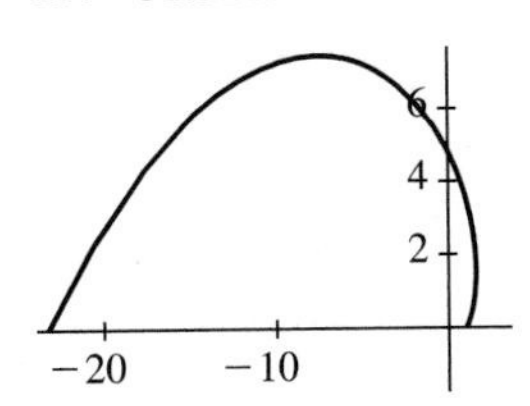

15. $\pi^2/2 \approx 4.9348$ meters

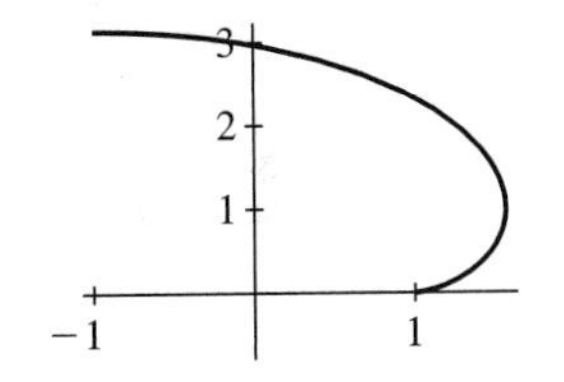

17. 1.14779 meters

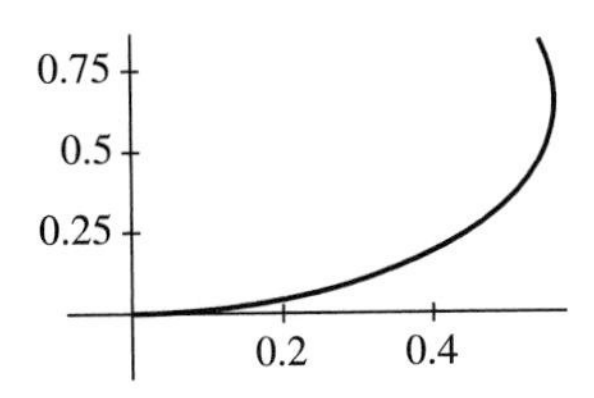

19. 3.96636 meters

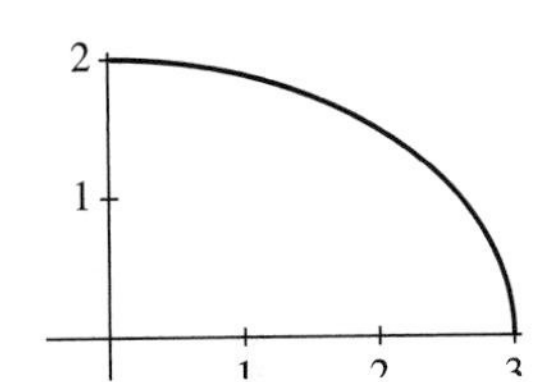

21. 1.54 meters

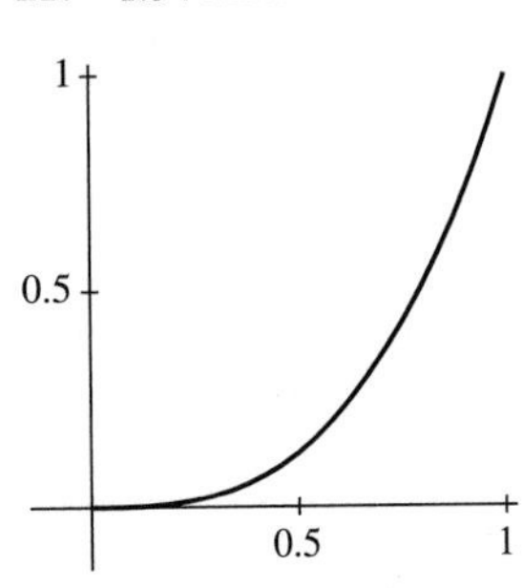

23. $\mathbf{T}(\pi/4) = \langle -3/\sqrt{13}, 2/\sqrt{13} \rangle$

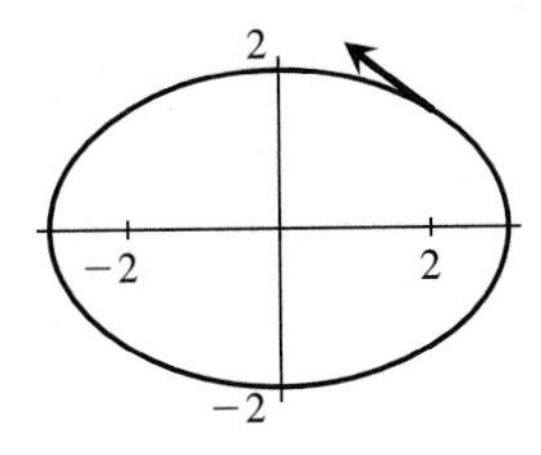

25. $\mathbf{T}(1) = \langle 1/\sqrt{10}, 3/\sqrt{10} \rangle$

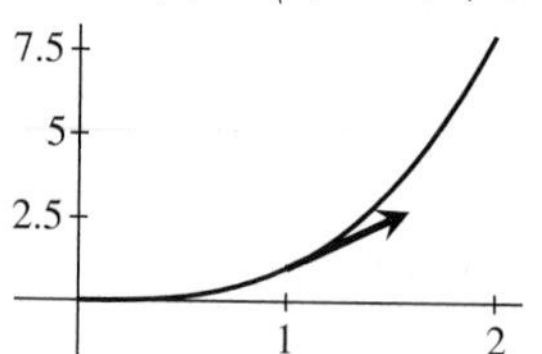

27. $s = 4 \cdot (19.27 \pm 0.01)$AU

29. From (3), $s \approx 2.27322$. Other things remaining the same, the approximation would improve on the interval $[3/2, 3]$.

33. The second equation describes the motion of the hare. At $t = 0.2$ the hare's speed is 6.78584 units per time unit, while the tortoise is going 2π units per time unit. At $t = 0.5$ the hare is resting and the tortoise is still going at 2π units per time unit.

35. No, because $\mathbf{r}'(t) = \langle 0, 0\rangle$ at $t = 0, 2\pi$. However, one might argue that unit tangents can be defined at $\mathbf{r}(0)$ and $\mathbf{r}(2\pi)$ by taking a limit of $\mathbf{T}(t)$ as $t \to 0^+$ or $t \to 2\pi$.

6.5 Areas of Regions Described by Polar Equations

1. $19\pi^3/1296$ square units

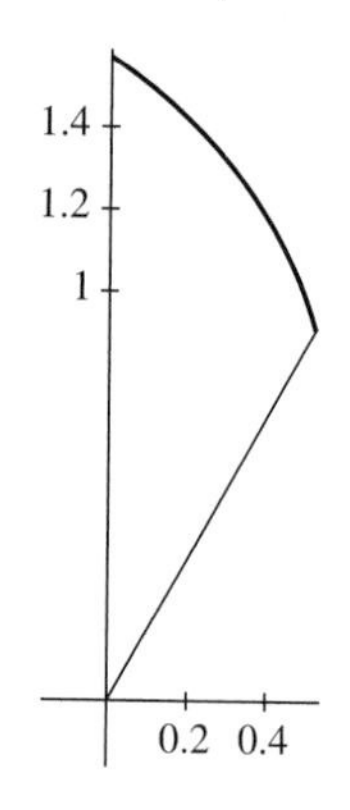

3. $\frac{1}{4}(e^\pi - 1)$ square units

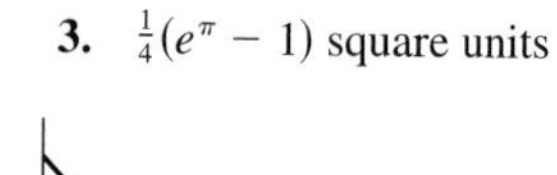
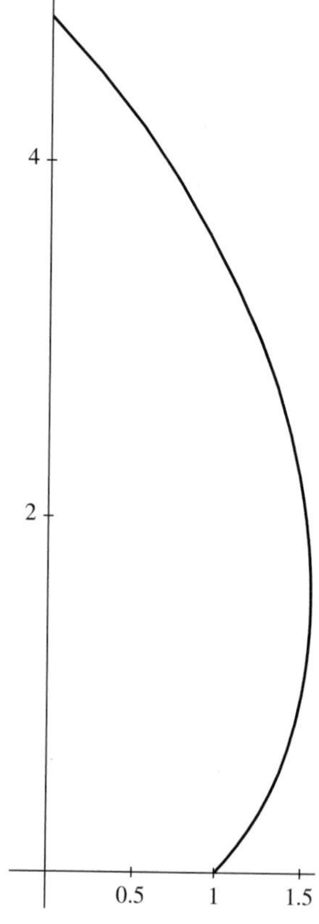

5. $\frac{1}{48}(3\sqrt{3} + 2\pi)$ square units

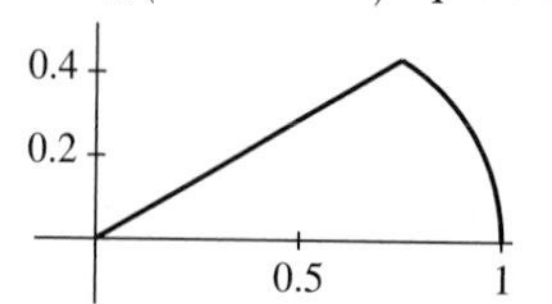

7. $3\pi/2$ square units

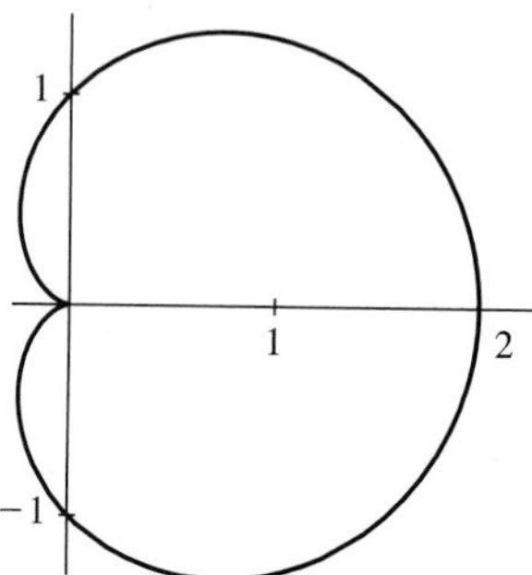

9. π square units

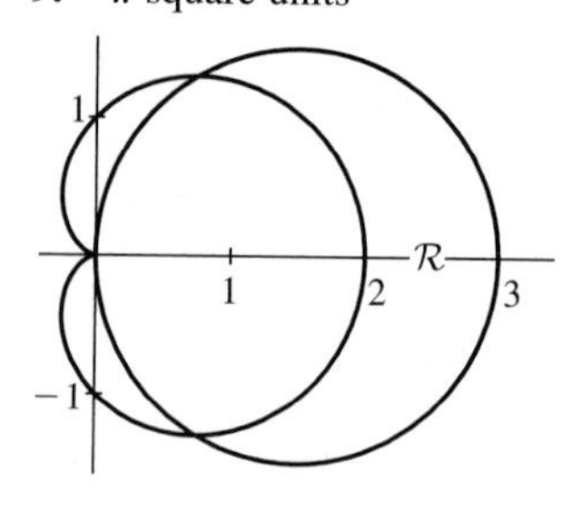

11. $\pi/4$ square units

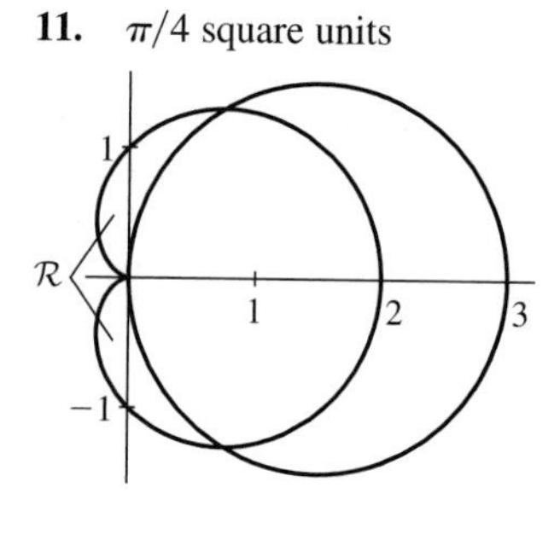

13. $\pi/8 - 1/4$ square units

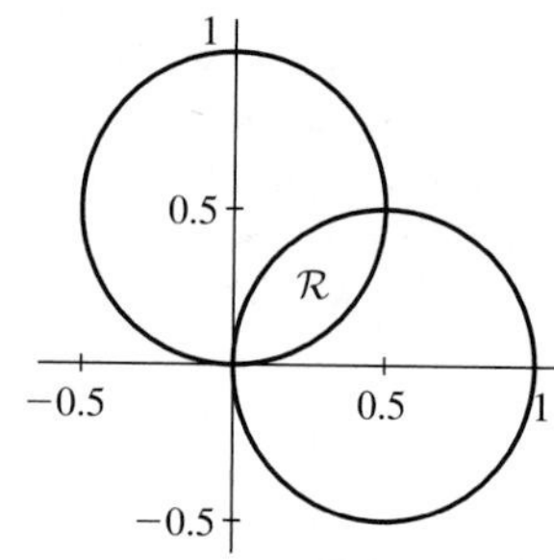

15. $\pi - 3\sqrt{3}/2$ square units

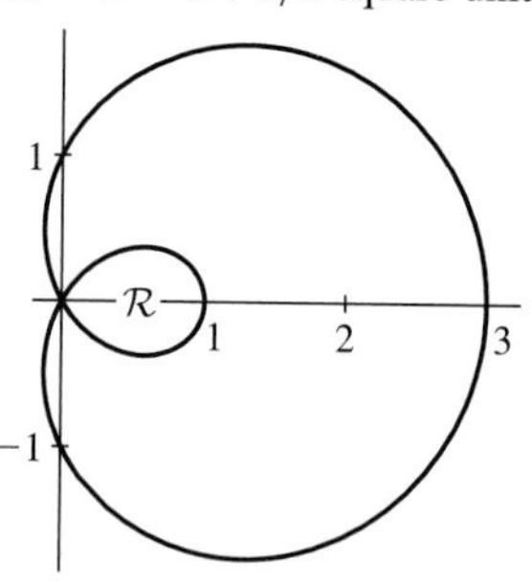

17. $4\pi/3 - \sqrt{3}$ square units

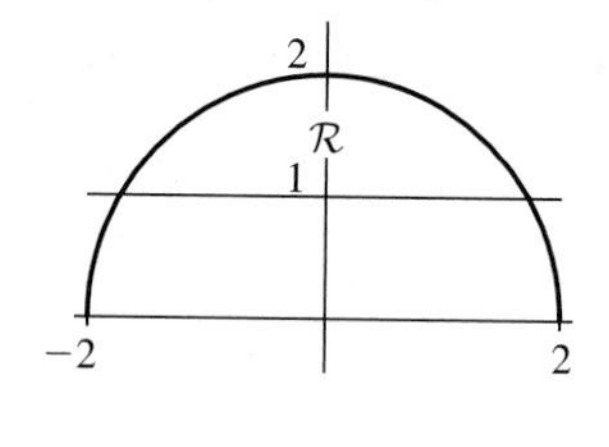

19. 1 square unit

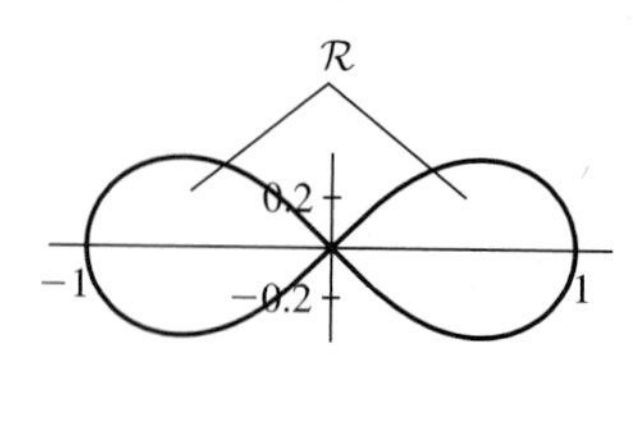

21. $3\pi/16 - \sqrt{2}/8$ square units

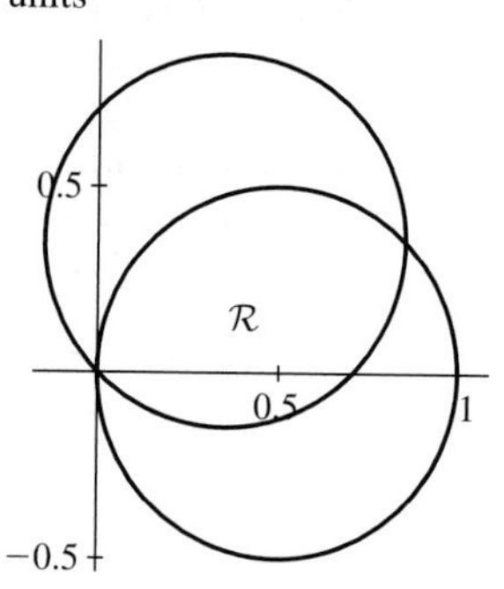

23. About 11 g

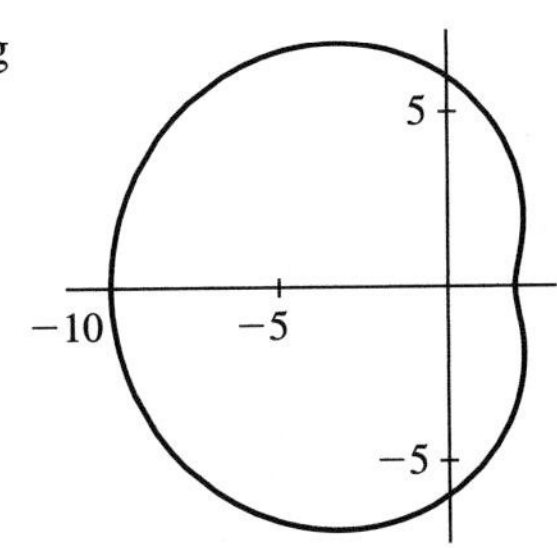

25. 0.150751 square units **27.** 0.0778954 square units
29. 0.0167804 square units

6.6 Work

1. 150 J **3.** 22.5 J **5.** 22.5 J
7. $2k$ J, where k is the proportionality constant
9. 2.04×10^{10} J **11.** $5mg$ J **13.** 62500 J **15.** 7400 J
17. 7.6×10^{10} J **19.** 1.8×10^{7} J **21.** 2.6×10^{7} J
23. 3400 J **25.** 2.8×10^{8} J **27.** 1.1×10^{7} J
29. 32/31 **31.** 5.7×10^{6} J **33.** 2.3×10^{12} J
35. The speed is about 28 m/s. As shown in the text, the work done by the gravitational force is $\frac{1}{2}mv_b^2$ if we assume that the car started from rest. Because the Earth's gravitational field is conservative, the work done by the gravitational force in moving the car from A to B is $mg \cdot 40$. Hence, $v_b = \sqrt{2g \cdot 40} = 28$ m/s.

6.7 Center of Mass

1. 7/9 m **3.** 64/11 m from the left end
5. The center of mass is at the point with position vector $\langle 7/26, -9/26 \rangle$.
7. $m = 1.675$ kg and the center of mass is at the 0.258706 mark from the less-dense end.
9. $\langle 13a/10, 9a/10 \rangle$-length units
11. $m = \delta(e-1)$; $\mathbf{R} = \langle 1/(e-1), (e+1)/4 \rangle$
17. In polar coordinates, $(3.63636, \pi)$
19. In polar coordinates, $(1.38021, 2.53735)$
21. $(X, Y) = (0, 52/5)$ m
23. $(X, Y) = (0.409429, 0.687985)$
25. $(X, Y) = (1, 1/4)$
29. Set up a coordinate system at the center of the circular lamina. Let $m_1 = \delta\pi a^2$ be the mass of the to-be-punched circular region of radius a and $\mathbf{R}_1 = \langle h, k \rangle$ its center/center of mass; let $m_2 = \delta\pi A^2$ be the mass of the unpunched circular lamina of radius A and $\mathbf{R}_2 = \langle 0, 0 \rangle$ its center/center of mass; and let $m_3 = m_2 - m_1$ be the mass of the punched lamina and $\mathbf{R}_3 = \langle X, Y \rangle$ its center of mass. Applying the Mass Subdivision Theorem to the two systems consisting of the punched lamina and the circular region to be punched,

$$\mathbf{R}_2 = \frac{1}{m_1 + m_3}(m_1\mathbf{R}_1 + m_3\mathbf{R}_3).$$

From this equation, we can solve for X and Y.

31. Mass is 241.925 kg; center of mass is at the 5.4386-meter mark, measuring from the thinner end.
33. $\mathbf{R} = \langle \pi, 4/3 \rangle$-length units
35. The center of mass of the hammer is on the center line of the handle, down approximately 3.8 cm from the end of the handle flush with a face.

6.8 Curvature, Acceleration, and Kepler's Second Law

1. $\mathbf{T}(0) = \langle 1, 0 \rangle$; $\mathbf{N}(0) = \langle 0, 1 \rangle$; $\kappa(0) = 1/2$; curvature is least at $\mathbf{r}(2)$ and greatest at $\mathbf{r}(0)$.

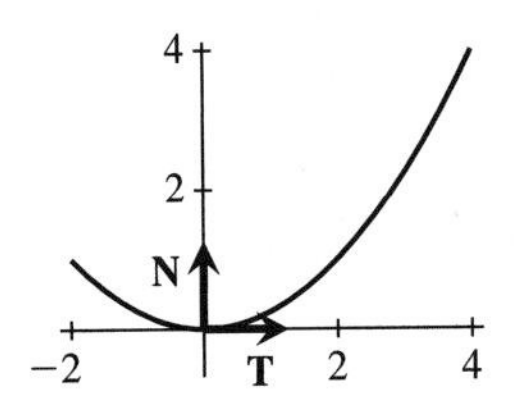

3. $\mathbf{T}(\pi/4) = \left\langle -3/\sqrt{13}, 2/\sqrt{13} \right\rangle$;
$\mathbf{N}(\pi/4) = \left\langle -2/\sqrt{13}, -3/\sqrt{13} \right\rangle$; $\kappa(\pi/4) = 12\sqrt{26}/169$; curvature is least at $\mathbf{r}(\pi/2)$ and greatest at $\mathbf{r}(0)$.

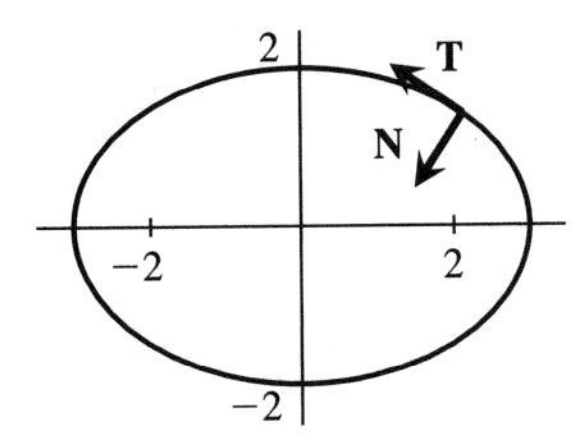

5. $\mathbf{T}(1) = \left\langle 1/\sqrt{10}, 3/\sqrt{10} \right\rangle$; $\mathbf{N}(1) = \left\langle -3/\sqrt{10}, 1/\sqrt{10} \right\rangle$; $\kappa(1) = 3/(5\sqrt{10})$; curvature is least at $(0, 0)$ and greatest near $(0.4, 0.06)$.

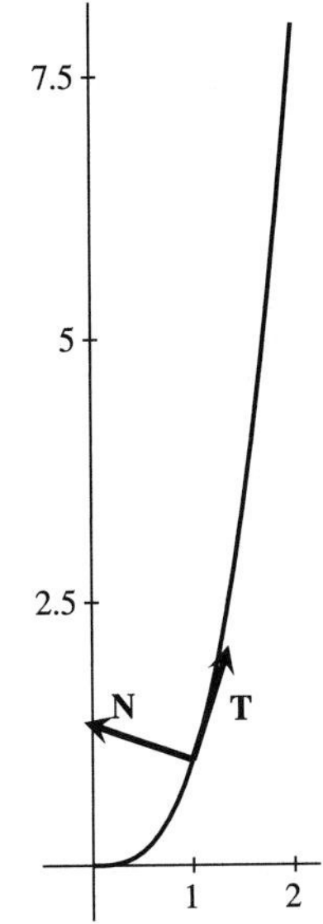

7. $\mathbf{T}(\pi/4) = \langle -1, 0\rangle$; $\mathbf{N}(\pi/4) = \langle 0, -1\rangle$; $\kappa(\pi/4) = 1/2$; curvature is everywhere the same.

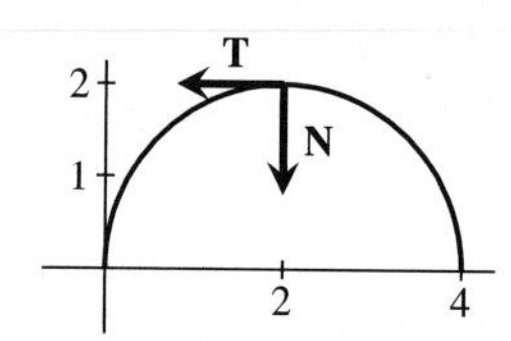

9. $\mathbf{T}(\pi/6) = \langle 2/\sqrt{7}, \sqrt{3/7}\rangle$; $\mathbf{N}(\pi/6) = \langle \sqrt{3/7}, -2/\sqrt{7}\rangle$; $\kappa(\pi/6) = 4/(7\sqrt{7})$; curvature is least at $(0, 0)$ and greatest at $(\pi/2, 1)$.

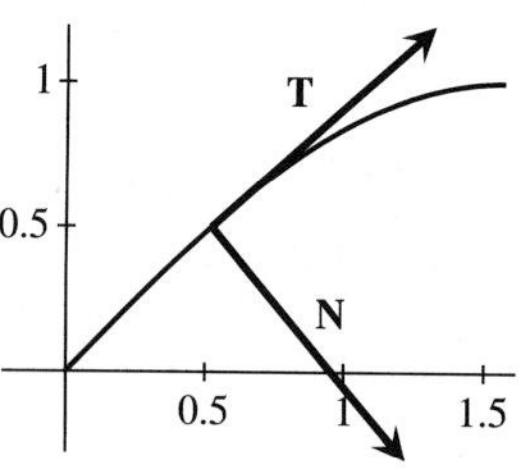

11. $\mathbf{T}(1) = \langle 1/\sqrt{1+e^2}, e/\sqrt{1+e^2}\rangle$; $\mathbf{N}(1) = \langle -e/\sqrt{1+e^2}, 1/\sqrt{1+e^2}\rangle$; $\kappa(1) = e/(1+e^2)^{3/2}$; curvature is least at $(1.5, e^{1.5})$ and greatest at $(0, 1)$.

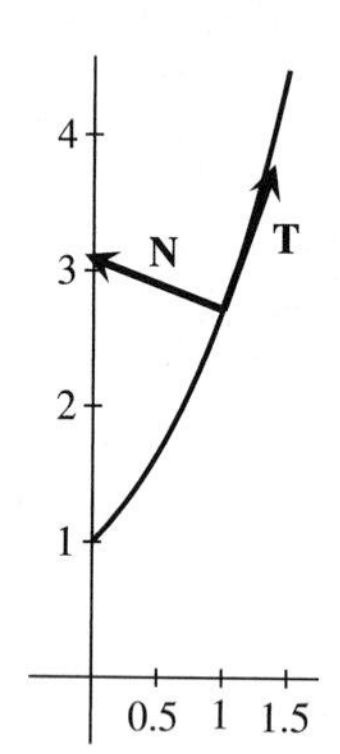

15. $\mathbf{a}(\pi/18) = \langle -9\sqrt{3}, 9\rangle$

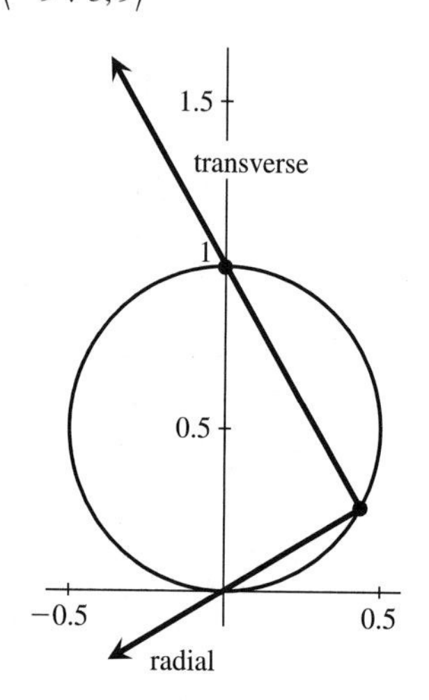

17. The maximum curvature is $2/(3\sqrt{3})$ and it occurs at $t = -\ln(2)/2$.

21. $\kappa(\theta) = \dfrac{1}{2^{\theta}\sqrt{1 + (\ln 2)^2}}$ **23.** $\kappa(\theta) = \dfrac{\theta^2 + 6}{\theta(\theta^2 + 4)^{3/2}}$

25.

27.

29. Take $r = 0.5$ units. Curve A can be described by

$$\mathbf{r}(x) = \left\langle x - \frac{x}{\sqrt{1 + 4x^2}}, x^2 + \frac{1}{2\sqrt{1 + 4x^2}}\right\rangle, \qquad -1 \le x \le 1.$$

33. The coordinates of P are approximately $(0.15, -1.08)$-length units.

6.9 Improper Integrals

1. Divergent **3.** Convergent; 2 **5.** Divergent
7. Divergent **9.** Convergent; $\pi/2$ **11.** Convergent; $\pi/3$
13. Divergent **15.** Convergent; ln 2 **17.** Convergent; -1
19. Convergent; π **23.** 1 **25.** Divergent
27. Convergent **29.** 3/13 **31.** Convergent
33. Convergent **35.** Convergent
37. Approximately $GM/(R_E)$ J

Chapter Review Exercises

1. $4000/\sqrt{3}$ cubic meters **3.** $\dfrac{4}{3}\pi a^3 \sin^2\left(\dfrac{1}{2}\theta\right)$ cubic meters
5. $8\pi/3$ cubic units **7.** $(1 - \pi/4)\pi$ cubic units
9. $\displaystyle\int_2^3 \sqrt{\frac{2x^2 - 1}{x^2 - 1}}\,dx$ **11.** $2a\sinh(b/a)$ units
13. 1 square unit **15.** 1.61729 square units
17. 1.44281 square units

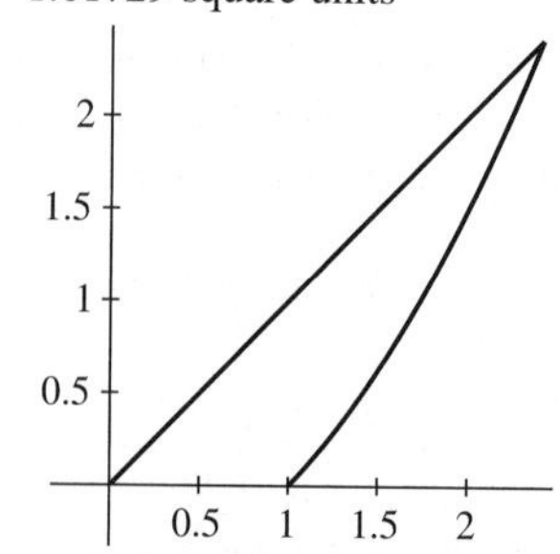

19. 4.38×10^{10} J **21.** 239/38 m from the 0-m mark
23. Mass of rod is 6 kg; center of mass at 1.02222 m.

25. $2\pi \int_0^{100} \frac{7r}{1+r^2}\,dr = 7\pi \ln(10{,}001)$ tons

27. Mass is $\frac{1}{2}a^2\delta\theta$; center of mass is $\frac{2a}{3\theta}\langle \sin\theta, 1 - \cos\theta\rangle$.

29.

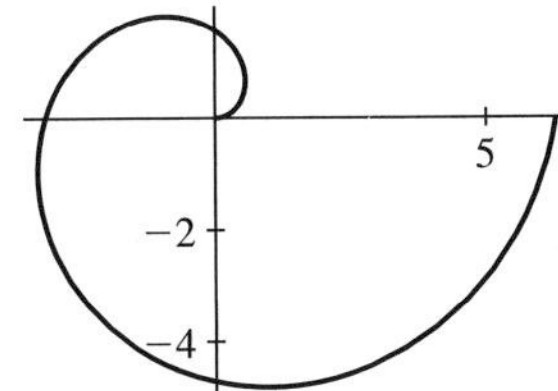

31. Perhaps by drawing a horizontal row of squares, with the sides of these squares $1, 1\sqrt{2^1}, 1\sqrt{2^2}, 1/\sqrt{2^3}, \ldots$, with the idea that these squares would have infinite extent. Now all you have to do is to convince her or him that $1 + 1/2^1 + 1/2^2 + \cdots = 2$.

33. The first integral is divergent; the second converges to 2/3.

35. $2/a^2$

CHAPTER 7—INFINITE SERIES, SEQUENCES, AND APPROXIMATIONS

7.1 Taylor Polynomials

1. $T_0(x;0) = 1$
$T_1(x;0) = 1 - x$
$T_2(x;0) = 1 - x + x^2$
$T_3(x;0) = 1 - x + x^2 - x^3$
$f(0.1) = 1/1.1 \approx 0.909091$, $T_3(0.1;0) = 0.909$

3. $T_0(x;0) = 0$
$T_1(x;0) = T_2(x;0) = x$
$T_3(x;0) = x - x^3/6$
$f(0.1) = \sin(0.1) \approx 0.0998334$, $T_3(0.1;0) \approx 0.0998333$

5. $T_0(x;0) = 0$
$T_1(x;0) = T_2(x;0) = x$
$T_3(x;0) = x - x^3/3$
$f(0.1) = \arctan(0.1) \approx 0.0996687$, $T_3(0.1;0) \approx 0.099667$

7. $T_0(x;-2) = -1$
$T_1(x;-2) = -1 - (x+2)$
$T_2(x;-2) = -1 - (x+2) - (x+2)^2$
$T_3(x;-2) = -1 - (x+2) - (x+2)^2 - (x+2)^3$
$f(-1.95) = 1/(-0.95) \approx -1.05263$, $T_3(-1.95;-2) = -1.052625$

9. $T_0(x;3) = 9\ln 3$
$T_1(x;3) = 9\ln 3 + (3 + 6\ln 3)(x-3)$
$T_2(x;3) = 9\ln 3 + (3 + 6\ln 3)(x-3) + (3/2 + \ln 3)(x-3)^2$
$T_3(x;3) = 9\ln 3 + (3 + 6\ln 3)(x-3) + (3/2 + \ln 3)(x-3)^2 + (x-3)^3/9$
$f(2.95) = 2.95^2 \ln 2.95 \approx 9.414409$, $T_3(2.95;3) \approx 9.414410$

11. $T_0(x;1) = \pi/4$
$T_1(x;1) = \pi/4 + (x-1)/2$
$T_2(x;1) = \pi/4 + (x-1)/2 - (x-1)^2/4$
$T_3(x;1) = \pi/4 + (x-1)/2 - (x-1)^2/4 + (x-1)^3/12$
$f(1.05) = \arctan(1.05) \approx 0.809784$, $T_3(1.05;1) \approx 0.809784$

13. the top graph

17. $-0.087 < x < 0.091$ **19.** $\frac{8!}{3}$

21. $k!\frac{k^k}{(3(k+1))^{k+1}}$ **23.** infinitely many

25. a. $T_6(x;0) = x - x^3/6 + x^5/120$

b.

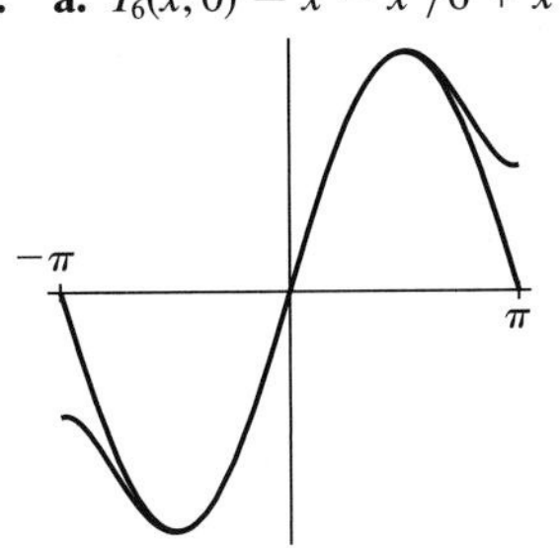

c.

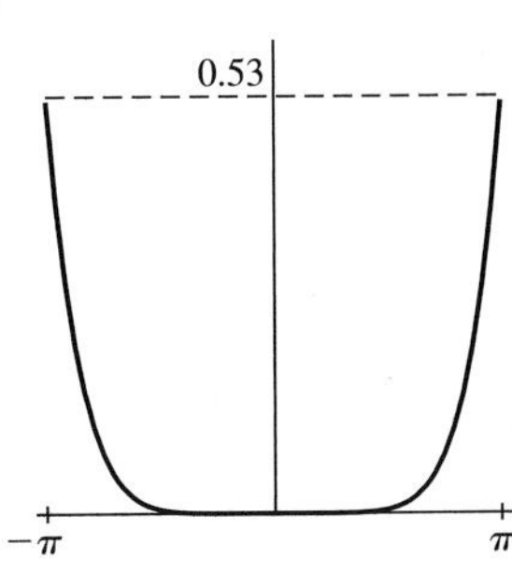

d. ≈ 0.53 **e.** $\sin(-0.2) \approx -0.198669$, $T_6(-0.2;0) = -0.198669$, $\sin 0 = 0$, $T_6(0;0) = 0$, $\sin(0.5) \approx 0.479426$, $T_6(0.5;0) \approx 0.479427$, $\sin 1 \approx 0.841471$, $T_6(1;0) \approx 0.841667$

27. a. $T_6(x;0) = 1 + x + x^2 + x^3 + x^4 + x^5 + x^6$

b.

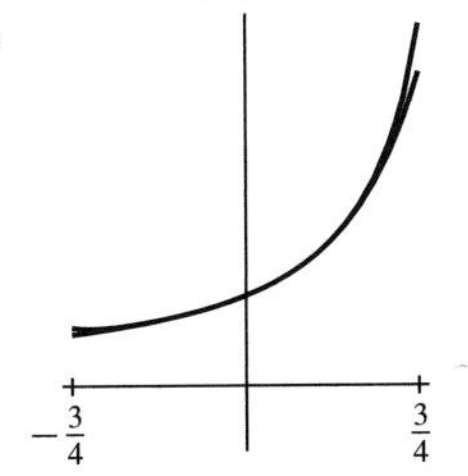

c.

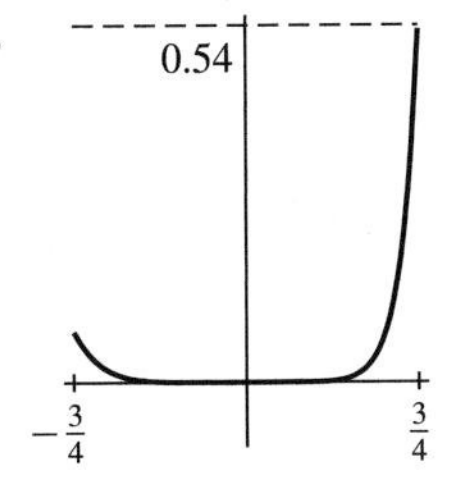

d. ≈ 0.54

e. $1/(1-(-0.9)) \approx 0.526316$, $T_6(-0.9;0) = 0.778051$,
$1/(1-(-0.05)) \approx 0.952381$, $T_6(-0.05;0) \approx 0.952381$,
$1/(1-(0.5)) = 2$, $T_6(0.5;0) = 1.984375$,
$1/(1-(0.7)) \approx 3.333333$, $T_6(0.7;0) = 3.058819$

29. $\sum_{k=0}^{n} x^k$ **31.** $\sum_{k=0}^{n} (-1)^k \left(\frac{x+1}{2}\right)^k$

33. does not exist

35. $T_0(x;c) = \sqrt{c}$
For $n \geq 1$,
$$\sqrt{c} + \sum_{k=1}^{n} \frac{1}{2}\left(\frac{1}{2} - 1\right)\left(\frac{1}{2} - 2\right)\cdots \left(\frac{1}{2} - (k-1)\right)\frac{c^{1/2-k}}{k!}(x-c)^k.$$

43. The largest n is 4.

45. $T_2(x;c)$ exists for all c.
$T_5(x;c)$ exists for $c \neq 0$.

49. $T_0(x;1) = 1$
$T_1(x;1) = 1 + (x-1)$
$T_2(x;1) = 1 + (x-1) + (x-1)^2$

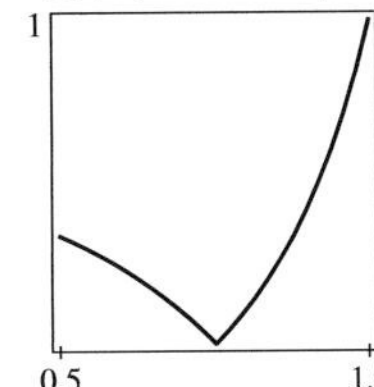

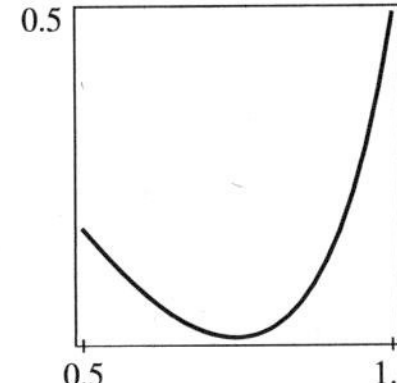

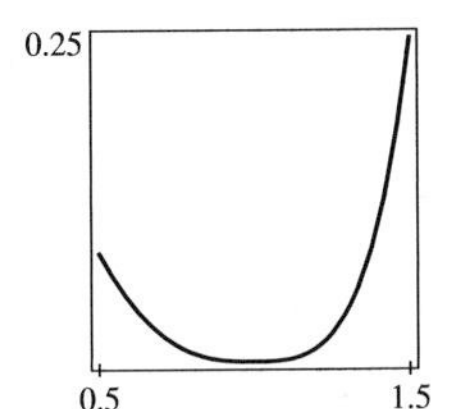

$E_0 = 1$, $E_1 = 0.5$, $E_2 = 0.25$

51. $T_0\left(x; \frac{1}{2}\right) = \frac{\pi}{6}$

$T_1\left(x; \frac{1}{2}\right) = \frac{\pi}{6} + \frac{2}{\sqrt{3}}\left(x - \frac{1}{2}\right)$

$T_2\left(x; \frac{1}{2}\right) = \frac{\pi}{6} + \frac{2}{\sqrt{3}}\left(x - \frac{1}{2}\right) + \frac{2}{3\sqrt{3}}\left(x - \frac{1}{2}\right)^2$

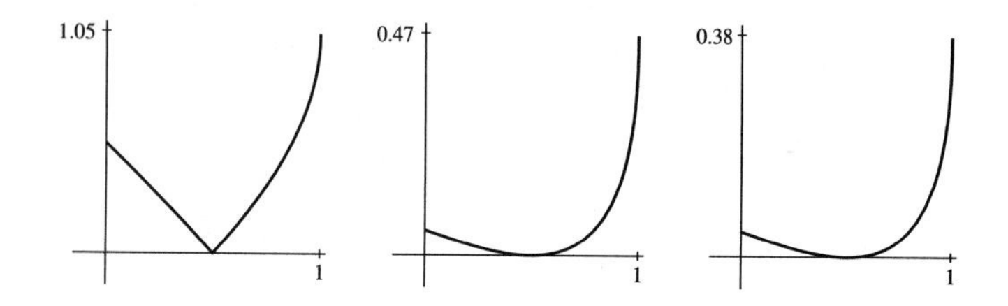

$E_0 \approx 1.05$, $E_1 \approx 0.47$, $E_2 \approx 0.38$

7.2 Approximations and Error

1. $T_2(x; 0) = 1 - 2x + 2x^2$,
$|R_2(x; 0)| \leq 8e^{0.6}(0.3)^3/3! \approx 0.0655963$

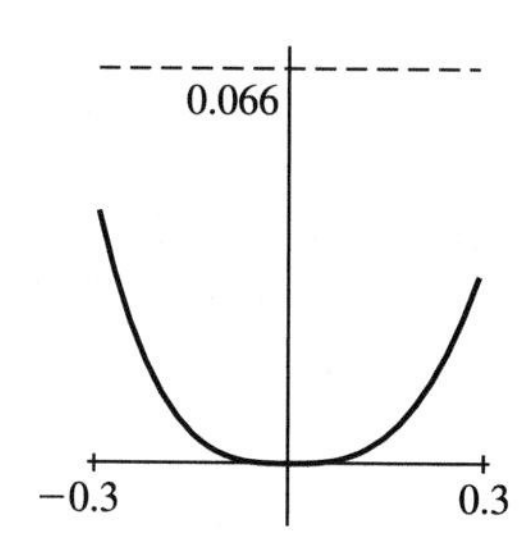

3. $T_2(x; 0) = x$,
$|R_2(x; 0)| \leq 1(\pi/6)^3/3! \approx 0.024$

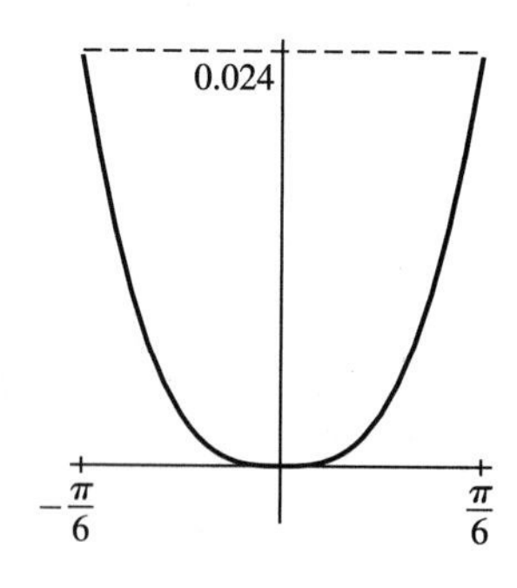

5. $T_2(x; 0) = 1 - x + x^2$,
$|R_2(x; 0)| \leq 30.375(1/3)^3/3! = 0.1875$

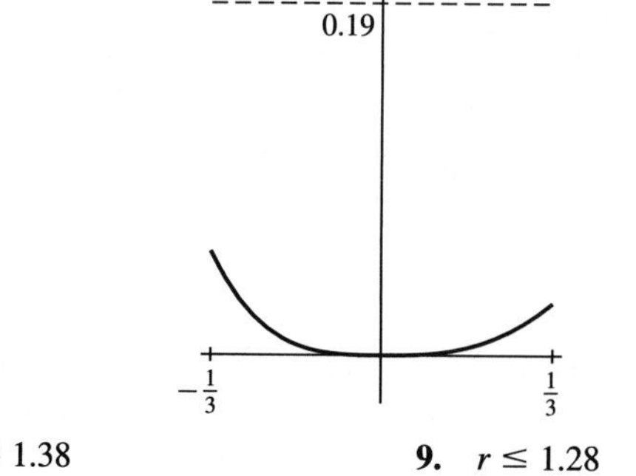

7. $r \leq 1.38$ **9.** $r \leq 1.28$

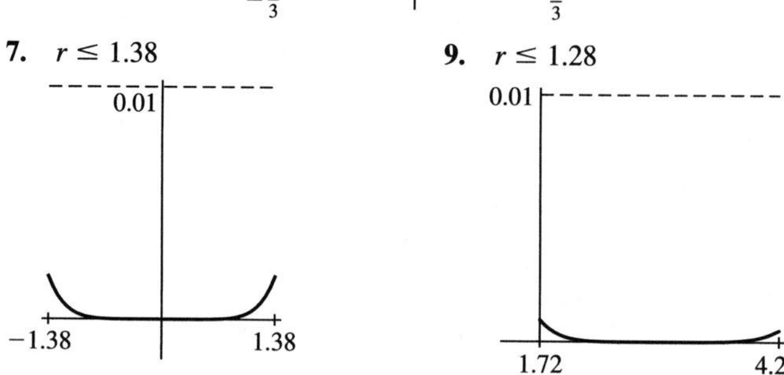

11. $f(x) = T_5(x; \sqrt{2})$ for all real x. Hence the error in the estimate $f(x) \approx T_5(x; \sqrt{2})$ is 0. r can be any positive number. The graph of $y = |R_5(x; \sqrt{2})| = 0$ is the x-axis.

15. $\sqrt{105} \approx 10.2469507659595983832$
$T_7(105; 100) \approx 10.24695076596450805 66$
$T_7(105; 100) - \sqrt{105} \approx 4.9096734 \times 10^{-12}$

17. $n \geq 7$ **19.** $n \geq 2$ **21.** $n \geq 6$

33. no

7.3 Sequences

1. 2, 4, 8, 16, 32 **3.** 3, 3, 3, 3, 3

5. $\sqrt{2}/2, 0, -\sqrt{2}/2, -1, -\sqrt{2}/2$

7. 1, 3/4, 85/108, 5413/6912, 16,922,537/21,600,000

9. 3.1415, 3.14159, 3.141592
10th term = 3.141592653
nth term = π to n digits

11. 3125, 46,656, 823,543
10th term = 10^{10}
nth term = n^n

13. $1 + x + x^2 + x^3 + x^4$, $1 + x + x^2 + x^3 + x^4 + x^5$, $1 + x + x^2 + x^3 + x^4 + x^5 + x^6$
10th term = $1 + x + x^2 + \cdots + x^9 = \Sigma_{k=0}^{9} x^k$
nth term = $1 + x + x^2 + \cdots + x^{n-1} = \Sigma_{k=0}^{n-1} x^k$

15. 1, 1, 2, 6, 24 **17.** 1, 3/4, 3/5, 1/2, 3/7

19. 1, 2, 3/2, 7/4, 13/8

21. $s_1 = 1$

$s_n = s_{n-1} + \frac{1}{n}, n \geq 2$

23. $s_1 = 2$
$s_n = 2^{s_{n-1}}, n \geq 2$
25. converges, limit is 1 **27.** diverges
29. converges, limit is $\pi/2$ **31.** diverges **33.** diverges
35. converges, limit is e^2
37. The lower of the two graphs (look very carefully) is the graph of $y = \ln x$.
39. The limit is $1/e$. **45.** Converges for $1/e^e < c \leq e^{1/e}$
47. The limit is π.

7.4 Infinite Series

1. terms: 1, 2, 3, 4, 5
partial sums: 1, 3, 6, 10, 15
3. terms: $1, x, 3x^2/4, x^3/2, 5x^4/16$
partial sums: $1, 1 + x, 1 + x + 3x^2/4,$
$1 + x + 3x^2/4 + x^3/2, 1 + x + 3x^2/4 + x^3/2 + 5x^4/16$
5. $s_n = \pi n, \lim_{n\to\infty} s_n = \infty$
7. $s_n = \sqrt{n}, \lim_{n\to\infty} s_n = \infty$
9. $s_n = \dfrac{1}{2} - \dfrac{2}{n+4}, \lim_{n\to\infty} s_n = \dfrac{1}{2}$
11. $s_n = (1 - e^{-2n})/(1 - e^{-2}), \lim_{n\to\infty} s_n = e^2/(e^2 - 1)$
13. $s_n = 16(1 - 0.8^n), \lim_{n\to\infty} s_n = 16$ **15.** converges
17. diverges **19.** diverges **21.** converges
23. 5/33 **25.** 811/990 **27.** 9/20 **29.** 4/5
31. no solution
33. $a_1 = 5/4, a_2 = -1/20, a_3 = -1/30, a_k = \dfrac{-1}{(k+2)(k+3)}$
35. $a_1 = \sin 1, a_2 = \sin 2 - \sin 1, a_3 = \sin 3 - \sin 2$
$a_k = \sin k - \sin(k-1)$
37. $2 - \sqrt{10} < x < 2 - 2\sqrt{2}$ and $2 + 2\sqrt{2} < x < 2 + \sqrt{10}$
39. $-\dfrac{\pi}{6} + \pi\ell \leq x \leq \dfrac{\pi}{6} + \pi\ell, \ell = 0, \pm 1, \pm 2, \ldots$
41. $2h$ **43.** $\approx$34.44 ft $\approx$5.28 s
45. $\approx$26.9 billion dollars **47.** **c.** The series converges.
53. No, the statement is not always true.

7.5 Tests for Convergence

1. diverges **3.** diverges **5.** diverges **7.** converges
9. converges **11.** converges **13.** converges
15. converges **17.** conditionally convergent
19. neither conditionally convergent nor absolutely convergent
21. absolutely convergent
23. 99 terms, $\sum_{r=1}^{99} (-1)^{r-1} \dfrac{1}{\sqrt{r^2+1}} \approx 0.446$
25. 100 terms, $\sum_{k=1}^{100} (-1)^{k-1} \dfrac{1}{2k-1} \approx 0.783$
27. 10 terms, $\sum_{j=0}^{9} (-1)^j \dfrac{1}{j!} \approx 0.3678792$ **29.** $x = \pm 3$
31. all real x **41.** $n \geq 10^{500}$ **43.** $n \geq e^{(10^{500})}$
45. $n > 36, n > 251{,}000$ **51.** yes

7.6 Power Series and Taylor Series

1. 1 **3.** ∞ **5.** $2e$ **7.** $\sum_{k=0}^{\infty} \dfrac{x^k}{k!}, R = \infty$
9. $\sum_{k=0}^{\infty} \dfrac{2 \cdot 3^k x^k}{k!}, R = \infty$
11. $50 - 47(x+2) + 16(x+2)^2 - 2(x+2)^3, R = \infty$
13. $\sum_{k=0}^{\infty} (-1)^k \left(\dfrac{1}{8^{k+1}} - \dfrac{1}{3^{k+1}}\right)(x-5)^k, R = 3$
15. $1 + \dfrac{x}{2} + \sum_{k=2}^{\infty} (-1)^{k-1} \dfrac{1 \cdot 3 \cdots (2k-3)}{2^k k!} x^k, -1 < x \leq 1$
17. $n \geq 10$ **19.** $n \geq 10$ **21.** $n \geq 7$
23. For $x = 1$, diverges
For $x = -1$, converges
27. $\sum_{k=0}^{\infty} \dfrac{(x-5)^k}{2^k}$; other answers are possible.
29. $1 - c$ **31.** ∞ **33.** $|x - 2| > 3$
35. all real x

7.7 Working with Power Series

1. $\sum_{k=0}^{\infty} (-1)^k \dfrac{3^k x^k}{k!}, -\infty < x < \infty$
3. $\sum_{k=0}^{\infty} (-1)^k \left(\dfrac{2x^{2k}}{(2k)!} - \dfrac{3x^{2k+1}}{(2k+1)!}\right), -\infty < x < \infty$
5. $1 + \sum_{k=1}^{\infty} (-1)^k \dfrac{2^{2k-1} x^{2k}}{(2k)!}, -\infty < x < \infty$
7. $\dfrac{\sqrt{2}}{2} \sum_{k=0}^{\infty} (-1)^k \left(\dfrac{(x - \pi/4)^{2k}}{(2k)!} - \dfrac{(x - \pi/4)^{2k+1}}{(2k+1)!}\right), -\infty < x < \infty$
9. $1 + (x+1) + \sum_{k=2}^{\infty} (-1)^{k-1} \dfrac{1 \cdot 3 \cdot 5 \cdots (2k-3)}{k!} (x+1)^k,$
$-\dfrac{3}{2} < x \leq -\dfrac{1}{2}$
11. $\ln 2 + \sum_{k=1}^{\infty} (-1)^{k-1} \dfrac{x^k}{k \cdot 2^k}, -2 < x \leq 2$
13. $1 + 2x + 5x^2/2 + 8x^3/3 + \cdots$
15. $x + x^3/6 + 3x^5/40 + 5x^7/112 + \cdots$
19. 0.242 **21.** 0.8274 **23.** $-e^{-(x+2)^2}$
25. $-\ln\left(2 - \dfrac{x}{2}\right)$ **29.** $x + \dfrac{1}{3}x^3 + \dfrac{2}{15}x^5 + \dfrac{17}{315}x^7 + \cdots$
33. a_m **35.** 0 **37.** 1/2 **39.** 1/6

Chapter Review Exercises

1. 0 **3.** does not exist **5.** does not exist **7.** $1/2\sqrt{2}$
9. e **11.** converges **13.** converges **15.** converges
17. converges **19.** 1 **21.** 3/2
23. $\sum_{k=0}^{\infty} (-1)^k \dfrac{2^{2k} x^{2k}}{(2k)!}$, radius of convergence $= \infty$
25. $\sum_{k=0}^{\infty} (-1)^k \left(\dfrac{3}{2^{k+1}} - \dfrac{2}{5^{k+1}}\right)(x-1)^k$, radius of convergence $= 2$

27. $\displaystyle\sum_{k=0}^{\infty}(-1)^k\frac{x^{4k+3}}{(4k+3)(2k+1)!}$, radius of convergence $= \infty$

29. 10 **31.** $e-1$ **33.** $\ln(3/2)$ **35.** $n \geq 17{,}252$

37. $n \geq 15$ **39.** $n \geq 10^{18}-1$ **41.** $n \geq 6$ **43.** $n \geq 9$

47. $\displaystyle\sum_{k=0}^{\infty}\frac{x^{2k+1}}{(2k+1)!}$

CHAPTER 8—VECTORS AND LINEAR FUNCTIONS

8.1 Vectors in Three Dimensions

1.

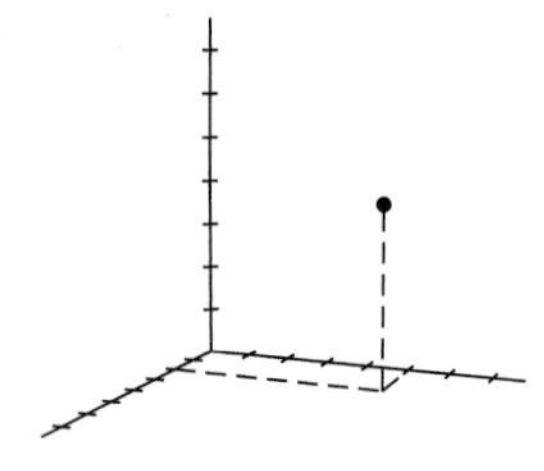

3.

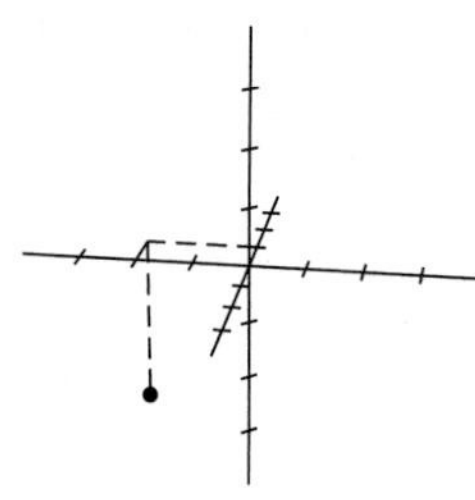

5.

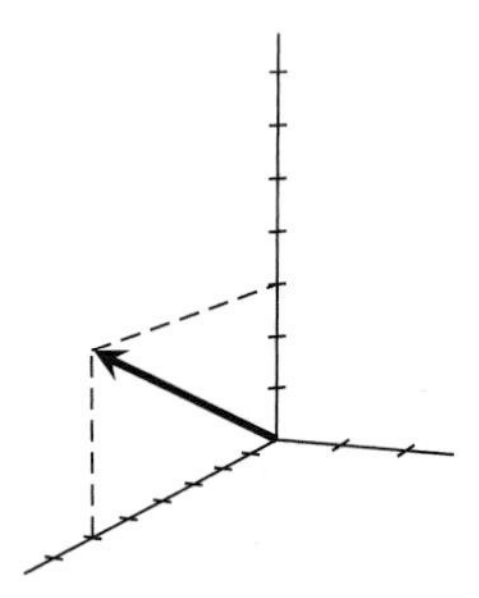

7.

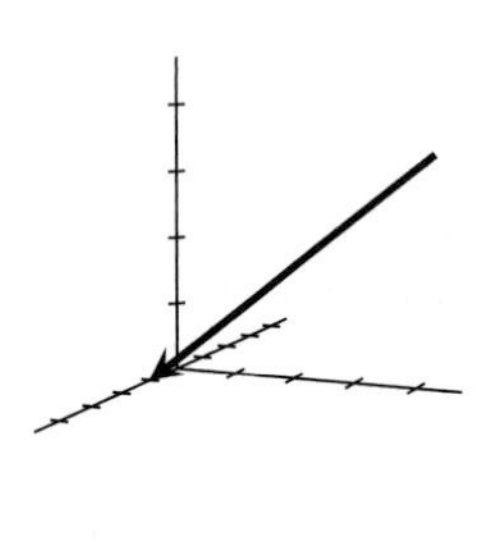

9.

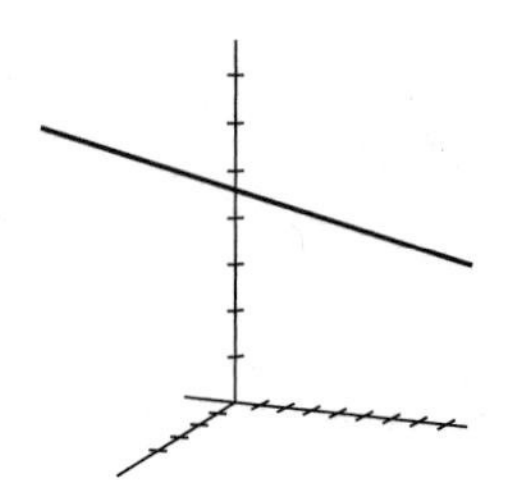

11. $\sqrt{34}$ **13.** $\langle 12, -1, -5\rangle$ **15.** 7

17. $\left\langle -5/\sqrt{34}, 0, 3/\sqrt{34}\right\rangle$

19. $\arccos\left(7/\sqrt{174}\right) \approx 1.011 \approx 57.95°$

21. $\langle -56/29, -42/29, -28/29\rangle$ **23.** $\sqrt{293}/2$

25. $7\mathbf{i} - 18\mathbf{j} + 3\mathbf{k}$ **27.** 2

29. $-\dfrac{6}{\sqrt{293}}\mathbf{i} + \dfrac{16}{\sqrt{293}}\mathbf{j} - \dfrac{1}{\sqrt{293}}\mathbf{k}$

31. $\arccos(2/3) \approx 0.841 \approx 48.19°$ **33.** $-\dfrac{1}{2}\mathbf{k}$

35. $2100\mathbf{i} - 500\mathbf{j} - 1500\mathbf{k}$ **37.** $20g/\sqrt{3}$ N

39. $\mathbf{r}(t) = \langle -3-2t, 5+4t, 5t\rangle$

$$\frac{x+3}{-2} = \frac{y-5}{4} = \frac{z}{5}$$

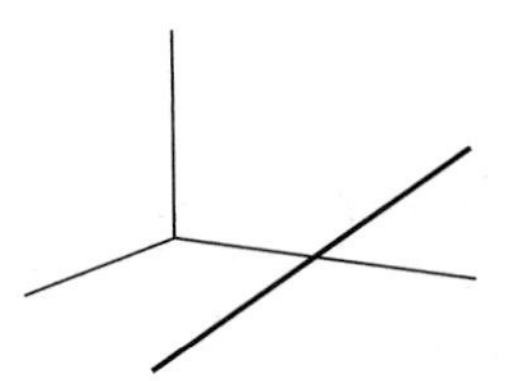

41. $\mathbf{r}(t) = \langle 101+101t, 201+201t, 301+301t\rangle$

$$\frac{x-101}{101} = \frac{y-201}{201} = \frac{z-301}{301}$$

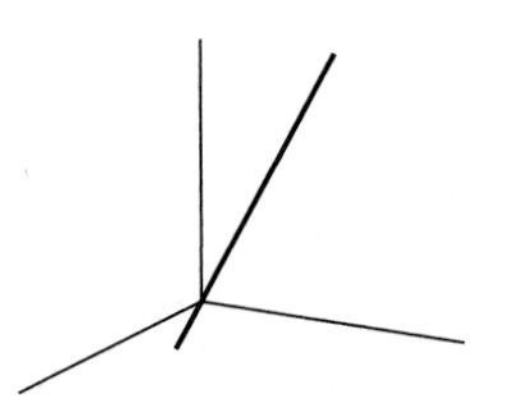

43. $\mathbf{r}(t) = \langle 1-4t, -1+6t, 2+3t\rangle$

$$\frac{x-1}{-4} = \frac{y+1}{6} = \frac{z-2}{3}$$

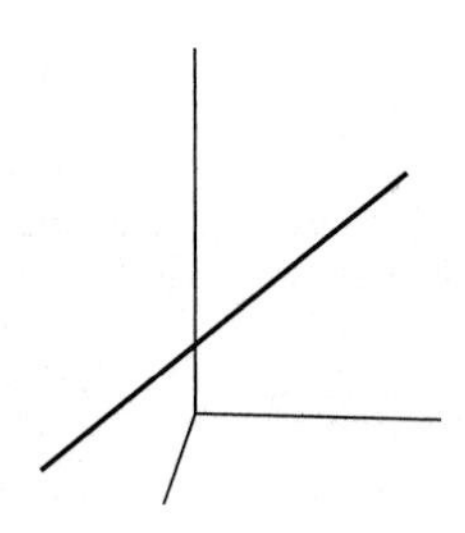

45. A is not on the line. B is on the line.

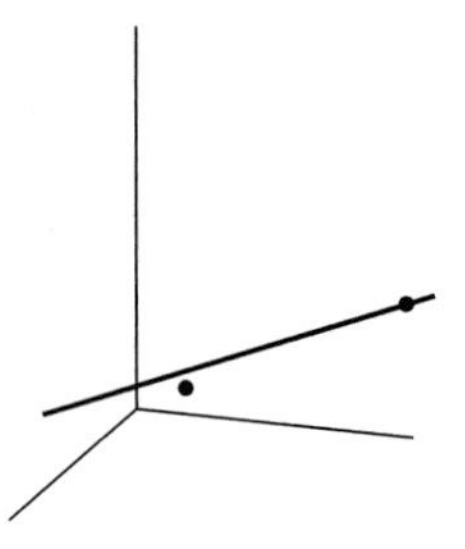

49. the same line
51. $\mathbf{r}(t) = \langle 4 - 2t, 5t, 7\rangle$
$\mathbf{r}(t) = \langle 4t, 10 - 10t, 7\rangle$, other answers are possible
53. no **55.** yes **57.** yes **59.** no
63. $\mathbf{r}(t) = (1 - t)\langle 0, 4, -4\rangle + t\langle 7, 9, -5\rangle, 0 \leq t \leq 1$

8.2 Matrices and Determinants

1. $\begin{pmatrix} 6 & -9 \\ -12 & -3 \\ 0 & -24 \end{pmatrix}$ **3.** undefined **5.** (7)
7. undefined **9.** undefined **11.** undefined
13. **a.** 3 **b.** 3 **c.** -2 **d.** undefined **e.** undefined
19. $ad - bc$ **21.** undefined **23.** 0 **29.** $x = 2, y = -1$
31. $x = 4, y = 0, z = -3$

8.3 The Cross Product

1. $\langle -3, -11, -5\rangle$ **3.** $\langle -14, 14, 7\rangle$ **5.** 0
7. $\langle 12/\sqrt{545}, 1/\sqrt{545}, 20/\sqrt{545}\rangle$ **9.** $\langle -2/\sqrt{5}, 0, 4/\sqrt{5}\rangle$
11. $-15\mathbf{i} - \frac{11}{2}\mathbf{j} + 2\mathbf{k}$ **13.** $-2\mathbf{j}$ **15.** 0
17. $\frac{3}{\sqrt{14}}\mathbf{i} + \frac{1}{\sqrt{14}}\mathbf{j} - \frac{2}{\sqrt{14}}\mathbf{k}$ **19.** $-\frac{4}{\sqrt{5}}\mathbf{i} - \frac{2}{\sqrt{5}}\mathbf{j}$
21. 14

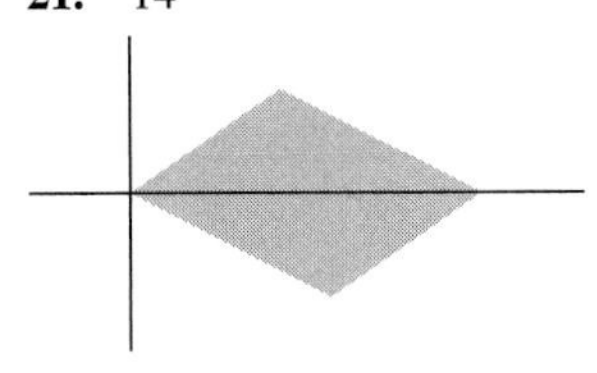

23. $2\sqrt{6}$

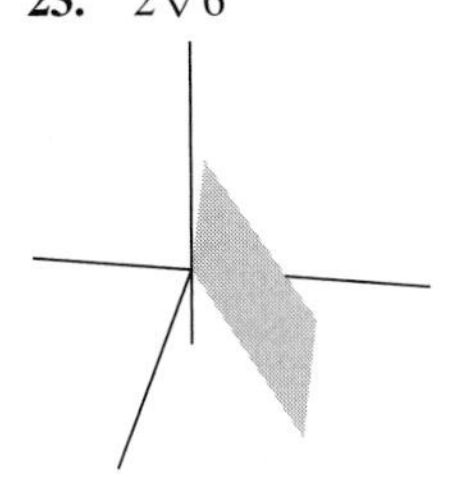

25. $\sqrt{381}$

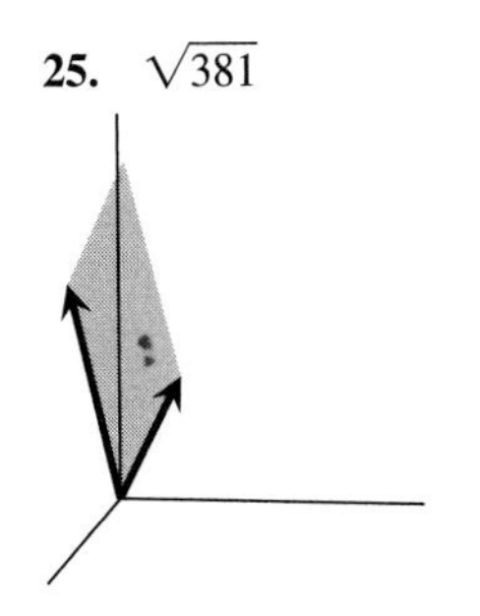

27. 150

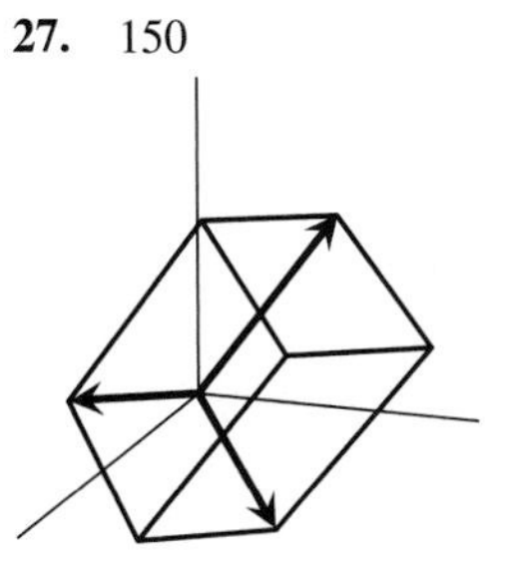

29. $\mathbf{r}(t) = \langle 4, -3, 1\rangle + t\langle 1, -1, 1\rangle$

31.

33. **a.** 20 joules **b.** 10 joules **c.** $10\sqrt{2}$ joules
35. $5\sqrt{2}$ joules
37. distance $= \sqrt{170}/3$
closest point: $(-38/9, 17/9, 29/9)$
39. distance $= \dfrac{\sqrt{(a_2b_1 - a_1b_2)^2 + (a_1b_3 - a_3b_1)^2 + (a_3b_2 - a_2b_3)^2}}{\sqrt{b_1^2 + b_2^2 + b_3^2}}$
closest point: $\left(a_1 - \dfrac{b_1(a_1b_1 + a_2b_2 + a_3b_3)}{b_1^2 + b_2^2 + b_3^2},\ a_2 - \dfrac{b_2(a_1b_1 + a_2b_2 + a_3b_3)}{b_1^2 + b_2^2 + b_3^2}, a_3 - \dfrac{b_3(a_1b_1 + a_2b_2 + a_3b_3)}{b_1^2 + b_2^2 + b_3^2}\right)$
41. distance $= \sqrt{209/13}$
closest point: $(-42/13, 5, 11/13)$
43. **a.** $d(t) = \sqrt{(2 - 3t)^2 + 1^2 + (t + 2)^2}$ **b.** $\sqrt{37/5}$
c. $(-6/5, 5, 17/5)$
45. $\mathbf{v} \times \mathbf{v} = \mathbf{0}$ **47.** yes, infinitely many such vectors
49. $(3, 4, 0)$, $(-9, 4, 0)$, or $(-1, 0, 2)$
51. $(3, 3, 5)$, $(-2, 8, 5)$, $(-2, 8, 10)$, $(-2, 3, 10)$, $(3, 3, 10)$
53. Base can be taken to be the parallelogram determined by **a** and **b** and corresponding height $\|\mathbf{a} \times \mathbf{b}\|$. Volume is $\|\mathbf{a} \times \mathbf{b}\|^2$.
55. The parallelogram is degenerate; all sides lie in the same plane. The volume is 0.
61. $\mathbf{r}(t) = \langle 2, -1, 1\rangle + t\langle 15, 3, -1\rangle$

8.4 Linear Functions

1. $\begin{pmatrix} -7 \\ -10 \end{pmatrix}, \begin{pmatrix} 1 \\ 4 \end{pmatrix}$ **3.** $(-5/2), (1001)$ **5.** $\begin{pmatrix} -4 \\ -48 \end{pmatrix}, \begin{pmatrix} 4 \\ 9 \end{pmatrix}$
7. $L\begin{pmatrix} x \\ y \end{pmatrix} = \begin{pmatrix} -3 & 1 \\ 4 & 7 \end{pmatrix}\begin{pmatrix} x \\ y \end{pmatrix}$, 2×2 matrix, domain R^2, range in R^2
9. $L\begin{pmatrix} x \\ y \\ z \end{pmatrix} = \begin{pmatrix} 1 & 0 & 0 \\ 0 & 1 & 0 \\ 0 & 0 & 1 \end{pmatrix}\begin{pmatrix} x \\ y \\ z \end{pmatrix}$, 3×3 matrix, domain R^3, range in R^3
11. $L\begin{pmatrix} x \\ y \\ z \end{pmatrix} = (-2 \quad 3 \quad 4)\begin{pmatrix} x \\ y \\ z \end{pmatrix}$, 1×3 matrix, domain R^3, range in R
13. $L\begin{pmatrix} x \\ y \end{pmatrix} = \begin{pmatrix} -2x + 3y \\ 6x \end{pmatrix}$
domain: R^2, range in R^2
15. $L\begin{pmatrix} x \\ y \\ z \end{pmatrix} = (ax + by + cz)$
domain: R^3, range in R^1

17. $L(x) = (14.92x)$
domain: R^1, range in R^1

19. $M \circ L: \begin{pmatrix} 2 & -1 \\ 10 & 9 \end{pmatrix}$, $L \circ M: \begin{pmatrix} 3 & 4 \\ -1 & 8 \end{pmatrix}$

21. $M \circ L: \begin{pmatrix} 1 & 0 \\ 0 & 1 \end{pmatrix}$, $L \circ M: \begin{pmatrix} 1 & 0 \\ 0 & 1 \end{pmatrix}$

For any vector **v** in R^2, we have $L \circ M(\mathbf{v}) = M \circ L(\mathbf{v}) = \mathbf{v}$.

27. $\begin{pmatrix} x_1 \\ y_1 \end{pmatrix} = \begin{pmatrix} 179,169,000 \\ 71,483,000 \end{pmatrix}$

To find the 1992 populations, multiply $\begin{pmatrix} x_1 \\ y_1 \end{pmatrix}$ by $\begin{pmatrix} 0.99 & 0.055 \\ 0.023 & 0.94 \end{pmatrix}$.

29. $\begin{pmatrix} 6/17 \\ 4/17 \end{pmatrix}$

31. $\begin{pmatrix} 1 \\ 1 \\ 1 \end{pmatrix}, \begin{pmatrix} -1 \\ 6 \\ -2 \end{pmatrix}, \begin{pmatrix} 3 \\ -4 \\ 4 \end{pmatrix}$; other answers are possible

35. $\begin{pmatrix} 9k \\ 6k \\ 7k \end{pmatrix}$, where k is any real number

37. $\begin{pmatrix} x \\ y \\ z \end{pmatrix} = \begin{pmatrix} -6k \\ 8k \\ 31k \end{pmatrix}$, where k is any real number

8.5 The Geometry of Linear Functions

1. line segment

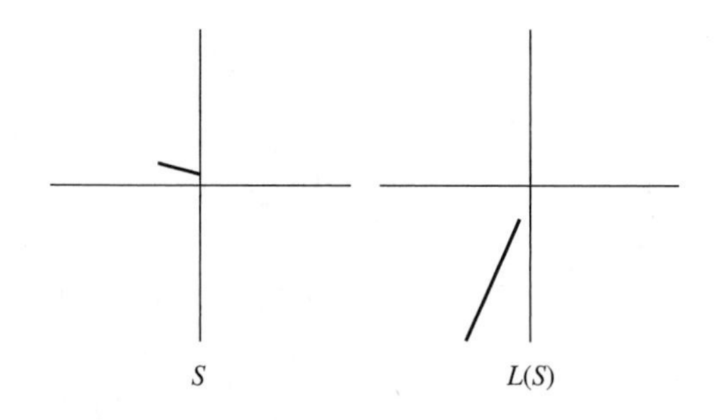

3. line segment

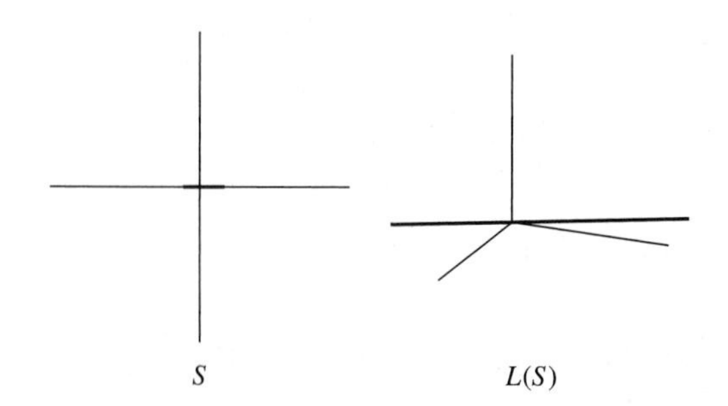

5. parallelogram

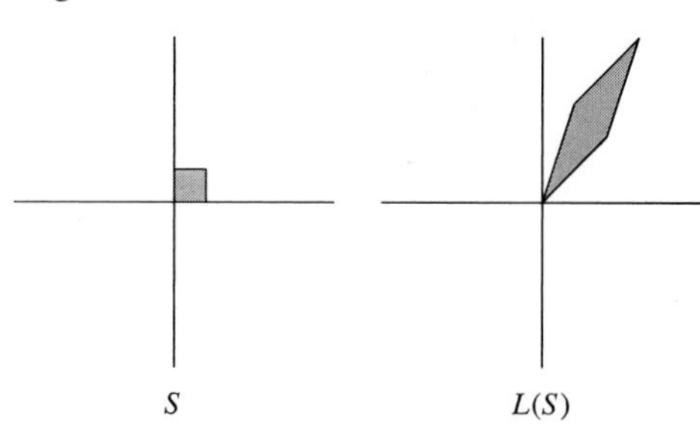

7. parallelogram

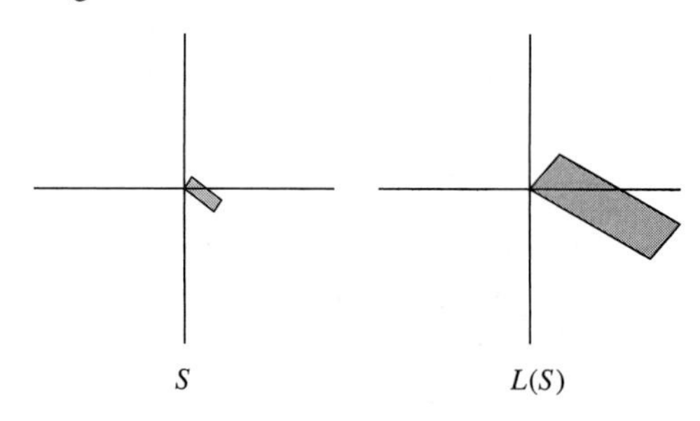

9. line segment

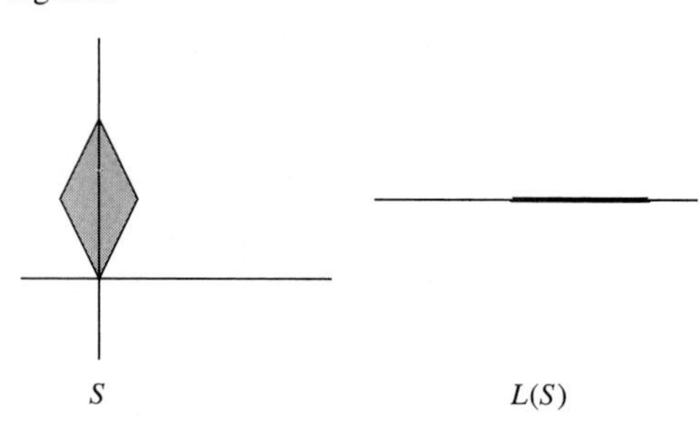

11. line segment

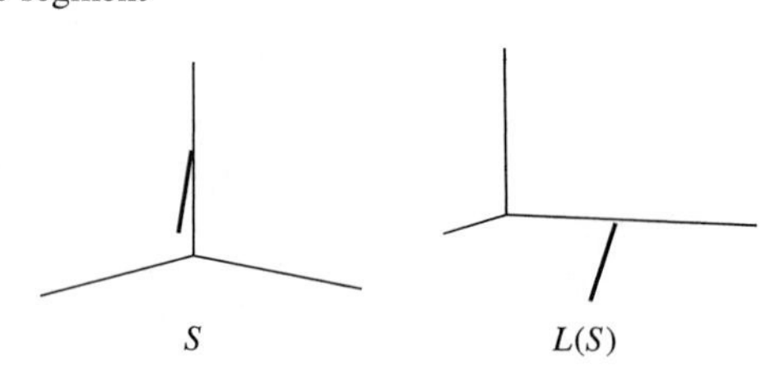

13. parallelepiped

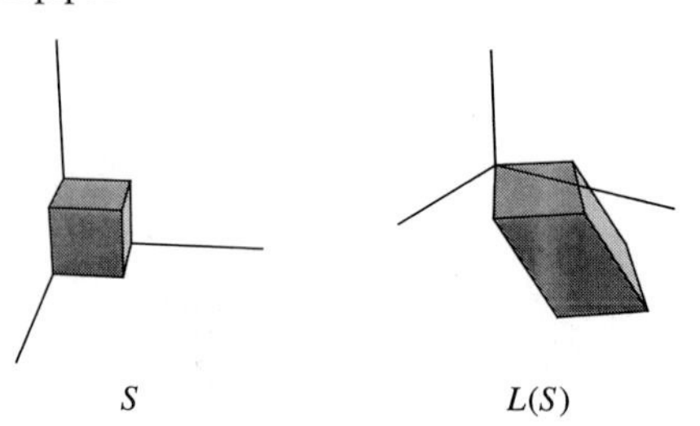

15. hexagon

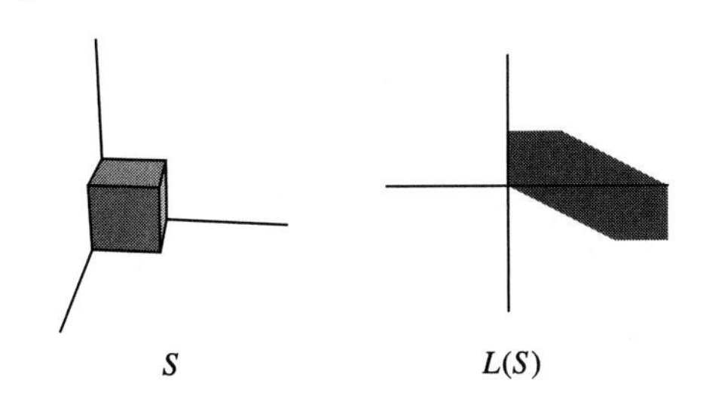

17. 300

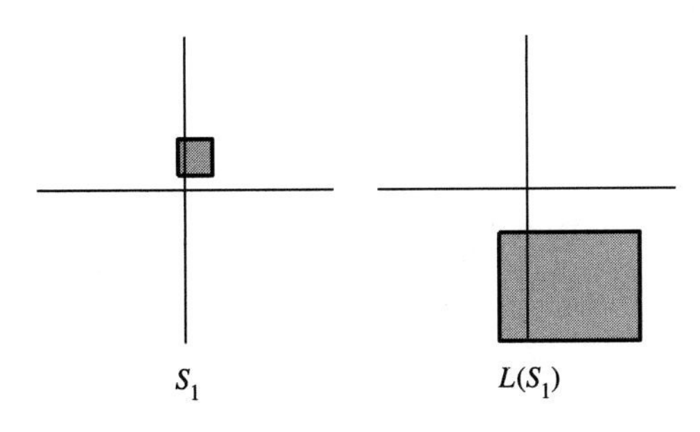

19. 176

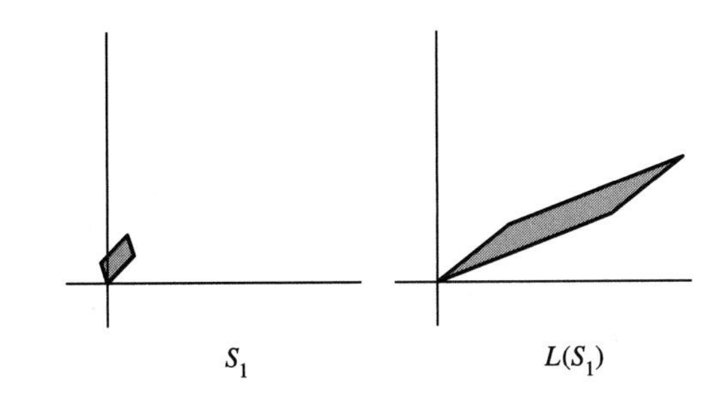

21. 324

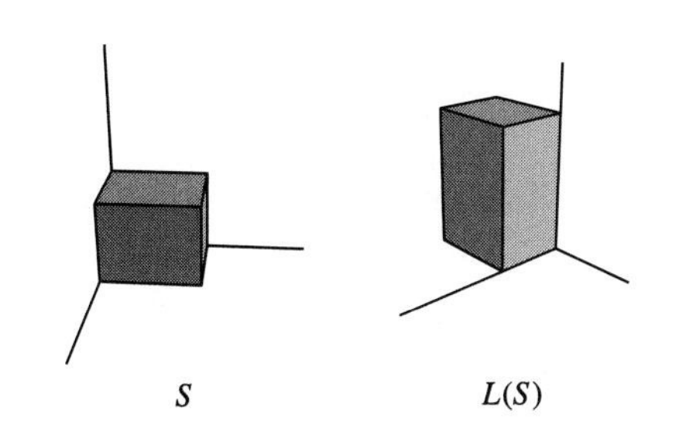

23. 1600

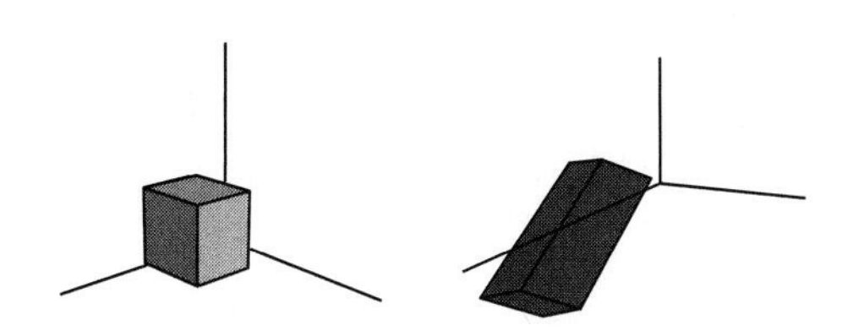

25. $T(\mathbf{v}) = A\mathbf{v} + \mathbf{b}$, where $A = \begin{pmatrix} -1 & 3 \\ 4 & 0 \end{pmatrix}$, $\mathbf{b} = \begin{pmatrix} -4 \\ 1 \end{pmatrix}$

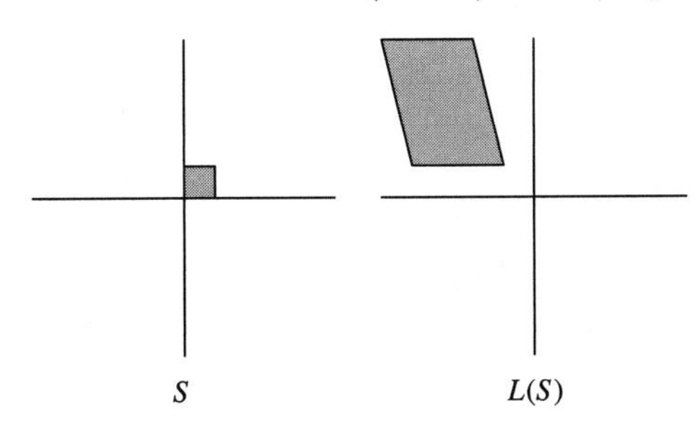

27. $T(\mathbf{v}) = A\mathbf{v} + \mathbf{b}$, where $A = \begin{pmatrix} 1 & 3 \\ 2 & -1 \end{pmatrix}$, $\mathbf{b} = \begin{pmatrix} -8 \\ 0 \end{pmatrix}$

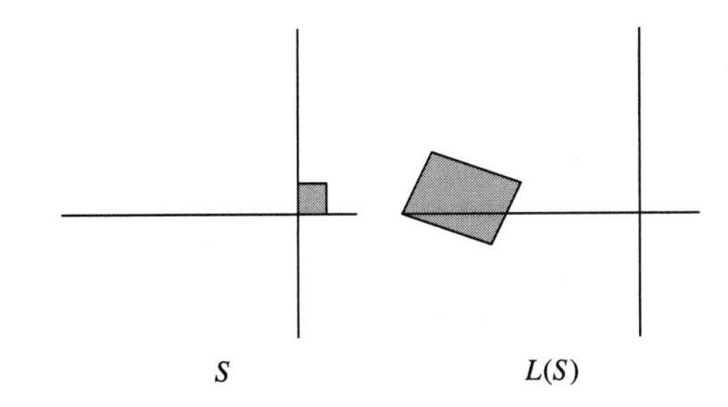

29. **a.** $L\begin{pmatrix} x \\ y \end{pmatrix} = \begin{pmatrix} -2 & 6 \\ 5 & -1 \end{pmatrix}\begin{pmatrix} x \\ y \end{pmatrix}$ **b.** 2

c. $K\begin{pmatrix} x \\ y \end{pmatrix} = \begin{pmatrix} 1/28 & 3/14 \\ 5/28 & 1/14 \end{pmatrix}\begin{pmatrix} x \\ y \end{pmatrix}$ **d.** 2

31. **a.** none **b.** none

33. **a.** $L\begin{pmatrix} x \\ y \\ z \end{pmatrix} = \begin{pmatrix} 1 & -1 & 3 \\ 2 & 0 & -4 \\ 1 & 2 & 5 \end{pmatrix}\begin{pmatrix} x \\ y \\ z \end{pmatrix}$ **b.** 6

c. $K\begin{pmatrix} x \\ y \\ z \end{pmatrix} = \begin{pmatrix} 4/17 & 11/34 & 2/17 \\ -7/17 & 1/17 & 5/17 \\ 2/17 & -3/34 & 1/17 \end{pmatrix}\begin{pmatrix} x \\ y \\ z \end{pmatrix}$ **d.** 6

35. $L\begin{pmatrix} x \\ y \end{pmatrix} = \begin{pmatrix} 1/2 & -\sqrt{3}/2 \\ \sqrt{3}/2 & 1/2 \end{pmatrix}\begin{pmatrix} x \\ y \end{pmatrix}$

37. $L\begin{pmatrix} x \\ y \end{pmatrix} = \begin{pmatrix} 2\cos(100°) & 2\sin(100°) \\ -2\sin(100°) & 2\cos(100°) \end{pmatrix}\begin{pmatrix} x \\ y \end{pmatrix}$

39. $L\begin{pmatrix} x \\ y \end{pmatrix} = \begin{pmatrix} 1/2 & \sqrt{3}/2 \\ -\sqrt{3}/2 & 1/2 \end{pmatrix}\begin{pmatrix} x \\ y \end{pmatrix}$

41. $L\begin{pmatrix} x \\ y \end{pmatrix} = \begin{pmatrix} 23/13 & 11/13 \\ -11/13 & 23/13 \end{pmatrix}\begin{pmatrix} x \\ y \end{pmatrix}$

49. yes **55.** no

57. a point, a line segment, or a parallelogram

8.6 Planes

1. $-x + 3y + 5z = -27$

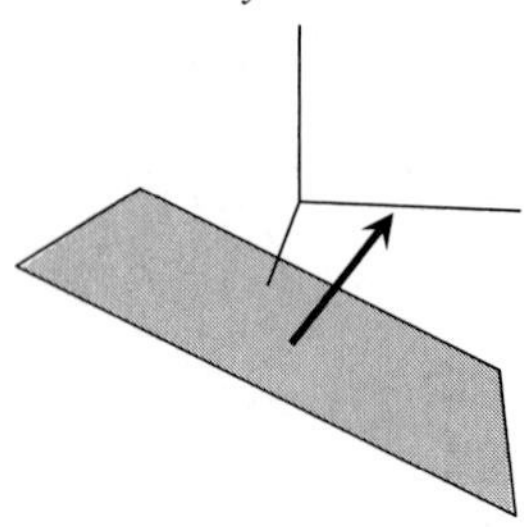

3. $z = 0$

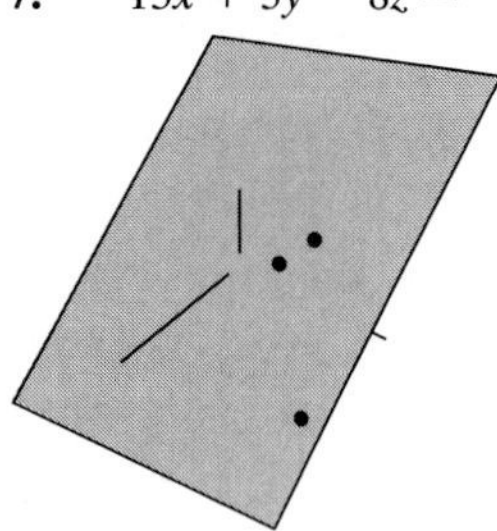

5. $3x + y - 7z = 15$

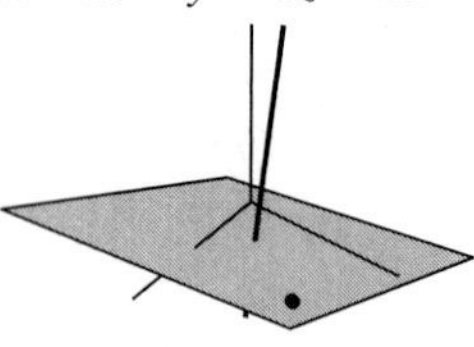

7. $-13x + 3y - 8z = -10$

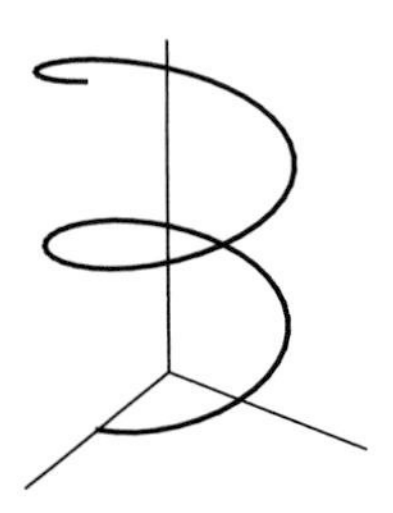

9. $9x + 22y + 12z = -42$

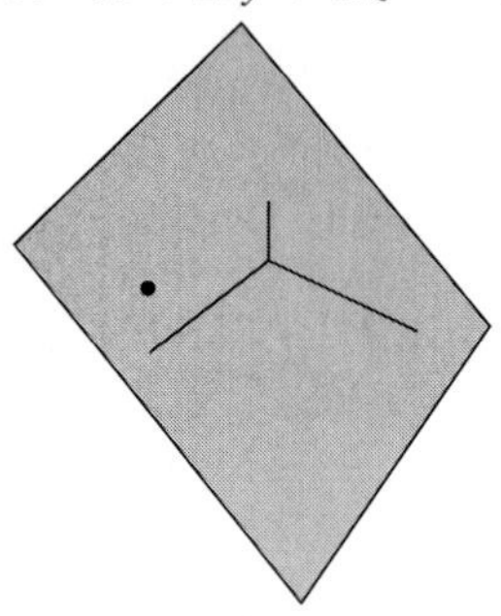

11. $\arccos(9/\sqrt{170}) \approx 0.808936$
13. $\arccos(17/\sqrt{990}) \approx 1.00001$
15. $(3 + 3\sqrt{11})x + (9 - \sqrt{11})y = 0$
17. $1.71895x - 0.212656y = 0$ **19.** $y = 0$
21. $4x + y + 5 = 0$ **23.** **b.** $(12/5, 3, 3/5)$
25. **b.** $11(x + 2) - 5y + 13(z + 1) = 0$
27. **b.** $17x - 12y - 4z = 51$ **33.** $\sqrt{3}$ **35.** $\dfrac{|d|}{\sqrt{a^2 + b^2 + c^2}}$
39. normal vector: $\langle 4, 0, 7\rangle$, point: $(0, 0, 0)$
41. normal vector: $\langle 0, 4, 0\rangle$, point: $(0, (3 + \sqrt{3})/4, 0)$

8.7 Motion in Three Dimensions

1. line segment

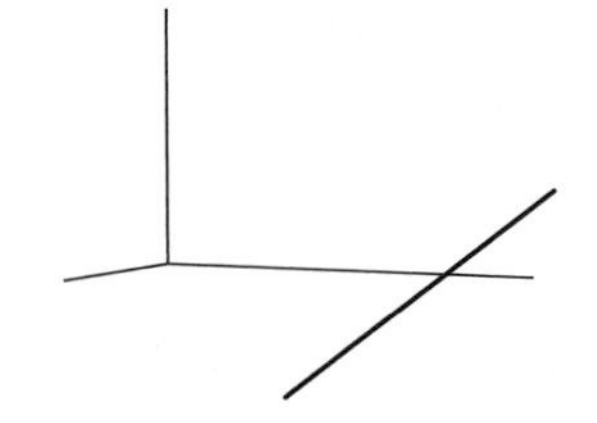

3. helix wrapping around z-axis

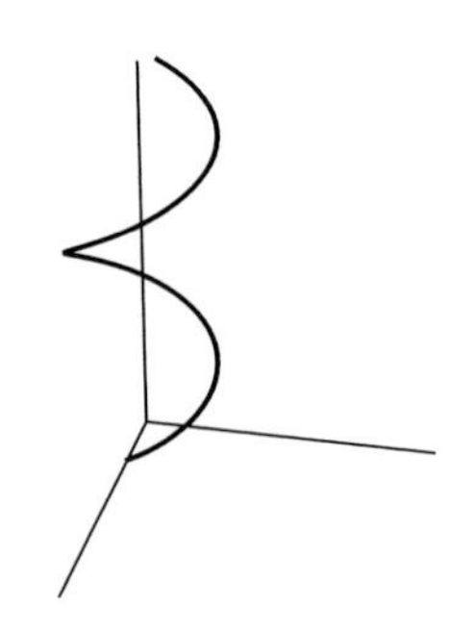

5. helix wrapping around z-axis

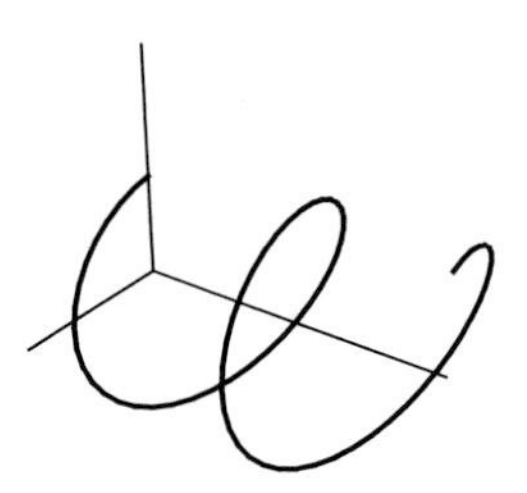

7. elliptical helix wrapping around y-axis

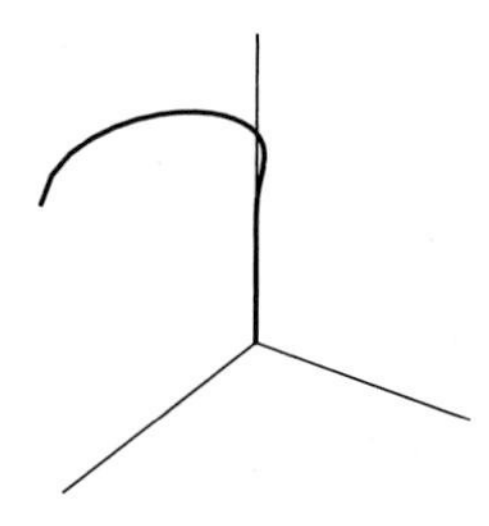

9. spring shape with cycles of increasing size, wrapping around z-axis

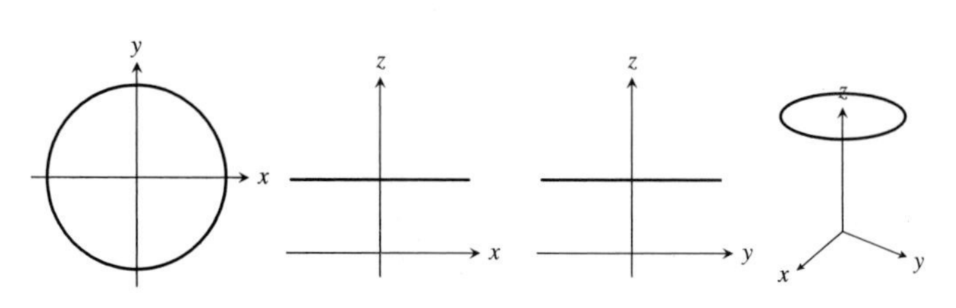

11. a circle in the (x, y)-plane, segments in the (x, z) and (y, z)-planes

13. an exponential graph in the (x, y)-plane, part of a parabola in the (x, z)-plane, part of a logarithm graph in the (y, z)-plane

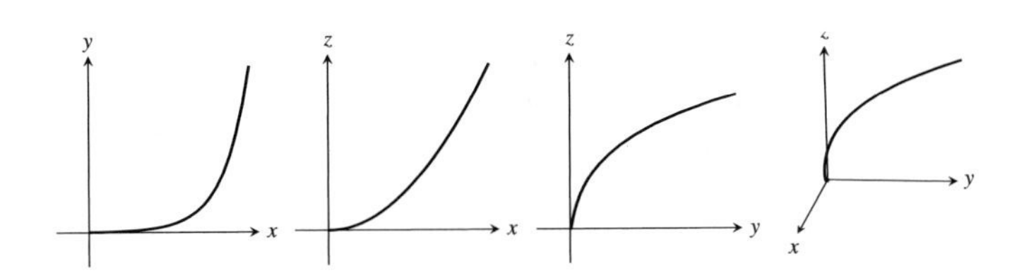

15. $\mathbf{r}(1) = \langle -2, 7, -2\rangle$, $\mathbf{v}(1) = \langle -2, 3, 2\rangle$, $\mathbf{a}(1) = \langle 0, 0, 0\rangle$

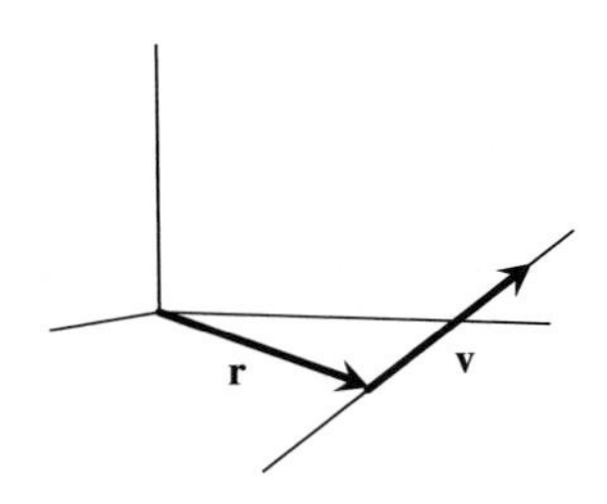

17. $\mathbf{r}(1) = \langle 1, 3\cos 1, \sin 1\rangle$, $\mathbf{v}(1) = \langle 1, -3\sin 1, \cos 1\rangle$, $\mathbf{a}(1) = \langle 0, -3\cos 1, -\sin 1\rangle$

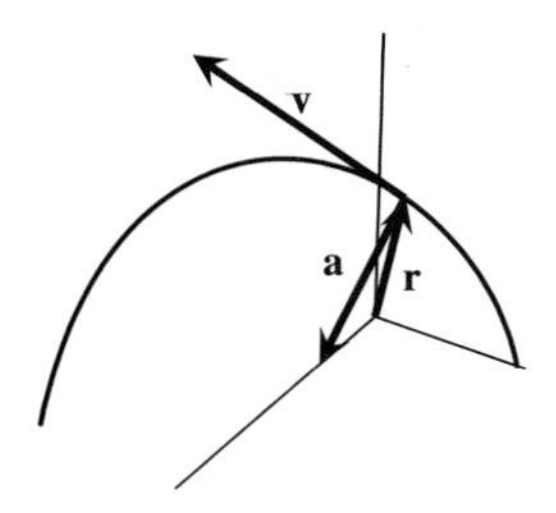

19. $\mathbf{T}(2) = \left\langle \frac{2}{\sqrt{5}}, \frac{1}{\sqrt{5}}, 0 \right\rangle \approx \langle 0.894, 0.447, 0\rangle$,
$\mathbf{N}(2)$ is undefined
$\kappa(2) = 0$
$a_T(2) = 2\sqrt{5} \approx 4.472$
There is no normal component to $\mathbf{a}(2)$.

21. $\mathbf{T}(\pi) = \left\langle \frac{\sqrt{2}}{2}, 0, -\frac{\sqrt{2}}{2} \right\rangle \approx \langle 0.707, 0, -0.707\rangle$
$\mathbf{N}(\pi) = \langle 0, 1, 0\rangle$
$\kappa(\pi) = \frac{3}{2}$
$a_T(\pi) = 0$
$a_N(\pi) = 3$

23. $\mathbf{T}(2) = \left\langle \frac{1}{\sqrt{146}}, \frac{1}{\sqrt{146}}, \frac{12}{\sqrt{146}} \right\rangle \approx \langle 0.083, 0.083, 0.993\rangle$
$\mathbf{N}(2) = \left\langle -\frac{6}{\sqrt{73}}, -\frac{6}{\sqrt{73}}, \frac{1}{\sqrt{73}} \right\rangle \approx \langle -0.702, -0.702, 0.117\rangle$
$\kappa(2) = \frac{6}{73\sqrt{73}} \approx 0.010$
$a_T(2) = \frac{144}{\sqrt{146}} \approx 11.918$
$a_N(2) = \frac{12}{\sqrt{73}} \approx 1.404$

29. ≈ 29.7807 **33.** $3\sqrt{2} \approx 4.243$ **35.** ≈ 1.12
37. $\mathbf{r}(t) = (1 - t)\langle -2, 4, 3\rangle + t\langle 2, 8, 5\rangle$
39. $\mathbf{r}(t) = \frac{ts}{\sqrt{a_1^2 + a_2^2 + a_3^2}}\langle a_1, a_2, a_3\rangle$
41. $\mathbf{r}(t) = \langle 5\sin(2t), 5\cos(2t), (10\sqrt{3})t\rangle$
$\mathbf{r}(t) = \langle 5\sin(\sqrt{12}t), 5\cos(\sqrt{12}t), 10t\rangle$
$\mathbf{r}(t) = \langle 5\sin(t), 5\cos(t), 5\sqrt{15}t\rangle$
$\mathbf{r}(t) = \langle -5\sin(t), 5\cos(t), 5\sqrt{15}t\rangle$
43. $\langle 0, -216{,}000\pi^2, 0\rangle$ newtons
51. a. $\mathbf{r}(s) = \langle -2\cos(3s/\sqrt{61}), 2\sin(3s/\sqrt{61}), 5s/\sqrt{61}\rangle$
b. $18/61$
53. $\mathbf{B} = \langle 1/\sqrt{2}, -1/\sqrt{2}, 0\rangle$
55. $\mathbf{B} = \langle -\sqrt{6}/3, 1/\sqrt{6}, 1/\sqrt{6}\rangle$
57. $\mathbf{r}(t) = \langle 1 + 2t, 3 + 3t, 3t\rangle$ **59.** $\mathbf{r}(t) = \langle t, 0, t\rangle$
61. a. $\mathbf{l}(t) = \langle t, t, -4\rangle$ **b.** $\mathbf{r}(0.1) \approx \mathbf{l}(0.1) = \langle 0.1, 0.1, -4\rangle$
c. perhaps by calculating $\|\mathbf{r}(0.1) - \mathbf{l}(0.1)\|$

Chapter Review Exercises

1. $\langle -17, 4, -29\rangle$ **3.** $\langle -44, 2, -16\rangle$ **5.** $48\sqrt{3}$ **7.** 288
9. $\langle 2/7, -4/7, -6/7\rangle$ **11.** 192 **13.** $\begin{pmatrix} 8 & -14 & -6 \\ 0 & -4 & 12 \end{pmatrix}$
15. $\begin{pmatrix} 13 & 13 \\ -25 & 39 \end{pmatrix}$ **17.** 2 **19.** undefined
21. $\begin{pmatrix} 31 & 49 \\ -22 & -66 \end{pmatrix}$ **23.** undefined
25. b. $\sqrt{1066}/2 \approx 16.3248$
27. $L(S)$ is the parallelogram with vertices $(4, 5)$, $(30, 13)$, $(56, 28)$, and $(82, 36)$. Area$(L(S)) = 182$.

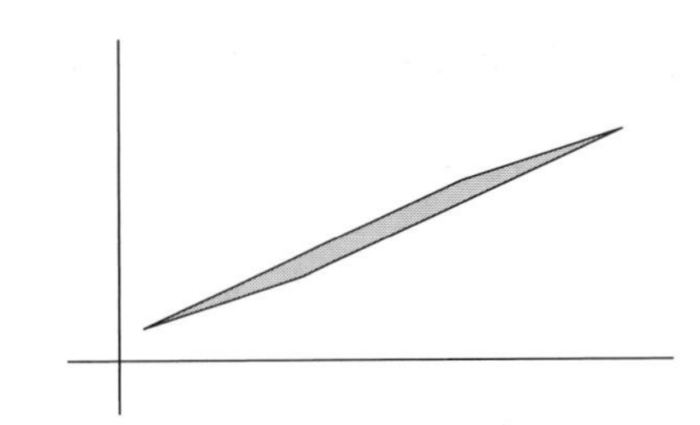

31. $6x - 15y + 8z = 0$ and $7x - 17y + 9z + 1 = 0$; other answers are possible.
33. $14x + 12y - 61z = 0$
35. $\mathbf{r}(t) = \langle -3t - 10, 15t + 8, 20t + 17 \rangle$
37. **a.** 264 **b.** 264 **c.** 1936 **d.** 1584
39. domain: R^2, range R
matrix: (4, 2)
43. $a = -3$ or $a = 2$
45. **a.** a spiral of increasing radius winding around the z-axis

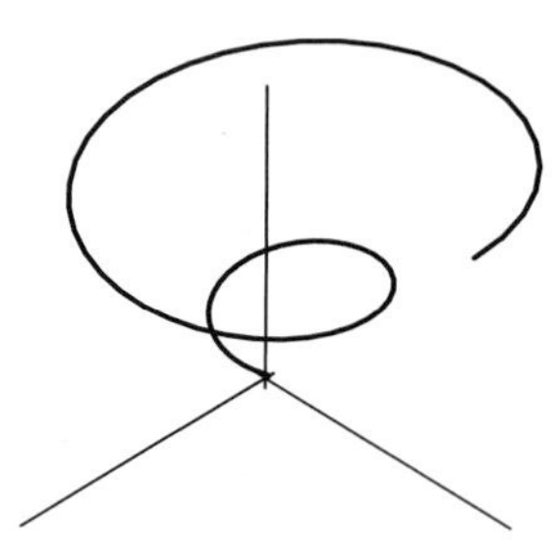

b. $\sqrt{6} + \ln(\sqrt{2} + \sqrt{3}) \approx 3.59571$
c. $\mathbf{T} = \langle 0, 1/\sqrt{2}, 1/\sqrt{2} \rangle$, $\mathbf{N} = \langle 1, 0, 0 \rangle$
d. $a_T(0) = 0$, $a_N(0) = 2$

CHAPTER 9—FUNCTIONS OF SEVERAL VARIABLES

9.1 Conic Sections

1. vertex: $(0, 0)$, focus: $\left(0, \frac{1}{4}\right)$, directrix: $y = -\frac{1}{4}$

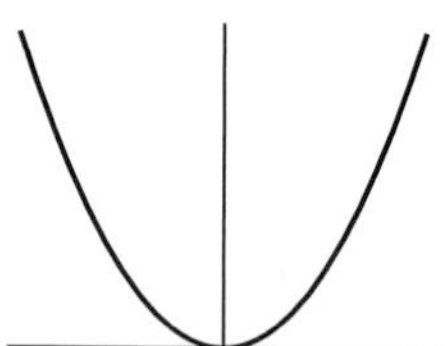

3. vertex: $(-1, -2)$, focus: $\left(-1, -\frac{23}{12}\right)$, directrix: $y = -\frac{25}{12}$

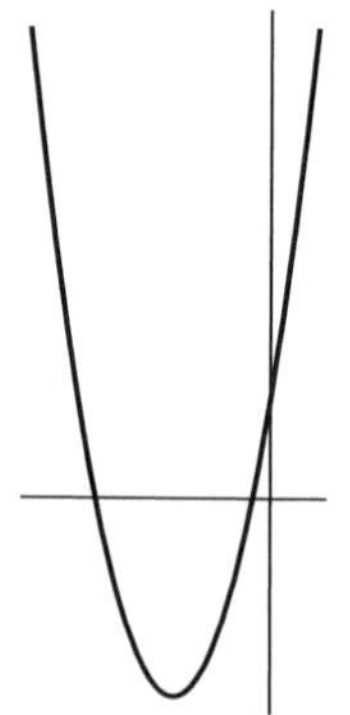

5. vertex: $(-1, -1)$, focus: $\left(-1, -\frac{3}{4}\right)$ directrix: $y = -\frac{5}{4}$,

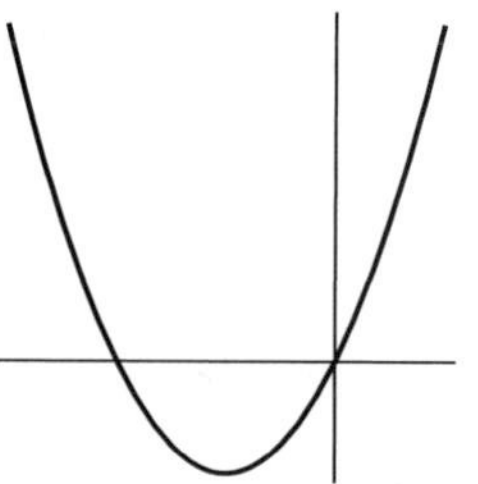

7. vertex: $(3, -1)$, focus: $(5, -1)$, directrix: $x = 1$

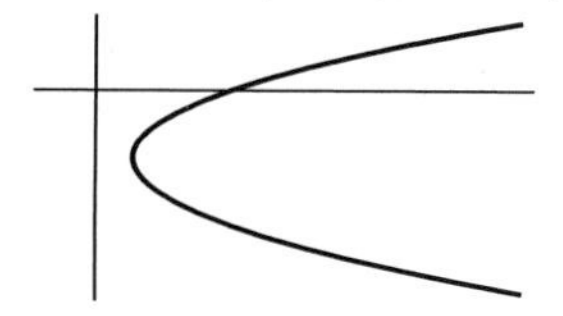

9. ellipse, center: $(0, 0)$, foci: $(0, \pm\sqrt{7})$, vertices: $(\pm 3, 0)$, $(0, \pm 4)$

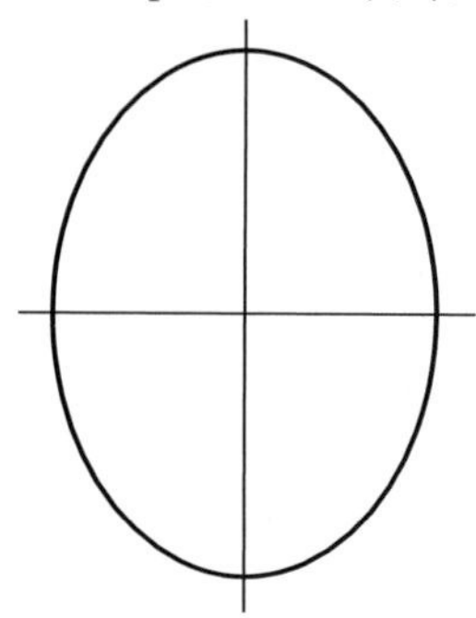

11. hyperbola, center: $(-1, 1)$, foci: $(-1, 1 \pm \sqrt{3})$,
asymptotes: $y = \pm\frac{1}{\sqrt{2}}(x + 1) + 1$, vertices: $(-1, 2)$, $(-1, 0)$

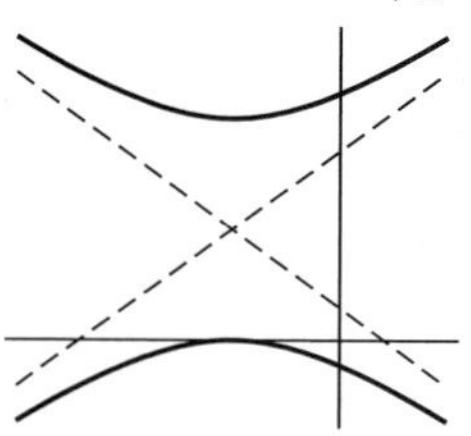

13. ellipse (circle), center: $\left(\frac{1}{4}, 0\right)$, foci coincide at the center

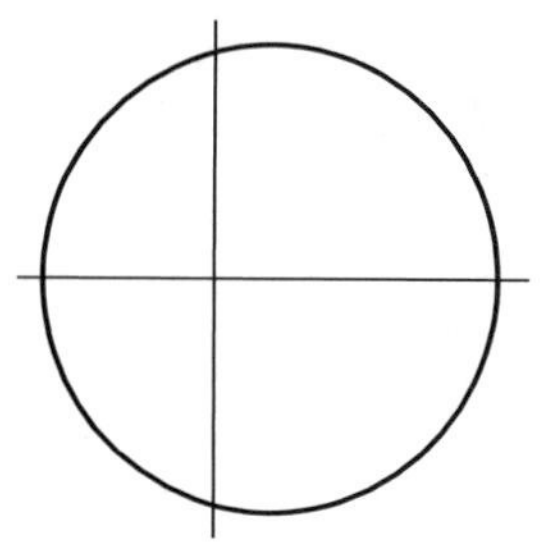

15. hyperbola, center: $(-3,-2)$, foci: $\left(-3,-2 \pm \sqrt{5}\right)$, asymptotes: $y = \pm 2(x+3) - 2$, vertices: $(-3,0)$, $(-3,-4)$

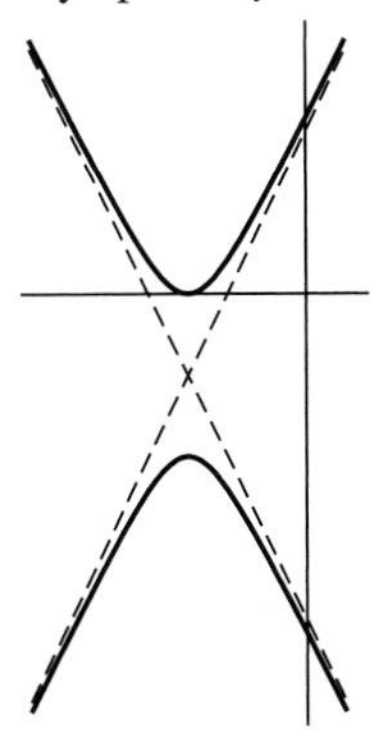

17. $\dfrac{x^2}{12} + \dfrac{y^2}{16} = 1$ **19.** $y + 2 = \dfrac{1}{16}(x-2)^2$

21. $(x-2)^2 - \dfrac{(y-1)^2}{15} = 1$

23. hyperbola: $\dfrac{(x+4)^2}{4} - \dfrac{y^2}{12} = 1$

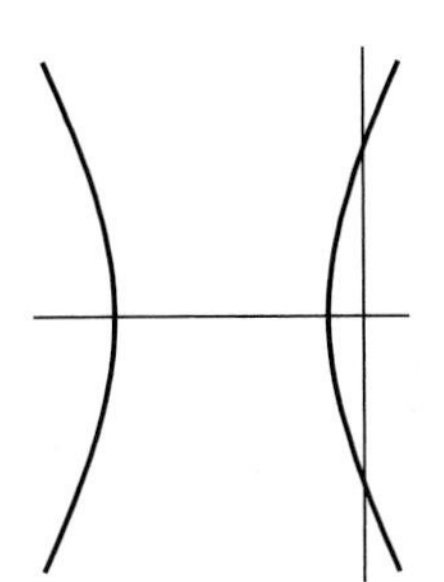

25. parabola: $x + \dfrac{3}{2} = \dfrac{1}{6}y^2$

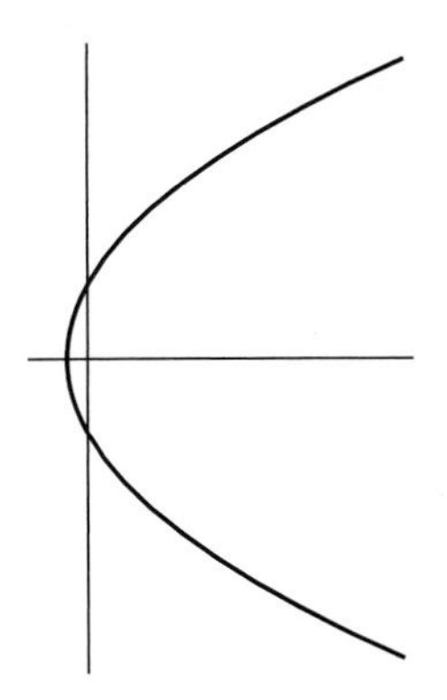

27. ellipse: $\dfrac{(x+2)^2}{3} + \dfrac{(y-4)^2}{4} = 1$

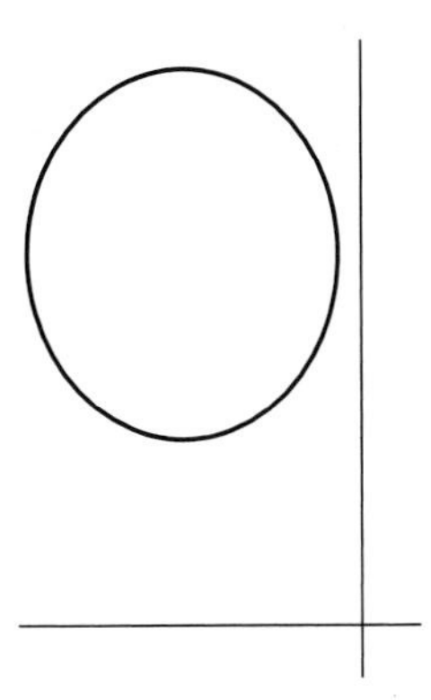

35. hyperbola, $\theta = 45°$ **37.** hyperbola, $\theta = \frac{1}{2}\operatorname{arccot}(3/4)$
39. parabola, $\theta = \frac{1}{2}\operatorname{arccot}\left(-1/\sqrt{8}\right)$
41. $x^2 + 2\sqrt{3}xy + 3y^2 + 2\sqrt{3}x - 2y + 4 = 0$
43. $2xy - 3 = 0$ **45.** $x^2 - 2xy + y^2 + 4x + 4y - 4 = 0$
47. $x^2 - 4xy + y^2 - 8x + 8y + 8 = 0$
49. $12x^2 + 32\sqrt{3}xy + 44y^2 - 15 = 0$
53. the lines $x = -2$ and $y = 3$
55. the lines $x + 2y = 3$ and $x + 2y = -4$
57. $45°$

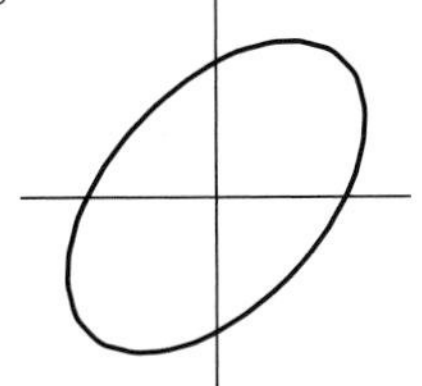

59. $-15°$

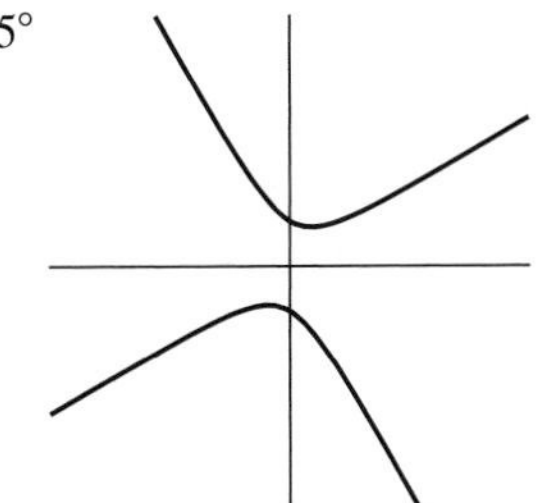

9.2 Real World Functions

1. domain $= R^2$

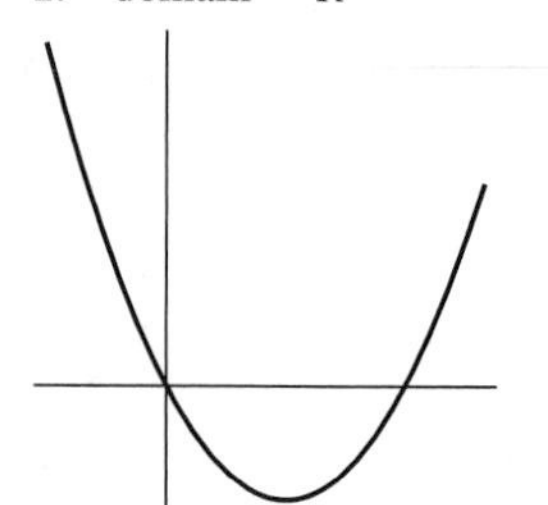

3. domain $= R^2$

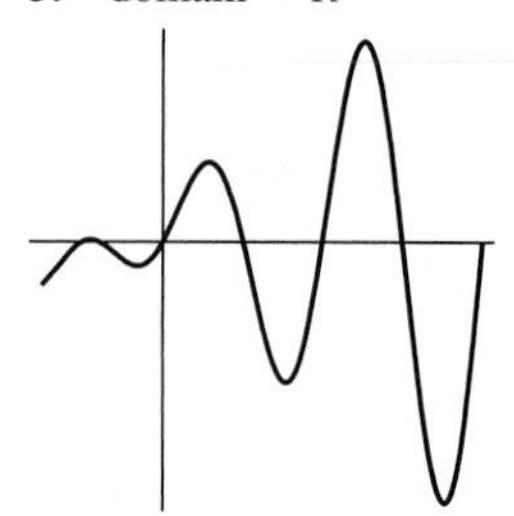

5. domain $= R^3$

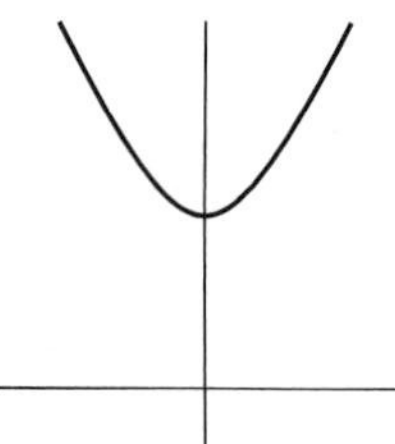

7. domain $= \{(x, y, z) : x + y + z > 0\}$

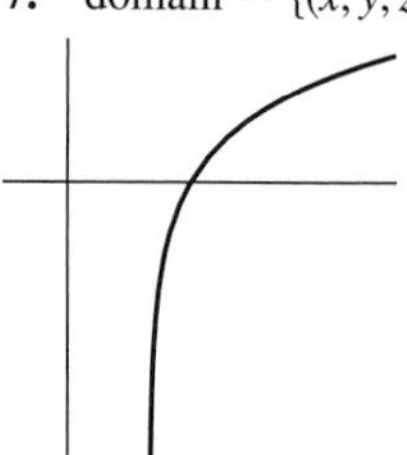

9. domain $= \{(x, y, z) : x^2 + y^2 + z^2 \leq 1\}$

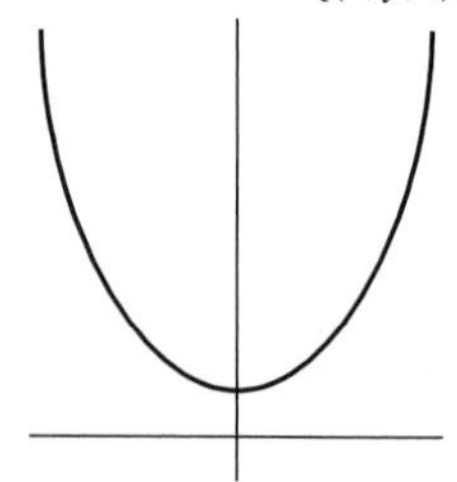

11. When the mass M and the response height h of the block are constant, the quantity $v - \sqrt{2gh}$ is inversely proportional to m, the mass of the bullet. When m is small, v is large and, as m grows, v decreases.

13. $p = p(0.9, f, 0.3) = 0.9 - 0.63f$, so p is a linear function of f. The probability p of being eaten decreases as the probability f of the mimic fooling the predator increases, but p can never be lower than 0.27.

15. $t \approx 57.19$ years (Use Newton's method to find an approximation to the root of $(d/dt)P(0.6, t) = 0$ near $t = 57$.)

17. $P(0.06, t) = \dfrac{8.7t + 43}{e^{0.06t} - 1}$, $100 \leq t \leq 110$

19. $95{,}000e^{-1.44} \approx 22{,}508$ dollars

21. If an object starts from rest, it has velocity $v = \sqrt{2gh}$ after falling height h. At this time the kinetic energy is $\text{KE} = (1/2)mv^2 = mgh$. Hence the gravitational potential energy of the object at height h is mgh. The bullet-block system of Example 2 has mass $M + m$. If it rises to height h, then at this instant the gravitational potential energy is $(M + m)gh$. Equate this to the kinetic energy expression in (7) to get

$$(M + m)gh = \frac{1}{2}(M + m)v_1^2.$$

Solve to obtain $v_1 = \sqrt{2gh}$. Substitute this into (6) to obtain (5).

23. **a.** domain $= \{(r, m, M) : r \geq 0, M \geq m > 0\}$

b. s is a linear function of r.

c. s is a logarithmic function of M.

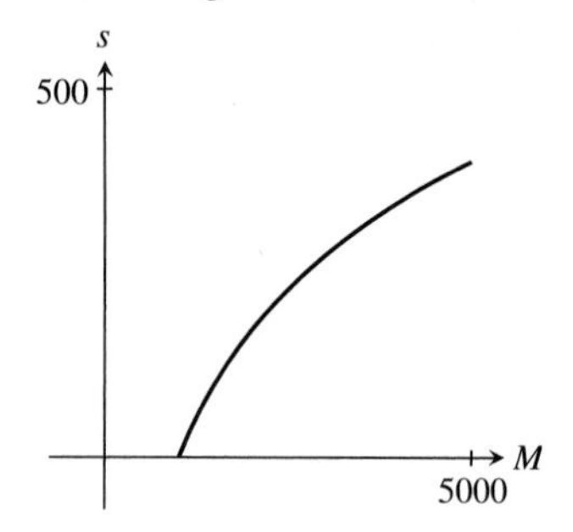

d. s is a logarithmic function of $1/m$.

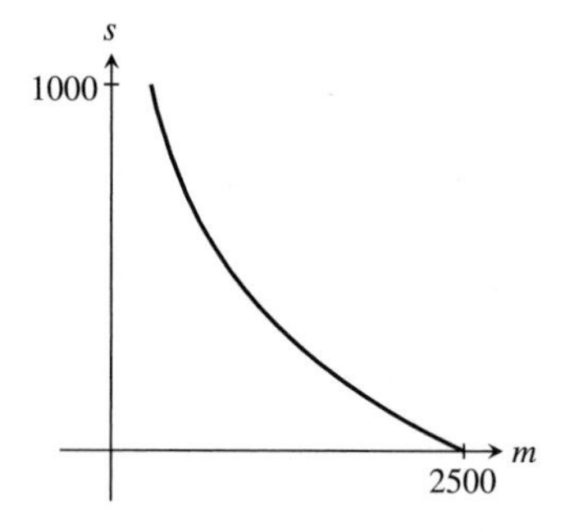

25. **a.**

b.

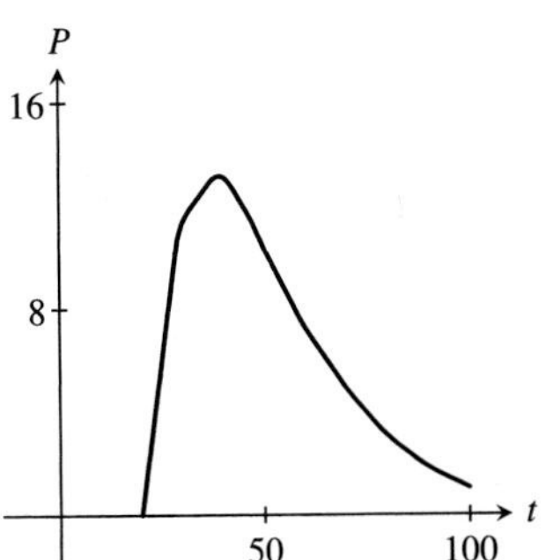

c. $t \approx 36.46$ years

27. **a.** domain $= \{(s, d) : 0 \le s \le 20, 0 \le d \le 20\}$
b. In this figure waterline w is on the vertical axis and distance d from center is on the horizontal axis. The graph shows the cross section of half the hull if a vertical cut is made at Station 11. For the full cross section, reflect across the w-axis. The dotted segment does not correspond to given data, but was filled in assuming that at Station 11 the center of the hull is at the lowest (i.e., 0) level.

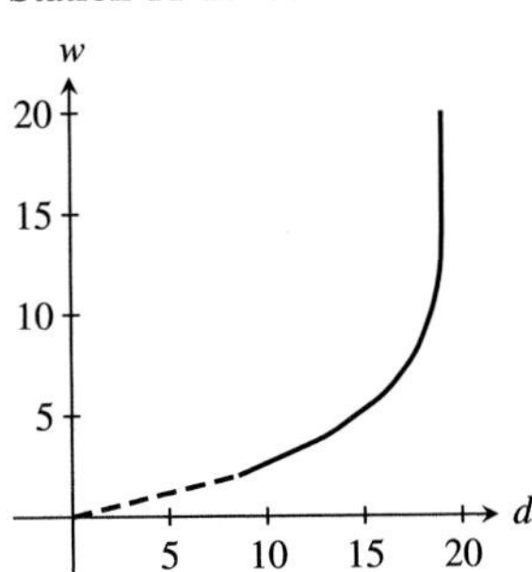

c. See the explanation for part b.

d. The vertical axis records stations, the horizontal axis the distance from center. The graph represents what a horizontal cross section of the hull at height 8 would look like. To get the full cross section, reflect about the s-axis. Remember that each station is 13.5 feet. Hence, for a true picture of the cross section, the illustration should be magnified by a factor of 13.5 in the vertical direction.

9.3 Graphing: Surfaces and Level Curves

1. The (x, y)-plane cross section is $(0, 0)$.
The (x, z)-plane cross section is a parabola.
The (y, z)-plane cross section is a parabola.

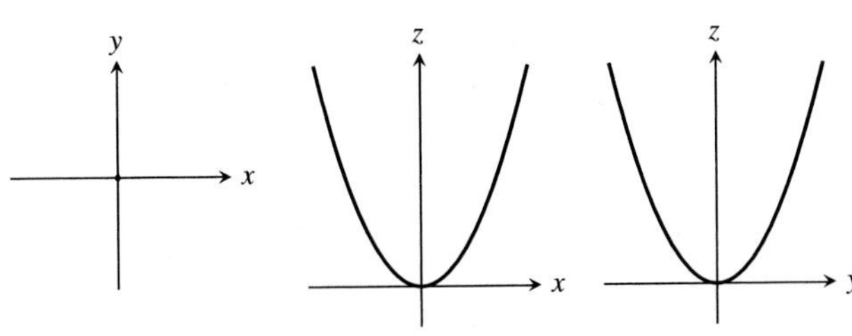

3. The (x, y)-plane cross section is a line.
The (x, z)-plane cross section is a line.
The (y, z)-plane cross section is a line.

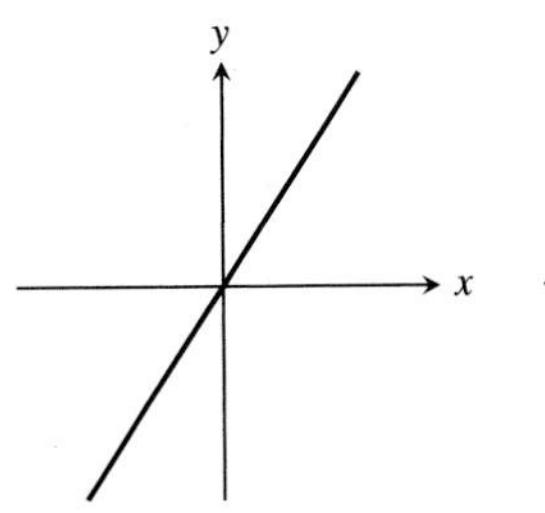

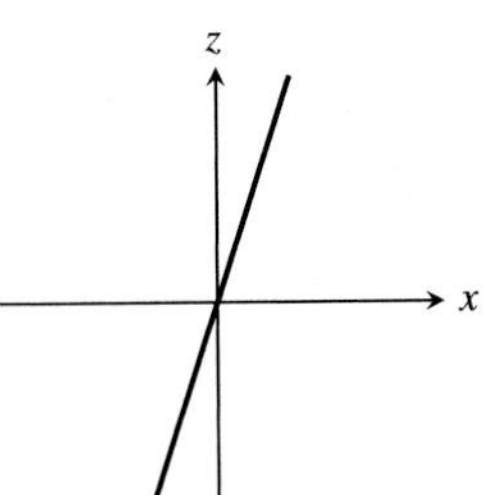

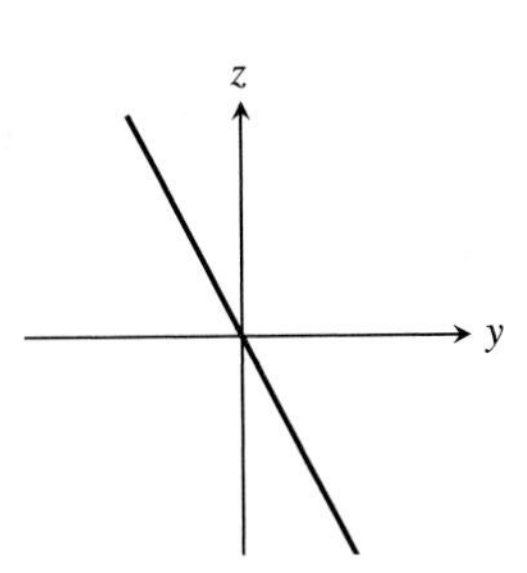

5. The (x, y)-plane cross section is two lines.
The (x, z)-plane cross section is a parabola.
The (y, z)-plane cross section is a parabola.

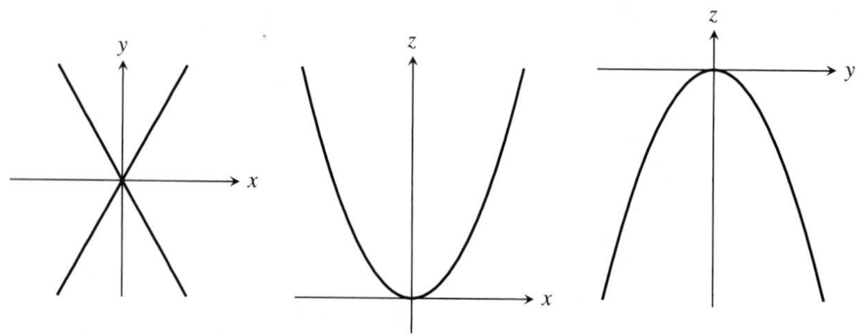

7. There is no (x, y)-plane cross section.
The (x, z)-plane cross section is a line.
The (y, z)-plane cross section is a line.

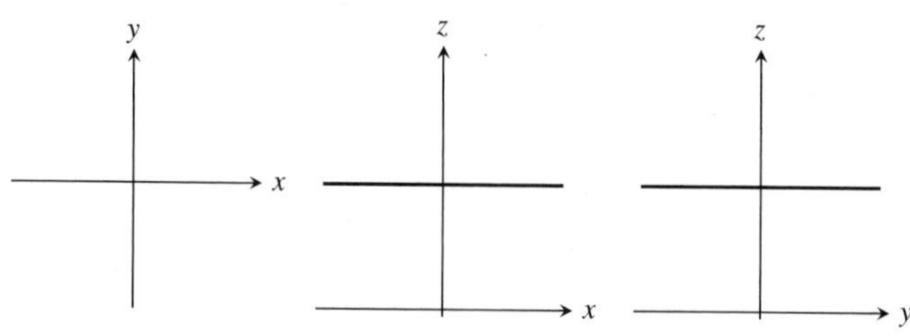

9. lines

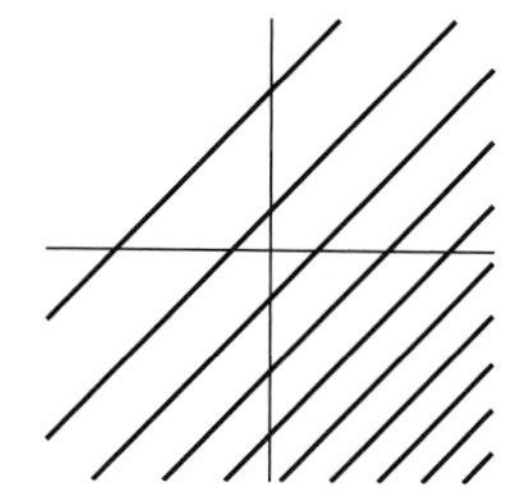

11. parabolas

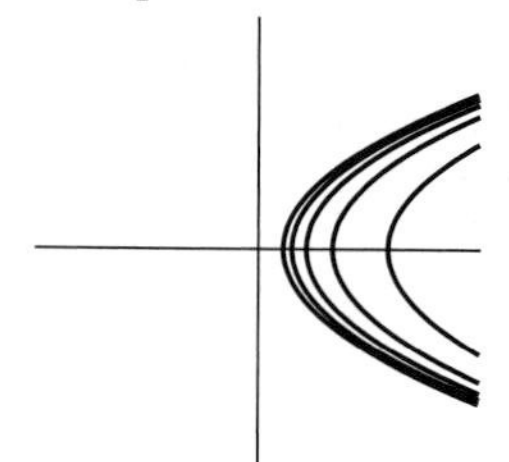

13. hyperbolas

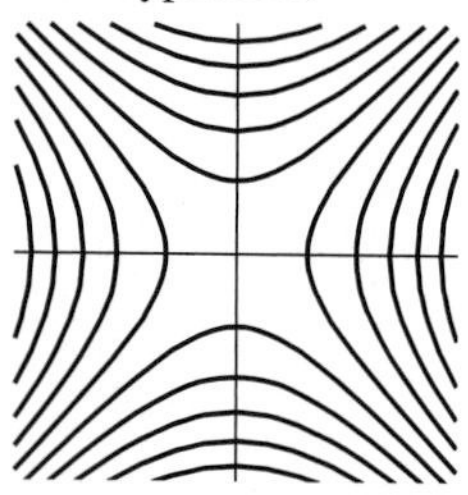

15. plane

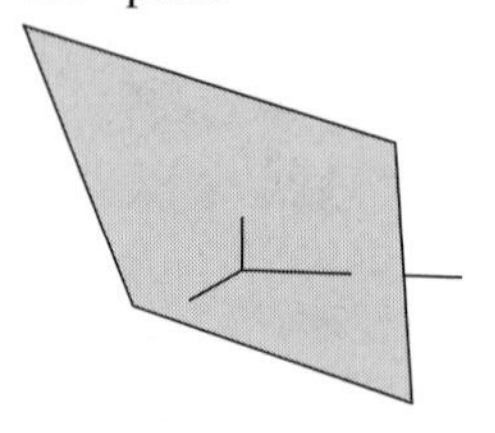

17. graph only above first quadrant of (x, y)-plane, rising as x or y grows

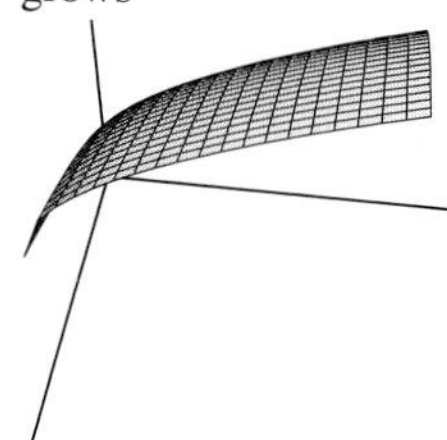

19. symmetric about z-axis, approaches (x, y)-plane as $|x|$ or $|y|$ grows large

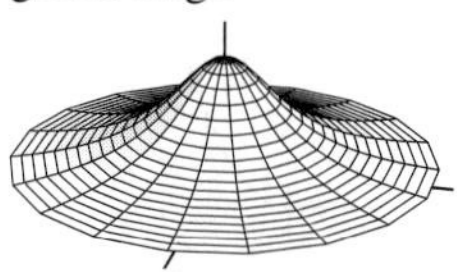

21. Because m is positive and v is nonnegative, the level curves are first-quadrant portions of graphs of the form $m = k/v^2$.

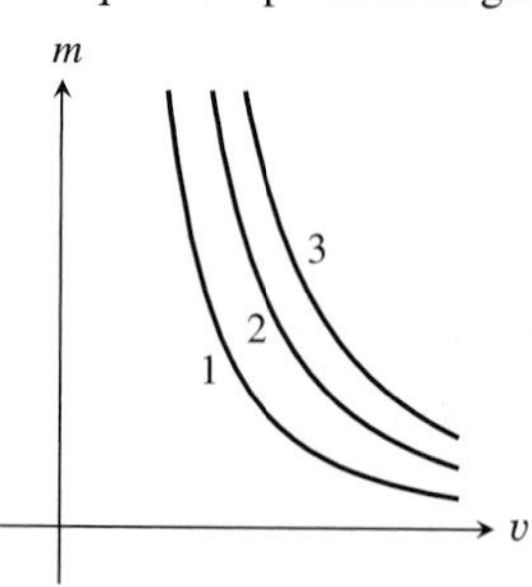

23. Taking s_1 and s_2 to be positive, the level curve for $f = c$ is the first-quadrant portion of the hyperbola with equation $s_2 = s_1/(cs_1 - 1)$.

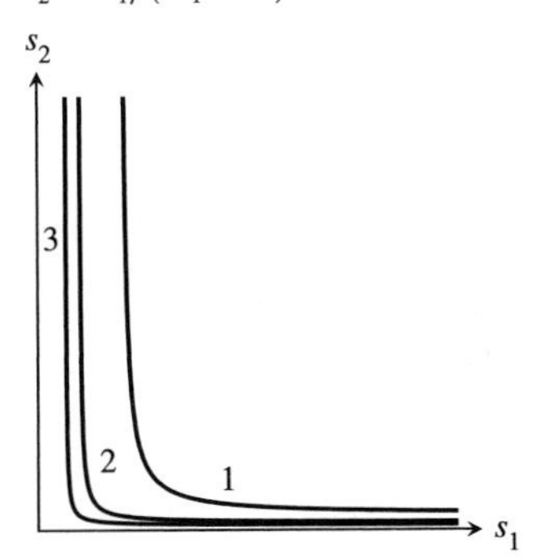

25. The function has as its domains locations in the United States. The range is the set of average wind powers for all locations. Given a location as input, the function returns the average wind power available at that location.

27. ellipsoids

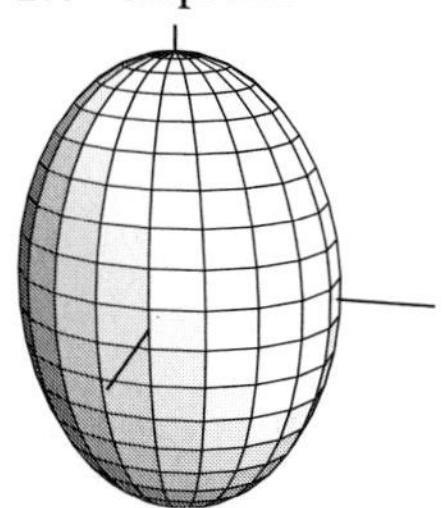

29. paraboloids with circular cross sections

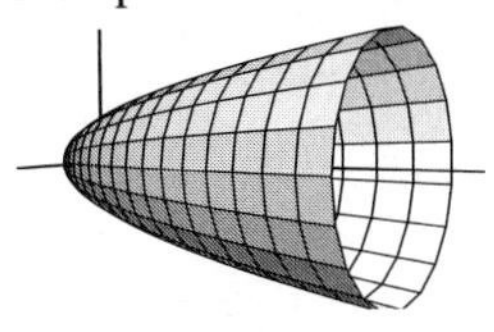

31. surfaces with both hyperbolas and parabolas as cross sections

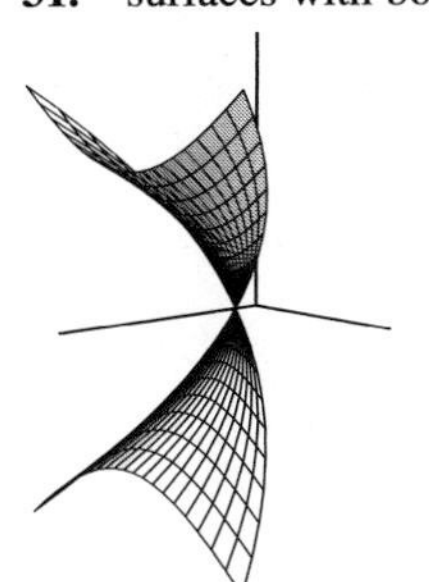

33. B, v **35.** F, iii **37.** C, iv

9.4 Graphing: Parametric Representations of Surfaces

1. A is on S, B is not on S. **3.** A is on S, B is on S.
5. A is on S, B is on S.

7. $2x + 7y - z = 0$

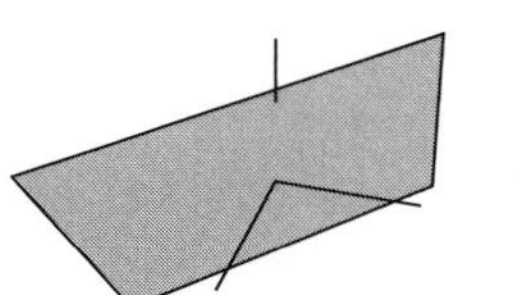

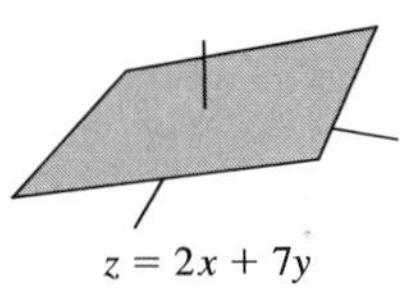

$z = 2x + 7y$

9. $z = x^2 + y^2 - 2$

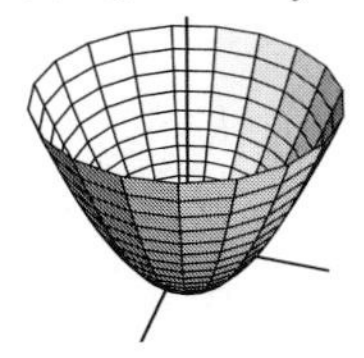

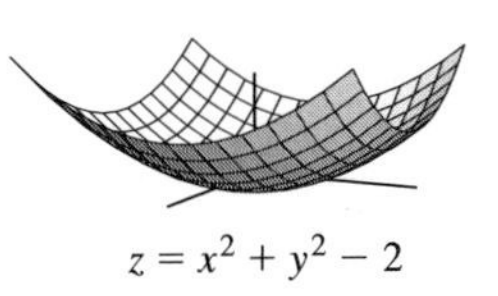

$z = x^2 + y^2 - 2$

11. $x = \dfrac{y^2}{16} + z^2 + 2$

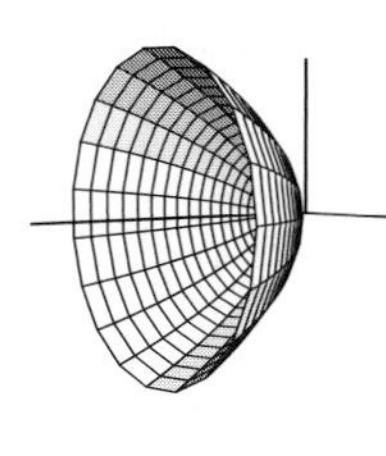

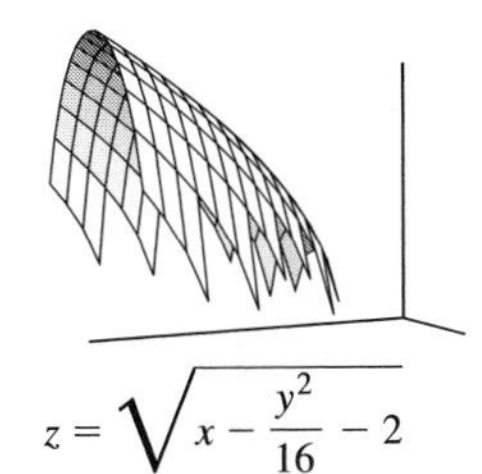

$z = \sqrt{x - \dfrac{y^2}{16} - 2}$

13. $y = \dfrac{x^2}{4} - \dfrac{z^2}{16} + 1$

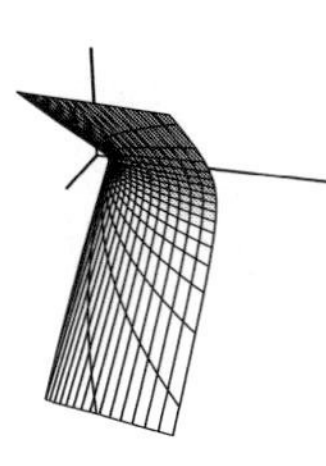

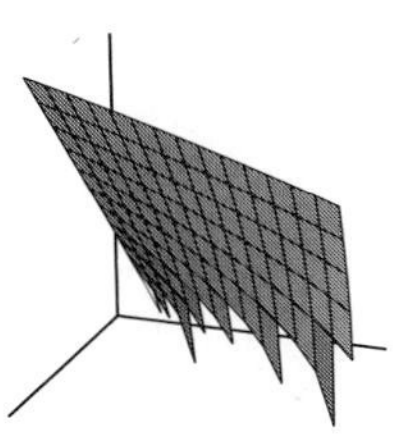

$z = \sqrt{4x^2 - 16y + 16}$

15. $z = e^{x+y}$

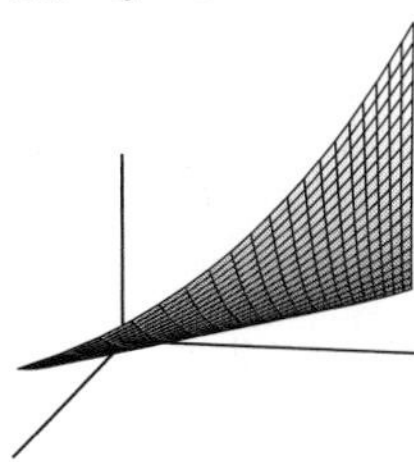

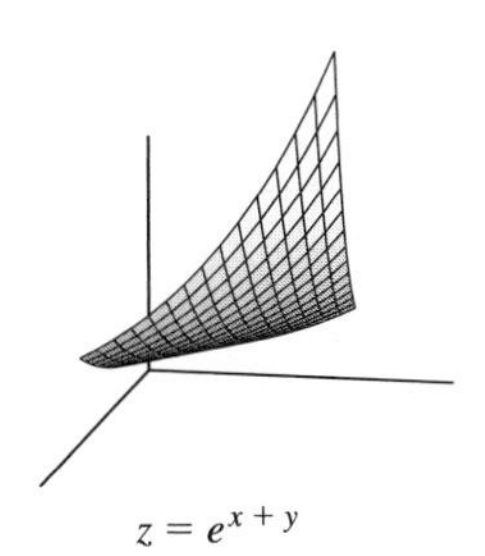

$z = e^{x+y}$

17. $\mathbf{f}(r, \theta) = \langle r\cos\theta, r\sin\theta, 2r^2\rangle$, $r \geq 0$, $0 \leq \theta \leq 2\pi$

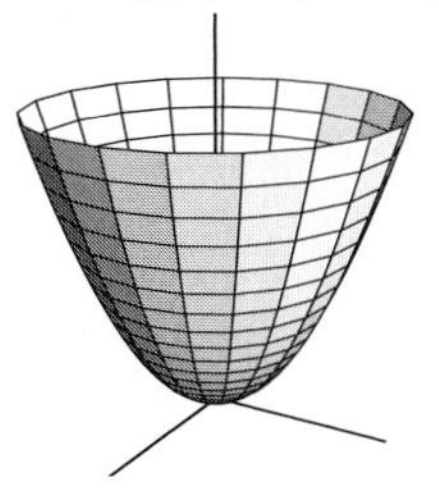

19. using level x-curves:
$\mathbf{r}(r,\theta) = \left\langle r^2 + 1, \sqrt{2}r\cos\theta, \dfrac{1}{\sqrt{2}}r\sin\theta \right\rangle, r \geq 0, 0 \leq \theta \leq 2\pi$

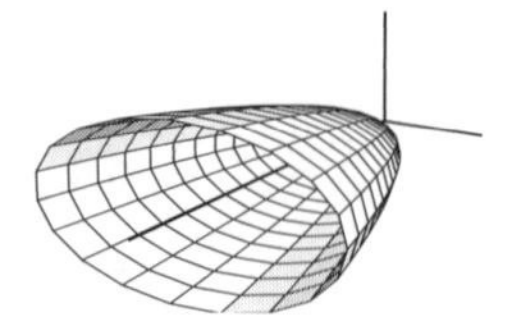

21. using level y-curves: $\mathbf{f}(r,\theta) = \left\langle \dfrac{1}{\sqrt{3}}r\tan\theta, r^2, r\sec\theta \right\rangle$,
$r \geq 0, -\pi/2 < \theta < \pi/2$

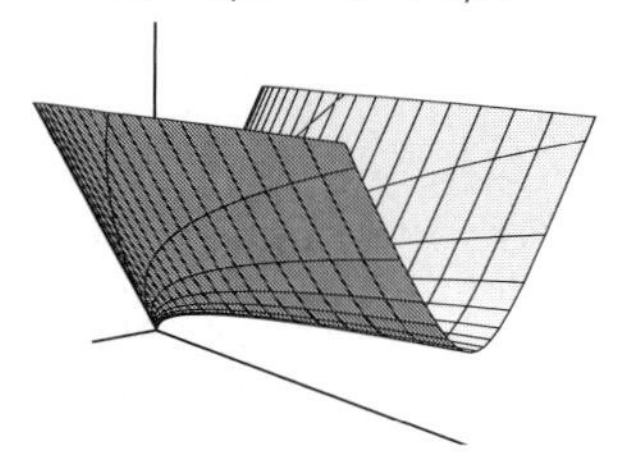

23. $\mathbf{r}(u,v) = \left\langle u, -2 + \dfrac{2}{3}u + \dfrac{5}{3}v, v \right\rangle, -\infty < u, v < \infty$

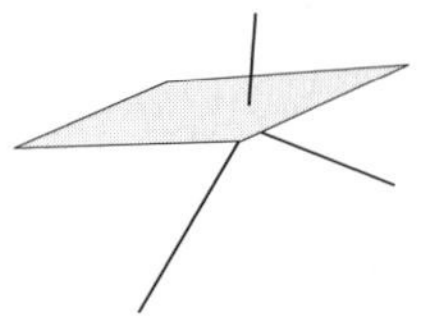

25. using level y-curves: $\mathbf{f}(r,\theta) = \left\langle \dfrac{1}{\sqrt{2}}r\cos\theta, 2\ln r, \dfrac{1}{2}r\sin\theta \right\rangle$,
$r > 0, 0 \leq \theta \leq 2\pi$

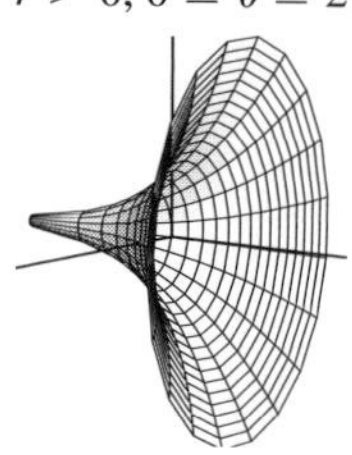

27. $\mathbf{r}(u,v) = \left\langle u, v, \dfrac{1}{2}(5 - 3u + v) \right\rangle, -\infty < u, v < \infty$
$\mathbf{r}(u,v) = \left\langle \dfrac{1}{3}(5 + u - 2v), u, v \right\rangle, -\infty < u, v < \infty$
$\mathbf{r}(u,v) = \langle u, 3u + 2v - 5, v\rangle, -\infty < u, v < \infty$; other answers are possible

29. $\mathbf{r}(u,v) = \langle u, e^{u+2v}, v\rangle, -\infty < u, v < \infty$
$\mathbf{r}(u,v) = \left\langle u, v, \dfrac{1}{2}(\ln v - u) \right\rangle, -\infty < u < \infty, v > 0$
$\mathbf{r}(u,v) = \langle \ln u - 2v, u, v\rangle, u > 0, -\infty < v < \infty$; other answers are possible

31. no

33. Surface S_2 is the reflection of S_1 across the plane $y = z$.
Surface S_3 is the reflection of S_1 across the plane $x = y$.
Surface S_3 is the reflection of S_2 across the plane $x = y$, followed by reflection across the plane $x = z$.

35. Surface S_1, represented by $\mathbf{r}_1$, is the reflection of S across the plane $y = z$.
Surface S_2, represented by $\mathbf{r}_2$, is the reflection of S across the plane $x = y$.
Surface S_3, represented by $\mathbf{r}_3$, is the reflection of S across the plane $x = y$, followed by reflection across the plane $y = z$.
Surface S_4, represented by $\mathbf{r}_4$, is the reflection of S across the plane $x = y$, followed by reflection across the plane $x = z$.
Surface S_5, represented by $\mathbf{r}_5$, is the reflection of S across the plane $x = z$.

37. **b.**

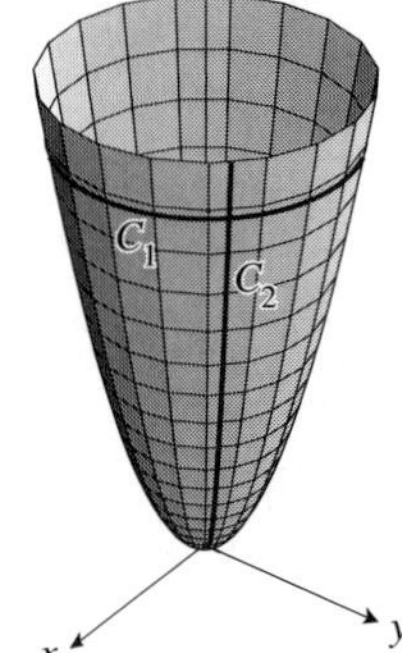

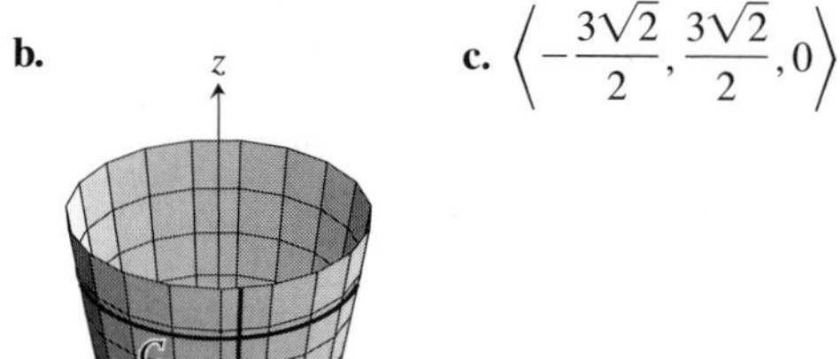

c. $\left\langle -\dfrac{3\sqrt{2}}{2}, \dfrac{3\sqrt{2}}{2}, 0 \right\rangle$

d. $\left\langle \dfrac{\sqrt{2}}{2}, \dfrac{\sqrt{2}}{2}, 6 \right\rangle$ **e.** $3\sqrt{2}x + 3\sqrt{2}y - z = 9$

39. **b.**

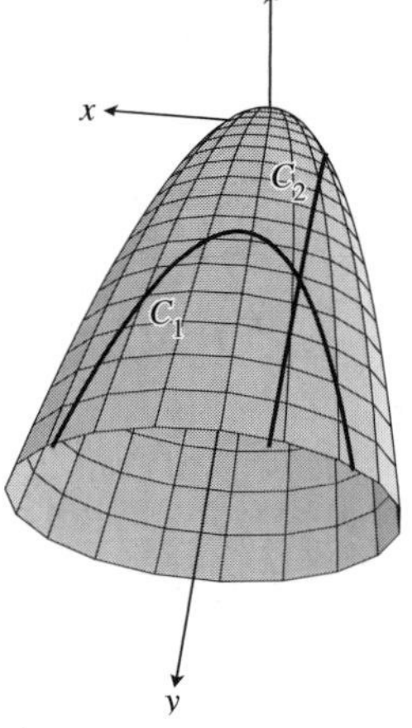

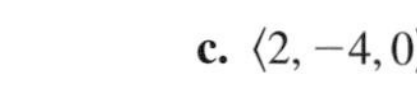

c. $\langle 2, -4, 0\rangle$

d. $\langle 0, 8, 1\rangle$ **e.** $-2x - y + 8z = 28$

41. Let the base square have side $\sqrt{2}$. Place the square in the plane so its center is at the origin and the diameter for the cross sections is

on the x-axis. Then the surface is described by
$\mathbf{r}(x,y) = \langle x, (1-|x|)y, (1-|x|)(1-|y|)\rangle$, for $-1 \le x, y \le 1$.

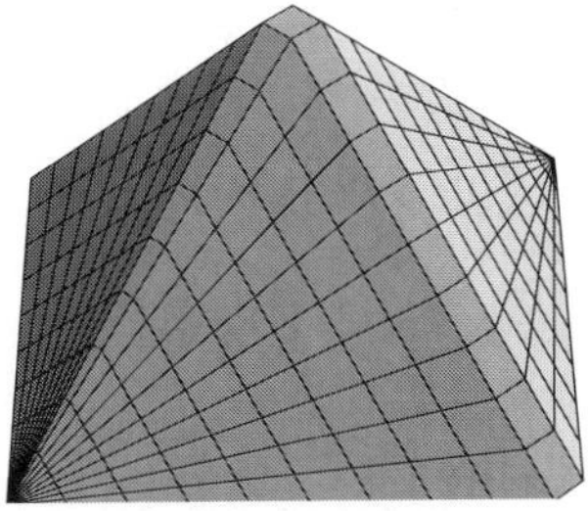

43. Let the base triangle be the plane with vertices $(0,\sqrt{3})$, $(-1,0)$, and $(1,0)$. Then the surface is described by
$$\mathbf{r}(c,\theta) = \left\langle \left(1 - \frac{c}{\sqrt{3}}\right)\cos\theta, c, \left(1 - \frac{c}{\sqrt{3}}\right)\sin\theta \right\rangle,$$
for $0 \le c \le \sqrt{3}$ and $0 \le \theta \le \pi$.

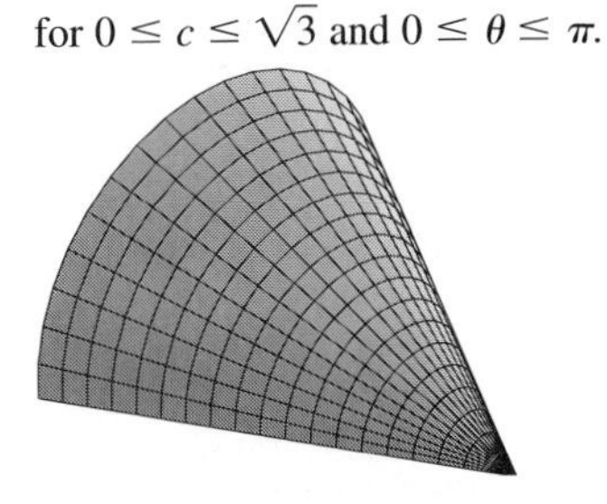

9.5 Cylindrical and Spherical Coordinates

1. cylindrical: $(\sqrt{2}, \pi/4, 1)$
spherical: $(\sqrt{3}, \pi/4, \arccos(1/\sqrt{3}))$
3. cylindrical: $(\sqrt{2}, -\pi/4, -\sqrt{2})$
spherical: $(2, 7\pi/4, 3\pi/4)$
5. Cartesian: $(42, 0, 0)$
spherical: $(42, 0, \pi/2)$
7. Cartesian: $(\sqrt{3}, 1, -2)$
spherical: $(2\sqrt{2}, \pi/6, 3\pi/4)$
9. Cartesian: $(0, 0, 42)$
cylindrical: $(0, 0, 42)$
11. Cartesian: $(0, 0, 0)$
cylindrical: $(0, 0, 0)$
13. cylindrical: $r^2 + z^2 = 16$
spherical: $\rho = 4$
15. cylindrical: $r = 5\csc\theta$
spherical: $\rho = 5\csc\phi\csc\theta$
17. Cartesian: $x^2 + y^2 + z^2 = 9$
spherical: $\rho = 3$
19. Cartesian: $x^2 + y^2 = 64$
spherical: $\rho = 8\csc\phi$
21. Cartesian: $x^2 + y^2 + z^2 = 25$
cylindrical: $r^2 + z^2 = 25$
23. Cartesian: $z = 3$
cylindrical: $z = 3$

25. $\mathbf{f}(\theta, t) = \langle\sqrt{5}\cos\theta, \sqrt{5}\sin\theta, t\rangle$, $0 \le \theta \le 2\pi$, $-\infty < t < \infty$

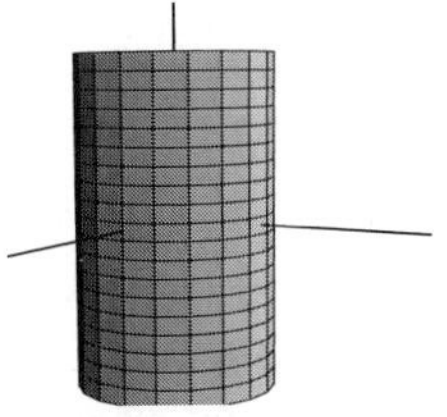

27. $\mathbf{r}(\phi,\theta) = \langle 4\sin^2\phi\sin\theta\cos\theta, 4\sin^2\phi\sin^2\theta, 4\sin\phi\sin\theta\cos\phi\rangle$, $0 \le \phi \le \pi$, $0 \le \theta \le 2\pi$

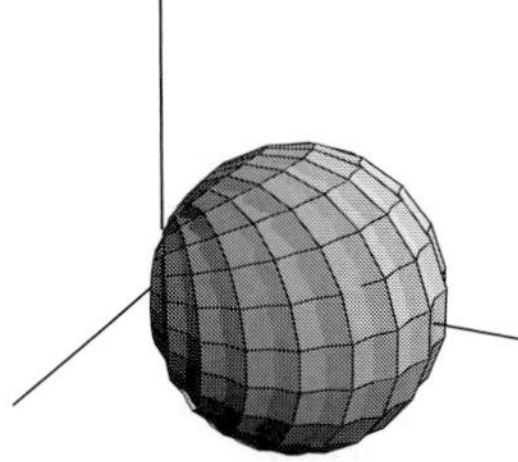

29. $\mathbf{f}(r,\theta) = \left\langle \frac{1}{2}r\cos\theta, r\sin\theta, e^{-r^2}\right\rangle$, $0 \le \theta \le 2\pi$, $r \ge 0$

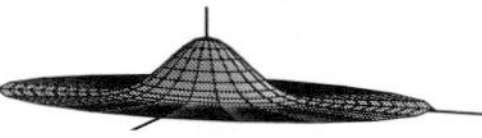

31. $\mathbf{r}(\phi,\theta) = \left\langle \frac{1}{2}\sin\phi\cos\theta, \frac{1}{3}\sin\phi\sin\theta, 2\cos\phi\right\rangle$,
$0 \le \phi \le \pi$, $0 \le \theta \le 2\pi$

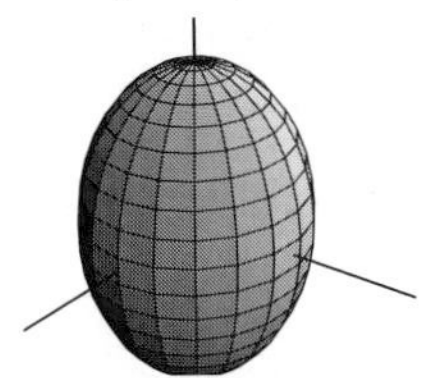

33. f, III **35.** b, V **37.** c, VI
39. The surface is shaped like a double saddle.
43. **a.** $x^2 + 2\sqrt{3}x + y^2 - 6y + z^2 + 4z = 0$
b. $r^2 + 2\sqrt{3}r\cos\theta - 6r\sin\theta + z^2 + 4z = 0$
c. $\rho = -2\sqrt{3}\sin\phi\cos\theta + 6\sin\phi\sin\theta - 4\cos\phi$
45. **a.** $x^2 + y^2 + z^2 + 10\sqrt{2}x - 10\sqrt{2}y = 0$
b. $r^2 + z^2 + 10\sqrt{2}r\cos\theta - 10\sqrt{2}r\sin\theta = 0$
c. $\rho = 10\sqrt{2}\sin\phi\sin\theta - 10\sqrt{2}\sin\phi\cos\theta$
47. S^* can be viewed as S with a cylinder of radius 1 about the z-axis cut out and moved in radially by one unit.
49. S^* is obtained by moving each point of S radially one unit farther from the origin.
51. S^* is obtained from S by removing the $0 < \phi < \frac{\pi}{6}$ portion of S, then for the remaining points decreasing the ϕ angle by $\pi/6$ for each point on S.
55. **a.** $\{-4, 2, 4\}$ **b.** $(10, -2, 5)$
c. If the vector coordinates are $\{a, b, c\}$, then the Cartesian coordinates are $(a + 4b, a, a + b + c)$.

d. If the Cartesian coordinates are (x, y, z), then the vector coordinates are $\{y, x/4 - y/4, -x/4 - 3y/4 + z\}$.

9.6 Limits

1. 96 **3.** −9765625 **5.** does not exist **7.** does not exist **9.** does not exist **11.** 1 **13.** does not exist **15.** 0

17. The boundary is the triangular edge. The interior is the region inside of this triangle.

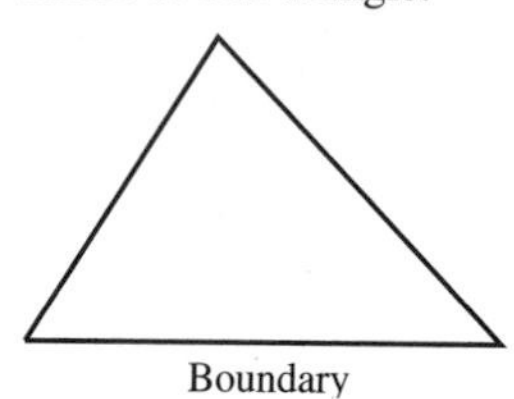

Boundary

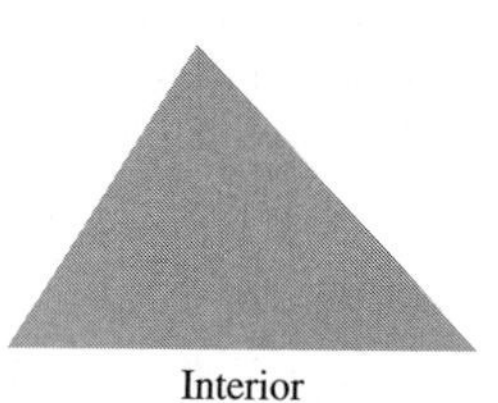

Interior

19. The boundary is the edge around the flower. The interior is the shaded region and does not include any boundary points.

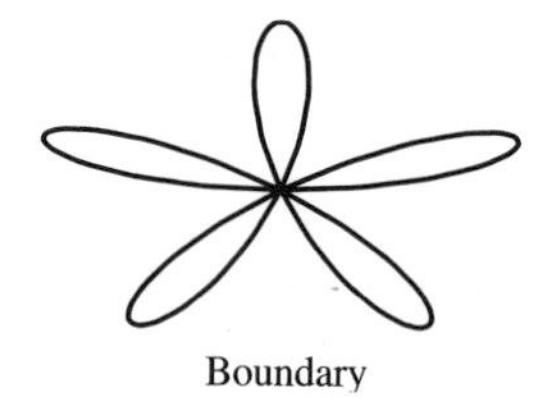

Boundary

Interior

21. The boundary is the circle of center $(0, 0)$ and radius 1. The interior is the set of all points inside of the boundary circle. The set is closed.

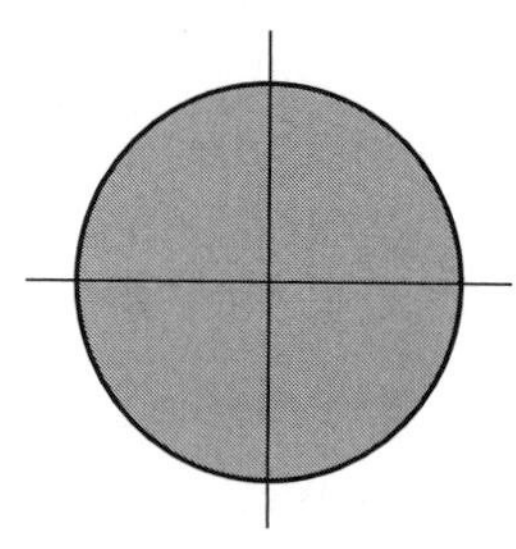

23. The boundary is the triangle with vertices $(0, 0)$, $(2, 0)$, and $(0, 2)$.
The interior is the set of all points inside of the boundary triangle. The set is closed.

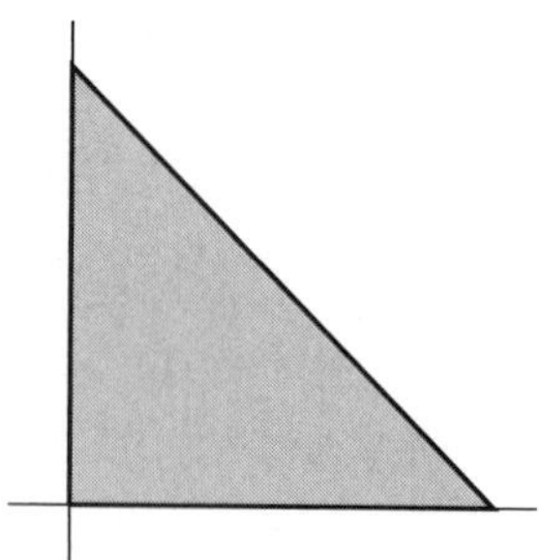

25. $\delta = 0.001$. Any smaller positive value is also correct.

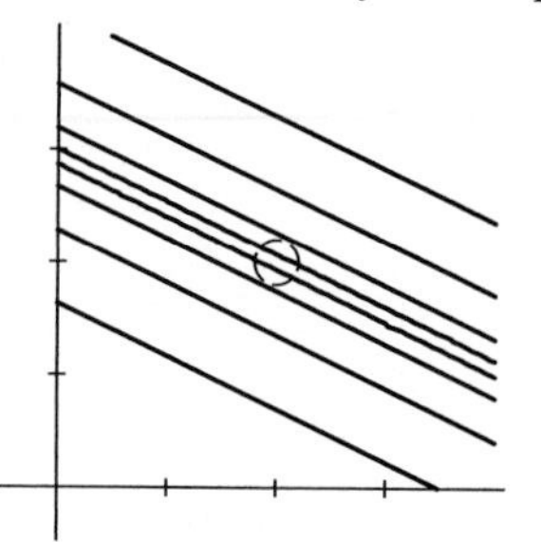

27. f is not continuous at points with positive x-coordinate on the line with equation $3x - 2y + 1 = 0$.

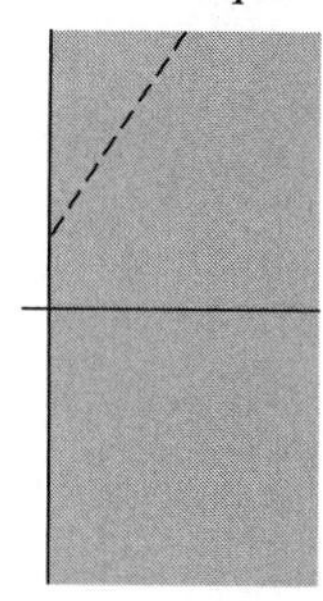

29. f is not continuous at points of D on or to the left of the line with equation $4x - 2y + 1 = 0$.

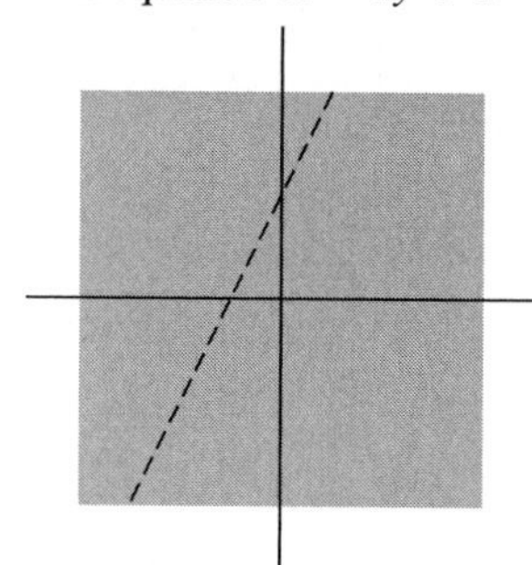

31. $\delta = 0.0001$ **33.** $\delta = 0.0025$ **35.** **b.** $\delta = 0.00002$ **37.** **b.** $\delta = 0.000004$ **43.** **a.** $\delta = 0.1$

9.7 Derivatives

1. $f_x(x, y) = -3$
$f_y(x, y) = \sqrt{2}$

3. $$f_x(x, y) = \frac{5y}{(-2x + y)^2(1 + y^2)}$$
$$f_y(x, y) = \frac{4x^2y + xy^2 - 4y^3 - 5x}{(-2x + y)^2(1 + y^2)^2}$$

5. $$f_x(x, y) = \frac{e^y(xy^2e^x - y^2e^x - 2e^x + e^y)}{(e^y - y^2e^x)^2}$$
$$f_y(x, y) = \frac{e^x(-xy^2e^y + 2xye^y - 4ye^x + 2e^y)}{(e^y - y^2e^x)^2}$$

7. $g_x(x,y,z) = y^2z^3$
$g_y(x,y,z) = 2xyz^3$
$g_z(x,y,z) = 3xy^2z^2$

9. $g_x(x,y,z) = \dfrac{4x(-y^2+z^2)}{(x^2-y^2+z^2)^2}$
$g_y(x,y,z) = \dfrac{4x^2y}{(x^2-y^2+z^2)^2}$
$g_z(x,y,z) = \dfrac{-4x^2z}{(x^2-y^2+z^2)^2}$

11. $g_x(x,y,z) = \cos y \sin 2y$
$g_y(x,y,z) = (x+z^2)(-\sin y \sin 2y + 2\cos y \cos 2y)$
$g_z(x,y,z) = 2z \cos y \sin 2y$

13. $\langle -3, \sqrt{2} \rangle$ **15.** $\left\langle \dfrac{2x+3y}{x^2+3xy+y^3}, \dfrac{3x+3y^2}{x^2+3xy+y^3} \right\rangle$

17. $\langle 0,0,0 \rangle$

19.
$$\left\langle \frac{2t_1^2t_2^{3/2}}{1+4t_1^2t_2^2} + 2t_1\sqrt{t_2}\arctan(2t_1t_2), \frac{2t_1^3\sqrt{t_2}}{1+4t_1^2t_2^2} + \frac{t_1^2}{2\sqrt{t_2}}\arctan(2t_1t_2), 0 \right\rangle$$

21. $-\dfrac{9}{5} - \dfrac{4\sqrt{2}}{5}$ **23.** $\dfrac{2a+3b}{a^2+3ab+b^3}$ **25.** 0 **27.** 0

29. $\mathbf{u} = \left\langle \dfrac{1}{\sqrt{2}}, -\dfrac{1}{\sqrt{2}} \right\rangle$, $D_{\mathbf{u}}f(P) = 11\sqrt{2}$
$\mathbf{v} = \left\langle -\dfrac{1}{\sqrt{2}}, \dfrac{1}{\sqrt{2}} \right\rangle$, $D_{\mathbf{v}}f(P) = -11\sqrt{2}$

31. $\mathbf{u} = \langle 1,0 \rangle$, $D_{\mathbf{u}}f(P) = 3e$
$\mathbf{v} = \langle -1,0 \rangle$, $D_{\mathbf{v}}f(P) = -3e$

33. $\pm\left\langle \dfrac{1}{\sqrt{2}}, \dfrac{1}{\sqrt{2}} \right\rangle$

35. $\pm\langle 0,1 \rangle$

37. ≈ 0.0001477 **39.** $\dfrac{\sqrt{2}}{2y}$ **41.** $\dfrac{\sqrt{5}}{18}$

43. $\langle 9/\sqrt{2}, 7/\sqrt{2} \rangle$

45. a.

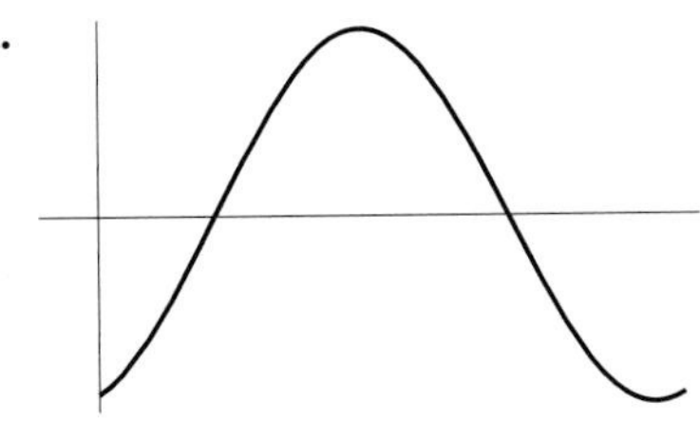

b. maximum at $\theta = \pi + \arctan(-1/3) \approx 2.82$
minimum at $\theta = 2\pi + \arctan(-1/3) \approx 5.961$
$d(\theta) = 0$ when $\theta = \arctan 3 \approx 1.249$ and $\theta = \pi + \arctan 3 \approx 4.391$
c. From $(1,-1)$, the direction of maximal increase for f is $\langle -3/\sqrt{10}, 1/\sqrt{10} \rangle$ and the direction of greatest decrease is $\langle 3/\sqrt{10}, -1/\sqrt{10} \rangle$.

47. a. $\mathbf{r}(t) = \langle 1, 2+t, 3+t \rangle$ **b.** $\mathbf{r}(t) = \langle 1+t, 2, 3+2t \rangle$
c. $-2x - y + z + 1 = 0$

49. a. $\mathbf{r}(t) = \langle 0, 1+t, 2t \rangle$ **b.** $\mathbf{r}(t) = \langle t, 1, 0 \rangle$
c. $2y - z - 2 = 0$

51. a. ≈ 0.707 **b.** ≈ 0.424
c. $g_x(0.5, 0.5) \approx 0.1$, $g_y(0.5, 0.5) \approx 0.8$

53. a. ≈ 1 **b.** ≈ 1.2 **c.** $\approx \langle -0.22, -6.4 \rangle$
d. For the direction for a maximum directional derivative, draw at $(-1,-1)$ a vector in the same direction as that of $\nabla g(-1,-1)$. This vector should be perpendicular to the level curve through $(-1,-1)$. For the direction for a minimum directional derivative, draw at $(-1,-1)$ a vector in the direction opposite that of $\nabla g(-1,-1)$. This vector should be perpendicular to the level curve through $(-1,-1)$.
e. The directions in which the directional derivative are 0 are tangent to the level curve at $(-1,-1)$.

55. a. The directional derivative is maximum in the direction of $\overrightarrow{P_0Q}$, minimum in the direction of $\overrightarrow{QP_0}$.
b. The directional derivative is 0 in directions tangent to the sphere of center P_0 that passes through Q; that is, in directions perpendicular to $\overrightarrow{P_0Q}$.
c. Although $\nabla d(P_0)$ is undefined, it can be shown, using the definition of the directional derivative, that the directional derivative is 1 in every direction from P_0.

57. a. The directional derivative is maximum in the direction perpendicular to the plane and pointing away from Π. The directional derivative is minimum in the direction perpendicular to the plane and pointing towards Π.
b. The directional derivative is 0 in directions parallel to Π.
c. Although ∇d is undefined on Π, the directional derivatives at a point on Π can be calculated using the definition of the directional derivative. The directional derivative is maximal in directions perpendicular to Π and minimal in directions parallel to Π.

59. a. in the direction $\langle -52, 50 \rangle$ as viewed from above
b. in the direction $\langle 52, -50 \rangle$ as viewed from above

61. in the direction $\langle -1.5, -0.6 \rangle$

Chapter Review Exercises

1. $y - 2 = -\dfrac{1}{4}(x-3)^2$ **3.** $-\dfrac{(x-2)^2}{35/4} + \dfrac{(y-3)^2}{1/4} = 1$

5. $x - 2 = -\dfrac{1}{8}y^2$

7. $f_x(x,y) = y^3\sec^2(xy^2)$, $f_y(x,y) = \tan(xy^2) + 2xy^2\sec^2(xy^2)$

9. $\dfrac{d\rho}{d\phi} = \dfrac{\cos\phi\sin\theta}{1-\cos\phi\cos\theta} - \dfrac{\cos\theta\sin\phi(1+\sin\phi\sin\theta)}{(1-\cos\phi\cos\theta)^2}$
$\dfrac{d\rho}{d\theta} = \dfrac{\cos\theta\sin\phi}{1-\cos\phi\cos\theta} - \dfrac{\cos\phi\sin\theta(1+\sin\phi\sin\theta)}{(1-\cos\phi\cos\theta)^2}$

11. $h_u(u,v,w) = \dfrac{2\cos(u)\sin(u)}{1+\cos(2w)^2+\sin(u)^2+\sin(v)^6}$
$h_v(u,v,w) = \dfrac{6\cos(v)\sin(v)^5}{1+\cos(2w)^2+\sin(u)^2+\sin(v)^6}$
$h_w(u,v,w) = \dfrac{-4\cos(2w)\sin(2w)}{1+\cos(2w)^2+\sin(u)^2+\sin(v)^6}$

13. $-25/\sqrt{2}$ **15.** $-4/9$ **17.** $(-3\sqrt{2} + 2\sqrt{3} + \sqrt{6})/24$

19. direction: $\mathbf{u} = \left\langle -\dfrac{14}{\sqrt{317}}, \dfrac{11}{\sqrt{317}} \right\rangle$
$D_{\mathbf{u}}f(-2,1) = \sqrt{317}$

21. direction: $\mathbf{u} = \left\langle -\dfrac{1}{\sqrt{5}}, -\dfrac{2}{\sqrt{5}}, 0 \right\rangle$

$D_{\mathbf{u}}f(1, 1, -1) = \sqrt{5}$

23. Cross sections perpendicular to the x- and y-axes are parabolas. Cross sections perpendicular to the z-axis are circles.

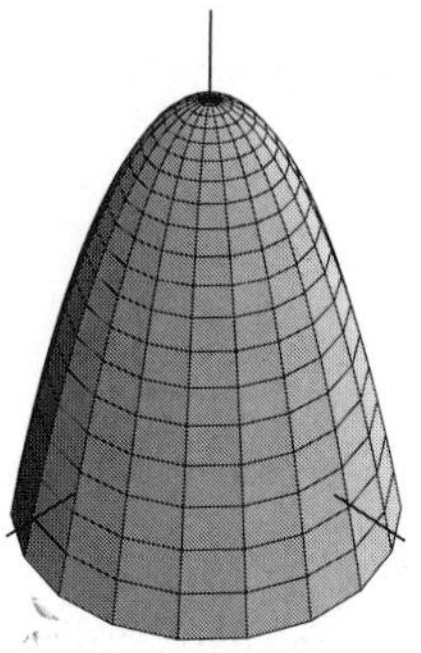

25. Cross sections perpendicular to the x- and z-axes are hyperbolas. Cross sections perpendicular to the y-axis are circles.

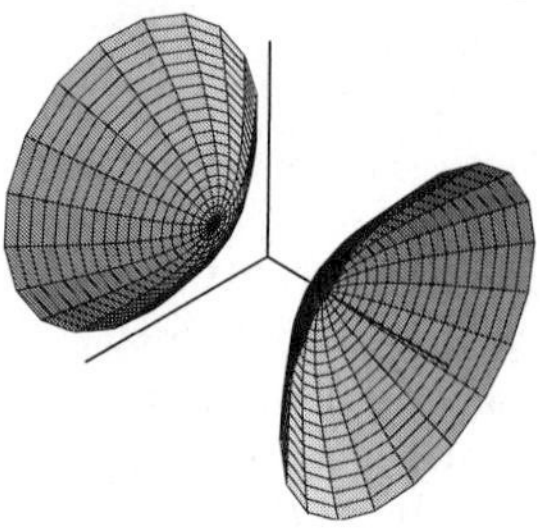

27. $\mathbf{r}(u, v) = \langle u \cos v, u \sin v, 4 - u^2 \rangle$, $u \geq 0$, $0 \leq v \leq 2\pi$

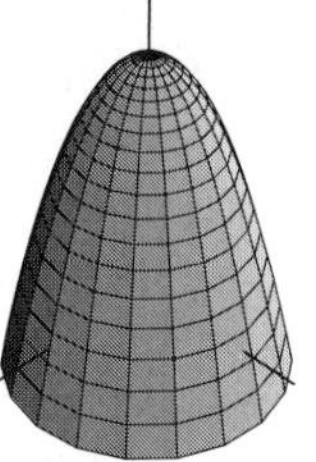

29. $\mathbf{r}(t, c) = \left\langle \dfrac{1}{2}t^2 + \dfrac{1}{2}c^2, c, t \right\rangle$, $c \geq 0$, $-\infty < t < \infty$

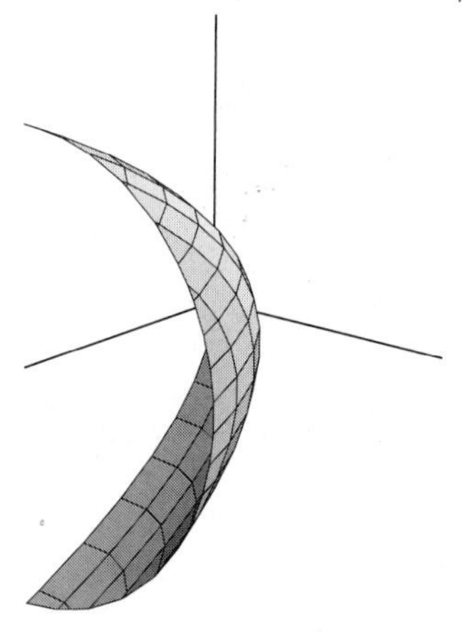

31. $z = 3 + r^2$

$\mathbf{w}(r, \theta) = \langle r \cos \theta, r \sin \theta, 3 + r^2 \rangle$

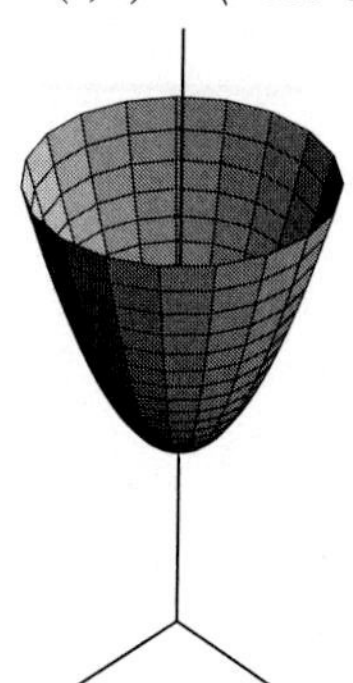

33. $\rho = 2$

$\mathbf{r}(\theta, \phi) = \langle 2 \cos \theta \sin \phi, 2 \sin \theta \sin \phi, 2 \cos \phi \rangle$

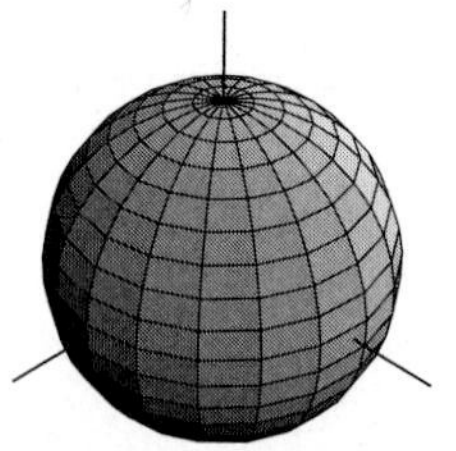

35. $\rho = 4 \sec \theta \csc \phi$

$\mathbf{r}(\theta, \phi) = \langle 4, 4 \tan \theta, 4 \sec \theta \cot \phi \rangle$

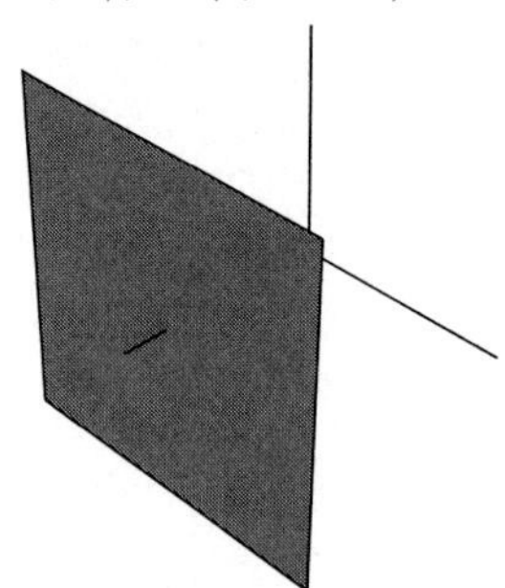

37. 17/2 **39.** 0 **41.** $\delta = 0.3$

43. Assume distances are in centimeters, time in seconds.

$$\frac{\partial s}{\partial x} = \frac{\pi}{120} \cos(20\pi t) \cos(\pi x/30) + \frac{\pi}{240} \cos(40\pi t) \cos(\pi x/15)$$

with units of cm/cm.

$$\frac{\partial s}{\partial t} = -5\pi \sin(20\pi t) \sin(\pi x/30) - \frac{5\pi}{2} \sin(40\pi t) \sin(\pi x/15)$$

with units of cm/s.

45. Assume distances are in meters, time in seconds, and mass in kilograms.

$\dfrac{\partial F}{\partial x} = -\dfrac{2xGmM}{(x^2 + y^2 + z^2)^2}$ with units of newtons/m.

$\dfrac{\partial F}{\partial y} = -\dfrac{2yGmM}{(x^2 + y^2 + z^2)^2}$ with units of newtons/m.

$\dfrac{\partial F}{\partial z} = -\dfrac{2zGmM}{(x^2 + y^2 + z^2)^2}$ with units of newtons/m.

$$\frac{\partial F}{\partial M} = \frac{Gm}{x^2 + y^2 + z^2}$$ with units of newtons/kg.

$$\frac{\partial F}{\partial m} = \frac{GM}{x^2 + y^2 + z^2}$$ with units of newtons/kg.

47. Assume distances are in meters, time in seconds, and mass in kilograms.

$$\frac{\partial s}{\partial r} = \ln\left(\frac{M}{m}\right)$$ with units of (m/s)/(m/s).

$$\frac{\partial s}{\partial M} = \frac{r}{M}$$ with units of (m/s)/kg.

$$\frac{\partial s}{\partial m} = -\frac{r}{m}$$ with units of (m/s)/kg.

CHAPTER 10—DIFFERENTIABLE FUNCTIONS OF SEVERAL VARIABLES

10.1 Differentiability

1. 3.50400; 3.30000; 0.204000
3. 0.322083; 0.350000; 0.0279165
5. 0.128800; 0.100000; 0.0288000
7. $(21 \quad 13)$; $21(x-1) + 13(y-1)$
9. $(\sqrt{3}/2 \quad -\pi\sqrt{3}/12)$; $(\sqrt{3}/2)(x - \pi/6) - (\pi\sqrt{3}/12)(y-1)$
11. $(-2/e \quad 1/e)$; $(-2/e)(x-1) + (1/e)y$
13. $(1/\sqrt{3} \quad 2\sqrt{2/3})$; $(1/\sqrt{3})(x-1) + 2\sqrt{2/3}(y - 1/\sqrt{2})$
15. $(-1 \quad 2 \quad -1)$; $-(x-1) + 2(y+2) - (z-1)$
17. $\mathbf{n} = \langle -8, 6, 1\rangle$; $z = 5 + 8(x-2) - 6(y-1)$
19. $\mathbf{n} = \langle -e^2, 2e^2, 1\rangle$; $z = e^2 + e^2(x-2) - 2e^2(y-1)$
21. $\mathbf{n} = \langle 6\cos(13), -4\cos(13), 1\rangle$;
$z = \sin(13) - 6\cos(13)(x+3) + 4\cos(13)(y-2)$
23. Approximate percentage error would be 6%.
25. 1.885 ft^3
27. $x = (1/\sqrt{2})(1 + (u-1) - (v - \pi/4))$
$y = (1/\sqrt{2})(1 + (u-1) + (v - \pi/4))$
$z = 1 + 2(u-1)$
From these equations, $\mathbf{r}(u_1, v_1) \approx \langle 0.767492, 0.788143, 1.20000\rangle$, while $\mathbf{r}(u_1, v_1) = \langle 0.766377, 0.789092, 1.21000\rangle$.
29. $x = 1 + (1/e)(u-e)$
$y = 1 + (1/e)(v-e)$
$z = e^2 + e(u-e) + e(v-e)$
From these equations, $\mathbf{r}(u_1, v_1) \approx \langle 0.956487, 1.03006, 7.28967\rangle$, while $\mathbf{r}(u_1, v_1) = \langle 0.955511, 1.02962, 7.28000\rangle$.
31. $x = u + 4(v - \pi/4)$
$y = \sqrt{2}(2 + (u-2) + 2(v - \pi/4)$
$z = 4 + 4(u-2)$
From these equations, $\mathbf{r}(u_1, v_1) \approx \langle 2.25841, 3.15257, 4.8\rangle$, while $\mathbf{r}(u_1, v_1) = \langle 2.26520, 3.15771, 4.84000\rangle$.
33. $\mathbf{f}$ is differentiable because all coordinate functions have continuous partials;

$$\mathbf{Df}(1, \pi/6) = \begin{pmatrix} \sqrt{3}/2 & -1/2 \\ 1/2 & \sqrt{3}/2 \end{pmatrix}$$

$$\mathbf{df}(\mathbf{x} - \mathbf{a}) = \begin{pmatrix} 0.359808 \\ -0.0232051 \end{pmatrix}.$$

35. The linearization is $g(x,y) = e + e(x-1) - e(y-1)$. The absolute values of the difference between f and its linearization at each of the points $(1.1, 1.1)$, $(1.1, 0.9)$, $(0.9, 1.1)$, and $(0.9, 0.9)$ are 0, 0.132785, 0.0917499, and 0. The error is largest at the point $(1.1, 0.9)$.
37. Using the differential for the calculation, the worst error is about 0.004 s.
39. The values of $T(L,g) - g(L,g)$ at $(1.1, 9.82)$, $(1.1, 9.8)$, $(0.9, 9.82)$, and $(0.9, 9.8)$ are -0.0024, -0.0023, -0.0026, and -0.0027.

10.2 The Chain Rule

1. $dy/dt = 2e^{2t} - 2e^{-2t}$
3. $\frac{d}{dt}f(\mathbf{g}(t)) = (\cos\sqrt{t})/(2\sqrt{t})$
5. $dy/dt = 1/(t^2+1)$ **7.** $d(f(\mathbf{g}(t)))/dt = 3e^{3t}$
9. $dy/dt = -3t/\sqrt{1-t^2}$ **11.** $z_u = 0$; $z_v = 1 + \tan^2 v$
13. $z_u = uf_x + vf_y$; $z_v = -vf_x + uf_y$
15. $\nabla z(0, 1/2) = \langle 0, -\sqrt{2}\rangle$
17. $$\begin{pmatrix} 2t\cos(t^2)\cos(t^3) - 3t^2\sin(t^2)\sin(t^3) \\ -3t^2\cos(t^3)\sin(t^2) - 2t\cos(t^2)\sin(t^3) \end{pmatrix}$$
19. $$\begin{pmatrix} 1 + 2t \\ -1/t^2 \end{pmatrix}$$ **21.** $$\begin{pmatrix} (5/2)t^{3/2} \\ (3/2)t^{1/2} \\ 2t \end{pmatrix}$$
23. $$e^t\begin{pmatrix} \cos(2t^2) - 4t\sin(2t^2) \\ 4t\cos(2t^2) + \sin(2t^2) \\ -4t - 2t^2 \end{pmatrix}$$
25. Because $u(\mathbf{r}(t)) = 6$ for all t, C is a subset of a level curve of u; $\mathbf{r}(\pi/4) = \mathbf{a}$; $\nabla u(\mathbf{a}) \cdot \mathbf{r}'(\pi/4) = 0$.
27. Because $u(\mathbf{r}(t)) = 1$ for all t, C is a subset of a level curve of u; $\mathbf{r}(\ln(\sqrt{3}+2)) = \mathbf{a}$; $\nabla u(\mathbf{a}) \cdot \mathbf{r}'(\ln(\sqrt{3}+2)) = 0$.
29. $T'(\pi/2) \approx 18.28$ degrees Fahrenheit per length unit; $T(1.5) \approx 148.7$ degrees Fahrenheit
31. $\langle -1/4, 0, 3/4\rangle$
33. $$\frac{\partial f_1}{\partial x} = \cos\psi\frac{\partial f_1}{\partial X} + \sin\psi\frac{\partial f_1}{\partial Y}$$
$$\frac{\partial f_1}{\partial y} = -\sin\psi\frac{\partial f_1}{\partial X} + \cos\psi\frac{\partial f_1}{\partial Y}$$
$$\frac{\partial f_1}{\partial z} = \frac{\partial f_1}{\partial Z}$$
35. $$\mathbf{Df}(1,1) = \begin{pmatrix} e & e \\ e & e \end{pmatrix}$$
$$\mathbf{Dg}(1,0) = \begin{pmatrix} 1 & 0 \\ 0 & 1 \end{pmatrix}$$
$$\mathbf{Dz}(1,0) = \begin{pmatrix} e & e \\ e & e \end{pmatrix}$$
$\mathbf{z}(1.05, 0.03) \approx \langle 2.93574, 2.93574\rangle$; the matrices associated with the linear transformations $\mathbf{df}(1,1)$, $\mathbf{dg}(1,0)$, and $\mathbf{dz}(1,0)$ are $\mathbf{Df}(1,1)$, $\mathbf{Dg}(1,0)$, and $\mathbf{Dz}(1,0)$, which are listed above.

10.3 Applications of the Chain Rule

1. $\nabla f(x_0, y_0) = \langle 2, 3\rangle$; solving $f(x,y) = k$ for y and calculating the slope of this curve at (x_0, y_0) we find $y' = -2/3$, from which a tangent vector to the curve is $\langle 3, -2\rangle$; this tangent vector is perpendicular to $\nabla f(x_0, y_0)$.

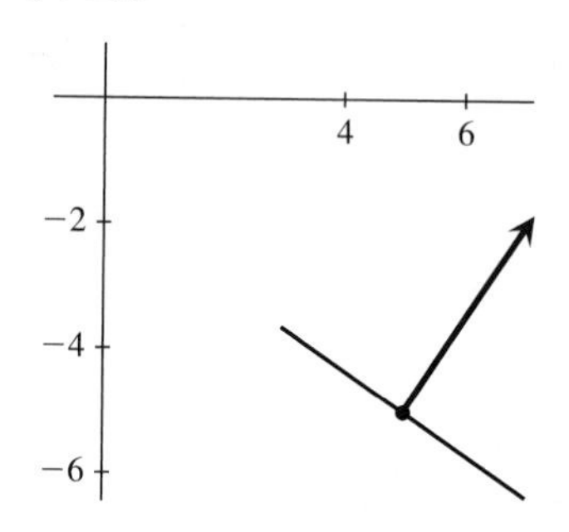

3. $\nabla f(x_0, y_0) = \langle -4, 1\rangle$; solving $f(x, y) = k$ for y and calculating the slope of this curve at (x_0, y_0) we find $y' = 4$, from which a tangent vector to the curve is $\langle 1, 4\rangle$; this tangent vector is perpendicular to $\nabla f(x_0, y_0)$.

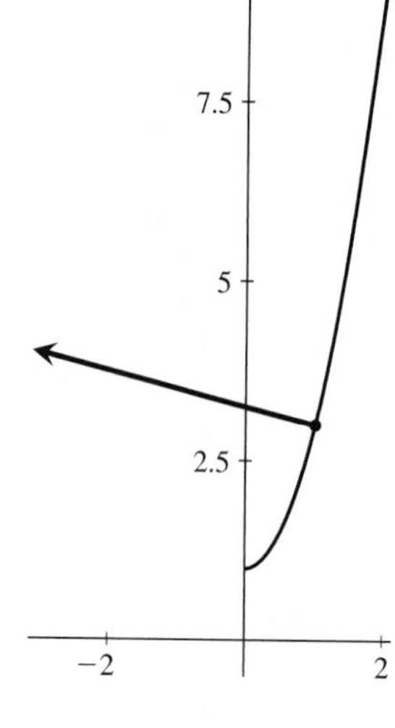

5. $\nabla f(x_0, y_0) = \langle 4, 8\rangle$; solving $f(x, y) = k$ for y and calculating the slope of this curve at (x_0, y_0) we find $y' = -1/2$, from which a tangent vector to the curve is $\langle 1, -1/2\rangle$; this tangent vector is perpendicular to $\nabla f(x_0, y_0)$.

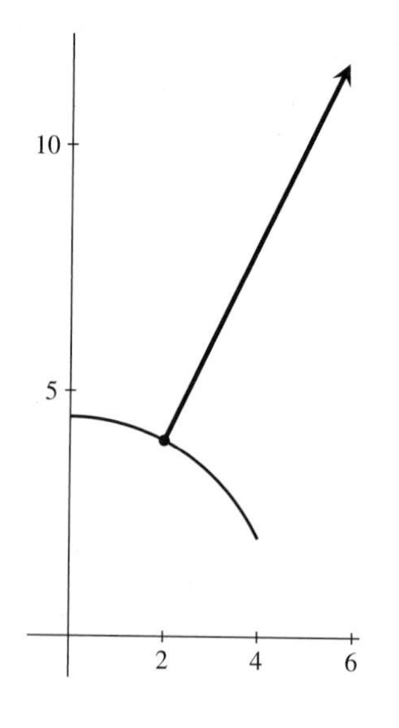

7. $\langle -f_x(x_0, y_0), -f_y(x_0, y_0), 1\rangle = \langle 1/3, -4/3, 1\rangle$; $\nabla F(x_0, y_0, z_0) = \langle -1, 4, -3\rangle$; yes, they are parallel; $-(x-1) + 4(y-1) - 3(z-1) = 0$
9. $\langle -f_x(x_0, y_0), -f_y(x_0, y_0), 1\rangle = \langle -\sqrt{3}, -\sqrt{3}, 1\rangle$; $\nabla F(x_0, y_0, z_0) = \langle -\sqrt{3}/2, -\sqrt{3}/2, 1/2\rangle$; yes, they are parallel; $-\sqrt{3}(x - \pi/6) - \sqrt{3}(y - \pi/6) + z = 0$
11. $\nabla F(3, 1, 1) = \langle 6, 2, 2\rangle$, which is perpendicular to the vectors $\mathbf{a} = \langle 3, 1, -10\rangle$ and $\mathbf{b} = \langle -1, 3, 0\rangle$ tangent, respectively, to the meridian and parallel of latitude through $(3, 1, 1)$.

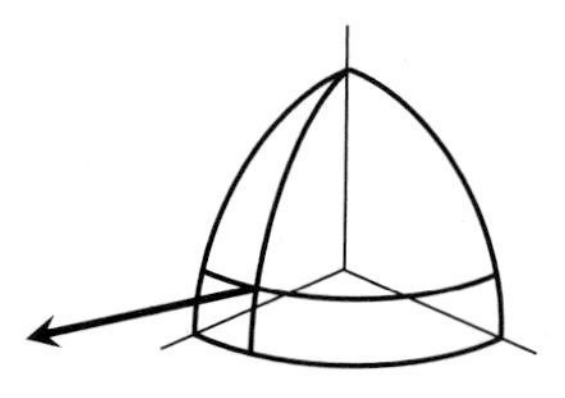

13. With $F(x, y, z) = x^2 - y^2 - z$, $\nabla F(2, 1, 3) = \langle 4, -2, -1\rangle$; an equation of the tangent plane at $(2, 1, 3)$ is $4(x-2) + (-2)(y-1) + (-1)(z-3) = 0$.
15. $y' = 1$ **17.** $y' = 1$
19. $f_x(x_0, y_0) = -6/(\pi - 6)$; $f_y(x_0, y_0) = -\pi/(\pi - 6)$; $\langle 6/(\pi - 6), \pi/(\pi - 6), 1\rangle$
21. $f_x(x_0, y_0) = 0$; $f_y(x_0, y_0) = 1/3$; $\langle 0, -1/3, 1\rangle$
23. $f_{xx} = 6xy^2$ **25.** $f_{yy} = x^5 e^{xy}$
27. $f_{xxy} = 2\cos(xy) - x^2y^2\cos(xy) - 4xy\sin(xy)$
33. $1 + \frac{1}{2}(x-1) + \frac{1}{2}(y-1) - \frac{1}{8}(x-1)^2 + \frac{1}{4}(x-1)(y-1) - \frac{1}{8}(y-1)^2$
35. $1 + x + \frac{1}{2}x^2 - \frac{1}{2}y^2$
37. Error is less than 0.0644 for $(x, y) \in S_{0.5}$.
41. $y' = \dfrac{x^2yf_u - y^3f_v}{x^3f_u - xy^2f_v}$
43. Recalling that $y' = -F_x/F_y$, it may happen that $F_y \to 0$ as we approach a vertical asymptote.
51. No, because if the second-order partial derivatives of f were continuous at $(0, 0)$, then $f_{xy}(0, 0) = f_{yx}(0, 0)$.

10.4 Further Applications of the Chain Rule

13. $U_{rr} + U_{ss} = 0$ **15.** $U_{rr} + U_{ss} = 0$
17. $-\frac{1}{4}(s^2 - r^2) - U_s + U_{rr} - U_r - U_{rr} = 0$
19. $-(r\cos\theta - s\sin\theta)(r\sin\theta + s\cos\theta) + \sin(\theta)U_s + \cos(2\theta)U_{rs} - \cos(\theta)U_r - \frac{1}{2}\sin(2\theta)(U_{ss} - U_{rr}) = 0$
23. No, the second derivative fails to exist at $x = \pm 1$. In the derivation leading to (21), it was assumed that u has continuous second partial derivatives. If this is true, then the initial condition $u(x, 0) = f(x)$ cannot hold everywhere.

29. Points on the 0.2 contour, as calculated by Newton's method and following the above outline, are

$(0, 3, 0.38), (0.6, 1.04), (0.9, 1.37), (1.2, 1.54), (1.5, 1.61), (1.8, 1.58)$

The curve containing these points will resemble the 0.2 contour in the figure for this problem.

10.5 Optimization

1. $(1/3, 0)$ and $(1, 0)$ **3.** $(-2, -1)$
5. $(0, \pi/2)$ and $(0, 3\pi/2)$
7. $(0, 0)$ and $(1, 1)$ **9.** $(1, 1), (-1, -1)$, and $(0, 0)$
11. $(-2, -2, 2), (-2, 2, -2), (0, 0, 0), (2, -2, -2)$, and $(2, 2, 2)$
13. Absolute minimum is 0 at $(0, 0)$; absolute maximum is 6 at $(1, 1)$
15. Absolute minimum of 0 at $(0, 0)$; absolute maximum of $4e^{-3}$ at $(2, 1)$
17. Absolute minimum of 0 at $(0, 1)$; absolute maximum of 4 at $(1, 0)$
19. Absolute minimum of -10 at $(0, 10)$; absolute maximum of 27 at $(6, 9)$
21. Absolute minimum of $-17/8$ at $(-1/4, 0)$; absolute maximum of 8 at $(2, 0)$
23. 4 ft^3. We assume that there is in fact an absolute maximum volume. If the depth h is eliminated through the glass side condition, the domain of the volume function (of w and ℓ) is in the first quadrant, beneath the hyperbola described by $w\ell = 12$. The volume function is everywhere positive and tends towards 0 on boundaries. There is just one candidate on the interior of the region, $(w, \ell) = (2, 2)$, which gives an absolute maximum.
25. We assume that there is a point closest to the origin. The point closest to the origin is $\left(2^{1/8}, 2^{-3/8}, 2^{-3/8}\sqrt{5}\right)$. By symmetry, we may ignore all but the first quadrant of the (x, y)-plane. The square of the distance function is everywhere positive and becomes large along the axes and far away from the origin. There is just one candidate on the interior of the region, $(2^{1/8}, 2^{-3/8})$, which gives an absolute minimum.
27. The dimensions are $a/3$ by $b/3$ by $c/3$. We may take the domain of the volume function to be a triangle and its interior. Hence, the volume function has a maximum. On the boundaries of the domain the volume function is 0; the only candidate in the interior of the domain is $(a/3, b/3)$ (if the domain lies in the (x, y)-plane). This will give the absolute maximum.
29. Absolute minimum of 0 at $(0, 0)$; absolute maximum of $3\sqrt{3}/2$ at $(\pi/3, \pi/3)$
31. If the sides are a, b, and c, then the triangle with maximum area has sides of length $p/3$; its area is $p^2/(12\sqrt{3})$. Because the area may be as close to 0 as desired, there is no triangle with minimum area.
35. $\mathbf{a}_2 = \langle 1/3, 0\rangle$; $\langle 1/3, 1/30\rangle$ is a point near $\mathbf{a}_2$ at which f is smaller.
37. $\mathbf{a}_2 = \langle -3/14, 3/14, -11/14\rangle$; $\langle -29/140, 29/140, -27/35\rangle$ is a point near $\mathbf{a}_2$ at which f is smaller.
39. The absolute maximum of f is $3\sqrt{3}a^2/32$ and it occurs at $(2/\sqrt{3}, 0)$.

10.6 Second Derivatives Test

1.

3. There is one stationary point, $(0, 0)$; it is a saddle point.
5. There is one stationary point, $(0, -3/2)$; it is a saddle point.
7. There are two stationary points, $(0, 0)$ and $(1, 1)$; the first is a saddle point and the second is a local minimum.
9. There are three stationary points, $(0, 0)$, $(-1, -1)$, and $(1, 1)$; the first is a saddle point and the second and third are local minima.
11. There is one stationary point, $(1, 1)$; it is a saddle point.
17. There is one stationary point, approximately $(1.98124, 1.87345)$, which is a saddle point.
23. The point near $(0, 0)$ is a local minimum and the points near $(0.4, 0.8)$ and $(-0.4, -0.8)$ are saddle points.
25. For $k > -1/4$ the graph will have one stationary point, $(-2/(1 + 4k), 4/(1 + 4k))$, which is a saddle point; for $k < -1/4$ the graph will have one stationary point, $(-2/(1 + 4k), 4/(1 + 4k))$, which is a local minimum; for $k = -1/4$, the graph has no stationary points.
29. The equation of the least squares line is $y = 0.764x + 39.052$.

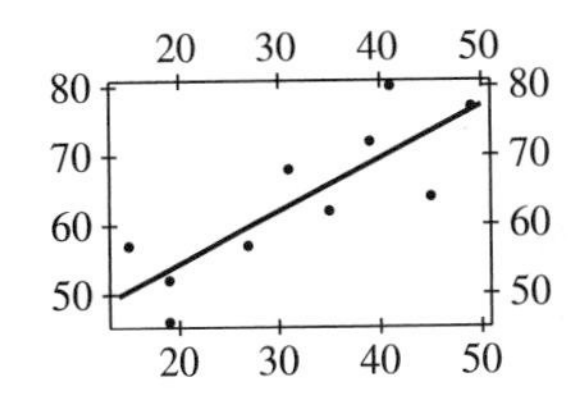

33. The surface has an absolute minimum at $(0, 0, 0)$.

10.7 Optimization with Constraints

1. $\left(\pm 2\sqrt{2}, 2\right)$ **3.** $\left(2, \sqrt{2}\right)$ **5.** $(\pm 2^{1/10}, \pm 2^{1/10}, 2^{-2/5})$
7. $\left(6/\sqrt{11}, 2/\sqrt{11}, -2/\sqrt{11}\right)$
9. $(1/3, 1/3, 1/3)$
11. Height 36 inches and radius $36/\pi$ inches
13. $500 \cdot 6^{2/3}$ units of labor and $2000(4/3)^{1/3}$ units of capital
15. $\left(\frac{bc}{P}, \frac{ac}{P}, \frac{ab}{P}\right)$ where $P = \sqrt{b^2c^2 + a^2c^2 + a^2b^2}$
17. The dimensions are 1 by $2/3$ by $1/2$.
21. They make f a minimum on the sphere.
23. The distance is $\dfrac{\sqrt{d^2(a^2 + b^2 + c^2)}}{a^2 + b^2 + c^2}$.
25. $\sqrt{30 + 20\sqrt{2}}$ and $\sqrt{30 - 20\sqrt{2}}$

Chapter Review Exercises

3. $z = 2 + \frac{1}{2}(x-1) + \frac{1}{2}(y-1)$

5. $f(x,y) \approx \frac{1}{2} - \frac{4}{3}\left(x - \frac{3}{4}\right) - \left(y - \frac{1}{2}\right)$

The percentage error is about 7.9 percent.

7. It crosses the sphere $x^2 + y^2 + z^2 = 10$ at $t \approx 0.149021$ seconds. A normal to the surface with equation $F(x,y,z) = x^2 + 2y^2 + 3z^2 - 6 = 0$ is $\nabla F(x,y,z)$. At $(1,1,1)$ the normal is $\langle 2,4,6\rangle$. The motion of the particle is described by $\mathbf{r} = \mathbf{r}(t) = \langle 1,1,1\rangle + \left(10/\sqrt{14}\right)t\langle 1,2,3\rangle$, for $t \geq 0$. Substituting into the equation $x^2 + y^2 + z^2 = 10$ gives an equation from which $t \approx 0.149021$ can be calculated.

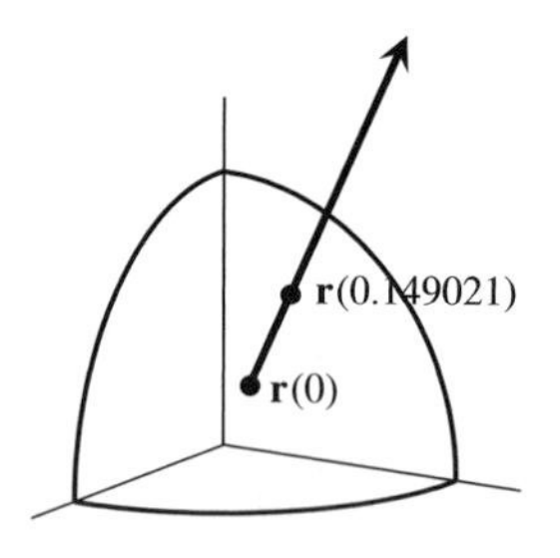

9. $H_s(-1,2) = -12$

11. $Df(1,2) = (4 \quad 13)$
$f(x,y) \approx 10 + 4(x-1) + 13(y-2)$

13. $w_t = 25$ and $w_{tu} = -9$

15. $z = 5 + 8(x-1) + 2(y-1)$; $\mathbf{N} = \left(1/\sqrt{69}\right)\langle -8,-2,1\rangle$

17. $dz/dt = f_x\left(1/t, \sqrt{t}\right)\left(-1/t^2\right) + f_y\left(1/t, \sqrt{t}\right)\left(1/\left(2\sqrt{t}\right)\right)$

19. $\mathbf{v}(\pi) = \langle 0,-2,3.89\rangle$ length units per time unit

21. $rU_r - sU_s = 2r$

23. $\begin{pmatrix} \frac{\partial f_1}{\partial x_1} & \frac{\partial f_1}{\partial x_2} \\ \frac{\partial f_2}{\partial x_1} & \frac{\partial f_2}{\partial x_2} \end{pmatrix} \begin{pmatrix} 2t_1 & 2t_2 \\ t_2 & t_1 \end{pmatrix}$

25. Thinking of $H(u,v) = f(x,y)$, where $x = uv$ and $y = u/v$,

$$\partial^2 H/\partial v^2 = \frac{u^2v^4 f_{xx} - 2u^2v^2 f_{xy} + 2uvf_y + u^2 f_{yy}}{v^4},$$

where the partials of f are evaluated at $(uv, u/v)$.

27. $f'(x) = 1$

29. $(-1/3,-1/3)$ and $(1,1)$ are stationary points; the first is a saddle point, the second is a local minimum.

31. f takes on its maximum value of 1 at $(1,0)$ and $(1,3)$; f takes on its minimum value of $-5/4$ at $(1,3/2)$.

33. The function f has two stationary points, $\left(-\sqrt{3}/3, \sqrt{3}\right)$ and $\left(\sqrt{3}/3, -\sqrt{3}\right)$, both of which are saddle points. If f has a maximum or a minimum, it would appear among the stationary points; hence, f has neither a maximum nor a minimum on R^2.

35. $x = 100a/(a+b+c)$, $y = 100b/(a+b+c)$, $z = 100c/(a+b+c)$

37. The hottest point is in the middle, at $(0,0)$; the coldest point is at $(1,1)$.

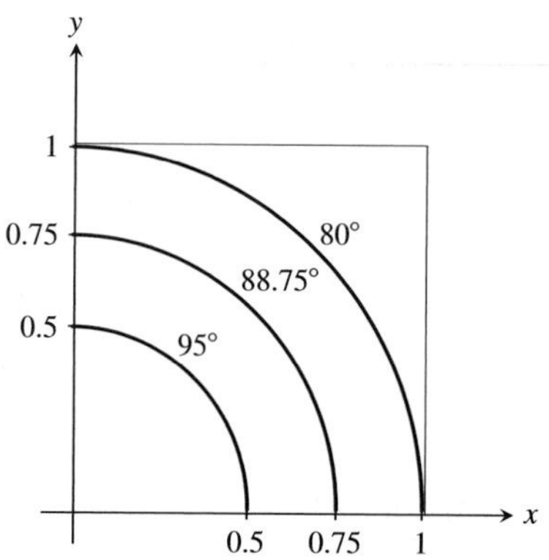

CHAPTER 11—MULTIPLE INTEGRALS

11.1 The Double Integral on Rectangles

1. $R_2 = 3.5$ **3.** $R_3 = 39.1111$ **5.** $R_2 = 1.01295$

7. From Exercise 1, $R_2 = 3.5$; volume is 3.5 cubic units; percentage error is 0

9. From Exercise 3, $R_3 = 39.1111$; volume is 48 cubic units; percentage error is about 19%

11. From Exercise 5, $R_2 = 1.01295$; volume is 1 cubic unit; percentage error is about 1%

13. 29.62×10^4 square-mile inches

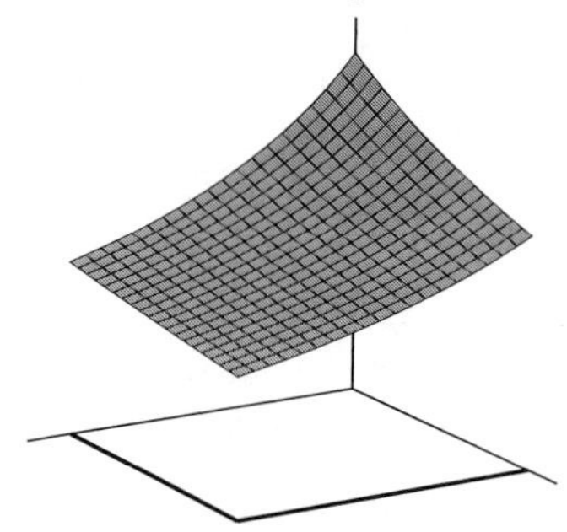

15. If we believe that the volume of the region beneath the graph of r and above $\mathcal{D}$ is equal to the amount of rain received by $\mathcal{D}$, then the volume $100^2 w$ of the pool is equal to V. The value of w is equal to the average value of the rainfall function on $\mathcal{D}$ because the product of the area of $\mathcal{D}$ and w is equal to the amount of rain received by $\mathcal{D}$. Moreover, this fits with the definition of average value given for one-dimensional integrals, that is, $f_{\text{ave}} = (1/(b-a))\int_a^b f(x)\,dx$. We may take $(x^*, y^*) \approx (42.3250, 50)$.

17. $2\cdot 6\cdot \frac{12}{5} + \frac{1}{2}\left(2\cdot 6\cdot\left(8 - \frac{12}{5}\right)\right) = 62.4$ cubic meters

19.

$$U_n = L_n - \frac{1}{n^2}\left(f(0,0) + \sum_{j=1}^{n-1} f(0,j/n) + \sum_{i=1}^{n-1} f(i/n,0)\right) + \frac{1}{n^2}\left(f(1,1) + \sum_{j=1}^{n-1} f(1,j/n) + \sum_{i=1}^{n-1} f(i/n,1)\right)$$

If one were using the sums L_n and U_n to approximate an integral, a significant savings could be realized by using the above formula for U_n, particularly for large n.

11.2 Extending the Double Integral and Applications

1. $1/6$

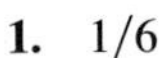

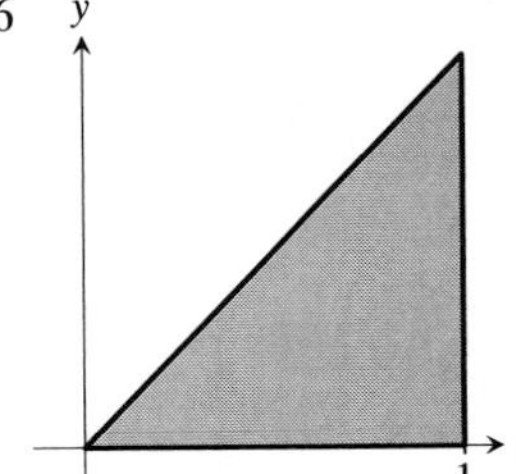

3. $-\frac{3}{4} + 2\ln 2$

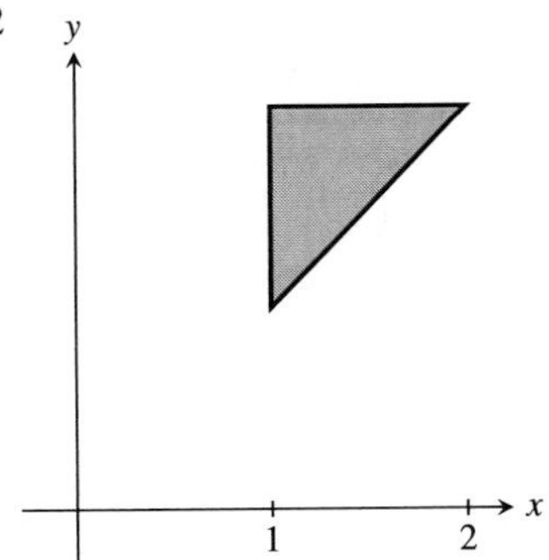

5. $\frac{1}{2} - \frac{1}{2}\cos 1$

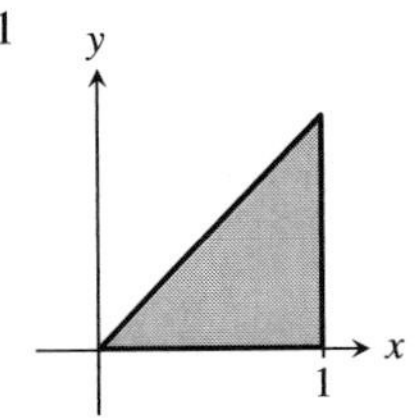

7. $\int_1^2 \int_0^{2-x} f(x, y)\, dy\, dx$ **9.** $\int_{-1/\sqrt{2}}^{\sqrt{2}} \int_{y^2}^{1+y/\sqrt{2}} f(x, y)\, dx\, dy$

11. $\int_{-1/\sqrt{2}}^{1/\sqrt{2}} \int_{x^2}^{1-x^2} f(x, y)\, dx\, dy$

13. $\int_0^2 \int_0^{y/2} xy^2\, dx\, dy = 4/5$

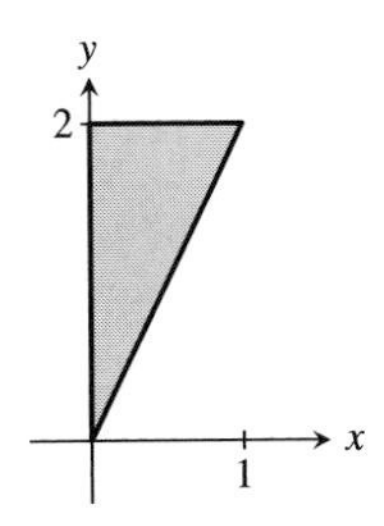

15. $\int_0^{\pi/4} \int_0^{\tan x} y\, dy\, dx = \frac{1}{8}(4 - \pi)$

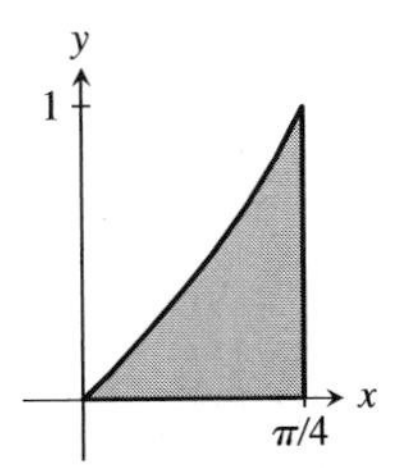

17. $1/3$ **19.** 4 **21.** $7/20$ **23.** 3π **25.** $16a^3/3$

29. $\mathbf{R} = \langle 163/185, 153/74 \rangle$; $m = 37k/12$

31. $\mathbf{R} = \langle 0, 3/(2\pi) \rangle$

33. $\int_0^1 \int_{-\sqrt{x}}^{\sqrt{x}} (4 - x)\, dy\, dx + \int_1^4 \int_{x-2}^{\sqrt{x}} (4 - x)\, dy\, dx$

35. It appears that all choices of domains in the (x, y)-plane give regions requiring at least two separate iterated integrals, because of the combination of straight and curved boundaries.

37. $k\int_0^2 \int_y^{5y/2} x\, dx\, dy + k\int_2^3 \int_y^{9-2y} x\, dx\, dy$

43. $\pi/2$ **45.** $2\pi^2 r^2 b$

47. The height of the element is $z = 4\sqrt{89}/15$. The curve in the plane with equation $y = 2$ can be described by

$$\mathbf{r}(\theta) = \left\langle \left(3\sqrt{21}/5\right)\cos\theta, 2, \left(4\sqrt{21}/5\right)\sin\theta \right\rangle,$$

where $0 \le \theta \le 2\pi$. The curve in the plane with equation $x = 2$ can be described by

$$\mathbf{r}(\theta) = \left\langle 2, \left(5\sqrt{5}/3\right)\cos\theta, \left(4\sqrt{5}/3\right)\sin\theta \right\rangle,$$

where $0 \le \theta \le 2\pi$.

11.3 Surface Area

1. 14 square units

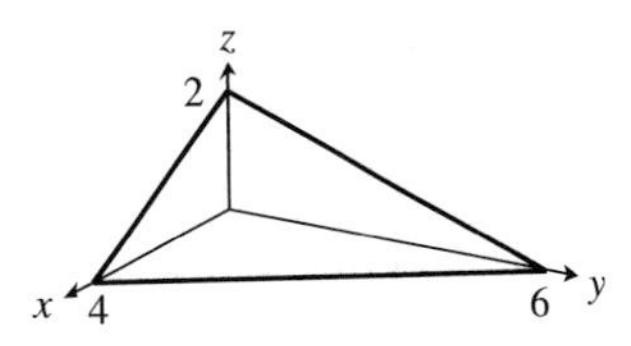

3. 6.88676 square units

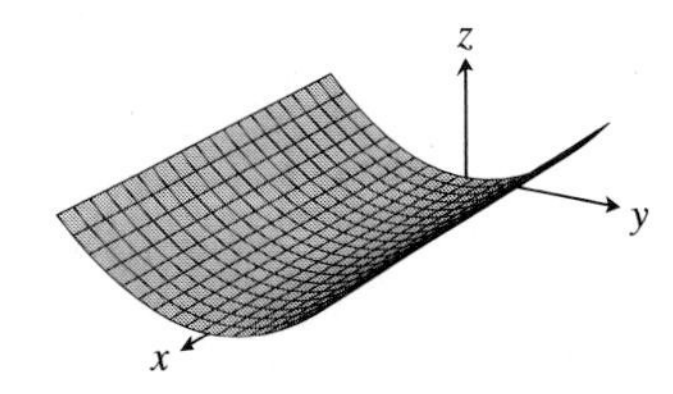

5. $\sqrt{3}$ square units

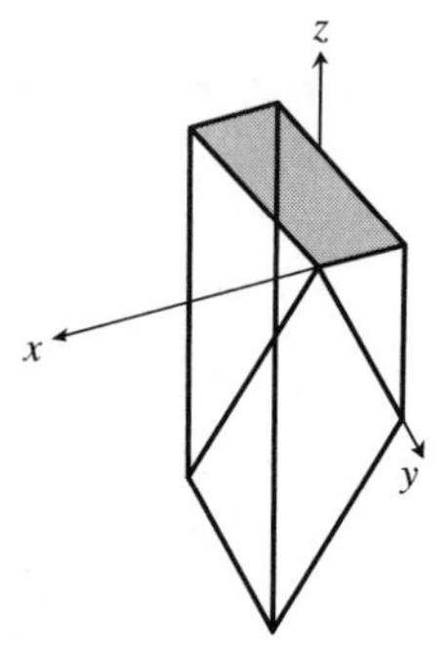

7. 10.4670 square units

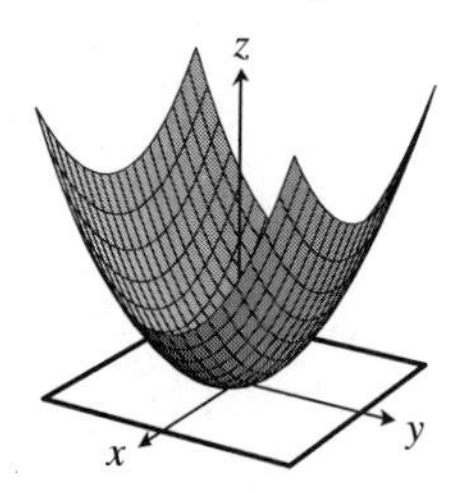

9. $\int_0^{2\pi}\int_{1/2}^{1}\sqrt{u^2+1/25}\,du\,dv$

11. $2\int_{0.9}^{1.1}\int_0^{\sqrt{0.1^2-(x-1)^2}}\sqrt{1+4x^2+4y^2}\,dy\,dx$

13. $64\pi a^2/3$ square units **15.** 22.9430 square units

17. $m = a\delta\pi\sqrt{a^2+b^2}$ kg; $Z = 2b/3$ m

19. $m = 16.034\delta$ kg; $Z = 1.29408$ m

23. $A = 4\pi^2 ab$ **25.** $A(S) = 2\pi\int_a^b f(x)\sqrt{1+f'(x)^2}\,dx$

27. 0.77%

31. approximately 2314 cubic feet of cement/gravel aggregate

11.4 Change-of-Variables Formula for Double Integrals

1. $2u^2 + 2v^2$ **3.** $2u^3$ **5.** $2u/(3v^2)$

7.

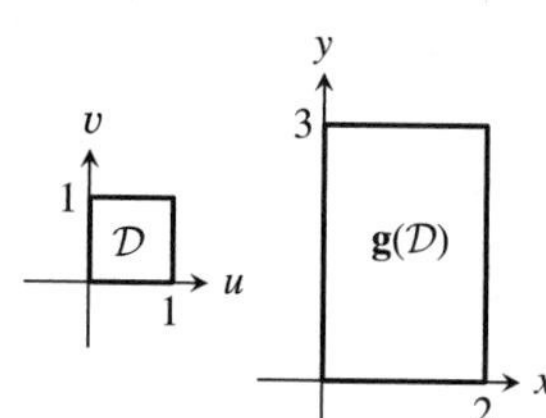

9.

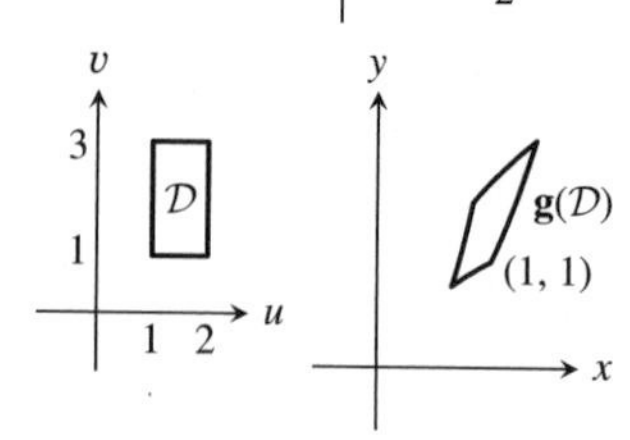

11.

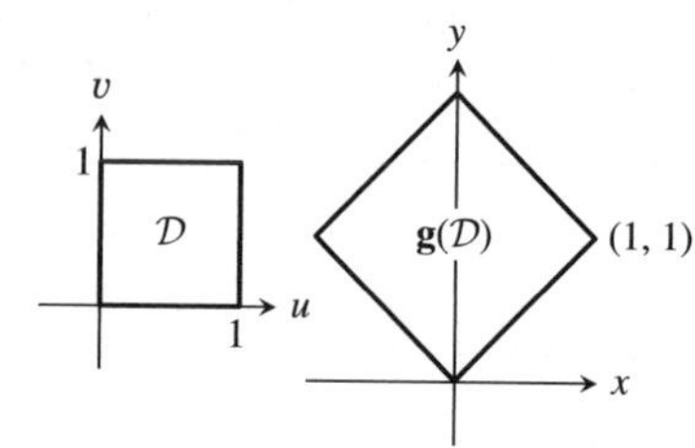

13. $2/3$

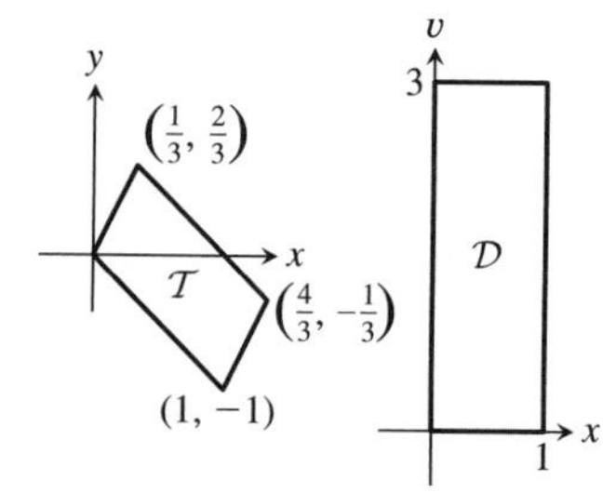

15. $56 \ln(3)/9$

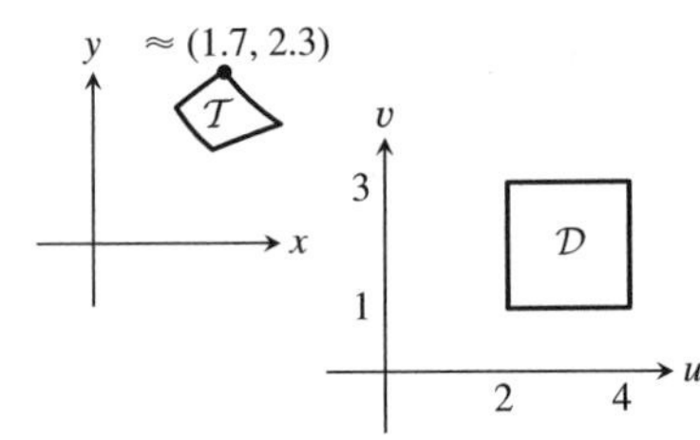

17. $-\frac{1}{4}\pi\sqrt{3/2}$

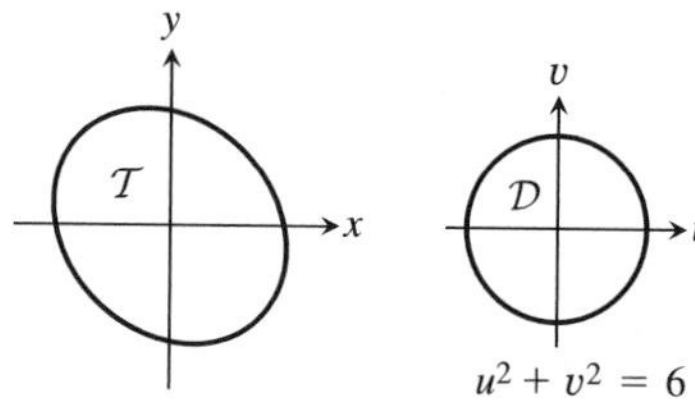

19. $\pi/12$

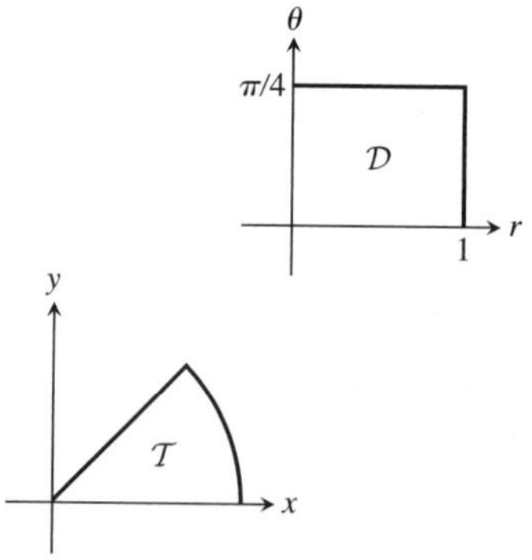

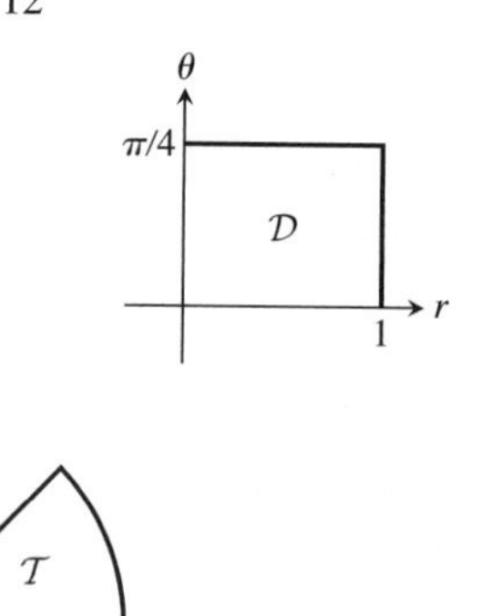

21. $2\pi\left(1 - 3\sqrt{11}/10\right)$ square units

23. $64/9$ cubic units **25.** $\frac{1}{24}\pi a^3 b^3$

27. One change of variable is given by the equations $X = r\cos\theta$ and $Y = r\sin\theta$ and the other is given by $x = aX$ and $y = bY$. Combining these changes of variable gives the change of variable $x = ar\cos\theta$ and $y = br\sin\theta$. The Jacobian of the composition, which is abr, is the product of the Jacobians of the individual changes of variable.
31. 72π mm^2 **33.** approximately 0.217605
35. $-1 + \sqrt{2}$ **37.** $\sqrt{5}\pi$ square units

11.5 Triple Integrals

1. $R_2 = 19.125$ mass units; the true mass is 19.5 mass units.
3. A region $\mathcal{B}$ in R^3 is a region of Type II if it lies between the graphs of bounded continuous functions p and q defined on a bounded subset $\mathcal{D}$ of the (x, y)-plane. If f is continuous and is defined on such a region $\mathcal{B}$, then

$$\iiint_{\mathcal{B}} f(x, y, z)\,dV = \iint_{\mathcal{D}} \left(\int_{p(x,z)}^{q(x,z)} f(x, y, z)\,dy \right) dx\,dz$$

5. $43/105$ cubic units

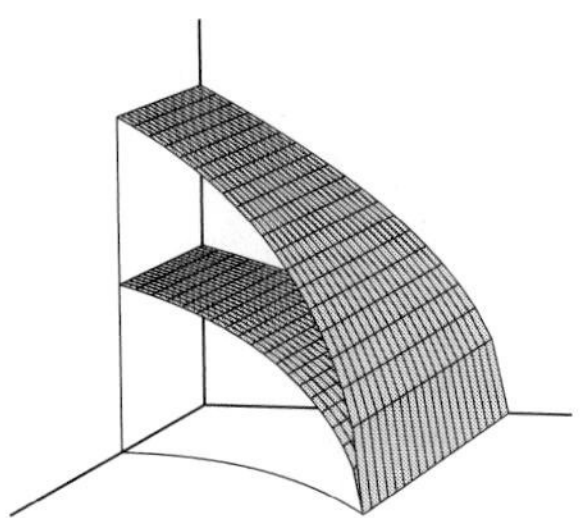

9. $2/3$ cubic units **11.** $14/15$ cubic units
13. $1/4$ cubic units **15.** 2 cubic units
17. $64/15$ cubic units
19. $m = \frac{1}{8}kabc^2$; $\mathbf{R} = \frac{1}{15}\langle 4a, 4b, 3c\rangle$ meters
21. $1/120$ **23.** 3 **25.** π **27.** $\mathbf{R} = \left\langle 0, 0, \frac{2}{5}h \right\rangle$
29. $32/3$ cubic units **31.** $\frac{1}{3}\pi kgh^2a^3$ J

35. $$\int_0^1 \int_0^1 \int_0^{1-z^2} 1\,dy\,dz\,dx + \int_1^2 \int_0^{\sqrt{1-\sqrt{x-1}}} \int_{\sqrt{x-1}}^{1-z^2} 1\,dy\,dz\,dx$$

11.6 Change-of-Variables Formula for Triple Integrals

1. 1 cubic unit either way. The first comes from a direct, geometrically based calculation with vectors; the second is the value of the integral

$$\iiint_{g(\mathcal{B})} 1\,dx\,dy\,dz = \iiint_{\mathcal{D}} \frac{\partial(x, y, z)}{\partial(u, v, w)}\,du\,du\,dw.$$

3. $$\int_0^1 \int_0^{r\pi^2/9} \int_{\sqrt{z/r}}^{\pi/3} r\,d\theta\,dz\,dr = \pi^3/243 \text{ cubic units}$$

5. $\frac{2}{9}a^3(3\pi - 4)$ cubic units **7.** 16π cubic units
9. $\frac{2}{3}\pi ka^3h$ mass units **11.** $Z = \frac{3}{8}a(1 + \cos b)$ length units
13. $\frac{4}{3}\pi(1 - 1/e)$ **15.** $\pi/16$
17. $\frac{1}{6}abc(\pi - 2\arctan(a/b))$ cubic units

19. $\frac{1}{32}\pi$ cubic units **21.** $\frac{2}{3}\pi\left(8\pi - 9\sqrt{3}\right)$ cubic units
23. $Z = \frac{3}{8}a$ units **29.** $3.00619k$ mass units

Chapter Review Exercises

1. $\iint_{\mathcal{D}} f(x, y)\,dx\,dy = 7/6 \approx 1.16667$; the Riemann sum is 1.15625; the percentage error is approximately 0.89%.
3. $\frac{1}{2}(e - 1)$

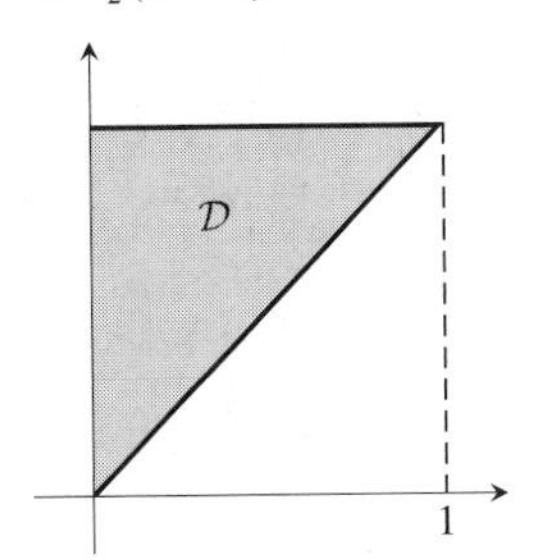

5. 8 cubic units
7. $m = \int_0^1 \int_0^2 \int_0^{1+x^2+2y^2} (2 + z)\,dz\,dx\,dy$ grams
9.

$$m = k\int_0^{\pi/2} \int_0^{\theta} r^2\,dr\,d\theta$$

$$X = (k/m)\int_0^{\pi/2} \int_0^{\theta} (r\cos\theta)r^2\,dr\,d\theta$$

$$Y = (k/m)\int_0^{\pi/2} \int_0^{\theta} (r\sin\theta)r^2\,dr\,d\theta$$

11. $629\sqrt{6}/10$ kg
13. $\frac{1}{2}(1 - \cos(1))$; as given, an antiderivative is not forthcoming, but with the order of integration reversed, the integrations are easy.

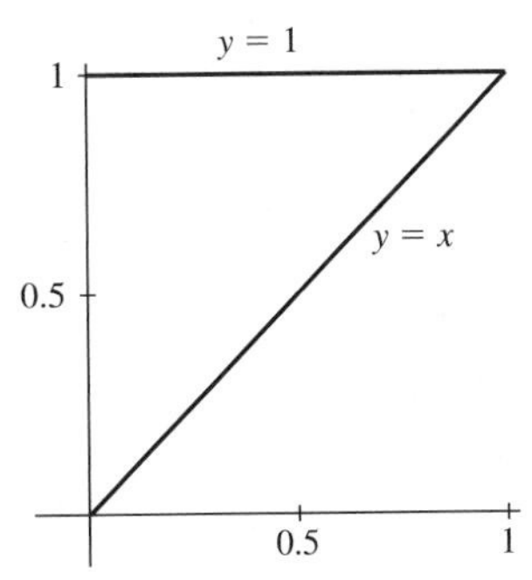

15. $19/24$ cubic units
17. $\mathbf{N} = \left(1/\sqrt{21}\right)\langle -1, -2, 4\rangle$;

$$\int_0^4 \int_0^4 2\sqrt{4u^2 + v^2 + 4u^2v^2}\,du\,dv \text{ square units}$$

19. $26/3$ square units
21. $\int_0^{2\pi} \int_0^2 r\sqrt{1 + 4r^2}\,dr\,d\theta$ square units
23. $\frac{1}{2}\int_0^a \int_0^{\sqrt{a^2-z^2}} \sqrt{\frac{1 + 8x^2 + 4z}{x^2 + z}}\,dx\,dz$ square units

25. $4\pi^2 R$ square units **27.** $\frac{1}{8}\pi \ln 5$

29. $\frac{1}{3}\pi \cdot 20^2 \cdot 12 - 2\int_0^{\pi/2}\int_0^{2\cos\theta} 12r(1 - r/20)\,dr\,d\theta$ cubic units

31. $1/3$ **33.** $\int_0^1\int_0^{3-3y}\int_{3-x-y}^{4(1-x/3-y/2)} dz\,dx\,dy$ cubic units

35. $\int_0^{(-1+\sqrt{5})/2}\int_{x^2}^{1-x}\int_0^{1-x-y} dz\,dy\,dx$ cubic units

37. $\int_{-R}^{R}\int_{-\sqrt{R^2-x^2}}^{\sqrt{R^2-x^2}}\int_0^H \frac{\sqrt{x^2+y^2}}{1+x^2+y^2}\,dz\,dy\,dx$

39. $\frac{1}{3}\pi(b^3 - a^3)(2 - \sqrt{2})$ cubic units

41. In cylindrical coordinates,

$$\int_0^{2\pi}\int_0^1\int_{\sqrt{3}r}^{\sqrt{4-r^2}} rz\,dz\,dr\,d\theta.$$

In spherical coordinates,

$$\int_0^{2\pi}\int_0^{\pi/6}\int_0^2 \rho^2\sin\phi(\rho\cos\phi)\,d\rho\,d\phi\,d\theta.$$

43. $\int_0^{2\pi}\int_0^{3/4}\int_r^{3(1-r)} r\,dz\,dr\,d\theta$ cubic units

CHAPTER 12 — LINE AND SURFACE INTEGRALS

12.1 The Line Integral

1. $28\sqrt{14} \approx 104.77$

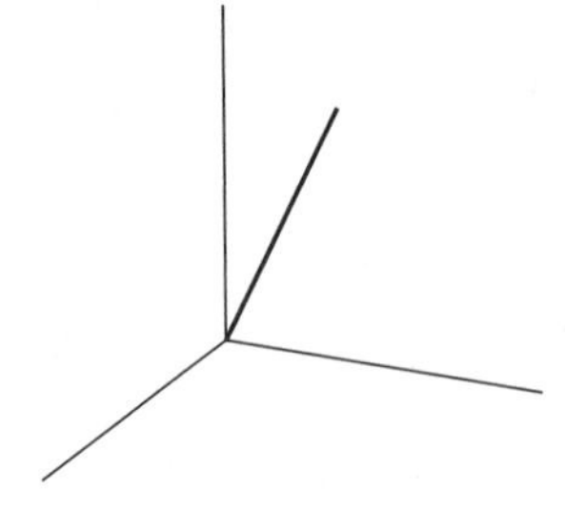

3. $4\pi^2\sqrt{5} \approx 88.28$

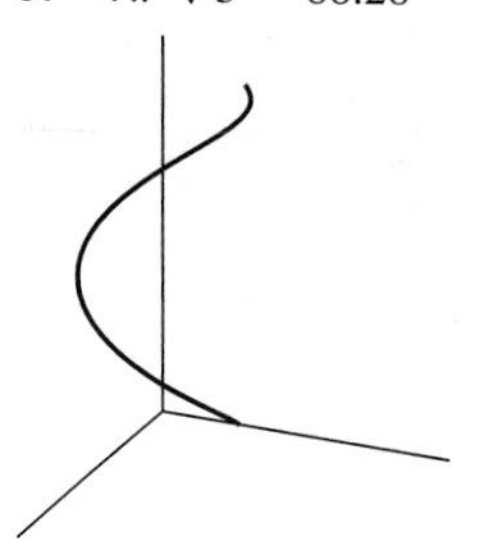

5. $33\sqrt{11}/2 \approx 54.72$

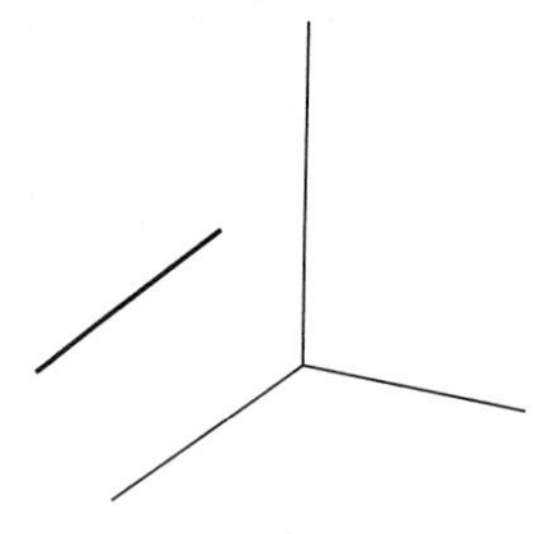

7. $-4\pi \approx -12.57$

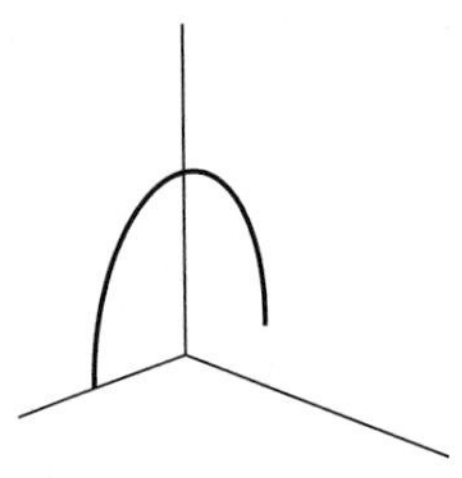

9. $\frac{\pi(108 + 7\pi^2)}{324\sqrt{2}} \approx 1.21$

11. $25.0848 \le \sqrt{10}\int_0^{\pi/2}(1 + 9\cos^2 t)\,dt \le 29.5554$

13. $\int_0^1 (t^2 + t^6 + 1)\sqrt{1 + 4t^2 + 9t^4}\,dt \approx 3.17$, by numerical integration

15. $\frac{5}{3} + 4\sqrt{2} \approx 7.32$

19. $\frac{1}{25}\int_{-50}^{50}\sqrt{t^2 + 625}\left(2 + \sin\left(\frac{t^2 + 50t}{200}\right)\right)dt \approx 258.23$ square units

23. $\frac{1}{M}\left\langle \frac{144e^5 + 100e^3 + 86}{225}, \frac{4e^6 + 3e^4 - 7}{3}, \frac{20e^7 + 14e^5 - 34}{35}\right\rangle \approx \langle 0.798, 4.517, 5.243\rangle$, where $M = \frac{12e^5 + 10e^3 - 22}{15}$

25. $\left\langle \pi - \frac{1}{4}, 0, 0\right\rangle$

12.2 Vector Fields, Work, and Flows

1.

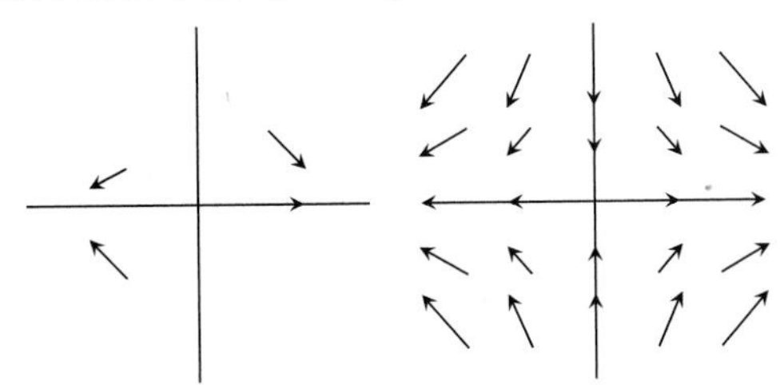

3.

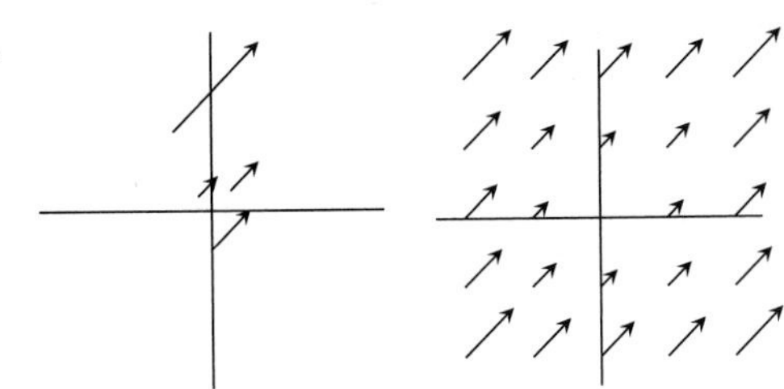

5.

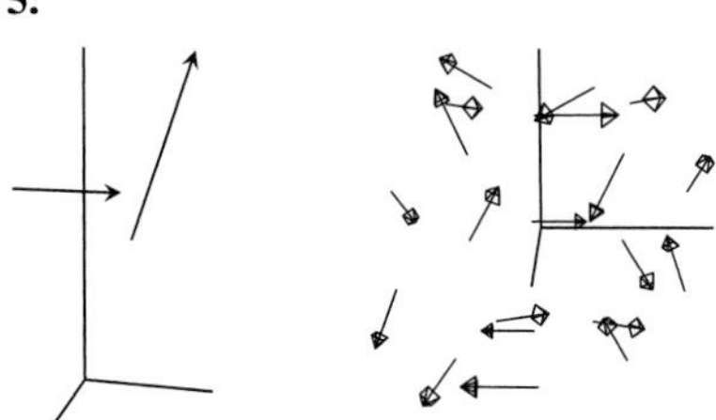

7.

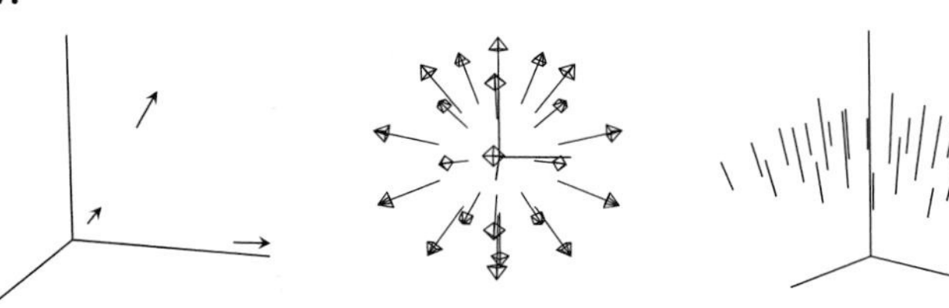

9. $17/2$ **11.** $-\frac{1}{3}e^3 + \frac{1}{3}e^{-3} - 4 \approx -10.68$ **13.** -5

15. 128 **17.** $40\sqrt{5}/3 \approx 29.81$ **21.** -33 g/s **23.** 0

25. a. There appears to be neither a source nor a sink inside the triangle. **b.** 0

c.

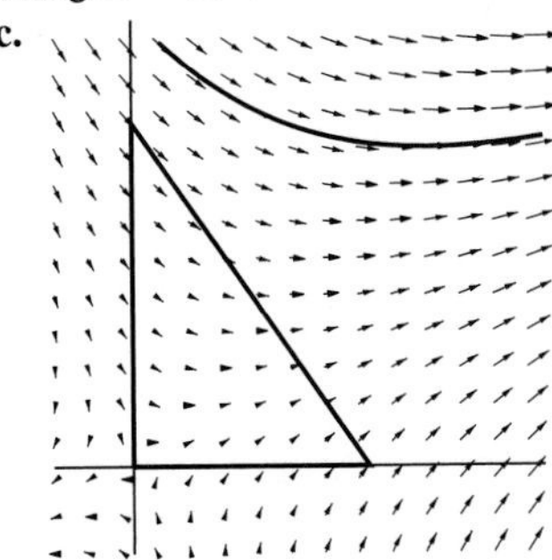

29. Let $f(x, y, z) = xy \ln z$. The value of the integral is $f(B) - f(A)$.

12.3 The Fundamental Theorem of Line Integrals

1. $782 + 2\sqrt{14} \approx 789.48$ **3.** 0

5. $f(x, y) = \frac{1}{2}x^2 + xy - \frac{3}{2}y^2 + 2x - 4y$

7. $f(x, y) = y \tan(xy)$ **9.** $f(x, y, z) = y^2z - yz^2$

11. no potential function **13.** $\sqrt{5} - 1 \approx 1.24$

15. 0 **17.** $f(x, y) = xy - 2x + y$

19. $f(x, y, z) = x \sin(yz) - \frac{1}{2}y^2 - \frac{9}{2}$

23. domain $= \{(x, y, z) : x^2 + y^2 + z^2 > R^2\}$, where R is the radius of the Earth

25. 6

29. a. $\frac{\partial}{\partial y}\frac{-x}{(x^2 + y^2)^2} = \frac{4xy}{(x^2 + y^2)^3} = \frac{\partial}{\partial x}\frac{-y}{(x^2 + y^2)^2}$,

potential function: $f(x, y) = \frac{1}{2(x^2 + y^2)}$

b. There is no contradiction because the theorem makes no statement about the behavior of functions on non–simply connected sets.

31. $\mathbf{G}(x, y) = \mathbf{F}(x, y) + \langle 0, e^{xy}\rangle$; value of integral is $2e$

33. $\mathbf{G}(x, y, z) = \mathbf{F}(x, y, z) + \langle 0, 0, 2xz\rangle$; value of integral is $16\pi^2$

12.4 Green's Theorem

1. -2π **3.** 0 **5.** 20 **7.** πa^2 **9.** $3\pi/2$ **11.** 0

13. 0 **15.** 0 **17.** 2π

25. $P(x, y) = y$ and $Q(x, y) = 2x$, $P(x, y) = y + \sqrt{x^3 + 1}$ and $Q(x, y) = 2x - e^{\sin y}$; other answers are possible

27. The value of the integral is 0.

12.5 Divergence and Curl

1. $\text{div}(\mathbf{F}) = 0$
$\text{curl}(\mathbf{F}) = \langle 0, 0, 0\rangle$

3. $\text{div}(\mathbf{v}) = 4x$
$\text{curl}(\mathbf{v}) = \langle 0, 0, 4y\rangle$

5. $\text{div}(\mathbf{H}) = 28$
$\text{curl}(\mathbf{H}) = \langle 0, 0, 0\rangle$

7. $\nabla\mathbf{F}$, undefined **9.** $\nabla \times (\nabla \cdot \mathbf{F})$, undefined

11. $\nabla(\nabla \cdot f)$, undefined **13.** $\nabla(\nabla \cdot \mathbf{F})$, defined

15. $\nabla \cdot (\nabla \times (\nabla f))$, defined

17. The property suggested is the distribution of the dot product over the vector sum.

12.6 Surface Integrals

1. $64\pi/3$ **3.** $38\pi/3$ **5.** $-21\sqrt{14}$ **7.** $4\pi\sqrt{5}$ g

9. $999\pi/2$ g **11.** $8/3$ **13.** 8π **15.** 0 **17.** $4\pi^3/3$

19. The integral is equal to the area of S.

21.

$$\mathbf{N}(x, y, z) = \frac{1}{\sqrt{(\arctan y)^2 + x^2/(1 + y^2)^2 + 1}}\left\langle \arctan y, \frac{x}{1 + y^2}, -1 \right\rangle$$

23. $\mathbf{N}(x, y, z) = \dfrac{1}{\sqrt{x^2 + y^2 + z^2}}\langle x, -y, z\rangle$

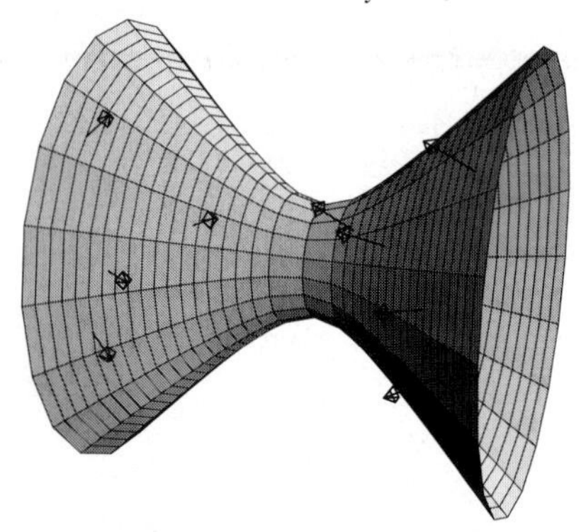

27. $(0, 0, 1/2)$ **29.** $(26/9, 0, 0)$
31. If $\mathbf{F}: R^3 \to R^3$ is conservative and has a continuous derivative, then curl $\mathbf{F} = \mathbf{0}$.

12.7 The Divergence Theorem

1. 4π **3.** $-\pi/2$ **5.** $8\pi/3$ **7.** 27π **9.** $(8\pi/3)$ g/s
11. 12 g/s
13. Source at all points $(x, y, z) \neq (0, 0, 0)$. Neither a source nor a sink at $(0, 0, 0)$.
15. Source at all points $(x, y, z) \neq (0, 0, 0)$. Neither a source nor a sink at $(0, 0, 0)$.
17. 0 cal/s **19.** -0.32π cal/s
21. Sample answers (other answers are possible):
$\mathbf{F}(x, y, z) = \langle x, 0, 0\rangle$
$\mathbf{F}(x, y, z) = \langle 0, y, 0\rangle$
25. **a.** $10\pi/3$ **b.** 0 **c.** $10\pi/3$ **27.** 40π

12.8 Stokes' Theorem

9. $4/3$ **11.** -1 **13.** 0 **15.** $\pi\sqrt{2}$ **17.** 2π
19. $\pi/8$ **21.** Yes, $\mathbf{F}$ is conservative on $\mathcal{D}$. **23.** 8π

Chapter Review Exercises

1. $-\dfrac{5}{12}\sqrt{58} \approx -3.1732$ **3.** $4/e \approx 1.47152$ **5.** 50
7. $8\pi/3$ **9.** 2π **11.** 0 **13.** 45 **15.** 0
17. no potential function
19. potential function: $f(x, y) = \sqrt{x^2 + 2y} + C$
21. no potential function **23.** **a.** 0 **b.** $3\pi/4$
27. **a.** $E(\mathbf{r}) = E(x, y, z) = \dfrac{kc}{(x^2 + y^2 + z^2)^{3/2}}\langle x, y, z\rangle$ where c is the charge of the particle at the origin and k is a positive constant of proportionality.
b.

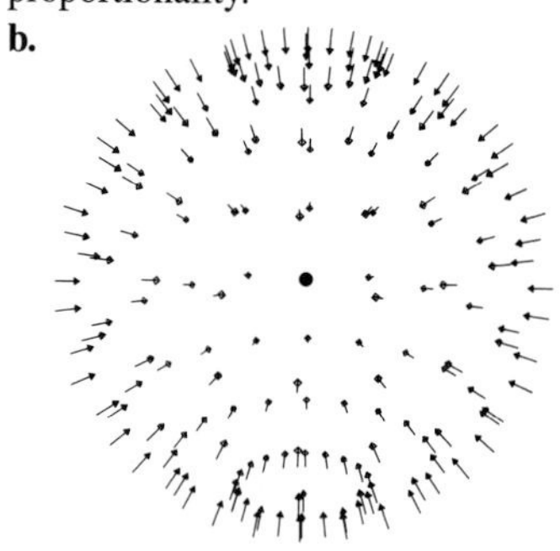

c.

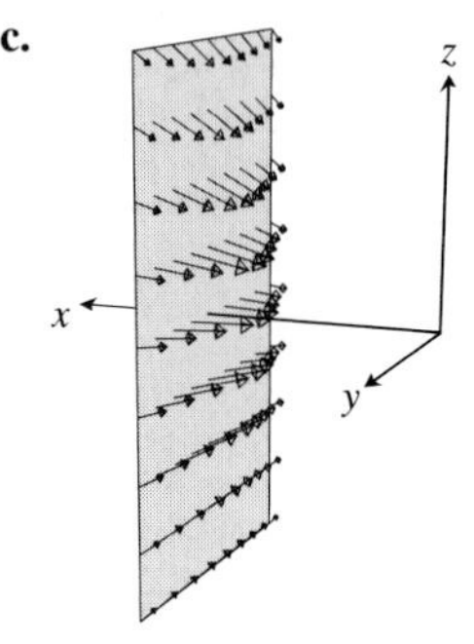

Index

DERIVATIVE FORMULAS

General Formulas

If c is a constant,

$$(cf(x))' = cf'(x)$$

Sums

$$(f(x) + g(x))' = f'(x) + g'(x)$$

The Product Rule

$$(u(x)v(x))' = u'(x)v(x) + u(x)v'(x)$$

The Quotient Rule

$$\left(\frac{u(x)}{v(x)}\right)' = \frac{u'(x)v(x) - u(x)v'(x)}{(v(x))^2}$$

The Chain Rule (Version I)

$$(f(g(x)))' = f'(g(x))g'(x)$$

The Chain Rule (Version II)

If $y = f(u)$ and $u = g(x)$, then $\dfrac{dy}{dx} = \dfrac{dy}{du}\dfrac{du}{dx}$

Specific Formulas

Simple Derivative

$$\frac{d}{dx}x^n = nx^{n-1}$$
$$\frac{d}{dx}\sin x = \cos x$$
$$\frac{d}{dx}\cos x = -\sin x$$
$$\frac{d}{dx}\tan x = \sec^2 x$$
$$\frac{d}{dx}\sec x = \sec x \tan x$$
$$\frac{d}{dx}\cot x = -\csc^2 x$$
$$\frac{d}{dx}\csc x = -\csc x \cot x$$
$$\frac{d}{dx}e^x = e^x$$
$$\frac{d}{dx}a^x = (\ln a)a^x$$
$$\frac{d}{dx}\ln x = \frac{1}{x}$$
$$\frac{d}{dx}\sinh x = \cosh x$$
$$\frac{d}{dx}\cosh x = \sinh x$$
$$\frac{d}{dx}\tanh x = \operatorname{sech}^2 x$$
$$\frac{d}{dx}\operatorname{sech} x = -\operatorname{sech} x \tanh x$$
$$\frac{d}{dx}\coth x = -\operatorname{csch}^2 x$$
$$\frac{d}{dx}\operatorname{csch} x = -\operatorname{csch} x \coth x$$
$$\frac{d}{dx}\arcsin x = \frac{1}{\sqrt{1-x^2}}$$
$$\frac{d}{dx}\arccos x = -\frac{1}{\sqrt{1-x^2}}$$
$$\frac{d}{dx}\arctan x = \frac{1}{1+x^2}$$
$$\frac{d}{dx}\operatorname{arcsec} x = \frac{1}{|x|\sqrt{x^2-1}}$$

Chain Rule Version: u a function of x

$$\frac{d}{dx}u^n = nu^{n-1}\frac{du}{dx}$$
$$\frac{d}{dx}\sin u = \cos u\frac{du}{dx}$$
$$\frac{d}{dx}\cos u = -\sin u\frac{du}{dx}$$
$$\frac{d}{dx}\tan u = \sec^2 u\frac{du}{dx}$$
$$\frac{d}{dx}\sec u = \sec u \tan u\frac{du}{dx}$$
$$\frac{d}{dx}\cot u = -\csc^2 u\frac{du}{dx}$$
$$\frac{d}{dx}\csc u = -\csc u \cot u\frac{du}{dx}$$
$$\frac{d}{dx}e^u = e^u\frac{du}{dx}$$
$$\frac{d}{dx}a^u = (\ln a)a^u\frac{du}{dx}$$
$$\frac{d}{dx}\ln u = \frac{1}{u}\frac{du}{dx}$$
$$\frac{d}{dx}\sinh u = \cosh u\frac{du}{dx}$$
$$\frac{d}{dx}\cosh u = \sinh u\frac{du}{dx}$$
$$\frac{d}{dx}\tanh u = \operatorname{sech}^2 u\frac{du}{dx}$$
$$\frac{d}{dx}\operatorname{sech} u = -\operatorname{sech} u \tanh u\frac{du}{dx}$$
$$\frac{d}{dx}\coth u = -\operatorname{csch}^2 u\frac{du}{dx}$$
$$\frac{d}{dx}\operatorname{csch} u = -\operatorname{csch} u \coth u\frac{du}{dx}$$
$$\frac{d}{dx}\arcsin u = \frac{1}{\sqrt{1-u^2}}\frac{du}{dx}$$
$$\frac{d}{dx}\arccos u = -\frac{1}{\sqrt{1-u^2}}\frac{du}{dx}$$
$$\frac{d}{dx}\arctan u = \frac{1}{1+u^2}\frac{du}{dx}$$
$$\frac{d}{dx}\operatorname{arcsec} u = \frac{1}{|u|\sqrt{u^2-1}}\frac{du}{dx}$$

TABLE OF INTEGRALS

Basic

(A) $\int x^n\,dx = \dfrac{x^{n+1}}{n+1} + C, \qquad n \neq -1$

(B) $\int \dfrac{1}{x}\,dx = \ln|x| + C$

(C) $\int \sin x\,dx = -\cos x + C$

(D) $\int \cos x\,dx = \sin x + C$

(E) $\int e^x\,dx = e^x + C$

(F) $\int a^x\,dx = \dfrac{a^x}{\ln a} + C$

(G) $\int \dfrac{dx}{x^2 + a^2} = \dfrac{1}{a}\arctan\dfrac{x}{a} + C$

(H) $\int \dfrac{dx}{\sqrt{a^2 - x^2}} = \arcsin\dfrac{x}{a} + C$

Algebraic

(1) $\int \dfrac{dx}{\sqrt{x^2 \pm a^2}} = \ln\left|x + \sqrt{x^2 \pm a^2}\right| + C$

(2) $\int \dfrac{dx}{x^2 - a^2} = \dfrac{1}{2a}\ln\left|\dfrac{x-a}{x+a}\right| + C$

(3) $\int \dfrac{dx}{x\sqrt{a^2 \pm x^2}} = -\dfrac{1}{a}\ln\left|\dfrac{a + \sqrt{a^2 \pm x^2}}{x}\right| + C$

(4) $\int \sqrt{x^2 \pm a^2}\,dx = \dfrac{x}{2}\sqrt{x^2 \pm a^2} \pm \dfrac{a^2}{2}\ln\left|x + \sqrt{x^2 \pm a^2}\right| + C$

(5) $\int \dfrac{\sqrt{x^2 \pm a^2}}{x^2}\,dx = -\dfrac{\sqrt{x^2 \pm a^2}}{x} + \ln\left|x + \sqrt{x^2 \pm a^2}\right| + C$

(6) $\int \dfrac{dx}{x\sqrt{x^2 - a^2}} = \dfrac{1}{a}\,\text{arcsec}\,\dfrac{x}{a} + C$

(7) $\int \sqrt{a^2 - x^2}\,dx = \dfrac{x}{2}\sqrt{a^2 - x^2} + \dfrac{a^2}{2}\arcsin\dfrac{x}{a} + C$

(8) $\int x^2\sqrt{a^2 - x^2}\,dx = \dfrac{x}{8}(2x^2 - a^2)\sqrt{a^2 - x^2} + \dfrac{a^4}{8}\arcsin\dfrac{x}{a} + C$

(9) $\int \dfrac{\sqrt{a^2 - x^2}}{x^2}\,dx = -\dfrac{1}{x}\sqrt{a^2 - x^2} - \arcsin\dfrac{x}{a} + C$

(10) $\int \dfrac{x^2}{\sqrt{a^2 - x^2}}\,dx = -\dfrac{x}{2}\sqrt{a^2 - x^2} + \dfrac{a^2}{2}\arcsin\dfrac{x}{a} + C$

(11) $\int \dfrac{\sqrt{a^2 \pm x^2}}{x}\,dx = \sqrt{a^2 \pm x^2} - a\ln\left|\dfrac{a + \sqrt{a^2 \pm x^2}}{x}\right| + C$

(12) $\int \dfrac{x}{(ax+b)^2}\,dx = \dfrac{b}{a^2(ax+b)} + \dfrac{1}{a^2}\ln|ax + b| + C$

(13) $\int \dfrac{dx}{x(ax+b)} = \dfrac{1}{b}\ln\left|\dfrac{x}{ax+b}\right| + C$

(14) $\int \dfrac{dx}{x(ax+b)^2} = \dfrac{1}{b(ax+b)} + \dfrac{1}{b^2}\ln\left|\dfrac{x}{ax+b}\right| + C$

Trigonometric

(15) $\int \sec^2 x\,dx = \tan x + C$

(16) $\int \sec x \tan x\,dx = \sec x + C$

(17) $\int \tan x\,dx = -\ln|\cos x| + C$

(18) $\int \sec x\,dx = \ln|\sec x + \tan x| + C$

(19) $\int \sec^3 x\,dx = \frac{1}{2}\sec x \tan x + \frac{1}{2}\ln|\sec x + \tan x| + C$

(20) $\int \sin ax \sin bx\,dx = \dfrac{\sin(a-b)x}{2(a-b)} - \dfrac{\sin(a+b)x}{2(a+b)} + C$